Biomedical Engineering Desk Reference

Note from the Publisher

This book has been compiled using extracts from the following books within the range of Biomedical Engineering books in the Elsevier collection:

Dyro (2004) *Clinical Engineering Handbook*, 9780122265709
Dunn *et al.* (2006) *Numerical Methods in Biomedical Engineering*, 9780121860318
Semmlow (2005) *Circuits, Signals, and Systems for Bioengineers*, 9780120884933
Ratner *et al.* (2004) *Biomaterials Science*, 9780125824637
Grimnes and Martinsen (2008) *Bioimpedance and Bioelectricity*, 9780123032607
Kyle and Murray (2008) *Clinical Simulation*, 9780123725318
Perez (2002) *Design of Medical Electronic Devices*, 9780125507110
Bankman (2000) *Handbook of Medical Imaging*, 9780120777907
Ikada (2006) *Tissue Engineering: Interface Science and Technology*, 9780123705822
Vallero (2007) *Biomedical Ethics for Engineers*, 9780750682275

The extracts have been taken directly from the above source books, with some small editorial changes. These changes have entailed the re-numbering of Sections and Figures. In view of the breadth of content and style of the source books, there is some overlap and repetition of material between chapters and significant differences in style, but these features have been left in order to retain the flavour and readability of the individual chapters.

End of chapter questions
Within the book, several chapters end with a set of questions; please note that these questions are for reference only. Solutions are not always provided for these questions.

Units of measure
Units are provided in either SI or IP units. A conversion table for these units is provided at the front of the book.

Upgrade to an Electronic Version
An electronic version of the Desk reference, the *Biomedical Engineering e-Mega Reference*, 9780123746474

- A fully searchable Mega Reference eBook, providing all the essential material needed by Biomedical Engineers on a day-to-day basis.
- Fundamentals, key techniques, engineering best practice and rules-of-thumb at one quick click of a button
- Over 1,500 pages of reference material, including over 1,000 pages not included in the print edition

Go to http://www.elsevierdirect.com/9780123746467 and click on **Ebook Available**

Biomedical Engineering Desk Reference

Amsterdam · Boston · Heidelberg · London · New York · Oxford
Paris · San Diego · San Francisco · Sydney · Tokyo
Academic Press is an imprint of Elsevier
ELSEVIER

Academic Press is an imprint of Elsevier
Linacre House, Jordan Hill, Oxford OX2 8DP, UK
525 B Street, Suite 1900, San Diego, CA 92101-4495, USA

First edition 2009

British Library Cataloguing in Publication Data
A catalogue record for this book is available from the British Library

Library of Congress Cataloging-in-Publication Data
A catalog record for this book is available from the Library of Congress

ISBN: 978-0-12-374646-7

For information on all Academic Press publications visit our web site at elsevierdirect.com

Printed and bound in the United States of America

09 10 11 11 10 9 8 7 6 5 4 3 2 1

Contents

Author Biographies

Dr. Isaac N. Bankman is a member of the Electro-Optical Systems Group in the Air Defense Systems Department at the Applied Physics Laboratory. He worked on signal and image processing projects as a Postdoctoral Fellow and Research Associate at the Johns Hopkins Biomedical Engineering Department before joining APL, where he serves as principal investigator in sponsored programs and IR&D projects on laser radar remote sensing and imaging. He has authored more than 30 scientific publications and is a member of SPIE, OSA, and IEEE.

Alkis Constantinides is a Professor of Chemical and Biochemical Engineering at Rutgers, The State University of New Jersey, with over forty years of academic and industrial experience. He has served as Chairman of the Department, Director of the Graduate Program, Director of the Undergraduate Program, and Director of Alumni Relations. He is the recipient of the prestigious Warren I. Susman Award for Excellence in Teaching, and the recipient of the Best Teacher of the Year Award several times in recent years, as chosen by the Graduating Senior Class of the Department of Chemical and Biochemical Engineering.

Prabhas Moghe is jointly appointed as Professor of Biomedical Engineering and Professor of Chemical and Biochemical Engineering at Rutgers University in New Jersey. An elected Fellow of the American Institute of Medical and Biological Engineering (AIMBE) and the American Academy of Nanomedicine (AANM), Dr. Moghe has received several awards/honours for his research and accomplishments, including the NSF CAREER Award and the Integra LifeSciences Excellence Award. An author of fifty publications and over 100 presentations, he currently directs an NSF IGERT training program on Integratively Engineered Biointerfaces.

Stanley M. Dunn is Faculty Scholar and Associate Dean at the School of Engineering, Rutgers University.

Dr. Joseph F. Dyro is President of the Biomedical Resource Group in Setauket, NY and Editor of the Journal of Clinical Engineering. He is also past President and Founding Fellow of the American College of Clinical Engineering (ACCE), a Founding Fellow of the American Institute of Medical and Biological Engineering (AIMBE) and a Senior Member of IEEE.

Dr. Sverre Grimnes is a full time research scientist at the Department of Biomedical and Clinical Engineering at Rikshospitalet, Oslo, Norway, where he was Head of Department for 27 years. Since 1984 he has also been professor with the Department of Physics. His main research activities are in bioimpedance and bioelectricity.

Professor Allan Hoffman taught in the faculty of the Chemical Engineering Department at M.I.T. for ten years. Since 1970 he has been Professor of Bioengineering and Chemical Engineering at the University of Washington in Seattle. He is currently a Visiting Professor at Shanghai University and Wuhan University of Technology in China. He was formerly President of the Society for Biomaterials, USA and was elected to the U.S. National Academy of Engineering. He has won numerous awards in the USA, including the Founders' Award of the Society for Biomaterials, and in Japan, such as the Biomaterials Science Prize from the Japan Biomaterials Society.

Dr. Yoshito Ikada has over forty year's experience as a professor at a number of universities in Japan and China in the field of biomedicine. Currently at Nara Medical University, he is also Emeritus Professor at Kyoto University and Guest Professor at Peking University. He has received the Prize of Japanese Society for Biomaterials and was President of the Japanese Society for Biomaterials for four years. He was also Chairman of National project, "Tissue Engineering" by Japan Society for the Promotion of Science.

Richard Kyle co-founded the Patient Simulation Laboratory at the Uniformed Services University with CAE-Eagle PatientSim #47. Since then, he has ridden the expanding wave of clinical simulation program creation and development while making new friends around the globe. He now mentors clinical faculty in integrating the principle of simulation – learning through doing, safely – into their daily teachings.

Jack E. Lemons is a professor of dentistry in the Department of Prosthodontics at the UAB School of Dentistry. His interests include dental biomaterials, biological tissue reaction to synthetic materials and biomechanics. He has authored hundreds of publications and more than 400 presentations and abstracts. He is a member of numerous academic and scientific organizations. Throughout his career, Lemons has received many awards and honors, including the UAB Distinguished Faculty Award, the W.T. Cavanaugh Memorial Award and the Distinguished Alumni Award from the University of Florida.

Ørjan G. Martinsen is a professor and Deputy Head with the Department of Physics, University of Oslo, where his research activities are mainly focused on electrical bioimpedance.

Dr. W. Bosseau Murray has been (and still is) a practicing anaesthesiologist for over 30 years. Through fourteen years experience teaching in the Pennsylvania State University Simulation Lab, he has developed a strong teaching background with multiple "Teacher of the Year" awards from medical students as well as anaesthesiology residents. He taught physics, clinical measurement and statistics to anaesthesia registrars, and was also a certified flight surgeon and diving medical officer.

Reinaldo Perez, M.R. Research, Littleton, Colorado, U.S.A.

Dr. Buddy D. Ratner is the Director of UWEB Engineering Research Center and the Michael L. and Myrna Darland Endowed Chair in Technology Commercialization. He is Professor of Bioengineering and Chemical Engineering at the University of Washington. He is the editor of the *Journal of Undergraduate Research in Bio-Engineering*, an Associate Editor of *Journal of Biomedical Materials Research*, on the advisory board of Biointerphases and serves on the editorial boards of ten other journals. He has authored over 400 scholarly works and has seventeen issued patents. He has won numerous awards including the C. William Hall Award from the Society for Biomaterials and the BMES Pritzker Distinguished Lecturer Award.

Professor Frederick J. Schoen is Professor of Pathology and Health Sciences and Technology, Harvard Medical School, and Director of Cardiac Pathology and Executive Vice-Chairman in the Department of Pathology at the Brigham and Women's Hospital in Boston. He is a Director of the newly established BWH Biomedical Research Institute Technology in Medicine Initiative and Site Miner for the Center for Integration of Medicine and Innovative Technology. He is author or co-author of over 400 manuscripts in journals and books. He is Past-President of the Society For Biomaterials and the Society for Cardiovascular Pathology, and was Founding Fellow of the American Institute of Medical and Biological Engineering.

John Semmlow has held faculty positions at the University of California, Berkeley, and the University of Illinois, Chicago, and currently holds a joint position as Professor of Surgery, UMDNJ- Robert Wood Johnson Medical School and Professor of Biomedical Engineering at Rutgers University, New Jersey. He was a NSF/CNRS Fellow in the Sensorimotor Control Laboratory of the University of Provence, Marseille, France, and was appointed a Fellow of the IEEE in recognition of his work in acoustic detection of coronary artery disease. He is also a Fellow of AIMBE and of BMES.

Dr. Daniel Vallero has led the establishment of the Engineering Ethics program at Duke University. He co-directs Duke's first-year Engineering Frontiers Focus program and teaches courses in ethics, sustainable design and green engineering. He is an authority on societal aspects of emerging technologies, such as nanomaterials and the macroethical issues associated with public health and sustainability. Dr. Vallero is the author of seven engineering textbooks and consulting editor (environmental engineering) to the *McGraw-Hill Yearbook of Science and Technology* and the *Encyclopedia of Science and Technology*, with his articles appearing in both.

Section One

Introduction

Chapter 1.0

Introduction

Joseph D. Bronzino

Technological innovation has continually reshaped the field of medicine and the delivery of health care services. Throughout history, advances in medical technology have provided a wide range of positive diagnostic, therapeutic, and rehabilitative tools. With the dramatic role that technology has played in shaping medical care during the latter part of the 20th century, engineering professionals have become intimately involved in many medical ventures. As a result, the discipline of biomedical engineering has emerged as an integrating medium for two dynamic professions: medicine and engineering. Today, biomedical engineers assist in the struggle against illness and disease by providing materials, tools, and techniques (such as medical imaging and artificial intelligence) that can be utilized for research, diagnosis, and treatment by health care professionals. In addition, one subset of the biomedical engineering community, namely clinical engineers, has become an integral part of the health care delivery team by managing the use of medical equipment within the hospital environment. The purpose of this chapter is to discuss the evolution of clinical engineering, to define the role played by clinical engineers, and to present the status of the professionalization of the discipline and to reflect upon its future.

What is clinical engineering?

Many of the problems confronting health care professionals today are of extreme interest to engineers because they involve the design and practical application of medical devices and systems—processes that are fundamental to engineering practice. These medically related problems can range from very complex, large-scale constructs, such as the design and implementation of automated clinical laboratories, multiphasic screening facilities, and hospital information systems, to the creation of relatively small and "simple" devices, such as recording electrodes and biosensors that are used to monitor the activity of specific physiological processes in a clinical setting. Furthermore, these problems often involve addressing the many complexities found in specific clinical areas, such as emergency vehicles, ORs, and intensive care units.

The field of biomedical engineering, as it has evolved, now involves applying the concepts, knowledge, and approaches of virtually all engineering disciplines (e.g., electrical, mechanical, and chemical engineering) to solve specific health care-related problems (Bronzino, 1995; 2000). When biomedical engineers work within a hospital or clinical environment, they are more properly called clinical engineers.

But what exactly is the definition of a "clinical engineer"? Over the years, a number of organizations have attempted to provide an appropriate definition (Schaffer and Schaffer, 1992). For example, the AHA defines a clinical engineer as:

- *"a person who adapts, maintains, and improves the safe use of equipment and instruments in the hospital,"* (AHA, 1986).

The American College of Clinical Engineering defines a clinical engineer as:

- *"a professional who supports and advances patient care by applying engineering and managerial skills to health care technology,"* (Bauld, 1991).

The definition that the AAMI originally applied to board certified practitioners describes a clinical engineer as:

- *"a professional who brings to health care facilities a level of education, experience, and accomplishment which will enable him to responsibly, effectively, and safety manage and interface with medical devices, instruments, and systems and the user thereof during patient care…,"* (Goodman, 1989).

Biomedical Engineering Desk Reference; ISBN: 9780123746467

For the purpose of certification, the Board of Examiners for Clinical Engineering Certification considers a clinical engineer to be:

- *"an engineer whose professional focus is on patient-device interfacing; one who applies engineering principles in managing medical systems and devices in the patient setting,* (ICC, 1991).

The *Journal of Clinical Engineering* has defined the distinction between a biomedical engineer and a clinical engineer by suggesting that the biomedical engineer:

- *"applies a wide spectrum of engineering level knowledge and principles to the understanding, modification or control of human or animal biological systems* (Pacela, 1991).

Finally, in the book "Management of Medical Technology," a clinical engineer was defined as:

- *"an engineer who has graduated from an accredited academic program in engineering or who is licensed as a professional engineer or engineer-in-training and is engaged in the application of scientific and technological knowledge developed through engineering education and subsequent professional experience within the health care environment in support of clinical activities* (Bronzino, 1992).

It is important to emphasize that one of the major features of these definitions is the clinical environment (i.e., that portion of the health care system in which patient care is delivered). Clinical activities include direct patient care, research, teaching, and management activities that are intended to enhance patient care.

Engineers were first encouraged to enter the clinical scene during the late 1960s in response to concerns about patient safety as well as the rapid proliferation of clinical equipment, especially in academic medical centers. In the process, a new engineering discipline—clinical engineering—evolved to provide the technological support that was necessary to meet these new needs. During the 1970s, a major expansion of clinical engineering occurred, primarily due to the following events (Bronzino and Hayes, 1988; Bronzino, 1992):

- The Veterans' Administration (VA), convinced that clinical engineers were vital to the overall operation of the VA hospital system, divided the country into biomedical engineering districts, with a chief biomedical engineer overseeing all engineering activities in the hospitals in a district.
- Throughout the United States, clinical engineering departments were established in most large medical centers and hospitals and in some smaller clinical facilities with at least 300 beds.
- Clinical engineers were hired in increasing numbers to help these facilities to use existing technology and to incorporate new technology.

Having entered the hospital environment, routine electrical safety inspections exposed the clinical engineer to all types of patient equipment that were not being properly maintained. It soon became obvious that electrical safety failures represented only a small part of the overall problem posed by the presence of medical equipment in the clinical environment. This equipment was neither totally understood nor properly maintained. Simple visual inspections often revealed broken knobs, frayed wires, and even evidence of liquid spills. Investigating further, it was found that many devices did not perform in accordance with manufacturers' specifications and were not maintained in accordance with manufacturers' recommendations. In short, electrical safety problems were only the tip of the iceberg. By the mid-1970s, complete performance inspections before and after use became the norm, and sensible inspection procedures were developed (Newhouse et al., 1989). Clinical engineering departments became the logical support center for all medical technologies. As a result, clinical engineers assumed additional responsibilities, including the management of high-technology instruments and systems used in hospitals, the training of medical personnel in equipment use and safety, and the design, selection, and use of technology to deliver safe and effective health care.

In the process, hospitals and major medical centers formally established clinical engineering departments to address these new technical responsibilities and to train and supervise biomedical engineering technicians to carry out these tasks. Hospitals that established centralized clinical engineering departments to meet these responsibilities used clinical engineers to provide the hospital administration with an objective opinion of equipment function, purchase, application, overall system analysis, and preventive maintenance policies. With the in-house availability of such talent and expertise, the hospital was in a far better position to make more effective use of its technological resources (Jacobs, 1975; Bronzino, 1977; 1986; 1992). It is also important to note that competent clinical engineers, as part of the health care system, also created a more unified and predictable market for biomedical equipment. By providing health care professionals with needed assurance of safety, reliability, and efficiency in using new and innovative equipment, clinical engineers identified poor quality and ineffective equipment much more readily. These activities, in turn, led to a faster, more appropriate utilization of new medical equipment and provided a natural incentive for greater industrial involvement—a step that is an essential prerequisite to widespread use of any technology (Newhouse et al., 1989; Bronzino, 1992). Thus, the presence of clinical engineers not only

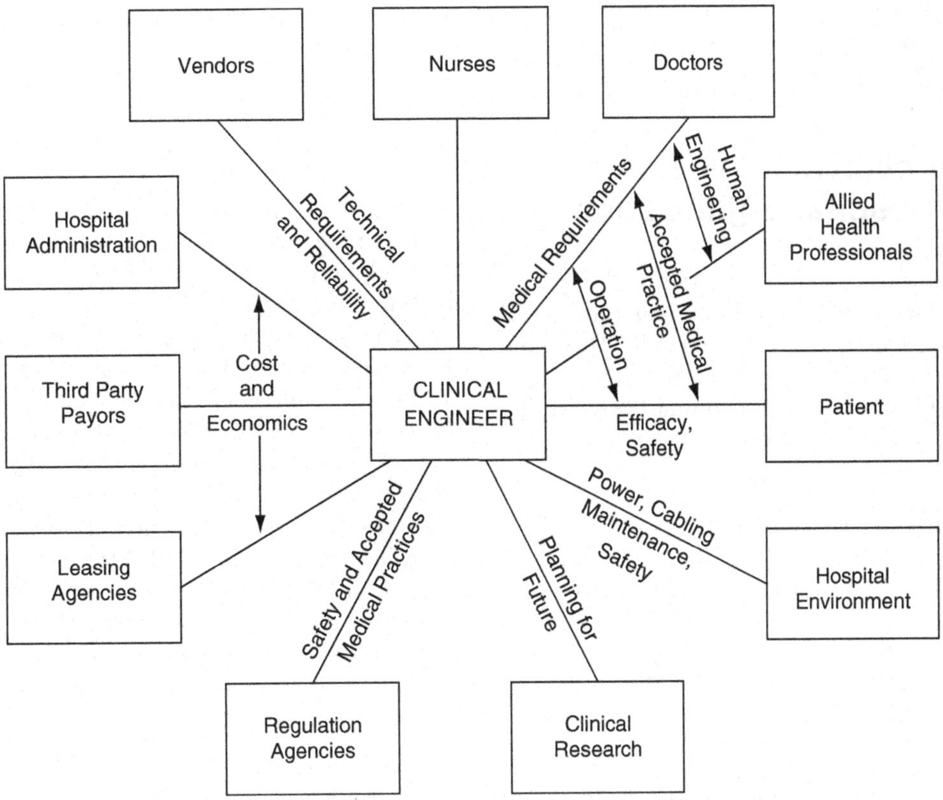

Figure 1.0-1 Diagram illustrating the range of interactions in which a clinical engineer might be required to engage in a hospital setting.

ensured the establishment of a safer environment, but also facilitated the use of modern medical technology to make patient care more efficient and effective.

Today, clinical engineers are an integral part of the health care delivery team. In fact, their role is multifaceted. Figure 1.0-1 illustrates the multifaceted role played by clinical engineers. They must successfully interface with clinical staff, hospital administrators and regulatory agencies to ensure that the medical equipment within the hospital is safely and effectively used.

To further illustrate the diversity of their tasks, some typical pursuits of clinical engineers are provided in the following:

- Supervision of a hospital clinical engineering department that includes clinical engineers and biomedical equipment technicians (BMETs)
- Pre-purchase evaluation and planning for new medical technology
- Design, modification, or repair of sophisticated medical instruments or systems
- Cost-effective management of a medical equipment calibration and repair service
- Safety and performance testing of medical equipment by BMETs
- Inspection of all incoming equipment (new and returning repairs)
- Establishment of performance benchmarks for all equipment
- Medical equipment inventory control
- Coordination of outside services and vendors
- Training of medical personnel in the safe and effective use of medical devices and systems
- Clinical applications engineering, such as custom modification of medical devices for clinical research or evaluation of new noninvasive monitoring systems
- Biomedical computer support
- Input to the design of clinical facilities where medical technology is used (e.g., operating rooms (ORs) or intensive-care units).
- Development and implementation of documentation protocols required by external accreditation and licensing agencies

Clinical engineers thereby provide extensive engineering services for the clinical staff and, in recent years, physicians, nurses, and other clinical professionals have increasingly accepted them as valuable team members. The acceptance of clinical engineers in the hospital setting has led to different types of engineering–medicine interactions, which in turn have improved health care delivery. Furthermore, clinical engineers serve as a significant resource for the entire hospital. Since they possess in-depth knowledge regarding available in-house technological capabilities and the technical resources available from outside firms, the hospital is able to make more effective and efficient use of all of its technological resources.

The role of clinical engineering within the hospital organization

Over the years, management organization within hospitals has evolved into a diffuse authority structure that is commonly referred to as the "triad model." The three primary components are the governing board (trustees), hospital administration (CEO and administrative staff), and the medical staff organization.

Clinical engineering program

In many hospitals, administrators have established clinical engineering departments to manage effectively all the technological resources, especially those relating to medical equipment, that are necessary for providing patient care. The primary objective of these departments is to provide a broad-based engineering program that addresses all aspects of medical instrumentation and systems support.

Figure 1.0-2 illustrates the organizational chart of the medical support services division of a typical major medical facility. Note that within this organizational structure, the director of clinical engineering reports directly to the vice president of medical support services. This administrative relationship is extremely important because it recognizes the important role that clinical engineering departments play in delivering quality care. It should be noted, however, that in other organizational structures, clinical engineering services fall under the category of "facilities," "materials management," or even simply "support services."

In practice, there is an alternative capacity in which clinical engineers can function. They can work directly with clinical departments, thereby bypassing much of the hospital hierarchy. In this situation, clinical departments can offer the clinical engineer both the chance for intense specialization and, at the same time, the opportunity to develop a personal relationship with specific clinicians based on mutual concerns and interests (Wald, 1989).

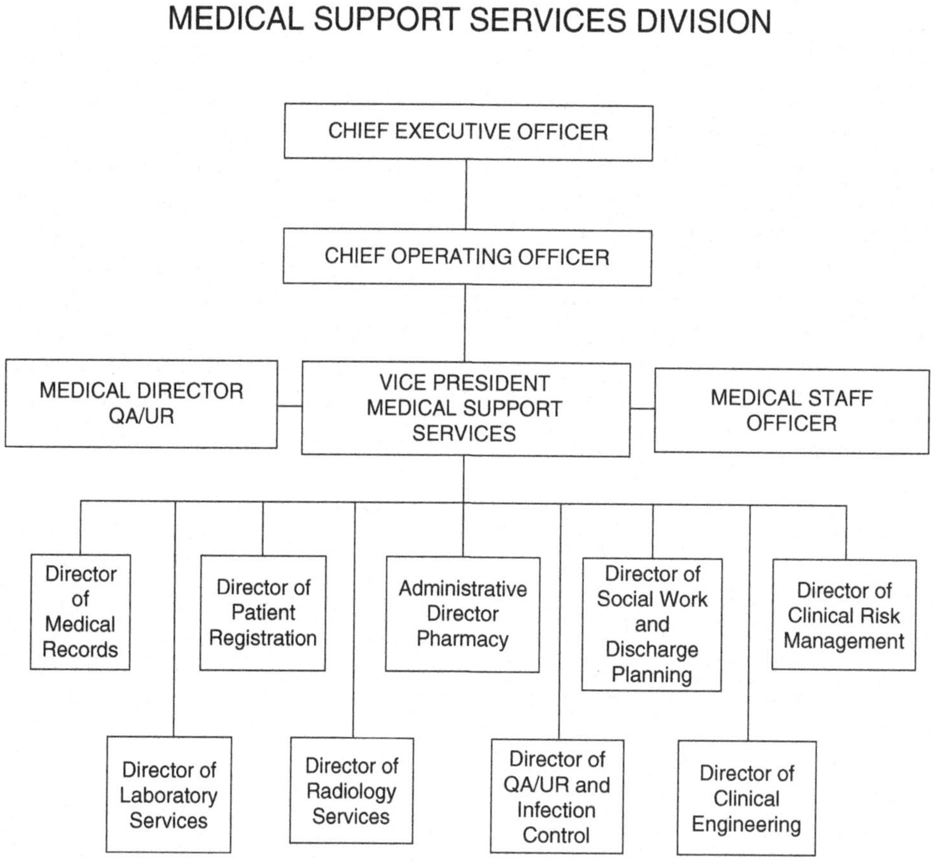

Figure 1.0-2 Organizational chart of medical support services division for a typical, major medical facility. This organizational structure points out the critical interrelationship between the clinical engineering department and the other primary services provided by the medical facility.

What is important today is the presence of clinical engineering at the appropriate point in the organizational structure for it to have a maximum impact on the proper use and management of modern medical technology (Bronzino, 1992; 1995; 2000).

Major functions of a clinical engineering department

The role of the clinical engineer in today's hospital can be both challenging and gratifying because the care of patients requires a greater partnership between medical staff and modem technology. As previously discussed, this interchange has led to a close working relationship between the clinical engineer and many members of the medical and hospital staff. The team approach is key to the successful operation of any clinical engineering program. Figure 1.0-3 illustrates the degree of teamwork and interdependence that is required in order to maintain constructive interrelationships. In this matrix presentation, it is important to note that the health care team approach to the delivery of patient care creates both vertical and lateral reporting relationships. Although clinical engineers report hierarchically to their hospital administrator, they also interact with hospital staff to meet patients' requirements.

As a result of the wide-ranging scope of interrelationships within the medical setting, the duties and responsibilities of clinical engineering directors continue to be extremely diversified. Yet, a common thread is provided by the very nature of the technology that they manage. Directors of clinical engineering departments are usually involved in the following areas:

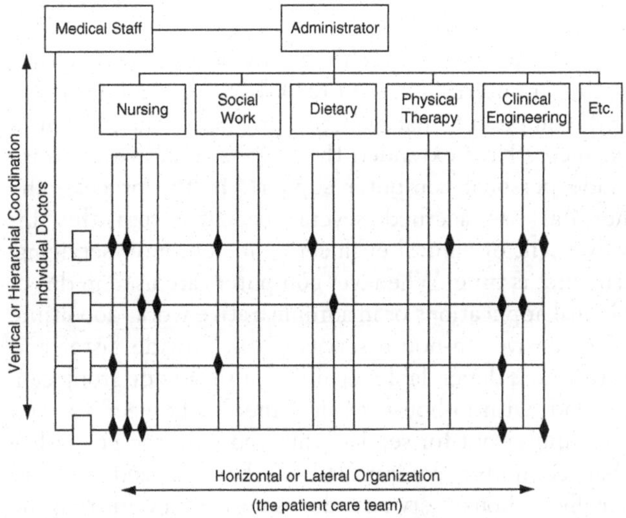

Figure 1.0-3 Matrix diagram illustrating the bi-directional interdependence and degree of teamwork required to maintain effective interaction between the members of the health care delivery team.

- Developing, implementing, and directing equipment management programs. Specific tasks include evaluating and selecting new technology, accepting and installing new equipment, and managing the inventory of medical instrumentation, all in keeping with the responsibilities and duties defined by the hospital administration. The clinical engineering director advises the administrator of the budgetary, personnel, space, and test equipment requirements that are necessary to support this equipment management program.
- Advising administration and medical and nursing staffs in areas such as safety, the purchase of new medical instrumentation and equipment, and the design of new clinical facilities
- Evaluating and taking appropriate action on incidents attributed to equipment malfunction or misuse. The clinical engineering director summarizes the technological significance of each incident and documents the findings of the investigation, subsequently submitting a report to the appropriate hospital authority and, according to the Safe Medical Devices Act of 1990, to the device manufacturer and/or the Food and Drug Administration (FDA).
- Selecting departmental staff and training them to perform their functions in a professional manner
- Establishing departmental priorities, developing and enforcing departmental policies and procedures, and supervising and directing departmental activities. The clinical engineering director takes an active role in leading the department to achieve its overall technical goals.

Therefore, the core functions of a clinical engineering department can be summarized as follows:

1. Technology management
2. Risk management
3. Technology assessment
4. Facilities design and project management
5. Quality assurance
6. Training

Professional status of clinical engineering

Upon careful review of our definition of clinical engineering and the responsibilities and functions that clinical engineers assume within the hospital, it is clear that the term *clinical engineer* must be associated with individuals who can provide engineering services, not simply technical services. Clinical engineers, therefore, must be individuals who have a minimum of a four-year bachelor's degree in an engineering discipline. They must be well versed in the design, modification, and testing of medical instrumentation—skills that fall predominantly in the field of engineering practice. Only with an engineering background can clinical engineers assume their proper role working with other health professionals to use available technological resources effectively and to improve health care delivery.

By clearly linking clinical engineering to the engineering profession, a number of important objectives are achieved. First, it enables hospital administrators to identify qualified individuals to serve as clinical engineers within their institutions and to understand better the wide range of functions that clinical engineers can perform, while making it clear that technicians cannot assume this role. Second, from this foundation, it is possible for the profession of clinical engineering to continue to mature. It has been pointed out that professional activities exist if "a cluster of roles in which the incumbents perform certain functions valued in the society in general" can be identified (Parsons, 1954; Courter, 1980; Goodman, 1989). Clearly, this goal has been achieved for the clinical engineer.

One can determine the status of professionalization by noting the occurrence of six crucial events: (1) the first training school; (2) the first university school; (3) the first local professional association; (4) the first national professional association; (5) the first state license law; and (6) the first formal code of ethics (Wilensky, 1964; Goodman, 1989).

Now let us consider the present status of professionalism of clinical engineering. Consider the following: (1) there is continued discussion about the educational needs of existing, as well as beginning, professionals; (2) there is a professional society, the American College of Clinical Engineering, which is effective in establishing the knowledge base on which the profession is to develop; (3) there exists a credentials process that reflects the needs of this new profession; and (4) there is a code of ethics, which the American College of Clinical Engineering has developed for use by all in the profession.

This process toward professionalization will certainly continue in the years ahead as this new professional society continues to seek to define and control certification activities, to define the educational process that is required for these new professionals, and to promote the status of clinical engineering to hospital administrators and society as a whole.

Future of clinical engineering

From its early days—when electrical safety testing and basic preventive maintenance were the primary concerns—to the present, the practice of clinical engineering has changed enormously. Yet, it is appropriate to use the cliché, "The more things change, the more they stay the same." Today, hospital-based clinical engineers still have the following as their primary concerns: patient safety and good hospital equipment management. However, these basic concerns are being supplemented by new areas of responsibility, making the clinical engineer not only the chief technology officer but also an integral part of the hospital management team.

In large part, these demands are due to the economic pressures that hospitals face. State-of-the-art, highly complex instruments, such as MRI systems, surgical lasers, and other sophisticated devices, are now used as a matter of course in patient care. Because of the high cost and complexity of such instrumentation, the institution needs to plan carefully—at both a technical and a managerial level—for the assessment, acquisition, and use of this new technology.

With these needs in mind, hospital administrators have begun to turn to their clinical engineering staffs for assistance in operational areas. Clinical engineers now provide assistance in the application and management of many other technologies that support patient care (e.g., computer support, telecommunications, facilities operations, and strategic planning).

Computer support

The use of personal computers (PCs) has grown enormously in the past decade. PCs are now commonplace in every facet of hospital operations, including data analysis for research, use as a teaching tool, and many administrative tasks. PCs are also increasingly used as integral parts of local area networks (LANs) and hospital information systems.

Because of their technical training and experience with computerized patient record systems and inventory and equipment management programs, many clinical engineers have extended their scope of activities to include personal computer support. In the process, the hospital has accrued several benefits from this involvement of clinical engineering in computer servicing. The first is time: Whenever computers are used in direct clinical applications or in administrative work, downtime is expensive. In-house servicing can provide faster and often more dependable repairs than an outside group can. Second, with in-house service, there is no need to send a computer out for service, thus reducing the possibility that computer equipment will be damaged or lost. Finally, in-house service reduces costs by permitting the hospital to avoid expensive service contracts for computers and peripheral equipment.

With all of these benefits, it might seem that every clinical engineering department should carry out computer servicing. However, the picture is not so simple. At the most basic level, the clinical engineering program must be sure that it has the staff, money, and space to do the job well. To assist in making this determination, several questions should be asked: Will computer repair take too much time away from the department's primary goal of patient care instrumentation? Is there enough money and space to stock needed parts, replacement boards, diagnostic software, and peripheral devices? For those hospitals that do commit the resources needed to support computer repair, clinical engineers have found that their departments can provide these services very efficiently, and they subsequently receive added recognition and visibility within the hospital.

Telecommunications

Another area of increased clinical engineering involvement is hospital-based telecommunications. In the modern health care institution, telecommunications covers many important activities, the most visible of which is telephone service. Broadly speaking, however, telecommunications today includes many other capabilities.

Until the 1970s, telephone systems were basically electromechanical, using switches, relays, and other analog circuitry. During the 1970s, digital equipment began to appear. This development allowed the introduction of innovations such as touch tone dialing, call forwarding, conference calling, improved call transferring, and other advanced services. The breakup of the Bell System also changed the telecommunications field enormously by opening it to competition and diversification of services.

Data transmission capability allows the hospital to send scans and reports to physicians at their offices or at other remote locations. Data, such as patient ECGs, can be transmitted from the hospital to a data analysis system at another location, and the results can be transmitted back. Hospitals are also making increased use of facsimile (fax) transmission. This equipment allows documents, such as patient charts, to be sent via telephone line from a remote location and reconstructed at the receiving site in a matter of minutes.

Modern telecommunications equipment also allows the hospital to conduct educational conferences through microwave links that allow video transmission of a conference taking place at a separate site. Some newer equipment allows pictorial information, such as patient slides, to be digitally transmitted via a phone line and then electronically reassembled to produce a video image.

Clinical engineers can play an important role in helping the hospital administrators to develop plans for a continually evolving telecommunications system. They can provide technical support during the planning stage, can assist in the development of requests for proposals for a new system, and can help to resolve any physical plant issues associated with installing the new system. Thereafter, the clinical engineer can assist in reviewing responses to the hospital's requests for quotations and can lend a hand during installation of the system.

Facilities operations

Recently, some hospital administrators have begun to tap the high-level technical skills that are available within their clinical engineering departments to assist in other operational areas. One of these is facilities operations, which includes heating, ventilation, and air conditioning (HVAC), electrical supply and distribution (including isolated power), central gas supply and vacuum systems, and other physical plant equipment (Newhouse et al., 1989). This trend has occurred for a number of reasons. First, the modem physical plant contains microprocessor-driven control circuits, sophisticated circuitry, and other high-level technology. In many instances, facilities personnel lack the necessary training to understand the engineering theory underlying these systems. Thus, clinical engineers can effectively work in a consultation role to help correct malfunctions in the hospital's physical plant systems. Another reason is cost; the hospital might be able to avoid expensive service contracts by performing this work in-house.

Clinical engineers can assist in facilities operations in other ways. For example, they can lend assistance when compressed gas or vacuum delivery systems need to be upgraded or replaced. While the work is taking place, the clinical engineer can serve as the administration's technical arm, ensuring that the job is done properly, that it is in compliance with code, and that it is done with minimal disruption.

Strategic planning

Today's emphasis on health care cost control requires that clinical engineers assist in containing costs associated with the use of modern medical technology. To accomplish this objective, clinical engineers are increasingly becoming involved in strategic planning, technology assessment, and purchase review. During assessment and purchase review, the clinical engineer studies a request to buy a new system or device and ensures that the purchase request includes (1) needed accessories; (2) warranty information; and (3) user and service training. This review process ensures that the device is, in fact, needed (or whether there exists a less costly unit that meets the clinician's requirements) and that it will be properly integrated into the hospital's existing equipment and physical environment.

Clinical engineers can provide valuable assistance during the planning for, and financial analysis of, potential new services. If one considers the steps that are involved in planning for a new patient-care area, many design questions immediately come to mind: What is the best layout for the new area? What equipment will be used there? How many suction, air, oxygen, and electrical outlets will be needed? Is there a need to provide for special facilities, such as those that are required for hemodialysis treatment at the bedside? With their knowledge of instrumentation and user needs, clinical engineers can help select reliable, cost-effective equipment to help ensure that the hospital obtains the best possible plan.

In the future, clinical engineering departments will need to concentrate even more heavily on management issues by emphasizing the goals of increased productivity and reduced costs. By continually expanding their horizons, clinical engineers can be major players in ensuring high-quality patient care at a reasonable cost.

References

AHA. *Hospital Administration Terminology, ed* 2, American Hospital Publishing, Washington, DC, 1986.

Bauld TJ. The Definition of a Clinical Engineer. *J Clin Engin* 16:403–405, 1991.

Bronzino JD. *Technology for Patient Care.* St. Louis, C.V. Mosby, 1977.

Bronzino JD et al. A Regional Model for a Hospital-Based Clinical Engineering Internship Program. *J Clin Engin* 7:34–37, 1979.

Bronzino JD. Clinical Engineering Internships: A Regional Hospital-Based Approach. *J Clin Engin* 10:239, 1985.

Bronzino JD. *Biomedical Engineering and Instrumentation: Basic Concepts and Applications*. Boston, PWS Publishing Co, 1986.

Bronzino JD. Biomedical Engineering. In Trigg G (ed): *Encyclopedia of Applied Physics*. New York, VCH Publishers, 1991.

Bronzino JD, Hayes TP. Hospital-Based Clinical Engineering Programs. In *Handbook for Biomedical Engineering*. New York, Academic Press, 1988.

Bronzino JD, Smith V, Wade M. *Medical Technology and Society.* MIT Press, Cambridge, 1990.

Bronzino JD. *Management of Medical Technology: A Primer for Clinical Engineers*. Philadelphia, Butterworth-Heinemann, 1992.

Bronzino JD. Clinical Engineering: Evolution of a Discipline. In *Biomedical Engineering Handbook, ed 1, 2.* Boca Raton, FL, CRC Press, 1995, 2000.

Courter SS. The Professional Development Degree for Biomedical Engineers. *J Clin Engin* 5:299–302, 1980.

Goodman G. The Profession of Clinical Engineering *J Clin Engin* 14:27–37, 1989.

International Certification Commission's (ICC) Definition of a Clinical Engineer, International Certification Commission Fact Sheet, Arlington, VA, ICC, 1991.

Jacobs JE. The Biomedical Engineering Quandary. *IEEE Trans Biomed Engin* 22: 1106, 1975.

Newhouse V et al. The Future of Clinical Engineering in the 1990s. *J Clin Engin 1989* 14:417–430, 1989.

Pacela A. *Bioengineering Education Directory,* Brea, CA, Quest Publishing Co, 1990.

Pacela A. Career "Fact Sheets" for Clinical Engineering and Biomedical Technology. *J Clin Engin* 16:407–416, 1991.

Painter FR. *Clinical Engineering and Biomedical Equipment Technology Certification.* World Health Organization Meeting on Manpower Development and Training for Health Care Equipment, Management, Maintenance and Repair. WHO/SHS/HHP/90.4, 1989, pp 130–185.

Parsons T. *Essays in Sociological Theory, rev ed.* Glencoe, IL, Free Press, 1954.

Schaffer MJ, Schaffer MD. The Professionalization of Clinical Engineering. *Biomed Instr Technol* 23: 370–374, 1989.

Schaffer MJ, Schaffer MD. What Is a Clinical Engineer? Issues in Definition. *Biomed Instr Technol* 277–282, 1992.

Wald A. Clinical Engineering in Clinical Departments: A Different Point of View. *Biomed Instr Technol* 23:58–63, 1989.

Wilensky HL: The Professionalization of Everyone. *Am J Sociol* 69:137–158, 1964.

Section Two

Modeling biosystems

Chapter 2.1a

2.1a

Modeling biosystems

Stanley M. Dunn, Alkis Constantinides, and Prabhas Moghe

2.1a.1 Biomedical engineering

Biomedical Engineering (BME) is the branch of engineering that is concerned with solving problems in biology and medicine. This text is an introduction to *Numerical Methods* for biomedical engineers. Numerical methods are mathematical techniques for performing accurate, efficient and stable computation, by computer, to solve mathematical models of biomedical systems. Numerical methods are the tools engineers use to realize computer implementation of analytic models of system behavior.

Biomedical engineers use principles, methods, and approaches drawn from the more traditional branches of electrical, mechanical, chemical, materials, and computer engineering to solve this wide range of problems. These methods include: principles of Electrical Engineering, such as circuits and systems; imaging and image processing; instrumentation and measurements; and sensors. The principles from Mechanical Engineering include fluid and solid mechanics; heat transfer; robotics and automation; and thermodynamics. Principles from Chemical Engineering include transport phenomena; polymers and materials; biotechnology; drug design; and pharmaceutical manufacturing.

Biomedical engineers apply these and other principles to problems in the life sciences and healthcare fields. That means the biomedical engineer must also be familiar with biological concepts of anatomy and physiology at the system, cellular, and molecular levels. Working in healthcare requires familiarity with the cardiovascular system, the nervous system, respiration, circulation, kidneys, and body fluids.

Other terms, such as bioengineering, clinical engineering, and tissue engineering, are used to identify subgroups of biomedical engineers. Bioengineering refers to biomedical engineers who focus on problems in biology and the relationship between biological and physiological systems. Clinical engineering is a term used to refer to biomedical engineers who solve problems related to the clinical aspects of healthcare delivery systems and patient care. Tissue engineering is the subspecialty in which engineering is used to design and create tissues and devices to replace structures with lost or impaired function. Tissue engineers use a combination of cells, engineering materials, and suitable biochemical factors to improve or replace biological functions in an effort to effect the advancement of medicine.

Although the most visible contributions of BME to clinical practice involve instrumentation for diagnosis, therapy, and rehabilitation, there are examples in this text drawn from both the biological and the medical arenas to show the wide variety of problems that biomedical engineers can and do work on.

The field of BME is rapidly expanding. Biomedical engineers will play a major role in research in the life sciences and development of devices for efficient delivery of healthcare. The scope of BME ranges from bionanotechnology to assistive devices, from molecular and cellular engineering to surgical robotics, and from neuromuscular systems to artificial lungs. The principles presented in this text will help prepare biomedical engineers to work in this diverse field.

There are a number of good histories of BME each has its own beginning date of the field and the significant milestones. The beginning of BME can be traced to either the 17th, 18th, or 19th century, the choice depending on the definition used for BME. The reader is referred to Nebeker (2002) for a comprehensive history of how biomedical engineers build diagnostic (data that characterizes the system) or therapeutic (replace or enhance

Biomedical Engineering Desk Reference; ISBN: 9780123746467

lost function) devices as solutions to problems in healthcare or the life sciences.

2.1a.2 Fundamental aspects of BME

Any BME device includes one or more *measurement*, *modeling*, or *manipulation* tasks. By measurement, we mean sensing properties of the physical, chemical, or biological system under consideration. Realizing that no property can be measured exactly is an important concept for biomedical engineers to understand. For this reason, an appreciation of statistics is required and will be covered later in this text. Principles of measurement also include an understanding of variability and sources of noise; sensing instruments and accuracy; and resolution and reproducibility as characterizations of measurements.

Manipulation in this engineering sense means interacting with a system in some way. For the most part, biomedical engineers will interact with the human body or biological system by constructing diagnostic or therapeutic interventions. The process of developing an intervention system starts with requirements and constraints, which, in turn, lead to specifications for the particular intervention and then to design, fabrication, and testing.

The task of engineering modeling is the process by which a biomedical engineer expresses the principles of physics, chemistry, and biology in a mathematical statement that describes the phenomena or system under consideration. The mathematical model is a precise statement of how the system interacts with its environment. *A model is a tool that allows the engineer to predict how the system will react to changes in one of the system parameters.*

Engineering models are mathematical statements using one or more of four areas of continuous mathematics: algebra, calculus, differential equations and statistics. Other than the pure modeling tasks, any of the other significant BME projects highlighted above use models, signal processing, or control systems that are implemented in a computer system. In these cases, one must describe the behavior algebraically, as an integral, differential equation or as an expression of the variability in analogs of these continuous mathematical models; the challenge is to preserve the accuracy and resolution to the highest degree possible and perform stable computation. This is the purpose of numerical methods.

2.1a.3 Constructing engineering models

Practicing engineers are asked to solve problems based on physical relationships between something of interest and the surrounding environment. Because there is quite a wide range of phenomena that will occur in biosystems, it is important to have a common language that can be used to describe and model bioelectric, biomechanical, and biochemical phenomena.

2.1a.3.1 A framework for problem solving

The problem-solving framework is based on first identifying the conservation law that governs the observed behavior. There are four steps to developing the solution to a problem in BME

1. **Identify the system to be analyzed:** A *system* is any region in space or quantity of matter set aside for analysis; it's the part of the universe in which we are interested. The *environment* is everything not inside the system. The system *boundary* is an infinitesimally thin surface, real or imagined, that separates the system from its environment. The boundary has no mass, and merely serves as a delineator of the extent of the system.
2. **Determine the extensive property to be accounted for:** An *extensive property* does not have a value at a point, and its value depends on the extent or size of the system; e.g., it is proportional to the mass of the system. The amount of an extensive property for a system can be determined by summing the amount of extensive property for each subsystem that comprises the system. The value of an extensive property for a system only depends upon time. Some examples that we are familiar with are mass and volume.

   There is scientific evidence that suggests that the property can neither be created nor destroyed. An extensive property that satisfies this requirement is called a *conserved property.*

   There are many experiments reported in the literature that support the idea that charge, linear momentum and angular momentum are conserved. Mass and energy, on the other hand, are conserved under some restrictions:

   a. If moving, the speed of the system is significantly less than the speed of light.
   b. The time interval is long when compared to the time intervals of quantum mechanics.
   c. There are no nuclear reactions.

It is very unlikely that conditions a and b will be violated by any biomedical system that will be studied. In the time scale of biology, the shortest event is on the order of 10^{-8}s which is still longer than nuclear events. However, emission tomography and nuclear medicine are systems where nuclear reactions do occur and one will have to be careful about assuming that mass or energy is conserved in these systems.

Conservation of mass, charge, energy, and momentum can be very useful in developing mathematical models for analysis of engineering artifacts. In addition to conserved properties, there are other extensive properties for which we know limits on the generation/consumption terms. The classic example of this is the *Second Law of Thermodynamics* and its associated property, *entropy*. When written as a conservation equation, it is easy to see that entropy can only be produced within a system. Furthermore, in the limit of an internally reversible process, the entropy production rate reduces to zero.

3. **Determine the time period to be analyzed:** When a system undergoes a change in state, we say that the system has undergone a *process*. It is frequently the goal of engineering analysis to predict the behavior of a system, i.e., the path of states that result, when it undergoes a specified process. Processes can be classified in three ways based on the time interval involved: steady-state, finite-time, and transient processes.

4. **Formulate a mathematical expression of the conservation law:** Experience has taught us that the amount of an extensive property within a system may change with time. This change can only occur by two mechanisms:
   - Transport of the extensive property across the system boundary
   - Generation (production) or consumption (destruction) of the extensive property inside the system

   Thus, the change of an extensive property within a system can be related to the amount of the extensive property transported across the boundary and the amount of the extensive property generated (and/or consumed) within the system. This "accounting principle" for an extensive property is known to engineers as a *balance equation*. Although this principle can be applied to a system for any extensive property, it will be especially useful for properties that are conserved. There are two forms of a balance equation: the accumulation form and the rate form.

2.1a.3.2 Formulating the mathematical expression of conservation

The accumulation form

In the accumulation form, the time period used in the analysis is finite and is therefore used to formulate equations that model steady-state or finite-time processes. When accounting for the input and output, the total amount that enters the system (during the time period) is computed and the total amount that exits in the same time is subtracted. The accounting statement is:

$$\left\langle \begin{array}{c}\text{Net amount}\\ \text{accumulated}\\ \text{inside the system}\end{array}\right\rangle = \left\langle \begin{array}{c}\text{Net amount}\\ \text{transported}\\ \text{into the system}\end{array}\right\rangle + \left\langle \begin{array}{c}\text{Net amount}\\ \text{generated}\\ \text{inside the system}\end{array}\right\rangle$$

or, for a conserved property P:

$$P^{final}_{inside} - P^{initial}_{inside} = (P_i - P_o) + (P_G - P_C)$$

where the left-hand side is the net amount accumulated inside; the first difference term on the right-hand side is the net amount transported into the system (input – output); and the second difference term on the right is the net amount of the property generated (generated – consumed).

The advantage of using an accumulation form of the conservation or accounting laws is that the mathematical expression is in the form of either algebraic or integral equations. The disadvantage of the accumulation form of the law is that it is not always possible to determine the amount of the property of interest entering or exiting from the system.

The rate form

The rate form of a balance equation is similar to the accumulation form, except that the time period is infinitesimally small, so in the limit the net amounts become rates of change. The mathematical relationship between the accumulation form and the rate form can easily be developed by dividing the accumulation form through by the time interval Δt and taking the limit as $\Delta t \to 0$. That is:

$$\left\langle \begin{array}{c}\text{Rate of change}\\ \text{inside the system}\\ \text{at time } t\end{array}\right\rangle = \left\langle \begin{array}{c}\text{Transport rate}\\ \text{into the system}\\ \text{at time } t\end{array}\right\rangle + \left\langle \begin{array}{c}\text{Generation rate}\\ \text{into the system}\\ \text{at time } t\end{array}\right\rangle$$

or, for a conserved property P:

$$\frac{dP}{dt} = (\dot{P}_i - \dot{P}_o) + (\dot{P}_G - \dot{P}_C)$$

where the first term on the right-hand side is the difference in transport rates across the system boundary, and the second term is the difference in the rates at which the property is generated inside the system. The sum of these two terms is the rate of change of the property, inside the system, with respect to time. The advantage of the rate form of the law is that the laws of physics generally make it easy to find the rates at which things are happening. The disadvantage of the rate form is that it generates differential equations.

Although the accounting principle can be applied for any extensive property, it is most useful when the transport and generation/consumption terms have physical significance. The most useful applications of this principle occur when something is known *a priori* about the generation/consumption term.

For conserved extensive properties the equations that apply the accounting principle are significantly simpler. In the accumulation form, the equations become:

$$P^{final}_{inside} - P^{initial}_{inside} = (P_i - P_o).$$

2.1a.3.3 Using balance equations

The mathematical formulation of problems that biomedical engineers solve is based on one or more of the conservation laws from chemistry and physics. The solution to a problem becomes obvious when one has the right formulation. The approach used herein is that the solution to a problem can be formulated using one or more of the forms of the following conservation laws.

The model formulation and problem-solving framework presented in Section 2.1a.3.1 shows that there are two forms for each of the conservation laws: the accumulation (or sum) form and the rate form. The former is used in solutions of steady-state or finite-time problems, whereas the latter is used in solving problems with transient behavior.

Example 2.1a.1 How conservation laws lead to the Nernst equation.

Show how Fick's law, Ohm's law and the Einstein relationship can be derived from the conservation laws in Section 2.1a.3.1. Show how these three conservation models lead to the Nernst equation. This problem is derived from section 3.4 of Enderle et al. (2000).

The rate form of the principle of conservation of mass is:

$$\frac{dm_{sys}}{dt} = \sum_{inlets} \dot{m}_i - \sum_{outlets} \dot{m}_e$$

meaning that the overall time rate of change of mass in the system is equal to the difference between the sum of the rates of change of mass into the system and the sum of the rates of change of mass out of the system.

For the derivation of the Nernst equation, the system being considered is a cell membrane surrounded by extracellular fluid, with ion flow across the membrane. Assume that one of K^+, Na^+, or Cl^- ions flow across the membrane, with the positive direction taken to be from the outside, across the membrane, to inside the cell. The rate form above leads to the following three relationships:

1. *Fick's Law* is a model that relates the flow of ions due to diffusion and the ion concentration gradient across the cell membrane. It is a law that expresses the influence of chemical force on the conservation of mass in the system. The left-hand side of the conservation law is the flow rate of mass, that is, the mass flux in the system, which is the flow due to diffusion. The right-hand side is the sum of the rates of change of mass through the system inlets and outlets, that is, the flow through the ion channels. The rate form of the principle of conservation of mass leads to:

$$J_{diffusion} = -D\frac{dI}{dx}$$

that is, the flow of ions due to diffusion is equal to the ion concentration gradient across the membrane scaled by the diffusivity constant D. That is, there is ion flow in response to an ion concentration gradient. The negative sign indicates that the ion flow is in the opposite direction of the gradient.

2. *Ohm's Law* is a model for the influence of electrical force on ion flow across the cell membrane. It is still based on conservation of mass, but the driving force on the right-hand side is mass flow due to an electric field induced by other charged particles. An electric field $\vec{E}$ is applied, creating a rate of change of potential, dv/dx. The right-hand side is the ion flow due to this potential, and leads to:

$$J_{drift} = -\mu Z[1]\frac{dv}{dx}$$

in which J_{drift} is the ion flux due to the electric field, μ is the mobility, Z is the ion valence, $[I]$ is the ion concentration, and v is the voltage.

3. *The Einstein relationship* is a form of conservation of momentum that expresses a relationship between diffusion and ion mobility. The electric field induces a force on the ions that causes a flow that is balanced by osmotic pressure. The conservation law,

$$\frac{dP_{sys}}{dt} = \sum F_{ext} + \sum_{inlets} \dot{P}_i - \sum_{outlets} \dot{P}_o$$

leads to the form:

$$D = \frac{KT\mu}{q}$$

in which K is Boltzmann's constant (1.38×10^{-23} J/K), T is absolute temperature in Kelvin, and q is the magnitude of the electric charge (1.61×10^{-19} C).

Both Fick's Law and Ohm's Law express conservation of mass and can be combined by the engineering principle of superposition, yielding an expression for total flow:

$$J = J_{diffusion} + J_{drift} = -D\frac{d[K^+]}{dx} - \mu Z[K^+]\frac{dv}{dx}$$

which, in the steady state, is 0 (that is, no net transport). The conservation of momentum principle, Einstein's relationship, relates the two rates on the right-hand side:

$$0 = J_{diffusion} + J_{drift} = -\frac{KT}{q}\mu\frac{d[K^+]}{dx} - \mu Z[K^+]\frac{dv}{dx}$$

for K^+, Z=1 and after integrating the potential across the cell boundary:

$$\int_{v_o}^{v_i} dv = -\frac{KT}{q}\int_{K_o^+}^{K_i^+}\frac{d[K^+]}{[K^+]}$$

which yields:

$$v_i - v_o = -\frac{KT}{q}\ln\frac{[K^+]_i}{[K^+]_o}$$

known as the Nernst equation, expressing the potential difference across a cell membrane as a function of chemical and electrical forces driving ion transport.

2.1a.4 Examples of solving BME models by computer

The materials presented in the previous section are the basic tools that a biomedical engineer will use to solve a problem or design a new diagnostic or therapeutic device. However, an analytic solution is not sufficient; the device will be designed and/or controlled using a computer. This means that one must know how to transform the model from continuous mathematics to discrete mathematics (numerical analysis) and then write a computer program to implement the new equation (numerical methods). This section shows some examples of how biomedical engineers use numerical analysis and numerical methods.

2.1a.4.1 Modeling rtPCR efficiency

PCR is an acronym that stands for polymerase chain reaction, a technique that will allow a short stretch of DNA (usually fewer than 3,000 base pairs) to be amplified about a million fold so that one can determine its size and nucleotide sequence.

The particular stretch of DNA to be amplified, called the *target sequence*, is identified by a specific pair of DNA primers, which are usually about 20 *nucleotides* in length. For convenience, these four nucleotides are called dNTPs.

There are three major steps in a PCR reaction, which are repeated for 30 or 40 cycles. This process is done on an automated cycler, which can heat and cool the tubes with the reaction mixture in a very short time. The three steps of a PCR reaction (illustrated in Fig. 2.1a-1) are:

1. **Denaturation** at 94°C: During the denaturation, the double strand melts open to single-stranded DNA and all enzymatic reactions stop.
2. **Annealing** at 54°C: Bonds are constantly formed and broken between the single-stranded primer and the single-stranded template. When primers fit exactly, the bonds are more stable and last a bit longer. On that piece of double-stranded DNA (template and primer), the polymerase can attach and starts copying the template. Once there are a few bases built in, the ionic bond is so strong between the template and the primer that it does not break.
3. **Extension** at 72°C: The primers, where there are a few bases built in, already have a strong attraction to the template. Primers that are on positions with no exact match get loose again and don't give an extension of the fragment. The bases complementary to the template are coupled to the primer on the 3′ side.

Because both strands are copied during PCR, there is an **exponential** increase of the number of copies of the gene.

Real-time PCR (rtPCR) is used to determine gene expression over and above size and sequence information. In rtPCR, it is assumed that the efficiency is constant; however, analysis of data shows that the efficiency is *not* constant, but is instead a function of cycle number (Gevertz et al., 2005). Based on this observation, it seems reasonable that rtPCR quantification techniques can be improved upon by understanding the behavior of rtPCR efficiency. A mathematical model of rtPCR will provide this understanding and will lead to more accurate methods to quantify gene expression levels from rtPCR data.

The mathematical model of annealing and extension is as follows: after the double-stranded DNA is denatured,

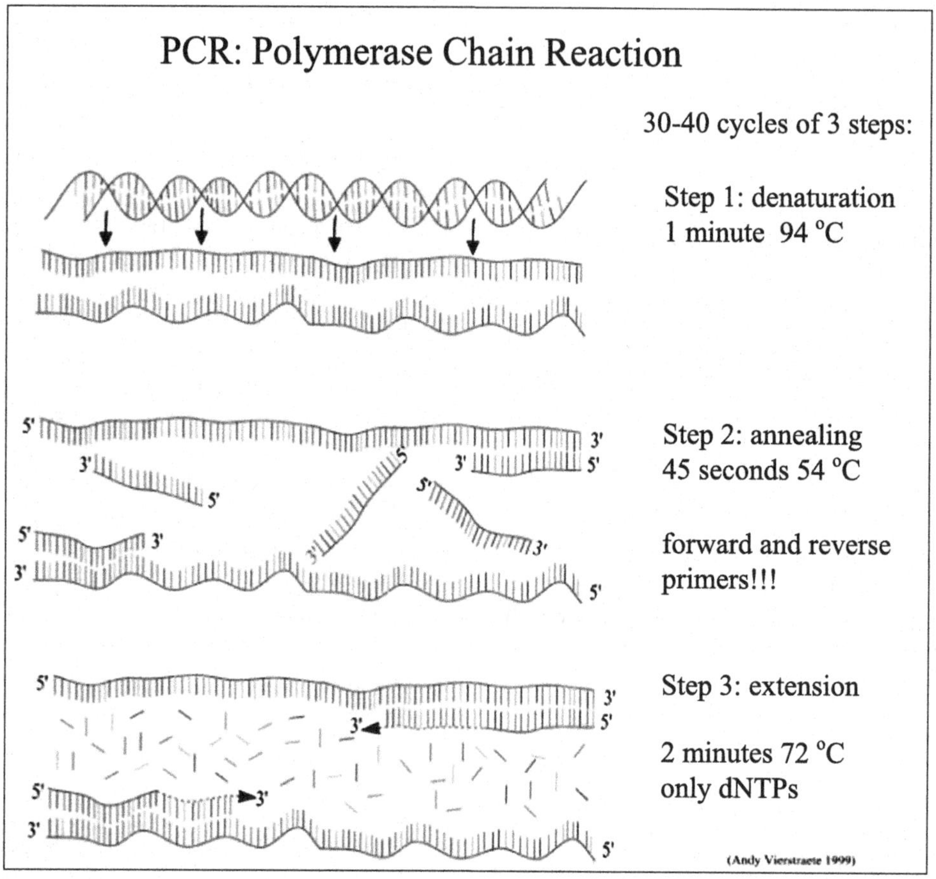

Figure 2.1a-1 The exponential amplification of the gene in PCR. From Vierstraete (1999).

there are two single strands of DNA, called T1 and T2, for template 1 and template 2, respectively. Once the temperature is cooled to 54°C (signifying the starting of the annealing stage), we expect that primer 1 (P1) will anneal with T1 to form a hybrid (H1) and we also expect that P2 and T2 will anneal to form a hybrid (H2). These two reactions can be represented by the chemical equations:

$$P1 + T1 \leftrightarrow H1 \tag{2.1a.1}$$

$$P2 + T2 \leftrightarrow H2 \tag{2.1a.2}$$

Unfortunately, other reactions can occur during the annealing stage of PCR. The template strands, T1 and T2, are complementary and can re-anneal upon contact, forming a template hybrid (HT) as represented by the following reaction:

$$T1 + T2 \leftrightarrow HT \tag{2.1a.3}$$

Finally, depending on primer design, primer-dimers (D) can form as well:

$$P1 + P2 \leftrightarrow D \tag{2.1a.4}$$

Hence, four reactions can occur simultaneously during the primer annealing stage. These reactions can be modeled using either thermodynamic (steady state) or kinetic (transient) equations.

The steady-state thermodynamic model tracks the total concentration of all products performing active roles during the annealing stage of PCR, a mass balance is performed on reactions (2.1a.1) to (2.1a.4). This procedure results in the following four equations:

$$\begin{aligned}
[P1]_T &= [P1] + [H1] + [D] \\
[P2]_T &= [P2] + [H2] + [D] \\
[T1]_T &= [T1] + [H1] + [HT] \\
[T2]_T &= [T2] + [H2] + [HT]
\end{aligned} \tag{2.1a.5}$$

where $[X]_T$ is a parameter denoting the total concentration of product X.

To complete the thermodynamic model, we need to introduce four more parameters into the system—the equilibrium constants of each reaction. We define K_{H1}, K_{H2}, K_{HT}, and K_D to be the equilibrium constants of (2.1a.1) to (2.1a.4), respectively. Since the equilibrium constant is, by definition, the fixed ratio of reactant to product concentrations, these allow us to introduce four more equations into the model:

$$K_{H1} = \frac{[P1]\cdot[T1]}{[H1]}$$
$$K_{H2} = \frac{[P2]\cdot[T2]}{[H2]}$$
$$K_{HT} = \frac{[T1]\cdot[T2]}{[HT]} \qquad (2.1a.6)$$
$$K_D = \frac{[P1]\cdot[P2]}{[D]}$$

The efficiency of the annealing stage, $\varepsilon_{ann}(n)$, can be calculated by comparing the amount of hybrids after the nth annealing stage to the total amount of template present throughout the nth annealing stage:

$$\varepsilon_{ann}(n) = 0.5\left(\frac{[H1]}{[T1]_T} + \frac{[H2]}{[T2]_T}\right) \qquad (2.1a.7)$$

in which [H1] and [H2] can be found by solving the nonlinear system of eight equations with eight unknowns in terms of the eight free parameters.

This deceivingly simple system of eight equations can be solved analytically either by hand or using a computer program for solving the system of equations symbolically.

2.1a.4.2 Modeling transcranial magnetic stimulation

Transcranial magnetic stimulation (TMS), the stimulation of cortical tissue by magnetic induction, is potentially a new diagnostic and therapeutic tool in clinical neurophysiology. The magnetic fields are delivered to the tissue by placing coils on the surface of the skull; TMS shows the promise of being useful for brain mapping and appears to show potential for treating brain disorders (Hallett 2000). The advantage of techniques like TMS and EEG (electroencephalogram) (see Chapter 2.2) is high temporal resolution; a disadvantage of TMS is that the 3D spatial resolution and depth penetration is not very good. Poor spatial resolution is due to the fact that the magnetic fields cannot easily be focused at a particular point in the brain.

Norton (2003) proposed a different method of stimulating cortical tissue, which potentially may allow deeper penetration and better focusing in cortical tissue. His idea is to create an electrical current by propagating an ultrasonic wave in the presence of a strong DC magnetic field. His results (Norton 2003) show that the amplitude of the magnetic field generated in this fashion is less than that of a traditional TMS approach, but there are some distinct advantages in both spatial and temporal characteristics of the induced field.

As one can imagine, numerical methods play a large role in calculating the magnitude of an electric field throughout the neural tissue in the brain. Briefly, let the cortical tissue be represented in a cylindrical 3D coordinate system (r, ϕ, z) and assume that the ultrasonic beam is collimated and propagating in the z direction. Furthermore, assume that the profile is axially symmetric and will be represented by $p(r)$. Norton (2003) modeled the particle velocity by:

$$v(\mathbf{r}) = v_0 p(r) e^{ik_0 z}\hat{z} \qquad (2.1a.8)$$

where v_0 is the peak velocity of the particles, $\hat{z}$ is the unit vector in the z direction and the wave number $k_0 = \omega/c_0$, i.e., the frequency of the ultrasonic wave divided by the speed of the ultrasonic wave.

The components of the magnetic field $\mathbf{E}_s$ induced in the brain are given by:

$$E_r(r,\phi,z) = \mathbf{B}_0 v_0\left[\frac{d^2A(r)}{dr^2}\right]e^{ik_0 z}\sin\phi$$
$$E_\phi(r,\phi,z) = \mathbf{B}_0 v_0\left[\frac{1}{r}\frac{dA(r)}{dr}\right]e^{ik_0 z}\cos\phi \qquad (2.1a.9)$$
$$E_z(r,\phi,z) = \mathbf{B}_0 v_0\left[ik_0\frac{dA(r)}{dr}\right]e^{ik_0 z}\sin\phi$$

where

$$A(r) = K_0(k_0 r)\int_r^0 I_0(k_0 r')p(r')r'dr' + I_0(k_0 r)\int_r^\infty K_0(k_0 r')p(r')r'dr' \qquad (2.1a.10)$$

and both $I_0(\cdot)$ and $K_0(\cdot)$ are functions that model the ultrasonic wave propagation.

This mathematical model is used to predict the distribution of the induced electric field in the brain when generated by an ultrasonic wave. In order for this set of equations to be used, they have to be solved numerically. But, in order to solve the three equations, one has to solve Eq. (2.1a.10) for $A(r)$. Except for a very ideal case the integrals in Eq. (2.1a.10) and the differential equations in Eq. (2.1a.9) have to be evaluated using the numerical integration techniques and the differential equation solvers.

2.1a.4.3 Modeling cardiac electrophysiology

Cardiac arrythmias are a leading cause of death in the United States and elsewhere. Computer simulations are rapidly becoming a powerful tool for modeling the factors that cause these life-threatening conditions (see Chapter 2.2). High accuracy simulations require fine spatial sampling and time-step sizes at or below a microsecond. To further complicate matters, there are many factors such as heat transport, fluid flow and electrical activity to model. The heart does not have a simple geometry and is not composed of one type of tissue. All of these factors mean that a complete simulation will require fast processors and lots of memory.

Chapter 2.2 shows examples of modeling fluid flow and heat transport in the heart. Pormann et al. (2000) developed a simulation system for the flow of electrical current. This simulation is based on a set of partial differential equations called the *Bidomain Equations*, which are a widely used model of cardiac electrophysiology:

$$\begin{aligned}\nabla\cdot\sigma_i\nabla\Phi_i &= \beta C_m\frac{\partial V_m}{\partial t}+I_{ion}(V_m,q)\\ \nabla\cdot\sigma_e\nabla\Phi_e &= -\beta C_m\frac{\partial V_m}{\partial t}-I_{ion}(V_m,q)\\ \frac{dq}{dt} &= M(V_m q)\end{aligned} \quad (2.1a.11)$$

where Φ_i and Φ_e are the intra- and extra-cellular potentials respectively, V_m is the transmembrane potential ($V_m = \Phi_i - \Phi_e$), and q is a set of state variables which define the physiological state of the cellular structures. I_{ion} and M are functions that approximate the cellular membrane dynamics, C_m is a transmembrane capacitance and σ_i and σ_e are conductivities.

This system of equations can be used to model 1-D nerve fibers, 2-D sheets of tissue, or 3-D geometries of the heart. The conductivities σ_i may be inhomogeneous (to model dead or diseased tissue). Different I_{ion} and M functions simulate nerve, atrial, or ventricular dynamics. The model parameters can be varied spatially to simulate diseased tissue or to study the effects of a channel-blocking drug on electrical conductivity.

The Bidomain Equations are solved for the potential V_m at each point in the 1-D, 2-D, or 3-D space, depending on the problem to be solved. In Pormann et al. (2000), the user has a choice of ten numerical integration methods to solve this set of partial differential equations. Some of these methods are: explicit, semi- and fully implicit time integrators, adaptive time steppers and Runge–Kutta methods.

2.1a.4.4 Using numerical methods to model the response of the cardiovascular system to gravity

Since the beginning of the space program, understanding the response of the cardiovascular system to returning to a normal gravitational environment has been a problem. Astronauts returning to earth may experience *post-spaceflight orthostatic intolerance* (OI) – the inability to stand after returning to normal gravity. OI is an active area of modeling research, including: explaining observations seen during space flight, simulating the cardiovascular response from experiments in earth gravity and modeling interventions to the cardiovascular problems caused by return to gravity. A review of a number of models for OI is given by Melchior et al. (1992).

Modeling the response to gravity can best be illustrated with the work of Heldt et al. (2002), where a single cardiovascular model is used to simulate the steady-state and transient response to ground-based tests. The authors compared their modeling results with population-averaged hemodynamic data and found that their predicted results compared well with subject data. Their model provides a framework with which to interpret experimental observations and to study competing physiological hypotheses of the cause of OI.

The hemodynamics are modeled in terms of an electrical network; Fig 2.1a-2 shows the model for a single compartment. Assuming that the devices behave linearly, the model is a set of first-order differential equations. Although the equations are in terms of electrical units, the assumption of linearity allows one to use the model for hemodynamics. The flow rates, q, across the resistors, R, and capacitor, C, expressed in terms of the pressures, P, are given by:

$$\begin{aligned}q_1 &= (P_{n-1}-P_n)/R_n\\ q_2 &= (P_n-P_{n+1})/R_{n+1}\\ q_3 &= \frac{d}{dt}[C_n(P_n-P_{bias})]\end{aligned} \quad (2.1a.12)$$

The subscripts are defined in Fig. 2.1a-2.

Applying conservation of charge to the node at P_n yields $q_1 = q_2 + q_3$. The rate form of the conservation law yields:

$$\begin{aligned}\frac{dP_n}{dt} &= \frac{P_{n+1}-P_n}{C_nP_{n+1}}+\frac{P_{n-1}-P_n}{C_nR_n}+\frac{P_{bias}-P_n}{C_n}\cdot\left(\frac{dC_n}{dt}\right)\\ &\quad+\frac{dP_{bias}}{dt}\end{aligned} \quad (2.1a.13)$$

The entire compartmental model is shown in Fig. 2.1a-3. The peripheral circulation is divided into upper body, renal, splanchnic, and lower extremity sections; the intrathoracic superior and inferior vena cavae and extrathoracic vena

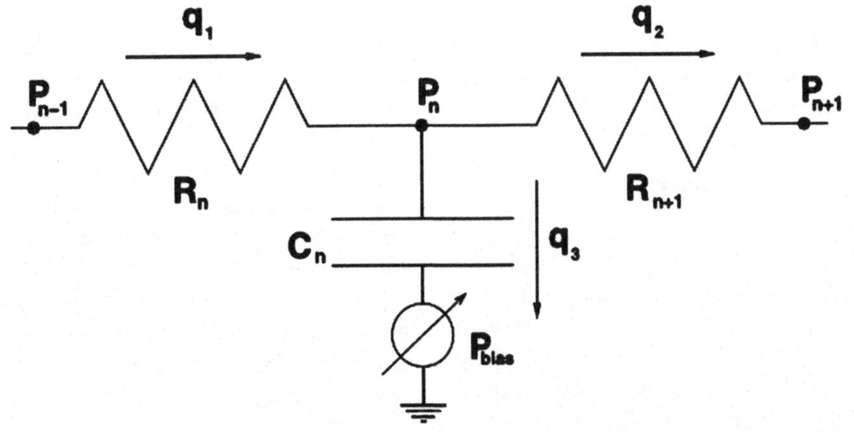

Figure 2.1a-2 Single-compartment circuit representation. *P* is pressure; *R* is resistance; *C* is vascular compliance; q_1, q_2, and q_3 are blood flow rates; *n*−1, *n*, *n*+1 are compartment indexes; P_{bias}, external pressure. From Heldt et al. (2002).

cava are separately identified. The model thus consists of 10 compartments, each of which is represented by a linear resistance, *R* and a capacitance, C. There are many similarities with the PHYSBE system in Chapter 2.2, but the reader will find PHYSBE to be a more detailed model of the cardiovascular system.

The model of Heldt et al. is described by 12 differential equations similar to Eq. (2.1a.13). The authors used an adaptive step-size fourth-order Runge–Kutta integration routine to solve the system of differential equations numerically. The results reported by Heldt et al. use integration steps that range from 6.1×10^{-4} to 0.01 s with

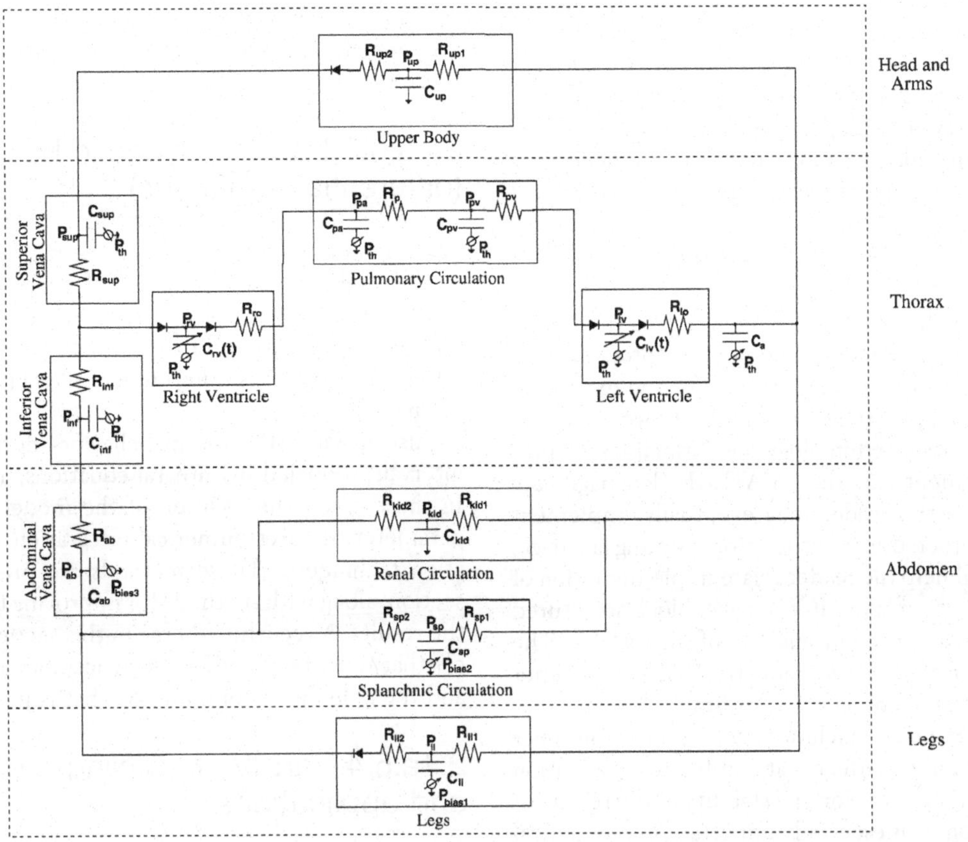

Figure 2.1a-3 Circuit diagram of the hemodynamic system. lv, left ventricle; a, arterial; up, upper body; kid, kidney; sp, splanchnic; ll, lower limbs; ab, abdominal vena cava; inf, inferior vena cava; sup, superior vena cava; rv, right ventricle; p, pulmonary; pa, pulmonary artery; pv, pulmonary vein; ro, right ventricular outflow; lo, left ventricular outflow; th, thoracic; bias, as defined in Fig. 2.1a-2. From Heldt et al. (2002).

a mean step size of 5.6×10^{-3} s. The initial pressures, required to start the solution, were computed by a linear algebraic solution of a hemodynamic system where all pressures are assumed constant.

2.1a.5 Overview of the text

The material presented in this text shows how to apply the principles and techniques of computing and numerical problem solving in a wide variety of problems that arise in BME. The aim here is to provide the student with a *working knowledge* of Numerical Methods, i.e., to be able to read, understand and use Numerical Methods to BME problems. A working knowledge has as its emphasis the understanding and application of the fundamentals. This approach provides the reader with exposure to a broad range of principles and techniques, but not theoretically rigorous derivations of the methods; the mathematical foundations are more appropriate for courses in Numerical Methods in a mathematics or computer science curriculum.

Our second aim is to give the reader examples of how to construct engineering models of biomedical systems. The *conservation laws' first* theme from this chapter is reinforced in the models presented in the examples. Thus, the text is organized into four sections around the physical principles: fundamentals, using models of steady-state behavior, using models of finite time behavior and using models of transient behavior. Throughout this text, examples from a variety of BME specialties, including biomedical instrumentation, imaging, bioinformatics, biomechanics, and biomaterials, are used to reinforce the concepts.

2.1a.5.1 Part I: Fundamentals

The first part of the text is an introduction to the fundamental principles of numerical methods. As programming is a necessary part of numerical methods, the examples, problems and applications in the text are given in MATLAB and basic terminology and principles of program development in the MATLAB language are reviewed. Since the reader may eventually implement a numerical method using another programming language, this section will help the reader relate implementation of common concepts of computer science: block structured design, data structures and analysis of algorithms. The emphasis is on design and the tradeoffs that a programmer makes when implementing an algorithm.

The introduction also includes a discussion of number representation and the effect that number representation has on the accuracy, precision and stability of the results of the computation. It is especially important in biomedical and healthcare applications that accuracy and stability be preserved and the system is as robust as possible. Whether or not MATLAB is used, keeping in mind concepts such as machine epsilon (the smallest number that can be represented by the computer) will help the programmer understand the importance of robust program and system design and controlling the propagation of error.

Lastly, this introduction includes a discussion of one of the most important concepts in numerical analysis, the role that a Taylor series approximation plays in mapping continuous models to their discrete analogs and methods for solving the discrete representation. The Taylor series plays an important role in deriving numerical algorithms and in characterizing the error introduced by the discrete approximation and also in the error propagated by performing a sequence of calculations.

2.1a.5.2 Part II: Steady-state behavior (algebraic models)

Part II is an overview of techniques used to analyze systems that are in steady state and whose models are formulated as algebraic equations that could be either linear or nonlinear. A single equation that is explicit in the unknown can easily be solved by methods from precollege algebra; if the equation is implicit in the unknown, then root-finding techniques must be used. If the model is a set of simultaneous equations then numerical algebraic methods are used. Of course, the case of simultaneous, implicit equations is also treated. Each of these techniques is presented in this part of the text.

2.1a.5.3 Part III: Dynamic behavior (differential equations)

Part III is of greatest interest to the biomedical engineer: modeling the transient behavior of dynamic systems and solving for the output of such systems. This section of the text includes methods for solving both ordinary and partial differential equations using numerical techniques as well as Simulink.

Also in Part III, system behaviors during finite-time intervals, modeled by integral equations, are considered. In these cases, the solution of the model for an output parameter requires numerical integration and differentiation techniques. These methods are presented in this section, along with methods for improving the accuracy of the results. A recurring theme in this section is the tradeoff that must be made between accuracy of the solution and the amount of computation that is performed.

2.1a.5.4 Part IV: Modeling tools and applications

Part IV is an introduction to developing models of complex systems and to tools and techniques for analyzing complex behaviors. Examples of multicompartmental models of the circulatory, respiratory, and nervous systems are presented along with MATLAB implementations of the computational models. Tools and techniques for statistical and time series analysis are also presented with applications of these methods.

As the physical scale of problems continues to decrease, the need for mathematical models of biomedical systems continues to increase. The biomedical engineer who masters the basic material such as presented here will be well prepared to implement methods to solve mathematical models of biomedical systems for the steady-state, finite-time, or transient behavior of the system. He or she will be able to combine his or her analytical skills, computational skills, and mastery of the link between the two, the numerical methods.

2.1a.6 Lessons learned in this chapter

After studying this chapter, the BME student will have learned the following:

- Mathematical models are tools that biomedical engineers use to predict the behavior of a system.
- Biomedical engineers model systems in one or more of three different states: steady-state behavior, behavior over a finite period of time, or transient behavior.
- Derivation of mathematical models begins with a conservation law.
- Numerical methods are the bridge between analytical formulation of the models (using algebra, calculus, or differential equations) and computer implementation.

2.1a.7 Problems

2.1a.1 List five applications of computers in BME and briefly describe each application.

2.1a.2 You have been assigned the task of designing a computerized patient-monitoring system for an intensive-care unit in a medium-sized community hospital. What parameters would you monitor? What role would you have the computer play?

2.1a.3 Why is a computer-averaged, EEG-evoked response signal easier to analyze than a raw signal?

2.1a.4 What advantage does a computer give in the automated clinical laboratory?

2.1a.5 List the types of monitoring equipment normally found in the ICU/CCU of a hospital.

2.1a.6 Are computer systems always applicable in biomedical equipment?

2.1a.7 Describe three computer applications in medical research.

2.1a.8 On a BME examination, a student named George computes in feet and inches the maximum distance that a certain artificial heart design could pump blood upward against gravity. Unfortunately, in recording this distance on his examination paper, he reverses the numbers for feet and inches. As a result, his recorded answer is only 30% of the computed length, which was less than 10 feet with no fractional feet or inches. What length did George compute in feet and inches?

References

Enderle, J. D., Blanchard, S. M., and Bronzino, J. D. 2000. *Introduction to Biomedical Engineering*. San Diego, CA: Academic Press.

Fournier, R. L. 1999. *Basic Transport Phenomena in Biomedical Engineering*. Philadelphia, PA: Taylor & Francis.

Gevertz, J. L., Dunn S. M., Roth, C. M. 2005. Mathematical model of real-time PCR kinetics. *Biotechnol. Bioeng.*, in press.

Hallett, M. (2000). Transcranial Magnetic Stimulation and the Brain. *Nature*, **406**: 147–150

Heldt, T., Shim, E. B., Kamm, R. D., Mark, R. G. 2002. Computational modeling of cardiovascular response to orthostatic stress. *J Appl Physiol.*, Mar:**92**(3): 1239-1254.

Melchior, F. M., Srinivasan, R. S., Charles, J. B. 1992. Mathematical modeling of human cardiovascular system for simulation of orthostatic response. *Am J Physiol.*, **262**(6 Pt 2):H1920–1933.

Norton, S. J. 2003. Can ultrasound be used to stimulate nerve tissue? *Biomed. Eng. Online*, **4**:2(1):6.

Nebeker, F. (2002). Golden accomplishments in biomedical engineering. *IEEE Engineering in Medicine and Biology Magazine*, 21:17-47.

Pormann, J. B., Henriquez, C.S., Board, J. A., Rose, D. J., Harrild, D.M., and Henriquez, A. P. 2000. Computer Simulations of Cardiac Electrophysiology. Article No. 24, *Proceedings of the 2000 ACM/IEEE Conference on Supercomputing*. Dallas, TX.

Thompson, W. J. 2000. *Introduction to Transport Phenomena*. Upper Saddle River, NJ: Prentice Hall PTR.

Vierstraete, A. 1999. http://users.ugent.be/~avierstr/principles/pcr.html. Last viewed August 24, 2005.

Chapter 2.1b

2.1b

Introduction to computing

Stanley M. Dunn, Alkis Constantinides, and Prabhas Moghe

2.1b.1 Introduction

The computer is a ubiquitous tool for all engineering disciplines, and biomedical engineering is no exception. In biomedical engineering, computers are used to control instrumentation and data collection, analyze images, simulate models, and perform statistical analysis, among many other tasks. Biomedical engineers with good computing skills will be able to apply their expertise to any problem in this diverse field.

In this chapter, we will introduce the core areas of computing: programming languages and program design, data structures, and the analysis of algorithms. There are many MATLAB examples that illustrate how to develop a computer program using good program design principles.

The material in this chapter will enable the student to accomplish the following:

- Identify functions performed by computers in biomedical engineering applications
- Differentiate between imperative, functional, and object-oriented programming languages
- Identify the common data structures used in computer programming languages
- Describe how numbers are represented in computer programming languages
- Describe a computer algorithm and its role in biomedical engineering applications
- Identify a block-structured programming language and describe good programming practice
- Write example computer programs in MATLAB

Biomedical Engineering Desk Reference; ISBN: 9780123746467

2.1b.2 The role of computers in biomedical engineering

The sphygmomanometer, a device for measuring blood pressure (BP), is one example of biomedical instrumentation that is available to the general consumer because of the widespread use of computers in medical devices.

The computer allows a consumer to operate the device by providing the:

1. primary user interface
2. primary control for the overall system
3. data storage for the system
4. primary signal processing functions for the system
5. safe and reliable operation of the overall system, including cuff inflation

The computer program in the BP monitor must do all of these functions and execute them correctly each and every time the monitor is used. The biomedical engineer who writes the computer program must keep in mind the following criteria:

1. The program must be correct and operate the same way each and every time.
2. The users must have the confidence that all measurements are equally reliable.
3. The monitor must be easy to use.

The sphygmomanometer for home use (Fig. 2.1b-1) is only one example of the impact that computers have had in biomedical engineering; other examples include medical imaging, especially computed tomography (CT) and magnetic resonance imaging (MRI); bioinformatics; hemodynamic and other simulation; and high-resolution microscopy.

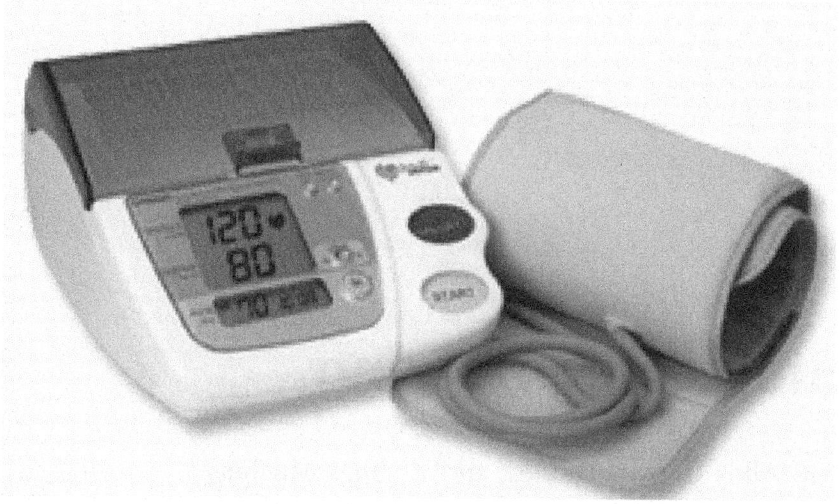

Figure 2.1b-1 An automatic sphygmomanometer for home use. Image courtesy of Omron Healthcare.

Instrumentation such as the BP monitor, CT or MRI imager, and the microscope have a common structure. In all cases, the computer controls the electronic interface (the user interface and data collection), which is connected to the subsystems that sense the tissue properties and control the sensor.

All of these functions could be performed by a trained user, but then instruments such as the BP monitor could not be put into the hands of the consumer for two reasons: First, the general consumer is untrained. Second, human performance changes over time. The computers embedded in the instruments have helped to standardize the operation, so that non-trained users can operate the BP monitor and other medical devices while being in better control of their healthcare.

Once a computer is given a standard set of instructions, the performance will be the same each time the device is used, unless there is a hardware failure. Some of the advantages of using computers are:

1. The actions directed by the computer and its program are *reproducible*: the operation will not change over time unless one or more parts (the electronics, actuator, or sensor) fail.
2. The precision and accuracy can be controlled: *precision* is the smallest possible resolvable event; it differs from *accuracy*, which is how close you can get to the truth (the real property). Precision and accuracy are dependent on both the sensors and the computation. The emphasis in this text will be on how we can best optimize the computer programs to maximize the precision and accuracy of the results.

Computers are not without their drawbacks. Some of the disadvantages of using computers instead of human control are:

1. Computers require precise and complete instructions. There is no room for ambiguity or assumptions.
2. All too often, program testing is not clear or complete. Testing a computer program, especially a large computer program, can be a full-time job by itself. You often have to worry about unanticipated input or conflicting input that is not expected. You may collect more data than you have accounted for. The number of possible combinations of inputs is exponential in the number of independent variables; for a moderate number of input variables, the number of possible inputs is too large to test.
3. There can be logic errors, or syntactic errors, that can later cause execution errors. In very large applications, like bioinformatics programs, special Web interfaces are built just to keep track of the bugs. A good example is http://www.biojava.org, the Web site for a set of Java-based tools for processing biomedical data, such as gene sequences.

Besides controlling or automating equipment, computers can make biomedical engineers more efficient by:

1. Providing design tools, including simulation (using tools such as Simulink, Pro-E, and ANSYS)
2. Keeping documentation (using tools such as MS Word or Excel)
3. Providing test procedures and results (using tools such as MATLAB and LabVIEW)

The emphasis here is to help the new biomedical engineer develop computer programs that control the precision and accuracy of the measurements and computations as an integral part of solutions to real-world bioengineering problems. First, the programmer must be familiar with the tools available in a programming environment such as MATLAB.

2.1b.3 Programming language tools and techniques

There are three classes of programming languages in use today: the imperative, or state transition class; the object-oriented class; and the functional class. In each class, there are a large number of programming languages. How can an engineer decide which class and which language to use?

The best examples of imperative or state-transition languages are C and Fortran. Programs written in these or other imperative languages are executed one statement at a time, in the order that the statements are written.

Object-oriented programming languages constitute the largest and most widely used class of languages today and include C++, Java, Python, and Smalltalk, the programming language that started the object-oriented revolution. Object-oriented programs execute much differently than imperative programs; the operations executed are dictated by the type of the data object and the desired action. The implementation of an action will depend on the type of the object (think about the difference between adding two integers and adding two vectors) although the *name* of the operation (e.g., *add*) may be the same in both cases. The examples in this chapter are written in MATLAB, an object-oriented programming language and environment. The user not familiar with MATLAB is strongly encouraged to be familiar with Appendix A, an introduction to the MATLAB programming language and environment.

Functional programming languages are written as a set of definitions of functions and a composition of these functions to be executed. The execution of this composition can be sequential, like the imperative languages, or parallel. Some common examples are Scheme, Sisal, or Caml. Functional languages often include object-oriented properties. This class of languages will not be considered here.

Both imperative and object-oriented languages have common programming language constructs that are designed to help the programmer develop readable and testable programs efficiently. These programming language constructs (statements) allow the programmer to manage the computation being performed and the flow of execution. Modern programming languages limit the control flow statements to one of three types: a sequence of statement; a statement for conditional execution; and a control structure for repeating a sequence of statement. These languages are called block-structured languages. All three of these statements have one path in and one path out; this standardizes program design and development.

2.1b.3.1 Sequences of statements

A sequence of instructions is delimited by special symbols or keywords. These delimiters indicate the single beginning and end to the block.

Example 2.1b.1 Programs that are sequences of statements.

Create a vector of the even whole numbers between 31 and 75.

Solution

The even whole numbers between 31 and 75 are the even numbers 32 to 74. There are two ways to create the vector: First, by typing the list of numbers or second, by using the MATLAB shorthand notation **first: increment:last.**

The solution using the first method is left to the reader. The solution using the second method is as follows:

1. The first number in the vector is 32.
2. The increment is 2, since the vector is to contain only the even numbers.
3. The last number in the vector is 74.

Therefore, the vector is specified as `[32:2:74]`. The MATLAB output is:

```
>> x = [32:2:74]
x =
  Columns 1 through 11
    32 34 36 38 40 42 44 46 48 50 52
  Columns 12 through 22
    54 56 58 60 62 64 66 68 70 72 74
```

2.1b.3.2 Conditional execution

A conditional statement is required if there is a need for an execution path that depends on the values of the data input, or external input. These conditional statements also have delimiters: generically, `if...then` or `if...then...else`. There is still only one entrance to the block at the `if` keyword and one exit where the branch(es) return to the main line of the control flow.

2.1b.3.2.1 If–then statements

The simplest form of conditional execution is the `if...then` statement, which in MATLAB has the form `if` *`expression statements`* `end`. If the *`expression`* is evaluated to be true, then the statement block is executed. The expression is always evaluated, and control flow always follows with the statement after the end.

2.1b.3.2.2 If–then–else

The if...then statement provides the ability to program one type of conditional execution control flow. The if...then...else construct provides the ability to implement a control flow where one of two blocks, but not both, are executed.

In MATLAB, the syntax of the if...then...else statement is

```
if expression
  statement1
else
  statement2
end
```

The expression is always evaluated; if the expression is true, then the block statement1 is executed; if the expression is evaluated to be false, then the block statement2 is executed.

There is another variant of conditional execution in MATLAB, the if...elseif...else...end statement with the syntax

```
if expression1
  statement1
elseif expression2
  statement2
else
  statement3
end
```

The elseif clause allows for the evaluation of an alternate expression; if the first expression is false and the second expression is true, then the block statement2 is executed. Here is an example:

Example 2.1b.2 Simple control flow using if...then...else

Write a MATLAB script to compute the following function:

$$f(x) = \begin{cases} -1 & x < 0 \\ 0 & x = 0 \\ 1 & x > 0 \end{cases}$$

Solution

In this example you are asked to create an m-file, a MATLAB script that computes the function f defined above. The value returned by the script is one of three values that depend on the input. The mathematics can be easily translated into MATLAB commands.

The valued to be returned by the m-file is -1 if the argument x is less than zero. The mathematical statement above is easily translated into the MATLAB statement

```
if x<0 f = −1;
```

where the variable f is the variable whose value is returned by the function.

If x is less than zero then the value to be returned has been determined (f=−1) and no more MATLAB statements are evaluated. If the value of x is not less than zero, then f=−1 is not executed and the final value of f must still be determined.

If the value of x is not less than zero, then either it is equal to zero or it is greater than zero. According to the mathematical definition above, if x is equal to zero, then the value of f should be 0; If x is greater than zero, then the value of f should be 1.

The MATLAB script that implements this function will have three *expression-statement* pairs. However, the three statements should not be executed as a sequence; if the first assignment statement is executed, then the evaluation of f should be stopped. Can you tell why?

To execute only those statements that should be executed and no others, use the structure to test multiple conditions and evaluate only the statements that satisfy the conditions. The control structure to implement this design requires the elseif clause in the conditional statement. The final code is:

```
if x < 0
   f = -1;
elseif x == 0
   f = 0;
elseif x > 0
   f = 1;
end
```

As a followup to this problem, we leave it to the reader to show that this MATLAB code performs the same function as the built-in function sign.

2.1b.3.2.3 Switch statement

In most block-structured programming languages, including MATLAB, there is a facility for conditionally executing a sequence of statements from an arbitrary number of alternatives. Of course, one could implement this control flow structure using a large if...elseif...end structure, but there is a more convenient shorthand notation in MATLAB, the switch...case...otherwise...end structure. The syntax of the MATLAB switch statement is

```
switch switch_expr
  case case_expr
    statement,...,statement
```

```
case
{case_expr1,case_expr2,case_expr3,...}
statement,...,statement
...
otherwise
statement,...,statement
end
```

Each alternative is called a case statement and has three parts: the `case` directive, one or more case expressions, and a sequence of one or more statements. When the `switch_expr` equals a `case_expr`, the corresponding sequence of statements is executed. If no `case_expr` equals the `switch_expr`, then the statements of the `otherwise` case are executed.

There is one important difference with the switch statement: The expressions must be either scalars or strings. MATLAB finds the right sequence of statements to execute by attempting to find the `case_expr` that matches the `switch_expr`.

Once the statements are executed, the control flow resumes with the statement that follows the switch statement, as with all other statements.

Example 2.1b.3 Use of the `switch` statement.

Write a MATLAB script that reads a character variable (s or c) and an integer variable (1, 2, or 3) and plots a function based on the following chart:

	1	2	3
s	$\sin\theta$	$e^{-\theta}\sin^2\theta$	$\sin 3\theta/e^{-\theta}$
c	$\cos\theta$	$\cos 2\theta/e^{-2\theta}$	$e^{-\theta}\cos^3\theta$

You may assume that there are 100 points over the interval $0 \leq \theta \leq 2\pi$.

Solution

Instead of beginning by immediately starting to write MATLAB code for this problem, it is best to begin by becoming familiar with using the plot function. The function plot can take a number of different forms. The function call `plot (Y)` will plot the vector Y against its index (or independent variable). The function call `plot (X, Y)` will plot the values of the dependent variable in the vector Y against the values of the independent variable in the vector X. For example, the MATLAB commands result in Fig. 2.1b-2.

```
>> x=1:100;
>> plot (sin(x/3));
```

Notice that the values of the **independent variable** x are plotted on the horizontal axis and the values of the **dependent variable** are plotted on the vertical axis. The independent variable for the given problem is the angle θ,

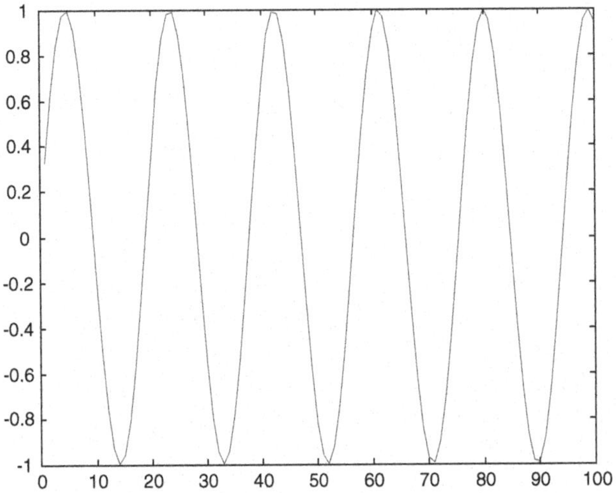

Figure 2.1b-2 sin(x/3) curve.

in radians. Since the problem specifies using 100 values over the range, the vector

```
x=0: (2*pi)/100:2*pi;
```

specifies the values of the independent variable for this problem.

The solution to this problem is a MATLAB script that reads two inputs. These two inputs identify a row (indicated by either the character 's' representing the function sine or the character 'c' representing the function cosine) and column (indicated by one of the integer 1, 2, or 3) indices of the table above. The MATLAB script must evaluate and plot the function in the cell identified.

It is easiest to use `switch` statements, rather than a large number of `if-then-else` statements, to control the flow of execution of the program. An easy strategy will be to have one switch statement that selects between the two rows and in each statement block, have another switch statement that will select one of the columns. To begin,

```
switch letter
case 's'
a switch statement for row s goes here
case 'c'
a different switch statement for row c
  goes here
end
```

With this strategy, the switch statements for each row can be independently written and independently tested, keeping with our principles of good programming practice. The switch statement for the row *s:*

```
switch col
  case 1
    plot(theta,sin(theta));
  case 2
    plot(theta,exp(-theta).*(sin(theta)).^2);
```

```
    case 3
      plot(theta,sin(3*theta)./exp(-theta));
    end
```

can be inserted as the block following the case 's'. The complete MATLAB script for this exercise is:

```
twopi=2*pi;
theta=0 : (twopi/100):twopi;
letter=input ('Please enter either s
  or c: ') ;
col=input ('Please enter either 1, 2
  or 3: ') ;
figure (1) ;
switch letter
case 's'
  switch col
  case 1
    plot (theta, sin (theta));
  case 2
    plot (theta,exp(-theta).* (sin
      (theta) ).^2);
  case 3
    plot (theta, sin (3* theta)./ exp
      (-theta) );
  end
case 'c'
  switch col
  case 1
    plot (theta, cos (theta) ) ;
  case 2
    plot (theta, cos (2* theta) . / exp
      (-2* theta) );
  case 3
    plot (theta, exp (-theta) .* (cos
      (theta) ).^3);
  end
end
```

A sample session using this script is below, and the results are shown in Fig. 2.1b-3. Why does the `plot` function use the scale from 0 to 7 for the independent variable?

```
Please enter either s or c: 's'
Please enter either 1, 2 or 3: 2
```

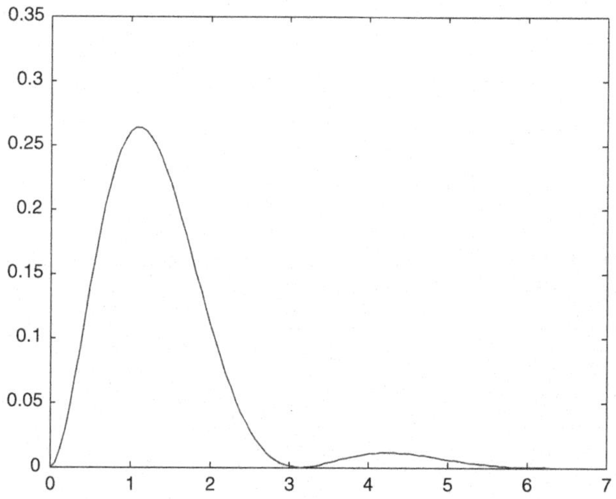

Figure 2.1b-3

Example 2.1b.4 The use of while loops.

Write a script that computes the number of random numbers required to add up to 20 (or more). You are to use the random-number generator `rand` in this exercise.

Solution

The random number generator in MATLAB returns a pseudo-random number that is uniformly distributed between 0 and 1. We refer to it as a pseudo-random number because the sequence of numbers generated depend on the state of the computer random number generator at the time that the function call is made. For example, the following call to the random number generator

```
>> rand
ans =
     0.9501
```

returns the result shown. The random number returned can be equal to or greater than zero, but is strictly less than 1.

The MATLAB script should stop executing once the sum of all the random numbers generated reaches 20. This means that the script will continue to **repeat** generating random numbers, adding each new one to the running sum, until the sum reaches at least 20. You also have to count the number of random numbers generated. Since it is not possible to predict exactly how many numbers need to be added, the program will have to repeat generating random numbers, until the condition is satisfied. The MATLAB `while...end` construct will be used to control the repetition.

The algorithm for solving this problem is:

1. The running sum of random numbers must start at 0 before the random numbers are generated. Set the count to be zero.
2. Check to see if the sum is greater than or equal to 20. If it is, then stop this script. If the sum is less than 20, then go to step 3.

2.1b.3.3 Iteration

2.1b.3.3.1 While loops

Repetition is where a sequence of instructions, or block, is repeated until a condition is satisfied. The condition can be expressed in terms of state of the data or an external input. A generic repetition structure is a `while...do loop`.

3. Generate a random number and add it to the sum. Add 1 to the count of random numbers generated. Go to step 2 to check the sum.

The MATLAB script, shown below, is saved as `whileloop.m`

```
% whileloop.m
add=0;
count=0;
while add<20
  add=add+rand;
  count=count+1;
end
display(count);
display(add);
```

and is executed from the command window to obtain the output:

```
>> whileloop
count =
     43
add =
    20.0364
```

2.1b.3.3.2 For loops

There are variations of iterative control structures, just as there were variations of the control structures for conditional execution. A *for loop* (`for...end`, in MATLAB) is a repetition structure where the block is repeated a fixed number of times, once for each value of a control variable. Typically, the execution of the block changes with the control variable—for example, accessing a different element of an array.

Example 2.1b.5 Using `for...end` loops.

Given the vector x = [1 8 3 9 0 1], create a MATLAB script to add the values of the elements (Check your result with the `sum` command.)

Solution

The statement of this problem gives the clues that the script need only add up the values in the given vector, not an arbitrary vector. You know that the length of the vector is 6; one way to write the script would be the commands

```
x=[1 8 3 9 0 1];
add=0;
add=add+x(1);
add=add+x(2);
add=add+x(3);
add=add+x(4);
add=add+x(5);
add=add+x(6);
add
sum(x)
```

Notice that this sequence has a common structure: each element of the array is added to the sum. If the vector was longer, writing out this sequence of statements would become very tedious; since the statements are similar, the complete sequence can be replaced with a shorthand description that uses a *for loop:*

```
% forloop.m
x=[1 8 3 9 0 1];
add=0 ;
for i=1:length(x)
  add=add+x(i);
end
add
sum (x)
```

where the `for...end` loop is the shorthand notation for the six statements that add each element of `x`, in turn. The result of executing `forloop.m` is:

```
>> forloop
add =
  22
ans =
  22
```

Note that the equivalent of all the statements in the script forloop.m may be accomplished with a single command: `sum (x)`. In a small way, this demonstrates the power of MATLAB.

2.1b.3.4 Encapsulation

The last key to writing well-structured MATLAB programs that are easy to read, understand and maintain is *encapsulation.* Encapsulation means to group together those instructions or statements that form one particular function. Encapsulation will improve readability and make the job of testing and debugging easier. It is hard to manage one large sequence of MATLAB statements.

Encapsulation is implemented in MATLAB by using *m-files*: text files that contain code in the MATLAB language. If large MATLAB programs are organized well, there will be one m-file for the main program and other m-files for the *functions* and *scripts* that are called by the main program or invoked at the MATLAB command prompt.

All modern programming languages have features for encapsulating sequences of instructions. There are typically two types of encapsulating programming constructs: *subroutines* and *functions.*

A *script* is a sequence of instructions that does not accept input arguments or return output arguments. A script operates on or modifies existing data in the workspace. In other block structured programming languages, the organizational equivalent of MATLAB scripts will be referred to as *subroutines.*

A *function* is a sequence of instructions like a script, but the function may accept input arguments and return output arguments. The function may also have MATLAB variables that can only be accessed inside the function.

Like all block-structured programming tools, subroutines and functions each have only one control flow entry point and one exit.

Example 2.1b.6 Using scripts and functions.

Write a *script* that asks for a temperature (in degrees Fahrenheit) and calls *a function* to compute the equivalent temperature in degrees Celsius. The script should keep running until no number is provided to convert (Note: The command `isempty` will be useful here).

Solution

The solution to this problem has two parts. The first is the function that converts a temperature in Fahrenheit into its equivalent temperature in Celsius. The second part of the solution is the script that calls this function. The best strategy is to write and test the function first, and only after it is determined to be correct, write the script that calls the function.

The MATLAB function needs to compute the temperature in Celsius using the following formula:

$$C = \frac{5}{9}(F - 32)$$

The temperature in Fahrenheit is the argument to the function, which returns the temperature in degrees Celsius. The MATLAB function is:

```
% Far2Cel.m
function C=Far2Cel (F)
C=(5/9) * (F-32)
end
```

which should be tested before going on to complete the exercise. The script that uses this function should take as input from the keyboard a temperature in degrees Fahrenheit, display the temperature converted to degrees Celsius and then wait for the next input.

The description should suggest that there is a `while` loop, not a `for` loop in this script since the number of iterations is uncertain and cannot be computed *a priori*.

First, get the keyboard input of temperature in Fahrenheit and check to see if the input is empty. If the input is empty, stop execution; otherwise, convert the temperature to Celsius and then get another input temperature and once again check to see if the input is empty. Iterate until there is no more keyboard input. The script `F2C.m` is:

```
% F2C.m
F = input ('The value of the temperature
    in degree Fahrenheit = ');
while ~isempty (F)
C = Far2Cel (F);
display(C);
F = input ('The value of the temperature
    in degree Fahrenheit =');
end
```

which is called from the MATLAB command line. In the sample execution below, note that the execution stops when there is no numeric input.

```
>> F2C
The value of the temperature in degree
Fahrenheit = 32
C = 0
The value of the temperature in degree
Fahrenheit = 100
C = 37.7778
The value of the temperature in degree
Fahrenheit =212
C = 100
The value of the temperature in degree
Fahrenheit = 77
C = 25
The value of the temperature is in degree
Fahrenheit. =
```

2.1b.4 Fundamentals of data structures for MATLAB

MATLAB programs have two parts: a sequence of instructions for the computation to be performed and the representation of the data that are being operated on. The previous section is an overview of the basic programming language constructs that underlie all MATLAB programs. This section is an overview of the techniques available in MATLAB for representing data.

Although the emphasis of this text is on numerical methods and quantitative results, there is still the need to include other information, such as names of variables, the units of the results, and information on the interpretation. The format of the numerical output is important as well.

All modern programming languages include a facility for creating and using different data types. There are fundamental data types and mechanisms for forming aggregates of the data. This section is an overview of the six fundamental data types in MATLAB: `double`, `char`, `sparse`, `uint8`, `cell`, and `struct`.

2.1b.4.1 Number representation

In calculus and courses on differential equations as well, the real numbers were used, whether this was made explicit or not. The real numbers have infinite precision, which is something that is not available in computer programs. Infinite precision would require infinite memory; no computer has infinite memory.

Thus, the real numbers have to be represented by an approximation—called the floating-point numbers. Floating-point numbers require only a finite amount of memory, but that means that there are real numbers that cannot be represented using a floating-point representation. Floating-point representation has a large impact on the accuracy and precision of the computation and is an important topic to master before studying numerical methods in detail.

A number written in a MATLAB script or function is written using the conventional notation. MATLAB does not distinguish between an integer, real, or complex number; they are all stored as the same type of MATLAB variable.

A number may have a leading plus or minus sign and an optional decimal point. Numbers written in scientific notation use the letter `e` to specify the exponent (base 10). Imaginary numbers use either `i` or `j` as a suffix to indicate an imaginary component. Some examples of numbers that can be represented in MATLAB are:

Example 2.1b.7 Number representation in MATLAB.

Positive integer	3
Negative integer	−45
Real number	0.00001
Scientific notation	2.71828e9
Complex number	3+5i
Imaginary number	−3.14159j

The amount of memory available to store each number is limited; there is an upper and lower limit to the real numbers that can be represented by the floating-point numbers in MATLAB. The MATLAB functions `realmin` and `realmax` return, respectively, the smallest and the largest real numbers that can be represented in MATLAB. The function `inf` returns the representation of infinity, and NaN returns a representation for Not-a-Number, the result of an undefined operation (such as 0.0/0.0) in MATLAB.

Complex numbers can be assigned directly to numeric variables or they can be created from two real components; the MATLAB command `c=complex (3,5)` creates a complex number `c` that is `3+5i`. The result of the command `c=complex(3, 0)` is a complex number with an imaginary component 0 and is not a real number.

Example 2.1.8 Complex numbers.

Compute the inverse of the complex number 3+5i and verify the result by hand calculation.

Solution

Recall that the inverse of a complex number $a+bi$ can be computed by

$$(a+bi)^{-1} = \frac{(a-bi)}{(a+bi)(a-bi)}$$

The MATLAB function `inv ()` computes the inverse of the argument; the command:

```
>> inv(3+5i)
ans =
   0.0882 - 0.1471i
```

can be verified by the calculation using the formula above, using the MATLAB function `conj ()` to compute the complex conjugate of the argument:

```
>> x=3+5i;
>> conj (x)/(x*conj(x))
ans =
     0.0882 - 0.1471i
```

2.1b.4.2 Arrays

An array is a data structure for representing a collection of objects that provides the ability to access each element of the collection in the same amount of time. You do not have to search all the way down a list, or through a matrix in order to get to a desired element.

All arrays in MATLAB are stored as column vectors, and each element can be accessed using only the single index into the column vector. However, if there is more than one index in the array reference, the location in the column vector is computed using the formula:

$$(j-1)*d+i$$

to reference the array element $A(i,j)$, where d is the length of one column. Subarrays of a multidimensional array can be referenced in their entirety as well.

Example 2.1b.9 Indexing arrays in MATLAB.

Create a 5x5 matrix with elements $A(i,j) = 1/(2^i + 3^j)$. Print the entire matrix, the first row and the second column. Give two ways of displaying the element $A(3,2)$.

Solution

Unlike many other programming languages, MATLAB does not have a dimension statement or method for declaring the space allocated for a variable; MATLAB automatically allocates storage for matrices and other arrays. For this problem, space for the matrix A is allocated at the time the elements are created. Since a formula is given for each element, the matrix can be created in a script that also includes the statements to print the elements (including the rows and columns). The size of the matrix is specified, so the elements can be computed using for loops, followed by the references to print the subelements. The MATLAB script is:

```
% ArraysRef.m
for i=1:5
  for j=1:5
    A(i, j)=1/(2^i+3^j);
  end
end
A
A (1,:)
A (:,2)
A (8)
A (3,2)
```

Notice that the first row is referred to as `A (1, :)`. The subscript 1 refers for the row and the second subscript is not a number, but rather the symbol ':' which refers to **all elements in this dimension.** The ':' in `A (:,2)` has the same meaning, but this time it refers to the row dimension and means all rows with the column being fixed at 2.

Remember that all MATLAB arrays are stored as column vectors, with the columns following each other in order. Therefore the element `A (3, 2)` can be referred to as an element in a vector or as the element in the matrix (which is logically how we would expect to refer to it). The above script produces the following results when executed from the command window:

```
>> ArrayRefs
A =
0.2000  0.0909  0.0345  0.0120  0.0041
0.1429  0.0769  0.0323  0.0118  0.0040
0.0909  0.0588  0.0286  0.0112  0.0040
0.0526  0.0400  0.0233  0.0103  0.0039
0.0286  0.0244  0.0169  0.0088  0.0036
ans =
0.2000  0.0909  0.0345  0.0120  0.0041
ans =
0.0909
0.0769
0.0588
0.0400
0.0244
ans =
0.0588
ans =
0.0588
```

Notice that the row is displayed as a row and the column is displayed as a column.

Chapter 2.2

2.2

Simulation and estimation

Stanley M. Dunn, Alkis Constantinides, and Prabhas Moghe

2.2.1 Numerical modeling of bioengineering systems

This chapter offers a number of examples designed to illustrate the range of applications of numerical methods and computing environments such as MATLAB and Simulink to biomedical engineering problems. Every attempt has been made to first state the clinical or physiological problem, introduce the modeling method and show how MATLAB and/or Simulink are used to create the computer solution using the numerical methods from this text.

A number of these examples are drawn from resources that are commonly available. Section 2.2.2 describes *PhysioNet*, a resource that includes a very large archive of bioelectric signals and software tools that can be used for research projects. The examples are a brief introduction as to how this resource can be used. The electroencephalogram (EEG) data used in Section 2.2.3 is available, as is the Simulink physiological simulation benchmark experiment (PHYSBE) used in Section 2.2.7. In all other cases, the problems were taken from the biomedical engineering literature.

There is no particular order in which to read or review these examples. As a guide, Table 2.2-1 is a summary of the examples presented in this chapter, in which the problems have been categorized by the underlying phenomena and the numerical method used. Simulink is indicated in the cases where it is used.

The material in this chapter will enable the student to accomplish the following:

- Learn how MATLAB can be used to solve biomedical engineering problems
- Learn how Simulink can be used to simulate models of physiological systems
- Write MATLAB programs to solve bioengineering problems
- Develop Simulink models of physiological systems
- Gain an understanding of the role that numerical methods play as the interface between mathematical model and computer solution

2.2.2 PhysioNet, PhysioBank, and PhysioToolkit

PhysioNet is an Internet resource for biomedical research and development sponsored by the National Center for Research Resources of the National Institute of Health (NIH). PhysioBank is an archive of contributed and standardized physiologic signals and annotations. It contains several examples to show how models, specifically frequency domain models, can be used to answer questions about the physiology.

Another component of PhysioNet is PhysioToolkit, which is a repository of software for analyzing PhysioBank data, visualizing data (including PhysioBank data), signal processing, software development, and simulation. Several of these software tools can be used with MATLAB; some will be shown here as examples.

2.2.2.1 ECG simulation

The MATLAB script given in Chapter 2.1b can be used to generate a very crude simulation of a single QRS complex. In PhysioToolkit, there is a MATLAB

Biomedical Engineering Desk Reference; ISBN: 9780123746467

Table 2.2-1 A summary of the application examples in this chapter

Section	Examples	Application	Phenomena	Models/Methods
10.2	10.1, 10.2	PhysioNet and PhysioBank	Bioelectric	Computing Model fitting ODEs*
10.3	10.3	Brain activation	Bioelectric	Model fitting Statistics
10.4	10.4	Diabetes and insulin regulation	Biochemical	Systems of ODEs Simulink
10.5	10.5	Renal clearance	Biochemical	ODEs
10.6	10.6	Gait and motion estimation	Biomechanical	Numerical linear algebra
10.7	10.7 to 10.12	PHYSBE	Biochemical	Systems of ODEs Simulink

* ODEs stands for Ordinary Differential Equations.

script for generating both normal complexes and arrhythmia. The script, called ECGwaveGen, was in part contributed by Floyd Harriott. In this script, the user can control the heart rate, signal duration, sampling frequency, QRS amplitude and duration, and T-wave amplitude of the synthesized signal.

Example 2.2.1 Using the MATLAB script ECGwaveGen to synthesize ECG data

Download the `ECGwaveGen.m` and the `QRSpulse.m` MATLAB scripts from http://www.physionet.org/physiotools/matlab/ECGwaveGen/, and use these programs to synthesize five seconds each of normal and sinus tachycardial rhythms.

Normal rhythm, sometimes called sinus rhythm, is synthesized using `ECGwaveGen` by default. The help for `ECGwaveGen` is:

```
[QRSwave] = ECGwaveGen(bpm,dur,fs,amp)
generates an artificial ECG/EKG waveform
  Heart rate (bpm) sets the qrs event
frequency (RR interval). Duration of the
entire waveform (dur) is in units of
seconds. Sample frequency (fs) sets the
sample frequency in Hertz. Amplitude (amp)
of the QRS event is measured in micro
Volts. The waveform consists of a QRS
complex and a T-wave. No attempt to repre-
sent a P-wave has been made.
  There are two additional parameters that
can be changed from within the function.
They are the parameters that set the QRS
width (default 0.1 secs) and the t-wave
amplitude (default 500 uV).
```

The MATLAB commands:

```
Norm = ECGwaveGen (60, 5, 64,1000);
time = [(1:length(Norm)) /64];
plot (time,Norm);
```

Create a vector of time values (not sample numbers), the synthesized QRS complexes, and plot the ECG against the time, as shown in Fig. 2.2-1, where the QRS complex peak is approximately .75 mV and the T-wave peak is .5mV.

Sinus tachycardia is characterized by a heart rate greater than 100 beats per minute (bpm). The MATLAB code to simulate sinus tachycardia at 110 bpm is reproduced below, and the results shown in Fig. 2.2-2.

```
figure(2);
Norm = ECGwaveGen(110, 5, 64, 1000);
time = [(1:length(Norm))/64];
plot (time, Norm);
title ('ECGwaveGen synthesis of Sinus
  Tachycardia');
xlabel ('Time in seconds');
ylabel ('Amplitude in microvolts');
```

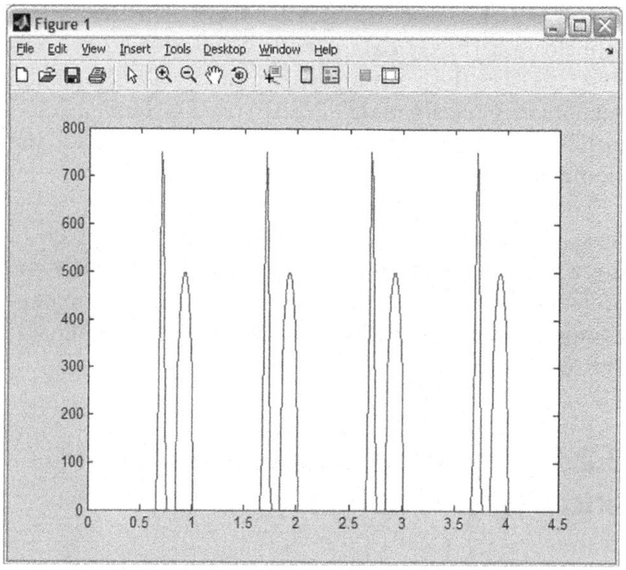

Figure 2.2-1 ECG profile.

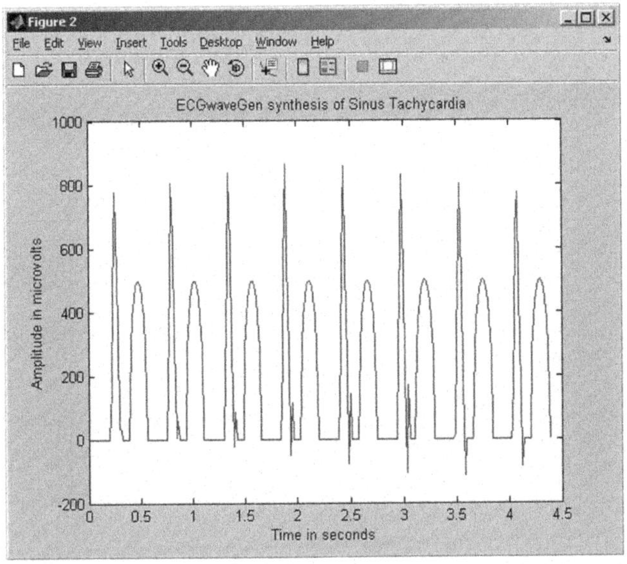

Figure 2.2-2 Simulation of sinus tachycardia.

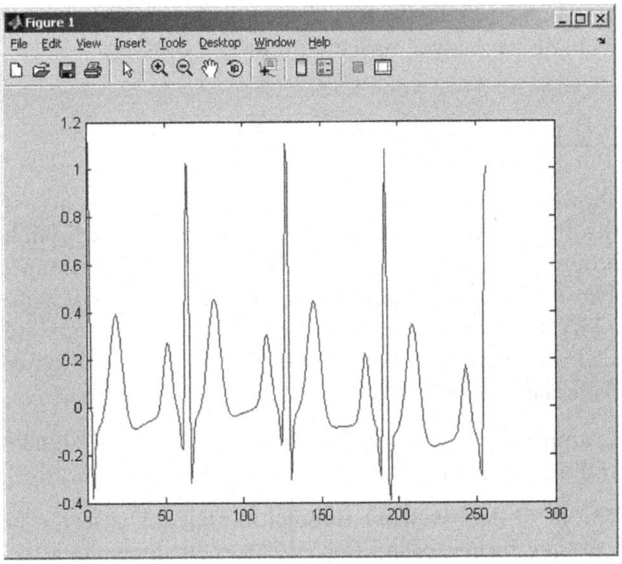

Figure 2.2-3 Normal sinus rhythm generated using `ecgsyn`.

Unfortunately, the `ECGwaveGen` script available at PhysioNet will not generate sinus tachycardia greater than 110 bpm. There is a second script at PhysioNet called `ecgsyn`. This script, based on the work of McSharry et al. (2003), allows the user to have more control over the morphology and frequency content. `ecgsyn` is a simulation of three coupled differential equations. The `ecgsyn.m` MATLAB script may be downloaded from http://www.physionet.org/physiotools/ecgsyn/.

With no arguments, the ecgsyn script will /generate synthetic ecg with the following parameters:

```
ECG sampled at 256 Hz
Approximate number of heart beats: 256
Measurement noise amplitude: 0
Heart rate mean: 60 bpm
Heart rate std: 1 bpm
LF/HF ratio: 0.5
Internal sampling frequency: 512
```

To generate sinus rhythm of 10 beats sampled at 64 Hz with no noise:

```
s = ecgsyn(64, 10, 0, 60);
```

Since the ECG is sampled at 64 Hz, the first 256 samples would cover 4 complexes (see Fig. 2.2-3).

```
plot (s(1:256))
```

The fourth argument is changed to vary the heart rate. To simulate tachycardia using `ecgsyn`:

```
s = ecgsyn(64,16, 0, 110);
```

and display the first 256 samples as before (Fig. 2.2-4).

```
plot(s (1:256)
```

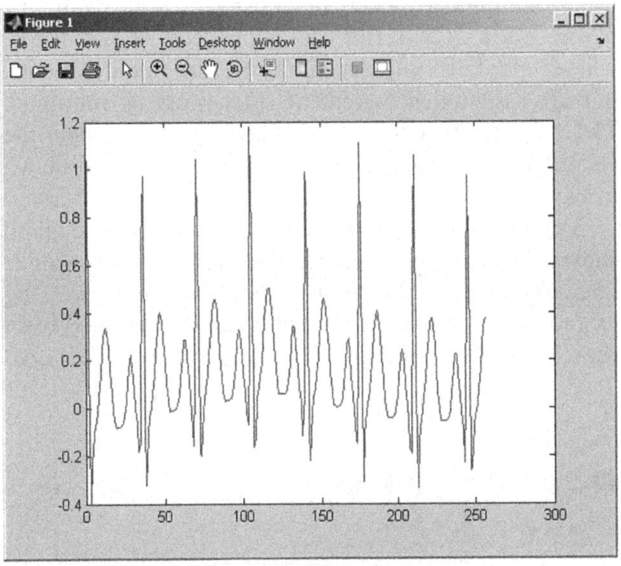

Figure 2.2-4 Tachycardia generated using `ecgsyn`.

There are roughly twice the number of PQRST complexes in Fig. 2.2-4 as there are in the same interval of time in Fig. 2.2-3

Synthesized ECG is not meant to be realistic, as it is normally used to test biomedical instruments, using an ECG-like signal with well-defined characteristics. Nevertheless, the difference between the two simulators is obvious by visually comparing their output: the `ecgsyn` output is much more realistic, as it includes the P waves and an S-T depression. The trade-off is obvious: a more realistic simulation is required for solving a system of three differential equations rather than a simulation based on curve fitting techniques. A more realistic signal will still have more realistic frequency content.

2.2.2.2 Reading PhysioBank data

To solve some of the problems, one may have to access the PhysioBank archives using the Chart-O-Matic Web browser (which uses the PhysioToolkit function rdsamp). The instructions are to download a text file and write a MATLAB script that would read the text file. There are also MATLAB scripts available to read the raw PhysioBank data files directly.

Example 2.2.2 Read and visualize PhysioBank signals and annotations.

The ECG data available through PhysioBank is stored in a unique format called the "212" format. Download the relevant sample problem and display it directly in MATLAB using the `rddata.m` script.

Throughout this text, we have stressed the value of using arguments to functions to preserve their flexibility. In the `rddata.m` script, the parameters of pathname, header, attribute, and data file names are not arguments, but are assignment statements that must be modified. The output, when using the MIT-BIH sample data file (in three parts 100.hea, 100.atr and 100.dat), is shown in Fig. 2.2-5.

Although difficult to dissect at first glance, the graph shows that the ECG recording is just over 80 seconds long. A small section of the data can be viewed by extracting the data (and attributes and annotations) from the MATLAB workspace variables. The MATLAB

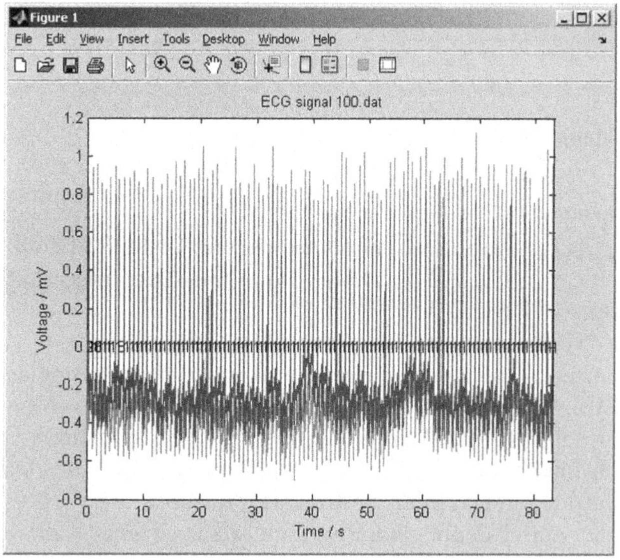

Figure 2.2-5 ECG sample data file from the MIT-BIH database.

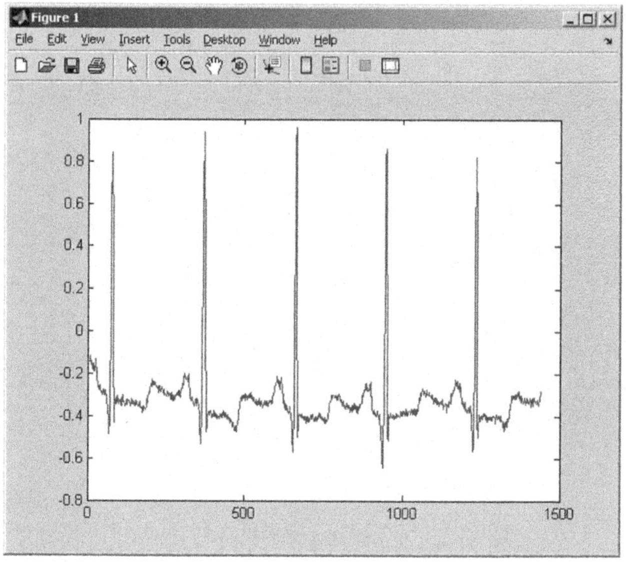

Figure 2.2-6 The first 1440 samples from the 100.dat file in the MIT-BIH database.

variable M contains the sampled data, that is, two signals, one in each column of the array.

The sampling frequency is stored in a separate MATLAB variable, `sfreq`. For the given sample, the sampling rate was 360. If the subject's heart rate was 60 bpm, at the given sample rate, 4 QRS complexes would be captured in 1440 samples.

Displaying the first 1440 samples of the first channel, as shown in Fig. 2.2–6,

```
plot(M(1:1440,1))
```

shows that the real heart rate is greater than 60 bpm. Now that the data is in a MATLAB variable, the true heart rate can be easily computed by finding the R waves (the peaks) and computing the R–R interval.

There are two QRS complexes in the first 512 samples: one in the first 256 samples, and the second somewhere between sample 256 and sample 512 (see the graph above). The R–R interval is computed by finding the maxima and their indices.

```
[p1, i1]=max(M(1:256,1));
[p2,i2]=max(M(256:512,1));
```

The amplitude and location of the peaks can be verified against the graph above. The R–R interval is:

```
RR=(i2+256)-i1;
```

Heart rate is the inverse of the R–R interval. Since the data is sampled at 360 samples per second, the heart rate can be estimated by

```
360*60/RR
ans =
   73.4694
```

This simple example illustrates how to read the raw PhysioBank data. Many of the processing and analysis techniques shown here can now be applied to determine the properties of this and other physiologic signals.

2.2.3 Signal processing: EEG data

EEG signals and their analysis are tools for (1) understanding the dynamic processes in the brain that are the bases of physical and mental behavior and (2) localizing the source of the brain activity associated with specific tasks or behaviors. The purpose of this example is to determine whether or not there is differential brain activity between the right and left hemispheres during certain cognitive tasks.

The standard convention for recording EEG data is called the 10-20 system. The name 10-20 refers to the percentage of arc length from nasion to inion through the vertex, as shown in Fig. 2.2-7.

The signals recorded from A1 and A2 are reference signals that record a signal called the electro-oculogram (EOG): muscle artifact in the EEG signals that is due to eye blinking.

The data for this project, from Keirn (1988), is publicly available from Professor Charles Anderson in the Department of Computer Science at Colorado State University. In addition to the EOG, signals were recorded at O1, O2, P3, P4, and C3, C4.

The subjects in this study underwent the following five tasks:

1. In the **resting** or **baseline** task, the subjects opened and closed their eyes and were asked to relax as much as possible. In a resting phase, alpha waves are produced and any left–right hemispheric asymmetries can be determined so that they can be subtracted from the measurements during cognitive tasks.
2. In the **arithmetic** task, subjects were asked to solve a complex multiplication problem without speaking or other muscle movements.
3. In the **geometric** task, subjects were given 30 seconds to study a drawing of a 3D block figure. After the figure was removed, the subjects were instructed to visualize the object rotating around an axis.
4. In the **letter composition** task, the subjects were instructed to mentally compose a letter to a friend or relative without speaking. In the given study, the tasks were repeated several times; during the repetitions of the letter composition task, the subjects were instructed to pick up where they left off in the previous trial.
5. In the **visual counting** task, subjects were asked to visualize numbers being written on a blackboard. The numbers were written sequentially, and each one was erased before the next was written. As with the other tasks, the subjects were asked not to speak, but simply to visualize the numbers. In the repetitions of the visual counting task, the subjects were told to pick up counting where they left off in the previous trial.

These five cognitive tests were administered to seven subjects as summarized in Table 2.2-2.

To the greatest extent possible, the tests were performed without vocalization or physical movement. Each task was repeated five times, in each session. From the chart above, there were a total of 13 sessions; thus, there are

$$13 \text{ (sessions)} \times 5 \text{ (tasks)} \times 5 \text{ (trials per task)} = 325 \text{ trials}$$

in the complete dataset.

The complete dataset is available as a MATLAB variable that is organized as a cell array. Recall from Chapter 2.1b that a cell array is a data structure in MATLAB used to organize data that is related but may be of different types (such as floating-point numbers,

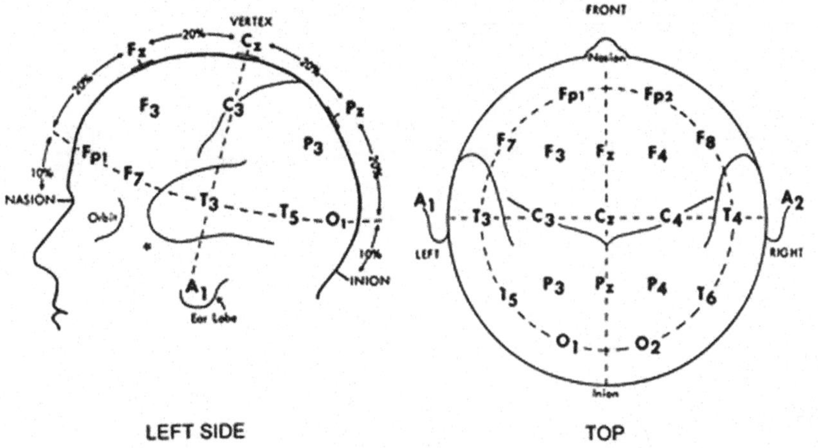

Figure 2.2-7 The International 10-20 system for lead placement in EEG recordings, from Jasper (1958).

Table 2.2-2 Demographic information on subjects tested in Keirn (1988).

Subject	Age	Handedness	Gender	Sessions
1	48	Left	Male	2
2	39	Right	Male	1
3	<30	Right	Male	2
4	<30	Right	Male	2
5	<30	Right	Female	3
6	<30	Right	Male	2
7	<30	Right	Male	1

integers, and character strings). The data set includes a set of trials of EEG measurement and some annotation. Each trial has four elements in the cell:

- Subject number
- Type of the cognitive task
- Trial number
- The digitized EEG data

Each digitized EEG sample is 10 seconds long and was sampled at 250 Hz. Data was recorded from six channels: C3, C4, O1, O2, P3 and P4. The EOG was measured at A1 and A2. Each trial has

$$250 \text{ samples/second} \times 10 \text{ seconds} \times 6 \text{ channels} = 15{,}000 \text{ data}$$

The sampled data has been bandpass filtered for 0.1 to 100 Hz. The frequency components of EEG data are:

- *delta* waves, from 0 to 2 Hz
- *theta* waves , from 2 to 7 Hz
- *alpha* waves, from 7 to 13 Hz
- *low beta* waves, from 14 to 20 Hz
- *high beta* waves, above 20 Hz, typically up to 64 Hz

```
% constants
TRIALS = 5;
EXPERIMENTS = 5;
SESSIONS = 13;
SR = 250; %sample rate
% subject/trial map
SUBJ{1} = 1:50;
SUBJ{2} = 51:75;
SUBJ{3} = 76:125;
SUBJ{4} = 126:175;
SUBJ{5} = 176:250;
SUBJ{6} = 251:300;
SUBJ{7} = 301:325;
load eegdata.mat
```

The problem is to analyze the data and determine whether there is interhemispheric differential activity that shows up in any or all of these cognitive tasks. The potential sites of differential activity for each task:

1. For the baseline/resting task there should be no differential activity.
2. For the verbal tasks, it is possible that there is differential activity in the left hemisphere in the occipital area (O1 or O2).
3. For the mathematical tasks, it is possible that there is differential activity in the left hemisphere, in the parietal area (P3 or P4).

How does one model brain activation? For the analysis, one needs to find a signal representation that is as small as possible, but it should contain the information necessary to differentiate different mental states. The MATLAB program in Example 2.2.3 was written to analyze this large set of data and determine whether there are differences in activation, using a mathematical (and numerical) model of brain activation as frequency content.

If the frequency content of a particular signal is primarily in the alpha band (7 to 13 Hz), then the subject must be resting and not performing any tasks. If the subject is performing a task then the majority of the power shifts to the beta frequency band. Of course, artifact due to 60 Hz noise has to be removed and muscle artifact due to eye blinking has to be removed. The 60 Hz noise is removed from high beta and the muscle artifact will show up in the delta and theta frequency bands.

Example 2.2.3 Differential brain activity in the left and right hemispheres.

Given the dataset described above, analyze the frequency content of the EEG signals to determine if there are differences in brain activity, between hemispheres, that can be attributed to cognitive task, frequency band, or any of the demographic characteristics.

```
for i = 1:length(data) data{i}{4}=double(data{i}{4}); end ;
datamag = data; % create copy of data
f = SR*(0:2500/2)/2500; % frequency data
for m = 1:length(data)
    kernel=fft(datamag{m}{4}(7,:)) ;
    kernel=kernel/max(kernel) ;
    datamag{1} ;
    for (n = 1:6)
        z = abs(fft(data{m}{4}(n,:)).*(1 - kernel)); % remove eog
        z = z (1:length(f)) ;
        z = bpf(59.5, 60.5, SR, z, 'f', 1); % apply bandstop
filter at 60Hz (see bpf.m)
        datamag{m}{5}(n,:) = abs(z); % power series for trial
    end
end
clear avgdata ;
% average over all experiments
for m = 1 : EXPERIMENTS
    trials = [ ] ;
  for n = 1:SESSIONS
        trials = [trials (m – 1)* TRIALS + (n – 1) * EXPERIMENTS * TRIALS + (1:TRIALS)] ;
    end
    avgdata{m} = averagedata(datamag, trials);
    %by subject
    for i = 1 : length(SUBJ)
        subjdata{i,m} = averagedata (datamag, intersect(trials, SUBJ{i})) ;
    end
    %by age
    for i = 1 : 3
        if (i == 1)
            z = [SUBJ{3} SUBJ{4} SUBJ{5} SUBJ{6} SUBJ{7}]; %age < 30
        elseif (i == 2)
            z = SUBJ{2}; % age 30-40
        else
            z = SUBJ{1}; % age 40 - 50
        end
        agedata{i,m} = processavg (averagedata (datamag, intersect (trials, z)), SR) ;
    end
    %by handedness
    for i = 1:2
        if (i == 1)
            z = SUBJ{1}; % LH
        else
        z = setdiff(1:EXPERIMENTS*SESSIONS*TRIALS, SUBJ{1}); % RH, all not in LH
        end
        handdata{i,m} = processavg(averagedata(datamag, intersect(trials, z)), SR) ;
    end
    % by gender
    for i = 1:2
        if (i == 1)
            z = SUBJ{5}; % female
        else
            z = setdiff(1:EXPERIMENTS*SESSIONS*TRIALS, SUBJ{5}); % male (setdiff ==> everyone
              else)
        end
```

```
            genderdata{i,m} = processavg(averagedata(datamag, intersect(trials, z)), SR);
        end
    end
    trialdiff = processavg(avgdata, SR);
    disp('Done');
    end
```

There are several functions, such as bandpass filtering, that are in separate MATLAB scripts. These scripts are not included here, but are available on the companion Web site for this text.

Notice that most of the references here are with respect to the cell array called data. The reader is encouraged to study Chapter 2.1b and the MATLAB help files for more information on how to use cell arrays.

The signal processing steps involved in the filtering are not shown here. The EOG is removed using a technique called a matched filter: the frequency spectra of the EOG signal is computed by the FFT of the A1-A2 difference signal. This frequency content is then subtracted from the frequency content of the other six signals in the recordings for a given trial.

The amount of data is too large to present or analyze here, but an example of the type of analysis is shown in Table 2.2-3, where positive differences are to the left hemisphere and negative differences to the right. The task numbers are references to the list above.

This data suggests that there is a large activation in the left occipital lobe during the geometry task (task 3) and a significant but less left occipital activation during the other mathematics tasks. Although these differences were expected in the parietal region, there were only six electrodes used in the study and it is possible that the signals measured at O1 and O2 had components from the parietal region.

2.2.4 Diabetes and insulin regulation

Most, but not all, diabetics are required to use insulin to manage their glucose (blood sugar) levels. The insulin, administered either as a tablet or injection, acts as a feedback control system to stabilize the blood sugar or keep it within an acceptable range.

Glucose is typically monitored with a glucose tolerance test (GTT). In a GTT, blood samples are taken from a fasting subject at regular intervals of time, following a single intravenous injection of glucose. The blood samples are then analyzed for glucose and insulin content.

To determine what is "normal," or to be expected, requires comparison of the measured data to a model of normality. Since glucose regulation is a dynamic process, this model will take the form of a set of differential equations that express how the concentrations of plasma glucose change over time. If the observed concentrations are comparable to those predicted by the model, then the results can be interpreted by the clinician as normal.

These models can vary in level of detail from the very simple (called minimal) to the complex. The level of detail changes with the number of compartments or systems: Either glucose alone or glucose and insulin concentrations in the blood can be modeled; and the abdomen, kidneys and pancreas can be modeled alone or as a combined system.

Van Riel (2004) describes a minimal model with three compartments:

- Plasma Glucose, $G(t)$, in units of mg/dL,
- Plasma Insulin, $I(t)$, in units of μU/mL, and
- Interstitial Insulin, $X(t)$ in units min^{-1}

The latter is a parameter of a single compartment that accounts for insulin in the abdomen, kidneys and pancreas. The variable $X(t)$ is not a physiological quantity, but is used to model insulin activity. The pair of differential equations that govern this system are

$$\begin{aligned}\frac{dG(t)}{dt} &= k_1(G_b - G(t)) - X(t)G(t)\\ \frac{dx(t)}{dt} &= k_2(I(t) - I_b) - k_3X(t)\end{aligned} \tag{2.2.1}$$

where $G(t_0) = G_0$, the initial concentration of plasma glucose, and $X(t_0) = 0$, that is, there is no interstitial insulin. The term G_b represents the basal level of glucose in the blood; if the glucose concentration is less than the basal level, glucose enters the blood and if the concentration rises above this level then glucose leaves the plasma compartment. The basal levels are typically measured before the administration of the glucose in a GTT.

The second equation expresses that insulin enters, or leaves, the interstitial tissues at rates k_2 or k_3, respectively. Insulin enters the interstitial tissues if the plasma insulin concentration rises above the basal level I_b, and insulin leaves the interstitial tissues if the plasma level falls below the basal level.

This system of equations can be easily solved in either MATLAB or Simulink.

Table 2.2-3 Summary results of hemispheric activation differences in 5 tests on 7 subjects from Keirn (1988). These quantities are expressed in percent differences between the left and right hemispheres.

	Central (C3, C4)					**Parietal (P3, P4)**					**Occipital (01, 02)**				
Freq band	δ	θ	α	βl	βh	δ	θ	α	βl	βh	δ	θ	α	βl	βh
Task	**Composite differences**														
1	−1.66	−0.07	1.64	−2.30	−6.12	−0.64	−0.05	1.24	−1.71	−2.70	−0.40	1.56	4.81	1.72	2.84
2	0.72	0.75	−3.86	−3.60	−0.35	0.97	0.39	−5.59	−7.35	1.67	0.70	1.71	2.09	0.76	4.60
3	−1.62	−1.33	−4.12	−3.44	−1.78	−0.62	−0.99	−1.44	−2.19	0.48	0.75	2.68	6.93	4.89	5.73
4	−0.83	1.13	−2.49	−2.64	−0.38	−0.60	0.54	−1.42	−4.87	0.32	1.89	2.09	3.33	1.27	3.74
5	0.06	0.16	−1.84	−2.66	−2.36	−0.14	0.65	−1.32	−3.72	−0.62	0.46	1.95	4.24	2.17	0.93

Example 2.2.4 Simulink model of glucose regulation.

Use Simulink to solve the system of equations (2.2.1) for glucose as a function of time. Use the constants k_1, k_2 and k_3 and insulin profile $I(t)$ from Van Riel (2004).

To use Eqs. (2.2.1), four unknowns must be determined before using MATLAB or Simulink to solve for $G(t)$. Van Riel (2004) rewrites the second of Eqs. (2.2.1) as

$$\frac{dX(t)}{dt} = k_3(S_I(I(t) - I_b) - X(t))$$

where $S_I = k_2/k_3$ is called the insulin sensitivity.

Van Riel solves this problem using the Runge–Kutta solver ode 4 5 in a MATLAB script. While there is flexibility in using the script to vary the parameters or the insulin profile, a set of MATLAB scripts and a Simulink implementation make for easier documentation of the different model systems. The MATLAB script for normal glucose kinetics is

```
% Set the workspace variables for the
% Simulink model VanRiel.mdl
% Basal Glucose and Insulin
Gb=92;
Ib=11;
G0=279;
% Model constants
SI=5e-4;
k3=0.025;
k1=2.6e-2;
```

where the constants and insulin profile are taken from Van Riel (2004). Notice that the constants G_b and I_b are MATLAB variables and the initial value of the integrator labeled dG/dt is the MATLAB variable `G0` . The gain boxes serve as multipliers by the constants k_1, S_I and k_3, respectively, rather than using a multiplier box and

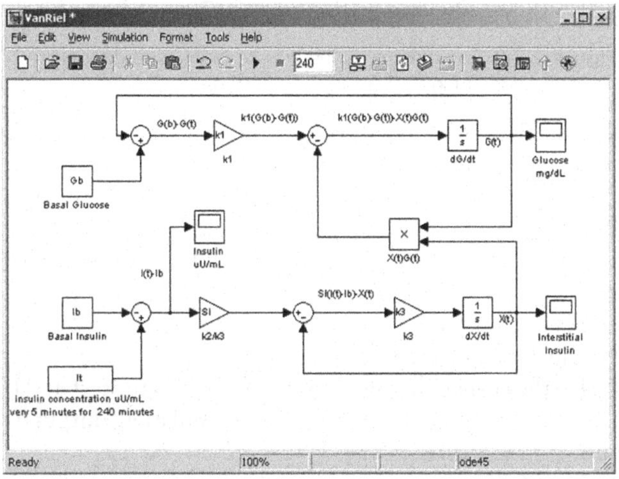

Figure 2.2-8 Simulink model of glucose regulation.

a constant for each. The Simulink model is shown in Fig. 2.2-8.

The key to making this simulation work is the "From Workspace" block, which is in source of the insulin profile $I[t]$. The source for a "From Workspace" block is a 2D array, with the time points in the first column and the values in the second. For this problem, the array is

```
%
% Time course of insulin I(t)
%
t_insu=[0 5 10 15 20 25 30 40 60 80 100 120 140
  160 180 240];
u=Ib+[0 100 100 100 100 00000000000];
It=[t_insu' u'];
```

Notice that the first seven samples are every five minutes and they are spaced further apart until 240 minutes. Since the time step of the simulator must be a constant, it is necessary to indicate to Simulink that it must interpolate the missing values. Also, in the event that the input is

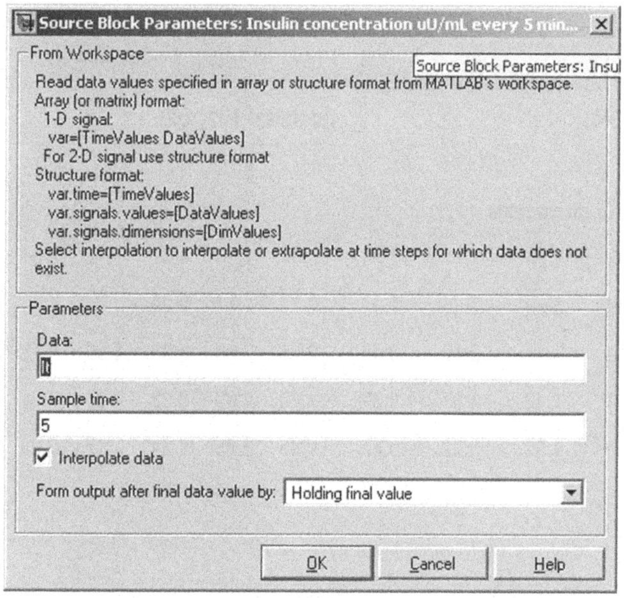

Figure 2.2-9 Parameters for the source block labelled insulin concentration in Fig. 2.2-8.

shorter than the simulation, the "From Workspace" block provides the ability to specify what to do for the missing data. These source sampling parameters are specified in the block parameters, shown in Fig. 2.2-9.

where the data is the MATLAB variable `It`, the sample time is 5 (meaning 5 minutes), the missing data is interpolated, and the final value is held as the output after the final data element.

Lastly, change the simulation parameters to start at 0.0 and end at 240.0 (minutes). The simulation output is shown in the three scopes (Figs. 2.2-10–2.2-12).

In a GTT, the samples measured are compared to a glucose profile that is similar to the simulation results

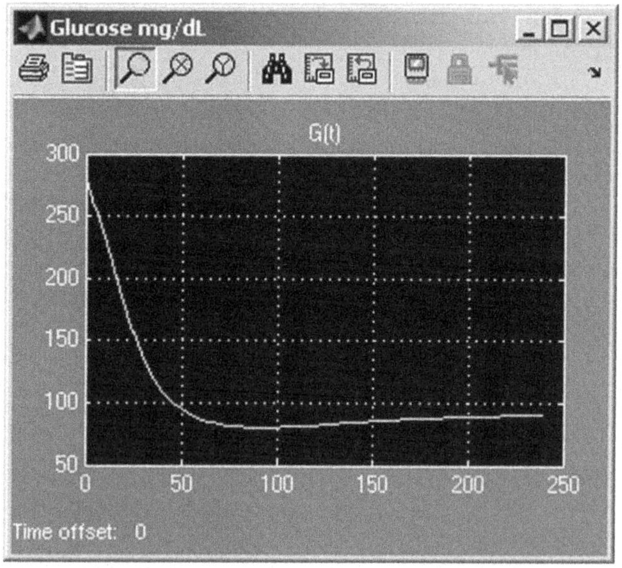

Figure 2.2-10 Glucose.

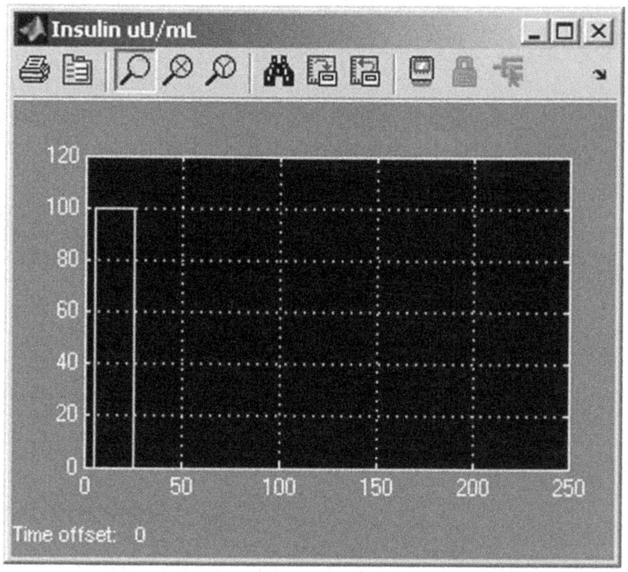

Figure 2.2-11 Plasma insulin.

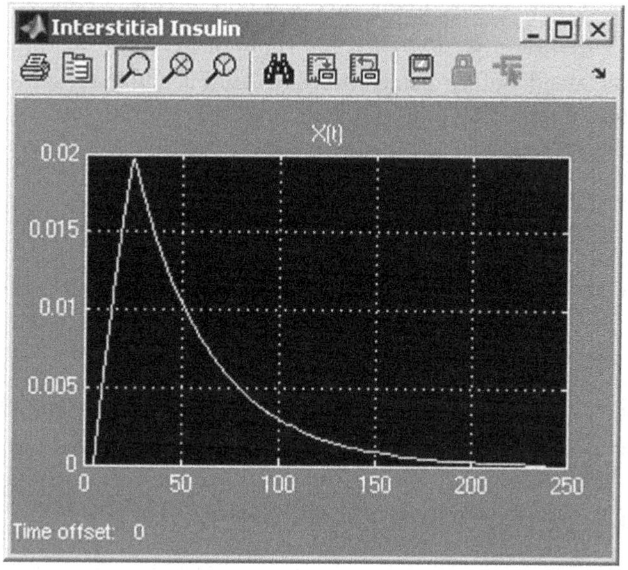

Figure 2.2-12 Interstitial insulin.

above. Notice that the glucose concentration dips below the basal level at approximately one hour after the glucose intake and subsequent insulin release. For this simulation of normal conditions, the minimum glucose concentration is 80 mg/dL.

In normal healthy adults, cognitive function is impaired when plasma glucose falls below 65 mg/dL. If the plasma glucose dips below this level and the patient has altered mood and/or impaired cognitive function, a diagnosis of reactive hypoglycemia is often made. If the glucose level is normal, but the patient exhibits these signs, then the condition is referred to as *postprandial syndrome*.

The same model can also be used to simulate glucose regulation in a diabetic. The MATLAB script for the model parameters is

```
% Set the workspace variables for the
% Simulink model VanRiel.mdl
% Basal Glucose and Insulin
Gb=92;
Ib=11;
G0=365;
% Model constants
SI=0.7e-4;
k3=0.01;
k1=1.7e-2;
```

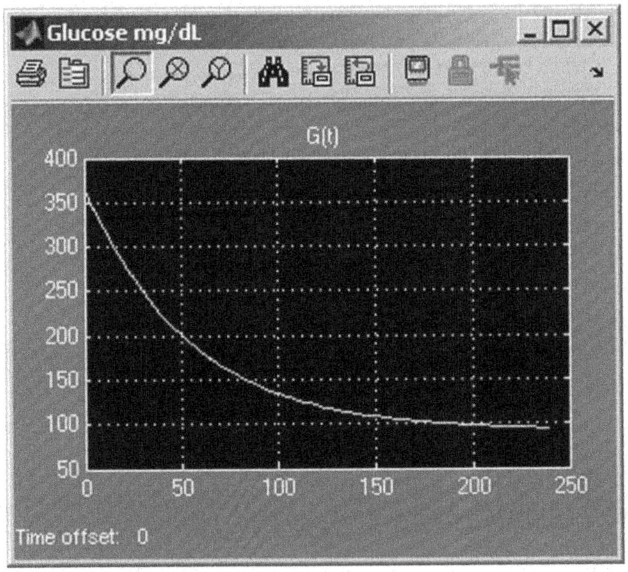

Figure 2.2-13 Glucose concentration.

and if the simulation is run for 240 minutes with the same insulin profile, the glucose concentration is shown in Fig. 2.2-13.

The insulin concentration is shown in Fig. 2.2-14.

The interstitial insulin concentration is shown in Fig. 2.2-15.

The NIH-NIDDK criterion for diabetes is a glucose concentration of 200 mg/dL or more, 2 hours after the glucose administration. This simulation shows a profile that would be diagnosed as pre-diabetes, or impaired glucose tolerance.

2.2.5 Renal clearance

Renal clearance (Cl_R) is a measure of kidney transport in units of volume of plasma per unit time. The volume of plasma measured is that volume for which a given substance (e.g., urea or drugs) is completely removed per minute. There are a number of specific forms of renal

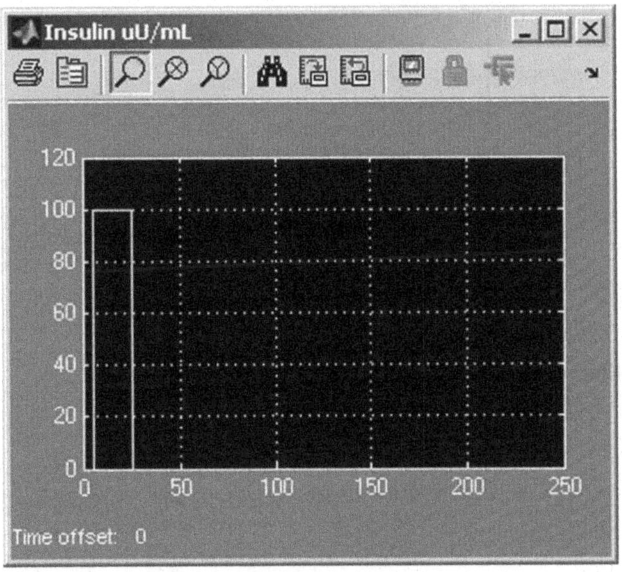

Figure 2.2-14 Insulin concentration.

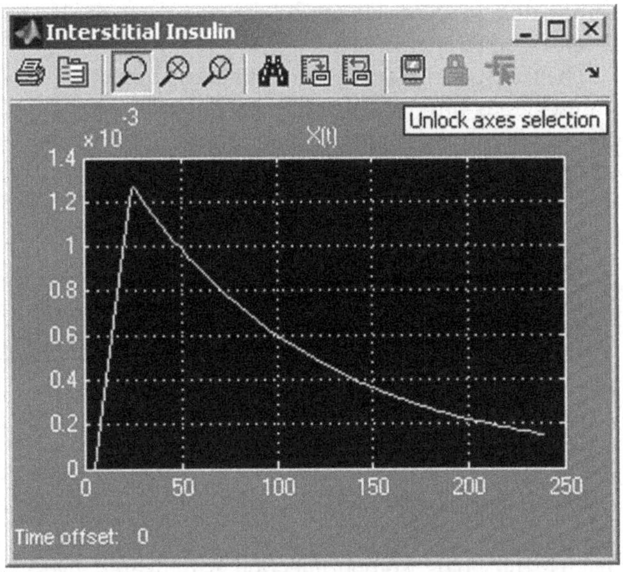

Figure 2.2-15 Interstitial insulin concentration.

clearance: Glomerular Filtration Rate, Effective Renal Plasma Flow, or Tubular Extraction Rate.

There are a number of techniques for evaluating renal kinetics, some of which are based on measuring radionuclide concentration over time from urine and plasma; plasma alone; or using a gamma camera to measure isotope concentration in the kidneys as the isotope is extracted from the plasma. Renal function is assessed by comparing clinical measurements to simulation results based on a model of normal kinetics. The last two of the three cases are easiest to model with only two compartments, central and peripheral.

One such two-compartment model was described by Estelberger and Popper (2002). In that model, there are two functionally separated spaces, a well-perfused

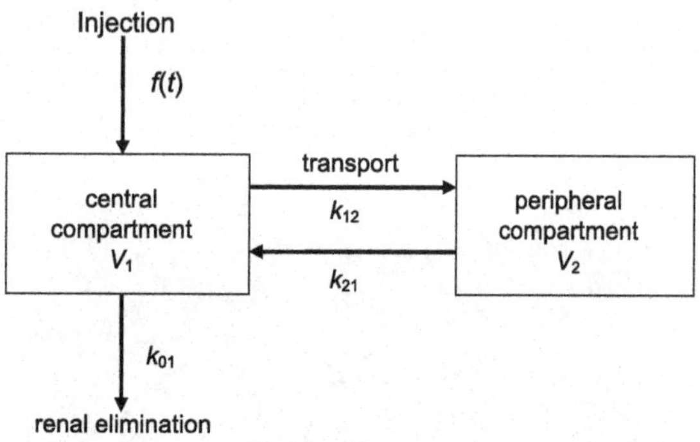

Figure 2.2-16 A two-compartment model of renal clearance, from Estelberger and Popper (2002).

central volume and a less-perfused peripheral compartment. The marker kinetics, measured indirectly from the time course of radionuclide concentration in the two compartments, is the result of the isotope injection, the transport between the two compartments, and the renal elimination process. Figure 2.2-16 is a diagram of the Estelberger/Popper model.

The transport model can be formulated by a set of two simultaneous differential equations describing the rates of change of radionuclide concentration in the two compartments:

$$\frac{dc}{dt} = d(t) - (k_{01} + k_{21})c + k_{12}p$$
$$\frac{dp}{dt} = k_{21}c - k_{12}p \tag{2.2.2}$$

where

$$d(t) = \frac{D}{T}, \quad 0 \le t < T \tag{2.2.3}$$

that is, the bolus isotope injection is delivered over T seconds. For purposes of this exercise, assume that the isotope is delivered uniformly during this period of time. Assume also that the central volume V_1 is unknown.

The brief paper by Estelberger and Popper shows how the group estimates the parameters of the model from empirical data. The optimization procedure is beyond the scope of this text and rather than detract from the example, the reader is referred to Estelberger and Popper (2002) for details.

In comparison to the simulation given in Section 2.2.3, this set of differential equations will be solved using MATLAB, rather than Simulink.

Example 2.2.5 Renal clearance.

Solve the system of two differential equations (2.2.2) and the initial value condition (2.2.3) for renal clearance using the constants $k_{01} = 0.0041$, $k_{12} = 0.0585$, $k_{21} = 0.0498$, and $V_1 = 7.3$, $c(0) = p(0) = 0$.

This problem is designed to illustrate the use of global variables in MATLAB. The mathematical problem is formulated as a conservation of mass with the isotope injection beginning at time zero; the length of the bolus injection varies with trial.

The parameters of the model and the parameters of the bolus injection are set as global variables, which can then be used by the script that generates the graph and the script that evaluates the system of differential equations, called `renal.m`.

First, the system of equations (2.2.2) is coded as a MATLAB function as per the MATLAB convention, with arguments time t and the vector of function values, y. All other model parameters are global variables.

```
function yprime=renal(t,y)
% separate the components of the state:
% central compartment concentration and
% peripheral compartment concentration
c=y(1);
p=y(2);
%
global D tau;
global k01 k12 k21 V1;
% let time t be in minutes
% now compute f(t)
%
if t <= tau
   f=D/tau;
else f=0;
end
%
dcdt=(f-(k01+k21)*c+k12*p)/V1;
dpdt=k21*c-k12*p;
yprime=[dcdt, dpdt]';
```

There is a single script that solves the system of equations three times—once for each of the different injections of isotopes. A single figure with the three concentrations in the central compartment is plotted with different symbols. The succeeding plots are added to the same figure by using the `hold` on command.

```
% Example 2.2.5 Renal Clearance
global D tau;
global k01 k12 k21 V1;
% parameters
k01=0.0041;
k12=0.0585;
k21=0.0498;
V1=7.3;
% initial value conditions
c0=0;
p0=0;
```

```
%
D=2500;
tau=0.5;
% use ode45
[t,y]=ode45(@renal, [0:0.5:240], [c0p0]);
plot(t,y(:,1),'b.');
xlabel('Time t (min)')
ylabel('Isotope concentration c(t)')
hold on
%
D=2500;
tau=10.0;
% use ode45
[t,y]=ode45(@renal, [0:0.5:240], [c0p0]);
plot(t,y(:,1), 'g:');
%
D=2500;
tau=240;
% use ode45
[t,y]=ode45(@renal,[0:0.5:240],[c0 p0]);
plot(t,y(:,1), 'r--');
hold off
```

In each case, the solution is run for 240 minutes in time steps of 0.5 minutes. The set of solutions is shown in the graph in Fig. 2.2-17. Notice that the first two solutions have very similar trajectories; the traces of isotope concentration are the impulse response of the system to the bolus radionuclide input, whereas the solution to the ramp input (the dashed line) does not exhibit the feedback control at 240 minutes.

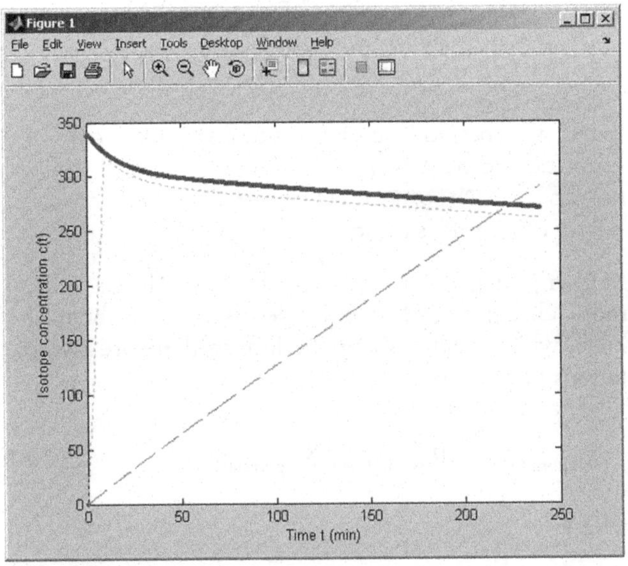

Figure 2.2-17 Solutions to renal clearance program. Solid line: bolus injection over 0.5 minutes; dotted line: injection over 10 minutes; dashed line: injection over 240 minutes.

2.2.6 Correspondence problems and motion estimation

A key concept in estimating motion from image sequences is that the relative motion between a pair of images can be estimated from the correspondence of a number of markers or features on the object(s). The features must be uniquely identifiable, and it is often assumed that the object is rigid or articulated.

Gait, the analysis of human body motion, is estimated by recording the location of features on the hips and legs. These features can be, for example, light emitting diodes (LEDs), and the motion is recorded by having the subject walk in a dark room. Many rehabilitation facilities have gait laboratories where such recordings take place, and the motion recorded can be used for diagnostic purposes.

This example, courtesy of Professor Charles Krousgrill at Purdue University, illustrates how rotation matrices can be estimated from the correspondence of markers on the hips and legs (see Fig. 2.2-18).

Example 2.2.6 Estimating motion from features on a rigid body.

Six markers have been attached to the hip, thigh, and lower leg, as shown in Fig. 2.2-18. The coordinates for these markers are given in Table 2.2-4.

Position data have been collected for these markers and the hip from one frame of motion. These coordinates are given in terms of global coordinates in 3D in Table 2.2-5.

Assuming a segment (thigh or lower leg) length of 18.13 inches, use the above data to determine the rotation matrices for the thigh and lower leg. Determine the 3D global coordinates for the position of the ankle B.

The position of the markers on the hip and legs are specified in *local* coordinates, 3D position with respect to one of the markers. For this exercise, the hip is the reference marker for the markers on the thigh, and the knee (position A) is the reference for the lower leg, all shown

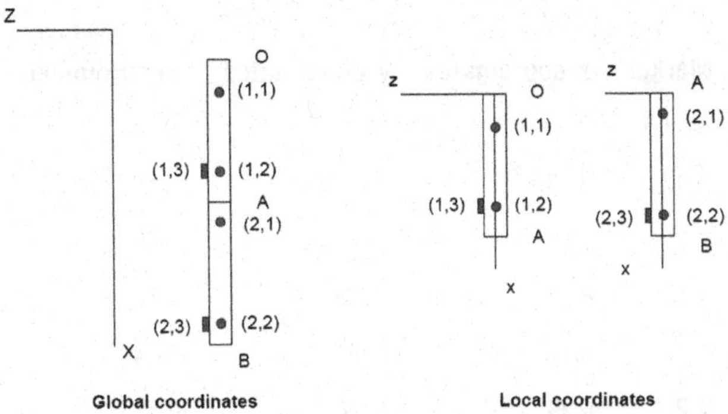

Figure 2.2-18 Coordinate systems for gait markers on the hip and legs, from Krousgrill (2005).

Table 2.2-4 Local coordinates for gait markers on the hip and legs, from Krousgrill (2005)

Marker	x-coordinates	y-coordinates	z-coordinates
(1,1)	6	4	0
(1,2)	12	4	0
(1,3)	12	0	4
(2,1)	6	3	0
(2,2)	12	3	0
(2,3)	12	0	4

in Table 2.2-5. The local coordinates are given in Table 2.2-4.

The motion cannot be recorded in a local coordinate system, but only in the *global* coordinate system of the gait laboratory. The cameras record images of the subject while walking, and the position of the markers in these images can be used to determine the motion *with respect to the position of the camera(s) in the laboratory, not with respect to the hip or leg.* The problem of determining the rotations of the thigh and lower leg then means switching between the local and global coordinates.

The general strategy to solve this problem is to (1) convert the recorded global coordinates to local coordinates for each segment, and (2) determine the rotation from the correspondence of the markers when referenced to the initial or resting position. The rotation of the thigh must be computed first, since it is required when solving for the rotation of the lower leg.

Bookkeeping is always a challenge when solving for motion. There are three sets of coordinates to keep track of: the local coordinates of each segment before motion (at rest), the global coordinates of each marker after motion, and the local coordinates of each segment, which are computed.

Table 2.2-5 Global coordinates of gait markers on the hip and legs after rigid body motion, from Krousgrill (2003)

Marker	x-coordinates	y-coordinates	z-coordinates
Hip (0)	2	1	0
(1,1)	3.18	7.13	3.60
(1,2)	7.16	10.47	6.60
(1,3)	8.89	5.88	9.41
(2,1)	9.11	15.45	10.44
(2,2)	6.99	20.66	12.52
(1,1)	9.36	19.88	16.86

The first step is to initialize some MATLAB variables with the data from Tables 2.2-4 and 2.2-5.

```
% Example 2.2.6
% At rest (in local coordinates)
% Each column vector is the 3D coordinates
% of a marker (initial conditions)
ThighRestLocal=[ 6 4 0; 12 4 0; 12 0 4]';
LegRestLocal=[6 3 0; 12 3 0; 12 0 4 ];
ThighFinalGlobal=[3.18  7.13  3.60;  7.16
  10.47 6.6; 8.89 5.88 9.41] ';
LegFinalGlobal=[9.11  15.45  10.44;  6.99
  20.66 12.52; 9.36 19.88 16.86]';
% Column vectors of positions of O, A and B
HipPos=[2 1 0]';
ThighLen=[18.13 0 0];
LegLen=[18.13 0 0]';
```

The final position, in global coordinates of the thigh after motion, is given by the equation

$$X_{thigh,motion} = R_{thigh} \cdot X_{thigh,initial} + O \tag{2.2.4}$$

where $X_{initial}$ and X_{motion} are matrices whose column vectors are the positions of the markers, R_{thigh} is the rotation about the hip, and O is the 3D position of the hip after motion. R_{thigh} can be computed from

$$R_{thigh} = (X_{thigh,motion} - O) \cdot (X_{thigh,initial})^{-1} \tag{2.2.5}$$

The MATLAB implementation is

```
% Find rotation of thigh about hip
% First, determine coordinates of thigh
  markers
% with respect to the hip
ThighFinalLocal=ThighFinalGlobal-
  [HipPos HipPos HipPos];
% Rotation computed by correspondence
  matching
% and is in the local coordinate system
  centered at the Hip
RotThigh=ThighFinalLocal*
  inv(ThighRestLocal);
```

Computing the rotation of the lower leg is a little more challenging. The lower leg is rotated about the knee, which rotates about the hip, so there are two rotations to the motion:

$$X_{leg,motion} = R_{thigh} \cdot R_{leg} \cdot X_{leg,initial} + A \tag{2.2.6}$$

where

$$A = O + R_{thigh} \cdot [18.13\ 0\ 0]^T \tag{2.2.7}$$

After substitution,

$$R_{leg} = R_{thigh} \cdot (X_{leg,motion} - A) \cdot (X_{leg,initial})^{-1} \quad (2.2.8)$$

For which the MATLAB implementation is

```
% Find final position of knee
KneePos=ThighLen;
FinalKneePos=(RotThigh*KneePos)+HipPos;
% Find rotation of leg about knee
% First determine local coordinates of leg
  markers
% with respect to the knee
% The rotation is in the local coordinate
  system of the Knee
LegFinalLocal=LegFinalGlobal-
  [FinalKneePos FinalKneePos
FinalKneePos];
CombinedRotLeg=LegFinalLocal*
  inv(LegRestLocal);
RotLeg=inv(RotThigh)*CombinedRotLeg;
```

Lastly, the position of the ankle has to be computed, since there is no marker at the ankle. The global position after motion is given by rotating the end of the lower leg (position B in Fig. 2.2-3) and translating it with respect to the position of the knee:

$$B = R_{thigh} \cdot R_{leg} \cdot [18.13\ 0\ 0]^T + A \quad (2.2.9)$$

which is, in MATLAB,

```
% Coordinates of the ankle are
% Ankle=Knee+Rot*Ankle at Rest
% Ankle is in the local coordinates of the
% Knee, which is the center of rotation
AnklePos=LegLen;
FinalAnklePos=FinalKneePos+
  CombinedRotLeg*AnklePos;
```

The rotation matrices are

```
>> RotThigh
RotThigh =
   0.6633  -0.7000  -0.2675
   0.5567   0.6975  -0.4500
   0.5000   0.1500   0.8525
>> RotLeg
RotLeg =
   0.4233  -0.8933   0.1581
   0.9063   0.4183  -0.0749
  -0.0011   0.1747   0.9864
```

and the position of the ankle is

```
>> FinalAnklePos
FinalAnklePos = 7.6203
  26.8353
  15.3501
```

Rotation matrices should be orthogonal, since the transformation does not change the magnitude or length of the thigh or lower leg. A matrix **R** is orthogonal if $\mathbf{R}^T\mathbf{R}$ is the identity matrix **I**

```
>> RotThigh'*RotThigh
ans =
   0.9999  -0.0011  -0.0017
  -0.0011   0.9990   0.0013
  -0.0017   0.0013   1.0008
>> RotLeg '* RotLeg
ans=
   1.0007   0.0007  -0.0020
   0.0007   1.0035  -0.0003
  -0.0020  -0.0003   1.0036
```

which are both the identity matrix I to within numerical error. Check the condition number of each matrix, and check how error is propagated when computing an inverse.

2.2.7 PHYSBE Simulations

PHYSBE is a model of the circulatory system for simulating the flow of oxygen, nutrients, heat, or chemical tracers within the bloodstream (McCleod, 1966, 1968). Although the fundamental work on PHYSBE dates from the 1960s, PHYSBE has since been implemented in Simulink, making it accessible for educational and research uses. The Simulink implementation of PHYSBE can be downloaded from the MathWorks MATLAB Central Web site. This section includes several examples of how to use PHYSBE to model cardiovascular system pathologies and predict the result on blood flow.

Example 2.2.7 Normal PHYSBE operation.

Use a procedure to install and run one simulation of PHYSBE.

Start the PHYSBE simulation by running the MATLAB script `pctrl2` to open the PHYSBE control panel (Fig. 2.2-19). Do not change any of the default values, but be sure to save the parameter file.

Next, start the PHYSBE Simulink model (Fig. 2.2-20) and start the simulation.

The simulation results which are displayed on a Simulink scope are the heart pressure (Fig. 2.2-21) and heart volume (Fig. 2.2-22).

Both are easiest to differentiate by color. There is also a floating scope, which can be tied to other system parameters of interest.

These normal results will be compared to the pressure and volume relationships in each of the examples below.

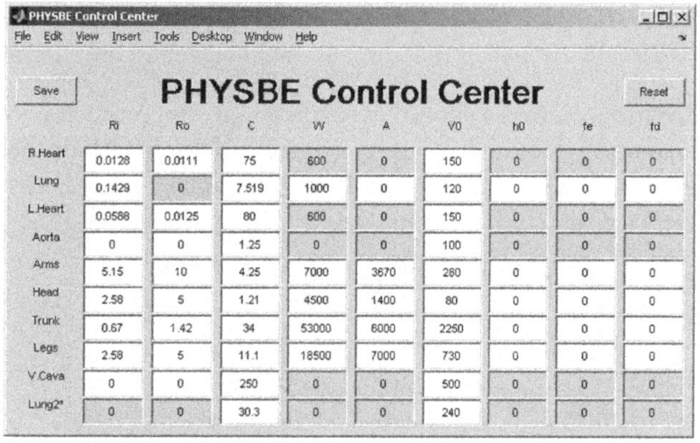

Figure 2.2-19 PHYSBE control panel.

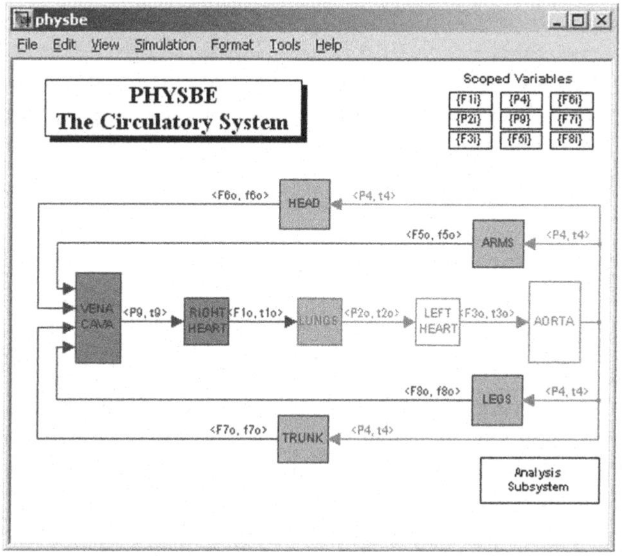

Figure 2.2-20 PHYSBE Simulink model.

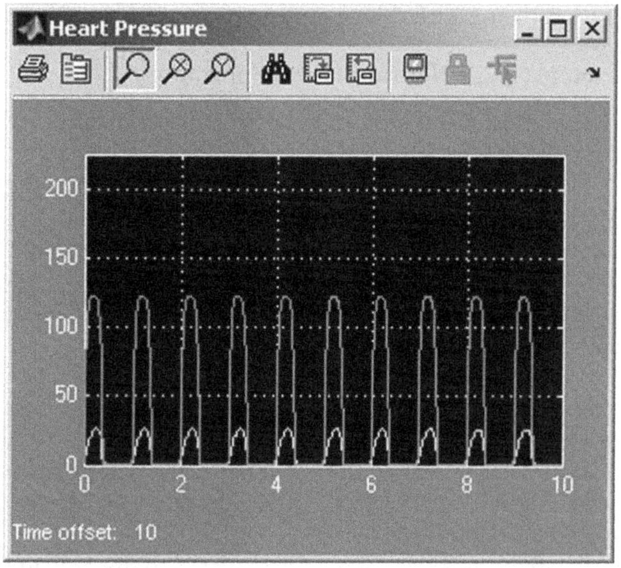

Figure 2.2-21 PHYSBE Simulink model: heart pressure.

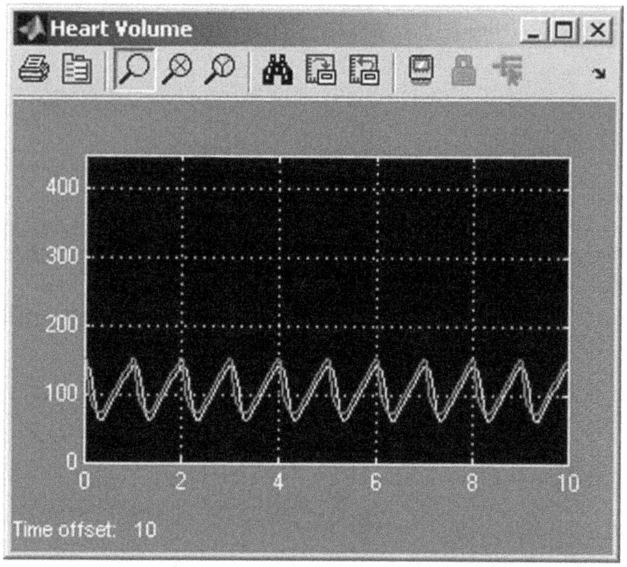

Figure 2.2-22 PHYSBE Simulink model: heart volume.

2.2.7.1 Coarctation of the aorta

The affliction characterized by a short constriction of the aorta is known as a *coarctation*. Such narrowing creates a pressure difference whereby abnormally high blood pressure exists before the point of coarctation, and abnormally low blood pressure exists after the point of coarctation. Commonly, this narrowing is situated such that there is a pressure differential between the upper body and arms (high blood pressure) and the lower body and legs (low blood pressure) (Suk, 2001). Fig. 2.2-23 shows where a coarctation occurs.

The untreated consequences of coarctation of the aorta are numerous and severe. Without repair, this condition results in high mortality from complications such as myocardial infarction, intracranial hemorrhage, infective endocarditis, coronary artery disease, and aortic rupture. There is a stunning mortality rate of 90% before the age of 50 when this condition is left untreated.

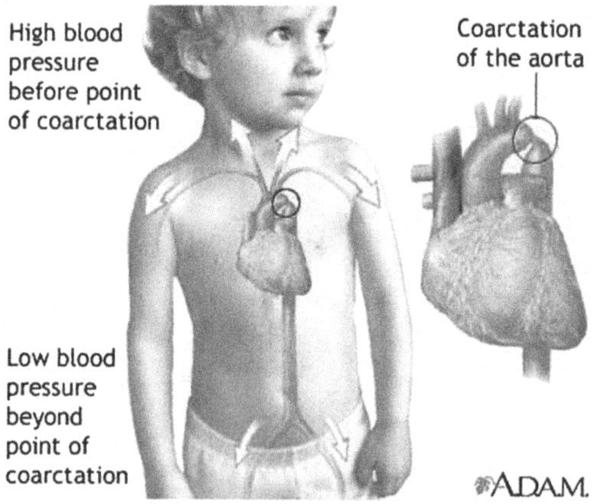

Figure 2.2-23 Coarctation of the aorta, from Suk (2001).

Optimistically, the dire consequences of this ailment can be almost completely eliminated with modern medical techniques.

Coarctation is a sort-of "pinch" in the aorta located between significant arterial branchings (Sokolow and McIlroy, 1981). This pinch translates into a significantly reduced aortic radius. Because blood flow rate immediately before, within, and after the pinch must remain constant, velocity must increase (since $A = \pi r^2$ will decrease) and this translates into a pressure gradient that *reduces* pressure to the lower extremities and trunk. A reduction in pressure to the lower extremities and trunk is implemented in Simulink with the addition of a resistance to the flow entering these regions. The target change in pressure is generated by this resistance according to Ohm's relation $Q = \Delta P/R$ (Li, 2004).

Example 2.2.8 Simulink model of coarctation of the aorta.

Modify the Simulink implementation of PHYSBE to model a coarctation of the aorta and predict the effect on blood pressure and volume in the heart.

The only direct modifications to the PHYSBE model that need to be made are the addition of a resistance and a pulse delay to the descending aorta.

In the Simulink model, the major modifications were implemented in a subsystem format (for simplicity and ease of use). The pulse delay was implemented via a pulse delay block. The time-of-delay parameter input to this block was 0.08 seconds. Data for the pulse delay was in this range, and cited as anywhere between 0.1 and 0.08 seconds.

The resistance is implemented by adding a gain block with a value equal to *1/R*, following the pulse delay and in series with the vessels extending to legs and trunk. Because the value of this resistance is dependent upon individual physiology, conditional geometry, and severity, it can vary from case to case.

The coarctation subsystem (Fig. 2.2-24) is between the ascending aorta and the legs (Fig. 2.2-25).

Lastly, the pressures on either side of the coarctation are output to MATLAB objects for plotting and quantification purposes. These new MATLAB objects, called PA and PB (pressure above and pressure below), are visible in the coarctation subsystem in Fig. 2.2-24. These objects were used by the PHYSBE Control and Analysis subsection and shown as the 10th and 11th columns in the modified PHYSBE Control Center.

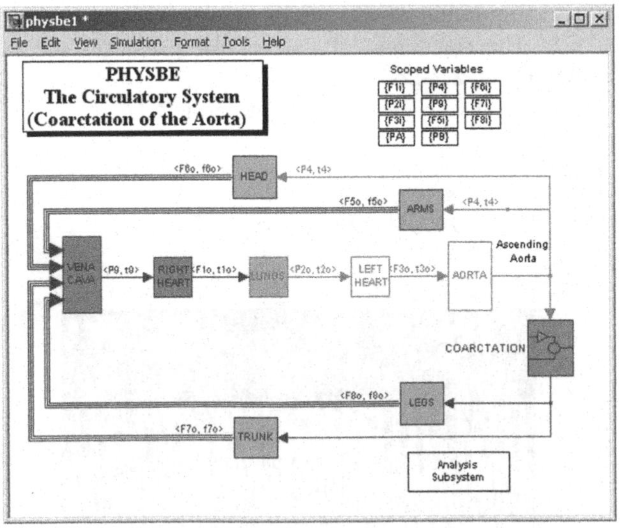

Figure 2.2-25 Location of coarctation subsystem in circulatory system.

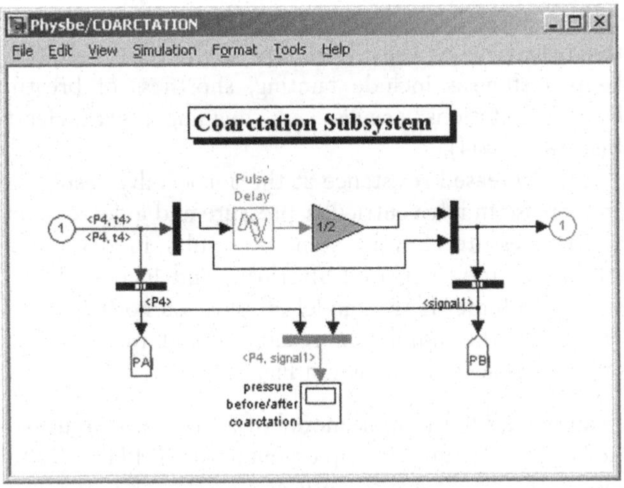

Figure 2.2-24 Coarctation subsystem.

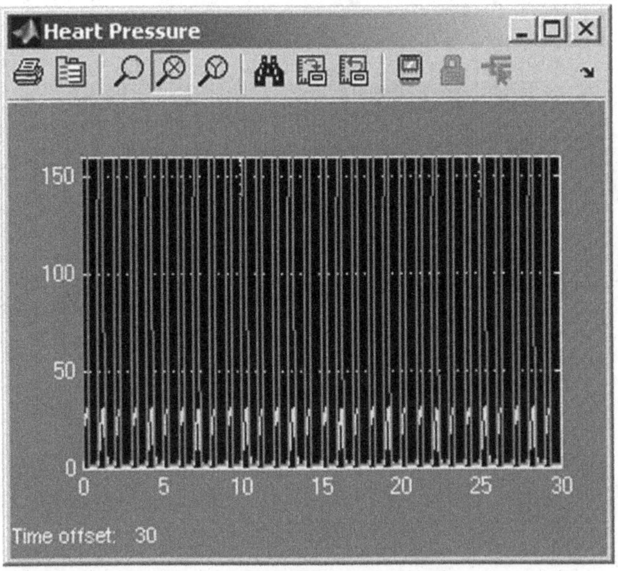

Figure 2.2-26 Heart pressure simulation results: right and left ventricular pressures.

The simulation results are as shown in Figs. 2.2-26–2.2-28. The right and left ventricular pressures are as shown in Fig. 2.2-26.

The blood volumes in the right and left heart:

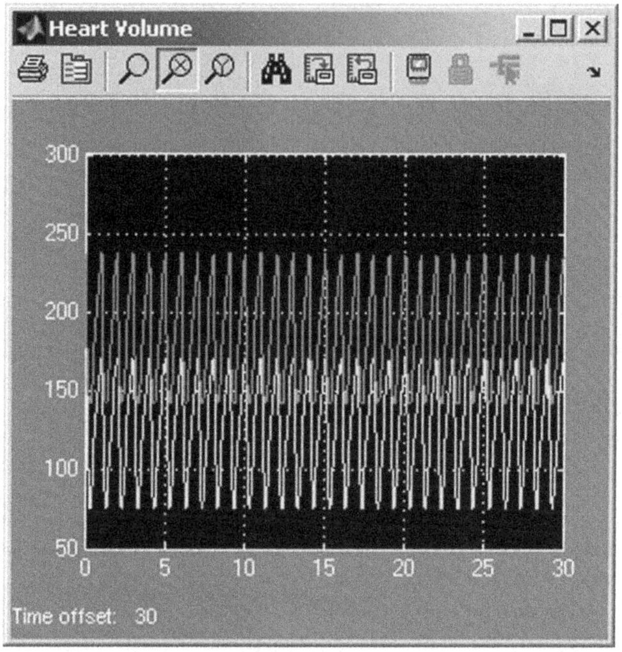

Figure 2.2-27 Blood volume simulation results, left and right heart.

The blood pressure before and after the coarctation:

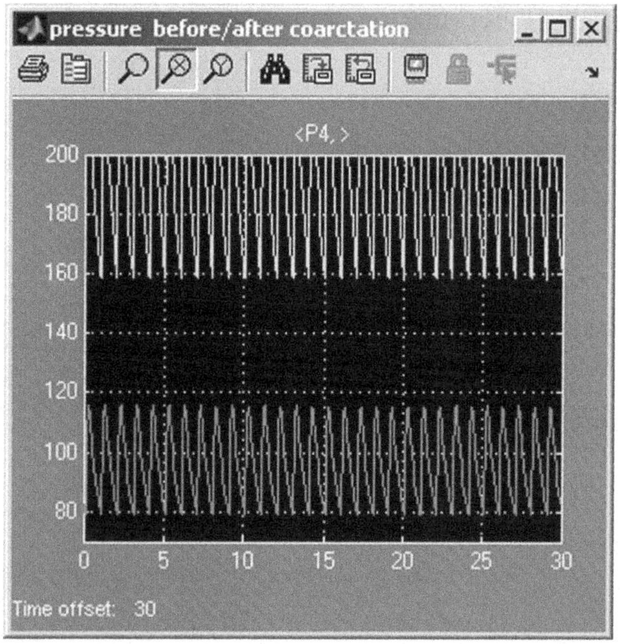

Figure 2.2-28 Blood pressure simulation results, before and after coarctation.

Notice that there is an increase in left ventricular pressure, while the right ventricular pressure is not affected. The coarctation causes an increase in left ventricular pressure as blood encounters greater resistance. By the time the blood has passed through the capillaries and pooled during venous return, the pressure drop caused by the coarctation is diminished. The same analysis could be traced in reverse over the pulmonary circuit.

These results suggest that a small change in resistance (that is, conductance) affected pressures throughout the body. However, the PHYSBE system is somewhat limited in that a pressure differential of 15 mmHg between the right and left arms suggests the narrowing of the artery (Braunwald, 1988). This incipient coarctation could not be modeled by PHYSBE, as the arms as well as the legs are lumped into one subsystem each.

2.2.7.2 Aortic stenosis

Aortic stenosis is a condition where deposits on the aortic valve cause it to become narrowed or blocked. As a result, the valve is unable to open properly, and blood flow out of the left ventricle into the aorta is reduced, causing the left ventricle to work harder to compensate and ensure that the body receives the necessary blood supply (Texas Heart Institute, 2004).

There are numerous causes of this condition, including congenital defects, rheumatic fever, and calcification on the aortic valve. A small percentage of people are born with two cusps on their aortic valve instead of three and, as a result of wear and tear over the years, the valve may become calcified or scarred, or its motility may be reduced. Rheumatic fever also damages the cusps of the aortic valve, causing the edges of the cusps to fuse together. As a person ages, the collagen in his/her body, including in the cusps of the aortic valve, is destroyed, and calcium deposits form. The calcium deposits on the valve reduce the cusps' motility and therefore increase the resistance of the blood flow. Symptoms resulting from a stenosis include fainting, shortness of breath, heart palpitations, angina, and coughing (Texas Heart Institute, 2004).

The increased resistance at the aortic valve results in an increase in left ventricular pressure and a decrease in aortic pressure. Severe stenosis results in decreased stroke volume, increased afterload, and increased end systolic volume. If the model of aortic stenosis is accurate, then the simulation results should mimic these pressure changes as shown in Fig. 2.2-29.

Example 2.2.9 Simulink model of aortic valve stenosis. Modify the Simulink implementation of PHYSBE to model an aortic valve stenosis and predict the effect on blood pressure and volume in the heart.

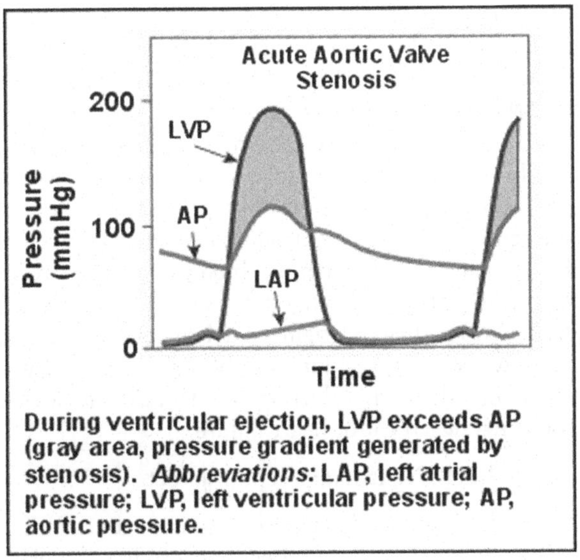

Figure 2.2-29 Pressure as a function of time during Aortic Valve Stenosis, from Klabunde (2004).

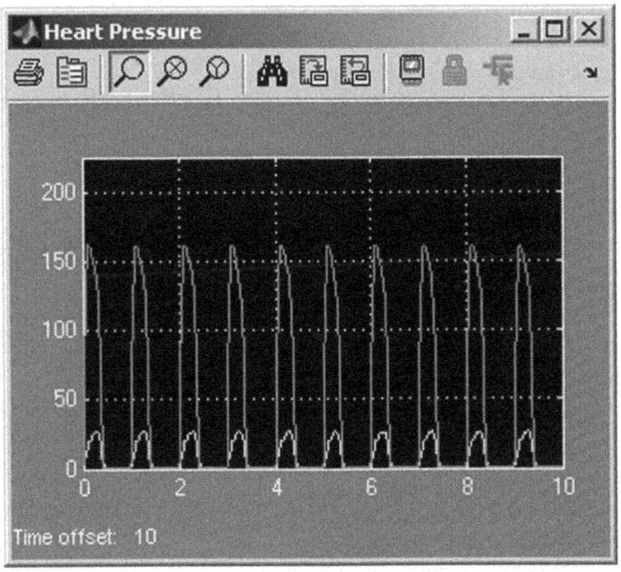

Figure 2.2-30 Aortic stenosis: heart pressure simulation results.

The PHYSBE model does not allow for changing physical parameters such as the radius of the aortic valve, which would be the most accurate model. Therefore, in order to model an aortic valve stenosis using PHYSBE, the resistance (R_0) of the left heart was increased. When stenosis occurs and the radius of the valve decreases, the resistance to blood flow increases, since resistance is inversely proportional to radius as given by Poiseuille's Law (Germann, 2005). The physiological effect of a stenosis is a decrease in blood flow due to the increase in resistance (due to the decrease in radius) as modeled by the relationship $F = \Delta P/R$, where F is blood flow, ΔP is the change in pressure, and R is the resistance. This decrease in blood flow was used to show whether or not an aortic valve stenosis was modeled.

The resistance was changed from 0.0125 mmHg/mL/s to 0.135 mmHg/mL/s, which is a physiologically relevant value for an aortic stenosis. This value was obtained by converting the value for severe aortic stenosis, 180 dyne s, which was reported by Mascherbauer (2004).

When there is an aortic stenosis, maximum heart pressure within the left ventricle increases to over 160 mmHg (Fig. 2.2-30).

When an aortic valve stenosis occurs, the end systolic volume, or the volume of blood left in the heart after a single pump, increases to about 85 mL (Fig. 2.2-31).

In a normal heart, maximum flow out of the left ventricle is about 925 mL/min (Fig. 2.2-32).

In the case of an aortic valve stenosis, the blood flow out of the heart is cut in half to 460 mL/min (Fig. 2.2-33).

In this simulation, as in the pathophysiology, blockage of the aortic valve results in an increased resistance to blood flow, due to the inverse relationship of radius and resistance. Because the relationship of blood flow to resistance is $F = \Delta P/R$, an increase in resistance should result in a decrease in blood flow out of the heart. If not enough blood is being ejected from the heart, and if blood flow is not high enough, the necessary oxygen and other nutrients, which the tissues need to survive, will not be efficiently delivered via the vascular system. The decrease in flow also triggers a homeostatic response within the body to compensate for the loss of blood flow. The left ventricle has to pump harder and thus fatigues at a faster rate than a normal heart.

In addition, as a result of the increased resistance, blood cannot escape the heart, specifically the left ventricle, as quickly as under normal conditions. Thus the

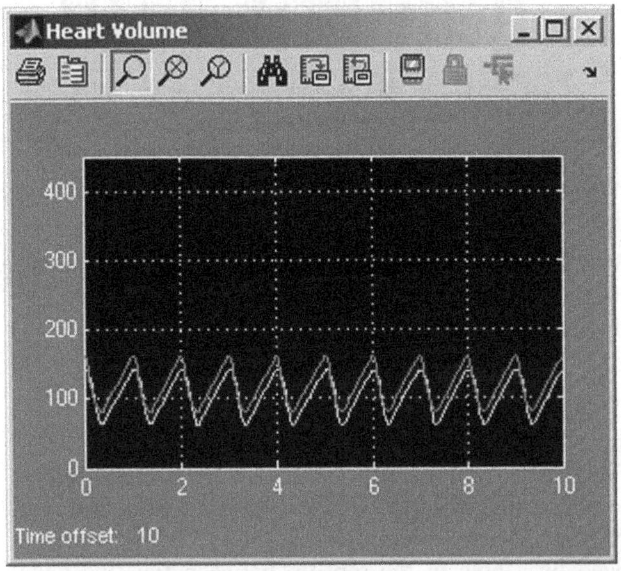

Figure 2.2-31 Aortic stenosis: heart volume simulation results.

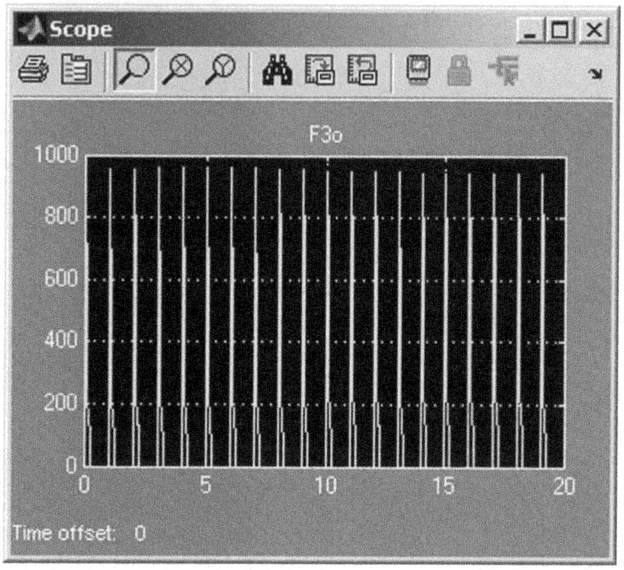

Figure 2.2-32 Normal heart: left ventrical outflow simulation results.

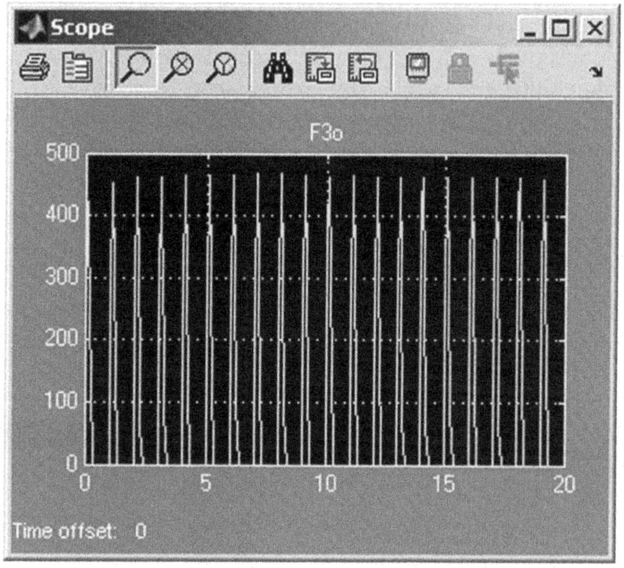

Figure 2.2-33 Aortic stenosis: heart blood outflow simulation results.

end systolic volume, or the amount of blood that remains in the heart after the heart has finished contracting, also increases.

2.2.7.3 Ventricular septal defect

Ventricular septal defect (VSD) is a heart malformation often arising at birth or in conjunction with other pathology, such as myocardial infarction. VSD is characterized by a hole in the septum between the two lower ventricles (Lue and Takao, 1986) and is the most common congenital heart defect. The significance of VSD varies with the size of the hole: the greater the hole, the higher the risk. The size of the hole can range from a pinpoint to almost the absence of the septum. In the case of a small hole, it can spontaneously close within the first three years of life. In the case of a large hole, the left ventricle shunts blood to the right ventricle via the septum hole. The extra blood causes the right ventricle to do more work and makes the lung receive too much blood, and this increases the pressure. If the problem is not resolved, it can result in a weakened or enlarged right ventricle caused by stress and overwork. Also, the lung can become crowded with the extra blood, causing clotting in the lungs that can lead to abnormal heartbeat or heart failure.

Example 2.2.10 Ventricular septal defect.

Modify the Simulink implementation of PHYSBE to model a ventricular septal defect and predict the effect on blood pressure and volume in the heart.

VSD can be crudely modeled in PHYSBE as blood flow from the left to right ventricles. The flow between the ventricles, stimulated by a pressure difference, can be modeled using Ohm's Law:

$$Q_s = \frac{P_{lv} - P_{rv}}{R} \quad \text{and} \quad R = \frac{8\eta L}{\pi r^4}$$

Blood has a viscosity, η, of 3 cP (Fournier, 1999), and the thickness of the ventricular wall is, in an adult, 4 mm (Lue and Takao, 1986). Here Q represents the blood flow, R the resistance, and r the radius of the hole.

A ventricular septal defect is modeled in PHYSBE in the left heart. First, the right ventricular pressure is connected to the left heart to allow modeling through a global "goto" tag (Fig. 2.2-34).

This is subtracted from the left ventricular pressure to measure the pressure change (Fig. 2.2-35).

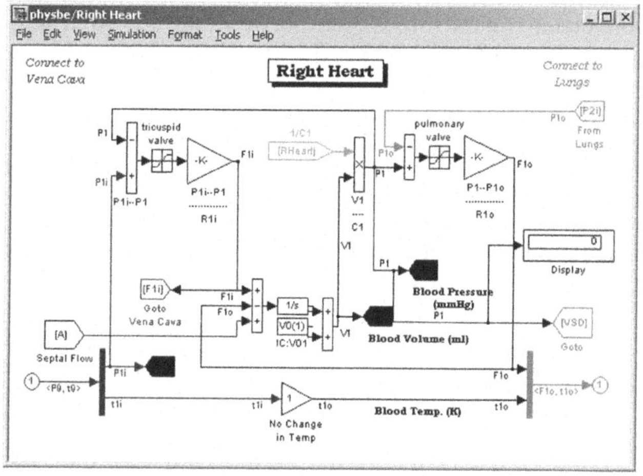

Figure 2.2-34 PHYSBE model of the right heart with VSD.

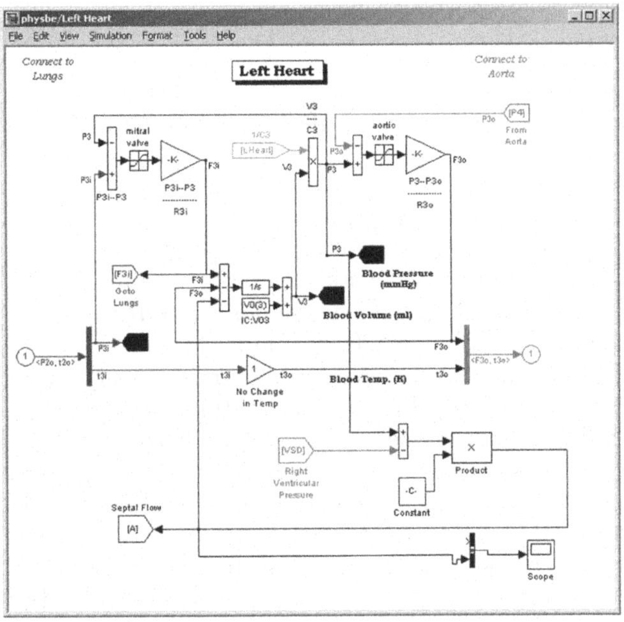

Figure 2.2-35 PHYSBE model of the left heart with VSD.

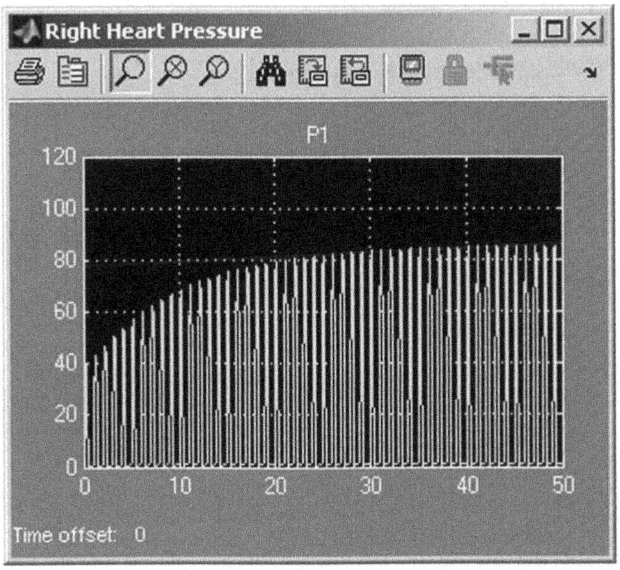

Figure 2.2-37 Right heart pressure.

This difference is divided by resistance, as determined by the Bernoulli equation:

$$\frac{8(3\text{cP})(4\ \text{mm})}{\pi r^4} = \frac{0.2292}{r^4}\left(\frac{\text{mmHg}\cdot\text{s}}{\text{mL}}\right)$$

with r in millimeters. The reciprocal (since the formula calls for division) represents the constant C in Fig. 2.2-34. This difference is subtracted from the left heart flow rate, and added to the right heart flow rate, shown in Fig. 2.2-34 for the left heart, and appears as a source block in the model of the right heart, above.

This simulation of VSD predicts that the blood pressure in the left heart reaches a steady state value that is less than the normal pressure, whereas the pressure in the right heart increases significantly. This pressure change is consistent with theory since the right ventricle has an increased flow from the left ventricle. (See Figs. 2.2-36 and 2.2-37.)

In this simulation, a hole greater than one square centimeter was modeled. The theory would predict that with a hole this size, a considerable amount of blood would be in the right ventricle and the blood volume in the pulmonary arteries would increase. The increased blood

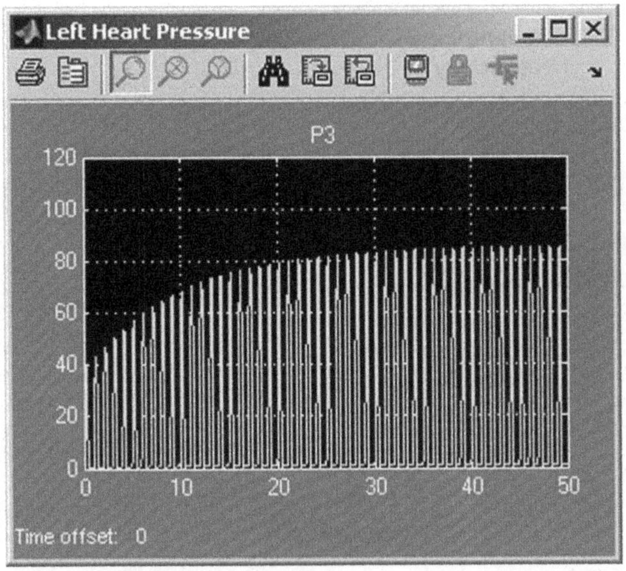

Figure 2.2-36 Left heart pressure.

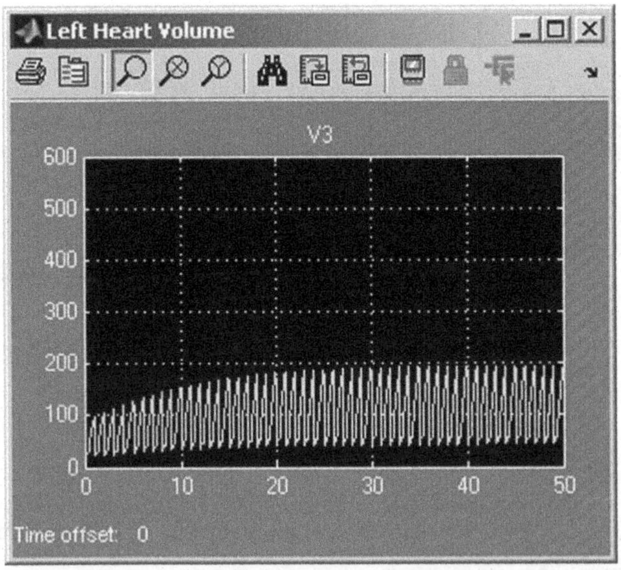

Figure 2.2-38 Left heart volume.

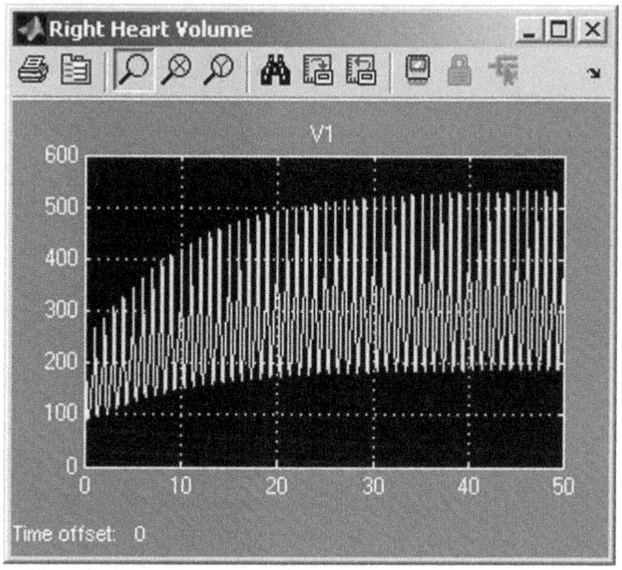

Figure 2.2-39 Right heart volume.

volume in the pulmonary arteries causes the pressure in the pulmonary circuit to increase—this increased pressure in the lung is called *hypertension.* The plots in Figs. 2.2-38 and 2.2-39 show that the blood volume in both the right and left heart are greater than normal.

VSD is a problem that affects the pressure in the heart because of the blood flow between the left and right ventricles through the septum. If the hole is large, the blood from the left ventricle blood flows into the right ventricle because of the pressure difference. The additional blood volume also causes more blood in the lung, causing extra pressure that sometimes results in hypertension. If VSD is not treated, the arteries in the lungs can thicken up under the extra pressure and permanent damage can be done to the lung, leading to the weakening of the valve.

2.2.7.4 Left ventricular hypertrophy

The heart responds to prolonged inadequate myocardial contraction by enlarging to keep pace with the requirements for increased cardiac output (Hori et al., 1989). This increased workload causes an increase in the wall thickness, a condition commonly referred to as *left ventricular hypertrophy* (shown in Fig. 2.2-40).

This inward expansion of the ventricular wall reduces both the rate and amount of relaxation during diastole, leading to a decrease in the wall compliance. The decrease in compliance (increase in wall stiffness), coupled with the concomitant increase in pressure, leads to a reduction in the size of the left chamber. If the left chamber is too small, it cannot fill efficiently, which leads to blood backing up into the vessels of the lungs and less blood circulating to the vital organs. At times, hypertrophy is itself the primary disease; but more commonly it is the consequence of another disorder.

Frequently, chronic hypertension is found to be the underlying cause of left ventricular dysfunction in humans. When operating correctly, the aortic and pulmonary valves are responsible for creating the pressure differences that lead to proper blood flow and circulation. Abnormal functioning of the valves is the underlying factor that produces either (1) pressure overloading due to restricted opening; or (2) volume overloading due to inadequate closure (Legato, 1987).

A narrowing of the aortic or pulmonary valves induces a decrease in blood flow (by Ohm's Law). Assuming constant resistance in the blood transport vessels, the

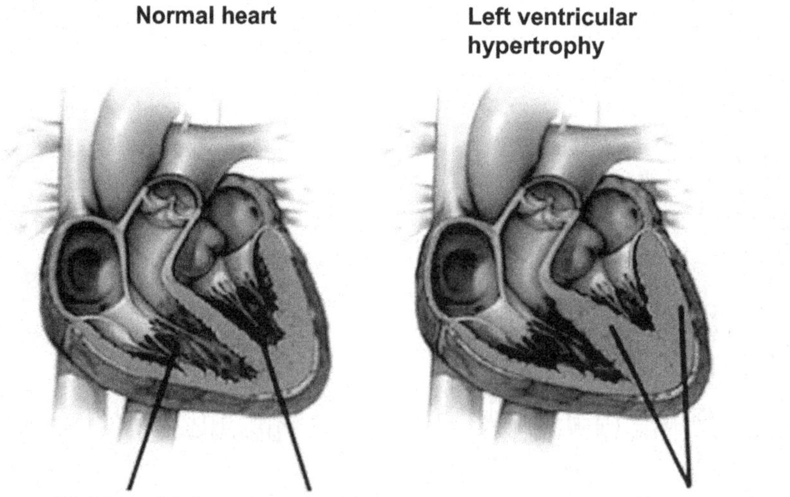

Figure 2.2-40 Overwork of the left ventricle causes an increase in wall thickness, a condition called left ventricular hypertrophy. Picture courtesy of the Mayo Educational Foundation.

narrowing of each valve leads to a decrease in the corresponding pressure gradient. The arterial compliance ($C = \Delta V/\Delta P$) will decrease in response to an impaired pressure gradient caused by valve dysfunction, reflecting the change in blood volume due to the decreasing arterial pressure. This model then predicts that the blood volume in the transport vessels must increase, leading to a decrease in left ventricular blood volume.

PHYSBE can be used to model left ventricular hypertrophy indirectly by first modeling inadequate aortic and pulmonary valves. The weakened valves induce left ventricular pressure overloading and a reduction in blood flow through the valves. This condition gives rise to left ventricular hypertrophy.

Example 2.2.11 Left ventricular hypertrophy.

Modify the Simulink implementation of PHYSBE to model left ventricular hypertrophy.

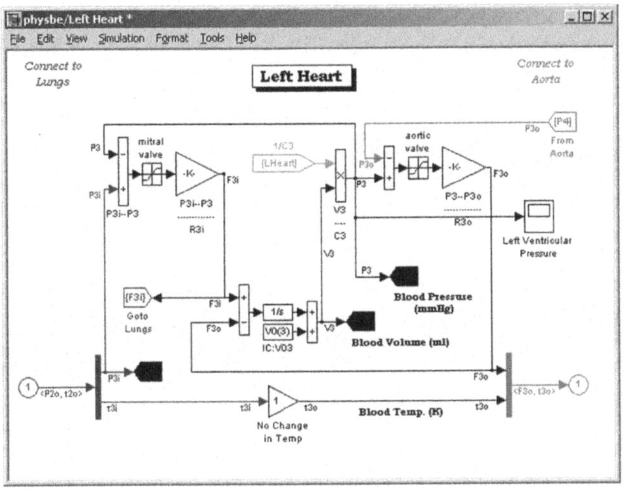

Figure 2.2-41 Modeling left ventricular hypertrophy: left heart.

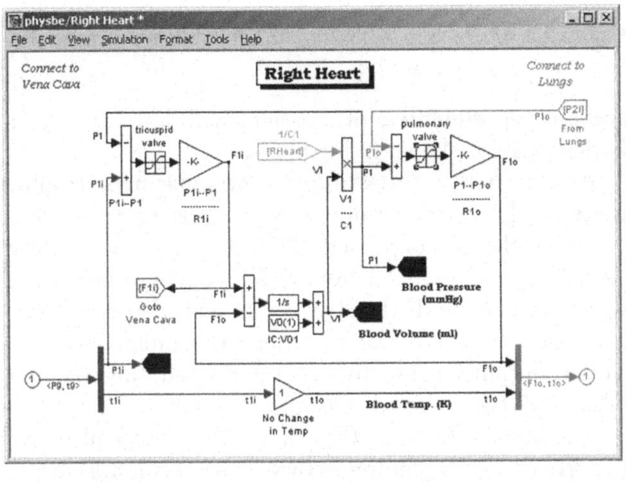

Figure 2.2-42 Modeling left ventricular hypertrophy: right heart.

Weakened aortic and pulmonary valves can be modeled by changing the upper limits of the aortic and pulmonary valves in the left heart (Fig. 2.2-41) and right heart (Fig. 2.2-42), respectively.

In addition to the scope output of the left ventricular pressure in the left heart, a scope should be added to the aorta to show the effect on aortic blood volume (Fig. 2.2-43).

Left ventricular hypertrophy is characterized by an increase in pressure in the left ventricle (Fig. 2.2-44).

The volume of blood in the aorta decreases commensurately with the increase in pressure in left ventricular hypertrophy (Fig. 2.2-45).

The Pressure–Volume (PV) loop is a graphical tool for assessing the interplay of ventricular function and circulation.

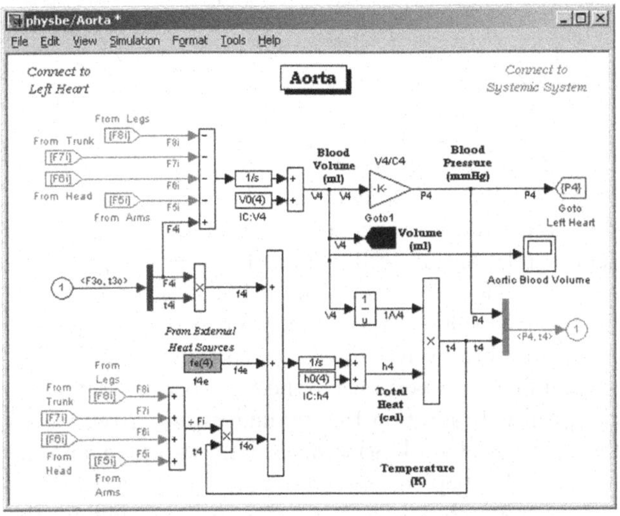

Figure 2.2-43 Modeling left ventricular hypertrophy: aorta.

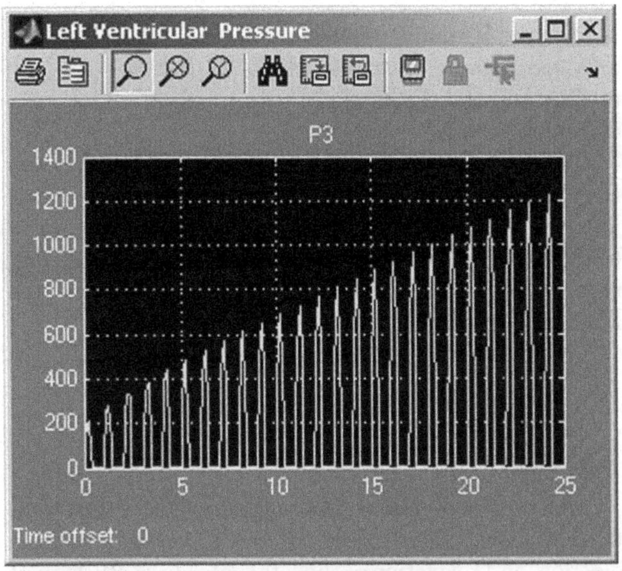

Figure 2.2-44 Left ventricular pressure.

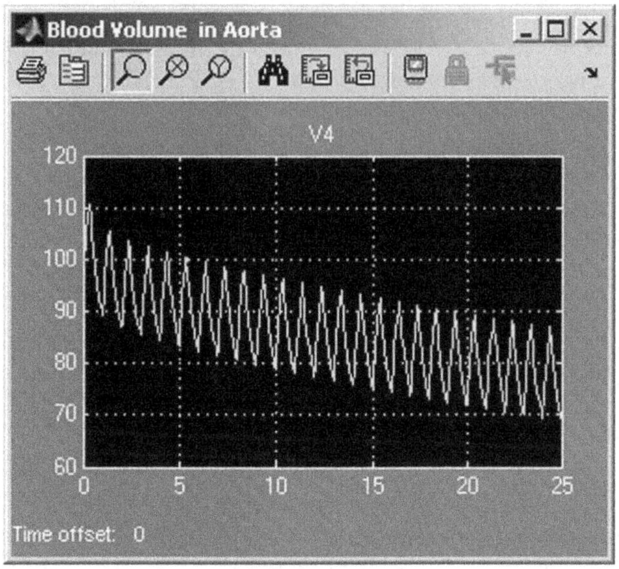

Figure 2.2-45 Blood volume in aorta.

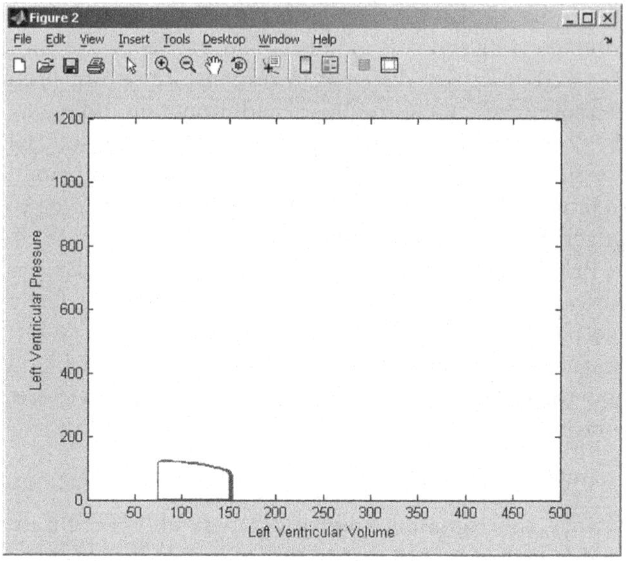

Figure 2.2-46 PV-loop of normal cardiovascular conditions.

Under normal conditions, the external work performed by the left ventricle is confined within the boundaries that show the interaction between the end-systolic pressure–volume relationship (ESPVR) and end-diastolic pressure–volume relationship (EDPVR). The place where the ESPVR is tangent to the PV-loop represents the point at which the aortic valve closes. The corresponding EDPVR tangent line is proportional to the reciprocal of ventricular compliance. In addition, the area under the loop represents the mechanical work performed by the ventricle, and the width of the loop corresponds to the difference between end diastolic volume and end systolic volume, defining the stroke volume (Li, 2004)

Example 2.2.12 Pressure–volume loops.

Modify the Simulink implementation of PHYSBE so that a PV-loop can be created in MATLAB. Use the PV-loop to illustrate the dynamic characteristics of left ventricular hypertrophy.

PV-loops are easily generated in MATLAB from PHYSBE data by plotting left ventricular pressure against left ventricular volume, measurements that were saved to the MATLAB variables LV_Volume and LV_Pressure, respectively. The PV-loops are generated with the MATLAB script:

```
plot(LV_Volume,LV_Pressure)
ylabel('Left Ventricular Pressure')
xlabel('Left Ventricular Volume')
axis([0 500 0 1200])
```

The pressure–volume loop diagram of the left ventricle is constructed based on the filling, contraction, ejection, and relaxation of the left ventricle. A PV-loop of normal cardiovascular conditions is shown in Fig. 2.2-46,

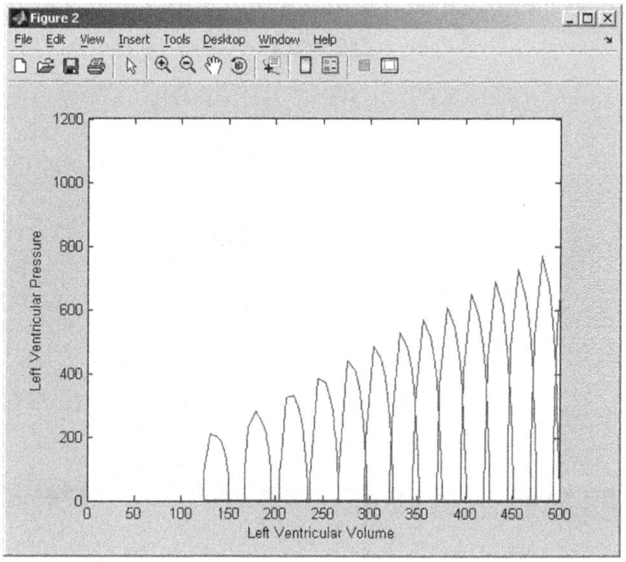

Figure 2.2-47 PV-loop of patient with left ventricular hypertrophy.

and one representative of a patient with left ventricular hypertrophy is shown in Fig. 2.2-47.

As expected in cases of left ventricular hypertrophy, there is a leftward and upward shift in the ESPVR, and a shift in the PV-loop itself. The total area enclosed by each loop was reduced, reflecting a decrease in stroke work and volume. If the stroke volume is decreased, the heart rate must increase in order to maintain a normal cardiac output. Eventually, if left untreated, this adaptive effect will fail and cardiac output will fall.

This PHYSBE simulation shows that left ventricular hypertrophy is a compensatory mechanism for maintaining normal cardiac output under abnormal increases in pressure and decreases in volume. A decrease in compliance and an increase in wall stiffness resulted from stimuli originating in the body's control center. Furthermore, reductions in stroke volume and work, computed from the PV-loops, are characteristic of the disease.

This set of examples is meant to illustrate to the reader the broad range of biomedical applications of numerical methods and show the reader how to use computing tools and numerical methods to solve problems in medicine and physiology.

References

Bhat, M. A. 2001. Fate of Hypertension after Repair of Coarctation of the Aorta in Adults. *British Journal of Surgery*, **88**: 536–538.

Braunwald, E. 1988. *Heart Disease: a Textbook of Cardiovascular Medicine*. Philadelphia, PA: W.B. Saunders Company.

Dunn, J. M., ed. 1988. *Cardiac Valve Disease In Children*. New York: Elsevier.

Estelberger, W., and Popper, N. 2002. Comparison 15: Clearance Identification. *Simulation News Europe* **35/36**:65–66.

Fournier, R. L. 1999. *Basic Transport Phenomena in Biomedical Engineering*. Philadelphia, PA: Taylor & Francis.

Germann, W. J. 2005. *Principles of Human Physiology*, 2nd ed. San Francisco, CA: Pearson Education.

Goldberger, A. L., Amaral, L. A. N., Glass, L., Hausdorff, J. M., Ivanov, P. Ch., Mark, R. G., Mietus, J. E., Moody, G. B., Peng, C. K., and Stanley, H. E. 2000. PhysioBank, PhysioToolkit, and Physionet: Components of a New Research Resource for Complex Physiologic Signals. *Circulation* **101**(23):e215–e220. (Circulation Electronic Pages; http://circ.ahajournals.org/cgi/content/full/101/23/e215/.)

Hori, M., Suga, J., and Yellin, E. L. 1989. *Cardiac Mechanics and Function in the Normal and Diseased Heart*. Tokyo, Japan: Springer-Verlag.

Jasper, H. H. 1958. The ten-twenty electrode system of the international federation. *Electroencephalography and Clinical Neurophysiology*, **10**:371–373.

Keirn, Z. 1988. *Alternative modes of communication between man and machine*. Master's thesis, Purdue University.

Klabunde, R. E. 2004. *Cardiovascular Physiology Concepts: Valvular Stenosis*. http://www.cvphysiology.com/HeartDisease/HD004.htm.

Krousgrill, C. M. 2005. Personal Communication.

Legato, M. J. 1987. *The Stressed Heart*. Boston, MA: Martinus Nijhoff Publishing.

Li, J. K-J. (2004). *Dynamics of the Vascular System. Boston*, MA: World Scientific Publishing.

Lue, H. C., and Takao A., eds. 1986. *Subpulmonic Ventricular Septal Defects*. Tokyo, Japan: Springer-Verlag.

Mascherbauer, J. 2004. Value and limitations of aortic valve resistance with particular consideration of low flow-low gradient aortic stenosis: an in vitro study. *Eur Heart J*. **25**(9):787–793.

McLeod, J. 1966. PHYSBE... a physiological simulation benchmark experiment. *SIMULATION*, **7**(6): 324–329.

McLeod, J. 1968. PHYSBE...a year later, *SIMULATION*, **10**(1):37–45.

McSharry, P. E., Clifford, G. D., Tarassenko, L., and Smith, L. 2003. A dynamical model for generating synthetic electrocardiagram signals. *IEEE Transactions on Biomedical Engineering*, **50**(3):289-294.

Mohiaddin, R. H. 1993. Magnetic Resonance Volume Flow and Jet Velocity Mapping in Aortic Coarctation. *JACC*, **22**:1515–1521.

Ruha, A., and Nissila, S. 1997. A real-time microprocessor QRS detector system with a 1-ms timing accuracy for the measurement of ambulatory HRV. *IEEE Trans Biomed Eng* **44**(3):159–167.

Sokolow, M., and McIlroy, M. B. 1981. *Clinical Cardiology*. Los Altos, CA: LANGE Medical Publications.

Suk, J., ed. 2001. Coarctation of the Aorta. *Yahoo Health*. http://health.yahoo.com/health/ency/adam/000191/i18128.

Texas Heart Institute. 2004. *Leading With the Heart*. http://www.tmc.edu/thi/vaortic.html.

Van Riel, N. 2004. *Minimal Models for Glucose and Insulin Kinetics* - A MATLAB implementation; version February, 5, 2004. Technical Report, Technische Universiteit Eindhoven.

Web site links

http://www.physionet.org/physiotools/matlab/ECGwaveGen/

http://www.physionet.org/physiotools/ecgsyn/

http://www.emedicine.com/ped/topic2543.htm

http://circ.ahajournals.org/cgi/content/full/101/23/e215

http://www.cvphysiology.com/HeartDisease/HD004.htm

http://health.yahoo.com/conditions/

http://www.tmc.edu/thi/vaortic.html

http://physionet.incor.usp.br/

http://www.physionet.org/cgi-bin/rdsamp

Chapter 2.3

2.3

Signals and systems

John Semmlow

2.3.1 Biological systems

A system is a collection of processes or components that interact for some common purpose, although that purpose may only be the invention of human intellect. Many systems of the human body are based on function. The cardiovascular system's function is to deliver oxygen-carrying blood to the peripheral tissues. The pulmonary system is responsible for the exchange of gases [primarily oxygen (O_2) and carbon dioxide (CO_2)] between the blood and air, whereas the renal system regulates water and ion balance and adjusts the concentration of other types of ions and molecules. Some systems are organized around mechanism rather than function. The endocrine system mediates a range of communication functions using complex molecules distributed through the blood stream. The nervous system performs an enormous number of tasks using neurons and axons to process and transmit information coded as electrical impulses.

The study of classical physiology and of many medical specialties is structured around human physiological systems. (The term *classical physiology* is used here to mean the study of whole organs or organ systems as opposed to newer molecular-based approaches.) For example, cardiologists specialize in the cardiovascular system, neurologists in the nervous system, ophthalmologists in the visual system, nephrologists in the kidneys, pulmonologists in the respiratory system, gastroenterologists in the digestive system, and endocrinologists in the endocrine system. There are medical specialties or subspecialties to cover most physiological systems. (Another set of medical specialties is based on common tools or approaches, including surgery, radiology, and anesthesiology, whereas one specialty, pediatrics, is based on the type of patient.)

Given this systems-based approach to physiology and medicine, it is not surprising that early bioengineers applied their engineering tools, especially those designed for the analysis of systems, to some of these physiological systems. Early applications in bioengineering research include the analysis of breathing patterns and the oscillatory movements of the iris muscle. Applications of basic science to medical research date from the eighteenth century. In the late nineteenth century, Einthoven recorded the electrical activity of the heart, and throughout that century, electrical stimulation was used therapeutically (largely to no avail). Although early researchers may not have considered themselves engineers, they did draw on the engineering tools of their day.

The nervous system, with its apparent similarity to early computers, was another favorite target of bioengineers, as was the cardiovascular system with its obvious links to hydraulics and fluid dynamics. Some of these early efforts are discussed in the sections on system and analog models (Sections 2.3.3.2 and 2.3.3.3). As bioengineering has expanded into areas of molecular biology, systems on the cellular, or even subcellular levels, have become of interest.

Regardless of the type of biological system, its scale, or its function, we must have some way of interacting with that system. Interaction or communication with a biological system is done through biosignals. The communication may only be one-way, such as when we attempt to infer the state of the system by measuring various biological or physiological variables to make a medical diagnosis. From a systems analytic point of view, changes in physiological variables constitute biosignals. Common signals measured in diagnostic medicine include electrical activity of the heart, muscles and brain; blood pressure; heart rate; blood gas concentrations and

Biomedical Engineering Desk Reference; ISBN: 9780123746467

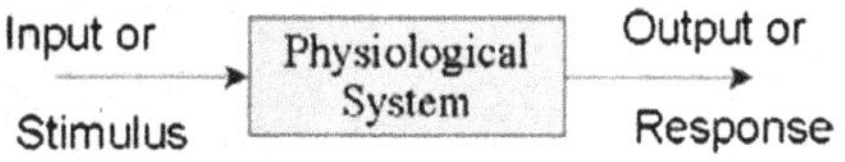

Figure 2.3-1 A classic systems view of a physiological system that receives an external input, or stimulus, that evokes an output, or response.

concentrations of other blood components; and sounds generated by the heart and its valves.

Often it is desirable to send signals into a biological system for purposes of experimentation or therapy. In a general sense, all drugs introduced into the body can be considered biosignals. We often use the term *stimulus* for signals directed into some physiological process, and if an output signal is evoked by these inputs we term it a *response*. (Terms shown in *italics* are an important part of a bioengineer's vocabulary.) In this scenario, the biological system is acting like an input–output system, a classic construct or model used in systems analysis (Figure 2.3-1).

Classical examples include the knee-jerk reflex, where the input is a mechanical force and the output is mechanical motion, and the pupillary light reflex, where the input is light and the output is a mechanical change in the iris muscles. Drug treatments can be included in this input–output description, where the input is the molecular configuration of the drug and the output is the therapeutic benefit (if any). Such representations are further explored in the sections on systems and analog modeling (Sections 2.3.3.2 and 2.3.3.3).

Systems that produce an output without the need for an input stimulus, for example the electrical activity of the heart, can be considered biosignal *sources*. (Although the electrical activity of the heart can be moderated by several different stimuli, exercise for example, the basic signal does not require a specific stimulus.) Input-only systems are not usually studied, because the purpose of any input signal is to produce some sort of response: even a placebo, which is designed to produce no physiological response, often produces substantive results.

Because all of our interactions with physiological systems are through biosignals, the characteristics of these signals are of great importance. Indeed, much of modern medical technology is devoted to extracting new physiological signals from the body or gaining more information from existing biosignals. The next section discusses some of the basic aspects of these signals.

2.3.2 Biosignals

Much of the activity in biomedical engineering, be it clinical or research, involves the measurement, processing, analysis, display, and/or generation of signals. Signals are variations in energy that carry information. The variable that carries the information (the specific energy fluctuation) depends upon the type of energy involved. Table 2.3-1 summarizes the different energy types that can be used to carry information, and the associated variables that encode this information. Table 2.3-1 also shows the physiological measurements that involve these energy forms as discussed later in the chapter.

Table 2.3-1 Energy forms and associated information-carrying variables

Energy	Variables (specific fluctuation)	Common measurements
Chemical	Chemical activity and/or concentration	Blood ion, oxygen, carbon dioxide, pH, hormonal concentrations, and other chemistry
Mechanical	Position Force, torque, or pressure	Muscle movement, cardiovascular pressures, muscle contractility Valve and other cardiac sounds
Electrical	Voltage (potential energy of charge carriers) Current (charge carrier flow)	EEG, ECG, EMG, EOG, ERG, EGG, GSR
Thermal	Temperature	Body temperature, thermography

ECG, electrocardiogram; EEG, electroencephalogram; EGG, electrogastrogram; EMG, electromyogram; EOG, electrooculogram; ERG, electroretinogram; GSR, galvinic skin response.

Biological signals are usually encoded into variations of electrical, chemical, or mechanical energy, although occasionally variations in thermal energy are of interest. For communication within the body, signals are primarily encoded as variations in electrical or chemical energy. When chemical energy is used, the encoding is usually done by varying the concentration of the chemical within a *physiological compartment*, for example, the concentration of a hormone in the blood. Bioelectric signals use the flow or concentration of ions, the primary charge carriers within the body, to transmit information. Speech, the primary form of communication between humans, encodes information as variations in air pressure.

Outside the body, information is commonly transmitted and processed as variations in electrical energy, although mechanical energy was used in the seventeenth and early eighteenth centuries to send messages. The semaphore telegraph used the position of one or more large arms placed on a tower or high point to encode letters of the alphabet. These arm positions could be observed at some distance (on a clear day), and relayed onward if necessary. Information processing can also be accomplished mechanically, as in the early numerical processors constructed by Babbage. More recently, mechanically based digital components have been attempted using variations in fluid flow. Modern electronics provides numerous techniques for modifying electrical signals at very high speeds. The body also uses electrical energy to carry information when speed is important. Since the body does not have many free electrons, it relies on ions, notably Na^+, K^+, and Cl^-, as the primary charge carriers. Outside the body, electrically based signals are so useful that signals carried by other energy forms are usually converted to electrical energy when significant transmission or processing tasks are required. The conversion of physiological energy to an electric signal is an important step, often the first step, in gathering information for clinical or research use. The energy conversion task is done by a device termed a *transducer*, specifically a *biotransducer*.

A transducer is a device that converts energy from one form to another. By this definition, a light bulb or a motor is a transducer. In signal processing applications, the purpose of energy conversion is to transfer information, not to transform energy as with a light bulb or a motor. In physiological measurement systems, all transducers are so-called *input transducers:* they convert nonelectrical energy into an electronic signal. An exception to this is the electrode, a transducer that converts electrical energy from ionic to electronic form. Usually, the output of a biotransducer is a voltage (or current) whose amplitude is proportional to the measured energy. Figure 2.3-2 shows a device to measure the movements of the

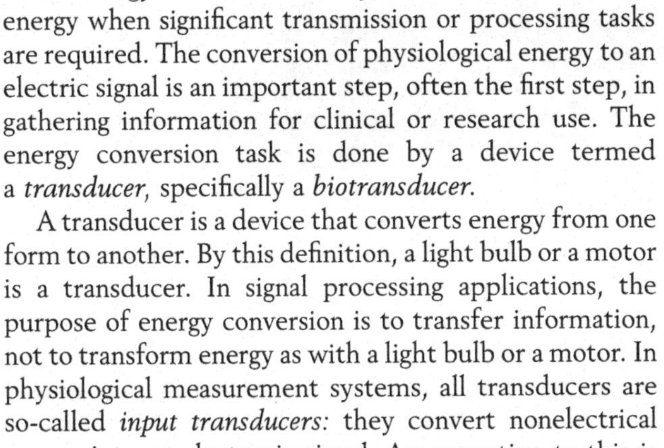

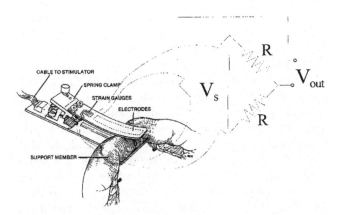

Figure 2.3-2 A device used to measure small movements of the intestine during surgery. These movements can be used to assess the viability of a segment of intestine. The device consists of an inflexible lower plate and a flexible upper plate. Movement of the upper plate is detected by two strain gages placed on its upper and lower surfaces. Strain gage transducers change their resistance in response to small changes in length. Subsequent electronics detect these resistance changes.

intestine during surgical procedures. The mechanical transducers used in the device are called *strain gages* and they change their electrical resistance when stretched even slightly.

The energy that is converted by the input transducer may be generated by the physiological process itself, may be energy that is indirectly related to the physiological process, or may be energy produced by an external source. In the latter case, the externally generated energy interacts with, and is modified by, the physiological process, and it is this alteration that produces the measurement. For example, when externally produced x-rays are transmitted through the body, they are absorbed by the intervening tissue, and a measurement of this absorption is used to construct an image. Most medical imaging systems are based on this external energy approach.

Images can also be constructed from energy sources internal to the body as in the case of radioactive emissions from radioisotopes injected into the body. These techniques make use of the fact that selected, or tagged, molecules will collect in specific tissue. The areas where these radioisotopes collect can be mapped using a gamma camera or, with certain short-lived isotopes, better localized using positron emission tomography (PET).

Many physiological processes produce energy that can be detected directly. For example, cardiac internal pressures are usually measured using a pressure transducer placed on the tip of a catheter introduced into the appropriate chamber of the heart. The measurement of electrical activity in the heart, muscles, or brain provides other examples of the direct measurement of physiological energy. For these measurements, the energy is already electrical and only needs to be converted from ionic to electronic current using an *electrode*. These sources are usually given the term *ExG*, where the *x* represents the physiological process that produces the electrical energy: ECG, electrocardiogram; EEG, electroencephalogram; EMG, electromyogram; EOG, electrooculogram; ERG, electroretinogram; and EGG, electrogastrogram. An exception to this terminology is the galvanic skin response, GSR, the electrical activity generated by the skin. Typical physiological measurements that involve the conversion of other energy forms to electrical energy are shown in Table 2.3-1. Figure 2.3-3 shows the early ECG machine where the interface between the body and the electrical monitoring equipment was buckets filled with saline (Figure 2.3-3E).

The *biotransducer* is often the most critical element in the system because it constitutes the interface between the subject or life process and the rest of the system. The transducer establishes the risk, or *invasiveness*, of the overall system. For example, an imaging system based on differential absorption of x-rays, such as a CT (computed tomography) scanner, is considered more invasive than

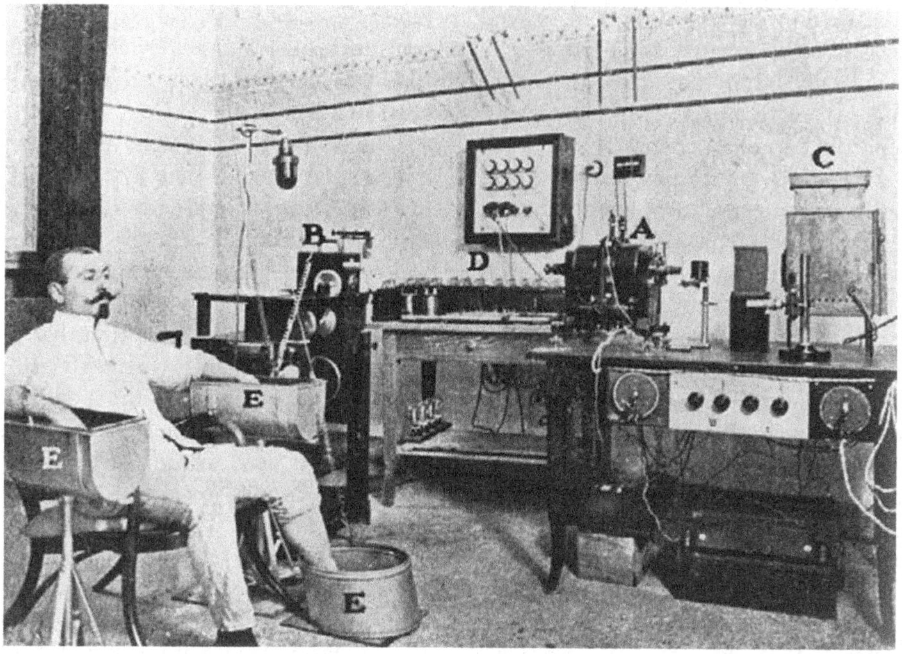

Figure 2.3-3 An early electrocardiogram machine.

an imaging system based on ultrasonic reflection, because CT uses ionizing radiation that may have an associated risk. (The actual risk of ionizing radiation is still an open question, and imaging systems based on x-ray absorption are considered *minimally invasive*.) Ultrasound and radiographic imaging would be considered less invasive than, for example, monitoring internal cardiac pressures through cardiac catheterization in which a small catheter is threaded into the heart chamber. Indeed, many of the outstanding problems in biomedical measurement, such as noninvasive measurement of internal cardiac pressures or intracranial pressure, await an appropriate (and undoubtedly clever) transducer mechanism.

2.3.2.1 Signal encoding

Given the importance of electrical signals in biomedical engineering, much of the discussion in this text is based on electrical or electronic signals. Nonetheless, many of the principles described are general and could be applied to signals carried by any energy form. Regardless of the energy form or specific variable used to carry information, some type of encoding scheme is necessary. Encoding schemes vary in complexity: human speech is so complex that automated decoding is still a challenge for voice-recognition computer programs. Yet, the exact same information could be encoded into the relatively simple series of long and short pulses known as Morse code.

Most encoding strategies can be divided into two broad categories or domains: continuous and discrete. The discrete domain is used almost exclusively in computer-based technology, because such signals are easier to manipulate electronically. Discrete signals are usually transmitted as a series of pulses at even (synchronous transmission) or uneven (asynchronous transmission) intervals. These pulses may be of equal duration, or the information can be encoded into the pulse length. Within the digital domain, many different encoding schemes can be used. For encoding alphanumeric characters, those featured on the keys of a computer keyboard, the ASCII (American Standard Code for Information Exchange) code is used. Here each letter, the numbers 0 through 9, and many characters are encoded into an 8-bit binary number. For example, the letters *a* though *z* are encoded as 97 (for *a*) through 122 (for *z*) whereas the capital letters *A* through Z are encoded by numbers 65 (*A*) through 90 (Z). The complete ASCII code can be found in some computer texts or on the Internet.

In the continuous domain, information is encoded in terms of signal amplitude, usually the intensity of the signal at any given time. For an electronic signal, this could be the value of the voltage or current at a given time. Note that all signals are by nature *time varying*, because a single constant value contains no information. (Modern information theory makes explicit the difference between information and meaning. The latter depends upon the receiver; that is, the device or person for which the information is intended. Many students have attended lectures with a considerable amount of information that, for them, had little meaning. This text strives valiantly for both information and meaning.) If the information is linearly encoded into signal amplitude, the signal is referred to as an *analog signal*. For example, the temperature in a room can be encoded so that 0 V represents 0.0°C, 5 V represents 10°C, 10 V represents 20°C, and so on, so that the encoding equation for temperature would be as follows:

Temperature = 2 × Voltage amplitude

Analog encoding is common in consumer electronics such as high-fidelity amplifiers and television receivers, although many applications that traditionally used analog encoding, such as sound and video recording, now use discrete or digital encoding. Nonetheless, analog encoding is likely to remain important to the biomedical engineer, if only because many physiological systems use analog encoding, and most biotransducers generate analog-encoded signals.

The typical analog signal is one whose amplitude varies in time as follows:

$$x(t) = f(t) \qquad \text{[Eq. 2.3.1]}$$

When a continuous analog signal is converted to the digital domain, it is represented by a series of numbers that are discrete samples of the analog signals at a specific point in time:

$$X[n] = x[1], x[2], x[3], \ldots, x[n] \qquad \text{[Eq. 2.3.2]}$$

Usually this series of numbers would be stored in sequential memory locations with x_1 followed by x_2, then x_3, and so forth. {It is common to use brackets to identify a discrete variable (i.e., $x[n]$); but note that the MATLAB® (MathWorks, Natick, MA) programming language used throughout this text also uses brackets in a different context.} Because digital numbers can only represent discrete or specific amplitudes, the analog signal must also be sliced up in amplitude. Hence, to *digitize* an analog signal requires slicing the signal in two ways: in time and in amplitude.

Slicing the signal into discrete points in time is termed *time sampling* or simply *sampling*. Time slicing *samples* the continuous waveform, $x(t)$, at discrete prints in time, nT_s, where T_s is the sample interval. Slicing the signal amplitude in discrete levels is termed *quantization* (Figure 2.3-4). The equivalent number can only approximate the level of the analog signal, and the degree of approximation will depend on the range of binary numbers and the amplitude of the analog signal. For example, if the signal is converted into an 8-bit binary number, this number is capable of 2^8 or 256 discrete values. If the analog signal amplitude ranges between 0.0 and 5.0 V, the quantization interval in volts will be 5/256 or 0.019 V. If, as is usually the case, the analog signal is time varying in a continuous manner, it must be approximated by a series of binary numbers representing the approximate analog signal level at discrete points in time (Figure 2.3-4).

Example 2.3.1: A 12-bit analog-to-digital converter (ADC) advertises an accuracy of ± the least significant bit (LSB). If the input range of the ADC is 0 to 10 V, what is the resolution of the ADC in analog volts?
Solution: If the input range were 10 V, the analog voltage represented by the LSB would be as follows:

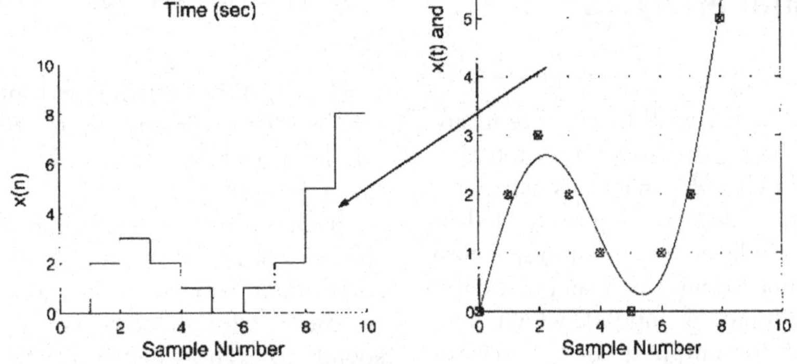

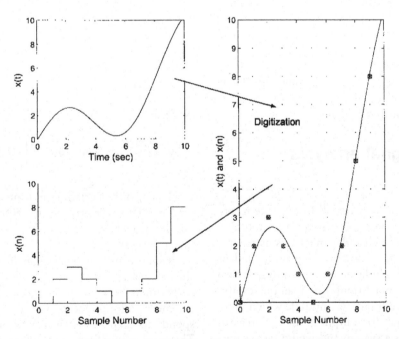

Figure 2.3-4 Digitizing a continuous signal (*upper left*) requires slicing the signal in time and amplitude (*right side*). The result is a series of discrete numbers (*x*'s) that approximate the original signal, and the resultant digitized signal (*lower left*) consists of a series of discrete steps in time and value.

$$V_{\text{LSB}} = \frac{V_{\max}}{2^{\text{Nu bits}}} = \frac{10}{2^{12}} = \frac{10}{4,096} = 0.0024 \text{ volts}$$

Hence, the resolution would be ± 0.0024 V.

It is relatively easy, and common, to convert between the analog and digital domains using electronic circuits specially designed for this purpose. Many medical devices acquire the physiological information as an analog signal but convert it to digital format using an ADC so that it can be processed using a computer. For example, the electrical activity produced by the heart can be detected using properly placed electrodes, and the resulting signal, the ECG, is an analog-encoded signal. This signal might undergo some *preprocessing* or *conditioning* using analog electronics before being converted to a digital signal using an ADC. The converted digital signal would be sent to a computer for more complex processing and storage. (In fact, conversion to digital format is usually done even if the data are only to be stored for later use.) Conversion from the digital to the analog domain is possible using a *digital-to-analog* converter (DAC). Most personal computers include both ADCs and DACs as part of a sound card. This circuitry is specifically designed for the conversion of audio signals, but can be used for other analog signals. Data transformation cards designed as general-purpose ADCs and DACs are readily available and offer greater flexibility in sampling rates and conversion gains. These cards provide multichannel ADCs (usually eight to 16 channels) and several channels of DAC.

Basic concepts that involve signals are often introduced or discussed in terms of analog signals, but most of these concepts apply equally well to the digital domain. In this text, the equivalent digital domain equation is often presented alongside the analog equation to emphasize the equivalence. Many of the problems and examples use a computer, so they obviously are being implemented in the digital domain even if they are presented as analog-domain problems.

2.3.3 Linear signal analysis: overview

From a mechanistic point of view, all living systems are composed of processes. These processes act, or interact, through manipulation of molecular mechanisms, chemical concentrations, ionic electrical current, and/or mechanical forces and displacements. A physiological process performs some operation(s) or manipulation(s) in response to a specific input (or inputs), which gives rise to a specific output (or outputs). In this regard, a process is the same as a system and would be systematically represented as shown in Figure 2.3-1. Sometimes the term *system* is reserved for larger structures composed of several processes, but the two terms are often used interchangeably, as they will be throughout this text. To study and quantify complex processes, we often impose rather severe simplifying constraints. The most common assumption is that the process and its components or subprocesses behave in a linear manner, and that their basic characteristics do not change over time. This assumption is referred to as the "linear time-invariant" (LTI) model. Such an assumption allows us to apply a powerful array of mathematical tools that are known collectively as linear systems analysis. Of course, most living systems change over time, are adaptive, and are often nonlinear. Nonetheless, the power of linear systems analysis is sufficiently seductive that assumptions or approximations are often made so that these tools can be used. Linearity can be approximated by using small-signal conditions where many systems behave more or less linearly. Alternatively, piecewise linear approaches can be used where the analysis is confined to operating ranges over which the system behaves linearly. One approach to dealing with a process that changes over time is to study that process within a short enough time frame that it can be considered time-invariant.

The concept of linearity has a rigorous definition, but the basic concept is one of proportionality of response. If you double the stimulus into a linear system, you will get twice the response. One way of stating this proportionality property mathematically is the following: if the independent variables of linear function are multiplied by a constant, k, the output of the function is simply multiplied by k:

$$y = f(x);$$

where f is a linear function, then:

$$ky = f(kx) \qquad \text{[Eq. 2.3.3]}$$

Note that:

$$ky = \frac{df(kx)}{dt} \quad \text{and} \quad ky = \int f(kx)dt \qquad \text{[Eq. 2.3.4]}$$

Hence, differentiation and integration are linear operations. The major transforms described in this text, the Fourier transform and the Laplace transform, are also linear processes.

Response proportionality, or linearity, is required for the application of an important concept known as *superposition*. Superposition states that if there are two (or more) stimuli acting on the system, the system responds to each as if it were the only stimulus present. The combined influence of the multiple stimuli is simply the addition of each stimulus acting alone. This allows complex stimuli to be broken down so that the problem of determining a system's response to such stimuli is greatly reduced.

2.3.3.1 Analysis of linear systems

A linear system is usually viewed as acting on a specific input signal to produce an output as shown in Figure 2.3-1. This is a very general concept: inputs can take many different energy forms (chemical, electrical, mechanical, or thermal), and outputs can be of the same or different energy forms. There are several ways to study a linear system. Here, two different approaches are developed and explored: *analog analysis* using *analog models*, and *systems analysis* using *systems models*. There is potential confusion in this terminology. Although analog analysis and systems analysis are two different approaches, both are included as tools of *linear systems analysis*. Hence, linear systems analysis includes both analog and systems analysis.

The primary difference between analog and systems analysis is the way the underlying physiological processes are represented. In analog analysis, individual components are represented by analogous elements. Often these elements show detailed structures and provide some insight into the way in which a given process is implemented, although they may also represent processes more globally. In systems analysis, a whole process can be represented by a single mathematical equation. The advantage of using analog analysis is that the model is often closer to the underlying physiological processes. Conversely, analyzing at the process level, as in systems analysis, provides a more succinct description and offers a better overall view of the system under study. In addition, the more abstract representation provided by systems models emphasizes behavioral characteristics and may aid in identifying behavioral similarities between processes that contain quite different elements.

2.3.3.2 Analog analysis and analog models

In analog analysis, there is a direct relationship between the physiological mechanism and the analog elements used in the model, although the elements may not necessarily be in the same energy modality as the physiological mechanism. For example, Figure 2.3-5 shows what appears to be an electric circuit. Of course it is an electric circuit, but it is also an early analog model of the cardiovascular system known as the *windkessel model*. In this circuit, voltage represents blood pressure, current represents blood flow, R_P and C_P are the resistance and compliance of the systemic arterial tree, and Z_o is the characteristic impedance of the proximal aorta. Later we will find that using an electrical network to represent

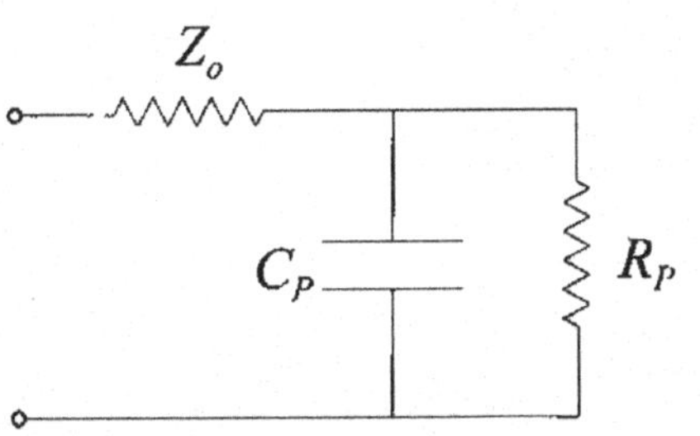

Figure 2.3-5 An early analog model of the cardiovascular system that used electrical elements to represent mechanical processes. In this model, voltage is equivalent to blood pressure and current to blood flow.

what is a mechanical system is mathematically appropriate. In this analog model, the elements are not very elemental, because they represent processes distributed throughout various segments of the cardiovascular system; however, the model can be expanded to represent the system at a more detailed level.

Figure 2.3-6 shows an analog model of the muscle skeletal muscle that uses mechanical elements. The muscle's force originates at the *contractile element*, but this force, F_o, is modified by the muscle mechanical processes before it appears at the output, *F*. The internal mechanical processes include the tissue viscosity, a sort of internal friction, the parallel elastic element which represents the elastic properties of the sarcolemma, and the series elastic element that reflects the elastic behavior of muscle tendons. In real muscle, these elements are nonlinear, but are often approximated as linear providing a linearized skeletal muscle model.

This basic model of skeletal muscle shown in Figure 2.3-6 has been used with additional mechanical elements to construct a mechanical model of the eye movement system including a pair of extraocular

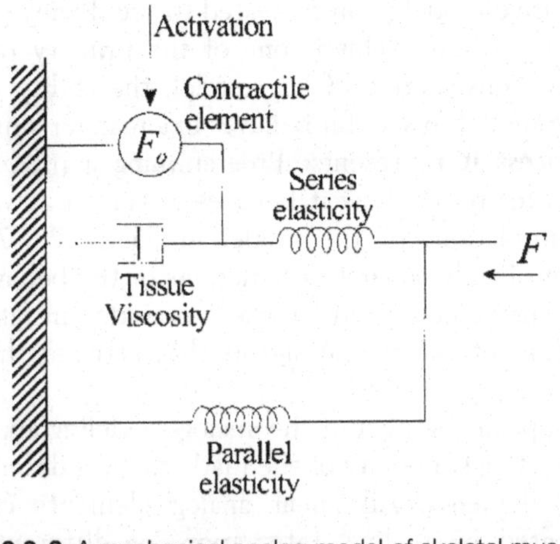

Figure 2.3-6 A mechanical analog model of skeletal muscle. The various elements correspond to specific properties of real muscle.

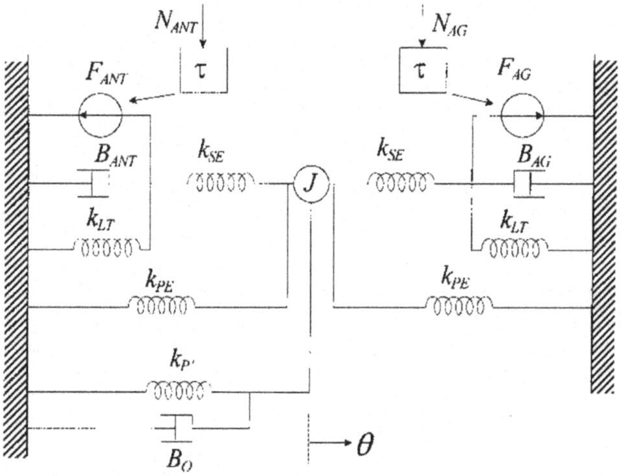

Figure 2.3-7 Analog model of the lateral and medial rectus muscles and associated mechanical involved in directing horizontal eye position. The neural signals, N_{ANT} and N_{AG}, are the inputs and the angular position, θ, is the output. The function of the various analog components is discussed in the text. (Adapted from Bahill and Stark, 1979.)

muscles, the lateral and medial rectus (Figure 2.3-7). These muscles are the mechanical elements involved in controlling the horizontal position of the eye. Each of the two extraocular muscles shown, the lateral and medial rectus, is represented by a force generator F_{ANT} (or F_{AG}), a viscous element B_{ANT} (or B_{AG}), a series elastic element, k_{SE}, and a parallel elastic element, k_{PE}. The two muscle representations also include an additional elastic element k_{LT}. Three other elements represent the mechanical properties of the eyeball and the orbit: an inertial component, J, representing the moment of inertia of the eyeball; a viscous element Bo, representing the friction between the eye and orbit; and a parallel elastic element, $k_{P¢}$, representing the elastic properties of the eye in the orbit. The neural signals, N_{ANT} and N_{AG}, are the inputs and the angular position, θ, is the output.

With the aid of a computer, this model, and all quantitative models, can be tested to see if they predict reasonable results. This is one of the primary motivations for construction of any model, the ability to 'try out' the model to see if it behaves in a manner similar to the process it represents. Programming a model into a computer to see how it behaves is known as *simulation*. Simulations of the model in Figure 2.3-7 have produced highly accurate predictions of the behavior of real eye movements and have also provided insight into the nature of the neural signals that activate the two muscles.

An appealing aspect of the analog-modeling approach is the relative simplicity of the mathematical description of the elements. All linear analog elements can be represented by scaling, integration, or differentiation operations between the associated variables:

$$v_1 = Av_2 \text{ Scaling (Dissipative element)}$$
$$v_1 = A\frac{dv_2}{dt} \text{ Time differentiation (Inertial elements)}$$
$$v_1 = A\int v_2 dt \text{ Time integration (Capacitive elements)} \quad \text{[Eq. 2.3.5]}$$

where v_1 and v_2 are the variables associated with the analog element. The specific variables depend on the type of elements. For electrical elements, they are voltage and current, whereas for mechanical elements they are force and velocity (Table 2.3-2).

One of the advantages of analog models is that although the systems they represent can be quite complicated, the individual analog components behave in a straightforward manner as noted in Eq. 2.3.5. Analog models become complicated because of the number of elements involved and their configuration, but the elements themselves are simple. Another advantage of analog analysis is that given the configuration of elements, a mathematical description of the overall model follows directly. We will find that by applying the conservation laws (conservation of charge, energy, and force) to a specific configuration of elements, a mathematical description follows in algorithmic fashion. One need merely follow a set of rules to obtain a mathematically complete description of the model.

It is possible to introduce nonlinear components into an analog model, but this complicates the analysis. For example, the model shown in Figure 2.3-5 is actually nonlinear because the capacitance changes its value as blood pressure changes. Often, a piecewise linear approach can be used where the model is analyzed over several different operating regions within which the nonlinear elements can be taken as linear.

Example 2.3.2: A constant force of 4 dyne is applied to a 2-g mass. Find the velocity of the mass after 5 seconds.

Table 2.3-2 Variables associated with analog elements and related conservation laws

Element type	Variable	Conservation law	Element (type)
Electrical	Voltage, *V* (volts)	Charge (Kirchhoff's current law)	Resistor (dissipative)
	Current, *i* (amps)	Energy (Kirchhoff's voltage law)	Inductor (inertial) Capacitor (capacitive)
Mechanical	Force, *F* (newtons) Velocity, *v* (cm/sec)	Force (Newton's law)	Friction (dissipative) Mass (inertial) Elasticity (capacitive)

Solution: Inertia is one property of mass that is defined as an integral relationship between the mechanical variables force (*F*) and velocity (*u*):

$$F = m(dv/dt)$$

a modification of Newton's equation, $F = ma$. To find the velocity of the mass, solve for v in the above equation by time integrating both sides of the equation:

$$\int Fdt = mv; \quad v = \frac{1}{m}\int Fdt = \frac{1}{2}\int 4dt$$

$$v = \frac{4}{2}t = 2(5) = 10 \text{ cm/sec}$$

2.3.3.3 Systems analysis and systems models

Systems models usually represent whole processes using so-called *black box* components. Each element of a systems model consists only of an input–output relationship defined by an equation and represented by a geometric shape, usually a rectangle. No effort is made to determine what is actually inside the box; hence, the term *black box*. The modeler pays no heed to what is the inside the box, only its overall input–output (or stimulus/response) characteristics. A typical element in a systems model is shown graphically as a box or sometimes as a circle when an arithmetic process is involved (Figure 2.3-8). The inputs and outputs of all elements are signals with a well-defined direction of flow or influence. These signals and their direction of influence are shown by lines and arrows connecting the system elements (Figure 2.3-8).

The letter G in the right-hand element of Figure 2.3-8 represents the mathematical operation that converts the input signal into an output signal, usually expressed as a ratio of output to input:

$$G = \frac{Output}{Input}; \quad Output = Input(G) \qquad \text{[Eq. 2.3.6]}$$

For many elements, the mathematical operation defined by the letter G in Figure 2.3-8 can be quite complex,

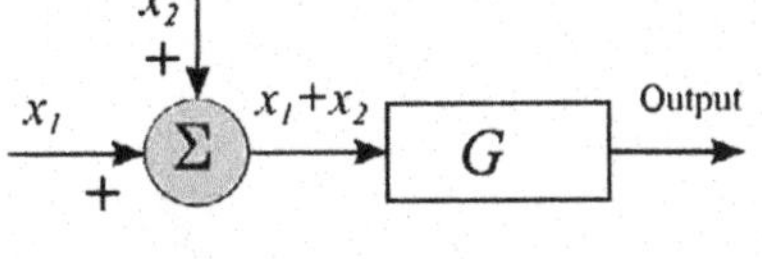

Figure 2.3-8 Typical elements in a system model. The left-hand element is an 'adder whose output signal is the sum of the two inputs x_1 and x_2. The right-hand element is a general element that takes the input ($x_1 + x_2$ in this case) and operates on it with the mathematical operation G to produce the output signal.

involving integral and differential operations, just as in an analog model. In fact, it is a straightforward task to convert from an analog model to a systems model, but not vice versa. It requires some mathematical tricks to structure calculus operations into a format that involves simply algebraic multiplication.

Occasionally, the operation performed by an element is simple, such as a scaling of the input; that is, the output is the same as the input, but multiplied by a constant *gain*. In such cases, the equation for G would be a simple constant defining the multiplying gain. Under static or *steady-state* conditions when the inputs have a constant value, and all internal signals have also settled to a constant value, the element equations, the G's, can usually be reduced to constants. If all the element functions are constants, the model can be solved, that is, the value of the output and all internal signals determined, using algebra. A steady-state solution of a systems model is given in Example 2.3.4 below.

One of the earliest physiological systems models, the pupil light reflex, is shown in Figure 2.3-9, and includes two processes. The pupil light reflex is the response of the iris to changes in light intensity falling on the retina. Increases or decreases in ambient light cause the muscles of the iris to change the size of the pupil in an effort to keep light falling on the retina constant. (This system was one of the first to be studied using engineering tools.) The two-component system receives light as the input and produces a movement of the iris muscles that changes pupil area, the aperture in the visual optics. The first box represents all of the neural processing associated with this reflex, including the light receptors in the eye. It generates a neural control signal, which is sent to the second box. The second box represents the iris musculature, including its geometric configuration. The input to this second box is the neural control signal from the first box and the output is pupil area.

The systems model shown in Figure 2.3-9 demonstrates the strengths and the weaknesses of systems

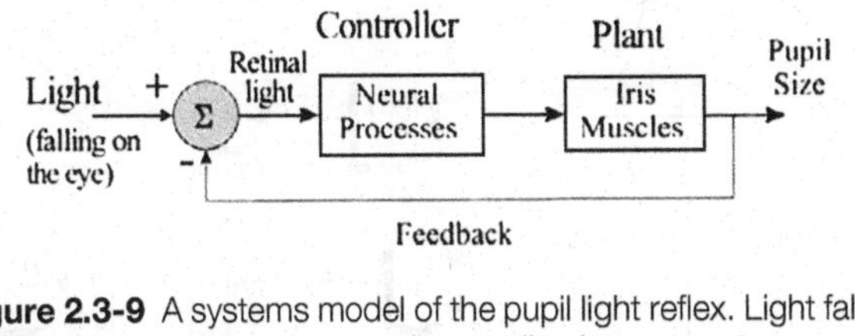

Figure 2.3-9 A systems model of the pupil light reflex. Light falling on the retina stimulates a neural *controller* that generates a neural signal that is sent to the iris muscles, the *plant* or *effector* apparatus. The system involves *feedback* because as the pupil (the hole in the iris) reduces in size, it reduces the light falling on the retina. This is considered a negative feedback system because a positive increase in the response (in this case a reduction in pupil size) leads to a decrease in the stimulus (i.e., the light falling on the retina).

analysis. By compressing a number of complex processes into a single black box, and representing these processes by a single input–output equation, a systems model can provide a concise, highly simplified representation of a very complex system. You need not understand how a biological process accomplishes a given task. As long as you can document some of its behavior quantitatively (which allows you to construct the input–output equation), you can usually construct a system representation. This will allow you to analyze the system's behavior over a large (perhaps nonphysiological) stimulus range or incorporate that process into the analysis of a larger system. However, this ability to reduce complex processes to a few elements, each represented by a single equation, means that these models do not provide much insight into how the process or processes are implemented by the underlying physiological mechanisms.

The effector apparatus in Figure 2.3-9, the iris musculature, is often termed the *plant* in systems models. This curious terminology comes from early applications of linear systems analysis to study the control of large chemical plants. These large systems were divided into control processes under the heading *controller,* and effector processes grouped under the heading *plant.* This terminology has been transferred to physiological control models, especially those involving motor control systems.

To complete this system description of the light reflex, note that changes in pupil area, the output of the system, will alter the light falling on the retina, the input to the system. Hence, the output feeds back to the input, creating a classic *feedback control system* (Figure 2.3-9). The feedback is negative because an increase in light stimulation will generate an increase in the response—in this case a decrease in pupil area—decreasing the light falling on the retina and offsetting, to some extent, the increase in light stimulus. In the pupil light reflex, the decrease in retinal light produced by the decrease in pupil size does not fully compensate for the increase in stimulus, so the feedback gain is less than 1 (unity).

Systems models often provide more detail than that given in the very basic structure of the model in Figure 2.3-9. Figure 2.3-10 shows a more detailed model of the neural pathways that mediate the vergence eye movement response, the processes used to turn the eyes inward to track visual targets at different depths. The model shows three neural paths converging on the elements representing the oculomotor plant (the two rightmost system elements). Neural processes in the upper two pathways provide a velocity-dependent signal to move the eyes quickly to approximately the right position. The lower pathway represents the processes that use visual feedback to more slowly fine-tune the position of the eyes and attain a very accurate final position. The error between the angle required to precisely image a stimulus in the two eyes and that actually attained by this neural controller is generally less than a tenth of a degree.

As with analog models, systems models can be evaluated by simulating their behavior on a computer. This not only provides a reality check—do they produce a response similar to that of the real system—but also permits evaluation of internal components and signals not available to the experimentalist. For example, what

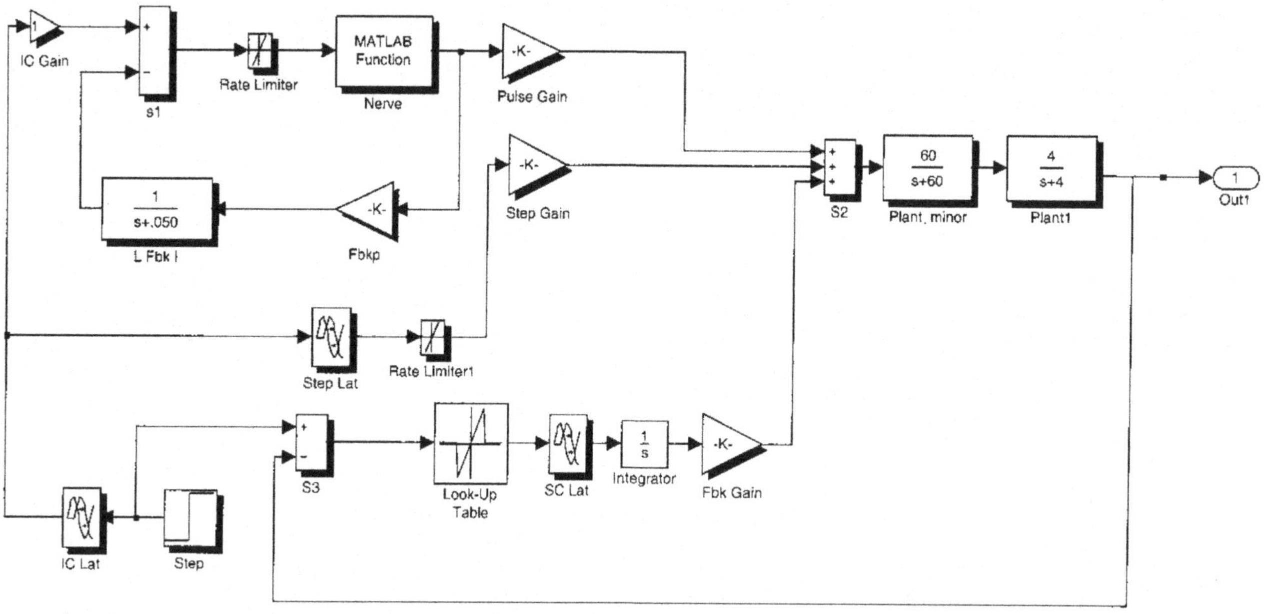

Figure 2.3-10 A model of the vergence eye movement neural control system showing more of the control details. This model can be simulated using MATLAB's Simulink program.

would happen in the vergence model if the neural components responsible for the pulse signal (upper pathway) were not functioning, or functioning erratically perhaps due to a brain tumor? Systems-type models are even easier to simulate than analog models because MATLAB provides a software package known as Simulink that uses graphics to convert a system model into computer code. Simulink simulations of the vergence model produce responses that are very close to actual vergence eye movements. Although Simulink is only developed for systems models, it is easy to convert analog models into system format so this software can be used to simulate analog models as well. Alternatively, there are programs such as pSpice specifically designed to simulate electronic circuits that can also be used for analog models.

Figures 2.3-9 and 2.3-10 show another important property of systems models. The influence of one process on another is explicitly stated and indicated by the line connecting two processes. This line has a direction usually indicated by an arrow, which implies that the influence or information flow travels only in that direction. If there is also a reverse flow of information, such as in the case in feedback systems, this must be explicitly stated in the form of an additional connecting line showing information flow in the reverse direction.

The next example has some of the flavor of the simulation approach, but will not require the use of Simulink.

Example 2.3.3: There is a MATLAB function on the disk that simulates some unknown process. The function is called `process_x` and takes an input variable, `x`, and generates a variable signal, `y`. (The `Courier` typeface is used to indicate a MATLAB variable, function, or code.) The function expects the input to be a signal represented by an array of numbers (as if x were a digitized signal), and produces an output signal that will be an array of number the same length as the input. We are to determine if `process_x` is a linear process over an input stimulus range of 0 to ±100. We can input to the process any signal we desire and examine the output.

Solution: Our basic strategy will be to input different signals having values within the desired range and see if the outputs are proportional. However, what is the best signal to use? The easiest might be to input two or three signals that have a constant value; for example, $x(t) = 1$, then $x(t) = 10$, then $x(t) = 100$, along with the negative values. The output should be proportional. However, what if the process contains a derivative operation? Although the derivative is a linear operation, the derivative of a constant is zero, and so we would get zero out for all three signals. Similarly, if the process contains integrations, the output to a constant could be difficult to interpret. Going back to basic calculus, recall that the derivative of a sine is a cosine, and the integral of a sine is a negative cosine. Thus, if the input signal were a sine, the output would still be sinusoidal even if the process contained integrations and/or differentiations. If the process contained derivative or integral operations, the sinusoidal output would be scaled (by the frequency), but this scaling would apply to all sine inputs.

Our strategy will be to input different sines having different amplitudes. If the output signals are proportional to the input amplitudes, we would guess that `process_x` is a linear process over the values tested. Because the work will be done on a computer, we can use any number of sine inputs, so let us try 100 different input signals ranging in amplitude from ±1 to ±100. If we plot the amplitude of the sinusoidal output, it should plot as straight line if the system is linear, and some other curve if it is not. The MATLAB code for this evaluation is as follows:

```
% Example 2.3.3 Example to evaluate an unknown
  process
% called 'process_x' to determine if it is
  linear.
%
t = 0:2*pi/500:2*pi;   % Sine wave argument
for k = 1:100          % Amplitudes will vary
                         from 1
                       % to 100
  x = k*sin(t);        % Generate a 1 cycle sine
                         wave
  y = process_x(x);    % Input sine to process
  output(k) = max(y);  % Save max value of
                         output
end
plot (output);         % Plot and label output
xlabel ('Input Amplitude');
ylabel ('Output Amplitude');
```

Analysis: Within the `for-loop`, the program generates a one-cycle sine wave having the desired amplitude. The amplitudes are incremented from 1 to 100 as the loop progresses. The sine wave, stored as variable `x`, becomes the input signal to `process_x`. The function produces an output signal, `y`. The maximum value of the output signal is found using MATLAB's *max* routine, and save in variable array, `output`. When the loop completes, the 100 values of `output` are plotted.

The figure below shows one of the input signals (solid curve) and the corresponding output (dashed curve) produced by `process_x`. The input signal looks like a sine function as expected, whereas the output looks like a cosine function. This suggests that the process contains a derivative operation. The output signal is also slightly larger than the input signal.

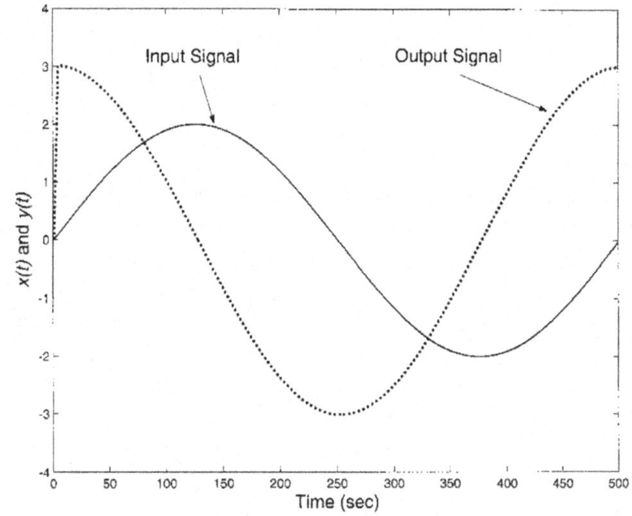

The plot of maximum output values (i.e., variable `output`) is a straight line indicating a linear relationship between the amplitude of output and input signal. Whatever the process really is, it appears to be linear.

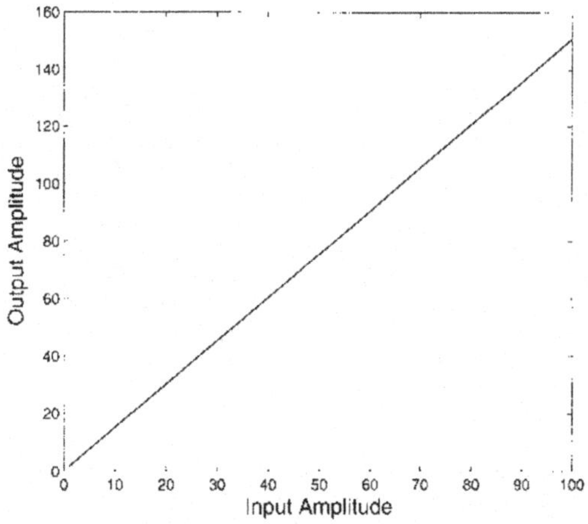

Figuring out exactly what the process is solely by testing it with external signals can be a major challenge. The field of *system identification* deals with approaches to obtain a mathematical representation for an unknown system by probing it with external signals. A few examples and problems later in this text show some of the techniques used to evaluate linear systems in this manner. Comparing the input and output sinusoids it looks like `process_x` contains a derivative and a multiplying factor that increases signal amplitude, but more input–output combinations would have to be evaluated to confirm this guess.

Example 2.3.4: Find the overall input–output relationship for the systems model below. Assume that the system is in steady-state condition so that all the signals have constant values and the two elements, represented by the equations G and *H*, are simply gain constants.

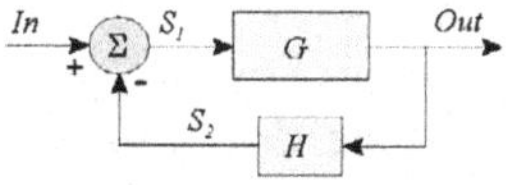

Solution: Generate an algebraic equation based on the configuration of the system and the fact that the output of each process is just the input multiplied by the associated gain term as stated in Eq. 2.3.6.

By definition: $G = Out/S1$; $H = S_3/Out$; and

$S_1 = In - S_3$ where G and H are constants

Hence $S_1 = Out/G$; $S_3 = Out(H)$

Since $S_1 = In - S_3$ substituting in the above:

$Out/G = In - Out(H)$

Rearranging: $Out = In(G) - Out(GH)$;

$Out(1 + GH) = In(G)$

$$\frac{Out}{In} = \frac{G}{1 + GH} \qquad \text{[Eq. 2.3.7]}$$

The solution is the classic *feedback equation*. Since the two elements, G and *H*, could be represented by simple gain constants, algebra alone can be used to work out the input–output equations. What of more complicated situations where the model is not in steady state and/or the processes must be defined using differential and integral operations?

2.3.3.4 Systems and analog analysis: summary

The basic differences and relative strengths and weaknesses of systems analysis versus analog analysis have already been described. Systems analysis only tries to represent the behavior of a process whereas analog analysis makes some effort to mimic the way in which the process produces that behavior. This is done in analog models by representing the process using elements that are, to some degree, analogous to those in the actual process. Analogous elements have the same general behavior as the physiological elements they represent; hence, analog models usually represent the system at a lower level, and in greater detail, than do systems models. However, not all analog models offer this detail. This can be seen in the windkessel cardiovascular model of Figure 2.3-5. The single capacitor, *Cp*, represents the combined elastic behavior, or springlike characteristics, of the entire arterial tree. Analog models often provide better representation of secondary features such as energy use, which is usually similar between analog elements and the actual components they represent.

Systems models provide better clarity, particularly with regard to information flow or influences. In analog models, all components may interact to some extent and this interaction may not be obvious from inspection of the model. Referring to the eye muscle model of Figure 2.3-7, a change in just one parallel elastic element, k_B, will modify the force on every element in the model. In systems models, all influences are explicitly shown and their interactions are immediately apparent from an inspection of the model. For example, in the model of Figure 2.3-9, the fact that the iris also influences the neural controller is explicitly shown by the feedback pathway. This can be of great benefit in clarifying the control structure of a complex system. Perhaps the most significant advantage of the systems approach is that it allows processes to be rigorously represented without requiring the modeler to know the details of the underlying physiological mechanism.

2.3.4 Noise and variability

Where there is signal, there is noise. Occasionally, the noise will be at such a low level that it is of little concern, but usually the noise limits the usefulness of the signal. This is especially true for physiological signals because they experience many potential sources of noise or variability. In most usages, *noise* is a general and relative term: noise is what you do not want, and signal is what you do want. This leads to a definition of noise as any form of unwanted variability. Noise is inherent in most measurement systems and is often the limiting factor in the performance of a medical instrument. Indeed, many medical instruments go to great lengths in terms of signal-conditioning circuitry and signal-processing algorithms to compensate for unwanted variability.

In biomedical measurements, noise or variability has four possible origins: (a) physiological variability; (b) environmental noise or interference; (c) measurement or transducer artifact; and (d) electronic noise. Physiological variability comes about because the information you desire is based on measurements subject to biological influences other than those of interest. For example, assessment of respiratory function based on the measurement of blood Po_2 could be confounded by other physiological mechanisms that alter blood Po_2. Measurement errors due to physiological variability can be very difficult to resolve, sometimes requiring a total redesign (or rethinking) of the approach.

Environmental noise can come from sources external or internal to the body. A classic example is the measurement of the fetal ECG signal where the desired signal is corrupted by the mother's ECG. Because it is not possible to describe the specific characteristics of environmental noise, typical noise reduction approaches such as filtering are not usually successful. Sometimes environmental noise can be reduced using *adaptive filters* or *noise cancellation* techniques that adjust their filtering properties on the basis of the current environment.

Measurement artifact is produced when the measurement device, or *transducer*, responds to energy modalities other than those desired. For example, recordings of electrical signals of the heart, the ECG, are made using electrodes placed on the skin. These electrodes are also sensitive to movement, so-called *motion artifact*, where the electrodes respond to mechanical movement as well as to the desired electrical activity of the heart. This artifact is not usually a problem when the ECG is recorded in the physician's office, but it can be if the recording is made during a patient's normal daily living, as in a Holter recording. Measurement artifacts can sometimes be successfully addressed by modifications in transducer design. Aerospace research has led to the development of electrodes that are relatively insensitive to motion artifact.

Unlike the other sources of variability, electronic noise has well-known sources and characteristics. Electronic noise falls into two broad classes: *thermal* or *Johnson* noise, and *shot* noise. The former is produced primarily in resistor or resistance materials whereas the latter is related to voltage barriers associated with semiconductors. Both sources produce noise that contains energy over a broad range of frequencies, often extending from DC to 10^{12} to 10^{13} Hz. Such broad-spectrum noise is referred to as *white noise* because it contains energy at all frequencies (or at least all the frequencies of interest to bioengineers) just as white light contains energy at all frequencies (or at least, all the frequencies we can see). Figure 2.3-11 shows a plot of the energy in a simulated white-noise waveform (actually, an array of random numbers) plotted against frequency. This is similar to a plot of the energy in a beam of light versus wavelength (or frequency) and, as with light, is also referred to as a *spectral plot* or *spectrum*. Note that the energy of the simulated noise is constant across the spectral range.

The various sources of noise or variability along with their causes and possible remedies are presented in Table 2.3-3. Note that in three of four instances, appropriate transducer design may aid in the reduction of the variability or noise. This demonstrates the important role of the transducer in the overall performance of the instrumentation system.

2.3.4.1 Electronic noise

Johnson or thermal noise is produced by resistance sources and the amount of noise generated is related to the resistance and to the temperature:

$$V_J = \sqrt{4kTRBW} \text{ volts} \quad \text{[Eq. 2.3.8]}$$

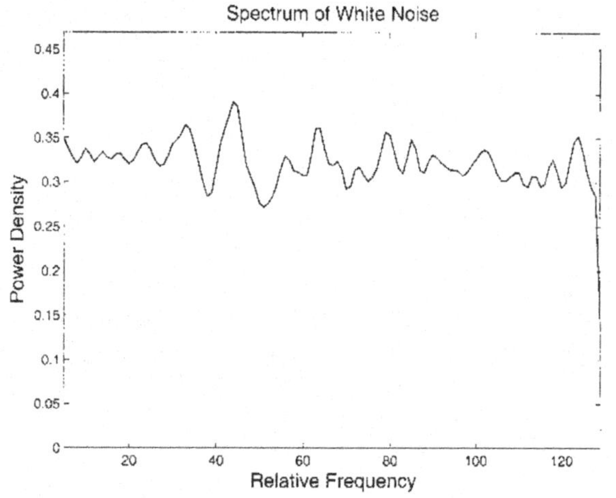

Figure 2.3-11 A plot of the energy in white noise as a function of frequency. The noise has a flat spectral characteristic showing similar energy levels over a wide range of all frequencies plotted. This equal-energy characteristic gives rise to the term *white noise.*

where R is the resistance in ohms, T the temperature in kelvin, and k Boltzmann's constant ($k = 1.38 \times 10^{-23}$ joules/kelvin). (A temperature of 310 K is often used as room temperature, in which case $4kT = 1.7 \times 10^{-20}$ J.) Here BW is the range of frequencies that is included in the signal. This range of frequencies is termed *bandwidth*. This frequency range is usually determined by the characteristics of the measurement system, often the filters used in the system. Because Johnson noise is spread over all frequencies, the greater the signal's bandwidth, the greater the noise in any given signal.

If noise current is of interest, the equation for Johnson noise current can be obtained from Eq. 2.3.8 in conjunction with Ohm's law:

$$I_J = \sqrt{4kTBW/R} \text{ amps} \qquad \text{[Eq. 2.3.9]}$$

Table 2.3-3 Sources of variability

Source	Cause	Potential remedy
Physiological variability	Measurement only indirectly related to variable of interest	Modify overall approach
Environmental (internal or external)	Other sources of similar energy form	Noise cancellation Transducer design
Artifact	Transducer responds to other energy sources	Transducer design
Electronic	Thermal or shot noise	Transducer or electronic design

In practice, there will be limits imposed on the frequencies present within any waveform (including noise waveforms), and these limits are used to determine band width. In the problems given here the bandwidth is simply stated. Bandwidth is usually specified in hertz with units of inverse seconds (i.e., 1/second). Because bandwidth is not always known in advance, it is common to describe a relative noise, specifically the noise that would occur if the bandwidth were 1.0 Hz. Such relative noise specification can be identified by the unusual units required: $V/\sqrt{Hz}$ or ampere $(A)/\sqrt{Hz}$. Shot noise is defined as a current noise and is proportional to the baseline current through a semiconductor junction:

$$I_s = \sqrt{2qI_dBW} \text{ amps} \qquad \text{[Eq. 2.3.10]}$$

where q is the charge on an electron (1.602×10^{-19} coulomb [coul]), and I_d is the baseline semiconductor current. (In photodetectors, the baseline current that generates shot noise is termed the *dark current*, hence the letter d in the current symbol, I_d in Eq. 2.3.10.) Again, the noise is spread across all frequencies so the bandwidth must be specified to obtain a specific value, or a relative noise can be specified in $A/\sqrt{Hz}$.

When multiple noise sources are present, as is often the case, their voltage or current contributions add to the total noise as the square root of the sum of the squares, assuming that the individual noise sources are independent. For voltages:

$$V_T = \sqrt{V_1^2 + V_2^2 + V_3^2 + \ldots + V_N^2} \qquad \text{[Eq. 2.3.11]}$$

A similar equation applies to current noise.

Example 2.3.5: A 20-mA current flows through a diode (i.e., a semiconductor) and a 200-Ω resistor. What is the net current noise, i_n? Assume a bandwidth of 1 MHz (1×10^6 Hz).

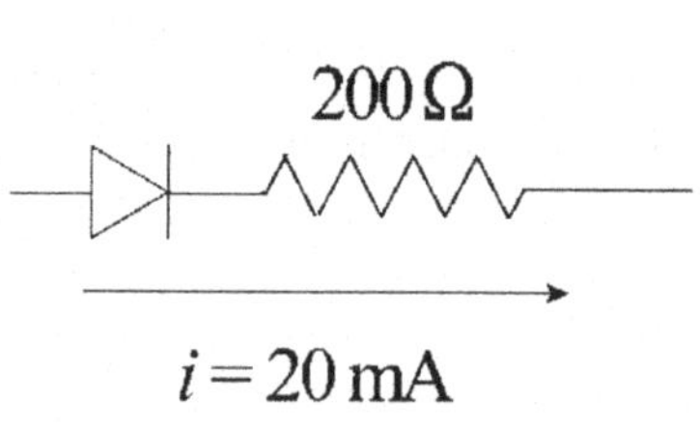

Solution: Find the noise contributed by the diode using Eq. 2.3.10, the noise contributed by the resistor using Eq. 2.3.9, then combine them using Eq. 2.3.11.

$$i_{nd} = \sqrt{2qI_dBW} = \sqrt{2(1.602\times10^{-19})(20\times10^{-3})10^6}$$

$$= 8.00\times10^{-8}\ \text{amps}$$

$$i_{nR} = \sqrt{4kTBW/R} = \sqrt{1.7\times10^{-20}(10^6/200)}$$

$$= 9.22\times10^{-9}\ \text{amps}$$

$$i_{nT} = \sqrt{i_{nd}^2 + i_{nR}^2} = \sqrt{6.4\times10^{-15} + 8.5\times10^{-17}}$$

$$= 8.1\times10^{-8}\ \text{amps}$$

Note that most of the current noise is coming from the diode, so the addition of the resistor's current noise does not contribute much to the diode noise current. The mathematics in this example could be simplified by calculating the square of the noise current (i.e., not taking the square roots) and using those values to get the total noise before taking the square roots.

2.3.4.2 Signal-to-noise ratio

Most waveforms consist of signal plus noise mixed together. As noted previously, signal and noise are relative terms, relative to the task: the signal is that portion of the waveform of interest whereas the noise is everything else. Often the goal of signal processing is to separate out a signal from noise, identify the presence of a signal buried in noise, or detect features of a signal buried in noise.

The relative amount of signal and noise present in a waveform is usually quantified by the *signal-to-noise ratio* (SNR). As the name implies, this is simply the ratio of signal to noise, both measured in RMS (root-mean-squared) amplitude. This measurement is rigorously defined in the next chapter. The SNR is often expressed in decibels (dB) where:

$$\text{SNR} = 20\log\left(\frac{\text{signal}}{\text{noise}}\right) \qquad \text{[Eq. 2.3.12]}$$

To convert from decibel scale to a linear scale:

$$\text{SNR}_{\text{Linear}} = 10^{\text{dB}/20} \qquad \text{[Eq. 2.3.13]}$$

For example, an SNR of 20 dB means that the RMS value of the signal is 10 times the RMS value of the noise ($10^{(20/20)} = 10$), +3 dB indicates a ratio of 1.414 ($10^{(3/20)} = 1.414$), 0 dB means the signal and noise are equal in RMS value, −3 dB means that the ratio is 1/1.414, and −20 dB means the signal is 1/10 of

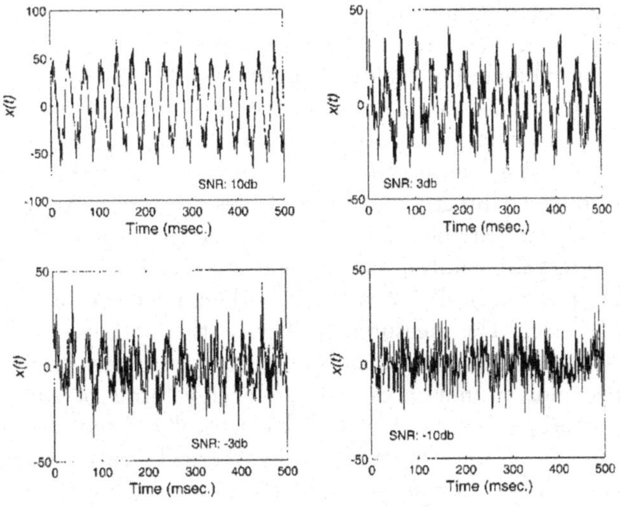

Figure 2.3-12 A 30-Hz sine wave with varying amounts of added noise. The sine wave is barely discernible when the signal-to-noise ratio is −3 dB and not visible when the signal-to-noise ratio is −10 dB.

the noise in RMS units. Figure 2.3-12 shows a sinusoidal signal with various amounts of white noise. Note that it is difficult to detect the presence of the signal visually when the SNR is −3 dB, and impossible when the SNR is −10 dB.

2.3.5 Summary

Biological systems include a variety of physiological processes ranging from the organ level through the cellular level to the molecular level. Classic physiology is structured around large-scale biological systems such as the cardiovascular system, the endocrine system, the gastrointestinal system, and others. All biological systems communicate with one another, and with themselves, via biosignals. Such signals are carried by electrical, chemical, mechanical, or thermal energy. All signals involve some form of coding process. For analog signals, the information is encoded into the amplitude of the signal at any given instant in time. If these analog signals are processed by digital computers, they must be converted to digital format, a process that involves slicing the signal in both amplitude and time. Amplitude slicing is known as quantization, whereas time slicing is known as sampling.

The field of linear systems analysis encompasses analog and systems processes. Both representations use linear elements so that this analysis only applies to linear processes or processes that can be taken, or approximated, as linear. Electrical analog models are used to analyze electric circuits and to represent physiological processes. Mechanical analog models can also be used from either of these two perspectives. Examples were given of the early use of an electric circuit model to represent the cardiovascular system and an early mechanical model of skeletal muscle.

All signals are contaminated by noise, variability, or other artifact. Efforts to obtain meaningful information from physiological processes are often thwarted by the inability to directly measure variables of interest. All too frequently variables only loosely related to the process of interest are readily available, and these may be altered by the physiological process itself or the influence of other processes. The measurement device, the biotransducer, is often a major source of measurement errors because it responds to influences of other energy forms or environmental factors. Finally, all electrically based measurements are contaminated by thermal and/or shot noise. These two noise processes are well defined and contain noise energy over a wide range of frequencies. This broad distribution of energy means that thermal and shot noise can always be improved by limiting the frequencies contained in the signal. A variety of filters exist in analog and digital forms to limit the frequencies in a signal to only those that carry the desired information.

Problems

1. An electrical inductor has a defining equation that is the same as a mass if the variable voltage and current are substituted for force and velocity (specifically, $V_L = L\ di/dt$). A constant voltage of 10 V is placed across a 1-H inductor. How long will it take for the current through the inductor to reach 1 A? (See Example 2.3.2.)
2. Assume that the feedback control system presented in Example 2.3.4 is in steady-state or static conditions. If G = 100 and $H = 1$ (i.e., a unity gain feedback control system), find the output if the input equals 1. Find the output if the input is increased to 10. [Note how the output is proportional to the input, which accounts for why a system (having this configuration) is sometimes termed a *proportional control system.*] Now find the output if the input is 10 and G is increased to 1,000. Note that the difference between the input and output values depends on the value of G.
3. In the system given in Problem 2.3.2, the input is changed to a signal that smoothly goes from 0.0 to 5.0 in 10 seconds [i.e., $In(t) = 0.5t$ seconds]. What will the output look like? (*Note:* G and *H* are simple constants, so Eq. 2.3.7 still holds.)
4. A resistor produces 10-μV noise when the room temperature is 310 K and the bandwidth is 1 kHz. What current noise would be produced by this resistor?
5. The noise voltage out of a 1-MΩ resistor is measured using a digital volt meter as 1.5 μV at a room temperature of 310 K. What is the effective bandwidth of the voltmeter?
6. If a signal is measured as 2.5 V and the noise is 28 mV (28×10^{-3} V), what is the SNR in decibels?
7. A single sinusoidal signal is found in a large amount of noise. (If the noise is larger than the signal, the signal is sometimes said to be 'buried in noise.') The RMS value of the noise is 0.5 V and the SNR is 10 dB. What is the RMS amplitude of the sinusoid?

MATLAB problems

8. Use the approach presented in Example 2.3.3 to determine if either of two processes, `process_y` or `process_z`, are linear. The two processes are found on the disk as MATLAB functions.
9. Write a MATLAB function that takes in two variables, and input variable `x` and gain variable `G`, and produces and output variable `y`. This function should implement the feedback equation (Eq. 2.3.7), with the variable `H` set to 1.0 (i.e., a unity gain feedback system). (Name the function '`fbk_system.`') Input a two-cycle sine wave as in Example 2.3.3 having an amplitude of 1. Plot the maximum values of the input–output relationship for this process [i.e., `max(y)/max(x)`] as a function of `G`, where `G` ranges between 1 and 1,000. (*Hint:* Put the process in a for-loop as in Example 2.3.3 and increment G. This will provide a more detailed demonstration of the relationship between the input–output ratio and the importance of the value of G in a feedback system.)

Chapter 2.4

2.4

Basic signal processing

John Semmlow

2.4.1 Basic signals: the sinusoidal waveform

Signals are the foundation of information processing, transmission, and storage. Signals also provide the interface with physiological systems and are the basis for communication between biological processes (Figure 2.4-1). Given the ubiquity of signals within and outside the body, it should be no surprise that understanding at least the basics of signals is fundamental to understanding, and interacting with, biological processes.

A few signals are simple and can be defined analytically, that is as mathematical functions. For example, a sinusoidal signal is defined by the equation:

$$x(t) = A\sin(\omega_p t) = A\sin(2\pi f_p t) = A\sin\left(\frac{2\pi t}{T}\right) \quad \text{[Eq. 2.4.1]}$$

where A is the signal amplitude, or more accurately the *peak-to-peak amplitude*, ω_p is the frequency in radians per second, f_p is the frequency in hertz, and T is the period in seconds, and t is time in seconds. Recall that frequency can be expressed in either radians or hertz (the units formerly known as cycles per second) and are related by 2π:

$$\omega_p = 2\pi f_p \quad \text{[Eq. 2.4.2]}$$

Both forms of frequency are used in the text, and the reader should be familiar with both. The frequency in Hz is also the inverse of the period, T:

$$f_p = \frac{1}{T} \quad \text{[Eq. 2.4.3]}$$

The signal presented in Eq. 2.4.1 is completely defined by A and f_p (or w_p, or T); once you specify these two terms, you have characterized the sine signal for all time. The sine wave signal is rather boring: if you have seen one cycle, you have seen them all. Moreover, because the signal is completely defined by A and f_p, if neither the amplitude, A, nor the frequency, f_p, changes over time, it is hard to see how this signal could carry much information. These limitations notwithstanding, sine waves (and cosine waves) are at the foundation of many signal analysis techniques. In part, their importance stems from their simplicity and the way they are treated by linear systems. Sine wave-like signals can also be represented by cosines, and the two are related.

$$A\cos(\omega t) = A\sin\left(\omega t + \frac{\pi}{2}\right) = A\sin(\omega t + 90 \text{ degrees})$$

$$A\sin(\omega t) = A\cos\left(\omega t - \frac{\pi}{2}\right) = A\cos(\omega t - 90 \text{ degrees}) \quad \text{[Eq. 2.4.4]}$$

Note that the second representations [i.e., $A\sin(\omega t + 90$ degrees) and $A\cos(\omega t - 90$ degrees)] have conflicting units: the first part of the sine argument, ωt, is in radians, whereas the second part is in degrees. Nonetheless, this is common usage and is the form that is used throughout this text.

A general *sinusoid* (as opposed to a pure sine wave or pure cosine wave) is a sine or cosine with a general phase term as shown in Eq. 2.4.5:

$$\begin{aligned} x(t) &= A\sin(\omega_p t + \theta) = A\sin(2\pi f_p t + \theta) \\ &= A\sin\left(\frac{2\pi t}{T} + \theta\right) \text{ or equivalently} \\ x(t) &= A\cos(\omega_p t - \theta) = A\cos(2\pi f_p t - \theta) \\ &= A\cos\left(\frac{2\pi t}{T} - \theta\right) \end{aligned} \quad \text{[Eq. 2.4.5]}$$

Biomedical Engineering Desk Reference; ISBN: 9780123746467

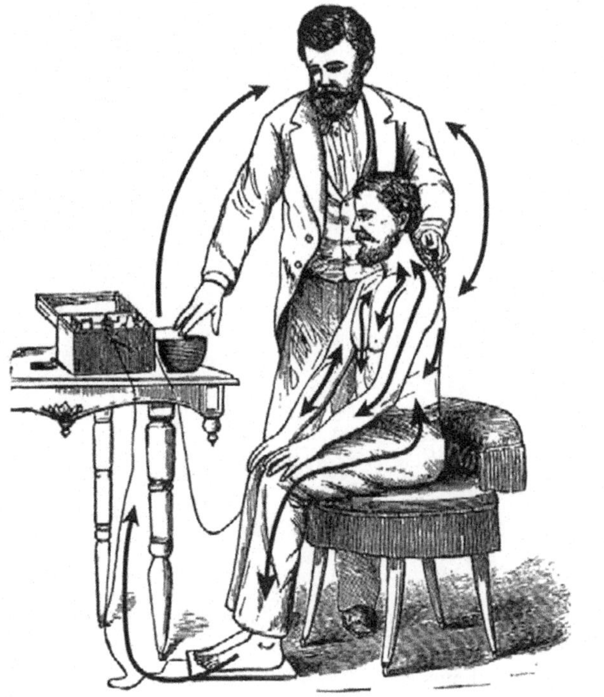

Figure 2.4-1 Signals continuously pass between various parts of the body. These *biosignals* are carried either by electrical energy, as in the nervous system, or by molecular signatures, as in the endocrine system and many other biological processes. Measurement of these biosignals is fundamental to diagnostic medicine and to bioengineering research.

where again the phase, θ, would be expressed in degrees even though the frequency descriptor ($\omega_p t$, or $2\pi f_p t$, or $2\pi t/T$) is expressed in radians or hertz. Many of the sinusoidal signals described in this text are expressed as in Eq. 2.4.5. Figure 2.4-2 shows two sinusoids that differ by 60 degrees.

To convert the difference in phase angle to a difference in time, note that the phase angle varies through 360 degrees during the course of one period, T seconds. To calculate the time difference or time delay between the two sinusoids, t_d, given the phase angle θ:

$$t_d = \frac{\theta}{360}T = \frac{\theta}{360f} \quad \text{or} \quad \theta = \frac{t_d}{T}360 = 360t_d f \qquad \text{[Eq. 2.4.6]}$$

where $f = f_p = 1/T$ For the 2-Hz sinusoids in Figure 2.4-2, $T = 1/f = 0.5$ seconds, so:

$$t_d = \frac{\theta}{360}T = \frac{60}{360}0.5$$
$$= 0.0833 \text{ seconds} = 83.3 \text{ milli seconds}$$

Example 2.4.1: Find the time difference or delay between two sinusoids:

$$x_1(t) = \cos(4t + 30) \quad \text{and} \quad x_2(t) = -2\sin(4t)$$

Solution: Convert both to a sine or cosine (here we convert to cosines):

$$x_2(t) = -2\cos(4t - 90)$$

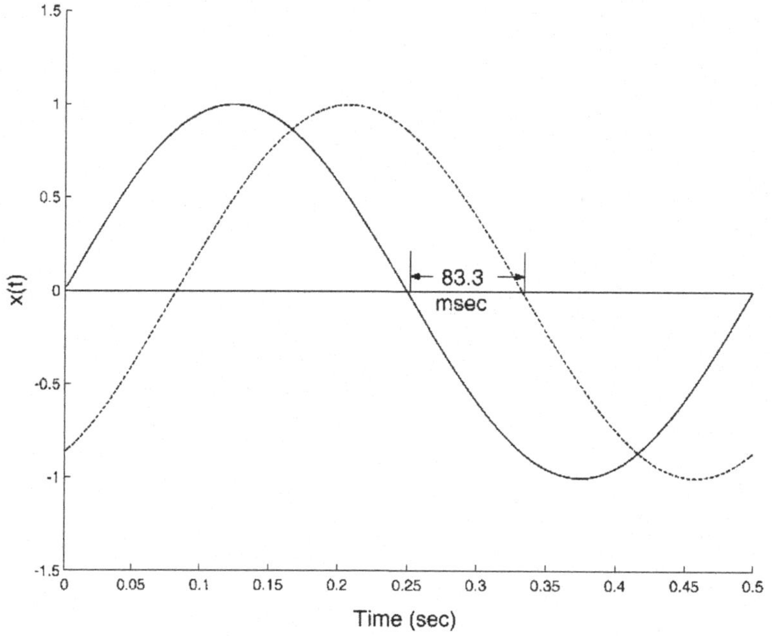

Figure 2.4-2 Two 2-Hz sinusoids that differ in phase by 60 degrees. This phase difference translates to a time difference or delay of 83.3 msec.

Thus, the angle between the two sinusoids is 120 degrees [30 – (–90)]. The period is given by:

$$T = \frac{1}{f} = \frac{1}{\omega/2\pi} = \frac{1}{4/2\pi} = 1.57 \text{ seconds}$$

and the time delay is:

$$t_d = \frac{\theta}{360}T = \frac{120}{360}1.57 = 0.523 \text{ seconds}$$

2.4.1.1 Sinusoidal arithmetic

Equation 2.4.5 describes an intuitive way of thinking about a sinusoid, as a sine wave with a phase shift. Alternatively, Eq. 2.4.5 shows that a cosine could just as well be used instead of the sine to represent a general sinusoid, and in this text, we use both. Sometimes it is mathematically convenient to represent a sinusoid as a combination of a pure sine and a pure cosine. This representation can be achieved using the well-known trigonometric identity for the sum of two arguments of a cosine function:

$$\cos(x-y) = \cos(x)\cos(y) + \sin(x)\sin(y)$$

[Eq. 2.4.7]

Based on this identity, the equation for a sinusoid can be written as:

$$\begin{aligned} &C\cos(2\pi ft - \theta) \\ &= C\cos(\theta)\cos(2\pi ft) + C\sin(\theta)\sin(2\pi ft) \\ &= a\cos(2\pi ft) + b\sin(2\pi ft) \end{aligned}$$

where:

$$a = C\cos(\theta); \quad b = C\sin(\theta)$$

[Eq. 2.4.8]

To convert from a sine and cosine to a single sinusoid with angle θ, start with Eq. 2.4.8.

If $a = C\cos(\theta)$ and $b = C\sin(\theta)$, then to determine C:

$$\begin{aligned} a^2 + b^2 &= C^2(\cos^2\theta + \sin^2\theta) = C^2 \\ C &= \sqrt{a^2+b^2} \end{aligned}$$

[Eq. 2.4.9]

Equation 2.4.10 shows the calculation for θ given a and b:

$$\frac{b}{a} = \frac{C\sin(\theta)}{C\cos(\theta)} = \tan(\theta), \ \theta = \tan^{-1}\left(\frac{b}{a}\right)$$

[Eq. 2.4.10]

Care must be taken in evaluating Eq. 2.4.10 to ensure that θ is determined to be in the correct quadrant on the basis of the signs of a and b. If both a and b are positive, θ must be between 0 and 90 degrees; if b is positive and a is negative, θ must be between 90 and 180 degrees (a calculator or MATLAB will not know this and will put any negative product in the fourth quadrant); if both a and b are negative, θ must be between 180 and 270 degrees (calculators and MATLAB put positive arguments in the first quadrant even if they result from two negative numbers); and finally, if b is negative and a is positive, θ must be between 270 and 360 degrees. Again, it is common to use degrees for phase angle.

To add sine waves, simply add their amplitudes. The same applies to cosine waves:

$$\begin{aligned} a_1\cos(\omega t) + a_2\cos(\omega t) &= (a_1 + a_2)\cos(\omega t) \\ a_1\sin(\omega t) + a_2\sin(\omega t) &= (a_1 + a_2)\sin(\omega t) \end{aligned}$$

[Eq. 2.4.11]

To add two sinusoids [i.e., $C\sin(\omega t + \theta)$ or $C\cos(\omega t - \theta)$], convert them to sines and cosines using Eq. 2.4.8, add sines to sines and cosines to cosines, and convert back to a single sinusoid if desired.

Example 2.4.2: Convert the sum of a sine and cosine wave, $x(t) = -5\cos(10t) - 3\sin(10t)$ into a single sinusoid.

Solution: Apply Eq. 2.4.9 and Eq. 2.4.10:

$$a = -5 \text{ and } b = -3$$

$$C = \sqrt{a^2+b^2} = \sqrt{5^2+3^2} = 5.83$$

$$\theta = \tan^{-1}\left(\frac{b}{a}\right) = \tan^{-1}\left(\frac{-3}{-5}\right) = 31 \text{ degrees,}$$

but θ must be in the third quadrant since both a and b are negative:

$$\theta = 31 + 180 = 211 \text{ degrees}$$

Therefore, the single sinusoid representation would be as follows:

$$x(t) = C\cos(\omega t - \theta) = 5.83\cos(10t - 211 \text{ degrees})$$

Analysis: Using Equations 2.4.8 through 2.4.11, any number of sines, cosines, or sinusoids can be combined into a single sinusoid if they are all at the same frequency. This is demonstrated in Example 2.4.3.

Example 2.4.3: Combine $x(t) = 4\cos(2t - 30 \text{ degrees}) + 3\sin(2t + 60 \text{ degrees})$ into a single sinusoid.

Solution: Expand each sinusoid into a sum of cosine and sine, algebraically add the cosines and sines, and recombine them into a single sinusoid. Be sure to convert the sine into a cosine [recall Eq. 2.4.4: $\sin(\omega) = \cos(\omega t - 90$ degrees$)$] before expanding this term.

$$4\cos(2t - 30) = a\cos(2t) + b\sin(2t)$$

where:

$$a = C\cos(\theta) = 4\cos(-30) = 3.5 \quad \text{and}$$
$$b = C\sin(\theta) = 4\sin(-30) = -2$$
$$4\cos(2t + 30) = 3.5\cos(2t) - 2\sin(2t)$$

Converting the sine to a cosine then decomposing the sine into a cosine plus a sine:

$$3\sin(2t + 60) = 3\cos(2t - 30)$$
$$= 2.6\cos(2t) - 1.5\sin(2t)$$

Combining cosine and sine terms algebraically:

$$4\cos(2t - 30) + 3\sin(2t + 60)$$
$$= (3.5 + 2.6)\cos(2t) + (-2 - 1.5)\sin(2t)$$
$$= 6.1\cos(2t) - 3.5\sin(2t)$$
$$= C\cos(2t + \theta) \quad \text{where} \quad C = \sqrt{6.1^2 + 3.5^2} \quad \text{and}$$
$$\theta = \tan^{-1}\left(\frac{-3.5}{6.1}\right)$$

$C = 7.0$; $\theta = 30$ (since b is negative, θ is in fourth quadrant) so $\theta = -30$ degrees

$$x(t) = 7.0\cos(2t - 30)$$

This approach can be extended to any number of sinusoids. An example involving three sinusoids is found in Problem 4.

2.4.1.2 Complex representation

An even more compact representation of a sinusoid is possible using complex notation. A *complex number* combines a *real number* and an *imaginary number.* Real numbers are commonly used, whereas imaginary numbers are the product of square roots and are represented by real numbers multiplied by the $\sqrt{-1}$. In mathematics, the $\sqrt{-1}$ is represented by the letter i, whereas engineers tend to use the letter j, the letter i being reserved for current. A *complex variable* simply combines a real and an imaginary variable: $z = x + jy$. Hence, although 5 is a real number, $j5$ is an imaginary number, and $5 + j5$ is a complex number. We will review here variables and arithmetic operations.

The beauty of complex numbers and complex variables is that the real and imaginary parts are *orthogonal.* One consequence of orthogonality is that the real and complex numbers (or variables) can be represented as if they are plotted on perpendicular axes (Figure 2.4-3).

Orthogonality is discussed in more detail later, but the importance of orthogonality with respect to complex numbers is that the real number or variable does not "interfere" with the imaginary number or variable and vice versa. Operations on one component do not affect the other. This means that a complex number behaves like two separate numbers rolled into one, and a complex variable like two variables in one. This feature comes in particularly handy when sinusoids are involved because a sinusoid at a given frequency can be uniquely defined by two variables: its magnitude and phase angle (or equivalently, using Eq. 2.4.8, its cosine and sine magnitudes, a and b). It follows that a sinusoid at a given frequency can be represented by a single complex number.

To find the complex representation, we will use the identity developed by the Swiss mathematician, Euler (Leonhard Euler's last name is pronounced "oiler". The use of the symbol e for the basis of the natural logarithmic system is a tribute to his extraordinary mathematical contributions):

$$e^{jx} = \cos x + j\sin x \qquad \text{[Eq. 2.4.12]}$$

This equation links sinusoids and exponentials, providing a definition of the sine and cosine in terms of complex exponentials. It also provides a concise representation of a sinusoid since a complex exponential contains both a sine and a cosine, although a few extra mathematical features are required to account for the fact that the second term is an imaginary sine term. This equation will prove very useful in two sinusoidally based analysis techniques: Fourier analysis and phasor analysis.

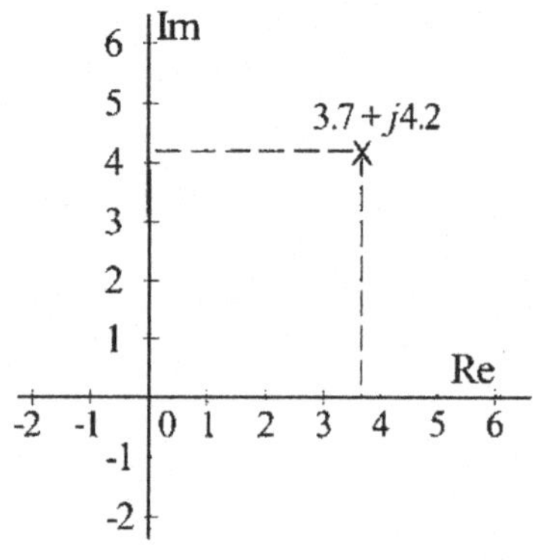

Figure 2.4-3 A complex number represented as an orthogonal combination of a real number on the horizontal axis and an imaginary number on the vertical axis. This graphic representation is useful for understanding complex numbers and aids in the interpretation of some arithmetic operations.

2.4.2 Signal properties: basic measurements

Biosignals and other information-bearing signals are often quite complicated and defy a straightforward analytical description. An archetype biomedical signal is the electrical activity of the brain as it is detected on the scalp by electrodes, the electroencephalogram (EEG) shown in Figure 2.4-4. Although a time display of this signal, as in Figure 2.4-4, constitutes a unique description, the information carried by this signal is not apparent from the time display, at least not to the untrained eye. Nonetheless, physicians and technicians are trained to extract useful diagnostic information by examining the time display of biomedical signals including the EEG. The time display of the electrocardiogram (ECG) signal is so medically useful that it is displayed continuously for patients undergoing surgery or those admitted to intensive care units (ICUs). This signal has become an indispensable image in television and movie medical dramas. Medical images, which can be thought of as two-dimensional signals, often need only visual inspection to provide information useful for diagnosis.

For some signals, a simple time display provides useful information, but many biomedical signals are not easy to interpret from their time characteristics alone. Nearly all signals will benefit from some additional signal processing. For example, the time display of the EEG signal in Figure 2.4-4 may have meaning for a trained neurologist, but it is likely to be uninterpretable to most readers. A number of basic measurements can be applied to a signal to extract more information, while other analyses can be used to probe the signal for specific features. Transformations can be used to provide a different view of the signal. In this section, basic measurements are described followed by more involved analyses.

One of the most straightforward of signal measurements is the assessment of its average value. Averaging is most easily described in the digital domain. To determine the average of a series of numbers, simply add the numbers together and divide by the length of the series (i.e., the number of terms in the series). This is mathematically stated as follows:

$$x_{avg} = \overline{x} = \frac{1}{N}\sum_{k=1}^{N} x_k \qquad \text{[Eq. 2.4.13]}$$

where k is an index number indicating a specific number in the series. The bar over the x in Eq. 2.4.13 stands for "the average of...". Equation 2.4.13 would be appropriate only for finding the average of a digital signal. An analog signal is a continuous function of time, $x(t)$, so the summation becomes an integration. The average or mean of a continuous signal, the continuous version of Eq. 2.4.13, is obtained by integrating the signal over time and dividing by the time length of the signal:

$$\overline{x}(t) = \frac{1}{T}\int_0^T x(t)dt \qquad \text{[Eq. 2.4.14]}$$

Note that the primary difference between digital and analog domain equations is the conversion of summation

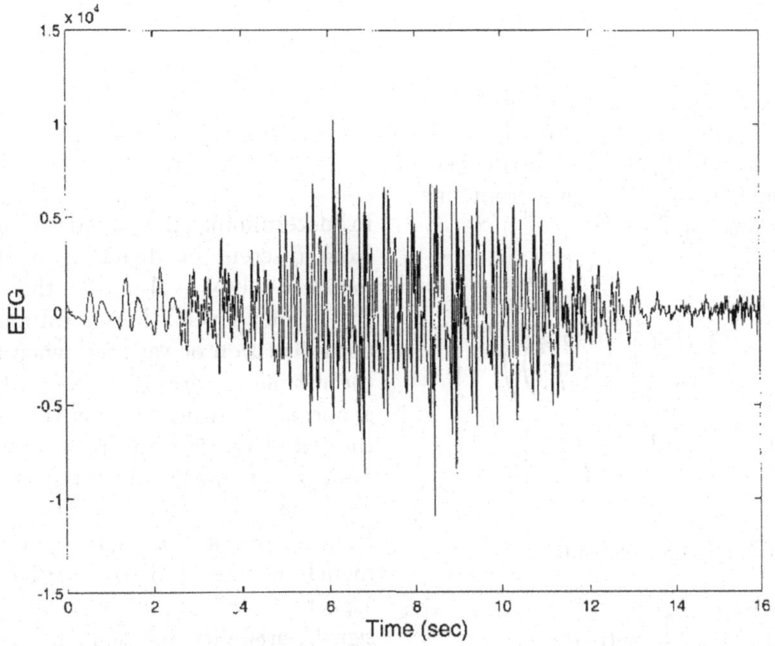

Figure 2.4-4 Segment of an electroencephalogram signal. (From the PhysioNet data bank, Goldberger et al., 2000.)

to integration and the use of a continuous variable, t, in place of the discrete integer, k. These conversion relationships are generally applicable, and most digital-domain equations can be transferred to continuous or analog equations in this manner. In this text, usually the reverse operation is used: the continuous-domain equation is developed first, then the corresponding digital-domain equation is derived by substitution of summation for integration and an integer variable for the continuous time variable.

Although the average value is a basic property of a signal, it does not provide any information about the variability of the signal. The root-mean-squared (RMS) value is a measurement that includes the signal's variability and its average. Obtaining the RMS value of a signal is just a matter of following the measurement's acronym in reverse: first squaring the signal, then taking its average, and finally taking the square root of this average:

$$x(t)_{rms} = \left[\frac{1}{T}\int_0^T x(t)^2 dt\right]^{1/2} \quad \text{[Eq. 2.4.15]}$$

The discrete form of the equation can be obtained by following the simple rules described above.

$$x_{rms} = \left[\frac{1}{N}\sum_{k=1}^{N} x_k^2\right]^{1/2} \quad \text{[Eq. 2.4.16]}$$

Example 2.4.4: Find the RMS value of the sinusoidal signal:

$$x(t) = A\sin(\omega_p t) = A\sin(2\pi t/T)$$

Solution: Because this signal is periodic, with each period the same as the previous one, it is sufficient to apply the RMS equation over a single period. (This is true for most operations on sinusoids.) Neither the RMS value nor anything else about the signal will change from one period to the next. Applying Eq. 2.4.15:

$$\overline{x}(t)_{rms}$$

$$= \left[\frac{1}{T}\int_0^T x(t)^2 dt\right]^{1/2} = \left[\frac{1}{T}\int_0^T \left(A\sin\left(\frac{2\pi t}{T_p}\right)\right)^2 dt\right]^{1/2}$$

$$= \left[\frac{1}{T}\frac{A^2}{2\pi}\left(-\cos\left(\frac{2\pi t}{T}\right)\sin\left(\frac{2\pi t}{T}\right)+\frac{\pi t}{T}\right)\Bigg|_0^T\right]^{1/2}$$

$$= \left[\frac{A^2}{2\pi}(-\cos(2\pi)\sin(2\pi)+\pi+\cos 0\sin 0)\right]^{1/2}$$

$$= \left[\frac{A^2\pi}{2\pi}\right]^{1/2} = \left[\frac{A^2}{2}\right]^{1/2} = \frac{A}{\sqrt{2}} \cong 0.707A$$

Hence, there is a proportional relationship between the peak-to-peak amplitude of a sinusoid (A in this example) and its RMS value: specifically, the RMS value is $1/\sqrt{2}$ times the peak-to-peak amplitude, rounded in this text to 0.707. This relationship is only true for sinusoids. For other waveforms, the application of Eq. 2.4.15 or Eq. 2.4.16 is required.

A statistical measure related to the RMS value is the *variance*, σ^2. The variance is a measure of signal variability regardless of its average. The calculation of variance for discrete and continuous signals is as follows:

$$\sigma^2 = \frac{1}{T}\int_0^T (x(t)-\overline{x})^2 dt \quad \text{[Eq. 2.4.17]}$$

$$\sigma^2 = \frac{1}{N-1}\sum_{k=1}^{N}(x_k-\overline{x})^2 \quad \text{[Eq. 2.4.18]}$$

where $\overline{x}$ is the mean or signal average. In statistics, the variance is defined in terms of an estimator known as the *expectation* operation applied to the probability distribution function of the data. Because the distribution of a signal is rarely known in advance, the equations given here are used to calculate variance in practical situations.

The *standard deviation* is another measure of a signal's variability and is simply the square root of the variance:

$$\sigma = \left[\frac{1}{T}\int_0^T (x(t)-\overline{x})^2 dt\right]^{1/2} \quad \text{[Eq. 2.4.19]}$$

$$\sigma = \left[\frac{1}{N-1}\sum_{k=1}^{N}(x_k-\overline{x})^2\right]^{1/2} \quad \text{[Eq. 2.4.20]}$$

In determining the standard deviation and variance from discrete or digital data, it is common to normalize by $1/N - 1$ rather than $1/N$. This is because the former gives a better estimate of the actual standard deviation or variance when the data being used in the calculation are samples of a larger data set that has a normal distribution (rarely the case for signals). If the data have zero mean, the standard deviation is the same as the RMS value except for the normalization factor in the digital calculation. Nonetheless, they are from different traditions (statistics versus measurement) and are used to describe conceptually different aspects of a signal: signal magnitude for RMS and signal variability for standard deviation. Figure 2.4-5 shows the EEG data in Figure 2.4-4 with positive and

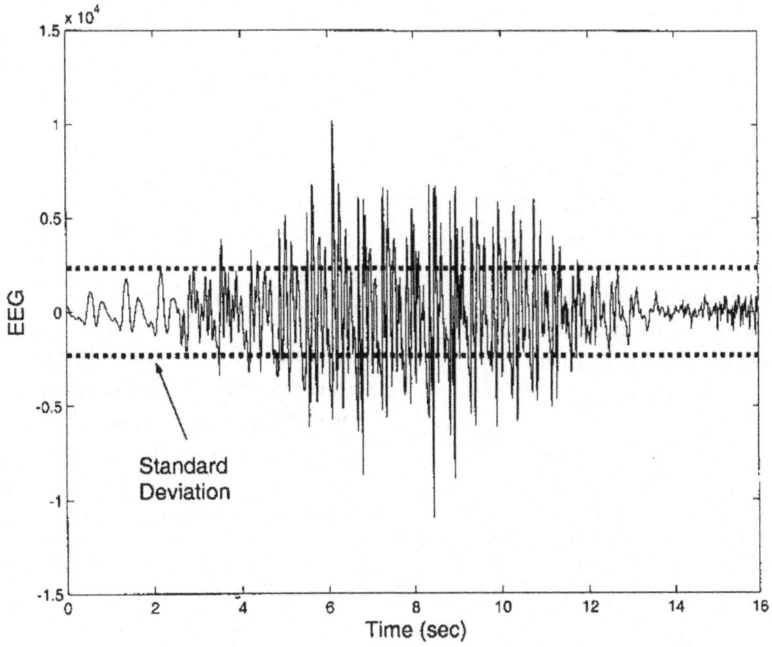

Figure 2.4-5 A segment of electroencephalogram signal shown in Figure 2.4-4 with the positive and negative standard deviation (*dotted horizontal line*).

negative values of standard deviation indicated by horizontal dotted lines.

When multiple measurements are made, multiple values or signals will be generated. If these measurements are combined or added together, the means add so that the combined value, or signal, has a mean that is the average of the individual means. The same is true for the variance: the variances add and the average variance of the combined measurement is the mean, or average, of the individual variances:

$$\bar{\sigma}^2 = \frac{1}{N}\sum_{k=1}^{N}\sigma_k^2 \qquad \text{[Eq. 2.4.21]}$$

where N is the number of signals being averaged. The standard deviation of the averaged signal is the square root of the variance so the standard deviations add as the $\sqrt{N}$ times the average standard deviation. Accordingly, the mean standard deviation is the average of the individual standard deviations divided by $\sqrt{N}$. Stated mathematically, from Eq. 2.4.21:

$$\sum_{k=1}^{N}\sigma_k^2 = N\bar{\sigma}^2$$

Taking the square root of both sides:

$$\sum_{k=1}^{N}\sigma_k = \sqrt{N\bar{\sigma}^2} = \sqrt{N}\bar{\sigma}$$

The mean standard deviation becomes:

$$\text{Mean } \sigma = \frac{1}{N}\sum_{k=1}^{N}\sigma_k = \frac{1}{N}\sqrt{N}\bar{\sigma} = \bar{\sigma}/\sqrt{N}$$

[Eq. 2.4.22]

In other words, averaging measurements from different sensors, or averaging multiple measurements from the same source, will reduce the standard deviation of the measurement's variability by the square root of the number of averages. For this reason, it is common to make multiple measurements whenever possible and average the results. This approach can also be applied to entire signals, a technique known as *ensemble averaging*. An example of ensemble averaging is given in the MATLAB implementation section of this chapter.

2.4.2.1 Decibels

It is common to compare the intensity of two signals using ratios, V_{Sig1}/V_{Sig2}, and to represent such ratios in units of *decibels*. Actually, decibels (dB) are not really units, but are simply a logarithmic scaling of ratios. The decibel has several advantageous features: (a) It provides a measurement of the effective power, or power ratio; (b) the log operation compresses the range of values (for example, a range of 1 to 1,000 becomes a range of 1 to 3 in log units); (c) when numbers or ratios are to be multiplied, they simply add if they are in log units; and (d) the logarithmic characteristic is similar to human perception. This latter feature motivated Alexander Graham Bell to develop the logarithmic unit called the *bel.* Audio power increments in logarithmic bels were perceived as equal increments by the human ear. The bel turned out to be inconveniently large, so it has been replaced by the decibel (1/10 bel). While originally defined only in terms of a ratio, decibel units are also used to express the intensity of a single signal. In this case, it has a dimension, the dimension of the signal (volts, amps, dynes, and so forth), but these units are often ignored.

When applied to a power measurement, the decibel is defined as 10 times the log of the power ratio:

$$P_{dB} = 10 \log\left(\frac{P_2}{P_1}\right) dB \qquad \text{[Eq. 2.4.23]}$$

When applied to a voltage ratio (or simply a voltage), the decibel is defined as 10 times the log of the RMS value squared, or voltage ratio squared. Because the log is taken, this is the same as 20 times the unsquared ratio or value. If a ratio of sinusoids is involved, then peak-to-peak voltages (or whatever units the signal is in) can also be used, because they are related to RMS values by a constant (0.707), and the constants will cancel in the ratio.

$$\begin{aligned} v_{dB} &= 10 \log(v_2^2/v_1^2) = 20 \log(v_2/v_1) \quad \text{or} \\ v_{dB} &= 10 \log v_{RMS}^2 = 20 \log v_{RMS} \end{aligned} \qquad \text{[Eq. 2.4.24]}$$

The logic behind taking the *square* of the RMS voltage value before taking the log is that the RMS voltage squared is proportional to signal power. Consider the case where the signal is a time-varying voltage, $v(t)$. To draw energy from this signal, it is necessary to feed it into a resistor, or a resistor-like element that consumes energy. (Recall from basic physics that resistors convert electrical energy into thermal energy, i.e., heat.) The power (energy per unit time) transferred from the signal to the resistor is given by the following equation:

$$P = v_{RMS}^2/R \qquad \text{[Eq. 2.4.25]}$$

where R is the resistance. This equation shows that the power imparted to a resistor by a given voltage depends, in part, on the value of the resistor. Assuming a nominal resistor value of 1 Ω, the power will be equal to the voltage squared; however, for any resistor value, the power transferred will be proportional to the voltage squared. When decibel units are used to describe a ratio of voltages, the value of the resistor is irrelevant, because the resistor values will cancel out:

$$v_{dB} = 10 \log\left(\frac{v_2^2/R}{v_1^2/R}\right) = 10 \log\left(\frac{v_2^2}{v_1^2}\right) = 20 \log\left(\frac{v_2}{v_1}\right) \qquad \text{[Eq. 2.4.26]}$$

If decibel units are used to express the intensity of a single signal, the units will be proportional to the log power in the signal.

To convert a voltage from decibel to RMS, use the inverse of the defining equation (Eq. 2.4.26):

$$v_{RMS} = 10^{X_{dB}/20} \qquad \text{[Eq. 2.4.27]}$$

Decibel units are particularly useful when comparing ratios of signal and noise.

Example 2.4.5: A sinusoidal signal is fed into an *attenuator* that reduces the intensity of the signal. The input signal has a peak-to-peak amplitude of 2.8 V and the output signal is measured at 2 V peak-to-peak. Find the ratio of output to input voltage in decibels. Compare the power-generating capabilities of the two signals in linear units.

Solution: Convert each peak-to-peak voltage to RMS, then apply Eq. 2.4.26 to the given ratio. Calculate the ratio without taking the log.

$$\begin{aligned} V_{RMSdB} &= 20 \log(V_{outRMS}/V_{inRMS}) \\ &= 20 \log\left(\frac{2.0 \times 0.707}{2.8 \times 0.707}\right) \\ V_{RMSdB} &= -3\,dB \end{aligned}$$

The power ratio is:

$$\text{Power ratio} = \frac{V_{outRMS}^2}{V_{inRMS}^2} = \frac{(2.0 \times 0.707)^2}{(2.8 \times 0.707)^2} = 0.5$$

Analysis: The ratio of the amplitude of a signal coming out of a process to that going into the process is known as the *gain,* and is often expressed in decibels. When the gain is less than 1, it means there is a loss, or reduction, in signal amplitude. In this case, the signal loss is 3 dB, so the "gain" of the attenuator is actually −3 dB. To add to the confusion, you can reverse the logic and say that the attenuator has an attenuation (i.e., loss) of +3 dB. In this example, the power ratio was 0.5, meaning that the signal coming out of the attenuator has half the power-generating capabilities of the signal that went in. A 3 dB attenuation is equivalent to a loss of half the signal's energy. Of course, it was not necessary to convert the peak-to-peak voltages to RMS because a ratio of these voltages was taken and the conversion factor (0.707) cancels out.

2.4.3 Advanced measurements: correlations and covariances

Applying the basic measurements we have just learned to the EEG data in Figure 2.4-4, we find the signal has a comparatively small mean of −29.8, an RMS value of 2,309 (or 67 dB), and a standard deviation of 2,310

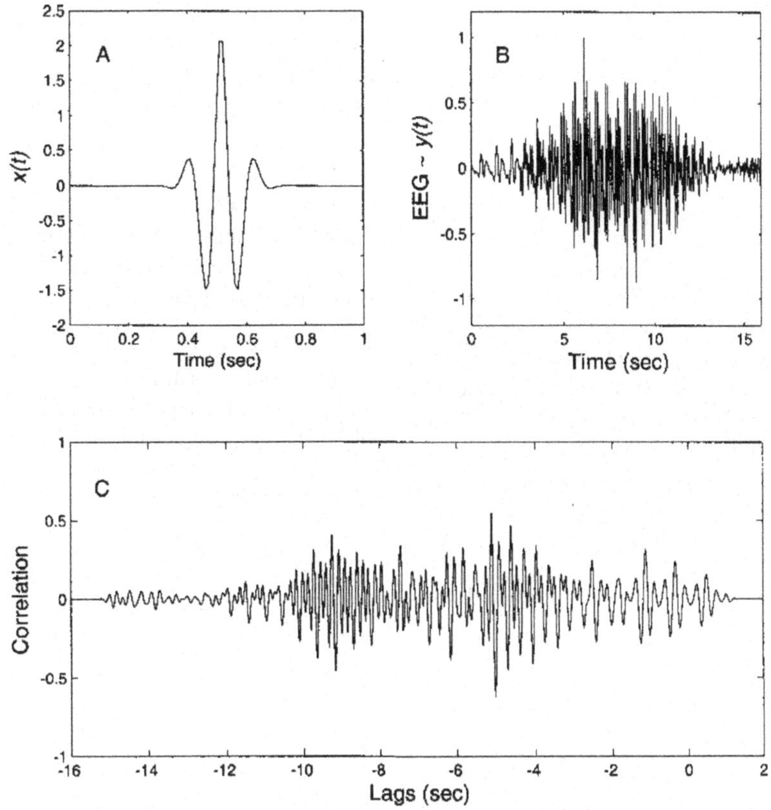

Figure 2.4-6 The reference waveform (A) is compared with the electroencephalogram signal (B) using a running correlation to determine to what extent the electroencephalogram signal contains this pattern. The running correlation between the two waveforms varies over time, but at maximum is only around 0.5. The running correlation operation is described and carried out in Section 2.4.3.2.

(Figure 2.4-5). (These numbers are all in relative units that relate to voltage in the brain by an unknown calibration factor.) These basic measurement numbers are not enlightening about the EEG signal or the processes that created it. More insight might be gained by comparing the EEG signal with one or more reference signals, or mathematical functions. For example, we might ask, "How much is the EEG signal like a 10-Hz sinusoid?" Or, "How much is it like a 12-Hz sinusoid, or a 12-Hz diamond-shaped wave, or any other function/waveform that might shed some light on the nature of the signal?" Such comparisons can be carried out using an operation known as *correlation.*

Another somewhat related question we might ask of an unknown waveform such as the EEG signal is whether the EEG signal contains anything like a brief waveform such as that shown in Figure 2.4-6A, or other short time period waveform.

2.4.3.1 Standard correlation and covariance

Although it is common in everyday language to take the word *uncorrelated* as meaning *unrelated* (and thus *independent)*, this is not the case in mathematical analysis, particularly if variables are related in a nonlinear manner. In the statistical sense, if two (or more) variables are independent, they are uncorrelated, but the reverse is not generally true. Moreover, signals that are very much alike can still have a mathematical correlation of zero. With these caveats in mind, correlation seeks to quantify (i.e., to assign a number to) how much one thing is like another. When comparing two mathematical functions, we use the technique of multiplying one by the other, then averaging the results. This average is often scaled by a normalizing factor. This gives us what is known as the *linear association* between two sets of variables. The same approach is used when correlation is applied to two signals. Given two functions, their average product will have the largest possible positive value when the two functions are identical. This process, since it is based on multiplication, will have the largest negative value when the two functions are exact opposites of one another (i.e., one function is the negative of the other). The average product will be zero when the two functions are, on average, completely dissimilar, again in a mathematical sense. Stated as an equation, the correlation between two signals, $x(t)$ and $y(t)$, over a time frame T is as follows:

$$Corr = \frac{1}{T}\int_0^T x(t)y(t)dt$$

or in discrete form

$$Corr = \frac{1}{N}\sum_{k=1}^{N} x(k)y(k) \qquad \text{[Eq. 2.4.28]}$$

The integration (or summation in the discrete form) and scaling (dividing by T or N) simply take the average of the product. It is common to modify Eq. 2.4.28 by dividing by the square root of the product of the variances of the two signals. This will make the correlation value equal to 1.0 when the two signals are identical and −1 if they are exact opposites:

$$Corr_{normalized} = \frac{Corr}{\sqrt{\sigma_1^2\sigma_2^2}} \qquad \text{[Eq. 2.4.29]}$$

where the variances, σ^2, are defined in Eq. 2.4.17 and Eq. 2.4.18. The term *correlation* implies this normalization.

Correlation between two signals is illustrated in Figure 2.4-7, which shows various pairs of waveforms and the correlation between them. Note that a sine and a cosine have no (zero) correlation even though the two are alike in the sense that they are both sinusoids (upper plot). Intuitively, we see that this is because any positive correlation between them over one portion of a cycle is canceled by negative correlation over the rest of the cycle. Mathematically, this is a demonstration that a sine and a cosine of the same frequency are *orthogonal* functions, which, by definition, are uncorrelated. Indeed, a good way to test if two functions are orthogonal is to assess their correlation. Correlation does not necessarily measure general similarity, so a sine and a cosine of the same frequency are, by this mathematical definition, as unalike as possible, even though they have very similar oscillatory patterns.

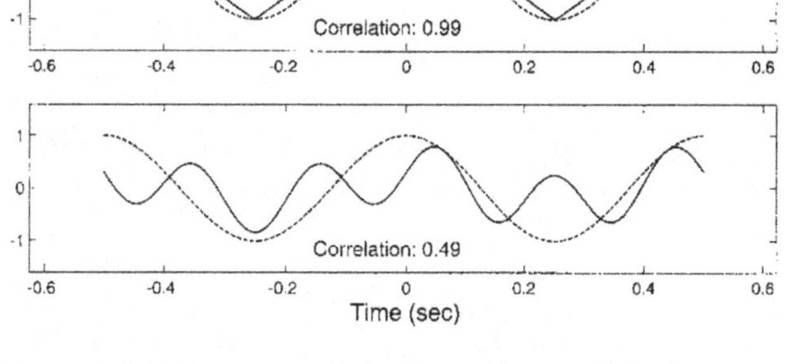

Figure 2.4-7 Three pairs of signals and the correlation between them as given by Equation 2.4.28 and normalized as in Equation 2.4.29. Note the high correlation between the sine and triangle wave (*center plot*) correctly expressing the general similarity between them. However, the correlation between a sine and cosine (*upper plot*) is zero, even though they are both sinusoids.

Example 2.4.6: Use Eq. 2.4.28 (continuous form) to find the correlation (unnormalized) between the sine wave and the square wave shown below. Both have an amplitude of 1.0 V (peak-to-peak) and a period of 1.0 second.

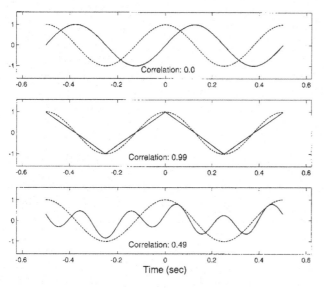

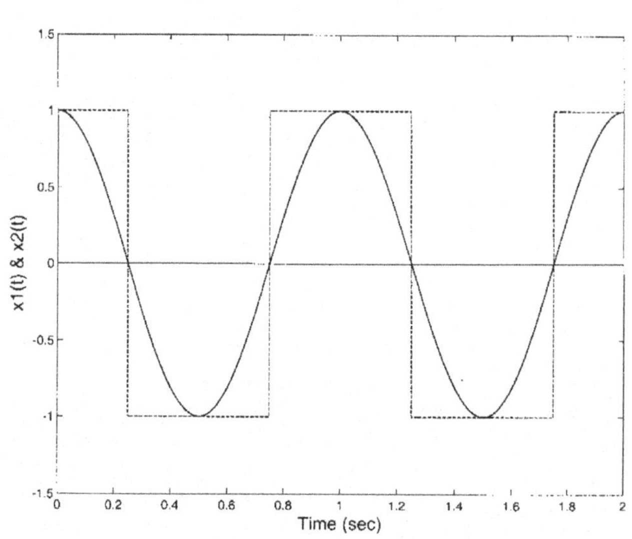

Solution: By symmetry, the correlation in the second half of the 1-second period equals the correlation in the first half, so it is only necessary to calculate the correlation period in the first half period.

$$Corr = \frac{1}{T}\int_0^T x(t)y(t)dt$$
$$= \frac{2}{T}\int_0^{T/2}(1)\sin\left(\frac{2\pi t}{T}\right)dt$$
$$= \frac{2}{T}\frac{T}{2\pi}\left(-\cos\left(\frac{2\pi t}{T}\right)\right)\Bigg|_0^{T/2}$$
$$Corr = \frac{1}{\pi}(-\cos(\pi) - (-\cos(0))) = \frac{2}{\pi}$$

Covariance computes the variance that is shared between two (or more) signals. Covariance is usually defined in discrete format as follows:

$$\sigma_{xy} = \frac{1}{N-1}\sum_{k=1}^{k}(x_k - \bar{x})(y_k - \bar{y}) \qquad \text{[Eq. 2.4.30]}$$

The equation for covariance is similar to the discrete form of correlation except that the average values of the signals have been removed. Of course, if the signals have average values of zero, the two discrete operations (unnormalized correlation and covariance) are the same. More extensive use of correlation is presented in the section on MATLAB implementation.

2.4.3.2 Autocorrelation and cross-correlation

The mathematical dissimilarity between a sine and a cosine is disconcerting and a real problem if you are trying to determine if a signal has general sinusoidal-like features. For example, a signal could be quite similar to a cosine, but if you are correlating using a sine wave reference, you would find only a small correlation. The same would be true if you were probing a sinelike signal with a cosine reference function. You might think that these signals are not sinusoidal when in fact they were very much like a sinusoid, just not the one you selected as a reference. To circumvent this problem, you could still use only a sine (or cosine) reference, but shift this reference signal in time, performing the correlation for many different time shifts. For example, comparing a cosine with a shifted sine shows increasing correlation with greater shifts. When the sine is shifted so that its phase is modified by 90 degrees, it will be identical to a cosine and will have a correlation of 1.0. Figure 2.4-8 shows the correlations between a sine and a cosine as the sine is shifted relative to the cosine. Figure 2.4-8 (lower right) plots a cosine/sine correlation against time shift for a 2-Hz sine. When the sine is shifted by 0.125 seconds, corresponding to a phase shift of 90 degrees, the correlation reaches a maximum value of 1.0, after which it begins to decrease to a minimum of −1.0 at 0.375 seconds, corresponding to a shift of 270 degrees.

The effect of shifting the reference waveform shown in Figure 2.4-8 suggests an approach for using correlation to search for general signal properties such as oscillatory behavior. Rather than correlate the signal with either a sine or a cosine, correlate the signal using a sine time-shifted by different amounts, performing the correlation operation (Eq. 2.4.28) at each time shift. The maximum correlation will describe how much the signal is like a sinusoid. (Alternatively, a cosine could be used as the reference with similar results, although the shift required for maximum correlation would be different.) This approach also provides information on how much time shifting is required to achieve the maximum correlation, which may be of interest in some applications. This approach is demonstrated in Example 2.4.11 at the end of this chapter.

When correlation is performed by time-shifting one waveform with respect to another, it is termed *cross-correlation*. This shifting correlation can be achieved by introducing a variable time delay, or time lag, or simply *lag*, into one of the two waveforms in the correlation. It

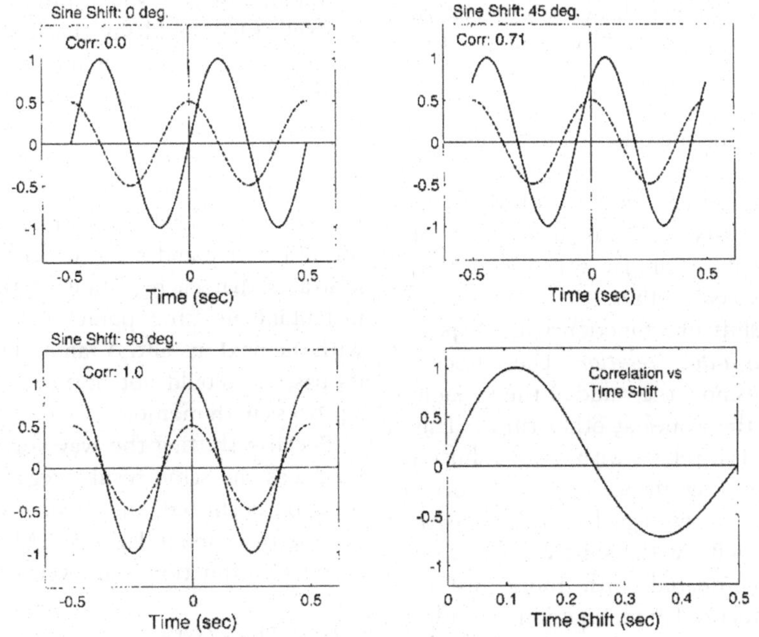

Figure 2.4-8 Upper left: The correlation between a 2-Hz cosine reference (*dashed line*) and an unshifted 2-Hz sine is 0.0. Upper right: When the sine time is shifted by the equivalent of 45 degrees, the correlation is 0.71. Lower left: When the sine is shifted by 90 degrees, the two functions are identical and the correlation is 1.0. Lower right: A plot of the correlation between cosine and sine as a function of the sine shift shows a peak value of 1.0 for a shift of 0.125 seconds corresponding to a shift of 90 degrees, a zero at 0.25 seconds corresponding to a shift of 180 degrees, and a correlation of −1.0 at a time shift of 0.375 seconds, corresponding to a shift of 270 degrees.

does not matter which function is shifted with respect to the other, although shifting the reference waveform is more common. The correlation operations of Eq. 2.4.28 then become a series of correlations over different time shifts or lags. For continuous signals, the time shifting can be continuous and the correlation becomes a continuous function of the time shift. This leads to an equation for cross-correlation that is an extension of Eq. 2.4.28 that adds a time shift variable, τ:

$$\textit{Cross-correlation} = r_{xy(\tau)} = \frac{1}{T}\int_0^T y(t)x(t+\tau)dt \qquad \text{[Eq. 2.4.31]}$$

The variable τ is a continuous variable of time used to shift $x(t)$ with respect to $y(t)$. The variable τ is *a* time variable, but not *the* time variable (which is t). To emphasize this τ is sometimes curiously referred to as a *dummy time variable*. The correlation is now a function of the time shift, τ, also known as *lag*. Cross-correlation is often abbreviated as r_{xy}, where x and y are the two functions being correlated. Again, this equation can be converted to a discrete form by substituting summation for integration and the integers i and k for the continuous variables t and τ:

$$r_{xy}(m) = \frac{1}{N}\sum_{k=1}^{N} y(k)x(k+m) \qquad \text{[Eq. 2.4.32]}$$

Figure 2.4-9A (lower plot) shows the cross-correlation function for a sinusoid and a triangle waveform. The cross-correlation shows that they are most similar (i.e., have the highest correlation) when one signal is shifted 0.18 seconds with respect to the other. This is demonstrated by shifting one of the functions by that amount in Figure 2.4-9B to provide a visual demonstration of this similarity. This also suggests a useful application of cross-correlation–alignment of similar waveforms that are shifted with respect to each other.

It is also possible to shift one function with respect to itself, a process called *autocorrelation*. The autocorrelation function describes how the value of the variable at one time depends on the values at other times. This will show how well a signal correlates with various shifted versions of itself. Another way of looking at autocorrelation is that it shows how the signal correlates with neighboring portions of itself. As the shift variable τ increases, the signal is compared with more distant neighbors. A signal's autocorrelation function provides some insight into how the signal was generated or altered by intervening processes. For example, a signal that remains highly correlated with itself over a long time shift must have been produced, or modified, by a process that took into account past values of the signal. Such a process can be described as having 'memory', because it must remember past values of the signal (or input) and use this information to shape the signal's current values. The longer the memory, the more the signal will remain partially correlated with shifted versions of itself. Just as memory tends to fade over time, the autocorrelation function usually goes to zero for large enough time shifts.

To perform an autocorrelation, simply substitute the same variable for x and y in Eq. 2.4.31 or Eq. 2.4.32:

$$\textit{Autocorrelation} = r_{xx(\tau)} = \frac{1}{T}\int_0^T x(t)x(t+\tau)dt \qquad \text{[Eq. 2.4.33]}$$

$$r_{xx}(m) = \frac{1}{N}\sum_{k=1}^{N} x(k)x(k+m) \qquad \text{[Eq. 2.4.34]}$$

Figure 2.4-10 shows the autocorrelation of several different waveforms. In all cases, the correlation has a maximum value of 1 at zero lag (i.e., no time shift) because when the lag (τ or n) is zero, this signal is being correlated with itself. The autocorrelation of a sine wave is another sinusoid (Figure 2.4-10A) because the correlation varies sinusoidally with the lag, or phase shift.

In Figure 2.4-10A, the sinusoidal pattern produced by autocorrelation falls off with increasing lags because this sinusoid had finite length. If the sinusoid were infinite in length, the autocorrelation function would be a constant amplitude cosine. A rapidly varying signal (Figure 2.4-10C) *decorrelates* quickly; that is, the self-correlation falls off rapidly for even small shifts of the signal with respect to itself. One could say that this signal has a very poor memory of its past values and was probably the product of a process with a short memory. For slowly varying signals, the correlation falls slowly (Figure 2.4-10B). Nonetheless, for all of these signals, there is some time shift for which the signal becomes completely decorrelated with itself. For a random signal, the correlation falls to zero instantly for all positive and negative lags (Figure 2.4-10D). This indicates that each instant of the random signal (each instantaneous time point) is completely uncorrelated with the next instant. A random signal has no memory of its past and could not be the product of, or altered by, a process with memory.

Because shifting the waveform with respect to itself produces the same results regardless of which way the function is shifted, the autocorrelation function will be symmetrical about lag zero. Mathematically, the autocorrelation function is an even function:

$$r_{xx}(-\tau) = r_{xx}(\tau) \qquad \text{[Eq. 2.4.35]}$$

In addition, the value of the function at lag zero, where the waveform is correlated with itself, will be as large, or larger than, any other value. If the autocorrelation is normalized by the variance, the value will be 1. (Because only one

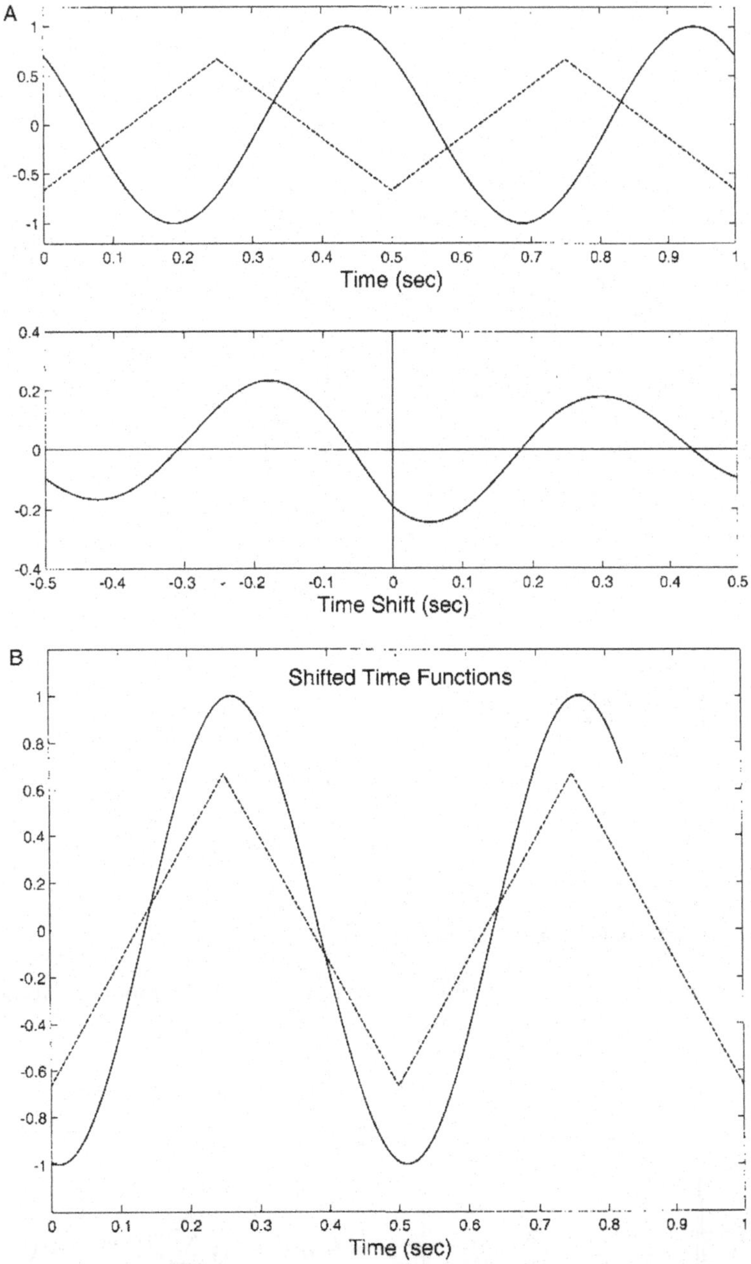

Figure 2.4-9 A (*upper plot*): a sinusoid and triangular wave at the same frequency, but not the same phase. *Lower plot:* The cross-correlation function for these two waveforms shows a peak at around −0.18 seconds when the functions are most alike. B: The two functions in A (*upper plot*) after shifting the sinusoid by an amount corresponding to the maximum cross-correlation given in A (*lower plot*).

function is involved in autocorrelation, the normalization equation given in Eq. 2.4.29 reduces to $1/\sigma^2$.)

Figure 2.4-11 shows the autocorrelation function of the EEG signal shown previously. The signal decorrelates quickly, reaching a value of zero correlation after a time shift of approximately 0.03 seconds. However, the EEG signal is likely to be contaminated with noise and the autocorrelation function of a signal plus noise is the sum of the autocorrelation function of the signal plus the autocorrelation of the noise. Because noise decorrelates instantly (Figure 2.4-10D), some of the rapid decorrelation seen in Figure 2.4-11 is due to the noise. A common approach to estimating the autocorrelation of the signal without the noise is to draw a smooth curve across the peaks and use that curve as the estimated autocorrelation function of signal without noise. From Figure 2.4-11, we see that such an estimated function would decorrelate at a longer time shift of 0.5 to 0.6 seconds.

Two operations closely related to autocorrelation and cross-correlation are autocovariance and cross-covariance. The relationship between correlation and covariance *functions* is similar to the relationship

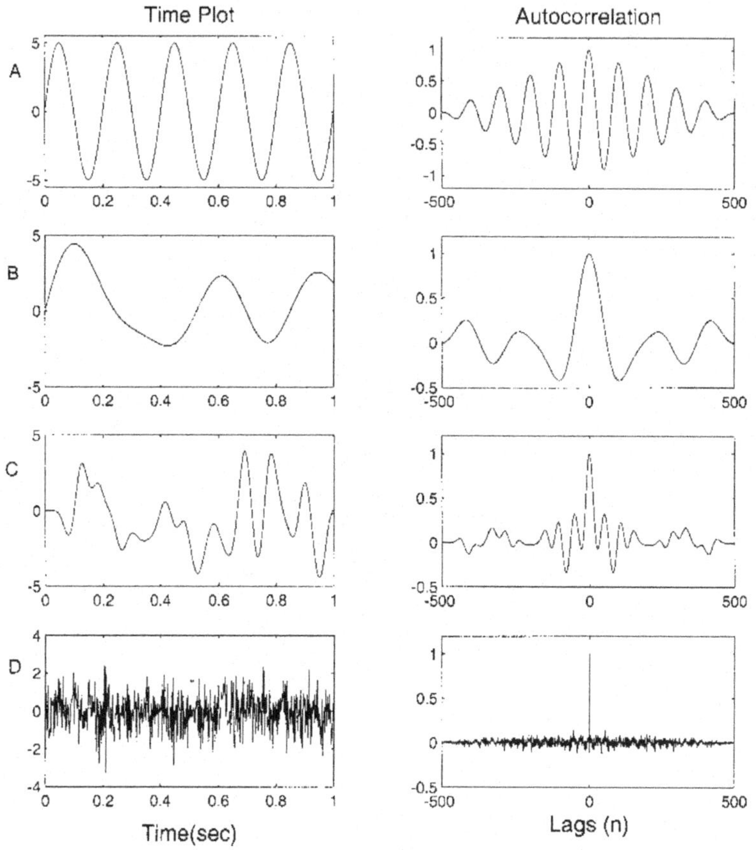

Figure 2.4-10 Four different signals (*left side*) and their autocorrelation functions (*right side*). A: A truncated sinusoid. The reduction in amplitude is due to the finite length of the signal. A true (i.e., infinite) sinusoid would have a nondiminishing cosine wave as its autocorrelation function. B: A slowly varying signal. C: A rapidly varying signal. D: A random signal.

between standard correlation and covariance given in the last section. Covariance and correlation functions are the same except that, in covariance, the means have been removed from the input signals, $x(t)$ and $y(t)$ [or just $x(t)$ in the case of autocovariance]:

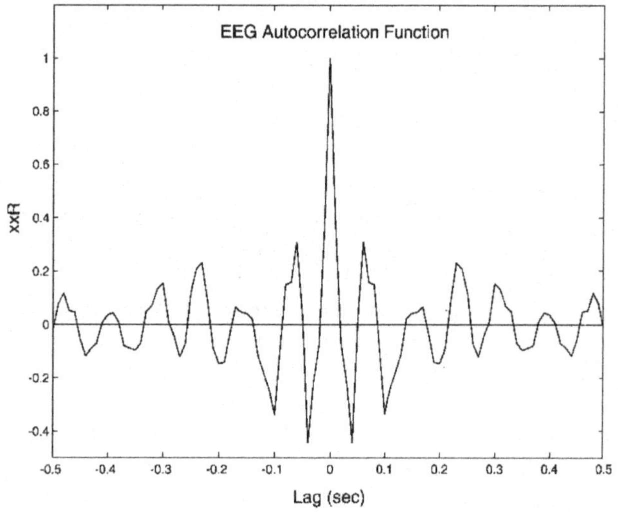

Figure 2.4-11 Autocorrelation function of the electroencephalogram signal in Figure 2.4-4. The autocorrelation function decorrelates rapidly probably due to the noise in the signal. Some correlation is seen out to 0.5 seconds.

$$Autocovariance \equiv C_{xx}(\tau) = \frac{1}{T}\int_0^T [x(t) - \overline{x(t)}] \times [x(t+\tau) - \overline{x(t)}]dt$$

$$C_{xx}[i] = \frac{1}{N}\sum_{k=1}^{N}[x(k) - \overline{x}][x(k+i) - \overline{x}] \quad \text{[Eq. 2.4.36]}$$

$$Crosscovariance \equiv C_{xy}(\tau) = \frac{1}{T}\int_0^T [y(t) - \overline{y(t)}] \times [x(t+\tau) - \overline{x(t)}]dt$$

$$Crosscovariance \equiv C_{xy[i]} = \frac{1}{N}\sum_{k=1}^{N}[y(k) - \overline{y(k)}] \times [x(k+i) - \overline{x(k)}]$$

[Eq. 2.4.37]

The autocovariance function can be thought of as measuring the memory or self-similarity of the *deviation* of a signal about its mean level. Similarly, the cross-covariance is a measure of the similarity of the deviation of two signals about their respective means. An example of the application of the autocovariance to the analysis of heart rate variability is given in the next section on MATLAB Implementation.

2.4.4 MATLAB implementation

All of the analyses described thus far are relatively easy to implement in MATLAB. In most cases, MATLAB has a function that will perform these operations.

2.4.4.1 Mean, variance, and standard deviation

Many of the techniques described in this chapter can be expeditiously, and conveniently, implemented in MATLAB. For example, the mean, variance, and standard deviations are implemented as shown in the three code lines below.

```
xm = mean (x);           % Evaluate mean of x
xvar = var (x);          % Variance of x
                           normalizing by N-1
xnorm = var(x,1); % Variance of x
                           normalizing by N
xstd = std(x);           % Evaluate the standard
                           deviation of x
```

If x is an array or series of numbers (also termed a *vector* for reasons given later) the output of these routines is a scalar representing the mean, variance, or standard deviation. If x is a matrix, the output is a row vector resulting from applying the appropriate calculation (mean, variance, or standard deviation) to each column of the matrix.

Example 2.4.7: Figure 2.4-12 shows heart rate variability for one subject under normal conditions (left side) and during a meditative state. Find the mean and standard deviation for the two conditions.

Solution: Apply the MATLAB routines `mean` and `std` (standard deviation) to the data. The program below loads the heart rate data from the .mat files `HR_pre` and `HR_med`. These files are assumed to be in workspace in this example, but are found on the accompanying CD. These files were originally obtained from the PhysioNet database (http://www.physionet.org) and contain approximately 500 seconds of heart rate data from a subject in a normal (`Hr_pre.mat`) and meditative state (`Hr_med.mat`). Each file contains a time variable (`t_pre` or `t_med`) and a heart rate variable (`hr_pre` or `hr_med`). The mean and standard deviation of the two heart rate variables will be determined using the appropriate MATLAB routines and the two variables plotted as functions of time.

Example 2.4.7: Plot the mean and standard deviation of the heart rate before and after meditation.

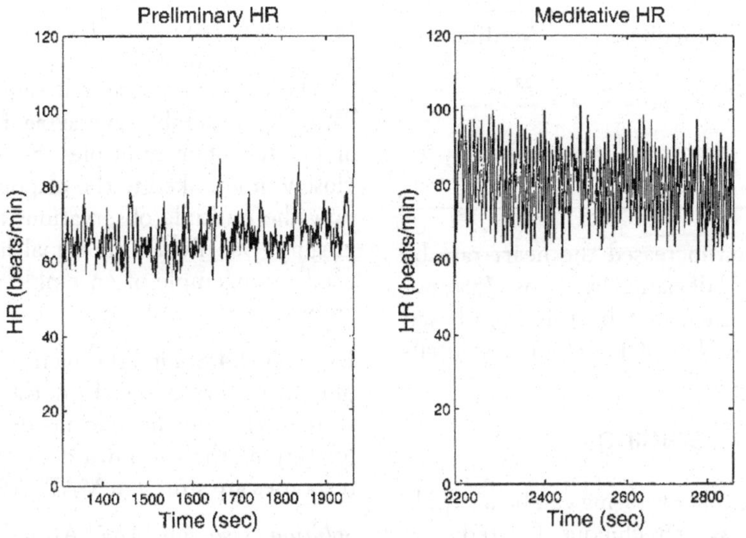

Figure 2.4-12 Heart rate over time during normal conditions (*left*) and during meditation (*right*). (From the PhysioNet database, Goldberger *et al.*, 2000.)

```
% Plots Figure 2.4.12
%
load Hr_pre                                     % Pre-meditative HR
load Hr_med                                     % Meditative HR
%
% Calculate the averages and standard deviations
Avg_pre = mean(hr_pre)                          % Average HR, normal
SD_pre = std(hr_pre)                            % Standard deviation, normal
Avg_med = mean(hr_med)                          % Average and std
SD_med = std(hr_med)                            % HR meditative
%
% Plot the heart rate data. Label axes
subplot(1,2,1);
  plot(t_pre,hr_pre,'k');                       % Plot normal HR data
  xlabel('Time (sec)');ylabel('HR (beats/min)');
  axis([t_pre(1) t_pre(end) 0 120]);
  title('Preliminay HR');
subplot(1,2,2);
  plot(t_med,hr_med,'k');                       % Plot meditative HR data
  xlabel('Time (sec)'); ylabel('HR (beats/min)');
  axis([t_med(1) t_med(end) 0 120]);
  title('Meditative HR')
```

Analysis: The program is a straightforward application of routines `mean` and `std`. The `var` routine could have been used if the variance was desired. In the plotting section, the `axis` routine was used to scale the vertical axis to be between 0.0 and 120 beats per minute. Because the time variables had different beginning and end times, the time limits were specified using the time array (`t_pre` or `t_med`) endpoints. (Recall that MATLAB is case sensitive.) The MATLAB files `Hr_pre` and `Hr_med` contain variables: `hr_pre`, `t_pre`, `hr_med`, and `t_med`.

Results:

	Premeditative	**Meditative**
Average heart rate (beats/min)	66.5	81.33
Standard deviation (beats/min)	5.36	9.35

In this subject, meditation increased the heart rate by about 22% and the standard deviation by almost 75%, not a result that might be anticipated by this Yoga-based meditation. (See the PhysioNet database for more details on the meditative conditions.)

2.4.4.2 Ensemble averaging

Equation 2.4.22 indicates that averaging can be a simple yet powerful signal-processing technique for reducing noise when multiple observations of the signal are possible. Such multiple observations could come from multiple sensors, but in many biomedical applications, the multiple observations come from repeated responses to the same stimulus. In *ensemble averaging,* a group, or ensemble, of time responses is averaged together on a point-by-point basis; that is, an *average signal* is constructed by taking the average, for each point in time, over all signals in the ensemble. A classic biomedical engineering example of the application of ensemble averaging is the visual evoked response (VER) in which a visual stimulus produces a small neural signal embedded in the EEG. Usually this signal cannot be detected in the EEG signal, but by averaging hundreds of observations of the EEG, time-locked to the visual stimulus, the visually evoked signal emerges.

There are two essential requirements for the application of ensemble averaging for noise reduction: the ability to obtain multiple observations and a reference closely time-linked to the response. The reference shows how the multiple observations are to be aligned for averaging. Usually a time signal linked to the stimulus is used. An example of ensemble averaging is given in Example 2.4.8.

Example 2.4.8: Find the average response given a number of individual responses from the vergence eye movement system. The vergence eye movement system is responsible for turning the eye inward to view a near target. These responses are stored in MATLAB file `vergence.mat`.

Solution: Use the MATLAB averaging routine `mean`. If this routine is given a matrix variable, it averages each column. Hence, if the various signals are arranged as rows in the matrix, the mean routine will produce the ensemble average.

Example 2.4.8: Load eye movement data, plot the data, then construct and plot the ensemble average.

```
close all; clear all;
load vergence;            % Get vergence eye
                            movement data
Ts = .005;                % Sample interval =
                            5 msec
[nu,N] = size             % Get data length
(data_out) ;                (N)
t = (1:N)*Ts;             % Generate time
                            vector (t = N Ts)
%
% Plot ensemble
data superimposed
plot(t,data_out,'k');
hold on;
%
% Construct and plot
  the ensemble average
avg = mean(data_out); % Calculate
                        ensemble
                        average and
plot(t,avg-3,'k');    % plot, separate
                        from the other
                      % data
  xlabel('Time        % Label axes
  (sec)');
  ylabel
  ('Eye Position');
plot([.43 .43],       % Plot horizontal
  [0 5];'-k');          line
text(1,1.2,'          % Label data
 Averaged ... Data');   average
```

The results are shown in Figure. 2.4-13.

2.4.4.3 Covariance and correlation

MATLAB has specific functions for determining the correlation and/or covariance between two or more signals. Correlation or covariance matrices are calculated using the `corrcoef` or `cov` functions, respectively. Again, the calls are similar for both functions:

```
Rxx = corrcoef(x);  % Signal correlation
S = cov(x);         % Signal covariance
```

where x is a matrix that contains the various signals to be compared in columns. Some options are available as explained in the associated MATLAB help file. The output, Rxx, of the `corrcoef` routine will be an *n-by-n* matrix where n is the number of signals (i.e., columns of x). The diagonals of this matrix represent the correlation of the signals with themselves, r_{xx} (and, hence, will be 1), and the off-diagonals represent the correlations of the various combinations. For example, r_{12} is the correlation between signals 1 and 2. Because the correlation of signal 1 with signal 2 is the same as signal 2 with signal 1, $r_{12} = r_{21}$, and in general $r_{m,n} = r_{n,m}$, so the matrix will be symmetrical about the diagonals:

$$r_{xx} = \begin{bmatrix} r_{1,1} & r_{1,2} & \cdots & r_{1,N} \\ r_{2,1} & r_{2,2} & \cdots & r_{2,N} \\ \vdots & \vdots & \ddots & \vdots \\ r_{N,1} & r_{N,2} & \cdots & r_{N,N} \end{bmatrix} \quad \text{[Eq. 2.4.38]}$$

The `cov` routine produces a similar output, except the diagonals are the variances of the various signals and the off-diagonals are the covariances as shown in Eq. 2.4.39 below.

$$S = \begin{bmatrix} \sigma_{1,1}^2 & \sigma_{1,2}^2 & \cdots & \sigma_{1,N}^2 \\ \sigma_{2,1}^2 & \sigma_{2,2}^2 & \cdots & \sigma_{2,N}^2 \\ \vdots & \vdots & \ddots & \vdots \\ \sigma_{N,1}^2 & \sigma_{N,2}^2 & \cdots & \sigma_{N,N}^2 \end{bmatrix} \quad \text{[Eq. 2.4.39]}$$

Example 2.4.8 uses covariance and correlation analysis to determine if sines and cosines of the same frequency and sine waves at multiple frequencies are orthogonal. Recall that two orthogonal signals will have zero correlation. Either covariance or correlation could be used to determine if signals are orthogonal. Example 2.4.9 uses both.

Example 2.4.9: Determine if a sine wave and a cosine wave at the same frequency are orthogonal and if sine waves at harmonically related frequencies are orthogonal. Include one sinusoid at a nonharmonic frequency.

Solution: Generate a data matrix where the columns consist of a 1.0-Hz sine and cosine, a 2.0-Hz sine and cosine, and a 3.0-Hz sine and a 3.5-Hz (i.e., nonharmonic) cosine. The six sinusoids should all be at different amplitudes. Apply the covariance (`cov`) and correlation (`corrcoef`) MATLAB functions. All of the sinusoids except the 3.5-Hz cosine are orthogonal and should show negligible correlation and covariance.

```
% Example 2.4.9: Application of the
  correlation and
% covariance matrices to sinusoids that
  are orthogonal and
% nonorthogonal
%
clear all; close all;
N = 256;                 % Number of points in
                           waveform
```

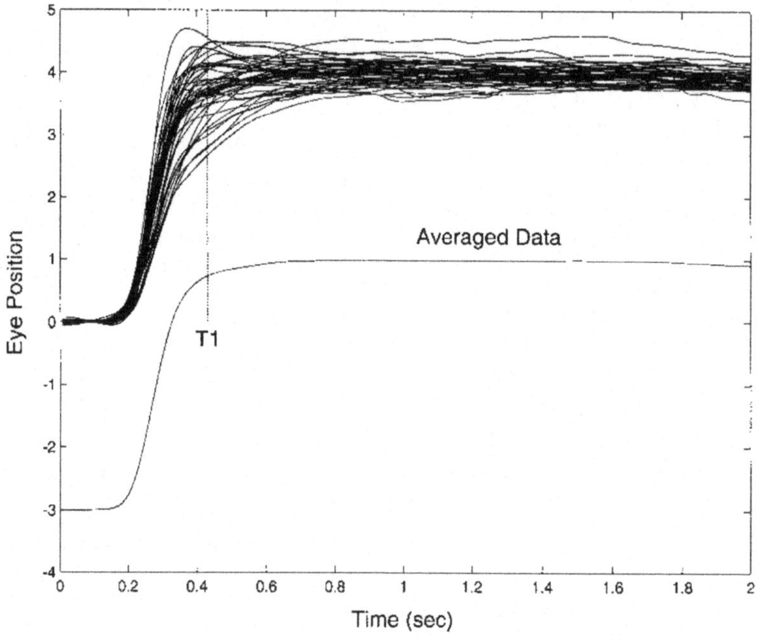

Figure 2.4-13 Upper traces: An ensemble of individual (vergence) eye movement responses to a step change in stimulus. Lower trace: The ensemble average, displaced downward for clarity. The ensemble average is constructed by averaging the individual responses at each point in time. Hence, the value of the average response at time T1 (*vertical line*) is the average of the individual responses at that time.

```
fs = 256;               % Assumed sample frequency
n = (1:N)/fs;           % Time vector: 1 sec of data
%
% Generate the sinusoids as columns of the matrix
x(:,1) = sin(2*pi*n)';        % Generate a 1 HZ sin
x(:,2)=2*cos(2*pi*n)';        % Generate a 1 HZ cos
x(:,3)=1.5*sin(4*pi*n)';      % Generate a 2 HZ sin
x(:,4)=3*cos(4*pi*n)';        % Generate a 2 HZ cos
x(:,5)=2.5*sin(6*pi*n)';      % Generate a 3 HZ sin
x(:,6)=1.75*cos(7*pi*n)';     % Generate a 3.5 HZ cos
%
S = cov(x)              % print covariance matrix
Rxx = corrcoef(x)       % and correlation matrix
```

Analysis: The program defines a time vector n that is 256 points long and achieves the proper time interval by dividing by the sampling frequency, fs (also 256). (Because MATLAB is case sensitive, n and N are different variables.) The program then generates the six sinusoids using this time vector in conjunction with sin and cos functions, arranging the signals as columns of x. The program then determines the covariance (cov) and correlation (corrcoef) matrices of x.

Results: The output from this program is a covariance and correlation matrix. The covariance matrix is as follows:

```
Correlation Matrix S =
 0.5020  0.0000   0.0000  0.0000  0.0000 -0.0497
 0.0000  2.0078  -0.0000 -0.0000 -0.0000 -0.0137
 0.0000 -0.0000   1.1294  0.0000 -0.0000 -0.2034
 0.0000 -0.0000   0.0000  4.5176 -0.0000 -0.0206
 0.0000 -0.0000  -0.0000 -0.0000  3.1373 -1.2907
-0.0497 -0.0137  -0.2034 -0.0206 -1.2907  1.5372
```

The diagonals of the covariance matrix give the variance of the six signals and these differ since the amplitudes of the signals are different. The correlation matrix shows similar results except that the diagonals are now 1.0 because these reflect the correlation of the signal with itself.

```
Correlation Matrix Rxx =
 1.0000  0.0000   0.0000  0.0000  0.0000 -0.0566
 0.0000  1.0000  -0.0000 -0.0000 -0.0000 -0.0078
 0.0000 -0.0000   1.0000  0.0000 -0.0000 -0.1544
 0.0000 -0.0000   0.0000  1.0000 -0.0000 -0.0078
 0.0000 -0.0000  -0.0000 -0.0000  1.0000 -0.5878
-0.0566 -0.0078  -0.1544 -0.0078 -0.5878  1.0000
```

The covariance and correlation between the various signals are given by the off-diagonals and are zero for all combinations between signals 1 and 5, demonstrating the orthogonality of all of these harmonic signals. Conversely, nonzero covariances and correlations are found between signals 1 through 5 and signal 6, the 3.5-Hz cosine. This shows that the nonharmonically related cosine is not orthogonal to any of the other sines or cosines. Note that the bottom row is the same as the last column, reflecting the symmetry of these matrices.

It may seem a little surprising that a 1-Hz sine wave and a 2-Hz sine wave are orthogonal, but it is easily demonstrated by sketching the two waveforms. Consider the product of two sine waves seen in Figure 2.4.14. The first half of the 1-Hz sine wave will be multiplied by a full cycle of the 2-Hz sine wave and the result will be 0.0. This would be true for any higher harmonic signal: if the 2-Hz sine wave were a 4-Hz sine wave, for example. The orthogonality of harmonically related sinusoids is a feature that will be used in the Fourier transform. It means that operations (such as correlation) involving a sinusoid do not interfere with operations that involve sinusoids at harmonically related frequencies.

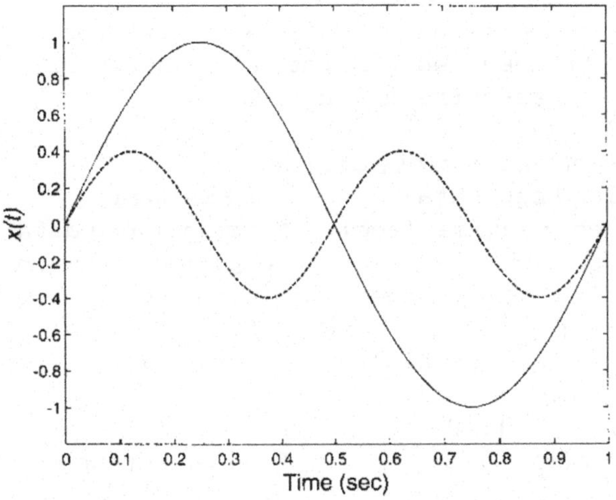

Figure 2.4-14 A 1-Hz sine wave plotted with a 2-Hz sine wave. The product of the two will clearly average to 0.0.

2.4.4.4 Autocorrelation and cross-correlation

The cross-correlation and autocorrelation operations are performed with the same MATLAB routine, with autocorrelation being treated as a special case. The program, `axcor`, is supplied on the accompanying CD:

```
[r, lags] = axcor (x,y);
```

Only the first input argument, x, is required. If no y variable is specified, autocorrelation is performed and the output is normalized to be 1.0 at zero lag. If both variables are given, the cross-correlation is normalized as in Eq. 2.4.29. The time shift extends over the entire range of the longer variable. If the MATLAB signal-processing toolbox is available, a MATLAB routine called `xcorr` is available that features a wider range of options. The `axcor` function produces an output argument, r, which is an array that is twice the length of the shortest input array. The optional output argument, `lags`, is simply an array containing the lag values, which is helpful in plotting the function.

Recall that auto- and cross-covariance are the same as auto- and cross-correlation if the data have zero means. Hence, autocovariance or cross-covariance can be determined using `axcor` simply by subtracting the variable means before calling the function.

```
[c, lags] = axcor(x-mean(x), y-mean(x));
```

The autocorrelation and autocovariance functions describe how one segment of data is correlated, *on average*, with adjacent segments. Such correlations could be due to memory-like properties in the process that generated the data. Many physiological processes are repetitive, such as respiration and heart rate, yet vary somewhat on a cycle-to-cycle basis. Autocorrelation and cross-correlation can be used to explore this variation. For example, considerable interest revolves around the heart rate and its beat-to-beat variations. Autocovariance can be used to tell us if these variations are completely random or if there is (again, on average) some correlation between beats or over several beats. In this instance, we use autocovariance, not autocorrelation, because we are interested in correlation of heart rate *variability*, not the correlation of heart rate per se. (Recall that autocovariance will remove the mean value of the heart rate from the data and analyze only the variation.) Example 2.4.10 analyzes the normal heart rate data presented in Figure 2.4-12 (Preliminary Heart Rate) to determine the correlation over successive beats.

Example 2.4.10: Determine if there is any correlation in the variation between the timing of successive heart beats under normal resting conditions.

Solution: Load the heart rate data taken during normal resting conditions (file Hr_pre.mat). Isolate the heart rate variable (the second column) and then take the autocovariance function. Plot this function to show potential correlation over approximately 30 successive beats.

```
% EXAMPLE 2.4.10 and Figure 2.4-15
% Use of autocovariance to determine
  the correlation
% of heart rate variation between
  heart beats
%
clear all; close all;
figure ;
load Hr_pre               % Load data
[c,lags]=axcor            % Autocovariance
(hr_pre-mean                (mean subtracted)
(hr_pre));
plot(lags,c, 'k');        % Plot
hold on;                    autocovariance
plot([lags(1)             % Plot zero
lags(end)], [0 0],'k')      line for
                          % reference
xlabel('Lags (N)');
ylabel
('Autocovariance');
grid on;
axis([-30 30 -.2 1.2]); % Limit plot range
                        % to ± 30 beats
```

Analysis: After loading the data file, the program calculates the autocovariance using routine axcor. The mean is subtracted from the data variable so that autocovariance will be performed. The data are then plotted along with a zero line and the axis is rescaled to show only the first ±30 lags. The plotting grid is enabled.

Results: The results in Figure 2.4-15 show there is high correlation for heart beats that are within a few seconds of each other (approximately 0.5 within ±4 seconds). This correlation falls off rapidly so there is little or no correlation between beats that are more than about 15 seconds apart. If the variability were completely random, the autocovariance function would be 1.0 for zero lag and 0.0 everywhere else (Figure 2.4-10D). Problem 15 applies this analysis to the heart rate data taken during the meditative state.

One of the most popular reference signals is the sinusoid. It is common to compare the signal of interest not just with one sinusoid but with a range of sinusoids having different frequencies. To ensure that we correlate with a sinusoid having the most appropriate phase shift, we use cross-correlation and take the maximum cross-correlation values as related to the amount of

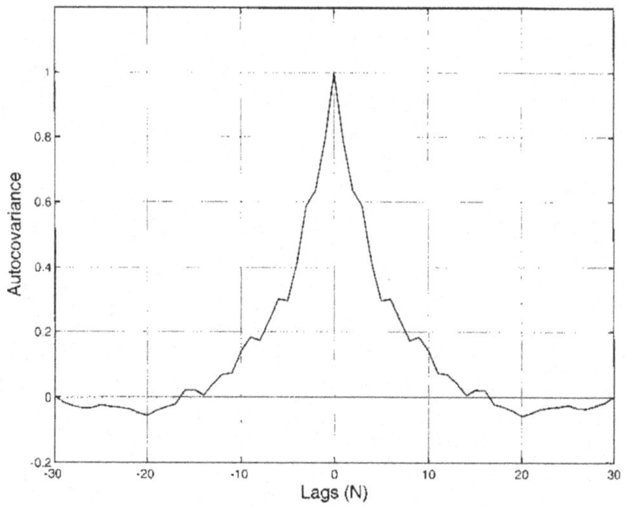

Figure 2.4-15 Autocovariance function of the heart rate from one subject under normal resting conditions. Some correlation is observed over approximately 10 successive heart beats.

sinusoid "in" the signal at a given frequency. This strategy is demonstrated in the next example.

Example 2.4.11: Find the sinusoidal content in the EEG signal over a range of frequencies from 0.5 to 50 Hz. The frequency resolution of the comparison should be 0.5 Hz.

Solution: Generate a series of sine waves from 0.5 to 50 Hz in 0.5-Hz increments. (Cosine waves would work just as well.) Cross-correlate these sine waves with the EEG signal and find the maximum cross-correlation. Plot this maximum correlation as a function of the sine wave frequency. This procedure is remarkably easy to program in MATLAB.

```
% Example 2.4.11 and Figure 2.4-16
% Comparison of sinusoids at different
  frequencies with the EEG signal using
  cross-correlation.
%
clear all; close all;
load eeg_data;            % Get EEG data
eeg = eeg/max(eeg);       % Normalize eeg data
fs =50;                   % Sampling frequency
t = (1:length(eeg))       % Time vector
/fs;
% Cross-correlate over a range of
  frequencies.
for i = 1:100
  f(i) = i/2;             % Frequency range:
                            0.5 - 50 Hz
  x = sin(2*pi*           % Generate sin
  f(i)*t);
```

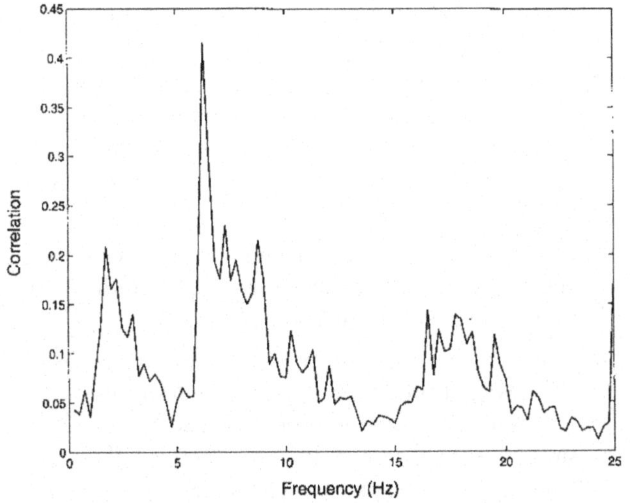

Figure 2.4-16 The maximum cross-correlation between sine waves and the electroencephalogram signal plotted as a function of the sine wave's frequency. A peak is seen between 7 and 9 Hz, which indicates the presence of an oscillatory pattern known as the 'alpha wave.'

```
    r = axcor(eeg,x); % Perform cross-
                        correlation
    rmax (i)= max(r); % Store max value
  end
......labels and plot......
```

Results: The result of the cross-correlations is seen in Figure 2.4.16 and an interesting structure is seen to emerge. Some frequencies show much higher correlation with sine and EEG, indicating more sine wave content at these frequencies. A particularly strong peak is seen in the region of 7 to 9 Hz, indicating the presence of an oscillatory pattern known as the *alpha wave.*

2.4.5 Summary

The sinusoidal waveform is arguably the single most important waveform in signal processing. Because of its importance, it is essential to know the mathematics associated with sines, cosines, and general sinusoids, including complex representations.

Several basic measurements apply to any signal including mean value, RMS value, and variance or standard deviation. Although these measurements provide some essential basic information, they do not provide much information on signal content or meaning. A common approach to obtaining more information is to probe a signal by correlating it with one or more reference waveforms. One of the most popular probing signals is the sinusoid.

Sometimes a signal will be correlated with another signal in its entirety, a process known as correlation, or the closely related covariance. If the correlation between the signal of interest and the reference is zero, it does not necessarily mean the two signals have nothing in common, but it does mean the signals are mathematically orthogonal.

If the probing signal is short, a running correlation known as cross-correlation may be appropriate. Cross-correlation not only shows the match between the probing signal and the signal of interest, but also where that match is greatest. A signal can also be correlated with shifted versions of itself, a process known as auto. The autocorrelation function describes the period for which a signal remains partially correlated with itself and this relates to the structure of the signal. For example, a signal consisting of random noise decorrelates immediately, whereas a slowly varying signal will remain correlated over for long period. Correlation, cross-correlation, autocorrelation, and the related covariances are all easy to imple in MATLAB.

Problems

1. Two 10-Hz sine waves have a relative phase shift of 30 degrees. What is the time difference between them? If the frequency of these sine waves doubles, but the time difference stays the same, what is the phase difference between them?
2. Convert $x(t) = 6\sin(5t) - 5\cos(5t)$ into a single sinusoid [i.e., $M\sin(5t + \theta)$].
3. Convert $x(t) = 30\sin(2t + 50)$ into sine and cosine components.
4. Convert $x(t) = 5\cos(10t + 30) + 2\sin(10t - 20) + 6\cos(10t + 80)$ into a single sinusoid as in Problem 2.
5. Find the delay between $x_1(t) = \cos(10t + 20)$ and $x_2(t) = \sin(10t - 20)$.
6. Equations 2.4.8, 2.4.9, and 2.4.10 were developed to convert a sinusoid such as $\cos(\omega t - \theta)$ into a sine and cosine wave and vice versa. Derive the equations to convert between a sinusoid based on the sine, $\sin(\omega t + \theta)$ and a sine and cosine wave.
7. Find the RMS value of the square wave with amplitude of 1.0 V and a period 0.2 seconds.
8. If a signal is measured as 2.5 V peak-to-peak and the noise is measured as 28 mV RMS, what is the SNR in decibels?
9. Use Eq. 2.4.28 to find the correlation (unnormalized) between $\sin(2\pi t)$ and $\cos(2\pi t)$.
10. Use Eq. 2.4.28 to find the correlation between a cosine and a square wave as shown below. This is

the same as Example 2.4.6 except that the sine has been replaced by a cosine.

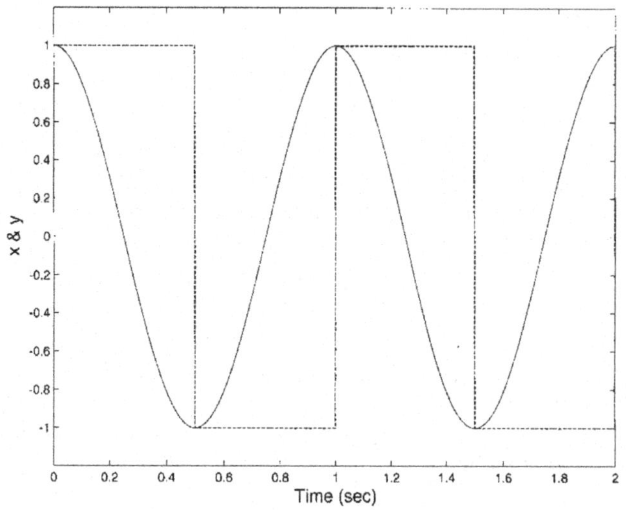

MATLAB problems

11. Load the data in `ensemble_data.mat` found on the CD. This file contains a data matrix labeled "data." The data matrix contains 100 responses of a signal in noise. Determine whether the responses are stored in the rows or columns of the matrix. Plot several randomly selected samples of these responses. Is it possible to identify the signal from any single record? Construct and plot the ensemble average for these data. (Be sure that the matrix is in the correct orientation.) Also construct and plot the ensemble standard deviation.

12. Demonstrate that 1-Hz and 4-Hz cosine waves are orthogonal by multiplying them together and averaging the product. (Note when you multiply be sure to use the point-by-point multiplication, `'.*'`.)

13. The file `two_var.mat` contains two variables `x` and `y`. Are either of these variables random? Are they orthogonal to each other?

14. The file `nerve.mat` contains two signals, `x` and `y`, along with a time vector, `t`. The two signals represent nerve action potentials taken simultaneously at two different sections of a nerve axon along with inevitable noise. Use cross-correlation to determine the average time delay between the two signals. Assume the sampling frequency of the nerve signals was 2 kHz.

15. Develop a program along the lines of Example 2.4.10 to determine the correlation in heart rate variability during meditation. Load file `Hr_med.mat`, which contains the heart rate (beats/min as a function of time) in variable `hr_med` and the time in variable `t_med`, then determine and plot the autocovariance. The result will show that the heart rate under meditative conditions contains some periodic elements. Can you determine the frequency of these periodic elements?

Section Three

Biomaterials science

Chapter 3.1

3.1

Properties of materials

Buddy D. Ratner, Allan Hoffman, Frederick J. Schoen, and Jack E. Lemons

3.1.1 Introduction

Jack E. Lemons

The bulk and surface properties of biomaterials used for medical implants have been shown to directly influence, and in some cases, control the dynamic interactions that take place at the tissue–implant interface. These interactions are included in the concept of compatibility, which should be viewed as a two-way process between the implanted materials and the host environment that is ongoing throughout the *in vivo* lifetime of the device.

It is critical to recognize that synthetic materials have specific bulk and surface properties or characteristics. These characteristics must be known prior to any medical application, but also must be known in terms of changes that may take place over time *in vivo*. That is, changes with time must be anticipated at the outset and accounted for through selection of biomaterials and/or design of the device.

Information related to basic properties is available from national and international standards, plus handbooks and professional journals of various types. However, this information must be evaluated within the context of the intended biomedical use, since applications and host tissue responses are quite specific for given areas, e.g., cardiovascular (flowing blood contact), orthopedic (functional load bearing), and dental (percutaneous).

In the following, we discuss basic information about bulk and surface properties of biomaterials based on metallic, polymeric, and ceramic substrates, the finite element (FE) modeling and analyses, and the role(s) of water and surface interaction with biomaterials. Also included are details about how some of these characteristics have been determined. The content of the various sections is intended to be relatively basic.

3.1.2 Bulk properties of materials

Francis W. Cooke

Introduction: the solid state

Solids are distinguished from the other states of matter (liquids and gases) by the fact that their constituent atoms are held together by strong interatomic forces (Pauling, 1960). The electronic and atomic structures, and almost all the physical properties, of solids depend on the nature and strength of the interatomic bonds. For a full account of the nature of these bonds one would have to resort to the modern theory of quantum mechanics. However, the mathematical complexities of this theory are much beyond the scope of this chapter and we will instead content ourselves with the earlier, classical model, which is still very adequate. According to the classical theory there are three different types of strong or primary interatomic bonds: ionic, covalent, and metallic.

Ionic bonding

In the ionic bond, electron donor (metallic) atoms transfer one or more electrons to an electron acceptor (nonmetallic) atom. The two atoms then become a cation (e.g.,

Biomedical Engineering Desk Reference; ISBN: 9780123746467

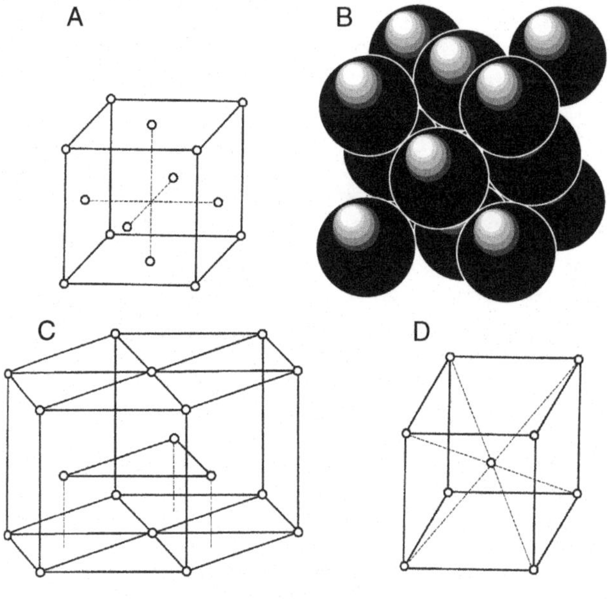

Fig. 3.1.2-1 Typical metal crystal structures (unit cells). (A) Face-centered cubic (FCC). (B) Full size atoms in FCC. (C) Hexagonal close-packed (HCP). (D) Body-centered cubic (BCC).

metal) and an anion (e.g., nonmetal), which are strongly attracted by the electrostatic or Coulomb effect. This attraction of cations and anions constitutes the ionic bond (Hummel, 1997).

In ionic solids composed of many ions, the ions are arranged so that each cation is surrounded by as many anions as possible to reduce the strong mutual repulsion of cations. This packing further reduces the overall energy of the assembly and leads to a highly ordered arrangement called a crystal structure (Fig. 3.1.2-1). Note that in such a crystal no discrete molecules exist, but only an orderly collection of cations and anions. The loosely bound electrons of the atoms are now tightly held in the locality of the ionic bond. These bound electrons are no longer available to serve as charge carriers and ionic solids are poor electrical conductors. Finally, the low overall energy state of these substances endows them with relatively low chemical reactivity. Sodium chloride (NaCl) and magnesium oxide (MgO) are examples of ionic solids.

Covalent bonding

Elements that fall along the boundary between metals and nonmetals, such as carbon and silicon, have atoms with four valence electrons and about equal tendencies to donate and accept electrons. For this reason, they do not form strong ionic bonds. Rather, stable electron structures are achieved by sharing valence electrons. For example, two carbon atoms can each contribute an electron to a shared pair. This shared pair of electrons constitutes the covalent bond $-\overset{|}{\underset{|}{C}}-\overset{|}{\underset{|}{C}}-$ (Hummel, 1997).

If a central carbon atom participates in four of these covalent bonds (two electrons per bond), it has achieved a stable outer shell of eight valence electrons. More carbon atoms can be added to the growing aggregate so that every atom has four nearest neighbors with which it shares one bond each. Thus, in a large grouping, every atom has a stable electron structure and four nearest neighbors. These neighbors often form a tetrahedron, and the tetrahedra in turn are assembled in an orderly repeating pattern (i.e., a crystal) (Fig. 3.1.2-2). This is the structure of both diamond and silicon. Diamond is the hardest of all materials, which shows that covalent bonds can be very strong. Once again, the bonding process results in a particular electronic structure (all valence electrons in pairs localized at the

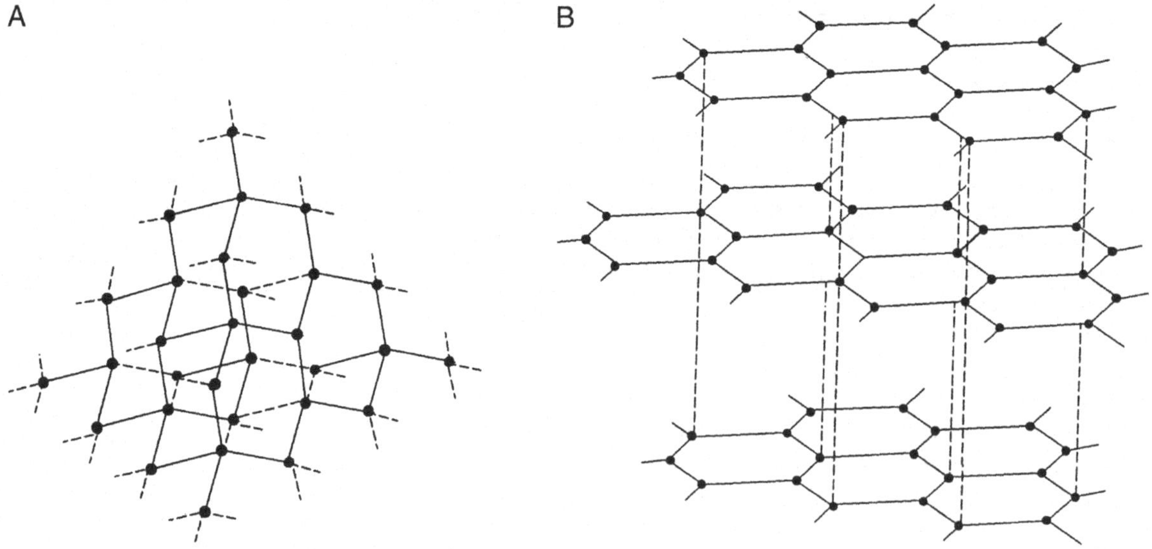

Fig. 3.1.2-2 Crystal structures of carbon. (A) Diamond (cubic). (B) Graphite (hexagonal).

covalent bonds) and a particular atomic arrangement or crystal structure. As with ionic solids, localization of the valence electrons in the covalent bond renders these materials poor electrical conductors.

Metallic bonding

The third, the least understood of the strong bonds, is the metallic bond. Metal atoms, being strong electron donors, do not bond by either ionic or covalent processes. Nevertheless, many metals are very strong (e.g., cobalt) and have high melting points (e.g., tungsten), suggesting that very strong interatomic bonds are at work here, too. The model that accounts for this bonding envisions the atoms arranged in an orderly, repeating, three-dimensional (3D) pattern, with the valence electrons migrating between the atoms like a gas.

It is helpful to imagine a metal crystal composed of positive ion cores, atoms without their valence electrons, about which the negative electrons circulate. On the average, all the electrical charges are neutralized throughout the crystal and bonding arises because the negative electrons act like a glue between the positive ion cores. This construct is called the free electron model of metallic bonding. Obviously, the bond strength increases as the ion cores and electron "gas" become more tightly packed (until the inner electron orbits of the ions begin to overlap). This leads to a condition of lowest energy when the ion cores are as close together as possible.

Once again, the bonding leads to a closely packed (atomic) crystal structure and a unique electronic configuration. In particular, the nonlocalized bonds within metal crystals permit plastic deformation (which strictly speaking does not occur in any nonmetals), and the electron gas accounts for the chemical reactivity and high electrical and thermal conductivity of metallic systems (Hummel, 1997).

Weak bonding

In addition to the three strong bonds, there are several weak secondary bonds that significantly influence the properties of some solids, especially polymers. The most important of these are van der Waals bonding and hydrogen bonding, which have strengths 3–10% that of the primary C–C covalent bond.

Atomic structure

The 3D arrangement of atoms or ions in a solid is one of the most important structural features that derives from the nature of the solid-state bond. In the majority of solids, this arrangement constitutes a crystal. A crystal is a solid whose atoms or ions are arranged in an orderly repeating pattern in three dimensions. These patterns allow the atoms to be closely packed [i.e., have the maximum possible number of near (contacting) neighbors] so that the number of primary bonds is maximized and the energy of the aggregate is minimized.

Crystal structures are often represented by repeating elements or subdivisions of the crystal called unit cells (Fig. 3.1.2-1). Unit cells have all the geometric properties of the whole crystal. A model of the whole crystal can be generated by simply stacking up unit cells like blocks or hexagonal tiles. Note that the representations of the unit cells in Fig. 3.1.2-1 are idealized in that atoms are shown as small circles located at the atomic centers. This is done so that the background of the structure can be understood. In fact, all nearest neighbors are in contact, as shown in Fig. 3.1.2-1B (Newey and Weaver, 1990).

Materials

The technical materials used to build most structures are divided into three classes, metals, ceramics (including glasses), and polymers. These classes may be identified only roughly with the three types of interatomic bonding.

Metals

Materials that exhibit metallic bonding in the solid state are metals. Mixtures or solutions of different metals are alloys.

About 85% of all metals have one of the crystal structures shown in Fig. 3.1.2-1. In both face-centered cubic (FCC) and hexagonal close-packed (HCP) structures, every atom or ion is surrounded by twelve touching neighbors, which is the closest packing possible for spheres of uniform size. In any enclosure filled with close-packed spheres, 74% of the volume will be occupied by the spheres. In the body-centered cubic (BCC) structure, each atom or ion has eight touching neighbors or eightfold coordination. Surprisingly, the density of packing is only reduced to 68% so that the BCC structure is nearly as densely packed as the FCC and HCP structures (Hummel, 1997).

Ceramics

Ceramic materials are usually solid inorganic compounds with various combinations of ionic and covalent bonding. They also have tightly packed structures, but with special requirements for bonding such as fourfold coordination for covalent solids and charge neutrality for ionic solids (i.e., each unit cell must be electrically neutral). As might be expected, these additional requirements lead to more open and complex crystal structures. Aluminum oxide or alumina (Al_2O_3) is an example of a ceramic that has found some use as an orthopedic implant material. (Kingery, 1976).

Carbon is often included with ceramics because of its many ceramic-like properties, even though it is not a compound and conducts electrons in its graphitic form. Carbon is an interesting material since it occurs with two different crystal structures. In the diamond form, the four valence electrons of carbon lead to four nearest neighbors in tetrahedral coordination. This gives rise to the diamond cubic structure (Fig. 3.1.2-2A). An interesting variant on this structure occurs when the tetrahedral arrangement is distorted into a nearly flat sheet. The carbon atoms in the sheet have a hexagonal arrangement, and stacking of the sheets (Fig. 3.1.2-2B) gives rise to the graphite form of carbon. The (covalent) bonding within the sheets is much stronger than the bonding between sheets.

The existence of an element with two different crystal structures provides a striking opportunity to see how physical properties depend on atomic and electronic structure (Table 3.1.2-1).

Inorganic glasses

Some ceramic materials can be melted and upon cooling do not develop a crystal structure. The individual atoms have nearly the ideal number of nearest neighbors, but an orderly repeating arrangement is not maintained over long distances throughout the 3D aggregates of atoms. Such noncrystals are called glasses or, more accurately, inorganic glasses and are said to be in the amorphous state. Silicates and phosphates, the two most common glass forming materials, have random 3D network structures.

Polymers

The third category of solid materials includes all the polymers. The constituent atoms of classic polymers are usually carbon and are joined in a linear chainlike structure by covalent bonds. The bonds within the chain require two of the valence electrons of each atom, leaving the other two bonds available for adding a great variety of atoms (e.g., hydrogen), molecules, functional groups, etc.

Based on the organization of these chains, there are two classes of polymers. In the first, the basic chains are all straight with little or no branching. Such "straight" chain or linear polymers can be melted and remelted without a basic change in structure (an advantage in fabrication) and are called thermoplastic polymers. If side chains are present and actually form (covalent) links between chains, a 3D network structure is formed. Such structures are often strong, but once formed by heating will not melt uniformly on reheating. These are thermosetting polymers.

Usually both thermoplastic and thermosetting polymers have intertwined chains so that the resulting structures are quite random and are also said to be amorphous like glass, although only the thermoset polymers have sufficient cross linking to form a 3D network with covalent bonds. In amorphous thermoplastic polymers, many atoms in a chain are in close proximity to the atoms of adjacent chains, and van der Waals and hydrogen bonding holds the chains together. It is these interchain bonds together with chain entanglement that are responsible for binding the substance together as a solid. Since these bonds are relatively weak, the resulting solid is relatively weak. Thermoplastic polymers generally have lower strengths and melting points than thermosetting polymers (Billmeyer, 1984).

Table 3.1.2-1 Relative physical properties of diamond and graphite[a]

Property	Diamond	Graphite
Hardness	Highest known	Very low
Color	Colorless	Black
Electrical conductivity	Low	High
Density (g/cm^3)	3.51	2.25
Specific heat (cal/gm atm/deg.C)	1.44	1.98

[a] Adapted from D. L. Cocke and A. Clearfield, eds., *Design of New Materials*, Plenum Publ., New York, 1987, with permission.

Microstructure

Structure in solids occurs in a hierarchy of sizes. The internal or electronic structures of atoms occur at the finest scale, less than 10^{-4} μm (which is beyond the resolving power of the most powerful direct observational techniques), and are responsible for the interatomic bonds. At the next higher size level, around 10^{-4} μm (which is detectable by X-ray diffraction, field ion microscopy, scanning tunneling microscopy (STM), etc.), the long-range, 3D arrangement of atoms in crystals and glasses can be observed.

At even larger sizes, 10^{-3} to 10^{2} μm (detectable by light and electron microscopy), another important type of structural organization exists. When the atoms of a molten sample are incorporated into crystals during freezing, many small crystals are formed initially and then grow until they impinge on each other and all the liquid is used up. At that point the sample is completely solid. Thus, most crystalline solids (metals and ceramics) are composed of many small crystals or crystallites called grains that are tightly packed and firmly bound together. This is the microstructure of the material that is observed at magnifications where the resolution is between 1–100 μm.

In pure elemental materials, all the crystals have the same structure and differ from each other only by virtue of their different orientations. In general, these crystallites or grains are too small to be seen except with a light microscope. Most solids are opaque, however, so the common transmission (biological) microscope cannot be used. Instead, a metallographic or ceramographic reflecting microscope is used. Incident light is reflected from the polished metal or ceramic surface. The grain structure is revealed by etching the surface with a mildly corrosive medium that preferentially attacks the grain boundaries. When this surface is viewed through the reflecting microscope the size and shape of the grains, i.e., the microstructure, is revealed.

Grain size is one of the most important features that can be evaluated by this technique because fine-grained samples are generally stronger than coarse-grained specimens of a given material. Another important feature that can be identified is the coexistence of two or more phases in some solid materials. The grains of a given phase will all have the same chemical composition and crystal structure, but the grains of a second phase will be different in both these respects. This never occurs in samples of pure elements, but does occur in mixtures of different elements or compounds where the atoms or molecules can be dissolved in each other in the solid state just as they are in a liquid or gas solution.

For example, some chromium atoms can substitute for iron atoms in the FCC crystal lattice of iron to produce stainless steel, a solid solution alloy. Like liquid solutions, solid solutions exhibit solubility limits; when this limit is exceeded, a second phase precipitates. For example, if more Cr atoms are added to stainless steel than the FCC lattice of the iron can accommodate, a second phase that is chromium rich precipitates. Many important biological and implant materials are multiphase (Hummel, 1997). These include the cobalt-based and titanium-based orthopedic implant alloys and the mercury-based dental restorative alloys, i.e., amalgams.

Mechanical properties of materials

Solid materials possess many kinds of properties (e.g., mechanical, chemical, thermal, acoustical, optical, electrical, magnetic). For most (but not all) biomedical applications, the two properties of greatest importance are strength (mechanical) and reactivity (chemical). The remainder of this section will be devoted to mechanical properties, their measurement, and their dependence on structure. It is significant to note that the dependence of mechanical properties on microstructure is so great that it is one of the fundamental objectives of materials science to control mechanical properties by modifying microstructure.

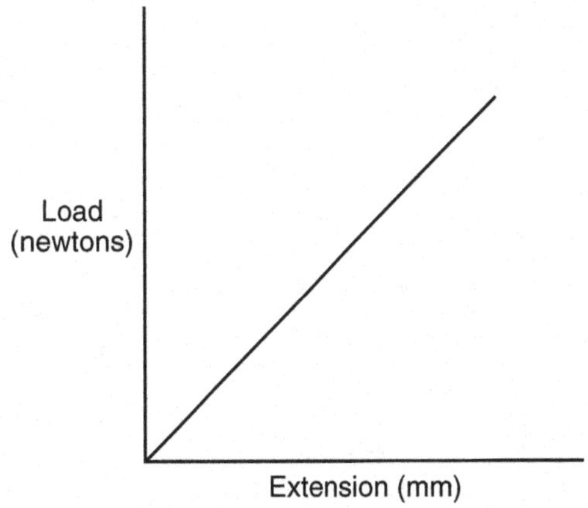

Fig. 3.1.2-3 Initial extension is proportional to load according to Hooke's law.

Elastic behavior

The basic experiment for determining mechanical properties is the tensile test. In 1678, Robert Hooke showed that a solid material subjected to a tensile (distraction) force would extend in the direction of traction by an amount that was proportional to the load (Fig. 3.1.2-3). This is known as Hooke's law and simply expresses the fact that most solids behave in an elastic manner (like a spring) if the loads are not too great.

Stress and strain

The extension for a given load varies with the geometry of the specimen as well as its composition. It is, therefore, difficult to compare the relative stiffness of different materials or to predict the load-carrying capacity of structures with complex shapes. To resolve this confusion, the load and deformation can be normalized. To do this, the load is divided by the cross-sectional area available to support the load, and the extension is divided by the original length of the specimen. The load can then be reported as load per unit of cross-sectional area, and the deformation can be reported as the elongation per unit of the original length over which the elongation occurred. In this way, the effects of specimen geometry can be normalized.

The normalized load (force/area) is stress (σ) and the normalized deformation (change in length/original length) is strain (ε) (Fig. 3.1.2-4).

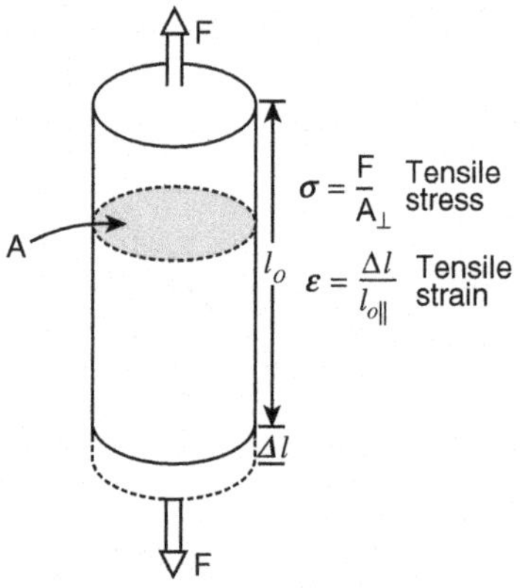

Fig. 3.1.2-4 Tensile stress and tensile strain.

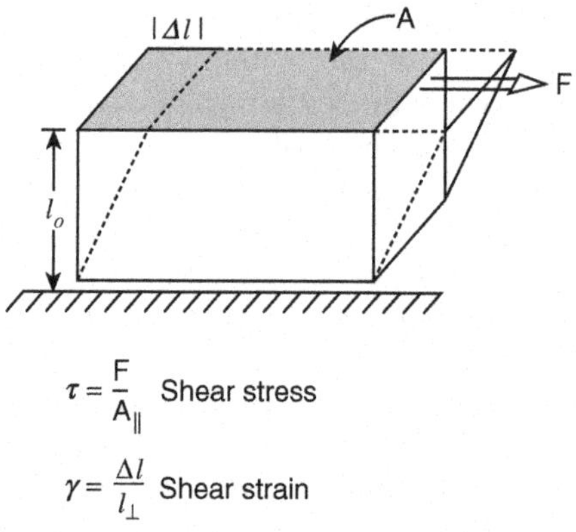

Fig. 3.1.2-5 Shear stress and shear strain.

Tension and compression

In tension and compression the area supporting the load is perpendicular to the loading direction (tensile stress), and the change in length is parallel to the original length (tensile strain).

If weights are used to provide the applied load, the stress is calculated by adding up the total number of pounds-force (lb) or newtons (N) used and dividing by the perpendicular cross-sectional area. For regular specimen geometries such as cyclindrical rods or rectangular bars, a measuring instrument, such as a micrometer, is used to determine the dimensions. The units of stress are pounds per inch squared (psi) or newtons per meter squared (N/m^2). The N/m^2 unit is also known as the pascal (Pa).

The measurement of strain is achieved, in the simplest case, by applying reference marks to the specimen and measuring the distance between with calipers. This is the original length, l_o. A load is then applied, and the distance between marks is measured again to determine the final length, l_f. The strain, ε, is then calculated by:

$$\varepsilon = \frac{l_f - l_o}{l_o} = \frac{\Delta l}{l_o}. \qquad (3.1.2.1)$$

This is essentially the technique used for flexible materials like rubbers, polymers, and soft tissues. For stiff materials like metals, ceramics, and bone, the deflections are so small that a more sensitive measuring method is needed (i.e., the electrical resistance strain gage).

Shear

For cases of shear, the applied load is parallel to the area supporting it (shear stress, τ), and the dimensional change is perpendicular to the reference dimension (shear strain, γ) (Fig. 3.1.2-5).

Elastic constants

By using these definitions of stress and strain, Hooke's law can be expressed in quantitative terms:

$$\sigma = E\varepsilon, \text{ tension or compression,} \qquad (3.1.2\text{-}2a)$$

$$\tau = G\gamma, \text{ shear.} \qquad (3.1.2\text{-}2b)$$

E and G are proportionality constants that may be likened to spring constants. The tensile constant, E, is the tensile (or (Young's) modulus and G is the shear modulus. These moduli are also the slopes of the elastic portion of the stress versus strain curve (Fig. 3.1.2-6). Since all geometric influences have been removed, E and G represent inherent properties of the material. These two moduli are direct macroscopic manifestations of the strengths of the interatomic bonds. Elastic strain is achieved by actually increasing the interatomic distances in the crystal (i.e., stretching the bonds). For materials with strong bonds (e.g., diamond, Al_2O_3, tungsten), the moduli are high and a given stress produces only a small strain. For materials with weaker bonds (e.g., polymers and gold), the moduli are lower (Hummel, 1997). The tensile elastic moduli for some important biomaterials are presented in Table 3.1.2-2.

Isotropy

The two constants, E and G, are all that are needed to fully characterize the stiffness of an isotropic material (i.e., a material whose properties are the same in all directions).

Single crystals are anisotropic (not isotropic) because the stiffness varies as the orientation of applied force changes relative to the interatomic bond directions in the crystal. In polycrystalline materials (e.g., most metallic and ceramic specimens), a great multitude of grains (crystallites) are aggregated with multiply distributed orientations. On the

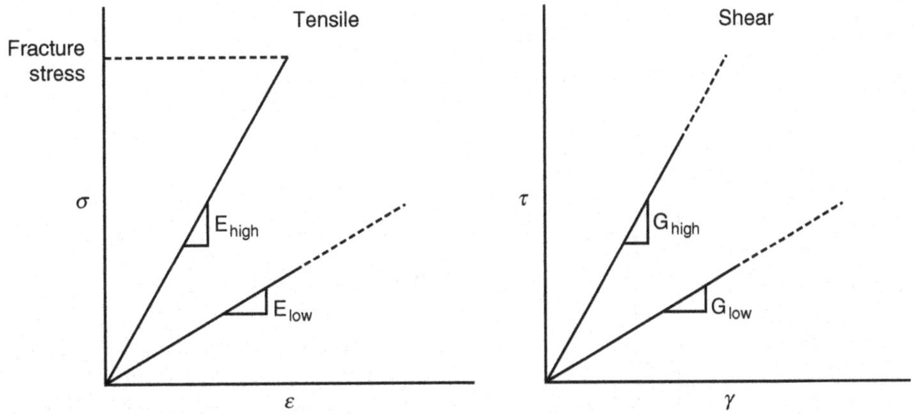

Fig. 3.1.2-6 Stress versus strain for elastic solids.

average, these aggregates exhibit isotropic behavior at the macroscopic level, and values of *E* and *G* are highly reproducible for all specimens of a given metal, alloy, or ceramic.

On the other hand, many polymeric materials and most tissue samples are anisotropic (not the same in all directions) even at the macroscopic level. Bone, ligament, and sutures are all stronger and stiffer in the fiber (longitudinal) direction than they are in the transverse direction. For such materials, more than two elastic constants are required to relate stress and strain properties.

Mechanical testing

To conduct controlled load-deflection (stress–strain) tests, a load frame is used that is much stiffer and stronger than the specimen to be tested (Fig. 3.1.2-7). One cross-bar or cross-head is moved up and down by a screw or a hydraulic piston. Jaws that provide attachment to the specimen are connected to the frame and to the movable cross-head. In addition, a load cell to monitor the force being applied is placed in series with the specimen. The load cell functions somewhat like a very stiff spring scale to measure the applied loads.

Tensile specimens usually have a reduced gage section over which strains are measured. For a valid determination of fracture properties, failure must also occur in this reduced section and not in the grips. For compression testing, the direction of cross-head movement is reversed and cylindrical or prismatic specimens are simply squeezed between flat anvils. Standardized specimens and procedures should be used for all mechanical testing to ensure

Table 3.1.2-2 Mechanical properties of some implant materials and tissues

	Elastic modulus (GPa)	Yield strength (MPa)	Tensile strength (MPa)	Elongation to failure (%)
Al_2O_3	350	–	1000 to 10,000	0
CoCr Alloy[a]	225	525	735	10
316 S.S.[b]	210	240 (800)[c]	600 (1000)[c]	55 (20)[c]
Ti–6Al–4V	120	830	900	18
Bone (cortical)	15 to 30	30 to 70	70 to 150	0–8
PMMA	3.0	–	35 to 50	0.5
Polyethylene[d]	0.6–1.8	–	23 to 40	200–400
Cartilage	[e]	–	7 to 15	20

[a] 28% Cr, 2% Ni, 7% Mo, 0.3% C (max), Co balance.
[b] Stainless steel, 18% Cr, 14% Ni, 2 to 4% Mo, 0.03 C (max), Fe balance.
[c] Values in parentheses are for the cold-worked state.
[d] High density polyethylene (HDPE) and ultrahigh molecular weight polyethylene (UHMWPE).
[e] Strongly viscoelastic.

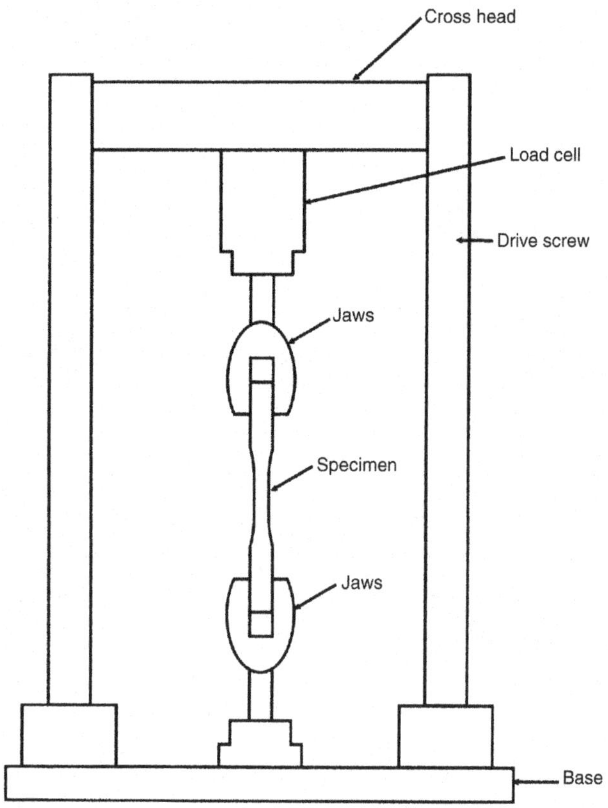

Fig. 3.1.2-7 Mechanical testing machine.

reproducibility of results (see the publications of the American Society for Testing and Materials, 100 Barr Harbor Dr., West Conshohocken, PA 19428-2959).

Another useful test that can be conducted in a mechanical testing machine is the bend test. In bend testing, the outside of the bowed specimen is in tension and the inside in compression. The outer fiber stresses can be calculated from the load and the specimen geometry (see any standard text on strength of materials; Meriam, 1996). Bend tests are useful because no special specimen shapes are required and no special grips are necessary. Strain gages can also be used to determine the outer fiber strains. The available formulas for the calculation of stress states are only valid for elastic behavior. Therefore, they cannot be used to describe any nonelastic strain behavior.

Some mechanical testing machines are also equipped to apply torsional (rotational) loads, in which case torque versus angular deflection can be determined and used to calculate the torsional properties of materials. This is usually an important consideration when dealing with biological materials, especially under shear loading conditions (Hummel, 1997).

Elasticity

The tensile elastic modulus, E (for an isotropic material), can be determined by the use of strain gages, an accurate load cell, and cyclic testing in a standard mechanical testing machine. To do so, Hooke's law is rearranged as follows:

$$E = \frac{\sigma}{\varepsilon}. \qquad (3.1.2.3)$$

Brittle fracture

In real materials, elastic behavior does not persist indefinitely. If nothing else intervenes, microscopic defects, which are present in all real materials, will eventually begin to grow rapidly under the influence of the applied tensile or shear stress, and the specimen will fail suddenly by brittle fracture. Until this brittle failure occurs, the stress–strain diagram does not deviate from a straight line, and the stress at which failure occurs is called the fracture stress (Fig. 3.1.2-6). This behavior is typical of many materials, including glass, ceramics, graphite, very hard alloys (scalpel blades), and some polymers like polymethyl-methacrylate (bone cement) and unmodified poly vinyl chloride (PVC). The number and size of defects, particularly pores, is the microstructural feature that most affects the strength of brittle materials.

Plastic deformation

For some materials, notably metals, alloys, and some polymers, the process of plastic deformation sets in after a certain stress level is reached but before fracture occurs. During a tensile test, the stress at which 0.2% plastic strain occurs is called the 0.2% offset yield strength. Once plastic deformation starts, the strains produced are very much greater than those during elastic deformation (Fig. 3.1.2-8); they are no longer proportional to the stress and they are not recovered when the stress is removed. This happens because whole arrays of atoms under the

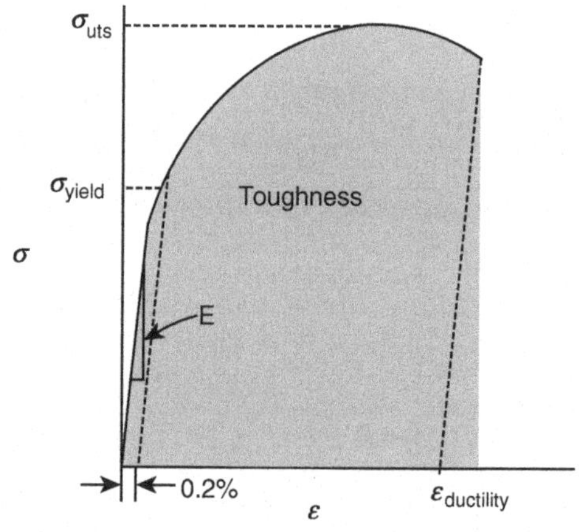

Fig. 3.1.2-8 Stress versus strain for a ductile material.

Table 3.1.2-3 Mechanical properties derivable from a tensile test

Property	Units		
	Fundamental[a]	International	English
1. Elastic modulus (E)	F/A	N/m^2 (Pa)	$lbf^b/in.^2$ (psi)
2. Yield strength (σ_{yield})	F/A	N/m^2 (Pa)	$lbf/in.^2$ (psi)
3. Ultimate tensile strength (σ_{uts})	F/A	N/m^2 (Pa)	$lbf/in.^2$ (psi)
4. Ductility ($\varepsilon_{ductility}$)	%	%	%
5. Toughness (work to fracture per unit volume)	$F \times d/V$	J/m^3	in $lbf/in.^3$

[a] *F*, force; *A*, area; *d*, length; *V*, volume.
[b] lbf, pounds force.

influence of an applied stress are forced to move, irreversibly, to new locations in the crystal structure. This is the microstructural basis of plastic deformation. During elastic straining, on the other hand, the atoms are displaced only slightly by reversible stretching of the interatomic bonds.

Large scale displacement of atoms without complete rupture of the material, i.e., plastic deformation, is only possible in the presence of the metallic bond so only metals and alloys exhibit true plastic deformation. Since long-distance rearrangement of atoms under the influence of an applied stress cannot occur in ionic or convolutely bonded materials, ceramics and many polymers do not undergo plastic deformation.

Plastic deformation is very useful for shaping metals and alloys and is called ductility or malleability. The total permanent (i.e., plastic) strain exhibited up to fracture by a material is a quantitative measure of its ductility (Fig. 3.1.2-8). The strength, particularly the 0.2% offset yield strength, can be increased significantly by reducing the grain size as well as by prior plastic deformation or cold work. The introductions of alloying elements and multiphase microstructures are also potent strengthening mechanisms.

Other properties can be derived from the tensile stress–strain curve. The tensile strength or the ultimate tensile stress (UTS) is the stress that is calculated from the maximum load experienced during the tensile test (Fig. 3.1.2-8).

The area under the tensile curve is proportional to the work required to deform a specimen until it fails. The area under the entire curve is proportional to the product of stress and strain, and has the units of energy (work) per unit volume of specimen. The work to fracture is a measure of toughness and reflects a material's resistance to crack propagation (Fig. 3.1.2-8) (Newey and Weaver, 1990). The important mechanical properties derived from a tensile test and their units are listed in Table 3.1.2-3. Representative values of these properties for some important biomaterials are listed in Table 3.1.2-2.

Creep and viscous flow

For all the mechanical behaviors considered to this point, it has been assumed that when a stress is applied, the strain response is instantaneous. For many important biomaterials, including polymers and tissues, this is not a valid assumption. If a weight is suspended from an excised ligament, the ligament elongates essentially instantaneously when the weight is applied. This is an elastic response. Thereafter the ligament continues to elongate for a considerable time even though the load is constant (Fig. 3.1.2-9A). This continuous, time-dependent extension under load is called "creep."

Similarly, if the ligament is extended in a tensile machine to a fixed elongation and held constant while the load is monitored, the load drops continuously with time (Fig. 3.1.2-9B). The continuous drop in load at constant extension is called stress relaxation. Both these responses are the result of viscous flow in the material. The mechanical analog of viscous flow is a dash-pot or cylinder and piston (Fig. 3.1.2-10A). Any small force is enough to keep the piston moving. If the load is increased, the rate of displacement will increase.

Despite this liquid-like behavior, these materials are functionally solids. To produce such a combined effect, they act as though they are composed of a spring (elastic element) in series with a dashpot (viscous element) (Fig. 3.1.2-10B). Thus, in the creep test, instantaneous strain is produced when the weight is first applied (Fig. 3.1.2-9A). This is the equivalent of stretching the spring to its equilibrium length (for that load). Thereafter, the additional time-dependent strain is modeled by the movement of the dashpot. Complex arrangements of springs and dashpots are often needed to adequately model the actual behavior of polymers and tissues.

Materials that behave approximately like a spring and dash-pot system are viscoelastic. One consequence of viscoelastic behavior can be seen in tensile testing where

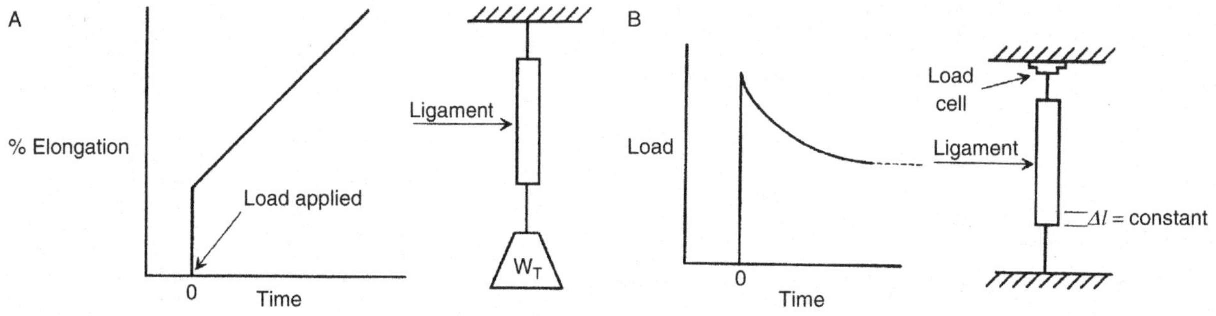

Fig. 3.1.2-9 (A) Elongation versus time at constant load (creep) of ligament. (B) Load versus time at constant elongation (stress relaxation) for ligament.

the load is applied at some finite rate. During the course of load application, there is time for some viscous flow to occur along with the elastic strain. Thus, the total strain will be greater than that due to the elastic response alone. If this total strain is used to estimate the Young's modulus of the material ($E = \sigma/\varepsilon$), the estimate will be low. If the test is conducted at a more rapid rate, there will be less time for viscous flow during the test and the apparent modulus will increase. If a series of such tests is conducted at ever higher loading rates, eventually a rate can be reached where no detectable viscous flow occurs and the modulus determined at this critical rate will be the true elastic modulus, i.e., the spring constant of the elastic component. Tests at even higher rates will produce no further increase in modulus. For all viscoelastic materials, moduli determined at rates less than the critical rate are "apparent" moduli and must be identified with the strain rate used. Failure to do this is one reason why values of tissue moduli reported in the literature may vary over wide ranges.

Finally, it should be noted that it may be difficult to distinguish between creep and plastic deformation in ordinary tensile tests of highly viscoelastic materials (e.g., tissues). For this reason, the total nonelastic deformation of tissues or polymers may at times be loosely referred to as plastic deformation even though viscous flow is involved.

Other important properties of materials

Fatigue

It is not uncommon for materials, including tough and ductile ones like 316L stainless steel, to fracture even though the service stresses imposed are well below the yield stress. This occurs when the loads are applied and removed for a great number of cycles, as happens to prosthetic heart valves and prosthetic joints. Such repetitive loading can produce microscopic cracks that then propagate by small steps at each load cycle.

The stresses at the tip of a crack, a surface scratch, or even a sharp corner are locally enhanced by the stress-raising effect. Under repetitive loading, these local high stresses actually exceed the strength of the material over a small region. This phenomenon is responsible for the stepwise propagation of the cracks. Eventually, the load-bearing cross-section becomes so small that the part finally fails completely.

Fatigue, then, is a process by which structures fail as a result of cyclic stresses that may be much less than the UTS. Fatigue failure plagues many dynamically loaded structures, from aircraft to bones (march- or stress-fractures) to cardiac pacemaker leads.

The susceptibility of specific materials to fatigue is determined by testing a group of identical specimens in cyclic tension or bending (Fig. 3.1.2-11A) at different maximum stresses. The number of cycles to failure is then plotted against the maximum applied stress (Fig. 3.1.2-11B). Since the number of cycles to failure is quite variable for a given stress level, the prediction of fatigue life is a matter of probabilities. For design purposes, the stress that will provide a low probability of failure after 10^6 to 10^8 cycles is often adopted as the fatigue strength or endurance limit of the material. This may be as little as one third or one fourth of the single-cycle yield strength. The fatigue strength is sensitive to environment, temperature, corrosion, deterioration (of tissue specimens), and cycle rate (especially for viscoelastic materials) (Newey and Weaver, 1990).

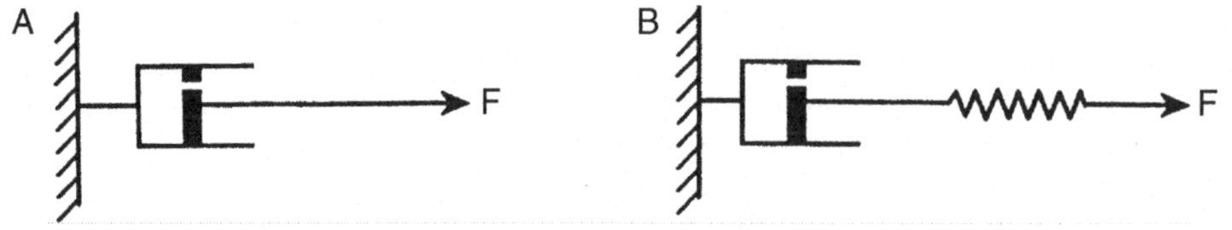

Fig. 3.1.2-10 (A) Dash pot or cylinder and piston model of viscous flow. (B) Dash pot and spring model of a viscoelastic material.

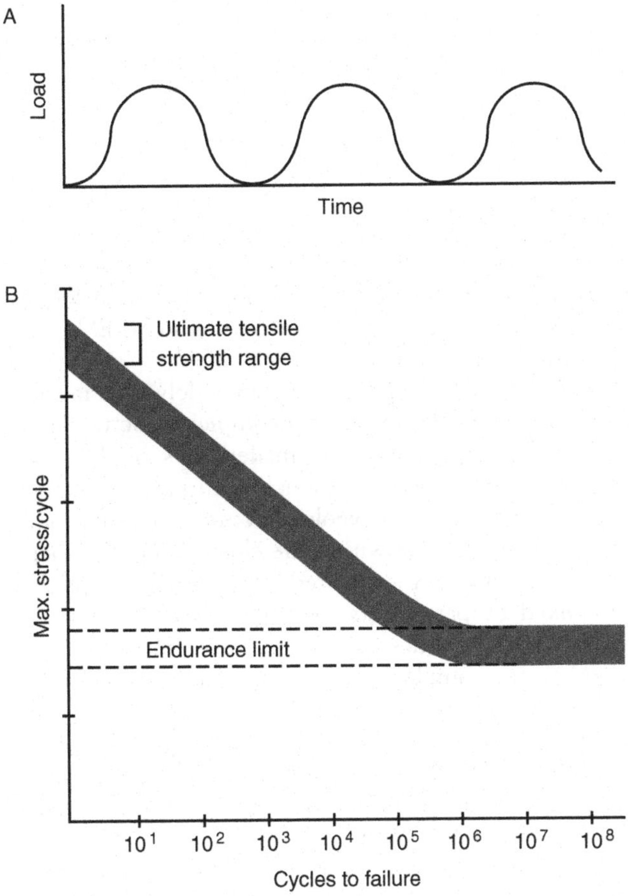

Fig. 3.1.2-11 (A) Stress versus time in a fatigue test. (B) Fatigue curve: fatigue stress versus cycles to failure.

Careful attention to these details is required if laboratory fatigue results are to be successfully transferred to biomedical applications.

Toughness

The ability of a material to plastically deform under the influence of the complex stress field that exists at the tip of a crack is a measure of its toughness. If plastic deformation does occur, it serves to blunt the crack and lower the locally enhanced stresses, thus hindering crack propagation. To design "failsafe" structures with brittle materials, it has become necessary to develop an entirely new system for evaluating service worthiness. This system is fracture toughness testing and requires the testing of specimens with sharp notches. The resulting fracture toughness parameter is a function of the apparent crack propagation stress and the crack depth and shape. It is called the critical stress intensity factor (K_{Ic}) and has units of Pa$\sqrt{m}$ or N $\cdot$ $m^{3/2}$ (Meyers and Chawla, 1984). For materials that exhibit extensive plastic deformation at the crack tip, an energy-based parameter, the J integral, can be used. The energy absorbed in impact fracture is also a measure of toughness, but at higher loading rates (Newey and Weaver, 1990).

Effect of fabrication on strength

A general concept to keep in mind when considering the strength of materials is that the process by which a material is produced has a major effect on its structure and hence its properties (Newey and Weaver, 1990). For example, plastic deformation of most metals at room temperature flattens the grains and produces strengthening while reducing ductility. Subsequent high-temperature treatment (annealing) can reverse this effect. Polymers drawn into fibers are much stronger in the drawing direction than are undrawn samples of the same material.

Because strength properties depend on fabrication history, it is important to realize that there is no unique set of strength properties of each generic material (e.g., 316L stainless steel, polyethylene, aluminum oxide). Rather, there is a range of properties that depends on the fabrication history and the microstructures produced.

Conclusion

The determination of mechanical properties is not only an exercise in basic materials science but is indispensable to the practical design and understanding of load-bearing structures. Designers must determine the service stresses in all structural members and be sure that at every point these stresses are safely below the yield strength of the material. If cyclic loads are involved (e.g., lower-limb prostheses, teeth, heart valves), the service stresses must be kept below the fatigue strength.

Subsequently where the properties and behavior of materials are discussed in detail, it is well to keep in mind that this information is indispensable to understanding the mechanical performance (i.e., function) of both biological and manmade structures.

Bibliography

Billmeyer, F.W. (1984). *Textbook of Polymer Science*. John Wiley and Sons Inc., New York.

Hummel, R.E. (1997). *Understanding Materials Science*. Springer-Verlag, New York.

Kingery, W.D. (1976). *Introduction to Ceramics*. John Wiley and Sons Inc., New York.

Meriam, J.L. (1996). *Engineering Mechanics*, Vol. 1, *Statics*, 14th ed. John Wiley and Sons Inc., New York.

Meyers, M.A., and Chawla, K.K. (1984). *Mechanical Metallurgy*. Prentice-Hall Inc., Upper Saddle River, NJ.

Newey, C., and Weaver, G. (1990). *Materials Principals and Practice*. Butterworth-Heinemann Ltd., Oxford, UK.

Pauling, L. (1960). *The Nature of the Chemical Bond and the Structure of Molecules and Crystals*. Cornell Univ. Press, Ithaca, NY.

3.1.3 Finite element analysis

Ivan Vesely and Evelyn Owen Carew

Introduction

The reader may be familiar with the concepts of elasticity, stress, and strain. Estimations of material stress and strain are necessary during the course of device design to minimize the chance of device failure. For example, artificial hip joints need to be designed to withstand the loads that they are expected to bear without fracture or fatigue. Stress analysis is therefore required to ensure that all components of the device operate below the fatigue limit. For deformable structures such as diaphragms for artificial hearts, an estimate of strains or deformations is required to ensure that during maximal deformation, components do not contact other structures, potentially causing interference and unexpected failure modes such as abrasion.

For simple calculations, such as the sizing of a bolt to connect two components that bear load, simple analytical calculations usually suffice. Often, these calculations are augmented by reference to engineering tables that can be used to refine the stress estimates based on local geometry, such as the pitch of the threads. Such analytical methods are preferred because they are exact and can be supported by a wealth of engineering experience. Unfortunately, analytical solutions are usually limited to linear problems and simple geometries governed by simple boundary conditions. The boundary conditions can be considered input data or constraints on the solution that are applied at the boundaries of the system. Most practical engineering problems involve some combination of material or geometrical nonlinearity, complex geometry, and mixed boundary conditions. In particular, all biological materials have nonlinear elastic behavior and most experience large strains when deformed. As a result, nonlinearities of one form or the other are usually present in the formulation of problems in biomechanics. These nonlinearities are described by the equations relating stress to strain and strain to displacement. Applying analytical methods to such problems would require so many assumptions and simplifications that the results would have poor accuracy and would thus be of little engineering value. There is therefore no alternative but to resort to approximate or numerical methods. The most popular numerical method for solving problems in continuum mechanics is the finite element method (FEM), also referred to as finite element analysis (FEA).

FEA is a computational approach widely used in solid and fluid mechanics in which a complex structure is divided into a large number of smaller parts, or elements, with interconnecting nodes, each with geometry much simpler than that of the whole structure. The behavior of the unknown variable within the element and the shape of the element are represented by simple functions that are linked by parameters that are shared between the elements at the nodes. By linking these simple elements together, the complexity of the original structure can be duplicated with good fidelity. After boundary conditions are taken into account, a large system of equations for the unknown nodal parameters always results; these equations are solved simultaneously by a computer, using indirect or iterative means.

FEA is extremely versatile. The size and configuration of the elements can be adjusted to best suit the problem; complex geometries can be discretized and solutions can be stepped through time to analyze dynamic systems. Very often, simple analytical methods are used to make a first approximation to the design of the device, and FEA is subsequently used to further refine the design and identify potential stress concentrations. FEA can be applied to both solids and fluids or, with additional complexity, to systems containing both. FEA software is very mature and computing power is now sufficiently cheap to allow FEMs to be applied to a wide range of problems. In fluid flow, FEA has been applied to weather forecasting and supersonic flow around aircraft and within engines, and in the medical field, to optimizing blood pumps and cannulas. In solids, FEA has been used to design, build, and crash automobiles, estimate the impact of earthquakes, and reconstruct crime scenes. In biomaterials, FEA has been applied to almost every implantable device, ranging from artificial joints to pacemaker leads. Although originally developed to help structural engineers analyze stress and strain, FEA has been adopted by basic scientists and biologists to study the dynamic environment within arteries, muscles and even cells.

We now hope to introduce the reader to FEMs without digressing into detailed discussion of some of the more difficult concepts that are often required to properly define and execute a real-world problem. For that, the reader is referred to the many excellent texts in the field, some of which are included in the bibliography.

Overview of the FEM

The essential steps in implementing the FEM follow:

(i) The region of interest (continuum) is discretized, that is, subdivided into a smaller number of regions called elements, interconnected at nodal points. Nodes may also be placed in the interior of an element. In one dimension, the elements are line segments; in two dimensions, they are usually triangles or quadrilaterals (Fig. 3.1.3-1); in three dimensions, they can be rectangular prisms (hexahedra) or triangular prisms (tetrahedra), for example (Fig. 3.1.3-2). Elements may be quite general with the possibility of non-planar faces and curvilinear sides or edges (Desai, 1979; Zienkiewicz and Taylor, 1994).

(ii) The unknown variables within the continuum (e.g., displacement, stress, or velocity components) are defined within each element by suitable interpolating functions. Interpolating functions are traditionally piece-wise polynomials and are also known as basis or shape functions. The order of the interpolating functions (i.e., first, second, or third order) is usually used to fix the number of nodes in the elements (Fig. 3.1.3-3).

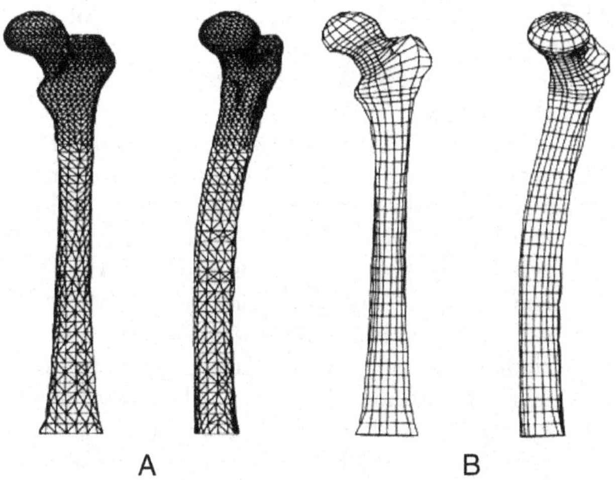

Fig. 3.1.3-2 3D FE representations of the human femur. (A) Tetrahedral elements; (B) hexahedral elements. (From Middleton *et al.*, 1996, p. 125. Reproduced with permission of Gordon and Breach Publishers, Overseas Publishers Assn., Amsterdam.)

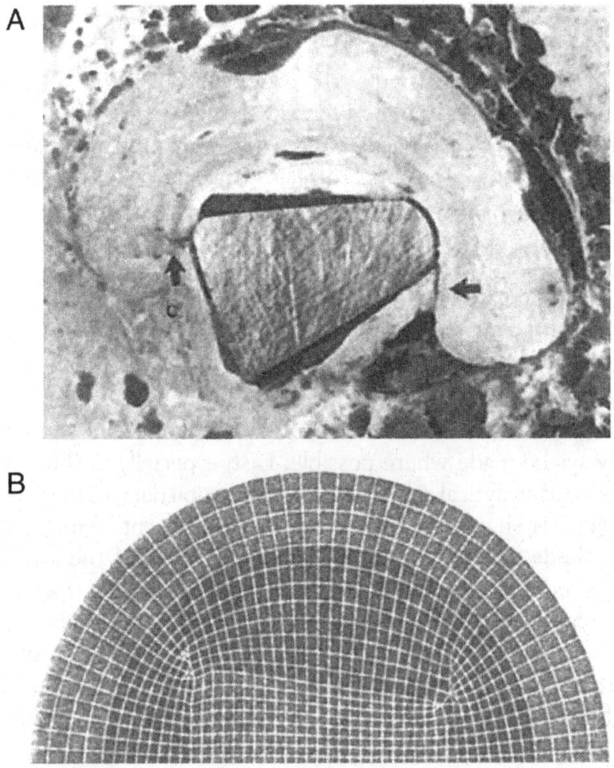

Fig. 3.1.3-1 (A) Cross-section of an autopsy-retrieved femur showing a cracked mantle (arrows). (B) Mixed planar quadrilateral/triangle FE representation of (A). (From Middleton *et al.*, 1996, p. 35. Reproduced with permission of Gordon and Breach Publishers, Overseas Publishers Assn., Amsterdam.)

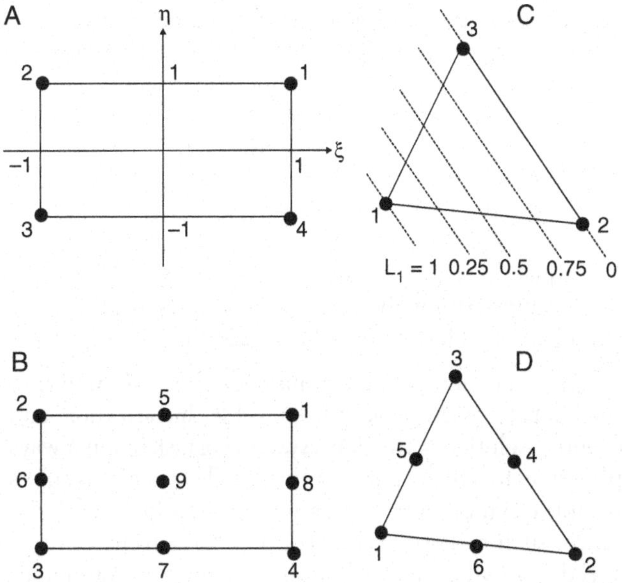

Fig. 3.1.3-3 Examples of two-dimensional elements and their corresponding local coordinate systems [embedded in (A) and (C)]. For the rectangles, the local coordinates (ξ, η) are referred to a cartesian system with $-1 \leq \xi, \eta \leq 1$; for the triangles, the local coordinates (L_1, L_2, L_3) are area coordinates satisfying $0 \leq L_1, L_2, L_3 \leq 1$. Elements with linear interpolating functions (first order) are shown in (A) and (C). Quadratic elements (second-order interpolating functions) are shown in (B) and (D).

(iii) The equations that define the behavior of the unknown variable, such as the equations of motion or the relationships between stress and strain or strain and displacement, are formulated for each element in the form of matrices. These element matrices are then assembled into a global system of equations for the entire discretized domain. This system is defined by a coefficient matrix, an unknown vector of nodal values, and a known "right-hand side" (RHS) vector. Boundary conditions in derivative form would already be included in the RHS vector at this stage, but those that set the unknown function to a known value at the boundary have to be incorporated into the system matrix and RHS vector by overwriting relevant rows and columns. Since the RHS vector contains information about the boundary conditions, it is sometimes called the "external load vector."

(iv) The final step in FEA involves solving the global system of equations for the unknown vector. In theory, this can be achieved by premultiplying the RHS vector by the inverse of the coefficient matrix. The result is the discrete (pointwise) solution to the original problem. If the problem is linear and isotropic, the elements of the matrix are constants and the required matrix inversion can be done. If the defining equations are nonlinear or the material is anisotropic, the coefficient matrix itself will be a function of the unknown variables and matrix inversion is not straightforward. Some kind of linearization is necessary before the matrix can be inverted (e.g., successive approximation or Newton's methods; see, for example, Harris and Stöcker, 1998). In practice, the global system matrix, whether linear or nonlinear, is seldom inverted directly, usually because it is too large. Some indirect method of solving the system of equations is preferred [i.e., lower-upper (LU) decomposition, Gaussian elimination; see, for example, Harris and Stöcker, 1998].

The evaluation of element matrices, their assembly into the global system, and the possible linearization and eventual solution of the global system is a task that is always passed on to a high-speed computer. This usually requires complex computer programs written in a high-level language, such as Fortran. Indeed, it is the advent of high-speed computers and workstations and the continuous improvements in processor speed, memory management, and disk storage that have enabled large-scale FE problems to be tackled with relative ease.

The modern-day FEA toolbox also includes facilities for data pre- and postprocessing. Data preprocessing usually involves input formatting and grid definition, the latter of which may require some ingenuity, because mesh design may affect the convergence and accuracy of the numerical solution. Element size is governed by local geometry and the rate of change of the solution in different parts of the domain. Mesh refinement (a gradation of element size) in the vicinity of sharp corners, boundary layers, high solution gradients, stress concentrations or vortices is done routinely to enhance the accuracy and convergence of the solution. Adaptive procedures that allow the mesh to change with the solution according to some error criteria are usually incorporated into the FE process (George, 1991; Brebbia and Aliabadi, 1993; Zienkiewicz and Taylor, 1994). Typically, this means that the mesh is refined in areas where the solution gradient is high, and elements are removed from regions where the solution is changing slowly. The result is usually a dramatic improvement in convergence, accuracy, and computational efficiency. Postprocessing of data involves the evaluation of *ad hoc* variables such as strains, strain rates, stresses; generating plots such as simple *xy*-plots, contour plots, and particle paths; and solution visualization and animation. All of the additional information facilitates the understanding and interpretation of the results.

The importance of checking and validating FE solutions cannot be overemphasized. The most basic validation involves a "patch test" (Zienkiewicz and Taylor, 1994) in which a few elements (i.e., a patch of the material) are analyzed to verify the formulation of interpolating functions and the consistency of the code itself. Second, a very simple problem with known analytical solution is simulated with a coarse grid to verify that the code reproduces the known solution with acceptable accuracy. For example, parabolic flow in a tube can be simulated with a very coarse grid and the result quickly compared against the analytical solution. We caution, however, that reproducing the solution in a simpler problem does not guarantee that the code will work in more realistic and complicated cases. It is also recommended that numerical solutions be obtained from at least three meshes with increasing degrees of mesh refinement. Such solutions should converge with mesh refinement (*h*-convergence, Strang and Fix, 1973). Comparison of numerical results to experimental data should always be made where possible. Last, especially in the absence of analytical solutions or experimental data, numerical solutions should be compared across different numerical methods, or across different numerical codes if the same method is used. There is no gold standard for the number of validation tests that are required for any particular problem. The greater the variety of test problems and checks, the greater the degree of confidence one can have in the results of the FEM.

The continuum equations

Whether we use FEA to compute the stress in a prosthetic limb or to simulate blood flow in bifurcating arteries, the first objective in setting up an FEA problem is to identify and specify the equations that define the behavior of unknown variables in the continuum. Such equations typically result from applying the universal laws of conservation of mass, momentum, and energy, as well as the constitutive equations that define the stress-strain or other relationships within the material. The resulting differential or integral equations must then be closed by specifying the appropriate boundary conditions.

A "well-behaved" solution to the continuum problem is guaranteed if the differential or integral equations and boundary conditions systems are "well posed." This means that a solution to the continuum problem should exist, be unique, and only change by a small amount when the input data change by a small amount. Under these circumstances the numerical solution is guaranteed to converge to the true solution. Proving in advance that a general continuum problem is "well posed" is not a trivial exercise. Fortunately, consistency and convergence of the numerical solution can usually be monitored by other means, for example, the already mentioned "patch test" (Zienkiewicz and Taylor, 1994).

The equations governing the description of a continuum can be formulated via a differential or variational approach. In the former, differential equations are used to describe the problem; in the latter, integral equations are used. In some cases, both formulations can be applied to a problem. As an illustration we present a case for which both formulations apply and later show that these lead to the same FE equations.

The differential formulation

Consider the function $u(x, y)$, defined in some two-dimensional domain Ω bounded by the curve Γ (Fig. 3.1.3-4), which satisfies the differential equation

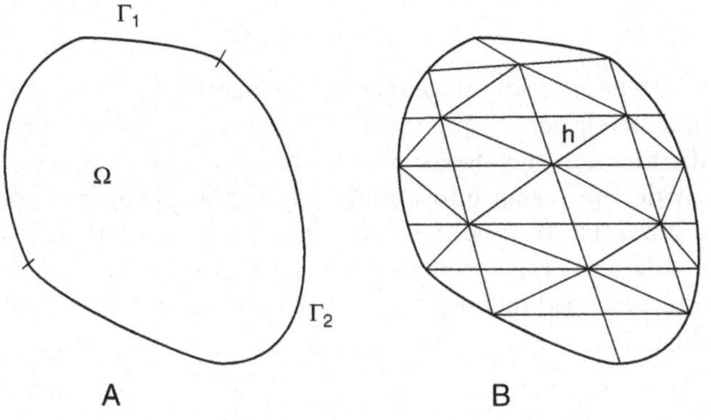

Fig. 3.1.3-4 (A) A continuum Ω enclosed by the boundary $\Gamma = \Gamma_1 \cup \Gamma_2$; the function itself is specified on Γ_1 and its derivative on Γ_2. (B) A finite element representation of the continuum. The domain has been discretized with general arbitrary triangles of size h, with the possibility of having curved sides.

$$-\nabla^2 u + qu = f \text{ in } \Omega \qquad (3.1.3.1a)$$

subject to the boundary conditions

$$u = g \text{ on } \Gamma_1 \qquad (3.1.3.1b)$$

$$\frac{\partial u}{\partial n} = 0 \text{ on } \Gamma_2 \qquad (3.1.3.1c)$$

where $\nabla^2 \equiv \partial^2/\partial x^2 + \partial^2/\partial y^2$ is the Laplacian operator in two dimensions, n is the unit outward normal to the boundary, and q, f, g are assumed to be constants for simplicity, with $q \geq 0$. Here, the boundary Γ is made up of two parts, Γ_1 and Γ_2, where different boundary conditions apply.

When $f = 0$, Eq. 3.1.3.1a means that the spatial change of the gradient of u at any point in the x–y space is proportional to u. The boundary condition 1b sets u to have a fixed value at one part of the boundary. On another part of the boundary, the rate of change of u in the normal direction is set to zero (boundary condition 3.1.3.1c). The system represented by Eqs. 3.1.3.1a–c can be used to describe the transverse deflection of a membrane, torsion in a shaft, potential flows, steady-state heat conduction, or groundwater flow (Desai, 1979; Zienkiewicz and Taylor, 1994).

The variational formulation

A variational equation can arise, for example, from the physical requirement that the total potential energy (TPE) of a mechanical system must be a minimum. Thus the TPE will be a function of a displacement function, for example, itself a function of spatial variables. A "function of a function" is referred to as a functional. We consider, as an example, the functional $I(v)$ of the function $v(x, y)$ of the spatial variables x and y, defined by:

$$I(v) = \iint_{\Omega} \left\{ (\nabla v)^2 + qv^2 - 2vf \right\} d\Omega \qquad (3.1.3.2)$$

(Strang and Fix, 1973; Zienkiewicz and Taylor, 1994). The relevant question is that of all the possible functions $v(x, y)$ that satisfy Eq. 3.1.3.2, what particular $v(x, y)$ minimizes $I(v)$? We get the answer by equating the first variation of $I(v)$, written $\delta I(v)$, to zero. To perform the variation of a functional, one uses the standard rules of differentiation. It can be shown that the variation of $I(v)$ over v results in Eq. 3.1.3.1a, provided Eqs. 3.1.3.1b and 3.1.3.1c hold and the variation of v is zero on Γ_1. Thus the function that minimizes the functional defined in Eq. 3.1.3.2 is the same function that solves the boundary value problem given by Eqs. 3.1.3.1a–c.

The finite element equations

There are four basic methods of formulating the equations of FEA. These are: (i) the direct or displacement method, (ii) the variational method, (iii) the weighted residual method, and (iv) the energy balance method. Only the more popular variational and weighted residual methods will be described here. The integral equation 2 will be used to illustrate the variational method, while the differential equation system 3.1.3.1a–c will be used to illustrate the weighted residual method.

The variational approach

The FEM is introduced in the following way. The region is divided into a finite number of elements of size h (Fig. 3.1.3-4). The h notation is to be interpreted as referring to the subdivided domain. Instead of seeking the function v that minimizes $I(v)$ in the continuous domain, i.e., the exact solution, we instead seek an approximate solution by looking for the function v^h that minimizes $I(v^h)$ in the discrete domain. The following trial functions are defined over the discretized domain:

$$v^h(x,y) = \sum_{i=1}^{n} v_i N_i(x,y) \qquad (3.1.3.3)$$

where N_i are global basis or shape functions and v_i are nodal parameters. The sum is over the total number of nodes n in the mesh. Using Eq. 3.1.3.3 in Eq. 3.1.3.2, the functional becomes

$$I(v^h) = \sum_{i,j} v_i v_j \iint_\Omega \nabla N_i \nabla N_j d\Omega + q \sum_{i,j} v_i v_j \times \iint_\Omega \nabla N_i N_j d\Omega - 2\sum_{i,j} v_i \iint_\Omega f N_i d\Omega \qquad (3.1.3.4)$$

which can be written in matrix notation as

$$I(v^h) = v^T K v + q v^T M v - 2 v^T F \qquad (3.1.3.5)$$

where

$$K = \iint_\Omega \nabla N_i \nabla N_j \, d\Omega, \qquad M = \iint_\Omega N_i N_j \, d\Omega,$$

$$F = \iint_\Omega f N_i \, d\Omega$$

v^T represents the *transpose* of the vector v; K is known as the stiffness matrix, M as the mass matrix, and F as the local load vector. The function v^h that minimizes Eq. 3.1.3.5 should satisfy $\delta I(v^h) = 0$. This gives

$$(K + qM)v = F \qquad (3.1.3.6)$$

that is, a set of simultaneous equations for the nodal parameters v.

Weighted residual approach

The weighted residual approach can be applied directly to any system of differential equations such as 3.1.3.1a–c and even to those problems for which a variational principle may not exist. This approach is therefore more general. The method assumes an approximation $u^h(x,y)$ for the real solution $u(x,y)$. Because u^h is approximate, its substitution into Eq. 3.1.3.1a will result in an error or residual R^h:

$$R^h = -\nabla^2 u^h + q u^h - f \qquad (3.1.3.7)$$

The weighted residual approach requires that some weighted average of the error due to nonsatisfaction of the differential equation by the approximate solution u^h (Eq. 3.1.3.7) vanish over the domain of interest:

$$\iint_\Omega R^h w \, d\Omega = \iint_\Omega (-\nabla^2 u^h + q u^h - f) w \, d\Omega = 0 \qquad (3.1.3.8)$$

where $w(x, y)$ is a weighting function. A function u^h that satisfies Eq. 3.1.3.8 for all possible w selected from a certain class of functions must necessarily satisfy the original differential equations 3.1.3.1a–c. It actually does so only in an average or "weak" sense. Equation 3.1.3.8 is therefore known as a "weak form" of the original equation 3.1.3.1a. The second-order derivatives of the ∇^2 term are usually reduced to first order derivatives by an integration by parts (Harris and Stocker, 1998). The result is another weak form:

$$\iint_\Omega \left\{ \nabla u^h \nabla w + q u^h w - f w \right\} d\Omega = 0 \qquad (3.1.3.9)$$

which has the advantage that approximating functions can now be chosen from a much larger space, a space where the function only needs be once-differentiable. Again, we divide the region into a finite number of elements and assume that the approximate solution can be represented by the sum of the product of unknown nodal values v_j and interpolating functions $N_j(x, y)$, defined at each node j of the mesh:

$$u^h = \sum_{j=1}^{n} v_j N_j \qquad (3.1.3.10)$$

When Eq. 3.1.3.10 is substituted into Eq. 3.1.3.9 with $w = N_i$, Eq. 3.1.3.6 results as before, proving the equivalence of the weighted residual and variational methods for this particular example. We note that the weighting function is required to be zero on those parts of the boundary where the unknown function is specified (Γ_1, in our example) and that there can be other choices of the weighting function w. Choosing weighting functions to be the same as interpolating functions defines the Galerkin FEM (Strang and Fix, 1973; Zienkiewicz and Taylor, 1994).

Properties of interpolating functions

The process of discretizing the continuum into smaller regions means that the global shape functions $N_j(x, y)$ are replaced by local shape functions $N_j^e(\xi, \eta)$, defined within each element e, where ξ, η are the local coordinates within the element (Fig. 3.1.3-3). In the FEM, interpolating functions are usually piecewise polynomials that are required to have (a) the minimum degree of smoothness, (b) continuity between elements, and (c) "local support."

The minimum degree of smoothness is dictated by the highest derivative of the unknown function that occurs in the "weak" or variational form of the continuum problem. The requirement for continuity between elements can always be satisfied by an appropriate choice of the approximating polynomial and number of boundary

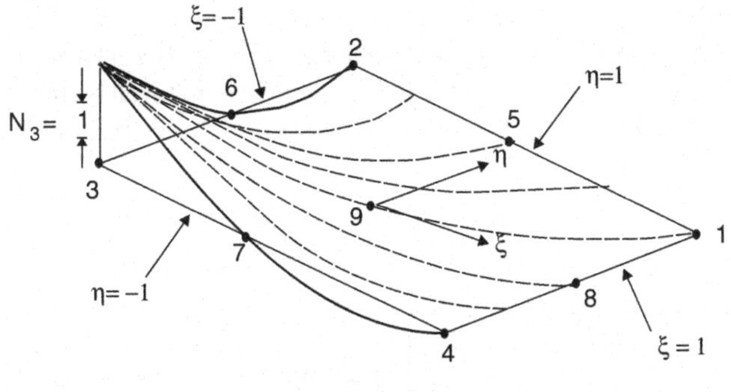

$$N_1^e(\xi, \eta) = \xi\eta\,(1+\xi)\,(1+\eta)/4$$
$$N_2^e(\xi, \eta) = -\xi\eta\,(1-\xi)\,(1+\eta)/4$$
$$N_3^e(\xi, \eta) = \xi\eta\,(1-\xi)\,(1-\eta)/4$$
$$N_4^e(\xi, \eta) = -\xi\eta\,(1+\xi)\,(1-\eta)/4$$
$$N_5^e(\xi, \eta) = \eta\,(1-\xi^2)\,(1+\eta)/2$$
$$N_6^e(\xi, \eta) = -\xi(1-\xi^2)\,(1-\eta^2)/2$$
$$N_7^e(\xi, \eta) = -\eta(1-\xi^2)\,(1-\eta)/2$$
$$N_8^e(\xi, \eta) = \xi(1+\xi^2)\,(1-\eta^2)/2$$
$$N_9^e(\xi, \eta) = \xi(1-\xi^2)\,(1-\eta^2)$$

Fig. 3.1.3-5 Sample shape functions for a nine-noded rectangular element. Shape functions are defined in terms of local coordinates ξ and η where $-1 \leq \xi, \eta \leq 1$; $N_3^e(\xi, \eta)$ is shown in the plot. It can be checked that $N_i = 1$ at node i and zero at all other nodes (compact support) as required.

nodes that define the element. The requirement for "local support" means that within an element,

$$N_i^e(\xi, \eta) = \begin{cases} 1 & \text{at node } i \\ 0 & \text{at all other nodes} \end{cases} \tag{3.1.3.11}$$

as shown in Fig. 3.1.3-5. This is the single most important property of the interpolating functions. This property makes it possible for the contributions of all the elements to be summed up to give the response of the whole domain.

The notation P^m is conventionally used to indicate the degree m of the interpolating polynomial. The notation C^n is used to indicate that all derivatives of the interpolating function, up to and including $n - 1$, exist and are continuous. By convention, the notation $P^m - C^n$ is therefore used to indicate the order and smoothness properties of the interpolating polynomials.

Examples from biomechanics

The following are examples of FEA applications in biomaterials science and biomechanics.

Analysis of commonplace maneuvers at risk for total hip dislocation

Dislocation is a frequent complication of total hip arthroplasty (THA). In this FE study (Nadzadi *et al.*, 2003), a motion tracking system and a recessed force plate were used to capture the kinematics and ground reaction forces from several trials of realistic dislocation-prone maneuvers performed by actual subjects. Kinematics and kinetic data associated with the experiments were imported into a FE model of THA dislocation. The FE model was used to compute stresses developed within the implant, given the observed angular motion of the hip and contact force inferred from inverse dynamics. The FE mesh (Fig. 3.1.3-6A) was created using PATRAN version 8.5 and the simulations were executed with ABAQUS version 5.8. In the FEA, the resultant resisting moment developed around the hip-cup center was tracked, as a function of hip angle. The peak of this resistive moment was a key outcome measure used to estimate the relative risk of dislocations from the motions. All seven maneuvers studied led to frequent instances of computationally predicted dislocation (Fig. 3.1.3-6B). The authors conclude that this library of dislocation-prone maneuvers appear to substantially extend the information base previously available to study this important complication of THA. Additionally the hope is that their results will contribute to improvements in implant design and surgical technique and reduce *in vivo* incidence.

A

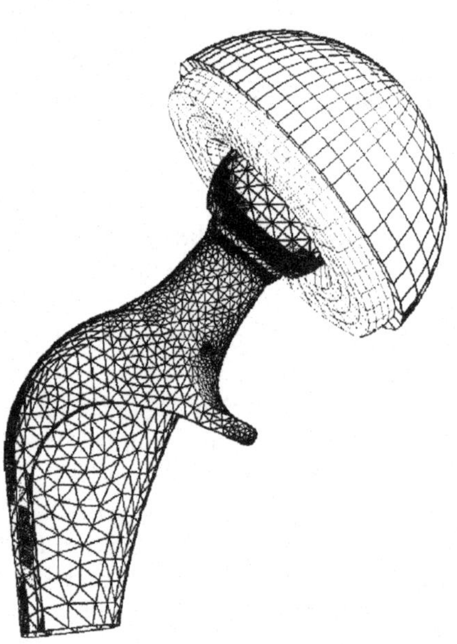

B

Maneuver	No. of trials	No. of dislocations	% of trials dislocating
Low sit-to-stand	47	41	87
Normal sit-to-stand	55	33	64
Tie	69	31	45
Leg cross	64	22	34
Stoop	42	6	14
Post. disloc. maneuvers	277	133	48
Pivot	58	23	40
Roll	19	12	63
Ant. disloc. maneuvers	77	35	45
Overall series	353	168	47

Fig. 3.1.3-6 (A) Finite element model of a contemporary 22-mm modular THA system. (B) Table of FE dislocation predictions of the seven challenge maneuvers simulated. (Reproduced with permission from Nadzadi *et al.*, 2003.)

A finite element model for the lower cervical spine

A parametric study was conducted to determine the variations in the biomechanical responses of the spinal components in the lower cervical spine (Yoganandan *et al.*, 1997). Axial compressive load was imposed uniformly on the superior surface of the C4–C6 unit. The various components were assumed to have linear isotropic and homogeneous elastic behavior and appropriate material parameters were taken from the literature. A detailed 3D FE model was reconstructed from 1.0-mm CT scans of a human cadaver, resulting in a total of 10,371 elements (Fig. 3.1.3-7A). The results show that an increase in elastic moduli of the disks resulted in an increase in endplate stresses and that the middle C5 vertebral body produced the highest compressive stresses (Fig. 3.1.3-7B). The model appears to confirm clinical experience that cervical fractures are induced by external compressive forces.

FEA of indentation tests on pyrolytic carbon

Pyrolytic carbon (PyC) heart valves are known to fail through cracks initiated at the contact areas between leaflets and their housing. In Gilpin *et al.* (1996), this phenomenon is simulated with a 5.1-mm steel ball indenting a graphite sheet coated on each side with PyC, similar to the makeup of real heart valves. Two types of contacts were analyzed: when the surface material is thick (rigid backing) and when it is fairly thin (flexible backing).

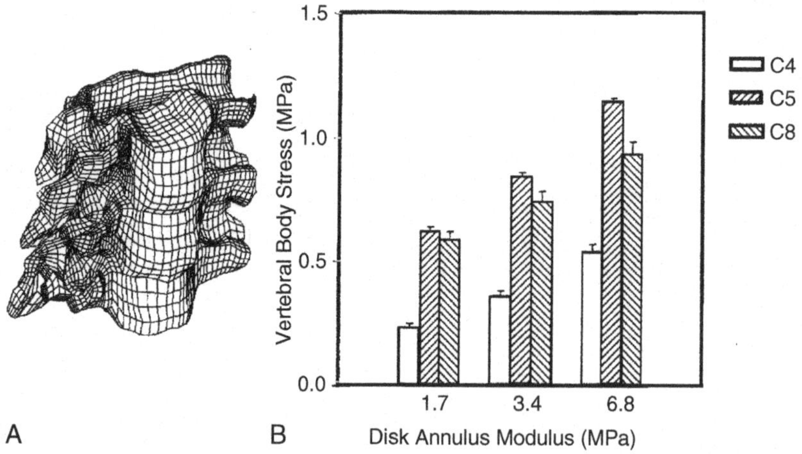

Fig. 3.1.3-7 Finite element model of the C4–C6 unit of the lower cervical spine: (A) mesh showing 3D solid elements and (B) plot of vertebral body stress as a function of disk annulus moduli. (Reproduced with permission from Yoganandan *et al.*, 1997.)

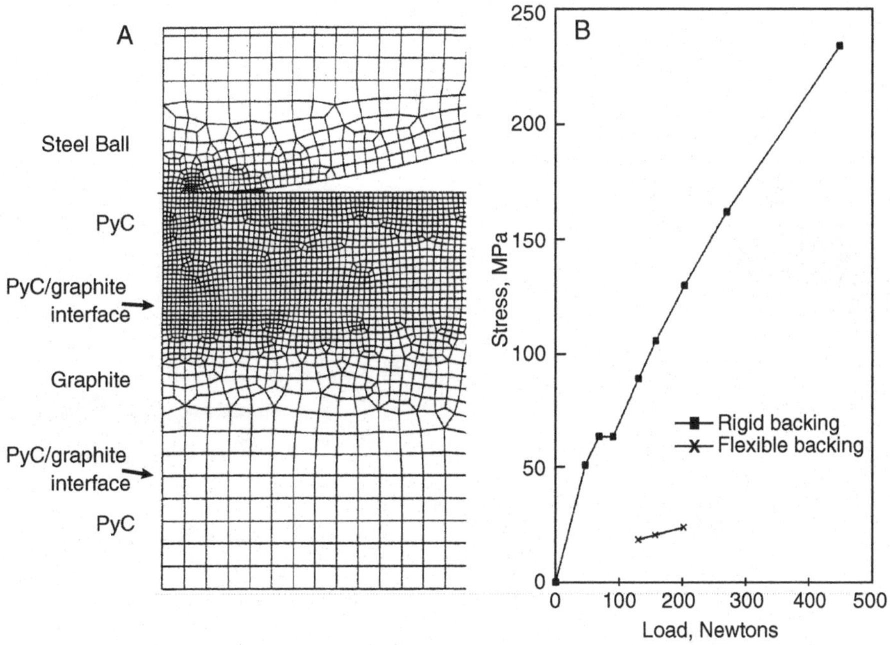

Fig. 3.1.3-8 FEA of indentation tests on PyC. (A) Part of the FE mesh showing a steel ball in contact with a PyC/graphite material. (B) Maximum principal stress on the PyC surface adjacent to ball contact radius. (Reproduced with permission from Gilpin *et al.*, 1996.)

FEA was used to evaluate the stresses resulting from a range of loads. The geometry was taken to be axisymmetric, PyC was assumed to be an elastic material and quadrilateral solid elements were used. Figure 3.1.3-8A shows part of the FE mesh. Note that the mesh is refined in the contact areas but gets progressively coarser toward the noncontact areas. Figure 3.1.3-8B shows the maximum principal stress on the PyC surface adjacent to ball contact, as a function of the indentation load. "Flexible backing" is seen to greatly reduce the maximum principal stress in this area. The FE results were correlated with data from experiments and used to develop failure criteria for contact stresses. This in turn provided criteria for designing contact regions in pyrolytic heart valves.

Numerical analysis of 3D flow in an aorta through an artificial heart valve

Three-dimensional transient flow past a Björk-Shiley valve in the aorta is simulated by the FEM combined with a time-stepping algorithm (Shim and Chang, 1997). The FE mesh is shown in Fig. 3.1.3-9A, comprising some 32,880 elements and 36,110 nodes. The results indicate that the flow is split into two major jet flows by the valve, which later merge downstream. A 3D plot of velocity vectors show large velocities in the upper and lower jet flow regions in the sinus region, large velocities only in the upper part of the merged jet, and an almost uniform paraboloid distribution near the outflow region (Fig. 3.1.3-9B). Twin spiral vortices are generated immediately downstream of the valve, in the sinus region (Fig. 3.1.3-9C) and are convected downstream, where they quickly die away by diffusion. Shear stress along the surface of the valve is shown to be a maximum in the vicinity of its leading edge (Fig. 3.1.3-9D). A study such as this provides useful information on the function of the valve *in vivo*.

Conclusion

The FEM is an approximate, numerical method for solving boundary-value problems of continuum mechanics that are posed in differential or variational form. The main advantages of the FEM over other numerical methods lie in its generalization to three dimensions and the relative ease in which arbitrary geometries, boundary conditions, and material anisotropy can be incorporated into the solution process. The same FE code can be applied to solve a wide range of nonrelated problems. Its main disadvantage has been its complexity to implement. Fortunately, the abundance and availability of commercial codes in recent years and an emphasis on a "black box" approach with minimum user interaction have reduced the level of expertise required in the implementation of FEA to most engineering problems.

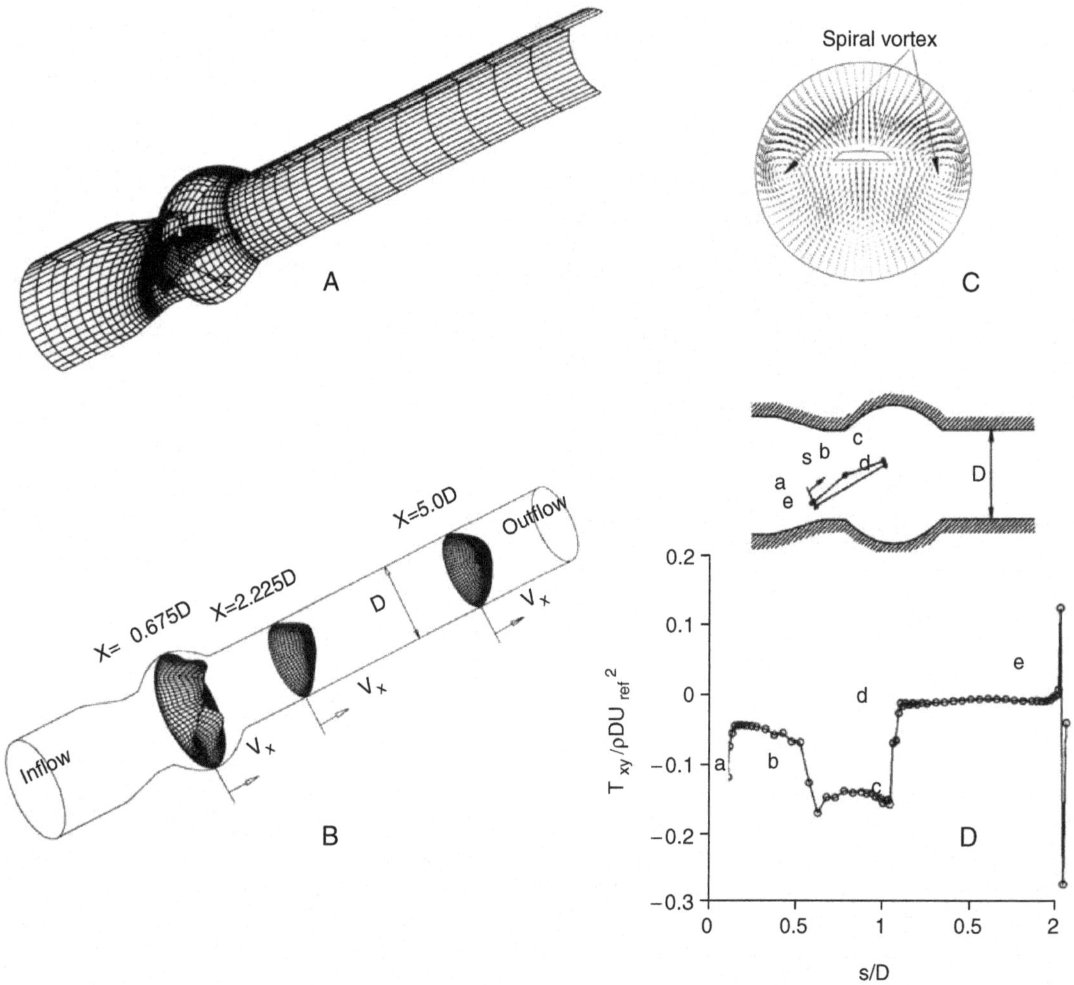

Fig. 3.1.3-9 FEA of transient 3D flow past a Bjork–Shiley valve in the aorta: (A) surface grid of aorta and fully opened Bjork–Shiley valve prosthesis, (B) carpet plot of axial velocity vectors, (C) secondary flow vector plot showing spiral vortices, and (D) shear stress along the valve surface in the symmetric mid-plane. (Reproduced with permission from Shim and Chang, 1997.)

Bibliography

Brebbia, C.A., and Aliabadi, M.H., eds. (1993). *Adaptive Finite and Boundary Element Methods*. Elsevier, New York.

Desai, C.S. (1979). *Elementary Finite Element Method*. Prentice-Hall, Upper Saddle River, NJ.

George, P.L. (1991). *Automatic Mesh Generation: Application to Finite Element Methods*. Wiley, New York.

Gilpin, C.B., Haubold, A.D., and Ely, J.L. (1996). Finite element analysis of indentation tests on pyrolytic carbon. *J. Heart Valve Dis.* **5(Suppl. 1)**: S72.

Harris, J.W., and Stöcker, H. (1998). *Handbook of Mathematics and Computational Science*. Springer, New York.

Middleton, J., Jones, M.L., and Pande, G.N., eds. (1996). *Computer Methods in Biomechanics and Biomedical Engineering*. Gordon and Breach, Amsterdam.

Nadzadi, M.E., Pedersen, D.R., Yack, H.J., Callaghan, J.J., and Brown, T.D. (2003). Kinematics, kinetics, and finite elements analysis of commonplace maneuvers at risk for total hip dislocation. *J. Biomech.* **36**: 577.

Shim, E.B., and Chang, K.S. (1997). Numerical analysis of three-dimensional Björk-Shiley valvular flow in an aorta. *J. Biomech. Eng.* **119**: 45.

Strang, G., and Fix, G.J. (1973). *An Analysis of the Finite Element Method*. Prentice-Hall, Upper Saddle River, NJ.

Yoganandan, N., Kumaresan, S., Voo, L., and Pintar, F.A. (1997). Finite element model of the human lower cervical spine: parametric analysis of the C4-C6 unit. *J. Biomech. Eng.* **119**: 87.

Zienkiewicz, O.C., and Taylor, R.L. (1991, 1994). *The Finite Element Method*, 4th ed., 2 vols. McGraw-Hill, London.

3.1.4 Surface properties and surface characterization of materials

Buddy D. Ratner

Introduction

Consider the atoms that make up the outermost surface of a biomaterial. As we shall discuss in this section, these atoms that reside at the surface have a special organization and reactivity. They require special methods to characterize them and novel methods to tailor them, and they drive many of the biological reactions that occur in response to the biomaterial (protein adsorption, cell adhesion, cell growth, blood compatibility, etc.). The importance of surfaces for biomaterials science has been appreciated since the 1960s. Almost every biomaterials meeting will have sessions addressing surfaces and interfaces. Here we focus on the special properties of surfaces, definitions of terms, methods to characterize surfaces, and some implications of surfaces for bioreaction to biomaterials.

In developing biomedical implant devices and materials, we are concerned with function, durability, and biocompatibility. In order to function, the implant must have appropriate properties such as mechanical strength, permeability, or elasticity, just to name a few. Well-developed methods typically exist to measure these bulk properties — often these are the classic methodologies of engineers and materials scientists. Durability, particularly in a biological environment, is less well understood. Still, the tests we need to evaluate durability have been developed over the past 20 years. Biocompatibility represents a frontier of knowledge in this field, and its study is often assigned to the biochemist, biologist, and physician. However, an important question in biocompatibility is how the device or material "transduces" its structural makeup to direct or influence the response of proteins, cells, and the organism to it. For devices and materials that do not leach undesirable substances in sufficient quantities to influence cells and tissues (i.e., that have passed routine toxicological evaluation), this transduction occurs through the surface structure – the body "reads" the surface structure and responds. For this reason we must understand the surface structure of biomaterials.

General surface considerations and definitions

This is the appropriate point to highlight general ideas about surfaces, especially solid surfaces. First, the surface region of a material is known to be of unique reactivity (Fig. 3.1.4-1A). Catalysis (for example, as used in petrochemical processing) and microelectronics both capitalize on special surface reactivity—thus, it would be surprising if biology did not also use surfaces to do its work. This

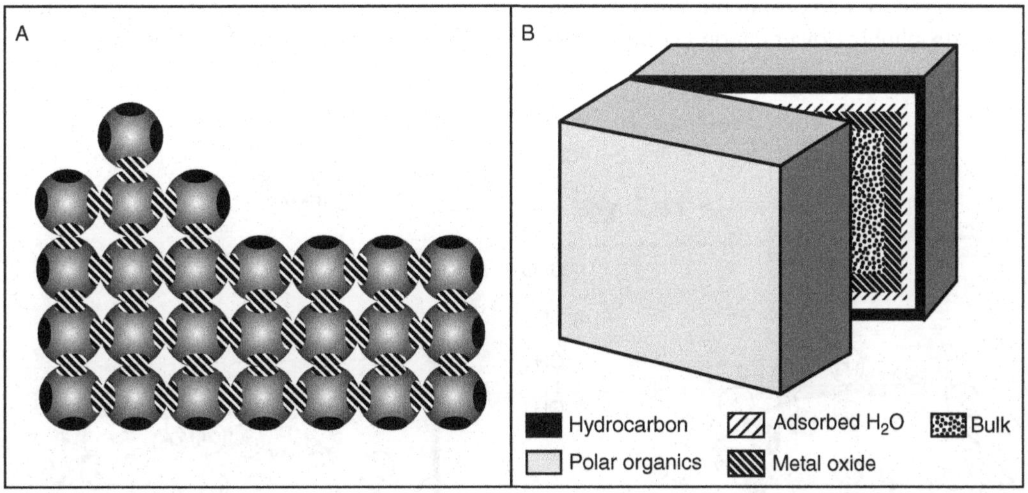

Fig. 3.1.4-1 (A) A two-dimensional representation of a crystal lattice illustrating bonding orbitals (black or crosshatched ovals). For atoms in the center (bulk) of the crystal (crosshatched ovals), all binding sites are associated. At planar exterior surfaces, one of the bonding sites is unfulfilled (black oval). At corners, two bonding sites are unfulfilled. The single atom on top of the crystal (an adatom) has three unfulfilled valencies. Energy is minimized where more of these unfulfilled valencies can interact. (B) In a "real world" material (e.g., a block of metal from an orthopedic device), if we cleave the block (under ultrahigh vacuum to prevent recontamination) we should find hydrocarbon on the outermost layer (perhaps 3 nm, surface energy ~22 ergs/cm^2), polar organic molecules (>1 nm, surface energy ~45 ergs/cm^2), adsorbed water (<1 nm, surface energy ~72 ergs/cm^2), metal oxide (approximately 5 nm, surface energy ~200 ergs/cm^2), and finally, the uniform bulk interior (surface energy ~1000 ergs/cm^2). The interface between air and material has the lowest interfacial energy (~22 ergs/cm^2). The layers are not drawn to scale.

reactivity also leads to surface oxidation and other surface chemical reactions. Second, the surface of a material is inevitably different from the bulk. The traditional techniques used to analyze the bulk structure of materials are not suitable for surface determination because they typically do not have the sensitivity to observe the small amount of material comprising the unique surface chemistry/structure. Third, there is not much total mass of material at a surface. An example may help us to appreciate this—on a 1 cm^3 cube of titanium, the 100 Å oxide surrounding the cube is in the same proportion as a 5-m wide beach on each coast of the United States is to the roughly 5,000,000 m distance from coast to coast. Fourth, surfaces readily contaminate with components from the vapor phase (some common examples are hydrocarbons, silicones, thiols, iodine). Under ultrahigh vacuum conditions (pressures $<10^{-7}$ Pa) we can retard this contamination. However, in view of the atmospheric pressure conditions under which all biomedical devices are used, we must learn to live with some contamination. The key questions here are whether we can make devices with controlled and acceptable levels of contamination and also avoid undesirable contaminants. This is critical so that a laboratory experiment on a biomaterial generates the same results when repeated after 1 day, 1 week, or 1 year, and so that the biomedical device is dependable and has a reasonable shelf life. Finally, the surface structure of a material is often mobile. A modern view of what might be seen at the surface of a real-world material is illustrated in Fig. 3.1.4-1B.

The movement of atoms and molecules near the surface in response to the outside environment is often highly significant. In response to a hydrophobic environment (e.g., air), more hydrophobic (lower energy) components may migrate to the surface of a material—a process that reduces interfacial energy (Fig. 3.1.4-1B). Responding to an aqueous environment, the surface may reverse its structure and point polar (hydrophilic) groups outward to interact with the polar water molecules. Again, energy minimization drives this process. An example of this is schematically illustrated in Fig. 3.1.4-2. For metal alloys, one metal tends to dominate the surface, for example, silver in a silver–gold alloy or chromium in stainless steel.

The nature of surfaces is complex and the subject of much independent investigation. The reader is referred to one of many excellent monographs on this important subject for a complete and rigorous introduction (see Somorjai, 1981, 1994; Adamson and Gast, 1997; Andrade, 1985). For overviews of the relationship between surface science, biology and biomaterials, see Castner and Ratner (2002), Tirrell *et al.* (2000), Ratner (1988).

When we say "surface," a question that immediately comes to mind is, "how deep into the material does it extend?" Although formal definitions are available, for all practical purposes, the surface is the zone where the structure and composition, influenced by the interface, differs from the average (bulk) composition and structure. This value often scales with the size of the molecules making up the surface. For an "atomic" material, for example gold, after penetration of about five atomic layers (0.5–1 nm), the composition becomes uniform from layer to layer (i.e., you are seeing the bulk structure). At the outermost atomic layer, the organization of the gold atoms at the surface (and their reactivity) can be substantially different from the organization in the averaged bulk. The gold, in air, will always have a contaminant overlayer, largely hydrocarbon, that may be roughly 2 nm thick. This

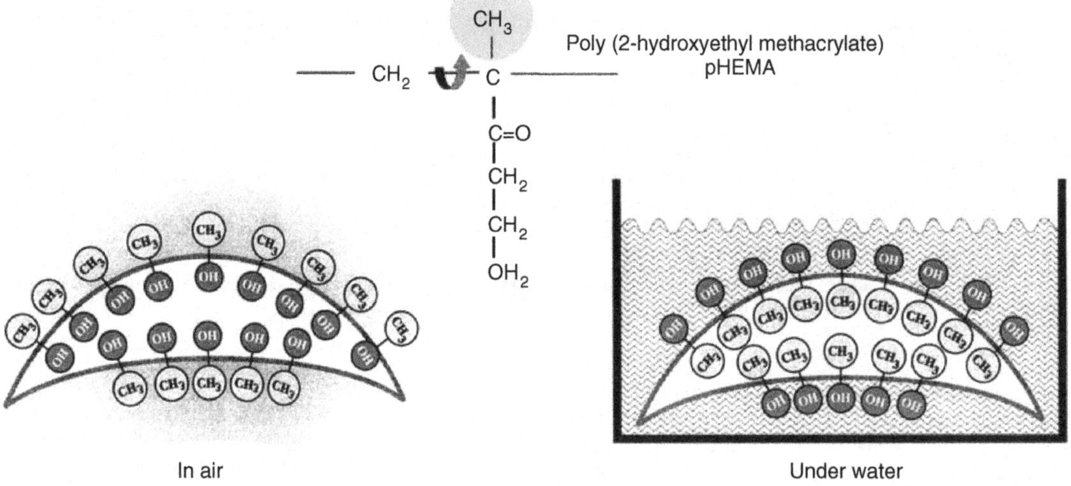

Fig. 3.1.4-2 Many materials can undergo a reversal of surface structure when transferred from air into a water environment. In this schematic illustration, a hydroxylated polymer (for example, a pHEMA contact lens) exhibits a surface rich in methyl groups (from the polymer chain backbone) in air, and a surface rich in hydroxyl groups under water. This has been observed experimentally (Ratner *et al.*, 1978, J. *Appl. Polym. Sci.* 22: 643; Chen *et al.*, 1999, J. *Am. Chem. Soc.* 121(2): 446).

is also a difference in composition between bulk and surface, but it is not the atomic/molecular rearrangements we are discussing here. For a polymer, the unique surface zone may extend from 10 nm to 100 nm (depending on the polymeric system and the chain molecular weight). Figure 3.1.4-1B addresses some of these issues about surface definitions. Two more definitions must be considered. An interface is the transition between two phases, in principle an infinitely thin separation plane. An interphase is the unique compositional zone between two phases. For the example, for gold, we might say that the interphase between gold and air is 3 nm thick (the structurally rearranged gold atoms + the contaminant layer).

Parameters to be measured

Many parameters describe a surface, as shown in Fig. 3.1.4-3. The more of these parameters we measure, the better we can piece together a complete description of the surface. A complete characterization requires the use of many techniques to compile all the information needed. Unfortunately, we cannot yet specify which parameters are most important for understanding biological responses to surfaces. Studies have been published on the importance of roughness, wettability, surface mobility, chemical composition, electrical charge, crystallinity, and heterogeneity to biological reaction. Since we cannot be certain which surface factors are predominant in each situation, the controlling variable or variables must be independently ascertained.

Surface analysis techniques

General principles

A number of general ideas can be applied to all surface analysis. They can be divided into the categories of sample preparation and analysis described in the following paragraphs.

Sample preparation

A key consideration for sample preparation is that the sample should resemble, as closely as possible, the material or device being subjected to biological testing or implantation. Needless to say, fingerprints on the surface of the sample will cover up things that might be of interest. If the sample is placed in a package for shipping or storage prior to surface analysis, it is critical to know whether the packaging material can induce surface contamination. Plain paper in contact with most specimens will transfer material (often metal ions) to the surface of the material. Many plastics are processed with silicone oils or other additives that can be transferred to the specimen. The packaging material used should be examined by surface analysis methods to ascertain its purity. Samples can be surface analyzed prior to and after storage or shipping in containers to ensure that the surface composition measured is not due to the container. As a general rule, the polyethylene press-close bags used in electron microscopy and cell culture plasticware are clean storage containers. However, abrasive contact must be avoided and each brand must be evaluated so that a meticulous specimen preparation is not ruined by contamination. Many brands of aluminum foil are useful for packing specimens, but some are treated with a surface layer of stearic acid that can surface contaminate wrapped biomaterials, implants or medical devices. Aluminum foil should be checked for cleanliness by surface analysis methods before it is used to wrap important specimens.

Surface analysis general comments

Two general principles guide sample analysis. First, all methods used to analyze surfaces also have the potential to alter the surface. It is essential that the analyst be aware of the damage potential of the method used. Second, because of the potential for artifacts and the need for many pieces of information to construct a complete picture of the surface (Fig. 3.1.4-3), more than one method should be used whenever possible. The data derived from two or more methods should always be corroborative. When data are contradictory, be suspicious and question why. A third or fourth method may then be necessary to allow confident conclusions to be drawn about the nature of a surface.

These general principles are applicable to all materials. There are properties (only a few of which will be presented here) that are specific to specific classes of materials. Compared with metals, ceramics, glasses, and carbons, organic and polymeric materials are more easily damaged by surface analysis methods. Polymeric systems also exhibit greater surface molecular mobility than inorganic systems. The surfaces of inorganic materials are contaminated more rapidly than polymeric materials because of their higher surface energy. Electrically conductive metals and carbons will often be easier to characterize than insulators using the electron, X-ray, and ion interaction methods. Insulators accumulate a surface electrical charge that requires special methods (e.g., a low-energy electron beam) to neutralize. To learn about other concerns in surface analysis that are specific to specific classes of materials, published papers become a valuable resource for understanding the pitfalls that can lead to an artifact or inaccurate results.

Table 3.1.4-1 summarizes the characteristics of many common surface analysis methods, including their depth of analysis and their spatial resolution (spot size analyzed). A few of the more frequently used techniques are described in the next section. However, space limitations prevent a developed discussion of these methods. The reader is referred to many comprehensive books on the general subject of surface analysis (Andrade, 1985; Briggs and Seah, 1983; Feldman and Mayer, 1986; Vickerman, 1997). References on specific surface analysis methods will be presented in sections on each of the key methods.

Contact angle methods

A drop of liquid sitting on a solid surface represents a powerful, but simple, method to probe surface properties. Experience tells us that a drop of water on highly polished automobile body surfaces will stand up (bead up),

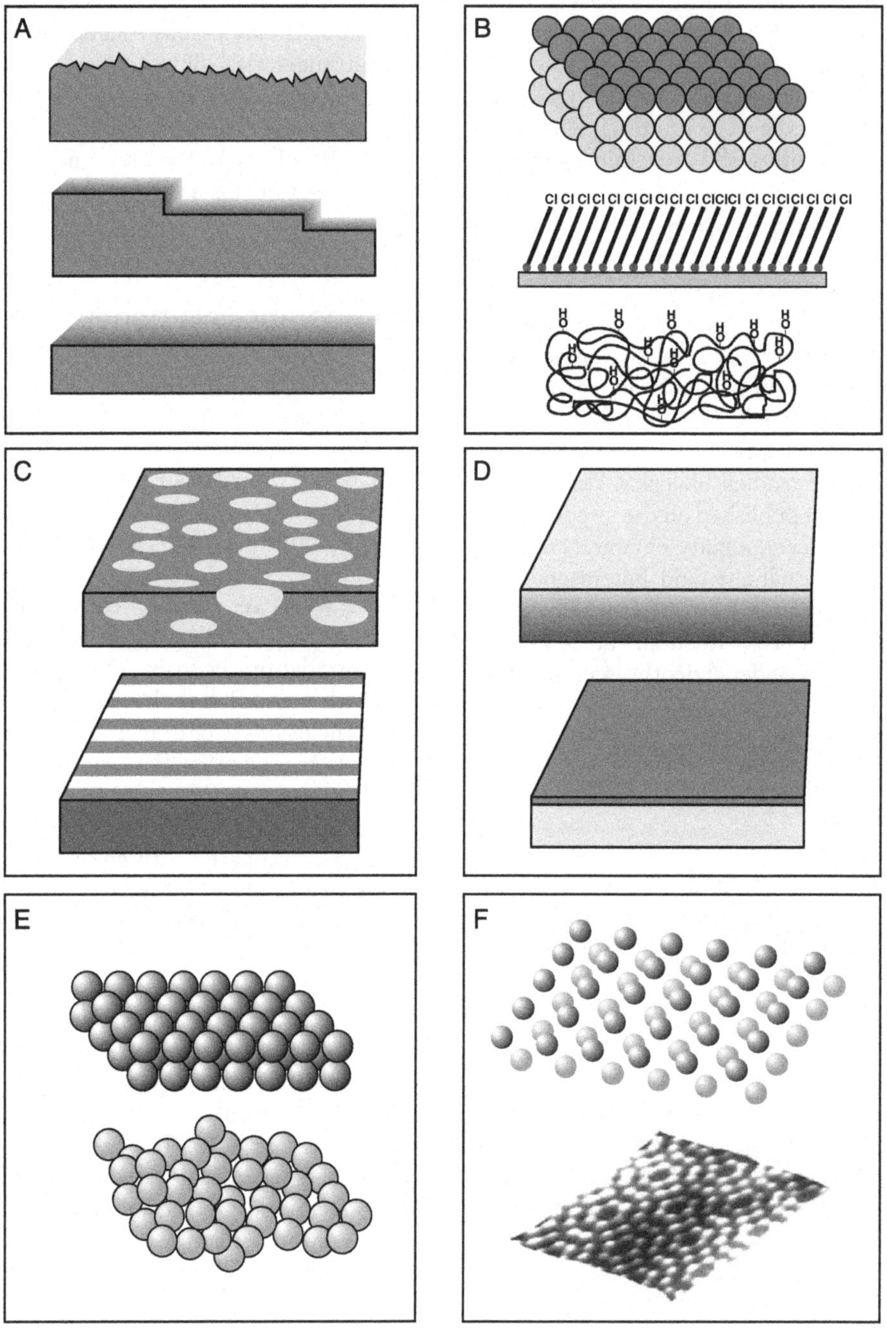

Fig. 3.1.4-3 What might be measured to define surface structure? (A) Surfaces can be rough, stepped or smooth. (B) Surfaces can be composed of different chemistries (atomic, supramolecular, macromolecular). (C) Surfaces may be structurally or compositionally inhomogeneous in the plane of the surface such as phase-separated domains or microcontact printed lanes. (D) Surfaces may be inhomogeneous with depth into the specimen or simply overlayered with a thin film. (E) Surfaces may be highly crystalline or disordered. (F) Crystalline surfaces are found with many organizations such as a silicon (100) unreconstructed surface or a silicon (111)(7 × 7) reconstructed surface.

Table 3.1.4-1 Common methods to characterize biomaterial surfaces

Method	Principle	Depth analyzed	Spatial resolution	Analytical sensitivity	Cost[c]
Contact angles	Liquid wetting of surfaces is used to estimate the energy of surfaces	3–20 Å	1 mm	Low or high depending on the chemistry	$
ESCA (XPS)	X-rays induce the emission of electrons of characteristic energy	10–250 Å	10–150 μm	0.1 at%	$$$
Auger electron spectroscopy[a]	A focused electron beam stimulates the emission of Auger electrons	50–100 Å	100 Å	0.1 atom%	$$$
SIMS	Ion bombardment sputters secondary ions from the surface	10 Å–1 μm[b]	100 Å	Very high	$$$
FTIR-ATR	IR radiation is adsorbed and excites molecular vibrations	1–5 μm	10 μm	1 mol%	$$
STM	Measurement of the quantum tunneling current between a metal tip and a conductive surface	5 Å	1 Å	Single atoms	$$
SEM	Secondary electron emission induced by a focused electron beam is spatially imaged	5 Å	40 Å, typically	High, but not quantitative	$$

[a] Auger electron spectroscopy is damaging to organic materials and is best used for inorganics.
[b] Static SIMS ≈ 10 Å, dynamic SIMS to 1 μm
[c] $, up to $5000; $$, $5000–$100,000; $$$, >$100,000.

whereas if that car has not been polished in a long time, the liquid will flow evenly over the surface. This observation, with some understanding of the method, tells us that the highly polished car probably has silicones or hydrocarbons at its surface, while the unpolished car surface consists of oxidized material. This type of observation, backed up with a quantitative measurement of the drop angle with the surface, has been used in biomaterials science to predict the performance of vascular grafts and the adhesion of cells to surfaces.

The phenomenon of the contact angle can be explained as a balance between the force with which the molecules of the liquid (in the drop) are being attracted to each other (a cohesive force) and the attraction of the liquid molecules for the molecules that make up the surface (an adhesive force). An equilibrium is established between these forces, the energy minimum. The force balance between the liquid–vapor surface tension (γ_{lv}) of a liquid drop and the interfacial tension between a solid and the drop (γ_{sl}), manifested through the contact angle (θ) of the drop with the surface, can be used to quantitatively characterize the energy of the surface (γ_{sv}). The basic relationship describing this force balance is:

$$\gamma_{sv} = \gamma_{sl} + \gamma_{lv}\cos\theta$$

The energy of the surface, which is directly related to its wettability, is a useful parameter that has often correlated strongly with biological interaction. Unfortunately, γ_{sv} cannot be directly obtained since this equation contains two unknowns, γ_{sl} and γ_{sv}. Therefore, the γ_{sv} is usually approximated by the Zisman method for obtaining the critical surface tension (Fig. 3.1.4-4), or calculated by solving simultaneous equations with data from liquids of different surface tensions. Some critical surface tensions for common materials are listed in Table 3.1.4-2.

Experimentally, there are a number of ways to measure the contact angle, and some of these are illustrated in Fig. 3.1.4-5. Contact angle methods are inexpensive, and, with some practice, easy to perform. They provide a "first line" characterization of materials and can be performed in any laboratory. Contact angle measurements provide

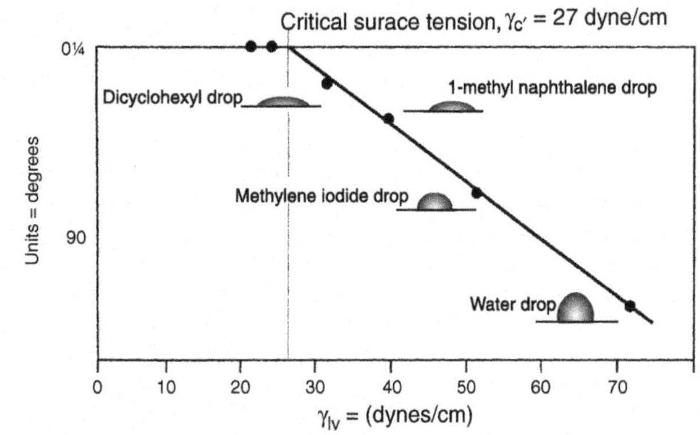

Fig. 3.1.4-4 The Zisman method permits a critical surface tension value, an approximation to the solid surface tension, to be measured. Drops of liquids of different surface tensions are placed on the solid, and the contact angles of the drops are measured. The plot of liquid surface tension versus angle is extrapolated to zero contact angle to give the critical surface tension value.

unique insight into how the surface will interact with the external world. However, in performing such measurements, a number of concerns must be addressed to obtain meaningful data (Table 3.1.4-3). Review articles are available on contact angle measurement for surface characterization (Andrade, 1985; Good, 1993; Zisman, 1964; McIntire, *et al.*, 1985).

Table 3.1.4-2 Critical surface tension values for common materials calculated from contact angle measurements

Material	Critical surface tension (dyn/cm)
Polytetrafluoroethylene	19
Poly(dimethyl siloxane)	24
Poly(vinylidine fluoride)	25
Poly(vinyl fluoride)	28
Polyethylene	31
Polystyrene	33
Poly(2-hydroxyethyl methacrylate)	37
Poly(vinyl alcohol)	37
Poly(methyl methacrylate)	39
Poly(vinyl chloride)	39
Polycaproamide (nylon 6)	42
Poly(ethylene oxide)-diol	43
Poly(ethylene terephthalate)	43
Polyacrylonitrile	50

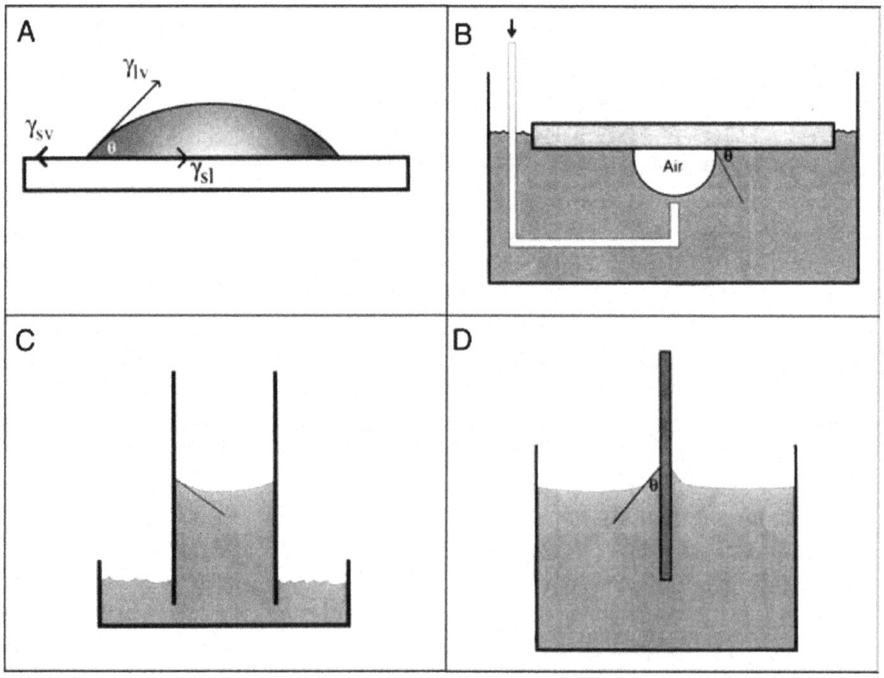

Fig. 3.1.4-5 Four possibilities for contact angle measurement: (A) sessile drop, (B) captive air bubble method, (C) capillary rise method, (D) Wilhelmy plate method.

Table 3.1.4-3 Concerns in contact angle measurement

The measurement is operator dependent.
Surface roughness influences the results.
Surface heterogeneity influences the results.
The liquids used are easily contaminated (typically reducing their V_{1v}).
The liquids used can reorient the surface structure.
The liquids used can absorb into the surface, leading to swelling.
The liquids used can dissolve the surface.
Few sample geometries can be used.
Information on surface structure must be inferred from the data obtained.

Electron spectroscopy for chemical analysis

Electron spectroscopy for chemical analysis (ESCA) provides unique information about a surface that cannot be obtained by other means (Andrade, 1985; Ratner, 1988; Dilks, 1981; Ratner and McElroy, 1986; Ratner and Castner, 1997; Watts and Wolstenholme, 2003). In contrast to the contact angle technique, ESCA requires complex, expensive apparatus (Fig. 3.1.4-6A) and demands considerable training to perform the measurements. However, since ESCA is available from commercial laboratories, university analytical facilities, national centers (for example, NESAC/BIO at the University of Washington), and specialized research laboratories, most biomaterials scientists can get access to it to have their samples analyzed. The data can be interpreted in a simple but still useful fashion, or more rigorously. ESCA has contributed significantly to the development of biomaterials and medical devices, and to understanding the fundamentals of biointeraction.

The ESCA method (also called X-ray photoelectron spectroscopy, XPS) is based upon the photoelectric effect, properly described by Einstein in 1905. X-rays are focused upon a specimen. The interaction of the X-rays with the atoms in the specimen causes the emission of a core level (inner shell) electron. The energy of this electron is measured and its value provides information about the nature and environment of the atom from which it came. The basic energy balance describing this process is given by the simple relationship:

$$\mathrm{BE} = h\nu - \mathrm{KE}$$

where BE is the energy binding the electron to an atom (the value desired), KE is the kinetic energy of the emitted electron (the value measured in the ESCA spectrometer), and $h\nu$ is the energy of the X-rays, a known value. A simple schematic diagram illustrating an ESCA instrument is shown in Fig. 3.1.4-6B. Table 3.1.4-4 lists some of the types of information about the nature of a surface that can be obtained by using ESCA. The origin of the surface sensitivity of ESCA is described in Fig. 3.1.4-7.

ESCA has many advantages, and a few disadvantages, for studying biomaterials. The advantages include the high information content, the surface localization of the measurement, the speed of analysis, the low damage potential, and the ability to analyze most samples with no specimen preparation. The last advantage is particularly important because it means that many biomedical devices (or parts of devices) can be inserted, as fabricated and sterilized,

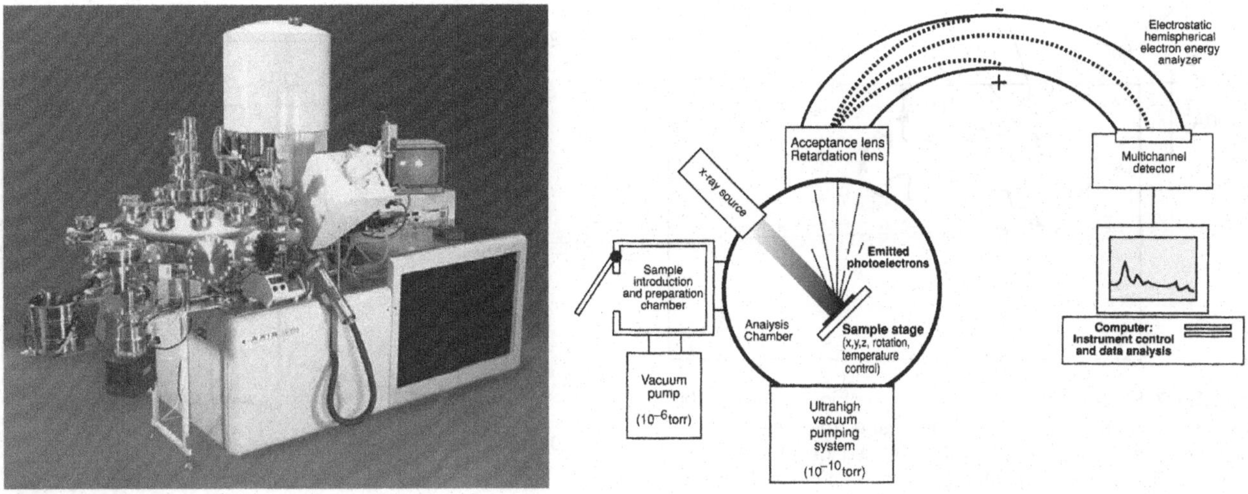

A B

Fig. 3.1.4-6 (A) Photograph of a contemporary ESCA instrument (photo by Kratos Corp.). (B) Schematic diagram of a monochromatized ESCA instrument.

Table 3.1.4-4 Information derived from an ESCA experiment

In the outermost 100 Å of a surface, ESCA can provide:
Identification of all elements (except H and He) present at concentrations >0.1 at %
Semiquantitative determination of the approximate elemental surface composition (±10%)
Information about the molecular environment (oxidation state, bonding atoms, etc.)
Information about aromatic or unsaturated structures from shake-up $\pi^* \leftarrow \pi$ transitions
Identification of organic groups using derivatization reactions
Nondestructive elemental depth profiles 100 Å into the sample and surface heterogeneity assessment using angular-dependent ESCA studies and photoelectrons with differing escape depths
Destructive elemental depth profiles several thousand angstroms into the sample using argon etching (for inorganics)
Lateral variations in surface composition (spatial resolution 8–150 μm, depending upon the instrument)
"Fingerprinting" of materials using valence band spectra and identification of bonding orbitals
Studies on hydrated (frozen) surfaces

directly in the analysis chamber for study. The disadvantages include the need for vacuum compatibility (i.e., no outgassing of volatile components), the vacuum environment (particularly for hydrated specimens), the possibility of sample damage by X-rays if long analysis times are used, the need for experienced operators, and the cost associated with this complex instrumentation. The vacuum limitations can be sidestepped by using an ESCA system with a cryogenic sample stage. At liquid nitrogen temperatures, samples with volatile components, or even wet, hydrated samples, can be analyzed.

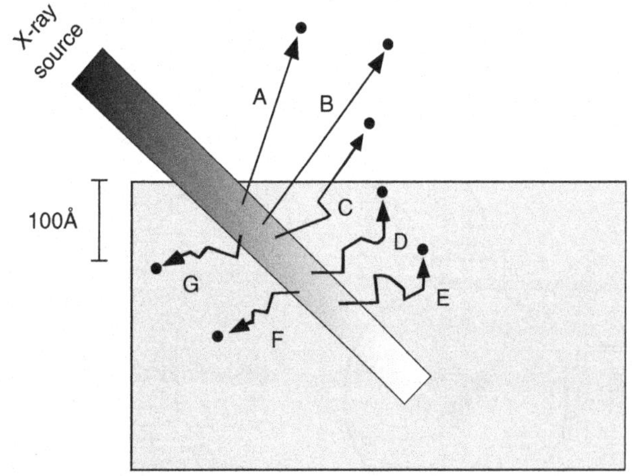

Fig. 3.1.4-7 ESCA is a surface-sensitive method. Although the X-ray beam can penetrate deeply into a specimen, electrons emitted deep in the specimen (D, E, F, G) will lose their energy in inelastic collisions and never emerge from the surface. Only those electrons emitted near the surface that lose no energy (A, B) will contribute to the ESCA signal used analytically. Electrons that lose some energy but still have sufficient energy to emerge from the surface (C) contribute to the background signal.

The use of ESCA is best illustrated with a brief example. A poly(methyl methacrylate) (PMMA) ophthalmologic device is to be examined. Taking care not to touch or damage the surface of interest, the device is inserted into the ESCA instrument introduction chamber. The introduction chamber is then pumped down to 10^{-6}torr (1.33×10^{-4}Pa) pressure. A gate valve between the introduction chamber and the analytical chamber is opened and the specimen is moved into the analysis chamber. In the analysis chamber, at 10^{-9}torr (1.33×10^{-7}Pa) pressure, the specimen is positioned (on contemporary instruments, using a microscope or TV camera) and the X-ray source is turned on. The ranges of electron energies to be observed are controlled (by computer) with the retardation lens on the spectrometer. First, a wide scan is made in which the energies of all emitted electrons over, typically, a 1000 eV range are detected (Fig. 3.1.4-8). Then, narrow scans are made in which each of the elements detected in the wide scan is examined in higher resolution (Fig. 3.1.4-9).

From the wide scan, we learn that the specimen contains carbon, oxygen, nitrogen, and sulfur. The presence of sulfur and nitrogen is unexpected for PMMA. We can calculate elemental ratios from the wide scan. The sample surface contains 58.2% carbon, 27.7% oxygen, 9.5% nitrogen, and 4.5% sulfur. The narrow scan for the carbon region (C1s spectrum) suggests four species: hydrocarbon, carbons singly bonded to oxygen (the predominant species), carbons in amide-like molecular environments, and carbons in acid or ester environments. This is different from the spectrum expected for pure PMMA. An examination of

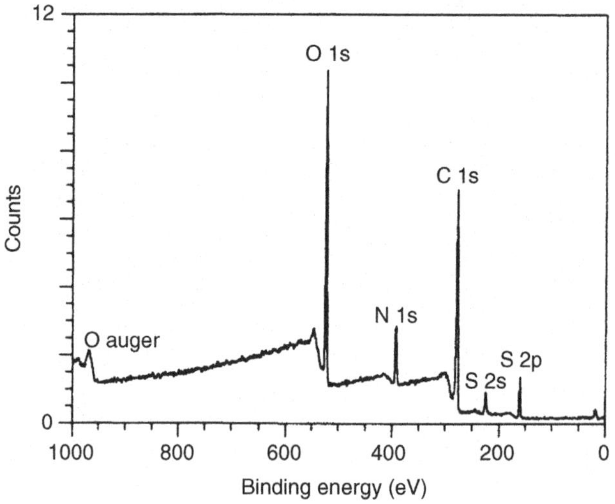

Fig. 3.1.4-8 ESCA wide scan of a surface-modified PMMA ophthalmologic device.

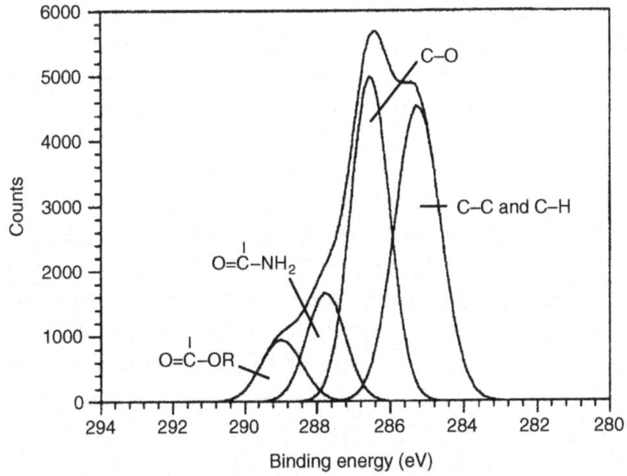

Fig. 3.1.4-9 The carbon 1s narrow scan ESCA spectrum of a surface-modified PMMA ophthalmologic device. Narrow scan spectra can be generated for each element seen in low-energy resolution mode in Fig. 3.1.4-8.

Table 3.1.4-5 Analytical capabilities of SIMS

	Static SIMS	Dynamic SIMS
Identify hydrogen and deuterium	✓	✓
Identify other elements (often must be inferred from the data)	✓	✓
Suggest molecular structures (inferred from the data)	✓	–
Observe extremely high mass fragments (proteins, polymers)	✓	–
Detection of extremely low concentrations	✓	✓
Depth profile to 1 μm into the sample	*	✓
Observe the outermost 1–2 atomic layers	✓	✓
High spatial resolution (features as small as ~400 Å)	✓	✓
Semiquantitative analysis (for limited sets of specimens)	✓	✓
Useful for polymers	✓	–
Useful for inorganics (metals, ceramics, etc.)	✓	✓
Useful for powders, films, fibers, etc.	✓	✓

* Cluster ion sources may allow depth profiling with static SIMS-like information content

the peak position in the narrow scan of the sulfur region (S2p spectrum) suggests sulfonate-type groups. The shape of the C1s spectrum, the position of the sulfur peak, and the presence of nitrogen all suggest that heparin was immobilized to the surface of the PMMA device. Since the stoichiometry of the lens surface does not match that for pure heparin, this suggests that we are seeing either some of the PMMA substrate through a <100 Å layer of heparin, or we are seeing some of the bonding links used to immobilize the heparin to the lens surface. Further ESCA analysis will permit the extraction of more details about this surface-modified device, including an estimate of surface modification thickness, further confirmation that the coating is indeed heparin, and additional information about the nature of the immobilization chemistry.

Secondary ion mass spectrometry

Secondary ion mass spectrometry (SIMS) is an important addition to the armamentarium of tools that the surface analyst can bring to bear on a biomedical problem. SIMS produces a mass spectrum of the outermost 10 Å of a surface. Like ESCA, it requires complex instrumentation and an ultrahigh vacuum chamber for the analysis. However, it provides unique information that is complementary to ESCA and greatly aids in understanding surface composition. Some of the analytical capabilities of SIMS are summarized in Table 3.1.4-5. Review articles on SIMS are available (Ratner, 1983; Scheutzle *et al.*, 1984; Briggs, 1986; Davies and Lynn, 1990; Vickerman *et al.*, 1989; Benninghoven, 1983; Van Vaeck *et al.*, 1999; Belu *et al.*, 2003).

In SIMS analysis, a surface is bombarded with a beam of accelerated ions. The collision of these ions with the atoms and molecules in the surface zone can transfer enough energy to them so they sputter from the surface into the vacuum phase. The process is analogous to racked pool balls that are ejected from the cluster by the impact of the cue ball; the harder the cue ball hits the rack of balls, the more balls are emitted from the rack. In SIMS, the cue balls are ions (xenon, argon, cesium, and gallium ions are commonly used) that are accelerated to energies of 5000–20,000 eV. The particles ejected from the surface are positive and negative ions (secondary ions), radicals, excited states, and neutrals. Only the secondary ions are measured in SIMS. In ESCA, the energy of emitted particles (electrons) is measured. SIMS measures the mass of emitted ions (more rigorously, the ratio of mass to charge, m/z) using a time-of-flight (TOF) mass analyzer or a quadrupole mass analyzer.

There are two modes for SIMS analysis, depending on the ion flux: dynamic and static. Dynamic SIMS uses high ion doses in a given analysis time. The primary ion beam sputters so much material from the surface that the surface erodes at an appreciable rate. We can capitalize on this to do a depth profile into a specimen. The intensity of the m/z peak of a species of interest (e.g., sodium ion, $m/z = 23$)

might be followed as a function of time. If the ion beam is well controlled and the sputtering rate is constant, the sodium ion signal intensity measured at any time will be directly related to its concentration at the erosion depth of the ion beam into the specimen. A concentration depth profile (sodium concentration versus depth) can be constructed over a range from the outermost atoms to a micron or more into the specimen. However, owing to the damaging nature of the high-flux ion beam, only atomic fragments can be detected. Also, as the beam erodes deeper into the specimen, more artifacts are introduced in the data by "knock-in" and scrambling of atoms.

Static SIMS, in comparison, induces minimal surface destruction. The ion dose is adjusted so that during the period of analysis less than 10% of one monolayer of surface atoms is sputtered. Since there are typically 10^{13}–10^{15} atoms in 1 cm^2 of surface, a total ion dose of less than 10^{13} ions/cm^2 during the analysis period is appropriate. Under these conditions, extensive degradation and rearrangement of the chemistry at the surface does not take place, and large, relatively intact molecular fragments can be ejected into the vacuum for measurement. Examples of molecular fragments are shown in Fig. 3.1.4-10. This figure also introduces some of the ideas behind SIMS

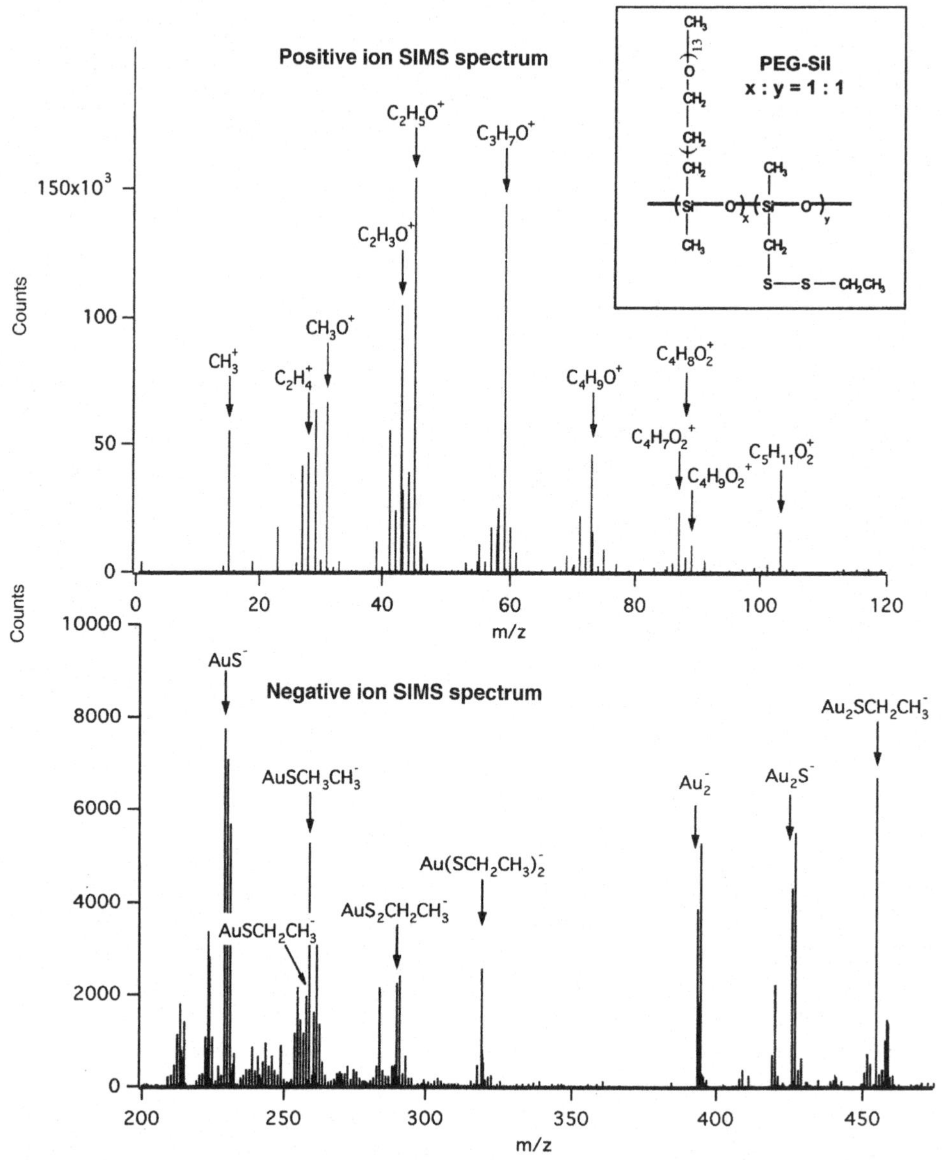

Fig. 3.1.4-10 Static positive and negative ion SIMS spectra of a poly(ethylene glycol)–poly(dimethyl siloxane) copolymer containing disulfide side groups on a gold surface. The primary peaks are identified. The low-mass region of the negative ion spectrum offers little insight into the polymer structure, but the high-mass region is rich in information. In this case, the low-mass positive spectrum is rich in information. Further details on this class of polymers can be found in *Macromolecules* 27: 3053 (1994). (Figure supplied by D. Castner.)

spectral interpretation. A more complete introduction to the concepts behind static SIMS spectral interpretation can be found in Van Vaeck *et al.*, (1999) or in standard texts on mass spectrometry.

Magnetically or electrostatically focusing the primary ion beam permits the SIMS technique to have high spatial resolution in the *x,y* plane. In fact, SIMS analysis can be performed in surface regions of 10 nm diameter or smaller. For static SIMS analysis, less than 10% of the atoms in any area are sampled. Thus, as the spot size gets smaller, the challenge to achieve high analytical sensitivity increases sharply. Still, static SIMS measurements have been performed in areas as small as 40 nm. Newly developed cluster ion sources (for example, using gold molecular clusters, Au_3, or C_{60} as the impacting primary particles) show high secondary ion yields and relatively low surface damage. These may improve spatial resolution and also permit depth profiling of organic surfaces by sputtering down into a surface while monitoring secondary ion emission as a function of time.

If the focused primary ion beam is rastered over the surface and the *x,y* position of the beam correlated with the signal emitted from a given spot, the SIMS data can be converted into an elemental image. Patterning and spatial control of chemistry is becoming increasingly important in biomaterials surface design. For example, microcontact printing allows patterned chemistry to be transferred to surfaces at the micron level using a relatively simple rubber stamp. Imaging SIMS is well suited to studying and monitoring such spatially defined chemistry. An example is presented in Fig. 3.1.4-11. Imaging SIMS is also valuable for observing defects in thin films (pinholes), analyzing the chemistry of fine particulates or assessing causes of implant failure.

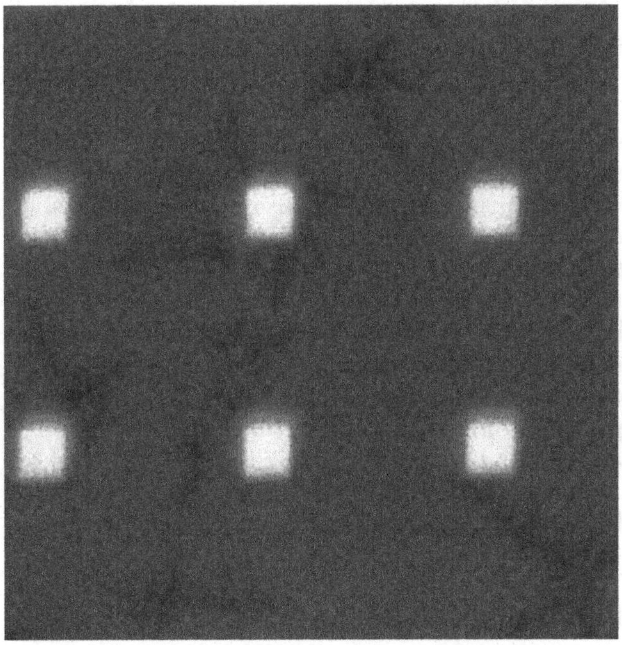

Fig. 3.1.4-11 Static SIMS image of protein islands on a poly(ethylene oxide) background. (For details, see Veiseh, M., Wickes, B. T., Castner, D. G., Zhang, M., "Guided cell patterning by surface molecular engineering." In press.)

Scanning electron microscopy

Scanning electron microscopy (SEM) images of surfaces have great resolution and depth of field, with a three-dimensional quality that offers a visual perspective familiar to most users. SEM images are widely used and much has been written about the technique. The comments here are primarily oriented toward SEM as a surface analysis tool.

SEM functions by focusing and rastering a relatively high-energy electron beam (typically, 5–100 keV) on a specimen. Low-energy secondary electrons are emitted from each spot where the focused electron beam impacts. The measured intensity of the secondary electron emission is a function of the atomic composition of the sample and the geometry of the features under observation. SEM image surfaces by spatially reconstructing on a phosphor screen [or charged coupled device (CCD) detector] the intensity of the secondary electron emission. Because of the shallow penetration depth of the low-energy secondary electrons produced by the primary electron beam, only secondary electrons generated near the surface can escape from the bulk and be detected (this is analogous to the surface sensitivity described in Fig. 3.1.4-7). Consequently, SEM is a surface analysis method.

Nonconductive materials observed in the SEM are typically coated with a thin, electrically grounded layer of metal to minimize negative charge accumulation from the electron beam. However, this metal layer is always so thick (>200 Å) that the electrons emitted from the sample beneath cannot penetrate. Therefore, in SEM analysis of nonconductors, the surface of the metal coating is, in effect, being monitored. If the metal coat is truly conformal, a good representation of the surface geometry will be conveyed. However, the specimen surface chemistry no longer influences secondary electron emission. Also, at very high magnifications, the texture of the metal coat and not the surface may be under observation.

SEM, in spite of these limitations in providing true surface information, is an important corroborative method to use in conjunction with other surface analysis methods. Surface roughness and texture can have a profound influence on data from ESCA, SIMS, and contact angle determinations. Therefore, SEM provides important information in the interpretation of data from these methods.

The development of low-voltage SEM offers a technique to truly study the surface chemistry (and geometry) of non-conductors. With the electron accelerating voltage lowered to approximately 1 keV, charge accumulation is not as critical and metallization is not required. Low-voltage SEM has been used to study platelets and phase separation in polymers. Also, the environmental SEM (ESEM) permits wet, uncoated specimens to be studied.

The primary electron beam also results in the emission of X-rays. These X-rays are used to identify elements with the technique called energy-dispersive X-ray analysis (EDXA). However, the high-energy primary electron beam penetrates deeply into a specimen (a micron or more). The X-rays produced from the interaction of these electrons with atoms deep in the bulk of the specimen can penetrate through the material and be detected. Therefore, EDXA is not a surface analysis method.

The primary use of SEM is in image topography. SEM for this application is well elaborated in the literature.

Infrared spectroscopy

Infrared spectroscopy (IRS) provides information on the vibrations of atomic and molecular species. It is a standard analytical method that can reveal information on specific chemistries and the orientation of structures. Fourier transform infrared (FTIR) spectrometry offers outstanding signal-to-noise ratio (S/N) and spectral accuracy. However, even with this high S/N, the small absorption signal associated with the minute mass of material in a surface region can challenge the sensitivity of the spectrometer. Also, the problem of separating the vastly larger bulk absorption signal from the surface signal must be addressed.

Surface FTIR methods couple the infrared radiation to the sample surface to increase the intensity of the surface signal and reduce the bulk signal (Allara, 1982; Leyden and Murthy, 1987; Urban, 1993; Dumas *et al.*, 1999). Some of these sampling modes, and their characteristics, are illustrated in Fig. 3.1.4-12.

The attenuated total reflectance (ATR) mode of sampling has been used most often in biomaterials studies. The penetration depth into the sample is 1–5 μm. Therefore, ATR is not highly surface sensitive, but observes a broad region near the surface. However, it does offer the wealth of rich structural information common to infrared spectra. With extremely high S/N FTIR instruments, ATR studies of proteins and polymers under water have been performed. In these experiments, the water signal (which is typically 99% or more of the total signal) is subtracted from the spectrum to leave only the surface material (e.g., adsorbed protein) under observation.

Another infrared method that has proven immensely valuable for observing extremely thin films on reflective surfaces is infrared reflection absorption spectroscopy (IRAS), Fig. 3.1.4-12. This method has been widely applied to self-assembled monolayers (SAMs), but is

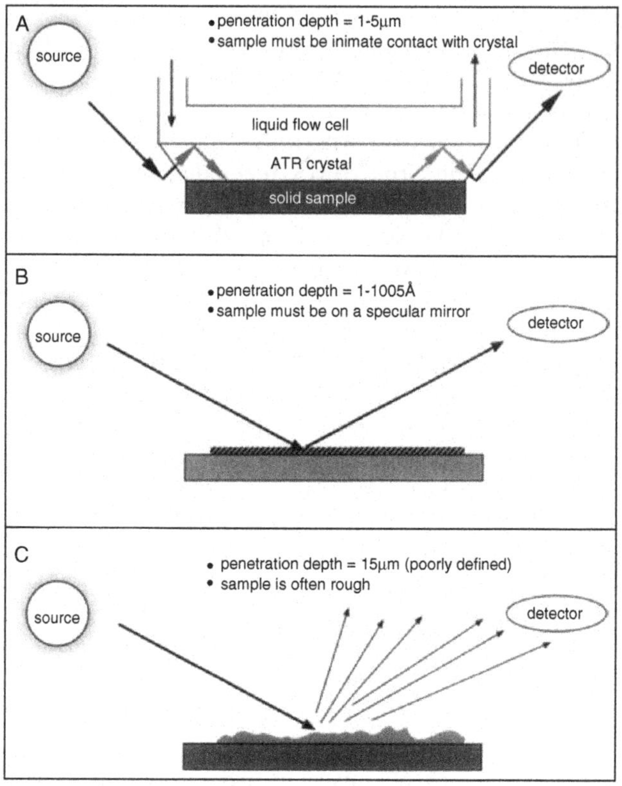

Fig. 3.1.4-12 Three surface-sensitive infrared sampling modes: (A) ATR-IR, (B) IRAS, (C) diffuse reflectance.

applicable to many surface films that are less than 10 nm in thickness. The surface upon which the thin film resides must be highly reflective and metal surfaces work best, though a silicon wafer can be used. IRAS gives information about composition, crystallinity and molecular orientation. IRS is one member of a family of methods called vibrational spectroscopies. Two other vibrational spectroscopies, sum frequency generation (SFG) and Raman, will be mentioned later in the section on newer methods.

Scanning tunneling microscopy, atomic force microscopy, and the scanning probe microscopies

STM and atomic force microscopy (AFM) have developed from novel research tools to key methods for biomaterials characterization. AFM has become more widely used than STM because oxide-free, electrically conductive surfaces are not needed with AFM. General review articles (Binnig and Rohrer, 1986; Avouris, 1990; Albrecht *et al.*, 1988) and articles oriented toward biological studies with these methods (Hansma *et al.*, 1988; Miles *et al.*, 1990; Rugar and Hansma, 1990; Jandt, 2001) are available.

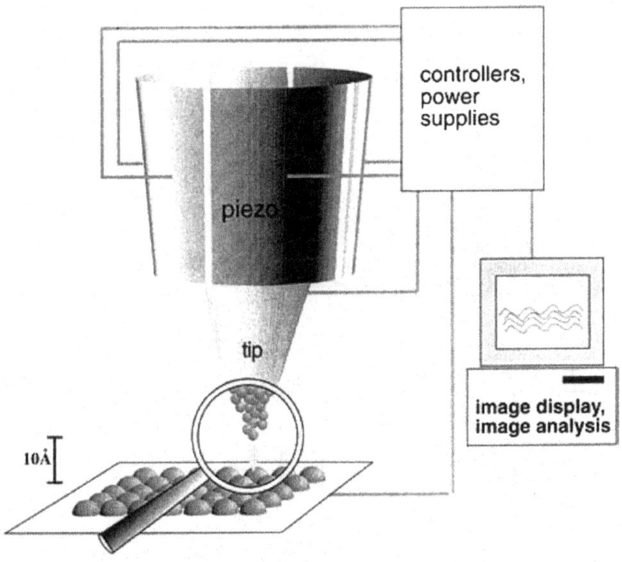

Fig. 3.1.4-13 Schematic diagram illustrating the principle of the STM—a tip terminating in a single atom permits localized quantum tunneling current from surface features (or atoms) to tip. This tunneling current can be spatially reconstructed to form an image.

The STM was invented in 1981 and led to a Nobel Prize for Binnig and Rohrer in 1986. The STM uses quantum tunneling to generate an atom-scale electron density image of a surface. A metal tip terminating in a single atom is brought within 5–10 Å of an electrically conducting surface. At these distances, the electron cloud of the atom at the "tip of the tip" will significantly overlap the electron cloud of an atom on the surface. If a potential is applied between the tip and the surface, an electron tunneling current will be established whose magnitude, J, follows the proportionality:

$$J \propto e^{(-Ak_0 S)}$$

where A is a constant, k_0 is an average inverse decay length (related to the electron affinity of the metals), and S is the separation distance in angstrom units. For most metals, a 1 Å change in the distance of the tip from the surface results in an order of magnitude change in tunneling current. Even though this current is small, it can be measured with good accuracy.

To image a surface, this quantum tunneling current is used in one of two ways. In constant current mode, a piezoelectric driver scans a tip over a surface. When the tip approaches an atom protruding above the plane of the surface, the current rapidly increases, and a feedback circuit moves the tip up to keep the current constant. Then, a plot is made of the tip height required to maintain constant current versus distance along the plane. In constant height mode, the tip is moved over the surface and the change in current with distance traveled along the plane of the surface is directly recorded. A schematic diagram of a scanning tunneling microscope is presented in Fig. 3.1.4-13. Two STM scanning modes are illustrated in Fig. 3.1.4-14.

The STM measures electrical current and therefore is well suited for conductive and semiconductive surfaces. However, biomolecules (even proteins) on conductive substrates appear amenable to imaging. It must be remembered that STM does not "see" atoms, but monitors electron density. The conductive and imaging mechanism for proteins is not well understood. Still, Fig. 3.1.4-15 suggests that valuable images of biomolecules on conductive surfaces can be obtained.

The AFM uses a similar piezo drive mechanism. However, instead of recording tunneling current, the deflection of a tip mounted on a flexible cantilever arm due to van der Waals and electrostatic repulsion and attraction between an atom at the tip and an atom on the surface is measured. Atomic-scale measurements of cantilever arm movements can be made by reflecting a laser beam off a mirror on the cantilever arm (an optical lever). A one-atom deflection of the cantilever arm can easily be magnified by monitoring the position of the laser reflection on a spatially resolved photosensitive detector. Other principles are also used to measure the deflection of the tip. These include capacitance measurements and interferometry. A diagram of a typical AFM is presented in Fig. 3.1.4-16.

Tips are important in AFM as the spatial resolution of the method is significantly associated with tip terminal diameter and shape. Tips are made from microlithographically fabricated silicon or silicon nitride. Also carbon whiskers, nanotubes, and a variety of nanospherical particles have been mounted on AFM tips to increase their sharpness or improve the ability to precisely define tip geometry. Tips are also surface-modified to alter the strength and types of interactions with surfaces (static SIMS can be used to image these surface modifications). Finally, cantilevers are sold in a range of stiffnesses so the analysis modes can be tuned to needs of the sample and the type of data being acquired. The forces associated with the interaction of an AFM tip with a surface as it approaches and is retracted are illustrated in Fig. 3.1.4-16. Since force is being measured and Hooke's law applies to the deformation of an elastic cantilever, the AFM can be used to quantify the forces between surface and tip. An exciting application of AFM is to measure the strength of interaction between two biomolecules (for example, biotin and streptavidin; see Chilkoti *et al.*, 1995).

AFM instruments are commonly applied to surface problems using one of two modes, contact mode and tapping mode. In contact mode, the tip is in contact with the surface (or at least the electron clouds of tip and surface essentially overlap). The pressures resulting from the force of the cantilever delivered through the extremely small

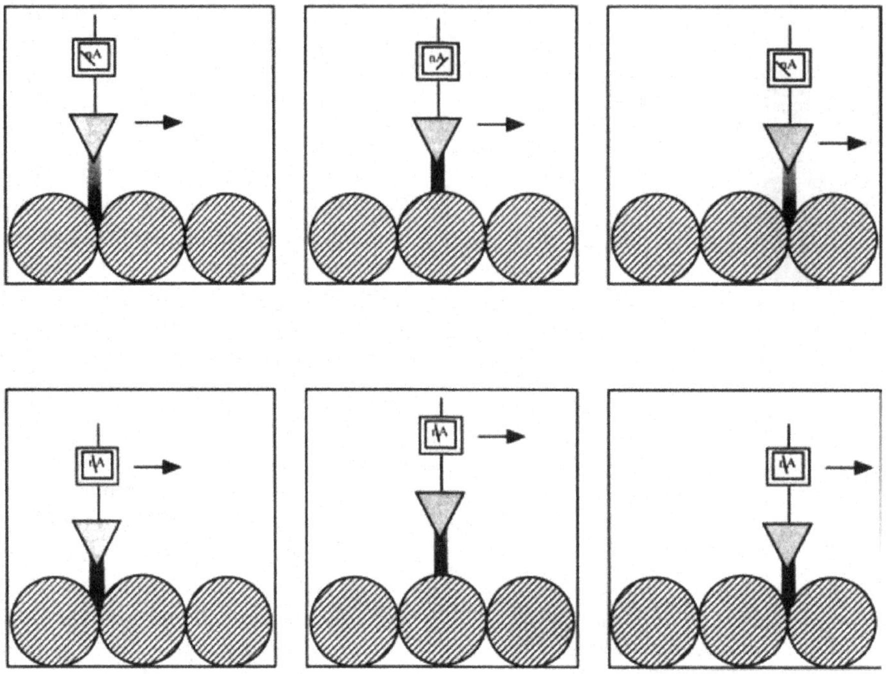

Fig. 3.1.4-14 STM can be performed in two modes. In constant height mode, the tip is scanned a constant distance from the surface (typically 5–10Å) and the change in tunneling current is recorded. In constant current mode, the tip height is adjusted so that the tunneling current is always constant, and the tip distance from the surface is recorded as a function of distance traveled in the plane of the surface.

surface area of the tip can be damaging to soft specimens (proteins, polymers, etc). However, for more rigid specimens, excellent topographical imaging can be achieved in contact mode. In tapping mode, the tip is oscillated at a frequency near the resonant frequency of the cantilever. The tip barely grazes the surface. The force interaction of tip and surface can affect the amplitude of oscillation and the oscillating frequency of the tip. In standard tapping mode, the amplitude change is translated into topographic spatial information. Many variants of tapping mode have been developed allowing imaging under different conditions and using the phase shift between the applied oscillation to the tip and the actual tip oscillation in the force field of the surface to provide information of the mechanical properties of the surface (in essence, the viscoelasticity of the surface can be appreciated).

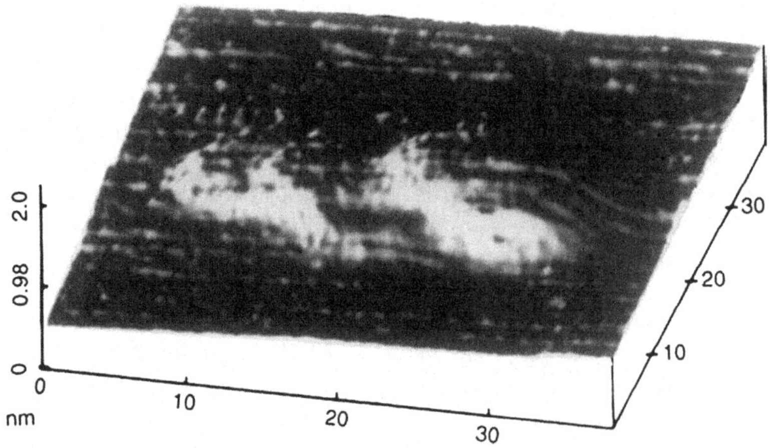

Fig. 3.1.4-15 Scanning tunneling micrograph image of a fibrinogen molecule on a gold surface, under buffer solution (image by Dr. K. Lewis).

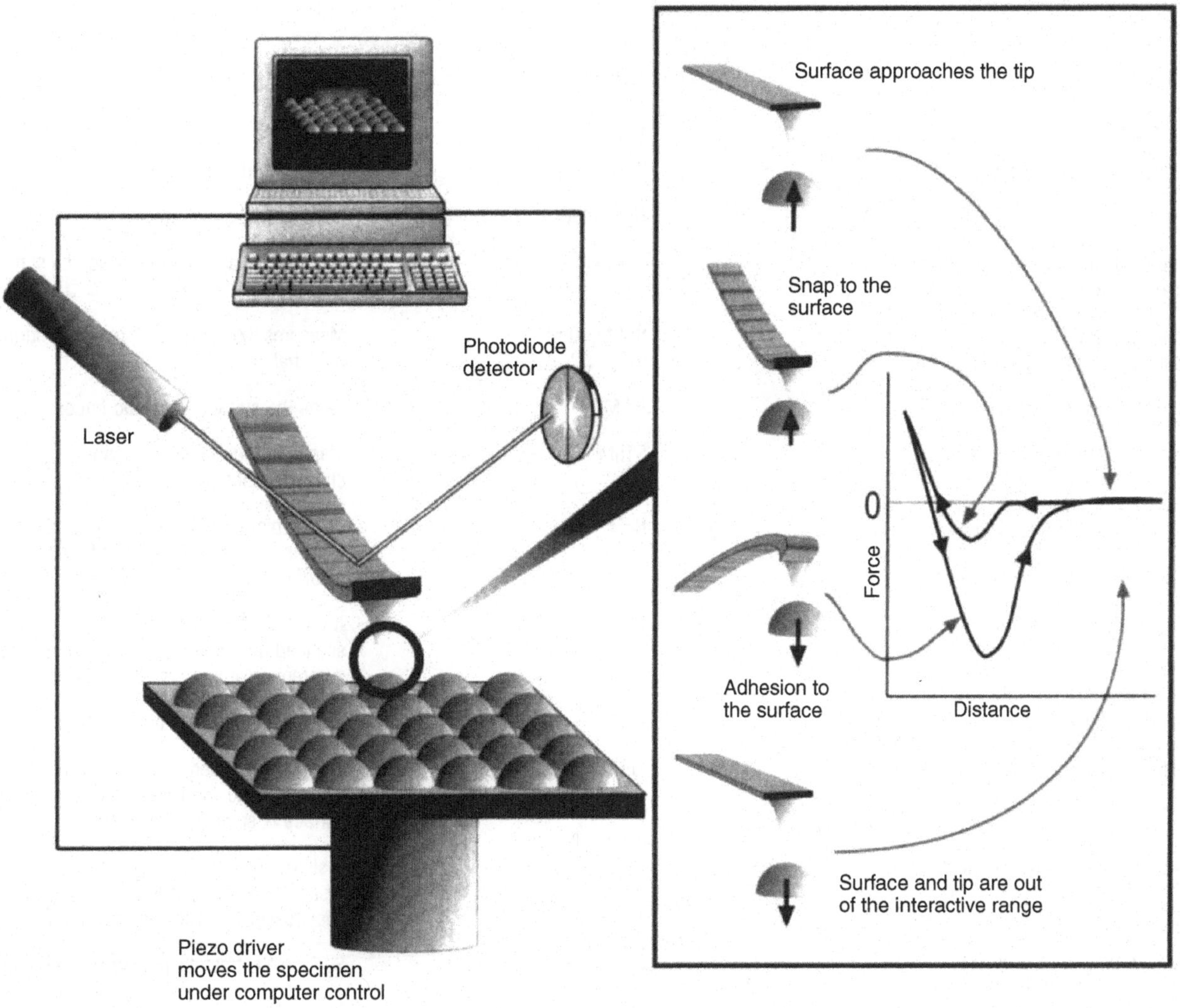

Fig. 3.1.4-16 Schematic diagram illustrating the principle of the atomic force microscope.

The potential of the AFM to explore surface problems has been greatly expanded by ingenious variants of the technique. In fact, the term "atomic force microscopy" has been generalized to "scanning probe microscopy (SPM)." Table 3.1.4-6 lists many of these creative applications of the AFM/STM idea.

Since the AFM measures force, it can be used with both conductive and nonconductive specimens. Force must be applied to bend a cantilever, so AFM is subject to artifacts caused by damage to fragile structures on the surface. Both AFM and STM can function well for specimens under water, in air, or under vacuum. For exploring biomolecules or mobile organic surfaces, the "pushing around" of structures by the tip is a significant concern. This surface artifact can be capitalized upon to write and fabricate surface structures at the nanometer scale (Fig. 3.1.4-17) (Boland *et al.*, 1998; Quate, 1997; Wilson *et al.*, 2001).

Newer methods

There are many other surface characterization methods that have the potential to become important in future years. Some of these are listed in Table 3.1.4-7. A few of these evolving techniques that will be specifically mentioned here include SFG Raman, and synchrotron methods.

SFG uses two high-intensity, pulsed laser beams, one in the visible range (frequency = $\omega_{visible}$) and one in the infrared (frequency = ω_{ir}), to illuminate a specimen. The light emitted from the specimen by a non-linear optical process, $\omega_{sum} = \omega_{visible} + \omega_{ir}$, is detected and quantified (Fig. 3.1.4-18). The intensity of the light at ω_{sum} is proportional to the square of the sample's second-order nonlinear susceptibility ($\chi^{(2)}$). The term susceptibility refers to the effect of the light field strength on the molecular polarizability. The ω_{sum} light intensity vanishes when

Table 3.1.4-6 Scanning probe microscopy (SPM) modes

Name	Acronym	Use
Contact mode	CM-AFM	Topographic imaging of harder specimens
Tapping (intermittent force) mode	IF-AFM	Imaging softer specimens
Noncontact mode	NCM-AFM	Imaging soft structures
Force modulation (allows slope of force–distance curve to be measured)	FM-AFM	Enhances image contrast based on surface mechanics
Scanning surface potential microscopy (Kelvin probe microscopy)	SSPM, KPM	Measures the spatial distribution of surface potential
Magnetic force microscopy	MFM	Maps the surface magnetic forces
Scanning thermal microscopy	SThM	Maps the thermal conductivity characteristics of a surface
Recognition force microscopy	RFM	Uses a biomolecule on a tip to probe for regions of specific biorecognition on a surface
Chemical force microscopy	CFM	A tip derivatized with a given chemistry is scanned on a surface to spatially measure differences of interaction strength
Lateral force microscopy	LFM	Maps frictional force on a surface
Electrochemical force microscopy	EFM	The tip is scanned under water and the electrochemical potential between tip and surface is spatially measured
Nearfield scanning optical microscopy	NSOM	A sharp optical fiber is scanned over a surface allowing optical microscopy or spectroscopy at 100-nm resolution
Electrostatic force microscopy	EFM	Surface electrostatic potentials are mapped
Scanning capacitance microscopy	SCM	Surface capacitance is mapped
Conductive atomic force microscopy	CAFM	Surface conductivity is mapped with an AFM instrument
Nanolithographic AFM		An AFM tip etches, oxidizes, or reacts a space permitting pattern fabrication at 10 nm or better resolution
Dip-pen nanolithography	DPN	An AFM tip, inked with a thiol or other molecule, writes on a surface at the nanometer scale

a material has inversion symmetry, i.e., in the bulk of the material. At an interface, the inversion symmetry is broken and an SFG signal is generated. Thus, SFG is exquisitely sensitive to the plane of the interface. In practice, ω_{ir} is scanned over a vibrational frequency range — where vibrational interactions occur with interface molecules, then the SFG signal is resonantly enhanced and we see a vibrational spectrum. The advantages are the superb surface sensitivity, the cancellation of bulk spectral intensity (for example, this allows measurements at a water/solid interface), the richness of information from vibrational spectra, and the ability to study molecular orientation due to the polarization of the light. SFG is not yet a routine method. The lasers and optical components are expensive and require precision alignment. Also, the range in the infrared over which lasers can scan is limited (though it has slowly expanded with improved equipment). However, the power of SFG for biomaterials studies has already been

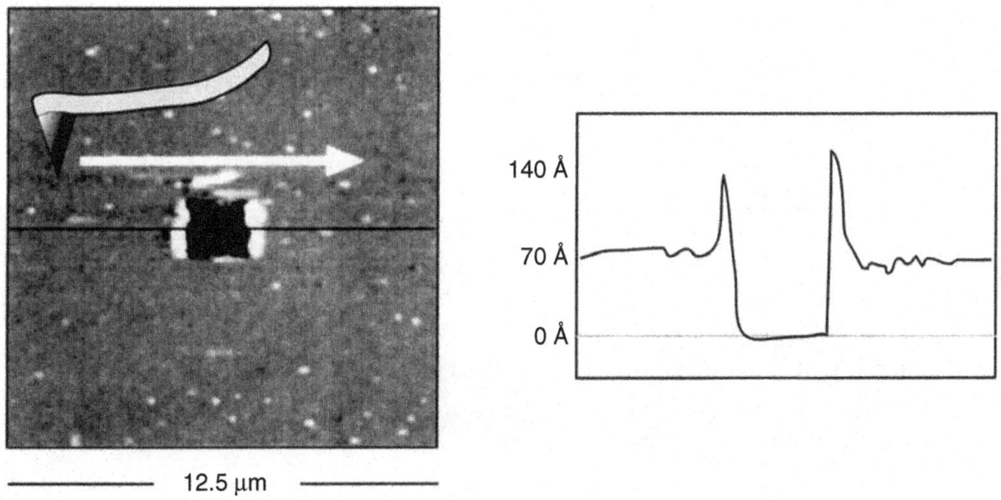

Fig. 3.1.4-17 An AFM tip, using relatively high force, was used to scratch a rectangular feature into a thin (70 Å) plasma deposited film. The AFM could also characterize the feature created.

proven with studies on hydrated hydrogels, polyurethanes, surface active polymer additives, and proteins (Shen, 1989; Chen *et al.*,2002).

In Raman spectroscopy a bright light is shined on a specimen. Most of the light scatters back at the same frequency as the incident beam. However, a tiny fraction of this light excites vibrations in the specimen and thereby loses or gains energy. The frequency shift of the light corresponds to vibrational bands indicative of the molecular structure of the specimen. The Raman

Table 3.1.4-7 Methods that may have applicability for the surface characterization of biomaterials

Method	Information obtained
Second-harmonic generation (SHG)	Detect submolayer amounts of adsorbate at any light accessible interface (air–liquid, solid–liquid, solid–gas)
Surface-enhanced Raman spectroscopy (SERS)	High-sensitivity Raman at rough metal interfaces
Ion scattering spectroscopy (ISS)	Elastically reflected ions probe only the outermost atomic layer
Laser desorption mass spectrometry (LDMS)	Mass spectra of adsorbates at surfaces
Matrix assisted laser desorption ionization (MALDI)	Though generally a bulk mass spectrometry method, MALDI has been used to analyze large adsorbed proteins
IR photoacoustic spectroscopy (IR-PAS)	IR spectra of surfaces with no sample preparation based on wavelength-dependent thermal response
High-resolution electron energy loss spectroscopy (HREELS)	Vibrational spectroscopy of a highly surface-localized region, under ultrahigh vacuum
X-ray reflection	Structural information about order at surfaces and interfaces
Neutron reflection	Thickness and refractive index information about interfaces from scattered neutrons—where H and D are used, unique information on interface organization can be obtained
Extended X-ray absorption fine structure (EXAFS)	Atomic-level chemical and nearest-neighbor (morphological) information
Scanning Auger microprobe (SAM)	Spatially defined Auger analysis at the nanometer scale
Surface plasmon resonance (SPR)	Study aqueous adsorption events in real time by monitoring changes in surface refractive index
Rutherford backscattering spectroscopy (RBS)	Depth profiling of complex, multiplayer interfacial systems

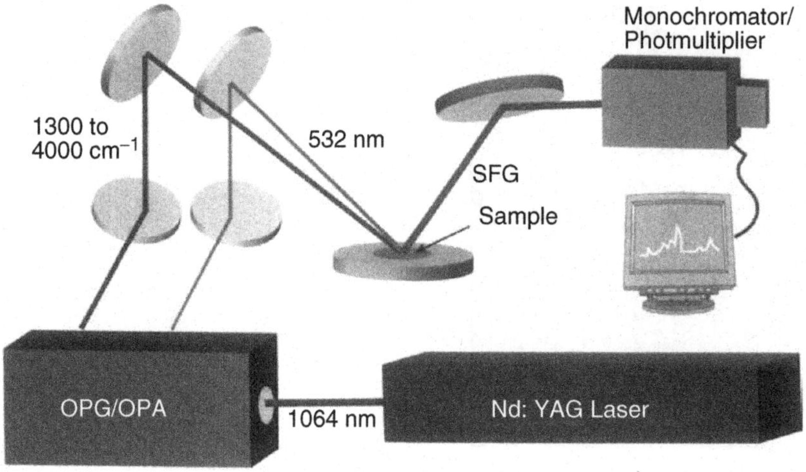

Fig. 3.1.4-18 Schematic diagram of a SFG apparatus (based upon a diagram developed by Polymer Technology Group, Inc.).

spectroscopic technique has been severely limited for surface studies because of its low signal level. However, in recent years, great strides in detector sensitivity have allowed Raman to be applied for studying the minute mass of material at a surface. Also, surface enhanced Raman spectroscopy (SERS), Raman spectra taken from molecules on a roughened metal surface, can enhance Raman signal intensity by 10^6 or more. Raman spectra will be valuable for biomedical surface studies because water, which absorbs radiation very strongly in the infrared range, has little effect on Raman spectra that are often acquired with visible light (Storey *et al.*, 1995).

Synchrotron sources of energetic radiation that can be used to probe matter were originally confined to the physics community for fundamental studies. However, there are now more synchrotron sources, better instrumentation, and improved data interpretation. Synchrotron sources are typically national facilities costing >$100M and often occupying hundreds of acres (Fig. 3.1.4-19). By accelerating electrons to near the speed of light in a large, circular ring, energies covering a broad swath of the electromagnetic spectrum (IR to energetic X-rays) are emitted. A synchrotron source (and ancillary equipment) permits a desired energy of the probe beam to be "dialed in" or scanned through a frequency range. Other advantages include high source intensity (bright light) and polarized light. Some of the experimental methods that can be performed with great success at synchrotron sources include crystallography, scattering, spectroscopy, microimaging, and nanofabrication. Specific surface spectroscopic methods include scanning photoemission microscopy (SPEM, 100 nm spatial resolution), ultraESCA (100 μm spatial resolution, high energy resolution), and near edge X-ray absorption spectrometry (NEXAFS).

Fig. 3.1.4-19 The Advanced Photon Source, Argonne National Laboratories, a modern synchrotron source.

Studies with surface methods

Hundreds of studies have appeared in the literature in which surface methods have been used to enhance the understanding of biomaterial systems. A few studies that demonstrate the power of surface analytical methods for biomaterials science are briefly described here.

Platelet consumption and surface composition

Using a baboon arteriovenous shunt model of platelet interaction with surfaces, a first-order rate constant of reaction of platelets with a series of polyurethanes was measured. This rate constant, the platelet consumption by the material, correlated in an inverse linear fashion with the fraction of hydrocarbon-type groups in the ESCA C1s spectra of the polyurethanes (Hanson *et al.*,

1982). Thus, surface analysis revealed a chemical parameter about the surface that could be used to predict long-term biological reactivity of materials in a complex *ex vivo* environment.

Contact-angle correlations

The adhesion of a number of cell types, including bacteria, granulocytes, and erythrocytes, has been shown, under certain conditions, to correlate with solid-vapor surface tension as determined from contact-angle measurements. In addition, immunoglobulin G adsorption is related to ν_{sv} (Neumann *et al.*, 1983).

Contamination of intraocular lenses

Commercial intraocular lenses were examined by ESCA. The presence of sulfur, sodium, and excess hydrocarbon at their surfaces suggested contamination by sodium dodecyl sulfate (SDS) during the manufacture of the lenses (Ratner, 1983). A cleaning protocol was developed using ESCA to monitor results that produced a lens surface of clean PMMA.

Titanium

The discoloration sometimes observed on titanium implants after autoclaving was examined by ESCA and SIMS (Lausmaa *et al.*, 1985). The discoloration was found to be related to accelerated oxide growth, with oxide thicknesses to 650 Å. The oxide contained considerable fluorine, along with alkali metals and silicon. The source of the oxide was the cloth used to wrap the implant storage box during autoclaving. Since fluorine strongly affects oxide growth, and since the oxide layer has been associated with the biocompatibility of titanium implants, the authors advise avoiding fluorinated materials during sterilization of samples. A newer paper contains detailed surface characterization of titanium using a battery of surface methods and addresses surface preparation, contamination, and cleaning (Lausmaa, 1996).

SIMS for adsorbed protein identification and quantification

All proteins are made up of the same 20 amino acids and thus, on the average, are compositionally similar. Surface analysis methods have shown the ability to detect and quantify surface-bound protein, but biological tools have, until recently, been needed to identify specific proteins. Modern static SIMS instrumentation, using a multivariate statistical analysis of the data, has demonstrated the ability to distinguish between more than 13 different proteins adsorbed on surfaces (Wagner and Castner, 2001). Also, the limits of detection for adsorbed proteins on various surfaces were compared by ESCA and SIMS (Wagner *et al.*, 2002).

Poly(glycolic acid) degradation studied by SIMS

The degradation of an important polymer for tissue engineering, poly(glycolic acid), has been studied by static SIMS. As well as providing useful information on this degradation process, the study illustrates the power of SIMS for characterizing synthetic polymers and their molecular weight distributions (Chen *et al.*, 2000).

Conclusions

The instrumentation of surface analysis steadily advances and newer instruments and techniques can provide invaluable information about biomaterials and medical devices. The information obtained can be used to monitor contamination, ensure surface reproducibility, and explore fundamental aspects of the interaction of biological systems with living systems. Considering that biomedical experiments are typically expensive to perform, the costs for surface analysis are modest to ensure that the surface is as expected, stable and identical from experiment to experiment. Surface analysis can also contribute to the understanding of medical device failure (and success). Myriad applications for surface methods are found in device optimization, manufacture and quality control. Predicting biological reaction based on measured surface structure is a frontier area for surface analysis.

Acknowledgment

Support was received from the UWEB NSF Engineering Research Center and the NESAC/BIO National Resource, NIH grant EB-002027, during the preparation of this work and for some of the studies described herein.

Questions

1. Scan the table of contents and abstracts from the last three issues of the *Journal of Biomedical Materials Research* or *Biomaterials*. List all the surface analysis

methods used in the articles therein and briefly describe what was learned by using them.

2. How is critical surface tension related to wettability? For the polymers in Table 3.1.4-2, draw the chemical formulas of the chain repeat units and attempt to relate the structures to the wettability. Where inconsistencies are noted, explain those inconsistencies using Table 3.1.4-3.
3. A titanium dental implant was being manufactured by the Biomatter Company for the past 8 years. It performed well clinically. For economic reasons, manufacturing of the titanium device was outsourced to Metalsmed, Inc. Early clinical results on this Metalsmed implant, supposedly identical to the Biomatter implant, suggested increased inflammation. How would you compare the surface chemistry and structure of these two devices to see if a difference that might account for the difference in clinical performance could be identified?

Bibliography

Adamson, A.W., and Gast, A. (1997). *Physical Chemistry of Surfaces*, 6th ed. Wiley-Interscience, New York.

Albrecht, T.R., Dovek, M.M., Lang, C.A., Grutter, P., Quate, C.F., Kuan, S.W.J., Frank, C.W., and Pease, R.F.W. (1988). Imaging and modification of polymers by scanning tunneling and atomic force microscopy. *J. Appl. Phys.* **64:** 1178–1184.

Allara, D.L. (1982). Analysis of surfaces and thin films by IR, Raman, and optical spectroscopy. *ACS Symp. Ser.* **199:** 33—47.

Andrade, J.D. (1985). *Surface and Interfacial Aspects of Biomedical Polymers*, Vol. 1: *Surface Chemistry and Physics*. Plenum Publishers, New York.

Avouris, P. (1990). Atom-resolved surface chemistry using the scanning tunneling microscope. *J. Phys. Chem.* **94:** 2246–2256.

Belu, A.M., Graham, D.J., and Castner, D.G. (2003). Time-of-flight secondary ion mass spectrometry: techniques and applications for the characterization of biomaterial surfaces. *Biomaterials* **24:** 3635–3653.

Benninghoven, A. (1983). Secondary ion mass spectrometry of organic compounds (review). in *Springer Series of Chemical Physics: Ion Formation from Organic Solids*, Vol. 25, A. Benninghoven, ed. Springer-Verlag, Berlin, pp. 64–89.

Binnig, G., and Rohrer, H. (1986). Scanning tunneling microscopy. *IBM J. Res. Develop.* **30:** 355–369.

Boland, T., Johnston, E.E., Huber, A., and Ratner, B.D. (1998). Recognition and nanolithography with the atomic force microscope, in *Scanning Probe Microscopy of Polymers*, Vol. 694, B.D. Ratner and V.V. Tsukruk, eds. American Chemical Society, Washington, D.C., pp. 342–350.

Briggs, D. (1986). SIMS for the study of polymer surfaces: a review. *Surf. Interface Anal.* **9:** 391–404.

Briggs, D., and Seah, M.P. (1983). *Practical Surface Analysis*. Wiley, Chichester, UK.

Castner, D.G., and Ratner, B.D. (2002). Biomedical surface science: foundations to frontiers. *Surf. Sci.* **500:** 28–60.

Chen, J., Lee, J.-W., Hernandez, N.L., Burkhardt, C.A., Hercules, D.M., and Gardella, J.A. (2000). Time-of-flight secondary ion mass spectrometry studies of hydrolytic degradation kinetics at the surface of poly(glycolic acid). *Macromolecules* **33:** 4726–4732.

Chen, Z., Ward, R., Tian, Y., Malizia, F., Gracias, D.H., Shen, Y.R., and Somorjai, G.A. (2002). Interaction of fibrinogen with surfaces of end-group-modified polyurethanes: A surface-specific sum-frequency-generation vibrational spectroscopy study. *J. Biomed. Mater. Res.* **62:** 254–264.

Chilkoti, A., Boland, T., Ratner, B.D., and Stayton, P.S. (1995). The relationship between ligand-binding thermodynamics and protein-ligand interaction forces measured by atomic force microscopy. *Biophys. J.* **69:** 2125–2130.

Davies, M.C., and Lynn, R.A.P. (1990). Static secondary ion mass spectrometry of polymeric biomaterials. *CRC Crit. Rev. Biocompat.* **5:** 297–341.

Dilks, A. (1981). X-ray photoelectron spectroscopy for the investigation of polymeric materials. in *Electron Spectroscopy: Theory, Techniques, and Applications*, Vol. 4, A.D. Baker and C.R. Brundle, eds. Academic Press, London, pp. 277–359.

Dumas, P., Weldon, M.K., Chabal, Y.J. and Williams, G.P. (1999). Molecules at surfaces and interfaces studied using vibrational spectroscopies and related techniques. *Surf. Rev. Lett.*, **6**(2): 225–255.

Feldman, L.C., and Mayer, J.W. (1986). *Fundamentals of Surface and Thin Film Analysis*. North-Holland, New York.

Good, R.J. (1993). Contact angle, wetting, and adhesion: a critical review. in *Contact Angle, Wettability and Adhesion*, K.L. Mittal, ed. VSP Publishers, The Netherlands.

Hansma, P.K., Elings, V.B., Marti, O., and Bracker, C.E. (1988). Scanning tunneling microscopy and atomic force microscopy: application to biology and technology. *Science* **242:** 209–216.

Hanson, S.R., Harker, L.A., Ratner, B.D., and Hoffman, A.S. (1982). Evaluation of artificial surfaces using baboon arteriove-nous shunt model. in *Biomaterials 1980, Advances in Biomaterials*, G.D. Winter, D.F. Gibbons, and H. Plenk, eds., Vol. 3, Wiley, Chichester, UK, pp. 519–530.

Jandt, K.D. (2001). Atomic force microscopy of biomaterials surfaces and interfaces. *Surf. Sci.* **491:** 303–332.

Lausmaa, J. (1996). Surface spectroscopic characterization of titanium implant materials. *J. Electron Spectrosc. Related Phenom.* **81:** 343–361.

Lausmaa, J., Kasemo, B., and Hansson S. (1985). Accelerated oxide growth on titanium implants during autoclaving caused by fluorine contamination. *Biomaterials* **6:** 23–27.

Leyden, D.E., and Murthy, R.S.S. (1987). Surface-selective sampling techniques in Fourier transform infrared spectroscopy. *Spectroscopy* **2:** 28–36.

McIntire, L., Addonizio, V.P., Coleman, D.L. , Eskin, S.G., Harker, L.A., Kardos, J.L., Ratner, B.D., Schoen, F.J., Sefton, M.V., and Pitlick, F.A. (1985). *Guidelines for Blood-Material Interactions—Devices and Technology Branch, Division of*

Heart and Vascular Diseases, National Heart, Lung, and Blood Institute, NIH Publication No. 85–2185, revised July 1985, U.S. Department of Health and Human Services.

Miles, M.J., McMaster, T., Carr, H.J., Tatham, A.S., Shewry, P.R., Field, J.M., Belton, P.S., Jeenes, D., Hanley, B., Whittam, M., Cairns, P., Morris, V.J., and Lambert, N. (1990). Scanning tunneling microscopy of biomolecules. *J. Vac. Sci. Technol. A* **8:** 698–702.

Neumann, A.W., Absolom, D.R., Francis, D.W., Omenyi, S.N., Spelt, J.K., Policova, Z., Thomson, C., Zingg, W., and Van Oss, C.J. (1983). Measurement of surface tensions of blood cells and proteins. *Ann. N.Y. Acad. Sci.* **416:** 276–298.

Quate, C.F. (1997). Scanning probes as a lithography tool for nanostructures. *Surf. Sci.* **386:** 259–264.

Ratner, B.D. (1983). Analysis of surface contaminants on intraocular lenses. *Arch. Ophthal.* **101:** 1434–1438.

Ratner, B.D. (1988). *Surface Characterization of Biomaterials.* Elsevier, Amsterdam.

Ratner, B.D., and Castner, D.G. (1997). Electron spectroscopy for chemical analysis. in *Surface Analysis—The Principal Techniques.* J.C. Vickerman, ed. John Wiley and Sons, Ltd., Chichester, UK, pp. 43–98.

Ratner, B.D., and McElroy, B.J. (1986). Electron spectroscopy for chemical analysis: applications in the biomedical sciences. in *Spectroscopy in the Biomedical Sciences*, R.M. Gendreau, ed. CRC Press, Boca Raton, FL, pp. 107–140.

Rugar, D., and Hansma, P. (1990). Atomic force microscopy. *Physics Today* **43:** 23–30.

Scheutzle, D., Riley, T.L., deVries, J.E., and Prater, T.J. (1984). Applications of high-performance mass spectrometry to the surface analysis of materials. *Mass Spectrom.* **3:** 527–585.

Shen, Y.R. (1989). Surface properties probed by second-harmonic and sum-frequency generation. *Nature* **337**(6207): 519–525.

Somorjai, G.A. (1981). *Chemistry in Two Dimensions: Surfaces.* Cornell Univ. Press, Ithaca, NY.

Somorjai, G.A. (1994). *Introduction to Surface Chemistry and Catalysis.* John Wiley and Sons, New York.

Storey, J.M.E., Barber, T.E., Shelton, R.D., Wachter, E.A., Carron, K.T., and Jiang, Y. (1995). Applications of surface-enhanced Raman scattering (SERS) to chemical detection. *Spectroscopy* **10**(3): 20–25.

Tirrell, M., Kokkoli, E., and Biesalski, M. (2000). The role of surface science in bioengineered materials. *Surf. Sci.* **500:** 61–83.

Urban, M.W. (1993). *Vibrational Spectroscopy of Molecules and Macromolecules on Surfaces.* Wiley-Interscience, New York.

Van Vaeck, L., Adriaens, A., and Gijbels, R. (1999). Static secondary ion mass spectrometry (S-SIMS): part I. Methodology and structural interpretation. *Mass Spectrom. Rev.* **18:** 1—47.

Vickerman, J.C. (1997). *Surface Analysis: The Principal Techniques.* John Wiley and Sons, Chichester, UK.

Vickerman, J.C., Brown, A., and Reed, N. M. (1989). *Secondary Ion Mass Spectrometry, Principles and Applications.* Clarendon Press, Oxford.

Wagner, M.S., and Castner, D.G. (2001). Characterization of adsorbed protein films by time-of-flight secondary ion mass spectrometry with principal component analysis. *Langmuir* 17: 4649–4660.

Wagner, M.S., McArthur, S.L., Shen, M., Horbett, T.A., and Castner, D.G. (2002). Limits of detection for time of flight secondary ion mass spectrometry (ToF-SIMS) and X-ray photo-electron spectroscopy (XPS): detection of low amounts of adsorbed protein. *J. Biomater. Sci. Polymer Ed.* **13**(4): 407–428.

Watts, J.F., and Wolstenholme, J. (2003). *An Introduction to Surface Analysis by XPS and AES.* John Wiley & Sons, Chichester, UK.

Wilson, D.L., Martin, R., Hong, S.I., Cronin-Golomb, M., Mirkin, C.A., and Kaplan, D.L. (2001). Surface organization and nanopat-terning of collagen by dip-pen nanolithography. *Proc. Natl. Acad. Sci. USA* **98**(24): 13,360–13,664.

Zisman, W.A. (1964). Relation of the equilibrium contact angle to liquid and solid constitution. in *Contact Angle, Wettability and Adhesion, ACS Advances in Chemistry Series*, Vol. 43, F.M. Fowkes, ed. American Chemical Society, Washington, D.C., pp. 1–51.

3.1.5 Role of water in biomaterials

Erwin A. Vogler

The primary role water plays in biomaterials is as a solvent system. Water, the "universal ether" (Baier and Meyer, 1996), dissolves inorganic salts and large organic macromolecules such as proteins or carbohydrates (solutes) with nearly equal efficiency (Pain, 1982). Water suspends living cells, as in blood, for example, and is the principal constituent of the interstitial fluid that bathes tissues. Water is not just a bland, neutral carrier system for biochemical processes, however. Far from this, water is an active participant in biology, which simply could not and would not work the way it does without the special mediating properties of water. Moreover, it is widely believed that water is the first molecule to contact biomaterials in any clinical application (Andrade *et al.*, 1981; Baier, 1978). This is because water is the majority molecule in any biological mixture, constituting 70 wt% or more of most living organisms, and because water is such a small and agile molecule, only about 0.25 nm in the longest dimension. Consequently, behavior of water near surfaces and the role of water in biology are very important subjects in biomaterials science. Some of these important aspects of water are discussed here.

Water solvent properties

Figures 3.1.5-1A-3.1.5-1D collect various diagrams of water illustrating the familiar atomic structure and how this arrangement leads to the ability to form a network of self-associated molecules through hydrogen bonding. Self-association confers unique properties on water, many of which are still active areas of scientific investigation even after more than 200 years of chemical and physical research applied to water (Franks, 1972).

Hydrogen bonds in water are relatively weak 3–5 kcal/mole associations with little covalent character (Iassacs *et al.*, 1999; Marshall, 1999). As it turns out, hydrogen bond strength is approximately the same as the energy transferred from one molecule to another by collisions at room temperature (Vinogradov and Linnell, 1971). So hydrogen bonds are quite transient in nature, persisting only for a few tens of picoseconds (Berendsen, 1967; Luzar and Chandler, 1996). Modern molecular simulations suggest, however, that more than 75% of liquid-water molecules are interconnected in a 3D network of three or four nearest neighbors at any particular instant in time (Robinson *et al.*, 1996). This stabilizing network of self-associated water formed from repeat units illustrated in Fig. 3.1.5-1D is so extensive, that it is frequently termed "water structure," especially in the older literature (Narten and Levy, 1969). These somewhat dated water-structure concepts will not be discussed further here, other than to caution the reader that the transient nature of hydrogen bonding greatly weakens the notion of a "structure" as it might be practically applied by a chemist for example (Berendsen, 1967) and that reference to water structure near solutes and surfaces in terms of "icebergs" or "melting" should not be taken too literally, as will be discussed further subsequently.

A very important chemical outcome of this propensity of water to self-associate is the dramatic effect on water solvent properties. One view of self-association is from the standpoint of Lewis acidity and basicity. It may be recalled from general chemistry that a Lewis acid is a molecule that can accept electrons or, more generally, electron density from a molecular orbital of a donor molecule. An electron-density donor molecule is termed a Lewis base. Water is amphoteric in this sense because, as illustrated in Figs. 3.1.5-1A and 3.1.5-1D, it can simultaneously share and donate electron density. Hydrogen atoms (the Lewis acids) on one or more adjacent water molecules can accept electron density from the unshared electron pairs on the oxygen atom (the Lewis bases) of another water molecule. In this manner, water forms a 3D network through Lewis acid–base self-association reactions.

If the self-associated network is more complete than some arbitrary reference state, then there must be

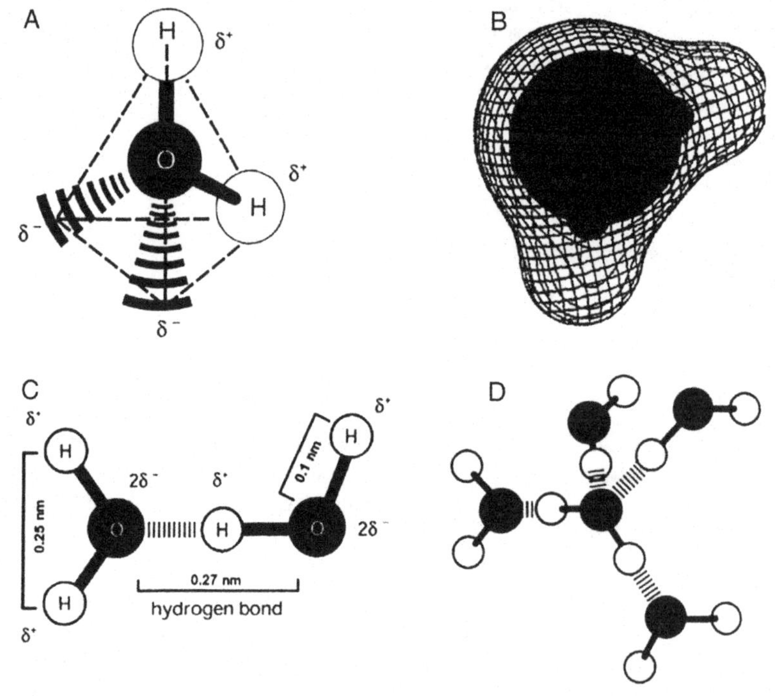

Fig. 3.1.5-1 Atomic structure of water illustrating (A) tetrahedral bonding arrangement wherein hydrogen atoms (H, light-colored spheres) are Lewis acid centers and the two lone-pair electrons on oxygen (O, dark-colored spheres) are Lewis base centers that permit water to hydrogen bond with four nearest-neighbor water molecules; (B) electron density map superimposed on an atomic-radius sphere model of water providing a more authentic representation of molecular water; (C) approximate molecular dimensions; and (D) five water molecules participating in a portion of a hydrogen-bond network.

relatively fewer unmatched Lewis acid–base pairings than in this reference state. Conversely, in less-associated water, the network is relatively incomplete and there are more unmatched Lewis acid–base pairings than in the reference state. These unmatched pairings in less associated water are readily available to participate in other chemical reactions, such as dissolving a solute molecule or hydrating a water-contacting surface. Therefore, it can be generally concluded that less-associated water is a stronger solvent than more-associated water because it has a greater potential to engage in reactions other than self-association. In chemical terminology, less self-associated water has a greater chemical potential than more self-associated water. Interestingly, more self-associated water with a relatively more complete 3D network of hydrogen bonds must be less dense (greater partial molar volume) than less self-associated water because formation of linearly directed hydrogen bonds takes up space (Fig. 3.1.5-1C), increasing free volume in the liquid. This is why water ice with a complete crystalline network is less dense than liquid water and floats upon unfrozen water, a phenomenon with profound environmental impact. Thus, less associated water is not only more reactive but also more dense. These inferred relationships between water structure and reactivity are summarized in Table 3.1.5-1, which will be a useful aid to subsequent discussion.

A variety of lines of evidence ranging from molecular simulations (Lum *et al.*, 1999; Robinson *et al.*, 1996) to experimental studies of water's solvent properties in porous media (Qi and Soka, 1998; Wiggins, 1988) suggest that water expands and contracts in density (molar volume) with commensurate changes in chemical potential to accommodate presence of imposed solutes and surfaces. The word "imposed" is specifically chosen here to emphasize that a solute (e.g., an ion or a macromolecule) or an extended surface (e.g., the outer region of a biomaterial) must in some way interfere with self-association. Simply stated, the solute or surface gets in the way and water molecules must reorient to maintain as many hydrogen bonds with neighbors as is possible in this imposed presence of solute or surface. Water may not be able to maintain an extensive hydrogen-bond network in certain cases and this has important and measurable effects on water solvency. The next sections will first consider "hydrophobic" and "hydrophilic" solute molecules and then extend the discussion to hydrophobic and hydrophilic biomaterial surfaces, at least to the extent possible within the current scientific knowledge base.

Table 3.1.5-1 Relationships among water structure and solvent properties

Extent of water self-association	Density	Partial molar volume	Chemical potential (number of available hydrogen bonds)
More	Less	More	Less
Less	More	Less	More

The hydrophobic effect

The hydrophobic effect is related to the insolubility of hydrocarbons in water and is fundamental to the organization of lipids into bilayers, the structural elements of life as we know it (Tanford, 1973). Clearly then, the hydrophobic effect is among the more fundamental, life-giving phenomena attributable to water. Hydrocarbons are sparingly soluble in water because of the strong self-association of water, not the strong self-association of hydrocarbons as is sometimes thought. Thus water structure is seen to be directly related to solvent properties in this very well-known case.

The so-called "entropy of hydrophobic hydration" (ΔS) has received a great deal of research attention from the molecular-simulation community because it dominates the overall (positive) free energy of hydrophobic hydration (ΔG) at ambient temperatures and pressures. The rather highly negative entropy of hydration of small hydrocarbons ($\Delta S \approx -20$ e.u.; see Kauzmann, 1959, for discussion related to lipids and proteins) turns out to be substantially due to constraints imposed on water-molecule orientation and translation as water attempts to maintain hydrogen-bond neighbors near the solute molecule (Paulaitis *et al*, 1996). Apparently, there are no structural "icebergs" with enhanced self-association around small hydrocarbons (Besseling and Lyklema, 1995) as has been invoked in the past to account for ΔS (Berendsen, 1967). Instead, water surrounding small solutes such as methane or ethane may be viewed as spatially constrained by a "solute-straddling" effect that maximizes as many hydrogen-bonded neighbors as possible at the expense of orientational flexibility. Interestingly, while these constraints on water-molecule orientation do not significantly promote local self-association (i.e., increase structure), this lack of flexibility does have the effect of reducing repulsive, non-hydrogen-bonding interactions between water-molecule neighbors, accounting for a somewhat surprisingly exothermic (≈ -2 kcal/mol) enthalpy of hydration (ΔH) of small hydrocarbons (Besseling and Lyklema, 1995). The strong temperature sensitivities of these entropic and enthalpic effects are nearly equal and opposite and compensate in a way that causes the overall free energy of hydration ($\Delta G = \Delta H - T\Delta S$) to be essentially temperature insensitive. Increasing temperature expands the self-associated network of water, creating more space for a hydrophobic solute to occupy, and ΔS becomes more positive ($-T\Delta S$

more negative). On the other hand, nonbonding (repulsive) contacts between water molecules increase with temperature, causing ΔH to become more positive.

As one might imagine, difficulties in maintaining a hydrogen-bonded network are exacerbated near very large hydrophobic solutes where no orientations can prevent separation of water-molecule neighbors. Another water-driven mechanism comes into play in some of these cases wherein hydrophobic patches or domains on a solute such as a protein aggregate, exclude water, and participate in what has been termed "hydrophobic bonding" (DeVoe, 1969; Dunhill, 1965; Kauzmann, 1959; Tanford, 1966). This aspect of the hydrophobic effect is very important in biomaterials because it controls the folding of proteins and is thus involved in protein reactions at surfaces, especially denaturation of proteins at biomaterial surfaces induced by unfolding reactions in the adsorbed state. Water near large hydrophobic patches will be further considered in relation to hydrophobic surfaces that present analogous physical circumstances to water.

The hydrophilic effect

There is no broadly recognized "hydrophilic effect" in science that is the antithesis to the well-known hydrophobic effect just discussed. But generally speaking, the behavior of water near hydrophilic solutes is so substantially different from that occurring near hydrophobic solutes that hydrophilicity may well be granted a distinguishing title of its own. The terms hydrophilic and hydrophobic are poorly defined in biomaterials and surface science (Hoffman, 1986; Oss and Giese, 1995; Vogler, 1998), requiring some clarification at this juncture since a distinction between hydrophilic and hydrophobic needs to be made. For the current purposes, let the term hydrophilic be applied to those solutes that compete with water for hydrogen bonds. That is to say, hydrophilic solutes exhibit Lewis acid or base strength comparable to or exceeding that of water, so that it is energetically favorable for water to donate electron density to or accept electron density from hydrophilic solutes instead of, or at least in competition with, other water molecules. For the sake of clarity, let it be added that there is no chemistry or other energetic reason for water to hydrogen bond with a hydrophobic solute as defined herein. Generally speaking, free energies of hydrophilic hydration are greater than that of hydrophobic hydration since acid–base chemistry is more energetic than the non-bonding "hydrophobic" reactions previously considered, and this frequently manifests itself in large enthalpic contributions to ΔG.

Familiar examples of hydrophilic solutes with biomedical relevance would include cations such as Na^+, K^+, Ca^{2+}, and Mg^{2+} or anions such as Cl^-, HCO_3^{-1} and HPO_4^{-2}. These ions are surrounded by a hydration sphere of water directing oxygen atoms toward the cations or hydrogen atoms toward the anions. Water structuring near ions is induced by a strong electric field surrounding the ion that orients water dipole moments in a manner that depends on ionic size and extent of hydration (Marcus, 1985). Some ions are designated "structure promoting" and others "structure breaking." Structure-promoting ions are those that impose more local order in surrounding water than occurs distant from the ion whereas structure-breaking or "chaotropic" ions increase local disorder and mobility of adjacent water molecules (Wiggins, 1971). Another feature of ion solvation important in biomaterials is that certain ions such as Ca^{2+} and Mg^{2+} are more hydrated (surrounded by more water molecules) than K^+ and Na^+ in the order of the so-called Hofmeister or lyotropic series. This implies that highly hydrated ions will partition into less associated water with more available hydrogen bonds for solvation (see Table 3.1.5-1) preferentially over more associated water with fewer available hydrogen bonds (Christenson *et al.*, 1990; Vogler, 1998; Wiggins, 1990; Wiggins and Ryn, 1990). This ion partitioning can have dramatic consequences on biology near surfaces since Ca^{2+} and Mg^{2+} have strong allosteric effects on enzyme reactions, a point that will be raised again in the final section of this chapter.

As in the hydrophobic effect, size plays a big role in the solvation of hydrophilic ions. Small inorganic ions are completely ionized and lead to separately hydrated ions in the manner just discussed above. Hydration of a polyelectrolyte such as hyaluronic acid or a single strand of DNA is more complicated because a counterion "atmosphere" surrounds the dissolved polyelectrolyte. The countercharge distribution within this atmosphere is not uniform in space but instead diminishes in concentration with distance from the polyelectrolyte. This means that water in a hypothetical compartment near the polyelectrolyte is enriched in countercharges (higher ionic strength, lower water chemical potential) relative to that of an identical compartment distant from the polyelectrolyte (lower ionic strength, higher water chemical potential). Since concentration (chemical potential) gradients cannot persist at equilibrium, there must be some route to making chemical potentials uniform throughout solution. Wiggins has argued that the only means available to such a system of dissolved polyelectrolytes at constant temperature, pressure, and fixed composition (including water) is adjustment of water density or, more precisely, partial molar volume (Wiggins, 1990). That is to say, in order to increase water chemical potential in the near compartment relative to that of water in the distant compartment, water density must increase (see row 2 of Table 3.1.5-1, more molecules/unit volume available for chemical work). At the same time, in order to decrease water chemical potential in the distant compartment relative to that in the closer one; water density must decrease (see row 2 of Table 3.1.5-1, fewer molecules/unit volume available for chemical work). This thinking gives rise to the notion of contiguous regions of variable water density within a polyelectrolyte solution. Here again, it is evident that adjustment of water chemical potential to accommodate the presence of a large solute molecule appears to be a necessary mechanism to account for commonly observed hydration effects. The next section will explore how these same effects might account for surface wetting effects.

The surface wetting effect

It is a very common observation that water wets certain kinds of surfaces whereas water beads up on others, forming droplets with a finite "contact angle." This and related wetting phenomena have intrigued scientists for almost three centuries, and the molecular mechanisms of wetting are still an important area of research to this day. The reason for such continued interest is that wetting phenomena probe the various intermolecular forces and interactions responsible for much of the chemistry and physics of everyday life. Some of the remaining open questions are related to water structure and solvent properties near different kinds of surfaces.

Although surfaces on which water spreads are commonly termed hydrophilic and those on which water droplets form hydrophobic, the definitions employed in preceding sections based on presence or absence of Lewis acid/base groups that can hydrogen bond with water will continue to be used here, as this is a somewhat more precise way of categorizing biomaterials. Thus, hydrophobic surfaces are distinguished from hydrophilic by virtue of having no Lewis acid or base functional groups available for water interaction.

Water near hydrophobic surfaces finds itself in a predicament similar to that briefly mentioned in the preceding section on the hydration of large hydrophobic solutes in that there are no configurational options available to water molecules closest to the surface that allow maintenance of nearest-neighbor hydrogen bonds. These surface-contacting water molecules are consequently in a less self-associated state and, according to row 2 of Table 3.1.5-1, must temporarily be in a state of higher chemical potential than bulk water. The key word here is temporarily, because chemical potential gradients cannot exist at equilibrium. At constant temperature and pressure, the only recourse available to the system toward establishment of equilibrium is decreasing local water density by increasing the extent of water self-association (row 1, Table 3.1.5-1). Thus it is reasoned that water in direct contact with a hydrophobic surface is less dense than bulk water some distance away from the hydrophobic surface.

This reasoning has been recently corroborated theoretically through molecular simulations of water near hydrophobic surfaces (Besseling and Lyklema, 1995; Lum *et al.*, 1999; Silverstein *et al.*, 1998) and experimentally by application of sophisticated vibrational spectroscopies (Du *et al.*, 1994; Gragson and Richmond, 1997). Although there is no precise uniformity among all investigators using a variety of different computational and experimental approaches, it appears that density variations propagate something of the order of 5 nm from a hydrophobic surface, or about 20 water layers.

There are at least two classes of hydrophilic surfaces that deserve separate mention here because these represent important categories of biomaterials as well (Hoffman, 1986). One class includes surfaces that *ad*sorb water through the interaction with surface-resident Lewis acid or base groups. These water–surface interactions are constrained to the outermost surface layer, say the upper 1 nm or so. Examples of these biomaterials might include polymers that have been surface treated by exposure to gas discharges, use of flames, or reaction with oxidative reagents as well as ceramics, metals, and glass. Another category of hydrophilic surfaces embraces those that significantly *ab*sorb water. Examples here are hydrogel polymers such as poly(vinyl alcohol) (PVA), poly(ethylene oxide) (PEO), or hydroxyethylmethacrylate (HEMA) that can visibly swell or even go into water solution, depending on the molecular weight and extent of crosslinking. Modern surface engineering can create materials that fall somewhere between water-adsorbent and -absorbent by depositing very thin films using self-assembly techniques (P.-Grosdemange *et al.*, 1991; Prime and Whitesides, 1993), reactive gas plasma deposition (Lopez *et al.*, 1992), or radiation grafting (Hoffman and Harris, 1972; Hoffman and Kraft, 1972; Ratner and Hoffman, 1980) as examples. Here, oligomers that would otherwise dissolve in water form a thin-film surface that cannot swell in the usual, macroscopic application of the word. In all of the mentioned cases, however, water hydrogen bonds with functional groups that may be characterized as either Lewis acid or base. In the limit of very strong (energetic) surface acidity or basicity, water can become ionized through proton or hydroxyl abstraction.

The subject of water structure near hydrophilic surfaces is considerably more complex than water structuring at hydrophobic surfaces just discussed, which itself is no trivial matter. This extra complexity is due to three related features of hydrophilic surfaces. First, each hydrophilic surface is a unique combination of both type and surface concentration of water-interactive Lewis acid or base functional groups (amine, carboxyl, ether, hydroxyl, etc.). Second, hydrophilic surfaces interact with water through both dispersion forces and Lewis acid–base interactions. This is to be contrasted to hydrophobic surfaces that interact with water only through dispersion forces (dispersion forces being a class of intermolecular interactions between the momentary dipoles in matter that arise from rapid fluctuations of electron density within molecular orbitals). Third, as a direct result of these two features, the number of possible water interactions and configurations is very large, especially if the hydrophilic surface is heterogeneous on a microscopic scale. These features make the problem of water behavior at hydrophilic surfaces both computationally and experimentally challenging.

In spite of this complexity, the reasoning and rationale applied to large polyelectrolytes in the preceding section should apply in an approximate way to extended hydrophilic surfaces, especially the more water-wettable types where acid–base interactions with water predominate over weaker dispersion interactions. This would suggest, then, that water near hydrophilic surfaces is more dense than bulk water, with a correspondingly less extensive self-associated water network (row 2, Table 3.1.5-1). There is some support for this general conclusion from simplified molecular models (Besseling, 1997; Silverstein *et al.*, 1998).

Thickness of this putative denser-water layer must depend in some way on the surface concentration (number) of Lewis acid/base sites and on whether the surface is predominately acid or predominately basic, but these relationships are far from worked out in detail. One set of experimental results suggesting that hydration layers near water-wettable surfaces can be quite thick comes from the rather startling finding by Pashley and Kitchener (1979) of 150-nm-thick, free-standing water films formed on fully water-wettable quartz surfaces from water vapor. These so-called condensate water films would comprise some 600 water molecules organized in a layer through unknown mechanisms. Perhaps these condensate films are formed from water-molecule layers with alternating oriented dipoles similar to the water layers around ions briefly discussed in the previous section. Note that this hypothetical arrangement defeats water self-association throughout the condensate-film layer in a manner consistent with the inferred less self-associated, high-density nature of water near hydrophilic surfaces.

Stepping back and viewing the full range of surface wetting behaviors discussed herein, it is apparent that water solvent properties (structure) near surfaces can be thought of as a sort of continuum or spectrum. At one end of the spectrum lie perfectly hydrophobic surfaces with no surface-resident Lewis acid or base sites. Water interacts with these hydrophobic surfaces only through dispersion forces mentioned above. At the other end of the spectrum, surfaces bear a sufficient surface concentration of Lewis sites to completely disrupt bulk water structure through a competition for hydrogen bonds, leading to complete water wetting (0° contact angle). Structure and solvent properties of water in contact with surfaces between these extremes must then exhibit some kind of graded properties associated with the graded wettability observed with contact angles. If the surface region is composed of molecules that hydrate to a significant degree, as in the case of hydrogel materials, then the surface can adsorb water and swell or dissolve. At the extreme of water–surface interactions, surface acid or base groups can abstract hydroxyls or protons from water, respectively, leading to water ionization at the surface.

Finally, in closing this section on water properties near surfaces, it is worthwhile to note that whereas insights gained from computational models employing hypothetical surfaces and experimental systems using atomically smooth mica and highly polished semiconductor-grade silicon wafers provide very important scientific insights, these results have limited direct biomedical relevance because practical biomaterial surfaces are generally quite rough relative to the dimensions of water (Fig. 3.1.5-1C). At the 0.25-nm scale, water structure near a hydrophobic polymer such as polyethylene, for example, might better be envisioned as a result of hydrating molecular-scale domains where methyl- and methylene-group protrusions from a "fractal" surface solvate in water rather than a sea of close-packed groups disposed erectly on an infinitely flat plane that one might construct in molecular modeling. Surfaces of functionalized polymers such as poly(ethylene terephthalate) (PET) would be even more complex. Both surface topography and composition will play a role in determining water structure near surfaces.

Water and the biological response to materials

It has long been assumed that the observed biological response to materials is initiated or catalyzed by interactions with material residing in the same thin surface region that affects water wettability, arguably no thicker than about 1 nm. In particular, it is frequently assumed that biological responses begin with protein adsorption. These assumptions are based on the observations that cells and proteins interact only at the aqueous interface of a material, that this interaction seemingly does not depend on the macroscopic thickness of a rigid material, and that water does not penetrate deeply into the bulk of many materials (excluding those that absorb water). Thus, one may conclude that biology does not "sense" or "see" bulk properties of a contacting material, only the outermost molecular groups protruding from a surface. Over the past decade or so, the validity of this assumption seems to have been confirmed through numerous studies employing SAMs supported on glass, gold, and silicon in which variation of the outermost surface functional groups exposed to blood plasma, purified proteins, and cells indeed induces different outcomes (Fragneto *et al.*, 1995; Liebmann-Vinson *et al.*, 1996; Margel *et al.*, 1993; Mooney *et al.*, 1996; Owens *et al.*, 1988; Petrash *et al.*, 1997; Prime and Whitesides, 1993; Scotchford *et al.*, 1998; Singhvi *et al.*, 1994; Sukenik *et al.*, 1990; Tidwell *et al.*, 1997; Vogler *et al.*, 1995a, b). But exactly how surfaces influence "biocompatibility" of a material is still not well understood.

Theories attempting to explain the role of surfaces in the biological response fall into two basic categories. One asserts that surface energy is the primary correlating surface property (Akers *et al.*, 1977; Baier, 1972; Baier *et al.*, 1969), the other states that water solvent properties near surfaces are the primary causative agent (Andrade *et al.*, 1981; Andrade and Hlady, 1986; Vogler, 1998). The former attempts to correlate surface energy factors such as critical surface energy σ_c or various interfacial tension components while the latter attempts correlations with water contact angle θ or some variant thereof such as water adhesion tension $\tau = \sigma_{lv} \cos \theta$, where σ_{lv} is the interfacial tension of water. Both approaches attempt to infer structure–property relationships between surface energy/wetting and some measure of the biological response. These two ideas would be functionally equivalent if water structure and solvent properties were directly related to surface energy in a straightforward way (e.g., linear), but this appears not to be the case (Vogler, 1998) because of water structuring in response to surface (adsorption) energetics, as described in preceding sections.

Both surface energy and water theories acknowledge that the principle interfacial events surfaces can promote or catalyze are adsorption and adhesion. Adsorption of proteins and/or adhesion of cells/tissues is known (or at least strongly suspected) to be involved in the primary interactions of biology with materials. Therefore, it is reasonable to anticipate that surfaces induce a biological response through adsorption and/or adhesion mechanisms. The surface energy theory acknowledges this connection by noting that surface energy is the engine that drives adsorption and adhesion. The water theory recognizes the same but in a quite different way. Instead, water theory asserts that surface energetics is the engine that drives adsorption of water and then, in subsequent steps, proteins and cells interact with the resulting hydrated interface either through or by displacing a so-called vicinal water layer that is more or less bound to the surface, depending on the energetics of the original water–surface interaction. Furthermore, water theory suggests that the ionic composition of vicinal water may be quite different than that of bulk water, with highly hydrated ions such as Ca^{2+} and Mg^{2+} preferentially concentrating in water near hydrophilic surfaces and less hydrated ions such as Na^{+} and K^{+} preferentially concentrating in water near hydrophobic surfaces. It is possible that the ionic composition of vicinal water layers further accounts for differences in the biological response to hydrophilic and hydrophobic materials on the basis that divalent ions have allosteric effects on enzyme reactions and participate in adhesion through divalent ion bridging.

Water is a very small, but very special, molecule. Properties of this universal biological solvent, this essential medium of life as we understand it, remain more mysterious in this century of science than those of the very atoms that compose it. Self-association of water through hydrogen bonding is the essential mechanism behind water solvent properties, and understanding self-association effects near surfaces is a key to understanding water properties in contact with biomaterials. It seems safe to conclude that no theory explaining the biological response to materials can be complete without accounting for water properties near surfaces and that this remains an exciting topic in biomaterials surface science.

Acknowledgments

The author is indebted to the editors for helpful and detailed discussion of the manuscript and to Professor J. Kubicki for molecular models used in construction of figures. Mr. Brian J. Mulhollem is thanked for reading the manuscript for typographical errors.

Bibliography

Akers, C. K., Dardik, I., Dardik, H., and Wodka, M. (1977). Computational methods comparing the surface properties of the inner walls of isolated human veins and synthetic biomaterials. *J. Colloid Interface Sci.* 59: 461–467.

Andrade, J. D., and Hlady, V. (1986). Protein adsorption and materials biocompatibility: a tutorial review and suggested mechanisms. *Adv. Polym. Sci.* 79: 3–63.

Andrade, J. D., Gregonis, D. E., and Smith, L. M. (1981). Polymer–water interface dynamics. in *Physicochemical aspects of polymer surfaces*, K. L. Mittal, ed. Plenum Press, New York, pp. 911–922.

Baier, R. E. (1972). The role of surface energy in thrombogenesis. *Bull. N. Y. Acad. Med.* 48: 257–272.

Baier, R. E. (1978). Key events in blood interactions at nonphysiologic interfaces—a personal primer. *Artificial Organs* **2**: 422–426.

Baier, R. E., and Meyer, A. E. (1996). Physics of solid surfaces. in *Interfacial Phenomena and Bioproducts*, J. L. Brash and P. W. Wojciechowski, eds. Marcel Dekker, New York, pp. 85–121.

Baier, R. E., Dutton, R. C., and Gott, V. L. (1969). Surface chemical features of blood vessel walls and of synthetic materials exhibiting thromboresistance. in *Surface Chemistry of Biological Systems*, M. Blank, ed. Plenum Publishers, New York, pp. 235–260.

Berendsen, H. J. C. (1967). Water structure. in *Theoretical and Experimental Biophysics*, A. Cole, ed. Marcel Dekker, New York, pp. 1–74.

Besseling, N. A. M. (1997). Theory of hydration forces between surfaces. *Langmuir* **13**: 2113–2122.

Besseling, N. A. M., and Lyklema, J. (1995). Hydrophobic hydration of small apolar molecules and extended surfaces: a molecular model. *Pure Appl. Chem.* **67**: 881–888.

Christenson, H. K., Fang, J., Ninham, B. W., and Parker, J. L. (1990). Effect of divalent electrolyte on the hydrophobic attraction. *J. Phys. Chem.* **94**: 8004–8006.

DeVoe, H. (1969). Theory of the conformations of biological macro-molecules in solution. in *Structure and Stability of Biological Molecules*, S. N. Timasheff and G. D. Fasman, eds. Marcel Dekker, New York, pp. 1–59.

Du, Q., Freysz, E., and Shen, Y. R. (1994). Surface vibrational spectroscopic studies of hydrogen bonding and hydrophobicity. *Science* **264**: 826–828.

Dunhill, P. (1965). How proteins acquire their structure. *Sci. Progr.* **53**: 609–619.

Fragneto, G., Thomas, R. K., Rennie, A. R., and Penfold, J. (1995). Neutron reflection study of bovine casein adsorbed on OTS self-assembled monolayers. *Science* **267**: 657–660.

Franks, F. (1972). Introduction — water, the unique chemical. in *Water: A Comprehensive Treatise*, F. Franks, ed. Plenum Publishers, New York, pp. 1–17.

Gragson, D. E., and Richmond, G. L. (1997). Comparisons of the structure of water at near oil/water and air/water interfaces as determined by vibrational sum frequency generation. *Langmuir* **13**: 4804–4806.

Hoffman, A. S. (1986). A general classification scheme for "hydrophilic" and "hydrophobic" biomaterial surfaces. *J. Biomed. Mater. Res.* **20**: ix–xi.

Hoffman, A. S., and Harris, C. (1972). Radiation-grafted hydrogels on silicone rubber surfaces — a new biomaterial. *Polym. Preprints* **13**: 740–746.

Hoffman, A. S., and Kraft, W. G. (1972). Radiation-grafted hydrogels on polyurethane surfaces—a new biomaterial. *Polym. Preprints* **13**: 723–728.

Iassacs, E. D., Shukla, A., Platzman, P. M., Hamann, D. R., Bariellini, B., and Tulk, C. A. (1999). Covalency of the hydrogen bond in ice: a direct X-ray measurement. *Phys. Rev. Lett.* **82**: 600–603.

Kauzmann, W. (1959). Some factors in the interpretation of protein denaturation. *Adv. Protein Chem.* **14**: 1–63.

Liebmann-Vinson, A., Lander, L. M., Foster, M. D., Brittain, W. J., Vogler, E. A., Majkrak, C. F., and Satija, S. (1996). A neutron reflectometry study of human serum albumin adsorption in Situ. *Langmuir* **12**: 2256–2262.

Lopez, G. P., Ratner, B. D., Tidwell, C. D., Haycox, C. L., Rapoza, R. J., and Horbett, T. A. (1992). Glow discharge plasma deposition of tetraethylene glycol dimethyl ether for fouling-resistant biomaterial surfaces. *J. Biomed. Mater. Res.* **26**: 415–439.

Lum, K., Chandler, D., and Weeks, J. D. (1999). Hydrophobicity at small and large length scales. *J. Phys. Chem. B* **103**: 4570–4577.

Luzar, A., and Chandler, D. (1996). Hydrogen-bond kinetics in liquid water. *Nature* **379**: 55–57.

Marcus, Y. (1985). *Ion Solvation*. Wiley, New York.

Margel, S., Vogler, E. A., Firment, L., Watt, T., Haynie, S., and Sogah, D. Y. (1993). Peptide, protein, and cellular interactions with self-assembled monolayer model surfaces. *J. Biomed. Mater. Res.* **27**: 1463–1476.

Marshall, E. (1999). Getting to the bottom of water. *Science* **283**: 614–615.

Mooney, J. F., Hunt, A. J., McIntosh, J. R., Leberko, C. A., Walba, D. M., and Rogers, C. T. (1996). Patterning of functional antibodies and other proteins by photolithograpy of silane monolayers. *Proc. Natl. Acad. Sci. USA* **93**: 12287–12291.

Narten, A. H., and Levy, H. A. (1969). Observed diffraction pattern and proposed models of liquid water. *Science* **165**: 447–454.

Oss, C. J. v., and Giese, R. F. (1995). The hydrophilicity and hydrophobicity of clay minerals. *Clays Clay Miner.* **43**: 474–477.

Owens, N. F., Gingell, D., and Trommler, A. (1988). Cell adhesion to hydroxyl groups of a monolayer film. *J. Cell Sci.* **91**: 269–279.

P.-Grosdemange, C., Simon, E. S., Prime, K. L., and Whitesides, G. M. (1991). Formation of self-assembled monolayers by chemisorption of derivatives of oligo(ethylene glycol) on gold. *J. Am. Chem. Soc.* **113**: 13–20.

Pain, R. H. (1982). Molecular hydration and biological function. in *Biophysics of Water*, F. Franks and S. Mathias, eds. John Wiley and Sons, Chichester, UK, pp. 3–14.

Pashley, R. M., and Kitchener, J. A. (1979). Surface forces in adsorbed multilayers of water on quartz. *J. Colloid Interface Sci.* **71**: 491–500.

Paulaitis, M. E., Garde, S., and Ashbaugh, H. S. (1996). The hydrophobic effect. *Curr. Opin. Colloid Interface Sci.* **1**: 376–383.

Petrash, S., Sheller, N. B., Dando, W., and Foster, M. D. (1997). Variation in tenacity of protein adsorption on self-assembled monolayers with monolayer order as observed by X-ray reflectivity. *Langmuir* **13**: 1881–1883.

Prime, K. L., and Whitesides, G. M. (1993). Adsorption of proteins onto surfaces containing end-attached oligo ethylene oxide: a model system using self-assembled monolayers. *J. Am. Chem. Soc.* **115**: 10714–10721.

Qi, Z., and Soka, M. (1998). Dynamic properties of individual water molecules in a hydrophobic pore lined with acyl chains: a molecular dynamics study. *Biophys. Chem.* **71**: 35–50.

Ratner, B. D., and Hoffman, A. S. (1980). Surface characterization of hydrophilic–hydrophobic copolymer model systems I. A preliminary study. *Polymer. Sci. Technol.* **12B**: 691–706.

Robinson, G. W., Zhu, S.-B., Singh, S., and Evans, M. W. (1996). *Water in Biology, Chemistry, and Physics*. World Scientific, London.

Scotchford, C. A., Cooper, E., Leggett, G. J., and Downes, S. (1998). Growth of human osteoblast-like cells on alkanethiol on gold self-assembled monolayers: the effect of surface chemistry. *J. Biomed. Mater. Res.* **41**: 431–442.

Silverstein, K.A.T., Haymet, A.D.J., and Dill, K. A. (1998). A simple model of water and the hydrophobic effect. *J. Am. Chem. Soc.* **120**: 3166–3175.

Singhvi, R., Kumar, A., Lopez, G. P., Stephanopoulos, G. N., Wang, D.I.C., Whitesides, G. M., and Ingber, D. E. (1994). Engineering cell shape and function. *Science* **264**: 696–698.

Sukenik, C. N., Balachander, N., Culp, L. A., Lewandowska, K., and Merritt, K. (1990). Modulation of cell adhesion by modification of titanium surfaces with covalently attached self-assembled monolayers. *J. Biomed. Mater. Res.* **24**: 1307–1323.

Tanford, C. (1966). *Physical Chemistry of Macromolecules*. John Wiley and Sons, New York.

Tanford, C. (1973). *The Hydrophobic Effect: Formation of Micelles and Biological Membranes*. John Wiley and Sons, New York.

Tidwell, C. D., Ertel, S. I., Ratner, B. D., Tarasevich, B. J., Atre, S., and Allara, D. L. (1997). Endothelial cell growth and protein adsorption on terminally functionalized, self-assembled monolayers of alkanethiols on gold. *Langmuir* **13**: 3404–3413.

Vinogradov, S. N., and Linnell, R. H. (1971). *Hydrogen Bonding*. Van Nostrand Reinhold, New York.

Vogler, E. A. (1998). Structure and reactivity of water at biomaterial surfaces. *Adv. Colloid Interface Sci.* **74**: 69–117.

Vogler, E. A., Graper, J. C., Harper, G. R., Lander, L. M., and Brittain, W. J. (1995a). Contact activation of the plasma coagulation cascade. 1. Procoagulant surface energy and chemistry. *J. Biomed. Mater. Res.* **29**: 1005–1016.

Vogler, E. A., Graper, J. C., Sugg, H. W., Lander, L. M., and Brittain, W. J. (1995b). Contact activation of the plasma coagulation cascade. 2. Protein adsorption on procoagulant surfaces. *J. Biomed. Mater. Res.* **29**: 1017–1028.

Wiggins, P. M. (1971). Water structure as a determinant of ion distribution in living tissue. *J. Theor. Biol.* **32**: 131–146.

Wiggins, P. M. (1988). Water structure in polymer membranes. Prog. Polym. Sci. **13**: 1–35.

Wiggins, P. M. (1990). Role of water in some biological processes. *Microbiol. Rev.* **54**: 432–449.

Wiggins, P. M., and Ryn, R. T. v. (1990). Changes in ionic selectivity with changes in density of water in gels and cells. *Biophys. J.* **58**: 585–596.

Chapter 3.2

3.2

Classes of materials used in medicine

Buddy D. Ratner, Allan Hoffman, Frederick J. Schoen, and Jack E. Lemons

3.2.1 Introduction

Allan S. Hoffman

Biomaterials can be divided into four major classes of materials: polymers, metals, ceramics (including carbons, glass-ceramics, and glasses), and natural materials (including those from both plants and animals). Sometimes two different classes of materials are combined together into a composite material, such as silica-reinforced silicone rubber (SR) or carbon fiber- or hydroxyapatite (HA) particle-reinforced poly (lactic acid). Such composites are a fifth class of biomaterials. What is the history behind the development and application of such diverse materials for implants and medical devices, what are the compositions and properties of these materials, and what are the principles governing their many uses as components of implants and medical devices? This chapter critically reviews this important literature of biomaterials.

The wide diversity and sophistication of materials currently used in medicine and biotechnology is testimony to the significant scientific and technological advances that have occurred over the past 50 years. From World War II to the early 1960s, relatively few pioneering surgeons were taking commercially available polymers and metals, fabricating implants and components of medical devices from them, and applying them clinically. There was little government regulation of this activity, and yet these earliest implants and devices had a remarkable success. However, there were also some dramatic failures. This led the surgeons to enlist the aid of physical, biological, and materials scientists and engineers, and the earliest interdisciplinary "bioengineering" collaborations were born.

These teams of physicians and scientists and engineers not only recognized the need to control the composition, purity, and physical properties of the materials they were using, but they also recognized the need for new materials with new and special properties. This stimulated the development of many new materials in the 1970s. New materials were designed *de novo* specifically for medical use, such as biodegradable polymers and bioactive ceramics. Some were derived from existing materials fabricated with new technologies, such as polyester fibers that were knit or woven in the form of tubes for use as vascular grafts, or cellulose acetate (CA) plastic that was processed as bundles of hollow fibers for use in artificial kidney dialysers. Some materials were "borrowed" from unexpected sources such as pyrolytic carbons or titanium alloys that had been developed for use in air and space technology. And other materials were modified to provide special biological properties, such as immobilization of heparin for anti-coagulant surfaces. More recently biomaterials scientists and engineers have developed a growing interest in natural tissues and polymers in combination with living cells. This is particularly evident in the field of tissue engineering, which focuses on the repair or regeneration of natural tissues and organs. This interest has stimulated the isolation, purification, and application of many different natural materials. The principles and applications of all of these biomaterials and modified biomaterials are critically reviewed in this chapter.

3.2.2 Polymers

Stuart L. Cooper, Susan A. Visser, Robert W. Hergenrother, and Nina M. K. Lamba

Many types of polymers are widely used in biomedical devices that include orthopedic, dental, soft tissue, and

Biomedical Engineering Desk Reference; ISBN: 9780123746467

cardiovascular implants. Polymers represent the largest class of biomaterials. In this section, we will consider the main types of polymers, their characterization, and common medical applications. Polymers may be derived from natural sources, or from synthetic organic processes.

The wide variety of natural polymers relevant to the field of biomaterials includes plant materials such as cellulose, sodium alginate, and natural rubber, animal materials such as tissue-based heart valves and sutures, collagen, glycosaminoglycans (GAGs), heparin, and hyaluronic acid, and other natural materials such as deoxyribonucleic acid (DNA), the genetic material of all living creatures. Although these polymers are undoubtedly important and have seen widespread use in numerous applications, they are sometimes eclipsed by the seemingly endless variety of synthetic polymers that are available today. Synthetic polymeric biomaterials range from hydrophobic, non-water-absorbing materials such as SR, polyethylene (PE), polypropylene (PP), poly(ethylene terephthalate) (PET), polytetrafluoroethylene (PTFE), and poly(methyl methacrylate) (PMMA) to somewhat more polar materials such as poly(vinyl chloride) (PVC), copoly(lactic–glycolic acid) (PLGA), and nylons, to water-swelling materials such as poly(hydroxyethyl methacrylate) (PHEMA) and beyond, to water-soluble materials such as poly(ethylene glycol) (PEG or PEO). Some are hydrolytically unstable and degrade in the body while others may remain essentially unchanged for the lifetime of the patient.

Both natural and synthetic polymers are long-chain molecules that consist of a large number of small repeating units. In synthetic polymers, the chemistry of the repeat units differs from the small molecules (monomers) that were used in the original synthesis procedures, resulting from either a loss of unsaturation or the elimination of a small molecule such as water or HCl during polymerization. The exact difference between the monomer and the repeat unit depends on the mode of polymerization, as discussed later.

The task of the biomedical engineer is to select a biomaterial with properties that most closely match those required for a particular application. Because polymers are long-chain molecules, their properties tend to be more complex than those of their short-chain precursors. Thus, in order to choose a polymer type for a particular application, the unusual properties of polymers must be understood.

This section introduces the concepts of polymer synthesis, characterization, and property testing as they are relevant to the eventual application of a polymer as a biomaterial. Following this, examples of polymeric biomaterials currently used by the medical community are cited and their properties and uses are discussed.

Molecular weight

In polymer synthesis, a polymer is usually produced with a distribution of molecular weights. To compare the molecular weights of two different batches of polymer, it is useful to define an average molecular weight. Two statistically useful definitions of molecular weight are the number average and weight average molecular weights. The number average molecular weight (M_n) is the first moment of the molecular weight distribution and is an average over the number of molecules. The weight average molecular weight (M_w) is the second moment of the molecular weight distribution and is an average over the weight of each polymer chain. Equations 3.2.2.1 and 3.2.2.2 define the two averages:

$$M_n = \frac{\sum N_i M_i}{\sum N_i} \quad (3.2.2.1)$$

$$M_w = \frac{\sum N_i M_i^2}{\sum N_i M_i} \quad (3.2.2.2)$$

where N_i is the number of moles of species i, and M_i is the molecular weight of species i.

The ratio of M_w to M_n is known as the polydispersity index (PI) and is used as a measure of the breadth of the molecular weight distribution. Typical commercial polymers have polydispersity indices of 1.5–50, although polymers with polydispersity indices of less than 1.1 can be synthesized with special techniques. A molecular weight distribution for a typical polymer is shown in Fig. 3.2.2-1.

Linear polymers used for biomedical applications generally have M_n in the range of 25,000 to 100,000 and M_w from 50,000 to 300,000, and in exceptional cases, such as the PE used in the hip joint, the M_w may range up to a million. Higher or lower molecular weights may be necessary, depending on the ability of the polymer chains

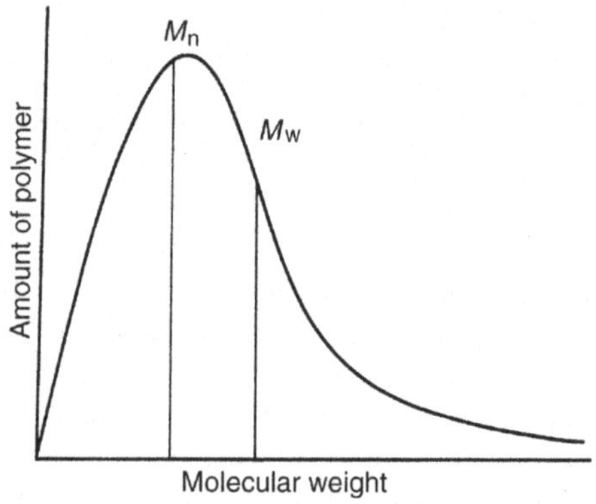

Fig. 3.2.2-1 Typical molecular weight distribution of a polymer.

Table 3.2.2-1 Mechanical properties of biomedical polymers

Polymer	Water absorption (%)	Bulk modulus (GPa)	Tensile strength (MPa)	Elongation at break (%)	T_g (K)	T_m (K)
Polyethylene	0.001–0.02	0.8–2.2	30–40	130–500	160–170	398–408
Polypropylene	0.01–0.035	1.6–2.5	21–40	100–300	243–270	433–453
Polydimethyl–siloxane	0.08–0.1		3–10	50–800	148	233
Polyurethane	0.1–0.9	1.5–2	28–40	600–720	200–250	453–523*
Polytetrafluoro–ethylene	0.01–0.05	1–2	15–40	250–550	293–295	595–600
Polyvinyl–chloride	0.04–0.75	3–4	10–75	10–400	250–363	423*
Polyamides	0.25–3.5	2.4–3.3	44–90	40–250	293–365	493–540
Polymethyl–methacrylate	0.1–0.4	3–4.8	38–80	2.5–6	379–388	443*
Polycarbonate	0.15–0.7	2.8–4.6	56–75	8–130	418	498–523
Polyethylene–terephthalate	0.06–0.3	3–4.9	42–80	50–500	340–400	518–528

* = decomposition temperature

to crystallize or to exhibit secondary interactions such as hydrogen bonding. The crystallinity and secondary interactions can give polymers additional strength. In general, increasing molecular weight corresponds to increasing physical properties; however, since melt viscosity also increases with molecular weight, processability will decrease and an upper limit of useful molecular weights is usually reached. Mechanical properties of some polymeric biomaterials are presented in Table 3.2.2-1.

Synthesis

Methods of synthetic polymer preparation fall into two categories: addition polymerization (chain reaction) and condensation polymerization (stepwise growth) (Fig. 3.2.2-2). (Ring opening is another type of polymerization and is discussed in more detail later in the section on degradable polymers.) In addition polymerization, unsaturated monomers react through the stages of

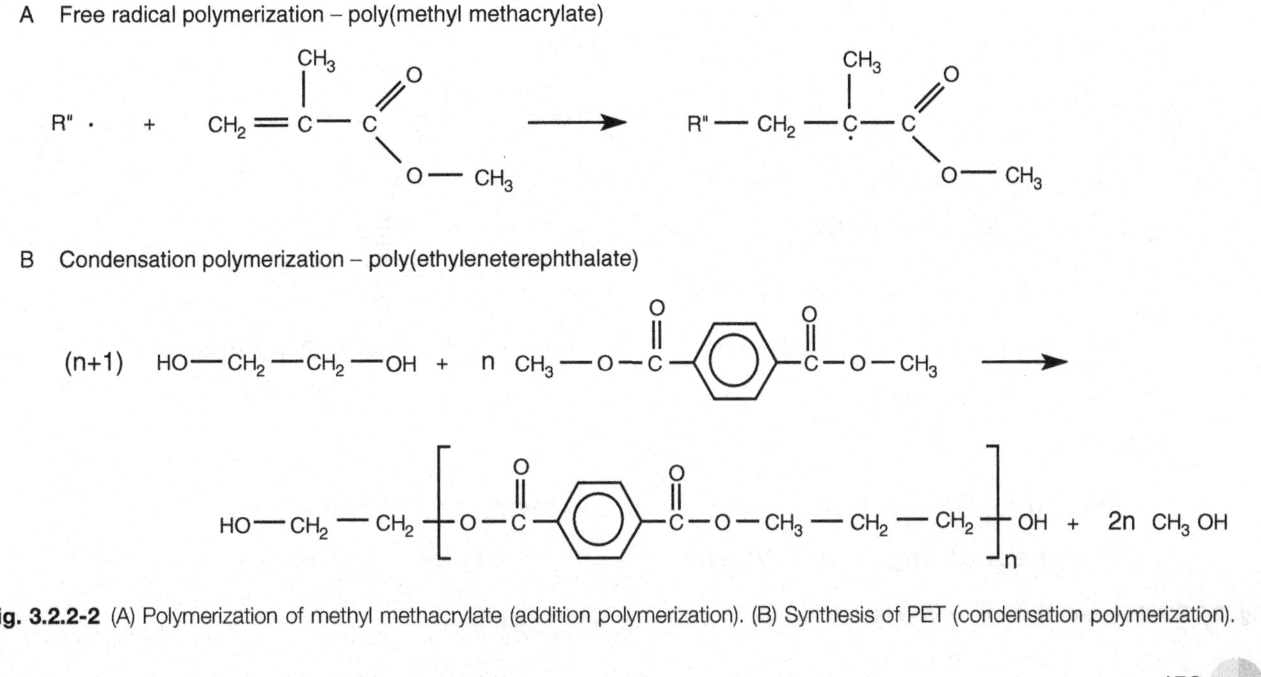

Fig. 3.2.2-2 (A) Polymerization of methyl methacrylate (addition polymerization). (B) Synthesis of PET (condensation polymerization).

initiation, propagation, and termination to give the final polymer product. The initiators can be free radicals, cations, anions, or stereospecific catalysts. The initiator opens the double bond of the monomer, presenting another "initiation" site on the opposite side of the monomer bond for continuing growth. Rapid chain growth ensues during the propagation step until the reaction is terminated by reaction with another radical, a solvent molecule, another polymer molecule, an initiator, or an added chain transfer agent. PVC, PE, and PMMA are relevant examples of addition polymers used as biomaterials. The polymerization of MMA to form PMMA is shown in Fig. 3.2.2-2A.

Condensation polymerization is completely analogous to condensation reactions of low-molecular-weight molecules. Two monomers react to form a covalent bond, usually with elimination of a small molecule such as water, hydrochloric acid, methanol, or carbon dioxide. Nylon and PET (Fig. 3.2.2-2B) are typical condensation polymers and are used in fiber or fabric form as biomaterials. The reaction continues until almost all of one reactant is used up. There are also polymerizations that resemble the stepwise growth of condensation polymers, although no small molecule is eliminated. Polyurethane synthesis bears these characteristics, which is sometimes referred to as polyaddition or rearrangement polymerization (Brydson, 1995).

The choice of polymerization method strongly affects the polymer obtained. In free radical polymerization, a type of addition polymerization, the molecular weights of the polymer chains are difficult to control with precision. Added chain transfer agents are used to control the average molecular weights, but molecular weight distributions are usually broad. In addition, chain transfer reactions with other polymer molecules can produce undesirable branched products (Fig. 3.2.2-3) that affect the ultimate properties of the polymeric material. In contrast, molecular architecture can be controlled very precisely in anionic polymerization. Regular linear chains with PI indices close to unity can be obtained. More recent methods of living free radical polymerizations called ATRP and RAFT may also yield low PIs.

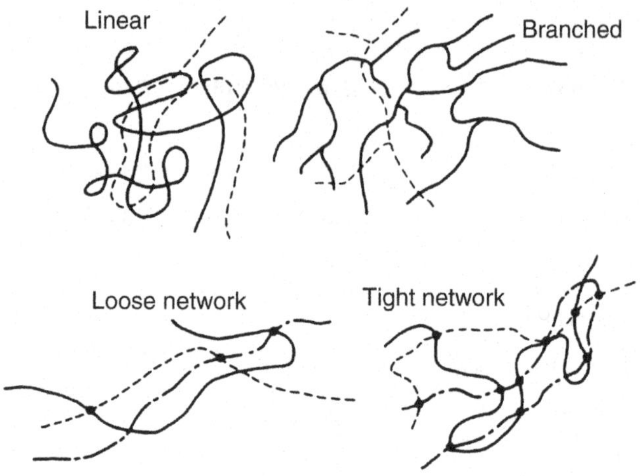

Fig. 3.2.2-3 Polymer arrangements. (From F. Rodriguez, *Principles of Polymer Systems*, Hemisphere Publ., 1982, p. 21, with permission.)

Polymers produced by addition polymerization can be homopolymers, i.e., polymers containing only one type of repeat unit, or copolymers with two or more types of repeat units. Depending on the reaction conditions and the reactivity of each monomer type, the copolymers can be random, alternating, graft, or block copolymers, as illustrated in Fig. 3.2.2-4. Random copolymers exhibit properties that approximate the weighted average of those of the two types of monomer units, whereas block copolymers tend to phase separate into a monomer-A-rich phase and a monomer-B-rich phase, displaying properties unique to each of the homopolymers. Figure 3.2.2-5 shows the repeat units of many of the homopolymers used in medicine.

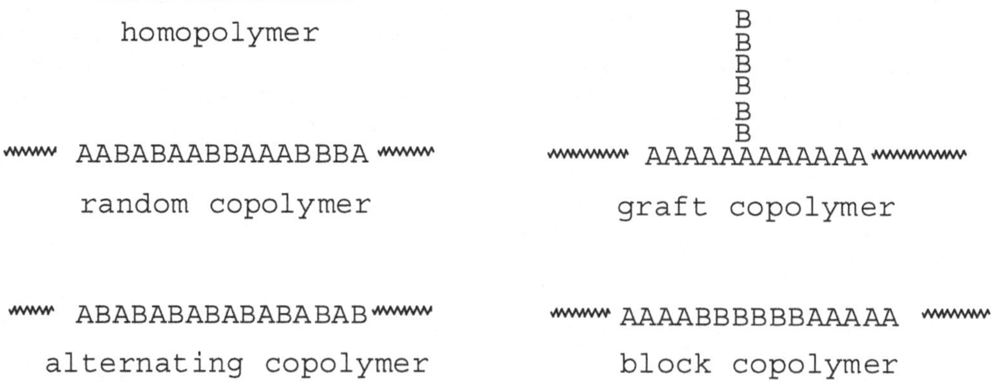

Fig. 3.2.2-4 Possible structures of polymer chains.

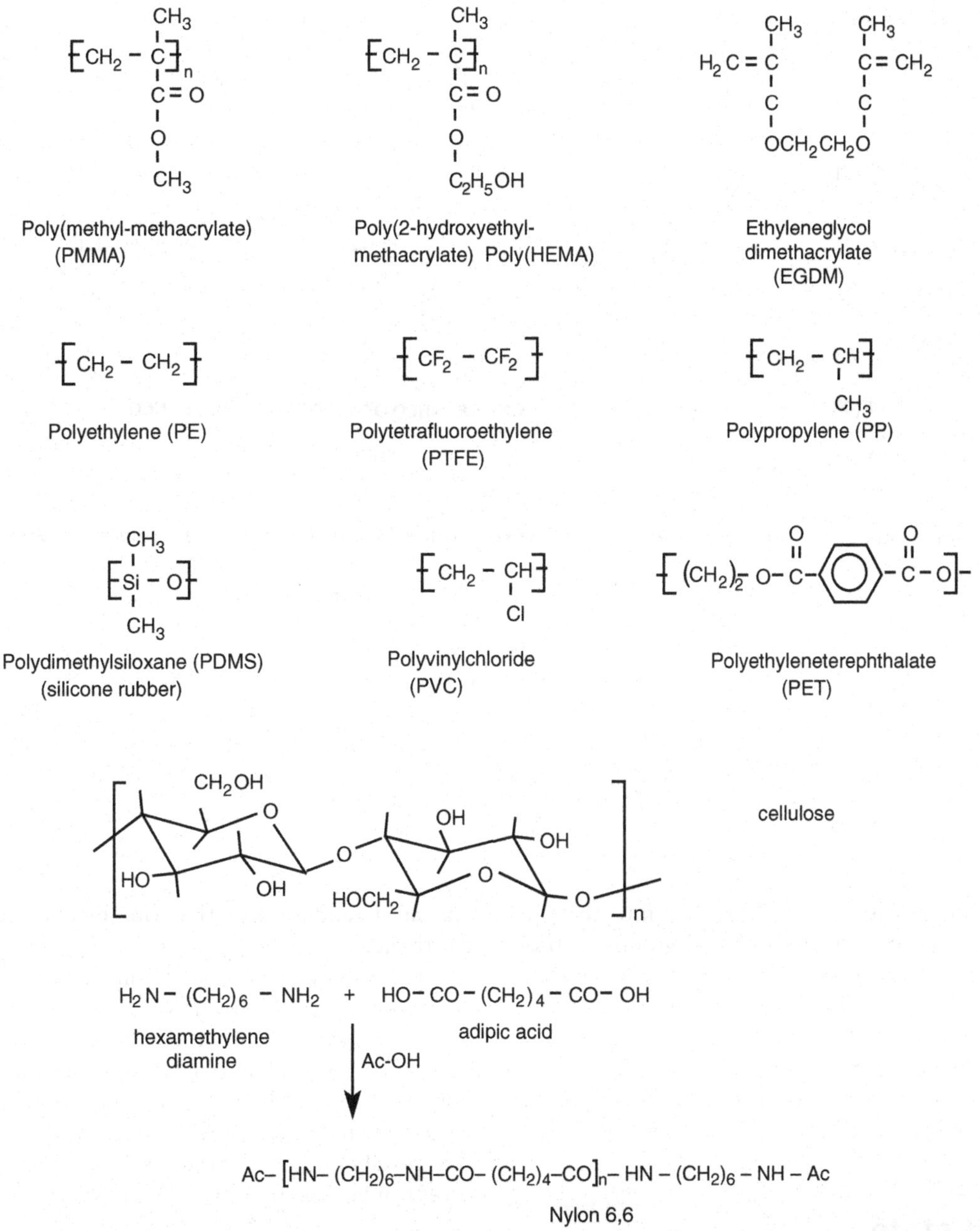

Fig. 3.2.2-5 Homopolymers used as biomaterials.

Condensation polymerization can also result in copolymer formation. The properties of the condensation copolymer depend on three factors: the chemistry of monomer units; the molecular weight of the polymer product, which can be controlled by the ratio of one reactant to another and by the time of polymerization; and the final distribution of the molecular weight of the copolymer chains. The use of bifunctional monomers gives rise to linear polymers, while multifunctional monomers may be used to form covalently cross-linked networks. Figure 3.2.2-6 shows the reactant monomers and polymer products of some biomedical copolymers.

Postpolymerization cross-linking of addition or condensation polymers is also possible. Natural rubber, for example, consists mostly of linear molecules that can be cross-linked to a loose network with 1–3% sulfur (vulcanization) or to a hard rubber with 40–50% sulfur (Fig. 3.2.2-3). In addition, physical, rather than chemical, cross-linking of polymers can occur in microcrystalline regions, that are present in nylon (Fig. 3.2.2-7A). Alternatively, physical cross-linking can be achieved through incorporation of ionic groups in the polymer (Fig. 3.2.2-7B). This is used in acrylic acid cement systems (e.g., for dental cements) where divalent cations

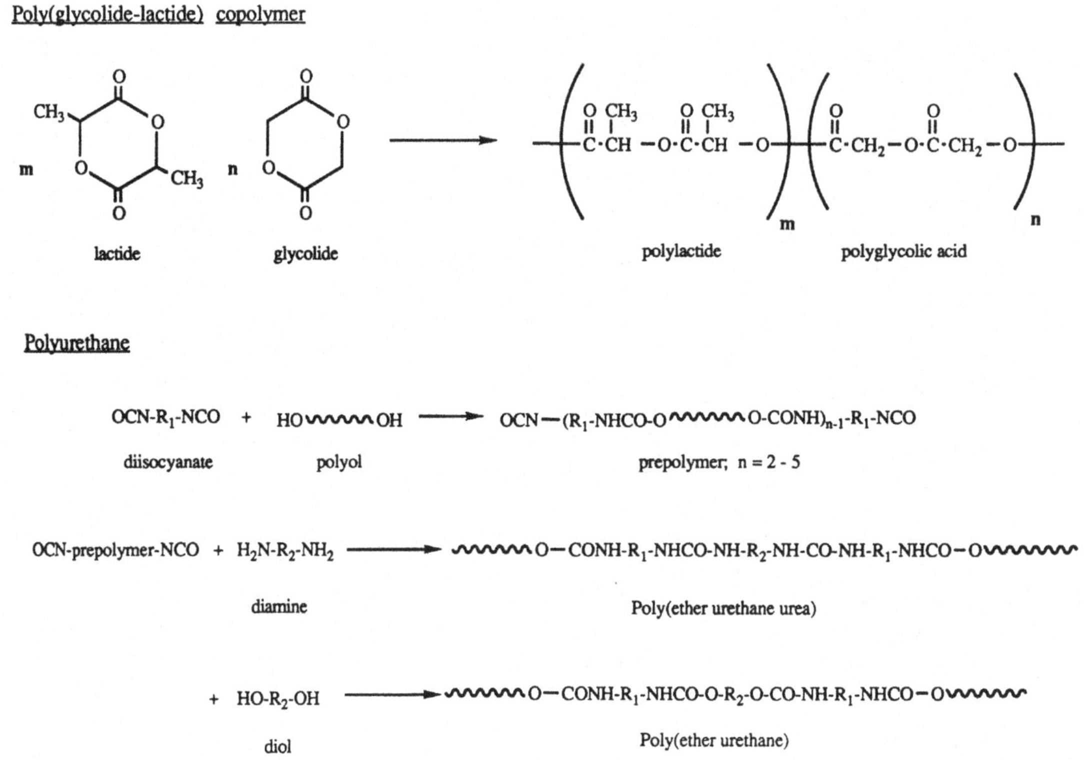

Fig. 3.2.2-6 Copolymers used in medicine and their base monomers.

such as zinc and calcium are incorporated into the formulation and interact with the carboxyl groups to produce a strong, hard material. The alginates, which are polysaccharides derived from brown seaweed, also contain anionic residues that will interact with cations and water to form a gel. The alginates are used successfully to dress deep wounds and are also being studied as tissue engineering matrices.

The solid state

Tacticity

Polymers are long-chain molecules and, as such, are capable of assuming many conformations through rotation of valence bonds. The extended chain or planar zigzag conformation of PP is shown in Fig. 3.2.2-8. This figure illustrates the concept of tacticity. Tacticity refers to the arrangement of substituents (methyl groups in the case of PP) around the extended polymer chain. Chains in which all substituents are located on the same side of the zigzag plane are isotactic, whereas syndiotactic chains have substituents alternating from side to side. In the atactic arrangement, the substituent groups appear at random on either side of the extended chain backbone.

Atactic polymers usually cannot crystallize, and an amorphous polymer results. Isotactic and syndiotactic polymers may crystallize if conditions are favorable. PP is an isotactic crystalline polymer used as sutures. Crystalline polymers, such as PE, also possess a higher level of structure characterized by folded chain lamellar growth that results in the formation of spherulites. These structures can be visualized in a polarized light microscope.

Crystallinity

Polymers can be either amorphous or semicrystalline. They can never be completely crystalline owing to lattice defects that form disordered, amorphous regions. The tendency of a polymer to crystallize is enhanced by the small side groups and chain regularity. The presence of crystallites in the polymer usually leads to enhanced mechanical properties, unique thermal behavior, and increased fatigue strength. These properties make semicrystalline polymers (often referred to simply as crystalline polymers) desirable materials for biomedical applications. Examples of crystalline polymers used as biomaterials are PE, PP, PTFE, and PET.

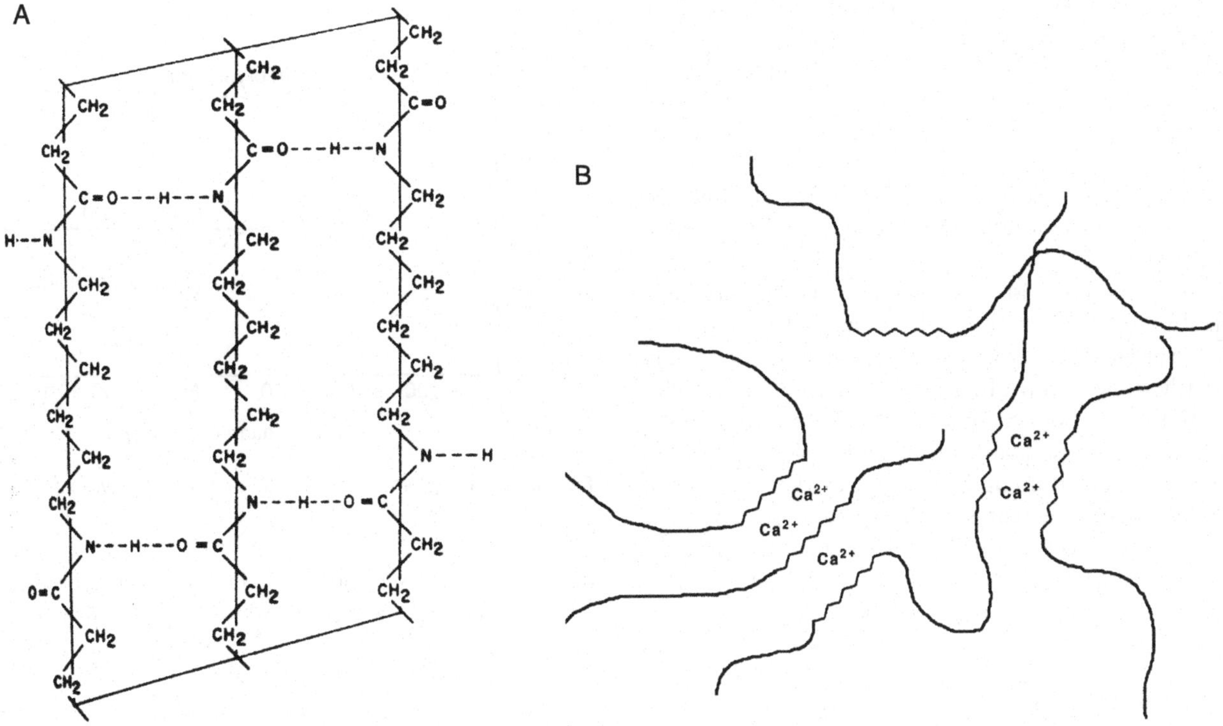

Fig. 3.2.2-7 (A) Hydrogen bonding in nylon-6,6 molecules in a triclinic unit cell: σ form. (From L. Mandelkern, *An Introduction to Macromolecules*, Springer-Verlag, 1983, p. 43, with permission.) (B) Ionic aggregation giving rise to physical cross-links in copolymers.

Mechanical properties

The tensile properties of polymers can be characterized by their deformation behavior (stress-strain response (Fig. 3.2.2-9). Amorphous, rubbery polymers are soft and reversibly extensible. The freedom of motion of the polymer chain is retained at a local level while a network structure resulting from chemical cross-links and chain entanglements prevents large-scale movement or flow. Thus, rubbery polymers tend to exhibit a lower modulus, or stiffness, and extensibilities of several hundred percent, as shown in Table 3.2.2-1. Rubbery materials may also exhibit an increase of stress prior to breakage as a result of strain-induced crystallization assisted by molecular orientation in the direction of stress. Glassy and semicrystalline polymers have higher moduli and lower extensibilities.

The ultimate mechanical properties of polymers at large deformations are important in selecting particular

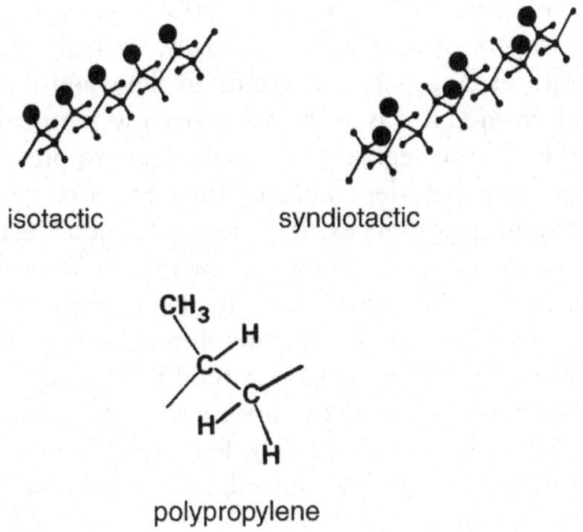

Fig. 3.2.2-8 Schematic of stereoisomers of PP. (From F. Rodriguez, *Principles of Polymer Systems*, Hemisphere Publ., 1982, p. 22, with permission.)

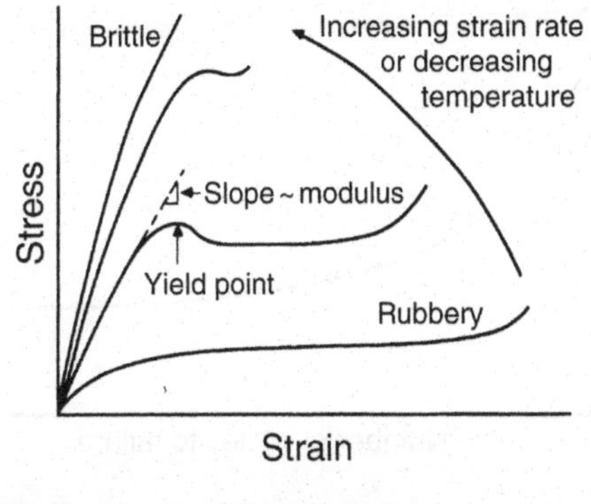

Fig. 3.2.2-9 Tensile properties of polymers.

polymers for biomedical applications. The ultimate strength of polymers is the stress at or near failure. For most materials, failure is catastrophic (complete breakage). However, for some semicrystalline materials, the failure point may be defined by the stress point where large inelastic deformation starts (yielding). The toughness of a polymer is related to the energy absorbed at failure and is proportional to the area under the stress-strain curve.

The fatigue behavior of polymers is also important in evaluating materials for applications where dynamic strain is applied. For example, polymers that are used in the artificial heart must be able to withstand many cycles of pulsating motion. Samples that are subjected to repeated cycles of stress and release, as in a flexing test, fail (break) after a certain number of cycles. The number of cycles to failure decreases as the applied stress level is increased, as shown in Fig. 3.2.2-10. For some materials, a minimum stress exists below which failure does not occur in a measurable number of cycles.

Thermal properties

In the liquid or melt state, a noncrystalline polymer possesses enough thermal energy for long segments of each polymer to move randomly (Brownian motion). As the melt is cooled, a temperature is eventually reached at which all long-range segmental motions cease. This is the glass transition temperature (T_g), and it varies from polymer to polymer. Polymers used below their T_g, such as PMMA, tend to be hard and glassy, while polymers used above their T_g, such as SR, are rubbery. Polymers with any crystallinity will also exhibit a melting temperature (T_m) owing to melting of the crystalline phase. These polymers, such as PET, PP, and nylon, will be relatively hard and strong below T_g, and tough and strong above T_g. Thermal transitions in polymers can be measured by differential scanning calorimetry (DSC), as

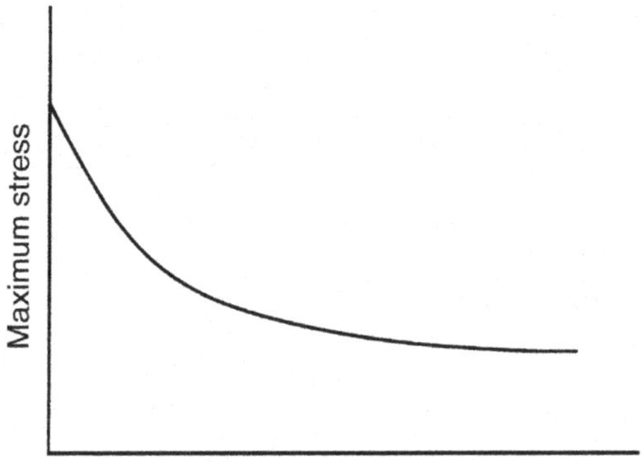

Fig. 3.2.2-10 Fatigue properties of polymers.

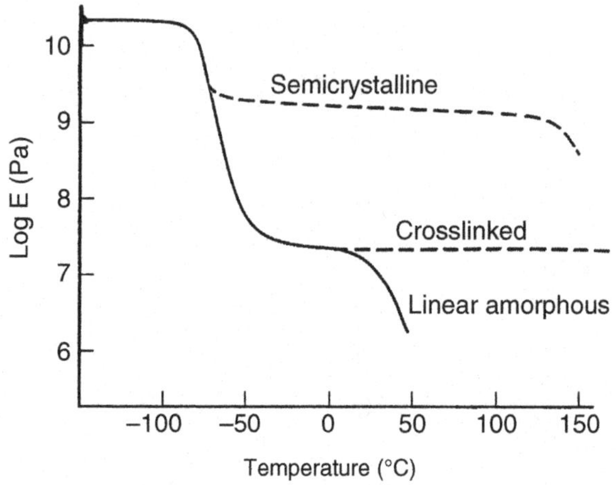

Fig. 3.2.2-11 Dynamic mechanical behavior of polymers.

discussed in the section on characterization techniques. All polymers have a T_g, but only polymers with regular chain architecture can pack well, crystallize, and exhibit a T_m. The T_g is always below the T_m.

The viscoelastic responses of polymers can also be used to classify their thermal behavior. The modulus versus temperature curves shown in Fig. 3.2.2-11 illustrate behaviors typical of linear amorphous, cross-linked, and semicrystalline polymers. The response curves are characterized by a glassy modulus below T_g of approximately 3×10^9 Pa. For linear amorphous polymers, increasing temperature induces the onset of the glass transition region where, in a 5–10°C temperature span (depending on heating rate), the modulus drops by three orders of magnitude, and the polymer is transformed from a stiff glass to a leathery material. The relatively constant modulus region above T_g is the rubbery plateau region where long-range segmental motion is occurring but thermal energy is insufficient to overcome entanglement interactions that inhibit flow. This is the target region for many biomedical applications. Finally, at high enough temperatures, the polymer begins to flow, and a sharp decrease in modulus is seen over a narrow temperature range. This is the region where polymers are processed into various shapes, depending on their end use.

Crystalline polymers exhibit the same general features in modulus versus temperature curves as amorphous polymers; however, crystalline polymers possess a higher plateau modulus owing to the reinforcing effect of the crystallites. Crystalline polymers tend to be tough, ductile plastics whose properties are sensitive to processing history. When heated above their flow point, they can be melt processed and will crystallize and become rigid again upon cooling.

Chemically cross-linked polymers exhibit modulus versus temperature behavior analogous to that of linear amorphous polymers until the flow regime is approached.

Unlike linear polymers, chemically cross-linked polymers do not display flow behavior; the cross links inhibit flow at all temperatures below the degradation temperature. Thus, chemically cross-linked polymers cannot be melt processed. Instead, these materials are processed as reactive liquids or high-molecular-weight amorphous gums that are cross-linked during molding to give the desired product. SR is an example of this type of polymer. Some cross-linked polymers are formed as networks during polymerization, and then must be machined to be formed into useful shapes. The soft contact lens, PHEMA, is an example of this type of network polymer; it is shaped in the dry state, and used when swollen with water.

Copolymers

In contrast to the thermal behavior of homopolymers discussed earlier, copolymers can exhibit a number of additional thermal transitions. If the copolymer is random, it will exhibit a T_g that approximates the weighted average of the T_g values of the two homopolymers. Block copolymers of sufficient size and incompatible block types, such as the polyurethanes, will exhibit two individual transitions, each one characteristic of the homopolymer of one of the component blocks (in addition to other thermal transitions) but slightly shifted owing to incomplete phase separation.

Characterization techniques

Determination of molecular weight

Gel permeation chromatography (GPC), a type of size exclusion chromatography, involves passage of a dilute polymer solution over a column of porous beads. High-molecular-weight polymers are excluded from the beads and elute first, whereas lower molecular-weight molecules pass through the pores of the bead, increasing their elution time. By monitoring the effluent of the column as a function of time using an ultraviolet or refractive index detector, the amount of polymer eluted during each time interval can be determined. Comparison of the elution time of the samples with those of monodisperse samples of known molecular weight allows the entire molecular weight distribution to be determined. A typical GPC trace is shown in Fig. 3.2.2-12.

Osmotic pressure measurements can be used to measure M_n. A semipermeable membrane is placed between two chambers.

Only solvent molecules flow freely through the membrane. Pure solvent is placed in one chamber, and a dilute polymer solution of known concentration is placed in the other chamber. The lowering of the activity

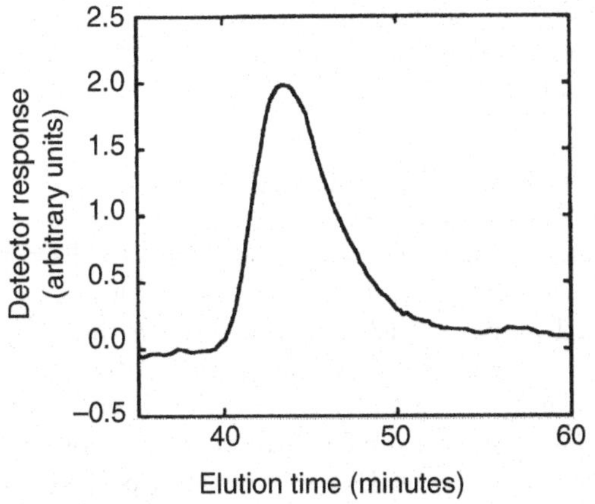

Fig. 3.2.2-12 A typical trace from a GPC run for a poly(tetramethylene oxide)/toluene diisocyanate-based polyurethane. The response of the ultraviolet detector is directly proportional to the amount of polymer eluted at each time point.

of the solvent in solution with respect to that of the pure solvent is compensated by applying a pressure π on the solution. π is the osmotic pressure and is related to $M_n$by:

$$\frac{\pi}{c} = RT\left[\frac{1}{M_n} + A_2c + A_3c^2 + \ldots\right] \tag{3.2.2.3}$$

where c is the concentration of the polymer in solution, R is the gas constant, T is temperature, and A_2 and A_3 are virial coefficients relating to pairwise and triplet interactions of the molecules in solution. In general, a number of polymer solutions of decreasing concentration are prepared, and the osmotic pressure is extrapolated to zero:

$$\lim_{c\to 0}\frac{\pi}{c} = \frac{RT}{M_n} \tag{3.2.2.4}$$

A plot of π/c versus c then gives as its intercept the number average molecular weight.

A number of other techniques, including vapor pressure osmometry, ebulliometry, cryoscopy, and end-group analysis, can be used to determine the M_n of polymers up to molecular weights of about 40,000.

Light-scattering techniques are used to determine M_w. In dilute solution, the scattering of light is directly proportional to the number of molecules. The scattered intensity is observed at a distance r and an angle θ from the incident beam I_o is characterized by Rayleigh's ratio R_θ:

$$R_\theta = \frac{i_o r^2}{I_o} \tag{3.2.2.5}$$

The Rayleigh ratio is related to M_w by:

$$\frac{K_c}{R_\theta} = \frac{1}{M_w} + 2A_2c + 3A_2c^2 + \ldots \tag{3.2.2.6}$$

A number of solutions of varying concentrations are measured, and the data are extrapolated to zero concentration to determine M_w.

Determination of structure

Infrared (IR) spectroscopy is often used to characterize the chemical structure of polymers. IR spectra are obtained by passing IR radiation through the sample of interest and observing the wavelength of the absorption peaks. These peaks are caused by the absorption of the radiation and its conversion into specific motions, such as C–H stretching. The IR spectrum of a polyurethane is shown in Fig. 3.2.2-13, with a few of the bands of interest marked.

Nuclear magnetic resonance (NMR), in which the magnetic spin energy levels of nuclei of spin 1/2 or greater are probed, may also be used to analyze chemical composition. ^{1}H and ^{13}C NMR are the most frequently studied isotopes. Polymer chemistry can be determined in solution or in the solid state. Figure 3.2.2-14 shows a ^{13}C NMR spectrum of a polyurethane with a table assigning the peaks to specific chemical groups. NMR is also used in a number of more specialized applications relating to local motions and intermolecular interactions of polymers.

Wide-angle X-ray scattering (WAXS) techniques are useful for probing the local structure of a semicrystalline polymeric solid. Under appropriate conditions, crystalline materials diffract X-rays, giving rise to spots or rings. According to Bragg's law, these can be interpreted as interplanar spacings. The interplanar spacings can be used without further manipulation or the data can be fit to a model such as a disordered helix or an extended chain. The crystalline chain conformation and atomic placements can then be accurately inferred.

Small-angle X-ray scattering (SAXS) is used in determining the structure of many multiphase materials. This technique requires an electron density difference to be present between two components in the solid and has been widely applied to morphological studies of copolymers and ionomers. It can probe features of 10–1000 Å in size. With appropriate modeling of the data, SAXS can give detailed structural information unavailable with other techniques.

Electron microscopy of thin sections of a polymeric solid can also give direct morphological data on a polymer of interest, assuming that (1) the polymer possesses sufficient electron density contrast or can be appropriately stained without changing the morphology and (2) the structures of interest are sufficiently large.

Mechanical and thermal property studies

In stress-strain or tensile testing, a dog-bone-shaped polymer sample is subjected to a constant elongation, or strain, rate, and the force required to maintain the constant elongation rate is monitored. As discussed earlier, tensile testing gives information about modulus, yield point, and ultimate strength of the sample of interest.

Dynamic mechanical analysis (DMA) provides information about the small deformation behavior of

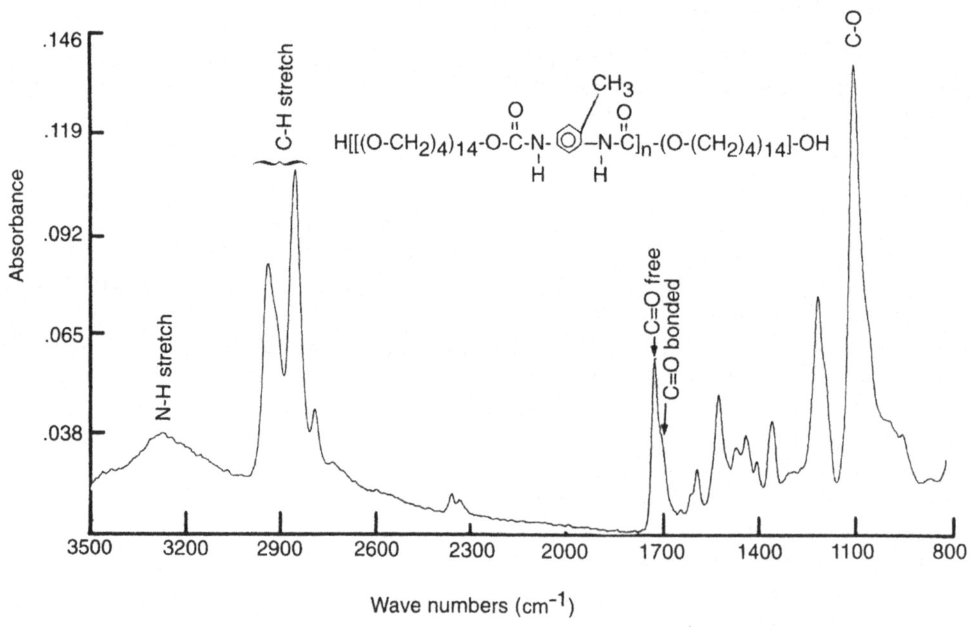

Fig. 3.2.2-13 IR spectrum of a poly(tetramethylene oxide)/toluene diisocyanate-based polyurethane.

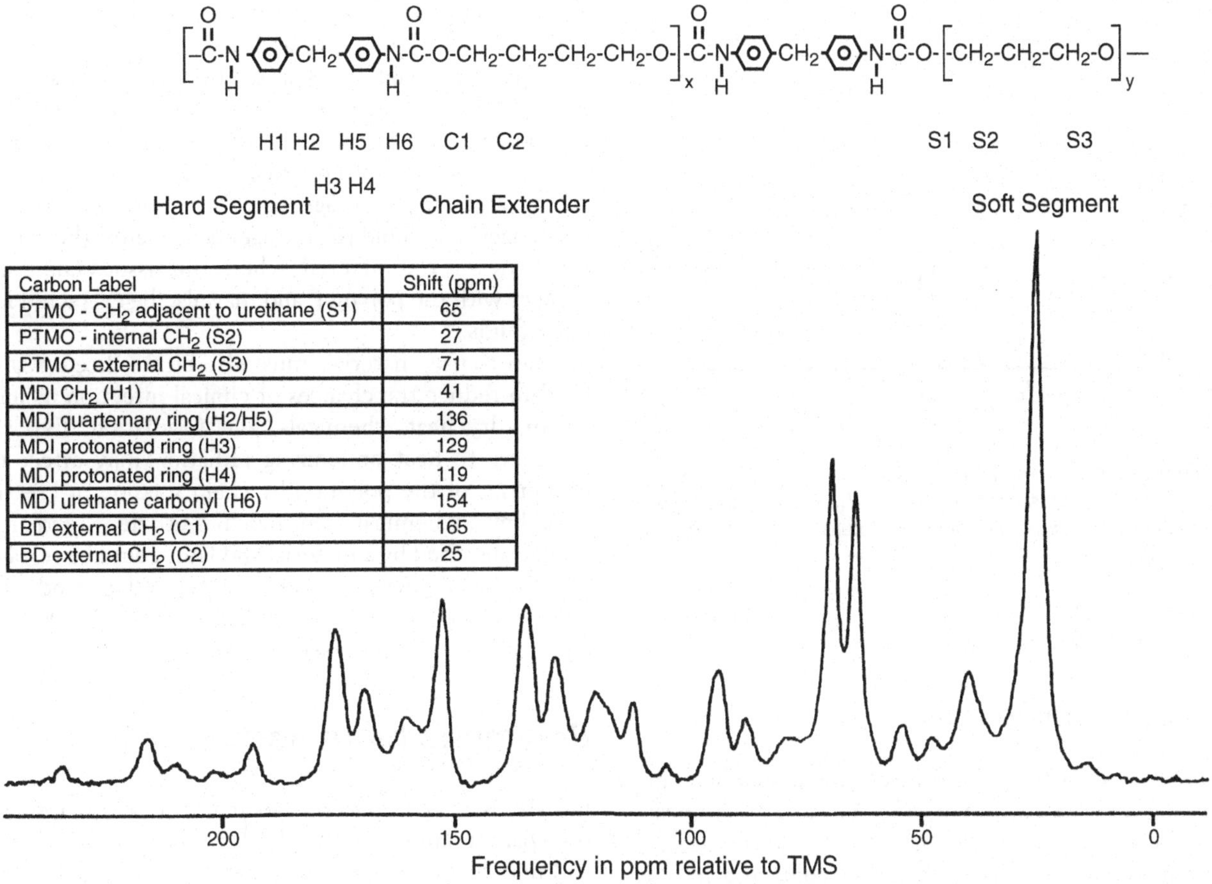

Carbon Label	Shift (ppm)
PTMO - CH_2 adjacent to urethane (S1)	65
PTMO - internal CH_2 (S2)	27
PTMO - external CH_2 (S3)	71
MDI CH_2 (H1)	41
MDI quarternary ring (H2/H5)	136
MDI protonated ring (H3)	129
MDI protonated ring (H4)	119
MDI urethane carbonyl (H6)	154
BD external CH_2 (C1)	165
BD external CH_2 (C2)	25

Fig. 3.2.2-14 ^{13}C NMR spectrum and peak assignation of a polyurethane [diphenylmethane diisocyanate (MDI, hard segment), polytetramethylene oxide (PTMO, soft segment), butanediol (BD, chain extender)]. Obtained by cross-polarization magic angle spinning of the solid polymer. (From Okamoto, D. T., Ph.D. thesis, University of Wisconsin, 1991. Reproduced with permission.)

polymers. Samples are subjected to cyclic deformation at a fixed frequency in the range of 1–1000 Hz. The stress response is measured while the cyclic strain is applied and the temperature is slowly increased (typically at 2–3 degrees/min). If the strain is a sinusoidal function of time given by:

$$\varepsilon(\omega) = \varepsilon_o \sin(\omega t) \quad (3.2.2.7)$$

where ε is the time-dependent strain, ε_o is the strain amplitude, ω is the frequency of oscillation, and t is time, the resulting stress can be expressed by:

$$\sigma(\omega) = \sigma_o \sin(\omega t + \delta) \quad (3.2.2.8)$$

where σ is the time-dependent stress, σ_o is the amplitude of stress response, and δ is the phase angle between stress and strain. For Hookean solids, the stress and strain are completely in phase ($\delta = 0$), while for purely viscous liquids, the stress response lags by 90°. Real materials demonstrate viscoelastic behavior where δ has a value between 0° and 90°.

A typical plot of tan δ versus temperature will display maxima at T_g and at lower temperatures where small-scale motions (secondary relaxations) can occur. Additional peaks above T_g, corresponding to motions in the crystalline phase and melting, are seen in semi-crystalline materials. DMA is a sensitive tool for characterizing polymers of similar chemical composition or for detecting the presence of moderate quantities of additives.

DSC is another method for probing thermal transitions of polymers. A sample cell and a reference cell are supplied energy at varying rates so that the temperatures of the two cells remain equal. The temperature is increased, typically at a rate of 10-20 degrees/min over the range of interest, and the energy input required to maintain equality of temperature in the two cells is recorded. Plots of energy supplied versus average temperature allow determination of T_g, crystallization temperature (T_c), and T_m. T_g is taken as the temperature at which one half the change in heat capacity, ΔC_p, has occurred. The T_c and T_m are easily identified, as shown in

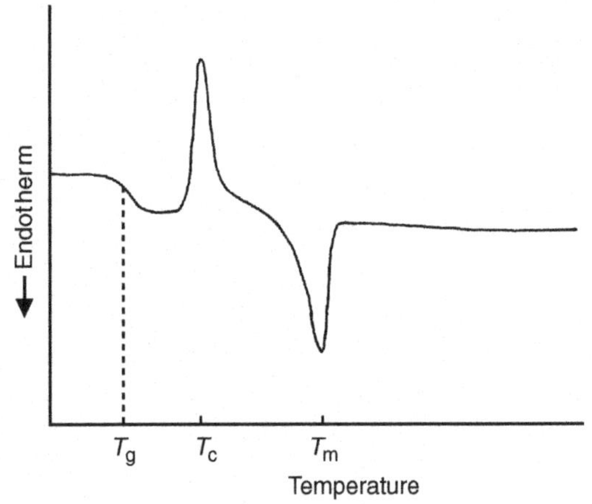

Fig. 3.2.2-15 DSC thermogram of a semicrystalline polymer, showing the glass transition temperature (T_g), the crystallization temperature (T_c), and the melting temperature (T_m) of the polymer sample.

Fig. 3.2.2-15. The areas under the peaks can be quantitatively related to enthalpic changes.

Surface characterization

Surface characteristics of polymers for biomedical applications are critically important. The surface composition is inevitably different from the bulk, and the surface of the material is generally all that is contacted by the body. The main surface characterization techniques for polymers are X-ray photoelectron spectroscopy (XPS), contact angle measurements, attenuated total reflectance Fourier transform infrared (ATR-FTIR) spectroscopy, and scanning electron microscopy (SEM). The techniques are discussed in detail in Section 3.1.4.

Fabrication and processing

Before a polymer can be employed usefully in a medical device, the material must be manipulated physically, thermally, or mechanically into the desired shape. This can be achieved using the high-molecular-weight polymer at the start of the process and may require additives in the material to aid processing, or the end use. Such additives can include antioxidants, UV stabilizers, reinforcing fillers, lubricants, mold release agents, and plasticizers.

Alternatively, polymer products can be fabricated into end-use shapes starting from the monomers or low-molecular-weight prepolymers. In such processes, the final polymerization step is carried out once the precursors are in a casting or molding device, yielding a solid, shaped end product. A typical example is PMMA dental or bone cement, which is cured *in situ* in the body. Polymers can be fabricated into sheets, films, rods, tubes, and fibers, as coatings on another substrate, and into more complex geometries and foams.

It is important to realize that the presence of processing and functional aids can affect other properties of a polymer. For example, plasticisers are added to rigid PVC to produce a softer material, e.g., for use as dialysis tubing and blood storage bags. But additives such as plasticizers and mold release agents may alter the surface properties of the material, where the tissues come into contact with the polymer, and may also be extracted into body fluids.

Prior to use, materials must also be sterilized. Agents used to reduce the chances of clinical infection include, steam, dry heat, chemicals, and irradiation. Exposing polymers to heat or ionizing radiation may affect the properties of the polymer, by chain scission or creating cross-links. Chemical agents such as ethylene oxide may also be absorbed by a material and later could be released into the body. Therefore, devices sterilized with ethylene oxide require a period of time following sterilization for any residues to be released before use.

Polymeric biomaterials

PMMA is a hydrophobic, linear chain polymer that is transparent, amorphous, and glassy at room temperature and may be more easily recognized by such trade names as Lucite or Plexiglas. It is a major ingredient in bone cement for orthopedic implants. In addition to toughness and stability, it has excellent light transmittance, making it a good material for intraocular lenses (IOLs) and hard contact lenses. The monomers are polymerized in the shape of a rod from which buttons are cut. The button or disk is then mounted on a lathe, and the posterior and anterior surfaces machined to produce a lens with defined optical power. Lenses can also be fabricated by melt processing, compression molding, or casting, but lathe machining methods are most commonly used.

Soft contact lenses are made from the same methacrylate family of polymers. The substitution of the methyl ester group in methylmethacrylate with a hydroxyethyl group (2-hydroxyethyl methacrylate or HEMA) produces a very hydrophilic polymer. For soft contact lenses, the poly(HEMA) is slightly cross-linked with ethylene glycol dimethyacrylate (EGDMA) to retain dimensional stability for its use as a lens. Fully hydrated, it is a swollen hydrogel. PHEMA is glassy when dried, and therefore, soft lenses are manufactured in the same way as hard lenses; however, for the soft lens a swelling factor must be included when defining the optical specifications. This class of hydrogel polymers is discussed in more detail in Section 3.2.5.

Polyacrylamide is another hydrogel polymer that is used in biomedical separations (e.g., polyacrylamide gel electrophoresis, or PAGE). The mechanical properties and the degree of swelling can be controlled by cross-linking with methylene-bis-acrylamide (MBA). Poly(N-alkylacrylamides) are environmentally sensitive, and the degree of swelling can be altered by changes in temperature and acidity. These polymers are discussed in more detail in Section 3.2.6 and; see also Hoffman (1997).

Polyacrylic acids also have applications in medicine. They are used as dental cements, e.g., as glass ionomers. In this use, they are usually mixed with inorganic salts, where the cation interacts with the carboxyl groups of the acid to form physical cross-links. Polyacrylic acid is also used in a covalently cross-linked form as a mucoadhesive additive to mucosal drug delivery formulations. Polymethacrylic acid may also be incorporated in small quantities into contact lens polymer formulations to improve wettability.

PE is used in its high-density form in biomedical applications because low-density material cannot withstand sterilization temperatures. It is used as tubing for drains and catheters, and in ultrahigh-MW form as the acetabular component in artificial hips and other prosthetic joints. The material has good toughness and wear resistance and is also resistant to lipid absorption. Radiation sterilization in an inert atmosphere may also provide some covalent cross-linking that strengthens the PE.

PP is an isotactic crystalline polymer with high rigidity, good chemical resistance, and good tensile strength. Its stress cracking resistance is excellent, and it is used for sutures and hernia repair.

PTFE, also known as PTFE Teflon, has the same structure as PE, except that the four hydrogens in the repeat unit of PE are replaced by fluorines. PTFE is a very high melting polymer ($T_m = 327°C$) and as a result it is very difficult to process. It is very hydrophobic, has excellent lubricity, and is used to make catheters. In microporous form, known generically as e-PTFE or most commonly as the commercial product Gore-Tex, it is used in vascular grafts. Because of its low friction, it was the original choice by Dr. John Charnley for the acetabular component of the first hip joint prosthesis, but it failed because of its low wear resistance and the resultant inflammation caused by the PTFE wear particles.

PVC is used mainly as tubing and blood storage bags in biomedical applications. Typical tubing uses include blood transfusion, feeding, and dialysis. Pure PVC is a hard, brittle material, but with the addition of plasticizers, it can be made flexible and soft. PVC can pose problems for long-term applications because the plasticizers can be extracted by the body. While these plasticizers have low toxicities, their loss also makes the PVC less flexible.

Poly(dimethyl siloxane) (PDMS) or SR is an extremely versatile polymer, although its use is often limited by its relatively poor mechanical strength. It is unique in that it has a silicon–oxygen backbone instead of a carbon backbone. Its properties are less temperature sensitive than other rubbers because of its very low T_g. In order to improve mechanical properties, SR is usually formulated with reinforcing silica filler, and sometimes the polysiloxane backbone is also modified with aromatic rings that can toughen it. Because of its excellent flexibility and stability, SR is used in a variety of prostheses such as finger joints, heart valves, and breast implants, and in ear, chin, and nose reconstruction. It is also used as catheter and drainage tubing and in insulation for pacemaker leads. It has also been used in membrane oxygenators because of its high oxygen permeability, although porous PP or polysulfone polymers have recently become more used as oxygenator membranes. Silicones are so important in medicine that details on their chemistry are provided in Section 3.2.3.

PET is one of the highest volume polymeric biomaterials. It is a polyester, containing rigid aromatic rings in a "regular" polymer backbone, which produces a high-melting ($T_m = 267°C$) crystalline polymer with very high tensile strength. It may be fabricated in the forms of knit, velour, or woven fabrics and fabric tubes, and also as nonwoven felts. Dacron is a common commercial form of PET used in large-diameter knit, velour, or woven arterial grafts. Other uses of PET fabrics are for the fixation of implants and hernia repair. PET can also be used in ligament reconstruction and as a reinforcing fabric for tissue reconstruction with soft polymers such as SR. It is used in a nonwoven felt coating on the peritoneal dialysis shunt (where it enters the body and traverses the skin) to enhance ingrowth and thereby reduce the possibility of infection.

PEG is used in drug delivery as conjugates with low solubility drugs and with immunogenic or fairly unstable protein drugs, to enhance the circulation times and stabilities of the drugs. It is also used as PEG–phospholipid conjugates to enhance the stability and circulation time of drug-containing liposomes. In both cases it serves to "hide" the circulating drug system from immune recognition, especially in the liver. PEG has also been immobilized on polymeric biomaterial surfaces to make them "nonfouling." PEGs usually exist in a highly hydrated state on the polymer surfaces, where they can exhibit steric repulsion based on an osmotic or entropic mechanism. This phenomenon contributes to the protein- and cell-resistant properties of surfaces containing PEGs (See Section 3.2.13).

Regenerated cellulose, for many years, was the most widely used dialysis membrane. Derivatives of cellulose, such as CA, are also used, since CA can be melt processed as hollow fibers for the hollow fiber kidney dialyser. CA is also used in osmotic drug delivery devices.

Polymerization of bisphenol A and phosgene produces polycarbonate, a clear, tough material. Its high impact strength dictates its use as lenses for eyeglasses and safety glasses, and housings for oxygenators and heart–lung bypass machine. Polycarbonate macrodiols have also been used to prepare copolymers such as polyurethanes. Polycarbonate segments may confer enhanced biological stability to a material.

Nylon is the name originally given by Du Pont to a family of polyamides; the name is now generic, and many other companies make nylons. Nylons are formed by the reaction of diamines with dibasic acids or by the ring opening polymerization of lactams. Nylons are used as surgical sutures (see also Section 3.2.4).

Biodegradable polymers

PLGA is a random copolymer used in resorbable surgical sutures, drug delivery systems, and orthopedic appliances such as fixation devices. The degradation products are endogenous compounds (lactic and glycolic acids) and as such are nontoxic. PLGA polymerization occurs via a ring-opening reaction of a glycolide and a lactide, as illustrated in Fig. 3.2.2-6. The presence of ester linkages in the polymer backbone allows gradual hydrolytic degradation (resorption). The rate of degradation can be controlled by the ratio of polyglycolic acid (PGA) to PLA.

Copolymers

Copolymers are another important class of biomedical materials. A copolymer of tetrafluoroethylene with a small amount of hexafluoropropylene (FEP Teflon) is used as a tubing connector and catheter. FEP has a crystalline melting point near 265°C compared with 327°C for PTFE. This enhances the processability of FEP compared with PTFE while maintaining the excellent chemical inertness and low friction characteristic of PTFE.

Polyurethanes are block copolymers containing "hard" and "soft" blocks. The "hard" blocks, having T_g values above room temperature and acting as glassy or semicrystalline reinforcing blocks, are composed of a diisocyanate and a chain extender. The diisocyanates most commonly used are 2,4-toluene diisocyanate (TDI) and methylene di(4-phenyl isocyanate) (MDI), with MDI being used in most biomaterials. The chain extenders are usually shorter aliphatic glycol or diamine materials with two to six carbon atoms. The "soft" blocks in polyurethanes are typically polyether or polyester polyols whose T_g values are much lower than room temperature, allowing them to give a rubbery character to the materials. Polyether polyols are more commonly used for implantable devices because they are stable to hydrolysis. The polyol molecular weights tend to be on the order of 1000 to 2000.

Polyurethanes are tough elastomers with good fatigue and blood-containing properties. They are used in pacemaker lead insulation, catheters, vascular grafts, heart assist balloon pumps, artificial heart bladders, and wound dressings.

Final remarks

The chemistry, physics, and mechanics of polymeric materials are highly relevant to the performance of many devices employed in the clinic today. Polymers represent a broad, diverse family of materials, with mechanical properties that make them useful in applications relating to both soft and hard tissues. The presence of functional groups on the backbone or side chains of a polymer also means that they are readily modified chemically or biochemically, especially at their surfaces. Many researchers have successfully altered the chemical and biological properties of polymers, by immobilizing anticoagulants such as heparin, proteins such as albumin for passivation and fibronectin for cell adhesion, and cell-receptor peptide ligands to enhance cell adhesion, greatly expanding their range of applications (see Section 3.2.16).

Bibliography

Billmeyer, F.W., Jr. (1984). *Textbook of Polymer Science*, 3rd ed. Wiley-Interscience, New York.

Black, J., and Hastings, G. (1998). *Handbook of Biomaterial Properties*. Chapman and Hall, London.

Brydson, J.A. (1995). *Plastics Materials*, 3rd ed. Butterworth Scientific, London.

Flory, P.J. (1953). *Principles of Polymer Chemistry.* Cornell Univ. Press, Ithaca, NY.

Hoffman, A.S. (1997). Intelligent Polymers. in *Controlled Drug Delivery*, K. Park, ed. ACS Publications, ACS, Washington, D.C.

Lamba, N.M.K., Woodhouse, K.A. and Cooper, S.L. (1998). *Polyurethanes in Biomedical Applications.* CRC Press, Boca Raton, FL.

Mandelkern, L. (1983). *An Introduction to Macromolecules.* Springer-Verlag, New York.

Rodriguez, F. (1996). *Principles of Polymer Systems*, 4th ed. Hemisphere Publishing, New York.

Sperling, L.H. (1992). *Introduction to Physical Polymer Science*, 2nd ed. Wiley-Interscience, New York.

Stokes, K., McVenes, R., and Anderson, J. M. (1995). Polyurethane elastomer biostability. *J. Biomater. Appl.* 9: 321–354.

Szycher, M. (ed.) *High Performance Biomaterials.* Technomic, Lancaster, PA.

3.2.3 Silicone biomaterials: history and chemistry

André Colas and Jim Curtis

Chemical structure and nomenclature

Silicones are a general category of synthetic polymers whose backbone is made of repeating silicon to oxygen bonds. In addition to their links to oxygen to form the polymeric chain, the silicon atoms are also bonded to organic groups, typically methyl groups. This is the basis for the name "silicones," which was assigned by Kipping based on their similarity with ketones, because in most cases, there is on average one silicone atom for one oxygen and two methyl groups (Kipping, 1904). Later, as these materials and their applications flourished, more specific nomenclature was developed. The basic repeating unit became known as "siloxane" and the most common silicone is PDMS.

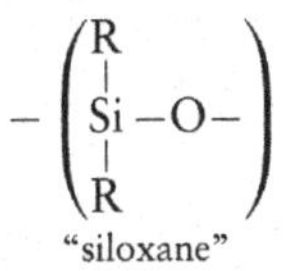

$$-\left(\begin{array}{c}R\\|\\Si-O-\\|\\R\end{array}\right) \quad \text{and if R is } CH_3, \quad -\left(\begin{array}{c}CH_3\\|\\Si-O-\\|\\CH_3\end{array}\right)_n$$

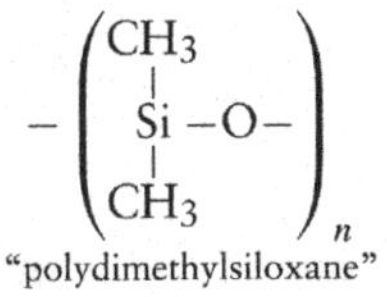

"siloxane" "polydimethylsiloxane"

Many other groups, e.g., phenyl, vinyl and trifluoropropyl, can be substituted for the methyl groups along the chain. The simultaneous presence of "organic" groups attached to an "inorganic" backbone gives silicones a combination of unique properties, making possible their use as fluids, emulsions, compounds, resins, and elastomers in numerous applications and diverse fields. For example, silicones are common in the aerospace industry, due principally to their low and high temperature performance. In the electronics field, silicones are used as electrical insulation, potting compounds and other applications specific to semiconductor manufacture. Their long-term durability has made silicone sealants, adhesives and waterproof coatings commonplace in the construction industry. Their excellent biocompatibility makes many silicones well suited for use in numerous personal care, pharmaceutical, and medical device applications.

Historical milestones in silicone chemistry

Key milestones in the development of silicone chemistry— thoroughly described elsewhere by Lane and Burns (1996), Rochow (1987), and Noll (1968)—are summarized in Table 3.2.3-1.

Table 3.2.3-1 Key milestones in the development of silicone chemistry

1824	Berzelius discovers silicon by the reduction of potassium fluorosilicate with potassium: $4K + K_2SiF_6 \rightarrow Si + 6KF$. Reacting silicon with chlorine gives a volatile compound later identified as tetrachlorosilane, $SiCl_4$: $Si + 2Cl_2 \rightarrow SiCl_4$.
1863	Friedel and Craft synthesize the first silicon organic compound, tetraethylsilane: $2Zn(C_2H_5)_2 + SiCl_4 \rightarrow Si(C_2H_5)_4 + 2ZnCl_2$.
1871	Ladenburg observes that diethyldiethoxysilane, $(C_2H_5)_2Si(OC_2H_5)_2$, in the presence of a diluted acid gives an oil that decomposes only at a "very high temperature."
1901–1930s	Kipping lays the foundation of organosilicon chemistry with the preparation of various silanes by means of Grignard reactions and the hydrolysis of chlorosilanes to yield "large molecules." The polymeric nature of the silicones is confirmed by the work of Stock.
1940s	In the 1940s, silicones become commercial materials after Hyde of Dow Corning demonstrates the thermal stability and high electrical resistance of silicone resins, and Rochow of General Electric finds a direct method to prepare silicones from silicon and methylchloride.

Nomenclature

The most common silicones are the polydimethylsiloxanes trimethylsilyloxy terminated, with the following structure:

$$CH_3-\underset{\underset{CH_3}{|}}{\overset{\overset{CH_3}{|}}{Si}}-O-\left(\underset{\underset{CH_3}{|}}{\overset{\overset{CH_3}{|}}{Si}}-O\right)_n-\underset{\underset{CH_3}{|}}{\overset{\overset{CH_3}{|}}{Si}}-CH_3,$$

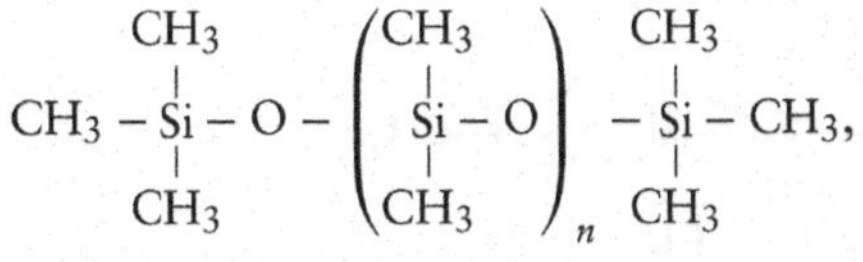

$$(n = 0, 1, \ldots)$$

These are linear polymers and liquids, even for large values of n. The main chain unit, $-(Si(CH_3)_2 O)_n-$, is often represented by the letter D because, as the silicon atom is connected with two oxygen atoms, this unit is capable of expanding within the polymer in two directions. M, T and Q units are defined in a similar manner, as shown in Table 3.2.3-2.

The system is sometimes modified by the use of superscript letters designating nonmethyl substituents, for example, $D^H = H(CH_3)SiO_{2/2}$ and M^ϕ or $M^{Ph} = (CH_3)_2(C_6H_5)SiO_{1/2}$ (Smith, 1991). Further examples are shown in Table 3.2.3-3.

Table 3.2.3-2 Shorthand notation for siloxane polymer units

M	D	T	Q
CH₃–Si(CH₃)(CH₃)–O–	–O–Si(CH₃)(CH₃)–O–	–O–Si(CH₃)(O–)–O–	–O–Si(O–)(O–)–O–
$(CH_3)_3 SiO_{1/2}$	$(CH_3)_2 SiO_{2/2}$	$CH_3 SiO_{3/2}$	$SiO_{4/2}$

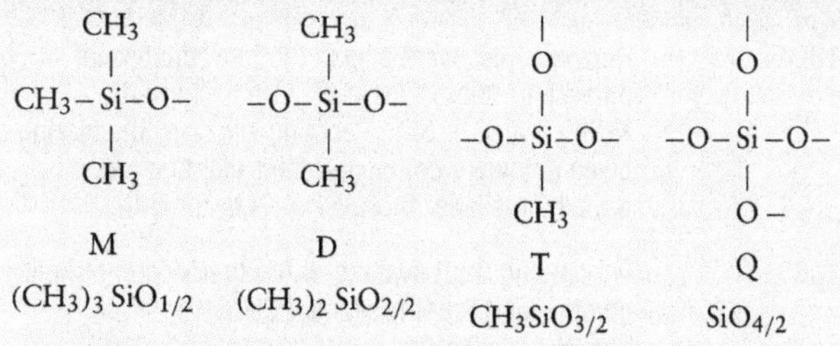

Table 3.2.3-3 Examples of silicone shorthand notation

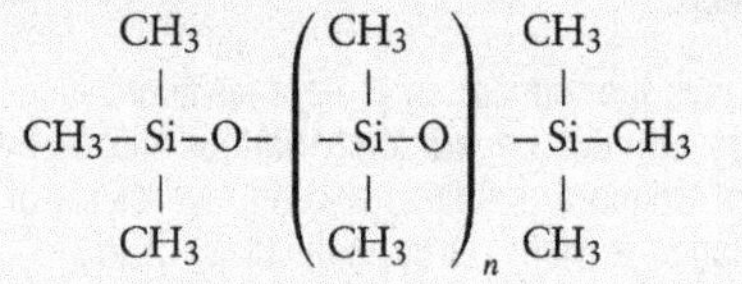

MD_nM

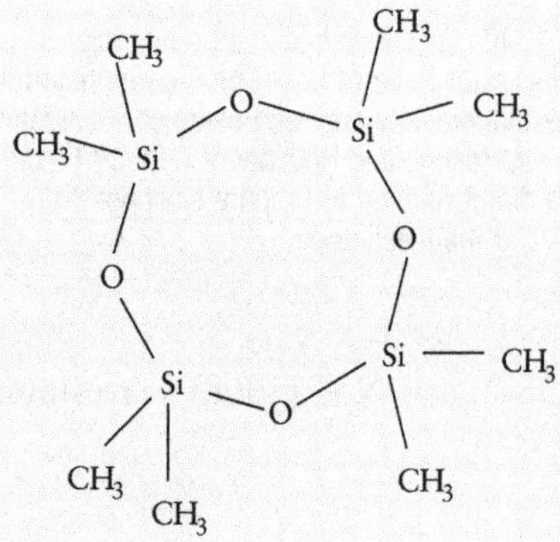

D_4

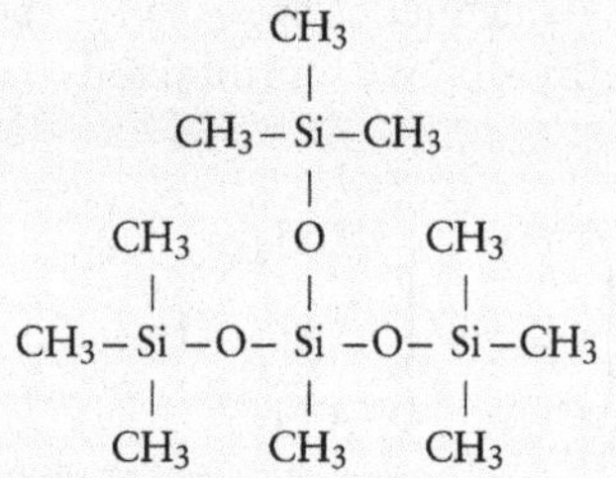

TM_3

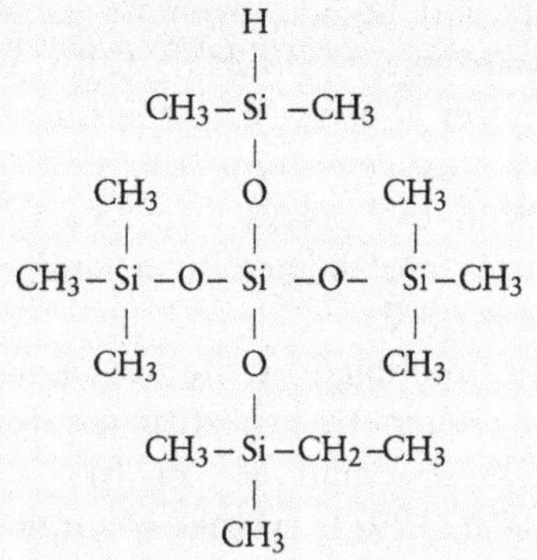

$QM_2M^HM^{C_2H_5}$ or $QM_2M^HM^{Et}$

Preparation

Silicone polymers

The modern synthesis of silicone polymers is multifaceted. It usually involves the four basic steps described in Table 3.2.3-4. Only step 4 in this table will be elaborated upon here.

Polymerization and Polycondensation. The linear [4] and cyclic [5] oligomers resulting from dimethyldichlorosilane hydrolysis have chain lengths too short for most applications. The cyclics must be polymerized, and the linears condensed, to give macromolecules of sufficient length (Noll, 1968).

Catalyzed by acids or bases, cyclosiloxanes $(R_2SiO)_m$ are ring-opened and polymerized to form long linear chains. At equilibrium, the reaction results in a mixture of cyclic oligomers plus a distribution of linear polymers. The proportion of cyclics will depend on the substituents along the Si–O chain, the temperature, and the presence of a solvent. Polymer chain length will depend on the presence and concentration of substances capable of giving chain ends. For example, in the KOH-catalyzed polymerization of the cyclic tetramer octamethylcyclotetrasiloxane $(Me_2SiO)_4$ (or D_4 in shorthand notation), the average length of the polymer chains will depend on the KOH concentration:

$$x(Me_2SiO)_4 + KOH \rightarrow (Me_2SiO)_y + KO(Me_2SiO)_zH$$

A stable hydroxy-terminated polymer, $HO(Me_2SiO)_zH$, can be isolated after neutralization and removal of the remaining cyclics by stripping the mixture under vacuum at elevated temperature. A distribution of chains with different lengths is obtained.

The reaction can also be made in the presence of $Me_3SiOSiMe_3$, which will act as a chain end blocker:

$$\cdots Me_2SiOK + Me_2SiOSiMe_3$$
$$\rightarrow \cdots Me_2SiOSiMe_3 + Me_3SiOK$$

where $\cdots$ represents the main chain.

The Me_3SiOK formed will attack another chain to reduce the average molecular weight of the linear polymer formed.

The copolymerization of $(Me_2SiO)_4$ in the presence of $Me_3SiOSiMe_3$ with Me_4NOH as catalyst displays a surprising viscosity change over time (Noll, 1968). First a peak or viscosity maximum is observed at the beginning of the reaction. The presence of two oxygen atoms on each silicon in the cyclics makes them more susceptible to a nucleophilic attack by the base catalyst than the silicon of the endblocker, which is substituted by only one oxygen atom. The cyclics are polymerized first into

Table 3.2.3-4 The Basic Steps in Silicone Polymer Synthesis

1. Silica reduction to silicon $\quad SiO_2 + 2C \rightarrow Si + 2CO$

2. Chlorosilanes synthesis $\quad Si + 2CH_3Cl \rightarrow \underset{[1]}{(CH_3)_2SiCl_2} + \underset{[2]}{CH_3SiCl_3} + \underset{[3]}{(CH_3)_3SiCl} + CH_3HSiCl_2 + \cdots$

3. Chlorosilanes hydrolysis

$$\underset{[1]}{\begin{matrix} CH_3 \\ | \\ Cl-Si-Cl \\ | \\ CH_3 \end{matrix}} + 2H_2O \rightarrow \underset{\substack{\text{linears} \\ [4]}}{HO-\left(\begin{matrix} CH_3 \\ | \\ -Si-O- \\ | \\ CH_3 \end{matrix}\right)_x - H} + \underset{\substack{\text{cyclics} \\ [5]}}{\left(\begin{matrix} CH_3 \\ | \\ Si-O \\ | \\ CH_3 \end{matrix}\right)_{3,4,5}} + HCl$$

4. Polymerization and polycondensation

$$\underset{\substack{\text{cyclics}\\ [5]}}{\left(\begin{matrix} CH_3 \\ | \\ Si-O \\ | \\ CH_3 \end{matrix}\right)_{3,4,5}} \rightarrow \underset{\text{polymer}}{-\left(\begin{matrix} CH_3 \\ | \\ -Si-O- \\ | \\ CH_3 \end{matrix}\right)_y -}$$

$$\underset{\substack{\text{linears}\\ [4]}}{HO-\left(\begin{matrix} CH_3 \\ | \\ -Si-O- \\ | \\ CH_3 \end{matrix}\right)_x - H} \rightarrow \underset{\text{polymer}}{-\left(\begin{matrix} CH_3 \\ | \\ -Si-O- \\ | \\ CH_3 \end{matrix}\right)_z -} + z\,H_2O$$

---

very long, viscous chains that are subsequently reduced in length by the addition of terminal groups provided by the endblocker, which is slower to react. This reaction can be described as follows:

$$Me_3SiOSiMe_3 + x(Me_2SiO)_4 \xrightarrow{cat} Me_3SiO(Me_2SiO)_nSiMe_3$$

or, in shorthand notation:

$$MM + xD_4 \xrightarrow{cat} MD_nM$$

where $n = 4x$ (theoretically).

The ratio between D and M units will define the average molecular weight of the polymer formed.

Catalyst removal (or neutralization) is always an important step in silicone preparation. Most catalysts used to prepare silicones can also catalyze the depolymerization (attack along the chain), particularly at elevated temperatures in the presence of traces of water.

$$\cdots(Me_2SiO)_n\cdots + H_2O$$
$$\xrightarrow{cat} \cdots(Me_2SiO)_yH + HO(Me_2SiO)_z\cdots$$

It is therefore essential to remove all remaining traces of the catalyst, providing the silicone optimal thermal stability. Labile catalysts have been developed. These decompose or are volatilized above the optimum polymerization temperature and consequently can be eliminated by a brief overheating. In this way, catalyst neutralization or filtration can be avoided (Noll, 1968).

The cyclic trimer $(Me_2SiO)_3$ has an internal ring tension and can be polymerized without reequilibration of the resulting polymers. With this cyclic, polymers with narrow molecular-weight distribution can be prepared, but also polymers only carrying one terminal reactive function (living polymerization). Starting from a mixture of different "tense" cyclics also allows the preparation of block or sequential polymers (Noll, 1968).

Linears can combine when catalyzed by many acids or bases to give long chains by intermolecular condensation of silanol terminals (Noll, 1968; Stark et al., 1982).

$$\cdots O-\underset{\underset{Me}{|}}{\overset{\overset{Me}{|}}{Si}}-OH + HO-\underset{\underset{Me}{|}}{\overset{\overset{Me}{|}}{Si}}-O\cdots$$

$$\underset{+H_2O}{\overset{-H_2O}{\rightleftharpoons}} \cdots O-\underset{\underset{Me}{|}}{\overset{\overset{Me}{|}}{Si}}-O-\underset{\underset{Me}{|}}{\overset{\overset{Me}{|}}{Si}}-O\cdots$$

A distribution of chain lengths is obtained. Longer chains are favored when working under vacuum and/or at elevated temperatures to reduce the residual water concentration. In addition to the polymers described above, reactive polymers can also be prepared. This can be achieved when reequilibrating oligomers or existing polymers to obtain a polydimethyl-methylhydrogenosiloxane, $MD_zD^H_wM$.

$$Me_3SiOSiMe_3 + x(Me_2SiO)_4 + Me_3SiO(MeHSiO)_ySiMe_3$$
$$\xrightarrow{cat} cyclics + Me_3SiO(Me_2SiO)_z(MeHSiO)_wSiMe_3$$

[6]

Additional functional groups can be attached to this polymer using an addition reaction.

$$Me_3SiO(Me_2SiO)_z(MeHSiO)_wSiMe_3 + H_2C = CHR$$

[6]

$$\xrightarrow{Pt\ cat} Me_3SiO(Me_2SiO)_z \underset{\underset{CH_2CH_2R}{|}}{(Me\ Si\ O)_w} SiMe_3$$

The polymers shown are all linear or cyclic, comprising difunctional units, D. In addition to these, branched polymers or resins can be prepared if, during hydrolysis, a certain amount of T or Q units are included, which will allow molecular expansion, in three or four directions, as opposed to just two. For example, consider the hydrolysis of methyltrichlorosilane in the presence of trimethylchlorosilane, which leads to a branched polymer as shown next:

$$x\ Me-\underset{\underset{Me}{|}}{\overset{\overset{Me}{|}}{Si}}-Cl + y\ Me-\underset{\underset{Cl}{|}}{\overset{\overset{Cl}{|}}{Si}}-Cl \xrightarrow[-HCl]{+H_2O} z$$
[3]　　　　[2]

$$Me-\underset{\underset{Me}{|}}{\overset{\overset{Me}{|}}{Si}}-O-\underset{\underset{O}{|}}{\overset{\overset{Me}{|}}{Si}}-O-\underset{\underset{OH}{|}}{\overset{\overset{Me}{|}}{Si}}-O\ \cdots$$

$$Me-\underset{\underset{O}{|}}{\overset{}{Si}}-O\ \cdots$$

$$Me-\underset{\underset{Me}{|}}{\overset{}{Si}}-Me$$

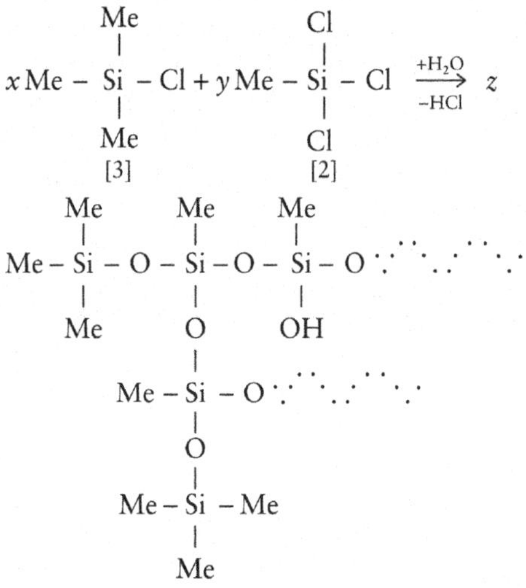

The resulting polymer can be described as $((Me_3SiO_{1/2})_x (MeSiO_{3/2})_y$ or M_xT_y, using shorthand notation. The formation of three silanols on the $MeSiCl_3$ by hydrolysis yields a three-dimensional structure or resin, rather than a linear polymer. The average molecular weight depends upon the number of M units that come from the trimethylchlorosilane, which limits the growth of the resin molecule. Most of these resins are prepared in a solvent and usually contain some residual hydroxyl groups. These could subsequently be used to cross-link the resin and form a continuous network.

Silicone elastomers

Silicone polymers can be easily transformed into a three-dimensional network by way of a cross-linking reaction, which allows the formation of chemical bonds between adjacent chains. The majority of silicone elastomers are cross-linked according to one of the following three reactions.

1. Cross-Linking with Radicals Efficient cross-linking with radicals is only achieved when some vinyl groups are present on the polymer chains. The following mechanism has been proposed for the cross-linking reaction associated with radicals generated from an organic peroxide (Stark et al., 1982):

$$R^{\cdot} + CH_2 = CH - Si \equiv\ \rightarrow R - CH_2 - CH^{\cdot} - Si \equiv$$

$$RCH_2 - CH^{\cdot} - Si \equiv\ + CH_3 - Si \equiv$$
$$\rightarrow RCH_2 - CH_2 - Si \equiv\ +\ \equiv Si - CH_2^{\cdot}$$

$$\equiv Si - CH_2^{\cdot} + CH_2 = CH - Si \equiv$$
$$\rightarrow\ \equiv Si - CH_2 - CH_2 - CH^{\cdot} - Si \equiv$$

$$\equiv Si - CH_2 - CH_2 - CH^{\cdot} - Si \equiv\ +\ \equiv Si - CH_3$$
$$\rightarrow\ \equiv Si - CH_2 - CH_2 - CH_2 - Si \equiv\ +\ \equiv Si - CH_2^{\cdot}$$

$$2 \equiv Si - CH_2^{\cdot} \rightarrow\ \equiv Si - CH_2 - CH_2 - Si \equiv$$

where $\equiv$ represents two methyl groups and the rest of the polymer chain.

This reaction has been used for high-consistency silicone rubbers (HCRs) such as those used in extrusion or injection molding, as well as those that are cross-linked at elevated temperatures. The peroxide is added before processing. During cure, some precautions are needed to avoid the formation of voids by the peroxide's volatile residues. Postcure may also be necessary to remove these volatiles, which can catalyze depolymerization at high temperatures.

2. Cross-Linking by Condensation Although mostly used in silicone caulks and sealants for the construction industry and do-it-yourselfer, this method has also found utility for medical devices as silicone adhesives facilitating the adherence of materials to silicone elastomers, as an encapsulant and as sealants such as around the connection of a pacemaker lead to the pulse generator (Fig. 3.2.3-1 shows Silastic Medical Adhesive, type A).

These products are ready to apply and require no mixing. Cross-linking starts when the product is squeezed from the cartridge or tube and comes into contact with moisture, typically from humidity in the ambient air. These materials are formulated from a reactive polymer prepared from a hydroxy end-blocked polydimethylsiloxane and a large excess of methyltriacetoxysilane.

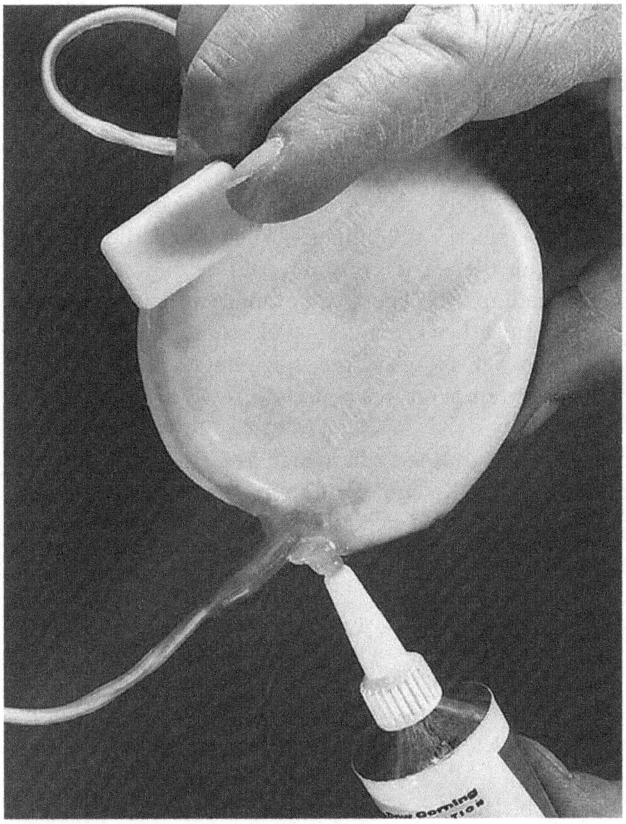

Fig. 3.2.3-1 RTV silicone adhesive.

$$HO-(Me_2SiO)_x-H + excess\ MeSi(OAc)_3$$
$$\xrightarrow[-2AcOH]{} (AcO)_2MeSiO(Me_2SiO)_xOSiMe(OAc)_2$$
[7]

where $Ac = -\overset{\underset{|}{CH_3}}{C} = O$

Because a large excess of silane is used, the probability of two different chains reacting with the same silane molecule is remote. Consequently, all the chains are end-blocked with two acetoxy functional groups. The resulting product is still liquid and can be packaged in sealed tubes and cartridges. Upon opening the acetoxy groups are hydrolyzed by the ambient moisture to give silanols, which allow further condensation to occur.

$$\cdots O-\underset{\underset{OAc}{|}}{\overset{\overset{Me}{|}}{Si}}-OAc \xrightarrow[-AcOH]{+H_2O} \cdots O-\underset{\underset{OAc}{|}}{\overset{\overset{Me}{|}}{Si}}-OH$$
[7] [8]

$$\cdots O-\underset{\underset{OAc}{|}}{\overset{\overset{Me}{|}}{Si}}-OH + AcO-\underset{\underset{OAc}{|}}{\overset{\overset{Me}{|}}{Si}}-O\cdots$$
[8] [7]

$$\xrightarrow[-AcOH]{} \cdots O-\underset{\underset{OAc}{|}}{\overset{\overset{Me}{|}}{Si}}-O-\underset{\underset{OAc}{|}}{\overset{\overset{Me}{|}}{Si}}-O\cdots$$

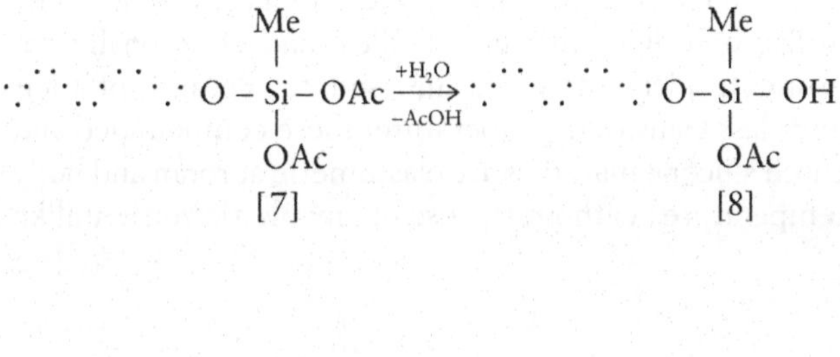

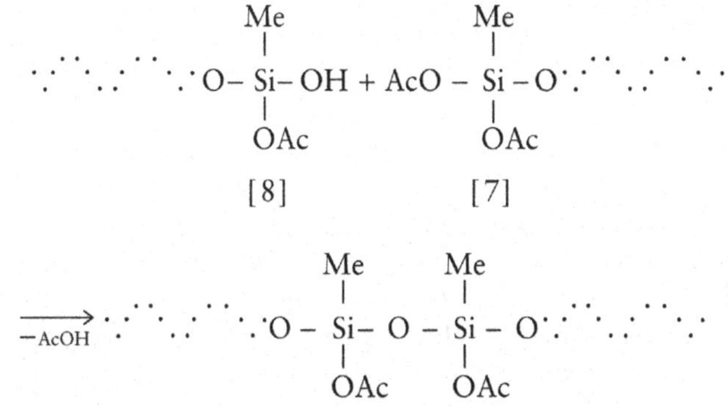

In this way, two chains have been linked, and the reaction will proceed further from the remaining acetoxy groups. An organometallic tin catalyst is normally used. The cross-linking reaction requires moisture to diffuse into the material. Accordingly cure will proceed from the outside surface inward. These materials are called one-part room temperature vulcanization (RTV) sealants, but actually require moisture as a second component. Acetic acid is released as a by-product of the reaction. Problems resulting from the acid can be overcome using other cure (cross-linking) systems that have been developed by replacing the acetoxysilane $RSi(OAc)_3$ with oximosilane $RSi(ON=CR'_2)_3$ or alkoxysilane $RSi(OR')_3$.

Condensation curing is also used in some two-part systems where cross-linking starts upon mixing the two components, e.g., a hydroxy end-blocked polymer and an alkoxysilane such as tetra-n-propoxysilane (Noll, 1968):

$$4\cdots\underset{\underset{Me}{|}}{\overset{\overset{Me}{|}}{Si}}-OH + nPrO-\underset{\underset{O}{|}}{\overset{\overset{nPr}{\overset{|}{O}}}{Si}}-OnPr$$

$$\xrightarrow[-4nPrOH]{cat} \cdots \underset{\underset{Me}{|}}{\overset{\overset{Me-Si\cdots}{|}}{Si}}-O-\underset{\underset{Me}{|}}{\overset{\overset{O}{|}}{Si}}-O-\underset{\underset{|}{Me-Si\cdots}}{\overset{\overset{Me}{|}}{Si}}\cdots$$

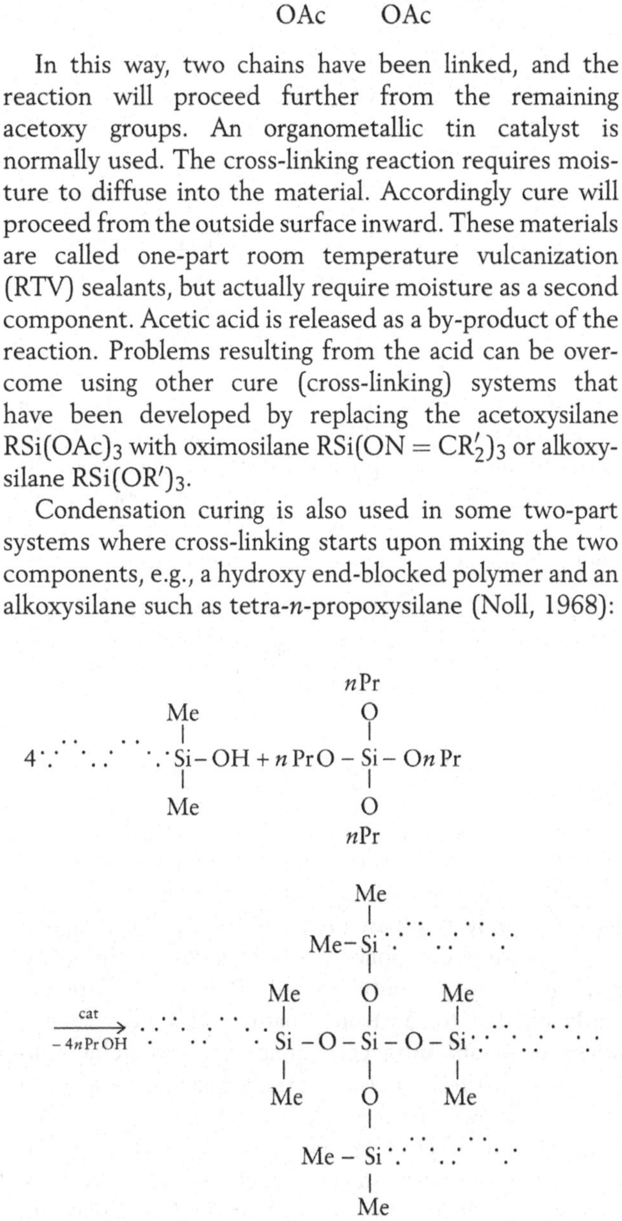

Here, no atmospheric moisture is needed. Usually an organotin salt is used as catalyst, but it also limits the stability of the resulting elastomer at high temperatures. Alcohol is released as a by-product of the reaction, leading to a slight shrinkage upon cure (0.5 to 1% linear shrinkage). Silicones with this cure system are therefore not suitable for the fabrication of parts with precise tolerances.

3. Cross-linking by Addition Use of an addition-cure reaction for cross-linking can eliminate the shrinkage problem mentioned above. In addition cure, cross-linking is achieved by reacting vinyl endblocked polymers with Si–H groups carried by a functional oligomer such as described above [6]. A few polymers can be bonded to this functional oligomer [6], as follows (Stark *et al.*, 1982):

$$\cdots O-\underset{Me}{\overset{Me}{\mid}}\!\!\!\!\!\underset{\mid}{Si}-CH=CH_2+H-Si\equiv \qquad [5]$$

$$\xrightarrow{cat} \cdots O-\underset{Me}{\overset{Me}{\mid}}\!\!\!\!\!\underset{\mid}{Si}-CH_2-CH_2-Si\equiv$$

where ≡ represents the remaining valences of the Si in [6].

The addition occurs mainly on the terminal carbon and is catalyzed by Pt or Rh metal complexes, preferably as organometallic compounds to enhance their compatibility. The following mechanism has been proposed (oxidative addition of the ≡Si to the Pt, H transfer to the double bond, and reductive elimination of the product):

$$\equiv Si{-}Pt{-}H \quad a+\equiv Si{-}CH{=}CH_2$$

$$\rightleftarrows\ \equiv Si{-}CH_2{-}CH_2{-}Pt{-}Si\equiv$$

$$\underset{-Pt}{\rightarrow}\ \equiv Si{-}CH_2{-}CH_2{-}Si\equiv$$

where, to simplify, other Pt ligands and other Si substituents are omitted.

There are no by-products with this reaction. Molded pieces made with silicone using this addition-cure mechanism are very accurate (no shrinkage). However, handling these two-part products (i.e., polymer and Pt catalyst in one component, SiH oligomer in the other) requires some precautions. The Pt in the complex is easily bonded to electron-donating substances such as amine or organosulfur compounds to form stable complexes with these "poisons," rendering the catalyst inactive and inhibiting the cure.

The preferred cure system can vary by application. For example, silicone-to-silicone medical adhesives use acetoxy cure (condensation cross-linking), and platinum cure (cross-linking by addition) is used for precise silicone parts with no by-products.

4. Elastomer Filler In addition to the silicone polymers described above, the majority of silicone elastomers incorporate "filler." Besides acting as a material extender, as the name implies, filler acts to reinforce the cross-linked matrix. The strength of silicone polymers without filler is generally unsatisfactory for most applications (Noll, 1968). Like most other noncrystallizing synthetic elastomers, the addition of reinforcing fillers reduces silicone's stickiness, increases its hardness and enhances its mechanical strength. Fillers might also be employed to affect other properties; for example, carbon black is added for electrical conductivity, titanium dioxide improves the dielectric constant, and barium sulfate increases radiopacity. These and other materials are used to pigment the otherwise colorless elastomer; however, care must be taken to select only pigments suitable for the processing temperatures and end-use application.

Generally, the most favorable reinforcement is obtained using fumed silica, such as Cab–O–Sil, Aerosil, or Wacker HDK. Fumed silica is produced by the hydrolysis of silicon tetrachloride vapor in a hydrogen flame:

$$SiCl_4 + 2H_2 + O_2 \xrightarrow{1800^\circ C} SiO_2 + 4HCl$$

Unlike many naturally occurring forms of crystalline silica, fumed silica is amorphous. The very small spheroid silica particles (on the order of 10 nm diameter) fuse irreversibly while still semimolten, creating aggregates. When cool, these aggregates become physically entangled to form agglomerates. Silica produced in this way possesses remarkably high surface area, 100 to 400 m^2/g as measured by the BET method developed by Brunauer, Emmett, and Teller (Brunauer *et al.*, 1938; Noll, 1968; Cabot Corporation, 1990).

The incorporation of silica filler into silicone polymers is called "compounding." This is accomplished prior to cross-linking, by mixing the silica into the silicone polymers on a two-roll mill, in a twin-screw extruder, or in a Z-blade mixer capable of processing materials with this rheology.

Reinforcement occurs with polymer adsorption encouraged by the silica's large surface area and when hydroxyl groups on the filler's surface lead to hydrogen bonds between the filler and the silicone polymer, thereby contributing to the production of SRs with high tensile strength and elongation capability (Lynch, 1978). The addition of filler increases the polymer's already high viscosity. Chemical treatment of the silica filler with silanes further enhances its incorporation in, and reinforcement of, the silicone elastomer, resulting in increased material strength and tear resistance (Lane and Burns, 1996) (Fig. 3.2.3-2).

Silicone elastomers for medical applications normally utilize only fillers of fumed silica, and occasionally appropriate pigments or barium sulfate. Because of their low glass transition temperature, these compounded and cured silicone materials are elastomeric at room and body temperatures without the use of any plasticizers—unlike

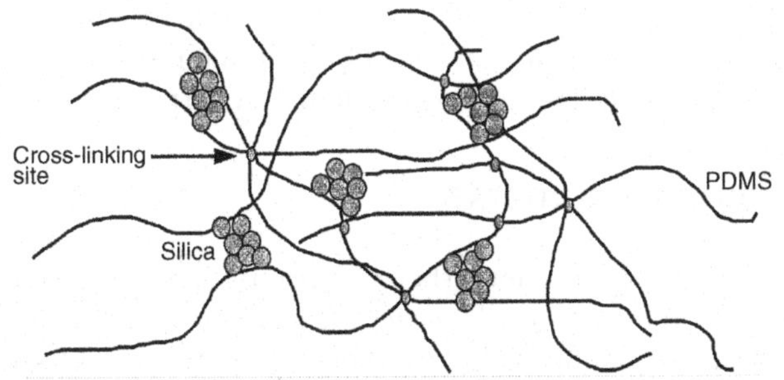

Fig. 3.2.3-2 Silicone elastomer matrix.

other medical materials such as PVC, which might contain phthalate additives.

Physicochemical properties

Silicon's position just under carbon in the periodic table led to a belief in the existence of analog compounds where silicon would replace carbon. Most of these analog compounds do not exist, or behave very differently. There are few similarities between Si–X bonds in silicones and C–X bonds (Corey, 1989; Hardman, 1989; Lane and Burns, 1996; Stark *et al.*, 1982).

Between any given element and Si, bond lengths are longer than for C with this element. The lower electronegativity of silicon ($\chi^{Si} \approx 1.80$, $\chi^{C} \approx 2.55$) leads to more polar bonds compared to carbon. This bond polarity also contributes to strong silicon bonding; for example, the SiO bond is highly ionic and has large bond energy. To some extent, these values explain the stability of silicones. The SiO bond is highly resistant to homolytic scission. On the other hand, heterolytic scissions are easy, as demonstrated by the reequilibration reactions occurring during polymerizations catalyzed by acids or bases (see earlier discussion).

Silicones exhibit the unusual combination of an inorganic chain similar to silicates and often associated with high surface energy, but with side methyl groups that are very organic and often associated with low surface energy (Owen, 1981). The Si–O bonds are quite polar and without protection would lead to strong intermolecular interactions (Stark *et al.*, 1982). Yet, the methyl groups, only weakly interacting with each other, shield the main chain (see Fig. 3.2.3-3).

This is made easier by the high flexibility of the siloxane chain. Barriers to rotation are low and the siloxane chain can adopt many configurations. Rotation energy around a $H_2C–CH_2$ bond in PE is 13.8 kJ/mol but only 3.3 kJ/mol around a $Me_2Si–O$ bond, corresponding to a nearly free rotation. In general, the siloxane chain adopts a configuration such that the chain exposes a maximum number of methyl groups to the outside, whereas in hydrocarbon polymers, the relative rigidity of the polymer backbone does not allow a "selective" exposure of the most organic or hydrophobic methyl groups. Chain-to-chain interactions are low, and the distance between adjacent chains is also greater in silicones. Despite a very polar chain, silicones can be compared to paraffin, with a low critical surface tension of wetting (Owen, 1981).

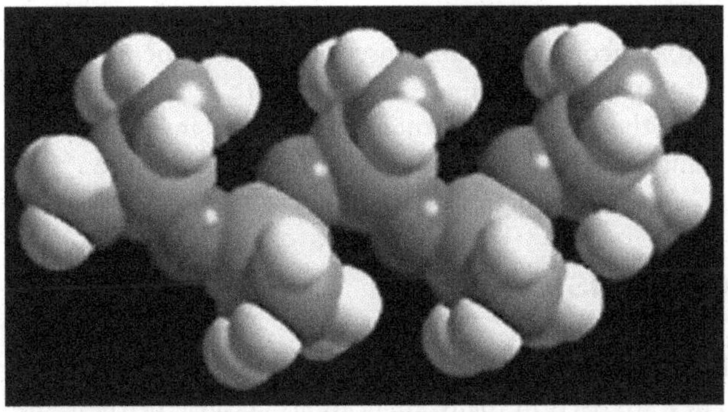

Fig. 3.2.3-3 Three-dimensional representation of dodecamethylpentasiloxane, Me_3SiO-$(SiMe_2O)_3SiMe_3$ or MD_3M. (Courtesy S. Grigoras, Dow Corning.)

The surface activity of silicones is evident in many ways (Owen, 1981):

- The polydimethylsiloxanes have a low surface tension (20.4 mN/m) and are capable of wetting most surfaces. With the methyl groups pointing to the outside, this gives very hydrophobic films and a surface with good release properties, particularly if the film is cured after application. Silicone surface tension is also in the most promising range considered for biocompatible elastomers (20–30 mN/m).
- Silicones have a critical surface tension of wetting (24 mN/m) higher than their own surface tension. This means that silicones are capable of wetting themselves, which promotes good film formation and good surface covering.
- Silicone organic copolymers can be prepared with surfactant properties, with the silicone as the hydrophobic part, e.g., in silicone glycols copolymers.

The low intermolecular interactions in silicones have other consequences (Owen, 1981):

- Glass transition temperatures are very low, e.g., 146 K for a polydimethylsiloxane compared to 200 K for polyisobutylene, the analog hydrocarbon.
- The presence of a high free volume compared to hydrocarbons explains the high solubility and high diffusion coefficient of gas into silicones. Silicones have a high permeability to oxygen, nitrogen, or water vapor, even though liquid water is not capable of wetting a silicone surface. As expected, silicone compressibility is also high.

- The viscous movement activation energy is very low for silicones, and their viscosity is less dependent on temperature compared to hydrocarbon polymers. Furthermore, chain entanglements are involved at higher temperature and contribute to limit the viscosity reduction (Stark *et al.*, 1982).

Conclusion

Polydimethylsiloxanes are often referred to as silicones. They are used in many applications because of their stability, low surface tension, and lack of toxicity. Methyl group substitution or introduction of tri- or tetra-functional siloxane units leads to a wide range of structures. Polymers are easily cross-linked at room or elevated temperature to elastomers, without loosing the above properties.

Acknowledgments

Part of this section (here revised) was originally published in *Chimie Nouvelle*, the journal of the Société Royale de Chimie (Belgium), Vol. 8 (30), 847 (1990) by A. Colas and are reproduced here with the permission of the editor. The authors thank S. Hoshaw and P. Klein, both from Dow Corning, for their contribution regarding breast implant epidemiology.

Bibliography

Brunauer, S., Emmett, P. H., and Teller, E. (1938). Adsorption of gases in multimolecular layers. *J. Am. Chem. Soc.* 60: 309.
Cabot Corporation (1990). *CAB-O-SIL Fumed Silica Properties and Functions*. Tuscola, IL.
Corey, J. Y. (1989). Historical overview and comparison of silicone with carbon. in *The Chemistry of Organic Silicon Compounds*, Part 1, S. Patai and Z. Rappoport eds. John Wiley & Sons, New York.
Hardman, B. (1989). Silicones. *Encyclopedia of Polymer Science and Engineering*. John Wiley & Sons, New York, Vol. 15, p. 204.
Kipping, F. S. (1904). Organic derivative of silicon. Preparation of alkylsilicon chlorides. *Proc. Chem. Soc.* 20: 15.
Lane, T. H., and Burns, S. A. (1996). Silica, silicon and silicones ... unraveling the mystery. *Curr. Top. Microbiol. Immunol.* 210: 3-12.
Lynch, W. (1978). *Handbook of Silicone Rubber Fabrication*. Van Nostrand Reinhold, New York.
Noll, W. (1968). *Chemistry and Technology of Silicones*. Academic Press, New York.
Owen, M. J. (1981). Why silicones behave funny. *Chemtech* 11: 288.
Rochow, E. G. (1987). *Silicon and Silicones*. Springler-Verlag, New York.
Smith, A. L. (1991). Introduction to silicones. *The Analytical Chemistry of Silicones*. John Wiley & Sons, New York.
Stark, F. O., Falender, J. R., and Wright, A. P. (1982). Silicones. In Comprehensive Organometallic Chemistry, G. Wikinson, F. G. A. Sone, and E. W. Ebel, eds. Pergamon Press, Oxford, Vol. 2, pp. 288-297.

3.2.4 Medical fibers and biotextiles

Steven Weinberg and Martin W. King

The term "medical textiles" encompasses medical products and devices ranging from wound dressings and bandages to high-technology applications such as biotextiles, tissue engineered scaffolds, and vascular implants (King, 1991). The use of textiles in medicine goes back to the Egyptians and the Native Americans who used textiles as bandages to cover and draw wound edges together after injury (Shalaby, 1985). Over the past several decades, the use of fibers and textiles in medicine has grown dramatically as new and innovative fibers, structures, and therapies have been developed. Advances in fabrication techniques, fiber technology, and composition have led to numerous new concepts for both products and therapies, some of which are still in development or in clinical trials. In this section, an introduction to fiber and textile fabric technology will be presented along with discussion of both old and new application areas. Traditional and nontraditional fiber and fabric constructions, processing issues, and fabric testing will be included in order to offer an overview of the technology associated with the use of textiles in medicine. Table 3.2.4-1 illustrates some of the more common application areas for textiles in medicine. As can be seen from this table, the products range from the simplest products (i.e., gauze bandages) to the most complex textile products such as vascular grafts and tissue scaffolds.

Medical fibers

All textile-based medical devices are composed of structures fabricated from monofilament; multifilaments; or staple fibers formulated from synthetic polymers, natural polymers (biopolymers), or genetically engineered polymers. When choosing the appropriate fiber configuration and polymer for a specific application,

Table 3.2.4-1 Textile structures and applications (Ko, 1990)

Application	Material	Yarn structure	Fabric structure
Arteries	Dacron T56 Teflon	Textured Multifilament	Weft/warp knit Straight/ bifurcations Woven/non-woven
Tendons	Dacron T56 Dacron T55 Kevlar	Low-twist filament Multifilament	Coated woven tape
Hernia repair	Polypropylene	Monofilament	Tricot knit
Esophagus	Regenerated collagen	Monofilament	Plain weave Knit
Patches	Dacron T56	Monofilament Multifilament	Woven Knit/knit velour
Sutures	Polyester Nylon Regenerated collagen Silk	Monofilament Multifilament	Braid Woven tapes
Ligaments	Polyester Teflon Polyethylene	Monofilament Multifilament	Braid
Bones and joints	Carbon in thermoset or thermoplastic Matrix	Monofilament	Woven tapes Knits/braids

consideration must be given to the device design requirements and the manner in which the fiber is to be used. For example, collagen-based implantable hemostatic wound dressings are available in multiple configurations including loose powder (Avitine), nonwoven mats (Helistat and Surgicel Fibrillar Hemostat), and knitted collagen fibrils (Surgicel Nu-Knit). In addition, other materials are also available for the same purpose (e.g., Surgicel Absorbable Hemostat is knitted from regenerated cellulose). Fibers can be fabricated from nonabsorbable synthetic polymers such as PET or polyester (e.g., Dacron) and PTFE (e.g., Teflon), or absorbable synthetic materials such as polylactide (PLA) and polyglycolide (PGA) (Hoffman, 1977). Natural materials (biopolymers), such as collagen or polysaccharides like alginates, have also been used to fabricate medical devices (Keys, 1996). And there are recent reports that biomimetic polymers have been synthesized in experimental quantities by genetic engineering of peptide sequences from elastin, collagen, and spider dragline silk protein, and expressed in *Escherichia coli* and yeast using plasmid vectors (Huang, 2000; Teule *et al.*, 2003).

Cotton was and still is commonly being used for bandages, surgical sponges, drapes, and surgical apparel, and in surgical gowns. In current practice, cotton has been replaced in many applications by coated nonwoven disposable fabrics, especially in cases when nonabsorbency is critical.

It is important to note that most synthetic polymers currently used in medicine were originally developed as commercial polymers for nonmedical applications and usually contain additives such as dyes, delustrants, stabilizers, antioxidants, and antistatic agents. Some of these chemicals may not be desirable for medical applications, and so must be removed prior to use. To illustrate this point, PET, formerly Dacron, which at present is the material of choice for most large-caliber textile vascular grafts, was originally developed for apparel use. A complex cleaning process is required before the material can be used in an implant application. Additional reading relating to this point can be found in Goswami *et al.* (1977) and Piller (1973).

Synthetic fibers

Various synthetic fibers have been used to fabricate medical devices over the past 25 years. Starting in the 1950s, various materials were evaluated for use in vascular grafts, such as Vinyon (PVC copolymer), acrylic polymers, poly(vinyl alcohol), nylon, PTFE, and polyester (PET) (King *et al.*, 1983). Today, only PTFE and PET are still used for vascular graft applications since they are reasonably inert, flexible, resilient, durable, and resistant to biological degradation. They have withstood the test of time, whereas other materials have not proven to be durable when used in an implant application. Table 3.2.4-2 shows a partial list of synthetic polymers that have been prepared as fibers, their method of fabrication, and how they are used in the medical field.

Most synthetic fibers are formed either by a melt spinning or a wet spinning process.

Melt spinning

With melt spinning the polymer resin is heated above its melting temperature and extruded through a spinneret. The number of holes in the spinneret defines the number of filaments in the fiber being produced. For example, a spinneret for a monofilament fiber contains one hole, whereas 54 holes are required to produce the 54-multifilament yarn that is commonly used in vascular graft construction. Once the monofilament or multifilament yarn is extruded, it is then drawn and cooled prior to being wound onto spools. The yarn can also be further processed to form the final configuration. For example, most yarns used for application in vascular grafts are texturized to improve the handling characterizes of the

Table 3.2.4-2 Synthetic polymers (Shalaby, 1996)

Type	Chemical and physical aspects	Construction/useful forms	Comments/applications
Polyethylene (PE)	High-density PE (HDPE): melting temperature $T_m = 125°C$ Low-density PE (LDPE): $T_m = 110°C$, Linear low-density (LLDPE) Ultrahigh molecular weight PE (UHMWPE) ($T_m = 140–150°C$), exceptional tensile strength and modulus	Melt spun into continuous yarns for woven fabric and/or melt blown onto nonwoven fabric Converted to very high tenacity yarn by gel spinning	The HDPE, LDPE and LLDP are used in a broad range of health care products Used experimentally as reinforced fabrics in lightweight orthopedic casts, ligament prostheses, and load-bearing composites
Polypropylene (PP)	Predominantly isotactic, $T_m = 165–175°C$; higher fracture toughness than HDPE	Melt spun to monofilaments and melt blown to nonwoven fabrics Hollow fibers	Sutures, hernia repair meshes, surgical drapes, and gowns Plasma filtration
Poly(tetrafluoroethylene) (PTFE)	High melting ($T_m = 325°C$) and high crystallinity polymer (50–75% for processed material)	Melt extruded	Vascular fabrics, heart valve sewing rings, orthopedic ligaments
Nylon 6	$T_g = 45°C$, $T_m = 220°C$, thermoplastic, hydrophilic	Monofilaments, braids	Sutures
Nylon 66	$T_g = 50°C$, $T_m = 265°C$, thermoplastic, hydrophilic	Monofilaments, braids	Sutures
Poly(ethylene terephthalate) (PET)	Excellent fiber-forming properties, $T_m = 265°C$, $T_g = 65–105°C$	Multifilament yarn for weaving, knitting, and braiding	Sutures, hernia repair meshes, and vascular grafts

final product. In contrast to flat or untexturized yarn, texturization results in a yarn that imparts bulk to the fabric for improved "hand" or feel, flexibility, ease of handling and suturing, and more pores for tissue ingrowth. Melt spinning is typically used with thermoplastic polymers that are not affected by the elevated temperatures required in the melt spinning process. Figure 3.2.4-1 is a schematic representation of a melt spinning process.

In this process, the molten resin is extruded through the spinning head containing one (monofilament) or multiple holes (multifilament). Air is typically used to cool and solidify the continuous threadline prior to lubricating, twisting, and winding up on a bobbin.

Wet spinning

If the polymer system experiences thermal degradation at elevated temperatures, as is the case with a polymer containing a drug, a low-temperature wet solution spinning process can be used. In this process the polymer is dissolved in a solvent and then extruded through a spinneret into a nonsolvent in a spin bath. Because the solvent is soluble in the spin bath, but the polymer is not, the continuous polymer stream precipitates into a solid filament, which is then washed to remove all solvents and nonsolvents, drawn, and dried before winding up (Adanur, 1995). Figure 3.2.4-2 presents a schematic of a typical wet solution spinning process.

Electrospinning

The diameters of fibers spun by melt spinning and wet solution spinning are controlled by the size of the hole in the spinneret and the amount of draw or stretch applied

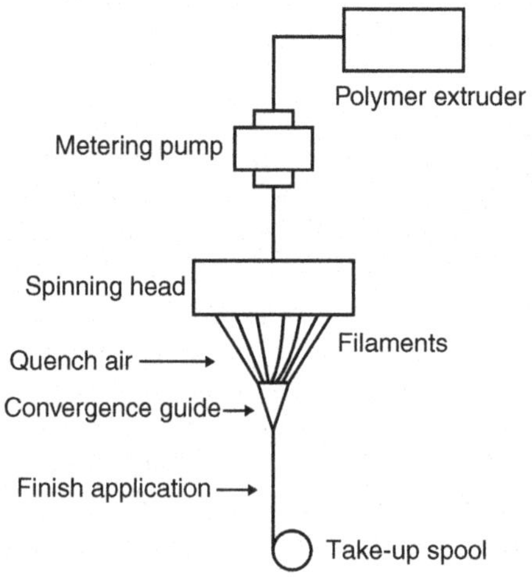

Fig. 3.2.4-1 Melt spinning process.

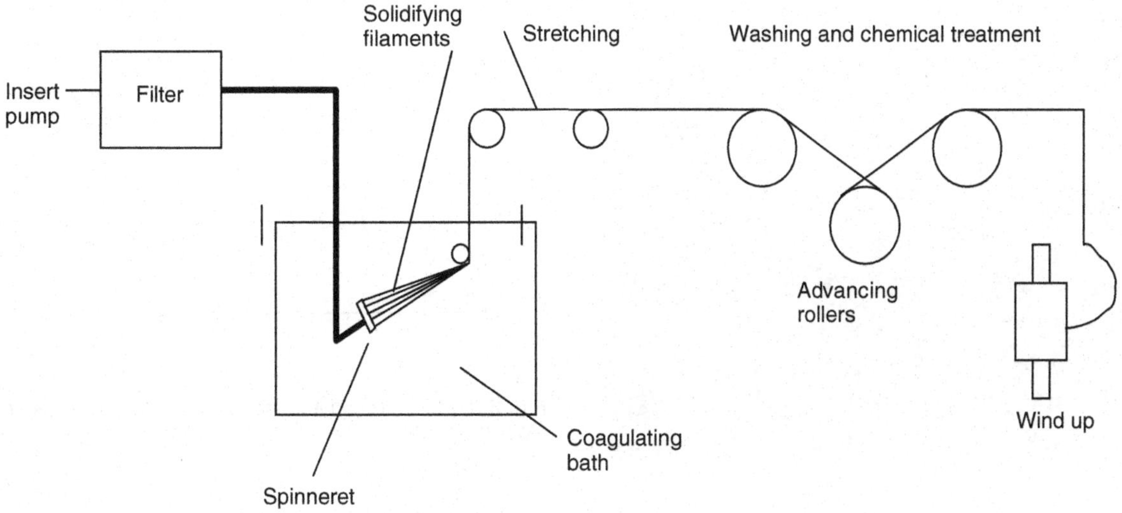

Fig. 3.2.4-2 Wet solution spinning process.

to the filament prior to wind-up. So the diameters of conventional spun fibers fall in a range from about 10 μm for multifilament yarns to 500 μm or thicker for monofilaments. To obtain finer fiber diameters it is necessary to employ alternative spinning technologies such as the bicomponent fiber (BCF) approach (see later section entitled "Hybrid BCFs"), or an electrospinning technique. This method of manufacturing microfibers and nanofibers has been known since 1934 when the first patent was filed (Formhals, 1934). Since then Freudenberg Inc. has used this process for the commercial production of ultrahigh-efficiency filters (Groitzsch and Fahrbach, 1986).

Electrospinning occurs when a polymer solution or melt is exposed to an electrostatic field by the application of a high voltage (5–30 kV), which overcomes the surface tension of the polymer and accelerates fine jets of the liquid polymer towards a grounded target (Reneker *et al.*, 2000). As the polymer jets cool or lose solvent they are drawn in a series of unstable loops, solidified, and collected as an interconnected web of fine fibers on a grounded rotating drum or other specially shaped target (Fig. 3.2.4-3).

The fineness of the fibers produced depends on the polymer chemistry, its solution or melt viscosity, the strength and uniformity of the applied electric field, and the geometry and operating conditions of the spinning system. Fiber diameters in the range of 1 μm down to 100 nm or less have been reported.

In addition to being used to fabricate ultrathin filtration membranes, electrospinning techniques have also been applied to the production of nonwoven mats for wound dressings (Martin and Cockshott, 1977), and there is currently much interest in making scaffolds for tissue engineering applications. Nonwoven scaffolds spun from Type I collagen and synthetic polymers such as poly(L-lactide), poly(lactide-*co*-glycolide), poly(vinyl alcohol), poly(ethylene-*co*-vinyl acetate), PEO, polyurethanes, and polycarbonates have been reported (Stitzel *et al.*, 2001; Matthews *et al.*, 2002; Kenawy *et al.*, 2002; Theron *et al.*, 2001; Schreuder-Gibson *et al.*, 2002). In addition genetic engineering has been used to synthesize an elastin–biomimetic peptide polymer based on the elastomeric peptide sequence of elastin and expressed from recombinant plasmid pRAM1 in

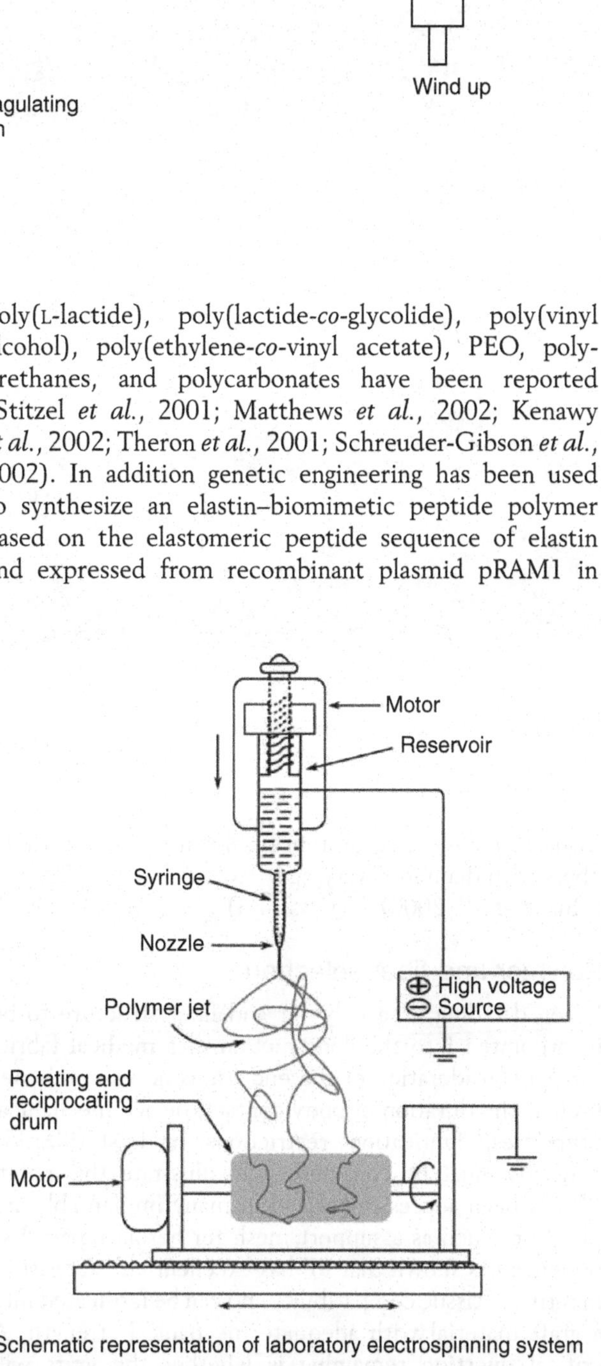

Fig. 3.2.4-3 Electrospinning system.

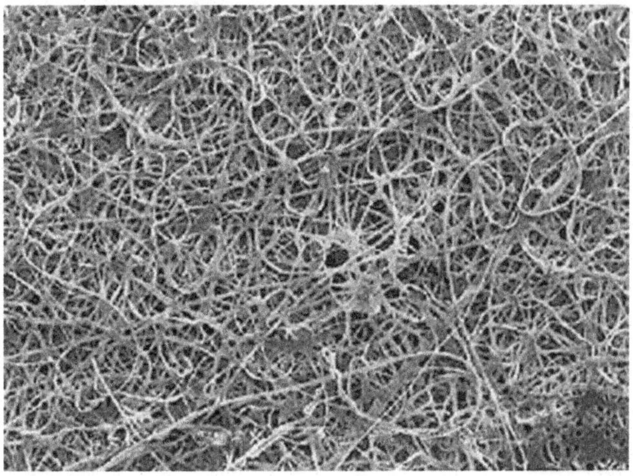

A

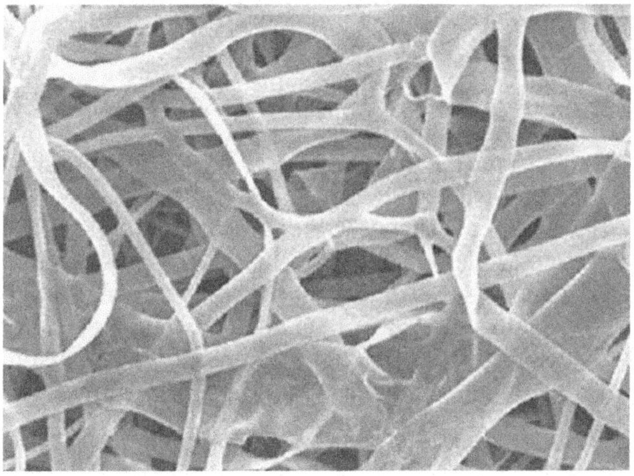

B

Fig. 3.2.4-4 Electrospun fibers from biomimetic-elastin peptide.

Escherichia coli. The protein has been electrospun into fibers with diameters varying between 3 nm and 200 nm (Huang *et al.*, 2000) (Fig. 3.2.4-4).

Polymer and fiber selection

When deciding on a polymer and fiber structure to be incorporated into the construction of a medical fabric, careful consideration of the end use is necessary. Issues such as the duration of body contact, device mechanical properties, fabrication restrictions, and sterilization methods must be considered. To illustrate this point, PP has been successfully used in many implantable applications such as a support mesh for hernia repair. Experience has shown that PP has excellent characteristics in terms of tissue compatibility and can be fabricated into a graft material with adequate mechanical strength. A critical question remaining is whether the graft will remain stable and survive as a long-term implant.

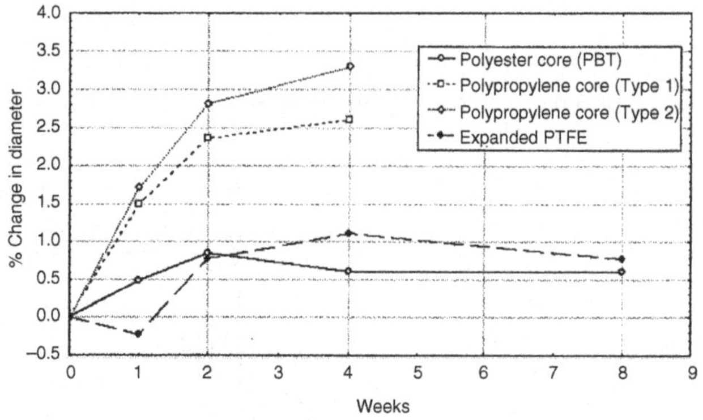

Fig. 3.2.4-5 Creep characteristics of various graft materials (Weinberg, 1998).

Figure 3.2.4-5 demonstrates the creep characteristics of grafts fabricated from expanded polytetrafluoroethylene (e-PTFE), polyester, and a BCF containing PP yarns. In the case of the first BCF design (see later section), the PP was used as the nonabsorbable core material and the main structural component of the fiber. Figure 3.2.4-5 represents the outer diameter of a series of pressurized graft materials as a function of time. Classical graft materials such as PET and e-PTFE show no creep over time, whereas the PP-based materials continue to creep over time, making them unacceptable for long-term vascular implants. However, in other applications such as for hernia repair meshes and sutures, PP has been used very successfully. It should be noted that in the second-generation BCF design, the core material was changed to poly(butylene terephthalate) (King *et al.*, 2000).

Absorbable synthetic fibers

Another series of synthetic fibers used in clinical applications are constructed from polymers that are designed to be absorbed over time when placed in the body. They classically have been used as sutures, but have also been used experimentally for neurological, vascular graft, and tissue scaffold applications. Table 3.2.4-3 is a list of bioabsorbable polymers that have been used in the past to fabricate medical devices. When in contact with the body, these polymers degrade either by hydrolysis or by enzymatic degradation into nontoxic by products. They break down or degrade either through an erosion process that starts on the exterior surface of the fiber and continues until the fiber has been totally absorbed, or by a bulk erosion mechanism in which the process is autocatalytic and starts in the center of the fiber. Caution should be exercised when using these types of materials. In vascular applications, the risk of distal embolization to the microvasculature may occur if small pieces of the polymer break off during the erosion or absorption process.

Table 3.2.4-3 Absorbable synthetic polymers

Type	Chemical and physical aspects	Construction/useful forms	Comments/applications
Poly(glycolide) (PGA)	Thermoplastic crystalline polymer ($T_m = 225°C$, $T_g = 40–45°C$)	Multifilament yarns, for weaving, knitting and braiding, sterilized by ethylene oxide	Absorbable sutures and meshes (for defect repairs and periodontal inserts)
10/90 Poly(L-lactide-*co*-glycolide) (Polyglactin 910)	Thermoplastic crystalline co-polymer ($T_m = 205°C$, $T_g = 43°C$)	Multifilament yarns, for weaving, knitting and braiding, sterilized by ethylene oxide	Absorbable sutures and meshes
Poly(p-dioxanone) (PDS)	Thermoplastic crystalline co-polymer ($T_m = 110–115°C$, $T_g = 10°C$)	Melt spun to monofilament yarn	Sutures, intramedullary pins and ligating clips
Poly(alkylene oxalates)	A family of absorbable polymers with T_m between 64 and 104°C	Can be spun to monofilament and multifilament yarns	Experimental sutures
Isomorphic poly(hexamethylene-*co-trans*-1, 4-cyclohexane dimethylene oxalates)	A family of crystalline polymers with T_m between 64 and 225°C	Can be spun to monofilament and multifilament yarns	Experimental sutures

Modified natural fibers

In addition to synthetic polymers, a class of fibers exists that is composed of natural biopolymer based materials. In contrast to synthetic fibers that have been adapted for medical use, natural fibers have evolved naturally and so can be particularly suited for medical applications. Cellulose, which is obtained from processed cotton or wood pulp, is one of the most common fiber-forming biopolymers. Because of the highly absorbent nature of cellulose fibers, they are commonly used in feminine hygiene products, diapers, and other absorbable applications, but typically are not used *in vivo* because of the highly inflammatory reactions associated with these materials. In certain cases, these properties can be used to advantage such as in the aforementioned hemostat Surgicel. In this application, the thrombogenicity and hydration characteristics of the regenerated cellulose are used in stopping internal bleeding from blood vessels and the surface of internal organs. Also of growing interest are fibers created from modified polysaccha-rides including alginates, xanthan gum, chitosan, dextran, and reticulated cellulose (Shalaby and Shah, 1991; Keys, 1996). These materials are obtained from algae, crustacean shells, and through bacterial fermentation. A list of several forms of alginates and their proposed uses is presented in Table 3.2.4-4 (Keys, 1996). Another natural material, chitosan, has been used to fabricate surgical sutures and meshes, and it is currently under investigation for use as a substrate or scaffold for tissue-engineered materials (Skjak-Braek and Sanford, 1989). Chitosan and alginate fibers are formed when the polymer is coagulated in a wet solution spinning process.

Silk and collagen are two natural fibers that have been widely used in medicine for multiple applications. Silk from the silkworm, *Bombyx mori*, has been used for decades as a suture. Because of the fineness of individual silk fibers, it is necessary to braid the individual fibers or brins together into thicker yarn bundles. Collagen has been used either in a reconstituted form or in its natural state. Reconstituted collagen is obtained from enzymatic chemical treatment of either bovine skin or tendon followed by reconstitution into fibrils. These fibrils can then be spun into fibers and fabricated into textile structures or can be left in their native fibrillar form for use in hemostatic mats and tissue-engineered substrates. "Catgut," a natural collagen-based suture material obtained from ovine intestine, which is cross-linked and

Table 3.2.4-4 Potential uses of alginates (Keys, 1996)

Type	Current use
Ca alginate (non-woven)	Absorbent wound dressings Pledgets Scaffold for cell culture Surgical hemostats
Ca alginate (particulate)	Acid-labile conjugates of alginate and doxirubicin Sequestration of 90Sr from ingested contaminated food and water
Na alginate (ultra pure)	Microencapsulation Bioreactors
Ca/Na alginate (hydrogel)	Wound management

cut into narrow strips, was one of the first bioabsorbable fibers used in surgery.

Hybrid BCFs

Hybrid BCF technology is a novel fiber concept that has been under development for a number of years for use in vascular grafts and other cardiovascular applications. One of the configurations of a BCF is a sheath of an absorbable polymer around an inner core of a second nonabsorbable or less absorbable polymer. With a multifilament BCF yarn, each of the filaments of the yarn bundle is identical and contains an identical inner core and outer sheath. Prior to the development of the BCF yarn, when a bicomponent fabric was to be produced it was fabricated by weaving, braiding, or knitting together two (or more) homogenous yarns (e.g., a polyester yarn and a PLA yarn). With such constructions, the tissue or blood sees multiple polymers at the same time. In contrast, with BCF technology only one polymer in the sheath makes initial contact with the tissue. If the outer sheath of a BCF fiber is composed of a bioabsorbable material such as PGA, the inner core polymer is only exposed when the sheath is absorbed. The composition and molecular weight of the polymer and the thickness of the sheath regulate its absorption rate. The hypothesis relating to the BCF concept is that the healing process can be modulated by slowly exposing the less biocompatible inner core material. Preliminary data has shown that the absorption rate can be regulated and will affect the healing process (King *et al.*, 1999). By constructing the inner core from a nonabsorbable biostable polymer such as PET, or a slower absorbing polymer such as PLA, the strength of the fiber will be maintained even as the outer sheath dissolved.

Additionally, drugs can be incorporated into the outer absorbable sheath and delivered at predefined rates depending on the choice and thickness of the outer polymer. By using this BCF technology, both the material strength profile and the biological properties can be engineered into the fiber to meet specific medical requirements.

Construction

After a fiber or yarn is produced, it is then fabricated into a textile structure in order to obtain the desired mechanical and biological properties. Typical biotextile structures used for medical applications include nonwovens, wovens, knits, and braids. Within each of these configurations, many variations exist. Each type of construction has positive and negative attributes, and in most cases, the final choice represents a compromise between desired and actual fabricated properties. For example, woven fabrics typically are stronger and can be fabricated with lower porosities or water/blood permeability as compared to knits, but are stiffer, less flexible, and more difficult to handle and suture. Knits have higher permeability than woven designs and are easier to suture, but may dilate after implantation. Braids have great flexibility, but can be unstable except when subject to longitudinal load, as in the case of a suture. Multilayer braids are more stable, but are also thicker and less flexible than unidimensional braids. Each construction is a compromise.

Nonwovens

By definition, a nonwoven is a textile structure produced directly from fibers without the intermediate step of yarn production. The fibers are either bonded or interlocked together by means of mechanical or thermal action, or by using an adhesive or solvent or a combination these approaches. Figure 3.2.4-6 is a representation of both wet and dry nonwoven forming processes. The fibers may be oriented randomly or preferentially in one or more directions, and by combining multiple layers one can engineer the mechanical properties independently in the machine (lengthwise) and cross directions. The average pore size of a nonwoven web depends on the density of fibers, as well as the average fiber diameter, and falls under a single distribution (Krcma, 1971). For this reason some tissue-engineered substrates under development use nonwovens to form the underlying tissue scaffold (Chu, 2002).

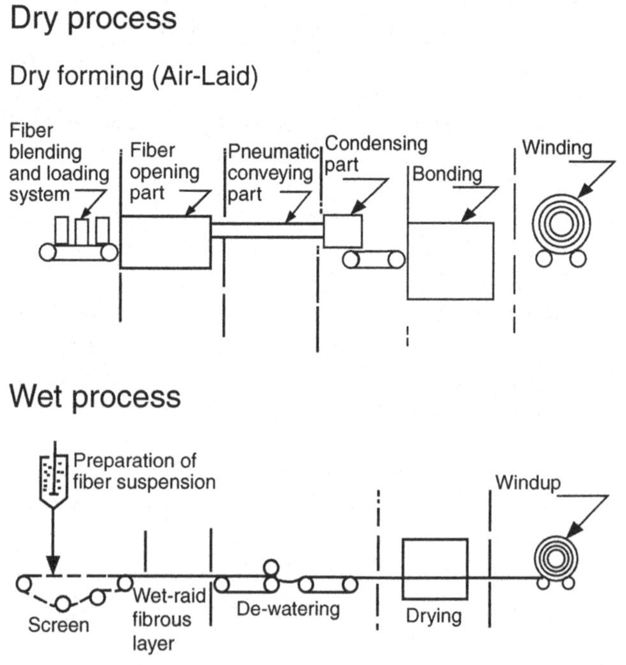

Fig. 3.2.4-6 Wet and dry nonwoven processes.

Woven fabrics

The term "woven" is used to describe a textile configuration where the primary structural yarns are oriented at 90° to each other. The machine direction is called the warp direction and the cross direction is identified as the filling or weft direction. Because of the orthogonal relationship between the warp and filling yarns, woven structures display low elongation and high breaking strength in both directions. There are many types of woven constructions including plain, twill, and satin weaves (Robinson and Marks, 1967). Figure 3.2.4-7 is a sketch showing several weave designs commonly used in vascular graft fabrications. Water permeability is one critical parameter used in the assessment of textile structures for vascular implants. Water permeability is a measure of the water flux through a fabric under controlled conditions and is given in units of ml cm^{-2} min^{-1}. It is measured by placing fabric into a test fixture having a fixed orifice size and applying a pressure of 120 mmHg across the fabric. The water passing through the fabric is collected and measured over time and water permeability is calculated (ISO 7198, section 8.2.2, Water Permeability). Surgeons use this parameter as a guide to determine if "pre-clotting" of a graft material is necessary prior to implantation. "Pre-clotting" is a process where a graft material is clotted with a patient's blood prior to implantation, rendering the fabric nonpermeable to blood after implantation. Fabric grafts with water permeability values less than 50 ml cm^{-2} min^{-1} usually do not require pre-clotting prior to implantation. The water permeability of the woven graft fabrics can be controlled through the weaving and finishing process and can range from a low of 50 ml cm^{-2} min^{-1} up to about 350 ml cm^{-2} min^{-1}. Above this range, a woven fabric starts becoming mechanically unstable.

Table 3.2.4-5 offers a list of a number of commercial woven graft designs with their respective mechanical properties. As can be seen, many variations in design are possible, presenting a difficult selection process for the surgeon. It is interesting to note that the choice of a graft by a surgeon is often based on the graft's "ease of handling" or "ease of suturing" rather than on its reported long-term performance. Plain weaves, in contrast to knits, can be made very thin (< 0.004 in.) and have thus become the material of choice for many endovascular graft designs.

Knits

Knitted constructions are made by interloping yarns in horizontal rows and vertical columns of stitches. They are softer, more flexible and easily conformable, and have better handling characteristics than woven graft designs. Knit fabrics can be built with water permeability values as high as 5,000 ml cm^{-2} min^{-1} and still maintain structural stability. Currently, highly porous grafts materials are usually coated or impregnated with collagen or gelatin so that the surgeon does not have to perform the time consuming pre-clotting process at the time of surgery. The water permeability values for non-coated knitted grafts range from about 1200 ml cm^{-2} min^{-1} up to about 3500 ml cm^{-2} min^{-1}. When knits are produced, the fabric is typically very open and requires special processing to tighten the looped structure and lower its permeability. This compaction process is usually done using a chemical shrinking agent such as methylene chloride or by thermal shrinking. Because of their open structure, knits are typically easier to suture and have better handling characteristics; however, in vascular graft

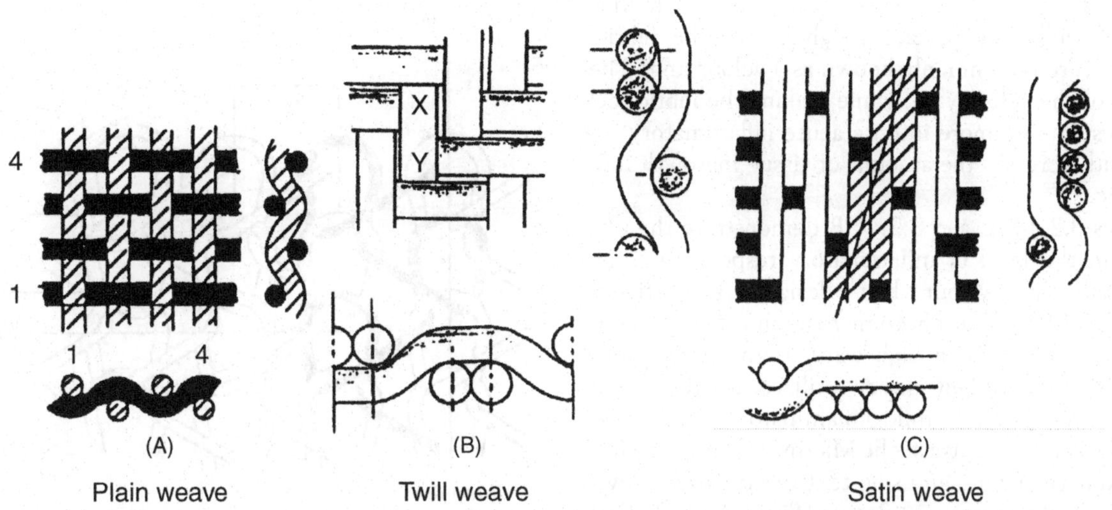

Fig. 3.2.4-7 Examples of woven graft designs.

Table 3.2.4-5 Woven graft properties and construction (King *et al.*, 1991)

Prosthesis	Type of weave	Ends per inch	Picks per inch	Bursting strength (N)	Water permeability	Suture retention strength (N)	Dilatation at 120 mm Hg (%)
Twill woven	1/1 Plain with float	42p22f	48	280	330	25	0
Debakey soft woven	1/1 Plain	52	32	366	220	35	0.2
Debaky extra low porosity	1/1 Plain	55	40	439	50	40	—
Vascutek woven	1/1 Plain	56	30	227	80	30	0.5
Meadox woven double velour	6/4 Satin+ 1/1 plain	36s36p	38	310	310	48	1.2
Meadox cooley verisoft	1/1 Plain	58	35	211	180	30	0.2
Intervascular oshner 200	1/1 Plain with leno	42p14L	21	268	250	22	0.5
Intervascular oshner 500	1/1 Plain with leno	42p14L	21	259	530	26	1.2

applications, some ultralightweight designs have been known to continuously dilate or expand when implanted in hypertensive patients. It is not uncommon to have lighter weight weft knitted grafts increase up to 20% in diameter shortly after implantation.

As is the case with woven structures, there are several variations in knits; the most common are the weft knit and warp knit constructions (see Fig. 3.2.4-8). Warp knitted structures have less stretch than weft knits, and therefore are inherently more dimensionally stable, being associated with less dilation *in vivo*. Warp knits do not run and ravel when cut at an angle (King *et al.*, 1991). Warp knits can be further modified by the addition of an extra yarn in the structure, which adds thickness, bulk, and surface roughness to the fabric. This structure is commonly known as a velour knit. The addition of the velour yarn, while making the fabric feel softer, results in a more intense acute inflammatory reaction and increases the amount of tissue ingrowth into the fabric.

Figures 3.2.4-9A and 3.2.4-9B demonstrate the difference in the level of inflammatory response as seen with plain and velour knit designs, respectively. Figure 3.2.4-9A is a photomicrograph of a Golaski Microkit weft knit with high water permeability. This high porosity weft knit design utilized nontexturized yarns that resulted in a mild inflammatory response as seen at 4 weeks. In contrast, the Microvel fabric, which is a warp knit velour design using texturized yarns, shows an intense acute response at 3 days (Fig. 3.2.4-9B). This more intense acute reaction was designed intentionally so

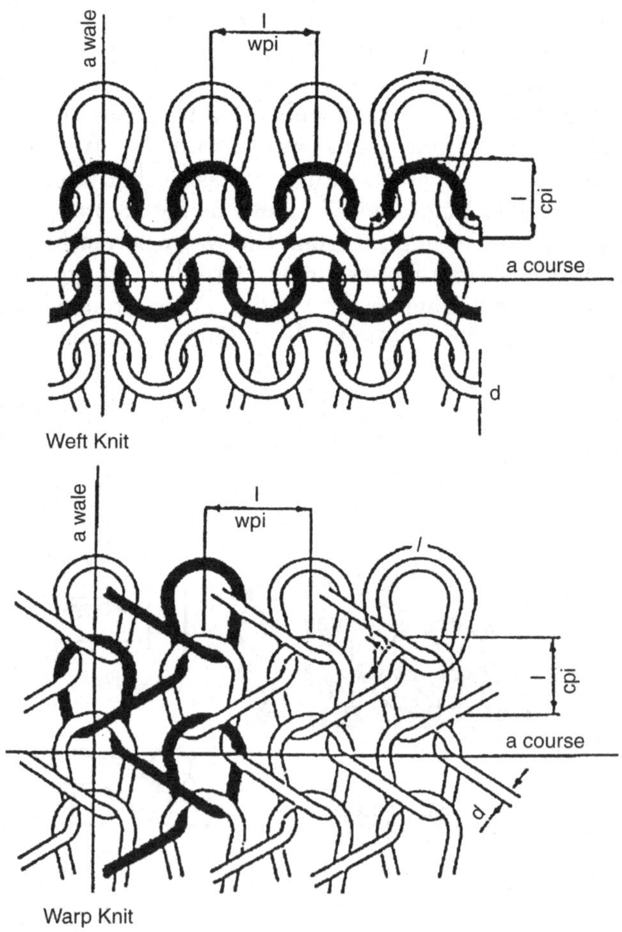

Fig. 3.2.4-8 Types of knit fabrics (Spencer, 1983).

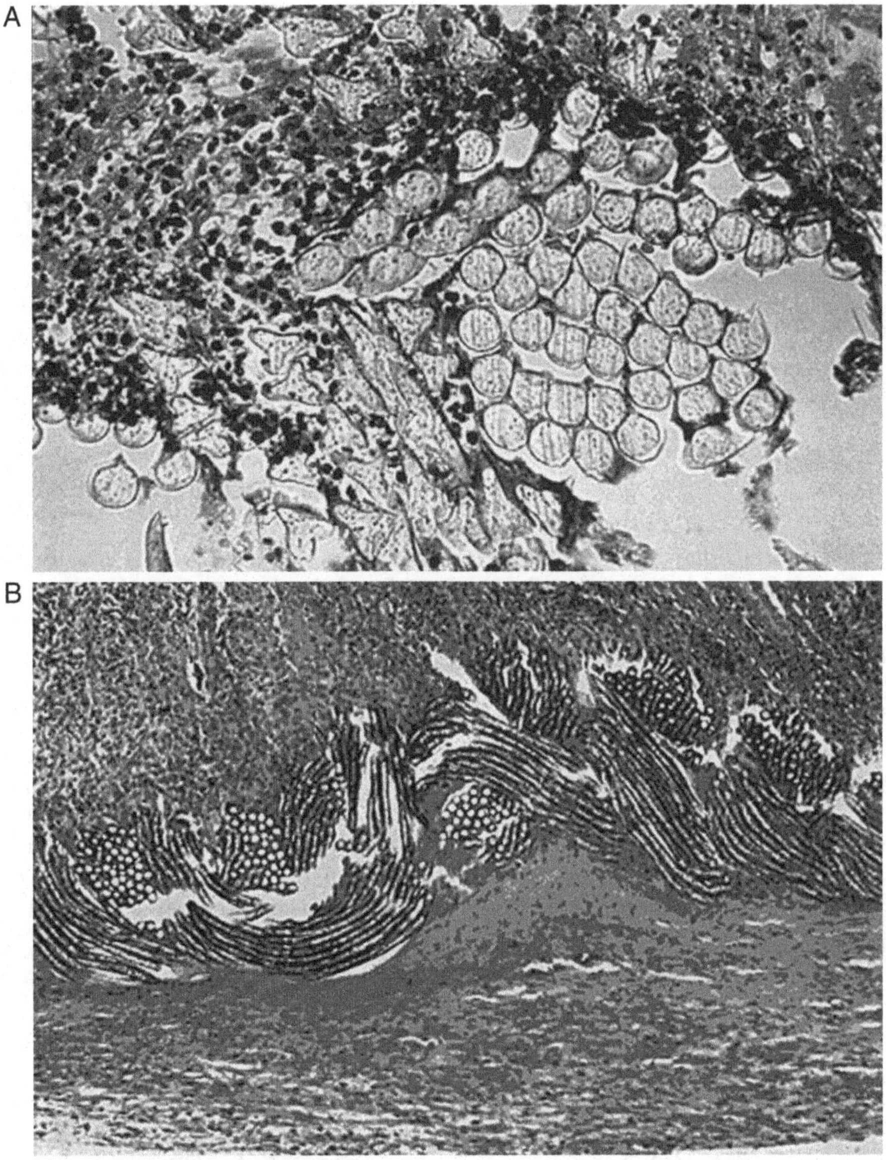

Fig. 3.2.4-9 (A) Weft knit inflammatory response at 4 weeks (Golaski Microkit); (B) Warp knit inflammatory response at 3 days (Microvel).

as to make the graft easier to preclot and to increase the extent of tissue incorporation into the graft wall.

Braids

Braids have found their way into medical use primarily in the manufacture of suture materials and anterior cruciate ligament (ACL) prostheses. Common braided structures involve the interlacing of an even number of yarns, leading to diamond, regular, and Hercules structures that can be either two-or three-dimensional (see Fig 3.2.4-10). A myriad of structural forms can be achieved with 3D braiding, such as "I" beams, channels, and solid tubes. A sketch of a flat braiding machine is included in Fig. 3.2.4-11.

Processing and finishing

Once a fabric has been produced from yarn, the subsequent processing steps are known as finishing. As mentioned previously, the starting yarn may contain additives that can result in cytotoxicity and adverse reactions when in contact with tissue. Some of these additives, such as titanium dioxide, which is used as a delusterant to increase the amount of light reflected, are inside the spun fiber and cannot be removed in the

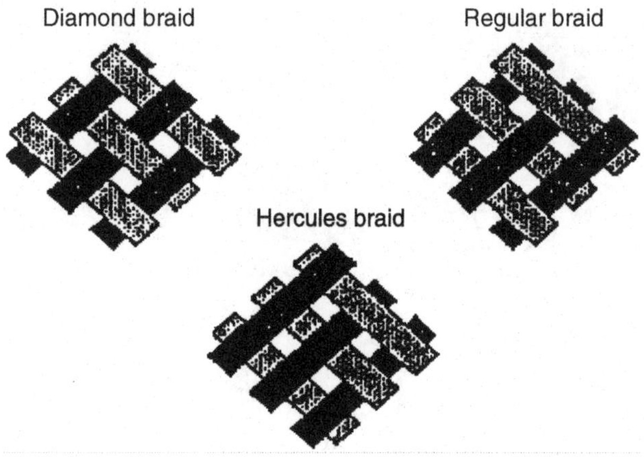

Fig. 3.2.4-10 Braided constructions.

finishing operation. Other surface finishes, on the other hand, such as yarn lubricants, can be removed with the proper cleaning and scouring operations. Typically such surface additives are mineral oil based and demand specially designed aqueous-based washing procedures or dry-cleaning techniques with organic solvents to ensure complete removal. In addition to such surface lubricants, the warp yarns may be coated with a sizing agent prior to weaving. This sizing protects the yarns from surface abrasion and filament breakage during weaving. Since each polymer and fabrication process is different, the finishing operation must be material and device specific.

Finishing includes such steps as cleaning, heat setting, bleaching, shrinking (compaction), inspection, packaging, and sterilization and will influence the ultimate properties of the biotextile fabric. Figure 3.2.4-12 represents a schematic of a typical finishing operation used in vascular graft manufacturing. The chemicals used in the finishing operation may differ among manufacturers and are usually considered proprietary. If the cleaning process is properly designed, all surface finishes are removed during the finishing process. Testing of the finished product for cytotoxicity and residual extractables is typically used to ensure all the surface additives are removed from the product's surface prior to packaging and sterilization.

Testing and evaluation

Once the biotextile is in its final form, it must be tested and evaluated to confirm that it meets published standards and its intended end use. The testing will include component testing on each component including the textile as well as final functional testing of the entire device. When developing and implementing a testing program, various pieces of reference information may apply, including ASTM standards, AAMI/ISO standards, FDA documents, prior regulatory submissions, and the results of failure analyses. In setting up the test plan a fine balance is needed so as to minimize the scope of the testing program while still ensuring that the

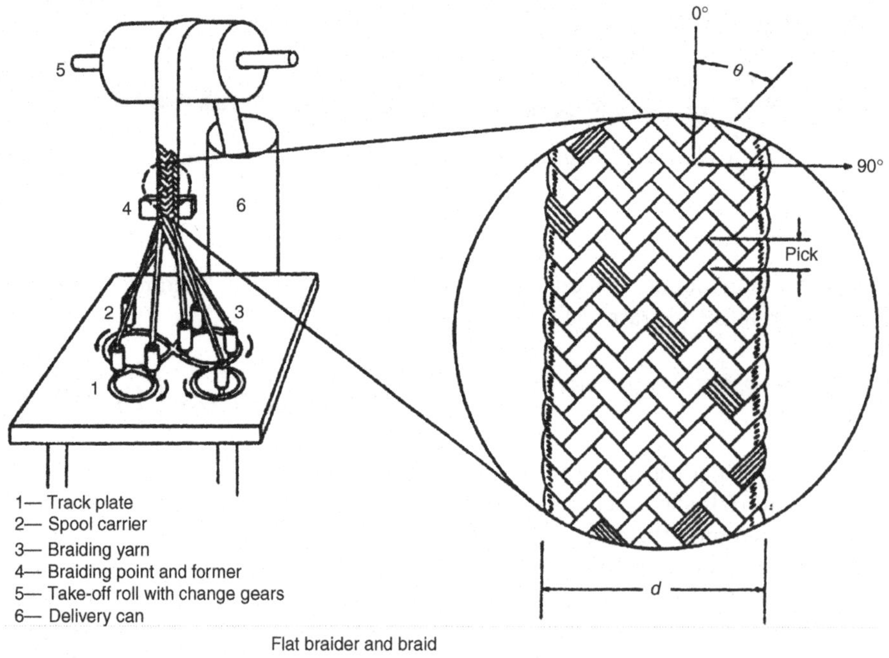

Fig. 3.2.4-11 Sketch of flat braider.

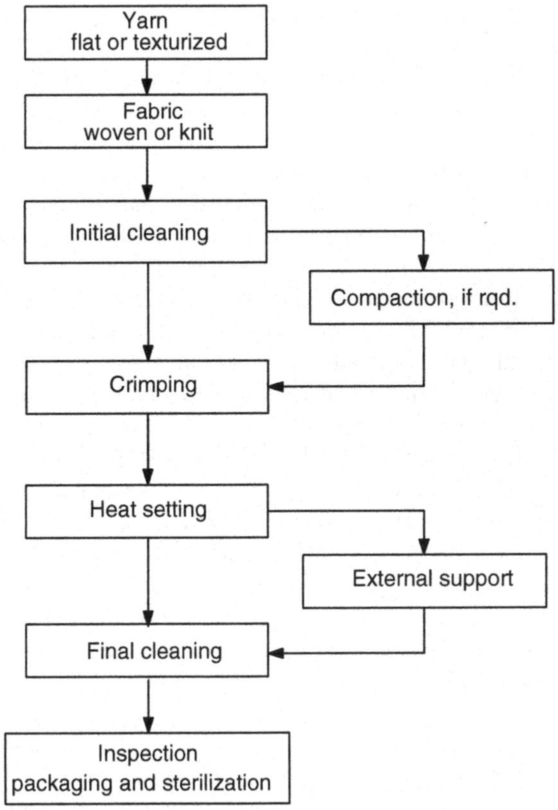

Fig. 3.2.4-12 Typical graft finishing operation.

Table 3.2.4-6 Sample test methods for large-diameter textile grafts

Test	Required regulatory testing	Routine quality testing
Visual inspection for defects	X	X
Water permeability	X	X
Longitudinal tensile strength	X	
Burst strength	X	X
Usable length	X	X
Relaxed internal diameter	X	X
Pressurized internal diameter	X	
Wall thickness	X	
Suture retention strength	X	
Kink diameter/radius	X	
Dynamic compliance	X	
Animal trials	X	
Shelf life	X	
Sterility	X	X
Biomaterials/toxicity and pyrogen testing	X	X

polymer, textile, and final product will be safe and efficacious. Table 3.2.4-6 is a list of the suggested test methods used in the development of a textile-based vascular graft for large vessel replacement (ANSI/AAMI/ISO, 2001).

Applications

The application of fibers and biotextiles as components for implant devices is widespread and covers all aspects of medicine and health care. Textiles are used as basic care items such as drapes, protective apparel, wound dressings, and diapers and in complex devices such as heart valve sewing rings, vascular grafts, hernia repair meshes, and percutaneous access devices.

Drapes and protective apparel

The most common nonimplantable medical use of textiles is for protective surgical gowns, operating room drapes, masks, and shoe covers. Nonwovens and wovens are most frequently used for these applications, with nonwovens being the material of choice for single-use (disposable) products, and wovens for reusable items. Most of these barrier-type fabrics are made from cellulose (cotton, viscose rayon, and wood pulp), P, and PP fibers. Many fabrics contain finishes that render them water repellent depending on the clinical need. Additionally, such fabrics must generally be fire retardant because of the risk of explosions due to exposure to flammable gases used for anesthesia. In applications such as facemasks, the fabric must minimize the passage of bacteria through the mask. This can be ensured by engineering the appropriate pore size distribution in the filtration fabric (Schreuder-Gibson, 2002). Antibacterial coatings are also placed on surgical drapes to minimize the risk of wound contamination. Drapes and protective apparel typically require some assembly that can be done either through conventional sewing or by ultrasonic seaming methods. The latter method is preferred for those products used in sterile fields since the holes created by conventional sewing needles can render the fabric permeable to liquids and liquid-borne pathogens. Drapes are usually constructed of a nonwoven fabric laminated to a plastic film to ensure that they are impervious to blood and other fluids. Another common use of textiles is in the fabrication of adhesive tapes. These tapes generally consist of an adhesive layer that is laminated onto a woven, knitted, or nonwoven fabric substrate.

Topical and percutaneous applications

Textiles have been used for many years as bandages, wound coverings, and diapers. Gauze, which is basically an open woven structure made from cotton fiber, is manufactured in many forms and sold by many companies worldwide. Elastic bandages are basically woven tapes where an expandable yarn, such as spandex polyurethane, is placed in the warp direction to allow for longitudinal stretch and recovery. Development continues to improve wound dressing products by the addition of antibiotics, barrier fabrics, growth factors, and modification of the basic underlining bandage construction. One example of the latter is the work of Karamuk *et al.* (2001), in which a three-layered laminate was formed from a nonwoven polyester/PP/cotton outer layer, a monofilament polyester middle layer, and a three-dimensional embroidered polyester inner layer with large pores to promote angiogenesis.

Blood access devices are a class of medical devices where tubes, wires, or other components pass through the skin. These include percutaneous drug delivery devices, blood access shunts, air or power lines for heart and left ventricular assist devices, and many types of leads. All of these devices suffer from the same basic problem, a high risk of infection at the skin-device interface due to the migration of bacteria along the surface of the percutaneous lead. If a textile cuff is placed around the tube, at the point of entry through the skin, aggressive tissue ingrowth into the fabric reduces the risk of infection at the percutaneous site. These cuffs are usually made from knits, nonwoven felts, and velour materials. Once a device is infected, it must be removed to prevent further spreading of the infection. Surface additives, such as silver or antibiotics, are sometimes coated on the fabric to reduce infection rates (Butany *et al.*, 2002; Takai *et al.*, 2002).

In vivo applications

Cardiovascular devices

Biotextiles developed for cardiovascular use include applications such as heart valve sewing rings, angioplasty rings,vascular grafts, valved conduits, endovascular stent grafts, and the components of left ventricular assist devices. One of the most important uses of textile fabrics in medicine is in the fabrication of large diameter vascular grafts (10–40 mm in diameter). As previously noted, polyester [PET] is the principal polymer used to fabricate vascular grafts. These grafts can either be woven or knitted and are produced in straight or bifurcated configurations. Within each type of construction, various properties can be incorporated into the product as illustrated in Table 3.2.4-5. Manufacturers recommend that all woven and knit grafts with water permeability rates over 50 ml cm^{-2} min^{-1} be pre-clotted to prevent blood loss through the fabric at the time of implantation. To eliminate the need for this pre-clotting procedure, textile-based vascular grafts are usually manufactured with a coating or sealant of collagen or gelatin.

Today a substantial amount of research activity is being directed toward the development of a small vessel prosthesis with diameters less than 6 mm for coronary artery bypass and tibial/popliteal artery replacement. Currently, no successful commercial products exist to meet this market need. The question still remains as to whether a biotextile will work as a small vessel prosthesis if it is fabricated to have the required compliance and mechanical properties and its surface is modified with surface coatings, growth factors, and other bioactive agents to prevent thrombosis and thrombo-embolic events. Current development activities are directed toward tissue-engineered grafts (Teebken and Haverich, 2002; Huang, 2000), coated or surface-modified synthetic and textile grafts (Chinn *et al.*, 1998), and biologically based grafts (Weinberg and Abbott, 1995).

During the past 10 years, large amounts of financial and personnel resources have gone into the development of endovascular stent grafts (Makaroun *et al.*, 2002). These grafts have been used for aortic aneurysm repair, occlusive disease, and vascular trauma. Endovascular prostheses or stent grafts are tubular grafts with an internal or external stent or rigid scaffold. The stent grafts range in size from about 20 mm up to 40 mm ID and are collapsed and folded into catheters and inserted through the femoral artery, thus avoiding the need for open surgery. The stents are typically made from nitinol, stainless steel, and Elgiloy wires and are similar to the coronary stents, however, much larger in diameter (e.g., 24 mm versus 4 mm, respectively). There are balloon expandable or self-expanding stents, which are manufactured in straight or bifurcated configurations. The stents are then covered in either ultrathin ePTFE (Cartes-Zumelzu *et al.*, 2002) or woven polyester (Areydi, 2003). Most of the endovascular graft designs incorporate an ultra-thin woven polyester tube. Most biotextile tubes are plain woven structures with water permeabilities ranging from 150 to 300 ml cm^{-2} min^{-1} depending on the manufacturer. They have been woven from 40 or 50 denier untexturized polyester yarn so as to minimize the overall wall thickness of the device.

General surgery

Three key applications of biotextiles in general surgery are sutures, hemostatic devices, and hernia repair meshes. Commercial sutures are typically monofilament or braided; they can be constructed of natural materials such as silk or collagen (catgut), or synthetic materials such as nylon, PP, and polyester. Sutures can be further classified into absorbable and nonabsorbable types. For obvious reasons, when blood vessels are ligated, only nonabsorbable sutures are used, and these are typically constructed of either braided polyester or pp monofilaments. On the other hand, when ligating soft tissue or closing wounds subcutaneously, absorbable sutures are preferred. Absorbable sutures do not create a chronic inflammatory response and do not require removal. These are typically made from PGA or poly(glycolide-*co*-lactide) copolymers.

Another common application of biotextiles and fiber technology in general surgery is the use of absorbable hemostatic agents, including those constructed of collagen and oxidized regenerated cellulose. As mentioned previously, these can be fabricated as nonwoven mats or woven and knitted fabrics, or they can be left in fibrillar form. Table 3.2.4-7 highlights some commercially available hemostatic agents and their representative properties. As can be seen in Table 3.2.4-7, collagen-based hemostatic devices are available in layered fibril, foam, and powdered forms. The regenerated cellulose pad is also available as a knitted fabric and is sold under the trade name of Surgicel. This material is commonly used to control suture line bleeding. The nonwoven and powdered forms are generally used to stop diffuse bleeding that results from trauma to the liver and spleen. Experience has shown that the loose fibril form is more difficult to use, so most surgeons prefer the more structured form of the product.

Various forms of open mesh fabrics are used as secondary support material in hernia repair. Traditional constructions are warp knitted from PP monofilaments, and some forms of the mesh are preshaped for easy nstallation. More recently three-dimensional Raschel knits using polyester multifilament yarns have been found to be more flexible and therefore can be implanted endoscopically. As with other textile structures, various properties can be engineered into the mesh to meet design goals that may include added flexibility, increased strength, reduced thickness, improved handling, and better suture holding strength. Some designs include a protein or microporous PTFE layer on one side only, which reduces the risk of unwanted adhesions *in vivo*.

Orthopedics

Attempts have been made to construct replacement ligaments and tendons using woven and braided fabrications. One design, which has had some limited clinical success, is a prestretched knitted graft, material used to repair separated shoulder joints. A similar design, using a high-tenacity polyester woven web inside of a prestretched knitted graft, was evaluated for ACL repair in the knee joint with limited success. In general, biotextiles have had limited success in orthopedic ligament and tendon applications as a result of abrasion wear problems, inadequate strength, and poor bone attachment (Guidoin *et al.*, 2000). An attempt was made to use a braided PTFE structure for ACL repair, but early failures occurred as a result of creep problems associated with the PTFE polymer. Roolker *et al.* (2000) recently reported on using the e-PTFE ligament prosthesis on 52 patients. However, during the follow-up they experienced increasing knee instability over time indicating prosthesis failure. Cooper (2000) and Lu (2001) have reported the development of a three-dimensional bioabsorbable braid using poly(glycolide-*co*-lactide) fibers for ligament replacement. They were able to modify the scaffold porosity, mechanical properties and matrix design using a three-dimensional braiding technique. A successful ACL ligament replacement would be a significant advance for orthopedic surgery, but at present, no biotextile or other type of prosthesis has shown clinical promise.

Table 3.2.4-7 Comparison between commercial hemostats (Ethicon, 1998)

	Surgicel fibrillar hemostat	Oxycel	Collagen power	Gelfoam
Bacterial activity	Inhibits bacterial growth	No antibacterial activity	No antibacterial activity	No antibacterial activity
Hemostasis time	3.5 to 4.5 minutes	2 to 8 minutes	2 to 4 minutes	Not specified
Bioresorbability	7 to 14 days	3 to 4 weeks	8 to 10 weeks	4 to 6 weeks
Packaging	Foil/Tyvek sterile	Glass vials	Glass jars	Peel envelope
Preparation	Packaged for use	Packaged for use	Packaged for use	Must be cut/soaked

Tissue engineering scaffolds and the future

One key area of research gaining significant attention over the past several years is tissue engineered scaffolds. This technology combines an engineered scaffold, or three-dimensional structure, with living cells. These scaffolds can be constructed of various materials and into various shapes depending on the desired application. One such concept is the use of the biodegradable hydrogel–textile substrate (Chu *et al.*, 2002). Their concept uses a 3D porous biodegradable hydrogel on a non-woven fabric structure. An alternate concept developed by Karamuk *et al.* (2000, 2001) uses a 3D embroidered scaffold to form a tissue-engineered substrate. With this concept, polyester yarns were used to form a complex textile structure, which allowed for easy deformation that they believe will enhance cellular attachment and cell growth. Risbud *et al.* (2002) reported on the development of 3D chitosan–collagen hydrogel coating for fabric meshes to support endothelial cell growth. They are directing their research toward the development of liver bioreactors.

Further in the future, various novel concepts will be undergoing development. Heim *et al.* (2002) reported on the development of a textile-based tissue engineered heart valve.

Using microfiber woven technology, Heim *et al.* (2002) hypothesized that the filaments could be oriented along the stress lines and the fabric based leaflet structure would have good fatigue resistance with minimal bending stiffness. Significant development is required before this concept can be used *in vivo*. Coatings on textile based vascular grafts continue to be an area of interest. Coury *et al.*, (2000) reported on the use of a synthetic hydrogel coating based on PEG to replace collagen. If successful, the use of a synthetic coating would be preferable to use of a collagen one since it will reduce manufacturing costs and graft-to-graft variability that typically occurs with naturally derived collagen materials. As mentioned earlier, even silk is undergoing modifications to enhance its biocompatibility for cardiovascular applications by sulfation and copolymerization with various monomers (Tamada *et al.*, 2000). These concepts will provide new and novel implantable products for advancing medical treatments and therapies in the future.

Summary

In summary, it can be stated that the use of biotextiles in medicine will continue to grow as new polymers, coatings, constructions, and finishing processes are introduced to meet the device needs of the future. In particular, advances in genetic engineering, fiber spinning, and surface modification technologies will provide a new generation of biopolymers and fibrous materials with unique chemical, mechanical, biological, and surface properties that will be responsible for achieving the previously unobtainable goal of tissue-engineered organs.

Acknowledgments

The authors thank Ruwan Sumansinghe and Henry Sun for their technical assistance in preparing this manuscript.

Bibliography

Adanur, S. (1995). *Wellington Sears Handbook of Industrial Textiles.* Technomic Publishing Company, Lancaster, PA, pp. 57–65.

Ayerdi, J., McLafferty, R. B., Markwell, S. J., Solis, M. M., Parra, J. R., Gruneiro, L. A., Ramsey, D. E., and Hodgson, K. J. (2003). Indications and outcomes of AneuRx phase III trial versus use of commercial AneuRx stent graft (In Process Citation). *J. Vascular Surgery* **37**(4): 739–743.

ANSI/AAMI/ISO 7198: 1998/2001. *Cardiovascular Implants—Vascular Prostheses*, 2001. Association for the Advancement of Medical Instrumentation.

Butany, J., Scully, H. E., Van Arsdell, G., and Leask, R. (2002). Prosthetic heart valves with silver-coated sewing cuff fabric: Early morphological features in two patients. *Can. J. Cardiol.* **18**(7): 733–738.

Cartes-Zumelzu, F., Lammer, J., Hoelzenbein, T., Cejna, M., Schoder, M., Thurnher, S., and Kreschmer, G. (2002). Endovascular placement of a nitinol-ePTFE stent-graft for abdominal aortic aneurysms: Initial and midterm results. *J. Vasc. Interv. Radiol.* **13**(5): 465–473.

Chinn, J. A., Sauter, J. A., Phillips, R. E., Kao, W. J., Anderson, J. M., Hanson, S. R., and Ashton, T. R. (1998). Blood and tissue compatability of modified polyester: Thrombosis, inflammation, and healing. *J. Biomed. Mater. Res.* **39**(1): 130–140.

Chu, C., Zhang, X. Z., and Van Buskirk, R. (2002). Biodegradable hydrogel-textile hybrid for tissue engineering. National Textile Center Research Briefs—Materials Competency: June 2002 (NTC Project: M01-B01).

Cooper, J. A., Lu, H. H., Ko, F. K., and Laurencin, C. T. (2000). Fiber-based tissue engineered scaffold for ligament replacement: Design considerations and in vitro evaluation, 208. Society for Biomaterials, Sixth World Biomaterials Congress Transactions.

Coury, A., Barrows, T., Azadeh, F., Roth, L., Poff, B., VanLue, S., Warnock, D., Jarrett, P., Bassett, M., and Doherty, E. (2000). Development of synthetic coatings for textile vascular prostheses, 1497. Society for

Biomaterials, Sixth World Biomaterials Congress Transactions.

Ethicon, Inc. (1998). Surgicel Fibrillar, Absorbable Hemostat. Somerville, NJ.

Formhals A. (1934). Process and apparatus for preparing artificial threads. US Patent 1,975,504.

Goswami, B. C., Martindale, J. G., and Scardono, F. L. (1977). *Textile Yarns: Technology, Structure and Applications.* John Wiley and Sons, New York.

Groitzsch D., and Fahrbach, E. (1986). Microporous multiplayer nonwoven material for medical applications. US Patent 4,618,524.

Guidoin, M. F., Marois, Y., Bejui, J., Poddevin, N., King, M. W., and Guidoin, R. (2000). Analysis of retrieved polymer fiber based replacements for the ACL. *Biomaterials* **21**(23): 2461–2474.

Heim, F., Chakfe, N., and Durand, B. (2002). A new concept of a flexible textile heart valve prosthesis, 665. Society for Biomaterials, 28th Annual Meeting Transactions.

Hoffman, A. S. (1977). Medical application of polymeric fibers. *J. Appl. Polym. Sci., Appl. Polym. Symp.* **31**: 313.

Huang, L., McMillan, R. A., Apkarian, R. P., Pourdeyhimi, B., Conticello, V. P., and Chaikof, E. L. (2000). Generation of synthetic elastin-mimetic small diameter fibers and fiber networks. *Macromolecules* **33**: 2989–2997.

Karamuk, E., Raeber, G., Mayer, J., Wagner, B., Bischoff, B., Billia, M., Seidl, R., and Wintermantel, E. (2000). Structual and mechanical aspects of embroidered textile scaffolds for tissue engineering, 4. Society for Biomaterials, Sixth World Biomaterials Congress Transactions.

Karamuk, E., Mayer, J., Selm, B., Bischoff, B., Ferrario, R., Heller, M., Billia, M., Seidel, R., Wanner, M., and Moser, R. (2001). Development of a structured wound dressing based on a textile composite funtionalised by embroidery technology. Tissupor, KTI. Projekt N-511.

Kenawy, E. R., Bowlin, G. L., Mansfield, K., Layman, J., Simpson, D. G., Sanders, E., and Wnek, G. E. (2002). Release of tetracycline hydrochloride from electrospun poly(ethylene-co-vinyl acetate), poly(l-lactic acid) and a blend. *J. Control Release* **81**(1-2): 57–64.

Keys, A. F. (1996). Presentation to the Texticeutical Meeting, 16 January.

King, M. W. (1991). Designing fabrics for blood vessel replacement. *Canadian Textile Journal* **108**(4): 24–30.

King, M. W., Guidoin, R. G., Gunasekera, K. R., and Gosselin, C. (1983). Designing polyester vascular prostheses for the future. Medical Progress Technology, Springer-Verlag.

King, M. W., Ornberg, R. L., Marois, Y., Marinov, G. R., Cadi, R., Roy, R., Cossette, F., Southern, J. H., Joardar, S. J., Weinberg, s. L., Shalaby, W.,and Guidon, R. (1999). Healing response of partially bioresorbably bicomponent fibers: A subcutaneous rat study. Society for Biomaterials, 25th Annual Meeting Transactions, Providence R.I.

King, M. W. (1991). Designing fabrics for blood vessel replacement. *Canadian Textile Journal* **108**(4): 24–30.

King, M. W., Ornberg, R. L., Marois, Y., Marinov, G. R., Cadi, R., Southern, J. H., Joardar, S. J., Weinberg, S. L., Shalaby, S. W., and Guidoin, R. (2000). Partially bioresorbable bicomponent fibers for tissue engineering: mechanical stability of core polymers, 533. Sixth World Biomaterials Congress, May 15–20, Kamuela, Hawaii.

Krcma, R. (1971). Manual of Nonwovens. Textile Trade Press, Manchester, England.

Ko, F. K. (1990). Presentation on fabrication, structure and properties of fibrous assemblies for medical applications, Drexel University and Medical Textiles, Inc. Philadelphia, PA. Workshop on Medical Textiles, Society for Biomaterials 16th Annual Meeting, Charleston, South Carolina, May 19.

Lu, H. H., Cooper, J. A., Ko, F. K, Attawia, M. A., and Laurencin, C. T. (2001). Effect of polymer scaffold composition on the morphology and growth of anterior cruciate ligament cells, 140. Society of Biomaterials, 27th Annual Meeting Transactions.

M. S., Chaikof, E., Naslund, T., and Matsumura, J. S. (2002). Efficacy of a bifurcated endograft versus open repair of abdominal aortic aneurysms: A reappraisal. *J. Vascular Surg.* **35**: 203–210.

Martin, C. E., and Cockshott, I. D. (1977). US Patent 4,043,331.

Matthews, J. A., Wnek, G. E., Simpson, D. G., and Bowlin, G. L. (2002). Electrospinning of collagen nanofibers. *Biomacromolecules* 3: 232–239.

Piller,B. (1973). *Bulked Yarns.* SNTL/ Textile Trade Press, Manchester, England.

Reneker, D. H., Yarin, A. L., Fong, H., and Koombhongse, S. (2000) Bending instability of electrically charged liquid jets of polymer solutions in electrospinning. *J. Appl. Phys., Part 1* **87**: 4531.

Risbud, M. V., Karamuk, E., Moser, R., and Mayer, J. (2002). Hydrogen-coated textile scaffolds as three-dimensional growth support for human umbilical vein endothelial cells (HUVECs): Possibilities as coculture system in liver tissue engineering. *Cell Transplant* **11**(**4**): 369–377.

Robinson, A. T. C., and Marks, R. (1967). *Woven Cloth Construction.* Plenum Press, New York.

Roolker, W., Patt, T. W., Van Dijk, C. N., Vegter, M., and Marti, R. K. (2000). The Gore-Tex Prosthetic Ligament as a Salvage Procedure in Deficient Knees. *Knee. Surg. Sports Taumatol. Arthrosc.* 8(1): 20–25.

Schreuder-Gibson, H., Gibson, P., Senecal, K., Sennett, M., Walker, J., Yeoman, W., Ziegler D., and Tsai, P. P. (2002). Protective textile materials based on electrospun nanofibers. *J. Adv. Maters.* **34**(3): 44–55.

Shalaby, S. W. (1985). Fibrous materials for biomedical applications, in *High Technology Fibers, Part A*, M. Lewin and J. Preson, eds. Marcel Dekker, New York.

Shalaby, S. W. (1996). Fabrics. in *Biomaterials Science: An Introduction to Materials in Medicine.* Hoffman, Lemons, Ratner & Schoen, eds., 118–124. Academic Press, Boston.

Shalaby S. W., and Shah, K. R., (1991). Chemical modification of natural polymers and their technological relevance. in *Water-Soluable Polymers: Chemisty and Applications*, S. W. Shalaby, G. B. Butler, and C. L. McCormick, eds., 74. ACS Symposium Series, American Chemical Society, Washington, D.C.

Skjak-Braek, G., and Sanford, P. A. eds. (1989). *Chitin and Chitosan: Sources, Chemistry, Biochemistry, Physical Properties, and Applications.* Elsevier, New York.

Spencer, D. J. (1983). *Knitting Technology.* Pergamon Press, Oxford.

Stitzel, J. D., Pawlowski, K. J., Wnek, G. E., Simpson, D. G., and Bowlin, G. L. (2001). Arterial smooth muscle cell proliferation on a novel biomimiking, biodegradable vascular graft scaffold. *J. Biomaterials Applications* **15**: 1.

Takai, K., Ohtsuka, T., Senda, Y., Nakao, M., Yamamoto, K., Matsuoka, J., and Hiari, Y. (2002). Antibacterial properties of antimicrobial-finished textile products. *Microbiol. Immunol.* **46**(2): 75–81.

Tamada, Y., Furuzono, T., Ishihara, K., and Nakabayashi, N. (2000). Chemical modification of silk to utilize as a new biomaterial. Society for Biomaterials, Sixth World Biomaterials Congress Transactions.

Teebken, O. E., and Haverich, A. (2002). Tissue engineering of small diameter vascular graft. *Eur. J. Vasc. Endovasc. Surg.* **23**(6): 475–487.

Teule, F., Aube, C., Ellison, M., and Abbott, A. (2003). Biomimetic manufacturing of customized novel fiber proteins for specialized applications, 38–43. Proceedings 3rd Autex Conference, Gdansk, Poland.

Theron, A., Zussman, E., and Yarin, A. L. (2001). Electrostatic field assisted alignment of electrospun nanofibers. *Nanotechnology* **12**: 384–390.

Weinberg, S. L. (1998). Biomedical Device Consultants Laboratory Data.

Weinberg, S., Abbott, W. M., (1995). Biological vascular grafts: Current and emerging technologies. in *Vascular Surgery: Theory and Practice*, A. D. Callow and C. B. Ernst, eds., 1213–1220. McGraw-Hill, New York.

3.2.5 Hydrogels

Nicholas A. Peppas

Hydrogels are water-swollen, cross-linked polymeric structures containing either covalent bonds produced by the simple reaction of one or more comonomers, physical cross-links from entanglements, association bonds such as hydrogen bonds or strong van der Waals interactions between chains (Peppas, 1987), or crystallites bringing together two or more macromolecular chains (Hickey and Peppas, 1995). Hydrogels have received significant attention because of their exceptional promise in biomedical applications. The classic book by Andrade (1976) offers some of the best work that was available prior to 1975. The more recent book and other reviews by Peppas (1987, 2001) addresses the preparation, structure, and characterization of hydrogels.

Here, we concentrate on some features of the preparation of hydrogels, as well as characteristics of their structure and chemical and physical properties.

Classification and basic structure

Depending on their method of preparation, ionic charge, or physical structure features, hydrogels maybe classified in several categories. Based on the method of preparation, they may be (i) homopolymer hydrogels, (ii) copolymer hydrogels, (iii) multipolymer hydrogels, or (iv) interpenetrating polymeric hydrogels. Homopolymer hydrogels are cross-linked networks of one type of hydrophilic monomer unit, whereas copolymer hydrogels are produced by cross-linking of two comonomer units, at least one of which must be hydrophilic to render them swellable. Multipolymer hydrogels are produced from three or more comonomers reacting together (see e.g., Lowman and Peppas, 1997, 1999). Finally, interpenetrating polymeric hydrogels are produced by preparing a first network that is then swollen in a monomer. The latter reacts to form a second intermeshing network structure. Based on their ionic charges, hydrogels may be classified (Ratner and Hoffman, 1976; Brannon-Peppas and Harland, 1990) as (i) neutral hydrogels, (ii) anionic hydrogels, (iii) cationic hydrogels, or (iv) ampholytic hydrogels. Based on physical structural features of the system, they can be classified as (i) amorphous hydrogels, (ii) semicrystalline hydrogels, or (iii) hydrogen-bonded or complexation structures. In amorphous hydrogels, the macromolecular chains are arranged randomly. Semicrystalline hydrogels are characterized by dense regions of ordered macromolecular chains (crystallites). Finally, hydrogen bonds and complexation structures may be responsible for the three-dimensional structure formed.

Structural evaluation of hydrogels reveals that ideal networks are only rarely observed. Figure 3.2.5-1A shows an ideal macromolecular network (hydrogel) indicating tetrafunctional cross-links (junctions) produced by covalent bonds. However, in real networks it is possible to encounter multifunctional junctions (Fig. 3.2.5-1B) or physical molecular entanglements (Fig. 3.2.5-1C) playing the role of semipermanent junctions. Hydrogels with molecular defects are always possible. Figures 3.2.5-1D and 3.2.5-1E indicate two such effects: unreacted functionalities with partial entanglements (Fig. 3.2.5-1D) and chain loops (Fig. 3.2.5-1E). Neither of these effects contributes to the mechanical or physical properties of a polymer network.

The terms "cross-link," "junction," or "tie-point" (an open circle symbol in Fig. 3.2.5-1D) indicate the connection points of several chains. These junctions may be carbon atoms, but they are usually small chemical bridges [e.g., an acetal bridge in the case of cross-linked poly(vinyl alcohol)] with molecular weights much smaller than those of the cross-linked polymer chains. In other situations, a junction may be an association of macromolecular chains caused by van der Waals forces, as in the case of the glycoproteinic network structure of natural mucus, or an aggregate formed by hydrogen bonds, as in the case of aged microgels formed in polymer solutions.

Finally, the network structure may include effective junctions that can be either simple physical entanglements of permanent or semipermanent nature, or ordered chains forming crystallites. Thus, the junctions

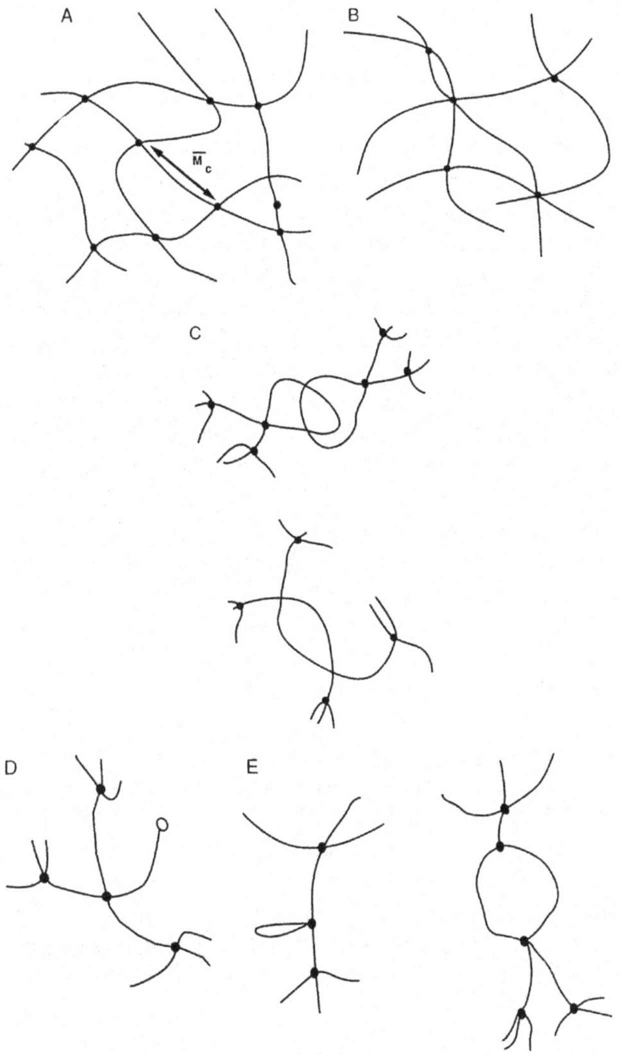

Fig. 3.2.5-1 (A) Ideal macromolecular network of a hydrogel. (B) Network with multifunctional junctions. (C) Physical entanglements in a hydrogel. (D) Unreacted functionality in a hydrogel. (E) Chain loops in a hydrogel.

should never be considered as points without volume, which is the usual assumption made when developing structural models for analysis of the cross-linked structure of hydrogels (Flory, 1953). Instead, they have a finite size and contribute to the deformational distribution during biomedical applications.

Preparation

Hydrogels are prepared by swelling cross-linked structures in water or biological fluids. Water or aqueous solutions may be present during the initial preparation of the cross-linked structure. Methods of preparation of the initial networks include chemical cross-linking, photopolymerization, or irradiative cross-linking (Peppas *et al.*, 2000).

Chemical cross-linking calls for direct reaction of a linear or branched polymer with at least one difunctional, small molecular weight, cross-linking agent. This agent usually links two longer molecular weight chains through its di- or multifunctional groups. A second method involves a copolymerisation-cross-linking reaction between one or more abundant monomers and one multifunctional monomer that is present in relatively small quantities. A third method involves using a combination of monomer and linear polymeric chains that are cross-linked by means of an interlinking agent, as in the production of polyurethanes. Several of these techniques can be performed in the presence of UV light leading to rapid formation of a three-dimensional network. Ionizing radiation cross-linking (Chapiro, 1962) utilizes electron beams, gamma rays, or X-rays to excite a polymer and produce a cross-linked structure via free radical reactions.

Swelling behavior

The physical behavior of hydrogels is dependent on their equilibrium and dynamic swelling behavior in water, since upon preparation they must be brought in contact with water to yield the final, swollen network structure. Figure 3.2.5-2 shows one of two possible processes of swelling. A dry, hydrophilic cross-linked network is placed in water. Then, the macromolecular chains interact with the solvent molecules owing to the relatively

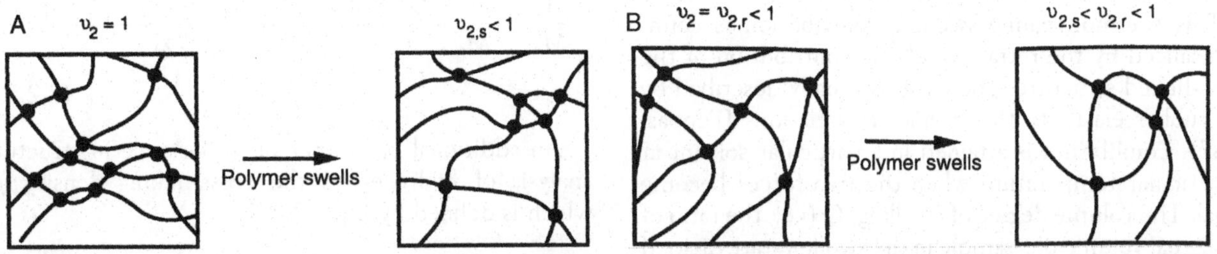

Fig. 3.2.5-2 (A) Swelling of a network prepared by cross-linking in dry state. (B) Swelling of a network prepared by cross-linking in solution.

good thermodynamic compatibility. Thus, the network expands to the solvated state.

The Flory-Huggins theory can be used to calculate thermodynamic quantities related to that mixing process. Flory (1953) developed the initial theory of the swelling of cross-linked polymer gels using a Gaussian distribution of the polymer chains. His model describing the equilibrium degree of cross-linked polymers postulated that the degree to which a polymer network swelled was governed by the elastic retractive forces of the polymer chains and the thermodynamic compatibility of the polymer and the solvent molecules. In terms of the free energy of the system, the total free energy change upon swelling was written as:

$$\Delta G = \Delta G_{elastic} + \Delta G_{mix} \qquad (3.2.5.1)$$

Here, $\Delta G_{elastic}$ is the contribution due to the elastic retractive forces and ΔG_{mix} represents the thermodynamic compatibility of the polymer and the swelling agent (water).

Upon differentiation of Eq. 3.2.5.1 with respect to the water molecules in the system, an expression can be derived for the chemical potential change of water in terms of the elastic and mixing contributions due to swelling.

$$\mu_1 - \mu_{1,0} = \Delta\mu_{elastic} + \Delta\mu_{mix} \qquad (3.2.5.2)$$

Here, μ_1 is the chemical potential of water within the gel and $\mu_{1,0}$ is the chemical potential of pure water.

At equilibrium, the chemical potentials of water inside and outside of the gel must be equal. Therefore, the elastic and mixing contributions to the chemical potential will balance one another at equilibrium. The chemical potential change upon mixing can be determined from the heat of mixing and the entropy of mixing. Using the Flory–Huggins theory, the chemical potential of mixing can be expressed as:

$$\Delta\mu_{mix} = RT(\ln(1 - 2\upsilon_{2,s}) + \upsilon_{2,s} + X_1\upsilon_{2,s}^2) \qquad (3.2.5.3)$$

where χ_1 is the polymer-water interaction parameter, $\upsilon_{2,s}$ is the polymer volume fraction of the gel, T is absolute temperature, and R is the gas constant.

This thermodynamic swelling contribution is counterbalanced by the retractive elastic contribution of the cross-linked structure. The latter is usually described by the rubber elasticity theory and its variations (Peppas, 1987). Equilibrium is attained in a particular solvent at a particular temperature when the two forces become equal. The volume degree of swelling, Q (i.e., the ratio of the actual volume of a sample in the swollen state divided by its volume in the dry state), can then be determined from Eq. 3.2.5.4.

$$\upsilon_{2,s} = \frac{\text{Volume of polymer}}{\text{Volume of swollen gel}} = \frac{V_p}{V_{gel}} = 1/Q \qquad (3.2.5.4)$$

Researchers working with hydrogels for biomedical applications prefer to use other parameters in order to define the equilibrium-swelling behavior. For example, Yasuda *et al.* (1969) introduced the use of the so-called hydration ratio, H, which has been accepted by those researchers who use hydrogels for contact lens applications (Peppas and Yang, 1981). Another definition is that of the weight degree of swelling, q, which is the ratio of the weight of the swollen sample to that of the dry sample.

In general, highly swollen hydrogels include those of cellulose derivatives, poly(vinyl alcohol), poly(*N*-vinyl-2-pyrrolidone) (PNVP), and PEG, among others. Moderately and poorly swollen hydrogels are those of PHEMA and many of its derivatives. In general, a basic hydrophilic monomer can be copolymerized with other more or less hydrophilic monomers to achieve desired swelling properties. Such processes have led to a wide range of swellable hydrogels, as Gregonis *et al.* (1976), Peppas (1987, 1997), and others have pointed out. Knowledge of the swelling characteristics of a polymer is of utmost importance in biomedical and pharmaceutical applications since the equilibrium degree of swelling influences (i) the solute diffusion coefficient through these hydrogels, (ii) the surface properties and surface mobility, (iii) the optical properties, especially in relation to contact lens applications, and (iv) the mechanical properties.

Determination of structural characteristics

The parameter that describes the basic structure of the hydrogel is the molecular weight between cross-links, $\overline{M}_c$ (as shown in Fig. 3.2.5-1A). This parameter defines the average molecular size between two consecutive junctions regardless of the nature of those junctions and can be calculated by Eq. 3.2.5.5.

$$\frac{1}{\overline{M}_c} = \frac{2}{\overline{M}_c} - \frac{(\upsilon/V_1)\left[\ln(1 - \upsilon_{2,s}) + \upsilon_{2,s} + X_1\upsilon_{2,s}^2\right]}{(\upsilon_{2,s}^{1/3} - \upsilon_{2,s}/2)} \qquad (3.2.5.5)$$

An additional parameter of importance in structural analysis of hydrogels is the cross-linking density, ρ_x, which is defined by Eq. 3.2.5.6.

$$\rho_x = \frac{1}{\bar{\upsilon}\overline{M}_c} \qquad (3.2.5.6)$$

In these equations, v is the specific volume of the polymer (i.e., the reciprocal of the amorphous density of the polymer), and $\overline{M}_n$ is the initial molecular weight of the un-cross-linked polymer.

Properties of important biomedical hydrogels

The multitude of hydrogels available leaves numerous choices for polymeric formulations. The best approach for developing a hydrogel with the desired characteristics for biomedical application is to correlate the macromolecular structures of the polymers available with the swelling and mechanical characteristics desired (Peppas *et al.*, 2000; Peppas, 2001).

The most widely used hydrogel is water-swollen, cross-linked PHEMA, which was introduced as a biological material by Wichterle and Lim (1960). The hydrogel is inert to normal biological processes, shows resistance to degradation, is permeable to metabolites, is not absorbed by the body, is biocompatible, withstands heat sterilization without damage, and can be prepared in a variety of shapes and forms.

The swelling, mechanical, diffusional, and biomedical characteristics of PHEMA gels have been studied extensively. The properties of these hydrogels are dependent upon their method of preparation, polymer volume fraction, degree of cross-linking, temperature, and swelling agent.

Other hydrogels of biomedical interest include polyacrylamides. Tanaka (1979) has done extensive studies on the abrupt swelling and deswelling of partially hydrolyzed acrylamide gels with changes in swelling agent composition, curing time, degree of cross-linking, degree of hydrolysis, and temperature. These studies have shown that the ionic groups produced in an acrylamide gel upon hydrolysis give the gel a structure that shows a discrete transition in equilibrium-swollen volume with environmental changes.

Discontinuous swelling in partially hydrolyzed polyacrylamide gels has been studied by Gehrke *et al.* (1986). Besides HEMA and acrylamides, *N*-vinyl-2-pyrrolidone (NVP), methacrylic acid (MAA), methyl methacrylate (MMA), and maleic anhydride (MAH) have all been proven useful as monomers for hydrogels in biomedical applications. For instance, cross-linked PNVP is used in soft contact lenses. Small amounts of MAA as a comonomer have been shown to dramatically increase the swelling of PHEMA polymers. Owing to the hydrophobic nature of MMA, copolymers of MMA and HEMA have a lower degree of swelling than pure PHEMA (Brannon-Peppas and Peppas, 1991). All of these materials have potential use in advanced technology applications, including biomedical separations, and biomedical and pharmaceutical devices.

Intelligent or smart hydrogels

Hydrogels may exhibit swelling behavior dependent on the external environment. Over the past 30 years there has been a significant interest in the development and analysis of environmentally or physiologically responsive hydrogels (Peppas, 1991). Environmentally responsive materials show drastic changes in their swelling ratio due to changes in their external pH, temperature, ionic strength, nature and composition of the swelling agent, enzymatic or chemical reaction, and electrical or magnetic stimuli (Peppas, 1993). In most responsive networks, a critical point exists at which this transition occurs.

An interesting characteristic of numerous responsive gels is that the mechanism causing the network structural changes can be entirely reversible in nature. The ability of pH- or temperature-responsive gels to exhibit rapid changes in their swelling behavior and pore structure in response to changes in environmental conditions lend these materials favorable characteristics as carriers for bioactive agents, including peptides and proteins. This type of behavior may allow these materials to serve as self-regulated, pulsatile drug delivery systems.

pH-Sensitive hydrogels

One of the most widely studied types of physiologically responsive hydrogels is pH-responsive hydrogels. These hydrogels are swollen ionic networks containing either acidic or basic pendant groups. In aqueous media of appropriate pH and ionic strength, the pendant groups can ionize developing fixed charges on the gel. All ionic materials exhibit a pH and ionic strength sensitivity. The swelling forces developed in these systems are increased over those of nonionic materials. This increase in swelling force is due to the localization of fixed charges on the pendant groups. As a result, the mesh size of the polymeric networks can change significantly with small pH changes.

Temperature sensitive hydrogels

Another class of environmentally sensitive gels exhibits temperature-sensitive swelling behavior due to a change in the polymer/swelling agent compatibility over the temperature range of interest. Temperature-sensitive polymers typically exhibit a lower critical solution temperature (LCST), below which the polymer is soluble. Above this temperature, the polymers are typically hydrophobic and do not swell significantly in water (Kim,

1996). However, below the LCST, the cross-linked gel swells to significantly higher degrees because of the increased compatibility with water.

Complexing hydrogels

Some hydrogels may exhibit environmental sensitivity due to the formation of polymer complexes. Polymer complexes are insoluble, macromolecular structures formed by the non-covalent association of polymers with affinity for one another. The complexes form as a result of the association of repeating units on different chains (interpolymer complexes) or on separate regions of the same chain (intrapolymer complexes). Polymer complexes are classified by the nature of the association as stereocomplexes, polyelectrolyte complexes, or hydrogen-bonded complexes. The stability of the associations is dependent on such factors as the nature of the swelling agent, temperature, type of dissolution medium, pH and ionic strength, network composition and structure, and length of the interacting polymer chains.

In this type of gel, complex formation results in the formation of physical cross-links in the gel. As the degree of effective cross-linking is increased, the network mesh size and degree of swelling is significantly reduced. As a result, if hydrogels are used as drug carriers, the rate of drug release will decrease dramatically upon the formation of interpolymer complexes.

Applications

Biomedical applications

The physical properties of hydrogels make them attractive for a variety of biomedical and pharmaceutical applications. Their biocompatibility allows them to be considered for medical applications, whereas their hydrophilicity can impart desirable release characteristics to controlled and sustained release formulations.

Hydrogels exhibit properties that make them desirable candidates for biocompatible and blood-compatible biomaterials (Merrill *et al.*, 1987). Nonionic hydrogels for blood contact applications have been prepared from poly(vinyl alcohol), polyacrylamides, PNVP, PHEMA, and poly(ethylene oxide) (PEO) (Peppas *et al.*, 1999). Heparinized polymer hydrogels also show promise as materials for blood-compatible applications (Sefton, 1987).

One of the earliest biomedical applications of hydrogels was in contact lenses (Tighe, 1976; Peppas and Yang, 1981) because of their relatively good mechanical stability, favorable refractive index, and high oxygen permeability.

Other potential applications of hydrogels include (Peppas, 1987) artificial tendon materials, wound-healing bioadhesives, artificial kidney membranes, articular cartilage, artificial skin, maxillofacial and sexual organ reconstruction materials, and vocal cord replacement materials (Byrne *et al.*, 2002).

Pharmaceutical applications

Pharmaceutical hydrogel applications have become very popular in recent years. Pharmaceutical hydrogel systems include equilibrium-swollen hydrogels, i.e., matrices that have a drug incorporated in them and are swollen to equilibrium. The category of solvent-activated, matrix-type, controlled-release devices comprises two important types of systems: swellable and swelling-controlled devices. In general, a system prepared by incorporating a drug into a hydrophilic, glassy polymer can be swollen when brought in contact with water or a simulant of biological fluids. This swelling process may or may not be the controlling mechanism for diffusional release, depending on the relative rates of the macromolecular relaxation of the polymer and drug diffusion from the gel.

In swelling-controlled release systems, the bioactive agent is dispersed into the polymer to form nonporous films, disks, or spheres. Upon contact with an aqueous dissolution medium, a distinct front (interface) is observed that corresponds to the water penetration front into the polymer and separates the glassy from the rubbery (gel-like) state of the material. Under these conditions, the macromolecular relaxations of the polymer influence the diffusion mechanism of the drug through the rubbery state. This water uptake can lead to considerable swelling of the polymer with a thickness that depends on time. The swelling process proceeds toward equilibrium at a rate determined by the water activity in the system and the structure of the polymer. If the polymer is cross-linked or if it is of sufficiently high molecular weight (so that chain entanglements can maintain structural integrity), the equilibrium state is a water-swollen gel. The equilibrium water content of such hydrogels can vary from 30% to 90%. If the dry hydrogel contains a water-soluble drug, the drug is essentially immobile in the glassy matrix, but begins to diffuse out as the polymer swells with water. Drug release thus depends on two simultaneous rate processes: water migration into the device and drug diffusion outward through the swollen gel. Since some water uptake must occur before the drug can be released, the initial burst effect frequently observed in matrix devices is moderated, although it may still be present. The continued swelling of the matrix causes the drug to diffuse increasingly easily, ameliorating the slow tailing off of the release curve. The net effect of the swelling process is to prolong and linearize the release curve. Details of hydrogels for medical and pharmaceutical applications have been presented by Korsmeyer and Peppas in 1981 for poly(vinyl alcohol) (PVA) systems, and by Peppas and Yang (1981) for PHEMA systems and their copolymers. One of numerous examples of such swelling-controlled systems was reported by Franson and Peppas (1983), who prepared cross-linked copolymer gels of poly-(HEMA-*co*-MAA) of varying compositions. Theophylline release was studied and it was found that near zero-order release could be achieved using copolymers containing 90% PHEMA.

PVA

Another hydrophilic polymer that has received attention is PVA. This material holds tremendous promise as a biological drug delivery device because it is nontoxic, is hydrophilic, and exhibits good mucoadhesive properties. Two methods exist for the preparation of PVA gels. In the first method, linear PVA chains are cross-linked using glyoxal, glutaraldehyde, or borate. In the second method, Hassan and Peppas (2000), semicrystalline gels were prepared by exposing aqueous solutions of PVA to repeating freezing and thawing. The freezing and thawing induced crystal formation in the materials and allowed for the formation of a network structure cross-linked with the quasi-permanent crystallites. The latter method is the preferred method for preparation as it allows for the formation of an "ultrapure" network without the use of toxic cross-linking agents. Ficek and Peppas (1993) used PVA gels for the release of bovine serum albumin using novel PVA microparticles.

PEG

Hydrogels of PEO and PEG have received significant attention for biomedical applications in the past few years (Graham, 1992). Three major preparation techniques exist for the preparation of cross-linked PEG networks: (i) chemical cross-linking between PEG chains, (ii) radiation cross-linking of PEG chains, and (iii) chemical reaction of mono- and difunctional PEGs. The advantage of using radiation-cross-linked PEO networks is that no toxic cross-linking agents are required. However, it is difficult to control the network structure of these materials. Stringer and Peppas (1996) have prepared PEO hydrogels by radiation cross-linking. In this work, they analyzed the network structure in detail. Additionally, they investigated the diffusional behavior of smaller molecular weight drugs, such as theophylline, in these gels. Kofinas *et al.* (1996) have prepared PEO hydrogels by a similar technique. In this work, they studied the diffusional behavior of various macromolecules in these gels. They noted an interesting, yet previously unreported dependence between the cross-link density and protein diffusion coefficient and the initial molecular weight of the linear PEGs.

Lowman *et al.* (1997) have presented an exciting new method for the preparation of PEG gels with controllable structures. In this work, highly cross-linked and tethered PEG gels were prepared from PEG dimethacrylates and PEG monomethacrylates. The diffusional behavior of diltiazem and theophylline in these networks was studied. The technique presented in this work is promising for the development of a new class of functionalized PEG-containing gels that may be of use in a wide variety of drug delivery applications.

pH-Sensitive hydrogels

Hydrogels that have the ability to respond to pH changes have been studied extensively over the years. These gels typically contain side ionizable side groups such as carboxylic acids or amine groups. The most commonly studied ionic polymers include poly(acrylamide) (PAAm), poly(acrylic acid) (PAA), poly(methacrylic acid) (PMAA), poly(diethylaminoethyl methacrylate) (PDEAEMA), and poly(dimethylaminoethyl methacrylate) (PDMAEMA). The swelling and release characteristics of anionic copolymers of PMAA and PHEMA (PHEMA-*co*-MAA) have been investigated. In acidic media, the gels did not swell significantly; however, in neutral or basic media, the gels swelled to a high degree because of ionization of the pendant acid group. Brannon-Peppas and Peppas (1991) have also studied the oscillatory swelling behavior of these gels.

Temperature-sensitive hydrogels

Some of the earliest work with temperature-sensitive hydrogels was done by Hirotsu *et al.* (1987). They synthesized cross-linked poly(*N*-isopropyl acrylamide) (PNIPAAm) and determined that the LCST of the PNIPAAm gels was 34.3°C. Below this temperature, significant gel swelling occurred. The transition about this point was reversible. They discovered that the transition temperature was raised by copolymerizing PNIPAAm with small amounts of ionic monomers. Dong and Hoffman (1991) prepared heterogeneous gels containing PNIPAAm that collapsed at significantly faster rates than homopolymers of PNIPAAm. Yoshida *et al.* (1995) and Kaneko *et al.* (1996) developed an ingenious method to prepare comb-type graft hydrogels of PNIPAAm. The main chain of the cross-linked PNIPAAm contained small-molecular-weight grafts of PNIPAAm. Under conditions of gel collapse (above the LCST), hydrophobic regions were developed in the pores of the gel resulting in a rapid collapse. These materials had the ability to collapse from a fully swollen conformation in less than 20 minutes, whereas comparable gels that did not contain graft chains required up to a month to fully collapse. Such systems show major promise for rapid and abrupt or oscillatory release of drugs, peptides, or proteins.

Complexation hydrogels

Another promising class of hydrogels that exhibit responsive behavior is complexing hydrogels. Bell and Peppas (1995) have discussed a class of graft copolymer gels of PMAA grafted with PEG, poly(MAA-*g*-EG). These gels exhibited pH-dependent swelling behavior due to the presence of acidic pendant groups and the formation of interpolymer complexes between the ether groups on the graft chains and protonated pendant groups. In these covalently cross-linked, complexing poly(MAA-*g*-EG) hydrogels, complexation resulted in the formation of temporary physical cross-links due to hydrogen bonding between the PEG grafts and the PMAA pendant groups. The physical cross-links were reversible in nature and dependent on the pH and ionic strength of the environment. As a result, these complexing hydrogels exhibit drastic changes in their mesh size in response to small changes of pH.

Promising new methods for the delivery of chemotherapeutic agents using hydrogels have been recently reported. Novel biorecognizable sugar-containing copolymers have been investigated for the use in targeted delivery of anti-cancer drugs. Peterson *et al.* (1996) have used poly(*N*-2-hydroxypropyl methacrylamide) carriers for the treatment of ovarian cancer.

Self-assembled structures

In the past few years there have been new, creative methods of preparation of novel hydrophilic polymers and hydrogels that may represent the future in drug delivery applications. The focus in these studies has been the development of polymeric structures with precise molecular architectures. Stupp *et al.* (1997) synthesized self-assembled triblock copolymer nanostructures that may have very promising biomedical applications.

Star polymers

Dendrimers and star polymers (Dvornik and Tomalia, 1996) are exciting new materials because of the large number of functional groups available in a very small volume. Such systems could have tremendous promise in drug targeting applications. Merrill (1993) has offered an exceptional review of PEO star polymers and applications of such systems in the biomedical and pharmaceutical fields. Griffith and Lopina (1995) have prepared gels of controlled structure and large biological functionality by irradiation of PEO star polymers. Such new structures could have particularly promising drug delivery applications when combined with emerging new technologies such as molecular imprinting.

Bibliography

Andrade, J. D. (1976). *Hydrogels for Medical and Related Applications.* ACS Symposium Series, Vol. 31, American Chemical Society, Washington, D.C.

Bell, C. L., and Peppas, N. A. (1995). Biomedical membranes from hydrogels and interpolymer complexes. *Adv. Polym. Sci.* **122**: 125–175.

Brannon-Peppas, L., and Harland, R. S. (1990). *Absorbent Polymer Technology.* Elsevier, Amsterdam.

Brannon-Peppas, L., and Peppas, N. A. (1991). Equilibrium swelling behavior of dilute ionic hydrogels in electrolytic solutions. J. *Controlled Release* **16**: 319–330.

Brannon-Peppas, L., and Peppas, N. A. (1991). Time-dependent response of ionic polymer networks to pH and ionic strength changes. *Int. J. Pharm.* **70**: 53–57.

Byrne, M. E., Henthorn, D. B., Huang, Y., and Peppas, N. A. (2002). Micropatterning biomimetic materials for bioadhesion and drug delivery. in *Biomimetic Materials and Design: Biointerfacial Strategies Tissue Enginering and Targeted Drug Delivery,* A. K. Dillow and A. M. Lowman, eds. Dekker, New York, pp. 443–470.

Chapiro, A. (1962). *Radiation Chemistry of Polymeric Systems.* Interscience, New York.

Dong, L. C., and Hoffman, A. S. (1991). A novel approach for preparation of pH-sensitive hydrogels for enteric drug delivery. *J. Controlled Release* **15**: 141–152.

Dvornik, P. R., and Tomalia, D. A. (1996). Recent advances in dendritic polymers. *Curr. Opin. Colloid Interface Sci.* **1**: 221–235.

Ficek B. J., and Peppas, N. A. (1993). Novel preparation of poly(vinyl alcohol) microparticles without crosslinking agent. J. *Controlled Rel.* **27**: 259–264.

Flory, P. J. (1953). *Principles of Polymer Chemistry.* Cornell Univ. Press, Ithaca, NY.

Franson, N. M., and Peppas, N. A. (1983). Influence of copolymer composition on water transport through glassy copolymers. *J. Appl. Polym. Sci.* **28**: 1299–1310.

Gehrke, S. H., Andrews, G. P., and Cussler, E. L. (1986). Chemical aspects of gel extraction. *Chem. Eng. Sci.* **41**: 2153–2160.

Graham, N. B. (1992). Poly(ethylene glycol) gels and drug delivery, in *Poly(ethylene glycol) Chemistry, Biotechnical and Biomedical Applications*, J. M. Harris, ed. Plenum Press, New York, pp. 263–281.

Gregonis, D. E., Chen, C. M., and Andrade, J. D. (1976). The chemistry of some selected methacrylate hydrogels. in *Hydrogels for Medical and Related Applications*, J. D. Andrade, ed. ACS Symposium Series, Vol. 31. American Chemical Society, Washington, D.C., pp. 88–104.

Griffith, L., and Lopina, S. T. (1995). Network structures of radiation cross-linked star polymer gels. *Macromolecules* **28**: 6787–6794.

Hassan, C. M., and Peppas, N. A. (2000). Structure and morphology or freeze/thawed PVA hydrogels. *Macromolecules* **33**: 2472–2479.

Hickey, A. S., and Peppas, N. A. (1995). Mesh size and diffusive characteristics of semicrystalline poly(vinyl alcohol) membranes. *J. Membr. Sci.* **107:** 229–237.

Hirotsu, S., Hirokawa, Y., and Tanaka, T. (1987). Swelling of gels. *J. Chem. Phys.* **87**: 1392–1395.

Kaneko,Y., Saki, K., Kikuchi, A., Sakurai, Y., and Okano, T. (1996). Fast swelling/deswelling kinetics of comb-type grafted poly(N-isopropyl acrylamide) hydrogels. *Macromol. Symp.* **109:** 41–53.

Kim, S. W. (1996). Temperature sensitive polymers for delivery of macromolecular drugs. in *Advanced Biomaterials in Biomedical Engineering and Drug Delivery Systems*, N. Ogata, S. W. Kim, J. Feijen, and T. Okano, eds. Springer, Tokyo, pp. 125–133.

Kofinas, P., Athanassiou, V. and Merrill, E. W. (1996). Hydrogels prepared by electron beam irradiation of poly(ethylene oxide) in water solution: unexpected dependence of cross-link density and protein diffusion coefficients on initial PEO molecular weight. *Biomaterials* **17**: 1547–1550.

Korsmeyer, R. W., and Peppas, N. A. (1981). Effects of the morphology of hydrophilic polymeric matrices on the diffusion and release of water soluble drugs. *J. Membr. Sci.* **9**: 211–227.

Lowman, A. M., and Peppas, N. A. (1997). Analysis of the complexation/decomplexation phenomena in graft copolymer networks. *Macromolecules* **30**: 4959–4965.

Lowman, A. M., and Peppas, N. A. (1999). Hydrogels. in *Encyclopedia of Controlled Drug Delivery*, E. Mathiowitz, ed. Wiley, New York, pp. 397–418.

Lowman, A. M., Dziubla, T. D., and Peppas, N. A. (1997). Novel networks and gels containing increased amounts of grafted and crosslinked poly(ethylene glycol). *Polymer Preprints* **38**: 622–623.

Merrill, E. W. (1993). Poly(ethylene oxide) star molecules: synthesis, characterization, and applications in medicine and biology. *J. Biomater. Sci. Polym. Edn.* 5: 1—11.

Merrill, E. W., Pekala, P. W., and Mahmud, N. A. (1987). Hydrogels for blood contact. in *Hydrogels in Medicine and Pharmacy*, N. A. Peppas, ed. CRC Press, Boca Raton, FL, Vol. 3, pp. 1–16.

Peppas, N. A. (1987). *Hydrogels in Medicine and Pharmacy.* CRC Press, Boca Raton, FL.

Peppas, N. A. (1991). Physiologically responsive hydrogels. *J. Bioact. Compat. Polym.* **6**: 241–246.

Peppas, N. A. (1993). Fundamentals of pH- and temperature-sensitive delivery systems. in *Pulsatile Drug Delivery*, R. Gurny, H. E. Juninger, and N. A. Peppas, eds. Wissenschaftliche Verlagsgesellschaft, Stuttgart, pp. 41–56.

Peppas, N. A. (1997). Hydrogels and drug delivery. *Crit. Opin. Colloid Interface Sci.* **2**: 531–537.

Peppas, N. A. (2001). Gels for drug delivery. in *Encyclopedia of Materials: Science and Technology.* Elsevier, Amsterdam, pp. 3492–3495.

Peppas, N. A., and Yang, W. H. M. (1981). Properties-based optimization of the structure of polymers for contact lens applications. *Contact Intraocular Lens Med. J.* **7**: 300–321.

Peppas, N. A., Huang, Y., Torres-Lugo, M., Ward, J. H., and Zhang, J. (2000). Physicochemical foundations and structural design of hydrogels in medicine and biology. *Ann. Rev. Biomed. Eng.* **2**: 9–29.

Peppas, N. A., Keys, K. B., Torres-Lugo, M., and Lowman, A. M. (1999). Poly(ethylene glycol)-Containing Hydrogels in Drug Delivery. *J. Controlled Release* **62:** 81–87.

Peterson, C. M., Lu, J. M., Sun, Y., Peterson, C. A., Shiah, J. G., Straight, R. C., and Kopecek, J. (1996). *Cancer Res.* **56:** 3980–3985.

Ratner, B. D., and Hoffman, A. S. (1976). Synthetic hydrogels for biomedical applications. in *Hydrogels for Medical and Related Applications*, J. D. Andrade, ed. ACS Symposium Series, American Chemical Society, Washington, D.C., Vol. 31, pp. 1–36.

Sefton, M. V. (1987). Heparinized hydrogels. in *Hydrogels in Medicine and Pharmacy*, N. A. Peppas, ed. CRC Press, Boca Raton, FL, Vol. 3, pp. 17–52.

Stringer, J. L., and Peppas, N. A. (1996). Diffusion in radiation-crosslinked poly(ethylene oxide) hydrogels. *J. Controlled Rel.* **42**: 195–202.

Stupp, S. I., LeBonheur, V., Walker, K., Li, L. S., Huggins, K. E., Keser M., and Amstutz, A. (1997). *Science* **276**: 384–389.

Tanaka, T. (1979). Phase transitions in gels and a single polymer. *Polymer* **20**: 1404–1412.

Tighe, B. J. (1976). The design of polymers for contact lens applications. *Brit. Polym. J.* **8**: 71–90.

Wichterle, O., and Lim, D. (1960). Hydrophilic gels for biological use. *Nature* **185:** 117–118.

Yasuda, H., Peterlin, A., Colton, C. K., Smith, K. A., and Merrill, E. W. (1969). Permeability of solutes through hydrated polymer membranes. III. Theoretical background for the selectivity of dialysis membranes. *Makromol. Chem.* **126:** 177–186.

Yoshida, R., Uchida, K., Kaneko, Y., Sakai, K., Kikcuhi, A., Sakurai, Y., and Okano, T. (1995). Comb-type grafted hydrogels with rapid deswelling response to temperature changes. *Nature* **374**: 240–242.

3.2.6 Applications of "smart polymers" as biomaterials

Allan S. Hoffman

Introduction

Stimulus-responsive, "intelligent" polymers are polymers that respond with sharp, large property changes to small changes in physical or chemical conditions. They are also known as "smart" or "environmentally sensitive" polymers. These polymers can take many forms; they may be dissolved in aqueous solution, adsorbed or grafted on aqueous–solid interfaces, or cross-linked in the form of hydrogels.

Many different stimuli have been investigated, and they are listed in Table 3.2.6-1. Typically, when the polymer's critical response is stimulated, the behavior will be as follows (Fig. 3.2.6-1):

- The smart polymer that is dissolved in an aqueous solution will show a sudden onset of turbidity as it phase separates, and if its concentration is high enough, it will convert from a solution to a gel.
- The smart polymer that is chemically grafted to a surface and is stimulated to phase separate will collapse, converting that interface from a hydrophilic to a hydrophobic interface. If the smart polymer is in solution and it is stimulated to phase separate, it may physically adsorb to a hydrophobic surface whose composition has a balance of hydrophobic and polar groups that is similar to the phase-separated smart polymer.
- The smart polymer that is cross-linked in the form of a hydrogel will exhibit a sharp collapse, and release much of its swelling solution.

These phenomena are reversed when the stimulus is reversed. Sometimes the rate of reversion is slower when the polymer has to redissolve or the gel has to reswell in aqueous media. The rate of collapse or reversal of smart polymer systems is sensitive to the dimensions of the smart polymer system, and it will be much more rapid for systems with nanoscale dimensions.

Smart polymers may be physically mixed with or chemically conjugated to biomolecules to yield a large and diverse family of polymerbiomolecule hybrid systems that can respond to biological as well as to physical and chemical stimuli. Biomolecules that may be combined with smart polymer systems include proteins and oligopeptides, sugars and polysaccharides, single and double-stranded oligonucleotides, RNA and DNA, simple lipids and phospholipids, and a wide spectrum of recognition ligands and synthetic drug molecules. In addition, PEG, which is also a smart polymer, may be

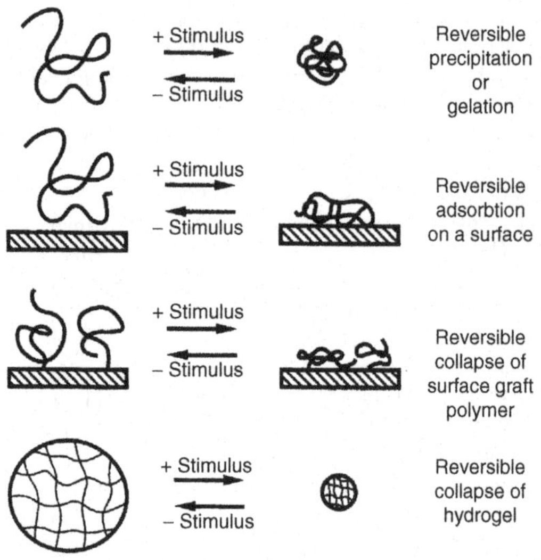

Fig. 3.2.6-1 Schematic illustration showing the different types of responses of "intelligent" polymer systems to environmental stimuli. Note that all systems are reversible when the stimulus is reversed (Hoffman *et al.*, *Journal of Biomedical Materials Research* © 2000).

Table 3.2.6-1 Environmental stimuli

Physical
Temperature
Ionic strength
Solvents
Radiation (UV, visible)
Electric fields
Mechanical stress
High pressure
Sonic radiation
Magnetic fields
Chemical
pH
Specific ions
Chemical agents
Biochemical
Enzyme substrates
Affinity ligands
Other biological agents

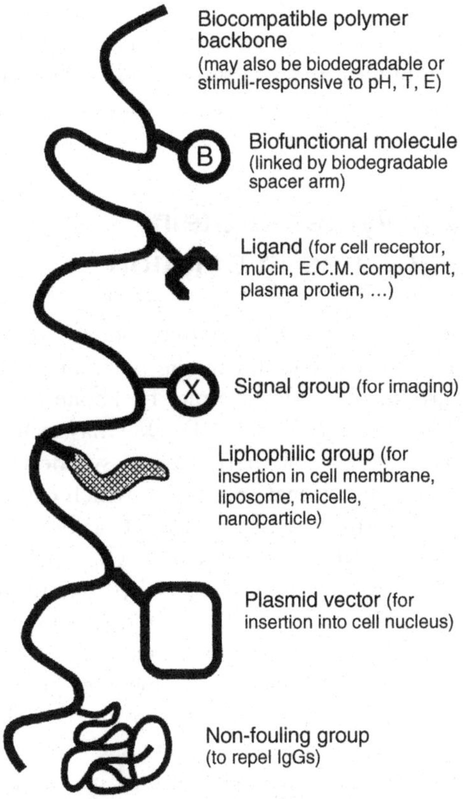

Fig. 3.2.6-2 Schematic illustration showing the variety of natural or synthetic biomolecules which may be conjugated to a smart polymer. In some cases, only one molecule may be conjugated, such as a recognition protein, which may be linked to the protein at a reactive terminal group of the polymer, or it may be linked at a reactive pendant group along the polymer backbone. In other cases more than one molecule may be conjugated along the polymer backbone, such as a targeting ligand along with many drug molecules. (Hoffman *et al.*, *Journal of Biomedical Materials Research* © 2000).

conjugated to the smart polymer backbone to provide it with "stealth" properties (Fig. 3.2.6-2).

Combining a smart polymer and a biomolecule produces a new, smart "biohybrid" system that can synergistically combine the individual properties of the two components to yield new and unusual properties. One could say that these biohybrids are "doubly smart." Among the most important of these systems are the smart polymer–biomolecule conjugates, especially the polymer–drug and polymer–protein conjugates. Such smart bioconjugates, and even a physical mixture of the individual smart polymers and biomolecules, may be physically adsorbed or chemically immobilized on solid surfaces. The biomolecule may also be physically or chemically entrapped in smart hydrogels. All of these hybrid systems have been extensively studied and this section reviews these studies. There have been a number of successful applications in both medicine and biotechnology for such smart polymer–biomolecule systems, and as such, they represent an important extension of polymeric biomaterials beyond their well-known uses in implants and medical devices. Several review articles are available on these interesting smart hybrid biomaterials (Hoffman, 1987, 1995, 1997; Hoffman *et al.*, 1999, 2000; Okano *et al.*, 2000).

Smart polymers in solution

There are many polymers that exhibit thermally induced precipitation (Table 3.2.6-2), and the polymer that has been studied most extensively is PNIPAAm. This polymer is soluble in water below 32°C, and it precipitates sharply as temperature is raised above 32°C (Heskins and Guillet, 1968). The precipitation temperature is called the LCST. If the solution contains buffer and salts the LCST will be reduced several degrees. If NIPAAm

Table 3.2.6-2 Some polymers and surfactants that exhibit thermally-induced phase separation in aqueous solutions
Polymers/Surfactants with Ether Groups
Poly(ethylene oxide) (PEO)
Poly(ethylene oxide/propylene oxide) random copolymers [poly(EO/PO)]
PEO-PPO-PEO triblock surfactants (Polyoxamers or Pluronics)
PLGA-PEO-PLGA triblock polymers
Alkyl-PEO block surfactants (Brij)
Poly(vinyl methyl ether)
Polymers with Alcohol Groups
Poly(hydropropyl acrylate)
Hydroxypropyl cellulose
Methylcellulose
Hydroxypropyl methylcellulose
Poly(vinyl alcohol) derivatives
Polymers with Substituted Amide Groups
Poly(N-substituted acrylamides)
Poly(N-acryloyl pyrrolidine)
Poly(N-acryloyl piperidine)
Poly(acryl-L-amino acid amides)
Others
Poly(methacrylic acid)

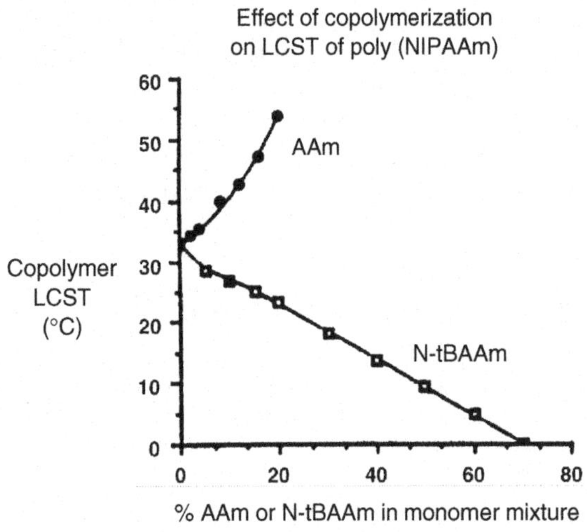

Fig. 3.2.6-3 Copolymerization of a thermally sensitive polymer, PNIPAAm, with a more hydrophilic comonomer, AAm, raises the LCST of the copolymer, whereas copolymerization with a more hydrophobic comonomer, N-tBAAm, lowers the LCST. (Hoffman *et al.*, *Journal of Biomedical Materials Research* © 2000).

monomer is copolymerized with more hydrophilic monomers such as acrylamide, the LCST increases and may even disappear. If NIPAAm monomer is copolymerized with more hydrophobic monomers, such as n-butylacrylamide, the LCST decreases (Fig. 3.2.6-3) (Priest *et al.*, 1987). NIPAAm may also be copolymerized with pH-sensitive monomers, leading to random copolymers with temperature- and pH-responsive components (Dong and Hoffman, 1987; Zareie *et al.*, 2000). NIPAAm has been copolymerized with pH-responsive macromonomers, leading to graft copolymers that independently exhibit two separate stimulus-responsive behaviors (Chen and Hoffman, 1995).

A family of thermally gelling, biodegradable triblock copolymers has been developed for injectable drug delivery formulations (Vernon *et al.*, 2000; Lee *et al.*, 2001; Jeong *et al.*, 2002). They form a medium viscosity, injectable solution at room temperature and a solid hydrogel at 37°C. These polymers are based on compositions of hydrophobic, degradable polyesters combined with PEO. The copolymers are triblocks with varying MW segments of PLGA and PEO. Typical compositions are PEO-PLGA-PEO and PLGA-PEO-PLGA.

Tirrell (1987) and more recently, Stayton, Hoffman, and co-workers have studied the behavior of pH-sensitive alpha-alkylacrylic acid polymers in solution (Lackey *et al.*, 1999; Murthy *et al.*, 1999; Stayton *et al.*, 2000). As pH is lowered, these polymers become increasingly protonated and hydrophobic, and eventually phase separate; this transition can be sharp, resembling the phase transition at the LCST. If a polymer such as poly(ethylacrylic acid) or poly(propylacrylic acid) is in the vicinity of a lipid bilayer membrane as pH is lowered, the polymer will interact with the membrane and disrupt it. These polymers have been used in intracellular drug delivery to disrupt endosomal membranes as pH drops in the endosome, enhancing the cytosolic delivery of drugs, and avoiding exposure to lysosomal enzymes.

Smart polymer–protein bioconjugates in solution

Smart polymers may be conjugated randomly to proteins by binding the reactive end of the polymer or reactive pendant groups along the polymer backbone to reactive sites on the protein (Fig. 3.2.6-4). One may utilize chain transfer free radical polymerization to synthesize oligomers with one functional end group, which can then be derivatized to form a reactive group that can be conjugated to the protein. NIPAAm has also been copolymerized with reactive comonomers (e.g., *N*-hydroxysuccinimide acrylate, or NHS acrylate) to yield a random copolymer with reactive pendant groups, which have then been conjugated to the protein. Vinyl monomer groups have been conjugated to proteins to provide sites for copolymerization with free monomers such as NIPAAm. These synthesis methods are described in several publications (Cole *et al.*, 1987; Monji and Hoffman, 1987; Shoemaker *et al.*, 1987; Chen *et al.*, 1990; Chen and Hoffman, 1990, 1994; Yang *et al.*, 1990; Takei *et al.*, 1993a; Monji *et al.*, 1994; Ding *et al.*, 1996) (see also Section 3.2.16 on biologically functional materials).

Normally the lysine amino groups are the most reactive protein sites for random polymer conjugation to

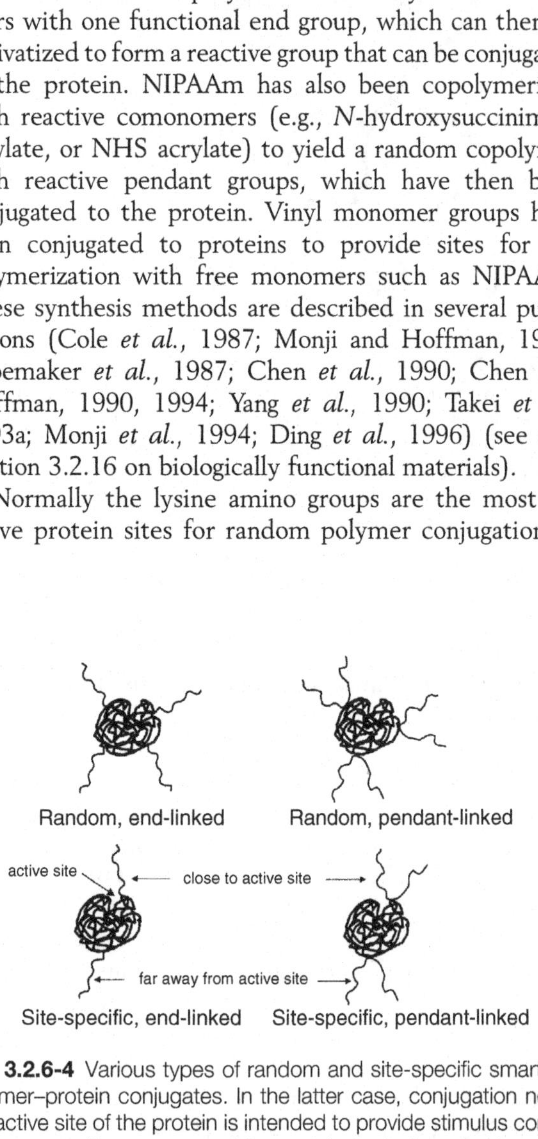

Fig. 3.2.6-4 Various types of random and site-specific smart polymer–protein conjugates. In the latter case, conjugation near the active site of the protein is intended to provide stimulus control of the recognition process of the protein for its ligand, whereas conjugation far away from the active site should avoid any interference of the polymer with the protein's natural activity (Hoffman *et al.*, *Journal of Biomedical Materials Research* © 2000).

proteins, and NHS attachment chemistry is most often utilized. Other possible sites include –COOH groups of aspartic or glutamic acid, –OH groups of serine or tyrosine, and –SH groups of cysteine residues. The most likely attachment site will be determined by the reactive group on the polymer and the reaction conditions, especially the pH. Because these conjugations are generally carried out in a nonspecific way, the conjugated polymer can interfere sterically with the protein's active site or modify its microenvironment, typically reducing the bioactivity of the protein. On rare occasions the conjugation of a polymer increases the activity of the protein. (e.g., Ding *et al.*, 1998).

Biomedical uses of smart polymers in solution have mainly been as conjugates with proteins. Random conjugation of temperature-sensitive (mainly) and pH-sensitive (occasionally) polymers to proteins has been extensively investigated, and applications of these conjugates have been focused on immunoassays, affinity separations, enzyme recovery, and drug delivery (Schneider *et al.*, 1981; Okamura *et al.*, 1984; Nguyen and Luong, 1989; Taniguchi *et al.*, 1989, 1992; Chen and Hoffman, 1990; Monji *et al.*, 1990; Pecs *et al.*, 1991; Park and Hoffman, 1992; Takei *et al.*, 1993b, 1994a; Galaev and Mattiasson, 1993; Fong *et al.*, 1999; Anastase-Ravion *et al.*, 2001). In some cases the "smart" polymer is a polyligand, such as polybiotin or poly(glycosyl methacrylate), which is used to phase separate target molecules by complexation to multiple binding sites on target proteins, such as streptavidin and Concanavalin A, respectively (Larsson and Mosbach, 1979; Morris *et al.*, 1993; Nakamae *et al.*, 1994). Wu *et al.* (1992, 1993) have synthesized PNIPAAm–phospholipid conjugates for use in drug delivery formulations as components of thermally sensitive composites and liposomes.

Smart polymers on surfaces

One may covalently graft a polymer to a surface by exposing the surface to ionizing radiation in the presence of the monomer (and in the absence of air), or by preirradiating the polymer surface in air, and later contacting the surface with the monomer solution and heating in the absence of air. (See also Section 3.1.4 on surface properties of materials.) These surfaces exhibit stimulus-responsive changes in wettability (Uenoyama and Hoffman, 1988; Takei *et al.*, 1994b; Kidoaki *et al.*, 2001). Ratner and co-workers have used a gas plasma discharge to deposit temperature-responsive coatings from a NIPAAm monomer vapor plasma (Pan *et al.*, 2001). Okano and Yamato and co-workers have utilized the radiation grafting technique to form cell culture surfaces having a surface layer of grafted PNIPAAm. (Yamato and Okano, 2001; Shimizu *et al.*, 2003). They have cultured cells to confluent sheets on these surfaces at 37°C, which is above the LCST of the polymer. When the PNIPAAm collapses, the interface becomes hydrophobic and leads to adsorption of cell adhesion proteins, enhancing the cell culture process. Then when the temperature is lowered, the interface becomes hydrophilic as the PNIPAAm chains rehydrate, and the cell sheets release from the surface (along with the cell adhesion proteins). The cell sheet can be recovered and used in tissue engineering, e.g., for artificial cornea and other tissues. Patterned surfaces have also been prepared (Yamato *et al.*, 2001). Smart polymers may also be grafted to surfaces to provide surfaces of gradually varying hydrophilicity and hydrophobicity as a function of the polymer composition and conditions. This phenomenon has been applied by Okano, Kikuchi, and co-workers to prepare chromatographic column packing, leading to eluate-free ("green") chromatographic separations (Kobayashi *et al.*, 2001; Kikuchi and Okano, 2002). Ishihara *et al.* (1982, 1984b) developed photo-responsive coatings and membranes that reversibly changed surface wettability or swelling, respectively, due to the photoinduced isomerization of an azobenzene-containing polymer.

Site-specific smart polymer bioconjugates on surfaces

Conjugation of a responsive polymer to a specific site near the ligand-binding pocket of a genetically engineered protein is a powerful new concept. Such site-specific protein–smart polymer conjugates can permit sensitive environmental control of the protein's recognition process, which controls all living systems. Stayton and Hoffman *et al.* (Stayton *et al.*, 2000) have designed and synthesized smart polymer–protein conjugates where the polymer is conjugated to a specific site on the protein, usually a reactive –SH thiol group from cysteine that has been inserted at the selected site (Fig. 3.2.6-5). This is accomplished by utilizing cassette mutagenesis to insert a site-specific mutation into the DNA sequence of the protein, and then cloning the mutant in cell culture. This method is applicable only to proteins whose complete peptide sequence is known. The preparation of the reactive smart polymer is similar to the method described above, but now the reactive end or pendant groups and the reaction conditions are specifically designed to favor conjugation to –SH groups rather than to $-NH_2$ groups. Typical SH-reactive polymer end groups include maleimide and vinyl sulfone groups.

The specific site for polymer conjugation can be located far away from the active site (Chilkoti *et al.*, 1994), in order to avoid interference with the biological

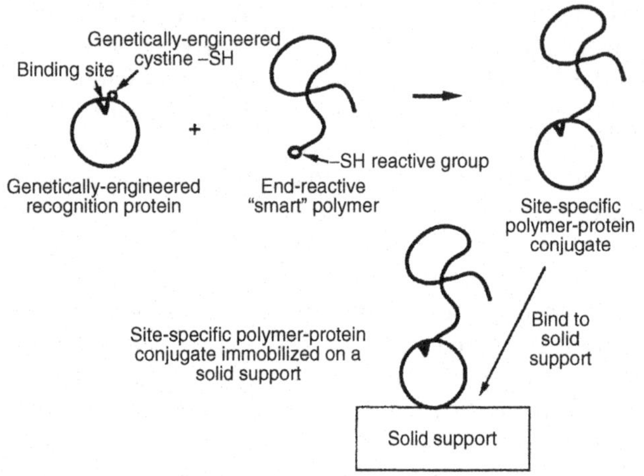

Fig. 3.2.6-5 Schematic illustration of the process for preparing an immobilized, site-specific conjugate of a smart polymer with a genetically-engineered, mutant protein (Hoffman *et al.*, *Journal of Biomedical Materials Research* © 2000).

functioning of the protein, or nearby or even within the active site, in order to control the ligand–protein recognition process and the biological activity of the protein (Fig. 3.2.6-4) (Ding *et al.*, 1999, 2001; Bulmus *et al.*, 1999; Stayton *et al.*, 2000; Shimoboji *et al.*, 2001, 2002a, b, 2003). The latter has been most studied by the Stayton/Hoffman group. Temperature-, pH-, and light-sensitive smart polymers have been used to form such novel, "doubly smart" bioconjugates. Since the objective is to control the activity of the protein, and not to phase separate it, these smart polymer–engineered protein bioconjugates have usually been immobilized on the surfaces of microbeads or nanobeads. Stayton, Hoffman, and co-workers have used such beads in microfluidic devices for immunoassays (Malmstadt *et al.*, 2003). Earlier work by Hoffman and co-workers established the importance of matching the smart polymer composition with the surface composition in order to enhance the stimulus-driven adsorption of the smart polymer on the surface (Miura *et al.*, 1994). Others have also recently utilized this phenomenon in microfluidic devices (Huber *et al.*, 2003).

The proteins that have been most studied by the Stayton/Hoffman group to date include streptavidin and the enzyme cellulase. PNIPAAm–streptavidin site-specific bioconjugates have been used to control access of biotin to its binding site on streptavidin, and have enabled separation of biotinylated proteins according to the size of the protein (Ding *et al.*, 2001). Ding, Stayton, and Hoffman *et al.* (1999) also found that raising the temperature to thermally induce the collapse of the polymer "triggered" the release of the bound biotin molecules (Ding *et al.*, 1999). For the site-specific enzyme conjugates, a combined temperature- and light-sensitive polymer was conjugated to specific sites on an endo-cellulase, which provided on–off control of the enzyme activity with either light or temperature (Shimoboji *et al.*, 2001, 2002a, b, 2003).

Triggered release of bound ligands by the smart polymer–engineered protein bioconjugates could be used to release therapeutics, such as for topical drug delivery to the skin or mucosal surfaces of the body, and also for localized delivery of drugs within the body by stimulated release at pretargeted sites using noninvasive, focused stimuli, or delivery of stimuli from catheters. Triggered release could also be used to release and recover affinity-bound ligands from chromatographic and other supports in eluate-free conditions, including capture and release of specific cell populations to be used in stem cell and bone marrow transplantation. These processes could involve two different stimulus-responsive polymers with sensitivities to the same or different stimuli. For delicate target ligands such as peptides and proteins, recovery could be affected without the need for time-consuming and harsh elution conditions. Triggered release could also be used to remove inhibitors, toxins, or fouling agents from the recognition sites of immobilized or free enzymes and affinity molecules, such as those used in biosensors, diagnostic assays, or affinity separations. This could be used to "regenerate" such recognition proteins for extended process use. Light-controlled binding and release of site-specific protein conjugates may be utilized as a molecular switch for various applications in biotechnology, medicine, and bioelectronics, including hand-held diagnostic devices, biochips, and lab-on-a-chip devices.

Fong, Stayton, and Hoffman (Fong *et al.*, 1999) have developed an interesting construct to control the distance of the PNIPAAm from the active site. For this purpose, they conjugated one sequence of complementary nucleotides to a specific site near the binding pocket of streptavidin, and a second sequence to the end of a PNIPAAm chain. Then, by controlling the location and length of the complementary sequence, the self-assembly via hybridization of the two single-chain DNA sequences could be used to control the distance of the polymer from the streptavidin binding site.

Smart polymer hydrogels

When a smart polymer is cross-linked to form a gel, it will collapse and re-swell in water as a stimulus raises or lowers it through its critical condition. PNIPAAm gels have been extensively studied, starting with the pioneering work of Toyoichi Tanaka in 1981 (Tanaka, 1981). Since then, the properties of PNIPAAm hydrogels have been widely investigated in the form of beads, slabs, and multilamellar laminates (Park and Hoffman,

1992a, b, 1994; Hu *et al.*, 1995, 1998; Mitsumata *et al.*, 2001; Kaneko *et al.*, 2002; Gao and Hu, 2002). Okano and co-workers have developed smart gels that collapse very rapidly, by grafting PNIPAAm chains to the PNIPAAm backbone in a cross-linked PNIPAAm hydrogel (Yoshida *et al.*, 1995; Masahiko *et al.*, 2003). Smart hydrogel compositions have been developed that are both thermally gelling and biodegradable (Zhong *et al.*, 2002; Yoshida *et al.*, 2003). These sol-gel systems have been used to deliver drugs by *in vivo* injections and are discussed in the section on smart polymers in solution.

Hoffman and co-workers were among the first to recognize the potential of PNIPAAm hydrogels as biomaterials; they showed that the smart gels could be used (a) to entrap enzymes and cells, and then turn them on and off by inducing cyclic collapse and swelling of the gel, and (b) to deliver or remove biomolecules, such as drugs or toxins, respectively, by stimulus-induced collapse or swelling (Dong and Hoffman, 1986, 1987, 1990; Park and Hoffman, 1988, 1990a, b, c) (Fig. 3.2.6-6). One unique hydrogel was developed by Dong and Hoffman (1991). This pH- and temperature-sensitive hydrogel was based on a random copolymer of NIPAAm and AAc, and it was shown to release a model drug linearly over a 4-hour period as the pH went from gastric to enteric conditions at 37°C. At body temperature the NIPAAm component was trying to maintain the gel in the collapsed state, while as the pH went from acidic to neutral conditions, the AAc component was becoming ionized, forcing the gel to swell and slowly release the drug (see Fig. 3.2.6-6B).

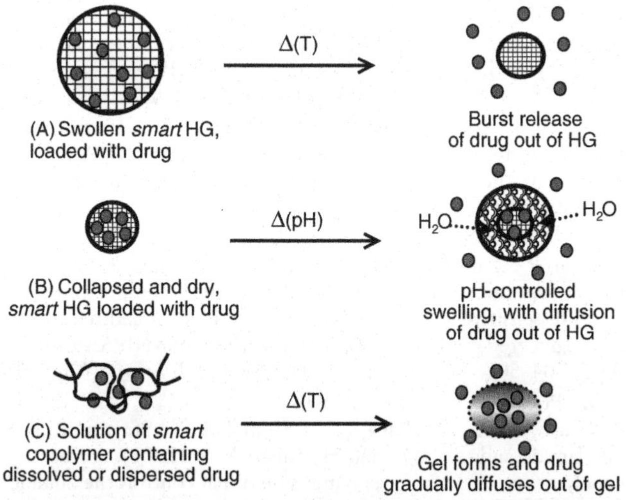

Fig. 3.2.6-6 Schematic illustration showing three ways that smart gel formulations may be stimulated to release bioactive agents: (A) thermally induced collapse, which is relevant to skin or mucosal drug delivery; (B) pH-induced swelling, which is relevant to oral drug delivery, where the swelling is induced by the increase in pH in going from the gastric to enteric regions; and (C) sol-to-gel formation, which is relevant to injectable or topical formulations of a triblock copolymer solution that are thermally gelled at body temperature. For *in vivo* uses, the block copolymer is designed to be degradable. The first two apply to cross-linked gels applied topically or orally, and the third is relevant to thermally induced formation of gels from polymer solutions that are delivered topically or by injection.

Kim, Bae, and co-workers have investigated smart gels containing entrapped cells that could be used as artificial organs (Vernon *et al.*, 2000). Matsuda and co-workers have incorporated PNIPAAm into physical mixtures with natural polymers such as hyaluronic acid and gelatin, for use as tissue engineering scaffolds (Ohya *et al.*, 2001a, b).

Peppas and co-workers (Robinson and Peppas, 2002) have studied pH-sensitive gels in the form of nanospheres. Nakamae, Hoffman, and co-workers developed novel compositions of smart gels containing phosphate groups that were used to bind cationic proteins as model drugs and then to release them by a combination of thermal stimuli and ion exchange (Nakamae *et al.*, 1992, 1997; Miyata *et al.*, 1994).

Smart gels that respond to biological stimuli

A number of drug delivery devices have been designed to respond to biologic signals in a feedback manner. Most of these gels contain an immobilized enzyme. Heller and Trescony (1979) were among the first to work with smart enzyme gels. In this early example, urease was immobilized in a gel, and urea was metabolized to produce ammonia, which caused a local pH change, leading to a permeability change in the surrounding gel. Ishihara *et al.* (1985) also developed a urea-responsive gel containing immobilized urease. Smart enzyme gels containing glucose oxidase (GOD) were designed to respond to a more relevant signal, that of increasing glucose concentration. In a typical device, when glucose concentration increases, the entrapped GOD converts the glucose in the presence of oxygen to gluconic acid and hydrogen peroxide. The former lowers pH, and the latter is an oxidizing agent. Each of these byproduct signals has been used in various smart hydrogel systems to increase the permeability of the gel barrier to insulin delivery (Horbett *et al.*, 1984; Albin *et al.*, 1985; Ishihara *et al.*, 1983, 1984a; Ishihara and Matsui, 1986; Ito *et al.*, 1989; Iwata and Matsuda, 1988).

In one case, the lowered pH due to the GOD byproduct, gluconic acid, accelerated hydrolytic erosion of the polymer matrix that also contained entrapped insulin, releasing the insulin (Heller *et al.*, 1990). Siegel and co-workers have used the glucose-stimulated swelling and collapse of hydrogels containing entrapped GOD to drive a hydrogel piston in an oscillating manner, for release of insulin in a glucose-driven, feedback manner (Dhanarajan *et al.*, 2002). Other smart enzyme gels for drug delivery have been developed based on activation of an inactivated enzyme by a biologic signal (Schneider *et al.*, 1973; Roskos *et al.*, 1993).

Smart gels have also been developed that are based on affinity recognition of a biologic signal. Makino *et al.* (1990) developed a smart system that contained glycosylated insulin bound by affinity of its glucose groups to an immobilized Concanavalin A in a gel. When glucose concentration increases, the free glucose competes off the insulin, which is then free to diffuse out of the gel. Nakamae *et al.* (1994) developed a gel based on a similar concept, using a cross-linked poly(glycosylethyl methacrylate) hydrogel containing physically or chemically entrapped Concanavalin A. In this case, the ConA is bound by affinity to the pendant glucose groups on the polymer backbone, acting as a cross-linker because of its four affinity binding sites for glucose; when free glucose concentration increases, the ConA is competed off the polymer backbone. This leads to swelling of the gel, which acts to increase permeation of insulin through the gel. Miyata and co-workers have designed and synthesized smart affinity hydrogels that are stimulated to swell or collapse by the binding of affinity biomolecules (Miyata *et al.*, 1999, 2002).

Conclusions

Smart polymers in solution, on surfaces, and as hydrogels have been utilized in many interesting ways, especially in combination with biomolecules such as proteins and drugs. Important applications include affinity separations, enzyme processes, immunoassays, drug delivery, and toxin removal. These smart polymer–biomolecule systems represent an important extension of polymeric biomaterials beyond their well-known uses in implants and medical devices.

Bibliography

Albin, G., Horbett, T. A., and Ratner, B. D. (1985). Glucose sensitive membranes for controlled delivery of insulin: insulin transport studies. *J. Controlled Release* **2**: 153–164.

Anastase-Ravion, S., Ding, Z., Pelle, A., Hoffman, A. S., and Letourneur, D. (2001). New antibody purification procedure using a thermally-responsive polyNIPAAm-dextran derivative conjugate. *J. Chromatogr. B.* **761**: 247–254.

Bulmus, V., Ding, Z., Long, C. J., Stayton, P. S., and Hoffman, A. S. (1999). Design, synthesis and site-specific conjugation of a pH- and temperature-sensitive polymer to streptavidin for pH-controlled binding and triggered release of biotin. *Bioconj. Chem.* **11**: 78–83.

Chen, J. P., and Hoffman, A. S. (1990). Polymer–protein conjugates. II. Affinity precipitation of human IgG by poly(*N*-isopropyl acrylamide) -protein A conjugates. *Biomaterials* **11**: 631–634.

Chen, G., and Hoffman, A. S. (1994). Synthesis of carboxylated poly(NIPAAm) oligomers and their application to form thermo-reversible polymer-enzyme conjugates. *J. Biomater. Sci. Polymer Edn.* **5**: 371–382.

Chen, G., and Hoffman, A. S. (1995). Graft copolymer compositions that exhibit temperature-induced transitions over a wide range of pH. *Nature* **373**: 49–52.

Chen, J. P., Yang, H. J., and Hoffman, A. S. (1990). Polymer–protein conjugates. I. Effect of protein conjugation on the cloud point of poly(*N*-isopropyl acrylamide). *Biomaterials* **11**: 625–630.

Chilkoti, A., Chen, G., Stayton, P. S., and Hoffman, A. S. (1994). Site-specific conjugation of a temperature-sensitive polymer to a genetically-engineered protein. *Bioconj. Chem.* **5**: 504–507.

Cole, C. A., Schreiner, S. M., Priest, J. H., Monji, N. and Hoffman, A. S. (1987). *N*-Isopropyl acrylamide and *N*-acryl succinimide copolymers: a thermally reversible water soluble activated polymer for protein conjugation. *Reversible Polymeric Gels and Related Systems,* ACS Symposium Series, Vol. 350, P. Russo, ed. ACS, Washington, D.C., pp. 245–254.

Dhanarajan, A. P., Misra, G. P., and Siegel, R. A. (2002). Autonomous chemomechanical oscillations in a hydrogel/enzyme system driven by glucose. *J. Phys. Chem.* **106**: 8835–8838.

Ding, Z. L., Chen, G., and Hoffman, A. S. (1996). Synthesis and purification of thermally-sensitive oligomer–enzyme conjugates of poly(NIPAAm)–trypsin. *Bioconj. Chem.* **7**: 121–125.

Ding, Z. L., Chen, G., and Hoffman, A. S. (1998). Properties of polyNIPAAm–trypsin conjugates. *J. Biomed. Mater. Res.* **39**: 498–505.

Ding, Z. L., Long, C. J., Hayashi, Y., Bulmu, E. V., Hoffman, A. S., and Stayton, P. S. (1999). Temperature control of biotin binding and release with a streptavidin–polyNIPAAm site-specific conjugate. *J. Bioconj. Chem.* **10**: 395–400.

Ding, Z. L., Shimoboji, T., Stayton, P. S., and Hoffman, A. S. (2001). A smart polymer shield that controls the binding of different size biotinylated proteins to streptavidin. *Nature* **411**: 59–62.

Dong, L. C., and Hoffman, A. S. (1986). Thermally reversible hydrogels: III. Immobilization of enzymes for feedback reaction control. *J. Contr. Rel.* **4**: 223–227.

Dong, L. C., and Hoffman, A. S. (1987). Thermally reversible hydrogels: swelling characteristics and activities of copoly(NIPAAm-AAm) gels containing immobilized asparaginase. in *Reversible Polymeric Gels and Related Systems,*

ACS Symposium Series, Vol 350, P. Russo, ed. ACS, Washington, D.C., pp. 236–244.

Dong, L. C., and Hoffman, A. S. (1990). Synthesis and application of thermally-reversible heterogels for drug delivery. *J. Contr. Release* 13: 21–32.

Dong, L. C., and Hoffman, A. S. (1991). A novel approach for preparation of pH- and temperature-sensitive hydrogels for enteric drug delivery. *J. Contr. Release* 15: 141–152.

Fong, R. B., Ding, Z. L., Long, C. J., Hoffman, A. S., and Stayton, P. S. (1999). Thermoprecipitation of streptavidin via oligonucleotide-mediated self-assembly with poly(NIPAAm). *Bioconj. Chem.* **10:** 720–725.

Galaev, I. Y., and Mattiasson, B. (1993). Affinity thermoprecipitation: Contribution of the efficiency and access of the ligand. *Biotechnol. Bioeng.* **41:** 1101–1106.

Gao, J., and Hu, Z. B. (2002). Optical properties of *N*-iso-propylacrylamide microgel spheres in water. *Langmuir* 18: 1360–1367.

Heller, J., and Trescony, P. V. (1979). Controlled drug release by polymer dissolution II. Enzyme-mediated delivery device. *J. Pharm. Sci.* **68:** 919–921.

Heller, J., Chang, A. C., Rodd, G., and Grodsky, G. M. (1990). Release of insulin from a pH-sensitive poly (ortho ester). *J. Controlled Release* **14:** 295–304.

Heskins, H., and Guillet, J. E. (1968). Solution properties or poly(*N*-isopropyl acrylamide). *J. Macromol. Sci. Chem. A2* **6:** 1209.

Hoffman, A. S. (1987). Applications of thermally reversible polymers and hydrogels in therapeutics and diagnostics. *J. Contr. Rel.* **6**: 297–305.

Hoffman, A. S. (1995). Intelligent polymers in medicine and biotechnology. *Macromol. Symp.* 98: 645–664.

Hoffman, A. S. (1997). Intelligent polymers in medicine and biotechnology, in *Controlled Drug Delivery,* K. Park, ed. ACS Publications, Washington, D.C.

Hoffman, A. S., *et al.* (2000). Really smart bioconjugates of smart polymers and receptor proteins. J. *Biomed. Mater. Res.* **52**: 577–586.

Hoffman, A. S., Chen, G., Wu, X., Ding, Z., Matsuura, J. E., and Gombotz, W. R. (1999). Stimuli-responsive polymers grafted onto polyacrylic acid and chitosan backbones as bioadhesive carriers for mucosal drug delivery. in *Frontiers in Biomedical Polymer Applications,* R. M. Ottenbrite, ed. Technomic Publ., Lancaster, UK, pp. 17–29.

Horbett, T. A., Kost, J., and Ratner, B. D. (1984). Swelling behavior of glucose sensitive membranes. in *Polymers as Biomaterials,* S. Shalaby, A. S. Hoffman, B. D. Ratner, and T. A., Horbett, eds. Plenum Press, New York, pp. 193–207.

Hu, Z. B., Zhang, X. M., and Li, Y. (1995). Synthesis and application of modulated polymer gels. *Science* **269**: 525.

Hu, Z. B., Chen, Y. Y., Wang, C. J., Zheng, Y. Y. and Li, Y. (1998). Polymer gels with engineered environmentally responsive surface patterns. *Nature* **393**: 149.

Huber, D. L., Manginell, R. P., Samara, M. A., Kim, B. I., and Bunkar, B. C. (2003). Programmed adsorption and release of proteins in a microfluidic device. *Science* **301:** 352.

Ishihara, K., and Matsui, K. (1986). Glucose-responsive insulin release from polymer capsule. *J. Polymer Sci., Polymer Lett. Ed.* **24**: 413–417.

Ishihara, K., Okazaki, A., Negishi, N., Shinohara, I., Okano, T., Kataoka, K., and Sakurai, Y. (1982). Photo-induced change in wettability and binding ability of azoaromatic polymer. *J. Appl. Polymer Sci.* **27**: 239–245.

Ishihara, K., Kobayashi, M., and Shinohara, I. (1983). Control of insulin permeation through a polymer membrane with responsive function for glucose. *Makromol. Chem. Rapid Commun.* **4**: 327–331.

Ishihara, K., Kobayashi, M., Ishimaru, N., and Shinohara, I. (1984a). Glucose-induced permeation control of insulin through a complex membrane consisting of immobilized glucose oxidase and a poly(amine). *Polymer J.* **16:** 625–631.

Ishihara, K., Hamada, N., Kato, S., and Shinohara, I. (1984b). Photo-induced swelling control of amphiphilic azoaromatic polymer membrane. *Polymer Sci., Polymer Chem. Ed.* **22**: 21–128.

Ishihara, K., Muramoto, N., Fujii, H., and Shinohara, I. (1985). Preparation and permeability of urea-responsive polymer membrane consisting of immobilized urease and a poly(aromatic carboxylic acid). *J. Polymer Sci., Polymer Lett. Ed.* **23**: 531–535.

Ito, Y., Casolaro, M., Kono, K., and Imanishi, Y. (1989). An insulin-releasing system that is responsive to glucose. *J. Controlled Release* **10:** 195–203.

Iwata, H., and Matsuda,T. (1988). Preparation and properties of novel environment-sensitive membranes prepared by graft polymerization onto a porous substrate. *J. Membrane Sci.* **38**: 185–199.

Jeong, B., Kim, S. W., and Bae, Y. H. (2002). Thermosensitive sol–gel reversible hydrogels. *Adv. Drug Delivery Rev.* **54**: 37–51.

Kaneko, D., Gong, J. P., and Osada, Y. (2002). Polymer gels as soft and wet chemomechanical systems—an approach to artifical muscles. *J. Mater. Chem.* **12**: 2169–2177.

Kawaguchi H., Kisara K., Takahashi T., Achiha K., Yasui M., and Fujimoto K. (2000). Versatility of thermosensitive particles. *Macromol. Symp.* **151:** 591–598.

Kawaguchi H., Isono Y., and Tsuji S. (2002). Hairy particles prepared by living radical graft polymerization. *Macromol. Symp.* **179**: 191–206.

Kidoaki, S., Ohya, S., Nakayama, Y., and Matsuda T. (2001). Thermoresponsive structural change of a PNIPAAm graft layer measured with AFM. *Langmuir* **17**: 2402–2407.

Kikuchi, A., and Okano, T. (2002). Intelligent thermoresponsive polymeric stationary phases for aqueous chromatography of biological compounds. *Progr. Polymer Sci.* **27**: 1165–1193.

Kobayashi, J., Kikuchi, A., Sakai, K., and Okano, T. (2001). Aqueous chromatography utilizing pH-/temperature-responsive polymers as column matrix surfaces for separation of ionic bioactive compounds. *Anal. Chem.* **73**(9): 2027–2033.

Lackey, C. A., Murthy, N., Press, O. W., Tirrell, D. A., Hoffman, A. S., and Stayton, P. S. (1999). Hemolytic activity of pH-responsive polymer–streptavidin bioconjugates. *Bioconj. Chem.* **10**: 401–405.

Larsson, P. O., and Mosbach, K. (1979). Affinity precipitation of enzymes. *FEBS Lett.* **98**: 333–338.

Lee, D. S., Shim, M. S., Kim, S. W., Lee, H., Park, I., and Chang, T. (2001). Novel thermoreversible gelation of biodegradable PLGA-*block-PEO-block*-PLGA triblock copolymers in aqueous solution. *Macromol. Rapid Commun.* **22:** 587–592.

Makino, K., Mack, E. J., Okano, T., and Kim, S. W. (1990). A microcapsule self-regulating delivery system for insulin. *J. Controlled Release* **12**: 235–239.

Malmstadt, N., Yager, P., Hoffman, A. S., and Stayton, P. S. (2003). A smart microfluidic affinity chromatography matrix composed of poly(*N*-isopropylacrylamide)-coated beads. *Anal. Chem.* **75**: 2943–2949.

Masahiko, A., Matsuura, T., Kasai, M., Nakahira, T., Hara, Y., and Okano, T. (2003). Preparation of comb-type *N*-isopropylacrylamide hydrogel beads and their application for size-selective separation media. *Biomacromolecules* **4**: 395–403.

Mitsumata, T., Gong, J. P., and Osada, Y. (2001). Shape memory functions and motility of amphiphilic polymer gels. *Polymer Adv. Technol.* **12**: 136–150.

Miura, M., Cole, C. A., Monji, N., and Hoffman, A. S. (1994). Temperature-dependent adsorption/desorption behavior of LCST polymers on various substrates. *J. Biomater. Sci. Polymer Ed.* **5**: 555–568.

Miyata, T., Nakamae, K., Hoffman, A. S., and Kanzaki, Y. (1994). Stimuli-sensitivities of hydrogels containing phosphate groups. *Macromol. Chem. Phys.* **195**: 1111–1120.

Miyata, T., Asami, N., and Uragami, T. (1999). A reversibly antigen-responsive hydrogel. *Nature* **399**: 766–769.

Miyata, T., Uragami, T., and Nakamae, K. (2002). Biomolecule-sensitive hydrogels. *Adv. Drug Delivery Rev.* **54**: 79–98.

Monji, N. and Hoffman, A. S. (1987). A novel immunoassay system and bioseparation process based on thermal phase separating polymers. *Appl. Biochem. Biotechnol.* **14**: 107–120.

Monji, N., Cole, C. A., and Hoffman, A. S. (1994). Activated, N-substituted acrylamide polymers for antibody coupling: application to a novel membrane-based immunoassay. *J. Biomater. Sci. Polymer Ed.* **5**: 407–420.

Monji, N., Cole, C. A., Tam, M., Goldstein, L., Nowinski, R. C., and Hoffman, A. S. (1990). Application of a thermally-reversible polymer–antibody conjugate in a novel membrane-based immunoassay. *Biochem. Biophys. Res. Commun.* **172**: 652–660.

Morris, J. E., Hoffman, A. S., and Fisher, R. R. (1993). Affinity precipitation of proteins by polyligands. *Biotechnol. Bioeng.* **41**: 991–997.

Murthy, N., Stayton, P. S., and Hoffman, A. S. (1999). The design and synthesis of polymers for eukaryotic membrane disruption. *J. Controlled Release* **61**: 137–143.

Nakamae, K., Miyata, T., and Hoffman, A. S. (1992). Swelling behavior of hydrogels containing phosphate groups. *Macromol. Chem.* **193**: 983–990.

Nakamae, K., Miyata, T., Jikihara, A., and Hoffman, A. S. (1994). Formation of poly(glucosyloxyethyl methacrylate)–Concanavalin A complex and its glucose sensitivity. *J. Biomater. Sci. Polymer Ed.* **6**: 79–90.

Nakamae, K., Nizuka, T., Miyata, T., Furukawa, M., Nishino, T., Kato, K., Inoue, T., Hoffman, A. S., and Kanzaki, Y. (1997). Lysozyme loading and release from hydrogels carrying pendant phosphate groups. *J. Biomater. Sci., Polymer Ed.* **9**: 43–53.

Nguyen, A. L., and Luong, J. H. T. (1989). Syntheses and application of water soluble reactive polymers for purification and immobilization of biomolecules. *Biotechnol. Bioeng.* 34: 1186–1190.

Ohya, S., Nakayama, Y., and Matsuda, T. (2001a). Thermoresponsive artificial extracellular matrix for tissue engineering: hyaluronic acid bioconjugated with poly(*N*-isopropylacrylamide) grafts. *Biomacromolecules* **2**: 856–863.

Ohya, S., Nakayama, Y., and Matsuda, T. (2001b). Material design for an artificial extracellular matrix: cell entrapment in poly(*N*-isopropylacrylamide) (PNIPAM)-grafted gelatin hydrogel. *J. Artif. Organs* **4**: 308–314.

Okamura, K., Ikura, K., Yoshikawa, M., Sakaki, R., and Chiba, H. (1984). Soluble–insoluble interconvertible enzymes. *Agric. Biol. Chem.* **48**: 2435–2440.

Okano, T., Kikuchi, A., and Yamato, M. (2000). Intelligent hydrogels and new biomedical applications. in *Biomaterials and Drug Delivery toward the New Millennium.* Han Rim Won Publishing Co., Seoul, Korea, pp. 77–86.

Pan, Y. V., Wesley, R. A., Luginbuhl, R., Denton, D. D., and Ratner, B. D. (2001). Plasma-polymerized *N*-isopropylacylamide: synthesis and characterization of a smart thermally responsive coating. *Biomacromolecules* **2**: 32–36.

Park, T. G., and Hoffman, A. S. (1988). Effect of temperature cycling on the activity and productivity of immobilized β-galactosidase in a thermally reversible hydrogel bead reactor. *Appl. Biochem. Biotechnol.* **19**: 1–9.

Park, T. G., and Hoffman, A. S. (1990a). Immobilization and characterization of β-galactosidase in thermally reversible hydrogel beads. *J. Biomed. Mater. Res.* 24: 21–38.

Park, T. G., and Hoffman, A. S. (1990b). Immobilization of A. *simplex* cells in a thermally-reversible hydrogel: effect of temperature cycling on steroid conversion. *Biotech. Bioeng.* **35**: 52–159.

Park, T. G., and Hoffman, A. S. (1990c). Immobilized biocatalysts in reversible hydrogels. in *Enzyme Engineering X*, A. Tanaka, ed. *Ann. N.Y. Acad. Sci.*, Vol. 613, pp. 588–593.

Park, T. G., and Hoffman, A. S. (1992a). Preparation of large, uniform size temperature-sensitive hydrogel beads. *J. Polymer Sci. A Polymer Chem.* **30**: 505–507.

Park, T. G., and Hoffman, A. S. (1992b). Synthesis and characterization of pH- and/or temperature-sensitive hydrogels. *J. Appl. Polymer Sci.* **46**: 659–671.

Park, T. G., and Hoffman, A. S. (1993). Synthesis and characterization of a soluble, temperature-sensitive polymer-conjugated enzyme. *J. Biomater. Sci. Polymer Ed.* **4**: 493–504.

Park, T. G., and Hoffman, A. S. (1994). Estimation of temperature-dependent pore sizes in poly(NIPAAm) hydrogel beads. *Biotechnol. Progr.* **10**: 82–86.

Pecs, M., Eggert, M., and Schügerl, K. (1991). Affinity precipitation of extracellular microbial enzymes. *J. Biotechnol.* **21**: 137–142.

Priest, J. H., Murray, S., Nelson, R. G., and Hoffman, A. S. (1987). LCSTs of aqueous copolymers of *N*-isopropyl acrylamide and other *N*-substituted acrylamides. *Reversible Polymeric Gels and Related Systems*, ACS Symposium Series, Vol. 350, P. Russo, ed. ACS, Washington, D.C., pp. 255–264.

Robinson, D. N., and Peppas, N. A. (2002). Preparation and characterization of pH-responsive poly(methacrylic acid-g-ethylene glycol) nanospheres. *Macromolecules* **35**: 3668–3674.

Roskos, K. V., Tefft, J. A. and Heller, J. (1993). A morphine-triggered delivery system useful in the treatment of heroin addiction. *Clin. Mater.* **13**: 109–119.

Schneider, M., Guillot, C., and Lamy, B. (1981). The affinity precipitation technique: application to the isolation and purification of trypsin from bovine pancreas. *Ann. N.Y. Acad. Sci.* **369**: 257–263.

Schneider, R. S., Lidquist, P., Wong, E. T., Rubenstein, K. E., and Ullman, E. F. (1973). Homogeneous enzyme immunoassay for opiates in urine. *Clin. Chem.* **19**: 821–825.

Shimizu, T., Yamato, M., Kikuchi, A., and Okano, T. (2003). Cell sheet engineering for myocardial tissue reconstruction. *Biomaterials* **24**: 2309–2316.

Shimoboji, T., Ding, Z., Stayton, P. S., and Hoffman, A. S. (2001). Mechanistic investigation of smart polymer–protein conjugates. *Bioconj. Chem.* **12**: 314–319.

Shimoboji, T., Ding, Z. L., Stayton, P. S., and Hoffman, A. S. (2002a). Photoswitching of ligand association with a photoresponsive polymer–protein conjugate. *Bioconj. Chem.* **13**: 915–919.

Shimoboji, T., Larenas, E., Fowler, T., Kulkarni, S., Hoffman, A. S., and Stayton, P. S. (2002b). Photoresponsive polymer–enzyme switches. *Proc. Natl. Acad. Sci. USA* **99**: 16592–16596.

Shimoboji, T., Larenas, E., Fowler, T., Hoffman, A. S., and Stayton, P. S. (2003). Temperature-induced switching of enzyme activity with smart polymer–enzyme conjugates. *Bioconj. Chem.* **14**: 517–525.

Shoemaker, S., Hoffman, A. S., and Priest, J. H. (1987). Synthesis of vinyl monomer–enzyme conjugates. *Appl. Biochem. Biotechnol.* **15**: 11.

Stayton, P. S., Hoffman, A. S., Murthy, N., Lackey, C., Cheung, C., Tan, P., Klumb, L. A., Chilkoti, A., Wilbur, F. S., and Press, O. W. (2000). Molecular engineering of proteins and polymers for targeting and intracellular delivery of therapeutics. *J. Contr. Rel.* **65**: 203–220.

Stayton, P. S., Shimoboji, T., Long, C., Chilkoti, A., Chen, G., Harris, J. M., and Hoffman, A. S. (1995). Control of protein–ligand recognition using a stimuli-responsive polymer. *Nature* **378**: 472–474.

Takei, Y. G., Aoki, T., Sanui, K., Ogata, N., Okano, T., and Sakurai, Y. (1993a). Temperature-responsive bioconjugates. 1. Synthesis of temperature-responsive oligomers with reactive end groups and their coupling to biomolecules. *Bioconj. Chem.* **4**: 42–46.

Takei, Y. G., Aoki, T., Sanui, K., Ogata, N., Okano, T., and Sakurai, Y. (1993b). Temperature-responsive bioconjugates. 2. Molecular design for temperature-modulated bioseparations. *Bioconj. Chem.* **4**: 341–346.

Takei, Y. G., Matsukata, M., Aoki, T., Sanui, K., Ogata, N., Kikuchi, A., Sakurai, Y., and Okano, T. (1994a). Temperature-responsive bioconjugates. 3. Antibody-poly(*N*-isopropylacrylamide) conjugates for temperature-modulated precipitations and affinity bioseparations. *Bioconj. Chem.* **5**: 577–582.

Takei, Y. G., Aoki, T., Sanui, K., Ogata, N., Sakurai, Y., and Okano, T. (1994b). Dynamic contact angle measurements of temperature-responsive properties for PNIPAAm grafted surfaces. *Macromolecules* **27**: 6163–6166.

Tanaka, T. (1981). Gels. *Sci. Am.* **244**: 124.

Taniguchi, M., Kobayashi, M., and Fujii, M. (1989). Properties of a reversible soluble–insoluble cellulase and its application to repeated hydrolysis of crystalline cellulose. *Biotechnol. Bioeng.* **34**: 1092–1097.

Taniguchi, M., Hoshino, K., Watanabe, K., Sugai, K., and Fujii, M. (1992). Production of soluble sugar from cellulosic materials by repeated use of a reversibly soluble-autoprecipitating cellulase. *Biotechnol. Bioeng.* **39**: 287–292.

Tirrell, D. (1987). Macromolecular switches for bilayer membranes. *J. Contr. Rel.* **6**: 15–21.

Uenoyama, S., and Hoffman, A. S. (1988). Synthesis and characterization of AAm/NIPAAm grafts on silicone rubber substrates. *Radiat. Phys. Chem.* **32**: 605–608.

Vernon, B., Kim, S. W., and Bae, Y. H. (2000). Thermoreversible copolymer gels for extracellular matrix. *J. Biomed. Mater. Res.* **51**: 69–79.

Wu. X. S., Hoffman, A. S., and Yager, P. (1992). Conjugation of phosphatidylethanolamine to poly(NIPAAm) for potential use in liposomal drug delivery systems. *Polymer* **33**: 4659–4662.

Wu, X. S., Hoffman, A. S., and Yager, P. (1993). Synthesis of and insulin release from erodible polyNIPAAm-phospholipid composites. *J. Intell. Mater. Syst. Struct.* **4**: 202–209.

Yamato, M., and Okano, T. (2001). Cell sheet engineering for regenerative medicine. *Macromol. Chem. Symp.* **14**(2): 21–29.

Yamato, M., Kwon, O. H., Hirose, M., Kikuchi, A., and Okano, T. (2001). Novel patterned cell co-culture utilizing thermally responsive grafted polymer surfaces. *J. Biomed. Mater. Res.* **55**: 137–140.

Yang, H. J., Cole, C. A., Monji, N., and Hoffman, A. S. (1990). Preparation of a thermally phase-separating copolymer with a controlled number of active ester groups per polymer chain. *J. Polymer Sci. A., Polymer Chem.* **28**: 219–226.

Yoshida, R., Uchida, K., Kaneko, Y., Sakai, K., Kikuchi, A., Sakurai, Y., and Okano, T. (1995). Comb-type grafted hydrogels with rapid de-swelling response to temperature changes. *Nature* **374**: 240–242.

Yoshida, T., Aoyagi, T., Kokufuta, E., and Okano, T. (2003). Newly designed hydrogel with both sensitive thermoresponse and biodegradability. *J. Polymer Sci. A: Polymer Chem.* **41**: 779–787.

Zareie, H. M., Bulmus, V., Gunning, P. A., Hoffman, A. S., Piskin, E., and Morris, V. J. (2000). Investigation of a pH- and temperature-sensitive polymer by AFM. *Polymer* **41**: 6723–6727.

Zhong, Z., Dijkstra, P. J., Feijen, J., Kwon, Y.-Mi., Bae, Y. H., and Kim, S. W. (2002). Synthesis and aqueous phase behavior of thermoresponsive biodegradable poly(D, L-3-methyl glycolide)-b-poly(ethylene glycol)-b-poly(D, L-3-methyl glycolide) triblock copolymers. *Macromol. Chem. Phys.* **203**: 1797–1803.

3.2.7 Bioresorbable and bioerodible materials

Joachim Kohn, Sascha Abramson, and Robert Langer

Introduction

Since a degradable implant does not have to be removed surgically once it is no longer needed, degradable polymers are of value in short-term applications that require only the temporary presence of a device. An additional advantage is that the use of degradable implants can circumvent some of the problems related to the long-term safety of permanently implanted devices. A potential concern relating to the use of degradable implants is the toxicity of the implant's degradation products. Since all of the implant's degradation products are released into the body of the patient, the design of a degradable implant requires careful attention to testing for potential toxicity of the degradation products. This section covers basic definitions relating to the process of degradation and/or erosion, the most important types of *synthetic*, degradable polymers available today, a classification of degradable medical implants, and a number of considerations specific for the design and use of degradable medical polymers (shelf life, sterilization, etc.).

Definitions relating to the process of erosion and/or degradation

Currently four different terms (biodegradation, bioerosion, bioabsorption, and bioresorption) are being used to indicate that a given material or device will eventually disappear after having been introduced into a living organism. However, when reviewing the literature, no clear distinctions in the meaning of these four terms are evident. Likewise, the meaning of the prefix "bio" is not well established, leading to the often-interchangeable use of the terms "degradation" and "biodegradation," or "erosion" and "bioerosion." Although efforts have been made to establish generally applicable and widely accepted definitions for all aspects of biomaterials research (Williams, 1987), there is still significant confusion even among experienced researchers in the field as to the correct terminology of various degradation processes.

Generally speaking, the term "degradation" refers to a chemical process resulting in the cleavage of covalent bonds. Hydrolysis is the most common chemical process by which polymers degrade, but degradation can also occur via oxidative and enzymatic mechanisms. In contrast, the term "erosion" refers often to physical changes in size, shape, or mass of a device, which could be the consequence of either degradation or simply dissolution. Thus, it is important to realize that erosion can occur in the absence of degradation, and degradation can occur in the absence of erosion. A sugar cube placed in water erodes, but the sugar does not chemically degrade. Likewise, the embrittlement of plastic when exposed to UV light is due to the degradation of the chemical structure of the polymer and takes place before any physical erosion occurs.

In the context of this section, we follow the usage suggested by the Consensus Conference of the European Society for Biomaterials (Williams, 1987) and refer to "biodegradation" only when we wish to emphasize that a biological agent (enzyme, cell, or microorganism) is causing the chemical degradation of the implanted device. After extensive discussion in the literature, it is now widely believed that the chemical degradation of the polymeric backbone of poly(lactic acid) (PLA) is predominantly controlled by simple hydrolysis and occurs independently of any biological agent (Vert *et al.*, 1991). Consequently, the degradation of PLA to lactic acid should not be described as "biodegradation." In agreement with Heller's suggestion (Heller, 1987), we define a "bioerodible polymer" as a water-insoluble polymer that is converted under physiological conditions into water-soluble material(s) without regard to the specific mechanism involved in the erosion process. "Bioerosion" includes therefore both physical processes (such as dissolution) and chemical processes (such as backbone cleavage). Here the prefix "bio" indicates that the erosion occurs under physiological conditions, as opposed to other erosion processes, caused for example by high temperature, strong acids or bases, UV light, or weather conditions. The terms "bioresorption" and "bioabsorption" are used interchangeably and often imply that the polymer or its degradation products are removed by cellular activity (e.g., phagocytosis) in a biological environment. These terms are somewhat superfluous and have not been clearly defined.

Overview of currently available degradable polymers

From the beginnings of the material sciences, the development of highly stable materials has been a major research challenge. Today, many polymers are available that are virtually nondestructible in biological systems, e.g., Teflon, Kevlar, or poly(ether ether ketone) (PEEK). On the other hand, the development of degradable biomaterials is a relatively new area of research. The variety of available, degradable biomaterials is still too limited to cover a wide enough range of diverse material properties.

Thus, the design and synthesis of new, degradable biomaterials is currently an important research challenge, in particular within the context of tissue engineering where the development of new biomaterials that can provide predetermined and controlled cellular responses is a critically needed component of most practical applications of tissue engineering (James and Kohn, 1996).

Degradable materials must fulfill more stringent requirements in terms of their biocompatibility than nondegradable materials. In addition to the potential problem of toxic contaminants leaching from the implant (residual monomers, stabilizers, polymerization initiators, emulsifiers, sterilization by-products), one must also consider the potential toxicity of the degradation products and subsequent metabolites. The practical consequence of this consideration is that only a limited number of nontoxic, monomeric starting materials have been successfully applied to the preparation of degradable biomaterials.

Over the past decade dozens of hydrolytically unstable polymers have been suggested as degradable biomaterials; however, in most cases no attempts have been made to develop these new materials for specific medical applications. Thus, detailed toxicological studies *in vivo*, investigations of degradation rate and mechanism, and careful evaluations of the physicomechanical properties have so far been published for only a very small fraction of those polymers. An even smaller number of synthetic, degradable polymers have so far been used in medical implants and devices that gained approval by the Food and Drug Administration (FDA) for use in patients. Note that the FDA does not approve polymers or materials per se, but only specific devices or implants. As of 1999, only five distinct synthetic, degradable polymers have been approved for use in a narrow range of clinical applications. These polymers are PLA, PGA, polydioxanone (PDS), polycaprolactone (PCL), and a poly(PCPP-SA anhydride) (see later discussion). A variety of other synthetic, degradable biomaterials currently in clinical use are blends or copolymers of these base materials such as a wide range of copolymers of lactic and glycolic acid. Note that this listing does not include polymers derived from animal sources such as collagen, gelatin, or hyaloronic acid.

Recent research has led to a number of well-established investigational polymers that may find practical applications as degradable implants within the next decade. It is beyond the scope of this section to fully introduce all of the polymers and their applications under investigation, thus only representative examples of these polymers are described here. This section will concern itself mostly with *synthetic* degradable polymers, as *natural* polymers (e.g., polymers derived from animal or plant sources) are described elsewhere. Table 3.2.7-1 provides an overview of some representative degradable polymers.

Table 3.2.7-1 Degradable polymers and representative applications under investigation

Degradable polymer	Current major research applications
Synthetic degradable polyesters	
Poly(glycolic acid), poly(lactic acid), and copolymers	Barrier membranes, drug delivery, guided tissue regeneration (in dental applications), orthopedic applications, stents, staples, sutures, tissue engineering
Polyhydroxybutyrate (PHB), polyhydroxyvalerate (PHV), and copolymers thereof	Long-term drug delivery, orthopedic applications, stents, sutures
Polycaprolactone	Long-term drug delivery, orthopedic applications, staples, stents
Polydioxanone	Fracture fixation in non-load-bearing bones, sutures, wound clip
Other synthetic degradable polymers	
Polyanhydrides	Drug delivery
Polycyanoacrylates	Adhesives, drug delivery
Poly(amino acids) and "pseudo"-Poly(amino acids)	Drug delivery, tissue engineering, orthopedic applications
Poly(ortho ester)	Drug delivery, stents
Polyphosphazenes	Blood contacting devices, drug delivery, skeletal reconstruction
Poly(propylene fumarate)	Orthopedic applications
Some natural resorbable polymers	
Collagen	Artificial skin, coatings to improve cellular adhesion, drug delivery, guided tissue regeneration in dental applications, orthopedic applications, soft tissue augmentation, tissue engineering, scaffold for reconstruction of blood vessels, wound closure
Fibrinogen and fibrin	Tissue sealant
Gelatin	Capsule coating for oral drug delivery, hemorrhage arrester
Cellulose	Adhesion barrier, hemostat
Various polysaccharides such as chitosan, alginate	Drug delivery, encapsulation of cells, sutures, wound dressings
Starch and amylose	Drug delivery

For completeness, some of the natural polymers have also been included here. Structural formulas of commonly investigated synthetic degradable polymers are provided in Fig. 3.2.7-1. It is an interesting observation that a large proportion of the currently investigated, *synthetic*, degradable polymers are polyesters. It remains to be seen whether some of the alternative backbone structures such as polyanhydrides, polyphosphazenes, polyphosphonates, polyamides, or polycarbonates will be able to challenge the predominant position of the polyesters in the future.

PDS is a poly(ether ester) made by a ring-opening polymerization of *p*-dioxanone monomer. PDS has gained increasing interest in the medical field and pharmaceutical field due to its degradation to low-toxicity monomers *in vivo*. PDS has a lower modulus than PLA or PGA. It became the first degradable polymer to be used to make a monofilament suture. PDS has also been

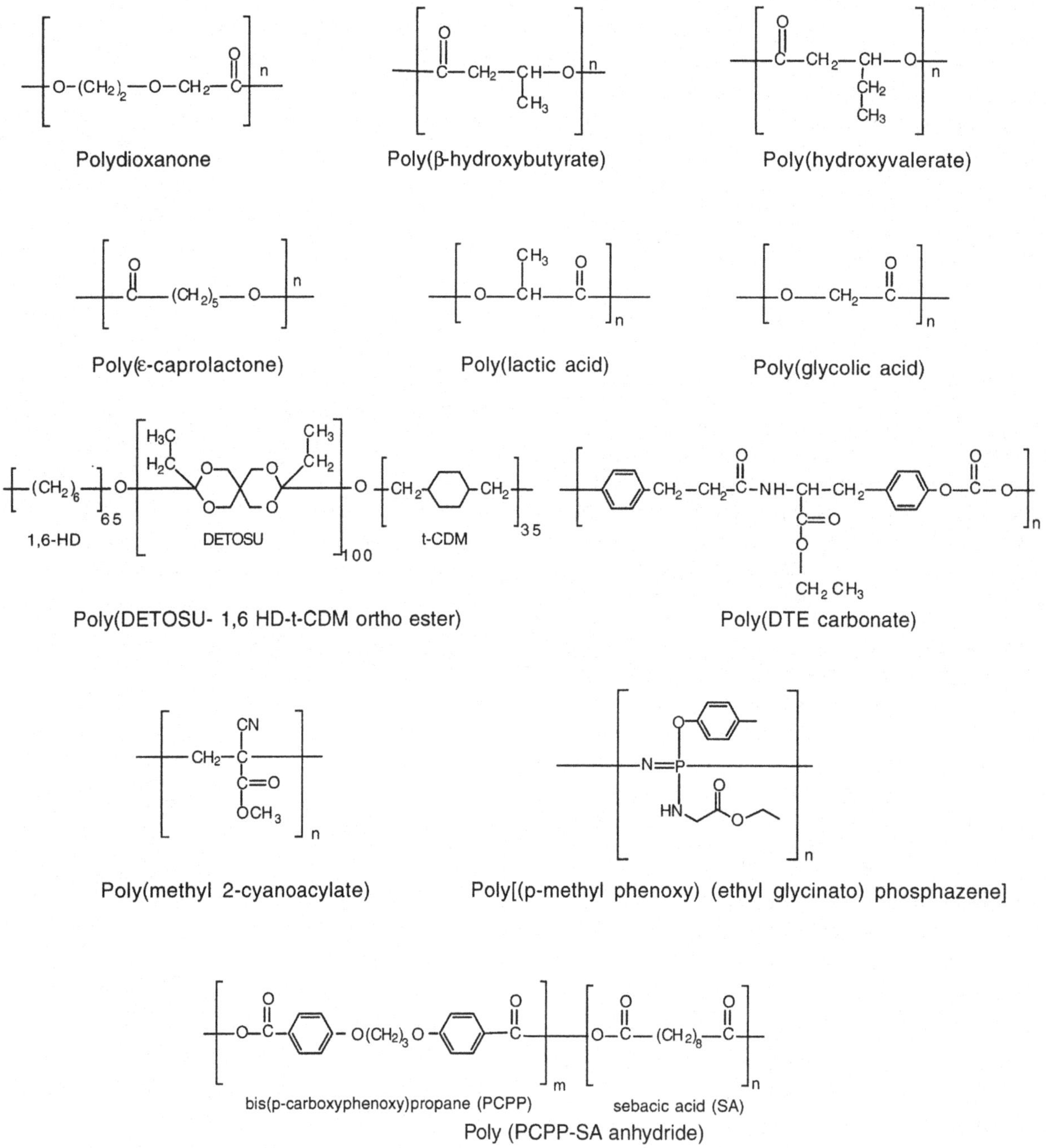

Fig. 3.2.7-1 Chemical structures of widely investigated degradable polymers.

introduced to the market as a suture clip as well as a bone pin marketed under the name OrthoSorb in the USA and Ethipin in Europe.

Poly(hydroxybutyrate) (PHB), poly(hydroxyvalerate) (PHV), and their copolymers represent examples of bioresorbable polyesters that are derived from microorganisms. Although this class of polymers are examples of *natural* materials (as opposed to *synthetic* materials), they are included here because they have similar properties and similar areas of application as the widely investigated PLA. PHB and its copolymers with up to 30% of 3-hydroxyvaleric acid are now commercially available under the trade name "Biopol" (Miller and Williams, 1987). PHB and PHV are intracellular storage polymers providing a reserve of carbon and energy to microorganisms similar to the role of starch in plant metabolism. The polymers can be degraded by soil bacteria (Senior *et al.*, 1972) but are relatively stable under physiological conditions (pH 7, 37°C). Within a relatively narrow window, the rate of degradation can be modified slightly by varying the copolymer composition; however, all members of this family of polymers require several years for complete resorption *in vivo. In vivo,* PHB degrades to D-3-hydroxybutyric acid, which is a normal constituent of human blood (Miller and Williams, 1987). The low toxicity of PHB may at least in part be due to this fact.

PHB homopolymer is very crystalline and brittle, whereas the copolymers of PHB with hydroxyvaleric acid are less crystalline, more flexible, and more readily processible (Barham *et al.*, 1984). The polymers have been considered in several biomedical applications such as controlled drug release, sutures, artificial skin, and vascular grafts, as well as industrial applications such as medical disposables. PHB is especially attractive for orthopedic applications because of its slow degradation time. The polymer typically retained 80% of its original stiffness over 500 days on *in vivo* degradation (Knowles, 1993).

PCL became available commercially following efforts at Union Carbide to identify synthetic polymers that could be degraded by microorganisms (Huang, 1985). It is a semicrystalline polymer. The high solubility of PCL, its low melting point (59–64°C), and its exceptional ability to form blends has stimulated research on its application as a biomaterial. PCL degrades at a slower pace than PLA and can therefore be used in drug delivery devices that remain active for over 1 year. The release characteristics of PCL have been investigated in detail by Pitt and his co-workers (Pitt *et al.*, 1979). The Capronor system, a 1-year implantable contraceptive device (Pitt, 1990), has become commercially available in Europe and the United States. The toxicology of PCL has been extensively studied as part of the evaluation of Capronor. Based on a large number of tests, ε-caprolactone and PCL are currently regarded as nontoxic and tissue-compatible materials. PCL is currently being researched as part of wound dressings, and in Europe, it is already in clinical use as a degradable staple (for wound closure).

Polyanhydrides were explored as possible substitutes for polyesters in textile applications but failed ultimately because of their pronounced hydrolytic instability. It was this property that prompted Langer and his co-workers to explore polyanhydrides as degradable implant materials (Tamada and Langer, 1993). Aliphatic polyanhydrides degrade within days, whereas some aromatic polyanhydrides degrade over several years. Thus aliphatic-aromatic copolymers are usually employed which show intermediate rates of degradation depending on the monomer composition.

Polyanhydrides are among the most reactive and hydrolytically unstable polymers currently used as biomaterials. The high chemical reactivity is both an advantage and a limitation of polyanhydrides. Because of their high rate of degradation, many polyanhydrides degrade by surface erosion without the need to incorporate various catalysts or excipients into the device formulation. On the other hand, polyanhydrides will react with drugs containing free amino groups or other nucleophilic functional groups, especially during high-temperature processing (Leong *et al.*, 1986). The potential reactivity of the polymer matrix toward nucleophiles limits the type of drugs that can be successfully incorporated into a polyanhydride matrix by melt processing techniques. Along the same line of reasoning, it has been questioned whether amine-containing biomolecules present in the interstitial fluid around an implant could react with anhydride bonds present at the implant surface.

A comprehensive evaluation of the toxicity of the polyanhydrides showed that, in general, the polyanhydrides possess excellent *in vivo* biocompatibility (Attawia *et al.*, 1995). The most immediate applications for polyanhydrides are in the field of drug delivery. Drug-loaded devices made of polyanhydrides can be prepared by compression molding or microencapsulation (Chasin *et al.*, 1990). A wide variety of drugs and proteins including insulin, bovine growth factors, angiogenesis inhibitors (e.g., heparin and cortisone), enzymes (e.g., alkaline phosphatase and β-galactosidase), and anesthetics have been incorporated into polyanhydride matrices, and their *in vitro* and *in vivo* release characteristics have been evaluated (Park *et al.*, 1998; Chasin *et al.*, 1990). Additionally, polyanhydrides have been investigated for use as nonviral vectors of delivering DNA in gene therapy (Shea and Mooney, 2001). The first polyanhydride-based drug delivery system to enter clinical use is for the delivery of chemotherapeutic agents. An example of this application is the delivery of bischloroethylnitrosourea (BCNU) to the brain for the treatment of glioblastoma multiformae, a universally fatal brain cancer (Madrid *et al.*, 1991). For this application, BCNU-loaded implants made of the polyanhydride derived from bis-*p*-carboxyphenoxypropane and sebacic acid received FDA regulatory approval in the fall of 1996 and are currently being marketed under the name Gliadel.

Poly(ortho esters) are a family of synthetic, degradable polymers that have been under development for a number of years (Heller *et al.*, 1990). Devices made of poly(ortho esters) can erode by "surface erosion" if appropriate excipients are incorporated into the polymeric matrix. Since surface eroding, slab-like devices tend to release drugs embedded within the polymer at a constant rate, poly(ortho esters) appear to be particularly useful for controlled-release drug delivery applications. For example, poly(ortho esters) have been used for the controlled delivery of cyclobenzaprine and steroids and a significant number of publications describe the use of poly(ortho esters) for various drug delivery applications (Heller, 1993). Poly(ortho esters) have also been investigated for the treatment of postsurgical pain, ostearthritis, and ophthalmic diseases (Heller *et al.*, 2002). Since the ortho ester linkage is far more stable in base than in acid, Heller and his co-workers controlled the rate of polymer degradation by incorporating acidic or basic excipients into the polymer matrix. One concern about the "surface erodability" of poly(ortho esters) is that the incorporation of highly water-soluble drugs into the polymeric matrix can result in swelling of the polymer matrix. The increased amount of water imbibed into the matrix can then cause the polymeric device to exhibit "bulk erosion" instead of "surface erosion" (see below for a more detailed explanation of these erosion mechanisms) (Okada and Toguchi, 1995).

By now, there are three major types of poly(ortho esters). First, Choi and Heller prepared the polymers by the trans-esterification of 2,2′-dimethoxyfuran with a diol. The next generation of poly (ortho esters) was based on an acid-catalyzed addition reaction of diols with diketeneacetals (Heller *et al.*, 1980). The properties of the polymers can be controlled to a large extent by the choice of the diols used in the synthesis. Recently, a third generation of poly(ortho esters) have been prepared. These materials are very soft and can even be viscous liquids at room temperature. Third-generation poly (ortho esters) can be used in the formulation of drug delivery systems that can be injected rather than implanted into the body.

Poly(amino acids) and "Pseudo"-Poly(amino acids) Since proteins are composed of amino acids, it is an obvious idea to explore the possible use of poly(amino acids) in biomedical applications (Anderson *et al.*, 1985). Poly(amino acids) were regarded as promising candidates since the amino acid side chains offer sites for the attachment of drugs, cross-linking agents, or pendent groups that can be used to modify the physicomechanical properties of the polymer. In addition, poly(amino acids) usually show a low level of systemic toxicity, due to their degradation to naturally occurring amino acids.

Early investigations of poly (amino acids) focused on their use as suture materials (Miyamae *et al.*, 1968), as artificial skin substitutes (Spira *et al.*, 1969), and as drug delivery systems (McCormick-Thomson and Duncan, 1989). Various drugs have been attached to the side chains of poly(amino acids), usually via a spacer unit that distances the drug from the backbone. Poly(amino acid)–drug combinations investigated include poly(L-lysine) with methotrexate and pepstatin (Campbell *et al.*, 1980), and poly(glutamic acid) with adriamycin, a widely used chemotherapeutic agent (van Heeswijk *et al.*, 1985).

Despite their apparent potential as biomaterials, poly(amino acids) have actually found few practical applications. Most poly(amino acids) are highly insoluble and nonprocessible materials. Since poly(amino acids) have a pronounced tendency to swell in aqueous media, it can be difficult to predict drug release rates. Furthermore, the antigenicity of polymers containing three or more amino acids limits their use in biomedical applications (Anderson *et al.*, 1985). Because of these difficulties, only a few poly(amino acids), usually derivatives of poly(glutamic acid) carrying various pendent chains at the γ-carboxylic acid group, have been investigated as implant materials (Lescure *et al.*, 1989). So far, no implantable devices made of a poly(amino acid) have been approved for clinical use in the United States.

In an attempt to circumvent the problems associated with conventional poly(amino acids), backbone-modified "pseudo"-poly(amino acids) were introduced in 1984 (Kohn and Langer, 1984, 1987). The first "pseudo"-poly(amino acids) investigated were a polyester from N-protected *trans*-4-hydroxy-L-proline, and a polyiminocarbonate derived from tyrosine dipeptide. The tyrosine-derived "pseudo"-poly(amino acids) are easily processed by solvent or heat methods and exhibit a high degree of biocompatibility. Recent studies indicate that the backbone modification of poly(amino acids) may be a generally applicable approach for the improvement of the physicomechanical properties of conventional poly(amino acids). For example, tyrosine-derived polycarbonates (Nathan and Kohn, 1996) are high-strength materials that may be useful in the formulation of degradable orthopedic implants. One of the tyrosine-derived pseudo-poly(amino acids), poly(DTE carbonate) exhibits a high degree of bone conductivity (e.g., bone tissue will grow directly along the polymeric implant) (Choueka *et al.*, 1996; James and Kohn, 1997).

The reason for the improved physicomechanical properties of "pseudo"-poly(amino acids) relative to conventional poly(amino acids) can be traced to the reduction in the number of interchain hydrogen bonds: In conventional poly(amino acids), individual amino acids are polymerized via repeated amide bonds leading to strong interchain hydrogen bonding. In natural peptides, hydrogen bonding is one of the interactions leading to the spontaneous formation of secondary structures such as α-helices or β-pleated sheets. Strong hydrogen bonding also results in high processing temperatures and low solubility in organic solvents which tends to lead to intractable polymers with limited applications. In "pseudo"-poly(amino acids), half on the amide bonds are replaced by other linkages (such as carbonate, ester, or iminocarbonate bonds) that have a much lower tendency to form interchain hydrogen bonds, leading to better processibility and, generally, a loss of crystallinity.

Polycyanoacrylates are used as bioadhesives. Methyl cyanoacrylates are more commonly used as general-purpose glues and are commercially available as "Crazy Glue." Methyl cyanoacrylate was used during the Vietnam war as an emergency tissue adhesive, but is no longer used today. Butyl cyanoacrylate is approved in Canada and Europe as a dental adhesive. Cyanoacrylates undergo spontaneous polymerization at room temperature in the presence of water, and their toxicity and erosion rate after polymerization differ with the length of their alkyl chains (Gombotz and Pettit, 1995). All polycyanoacrylates have several limiting properties: First, the monomers (cyanoacrylates) are very reactive compounds that often have significant toxicity. Second, upon degradation polycyanoacrylates release formaldehyde resulting in intense inflammation in the surrounding tissue. In spite of these inherent limitations, polycyanoacrylates have been investigated as potential drug delivery matrices and have been suggested for use in ocular drug delivery (Deshpande *et al.*, 1998).

Polyphosphazenes are very unusual polymers, whose backbone consists of nitrogen-phosphorus bonds. These polymers are at the interface between inorganic and organic polymers and have unusual material properties. Polyphosphazenes have found industrial applications, mainly because of their high thermal stability. They have also been used in investigations for the formulation of controlled drug delivery systems (Allcock, 1990). Polyphosphazenes are interesting biomaterials, in many respects. They have been claimed to be biocompatible and their chemical structure provides a readily accessible "pendant chain" to which various drugs, peptides, or other biological compounds can be attached and later released via hydolysis. Polyphosphazenes have been examined for use in skeletal tissue regeneration (Laurencin *et al.*, 1993). Another novel use of polyphosphazenes is in the area of vaccine design where these materials were used as immunological adjuvants (Andrianov *et al.*, 1998).

PGA and PLA and their copolymers are currently the most widely investigated, and most commonly used synthetic, bioerodible polymers. In view of their importance in the field of biomaterials, their properties and applications will be described in more detail.

PGA is the simplest linear, aliphatic polyester (Fig. 3.2.7-1). Since PGA is highly crystalline, it has a high melting point and low solubility in organic solvents. PGA was used in the development of the first totally synthetic, absorbable suture. PGA sutures have been commercially available under the trade name "Dexon" since 1970. A practical limitation of Dexon sutures is that they tend to lose their mechanical strength rapidly, typically over a period of 2 to 4 weeks after implantation. PGA has also been used in the design of internal bone fixation devices (bone pins). These pins have become commercially available under the trade name "Biofix."

In order to adapt the materials properties of PGA to a wider range of possible applications, copolymers of PGA with the more hydrophobic PLA were intensively investigated (Gilding and Reed, 1979, 1981). The hydrophobicity of PLA limits the water uptake of thin films to about 2% and reduces the rate of backbone hydrolysis as compared to PGA. Copolymers of glycolic acid and lactic acid have been developed as alternative sutures (trade names "Vicryl" and "Polyglactin 910").

It is noteworthy that there is no linear relationship between the ratio of glycolic acid to lactic acid and the physicomechanical properties of the corresponding copolymers. Whereas PGA is highly crystalline, crystallinity is rapidly lost in copolymers of glycolic acid and lactic acid. These morphological changes lead to an increase in the rates of hydration and hydrolysis. Thus, 50:50 copolymers degrade more rapidly than either PGA or PLA.

Since lactic acid is a chiral molecule, it exists in two steroisomeric forms that give rise to four morphologically distinct polymers: the two stereoregular polymers, D-PLA and L-PLA, and the racemic form D, L-PLA. A fourth morphological form, *meso*-PLA can be obtained from D, L-lactide but is rarely used in practice.

The polymers derived from the optically active D and L monomers are semicrystalline materials, while the optically inactive D, L-PLA is always amorphous. Generally, L-PLA is more frequently employed than D-PLA, since the hydrolysis of L-PLA yields L(+)-lactic acid, which is the naturally occurring stereoisomer of lactic acid.

The differences in the crystallinity of D, L-PLA and L-PLA have important practical ramifications: Since D, L-PLA is an amorphous polymer, it is usually considered for applications such as drug delivery, where it is important to have a homogeneous dispersion of the active species within the carrier matrix. On the other hand, the semicrystalline L-PLA is preferred in applications where high mechanical strength and toughness are required, such as sutures and orthopedic devices.

PLA and PGA and their copolymers have been investigated for more applications than any other degradable polymer. The high interest in these materials is based, not on their superior materials properties, but mostly on the fact that these polymers have already been used successfully in a number of approved medical implants and are considered safe, nontoxic, and bio-compatible by regulatory agencies in virtually all developed countries. Therefore, implantable devices prepared from PLA, PGA, or their copolymers can be brought to market in less time and for a lower cost than similar devices prepared from novel polymers whose biocompatibility is still unproven.

Currently available and approved products include sutures, GTR membranes for dentistry, bone pins, and implantable drug delivery systems. The polymers are also being widely investigated in the design of vascular and urological stents and skin substitutes, and as scaffolds for tissue engineering and tissue reconstruction. In many of these applications, PLA, PGA, and their copolymers have performed with moderate to high degrees of success. However, there are still unresolved issues: First, in tissue culture experiments, most cells do not attach to PLA or PGA surfaces and do not grow as vigorously as on the surface of other materials, indicating that these polymers are actually poor substrates for cell growth *in vitro*. The significance of this finding for the use of PLA and PGA as tissue engineering scaffolds *in vivo* is currently a topic of debate. Second, the degradation products of PLA and PGA are relatively strong acids (lactic acid and glycolic acid). When these degradation products accumulate at the implant site, a delayed inflammatory response is often observed months to years after implantation (Bergsma *et al.*, 1995; Athanasiou *et al.*, 1998; Törmälä *et al.*, 1998).

Table 3.2.7-2 Some "short-term" medical applications of degradable polymeric biomaterials

Application	Comment
Sutures	The earliest, successful application of synthetic degradable polymers in human medicine.
Drug delivery devices	One of the most widely investigated medical applications for degradable polymers.
Orthopedic fixation devices	Requires polymers of exceptionally high mechanical strength and stiffness.
Adhesion prevention	Requires polymers that can form soft membranes or films.
Temporary vascular grafts and stents made of degradable polymers	Only investigational devices are presently available. Blood compatibility is a major concern.
Tissue engineering or guided tissue regeneration scaffold	Attempts to recreate or improve native tissue function using degradable scaffolds

Applications of synthetic, degradable polymers as biomaterials

Classification of degradable medical implants

Some typical short-term applications of biodegradable polymers are listed in Table 3.2.7-2. From a practical perspective, it is convenient to distinguish between five main types of degradable implants: the temporary support device, the temporary barrier, the drug delivery device, the tissue engineering scaffold, and the multifunctional implant.

A *temporary support device* is used in those circumstances in which the natural tissue bed has been weakened by disease, injury, or surgery and requires some artificial support. A healing wound, a broken bone, or a damaged blood vessel are examples of such situations. Sutures, bone fixation devices (e.g., bone nails, screws, or plates), and vascular grafts would be examples of the corresponding support devices. In all of these instances, the degradable implant would provide temporary, mechanical support until the natural tissue heals and regains its strength. In order for a temporary support device to work properly, a gradual stress transfer should occur: As the natural tissue heals, the degradable implant should gradually weaken. The need to adjust the degradation rate of the temporary support device to the healing of the surrounding tissue represents one of the major challenges in the design of such devices.

Currently, sutures represent the most successful example of a temporary support device. The first synthetic, degradable sutures were made of PGA and became available under the trade name "Dexon" in 1970. This represented the first routine use of a degradable polymer in a major clinical application (Frazza and Schmitt, 1971). Later copolymers of PGA and PLA were developed. The widely used "Vicryl" suture, for example, is a 90:10 copolymer of PGA/PLA, introduced into the market in 1974. Sutures made of PDS became available in the United States in 1981. In spite of extensive research efforts in many laboratories, no other degradable polymers are currently used to any significant extent in the formulation of degradable sutures.

A ***temporary barrier*** has its major medical use in adhesion prevention. Adhesions are formed between two tissue sections by clotting of blood in the extravascular tissue space followed by inflammation and fibrosis. If this natural healing process occurs between surfaces that were not meant to bond together, the resulting adhesion can cause pain, functional impairment, and problems during subsequent surgery. Surgical adhesions are a significant cause of morbidity and represent one of the most significant complications of a wide range of surgical procedures such as cardiac, spinal, and tendon surgery. A temporary barrier could take the form of a thin polymeric film or a meshlike device that would be placed between adhesion-prone tissues at the time of surgery. To be useful, such as temporary barrier would have to prevent the formation of scar tissue connecting adjacent tissue sections, followed by the slow resorption of the barrier material (Hill *et al.*, 1993). This sort of barrier has also been investigated for the sealing of breaches of the lung tissue that cause air leakage.

Another important example of a temporary barrier is in the field of skin reconstruction. Several products are available that are generally referred to as "artificial skin" (Beele, 2002). The first such product consists of an artificial, degradable collagen/glycosaminoglycan matrix that is placed on top of the skin lesion to stimulate the regrowth of a functional dermis. Another product consists of a degradable collagen matrix with preseeded human fibroblasts. Again, the goal is to stimulate the regrowth of a functional dermis. These products are used in the treatment of burns and other deep skin lesions and represent an important application for temporary barrier type devices.

An implantable drug delivery device is by necessity a temporary device, as the device will eventually run out of drug or the need for the delivery of a specific drug is eliminated once the disease is treated. The development of implantable drug delivery systems is probably the most widely investigated application of degradable polymers (Langer, 1990). One can expect that the future acceptance of implantable drug delivery devices by physicians and patients alike will depend on the availability of degradable systems that do not have to be explanted surgically.

Since PLA and PGA have an extensive safety profile based on their use as sutures, these polymers have been very widely investigated in the formulation of implantable controlled release devices. Several implantable, controlled release formulations based on copolymers of lactic and glycolic acid have already become available. However, a very wide range of other degradable polymers have been investigated as well. Particularly noteworthy is the use of a type of polyanhydride in the formulation of an intracranial, implantable device for the administration of BCNU (a chemotherapeutic agent) to patients suffering from glioblastoma multiformae, a usually lethal form of brain cancer (Chasin *et al.*, 1990).

The term ***tissue engineering scaffold*** will be used in this section to describe a degradable implant that is designed to act as an artificial extracellular matrix (ECM) by providing space for cells to grow into and to reorganize into functional tissue (James and Kohn, 1996).

It has become increasingly obvious that manmade implantable prostheses do not function as well as the native tissue or maintain the functionality of native tissue over long periods of time. Therefore, tissue engineering has emerged as an interdisciplinary field that utilizes degradable polymers, among other substrates and biologics, to develop treatments that will allow the body to heal itself without the need for permanently implanted, artificial prosthetic devices. In the ideal case, a tissue engineering scaffold is implanted to restore lost tissue function, maintain tissue function, or enhance existing tissue function (Langer and Vacanti, 1993). These scaffolds can take the form of a feltlike material obtained from knitted or woven fibers or from fiber meshes. Alternatively, various processing techniques can be used to obtain foams or sponges. For all tissue engineering scaffolds, pore interconnectivity is a key property, as cells need to be able to migrate and grow throughout the entire scaffold. Thus, industrial foaming techniques, used for example in the fabrication of furniture cushions, are not applicable to the fabrication of tissue engineering scaffolds, as these industrial foams are designed contain "closed pores," whereas tissue engineering scaffolds require an "open pore" structure. Tissue engineering scaffolds may be preseeded with cells *in vitro* prior to implantation. Alternatively, tissue engineering scaffolds may consist of a cell-free structure that is invaded and "colonized" by cells only after its implantation. In either case, the tissue engineering scaffold must allow the formation of functional tissue *in vivo*, followed by the safe resorption of the scaffold material.

There has been some debate in the literature as to the exact definition of the related term "guided tissue regeneration" (GTR). GTR is a term traditionally used in dentistry. This term sometimes implies that the scaffold encourages the growth of specific types of tissue. For example, in the treatment of periodontal disease, periodontists use the term "guided tissue regeneration" when using implants that favor new bone growth in the periodontal pocket over soft-tissue ingrowth (scar formation).

One of the major challenges in the design of tissue engineering scaffolds is the need to adjust the rate of scaffold degradation to the rate of tissue healing. Depending upon the application the scaffold, the polymer may need to function on the order of days to months. Scaffolds intended for the reconstruction of bone illustrate this point: In most applications, the scaffold must maintain some mechanical strength to support the bone structure while new bone is formed. Premature degradation of the scaffold material can be as detrimental to the healing process as a scaffold that remains intact for excessive periods of time. The future use of tissue engineering scaffolds has the potential to revolutionize the way aging-, trauma-, and disease-related loss of tissue function can be treated.

Multifunctional devices, as the name implies, combine several of the functions just mentioned within one single device. Over the past few years, there has been a trend toward increasingly sophisticated applications for degradable biomaterials. Usually, these applications envision the combination of several functions within the same device and require the design of custom-made materials with a narrow range of predetermined materials properties. For example, the availability of biodegradable bone nails and bone screws made of ultrahigh-strength poly (lactic acid) opens the possibility of combining the "mechanical support" function of the device with a "site-specific drug delivery" function: a biodegradable bone nail that holds the fractured bone in place can simultaneously stimulate the growth of new bone tissue at the fracture site by slowly releasing bone growth factors (e.g., bone morphogenic protein (BMP) or transforming growth factor β) throughout its degradation process.

Likewise, biodegradable stents for implantation into coronary arteries are currently being investigated (Agrawal *et al.*, 1992). The stents are designed to mechanically prevent the collapse and restenosis (reblocking) of arteries that have been opened by balloon angioplasty. Ultimately, the stents could deliver an antiinflammatory or antithrombogenic agent directly to the site of vascular injury. Again, it would potentially be possible to combine a mechanical support function with site specific drug delivery.

Various functional combinations involve the tissue engineering scaffold. Perhaps the most important multifunctional device for future applications is a tissue engineering scaffold that also serves as a drug delivery system for cytokines, growth hormones, or other agents that directly affect cells and tissue within the vicinity of the implanted scaffold. An excellent example for this concept is a bone regeneration scaffold that is placed within a bone defect to allow the regeneration of bone while releasing BMP at the implant site. The release of BMP has been reported to stimulate bone growth and therefore has the potential to accelerate the healing rate. This is particularly important in older patients whose natural ability to regenerate tissues may have declined.

The process of bioerosion

One of the most important prerequisites for the successful use of a degradable polymer for any medical application is a thorough understanding of the way the device will degrade/erode and ultimately resorb from the implant site. Within the context of this section, we are limiting our discussion to the case of a solid, polymeric implant. The transformation of such an implant into water-soluble material(s) is best described by the term "bioerosion." The bioerosion process of a solid, polymeric implant is associated with macroscopic changes in the appearance of the device, changes in its physicomechanical properties and in physical processes such as swelling, deformation, or structural disintegration, weight loss, and the eventual depletion of drug or loss of function.

All of these phenomena represent distinct and often independent aspects of the complex bioerosion behavior of a specific polymeric device. It is important to note that the bioerosion of a solid device is not necessarily due to the chemical cleavage of the polymer backbone or the chemical cleavage of cross-links or side chains. Rather, simple solubilization of the intact polymer, for instance, due to changes in pH, may also lead to the erosion of a solid device.

Two distinct modes of bioerosion have been described in the literature. In "bulk erosion," the rate of water penetration into the solid device exceeds the rate at which the polymer is transformed into water-soluble material(s). Consequently, the uptake of water is followed by an erosion process that occurs throughout the entire volume of the solid device. Because of the rapid penetration of water into the matrix of hydrophilic polymers, most of the currently available polymers will give rise to bulk eroding devices. In a typical "bulk erosion" process, cracks and crevices will form throughout the device that may rapidly crumble into pieces. A good illustration for a typical bulk erosion process is the disintegration of an aspirin tablet that has been placed into water. Depending on the specific application, the often uncontrollable tendency of bulk eroding devices to crumble into little pieces can be a disadvantage.

Alternatively, in "surface erosion," the bioerosion process is limited to the surface of the device. Therefore, the device will become thinner with time, while maintaining its structural integrity throughout much of the erosion process. In order to observe surface erosion, the polymer must be hydrophobic to impede the rapid imbibition of water into the interior of the device. In addition, the rate at which the polymer is transformed into water-soluble material(s) has to be fast relative to the rate of water penetration into the device. Under these conditions, scanning electron microscopic evaluation of surface eroding devices has sometimes shown a sharp border between the eroding surface layer and the intact polymer in the core of the device (Mathiowitz *et al.*, 1990).

Surface eroding devices have so far been obtained only from a small number of hydrophobic polymers containing hydrolytically highly reactive linkages in the backbone. A possible exception to this general rule is enzymatic surface erosion. The inability of enzymes to penetrate into the interior of a solid, polymeric device may result in an enzyme-mediated surface erosion mechanism. Currently, polyanhydrides and poly(ortho esters) are the best known examples of polymers that can be fabricated into surface eroding devices.

Mechanisms of chemical degradation

Although bioerosion can be caused by the solubilization of an intact polymer, chemical degradation of the polymer is usually the underlying cause for the bioerosion of a solid, polymeric device. Several distinct types of chemical degradation mechanisms have been identified (Fig. 3.2.7-2) (Rosen *et al.*, 1988). Chemical reactions can lead to cleavage of crosslinks between water-soluble polymer chains (mechanism I), to the cleavage of polymer side chains resulting in the formation of polar or charged groups (mechanism II), or to the cleavage of the polymer backbone (mechanism III). Obviously, combinations of these mechanisms are possible: for instance, a cross-linked polymer may first be partially solubilized by the cleavage of crosslinks (mechanism I), followed by the cleavage of the backbone itself (mechanism III). It should be noted that water is key to all of these degradation schemes. Even enzymatic degradation occurs in aqueous environment.

Since the chemical cleavage reactions described above can be mediated by water or by biological agents such as enzymes and microorganisms, it is possible to distinguish between hydrolytic degradation and biodegradation, respectively. It has often been stated that the availability of water is virtually constant in all soft tissues and varies little

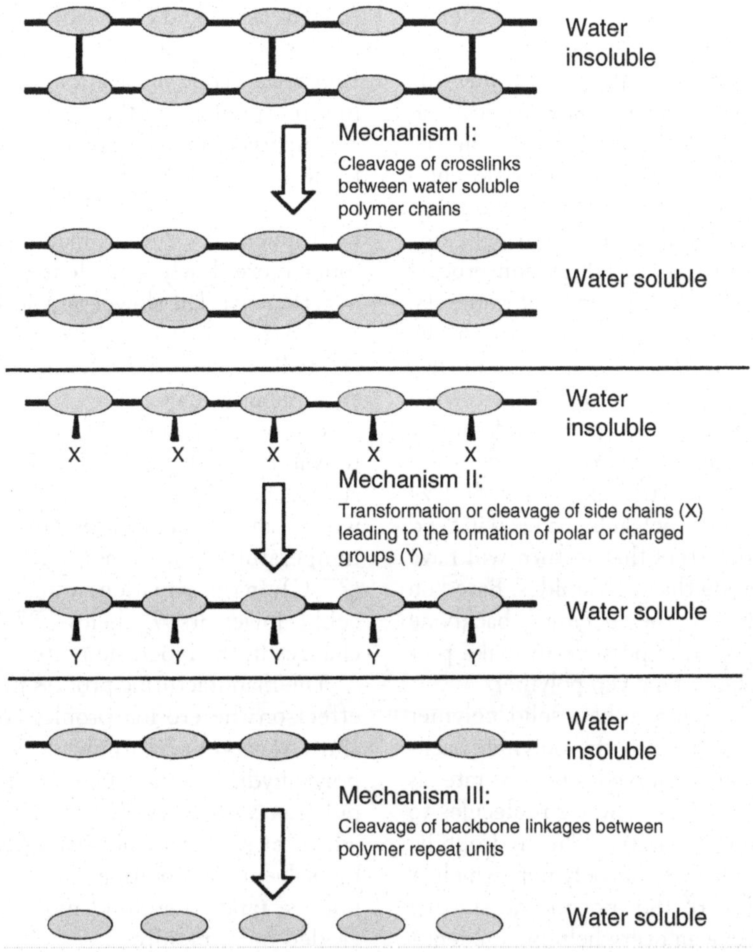

Fig. 3.2.7-2 Mechanisms of chemical degradation. Mechanism I involves the cleavage of degradable cross-links between water-soluble polymer chains. Mechanism II involves the cleavage or chemical transformation of polymer side chains, resulting in the formation of charged or polar groups. The presence of charged or polar groups leads then to the solubilization of the intact polymer chain. Mechanism III involves the cleavage of unstable linkages in the polymer backbone, followed by solubilization of the low-molecular-weight fragments.

from patient to patient. On the other hand, the levels of enzymatic activity may vary widely not only from patient to patient but also between different tissue sites in the same patient. Thus polymers that undergo hydrolytic cleavage tend to have more predictable *in vivo* erosion rates than polymers whose degradation is mediated predominantly by enzymes. The latter polymers tend to be generally less useful as degradable medical implants.

Factors that influence the rate of bioerosion

Although the solubilization of intact polymer as well as several distinct mechanisms of chemical degradation have been recognized as possible causes for the observed bioerosion of a solid, polymeric implant, virtually all currently available implant materials erode because of the hydrolytic cleavage of the polymer backbone (mechanism III in Fig. 3.2.7-2). We therefore limit the following discussion to solid devices that bioerode because of the hydrolytic cleavage of the polymer backbone.

In this case, the main factors that determine the overall rate of the erosion process are the chemical stability of the hydrolytically susceptible groups in the polymer backbone, the hydrophilic/hydrophobic character of the repeat units, the morphology of the polymer, the initial molecular weight and molecular weight distribution of the polymer, the device fabrication process used to prepare the device, the presence of catalysts, additives, or plasticizers, and the geometry (specifically the surface area to volume ratio) of the implanted device.

The susceptibility of the polymer backbone toward hydrolytic cleavage is probably the most fundamental parameter. Generally speaking, anhydrides tend to hydrolyze faster than ester bonds that in turn hydrolyze faster than amide bonds. Thus, polyanhydrides will tend to degrade faster than polyesters that in turn will have a higher tendency to bioerode than polyamides. Based on the known susceptibility of the polymer backbone structure toward hydrolysis, it is possible to make predictions about the bioerosion of a given polymer.

However, the actual erosion rate of a solid polymer cannot be predicted on the basis of the polymer backbone structure alone. The observed erosion rate is strongly dependent on the ability of water molecules to penetrate into the polymeric matrix. The hydrophilic versus hydrophobic character of the polymer, which is a function of the structure of the monomeric starting materials, can therefore have an overwhelming influence on the observed bioerosion rate. For instance, the erosion rate of polyanhydrides can be slowed by about three orders of magnitude when the less hydrophobic sebacic acid is replaced by the more hydrophobic bis(carboxy phenoxy)propane as the monomeric starting material. Likewise, devices made of PGA erode faster than identical devices made of the more hydrophobic PLA, although the ester bonds have about the same chemical reactivity toward water in both polymers.

The observed bioerosion rate is further influenced by the morphology of the polymer. Polymers can be classified as either semicrystalline or amorphous. At body temperature (37°C) amorphous polymers with T_g above 37°C will be in a glassy state, and polymers with a T_g below 37°C will in a rubbery state. In this discussion it is therefore necessary to consider three distinct morphological states: semicrystalline, amorphous-glassy, and amorphous-rubbery.

In the crystalline state, the polymer chains are densely packed and organized into crystalline domains that resist the penetration of water. Consequently, backbone hydrolysis tends to occur in the amorphous regions of a semicrystalline polymer and at the surface of the crystalline regions. This phenomenon is of particular importance to the erosion of devices made of poly (L-lactic acid) and PGA which tend to have high degrees of crystallinity around 50%.

Another good illustration of the influence of the polymer morphology on the rate of bioerosion is provided by a comparison of poly(L-lactic acid) (PLLA) and poly(D, L-lactic acid): Although these two polymers have chemically identical backbone structures and an identical degree of hydrophobicity, devices made of PLLA tend to degrade much more slowly than identical devices made of poly(D, L-lactic acid). The slower rate of bioerosion of PLLA is due to the fact that this stereoregular polymer is semicrystalline, while the racemic poly(D, L-lactic acid) is an amorphous polymer.

Likewise, a polymer in its glassy state is less permeable to water than the same polymer when it is in its rubbery state. This observation could be of importance in cases where an amorphous polymer has a glass transition temperature that is not far above body temperature (37°C). In this situation, water sorption into the polymer could lower its T_g below 37°C, resulting in abrupt changes in the bioerosion rate.

The manufacturing process may also have a significant effect on the erosion profile. For example, Mathiowitz and co-workers (Mathiowitz *et al*, 1990) showed that polyanhydride microspheres produced by melt encapsulation were very dense and eroded slowly, whereas when the same polymers were formed into microspheres by solvent evaporation, the microspheres were very porous (and therefore more water permeable) and eroded more rapidly.

The preceding examples illustrate an important technological principle in the design of bioeroding devices: The bioerosion rate of a given polymer is not an unchangeable property, but depends to a very large degree on readily controllable factors such as the presence of plasticizers or additives, the manufacturing process, the initial molecular weight of the polymer, and the geometry of the device.

Storage stability, sterilization, and packaging

It is important to minimize premature polymer degradation during fabrication and storage. Traces of moisture can seriously degrade even relatively stable polymers such as poly (bisphenol A carbonate) during injection molding or extrusion. Degradable polymers are particularly sensitive to hydrolytic degradation during high-temperature processing. The industrial production of degradable implants therefore often requires the construction of "controlled atmosphere" facilities where the moisture content of the polymer and the ambient humidity can be strictly controlled.

After fabrication, γ -irradiation or exposure to ethylene oxide may be used for the sterilization of degradable implants. Both methods have disadvantages and as a general rule, the choice is between the lesser of two evils. γ-Irradiation at a dose of 2 to 3 Mrad can result in significant backbone degradation. Since the aliphatic polyesters PLA, PGA, and PDS are particularly sensitive to radiation damage, these materials are usually sterilized by exposure to ethylene oxide and not by γ-irradiation. Unfortunately, the use of the highly dangerous ethylene oxide gas represents a serious safety hazard as well as potentially leaving residual traces in the polymeric device. Polymers sterilized with ethylene oxide must be degassed for extended periods of time.

Additionally, for applications in tissue engineering, biodegradable scaffolds may be preseeded with viable cells or may be impregnated with growth factors or other biologics. There is currently no method that could be used to sterilize scaffolds that contain viable cells without damaging the cells. Therefore, such products must be manufactured under sterile conditions and must be used within a very short time after manufacture. Currently, a small number of products containing preseeded, living cells are in clinical use. These products are extremely expensive, are shipped in special containers, and have little or no shelf life.

Likewise, it has been shown that sterilization of scaffolds containing osteoinductive or chondroinductive agents leads to significant losses in bioactivity, depending on the sterilization method used (Athanasiou *et al.*, 1998). The challenge of producing tissue engineering scaffolds that are preseeded with viable cells or that contain sensitive biological agents has not yet been fully solved.

After sterilization, degradable implants are usually packaged in air-tight aluminum-backed plastic-foil pouches. In some cases, refrigeration may also be required to prevent backbone degradation during storage.

Bibliography

Agrawal, C. M., Hass, K. F., Leopold, D. A., and Clark, H. G. (1992). Evaluation of poly(L-lactic acid) as a material for intravascular polymeric stents. *Biomaterials* **13:** 176–182.

Allcock, H. R. (1990). In *Biodegradable Polymers as Drug Delivery Systems*, M. Chasin, and R. Langer, eds. Marcel Dekker, New York, pp. 163–193.

Anderson, J. M., Spilizewski, K. L.,and Hiltner, A. (1985). In *Biocom-patibility of Tissue Analogs*, Vol. 1, D. F. Williams, ed. CRC Press Inc., Boca Raton, FL, pp. 67–88.

Andrianov, A. K., Sargent, J. R., Sule, S. S., LeGolvan, M. P., Woods, A. L., Jenkins, S. A., and Payne, L. G. (1998). Synthesis, physico-chemical properties and immunoadjuvant activity of water-soluble polyphosphazene polyacids. *J. Bioactive Comp. Polym.* **13:** 243–256.

Athanasiou, K., Agrawal, M., Barber, A., and Burkhart, S. (1998). Orthopaedic applications for PLA–PGA biodegradable polymers. *Arthroscopy* **14:** 726–737.

Attawia, M. A., Uhrich, K. E., Botchwey, E. , Fan, M., Langer, R., and Laurencin, C. T. (1995). Cytotoxicity testing of poly(anhydride-co-imides) for orthopedic applications. *J. Biomed. Mater. Res.* **29:** 1233–1240.

Barham, P. J., Keller, A., Otun, E. L., and Holmes, P. A. (1984). Crystallization and morphology of a bacterial thermoplastic: poly-3-hydroxybutyrate. *J. Mater. Sci.* **19:** 2781–2794.

Beele, H. (2002). Artificial skin: Past, present and future. *Int. J. Artificial Organs* 25: 163–173.

Bergsma, J. E., de Bruijn, W. C., Rozema, F. R., Bos, R. R. M., and Boering, G. (1995). Late degradation tissue response to poly(l-lactic) bone plates and screws. *Biomaterials* **16:** 25–31.

Campbell, P., Glover, G. I., and Gunn, J. M. (1980). Inhibition of intracellular protein degradation by pepstatin, poly(L-lysine) and pepstatinyl-poly (L-lysine). *Arch. Biochem. Biophys.* **203:** 676–680.

Chasin, M., Domb, A., Ron, E., Mathiowitz, E., Langer, R., Leong, K., Laurencin, C., Brem, H., and Grossman, S. (1990). In *Biodegradable Polymers as Drug Delivery Systems*, M. Chasin and R. Langer, eds. Marcel Dekker, New York, pp. 43–70.

Choueka, J., Charvet, J. L., Koval, K. J., Alexander, H., James, K. S., Hooper, K. A., and Kohn, J. (1996). Canine bone response to tyrosine-derived polycarbonates and poly(L-lactic acid). *J. Biomed. Mater. Res.* **31:** 35–41.

Deshpande, A. A., Heller, J., and Gurny, R. (1998). Bioerodible polymers for ocular drug delivery. *Crit. Rev. Thera. Drug Carrier Syst.* **15**: 381–420.

Frazza, E. J., and Schmitt, E. E. (1971). A new absorbable suture. *J. Biomed. Mater. Res.* **1:** 43–58.

Gilding, D. K., and Reed, A. M. (1979). Biodegradable polymers for use in surgery—poly(glycolic)/poly(lactic acid) homo- and copolymers: 1. *Polymer* **20:** 1459–1464.

Gilding, D. K., and Reed, A. M. (1981). Biodegradable polymers for use in surgery—poly(glycolic)/poly(lactic acid) homo- and copolymers: 2: In vitro degradation. *Polymer* **22:** 494–498.

Gombotz, W. R., and Pettit, D. K. (1995). Biodegradable polymers for protein and peptide drug delivery. *Bioconjugate Chem.* **6:** 332–351.

Heller, J. (1987). in *Controlled Drug Delivery, Fundamentals and Applications*, 2nd ed. J. R. Robinson and V. H. L. Lee, eds. Marcel Dekker, New York, pp. 180–210.

Heller, J. (1993). Polymers for controlled parenteral delivery of peptides and proteins. *Adv. Drug Delivery Rev.* **10:** 163–204.

Heller, J., Barr, J., Ng, S. Y., Abdellauoi, K. S., and Gurny, R. (2002). Poly(ortho esters): synthesis, characterization, properties and uses. *Adv. Drug Delivery Rev.* **54:** 1015–1039.

Heller, J., Penhale, D. W. H., and Helwing, R. F. (1980). Preparation of poly(ortho esters) by the reaction of diketene acetals and polyols. *J. Polymer Sci. (Polymer Lett. Ed.)* **18:** 619–624.

Heller, J., Sparer, R. V., and Zentner, G. M. (1990). in *Biodegradable Polymers as Drug Delivery Systems*, M. Chasin and R. Langer, eds. Marcel Dekker, New York, pp. 121–162.

Hill, J. L., Sawhney, A. S., Pathak, C. P. and Hubbell, J. A. (1993). Prevention of post-operative adhesions using biodegradable hydrogels. 19th Annual Meeting of the Society of Biomaterials, Minneapolis, MN, p. 199.

Huang, S. (1985) in *Encyclopedia of Polymer Science and Engineering*, Vol. 2. F. H. Mark, N. M. Bikales, C. G. Overberger, G. Menges, and J. I. Kroshwitz, eds. John Wiley, New York, pp. 220–243.

James, K., and Kohn, J. (1996). New biomaterials for tissue engineering. *MRS Bull.* **21:** 22–26.

James, K., and Kohn, J. (1997). in *Controlled Drug Delivery: Challenges and Strategies*, K. Park, ed. American Chemical Society, Washington, D.C., pp. 389–403.

Knowles, J. C. (1993). Development of a natural degradable polymer for orthopedic use. *J. Med. Eng. Technol.* **17:** 129–137.

Kohn, J., and Langer, R. (1984). in *Polymeric Materials, Science and Engineering*, Vol. 51. American Chemical Society, Washington, D.C., pp. 119–121.

Kohn, J., and Langer, R. (1987). Polymerization reactions involving the side chains of α-L-amino acids. *J. Am. Chem. Soc.* **109:** 817–820.

Langer, R. (1990). New methods of drug delivery. *Science* **249:** 1527–1533.

Langer, R., and Vacanti, J. P. (1993). Tissue engineering. *Science* **260:** 920–926.

Laurencin, C. T., Norman, M. E., Elgendy, H. M., El-Amin, S. F., Allcock, H. R., Pucher, S. R., and Ambrosio, A. A. (1993). Use of polyphosphazenes for skeletal tissue regeneration. *J. Biomed. Mater. Res.* **27:** 963–973.

Leong, K. W., D' Amore, P. D., Marletta, M., and Langer, R. (1986). Bioerodible polyanhydrides as drug-carrier matrices. II: Biocompatibility and chemical reactivity.J. *Biomed. Mater. Res.* **20:** 51–64.

Lescure, F., Gurney, R., Doelker, E., Pelaprat, M. L., Bichon, D., and Anderson, J. M. (1989). Acute histopathological response to a new biodegradable, polypeptidic polymer for implantable drug delivery system. *J. Biomed. Mater. Res.* **23:** 1299–1313.

Madrid, Y., Langer, L. F., Brem, H., and Langer, R. (1991). New directions in the delivery of drugs and other substances to the central nervous system. *Adv. Pharmacol.* **22:** 299–324.

Mathiowitz, E., Kline, D., and Langer, R. (1990). Morphology of polyanhydride microsphere delivery systems. *J. Scanning Microsc.* **4:** 329–340.

McCormick-Thomson, L. A., and Duncan, R. (1989). Poly(amino acid) copolymers as a potential soluble drug delivery system. 1. Pinocytic uptake and lysosomal degradation measured *in vitro*. *J. Bioact. Biocompat. Polymer* **4:** 242–251.

Miller, N. D., and Williams, D. F. (1987). On the biodegradation of poly-β-hydroxybutyrate (PHB) homopolymer and poly-β-hydroxybutyrate-hydroxyvalerate copolymers. *Biomaterials* **8:** 129–137.

Miyamae, T., Mori, S. and Takeda, Y. (1968). Poly-L-glutamic acid surgical sutures. US Patent 3,371,069.

Nathan, A. and Kohn, J. (1996). in *Protein Engineering and Design*, P. Carey, ed. Academic Press, New York, pp. 265–287.

Okada, H., and Toguchi, H. (1995). Biodegradable microspheres in drug delivery. *Crit. Rev. Ther. Drug Carrier Syst.* **12:** 1–99.

Park, E.-S., Maniar, M., and Shah, J. (1998). Biodegradable polyanhydride devices of cefazolin sodium, bupivacaine, and taxol for local drug delivery: preparation, kinetics and mechanism of in vitro release. *J. ControlRel.* **52:** 179–189.

Pitt, C. G. (1990). in *Biodegradable Polymers as Drug Delivery Systems*, M. Chasin and R. Langer, eds. Marcel Dekker, New York, pp. 71–120.

Pitt, C. G., Gratzl, M. M., Jeffcoat, A. R., Zweidinger, R., and Schindler, A. (1979). Sustained drug delivery systems II: Factors affecting release rates from poly(ε-caprolactone) and related biodegradable polyesters. *J. Pharm. Sci.* **68:** 1534–1538.

Rosen, H., Kohn, J., Leong, K., and Langer, R. (1988). in *Controlled Release Systems: Fabrication Technology*, D. Hsieh, eds. CRC Press, Boca Raton, FL, pp. 83–110.

Senior, P. J., Beech, G. A., Ritchie, G. A. and Dawes, E. A. (1972). The role of oxygen limitation in the formation of poly-β-hydroxybutyrate during batch and continuous culture of *Azotobacter beijerinckii. Biochem.J.* **128:** 1193–1201.

Shea, L. D., and Mooney, D. J. (2001). Nonviral DNA delivery from polymeric systems. *Methods Mol. Med.* **65:** 195–207.

Spira, M., Fissette, J., Hall, C. W., Hardy, S. B., and Gerow, F. J. (1969). Evaluation of synthetic fabrics as artificial skin grafts to experimental burn wounds. *J. Biomed. Mater. Res.* **3:** 213–234.

Tamada, J. A., and Langer, R. (1993). Erosion kinetics of hydrolytically degradable polymers. *Proc. Natl. Acad. Sci. USA* **90:** 552–556.

Törmälä, P., Pohjonen, T. and Rokkanen, P. (1998). Bioabsorbable polymers: materials technology and surgical applications. *Proc. Inst. Mech. Engr.* **212:** 101–111.

van Heeswijl, W. A. R., Hoes, C. J. T., Stoffer, T., Eenink, M. J. D., Potman, W., and Feijen, J. (1985). The synthesis and characterization of polypeptide–adriamycin conjugates and its complexes with adriamycin. Part 1. *J. Control Rel.* **1:** 301–315.

Vert, M., Li, S., and Garreau, H. (1991). More about the degradation of LA/GA-derived matrices in aqueous media. *J. Control Release* **16:** 15–26.

Williams, D. F. (1987). *Definitions in Biomaterials—Proceedings of a Consensus Conference of the European Society for Biomaterials.* Elsevier, New York.

3.2.8 Natural materials

Ioannis V. Yannas

Natural polymers offer the advantage of being very similar, often identical, to macromolecular substances which the biological environment is prepared to recognize and to deal with metabolically (Table 3.2.8-1). The problems of toxicity and stimulation of a chronic inflammatory reaction, as well as lack of recognition by cells, which are frequently provoked by many synthetic polymers, may thereby be suppressed. Furthermore, the similarity to naturally occurring substances introduces the interesting capability of designing biomaterials that function biologically at the molecular, rather than the macroscopic, level. On the other hand, natural polymers are frequently quite immunogenic. Furthermore, because they are structurally much more complex than most synthetic polymers, their technological manipulation is quite a bit more elaborate. On balance, however, these opposing factors have conspired to lead to a substantial number of biomaterials applications in which naturally occurring polymers, or their chemically modified versions, have provided unprecedented solutions.

An intriguing characteristic of natural polymers is their ability to be degraded by naturally occurring enzymes, a virtual guarantee that the implant will be eventually metabolized by physiological mechanisms. This property may, at first glance, appear as a disadvantage since it detracts from the durability of the implant. However, it has been used to advantage in biomaterials applications in which it is desired to deliver a specific function for a temporary period of time, following which the implant is expected to degrade completely and to be disposed of by largely normal metabolic processes. Since, furthermore, it is possible to control the degradation rate of the implanted polymer by chemical cross-linking or other chemical modifications, the designer is offered the opportunity to control the lifetime of the implant.

A potential problem to be dealt with when proteins are used as biomaterials is their frequently significant immunogenicity, which, of course, derives precisely from their similarity to naturally occurring substances. The immunological reaction of the host to the implant is directed against selected sites (antigenic determinants) in the protein molecule. This reaction can be mediated by molecules in solution in body fluids (immunoglobulins). A single such molecule (antibody) binds to single or multiple determinants on an antigen. The immunological reaction can also be mediated by molecules that are held

Table 3.2.8-1 General properties of certain natural polymers

	Polymer	Incidence	Physiological function
A. Proteins	Silk	Synthesized by arthropods	Protective cocoon
	Keratin	Hair	Thermal insulation
	Collagen	Connective tissues (tendon, skin, etc.)	Mechanical support
	Gelatin	Partly amorphous collagen	(Industrial product)
	Fibrinogen	Blood	Blood clotting
	Elastin	Neck ligament	Mechanical support
	Actin	Muscle	Contraction, motility
	Myosin	Muscle	Contraction, motility
B. Polysaccharides	Cellulose (cotton)	Plants	Mechanical support
	Amylose	Plants	Energy reservoir
	Dextran	Synthesized by bacteria	Matrix for growth of organism
	Chitin	Insects, crustaceans	Provides shape and form
	Glycosaminoglycans	Connective tissues	Contributes to mechanical support
C. Polynucleotides	Deoxyribonucleic acids (DNA)	Cell nucleus	Direct protein biosynthesis
	Ribonucleic acids (RNA)	Cell nucleus	Direct protein biosynthesis

tightly to the surface of immune cells (lymphocytes). The implant is eventually degraded. The reaction can be virtually eliminated provided that the antigenic determinants have been previously modified chemically. The immunogenicity of polysaccharides is typically far lower than that of proteins. The collagens are generally weak immunogens relative to the majority of proteins.

Another potential problem in the use of natural polymers as biomaterials derives from the fact that these polymers typically decompose or undergo pyrolytic modification at temperatures below the melting point, thereby precluding the convenience of high-temperature thermoplastics processing methods, such as melt extrusion, during the manufacturing of the implant. However, processes for extruding these temperature-sensitive polymers at room temperature have been developed. Another serious disadvantage is the natural variability in structure of macromolecular substances which are derived from animal sources. Each of these polymers appears as a chemically distinct entity not only from one species to another (species specificity) but also from one tissue to the next (tissue specificity). This testimonial to the elegance of the naturally evolved design of the mammalian body becomes a problem for the manufacturer of implants, which are typically required to adhere to rigid specifications from one batch to the next. Consequently, relatively stringent methods of control of the raw material must be used.

Most of the natural polymers in use as biomaterials today are constituents of the ECM of connective tissues such as tendons, ligaments, skin, blood vessels, and bone. These tissues are deformable, fiber-reinforced composite materials of organ shape as well as of the organism itself. In the relatively crude description of these tissues as if they were manmade composites, collagen and elastin fibers mechanically reinforce a "matrix" that primarily consists of protein polysaccharides (proteoglycans) highly swollen in water. Extensive chemical bonding connects these macromolecules to each other, rendering these tissues insoluble and, therefore, impossible to characterize with dilute solution methods unless the tissue is chemically and physically degraded. In the latter case, the solubilized components are subsequently extracted and characterized by biochemical and physicochemical methods. Of the various components of extracellular materials that have been used to fashion biomaterials, collagen is the one most frequently used. Other important components, to be discussed later, include the proteoglycans and elastin.

Almost inevitably, the physicochemical processes used to extract the individual polymer from tissues, as well as subsequent deliberate modifications, alter the native structure, sometimes significantly. The description in this section emphasizes the features of the naturally occurring, or native, macromolecular structures. Certain modified forms of these polymers are also described.

Structure of native collagen

Structural order in collagen, as in other proteins, occurs at several discrete levels of the structural hierarchy. The collagen in the tissues of a vertebrate occurs in at least 10 different forms, each of these being predominant in a specific tissue. Structurally, these collagens share the characteristic triple helix, and variations among them are restricted to the length of the nonhelical fraction, as well as the length of the helix itself and the number and nature of carbohydrates attached on the triple helix. The collagen in skin, tendon, and bone is mostly type I collagen. Type II collagen is predominant in cartilage, while type III collagen is a major constituent of the blood vessel wall as well as being a minor contaminant of type I collagen in skin. In contrast to these collagens, all of which form fibrils with the distinct collagen periodicity, type IV collagen, a constituent of the basement membrane that separates epithelial tissues from mesodermal tissues, is largely nonhelical and does not form fibrils. We follow here the nomenclature that was proposed by W. Kauzmann (1959) to describe in a general way the structural order in proteins, and we specialize it to the case of type I collagen (Fig. 3.2.8-1).

The primary structure denotes the complete sequence of amino acids along each of three polypeptide chains as well as the location of interchain cross-links in relation to this sequence. Approximately one-third of the residues are glycine and another quarter or so are proline or hydroxyproline. The structure of the bifunctional interchain cross-link is the relatively complex condensation product of a reaction involving lysine and hydroxylysine residues; this reaction continues as the organism matures, thereby rendering the collagens of older animals more difficult to extract from tissues.

The secondary structure is the local configuration of a polypeptide chain that results from satisfaction of stereochemical angles and hydrogen-bonding potential of peptide residues. In collagen, the abundance of glycine residues (Gly) plays a key configurational role in the triplet Gly–X–Y, where X and Y are frequently proline or hydroxyproline, respectively, the two amino acids that control the chain configuration locally by the very rigidity of their ring structures. On the other hand, the absence of a side chain in glycine permits close approach of polypeptide chains in the collagen triple helix.

Tertiary structure refers to the global configuration of the polypeptide chains; it represents the pattern according to which the secondary structure is packed within the complete macromolecule and it constitutes the structural unit that can exist as a physicochemically stable entity in solution, namely, the triple helical collagen molecule.

In type I collagen, two of the three polypeptide chains have identical amino acid composition, consisting of 1056 residues and are termed a1(I) chains, while the third has a different composition, it consists of 1038 residues and is termed a2(I). The three polypeptide chains fold to produce a left-handed helix, whereas the three-chain supercoil is actually right-handed with an estimated pitch of about 100 nm (30–40 residues). The helical structure extends over 1014 of the residues in each of the three chains, leaving the remaining residues at the ends (telopeptides) in a nonhelical configuration. The residue spacing is 0.286 nm and the length of the helical portion of the molecule is, therefore, 1014 × 0.286 or 290 nm long.

The fourth-order or quaternary structure denotes the repeating supermolecular unit structure, comprising several molecules packed in a specific lattice, which constitutes the basic element of the solid state (microfibril). Collagen molecules are packed in a quasi-hexagonal lattice at an interchain distance of about 1.3 nm, which shrinks considerably when the microfibril is dehydrated. Adjacent molecules in the microfibril are approximately parallel to the fibril axis; they all point in the same direction along the fibril and are staggered regularly, giving rise to the well-known D-period of collagen, about 64 nm, which is visible in the electron microscope. Higher levels of order, eventually leading to gross anatomical features that can be readily seen with the naked eye, have been proposed, but there is no general agreement on their definition.

Biological effects of physical modifications of the native structure of collagen

Crystallinity in collagen can be detected at two discrete levels of structural order: the tertiary (triple helix) (Fig. 3.2.8-1C) and the quaternary (lattice of triple helices) (Fig. 3.2.8-1D). Each of these levels of order corresponds, interestingly enough, to a separate melting transformation. A solution of collagen triple helices is thus converted to the randomly coiled gelatin by heating above the helix–coil transition temperature, which is approximately 37°C for bovine collagen, or by exceeding a critical concentration of certain highly polarizable anions, e.g., bromide or thiocyanate, in the solution of collagen molecules. IR spectroscopic procedures, based on helical marker bands in the mid- and far IR, have been developed to assay the gelatin content of collagen in the solid or semisolid states in which collagen is commonly used as an implant. Since implanted gelatin is much more rapidly degradable than collagen, a characteristic that can seriously affect implant performance, these assays are essential tools for quality control of collagen-based biomaterials. Frequently, such biomaterials have been processed under manufacturing conditions that may threaten the integrity of the triple helix.

Collagen fibers also exhibit a characteristic banding pattern with a period of about 65 nm (quaternary structure). This pattern is lost reversibly when the pH of a suspension of collagen fibers in acetic acid is lowered below 4.25 ± 0.30. Transmission electron microscopy or small-angle X-ray diffraction can be used to determine the fraction of fibrils that possess banding as the pH of the system is altered. During this transformation, which appears to be a first-order thermodynamic transition, the triple helical structure remains unchanged. Changes in pH can, therefore, be used to selectively abolish the quaternary structure while maintaining the tertiary structure intact.

This experimental strategy has made it possible to show that the well-known phenomenon of blood platelet aggregation by collagen fibers (the reason for use of collagen sponges as hemostatic devices) is a specific property of the quaternary rather than of the tertiary structure. Thus collagen that is thrombo-resistant *in vitro* has been prepared by selectively "melting out" the packing order of helices while preserving the triple helices themselves. Figure 3.2.8-2 illustrates the banding pattern of such collagen fibers. Notice that short segments of banded fibrils persist even after very long treatment at low pH, occasionally interrupting long segments of nonbanded fibrils (Fig. 3.2.8-2, inset).

The porosity of a collagenous implant normally makes an indispensable contribution to its performance. A porous structure provides an implant with two critical functions. First, pore channels are ports of entry for cells migrating from adjacent tissues into the bulk of the implant for tissue serum (exudate) that enters via capillary suction or of blood from a hemorrhaging blood vessel nearby. Second, pores endow a material with a frequently enormous specific surface that is made available either for specific interactions with invading cells (e.g., myofibroblasts bind extensively on the surface of porous collagen–glycosaminoglycan copolymer structures that induce regeneration of skin in burned patients) or for interaction with coagulation factors in blood flowing into the device (e.g., hemostatic sponges).

Pores can be incorporated by first freezing a dilute suspension of collagen fibers and then inducing sublimation of the ice crystals by exposing the suspension to low-temperature vacuum. The resulting pore structure is a negative replica of the network of ice crystals (primarily dendrites). It follows that control of the conditions for

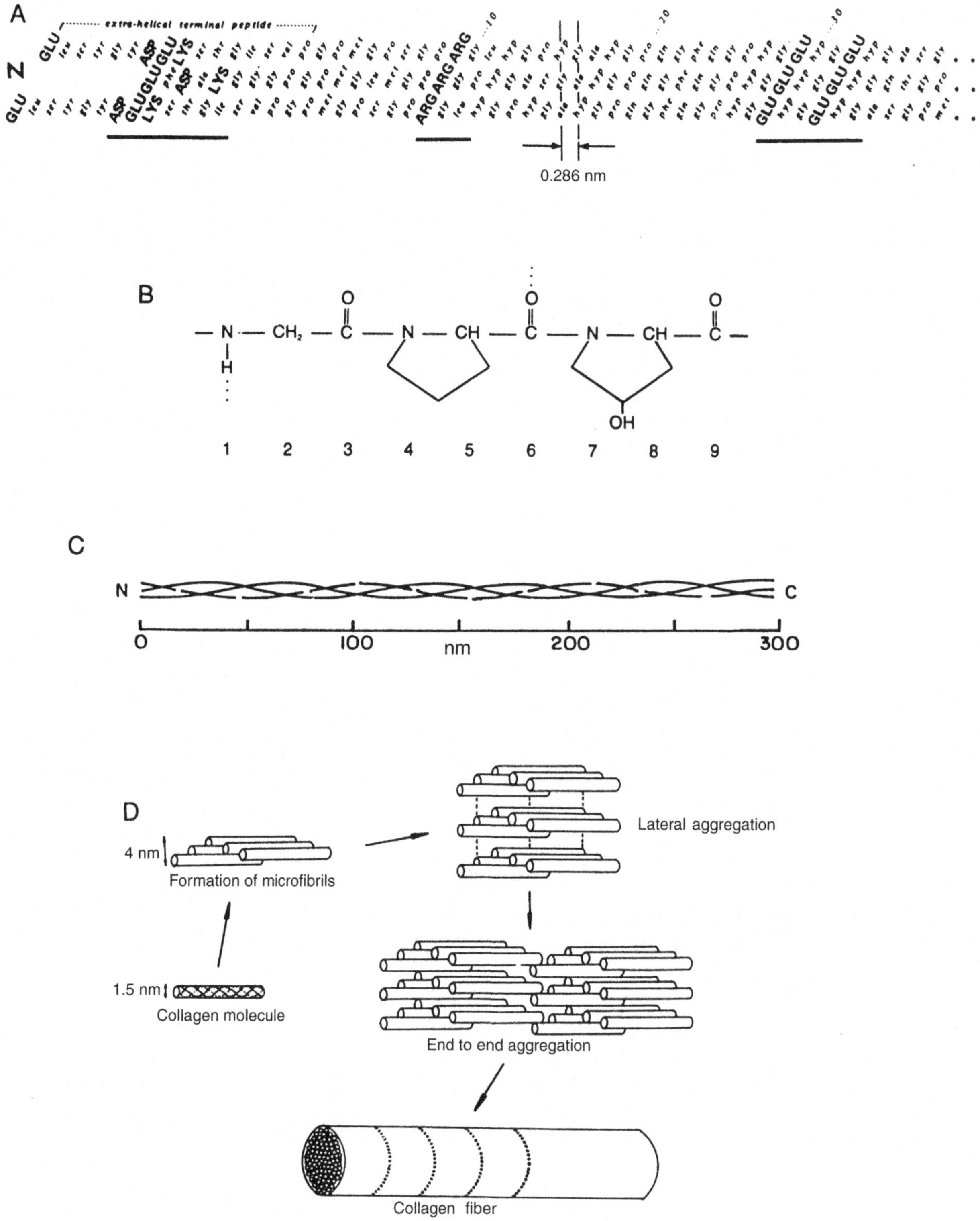

Fig. 3.2.8-1 Collagen, like other proteins, is distinguished by several levels of structural order. (A) Primary structure—the complete sequence of amino acids along each polypeptide chain. An example is the triple chain sequence of type I calf skin collagen at the N-end of the molecule. Roughly 5% of a complete molecule is shown above. No attempt has been made to indicate the coiling of the chains. Amino acid residues participating in the triple helix are numbered, and the residue-to-residue spacing (0.286 nm) is shown as a constant within the triple helical domain, but not outside it. Bold capitals indicate charged residues which occur in groups (underlined). (Reprinted from J. A. Chapman and D. J. S. Hulmes (1984). In *Ultrastructure of the Connective Tissue Matrix*, A. Ruggeri and P. M. Motta, eds. Martinus Nijhoff, Boston, Chap. 1, Fig. 3.2.8-1, p. 2, with permission.) (B) Secondary structure—the local configuration of a polypeptide chain. The

ice nucleation and growth can lead to a large variety of pore structures (Fig. 3.2.8-3).

In practice, the average pore diameter decreases with decreasing temperature of freezing while the orientation of pore channel axes depends on the magnitude of the major heat flux vector during freezing. In experimental implants the mean pore diameter has ranged between about 1 and 800 μm; pore volume fractions have ranged up to 0.995; the specific surface has been varied between about 0.01 and 100 m^2/g dry matrix; and the orientation of axes of pore channels has ranged from strongly uniaxial to almost random. The ability of collagen–glycosaminoglycan copolymers to induce regeneration of tissues such as skin, the conjunctiva and peripheral nerves depends critically, among other factors, on the adjustment of the pore structure to desired levels, e.g., a pore size range of about 20–125 μm for skin regeneration and less than 10 μm for sciatic nerve regeneration appear to be mandatory. Determination of pore structure is based on principles of stereology, the discipline which allows the quantitative statistical properties of three-dimensional structures of implants to be related to those of two-dimensional projections, e.g., sections used for histological analysis.

Chemical modification of collagen and its biological consequences

The primary structure of collagen is made up of long sequences of some 20 different amino acids. Since each amino acid has its own chemical identity, there are 20 types of pendant side groups, each with its own chemical reactivity, attached to the polypeptide chain backbone. As examples, there are carboxylic side groups (from glutamic acid and aspartic acid residues), primary amino groups (lysine, hydroxylysine, and arginine residues), and hydroxylic groups (tyrosine and hydroxylysine). The collagen molecule is therefore subject to modification by a large variety of chemical reagents. Such versatility comes with a price: Even though the choice of reagents is large, it is important to ascertain that use of a given reagent has actually led to modification of a substantial fraction of the residues of an amino acid in the molecule. This is equivalent to proof that a reaction has proceeded to a desired "yield." Furthermore, proof that a given reagent has attacked only a specific type of amino acid, rather than all amino acid residue types carrying the same functional group, also requires chemical analysis.

Historically, the chemical modification of collagen has been practiced in the leather industry (since about 50% of the protein content of cowhide is collagen) and in the photographic gelatin industry. Today, the increasing use of collagen in biomaterials applications has provided renewed incentive for novel chemical modification, primarily in two areas. First, implanted collagen is subject to degradative attack by collagenases, and chemical cross-linking is a well-known means of decelerating the degradation rate. Second, collagen extracted from an animal source elicits production of antibodies (immunogenicity) and chemical modification of antigenic sites may potentially be a useful way to control the immunogenic response. Although it is widely accepted that implanted collagen elicits synthesis of antibodies at a far smaller concentration than is true of most other implanted proteins, treatment with specific reagents, including enzymatic treatment, or cross-linking, is occasionally used to reduce significantly the immunogenicity of collagen.

Collagen-based implants are normally degraded by mammalian collagenases, naturally occurring enzymes that attack the triple helical molecule at a specific location. Two characteristic products result, namely, the N-terminal fragment, which amounts to about two-thirds of the molecule, and the C-terminal fragment. Both of these fragments become spontaneously transformed (denatured) to gelatin at physiological temperatures via the helix–coil transition and the gelatinized fragments are then cleaved to oligopeptides by naturally occurring enzymes that degrade several other tissue proteins (nonspecific proteases).

Collagenases are naturally present in healing wounds and are credited with a major role in the degradation of collagen fibers at the site of trauma. At about the same time that degradation of collagen and of other ECM

triplet sequence Gly-Pro-Hyp illustrates elements of collagen triple-helix stabilization. The numbers identify peptide backbone atoms. The conformation is determined by trans peptide bonds (3-4, 6-7, and 9-1); fixed rotation angle of bond in proline ring (4-5); limited rotation of proline past the C=O group (bond 5-6); interchain hydrogen bonds (dots) involving the NH hydrogen at position 1 and the C=O at position 6 in adjacent chains; and the hydroxy group of hydroxyproline, possibly through water-bridged hydrogen bonds. (Reprinted from K. A. Piez and A. H. Reddi, editors (1984). *Extracellular Matrix Biochemistry*. Elsevier, New York, Chap. 1, Fig 1.6. p. 7, with permission.) (C) Tertiary structure—the global configuration of polypeptide chains, representing the pattern according to which the secondary structures are packed together within the unit substructure. A schematic view of the type I collagen molecule, a triple helix 300 nm long. (Reprinted fromK. A. Piez and A. H. Reddi, editors (1984). *Extracellular Matrix Biochemistry*. Elsevier, New York, Chap. 1, Fig. 1.22, p. 29, with permission.) (D) Quaternary structure—the unit supermolecular structure. The most widely accepted unit is one involving five collagen molecules (microfibril). Several microfibrils aggregate end to end and also laterally to form a collagen fiber that exhibits a regular banding pattern in the electron microscope with a period of about 65 nm. (Reprinted from E. Nimni, editor (1988). *Collagen*, Vol. I, Biochemistry, CRC Press, Boca Raton, FL Chap. 1, Fig. 10, p. 14, with permission.)

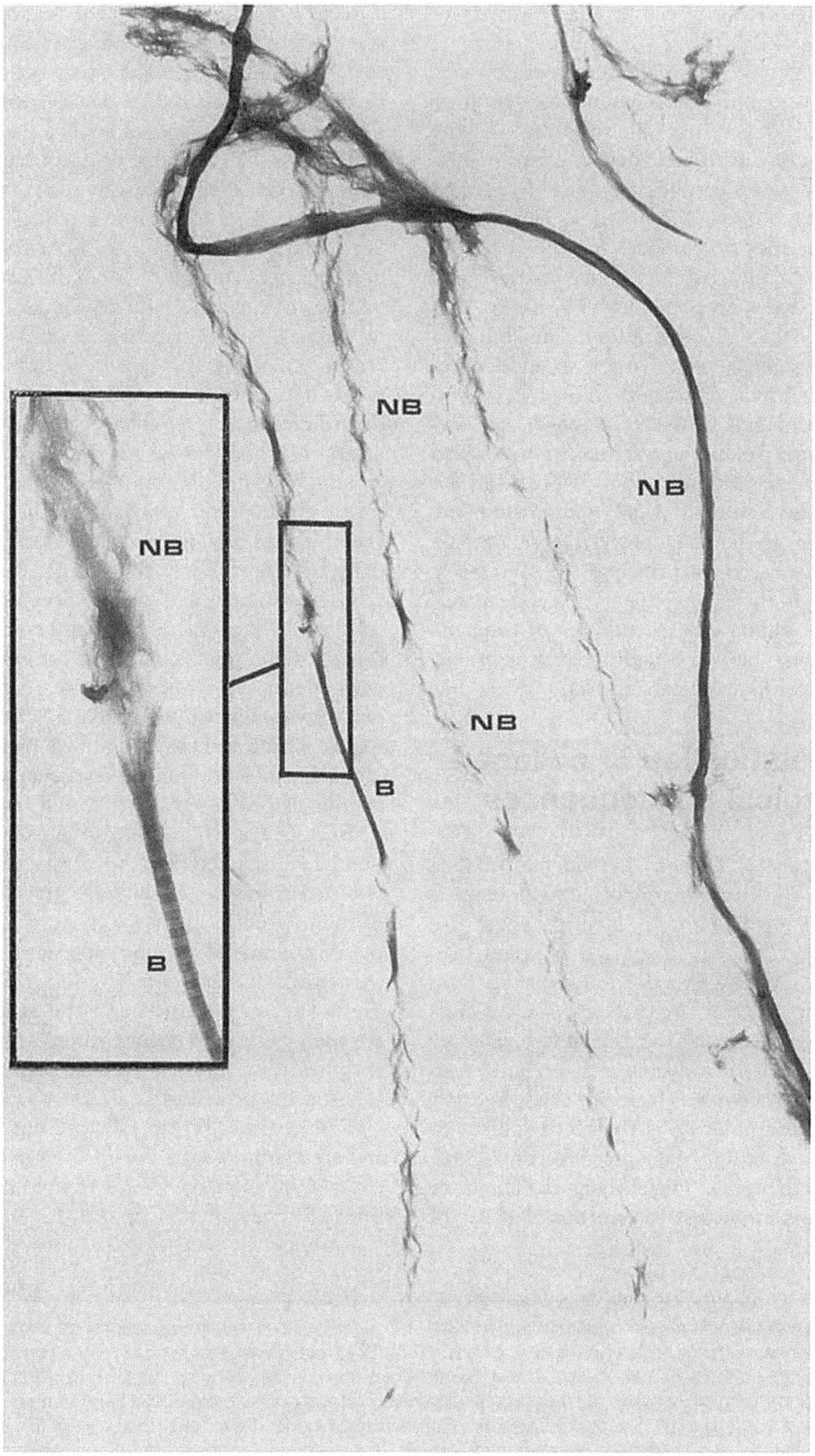

Fig. 3.2.8-2 Following exposure to pH below 4.25 ± 0.30 the banding pattern of type I bovine hide collagen practically disappears. Short lengths of banded collagen (B) do, however, persist next to very long lengths of nonbanded collagen (NB), which has tertiary but not quaternary structure. This preparation does not induce platelet aggregation provided that the fibers are prevented from recrystallizing to form banded structures when the pH is adjusted to neutral in order to perform the platelet assay. Stained with 0.5 wt.% phosphotungstic acid. Banded collagen period, about 65 nm. Original magnification: 15,000×. Inset original mag.: 75,000×. (Reprinted from M. J. Forbes, M.S. dissertation, Massachusetts Institute of Technology, 1980, courtesy of MIT.)

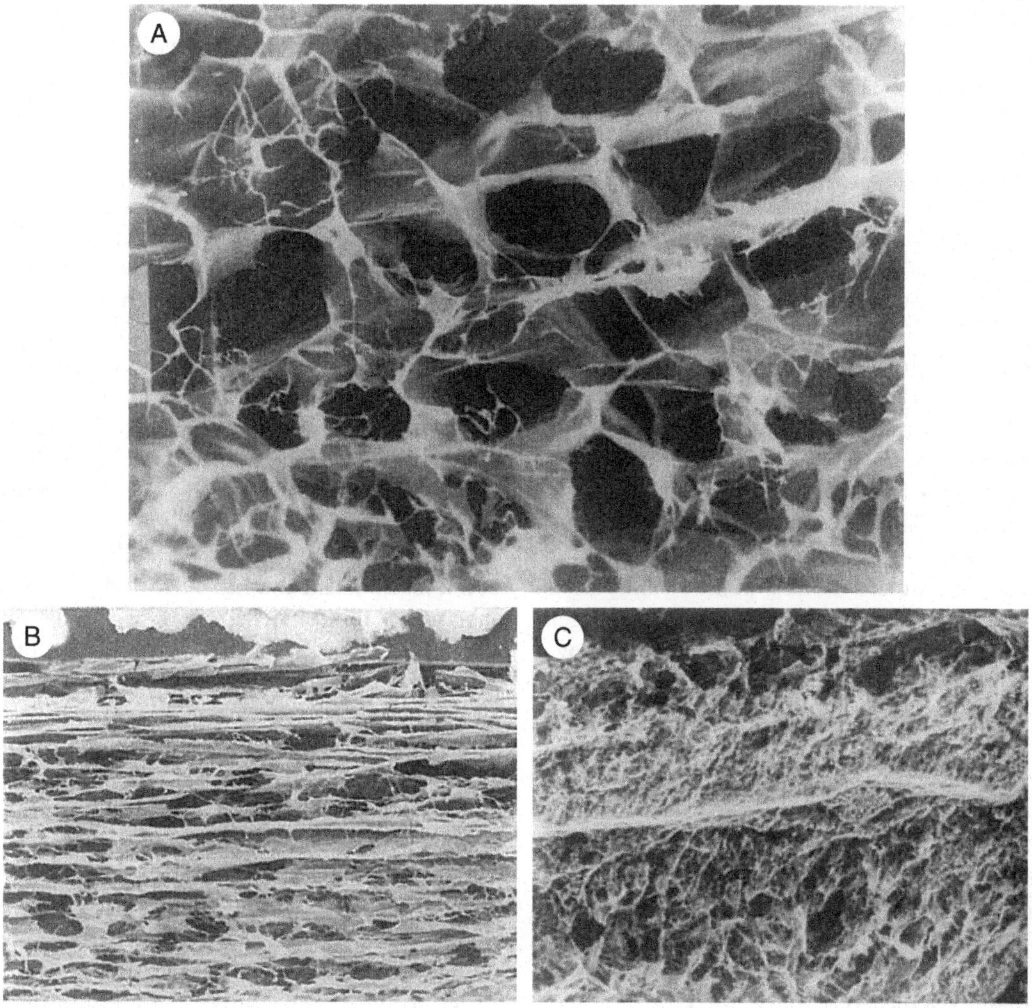

Fig. 3.2.8-3 Illustration of the variety of porous structures that can be obtained with collagen–GAG copolymers by adjusting the kinetics of crystallizaton of ice to the appropriate magnitude and direction. Pores form when the ice dendrites are eventually sublimed. SEM. (Courtesy of MIT.)

components proceeds in the wound bed, these components are being synthesized *de novo* by cells at the same anatomical site. Eventually, new architectural arrangements of collagen fibers, such as scar tissue, are synthesized. Although it is not a replica of the intact tissue, scar tissue forms a stable endpoint to the healing process and acts as a tissue barrier that allows the healed organ to continue functioning at a nearly physiological level. One of the frequent challenges in the design of collagen implants is to modify collagen chemically in a way that the rate of its degradation at the implantation site is either accelerated or slowed down to a desired level.

An effective method for reducing the rate of degradation of collagen by naturally occurring enzymes is by chemical cross-linking. A very simple self-cross-linking procedure, dehydrative cross-linking, is based on the fact that removal of water below ca. 1 wt.% insolubilizes collagen as well as gelatin by inducing formation of interchain peptide bonds. The nature of cross-links formed can be inferred from results of studies using chemically modified gelatins. Gelatin that had been modified either by esterification of the carboxylic groups of aspartyl and glutamyl residues, or by acetylation of the ε-amino groups of lysyl residues, remained soluble in aqueous solvents after exposure of the solid protein to high temperature, while unmodified gelatins lost their solubility. Insolubilization of collagen and gelatin following severe dehydration has been, accordingly, interpreted as the result of drastic removal of the aqueous product of a condensation reaction that led to formation of interchain amide links. The proposed mechanism is consistent with results, obtained by titration, showing that the number of free carboxylic groups and free amino groups in collagen are both significantly decreased following high-temperature treatment.

Removal of water to the extent necessary to achieve a density of cross-links in excess of 10^{-5} mol cross-links/g dry protein, which corresponds to an average molecular weight between crosslinks, M_c, of about 70 kDa, can be achieved within hours by exposure to temperatures in excess of 105°C under atmospheric pressure. The possibility that cross-linking achieved under these conditions is caused by a pyrolytic reaction has been ruled out. Furthermore, chromatographic data have shown that the amino acid composition of collagen remains intact after exposure to 105°C for several days. In fact, it has been observed that gelatin can be cross-linked by exposure to temperatures as low as 25°C provided that a sufficiently high vacuum is present to achieve the drastic moisture removal that drives the cross-linking reaction.

Exposure of highly hydrated collagen to temperatures in excess of ca. 37°C is known to cause reversible melting of the triple helical structure, as described earlier. The melting point of the triple helix increases with the collagen–diluent ratio from 37°C, the helix-coil transition of the infinitely dilute solution, to about 120°C for collagen swollen with as little as 20 wt.% diluent and up to 210°C, the approximate melting point of anhydrous collagen. Accordingly, it is possible to cross-link collagen using the drastic dehydration procedure described above without loss of the triple helical structure. It is simply sufficient to adjust the moisture content of collagen to a low enough level prior to exposure to the high temperature levels required for rapid dehydration.

Dialdehydes have been long known in the leather industry as effective tanning agents and in histological laboratories as useful fixatives. Both of these applications are based on the reaction between the dialdehyde and the ε-amino group of lysyl residues in the protein, which induces formation of interchain cross-links. Glutaraldehyde cross-linking is a relatively widely used procedure in the preparation of implantable biomaterials. Free glutaraldehyde is a toxic substance for cells; it cross-links vital cell proteins. However, clinical studies and extensive clinical use of implants have shown that the toxicity of glutaraldehyde becomes effectively negligible after the unreacted glutaraldehyde has been carefully rinsed out following reaction with an implant, e.g., one based on collagen. The nature of the cross-link formed has been the subject of controversy, primarily due to the complex, apparently polymeric, character of this reagent. Considerable evidence supports a proposed anabilysine structure, which is derived from two lysine side chains and two molecules of glutaraldehyde.

Evidence for other mechanisms has been presented. By comparison with other aldehydes, glutaraldehyde has shown itself to be a particularly effective cross-linking agent, as judged, for example, by its ability to increase the crosslink density to very high levels. Values of the average molecule weight between cross-links (M_c) provide the experimenter with a series of collagens in which the enzymatic degradation rate can be studied over a wide range, thereby affording implants that effectively disappear from tissue between a few days and several months following implantation. The mechanism of the reaction between glutaraldehyde and collagen at neutral pH is understood in part; however, the reaction in acidic media has not been studied extensively. Evidence that covalent cross-linking is involved comes from measurements of the equilibrium tensile modulus of films that have been treated to induce cross-linking and have subsequently been gelatinized by treatment in 1 M NaCl at 70°C. Under such conditions, only gelatin films that have been converted into a three-dimensional network by cross-linking support an equilibrium tensile force; by contrast, un-cross-linked specimens dissolve readily in the hot medium.

Several other methods for cross-linking collagen have been studied, including hexamethylene diisocynate, acyl azide, and a carbodiimide, 1-ethyl-3-(3-dimethlyaminopropyl) carbodiimide (EDAC).

The immunogenicity of the collagen used in implants is not insignificant and has been studied assiduously using laboratory preparations. However, the clinical significance of such immunogenicity has been shown to be very low and is often considered to be negligible. The validity of this simple approach to using collagen as a biomaterial was long ago recognized by manufacturers of collagen-based sutures. The apparent reason for the low antigenicity of type I collagen mostly stems from the small species difference among type I collagens (e.g., cow versus human). Such similarity is, in turn, probably understandable in terms of the inability of the triple helical configuration to incorporate the substantial amino acid substitutions that characterize species differences with other proteins. The relative constancy of the structure of the triple helix among the various species is, in fact, the reason why collagen is sometimes referred to as a "successful" protein in terms of its evolution or, rather, the relative lack of it.

In order to reduce the immunogenicity of collagen it is useful to consider the location of its antigenic determinants, i.e., the specific chemical groups that are recognized as foreign by the immunological system of the host animal. The configurational (or conformational) determinants of collagen depend on the presence of the intact triple helix and, consequently, are abolished when collagen is denatured into gelatin; the latter event (see earlier discussion) occurs spontaneously after the triple helix is cleaved by a collagenase. Gelatinization exposes effectively the sequential determinants of collagen over the short period during which gelatin retains its macromolecular character, before it is cleared away following attack by one of several nonspecific proteases. Control of the stability of the triple helix during processing of collagen, therefore, partially prevents the display of the sequential determinants.

Sequential determinants also exist in the nonhelical end (telopeptide region) of the collagen molecule, and

Table 3.2.8-2 Certain applications of collagen-based biomaterials

Application	Physical state
Sutures	Extruded tape (Schmitt, 1985)
Hemostatic agents	Powder, sponge, fleece (Stenzel *et al.* 1974; Chvapil, 1979)
Blood vessels	Extruded collagen tube, processed human or animal blood vessel (Nimni, 1988)
Heart valves	Processed porcine heart valve (Nimni, 1988)
Tendon, ligaments	Processed tendon (Piez, 1985)
Burn treatment (dermal regeneration)	Porous collagen–glycosaminoglycan (GAG) copolymers (Yannas *et al.*, 1982, 1989: Burke *et al.*, 1981; Heimbach *et al.*, 1988)
Peripheral nerve regeneration	Porous collagen–GAG copolymers (Chang and Yannas, 1992)
Meniscus regeneration	Porous collagen–GAG copolymers (Stone *et al.*, 1997)
Skin regeneration (plastic surgery)	Porous collagen–GAG copolymers
Intradermal augmentation	Injectable suspension of collagen particles (Piez, 1985)
Gynecological applications	Sponges (Chvapil, 1979)
Drug-delivery systems	Various forms (Stenzel *et al.*, 1974; Chvapil, 1979)

this region has been associated with most of the immunogenicity of collagen-based implants. Several enzymatic treatments have been devised to cleave the telopeptide region without destroying the triple helix. Treatment of collagen with glutaraldehyde not only reduces its degradation rate by collagenase but also appears to reduce its antigenicity. The mechanism of this effect is not well understood.

Certain applications of collagen-based biomaterials are shown in Table 3.2.8-2.

Proteoglycans and GAG

GAGs occur naturally as polysaccharide branches of a protein chain, or protein core, to which they are covalently attached via a specific oligosaccharide linkage. The entire branched macromolecule, which has been described as having a "bottle brush" configuration, is known as a proteoglycan and typically has a molecular weight of about 10^3 kDa.

The structure of GAGs can be generically described as that of an alternating copolymer, the repeat unit consisting of a hexosamine (glucosamine or galactosamine) and of another sugar (galactose, glucuronic acid, or iduronic acid). Individual GAG chains are known to contain occasional substitutions of one uronic acid for another; however, the nature of the hexosamine component remains invariant along the chain. There are other deviations from the model of a flawless alternating copolymer, such as variations in sulfate content along the chain. It is, nevertheless, useful for the purpose of getting acquainted with the GAGs to show their typical (rather, typified) repeat unit structure, as in Fig. 3.2.8-4. The molecular weight of many GAGs is in the range 5–60 kDa with the exception of hyaluronic acid, the only GAG which is not sulfated; it exhibits molecular weights in the range 50–500 kDa. Sugar units along GAG chains are linked by α or β glycosidic bonds that are 1,3 or 1,4 (Fig. 3.2.8-4). There are several naturally occurring enzymes which degrade specific GAGs, such as hyaluronidase and chondroitinase. These enzymes are primarily responsible for the physiological turnover rate of GAGs, which is in the range 2–14 days.

The nature of the oligosaccharide linkage appears to be identical for the GAGs, except for keratan sulfate, and is a galactosyl–galactosyl–xylose, with the last glycosidically linked to the hydroxyl group of serine in the protein core.

The very high molecular weight of hyaluronic acid is the basis of most uses of this GAG as a biomaterial: Almost all make use of the exceptionally high viscosity and the facility to form gels that characterize this polysaccharide. Hyaluronic acid gels have found considerable use in ophthalmology because they facilitate cataract surgery as well as retinal reattachment. Other reported uses of GAGs are in the treatment of degenerative joint dysfunction in horses and in the treatment of certain orthopedic dysfunctions in humans. On the other hand, sulfated GAGs are anionically charged and can induce precipitation of collagen at acidic pH levels, a process that yields collagen–GAG coprecipitates that can be subsequently freeze-dried and covalently cross-linked to yield biomaterials that have been shown capable of inducing regeneration of skin (dermis), peripheral nerves, and the conjunctiva (Table 3.2.8-2).

Elastin

Elastin is one of the least soluble protein in the body, consisting as it does of a three-dimensional cross-linked network. It can be extracted from tissues by dissolving and degrading all adjacent substances. It appears to be highly amorphous and thus has eluded elucidation of its structure by crystallographic methods. Fortunately, it exhibits ideal rubber elasticity and it thus becomes possible to study certain features of the macromolecular network. For example, mechanical measurements have shown that the average number of amino acid units between cross-links is 71–84. Insoluble elastin preparations can be degraded by the enzyme elastase. The soluble preparations prepared thereby have not yet been applied extensively as biomaterials.

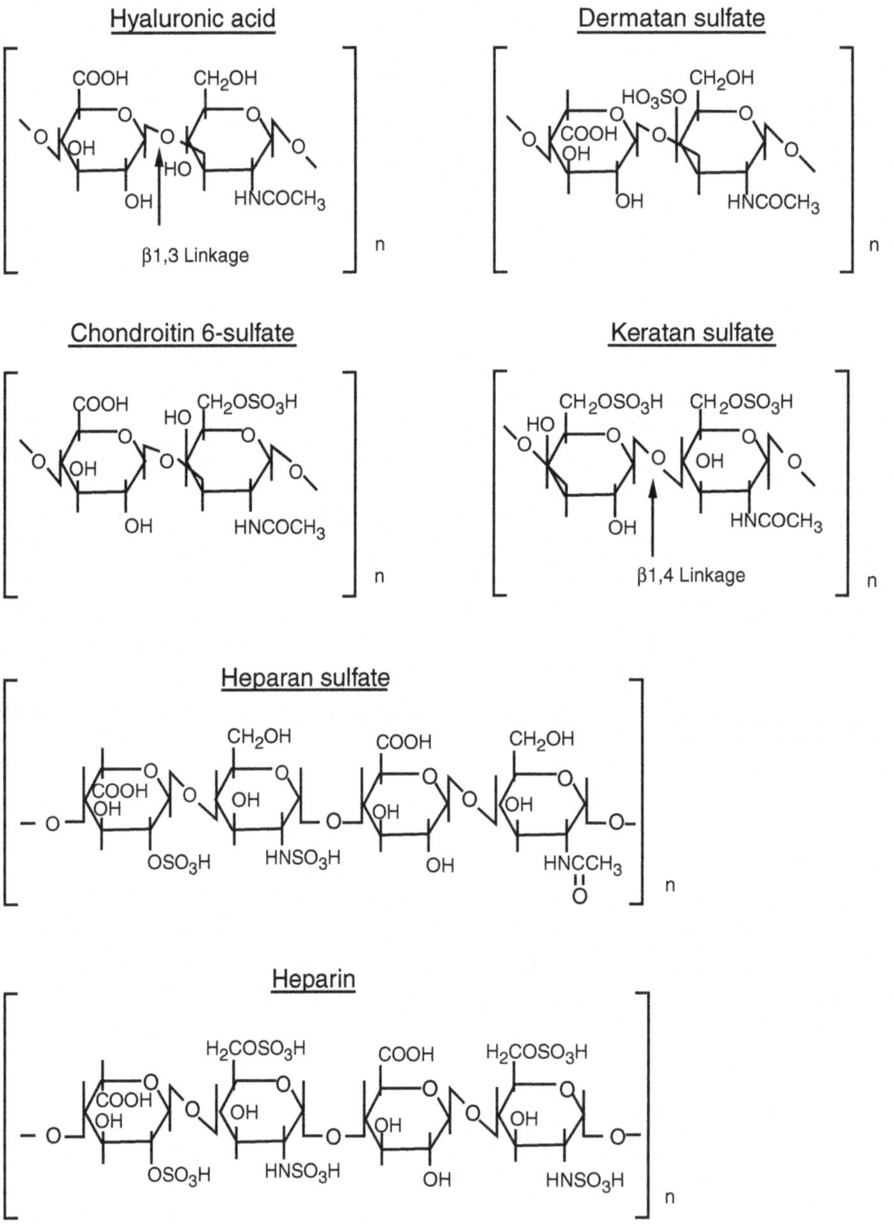

Fig. 3.2.8-4 Repeat units of GAGs. (Reprinted from J. Uitto and A. J. Perejda, editors (1987). *Connective Tissue Disease, Molecular Pathology of the Extracellular Matrix*, Vol. 12 in the series The Biochemistry of Disease. Marcel Dekker, New York, Chap. 4, Figs. 3.2.1 and 3.2.2, p. 85, with permission.)

Graft copolymers of collagen and GAGs

The preceding discussion in this chapter has focused on the individual macromolecular components of ECMs. Naturally occurring ECMs are insoluble networks comprising several macromolecular components. Several types of ECMs are known to play critical roles during organ development. During the past several years certain analogs of ECMs have been synthesized and have been studied as implants. This section summarizes the evidence for the unusual biological activity of a small number of ECM analogs.

In the 1970s it was discovered that a highly porous graft copolymer of type I collagen and chondroitin 6-sulfate was capable of modifying dramatically the kinetics and mechanism of healing of full-thickness skin wounds in rodents. In the adult mammal, full-thickness skin wounds represent anatomical sites that are demonstrably devoid of both epidermis and dermis, the two main tissues that comprise skin, respectively. Such wounds normally close by contraction of wound edges and by synthesis of scar tissue. Previously, collagen and various GAGs, each prepared in various forms such as powder and films, had been used to cover such deep wounds without observation of a significant modification in the outcome of the wound healing process.

Surprisingly, grafting of these wounds with the porous CG copolymer on guinea pig wounds blocked the onset of wound contraction by several days and led to synthesis of new connective tissue within about 3 weeks in the space occupied by the copolymer. The copolymer underwent substantial degradation during the 3-week period, at the end of which it had degraded completely at the wound site. Studies of the connective tissue synthesized in place of the degraded copolymer eventually showed that the new tissue was distinctly different from scar and was very similar, though not identical, to physiological dermis. In particular, new hair follicles and new sweat glands had not been synthesized. This marked the first instance where scar synthesis was blocked in a full-thickness skin wound of an adult mammal and, in its place, a nearly physiological dermis had been synthesized. That this result was not confined to guinea pigs was confirmed by grafting the same copolymer on full-thickness skin wounds in other adult mammals, including swine and, most importantly, human victims of massive burns as well as humans who underwent reconstructive surgery of the skin.

Although a large number of CG copolymers were synthesized and studied as grafts, it was observed that only one possessed the requisite activity to dramatically modify the wound healing process in skin. In view of the nature of its unique regenerative activity this biologically active macromolecular network has been referred to as dermis regeneration template (DRT). The structure of DRT required specification at two scales: At the nanoscale, the average molecular weight of the cross-linked network that was required to induce regeneration of the dermis was $12,500 \pm 5000$; at the microscale, the average pore diameter was between 20 and 120 μm. Relatively small deviations from these structural features led to loss of activity.

The regeneration of dermis was followed by regeneration of a quite different organ, the peripheral nerve. This was accomplished using a distinctly different ECM analog, termed nerve regeneration template (NRT). Although the chemical composition of the two templates was nearly identical, there were significant differences in other structural features. NRT degrades considerably more slowly than DRT (half-life of about 6 weeks for NRT compared to about 2 weeks for DRT) and is also characterized by a much smaller average pore diameter (about 5 μm compared to 20–120 μm for DRT). DRT was also shown capable of inducing regeneration of the conjunctiva, a specialized structure underneath the eyelid that provides for tearing and other functions that preserve normal vision. The mechanism of induced organ regeneration by templates appears to consist primarily of blocking of contraction of the injured site followed by synthesis of new physiological tissue at about the same rate that the tissue originally present is degraded (synchronous isomorphous replacement). These combined findings suggest that other ECM analogs, still to be discovered, could induce regeneration of organs such as a kidney or the pancreas.

Bibliography

Burke, J. F., Yannas, I. V., Quimby, W. C., Jr., Bondoc, C. C., and Jung, W. K. (1981). Successful use of a physiologically acceptable artificial skin in the treatment of extensive burn injury. *Ann. Surg.* **194:** 413–428.

Chamberlain, L. J., Yannas, I. V., Hsu, H-P., Strichartz, G., and Spector, M. (1998). Collagen-GAG substrate enhances the quality of nerve regeneration through collagen tubes up to level of autograft. *Exp. Neurol.* **154:** 315–329.

Chang, A. S., and Yannas, I. V. (1992). Peripheral nerve regeneration. in *Neuroscience Year* (Suppl. 2 to *The Encyclopedia of Neuroscience*), B. Smith and G. Adelman, eds. Birkhauser, Boston, pp. 125–126.

Chvapil, M. (1979). Industrial uses of collagen. in *Fibrous Proteins: Scientific, Industrial and Medical Aspects*, D. A. D. Parry and L. K. Creamer, eds. Academic Press, London, Vol. 1, pp.247–269.

Compton, C. C., Butler, C. E., Yannas, I. V., Warland, G., and Orgill, D. P. (1998). Organized skin structure is regenerated *in vivo* from collagen-GAG matrices seeded with autologous keratinocytes. *J. Invest. Dermatol.* **110**: 908–916.

Davidson, J. M. (1987). Elastin, structure and biology. in *Connective Tissue Disease*, J. Uitto and A. J. Perejda, eds. Marcel Dekker, New York, Chap. 2, pp. 29–54.

Heimbach, D., Luterman, A., Burke, J., Cram, A., Herndon, D., Hunt, J., Jordan, M., McManus, W., Solem, L., Warden, G., and Zawacki, B. (1988). Artificial dermis for major burns. *Ann. Surg.* **208**: 313–320.

Hsu, W-C., Spilker M. H., Yannas I. V., and Rubin P. A. D. (2000). Inhibition of conjunctival scarring and contraction by a porous collagen-GAG implant. *Invest. Ophthalmol. Vis. Sci.* 41: 2404–2411.

Kauzmann, W. (1959). Some factors in the interpretation of protein denaturation. *Adv. Protein Chem.* **14**: 1–63.

Li, S.-T. (1995). Biologic biomaterials: tissue-derived biomaterials (collagen). in *The Biomedical Engineering Handbook*, J. D.Bronzino, ed. CRC Press, Boca Raton, FL, Chap. 45, pp. 627–647.

Nimni, M. E., editor. (1988). *Collagen*, Vol. III, *Biotechnology*. CRC Press, Boca Raton, FL.

Piez, K. A. (1985). Collagen. in *Encyclopedia of Polymer Science and Technology*, Vol. 3, pp. 699–727.

Schmitt, F. O. (1985). Adventures in molecular biology. *Ann. Rev.Biophys. Biophys. Chem.* **14**: 1–22.

Shalaby, S. W. (1995). Non-blood-interfacing implants for soft tissues in *The Biomedical Engineering Handbook*, J. D. Bronzino, ed. CRC Press, Boca Raton, FL, Chap. 46.2, pp. 665–671.

Silbert, J. E. (1987). Advances in the biochemistry of proteoglycans. in *Connective Tissue Disease*, J. Uitto and A. J. Perejda, eds. Marcel Dekker, New York, Chap. 4, pp. 83–98.

Stenzel, K. H., Miyata, T., and Rubin, A. L. (1974). Collagen as a biomaterial. in *Annual Review of Biophysics and Bioengineering*, L. J. Mullins, ed. Annual Reviews Inc., Palo Alto, CA, Vol. 3, pp. 231–252.

Stone, K. R., Steadman, R., Rodkey, W. G., and Li, S.-T. (1997). Regeneration of meniscal cartilage with use of a collagen scaffold. *J. Bone Joint Surg.* 79-A: 1770–1777.

Yannas, I. V. (1972). Collagen and gelatin in the solid state. *J. Macromol. Sci.-Revs. Macromol. Chem.* C7(1): 49–104.

Yannas, I. V., Burke, J. F., Orgill, D. P., and Skrabut, E. M. (1982). Wound tissue can utilize a polymeric template to synthesize a functional extension of skin. *Science* **215**: 174–176.

Yannas, I. V., Lee, E., Orgill, D. P., Skrabut, E. M., and Murphy, G. F. (1989). Synthesis and characterization of a model extracellular matrix which induces partial regeneration of adult mammalian skin. *Proc. Natl. Acad. Sci. USA* **86**: 933–937.

Yannas, I. V. (1990). Biologically active analogs of the extracellular matrix. *Angew. Chem. Int. Ed.* **29**: 20–35.

Yannas, I. V. (1997). In vivo synthesis of tissue and organs. in *Principles of Tissue Engineering*, R. P. Lanza, R. Langer, and W. L. Chick, eds. R. G. Landes, Austin, Chap. 12, pp. 169–178.

Yannas, I. V. (2004). Synthesis of tissues and organs. *Chembiochem.* **5**(1): 26–39.

Yannas, I. V. (2001). Tissue and Organ Regeneration in Adults. New York: Springer.

Yannas, I. V., and Hill, B. J. (2004). Selection of biomaterials for peripheral nerve regeneration using data from the nerve chamber model. *Biomaterials.* **25**(9): 1593–1600.

3.2.9 Metals

John B. Brunski

Introduction

Implant materials in general, and metallic implant materials in particular, have a significant economic and clinical impact on the biomaterials field. The worldwide market for all types of biomaterials was estimated at over $5 billion in the late 1980s, but grew to about $20 billion in 2000 and is likely to exceed $23 billion by 2005. With the recent emergence of the field known as tissue engineering, including its strong biomaterials segment, the rate of market growth has been estimated at about 12–20% per year.

For the United States, the biomaterials market has been estimated at about $9 billion as of the year 2000, with a growth rate of about 20% per year. The division of this market into various submarkets is illustrated by older data: in 1991 the total orthopedic implant and instrument market was about $2 billion and was made up of joint prostheses made primarily of metallic materials ($1.4 billion), together with a wide variety of trauma products ($0.340 billion), instrumentation devices ($0.266 billion), bone cement accessories ($0.066 billion), and bone replacement materials ($0.029 billion). Estimates for other parts of the biomaterials market include $0.425 billion for oral and maxillofacial implants and $0.014 billion for periodontal treatments, and materials for alveolar ridge augmentation or maintenance.

Estimates of the size of the total global biomaterials market are substantiated by the statistics on clinical procedures. For example, of the approximately 3.6 million orthopedic operations per year in the United States, four of the 10 most frequent involve metallic implants: open reduction of a fracture and internal fixation (1 on the list); placement or removal of an internal fixation device without reduction of a fracture (6); arthroplasty of the knee or ankle (7), and total hip replacement or arthroplasty of the hip (8). Moreover, 1988 statistics show that although reduction of fractures was first on the list of inpatient procedures (631,000 procedures), second on the list was excision or destruction of an intervertebral disk (250,000 procedures). Since the latter often involves a bone graft of some kind (from the same patient of from a bone bank) and internal fixation with plates and screws, this represents yet another clinical procedure involving significant use of biomaterials. Overall, including all clinical specialties in 1988, statistics showed that about 11 million Americans (about 4.6% of the civilian population) had at least one implant (Moss *et al.*, 1990).

In view of this wide utilization of implants, many of which are metallic, the objective of this section is to describe the composition, structure, and properties of current metallic implant alloys. Major themes are the metallurgical principles underlying structure-property relationships, and the role that biomaterials play in the larger problem of design, production, and proper utilization of medical devices.

Steps in the fabrication of implants

Understanding the structure and properties of metallic implant materials requires an appreciation of the metallurgical significance of the material's processing history. Since each metallic device will ordinarily differ in exactly how it is manufactured, generic processing steps are outlined in Fig. 3.2.9-1A.

Metal-containing ore to raw metal product

With the exception of the noble metals (which do not represent a major fraction of implant metals), metals exist in the Earth's crust in mineral form wherein the metal is chemically combined with other elements, as in the case of metal oxides. These mineral deposits (ore) must be located and mined, and then separated and

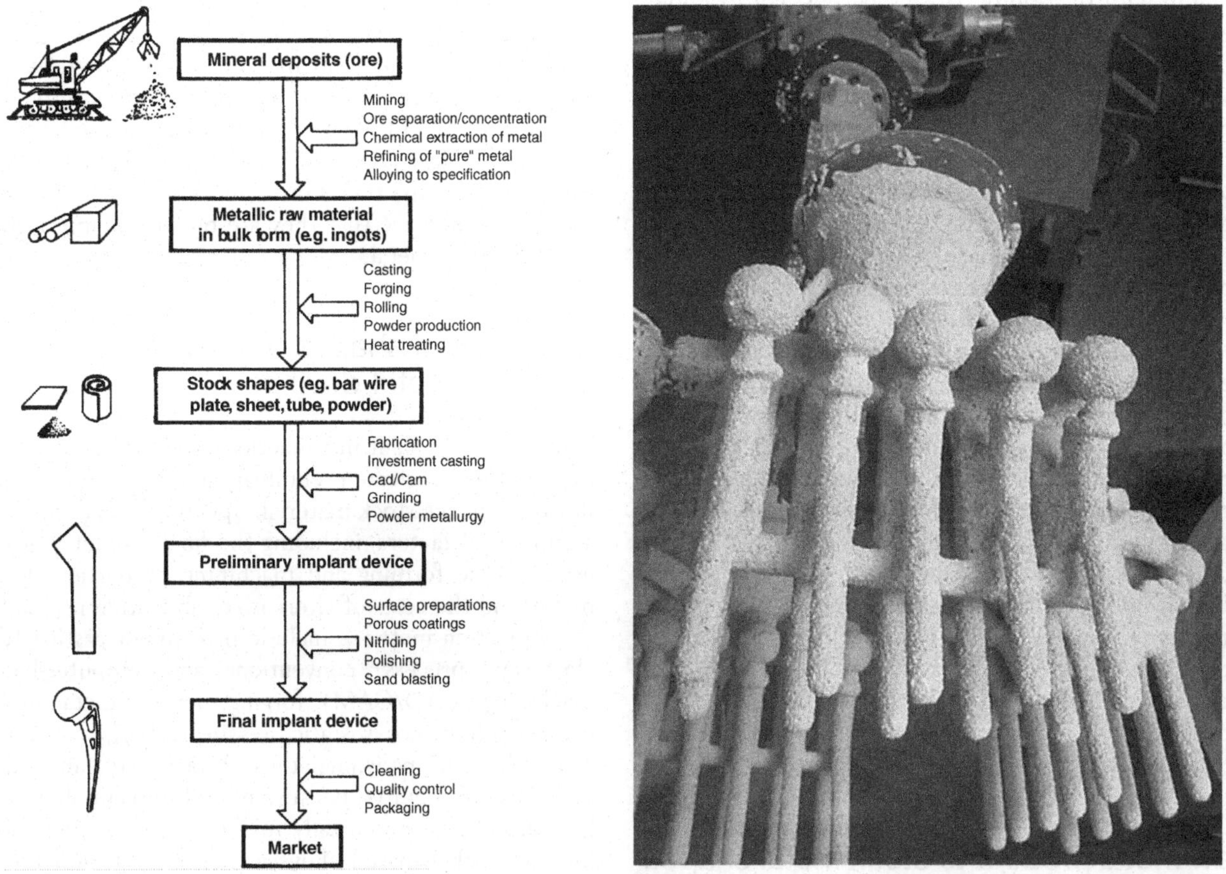

A B

Fig. 3.2.9-1 (A) Generic processing history of a typical metallic implant device, in this case a hip implant. (B) Image of one step during the investment casting ("lost wax") process of manufacturing hip stems; a rack of hip stems can be seen attached to a system of sprues through which molten metal can flow. At this point, ceramic investment material composes the mold into which the molten metal will flow and solidify during casting, thereby replicating the intended shape of a hip stem.

enriched to provide ore suitable for further processing into pure metal and/or various alloys.

For example, with titanium, certain mines in the southeastern United States yield sands containing primarily common quartz but also mineral deposits of zircon, titanium, iron, and rare earth elements. The sandy mixture can be concentrated by using water flow and gravity to separate out the metal-containing sands into titanium-containing compounds such as rutile (TiO_2) and ilmenite ($FeTiO_3$). To obtain rutile, which is particularly good for making metallic titanium, further processing typically involves electrostatic separations. Then, to extract titanium metal from the rutile, one method involves treating the ore with chlorine to make titanium tetrachloride liquid, which in turn is treated with magnesium or sodium to produce chlorides of the latter metals and bulk titanium "sponge" according to the Kroll process. At this stage, the titanium sponge is not of controlled purity. So, depending on the purity grade desired in the final titanium product, it is necessary to refine it further by using vacuum furnaces, remelting, and additional steps. All of this can be critical in producing titanium with the appropriate properties. For example, the four most common grades of commercially pure (CP) titanium differ in oxygen content by only tenths of a percent, but these small differences in oxygen content can make major differences in mechanical properties such as yield and tensile and fatigue strength of titanium, as discussed later in this section. In any case, from the preceding extraction steps, the resulting raw metal product eventually emerges in some type of bulk form, such as ingots, which can be supplied to raw materials vendors or metal manufacturers.

In the case of multicomponent metallic implant alloys, the raw metal product will usually have to be processed further both chemically and physically. Processing steps include remelting, the addition of alloying elements, and controlled solidification to produce an alloy that meets certain chemical and metallurgical specifications. For example, to make ASTM (American Society for Testing and Materials) F138 316L stainless steel, iron is alloyed with specific amounts of carbon, silicon, nickel, and chromium. To make ASTM F75 or F90 alloy, cobalt is alloyed with specific amounts of chromium, molybdenum, carbon, nickel, and other elements. Table 3.2.9-1 lists the chemical compositions of some metallic alloys for surgical implants.

Raw metal product to stock metal shapes

A metal supplier further processes the bulk raw metal product (metal or alloy) into "stock" bulk shapes, such as bars, wire, sheet, rods, plates, tubes, or powders. These stock shapes may then be sold to specialty companies (e.g., implant manufacturers) who need stock metal that is closer to the final form of the implant. For example, a maker of screw-shaped dental implants might want to buy rods of the appropriate metal to simplify the machining of the screws from the rod stock.

The metal supplier might transform the metal product into stock shapes by a variety of processes, including remelting and continuous casting, hot rolling, forging, and cold drawing through dies. Depending on the metal, there may also be heat-treating steps (carefully controlled heating and cooling cycles) designed to facilitate further working or shaping of the stock; relieve the effects of prior plastic deformation (e.g., as in annealing); or produce a specific microstructure and properties in the stock material. Because of the high chemical reactivity of some metals at elevated temperatures, high-temperature processes may require vacuum conditions or inert atmospheres to prevent unwanted uptake of oxygen by the metal, all of which adds to cost. For instance, in the production of fine powders of ASTM F75 Co–Cr–Mo alloy, molten metal is often ejected through a small nozzle to produce a fine spray of atomized droplets that solidify while cooling in an inert argon atmosphere.

For metallic implant materials in general, stock shapes are often chemically and metallurgically tested at this early stage to ensure that the chemical composition and microstructure of the metal meet industry standards for surgical implants (ASTM Standards), as discussed later in this section. In other words, an implant manufacturer will want assurance that they are buying an appropriate grade of stock metal.

Stock metal shapes to preliminary and final metal devices

Typically, an implant manufacturer will buy stock material and then fabricate preliminary and final forms of the device from the stock material. Specific steps depend on a number of factors, including the final geometry of the implant, the forming and machining properties of the metal, and the costs of alternative fabrication methods.

Fabrication methods include investment casting (the "lost wax" process), conventional and computer-based machining (CAD/CAM), forging, powder metallurgical processes (e.g., hot isostatic pressing (HIP)), and a range of grinding and polishing steps. A variety of fabrication methods are required because not all implant alloys can be feasibly or economically made in the same way. For instance, cobalt-based alloys are extremely difficult to machine by conventional methods into the complicated shapes of some implants. Therefore, many cobalt-based alloys are frequently shaped into implant forms by investment casting (e.g., Fig. 3.2.9-1B) or powder metallurgy. On the other hand, titanium is relatively difficult to

Table 3.2.9-1 Chemical compositions of stainless steels used for implants

Material	ASTM designation	Common/trade names	Composition (wt.%)	Notes
Stainless steel	F55 (bar, wire)	AISI 316 LVM	60–65 Fe	F55, F56 specify 0.03 max for P,S.
	F56 (sheet, strip) F138 (bar, wire)	316L 316L	17.00–20.00 Cr 12.00–14.00 Ni	F138, F139 specify 0.025 max for P and 0.010 max for S.
	F139 (sheet, strip)	316L	2.00–3.00 Mo	LVM = low vacuum melt.
			max 2.0 Mn	
			max 0.5 Cu	
			max 0.03 C	
			max 0.1 N	
			max 0.025 P	
			max 0.75 Si	
			max 0.01 S	
Stainless steel	F745	Cast stainless steel cast 316L	60–69 Fe 17.00–20.00 Cr	
			11.00–14.00 Ni	
			2.00–3.00 Mo	
			max 0.06 C	
			max 2.0 Mn	
			max 0.045 P	
			max 1.00 Si	
			max 0.030 S	

cast, and therefore is frequently machined even though titanium in general is not considered to be an easily machinable metal.

Another aspect of fabrication, which comes under the heading of surface treatment, involves the application of macro- or microporous coatings on implants, or the deliberate production of certain degrees of surface roughness. Such surface modifications have become popular in recent years as a means to improve fixation of implants in bone. The surface coating or roughening can take various forms and require different fabrication technologies. In some cases, this step of the processing history can contribute to metallurgical properties of the final implant device. For example, in the case of alloy beads or "fiber metal" coatings, the manufacturer applies the coating only over specific regions of the implant surface (e.g., on the proximal portion of a femoral stem), and the means by which such a coating is attached to the bulk substrate may involve a process such as high-temperature sintering. Generally, sintering involves heating the coating and substrate to about one-half or more of the alloy's melting temperature, which is meant to enable diffusive mechanisms to form necks that join the beads in the coating to one another and to the implant's surface (Fig. 3.2.9-2). Such temperatures can also modify the underlying metallic substrate.

An alternative surface treatment to sintering is plasma or flame spraying a metal onto an implant's surface. Hot, high-velocity gas plasma is charged with a metallic powder and directed at appropriate regions of an implant surface. The powder particles fully or partially melt and then fall onto the substrate surface, where they solidify rapidly to form a rough coating (Fig. 3.2.9-3).

Other surface treatments are also available, including ion implantation (to produce better surface properties), nitriding, and coating with a thin diamond film. In nitriding, a high-energy beam of nitrogen ions is directed at the implant under vacuum. Nitrogen atoms penetrate the surface and come to rest at sites in the substrate. Depending on the alloy, this process can produce

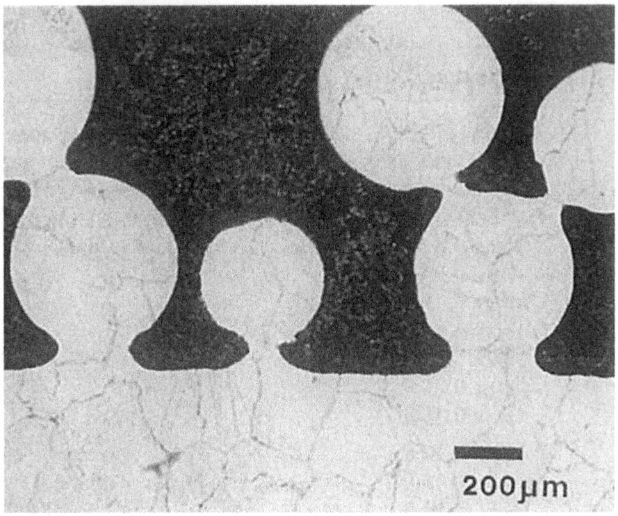

Fig. 3.2.9-2 Low-power view of the interface between a porous coating and solid substrate in the ASTM F75 Co–Cr–Mo alloy system. Note the structure and geometry of the necks joining the beads to one another and to the substrate. Metallographic cross section cut perpendicular to the interface; lightly etched to show the microstructure. (Photo courtesy of Smith & Nephew Richards, Inc. Memphis, TN.)

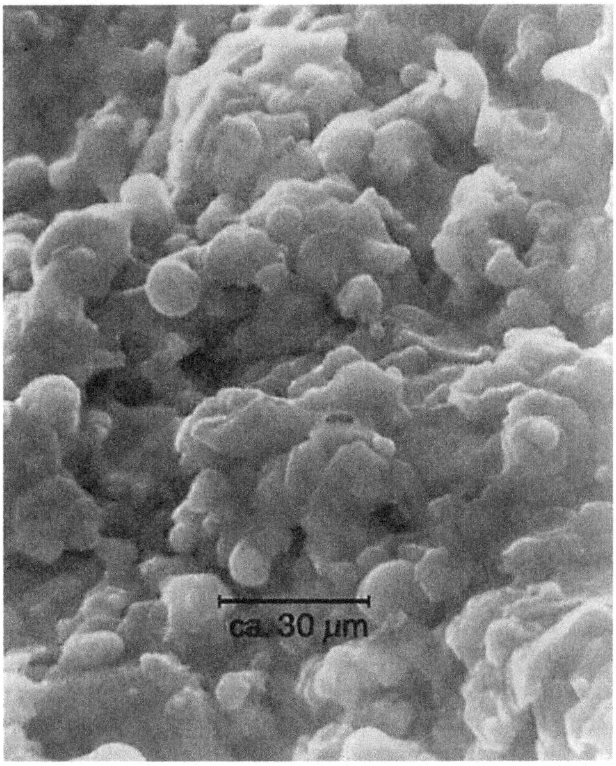

Fig. 3.2.9-3 Scanning electron micrograph of a titanium plasma spray coating on an oral implant. (Photo courtesy of A. Schroeder, E. Van der Zypen, H. Stich, and F. Sutter, *Int. J. Oral Maxillofacial Surg.* 9:15,1981.)

enhanced properties. These treatments are commonly used to increase surface hardness and wear properties.

Finally, the manufacturer of a metallic implant device will normally perform a set of finishing steps. These vary with the metal and manufacturer, but typically include chemical cleaning and passivation (i.e., rendering the metal inactive) in appropriate acid, or electrolytically controlled treatments to remove machining chips or impurities that may have become embedded in the implant's surface. As a rule, these steps are conducted according to good manufacturing practice (GMP) and ASTM specifications for cleaning and finishing implants.

It is worth emphasizing that these steps can be extremely important to the overall biological performance of the implant because they can affect the surface properties of the medical device, which is the surface that comes in direct contact with the blood and other tissues at the implant site.

Microstructures and properties of implant metals

In order to understand the properties of each alloy system in terms of microstructure and processing history, it is essential to know (1) the chemical and crystallographic identities of the phases present in the microstructure; (2) the relative amounts, distribution, and orientation of these phases; and (3) the effects of the phases on properties. This section of the chapter emphasizes mechanical properties of metals used in implant devices even though other properties, such as surface properties and wear properties, must also be considered and may actually be more critical to control in certain medical device applications. (Surface properties of materials are reviewed in more depth in Section 3.1.4 of this book.) The following discussion of implant alloys is divided into the stainless steels, cobalt-based alloys, and titanium-based alloys, since these are the most commonly used metals in medical devices.

Stainless steels

Composition

Although several types of stainless steels are available for implant use (Table 3.2.9-1), in practice the most common is 316L (ASTM F138, F139), grade 2. This steel has less than 0.030% (wt.%) carbon in order to reduce the possibility of *in vivo* corrosion. The "L" in the designation 316L denotes low carbon content. The 316L alloy is predominantly iron (60–65%) with significant alloying additions of chromium (17–20%) and nickel (12–14%), plus minor amounts of nitrogen, manganese, molybdenum, phosphorus, silicon, and sulfur.

With 316L, the main rationale for the alloying additions involves the metal's surface and bulk microstructure. The key function of chromium is to permit the development of corrosion-resistant steel by forming a strongly adherent surface oxide (Cr_2O_3). However, the downside to adding Cr is that it tends to stabilize the ferritic (body-centered cubic (BCC)) phase of iron and steel, which is weaker than the austenitic (face-centered cubic (FCC)) phase. Moreover, molybdenum and silicon are also ferrite stabilizers. So to counter this tendency to form weaker ferrite, nickel is added to stabilize the stronger austenitic phase.

The main reason for the low carbon content in 316L is to improve corrosion resistance. If the carbon content of the steel significantly exceeds 0.03%, there is increased danger of formation of carbides such as $Cr_{23}C_6$. Such carbides have the bad habit of tending to precipitate at grain boundaries when the carbon concentration and thermal history are favorable to the kinetics of carbide growth. The negative effect of carbide precipitation is that it depletes the adjacent grain boundary regions of chromium, which in turn has the effect of diminishing formation of the protective, chromium-based oxide Cr_2O_3. Steels in which such grain-boundary carbides have formed are called "sensitized" and are prone to fail through corrosion-assisted fractures that originate at the sensitized (weakened) grain boundaries.

Microstructure and mechanical properties

Under ASTM specifications, the desirable form of 316L is single-phase austenite (FCC); there should be no free ferritic (BCC) or carbide phases in the microstructure. Also, the steel should be free of inclusions or impurity phases such as sulfide stringers, which can arise primarily from unclean steel-making practices and predispose the steel to pitting-type corrosion at the metal-inclusion interfaces.

The recommended grain size for 316L is ASTM #6 or finer. The ASTM grain size number n is defined by the formula:

$$N = 2^{n-1} \quad (3.2.9.1)$$

where N is the number of grains counted in 1 square inch at 100-times magnification (0.0645 mm^2 actual area). As an example, when $n = 6$, the grain size is about 100 microns or less. Furthermore, the grain size should be relatively uniform throughout (Fig. 3.2.9-4A). The emphasis on a fine grain size is explained by a Hall–Petch-type relationship (Hall, 1951; Petch, 1953) between mechanical yield stress and grain diameter:

$$t_y = t_i + kd^{-m} \quad (3.2.9.2)$$

Here t_y and t_i are the yield and friction stress, respectively; d is the grain diameter; k is a constant associated with propagation of deformation across grain boundaries; and m is approximately 0.5. From this equation it follows that higher yield stresses may be achieved by a metal with a smaller grain diameter d, all other things being equal. A key determinant of grain size is manufacturing history, including details on solidification conditions, cold-working, annealing cycles, and recrystallization.

Another notable microstructural feature of 316L as used in typical implants is plastic deformation within grains (Fig. 3.2.9-4B). The metal is often used in a 30% cold-worked state because cold-worked metal has a markedly increased yield, ultimate tensile, and fatigue strength relative to the annealed state (Table 3.2.9-2). The trade-off is decreased ductility, but ordinarily this is not a major concern in implant products.

In specific orthopedic devices such as bone screws made of 316L, texture may also be a notable feature in the microstructure. Texture means a preferred orientation of deformed grains. Stainless steel bone screws show elongated grains in metallographic sections taken parallel to the long axis of the screws (Fig. 3.2.9-5). Texture arises as a result of the cold drawing or similar cold-working operations inherent in the manufacture of bar rod stock from which screws are usually machined. In metallographic sections taken perpendicular to the screw's long axis, the grains appear more equiaxed. A summary of representative mechanical properties of 316L stainless is provided in Table 3.2.9-2, but this should only be taken as a general guide, given that final production steps specific to a given implant may often affect properties of the final device.

Cobalt-based alloys

Composition

Cobalt-based alloys include Haynes-Stellite 21 and 25 (ASTM F75 and F90, respectively), forged Co–Cr–Mo alloy (ASTM F799), and multiphase (MP) alloy MP35N (ASTM F562). The F75 and F799 alloys are virtually identical in composition (Table 3.2.9-3), each being about 58–70% Co and 26–30% Cr. The key difference is their processing history, as discussed later. The other two alloys, F90 and F562, have slightly less Co and Cr, but more Ni in the case of F562, and more tungsten in the case of F90.

Microstructures and properties

ASTM F75

The main attribute of this alloy is corrosion resistance in chloride environments, which is related to its bulk composition and surface oxide (nominally Cr_2O_3). This alloy has a long history in both the aerospace and biomedical implant industries.

When F75 is cast into shape by investment casting ("lost wax" process), the alloy is melted at 1350–1450°C and then poured or pressurized into ceramic molds of the

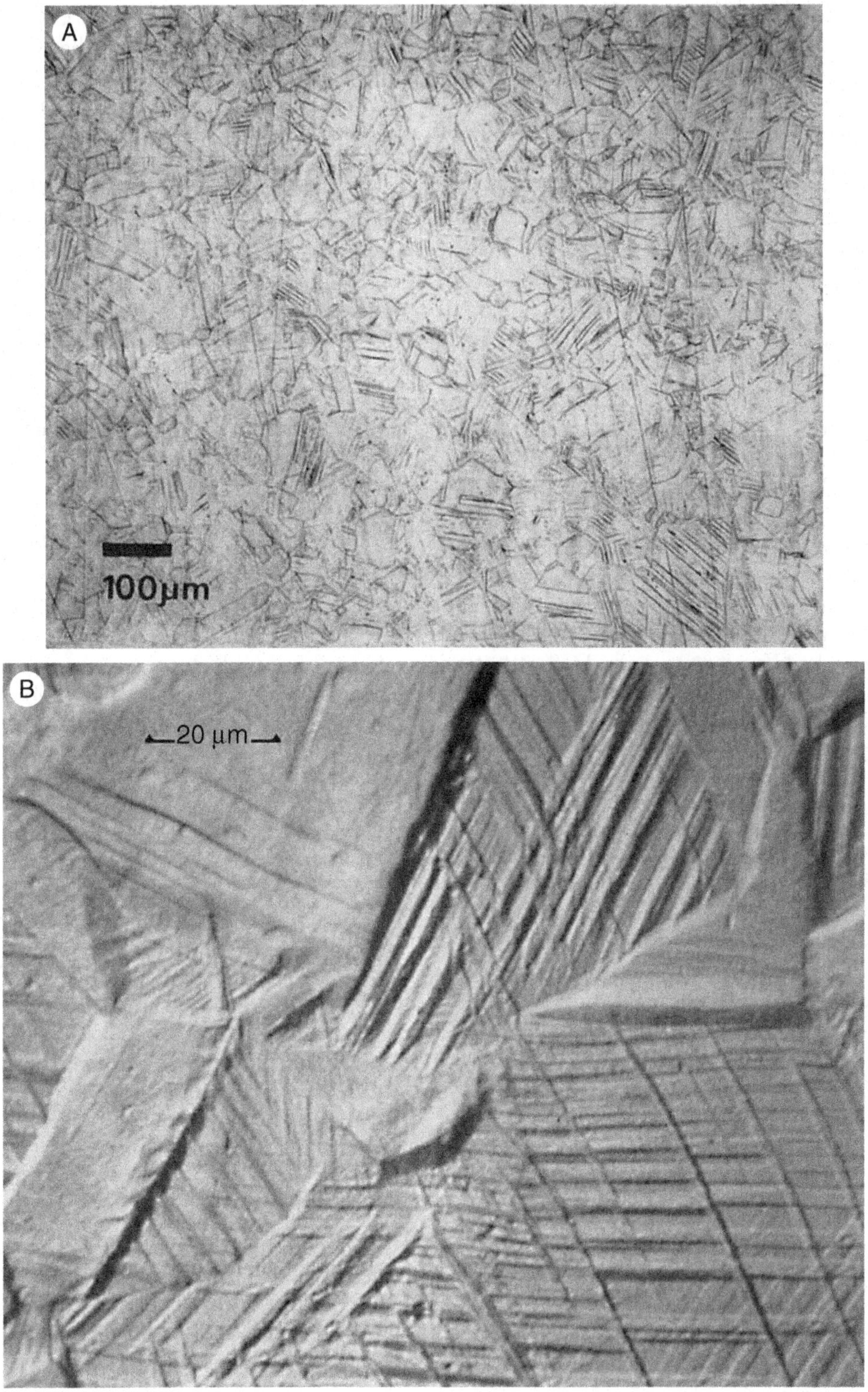

Fig. 3.2.9-4 (A) Typical microstructure of cold-worked 316L stainless steel, ASTM F138, in a transverse section taken through a spinal distraction rod. (B) Detail of grains in cold-worked 316L stainless steel showing evidence of plastic deformation. (Photo in B courtesy of Zimmer USA, Warsaw, IN.)

Table 3.2.9-2 Typical mechanical properties of implant metals[a]

Material	ASTM designation	Condition	Young's modulus (GPa)	Yield strength (MPa)	Tensile strength (MPa)	Fatigue endurance limit (at 10^7 cycles, $R = -1$[c]) (MPa)
Stainless steel	F745	Annealed	190	221	483	221–280
	F55, F56, F138, F139	Annealed	190	331	586	241–276
		30% Cold worked	190	792	930	310–448
		Cold forged	190	1213	1351	820
Co–Cr alloys	F75	As-cast/annealed	210	448–517	655–889	207–310
		P/M HIP[b]	253	841	1277	725–950
	F799	Hot forged	210	896–1200	1399–1586	600–896
	F90	Annealed	210	448–648	951–1220	Not available
		44% Cold worked	210	1606	1896	586
	F562	Hot forged	232	965–1000	1206	500
		Cold worked, aged	232	1500	1795	689–793 (axial tension $R = 0.05$, 30 Hz)
Ti alloys	F67	30% Cold-worked Grade 4	110	485	760	300
	F136	Forged annealed	116	896	965	620
		Forged, heat treated	116	1034	1103	620–689

[a] Data collected from references noted at the end of this section, especially table 1 in Davidson and Georgette (1986).
[b] P/M HIP; Powder metallurgy product, hot-isostatically pressed.
[c] R is defined as $\sigma_{min}/\sigma_{max}$.

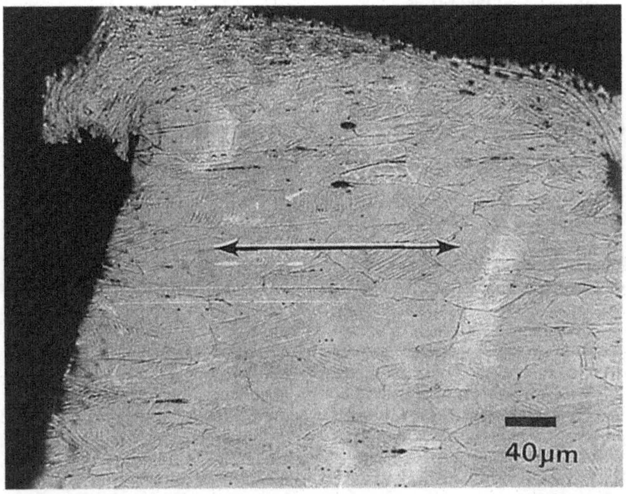

Fig. 3.2.9-5 Evidence of textured grain structure in 316L stainless steel ASTM F138, as seen in a longitudinal section through a cold-worked bone screw. The long axis of the screw is indicated by the arrow.

desired shape (e.g., femoral stems for artificial hips, oral implants, dental partial bridgework). The sometimes intricately shaped molds are made by fabricating a wax pattern to near-final dimensions of the implant and then coating (or investing) the pattern with a special ceramic, which then holds its shape after the wax is burned out prior to casting—hence the "lost wax" name of the process. Molten metal is poured into the ceramic mold through sprues, or pathways. Then, once the metal has solidified into the shape of the mold, the ceramic mold is cracked away and processing of the metal continues toward the final device.

Depending on the exact casting details, this process can produce at least three microstructural features that can strongly influence implant properties, often negatively.

First, as-cast F75 alloy (Figs. 3.2.9-6 and 3.2.9-7A) typically consists of a Co-rich matrix (alpha phase) plus interdendritic and grain-boundary carbides (primarily $M_{23}C_6$, where M represents Co, Cr, or Mo).

There can also be interdendritic Co and Mo-rich sigma intermetallic, and Co-based gamma phases. Overall, the

Table 3.2.9-3 Chemical compositions of co-based alloys for implants

Material	ASTM designation	Common trade names	Composition (wt.%)	Notes
Co–Cr–Mo	F75	Vitallium Haynes-Stellite 21 Protasul-2 Micrograin-Zimaloy	58.9–69.5 Co 27.0–30.0 Cr 5.0–7.0 Mo max 1.0 Mn max 1.0 Si max 2.5 Ni max 0.75 Fe max 0.35 C	Vitallium is a trade mark of Howmedica, Inc. Hayness-Stellite 21 (HS 21) is a trademark of Cabot Corp. Protasul-2 is a trademark of Sulzer AG, Switzerland. Zimaloy is a trademark of Zimmer USA.
Co–Cr–Mo	F799	Forged Co–Cr–Mo Thermomechanical Co–Cr–Mo FHS	58–59 Co 26.0–30.0 Cr 5.0–7.00 Mo max 1.00 Mn max 1.00 Si max 1.00 Ni max 1.5 Fe max 0.35 C max 0.25 N	FHS means, "forged high strength" and is a trademark of Howmedica, Inc.
Co–Cr–W–Ni	F90	Haynes-Stellite 25 Wrought Co–Cr	45.5–56.2 Co 19.0–21.0 Cr 14.0–16.0 W 9.0–11.0 Ni max 3.00 Fe 1.00–2.00 Mn 0.05–0.15 C max 0.04 P max 0.40 Si max 0.03 S	Haynes-Stellite 25 (HS25) is a trademark of Cabot Corp.
Co–Ni–Cr–Mo–Ti	F562	MP 35N Biophase Protasul-1( )	29–38.8 Co 33.0–37.0 Ni 19.0–21.0 Cr 9.0–10.5 Mo max 1.0 Ti max 0.15 Si max 0.010 S max 1.0 Fe max 0.15 Mn	MP35 N is a trademark of SPS Technologies, Inc. Biophase is a trademark of Richards Medical Co. Protasul-10 is a trademark of Sulzer AG, Switzerland.

relative amounts of the alpha and carbide phases should be approximately 85% and 15%, respectively. However, because of nonequilibrium cooling, a "cored" microstructure can develop. In this situation, the interdendritic regions become solute (Cr, Mo, C) rich and contain carbides, while the dendrites become depleted in Cr and richer in Co. This is an unfavorable electrochemical situation, with the Cr-depleted regions being anodic with respect to the rest of the microstructure. (This is also an unfavorable situation if a porous coating will subsequently be applied by sintering to this bulk metal.) Subsequent solution-anneal heat treatments at 1225°C for 1 hour can help alleviate this situation.

Second, the solidification during the casting process results not only in dendrite formation, but also in a relatively large grain size. This is generally undesirable because it decreases the yield strength via a Hall–Petch relationship between yield strength and grain diameter (see Eq. 3.2.9.2 in the section on stainless steel). The dendritic growth patterns and large grain diameter (~4 mm) can be easily seen in Fig. 3.2.9-7A, which shows a hip stem manufactured by investment casting.

Third, casting defects may arise. Figure 3.2.9-7B shows an inclusion in the middle of a femoral hip stem. The inclusion was a particle of the ceramic mold (investment) material, which presumably broke off and became entrapped within the interior of the mold while the metal was solidifying. This contributed to a fatigue fracture of the implant device *in vivo*, most likely because of stress concentrations and crack initiation sites associated with the ceramic inclusion. For similar reasons, it is also desirable to avoid macro- and microporosity arising

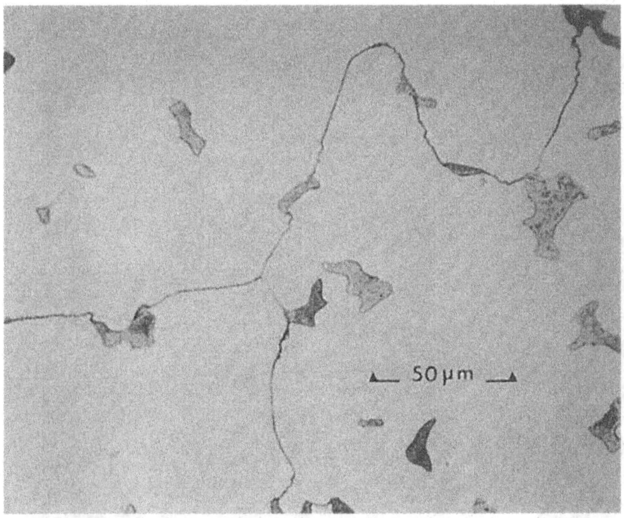

Fig. 3.2.9-6 Microstructure of as-cast Co–Cr–Mo ASTM F75 alloy, showing a large grain size plus grain boundary and matrix carbides. (Photo courtesy of Zimmer USA, Warsaw, IN.)

from metal shrinkage upon solidification of castings. Figures 3.2.9-7C and 3.2.9-7D exemplify a markedly dendritic microstructure, large grain size, and evidence of microporosity at the fracture surface of a ASTM F75 dental device fabricated by investment casting.

To avoid problems such as the above with cast F75, and to improve the alloy's microstructure and mechanical properties, powder metallurgical techniques have been used. For example, in HIP, a fine powder of F75 alloy is compacted and sintered together under appropriate pressure and temperature conditions (about 100 MPa at 1100°C for 1 hour) and then forged to final shape. The typical microstructure (Fig. 3.2.9-8) shows a much smaller grain size (~8 µm) than the as-cast material. Again, according to a Hall–Petch relationship, this microstructure gives the alloy higher yield strength and better ultimate and fatigue properties than the as-cast alloy (Table 3.2.9-2). Generally speaking, the improved properties of the HIP versus cast F75 result from both the finer grain size and a finer distribution of carbides, which has a hardening effect as well.

In porous-coated prosthetic devices based on F75 alloy, the microstructure will depend on the prior manufacturing history of the beads and substrate metal as well as on the sintering process used to join the beads together and to the underlying bulk substrate. With Co–Cr–Mo alloys, for instance, sintering can be difficult, requiring temperatures near the melting point (1225°C). Unfortunately, these high temperatures can decrease the fatigue strength of the substrate alloy. For example, cast-solution-treated F75 has a fatigue strength of about 200–250 MPa, but it can decrease to about 150 MPa after porous coating treatments. The reason for this decrease probably relates to further phase changes in the non-equilibrium cored microstructure in the original cast F75 alloy. However, it has been found that a modified sintering treatment can return the fatigue strength back up to about 200 MPa (Table 3.2.9-2).

Beyond these metallurgical issues, a related concern with porous-coated devices is the potential for decreased fatigue performance due to stress concentrations inherent in the geometrical features where particles are joined to the substrate (e.g., Fig. 3.2.9-2).

ASTM F799

The F799 alloy is basically a modified F75 alloy that has been mechanically processed by hot forging (at about 800° C) after casting. It is sometimes known as thermo-mechanical Co–Cr–Mo alloy and has a composition slightly different from that of ASTM F75. The microstructure reveals a more worked grain structure than as-cast F75 and a hexagonal close-packed (HCP) phase that forms via a shear-induced transformation of FCC matrix to HCP platelets. This microstructure is not unlike that which occurs in MP35N (see ASTM F562).

The fatigue, yield, and ultimate tensile strengths of this alloy are approximately twice those of as-cast F75 (Table 3.2.9-2).

ASTM F90

Also known as Haynes Stellite 25 (HS-25), F90 alloy is based on Co–Cr–W–Ni. Tungsten and nickel are added to improve machinability and fabrication. In the annealed state, its mechanical properties are about the same as those of F75 alloy, but when cold worked to 44%, the properties more than double (Table 3.2.9-2).

ASTM F562

Known as MP35N, F562 alloy is primarily Co (29–38.8%) and Ni (33–37%), with significant amounts of Cr and Mo. The "MP" in the name refers to the multiple phases in its microstructure. The alloy can be processed by thermal treatments and cold working to produce a controlled microstructure and a high-strength alloy, as follows.

To start with, under equilibrium conditions pure solid cobalt has an FCC Bravais lattice above 419°C and a HCP structure below 419°C. However, the solid-state transformation from FCC to HCP is sluggish and occurs by a martensitic-type shear reaction in which the HCP phase forms with its basal planes (0001) parallel to the close-packed (111) planes in FCC. The ease of this transformation is affected by the stability of the FCC phase, which in turn is affected by both plastic deformation and alloying additions. Now, when cobalt is alloyed to make MP35N, the processing includes 50% cold work, which increases the driving force for the transformation of the FCC to the HCP phase. The HCP phase emerges as fine platelets within FCC grains. Because the FCC grains are small (0.01–0.1 µm,

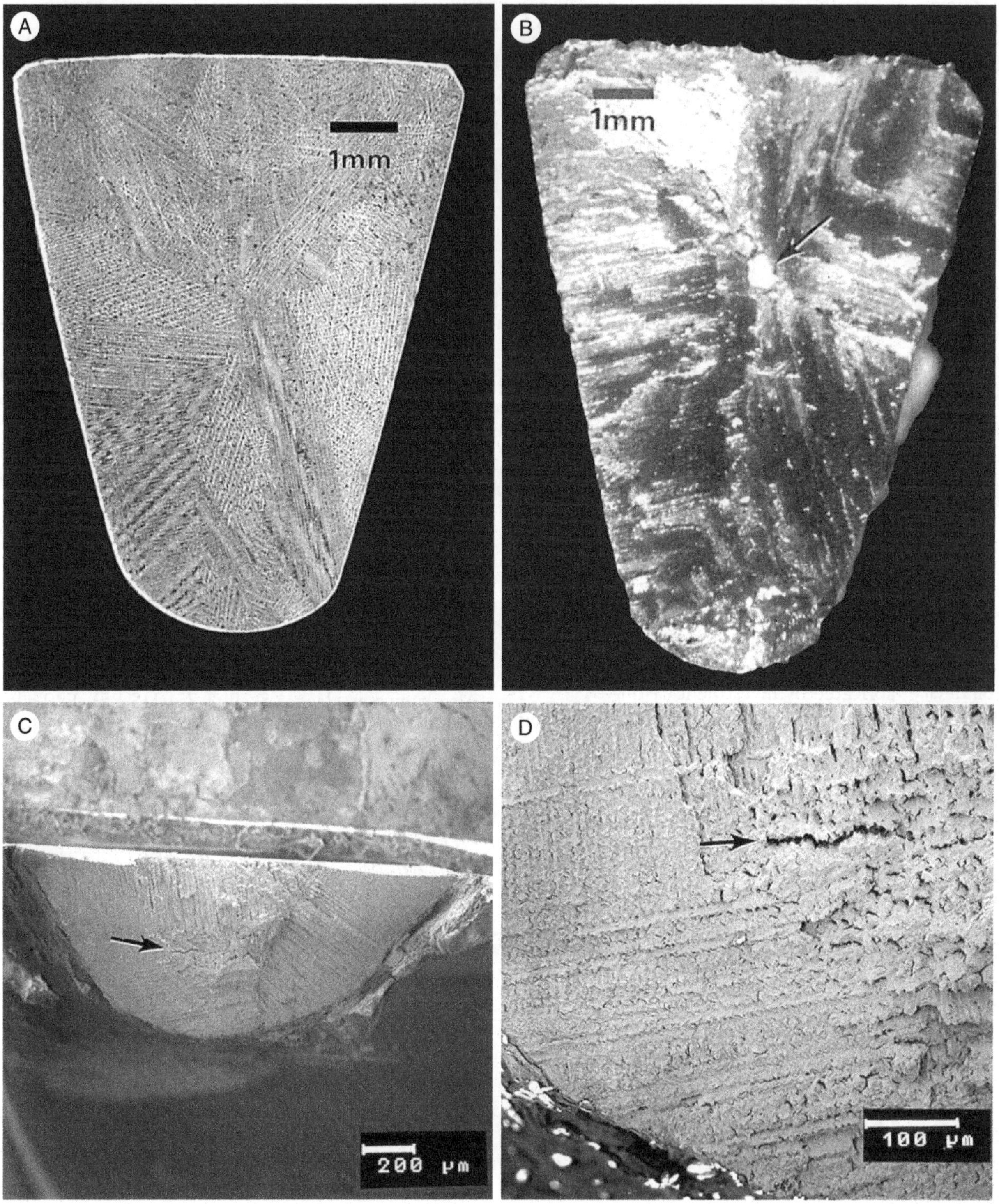

Fig. 3.2.9-7 (A) Macrophoto of a metallographically polished and etched cross section of a cast Co–Cr–Mo ASTM F75 femoral hip stem, showing dendritic structure and large grain size. (B) Macrophoto of the fracture surface of the same Co–Cr–Mo ASTM F75 hip stem as in (A). Arrow indicates large inclusion within the central region of the cross section. Fracture of this hip stem occurred *in vivo*. (C, D) Scanning electron micrographs of the fracture surface from a cast F75 subperiosteal dental implant. Note the large grain size, dendritic microstructure, and interdendritic microporosity (arrows).

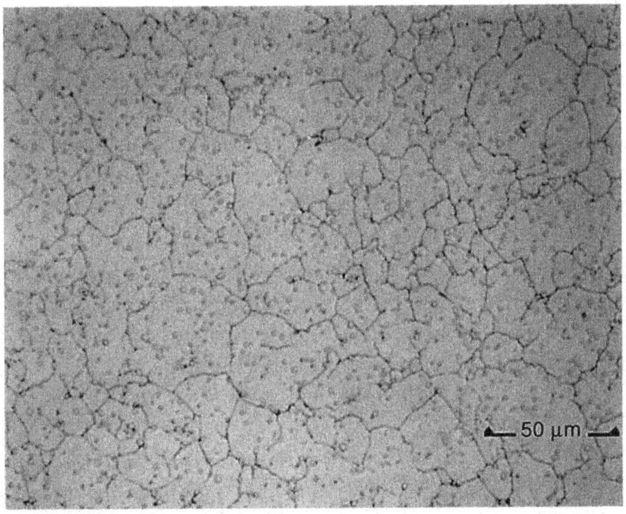

Fig. 3.2.9-8 Microstructure of the Co–Cr–Mo ASTM F75 alloy made via HIP, showing the much smaller grain size relative to that in Fig. 6. (Photo courtesy of Zimmer USA, Warsaw, IN.)

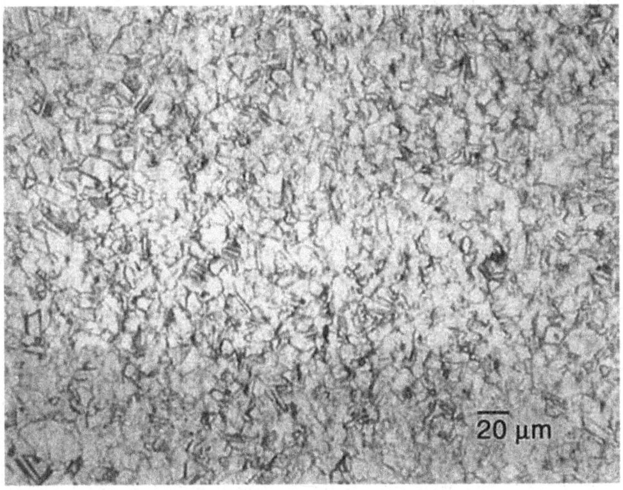

Fig. 3.2.9-9 Microstructure of Co-based MP35N, ASTM F562, Biophase. (Photo courtesy of Smith & Nephew Richards, Inc., Memphis, TN.)

Fig. 3.2.9-9) and the HCP platelets further impede dislocation motion, the resulting structure is significantly strengthened (Table 3.2.9-2). It can be strengthened even further (as in the case of Richards Biophase) by an aging treatment at 430–650°C. This produces Co_3Mo precipitates on the HCP platelets. Hence, the alloy is truly multiphasic and derives strength from the combination of a cold-worked matrix phase, solid solution strengthening, and precipitation hardening. The resulting mechanical properties make the family of MP35N alloys among the strongest available for implant applications.

Titanium-based alloys

Composition

CP titanium (ASTM F67) and extra-low interstitial (ELI) Ti–6Al–4Valloy (ASTM F136) are the two most common titanium-based implant biomaterials. The F67 CP Ti is 98.9–99.6% Ti (Table 3.2.9-4). The oxygen content of CP Ti (and other interstitial elements such as C and N) affects its yield and its tensile and fatigue strengths significantly, as discussed shortly.

With Ti–6Al–4V ELI alloy, the individual Ti–Al and Ti–V phase diagrams suggest the effects of the alloying

Table 3.2.9-4 Chemical compositions of Ti-based alloys for implants

Material	ASTM designation	Common/ trade names	Composition (wt.%)	Notes
Pure Ti, grade 4	F67	CP Ti	Balance Ti max 0.10 C max 0.5 Fe max 0.0125–0.015 H max 0.05 N max 0.40 O	CP Ti comes in four grades according to oxygen content– Grade 1 has 0.18% max O Grade 2 has 0.25% max O Grade 3 has 0.35% max O Grade 4 has 0.40% max O
Ti–6Al–4V ELI*	F136	Ti–6Al–4V	88.3–90.8 Ti 5.5–6.5 Al 3.5–4.5 V max 0.08 C max 0.0125 H max 0.25 Fe max 0.05 N max 0.13 O	

* A more recent specification can be found from ASTM, the American Society for Testing and Materials, under *F136-98e1 Standard Specification for Wrought Titanium-6 Aluminium-4 ELI (Extra Low Intersitial) Alloy (R56401) for Surgical Implant Applications.*

additions in the ternary alloy. That is, since Al is an alpha (HCP) phase stabilizer while V is a beta (BCC) phase stabilizer, it turns out that the Ti–6Al–4V alloy used for implants is an alpha-beta alloy. The alloy's properties depend on prior treatments.

Microstructure and properties

ASTM F67

For relatively pure titanium implants, as exemplified by many current dental implants, typical microstructures are single-phase alpha (HCP), showing evidence of mild (30%) cold work and grain diameters in the range of 10–150 μm (Fig. 3.2.9-10), depending on manufacturing. The nominal mechanical properties are listed in Table 3.2.9-2. Interstitial elements (O, C, N) in both pure titanium and the Ti–6Al–4V alloy strengthen the metal through interstitial solid solution strengthening mechanisms, with nitrogen having approximately twice the hardening effect (per atom) of either carbon or oxygen.

As noted, it is clear that the oxygen content of CP Ti (and the interstitial content generally) will affect its yield and its tensile and fatigue strengths significantly. For example, data available in the ASTM standard show that at 0.18% oxygen (grade 1), the yield strength is about 170 MPa, whereas at 0.40% (grade 4) the yield strength is about 485 MPa. Likewise, the ASTM standard shows that the tensile strength increases with oxygen content.

The literature establishes that the fatigue limit of unalloyed CP Ti is typically increased by interstitial content, in particular the oxygen content. For example, Fig. 3.2.9-11A shows data from Beevers and Robinson (1969), who tested vacuum-annealed CP Ti having a grain size in the range 200–300 μm in tension–compression at a mean stress of zero, at 100 cycles/sec. The 10^7cycle endurance limit, or fatigue limit, for Ti 115 (0.085 wt.% O, grade 1), Ti 130 (0.125 wt.% O, grade 1), and Ti 160 (0.27 wt.% O, grade 3) was 88.3, 142, and 216 MPa, respectively. Figure 3.2.9-11B shows similar results from Turner and Roberts' (1968a) fatigue study on CP Ti (tension–compression, 160 cycles/sec, mean stress = zero) having a grain size in the range 26–32 μm. Here the fatigue limit for "H. P. Ti" (0.072 wt.% O, grade 1),Ti 120 (0.087 wt.% O, grade 1),and Ti 160 (0.32 wt.% O, grade 3) was 142, 172, and 295 MPa, respectively—again increasing with increasing oxygen content. Also, for grade 4 Ti in the cold-worked state, Steinemann *et al.* (1993) reported a 10^7 endurance limit of 430 MPa.

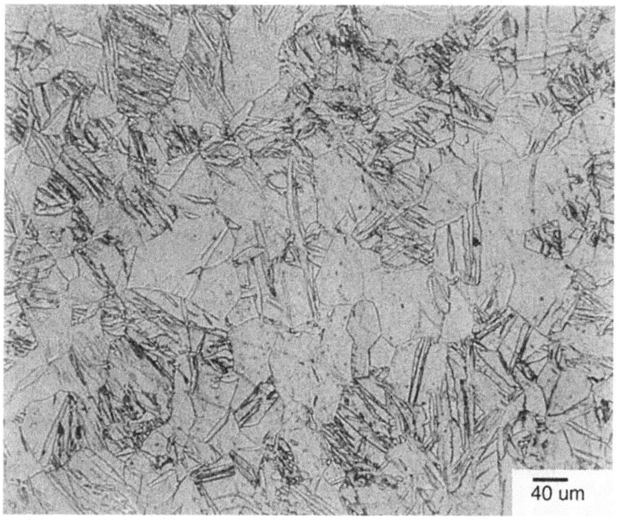

Fig. 3.2.9-10 Microstructure of moderately cold-worked commercial purity titanium, ASTM F67, used in an oral implant.

Figure 3.2.9-11C, from Conrad *et al.* (1973), summarizes data from several fatigue studies on CP Ti at 300K. Note that the ratio of fatigue limit to yield stress is relatively constant at about 0.65, independent of interstitial content and grain size. Conrad *et al.* suggest that this provides evidence that "the high cycle fatigue strength is controlled by the same dislocation mechanisms as the flow [yield] stress" (p. 996). The work of Turner and Roberts also reported that the ratio *f* (fatigue limit/ultimate tensile strength)—which is also called the "fatigue ratio" in materials design textbooks (e.g., Charles and Crane, 1989, p. 106)—was 0.43 for the high-purity Ti (0.072 wt.% O), 0.50 for Ti 120 (0.087 wt.% O), and 0.53 for Ti 160 (0.32 wt.% O). It seems clear that interstitial content affects the yield and tensile and fatigue strengths in CP Ti.

Also, cold work appears to increase the fatigue properties of CP Ti. For example, Disegi (1990) quoted bending fatigue data from for annealed versus cold-worked CP Ti in the form of unnotched 1.0 mm-thick sheet (Table 3.2.9-5); there was a moderate increase in UTS and "plane bending fatigue strength" when comparing annealed versus cold-rolled Ti samples. In these data, the ratio of fatigue strength to ultimate tensile strength ("endurance ratio" or "fatigue ratio", see paragraph above) varied between 0.45 and 0.66. Whereas the ASM *Metals Handbook* (Wagner, 1996) noted that the fatigue limit for high-purity Ti was only about 10% larger for cold-worked versus annealed material, Desegi's data shows that the fatigue strength increased by about 28%, on average.

In recent years there has been increasing interest in the chemical and physical nature of the oxide on the surface of titanium and its 6Al-4V alloy and its biological significance. The nominal composition of the oxide is TiO_2 for both metals, although there is some disagreement about exact oxide chemistry in pure versus alloyed Ti. Although there is no dispute that the oxide provides corrosion resistance, there is some controversy about exactly how it influences the biological performance of titanium at molecular and tissue levels, as suggested in literature on osseointegrated oral and maxillofacial implants by Brånemark and co-workers in Sweden (e.g., Kasemo and Lausmaa, 1988).

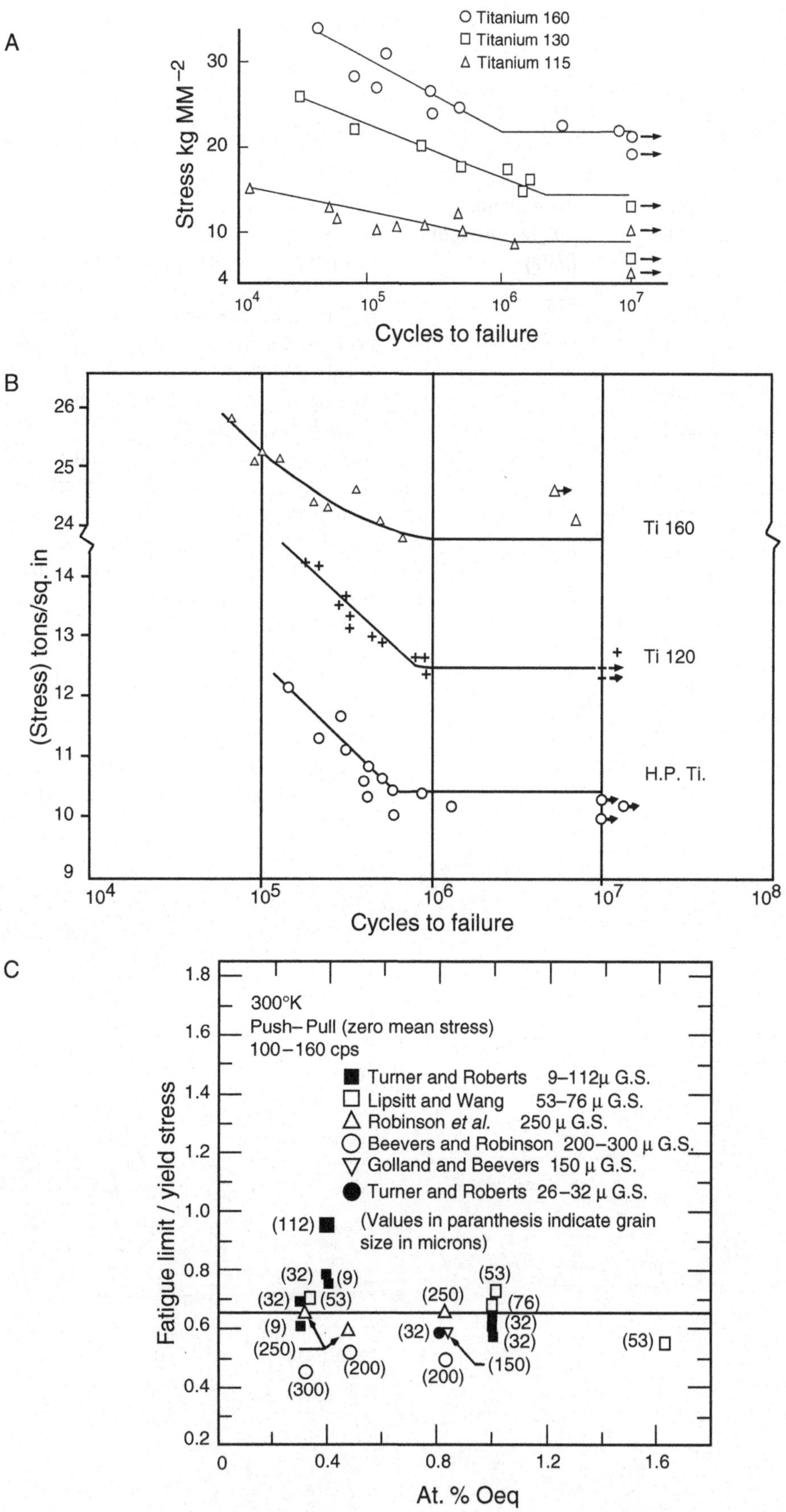

Fig. 3.2.9-11 (A) S–N curves (stress amplitude-number of cycles to failure) at room temperature for CP Ti with varying oxygen content (see text for O content of Ti 160, 130, and 115), from Beevers and Robinson (1969). (B) S–N curves at room temperature for CP Ti with varying oxygen content (see text), from Turner and Roberts (1968a). (C) Ratio of fatigue limit to yield stress in unalloyed Ti at 300 K as a function of at.% oxygen and grain size, from Conrad *et al.* (1973).

Table 3.2.9-5 Plane bending fatigue data for unnotched 1.0-mm-thick unalloyed titanium sheet, tested at 58 cycles/sec in air (from Disegi, 1990)

Ultimate tensile strength (MPa)	Sample condition	Plane bending fatigue strength (MPa)
371	Annealed	246
402	Annealed	235
432	Annealed	284
468	Annealed	284
510	Cold rolled	265
667	Cold rolled	314
667	Cold rolled	343
745	Cold rolled	334
766	Cold rolled	343
772	Cold rolled	383
820	Cold rolled	383

ASTM F136

This alloy is an alpha–beta alloy, the microstructure of which depends upon heat treating and mechanical working. If the alloy is heated into the beta phase field (e.g., above 1000°C, the region where only BCC beta is thermodynamically stable) and then cooled slowly to room temperature, a two-phase Widmanstätten structure is produced (Fig. 3.2.9-12). The HCP alpha phase (which is rich in Al and depleted in V) precipitates out as plates or needles having a specific crystallographic orientation within grains of the beta (BCC) matrix. Alternatively, if cooling from the beta phase field is very fast (as in oil quenching), a "basketweave" microstructure will develop, owing to martensitic or bainitic (nondiffusional shear) solid-state transformations. Most commonly, the F136 alloy is heated and worked at temperatures near but not exceeding the beta transus, and then annealed to give a microstructure of fine-grained alpha with beta as isolated particles at grain boundaries (mill annealed, Fig. 3.2.9-13).

Interestingly, all three of the just-noted microstructures in Ti–6Al–4V alloy lead to about the same yield and ultimate tensile strengths, but the mill-annealed condition is superior in high-cycle fatigue (Table 3.2.9-2), which is a significant consideration.

Like the Co-based alloys, the above microstructural aspects for the Ti systems need to be considered when

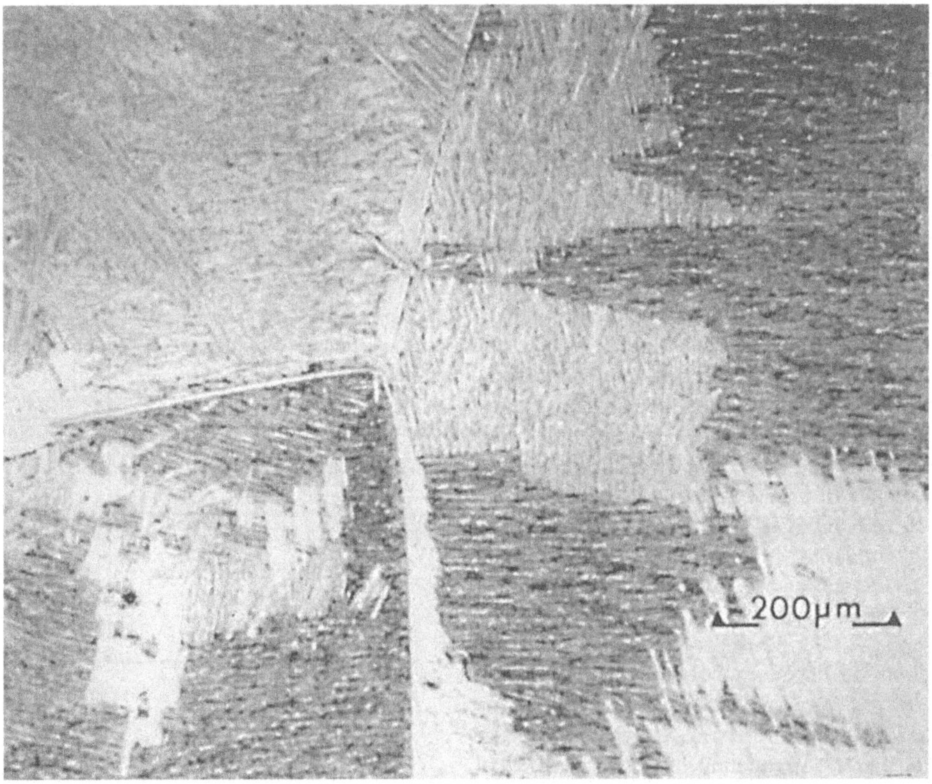

Fig. 3.2.9-12 Widmanstätten structure in cast Ti—Al—4V, ASTM F136. Note prior beta grains (three large grains are shown in the photo) and platelet alpha structure within grains. (Photo courtesy of Zimmer USA, Warsaw, IN.)

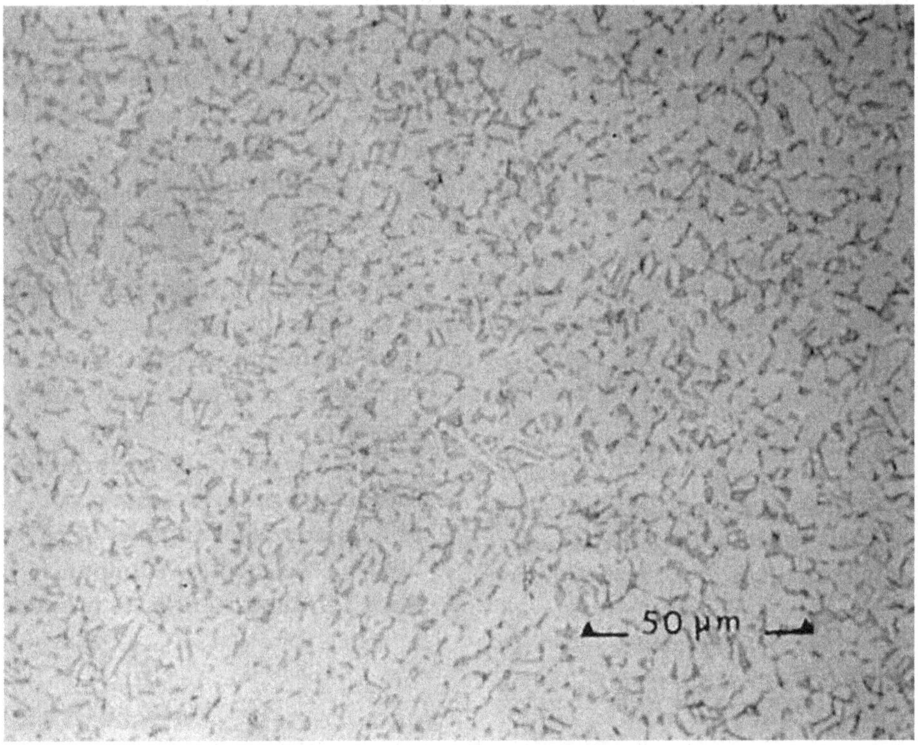

Fig. 3.2.9-13 Microstructure of wrought and mill-annealed Ti–6Al–4V, showing small grains of alpha (light) and beta (dark). (Photo courtesy of Zimmer USA, Warsaw, IN.)

evaluating the structure–property relationships of porous-coated or plasma-sprayed implants. Again, as in the case of the cobalt-based alloys, there is the technical problem of successfully attaching the coating onto the metallic substrate while maintaining adequate properties of both coating and substrate. Optimizing the fatigue properties of Ti–6Al–4V porous-coated implants becomes an interdisciplinary design problem involving not only metallurgy but also surface properties and fracture mechanics.

Concluding remarks

It should be evident that metallurgical principles guide understanding of structure–property relationships and inform judgments about implant design, just as they would in the design process for any well-engineered product. Although this section's emphasis has been on mechanical properties (for the sake of specificity), other properties, in particular surface texture, are receiving increasing attention in relation to biological performance of implants. Timely examples of this are (a) efforts to attach relevant biomolecules to metallic implant surfaces to promote certain desired interfacial activities; and (b) efforts to texture implant surfaces to optimize molecular and cellular reactions.

Another point to remember is that the intrinsic material properties of metallic implants—such as elastic modulus, yield strength, or fatigue strength—are not the sole determinant of implant performance and success. Certainly it is true that inadequate attention to material properties can doom a device to failure. However, it is also true that even with the best material, a device can fail because of faulty structural properties, inappropriate use of the implant, surgical error, or inadequate mechanical design of the implant in the first place. As an illustration of this point, Fig. 3.2.9-14 shows a plastically deformed 316L stainless steel Harrington spinal distraction rod that failed *in vivo* by metallurgical fatigue. An investigation of this case concluded that failure occurred not because 316L cold-worked stainless steel had poor fatigue properties per se, but rather due to a combination of factors: (a) the surgeon bent the rod to make it fit a bit better in the patent, but this increased the bending moment and bending stresses on the rod at the first ratchet junction, which was a known problem area; (b) the stress concentrations at the ratchet end of the rod were severe enough to significantly increase stresses at the first ratchet junction, which was indeed the eventual site of the fatigue fracture; and (c) spinal fusion did not occur in the patient, which contributed to relatively persistent loading of the rod over several months post-implantation. Here the point is that all three of these factors could have been anticipated and addressed during the initial design of the rod, during which both structural and material properties would be considered in various stress analyses related to possible failure modes. It must always be recalled that implant design is a multifaceted problem in which materials selection is only a part of the problem.

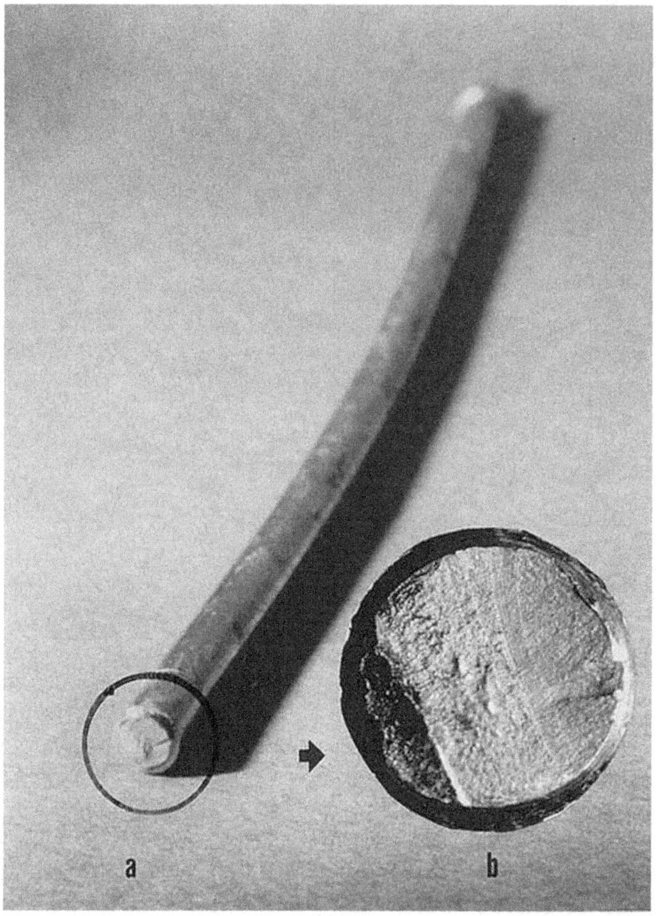

Fig. 3.2.9-14 The smooth part of a 316L stainless steel Harrington spinal distraction rod that fractured by fatigue *in vivo*. Note the bend inthe rod (the rod was originally straight) and (insert) the relationship of the crack initiation zone of the fracture surface to the bend. The inserted photo shows the nature of the fatigue fracture surface, which is characterized by a region of "beach marks" and a region of sudden overload failure. (Photo courtesy of J. B. Brunski, D. C. Hill, and A. Moskowitz, 1983. Stresses in a Harrington distraction rod: their origin and relationship to fatigue fractures *in vivo*. *J. Biomech. Eng.* **105:** 101–107.)

Bibliography

American Society for Testing and Materials (1978). ASTM Standards for Medical and Surgical Materials and Devices. Authorized Reprint from *Annual Book of ASTM Standards*, ASTM, Philadelphia, PA.

Beevers, C. J., and Robinson, J. L. (1969). Some observations on the influence of oxygen content on the fatigue behavior of α-titanium. J. *Less-Common Metals* **17:** 345–352.

Brunski, J. B., Hill, D. C., and Moskowitz, A. (1983). Stresses in a Harrington distraction rod: their origin and relationship to fatigue fractures in vivo. J. *Biomech. Eng.* **105:** 101–107.

Charles, J. A., and Crane, F. A. A. (1989). *Selection and Use of Engineering Materials*. 2nd ed. Butterworth–Heinemann Ltd., Halley Court, Oxford.

Compte, P. (1984). Metallurgical observations of biomaterials. in *Contemporary Biomaterials*, J. W. Boretos and M. Eden, eds. Noyes Publ., Park Ridge, NJ, pp. 66–91.

Conrad, H., Doner, M., and de Meester, B. (1973). Critical review: deformation and fracture. in *Titanium Science and Technology*, Vol. 2, R. I. Jaffee and H. M. Burte, eds. Plenum Press, New York, pp. 969–1005.

Cox, D. O. (1977). The fatigue and fracture behavior of a low stacking fault energy cobalt–chromium–molybdenum–carbon casting alloy used for prosthetic devices. Ph.D. dissertation, Engineering, University of California at Los Angeles.

Davidson, J. A., and Georgette, F. S. (1986). State-of-the-art materials for orthopaedic prosthetic devices. in *Implant Manufacturing and Material Technology*. Proc. Soc. of Manufacturing Engineering, Itasca, IL.

Disegi, J. (1990). AO/ASIF Unalloyed Titanium Implant Material. Technical Brochure available from Synthes (USA), P.O. Box 1766, 1690 Russell Road, Paoli, PA, 19301–1222.

Golland, D. I., and Beevers, C. J. (1971). Some effects of prior deformation and annealing on the fatigue response of α-titanium. J. *Less-Common Metals* **23:** 174.

Golland, D. I., and Beevers, C. J. (1971). The effect of temperature on the fatigue response of alpha-titanium. *Met. Sci. J.* **5:** 174.

Gomez, M., Mancha, H., Salinas, A., Rodríguez, J. L., Escobedo, J., Castro, M., and Méndez, M. (1997).

Relationship between microstructure and ductility of investment cast ASTM F-75 implant alloy.J. *Biomed. Mater. Res.* **34:** 157–163.

Hall, E. O. (1951). The deformation and ageing of mild steel: Discussion of results. *Proc. Phys. Soc. (London)* **64B:** 747–753.

Hamman, G., and Bardos, D. I. (1980). Metallographic quality control of orthopaedic implants. in *Metallography as a Quality Control Tool*, J. L. McCall and P. M. French, eds. Plenum Publishers, New York, pp. 221–245.

Honeycombe, R. W. K. (1968). *The Plastic Deformation of Metals* St. Martin's Press, New York, p. 234.

Kasemo, B., and Lausmaa, J. (1988). Biomaterials from a surface science perspective. in *Surface Characterization of Biomaterials*, B. D. Ratner, ed. Elsevier, New York, Ch. 1, pp. 1-12.

Lipsitt, H. A., and Wang, D. Y. (1961). The effects of interstitial solute atoms on the fatigue limit behavior of titanium. *Trans. AIME* **221:** 918.

Moss, A. J., Hamburger, S., Moore, R. M. Jr., Jeng, L. L., and Howie, L. J. (1990). Use of selected medical device implants in the United States, 1988. *Adv. Data* (191): 1–24.

Nanci, A., Wuest, J. D., Peru, L., Brunet, P., Sharma, V., Zalzal, S., and McKee, M. D. (1998). Chemical modification of titanium surfaces for covalent attachment of biological molecules. *J. Biomed. Mater. Res.* **40:** 324–335.

Petch, N. J. (1953). The cleavage strength of polycrystals.J. *Iron Steel Inst. (London)* **173:** 25.

Pilliar, R. M., and Weatherly, G. C. (1984). Developments in implant alloys. *CRC Crit. Rev. Biocompatibility* **1**(4): 371–403.

Richards Medical Company (1985). Medical Metals. Richards Medical Company Publication No. 3922, Richards Medical Co., Memphis,

TN. [Note: This company is now known as Smith & Nephew Richards, Inc.]

Robinson, S. L., Warren, M. R., and Beevers, C. J. (1969). The influence of internal defects on the fatigue behavior of α-titanium. J. *Less-Common Metals* **19:** 73–82.

Steinemann, S. G., Mäusli, P.-A., Szmuckler-Moncler, S., Semlitsch, M., Pohler, O., Hintermann, H.-E., and Perren, S. M. (1993). Beta-titanium alloy for surgical implants. In *Titanium '92 Science and Technology*, F. H. Froes and I. Caplan, eds. The Minerals, Metals & Materials Society, pp. 2689–2698.

Turner, N. G., and Roberts, W. T. (1968a). Fatigue behavior of titanium. *Trans. Met. Soc. AIME* **242:** 1223–1230.

Turner, N. G., and Roberts, W. T. (1968b). Dynamic strain ageing in titanium. *J. Less-Common Metals* **16:** 37.

www.biomateria.com/media_briefing.htm.

www.sric-bi.com/Explorer/BM.shtml

Wagner, L. (1996). Fatigue life behavior. in *ASM Handbook*, Vol. 19, *Fatigue and Fracture*, S. Lampman, G. M. Davidson, F. Reidenbach, R. L. Boring, A. Hammel, S. D. Henry, and W. W. Scott, Jr., eds., ASM International, pp. 837–853.

Zimmer USA (1984a). *Fatigue and Porous Coated Implants*. Zimmer Technical Monograph, Zimmer USA, Warsaw, IN.

Zimmer USA (1984b). *Metal Forming Techniques in Orthopaedics*.Zimmer Technical Monograph, Zimmer USA, Warsaw, IN.

Zimmer USA (1984c). *Physical and Mechanical Properties of Orthopaedic Alloys*. Zimmer Technical Monograph, Zimmer USA, Warsaw, IN.

Zimmer USA (1984d). *Physical Metallurgy of Titanium Alloy.* Zimmer Technical Monograph, Zimmer USA, Warsaw, IN.

3.2.10 Ceramics, glasses, and glass-ceramics

Larry L. Hench and Serena Best

Ceramics, glasses, and glass-ceramics include a broad range of inorganic/nonmetallic compositions. In the medical industry, these materials have been essential for eyeglasses, diagnostic instruments, chemical ware, thermometers, tissue culture flasks, and fiber optics for endoscopy. Insoluble porous glasses have been used as carriers for enzymes, antibodies, and antigens, offering the advantages of resistance to microbial attack, pH changes, solvent conditions, temperature, and packing under high pressure required for rapid flow (Hench and Ethridge, 1982).

Ceramics are also widely used in dentistry as restorative materials such as in gold–porcelain crowns, glass-filled ionomer cements, and dentures. These dental ceramics are discussed by Phillips (1991).

This section focuses on ceramics, glasses, and glass-ceramics used as implants. Although dozens of compositions have been explored in the past, relatively few have achieved clinical success. This section examines differences in processing and structure, describes the chemical and microstructural basis for their differences in physical properties, and relates properties and tissue response to particular clinical applications. For a historical review of these biomaterials, see Hulbert *et al.* (1987).

Types of bioceramics—tissue attachment

It is essential to recognize that no one material is suitable for all biomaterial applications. As a class of biomaterials, ceramics, glasses, and glass-ceramics are generally used to repair or replace skeletal hard connective tissues. Their success depends upon achieving a stable attachment to connective tissue.

The mechanism of tissue attachment is directly related to the type of tissue response at the implant–tissue interface. No material implanted in living tissue is inert because all materials elicit a response from living tissues. There are four types of tissue response (Table 3.2.10-1) and four different means of attaching prostheses to the skeletal system (Table 3.2.10-2).

Table 3.2.10-1 Types of implant–tissue response

If the material is toxic, the surrounding tissue dies.
If the material is nontoxic and biologically inactive (nearly inert), a fibrous tissue of variable thickness forms.
If the material is nontoxic and biologically active (bioactive), an interfacial bond forms.
If the material is nontoxic and dissolves, the surrounding tissue replaces it.

A comparison of the relative chemical activity of the different types of bioceramics, glasses, and glass-ceramics is shown in Fig. 3.2.10-1. The relative reactivity shown in Fig. 3.2.10-1A correlates very closely with the rate of formation of an interfacial bond of ceramic, glass, or glass-ceramic implants with bone (Fig. 3.2.10-1B). Figure 3.2.10-1B is discussed in more detail in the subsection on bioactive glasses and glass-ceramics in this section.

The relative level of reactivity of an implant influences the thickness of the interfacial zone or layer between the material and tissue. Analyses of implant material failures during the past 20 years generally show failure originating at the biomaterial–tissue interface. When biomaterials are nearly inert (type 1 in Table 3.2.10-2 and Fig. 3.2.10-1) and the interface is not chemically or biologically

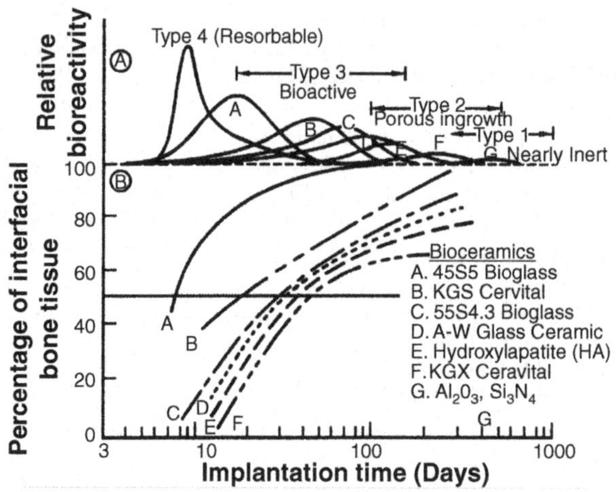

Fig. 3.2.10-1 Bioactivity spectra for various bioceramic implants: (A) relative rate of bioreactivity, (B) time dependence of formation of bone bonding at an implant interface.

bonded, there is relative movement and progressive development of a fibrous capsule in soft and hard tissues. The presence of movement at the biomaterial—tissue interface eventually leads to deterioration in function of the implant or the tissue at the interface, or both. The thickness of the non-adherent capsule varies, depending upon both material (Fig. 3.2.10-2) and extent of relative motion.

The fibrous tissue at the interface of dense Al_2O_3 (alumina) implants is very thin. Consequently, as discussed later, if alumina devices are implanted with a very tight mechanical fit and are loaded primarily in compression, they are very successful. In contrast, if a type 1 nearly inert implant is loaded so that interfacial movement can occur, the fibrous capsule can become several hundred micrometers thick, and the implant can loosen very quickly.

The mechanism behind the use of nearly inert microporous materials (type 2 in Table 3.2.10-2 and

Table 3.2.10-2 Types of bioceramic–tissue attachment and their classification

Type of attachment	Example
1. Dense, nonporous, nearly inert ceramics attach by bone growth into surface irregularities by cementing the device into the tissues or by press-fitting into a defect (termed "morphological fixation").	Al_2O_3 (single crystal and polycrystalline)
2. For porous inert implants, bone ingrowth occurs that mechanically attaches the bone to the material (termed "biological fixation").	Al_2O_3 (polycrystalline) Hydroxyapatite-coated porous metals
3. Dense, nonporous surface-reactive ceramics, glasses, and glass-ceramics attach directly by chemical bonding with the bone (termed "bioactive fixation").	Bioactive glasses Bioactive glass-ceramics Hydroxyapatite
4. Dense, nonporous (or porous) resorbable ceramics are designed to be slowly replaced by bone.	Calcium sulfate (plaster of Paris) Tricalcium phosphate Calcium-phosphate salts

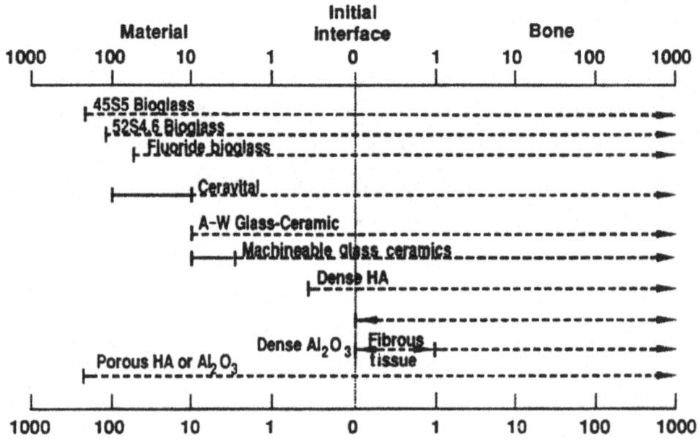

Fig. 3.2.10–2 Comparison of interfacial thickness (μm) of reaction layer of bioactive implants of fibrous tissue of inactive bioceramics in bone.

Fig. 3.2.10-1) is the ingrowth of tissue into pores on the surface or throughout the implant. The increased interfacial area between the implant and the tissues results in an increased resistance to movement of the device in the tissue. The interface is established by the living tissue in the pores. Consequently, this method of attachment is often termed "biological fixation". It is capable of withstanding more complex stress states than type 1 implants with "morphological fixation". The limitation with type 2 porous implants, however, is that for the tissue to remain viable and healthy, it is necessary for the pores to be greater than 50 to 150 μm (Fig. 3.2.10-2). The large interfacial area required for the porosity is due to the need to provide a blood supply to the ingrown connective tissue (vascular tissue does not appear in pore sizes less than 100 μm). Also, if micromovement occurs at the interface of a porous implant and tissue is damaged, the blood supply may be cut off, the tissues will die, inflammation will ensue, and the interfacial stability will be destroyed. When the material is a porous metal, the large increase in surface area can provide a focus for corrosion of the implant and loss of metal ions into the tissues. This can be mediated by using a bioactive ceramic material such as HA as a coating on the metal. The fraction of large porosity in any material also degrades the strength of the material proportional to the volume fraction of porosity. Consequently, this approach to solving interfacial stability works best when materials are used as coatings or as unloaded space fillers in tissues.

Resorbable biomaterials (type 4 in Table 3.2.10-2 and Fig. 3.2.10-1) are designed to degrade gradually over a period of time and be replaced by the natural host tissue. This leads to a very thin or nonexistent interfacial thickness (Fig. 3.2.10-2). This is the optimal biomaterial solution, if the requirements of strength and short-term performance can be met, since natural tissues can repair and replace themselves throughout life. Thus, resorbable biomaterials are based on biological principles of repair that have evolved over millions of years. Complications in the development of resorbable bioceramics are (1) maintenance of strength and the stability of the interface during the degradation period and replacement by the natural host tissue, and (2) matching resorption rates to the repair rates of body tissues (Fig. 3.2.10-1A) (e.g., some materials dissolve too rapidly and some too slowly). Because large quantities of material may be replaced, it is also essential that a resorbable biomaterial consist only of metabolically acceptable substances. This criterion imposes considerable limitations on the compositional design of resorbable biomaterials. Successful examples of resorbable polymers include PLA and PGA used for sutures, which are metabolized to CO_2 and H_2O and therefore are able to function for an appropriate time and then dissolve and disappear. Porous or particulate calcium phosphate ceramic materials such as tricalcium phosphate (TCP) have proved successful for resorbable hard tissue replacements when low loads are applied to the material.

Another approach to solving problems of interfacial attachment is the use of bioactive materials (type 3 in Table 3.2.10-2 and Fig. 3.2.10-1). Bioactive materials are intermediate between resorbable and bioinert. A bioactive material is one that elicits a specific biological response at the interface of the material, resulting in the formation of a bond between the tissues and the material. This concept has now been expanded to include a large number of bioactive materials with a wide range of rates of bonding and thicknesses of interfacial bonding layers (Figs. 3.2.10-1 and 3.2.10-2). They include bioactive glasses such as Bioglass; bioactive glass-ceramics such as Ceravital, A-W glass-ceramic, or machinable glass-ceramics; dense HA such as Durapatite or Calcitite; and bioactive composites such as HA-PE, HA-Bioglass, Palavital, and stainless steel fiber-reinforced Bioglass. All of these materials form an interfacial bond with adjacent tissue. However, the time dependence of bonding, the strength of bond, the mechanism of bonding, and the thickness of the bonding zone differ for the various materials.

It is important to recognize that relatively small changes in the composition of a biomaterial can dramatically affect whether it is bioinert, resorbable, or bioactive. These compositional effects on surface reactions are discussed in the section on bioactive glasses and glass-ceramics.

Characteristics and processing of bioceramics

The types of implants listed in Table 3.2.10-2 are made using different processing methods. The characteristics and properties of the materials, summarized in Table 3.2.10-3, differ greatly, depending upon the processing method used.

The primary methods of processing ceramics, glasses, and glass-ceramics are summarized in Fig. 3.2.10-3. These methods yield five categories of microstructures:

1. Glass
2. Cast or plasma-sprayed polycrystalline ceramic
3. Liquid-phase sintered (vitrified) ceramic
4. Solid-state sintered ceramic
5. Polycrystalline glass-ceramic

Differences in the microstructures of the five categories are primarily a result of the different thermal processing steps required to produce them. Alumina and calcium phosphate bioceramics are made by fabricating the product from fine-grained particulate solids. For example, a desired shape may be obtained by mixing the

Table 3.2.10-3 Bioceramic material characteristics and properties

Composition
Microstructure
Number of phases
Percentage of phases
Distribution of phases
Size of phases
Connectivity of phases
Phase state
Crystal structure
Defect structure
Amorphous structure
Pore structure
Surface
Flatness
Finish
Composition
Second phase
Porosity
Shape

particulates with water and an organic binder, then pressing them in a mold. This is termed "forming." The formed piece is called green ware. Subsequently, the temperature is raised to evaporate the water (i.e., drying) and the binder is burned out, resulting in bisque ware. At a very much higher temperature, the part is densified during firing. After cooling to ambient temperature, one or more finishing steps may be applied, such as polishing. Porous ceramics are produced by adding a second phase that decomposes prior to densification, leaving behind holes or pores (Schors and Holmes, 1993), or transforming natural porous organisms, such as coral, to porous HA by hydrothermal processing (Roy and Linnehan, 1974).

The interrelation between microstructure and thermal processing of various bioceramics is shown in Fig. 3.2.10-3, which is a binary phase diagram consisting of a network-forming oxide such as SiO_2 (silica), and some arbitrary network modifier oxide (MO) such as CaO. When a powdered mixture of MO and SiO_2 is heated to the melting temperature T_m, the entire mass will become liquid (L). The liquid will become homogeneous when held at this temperature for a sufficient length of time. When the liquid is cast (paths 1B, 2, 5), forming the shape of the object during the casting, either a glass or a polycrystalline microstructure will result. Plasma spray coating follows path 1A. However, a network-forming oxide is not necessary to produce plasma-sprayed coatings such as HAs, which are polycrystalline (Lacefield, 1993).

If the starting composition contains a sufficient quantity of network former (SiO_2), and the casting rate is sufficiently slow, a glass will result (path 1B). The viscosity of the melt increases greatly as it is cooled, until at approximately T_1, the glass transition point, the material is transformed into a solid.

If either of these conditions is not met, a polycrystalline microstructure will result. The crystals begin growing at T_L and complete growth at T_2. The final material consists of the equilibrium crystalline phases predicted by the

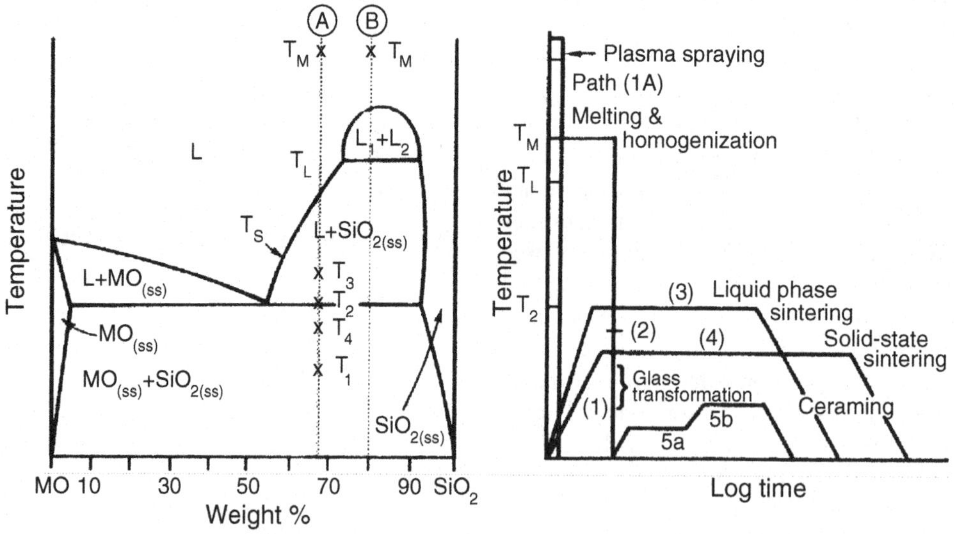

Fig. 3.2.10-3 Relation of thermal processing schedules of various bioceramics to equilibrium phase diagram.

phase diagram. This type of cast object is not often used commercially because the large shrinkage cavity and large grains produced during cooling make the material weak and subject to environmental attack.

If the MO and SiO_2 powders are first formed into the shape of the desired object and fired at a temperature T_3, a liquid-phase sintered structure will result (path 3). Before firing, the composition will contain approximately 10–40% porosity, depending upon the forming process used. A liquid will be formed first at grain boundaries at the eutectic temperature, T_2. The liquid will penetrate between the grains, filling the pores, and will draw the grains together by capillary attraction. These effects decrease the volume of the powdered compact. Since the mass remains unchanged and is only rearranged, an increased density results. Should the compact be heated for a sufficient length of time, the liquid content can be predicted from the phase diagram. However, in most ceramic processes, liquid formation does not usually proceed to equilibrium owing to the slowness of the reaction and the expense of long-term heat treatments.

The microstructure resulting from liquid-phase sintering, or vitrification as it is commonly called, will consist of small grains from the original powder compact surrounded by a liquid phase. As the compact is cooled from T_3 to T_2, the liquid phase will crystallize into a fine-grained matrix surrounding the original grains. If the liquid contains a sufficient concentration of network formers, it can be quenched into a glassy matrix surrounding the original grains.

A powder compact can be densified without the presence of a liquid phase by a process called solid-state sintering. This is the process usually used for manufacturing alumina and dense HA bioceramics. Under the driving force of surface energy gradients, atoms diffuse to areas of contact between particles. The material may be transported by either grain boundary diffusion, volume diffusion, creep, or any combination of these, depending upon the temperature or material involved. Because long-range migration of atoms is necessary, sintering temperatures are usually in excess of one-half of the melting point of the material: $T > T_L/2$ (path 4).

The atoms move so as to fill up the pores and open channels between the grains of the powder. As the pores and open channels are closed during the heat treatment, the crystals become tightly bonded together, and the density, strength, and fatigue resistance of the object improve greatly. The microstructure of a material that is prepared by sintering consists of crystals bonded together by ionic–covalent bonds with a very small amount of remaining porosity.

The relative rate of densification during solid-state sintering is slower than that of liquid-phase sintering because material transport is slower in a solid than in a liquid. However, it is possible to solid-state sinter individual component materials such as pure oxides since liquid development is not necessary. Consequently, when high purity and uniform fine-grained microstructures are required (e.g., for bioceramics) solid-state sintering is essential.

The fifth class of microstructures is called glass-ceramics because the object starts as a glass and ends up as a polycrystalline ceramic. This is accomplished by first quenching a melt to form the glass object. The glass is transformed into a glass-ceramic in two steps. First, the glass is heat treated at a temperature range of 500–700°C (path 5a) to produce a large concentration of nuclei from which crystals can grow. When sufficient nuclei are present to ensure that a fine-grained structure will be obtained, the temperature of the object is raised to a range of 600–900°C, which promotes crystal growth (path 5b). Crystals grow from the nuclei until they impinge and up to 100% crystallization is achieved. The resulting microstructure is nonporous and contains fine-grained, randomly oriented crystals that may or may not correspond to the equilibrium crystal phases predicted by the phase diagram. There may also be a residual glassy matrix, depending on the duration of the ceraming heat treatment. When phase separation occurs (composition B in Fig. 3.2.10-3), a nonporous, phase-separated, glass-in-glass microstructure can be produced. Crystallization of phase-separated glasses results in very complex microstructures. Glass-ceramics can also be made by pressing powders and a grain boundary glassy phase (Kokubo, 1993). For additional details on the processing of ceramics, see Reed (1988) or Onoda and Hench (1978), and for processing of glass-ceramics, see McMillan (1979).

Nearly inert crystalline ceramics

High-density, high-purity (>199.5%) alumina is used in load-bearing hip prostheses and dental implants because of its excellent corrosion resistance, good biocompatibility, high wear resistance, and high strength (Christel *et al.*, 1988; Hulbert, 1993; Hulbert *et al.*, 1987; Miller *et al.*, 1996). Although some dental implants are single-crystal sapphires (McKinney and Lemons, 1985), most Al_2O_3 devices are very fine-grained polycrystalline $< \alpha\text{-}Al_2O_3$ produced by pressing and sintering at $T = 1600\text{–}1700°C$. A very small amount of MgO (<0.5%) is used to aid sintering and limit grain growth during sintering.

Strength, fatigue resistance, and fracture toughness of polycrystalline $< \alpha\text{-}Al_2O_3$ are a function of grain size and percentage of sintering aid (i.e., purity). Al_2O_3 with an average grain size of < 4 μm and $> 99.7\%$ purity exhibits good flexural strength and excellent compressive strength. These and other physical properties are summarized in Table 3.2.10-4, along with the International Standards

Organization (ISO) requirements for alumina implants. Extensive testing has shown that alumina implants that meet or exceed ISO standards have excellent resistance to dynamic and impact fatigue and also resist subcritical crack growth (Drre and Dawihl, 1980). An increase in average grain size to >17 μm can decrease mechanical properties by about 20%. High concentrations of sintering aids must be avoided because they remain in the grain boundaries and degrade fatigue resistance.

Methods exist for lifetime predictions and statistical design of proof tests for load-bearing ceramics. Applications of these techniques show that load limits for specific prostheses can be set for an Al_2O_3 device based upon the flexural strength of the material and its use environment (Ritter *et al.*, 1979). Load-bearing lifetimes of 30 years at 12,000-N loads have been predicted (Christel *et al.*, 1988). Results from aging and fatigue studies show that it is essential that Al_2O_3 implants be produced at the highest possible standards of quality assurance, especially if they are to be used as orthopedic prostheses in younger patients.

Alumina has been used in orthopedic surgery for nearly 20 years (Miller *et al.*, 1996). Its use has been motivated largely by two factors: its excellent type 1 biocompatibility and very thin capsule formation (Fig. 3.2.10-2), which permits cementless fixation of prostheses; and its exceptionally low coefficients of friction and wear rates.

Table 3.2.10-4 Physical characteristics of Al_2O_3 bioceramics

	High alumina ceramics	ISO Standard 6474
Alumina content (% by weight)	>99.8	≥99.50
Density (g/cm^3)	>3.93	≥3.90
Average grain size (μm)	3–6	<7
Ra (μm)[a]	0.02	
Hardness (Vickers hardness number, VHN)	2300	>2000
Compressive strength (MPa)	4500	
Bending strength (MPa) (after testing in Ringer's solution)	550	400
Young's modulus (GPa)	380	
Fracture toughness (K1C) (MPa^{12})	5–6	
Slow crack growth	10–52	

[a] Surface roughness value.

The superb tribiologic properties (friction and wear) of alumina occur only when the grains are very small (< 4 μm) and have a very narrow size distribution. These conditions lead to very low surface roughness values (Ra < 40.02 μm, Table 3.2.10-4). If large grains are present, they can pull out and lead to very rapid wear of bearing surfaces owing to local dry friction.

Alumina on load-bearing, wearing surfaces, such as in hip prostheses, must have a very high degree of sphericity, which is produced by grinding and polishing the two mating surfaces together. For example, the alumina ball and socket in a hip prosthesis are polished together and used as a pair. The long-term coefficient of friction of an alumina–alumina joint decreases with time and approaches the values of a normal joint. This leads to wear on alumina-articulating surfaces being nearly 10 times lower than metal–PE surfaces (Fig. 3.2.10-4).

Low wear rates have led to widespread use in Europe of alumina noncemented cups press-fitted into the acetabulum of the hip. The cups are stabilized by the growth of bone into grooves or around pegs. The mating femoral ball surface is also made of alumina, which is bonded to a metallic stem. Long-term results in general are good, especially for younger patients. However, Christel *et al.* (1988) caution that stress shielding, owing to the high elastic modulus of alumina, may be responsible for cancellous bone atrophy and loosening of the acetabular cup in old patients with senile osteoporosis or rheumatoid arthritis. Consequently, it is essential that the age of the patient, nature of the disease of the joint, and biomechanics of the repair be considered carefully before any prosthesis is used, including alumina ceramics.

Zirconia (ZrO_2) is also used as the articulating ball in total hip prostheses. The potential advantages of zirconia in load-bearing prostheses are its lower modulus of elasticity and higher strength (Hench and Wilson, 1993). There are insufficient data to determine whether these properties will result in higher clinical success rates over long times (>15 years).

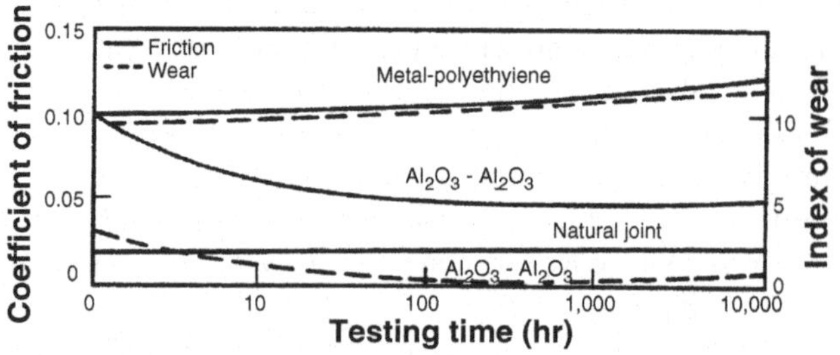

Fig. 3.2.10-4 Time dependence of coefficient of friction and wear of alumina–alumina versus metal–PE hip joint (*in vitro* testing).

Other clinical applications of alumina prostheses reviewed by Hulbert *et al.* (1987) include knee prostheses; bone screws; alveolar ridge and maxillofacial reconstruction; ossicular bone substitutes; keratoprostheses (corneal replacements); segmental bone replacements; and blade, screw, and post dental implants.

Porous ceramics

The potential advantage offered by a porous ceramic implant (type 2, Table 3.2.10-2, Figs. 3.2.10-1 and 3.2.10-2) is its inertness combined with the mechanical stability of the highly convoluted interface that develops when bone grows into the pores of the ceramic. The mechanical requirements of prostheses, however, severely restrict the use of low-strength porous ceramics to nonload-bearing applications. Studies reviewed by Hench and Ethridge (1982), Hulbert *et al.* (1987), and Schors and Holmes (1993) have shown that when load-bearing is not a primary requirement, porous ceramics can provide a functional implant. When pore sizes exceed 100 μm, bone will grow within the interconnecting pore channels near the surface and maintain its vascularity and long-term viability. In this manner, the implant serves as a structural bridge or scaffold for bone formation.

Commercially available porous products originate from two sources: HA converted from coral (e.g., Pro Osteon) or animal bone (e.g., Endobon). Other production routes; e.g., burnout techniques and decomposition of hydrogen peroxide (Peelen *et al.*, 1978; Driessen *et al.*, 1982) are not yet used commercially. The optimal type of porosity is still uncertain. The degree of inter–connectivity of pores may be more critical than the pore size. Eggli *et al.* (1988) demonstrated improved integration in interconnected 50–100 μm pores compared with less connected pores with a size of 200–400 μm. Similarly Kühne *et al.* (1994) compared two grades of 25–35% porous coralline apatite with average pore sizes of 200 and 500 μm and reported bone ingrowth to be improved in the 500 μm pore sized ceramic. Holmes (1979) suggests that porous coralline apatite when implanted in cortical bone requires interconnections of osteonic diameter for transport of nutrients to maintain bone ingrowth. The findings clearly indicate the importance of thorough characterisation of porous materials before implantation, and Hing *et al.* (1999) has recommended a range of techniques that should be employed.

Porous materials are weaker than the equivalent bulk form in proportion to the percentage of porosity, so that as the porosity increases, the strength of the material decreases rapidly. Much surface area is also exposed, so that the effects of the environment on decreasing the strength become much more important than for dense, nonporous materials. The aging of porous ceramics, with their subsequent decrease in strength, requires bone ingrowth to stabilize the structure of the implant. Clinical results for non-load-bearing implants are good (Schors and Holmes, 1993).

Bioactive glasses and glass-ceramics

Certain compositions of glasses, ceramics, glass-ceramics, and composites have been shown to bond to bone (Hench and Ethridge, 1982; Gross *et al.*, 1988; Yamamuro *et al.*, 1990; Hench, 1991; Hench and Wilson, 1993). These materials have become known as bioactive ceramics. Some even more specialized compositions of bioactive glasses will bond to soft tissues as well as bone (Wilson *et al.*, 1981). A common characteristic of bioactive glasses and bioactive ceramics is a time-dependent, kinetic modification of the surface that occurs upon implantation. The surface forms a biologically active carbonated HA layer that provides the bonding interface with tissues.

Materials that are bioactive develop an adherent interface with tissues that resist substantial mechanical forces. In many cases, the interfacial strength of adhesion is equivalent to or greater than the cohesive strength of the implant material or the tissue bonded to the bioactive implant.

Bonding to bone was first demonstrated for a compositional range of bioactive glasses that contained SiO_2, Na_2O, CaO, andP_2O_5 in specific proportions (Hench *et al.*, 1972) (Table 3.2.10-5). There are three key compositional features to these bioactive glasses that distinguish them from traditional soda–lime–silica glasses: (1) less than 60 mol% SiO_2, (2) high Na_2O and CaO content, and (3) a high CaO/P_2O_5 ratio. These features make the surface highly reactive when it is exposed to an aqueous medium.

Many bioactive silica glasses are based upon the formula called 45S5, signifying 45 wt.% SiO_2 (S = the network former) and 5 :1 ratio of CaO to P_2O_5. Glasses with lower ratios of CaO to P_2O_5 do not bond to bone. However, substitutions in the 45S5 formula of 5–15 wt.% B_2O_3 for SiO_2 or 12.5 wt.% CaF_2 for CaO or heat treating the bioactive glass compositions to form glass-ceramics has no measurable effect on the ability of the material to form a bone bond. However, adding as little as 3 wt.% Al_2O_3 to the 45S5 formula prevents bonding to bone.

The compositional dependence of bone and soft tissue bonding on the Na_2O–CaO–P_2O_5–SiO_2 glasses is illustrated in Fig. 3.2.10-5. All the glasses in Fig. 3.2.10-5 contain a constant 6 wt.% of P_2O_5. Compositions in the middle of the diagram (region A) form a bond with bone. Consequently, region A is termed the bioactive bone-bonding boundary. Silicate glasses within region B

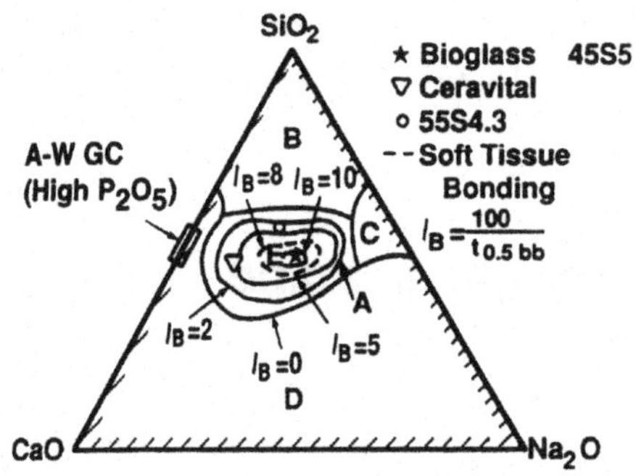

Fig. 3.2.10-5 Compositional dependence (in wt.%) of bone bonding and soft tissue bonding of bioactive glasses and glass-ceramics. All compositions in region A have a constant 6 wt.% of P_2O_5. A-W glass ceramic has higher P_2O_5 content (see Table 3.2.10-5 for details). I_B, Index of bioactivity.

(e.g., window or bottle glass, or microscope slides) behave as nearly inert materials and elicit a fibrous capsule at the implant–tissue interface. Glasses within region C are resorbable and disappear within 10 to 30 days of implantation. Glasses within region D are not technically practical and therefore have not been tested as implants.

The collagenous constituent of soft tissues can strongly adhere to the bioactive silicate glasses that lie within the dashed line region in Fig. 3.2.10-5. The interfacial thicknesses of the hard tissue–bioactive glasses are shown in Fig. 3.2.10-2 for several compositions. The thickness decreases as the bone-bonding boundary is approached.

Gross *et al.* (1988) and Gross and Strunz (1985) have shown that a range of low-alkali (0 to 5 wt.%) bioactive silica glass-ceramics (Ceravital) also bond to bone. They found that small additions of Al_2O_3, Ta_2O_5, TiO_2, Sb_2O_3, or ZrO_2 inhibit bone bonding (Table 3.2.10-5, Fig. 3.2.10-1). A two-phase silica–phosphate glass-ceramic composed of apatite [$Ca_{10}(PO_4)_6(OH_1F_2)$] and wollastonite [CaO, SiO_2] crystals and a residual silica glassy matrix, termed A-W glass-ceramic (A-WGC) (Nakamura *et al.*, 1985; Yamamuro *et al.*, 1990; Kokubo, 1993), also bonds with bone. The addition of Al_2O_3 or TiO_2 to the A-W glass-ceramic also inhibits bone bonding, whereas incorporation of a second phosphate phase, B-whitlockite (3CaO, P_2O_5), does not.

Another multiphase bioactive phosphosilicate containing phlogopite (Na, K) $Mg_3[AlSi_3O_{10}]F_2$ and apatite crystals bonds to bone even though Al is present in the composition (Höhland and Vogel, 1993). However, the Al^{3+} ions are incorporated within the crystal phase and do not alter the surface reaction kinetics of the material. The compositions of these various bioactive glasses and glass-ceramics are compared in Table 3.2.10-5.

Table 3.2.10-5 Composition of bioactive glasses and glass-ceramics (in weight percent)

	45S5 Bioglass	45S5F Bioglass	45S5.4F Bioglass	40S5B5 Bioglass	52S4.6 Bioglass	55S4.3 Bioglass	KGC Ceravital	KGS Ceravital	KGy213 Ceravital	A-W GC	MB GC
SiO_2	45	45	45	40	52	55	46.2	46	38	34.2	19–52
P_2O_5	6	6	6	6	6	6				16.3	4–24
CaO	24.5	12.25	14.7	24.5	21	19.5	20.2	33	31	44.9	9–3
$Ca(PO_3)_2$							25.5	16	13.5		
CaF_2		12.25	9.8							0.5	
MgO							2.9			4.6	5–15
MgF_2											
Na_2O	24.5	24.5	24.5	24.5	21	19.5	4.8	5	4		3–5
K_2O							0.4				3–5
Al_2O_3									7		12–33
B_2O_3				5							
Ta_2O_5/TiO_2									6.5		
Structure	Glass	Glass	Glass	Glass	Glass		Glass-ceramic	Glass-ceramic		Glass-ceramic	Glass-ceramic
Reference	Hench *et al.* (1972)	Hench *et al.* (1972)	Hench *et al.* (1972)	Hench *et al.* (1972)	Hench *et al.* (1972)	Hench *et al.* (1972)	Gross *et al.* (1988)	Gross *et al.* (1988)		Nakamura *et al.* (1985)	Höhland and Vogel (1993)

The surface chemistry of bioactive glass and glass-ceramic implants is best understood in terms of six possible types of surface reactions (Hench and Clark, 1978). A high-silica glass may react with its environment by developing only a surface hydration layer. This is called a type I response (Fig. 3.2.10-6). Vitreous silica (SiO_2) and some inert glasses at the apex of region B (Fig. 3.2.10-5) behave in this manner when exposed to a physiological environment.

When sufficient SiO_2 is present in the glass network, the surface layer that forms from alkali–proton exchange can repolymerize into a dense SiO_2-rich film that protects the glass from further attack. This type II surface (Fig. 3.2.10-6) is characteristic of most commercial silicate glasses, and their biological response of fibrous capsule formation is typical of many within region B in Fig. 3.2.10-5.

At the other extreme of the reactivity range, a silicate glass or crystal may undergo rapid, selective ion exchange of alkali ions, with protons or hydronium ions leaving a thick but highly porous and nonprotective SiO_2-rich film on the surface (a type IV surface) (Fig. 3.2.10-6). Under static or slow flow conditions, the local pH becomes sufficiently alkaline ($pH > 19$) that the surface silica layer is dissolved at a high rate, leading to uniform bulk network or stoichiometric dissolution (a type V surface). Both type IV and V surfaces fall into region C of Fig. 3.2.10-5.

Type IIIA surfaces are characteristic of bioactive silicates (Fig. 3.2.10-6). A dual protective film rich in CaO and P_2O_5 forms on top of the alkali-depleted SiO_2-rich film. When multivalent cations such as Al^{3+}, Fe^{3+}, and Ti^{4+} are present in the glass or solution, multiple layers form on the glass as the saturation of each cationic complex is exceeded, resulting in a type IIIB surface (Fig. 3.2.10-6), which does not bond to tissue.

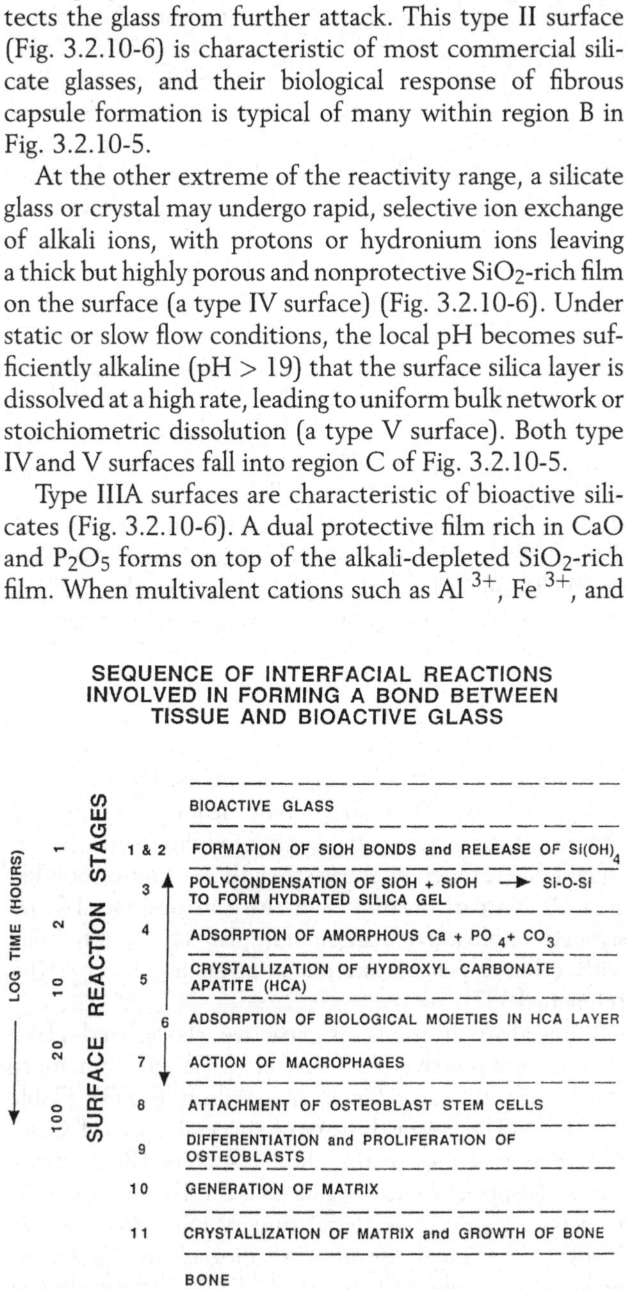

Fig. 3.2.10-6 Types of silicate glass interfaces with aqueous or physiological solutions.

A general equation describes the overall rate of change of glass surfaces and gives rise to the interfacial reaction profiles shown in Fig. 3.2.10-6. The reaction rate (R) depends on at least four terms (for a single-phase glass). For glass-ceramics, which have several phases in their microstructures, each phase will have a characteristic reaction rate, R_i.

$$R = -k_1t^{0.5} - k_2t^{1.0} - k_3t^{1.0} + k_4t^y + k_nt^z \quad (3.2.10\text{-}1)$$

The first term describes the rate of alkali extraction from the glass and is called a stage 1 reaction. A type II nonbonding glass surface (region B in Fig. 3.2.10-6) is primarily undergoing stage 1 attack. Stage 1, the initial or primary stage of attack, is a process that involves an exchange between alkali ions from the glass and hydrogen ions from the solution, during which the remaining constituents of the glass are not altered. During stage 1, the rate of alkali extraction from the glass is parabolic ($t^{1/2}$) character.

The second term describes the rate of interfacial network dissolution that is associated with a stage 2 reaction. A type IV surface is a resorbable glass (region C in Fig. 3.2.10-5) and is experiencing a combination of stage 1 and stage 2 reactions. A type V surface is dominated by a stage 2 reaction. Stage 2, the second stage of attack, is a process by which the silica structure breaks down and the glass totally dissolves at the interface. Stage 2 kinetics are linear ($t^{1.0}$).

A glass surface with a dual protective film is designated type IIIA (Fig. 3.2.10-6). The thickness of the secondary films can vary considerably—from as little as 0.01 μm for Al_2O_3–SiO_2-rich layers on inactive glasses, to as much as 30 μm for CaO–P_2O_5-rich layers on bioactive glasses.

A type III surface forms as a result of the repolymerization of SiO_2 on the glass surface by the condensation of the silanols (Si<pisbOH) formed from the stage 1 reaction. For example:

$$Si - OH + OH - Si \rightarrow Si - O - Si + H_2O \quad (3.2.10\text{-}2)$$

Stage 3 protects the glass surface. The SiO_2 polymerization reaction contributes to the enrichment of surface SiO_2 that is characteristic of type II, III, and IV surface profiles (Fig. 3.2.10-6). It is described by the third term in Eq. 3.2.10-1. This reaction is interface controlled with a time dependence of $+k_3t^{1.0}$. The interfacial thickness of the most reactive bioactive glasses shown in Fig. 3.2.10-2 is largely due to this reaction.

The fourth term in Eq. 3.2.10-1, $+k_4t^y$ (stage 4), describes the precipitation reactions that result in the multiple films characteristic of type III glasses. When only one secondary film forms, the surface is type IIIA. When several additional films form, the surface is type IIIB.

In stage 4, an amorphous calcium phosphate film precipitates on the silica-rich layer and is followed by crystallization to form carbonated HA crystals. The calcium and phosphate ions in the glass or glass-ceramic provide the nucleation sites for crystallization. Carbonate anions (CO_3^{2-}) substitute for OH^- in the apatite crystal structure to form a carbonate HA similar to that found in living bone. Incorporation of CaF_2 in the glass results in incorporation of fluoride ions in the apatite crystal lattice. Crystallization of carbonate HA occurs around collagen fibrils present at the implant interface and results in interfacial bonding.

In order for the material to be bioactive and form an interfacial bond, the kinetics of reaction in Eq. 3.2.10-1, and especially the rate of stage 4, must match the rate of biomineralization that normally occurs *in vivo*. If the rates in Eq. 3.2.10-1 are too rapid, the implant is resorbable, and if the rates are too slow, the implant is not bioactive.

By changing the compositionally controlled reaction kinetics (Eq. 3.2.10-1), the rates of formation of hard tissue at an implant interface can be altered, as shown in Fig. 3.2.10-1. Thus, the level of bioactivity of a material can be related to the time for more than 50% of the interface to be bonded ($t_{0.5bb}$) [e.g., I_B index of bioactivity: $= (100/t_{0.5bb})$] (Hench, 1988). It is necessary to impose a 50% bonding criterion for an I_B since the interface between an implant and bone is irregular (Gross *et al.*, 1988). The initial concentration of macrophages, osteoblasts, chondroblasts, and fibroblasts varies as a function of the fit of the implant and the condition of the bony defect. Consequently, all bioactive implants require an incubation period before bone proliferates and bonds. The length of this incubation period varies widely, depending on the composition of the implant.

The compositional dependence of I_B indicates that there are iso I_B contours within the bioactivity boundary, as shown in Fig. 3.2.10-5 (Hench, 1988). The change of I_B with the $SiO_2/(Na_2O + CaO)$ ratio is very large as the bioactivity boundary is approached. The addition of multivalent ions to a bioactive glass or glass-ceramic shrinks the isoI_B contours, which will contract to zero as the percentage of Al_2O_3, Ta_2O_5, ZrO_2, or other multivalent cations increases in the material. Consequently, the isoI_B boundary shown in Fig. 3.2.10-5 indicates the contamination limit for bioactive glasses and glass-ceramics. If the composition of a starting implant is near the I_B boundary, it may take only a few parts per million of multivalent cations to shrink the I_B boundary to zero and eliminate bioactivity. Also, the sensitivity of fit of a bioactive implant and length of time of immobilization postoperatively depends on the I_B value and closeness to the $I_B = 0$ boundary. Implants near the I_B boundary require more precise surgical fit and longer fixation times before they can bear loads. In contrast, increasing the surface area of a bioactive implant by using them in particulate form for bone augmentation expands the bioactive boundary. Small (<200 μm) bioactive glass granules behave as a partially resorbable implant and stimulate new bone formation (Hench, 1994).

Bioactive implants with intermediate I_B values do not develop a stable soft tissue bond; instead, the fibrous interface progressively mineralizes to form bone. Consequently, there appears to be a critical isoI_B boundary beyond which bioactivity is restricted to stable bone bonding. Inside the critical isoI_B boundary, the bioactivity includes both stable bone and soft-tissue bonding, depending on the progenitor stem cells in contact with the implant. This soft tissue–critical isoI_B limit is shown by the dashed contour in Fig. 3.2.10-5.

The thickness of the bonding zone between a bioactive implant and bone is proportional to its I_B (compare Fig. 3.2.10-1 with Fig. 3.2.10-2). The failure strength of a bioactively fixed bond appears to be inversely dependent on the thickness of the bonding zone. For example, 45S5 Bioglass with a very high I_B develops a gel bonding layer of 200 μm, which has a relatively low shear strength. In contrast, A-W glass-ceramic, with an intermediate I_B value, has a bonding interface in the range of 10–20 μm and a very high resistance to shear. Thus, the interfacial bonding strength appears to be optimum for I_B values of ~4. However, it is important to recognize that the interfacial area for bonding is time dependent, as shown in Fig. 3.2.10-1. Therefore, interfacial strength is time dependent and is a function of such morphological factors as the change in interfacial area with time, progressive mineralization of the interfacial tissues, and resulting increase of elastic modulus of the interfacial bond, as well as shear strength per unit of bonded area. A comparison of the increase in interfacial bond strength of bioactive fixation of implants bonded to bone with other types of fixation is given in Fig. 3.2.10-7 (Hench, 1987).

Clinical applications of bioactive glasses and glass-ceramics are reviewed by Gross *et al.* (1988), Yamamuro *et al.* (1990), and Hench and Wilson (1993) (Table 3.2.10-6). The 8-year history of successful use of Ceravital glass-ceramics in middle ear surgery (Reck *et al.*, 1988) is especially encouraging, as is the 10-year use of A-W glass-ceramic in vertebral surgery (Yamamuro *et al.*, 1990), the 10-year use of 45S5 Bioglass in endosseous ridge maintenance (Stanley *et al.*, 1996) and middle-ear replacement, and the 6-year success in repair of periodontal defects (Hench and Wilson, 1996; Wilson, 1994).

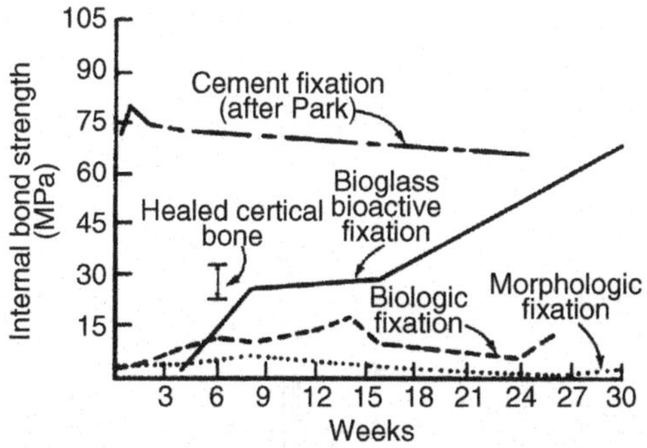

Fig. 3.2.10-7 Time dependence of interfacial bond strength of various fixation systems in bone. (After Hench, 1987.)

Calcium phosphate ceramics

Bone typically consists, by weight of 25% water, 15% organic materials and 60% mineral phases. The mineral phase consists primarily of calcium and phosphate ions, with traces of magnesium, carbonate, hydroxyl, chloride, fluoride, and citrate ions. Hence calcium phosphates occur naturally in the body, but they occur also within nature as mineral rocks, and certain compounds can be synthesized in the laboratory. Table 3.2.10-7 summarizes the mineral name, chemical name and composition of various phases of calcium phosphates.

Within the past 20–30 years interest has intensified in the use of calcium phosphates as biomaterials, but only certain compounds are useful for implantation in the body since both their solubility and speed of hydrolysis increase with a decreasing calcium-to-phosphorus ratio. Driessens (1983) stated that those compounds with a Ca/P ratio of less than 1:1 are not suitable for biological implantation.

The main crystalline component of the mineral phase of bone is a calcium-deficient carbonate HA, and various methods have been investigated to produce synthetic HA. The commercial routes are based on aqueous precipitation or conversion from other calcium compounds. Aqueous precipitation is most often performed in one of two ways: a reaction between a calcium salt and an alkaline phosphate (Collin, Hayek and Newsley, 1963; Eanes *et al.*, 1965; Bonel *et al.*, 1987; Young and Holcomb, 1982; Denissen *et al.*, 1980b; Jarcho *et al.*, 1976; Kijima and Tsutsumi, 1979) or a reaction between calcium hydroxide or calcium carbonate and phosphoric acid (Mooney and Aia, 1961; Irvine, 1981; Rao and Boehm, 1974; McDowell *et al.*, 1977; Akao *et al.*, 1981).

Other routes include solid-state processing (Monma *et al.*, 1981; Fowler, 1974; Young and Holcomb, 1982; Rootare *et al.*, 1978; Lehr *et al.*, 1967); hydrolysis (Schleede *et al.*, 1932; Morancho *et al.*, 1981; Young and Holcomb, 1982); hydrothermal synthesis (Young and Holcomb, 1982; Fowler, 1974; Roy, 1971; Skinner, 1973; Arends *et al.*, 1979).

Table 3.2.10-6 Present uses of bioceramics

Orthopedic load-bearing applications Al_2O_3	Coatings for tissue ingrowth (cardiovascular, orthopedic, dental and maxillofacial prosthetics) Al_2O_3
Coatings for chemical bonding (Orthopedic, dental and maxillofacial prosthetics) HA Bioactive glasses Bioactive glass-ceramics	Temporary bone space fillers Tricalcium phosphate Calcium and phosphate salts
Dental implants Al_2O_3 HA Bioactive glasses	Periodontal pocket obliteration HA HA–PLA composite Trisodium phosphate Calcium and phosphate salts Bioactive glasses
Alveolar ridge augmentations Al_2O_3 HA HA-autogenous bone composite HA-PLA composite Bioactive glasses	Maxillofacial reconstruction Al_2O HA HA–PLA composite Bioactive glasses
Otolaryngological Al_2O_3 HA Bioactive glasses Bioactive glass-ceramics	Percutaneous access devices Bioactive glasses Bioactive composites
Artifical tendon and ligament PLA–carbon fiber composite	Orthopedic fixation devices PLA–carbon fibers PLA–calcium/phosphorus–base glass fibers

Table 3.2.10-7 Calcium phosphates

Ca:P	Mineral name	Formula	Chemical name
1.0	Monetite	$CaHPO_4$	Dicalcium phosphate (DCP)
1.0	Brushite	$CaHPO_4 \cdot 2H_2O$	Dicalcium phosphate dihydrate (DCPD)
1.33	—	$Ca_8(HPO_4)_2(PO_4)_4 \cdot 5H_2O$	Octocalcium phosphate (OCP)
1.43	Whitlockite	$Ca_{10}(HPO_4)(PO_4)_6$	
1.5	—	$Ca_3(PO_4)_2$	Tricalcium phosphate (TCP)
1.67	Hydroxyapatite	$Ca_{10}(PO_4)_6(OH)_2$	
2.0		$Ca_4P_2O_9$	Tetracalcium phosphate

The route and conditions under which synthetic HA is produced will greatly influence its physical and chemical characteristics. Factors that affect the rate of resorption of the implant include physical factors such as the physical features of the material (e.g., surface area, crystallite size), chemical factors such as atomic and ionic substitutions in the lattice, and biological factors such as the types of cells surrounding the implant and location, age, species, sex, and hormone levels.

The thermodynamic stability of the various calcium phosphates is summarized in the form of the phase diagram shown in Fig. 3.2.10-8. The binary equilibrium phase diagram between CaO and P_2O_5 gives an indication of the compounds formed between the two oxides, and by comparing this with Table 3.2.10-7 it is possible to identify the naturally occurring calcium phosphate minerals. The diagram does not indicate the phase boundaries of apatite due to the absence of hydroxyl groups.

However, from the binary diagram an indication may be obtained of the stability of other calcium phosphates with temperature.

The stoichiometry of HA is highly significant where thermal processing of the material is required. Slight imbalances in the stoichiometric ratio of calcium and phosphorus in HA (from the standard molar ratio of 1.67), can lead to the appearance of either α or β-TCP on heat treatment. Many early papers concerning the production and processing of HA powders reported problems in avoiding the formation of these extraneous phases (Jarcho *et al.*, 1976; De With *et al.*, 1981a, b; Peelen *et al.*, 1978). However, using stoichiometric HA it should be possible to sinter, without phase purity problems, at temperatures in excess of 1300°C.

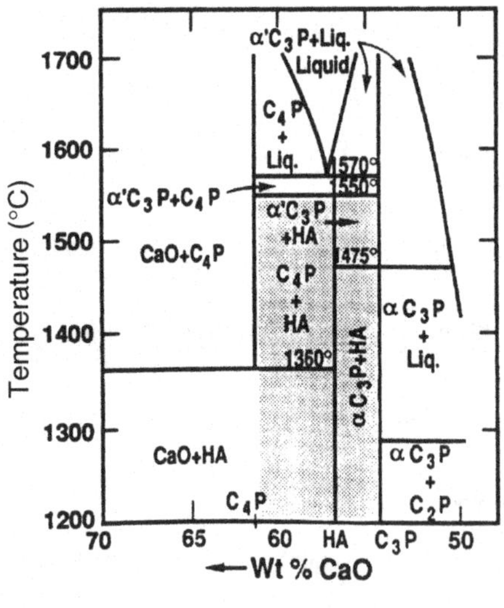

Fig. 3.2.10-8 Phase equilibrium diagram of calcium phosphates in a water atmosphere. Shaded area is processing range to yield HA-containing implants. (After K. de Groot. 1988. *Ann.* N. Y. *Acad. Sci.* **523:** 227.)

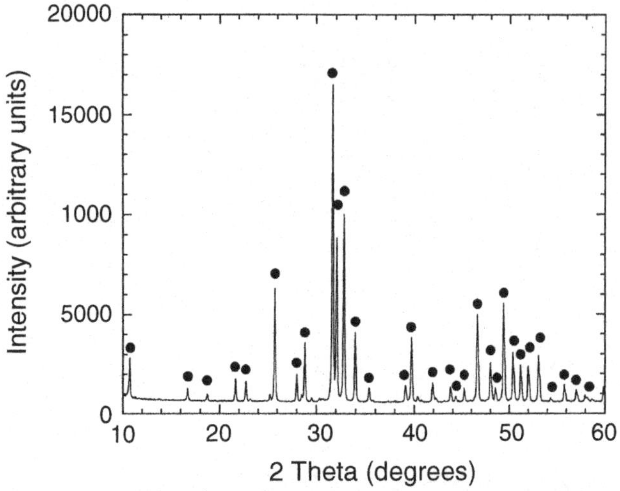

Fig. 3.2.10-9 X-ray diffraction of HA.

X-ray diffraction (Fig. 3.2.10-9) and infra-red spectroscopy (Fig. 3.2.10-10) should be used to reveal the phase purity and level of hydroxylation of HA. Kijima and Tsutsumi (1979) used these techniques to study HA sintered at different temperatures and reported that after sintering at 900°C, the material was fully hydroxylated, but after sintering at temperatures higher than this, dehydroxylation occurred. Dehydration of HA, produced by processes such as high temperature solid state reaction, result in the formation of oxyhydroxyapatite: $Ca_{10}(PO_4)_6(OH)_{2-2x}O_xV_x$ (where V is a hydroxyl vacancy). HA has a $P6_3/m$ space group: This signifies that the lattice is primitive Bravais, there is a sixfold axis parallel to the *c* axis and *a* 1/2 (3/6) translation along the length of the *c* axis (a screw axis) with a mirror plane situated perpendicular to the screw axis and the *c* axis. The *a* and *c* parameters for HA are 0.9418 nm and 0.6884 nm, respectively.

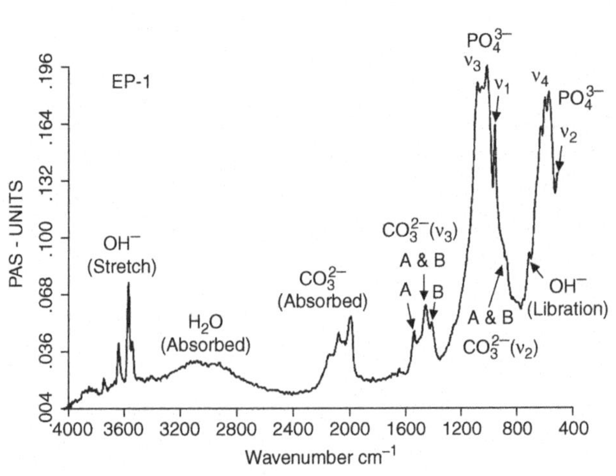

Fig. 3.2.10-10 Typical FT-IR spectrum for HA.

The structure assumed by any solid is such that, on an atomic level, the configuration of the constituents is of the lowest possible energy. In phosphates, this energy requirement results in the formation of discrete subunits within the structure and the PO_4^{3-} group forms a regular tetrahedron with a central P^{5+} ion and O^{2-} ions at the four corners. In a similar manner, the $(OH)^-$ groups are also ionically bonded. In terms of the volume occupied, the oxygen ions exceed all other elements in phosphates. Any other elements present may therefore be considered as filling the interstices, with the exact position being determined by atomic radius and charge (See Fig. 3.2.10-11).

The HA lattice contains two kinds of calcium positions; columnar and hexagonal. There is a net total of four "columnar calcium" ions that occupy the [1/3, 2/3, 0] and [1/3, 2/3, 1/2] lattice points. The "hexagonal calcium" ions are located on planes parallel to the basal plane at $c = 1/4$ and $c = 3/4$ and the six PO_4^{3-} tetrahedra are also located on these planes. The $(OH)^-$ groups are located in columns parallel to the c axis, at the corners of the unit cell, which may be viewed as passing through the centers of the triangles formed by the "hexagonal calcium" ions. Successive hexagonal calcium triangles are rotated through 60° (See Fig. 3.2.10-12).

Defects and impurities in HA may be identified as either substitutional or as discrete, extraneous crystalline phases (as discussed above). Methods of detection of impurities include X-ray diffraction, IR spectroscopy, and spectrochemical analysis. It is important to make a full spectrochemical analysis of HA since contact with any metal ions during production can lead to high levels of impurties in the product. Typical data for one commercial HA powder are shown in Table 3.2.10-8.

Ions that may be incorporated into the HA structure, either intentionally or unintentionally, include carbonate ions (substituting for hydroxyl or phosphate groups), fluoride ions (substituting for hydroxyl groups), silicon, or silicate ions (substituting for phosphorus or phosphate groups) and magnesium ions substituting for calcium, e.g., Newsley, 1963; Le Geros, 1965; Jha *et al.*, 1997; Gibson *et al.*, 1999).

The presence of carbonate may be observed directly, using IR spectroscopy, in the form of weak peaks at between 870 and 880 cm^{-1} and a stronger doublet between 1460 and 1530 cm^{-1}, and also through alterations in the HA lattice parameters from X-ray diffraction

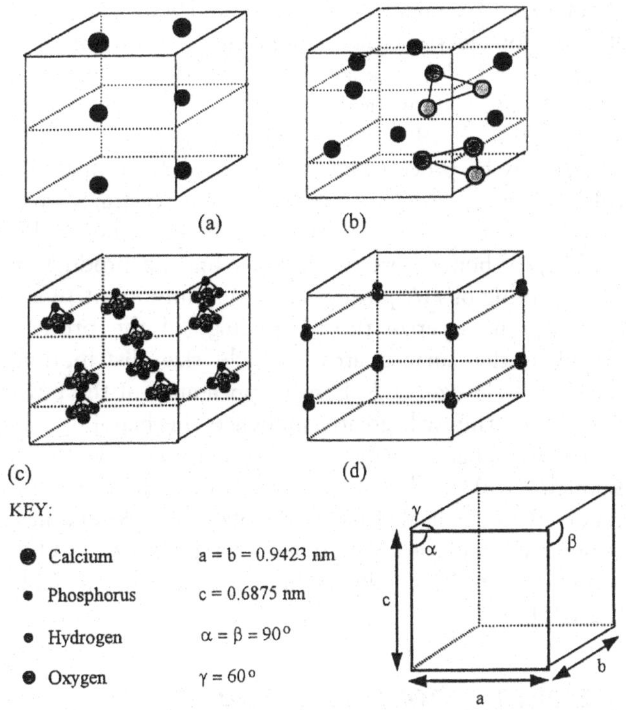

Fig. 3.2.10-12 Theoretical positions of the ionic species within the unit cell of HA (Hing, 1995).

Trace element	ppm
Al	600
Cu	1
Fe	1000
Ge	100
Mg	2000
Mn	300
Na	3000
Pb	4
Si	500
Ti	30

Table 3.2.10-8 Trace elements in a commercial HA

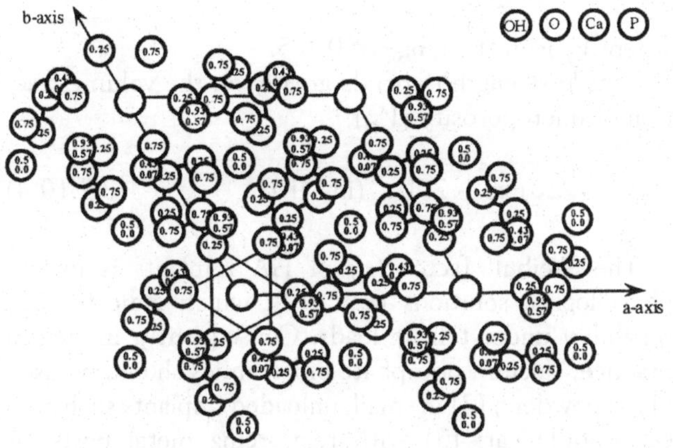

Fig. 3.2.10-11 Structure of HA projected on the *x,y* Plane, adapted from Kay, M. I., Young, R. A., and Posner, A. S. (1964). Crystal structure of hydroxyapatite. *Nature* 12: 1050. (Hing, 1995).

(Rootare and Craig, 1978; LeGeros, 1965). The substitution of electronegative anions such as fluorine and chlorine for $(OH)^-$ has also been reported to alter the lattice parameters of the material (e.g., Young and Elliot, 1966; Kay *et al.*, 1964; see also Elliot, 1994).

HA may be processed as a ceramic using compaction (die pressing, isostatic pressing, slip casting, etc.) followed by solid-state sintering (discussed earlier in this section). When reporting methods for the production and sintering of HA powders it is very important to adequately characterize the morphology of the product including the surface area, particle size distribution, mean particle size, and physical appearance of the powders, since this will greatly influence the handling and processing characteristics of the material (Best and Bonfield, 1994). There is great deal of variation in the reported mechanical performance of dense HA ceramics, dependent on phase purity, density and grain size, but the properties cited generally fall in the range shown in Table 3.2.10-9.

Calcium phosphate coatings

The clinical application of calcium phosphate ceramics is largely limited to non-major-load-bearing parts of the skeleton because of their inferior mechanical properties, and it was partly for this reason that interest was directed toward the use of calcium phosphate coatings on metallic implant subtrates. A very good review of techniques for the production of calcium phosphates was given by Wolke *et al.* in 1998. Many techniques are available for the deposition of HA coatings, including electrophoresis, sol–gel routes, electrochemical routes, biomimetic routes, and sputter techniques, but the most popular commercial routes are those based on plasma spraying. In plasma spraying, an electric arc is struck between two electrodes and stream of gases is passed through the arc. The arc converts the gases into a plasma with a speed of up to 400 m/sec and a temperature within the arc of 20,000 K. The ceramic powder is suspended in the carrier gas and fed into the plasma where it can be fired at a substrate. There are many variables in the process including the gases used, the electrical settings, the nozzle/substrate separation and the morphology, particle size, and particle size distribution of the powder. Because of the very high temperatures but very short times involved, the behavior of the HA powder particle is somewhat different than might be predicted in an equilibrium phase diagram. However, according to the particular conditions used, it is likely that at least a thin outer layer of the powder particle will be in a molten state and will undergo some form of phase transformation, but by careful control of the operating variables the transformed material should represent a relatively small volume fraction of the coating and the product should maintain the required phase purity and crystallinity (Cook *et al.*, 1998; Wolke *et al.*, 1992).

Table 3.2.10-9 Typical mechanical properties of dense HA ceramics

Theoretical density	3.156 g cm^3
Hardness	500–800 HV, 2000–3500 Knoop
Tensile strength	40–100 MPa
Bend strength	20–80 MPa
Compressive strength	100–900 MPa
Fracture toughness	Approx. 1 MPa $m^{0.5}$
Young's modulus	70–120 GPa

A number of factors influence the properties of the resulting coating, including coating thickness (this will influence coating adhesion and fixation—the agreed optimum now seems to be 50–100 μm; Soballe *et al.*, 1993 and de Groot *et al.*, 1987), crystallinity (this affects the dissolution and biological behavior; Klein *et al.*, 1994a, b; Clemens, 1995; Le Geros *et al.*, 1992), biodegradation (affected by phase purity, chemical purity, porosity, crystallinity), and adhesion strength (these may range between 5 and 65 MPa (97)).

The mechanical behavior of calcium phosphate ceramics strongly influences their application as implants. Tensile and compressive strength and fatigue resistance depend on the total volume of porosity. Porosity can be in the form of micropores (< 1 μm diameter, due to incomplete sintering) or macropores (>100 μm diameter, created to permit bone growth). The dependence of compressive strength (σ_c) and total pore volume (V_p) is described in de Groot *et al.* (1990) by:

$$\sigma_c = 700 \exp - 5V_p \quad \text{(in Mpa)} \qquad (3.2.10\text{-}3)$$

where V_p is in the range of 0–0.5.

Tensile strength depends greatly on the volume fraction of microporosity (V_m):

$$\sigma_t = 220 \exp - 20V_m \quad \text{(in MPa)} \qquad (3.2.10\text{-}4)$$

The Weibull factor (n) of HA implants is low in physiological solutions ($n = 12$), which indicates low reliability under tensile loads. Consequently, in clinical practice, calcium phosphate bioceramics should be used (1) as powders; (2) in small, unloaded implants such as in the middle ear; (3) with reinforcing metal posts, as in dental implants; (4) as coatings (e.g., composites); or (5) in low-loaded porous implants where bone growth acts as a reinforcing phase.

The bonding mechanism of dense HA implants appears to be very different from that described above for bioactive glasses. The bonding process for HA implants is described by Jarcho (1981). A cellular bone matrix from differentiated osteoblasts appears at the surface, producing a narrow amorphous electron-dense band only 3–5 μm wide. Between this area and the cells, collagen bundles are seen. Bone mineral crystals have been identified in this amorphous area. As the site matures, the bonding zone shrinks to a depth of only 0.05–0.2 μm (Fig. 3.2.10-2). The result is normal bone attached through a thin epitaxial bonding layer to the bulk implant. TEM image analysis of dense HA bone interfaces show an almost perfect epitaxial alignment of some of the growing bone crystallites with the apatite crystals in the implant. A consequence of this ultra-thin bonding zone is a very high gradient in elastic modulus at the bonding interface between HA and bone. This is one of the major differences between the bioactive apatites and the bioactive glasses and glass-ceramics. The implications of this difference for the implant interfacial response to Wolff's law is discussed in Hench and Ethridge (1982, Chap. 14).

Resorbable calcium phosphates

Resorption or biodegradation of calcium phosphate ceramics is caused by three factors:

1. Physiochemical dissolution, which depends on the solubility product of the material and local pH of its environment. New surface phases may be formed, such as amorphous calcium phosphate, dicalcium phosphate dihydrate, octacalcium phosphate, and anionic-substituted HA.
2. Physical disintegration into small particles as a result of preferential chemical attack of grain boundaries.
3. Biological factors, such as phagocytosis, which causes a decrease in local pH concentrations (de Groot and Le Geros, 1988).

Ideally, one would wish for a replacement material to be slowly resorbed by the body once its task of acting as a scaffold for new bone has been completed. Degradation or resorption of calcium phosphates *in vivo* occurs by a combination of phagocytosis of particles and the production of acids. However, when selecting a resorbable material for implantation, care must be taken to match the rate of resorption with that of the expected bone tissue regeneration. Where the solubility of calcium phosphates is higher than the rate of tissue regeneration, they will not be of use to fill bone defects. As mentioned previously, the rate of dissolution increases with decreasing calcium-to-phosphorus ratio, and consequently, TCP, with a Ca:P ratio of 1.5, is more rapidly resorbed than HA. TCP has four polymorphs, α β γ and super α. The γ polymorph is a high pressure phase and the super α polymorph is observed at temperatures above approximately 1500°C. Therefore the most frequently observed polymorphs in bioceramics are α and β-TCP. X-ray diffraction studies indicate that the β polymorph transforms to the α polymorph at temperatures between 1120°C and 1290°C (Gibson *et al.*, 1996).

In the 1980s, the idea of a new bone susbtitute material was introduced and the materials were referred to as calcium phosphate bone cements. These materials offer the potential for *in situ* molding and injectability. There are a variety of different combinations of calcium compounds (e.g., α-TCP and dicalcium phosphate) that are used in the formulation of these bone cements, but the end product is normally based on a calcium-deficient HA (Fernandez *et al.*, 1998,1999a, b).

All calcium phosphate ceramics biodegrade to varying degrees in the following order: increasing rate HA.

The rate of biodegradation increases as:

1. Surface area increases (powders > porous solid > dense solid)
2. Crystallinity decreases
3. Crystal perfection decreases
4. Crystal and grain size decrease
5. There are ionic substitutions of CO_2^{-3}, Mg^{2+}, and Sr^{2+} in HA

Factors that tend to decrease the rate of biodegradation include (1) F^- substitution in HA, (2) Mg^{2+} substitution in β-TCP, and (3) lower β-TCP/HA ratios in biphasic calcium phosphates.

Clinical applications of HA

Calcium phosphate-based bioceramics have been used in medicine and dentistry for nearly 20 years (Hulbert *et al.*, 1987; de Groot, 1983, 1988; de Groot *et al.*, 1990; Jarcho, 1981; Le Geros, 1988; Le Geros and Le Geros, 1993). Applications include dental implants, periodontal treatment, alveolar ridge augmentation, orthopedics, maxillofacial surgery, and otolaryngology (Table 3.2.10-6).

Most authors agree that HA is bioactive, and it is generally agreed that the material is osseoconductive, where osseoconduction is the ability of a material to encourage bone growth along its surface when placed in the vicinity of viable bone or differentiated bone-forming cells. A good recent review of *in vitro* and *in vivo* data for calcium phosphates has been prepared by Hing *et al.* (1998), who observed that there are a large number of "experimental parameters," including specimen, host, and test parameters, which need to be carefully controlled in order to allow adequate interpretation of data.

HA has been used clinically in a range of different forms and applications. It has been utilised as a dense, sintered ceramic for middle ear implant applications (van Blitterswijk *et al.*, 1990) and alveolar ridge reconstruction and augmentation (Quin and Kent, 1984; Cranin *et al.*, 1987), in porous form (Smiler and Holmes, 1987; Bucholz *et al.*, 1987), as granules for filling bony defects in dental and orthopaedic surgery (Aoki, 1994; Fujishiro *et al.*, 1997; Oonishi *et al.*, 1990; Froum *et al.*, 1986; Galgut *et al.*, 1990; Wilson and Low, 1992), and as a coating on metal implants (Cook *et al.*, 1992a, b; De Groot, 1987).

Another successful clinical application for HA has been in the form of a filler in a polymer matrix. The original concept of a bioceramic polymer composite was introduced by Bonfield and co-workers and the idea was based on the concept that cortical bone itself comprises an organic matrix reinforced with a mineral component. The material developed by Bonfield and co-workers contains up to 50 vol % HA in a PE matrix, has a stiffness similar to that of cortical bone, has high toughness, and has been found to exhibit bone bonding *in vivo.* The material has been used as an orbit implant for orbital floor fractures and volume augmentation (Tanner *et al.*, 1994) and is now used in middle ear implants, commercialized under the trade name HAPEX (Bonfield, 1996).

Bibliography

Akao, M., Aoki, H., and Kato, K. (1981). Mechanical Properties of Sintered Hydroxyapatite for Prosthetic Applications. *J. Mat. Sci.* **16**: 809.

Aoki, H. (1994). *Medical Applications of Hydroxyapatite.* Ishiyaku EuroAmerica Inc. Tokyo.

Levin, E. M., Robbins, C. R., and McMurdie, H. F. (eds.) (1964). *American Ceramic Society, Phase Diagrams for Ceramists.* American Ceramics Society, Ohio, p. 107.

Arends, J., Schutof, J., van der Linden, W. H., Bennema, P., and van den Berg, P. J. (1979). Preparation of Pure Hydroxyapatite Single Crystals by Hydrothermal Recrystallisation. *J. Cryst Growth* **46**: 213.

Barralet, J. E., Best, S. M., and Bonfield, W. (1998). Carbonate substitution in precipitated hydroxyapatite: An investigation into the effects of reaction temperature and bicarbonate Ion concentration. *J. Biomed. Mater. Res.* **41**: 79-86.

Berndt, C. C., Haddad, G. N., and Gross, K. A. (1989). Thermal spraying for bioceramics application. in *Bioceramics 2.* Proceedings of the 2nd International Symposium on Ceramics in Medicine, Heidelberg, pp. 201–206.

Best, S. M., and Bonfield, W. (1994). Processing behaviour of hydroxyapatite powders of contrasting morphology. *J. Mater. Sci. Mat. Med.* **5**: 516.

Bocholz, R. W., Carlton, A., and Holme, R. E. (1987). Hydroxyapatite and tricalcium phosphate bone graft substitute. *Orthop. Clin. North Am.* **18**: 323–334.

Bonel, G., Heughebeart, J-C., Heughebaert, M., Lacout, J. L., and Lebugle, A. (1987). Apatitic calcium orthophosphates and related compounds for biomaterials preparation. *Annals of the New York Academy of Science* 115.

Bonfield, W., Grynpas, M. D., Tully, A. E., Bowman, J., and Abram, J. (1989). Hydroxyapatite reinforced polyethylene—A mechanically compatible implant. *Biomaterials* **2**: 185–186.

Bonfield, W. (1996). Composite biomaterials. in *Bioceramics 9*, T. Kukubo, T. Nakamura, and F. Miyaji, eds. Proceedings of the 9th International Symposium on Ceramics in Medicine, Pergamon.

Christel, P., Meunier, A., Dorlot, J. M., Crolet, J. M., Witvolet, J., Sedel, L., and Boritin, P. (1988). Biomechanical compatability and design of ceramic implants for orthopedic surgery. in *Bioceramics: Material Characteristics versus In-Vivo Behavior,* P. Ducheyne and J. Lemons, eds. *Ann. N. Y. Acad. Sci.* **523**: 234.

Clemens, J. A. M. (1995). Fluorapatite Coatings for the Osseointegration of Orthopaedic Implants, Thesis, University of Leiden, Leiden, The Netherlands.

Clemens J. A. M., Klein C. P. A. T., Vriesde R. C., de Groot K., and Rozing P. M. (1995). Large gaps around calcium phosphate coated bone implants: Deficient bone apposition despite use of allograft bone. in *Proceedings of 21st Annual Meeting.* Society for Biomaterials, San Francisco.

Collin, R. L. (1959). Strontium—Calcium hydroxyapatite solid solutions: Preparations and lattice constant measurements. *J. Am. Chem. Soc.* **81**: 5275.

Cook, S. D., Thomas, K. A., Kay, J. F., and Jarcho, M. (1998). Hydroxylapatite coated titanium for orthopaedic implant applications. *Clin. Orthop. Rel. Res.* **232**: 225–243.

Cook, S. D., Thomas, K. A., Dalton, J. E., Volkman, R. K., White-cloud, T. S., and Kay, J. E. (1999). Hydroxylapatite coating of porous implants improves bone ingrowth and interface attachment strength. *J. Biomed. Mater. Res.* **26**: 989–1001.

Cranin, A. N., Tobin, G. P., and Gelbman, J. (1987). Applications of hydroxyapatite in oral and maxilofacial surgery, part II: Ridge augmentation and repair of major oral defects. *Compend. Contin. Educ. Dent.* 8: 334–345.

DeGroot, K. (1984). Surface chemistry of sintered hydroxyapatite: On possible relations with biodegredation and slow crack propagation, in *Adsorption and Surface Chemistry of Hydroxyapatite.* Ed. Misra, D. N. Plenum Publishing, New York.

DeGroot, K. (1987). Hydroxylapatite coatings for implants in surgery. in *High Tech Ceramics.* Ed. Vincenzini, P. Elsevier, Amsterdam.

DeGroot, K., Geesink, R. G. T., Klein, C. P. A. T., and Serekian, P. (1987). Plasma sprayed coatings of hydroxylapatite. *J. Biomed. Mater. Res.* **21**: 1375–1381.

DeGroot, K., Wolke, J. G. C., and Jansen, J. A. (1998). Calcium phosphate coatings for medical implants. *Proc. Inst. Mech. Engrs.* **212** (part H): 437.

Denissen, H. W., DeGroot, K., Driessen, A. A., Wolke, J. G. C., Peelen, J. G. J., van Dijk, H. J. A., Gehring, A. P., and Klopper, P. J. (1980a). Hydroxylapatite implants: Preparation, properties and use in alveolar ridge preservation. *Sci Ceram.***10**: 63.

Denissen, H. W., DeGroot, K., Klopper, P. J., van Dijk, H. J. A., Vermeiden, J. W. P., and Gehring, A. P. (1980b). Biological and mechanical evaluation of calcium hydroxypatite made by hot pressing. in *Mechanical Properties of Biomaterials*, Chapter 40, Eds. Hastings, G. W., and Williams, D. F. John Wiley and Sons, New York.

De With, G., van Dijk, H. J. A., and Hattu, N. (1981a). Mechanical behaviour of biocompatible hydroxyapatite ceramics. *Proc. Brit. Ceram. Soc.* **31**: 181.

De With, G., van Dijk, H. J. A., Hattu, N., and Prijs, K. (1981b). Preparation, microstructure and mechanical properties of dense polycrystalline hydroxyapatite. *J. Mat. Sci.* **16**: 1592.

Dickens, B., Schroeder, L. W., and Brown, W. E. (1974). Crystallographic studies of the role of Mg as a stabilising impurity in β-tricalcium phosphate. *J. Solid State Chem.* **10**: 232.

Driessen, A. A., Klein, C. P. A. T., and de Groot, K. (1982). Preparation and some properties of sintered ß-Whitlockite. *Biomaterials* **3**:113–116.

Driessens, F. C. M. (1983). Formation and stability of calcium phopshate in relation to the phase composition of the mineral in calcified tissue. in *Bioceramics of Calcium Phosphate*, Ed. DeGroot, K. CRC Press, Boca Raton, Florida.

de Groot, K. (1983). *Bioceramics of Calcium-Phosphate*. CRC Press, Boca Raton, FL.

de Groot, K. (1988). Effect of porosity and physicochemical properties on the stability, resorption, and strength of calcium phosphate ceramics. in *Bioceramics: Material Characteristics versus In-Vivo Behavior, Ann. N. Y. Acad. Sci.* **523**: 227.

de Groot, K., and Le Geros, R. (1988). in *Position Papers in Bioceramics: Materials Characteristics versus In-Vivo Behavior,* P. Ducheyne and J. Lemons, eds. *Ann. N. Y. Acad. Sci.* **523**: 227, 268, 272.

de Groot, K., Klein, C. P. A. T., Wolke, J. G. C., and de Blieck-Hogervorst, J. (1990). Chemistry of calcium phosphate bioceramics. in *Handbook on Bioactive Ceramics,* T. Yamamuro, L. L. Hench, and J. Wilson, eds. CRC Press, Boca Raton, FL, Vol. II, Ch. 1.

Drre, E., and Dawihl, W. (1980). Ceramic hip endoprostheses. in *Mechanical Properties of Biomaterials,* G. W. Hastings and D. Williams, eds. Wiley, New York, pp. 113–127.

Eanes, E. D., Gillessen, J. H., and Posner, A.L. (1965). Intermediate states in the precipitation of hydroxyapatite. *Nature* **208**: 365.

Eggli, P.S., Muller, W., and Schenk, R.K. (1988). Porous hydroxyapatite and tricalcium phosphate cylinders with two different pore size ranges implanted in the cancellous bone of rabbits. *Clin. Orthop. Relat. Res.* **232**: 127–138.

Elliot, J. R. (1994). *Structure and Chemistry of Apatites and Other Calcium Orthophosphate.* Elsevier, Amsterdam.

Fang, Y., Agrawal, D. K., Roy, D. M., and Roy, R. (1992). Fabrication of Porous Hydroxyapatite by Microwave Processing. *J. Mater. Res.* **7**(2): 490–494.

Fernandez, E., Gil, F. X., Ginebra, M. P., Driessens, F. C. M., Planell, J. A, and Best, S. M. (1999a). Calcium Phosphate Bone Cements for Clinical Applications, Part I, Solution Chemistry. *J. Mater. Sci. Mater. in Med.* **10**: 169–176.

Fernandez, E., Gil, F. X., Ginebra, M. P., Driessens, F. C. M., Planell, J. A, and Best, S. M. (1999b). Calcium Phosphate Bone Cements for Clinical Applications, Part II, Precipitate Formation during Setting Reactions. *J. Mater. Sci. Mater. in Med.* **10**: 177–184.

Fernandez, E., Planell, J. A., Best, S. M., and Bonfield W. (1998). Synthesis of Dahllite Through a Cement Setting Reaction, J. *Mater. Sci. Mater. in Med.* **9**: 789–792.

Fowler, B.O. (1974). Infrared studies of apatites, part II: Preparation of normal and isotopically substituted calcium, strontium and barium hydroxyapatites and spectra-structure correlations. *Inorg. Chem.* **13**: 207.

Froum, S. J., Kushner, J., Scopp, L., and Stahl, S. S. (1986). Human clinical and histologic responses to durapatite implants in intraosseous lesions: Case reports. *J. Periodontol.* **53**: 719–725.

Fujishiro, Y., Hench, L. L., and Oonishi, H. (1997). Quantitative rates of in-vivo bone generation for bioglass and hydroxyapatite particels as bone graft substitute, *J. Mater. Sci. Mater. in Med.* **8**: 649–652.

Galgut, P. N., Waite, I. M., and Tinkler, S. M. B. (1990). Histological investigation of the tissue response to hydroxyapatite used as an implant material in periodontal treatment. *Clin. Mater.* **6**: 105–121.

Gibson, I. R., Best, S. M., and Bonfield, W. (1996). Phase transformations of tricalcium phosphates using high temperature x-ray diffraction, Bioceramics **9**, Eds. Kokubo, Nakamura, and Miyaji, Publ. Pergamon Press, Oxford, pp. 173–176.

Gibson, I. R., Best, S. M., and Bonfield, W. (1999). Chemical characterisation of silicon-substituted hydroxyapatite. *J. Biomed. Mater. Res.* **44**: 422–428.

Gross, V., and Strunz, V. (1985). The interface of various glasses and glass-ceramics with a bony implantation bed. *J. Biomed. Mater. Res.* **19**: 251.

Gross, V., Kinne, R., Schmitz, H. J., and Strunz, V. (1988). The response of bone to surface active glass/glass-ceramics. CRC *Crit. Rev. Biocompatibility* **4**: 2.

Hayek, E., and Newsley, H. (1963). Pentacalcium monohydroxy-orthophosphate. *Inorganic Synthesis* **7**: 63.

Hench, L. L., and Wilson, J. (1993). *An Introduction to Bioceramics.* World Scientific, Singapore.

Hing, K. A. (1995). Assessment of porous hydroxyapatite for bone replacement. PhD Thesis, University of London.

Hing, K. A., Best, S. M., and Bonfield, W. (1999). Characterisation of porous hydroxyapatite. *J. Mater. Sci. Mater. in Med.* **10**: 135–160.

Hing, K. A., Best, S. M., Tanner, K. E., Revell, P. A., and Bonfield, W. (1998). Histomorphological and biomechamical characterisation of calcium phosphates in the osseous environment. *Proc. Inst. Mech. Engrs.* **212** (part H): 437.

Holmes, R. E. (1979). Bone regeneration within a coralline hydroxyapatite implant. *Plast. Reconstr. Surg.* **63**: 626–633.

Hulbert, S. F., Young, F. A., Mathews, R. S., Klawitter, J. J., Talbert, C. D., and Stelling, F. H. (1970). Potential of ceramic materials as permenantly implantable skeletal prosthesis. *J. Biomed. Mater. Res.* **4**: 433–456.

Hench, L. L. (1987). Cementless fixation. in *Biomaterials and Clinical Applications,* A. Pizzoferrato, P. G. Marchetti, A. Ravaglioli, and A. J. C. Lee, eds. Elsevier, Amsterdam, p. 23.

Hench, L. L. (1988). Bioactive ceramics. in *Bioceramics: Materials Characteristics versus In-Vivo Behavior,* P. Ducheyne and J. Lemons, eds. *Ann. N. Y. Acad. Sci.* **523**: 54.

Hench, L. L. (1991). Bioceramics: From concept to clinic. *J. Am. Ceram. Soc.* **74**: 1487–1510.

Hench, L. L. (1994). *Bioactive Ceramics: Theory and Clinical Applica,* O. H. Anderson and A. Yli-Urpo, eds. Butterworth–Heinemann, Oxford, England, pp. 3–14.

Hench, L. L., and Clark, D. E. (1978). Physical chemistry of glass surfaces. *J. Non-Cryst. Solids* **28**(1): 83–105.

Hench, L. L., and Ethridge, E. C. (1982). *Biomaterials: An Interfacial Approach.* Academic Press, New York.

Hench, L. L., and Wilson, J. W. (1993). *An Introduction to Bioceramics.* World Scientific, Singapore.

Hench, L. L., and Wilson, J. W. (1996). *Clinical Performance of Skeletal Prostheses.* Chapman and Hall, London.

Hench, L. L., Splinter, R. J., Allen, W. C., and Greenlec, T. K., Jr. (1972). Bonding mechanisms at the interface of ceramic prosthetic materials. *J. Biomed. Res. Symp. No. 2.* Interscience, New York, p. 117.

Höhland, W., and Vogel, V. (1993). Machineable and phosphate glass-ceramics, in *An Introduction to Bioceramics,* L. L. Hench and J. Wilson, eds. World Scientific, Singapore, pp. 125–138.

Hulbert, S. (1993). The use of alumina and zirconia in surgical implants. in *An Introduction to Bioceramics,* L. L. Hench and J. Wilson, eds. World Scientific, Singapore, pp. 25–40.

Hulbert, S. F., Bokros, J. C., Hench, L. L., Wilson, J., and Heimke, G. (1987). Ceramics in clinical applications: past, present, and future, in *High Tech Ceramics,* P. Vincenzini, ed. Elsevier, Amsterdam, pp. 189–213.

Irvine, G. D. (1981). Synthetic bone ash. British Patent number 1 586 915.

Irwin, G. R. (1959). Analysis of stresses and strains near the end of a crack traversing a plate. *J. Appl. Mechanics* **24**: 361.

Jarcho, M., Bolen, C. H., Thomas, M. B., Bobick, J., Kay, J. F., and Doremus, R. H. (1976). Hydroxylapatite synthesis in dense polycrystalline form. *J. Mat. Sci.* 11: 2027.

Jha, L., Best, S. M., Knowles, J., Rehman, I., Santos, J., and Bonfield, W. (1997). Preparation and characterisation of fluoride-substituted apatites. *J. Mater. Sci. Mater. in Med.* **8**: 185–191.

Johnson, P. D.,Prener, J. S., and Kingsley, J. D. (1963). Apatite: Origin of blue color. *Science* **141**: 1179.

Jarcho, M. (1981). Calcium phosphate ceramics as hard tissue prosthetics. *Clin. Orthop. Relat. Res.* **157**: 259.

Kay, M. I., Young, R. A., and Posner, A. S. (1964). Crystal structure of hydroxyapatite. *Nature* **12**: 1050.

Kijima, T., and Tsutsumi, M. (1979). Preparation and thermal properties of dense polycrystalline oxyhydroxyapatite. J. *Am. Ceram. Soc.* **62**(9): 455.

Klawitter, J. J., Bagwell, J. G., Weinstein, A. M., Sauer, B. W., and Pruitt, J. R. (1976). An evaluation of bone growth into porous high density polyethylene. *J. Biomed. Mater. Res.* **10**: 311–323.

Klein, C. P. A. T., Driessen, A.A., and de Groot, K. (1983a). Biodegredation of calciumphosphate ceramics–ultrastructural geometry and dissolubility of different calcium phosphate ceramics. in Proceedings of the 1st International Symposium on Biomaterials in Otology, 84–92.

Klein, C. P. A. T., Driessen, A. A., de Groot, K., and van den Hooff, A. (1983b). Biodegration behaviour of various calcium phosphate materials in bone tissue. *J. Biomed. Mater. Res.* **17**: 769–784.

Klein, C. P. A. T., Wolke, J. G. C., de Blieck-Hogervorst, J. M. A., and DeGroot, K. (1994a). Features of calcium phosphate coatings: an in-vitro study. *J. Biomed. Mater. Res.* **28**: 961–967.

Klein, C. P. A. T., Wolke, J. G. C., de Blieck-Hogervorst, J. M. A., and DeGroot, K. (1994b). Features of calcium phosphate coatings: An in-vivo study. *J. Biomed. Mater. Res.* **28**: 909–917.

Kühne, J.H., Bartl, R., Frish, B., Hanmer, C., Jansson, V., Zimmer, M. (1994). Bone formation in coralline hydroxyapatite: Effects of pore size studied in rabbits. *Acta Orthop. Scand.* **65** (3): 246–252.

Kokubo, T. (1993). A/W glass-ceramics: Processing and properties. in *An Introduction to Bioceramics,* L. L. Hench and J. Wilson, eds. World Scientific, Singapore, pp. 75–88.

LeGeros, R. Z. (1965). Effect of carbonate on the lattice parameters of apatite. *Nature* **206**: 403.

LeGeros, R. Z. (1988). Calcium phosphate materials in restorative dentistry: A review. *Adv. Dent Res.* **2** (1): 164.

LeGeros, R. Z., Daculsi, G., Orly, I., Gregoire, M., Heughebeart, M., Gineste, M., and Kijkowska, R. (1992). Formation of carbonate apatite on calcium phosphate materials: Dissolution/precipitation processes. in *Bone Bonding Biomaterials.* Eds. Ducheynes, P., Kukubo, T., and van Blitterswijk, C. A.. Reed Healthcare Communications, Leiderdorp, The Netherlands, pp. 78088.

Lehr, J. R, Brown, E. T., Frazier, A.W., Smith, J. P., and Thrasher, R. D. (1967). *Crystallographic Properties of Fertilizer Compounds.* Tennessee Valley Authority Chemical Engineering Bulletin, 6.

Lacefield, W. R. (1993). Hydroxylapatite coatings. in *An Introduction to Bioceramics,* L. L. Hench and J. Wilson, eds. World Scientific, Singapore, pp. 223–238.

Le Geros, R. Z. (1988). Calcium phosphate materials in restorative density: a review. *Adv. Dent. Res.* **2**: 164–180.

Le Geros, R. Z., and Le Geros, J. P. (1993). Dense hydroxyapatite. in *An Introduction to Bioceramics,* L. L. Hench and J. Wilson, eds. World Scientific, Singapore, pp. 139–180.

Maxian, S. H., Zawaddsky, J. P., and Dunn, M. G. (1994). Effect of calcium phosphate coating resorption and surgical fit on the bone/implant interface. *J. Biomed. Mater. Res.* **28**: 1311–1319.

Maxian, S.H., Zawaddsky, J.P., and Dunn, M.G. (1993). Mechanical and histological evaluation of amorphous calcium phosphate and poorly crystalised hydroxyapatite coatings on titanium implants. *J. Biomed. Mater. Res.* **27**: 717–728.

McDowell, H., Gregory, T. M., and Brown, W. E. (1977). Solubility of $Ca_5(PO_4)_3OH$ in the system $Ca(OH)_2$ - H3PO4 - H_2O at 5, 15, 25 and 37°C. J. *Res. Natl. Bureau Standards* **81A**: 273.

Monma, H., Ueno, S., and Kanazawa, T. T. (1981). Properties of hydroxyapatite prepared by the hydrolysis of tricalcium phosphate. J. *Chem Tech. Biotechnol.* **31**: 15.

Mooney, R. W., and Aia, M. A. (1961). Alkaline earth phosphates. *Chem. Rev.* **61**: 433.

Morancho, R., Ghommidh, J., and Buttazoni, B. G. (1981). Constant, thin films of several calcium phosphates obtained by chemical spray of aque ous calcium hydrogen phosphate solution: A route to hydroxyapatite films. *Proceedings of the 8th International Conference on Chemical Vapour Deposition.* Electrochemical Society, New York.

Moroni, A., Caja, V. J., Egger, E. L., Trinchese, L., and Chao, E. Y. S. (1994). Histomorphometry of hydroxyapatite coated and uncoated porous titanium bone implants. *Biomaterials* **15** (11): 926–930.

McKinney, Jr., R. V., and Lemons, J. (1985). *The Dental Implant.* PSG Publ., Littleton, MA.

McMillan, P. W. (1979). *Glass-Ceramics.* Academic Press, New York.

Miller, J. A., Talton, J. D., and Bhatia, S. (1996). in *Clinical Performance of Skeletal Prostheses,* L. L. Hench and J. Wilson, eds. Chapman and Hall, London, pp. 41–56.

Nagai, H., and Nishimura, Y. (1985). Hydroxyapatite ceramic material and process for the preparation thereof. US Patent no. 4,548,59.

Newsley, H. (1963). Crystallographic and morphological study of carbonate-apatite, M.h.f. *Chem.* **95**: 270.

Nakamura, T., Yamumuro, T., Higashi, S., Kokubo, T., and Itoo, S. (1985). A new glass-ceramic for bone replacement: Evaluation of its bonding to bone tissue. *J. Biomed. Mater. Res.* **19**: 685.

Oonishi, H., Tsuji, E., Ishimaru, H., Yamamoto, M., and Delecrin, J. (1990). Clinical sginificance of chemical bonds and bone in orthopaedic surgery. in *Bioceramics 2: Proceedings of the 2nd International Symposium on Ceramics in Medicine.* Ed, Heimke, G. German Ceramic Society, Cologne.

Onoda, G., and Hench, L. L. (1978). *Ceramic Processing before Firing.* Wiley, New York.

PDF card 9-432. ICDD, Newton Square, PA.

Peelen, J. G. J., Rejda, B. V., and DeGroot, K. (1978). Preparation and properties of sintered hydroxylapatite. *Ceramurgica International* **4**(2): 71.

Perloff, A., and Posner, A. S. (1956). Preparation of pure hydroxyapatite crystals. *Science* **124**: 583.

Phillips, R. W. (1991). *Skinners Science of Dental Materials,* 9th ed., Ralph W. Phillips, ed. Saunders, Philadelphia.

Quinn, J. H., and Kent, J .N. (1984). Alveolar ridge maintenance with solid non-porous hydroxylapatite root implants. *Oral Surg.* **58**: 511–516.

Rao, R. W., and Boehm, J. (1974). A study of sintered apatites. *J. Dent Res.* 1351.

Rahn, B. A., Neff, J., Leutenegger, A., Mathys, R., and Perren, S. M. (1986). Integration of synthetic apatite of various pore size and density in bone. in *Biological and Biomechanical Performance of Biomaterials.* Eds, Christel, P., Meunier, A., and Lee, A. J. C. Elsevier Science Publishers, Amsterdam.

Rootare, H., and Craig, R. G. (1978). Characterisation of hydroxyapatite powders and compacts at room temperature after sintering at 1200°C. *J. Oral Rehab.* **5**: 293.

Roy, D. M. (1971). Crystal growth of hydroxyapatite. *Mater. Res. Bull.* **6**: 1337.

Reck, R., Storkel, S., and Meyer, A. (1988). Bioactive glass-ceramics in middle ear surgery: an 8-year review. in *Bioceramics: Materials Characteristics versus In-Vivo Behavior,* P. Ducheyne and J. Lemons, eds. *Ann. N. Y. Acad. Sci.* **523**: 100.

Reed, J. S. (1988). *Introduction to Ceramic Processing.* Wiley, New York.

Ritter, J. E., Jr., Greenspan, D. C., Palmer, R. A., and Hench, L. L. (1979). Use of fracture mechanics theory in lifetime predictions for alumina and bioglass-coated alumina. *J. Biomed. Mater. Res.* **13**: 251–263.

Roy, D. M., and Linnehan, S. K. (1974). Hydroxyapatite formed from coral skeletal carbonate by hydrothermal exchange. *Nature* **247**: 220–222.

Schleede, A., Schmidt, W., and Kindt, H. (1932). Zu kenntnisder calciumphosphate und apatite. Z. *Elektrochem.* **38**: 633.

Skinner, H. C. W. (1973). Phase relations in the CaO-P_2O_5–H_2O system from 300 to 600°C at 2kb H_2O pressure. *J. Am. Sci.* **273**: 545.

Smiler, D. G., and Holmes, R. E. (1987). Sinus life procedure using prous hydroxyapaitte: A preliminary clinical report. *J. Oral. Implantology* **13**: 17–32.

Soballe, K., Hansen, E. S., Brockstedt-Rasmussen, H. B., and Bunger, C. (1993). Hydroxyapatite coating converts fibrous tissue to bone around loaded implants. *J. Bone Jt Surgery* **75B**: 270–278.

Stephenson, P. K., Freeman, M. A. R. F., Revell, P. A., German, J., Tuke, M., Pirie, C. J. (1991). The effect of hydroxyapatite coating on ingrowth of bone into cavities in an implant. *J. Arthroplasty* **6**(1): 51–58.

Schors, E. C., and Holmes, R. E. (1993). Porous hydroxyapatite. in *An Introduction to Bioceramics,* L. L. Hench and J. Wilson, eds. World Scientific, Singapore, pp. 181–198.

Stanley, H. R., Clark, A. E., and Hench, L. L. (1996). Alveolar ridge maintenance implants. in *Clinical Performance of Skeletal Prostheses.* Chapman and Hall, London, pp. 237–254.

Tanner, K. E., Downes, R. N., and Bonfield, W. (1994). Clinical applications of hydroxyapatite reinforced materials. *Brit. Ceram. Trans.* **4** (93): 104–107.

van Blitterswijk, C. A., Hessling, S. C., Grote, J. J., Korerte, H. K., and DeGroot, K. (1990). The biocompatibility of hydroxyapatite ceramic: A study of retrieved human middle ear implants. *J. Biomed. Mater. Res.* **24**: 433–453

Wilson, J., and Low, S. B. (1992). Bioactive ceramics for periodontal treatment: Comparative study in Patus monkey. *J. App. Biomat.* **2**: 123–129.

Wolke, J. G. C., de Blieck-Hogervorst, J. M. A., Dhert, W. J. A., Klein, C. P. A. T., and DeGroot, K. (1992). Studies on thermal spraying of apatite bioceramics. *J. Thermal Spray Technology* **1**: 75–82.

White, E., and Schors, E. C. (1986). Biomaterials aspects of interpore-200 porous hydroxyapatite. *Dent. Clin. North Am.* **30**: 49–67.

Wilson, J. (1994). *Clinical applications of bioglass implants.* in *Bioceramics-7,* O. H. Andersson, ed. Butterworth-Heinemann, Oxford, England.

Wilson, J., Pigott, G. H., Schoen, F. J., and Hench, L. L. (1981). Toxicology and biocompatibility of bioglass. J. *Biomed. Mater. Res.* **15**: 805.

Yamamuro, T., Hench, L. L., Wilson, J. (1990). *Handbook on Bioactive Ceramics,* Vol. I: *Bioactive Glasses and Glass-Ceramics,* Vol. II: *Calcium-Phosphate Ceramics.* CRC Press, Boca Raton, FL.

Young, R. A., and Elliot, J. C. (1966). Atomic scale bases for several properties of apatites. *Arch. Oral Biol.* **11**: 699.

Young, R. A., and Holcomb, D. W. (1982). Variability of hydroxyapatite preparations. *Calcif. Tiss Int.* **34**: S17.

3.2.11 Pyrolytic carbon for long-term medical implants

Robert B. More, Axel D. Haubold, and Jack C. Bokros

Introduction

Carbon materials are ubiquitous and of great interest because the majority of substances that make up living organisms are carbon compounds. Although many engineering materials and biomaterials are based on carbon or contain carbon in some form, elemental carbon itself is also an important and very successful biomaterial. Furthermore, there exists enough diversity in structure and properties for elemental carbons to be considered as a unique class of materials beyond the traditional molecular carbon focus of organic chemistry, polymer chemistry, and biochemistry. Through a serendipitous interaction between researchers during the late 1960s the outstanding blood compatibility of a special form of elemental pyrolytic carbon deposited at high temperature in a fluidized bed was discovered. The material was found to have not only remarkable blood compatibility but also the structural properties needed for long-term use in artificial heart valves (LaGrange *et al.*, 1969). The blood compatibility of pyrolytic carbon was recognized empirically using the Gott vena cava ring test. This test involved implanting a small tube made of a candidate material in a canine vena cava and observing the development of thrombosis within the tube in time. Prior to pyrolytic carbon, only surfaces coated with graphite, benzylalkonium chloride, and heparin would resist thrombus formation when exposed to blood for long periods. The incorporation of pyrolytic carbon in mechanical heart valve implants was declared "an exceptional event" (Sadeghi, 1987) because it added the durability and stability needed for heart valve prostheses to endure for a patient's lifetime. The objective of this section is to present the elemental pyrolytic carbon materials currently is used in the fabrication of medical devices and to describe their manufacture, characterization, and properties.

Elemental carbon

Elemental carbon is found in nature as two crystalline allotrophic forms: graphite and diamond. Elemental carbon also occurs as a spectrum of imperfect, turbostratic crystalline forms that range in degree of crystallinity from amorphous to the perfectly crystalline allotropes.

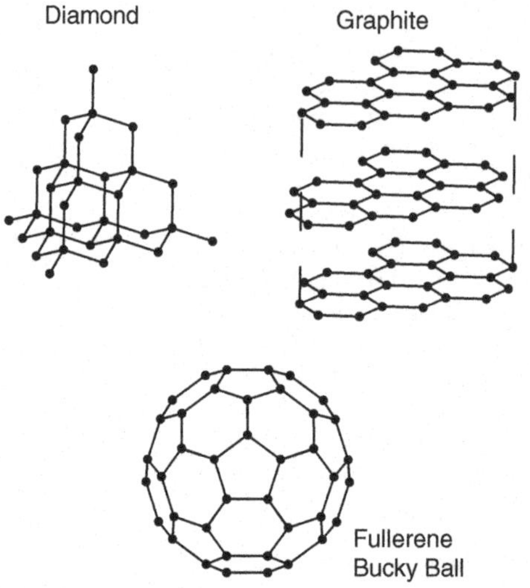

Fig. 3.2.11-1 Allotropic crystalline forms of carbon: diamond, graphite, and fullerene.

Recently a third crystalline form of elemental carbon, the fullerene structure, has been discovered. The crystalline polymorphs of elemental carbon are shown in Fig. 3.2.11-1.

The properties of the elemental carbon crystalline forms vary widely according to their structure. Diamond with its tetrahedral sp^3 covalent bonding is one of the hardest materials known. In the diamond crystal structure, covalent bonds of length 1.54 Å connect each carbon atom with its four nearest neighbors. This tetrahedral symmetry repeats in three dimensions throughout the crystal (Pauling, 1964). In effect, the crystal is a giant isotropic covalently bonded molecule; therefore, diamond is very hard.

Graphite with its anisotropic layered in-plane hexagonal covalent bonding and interplane van der Waals bonding structure is a soft material. Within each planar layer, each carbon atom forms two single bonds and one double bond with its three nearest neighbors. This bonding repeats in-plane to form a giant molecular (graphene) sheet. The in-plane atomic bond length is 1.42 Å, which is a resonant intermediate (Pauling, 1964) between the single-bond length of 1.54 Å and the double-bond length of 1.33 Å. The planer layers are held together by relatively weak van der Waals bonding at a distance of 3.4 Å, which is more than twice the 1.42-Å bond length (Pauling, 1964). Graphite has low hardness and a lubricating property because the giant molecular sheets can readily slip past one another against the van der Waals bonding. Nevertheless, although large-crystallite-size natural graphite is used as a lubricant, some artificially produced graphites can be very abrasive if the crystallite sizes are small and randomly oriented.

Fullerenes have yet to be produced in bulk, but their properties on a microscale are entirely different from those of their crystalline counterparts. Fullerenes and nanotubes consist of a graphene layer that is rolled up or folded (Sattler, 1995) to form a tube or ball. These large molecules, C_{60} and C_{70} fullerenes and (C_{60+18j}) nanotubes, are often mentioned in the literature (Sattler, 1995) along with more complex multilayer "onion skin" structures.

There exist many possible forms of elemental carbon that are intermediate in structure and properties between those of the allotropes diamond and graphite. Such "turbostratic" carbons occur as a spectrum of amorphous through mixed amorphous, graphite-like and diamond-like to the perfectly crystalline allotropes (Bokros, 1969). Because of the dependence of properties upon structure, there can be considerable variability in properties for the turbostratic carbons. Glassy carbons and pyrolytic carbons, for example, are two turbostratic carbons with considerable differences in structure and properties. Consequently, it is not surprising that carbon materials are often misunderstood through oversimplification. Properties found in one type of carbon structure can be totally different in another type of structure. Therefore it is very important to specify the exact nature and structure when discussing carbon.

Pyrolytic carbon (Pyc)

The biomaterial known as pyrolytic carbon is not found in nature: it is manmade. The successful pyrolytic carbon biomaterial was developed at General Atomic during the late 1960s using a fluidized-bed reactor (Bokros, 1969). In the original terminology, this material was considered a low-temperature isotropic carbon (LTI carbon). Since the initial clinical implant of a pyrolytic carbon component in the DeBakey–Surgitool mechanical valve in 1968, 95% of the mechanical heart valves implanted worldwide have at least one structural component made of pyrolytic carbon. On an annual basis this translates into approximately 500,000 components (Haubold, 1994). Pyrolytic carbon components have been used in more than 25 different prosthetic heart valve designs since the late 1960s and have accumulated a clinical experience of the order of 16 million patient-years. Clearly, pyrolytic carbon is one of the most successful, critical biomaterials both in function and application. Among the materials available for mechanical heart valve prostheses, pyrolytic carbon has the best combination of blood compatibility, physical and mechanical properties, and durability. However, the blood compatibility of pyrolytic carbon in heart-valve applications is not perfect; chronic anticoagulant therapy is needed for patients with mechanical heart valves. Whether the need for anticoagulant therapy arises from the biocompatible properties of the material itself or from the particular hydrodynamic interaction of a given device and the blood remains to be resolved.

The term "pyrolytic" is derived from "pyrolysis," which is thermal decomposition. Pyrolytic carbon is formed from the thermal decomposition of hydrocarbons such as propane, propylene, acetylene, and methane, in the absence of oxygen. Without oxygen the typical decomposition of the hydrocarbon to carbon dioxide and water cannot take place; instead a more complex cascade of decomposition products occurs that ultimately results in a "polymerization" of the individual carbon atoms into large macroatomic arrays.

Pyrolysis of the hydrocarbon is normally carried out in a fluidized-bed reactor such as the one shown in Fig. 3.2.11-2. A fluidized-bed reactor typically consists of a vertical tube furnace that may be induction or resistance heated to temperatures of 1000–2000°C (Bokros, 1969). Reactor diameters ranging from 2 cm to 25 cm have been used; however, the most common size used for medical devices has a diameter of about 10 cm. These high-temperature reactors are expensive to operate, and the reactor size limits the size of device components to be produced.

Small refractory ceramic particles are placed into the vertical tube furnace. When a gas is introduced into the bottom of the tube furnace, the gas causes the particle bed to expand: Interparticle spacing increases to allow for the flow of the gas. Particle mixing occurs and the bed of particles begins to "flow" like a fluid. Hence the term "fluidized bed." Depending upon the gas flow rate and volume, this expansion and mixing can be varied from a gentle bubbling bed to a violent spouting bed. An oxygen-free, inert gas such as nitrogen or helium is used

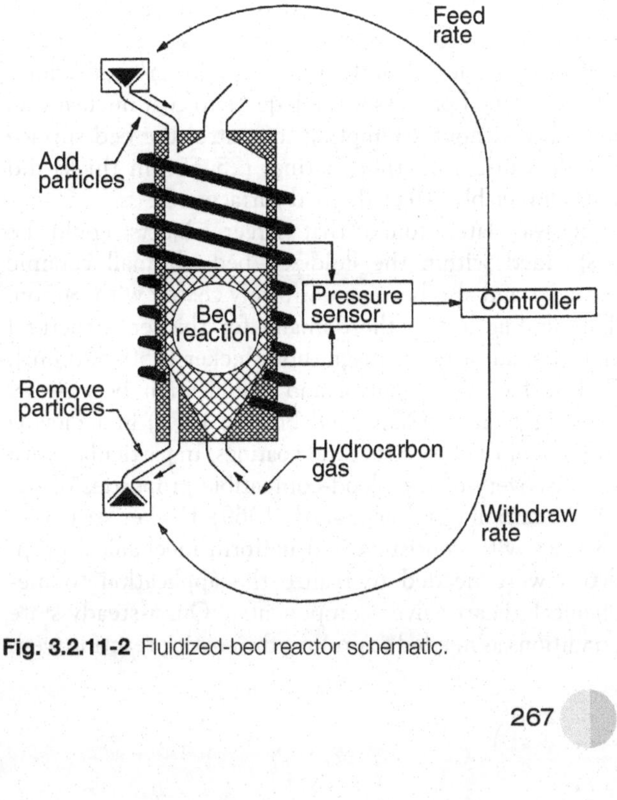

Fig. 3.2.11-2 Fluidized-bed reactor schematic.

to fluidize the bed, and the source hydrocarbon is added to the gas stream when needed.

At a sufficiently high temperature, pyrolysis or thermal decomposition of the hydrocarbon can take place. Pyrolysis products range from free carbon and gaseous hydrogen to a mixture of C_xH_y decomposition species. The pyrolysis reaction is complex and is affected by the gas flow rate, composition, temperature and bed surface area. Decomposition products, under the appropriate conditions, can form gas-phase nucleated droplets of carbon/hydrogen, which condense and deposit on the surfaces of the wall and bed particles within the reactor (Bokros, 1969). Indeed, the fluidized-bed process was originally developed to coat small (200–500 micrometer) diameter spherical particles of uranium/thorium carbide or oxide with pyrolytic carbon. These coated particles were used as the fuel in the high temperature gas-cooled nuclear reactor (Bokros, 1969).

Pyrolytic carbon coatings produced in vertical-tube reactors can have a variety of structures such as laminar or isotropic, granular, or columnar (Bokros, 1969). The structure of the coating is controlled by the gas flow rate (residence time in the bed), hydrocarbon species, temperature and bed surface area. For example, an inadequately fluidized or static bed will produce a highly anisotropic, laminar pyrolytic carbon (Bokros, 1969).

Control of the first three parameters (gas flow rate, hydrocarbon species, and temperature) is relatively easy. However, until recently, it was not possible to measure the bed surface area while the reactions were taking place. As carbon deposits on the particles in the fluidized bed, the diameter of the particles increases. Hence the surface area of the bed changes, which in turn influences the subsequent rate of carbon deposition. As surface area increases, the coating rate decreases since a larger surface area now has to be coated with the same amount of carbon available. Thus the process is not in equilibrium. The static-bed process was adequate to coat nuclear fuel particles without attempting to control the bed surface area, because such thin coatings (25–50 μm thick) did not appreciably affect the bed surface area.

It was later found that larger objects could be suspended within the fluidized bed of small ceramic particles and also become uniformly coated with carbon. This finding led to the demand for thicker, structural coatings, an order of magnitude thicker (250–500 μm). Bed surface area control and stabilization became an important factor (Akins and Bokros, 1974) in achieving the goal of thicker, structural coatings. In particular, with the discovery of the blood-compatible properties of pyrolytic carbon (LaGrange *et al.*, 1969), thicker structural coatings with consistent and uniform mechanical properties were needed to realize the application to mechanical heart-valve components. Quasi-steady-state conditions as needed to prolong the coating reaction were achieved empirically by removing coated particles and adding uncoated particles to the bed while the pyrolysis reaction was taking place (Akins and Bokros, 1974). However, the rates of particle addition and removal were based upon little more than good guesses.

Three of the four parameters that control carbon deposition could be accurately measured and controlled, but a method to measure and control bed surface area was lacking. Thus, the quasi-steady-state process was more of an art than a science. If too many coated particles were removed, the bed became too small to support the larger components within it and the bed collapsed. If too few particles were removed, the rate of deposition decreased, and the desired amount of coating was not achieved in the anticipated time. Furthermore, there were considerable variations in the mechanical properties of the coating from batch to batch. It was found that in order to consistently achieve the hardness needed for wear resistance in prosthetic heart valve applications, it was necessary to add a small amount of β-silicon carbide to the carbon coating. The dispersed silicon carbide particles within the pyrolytic carbon matrix added sufficient hardness to compensate for potential variations in the properties of the pyrolytic carbon matrix. The β-silicon carbide was obtained from the pyrolysis of methyltrichlorosilane, CH_3SiCl_3. For each mole of silicon carbide produced, the pyrolysis of methyltrichlorosilane also produces 3 moles of hydrogen chloride gas. Handling and neutralization of this corrosive gas added substantial complexity and cost to an already complex process. Nevertheless, this process allowed consistency for the successful production of several million components for use in mechanical heart valves.

A process has been developed and patented that allows precise measurement and control of the bed surface area. A description of this process is given in the patent literature and elsewhere (Emken *et al.*, 1993, 1994; Ely *et al.*, 1998). With precise control of the bed surface area it is no longer necessary to include the silicon carbide. Elimination of the silicon carbide has produced a stronger, tougher, and more deformable pure pyrolytic carbon. Historically, pure carbon was the original objective of the development program because of the potential for superior biocompatibility (LaGrange *et al.*, 1969). Furthermore, the enhanced mechanical and physical properties of the pure pyrolytic carbon now possible with the improved process control allows prosthesis design improvements in the hemodynamic contribution to thromboresistance (Ely *et al.*, 1998).

Structure of pyrolytic carbons

X-ray diffraction patterns of the biomedical-grade fluidized-bed pyrolytic carbons are broad and diffuse because of the small crystallite size and imperfections. In silicon-alloyed pyrolytic carbon, a diffraction pattern characteristic of the β form of silicon carbide also appears in the diffraction pattern along with the carbon bands. The carbon diffraction pattern indicates a turbostratic structure (Kaae and Wall, 1996) in which there is order within carbon layer planes, as in graphite; but, unlike graphite, there is no order between planes. This type of turbostratic structure is shown in Fig. 3.2.11-3 compared to that of graphite. In the disordered crystalline structure, there may be lattice vacancies and the layer planes are curved or kinked. The ability of the graphite layer planes to slip is inhibited, which greatly increases the strength and hardness of the pyrolytic carbon relative to that of graphite. From the Bragg equation, the pyrolytic carbon layer spacing is reported to be 3.48 Å, which is larger than the 3.35 Å graphite layer spacing (Kaae and Wall, 1996). The increase in layer spacing relative to graphite is due to both the layer distortion and the small crystallite size, and is common feature for turbostratic carbons. From the Scherrer equation the crystallite size is typically 25–40 Å (Kaae and Wall, 1996).

During the coating reaction, gas-phase nucleated droplets of carbon/hydrogen form that condense and deposit on the surfaces of the reactor wall and bed particles within the reactor. These droplets aggregate, grow, and form the coating. When viewed with high-resolution transmission electron microscopy, a multitude of near-spherical polycrystalline growth features are evident as shown in Fig. 3.2.11-4 (Kaae and Wall, 1996). These growth features are considered to be the basic building blocks of the material, and the shape and size are related to the deposition mechanism. In the silicon-alloyed carbon small silicon carbide particles are present within the growth features as shown in Fig. 3.2.11-5. Based on a crystallite size of 33 Å, each growth feature contains about 3×10^9 crystallites. Although the material is quasi-crystalline on a fine level, the crystallites are very small and randomly oriented in the fluidized bed pyrolytic

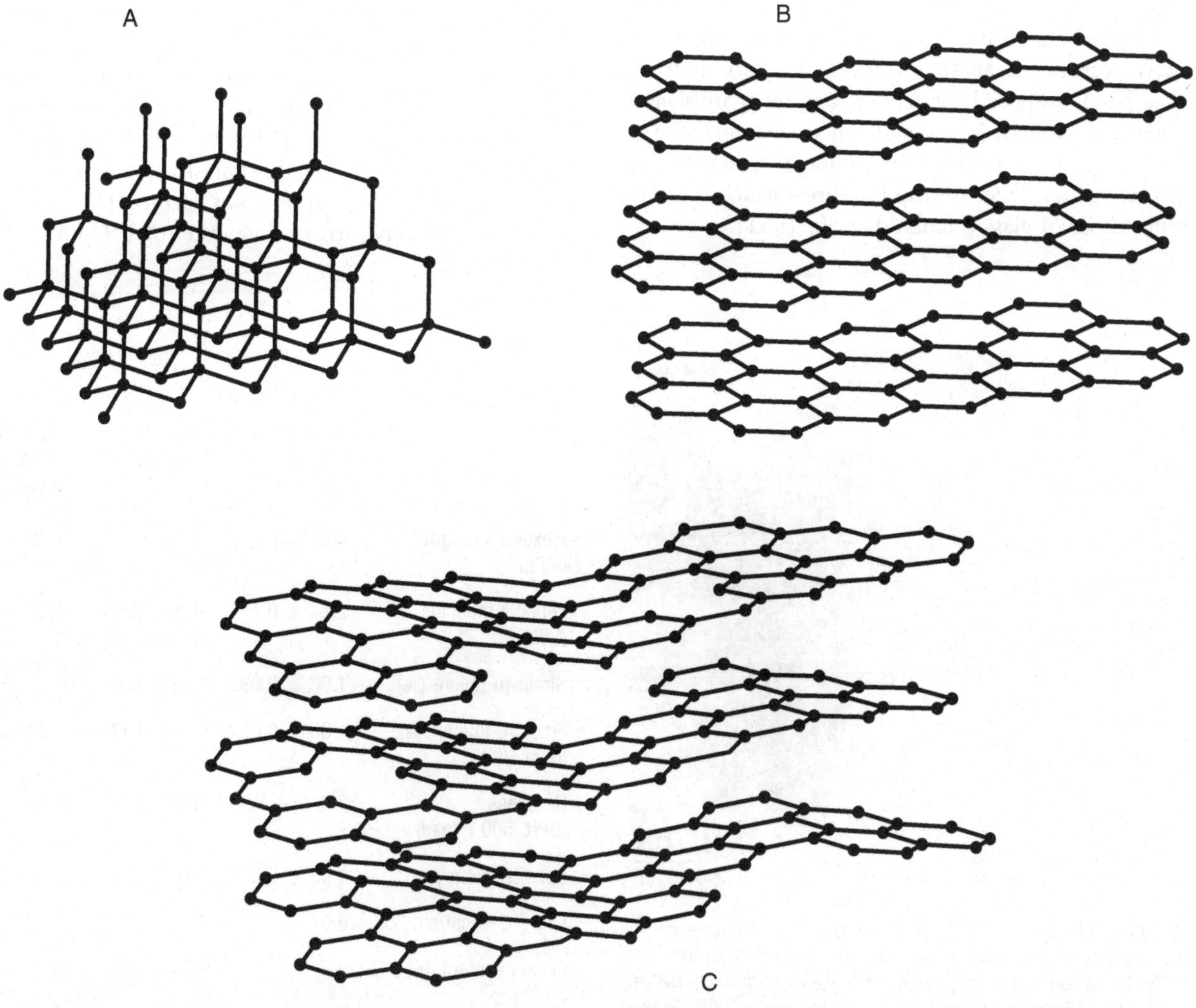

Fig. 3.2.11-3 Structures of (A) diamond, (B) graphite, and (C) turbostratic pyrolytic carbon.

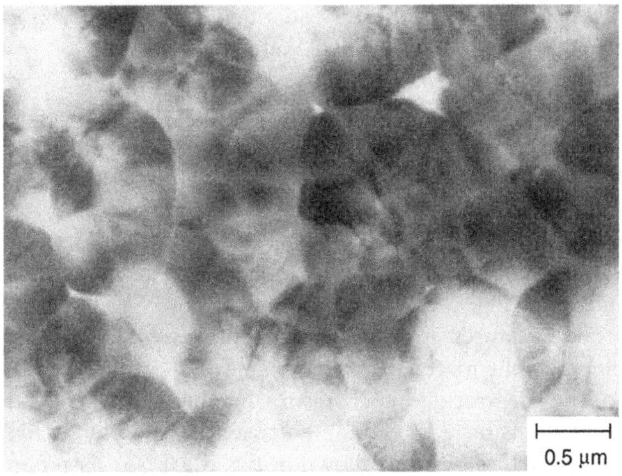

Fig. 3.2.11-4 Electron micrograph of pure pyrolytic carbon microstructure showing near-spherical polycrystalline growth features formed during deposition (Kaae and Wall, 1996).

carbons so that overall the material exhibits isotropic behavior.

Glassy carbon, also known as vitreous carbon or polymeric carbon, is another turbostratic carbon form that has been proposed for use in long-term implants. However, its strength is low and the wear resistance is poor. Glassy carbons are quasi crystalline in structure and are named "glassy" because the fracture surfaces closely resemble those of glass (Haubold *et al.*, 1981).

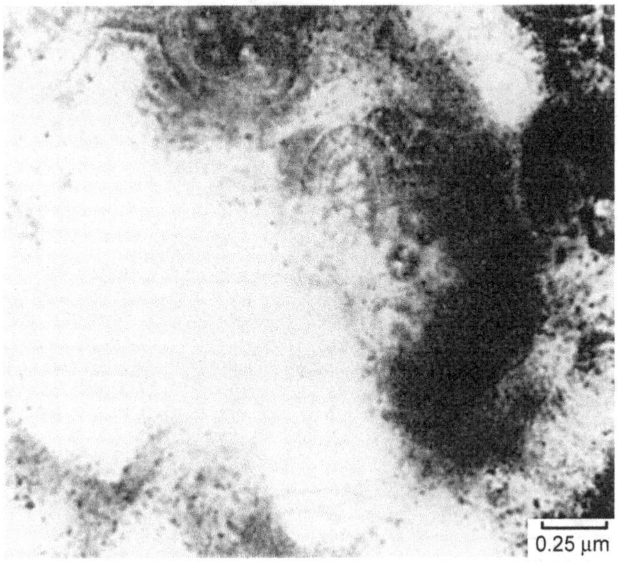

Fig. 3.2.11-5 Electron micrograph of silicon-alloyed pyrolytic carbon microstructure showing near-spherical polycrystalline growth features formed during deposition (Kaae and Wall, 1996). Small silicon carbide particles are shown in concentric rings in the growth features.

Vapor-deposited carbons are also used in heart-valve applications. Typically, the coatings are thin (< 1 μm) and may be applied to a variety of materials in order to confer the biochemical characteristics of turbostratic carbon. Some examples are vapor-deposited carbon coatings on heart-valve sewing cuffs and metallic orifice components (Haubold *et al.*, 1981).

Mechanical properties

Mechanical properties of pure pyrolytic carbon, silicon-alloyed pyrolytic carbons and glassy carbon are given in Table 3.2.11-1. Pyrolytic carbon flexural strength is high enough to provide the necessary structural stability for a variety of implant applications and the density is low enough to allow for components to move easily under the applied forces of circulating blood. With respect to orthopedic applications, Young's modulus is in the range reported for bone (Reilly and Burstein, 1974; Reilly *et al.*, 1974), which allows for compliance matching. Relative to metals and polymers, the pyrolytic carbon strain-to-failure is low; it is a nearly ideal linear elastic material and requires consideration of brittle material principles in component design. Strength levels vary with the effective stressed volume or stressed area as predicted by classical Weibull statistics (De Salvo, 1970). The flexural strengths cited in Table 3.2.11-1 are for specimens tested in four-point bending, third-point loading (More *et al.*, 1993) with effective stressed volumes of 1.93 mm^3. The

Table 3.2.11-1 Mechanical properties of biomedical carbons

Property	Pure PyC	Typical Si-alloyed PyC	Typical glassy carbon
Flexural strength (MPa)	493.7 ± 12	407.7 ± 14.1	175
Young's modulus (GPa)	29.4 ± 0.4	30.5 ± 0.65	21
Strain-to-failure (%)	1.58 ± 0.03	1.28 ± 0.03	
Fracture toughness (MPa √m)	1.68 ± 0.05	1.17 ±0.17	0.5–0.7
Hardness (DPH, 500 g load)	235.9 ± 3.3	287 ± 10	150
Density (g/cm^3)	1.93 ± 0.01	2.12 ± 0.01	<1.54
CTE (10–6 cm/cm °C)	6.5	6.1	
Silicon content (%)	0	6.58 ± 0.32	0
Wear resistance	Excellent	Excellent	Poor

pyrolytic carbon material Weibull modulus is approximately 10 (More *et al.*, 1993).

Fracture toughness levels reflect the brittle nature of the material, but the fluidized-bed isotropic pyrolytic carbons are remarkably fatigue resistant. Recent fatigue studies indicate the existence of a fatigue threshold that is very nearly the single-cycle fracture strength (Gilpin *et al.*, 1993; Ma and Sines, 1996, 1999, 2000). Fatigue-crack propagation studies indicate very high Paris-Law fatigue exponents, on the order of 80, and display clear evidence of a fatigue-crack propagation threshold (Ritchie *et al.*, 1990; Beavan *et al.*, 1993; Cao, 1996).

Crystallographic mechanisms for fatigue-crack initiation, as occur in metals, do not exist in the pyrolytic carbons (Haubold *et al.*, 1981). In properly designed and manufactured components, and in the absence of externally induced damage, fatigue does not occur in pyrolytic-carbon mechanical heart-valve components. In the 30 years of clinical experience, there have been no clear instances of fatigue failure. Few pyrolytic carbon component fractures have occurred, less than 60 out of more than 4 million implanted components (Haubold, 1994), and most are attributable to induced damage from handling or cavitation (Kelpetko *et al.*, 1989; Kafesjian *et al.*, 1994).

Wear resistance of the fluidized-bed pyrolytic carbons is excellent. The strength, stability, and durability of pyrolytic carbon are responsible for the extension of mechanical-valve lifetimes from less than 20 years to more than the recipient's expected lifetime (Wieting, 1996; More and Silver, 1990; Schoen, 1983; Schoen *et al.*, 1982).

Pyrolytic carbon in heart-valve prostheses is often used in contact with metals, either as a carbon disk in a metallic valve orifice or as a carbon orifice stiffened with a metallic ring. Carbon falls with the noble metals in the galvanic series (Haubold *et al.*, 1981), the sequence being silver, titanium, graphite, gold, and platinum. Carbon can accelerate corrosion when coupled to less noble metals *in vivo*. However, testing using mixed potential corrosion theory and potentiostatic polarization has determined that no detrimental effects occur when carbon is coupled with titanium or cobalt–chrome alloys (Griffin *et al.*, 1983; Thompson *et al.*, 1979). Carbon couples with stainless steel alloys are not recommended.

Steps in the fabrication of pyrolytic carbon components

To convert a gaseous hydrocarbon into a shiny, polished black component for use in the biological environment is not a trivial undertaking. Furthermore, because of the critical importance of long-term implants to a recipient's health, all manufacturing operations are performed to stringent levels of quality assurance under the auspices of U.S. Food and Drug Administration GMP and International Standards Organization ISO-9000 regulations. As in the case of fabrication of metallic implants, numerous steps are involved. Pyrolytic carbon is not machined from a block of material, as is the case with most metallic implants, nor is it injection or reactive molded, as are many polymeric devices. An overview of the processing steps leading to a finished pyrolytic carbon coated component for use in a medical device is shown in Fig. 3.2.11-6 and is further described in the following sections.

Substrate material

Since pyrolytic carbon is a coating, it must be deposited on an appropriately shaped, preformed substrate (preform). Because the pyrolysis process takes place at high temperatures, the choice of substrates is severely limited. Only a few of the refractory materials such as tantalum or molybdenum/rhenium alloys and graphite can withstand the conditions at which the pyrolytic carbon coating is produced. Some refractory metals have been used in heart-valve components; for example, Mo/Re preforms were coated to make the struts for the Beall-Surgitool mitral valve. It is important for the thermal expansion characteristics of the substrate to closely match those of the applied coating. Otherwise, upon cooling of the coated part to room temperature the coating will be highly stressed and can spontaneously crack. For contemporary

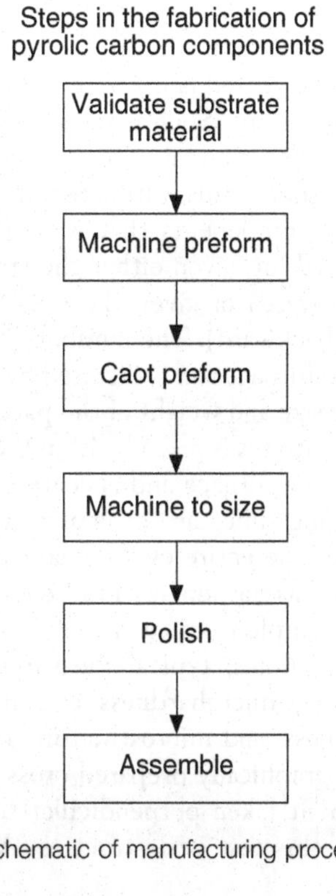

Fig. 3.2.11-6 Schematic of manufacturing processing steps.

heart-valve applications, fine-grained isotropic graphite is the most commonly used substrate. This substrate graphite can be doped with tungsten in order to provide radioopacity for X-ray visualizations of the implants. The graphite substrate does not impart structural strength. Rather, it provides a dimensionally stable platform for the pyrolytic carbon coating both at the reaction temperature and at room temperature.

Preform

Once the appropriate substrate material has been selected and prior to making a preform, it must be inspected to ensure that the material meets the desired specifications. Typically, the strength and density of the starting material are measured. Thermal expansion is ordinarily validated and monitored through process control. The preform, which is an undersized replica of the finished component, is normally machined using conventional machining methods. Because the fine-grained isotropic graphite is very abrasive, standard machine tools have given way to diamond-plated or single-point diamond tools. In the case of heart valves, numerical control machining methods are often required to maintain critical component dimensional tolerances. After the preform is completed, it is inspected to ensure that its dimensions fall within the specified tolerances and that it contains no visible flaws or voids.

Coating

Generally numerous preforms are coated in one furnace run. A batch to be coated is made up of substrates from a single lot of preforms. Such batch processing by lot is required in order to maintain "forward and backward" traceability. In other words, ultimately it is necessary to know all of the components that were prepared using a specific material lot, given either the starting material lot number (forward) or given the specific component serial number (backward). The number of parts that can be coated in one furnace run is dictated by the size of the furnace and the size and weight of the parts to be coated. The batch of substrates is placed within the fluidized bed in the vertical tube furnace and is coated to the desired thickness. Coating times are generally on the order of a few hours, but the entire cycle (heat-up, coating, and cool-down) may take as long as a full day.

A statistical sample from each coating lot is taken for analysis. At this point, typical measurements include coating thickness, microhardness and microstructure. The microhardness, and microstructure are determined from a metallo-graphically prepared cross section of the coated component taken perpendicular to plane of deposition. Thus, this test is destructive. An example of a metallographically prepared cross section of a pyrolytic carbon component is shown in Fig. 3.2.11-7.

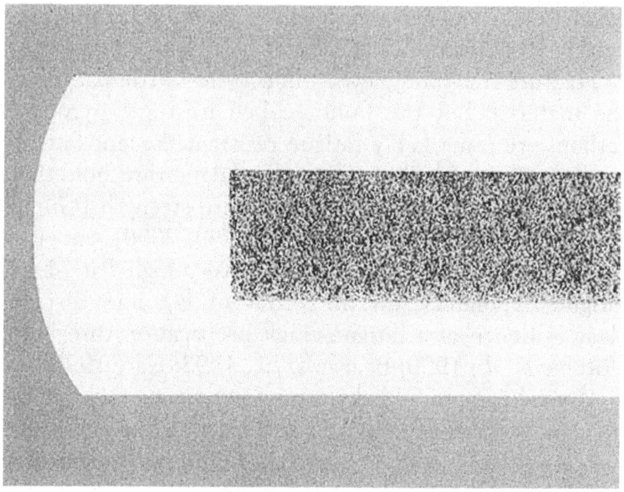

Fig. 3.2.11-7 Metallographic mount cross section of heat valve component. The light-colored pyrolytic carbon layer is coated over the interior, darker colored granular-appearing graphite substrate.

Machine to size

The components used to manufacture medical devices have strict dimensional requirements. Because of the inability, until recently, to precisely measure and control bed size and indirectly coating thickness, the preforms were generally coated more thickly than necessary to ensure adequate pyrolytic carbon coating thickness on the finished part. The strict dimensional requirements were then achieved through precision grinding or other machining operations. Because pyrolytic carbon is very hard, conventional machine tools again cannot be used. Diamond-plated grinding wheels and other diamond tooling are required. The dimensions of final machined parts are again verified.

Polish

The surface of as-deposited, machined and polished components is shown in Fig. 3.2.11-8. It was found early on in experiments (LaGrange *et al.*, 1969; Haubold *et al.*, 1981; Sawyer *et al.*, 1975) that clean polished pyrolytic carbon surfaces of tubes when placed within the vasculature of experimental animals accumulated minimal if any thrombus and certainly less than pyrolytic carbon tubes with the as-deposited surface. Consequently, the surfaces of pyrolytic carbon have historically been polished, either manually or mechanically, using fine diamond or aluminum oxide pastes and slurries. The surface finish achieved has roughness measured on the scale of nanometers. As can be seen from Table 3.2.11-2 (More

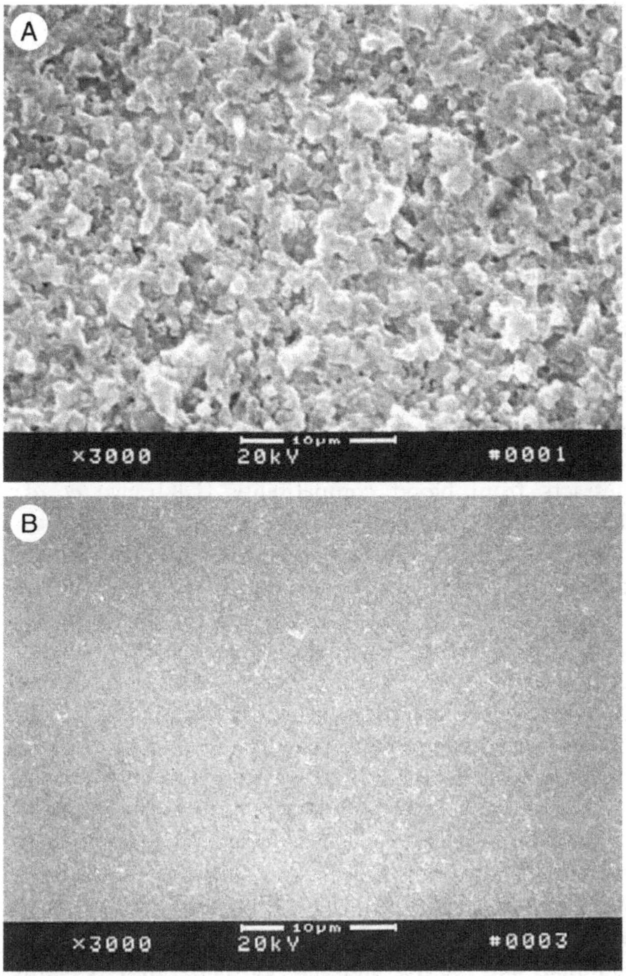

Fig. 3.2.11-8 Scanning electron microscope micrographs of (A) as-coated and (B) as-polished surfaces.

and Haubold, 1996), the surfaces of polished pyrolytic carbon (30–50 nm) are an order of magnitude smoother than the as-deposited surfaces (300–500 nm).

Once the desired surface quality is achieved, components are again inspected. The final component inspection may include measurement of dimensions, X-ray inspection in two orientations to verify coating thickness, and visual inspection for surface quality and flaws. In many cases, automated inspection methods with computer-controlled coordinate measurement machines are used. X-ray inspection can be used to ensure that minimum coating thickness requirements are met. Two orthogonal views ensure that machining and grinding of the coating was achieved uniformly and that the coating is symmetrical. The machining and grinding operation after coating is not without the risk of inducing cracks or flaws in the coating, which may subsequently affect the service life of the component. Such surface flaws are detected visually or with the aid of dye-penetrant techniques. Components may also be proof-tested to detect and

Table 3.2.11-2 Surface finish (R_a, average, and R_q, root mean square) of pyrolytic carbon heat valve components[a]

Specimen	R_a (nm)	R_q (nm)	Comments
Glass microscope slide	17.14	26.80	
On-X leaflet	33.95	42.12	Clinical
Sorin Bicarbon leaflet	40.12	50.63	Nonclinical
SJM leaflet	49.71	62.74	Clinical
CMI (SJM) leaflet	67.98	85.56	Nonclinical
Sorin Monoleaflet	99.59	128.10	Clinical
DeBakey-Surgitool ball	129.78	157.93	Nonclinical
As-coated slab	389.07	503.72	

[a] Components/prepared by: On-X/Medical Carbon Research Institute, Austin, TX; Sorin/Sorin Biomedica, Saluggia, Italy; SJM/Saint Jude Medical, Saint Paul, MN; CMI (SJM)/CarboMedics, Austin, TX; DeBakey-S/CarboMedics, San Diego (circa 1968). "Clinical" was from as-packaged valve; "nonclinical" lacks component traceability.

eliminate components with subsurface flaws. With the advent of bed size control, which allows coating to exact final dimensions, the concerns about flaws introduced during the machining and grinding operation have been eliminated.

The polished and inspected components, thus prepared, are now ready for assembly into devices, or are packaged and sterilized in the case of stand-alone devices. Shown in Fig. 3.2.11-9 are the three pyrolytic carbon components for a bileaflet mechanical heart valve. The

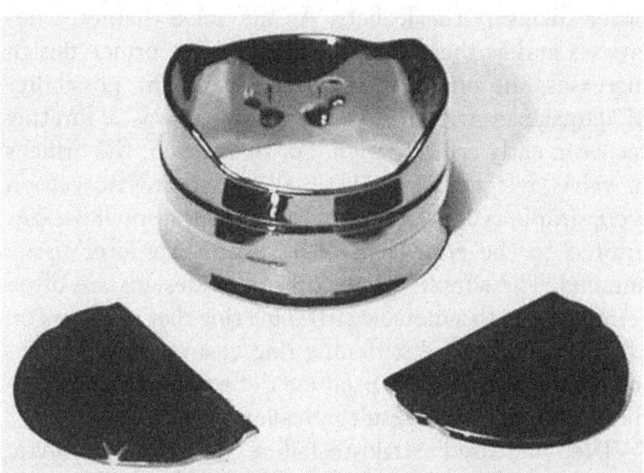

Fig. 3.2.11-9 Components for On-X bileaflet heart valve.

Fig. 3.2.11-10 Replacement metacarpophalangeal total joint prosthesis components, Ascension Orthopedics, Austin TX.

components were selected and matched for assembly using the data generated from the final dimensional inspection to achieve the dimensional requirements specified in the device design. In Fig. 3.2.11-10, the pyrolytic carbon components for a replacement metacarpophalangeal total joint prosthesis are shown.

Assembly

The multiple components of a mechanical heart must be assembled. The brittleness of pyrolytic carbon poses a significant assembly problem. Because the strain-to-failure is on the order of 1.28–1.58%, there is a limited range of deformation that can be applied in order to achieve a proper fit. Relative fit between the components defines the capture and the range of motion for components that move to actuate valve opening and closing. Furthermore, component obstructive bulk and tolerance gaps are critical to hemodynamic performance.

In designs that use a metallic orifice, the metallic components are typically deformed in order to insert the pyrolytic carbon occluder disk. For the all-carbon bileaflet designs, the carbon orifice must be deflected in order to insert the leaflets. As the valve diameter decreases and as the section modulus of the orifice design increases, the orifice stiffness increases. The possibility of damage or fracture during assembly was a limiting factor in early orifice design. For this reason, the orifices in valve designs using silicon-alloyed pyrolytic carbon were simple cylindrical geometries, and the smallest sizes limited to the equivalent of a 19-mm-diameter tissue annulus. The simple cylindrical orifice designs are often reinforced with a metallic stiffening ring that is shrunk on after assembly. The stiffening ring ensures that physiological loading will not produce deflections that can inhibit valve action or result in leaflet escape.

The increased strain-to-failure of pure pyrolytic carbon, relative to the silicon-alloyed carbon, allows designs with more complex orifice section moduli. This allows designers to utilize hydrodynamically efficient shapes such as flared inlets and to incorporate external stiffening bands that eliminate the need for a metallic stiffening ring. The increased strain-to-failure of On-X carbon has been used to advantage in the On-X mechanical heart valve design (Ely *et al.*, 1998).

Cleaning and surface chemistry

Pyrolytic carbon surface chemistry is important because the manufacturing and cleaning operations to which a component is subjected can change and redefine the surface that is presented to the blood. Oxidation of carbon surfaces can produce surface contamination that detracts from blood compatibility (LaGrange *et al.*, 1969; Bokros *et al.*, 1969). Historically, the initial examinations of pyrolytic carbon biocompatibility assumed de facto that the surface needed to be treated with a thromboresistant agent such as heparin (Bokros *et al.*, 1969). It was found, however, that the non-heparin-coated surface was actually more blood compatible than the treated surface. Hence, the efforts toward surface coating with heparin were abandoned.

In general it is desired to minimize the surface oxygen and any other non-carbon surface contaminants. From XPS analyses, a typical heart valve component surface has 76–86% C, 12–21% O, 0–2% Si, and 1–2% Al (More and Haubold, 1996; King *et al.*, 1981; Smith and Black, 1984). Polishing compounds tend to contain alumina and some alumina particles may become imbedded in the carbon surface. Other contaminants that may be introduced at low levels < 2% each are Na, B, Cl, S, Mg, Ca, Zn, and N. The XPS carbon 1s peak when scanned at high resolution can be deconvoluted to determine carbon oxidations states. The carbon 1s peak will typically consist primarily of hydrocarbon-like carbon (60–81%), ether alcohol/ester-like carbon (10–24%), ketone-like carbon (0–6%), and ester/acid-like carbon (1–12%) (More and Haubold, 1996). Each manufacturing, cleaning and sterilizing operation potentially redefines the surface. The effect of modified surface chemistry on blood compatibility is not well characterized, so this adds a level of uncontrolled variability when considering the blood compatibility of pyrolytic-carbon heart-valve materials from different manufacturing sources and different investigators. In general, the presence of oxygen and surface contaminants should be eliminated.

Biocompatibility of pyrolytic carbon

The suitability of a material for use in an implant is a complex issue. Biocompatibility testing is not the focus of this section. In the case of pyrolytic carbon, its successful history interfacing with blood in mechanical heart valves attests to its suitability for this application. A note of caution, however, is in order. Until about a decade ago, the pyrolytic carbon used so successfully in mechanical heart valves was produced by a single manufacturer. The material, its many applications in the biological environment, and the processes for producing the material were all patented. Since the expiration of the last of these patents in 1989, other sources for pyrolytic carbon have appeared that are copies of the original General Atomic material. When considering alternative carbon materials, it is important to recognize that the proper combination of physical, mechanical, and blood-compatible properties is required for the success of the implant application. Furthermore, because there are a number of different possible pyrolysis processes, it should be recognized that each can result in different microstructures with different properties. Just because a material is carbon, a turbostratic carbon, or a pyrolytic carbon does not qualify its use in a long-term human implant (Haubold and Ely, 1995). For example, pyrolytic carbons prepared by chemical vapor deposition processes, other than the fluidized-bed process, are known to exhibit anisotropy, nonhomogeneity, and considerable variability in mechanical properties (Agafonov *et al.*, 1999). Although these materials may exhibit biocompatibility, the potential for variability in structural stability and durability may lead to valve dysfunction.

The original General Atomic–type fluidized-bed pyrolytic carbons all demonstrate negligible reactions in the standard Tripartite and ISO 10993-1 type biocompatibility tests. Results from such tests are given in Table 3.2.11-3 (Ely *et al.*, 1998). Pure pyrolytic carbon is so non-reactive that it can serve as a negative control for these tests.

Table 3.2.11-3 Biological testing of pure PyC

Test description	Protocol	Results
Klingman maximization	ISO/CD 10993–10	Grade 1; not significant
Rabbit pyrogen	ISO/DIS 10993–11	Nonpyrogenic
Intracutaneous injection	ISO 10993–10	Negligible irritant
Systemic injection	ANSI/AAMI/ISO 10993–11	Negative—same as controls
Salmonella typhimurium reverse mutation assay	ISO 10993–3	Nonmutagenic
Physicochemical	USP XXIII, 1995	Exceeds standards
Hemolysis—rabbit blood	ISO 10993–4/NIH 77–1294	Nonhemolytic
Elution test (L929 mammalian cell culture)	ISO 10993–5, USP XXIII, 1995	Noncytotoxic

It is believed that pyrolytic carbon owes its demonstrated blood compatibility to its inertness and to its ability to quickly absorb proteins from blood without triggering a protein denaturing reaction. Ultimately, the blood compatibility is thought to be a result of the protein layer formed upon the carbon surface. Baier observed that pyrolytic carbon surfaces have a relatively high critical surface tension of 50 dyn/cm, which immediately drops to 28–30 dyn/cm following exposure to blood (Baier *et al.*, 1970). The quantity of sorbed protein was thought to be an important factor for blood compatibility. Lee and Kim (1974) quantified the amount of radiolabeled proteins sorbed from solutions of mixture proteins (albumin, fibrinogen, and gamma-globulin). While pyrolytic carbon does absorb albumin, it also absorbs considerable quantity of fibrinogen as shown in Fig. 3.2.11-11. As can be seen in Fig. 3.2.11-11, the amount of fibrinogen absorbed on pyrolytic carbon surfaces is far greater than the amount of albumin on these surfaces and is comparable to the amount of fibrinogen that sorbed on SR. The mode of albumin absorption, however, appears to be drastically different for these two materials. Albumin sorbs immediately on the pyrolytic carbon surfaces, whereas the buildup of fibrinogen is much slower. In the case of SR, both proteins sorb at a much slower rate. It appears that the mode of protein absorption is important and not the total amount sorbed.

Nyilas and Chiu (1978) studied the interaction of plasma proteins with foreign surfaces by measuring directly the heats of absorption of selected proteins onto such surfaces using microcalorimetric techniques. They found that the heats of absorption of fibrinogen, up to the completion of first monolayer coverage, are a factor of 8 smaller on pyrolytic carbon surfaces than on the known thrombogenic control (glass) surface as shown in Fig. 3.2.11-12. Furthermore, the measured net heats of absorption of gamma globulin on pyrolytic carbon were about 15 times smaller than those on glass. They concluded that low heats of absorption onto a foreign surface imply small interaction forces with no conformational changes of the proteins that might trigger the clotting cascade. It appears that a layer of continuously exchanging blood proteins in their unaltered state "masks" the pyrolytic carbon surfaces from appearing as a foreign body.

There is further evidence that the minimally altered sorbed protein layers on pyrolytic carbon condition blood compatibility. Salzman *et al.* (1977), for example,

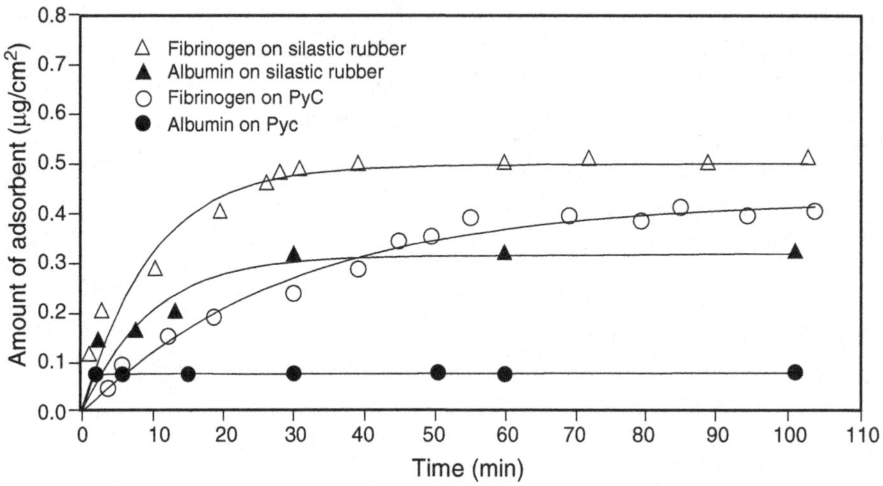

Fig. 3.2.11-11 Fibrinogen and albumin adsorption on pyrolytic carbon (PyC) and Silastic SR.

observed a significant difference in platelet reaction with pyrolytic carbon beads in packed columns prior to and after pretreatment with albumin. With no albumin preconditioning treatment, platelet retention by the columns was high, but the release of platelet constituents was low. However, with albumin pretreatment, platelet retention and the release of constituents was minimal.

The foregoing observations led to the view that pyrolytic carbon owes its demonstrated blood compatibility to its inertness and to its ability to quickly absorb proteins from blood without triggering a protein-denaturing reaction (Haubold *et al.*, 1981; Nyilas and Chiu, 1978). However, the assertion that pyrolytic carbon is an inert material and induces minimal conformational changes in adsorbed protein was reexamined by Feng and Andrade (1994). Using DSC and a variety of proteins and buffers, they found that pyrolytic carbon surfaces denatured all of the proteins studied. They concluded that whether or not a surface denatures protein cannot be the sole criteria for blood compatibility. Their suggestion was that the specific proteins and the sequence in which they are denatured may be important. For example, it was suggested that pyrolytic carbon may first adsorb and denature albumin, which forms a layer that subsequently passivates the surface and inhibits thrombosis.

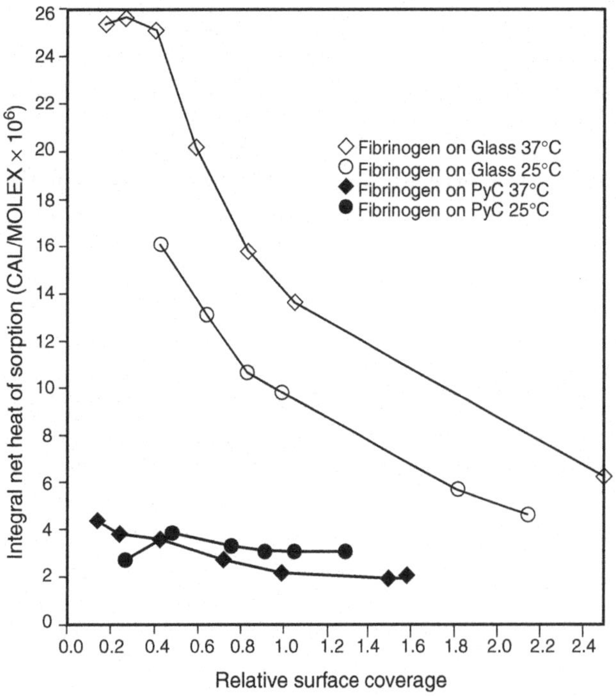

Fig. 3.2.11-12 Integral heat of sorption for fibrinogen on glass and fibrinogen on PyC at two different temperatures (Nyilas and Chiu, 1978).

Chinn *et al.* (1994) reexamined the adsorption of albumin and fibrinogen on pyrolytic carbon surfaces and noted that relatively large amounts of fibrinogen were adsorbed and speculated that the adsorbed fibrinogen was rapidly converted to a non-elutable form. If the elutable form is more reactive to platelets than the nonelutable form, then the nonelutable protein layer may contribute to the passivating effect.

Work on visualizing the carbon surface and platelet adhesion done by Goodman *et al.* (1995) using low-accelerating-voltage SEM, along with critical-point drying techniques, has discovered that the platelet spreading on pyrolytic carbon surfaces is more extensive than previously observed (Haubold *et al.*, 1981). However, platelet loading was in a static flow situation that does not model the physiological flow that a heart valve is subjected to. Hence, this approach cannot resolve kinetic effects on platelet adhesion. However, Okazaki, Tweden, and co-workers observed adherent platelets on valves following implantation in sheep that were not treated with anticoagulants (Okazaki *et al.*, 1997). There were no instances of valve thrombosis even though platelets were present on some of the valve surfaces. But, the relevance of this observation to clinical valve thromboses is not clear because human patients with mechanical heart valves undergo chronic anticoagulant therapy (Edmunds, 1987) and have a hemostatic system different from that of sheep.

A more contemporary version of the mechanism of pyrolytic carbon blood compatibility might be to reject the assumption that the surface is inert, as it is now thought by some that no material is totally inert in the body (Williaims, 1998), and to accept that the blood–material interaction is preceded by a complex, interdependent, and time-dependent series of interactions between the plasma proteins and the surface (Hanson, 1998) that is as yet poorly understood. To add to the confusion, it must also be recognized that much of the forementioned conjecture depends on the assumption that all of the carbon surfaces studied were in fact pure and comparable to one another.

Conclusion

Because the blood compatibility of pyrolytic carbon in mechanical heart valves is not perfect, anticoagulant therapy is required for mechanical heart valve patients. However, pyrolytic carbon has been the most successful material in heart valve applications because it offers excellent blood and tissue compatibility which, combined with the appropriate set of physical and mechanical properties and durability, allows for practical implant device design and manufacture. Improvements in biocompatibility are desired, of course, because when heart valves and other implants are used, a deadly or disabling disease is often treated by replacing it with a less pathological, more manageable chronic condition. Ideally, an implant should not lead to a chronic condition.

It is important to recognize that the mechanism for the blood compatibility of pyrolytic carbons is not fully understood, nor is the interplay between the biomaterial itself, design-related hemodynamic stresses, and the ultimate biological reaction. The elucidation of the mechanism for blood and tissue compatibility of pyrolytic carbon remains a challenge.

It is also worth restating that the suitability of carbon materials from new sources for long-term implants is not assured simply because the material is carbon. Elemental carbon encompasses a broad spectrum of possible structures and mechanical properties. Each new candidate carbon material requires a specific assessment of biocompatibility based on its own merits and not by reference to the historically successful General Atomic–type pyrolytic carbons.

Bibliography

Agafonov, A., Kouznetsova, E., Kouznetsova, V., and Reif, T. (1999). TRI carbon strength and macroscopic isotropy of boron carbide alloyed pyrolytic carbon. *Artif. Organs.* **23**(7): 80.

Akins, R. J., and Bokros, J. C. (1974). The deposition of pure and alloyed isotropic carbons and steady state fluidized beds. *Carbon* **12**: 439–452.

Baier, R. E., Gott, V. L., and Feruse, A. (1970). Surface chemical evaluation of thromboresistant materials before and after venous implantation. *Trans. Am. Soc. Artif. Intern. Organs* **16**: 50–57.

Beavan, L. A., James, D. W., and Kepner, J. L. (1993). Evaluation of fatigue in pyrolite carbon. in *Bioceramics*, Vol. 6, P. Ducheyne and D. Christiansen, eds. Butterworth–Heinemann, Oxford, pp. 205–210.

Bokros, J. C. (1969). Deposition, structure and properties of pyrolytic carbon. in *Chemistry and Physics of Carbon*, Vol. 5, P. L. Walker, ed. Marcel Dekker, New York, pp. 1–118.

Bokros, J. C., Gott, V. L., LaGrange, L. D., Fadall, A. M., Vos, K. D., and Ramos, M. D. (1969). Correlations between blood compatibility and heparin adsorptivity for an impermeable isotropic pyrolytic carbon. *J. Biomed. Mater. Res.* 3: 497–528.

Cao, H. (1996). Mechanical performance of pyrolytic carbon in prosthetic heart valve applications. *J. Heart Valve Dis.* 5(Suppl. I): S32–S49.

Chinn, J. A., Phillips, R. E., Lew, K. R., and Horbett, T. A. (1994). Fibrinogen and albumin adsorption to pyrolite carbon. *Trans. Soc. Biomater.* 17: 250.

De Salvo, G. (1970). *Theory and Structural Design Applications of Weibull Statistics*, Report WANL-TME-2688, Westinghouse Electric Corporation.

Dillard, J. G. (1995). X-ray photoelectron spectroscopy (XPS) and electron spectroscopy for chemical analysis (ESCA). in *Characterization of Composite Materials*, Vol. 1, H. Ishida and L. E. Fitzpatrick, eds. Butterworth–Heinemann, Boston, pp. 22.

Edmunds, L. H. (1987). Thrombotic and bleeding complications of prosthetic heart valves. *Ann. Thorac. Surg.* 44: 430–445.

Ely, J. L., Emken, M. R., Accuntius, J. A., Wilde, D. S., Haubold, A. D., More, R. B., and Bokros, J. C. (1998). Pure pyrolytic carbon: preparation and properties of a new material, on-X carbon for mechanical heart valve prostheses. *J. Heart Valve Dis.* 7: 626–632.

Emken, M. R., Bokros, J. C., Accuntius, J. A., and Wilde, D. S. (1993). Precise control of pyrolytic carbon coating. Presented at the 21st Biennial Conference on Carbon, Buffalo, New York, June 13-18, 1993, Extended Abstracts and Program Proceedings, pp. 531–532.

Emken, M. R., Bokros, J. C., Accuntius, J. A., and Wilde, D. S. (1994). U.S. Patent No. 5,284,676, Pyrolytic deposition in a fluidized bed, Feb. 8, 1994.

Feng, L., and Andrade, J. D. (1994). Protein adsorption on low-temperature isotropic carbon: I. Protein conformational change probed by differential scanning calorimetry. *J. Biomed. Mater. Res.* **28**: 735–743.

Gilpin, C. B., Haubold, A. D., and Ely, J. L. (1993). Fatigue crack growth and fracture of pyrolytic carbon composites. in *Bioceramics*, Vol. 6, P. Ducheyne and D. Christiansen, ed. Butterworth–Heinemann, Oxford, pp. 217–223.

Goodman, S. L., Tweden, K. S., and Albrecht, R. M. (1995). Three-dimensional morphology and platelet adhesion on pyrolytic carbon heart valve materials. *Cells Mater.* 5(1): 15–30.

Griffin, C. D., Buchanan, R. A., and Lemons, J. E. (1983). In vitro electrochemical corrosion study of coupled surgical implant materials. *J. Biomed. Mater. Res.* **17**: 489–500.

Hanson, S. R. (1998). Blood-material interactions. in *Handbook of Biomaterial Properties*, J. Black and G. Hastings, eds. Chapman and Hall, London, pp. 545–555.

Haubold, A. D. (1994). On the durability of pyrolytic carbon in vivo. *Medi. Prog. Technol.* **20**: 201–208.

Haubold, A. D., and Ely, J. L. (1995). Carbons used in mechanical heart valves. *Transactions Society for Biomaterials, 21st Annual Meeting, San Francisco*, p. 275.

Haubold, A. D., Shim, H. S., and Bokros, J. C. (1981). Carbon in medical devices. in *Biocompatibility of Clinical Implant Materials.* David, P. Williams, ed. CRC Press, Boca Raton, Florida, pp. 3–42.

Kaae, J. L., and Wall, D. R. (1996). Microstructural characterization of pyrolytic carbon for heart valves. *Cells Mater.* 4: 281–290.

Kafesjian, R., Howanec, M., Ward, G. D., Diep, L., Wagstaff, L., and Rhee, R. (1994). Cavitation damage of pyrolytic carbon in mechanical heart valves. *J. Heart Valve Dis.* 3(Suppl. I): S2–S7.

Kelpetko, V., Moritz, A., Mlczoch, J., Schurawitzki, H., Domanig, E., and Wolner, E. (1989). Leaflet fracture in Edwards-Duromedics bileaflet valves. *J. Thorac. Cardiovasc. Surg.* 97: 90–94.

King, R. N., Andrade, J. D., Haubold, A. D., and Shim, H. S. (1981). Surface analysis of silicon: alloyed and unalloyed LTI pyrolytic carbon, in *Photon, Electron and Ion Probes of Polymer Structure and Properties*, ACS Symposium Series 162, D. W. Dwight, T. J. Fabish, and H. R. Thomas, eds. American Chemical Society, Washington, D.C., pp. 383–404.

LaGrange, L. D., Gott, V. L., Bokros, J. C., and Ramos, M. D. (1969). Compatibility of carbon and blood. in *Artificial Heart Program Conference Proceedings*, R. J. Hegyeli, ed. U.S. Government Printing Office, Washington, D.C., pp. 47–58.

Lee, R. G., and Kim, S. W. (1974). Adsorption of proteins onto hydrophobic polymer surfaces: adsorption isotherms and kinetics. *J. Biomed Mater. Res.* **8**: 251.

Ma, L., and Sines, G. (1996). Fatigue of isotropic pyrolytic carbon used in mechanical heart valves. *J. Heart Valve Dis.* 5(Suppl. I): S59–S64.

Ma, L., and Sines, G. (1999). Unalloyed pyrolytic carbon for implanted heart valves. *J. Heart Valve Dis.* 8(5): 578–585.

Ma, L., and Sines, G. (2000). Fatigue behavior of pyrolytic carbon. *J. Biomed. Mater. Res.* **51**: 61–68.

More, R. B., and Haubold, A. D. (1996). Surface chemistry and surface roughness of clinical pyrocarbons. *Cells Mater.* **6**: 273–279.

More, R. B., and Silver, M. D. (1990). Pyrolytic carbon prosthetic heart valve occluder wear: in vivo vs. in vitro results for the Björk–Shiley prosthesis. *J. Appl. Biomater.* 1: 267–278.

More, R. B., Kepner, J. L., and Strzepa, P. (1993). Hertzian fracture in pyrolite carbon. in *Bioceramics*, Vol. 6, P. Ducheyne and D. Christiansen, eds. Butterworth–Heinemann, Oxford, pp. 225–228.

Nyilas, E., and Chiu, T. H. (1978). Artificial surface/sorbed protein structure/ hemocompatibility correlations. *Artif. Organs* **2**(Suppl): 56–62.

Okazaki, Y., Wika, K. E., Matsuyoshi, T., Fukamachi, K., Kunitomo, R., Tweeden, K. S., and Harasaki, H. (1997). Platelets were early postoperative depositions on the leaflet of a mechanical heart valve in sheep without postoperative anticoagulants or antiplatelet agents. *ASAIO J.* **42**: M750–M754.

Pauling, L. (1964). *College Chemistry*, 3rd ed. W. H. Freeman and Company, San Francisco.

Reilly, D. T., and Burstein, A. H. (1974). The mechanical properties of bone. *J. Bone Joint Surg. Am.* **56**: 1001.

Reilly, D. T., Burstein, A. H., and Frankel, V. H. (1974). The elastic modulus for bone. *J. Biomech.* 7: 271.

Richard, G., and Cao, H. (1996). Structural failure of pyrolytic carbon heart valves. *J. Heart Valve Dis.* 5(Suppl. I): S79–S85.

Ritchie, R. O., Dauskardt, R. H., Yu, W., and Brendzel, A. M. (1990). Cyclic fatigue-crack propagation, stress corrosion and fracture toughness behavior in pyrolite carbon coated graphite for prosthetic heart valve applications. *J. Biomed. Mat. Res.* **24**: 189–206.

Sadeghi, H. (1987). Dysfonctions des prostheses valvulaires cardaques et leur traitment chirgical. *Schwiez. Med. Wochenschr.* **117**: 1665–1670.

Salzman, E. W., Lindon, J., Baier, D., and Merril, E. W. (1977). Surface-induced platelet adhesion, aggregation and release. *Ann. N.Y. Acad. Sci.* **283**: 114.

Sattler, K. (1995). Scanning tunneling microscopy of carbon nanotubes and nanocones. *Carbon* 7: 915–920.

Sawyer, P. N., Lucas, L., Stanczewski, B., Ramasamy, N., Kammlott, G. W., and Goodenough, S. H. (1975). Evaluation techniques for potential cardiovascular prosthetic alloys experience with titanium aluminum 6–4 ELI tubes. *Proceedings of the San Diego Biomedical Symposium*, Vol. 14, pp. 423–427.

Schoen, F. J. (1983). Carbons in heart valve prostheses: foundations and clinical performance. in *Biocompatible Polymers, Metals and Composites*, M. Zycher, ed. Technomic, Lancaster, PA, pp. 240–261.

Schoen, F. J., Titus, J. L., and Lawrie, G. M. (1982). Durability of pyrolytic carbon-containing heart valve prostheses. *J. Biomed. Mater. Res.* **16**: 559–570.

Smith, K. L., and Black, K. M. (1984). Characterization of the treated surfaces of silicon alloyed pyrolytic carbon and SiC. *J. Vac. Sci. Technol.* A2: 744–747.

Thompson, N. G., Buchanan, R. A., and Lemons, J. E. (1979). In vitro corrosion of Ti-6Al-4V and Type 316L stainless steel when galvanically coupled with carbon. *J. Biomed. Mater. Res.* **13**: 35–44.

Wieting, D. W. (1996). The Björk–Shiley Delrin tilting disc heart valve: historical perspective, design and need for scientific analyses after 25 years. *J. Heart Valve Dis.* 5(Suppl. I): S157–S168.

Williams, D. F. (1998). General concepts of biocompatibility. in *Handbook of Biomaterial Properties*, J. Black, and G. Hastings, eds. Chapman and Hall, London, pp. 481–489.

3.2.12 Composites

Claudio Migliaresi and Harold Alexander

Introduction

The word *composite* means "consisting of two or more distinct parts." At the atomic level, materials such as metal alloys and polymeric materials could be called composite materials in that they consist of different and distinct atomic groupings. At the microstructural level (about 10^{-4}–10^{-2} cm), a metal alloy such as a plain-carbon steel containing ferrite and pearlite could be called a composite material since the ferrite and pearlite are distinctly visible constituents as observed in the optical microscope. At the molecular and microstructural level, tissues such as bone and tendon are certainly composites with a number of levels of hierarchy.

In engineering design a composite material usually refers to a material consisting of constituents in the micro- to macro-size range, favoring the macrosize range. For the purpose of discussion in this section, composites can be considered materials consisting of two or more chemically distinct constituents, on a macroscale, having a distinct interface separating them. This definition encompasses the fiber and particulate composite materials of primary interest as biomaterials. Such composites consist of one or more discontinuous phases embedded within a continuous phase. The discontinuous phase is usually harder and stronger than the continuous phase and is called the *reinforcement* or *reinforcing material*, whereas the continuous phase is termed the *matrix*.

Properties of composites are strongly influenced by the properties of their constituent materials, their distribution and content, and the interaction among them. The composite properties may be the volume fraction sum of the properties of the constituents, or the constituents may interact in a synergistic way due to geometrical orientation so as to provide properties in the composite that are not accounted for by a simple volume fraction sum. Thus in describing a composite material, besides specifying the constituent materials and their properties, one needs to specify the geometry of the reinforcement, their concentration, distribution, and orientation.

Most composite materials are fabricated to provide desired mechanical properties such as strength, stiffness, toughness, and fatigue resistance. Therefore, it is natural to study together the composites that have a common strengthening mechanism. The strengthening mechanism strongly depends upon the geometry of the reinforcement. Therefore, it is quite convenient to classify composite materials on the basis of the geometry of a representative unit of reinforcement. Figure 3.2.12-1 shows a commonly accepted classification scheme.

With regard to this classification, the distinguishing characteristic of a particle is that it is nonfibrous in nature. It may be spherical, cubic, tetragonal, or of other regular or irregular shape, but it is approximately equiaxial. A fiber is characterized by its length being much

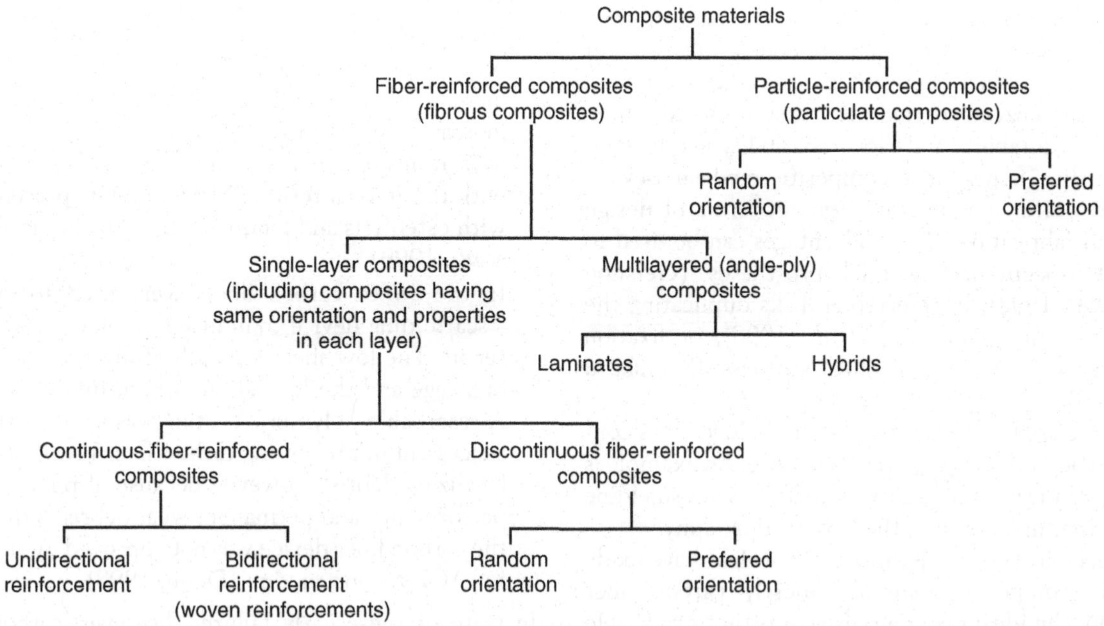

Fig. 3.2.12-1 Classification of composite materials. (From Agarwal, B. D., and Broutman, L. J. (1980). *Analysis and Performance of Fiber Composites*. Wiley-Interscience, New York.)

greater than its cross-sectional dimensions. Particle-reinforced composites are sometimes referred to as *particulate composites*. Fiber-reinforced composites are, understandably, called *fibrous composites*. *Laminates* are composite structures made by stacking laminae of fiber composites oriented to produce a structural element. Characteristics, number, and orientation of laminae are such as to match specific design requirements.

Fibers are much more mechanically effective than particles, and polymer-fiber composites can reach stiffness and strength comparable to those of metals and even higher.

Moreover, whereas particle-reinforced composites are isotropic, fiber-reinforced composites are basically anisotropic. Properties in different directions can be in most cases designed to match specific requirements.

At the molecular and microstructural level, tissues such as bone and tendon or vessels are certainly composites with a number of levels of hierarchy. Their properties are highly anisotropic, and the only possibility to mimic them is to use composites.

Failure of a composite material implant can expose fibers or particles to the surrounding biological environment. In many cases failure in composites is preceded by the failure of the interface between filler and matrix, this being due to idrothermal aging or stresses exceeding the interface strength. Sterilization methods or conditions can play an important role.

As with all biomaterials, the question of biocompatibility (tissue response to the composite) is paramount. Being composed of two or more materials, composites provide enhanced probability of causing adverse tissue reactions. Also, the fact that one constituent (the reinforcement) usually has dimensions on the cellular scale always leaves open the possibility of cellular ingestion of particulate debris that can result in either the production of tissue-lysing enzymes or transport into the lymph system.

Although durability and biocompatibility can be considered major issues in a composite medical device, composites offer unique advantages in terms of design ability and fabrication. These advantages can be used to construct isocompliant arterial prostheses (Gershon *et al.*, 1990, 1992), intervertebral disks duplicating the natural structure (Ambrosio *et al.*, 1996), or fixation plates and nails with controlled stiffness (Veerabagu *et al.*, 2003).

For some applications, moreover, radiolucency is considered to be a further potential advantage. An example is external or internal fracture fixation devices not shielding the bone fracture site from the X-ray radiography.

Design flexibility, strength, and lightweight have made polymeric composite materials, mostly carbon fiber reinforced, the ideal materials also for orthotic aids able to return walking and even athletic performances to impaired people (Dawson, 2000).

Reinforcing systems

The main reinforcing materials that have been used in biomedical composites are carbon fibers, polymer fibers, ceramics, and glasses. Depending upon the application, the reinforcements have been either inert or absorbable.

Carbon fiber

Carbon fiber is a lightweight, flexible, high-strength, high-tensile-modulus material produced by the pyrolysis of organic precursor fibers, such as rayon, polyacrylonitrile (PAN), and pitch in an inert environment. The term carbon is often indifferently interchanged with the term graphite, but carbon and graphite fibers differ in the temperature of fabrication, thermal treatment, and the content of carbon (93–95% for carbon fibers and more than 99% for graphite fibers). Because of their low density (depending on the precursor from 1.7 to 2.1 g/cm^3) and high mechanical properties (elastic modulus up to 900 GPa and strength up to 4.5 GPa, depending on the precursor and on the fabrication process—hence they can be much stiffer and stronger than steel!) these fibers are used in composites in a variety of applications demanding lightness and high mechanical properties. Their disadvantage is that carbon fibers have poor shear strength.

In medicine, several commercial products have used carbon fibers. Some of the first devices, however, have experienced severe negative effects and have been recalled from the market. Two examples are:

- Short carbon fiber reinforced UHMWPE for orthopedic applications. The assumption was that increase of strength and decrease of creep would increase the bearing longevity. The favorable indications of the laboratory wear tests contrasted with the *in vivo* results: Many patients presented with osteolysis and failure of the tibial inserts (Kurtz *et al.*, 1999).
- In the 1980s carbon fibers were used to develop a scaffolding device to induce tendon or ligament repair. The low shear strength of fibers caused fiber breakage and the formation of harmful debris. A resorbable polymeric coating was somewhat successful in preventing carbon fiber breakage and localizing debris. However, because of poor performance and permanent wear debris in the joint, the carbon fiber device was not approved by the FDA for ACL reconstruction (Dunn, 1998).

In spite of these early failures, however, carbon fibers display unique properties for the fabrication of load-bearing medical devices.

Polymer fibers

Whereas carbon fibers have been used for their superior mechanical properties, polymer fibers are not comparably strong or stiff as reinforcements for other polymers, with the possible exceptions of aramid fibers or ultrahigh-molecular-weight polyethylene (UHMWPE) fibers. For biomedical applications, biocompatibility, of course, and high strength and fatigue resistance are compulsory, while stiffness is a design parameter to be adapted to the specific conditions. This is why for some applications PET fibers have been used. In addition, thanks to their absorbability, not to their mechanical superiority, certain absorbable fibers have been employed.

- *Aramid* is the generic name for aromatic polyamide fibers. The most well known aramids are Kevlar and Nomex (DuPont trademarks), and Twaron (made by Teijin/Twaron of Japan). Kevlar is produced by spinning a sulfuric acid/poly(*p*-phenylene terephthalamide) solution through an air layer into a coagulating water bath. Aramid fibers are light (density = 1.44 g/cm^3), stiff (the modulus can go up to 190 GPa), and strong (tensile strength about 3.6 GPa); moreover, they resist impact and abrasion damage. A negative point that can be relevant for biomedical applications is that aramid fibers absorb moisture, and also worth noting is their poor compressive strength, about 1/8 of the tensile strength. Aramid fiber composites are used commercially where high tensile strength and stiffness, damage resistance, and resistance to fatigue and stress rupture are important. In medicine, these composites have not seen extensive use, due perhaps to some concerns about their biocompatibility or long-term fate. Main applications have been in dentistry (Pourdeyhimi *et al.*, 1986; Vallittu, 1996) and ligament prostheses (Wening *et al.*, 1994).
- Commercially available high-strength, high-modulus PE fibers include Spectra from Honeywell Performance Fibers (Colonial Heights, VA), Dyneema from DSM (Heerlen, The Netherlands), and Toyobo fibers from Toyobo (Shiga, Japan). UHMWPE fibers are produced by a gel-spinning technique starting from an approximately 2–8 wt.% solution of the ultrahigh-molecular-weight polymer ($M_w > 10^6$) in a common solvent, such as decalin. Spinning at 130–140°C and hot drawing at very high draw ratios produces fibers with the highest specific strength of all commercial fibers available to date. UHMWPE fibers possess high modulus and strength, besides displaying light weight (density about 0.97 g/cm^3) and high energy dissipation ability, compared to other fibers. In addition PE fibers resist abrasion and do not absorb water. However, the chemical properties of UHMWPE fibers are such that few resins bond well to the fiber surfaces, and so the structural properties expected from the fiber are often not fully realized in a composite. The low melting point of the fibers (about 147°C) impedes high-temperature fabrication. Bulk UHMWPE has extensive applications in medicine for the fabrication of bearings for joint prostheses, displaying excellent biocompatibility but with lifetime restricted by its wear resistance. PE fibers are used to reinforce acrylic resins for application in dentistry (Ladizesky *et al.*, 1994; Karaman *et al.*, 2002; Brown, 2000), or to make intervertebral disk prostheses (Kotani *et al.*, 2002). They have been also used for the fabrication of ligament augmentation devices (Guidoin *et al.*, 2000).
- Dacron is the name commonly used to indicate PET fibers. These fibers have several biomedical uses, most in cardiovascular surgery for arterial grafts. PET fibers, however, have been proposed in orthopedics for the fabrication of artificial tendons or ligaments (Kolarik *et al.*, 1981) and ligament augmentation devices, as fibers or fabrics alone, or imbedded in different matrices in composites. Other proposed applications include soft-tissue prostheses, intervertebral disks (Ambrosio *et al.*, 1996), and plastic surgery applications.
- PLA and PGA and their copolymers are the principal biodegradable polymers used for the fabrication of biodegradable fibers. These fibers have been used for a number of years in absorbable sutures. Properties of these fibers depend upon several factors, such as crystallinity degree, molecular weight, and purity (Migliaresi and Fambri, 1997). Fibers and tissues have been proposed for ligament reconstruction (Durselen *et al.*, 2001) or as scaffolds for tissue engineering applications (Lu and Mikos, 1996). They also have been employed in composites, in combination with parent biodegradable matrices. Examples are the intramedullary biodegradable pins and plates (Vert *et al.*, 1986, Middleton and Tipton, 2000) and biodegradable scaffolds for bone regeneration (Vacanti *et al.*, 1991, Kellomaki *et al.*, 2000).

Ceramics

A number of different ceramic materials have been used to reinforce biomedical composites. Since most biocompatible ceramics, when loaded in tension or shear, are relatively weak and brittle materials compared to metals, the preferred form for this reinforcement has usually been particulate. These reinforcements have included various calcium phosphates, aluminum- and zinc-based phosphates, glass and glass-ceramics, and bone mineral. Minerals in bone are numerous. In the past, bone has been defatted, ground, and calcined or heated to yield a relatively pure mix of the naturally occurring bone minerals. It was recognized early that this mixture of natural bone mineral was poorly defined and extremely variable. Consequently, its use as an implant material was limited.

The calcium phosphate ceramic system has been the most intensely studied ceramic system. Of particular interest are the calcium phosphates having calcium-to-phosphorus ratios of 1.5–1.67. TCP and HA form the boundaries of this compositional range. At present, these two materials are used clinically for dental and orthopedic applications. TCP has a nominal composition of $Ca_3(PO_4)_2$. The common mineral name for this material is whitlockite. It exists in two crystographic forms, α- and β-whitlockite. In general, it has been used in the β-form.

The ceramic HA has received a great deal of attention. HA is, of course, the major mineral component of bone. The nominal composition of this material is $Ca_{10}(PO_4)_6(OH)_2$.

TCPs and HA are commonly referred as bioceramics, i.e., bioactive ceramics. The definition refers to their ability to elicit a specific biological response that results in the formation of bond between the tissues and material (Hench *et al.*, 1971). HA ceramic and TCPs are used in orthopedics and dentistry alone or in combination with other substances, or also as coating of metal implants. The rationale behind the use of bioceramics in combination with polymeric matrix for composites is in their ability to enhance the integration in bone, while improving the device mechanical properties. An example are the HA–PE composites developed by Bonfield (Bonfield, 1988; Bonfield *et al.*, 1998), and today commercialized with the name of HAPEX (Smith & Nephew ENT, Memphis, TN).

Glasses

Glass fibers are used to reinforce plastic matrices to form structural composites and molding compounds. Commercial glass fiber plastic composite materials have the following favorable characteristics: high strength-to-weight ratio; good dimensional stability; good resistance to heat, cold, moisture, and corrosion; good electrical insulation properties; ease of fabrication; and relatively low cost. De Santis *et al.* (2000) have stacked glass and carbon/PEI laminae to manufacture a hip prosthesis with constant tensile modulus but with bending modulus increasing in the tip–head direction. An isoelastic intramedullary nail made of PEEK and chopped glass fibers has been evaluated by Lin *et al.* (1997), and glass fibers have been used to increase the mechanical properties of acrylic resins for applications in dentistry (Chen *et al.*, 2001).

Zimmerman *et al.* (1991) and Lin (1986) introduced an absorbable polymer composite reinforced with an absorbable calcium phosphate glass fiber. This allowed for the fabrication of a completely absorbable composite implant material. Commercial glass fiber produced from a lime–aluminum–borosilicate glass typically has a tensile strength of about 3 GPa and a modulus of elasticity of 72 GPa. Lin (1986) estimates the absorbable glass fiber to have a modulus of 48 GPa, comparing favorably with the commercial fiber. The tensile strength, however, was significantly lower, approximately 500 MPa.

Matrix systems

Ceramic matrix or metal matrix composites have important technological applications, but their use is restricted to specific cases (e.g., cutting tools, power generation equipment, process industries, aerospace), with just a few examples for biomedical applications (e.g., calcium phosphate bone cements).

Most biomedical composites have polymeric matrices, mostly thermoplastic, bioabsorbable or not.

The most common matrices are synthetic nonabsorbable polymers. By far the largest literature exists for the use of polysulfone, PEEK, UHMWPE, PTFE, PMMA, and hydrogels. These matrices, reinforced with carbon fibers, PE fibers, and ceramics, have been used as prosthetic hip stems, fracture fixation devices, artificial joint bearing surfaces, artificial tooth roots, and bone cements. Also, epoxy composite materials have been used. However, because of concerns about the toxicity of monomers (Morrison *et al.*, 1995) the research activity on epoxy composite for implantable devices gradually decreased.

Materials used and some examples of proposed applications are reported in Table 3.2.12-1.

Not all the proposed systems underwent clinical trial and only some of them are today regularly commercialized.

A review on biomedical applications of composites is in Ramakrishna *et al.* (2001).

Absorbable composite implants can be produced from absorbable α-polyester materials such as polylactic and polyglycolic polymers. Previous work has demonstrated that for most applications, it is necessary to reinforce these polymers to obtain adequate mechanical strength. PGA was the first biodegradable polymer synthesized (Frazza and Schmitt, 1971). It was followed by PLA and copolymers of the two (Gilding and Reed, 1979). These α-polyesters have been investigated for use as sutures and as implant materials for the repair of a variety of osseous and soft tissues. Important biodegradable polymers include poly(ortho esters), synthesized by Heller and coworkers (Heller *et al.*, 1980), and a class of bioerodable dimethyl-trimethylene carbonates (DMTMCs) (Tang

Table 3.2.12-1 Some examples of biomedical composite systems

Applications	Matrix/reinforcement	Reference
External fixator	Epoxy resins/CF	Migliaresi *et al.*, 2004; Baidya *et al.*, 2001
Bone fracture fixation plates, pins, screws	Epoxy resins/CF PMMA/CF PSU/CF PP/CF PE/CF PBT/CF PEEK/CF PEEK/GF PLLA/HA PLLA/PLLA fibers PGA/PGA fibers	Ali *et al.*, 1990; Veerabagu *et al.*, 2003 Woo *et al.*, 1974 Claes *et al.*, 1997 Christel *et al.*, 1980 Rushton and Rae, 1984 Gillett *et al.*, 1986 Fujihara *et al.*, 2001 Lin *et al.*, 1997 Furukawa *et al.*, 2000a Tormala, 1992; Rokkanen *et al.*, 2000 Tormala, 1992; Rokkanen *et al.*, 2000
Spine surgery	PU/bioglass PSU/bioglass PEEK/CF Hydrogels/PET fibers	Claes *et al.*, 1999 Marcolongo *et al.*, 1998 Ciappetta *et al.*, 1997 Ambrosio *et al.*, 1996
Bone cement	PMMA/HA particles PMMA/glass beads Calcium phosphate/aramid fibers, CF, GF, PLGA fibers PMMA/UHMWPE fibers	Morita *et al.*, 1998 Shinzato *et al.*, 2000 Xu *et al.*, 2000 Yang *et al.*, 1997
Dental cements and other dental applications	Bis-GMA/inorganic particles PMMA/KF	Moszner and Salz, 2001 Pourdeyhimi *et al.*, 1986; Vallittu, 1996
Acetabular cups	PEEK/CF	Wang *et al.*, 1998
Hip prostheses stem	PEI/CF-GF PEEK/CF	De Santis *et al.*, 2000 Akay and Aslan, 1996; Kwarteng, 1990
Bone replacement, substitute	PE/HA particles	Bonfield, 1988; Bonfield *et al.*, 1998
Bone filling, regeneration	Poly(propylene fumarate)/TCP PEG-PBT/HA PLGA/HA fibers P(DLLA-CL)/HA particles Starch/HA particles	Yaszemski *et al.*, 1996 Qing *et al.*, 1997 Thomson *et al.*, 1998 Ural *et al.*, 2000 Reis and Cunha, 2000; Leonor *et al.*, 2003
Tendons and ligaments	Hydrogels/PET Polyolefins/UHMWPE fibers	Kolarik *et al.*, 1981; Iannace *et al.*, 1995 Kazanci *et al.*, 2002
Vascular grafts	PELA/Polyurethane fibers	Gershon *et al.*, 1990; Gershon *et al.*, 1992
Prosthetic limbs	Epoxy resins/CF, GF, KF	Dawson, 2000

Legenda: PMMA, polymethylmethacrylate; PSU, polysulfone; PP, polypropylene; PE, polyethylene; PBT, poly(butylene terephthalate); PEEK, poly(ether ether ketone); PLLA, poly(L-lactic acid); PGA, poly(glycolic acid); PU, polyurethane; PET, poly(ethylene terephthalate); Bis-GMA, bis-glycidil dimethacrylate; PEI, poly(ether-imide); PEG, poly(ethylene glycol); PLGA, lactic acid–glycolic acid copolymer; PDLLA, poly(D,L-lactic acid); CL, poly(ε-caprolactone acid); PELA, ethylene oxide/lactic acid copolymer; CF, carbon fibers; GF, glass fibers; HA, hydroxyapatite; UHMWPE, ultrahigh-molecular-weight polyethylene; TCP, tricalcium phosphate; KF, Kevlar fibers.

et al., 1990). A good review of absorbable polymers by Barrows (1986) included PLA, PGA, poly(lactide-*co*-glycolide), PDS, poly(glycolide-*co*-trimethylene carbonate), poly(ethylene carbonate), poly(iminocarbonates), PCL, polyhydroxybutyrate, poly(amino acids), poly(ester amides), poly(ortho esters), poly(anhydrides), and cyanoacrylates. The more recent review by Middleton and Tipton (2000) focused on biodegradable polymers suited for orthopedic applications, mainly PGA and PLA. The authors examined chemistry, fabrication, mechanisms, degradation, and biocompatibility of different polymers and devices.

Natural-origin absorbable polymers have also been utilized in biomedical composites. Purified bovine collagen, because of its biocompatibility, resorbability, and availability in a well-characterized implant form, has been used as a composite matrix, mainly as a ceramic composite binder (Lemons *et al.*, 1984). A commercially available fibrin adhesive (Bochlogyros *et al.*, 1985) and calcium sulfate (Alexander *et al.*, 1987) have similarly been used for this purpose.

Reis *et al.* (1998) proposed alternative biodegradable systems to be used in temporary medical applications. These systems are blends of starch with various thermoplastic polymers. They were proposed for a large range of applications such as temporary hard-tissue replacement, bone fracture fixation, drug delivery devices, or tissue engineering scaffolds.

Fabrication of composites

Composite materials can be fabricated with different technologies. Some of them are peculiar for the type of filler (particle, short or long fiber) and matrix (thermoplastic or thermosetting). Some make use of solvents whose residues could affect the material biocompatibility, hence not being applicable for the fabrication of biomedical composites. The selection of the most appropriate manufacturing technology is also influenced by the relatively low volumes of the production, compared to other applications, and by the relatively low dominance of the manufacturing cost over the overall cost of the device.

Some biomedical composites, moreover, are fabricated *"in situ."* This is the case of composite bone cements.

The most common fabrication technologies for composites are:

1. Hand lay up
2. Spray up
3. Compression molding
4. Resin transfer molding
5. Injection molding
6. Filament winding
7. Pultrusion

In principle all of the listed technologies could be used for the fabrication of biomedical composites. Only some of them, however, have found practical use.

Fabrication of particle-reinforced composites

Injection molding, compression molding, and extrusion are the most common fabrication technologies for biomedical particulate composites. In some applications composites are manufactured *in situ.* This is the case of dental restorative composites and particle-reinforced bone cements.

Fabrication of fiber-reinforced composites

Fiber-reinforced composites are produced commercially by one of two classes of fabrication techniques: open or closed molding. Most of the open-molding techniques are not appropriate to biomedical composites because of the character of the matrices used (mainly thermoplastics) and the need to produce materials that are resistant to water intrusion.

Consequently, the simplest techniques, the hand lay-up and spray-up procedures, are seldom, if ever, used to produce biomedical composites. The two open-molding techniques that may find application in biomedical composites are the vacuum bag–autoclave process and the filament-winding process.

Vacuum bag-autoclave process

This process is used to produce high-performance laminates, usually of fiber-reinforced epoxy. Composite materials produced by this method are currently used in aircraft and aerospace applications. The first step in this process, and indeed many other processes, is the production of a "prepreg." This basic structure is a thin sheet of matrix imbedded with uniaxially oriented reinforcing fibers. When the matrix is epoxy, it is prepared in the partially cured state. Pieces of the prepreg sheet are cut out and placed on top of each other on a shaped tool to form a laminate. The layers, or plies, may be placed in different directions to produce the desired strength and stiffness.

After the laminate is constructed, the tooling and attached laminate are vacuum-bagged, with a vacuum being applied to remove entrapped air from the laminated part. Finally, the vacuum bag enclosing the laminate and the tooling is put into an autoclave for the final curing of the epoxy resin. The conditions for curing vary depending upon the material, but the carbon fiber–epoxy composite material is usually heated at about 190°C at a pressure of about 700 kPa. After being removed from the autoclave, the composite part is stripped from its tooling and is ready for further finishing operations. This procedure is potentially useful for the production of fracture fixation devices and total hip stems.

Filament-winding process

Another important open-mold process to produce high-strength hollow cylinders is the filament-winding process.

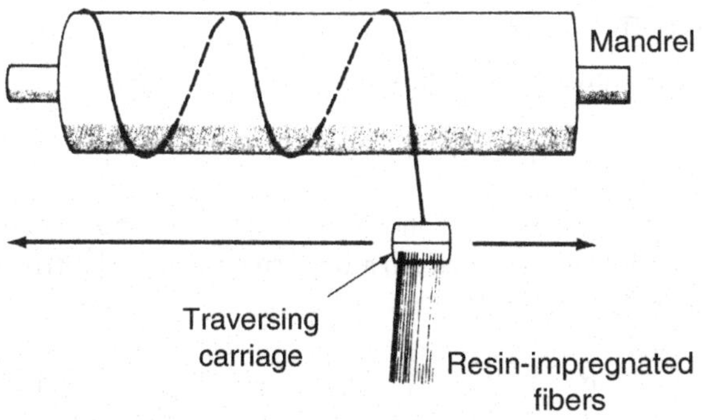

Fig. 3.2.12-2 Filament-winding process for producing fiber-reinforced composite materials.

In this process, the fiber reinforcement is fed through a resin bath and then wound on a suitable mandrel (Fig. 3.2.12-2). When sufficient layers have been applied, the wound mandrel is cured. The molded part is then stripped from the mandrel. The high degree of fiber orientation and high fiber loading with this method produce extremely high tensile strengths. Biomedical applications for this process include intramedullary rods for fracture fixation, prosthetic hip stems, ligament prostheses, intervertebral disks, and arterial grafts.

Closed-mold processes

There are many closed-mold methods used for producing fiber-reinforced plastic materials. The methods of most importance to biomedical composites are compression and injection molding and continuous pultrusion. In compression molding, the previously described prepregs are arranged in a two-piece mold that is then heated under pressure to produce the laminated part. This method is particularly useful for use with thermoplastic matrices. In injection molding the fiber–matrix mix is injected into a mold at elevated temperature and pressure. The finished part is removed after cooling. This is an extremely fast and inexpensive technique that has application to chopped fiber–reinforced thermoplastic composites. It offers the possiblity to produce composite devices, such as bone plates and screws, at much lower cost than comparable metallic devices.

Continuous pultrusion is a process used for the manufacture of fiber-reinforced plastics of constant cross section such as structural shapes, beams, channels, pipe, and tubing. In this process, continuous-strand fibers are impregnated in a resin bath and then are drawn through a heated die, which determines the shape of the finished stock (Fig. 3.2.12-3). Highly oriented parts cut from this stock can then be used in other structures or they can be used alone in such applications as intramedullary rodding or pin fixation of bone fragments.

Mechanical and physical properties of composites

Continuous fiber composites

Laminated continuous fiber-reinforced composites are described from either a micro- or macromechanical point of view. Micromechanics is the study of composite material behavior wherein the interaction of the constituent materials is examined on a local basis. Macromechanics is the study of composite material behavior wherein the material is presumed homogeneous and the effects of the constituent materials are detected only as averaged apparent properties of the composite. Both the micromechanics and macromechanics of experimental laminated composites will be discussed.

Micromechanics

There are two basic approaches to the micromechanics of composite materials: the mechanics of materials and the elasticity approach. The mechanics-of-materials approach embodies the concept of simplifying assumptions regarding the hypothesized behavior of the mechanical system. It is the simpler of the two and the traditional choice for micromechanical evaluation.

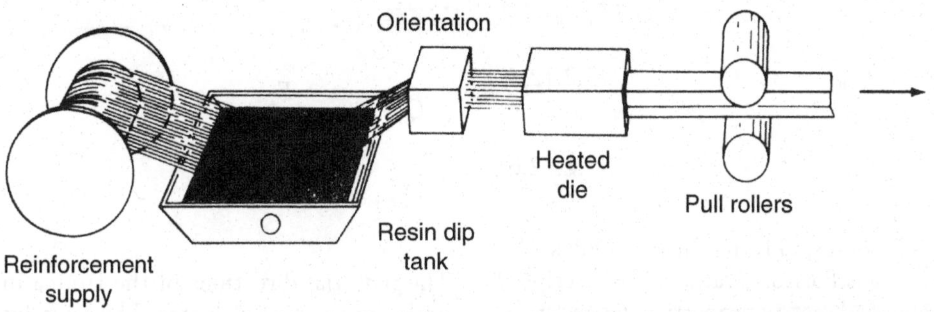

Fig. 3.2.12-3 The pultrusion process for producing fiber-reinforced polymer composite materials. Fibers impregnated with polymer are fed into a heated die and then are slowly drawn out as a cured composite material with a constant cross-sectional shape.

The most prominent assumption made in the mechanics-of-materials approach is that strains in the fiber direction of a unidirectional fibrous composite are the same in the fibers and the matrix. This assumption allows the planes to remain parallel to the fiber direction. It also allows the longitudinal normal strain to vary linearly throughout the member with the distance from the neutral axis. Accordingly, the stress will also have a linear distribution.

Some other important assumptions are as follows:

1. The lamina is macroscopically homogeneous, linearly elastic, orthotropic, and initially stress-free.
2. The fibers are homogeneous, linearly elastic, isotropic, regularly spaced, and perfectly aligned.
3. The matrix is homogeneous, linearly elastic, and isotropic.

In addition, no voids are modeled in the fibers, the matrix or between them.

The mechanical properties of a lamina are determined by fiber orientation. The most often used laminate coordinate system has the length of the laminate in the x direction and the width in the y direction. The principal fiber direction is the 1 direction, and the 2 direction is normal to that. The angle between the x and 1 directions is ϕ. A counterclockwise rotation of the 1–2 system yields a positive ϕ.

The mechanical properties of the lamina are dependent on the material properties and the volume content of the constituent materials. The equations for the mechanical properties of a lamina in the 1–2 directions are:

$$E_1 = E_f V_f + E_m V_m \qquad (3.2.12.1)$$

$$E_2 = \frac{E_f E_m}{V_m E_f} + V_f E_m \qquad (3.2.12.2)$$

$$\upsilon_{12} = V_m \upsilon_m + V_f \upsilon_f \qquad (3.2.12.3)$$

$$\upsilon_{21} E_1 = \upsilon_{12} E_2 \qquad (3.2.12.4)$$

$$G_{12} = \frac{G_f G_m}{V_m G_f} + V_f G_m \qquad (3.2.12.5)$$

$$V_m = 1 - V_f \qquad (3.2.12.6)$$

where E is Young's modulus, G is the shear modulus, V is the volume fraction, ν is Poisson's ratio, and subscripts f and m represent fiber and matrix properties, respectively. These equations are based on the law of mixtures for composite materials.

Macromechanics of a lamina

The generalized Hooke's law relating stresses to strains is

$$\sigma_i = C_{ij}\varepsilon_j \quad ij = 1,2,\ldots,6 \qquad (3.2.12.7)$$

where s_i = stress components, C_{ij} = stiffness matrix, and ϵ_j = strain components. An alternative form of the stress–strain relationship is

$$\varepsilon_{ij} = S_{ij}\sigma_i \quad ij = 1,2,\ldots,6 \qquad (3.2.12.8)$$

where S_{ij} = compliance matrix.

Given that $C_{ij} = C_{ji}$, the stiffness matrix is symmetric, thus reducing its population of 36 elements to 21 independent constants. We can further reduce the matrix size by assuming the laminae are orthotropic. There are nine independent constants for orthotropic laminae. In order to reduce this three-dimensional situation to a two-dimensional situation for plane stress, we have

$$\tau_3 = 0 = \sigma_{23} = \sigma_{13} \qquad (3.2.12.9)$$

thus reducing the stress–strain relationship to

$$\begin{vmatrix} \varepsilon_1 \\ \varepsilon_2 \\ \gamma_{12} \end{vmatrix} = \begin{vmatrix} S_{11} & S_{12} & 0 \\ S_{21} & S_{22} & 0 \\ 0 & 0 & S_{66} \end{vmatrix} \begin{vmatrix} \sigma_1 \\ \sigma_2 \\ \tau_{12} \end{vmatrix} \qquad (3.2.12.10)$$

The stress–strain relation can be inverted to obtain

$$\begin{vmatrix} \sigma_1 \\ \sigma_2 \\ \tau_{12} \end{vmatrix} = \begin{vmatrix} Q_{11} & Q_{12} & 0 \\ Q_{21} & Q_{22} & 0 \\ 0 & 0 & Q_{66} \end{vmatrix} \begin{vmatrix} \varepsilon_1 \\ \varepsilon_2 \\ \gamma_{12} \end{vmatrix} \qquad (3.2.12.11)$$

where Q_{ij} are the reduced stiffnesses. The equations for these stiffnesses are

$$Q_{11} = \frac{E_1}{1 - \upsilon_{21}\upsilon_{12}} \qquad (3.2.12.12)$$

$$Q_{12} = \frac{\upsilon_{12}E_2}{1 - \upsilon_{12}\upsilon_{21}} = \frac{\upsilon_{21}E_l}{1 - \upsilon_{12}\upsilon_{21}} = Q_{21} \qquad (3.2.12.13)$$

$$Q_{22} = \frac{E_2}{1 - \upsilon_{21}\upsilon_{21}} \qquad (3.2.12.14)$$

$$Q_{66} = G_{12} \qquad (3.2.12.15)$$

The material directions of the lamina may not coincide with the body coordinates. The equations for the transformation of stresses in the 1–2 direction to the x–y direction are

$$\begin{vmatrix} \sigma_x \\ \sigma_y \\ \tau_{xy} \end{vmatrix} = [T^{-1}].\begin{vmatrix} \sigma_1 \\ \sigma_2 \\ \tau_{12} \end{vmatrix} \quad (3.2.12.16)$$

where $[T^{-1}]$ is

$$[T^{-1}] = \begin{vmatrix} \cos^2\Phi & \sin^2\Phi & -2\sin\Phi\cos\Phi \\ \sin^2\Phi & \cos^2\Phi & 2\sin\Phi\cos\Phi \\ \sin\Phi\cos\Phi & -\sin\Phi\cos\Phi & \cos^2\Phi - \sin^2\Phi \end{vmatrix} \quad (3.2.12.17)$$

The x and 1 axes form angle Φ. This matrix is also valid for the transformation of strains,

$$\begin{vmatrix} \varepsilon_x \\ \varepsilon_y \\ \frac{1}{2}\gamma_{xy} \end{vmatrix} = [T^{-1}].\begin{vmatrix} \varepsilon_1 \\ \varepsilon_2 \\ \frac{1}{2}\gamma_{12} \end{vmatrix} \quad (3.2.12.18)$$

Finally, it can be demonstrated that

$$\begin{vmatrix} \sigma_x \\ \sigma_y \\ \tau_{xy} \end{vmatrix} = [\overline{Q}_{ij}].\begin{vmatrix} \varepsilon_x \\ \varepsilon_y \\ \gamma_{xy} \end{vmatrix} \quad (3.2.12.19)$$

where $[\overline{Q}_{ij}]$ is the transformed reduced stiffness. The transformed reduced stiffness matrix is

$$[\overline{Q}_{ij}] = \begin{bmatrix} \overline{Q}_{11} & \overline{Q}_{12} & \overline{Q}_{16} \\ \overline{Q}_{21} & \overline{Q}_{22} & \overline{Q}_{26} \\ \overline{Q}_{16} & \overline{Q}_{26} & \overline{Q}_{66} \end{bmatrix} \quad (3.2.12.20)$$

where,

$$\overline{Q}_{11} = Q_{11}\cos^4\Phi + Q_{22}\sin^4\Phi + 2(Q_{12} + 2Q_{66})\sin^2\Phi\cos^2\Phi \quad (3.2.12.21)$$

$$\overline{Q}_{22} = Q_{11}\sin^4\Phi + Q_{22}\cos^4\Phi + 2(Q_{12} + 2Q_{66})\sin^2\Phi\cos^2\Phi \quad (3.2.12.22)$$

$$\overline{Q}_{12} = (Q_{11} + Q_{22} - 4Q_{66})\sin^2\Phi\cos^2\Phi + Q_{12}(\sin^4\Phi\cos^4\Phi) \quad (3.2.12.23)$$

$$\overline{Q}_{66} = (Q_{11} + Q_{22} - 2Q_{12} - 2Q_{66})\sin^2\Phi\cos^2\Phi + Q_{66}(\sin^4\Phi\cos^4\Phi) \quad (3.2.12.24)$$

$$\overline{Q}_{16} = (Q_{11} + Q_{12} - 2Q_{66})\sin\Phi\cos^3\Phi - (Q_{22} - Q_{12} - 2Q_{66})\sin^3\Phi\cos\Phi \quad (3.2.12.25)$$

$$\overline{Q}_{26} = (Q_{11} - Q_{12} - 2Q_{66})\sin^3\Phi\cos\Phi - (Q_{22} - Q_{12} - 2Q_{66})\sin\Phi\cos^3\Phi \quad (3.2.12.26)$$

$\overline{Q}_{16} = \overline{Q}_{26} = 0$ for a laminated symmetric composite.

The transformation matrix$[T^{-1}]$ and the transformed reduced stiffness matrix $[\overline{Q}_{ij}]$ are very important matrices in the macromechanical analysis of bothe the laminae and laminates. These matrices play a key role in detemining the effective in-plane and bending properties and how a laminate will perform when subjected to different combinations of forces and moments.

Macromechanics of a laminate

The development of the A, B, and D matrices for laminate analysis is important for evaluating the forces and moments to which the laminate will be exposed and in determining the stresses and strains of the laminae. As given in Eq. (3.2.12.19),

$$(\sigma_K) = [\overline{Q}_{ij}](\varepsilon_K) \quad (3.2.12.27)$$

where σ = normal stresses, ε = normal strains, and $[\overline{Q}_{ij}]$ = stiffness matrix. The A, B, and D matrices are equivalent to the following:

$$[A_{ij}] = \sum_{k=1}^{n}(\overline{Q}_{ij})_k(h_k - h_{k-1}) \quad (3.2.12.28)$$

$$[B_{ij}] = \frac{1}{2}\sum_{k=1}^{n}(\overline{Q}_{ij})_k(h_k^2 - h_{k-1}^2) \quad (3.2.12.29)$$

$$[D_{ij}] = \frac{1}{3}\sum_{k=1}^{n}(\overline{Q}_{ij})_k(h_k^3 - h_{k-1}^3) \quad (3.2.12.30)$$

The matrix $[A]$ is called the *extensional stiffness matrix* because it relates the resultant forces to the midplane strains, while matrix $[D]$ is called the *bending stiffness matrix* because it relates the resultant moments to the laminate curvature. The so called *coupling stiffness matrix*, $[B]$, accounts for coupling between bending and extension, which means that normal and shear forces acting at the laminate midplane are causing laminate curvature or that bending and twisting moments are accompanied by midplane strain.

The letter k denotes the number of laminae in the laminate with a maximum number (N). The letter h represents the distances from the neutral axis to the edges of the respective laminae. A standard procedure for numbering laminae is used where the 0 lamina is at the bottom of a plate and the Kth lamina is at the top.

The resultant laminate forces and moments are:

$$\begin{vmatrix} N_x \\ N_y \\ N_{xy} \end{vmatrix} = [A_{ij}].\begin{vmatrix} \varepsilon_x \\ \varepsilon_y \\ \gamma_{xy} \end{vmatrix} + [B_{ij}].\begin{vmatrix} k_x \\ k_y \\ k_{xy} \end{vmatrix} \quad (3.2.12.31)$$

$$\begin{vmatrix} M_x \\ M_y \\ M_{xy} \end{vmatrix} = [B_{ij}].\begin{vmatrix} \varepsilon_x \\ \varepsilon_y \\ \gamma_{xy} \end{vmatrix} + [D_{ij}].\begin{vmatrix} k_x \\ k_y \\ k_{xy} \end{vmatrix} \quad (3.2.12.32)$$

The k vector represents the respective curvatures of the various planes. The resultant forces and moments of a loaded composite can be analyzed given the *ABD* matrices. If the laminate is assumed symmetric, the force equation reduces to

$$\begin{vmatrix} N_x \\ N_y \\ N_{xy} \end{vmatrix} = [A_{ij}].\begin{vmatrix} \varepsilon_x \\ \varepsilon_y \\ \gamma_{xy} \end{vmatrix} \quad (3.2.12.33)$$

Once the laminate strains are determined, the stresses in the *xy* direction for each lamina can be calculated. The most useful information gained from the *ABD* matrices involves the determination of generalized in-plane and bending properties of the laminate.

In a generic laminate, normal stresses N_x and/or N_y (or thermal stresses or liquid sorption) will cause deformations in the directions x and/or y, but also shear strains, unless A_{16} and A_{26} of the extensional stiffness matrix are equal to 0. These coefficient become 0 if the laminate is balanced, i.e., has the same number of laminae oriented at Φ and $-\Phi$.

Moreover, in a generic laminate, normal or shear stresses will produce bending, and bending or twisting will cause mid-plane strains. The coupling between bending and extension can be eliminated if the coefficients of the B_{ij} matrix are equal to zero, that is, if the laminate is fabricated symmetric with respect to its midplane.

The equivalent elastic constants (E_x, E_y, G_{xy}, ν_{xy}) of a symmetric and balanced laminate can be easily evaluated from the A_{ij} coefficients (Barbero, 1998):

$$E_x = \frac{1}{h}\frac{A_{11}A_{22} - A_{12}^2}{A_{22}} \quad (3.2.12.34)$$

$$E_y = \frac{1}{h}\frac{A_{11}A_{22} - A_{12}^2}{A_{11}} \quad (3.2.12.35)$$

$$\nu_{xy} = \frac{A_{12}}{A_{22}} \quad (3.2.12.36)$$

$$G_{xy} = \frac{1}{h}A_{66} \quad (3.2.12.37)$$

In the equations above h is the total thickness of the laminate.

Short-fiber composites

A distinguishing feature of the unidirectional laminated composites discussed above is that they have higher strength and modulus in the fiber direction, and thus their properties are amenable to alteration to produce specialized laminates. However, in some applications, unidirectional multiple-ply laminates may not be required. It may be advantageous to have isotropic laminae. An effective way of producing an isotropic lamina is to use randomly oriented short fibers as the reinforcement. Of course, molding compounds consisting of short fibers that can be easily molded by injection or compression molding may be used to produce generally isotropic composites. The theory of stress transfer between fibers and matrix in short-fiber composites goes beyond this text; it is covered in detail by Agarwal and Broutman (1980). However, the longitudinal and transverse moduli (E_L and E_T, respectively) for an aligned short-fiber lamina can be derived from the generalized Halpin-Tsai equations (Halpin and Kardos, 1976), as:

$$\frac{E_L}{E_m} = \frac{1 + ((2l/d)\eta_L V_f)}{1 - \eta_L V_f} \quad (3.2.12.38)$$

$$\frac{E_T}{E_m} = \frac{1 + 2\eta_T V_f}{1 - \eta_T V_f} \quad (3.2.12.39)$$

$$\eta_L = \frac{E_f/E_m - 1}{E_f/E_m + 2(l/d)} \quad (3.2.12.40)$$

$$\eta_T = \frac{E_f/E_m - 1}{E_f/E_m + 2} \quad (3.2.12.41)$$

In the previous equations E_m is the elastic modulus of the matrix, l and d are the fiber length and diameter respectively, and V_f is the fiber volume fraction.

For a ratio of fiber to matrix modulus of 20, the variation of longitudinal modulus of an *aligned* short-fiber lamina as a function of fiber aspect ratio, l/d, for different fiber volume fractions is shown in Fig. 3.2.12-4. It can be seen that approximately 85% of the modulus obtainable from a continuous fiber lamina is attainable with an aspect ratio of 20.

The problem of predicting properties of *randomly oriented* short-fiber composites is more complex. The following empirical equation can be used to predict the modulus of composites containing fibers that are randomly oriented in a plane:

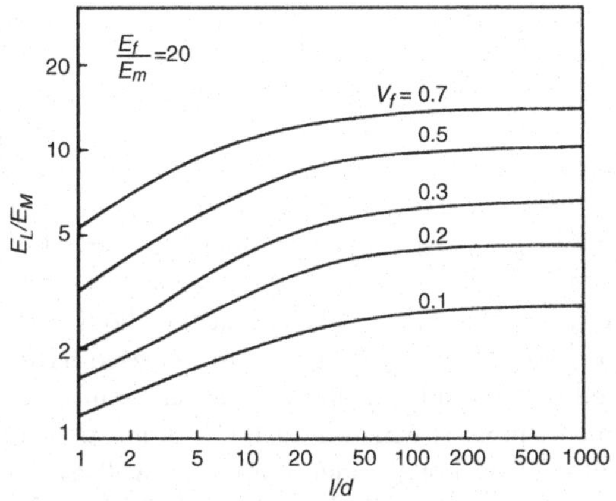

Fig. 3.2.12-4 Variations of longitudinal modulus of short-fiber composites against aspect ratio for different fiber volume fractions ($E_f/E_m = 20$).

$$E_{\text{random}} = \frac{3}{8}E_L + \frac{5}{8}E_T \qquad (3.2.12.42)$$

where E_L and E_T are respectively the longitudinal and transverse moduli of an aligned short-fiber composite having the same fiber aspect ratio and fiber volume fraction as the composite under consideration. Moduli E_L and E_T can either be determined experimentally or calculated using Eqs. 3.2.12.38 and 3.2.12.39.

Particulate composites

The reinforcing effect of particles on polymers was first recognized for rubbery matrices during studies of the effect of carbon black on the properties of natural rubber.

Several models have been introduced to predict the effect of the addition of particles to a polymeric matrix, starting from the equation developed by Einstein in 1956 to predict the viscosity of suspensions of rigid spherical inclusions. The paper by Ahmed and Jones (1990) well reviews theories developed to predict strength and modulus of particulate composites. One of the most versatile equation predicting the shear modulus of composites of polymers and spherical fillers is due to Kerner (1956):

$$G_c = G_m\left(1 + \frac{V_f}{V_m}\frac{15(1 - \upsilon_m)}{(8 - 10\upsilon_m)}\right) \qquad (3.2.12.43)$$

A more generalized form was developed by Nielsen (1974),

$$M_c = M_m\frac{1 + ABV_f}{1 - B\psi V_f} \qquad (3.2.12.44)$$

where M_c is any modulus—shear, Young's or bulk- of the composite, the constant A takes into account for the filler geometry and the Poisson's ratio of the matrix and the

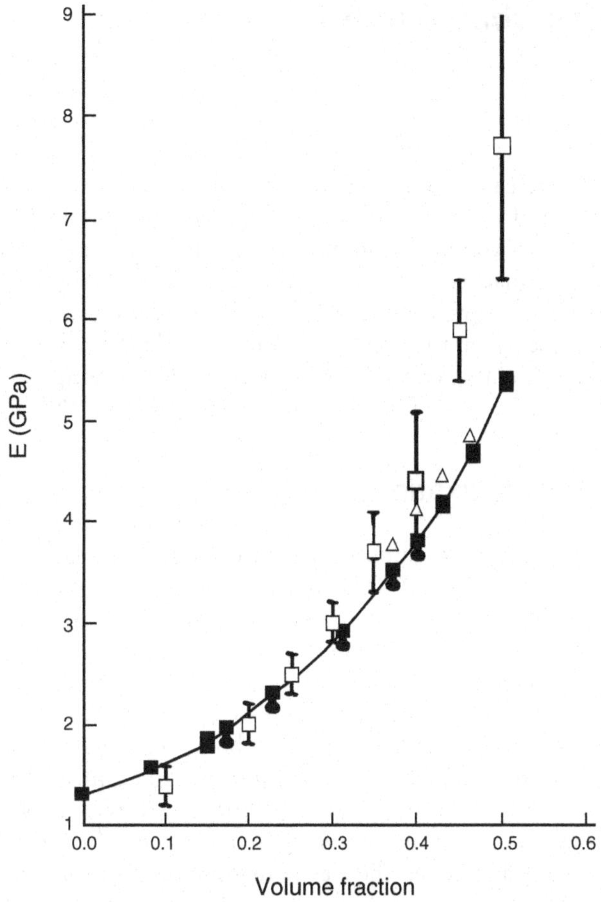

Fig. 3.2.12-5 Variation of the Young's modulus of HA–PE composites modulus with volume fraction: experimental values, □, and predicted values before and after the application of the statistical model; ■, primary; △, equal strain; ▲, equal stress (from Guild and Bonfield, 1993).

constant B depends on the relative moduli of the filler (M_f) and the matrix (M_m). The function ψ depends on the particle packing fraction.

By using a finite element analysis method Guild and Bonfield (1993) predicted the elastic modulus of hydroxyapatite–PE reinforced composites for various filler content. Their result (Fig. 3.2.12-5) indicated a good agreement between theoretical and experimental data, except at higher HA volume fraction.

While elastic modulus of a particulate composites increases with the filler content, strength decreases in tension and increases in compression. Size and shape of the inclusion play an important role, with a higher stress concentration cause by irregularly shaped inclusions. For spherical particles, tensile strength can be predicted by the equation (Nicolais and Narkis, 1971):

$$\sigma_{\text{cu}} = \sigma_{\text{mu}}\left(1 - 1.21V_f^{2/3}\right) \qquad (3.2.12.45)$$

where σ_{cu} and σ_{mu} are tensile strength of composite and matrix, respectively.

Absorbable matrix composites

Absorbable matrix composites have been used in situations where absorption of the matrix is desired. Matrix absorption may be desired to expose surfaces to tissue or to release admixed materials such as antibiotics or growth factors (drug release) (Yasko *et al.*, 1992). However, the most common reasons for the use of this class of matrices for composites has been to accomplish time-varying mechanical properties and assure complete dissolution of the implant, eliminating long-term biocompatibility concerns. A typical clinical example is fracture fixation (Daniels *et al.*, 1990; Tormala, 1992).

Fracture fixation

Rigid internal fixation of fractures has conventionally been accomplished with metallic plates, screws, and rods. During the early stages of fracture healing, rigid internal fixation maintains alignment and promotes primary osseous union by stabilization and compression. Unfortunately, as healing progresses, or after healing is complete, rigid fixation may cause bone to undergo stress protection atrophy. This can result in significant loss of bone mass and osteoporosis. Additionally, there may be a basic mechanical incompatibility between the metal implants and bone. The elastic modulus of cortical bone ranges from 17 to 24 GPa, depending upon age and location of the specimen, while the commonly used alloys have moduli ranging from 110 GPa (titanium alloys) to 210 GPa (316L steel). This large difference in stiffness can result in disproportionate load sharing, relative motion between the implant and bone upon loading, as well as high stress concentrations at bone–implant junctions.

Another potential problem is that the alloys currently used corrode to some degree. Ions so released have been reported to cause adverse local tissue reactions as well as allogenic responses, which in turn raises questions of adverse effects on bone mineralization as well as adverse systemic responses such as local tumor formation (Martin *et al.*, 1988). Consequently, it is usually recommended that a second operation be performed to remove hardware.

The advantages of absorbable devices are thus twofold. First, the devices degrade mechanically with time, reducing stress protection and the accompanying osteoporosis. Second, there is no need for secondary surgical procedures to remove absorbable devices. The state of stress at the fracture site gradually returns to normal, allowing normal bone remodeling.

Absorbable fracture fixation devices have been produced from PLLA polymer, PGA polymer, and PDS. An excellent review of the mechanical properties of biodegradable polymers was prepared by Daniels and co-workers (Daniels *et al.*, 1990; see Figs. 3.2.12-6 and 3.2.12-7).

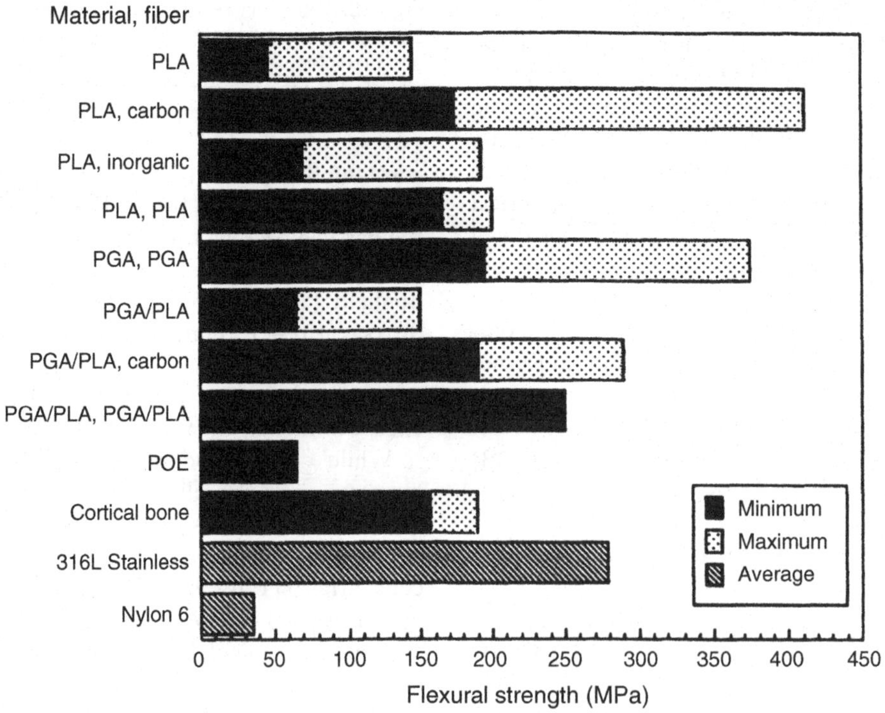

Fig. 3.2.12-6 Representative flexural strengths of absorbable polymer composites (from Daniels, A. U., Melissa, K. O., and Andriano, K. P. (1990). Mechanical properties of biodegradable polymers and composites proposed for internal fixation of bone. *J. Appl. Biomater.* 1(1): 57-78.).

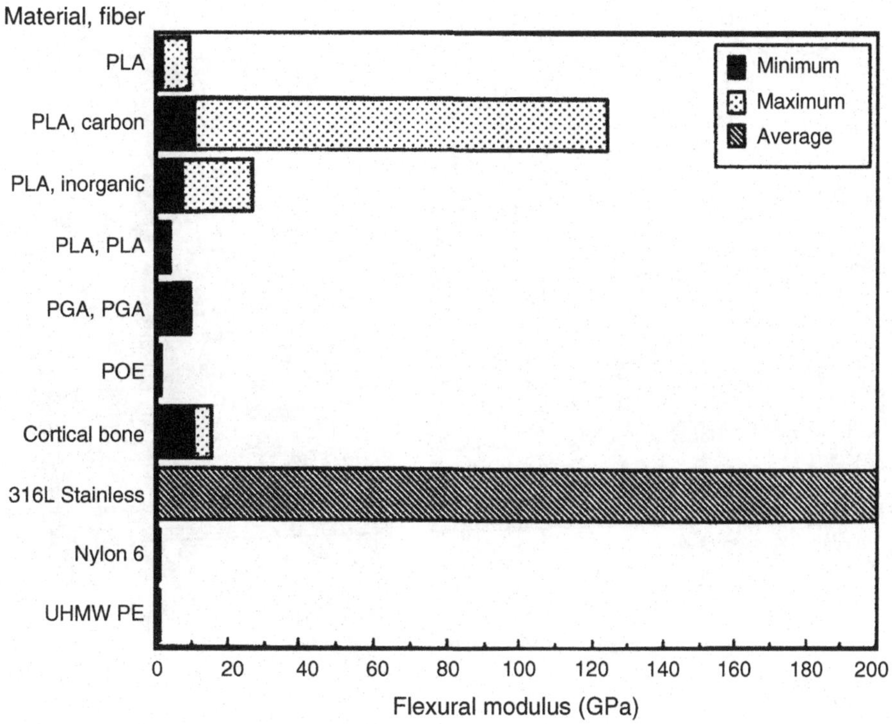

Fig. 3.2.12-7 Representative flexural moduli of absorbable polymer composites (from Daniels, A. U., Melissa, K. O., and Andriano, K. P. (1990). Mechanical properties of biodegradable polymers and composites proposed for internal fixation of bone. *J. Appl. Biomater.* 1(1): 57–78.).

Their review revealed that unreinforced biodegradable polymers are initially 36% as strong in tension as annealed stainless steel, and 54% in bending, but only 3% as stiff in either test mode. With fiber reinforcement, highest initial strengths exceeded those of stainless steel. Stiffness reached 62% of stainless steel with non-degradable carbon fibers, 15% with degradable inorganic fibers, but only 5% with degradable polymeric fibers.

Most previous work on absorbable composite fracture fixation has been performed with PLLA polymer. PLLA possesses three major characteristics that make it a potentially attractive biomaterial:

1. It degrades in the body at a rate that can be controlled.
2. Its degradation products are nontoxic, biocompatible, easily excreted entities. PLA undergoes hydrolytic deesterification to lactic acid, which enters the lactic acid cycle of metabolites. Ultimately it is metabolized to carbon dioxide and water and is excreted.
3. Its rate of degradation can be controlled by mixing it with PGA polymer.

PLLA polymer reinforced with randomly oriented chopped carbon fiber was used to produce partially degradable bone plates (Corcoran *et al.*, 1981). It was demonstrated that the plates, by virtue of the fiber reinforcement, exhibited mechanical properties superior to those of pure polymer plates. *In vivo*, the matrix degraded and the plates lost rigidity, gradually transferring load to the healing bone. However, the mechanical properties of such chopped fiber plates were relatively low; consequently, the plates were only adequate for low-load situations. Zimmerman *et al.* (1987) used composite theory to determine an optimum fiber layup for a long fiber composite bone plate. Composite analysis suggested the mechanical superiority of a 0°/±45° laminae layup. Although the 0°/±45° carbon/PLA composite possessed adequate initial mechanical properties, water absorption and subsequent delamination degraded the properties rapidly in an aqueous environment (Fig. 3.2.12-8). The fibers did not chemically bond to the matrix.

In an attempt to develop a totally absorbable composite material, a calcium-phosphate-based glass fiber has been used to reinforce PLA. Experiments were pursued to determine the biocompatibility and in vitro degradation properties of the composite (Zimmerman *et al.*, 1991). These studies showed that the glass fiber-PLA composite was biocompatible, but its degradation rate was too high for use as an orthopedic implant.

Shikinami and Okuno (2001), have produced mini-plates, rods, and screws made of HA poly(L-lactide). These composites have been principally applied for indications such as repair of bone fracture in osteosynthesis and fixation of bony fragments in bone grafting and osteotomy, exhibiting total resorbability and osteological

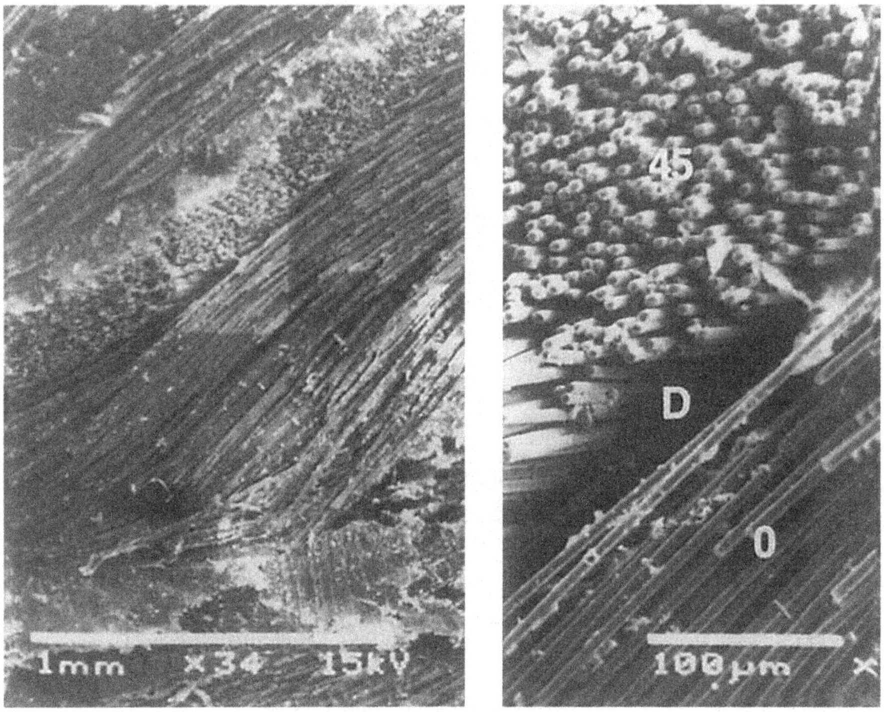

Fig. 3.2.12-8 Scanning electron micrograph of laminae buckling and delamination (D) between lamina in a carbon fiber-reinforced PLA fracture fixation plate (from Zimmerman, M. C., Parsons, J. R., and Alexander, H. (1987). The design and analysis of a laminated partially degradable composite bone plate for fracture fixation. *J. Biomed. Mater. Res. Appl. Biomater.* 21A(3): 345.).

bioactivity while retaining sufficient stiffness high stiffness retainable for a long period of time to achieve bony union. These plates are commercialized with the name of Fixsorb-MX.

Furukawa *et al.* (2000b) have investigated the *in vivo* biodegradation behavior of HA/poly(L-lactide) composite rods implanted *sub cutem* and in the intramedullary cavities of rabbits, showing that after 25 weeks of *sub cutem* implantation rods maintained a bending strength higher than 200 MPa. Their conclusion was that such a strength was sufficient for application of the rods in the fixation of human bone fractures.

By using a sintering technique, Tormala *et al.* (1988) have produced self-reinforced PGA (SR-PGA) rods that have been used in the treatment of fractures and osteotomies. Afterwards, by using the same technique, self-reinforced PLLA (SR-PLLA) pins and screws have been produced. The higher initial mechanical properties of SR-PLGA are counterbalanced by their faster decrease with respect to the SR-PLLA material, which has a slower degradation rate and is reabsorbed in 12–16 months. These products are commercially available.

Nonabsorbable matrix composites

Nonabsorbable matrix composites are generally used as biomaterials to provide specific mechanical properties unattainable with homogeneous materials. Particulate and chopped-fiber reinforcement has been used in bone cements and bearing surfaces to stiffen and strengthen these structures.

For fracture fixation, reduced-stiffness carbon-fiber-reinforced epoxy bone plates to reduce stress-protection osteoporosis have been made. These plates have also been entered into clinical use, but were found to not be as reliable or biocompatible as stainless steel plates. Consequently, they have not generally been accepted in clinical use. By far the most studied, and potentially most valuable use of nonabsorbable composites has been in total joint replacement.

Total joint replacement

Bone resorption in the proximal femur leading to aseptic loosening is an all-too-common occurrence associated with the implantation of metallic femoral hip replacement components. It has been suggested that proximal bone loss may be related to the state of stress and strain in the femoral cortex. It has long been recognized that bone adapts to functional stress by remodeling to reestablish a stable mechanical environment. When applied to the phenomenon of bone loss around implants, one can postulate that the relative stiffness of the metallic component is depriving bone of its accustomed load. Clinical and experimental results have shown the significant role that implant elastic characteristics play in allowing the femur to attain a physiologically acceptable

Table 3.2.12-2 Typical mechanical properties of polymer-carbon composites (three-point bending)

Polymer	Ultimate strength (MPa)	Modulus (GPa)
PMMA	772	55
Polysufone	938	76
Epoxy		
Stycast	535	30
Hysol	207	24
Polyurethane	289	18

stress state. Femoral stem stiffness has been indicated as an important determinant of cortical bone remodeling (Cheal *et al.*, 1992). Composite materials technology offers the ability to alter the elastic characteristics of an implant and provide a better mechanical match with the host bone, potentially leading to a more favorable bone remodeling response.

Using different polymer matrices reinforced with carbon fiber, a large range of mechanical properties is possible. St. John (1983) reported properties for ±15° laminated test specimens (Table 3.2.12-2) with moduli ranging from 18 to 76 GPa. However, the best reported study involved a novel press-fit device constructed of carbon fiber/polysulfone composite (Magee *et al.*, 1988). The femoral component designed and used in this study utilized composite materials with documented biologic profiles. These materials demonstrated strength commensurate with a totally unsupported implant region and elastic properties commensurate with a fully bone-supported implant region. These properties were designed to produce constructive bone remodeling. The component contained a core of unidirectional carbon/polysufone composite enveloped with a bidirectional braided layer composed of carbon/polysulfone composite covering the core. These regions were encased in an outer coating of pure polysulfone (Fig. 3.2.12-9). Finite-element stress analysis predicted that this construction would cause minimal disruption of the normal stresses in the intact cortical bone. Canine studies carried out to 4 years showed a favorable bone remodeling response. The authors proposed that implants fabricated from carbon/polysulfone composites should have the potential for use in load-bearing applications. An implant with appropriate elastic properties provides the opportunity for the natural bone remodeling response to enhance implant stability.

Adam *et al.* (2002) reported on the revision of 51 epoxy resin/carbon fiber composite press fit-hip prostheses implanted in humans. Their result showed that within 6 years 92% of the prostheses displayed aseptic loosening, i.e., did not induce bone ongrowth. Authors attributed the failure to the smoothness of the stem surface. No osteolysis or wear or inflammatory reaction were, however, observed.

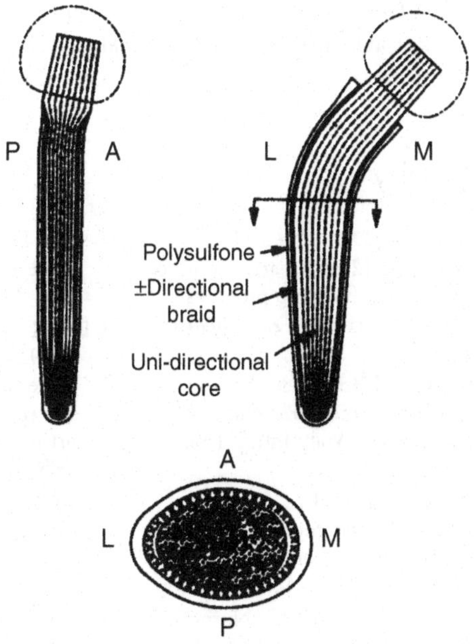

Fig. 3.2.12-9 Construction details of a femoral stem of a composite total hip prosthesis. (From Magee, F. P., Weinstein, A. M., Longo, J. A., Koeneman, J. B., and Yapp, R. A. (1988). A canine composite femoralstem. *Clin. Orthop. Rel. Res.* 235: 237.)

Different fibers matrices and fabrication technologies have been proposed for the fabrication of hip prostheses. Reviews of materials and methods are in Ramakrishna *et al.* (2001) and in de Oliveira Simopes and Marques (2001).

Conclusions

Biomedical composites have demanding properties that allow few, if any, "off the shelf" materials to be used. The designer must almost start from scratch. Consequently, few biomedical composites are yet in general clinical use. Those that have been developed to date have been fabricated from fairly primitive materials with simple designs. They are simple laminates, chopped fiber, or particulate reinforced systems with no attempts made to react or bond the phases together. Such bonding may be accomplished by altering the surface texture of the filler or by the introduction of coupling agents: molecules that can react with both filler and matrix. However, concerns about the biocompatibility of coupling agents and the high development costs of surface texture alteration procedures have curtailed major developments in this area. It is also possible to provide three-dimensional reinforcement with complex fiber weaving and impregnation procedures now regularly used in high-performance aerospace composites. Unfortunately, the high development costs associated with these techniques have restricted their application to biomedical composites.

Because of the high development costs and the small-volume market available, few biomedical materials have, to date, been designed specifically for biomedical use. Biomedical composites, because of their unique requirements, are probably be the first general class of materials developed exclusively for implantation purposes.

Bibliography

Adam, F., Hammer, D. S., Pfautsch, S., and Westermann, K. (2002). Early failure of a press-fit carbon fiber hip prosthesis with a smooth surface. *J. Arthroplasty* **17**: 217–223.

Agarwal, B. D., and Broutman, L. J. (1980). *Analysis and Performance of Fiber Composites.* Wiley-Interscience, New York.

Ahmed, S., and Jones F. R. (1990). A review of particulate reinforcement theories for polymer composites. *J. Mater. Sci.* **25**: 4933–42.

Akay, M., and Aslan, N. (1996). Numerical and experimental stress analysis of a polymeric composite hip joint prosthesis. *J. Biomed Mater. Res.* **31**: 167–82.

Alexander, H., Parsons, J. R., Ricci, J. L., Bajpai, P. K., and Weiss, A. B. (1987). Calcium-based ceramics and composites in bone reconstruction. *CRC Crit. Rev. Biocompat.* **4**(1): 43–77.

Ali, M. S., Hastings, G. W., Rushton, N., Ross, E. R. S., and Wynn-Jones, C. H. (1990). Carbon fiber composite plates. *J. Bone Joint Surg.* **72-B**: 586–591.

Ambrosio, L., Netti, P., Iannace, S., Huang, S. J., and Nicolais, L. (1996). Composite hydrogels for intervertebral disc prostheses. *J. Mater. Sci. Mater. Med.* 7: 251–4.

Baidya, K. P., Ramakrishna, S., Rahman, M., and Ritchie A. (2001). Advanced textile composite ring for Ilizarov external fixator system. *Proc. Inst. Mech. Eng. Part H, J. Eng. Med.* **215**: 11–23.

Barbero, E. J. (1998). *Introduction to Composite Materials Design.* Taylor and Francis, Philadelphia.

Barrows, T. H. (1986). Degradable implant materials: a review of synthetic absorbable polymers and their applications. *Clin. Mater.* **1**: 233.

Bochlogyros, P. M., Hensher, R., Becker, R., and Zimmerman, E. (1985). A modified hydroxyapatite implant material. *J. Maxillofac. Surg.* **13**(5): 213.

Bonfield, W. (1988). Composites for bone replacement. *J. Biomed. Eng.* **10**: 522.

Bonfield, W., Wang, M., and Tanner K. E. (1998). Interfaces in analogue biomaterials. *Acta Mater.* 7: 2509–2518.

Brown, D. (2000). Fibre-reinforced materials. *Dent. Update* **27**(9): 442–448.

Cheal, E. J., Spector, M., and Hayes, W. C. (1992). Role of loads and prosthesis material properties on the mechanics of the proximal femur after total hip arthroplasty. *J. Orthop. Res.* **10**: 405–422.

Chen, S. Y., Liang, W. M., and Yen, P. S. (2001). Reinforcement of acrylic denture base resin by incorporation of various fibers. *J. Biomed. Mater. Res.* **58**(2): 203–208.

Christel, P., Leray, J., Sedel, L., and Morel, E. (1980). Mechanical evaluation and tissue compatibility of materials for composite bone plates. in *Mechanical Properties of Biomaterials,* G. Hasting and D. F. Williams, eds. Wiley, New York, pp. 367–377.

Ciappetta, P., Boriani, S., and Fava, G. P. (1997). A carbon fiber reinforced polymer cage for vertebral body replacement: technical note. *Neurosurgery* **4**(5): 1203–1206.

Claes, L., Hutter, W., and Weiss, R. (1997). Mechanical properties of carbon reinforced polysulfone plates for internal fixation, in *Biological and Biomechanical Performance of Biomaterials,* P. Christel, A. Meunier, and A. J. C. Lee, eds. Elsevier, Amsterdam, pp. 81–86.

Claes, L., Schultheiss, M., Wolf, S., Wilke, H. J., Arand, M., and Kinzl, L. (1999). A new radiolucent system for vertebral body replacement: its stability in comparison to other systems. *J. Biomed. Mater. Res. Appl. Biomater.* **48**(1): 82–89.

Corcoran, S., Koroluk, J., Parsons, J. R., Alexander, H., and Weiss, A. B. (1981). The development of a variable stiffness, absorbable composite bone plate. in *Current Concepts for Internal Fixation of Fractures,* H. K. Uhthoff, ed. Springer-Verlag, New York, pp. 136.

Daniels, A. U., Melissa, K. O., and Andriano, K. P. (1990). Mechanical properties of biodegradable polymers and composites proposed for internal fixation of bone. *J. Appl. Biomater.* **1**(1): 57–78.

Dawson, D. K. (2000). Medical devices. in *Comprehensive Composite Materials,* Vol. 6, A. Kelly, ed. Elsevier, pp. 1–32.

de Oliveira Simoes, J. A., and Marques, A. T. (2001). Determination of stiffness properties of braided composites for the design of hip prostheses. *Composites, Part A* **32**: 655–662.

De Santis, R., Ambrosio, L., and Nicolais, L. (2000). Polymer-based composite hip prosthesis. *J. Inorg. Biochem.* **79**: 97–102.

Dunn, M. G. (1998). Anterior cruciate ligament prostheses. in *Encyclopedia of Sports Medicine and Science,* T. D. Fahey, ed. Internal Society for Sports Science, http://sportsci.org.

Durselen, L., Dauner, M., Hierlemann, H., Planck, H., Claes, L. E., and Ignatius, A. (2001) Resorbable polymer fibers for ligament augmentation. *J. Biomed. Mater. Res.* **58**(6): 666–672.

Einstein, A. (1956). in *Investigation of Theory of Brownian Motion.* Dover, New York.

Frazza, E. J., and Schmitt, E. E. (1971). A new absorbable suture. *J. Biomed. Mater. Res.* **10**: 43.

Furukawa, T., Matsusue, Y., Yasunaga, T., Nakagawa, Y., Shikinami, Y., Okuno, M., and Nakamura, T. (2000a). Bone bonding ability of a new biodegradable composite for internal fixation of bone fractures. *Clin. Orthop.* **379**: 247–258.

Furukawa, T., Matsusue, Y., Yasunaga, T., Shikinami, Y., Okuno, M., and Nakamura, T. (2000b). Biodegradation behavior of ultra-high-strength hydroxyapatite/poly(L-lactide) composite rods for internal fixation of bone fractures. *Biomaterials* **21**: 889–898.

Gershon, B., Cohn, D., and Marom, G. (1992). Compliance and ultimate strength of composite arterial prostheses. *Biomaterials* **13**: 38–43.

Gershon, B., Cohn, D., and Marom, G. (1990). The utilization of composite laminate theory in the design of synthetic soft tissue for biomedical prostheses. *Biomaterials* **11**: 548–552.

Gilding, D. K., and Reed, A. M. (1979). Biodegradable polymers for use in surgery: PGA/PLA homo- and copolymers. 1. *Polymer* **20**: 1459.

Gillett, N., Brown, S. A., Dumbleton, J. H., and Pool, R. P. (1986). The use of short carbon fiber reinforced thermoplastic plates for fracture fixation. *Biomaterials* **6**: 113–21.

Guidoin, M. F., Marois, Y., Bejui, J., Poddevin, N., King, M. W., and Guidoin R. (2000). Analysis of retrieved polymer fiber based replacements for the ACL. *Biomaterials* **21**: 2461–2474.

Guild, F. J., and Bonfield, W. (1993). Predictive modeling of hydroxyapatite–polyethylene composite. *Biomaterials* **14**(13): 985–993.

Halpin J. C., and Kardos J. L. (1976). The Halpin-Tsai equations: a review. *Poly. Eng. Sci.* **16**(5): 344–335.

Heller, J., Penhale, D. W. H., and Helwing, R. F. (1980). Preparation of poly(ortho esters) by the reaction of diketene acetals and polyols. *J. Polymer Sci. Polymer Lett. Ed.* **18**: 619.

Hench, L. L., Splinter, R. J., Allen, W. C., and Greenlee, T. K. (1971). Bonding mechanisms at the interface of ceramic prosthetic materials. *J. Biomed. Mater. Res.* **72**: 117–141.

Iannace, S., Sabatini, G., Ambrosio, L., and Nicolais, L. (1995). Mechanical behavior of composite artificial tendons and ligaments. *Biomaterials* **16**: 675–680.

Jarcho, M., Kay, J. F., Gumaer, K. I., Domerus, R. H., and Droback, H. P. (1977). Tissue cellular and subcellular events at a bone-ceramic hydroxyapatite interface. *J. Bioeng.* **1**: 79–92.

Karaman, A. I., Kir, N., and Belli, S. (2002). Four applications of reinforced polyethylene fiber material in orthodontic practice. *Am. J. Orthod. Dentofacial. Orthop.* **121**(6): 650–654.

Kazanci, M., Cohn, D., Marom, G., Migliaresi, C., and Pegoretti, A. (2002). Fatigue characterization of polyethylene fiber reinforced polyolefin biomedical composites. *Composites, Part A* **33**: 453–458.

Kellomaki, M., Niiranen, H., Puumanen, K., Ashammakhi, N., Waris, T., and Tormala, P. (2000). Bioabsorbable scaffolds for guided bone regeneration and generation. *Biomaterials* **21**: 2495–2505.

Kerner, E. H. (1956). The elastic and thermo-elastic properties of composite media. *Proc. Phys. Soc. B* **69**: 808–13.

Kolarik, J., Migliaresi, C., Stol, M., and Nicolais, L. (1981). Mechanical properties of model synthetic tendons, *J. Biomed. Mater. Res.* **15**: 147.

Kotani, Y., Abumi, K., Shikinami, Y., Takada, T., Kadoya, K., Shimamoto, N., Ito, M., Kadosawa, T., Fujinaga, T., and Kaneda, K. (2002). Artificial intervertebral disc replacement using bioactive three-dimensional fabric: design, development, and preliminary animal study. *Spine* **27**(9): 929–935.

Kurtz, S. M., Muratoglu, O. K., Evans, M., and Edidin, A. A. (1999). Advances in the processing, sterilization, and crosslinking of ultra-high molecular weight polyethylene for total joint arthroplasty. *Biomaterials* **20**: 1659–1688.

Kwarteng, K. B. (1990). Carbon fiber reinforced PEEK (APC-2/AS-4) composites for orthopaedic implants. *SAMPE Quart.* **21**(2): 10–14.

Ladizesky, N. H., Chow, T. W., and Cheng, Y. Y. (1994). Denture base reinforcement using woven polyethylene fiber. *Int. J. Prosthodont.* **7**(4): 307–314.

Lemons, J. E., Matukas, V. J., Nieman, K. M. W., Henson, P. G., and Harvey, W. K. (1984). Synthetic hydroxylapatite and collagen combinations for the surgical treatment of bone. in *Biomedical Engineering,* Vol. 3, Sheppard, L. C., ed. Pergamon, New York, p. 13.

Leonor, I. B., Ito, A., Kanzaki, N., and Reis R. L. (2003). In vitro bioactivity of starch thermoplastic/hydroxyapatite composite biomaterials: an in situ study using atomic force microscopy. *Biomaterials* **24**: 579–585.

Lin, T. C. (1986). Totally absorbable fiber reinforced composite from internal fracture fixation devices. *Trans. Soc. Biomater.* **9**: 166.

Lin, T. W., Corvelli, A. A., Frondoza, C. G., Roberts, J. C., and Hungerford D. S. (1997). Glass PEEK composite promotes proliferation and osteocalcin production of human osteoblastic cells. *J. Biomed. Mater. Res.* **36**(2): 137–144.

Lu, L., and Mikos A.G. (1996). The importance of new processing techniques in tissue engineering. *MRS Bull.* **21**(11): 28–32.

Magee, F. P., Weinstein, A. M., Longo, J. A., Koeneman, J. B., and Yapp, R. A. (1988). A canine composite femoral stem. *Clin. Orthop. Rel. Res.* **235**: 237.

Marcolongo, M., Ducheyne, P., Garino, J., and Schepers, E. (1998). Bioactive glass fiber/polymeric composites bond to bone tissue. *J. Biomed. Mater. Res.* **39**(1): 161–170.

Martin, A., Bauer, T. W., Manley, M. T., and Marks, K. E. (1988). Osteosarcoma at the site of total hip replacement. A case report. *J. Bone Joint Surg. Am.* **70**: 1561–1567.

Middleton, J. C., and Tipton, A. J. (2000). Synthetic biodegradable polymers as orthopedic devices. *Biomaterials* **21**: 2335–2346.

Migliaresi, C., and Fambri, L. (1997). Processing and degradation of poly (L-lactic acid) fibres. *Macromol. Symp.* **123**: 155–161.

Migliaresi, C., Nicoli, F., Rossi, S., and Pegoretti, A. (2004). Novel uses of carbon composites for the fabrication of external fixators. *Comp. Sci. Tech.* **64**(6): 873–883.

Morita, S., Furuya, K., Ishihara, K., and Nakabayashi, N. (1998). Performance of adhesive bone cement containing hydroxyapatite particles. *Biomaterials* **19**(17): 1601–1606.

Morrison, C., Macnair, R., MacDonald, C., Wykman, A., Goldie, I., and Grant, M. H. (1995). In vitro biocompatibility testing of polymers for orthopaedic implants using cultured fibroblasts and osteoblasts. *Biomaterials* **16**(3): 987–992.

Moszner, N., and Salz, U. (2001). New developments of polymeric dental composites. *Prog. Polymer Sci.* **26**(4): 535–576.

Nicolais, L., and Narkis, M. (1971). Stress-strain behavior of SAN/glass bead composites in the glassy region. *Polym. Eng. Sci.* 194–199.

Nielsen, L. E. (1974). Mechanical properties of polymers and composites, Marcel Dekker, Inc., New York.

Pourdeyhimi, B., Robinson, H. H., Schwartz, P., and Wagner, H. D. (1986). Fracture toughness of Kevlar 29/ poly(methyl methacrylate) composite materials for surgical implantations. *Ann. Biomed. Eng.* 14(3): 277–294.

Qing, L., de Wijn, J. R., and van Blitterswijk, C. A. (1997). Nano-apatite/polymer composites: mechanical and physicochemical characteristics. *Biomaterials* **18**: 1263–1270.

Ramakrishna, S., Mayer, J., Wintermantel, E., and Leong, K. W. (2001). Biomedical applications of polymer composite materials: a review. *Composites Sci. Technol.* **61**: 1189–1224.

Reis, R. L., and Cunha, A. M. (2000). New degradable load-bearing biomaterials based on reinforced thermoplastic starch incorporatin blends. *J. Appl. Med. Polymer.* **4**: 1–5.

Reis, R. L., Cunha, A. M., and Bevis, M. J. (1998). Shear controlled orientation injection molding of polymeric composites with enhanced properties. *SPE Proceedings*, 57th Annual Technical Conference, Atlanta, USA, pp. 487–493.

Rokkanen, P. U., Bostman, O., Hirvensalo, E., Makela, E. A., Partio, E. K., Patiala, H., Vainionpaa, S., Vihtonen, K., and Tormala, P. (2000). Bioabsorbable fixation in orthopaedic surgery and traumatology. *Biomaterials* **21**: 2607–2613.

Rushton, N., and Rae, T. (1984). The intra-articular response to particulate carbon fiber reinforced high density polyethylene and its constituents: an experimental study in mice. *Biomaterials* **5**: 352–356.

Shikinami, Y., and Okuno, M. (2001). Bioresorbable devices made of forged composites of hydroxyapatite (HA) particles and poly L-lactide (PLLA). Part II: practical properties of miniscrews and miniplates. *Biomaterials* **22**: 3197–3211.

Shinzato, S., Kobayashi, M., Farid Mousa, W., Kamimura, M., Neo, M., Kitamura, Y., Kokubo, T., and Nakamura, T. (2000). Bioactive polymethyl methacrylate-based bone cement: Comparison of glass beads, apatite- and wollastonite-containing glass-ceramic, and hydroxyapatite fillers on mechanical and biological properties. *J. Biomed. Mater. Res.* **51**(2): 258–272.

St. John, K. R. (1983). Applications of advanced composites in orthopaedic implants. in *Biocompatible Polymers, Metals, and Composites*, M. Szycher, ed. Technomic, Lancaster, PA, p. 861.

Tang, R., Boyle, Jr., W. J., Mares, F., and Chiu, T.-H. (1990). Novel bioresorbable polymers and medical devices. *Trans. 16th Ann. Mtg. Soc. Biomater.* **13**: 191.

Thomson, R. C., Yaszemski, M. J., Powers, J. M., and Mikos, A. G. (1998). Hydroxyapatite fiber reinforced poly (α-hydroxy ester) foams for bone regeneration. *Biomaterials* **19**: 1935–1943.

Tormala, P. (1992). Biodegradable self-reinforced composite materials: manufacturing structure and mechanical properties. *Clin. Mater.* 10: 29–34.

Tormala, P., Rokkanen, P., Laiho, J., Tamminmaki, M., and Vainionpaa, S. (1988). Material for osteosynthesis devices. U.S. Patent No. 4,734,257.

Ural, E., Kesenci, K., Fambri, L., Migliaresi, C., and Piskin, E. (2000). Poly(D,L-lactide/ε-caprolactone)/hydroxyapatite composites. *Biomaterials* **21**: 2147–2154.

Vacanti, C. A., Langer, R, Schloo, B., and Vacanti, J. P. (1991). Synthetic polymers seeded with chondrocytes provide a template for new cartilage formation. *Plast. Reconstr. Surg.* **88**(5): 753–759.

Vallittu, P.K. (1996). A review of fiber-reinforced denture base resins. *J. Prosthodont.* **5**(4): 270–276.

Veerabagu, S., Fujihara, K., Dasari, G. R., and Ramakrishna, S. (2003). Strain distribution analysis of braided composite bone plates. *Composites Sci. Technol.* **61**: 427–435.

Vert, M., Christel, P., Garreau, H., Audion, M., Chanavax, M., and Chabot, F. (1986). Totally bioresorbable composites systems for internal fixation of bone fractures. in *Polymers in Medicine*, Vol. 2, C. Migliaresi and L. Nicolais, eds. Plenum, New York, pp. 263–275.

Wang, A., Lin, R., Polineni, V. K., Essner, A., Stark, C., and Duble-ton, J. H. (1998). Carbon fiber reinforced polyether ether ketone composite as a bearing surface for total hip replacement. *Tribology Int.* **31**: 661–667.

Wening, J. V., Katzer, A., Nicolas, V., Hahn, M., Jungbluth, K. H., and Kratzer, A. (1994). Imaging of alloplastic ligament implant. An in vivo and in vitro study exemplified by Kevlar. *Unfallchirurgie* **20**(2): 61–65.

Woo, S. L. Y., Akeson, W. H., Levenetz, B., Coutts, R. D., Matthews, J. V., and Amiel, D. (1974). Potential application of graphite fiber and methylmethacrylate resin composites as internal fixation plates. *J. Biomed. Mater. Res.* **8**: 321–328.

Xu, H. K., Eichmiller, F. C., and Giuseppetti, A. A. (2000). Reinforcement of a self-setting calcium phosphate cement with different fibers. *J. Biomed. Mater. Res.* **52**(1): 107–114.

Yang, J. M., Huang, P. Y., Yang, M. C., and Lo, S. K. (1997). Effect of MMA-g-UHMWPE grafted fiber on mechanical properties of acrylic bone cement. *J. Biomed. Mater. Res.* **38**(4): 361–369.

Yasko, A., Fellinger, E., Waller, S., Tomin, A., Peterson, M., Wang, E., and Lane, J. (1992). Comparison of biological and synthetic carriers for recombinant human BMP induced bone formation. *Trans. Orth. Res. Soc.* **17**: 71.

Yaszemski, M. J., Paune, R. G., Hayes, W. C., Langer, R., and Mikos, A. G. (1996). In vitro degradation of a poly(propylene fumavate)-based composite materials. *Biomaterials* **17**: 2127–2130.

Zimmerman, M. C., Alexander, H., Parsons, J. R., and Bajpai, P. K. (1991). The design and analysis of laminated degradable composite bone plates for fracture fixation. in *Hi-Tech Textiles*, T. Vigo, ed. ACS Publications, Washington, D.C.

Zimmerman, M. C., Parsons, J. R., and Alexander, H. (1987). The design and analysis of a laminated partially degradable composite bone plate for fracture fixation. *J. Biomed. Mater. Res. Appl. Biomater.* **21A**(3): 345.

3.2.13 Nonfouling surfaces

Allan S. Hoffman and Buddy D. Ratner

Introduction

"Nonfouling" surfaces (NFSs) refer to surfaces that resist the adsorption of proteins and/or adhesion of cells. They are also loosely referred to as protein-resistant surfaces and "stealth" surfaces. It is generally acknowledged that surfaces that strongly adsorb proteins will generally bind cells, and that surfaces that resist protein adsorption will also resist cell adhesion. It is also generally recognized that hydrophilic surfaces are more likely to resist protein adsorption, and that hydrophobic surfaces usually will adsorb a monolayer of tightly adsorbed protein. Exceptions to these generalizations exist, but, overall, they are accurate statements.

An important area for NFSs focuses on bacterial biofilms. Bacteria are believed to adhere to surfaces via a "conditioning film" of molecules (often proteins) that adsorbs first to the surface. The bacteria stick to this conditioning film and begin to exude a gelatinous slime layer (the biofilm) that aids in their protection from external agents (for example, antibiotics). Such layers are particularly troublesome in devices such as urinary catheters and endotracheal tubes. However, they also form on vascular grafts, hip joint prostheses, heart valves, and other long-term implants where they can stimulate significant inflammatory reaction to the infected device. If the conditioning film can be inhibited, bacterial adhesion and biofilm formation can also be reduced. NFSs offer this possibility.

NFSs have medical and biotechnology uses as blood-compatible materials (where they may resist fibrinogen adsorption and platelet attachment), implanted devices, urinary catheters, diagnostic assays, biosensors, affinity separations, microchannel flow devices, intravenous syringes and tubing, and nonmedical uses as biofouling-resistant heat exchangers and ship bottoms. It is important to note that many of these uses involve *in vivo* implants or extracorporeal devices, and many others involve *in vitro* diagnostic assays, sensors, and affinity separations. As well as having considerable medical and economic importance, NFSs offer important experimental and theoretical insights into one of the important phenomena in biomaterials science, protein adsorption. Hence, they have been the subject of many investigations. Aspects of NFSs are addressed in Section 3.1.5 and 3.2.14.

The majority of the literature on non-fouling surfaces focuses on surfaces containing the relatively simple polymer PEG:

$$(-CH_2CH_2O-)_n$$

When n is in the range of 15–3500 (molecular weights of approximately 400–100,000), the PEG designation is used. When molecular weights are greater than 100,000, the molecule is commonly referred to as PEO. Where n is in the range of 2–15, the term oligo(ethylene glycol) (oEG) is often used. An interesting article on the origins of the use of PEG to enhance the circulation time of proteins in the body has recently been published by Davis (2002). Other natural and synthetic polymers besides PEG show nonfouling behavior, and they will also be discussed in this section.

Background

The published literature on protein and cell interactions with biomaterial surfaces has grown significantly in the past 30 years, and the following concepts have emerged:

- It is well established that hydrophobic surfaces have a strong tendency to adsorb proteins irreversibly (Horbett and Brash, 1987, 1995; Hoffman, 1986). The driving force for this action is most likely the unfolding of the protein on the surface, accompanied by release of many hydrophobically structured water molecules from the interface, leading to a large entropy gain for the system (Hoffman, 1999). Note that adsorbed proteins can be displaced from the surface by solution phase proteins (Brash *et al.*, 1974).
- It is also well known that at low ionic strengths cationic proteins bind to anionic surfaces and anionic proteins bind to cationic surfaces (Hoffman, 1999; Horbett and Hoffman, 1975). The major thermodynamic driving force for these actions is a combination of ion-ion coulombic interactions, accompanied by an entropy gain due to the release of counterions along with their waters of hydration. However, these interactions are diminished at physiologic conditions by shielding of the protein ionic groups at the 0.15 N ionic strength (Horbett and Hoffman, 1975). Still, lysozyme, a highly charged cationic protein at physiologic pH, strongly binds to hydrogel contact lenses containing anionic monomers (see Bohnert *et al.*, 1988 for discussion of class IV contact lenses).
- It has been a common observation that proteins tend to adsorb in monolayers, i.e., proteins do not adsorb nonspecifically onto their own monolayers (Horbett, 1993). This is probably due to retention of hydration water by the adsorbed protein molecules, preventing close interactions of the protein molecules in solution with the adsorbed protein molecules. In fact, adsorbed protein films are, in themselves, reasonable NFSs with regard to other proteins (but not necessarily to cells).
- Many studies have been carried out on surfaces coated with physically or chemically immobilized PEG, and a conclusion was reached that the PEG molecular weight should be above a minimum of ca.2000 in order to provide good protein repulsion (Mori *et al.*, 1983; Gombotz *et al.*, 1991; Merrill, 1992). This seems to be the case whether PEG is chemically bound as a side chain of a polymer that is grafted to the surface (Mori *et al.*, 1983), is bound by one end to the surface (Gombotz *et al.*, 1991; Merrill, 1992), or is incorporated as segments in a cross-linked network (Merrill, 1992). The minimum MW was found to be ca. 500–2000, depending on packing density (Mori *et al.*, 1983; Gombotz *et al.*, 1991; Merrill, 1992).
- The mechanism of protein resistance by the PEG surfaces may due to be a combination of factors, including the resistance of the polymer coil to compression due to its desire to retain the volume of a random coil (called "entropic repulsion" or

"elastic network" resistance) plus the resistance of the PEG molecule to release both bound and free water from within the hydrated coil (called "osmotic repulsion") (Gombotz *et al.*, 1991; Antonsen and Hoffman, 1992). The size of the adsorbing protein and its resistance to unfolding may also be an important factor determining the extent of adsorption on any surface (Lim and Herron, 1992). The thermodynamic principles governing the adsorption of proteins onto surfaces involve a number of enthalpic and entropic terms favoring or resisting adsorption. These terms are summarized in Table 3.2.13-1. The major factors favoring adsorption will be the entropic gain of released water and the enthalpy loss due to cation-anion attractive interactions between ionic protein groups and surface groups. The major factors favoring resistance to protein adsorption will be the retention of bound water, plus, in the case of an immobilized hydrophilic polymer, entropic and osmotic repulsion of the polymer coils.

- In spite of the evidence for a PEG molecular weight effect, excellent protein resistance can be achieved with very short chain PEGs (OEGs) and PEG-like surfaces (Lopez *et al.*, 1992; *Sheu et al.*, 1993).
- Surface-assembled monolayers (SAMs) of lipid-oligoEG molecules have been studied, and it has been found that at least about 50% of the surface should be covered before significant resistance to protein adsorption is observed (Prime and Whitesides, 1993). This suggests that protein resistance by OEG-coated surfaces may be related to a "cooperativity" between the hydrated, short OEG chains in the "plane of the surface," wherein the OEG chains interact together to bind water to the surface, in a way that is similar to the hydrated coil and its osmotic repulsion, as described above. It has also been observed that a minimum of 3 EG units are needed for highly effective protein repulsion (Harder *et al.*, 1998). Based on all of these observations, one may describe the mechanism as being related to the conformation of the individual oligoEG chains, along with their packing density in the SAM. It has been proposed that helical or amorphous oligoEG conformations lead to stronger water-oligoEG interactions than an all-trans oligoEG conformation (Harder *et al.*, 1998).

Table 3.2.13-1 Thermodynamics of protein adsorption

		Favoring adsorption
ΔH_{ads}	(−)	VdW interactions (short-range)
	(−)	Ion–ion interactions (long-range)
ΔS_{ads}	(+)	Desorption of many H_2Os
	(+)	Unfolding of protein
		Opposing adsorption
ΔH_{ads}	(+)	Dehydration (interface between surface and protein)
	(+)	Unfolding of protein
	(+)	Chain compression (PEO)
ΔS_{ads}	(−)	Adsorption of protein
	(−)	Protein hydrophobic exposure
	(−)	Chain compression (PEO)
	(−)	Osmotic repulsion (PEO)

Packing density of the nonfouling groups on the surface is difficult to measure and often overlooked as an important factor in preparing NFSs. Nevertheless, one may conclude that the one common factor connecting all NFSs is their resistance to release of bound water molecules from the surface. Water may be bound to surface groups by both hydrophobic (structured water) and hydrophilic (primarily via hydrogen bonds) interactions, and in the latter case, the water may be H-bonded to neutral polar groups, such as hydroxyl (−OH) or ether (−C−O−C−) groups, or it may be polarized by ionic groups, such as $-COO^-$ or $-NH^+{}_3$. *The overall conclusion from all of the above observations is that resistance to protein adsorption at biomaterial interfaces is directly related to resistance of interfacial groups to the release of their bound waters of hydration.*

Based on these conclusions, it is obvious why the most common approaches to reducing protein and cell binding to biomaterial surfaces have been to make them more hydrophilic. This has been accomplished most often by chemical immobilization of a hydrophilic polymer (such as PEG) on the biomaterial surface by one of the following methods: (a) using UV or ionizing radiation to graft copolymerize a hydrophilic monomer onto surface groups; (b) depositing such a polymer from the vapor of a precursor monomer in a gas discharge process; or (c) directly immobilizing a preformed hydrophilic polymer on the surface using radiation or gas discharge processes. Other approaches to make surfaces more hydrophilic have included the physical adsorption of surfactants or chemical derivatization of surface groups with neutral polar groups such as hydroxyls, or with negatively charged groups (especially since most proteins and cells are negatively charged) such as carboxylic acids or their salts, or sulfonates. Gas discharge has been used to covalently bind nonfouling surfactants such as Pluronic polyols to polymer surfaces (Sheu *et al.*, 1993), and it has also been used to deposit an "oligoEG-like" coating from vapors of triglyme or tetraglyme (Lopez *et al.*, 1992). More recently, a hydrophilic polymer containing phosphorylcholine zwitterionic groups along its backbone has been extensively studied for its nonfouling properties (Iwasaki *et al.*, 1999). Coatings of many hydrogels including poly(2-hydroxyethyl methacrylate) and polyacrylamide show reasonable nonfouling behavior. There have also been a number of naturally occuring biomolecules such as albumin, casein, hyaluronic acid, and mucin that have been coated on surfaces and have exhibited resistance to nonspecific adsorption of proteins. Naturally occurring ganglioside lipid surfactants having saccharide head groups have been used to make "stealth" liposomes (Lasic and Needham, 1995). One paper even suggested that the protein resistance of PEGylated surfaces is related to the "partitioning" of albumin into the PEG layers, causing those surfaces to "look like native albumin" (Vert and Domurado, 2000).

Recently, SAMs presenting an interesting series of head-group molecules that can act as H-bond acceptors but not as H-bond donors have been shown to yield surfaces with unexpected protein resistance (Chapman *et al.*, 2000; Ostuni *et al.*, 2001; Kane *et al.*, 2003). Interestingly, PEG also fits in this category of H-bond acceptors but not donors. However, this generalization does not explain all NFSs, especially a report in which mannitol groups with H-bond donor –OH groups were found to be nonfouling (Luk *et al.*, 2000). Another hypothesis proposes that the functional groups that impart a nonfouling property are kosmotropes, order-inducing molecules (Kane *et al.*, 2003). Perhaps because of the ordered water surrounding these molecules, they cannot penetrate the ordered water shell surrounding proteins so strong intermolecular interactions between surface group and protein cannot occur. An interesting kosmotrope molecule with good nonfouling ability described in this paper is taurine, $H_3N^+(CH_2)_2SO_3^-$. Table 3.2.13-2 summarizes some of the different compositions that have been applied as NFSs.

It is worthwhile to mention some computational papers (supported by some experiments) that offer new insights and ideas on NFSs (Lim and Herron, 1992; Pertsin *et al.*, 2002; Pertsin and Grunze, 2000). Also, many new experimental methods have been applied to study the mechanism of NFSs including neutron reflectivity to measure the water density in the interfacial region (Schwendel *et al.*, 2003), scanning force microscopy (Feldman *et al.*, 1999), and sum frequency generation (Zolk *et al.*, 2000).

Finally, it should be noted that bacteria tend to adhere and colonize almost any type of surface, perhaps even many protein-resistant NFSs. However, the best NFSs can provide acute resistance to bacteria and biofilm build-up better than most surfaces (Johnston *et al.*, 1997). Resistance to bacterial adhesion remains an unsolved problem in surface science. Also, it has been pointed out that susceptibility of PEGs to oxidative damage may reduce their utility as NFSs in real-world situations (Kane *et al.*, 2003).

Table 3.2.13-2 NFS compositions

Synthetic hydrophilic surfaces
• PEG polymers and surfactants
• Neutral polymers
Poly(2-hydroxyethyl methacrylate) Polyacrylamide Poly(*N*-vinyl-2-pyrrolidone) Poly(*N*-isopropyl acrylamide) (below 31°C)
• Anionic polymers
• Phosphoryl choline polymers
• Gas discharge-deposited coatings (especially from PEG-like monomers)
• Self-assembled *n*-alkyl molecules with oligo-PEG head groups
• Self-assembled *n*-alkyl molecules with other polar head groups
Natural hydrophilic surfaces
• Passivating proteins (e.g., albumin and casein)
• Polysaccharides (e.g., hyaluronic acid)
• Liposaccharides
• Phospholipid bilayers
• Glycoproteins (e.g., mucin)

Conclusions and perspectives

It is remarkable how many different surface compositions appear to be nonfouling. Although it is difficult to be sure about the existence of a unifying mechanism for this action, it appears that the major factor favoring resistance to protein adsorption will be the retention of bound water by the surface molecules, plus, in the case of an immobilized hydrophilic polymer, entropic and osmotic repulsion by the polymer coils. Little is known about how long a NFS will remain nonfouling *in vivo*. Longevity and stability for nonfouling biomaterials remains an uncharted frontier. Defects (e.g., pits, uncoated areas) in NFSs may provide "footholds" for bacteria and cells to begin colonization. Enhanced understanding of how to optimize the surface density and composition of NFSs will lead to improvements in quality and fewer microdefects. Finally, it is important to note that a clean, "nonfouled" surface may not always be desirable. In the case of cardiovascular implants or devices, emboli may be shed when such a surface is exposed to flowing blood (Hoffman *et al.*, 1982). This can lead to undesirable consequences, even though (or perhaps especially because) the surface is an effective NFS. In the case of contact lenses, a protein-free lens may seem desirable, but there are concerns that such a lens will not be comfortable. Although biomaterials scientists can presently create surfaces that are nonfouling for a period of time, applying such surfaces must take into account the specific application, the biological environment, and the intended service life.

Bibliography

Antonsen, K. P., and Hoffman, A. S. (1992). Water structure of PEG solutions by DSC measurements. in *Polyethylene Glycol Chemistry: Biotechnical and Biomedical Applications*, J. M. Harris, ed. Plenum Press, New York, pp. 15–28.

Bohnert, J. L., Horbett, T. A., Ratner, B. D., and Royce, F. H. (1988). Adsorption of proteins from artificial tear solutions to contact lens materials. *Invest. Ophthalom. Vis. Sci.* **29**(3): 362–373.

Brash, J. L., Uniyal, S., and Samak, Q. (1974). Exchange of albumin adsorbed on polymer surfaces. *Trans. Am. Soc. Artif. Int. Organs* **20**: 69–76.

Chapman, R. G., Ostuni, E., Takayama, S., Holmlin, R. E., Yan, L., and Whitesides, G. M. (2000). Surveying for surfaces that resist the adsorption of proteins. *J. Am. Chem. Soc.* **122:** 8303–8304.

Davis, F. F. (2002). The origin of pegnology. *Adv. Drug. Del. Revs.* **54**: 457–458.

Feldman, K., Hahner, G., Spencer, N. D., Harder, P., and Grunze, M. (1999). Probing resistance to protein adsorption of oligo(ethylene glycol)-terminated self-assembled monolayers by scanning force microscopy. *J. Am. Chem. Soc.* **121**(43): 10134–10141.

Gombotz, W. R., Wang, G. H., Horbett, T. A., and Hoffman, A. S. (1991). Protein adsorption to PEO surfaces. *J. Biomed. Mater. Res.* **25**: 1547–1562.

Harder, P., Grunze, M., Dahint, R., Whitesides, G. M., and Laibinis, P. E. (1998). Molecular conformation and defect density in oligo(ethylene glycol)-terminated self-assembled monolayers on gold and silver surfaces determine their ability to resist protein adsorption. *J. Phys. Chem. B* **102**: 426–436.

Hoffman, A. S. (1986). A general classification scheme for hydrophilic and hydrophobic biomaterial surfaces. *J. Biomed. Mater. Res.* **20**: ix.

Hoffman, A. S. (1999). Non-fouling surface technologies. *J. Biomater. Sci., Polymer Ed.* **10:** 1011–1014.

Hoffman, A. S., Horbett, T. A., Ratner, B. D., Hanson, S. R., Harker, L. A., and Reynolds, L. O. (1982). Thrombotic events on grafted polyacrylamide-Silastic surfaces as studied in a baboon. ACS *Adv. Chem. Ser.* **199:** 59–80.

Horbett, T. A. (1993). Principles underlying the role of adsorbed plasma proteins in blood interactions with foreign materials. *Cardiovasc. Pathol.* **2**: 137S–148S.

Horbett, T. A., and Brash, J. L. (1987). Proteins at interfaces: current issues and future prospects. in *Proteins at Interfaces, Physicochemical and Biochemical Studies,* ACS Symposium Series, Vol. 343, T. A. Horbett and J. L. Brash, eds. American Chemical Society, Washington, D.C., pp. 1–33.

Horbett, T. A., and Brash, J. L. (1995). Proteins at interfaces: an overview. in *Proteins at Interfaces II: Fundamentals and Applications,* ACS Symposium Series, Vol. 602, T. A. Horbett and J. L. Brash, eds. American Chemical Society, Washington, D.C., pp. 1–25.

Horbett, T. A., and Hoffman, A. S. (1975). Bovine plasma protein adsorption to radiation grafted hydrogels based on hydroxyethyl-methacrylate and *N*-vinylpyrrolidone, Advances in Chemistry Series, Vol. 145, *Applied Chemistry at Protein Interfaces,* R. Baier, ed. American Chemical Society, Washington D.C., pp. 230–254.

Iwasaki, Y., *et al. (1999)*. Competitive adsorption between phospholipid and plasma protein on a phospholipid polymer surface. *J. Biomater. Sci. Polymer Ed.* **10**: 513–529.

Johnston, E. E., Ratner, B. D., and Bryers, J. D. (1997). RF plasma deposited PEO-like films: Surface characterization and inhibition of *Pseudomonas aeruginosa* accumulation. in *Plasma Processing of Polymers,* R. d'Agostino, P. Favia and F. Fracassi, eds. Kluwer Academic, Dordrecht, The Netherlands, pp. 465–476.

Kane, R. S., Deschatelets, P., and Whitesides, G. M. (2003). Kosmotropes form the basis of protein-resistant surfaces. *Langmuir* **19**: 2388–2391.

Lasic, D. D., and Needham, D. (1995). The "stealth" liposome: A prototypical biomaterial. *Chem. Rev.* **95**(8): 2601–2628.

Lim, K., and Herron, J. N. (1992). Molecular simulation of protein–PEG interaction. in *Polyethylene Glycol Chemistry: Biotechnical and Biomedical Applications* J. M. Harris, ed. Plenum Press, New York, p. 29.

Lopez, G. P., Ratner, B. D., Tidwell, C. D., Haycox, C. L., Rapoza, R. J., and Horbett, T. A. (1992). Glow discharge plasma deposition of tetraethylene glycol dimethyl ether for fouling-resistant biomaterial surfaces. *J. Biomed. Mater. Res.* **26**(4): 415–439.

Luk, Y., Kato, M., and Mrksich, M. (2000). Self-assembled monolayers of alkanethiolates presenting mannitol groups are inert to protein adsorption and cell attachment. *Langmuir* **16**: 9605.

Merrill, E. W. (1992). Poly(ethylene oxide) and blood contact: a chronicle of one laboratory. in *Polyethylene Glycol Chemistry: Biotechnical and Biomedical Applications,* J. M. Harris, ed. Plenum Press, New York, pp. 199–220.

Mori, Y., *et al.* (1983). Interactions between hydrogels containing PEO chains and platelets. *Biomaterials* **4**: 825–830.

Ostuni, E., Chapman, R. G., Holmlin, R. E., Takayama, S., and Whitesides, G. M. (2001). A survey of structure–property relationships of surfaces that resist the adsorption of protein. *Langmuir* **17**: 5605–5620.

Pertsin, A. J., and Grunze, M. (2000). Computer simulation of water near the surface of oligo(ethylene glycol)-terminated alkanethiol self-assembled monolayers. *Langmuir* **16**(23): 8829–8841.

Pertsin, A. J., Hayashi, T., and Grunze, M. (2002). Grand canonical monte carlo simultations of the hydration interaction between oligo(ethylene glycol)-terminated alkanethiol self-assembled monolayers. *J. Phys. Chem. B.* **106**(47): 12274–12281.

Prime, K. L., and Whitesides, G. M. (1993). Adsorption of proteins onto surfaces containing end-attached oligo(ethylene oxide): a model system using self-assembled monolayers. *J. Am. Chem. Soc.* **115**: 10715.

Schwendel, D., Hayashi, T., Dahint, R., Pertsin, A., Grunze, M., Steitz, R., and Schreiber, F. (2003). Interaction of water with self-assembled monolayers: neutron reflectivity measurements of the water density in the interface region. *Langmuir* **19**(6): 2284–2293.

Sheu, M.-S., Hoffman, A. S., Terlingen, J. G. A., and Feijen, J. (1993). A new gas discharge process for preparation of non-fouling surfaces on biomaterials. *Clin. Mater.* **13**: 41–45.

Vert, M., and Domurado, D. (2000). PEG: Protein-repulsive or albumin-compatible? *J. Biomater. Sci., Polymer Ed.* **11**: 1307–1317.

Zolk, M., Eisert, F., Pipper, J., Herrwerth, S., Eck, W., Buck, M., and Grunze, M. (2000). Solvation of oligo(ethylene glycol)-terminated self-assembled monolayers studied by vibrational sum frequency spectroscopy. *Langmuir* **16**(14): 5849–5852.

3.2.14 Physicochemical surface modification of materials used in medicine

Buddy D. Ratner and Allan S. Hoffman

Introduction

Much effort goes into the design, synthesis, and fabrication of biomaterials and devices to ensure that they have the appropriate mechanical properties, durability, and functionality. To cite a few examples, a hip joint should withstand high stresses, a hemodialyzer should have the requisite permeability characteristics, and the pumping bladder in an artificial heart should flex for millions of cycles without failure. The bulk structure of the materials governs these properties.

The biological response to biomaterials and devices, on the other hand, is controlled largely by their surface chemistry and structure (see Section 3.1.4). The rationale for the surface modification of biomaterials is therefore straightforward: to retain the key physical properties of a biomaterial while modifying only the outermost surface to influence the biointeraction. If such surface modification is properly effected, the mechanical properties and functionality of the device will be unaffected, but the bioresponse related to the tissue–device interface will be improved or modulated.

Materials can be surface-modified by using biological, mechanical, or physicochemical methods. Many biological surface modification schemes are covered in Section 3.2.16. Generalized examples of physicochemical surface modifications, the focus of this section, are illustrated schematically in Fig. 3.2.14-1. Surface modification with Langmuir–Blodgett (LB) films has elements of both biological modification and physicochemical modification. LB films will be discussed later in this section. Some applications for surface modified biomaterials are listed in Table 3.2.14-1. Physical and chemical surface modification methods, and the types of materials to which they can be applied, are listed in Table 3.2.14-2. Methods to modify or create surface texture or roughness will not be explicitly covered here, though chemical patterning of surfaces will be addressed.

General principles

Surface modifications fall into two categories: (1) chemically or physically altering the atoms, compounds, or molecules in the existing surface (chemical modification, etching, mechanically roughening), or (2) overcoating the existing surface with a material having a different composition (coating, grafting, thin film deposition) (Fig. 3.2.14-1). A few general principles provide guidance when undertaking surface modification:

Thin surface modifications

Thin surface modifications are desirable. The modified zone at the surface of the material should be as thin as possible. Modified surface layers that are too thick can change the mechanical and functional properties of the material. Thick coatings are also more subject to delamination and cracking. How thin should a surface modification be? Ideally, alteration of only the outermost molecular layer (3–10 Å) should be sufficient. In practice, thicker films than this will be necessary since it is difficult to ensure that the original surface is uniformly covered when coatings and treatments are so thin. Also, extremely thin layers may be more subject to surface reversal (see later discussion) and mechanical erosion. Some coatings intrinsically have a specific thickness. For example, the thickness of LB films is related to the length of the amphiphilic molecules that form them (25–50 Å). Other coatings, such as PEG protein-resistant layers, may require a minimum thickness (a dimension related to the molecular weight of chains) to function. In general, surface modifications should be the minimum thickness needed for uniformity, durability, and functionality, but no thicker. This is often experimentally defined for each system.

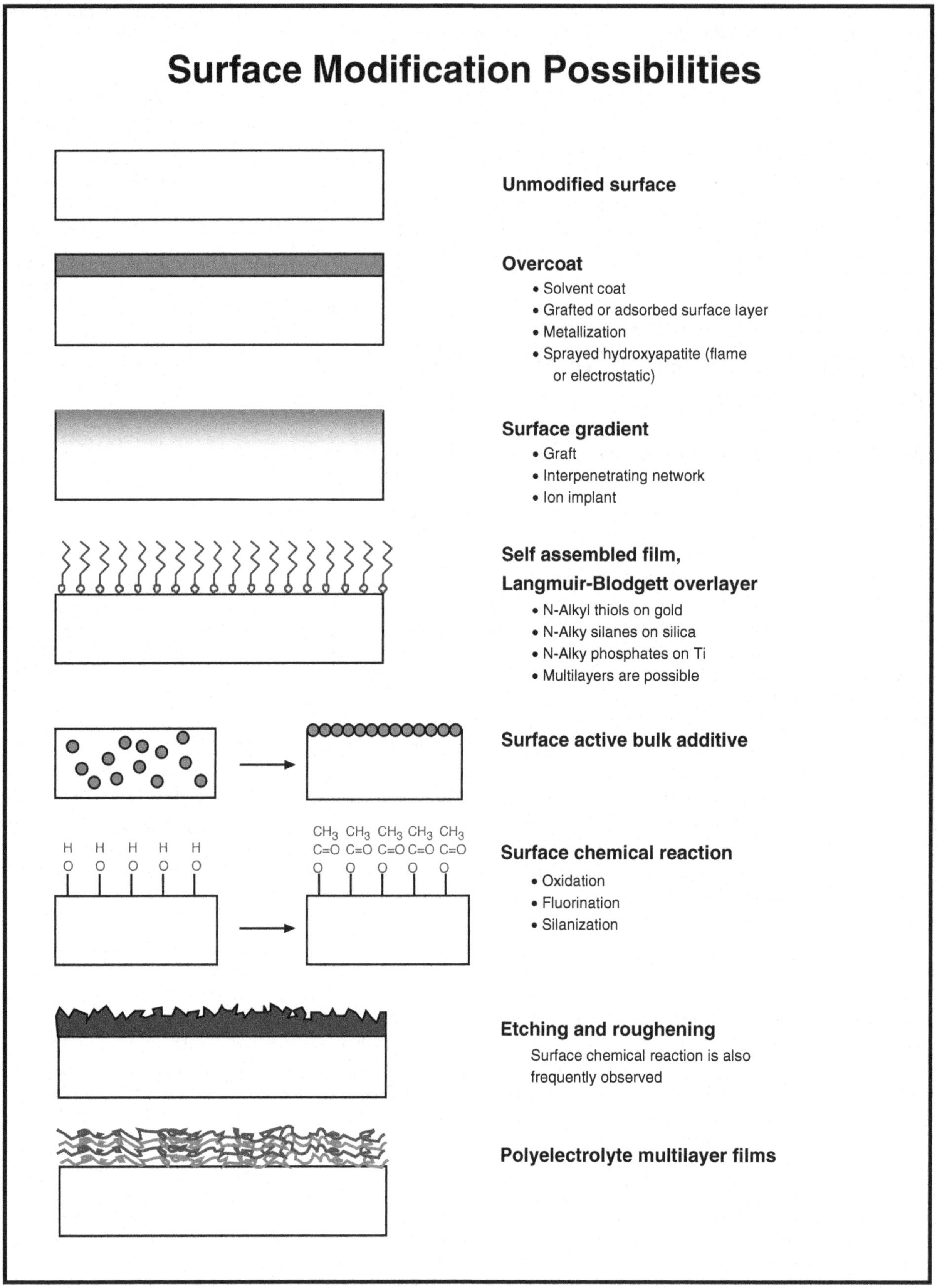

Fig. 3.2.14-1 Schematic representations of methods to modify surfaces.

Table 3.2.14-1 Some physicochemically surface-modified biomaterials

To modify blood compatibility
Octadecyl group attachment to surfaces (albumin affinity)
Silicone-containing block copolymer additive
Plasma fluoropolymer deposition
Plasma siloxane polymer deposition
Radiation grafted hydrogel
Chemically modified polystyrene for heparin-like activity
To influence cell adhesion and growth
Oxidized polystyrene surface
Ammonia plasma-treated surface
Plasma-deposited acetone or methanol film
Plasma fluoropolymer deposition (reduce endothelial adhesion to IOLs)
To control protein adsorption
Surface with immobilized poly(ethylene glycol) (reduce adsorption)
Treated ELISA dish surface (increase adsorption)
Affinity chromatography column
Surface cross-linked contact lens (reduce adsorption)
To improve lubricity
Plasma treatment
Radiation grafting (hydrogels)
Interpenetrating polymeric networks
To improve wear resistance and corrosion resistance
Ion implantation
Diamond deposition
Anodization
To alter transport properties
Polyelectrolyte grafting
To modify electrical characteristics
Polyelectrolyte grafting
Magnetron sputtering of titanium

Delamination resistance

The surface-modified layer should be resistant to delamination and cracking. Resistance to delamination is achieved by covalently bonding the modified region to the substrate, intermixing the components of the substrate and the surface film at an interfacial zone (for example, an interpenetrating network (IPN)), applying a compatibilizing ("primer") layer at the interface, or incorporating appropriate functional groups for strong intermolecular adhesion between a substrate and an overlayer (Wu, 1982).

Surface rearrangement

Surface rearrangement can readily occur. It is driven by a thermodynamic minimization of interfacial energy and enhanced by molecular mobility. Surface chemistries and structures can "switch" because of diffusion or translation of surface atoms or molecules in response to the external environment (see Section 3.1.4 and Fig. 3.1.4-2 in that section). A newly formed surface chemistry can migrate from the surface into the bulk, or molecules from the bulk can diffuse to cover the surface. Such reversals occur in metallic and other inorganic systems, as well as in polymeric systems. Terms such as "reconstruction," "relaxation," and "surface segregation" are often used to describe mobility-related alterations in surface structure and chemistry (Ratner and Yoon, 1988; Garbassi *et al.*, 1989; Somorjai, 1990, 1991). The driving force for these surface changes is a minimization of the interfacial energy. However, sufficient atomic or molecular mobility must exist for the surface changes to occur in reasonable periods of time. For a modified surface to remain as it was designed, surface reversal must be prevented or inhibited. This can be done by cross-linking, sterically blocking the ability of surface structures to move, or by incorporating a rigid, impermeable layer between the substrate material and the surface modification.

Surface analysis

Surface modification and surface analysis are complementary and sequential technologies. The surface-modified region is usually thin and consists of only minute amounts of material. Undesirable contamination can readily be introduced during modification reactions. The potential for surface reversal to occur during surface modification is also high. The surface reaction should be monitored to ensure that the intended surface is indeed being formed. Since conventional analytical methods are often insufficiently sensitive to detect surface modifications, special surface analytical tools are called for (Section 3.1.4).

Commercializability

The end products of biomaterials research are devices and materials that are manufactured to exacting

Table 3.2.14-2 Physical and chemical surface modification methods

	Polymer	Metal	Ceramic	Glass
Noncovalent coatings				
Solvent coating	✓	✓	✓	✓
Langmuir–Blodgett film deposition	✓	✓	✓	✓
Surface-active additives	✓	✓	✓	✓
Vapor deposition of carbons and metals[a]	✓	✓	✓	✓
Vapor deposition of parylene (*p*-xylylene)	✓	✓	✓	✓
Covalently attached coatings				
Radiation grafting (electron accelerator and gamma)	✓	—	—	—
Photografting (UV and visible sources)	✓	—	—	✓
Plasma (gas discharge) (RF, microwave, acoustic)	✓	✓	✓	✓
Gas-phase deposition				
• Ion beam sputtering	✓	✓	✓	✓
• Chemical vapor deposition (CVD)	—	✓	✓	✓
• Flame spray deposition	—	✓	✓	✓
Chemical grafting (e.g., ozonation + grafting)	✓	✓	✓	✓
Silanization	✓	✓	✓	✓
Biological modification (biomolecule immobilization)	✓	✓	✓	✓
Modifications of the original surface				
Ion beam etching (e.g., argon, xenon)	✓	✓	✓	✓
Ion beam implantation (e.g., nitrogen)	—	✓	✓	✓
Plasma etching (e.g., nitrogen, argon, oxygen, water vapor)	✓	✓	✓	✓
Corona discharge (in air)	✓	✓	✓	✓
Ion exchange	✓[b]	✓	✓	✓
UV irradiation	✓	✓	✓	✓
Chemical reaction				
• Nonspecific oxidation (e.g., ozone)	✓	✓	✓	✓
• Functional group modifications (oxidation, reduction)	✓	—	—	—
• Addition reactions (e.g., acetylation, chlorination)	✓	—	—	—
Conversion coatings (phosphating, anodization)	—	✓	—	—
Mechanical roughening and polishing	✓	✓	✓	✓

[a] Some covalent reaction may occur.
[b] For polymers with ionic groups.

specifications for use in humans. A surface modification that is too complex will be difficult and expensive to commercialize. It is best to minimize the number of steps in a surface modification process and to design each step to be relatively insensitive to small changes in reaction conditions.

Methods for modifying the surfaces of materials

General methods to modify the surfaces of materials are illustrated in Fig. 3.2.14-1, with many examples listed in Table 3.2.14-2. A few of the more widely used of these methods will be briefly described. Some of the conceptually simpler methods such as solution coating of a polymer onto a substrate or metallization by sputtering or thermal evaporation will not be elaborated upon here.

Chemical reaction

There are hundreds of chemical reactions that can be used to modify the chemistry of a surface. Chemical reactions, in the context of this article, are those performed with reagents that react with atoms or molecules at the surface, but do not overcoat those atoms or molecules with a new layer. Chemical reactions can be classified as nonspecific and specific.

Nonspecific reactions leave a distribution of different functional groups at the surface. An example of a nonspecific surface chemical modification is the chromic acid oxidation of PE surfaces. Other examples include the corona discharge modification of materials in air; radiofrequency glow discharge (RFGD) treatment of materials in oxygen, argon, nitrogen, carbon dioxide, or water vapor plasmas; and the oxidation of metal surfaces to a mixture of suboxides.

Specific chemical surface reactions change only one functional group into another with a high yield and few side reactions. Examples of specific chemical surface modifications for polymers are presented in Fig. 3.2.14-2. Detailed chemistries of biomolecule immobilization are described in Section 3.2.16.

Radiation grafting and photografting

Radiation grafting and related methods have been widely applied for the surface modification of biomaterials starting in the late 1960s (Hoffman *et al.*, 1972), and comprehensive review articles are available (Ratner, 1980; Hoffman, 1981; Hoffman *et al.*, 1983; Stannett, 1990; Safrany, 1997). The earliest applications, particularly for biomedical applications, focused on attaching chemically reactable groups (–OH, –COOH, $-NH_2$, etc) to the surfaces of relatively inert hydrophobic polymers. Within this category, three types of reactions can be distinguished: grafting using ionizing radiation sources (most commonly, a cobalt-60 or cesium-137 gamma radiation source) (Dargaville *et al.*, 2003), grafting using UV radiation (photografting) (Srinivasan and Lazare, 1985; Matsuda and Inoue, 1990; Dunkirk *et al.*, 1991; Swanson, 1996), and grafting using high-energy electron beams (Singh and Silverman, 1992). In all cases, similar processes occur. The radiation breaks chemical bonds in the material to be grafted, forming free radicals, peroxides, or other reactive species. These reactive surface groups are then exposed to a monomer. The monomer reacts with the free radicals at the surface and propagates as a free radical chain reaction incorporating other monomers into a surface grafted polymer. Electron beams and gamma radiation sources are also used for biomedical device sterilization.

These high-energy surface modification technologies are strongly dependent on the source energy, the radiation dose rate, and the amount of the dose absorbed. Gamma sources have energies of roughly 1 MeV (1 eV = 23.06 kcal/mol). Typical energies for electron beam processing are 5–10 MeV. UV radiation sources are of much lower energy (<6 eV). Radiation dose rates are low for UV and gamma and very high for electron beams. The amount of energy absorbed is measured in units of grays (Gy) where 1 kilogray (kGy) = 1000 joules/kilogram. Units of megarads (MR) are often used for gamma sources; $1\ MR = 1\times 10^6$ ergs/gram = 10 kGy.

Three distinct reaction modes can be described: (a) In the mutual irradiation method, the substrate material is immersed in an oxygen-free solution (monomer ± solvent) that is then exposed to the radiation source. (b) The substrate materials can also be exposed to the radiation under an inert atmosphere or at low temperatures (to stabilize free radicals). In this case, the materials are later contacted with a monomer solution to initiate the graft process. (c) Finally, the exposure to the radiation can take place in air or oxygen, leading to the formation of peroxide groups on the surface. Heating the material to be grafted in the presence of monomer, or addition of a redox reactant (e.g., Fe^{2+}) that will decompose the peroxide groups to form free radicals, can initiate the graft polymerization (in O_2-free conditions).

Graft layers formed by energetic irradiation of the substrate are often thick (>1 μm) and composed of relatively high-molecular-weight polymer chains. However, they are typically well-bonded to the substrate material. Since many polymerizable monomers are available, a wide range of surface chemistries can be created. Mixtures of monomers can form unique graft copolymers (Ratner and Hoffman, 1980). For example, the hydrophilic/hydrophobic ratio of surfaces can be controlled by varying the ratio of a hydrophilic and

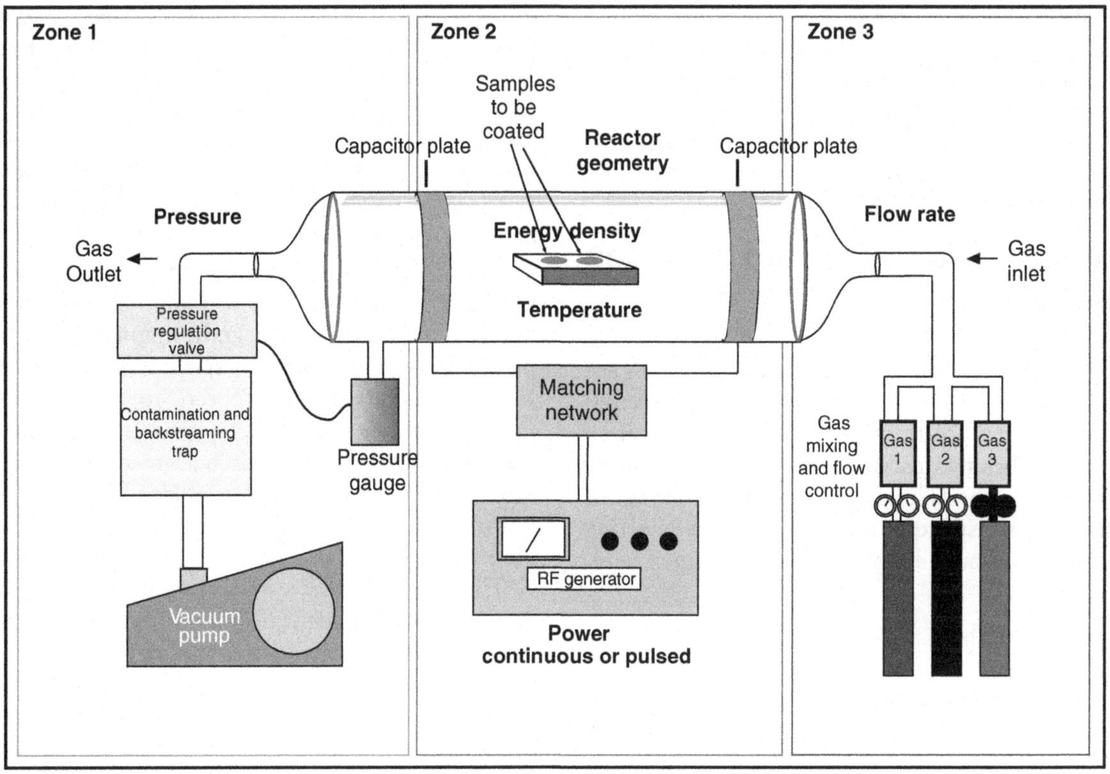

Fig. 3.2.14-2 A diagram of a capacitively coupled RF plasma reactor. Important experimental variables are indicated in bold typeface. Zone 1 shows gas storage and mixing. Zone 2 shows components that power the reactor. Zone 3 highlights components of the vacuum system.

a hydrophobic monomer in the grafting mixture (Ratner and Hoffman, 1980; Ratner *et al.*, 1979).

Photoinitiated grafting (usually with visible or UV light) represents a unique subcategory of surface modifications in which there is growing interest. There are many approaches to effect this photoinitiated covalent coupling. For example, a phenyl azide group can be converted to a highly reactive nitrene upon UV exposure. This nitrene will quickly react with many organic groups. If a synthetic polymer is prepared with phenyl azide side groups and this polymer is exposed simultaneously to UV light and a substrate polymer or polymeric medical device, the polymer containing the phenyl azide side groups will be immobilized to the substrate (Matsuda and Inoue, 1990). Another method involves the coupling of a benzophenone molecule to a hydrophilic polymer (Dunkirk *et al.*, 1991). In the presence of UV irradiation, the benzophenone is excited to a reactive triplet state that can abstract a hydrogen leading to radical cross-linking.

Radiation, electron beam, and photografting have frequently been used to bond hydrogels to the surfaces of hydrophobic polymers (Matsuda and Inoue, 1990; Dunkirk *et al.*, 1991). Electron beam grafting of *N*-isopropyl acrylamide to polystyrene has been used to create a new class of temperature-dependent surfaces for cell growth (Kwon *et al.*, 2000) (also see Section 3.2.6). The protein interactions (Horbett and Hoffman, 1975), cell interactions (Ratner *et al.*, 1975; Matsuda and Inoue, 1990), blood compatibility (Chapiro, 1983; Hoffman *et al.*, 1983), and tissue reactions (Greer *et al.*, 1979) of hydrogel graft surfaces have been investigated.

RFGD plasma deposition and other plasma gas processes

RFGD plasmas, as used for surface modification, are low-pressure ionized gas environments typically at ambient (or slightly above ambient) temperature. They are also referred to as glow discharge or gas discharge depositions or treatments. Plasmas can be used to modify existing surfaces by ablation or etching reactions or, in a deposition mode, to overcoat surfaces (Fig. 3.2.14-1). Good review articles on plasma deposition and its application to biomaterials are available (Yasuda and Gazicki, 1982; Hoffman, 1988; Ratner *et al.*, 1990; Chu *et al.*, 2002; Kitching *et al.*, 2003). Some biomedical applications of plasma-modified biomaterials are listed in Table 3.2.14-3.

Table 3.2.14-3 Biomedical applications of glow discharge plasma-induced surface modification processes

A. Plasma treatment (etching)
1. Clean
2. Sterilize
3. Cross-link surface molecules
B. Plasma treatment (etching) and plasma deposition
1. Form barrier films
a. Protective coating
b. Electrically insulating coating
c. Reduce absorption of material from the environment
d. Inhibit release of leachables
e. Control drug delivery rate
2. Modify cell and protein reactions
a. Improve biocompatibility
b. Promote selective protein adsorption
c. Enhance cell adhesion
d. Improve cell growth
e. Form nonfouling surfaces
f. Increase lubricity
3. Provide reactive sites
a. For grafting or polymerizing polymers
b. For immobilizing biomolecules

The application of RFGD plasma surface modification in biomaterials development is steadily increasing. Because such coatings and treatments have special promise for improved biomaterials, they will be emphasized in this section. The specific advantages of plasma-deposited films (and to some extent, plasma-treated surfaces) for biomedical applications are:

1. They are conformal. Because of the penetrating nature of a low-pressure gaseous environment in which mass transport is governed by both molecular (line-of-sight) diffusion and convective diffusion, complex geometric shapes can be treated.
2. They are free of voids and pinholes. This continuous barrier structure is suggested by transport studies and electrical property studies (Charlson *et al.*, 1984).
3. Plasma-deposited polymeric films can be placed upon almost any solid substrate, including metals, ceramics, and semiconductors. Other surface-grafting or surface-modification technologies are highly dependent upon the chemical nature of the substrate.
4. They exhibit good adhesion to the substrate. The energetic nature of the gas-phase species in the plasma reaction environment can induce mixing, implantation, penetration, and reaction between the overlayer film and the substrate.
5. Unique film chemistries can be produced. The chemical structure of the polymeric overlayer films generated from the plasma environment usually cannot be synthesized by conventional chemical methods.
6. They can serve as excellent barrier films because of their pinhole-free and dense, cross-linked nature.
7. Plasma-deposited layers generally show low levels of leachables. Because they are highly cross-linked, plasma-deposited films contain negligible amounts of low-molecular-weight components that might lead to an adverse biological reaction. They can also prevent leaching of low-molecular-weight material from the substrate.
8. These films are easily prepared. Once the apparatus is set up and optimized for a specific deposition, treatment of additional substrates is rapid and simple.
9. The production of plasma depositions is a mature technology. The microelectronics industry has made extensive use of inorganic plasma-deposited films for many years (Sawin and Reif, 1983; Nguyen, 1986).
10. Plasma surface modifications, although they are chemically complex, can be characterized by IR (Inagaki *et al.*, 1983; Haque and Ratner, 1988; Krishnamurthy *et al.*, 1989), NMR (Kaplan and Dilks, 1981), electron spectroscopy for chemical analysis (ESCA) (Chilkoti *et al.*, 1991a), chemical derivatization studies (Everhart and Reilley, 1981; Gombotz and Hoffman, 1988; Griesser and Chatelier, 1990; Chilkoti *et al.*, 1991a), and static secondary ion mass spectrometry (SIMS) (Chilkoti *et al.*, 1991b, 1992; Johnston and Ratner, 1996).
11. Plasma-treated surfaces are sterile when removed from the reactor, offering an additional advantage for cost-efficient production of medical devices.

It would be inappropriate to cite all these advantages without also discussing some of the disadvantages of plasma deposition and treatment for surface modification. First, the chemistry produced on a surface is often ill-defined (Fig. 3.2.14-3). For example, if tetrafluoroethylene gas is introduced into the reactor, PTFE will not be deposited on the surface. Rather, a complex, branched fluorocarbon polymer will be produced. This scrambling of monomer structure has been addressed in studies dealing with retention of monomer structure in the final film (Lopez and Ratner, 1991; Lopez *et al.*, 1993; Panchalingam *et al.*, 1993). Second, the apparatus used to produce plasma depositions can be expensive. A good laboratory-scale reactor will cost $10,000–30,000, and a production reactor can cost $100,000 or more. Third, uniform reaction within long, narrow pores can be difficult to achieve. Finally, contamination can be a problem and care must be exercised to prevent extraneous gases and pump oils from entering the reaction zone. However, the advantages of plasma reactions outweigh these potential disadvantages for many types of modifications that cannot be accomplished by other methods.

The nature of the plasma environment

Plasmas are atomically and molecularly dissociated gaseous environments. A plasma environment contains positive ions, negative ions, free radicals, electrons, atoms, molecules, and photons (visible and UV). Typical conditions within the plasma include an electron energy of 1–10 eV, a gas temperature of 25–60°C, an electron density of 10^{-9}–10^{-12}/cm^3, and an operating pressure of 0.025–1.0 torr.

A number of processes can occur on the substrate surface that lead to the observed surface modification or deposition. First, a competition takes place between deposition and etching by the high-energy gaseous species (ablation) (Yasuda, 1979). When ablation is more rapid than deposition, no deposition will be observed. Because of its energetic nature, the ablation or etching process can result in substantial chemical and morphological changes to the substrate.

A number of mechanisms have been postulated for the deposition process. The reactive gaseous environment and UV emission may create free radical and other reactive species on the substrate surface that react with and polymerize molecules from the gas phase. Alternately, reactive small molecules in the gas phase could combine to form higher-molecular-weight units or particulates that may settle or precipitate onto the surface. Most likely, the depositions observed are formed by some combination of these two processes.

Production of plasma environments for deposition

Many experimental variables relating both to reaction conditions and to the substrate onto which the deposition is placed affect the final outcome of the plasma deposition process (Fig. 3.2.14-2). A diagram of a typical inductively coupled radio frequency plasma reactor is presented in Fig. 3.2.14-2. The major subsystems that make up this apparatus are a gas introduction system (control of gas mixing, flow rate, and mass of gas entering the reactor), a vacuum system (measurement and control of reactor pressure and inhibition of backstreaming of molecules from the pumps), an energizing system to efficiently couple energy into the gas phase within the reactor, and a reactor zone in which the samples are treated. Radio-frequency, acoustic, or microwave energy can be coupled to the gas phase. Devices for monitoring the molecular weight of the gas-phase species (mass spectrometers), the optical emission from the glowing plasma (spectrometers), and the deposited film thickness (ellipsometers, vibrating quartz crystal microbalances) are also commonly found on plasma reactors. Technology has been developed permitting atmospheric-pressure plasma deposition (Massines *et al.*, 2000; Klages *et al.*, 2000). Another important development is "reel-to-reel" (continuous) plasma processing, opening the way to low-cost treatment of films, fibers, and tubes.

RFGD plasmas for the immobilization of molecules

Plasmas have often been used to introduce organic functional groups (e.g., amine, hydroxyl) on a surface that can be activated to attach biomolecules (see Section 3.2.16). Certain reactive gas environments can also be used for directly immobilizing organic molecules such as surfactants. For example, a PEG-*n*-alkyl surfactant will adsorb to PE via the propylene glycol block. If the PE surface with the adsorbed surfactant is briefly exposed to an argon plasma, the *n*-alkyl chain will be cross-linked, thereby leading to the covalent attachment of pendant PEG chains (Sheu *et al.*, 1992).

High-temperature and high-energy plasma treatments

The plasma environments described above are of relatively low energy and low temperature. Consequently, they can be used to deposit organic layers on polymeric or inorganic substrates. Under higher energy conditions, plasmas can effect unique and important inorganic surface modifications on inorganic substrates. For example, flame-spray deposition involves injecting a high-purity, relatively finely divided (~100 mesh) metal powder into a high-velocity plasma or flame. The melted or partially melted particles impact the surface and rapidly solidify

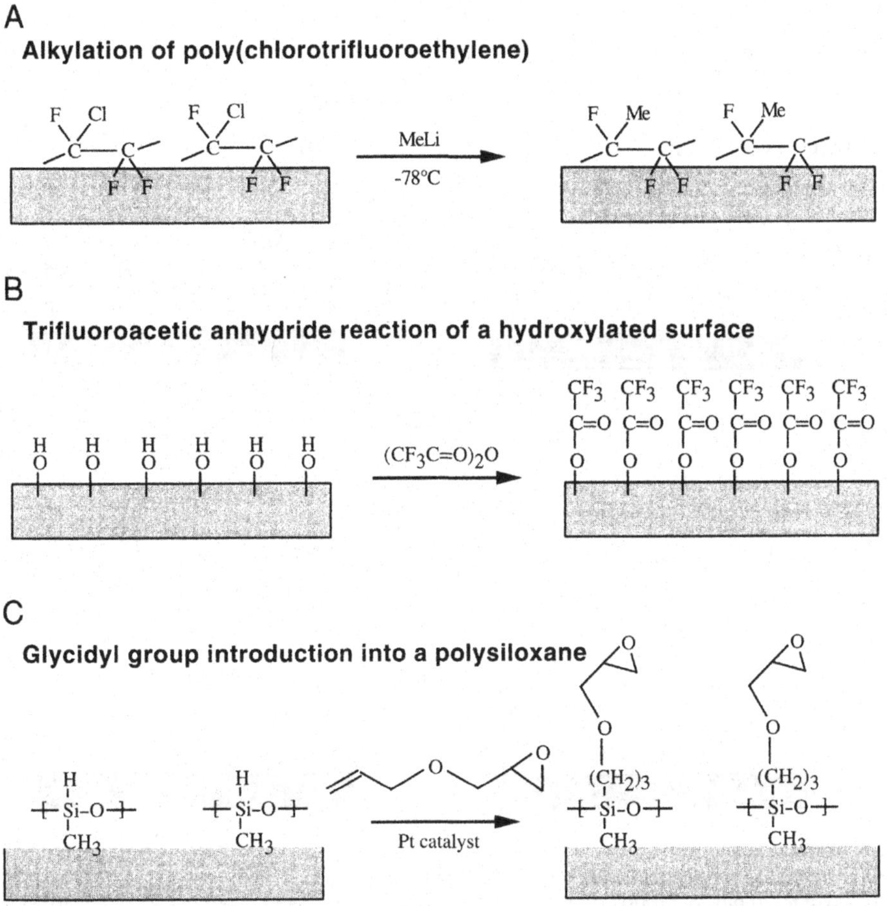

Fig. 3.2.14-3 Some specific chemical reactions to modify surfaces.

(see Section 3.2.9). An example of thermal spray coating on titanium is seen in Gruner (2001).

Silanization

Silane treatments of surfaces involve a liquid-phase chemical reaction and are straightforward to perform and low cost. A typical silane surface modification reaction is illustrated in Fig. 3.2.14-4. Silane reactions are most often used to modify hydroxylated surfaces. Since glass, silicon, germanium, alumina, and quartz surfaces, as well as many metal oxide surfaces, are rich in hydroxyl groups, silanes are particularly useful for modifying these materials. Numerous silane compounds are commercially available, permitting a broad range of chemical functionalities to be incorporated on surfaces (Table 3.2.14-4). The advantages of silane reactions are their simplicity and stability, attributed to their covalent, cross-linked structure. However, the linkage between a silane and an hydroxyl group is also readily subject to basic hydrolysis, and film breakdown under some conditions must be considered (Wasserman *et al.*, 1989).

Silanes can form two types of surface film structures. If only surface reaction occurs (perhaps catalyzed by traces of adsorbed surface water), a structure similar to that shown in Fig. 3.2.14-4 can be formed. However, if more water is present, a thicker silane layer can be formed consisting of both Si–O groups bonded to the surface and silane units participating in a "bulk", three-dimensional, polymerized network. The initial stages in the formation of a thicker silane film are suggested by the further reaction of the group at the right side of Fig. 3.2.14-4D with solution-phase silane molecules. Without careful control of silane liquid purity, water concentration, and reaction conditions, thicker silane films can be rough and inhomogeneous.

A new class of silane-modified surfaces based upon monolayer silane films and yielding self-assembled, highly ordered structures is of particular interest in precision engineering of surfaces (Pomerantz *et al.*, 1985; Maoz *et al.*, 1988; Heid *et al.*, 1996). These self-assembled monolayers (SAMs) are described in more detail later in this section.

Many general reviews and basic science studies on surface silanization are available (Arkles, 1977; Plueddemann, 1980; Rye *et al.*, 1997). Applications for silanized surface-modified biomaterials are on the increase and include cell attachment (Matsuzawa *et al.*, 1997; Hickman and

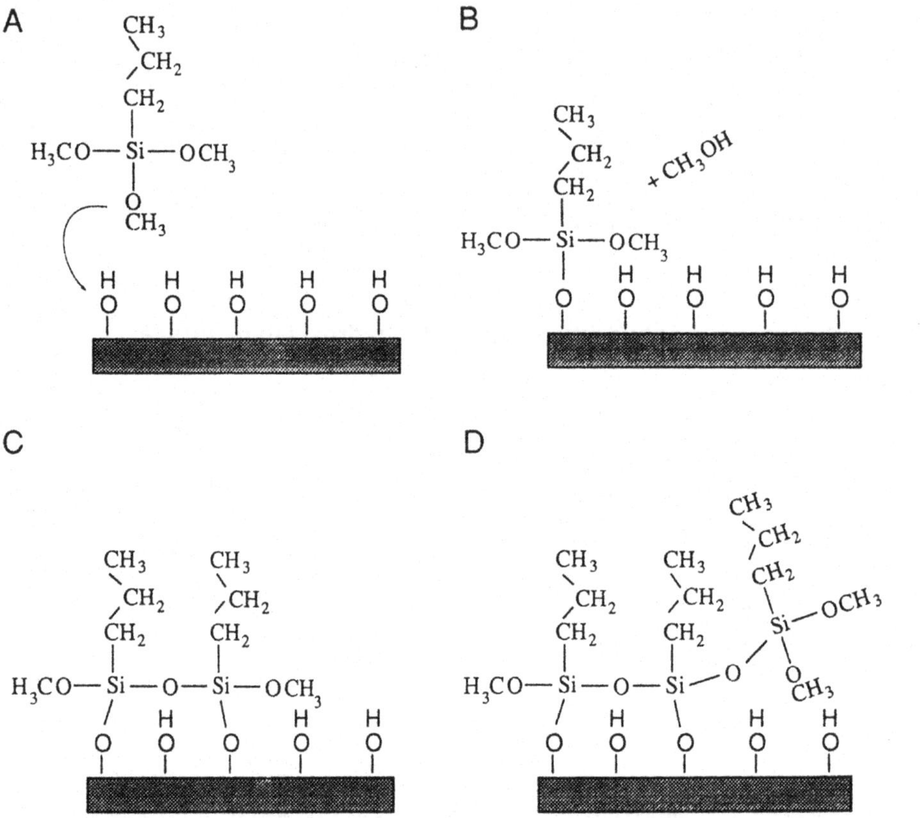

Fig. 3.2.14-4 The chemistry of a typical silane surface modification reaction. (A) A hydroxylated surface is immersed in a non-aqueous solution containing n-propyl trimethoxysilane (nPTMS). (B) One of the methoxy groups of the nPTMS couples with a hydroxyl group releasing methanol. (C) Two of the methoxy groups on another molecule of the nPTMS have reacted, one with a hydroxyl group and the other with a methoxy group from the first nPTMS molecule. (D) A third nPTMS molecule has reacted only with a methoxy group. This molecule is tied into the silane film network, but is not directly bound to the surface.

Stenger, 1994), biomolecule and polymer immobilization (Xiao et al., 1997; Mao et al., 1997), NFSs (Lee and Laibinis, 1998), surfaces for DNA studies (Hu et al., 1996), biomineralization (Archibald et al., 1996), and model surfaces for biointeraction studies (Jenney and Anderson, 1999).

Ion beam implantation

The ion-beam method injects accelerated ions with energies ranging from 10^1 to 10^6 eV (1 eV = 1.6×10^{-19} joules) into the surface zone of a material to alter its surface properties. It is largely, but not exclusively, used with metals and other inorganics such as ceramics, glasses, and semiconductors. Ions formed from most of the atoms in the periodic table can be implanted, but not all provide useful modifications to the surface properties. Important potential applications for biomaterial surfaces include modification of hardness (wear), lubricity, toughness, corrosion, conductivity, and bioreactivity.

If an ion with kinetic energy greater than a few electron volts impacts a surface, the probability that it will enter the surface is high. The impact transfers much energy to a localized surface zone in a very short time interval. Some considerations for the ion implantation process are illustrated in Fig. 3.2.14-5. These surface changes must be understood quantitatively for engineering of modified surface characteristics. Many review articles and books are available on ion implantation processes and their application for tailoring surface

Table 3.2.14-4 Silanes for surface modification of biomaterials

$$X-\underset{\underset{X}{|}}{\overset{\overset{X}{|}}{Si}}-R$$

X = leaving group	R = functional group
–Cl	–(CH$_2$)$_n$CH$_3$
–OCH$_3$	–(CH$_2$)$_3$NH$_2$
–OCH$_2$CH$_3$	–(CH$_2$)$_2$(CF$_2$)$_5$CF$_3$
	–(CH$_2$)$_3$O–C(=O)–C(CH$_3$)=CH$_2$
	–CH$_2$CH$_2$–C$_6$H$_5$

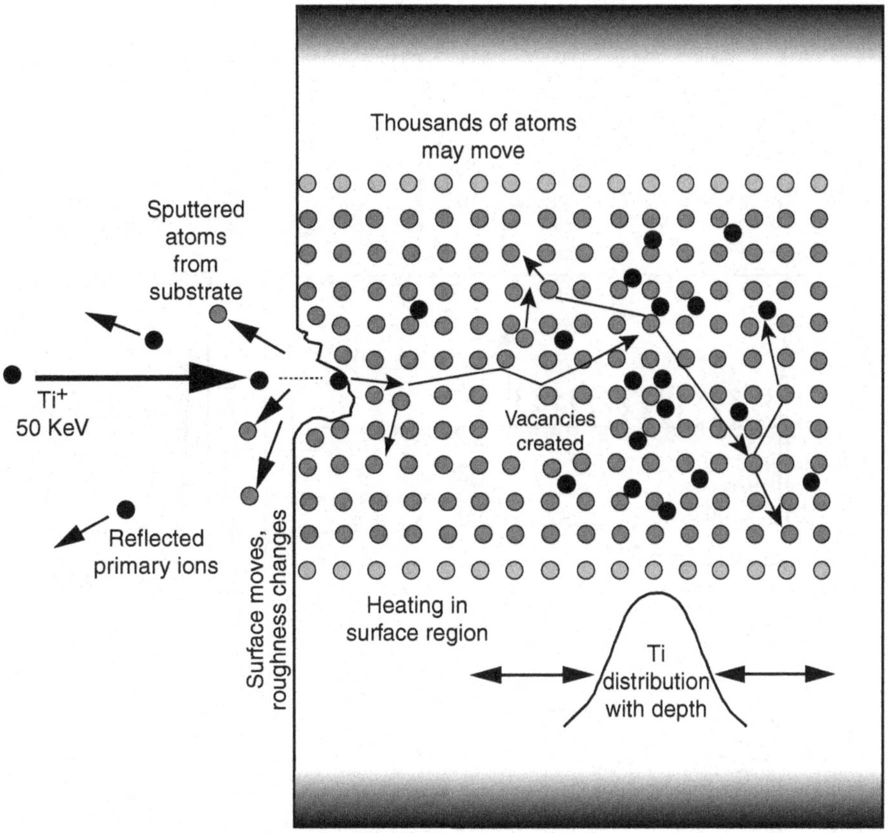

Fig. 3.2.14-5 Some considerations for the ion implantation process.

properties (Picraux and Pope, 1984; Colligon, 1986; Sioshansi, 1987; Nastasi *etal.*, 1996).

Specific examples of biomaterials that have been surface altered by ion implantation processes are plentiful. Iridium was ion implanted in a Ti–6Al–4V alloy to improve corrosion resistance (Buchanan *et al.*, 1990). Nitrogen implanted into titanium greatly reduces wear (Sioshansi, 1987). The ion implantation of boron and carbon into type 316L stainless steel improves the high cycle fatigue life of these alloys (Sioshansi, 1987). Silver ions implanted into polystyrene permit cell attachment (Tsuji *et al.*, 1998).

LB deposition

The (LB) deposition method overcoats a surface with one or more highly ordered layers of surfactant molecules. Each of the molecules that assemble into this layer contains a polar "head" group and a nonpolar "tail" group. The deposition of an LB film using an LB trough is illustrated schematically in Fig. 3.2.14-6. By withdrawing the vertical plate through the air–water interface, and then pushing the plate down through the interface, keeping the surface film at the air–water interface compressed at all times (as illustrated in Fig. 3.2.14-6), multilayer structures can be created. Some compounds that form organized LB layers are shown in Fig. 3.2.14-7. The advantages of films deposited on surfaces by this method are their high degree of order and uniformity. Also, since a wide range of chemical structures can form LB films, there are many options for incorporating new chemistries at surfaces. The stability of LB films can be improved by cross-linking or internally polymerizing the molecules after film formation, often through double bonds in the alkyl portion of the chains (Meller *et al.*, 1989). A number of research groups have investigated LB films for biomedical applications (Hayward and Chapman, 1984; Bird *et al.*, 1989; Cho *et al.*, 1990; Heens *et al.*, 1991). A unique cross between silane thin films and LB layers has been developed for biomedical surface modification (Takahara *et al.*, 2000). Many general reviews on these surface structures are available (Knobler, 1990; Ulman, 1991).

SAMs

SAMs are surface films that spontaneously form as highly ordered structures (two-dimensional crystals) on specific substrates (Maoz *et al.*, 1988; Ulman, 1990, 1991; Whitesides *et al.*, 1991; Knoll, 1996). In some

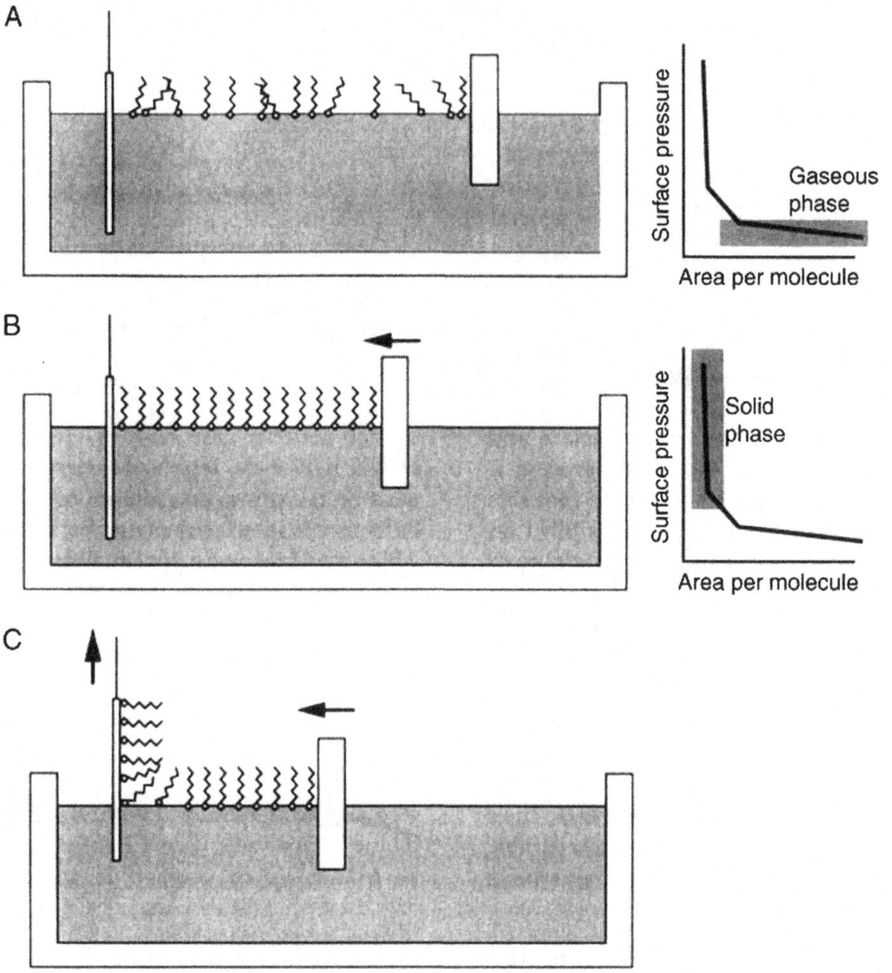

Fig. 3.2.14-6 Deposition of a lipid film onto a glass slide by the LB technique. (A) The lipid film is floated on the water layer. (B) The lipid film is compressed by a moveable barrier. (C) The vertical glass slide is withdrawn while pressure is maintained on the floating lipid film with the moveable barrier.

ways SAMs resemble LB films, but there are important differences, in particular their ease of formation. Examples of SAM films include *n*-alkyl silanes on hydroxylated surfaces (silica, glass, alumina), alkane thiols [e.g., $CH_3(CH_2)nSH$] and disulfides on coinage metals (gold, silver, copper), amines and alcohols on platinum, carboxylic acids on aluminum oxide, and silver and phosphates (phosphoric acid or phosphonate groups) on titanium or tantalum surfaces. Silane SAMs and thiols on gold are the most commonly used types. Most molecules that form SAMs have the general characteristics illustrated in Fig. 3.2.14-8. Two processes are particularly important for the formation of SAMs (Ulman, 1991): a moderate to strong adsorption of an anchoring chemical group to the surface (typically 30–100 kcal/mol), and van der Waals interaction of the alkyl chains. The bonding to the substrate (chemisorption) provides a driving force to fill every site on the surface and to displace contaminants from the surface. This process is analogous to the compression to the LB film by the movable barrier in the trough. Once adsorption sites are filled on the surface, the chains will be in sufficiently close proximity so that the weaker van der Waals interactive forces between chains can exert their influence and lead to a crystallization of the alkyl groups. Fewer than nine CH_2 groups do not provide sufficient interactive force to stabilize the 2D quasicrystal and are difficult to assemble. More than 24 CH_2 groups have too many options for defects in the crystal and are also difficult to assemble. Molecules with lengths between nine and 24 methylene groups will assemble well. Molecular mobility is an important consideration in this surface crystal formation process so that (1) the molecules have sufficient time to maneuver into position for tight packing of the binding end groups at the surface and (2) the chains can enter the quasicrystal.

The advantages of SAMs are their ease of formation, their chemical stability (often considerably higher than

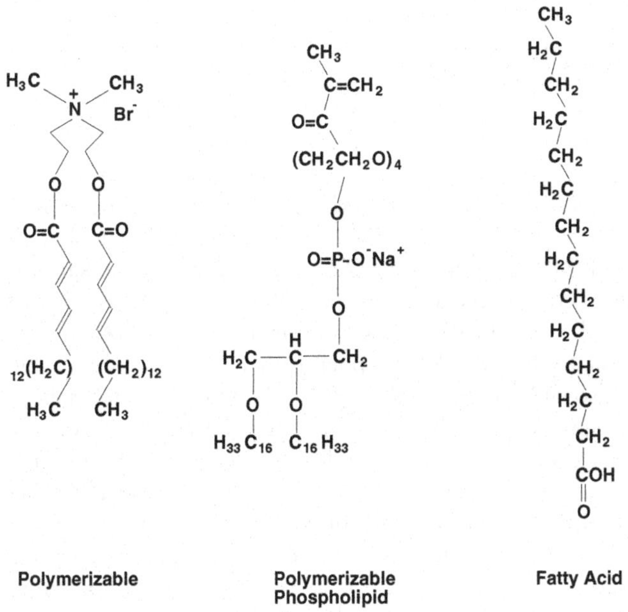

Fig. 3.2.14-7 Three examples of molecules that form organized LB films.

that of comparable LB films) and the many options for changing the outermost group that interfaces with the external environment. Many biomaterials applications have already been suggested for SAMs (Lewandowska *et al.*, 1989; Mrksich and Whitesides, 1996; Ferretti *et al.*, 2000). Useful SAMs for creating molecularly-engineered functional surfaces include headgroups of ethylene glycol oligomers, biotin, free radical initiators, N-HS esters, anhydrides, perfluoro groups, and amines, just to list a small sampling of the many possibilities. Though most SAMs are based on *n*-alkyl chain assembly, SAMs can form from other classes of molecules including proteins (Sara and Sleytr, 1996), porphyrins, nucleotide bases and aromatic ring hydrocarbons.

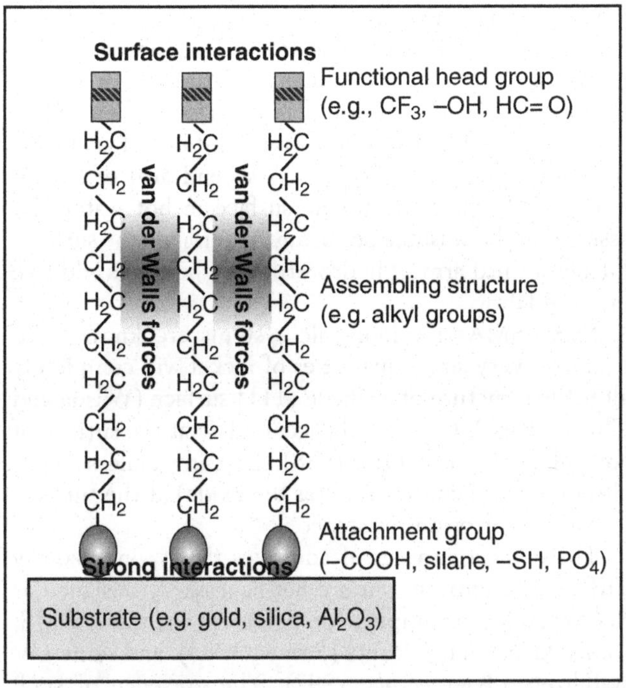

Fig. 3.2.14-8 General characteristics of molecules that form self-assembled monolayers.

Multilayer polyelectrolyte absorption

A new strategy for the surface modification of biomaterials has been developed within the past few years (Decher, 1996) and has already found application in biomaterials devices. Multilayer polyelectrolyte absorption requires a surface with either a fixed positive or a fixed negative charge. Some surfaces are intrinsically charged (for example, mica) and others can be modified with methods already described in this section. If the surface is negatively charged, it is dipped into an aqueous solution of a positively charged polyelectrolyte (e.g., polyethyleneimine). It is then rinsed in water and dipped in an aqueous solution of a negatively charged polyelectrolyte. This process is repeated as many times as desired to build up a polyelectrolyte complex multilayer of the appropriate thickness for a given application. Once a thin layer of a charged component adsorbs, it will repel additional adsorption thus tightly controlling the layer thickness and uniformity. The outermost layer can be the positively charged or negatively charged component. This strategy works well with charged biomolecules, for example hyaluronic acid (−) and chitosan (+). Layers formed are durable and assembly of these multiplayer structures is simple. The pH and ionic strength of polyelectrolyte solutions are important process variables. Such overlayer films are now being explored for application in contact lenses.

Surface-modifying additives

Specifically designed and synthesized surface-active compositions can be added in low concentrations to a material during fabrication and will spontaneously rise to and dominate the surface (Ward, 1989; Wen *et al.*, 1997). These surface-modifying additives (SMAs) are well known for both organic and inorganic systems. A driving force to minimize the interfacial energy causes the SMA to concentrate at the surface after blending homogeneously with a material. For efficient surface concentration, two factors must be taken into consideration. First, the magnitude of interfacial energy difference between the system without the additive and the same system with the SMA at the surface will determine the magnitude of the driving force leading to a SMA-dominated surface. Second, the molecular mobility of the bulk material and the SMA additive molecules within the bulk will determine the rate at which the SMA reaches the surface, or if it will get there at all. An additional concern is the durability and stability of the SMA at the surface.

A typical SMA designed to alter the surface properties of a polymeric material will be a relatively low molecular weight diblock or triblock copolymer (see Section 3.2.2). The "A" block will be soluble in, or compatible with, the bulk material into which the SMA is being added. The "B" block will be incompatible with the bulk material and have lower surface energy. Thus, the A block will anchor the B block into the material to be modified at the interface. This is suggested schematically in Fig. 3.2.14-9.

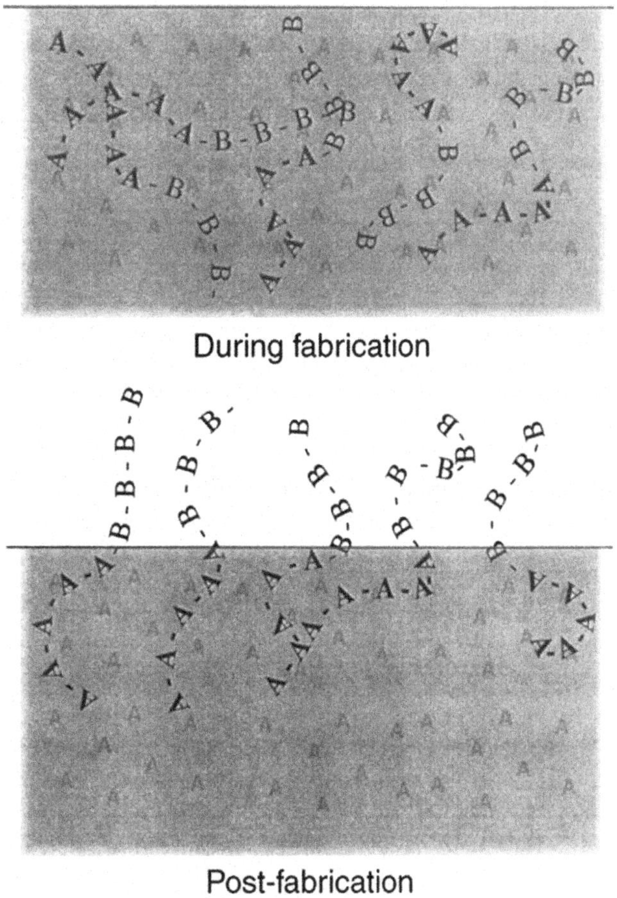

Fig. 3.2.14-9 A block copolymer SMA comprising an A block and a B block is blended into a support polymer (the bulk) with a chemistry similar to the A block. During fabrication, the block copolymer is randomly distributed throughout the support polymer. After curing or annealing, the A block anchors the SMA into the support, while the low-energy B block migrates to the air–polymer interface.

During initial fabrication, the SMA might be distributed uniformly throughout the bulk. After a period for curing or an annealing step, the SMA will migrate to the surface. Low-molecular-weight end groups on polymer chains can also provide the driving force to bring the end group to the surface.

As an example, on SMA for a polyurethane might have a low-molecular-weight polyurethane A block and a PDMS B block. The PDMS component on the surface may confer improved blood compatibility to the polyurethane. The A block will anchor the SMA in the polyurethane bulk (the polyurethane A block should be reasonably compatible with the bulk polyurethane), while the low-surface-energy, highly flexible silicone B block will be exposed at the air surface to lower the interfacial energy (note that air is "hydrophobic"). The A block anchor should confer stability to this system. However, consider that if the system is placed in an aqueous environment, a low-surface-energy polymer (the B block) is now in contact with water—a high interfacial energy situation. If the system, after fabrication, still exhibits sufficient chain mobility, it might phase-invert to bring the bulk polyurethane or the A block to the surface. Unless the system is specifically engineered to do such a surface phase reversal, this inversion is undesirable. Proper choice of the bulk polymer and the A block can impede surface phase inversion.

An example of a polymer additive that was developed by 3M specifically to take advantage of this surface chemical inversion phenomenon is a stain inhibitor for fabric. Though not a biomaterial, it illustrates design principles for this type of system. The compound has three "arms." A fluoropolymer arm, the lowest energy component, resides at the fabric surface in air. Fluoropolymers and hydrocarbons (typical stains) do not mix, so hydrocarbons are repelled. A second arm of hydrophilic PEO will come to the surface in hot water and assist with the washing out of any material on the surface. Finally, a third arm of hydrocarbon anchors this additive into the fabric.

Many SMAs for inorganic systems are known. For example, very small quantities of nickel will completely alter the structure of a silicon (111) surface (Wilson and Chiang, 1987). Copper will accumulate at the surface of gold alloys (Tanaka *et al.*, 1988). Also, in stainless steels, chromium will concentrate (as the oxide) at the surface, imparting corrosion resistance.

There are a number of additives that spontaneously surface-concentrate, but are not necessarily designed as SMAs. A few examples for polymers include PDMS, some extrusion lubricants (Ratner, 1983), and some UV stabilizers (Tyler *et al.*, 1992). The presence of such additives at the surface of a polymer may be unplanned and they will not necessarily form stable, durable surface layers. However, they can significantly contribute (either positively or negatively) to the bioresponse to the surface.

Conversion coatings

Conversion coatings modify the surface of a metal into a dense oxide-rich layer that imparts corrosion protection, enhanced adhesivity, altered appearance (e.g., color) and sometimes lubricity to the metal. For example, steel is frequently phosphated (treated with phosphoric acid) or chromated (with chromic acid). Aluminum is electrochemically anodized in chromic, oxalic, or sulfuric acid electrolytes. Electrochemical anodization may also be useful for surface-modifying titanium and Ti–Al alloys (Bardos, 1990; Kasemo and Lausmaa, 1985).

The conversion of metallic surfaces to "oxide-like," electrochemically passive states is a common practice for base-metal alloy systems used as biomaterials. Standard and recommended techniques have been published (e.g., ASTM F4-86) and are relevant for most musculoskeletal load-bearing surgical implant devices. The background literature supporting these types of surface passivation technologies has been summarized (von Recum, 1986).

Base-metal alloy systems, in general, are subject to electrochemical corrosion ($M \rightarrow M^+ + e-$) within saline environments. The rate of this corrosion process is reduced 10^3–10^6 times by the presence of a dense, uniform, minimally conductive, relatively inert oxide surface. For many metallic devices, exposure to a mineral acid (e.g., nitric acid in water) for times up to 30 minutes will provide a passivated surface. Plasma-enhanced surface passivation of metals, laser surface treatments, and mechanical treatments (shot peening) can also impart many of these characteristics to metallic systems.

The reason that many of these surface modifications are called "oxide-like" is that the structure is complex, including OH, H, and subgroups that may, or may not, be crystalline. Since most passive surfaces are thin films (5–500 nm) and are transparent or metallic in color, the surface appears similar before and after passivation. Further details on surfaces of this type can be found in Sections 3.1.4 and 3.2.9.

Parylene coating

Parylene (*para*-xylylene) coatings occupy a unique niche in the surface modification literature because of their wide application and the good quality of the thin film coatings formed (Loeb *et al.*, 1977a; Nichols *et al.*, 1984). The deposition method is also unique and involves the simultaneous evaporation, pyrolysis, deposition, and polymerization of the monomer, di-*para*-xylylene (DPX), according to the following reaction:

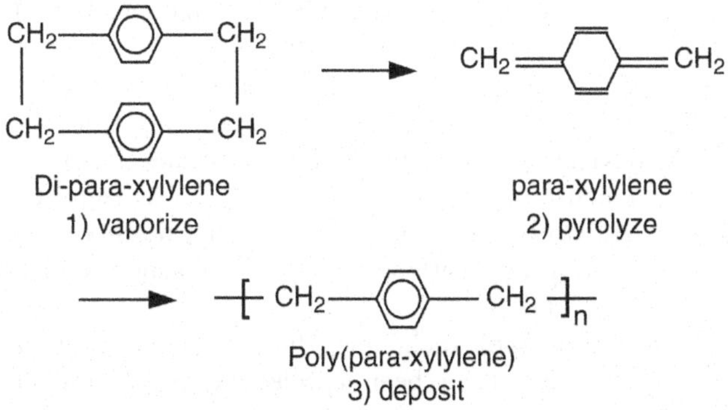

The DPX monomer is vaporized at 175°C and 1 torr, pyrolyzed at 700°C and 0.5 torr, and finally deposited on a substrate at 25°C and 0.1 torr. The coating has excellent electrical insulation and moisture barrier properties and has been used for protection of implant electrodes (Loeb *et al.*, 1977b; Nichols *et al.*, 1984) and implanted electronic circuitry (Spivack and Ferrante, 1969). Recently, a parylene coating has been used on stainless steel cardiovascular stents between the metal and a drug-eluting polymer layer.

Laser methods

Lasers can rapidly and specifically induce surface changes in organic and inorganic materials (Picraux and Pope, 1984; Dekumbis, 1987; Chrisey *et al*, 2003). The advantages of using lasers for such modification are the precise control of the frequency of the light, the wide range of frequencies available, the high energy density, the ability to focus and raster the light, the possibilities for using both heat and specific excitation to effect change, and the ability to pulse the source and control reaction time. Lasers commonly used for surface modification include ruby, neodymium: yttrium aluminum garnet (Nd: YAG), argon, and CO_2. Treatments are pulsed (100 nsec to picoseconds pulse times) and continuous wave (CW), with interaction times often less than 1 msec. Laser-induced surface alterations include annealing, etching, deposition, and polymerization. Polymers, metals, ceramics, and even tooth dentin have been effectively surface modified using laser energy. The major considerations in designing a laser surface treatment include the absorption (coupling) between the laser energy and the material, the penetration depth of the laser energy into the material, the interfacial reflection and scattering, and heating induced by the laser.

Patterning

Essentially all of the surface modification methods described in this section can be applied to biomaterial surfaces as a uniform surface treatment, or as patterns on the surface with length scales of millimeters, microns or even nanometers. There is much interest in deposition of proteins and cells in surface patterns and textures in order to control bioreactions (Section 3.2.16). Furthermore, devices "on a chip" frequently require patterning. Such devices include microfluidic systems ("lab on a chip"), neuronal circuits on a chip, and DNA diagnostic arrays. An overview of surface patterning methods for bioengineering applications has been published (Folch and Toner, 2000).

Photolithographic techniques that were developed for microelectronics have been applied to patterning of biomaterial surfaces when used in conjunction with methods described in this section. For example, plasma-deposited films were patterned using a photoresist (PR) lift-off method (Goessl *et al.*, 2001).

Microcontact printing (μCP) is a newer method permitting simple modification. Basically, a rubber stamp is made of the pattern that is desired on the biomaterial surface (Fig. 3.2.14-10). The stamp can be "inked" with thiols (to stamp gold), silanes (to stamp silicon), proteins (to stamp many types of surfaces) or polymer solutions (again, to stamp many types of surfaces). Spatial resolution of pattern features in the nanometer range has been demonstrated, though most patterns are applied in the micron range. Methods have been developed to accurately stamp curved surfaces. An example of cells on laminin-stamped lines is shown in Fig. 3.2.14-11. These laminin lines were durable for at least 2 weeks of cell contact. Durability remains a major consideration with patterns on surface generated by this relatively simple method.

There are many other options to pattern biomaterial surfaces. These include ion-beam etching, electron-beam lithography, laser methods, inkjet printers, and stochastic patterns made by phase separation of two components (Takahara *et al.*, 2000).

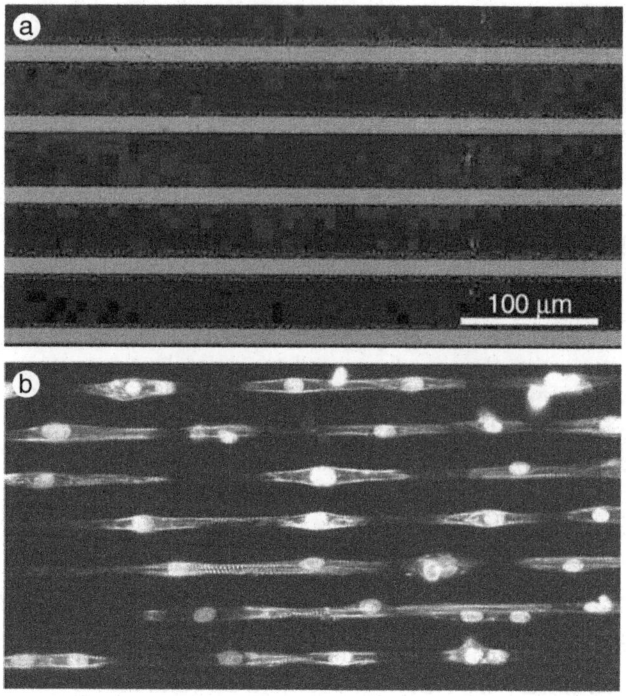

Fig. 3.2.14-11 (a) Microcontact printed lines of laminin protein (fluorescent labeled) on a cell-resistant background. (b) Cardiomyocyte cells adhering and aligning on the laminin printed lines (see *J. Biomed. Mater. Res. 60:472* for details) (used with the permission of P. Stayton, C. Murry, S. Hauschka, J. Angello and T. McDevitt).

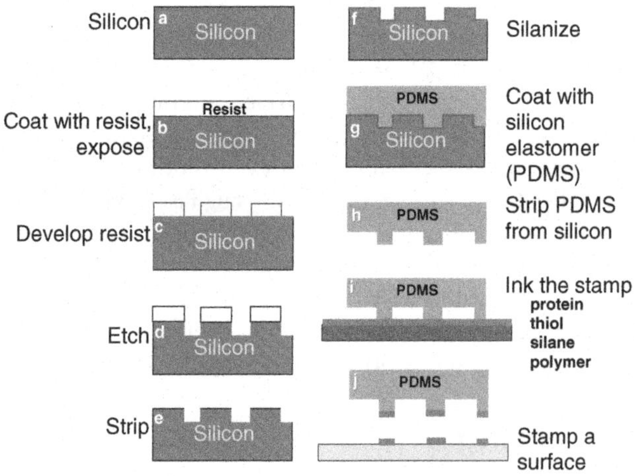

Fig. 3.2.14-10 Fabrication of a silicone elastomer stamp for μCP. The sequence of steps is a-j.

Conclusions

Surface modifications are being widely explored to enhance the biocompatibility of biomedical devices and improve other aspects of performance. Since a given medical device may already have appropriate performance characteristics and physical properties and be well understood in the clinic, surface modification provides a means to alter only the biocompatibility of the device without the need for redesign, retooling for manufacture, and retraining of medical personnel.

Acknowledgment

The suggestions and assistance of Professor J. Lemons have enhanced this section and are gratefully appreciated.

Questions

1. You are assigned the task of designing a proteomics array for cancer diagnostics. Six hundred and twenty-five proteins must be attached to the surface of

a standard, glass microscope slide in a 25 × 25 array. Design a scheme to make such a proteomic chip. What are the important surface issues? Which strategies might you apply to address each of the issues? You may find helpful ideas in Sections 3.1.4, 3.2.13, and 3.2.16.

2. A hydrogel surface must be put on a SR medical device. A viscous solution of the hydrogel polymer is used to spray-coat the device. When it is placed in aqueous buffer solution the hydrogel layer quickly delaminates from the silicone. How might you permanently attach a hydrogel layer to a silicone device? Briefly describe the method you would use and the general steps needed to produce a reliable coating.

3. List the molecular and design factors that can contribute to increasing the durability of an *n*-alkyl thiol SAM on gold.

Bibliography

Archibald, D. D., Qadri, S. B., and Gaber, B. P. (1996). Modified calcite deposition due to ultrathin organic films on silicon substrates. *Langmuir* **12**: 538–546.

Arkles, B. (1977). Tailoring surfaces with silanes. *Chemtech* **7**: 766–778.

Bardos, D. I. (1990). Titanium and titanium alloys in *Concise Encyclopedia of Medical and Dental Materials*, 1st ed., E. Williams, R. W. Cahn, and M. B. Bever, eds. Pergamon Press, Oxford, pp. 360–365.

Bird, R. R., Hall, B., Hobbs, K. E. F., and Chapman, D. (1989). New haemocompatible polymers assessed by thrombelastography. *J. Biomed. Eng.* **11**: 231–234.

Buchanan, R. A., Lee, I. S., and Williams, J. M. (1990). Surface modification of biomaterials through noble metal ion implantation. *J. Biomed. Mater. Res.* **24**: 309–318.

Chapiro, A. (1983). Radiation grafting of hydrogels to improve the thrombo-resistance of polymers. *Eur. Polym. J.* **19**: 859–861.

Charlson, E. J., Charlson, E. M., Sharma, A. K., and Yasuda, H. K. (1984). Electrical properties of glow-discharge polymers, parylenes, and composite films. *J. Appl. Polymer Sci. Appl. Polymer Symp.* **38**: 137–148.

Chilkoti, A., Ratner, B. D., and Briggs, D. (1991a). Plasma-deposited polymeric films prepared from carbonyl-containing volatile precursors: XPS chemical derivatization and static SIMS surface characterization. *Chem. Mater.* **3**: 51–61.

Chilkoti, A., Ratner, B. D., and Briggs, D. (1991b). A static secondary ion mass spectrometric investigation of the surface structure of organic plasma-deposited films prepared from stable isotope-labeled precursors. Part I. Carbonyl precursors. *Anal. Chem.* **63**: 1612–1620.

Chilkoti, A., Ratner, B. D., Briggs, D., and Reich, F. (1992). Static secondary ion mass spectrometry of organic plasma deposited films created from stable isotope-labeled precursors. Part II. Mixtures of acetone and oxygen. *J. Polymer Sci., Polymer Chem. Ed.* **30**: 1261–1278.

Cho, C. S., Takayama, T., Kunou, M., and Akaike, T. (1990). Platelet adhesion onto the Langmuir–Blodgett film of poly-(gamma-benzyl L-glutamate)-poly(ethylene oxide)-poly(gamma-benzyl L-glutamate) block copolymer. *J. Biomed. Mater. Res.* **24**: 1369–1375.

Chrisey, D. B., Piqué, A., McGill, R. A., Horowitz, J. S., Ringeisen, B. R., Bubb, D. M., and Wu, P. K. (2003). Laser deposition of polymer and biomaterial films. *Chem. Rev.* **103**: 553–576.

Chu, P. K., Chen, J. Y., Wang, L. P., and Huang, N. (2002). Plasma surface modification of biomaterials. *Mater. Sci. Eng. Rep.* **36**: 143–206.

Colligon, J. S. (1986). Surface modification by ion beams. *Vacuum* **36**: 413–418.

Dargaville, T. R., George, G. A., Hill, D. J. T., and Whittaker, A. K. (2003). High energy radiation grafting of fluoropolymers. *Prog. Polymer Sci.* **28**: 1355–1376.

Decher, G. (1996). Layered nanoarchitectures via directed assembly of anionic and cationic molecules. in *Comprehensive Supramolecular Chemistry*, Vol. 9, *Templating, Self-Assembly and Self-Organization*, J.-P. Sauvage and M. W. Hosseini, eds. Pergamon Press, Oxford, pp. 507-528.

Dekumbis, R. (1987). Surface treatment of materials by lasers. *Chem. Eng. Prog.* **83**: 23-29.

Dunkirk, S. G., Gregg, S. L., Duran, L. W., Monfils, J. D., Haapala, J. E., Marcy, J. A., Clapper, D. L., Amos, R. A., and Guire, P. E. (1991). Photochemical coatings for the prevention of bacterial colonization. *J. Biomater. Appl.* 131–156.

Everhart, D. S., and Reilley, C. N. (1981). Chemical derivatization in electron spectroscopy for chemical analysis of surface functional groups introduced on low-density polyethylene film. *Anal. Chem.* **53**: 665–676.

Ferretti, S., Paynter, S., Russell, D. A., and Sapsford, K. E. (2000). Self-assembled monolayers: a versatile tool for the formulation of bio-surfaces. *Trends Anal. Chem.* **19(9)**: 530–540.

Folch, A., and Toner, M. (2000). Microengineering of cellular interactions. *Ann. Rev. Bioeng.* **2**: 227–256.

Garbassi, F., Morra, M., Occhiello, E., Barino, L., and Scordamaglia, R. (1989). Dynamics of macromolecules: A challenge for surface analysis. *Surf. Interface Anal.* **14**: 585–589.

Goessl, A., Garrison, M. D., Lhoest, J., and Hoffman, A. (2001). Plasma lithography—thin-film patterning of polymeric+biomaterials by RF plasma polymerization I: surface preparation and analysis. J. *Biomater. Sci. Polymer Ed.* 12(7): 721-738.

Gombotz, W. R., and Hoffman, A. S. (1988). Functionalization of polymeric films by plasma polymerization of allyl alcohol and allylamine. *J. Appl. Polymer Sci. Appl. Polymer Symp.* **42**: 285–303.

Greer, R. T., Knoll, R. L., and Vale, B. H. (1979). Evaluation of tissue-response to hydrogel composite materials. *SEM* **2**: 871–878.

Griesser, H. J., and Chatelier, R. C. (1990). Surface characterization of plasma polymers from amine, amide and alcohol monomers. *J. Appl. Polymer Sci. Appl. Polymer Symp.* **46**: 361–384.

Gruner, H. (2001). Thermal spray coating on titanium. in *Titanium in Medicine*, D.M. Brunette, P. Tengvall, M. Textor and P. Thomsen, eds. Springer-Verlag, Berlin.

Haque, Y., and Ratner, B. D. (1988). Role of negative ions in the RF plasma deposition of fluoropolymer films from perfluoropropane. J. *Polymer Sci., Polymer Phys. Ed.* **26**: 1237–1249.

Hayward, J. A., and Chapman, D. (1984). Biomembrane surfaces as models for polymer design: The potential for haemocompatibility. *Biomaterials* **5**: 135–142.

Heens, B., Gregoire, C., Pireaux, J. J., Cornelio, P. A., and Gardella, J. A., Jr. (1991). On the stability and homogeneity of Langmuir–Blodgett films as models of polymers and biological materials for surface studies: An XPS study. *Appl. Surf. Sci.* **47**: 163–172.

Heid, S., Effenberger, F., Bierbaum, K., and Grunze, M. (1996). Self-assembled mono- and multilayers of terminally functionalized organosilyl compounds on silicon substrates. *Langmuir* **12**(8): 2118–2120.

Hickman, J. J., and Stenger, D. A. (1994). Interactions of cultured neurons with defined surfaces. in *Enabling Technologies for Cultured Neural Networks*, D. A. Stenger, T. N. McKenna, eds. Academic Press, San Diego, pp. 51–76.

Hoffman, A. S. (1981). A review of the use of radiation plus chemical and biochemical processing treatments to prepare novel biomaterials. *Radiat. Phys. Chem.* **18**: 323–342.

Hoffman, A. S. (1988). Biomedical applications of plasma gas discharge processes. *J. Appl. Polymer Sci. Appl. Polymer Symp.* **42**: 251–267.

Hoffman, A. S., Schmer, G., Harris, C., and Kraft, W. G. (1972). Covalent binding of biomolecules to radiation-grafted hydrogels on inert polymer surfaces. *Trans. Am. Soc. Artif. Int. Organs* **18**: 10–17.

Hoffman, A. S., Cohn, D. C., Hanson, S. R., Harker, L. A., Horbett, T. A., Ratner, B. D., and Reynolds, L. O. (1983). Application of radiation-grafted hydrogels as blood-contacting biomaterials. *Radiat. Phys. Chem.* **22**: 267–283.

Horbett, T. A., and Hoffman, A. S. (1975). Bovine plasma protein adsorption on radiation-grafted hydrogels based on hydroxyethyl methacrylate and *N*-vinyl-pyrrolidone. in *Applied Chemistry at Protein Interfaces*, Advances in Chemistry Series, R. E. Baier, ed. American Chemical Society, Washington, D.C., pp. 230–254.

Hu, J., Wang, M., Weier, U. G., Frantz, P., Kolbe, W., Ogletree, D. F., and Salmeron, M. (1996). Imaging of single extended DNA molecules on flat (aminopropyl)triethozysilane-mica by atomic force microscopy. *Langmuir* **12**(7): 1697–1700.

Inagaki, N., Nakanishi, T., and Katsuura, K. (1983). Glow discharge polymerizations of tetrafluoroethylene, perfluoromethylcyclohexane and perfluorotoluene investigated by infrared spectroscopy and ESCA. *Polymer Bull.* **9**: 502–506.

Jenney, C. R., and Anderson, J. M. (1999). Alkylsilane-modified surfaces: inhibition of human macrophage adhesion and foreign body giant cell formation. *J. Biomed. Mater. Res.* **46**: 11–21.

Johnston, E. E., and Ratner, B. D. (1996). XPS and SSIMS characterization of surfaces modified by plasma deposited oligo(glyme) films in *Surface Modification of Polymeric Biomaterials*, B. D. Ratner, D. G. Castner, eds. Plenum Press, New York, pp. 35–44.

Kaplan, S., and Dilks, A. (1981). A solid state nuclear magnetic resonance investigation of plasma-polymerized hydrocarbons. *Thin Solid Films* **84**: 419–424.

Kasemo, B., and Lausmaa, J. (1985). Metal selection and surface characteristics. in *Tissue-Integrated Prostheses*, P. I. Branemark, G. A. Zarb and T. Albrektsson, eds. Quintessence Publishing, Chicago, pp. 99–116.

Kitching, K. J., Pan, V., and Ratner, B. D. (2003). Biomedical applications of plasma-deposited thin films. in *Plasma Polymer Films*, H. Biederman, ed. Imperial College Press, London.

Klages, C. -P., Höpfner, K., and Kläke, N. (2000). Surface functionalization at atmospheric pressure by DBD-based pulsed plasma polymerization. *Plasmas Polymer* **5**: 79–89.

Knobler, C. M. (1990). Recent developments in the study of monolayers at the air–water interface. *Adv. Chem. Phys.* **77**: 397–449.

Knoll, W. (1996). Self-assembled microstructures at interfaces. *Curr. Opin. Colloid Interface Sci.* **1**: 137–143.

Krishnamurthy, V., Kamel, I. L., and Wei, Y. (1989). Analysis of plasma polymerization of allylamine by FTIR. *J. Polymer Sci., Polymer Chem. Ed.* **27**: 1211–1224.

Kwon, O. H., Kikuchi, A., Yamato, M., Sakurai, Y., and Okano, T. (2000). Rapid cell sheet detachment from poly(N-isopropylacrylamide)-grafted porous cell culture membranes. *J. Biomed. Mater. Res.* **50**: 82–89.

Lee, S. -W., and Laibinis, P. E. (1998). Protein-resistant coatings for glass and metal oxide surfaces derived from oligo(ethylene glycol)-terminated alkyltrichlorosilane. *Biomaterials* **19**: 1669–1675.

Lewandowska, K., Balachander, N., Sukenik, C. N., and Culp, L. A. (1989). Modulation of fibronectin adhesive functions for fibroblasts and neural cells by chemically derivatized substrata. *J. Cell. Physiol.* **141**: 334–345.

Loeb, G. E., Bak, M. J., Salcman, M., and Schmidt, E. M. (1977a). Parylene as a chronically stable, reproducible microelectrode insulator. *IEEE Trans. Biomed. Eng.* **BME-24**: 121–128.

Loeb, G. E., Walker, A. E., Uematsu, S., and Konigsmark, B. W. (1977b). Histological reaction to various conductive and dielectric films chronically implanted in the subdural space. *J. Biomed. Mater. Res.* **11**: 195–210.

Lopez, G. P., and Ratner, B. D. (1991). Substrate temperature effects of film chemistry in plasma deposition of organics. I. Nonpolymerizable precursors. *Langmuir* **7**: 766–773.

Lopez, G. P., Ratner, B. D., Rapoza, R. J., and Horbett, T. A. (1993). Plasma deposition of ultrathin films of poly-(2-hydroxyethyl methacrylate): Surface analysis and protein adsorption measurements. *Macromolecules* **26**: 3247–3253.

Mao, G., Castner, D. G., and Grainger, D. W. (1997). Polymer immobilization to alkylchlorosilane organic monolayer films using sequential derivatization reactions. *Chem. Mater.* **9**(8): 1741–1750.

Maoz, R., Netzer, L., Gun, J., and Sagiv, J. (1988). Self-assembling monolayers in the construction of planned supramolecular structures and as modifiers of surface properties. *J. Chim. Phys.* **85**: 1059-1064.

Massines, F., Gherardi, N., and Sommer, F. (2000). Silane-based coatings of polypropylene, deposited by atmospheric pressure glow discharge plasmas. *Plasmas Polymers* **5**: 151–172.

Matsuda, T., and Inoue, K. (1990). Novel photoreactive surface modification technology for fabricated devices. *Trans. Am. Soc. Artif. Internal Organs* **36**: M161-M164.

Matsuzawa, M., Umemura, K., Beyer, D., Sugioka, K., and Knoll, W. (1997). Micropatterning of neurons using organic substrates in culture. *Thin Solid Films* **305**: 74–79.

Meller, P., Peters, R., and Ringsdorf, H. (1989). Microstructure and lateral diffusion in monolayers of polymerizable amphiphiles. *Colloid Polymer Sci.* **267**: 97–107.

Mrksich, M., and Whitesides, G. M. (1996). Using self-assembled monolayers to understand the interactions of manmade surfaces with proteins and cells. *Annu. Rev. Biophys. Biomol. Struct.* **25**: 55–78.

Nastasi, M., Mayer, J., and Hirvonen, J. K. (1996). *Ion–solid interactions: fundamentals and applications.* Cambridge Univ. Press, Cambridge, UK.

Nguyen, S.V. (1986). Plasma assisted chemical vapor deposited thin films for microelectronic applications. J. *Vac. Sci. Technol. B* **4**: 1159–1167.

Nichols, M. F., Hahn, A. W., James, W. J., Sharma, A. K., and Yasuda, H. K. (1984). Evaluating the adhesion characteristics of glow-discharge plasma-polymerized films by a novel voltage cycling technique. *J. Appl. Polymer Sci. Appl. Polymer Symp.* **38**: 21–33.

Panchalingam, V., Poon, B., Huo, H. H., Savage, C. R., Timmons, R. B., and Eberhart, R. C. (1993). Molecular surface tailoring of biomaterials via pulsed RF plasma discharges. *J. Biomater. Sci. Polymer. Ed.* **5**(1/2): 131–145.

Picraux, S. T., and Pope, L. E. (1984). Tailored surface modification by ion implantation and laser treatment. *Science* **226**: 615–622.

Plueddemann, E. P. (1980). Chemistry of silane coupling agents. in *Silylated Surfaces*, D. E. Leyden, ed. Gordon and Breach Science Publishers, New York, pp. 31–53.

Pomerantz, M., Segmuller, A., Netzer, L., and Sagiv, J. (1985). Coverage of Si substrates by self-assembling monolayers and multilayers as measured by IR, wettability and x-ray diffraction. *Thin Solid Films* 132: 153–162.

Ratner, B. D. (1980). Characterization of graft polymers for biomedical applications. *J. Biomed. Mater. Res.* **14**: 665–687.

Ratner, B. D. (1983). ESCA studies of extracted polyurethanes and polyurethane extracts: Biomedical implications. in *Physicochemical Aspects of Polymer Surfaces*, K. L. Mittal, ed. Plenum Publishing, New York, pp. 969–983.

Ratner, B. D., and Hoffman, A. S. (1980). Surface grafted polymers for biomedical applications. in *Synthetic Biomedical Polymers. Concepts and Applications*, M. Szycher and W. J. Robinson, eds. Technomic Publishing, Westport, CT, pp. 133–151.

Ratner, B. D., and Yoon, S. C. (1988). Polyurethane surfaces: solvent and temperature induced structural rearrangements. in *Polymer Surface Dynamics*, J. D. Andrade, ed. Plenum Press, New York, pp. 137–152.

Ratner, B. D., Horbett, T. A., Hoffman, A. S., and Hauschka, S. D. (1975). Cell adhesion to polymeric materials: Implications with respect to biocompatibility. J. *Biomed. Mater. Res.* **9**: 407–422.

Ratner, B. D., Hoffman, A. S., Hanson, S. R., Harker, L. A., and Whiffen, J. D. (1979). Blood compatibility–water content relationships for radiation grafted hydrogels. *J. Polymer Sci. Polymer Symp.* **66**: 363–375.

Ratner, B. D., Chilkoti, A., and Lopez, G. P. (1990). Plasma deposition and treatment for biomaterial applications. in *Plasma Deposition, Treatment and Etching of Polymers*, R. D'Agostino, ed. Academic Press, San Diego, pp. 463–516.

Rye, R. R., Nelson, G. C., and Dugger, M. T. (1997). Mechanistic aspects of alkylchlorosilane coupling reactions. *Langmuir* **13**(11): 2965–2972.

Safrany, A. (1997). Radiation processing: synthesis and modification of biomaterials for medical use. *Nucl. Instrum. Methods Phys. Res., Sect. B* **131**: 376–381.

Sara, M., and Sleytr, U. B. (1996). Crystalline bacterial cell surface layers (S-layers): from cell structure to biomimetics. *Prog. Biophys. Mol. Biol.* **65**(1/2): 83–111.

Sawin, H. H., and Reif, R. (1983). A course on plasma processing in integrated circuit fabrication. *Chem. Eng. Ed.* **17**: 148–152.

Sheu, M.-S., Hoffman, A. S., and Feijen, J. (1992). A glow discharge process to immobilize PEO/PPO surfactants for wettable and non-fouling biomaterials. *J. Adhes. Sci. Technol.* **6**: 995–1101.

Singh, A., and Silverman, J., editors. (1992). *Radiation Processing of Polymers.* Oxford Univ. Press, New York.

Sioshansi, P. (1987). Surface modification of industrial components by ion implantation. *Mater. Sci. Eng.* **90**: 373–383.

Somorjai, G. A. (1990). Modern concepts in surface science and heterogeneous catalysis. *J. Phys. Chem.* **94**: 1013–1023.

Somorjai, G. A. (1991). The flexible surface. Correlation between reactivity and restructuring ability. *Langmuir* **7**: 3176–3182.

Spivack, M. A., and Ferrante, G. (1969). Determination of the water vapor permeability and continuity of ultrathin parylene membranes. *J. Electrochem. Soc.* **116**: 1592–1594.

Srinivasan, R., and Lazare, S. (1985). Modification of polymer surfaces by far-ultraviolet radiation of low and high (laser) intensities. *Polymer* **26**: 1297–1300.

Stannett, V. T. (1990). Radiation grafting—state-of-the-art. *Radiat. Phys. Chem.* **35**: 82–87.

Swanson, M. J. (1996). A unique photochemical approach for polymer surface modification. in *Polymer Surfaces and Interfaces: Characterization, Modification and Application*, K. L. Mittal and K. W. Lee eds. VSP, The Netherlands.

Takahara, A., Ge, S., Kojio, K., and Kajiyama, T. (2000). In situ atomic force mircroscopic observation of albumin adsorption onto phase-separated organosilane monolayer surface. *J. Biomater. Sci. Polymer. Ed.* **11**(1): 111–120.

Tanaka, T., Atsuta, M., Nakabayashi, N., and Masuhara, E. (1988). Surface treatment of gold alloys for adhesion. *J. Prosthet. Dent.* **60**: 271–279.

Tsuji, H., Satoh, H., Ikeda, S., Ikemoto, N., Gotoh, Y., and Ishikawa, J. (1998). Surface modification by silver-negative-ion implantation for controlling cell-adhesion properties of polystyrene. *Surf. Coat. Technol.* 103–104: 124–128.

Tyler, B. J., Ratner, B. D., Castner, D. G., and Briggs, D. (1992). Variations between Biomer lots. 1. Significant differences in the surface chemistry of two lots of a commercial polyetherurethane. *J. Biomed. Mater. Res.* **26**: 273–289.

Ulman, A. (1990). Self-assembled monolayers of alkyltrichlorosilanes: Building blocks for future organic materials. *Adv. Mater.* **2**: 573–582.

Ulman, A. (1991). *An Introduction to Ultrathin Organic Films.* Academic Press, Boston.

von Recum, A. F. (1986). *Handbook of Biomaterials Evaluation*, 1st ed. Macmillan, New York.

Ward, R. S. (1989). Surface modifying additives for biomedical polymers. *IEEE Eng. Med. Biol. Mag.* June, pp. 22–25.

Wasserman, S. R., Tao, Y. -T., and Whitesides, G. M. (1989). Structure and reactivity of alkylsiloxane monolayers formed by reaction of alkyltrichlorosilanes on silicon substrates. *Langmuir* **5**: 1074–1087.

Wen, J. M., Gabor, S., Lim, F., and Ward, R. (1997). XPS study of surface composition of a segmented polyurethane block copolymer modified by PDMS end groups and its blends with phenoxy. *Macromolecules* **30**: 7206–7213.

Whitesides, G. M., Mathias, J. P., and Seto, C. T. (1991). Molecular self-assembly and nanochemistry: a chemical strategy for the synthesis of nanostructures. *Science* **254**: 1312–1319.

Wilson, R. J., and Chiang, S. (1987). Surface modifications induced by adsorbates at low coverage: A scanning-tunneling-microscopy study of the Ni/Si(111) square-root-19 surface. *Phys. Rev. Lett.* **58**: 2575–2578.

Wu, S. (1982). *Polymer Interface and Adhesion.* Marcel Dekker, New York.

Xiao, S. J., Textor, M., Spencer, N. D., Wieland, M., Keller, B., and Sigrist, H. (1997). Immobilization of the cell-adhesive peptide arg-gly-asp-cys (RGDC) on titanium surfaces by covalent chemical attachment. *J. Mater. Sci.: Mater. Med.* **8**: 867–872.

Yasuda, H. K. (1979). Competitive ablation and polymerization (CAP) mechanisms of glow discharge polymerization. in *Plasma Polymerization*, ACS Symposium Series 108, M. Shen and A. T. Bell, eds. American Chemical Society, Washington, D.C., pp. 37–52.

Yasuda, H. K., and Gazicki, M. (1982). Biomedical applications of plasma polymerization and plasma treatment of polymer surfaces. *Biomaterials* **3**: 68–77.

3.2.15 Textured and porous materials

John A. Jansen and Andreas F. von Recum

Introduction

Surface irregularities on medical devices, such as grooves/ridges, hills, pores, and pillars, are expected to guide many types of cells (including immunological, epithelial, connective-tissue, neural, and muscle cells) and to aid tissue repair after injury. With the growing interest in tissue engineering, porous scaffold reactions *in vitro* and *in vivo* are assuming increasing importance. The final response to rough or porous materials is reflected in the organization of the cytoskeleton, the orientation of ECM components, the amount of produced ECM, and angiogenesis. Although significant progress has been made, the exact cellular and molecular events underlying cellular and matrix orientation are not yet completely understood.

This section will provide information about how surface roughness is defined, prepared, and measured. In addition, it will cover the biological effects of surface irregularities on cells.

Definition of surface irregularities

Surface irregularities can be considered as deviations from a geometrically ideal (flat) surface. They can be created accidentally by the production process or engineered for specific purposes. Surface irregularities can be classified according to their dimensions and the way they are achieved. In view of this, surface irregularities can be classified into six classes (Sander, 1991). The main distinctive characteristic is their horizontal pattern. Thus, Class 1 irregularities are associated with form errors of the substrate surface such as straightness, flatness, roundness, and cylindricity. Class 2 surface features deal with so-called waviness deviations. Waviness is considered to occur if the wave spacing is larger than the wave depth. Class 3, 4, and 5 irregularities all refer to surface roughness. Roughness is assumed if the space between two hills is about 5–100 times larger than the depth. Depending on the manufacturing process used, roughness can be periodic or random. A periodic surface roughness is also referred to as surface texture and represents a regular surface topography with well-defined dimensions and surface distribution. Further, distinction has to be made among macro, micro, and nano surface roughness. Microroughness deals with surface features sized in cellular and subcellular dimensions. Considering their appearance and morphological structure, class 3 surface roughness has a groove-type appearance; class 4 roughness deals with score marks, flakes, and protuberances, for example created by grit-blasting procedures; and class 5 surface roughness is the result of the crystal structure of a material.

Porosity

Besides the surface irregularities as mentioned earlier, porosity can also be considered as surface irregularity. Porosity can occur only at the substrate surface or can completely penetrate throughout a bulk material. It consists of individual openings and spacings or interconnecting pores. Porosity can be created intentionally by a specific production process, such as sintering of beads, leaching of salt, sugar, or starch crystals, or knitting and weaving of fibers. On the other hand, porosity can also arise as a manufacturing artifact, for example, in casting procedures.

For many biomedical applications, there is a need for porous implant materials. They can be used for artificial blood vessels, artificial skin, drug delivery, bone and cartilage reconstruction, periodontal repair, and tissue engineering (Lanza *et al.*, 1997). For each application, the porous materials have to fulfil a number of specific requirements. For example, for bone ingrowth the optimum pore size is in the range of 75–250 μm (Pilliar, 1987). On the other hand, for ingrowth of fibrocartilagenous tissue the recommended pore size ranges from 200 to 300 μm (Elema *et al.*, 1990). Besides pore size, other parameters play a role, such as compressibility, pore interconnectivity, pore interconnection throat size, and possibly degradibility of the porous material (de Groot *et al.*, 1990).

Although porosity can also be discerned as a different class of surface irregularity, the following sections will consider porosity as microtexture, much like other surface features. This choice is based on the many reports that emphasize the importance of this type of surface morphology for cell and tissue response.

Preparation of surface microtexture

For the production of microtextured implant surfaces, numerous techniques are available ranging from simple manual scratching to more controlled fabrication methods. For example, from semiconductor technology, photolithographic techniques used in conjunction with reactive plasma and ion-etching, LIGA and electroforming, have become available. Deep reactive ion etching (DRIE) enhances the depth of surface etched features and gives parallel sidewalls—it is especially well suited for microelectromechanical systems (MEMS) fabrication. μCP allows patterns to be transferred to biomaterial surfaces by a rubber stamp. Because these techniques are relatively fast and cheap, and also allow the texturing of surfaces of reasonable size, they appear to be promising for biomedical research and applications. Other methods that offer the ability to texture and pattern surfaces include UV laser machining, electron-beam etching, and ion-beam etching.

Reactive plasma and ion etching

For this method the material, usually silicon, is first cleaned and dried with filtered air (den Braber *et al.*, 1998a; Hoch *et al.*, 1996; Jansen *et al.*, 1996). Then it is coated with a primer and PR material. Photolithography is used to create a micropattern in the PR layer. Masks with predetermined dimensions are exposed with either UV light or electron beams depending on the size of the required surface configuration. Subsequently, the exposed resist is developed and rinsed off. Finally, this lithographically defined PR pattern is transferred into the underlying material by etching. This etching can be performed under wet or dry conditions. In the first situation, materials are placed in chemicals. Etch direction is along the crystal planes of the material. In the second situation, dry etching is performed using directed ions from a plasma or ion beam as etchants. This technique of physical etching allows a higher resolution than the wet technique. It is also applicable in noncrystalline materials because of the etch directionality without using crystal orientation. Finally, after the etching process, the remaining resist is removed. If a substrate is formed with microgrooves, the dimensions of the texture are usually described in pitch (or spacing), ridge width, and groove width (von Recum and van Kooten, 1995).

Plasma and ion etching techniques can be used to create micropatterns in a wide variety of biopolymers. The micropatterns can be prepared directly in the polymer surface or transferred into the polymer surface via solvent-casting or injection-molding methods, whereby a micropatterned silicon wafer is used as a template (Fig. 3.2.15-1).

Liga

Another technology suitable for creating surface microtextures is the so-called LIGA process (Rogner *et al.*, 1992). LIGA refers to the German "Lithographie, Galvanoformung, Abformung" (lithography, electroplating, molding). The LIGA technique differs completely from that described in the preceding section, since it is not based on etching. In the LIGA process a thick X-ray-resistant layer is exposed to synchrotron radiation using a special X-ray mask membrane. Subsequently, the exposed layer is developed, which results in the desired resist structure. Then, metal is deposited

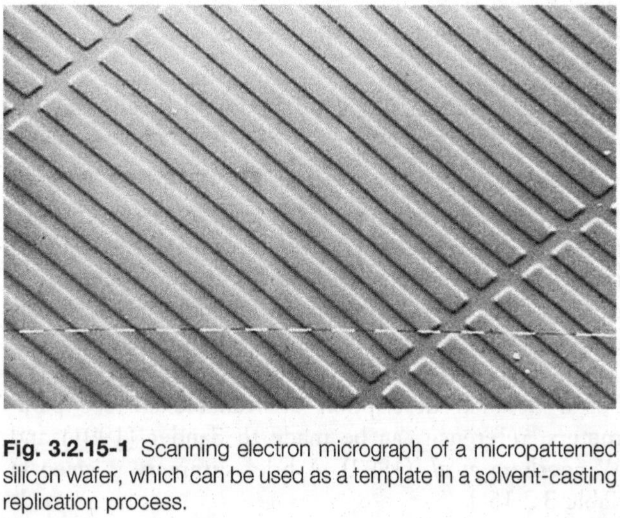

Fig. 3.2.15-1 Scanning electron micrograph of a micropatterned silicon wafer, which can be used as a template in a solvent-casting replication process.

onto the remaining resist structure by galvanization. After removal of the remaining resist either a metal structure or mold for subsequent cost-effective replication processes is achieved.

μCP

The μCP method, developed in the laboratory of George Whitesides, provides a simple method to create patterns over large surface areas at the micro and even nanoscale (Kumar *et al.*, 1994). A master silicon template or mold is formed by conventional photolithographic and etching methods generating the micron-scale pattern of interest. Onto that template, a curable silicone elastomer is poured. When the silicone polymer cures, it is peeled off and then serves as a rubber stamp. The stamp can be "inked" in thiols, silanes, proteins or other polymers (see Section 3.2.14). Flat and curved surfaces can be patterned with these μCP stamps.

Parameters for the assessment of surface microtexture

Since the final biological performance of a microtextured surface is determined by the size and dimensions of the surface features, specific surface parameters have to be provided to describe and define the surface structure.

The definition of surface parameters is mostly based on a two-dimensional profile section, Occasionally, three-dimensional profiles are created (see the next two sections).

In general, for the quantitative description of surface microtexture, three parameters can be used:

1. Amplitude parameters, to obtain information about height variations
2. Spacing parameters, to describe the spacing between features
3. Hybrid parameters, a combination of height and spacing parameters

These parameters are presented as Ra, Rq, Rt, Rz, Rsk, Rku (amplitude parameters), Scx, Scy, Sti (spacing parameters), and Δq and λq (hybrid parameters). The R-parameters are denominations for a two-dimensional description. The S-parameters stand for a three-dimensional evaluation. These S-denominations are generally accepted since the work of Stout *et al.* (1993). For a detailed description of available surface parameters, reference can be made to Sander (1991) and Wennerberg *et al.* (1992). A brief summary is given in Table 3.2.15-1.

Further, it has to be emphasized that for a correct assessment of surface parameters various requirements have to be met. A first condition is the provision of a reference line to which measurements can be related. Also, surface parameters have to be determined with a clear separation between roughness and waviness components. This separation has to be achieved by an electronic filtering procedure. In view of this, perhaps the most important measurement requirements are the parameters measuring length over the substrate surface and cutoff wavelength of the filter used. Measuring or tracing length has to be described in terms of real evaluation length (lm) and pre- and overtravel (lv and ln). The function of the electronic filter is to eliminate waviness and roughness frequencies out of the surface profile. As surface features differ in both their wavelength and surface profile depths, various filters are available. The filter type to be selected for a specific surface profile is defined in DIN standards. Use of the wrong filter will result in incorrect measurements (Sander, 1991).

Characterization of surface topography

Various methods are available to describe surface features. SEM can be used to obtain a qualitative image of the surface geometry. Contact and noncontact profilometry are methods to quantify the surface roughness.

Contact profilometry

The principle of contact profilometry is that a finely pointed stylus moves over the detected area. The vertical movements of the stylus are switched into numerical information. This method results in a two-dimensional description of the surface. The advantage of contact profilometry is that the method is inexpensive, direct, and reproducible. Contact profilometry can be applied on a wide variety of materials. The major disadvantage is that the diameter of the pointed stylus limits its use to surface features larger than the stylus point diameter. Another problem is that, because of the physical contact between the stylus and substrate surface, distortion of the surface profile can occur.

Noncontact profilometry

In this method, the pointed stylus is replaced by a light or laser spot. This spot never touches the substrate surface. The light or laser beam is focused on the surface and the light is reflected and finally converted to an electrical signal. In this way both two- and three-dimensional

Table 3.2.15-1 Definition of surface parameters

Parameter	Definition
lm	Evaluation length = the horizontal limitation for the assessment of surface parameters
lv	Pre-travel length = the distance traversed by the tracing system over the sample before the tracing (lt) starts
ln	Over-travel length = the distance traversed or area scanned by the tracing system over the sample after the tracing (lt)
lt	Tracing length = the distance traversed by the tracing system when taking a measurement. It comprises the pre- and overtravel, and the evaluation length
le	Sampling length = a standardized number of evaluation lengths/areas as required to obtain a proper surface characterization
Ra/Sa	Arithmetical mean roughness = the arithmetical average value of all vertical departures of the profile or surface from the mean line throughout the sampling length/area
Rq/Sq	Root-mean square roughness = the root-mean square value of the profile or surface departures within the sampling length/area
Rt/St	Maximum roughness depth = the distance between the highest and lowest points of the profile or surface within the evaluation length/area
Rz/Sz	Mean peak-to-valley height = the average of the single peak-to valley heights of five adjoining sampling lengths/areas
Rsk/Ssk	Skewness = measure of the symmetry of the amplitude density function (ADF)
ADF	Amplitude density function = the graphical representation of the material distribution within the evaluation length/area
Rku/Sku	Kurtosis = fourth central moment of the profile or surface amplitude density with the evaluation length/area. Kurtosis is the measure of the sharpness of the profile or surface
Rcx/Rcy Scx/Scy	= mean spacing between surface peaks of the surface/area profile along the X or Y direction
Sti	= surface texture index, i.e. min. (Rq/Sq divided by max. Rq/Sq + min. Rsk/Ssk divided by max Rsk/Ssk + min.(q divided by max.)q + min (Rc/Sc divided max. Rc/Sc) divided by 4
Δq	= the root mean square slope of the rough profile throughout the evaluation length/area
λq	= the root mean square of the spacings between local peaks and valleys, taking into account their relative amplitudes and individual spatial frequencies

surface profiles can be created (Fig. 3.2.15-2). Occasionally, techniques are used in which the reflected light is not directly translated to an electrical signal. In these so-called interferometers a surface profile is created by combining light reflecting off the surface with light reflecting off a reference substrate. When those two light bundles combine, the light waves interfere to produce a pattern of fringes, which are used to determine surface height differences.

The resolution of noncontact methods can be in the nanometer range. The limiting factor is the spot size. Several scans have to be taken to obtain a representative surface area. Occasionally, this is impossible or too laborious. In light beam interferometry, an additional disadvantage is that the substrate surface has to provide at least some reflectivity.

Atomic force microscopy

Atomic force microscopy (AFM) is a direct method for determining high-resolution surface patterns (Binnig *et al.*, 1986; van der Werf *et al.*, 1993) (also see Section 3.1.4). In AFM the substrate surface is brought close to a tip on a small cantilever which is attached to a piezo tube. The deflection of the cantilever, generated by interaction forces between tip and substrate surface, is detected and used as an input signal for a measuring

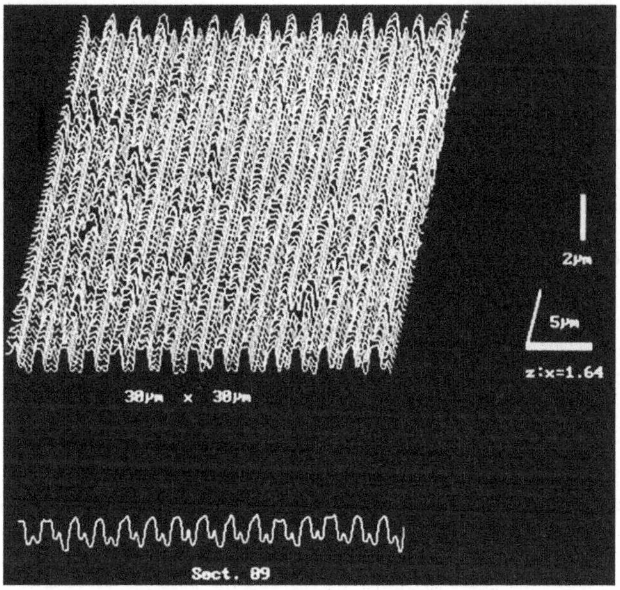

Fig. 3.2.15-2 Results of a confocal laser scanning microscope (CLSM) surface analysis of a microgrooved substratum. CLSM has to be considered as a noncontact technique. A three- and two-dimensional surface representation is obtained, composed from 256 optical *Z* sections. To the right of the 3D surface profile, the size of the scanned area (30 μm^2) and the difference in *X* versus *Z* enlargement can be found (Scale 1:1.64).

system. AFM is frequently used as a contact method. However, noncontact and transient contact modes of analysis are also available. The advantage of AFM above other contact techniques is that AFM is generally not as destructive. Considering resolution, a limiting factor in AFM is again the size of the used tip (Fig. 3.2.15-3). Still, a significantly smaller tip diameter is used compared with conventional contact methods such as profilometry.

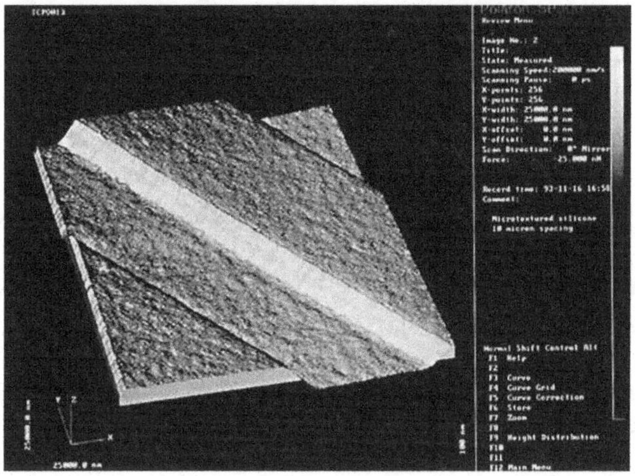

Fig. 3.2.15-3 Three-dimensional representation of an AFM measurement of a silicon wafer provided with 10-µm-wide and 0.5-µm-deep microgrooves. The raised wall of the edge shows a small inclination. This is a distortion due to the size and movement of the tip over the silicon surface.

Biological effects of surface microtexture

The role of standardized surface texture in inducing a specific cellular response is a field of active research. For example, various reports have suggested that a regular surface microtexture can benefit the clinical success of skin penetrating devices by preventing epithelial downgrowth (Brunette *et al.*, 1983; Chehroudi *et al.*, 1988) and reduce the inflammatory response (Campbell *et al.*, 1989) and fibrous encapsulation (Chehroudi *et al.*, 1991) of subcutaneous implants. Closely related to these studies, certain porosities have led to an increase of the vascularity of the healing response and a reduction of collagenous capsule density (Brauker *et al.*, 1995; Sharkawy *et al.*, 1998). The literature on the effect of surface texture on the healing of silicone breast implants is extensive (for example, see Pollock, 1992). Therefore, much current research has been focused on the effect of standardized surface roughness on the soft tissue reaction. Excellent reviews on the effect of surface microtexturing on cellular growth, migration, and attachment have been written by Singhvi *et al.* (1994), von Recum and van Kooten (1995), Brunette (1996), Curtis and Wilkinson (1997), and Folch and Toner (2000).

Hypotheses on contact guidance

Contact guidance is the phenomenon that cells adapt and orient to the substrate surface microtopography (Harrison, 1912). Early studies on contact guidance describe the alignment of cells and focal adhesions to microgrooves with dimensions 1.65–8.96 µm in width and 0.69 µm in depth. This cellular behavior was suggested to be due to the mechanical properties of the cytoskeleton (Dunn, 1982; Dunn and Brown, 1986). The relative inflexibility of cytoskeletal components was considered to prevent bending of cell protrusions over surface configurations with too large an angle.

Later studies and hypotheses focused on the relationships among cell contact site, deposited ECM, surface microtexture, and cell response. For example, a microtextured surface was supposed to possess local differences in surface free energy resulting in a specific deposition pattern of the substratum bound attachment proteins (Brunette, 1996; Maroudas, 1972; von Recum and van Kooten, 1995). The spatial arrangement of the adsorbed proteins and their conformational state were hypothesized to be affected. In addition to wettability properties, the specific geometric dimensions of the cell adhesion sites were suggested to induce a cell orientational effect (Dunn, 1982; Dunn and Brown, 1986; Ohara and Buck, 1979). A recent hypothesis suggests that contact guidance on microtextured surfaces is a part of the cellular efforts to achieve a biomechanical equlibrium condition with a resulting minimal net sum of forces. The signficance of this theory has been described extensively by Ingber (1993, 1994) in his tensegrity models. According to this model, the anisotropic geometry of substratum surface features establishes stress- and shear-free planes that influence the direction of cytokeletal elements in order to create a force economic situation (Oakley and Brunette, 1993, 1995; O'Neill *et al.*, 1990).

The *in vitro* effect of surface microtexturing

A considerable number of *in vitro* studies have been performed to determine which of the hypotheses mentioned in the preceding section can be experimentally supported. Up to this point, we have to emphasize that comparison of the obtained data is difficult because most of the studies had differences in the surface textures of the materials explored. In addition, different bulk materials were also applied. Modern surface feature fabrication methods have allowed more precise surfaces to be fabricated so studies from different groups might be compared.

In the experiments performed by Curtis *et al.* (Clark *et al.*, 1987, 1990, 1991; Curtis and Wilkinson, 1997) with fibroblasts and macrophages cultured on microgrooved glass substrates, groove depth was observed to be more important than groove width in the establishment of contact guidance. Therefore, these experiments believe that cytoskeletal flexibility and the possibility of making cellular protrusions are the determining cellular characteristics for contact guidance. As a consequence of these studies, other reseachers further explored the involvement of cytoskeletal elements in cell orientation processes. Also, the possibility of a relationship between cytoskeletal organization and cell–substrate contact sites was investigated (den Braber *et al.*, 1995, 1996, 1998b; Meyle *et al.*, 1991, 1993; Oakley and Brunette, 1993, 1995; Oakley *et al.*, 1997; Walboomers *et al.*, 1998a, 1999). Although these studies varied in cell type used, substrate surface feature dimensions and substrate bulk chemical composition, the results clearly confirmed that very fine microgrooves (≤ 2 μm) have an orientational effect on both cell body and cytoskeletal elements. Transmission electron microscopy observations showed that cells were only able to penetrate into very shallow (≤ 1 μm) or wide (≥ 5 μm) microgrooves. Cells were also observed to possess cell adhesion structures that were wrapped around the edge of a ridge or attached to the wall of the ridge. On the basis of these findings, these investigators suggested that the mechanical properties of cellular structures can never be the only determining factor in contact guidance.

Further, a mechanical model to explain contact guidance suggests that the "surface feature stimulus" is transduced to the cytoskeleton via cell contact sites and cell surface receptors. In this model, the cytoskeleton is considered as a static structure. This is incorrect. The cytoskeleton is a highly dynamic system (Lackie, 1986), which is constantly broken down and elongated in living cells. Consequently, if the mechanical theory is still true, the fundamentals should be derived from other processes than just the remodeling of the cytoskeleton (Walboomers *et al.*, 1998a). Studies on cell nuclear connections to the cytoskeleton may offer insights into the relationships between surface features and cell behavior (Maniotis *et al.*, 1997).

Apart from changes in cell size, shape, and orientation, surface microtopography has been reported to influence other cell processes. For example, several studies described changes in cellular differentiation, DNA/RNA transcription, cellular metabolism, and cellular protein production of cells cultured on microtextured surfaces (Chou *et al.*, 1995; Hong and Brunette 1987; Matsuzaka *et al.*, 1999; von Recum and van Kooten, 1995; Singhvi *et al.*, 1994; Wójciak-Stothard *et al.*, 1995). A study using μCP surfaces with square cell adhesive and nonadhesive domains has shown that where surface adhesive domains are small (<75 μm), apoptosis levels in endothelial cells is high (particularly so for 5 μm × 5 μm domains) and when cells are placed on larger domains, cell spreading and growth occurs (Chen *et al.*, 1997). Whether these additional effects have to be considered as independent phenomena is still a topic of discussion. According to Hong and Brunette (1987), the good news was that surface microtopography can enhance the production of specific, perhaps favorable proteins. On the other hand, the production or secretion of less favorable metabolic products can also be enhanced. If this occurs, this might have a deleterious effect on the overall cell response. For example, a rise in the production or release of proteinases may not be beneficial for connective tissue cell response. This example shows that, at least at the molecular level, the regulation of cell function by substrate surface microtexture may be a complex affair.

The *in vivo* effect of surface microtexturing

Based upon interesting results from *in vitro* experiments, *in vivo* studies with microtextured implants have been performed. Unfortunately, the results from the various studies are not consistent. For example, in some animal experiments it was demonstrated that silicone-coated filters and bulk SR implants provided with surface features of 1–3 μm showed a minimal inflammatory response with direct fibroblast attachment and a very reduced connective capsule (Campbell and von Recum, 1989; Schmidt and von Recum, 1991, 1992). In contrast, other animal studies suggested that implant surface microgrooves were unable to influence the wound healing process at all (den Braber *et al.*, 1997; Walboomers *et al.*, 1998b). These differing results may hint at multiple surface-texture-related factors that are not yet identified and controlled.

Besides the effect on wound healing, microtextured implants have also been used to inhibit epidermal downgrowth along skin penetrating devices (Chehroudi *et al.*, 1989, 1990, 1992). This downgrowth is considered as a major failure mode for this type of implant. Indeed, the experiments suggested that epidermal downgrowth can be prevented or delayed by percutaneous devices provided with surface microgrooves.

Directions for further developments

Considering the *in vitro* experiments, none of the earlier mentioned hypotheses to explain contact guidance has been fully supported. Therefore, based on various findings we suggest a new theory that is a refinement of the "mechanical" theory discussed earlier. The breakdown and formation of fibrous cellular components, especially in the filopodium, is influenced by the microgrooves. These microgrooves create a pattern of mechanical stress, which affects cell spreading and causes the alignment of cells. On the other hand, we must also notice that the ECM possesses mechanical properties. The ECM is not a rigid structure, but a dynamic mass of molecules. Many *in vitro* studies have already indicated that cell-generated forces of tension and traction can reorganize the ECM into structures that direct the behavior of single cells (Erickson, 1994; Choquet *et al.*, 1997; Janmey and Chaponnier, 1995; Janmey, 1998). As cells cannot penetrate very shallow or small grooves, we suppose that on those surfaces the forces as exerted by the cells will result in an enhanced reorganization of the deposited ECM proteins. Consequently, contact guidance and other cell behaviors are induced. No doubt, cell surface receptors and inside–outside cell signaling phenomena play an important role in this process. As far as *in vivo* applications of surface microtexturing, more research has to be done to learn and understand the full impact of surface microtexturing for medical devices. A first step is the development of techniques that enable the production of standardized microstructures on nonplanar surfaces. Evidently, this development will benefit not only biomaterial research, but also the production of microelectronic, mechanical, and optical devices and subsytems. As a second step, the relationship between the surface topographical design of an implant and histocompatibility has to be further documented. These studies must focus not only on the soft tissue response; they must also involve bone tissue behavior.

Bibliography

Binnig, G., Quate, C. F., and Gerber, C. (1986). Atomic force microscopy. *Phys. Rev. Lett.* **56**: 930–933.

Brauker, J. H., Carr-Brendel, V. E., Martinson, L. A., Crudele, J., Johnston, W. D., and Johnson, R. C. (1995). Neovascularization of synthetic membranes directed by membrane microarchitecture. *J. Biomed. Mater. Res.* **29**: 1517–1524.

Brunette, D. M. (1996). Effects of surface topography of implant materials on cell behavior in vitro and in vivo. in *Nanofabrication and Biosystems*, H. C. Hoch, L. W. Jelinski, and H. G. Craighead, eds. Cambridge University Press, Cambridge, UK, pp. 335–355.

Brunette, D. M., Kenner, G. S., and Gould, T. R. L. (1983). Grooved titanium surfaces orient growth and migration of cells from human gingival explants. *J. Dent. Res.* **62**: 1045–1048.

Campbell, C. E., and von Recum, A. F. (1989). Microtopography and soft tissue response. *J. Invest. Surg.* **2**: 51–74.

Chehroudi, B., Gould, T. R., and Brunette, D. M. (1988). Effects of a grooved epoxy substratum on epithelial cell behavior in vitro and in vivo. *J. Biomed. Mater. Res.* **22**: 459–473.

Chehroudi, B., Gould, T. R. L., and Brunette, D. M. (1989). Effects of a grooved titanium-coated implant surface on epithelial cell behavior *in vitro* and *in vivo*. *J. Biomed. Mater. Res.* **23**: 1067–1085.

Chehroudi, B., Gould, T. R., and Brunette, D. M. (1991). A light and electron microscope study of the effects of surface topography on the behavior of cells attached to titanium-coated percutaneous implants. *J. Biomed. Mater. Res.* **25**: 387–405.

Chehroudi, B., Gould, T. R. L., and Brunette, D. M. (1990). Titanium coated micromachined grooves of different dimensions affect epithelial and connective tissue cells differently *in vivo*. *J. Biomed. Mater. Res.* **24**: 1203–1219.

Chehroudi, B., Gould, T. R. L., and Brunette, D. M. (1992). The role of connective tissue in inhibiting epithelial downgrowth on titanium-coated percutaneous implants. *J. Biomed. Mater. Res.* **26**: 493–515.

Chen, C. S., Mrksich, M., Huang, S., Whitesides, G. M., and Ingber, D. E. (1997). Geometric control of cell life and death. *Science* **276**: 1425–1428.

Choquet, D., Felsenfeld D. P., and Sheetz, M. P. (1997). Extracellular matrix rigidity causes strengthening of integrin-cytoskeleton linkages. *Cell* **88**: 39–48.

Chou, L. S., Firth, J. D., Uitto, V. J., and Brunette, D. M. (1995). Substratum surface topography alters cell shape and regulates fibronectin mRNA level, mRNA stability, secretion and assembly in human fibroblasts. *J. Cell Sci.* **108**: 1563–1573.

Clark, P., Connoly, P., and Curtis, A. S. G. (1987). Topographical control of cell behavior I: simple step clues. *Development* **99**: 439–448.

Clark, P., Connoly, P., and Curtis, A. S. G. (1990). Topographical control of cell behavior II: multiple grooved substrata. *Development* **108**: 635–644.

Clark, P., Connoly, P., and Curtis, A. S. G. (1991). Cell guidance by ultrafine topography *in vitro*. *J. Cell Res.* **86**: 9–24.

Curtis, A. S. G., and Wilkinson, C. (1997). Topographical control of cells. *Biomaterials* **18**: 1573–1583.

den Braber, E. T., Ruijter, J. E., de Smits, H. T. J., Ginsel, L. A., Recum, A. F., and von Jansen, J. A. (1995). Effects of parallel surface microgrooves and surface energy on cell growth. *J. Biomed. Mat. Res.* **29**: 511–518.

den Braber, E. T., Ruijter, J. E., de Smits, H. T. J., Ginsel, L. A., Recum, A. F., and von Jansen, J. A. (1996). Quantitative analysis of cell proliferation and orientation on substrata with uniform parallel surface micro grooves. *Biomaterials* **17**: 1093–1099.

den Braber, E. T., Ruijter, J. E., and Jansen, J. A. (1997). The effect of a subcutaneous silicone rubber implant with shallow surface micro grooves on the surrounding tissue in rabbits. *J. Biomed. Mater. Res.* **37**: 539–547.

den Braber, E. T., Jansen, H. V., de Boer, M. J., Croes, H. J. E., Elwenspoek, M., Ginsel, L. A., and Jansen, J. A. (1998a). Scanning electron microscopic, transmission electron microscopic, and confocal laser scanning microscopic observation of fibroblasts cultured on microgrooved surfaces of bulk titanium substrata. *J. Biomed. Mater. Res.* **40**: 425–433.

den Braber, E. T., Ruijter, J. E., Ginsel, L. A., von Recum, A. F., and Jansen, J. A. (1998b). Orientation of ECM protein deposition, fibroblast cytoskeleton, and attachment complex components on silicone microgrooved surfaces. *J. Biomed. Mater. Res.* **40**: 291–300.

Dunn, G. A. (1982). Contact guidance of cultured tissue cells: a survey of potentially relevant properties of the substratum. in *Cell Behavior*, R. Bellairs, A. S. G. Curtis and G. A. Dunn, eds. Cambridge University Press, Cambridge, UK, pp. 247–280.

Dunn, G. A., and Brown, A. F. (1986). Alignment of fibroblasts on grooved surfaces described by a simple geometric transformation. *J. Cell Sci.* **83**: 313–340.

Elema. H., Groot, J. H., de Nijenhuis, A. J., Pennings, A. J., Veth, R. P. H., Klompmaker, J., and Jansen, H. W. B. (1990). Biological evaluation of porous biodegradable polymer implants in menisci. *Colloid Polymer Sci.* **268**: 1082–1088.

Erickson, H. P. (1994). Reversible unfolding of fibronectin type III and immunoglobulin domains provides the structural basis for stretch and elasticity of titin and fibronectin. *Proc. Natl. Acad. Sci. USA* **91**: 10114–10118.

Folch, A., and Toner, M. (2000). Microengineering of cellular interactions. *Annu. Rev. Biomed. Eng.* **2**: 227–256.

Groot, J. H., de Nijenhuis, A. J., Bruin, P., Pennings, A. J., Veth, R. P. H., Klompmaker, J., and Jansen, H. W. B. (1990). Preparation of porous biodegradable polyurethanes for the reconstruction of meniscal lesions. *Colloid Polymer Sci.* **268**: 1073–1081.

Harrison, R. G. (1912). The cultivation of tissues in extraneous media as a method of morphogenetic study. *Anat. Rec.* **6**: 181–193.

Hoch, H. C., Jelinski, L. W., and Craighead, H. G. (1996). *Nanofabrication and Biosystems.* Cambridge University Press, Cambridge, UK.

Hong, H. L., and Brunette, D. M. (1987). Effect of cell shape on proteinase secretion. *J. Cell Sci.* **87**: 259–267.

Ingber, D. E. (1993). Cellular tensegrity; defining new rules of biological design that govern the cytoskeleton. *J. Cell Sci.* **104**: 613–627.

Ingber, D. E. (1994). Cellular tensegrity and mechanochemical transduction. *in Cell Mechanics and Cellular Engineering*, V. C. Mow, F. Guilak, R. Tran-Son-Tay, and R. M. Hochmuth, eds. Springer-Verlag, New York, pp. 329–342.

Janmey, P. A., and Chaponnier, C. (1995). Medical aspects of the actin cytoskeleton. *Curr. Opin. Cell Biol.* **7**: 111–117.

Janmey, P. A. (1998). The cytoskeleton and cell signaling: component localization and mechanical coupling. *Physiol. Rev.* **78**: 763–781.

Jansen, H. V., Gardeniers, J. G. E., de Boer, M. J., Elwenspoek, M. E., and Fluitman, J. H. J. (1996). A survey on the reactive ion etching of silicon in microtechnology. *J. Micromech. Microeng.* **6**: 14–28.

Kumar, A., Biebuyck, H. A., and Whitesides, G. M. (1994). Patterning self-assembled monolayers: applications in materials science. *Langmuir* **10**(5): 1498–1511.

Lackie, J. M. (1986). *Cell Movement and Cell Behaviour.* Allen & Unwin, London.

Lanza, R. P., Langer, R., and Chick, W. L. (1997). *Principles of Tissue Engineering.* Academic Press, San Diego.

Maniotis, A. J.,Chen, C. S., and Ingber, D. E. (1997). Demonstration of mechanical connections between integrins, cytoskeletal filaments, and nucleoplasm that stablize nuclear structure. *Proc. Natl. Acad. Sci. USA* **94**: 849–854.

Maroudas, N. G. (1972). Anchorage dependence: correlation between amount of growth and diameter of bead, for single cells grown on individual glass beads. *Exp. Cell Res.* **74**: 337–342.

Matsuzaka, K., Walboomers, X. F., de Ruijter, J. E., and Jansen, J. A. (1999). The effect of poly-L-lactic acid with parallel surface micro groove on osteoblast-like cells *in vitro*. *Biomaterials* **20**(14): 1293–1301.

Meyle, J., von Recum, A. F., and Gibbesch, B. (1991). Fibroblast shape conformation to surface micromorphology. *J. Appl. Biomat.* **2**: 273–276.

Meyle, J., Gültig, K., Wolburg, H., and von Recum, A. F. (1993). Fibroblast anchorage to microtextured surfaces. *J. Biomed. Mater. Res.* **27**: 1553–1557.

Oakley, C., and Brunette, D. M. (1993). The sequence of alignment of microtubules, focal contacts and actin filaments in fibroblasts spreading on smooth and grooved titanium substrata. *J. Cell Sci.* **106**: 343–354.

Oakley, C., and Brunette, D. M. (1995). Topographic compensation: guidance and directed locomotion of fibroblasts on grooved micromachined substrata in the absence of microtubules. *Cell Motil. Cytoskeleton* **31**: 45–58.

Oakley, C., Jaeger, N. A., and Brunette, D. M. (1997). Sensitiv of fibroblasts and their cytoskeletons to substratum topographies: topographic guidance and topographic compensation by micromachined grooves of different dimensions. *Exp. Cell Res.* **234**: 413–424.

Ohara, P. T., and Buck, R. C. (1979). Contact guidance *in vitro*. A light, transmission, and scanning electron microscopic study. *Expl. Cell Res.* **121**: 235–249.

O'Neill, C., Jordan, P., and Riddle, P. (1990). Narrow linear strips of adhesive substratum are powerful inducers of both growth and total focal contact area. *J. Cell Sci.* **95**: 577–586.

Pilliar, R. M. (1987). Porous-surfaced metallic implants for orthopaedic applications. *J. Biomed. Mater. Res.* **21**: 1–33.

Pollock, H. (1992): Breast capsular contracture: A retrospective study of textured versus smooth silicone implants. *Plast. Reconstr. Surg.* **91**(3): 404–407.

Rogner, A., Eichner, J., Münchmeyer, D., Peters, R.-P., and Mohr, J. (1992). The LIGA technique—what are the opportunities? J. *Micromech. Microeng.* **2**: 133–140.

Sander, M. (1991). *A Practical Guide to the Assessment of Surface Texture.* Feinprüf Perthen Gmbh, Göttingen.

Schmidt, J. A., and von Recum, A. F. (1991). Texturing of polymer surfaces at the cellular level. *Biomaterials* **12**: 385–389.

Schmidt, J. A., and von Recum, A. F. (1992). Macrophage response to microtextured silicone. *Biomaterials* **13**: 1059–1069.

Sharkawy, A., Klitzman, B., Truskey, G. A., and Reichert, W. M. (1998). Engineering the tissue which encapsulates subcutaneous implants. II. Plasma–tissue exchange properties. *J. Biomed. Mater. Res.* **40**: 586–597

Singhvi, R., Stephanopoulos, G., and Wang, D. I. C. (1994). Review: effects of substratum morphology on cell physiology. *Biotechnol. Bioeng.* **43**: 764–771.

Stout K.-J., Sullivan, P. J., Dong, W. P., Mainsah, E., Luo, N., Mathia, T., and Zahouni, H. (1993). The devlopment of methods for the characterization of roughness in three dimensions. EUR 15178 EN of Commission of the European Communities, University of Birmingham, Birmingham, UK.

von Recum, A. F., and van Kooten, T. G. (1995). The influence of microtopography on cellular response and the implications for silicone implants. *J. Biomater. Sci. Polymer Ed.* **7**: 181–198.

Walboomers, X. F., Croes, H. J. E., Ginsel, L. A., and Jansen, J. A. (1998a). Growth behavior of fibroblasts on microgrooved polystyrene. *Biomaterials* **19**: 1861–1868.

Walboomers, X. F., Croes, H. J. E., Ginsel, L. A., and Jansen, J. A. (1998b). Microgrooved subcutaneous implants in the goat. *J. Biomed. Mater. Res.* **42**: 634–641.

Walboomers, X. F., Monaghan, W., Curtis, A. S. G., and Jansen, J. A. (1999). Attachment of fibroblasts on smooth and microgrooved polystyrene. *J. Biomed. Mater. Res.* **46**(2): 212–220.

Wennerberg, A., Albrektsson, T., Ulrich, H., and Krol, J. (1992). An optical three-dimensional technique for topographical descriptions of surgical implants. *J. Biomed. Eng.* **14**: 412–418.

Werf, K. O., van der, Putman, C. A. J., de Grooth, B. G., Segerink, F. B., Schipper, E. H., van Hulst, N. F., and Greve, J. (1993). Compact stand-alone atomic force microscope. *Rev. Sci. Instrum.* **64**: 2892–2897.

Wójciak-Stothard, B., Madeja, Z., Korohoda, W., Curtis, A., and Wilkinson, C. (1995). Activation of macrophage-like cells by multiple grooved substrata: topographical control of cell behaviour. *Cell Biol. Int.* **19**: 485–490.

3.2.16 Surface-immobilized biomolecules

Allan S. Hoffman and Jeffrey A. Hubbell

Biomolecules such as enzymes, antibodies, affinity proteins, cell receptor ligands, and drugs of all kinds have been chemically or physically immobilized on and within biomaterial supports for a wide range of therapeutic, diagnostic, separation, and bioprocess applications. Immobilization of heparin on polymer surfaces is one of the earliest examples of a biologically functional biomaterial. Living cells may also be combined with biomaterials, and the fields of cell culture, artificial organs, and tissue engineering are additional, important examples. These "hybrid" combinations of natural and synthetic materials confer "biological functionality" to the synthetic biomaterial. Since many sections in this text cover many aspects of this topic, including adsorption of proteins and adhesion of cells and bacteria on biomaterial surfaces, NFSs, cell culture, tissue engineering, artificial organs, drug delivery, and others, this section will focus on the methodology involving physical adsorption and chemical immobilization of biomolecules on biomaterial surfaces, especially for applications requiring bioactivity of the immobilized biomolecule.

Among the different classes of biomaterials that could be biologically modified, polymers are especially interesting because their surfaces may contain reactive groups *de novo*, or they may be readily derivatized with reactive groups that can be used to covalently link biomolecules. Another advantage of polymers as supports for biomolecules is that the polymers may be fabricated in many forms, including films, membranes, tubes, fibers, fabrics, particles, capsules, and porous structures. Furthermore, polymer compositions vary widely, and molecular structures include homopolymers, and random, alternating, block, and graft copolymers. Living anionic polymerization techniques, along with newer methods of living free radical polymerizations, now provide fine control of molecular weights with narrow distributions. The molecular forms of solid polymers include un-cross-linked chains that are insoluble at physiologic conditions, cross-linked networks, physical blends, and IPNs (e.g., Piskin and Hoffman, 1986; see also Section 3.2.2). When surfaces of metals or inorganic glasses or ceramics are involved, biological functionality can sometimes be added via a chemically immobilized or physisorbed polymeric or surfactant adlayer, or by use of techniques such as plasma gas discharge to deposit polymer compositions having functional groups (see also Section 3.2.14).

Patterned surfaces

Biomaterial surfaces may be functionalized uniformly or in geometric patterns (Bernard *et al.*, 1998; Blawas and

Reichert, 1998; James *et al.*, 1998; Kane *et al.*, 1999; Ito, 1999; Folch and Toner, 2000). Sometimes the patterned surfaces will have regions that repel proteins ("nonfouling" compositions) while others may contain covalently-linked cell receptor ligands (Neff *et al.*, 1999; Alsberg *et al.*, 2002; Csucs *et al.*, 2003; VandeVondele *et al.*, 2003), or may have physically adsorbed cell adhesion proteins (McDevitt *et al.*, 2002; Ostuni *et al.*, 2003). There has also evolved a huge industry based on "biochips" that contain microarrays of immobilized, single-stranded DNA (for genomic assays) or peptides or proteins (for proteomic assays) (Houseman and Mrksich, 2002; Lee and Mrksich, 2002). The majority of these microarrays utilize inorganic silica chips rather than polymer substrates directly, but it is possible to incorporate functionality through chemical modification with silane chemistries (Puleo, 1997) or adsorption of a polymeric adlayer (Scotchford *et al.*, 2003; Winkelmann *et al.*, 2003). A variety of methods have been used for the production of these patterned biochips, including photocontrolled synthesis (Ellman and Gallop, 1998; Folch and Toner, 2000), microfluidic fluid exposure (Ismagilov *et al.*, 2001), and protection with adhesive organic protecting layers that are lifted off after exposure to the biomolecular treatment (Jackman *et al.*, 1999).

Immoblized biomolecules and their uses

Many different biologically functional molecules can be chemically or physically immobilized on polymeric supports (Table 3.2.16-1) (Laskin, 1985; Tomlinson and Davis, 1986). When some of these solids are water-swollen they become hydrogels, and biomolecules may be immobilized on the outer gel surface as well as within the swollen polymer gel network. Examples of applications of these immobilized biological species are listed in Table 3.2.16-2. It can be seen that there are many diverse uses of such biofunctional systems in both the medical and biotechnology fields. For example, a number of immobilized enzyme supports and reactor systems (Table 3.2.16-3) have been developed for therapeutic uses in the clinic (Table 3.2.16-4) (De Myttenaere *et al.*, 1967; Kolff, 1979; Sparks *et al.*, 1969; Chang, 1972; Nose *et al.*, 1983, 1984; Schmer *et al.*, 1981; Callegaro and Denri, 1983; Lavin *et al.*, 1985; Sung *et al.*, 1986).

Table 3.2.16-1 Examples of biologically active molecules that may be immobilized on or within polymeric biomaterials

Proteins/peptides	Drugs
Enzymes	Antithrombogenic agents
Antibodies	Anticancer agents
Antigens	Antibiotics
Cell adhesion molecules	Contraceptives
"Blocking" proteins	Drug antagonists
	Peptide, protein drugs
Saccharides	Ligands
Sugars	Hormone receptors
Oligosaccharides	Cell surface receptors (peptides, saccharides)
Polysaccharides	Avidin, biotin
Lipids	Nucleic acids, nucleotides
Fatty acids	Single or double-stranded DNA, RNA (e.g., antisense oliogonucleotides)
Phospholipids	
Glycolipids	
Other	
Conjugates or mixtures of the above	

Table 3.2.16-2 Application of immobilized biomolecules and cells

Enzymes	Bioreactors (industrial, biomedical) Bioseparations Biosensors Diagnostic assays Biocompatible surfaces
Antibodies, peptides, and other affinity molecules	Biosensors Diagnostic assays Affinity separations Targeted drug delivery Cell culture
Drugs	Thrombo-resistant surfaces Drug delivery systems
Lipids	Thrombo-resistant surfaces Albuminated surfaces
Nucleic acid derivatives and nucleotides	DNA probes Gene therapy
Cells	Bioreactors (industrial) Bioartificial organs Biosensors

Immobilized cell ligands

Cell interactions with foreign materials are usually mediated by a biological intermediate, such as adsorbed

Table 3.2.16-3 Bioreactors supports and designs

"Artificial cell" suspensions (microcapsules, RBC ghosts, liposomes, reverse micelles [w/o] microspheres)
Biologic supports (membranes and tubes of collagen, fibrin ± GAGs)
Synthetic supports (porous or asymmetric hollow fibres, particulates, parallel plate devices)

Table 3.2.16-4 Examples of immobilized enzymes in therapeutic bioreactors

Medical application	Substrate	Substrate action
Cancer treatment		
L-Asparaginase	Asparagine	Cancer cell nutrient
L-Glutaminase	Glutamine	Cancer cell nutrient
L-Arginase	Arginine	Cancer cell nutrient
L-Phenylalanine lyase	Phenylalanine	Toxin
Indole-3-alkane α hydroxylase	Tryptophan	Cancer cell nutrient
Cytosine deaminase	5-Fluorocytosine	Toxin
Liver failure (detoxification)		
Bilirubin oxidase	Bilirubin	Toxin
UDP-gluceronyl transferase	Phenolics	Toxin
Other		
Heparinase	Heparin	Anticoagulant
Urease	Urea	Toxin

Table 3.2.16-5 Some advantages and disadvantages of immobilized enzymes

Advantages
Enhanced stability
Can modify enzyme microenvironment
Can separate and reuse enzyme
Enzyme-free product
Lower cost, higher purity product
No immunogenic response (therapeutics)
Disadvantages
Difficult to sterilize
Fouling by other biomolecules
Mass transfer resistances (substrate in and product out)
Adverse biological responses of enzyme support surfaces (*in vivo* or *ex vivo*)
Greater potential for product inhibition

proteins. An approach using biologically functional materials can be much more direct, by adsorbing or covalently grafting ligands for cell-surface adhesion receptors to the material surface. This has been accomplished with peptides grafted randomly over a substrate (Massia and Hubbell, 1991) as well as with peptides presented in a pre-clustered manner (Irvine *et al.*, 2001). The latter has important advantages: Cells normally cluster their adhesion receptors into assemblies referred to as focal contacts, and preassembly confers benefits in terms of both adhesion strength (Ward and Hammer, 1993) and cell signaling (Maheshwari *et al*, 2000). In addition to peptides, saccharides have also been grafted to polymer surfaces to confer biological functionality (Griffith and Lopina, 1998; Chang and Hammer, 2000).

Specific biomolecules can be immobilized in order to control cellular interactions; one important example is the polypeptide growth factor. Such molecules can be immobilized and retain their ability to provide biological cues that signal specific cellular behavior, such as support of liver-specific function in hepatocytes (Kuhl and Griffith-Cima, 1996), induction of neurite extension in neurons (Sakiyama-Elbert *et al.*, 2001), induction of angiogenesis (Zisch *et al.*, 2001), or the differentiation of mesenchymal stem cells into bone-forming osteoblasts (Lutolf *et al.*, 2003b). Other molecules may be immobilized that can partake in enzymatic reactions at the surface. McClung *et al.* (2001, 2003) have immobilized lysines, whose ε-amino groups may interact with preadsorbed tissue plasminogen activator (tPA) during coagulation, to enhance fibrin clot dissolution at that surface.

Some of the advantages and disadvantages of immobilized biomolecules are listed in Table 3.2.16-5, using enzymes as an example.

Immobilization methods

There are three major methods for immobilizing biomolecules (Table 3.2.16-6) (Stark, 1971; Zaborsky, 1973; Dunlap, 1974). It can be seen that two of them are physically based, while the third is based on covalent or "chemical" attachment to the support molecules. Thus, it is important to note that the term "immobilization" can refer either to a transient or to a long-term localization of the biomolecule on or within a support. In the case of a drug delivery system, the immobilized drug is supposed to be released from the support, while an immobilized

Table 3.2.16-6 Biomolecule immobilization methods

Physical adsorption
van der Waals
Electrostatic
Affinity
Adsorbed and cross-linked
Physical "entrapment"
Barrier systems
Hydrogels
Dispersed (matrix) systems
Covalent attachment
Soluble polymer conjugates
Solid surfaces
Hydrogels

enzyme or adhesion-promoting peptide in an artificial organ is designed to remain attached to or entrapped within the support over the duration of use. Either physical or chemical immobilization can lead to "permanent" or long-term retention on or within a solid support, the former being due to the large size of the biomolecule. If the polymer support is biodegradable, then the chemically immobilized biomolecule may be released as the matrix erodes or degrades away. The immobilized biomolecule may also be susceptible to enzymatic degradation *in vivo*, and this remains an interesting aspect that has received relatively little attention.

A large and diverse group of methods have been developed for covalent binding of biomolecules to soluble or solid polymeric supports (Weetall, 1975; Carr and Bowers, 1980; Dean *et al.*, 1985; Shoemaker *et al.*, 1987; Yang *et al.*, 1990; Park and Hoffman, 1990; Gombotz and Hoffman, 1986; Schense and Hubbell, 1999; Lutolf *et al.*, 2003a). Many of these methods are schematically illustrated in Fig 3.2.16-1. The same biomolecule may be immobilized by many different methods; specific examples of the most common chemical reactions utilized are shown in Fig. 3.2.16-2.

For covalent binding to an inert solid polymer surface, the surface must first be chemically modified to provide reactive groups (e.g., –OH, $-NH_2$, –COOH, –SH, or $-CH{=}CH_2$) for the subsequent immobilization step. If the polymer support does not contain such groups, then it is necessary to modify it in order to permit covalent immobilization of biomolecules to the surface. A wide number of solid surface modification techniques have been used, including ionizing radiation graft copolymerization, plasma gas discharge, photochemical grafting, chemical modification (e.g., ozone grafting), and chemical derivatization (Hoffman *et al.*, 1972,1986; Hoffman, 1987, 1988; Gombotz and Hoffman, 1986, 1987). (See also Section 3.2.14.)

A chemically immobilized biomolecule may also be attached via a spacer group, sometimes called an "arm" or a "tether" (Cuatrecasas and Anfinsen, 1971; Hoffman *et al.*, 1972; Hoffman, 1987). One of the most popular tethers is PEG that has been derivatized with different reactive end groups (Kim and Feijen, 1985), and some companies offer a variety of chemistries of heterobifunctional linkers having activated coupling end groups such as NHS, maleimide, pyridyl disulfide, and vinyl sulfone. Such spacer groups can provide greater steric freedom and thus greater specific activity for the immobilized biomolecule, especially in the case of smaller biomolecules. The spacer arm may also be either hydrolytically or enzymatically degradable, and therefore will release the immobilized biomolecule as it degrades (Kopecek, 1977; Hern and Hubbell, 1998).

Inert surfaces, whether polymeric, metal, or ceramic, can also be functionalized through modification of an polymeric adlayer. Such physisorbed or chemisorbed polymers can be bound to the surface via electrostatic interactions (VandeVondele *et al.*, 2003), hydrophobic interactions (Neff *et al.*, 1999), or specific chemical interactions, such as that between gold and sulfur atoms (Harder *et al.*, 1998; Bearinger *et al.*, 2003). Metal or ceramic surfaces may also be derivatized with functional groups using silane chemistry, such as with functionalized triethoxysilanes (Massia and Hubbell, 1991; Puleo, 1997). Plasma gas discharge has been used to deposit polymeric amino groups for conjugation of hyaluronic acid to a metal surface (Verheye *et al.*, 2000).

As noted earlier, hydrophobic interactions have been used to functionalize surfaces, utilizing ligands attached to hydrophobic sequences (e.g., Ista *et al.*, 1999; Nath and Chilkoti, 2003). Surfaces with hydrophobic gradients have also been prepared for this purpose (Detrait *et al.*, 1999). An interesting surface active product was developed several years ago that was designed to convert a hydrophobic surface to a cell adhesion surface by hydrophobic adsorption; it had an RGD cell adhesion peptide coupled at one end to a hydrophobic peptide sequence.

Sometimes more than one biomolecule may be immobilized to the same support. For example, a soluble polymer designed to "target" a drug molecule may have separately conjugated to it a targeting moiety such as an antibody, along with the drug molecule, which may be attached to the polymer backbone via a biodegradable spacer group (Ringsdorf, 1975; Kopecek, 1977; Goldberg, 1983). In another example, the wells in an immunodiagnostic microtiter plate usually will be coated first with an antibody and then with albumin or casein, each physically adsorbed to it, the latter acting to reduce nonspecific adsorption during the assay. In the case of affinity chromatography supports, the affinity ligand may be covalently coupled to the solid packing, and in some cases a "blocking" protein such as albumin or casein is then added to block nonspecific adsorption to the support.

It is evident that there are many different ways in which the same biomolecule can be immobilized to a polymeric support. Heparin and albumin are two common biomolecules that have been immobilized by a number of widely differing methods. These are illustrated schematically in Figs. 3.2.16-3 and 3.2.16-4.

Some of the major features of the different immobilization techniques are compared and contrasted in Table 3.2.16-7. The important molecular criteria for successful immobilization of a biomolecule are that a large fraction of the available biomolecules should be immobilized, and a large fraction of those immobilized biomolecules should retain an acceptable level of bioactivity over an economically and/or clinically appropriate time period.

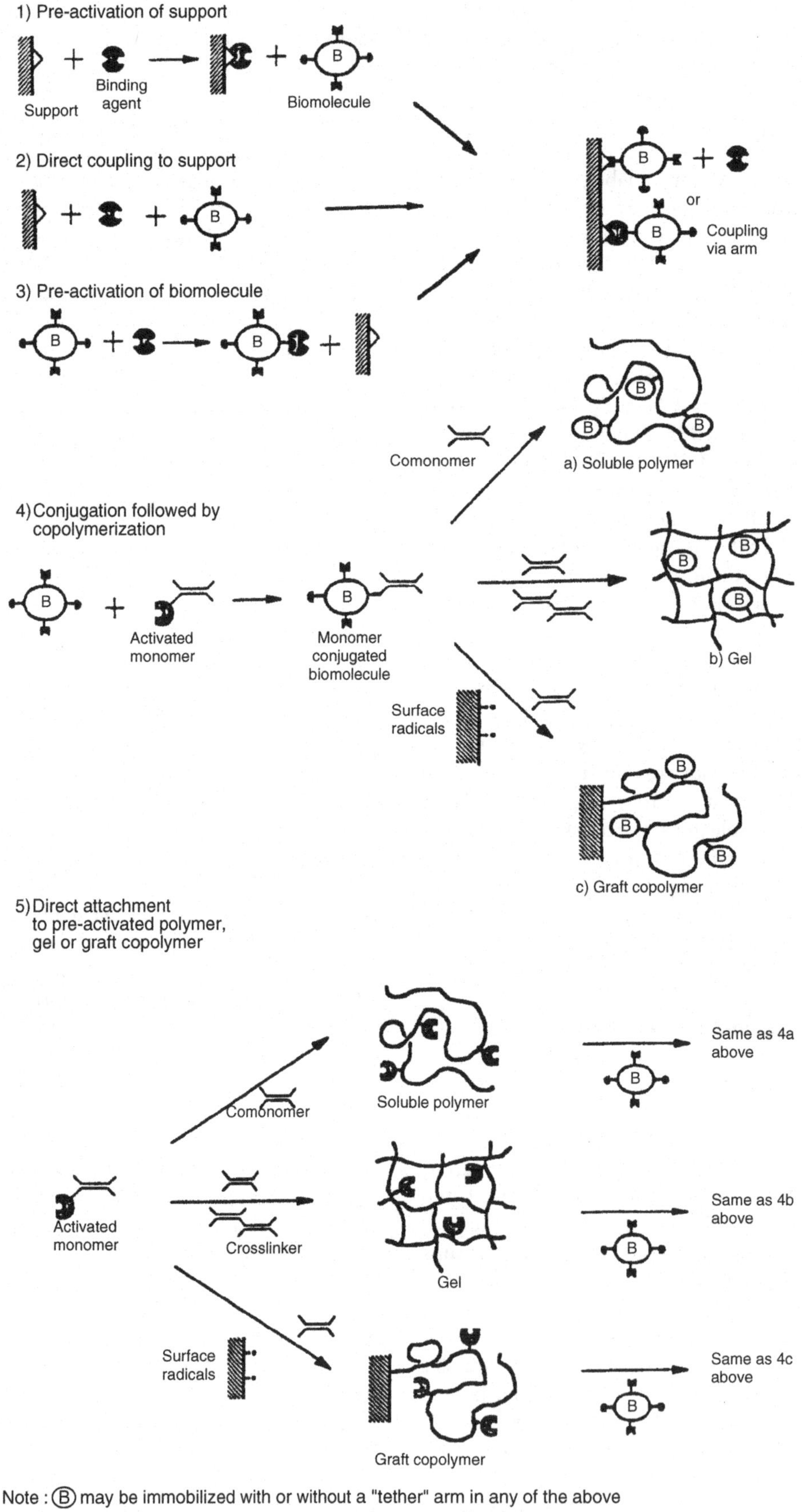

Fig. 3.2.16-1 Schematic cartoons showing various methods for covalent biomolecule immobilization.

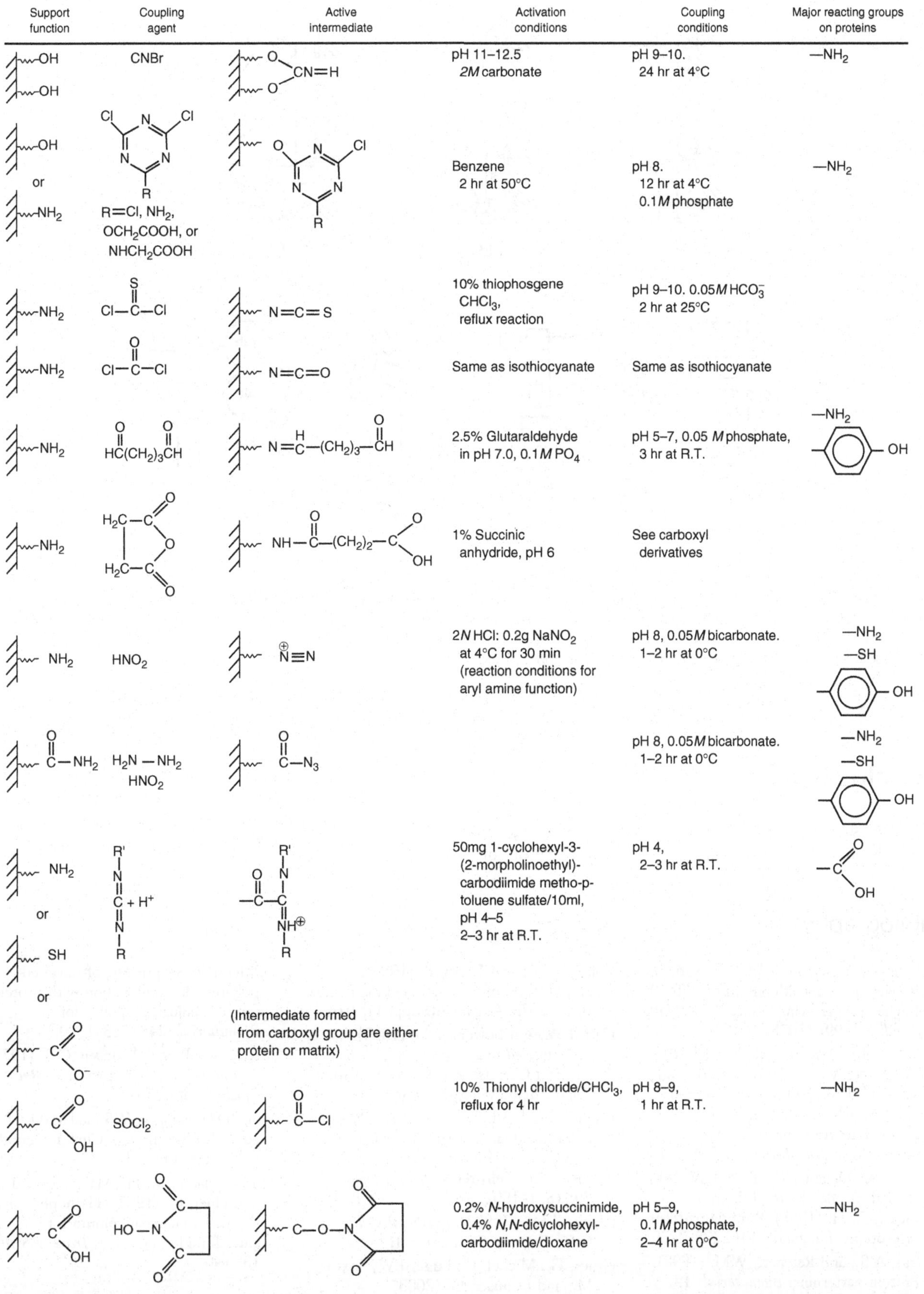
Support function
Coupling agent
Active intermediate
Activation conditions
Coupling conditions
Major reacting groups on proteins
CNBr
pH 11–12.5
2M carbonate
pH 9–10.
24 hr at 4°C
—NH2
or
R=Cl, NH2, OCH2COOH, or NHCH2COOH
Benzene
2 hr at 50°C
pH 8.
12 hr at 4°C
0.1M phosphate —NH_2
—NH_2	Cl—C(=S)—Cl	—N=C=S
10% thiophosgene
CHCl3,
reflux reaction
pH 9–10. 0.05M HCO3
2 hr at 25°C —NH_2
—NH_2	Cl—C(=O)—Cl	—N=C=O
Same as isothiocyanate
Same as isothiocyanate —NH_2
—NH_2	$HC(=O)(CH_2)_3CH(=O)$	—N=CH—$(CH_2)_3$—CH=O
2.5% Glutaraldehyde
in pH 7.0, 0.1M PO4
pH 5–7, 0.05 M phosphate,
3 hr at R.T.
1% Succinic
anhydride, pH 6
See carboxyl
derivatives
HNO2
2N HCl: 0.2g NaNO2
at 4°C for 30 min
(reaction conditions for
aryl amine function)
pH 8, 0.05M bicarbonate.
1–2 hr at 0°C —NH_2, —SH, tyrosyl (—OH)
—C(=O)—NH_2
H2N — NH2
HNO2 —C(=O)—N_3
pH 8, 0.05M bicarbonate.
1–2 hr at 0°C
50mg 1-cyclohexyl-3-
(2-morpholinoethyl)-
carbodiimide metho-p-
toluene sulfate/10ml,
pH 4–5
2–3 hr at R.T.
pH 4,
2–3 hr at R.T.
(Intermediate formed
from carboxyl group are either
protein or matrix)
SOCl2
10% Thionyl chloride/CHCl3,
reflux for 4 hr
pH 8–9,
1 hr at R.T.
0.2% N-hydroxysuccinimide,
0.4% N,N-dicyclohexyl-
carbodiimide/dioxane
pH 5–9,
0.1M phosphate,
2–4 hr at 0°C

Fig. 3.2.16-2 Examples of various chemical methods used to bond biomolecules directly to reactive supports (Carr and Bowers, 1980).

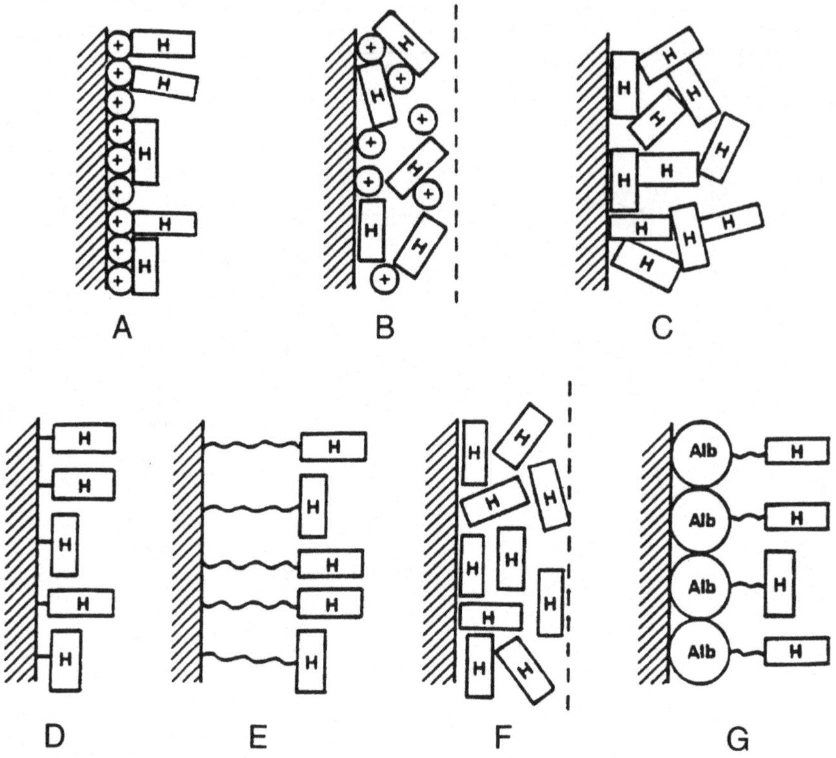

Fig. 3.2.16-3 Various methods for heparinization of surfaces: (A) heparin bound ionically on a positively charged surface; (B) heparin ionically complexed to a cationic polymer, physically coated on a surface; (C) heparin physically coated and self-cross-linked on a surface; (D) heparin covalently linked to a surface; (E) heparin covalently immobilized via spacer arms; (F) heparin dispersed into a hydrophobic polymer; (G) heparin–albumin conjugate immobilized on a surface (Kim and Feijen, 1985).

Conclusions

It can be seen that there is a wide and diverse range of materials and methods available for immobilization of biomolecules and cells on or within biomaterial supports. Combined with the great variety of possible biomedical and biotechnological applications, this represents a very exciting and fertile field for applied research in biomaterials.

Bibliography

Alsberg, E., Anderson, K.W., Albeiruti, A., Rowley, J.A., and Mooney, D.J. (2002). Engineering growing tissues. *Proc. Natl. Acad. Sci.* **99:** 12025.

Bearinger, J.P., Terrettaz, S., Michel, R., Tirelli, N., Vogel, H., Textor, M., and Hubbell, J.A. (2003). Chemisorbed poly(propylene sulphide)-based copolymers resist biomolecular interactions. *Nat. Mater.* **2:** 259–264.

Bernard, A., Delamarche, E., Schmid, H., Michel, B., Bosshard, H.R., and Biebuyck, H. (1998). Printing patterns of proteins. *Langmuir,* **14:** 2225–2229.

Blawas, A.S., and Reichert, W.M. (1998). Protein patterning. *Biomaterials* **19:** 595–609.

Callegaro, L., and Denri, E. (1983). Applications of bioreactors in medicine. *Int. J. Artif. Organs* 6(Suppl 1): 107.

Carr, P.W., and Bowers, L.D. (1980). *Immobilized Enzymes in Analytical and Clinical Chemistry: Fundamentals and Applications.* Wiley, New York.

Chang, K.C., and Hammer, D.A. (2000). Adhesive dynamics simulations of sialyl-Lewis(x)/E-selectin-mediated rolling in a cell-free system. *Biophys. J.* **79:** 1891–1902.

Chang, T.M.S. (1972). *Artificial Cells.* C.C. Thomas, Springfield, IL.

Csucs, G., Michel, R., Lussi, J.W., Textor, M., and Danuser, G. (2003). Microcontact printing of novel co-polymers in combination with proteins for cell-biological applications. *Biomaterials* **24:** 1713–1720.

Cuatrecasas, P., and Anfinsen, C.B. (1971). Affinity chromatography. *Ann. Rev. Biochem.* **40:** 259.

Dean, P.D.G., Johnson, W.S., and Middle, F.A., eds. (1985). *Affinity Chromatography.* IRL Press, Oxford.

De Myttenaere, M. H., Maher, J., and Schreiner, G. (1967). Hemoperfusion through a charcoal column for glutethimide poisoning. *Trans. ASAIO* **13:** 190.

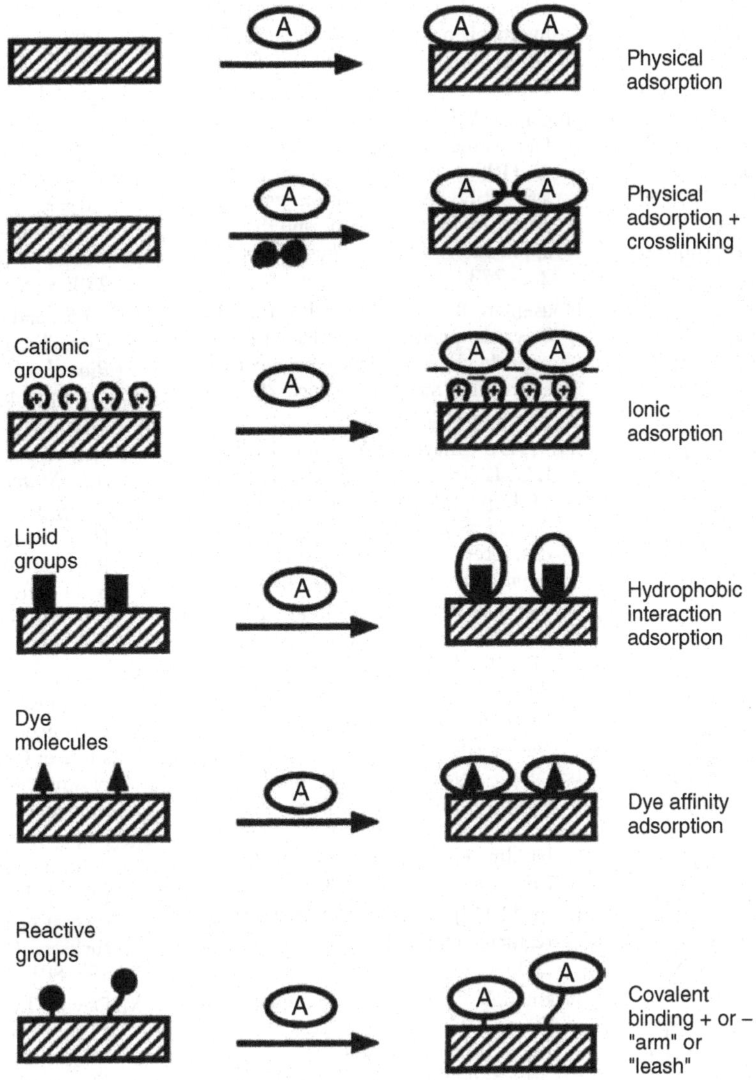

Fig. 3.2.16-4 Schematic of various ways that albumin may be immobilized on a surface. Albumin is often used as a "passivating" protein, to minimize adsorption of other proteins to a surface.

Table 3.2.16-7 Biomolecule immobilization methods

Method	Physical and electrostatic adsorption	Cross-linking (after physical adsorption)	Entrapment	Covalent binding
Ease	High	Moderate	Moderate to low	Low
Loading level possible	Low (unless high S/V)	Low (unless high S/V)	High	(depends on S/V and site density)
Leakage (loss)	Relatively high (sens. to △pH salts)	Relatively low	Low to none[a]	Low to none
Cost:	Low	Low to moderate	Moderate	High

[a] Except for drug delivery systems.

Detrait, E., Lhoest, J.B., Bertrand, P., and de Aguilar, V.B. (1999). Fibronectin–pluronic coadsorption on a polystyrene surface with increasing hydrophobicity: relationship to cell adhesion. *J. Biomed. Mater. Res.* **45:** 404–413.

Dunlap, B.R. ed. (1974). *Immobilized Biochemicals and Affinity Chromatography.* Plenum, New York.

Ellman, J.A., and Gallop, M.A. (1998). Combinatorial chemistry. *Curr. Opin. Chem. Biol.* **2:** 17–319.

Folch, A., and Toner, M. (2000). Microengineering of cellular interactions. *Annu. Rev. Biomed. Eng.* **2:** 227–256.

Goldberg, E., ed. (1983). *Targeted Drugs.* Wiley-Interscience, New York.

Gombotz, W.R., and Hoffman, A.S. (1986). Immobilization of biomolecules and cells on and within synthetic polymeric hydrogels, in *Hydrogels in Medicine and Pharmacy,* Vol. 1, N.A. Peppas, ed. CRC Press, Boca Raton, FL, pp. 95–126.

Gombotz, W.R., and Hoffman, A.S. (1987). Gas discharge techniques for modification of biomaterials. in *Critical Reviews in Biocompatibility,* Vol. 4, D. Williams, ed. CRC Press, Boca Raton, FL, pp. 1–42.

Griffith, L.G., and Lopina, S. (1998). Microdistribution of substratum-bound ligands affects cell function: Hepatocyte spreading on PEO-tethered galactose. *Biomaterials* **19:** 979–986.

Harder, P., Grunze, M., Dahint, R., Whitesides, G.M., and Laibinis, P.E. (1998). Molecular conformation and defect density in oligo (ethylene glycol)-terminated self-assembled monolayers on gold and silver surfaces determine their ability to resist protein adsoption. *J. Phys. Chem. B.* **102:** 426–436.

Hern, D.L., and Hubbell, J.A. (1998). Incorporation of adhesion peptides into nonadhesive hydrogels useful for tissue resurfacing. *J. Biomed. Mater. Res.* **39:** 266–276.

Hoffman, A.S. (1987). Modification of material surfaces to affect how they interact with blood. in *Blood in Contact with Natural and Artificial Surfaces,* E. Leonard, L. Vroman and V. Turitto, eds., *Ann. N.Y. Acad. Sci.* **516:** 96–101.

Hoffman, A.S. (1988). Applications of plasma gas discharge treatments for modification of biomaterial surfaces. *J. Appl. Polymer Sci. Symp.,* H. Yasuda and P. Kramer, eds. **42:** 251.

Hoffman, A.S., Schmer, G., Harris, C., and Kraft, W.G. (1972). Covalent binding of biomolecules to radiation-grafted hydrogels on inert polymer surfaces. *Trans. Am. Soc. Artif. Intenal. Organs* **18:** 10.

Hoffman, A.S., Gombotz, W.R., Uenoyama, S., Dong, L.C., and Schmer, G. (1986). Immobilization of enzymes and antibodies to radiation grafted polymers for therapeutic and diagnostic applications. *Radiat. Phys. Chem.* **27:** 265–273.

Houseman, B.T., and Mrksich, M. (2002). Towards quantitative assays with peptide chips: A surface engineering approach. *Trends Biotechnol.* **20:** 279–281.

Irvine, D.J., Mayes, A.M., and Griffith, L.G. (2001). Nanoscale clustering of RGD peptides at surfaces using comb polymers. 1. Synthesis and characterization of comb thin films. *Biomacromolecules* **2:** 85–94.

Ismagilov, R.F.,Ng, J.M.K.,Kenis, P.J.A., and Whitesides, G.M. (2001). Microfluidic arrays of fluid-fluid diffusional contacts as detection elements and combinatorial tools. *Anal. Chem.* **73:** 5207–5213.

Ista, L.K., Pérez-Luna, V.H., and López, G.P. (1999) Surface-grafted, environmentally sensitive polymers for biofilm release. *Appl. Environ. Microbiol.* **65:** 1603–1609.

Ito, Y. (1999). Surface micropatterning to regulate cell functions. *Biomaterials* **20:** 2333–2342.

Jackman, R.J., Duffy, D.C., Cherniavskaya, O., and Whitesides, G.M. (1999). Using elastomeric membranes as dry resists and for dry lift-off. *Langmuir* **15:** 2973–2984.

James, C.D., Davis, R.C., Kam, L., Craighead, H.G., Isaacson, M., Turner, J.N., and Shain, W. (1998). Patterned protein layers on solid substrates by thin stamp microcontact printing. *Langmuir* **14:** 741–744.

Kane, R.S., Takayama, S., Ostuni, E., Ingber, D.E., and Whitesides, G.M. (1999). Patterning proteins and cells using soft lithography. *Biomaterials* **20:** 2363–2376.

Kim, S.W., and Feijen, J. (1985). Methods for immobilization of Heparin. in *Critical Reviews in Biocompatibility.* D. Williams, ed. CRC Press, Boca Raton, FL, pp. 229–260.

Kolff, W.J. (1979). Artificial organs in the seventies. *Trans. ASAIO* **16:** 534.

Kopecek, J. (1977). Soluble biomedical polymers. *Polymer Med.* **7:** 191.

Kuhl, P.R., and Griffith-Cima, L.G. (1996). Tethered epidermal growth factor as a paradigm for growth factor-induced stimulation from the solid phase. *Nat. Med.* **2:** 1022–1027.

Laskin, A.I., ed. (1985). *Enzymes and Immobilized Cells in Biotechnology.* Benjamin/Cummings, Menlo Park, CA.

Lavin, A., Sung, C., Klibanov, A.M., and Langer, R. (1985). Enzymatic removal of bilirubin from blood: A potential treatment for neonatal jaundice. *Science* **230:** 543.

Lee, Y.S., and Mrksich, M. (2002). Protein chips: From concept to practice. *Trends Biotechnol.* **20:** S14–S18.

Lutolf, M.P., Raeber, G.P., Zisch, A.H., Tirelli, N., and Hubbell, J.A. (2003a). Cell-responsive synthetic hydrogels. *Adv. Mater.* **15:** 888–892.

Lutolf, M.R., Weber, F.E., Schmoekel, H.G., Schense, J.C., Kohler, T., Muller, R., and Hubbell, J.A. (2003b). Repair of bone defects using synthetic mimetics of collagenous extracellular matrices. Nat. *Biotechnol.* **21:** 513–518.

McClung, W.G., Clapper, D.L., Hu, S.P., and Brash, J.L. (2001). Lysine-derivatized polyurethane as a clot lysing surface: Conversion of plasminogen to plasmin and clot lysis in vitro. *Biomaterials* **22:** 1919–1924.

McClung, W.G., Clapper, D.L., Anderson, A.B., Babcock, D.E., and Brash, J.L. (2003). Interactions of fibrinolytic system proteins with lysine-containing surfaces. *J. Biomed. Mater. Res.* **66A:** 795–801.

McDevitt, T.C., Angelo, J.C., Whitney, M.L., Reinecke, H., Hauschka, S.D., Murry, C.E., and Stayton, P.S. (2002). In vitro generation of differentiated cardiac myofibers on micropatterned laminin surfaces. *J. Biomed. Mater. Res.* **60:** 472–479.

Maheshwari, G., Brown, G., Lauffenburger, D.A., Wells, A., and Griffith, L.G. (2000). Cell adhesion and motility depend on nanoscale RGD clustering. *J. Cell Sci.* **113:** 1677–1686.

Massia, S.P., and Hubbell, J.A. (1991). An RGD spacing of 440 nm is sufficient for integrin $\alpha_v \beta_3$ -mediated fibroblast spreading and 140 nm for focal contact and stress fiber formation. *J. Cell Biol.* **114:** 1089–1100.

Nath, N., and Chilkoti, A. (2003). Fabrication of reversible functional arrays of proteins directly from cells using a stimuli responsive polypeptide. *Anal. Chem.* **75:** 709–715.

Neff, J.A., Tresco, P.A., and Caldwell, K.D. (1999). Surface modification for controlled studies of cell–ligand interactions. *Biomaterials* **20:** 2377–2393.

Nose, Y., Malchesky, P.S., and Smith, J.W., eds. (1983). *Plasmapheresis: New Trends in Therapeutic Applications.* ISAO Press, Cleveland, OH.

Nose, Y., Malchesky, P.S., and Smith, J.W., eds. (1984). *Therapeutic Apheresis: A Critical Look.* ISAO Press, Cleveland, OH.

Ostuni, E., Grzybowski, B.A., Mrksich, M., Roberts, C.S., and Whitesides, G.M. (2003). Adsorption of proteins to hydrophobic sites on mixed self-assembled monolayers. *Langmuir* **19:** 1861–1872.

Park, T.G., and Hoffman, A.S., eds. (1990). Immobilizaiton of *Arthrobacter simplex* in a thermally reversible hydrogel: effect of temperature cycling on steroid conversion. *Biotech. Bioeng.* **35:** 152–159.

Piskin, E., and Hoffman, A.S., eds. (1986). *Polymeric Biomaterials.* M. Nijhoff, Dordrecht, The Netherlands.

Puleo, D.A. (1997). Retention of enzymatic activity immobilized on silanized Co–Cr–Mo and Ti-6Al-4V. *J. Biomed. Mater. Res.* **37:** 222–228.

Ringsdorf, H. (1975). Structure and properties of pharmacologically active polymers. *J. Polymer Sci.* **51:** 135.

Sakiyama-Elbert, S.E., Panitch, A., and Hubbell, J.A. (2001). Development of growth factor fusion proteins for cell-triggered drug delivery. *FASEB J.* **15:** 1300–1302.

Schense, J.C., and Hubbell, J.A. (1999). Cross-linking exogenous bifunctional peptides into fibrin gels with factor XIIIa. *Bioconjugate Chem.* **10:** 75–81.

Schmer, G., Rastelli, L., Newman, M.O., Dennis, M.B., and Holcenberg, J.S. (1981) The bioartificial organ: review and progress report. *Internat. J. Artif. Organs* **4:** 96.

Scotchford, C.A., Ball, M., Winkelmann, M., Voros, J., Csucs, C., Brunette, D.M., Danuser, G., and Textor, M. (2003). Chemically patterned, metal-oxide-based surfaces produced by photolithographic techniques for studying protein- and cell-interactions. II: Protein adsorption and early cell interactions. *Biomaterials* **24:** 1147–1158.

Shoemaker, S., Hoffman, A.S., and Priest, J.H. (1987). Synthesis and properties of vinyl monomer/enzyme conjugates: Conjugation of L-asparaginase with *N*-succinimidyl acrylate. *Appl. Biochem. Biotechnol.* **15:** 11.

Sparks, R.E., Solemme, R.M., Meier, P.M., Litt, M.H., and Lindan, O. (1969). Removal of waste metabolites in uremia by microencapsulated reactants. *Trans. ASAIO* **15:** 353.

Stark, G.R., ed. (1971). *Biochemical Aspects of Reactions on Solid Supports.* Academic Press, New York.

Sung, C., Lavin, A., Klibanov, A., and Langer, R. (1986). An immobilized enzyme reactor for the detoxification of bilirubin. *Biotech. Bioeng.* **28:** 1531.

Tomlinson, E., and Davis, S.S. (1986). *Site-Specific Drug Delivery: Cell Biology, Medical and Pharmaceutical Aspects.* Wiley, New York.

VandeVondele, S., Voros, J., and Hubbell, J.A. (2003). RGD-grafted poly-L-lysine-graft-(polyethylene glycol) copolymers block non-specific protein adsorption while promoting cell adhesion. *Biotechnol. Bioeng.* **82:** 784–790.

Verheye, S., Markou, C.P., Salame, M.Y., Wan, B., King III, S.B., Robinson, K.A., Chronos, N.A.F., and Hanson, S.R. (2000). Reduced thrombus formation by hyaluronic acid coating of endovascular devices. *Arterioscler. Thromb. Vasc. Biol.* **20:** 1168–1172.

Ward, M.D., and Hammer, D.A. (1993). A theoretical-analysis for the effect of focal contact formation on cell–substrate attachment strength. *Biophys. J.* **64:** 936–959.

Weetall, H.H., ed. (1975). *Immobilized Enzymes, Antigens, Antibodies, and Peptides: Preparation and Characterization.* Dekker, New York.

Winkelmann, M., Gold, J., Hauert, R., Kasemo, B., Spencer, N.D., Brunette, D.M., and Textor, M. (2003). Chemically patterned, metal oxide based surfaces produced by photolithographic techniques for studying protein– and cell–surface interactions I: Microfabrication and surface characterization. *Biomaterials* **24:** 1133–1145.

Yang, H.J., Cole, C.A., Monji, N., and Hoffman, A.S. (1990). Preparation of a thermally phase-separating copolymer, poly(*N*-isopropylacrylamide-*co*-*N*-acryloxysuccinimide) with a controlled number of active esters per polymer chain. *J. Polymer Sci. A. Polymer Chem.* **28:** 219–220.

Zaborsky, O. (1973). *Immobilized Enzymes.* CRC Press, Cleveland, OH.

Zisch, A.H., Schenk, U., Schense, J.C., Sakiyama-Elbert, S.E., and Hubbell, J.A. (2001). Covalently conjugated VEGF-fibrin matrices for endothelialization. *J. Controlled Release* **72:** 101–113.

Section Four

Clinical engineering

Chapter 4.1

4.1

Clinical applications of bioelectricity

Sverre Grimnes and Ørjan G. Martinsen

Bioimpedance research and development have been going on for a very long time, and the number of *clinical* applications does regularly increase but at a low rate. Bioimpedance is a general transducing mechanism for many physiological events, and the instrumentation is low cost. Basically the situation may have the aspect of a technology seeking a problem, but this must be reverted by taking another approach: go to the clinics and start with their problems and needs and see whether bioimpedance technology has a solution. Many of the transducing mechanisms can indeed offer solutions, but they must compete with the other solutions already in clinical use. Ultrasound imaging instrumentation for example can measure many of the parameters that bioimpedance can also measure. As the medical doctors already have the ultrasound probe in their hands as a multiparameter measuring device the bioimpedance technology must offer some definite advantages. The job is to find those clinical application areas.

Another job is to standardize the methods used by different models and companies. A problem related to standardization is when different models use non-scientific methods based on non-published algorithms. Another problem is that it is difficult to describe the clinical applications in detail because the more important they are the more patent rights and commercial interest complicate the picture.

4.1.1 Electrocardiography

Electrocardiography (ECG) is an important clinical examination routinely performed with 12 leads, 9 skin surface electrodes and a reference electrode. ECG is also important for long term monitoring in intensive care units. Also ECG control is extremely important during resuscitation and defibrillation.

During a routine ECG examination four electrodes are connected to the limbs. Three of these are for ECG signal pick up and one is a reference electrode (right leg (RL)) for noise reduction. Six electrodes are connected in the thorax region at well-defined positions near the heart.

4.1.1.1 Three limb electrodes (six limb leads)

The most basic ECG examination is with three skin surface electrodes on the limbs: one at the left arm (LA), one at the right arm (RA) and one at the left leg (LL). Figure 4.1-1 illustrates how six limb leads are derived from these three electrodes.

It is unusual that pick-up electrodes are placed so far away from the source organ; the main reasons in ECG are standardization and reproducibility. In our language, the limbs are salt bridges to the thorax. By this, the coupling to the thorax is well defined and electrocardiograms can be compared even when recorded in different hospitals and at long intervals. The position of the electrodes on each limb is uncritical because the distal part of each limb is isoelectric (with respect to ECG, not with respect to, for example, electromyography (EMG) sources of the arm muscles). Such is the reproducibility that the bipolar leads form the basis for determining the axis of the electric heart vector.

Three bipolar leads: I, II, III

Channel I is the voltage difference RA–LA; channel II: RA–LL; channel III: LA–LL. Einthoven found the

Biomedical Engineering Desk Reference; ISBN: 9780123746467

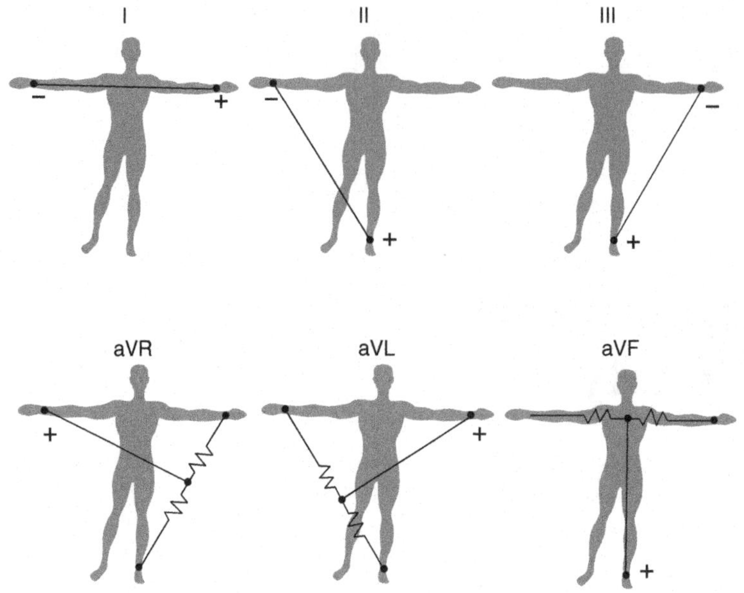

Figure 4.1-1 The six limb leads derived from three limb electrodes (both arms and LL). Here 'a' (e.g. in aVR) means augmented.

following relationship between the scalar lead voltages: $u_{II} = u_I + u_{III}$. Figure 4.1-2 shows an example of a typical ECG waveform.

The three unipolar limb leads: aVR, aVL, aVF

In the days of the first publications of Einthoven[1] no amplification tube had yet been invented, and the signal from the electrodes had to drive a galvanometer directly. Later the *unipolar augmented* leads were invented; they were called augmented because the signal amplitudes were higher. A "reference" voltage was obtained by summing the voltage from two of the limb electrodes (Fig. 4.1-1). More important, these vector leads have the interesting property of interlacing the vector lead angles of I, II and III (0°, 60° and 120°) so that each 30° is covered. This is shown in Fig. 4.1-6, where both an example of the heart vector **m** and the six lead directions are shown. Figure 4.1-3 shows the six limb leads arranged in a so-called Cabrera sequence. The text box in Fig. 4.1-3 contains a set of data extracted from the six ECG lead waveforms.

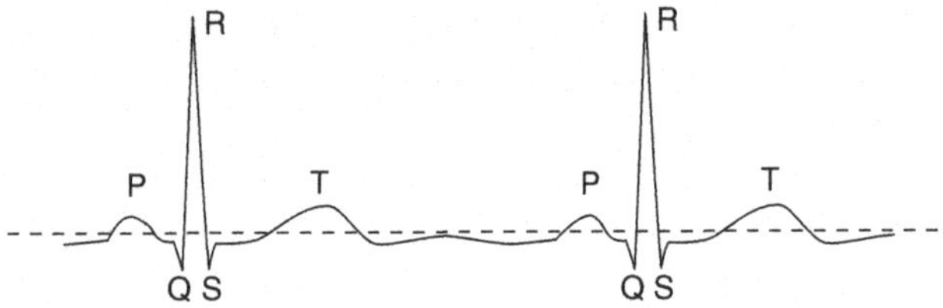

Figure 4.1-2 ECG waveform lead II. Dashed line is zero voltage determined by the electronic circuit high-pass filter functioning so that the negative and positive areas become equal.

Einthoven triangle and the lead vector

Einthoven (1913) proposed an equilateral triangle model with center in the heart center (Fig. 4.1-4). Each side of the triangle corresponded to each of the three bipolar leads I, II and III. He proposed the heart modeled as a bound dipole vector (he did not use that expression), and that a lead voltage was the projection of the heart vector on a corresponding triangle side (actually a dot product). He proposed that his model could be used for determining the heart vector axis. This means that he saw a solution to the inverse problem: from measured lead voltages to the heart vector, from measured surface potentials to source characterization. He regarded the triangle apexes as corresponding to right and left shoulder and the symphysis. The triangle is in the frontal plane of the patient, and the heart vector has no component perpendicular to the frontal plane. He considered the heart vector to be very short.

Burger and Milan (1946) formalized Einthoven's idea and introduced the concept of the *lead vector*. The lead vectors **H** of the I, II and III leads correspond to the sides of the Einthoven triangle with corners corresponding to the anatomy: the shoulders and the symphysis. The measured lead voltage u (volt) is a scalar and $u = \mathbf{H} \cdot \mathbf{m}$ so that u is the projection of **m** in the direction of a lead vector **H** (Fig. 4.1-4). **H** [Ω/meter] for each lead is

[1] Willem Einthoven (1860–1927), Dutch physician. Nobel prize laureate in medicine 1924 (on ECG).

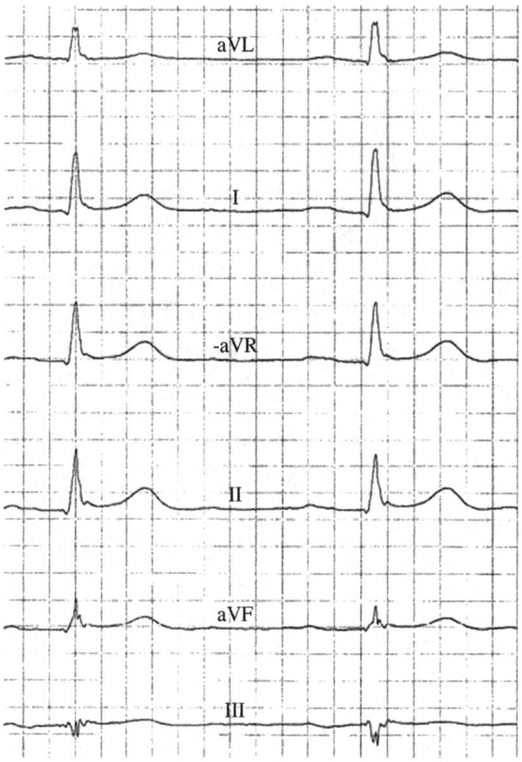

HR 52/minute	QRS 92ms	T axis 35°
RR 1150ms	QT 426ms	P(II) 0.08mV
P 146ms	P axis 40°	S (aVL) −1.12mV
PR 234ms	QRS axis 24°	R (aVF) 1.64mV

Figure 4.1-3 ECG waveforms of the six standard extremity leads shown with the augmented leads interlaced between the bipolar limb leads (Cabrera sequence).

obtained from the Einthoven triangle, and the heart vector **m** [Am] is the unknown. The magnitude and direction of **m** can be determined from measured potentials of at least two leads (Fig. 4.1-4).

m represents a bound vector with the fixed origin in the "center" of the heart, the lead vectors **H** are free vectors. Sampling the u values of the leads I, II and III at a given moment on Fig. 4.1-4 defines the *instantaneous* value and direction (angle α in Fig. 4.1-4) of the heart vector **m**. The mean direction during the QRS complex defines the *electrical axis* of the heart. Mean direction is used because the R tags are not completely concurrent for the I, II and III leads. The electrical axes for the P, QRS and T complexes have different directions (cf. the text box of Fig. 4.1-3). During a QRS complex the locus

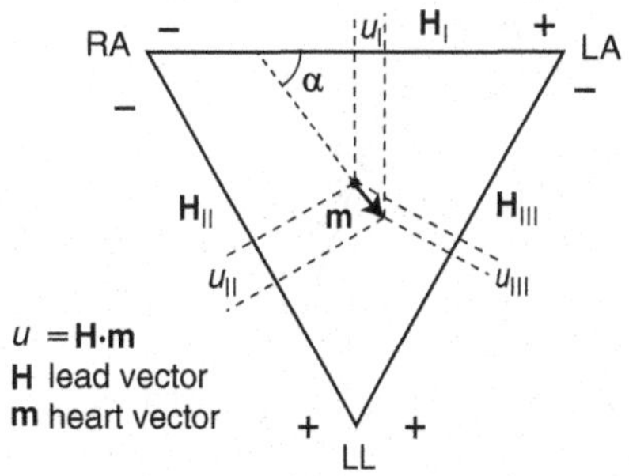

Figure 4.1-4 Einthoven's triangle. The triangle is in the frontal plane of the patient. **m** is the heart vector bound to the "center" of the heart; u is the instantaneous scalar voltage measured in a respective lead; α is the instantaneous angle of the electric axis of the heart.

of the vector arrow describes a closed loop in the frontal plane (Fig. 4.1-5).

The electrical axis of the heart can roughly be determined by looking at the net area of a QRS complex. A net positive area means that the heart vector has the same direction as the lead vector. A small net area means that the heart vector is perpendicular to the lead vector. A negative net area means that the heart vector has the opposite direction of the lead vector. It is interesting to compare this interpretation which is based on a bound heart vector with changing length and direction, with the model which is a moving current dipole with constant length and direction. With **m** and $\mathbf{H}_{II}$ parallel the waveform will be monophasic, with **m** and $\mathbf{H}_{III}$ perpendicular the waveform will be biphasic.

The *dipole* model in ECG has been a hot topic.[2] Einthoven (1913) did not use the dipole concept, but referred to the potential difference [V] between two close points in the center of the heart. He thus defined a current vector with the voltage difference between its poles. There is an ambiguity here: A potential difference may be regarded as a scalar [V], but may also be regarded as an electric field [V/m] vector. A bipolar lead may be regarded as an electric field transducer with scalar voltage [V] output. Einthoven used the concept of electromotive force (EMF) which may be considered to have direction when related to the transducing mechanism of the force exercised on a charge in an electric field, but may also be regarded as a scalar potential [V]. So the heart dipole has a vector moment [Am], the heart dipole moment has a direction from plus pole to minus pole, the resulting current density in the thorax is a vector field [A/m^2], the current density **J** and the electric field strength **E** are in

[2] "Unfortunately, the application of the long known and well understood principles of potential theory to electrocardiography was not in general well received. Many of the more or less theoretical and mathematical papers along these lines aroused a storm of opposition. Some of the criticism came from physicians who felt that electrocardiography was a purely empiric science and that progress in the field could come only from comparison of the electrographical findings with clinical and post mortem data. Much opposition came from physiologists, many eminent in their field, who not only discounted any article of theoretical nature but also regarded the dipole hypothesis as rank heresy" (Wilson, 1953).

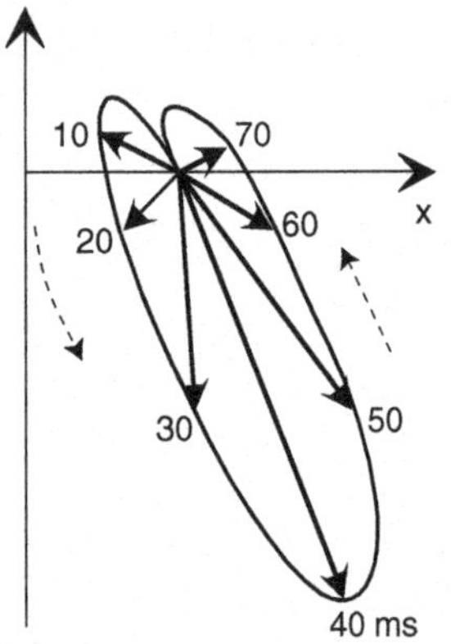

Figure 4.1-5 Locus of the heart vector **m** given each 10 ms in the QRS diastole. Derived from the Einthoven triangle (Fig. 4.1-4). The graph is in the frontal plane and the *x*-axis has the direction of the lead vector $\mathbf{H}_1$ ($\alpha = 0$).

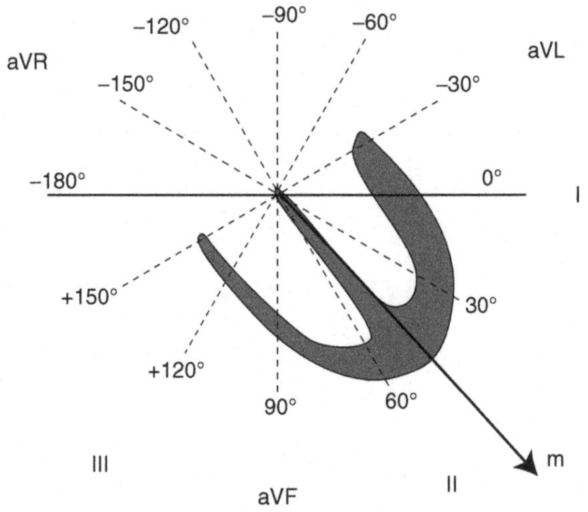

Figure 4.1-6 Myocard of left (thick) and right (thin) ventricles, the Einthoven triangle in the frontal plane. QRS heart vector **m**, normal angle values are between −30° and +90°.

isotropic media proportional and simply related by equation $\mathbf{J} = \sigma\mathbf{E}$. Einthoven did not use the vector concept nor the dipole concept. He referred to the direction of maximum potential difference, and this is in accordance with the Maxwell equation $\mathbf{E} = -\nabla\Phi$.

The Einthoven triangle gained more and more acceptance, but was also heavily criticized. In particular four aspects have been attacked:

1. *Redundancy:* Of course $u_{\mathrm{I}} + u_{\mathrm{II}} + u_{\mathrm{III}} = 0$ (Kirchhoff's law) in a linear system. Because the two input wires of the amplifier of lead II are swapped, lead II has changed sign so that: $u_{\mathrm{III}} = u_{\mathrm{II}} - u_{\mathrm{I}}$. Theoretically a third lead contains no new information, in practice however, it represents quality control and eases rapid waveform interpretations.
2. It is not a great surprise that an equivalent to the electrical activity of the whole myocard, the transmission from the sources to the different lead electrodes included, can not be modeled as one bound dipole vector alone. The surprise is that the accordance is so good. Refinements can only include expansion with spherical harmonic multipoles such as a quadrupole, not with moving dipoles, electromotive surfaces or multiple dipoles.
3. *Lack of 3D data:* The Einthoven triangle is flat as if all electrical activities in the heart occur in a thin vertical sheet. An additional back electrode would, for instance, represent a first primitive approach to a lead perpendicular to the classical Einthoven triangle.
4. The triangle is actually based on a model with two ideal dipoles. This presumption is broken because (1) the distance between the dipoles is not very much longer than the dipole lengths; (2) the medium is not infinite and homogeneous.

Torso models were built filled with electrolytes and with an artificial heart in the natural heart position. Burger and Milan in 1946 used two copper plates 2 cm in diameter and 2 cm apart with two wires supplying current as an artificial heart. The dipole concept was not used but the lead vector and an oblique triangle instead of the equilateral Einthoven triangle. The direction of the heart vector was defined as the direction in which a heart propagates the current flow; the dimension was given as [Vcm^2]. It is still easy to mix the dimensions, in Figs. 4.1-2 and 4.1-4 it is easy to believe that the heart vector is some strange voltage vector created by projections of the lead voltages. But the heart vector is a dipole vector according to $\mathbf{m} = i\mathbf{L}_{cc}$ [Am]. The direction of the vector is the electrical axis of the heart, and this is clinically used. The split components of **m**: the dipole current i and the distance vector $\mathbf{L}_{cc}$ are not used clinically.

Table 4.1-1 shows time intervals and heart vector axis is shown in Fig. 4.1-6

4.1.1.2 Six chest electrodes (six unipolar precordial leads)

A larger signal with higher information content is obtained by placing surface electrodes as near to the

Table 4.1-1 Normal values of time intervals and direction of the heart vector

	Normal values	
0.12 seconds < PQ < 0.25 seconds	QRS < 0.12 seconds	−30° < QRS axis < +90°

heart as possible. The drawback is that the exact electrode positions become critical, and it is difficult to obtain the same positions next time by another person or clinic. Six unipolar pick-up electrodes are positioned on the skin around the heart in a transverse plane. The lead vector of each of the six electrodes has the direction of the line from the heart center to the unipolar electrode in the transversal plane. The "reference" electrode system may be simply the RA electrode alone (CR lead). A more common "reference" is obtained (V lead) by summing the voltage from all three extremities, this is the Wilson reference.

Other unipolar leads

Invasive electrodes

During open heart surgery a net of 11 sterile electrodes are used for epicardial mapping. They have direct contact with the surface of the heart, and the spread of signal on the heart surface is examined by sampling at millisecond intervals in the systole. The signal amplitude is large, and the discriminative power also. The position of infarcted regions is revealed by abnormal epicardial potential spread.

Another invasive ECG recording is with intracardial catheter electrodes (e.g. for *His*[3]*-bundle* transmission). Both unipolar and bipolar leads are used, and the catheter is always advanced into the right atrium via the venous vessels. From there it is further advanced through the tricuspidal valve into the ventricle. Being so near to the source, only small position changes have a large influence on the recorded waveform. This is acceptable because usually the time intervals are of greatest interest. Since the electrodes are very near or on the myocard, the recorded signals have several millivolt of amplitude and with a frequency content up to above 500 Hz.

4.1.1.3 Standard 12 lead ECG

The 12 lead ECG clinical diagnostic test is composed of the six limb leads (see Fig. 4.1-3) and the six unipolar precordial leads.

4.1.1.4 Cardiac electrophysiology

The heart is a large muscle group (myocard) driven by a single firing unit followed by a special network for obtaining the optimal muscle squeeze and blood acceleration. It is a very important organ and a very well-defined signal source; therefore all sorts of recording electrodes have been taken into use: from invasive intracardiac catheter electrodes for local His-bundle recording, to multiple sterile electrodes placed directly on the heart surface during open heart surgery (epicardial mapping).

The cardiac cycle in the normal human heart is initiated by the sinoatrial (SA) node. The excitation wave spreads through the atria at a velocity of about 1 m/s. The electroanatomy of the heart separates the atria and ventricles (two chamber heart) so that the excitatory wave can be delayed before it reaches the ventricles. This delay of about 0.15 s is performed in the zone of the atrioventricular (AV) node, and permits a filling of the ventricles before their contraction. Beyond the AV node the excitation spreads rapidly (2–4 m/s) in the networks of His and Purkinje. These networks consist of specialized muscular (not nerve) tissue. From these networks the excitation wave spreads in the myocard at a much lower speed (0.3 m/s). The myocard cells communicate by channels that connect the intracellular electrolytes directly. Each muscle cell is therefore triggered by its neighbor myocard cell but guided by the Purkinje network. Myocard lacks direct nerve control, although the SA node is under nerve control. The intracellular volume is like one volume of cytoplasm with many nuclei (*syncytium*), and electrical models are based on the concept of the *bidomain*, the extracellular and intracellular domains. The tissue mass of the network is much smaller than the mass of the myocard and the electrical signal from myocard dominates the surface ECG.

Important clinical use of ECG data is in the study of beat-to-beat differences, both waveform and repetition rate (variation in the beat-to-beat Q–T interval is called *dispersion*), arrhythmias, blocks, detection of ischemia and infarcted muscle volumes and their positions.

4.1.1.5 Vector cardiography

Figure 4.1-5 actually shows a 2D vector cardiogram in the *frontal* plane. 3D vector cardiography according to Frank is based on the heart vector in a 3D Cartesian diagram. Five strategically positioned skin surface electrodes define the heart vector in the *transversal* plan, and two additional electrodes on the head and LL take care of the vertical vector component. The heart vector is calculated from the recorded lead voltages and projected into the three body planes. Three loci curves of the vector tip in the three planes are the basis for the doctor's description. In principle the 12 channel registration is reduced to three, even so this 3D data set contains more information than the 2D data set from lead I, II and III. However, the problem for vector cardiography is that it is very difficult to throw overboard the long tradition of interpreting curves obtained with the old standardized electrode positions. From long experience a clinical information bank has been assembled giving the relation between waveform and diagnosis. The reason for these relationships

[3] Wilhelm His Jr. (1863–1934), Swiss physician.

may be unknown, and is not of great concern as long as the empirical procedure furnishes precise diagnostic results. All doctors world over are trained according to this tradition (the interpretation of EEG waveforms is another striking example of such a practice). For special diagnostic problems the vector cardiograph has some distinct advantages. For example, the vector display of the QRS complex has a much better resolution than the narrow QRS waveform obtained with a standard 12 lead ECG. Also the vector cardiograph is valuable for training purposes to obtain a better understanding of the spatial distribution of the electrical activity of the heart.

4.1.1.6 Forward and inverse problem

The purpose of the ECG examination is to find the electrical properties of the heart by measuring potential differences on the skin surface. The EMFs of the heart produce currents in the surrounding tissue. The anatomy of the thorax and the conductivity distribution determine how the current flow spreads and which potential differences are to be found at the skin surface. It is generally accepted that the thorax tissue is linear with the usual endogenous current density/electrical field amplitudes generated by the heart. The different contributions of different myocard volumes can therefore simply be added. On the other hand the tissue may be anisotropic and the transfer impedance is frequency dependent. A rough demonstration of that can be obtained by measuring the transfer impedance between the arms and a bipolar skin surface pick-up electrode pair positioned over the apex and the sternum (Fig. 4.1-7). Because of reciprocity the transfer impedance is the same if the current is applied to the electrodes on the thorax and the potential difference between the hands is measured. From Fig. 4.1-7 the frequency components of the P and T waves are not much attenuated, but the frequency components of the QRS above 50 Hz are attenuated by a factor of about 10 or more.

To find the electrical properties of the heart is an *inverse* problem, and in principle it is unsolvable, as there are infinitely many source configurations which may result in the measured skin potentials.

We therefore start with the forward problem: from heart models in a conductive medium to surface potentials. Those problems are solvable, but it is difficult to model the heart and it is difficult to model the signal transmission from the heart up to the surface. The model may be an infinite homogeneous volume or a torso filled with saline or with a heterogeneous conductor mimicking the conductivity distribution of a thorax.

The most basic electrical model of the heart is a bound vector with the variable vector moment $\mathbf{m} = i\mathbf{L}_{cc}$. Plonsey (1966) showed that a model with more than one

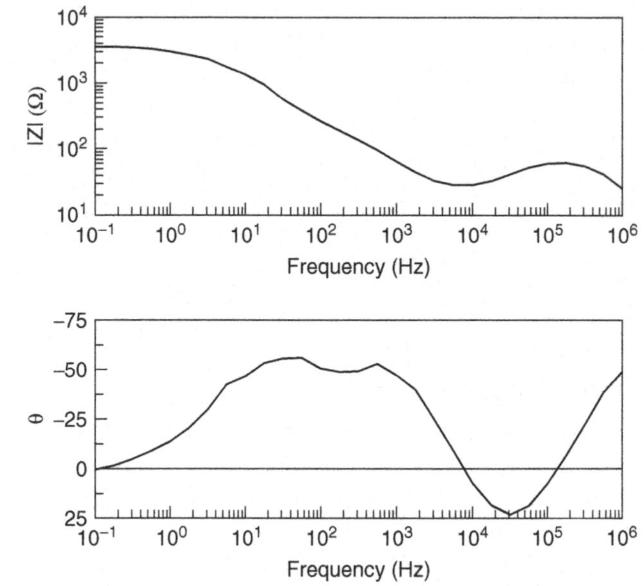

Figure 4.1-7 Transfer impedance between two chest surface electrodes (apex–sternum) and two electrodes in each of the hands.

dipole is of no use because it will not be possible from surface measurements to determine the contribution from each source. The only refinement is to let the single bound dipole be extended to a multipole of higher terms (e.g. with a quadrupole).

The Einthoven triangle was an early solution to the inverse problem: how to characterize the source from surface electrode derived data. It is astonishing how the original Einthoven triangle still is the basis for standard clinical ECG interpretations all over the world even if many improvements have been proposed. Actually the tradition of using a simple theoretical model with an ideal dipole in an infinite, homogeneous volume conductor is uninterrupted. The six limb leads have been heavily attacked for their redundancy. There must be some reason why they have endured all these attacks and reached the overwhelming global spread and acceptance they have today. The large amount of clinical data is one reason; perhaps the value of these data is based on the exceptional reproducibility of their lead vectors. The salt bridge principle assures a well-defined coupling to the shoulders and symphysis, and the position of the electrodes on the limbs is totally uncritical. The chest electrode positions are critical both for the precordial leads and the vector cardiographical leads.

4.1.1.7 Technology

Signal amplitude, limb lead II

The *R* voltage amplitude is usually around 1 mV. Amplitudes <0.5 mV are characterized as "low voltage" cases, and are pathological. Obesity does not change the ECG waveforms; the *R* amplitude may be somewhat reduced but not into the "low voltage" region.

Frequency spectrum

Routine bipolar diagnostic ECG is performed with a pass band of, for example, 0.05–60 Hz. The American Heart Association (AHA) recommends 150 Hz as minimum bandwidth and 500 Hz as minimum sampling rate for recording pediatric ECG. DC voltage is thus filtered out because it is related to the electrodes and not to electrophysiology. In monitoring situations the low frequency cut-off frequency may be increased to 0.5 Hz or more because the patient induced motion artefacts are reduced. With skin surface electrodes only the QRS complex has frequency components above 60 Hz, however in the QRS there are components up to and above 1000 Hz (Franke et al., 1962) which are not used clinically. With electrodes in direct connection with the heart the useful frequency spectrum extends to 500 Hz and more. With a pacemaker assisted heart the pacemaker pulses may have a width of <1 ms, and the frequency spectrum extends far above 1 kHz.

Zero line and isoelectric level

Because the ECG amplifier is AC coupled no true zero voltage is recorded. Instead the zero is a line defined by the low-pass filter of the ECG amplifier so that the positive and negative areas of the total ECG waveform are equal. The zero line can be determined by simply shorting the two input wires of a lead (no signal).

Noise

True physiological signals

The limb leads presuppose that the patient is lying down so that EMG signals from the limb and body muscles do not intervene. Slow baseline respiration waves are seen if the characteristic frequency of the low-pass filter is 0.05 Hz or lower.

Exogenous noise

- 50/60 Hz mains noise is electric or magnetic field coupled from power line wires to patient wires or patient body.
- Patient movement generates triboelectricity which may severely disturb the ECG waveform (electrostatic noise).

Solutions are shielding, active op-amp circuit clamping the body to ground, increased distance to the noise source, increased common mode rejection circuitry, averaging over several heart beats, wireless telemetry of electrode pick-up signals.

Indifferent electrode

With an ideal current dipole in an infinite volume, zero potential will exist in all directions if the distance to the dipole is large enough. Unfortunately an indifferent (neutral) ECG electrode does not exist because the human body is not large enough. If we go out in one direction along a limb, the limb proper is isoelectric with respect to the heart activity but not with respect to other sources (e.g. respiration). If we go out along a second limb that too will be isoelectric, however the potential will not be equal to that of the first limb. Therefore none of them represents a true indifferent electrode.

Since this is an unsolvable problem, the ideal unipolar lead does not exist. Non-ideal solutions are to use one electrode at some remote point (e.g. at a limb or an earlobe). Another solution is to add the voltages picked up by more than one electrode, such as two limb electrodes (augmented leads) or three limb electrodes (Wilson central terminal).

Ground, reference, indifferent electrode

In Einthoven's time without amplifiers there was no need for a "ground" electrode, the galvanometer is a floating input device. But mains supplied amplifiers needed a "ground" electrode to reduce noise, and it was the RL which was chosen for that purpose. Later safety philosophy advocated that the patient should be electrically floating with respect to ground (cf. Section 4.1.17.2). This implied that "ground" was not ground in the meaning the ground wire of the mains. Instead of "ground" this is the *reference* electrode, meaning that the amplifier and the patient are connected together so that they roughly are at the same potential. But then we also had the ECG tradition that in unipolar leads the reference may not only be one electrode, but for example, a Wilson terminal, used as "zero" or indifferent reference. Here *ground* means the protective ground electrically connected to the building and the room, supplied by the mains and having the symbol ⏚. The purpose of a *reference* electrode is to obtain potential equalization between the patient and the electronic circuitry, symbol ▽, ⏛ or ⊥. The reference electrode is usually an electrode on the RL. A *neutral* or *indifferent* electrode is a part of a unipolar electrode system, and is usually obtained by summing the potentials from more than one electrode (cf. the Wilson terminal).

4.1.2 Impedance plethysmography

Plethysmography is the measurement of volume. Dynamic plethysmography is usually associated with volume changes due to the heartbeats, but may also be related to, for example, respiration or peristaltic movements of the alimentary canal. During the heart systole with increased blood flow, the volume, for example, of a limb increases due to the inflow of blood (*swelling*). Impedance may in many cases be regarded as measuring both volume and flow, a volume change must be due to a flow. Measurements may be based on, for example, mechanical dimensional change (strain-gauge plethysmography, light absorption (photo-plethysmography), X-ray absorption or immittance change. Application areas are rather diversified, for example, heart stroke volume (SV), cardiac output (CO), respiration volume, fluid volume in pleural cavities, edema, urine bladder volume, uterine contractions, detection of vein thrombosis.

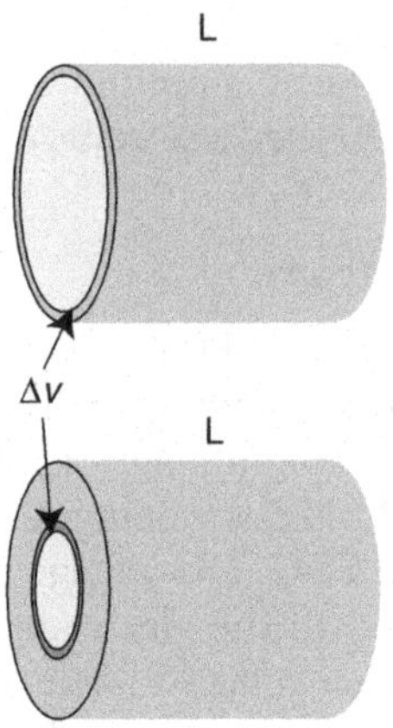

Figure 4.1-8 Cylinder models of length L and a small parallel volume increment Δv. Upper: one-compartment, lower: two-compartment model.

4.1.2.1 Ideal cylinder models

By ideal we mean that the biomaterial is considered incompressible and homogeneous. The cross sectional area of the cylinder may be circular, elliptic or have any plane form. Estimation of volume from immittance measurement is based on two effects:

1. A geometry-dependent effect illustrated by the cylinder model and the ratio A/L in the equation $G = \sigma A/L$. The resulting effect will be dependent on the constraints on the measured tissue volume: if the volume increase results in a swelling of length L, conductance will fall. If the volume increase results in a swelling of cross sectional area A, the conductance will increase. If the volume increase occurs outside the measured tissue volume, the measured conductance will not change with the geometrical volume increase.
2. A conductivity-dependent component. Of special interest is the flow dependence of the conductivity of blood.

For the further analysis of these effects it is useful to set up some simple cylinder models.

The geometry is shown in Fig. 4.1-8. For the single cylinder shown at the top of Fig. 4.1-8 the volume v is easily found from $G = \sigma A/L$ or $R = \rho L/A$:

$$v = G\rho L^2 = R\sigma A^2 = \frac{1}{R}\frac{L^2}{\sigma} \quad \text{(exact)} \tag{4.1.1}$$

Notice that with the presumption L = const, the volume v is proportional to G. If the presumption is that A = const, the volume v is proportional to R. *If swelling is longitudinal* the volume increase Δv is best modeled as a resistance increase in a series model. *If swelling is transverse (as supposed in many cases)*, the volume increase Δv is best modeled as a conductance increase in a parallel model. If it is not known whether the tissue swells in longitudinal or transverse direction, the conductance versions may be preferred because they lead to simpler and more exact expressions.

One-compartment model. Cylinder surrounded by air (cf. Fig. 4.1-8 top)

In many applications the absolute volume may remain unknown; the emphasis is instead on the *relative* volume change $\Delta v/v$. Also the relative conductance change AG/G is of special interest, because the ratio is related to the signal-to-noise ratio which should be as high as possible. From eq. (4.1.1):

$$\frac{\Delta G}{G} = \frac{\Delta v}{v} \quad \text{or} \quad \frac{\Delta G}{\Delta v} = \frac{\sigma}{L^2} \quad \text{or}$$
$$\Delta v = \Delta G \rho L^2 \quad \text{(exact)} \tag{4.1.2}$$

Thus it is clear that relative volume changes can be found without knowing the dimensions of the cylinder. In order to have a high sensitivity (large ΔG) for a given volume change Δv, the length L should be as short as possible.

Under the presumption that L = const. the conductance model is preferred. If we still use a resistance model we have from eq. (4.1.1):

$$\Delta v = \left(\frac{1}{R + \Delta R} - \frac{1}{R}\right)\rho L^2$$
$$= -\frac{\Delta R}{R}\frac{1}{R + \Delta R}\rho L^2 \text{(exact)} \tag{4.1.3}$$

Equation (4.1.3) becomes linear only if $\Delta R \ll R$:

$$\Delta v \cong -\Delta R \rho \left(\frac{L}{R}\right)^2 \quad (\Delta R \ll R) \tag{4.1.4}$$

The minus signs in eqs. (4.1.3) and (4.1.4) are because a resistance *increase* corresponds to a volume *decrease*.

Two-compartment model: an inner cylinder surrounded by an outer cylinder of same resistivity (Fig. 4.1-8 bottom)

$$\frac{\Delta G}{G} = \frac{\Delta v}{\Delta v + v_A + v_t} \quad \text{(exact)} \qquad (4.1.5)$$

In the two-compartment model the two cylinders are physically in parallel, and the conductance model is preferred with $L = \text{const.}$ $\Delta v + v_A$ is the volume of the inner cylinder, v_t is the volume of the outer cylinder and considered constant (implying that both the inner tube and outer tube swell when $\Delta v > 0$). Equation (4.1.5) shows that the sensitivity falls with a larger surrounding volume v_t. In plethysmography the measurement should be confined as much as possible to the volume where the volume change occurs. Thus the problem of high sensitivity plethysmography poses the same problem as in electrical impedance tomography (EIT): to selectively measure immittance in a selected volume.

4.1.2.2 The effect of different conductivities

In the two-compartment, constant length, parallel cylinder model analyzed above the conductivities in the inner and outer cylinders were considered equal. Different conductivities will of course also change G. With a conductivity σ_t of tissue outer cylinder and σ_b of blood in inner cylinder, and with constant geometry, the conductance is found from $G = \sigma A/L$:

$$G = (\sigma_A A_A + \sigma_t A_t)\frac{1}{L} \quad \text{(constant volume)} \qquad (4.1.6)$$

Equation (4.1.6) is not really a plethysmographic equation when the assumption is that no geometrical change shall occur. However, it relates to flow systems with a varying conductivity of the passing liquid, and flow with time is volume.

4.1.2.3 Models with any geometry and conductivity distribution

Figure 4.1-9 shows a dynamic system in a vessel where, for example, a blood bolus volume on its passage leads to a temporal local volume increase during heart systole. The measured zone in the inner cylinder is filled with blood (inflow phase). Later during the diastole the blood is transported further (outflow phase), but also returned via the venous system. Figure 4.1-9 shows a tetrapolar electrode system for the measurement of G.

In general a *tetrapolar* system is preferable; it may then be somewhat easier to confine the measured tissue volume to the zone of volume increase. The sensitivity for bolus detection with a tetrapolar electrode system

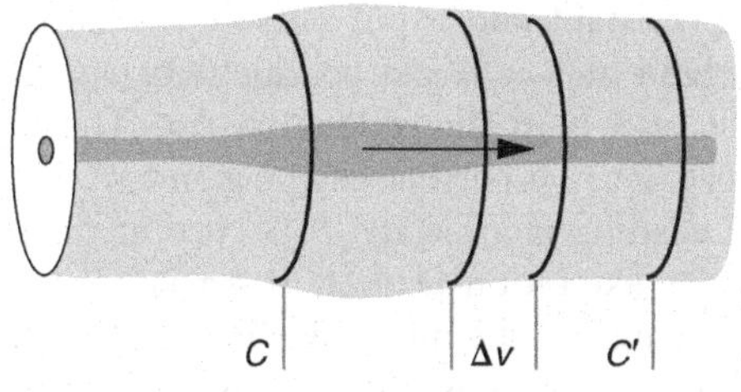

Figure 4.1-9 Tetrapolar electrode system and the effect of a bolus of blood passing the measured volume.

will be dependent on the bolus length with respect to the measured length.

To analyze the situation with a tetrapolar electrode system in contact with, for example, a human body, we must leave our simplified models and turn to lead field theory. The total measured transfer impedance is the ratio of recorded voltage to injected current. The impedance is the sum of the impedance contributions from each small volume dv in the measured volume. In each small volume the resistance contribution is the resistivity multiplied by the vector dot product of the space vectors $\mathbf{J}'_{reci}$ (the current density field due to a unit reciprocal current applied to the recording electrodes) and $\mathbf{J}'_{cc}$ (the current density field due to a unit current applied to the true current carrying electrodes). With disk-formed surface electrodes the constrictional resistance increase from the proximal zone of the electrodes may reduce sensitivity considerably. A prerequisite for two-electrode methods is therefore large band electrodes with minimal current constriction.

If the system is reciprocal the swapping of the recording and current carrying electrode pairs shall give the same transfer impedance. It is also possible to have the electrode system situated *into the volume* of interest (e.g. as needles or catheters). Such volume calculation (e.g. of CO) is used in some implantable heart pacemaker designs, cf. Section 4.1.11.

Changes in conductivity

Conductivity may change as a function of time (e.g. caused by flow). The special case of a changing conductivity with a general but constant geometry was analyzed by Geselowitz (1971) who developed an expression for ΔZ based on the potential field. Lehr (1972) proposed to use current density instead of potential in the development. Putting $\mathbf{E} = -\nabla\Phi$ and as $\mathbf{J} = \sigma\mathbf{E}$ in isotropic media, we have:

$$\Delta Z = \frac{-\Delta\sigma}{\sigma_0^2}\iiint \mathbf{J}'_0 \cdot \mathbf{J}'_{\Delta\sigma} dv \quad [\Omega] \qquad (4.1.7)$$

Here the integration is in the volume of conductivity change. The volume is homogeneous but with changing conductance, at a given moment of integration σ is therefore constant and is put outside the integral. $\mathbf{J}'_0$ is the lead field at $t = 0$ and conductivity σ_0, $\mathbf{J}'_{\Delta\sigma}$ is the reciprocal lead field dependent on $\Delta\sigma$. The choice of which port is to be current carrying and which is to be potential reading is arbitrary. J' is current density with unit current excitation $[1/m^2]$. If $\Delta\sigma$ is positive, ΔZ is negative because of the minus sign in eq. (4.1.7). If $\Delta\sigma$ is zero, ΔZ is zero, this is not the case for eq. (25.1) in Malmivuo and Plonsey (1995).

Sigman effect

Sigman et al. (1937) were the first to report that the resistivity of blood is flow dependent. They found that the resistivity fell about 7% when the blood velocity was increased from 10 to 40 cm/s. This is an application area for the Geselowitz (1971) equation, a change in measured conductance not related to volume and therefore not plethysmographic. It is a source of error in volume estimations, but not necessarily in flow estimations.

No Sigman effect is found in plasma or electrolytes (Geddes and Baker, 1989). The Sigman effect is due to the non-spherical bodies in the blood, in particular the erythrocytes. At higher velocities but still linear flow the erythrocytes reorient into the flow direction, and in a tube they also clump together around the central axis. The erythrocyte orientation means less hindrance to electric current flow and lower resistivity if resistance is measured in the axial direction. Kanai et al. (1976) reported that resistivity changes occurred at double the flow pulsation frequency, and that the magnitude became very small >3 Hz. This means that the orientation and clumping effects are rather slow.

4.1.2.4 Rheoencephalography

Rheoencephalography (REG) is a plethysmographic bioimpedance method widely used in countries like Russia and China, but not very well known in the USA and Europe. The ambition is to assess cerebral blood flow, (Geddes and Baker, 1989), however only a little part of the REG signal is caused by changes in brain conductivity, the rest relates to the pulsating blood flow of the scalp. REGs use has been limited because the reading is so highly contaminated by this scalp component. The anatomical background of REG is not clearly understood, and a multilayer spherical model of the head has been used so that the REG information is split into the extracranial and intracerebral flow signals.

The United States Food and Drug Administration definition states (Anonymous 1, 1997: (a) Identification). A rheoencephalograph is a device used to estimate a patient's cerebral circulation (blood flow in the brain) by electrical impedance methods with direct electrical connections to the scalp or neck area." In other words, the FDA definition includes the word "flow." On the basis of previous data REG is actually a reflection of volume rather than flow (Nyboer, 1960). REG and Cerebral Blood Flow (CBF) correlation have been described earlier (Hadjiev, 1968; Jacquy et al., 1974; Moskalenko, 1980; Jenkner, 1986). However, the correlation of global, local CBF and carotid flow was not investigated. More REG and related references can be found at www.isebi.org/.

REG pulse amplitude change reflects arteriolar, capillary and venular volume changes together rather than absolute brain blood flow. Early CBF–REG studies did not focus on this topic. It was previously described that the involvement of a vessel in CBF autoregulation is size-dependent: larger arteries are less involved than arteriola (Kontos et al., 1978). Consequently, the arteriolar change observed in brain by REG reflects arteriolar function more than it reflects functions in larger arteries (e.g. carotid). The clinical importance of these findings is that REG can be measured more conveniently and continuously in humans than Doppler ultrasound. Therefore, measurement of CBF autoregulation by REG has potential for use as a life sign monitoring modality.

REG is a potential method for cerebrovascular diagnostics as well. In order to reach the potential of widespread application of REG, there is a need for research to clarify the physiological and pathophysiological correlations and adequate data processing.

The physical basis of the REG measurement is based on the fact that blood and cerebrospinal fluid (CSF) are better conductors than the brain or other "dry" tissue. The REG signal reflects the impedance change: during blood inflow into the cranial cavity, electrical conductivity is increased (resistance decreased) represented by increasing REG pulse amplitude. The same electrical impedance change occurs generating pulse wave on peripheral site, as it was first described by Nyboer (1970) in the parallel-column model. In the skull the input is the volume of the arterial pulse, and the output is the venous outflow and the CSF together. The resulting impedance change – REG curve – is the result of the equation – involving all mentioned factors but not detailed individually. The measured electrical impedance value offers the basis of several volume calculations, detailed by Jenkner (1986).

A typical REG change is known to occur as a consequence of arteriosclerosis, expressed as elongation of REG pulse amplitude peak time or decreased slope of anacrotic part (Jenkner, 1986). The possible cause of this alteration is the decreased elasticity of arteriolar wall, which is the most sensitive indicator of disease progression.

Animal studies (Bodo et al., 2004, 2005a,b, 2007) show that REG can be measured more conveniently and continuously in humans than Doppler ultrasound.

Therefore, measurement of CBF autoregulation by REG has potential for use as a life sign monitoring modality.

Studies on humans have shown that reproducibility and sensitivity of the bioimpedance measurement – including REG – were comparable to the sensitivities of the pulse oximeter, laser Doppler and Doppler ultrasound. Results demonstrated that bioimpedance offers potential for use as a multifunctional, continuous, noninvasive life sign monitor for both military and civilian purposes (Bodo et al., 2006).

In a comparative population screening study (546 volunteers) REG measurements revealed symptoms of arteriosclerosis in 54% of the subjects; within the identical population the Doppler ultrasound measurements showed 30% with arteriosclerosis (Sipos et al., 1994; Bodo et al., 1995a).

REG may have potential for non-invasive continuous life sign monitoring and detection of early cerebrovascular changes. Since the early days of REG research, advances in the development of electronics, computation and signal processing techniques offer the possibility to reconsider the feasibility of implementing a portable or even wearable version of the REG monitoring technique to evaluate the adequacy of CBF reactivity.

REG is also used in cardiac applications and evaluation of edema in the legs.

In order to fulfill some of the expectations REG must reach a much higher level of standardization of both instrumentation and electrode geometry. The lack of electrode sensitivity analysis of the REG techniques has severely reduced the scientific soundness of the method.

4.1.3 Impedance cardiography

Impedance Cardiography (ICG) is impedance plethysmography based on the measurement of thoracic electrical bioimpedance (TEB). It also includes a component from the resistivity dependence on blood flow (Sigman effect). This is not a plethysmographic but a blood velocity component. Usually a measuring frequency of 50–100 kHz has been used. A TEB picks up both cardiac and respiration signals. The ambition is that the SV (L) and therefore CO (L/min) can be calculated with ICG, as well as the total thoracic fluid volume,for example, according to eq. (4.1.1).

Nyboer (1950) used two band electrodes around the neck, one band electrode corresponding to the apex of the heart and a fourth further in caudal direction. Nyboer regarded the thorax as a cylinder volume of length L and used the expression:

$$\Delta v = \Delta Z \rho \left(\frac{L}{Z_0}\right)^2 \; [\mathrm{m}^3] \quad \text{(Nyboer)} \qquad (4.1.8)$$

which is in accordance with eq. (4.1.4). The ICG tradition has since then been to use impedance Z instead of resistance R, and to use the series model. Equation (4.1.8) is surely valid as the condition $\Delta Z \ll Z_0$ is fulfilled in ICG, but the $\Delta Z/Z_0^2$ term is not so directly evident as the simple ΔG term in eq. (4.1.2). Also an increased volume corresponds to a conductance increase, but to an impedance decrease. Therefore a minus sign is often introduced as in eq. (4.1.4) (Geddes and Baker, 1989). Typically values for Z_0 is 25 Ω and ΔZ 0.2 Ω. The Z waveform is similar to the aorta blood pressure curve.

SV as developed by Kubicek et al. (1966) is still compatible with a basic physical model:

$$SV = \left(\frac{dZ}{dt}\right)_{\max} T \rho \left(\frac{L}{Z_0}\right)^2 \; [\mathrm{m}^2] \quad \text{(Kubicek)} \qquad (4.1.9)$$

The *first time derivative dZ/dt* is called the impedance cardiographic curve (ICG). T is the ventricular ejection time. As the pick-up electrodes are positioned near the heart they also pick-up the ECG signal, and this is used for the time estimation.

In the original Nyboer model the changes in conductance was associated with cylinders of different and changing cross sectional areas. The blood distribution process is of course much more complicated. With chest electrodes we have signals from the filling and emptying of the heart, aorta, lungs, muscles of the chest; as well as the Sigman effect.

Many electrode geometries have been used, in particular the old four-band technique and the newer spot electrode technique with four or eight or more electrodes. Some systems use two current carrying systems with four excitation and four recording electrodes, eight electrodes in total. Then four electrodes are connected around the neck, the others on the lower thorax. With two current sources the sensitivity field is complicated, and many algorithms are possible when weighing the results obtained in the two channels. Kauppinen et al. (1998) compared four different electrode systems using either band or spot electrodes. They used a 3D computer model with data from the US Library of Medicines Visible Human Project. They found that more than 55% of the measured transfer impedance was due to the skeletal muscle mass in the thorax and only about 15% originated from blood, heart and lungs. The sensitivity field for the four tested systems showed only small differences in total sensitivity, but all the same each was a complicated mixture of many factors.

Sramek (1981) and Bernstein (1986) further developed empirical equations also with biometrical data such as patient height, actual weight, ideal weight, body surface area, age and gender. Other transthoracic equations have also appeared (reviewed by Moshkovitz et al., 2004), partly with proprietary modifications making methods more accurate but perhaps less robust in the case of for instance critically ill patients or validity before and after surgery. Bernstein and Osypka (2003), and Bernstein and Lemmens (2005) introduced an index of transthoracic aberrant conduction (e.g. by excess extra-vascular lung water) in their equation. Suttner et al. (2006) tried it on critically ill patients with acceptable results.

Since ICG is a very simple and low cost technique very valuable applications may appear. It must also be remembered that a golden standard method for *SV* and CO does not exist; the intramethod variability may be larger than the intermethod variability. Reference methods have been thermodilution, dye dilution, oxygen Fick technique, radionuclear radiography and transoesophageal Doppler echocardiography.

4.1.4 Tissue characterization in urology

Bioimpedance techniques are interesting candidates for diagnosing disorders in the urinary system. There are many possible applications, ranging from measurements of uteric tone and contractility to assessment of prelabor ripening and detection of cervical cancer. Some examples are as follows.

Mudraya et al. (2007) used different instruments such as a multichannel impedance spectrometer for tetrapolar measurements on an array of nine electrodes, enabling simultaneous recording of six locations along the urinary tract. They found such measurements to be valuable for quantitative assessment of ureteric peristalsis, and they could also locate stricture regions by monitoring the ratio between low and high frequency impedance, Z_{low}/Z_{high}.

Abdul et al. (2005) developed a tetrapolar probe for impedance measurements in the frequency range 2–1200 kHz. They found their system to give similar sensitivity and specificity to currently used screening tests for cervical intraepithelial neoplasia, but with the evident advantage of providing instant results.

Gandhi et al. (2006) measured cervical stromal impedance (CSI) in non-pregnant women and women in different stages of pregnancy. They found the impedance to increase during pregnancy and their results suggest potential utility of CSI measurement for quantifying gestation-dependent changes in cervical stromal tissue.

4.1.5 EEG, ENG/ERG/EOG

The electrical activity of 10^{11} brain cells are recorded on the skin of the scull with a standardized electrode network of 21 electrodes. The leads may be bipolar or unipolar. The signal amplitude is only of the order of 50 μV, and the frequency content 1–50 Hz, so DC voltages are filtered out. The low frequency content is a clear sign that the scull has a detrimental effect on the signal transmission. Even so the number of electrodes indicate that the information content is sufficient to roughly localize a source. It is believed that the brain centers are less synchronized the higher the activity, resulting in smaller amplitude and more high frequency content of the EEG. At sleep the waves are with largest amplitude and lowest frequency content. The electrodes used are rather small, often made of tin/lead with collodium as contact/fixation medium.

The brain can be stimulated (e.g. by a sound or by looking at changing patterns). The EEG signal can be time averaged based on synchronization pulses from the stimulator. Electrical activity and the brain electric response can by this method be extracted from noise. In this way hearing sense of small children and babies can be examined.

Electrocorticography with electrodes placed directly on the cortex during surgery permits direct recording of high amplitude, high frequency EEG.

EOG (electro-oculography) is an electrophysiological method where DC potentials are utilized, and therefore AgCl-electrodes are used. The DC potential is dependent on the position of the eye, and is of particular interest, for example, when the eye lids are closed (REM sleep). As a DC recording method EOG tends to be prone to drift which makes the spatial localization of the point of gaze problematic. It is also sensitive to facial muscle activity and electrical interference. The signals are due to a potential between the cornea and the fundus of an eye with a functioning retina, and are not from the ocular muscles (Geddes and Baker, 1989).

ENG (electro-nystagmography) is also the recording of corneo-retinal potentials, usually used to confirm the presence of nystagmus (special eye movements). The electrodes are placed to the side (lateral), above and below each eye. A reference electrode is attached to the forehead. A special caloric stimulation test is performed, with cold and/or hot water brought into the canal of one ear. The electrodes record the duration and velocity of eye movements that occur when the ear is temperature stimulated.

ERG (electro-retinography) records the AC potentials from the retina. The electrode system is unipolar, with a gold foil or AgCl recording electrode embedded in a special saline filled contact lens in contact with the cornea. The eye may be considered as a fluid-filled sphere in contact with the retina as a thin, sheet-like bioelectric source. The ERG signal caused by a light flash is a very rapid wave with an initial rise time less than 0.1ms (early receptor potential) and an amplitude around 1 mV, followed by a late receptor potential lasting many milliseconds.

4.1.6 Electrogastrography

The typical electrogastrographical (EGG) signal due to stomach activity is recorded with a bipolar lead using a pair of standard ECG electrodes on the skin (e.g. 4 cm apart). The signal is typically of about 100 μV amplitude, and periodic with a period of about 20 seconds (0.05Hz fundamental). Best position for the EGG electrodes is along the projection of the stomach axis on the abdomen.

Internal electrodes are also used, but are in general not considered to provide more information than external EGG signals. Of course the internal electrodes are nearer to the source, implying higher amplitude signals with more high frequency content. Because of the very low frequency spectrum, external noise from, for example, slowly varying skin potentials tends to be greater problem than from internal recordings.

4.1.7 EMG and neurography

EMG

To record signals from muscles (EMG), both skin surface electrodes and invasive needles are used. The distance to the muscle is often short, and the muscle group large, so signals have high amplitude and high frequency content. But the signal is usually coming from many muscle groups, and the signal looks rather chaotic. The muscle activity is related to the rms value of the signal. If the muscle activity is low and controlled, single motor units become discernible with needle electrodes. EMG is often recorded in connection with active neurostimulation, involving both the muscle and the nervous system.

The frequency EMG spectrum covers 50–5000 Hz, and the amplitude may be several millivolt with skin surface electrodes.

ENeG

The sum of activities forming a nerve bundle may be picked up by skin surface electrodes (e.g. on the arm where the distance from the electrodes is only a few millimeters). Action potentials from single nerve fibers must be *measured* with invasive needle electrodes. They may be bipolar or unipolar. To find the right position, the signal is monitored during insertion, often by sounds in a loudspeaker.

Stimulation may be done with the same electrode designs: either transcutaneously right above the bundle of interest, or by needles. Muscles (e.g. in the hand) are stimulated by stimulating efferent nerves in the arm. EMG electrodes can pickup the result of this stimulation, for instance for nerve velocity determination. It can also be picked up by neurographic electrodes, but the signal is much smaller and must therefore be averaged with multiple stimuli. However, the method is more of interest because there is more information in the response waveform.

4.1.8 Electrical impedance myography

Electrical impedance myography (EIM) refers to a group of impedance-based methods for the clinical assessment of muscles. This includes primary disorders of muscle such as myopathic conditions (Rutkove et al., 2002; Tarulli et al., 2005) and the sarcopenia of aging (Aaron et al., 2006) as well as diseases that affect the nerve, such as localized neuropathies or nerve root injuries (Rutkove et al., 2005) and generalized problems, such as amyotrophic lateral sclerosis (Esper et al., 2005). This neuromuscular disease-focused application of bioimpedance is built on the earlier experimental and theoretical work of Shiffman and Aaron (see Aaron et al., 1997; Shiffman et al., 1999; Aaron and Shiffman, 2000). Importantly, the goal of EIM is not to image the muscle, but rather to assess quantitatively changes in its microscopic structure induced by neuromuscular disease states.

All methods utilize a tetrapolar technique and rely on the placement of voltage sensing electrodes along a muscle or muscle group of interest. Depending on the application, current-injecting electrodes can be placed in close proximity to or at a distance from the voltage electrode array. Both single frequency (50 kHz) and multifrequency (up to 2 MHz) methods have been studied with the former showing very high reproducibility of the major outcome variable, the spatially averaged phase θ_{avg} (Rutkove et al., 2006). The application of EIM in the setting of voluntary or stimulated muscle contraction represents another provocative and potentially important area of investigation that may allow assessment of the contractile apparatus (Shiffman et al., 2003).

Some limited animal work has also been performed. Nie et al. (2006) showed that consistent measurements could be obtained on the hamstring muscles of the rat and that substantial changes occur after experimental sciatic crush, including reductions in the measured phase and loss of the normal frequency dependence. Future animal work will be geared at disease differentiation and determining the relationship between muscle states and their impedance patterns.

The most straightforward application of the technique of EIM for clinical care is for its use as a quantifiable measure of muscle health, such that treatment or rehabilitation programs can be effectively monitored. Indeed, EIM has the potential to serve as a useful new outcome measure in clinical trials work (Tarulli et al., 2005) and studies are ongoing to verify the role of EIM as an outcome measure in amyotrophic lateral sclerosis, spinal muscular atrophy and exercise interventions in aging. It is uncertain whether the technique has the potential of supplanting standard neuromuscular diagnostic methods, most notably needle EMG, and this remains a subject of ongoing research.

4.1.9 Electrotherapy

Transcutaneous stimulation

Transcutaneous electrical nerve stimulation (TENS): is electrical stimulation through surface electrodes. The advantage of not using syringe injections is obvious, the electrical pulses stimulate the body's own mechanisms for obtaining pain relief. There are three theories as to how the pain relief is achieved.

Gating theory

Pain perception is controlled by a gate mechanism in the synapses, particularly in central nervous system (CNS) of the spine. This gate is controlled by separate nerve fibers, and by stimulating with pulses of high frequency (50–200 Hz), these fibers are stimulated and pain relief is obtained.

Endorphines

The body uses natural forms of morphine called endorphines for pain relief. The secretion of endorphines is obtained with low frequency (2–4 Hz) stimulation. These low frequencies correspond to the rhythmic movement of an *acupuncture* needle (in classical acupuncture it is also necessary to stimulate motor nerve fibers). The effect of endorphines is probably in the higher centers of the CNS.

Vasodilation

This effect is usually linked to pain in cold extremities. Increased blood flow may increase the temperature from the range 22–24°C to 31–34°C, including in the extremities not stimulated. The effect must therefore be elicited in higher centers of the CNS.

The afferent pain nerves have a higher threshold and rheobase than sensory and motor nerves. Thus it is possible to stimulate sensory and motor nerves without eliciting pain. Very short pulses of duration 10–400 μs are used, with a constant amplitude current up to about 50 mA and a treatment duration of 15 minutes or more. The skin electrodes may be bipolar or monopolar. The position is in the pain region, an electrode pair may, for example, be positioned on the skin on the back of the patient, or implanted with thin leads out through the skin. The electrode pair may also be positioned outside the pain area (e.g. at regions of high afferent nerve fiber densities in the hand).

Electroacupuncture

The secretion of endorphines is obtained with low frequency (2–4 Hz) stimulation, corresponding to the rhythmic movement of an *acupuncture* needle. Instead of, or in addition to the mechanical movement, the needle is used as monopolar electrode, pulsed by a low frequency in the same frequency range (1–4 Hz). This is not a TENS method strictly speaking, as the electrode is invasive and the current not transcutaneous.

4.1.9.1 Electrotherapy with DC

Applied DC through tissue for *long* duration (e.g. >10 seconds) is a method almost 200-years old, and is traditionally called *galvanization*. Today the DC effect is often ignored, even if it is quite clear that DC through tissue has some very special effects. Generally the physical/chemical effect of a DC through tissue is:

- electrolysis (local depletion or accumulation of ions)
- electrophoresis (e.g. protein and cell migration)
- iontophoresis (ion migration)
- electro-osmosis (volume transport)
- temperature rise.

Some of these effects are special for long duration (>10 seconds) DC or very low frequency AC, in particular electrolysis. Other effects are in common with AC.

Short term effects of DC through the skin is limited to the sweat ducts. The current density is much smaller in the stratum corneum, but long term currents may have an effect. Proximal to the electrode electrolytic effects influences the skin. Possible effects in deeper layers are erythema (skin reddening) and hyperemia (increased blood perfusion) due to the stimulation of vasomotor nerves. If the DC is applied transcutaneously, there is always a chance of unpleasant pricking, reddening and wound formation in the skin under the electrode.

Iontophoresis

Iontophoresis in the skin (cf. also Section 4.1.15): if a drug is in ionic form, the migration velocity and direction are determined by the polarity of the DC. It is a very famous experiment whereby strychnine was applied to one of the electrodes attached to a rabbit. With one polarity the rabbit died, with the reverse polarity nothing happened. The ions may be transported from the electrode, and thus have an effect both in the tissue during passage, and when assembled under the other electrode. Anesthethic agents may be introduced in the skin by iontophoresis, for minor surgery or the treatment of chronic pain. Also antibiotics and metallic silver has been introduced iontophoretically, as well as zinc for ischemic ulcers. The mechanism of iontophoresis is of course accompanied by a possible electrophoretic action (see below), but is not necessarily so pH-dependent as the latter. The advantage of iontophoretic instead of local syringe injection is not always obvious.

Electrotonus

Tonus is the natural and continuous slight contraction of a muscle. Electrotonus is the altered electrical state of nerve or muscle cells due to the passage of a DC. Subthreshold DC currents through nerves and muscles may make the tissue more (excitatory effect), or less (inhibitory effect) excitable. Making the outer nerve cell membrane less positive lowers the threshold and has an excitatory effect (at the cathode, *catelectrotonus*), the anode will have a certain inhibitory effect (*anelectrotonus*). This is used in muscle therapy with diadynamic currents, see below.

Wound healing

Many controlled studies have shown that a small microampere current as a long term treatment leads to accelerated healing. There are two classes of cases: accelerated healing of bone fractures, and of skin surface wounds. Ischemic dermal ulcers are treated with DC, and the healing rate is approximately doubled. It has been found that a monopolar cathodic application during the first days followed by an anodic application, gives the best results. In a skin wound it is believed that positive charge carriers (ions and proteins) are transported to the liquid wound zone by endogenic migration.

An increased rate of bone formation has also been found when small currents are applied to each side of a bone fracture (bipolar electrode system). This is of particular interest in cases of bone fractures that will not grow the natural way: so-called *non-union*). Nordenstram (1983) described the use of DC treatment by applied needles into tumors for the treatment of cancer.

DC ablation

At higher current densities the acid under the anode leads to coagulation, and the alkali under the cathode to liquidification, of the tissue. Warts can thus be treated, and with a needle cathode in the hair follicle, the local epidermis is destroyed and the hair removed.

DC shock pulses

DC shock pulses have also been used for destroying calculi in the urinary tract, *electrolithotrity*.

Hydrogen production

The application of a DC current to the inside of the eyeball by a needle electrode is used to produce bubbles of hydrogen in the aqueous humor, *electroparacentesis*.

4.1.9.2 Electrotherapy of muscles

Electrotherapy is a broad term, and should for instance include pacing, defibrillation and electroconvulsion. Here we will keep to the traditional meaning, however, which is more limited to methods stimulating *muscles*, either directly or via the nerves. Usually such stimulation also generates data of diagnostic value.

Figure 4.1-10 shows the minimum stimulus current to an efferent nerve fiber as a function of pulse duration for obtaining a certain muscle response. The coupling (synapses) between the nerve axon end plates and the muscle cells is an important part of this signal transmission line. It is not possible to lower the current under a certain minimum level, the *rheobase* value. The pulse length with current amplitude 2 × rheobase value is called the *chronaxie*.

Pflüger's law relates the muscle effect to the leading or trailing edge of the pulse, and to anodal or cathodal polarity:

Leading edge cathodal	Strongest effect
Leading edge anodal	↓
Trailing edge anodal	
Trailing edge cathodal	Weakest effect

If a linear triangular current pulse is used instead of a square wave pulse, an *accommodation* effect will appear, particularly at pulse duration >1 ms. A slowly increasing DC does not excite a nerve to the same extent as a DC step change, and the accommodation implies that the threshold current amplitude will be larger with triangular than with squared waveform. The current-time curves can be recorded for diagnostic purposes, the curves are quite different for degenerated muscles.

The pulses may be of unidirectional current (interrupted DC, *monophasic*), which implies that the current has a DC component. High voltage pulsed galvanic stimulation is also used, with pulse currents up to some amperes, but pulse duration only a few microseconds. If

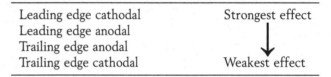

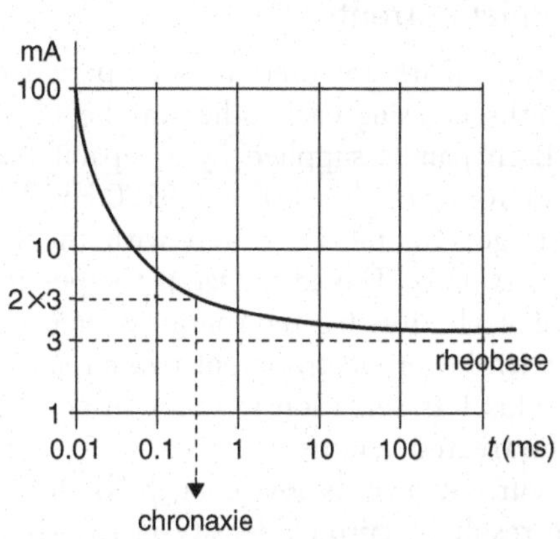

Figure 4.1-10 Minimum stimulus current to an efferent nerve fiber bundle for obtaining a certain muscle response, as a function of pulse duration.

DC effects are to be avoided (e.g. to reduce electrolytic effects or electrode metal corrosion), the current is *biphasic*. *Faradic* currents are biphasic currents of the type generated by an induction coil. If the pulses are slowly increased in amplitude, then reduced, and after a pause again increased, we have a *ramp* or *surged* current. As many effects are current controlled, it is often better to use a constant amplitude current mode than a constant amplitude voltage mode of the stimulator output.

Pulse waveform treatment of innervated muscles (faradization)

Short (0.5–5 ms) triangular pulses for tetanic muscular contractions, interrupted or with varying amplitude. Interval between pulses 10–25 ms. For muscle pain relief rectangular pulses of length 2 ms and interval 7 ms are used. TENS for pain relief is different. It is based on stimulating afferent nerve fibers with much shorter pulses (e.g. 0.2 ms).

Pulse waveform treatment of denervated muscles

As there are no innervation, the stimulation is directly to the muscle. The paralyzed part of the muscle mass can be stimulated selectively, because such muscles has a smaller accommodation at long pulse duration. Very long (1 second) triangular waveforms are used, with even longer intervals between the pulses.

Diadynamic currents for the treatment of pain and increase of blood perfusion

This is the summation of two currents: a pulse current superimposed on a DC. Each current is separately adjusted. Often the DC level is first increased slowly so that no perception occurs (*electrotonus*), and then with a constant DC flowing the pulse amplitude is increased until a weak vibration is felt. The pulse waveform may be a power line 50 Hz half-rectified (50) or fully rectified (100 Hz) current.

Interferential currents

Two-electrode pairs are used to set up two different current paths crossing each other in the target tissue volume. Each pair is supplied by a separate oscillator, adjusted to, for example, 5000 and 5100 Hz. The idea is that the target volume is treated with the frequency difference, 100 Hz. The advantage is the possible selective choice of a limited treated volume deep in the tissue, together with lower electrode polarization and skin impedance, plus less sensation in the skin and the tissue outside the treated volume.

If the current level is low enough for linear conditions, the resultant current density in the tissue is the linear summation of the two current densities. According to the superposition theorem in network theory and Fourier analysis, the new waveform $f(t)$ does not contain any new frequencies. The linear summation of two currents at two different frequencies remains a current with a frequency spectrum with just the two frequencies, no current at any new frequency is created.

The current density in the treated volume must be high enough to create non-linear effects. The process can then be described mathematically by a multiplication. If $\omega_1 \approx \omega_2 \approx \omega$, $f(t)$ is a signal of double frequency and half the amplitude, together with a signal of the low beat frequency $\omega_1 - \omega_2$ also of half the amplitude. The double frequency signal is not of interest, but the low beat frequency $\omega_1 - \omega_2$ is now present in the tissue.

A third electrode pair can be added, with additional flexibility of frequency and amplitude selection. Anyhow it must be taken into account that muscle impedance may be strongly anisotropic, with a possible 1:10 ratio between two directions.

4.1.10 Body composition analysis

The parameters of interest in body composition analysis (bioelectric impedance analysis BIA) are total body water (TBW), extracellular/intracellular fluid balance, muscle mass and fat mass. Application areas are as diversified as sports medicine, nutritional assessment and fluid balance in renal dialysis and transplantation.

One of the first to introduce BIA was Thomasset (1965), using a two-electrode method and 1 kHz signal frequency. With just two electrodes it is important to use large area band electrodes in order to reduce the contribution from the current constrictional zones near the electrodes. With a tetrapolar electrode system it is easier to select the preferred volume to be measured. The small circumference of the lower arm, wrist and fingers causes those body segments to dominate measured impedance in a so-called whole body measurement. With measuring electrodes, for example, on one hand and one foot the chest contribution is very small. The impedance of the chest segment is therefore the most difficult one to determine accurately, both because it is much lower than the impedance of the limbs and because it varies with respiration and heartbeats, as exploited in ICG (Section 4.1.3). Thoracic measurements were compared with whole body measurements by Nescolarde et al. (2006).

By using more than four electrodes it possible to measure more than one body segment. One method uses eight electrodes, two electrodes at each hand and foot. The body impedance is then modelled in five segments: arms, legs and chest (Fig. 4.1-11). One segment impedance is determined by letting two limbs be current carrying and use a third limb for zero current potential reading (five electrode lead, pentapolar). The leads are

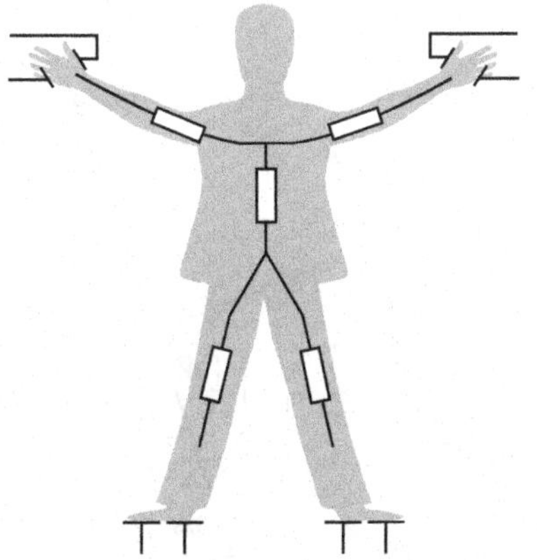

Figure 4.1-11 Human body divided into five impedance segments, octopolar electrodes.

then varied in succession. The system will be highly sensitive for the detection of asymmetrical limb bioimpedance. Standardization of the type of electrodes used and their placement is a major concern (Kyle et al., 2004). Cornish et al. (1999) provided a set of standard electrode sites for bioimpedance measurements.

Calibration is also a major concern in BIA. Calibration can be done with more accurate but cumbersome methods such as using deuterium, underwater weighing or dual energy X-ray absorption. However, dilution methods have their own errors (>2 l for TBW) and yield different results (e.g. 4% difference between the deuterium–TBW method and the ^{18}O-TBW method). Although body impedance reflects tissue hydration, soft tissue mass (lean and fat) can also be empirically derived by correlation in healthy subjects because the *compartments of soft tissue are correlated with each other through physiological constants*. However, physiological constants become flawed in patients with fluid disorders, which accounts for most conflicting results in the literature (Kyle et al,. 2004).

Body *position* is important because it influences the distribution both of blood and the fluids in the stomach/intestine tissue. Direct body segment to body segment skin contact must be avoided in order to obtain stable readings. The feet should therefore be kept at a distance from each other, and the arms should be held out from the chest. Scharfetter et al. (1998) also analyzed the artefacts produced by stray capacitance during whole body or segmental Bioimpedance Spectroscopy (BIS), and proposed a model for simulating the influence of stray capacitance on the measured data.

BIA as a tool for assessment of the hydration of soft tissue may be divided along *three methods of body fluid volume assessment*.

The first and the most validated method is prediction of TBW from whole body impedance measurements at a single frequency, often 50 kHz. Either the series impedance electrical model (often with the reactance component X neglected), or the parallel equivalent have been used. Several indexes have been introduced in order to increase the accuracy. Gender, age and anthropometric results such as total body weight and height are parameters used. An often used index is H^2/R_{segm}, where H is the body height and R_{segm} the resistance of a given segment. Because of the $1/R_{segm}$ term this is therefore actually a *conductance* index. Calibration can be done by determining the k-constants in the following equations:

$$TBW = k_1 \frac{H^2}{R_{segm}} + k_2 \tag{4.1.10}$$

$$TBW = k_1 \frac{H^2}{R_{segm}} + k_2 W + k_3 \tag{4.1.11}$$

Such equations are not directly derived from biophysical laws, but have been empirically selected because they give the best correlation. The correlation according to eq. (4.1.10) can be better than 0.95, and it can be slightly improved by also taking into account the body weight W, eq. (4.1.11). Hundreds of validation studies with isotope dilution have established a solid relation between whole-body impedance at 50 kHz and body fluid volume (Kyle et al., 2004). Complex impedance data can be given also as modulus and phase. Phase has been used as an index of *nutrition*. This is true only in comparison between vectors with the same modulus. For instance, short vectors with a small phase angle are associated with edema whereas long vectors with an increased phase angle indicate dehydration. *Fat free mass* is predicted either from TBW (TBW/0.73) or through specific regression equations including the same variables as TBW, with different partial regression coefficients (Sun et al., 2003; Kyle et al., 2004). Fat mass is calculated by difference. The prediction error of best equations while suitable for epidemiological studies is too high for the clinical use (standard error of the estimate in the order of 3–4 l for TBW and 3–4 kg for the fat free mass) (Sun et al., 2003).

The second method is the use of BIS following the Cole model approach (many groups call it a Cole-Cole model but that is a permittivity model), early used by Cornish et al. (1993). R values are extrapolated at extreme limit frequencies (0 and infinity) for prediction of TBW and extracellular water (ECW) respectively, and by difference intracellular water (ICW). *Body cell mass* is then predicted as a function of the ICW (De Lorenzo et al., 1997). The results of Lozano et al. (1995) showed that there is a sharp disequilibrium between the intracellular and extracellular compartments in the very first dialysis period and they stressed the importance of continuously monitoring segmental impedance during dialysis. BIS may be accurate with suspended cells, but unfortunately it is impossible to estimate the extracellular electric volume of tissues because of anisotropy and limited low frequency data. It is common practice not to measure below 1 kHz, and low frequency dispersions are therefore neglected. The measuring current is usually around 0.5 mA, higher current levels are difficult to use because the threshold of perception is reached at the lowest frequencies (cf. Section 4.1.17).

The third method is the prediction of ECW and TBW with low (1–5 kHz) and high (100–500 kHz) frequency impedance data (dual- or multifrequency BIA), with the ICW calculated by difference. Volume calibration is obtained with regression equations as in single-frequency BIA. Like BIS, multifrequency BIA relies on the wrong assumption of tissue isotropy with low frequency current only flowing around cells. If the hydration of the fat free mass is not fixed at 73% (e.g. in hemodialysis, in patients with edema or heart failure) or when body weight is meaningless for body compartments (e.g. in ascites, pregnancy and severe obesitiy); then BIA, multifrequency BIA and BIS prediction equations should not be used.

A more recent method is to use the complex impedance vector at 50 kHz in a probabilistic Wessel diagram. Such vector BIA (or BIVA) is based upon patterns of the impedance vector relating body impedance to body *hydration* (Piccoli et al., 1994). BIVA is a single frequency BIA that follows a black-box approach considering Z as a random output of a stochastic system (current flow through anisotropic tissues). The method consistently applies to whole body or segmental measurements normalized by the conductor length (height (H) for whole body, body segment length for segmental). BIVA only needs to take care of the measurement error (in the order of 3% at 50 kHz) and of the biological variability of subjects in any clinical condition. The intersubject variability of Z is represented with the bivariate normal distribution (i.e. with elliptical probability regions in the Wessel plane) which are confidence (95%) and tolerance ellipses (50%, 75%, 95%) for mean and individual vectors respectively (cf. Fig. 4.1-12). The intersubject variability of the impedance vector is represented with the bivariate normal distribution (i.e. with elliptical sex-specific probability regions (50%, 75%, and 95% tolerance ellipses) in the Wessel plane. Vector components are normalized by the subject's height (R/H, and X/H, in Ohm/m). Upper and lower poles of the 75% tolerance ellipses represent bioelectrical thresholds for dehydration and fluid overload, respectively. Vector components can also be transformed into dimensionless z-scores which allow comparisons of vector position between different analyzers (Piccoli et al., 2002). Clinical information on hydration is obtained through patterns of vector distribution with respect to the healthy population of the same

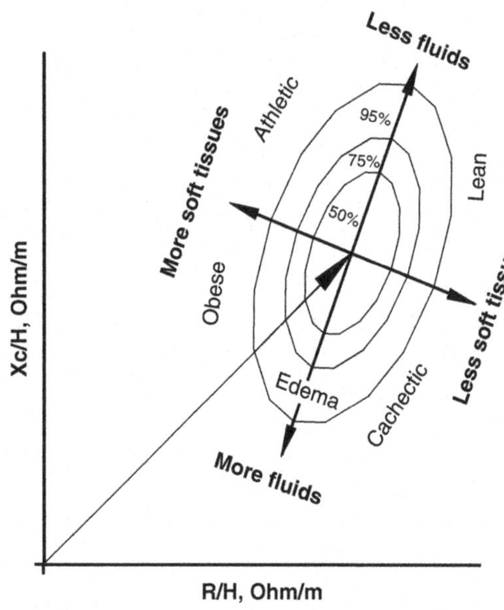

Figure 4.1-12 Z probability graph. Vector position and migration in the Wessel plane are interpreted and ranked according to directions: (a) Vector displacement parallel to the major axis of an ellipse is associated with a progressive change in soft tissue hydration (short term changes: hours, days). (b) A vector lying on the left or right side of the major axis of an ellipse is associated with more or less cell mass respectively (long term changes: weeks, months).

race, sex, and class of BMI (body mass index) and age (cf. Fig. 4.1-12). From clinical validation studies in adults, vectors falling out of the 75% tolerance ellipse indicate abnormal tissue impedance. Vector position is interpreted and ranked following two directions in the Wessel plane, as depicted in Fig. 4.1-12. The basic pattern has been recently validated with deuterium dilution (Lukaski et al., 2007).

4.1.11 Implanted active thoracic devices

Bioimpedance is especially attractive as a potentially useful transducing mechanism in implantable devices such as pacemakers (including pacemakers with cardiac resynchronization therapy (CRT) and internal cardioverter defibrillators (ICD's)) for several reasons: First, the device circuitry and electrode vector configurations to perform such measurements is relatively simple and already exists in many of the current implantable devices and lead configurations. However, sampling resolution is limited in many of these circuits preventing sufficient data for detecting changes in cardiac impedance waveforms. Therefore, low resolution impedance measurements that provide information on fluid status or respiratory impedance trends as a function oftime (day(s)/month(s)) are currently employed. Improved circuit designs planned for future generation devices will be ideally suited for implantable impedance applications where high resolution, real-time complex impedance waveform data is required. Second, impedance may be able to provide useful diagnostic information about multiple physiological parameters including heart rate, CO, respiratory rate, minute ventilation, thoracic fluid accumulation, myocardial contractility and ischemia detection. Third, impedance is a well established sensing means that it may provide relevant clinical diagnostic information used independently or in conjunction with other sensors such as pressure transducers or accelerometers.

4.1.11.1 Physiological impedance components

Impedance signals are acquired from selected implantable electrode vector configurations, defined in this context as the electric field generated by the injection current field electrodes and the voltage measured by the sense field electrodes. Signals acquired from each electrode vector configuration can be either bipolar where the injection current electrodes and sense field electrodes are the same, or tetrapolar, where the injection current electrodes and sense field electrodes are isolated from each other. Electrodes in implantable pacing devices consist of unipolar or bipolar electrodes positioned on the distal end of conventional pacing leads. Pacing and ICD leads can be implanted in the right atrium, right ventricle, superior vena cava and left ventricular cardiac vein.

Data acquired from the various possible electrode vectors typically contain three major physiological components that may provide useful information for diagnostics or implantable device control:

- A low frequency respiratory component; (fundamental <1 Hz).
- A higher frequency cardiac component; (fundamental = 1–3 Hz).
- A calculated mean impedance; (0Hz).

As shown in Fig. 4.1-13, the higher frequency cardiac component is superimposed on the low frequency respiratory component. The higher frequency cardiac component represents the impedance change during each cardiac cycle that occurs immediately following the QRS deflection on the electrocardiogram. The low frequency respiratory component represents the impedance change during respiration, due to the expansion of the lungs and thorax. Moreover, each component has a different fundamental frequency, typically 1.0–3.0 Hz for cardiac activity and 0.1–1.0 Hz for respiratory activity (Hettrick and Zielinski, 2006). This frequency differentiation allows extraction of each signal by specific filtering techniques. Mean impedance, is represented by a "DC shift" in impedance, changing according to the amount of static conductive fluid in the electrode vector lead field configuration as a function of time (hours/days).

More specifically, change in impedance waveform morphologies may be an indicator of change in blood volume, interstitial volume or tissue integrity. Deviations in the impedance waveform morphologies such as in the positive or negative slope, time duration between minimum and maximum magnitudes, delta between the minimum and maximum magnitudes, changes in the minimum and maximum first derivative, changes in the area under a specific waveform or other deviations in the waveform morphology of complex impedance may be indicative of a vector specific change in chamber or vessel blood volume, such as in heart disease, tissue degradation, such as in myocardial ischemia, or interstitial fluid accumulation, such as in peripheral or pulmonary edema, all secondary to cardiac, vascular or renal disease.

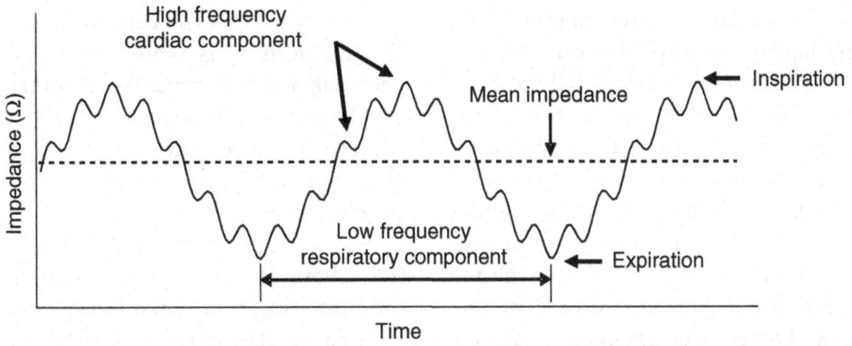

Figure 4.1-13 Simulated impedance waveform consisting of a higher frequency cardiac component superimposed on a low frequency respiratory component: the dotted line is the calculated mean impedance (measured during two respiratory cycles. An implantable impedance sensor may be able to leverage all three signal components in order to provide useful diagnostic or device control information.

4.1.11.2 Fluid status monitoring

Externally measured transthoracic impedance techniques have been shown to reflect alterations in intrathoracic fluid and pulmonary edema in both acute animal and human studies (Fein et al., 1979). The electrical conductivity and the value for transthoracic impedance are determined at any point in time by relative amounts of air and fluid within the thoracic cavity (Gotshall and Davrath, 1999). Additional studies have suggested that transthoracic impedance techniques provide an index of fluid volume in the thorax (Pomerantz et al., 1969; Ebert et al., 1986). Wang et al. (2005) employed a pacing-induced heart failure model to demonstrate that measurement of chronic impedance using an implantable device effectively revealed changes in left ventricular end-diastolic pressure in dogs with pacing-induced cardiomyopathy. Several factors were identified that may influence intrathoracic impedance with an implantable system. These include: (1) fluid accumulation in the lungs due to pulmonary vascular congestion, pulmonary interstitial congestion and pulmonary edema; (2) as heart failure worsens, heart chamber dilation and venous congestion occur and pleural effusion may develop; (3) after implant, the tissues near the pacemaker pocket swell and surgical trauma can cause fluid buildup (Wang et al., 2005).

Yu et al. (2005) also showed that sudden changes in thoracic impedance predicted imminent hospitalization in 33 patients with severe congestive heart failure (NYHA Class III–IV). During a mean follow-up of 20.7 ± 8.4 months, 10 patients had a total of 25 hospitalizations for worsening heart failure. Measured impedance gradually decreased before admission by an average of 12.3 ± 5.3% ($p < 0.001$) over a mean duration of 18.3 ± 10.1 days. The decline in impedance also preceded the symptom onset by a mean lead-time of 15.3 ± 10.6 days ($p < 0.001$). During hospitalization, impedance was inversely correlated with pulmonary wedge pressure (PWP) and volume status with $r = -0.61$ ($p < 0.001$) and $r = -0.70$ ($p < 0.001$), respectively. Automated detection of impedance decreases was 76.9% sensitive in detecting hospitalization for fluid overload with 1.5 false-positive (threshold crossing without hospitalization) detections per patient-year of follow-up. Thus, intrathoracic impedance from the implanted device correlated well with PWP and fluid status, and may predict imminent hospitalization with good sensitivity and low false alarm rate in patients with severe heart failure (Fig. 4.1-14). Figure 4.1-14a shows the results from an algorithm for detecting decrease in impedance over long time. Differences between measured impedance (bottom; circles) and reference impedance (solid line) are accumulated over time to produce fluid index (top). Threshold values are applied to fluid index to detect sustained decreases in impedance which may be indicative of acutely worsening thoracic congestion. Figure 4.1-14b shows an example of impedance reduction before heart failure hospitalization (arrow) for fluid overload and impedance increase during intensive diuresis during hospitalization. Label indicates reference baseline (initial reference impedance value when daily impedance value consistently falls below reference impedance line before hospital admission). Magnitude and duration of impedance reduction are also shown. Days in hospital are shaded.

Some commercially available implantable devices for the treatment of CHF and/ or ventricular tachyarrhythmias now continually monitor intrathoracic impedance and display fluid status trends. This information is then provided to the clinician via direct device interrogation or by remote telemetry. Recent reports based on actual clinical experience with this feature have attested to critical reliability and utility (Vollmann et al., 2007) and good correlation with other traditional tools (Luthje et al., 2007).

However, besides lung fluid, other physiological parameters might explain device measured changes in intrathoracic impedance. Some of these factors include ventricular dilation, atrial or pulmonary vascular dilation, anemia, hyper or hypovolemia, right and left ventricular preload, hematocrit, electrolyte balance, pocket infection, kidney dialysis, pneumonia, bronchitis; weight change (not related to fluid accumulation), lymphatic fluid changes, etc.

4.1.11.3 Cardiac pacemakers

Pacing of the heart may be done transcutaneously, but this is accompanied by pain. The usual method is with two epicardial electrodes and leads out through the chest to an external pacemaker, or with an implanted pacemaker.

The implanted pacemakers are of many models. Let us consider a demand pacemaker, with special recording ring electrodes on the catheter for the demand function. If QRS activity is registered, pacing is inhibited. The pacemaker housing may be of metal (titan) and function as a large neutral electrode. Pacing is done with a small catheter tip electrode, either unipolar with the neutral electrode or bipolar with a catheter ring proximal to the tip electrode.

As can be seen from Fig. 4.1-15, the chronaxie is less than about 500 μs, so there is an energy waste choosing the pulse duration much larger than 100 μs.

A pacemaker may be externally programmed by magnetic pulses. Also because of this a pacemaker is to certain degree vulnerable to external interference. Typical limits are: static magnetic field <1 gauss, 40 kV arcing >30 cm

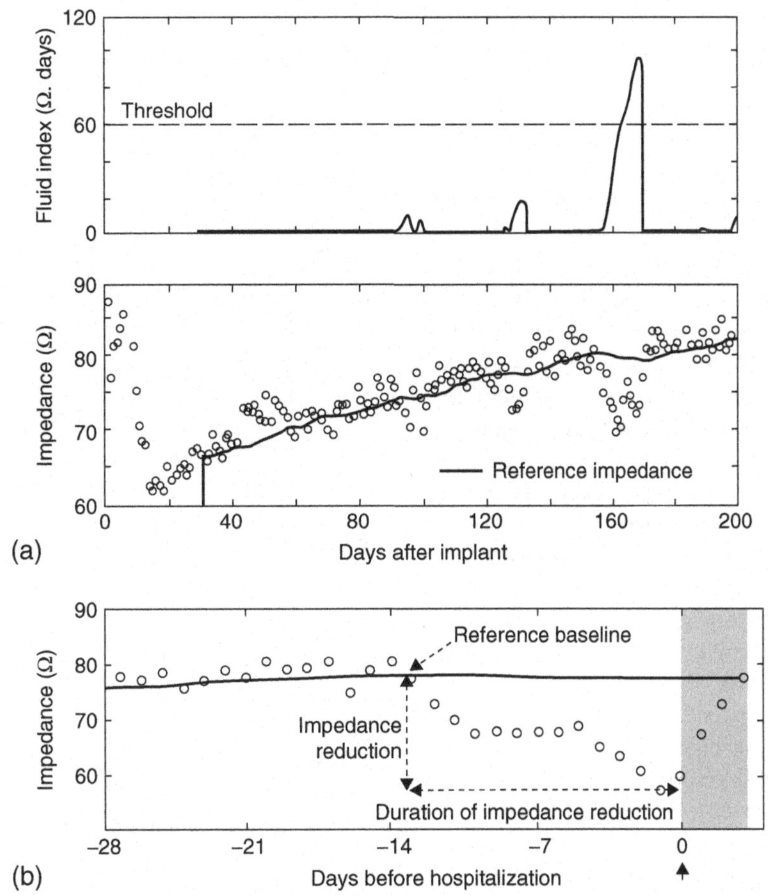

Figure 4.1-14 (a) Fluid index and impedance over 6 months; (b) example of impedance reduction before heart failure hospitalization (arrow) for fluid overload and impedance increase during intensive diuresis during hospitalization. *Source:* Reproduced with permission: Yu et al. (2005).

distance (car ignition system), radar 9 GHz E-field <1.2 kV/m. Typical pacemaker data are: pulse amplitude 5 mA, impedance monopolar electrode system 1 kΩ, load voltage 5 V, lithium battery 6.4 V with capacity 1800 mAh. The stimulus electrodes are AC coupled in the pacemaker's output stage, so that no DC can pass and unduly polarize the electrodes. The electrodes are made of noble metals to be biocompatible, and consequently they are highly polarizable. The monopolar electrode system impedance is not very dependent on faradic impedance because the admittance of the double layer capacitance is large at the frequencies used.

Pacemaker implant and the use of electrosurgery are treated in Section 4.1.13.

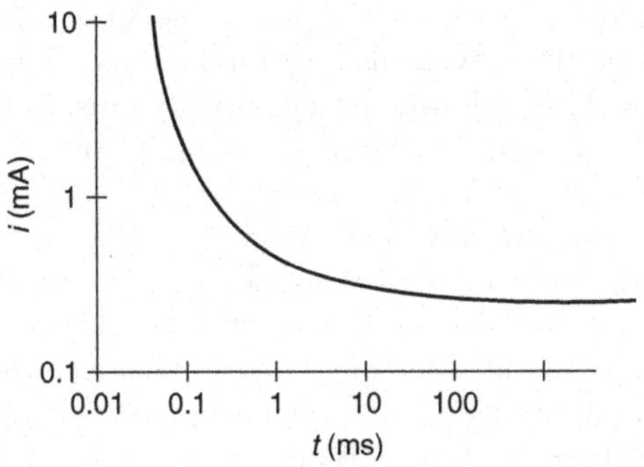

Figure 4.1-15 Current–time curves for heart pacing with a square wave pulse delivered during diastole with intracardial catheter electrode.

4.1.12 Defibrillation and electroshock

4.1.12.1 Defibrillator

Defibrillator shocks are the largest electric shocks used in clinical medicine, up to 50 A is applied for some milliseconds through the thorax, driven by approximately 5 kV. The electrode system is usually bipolar with two equal electrodes of surface *ca.* 50 cm^2 (adult, defibrillation of children is rare). They are positioned so that as much as possible of the current is passing the heart region. With a more unipolar system with one electrode under the shoulder the current path is more optimal, and this is used if the deflbrillation is planned (electroconversion).

Earlier conductive paste was used on the skin, today contact pads are used because they are quick to apply. They also make it possible to avoid usual contact paste that is easily smeared out on the skin surface and causes stray currents (either short-circuiting the shock energy, or representing a hazard for the personnel involved). The current density is so high that reddening of the skin often occurs, especially at the electrode edge. The large electrode and the large current cause an extremely low-ohmic system. Fifty ohm is the standardized resistance of the complete system with two electrodes and the tissue in between. The resistance is falling for each shock given, and this is attributed to tissue damage.

We must assume that it is the local current density in the heart that is the determining parameter for a successful conversion. As this is unknown, it is usual practice to characterize the shock in energy (wattseconds = joules). This refers to the capacitor used to store the energy (Fig. 4.1-16). Stored energy is $CV^2/2$, so a chock dose is simply chosen by choosing charging voltage. Maximum stored energy is usually 400 Ws. Not all the energy will be dissipated in the patient system, a part will be dissipated in the internal resistance R_i of the coil used to shape the waveform of the discharge current pulse. External shock is given transcutaneously, so the voltage must be high enough to break down the skin even at the lowest dose. For internal, direct epicardial application (internal shock), sterile electrode cups are used directly on the heart without any paste or pad. The necessary dose is usually less than 50 Ws.

There is a certain range of accepted current duration. Figure 4.1-17 shows some current discharge current waveforms. Note that some marks use biphasic waveforms, some use truly monophasic.

Defibrillators are also made as implanted types, using intracardial catheter electrodes. In order to reduce energy consumption new waveforms have been taken into use: the exponential truncated waveform. It may be monophasic or biphasic. The idea of the biphasic waveform is that the second pulse shall cancel the net charge caused by the first pulse and thereby reduce the chance of refibrillation.

Tissue impedance measurements with the defibrillator electrodes are used both in some external and internal defibrillator models. Measuring current and voltage during a shock gives a high current level, minimum value, non-linear region, peak voltage to peak current ratio. Between shocks the small signal, linear impedance is also monitored. The measured impedance value is used to customize both waveform and energy level for each shock given.

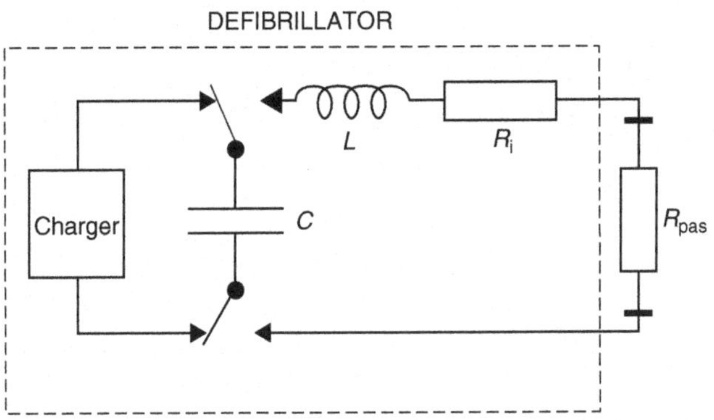

Figure 4.1-16 Classical defibrillator circuit. Typical values are $C = 20\ \mu F$, $L = 100$ mH, $R_i = 15\ \Omega$, $R_{pas} = 50\ \Omega$.

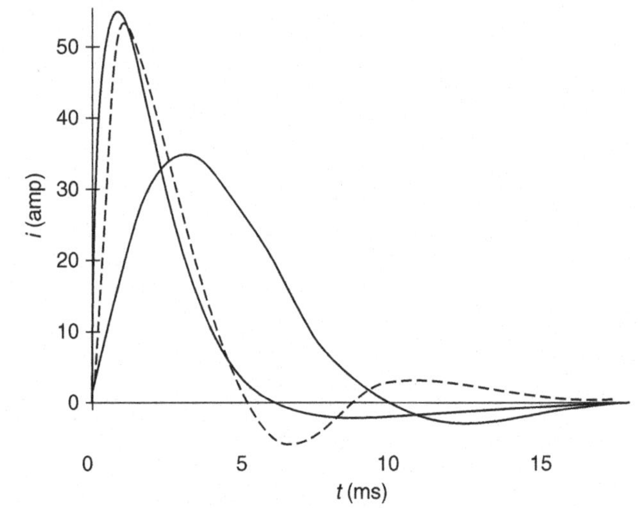

Figure 4.1-17 Some classical current discharge waveforms.

4.1.12.2 Electroshock (brain electroconvulsion)

Electroshock therapy is a somatic method in psychiatry, for the treatment of depressions. The traditional current waveform is a quarter-period power line 50 Hz sine wave, starting at the waveform maximum. Pulse duration is therefore 5 ms, followed by a pause of 15 ms. Automatic amplitude increase, or pulse grouping, is used. It is now often replaced by another waveform, a train of pulses of 1 ms duration with a total energy around 20 J. It is believed that with this waveform the memory problems are less. The corresponding voltage and current are several hundred volts and milliamps. Large bipolar electrodes are used on the temples. The positioning is usually bilateral, but ipsilateral positioning is also used.

ECT is a much discussed procedure, partly because it has been perceived as a brutal medical treatment. It is performed under anesthesia, and because of the heavy muscle contractions, muscle relaxants are given. The shock elicits a seizure not very different from a grand mal epileptic attack, the seizure is to last longer than 25 seconds. The effect is presumably due to the enormous synchronized activity of the whole CNS. The treatment is usually repeated several times within a few weeks span. The treatment is often followed by a loss of memory for recent events, and the therapeutic effect is not permanent.

4.1.13 Electrosurgery

High frequency (also called radio frequency, RF) current is used to cut or coagulate tissue. The method must not be confused with *electrocautery*. In *electrosurgery* the current is passing the tissue, with heat development in the tissue and cold electrodes (diathermy). With electrocautery the current is passed through a wire and not through tissue, and the wire is accordingly heated. Bipolar forceps are used for microsurgery, they represent a dipole current source in the tissue. A unipolar (in the field of surgery called monopolar) circuit is used in general surgery. The neutral electrode is a large flexible plate covered with sticky hydrogel for direct fixation to the skin. The neutral plate is often split into two, and a small current is passed between the two plates via the skin and tissue. Impedance is measured, and if this impedance is outside pre-set or memory set limits, the apparatus will warn of poor and dangerous plate contact.

The active electrode may be handheld, or endoscopic: long and thin types either flexible or rigid.

Figure 4.1-18 illustrates the monopolar circuit. The monopolar coagulation electrode is often in the form of a sphere.

The waveform used is more or less pure sinusoidal in *cut mode*, to highly pulsed with a crest factor of 10 or more for the *spread coagulation mode*. In spread coagulation tissue contact is not critical, the current is passed to the tissue mostly by *fulguration* (electric arc). The electromagnetic noise generated may be severe over a large frequency spectrum, and this causes trouble for medical instrumentation connected to the same patient.

Electrosurgery is based on the *heat effect* of the current, and this is proportional to the square of the *current density* (and the electric field) and tissue *conductivity*. The power volume density W_v is falling extremely rapidly with distance from the electrode, as shown by the equation for a voltage driven half sphere electrode at the surface of a half infinite homogeneous medium. With constant amplitude current the power volume density is:

$$W_v = \frac{i^2}{4\pi^2 \sigma r^4} \quad \text{(half sphere)} \qquad (4.1.12)$$

Tissue destruction therefore occurs only in the very vicinity of the electrode. Power dissipation is linked with conductance, not admittance, because the reactive part just stores the energy and sends it back later in the AC cycle. Heat is also linked with the rms values of voltage and current, ordinary instruments reading average values cannot be used. Because heat is so current *density* dependent, the effect is larger the smaller the cross sectional area of an electrode, or at a tissue zone constriction. This is an important reason for the many

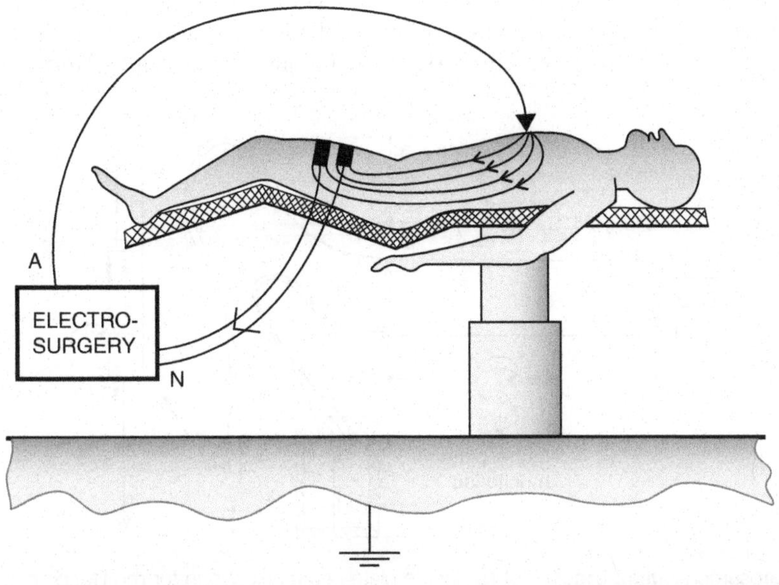

Figure 4.1-18 Monopolar electrosurgery.

hazard reports with the use of electrosurgery in hospitals. Another reason is that that the whole patient is electroactive in normal mode of electrosurgery use. The RF potentials of many body segments may easily attain some tenths of rms volt in normal mode operation, and insulation of these body segments is critical.

High frequencies have been chosen to avoid nerve and muscle stimulation (cf. the sensitivity curve of Fig. 4.1-31). The output is neither constant amplitude voltage nor constant amplitude current. The optimal output characteristic is linked with the very variable load resistance. Tissue resistance increases when coagulated, fat has higher resistance than muscles and blood, and the contact geometry is very dependent on the electrode chosen and the way it is held by the surgeon. If a constant amplitude current is chosen, power would be proportional to load resistance, and tissue would quickly be carbonized in high resistance situations. If constant amplitude voltage were chosen, power would be inversely proportional to load resistance, and when tissue layers around the electrode would coagulate, current would stop flowing. Modern instrumentation therefore measures both output voltage and current, and regulates for an isowatt characteristic.

Typical power levels in unipolar electrosurgery is about 80 W (500 Ω, 200 V, 400 mA rms), in bipolar work 15 W (100 Ω, 40 V, 400 mA rms). The frequency content of the sine wave is of course just the repetition frequency, usually around 500 kHz. In pulsed mode the frequency content is very broad, but most of the energy will be in the frequency band 0.5–5 MHz. In pulsed mode the peak voltage can reach 5000 V, so insulation in the very humid surroundings is a problem.

The arc formed particularly in coagulation mode, is a source both of noise and rectification. Rectification is strongly unwanted, because low frequency signal components may be generated which excites nerves and muscles. In order to avoid circulating rectified currents, the output circuit of a monopolar equipment always contain a safety blocking capacitor (Fig. 4.1-19). Low frequency voltage is generated, but it does not lead to low frequency current flow. Even so, there may be local low frequency current loops in multiple arc situations (Slager et al., 1993). The resulting nerve stimulation is a problem in certain surgical procedures.

Argon gas is sometimes used as an arc guiding medium. The argon gas flows out of the electrode mainly for two purposes: to facilitate and lead the formation of an arc between the electrode and tissue surface, and to impede oxygen in reaching the coagulation zone. In this mode of operation, no physical contact is made between the metal electrode and the tissue, the surgeon points the pen toward the tissue and coagulation is started just as if it was a laser beam (which it is often mixed up with). The gas jet also blows away liquids on the tissue surface, thus facilitating easy surface coagulation.

Implants

The use of electrosurgery on patients with metallic implants or cardiac pacemakers may pose problems. Metallic implants are usually considered not to be a problem if the form is round and not pointed (Etter et al., 1947). The pacemaker electrode tip is a small area electrode, where relative small currents may coagulate endocardial tissue. The pacemaker catheter positioning should therefore not be parallel with the electrosurgery current density lines. This is illustrated in Fig. 4.1-20 for a heart pacemaker implant.

Ablation

Through catheters it is possible to destroy tissue with RF currents in a minimal-invasive procedure. In cardiology this is called *ablation*. Both DC and RF current has been tried for this purpose. Choice of bipolar or monopolar technique is important (Anfmsen et al., 1998).

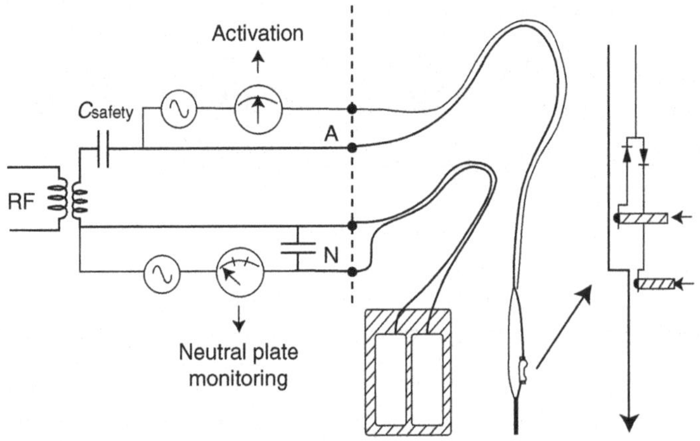

Figure 4.1-19 Typical electrosurgery output circuit. Notice double plate neutral electrode for monitoring of skin contact. Safety blocking capacitor shall prevent rectified *low frequency* currents in tissue.

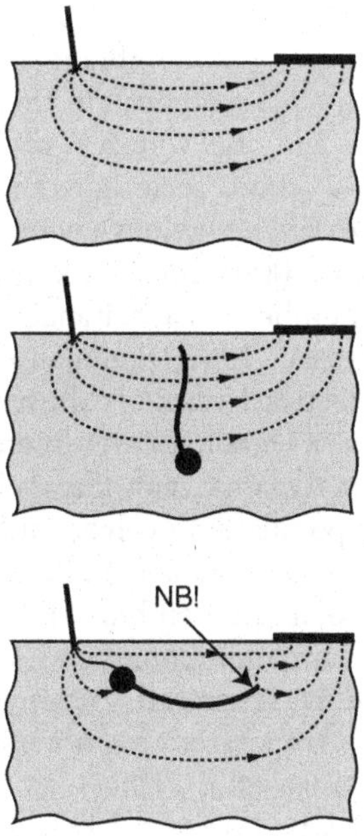

Figure 4.1-20 Monopolar electrosurgery and an implant, for example a pacemaker with intracardial catheter electrode. Importance of catheter direction with respect to current density direction.

4.1.14 Cell suspensions

4.1.14.1 Electroporation and electrofusion

Electroporation is the phenomenon in which cell membrane permeability to ions and macromolecules is increased by exposing the cell to short (microsecond to millisecond) high voltage electric pulses (Weaver, 2003). While the mechanism for electroporation is not yet completely understood experiments show that the application of electrical pulses can have different effects on the cell membrane, as a function of various pulse parameters; such as amplitude, duration, pulse shape and repetition rate (Mir, 2001). As a function of these parameters, the application of the electrical pulse can have no effect, can have a transient effect known as reversible electroporation or can cause permanent permeation known as irreversible electroporation (IRE) which leads to non-thermal cell death by necrosis (Weaver, 2003; Davalos et al., 2005). It is thought that the induced potential across the cell membrane causes instabilities in the polarized lipid bilayer (Weaver and Barnett, 1992; Weaver and Chizmadzhev, 1996; Edd et al., 2006). The unstable membrane then alters its shape forming aqueous pathways that possibly are nanoscale pores through the cell's plasma membrane (Neumann and Rosenheck, 1972; Neumann et al., 1989; Weaver, 1995). Irreversible behavior is attributed to bilayer rupture by uncontrolled pore growth and the outcome is governed by the local plasma membrane potential, $\Delta\psi_{PM}$, behavior and the relative large membrane tension (Esser et al., 2007).

IRE for tissue ablation

IRE has been studied extensively with in vitro cellular systems. IRE has also been studied as method to destroy prokaryotic (Sale and Hamilton, 1967) and eukaryotic cells in vitro and has gain momentum recently as a method to kill microorganisms (Vernhes et al., 1999), mammalian normal cells (Vernhes et al., 1999) as well as mammalian cancer cells (Miller et al., 2005) in vitro and tumors (Nuccitelli et al., 2006). These studies have demonstrated the ability of IRE, to completely eradicate an entire population of cells in vitro without inducing any thermal damage.

Lee et al. hypothesized that electrical injury is often characterized by the preferential death of large mammalian cells (skeletal muscle, nerves) in tissue regions where insignificant temperature rise occurs (Bhatt et al., 1990; Lee et al., 2000; Esser et al., 2007). With the key distinction between shock trauma and temperature change, this research group opened the door to the application of IRE as an alternate tissue ablation technique (Lee and Despa, 2005).

Davalos, Mir and Rubinsky recently postulated that IRE can be used as an independent drug-free tissue ablation modality for particular use in cancer therapy (Davalos et al., 2005). This minimally invasive procedure involves placing electrodes into or around the targeted area to deliver a series of short and intense electric pulses that induce the irreversible structural changes in the cell membrane (Edd and Davalos, 2007). This induced potential is dependent on a variety of conditions such as tissue type and cell size (Edd and Davalos, 2007). Due to the changes in the cell membrane resistance during electroporation the technique can be controlled and monitored with EIT, a real-time imaging method that maps the electrical impedance distribution inside the tissue (Davalos et al., 2004). Ivorra et al. concluded in (Ivorra and Rubinsky, 2007) that impedance measurements can be employed to detect and distinguish reversible and IRE in in vivo and in situ liver tissue. IRE produces a well-defined region of tissue ablation, without areas in which the extent of damage changes gradually as during thermal ablation (Rubinsky, 2007). A single cell is either destroyed by IRE or not (Rubinsky, 2007). The IRE pulses do not compromise the blood vessel matrix and appears to be safe and cause no complications as suggested in (Maor et al., 2007). In addition, it has been shown through mathematical modeling that the area ablated by irreversible tissue electroporation prior to the onset of thermal effects is substantial and comparable to that of other tissue ablation techniques, such as cryosurgery (Davalos et al., 2005). Thus, for certain medical applications IRE alone could be used as an effective technique for tissue ablation without the use of cytotoxic drugs like in chemotherapy (Davalos et al., 2005).

Electrical properties of tissue during electroporation

The electrical properties of any material, including biological tissue, can be broadly separated into two categories: conducting and insulating. In a conductor, the electric charges move freely in response to the application of an electric field, whereas in an insulator (dielectric), the charges are fixed and not free to move.

If a conductor is placed in an electric field, charges will move within the conductor until the interior field is zero. In the case of an insulator, no free charges exist, so net migration of charge does not occur. In polar materials, however, the positive and negative charge centers in the molecules do not coincide. An electric dipole moment, p, is said to exist. An applied field, E_0, tends to orient the dipoles and produces a field inside the dielectric, E_p, which opposes the applied field. This process is called polarization. Most materials contain a combination of orientable dipoles and relatively free charges so that the electric field is reduced in any material relative to its free-space value. The net field inside the material, E, is then

$$E = E_0 - E_p \tag{4.1.13}$$

The net field is lowered by a significant amount relative to the applied field if the material is an insulator and is essentially zero for a good conductor. This reduction is characterized by a factor ε_r, which is called the relative permittivity or dielectric constant, according to

$$E = \frac{E_0}{\varepsilon_r} \tag{4.1.14}$$

Biological systems are electrically heterogeneous (Gift and Weaver, 1995). Application of an electric field pulse results in rapid polarization changes that can deform mechanically unconstrained cell membranes (e.g., suspended vesicles and cells) followed by ionic charge redistribution governed by electrolyte conductivities and distributed capacitance (Weaver, 2000; Ivorra and Rubinsky, 2007). For most cells and tissues the latter charging times are of order $\tau_{CHG} \approx 10^{-6}$ seconds. Thus, if U_m is to exceed 0.5–1 V, much larger pulses must be used if the pulse is significantly shorter than τ_{CHG} (Weaver, 2000).

Electroporation is hypothesized to involve inhomogeneous nucleation of primary, hydrophilic pores based on transitions from much more numerous hydrophobic pores (Weaver, 2000). The basic idea is that a circular region of membrane is replaced with a pore. As primary pores appear in the membrane, its resistance drops, and the voltages within the system redistribute on a time scale governed by the instantaneous values of the various conductivities and capacitance (Weaver, 2000). Both experiment and theory show that the membrane capacitance change is small (Chernomordik et al., 1982; Freeman et al., 1994), so that the main electrical result is drastically decreased barrier resistance. Overall, bilayer membrane electroporation results in dynamic, non-linear changes as a heterogeneous pore population evolves rapidly in response to the local value of the transmembrane voltage U_{mlocal} along the surface of a cell membrane (Weaver, 2000). At the time of maximum membrane conductance, pores are nevertheless widely separated, occupying only about 0.1% of the electroporated membrane area (Hibino et al., 1991; Freeman et al., 1994). In this sense, electroporation is catalytic (Weaver, 1994). Not only is there the possibility of binding and lateral diffusion to the other side of the membrane as pores form and then vanish, but there is a tremendous increases in rate (of transport) due to small entities (pores) that occupy a small fraction of the membrane (Weaver, 2000).

Due to the changes in the cell membrane resistance during electroporation the technique can also be controlled and monitored with EIT IRE.

Single cell microelectroporation technology

There are different techniques to overcome the cell membrane barrier and introduce exogenous impermeable compounds, such as dyes, DNA, proteins and amino acids into the cell. Some of the methods include lipofection, fusion of cationic liposome, electroporation, microinjection, optoporation, electroinjection and biolistics. Electroporation has the advantage of being a non-contact method for transient permeabilization of cells (Olofsson et al., 2003). In contrast to microinjection techniques for single cells and single nuclei (Capecchi, 1980), the electroporation technique can be applied to biological containers of sub-femtoliter volumes, that are less than a few micrometers in diameter. Also, it can be extremely fast and well timed (Hibino et al., 1991; Kinosita et al., 1988), which is of importance in studying fast reaction phenomena (Ryttsen et al., 2000).

In addition to bulk electroporation methods, instrumentation has been developed that can be used for electroporation of a small number of cells in suspension (Kinosita and Tsong, 1979; Chang, 1989; Marszalek et al., 1997), and for a small number of adherent cells grown on a substratum (Zheng and Chang, 1991; Teruel and Meyer, 1997). These electroporation devices create homogeneous electric fields across fixed distances of 0.1–5 mm, several times larger than the size of a single mammalian cell (Ryttsen et al., 2000). Also there are numerous experimental methods for the biochemical and biophysical investigations of single cells. Such methods include (1) patch clamp techniques for measuring transmembrane currents through a single ion channel (Hamill et al., 1981), (2) scanning confocal and multi-photon microscopy for imaging and localizing bioactive components in single cells and single organelles (Maiti et al., 1997), (3) near-field optical probes for measuring pH in the cell interior (Song et al., 1997), (4) ultramicroelectrodes for monitoring the release of single catechol- and indol-amine-containing vesicles (Wightman et al., 1991; Chow et al., 1992), (5) optical trapping and capillary electrophoresis separations for analyzing the chemical composition of individual secretory vesicles (Chiu et al., 1998), (6) electroporation with solid microelectrodes (Lundqvist et al., 1998), (7) electroporation with capillaries and micropipettes (Haas et al., 2001; Nolkrantz et al., 2001, Rae and Levis, 2002), (8) microfabricated chips and multiplexed electroporation system (Huang and Rubinsky, 1999; Lin , 2001).

Rubinsky's group presented the first microfluidic device to electroporate a cell (Huang and Rubinsky, 1999; Davalos et al., 2000). Their device consisted of three silicon chips bonded together to form two chambers, separated by a 1 μm thick silicon nitride membrane with a 2–10 μm diameter hole. Since silicon nitride is non-conductive, any electrical current flowing from the top chamber to the bottom chamber must pass through this microhole. A cell suspension was introduced into the top chamber, followed by the immobilization of one cell in the hole by lowering the pressure in the bottom chamber. Since the trapped cell impedes the system's electrical path, only low voltage pulses are needed to induce large fields near the trapped cell and only the trapped cell is electroporated. With this chip, they were able to show the natural difference in electroporation behavior between human prostate adenocarcinoma and rat hepatocyte cells by studying the process using current-voltage measurements (Huang and Rubinsky, 1999). Davalos and colleagues advanced this technology by making the chambers and electrodes off-chip, simplifying it to one silicon chip. Such changes enabled ease of use, accessibility of the device and reusability (Lee et al., 2006, Robinson et al., 2007).

In recent years, several microfluidic electroporation designs for the analysis, transfection or pasteurization of biological cells have been reported (Fox et al., 2006a,b). The range of applications for microfluidic electroporation coupled with advances in microfabrication techniques, specifically the use of structural photoresist for soft lithography, has resulted in a variety of designs: microchips in which cells move through a treatment zone (Gao et al., 2004), microchips in which cells are trapped at a specific location (Huang and Rubinsky, 1999) and devices in which the cells are surface-bound (Lin and Huang, 2001; Fox et al., 2006a,b). Of all the types of designs created, only the few designs in which a cell is trapped at a specific location enables us to study the biophysics of electroporation at the single cell level. In addition to the original devices described in the previous paragraph, other designs have been developed in which a cell is trapped at a specific location and electroporated. For example, Huang and Rubinsky advanced their technology using structural photoresist to create microfluidic channels on top of their silicon wafer (Huang and Rubinsky, 2003). Khine et al. fabricated a device using soft lithography which was originally developed as a multiple patch clamp array. Their device contains a main channel with multiple perpendicular small side channels. The individual cells in the main channel are brought into contact with the opening of a side channel using pressure. The cell does pass the constriction because its diameter (12–17 μm) is approximately 4 times larger than the constriction (3.1 μm). The constriction enables potentials of less than 1V to deliver the high fields needed to induce electroporation, which is applied using an silver–silver chloride-electrode (Khine et al., 2005a,b). Such devices are useful to study the biophysical process of electroporation because the changes in electrical properties of an individual cell as well as the molecular transport into the cell can be tracked (Davalos et al., 2000).

Supraelectroporation

If the applied electric field is very high (>10 kV/cm) and the pulses are very short (nanaosecond range), not only the plasma membrane of a cell is rendered permeable but also intracellular structures (Schoenbach et al., 2004; Vasilkoski et al., 2006). This opens new perspectives for treatment of cells, especially involving intracellular structures.

Electroporation theory works well up to about 2 kV/cm applied electric field. At higher field strength some effects appear which are hardly explainable just by pore formation.

The cell membrane has a capacity on the order of 1 μF/cm^2. At 30 MHz its capacitive resistance is small compared to the resistance of the electrolytes (i.e. the membranes are de facto shortened). The conductivity at this frequency is therefore a good guide for the maximum extent of electroporation, which would be when all membranes contain 100% pores. Even if it is practically irrelevant, it gives an idea about the absolute maximum of conductance. As shown in Fig. 4.1-21, the conductance exceeds this maximum value considerably when the field exceeds about 5 kV/cm. An early explanation involved some kind of Wien effect. The first Wien effect

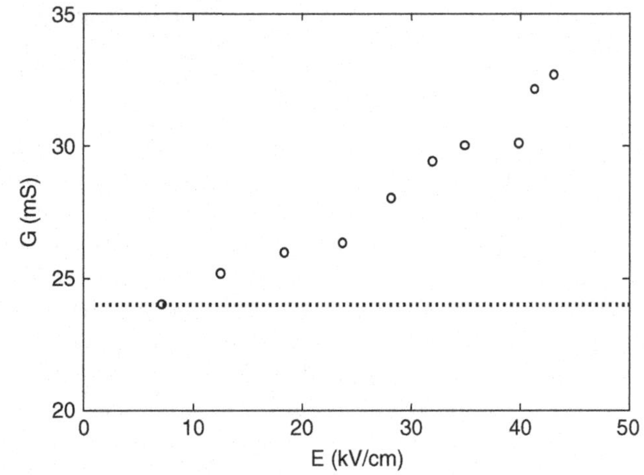

Figure 4.1-21 Measured conductance versus the field strength for a suspension of Jurkat-cells. The dashed line indicates the conductance at 30 MHz *Source:* From Pliquett et al. (2007).

is due to the liberation of ions from the counter-ion cloud around charged particles like proteins while the second one describes the creation of new charge carriers by field dissociation of week electrolytes. Both of these effects together can explain a conductivity increase by several percent but not by 140% as seen in Fig. 4.1-21. Moreover this dramatic conductivity increase is only found in solution containing aggregated amphiphiles like lipids.

The creation of very dense electropores (supra electroporation) is probably the initial step to a complete disintegration of the membrane. If the electric field is sufficiently high, micelles instead of membrane structures become stable. Because of the higher mobility of ions in the vicinity of the membrane, a significant increase in conductivity happens.

Electrofusion is the connection of two separate cell membranes into one by a similar pulse. It is believed that the process is based on the same field-induced restructuring of the bilayer lipid membranes (BLMs), a process which may be reversible or irreversible.

It is known that an ordinary cell membrane cannot withstand a prolonged DC potential difference ΔV_m more than about 150–300 mV without irreversible damage. For short pulses in the μs range, it has been found that at a threshold voltage Δv_m of about 1 V, the cell membrane becomes leaky and rather large macromolecules pass in and out of the cell (lysis). The following expression for the electric field in the membrane E_m is valid if the membrane thickness d is much less than cell radius r, and that the conductivity of the membrane material is much less than both the internal and external (σ_0) electrolyte conductivity:

$$E_m = 1.5(r/d)E(1 - e^{-Kt})\cos\theta \qquad (4.1.15)$$

E is the electric field in the external homogenous medium, $K = \sigma_0/3rC_m$, and θ the angle between the **E**-field and the cell radius r (cell center is origin).

For electroporation a threshold voltage of about 1V across the cell membrane has been found. The relationship between the cell membrane potential difference (corresponding to the order of 2–20 kV/cm in the suspension according to cell size, type, etc.), and electroporation may still be a reversible one as long as the latter is caused by a single pulse of a short time duration (e.g. of the order of 20 μs). If a train of such pulses is applied, the cell is killed because of the excessive material exchange. It is believed that a large part of the material exchange (lysis) is an *after-field* effect lasting up to 0.1 seconds or more. If the electroporation is reversible, the pores or cracks then reseal. Electrofusion is certainly an irreversible after-field effect.

The primary field effect shows threshold behavior, of about the same value for poration and fusion: The electric field effect in the cell membrane lipid bilayer is a molecular rearrangement with both hydrophobic or hydrophilic pore formation. Hydrophilic pores are considered to be water filled, with pore walls which may comprise embedded lipids. The threshold field strength has been found to be inversely proportional to the cell diameter. At the time of pulse application cell fusion may occur if two cells are in contact with each other, DNA uptake may occur if DNA is adsorbed to the cell surface. Cells may be brought in contact with each other by means of the pearl chain effect (see below). The electroporative cell transformation probability due to DNA entrance is low, typically 10^{-5}. Field values above threshold are believed to increase the pores in number and size, until a critical value is reached where complete membrane rupture occurs (irreversible non-thermal breakdown). The difference between the threshold level and the critical level is not large, so overdoses easily kill the cells. It is interesting to speculate whether electroporation is a mechanism in defibrillator chock treatment. The field strength used is lower (of the order of 500 V/cm), however the pulse duration is longer, of the order of some milliseconds.

The usual source for the electric field pulse is to discharge a charged capacitor (e.g. a 25 μF capacitor charged to 1500 V). The charge voltage and the distance between the capacitor plates determine the E-field strength, and the capacitance together with the system resistance determines the time constant of the discharge current waveform. The circuitry is very similar to the defibrillator circuit shown in Fig. 4.1-16, except that the inductor extending the time constant into the millisecond range, is not necessarily used. The pulse is accordingly a single exponentially decaying DC pulse, and the time constant is dependent on the liquid conductivity. With more complicated circuitry it is possible to make a square wave high voltage pulse generator. Because it is DC, there may be an appreciable electrolysis and change in pH near the electrodes. To keep the necessary voltage low, the distance between the electrode plates is small.

It is possible to use a RF pulse instead of DC. The RF causes mechanical vibrations in addition to the electrical effects, and this may increase the poration or fusion yield. As the effect is so dependent on the cell diameter, it may be difficult to fuse or porate cells of different sizes with DC pulses. The threshold level for the smallest cell will kill the largest.

4.1.14.2 Cell sorting and characterization by electrorotation and dielectrophoresis

The direction and rate of movement of bioparticles and cells due to electrorotation, dielectrophoresis and other electrokinetic effects depend on the dielectric properties of, for example, the cell. These dielectric properties may to some extent reflect the type of cell or the condition of the cell and there is consequently a significant potential in the utilization of these techniques for cell sorting or characterization.

Electrorotation was, for example, used to differentiate between viable and non-viable biofilms of bacteria. Because of their small size, determination of the dielectric properties of bacteria by means of electrorotation is impractical. By forming bacterial biofilms on polystyrene beads, however, Zhou et al. (1995) were able to investigate the effect of biocides on the biofilms.

Masuda et al. (1987) introduced the use of traveling wave configuration for the manipulation of particles. The frequency used was originally relatively low, so that electrophoresis rather than dielectrophoresis was predominant. The technique was later improved by, among others, Fuhr et al. (1991) and Talary et al. (1996), who used higher frequencies where dielectrophoresis dominates. Talary et al. (1996) used traveling wave dielectrophoresis to separate viable from non-viable yeast cells and the same group have used the technique to separate erythrocytes from white blood cells (Burt et al., 1998).

Hydrodynamic forces in combination with stationary electric fields have also been used for the separation of particles. Particles in a fluid flowing over the electrodes will to different extent be trapped to the electrodes by gravitational or dielectrophoretic forces. Separation is achieved by calibration of for example, the conductivity of the suspending medium or the frequency of the applied field. This approach has been used for separation between viable and non-viable yeast cells (Markx et al., 1994), different types of bacteria (Markx et al., 1996), leukemia and breast cancer cells from blood (Becker et al., 1994, 1995). Dielectrophoresis has also been successfully used for other types of bioparticles like DNA (Washizu and Kurosawa, 1990), proteins (Washizu et al., 1994) and viruses (Schnelle et al., 1996).

More recently, dielectrophoretic studies have for instance been reported on T-lymphocytes (Pethig et al., 2002; Pethig and Talary 2007) and on how cell destruction during dielectrophoresis can be minimized (or utilized) by appropriate choice of AC frequency and amplitude (Menachery and Pethig, 2005). Dielectrophoresis has also been used for measurement of membrane electrical properties such as capacitance and conductance for insulin secreting pancreatic cells (Pethig et al., 2005).

Another interesting approach to particle separation is called field-flow fractionation, and this technique can be used in combination with dielectrophoresis (Davis and Giddings, 1986). Particles are injected into a carrier flow and another force (e.g. by means of dielectrophoresis) is applied perpendicular to the flow. Dielectric and other properties of the particle will then influence the particle's distance from the chamber wall and hence its position in the parabolic velocity profile of the flow. Particles with different properties will consequently be released from the chamber at different rates and separation hence achieved. Washizu et al. (1994) used this technique for separating different sizes of plasmid DNA.

4.1.14.3 Cell-surface attachment and micromotion detection

Many types of mammalian cells are dependent on attachment to a surface in order to grow and multiply. Exceptions are the different cells of the blood and cancer cells which may spread aggressively (metastases). To study cell attachment a microelectrode is convenient: the half-cell impedance is more dominated by electrode polarization impedance the smaller the electrode surface is. Figure 4.1-22 shows the set up used by Giaever's group (Giaever and Keese, 1993).

A monopolar electrode system with two gold electrodes is used. A controlled current of 1 μA, 4 kHz is applied to a microelectrode <0.1 mm^2, and the corresponding voltage is measured by a lock-in amplifier. With cell attachment and spreading, both the in-phase and quadrature voltage increase as the result of cell-surface coverage. It is possible to follow cell motion on the surface, and the motion sensitivity is in the nm range. The method is very sensitive to subtle changes in the cells (e.g. by the addition of toxins, drugs and other chemical compounds). It is also possible to study the effect of high voltage shocks and electroporation.

Figure 4.1-23 shows an example of cell attachment and motion as measured with the electric cell-substrate impedance sensing (ECIS) instrument, which

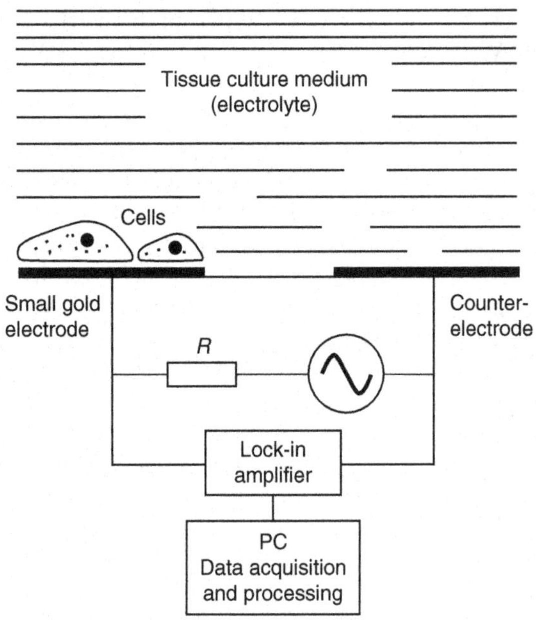

Figure 4.1-22 Impedance motion sensing with cells on a small gold electrode.

is a commercially available version of this system (Applied BioPhysics Inc., Troy, New York, USA).

4.1.14.4 Coulter counter

The principle is based on letting cells in suspension pass a narrow orifice. If a cell has different electrical properties than the liquid, the impedance of the pore will change at each cell passage. Cell counting is possible, and it is also possible to have information about each cells size, form or electrical properties. Figure 4.1-24 shows the basic set-up of the two-electrode conductance measuring cell. Typical dimensions (diameter, length) for a capillary is 50 and 60 μm (erythrocytes); 100 and 75 μm (leukocytes).

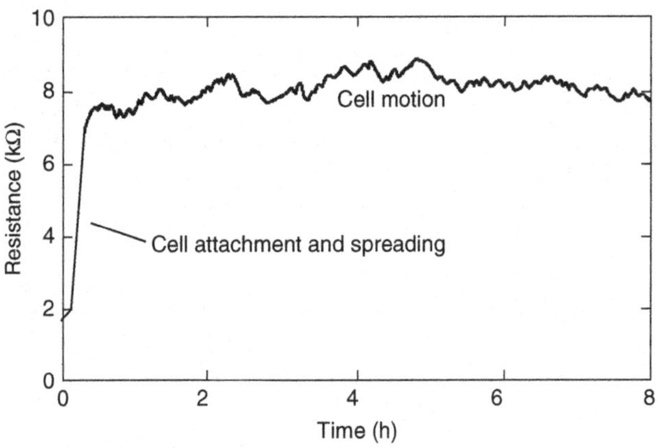

Figure 4.1-23 Changes in electrical resistance reflecting attachment and motion of cells on a small gold electrode.

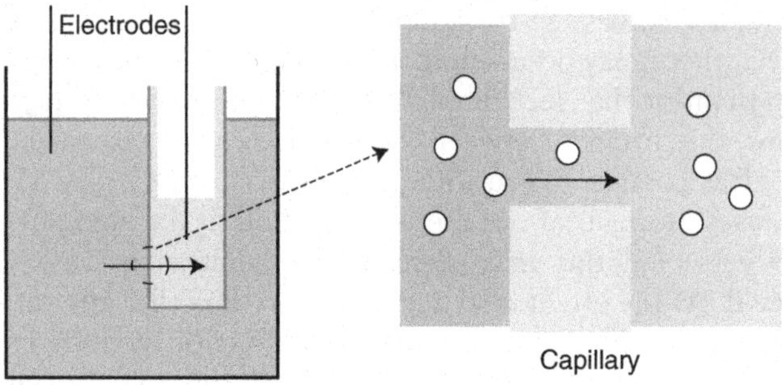

Figure 4.1-24 The measuring capillary cell of a Coulter counter.

4.1.15 Skin instrumentation

4.1.15.1 Fingerprint detection

Electronic fingerprint systems will in the near future eliminate the need for keys, pin-codes and access cards in a number of everyday products. While fingerprint recognition traditionally has been used only in high security applications, it is now gaining acceptance in mainstream consumer applications worldwide. One such large-scale application will be the need for secure mobile transactions when paying for the groceries with your mobile phone at the local supermarket. Several bioimpedance based fingerprint sensors have been developed, such as the electrode array based sensor from the company Idex ASA (www.idex.no). An array of electrodes is scanned as the fingertip is swept over the sensor stripe, giving a 500 dpi resolution impedance image of the fingerprint.

The fingerprint sensor market also demands systems for detecting fake fingers on the sensor, and a bioimpedance based solution for spoof detection was described by Martinsen et al. (2007). Their system is based on the simultaneous measurement of skin impedance at different depths, and the use of multivariate models to classify the fingers as living or fake.

4.1.15.2 Stratum corneum hydration

Stratum corneum hydration is essential for proper function and appearance of the skin. The moisture content can be measured in vitro by means of gravimetry or electron microscopy, or by magnetic resonance techniques in vivo. The resolution of the latter technique is, however, currently not sufficiently high to enable isolated measurements on the stratum corneum. Compared to these techniques, assessment of stratum corneum hydration by means of electrical measurements would represent a tremendous reduction in instrumental cost and complexity.

A prerequisite for using electrical measurements in this way is of course a detailed knowledge of how the different parts of the skin influence on the electrical impedance. Furthermore, the current and potential distribution in the skin will also be determined by the electrode geometry, which must be taken into account. The impedivities of the stratum corneum and the viable skin converge as the measuring frequency is increased. Measurements at high frequencies will hence normally be largely influenced by the deeper layers of the skin. The frequency must therefore be kept low in order to achieve isolated measurements on the stratum corneum. A frequency scan (i.e. impedance spectroscopy) cannot be utilized in stratum corneum hydration measurements, owing to the problems of interacting dispersion mechanisms. Contrary to certain opinions (Salter, 1998), the mere fact that the current distribution in the different skin layers will differ between different measuring frequencies, is enough to discard the multiple frequency approach on stratum corneum in vivo (Martinsen et al., 1999). Further complications are introduced by the dispersions of the electrode impedance and deeper skin layers, and also by the Maxwell–Wagner type of dispersion that is due to the interface between the dry stratum corneum and the viable epidermis.

Since the sweat ducts largely contribute to the DC conductance of the skin, the proper choice of electrical parameter for stratum corneum hydration assessment is consequently low frequency AC conductance (where DC conductance is removed) or susceptance.

There are a number of instruments for skin hydration assessment on the market. Most of them measure at rather high frequencies, which mean that they measure deep into the viable skin. Some instruments use closely spaced interdigitated microelectrodes. This reduces somewhat the contribution from viable skin layers, but the chance of only measuring in redundant moisture on the skin surface is obvious for such systems. Rationales for using a low frequency electrical susceptance method for skin hydration assessment and description of a method for absolute calibration of the measurements can be found in (Martinsen et al., 1998, 2008; Martinsen and Grimnes, 2001).

4.1.15.3 Skin irritation and skin diseases including skin cancer

Irritant contact dermatitis is a localized, superficial, non-immunological inflammation of the skin resulting from the contact with an external factor. The dermatitis may be acute, for example, if the influence from the external source was strong and of short duration, or of a more chronic kind if the influence is weaker but prolonged. The difference between irritant and allergic contact dermatitis is subtle, and depends mainly on whether the immune system is activated or not. Established signs of irritation are edema, erythema and heat, and any electrical parameter sensitive to these physiological changes could serve as a possible parameter for the assessment of skin irritation. As for other diagnostic bioimpedance measurements, the parameter should be immune to other, irrelevant changes in the skin. To eliminate the large variations in interpersonal electrical impedance baseline, normalization by means of indexes are often used rather than absolute impedance values.

A depth-selective skin electrical impedance spectrometer (formerly called SCIM) developed by S. Ollmar at the Karolinska Institute is an example of a commercial instrument intended for quantification and classification of skin irritation. It measures impedance at 31 logarithmically distributed frequencies from 1 kHz to 1 MHz, and the measurement depth can to some extent be controlled by electronically changing the virtual separation between two concentric surface electrodes (Ollmar, 1998).

Ollmar and Nicander (1995), Nicander et al. (1996), Nicander (1998) used the following indices:

$$\text{Magnitude index (MIX)} = |Z|_{20\ \text{kHz}}/|Z|_{500\ \text{kHz}}$$

$$\text{Phase index (PIX)} = \varphi_{20\ \text{kHz}} - \varphi_{500\ \text{kHz}}$$

$$\text{Real part index (RIX)} = R_{20\ \text{kHz}}/|Z|_{500\ \text{kHz}}$$

$$\text{Imaginary part index (IMIX)} = X_{20\ \text{kHz}}/|Z|_{500\ \text{kHz}}$$

where Z, R, X and φ have their usual meaning. The authors found significant changes in these indexes after treatment with sodium lauryl sulfate, nonanoic acid and benzalkonium chloride, and the measured changes correlated well with the results from subsequent histological examinations. The stratum corneum is soaked with saline before the measurements in order to provide good contact between the electrode system and the skin surface and to focus the measurements on the viable skin, although the barrier function of intact stratum corneum will still give a considerable contribution. The choice of frequencies for the indices hence seems reasonable. The group have also extended their impedance spectroscopy technology to further applications. Examples are the detection of other conditions and diseases in the skin or oral mucosa (Emtestam and Nyrén, 1997; Lindholm-Sethson, et al., 1998; Norlén et al., 1999), the early detection of transplanted organ complications (Ollmar, 1997; Halldorsson and Ollmar, 1998) and assessment of skin cancer (Emtestam et al., 1998).

After publication of a paper by Nicander et al. (1996) where it was demonstrated that skin reactions elicited by three irritants of different polarity created three different histopathological patterns and that each pattern could be correlated to corresponding patterns in the impedance indices, the Ollmar group has taken steps away from the data reduction technique based on the four indices in order to extract more information from the original impedance spectra. However, the indices are still useful for quantification of various aspects of responses to treatment or test substances, an example of which is given by Emtestam et al. (2007).

The finding of correlation between tissue structures (as seen in the microscope) with impedance properties (Nicander et al., 1996) triggered the idea of a potential diagnostic decision support tool intended to assist the doctor in a clinical environment. This idea is not new, for example, Fricke and Morse found a difference in capacity of tumors of the breast compared to normal breast tissue already in 1926. However, this finding is completely unspecific. Almost any tissue alteration can be detected by electrical impedance, but in order to be clinically useful, the method has to be able to differentiate benign alterations from malignant alterations, or be able to distinguish one disease from another in order to select a specific and adequate therapy. In other words: there is a difference between statistical significance and clinical significance. A p-value <0.001 may sound convincing in a statistical comparison, but may mean nothing in the clinic unless both sensitivity and specificity are good enough in the intended application. Thus, more information had to be extracted from the impedance spectra, and both the electrical impedance indices, which were sufficient for characterizing elicited skin reactions, and classical Cole-style models, were found inadequate to distinguish various disorders, according to the Ollmar group.

In skin testing, the central area of the volar forearm is very popular, mainly because of ease of access. It is also considered very homogeneous and stable, compared to other areas of the human body, and in extrapolation of this belief there have been studies without randomization of test sites within the volar forearm region. In search of suitable statistical tools to enhance discrimination power, this belief was challenged by Åberg et al. (2002), who used linear projection methods (in this case PARAFAC) to extract clinical information from the impedance spectra. The results showed systematic differences within the test area, which may not be important in comparison of strong reactions to, for example, detergents, but would have devastating impact on the outcome of a comparison of cosmetic preparations, where only small differences would be expected on normal skin, unless randomization of test sites is built into the study protocol.

It is known that diabetics are prone to develop ulcers difficult or impossible to heal, and therefore some difference in the skin properties might be present even when no clinical signs of ulceration, not even slight erythema, are present. Lindholm-Sethson et al. (1998) found such a difference in the skin between diagnosed diabetics without clinical signs in comparison with a healthy control group, but the difference was hidden behind more pronounced factors, such as age and sex, which were identified using PCA (principal component analysis). It seems that multivariate methods, such as PCA, make better use of the information inherent in electrical impedance spectra than simple indices or lumped parameter models, and that the extracted information sometimes reflects clinically interesting physiological or pathological conditions. In certain cases it might be possible to establish a strong correlation between specific principal components and well-defined physiological or pathological conditions.

To date, most skin studies involving electrical impedance are based on pure surface electrodes. Due to the extreme heterogeneity of the skin, such measurements (at least at low frequencies, as demonstrated in a simulation study by Martinsen et al., 1999), reflect mainly the conditions in the *stratum corneum* and the integrity of its inherent skin barrier. However, only living cells get irritated or sick, and if important information about an alteration in the living strata of skin resides in a relatively low frequency range, such information will be overshadowed or diluted by the intact *stratum corneum*. The dilution factor might be 1:100 or even 1:1000, and strongly frequency dependent! In several skin reactions or diseases, the skin barrier will be more or less destroyed by the chemical assault from the outside (fast event), or by sloppy maintenance of the *stratum corneum* provided by the damaged or sick living epidermis from the inside (slow event), and then surface electrodes would be sufficient. If not, something has to be done about the *stratum corneum*, and a number of methods have been tried, such as aggressive electrode gels (which will add to tissue damage), peeling creams or simply grinding or stripping off the outermost layer. The effect of tape stripping to various degrees of damage on skin impedance is illustrated in a book by Ollmar and Nicander (2005), and shows the dramatic overshadowing power of the intact *stratum corneum* on properties residing in the living strata of skin.

A new approach toward solving the *stratum corneum* dilemma has been presented by Griss et al. (2001), using electrodes furnished with micromachined conductive spikes, thin enough not to leave any damage after removal and short enough to only short circuit the *stratum corneum* without reaching blood vessels and nerve endings in deeper strata. This concept, originally developed as an improvement of ECG and EEG electrodes, has been further developed by the Ollmar group (Åberg et al., 2003a) to an electrode system intended to facilitate skin cancer detection even when the *stratum corneum* happens to be intact on top of any skin tumor.

For a clinical diagnostic decision support tool, it is not enough to detect an alteration; it must also reliably distinguish disease from other alterations. Key concepts in this context is sensitivity and specificity, as well as receiver operating characteristics (ROC) curves. In the skin context, reasonable clinical requirements on area under ROC curve have been published by Lee and Claridge (2005). A number of classifiers have been tried on data sets prepared by data decomposition techniques using indices as well as PCA, and using impedance data from both surface electrodes and spiked electrodes, by Åberg et al. (2003b, 2004, 2005). It seems that non-melanoma skin cancers (NMSCs), which are slow growing, degrade the skin barrier enough to make it more conductive and in this case there would be no need for demolition or short-circuiting of the *stratum corneum*, while, for example, malignant melanoma at an early stage the *stratum corneum* can be quite normal (electrically insulating) and in this case the spiked electrodes increase diagnostic power.

Figure 4.1-25 illustrates a PCA plot of NMSC and other lesions (neither normal skin nor malignant melanoma) using two principal components which in this case describe 84% of the variation in the data set that was obtained with non-invasive probes (without spikes). It is obvious that different ways to cut out the data volume including the cancer lesions would include different amounts of non-cancer data points, and that it is generally impossible to avoid classification of at least some harmless lesions as harmful. The choice of classifier and cut-off levels is a matter of risk assessment, common sense and data characteristics. In the case of malignant melanoma, missing a tumor might be assigned an unacceptable risk (life at stake), and therefore an almost 100% sensitivity required, which would entail a reduced specificity

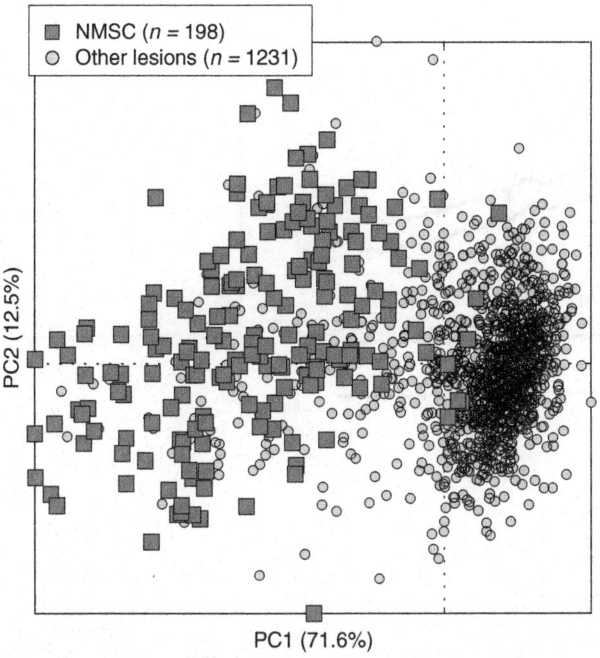

Figure 4.1-25 PCA plot of NMSC.

at a given area under ROC curve. In other cases it might be more important with high specificity to avoid painful and costly interventions, because a wrong diagnosis is not dangerous and the patient simply could come back if the complaint remains.

An update on this group's latest work on validation of their skin cancer detector in multi center clinical studies can be found in Ollmar et al. (2007).

Subcutaneous fibrosis is a common side effect of radiotherapy given, for example, to women with breast cancer. Nuutinen et al. (1998) measured the relative permittivity of the skin at 300 MHz with an open-ended coaxial probe, and found that the permittivity values were higher in fibrotic skin sites than in normal skin. Based on in vitro experiments with protein–water solutions indicating that the slope of the dielectric constant versus the electromagnetic frequency is a measure of the protein concentration, Lahtinen et al. (1999) demonstrated that skin fibrosis can also be measured with the slope technique. Both Nuutinen et al. (1998) and Lahtinen et al. (1999) found a significant correlation between the permittivity parameters and clinical score of subcutaneous fibrosis obtained by palpation. Finally, radiation-induced changes in the dielectric properties were also found in subcutaneous fat by modeling the skin as a three-layer dielectric structure (Alanen et al., 1998).

4.1.15.4 Electrodermal response

The sweat activity on palmar and plantar skin sites is very sensitive to psychological stimuli or conditions. One will however usually not be able to perceive these changes in sweat activity as a feeling of changes in skin hydration, except, for example, in stressing situations like speaking to a large audience. The changes are easily detected by means of electrical measurements, however, and since the sweat ducts are predominantly resistive, a low frequency conductance measurement is appropriate (Grimnes, 1982). Electrodermal response (EDR) measurements have during many years been based on DC voltage or current, and accordingly the method has been termed galvanic skin response (GSR).

The measured activity can be characterized as exosomatic or endosomatic. The *exosomatic* measurements are usually conducted as resistance or conductance measurements at DC or low frequency AC. Resistance and conductance will of course be inverse when using DC excitation, but when AC excitation is used, it is important to remember that resistance generally is part of a series equivalent of a resistor and a capacitor, while conductance is part of a parallel equivalent of these component. In this case it is obvious that resistance and conductance are no longer inverse, and conductance should be preferred to resistance since ionic conduction

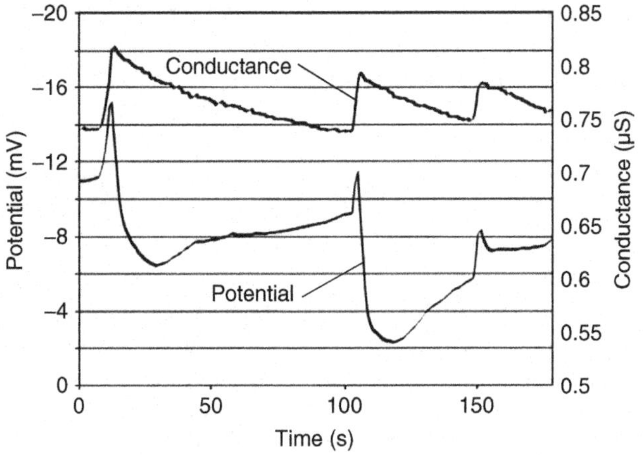

Figure 4.1-26 Simultaneous registration of exosomatic AC conductance and endosomatic DC voltage. *Source:* Courtesy of Azar Jabbari.

and polarization basically appear in parallel in biological tissue.

The *endosomatic* measurements are carried out as DC voltage measurements. The mechanisms behind the changes in skin potential during symphaticus activity are not known, but processes like sodium reabsorption across the duct walls and streaming potentials in the sweat ducts should be taken into account. Figure 4.1-26 shows that the origin of the endosomatic and exosomatic curves is not identical.

The so-called "lie detector" is perhaps the most well-known instrument in which the electrical detection of this activity is utilized. There are, however, several other applications for such measurements, mainly within the two categories; neurological diseases or psycho–physiological measurements. Examples of the first category are neuropathies (from diabetes), nerve lesions, depressions and anxiety. The latter category may include emotional disorders and lie detection. Qiao et al. (1987) developed a method to measure skin potential, skin electrical admittance, skin blood flow and skin temperature simultaneously at the same site of human palmar skin. This was done in order to be able to investigate a broader spectrum of responses to the activity of the efferent sympathetic nerve endings in palmar blood vessels and sweat glands.

Both *evoked* responses (e.g. to light, sound, questions, taking a deep breath) and *spontaneous* activity may be of interest. Measurement of spontaneous EDR is used in areas such as sleep research, the detection of the depth of anesthesia and in sudden infant death syndrome research.

Figure 4.1-27 shows detection of EDR by means of 88 Hz conductance (G) and susceptance (*B*) measurements (Martinsen et al., 1997a). The measurements were conducted on palmar sites in right and left hand. In Fig. 4.1-27 both hands show clear conductance waves, but no substantial susceptance waves. No time delay can be seen between the onset of the conductance waves in the two hands, but the almost undetectable susceptance waves appear a few seconds after the changes in conductance. This indicates that these changes have different causes. The rapid conductance change is presumably a sweat duct effect and the slower change in susceptance is most probably due to a resultant increased hydration of the stratum corneum itself. There are no susceptance waves that could indicate any significant capacitance in the sweat ducts.

Venables and Christie (1980) give a detailed suggestion on the analysis of EDR conductance waves based on the calculation of amplitude, latency, rise time and

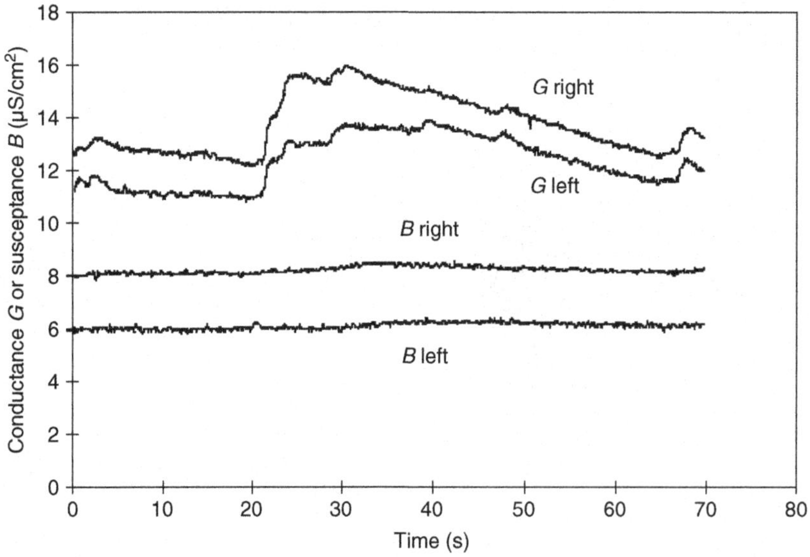

Figure 4.1-27 Measured 88 Hz admittance GSR activity on palmar skin sites. A deep breath at approximately 20 seconds on the time scale triggered the response.

recovery time. They also give extensive statistical data for these parameters in different age groups. The use of absolute values for the electrical properties of tissue is hazardous due to their liability to measurement error and their dependence on, for example, electrode size, gel composition and ambient environment. The use of indexes or other relative parameters would presumably prove beneficial also in EDR measurements. Mørkrid and Qiao (1988) analyzed the use of different parameters from the Cole admittance equation in EDR measurements and proposed a method for calculating these parameters from measurements at only two frequencies.

4.1.15.5 Sweat measurements

Quantitative assessment of sweat activity is of great importance also for other purposes than the measurement of EDR. Tronstad et al. (2007) have reported on the development of a portable multichannel instrument for long term logging of sweat activity. The instrument is based on conductance measurements and has been developed primarily for the assessment before and after treatment of patients with hyperhidrosis (see Section 4.1.15.6). Since the instrument is portable it can also be used in sports and during other kinds of physical activities, and with four independent channels it can monitor different body parts simultaneously. Their measurements indicate that the sweat activity of different skin sites behaves differently under physical stress. Hence, such measurements will be valuable for obtaining a better understanding of the physiology controlling thermal sweating.

4.1.15.6 Iontophoretic treatment of hyperhidrosis

Hyperhidrosis is a state of extreme sweat secretion in palmar, plantar or axilliary skin sites. The disorder can be treated with drugs (e.g. anticholinergica), tap water iontophoresis or surgical sympathectomy. Tap water iontophoresis has been in use at least since the beginning of this century, and represents a simple, effective, but somewhat painful cure. In its simplest form, the setup used comprises two water filled metal tubs and a DC supply. In case of palmar hyperhidrosis, the hands are placed in the two tubs and a DC current is driven from one hand to the other through the upper body. This treatment has now been stopped in Norway because the current is driven through the heart region. The Drionic is an example of an alternative device for the treatment of palmar hyperhidrosis where this problem is solved. When one hand is placed on the Drionic, half the hand is connected to one electrode through a wet sponge, and the other half to the other electrode through another wet sponge. The current is hence driven locally in the hand, and only one hand is treated at a time.

How tap water iontophoresis can impede excess sweating is still not fully understood. One theory suggests abnormal keratinization in the epidermis as a result of the current being shunted through the sweat ducts, leading to a plugging of the sweat orifices. This plugging cannot be found on micrographies of the skin, however, and a more plausible theory is presumably that the current leads to a reversible destruction of the sweat glands.

4.1.15.7 Iontophoresis and transdermal drug delivery

The transport of charged substances through the skin was shown early by the famous experiment by Munk (1873). He applied an aqueous solution of strychnine in HCl under two electrodes attached to the skin of a rabbit. Without current flow nothing happened to the rabbit; with application of a DC current for 45 minutes, the rabbit died.

Abramson and Gorin (1939, 1940) found that timothy pollen could be transported into the skin by electrophoresis. They studied the transport of dyes into human skin by electrophoresis. Without the application of electricity, no particular skin marks were seen after the dye had been in contact with the skin for some minutes. With an applied DC current, and after the superfluous dye had been wiped off, small dots were seen corresponding to the pores of the skin. Positively charged methylene blue was transported into the skin under an anode, and negatively charged eosin under a cathode. Some pores were colored with only one of the types.

Iontophoresis of pilocarpine is the classical method for obtaining sweat for the cystic fibrosis test (Gibson and Cooke, 1959). The penetration of pilocarpine in the skin enhances sweat production. The test is usually performed on children with both electrodes placed on the underarm (for safety reasons the current should not pass the thorax). A 0.5% solution of pilocarpine is placed under the positive electrode, and the DC current is slowly increased to a maximum of about 1.5 mA. The iontophoresis time is about 5 minutes.

A skin surface *negative* electrode attracts water from deeper layers, a positive electrode repels water. This is an *electro-osmotic* effect and not iontophoresis (Abramson and Gorin, 1939; Grimnes, 1983b).

The conductivity of human skin is very unevenly distributed. The current pathways have been found to be the pores of the skin, particularly the sweat ducts, only to a small extent through the hair follicles (Abramson and Gorin, 1940; Grimnes, 1984).

Transdermal drug delivery through iontophoresis has received widespread attention. A long term delivery with transdermal DC voltage of <5 V is used (Pliquett and

Weaver, 1996). High voltage pulses up to 200 V decaying in about 1 ms have also been used on human skin for enhancement of transport by electroporation (Pliquett and Weaver, 1996). The effect was found to be due to the creation of aqueous pathways in the stratum corneum.

4.1.16 Non-medical applications

- Fingerprint detection as described in Section 4.1.15.1 is a measurement on humans, but not with a medical purpose.
- The monitoring of fermentation processes in beer brewing or pharmaceutical industry is measuring on different sort of cell suspensions.
- Plant tissue is both strongly similar and very different from animal tissue, the cell membranes for instance are quite different.
- Meat quality can be estimated from bioimpedance measurements. See. e.g. Oliver et al., 2001 and Guerro et al., 2004.

In geophysics impedance measurements were used as early as in the 1920s for oil exploration (Schlumberger, 1920). Impedance measurements are also used for monitoring volcanic activity (e.g. on Iceland).

4.1.17 Electrical safety

4.1.17.1 Threshold of perception

The perception of a current through human skin is dependent on frequency, current density, effective electrode area (EEA) and skin site/condition. Current duration also is a factor, in the case of DC determining the quantity of electricity and thereby the electrolytic effects according to Faraday's law.

DC

If a DC source coupled to two skin surface electrodes is suddenly switched on, a transient sensation may be felt in the skin. The same thing happens when the DC current is switched off. This proves that many nerve endings are only sensitive to *changes* in a stimulus, and not to a *static* stimulus. At the moment a DC is switched on, it is not only a DC, it also contains an AC component. DC must therefore be applied with a slow increase from zero up to the desired level, if the threshold of DC perception is to be examined.

DC causes ion migration (iontophoresis) and cell/charged particle migration (electrophoresis). These charge carriers are depleted or accumulated at the electrodes, or when passing ion-selective membranes in the tissue. In particular, almost every organ in the body are encapsulated in a *macromembrane* of epithelia tissue. There are, for example, three membranes (meninges) around the brain and CNS (pia mater, arachnoidea, dura mater). There are membranes around the abdomen (peritoneum), fetus, heart (pericardium), lungs (pleura), inside the blood vessel (endothelium), around the nerves (myelin, neurolemma in the hand). At some tissue interfaces and at the electrodes the chemical composition will gradually change.

A sensation will start either under one of the electrodes (anode or cathode), or in the tissue between. The chemical reaction at an electrode is dependent on the electrode material and electrolyte, but also on the current level. A sensation around threshold current level is slowly developing and may be difficult to discern from other sensations, for example, the mechanical pressure or the cooling effect of the electrode. After the sensation is clear, and the current is slowly reduced to avoid AC excitation, the sensation remains for some time. That proves that the current does not trigger nerve ends directly, but that the sensation is of a chemical, electrolytic nature as described by the law of Faraday. The after current sensation period is dependent on the perfusion of the organ eliciting the sensation.

On palmar skin, with a surface electrode of varying area A, the current I_{th} or current density J_{th} at the threshold of perception follows the following equation for a sensation within 3 minutes after current onset (Martinsen et al., 2004).

$$J_{th} = J_0 A^{-0.83} \quad \text{or} \quad I_{th} = I_0 A^{0.17} \qquad (4.1.16)$$

The perception was only localized under the monopolar electrode, never in the tissue distal to the electrode. Surprisingly, according to eq. (4.1.16) the threshold as a function of electrode area A is more dependent on *current* than current density. There may be more than one reason for this:

1. A spatial summation effect in the nervous system. The current *density* is reduced when the same current is spread by a larger electrode, but at the same time a larger number of nerve endings are excited. Consequently the current threshold is not so much altered when electrode area is changed.
2. The DC current is not evenly distributed under a plate electrode, the current density is higher at the edge. The conductance of a surface sphere or plate electrode is proportional to radius or circumference, not to area.
3. The DC current is probably concentrated to the sweat ducts and the nerve endings there (Grimnes, 1984).

A practical use of DC perception is the old test of the condition of a battery by placing the poles at the tongue.

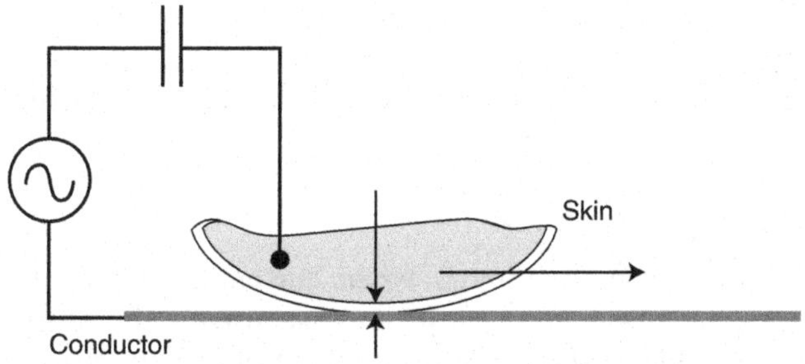

Figure 4.1-28 Electrovibration perception mechanism.

This test is actually also done clinically: *electrogustometry* is the testing of the sense of taste by applying a DC to the tongue.

Sine waves

The lowest level (<1 μA) of 50/60 Hz perception is caused by *electrovibration* (Grimnes,1983d; Fig. 4.1-28). It is perceived when the current carrying conductor slides on dry skin. Dry skin is a poor conductor, so that potential differences of several tenths of volts may exist across the dielectric which is the stratum corneum of the epidermis. With dry skin only a small microampere current flows. The electric field sets up an electrostatic compression force in the dielectric, pressing the stratum corneum to the metal plate. In the stratum corneum there are no nerve endings, and consequently no perception. However, if the skin is made to slide along the metal, the frictional force will be modulated by the electrostatic force and be felt as a lateral mechanical vibration synchronous with the double frequency of the AC voltage. Even if the voltage across the dielectric is >20 V at threshold, the corresponding (mainly capacitive) current may be <1 μA. If the skin is at rest, or if the skin is wet, no sensation is felt.

The second level (1 mA) is due to the *direct electric excitation* of nerve endings, which must be a function of the local current (density). The electric current threshold of perception with firm hand grip contact and contact area several square centimeters, is around 1 mA. Threshold current has a surprisingly small dependence on contact area. The reason for this is mentioned above in the section on DC perception. With a small area contact around 1 mm^2, the threshold of perception is around 0.1 mA, corresponding to 100 A/m^2.

Interpersonal variations and the dependence on age and sex are small. Skin condition is not important as long as there are no wounds. Skin site may be important. On the fingertips the density of nerve ending is large, but the stratum corneum is thick and the current will be rather uniformly distributed. Other skin sites may have much thinner skin and lower density of nerve endings, but conductive sweat ducts that canalize the current.

Frequency dependence

For sine waves the maximum sensitivity of our nervous system is roughly in the range 10–1000 Hz (cf. also Fig. 4.1-31). At lower frequencies each cycle begin to be discernible, and during each cycle it may be charge enough to give electrolytic effects. At frequencies >1000 Hz the sensitivity is strongly reduced, and at >100 kHz no perception remains, because the levels are so high that electric stimulation are shadowed by the heat effect of the current. That is the frequency range for electrosurgery.

A single pulse or a repetitive square wave may give both DC and AC effects. In both cases the duration of the pulse or square wave is an important variable (cf. rheobase and chronotaxi).

The exponential decaying discharge waveform is the case of electrostatic discharges (see below).

Electrostatic discharge pulse

The perception of an electrostatic discharge is an annoyance, and in some situations a hazard. It is particularly troublesome indoor during the winter with low relative humidity (RH). Low RH reduces the conductivity of most dielectrics (e.g. the stratum corneum) and also the conductivity of clothing, construction materials, tree, concrete, etc. A person may be charged up to more than 30 kV under such circumstances, and with a body capacitance to the room (ground) of about 300 pF, the electrical energy of the person is of the order of 0.1 J. A smaller discharge, near the threshold of perception, is typically with a time constant of a few μs, and the peak current around 100 mA (Fig. 4.1-29). It is obtained by discharging a capacitor of 100 pF charged to 1.4 kV. The point electrode is approached to the skin until an arc is formed, heard and perceived in the skin.

The maximum current is determined by the voltage drop in the arc (probable less than 100 V) and the resistance in the skin. The arc probably has a very small

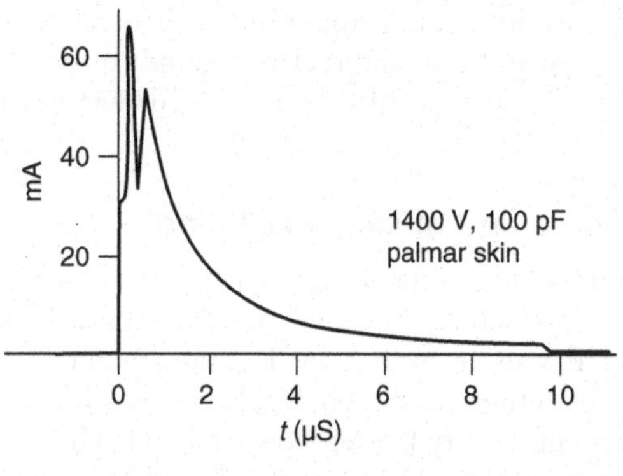

Figure 4.1-29 Capacitor discharge current flow through palmar skin. Monopolar electrode: 1.3 mm diameter pin of steel, sharpened at the tip. Indifferent electrode on the underarm.

cross sectional area, so most of the resistance is in the proximal zone in the stratum corneum. The current density is probably far out in the non-linear breakdown region of the skin, but because of the short pulse duration the impedance is presumably also determined by the capacitive properties of stratum corneum. A rough calculation with $\sigma = 0.1$ S/m and calculated resistance from measured current maximum: 20 kΩ, gives the arc contact diameter with the skin: $2a = 200$ μm.

The charge transferred around threshold level is of the order of 0.2 C. The threshold of perception as a function of stored energy is about 10 μJ. The formation of an arc in the air between the conductor and the skin is possible when the voltage difference is larger than about 400 V. The arc discharge can be heard as a click and felt as a prick in the skin.

4.1.17.2 Electrical hazards

Electromagnetic field effects

Coupling without galvanic tissue contact and electromagnetic hazards are outside the scope of this chapter. There is a vast amount of experimental data on this subject, and the interested reader is recommended the CRC handbook (Polk and Postow, 1986).

Continuous current

The risk of sudden death is related to stimulating the cells of three vital organs of the body: the heart, the lungs and the brain stem. Involuntary movements may indirectly lead to sudden deaths (loss of balance, falling). Heat and electrochemical effects may also be fatal by inducing injuries that develop during hours and days after the injury. In electrical injuries the question often arises as to whether the current is evenly distributed in the tissue, or follows certain high-conductance paths. Current marks and tissue destruction often reveal an uneven current distribution (cf. Ugland, 1967).

The current path is important, and organs without current flow are only indirectly affected: to be directly dangerous for the healthy heart, the current must pass the heart region.

Cell, nerve and muscle excitation

Heat effects are certainly related to current density in volume conductors, but this is not necessarily so for nerve and muscle excitation. Excitation under a plate electrode on the skin is more highly correlated to current than current density (cf. Section 4.1.17.1). The stimulus summation in the nerve system may reduce the current density dependence if the same current is spread out over a larger volume of the same organ. Therefore, and for practical reasons, safe and hazard levels are more often quoted as current, energy or quantity of current in the external circuit, and not current density in the tissue concerned.

Macro/microshock

A *macroshock* situation is when current is applied to tissue far from the organ of interest, usually the heart. The current is then spread out more or less uniformly, and rather large currents are needed in the external circuit (usually quoted >50 mA @ 50/60 Hz) in order to attain dangerous levels (Fig. 4.1-30).

The heart and the brain stem are particularly sensitive for small areas of high current density. Small area contacts occur, for example, with pacemaker electrodes, catheter electrodes and current carrying fluid-filled cardiac catheters. Small area contact implies a monopolar system with possible high local current densities at *low current levels* in the external circuit. This is called a *microshock* situation. The internationally accepted 50/60 Hz safety current limit for an applied part to the heart is therefore 10 μA in normal mode, and 50 μA under single fault condition (e.g. if the patient comes into contact with the mains voltage because of defective insulation). The macro- and microshock safety current levels differ by more than three orders of magnitude.

The heart is most vulnerable for an electric shock in the repolarization interval, that is in the T-wave of the ECG waveform. Therefore the probability of current passage during the approximate 100 ms duration of the T-wave is important. If the current lasts more than one heart cycle, the T-wave is certainly touched. For short current durations <1 seconds, the risk of heart stop is determined by the chance of coincidence with the T-wave.

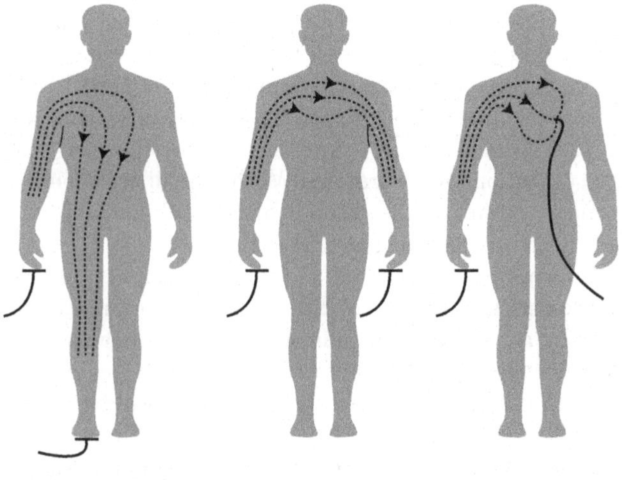

Figure 4.1-30 Macroshock (left) and microshock (right) situations.

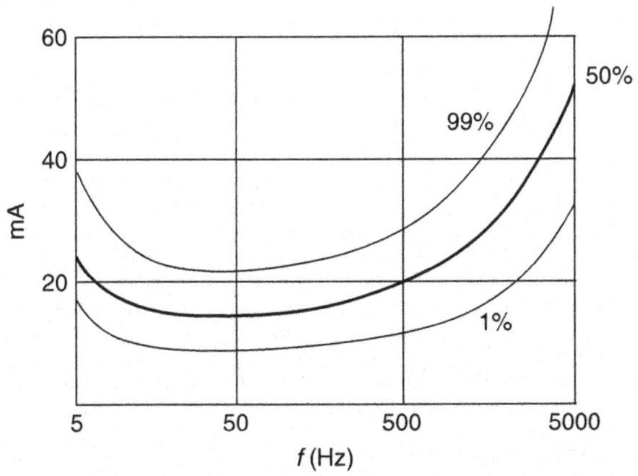

Figure 4.1-31 Frequency dependence of let-go currents. Statistic for 134 men and 28 women. *Source*: From Dalziel (1972).

Let-go current

Let-go current threshold (15 mA @ 50/60 Hz) is the current level when the current density in muscles and nerves is so large that the external current controls the muscles. As the grip muscles are stronger than the opening muscles of the hand, a grip around the current carrying conductor cannot be loosened by the person himself. Let-go current levels are therefore the most important data for safety analysis. The result in Fig. 4.1-31 shows that 1% of the population have a let-go threshold as low as 9 mA at power line frequencies.

Fatal levels are reached at current levels >50 mA @ 50/60 Hz if the current path is through vital organs: heart, lung or brain stem (cf. the electric chair, Table 4.1-2).

Heat effects

Jouleian heat is dependent on the in-phase components of potential difference and current density. The resulting temperature rise is dependent on the power density, the specific heat of the tissue and the cooling effect of the blood perfusion.

The tissue damage is very dependent on exposure time, cells can tolerate long time exposure of 43°C. Above about 45°C, the time duration becomes more and more critical. In high voltage accidents the heat effect may be very important, and patients are treated as thermal burn patients. In particular, special attention is paid to the fluid balance, because electrical burn patients tend to go into renal failure more readily than thermal burns of equal severity. As electric current disposes thermal energy directly into the tissue, the electric burn is often deeper than a thermal burn caused by thermal energy penetrating from the surface. The general experience is therefore that an electrical burn is more severe than it may look like the first hours after the injury.

Electrolytic effects

Electrolytic effects are related to DC, applied or rectified by non-linear effects at the electrodes or in the tissue. Also with very low frequency AC (e.g. <10 Hz), each half period may last so long as to cause considerable non-reversible electrolytic effects. With large quantities of electricity ($Q = It$) passed, the *electrolytic* effects may be systemic and dangerous (lightning and high voltage accidents). The risk of skin chemical burns is greater under the cathode (alkali formation) than the anode (acid formation), the natural skin pH is on the acidic side (pH < 5.5).

Nerve damage is often reported in high voltage accidents.

Current limiting body resistance

The most important current limiting resistance of the human body is the dry skin. This may be impaired by high-field electrical breakdown, skin moisturizing or a skin wound. Skin breakdown may occur under 10 V AC 50/60 Hz due to electro-osmotic breakdown (Grimnes, 1983b).

Without the protective action of the skin, the *internal body resistance* may be divided into a *constrictional zone resistance* with increased current density near an electrode, and *segmental resistances* of each body segment with rather uniform current density. With small area

Table 4.1-2 50/60 Hz threshold levels of perception and hazard

Current threshold	Voltage threshold (very approximate)	Organs affected	Type	Comments
0.3 μA	20 V	Skin	Perception threshold	Electrovibration, mechanical
10 μA	20 mV	Heart	Microshock hazard	Myocard excitation
1 mA	10 V	Skin	Perception threshold	Nerve excitation
15 mA	50 V	Muscles	Let-go	Loss of muscle control
50 mA	250 V	Heart, lung, brain stem	Macroshock hazard	Nerve excitation

electrode contact the constrictional zone resistance will dominate. The segmental resistance may be estimated from the equation $R_{sr} = L/\sigma A$. With constant σ the segmental resistance depends on the ratio L/A, and accordingly varies according to body or limb size.

Table of threshold values

From the threshold current levels, the corresponding voltages are found by estimating the minimum current limiting resistance. These worst case minima are found by assuming no protective action from the skin at all, only from the volume resistance of the living parts of the body. The levels are summarized in Table 4.1-2.

The question of current canalizing effects in tissue is one of the issues of our field. It is well known that the myelin sheet around the nerves serves electrical insulation and current canalization to the Ranvier nodes. But the extent of current canalization in many parts of the body is largely unknown. The blood has a high conductivity, but what are the electrical properties of the endothelium? Does electrosurgery current follow the bile duct? The current through skin is canalized through the sweat ducts. The high acid concentration in the stomach must have high electrical conductivity.

Single pulses

Trigger and safety levels have been examined by Zoll and Linenthal (1964) for external pacing of the heart. For TENS single monophasic pulses of duration <1 ms, FDA has set up the following values for the electric charge through the thorax (Table 4.1-3).

4.1.17.3 Lightning and electrocution

Lightning

An average lightning stroke may have a rise time of 3 μs and duration of 30 μs, energy dissipated 10^5J/m, length 3 km, peak current 50 000 A, power 10^{13} W. After the main stroke there are continuing currents of typically 100A and 200ms duration.

The mechanical hazards are due to the pressure rise in the lightning channel. The energy per meter is equivalent to about 22 g of TNT per meter. A direct hit from the main stroke is usually lethal, but often the current path is via a tree, the ground (current path from foot to foot, current through tissue determined by the *step voltage*), or from a part of the house or building. The current path is of vital importance, for humans a current from foot to foot does not pass vital organs, for a cow it may do.

It is believed that there are around 500 deaths caused by lightning per year in the USA.

Electrocution

The current path in an electric chair is from a scalp electrode to a calf electrode. The current is therefore passing the brain and the brain stem, the lung and the heart. It is believed that the person gets unconscious immediately after current onset, but it is well known that death is not immediate. The electric chair was used for the first time in 1890, and the first jolt was with 1400 V 60 Hz applied for 17 seconds, which proved insufficient. At present a voltage of about 2000 V applied for 30 seconds is common, followed by a lower voltage, for example, a minute. The initial 60 Hz AC current is about 5 A, and the total circuit resistance is therefore around 400 Ω. The power is around 10 kW and the temperature rise in the body, particularly in the regions of highest current densities in the head, neck and leg region, must be substantial. Because of the cranium the current distribution in the head may be very non-uniform. Temperature rise is proportional to time and the *square* of current density, so there are probably local high temperature zones in the head.

The scalp electrode is a concave metal device with a diameter about 7 cm and an area of about 30 cm^2. A sponge soaked with saline is used as contact medium.

Table 4.1-3 Electric charge values (microcoulomb) for single monophasic pulses through the chest

	μC
TENS threshold	3
TENS, max	7
Safe level	20
Hazard threshold	75
Heart pacing	100

4.1.17.4 Electric fence

The electric fence is used to control animals and livestock. There are two types of controllers: one type delivers a continuous controlled AC current of about 5 mA. The other delivers a capacitive discharge, like the working principle of a deflbrillator. The repetitive frequency is around 1 Hz, and the capacitor is charged to a DC voltage up to 10 kV. The large voltage secures that the shock will pass the animals' hair-covered skin. The shock is similar to an electrostatic discharge, even if the electric fence shock energy is higher and the duration longer. The capacitor is charged to an energy typically in the range 0.25–10 J.

4.1.17.5 Electrical safety of electromedical equipment

Special safety precautions are taken for electromedical equipment. Both patient and operator safety is considered (as well as damage of property). *Electromedical equipment* is equipment which is situated in the patient environment and is in physical contact with the patient, or which can deliver energy (electrical, mechanical or radiation) to the patient from a distance. Equipment for in vitro diagnosis is also important for patient safety with respect to correct diagnostic answers, but as long as it is not in the patient environment the safety aspects are different.

The basis for the national or international standards (IEC, UL, VDE, MDD (the European Medical Device Directive), etc.) is to reduce the risk of hazardous currents reaching the patient under normal conditions. Even under a *single fault condition* patient safety shall be secured.

The part of the equipment in physical contact with the patient is called the *applied part*. It may ground the patient (type B applied part), or keep the patient floating with respect to ground (BF – body floating, or CF – cardiac floating) by *galvanic separation* circuitry (magnetic or optical coupling, battery-operated equipment). In most situations higher safety is obtained by keeping the patient floating. If the patient by accident comes in contact with a live conductor the whole patient will be live, but *little current* will flow.

Figure 4.1-32 shows the most important parts of an electromedical device. The power line and earth connection are shown to the right. The signal connections to the outside world are shown at the upper part. Important safety aspects are linked with these signal input and output parts: they may be connected to recorders, printers, dataloggers, data networks, coaxial video cables, synchronization devices, etc. These devices may be remotely situated and outside the electrical control of the patient room. With a floating applied part hazardous currents from the outside do not reach the patient, the galvanic separation protects both ways.

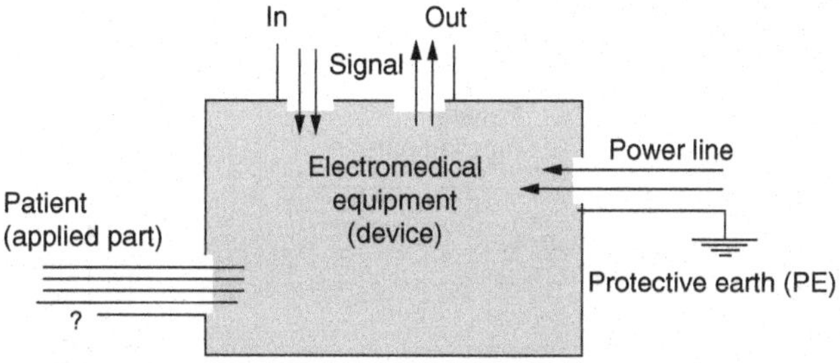

Figure 4.1-32 Basic parts of a grounded (class I) electromedical device.

The device may be grounded for safety reasons (safety class I, as shown in Fig. 4.1-32, maximum resistance in the *protective earth* (PE) conductor between power plug and chassis 0.2 Ω), or double insulated (safety class II).

Leakage currents are currents at power line frequency (50 or 60 Hz), they may be due to capacitive currents even with perfect insulation, and are thus difficult to avoid completely. *Patient leakage currents* are the leakage currents flowing to the patient via the applied part. Patient *auxiliary currents* are the functional currents flowing *between* leads of the applied part (e.g. for bioimpedance measurement). They are not leakage currents and therefore usually not at power line frequency. *Earth leakage current* is the current through the ground wire in the power line cord (not applicable for double insulated devices). According to IEC it shall be <500 μA during normal conditions for all types B, BF or CF. *Enclosure leakage current* is a possible current from a conductive accessible part of the device to earth. Grounded small devices have zero enclosure current during normal conditions, but if the ground wire is broken the enclosure leakage current is equal to the earth leakage current found under normal conditions.

The current limits according to IEC-60601-1 (2005) is shown in Table 4.1-4.

Table 4.1-4 Allowable values of continuous leakage and patient auxiliary currents (μA) according to IEC-60601 (N.C. = normal conditions)

Currents	Type B		Type BF		Type CF		
	N.C.	Single fault	N.C.	Single fault	N.C.	Single fault	
Earth leakage	500	1000	500	1000	500	1000	Higher values, for example, stationary equipment
Patient leakage	100	500	100	500	10	50	
Patient leakage				5000		50	Mains on applied part
Patient leakage		5000					Mains on signal part
Patient auxiliary	100	500	100	500	10	50	AC
Patient auxiliary	10	50	10	50	10	50	DC

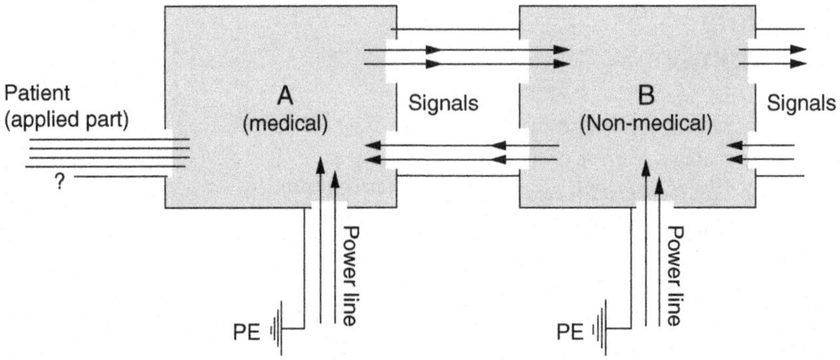

Figure 4.1-33 A non-electromedical device (B) within the patient environment. *Source:* From IEC-60601-1 (2005).

The insulation level is also specified. It is defined both in kV, creepage distances and air clearances in millimeters. Important additional specifications are related to maximum exposed surface temperature, protection against water penetration (drop/splash proof), cleaning–disinfection–sterilization procedures, technical and users documentation.

A non-medical device such as a PC may be situated within a patient environment (instrument B in Fig. 4.1-33), but in itself it must not have an applied part. An electromedical device (instrument A) must be inserted between the B and the patient. The connection between A and B is via the signal input/output of device A. If the instrument B has higher earth leakage current than 500 μA, an insulation power line transformer or extra ground must be provided. The reason for this is that a person can transfer the enclosure leakage current by touching the enclosure of B and the patient simultaneously. During the single fault condition of a broken ground wire to B, the earth leakage current of B may then be transferred to the patient. This would not happen with the interconnected signal ground wires as shown in Fig. 4.1-33, but could happen if A and B were in the same rack with one common power cord.

An electromedical device may have more than one applied part (Fig. 4.1-34).

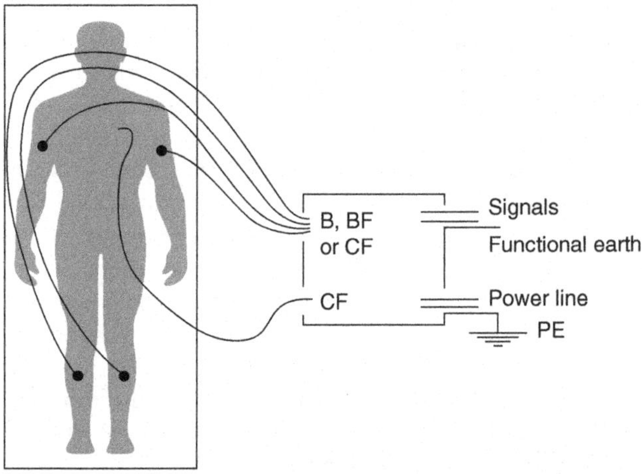

Figure 4.1-34 Electromedical equipment with two applied parts.

The producer must basically declare the intended use of his equipment. If an applied part is intended to be used in direct connection with the heart, it must be of type CF. The same instrument may have another applied part intended to be used with skin surface electrodes or sensors. That applied part may be of type B, BF or CF. A plug in the instrument may be marked with type B, but a box with a galvanic separation may be inserted in the cable so that the applied part is converted from type B to BF or CF.

Chapter 4.2

4.2

Intensive care facilities

Saul Miodownik

There has always been the perception that those patients closer to nursing stations fare somewhat better than their counterparts who are further away. The concept of the "intensive care unit" (ICU) evolved from a series of medical needs as varied as the treatment of large numbers of shock casualties during World War II, nursing shortages in the late 1940s, and the polio epidemics of 1947–1948. The care and monitoring of such patients was more efficient, and survival improved when patients were grouped in a single location. By 1958, approximately 25 percent of community hospitals with more than 300 beds reported having an ICU. By the late 1960s, most United States hospitals had at least one ICU. In 2002, there were approximately 6000 ICUs in the U.S., treating 55,000 patients daily, with an annual budget of approximately $180 billion. Several types of ICU exist, including surgical ICU, medical ICU, neonatal ICU, pediatric ICU, and Nero (burn) ICU. The training of the staff accommodates the specific needs of the each population and the specialized equipment used. Many of the techniques and interventions are common to all types of ICUs (Marino, 1991).

A patient's illness or trauma can rapidly become a life-threatening situation that demands immediate and continuous attention. ICUs are repositories of technology and expertise where patients can receive the treatment required for survival. In this arena, clinical engineers (CEs) are presented with their most complex and challenging environment. It is useful to categorize these technologies and equipment in the manner shown in Figure 4.2-1.

All four elements are required, to manage effectively any patient's care. However, a patient who has been admitted to an ICU presents with a set of abnormal physiological conditions of varying severity that can suddenly change and become life threatening. Consequently, physiological parameters such as blood pressure, heart rate, blood oxygen saturation, and respiration are continuously monitored; laboratory results (e.g., blood chemistry, gases) and other information are frequently, though intermittently, obtained, integrated, and interpreted; and a therapeutic intervention is put in place, adjusted, or removed. Although this "closed loop" process is by no means automatic or unique to the intensive care setting, it represents an algorithmic approach to patient management that requires tight control and rapid response to emergent conditions. Consequently, the technologies that are required to implement these functions are often specialized for the ICU setting.

Monitoring and diagnostics

Physiological monitoring in the ICU setting closely follows the patient's cardiac, hemodynamic, and respiratory statuses. While early physiological monitors could view one lead of the (electrocardiogram ECG) and, perhaps, an invasive blood pressure, today's systems integrate a vast array of parameters into a relatively small monitoring device. A typical, full-featured, physiological monitor might contain a set of built-in monitored parameters or might have the ability to add plug-in modules to configure the unit as the clinical situation demands. Monitors located at the bedside are typically connected to monitors at a central station where one may observe the physiological status of many patients at once (see Figure 4.2-2). ICU monitors commonly include the ability to display the waveforms, values, and trends for the following parameters:

- 12-lead ECG
- Arrhythmia monitoring and interpretation
- ST segment analysis

Biomedical Engineering Desk Reference; ISBN: 9780123746467

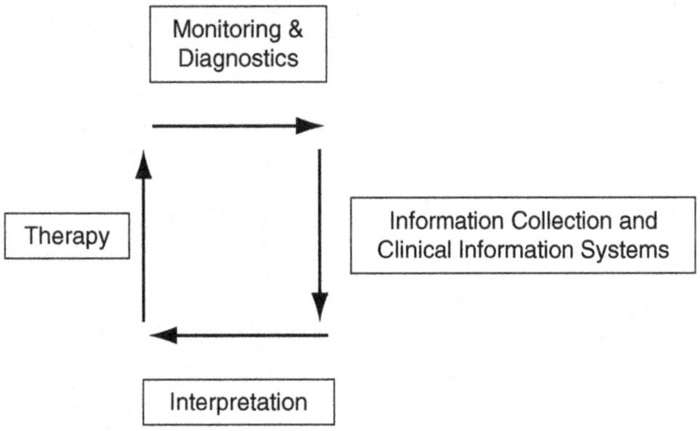

Figure 4.2-1 Four elements required for effective management of patient care.

- Impedance based respiration
- One to three invasive blood pressures
- Noninvasive blood pressure
- Pulse oximetry (SaO_2)
- Two temperatures
- Thermal dilution cardiac output

The following additional monitoring functions are also available as separate units or modules that can be relevant to the management of ICU patient:

- End tidal CO_2
- Continuous cardiac output (CCO)
- Impedance based cardiac output
- Metabolic monitoring
- Real-time blood gas monitoring
- Mass spectrometry

Although provided as part of the therapeutic aspect of the ICU, the life-support ventilator provides a full range of pulmonary monitoring parameters as well.

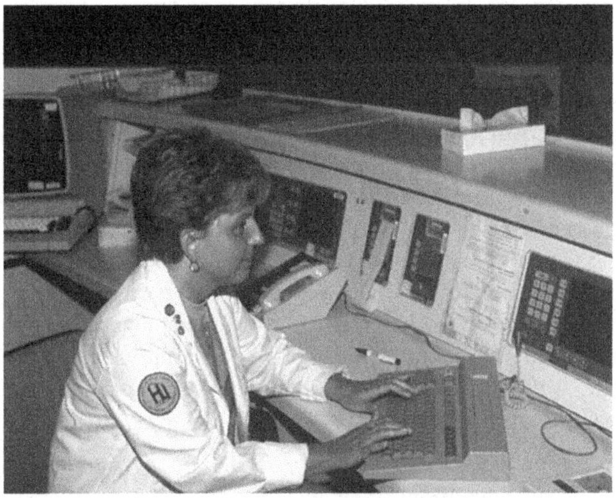

Figure 4.2-2 Physiological monitoring at the central station.

The admission of a patient to an ICU is often (but not always) prompted by compromises or threats to cardiac and/or respiratory function. Insufficiencies in these areas must be addressed rapidly. The delivery of oxygenated blood to body tissues is therefore the first concern of the ICU staff. ECG, pulse oximetry, and noninvasive blood pressure (NIBP) are physiological measurements that can be obtained immediately and noninvasively upon connection of the patient to the monitoring equipment.

Monitoring and therapy at this point become progressively more invasive with the addition of vascular access to measure internal blood pressures on both the arterial and venous side of the circulatory system. Catheters can be inserted into the pulmonary artery to obtain thermodilution cardiac output information (Trautman, 1988) as well as measurement of patient fluid load. Vascular access also provides a convenient site for obtaining blood samples for a variety of laboratory tests. Although attempted for many years, a practical and economical indwelling blood gas analysis catheter remains to be developed.

Additional diagnostic tools are often available in the ICU as well. It is common to find ultrasonic scanning and imaging systems as part of the unit's armamentarium. Although 12-lead ECG acquisition is often part of the bedside monitor, a separate ECG machine is also available. Depending on the ICU size and census, the ICU also can own items such as dedicated, portable X-ray machines and blood gas analyzers. Because of the volatility of the patient's medical condition, and the need to respond quickly to sudden changes, point-of-care blood chemistry machines, which are small enough to be positioned at a patient's bedside, have been developed.

Bedside physiological data is often routed to a central station for display, printing, and alarm monitoring. In recent years, the importance of the central station as a location to be carefully watched for adverse patient events has diminished. Staffing in the ICU environment usually involves a high nurse-to-patient ratio. Most alarm events are recognized at the bedside, with the central station serving more as workstation for archival retrieval of alarm history and analysis of trends. The availability of inexpensive computer hardware has made possible the implementation of full disclosure systems that permit the archiving of all waveforms for a period of up to several days for all of the beds in the ICU. These are often packaged as software within the central station monitor. A consequence of the capabilities has been a reduction in the need and use of strip-chart recorders at the bedside or central station. There is little need to sift through yards of paper strips to find the event of interest because the full disclosure or alarm history record can be printed at will. The laser printer has become the *de facto* hard-copy device of choice.

Information collection and clinical information systems

Most, if not all, health care facilities in the 21st century have installed an information systems (IS) infrastructure that permits the dissemination of all manner of information to the hospital staff. Computer workstations are inexpensive and widely available, and they tap into the institution's high-speed networks. Thus, ICU staff accessibility of such data as patient medical records and test results is assured.

As the patient is admitted to the ICU, information from his medical history and ongoing monitoring are collected and correlated. In today's modern facility, a vast array of information is available to assess the patient's condition. To this end (in addition or as part of the bedside monitoring), high-speed data networks route a large volume of information to the bedside for analysis. Past ECGs are retrievable from archived records and can be compared with the latest 12-lead ECG available at the bedside. Blood chemistry values are obtained, and relevant radiological and ultrasound images are retrieved electronically by way of a picture archiving and communications system (PACS) that is available for viewing near the patient's bedside.

The actual display devices for physiological monitoring and patient lab data are often identical. Increasingly, there has been a movement away from proprietary CRT monitors or thin screen displays in physiological monitoring. Many manufacturers have begun to use components and interfaces that are commonly and inexpensively available in the IS industry. Additionally, digital information is transacted using standard Ethernet protocols. This allows a certain degree of compatibility between physiologic monitoring systems and the patient's medical records. In many instances, parametric information from physiologic monitoring system passes to clinical information networks and workstations. To varying degrees, laboratory data are viewable on the physiologic monitors.

Interpretation

To address conditions that are life threatening and that require immediate intervention, several algorithms have been developed that, to a certain degree, address a given acute clinical situation. These algorithms can be taught to newer ICU staff and compensate to some extent for the experience that otherwise would be required. This distills the array of information into a simplified set of required parameters to treat a specific situation. A decision tree is followed with feedback information that is derived from the monitoring and diagnostic equipment.

Not all ICUs are managed the same way. There are two prevailing schools of thought regarding this issue. In some hospitals, when a patient is admitted to the ICU, most, if not all, of the patient's subsequent medical management is in the hands of a dedicated staff of intensive care specialists (IC specialists) whose responsibilities lie primarily in the ICU environment. This team of specialists includes physicians with specialties in such areas as pulmonary, cardio-vascular, infectious diseases, and internal medicine. Additional medical specialties are consulted as needed. Control of the patient's care is usually relinquished by the referring physician or surgeon. In other institutions, the referring physician might continue to monitor and adjust the therapy to the patient with the assistance of specialized nursing staff and technicians. The prevailing literature indicates that there may be benefits in terms of improved ICU survival rates, reduced length of stay, and a decrease in operating costs in an IC specialist-controlled ICU model.

The availability of this vast array of data and diagnostic tools does not necessarily translate into better outcomes for ICU patients. The complexity of a particular patient's condition makes the interpretation of the data a less-than-automatic or intuitive process. Additionally, clinical studies that challenge certain prevailing ICU practices are constantly emerging in the literature. The relatively wide latitude of patient response to, and tolerance of, interventional procedures further clouds the issue. It cannot be stated with any clear degree of certainty that the introduction of advanced monitoring, therapies, and data management technology has resulted in a definitive decline in patient mortality. What advanced information technology has provided, however, is a streamlining of the information assembly process. This simplifies and, to some extent, reduces the time required to assess and begin treating the patient.

Therapy

Patients who have been admitted to the ICU present with a host of problems that might or might not be life threatening. Severe dehydration, chest pain, and shortness of breath are common symptoms that dictate ICU entry. Upon admission, monitoring of basic vital signs is begun immediately and includes the measurement of ECG (Plonsey, 1988), NIBP (King, 1988), and arterial blood-oxygen saturation (S_aO_2) (Welch et al., 1990). The resulting parametric information points to the first therapeutic interventions. Disturbances in the heart rhythm and rate are immediately visualized, blood pressure in the abnormal range identified, and low levels of S_aO_2 addressed. A peripheral intravenous (IV) line is started for the administration of medications, fluid support, and withdrawal of blood samples for diagnostic evaluation.

Oxygen is applied by facemask to raise saturation levels. These are usual starting points, and many patients will require not much more therapy than this throughout their ICU stay. The patients are monitored, and the levels of therapeutic support are correspondingly adjusted, until the underlying disease process or condition is resolved. This type of patient represents a sizable percentage of the short-stay ICU population that is primarily being observed and requires minimal or moderate support during their stay. The more severely ill patients will undergo more radical and invasive therapies. Following are major categories of ICU therapeutic intervention.

Respiratory care: intubation, mechanical ventilation

Respiratory failure can be described as the inadequacy of the patient's intrinsic respiratory efforts to produce normal blood oxygenation and carbon dioxide clearance, as verified by arterial blood gas measurement. It can be caused by a host of factors, including infection, trauma, and paralysis. The primary consideration of mechanical ventilation has been to restore the patient's blood gas value to nominally normal levels. In spontaneous ventilation, the lungs inflate when negative pressure is created in the thoracic cavity by the downward displacement of the diaphragm. Until the mid-1950s, this was done with devices such as the iron lung, a container that surrounded the patient with his head protruding by way of an airtight collar (Mörch, 1985). This subjected the patient's body to a negative pressure, relative to the outside air, allowing the patient to inspire ambient air through his mouth as the iron lung cycled. This form of ventilation placed little stress on the patient's pulmonary system. He could eat, drink, and talk because his airways were otherwise unencumbered.

After the polio epidemics of the 1950s, positive-pressure mechanical ventilation became more widespread. In this mode, the lungs inflate when positive pressure is generated by the ventilator, forcing the lungs open and causing downward displacement of the diaphragm. Most often, this is done by delivering a volume of humidified, blended air and oxygen mixture by way of an endotracheal tube. Modern mechanical ventilators (see Figure 4.2-3) have a wide range of settings over a range of pressure limits such as O_2 percentage, tidal volumes respiratory rates, and positive-end expiratory pressure (PEEP) (Behbehani, 1995). This type of technology has been in place for more than 35 years, with various improvements in machine size, intelligence, display, modes of ventilation, interfacing, and data collection. However, there is a danger, long recognized, of ventilator-induced lung injury (VILI) or ventilator-associated pneumonia (VAP) with the

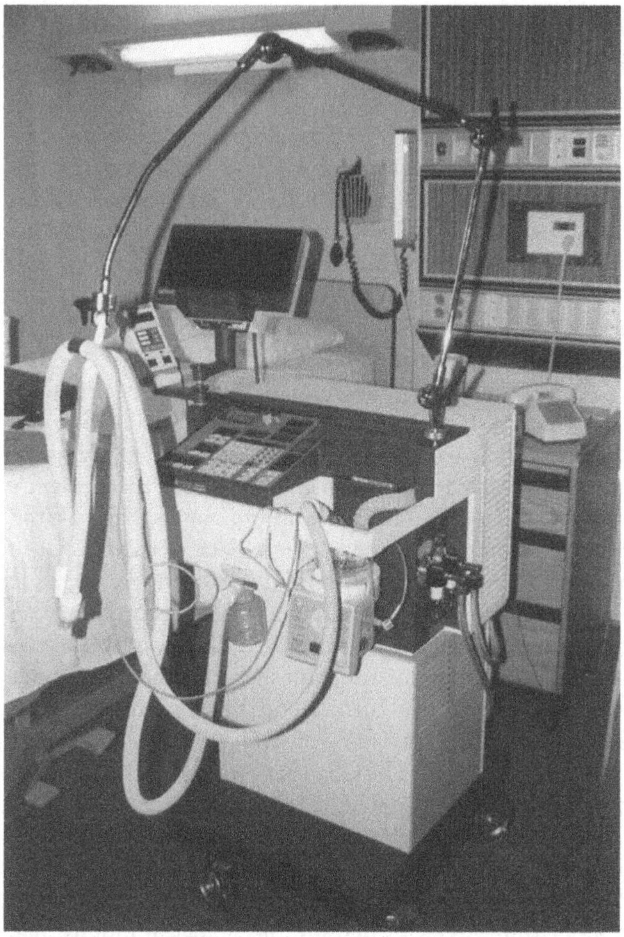

Figure 4.2-3 Puritan-Bennett model 7200 ventilator.

indiscriminate use of positive-pressure mechanical ventilators. Overdistension of the alveoli, either by excessive pressures or volumes, can cause dangerous lung damage especially in the presence of high-oxygen concentration. This is often evidenced in the longer-term patient requiring higher levels of respiratory support. In addition, injury to the lung can produce a cytokine cascade effect that can cause damage to organs throughout the rest of the body. One approach to alleviate this problem is the use of permissive hypercapnia. Smaller tidal volumes are used to prevent lung distension and high airway pressure. This produces less CO_2 clearance, and the blood pH will be lower (more acidic), but this method can protect the lungs from permanent injury.

As an alternative, other ventilator technologies are available. High-frequency jet ventilation (HFJV), high-frequency positive-pressure ventilation (HFPPV), and high-frequency oscillation (HFO) all limit the volume and, hence, the pressure provided to the patient on each breath by operating at supranormal respiratory rates (Hamilton, 1988). HFJV and HFPPV operate in the realm of 100–200 breaths per minute, and HFO in the domain of 10–30 Hz. This is far in excess of the 10–30 breaths per minute encountered in spontaneous or mechanical ventilation. These high-frequency breaths produce lower peak airway pressures, possibly reducing lung injury.

Cardiac care

Among some of the more frequently utilized devices in cardiac care are the defibrillator (see Figure 4.2-4), pacemaker, intra-aortic balloon pump (IABP) (Jaron and Moore, 1988), left ventricular assist device (LVAD) (Rosenberg, 1995), and extracorporeal membrane oxygenator (ECMO). For decades, life-threatening cardiac rhythm disturbances (e.g., ventricular fibrillation, ventricular tachycardia, and atrial fibrillation) have been successfully addressed utilizing defibrillators (Tacker, 1988). These devices have been standard in the ICU since their inception decades ago. Little has changed in their basic operation. They all deliver energy over a wide range of output energy and have had the ability to be synchronized (i.e., to perform ECG-synchronized cardioversion). Recently, however, the use of biphasic waveforms has been revisited. The assumption (not yet accepted in wide-ranging studies) is that more successful cardioversion can be achieved at lower energies, thus avoiding potential damage to the myocardium. Additionally, the refinement of arrhythmia detection has allowed these devices to be automated, detecting dangerous cardiac events and firing the defibrillator without the need for operator intervention. Additionally, many of these units can be configured to perform transthoracic pacing.

Because of myocardial infarction, or for other reasons, patients sometimes develop temporary or permanent heart block, preventing the heart's conduction system from beating at an appropriately consistent rate. A pacemaker is typically utilized to correct this condition. Depending on the acuity of a patient's condition, the first-order intervention probably would entail the use of a transthoracic pacemaker. Pacing also can be done by inserting a pacing catheter in the right heart, which is then connected to a small, external pacemaker.

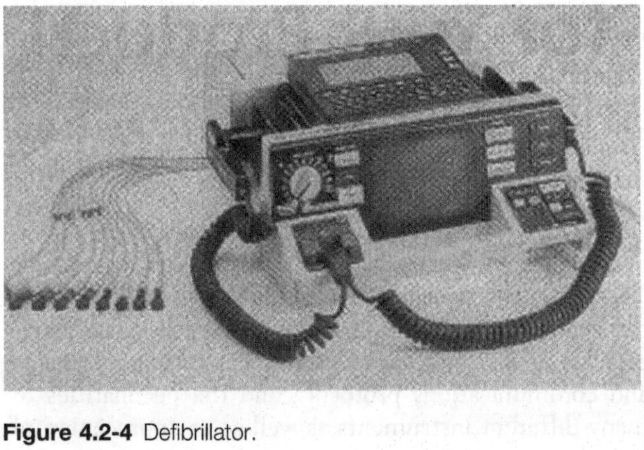

Figure 4.2-4 Defibrillator.

The aforementioned interventions effectively treat conduction or electrical defects that can arise spontaneously or in response to infection or myocardial infarction. The heart muscle is, for the most part, largely intact. When the heart muscle fails, additional measures must be taken, to sustain the patient's life. At the ICU bedside, this is accomplished by the use of the IABP (Jaron and Moore, 1988). A special catheter is inserted into the aortic arch containing a balloon that can be rapidly inflated and deflated by a low-viscosity gas, usually helium. The timing of the balloon inflation is synchronized to either the patient's ECG or blood pressure waveform. Its primary purpose is to improve perfusion of the heart muscle by inflating the balloon as the aortic valve opens, providing additional backpressure to the carotid arteries and rapidly deflating it so that the rest of the body can be perfused. This procedure sustains the patient as the myocardium can recover sufficiently to pump on its own. In the event that a replacement heart is required, the procedure will extend a patient's life until a donor heart is available.

Alternatively, there are a number of LVADs that enhance the pumping action of the left ventricle until a donor organ is found (Rosenberg, 1995). They are surgically implanted in the patient's chest or abdomen and are managed by the ICU staff. Less widely used, though available also, is the ECMO, or heart lung machine (Dorson and Loria, 1988). This device, used during coronary-artery bypass graft, can provide both respiratory and circulatory function, allowing the heart and lungs to be mechanically inactive for a time. Efforts such as this are extreme and require significant technical support during use.

Infusion devices

Most ICU patients arrive with some form of vascular access and usually receive additional IV lines during their stay. Much of the therapeutic intervention and monitoring occurs via vascular access. Fluid management and drug delivery can be accomplished using gravity as the source of positive pressure to push liquids into the patient. Infusion rates are monitored visually, and adjustments can be made to achieve the required goals. This approach is not viable when six to eight infusions are running simultaneously, with rates varying from 0.1 cc/hour, for certain pain medication, to hundreds of cc/hour, for rapid fluid replacement. This is handled now with a variety of programmable infusion pumps (Figure 4.2-5) and controllers (Voss and Butterfield, 1995).

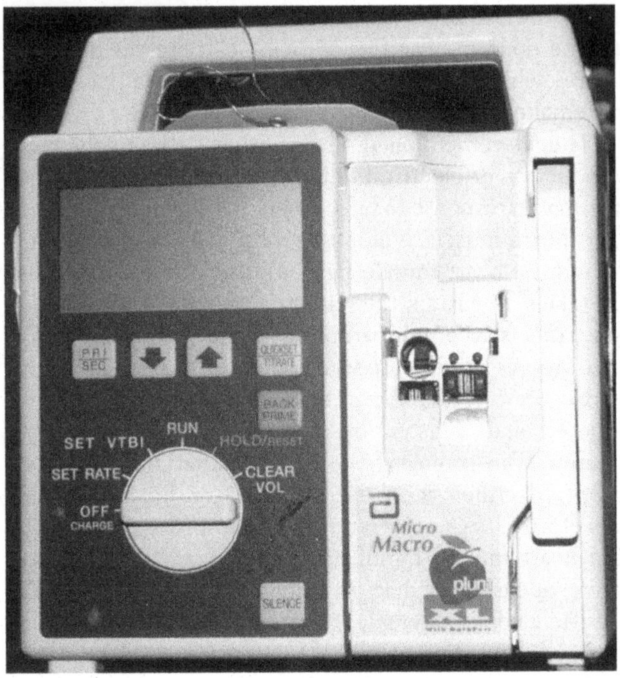

Figure 4.2-5 Abbot Plum XL infusion pump.

Infusion pumps and controllers employ a variety of electromechanical mechanisms to regulate fluid flow to the patient. These include peristaltic or diaphragm pumps, syringe devices, and variable pinch clamp mechanisms that control gravity-driven flow. Usually, the tubing sets used by these devices are dedicated and not interchangeable. Infusion pumps with a wide range of rates (1–999 cc/hour) have been available for decades. Modern infusion devices have a wide range of programmable infusion rates and extensive alarm capabilities. Features include proper set placement detection, pressure limits, proximal and distal occlusion alarms, end of infusion alarm, and programmable keep-vein-open (KVO) rates. Additional safety features to prevent unrestricted free flow of IV fluid to the patient are incorporated as well. In order to avoid medication errors, systems that will associate pharmacy requisitions, the infusion device, and patient identification are being developed. Additional intelligence in such systems will identify patient sensitivities, incompatibilities between infused medications, and violation of nominal drug concentrations and infusion rates.

Dialysis: kidney and organ support

The sickest ICU patient will begin to manifest failure in one or more organs. Acute renal failure (ARF) is not uncommon in the ICU setting. ARF occurs in approximately 5% of all hospitalized patients and, historically, has been associated with a high risk of mortality. Following the introduction of dialysis therapy more than 50 years ago, patient mortality dropped from 90% to 50%. However, mortality in the intervening years has not improved from the 50% level. In ICU patients presenting with other co-morbidities, the mortally rate is between 50% and 80%. Although the etiology of ARF varies, it is often associated with a reduction in urine output and a drop in systemic blood pressure. Waste products accumulate in the blood, electrolyte levels become abnormal, pH usually drops, and edema sets in. Initially, ARF was treated in the same manner as chronic renal failure, with intermittent hemodialysis. This approach quickly removes the metabolite, electrolyte, and fluid imbalance. However, these levels vary widely because the body continuously generates their production between dialysis sessions. Consequently, there has been a move toward continuous renal replacement therapy (CRRT) (Galletti et al., 1995). In this mode, renal function is supplied continuously while the patient recovers. There are several different CRRT modalities. Historically, arterial circulation was used to provide the force to move blood across a dialyzer cartridge. Because of the need for arterial access and inherent complications and the technical improvement in venovenous therapies, these methods have been largely abandoned. The venovenous modalities differ primarily in the methods of clearance. Slow continuous ultrafiltration (SCUF) is a method employed to remove volume by taking off fluid filtered by a dialysis membrane by convective force. Continuous venovenous hemodialysis (CVVHD) provides clearance using diffusive clearance by running dialysate across the membrane. Continuous venovenous hemofiltration (CVVHF) removes fluid by convection (as in SCUF) and then provides replacement fluid back to the patient. Finally, continuous venovenous hemodiafiltration combines the clearance properties of CVVHD and CVVHF. These approaches are provided by bedside machines incorporating the required pumps, control instrumentation, filters, and fluids.

Clinical engineering and the ICU

Perhaps the most complex and technically challenging hospital environment is the ICU. Traditionally, CEs have been involved in medical device evaluation, inspection, maintenance, layout, design, and integration of the various instrumentation systems. As the devices' and systems' complexity has increased over the years, CEs have been looked to as repositories of technical knowledge that is beyond that of the users (i.e., the nurses and physicians). This requires maintaining a substantial level of expertise in the physics and physiology of measurements and therapies; knowledge of computer operating systems; networks and communications protocols; and the peculiarities of many different instruments as well as an appreciation of ways to keep all of these systems in operation. The many interactions and safety issues of the various monitoring and therapeutic instruments must be understood and translated so that the CE can provide answers to questions posed by the clinical staff, whose responsibility is the care of the patient, not the technology.

References

Behbehani K. Mechanical Ventilation. In Bronzino JD (ed). *The Biomedical Engineering Handbook*, Boca Raton, FL, CRC Press, 1995.

Dorson WJ, Loria JB. Heart-Lung Machine. In Webster JG (ed). *Encyclopedia of Medical Devices and Instrumentation*, New York, Wiley, 1988.

Galletti PM, Colton CK, Lysaght MJ. Artificial Kidney. In Bronzino JD (ed). *The Biomedical Engineering Handbook*, Boca Raton, FL, CRC Press, 1995.

Hamilton LH. High-Frequency Ventilators. In Webster JG (ed). *Encyclopedia of Medical Devices and Instrumentation*. New York, Wiley, 1988.

Jaron D, Moore TW. Intraaortic Balloon Pump. In Webster JG (ed). *Encyclopedia of Medical Devices and Instrumentation*. New York, Wiley, 1988.

King GE. Blood Pressure Measurement. In Webster JG (ed). *Encyclopedia of Medical Devices and Instrumentation*, New York, Wiley, 1988.

Marino L. *The ICU Book*. Philadelphia, Lea and Febiger, 1991.

Mörch ET. History of Mechanical Ventilation. In Kirby RR, Smith RA, Desautels DA (eds). *Mechanical Ventilation*. New York, Churchill-Livingstone, 1985.

Plonsey R. Electrocardiography. In Webster JG (ed). *Encyclopedia of Medical Devices and Instrumentation*. New York, Wiley, 1988.

Roa RL. Clinical Laboratory: Separation and Spectral Methods. In Bronzino JD (ed). *The Biomedical Engineering Handbook*. Boca Raton, FL, CRC Press, 1995.

Roa RL. Clinical Laboratory: Nonspectral Methods and Automation. In Bronzino JD (ed). *The Biomedical Engineering Handbook*. Boca Raton, FL, CRC Press, 1995.

Rosenberg G. Artificial Heart and Circulatory Assist Devices. In Bronzino JD (ed). *The Biomedical Engineering Handbook*. Boca Raton, FL, CRC Press, 1995.

Tacker WA. Electrical Defibrillators. In Webster JG (ed). *Encyclopedia of Medical Devices and Instrumentation*. New York, Wiley, 1988.

Trautman ED. Thermodilution Measurement of Cardiac Output. In Webster JG (ed). *Encyclopedia of Medical Devices and Instrumentation*. New York, Wiley, 1988.

Voss GI, Butterfield RD. Parenteral Infusion Devices. In Bronzino JD (ed). *The Biomedical Engineering Handbook*. Boca Raton, FL, CRC Press, 1995.

Welch JP, DeCesare R, Hess D. Pulse Oximetry: Instrumentation and Clinical Applications. *Respir Care* 35(6): 584–601, 1990.

Further information

Kirby RR, Banner MJ, Downs JB. *Clinical Applications of Ventilatory Support*. New York, Churchill Livingstone, 1990.

Webster JG. *Medical Instrumentation: Application and Design*, 2nd ed. Boston, Houghton Mifflin, 1992.

Chapter 4.3

4.3

Operating theatre facilities

Chad J. Smith, Raj Rane, and Luis Melendez

The operating room (OR) is an exciting location in which to apply the principles of engineering. The surgical environment encompasses an extensive arrangement of medical technologies and processes. The clinical engineer must apply the principles of several engineering disciplines to support an efficient, productive, and safe OR.

This chapter describes the medical technology, surgical specialties, personnel, and physical setting of a conventional OR. The complex and dynamic interaction of these facets of the OR is thoroughly illustrated, thus enabling the clinical engineer to support this environment more effectively.

The role of the operating room

Surgery is the diagnosis and treatment of medical injuries, diseases, and deformities by manual or operative means. It is a fundamental service that the health care facility provides to a community. A vast assortment of surgical procedures is performed to treat the medical conditions within the scope of health care.

A surgical procedure is a delicate and complicated undertaking. It must be performed in a controlled environment with numerous technical resources and a staff of medical professionals and support personnel. Depending on the patient and the nature of the surgical procedure, surgical cases are characterized in one of three ways: Inpatient, outpatient, and same-day. Inpatient procedures are performed for patients who have been admitted to the hospital. Outpatient procedures are generally for those undergoing minor surgical procedures. In those cases, the patient typically will undergo a local anesthetic and will be admitted and released in the same day. Same-day surgery consists of more extensive cases than outpatient procedures and might include general anesthesia. However, the patient is also discharged in the same day. The OR is the location in a health care facility that is equipped for the performance of surgery. Patients who require invasive treatment are transported to this department to undergo induction of anesthesia, the surgical procedure, and resuscitation. It is a technologically advanced area filled with complex equipment, highly trained professionals, and stringent environmental controls. For the scope of this chapter, the term "surgical suite" will be used to refer to the individual room where a surgical procedure is performed. The "OR" will signify the entire department, within the health care facility, in which the operating suites and supporting areas are located.

A typical hospital comprise numerous departments that provide a wide range of services. The OR plays an important role in a hospital and in the overall process of providing patient care. The following lists outline the typical steps for a patient who is undergoing a scheduled surgical procedure, as well as an unscheduled procedure. It is meant to demonstrate methods in which a surgical procedure is arranged, and the relationships among hospital departments.

Scheduled surgery

- The patient consulted a primary care physician.
- The primary care physician made a diagnosis and directed the patient to consult a surgical specialist.
- A surgical specialist diagnosed the condition and confirmed the need for surgical treatment. Images from the radiology department and test result from the medical lab aided in the diagnosis.
- The surgery date was scheduled, and OR time was booked.

Biomedical Engineering Desk Reference; ISBN: 9780123746467

- The patient was admitted to a pre-admitting test center to review the medical history and to undergo various tests.
- An anesthesiologist studied the patient's medical history and planned the method of anesthesia.
- The patient was sent to the pre-operating area (induction room), where she was prepared for surgery, and anesthesia was initiated.
- Surgery was performed in a surgical suite.
- Following surgery, the patient was delivered to the post-anesthesia care unit (PACU) to recover from the effects of anesthesia.
- The patient was delivered to a care unit for complete recovery.
- The patient was discharged from the hospital.
- The patient made periodic visits to the hospital for physical rehabilitation.

Unscheduled (emergency) surgery

- The patient placed an emergency call, and an ambulance was dispatched.
- The patient was delivered to the hospital's emergency department.
- The ER physician ordered the patient be transported to the radiology department for imaging and sent samples to the medical lab for analysis.
- A surgical specialist was consulted to aid in the diagnosis.
- The patient was transported to the pre-operating area (induction room), where she was prepared for surgery, and anesthesia was initiated.
- Surgery was performed in a surgical suite reserved for emergency cases.
- Following surgery, the patient was delivered to the PACU to recover from the effects of anesthesia.
- The patient was delivered to the intensive care unit for complete recovery.
- The patient was discharged from the hospital.
- The patient made periodic visits to the hospital for physical rehabilitation.

A wide range of surgical procedures can be performed in an OR. The surgical specialties of a particular hospital are based on the resources and medical scope of that health care facility. The following is a list of some surgical specialties that are commonly found in a hospital's OR. A brief description of each specialty is included.

General surgery

This form of surgery includes a broad spectrum of surgical care that involves largely the surgical management of diseases of the bowel, gallbladder, stomach, and other digestive organs.

Thoracic and cardiovascular surgery

Thoracic and cardiovascular surgery concerns the diagnosis and treatment of disorders of the thorax. The thorax is the upper part of the trunk between the neck and the abdomen, containing the chief organs of the circulatory and respiratory systems, such as the lungs and heart. Two major types of thoracic surgery are classified as pulmonary (i.e., pertaining to the lungs) and cardiovascular. Cardiovascular surgery treats diseases and conditions of the heart and the blood vessels of the entire body. Common cardiovascular procedures include coronary bypass surgery, aortic or mitral valve replacement or repair, and aneurysm repair.

Neurosurgery

Neurosurgery is the specialty of surgery that addresses the diseases and disorders of the nervous system. Within the realm of neurosurgery are the brain, spinal cord, and associated vascular supply. Common procedures include cervical fusion. Disorders commonly treated by neurosurgeons include intracranial tumors, vascular malformations, carpal tunnel syndrome, spinal cord injury, and stroke.

Orthopedic surgery

Orthopedic surgery treats and corrects deformities, diseases, and injuries to the skeletal system, its articulations, and associated structures. Some examples of orthopedic surgery include hip and knee replacement, cartilage repair, and fracture repair.

Plastic surgery

Plastic surgery is concerned with the repair, restoration, or improvement of lost, injured, defective, or misshapen parts of the body, caused by congenital defects, developmental abnormalities, trauma, infection, tumors, or disease. It is generally performed to improve functions, but it also is done to approximate a normal appearance.

Gynecology

Gynecology involves the physiology and disorders primarily of the female genital tract, as well as female endocrinology and reproductive physiology.

Urology

Urology is the study, diagnosis, and treatment of diseases of the urinary tract in both sexes, and the genital tract in the male.

Burn treatment

Burn treatment is surgery performed to treat and manage burn injuries by reconstructing damaged tissues and improving functionality of the damaged area. A burn team might include plastic and general surgeons.

Pediatric surgery

Pediatric surgery is concerned with the health of infants, children, and adolescents.

For the purpose of this chapter, the setting for any of the following procedural areas can be characterized as a surgical suite. In many cases, the equipment and facility requirements that are present in these areas are similar to those found in a conventional OR environment.

- Labor and delivery
- Endoscopy
- Ophthalmology and otolaryngology procedure rooms
- Dental surgery
- Radiology suites, including interventional radiology and radiation oncology
- Interventional electrophysiology
- Treatment rooms in which minor procedures are performed

ORs, large and small, rely heavily on medical personnel and support staff to carry out a wide variety of tasks. The following is a list of some of the employees who might be found working in, or providing a service to, the OR.

- Physicians and anesthesiologists
- Registered nurses, circulating nurses, scrub nurses, nursing assistants, and surgical technicians
- Clinical engineers, biomedical engineering technicians, and anesthesia technicians
- Schedulers, record keepers, patient transporters, turnover team, purchasers, and administrators

Operating room floor plan

The floor plan of a well-organized and safe OR involves a great deal of research and planning. It is a complex environment with numerous design criteria. The layout of the facility directly effects user productivity and satisfaction. The OR must be arranged in a manner that is conducive to the flow of patients, staff, and equipment. It also must include space for functional support areas that are common to most ORs, such as instrument processing, technical support, and equipment storage areas. It is necessary for clinical engineers to have knowledge of the OR layout and design issues. Clinical engineers, as technical experts of medical technology and the environment in which it is used, also may serve as a valuable resource in the design of an OR.

The arrangement of surgical suites often reflects a hospital's available space and surgical requirements. Several OR floor plans have been developed to utilize available space and maximize productivity most ideally. The L- or T-shaped single-corridor layout is common in smaller ORs. Larger facilities typically utilize the multiple-corridor layout, with the clustering of surgical suites by surgical specialty (Bronzino, 1992). Grouping surgical specialties together aids in the efficiency of sharing common resources.

The location of functional support areas that serve the OR is another important consideration in a floor plan. The efficiency of these services is enhanced when the distances that must be traveled are minimized. The following is a list of some functional support areas that are commonly found in, or adjacent to, the OR.

Induction and pre-operative holding areas

Induction and pre-operative holding areas are utilized to prepare the patient for surgery, and the administration of anesthesia. These services are located either immediately adjacent to the OR or within the department. The procedures and services performed in these areas will vary from hospital to hospital. The holding areas at some facilities are used simply to prepare the patient for delivery to the surgical suite, and other hospitals might utilize this area to fully anesthetize the patient.

Scrub stations

Scrub stations used for hand washing contain sinks and cleaning supplies and are common to all ORs. Two scrub positions must be located adjacent to the entrance to each surgical suite.

Control desk

The control desk is the communication center of the OR. Arrangement of the case schedule is a primary function of the control desk. This is vital for the efficiency and productivity of the OR. Several things must be considered when planning the schedule. The scheduler must coordinate the urgency of surgical cases, surgical location,

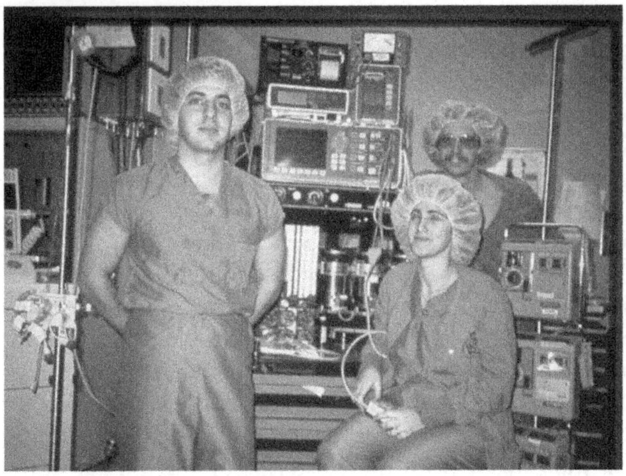

Figure 4.3-1 Clinical engineering department engineers and technicians support the OR.

time considerations, patients, surgeons, anesthesiologists, support staff, equipment and supply availability, and post-anesthesia care. The control desk also might be involved in patient billing and recordkeeping.

Technical support services

Technical support services are vital to the productivity of the OR. Support for OR technology must be immediately available to ensure the efficiency and safety of the surgical environment. Support groups serving the OR include anesthesia technical support and clinical engineering (see Figure 4.3-1). Anesthesia technical support is located within, or in the immediate vicinity of, the OR. The anesthesia workroom is primarily used for cleaning, testing, and storing anesthesia equipment. In larger facilities, clinical engineering support workrooms also might be situated within the OR. Later in this chapter, the responsibilities of the clinical engineering department will be discussed in detail.

Patient support service areas

Patient support service areas include personnel, equipment, and supplies used to provide assistance to a variety of OR functions. Space must be incorporated into the OR floor plan to accommodate services such as equipment or patient transporters, orderlies, and nurse's aides.

Housekeeping areas

Housekeeping areas are the workspaces for housekeeping and turnover team personnel, equipment, and supplies. Turnover team members are responsible for cleaning and setting up surgical suites between cases.

Pharmacy

The pharmacy is included in the overall design of a health care facility to provide medications for patient care. Hospitals of considerable size can include a pharmacy within the OR.

Instrument processing

Instrument processing can be managed in a variety of ways, depending on the resources and preferences of a particular institution. Regardless of the sterilization method, a process must be present for delivering contaminated instruments to the processing area and redistributing the appropriate instruments to the surgical location. Sterilization methods can be divided into two categories: centralized and decentralized.

A centralized processing department is designed to serve a large volume and can be located within the OR. Smaller hospitals sometimes utilize a centralized processing department that serves the entire facility. Centralized instrument processing can employ an assortment of sterilization processes. The two most common methods include steam sterilization and ethylene oxide (EtO) sterilization. Autoclaves are devices that sterilize instruments using steam, with temperature, exposure time, and pressure as variables. Variables in EtO sterilization are EtO gas concentration, moisture, temperature, and time. Dry-heat sterilization and chemical sterilization are other methods of instrument processing that can be utilized in health care facilities.

Decentralized instrument processing is the practice of sterilizing instruments inside, or within the immediate proximity of, a surgical suite. Perhaps the most common method involves the use of autoclave technology. An autoclave can be designated for each operating suite or shared among a cluster of suites. Table-top sterilizers are used for flash sterilization located in common areas in the OR. They are autoclavable at around 270°F for 3 to 5 minutes.

Materials management

A hospital also must allocate space for the delivery, storage, and distribution of supplies to the OR. The process of waste disposal also must be accommodated. Trash and biological waste must be properly stored, handled, and eliminated. In addition, the OR must establish a manner for processing surgical drapes, scrubs, and other linens used in the surgical environment.

Equipment storage areas

A common dilemma is the lack of sufficient storage space. The OR layout must include areas for the storage of medical devices. Large devices, such as stretchers, spare anesthesia machines, and X-ray equipment, must be stored in accessible locations where they will not clutter or obstruct hallways.

Other areas

Functional areas that also can be located in or near the OR include dark rooms, conference areas, and locker rooms.

Operating room facility infrastructure

The OR is a complex environment that requires precise climate control and numerous utility services. The following is a list of utilities services that are present in an OR. A description of each utility is included.

Climate control

The purpose of a heating, ventilating, and air-conditioning (HVAC) system is to provide and maintain environmental conditions, including proper airflow, heating, and cooling within a certain area or the entire hospital. Types of HVAC systems include variable air volume, multizone systems, displacement ventilation, and water-loop heat pumps. An HVAC system mainly consists of an air/water supply system, filters, heating and cooling coils, compressors, fans, motors, exhaust and evacuation systems, air ducts, vents, and control mechanisms. It can provide heat and ventilation to the entire hospital with provisions for controls for individual zones (e.g., the OR), or each zone can have their own HVAC system and associated control. The control mechanism is an integral part of a highly efficient HVAC system. The controls monitor, display, and allow the user to regulate various parameters, including temperature, pressure, and humidity. Precise control over temperature and humidity is important in the OR. For example, a clinician must be able to adjust the temperature of the operating suite for surgical specialties such as pediatrics and orthopedics, where higher room temperatures are desired in order to help maintain patient temperature. Some cases, such as burn treatments, require a higher humidity.

The OR requires specialized criteria for the ventilation system as well. The surgical area must be replenished with fresh air at regular time intervals time to maintain a safe and sterile environment. A minimum of three air changes per hour of outdoor air, and 15 total air changes per hour, is required (AIA, 1996). A positive pressure must exist between the OR and adjacent areas, to prevent the inflow of contaminated air. The ventilation system also should be designed so that there is no recirculation of air from room to room.

Adequate filtering is another important criterion of the ventilation system. A hospital-grade high efficiency particulate air (HEPA) filter is installed at the location where the air is drawn in, to filter debris and to control airborne infection. Hospital-grade HEPA filtration is constructed with an air-intake grill to catch larger debris such as leaves, seeds, and insects. Next is a layer composed of fibers to remove smaller objects like dirt and dust. Finally, a corrugated fine-mesh removes particles such as pollen, dust, and viruses.

Electrical power distribution

The prevalence of electrical devices for the diagnosis and treatment of patients has made electrical power distribution one of the more important systems in modern medical facilities. The nature of the OR environment demands a particularly elaborate set of requirements. Electrically powered devices surround the patient and perform a wide range of vital functions. Therefore, proper design and maintenance of the electrical power system are essential, to ensure the safety of both the patient and the clinician. As the need for a continuous power is essential throughout a hospital, electrical systems must comprise at least two sources of power. The typical primary power sources are public utility lines. To provide for the continuous supply of power to the critical areas of the hospital, an alternate, or emergency, source of power is necessary. In the event of an interruption, hospitals are required to have an in-house alternate source of power, such as a generator.

The OR is densely populated with electrically powered devices. Individual surgical suites must be supplied with enough electrical wall outlets to accommodate the equipment, and these outlets must be in locations that are conveniently accessible to the clinician. Wall outlets in a health care facility deliver power to equipment using a three-wire configuration of hot, neutral, and ground wires. The ground wire protects patients and staff from electrical hazards by providing a low resistance pathway to channel fault or leakage currents away from an electrically powered device to ground. Some surgical techniques and instrumentation bypass the patient's body resistance, thus increasing the patient's susceptibility to electrical energy from external sources. Minimizing the electrical energy to which a susceptible patient is exposed is prudent.

Although most equipment is designed to operate on the same power requirements, the facility also must be prepared to accommodate devices that require varying power consumption. While most wall receptacles provide 115 volts, some electrical devices, such as lasers and X-ray equipment, require 230 volts to operate.

In the past, the use of combustible anesthetic gases (e.g., cyclopropane and ether) created a volatile environment that demanded additional safety measures. It was essential to reduce the likelihood of a static discharge (spark) that could lead to an explosion in the operating suite. Although the risk of explosion is not typically an issue today, it has played an important role in the design of electrical power systems in many existing OR. For example, power outlets have typically been installed approximately 4 feet above floor level because volatile explosive agents tend to settle in the air. Special plugs, termed "explosion-proof," were also commonly installed on power cords that, when plugged in, created an airtight seal. Conductive flooring is also commonly found in ORs. This technology was installed to eliminate the electrical isolation of equipment and personnel in an operating suite. Electrical isolation between objects is conducive to the build-up and potential discharge of static electrical charge. Isolated ungrounded power systems, line isolation monitors (LIMs), and ground-fault circuit interrupters (GFCIs) are further examples of technologies employed in ORs to reduce the risk of explosion in the presence of combustible anesthetic agents. These technologies also serve to reduce the risk of explosion from electrical shocks resulting from stray (leakage) current.

An isolated ungrounded power system is an electrical power distribution system in which all of the current-carrying conductors are isolated from ground (and earth) by a high impedance (Feinberg, 1980). The most common and economical method of isolation is to use an isolating transformer. In a properly installed system, no hazardous current will flow from either conductor to ground, but the two conductors will function as though they were connected directly to a ground. This is especially useful in a "wet environment" such as the OR, where liquid spills and standing water increase the risk of electrical hazards.

An LIM displays the degree of electrical isolation in an isolated ungrounded power system. It provides an early warning system of possible leakage or fault currents from either of the current-carrying conductors to ground. A GFCI also serves to protect patients and staff members. This technology monitors the ground fault current and interrupts the power to the electrical receptacle when the current exceeds a preset limit. Although LIM technology is more expensive, one advantage over the GFCI is that the LIM will alarm when a fault is detected, but will not interrupt electrical service.

Gas pipelines

Pressurized gases must be distributed to locations throughout the health care facility. Medical gases such as oxygen, nitrous oxide, and medical air are particularly important for the administration of anesthesia in the OR. Medical gases are stored either in metal cylinders or in the reservoirs of bulk gas storage and central supply systems. The central supply system is the source of medical gases that are distributed via the facility pipeline system. The number and location of wall outlets in the OR varies, depending on facility design and surgical specialty. Figure 4.3-2 shows a wall outlet for nitrogen commonly used to power pneumatic tools such as saws and drills.

The method of storage for the central supply of oxygen depends on the demands of the facility. Smaller facilities tend to store oxygen in a series of cylinders connected by a manifold or high-pressure header systems. Larger facilities store their bulk oxygen in pressurized liquid form, which enables the hospital to store more oxygen in a smaller space. The central supply of medical air is most commonly generated by air compressors, stored in a reservoir, and delivered to the piping system. Other sources include pressurized cylinders or a proportioning system that mixes the appropriate amount of oxygen and nitrogen from central sources. The nitrous oxide supply systems include the cylinder manifold system or a bulk liquid storage system similar to the one used for oxygen system. All central supply systems for medical gases must be designed with a separate reserve system. Details concerning the use of medical gases in anesthesiology are discussed in Chapter 4.4.

Nitrogen and carbon dioxide can be piped into the OR, as well. Nitrogen, primarily used for gas-powered equipment, is stored as a series of pressurized tanks connected by a manifold. Carbon dioxide is also typically stored as a cylinder manifold system.

Vacuum

The central vacuum system is another essential utility that must be piped into the OR. The system is vital to both surgery and anesthesia. It provides pressure for drainage, aspiration, suction during surgical procedures,

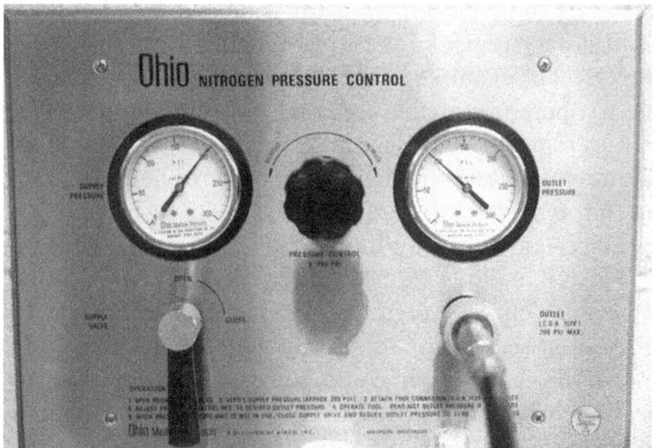

Figure 4.3-2 OR wall outlet for nitrogen gas.

and the scavenging of waste gases from the anesthesia machine breathing circuit. A health care facility must have two suction pumps, each individually capable of handling the overall demand.

Lighting

Lighting is an essential utility in a health care facility. The OR requires special attention to lighting because of the small-scale delicate tasks that must be performed. Although the choice of a surgical light is subjective, several characteristics must be addressed. The criteria include light intensity, light color, focusing capability and range, degree of shadow production, heat production and dissipation, choice of mounting, lamp head maneuverability, and ease of cleaning (Bronzino, 1992).

Communication

The speed and reliability of communication in the OR is vitally important. Communications systems are becoming increasingly complex with the continual advances in networking and wireless technology. Fundamental modes of communication such as telephones, paging systems, and intercoms are common in ORs.

Advances in computer technology and network systems have had a profound effect on the communications systems within the OR environment. E-mail, Internet access, online scheduling, and patient databases have vastly improved the quantity and quality of exchanged information.

Fire protection

All health care facilities must be equipped with a fire protection system. The fundamental elements of such a system include smoke detectors, alarms, a sprinkler system, and automatically closing doors to contain the fire. The hospital staff must have an established evacuation plan for patients and staff. Several OR fires occur each year. The combination of an oxygen-enriched atmosphere, fuel source (e.g., drapes, endotracheal tubes, and the patient) and ignition source (e.g., electrosurgical units and lasers) have produced catastrophic fires (de Richmond and Bruley, 1993).

Surgical suite layout

A well-designed surgical suite provides ample space for staff and equipment. It encourages a smooth flow of people and materials within the room. While the layout of an individual surgical suite varies by surgical service, all suites share some characteristics, as described in below.

The size of the operating suite should reflect the space required by the patient, staff, and equipment to carry out the surgical procedure. The standard surgical suite is at least 400 square feet. Surgical suites that require specialized equipment or additional personnel can be significantly larger. It is important to provide space so that sensitive medical equipment is not positioned in high-traffic areas. The shape of an operating suite can take on a variety of forms, (e.g., square, rectangular, round, or oval), depending on design preferences and overall OR layout.

The surgical table is typically oriented in the center of the room, with the surgeons positioned along the side of the patient. The scrub nurse is positioned in an area where both the instrument table and the surgeon's outstretched hand are within reach. The anesthesiologist and anesthesia equipment are generally located at the head end of the table. This allows the anesthesiologist to have access to the patient's airway. The anesthesia equipment also must be positioned near medical gas, vacuum, and power outlets that can be located on a wall or column. Other devices, such as defibrillators, electrosurgical units, video towers, and bypass machines, are located further outside the surgical field. Figure 4.3-3 shows a typical layout of an OR.

The surgical suite also must provide accommodations for storage. Commonly, cabinets, shelving, and carts will line the walls of a surgical suite. In addition, surgical suites are usually equipped with communication equipment. Computer, intercom, and telephone access are standard in modern surgical suite layouts.

Common operating room technologies

The following is a list of typical medical technologies commonly found in the OR. A brief description of each technology is included.

Furniture

Furniture is utilized extensively in the OR and serves a wide range of functions. Patient beds are employed for transporting patients and as an adaptable platform for performing surgery. Transport beds, or stretchers, are designed with side rails and large casters to safely transport patients within the hospital.

Surgical beds are valuable tools for patient positioning during surgery. In order to accommodate various procedures, surgical beds are more substantial and complex than transport beds. Modern beds are equipped with mechanical, electronic, and hydraulic systems that allow clinicians to position a patient in numerous orientations,

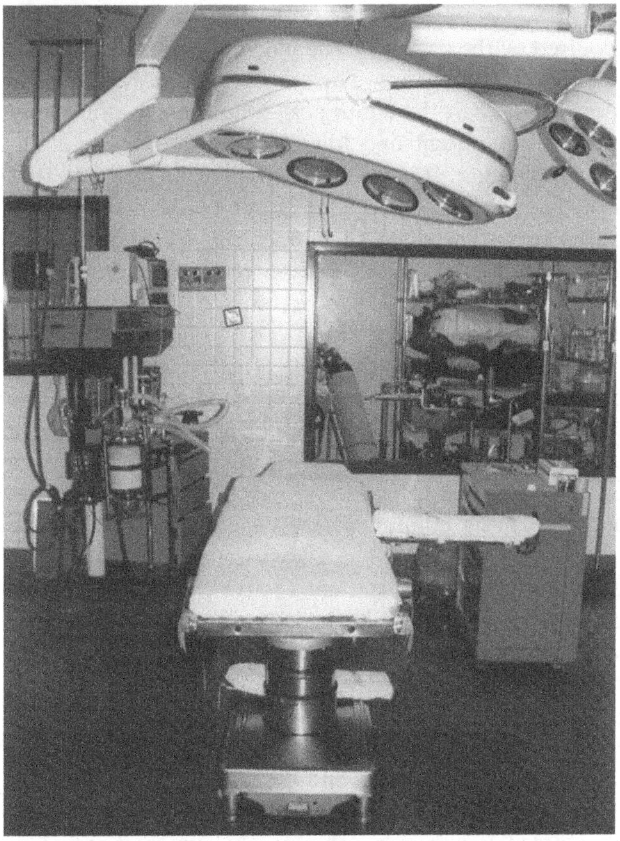

Figure 4.3-3 Typical OR layout showing OR table, overhead lights, anesthesia machine and monitors, medication and supplies cart, storage room, and wall outlets for gases.

such as raising or lowering the height, tilting to the left or right, Trendelenburg (head down, legs up), and reverse-Trendelenburg. Surgical bed accessories include arm support boards, leg supports, foot extensions, restraints, and padding. They also can be outfitted with a wide variety of specialized surgical positioning devices, such as neurosurgical headrests, radiolucent tops, and the Andrews Frame used for spinal surgery.

Other standard items of furniture in the OR include instrument tables, intravenous (IV) poles, waste storage bins, stools, and chairs. Racks, carts, and shelving, used to store medical supplies, surgical supplies, and instruments, are also common in a surgical suite.

Surgical drapes

Surgical drapes are used in the OR to protect the patient, clinicians, and equipment. Drapes can be made of cloth or paper, and reusable or disposable. Figure 4.3-4 shows a surgical procedure underway with the patient draped. Important characteristics include barrier protection effectiveness, resistance to ignition, and durability. Surgical drapes are employed to provide a physical barrier that

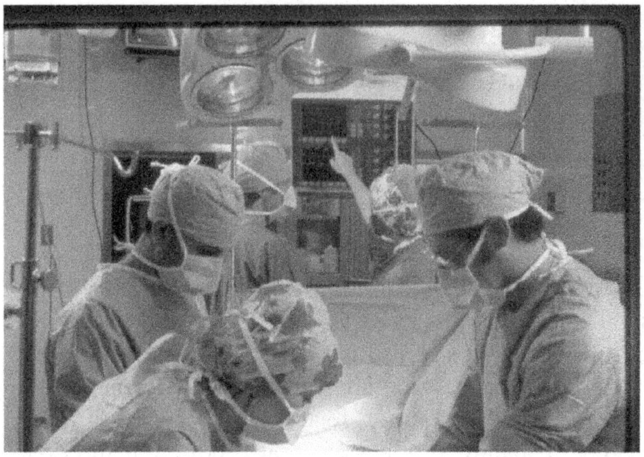

Figure 4.3-4 The surgical team at work on a draped patient.

protects the surgical field from contamination. An "ether screen" is the wall of drapes set up in order to provide a barrier between the anesthesia work area, at the head of the patient, and the surgical field. Drapes are also placed in the surgical field around the incision site to cover the patient and to collect fluids. They also can be used to wrap sterile surgical instruments and to cover equipment in the surgical suite.

Anesthesia equipment

The delivery of anesthesia requires a gas delivery system, as well as continuous and detailed monitoring of patient physiology. The anesthesia machine is used to deliver a known mixture of gases to the patient. Three main sections of the anesthesia machine are the gas supply and delivery system, the vaporizer, and the patient breathing circuit (see Chapter 4.4; Calkins, 1988 Dorsch and Dorsch, 1984).

A variety of technologies are employed by the anesthesiologist to monitor the physiology of the anesthetized patient. Physiological monitoring equipment includes electrocardiograph monitors, pulse oximetry, invasive blood pressure monitors, noninvasive blood pressure monitors, temperature monitors, respiratory gas monitors, and electroencephalograph monitors. Other devices utilized by the anesthesiology department include infusion devices and fluid warmers.

Surgical instruments

Surgical instruments are hand-held tools or implements used by clinicians for the performance of surgical tasks. A vast assortment of instruments can be found in an operating suite. Scalpels, forceps, scissors, retractors, and clamps are used extensively. The nature of certain surgical procedures requires a more specialized set of

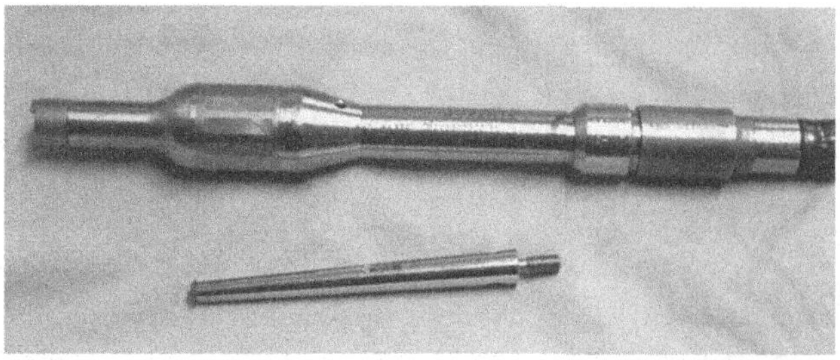

Figure 4.3-5 Surgical drill.

instruments. For example, bone saws, files, drills (Figure 4.3-5), and mallets are commonly utilized in orthopedic surgery. Surgical instruments are generally made of carbon steel, stainless steel, aluminum, or titanium, and are available in a range of sizes.

Electrosurgical units

The electrosurgical unit, or Bovie, is a surgical device used to incise tissue, destroy tissue through desiccation, and to control bleeding (hemostasis) by causing the coagulation of blood. This is accomplished with a high-powered and high-frequency generator that produces a radio frequency (RF) spark between a probe and the surgical site that causes localized heating and damage to the tissue (Gerhard GC, 1988). An electrosurgical generator (Figure 4.3-6) operates in two modes. In the monopolar mode, an active electrode concentrates the current to the surgical site and a dispersive (return) electrode channels the current away from the patient. In the bipolar mode, both the active and return electrodes are located at the surgical site.

Defibrillators

A defibrillator is a medical device that is used to deliver an electrical shock to the heart. The shock is intended to correct irregular electrical activity of the heart and to establish an organized rhythm. A shock of adequate power and duration will cause the cells of the heart to simultaneously repolarize and allow a normal rhythm to return. The defibrillator uses a capacitor to store the required energy, measured in joules (i.e., watts per second), to deliver the shock. A DC power supply charges the capacitor to the selected energy level.

Electrodes are used to deliver the electrical shock to the patient. Electrode types include reusable paddles and adhesive electrodes. External defibrillation is applied to the chest of the patient with external electrodes or paddles. An internal paddle set is used when defibrillation is delivered directly to the heart. Factors governing the set-up and performance of a defibrillator include patient

Figure 4.3-6 Aspen Labs MF 380 electrosurgical unit.

impedance, energy waveform shape, and electrode type and placement.

Most defibrillators are designed with technology to monitor the patient's ECG signal and allow for synchronized cardioversion. Synchronized cardioversion is the delivery of energy to the heart during ventricular depolarization, or upon the detection of the QRS complex. This feature serves to protect the patient by preventing the inadvertent delivery of energy during the ventricular refractory period.

Temperature regulation devices

Temperature monitoring and regulation are crucial to ensure the safety of a surgical patient (Vaughan, 1988). Although heating and air conditioning controls in the OR help to maintain a safe surgical environment, a patient is at risk of suffering from the effects of hyperthermia and, more commonly, hypothermia. Heat loss can be attributed to contact with conductive surfaces, exposed body cavities, cold irrigation solutions, and convective heat loss due to the considerable flow of air in operating suites. In addition, the effects of anesthesia can impair the body's natural mechanisms for maintaining proper temperature.

Several methods are employed to equalize a patient's body temperature. Blankets, sheets, and clothing are common methods of preventing heat loss. A blanket warmer is also utilized to prevent hypothermia by delivering warmed air to an inflatable blanket surrounding the patient. A hyper-hypothermia unit circulates heated or cooled water through a blanket, to raise or lower a patient's temperature. A fluid warmer is a device used for warming intravenous or irrigation fluids prior to contact with the patient.

Headlamps

Headlamps are used to supplement the facility lights in the surgical field. The headlamp apparatus, worn by the clinician, is connected to a light source by a fiber-optic cable. This versatile source of light is particularly useful when overhead lights are insufficient or obstructed.

Tourniquets

A tourniquet is a surgical device that is used primarily to temporarily occlude blood flow to a part of the body and to obtain a nearly bloodless operative field. A pneumatic tourniquet uses pressurized air to restrict blood flow and comprise an inflatable cuff, connective tubing, pressure source, pressure regulator, and a pressure display. Tourniquets are commonly used in amputations and various other orthopedic surgical procedures (see Figure 4.3-7).

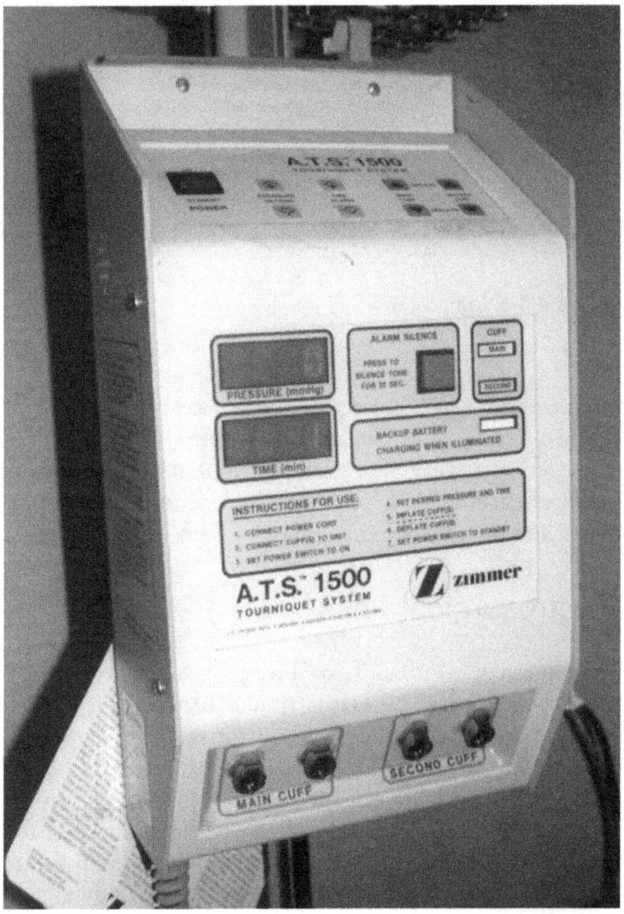

Figure 4.3-7 Zimmer ATS automated tourniquet system.

Specialized operating room technologies

Lasers

A laser is a device that directs an intense beam of radiation to the surgical site to cut, coagulate, or vaporize tissue (Judy, 1995). The major components of a laser are the lasing material, mirrors, a cooling system, an optical or electrical pump source, and a delivery system. The lasing material fixes the output wavelength of the laser and primarily comprises a gas-filled tubular cavity (gas) or a solid-state medium. Lasers with differing lasing material, wavelengths, beam shapes, and guidance method, are used in the OR for various physiological effects. Four laser types that are typically used in the OR to coagulate, ablate, or remove soft tissue are carbon dioxide (CO_2), argon ion (Ar-ion), neodymium-yttrium-aluminum-garnet (Nd:YAG), and gallium-aluminum arsenide (GaAlAs).

Robotics

The use of robotics has provided extensive benefits to many industries, and the medical field is no exception. Specifically, surgery is a field wherein the application of robotics can lead to advantages in providing health care. The inexhaustibility, repeatability, and precision of robotics are favorable in surgery.

Imaging technologies

Body organs, structures, and tissues are studied using imaging techniques. Several technologies are available to the clinician. Hard-tissue imaging is done using X-rays. Soft-tissue contrast and functional information cannot be sought using X-rays. There are many techniques, including computed tomography and magnetic resonance imaging, that are effective in fairly accurate determination of damaged cells or soft tissue.

X-rays

Radiography utilized electromagnetic waves to produce two-dimensional images of anatomy, captured on a photographic film. The intensity of the image on the film is determined by the intensity of the rays emerging from the anatomy. Tissues with different densities will exhibit varying degrees of X-ray absorption. This is the most common form of imaging technique used in clinical practice.

Angiography

Angiography is a radiographic technique where a radio-opaque (i.e., visible on an X-ray) contrast material is injected into a blood vessel for the purpose of identifying its anatomy on X-ray. This technique is used to image arteries in the brain, heart, kidneys, gastrointestinal tract, aorta, neck (carotids), chest, limb, and pulmonary circuit.

Fluoroscopy

Fluoroscopy is a form of diagnostic radiology that enables the radiologist, with the aid of a contrast agent, to visualize an organ or the area of concern in motion, via X-ray. This contrast agent allows the image to be viewed clearly on a television monitor or screen. Contrast agents, also known as "contrast media," can be introduced into the body by injection, swallowing, or enema.

Computed tomography

Computed tomography is a diagnostic technique that uses X-rays to acquire detailed information of soft tissue structures, muscles, bones, and organs. It uses an X-ray source that revolves around the object to be imaged. The detector captures the rays (raw data) that penetrate through the organs. The raw information is processed and reconstructed using a computer algorithm to form images. The images are in the form of cross-sectional slices. This technology is particularly useful for producing images of the brain.

MRI

Magnetic resonance imaging (MRI) is a noninvasive method of imaging structures and soft tissues inside the body. MRI imaging is very detailed and provides a high degree of diagnostic accuracy, as compared to other modes of imaging. It is based on the principle that hydrogen nuclei in a strong magnetic field absorb pulses of radiofrequency energy and emit them as radio waves, which can be reconstructed into computerized images. The images produced are of a high quality and give a good indication of the properties of internal body parts. MRI is widely utilized by many surgical specialties. It is particularly valuable for imaging soft tissue, the brain and spinal cord, joints, and the abdomen.

MRA

Magnetic resonance angiography (MRA) is a noninvasive method of vascular imaging and determination of internal anatomy without injection of contrast media or radiation exposure. The technique is used especially in cerebral angiography, the radiography of the vascular system of the brain, as well as for studies of other vascular structures.

Ultrasound

Ultrasound imaging is technique in which high-frequency sound waves are reflected from internal organs, and the echo pattern is converted into a two-dimensional picture of the structures beneath the transducer.

Minimally invasive surgical devices

For many surgical procedures, the method of choice has shifted from traditional open surgery to the use of less-invasive means. Minimally invasive surgery is performed with the aid of a viewing scope and specially designed surgical instruments (Garrett HMS, 1994). The scope allows the surgeon to perform major surgery through several tiny openings without making a large incision. These minimally invasive alternatives usually result in decreased pain, scarring, and recovery time for the patient, as well as reduced health care costs.

Endoscopy is an examination of the interior organs and body cavities, through a natural body opening or a surgical incision, using a light and a rigid or flexible viewing instrument called an endoscope. The viewing component of an endoscope is made up of hundreds of light-transmitting glass fibers bundled tightly together.

Laparoscopy has become a common surgical technique in the OR (Brooks, 1994). This procedure is the examination of the interior of the abdomen with a slender endoscope, called a laparoscope. The laparoscope is inserted through an incision in the abdominal wall in order to perform surgery. Laparoscopic techniques are performed to remove gall bladders, to perform antireflux operations on the esophagus, and to remove organs such as the adrenal glands. Although minimally invasive surgery is beneficial for the patient, it is technologically more demanding than traditional surgery. Special training is required for clinicians, and the associated medical technology is more advanced. A laparoscopic procedure is usually performed with a mobile cart that is outfitted with equipment for visualization, instruments for exposure and manipulation, and equipment and instruments for cutting and coagulation.

The equipment for viewing internal tissues and organs include a light source with fiber-optics to transmit the

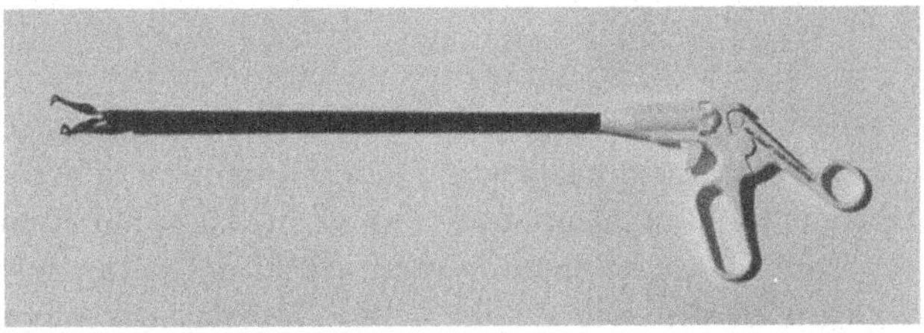

Figure 4.3-8 Endoscopic instrument: Autosuture Endo Babcock graspers.

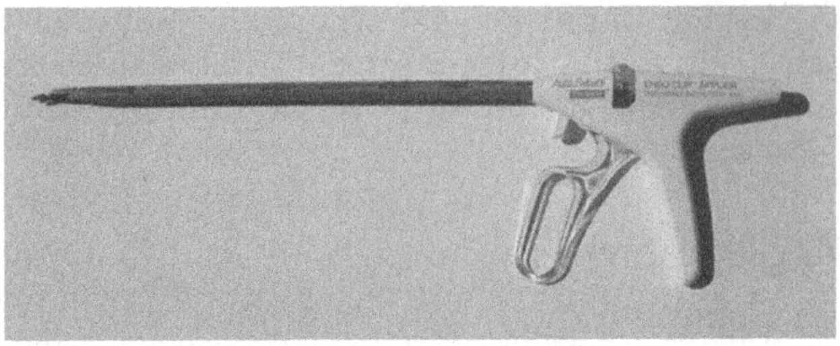

Figure 4.3-9 Endoscopic instrument: Autosuture Endoclip applier.

light, a high-resolution camera that can withstand sterilization, a video processor, a high-resolution monitor(s), a video recorder, and a printer. Several devices are employed for exposure and manipulation in laparoscopic procedures. Trocars are sharp-pointed surgical instruments, used with a cannula to puncture a body cavity and to provide intra-abdominal access. A high-flow carbon dioxide (CO_2) insufflator is used to expand the abdominal cavity, to make internal organs more accessible. Graspers are devices, generally with two movable and serrated jaws, that are used to grasp and retract organs (Figure 4.3-8). Figure 4.3-9 shows an endoscopic instrument for applying clips. Irrigators (or aspirators) are also commonly used. Instruments used for cutting and coagulation in laparoscopic procedures include microscissors, electrocoagulating dissectors and graspers, heater probes, and lasers.

Heart-lung machines (bypass)

A heart-lung machine (Figure 4.3-10) is an apparatus that does the work both of the heart (i.e., pumps blood) and the lungs (i.e., oxygenates the blood) during, for example, open-heart surgery (Galletti and Colton, 1995). The basic function of the machine is to oxygenate the body's venous supply of blood and then to pump it back into the arterial system. Blood returning to the heart is diverted through the machine before returning it to the arterial circulation. Some of the more important components of these machines include pumps, oxygenators, temperature regulators, and filters. The heart-lung machine also provides intracardiac suction, filtration, and temperature control.

Transesophageal echocardiogram

The transesophageal echocardiogram is a device that utilizes ultrasound technology for imaging cardiovascular anatomy and physiology. This device uses an ultrasound probe that is inserted in the esophagus and stomach to image the heart and its associated blood vessels. The internal perspective avoids the interference from the anatomy of the chest encountered with external echocardiograms. Because the esophagus is located just behind the heart this specialized view is clear. The transesophageal echocardiogram uses the same sound-wave technology as a regular echocardiogram. The ultrasound probe emits sound waves of a certain frequency upon the object to be imaged, and the return wave frequency is detected by a transducer. The characteristics of the structure are determined by analyzing the return frequency, and the signals are converted into computerized images.

Transesophageal echocardiograms are frequently used to monitor the heart during surgery. Other common uses include searching for an abnormality in the heart or major blood vessels that could be responsible for causing a stroke; looking for infections on the heart valves; and evaluating the aorta for a tear in its wall.

Cell savers

A cell saver is a device that collects and returns a patient's blood that otherwise would be lost during a surgical

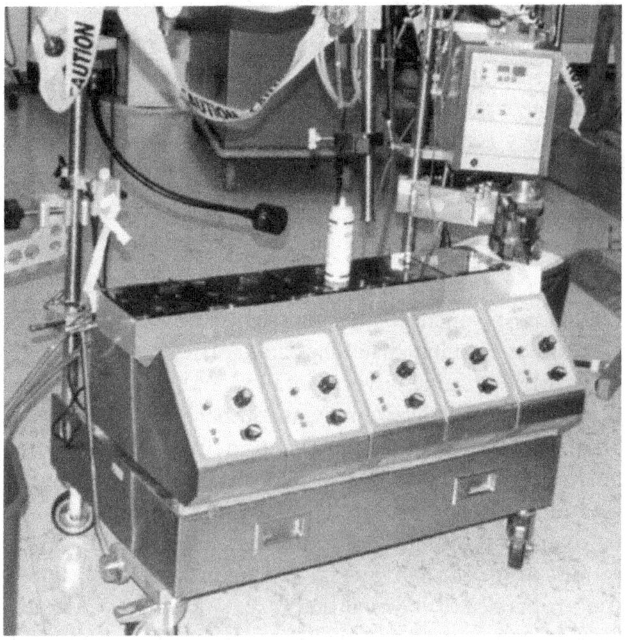

Figure 4.3-10 Stockert-Shiley heart-lung machine under laboratory testing.

procedure. This process is referred to as "auto transfusion." The cell saver receives salvaged blood that has been suctioned from the surgical site. The blood passes through filters to remove surgical and cellular debris. Centrifuge technology is employed to separate the heavier oxygen-carrying red blood cells from the other components of the blood. The red blood cells are processed and stored for auto transfusion.

Pacers and pacemakers

A pacemaker is an electrical pulse generator designed to support the electrical activity of the heart (Greatbatch and Seligman, 1988). A pacemaker can be implanted within a patient or used externally. External temporary pacemakers are commonly used in cardiovascular procedures to treat heart conditions such as bradycardia, atrial and/or ventricular arrhythmias, and cardiac arrest.

The basic pacemaker comprises a pulse generator, a programmer, and a lead. The pulse generator consists of the power source and circuitry, which senses the heart's electrical activity and generates the output. The programmer allows the clinician to adjust pacing variables such as pulse rate, amplitude, duration, and the sensitivity of pulse detection. The lead is an insulated wire that carries the stimulus from the pacemaker to the heart and delivers an **ECG** signal back to the pacemaker.

Ventricular assist devices

The primary aim of a ventricular assist device (VAD) is circulation support. When the myocardium is damaged, the heart is unable to maintain the required cardiac output and blood pressure to maintain blood flow. A VAD is implemented to relieve the workload of the myocardium. Another use of the VAD is ventricular assistance. Patients with heart failure, which can be reversed if the heart is given sufficient time to recover, are candidates for a VAD. Using a VAD is also common after patients undergo cardiopulmonary bypass or any other traumatic heart surgery. A VAD can assist in the recovery of the right or left ventricle (RVAD or LVAD) or both ventricles (BIVAD). An LVAD diverts blood from either the left atrium or left ventricle, sends the blood through a pump, and then returns the blood to the aorta. The RVAD operates in the same manner, with the blood diverted from the right atrium and returned into the pulmonary artery.

Microscopes

Due to the spread of microsurgical procedures through the various surgical disciplines, the surgical microscope has become a valuable tool in the OR. Microscopes are used to provide magnification and illumination in order to view objects during minute and intricate surgeries. These devices are generally employed when performing procedures, such as neurosurgery, where structures are often very small and cannot be viewed clearly with the naked eye. Surgical microscopes generally contain binocular lenses, light sources, optic fiber cables, and focusing mechanisms. Surgical microscopes must be equipped with a mounting or support system in order to be utilized in the surgical field. A variety of support systems are available, including mobile floor stands and ceiling mounts.

Safety in the operating room

The hazards associated with medical devices, clinical processes, and human error are of primary concern to the clinical engineer. A prevalence of medical equipment and staff in the OR, along with the vulnerability of the surgical patient, creates an environment that requires additional precautions. Potential hazards associated with OR devices and technologies are vast and can be encountered in many forms. Medical equipment, for example, can pose electrical and/or mechanical hazards to patients and staff. A patient whose natural resistance to current flow has been compromised (perhaps due to an invasive connection to a medical device) is particularly vulnerable to electrical hazards. Devices that are mechanical in nature, such as transport beds and surgical tables, also pose a risk to patients and staff if they are not used and maintained properly. It is a responsibility of the clinical engineering department to implement a program to ensure medical device safety and to manage risk. The program must include the preventive maintenance and inspection of medical devices and systems. It also must conform to codes standards set forth by regulatory agencies such as NFPA, Underwriters Laboratories (UL), Occupation Safety and Health Act (OSHA), and the Joint Commission on Accreditation of Healthcare Organizations (JCAHO).

Other hazards associated with technology in the OR include environmental hazards, such as the spread of infection due to poor filtration of OR air; biological hazards, resulting from poor sterilization practices; and radiation hazards from diagnostic machines and therapeutic devices that release radiation (Bronzino, 1992).

A lack of knowledge and communication within the clinical staff significantly increases the likelihood of injuries in the OR. Familiarity with medical devices, technical processes, and awareness of emergency procedures are key factors in avoiding injury. To meet these requirements, a clinical engineer must provide, or participate in, the in-servicing of new equipment, continual user training, and incident reporting and evaluation.

Adequate preparation requires that the medical staff be properly trained to respond to the following emergencies:

- External (community) disaster
- Fire
- Facility-system failure (e.g., power, medical gases, communication systems, elevators, sewer, steam supply, water loss or leak, or HVAC)
- Hazardous-material contamination (e.g., chemical, biohazard, or radiation)
- Infection control hazard

Clinical engineering roles

Clinical engineering is the application of technology to improve the quality of health care in health care facilities. It is the responsibility of a clinical engineer to apply engineering principles to understand, develop, control, and maintain medical technologies, systems, and processes. Ensuring the safety of patients and staff is also at the core of a clinical engineer's responsibilities. The role of clinical engineering departments can differ among institutions of varying resources and services. Some clinical engineering departments have an active role in the application of technology for patient care, while others are mainly a "fix-it" shop. Smaller institutions tend to have a centralized clinical engineering department that services the entire facility. On the other hand, larger institutions might allocate groups to specific departments within the hospital.

The OR is particularly challenging to a clinical engineer. It is an area of abundant and complex technologies. In order to provide proper support to the OR, the engineer must understand surgical processes and terminology, in addition to providing technological expertise. Ethics and professionalism are also vital characteristics in this field. Members of a clinical engineering department must be conscious of their impact on the patient care process and overall safety in the OR.

The evaluation and introduction of new technology is a major function of a clinical engineer. The process of medical equipment procurement includes identification of equipment needs, selecting the equipment specifications and vendor, demonstrating the equipment to the users, purchasing the equipment, preparing it for use, staff training, and installation. Medical devices and technologies are provided to promote quality patient care, yet ensure cost effectiveness. Hospitals generally operate with limited funds with a growing demand for medical services. Therefore, it is vital for clinical engineers to plan investments in medical technology carefully and to maintain the equipment following procurement. The clinical engineer also must be mindful of the codes and standards that regulatory agencies have applied to health care technology.

The OR's dependency on technology for patient care demands that medical equipment be reliable and available. The clinical engineer is responsible for ensuring that equipment is properly maintained. Preventative maintenance and scheduled inspections are necessary to minimize malfunctions, to verify functionality, and to prevent device-related injuries. Spare devices and replacement parts should be maintained for all vital equipment.

Inventory records must be kept in order, to manage the assets of the OR. A computer database is an essential tool for managing data such as equipment inventories, parts inventories, and inspection schedules. A device's individual performance and repair histories also should be stored for future reference. Other responsibilities of a clinical engineer can include device design, project management, budgeting, staff training, incident investigation, and facility design.

Conclusion

The OR is a complex department that serves an essential role within a health care facility. This environment comprises a wide range of clinical professionals, medical technologies, and complex systems. The clinical engineer, as a technical expert, is essential to the efficiency and safety of the health care facility. To support the OR effectively, the clinical engineer must maintain a thorough knowledge of medical devices, clinical practices and procedures, facility systems, and safety guidelines.

References

AIA. *Guidelines for Design and Construction of Hospital and Health Care Facilities*. Washington, DC, The American Institute of Architects Press, 1996.

BMET Certification Course. Mercer Island, WA, Morse Medical, Inc.

Bronzino JD. *Management of Medical Technology: A Primer for Clinical Engineers*. Boston, MA, Butterworth-Heinemann, 1992.

Bronzino JD. *The Biomedical Engineering Handbook*. Boca Raton, FL, CRC Press, 1995.

Brooks DC. *Current Techniques in Laparoscopy*. Philadelphia, PA, Current Medicine, 1994.

Calkins JM. Anesthesia Machines. In Bronzino JD (ed). *The Biomedical Engineering Handbook*. Boca Raton, FL, CRC Press, 1988.

De Richmond AL, Bruley ME. Head and Neck Surgical Fires. In Eisele DW (ed). *Complications in Head and Neck Surgery*. St. Louis, MO, Mosby, 1993.

Dorsch JA, Dorsch SE. *Understanding Anesthesia Equipment*. Baltimore, Williams & Wilkins, 1984.

Eichorn JH, Ehrenwerth J. Medical Gases: Storage and Supply. In Ehrenwerth J, Eisenkraft J (eds). *Anesthesia Equipment: Principle and Applications*. St. Louis, MO, Mosby, 1993.

Feinberg BN. *Handbook of Clinical Engineering Volume 1*. Boca Raton, FL, CRC Press, 1980.

Galletti PM, Colton CK. Artificial Lungs and Blood-Gas Exchange Devices. In Bronzino JD (ed). *The Biomedical Engineering Handbook*. Boca Raton, FL, CRC Press, 1995.

Gardner TW. *Health Care Facilities Handbook, 6th Ed.* Quincy, MA, National Fire Protection Association, 1999.

Garrett HMS. *Surgeon's Reference for Minimally Invasive Surgery Products.* Montvale, NJ, Medical Economics Data Production Co.

Gerhard GC. Electrosurgical Unit. In Webster JG (ed). *Encyclopedia of Medical Devices and Instrumentation*. New York, Wiley, 1988.

Greatbatch W, Seligman LJ. Pacemakers. In Webster JG (ed). *Encyclopedia of Medical Devices and Instrumentation*. New York, Wiley, 1988.

Judy MM. Biomedical Lasers. In Bronzino JD (ed). *The Biomedical Engineering Handbook*. Boca Raton, Florida, CRC Press, 1995.

Laufman H. Surgical Management: Developments in Operating Room Design and Instrumentation. In Ray CD (ed). *Medical Engineering Year Book*, Year Book Medical Publishers, 1974.

Chapter 4.4

4.4

Anesthesiology

Luis Melendez and Raj Rane

To excel as a clinical engineer (CE) or biomedical engineering technician supporting, the use of equipment in anesthesia requires equally strong understanding of both the clinical and technological components of the job. This is an important distinction that anyone must face when involved with anesthesia. Physiology and technology meet—the two cannot be separated. If you know the technology well but have less than able comprehension of medical terminology, human anatomy, and physiology, you will have difficulty making efficient decisions to correct a problem in the middle of surgery or to choose the best technology to be used for the next 15 years. This chapter is primarily dedicated to existing and present technology. Future machines are not discussed in detail, as it would be purely speculation as to what might work. Most discussions are limited to principles and not specific detail.

This chapter is best introduced and summarized with a case report. Events similar to the one described below have happened, and will happen, at almost any institution. Suppose that in the middle of a total hip replacement in an operating room (OR) where ultraviolet (UV) lights and laminar flow ventilation systems are used to keep infection rates to a minimum, the CE is asked to diagnose a large leak that has developed in the breathing circuit during the case. The resident and attending physicians are concerned about the course of events, and the CE can feel the tension upon entering the room. Alarms are sounding, and it appears that the doctors are having a difficult time ventilating the patient. The physiological monitor is indicating poor oxygen saturation; it tells the CE that the capnogram shows a poor waveform. The ventilator is flashing a number of alarms. The attending physician wants a new anesthesia machine, now. The CE is told that the patient is paralyzed and that they will use an Ambu bag (i.e., a manually operated resuscitator) while the CE gets that broken machine out of the room.

The CE wants to perform a couple of tests on the machine but is concerned that asking for the time to do it will add more tension. Wanting to solve the problem as fast as possible, the CE does as told. Quickly, the CE is out of the OR and off to find another machine. Meanwhile, they disconnect the patient from the machine, start manual ventilation, and rearrange things in the OR so that the CE can wheel the large device out and roll in another in its place. The CE returns and leaves the new machine just outside of the room, helps the physicians move all items and drugs from the existing machine, disconnects patient monitoring, gas-supply hoses, and electrical connections. The CE wheels the machine out of the room (careful not to touch sterile tables and drapes, or else there would be even more upset people) and moves the new machine in. With great care, things are reconnected, and every detail is put back in its place. The new machine and then the ventilator are turned on. The very same problem is still there. The bellows do not fill properly, even with a fresh gas-flow rate of 10 lpm. Tension is building. The CE has done everything asked of him, and there is nothing else that he can think of. The physicians are growing even more upset. Not wanting to be in the way, the CE decides that it is time to leave.

Back in the clinical engineering department, the CE finds a co-worker and discusses the recent occurrence. The co-worker convinces the CE that they both should go to the OR. The first machine is still in the hallway. They plug it in and turn on the oxygen cylinder. A quick breathing circuit leak test and a functional ventilator test indicate that nothing is wrong with that machine. In the room, everyone is near an uproar because they cannot ventilate adequately. The co-worker asks the physicians to turn off the ventilator and to manually ventilate the patient. They must use the oxygen-flush key just to keep enough volume in the breathing circuit. The CE notices

Biomedical Engineering Desk Reference; ISBN: 9780123746467

that the breathing circuit bag is actually collapsing between breaths. There is something pulling vacuum in the breathing circuit. The CE knows that there is now a different scavenger connected, so that cannot be the problem. Remembering the nasogastric suction catheter (used to empty the esophagus and stomach contents), the co-worker asks for it to be pulled out a little. The physician pulls back on the tube, and everything comes back to normal. The CE is amazed that a doctor missed the cause of such a leak.

Humans make mistakes. Safety is the responsibility of everyone involved. There must be redundancies to help prevent a number of events that would result in an adverse outcome. Multiple factors played into this case. Tunnel vision, a lack of comprehension in the application of technology, simply following orders without thinking or communicating, striving to be the one to solve a problem, and not seeking help from others all contributed to the confusion of both clinical and technical staff while putting the patient at greater risk and increasing the possibility of serious infection.

What is anesthesia?

Pre-operative assessment and plan

Anesthesia is more than the manual labor and technology involved in relieving patients of pain that they would otherwise endure during surgery. To determine whether a patient is ready to undergo anesthesia, anesthesiologists must assess a patient's current state with a physical examination, evaluate lab results and other relevant tests (e.g., EKG and renal and pulmonary function) and medical history while taking into consideration personal preferences, and even religious beliefs, on occasion. A perioperative management plan must be developed. The physician and patient will discuss options and risks of local, regional, and general anesthesia as well as backup plans in case unexpected events occur. In doing so, they must take into account any risks to which the patient will be subject. A neonate with poor lung function, a healthy adult, and a hemodynamically unstable elderly patient all pose different challenges. In anticipating what could happen during the surgical procedure and what the level of difficulty any one individual patient may be, the anesthesiologist must decide among many things, such as level of monitoring, vascular access (replenish fluids and sample blood gases) and airway management.

Amnesia and analgesia

Anesthesia is an induced, controlled state combining amnesia and analgesia during surgery, obstetric, therapeutic, and diagnostic procedures. Overall, the safest approach is to leave the patient as self-sufficient as possible. Under proper management, a patient who is spontaneously ventilating (breathing unassisted) is inherently at less risk than a patient requiring mechanical, or assisted, ventilation. For minor procedures, local anesthesia, as with the use of peripheral nerve blocks, is performed by an anesthesiologist and monitored by less-trained personnel during the procedure. This is known as monitored anesthesia care (MAC) and often involves nurses without extensive anesthesia training. For more invasive cases, spinals and epidurals (regional blocks), combined with sedation, can be used to perform many surgical procedures. There are significant risks involved. Regional anesthesia frequently requires the immediate availability of appropriately trained personnel and general anesthesia equipment as backup. Epidurals can be used for surgical needs and postoperative pain management.

Conscious sedation is often used in intensive care units (ICUs). It can be used as a means to restrict patient movement and to aid relief in highly stressful times. It is also used for imaging child patients or irritated individuals. There is a hazy, but fine, line (often misunderstood) between where conscious sedation ends and general anesthesia begins. Anesthesiologists tend to have a clear understanding of the boundaries but others who are not as experienced in the practice might not.

General anesthesia (GA) affects the entire body, rather than any one specific area. GA shuts down the body's reactions to noxious stimuli most often also involving paralysis. The MAC is the minimum anesthetic agent to prevent 50% of the population from moving because of noxious stimuli (i.e., incision). By definition, one MAC is not enough to treat patients, as half would react to an incision. A value of 1.3 MAC will prevent almost any patient from moving. One MAC varies by intravenous drug and inhalation agent. It is a measuring tool to standardize comparisons. For example, one MAC is accomplished at 1.68% of the volatile liquid anesthetic ethrane and at 1.15% of the anesthetic forane. Because GA affects the entire body, there is total loss of consciousness and the ability to communicate, thus requiring physiological monitoring to best determine the patient's condition. During GA, the patient is often paralyzed to the point where autonomic control of the diaphragm is shut down, thus requiring the use of positive-pressure (mechanical) ventilation and related monitoring.

Critical care

Many anesthesia departments are deeply rooted and involved in the day-to-day operations of ICUs. Training in anesthesia is a solid foundation for management of critically ill patients. Patient care during surgery is, in many ways, similar to that of other critical care areas, but changes happen in terms of minutes or hours, versus days or weeks. Because of their specialty and experience, anesthesiologists must understand intravenous administration of rapidly acting drugs, fluids, and volume control. Critical care calls for in-depth knowledge of mechanical ventilation, airway management, and cardiovascular monitoring, including invasive pressure lines of hemodynamically unstable patients. Anesthesiologists are also responsible for cardiac and pulmonary resuscitation (code response) in the ORs and ICUs.

Pain management

Anesthesia departments are also responsible for pain management during diagnostic, therapeutic, and obstetric procedures. Pain units that are staffed by anesthesiologists diagnose and treat painful syndromes when patients suffer from chronic issues. Obstetric floors often have anesthesiologists present, to help reduce the pain involved with labor and delivery.

Considering all of this, do not overestimate anesthesiologists. They are not superhuman. No individual will master all of these subspecialties. As a CE or biomedical equipment technician (BMET) supporting the needs of the department and its staff, it helps to understand the breadth of responsibilities and expectations of the physicians as a whole.

Where is anesthesia performed?

ORs, treatment rooms, and intervention radiology suites are beginning to look more similar than ever. Historical functions of the different areas and physician specialties are blending, making formerly clear lines quite hazy. Patient flow through the hospital's multifaceted care units must be carefully planned. Preparing and recovering a surgical day-care patient (rather than one from general surgery) will put significantly different demands on support systems and staffing. The hospital cannot afford bottlenecks that result with unnecessary delays in surgical, imaging, or other expensive care areas.

For various reasons, doctors and patients alike express interest in providing anesthesia at remote locations (out of the OR) of the hospital. This poses significant risks and unique challenges in planning, supporting, and actually meeting the demand. These remote sites most often were not designed with anesthetizing-location-facility requirements in mind. Details of facility criteria are discussed in Chapter 4.3. Reliable supplies of oxygen and suction, along with adequate electrical power and lighting, are key components. The lack of any of these can negate further discussion. If facility requirements are met, but physical space is an issue, the anesthesiologist, machine, and supplies are often crowded into an uncomfortable corner. From a user's perspective, they are away from the space to which they have grown accustomed. People must be extra vigilant when they are out of their normal element, as supplies and other things are not where they are expected to be. More importantly, support personnel are not immediately available to help in emergencies, as in they are in the OR. One example is the need of imaging pediatric patients. Small children tend not to stay still enough for adequate results, and therefore need sedation. Unless properly planned, pediatric supplies are not commonly found throughout the hospital. Delays in procuring critical care items could lead to disastrous results.

The special needs of magnetic resonance imaging (MRI) and radiation oncology (e.g., proton beam accelerators) present unique challenges to all involved in their use. The most significant are magnets and invisible forces. Both technologies involve strong forces that cannot be seen, and catastrophic events can result if mistakes occur. Specialized equipment is needed, to detect the energy field generated by both types of devices.

MRI requires highly specialized equipment: Non-ferrous magnetic materials can be used, CRTs are a problem, radio frequency (RF) sensitive environment, physiological monitors are less than optimum, the same level of monitoring is simply not available, application of MRI is expanding and it is being used more frequently during surgery. Radiation oncology requires leaving anesthetized patients by themselves, which is not something that people take lightly. Anesthesiologists still need to be able to monitor the patient and machine while they are out of the room.

Safety concerns

The single most significant difference in supporting anesthesia technology, as distinguished from any other clinical engineering function, is that there is a susceptible patient connected to life-support equipment, and they are given medications that bring them relatively close to death. Anesthesiologists are keenly aware of this. As a whole, stress levels of people in ORs are exacerbated because the patient has undergone two forms of injury when on an OR table. They have suffered the initial illness or trauma that has brought them to the OR, and they must undergo the trauma of the surgical procedure itself. It can be difficult to believe that the patient often must experience additional trauma in order to get better.

Patient safety is a team effort. No single department bears the responsibility for overall patient safety. Everyone must evaluate their individual responsibilities and must look for ways to prevent and improve their environment, to prevent mishaps from occurring. One example of a relatively simple sounding task, but often complicated by various factors for the OR team, is patient positioning. A number of injuries have occurred because of positioning. Different departments must take into account their own prospective needs. Surgeons must have access to certain areas for sterile prep and the surgical site. Nurses must make sure that the items for which they are responsible are accessible and do not cause potential pressure points, while anesthesiologists are looking out for potential nerve injuries. Because the patient is unconscious, he is unable to tell them that the position is uncomfortable, and staying that way for 6 hours can hurt.

Anesthesia safety is frequently compared to flying an airplane. On airplanes, most of the problems (and, therefore, most hazards) are associated with take-off and landing. In anesthesia, this is equivalent to induction and emergence. Both have multiple systems working in unison to maintain function. On a commercial plane, fuel is delivered precisely to engines that are attached via structural members to a fuselage, with a crew using RF communication and a global positioning system to move the plane from one location to another. During anesthesia, oxygen is mixed carefully with an inhaled anesthetic agent and delivered through a breathing circuit driven by a ventilator, controlled by a physician who watches a display to monitor electrical and physical changes in a patient who cannot communicate. As long as things are going well, everything appears to be relatively simple.

Unfortunately, things do not always go as planned. A number of factors can contribute to undesired events during anesthesia, including noise, fatigue, and boredom. Studies have shown that anesthesiologists can be idle 40–70% of the time during a surgical procedure (Drui et al., 1973; Boquet et al., 1980), which could further affect their vigilance at their primary responsibility of monitoring the patient, procedure, and equipment. When something out of the ordinary occurs and takes anesthesiologists off their planned course, it can lead to moments of high activity, where many things must be accomplished in little time, potentially in a state of concern. These periods are often associated with high task density, where people average less time on individual tasks in contrast with the time spent during less busy periods (Herndon et al., 1991). In both aviation and anesthesia, there are lengthy, intensive training periods and highly educated and skilled personnel. Each field relies on a person who has a widely varying workload, to maintain order. It does not take much time for one or two additional missed warning signs to result with significant demands on the individual at the controls. For a pilot or an anesthesiologist, an uneventful day is a good day. No one likes unpleasant surprises on an airplane or in anesthesia.

Redundant systems help to prevent surprises. Studies in the aviation industry show that adverse outcomes frequently happen when a number of undetected smaller events occur, involving different factors (e.g., human error, equipment failure, and supply mishaps) cumulating in an undesired result (Billings and Reynard, 1984). There always should be additional resources for people to turn to for help. This can be difficult for some, because they may feel that asking for help shows a weakness and that they should be able to figure it out on their own, lest they be seen as less than fully capable. Patient care environments must try to foster a setting in which looking for help when an individual is not fully comfortable with a condition is a perfectly respectable option and not cause for punitive action. Solutions can be relatively simple but not visible when an individual is fixated on something else. Another viewpoint might be all that is needed. Physicians frequently have enough to worry about and might not always see the solution; this state of affairs can be more prevalent at teaching institutions. CEs can back physicians up by understanding the demands placed on them and on the clinical environment.

To minimize the possibility of equipment failures during use, the aviation industry implemented a preflight inspection. Similarly, those who are concerned with anesthesia safety developed the United States Food and Drug Administration (FDA) Anesthesia Apparatus Check-out Recommendations (FDA, 1993). A team of people of varied backgrounds and interests created the FDA procedure. When followed correctly, this comprehensive procedure can identify almost any problem. Although it is a simple procedure, it does require the user to know proper technique. It can be simple to make a mistake (e.g., negative pressure leak test with the machine turned on) that results in false positives or negatives. Most anyone with a reasonable understanding of an anesthesia machine can perform the checkout. A technician or engineer with more in-depth knowledge of the components that are actually being tested can use it as a useful troubleshooting tool to easily identify system failures within the machine.

The aviation industry has used simulators for years. Their use is expanding medicine to put physicians in stressful situations without putting patients at risk. People need to be taught crisis management. One needs to conceptualize and understand how to work in a team when encountering stressful and challenging situations. The human mind does not function rationally when in a panic.

Safety at off-site locations is worth noting. A number of undesired events have prompted review of office-based anesthesia. The associated risk versus cost is an example of the many challenges faced in health care, particularly at large teaching hospitals. In an effort to stay price-competitive with the entire medical industry, procedures are being performed out of the OR and in doctors' offices. No one should ignore the fact that infrequent events with potentially catastrophic outcomes do happen. Key components can be as simple as proper and adequate supplies, additional personnel, and available telephones. Other components depend on the procedure and level of anesthesia performed. Even though some procedures appear to be relatively simple, the requirements for anesthesia can call for a machine to be present at all times. The emergent need for mechanical ventilation and use of volatile agents is quite possible. These machines might be underutilized, but they are subject to greater wear because they are moved more frequently than machines based in the OR.

There are supply concerns, as well. Additional stock is not as readily available when users are not in the main supply area, which normally is the OR. If items stored differently, individuals are out of their normal surrounding, and this contributes to disorientation and difficulty locating the items. Delays in supplies can have catastrophic results, as well. If the correct endotracheal tube is not readily available, minutes might as well be hours when there is a problem with an airway.

Infrastructure

Smith *et al.* described facilities' infrastructural needs (see Chapter 4.3). Much of an OR's evolution has involved the application of anesthesia and some of its historical limitations. Conductive flooring, other electrostatic discharge protection, and explosion-proof electrical connections were directly and indirectly employed because of former anesthetic technology. Currently, there is ongoing debate about the continued use of the isolated power supply (IPS) in new construction. The use of ground fault interrupts (GFIs) meets code requirements for wet locations at a lower cost than isolated power. Some facilities elect to continue with the more expensive option because it does not shut off power to an outlet at the first fault. If their application is poorly understood, GFIs could be hazardous when life-support equipment such as anesthesia or cardiopulmonary bypass machines is used. Shutting of life support equipment when one plugs a faulty device into an outlet on the same circuit is not a wise alternative.

Following are a number of other systems that are needed, to support specific needs for anesthesia:

- Immediate access to the hospital's pharmacy and a means to control narcotics with potentially high throughput for busy centers
- A well-organized means to purchase, receive, and stock single-use supplies
- Facilities and associated documentation to reprocess, decontaminate, and sterilize multiple-use devices
- Centralized storage of equipment that is not used everyday but must be available
- An area for preadmissions testing
- Space for an interview and simple testing

Patients and physicians need to interact preoperatively to assess and plan the anesthesia. This may also be done in general-care areas if the patient has been admitted in the hospital.

Induction (or holding) rooms offer a number of advantages and disadvantages. They require more patient movement and additional equipment but can reduce turnover time and can increase throughput. Primarily, they provide a location to help set up lines and epidurals without occupying time in the OR itself. They are no longer true induction rooms, as anesthesia is not induced in them, but they are helpful prep areas.

In the ORs, anesthesia machines must connect to existing infrastructure for gas supplies. It is useful to have standardized connections so that machines can be used in any location with adequate infrastructure. Individual surgical anesthetizing teams might have varying needs and preferences for machine and monitoring layout, depending on the surgical service that they support. There also are specialized sites like radiation oncology and MRI that will require the machine to be tailored to meet their requirements.

Post-anesthesia care units (PACUs) have established clinical personnel requirements for emerging patients (ASA, 1994). Patients can be unstable; so immediate critical care, airway management, and possibly anesthesia equipment must be available. Their workload varies. At one moment, the space could be used primarily for pre-op holding; at other times, things can be relatively calm. Then, suddenly, three or four cases finish at once. The PACU is an ICU and must have similar monitoring requirements with centralized alarms. It must accommodate varying demands for stretchers used in patient transport.

Airway management tools

One of the most critical tasks in caring for the ventilated patient is in securing an airway. If there is not a reliable means of moving gases in and out of the lungs, cardiac resuscitation follows respiratory arrest. Among the many tools available to anesthesiologists, none is more important for airway management than suction. Reliable and adequate sources of oxygen and suction are vital. Anesthesia cannot be performed safely without both. Yankaur catheters are the most typical device used to aspirate airways. Nasal-gastro tubes are used to empty stomach contents that otherwise could interfere with the airway.

Probably the device with which most people are familiar is a facemask. Masks are relatively simple devices that come in numerous sizes, both disposable and reusable. They can leak if not fitted correctly, and they require a hand or strap to hold them in place. Their greatest drawback is that they do not prevent possible aspiration of stomach contents. They are not the best option for longer cases, and excessive pressure can cause physical injury to the patient.

To use most any other airway management tool requires direct visualization with a laryngoscope. There are standard airway classifications, depending on patient anatomy. Laryngoscopes are available in various configurations to best meet the needs of different airway anatomy. Common blades that are used to obtain direct visualization of the vocal cords are straight, straight with curved tip, or curved (Jackson-Wisconsin, Miller, and MacIntosh, respectively). When patient anatomy or trauma is such that use of a laryngoscope is difficult or impossible, a fiber optic scope is used to help intubate the patient. The two services that most often require these tools are thoracic and plastic surgery. Thoracic teams use them to visualize airways more easily, evaluate tube placement, and aspirate secretions. Reconstructive plastic surgery requires the use of fiber optic scopes, as a significant percentage of these patients have disfigurement or trauma that has altered normal anatomy. Video equipment can help teach their proper use by enabling two people to visualize the same image at once.

Endotracheal (ET) tubes are the most common item used to maintain an airway. They are available in numerous sizes and in cuffed (Figure 4.4-1) and uncuffed configurations, although cuffed tubes are more common. They have a balloon-like outer section at the distal tip that inflates to seal with the inner walls of the trachea to prevent leaks and inhalation of gastric contents or other secretions. They are nearly always used on adults, and they pose other potential problems if the patient is intubated for periods over 48 hours. Uncuffed tubes do not put pressure on the inside of the trachea that can be more problematic with pediatric patients but can contribute to airway leaks. The most frequent problem associated with intubation is a sore throat from the pressure exerted on the inner tracheal mucosa. There are specialized ET tubes with two lumens used most frequently during lung surgery, enabling ventilation of one lung or both. Because they are in the immediate surgical vicinity, these tubes are subject to greater external forces and, therefore, are often reinforced. Another option is the laryngeal mask airway (LMA). Because of its seal design, its use is limited to ventilation pressures of about 20 cm H_2O, and it does not prevent aspiration of gastric contents. The LMA is most efficient in environments where surgical procedures are generally short, and it is a helpful tool for emergent needs.

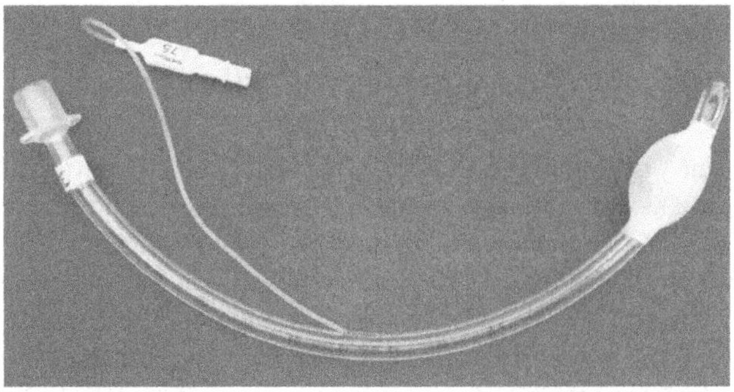

Figure 4.4-1 Cuffed endotracheal tube.

Services that pose unique challenges are pediatrics and oral surgery. Children are smaller, potentially making tasks more challenging. In oral surgery, scavenging can be challenging because surgery takes place in the immediate area where gases are flowing.

Anesthesia machines

Anesthesia machines (see Figure 4.4-2) are constructed of a number of systems assembled as one device. There are standards developed by the American Society for Testing and Materials (ASTM) (ASTM, 1989) for many of the subassemblies used on or with the machines. Its major systems can be broken down to gas delivery (frequently referred to as the "machine," itself), vaporizer(s), breathing circuit, ventilator (including related monitoring), physiological and CO_2 and agent monitors. One standard does not cover all aspects of the machine; for example, there are standards for machine, ventilator, oxygen monitor, and breathing circuit. Unfortunately, they can be vague and interpreted in different ways, making them somewhat difficult to read and understand.

Gas supplies

The machine's primary function is to reduce supply-line pressures, mix a number of gases (most typically oxygen, nitrous oxide, and air), and deliver a controlled output to the breathing circuit. Primary gas supplies feed the machine 50 psi. A pressure-relief valve opens above 75 psi in case of infrastructure system failure. Technicians and engineers need to be familiar with a number of pressure-measurement units. The most common are pounds per square inch, millimeter of mercury, and centimeters of water (psi, mmHg, and cmH_2O, respectively). A rough equivalent is that one psi is about 50

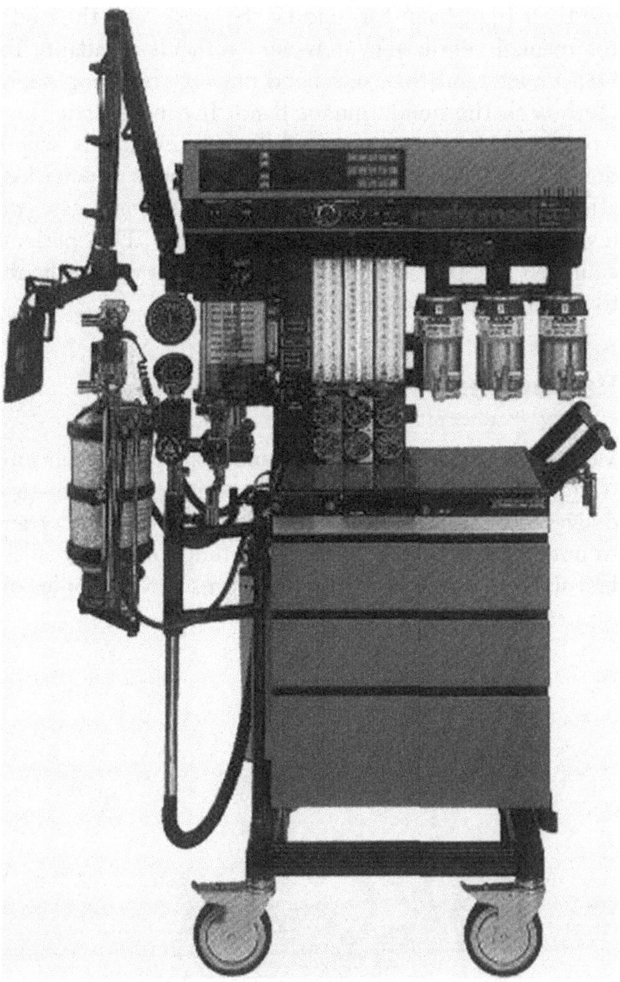

Figure 4.4-2 Anesthesia machine.

mmHg and about 70 cmH_2O. In the same units of measure, an anesthesia machine needs to safely reduce and control gases fed at 3500 cmH_2O and to supply them to the breathing circuit normally operating at about 35 cmH_2O. In other words, the supply pressures are 100 times that of the breathing circuit. Machines are constructed with a high-pressure side and a low-pressure side. The high-pressure side is primarily the supply, and the low-pressure side is any part operating near breathing-circuit pressure.

There are a number of items with designed incompatibility in anesthesia machines. Pipeline gas-supply connections are available in a few different configurations. The diameter indexing safety system (DISS) is a common configuration. The thread and inner diameters vary to prevent accidental connections to a wrong supply. Virtually all machines use E cylinders as backup supplies. Backup oxygen is vital for patient safety and must always be available. Cylinders use the pin index safety system (PISS) to prevent wrong connections. Holes in cylinders work in conjunction with pins in the yokes. For anesthesia-supply gases (oxygen, nitrous oxide, and air), one pin location is fixed, and the other varies depending on gas type. One inherent weakness in this system is that if the wrong pin pulls out or breaks, it could lead to a misconnection.

Oxygen pressure detecting system and distribution

To detect the presence of oxygen pressure, there is a pressure-controlled, normally off, valve called the "fail safe," which is intended to protect the patient from a hypoxic mixture. Its function is to shut off secondary and tertiary gases in case of oxygen loss. The name is a misnomer because the device does fail. It is ineffective when other gases are delivered in the oxygen pipeline or when flow-control valves or hypoxic mixture interlock systems are out of adjustment. Machines also have a pressure-sensing alarm that indicates when supply pressure has fallen below a threshold to inform the user to take evasive action (e.g., to turn on cylinder supply).

Oxygen-pipeline supply is plumbed to at least five locations at full pressure within the machine. The oxygen-flush valve is always active and capable of delivering 50–65 lpm. It incorporates a safety-pressure relief at about 120 mmHg for catastrophic failures that far exceed normal breathing-circuit pressures. The flush valve is dangerous to the patient in the hands of an uninformed user. Activating the flush for 1 second will increase the tidal volume delivered by 1 liter, which is easily enough to cause barotrauma to a susceptible patient. The fail-safe and pressure-sensing alarms, as described above, require line pressure to operate correctly. Flowmeters combined with the flow-control valves feed the oxygen supply to the patient via vaporizer(s) and breathing circuit. They are the interfaces between the high- and low-pressure sides of the machine. Ventilators need a working gas. The United States has standardized on oxygen, but other parts of the world could use air. Auxiliary flow meter and other power outlets are fed from accessory connections.

The anesthesia machine has two limits for the minimum oxygen-flow rate. One is fixed and sets at the absolute minimum deliverable, while the other varies in proportion to the nitrous oxide flow-rate setting. Most machines are designed for a minimum of 200 ml, but some go as low as 50 ml. This is intended to reduce the possibility of a hypoxic mixture because there is always the minimum patient-uptake requirement supplied. An average adult will metabolize about 2–300 ml/min of oxygen.

Nitrous oxide flow control is linked directly or indirectly to the flow of oxygen, also to prevent delivery of a hypoxic mixture. This introduces significant associated construction costs and complexity. As technology and practice evolve, people are questioning the long-term use of nitrous oxide and wonder whether it still should be made available in all machines. To compensate for potential supply pressure changes that could alter flow settings from one machine to the next connected to a common supply line, some manufactures use two-stage pressure reduction to minimize the effects.

Gas mixing

Traditionally constructed (mechanical) machines use a needle valve combined with a rotameter for each supply gas to control flow. The rotameters all feed a common output, thereby regulating gas composition fed to the vaporizers. After mixing with the inhalation agent delivered by the vaporizer, the combined mixture is called the "fresh gas flow" (FGF), delivered at the common gas outlet. Some electronic machines have settings for total gas flow and oxygen percentage that control gas-flow rates independently.

There is a check valve built into the common gas outlet on Datex-Ohmeda machines to prevent breathing-circuit pressures from back-pressurizing the vaporizers. This valve requires a negative pressure, to open and correctly perform a leak test on the low-pressure system. If a positive-pressure test is performed, it will not open this valve, and the machine will not be tested properly. Drager machines do not have this check valve.

Machine and space layout

Much of the machine consists simply of shelf space, essentially a convenient location to store monitoring equipment and supplies needed when performing anesthesia. A working surface and tabletop space is an important location to place medications and other items needed within hands' reach of the physician. There is also a chest of drawers to store supplies and other items needed for use with the machine.

People adjust the layout of almost anything they use regularly. They become accustomed to what they have, and grow to like where things are positioned. If the things with which one interfaces change radically, it affects one's comfort level. One is more likely to make mistakes by turning the wrong knob, leading into additional confusion because of rituals and habits of looking for things in one location. Consistency, particularly at large institutions, has its advantages and disadvantages.

The physical space layout and the way the machine is positioned in the room can be significant. If the user is forced, due to an awkward machine orientation, to use their right hand for holding the mask, and their left for manual ventilation, it is very difficult to adapt. To help understand this, one need only try brushing one's teeth with the nondominant hand. It can be done, but would one want a person's life in one's hands when doing it the first few times? The patient's orientation affects access to the endotracheal tube and ease of inspecting lines, leads, and electrodes. The patient could be prepped and draped so that it is very difficult to reach almost anything.

Vaporizers

Vaporizers (Figure 4.4-3) add and control the concentration of volatile inhalation anesthetic agent in the gas delivered to the machine's common gas outlet. In order to understand the way a vaporizer functions, the reader first must be familiar with a few terms and principles of

Figure 4.4-3 Vaporizer.

fluid dynamics. A fluid is anything (liquid, gas, or both) that takes the shape of its container. Materials have three phases: gas, liquid, and solid. Vapor is the gaseous phase of a liquid at room temperature and atmospheric pressure. Pressure, volume, and temperature are related. A gas is fully saturated when it contains the maximum amount of vapor possible without precipitating out to a liquid. If two or more gases are mixed in one container, the total pressure exerted is made up of the sum of partial pressures created by each individual gas.

Partial pressure

An understanding of partial pressure is required, to support anesthesia technology effectively. Partial pressures are an absolute measurement defining the total number of molecules where percentages are relative to total gas mixture. Partial pressure is used and applied in a number of technologies within anesthesia (e.g., vaporization, ventilation, and respiratory gas monitoring). One example to help clarify the term is dry (versus moist) air. Air is made of 21% oxygen. The remainder is mostly nitrogen and trace gases (negligible partial pressure for this example). Atmospheric pressure is approximately 760 mmHg. On a dry, warm day (98°F or 37°C) of 0% relative humidity, oxygen partial pressure is 21% of 760 (0.21×760), or 159.6 mmHg. The remaining pressure must be from nitrogen (760–159.6), or 600.4 mmHg. Then, on the following day, the temperature and atmospheric pressure are the same as the day before, but the relative humidity is 100% (the air is fully saturated with water vapor and cannot contain any more without rain-out). At 37°C, fully saturated water-vapor pressure is 47 mmHg. On the second day, atmospheric pressure contains three gases (nitrogen, oxygen, and water vapor). Oxygen concentration remains constant at 21%, but its partial pressure is a function of the three gases [$0.21 \times (760\text{–}47)$] or 149.7 mmHg. The partial pressure of nitrogen is $0.79 \times (760\text{–}47)$ mmHg, or 563.3 mmHg.

Because volatile inhalation agents have varying partial pressures that affect concentration output, modern vaporizers are made agent-specific. The most common agents currently in use are forane, ultane and suprane (manufacturer trade names are isoflurane, sevoflurane and desflurane, respectively). Vaporizers are available as either funnel- or key-filled. Funnel-filled vaporizers can be more convenient and reduce the possibility of vapor lock when filling, but they are susceptible to being filled with the wrong agent. Key-filled vaporizers are more cumbersome and require use of a filler, but they virtually eliminate the possibility of cross-contamination because the agent is not poured directly into the vaporizer as in the case of funnel-filled units.

Operating principles

Side-stream

All of the mixed gas flows through the inside of a side-stream vaporizer. The main stream of this gas (when flowing through a vaporizer it is also known as the "carrier gas") has a fraction diverted to the wick/sump assembly. The amount of flow of this side-stream diverted to the wick assembly is controlled by the setting on the vaporizer output dial. The higher the output setting, the greater the amount diverted. This side-stream flow becomes fully saturated with anesthetic vapor in the wick assembly and then returns to the mainstream controlling agent concentration output in the fresh gas flow.

Two physical principles cool this style of vaporizer, requiring temperature compensation. Forced convection (flow of gas) and latent heat of vaporization (energy required to vaporize a liquid) cool the vaporizer while in use. Cooling an anesthetic agent reduces its partial pressure, which in turn lowers the vaporizer's output. To compensate for this cooling, vaporizers incorporate a bimetallic temperature-sensitive diverter that adjusts the amount of flow fed to the side-stream from the main carrier gas. As temperature drops, the side-stream of gas increases to maintain a constant output over varying temperature during normal operation. Side-stream vaporizers, which are relatively simple in design, are the most commonly used and are available for all of the currently used agents, with the exception of Suprane.

Mainstream injection

Suprane is a volatile agent with high partial pressure that makes it difficult to deliver with a traditional side-stream vaporizer. It can be delivered only using a mainstream injection vaporizer. Suprane vaporizers are heated and pressurized to control more precisely the environment in which is it stored when in use. Mainstream vaporizers control output by injecting measured amounts of agent directly into the main stream of carrier gas.

Future machines might incorporate new designs that are more closely related to the mainstream vaporizers for other agents. Vaporizers are also used on heart-lung machines to control the anesthetic agent delivered through the oxygenator during cardiopulmonary bypass surgery.

Breathing circuits

Breathing circuits serve as the interface between patient and machine. Because people breathe in volumetric flow and pressure cycles and anesthesia machines deliver a unidirectional stream of gas (at a specific oxygen concentration and controlled anesthetic agent), an interface between a person and the anesthesia machine is required. Breathing circuits convert the machine's steady gas output to a flow and pressure cycle that is consonant with the human breathing cycle.

In keeping with the Ideal Gas Law, $PV = nRT$, to maintain a constant baseline pressure, one must also maintain a constant baseline volume. If there is excessive (or a growing) volume in a closed system, pressure will decrease. Respiratory baseline pressure is measured at end-tidal expiration when there is no longer significant flow. This measurement is known as "positive end-tidal expiratory pressure" (PEEP). A second ventilation indicator is "peak inspiratory pressure" (PIP), the maximum pressure attained during the inspiratory cycle.

When a patient is paralyzed to the point where they no longer spontaneously ventilate, positive-pressure ventilation is used. This involves delivering pressurized tidal volumes of gas to the patient's lungs for oxygen uptake supporting their required metabolism. Under normal circumstances (no leaks), an anesthesia machine's gas output has two places to go: Patient uptake or its scavenging system.

A primary component of breathing circuits is the tubing and its configuration to the patient. Some breathing circuits rely on high fresh gas flow rates to prevent rebreathing of respiratory gases while others involve absorbent to neutralize expired carbon dioxide. A thorough discussion of all the breathing circuit (open, semi-open, semi-closed, and closed) options is too great for this text. The two most common circuits used in the Unites States, the Bain Circuit and Circle System, are described below.

Bain circuit

A Bain Circuit is a semi-open breathing circuit that does not recirculate respiratory gases and relies on high fresh gas flow rates to prevent rebreathing. It is most often used for neonatal and thoracic applications. Its greatest advantages are that it creates little dead space, it is lightweight, and gases can be scavenged easily. The most significant drawbacks to this circuit are that it requires high fresh gas flow (delivering cold dry gases to the patient) and can be cost-prohibitive with costly volatile agents.

Circle system

By far, the most common is the circle system, where respiratory gases move through a housing incorporating unidirectional valves to recirculate gases, minimizing waste and maximizing warmth and humidity in the circuit. The system's greatest drawback is its level of complexity. However, nearly all anesthesia machines are designed to operate with a circle system.

Carbon dioxide absorbers

Recirculating breathing circuit gases require the patient's expired carbon dioxide to be removed, to prevent hypercarbia. This is accomplished by means of a CO_2 absorbent contained in a housing, called an "absorber." The breathing circuit absorber must allow the use of both automated (machine ventilator) and manual (breathing circuit bag) ventilation. Most present-day absorbers are stand-alone subassemblies that tie into related functions of the machine. They are modular and can be exchanged easily if needed. They are the union between patient and machine and are where respiratory gases mix with fresh gas flow from the machine. Delivered oxygen concentration is measured in the absorber's inspiratory limb that is most proximal to the patient and in a relatively safe location where it is unlikely to be damaged or disconnected. One advantage of modern absorbers is that they measure PIP after the inspiratory check valve. Older style absorbers require the use of a second gauge when a PEEP valve is used for clinical reasons (discussed below). Gages on older designs only display interior absorber pressure. When the inspiratory check valve closes, they do not display true airway pressure and do not reflect PEEP.

In an effort to address some of the limitations of the relatively simplistic time-cycled, volume-controlled ventilators that are typically found in anesthesia, manufacturers are introducing a new generation of absorbers that are more fully integrated into the machine. This design approach has advantages and disadvantages. New designs can be sterilized. While not required in the United States, parts of the European Community require that breathing circuit components be sterilized between patients.

One significant point to consider is that absorbers require daily service and therefore have greater potential for problems such as leaks. Although they are designed with this in mind, personnel who are not technically trained might partially disassemble the absorber. The system is subject to repeated changes of disposable breathing circuit tubing, spills, daily cleaning, and absorbent changes, and is otherwise exposed to a demanding environment.

Absorbent

Various commercially available absorbents are used to absorb carbon dioxide. For practical purposes, the CE and BMET must be aware and concerned with the absorbent's systematic application and its use. By-products of the chemical reactions between the absorbent and CO_2 are beneficial to the patient and to the proper function of the absorber. However, one reaction might prove to be a hindrance in excess. Removal of carbon dioxide is clearly beneficial and a primary function of the absorbent. In the process, a chemical reaction produces a color change in the absorbent, which acts as a visual indicator of the absorbent's activity. The color of absorbent changes from off-white to violet as its ability to absorb carbon dioxide diminishes. Under certain, but rare, conditions, the color change reverts to the original. A more common problem is that users overlook, or forget to notice, the color change. In addition, optical properties of the housing can change, thus making the color change difficult to see. Channeling occurs when respiratory gases take the path of least resistance and channel through a relatively narrow cross-section. This can result in unwanted inspired carbon dioxide.

Heat and water are by-products of CO_2 absorption. The heat generated is not excessive and is beneficial to the patient because it helps to warm cool supply gases, which could irritate the lung lining. Water produced is also beneficial because supply gases are dry, and added moisture reduces lung irritation. If water production is neglected, it will build to the point where it could contribute to problems in the breathing circuit. The most frequently noted problems are sticking expiratory check valves from surface tension between valve disc and housing, and water accumulation in hoses and tubing to the ventilator, affecting proper function by inhibiting proper pressure sensing and gas flow.

Gas scavengers

When the machine is set for an FGF rate of anything higher than the patient's uptake (normally 200–300 ml of oxygen per minute for most adults), any excess must be removed to avoid a build-up of volume as discussed above. Although PEEP is used for certain clinical indications, its uncontrolled application is disastrous for the patient. The scavenging system plays a vital role in conjunction with the breathing circuit to prevent this from happening. During automated ventilation, the bellows pop-off valve opens at the end of tidal expiration to divert excess volume to the scavenging system. During manual ventilation, the user must open and close (as necessary) an adjustable pressure-limiting (APL) valve to set the upper limit for breathing-circuit pressures. If the APL valve is open, or any pressure occurs in excess of its setting, gas flows to the scavenger. The physical construction of scavengers used on the machines varies and can require user interaction. The various designs offer their own risks and benefits; some put greater demands on the hospital infrastructure (vacuum pumps), while others require the user to make adjustments according to FGF rate.

PEEP valves

During spontaneous ventilation, the body maintains a residual volume in its lungs to prevent them from collapse. During positive-pressure ventilation, the ventilator maintains a slight PEEP (2–3 cm H_2O) to accomplish the same. Clinical indications (e.g., adult respiratory distress syndrome [ARDS]), PEEP valves are used to increase the resistance that a patient encounters during exhalation, increasing the residual volume in their lungs. Different models of PEEP valves can be permanently installed either on the machine/absorber or temporarily on the expiratory limb. For the latter, the design is a variation of a check valve, making it dangerous if installed backwards or in the inspiratory limb, as it would result in little gas flow and inadequate oxygenation (Figure 4.4-4).

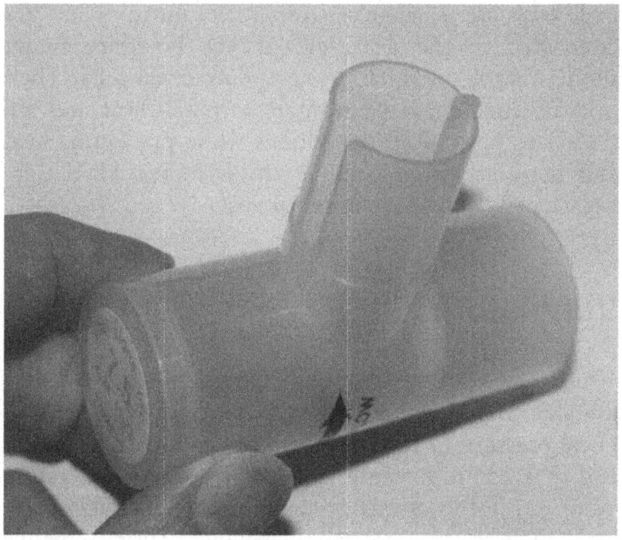

Figure 4.4-4 Example of a PEEP valve.

Humidification

As discussed previously, supply gases are dry and can irritate the lung's lining, particularly with susceptible patients who are on the machine for extended periods. Although the use of active humidifiers has declined over the years, they may still be used for certain patients. Two techniques are used to maintain moisture in the gas that the patient breathes: the heated active humidifier and the heat moisture exchanger (HME).

Active heated units are generally more effective in adding both humidity and heat to the breathing circuit. The operator must connect the unit to an electrical power source, must monitor its temperature, and must clean and fill them. Many units incorporate check valves to prevent inadvertent delivery of fluid to the patient if they are installed incorrectly. Backward installation can result in significant flow restriction and inadequate oxygenation. These units are installed in the inspiratory limb and have a predetermined "in" and "out."

HMEs are single-use, disposable units. They are connected to the distal end of the breathing-circuit tubing that is proximal to the endotracheal tube. Their primary function is to maintain as much heat and humidity as possible in the patient while preventing heat and humidity from entering the absorber. HMEs are occasionally known as "artificial noses."

Ventilators

Ventilators (Figure 4.4-5) free the users' hands so they are able to perform other tasks. It is a way to apply cyclical pressure to the equivalent of the breathing circuit bag at a controlled rate and frequency. An expiratory valve inside the ventilator closes during the inspiratory cycle to direct drive gas into the bellows housing, creating a positive pressure forcing breathing circuit gases to flow. At the end of inspiratory cycle, the expiratory valve opens, releasing drive gas from the bellows housing, and returns the patient to atmospheric pressure. The working gas that is used to drive the bellows varies by country, as discussed previously. Comparable components between manual and automatic ventilation are the bag and bellows and APL and bellows pop-off valve. The bag and bellows offer a means to buffer a volume as gases move in the breathing circuit. The APL and bellows pop-off valves control removal of excess gas from the breathing circuit to the scavenging system, as discussed previously. Both valves require a small amount of PEEP for preferential flow into the bellows (rather than the scavenger) but the bellows pop-off is set a fraction higher, at about 2–3 cm H_2O.

Pressure versus volume-controlled

Historically, anesthesia ventilators have been relatively simple, time-cycled, and volume-controlled which might or might not meet the changing health care environment. Patient care can be classified as acute or noncritical. Acute care is a growing population at larger teaching hospitals, as sicker patients are moved out of smaller community hospitals by modern health management organizations. This patient population is challenging to manage and requires equipment that is more sophisticated. Noncritical care machines should be simple to use, with relatively little user-interface complexity.

One example of the limitations of traditionally simplistic anesthesia ventilators is in the function of the bellows pop-off valve. It closes completely during inspiration so that FGF mixes directly with the set tidal volume delivered. Changes in FGF will affect actual volume delivered as a function of inspiratory time; the greater the flow, the larger the tidal volume without making any changes to settings. In response to the growing market of acute care ventilated patients, manufactures have begun to incorporate more complex pressure-support/controlled ventilators into their anesthesia machines.

Airway pressure monitoring

Most volume-controlled ventilators are pressure-limited and will not deliver settings if pressure limit is triggered. Some have an adjustable pressure limit, and others are preset. The transducer, which senses patient breathing-circuit pressures, is physically located in the ventilator fed from a tube taped into the breathing circuit. As

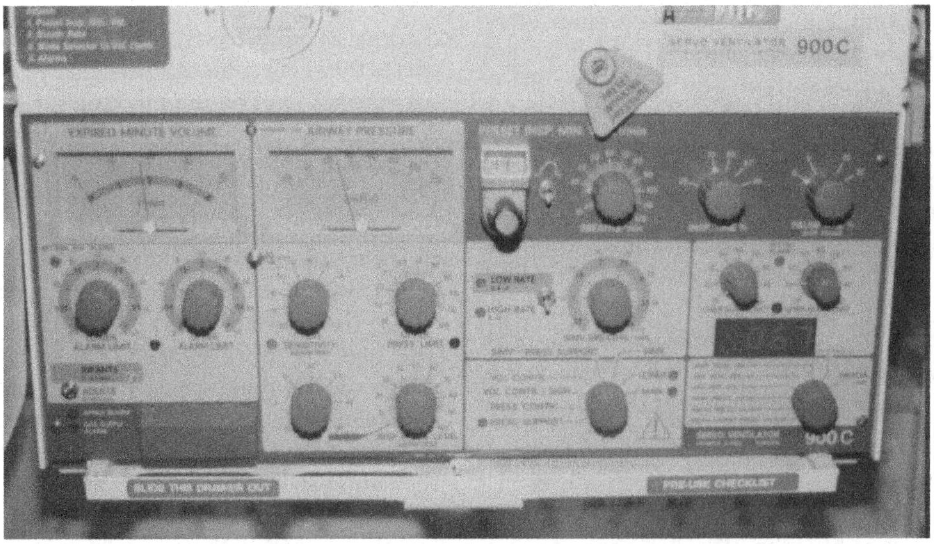

Figure 4.4-5 Siemens-Elema 900C ventilator.

mentioned above, this should be in the inspiratory limb after the check valve, but its location can vary depending on machine vintage (Figure 4.4-5).

Oxygen concentration

The ASTM standards require an oxygen concentration measured in the inspiratory limb that turns on with the machine and has a battery backup. Most utilize galvanic fuel-cell technology, which involves a chemical reaction between an electrolyte and two poles (anode and cathode), similar to those of a battery. The reaction produces an electrical potential relative to the oxygen concentration that is measured and displayed. It has a relatively long time constant and cannot provide breath-to-breath oxygen concentration.

Paramagnetic technology is incorporated in some CO_2 and agent monitors. It has a fast enough response to overcome the limitation of galvanic fuel cells but is more expensive. Using a switched magnetic field, it exploits oxygen's natural magnetic properties to create a pressure differential between gas streams (sample and reference), which is detected and displayed as a concentration (Ehrenwerth and Eisenkraft, 1993).

Volume measurements

Given the limitations of relatively simplistic volume-controlled ventilators, volume measurement is the most accurate device currently in use to provide users with feedback of what tidal and minute volume is actually delivered to the patient. Because of compliance of gas, breathing circuit tubing, and additional dead space in the circuit, the measurement is close, but not exact. A vane anemometer measures a volume of gas by means of monitoring the mechanical rotation of a vane or blade. These devices can be either analog or digital. Technology in newer machines includes ultrasound using Doppler acoustical properties between two points to measure gas flow. Hot wire systems measure energy requirements and changes of a heated wire within a housing as gas volume cools the surface. Pitot tubes in opposing directions working in conjunction with differential pressure transducers are also used to measure flow.

Associated alarms

Many of these measurements are, primarily, a means to inform the user of the possibility of a problem, such as disconnect or leak. Oxygen monitors identify potential hypoxic mixtures and catastrophic events and cannot be set below 18% without alarming. Pressure alarms can be used to indicate significant changes in the patient's condition or restriction in the airway. Subatmospheric alarms indicate the possibility of an active vacuum in the airway, or whether the patient is drawing a negative pressure by fighting the ventilator.

Monitoring

The American Society of Anesthesiologists (ASA) (ASA, 1998) has set minimum requirements for monitoring during anesthesia. Depending on the patient's acuity and other factors, such as the surgical procedure and anesthetic plan, the level of monitoring chosen by the anesthesiologist can exceed minimum requirements to include invasive pressures (arterial, pulmonary artery, and central venous), neuromuscular blockade, and consciousness.

The first requirement of the ASA monitoring standard is "qualified anesthesia personnel.... " The care provider is ultimately the most important monitor and must be able to concentrate on the patient without considerable distraction. Secondarily, but no less significantly, the standards specify monitoring for oxygenation, ventilation, circulation, and temperature. Continuous physiological monitoring is used during anesthesia, utilizing plethysmographs, electrocardiograms (ECGs), and oscillometric devices. Figure 4.4-6 shows a pulse oximeter used for monitoring the patient arterial blood-oxygen saturation, Sa_4O_2.

Studies have shown that although a large percentage of mishaps during anesthesia were due to human error, a significant number were related to equipment failures related to leaks, misconnects, gas-flow-control errors, and circuit disconnects (Cooper *et al.*, 1984). Two of the ventilation monitors (oxygen and volume) were discussed previously. The ASA standard requires that there be a device to detect a breathing-circuit disconnect and that it must have an audible alarm. Along with the use of pulse oximetry, the ASA specifies that there be adequate lighting and exposure to assess patient color and proper oxygenation. This serves as a reminder that one cannot rely solely on monitors. Taking care of a patient is ultimately the responsibility of the provider, not of the monitor.

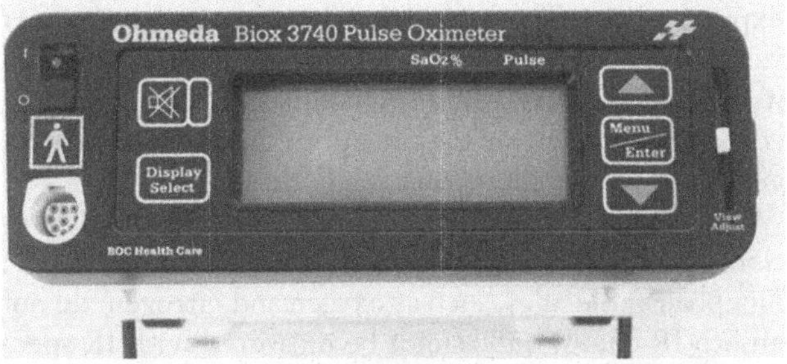

Figure 4.4-6 Ohmeda Biox model 3740 pulse oximeter.

Capnometry and agent analysis

ASA standards originally stated that capnometers would be used to verify the presence of carbon dioxide to ensure tracheal, versus esophageal, intubation. The latest revision specifies continuous CO_2 measurement. There are clinical indications that can be detected using capnometers. Apnea caused by disconnects, ventilator failure, and complete obstruction in the breathing circuit or the scavenging system can be detected, possibly early enough before arterial blood-oxygen desaturation occurs. CO_2 waveforms during controlled ventilation display characteristics that help the physician to identify potential patient-management issues. An abnormally slow rise can indicate a restricted airway or kinked endotracheal tube, while a baseline drift can indicate a faulty expiratory limb valve, consumed CO_2 absorbent, or channeling. BMETs and CEs should learn typical and atypical capnograms and their causes. This knowledge assists in field troubleshooting and in communication, especially when a physician states that the CO_2 waveform looks peculiar.

Monitors can be divided into two categories: side stream or mainstream measurement devices. Side stream monitors remove a small sample from the breathing-circuit gases and delivers it to a measurement chamber. The effluent can be either scavenged or returned to the expiratory limb to be recirculated. Mainstream monitors measure the patient's expired carbon dioxide concentration through an optical window in a tube connected to the breathing circuit, most often at the end of the endotracheal tube.

There are a number of different technologies employed in side-stream monitors, including mass spectrometry, infrared (IR) analyzers (single and dual beam), and Raman spectroscopy. The advantages of a side stream monitor are the simplicity of items connecting to the breathing circuit and the ability to read multiple gases. The ancillary items do not need to be reprocessed, are disposable (thus minimizing risk for cross-contamination), do not need optical properties or cleanliness, and therefore are easier to use on a daily basis. The components added to the breathing circuit are physically smaller—an added advantage when working with limited access and sterile drapes. Their largest drawback is in working with expired water. The sampling tube can condense water vapor, and if the monitor is unable to clear the droplets on its own, it requires user interaction. Many monitors incorporate a water trap that must be emptied on occasion, and hydrophobic filters that can occlude.

Molecular asymmetry is required for IR light absorption, resulting in vibration or rotation of dipole bonds. Nonpolar molecules, such as oxygen and nitrogen, do not absorb IR energy and cannot be measured with IR spectroscopy. Absorption correlates with the number of molecules, so it is an absolute, and not a relative, measurement. Carbon dioxide absorbs IR light at about 4.3 μm, nitrous oxide (N_2O) about 4.5 μm, and volatile anesthetic agents range from 9–12 μm. Because CO_2 and N_2O absorb IR closely to one another, there is signal bleeding between the two. As a result, some CO_2 monitors need N_2O compensation when calibrated. In the monitor, the sampled patient gas runs through water-permeable tubes to dry the moist sample prior to delivery to the measurement bench. Water vapor absorbs IR at the same spectrum and therefore is a contaminant for CO_2 measurements. To measure different gases, dual-beam infrared spectrometers use a spinning wheel with band pass filters to tune its spectrum for the particular gas of interest for maximum signal and a reference. The reference is a zero or minimal response point to which to compare the measurement. Absorption is detected and displayed as a waveform and number.

Volatile anesthetic agents can be measured using infrared spectroscopy and have individual signatures at different wavelengths. Some monitors require simple software updates to read newer agents, while others require more extensive replacement of hardware. Unfortunately, there also are models that are cost-prohibitive to update.

End-user calibration of side-stream monitors involves a zero and span setting. Newer-style monitors have automatic zeroing where room air sample is taken internally. Older monitors require the person completing the calibration to remove the sampling line from the breathing circuit, and then to complete a sequence of actions (i.e., turn knobs or activate keys). Span requires the person calibrating to use a known calibration gas that can be specific to the model of monitor. The calibration gas is sprayed into the sampling port, and the upper measurement range is established either by software or physically setting a potentiometer.

Existing mainstream monitors are more common for use outside of anesthesia due to technology constraints as they are limited to measuring only one gas. They are frequently found as options on physiological monitors.

Temperature

Temperature is the last physiological parameter mentioned in the ASA monitoring standard. Patients undergoing anesthesia frequently experience hypothermia caused by the mechanics of anesthesia and skin exposure to the cold environment of an OR. Induction of anesthesia suppresses the body's ability to regulate core temperature at the most fundamental state. An individual who is awake will make behavioral changes (e.g., in terms of dress or shivering) when sensing a change in environmental temperature. The hypothalamus responds to temperature changes and induces vasodilation or restriction as necessary to control and redistribute blood volume regulating core and peripheral body temperatures. Most volatile agents impair vasoconstriction, potentially altering blood-volume distribution. Greater blood volume in extremities increases heat loss, thus lowering core body temperature. Muscle relaxants and anesthesia otherwise clearly inhibit shivering and reduce heat production by resting muscles (Morley-Forster, 1986).

Accidental hyperthermia during anesthesia is not common in the United States, as most all ORs are air-conditioned. Hyperthermia is more common in tropical climates, in ORs without air conditioning. Malignant hyperthermia can be detected by monitoring the patient's temperature, but an increase in expired carbon dioxide is an earlier indicator.

Thermistors are most commonly used to measure the patient's temperature. Their use poses little risk to the patient; they are reliable and relatively simple to use, and they provide an accurate measurement when used correctly. The most frequent measurement location is nasopharyngeal. The most common problems are cooling from respiratory gases from leaking and misplacement. These devices generally consist of probes that connect to a temperature-sensing device, either a stand-alone box or integrated into a cardiovascular monitor. Thermistors are also used in pulmonary artery catheters, which are used for invasive pressure monitoring and provide a core-temperature measurement.

Infrared tympanic membrane measurement can provide an accurate core temperature but is susceptible to user error. An infrared scanning device with disposable protective cap is inserted in the ear canal, where the energy radiated from the tympanic membrane is measured, converted, and displayed as a temperature.

Liquid crystal thermometers are available to measure skin temperature. However, because skin temperature does not correlate well with core temperature, their use is not practical in anesthesia.

Peripheral nerve stimulators

Peripheral nerve stimulators are used to provide an indicator of neuromuscular blockade or otherwise general paralysis. During anesthesia, muscle relaxants are most frequently used to relieve natural muscle tension to facilitate intubation and surgery. In an ICU, muscle relaxants can be used to aid the use of mechanical ventilation. The drugs used are characterized as depolarizing or nondepolarizing. They work by blocking neuromuscular transmission, thereby reducing the muscle's ability to contract, and paralyzing the patient. Paralysis can be achieved also at high doses of volatile anesthetic agents. One important distinction is that a patient can be paralyzed but may not necessarily be anesthetized. Failure to render the patient unconscious results in a most undesirable situation in which the patient can experience excruciating pain of surgery yet, because of paralysis, is unable to cry out or otherwise signal to the surgeon.

Peripheral nerve stimulators (also known as "twitch monitors") are relatively simple pulse generators. Most stimulators are battery-powered devices that connect to the patient by means of a cable and electrodes. Most often, the electrodes are similar to those used for ECG monitoring but can be specific for the application. Most monitors provide information to the physician by pure observation. The user must feel and observe the patient's reaction to the electrical stimuli. There are units available that can monitor the patient's reaction electrically and/or mechanically and can display patient characteristics (electronic or paper chart recorders). These units are used less regularly, often for teaching purposes.

Most frequently, the electrodes are placed over the ulnar nerve near the wrist. The location preference is primarily related to muscle innervation. The ulnar nerve solely innervates the adductor pollicis muscle. Stimulating the ulnar nerve minimizes the possibility of crosstalk with other muscles, allowing the user to monitor for a response only on the patient's thumb. Less frequently, the electrodes can be placed on a facial nerve if access to extremities is difficult.

Twitch monitors generate a single-phase DC pulse, normally of a fixed duration and adjustable current. Preprogrammed units are available that deliver pulses in sets to help the physician to identify the state of neuromuscular blockade. The most common pulse sets are:

- Single twitch—upon demand, every 10 seconds or 1 per second
- Train of four—a set of four pulses at half-second intervals or can be set to repeat every 12 seconds
- Double burst—a set of three pulses, 20 milliseconds apart, followed by another similar burst 750 milliseconds later
- Tetanus—repetitive, single pulses at 50 Hz or greater

References

ASA. Standards for Basic Anesthetic Monitoring. Park Ridge, IL, American Society of Anesthesiologists, 1998.

ASA. Standards for Postanesthesia Care. Park Ridge, IL, American Society of Anesthesiologists, 1994.

ASTM. Minimum Performance and Safety Requirements for Components and Systems of Anesthesia Gas Machines, F1161-88. Philadelphia, PA, American Society for Testing and Materials, 1989.

Billings C, Reynard W. Human Factors in Aircraft Incidents: Results of a Seven-Year Study, *Aviat Space Environ Med* 55:960–965, 1984.

Boquet G, Bushman JA, Davenport HT. The Anesthesia Machine: A Study of Function and Design, *Br J Anaesth* 52: 61–67, 1980.

Cooper JB, Newbower RS, Kitz RJ. An Analysis of Major Errors and Equipment Failures in Anesthesia Management: Considerations for Prevention and Detection. *Anesthesiology* 60:34–42, 1984.

Drui AB, Behm RJ, Martin WE. Predesign Investigation of the Anesthesia Operational Environment. *Anesth Analg* 52:584–591, 1973.

Ehrenwerth J, Eisenkraft J, eds. *Anesthesia Equipment Principles and Applications.* St. Louis, MO, Mosby, 1993. FDA. Anesthesia Apparatus Check-Out Recommendations. Rockville, MD, United States Food and Drug Administration, 1993.

FDA. Anesthesia Apparatus Check-Out Recommendations. Rockville, MD, United States Food and Drug Administration, 1993.

Herndon O, Weinger M, Paulus M, et-al. Analysis of the Task of Administering Anesthesia: Additional Objective Measures. *Anesthesiology* 75:A487, 1991.

Morley-Forster PK. Unintentional Hypothermia in the Operating Room, *Can Anaesth Soc J* 33:515–527, 1986.

Chapter 4.5

4.5

Simulation facility design

Michael Seropian

4.5.1 The virtual hospital – a virtual fantasy?

I once was involved in a role-play. It went something like this. The hospital CEO walks in to announce that a major donor wants the hospital to build a simulation center. The monies are for start-up and construction but cannot be used for ongoing costs. The room is filled with excitement. Surgery has wanted to enter into this domain and sees the opportunity to have the center accredited by the American College of Surgeons. Nursing has long seen simulation as a mainstay for quality, workforce issues, and cost savings. Anesthesiology, Emergency Medicine, Obstetrics, and the Intensive care representatives are all equally enthusiastic to develop simulation programs for team, crisis management, and skills training. The CEO announces that the donor wants the hospital to build a simulation center that is a virtual hospital. It should look and feel like a hospital. Eyes around the table are big and clearly eager and pleased.

This scenario sounds great. Many consider it the utopian solution. Who wouldn't want a simulation facility that looked exactly like a hospital? The reality is that the notion of a virtual hospital may actually be candy for the mind. It is sweet and desirable. It is however a solution that is likely reserved for those with not only unlimited start-up funds but deep pockets for ongoing costs. The reality is that hospitals are built the way they are to deliver patient care as their primary focus. A hospital environment is filled with distraction and people with cross-purposes. Although dealing with this may be a course in itself, it is just a small part of the totality of clinical education. Education is a secondary or even tertiary focus in a hospital. On the other hand, a simulation center's primary focus must be education. The educational principles and environment must be optimized and must be pedagogically sound. Indeed, classrooms are built the way they are for a reason. The same consideration applies to the construction of a simulation center or facility. Although this point may seem subtle, it has profound implications. Mixing of students and participants undergoing distinct and different simulation experiences can be inherently distracting and artificial. They are in the same relative space for artificial reasons. This is potentially disruptive to the education process. As with most educational activities, the concept of contextual isolation should be a top priority. The notion that people need to learn in an environment that is filled with distractions is fallacious.

4.5.2 Design and build for the actual use

Unless one is able to design a virtual hospital that achieves true separation of activity, the risk of disruption of the immersive experience inherent to simulation is possible if not likely. As people exit a high-fidelity mannequin-based simulation, they are often still "in the moment" as they move to debriefing. Being distracted by passing colleagues having just completed a VR training session is not pedagogically desirable. So what is the answer? There is not one answer. It is important for people developing centers to carefully assess what is desired versus what is needed. They may not be the same thing. The seduction of reproducing the environment to the exacting detail is potent but likely a waste of money. As one considers what and how to build a center, it is important to consider how the activity (especially if it is immersive) allows a certain amount of leeway not possible in a hospital. Once a participant is immersed, do they really look at the wall details? Do the wall details immerse them or does the interactivity

Biomedical Engineering Desk Reference; ISBN: 9780123746467

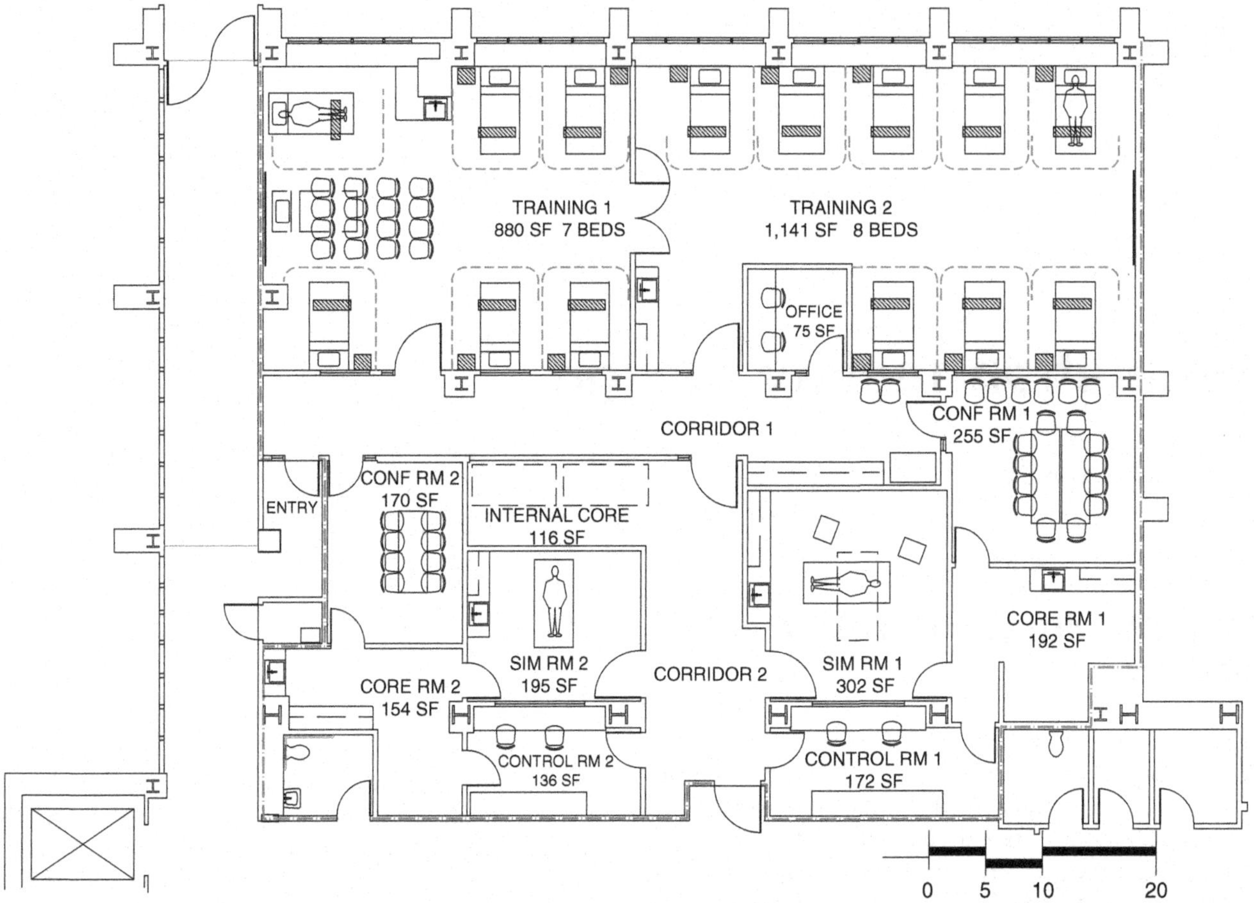

Figure 4.5-1 Floor plan for a 5000-square-foot simulation center. The apartment concept (also known as a suite) is composed of four distinct functional subsections: simulation, control, core, and conference. There are two such apartments shown, numbered 1 and 2, located in each lower quadrant. Look at # 1. Notice how the conference room is in the far corner, away from noise sources, like conversations in the control room. Trace the typical footpath of the students and their instructors. Locate # 2. Notice how the two control rooms are just a quiet shout away from each other. Note: Permission to use Figure 4.5-1 has been granted by Dahanukar Brandes Architects, Samuel Merritt College, and SimHealth Consultants LLC. The schematic was developed by Dahanukar Brandes Architects and SimHealth Consultants LLC (facility design consultants) for Samuel Merritt College in Oakland, California. All rights and copyright remain the property of Dahanukar Brandes Architects and SimHealth Consultants LLC.

of the whole environment achieve that? Some centers have made good attempts at creating hybrid versions of virtual hospitals and educational environments. These centers are worth looking at, not to replicate, but to be able to learn in which direction should the hybridization favor.

Should space be designed so specifically that its utility becomes limited to a specific activity or discipline alone? If a space is created to closely resemble an operating room (OR), then it may become difficult to use that space for anything else. Utilization of the space is therefore limited to the demand created by select specialties. The room lacks flexibility to be appealing to other disciplines and specialties. The common response to such a comment is "we can just put partitions up to hide the elements that label the room." The counter response to this is "why create the detail in the first place if you will ultimately want to hide it?" It is important to recognize that the amount of detail relates to the role it plays in the suspension of disbelief.

This chapter is intended to reveal issues rather than explain what your simulation center should look like. However, one floor plan that does meet most of one program's needs is offered as an example (Figure 4.5-1). The design of each center is unique but certain guiding principles around function, structure, and need are common in most instances. The topics that will be covered include:

- The center design team
- Function, flow, and utilization
- Simulation type
- Sound
- Gases
- Lighting
- Electrical
- Ventilation

A flow chart provides a graphical representation of the iterative process, that is, creation of a simulation program and its supporting facilities (Figure 4.5-2).

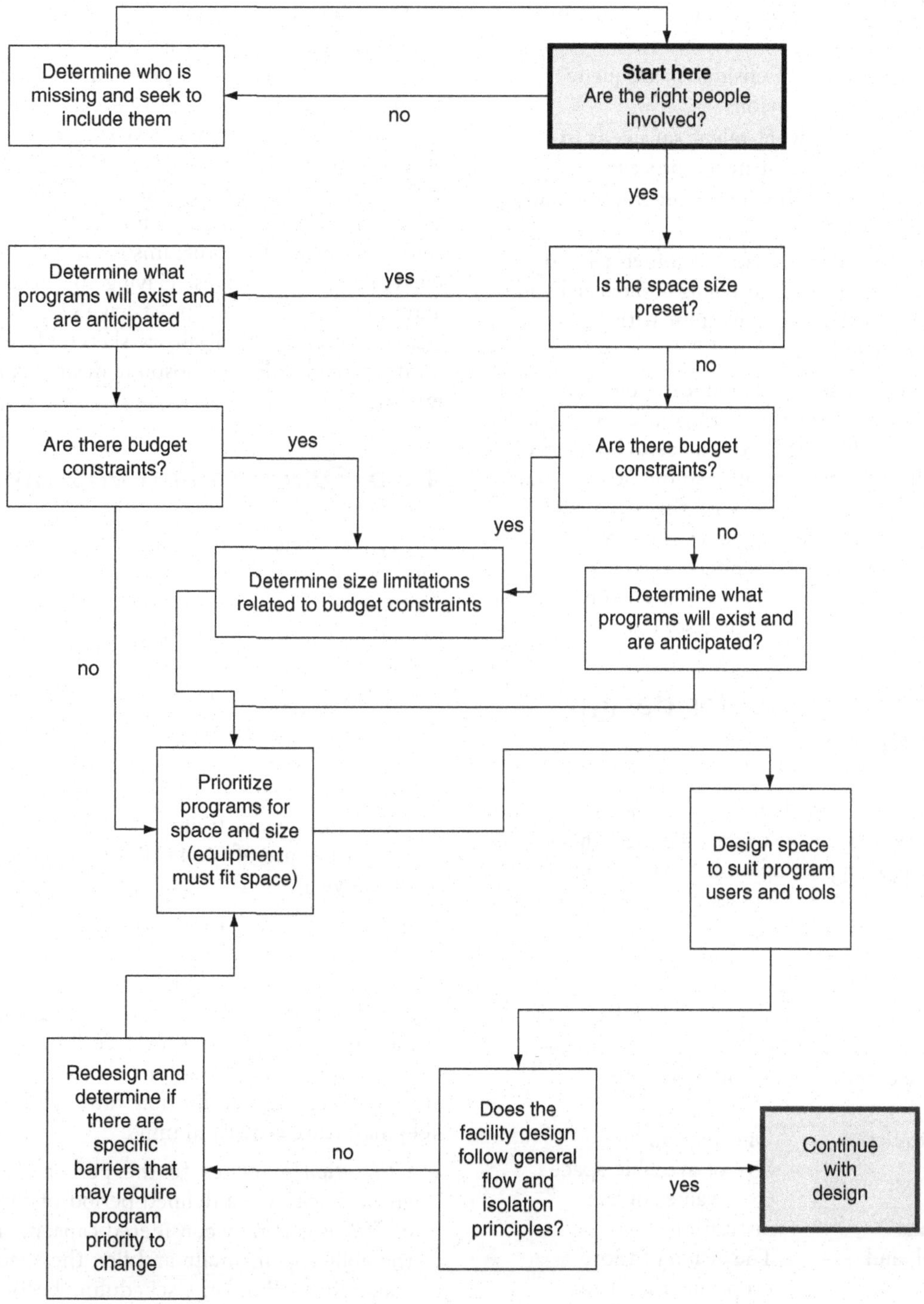

Figure 4.5-2 Flow chart: getting started on simulation facility design.

4.5.3 The center design team

The team that designs a facility should include an architect, the customer, the project manager, the contractor, AV and IT professionals, a simulation facility design consultant, and select faculty who will use the facility. The criterion of what makes a simulation design consultant skilled needs to be further matured and standardized. Some qualities include:

- Prior simulation experience
- Prior simulation facility construction experience
- Good spatial skills and qualities
- Ability to convert ideas into images

- Ability to convert customer vision into functional images
- Ability to preempt and consider consequences and limitations of design decisions
- Ability to understand what talent set needs to be involved and at what point in the process
- Ability to assess and understand construction and architectural drawings
- Ability to understand and guide audiovisual and software professionals in designing audiovisual and storage solutions that are consistent with the intended vision and application.

The team should be assembled or at least considered early on. The criteria for contracting with contractors or architects should not rely on their experience to build medical facilities but their ability to create a studio that will mimic one. The different components of the team will have relative importance at different times. The constant team members, however, are the architect, the customer, and the design consultant. Together, this core team will help the customer realize the vision they set out with.

4.5.4 Simulation center design considerations

The notion that simulation facility design differs from a hospital or classroom is just coming of age. Indeed, we see these concepts appearing on listservs. As alluded to earlier, the design of a simulation facility is not the recreation of a hospital but a studio designed for education that should provide adequate fidelity to immerse individuals to believe they are delivering patient care. In the case of skill trainers, the area should similarly embrace principles that provide an efficient and conducive learning environment. Center design should consider:

- Type of simulation
- Flow
- Budget
- Future plans
- Type, level, and discipline of participants
- Ongoing funding
- Size of available space
- Core values of the institution
- The vision of those proposing the facility

Missing just one of these elements could result in cost to the customer and limitations that were not previously foreseen. There are several examples of this throughout the world.

I often start the design of a facility with a blank sheet of paper. The outline of the space is drafted onto the paper and the interior only includes structures that are immovable or structural in nature (e.g., electrical panels and support columns, respectively). In creating this skeleton outline, the absence of any existing walls (in the case of a remodel) does not bias the creative process. Prior to any design work, detailed program information from the customer needs to be gathered. It is not uncommon that they may not have the program information straight away but this is where the facility design consultant can facilitate by asking contextually relevant questions. The facility is defined by the programs that reside within it. The programs are in turn defined by the curricula and objectives. Having this basic information and knowing what the "must-have's" are allows the designer to create a preliminary sketch. The sketch is not arbitrary and calls on hospital design only at specific junctures.

4.5.5 Function and utilization

The *type of simulation* to be done has a heavy influence on the design of a facility. The concepts for a VR training room versus a computer-based area versus a high-fidelity mannequin-based area are all different and unique. Having determined what programs will exist in a defined space, the design team can move to flow and volume.

- How many people will use each area in any given period of time?
- What concurrent sessions are likely to happen?
- What are the constraints of the users that may weight the use of the facility to a particular time of the year?

Knowing the *intended use* of the facility and the *relative utilization* requirements will help define the space requirements. In some instances, the space is predefined and will therefore be the constraining factor that will influence type and utilization (see flow chart). In either case, it is optimal to aim for a facility that is *functional* and achieves high utilization rates (70–75%) – while keeping future growth in mind.

- *Utilization* – relates to the percentage use of the given space over a defined period of time. A rate of 75% is generally considered capacity. Beyond this, the ability to maintain and flex the space becomes constrained. Similarly, scheduling becomes cumbersome and complex. The utilization rate is not only determined by the number of participants but also by the type of simulation. Scenario-based mannequin simulation may have a lower throughput of people per unit space per unit of time. The exercise of determining how many participants will likely be coming through the center per year (and for what activity), after 3–4 years of operations, is very valuable. In doing this, one can determine how many same use areas are needed and how much overall space is required (if space is not a constraining issue). If a space is to be used with a ratio of one participant per hour, then that would be dramatically different than a space that is used by three to four participants per hour. This is important for the design team and will also be beneficial for the customer in justifying space allocation to their respective executive.

- *Functionality* – relates to the ability of the space to comfortably achieve the desired function and to allow for good flow dynamics within the specified space. This includes the ability to give easy access and egress, easy access to washroom facilities, easy access to storage, and all necessary spaces for the defined activity. It also speaks to the ability of one space to not interfere with another. It is important to consider how people enter and leave a defined space. Do they need to traverse another teaching area? In considering functionality, the design team should consider if the functional integrity of the facility remains intact when in full use. I have often found it useful to draw footpaths on paper to represent groups of participants as they move in and out of the center and how they move within a center.
- *Type of Simulation* – It either determines the use of the space or is defined by the space (virtual hospital example). The type of simulation will define the requirements of the space (large room versus a set of rooms). This is discussed in considerable detail in the next section, as it is likely the most important factor. It can also be the most constraining.

4.5.6 Space by design

4.5.6.1 Virtual reality area

This skill setting may or may not require a single room for a single trainer or may, on the other hand, include many trainers in a single room. In this case, establishing the type of trainer and how it is to be used is of importance. For the purpose of this chapter, we will not touch on VR using "cave" technology. Placing trainers in different environments may affect the objectives of the session. As an example, interventional radiology trainers may be used in a realistic patient environment or in a generic classroom. In the case of the former, the trainee is not only exposed to the immersion of the technology but also to that of the environment as well. The environment is contextually closer to reality than a simple classroom. The objectives of a course can therefore be expanded from just procedural skill training to include cognitive skill training. The design of a more generic setting can range from a large room to a series of rooms with groupings of technology. The program requirements will dictate this. VR trainers may also have a significant footprint and utility requirements, therefore, will require more than just a cubicle.

4.5.6.2 Skills training area

This type of area embraces many of the same issues as with VR training. There is considerably more history with skills centers – especially within nursing. The old metaphor includes beds within a room often aligned along the wall with a central desk area in the middle. This arrangement has it benefits but requires a substantive footprint given the size of the beds and the space required around them. The important question to ask is for what purpose will these beds be used? Do the trainees really need all the beds for their activities? Do the training exercises always require a full-size mannequin or person in the bed? If not, then what percent of the time does it not? If the percent of time that actually truly requires a bed is small, then it is important for the customer to understand the trade-off – loss of functional space. Beds in particular are problematic as they cannot be easily moved and stored (especially with today's more specialized beds). Skill training areas are well suited for large "flexible rooms." In this environment, the instructor can address the entire room but also has the flexibility of setup not being hindered by beds. There are many activities that require tables rather than beds. This is especially true of partial-task trainers. Certain skill areas will require specialized smaller rooms (e.g., patient room). These rooms are useful for physical assessment that would inappropriate in a large room. These rooms can also be utilized for standardized patient encounters.

4.5.6.3 Computer-based learning area

Learning that leverages this type of technology has its unique characteristics. It is typically done in a larger area divided into cubicles or self-study areas. The question to be posed is whether a dedicated space for this type of training is required and if one already exists within the institution. The temptation is to make all that is simulation occur in the same facility. The demands of a computer-based environment are much more reliant on self-study than most other forms of simulation. In the case of computer-based training, there may already be sophisticated and well-developed space for this purpose. Careful consideration should therefore be given as to whether this space is required or if this activity can be done at another location. Recall that the simulation facility is premium space for very specialized activities.

4.5.6.4 Full-sized high-fidelity mannequin-based area

There have been many variations of design of this type of learning environment. There is no one good answer. The type of simulation, flow, and the participant type (and quantity) are important in the design of this setting.

The high-fidelity area is a special circumstance where more than one room is used for a single activity. The rooms include a conference room, control room, supply area, and simulation room. The challenge is in placing these rooms and areas in proximity to each other, so that flow within the space as a whole is self-contained and undisturbed by other learning activities in the center (revisit Figure 4.5-1).

This area is much like an apartment. In this case, the apartment contains a control room, a simulation theater, a debrief area, and a core area (where meds, patient supplies, etc., are housed). The apartment should be accessible in multiple ways and for different purposes. In one case, the purpose may be related to the participant and the other specific for the operator (to be able get equipment and deal with things behind the scenes without being seen or heard). If possible, the apartment should be designed in such a way that its rooms could be used together or for multiple activities. In this way, if a full-scale simulation session is not occurring, the debriefing room can be used as a conference room, and the simulation theater could be used for bedside training with the group in the room. This type of arrangement provides maximal flexibility while maintaining contextual isolation for the participant. A center can truly exist with multiple activities and little crossover of participants and faculty. A center that is built with a central corridor presents a specific dilemma when the space on either requires movement through the common corridor for activities excluding entering and leaving the facility. Careful arrangement/placement of rooms is paramount. A design that has its simulation rooms at one end and debriefing rooms at the other opens the center to considerable disruption of different courses as people move in and out of the different learning environments. Learning areas should have little or no overlap and the use of the corridor should be restricted for entrance, egress, bathroom needs, and other non-class-related issues.

4.5.7 Utilities

4.5.7.1 Gases and suction

As with a hospital, a simulation center will need to provide compressed gases for two purposes: (i) to deliver gases to virtual patients and (ii) to drive equipment. The plumbing for the gases does not need to be hospital grade but must meet the code for the facility. The ventilation must meet specific code requirements before certain gases such as pure oxygen and volatile anesthetics are used. There are several centers that do not use oxygen at all. This is usually satisfactory but may become problematic if the equipment that is in use requires the presence of pure oxygen to function properly (e.g., certain newer ventilators). The cost of decommissioning a ventilator by bypassing its built-in safety features may far outstrip the cost of effective ventilation. Similarly, the cost of ventilation to evacuate anesthetic gases may far outstrip the educational value of having real gases present. Ventilation must also consider heat, which is determined by the number of people in given space and all the heat generating equipment (e.g., an Audio/Video rack, or numerous video displays).

A functioning vacuum/suction system should also exist as participants expect it. This is an example of a detail that adds to a room's fidelity but does not restrict its functionality. The system may be small and self-contained, as are found in dental offices, or may tie into an existing system.

4.5.7.2 Sound

The design and structure of the center should involve the input of a sound specialist. It is important to consider the isolation of sound from one area to another. This includes considerations such as wall treatments, special ceiling materials, wall construction, etc. The ventilation system must also be considered as it can provide considerable background noise. The level of noise is determined by the size of the ducting and the velocity of the air flow. The reverberation and transmission of sound through materials and through hung ceilings can be considerable. These issues degrade sound quality and pose problems for the setup of the Audio/Video system for broadcasting and recording. The size and shape of a room will also change the sound dynamics more than one would expect. Building a room that is a perfect square will produce dead spots (nodes) where sound cancels itself out as it reflects off the walls. This may all sound like minutiae but considering this early will make a large difference in the quality and ease of setup of your AV equipment.

4.5.7.3 Lighting

Lighting affects the experience of the individual, the activity, and the quality of recording if any is to be done. The "temperature" of the lighting should be consistent with the lighting requirements of the video equipment in use. The placement of the lights should also be carefully thought out, lest they reside exactly where the ceiling-mounted LCD projector needs to go. Lighting must also accommodate for function. This is where one considers issues such as OR lights. These lights are costly and require expensive structural modifications to mount them. The design team should carefully evaluate whether they fall into the "candy for the mind" category or if they are truly educational. Beyond expense, they will label your room. This has been discussed earlier but is worthwhile reiterating.

4.5.7.4 Electrical and information technologies

Careful attention should be given to the electrical requirements (amount, type, and location) of a room within code limitations. This spans AC outlets to light switches and types of switches (dimmers versus a simple conventional binary switch). Also, consideration must be given to the need to electrically isolate a specific area should a scenario involve an electrical failure. The placement of electrical outlets should be frequent and at multiple heights along the walls. Power outlets for mounted projectors will need to be installed at the appropriate height or hung from the ceiling.

IT considerations are important and complex. A center may have its own dedicated intranet or may exist on an institutions network. Both have implications. New technologies exist that use IT infrastructure to not only deliver data but to stream audio and video as well. This may be of some importance for certain centers that wish to transmit to remote locations. An IT professional will be able to quickly determine where junction boxes are required and the amount of conduit and preinstalled dark fiber that would suit the application and its intended growth. In this way, the AV and IT professionals should not only work closely as much of their cabling may follow common pathways, but also share common cabling and network equipment.

4.5.7.5 Security

It would be great if this was not a consideration, but in most centers, theft is not unheard of. Not only is this pertinent for equipment, but also for materials that may be sensitive and confidential. It is therefore important to involve your institutional security to consult on what security measures would best suit the environment. This can be as simple as to what side of the door the key lock is on or the use of sophisticated card locks and tracking systems. For confidential information, one should consider redundant security measures as well as methods of encryption in the case of electronic media. Networked video systems originally designed for capturing the behaviors of students during sessions can be used to capture those of intruders after sessions.

4.5.7.6 Storage

This topic does not need much explanation, except to say it is almost always underestimated. All those beds and equipment have to go somewhere lest they line the halls of your new and beautiful center. As the design of a center evolves, storage space is usually the first to be decreased if more space is needed. Storage space is abstract and is not considered to be functional for the purpose of education. The exercise of creating a rough catalog of equipment and supplies and then placing them in the drawn storage areas can quickly make the point of how much storage is really needed.

We are all familiar with the use of storage areas in our own homes such as closets, basements, attics, and garages. The term *storage* connotes hidden and dormant. We are also familiar with the use of utility areas such as laundry, cleaning, and workshops. The term *utility* connotes accessible and active. Perhaps, relabeling all storage space as utility will help focus its essential value added to a simulation facility.

4.5.8 A walk-through

It is worthwhile taking a deep breath here and to take a stroll through a design example (Figure 4.5-1). Imagine that four students have come to the center. They are all here for four different courses:

- Pat needs to go to simulation area 1 at 09:00 for a course
- Jeff needs to go to training room 1 at 09:00
- Alex is scheduled for a course at 10:00
- Francis is scheduled for a course at 10:15 in sim area 2

Trace the steps of each of these individuals starting from point X on the schematic. Notice that the ONLY common area is the central hall and core corridor (which is faculty only). As each of these individuals enters separately or together, they do not disturb classes/courses in progress. Similarly, as individuals leave or need to go to the restroom, they cause minimal disruption within the center. All their activities are quarantined to their respective areas. In the event that you want a larger area, Training Rooms 1 and 2 can be opened up and Simulation Rooms 1 and 2 can also have some communication through the core corridor. The center design provides for easy access and egress and meets fire code for exits. Faculty and educators can meet with individuals in the office area if needed. The core corridor acts as a "faculty only" area where preparation for the day can occur and where supplies can be readily available given the natural variation found in full-size high-fidelity mannequin-based simulation. Training Rooms 1 and 2 may/maynot have beds. This has implications as to its function: a ward-like setting for training multiple participants skills at the bedside or an open area with multiple stations to use task trainers, VR trainers, etc. The key to this design is versatility and function, while maintaining the sanctity of educational isolation when needed.

4.5.9 Conclusion

The design of a simulation facility is deliberate and precise. It requires imagination and forethought. It is not a hospital or clinic; it is space dedicated to immersive interactive education. Attention to the selection of a design and project team is important. Who you include may be less important than who you do not include. The facility design should suit the program needs, which in turn have an obligation to curricular objectives. Flow, utilization, and functionality remain at the forefront of a good design specialist. There is not one single design but a set of guiding principles that have at least partially been covered in this chapter. If nothing else, this chapter should serve to spur the imagination *and* illustrate the complexity of thought that goes into effective design. In ending, summarizing this chapter, I find myself thinking of the virtual hospital again. The idea is truly exciting and hard to resist. I will however remain true to education and fight the desire – for the time being.

Section Five

Medical devices and instrumentation

Chapter 5.1

5.1

Evolution of medical device technology

Nandor Richter

The origins

In the early phase of civilization, instruments were used in medical treatment. Healers used sharpened obsidian stones for skull trepanation, and metal knives later became common tools. Thousands of years ago doctors sought to cast a glance into the human body. They tried to observe the inner structure and the functioning of the living organ through the mouth, the rectum, and the vagina. Observation through these parts caused minimum functional disorder, and minimal tools were needed. Around 400 BC, Hippocrates was the first to mention the endoscopy when he described a rectum examination. At that time, a light source was not yet available. Around AD 1000, the well-known Arab doctor Abu al-Quasim used a glass mirror for vaginoscopy. Because of inaccurate notions and the lack of measuring instruments, observations could lead to only rough conclusions.

The Enlightenment brought with it a significant impetus as the dogmatic, retarding forces of the Middle Ages influenced scientific activities less and less. During the seventeenth century, the exact sciences developed vigorously. The curious human examined and understood nature better and more objectively. Progress in healing required the increasing application of tools and equipment. For example, doctors realized rather early the importance of body temperature and its fluctuation. Again, it was Hippocrates who considered body temperature as the most important sign in the case of acute illnesses. However, it was not until around 1612 that the Italian doctor Santorio made the first clinical thermometer. He is also credited with the introduction of the weight balance and hygrometer. With these devices, the desire of Galen (AD 131–201), that not only the type of illness be identified but also the "quantity" of the illness be measured, could now be realized. Galileo stated "measure the measurable and make measurable what was not measurable till now." His words are equally important and basic even today. One can apply this to all diagnostics and to most medical equipment.

Temperature measurement

Thermometers have been enormously helpful to physicians. Time series of temperatures were made possible, providing information on the progress of an illness and the trend of the patient's condition. One good indicator of the importance of the thermometer is that over one billion temperature measurements are carried out yearly in the hospitals in the United States.

Physicians looked for symptoms that could increase their knowledge about the condition of the patient, the nature of illness, and its seriousness. Besides temperature, the color of the face, the color and odor of urine, and the rhythm of the heart were all peculiar and perceptible to the patient. It was desirable to measure these parameters quantitatively and not simply to observe the color of the patient's face. Quantifying color, heart rhythm, and chemical composition of urine had to wait for further developments in the chemical and technical sciences.

Bioelectrical signals

The discovery of bioelectrical signals was of paramount importance in that it led to diagnostic and therapeutic applications. Today, we know that bioelectrical phenomena are characteristic of all living organisms. Long

Biomedical Engineering Desk Reference; ISBN: 9780123746467

before bioelectricity was well understood, much was already known about the electrical activity of nerves and muscles. Plato (427–347 BC) mentioned electric rays (torpedinidae) found in the Mediterranean Sea. Aristotle (384–322 BC) also wrote about these marine creatures. He mentioned that touching these rays could cause deafness. In the eighteenth century, Aloisius Luigi Galvani (1737–1798), a professor at the University of Bologna, performed experiments on frog nerve-muscle preparations. He stimulated with electrical charges the nerve that led to the frog's femur and found that during stimulation the femur muscle contracted. In one of his experiments, he demonstrated the cell membrane potential, the polarizing potential of cells in rest. For his achievements, Galvani was named the father of electrophysiology. From that time onward, an increasing number of scientists started to study the action potentials generated in living organisms.

These electrical signals are primarily the cell membrane potentials, action potentials of heart and muscles, and the action potentials of the brain. Studying action potentials requires sensitive equipment and thorough knowledge in measuring technique. Depending on the source of their origin, the amplitude and frequency domains of these signals are significantly different. From the point of view of measurement technique, the heart action potentials, the ECG waves, can be considered as the easiest to measure. Voltage potentials of these signals are in the millivolt (mV) range, their useful frequency domain is 0.1–100 Hz. Brain action potentials are on the order of microvolt (μV) and the frequency domain is 0.1–10,000 Hz. Evoked potentials are even lower with a frequency range of approximately 10–10,000 Hz.

Around 1856, electrocardiographic signals of a frog's heart were measured for the first time. In 1903, Willem Einthoven (1860–1927) introduced his string-galvanometer and recorded ECG signals. Hans Berger started to study the electrical signals of the brain by electroencephalograph measurements in 1924. However, his first good-quality electroencephalogram was not ready until 1929. In 1943, Weddel succeeded in registering muscle action potentials with an electron beam oscilloscope. But the first commercial electromyograph did not enter the market until around 1960.

Blood pressure

Aware of the importance of blood circulation, studying blood pressure was a natural requirement. In one of the first experiments, a tube was inserted into the neck artery of a horse, and the pressure variation generated by the heart was measured. For human application, the bloodless measurement of blood pressure was a necessity. However, one had to give a measurable, practical definition of the blood pressure. There was no simple, bloodless way of measuring the instantaneous value of blood pressure until the introduction of the systolic–diastolic measurement. Still accepted in daily routine, this detection technique is based on the Korotkoff sounds. Digital technology made small blood pressure meters possible with a concomitant surge in their use in the home as well as the hospital.

Anesthesia and the relief of pain

Horace Wells, an American dentist, performed the first painless dental operation in 1844. He used nitrous oxide, known also as "laughing gas." Therefore, he is considered the inventor of narcosis. Anesthesia, resulting in a painless operation on patients, was an immense development in surgery. It gave enormous push to the development of surgical instruments and apparatuses used in operating rooms and beyond.

It is also interesting to draw attention to a different medical advance, acupuncture anesthesia. Western medicine is learning more and more about this practice of traditional Chinese medicine. While its mechanism is still not completely known, it does prove useful in certain areas of anesthesia.

X-ray and nuclear medicine

The desire of physicians for centuries to be able to look into the human body was finally realized in 1895 with Conrad Roentgen's discovery of X-rays. In the following century, the application of X-rays to diagnosis and therapy gave strong momentum to the advance of medicine. In hospitals and clinics, the medical applications of X-rays became a separate professional field. With the development of the engineering sciences and, later, informatics and computer techniques, application of the images generated by X-ray increased. Of course, this went along with the construction of more sophisticated equipment. Therefore, the demand for experts who are familiar with the operation and maintenance of these devices increased.

Imaging and image processing

X-ray images, with their increasingly fine details, provided increasingly more information. Image intensifiers and video monitors rendered possible the manifold applicability of X-ray equipment. High-power and fine-resolution X-ray tubes expanded further the application possibilities. X-ray equipment, such as the tomograph, angiograph, angio-cardiograph, and the urograph, was developed for special applications. Imaging and image-processing methods provided the information-enhancing possibility for physicians.

CT and MRI

In the field of imaging, the computer tomograph (conceived by William Oldendorf and developed by Godfrey Hounsfield and Allen Cormack) brought revolutionary change. The first successful clinical experiments occurred in 1972. The Nobel Prize was bestowed upon the two inventors in 1979, and Godfrey Hounsfield was knighted in 1981. The basic idea of computer tomography went beyond X-ray imaging. The principle is also becoming important in fields where the source is nonionizing radiation.

The basic phenomenon of magnetic resonance imaging was first observed in 1946. Certain atomic particles in strong magnetic fields absorb very high-frequency electromagnetic waves selectively. Particularly those nuclei in which at least one proton or one neutron is unpaired show this phenomenon. This absorption can be measured and evaluated. In the human body, water and lipid (fat) molecules, which contain hydrogen, show the effect in measurable magnitude. This noninvasive, relatively hazard-free diagnostic method is based on this selective absorption phenomenon. The first MRI equipment was put on the market during the early 1980s. Since then, its application has become broad including, among others, the study of the brain, breast, heart, kidneys, liver, pancreas, and spleen.

Nuclear medicine

The use of radioactive isotopes as tracers by George C. de Hevesy in 1912 was a great epoch-making discovery for which Hevesy was recognized with the Nobel Prize in 1943. Wide-ranging application of radioactive tracers became possible after World War II. Radioactive isotopes came to the market after the 1950s. Around this time, development of the equipment necessary for measurements was completed as well. Importance of manmade scintillation crystals was extraordinary. These crystals make possible the detection, counting, and measurement of the radiation emitted from disintegrating nuclei. The development of synthetic crystals, the electron multiplier, and spectrum analyzers altogether resulted in the general application of nuclear-measuring technique in health care. In the following decades, the evolution of imaging and image-processing technology further broadened the field of nuclear measuring techniques. The introduction of the gamma camera made possible the imaging of larger parts of the body.

Ultrasound

In 1880, Madame Maria Sklodowska Curie and her husband Pierre discovered radium and subsequently received the Nobel Prize. Pierre's brother Jacques discovered the piezoelectric effect. When high-frequency electric fields are imposed on a piezoelectric material, mechanical vibration results. Because the frequencies of these vibrations are above the audible frequency domain, the vibrations produced are termed "ultrasound." Mulwert and Voss reported the first ultrasound therapy intervention in 1928 when they tried to cure certain deafness with ultrasound irradiation. Pohlmann, in 1939, made the first ultrasound therapy equipment for treating humans. In 1942, K. Tr. Dussik was the first to attempt to apply ultrasound for diagnostics. After World War II, ultrasound was used in different fields of medicine. The one-dimensional method (A-scan) was followed by the two-dimensional method (B-scan). Essentially, these are imaging methods with which place and form of tissues and organs can be examined.

By the application of the Doppler effect, ultrasound was expanded to dynamic measurement areas. Reflection of ultrasound from certain moving parts of the body makes the measurement of radial velocity of the reflecting surface possible. This phenomenon is applied, for instance, in measurement of the blood flow velocity and in the examination of fetal heart movements.

Considerably higher power must be applied in the case of therapy. Accordingly, the problem of regulating and making accurate measurements of ultrasound dose and energy had to be solved. Today, medical application of ultrasound is expanding rapidly, and equipment with new features appears on the market continually. As a result, quantity and diversity of knowledge required for the application of ultrasound is growing.

Microscope and endoscope

Most likely, Zacharis Jansen, a Dutch optician, discovered the compound microscope in 1590. Its general use gave enormous impetus to the development of medical sciences. With its help, otherwise invisible elements of the body could be studied. Since this discovery, many varieties of microscopes have been developed. At the far end of the spectrum are sophisticated devices, such as the electron microscope and the atomic-force microscope.

For centuries, physicians wanted to directly visualize the inner parts of the functioning body. Various endoscopes were developed to assist in achieving this goal. The major improvement in endoscopes started with the invention of Bozzini, a physician in Frankfurt. Bozzini succeeded in introducing a beam of light into a hollow organ and directing the reflected light to the eye of the observer. The technician, Nitze, constructed the first truly practical endoscope. These instruments were used for rectoscopy and cystoscopy. From that time onward, the endoscopes have been continuously improved. Nowadays, endoscopes and laparoscopes furnished with flexible fiber optics or with cylindrically shaped lens systems (the Hopkins-optic) are in use. With the help of these modern devices, minimally invasive diagnostics and surgery progressed rapidly. The diagnosis and therapy of the digestive system, rectum, bladder, respiratory tract, joints, and abdominal cavity can be carried out with less trauma than with the conventional methods, which typically required large incisions. Introduction of catheters, biopsy forceps, electrocoagulation instruments, and lasers can be done with devices evolved from endoscope.

Laboratory devices

Together with the development of biology, medical science, and measurement techniques in chemistry and physics, the variety of laboratory devices has increased enormously. With the help of the modern laboratory automatons, the number of different measurements done today is in the order of hundreds of millions per day. Of course, this practice also raises the number of companies that supply the reagents and disposables.

Surgical instruments

These instruments are usually not of concern to clinical engineers. However, for the sake of completeness, they should be mentioned. The number of surgical instruments ranges in the thousands. Many of them are engineering marvels, the products of cutting-edge fabrication technologies and advances in metallurgy and materials science.

Patient monitoring devices for intensive therapy

Modern technology and engineering opened the door for the application of individual equipment in systems. Two good examples are the complex array of equipment in the operating theater or the installations in the intensive care units. Patient monitoring systems, respirators, defibrillators and others constitute an integrated system. The measurements of one device might affect the operating parameters of the other. The harmonized functioning of these devices has required the attentive presence and intervention of the physician or intensive therapy nurse.

One ongoing development is the automation of processes and the involvement of information technology. The patient will become a part of a feedback loop, in which the system automatically sets and maintains the optimum parameters. Such parameters might include pulse, respiration, ECG, blood pressure, and blood sugar level. This, to some extent, is the vision of the future; but considering the pace of development in the last few decades, one may safely say that this is not the illusion of the far future. The science of robotics is enabling machines to replace some of the operations of the surgeon. Will the human ever be completely replaced by machines?

Understanding the physical world

In the previous paragraphs the development of medical device technology was introduced in a nutshell, through examples. These brief episodes demonstrate well that, in the course of development, combination and interaction of different scientific disciplines have played an important role. Formerly, usually one person facilitated this interaction. In the time of Hippocrates, for example, scientifically oriented people had a wide interest and philosophical inclination. Because of the lesser quantity of the available scientific knowledge at that time, surveying it was easier for one person. This helped to develop those statements and discoveries that demanded multidisciplinary knowledge from the inventor. Such was the situation until the last few centuries.

Taking a great leap in historical terms to the Enlightenment, the progress in understanding the law's of nature was rapid and substantial. The seventeenth century is considered to have been the century of geniuses, and the eighteenth century is considered to have been the period of brightness and the mind. The Bernoulli brothers (Jacob and Johann), Herman von Helmholtz, Gottfried Wilhelm Leibnitz, and Isaac Newton towered above the geniuses. Newton, in his *Principia*, who also built upon the observations of Galileo, Copernicus, and their contemporaries, gave the best description of mechanics of their time. With the infinitesimal calculus, and with Newton's mechanics as a basis, Leonhard Euler, a Swiss mathematician, described the movement of solid bodies and developed the basic equations of hydrodynamics. Galvani and Volta laid the foundation for knowledge of contemporary electricity.

During the eighteenth century and beyond, the bulk of work went into observation and understanding natural phenomena, the qualitative recognition of the laws of nature, and their formulation in statements and equations. The growing amount of knowledge and the immersion into the details necessarily steered scientists toward specialization. The epoch of polymaths began to fade away. Due to the prompting effect of the industrial revolution, demand on practical application of sciences strengthened. A technical revolution followed the scientific revolution. Application of discoveries accelerated. The number of inventions grew. The steam engine and the principle of the dynamo, followed by the general use of electricity, had great effects on almost every aspect of life. Medical science provided its share, too. Physicians discovered the potential of electrical phenomena. For instance, they introduced electrical shock therapy for a variety of purposes. The rapid development of basic sciences, physics, chemistry, and biology, and the engineering sciences, together with the demands emerging from the medical side, gave impetus to the developers of medical devices. The pace of development during the nineteenth century became significant. Increasingly, humans began to understand the phenomena of their surroundings, the laws of nature, and their effects on the individual as well as on society. Discoveries and inventions followed each other.

During the first part of the twentieth century, significant milestones included the discovery of X-ray, the introduction of ECG, and the development of anesthetic and respiratory equipment. Then came World War II and the harnessing and release of atomic energy, with one of its practical applications being nuclear medicine. Discovery of special semiconductors and transistors in the 1950s also gave a large thrust to the development of electromedical equipment.

During the first half of the last century, pharmaceuticals dominated medical advances. But since the 1960s, developments in medical engineering have been unprecedented. In the second half of the last century, a great variety and number of medical devices entered hospitals and consulting rooms. Today, the number of different medical devices is around 10,000. If one considers the variations of the equipment manufactured by different companies for the same application, then this number increases to tens of thousands.

It is no exaggeration to state that health care and health care organizations are among the most complex and complicated structures of our society. Physicians must select from an extensive arsenal of methods, materials, instruments, and from combinations of them, and they must do so with the intention of providing the best care for the patient. Evidently, in such a complex system, various specialists must work together. In the course of time, pharmacists appeared in the hospital, joining the physician and the nurse. They helped with the proper handling and preparation of the vast number of pharmaceuticals available. As with pharmaceuticals, the introduction of a multitude of increasingly complicated medical devices urgently emphasized the importance of the support of specialists in physics and engineering. By the 1930s, medical physicists began to appear in hospitals. Their tasks were to handle and control X-ray equipment, to perform dose measurements, and, later, to work with other ionizing radiation sources. Dose planning and control and quality assurance also became their tasks. Nowadays, they are often found working with nonionizing radiation, such as ultrasound and lasers, as well.

Since the 1970s, complicated equipment such as patient monitoring systems and heart–lung machines have made the presence of clinical engineers in hospitals indispensable. Soon it became evident that besides requiring technical knowledge, these clinical engineers needed experience in other fields. Apart from becoming knowledgeable about the human body, these engineers needed to know how to communicate with the medical staff and to have experience in hospital management and administration. Step by step, the concept of clinical engineer was developed. In the present times, clinical engineering has become an accepted profession in many countries.

The number, variety, and complexity of medical devices, combined with the growth of informatics has made it important and possible to think and work with systems. Health informatics gradually gained ground all throughout health care. The large amount of patient data can be organized and managed only with the help of informatics. Informatics finds particular importance and applicability in acquisition, processing, storage, archiving, retrieving, transmission, and display of diagnostic images.

The future

The current trend will continue. The fields of medicine, engineering, and science will develop independently but will help each other, a phenomenon known in electronics as "bootstrapping." Progress in medical science makes new demands upon engineering, which in turn are reflected in the development of new methods and equipment. New engineering technologies provide fresh possibilities for physicians. Cellular and tissue engineering come closer to the routine application of their results. The introduction of nanotechnology will soon yield practical applications. Robotics will open up new opportunities in surgery. The combination of these technologies will result in spectacular achievements. As an example, new techniques in heart surgery combine robotics and endoscopy, thus enabling an operation without the need to fully open the thorax, without the need to use a heart–lung machine, and without having to lower the patient's temperature. The field of gene technology shows promise of progress and advancement.

Biomedical engineering was literally created and developed during the twentieth century. Considering the accelerated pace of development of the present, it is impossible to predict developments that will occur over the next few decades. However, it is possible to affirm with confidence that the physician–engineer relationship and the importance of clinical engineers in health care will remain and grow in prominence. Besides engineers who are directly involved in hospitals, clinical engineers will also play an important role in research and development. Equally important is the role of the engineers who are employed in medical equipment factories. In summary, relatively little is known about the forthcoming development in this century; however, even this modest foresight hints at enormous technical progress.

Social effects

What is the impact on the society of the rapid technical progress taking place in this century? Colossal resources of energy are at the disposal of mankind. The bulk of physical work has been taken over by machines. Comfort of life has increased. Initially, humans saw only the pleasant side of this progress, that living conditions improved. Only in the last few decades has mankind started to feel and get to know in more detail the harmful effects of the environmental hazards that accompanied the technical revolution and which affect the entire biosphere.

In addition to environmental damages, the technical revolution has affected the practice of medicine, which now must face new maladies. As illnesses shift in their nature and significance, adaptation of health care services requires continuous change and development.

The extended life span of humans brings into focus health problems of the aged. Attending to the health problems of the aged will impact medical device development in two ways: (1) requirements for innovative medical devices for home use and (2) increased number and variety of devices for hospital use. Those in the aged population are not necessarily ill, in the general sense, but they require special support structures and medical devices which may significantly differ from those found in a general hospital.

In the long run, man-made technical means influence the trend of human biological evolution. The so called "homo technomanipulatus" is coming into being. Abilities, once so important, like physical fitness, the subtle function of the sensory organs – the anatomy that was so well matched to extensive movement are all gradually diminishing. In place of these, new abilities required for intellectual work are being augmented. In the longer-term, negative effects of these will be significant in health care. For instance, the incidence of the diseases of the locomotor, cardiovascular, psychic, and neural systems will increase. The results of a recent research project demonstrated the likelihood of manipulation of the human genome. As a consequence, deliberate or accidental modification of the human genes will be possible. The "homo genomanipulatus" may come into being. Impact of this on humans is unimaginable today. The consequences of gene manipulation for our descendants are difficult to foresee. We do not know the future of the biosphere. However, genetic manipulation of humans will soon become reality. But we do not know to what extent the new features created in this way will be advantageous or rather disadvantageous in the biosphere that is also changing beyond our control and in an unknown way. Because of the different and long time constants of the processes, one cannot survey the whole system (man + biosphere).

It is clear that health care must be vigilant and ready to respond to new challenges. While some of the impact of advances in technology may be deleterious, it is some comfort to know that predicting and reacting positively to this impact is within mankind's reach. It is imperative that those engaged in health related technical activities must continuously progress by advancing to appropriate knowledge levels and developing technological means so as to enable the generation of solutions to complex and rapidly mutating emerging problems. Taking all of this into consideration, the knowledge and efforts of clinical engineers will be appreciated in the future, as they are in the present.

Further information

Borst C. Operating on a Beating Heart. *Scientific American* 283(4):46, 2000.

Csaba G. Quo Vadis Homine? *Természet Világa* (World of Nature) 125(1):12, 1994.

Csaba G. Homo Biomanipulatus. *Természet Világa* (World of Nature) 131(4):167, 2000.

Csaba G. Homo Technomanipulatus. *Természet Világa* (World of Nature) 131(8):357, 2000.

Csaba G. Homo Genomanipulatus. *Természet Világa* (World of Nature) 132(1):2, 2001.

Encyclopedia Britannica 1997.

Katona Z. Brief History of the Detection of Bioelectric Phenomena. *Kórház és Orvostechnika* (Hospital and Medical Engineering) 26(3):70, 1988.

Katona Z. History of Medical Technique. *Kórház és Orvostechnika* (Hospital and Medical Engineering) 26(4): 108, 1988.

Katona Z. Thermometer. *Kórház és Orvostechnika* (Hospital and Medical Engineering) 26(6): 161, 1988.

Katona Z. History of Instruments for Temperature Measurement. *Kórház és Orvostechnika* (Hospital and Medical Engineering) 27(1):19, 1989.

Katona Z. Short History of the Application of Ultrasounds in Medicine. *Kórház és Orvostechnika* (Hospital and Medical Engineering) 27(4):97, 1989.

Katona Z. Early History of Surgical Anesthesia. *Kórház és Orvostechnika* (Hospital and Medical Engineering) 29(4):97, 1991.

Katona Z. Short History of Endoscopy. *Kórház és Orvostechnika* (Hospital and Medical Engineering) 29(5): 141, 1991.

Olshansky SJ, Carnes BA, Butler RN. If Humans Were Built to Last. *Scientific American* 284(3)42, 2001.

Sorid D, Moore SK. The Virtual Surgeon. *IEEE Spectrum* 37(7):26, 2000.

Chapter 5.2

5.2

Medical device design and control in the hospital

Joel R. Canlas, Jay W. Hall, and Pam Shuck-Holmes

Clinical engineers (CEs) in the hospital setting use modern technical, scientific, engineering, and management training. Specialization, knowledge, and experience of CEs serve humankind by contributing significantly to the provision of safe and effective medical technology. This view is in harmony with the American College of Clinical Engineering (ACCE) definition of a CE as "a professional who supports and advances patient care by applying engineering and managerial skills to health care technology" (Bauld, 1991). In this chapter, the authors, reflecting on over 40 years of collective experience as CEs in a hospital setting, describe their role in designing, manufacturing, evaluating, and controlling medical devices to ensure their safe and effective application in health care.

The extent to which a CE is able to perform a responsible function in the hospital setting depends to a large degree on the acceptance and understanding of their roles and capabilities by the hospital administration and staff. Without the full support of the hospital, from the lowest to the highest levels, CEs cannot function to the best of their abilities. Thus, when empowered, CEs not only follow a course of action specified in good clinical engineering practice guidelines but also go beyond the call of duty. The CEs clients (i.e., doctors, nurses, respiratory therapists, imaging specialists, other hospital staff, patients and their families, health maintenance organizations [HMOs], and preferred provider organizations [PPOs]), are usually well informed about medical technologies, which is a result in part, of the Internet and the accessibility of up-to-date information that once lay only in the books and journals stored in the deep recesses of libraries.

As CEs, the authors have striven, consciously and subconsciously, to change the staff's perception of CEs as simply repairmen armed with a cell phone and three pagers dangling from their belts. At William Beaumont Hospital, CEs have focused on a systems approach to medical technology management. For example, hospital beds are not treated as simply a support surface for patients. The CE considers a hospital bed from the perspective of mechanical design, safety features, operating mechanisms, interfaces with the patient, intelligent nurse-call and alarm systems, and electrical power requirements. The CE analyzes the risks that the bed or its accessories pose to a patient, a nurse, or other equipment. Similarly, a robotic system includes the control device, all of the system components, the maintainers, the users, the storage environment, the physical-facility requirements, the specialized test equipment, and the training requirements. If all devices and systems were so perfect as to be predictable with no possibility of causing injury to the patient and user, then the need for CEs would diminish. As this is not the case, the CE must identify and utilize available resources, examples of which include the purchasing department, which can provide financial analyses and assess the manufacturers' financial health, test equipment, education and training in the scientific and engineering method, the manufacturers' engineers and technical staff, the Internet, and device-user experiences.

As in the health care industry at large, in the hospital there are reactive and proactive elements in managing medical technology. The proactive, or preventive, component in medicine staves off the possibilities that can cause disease in a patient; e.g., preventing obesity can prevent the various diseases known to be associated with it such as diabetes. The reactive, or treatment, component addresses the patient who is already afflicted with a disease and the means for making them well again. This similar approach works well with the hospital's overall goal of providing world-class customer service.

Biomedical Engineering Desk Reference; ISBN: 9780123746467

By describing clinical engineering experiences in this chapter, the authors provide practicing and prospective CEs with a window into the clinical environment, showing ways in which CEs can help the hospital and patients and can advance health care. Engineers are never satisfied with the status quo. They constantly explore various solutions to a problem, seeking the optimal outcome.

The following anecdote captures the sense of the engineer's fervent desire to seek engineering solutions to a problem. A long time ago, three folks were about to lose their heads on the guillotine. The first to go, a peasant, was asked by the executioner whether he wanted to have his head cut off while facing the sky, or downwards. He chose to look downwards. The blade was released but stopped just short of his head. Everyone took it as a miracle, and the peasant, to his extreme joy, was pardoned of his crime and was set free. The same thing happened to the second person, a lawyer, who chose to face the sky. The last person, an engineer, also choosing to face the sky, saw a knot in the rope and, just before the guillotine was released, cried out, "Hold everything—I see the problem!"

Design and modification of medical devices in the hospital

CEs thrive on change. They must, because medical technology is constantly evolving and changing. What seems to work well today could be superseded by something better tomorrow. Keeping up with the technology, and managing current and future needs are two services that CEs can provide adequately. This goes well with the hospital's goals of customer satisfaction by providing a holistic approach to medical care in a timely and effective manner. By helping to ensure an environment of safe and effective medical devices through good technology management, CEs enable clinicians to have the armamentarium that is needed to treat patients effectively.

Managing medical technology requires that both the medical device and the human factor component be understood. Ensuring that a medical device is safe and effective requires consideration of a host of factors, such as device design, construction materials, performance history in the clinical setting, the time in the market, FDA allowances and approvals, the manufacturer's reputation, parts availability, and the vendor's responsiveness to its clients' needs. The human factor component of technology management looks at the human interface, which includes the patient, the clinician, the clinical/biomedical engineer, the service provider(s), and the manufacturer's representatives. The hospital relies on the CE to facilitate communication among medical, nonmedical, and technical people. In addition to a working knowledge of the engineering process and specifications, the CE must be able to translate this information to such diverse professionals as doctors, nurses, financial managers, administrators, and technicians. At the same time, the CE needs to be able to transmit the clients' feedback on a medical product to the manufacturer in a constructive fashion in order to resolve product problems in a timely fashion. Such feedback often results in product improvements as well.

Ensuring medical device safety by clinical engineering

Various tools are available to the CE in the hospital setting to ensure the safety of medical devices. These tools, tried and tested by the authors, are listed below.

Equipment selection and evaluations

CEs conduct short equipment evaluations in-house to provide an educated estimation of the product's usefulness, safety, and effectiveness. The focus at this stage is on the product's suitability as a hospital standard. Such factors that must be considered include the manufacturer's financial health, which could determine whether the company will even exist in 10 years to support its products, and the product's record of performance.

The manufacturer's financial health can be assessed by analyzing information gleaned from annual reports, the Internet, and information services such as ECRI and the MDE Group. Purchasing and finance departments, by the nature of work that they do, are excellent resources. A product or a technology's track record, i.e., performance history, can be assessed through various sources, like interviewing user hospitals (vendors should be able to provide you with a list of hospitals and specific departments using their products), researching websites such as the FDA's Manufacturers And User Device Experience (MAUDE) database at www.fda.gov/, and ECRI at www.ecri.org.

Hospital technology management requires the monitoring of the life cycles of medical devices and systems (see Chapter 5.3). Systems comprise not merely the device, but also the support group of equipment and services that ensure its proper functioning and the competency of the user group. As a device approaches the end of its rated life expectancy, or if it has required inordinately high maintenance, this information must be conveyed to the user group. However, as experience shows, through constant communication, the clients (i.e., the hands-on staff members (users)) inform the CE of changes in a device such as deterioration, inaccurate measurement, or intermittent operation. When equipment has been identified for obsolescence in at least two years, work begins on selecting replacements.

At times, if the technology has not changed significantly, and the hospital is satisfied with the equipment, only a financial analysis of the pricing and service quotations from several vendors and manufacturers is required. Rapid changes in medical technology, however, dictate that even the most mature, most often used, and most accepted technologies, despite small generational improvements, should be compared with newer competing products and technologies through evaluation and, where necessary, clinical trials. Purchases that exceed a certain monetary value require a formal bid process entailing a request for proposal (RFP). The RFP is addressed later in this chapter.

A medical device becomes a hospital standard after the process of engineering evaluation, clinical trials, financial analysis (based in large part on the responses to the RFP), and selection as the most cost-effective product.

Clinical trials

Participation in clinical trials is often an enjoyable, rewarding, and challenging aspect of the CE's work. The rapid growth in the biomedical engineering and its application to medical device technology market results in new and varied devices coming to market. This, in turn, presents the CE with more devices to evaluate, and more clinical trials to manage. Clinical trials find CEs becoming increasingly involved with the clinical and medical staff. The following is a list of many of the steps taken by CEs in managing and running a clinical trial:

- Obtain user department cooperation. (This tends not to be a problem because the department usually has requested the trial.)
- Arrange with the vendors and manufacturers for use of their products on a clinical trial of a duration of a few days to several weeks. Vendors are generally quite helpful, even supplying necessary, and often costly, disposables. Occasionally, the hospital supplies the disposables, especially if these are stocked items.
- Arrange with the vendors and manufacturers for in-service training of the staff involved with the trial. Obtain user and service manuals and make them readily available to staff. Schedule the trial so that most staff members have an opportunity to use and evaluate the product.
- Prepare a survey instrument that lists the desired specifications and evaluation criteria (e.g., ease of use). It is recommended that the survey instrument be on only one page, that the scoring be succinctly explained, that all terms used be understandable by the evaluators, that it states where to return completed forms, and that it contain a space for the name and mailing address of the evaluator.
- Clearly explain and document test methodology to ensure scientific validity of the study.
- Facilitate the monitoring of the devices for proper operation. This monitoring is usually the responsibility of the vendor or manufacturer. Because the vendor cannot be at the hospital at all times, a person in the department who is properly in-serviced should be designated as the resource person.
- Prepare a schedule lasting over a reasonable time that covers every possible step needed for the clinical trial. This is essential to ensure timely completion of the trial.

Negotiating for safety features

Sometimes certain features that help to improve the safety of the device are incorporated or are even sold as options. One should ascertain that the products provided for the trial have all the desired specifications and that the competing brands have the same features.

Note the willingness of the manufacturer to provide training of in-house maintainers to the same level as that of the manufacturer's field-service technicians. While in most cases, the manufacturer is quite willing to provide this level of training, the manufacturer occasionally claims that the proprietary nature of its technology militates against this. In these cases, one should relinquish servicing responsibility to the manufacturer or should negotiate a plan for utilizing in-house maintainers. The reasons for this are threefold: (1) immediate response to any equipment problems; (2) control over the safety, turnaround times, and costs; and (3) an immediate familiarity with the equipment and user group.

Education

In-servicing of clinical users

The hospital should demand in-servicing for all new devices, including new standard devices that are upgrades of the old standard being replaced. Every department has an assigned education specialist, who will need to be trained well so that that specialist can train other staff members, i.e., a train-the-trainer program. (Preferably, every user should obtain hands-on training initially.) That department should provide periodic in-servicing to users, either on demand or on a scheduled basis. Training should entail the theory of operation. A knowledgeable user is a prerequisite for efficient and safe use of medical devices.

The operation of complex devices is not easily mastered by typical users. It is, therefore, imperative that staff members continuously receive in-servicing that includes theory of operation, device features, and available resources with the goal of achieving staff familiarity with, and respect for, devices.

Training the maintainers

Training and in-servicing should be provided to the servicing department. In particular, the in-house department should designate an education specialist, who will be the department's resource when questions arise.

Surveillance

Monitoring medical device reports for indicators

Researching the products' safety record can be done from several sources, including the hospital's medical equipment database; other hospitals; professional colleagues at local, regional or national conferences; the FDA; and ECRI. Maintaining a healthy communication line with every departmental client in-house helps tremendously because bits and pieces of information gleaned from users can prove invaluable in fine-tuning requirements to help keep the device in constant and safe operational condition.

Variance reporting process

Once the device has been purchased, has passed all safety inspections, and has been deployed in the hospital, surveillance of device performance and safety must occur. Any variance in device operation must be documented by means of a variance, incident, or adverse-occurrence report. (The title of such a report varies from institution to institution.) Any equipment that malfunctions or fails while in use on a patient must be investigated, and the variance, findings, and resolution must be recorded. The hospital has legal responsibilities under the Safe Medical Devices Act of 1990 (Alder, 1993; JCAHO, 1993; 21 CFR, 1995).

Cradle-to-grave management

This is a circular process of ensuring the continuity of making available at all times the necessary patient care equipment. Key to this process is the collaborative interaction among the client clinical staff, equipment maintainers, CEs, and purchasing and financial analysis departments. Having responsibility for certain steps in this process and interacting with members of the project team, the CE stands to achieve a relatively safe, reliable, and cost-effective medical equipment inventory. The various components in this process include the following:

- Identifying the equipment to be purchased based on (1) clinical need (newer technologies have a way of rendering some older technologies obsolete);
- Planned obsolescence (the device is nearing the end of its rated useful life);
- Maintenance cost (the cost of maintaining a device exceeds depreciated value).
- Conducting pre-purchase evaluation of equipment. This process consists of two components: (1) performing engineering evaluations of competing products, and (2) conducting clinical trials, after narrowing the field down to a few candidates, to rank and select the best fit. Where the cost of purchase exceeds a certain amount, a more formalized pre-purchase evaluation is conducted, the RFP that involves the competing vendors and manufacturers to submit competitive bids that include the desired hospital specifications.
- Justification of the selected product and vendor and manufacturer is based on results of the pre-purchase evaluation or the RFP, cost-benefit analysis, life cycle costing, and the anticipated cost of ownership that includes, among others, expected cost of maintenance, depreciation, and obsolescence.

CEs, bridging the communication gap between clinicians and design engineers, can positively influence product design. Frequently, the CE explains to the manufacturer those things that can improve a product's efficiency and safety. Consider an example that the authors encountered in which foot supports used on an operating room (OR) bed repeatedly loosened and fell, but only on one side. Clinical engineering analysis revealed that the clamps' unidirectional thread enabled them to tighten down on the bed rail when a load (the foot) was placed on it but to loosen on the other side because the load applied a loosening force. Working with the manufacturer produced a clamp that works on either the right or the left side of an OR bed, without loosening.

At some time in the life of a medical device, it is likely that failure will occur. If the clinical engineering department provides in-house service, the CEs and BMETs must be trained at the same level as that of the manufacturer's service providers. Based on the manufacturer's recommended preventive maintenance schedules, the equipment is included in the scheduled maintenance program.

Maintaining an up-to-date database is a prerequisite to being able to predict early on a device's maintenance schedule. Also, including the projected useful life span of every device in this database enables the CE to begin the replacement process before frequent breakdowns occur, which would force the hospital to take costly measures like renting similar equipment to make up for the shortage.

Ensuring efficacy and effectiveness

Specifying medical devices for cost-effectiveness

Cost-effectiveness means that the product meets all of the desired specifications at a price that is competitive with, or even lower than, that of other vendors. CEs must work closely with the client to prioritize specifications, listing them as (1) those that are necessary; (2) those that are desired but that the organization can live without; and (3) those that are not needed, based on current and anticipated use. It is also best to work closely with the manufacturers, who can recommend the best devices to suit the hospital's needs without including unwanted or nonessential features that would drive up the price.

Anticipating future needs during selection

Health care is a constantly evolving process, and, at times, it is even revolutionary. The instrumentation that is required to deliver current state-of-the-art technologies could become obsolete in a short time. For example, robotics is making great strides in the hospital setting. William Beaumont Hospital's medical laboratories are major users of robotic systems to perform various tests efficiently and cost- effectively, utilizing small specimens. It allowed the hospital laboratory to expand its capacity and capabilities without increasing skilled and educated labor. Robotic surgical systems have also revolutionized the way surgeries are performed in house. A robotic arm system was quickly superseded by a more sophisticated one that is a precursor of one enjoying more widespread use in surgical telemetry.

The CE, with various information sources available, is becoming an unofficial bio-informatics officer. With our participation in the institutional review board (IRB) process in-house; the technology acquisition process; hands-on familiarity with the standard equipment used in-house; communication on almost a daily basis with our clients regarding equipment performance; keeping up with medical research and current trends; consultations with clients, manufacturers, and their representatives, the in-house CE can provide technological guidance to the future.

An engineering professor's favorite question to his students is "When is software considered obsolete?" The answer is, "As soon as it is marketed." CEs need to be alert to the possibility that the product that is recommended to the hospital today will be obsolete in a short time. CEs do not need to have extrasensory perception to be able to look into the future. Being well-read and scientifically curious enables CEs to unearth information that, by itself or in various combinations with other technologies, can help to paint a picture of what is to come.

Evaluating designs for efficiency and reliability

When assessing efficiency and reliability, the CE should have the actual device, its user, service manuals, and an in-service from the manufacturer's representative. A block of time should be set aside to operate the device, while becoming acquainted with its various features and capabilities. This opportunity to learn new technology is quite often stimulating and enjoyable for the CE. Assessing a product's efficiency and reliability, long before it is used, need not be a guessing game. If the product's design and technology are relatively new and untested in the market, the CE looks at the manufacturer's reputation and record of accomplishment for similar products. The authors prefer devices that have been on the market for at least one year. Problems occurring in this one-year introductory period would appear in manufacturers' recalls, warning letters, ECRI Hazard Alerts, and the FDA MAUDE database.

The manufacturer's representatives usually can provide the CE with a list of user hospitals in the region or in the entire country. Checking with these users usually provides one a good insight into what to expect from the product, especially if one communicates with all levels of users and maintainers in the hospital. This is where membership at the local bio-medical engineering society or any similar professional groups can be of great help.

Minimizing user errors through ergonomics in device design

Elegant design often relies upon the KISS ("Keep It Simple, Sweetheart!") principle. The design should make the equipment intuitive, simple, and easy to use. The fewer controls for staff to manipulate, especially during procedures, the better. All critical controls should be on the front panel and not susceptible to fluid spills, which is a constant risk in the hospital setting. Operating the equipment should eliminate, as much as possible, steps that are unnecessary and consolidating those that are similar and that may require a modified sequence of operation. Going beyond the equipment itself, make sure that the supporting infrastructure is available in the areas in which it will be used. A simple oversight can lead to terrible delays, not to mention costs and frayed nerves, as well as having to install or customize the facility to accommodate the equipment's needs.

Criteria for involvement

Clinical engineering services are provided in response to the following conditions:

- Requests by human investigation committee (HIC) or IRB
- Requests by medical staff
- Requests to meet a "critical" patient need
- Humanitarian use devices (humanitarian device exemptions)

Requests by HIC or IRB

Investigational devices are evaluated for safety and efficacy prior to use in the hospital as part of a study, usually as part of the FDA premarket approval (PMA) process. Less frequently, the CE department assesses devices that are designed and constructed by in-house staff.

Requests by medical staff

Often, doctors return from conferences brimming with ideas and requests to introduce new products into the hospital. Where the products show promise of being more cost-effective without compromising on safety and efficacy, they become the subject of extensive CE evaluation and clinical trials. Through this mechanism, beneficial, innovative technology can be made available to the patient population.

Meet a "critical" patient need

At times, the caregiver or the CE conceives of ways to customize or to modify a device to expand its capabilities, thus providing more benefit to the patient and possibly to the user. The CE possesses the education, intellect, skill, and training to perform such customization or re-engineering of a device. Such modification typically requires that the CE expend time in research. The following example illustrates the way the CE does engineering. For certain procedures, OR patients must be positioned on their side, with their 'arms on the side in parallel fashion. Standard arm rests that clamp onto the OR table do not allow this positioning. Alerted to this clinical need, the CE working with the OR staff conceived of a double-decker design. Some research revealed that the local armrest manufacturer sold modular components. A double-decker armrest was created from modular components. In-house manufacture of a specialty device such as this would not have been cost-effective and would have required extensive time to design and prepare mechanical drawings. Even with the aid of computer-aided design (CAD) software, preparing the necessary drawings for an instrument maker to follow would have been prohibitively expensive. In this case, the use of standard modular hardware in a novel configuration accomplished the objective at a minimum expense.

Humanitarian use devices (humanitarian device exemptions)

Humanitarian use devices are exempt from normal FDA 510(k) and PMA processes. The FDA allows the use of such equipment on a limited number of patients where the product might be their best hope (hence the term "humanitarian device exemptions"). The hospital usually treats devices of this sort as investigational and processes the introduction of this technology by way of the IRB. Under the HDE, such devices need not go through the rigorous evaluation that the CE, as a technical consultant to the IRB, must perform. However, an assessment of its safety and efficacy must be weighed against the possible benefits that the product can bring to the patient. In most cases, the risk of patient injury is far outweighed by the alternatives; e.g., permanent disability or death of the patient if he is left untreated.

Redesigning or customizing medical devices

Improving ergonomics

Not all products come customized to a user's personal specifications. By helping to improve the ergonomics of equipment, the CE can help the clinician to perform a better job more safely. Because medical devices are regulated by the FDA, care must be exercised to ensure that specifications are not altered in such a way as to conflict with these regulations. Modifications are typically limited to helping improve a product's ergonomics and do not necessarily change its specifications or performance. Examples of such modifications include the following:

- Mounting an electrosurgical unit (ESU) on a wheeled cart at waist-level height, so that a 5′2″ tall doctor can operate comfortably
- Providing a padded armrest and a comfortable chair for a surgeon who is reconstructing a patient's middle ear
- Selection of foot-pedal controls to prevent accidentally stepping on two pedals at once
- Placing foam pads to protect a surgery patient from pressure necrosis and decubitus ulcer formation

Some in-house designs have been patented, but for the most part, the CE makes these adjustments based on need, without regard to commercialization (see Chapter 5.3).

Prototyping designs to fill specific and specialized needs

After producing mechanical drawings with CAD, the next step is to render the design into a working prototype. Many times, theory does not readily match reality. Engineers should be prepared to encounter the familiar *Murphy's Law*: Anything that can go wrong will go wrong. However, with a bit of persistence, a final design is rendered, and the finished product is executed in collaboration with the machine shop or instrumentation shop. After the device is deployed, care must be taken to monitor closely the product's safety and effectiveness. Close consulting with the clients greatly helps to perfect the finished product.

Typically, the cost of such device-design activity is built into the general operating fund for the hospital. Where the project involves a research or an activity with funding other than by the hospital, the CE department bills the grant or source of funding through the client department. This funding mechanism will differ among institutions.

Should a device design progress to the stage of patenting, the department could realize some financial advantage by selling the rights to manufacture the product. The transition from device design, to patent, to manufacturing, to marketing and sales is long and costly, requiring substantial financial resources and device development infrastructure. Nevertheless, patenting can be an effective way to preserve the rights of the inventor to some remuneration from the invention.

Design/ergonomics

Increased attention is being paid to ergonomics and human factors in medical device design. Increased efficiency and reduction in human error are achievable using an ergonomic approach in equipment design. For example, minimally invasive surgical procedures made possible by ergonomically designed endoscopic instrumentation has reduced (or, in many cases, eliminated) the need for an open incision, thus reducing surgical time and expense and improving patient outcome (e.g., less anesthesia and reduced chance of infection). Robotic arms now extend the surgeon's precision beyond unaided physical capabilities. Lengthy surgeries lead to physicians' tired muscles, but when translating motions into action, robot helpers can eliminate the tremors of a tired surgeon's hands. Years can be added to a surgeon's career by enhancing his senses. The patient population ultimately benefits from the prolonging of a skilled surgeon's longevity.

Documenting

A well-written report can be effective in impressing upon the hospital and clients the CE's thoroughness, knowledge of the subject, and grasp of the issues involved. Writing reports helps the CE to understand more thoroughly what was accomplished in the evaluation, clinical trial, or device design. Putting thoughts into words forces one to consolidate thoughts and to communicate in a logical, understandable fashion for the target audience. On occasion, in the process of writing, various issues, options, and alternate solutions emerge that were not considered during the progress of the work. The CE should never underestimate the power of the written word. A CE's responsibility in this endeavor is to be careful and deliberate in arriving at conclusions and recommendations and to be astute enough to modify them as new information becomes available. Preparing a well-written report can induce a customer-satisfaction response that exceeds financial reward. A satisfied customer will regard the engineer highly and will seek help in the future. Furthermore, the customer is likely to recommend the engineer to colleagues.

A CE's good work is only helpful when it is communicated properly to the correct audience, who can act on the findings and recommendations. Good communication skills require that the author know the intended audience and phrase words accordingly. In the hospital, the CE's reports are usually directed to the administrators, managers, the nursing staff, and the medical staff. These different audiences require slightly different approaches that match their expectations, training, education, and experience. A layperson's approach works well with nonmedical persons. The scientific approach appeals to the clinical professional. The nursing staff seems to respond well to reports that address protocol or standard procedures for its areas of responsibilities. Doctors, by virtue of their extensive scientific education and training, respond well to a reporting style used in medical textbooks and journals. With a mixed audience, if recommendations particularly affect one group, the reporting style should be slanted in that direction. Another option is simply to write two or three different reports to send to the different groups.

Product improvements

CEs are uniquely positioned and qualified to recommend to manufacturers device modifications that help to improve the safety and effectiveness of their products. Such modifications could be simply adding warning labels or changing existing warnings to make them more effective. Many product improvement ideas stem from work performed during engineering investigation of medical device variance reports, engineering evaluations, and clinical trials.

Clinical engineering in the hospital should not be limited to recommending and implementing changes in medical devices and systems. The CE has much to offer in helping clinical staff to modify protocols and other standard procedures to improve safety, efficiency, and the effectiveness of the equipment used. For example, recommending the use of split return pads when using ESUs and advocating for smoke evacuators in the OR where ESU is used, are two measures that enhance patient and staff safety.

Conclusion

The hospital-based CE is in a unique position to advance patient care through engineering skills. Engineering is the process of synthesis and design, the creation of something that was not there before—something that fills a need. The need for good clinical engineering abounds in today's health care organizations. The CE should not be content with the role so often cast by well-intentioned, but inadequately informed, fellow health care professionals. Although some engineers think of device modification and design as merely offshoots of the job of maintaining hospital medical equipment (a view more in keeping with the education), experience and skills of an engineer in that analysis and synthesis is the main stem. Caceres (1980) would be delighted to know that clinical engineering has emerged from the shadows of the distant repair shop to the bedside where engineering skills can be most effectively applied.

Acknowledgment

As staff CEs assigned to different hospital department clients, the authors did a lot of collaborative work, which helped to forge a warm friendship. We acknowledge each other's support and encouragement during this professional affiliation. We have supported each other through many rough bumps and have celebrated each other's career successes. As with almost everything else, all good things must eventually end. Although two of the group have left for positions in other institutions, we thought it fitting to assemble in this chapter some collective thoughts on the role that CEs can play in the hospital, with an emphasis on the design and redesign of medical devices.

References

21 CFR Parts 803 and 807; Medical Devices; Medical Device User Facility and Manufacturing Reporting, Certification and Registration. *Federal Register* 60(237):63578–63607, 1995.

Alder H. *Safe Medical Devices Act of 1990: Current Hospital Requirements and Recommended Actions; Health care Facilities Management Series*. Chicago, American Society for Hospital Engineering of the American Hospital Association, 1993.

American College of Clinical Engineering. *Enhancing Patient Safety: The Role of Clinical Engineering*. Plymouth Meeting, PA, American College of Clinical Engineering, 2001.

Bauld TJ. The Definition of a Clinical Engineer. *J Clin Eng* 16(5):403-405, 1991.

Caceres CA. *Management and Clinical Engineering*, Norwood, MA, Artech House Books, 1980.

JCAHO. Medical Equipment Safety: Meeting the Requirements of the Safe Medical Devices Act. *Plant, Technology & Safety Management Series* 2. Chicago, Joint Commission on Accreditation of Healthcare Organizations, 1993.

Chapter 5.3

5.3

Medical device research and design

Åke Öberg

Engineers who are active in the health care sector have great opportunities to contribute to long-term quality development by developing new techniques or improving existing ones. By working daily with commercially available medical devices and observing ways in which products are used in practice, the clinical engineer (CE) gains valuable knowledge that, when coupled with keen insight and creativity, can lead to ways to improve existing techniques or to solve long-standing problems. The CE has a perspective on developmental requirements, which is an extremely valuable attribute in order to function as an innovator. Furthermore, the CE's proximity to the point of health care delivery provides ample opportunities for the testing and trial of new products in the end-user environment. In the pursuit of improvements in health care, many with expertise in the medical sciences (e.g., doctors, nurses, and therapists) collaborate well with engineers, for they understand the engineer's gifted ability to analyze problems and synthesize solutions. The beneficial result of such analysis and synthesis, the warp and weft of the engineer's mantel, is invention, the creation of something useful; e.g., an object, a machine, or a technique, that did not exist before.

Clinical engineering was born from the concept that engineering attributes (i.e., the analysis and synthesis) are needed to improve health care (Caceres, 1980). Over the last two decades, however, the pendulum has swung and paused over the repair-shop and financial spreadsheets. Mercifully, it is swinging back to its original position, the bedside, the point of care delivery, with no small impetus imparted by recent revelations of inadequacies in the so-called health care delivery system (Kohn *et al.*, 2000). Regulator demands and financial pressures have absorbed and redirected time and talent, thus denying clinical engineering departments the resources for time-intensive product-development activities. Available personnel must resolve the immediate daily tasks such as keeping the hallway clear of equipment, fixing broken infusion pumps, and entering no-problem-found codes in the computer. Inventions and technical development work fell to the lowest rung on the departmental priority ladder. Sadly, hospital administrators underutilized or spurned the talents of even highly competent, well-educated, and skilled CEs, often out of ignorance of their enormous potential, but also in part because of the inability of CEs to articulate their value and to advocate their profession.

The need for improved or new products for health care is the major driving force for innovations and industrial production of medical devices. New medical products are born in the light of new clinical requirements and new technical possibilities. To be able to identify the new needs and possibilities requires a high degree of competence in both the engineering and medical fields.

Developing new products takes time. A 5–10-year period from inception to commercialization is not unrealistic. The interpretation of marketplace trends is important in successful product development. Two main questions arise: Where will health care be 10 years from now, and which new techniques will emerge in the meantime? The needs that one sees today will not necessarily be the same 10 years from now. To correctly assess future trends and needs is of the utmost importance in the development of new, commercially successful products. The world market for medical and health care products is worth approximately $150 billion, of which pharmaceuticals represent $100 billion; the remaining $50 billion represents instrumentation and medical devices (OECD, 1992).

Today health care and the biomedical industry use highly sophisticated technology in their products. It is reasonable to assume that technological progress will find

Biomedical Engineering Desk Reference; ISBN: 9780123746467

early and advanced applications in the health care field. Therefore, it is of interest to discuss personal qualities and knowledge that are prerequisites for the CE who wants to engage in, and to be dedicated to, innovation activities and product development.

From inventor to innovator

An innovation is defined as an invention that has been successfully marketed and has reached a stage of commercial success. Usually, a distinction is made between an inventor and an innovator. The role of an innovator is much more demanding than that of an inventor. An inventor is strongly focused on the technology and the technical development of the invention: the product. The inventor seldom finds invention commercialization exciting and avoids creating the contacts necessary for further business development of the product.

An inventor must learn to become a successful innovator and entrepreneur. Knowledge in the following fields is particularly important to gain:

- **Marketing**, to perform market assessments and to understand the essentials of marketing new products
- **Economy**, to do project budget plans
- **Law**, to the protection that a patent gives

The innovator as a person

Concepts like innovations, innovators, inventors, product ideas, and creativity are commonly used when the importance of product development is discussed. Some of these concepts are considered to be synonymous, but sometimes one needs a short and practically oriented discussion to define these concepts.

"Creativity," for example, has hundreds of definitions in the literature. In most cases, they can be summarized as "the ability to generate something new and/or useful." For others, the word "useful" can be exchanged for "interesting," "attracting" or even "selling." In the case of technical innovations, the "degree of utility" is defined as something that can be sold on the market or can be exploited financially. Within the Organization for Economic Cooperation and Development (OECD), the concept of innovation has been taken to mean a technical idea that has been realized in the form of a product, which has been successful on the market. Creativity is the willingness and ability to produce something new that is closely related to curiousness, an inherent human attribute. In addition, the inclination and the necessity to invent and to explore are deeply satisfying and rewarding activities for the inventor. Also satisfying are creating and developing new ideas and seeing solutions to problems. One reward is to see the ideas realized, a satisfaction that is similar to that of an artist over a finished painting or sculpture. Creativity is also a necessity, even a compulsion. The inventor is often obsessed by thoughts surrounding a problem. It can seem impossible not to come back to the problem repeatedly.

Creative persons might have many unique qualities, which are easily identified such as the following:

- Flexibility
- Sensitivity to problems
- Originality
- Motivation to create
- Endurance
- Concentration on the task

A visionary mind is crucial, e.g., a talent to foresee scenarios in the future. But a visionary mind is not enough. The abilities to get things done to control project development are important. The innovator must constantly draw knowledge from the wells of marketing, economics, visualization, and patent law, all of which are important in the process of developing an idea into a marketable product.

The life cycle of a product

Every product has a life cycle. It is born, matures, reaches a maximal sales figure, and finally disappears from the market. The life cycle curve of a product is a well-known concept within marketing, product development, and market research (Figure 5.3-1). It shows the sold quantity of a product as a function of time from its introduction on the market to the time when the product no longer is marketed, and it is usually divided into phases (Ohlsson, 1992):

- **Introduction**–when the product is first introduced on a market and the first sales occur.
- **Sales growth**–when the product has been on the market for some time and the awareness of its

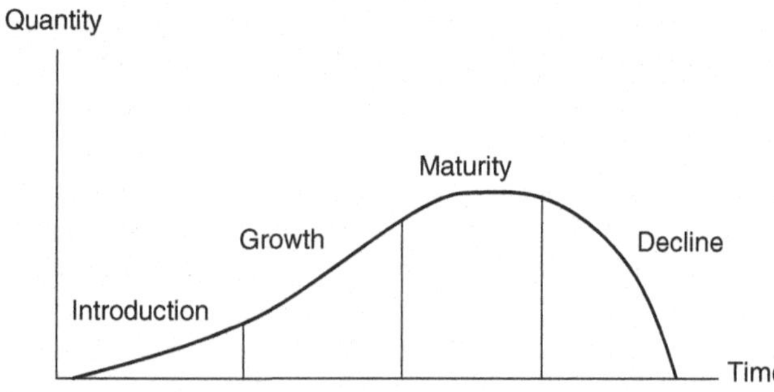

Figure 5.3-1 Product life cycle curve from introduction to obsolescence.

existence spreads and increasing numbers of the product are sold.

- **Maturity**–where the consumers interested in the product have already bought it and sales no longer increase.
- **Decline**–where the demand for the product decreases.

The duration of the life cycle can vary from a few months to many decades. The life cycle curve is a useful instrument by which the future market potential for a product can be analyzed. If the sales of a particular product are studied, one usually can tell which of the phases the demand will be in, at a given moment. Usually, the introduction and development phases are more interesting than the saturation and decline phases. The future total consumption can be calculated also by means of the life cycle curve. Using today's data, the future market potential can be forecast.

Patenting and publishing

Good patent protection is essential for anyone who wishes to exploit an invention industrially, to invest in it, and to manufacture it. The patent is a guarantee that no one else, without risking some form of legal action, can exploit the original idea for commercial profit. In order to secure a patent for an invention, it must be "new". If one places the idea before the public by means of a publication or a lecture, then patenting possibilities are voided, thus creating a high hurdle to jump in the path to product innovation.

At university hospitals and other research-oriented hospitals, there are often visible differences in outlook between the academic researchers who will publish, or give lectures on, their new results, and the innovators who will patent new ideas as a base for industrial development. Often, both groups can be satisfied by patenting first, and then publishing somewhat later. By this procedure, the publishing is marginally delayed while at the same time the great commercial value of such a patent can be utilized.

Strategic assessments for effective product development

A preliminary assessment of a new product idea must always include the following considerations:

- Market analysis
- Technology assessment
- Analyses

The results from these preliminary assessments are important cornerstones in the management of product development projects leading to a commercial success.

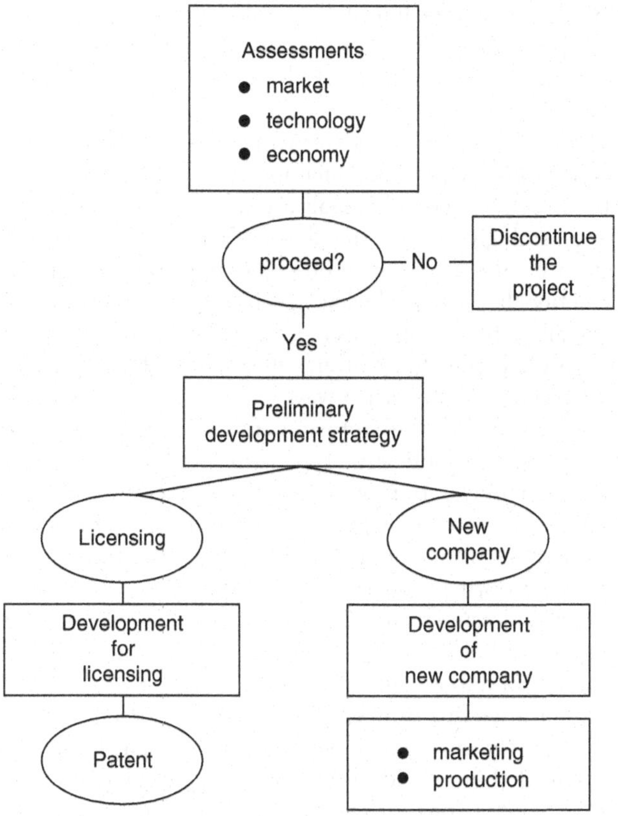

Figure 5.3-2 Product development flow diagram.

Figure 5.3-2 is a flow diagram showing the steps in product development, starting with assessment and moving toward patenting and marketing and production.

Market analysis

Early market assessments will answer two important questions (Committee of Science and Technology, 1980):

1. Is there a need (i.e., a "market") for the product that is under study?
2. Is the market large enough that an investment in a development project would be profitable?

Whether developing simple or complicated technology, one should aim to gather a good overview of the market. An early and preliminary view of a potential market helps the innovator to plan for the long-range development of the invention.

Technology assessment

Technology assessment includes analysis of problems relating to product technology as well as production technology. Product technology assessment is particularly important for medical products in the clinical environment. Of course, safety considerations are greatly important. The technical design of a medical device must comply with existing national and international standards and regulations. Accepted design principles must be used.

The involvement of external persons who are not familiar with the project is an important part of technology assessment. Such individuals are valuable for conducting tests and giving user feedback. Such an involvement often results in major improvements in the design and functionality of a device. Technology assessment of a small number of devices (e.g., 5–10) in "the real world" of hospitals and clinics often will affect the final design of the product in a decisive way.

Involvement of "external" persons in product assessment must include some type of protection in terms of a secrecy declaration or contract (nondisclosure or confidentiality agreement) in which the involved person is prohibited from revealing or utilizing the knowledge obtained during their evaluation work.

The assessment of production technology problems is important before mass production starts. The choice of proper production technologies can strongly affect the market price of a product. If new production equipment must be set up, then large investments might be necessary, and they would affect the economy of the whole project. If a long production run of many devices is contemplated, a production technology-oriented design will result in large savings, particularly for an inventor in a university or hospital environment, who usually does not have extensive experience in industrial production techniques. The initiation of close and early collaboration between the inventor and the final (industrial) producer of the device is strongly recommended. Such collaboration often shortens the route to the market considerably and can reduce production costs.

Economic assessment

According to its definition, an innovation must involve financial success on the market. Thus, it is essential that the project costs are monitored regularly and that, as early as possible, an assessment be made of the feasibility of the product being a technical and a commercial success. At an early stage, only rough calculations can be made. They should be based on the cost of product sales and an estimate of the size of the product's market share that is necessary to the product being a financial success.

Concept testing

Before starting to exploit an idea technically and commercially, an investigation should be launched to ascertain whether it really is new. A patent search often reveals that the same, or a similar, idea has been invented already. Concept testing is important for answering the following questions:

- Is the idea new?
- Does the invention infringe on another patent upon commercial exploitation?
- Can the idea be patented?
- What has been done in this area lately?

Concept testing can be carried out partly by screening patents in areas in the proximity of the current invention. A good conception of the "state-of-the-art" also can be gained by searching databases of different types. Libraries have large databases where correctly formulated questions in the form of search profiles can give adequate guidance on innovations in the current area of work. Published scientific literature can provide interesting information on the news value of an invention. The literature can suggest the diagnostic value that is inherent in the utilization of the invention. This, in turn, is an indicator of market size.

Prototype development

Prototype design is an important step of the innovation process. Prototype development enables one to test the strength of an idea from a technical and an economic point of view, to evaluate the response of the market, and to calculate the production cost. It is usually a long way from the prototype stage to mass production, and often it is practical to proceed in several steps:

- Sketches: Show the idea from a functional and technical point of view in terms of drawings or technical diagrams.
- Models: A three-dimensional presentation of the product idea in wood, metal, or plaster, without any functional demands. This form is especially good for marketing purposes.
- Functional model: A three-dimensional model, as the one mentioned above, but one that also can demonstrate how the idea will work.
- Prototype: The prototype is an exact model of the product as it will function and appear in mass production.
- O-series: A smaller series of the product, often used for demonstration to prospective customers, or for evaluation by experts
- Mass-produced products: Products from the production line

Forms of exploitation

Commercial exploitation of an invention can be accomplished in many different ways. The two most usual are

Table 5.3-1 Comparison of activities on exploiting an invention

Activity	Demand	
	New company	Licensing
Patenting	Less	More
Prototype design	Same	Same
Marketing	More	Less
Manufacturing	More	Less
Financing	Same	Same
Organization	More	Less
Time span	Better	Worse
Risks	Same	Same

licensing and the starting of new companies. Licensing entails transferring the rights of the invention to an entrepreneur, who is responsible for continued development and commercial introduction. Generally, when choosing the form of licensing, the inventor has strong patent protection as a basis for negotiations with the entrepreneur. Without strong patent protection, the licensing negotiations are difficult, as it is generally the patent that is, at this time, the only meaningful feature in the negotiation. In some cases, it can be advantageous to start a new company that conducts the development work and is responsible for market launching. If the invention to be exploited concerns completely new areas where few or no established companies are active, then a start-up company for commercialization can be a quicker route to the market. After successful product development and market launching, the company can be sold. Upon starting a company, the patent issue is generally less important, while matters concerning marketing and manufacturing are more important. After the invention has been assessed from market, technical, and economic perspectives, it is time to decide on the form of full-scale exploitation.

Often, it is the inventor's own disposition that determines the form of exploitation chosen. If the inventor is more interested in the continuous development of a new technique, then licensing (which amounts to transferring the marketing and financial problems to someone else) is the choice. If, however, the inventor finds it stimulating to work with the entire production process from technical development work through to sales, then starting a company is the choice. A CE who is employed by a hospital will find that the hospital's policy also influences the choice of form of exploitation. Table 5.3-1 describes differences between the two forms of exploitation.

References

Caceres CA. *Management and Clinical Engineering*. Artech House Books, 1980.

OECD: *Innovation Policy-Trends and Perspectives*. Paris, OECD, 1992.

Kohn LT, Corrigan JM, Donaldson MS. *To Err is Human: Building a Safer Health System*, National Academy Press, 2000.

Ohlsson L. *R&D for Swedish Industrial Renewal* (DS 1992:109). Stockholm, Utbildningsdepartementet, 1992.

Committee of Science and Technology. *Small, High Technology Firms and Innovation.* Washington, D.C., U.S. Government Printing Office, 1980.

Fölster S. *The Art of Encouraging Invention.* Stockholm, IUI, 1991.

Further information

Brown KA. *Inventors at Work: Interviews with 16 Notable American Inventors*, TEMPUS Books, 1988. This book gives interesting perspectives on the invention process.

In addition, the serials *Journal of Medical and Biological Engineering and Computing, Physiological Measurements, Medical Engineering*, and *Medical Physics* publish papers on new instrument ideas. National authorities give advice and recommend ways to start new companies.

Chapter 5.4

5.4

Medical device software development

Richard C. Fries and Andre E. Bloesch

Many kinds of medical devices are rapidly becoming software-intensive. Software controls their operation, collects and analyzes information to help make treatment decisions, and provides a way for users to interface with the medical device. In these devices, the software transforms a general purpose computer into a special-purpose medical device component. As in hardware design, specifying the software requirements, creating a sound software design and correctly implementing it are difficult intellectual challenges. Good software development is based on a combination of creativity and discipline. Creativity provides resolution to new technical hurdles and the challenges of new market and user needs. Discipline provides quality and reliability to the final product.

Software design and implementation is a multi-staged process in which system and software requirements are translated into a functional program that addresses each requirement. Software design begins with the work products of the Software Requirements Specification (SRS). The design itself is the system architecture, which addresses each of the requirements of the specification and any appropriate software standards or regulations. The design begins with the analysis of software-design alternatives and their trade-offs. The overall software architecture is then established, along with the design methodology to be used and the programming language to be implemented. A risk analysis is performed and then refined to ensure that malfunction of any software component will not cause harm to the patient, the user, or the system. Metrics are established to check for program effectiveness and reliability. The Requirements Traceability Matrix (RTM) is reviewed to ensure that all requirements have been addressed. Peers The review the software design for completeness.

The design continues with modularizing the software architecture, assigning specific functionality to each component and ensuring that internal and external interfaces are well defined. Coding style and techniques are chosen based on their proven value and the intended function and environment of the system. Peer reviews ensure the completeness and effectiveness of the design. The detailed design also establishes the basis for subsequent verification and validation activity. The use of automated tools throughout the development program is an effective method for streamlining the design and development process and assists in developing the necessary documentation.

The key to success in verification and validation is planning. Verification and validation planning encompasses the entire development life cycle, from requirements generation to product release. The initial planning of verification and validation is documented in a Software Verification and Validation Plan (SVVP). The SVVP describes the verification and validation life cycle, gives an overview of verification and validation, describes the verification and validation life cycle activities, defines the verification and validation documentation, and discusses the verification and validation administrative procedures. An excellent guideline for verification and validation planning can be found in IEEE Std 1012, with an explication of Std 1012 found in IEEE Std 1059.

Software standards and regulations

There are a myriad of software standards to assist the developer in designing and documenting a software program. IEEE standards cover documentation through all phases of design. Military standards describe the way that software is to be designed and developed for military use. There are also standards on software quality and

Biomedical Engineering Desk Reference; ISBN: 9780123746467

reliability to assist developers in preparing a quality program. The international community has produced standards that primarily address software safety. In each case, the standard is a voluntary document that has been developed to provide guidelines for designing, developing, testing, and documenting a software program.

In the United States, the FDA is responsible for ensuring that the device utilizing the software, or the software as a device, is safe and effective for its intended use. The FDA has produced several drafts of reviewer guidelines, auditor guidelines, software policy, and good manufacturing practices (GMP) regulations addressing device and process software. In addition, guidelines for FDA reviewers have been prepared, as have training programs for inspectors and reviewers. The new version of the GMP regulation addresses software as part of the design phase.

The United States is ahead of other countries in establishing guidelines for medical software development. However, there is movement within several international organizations to develop regulations and guidelines for software and software-controlled devices. For example, ISO 9000-3 specifically addresses software development in addition to what is contained in ISO 9001. CSA addresses software issues in four standards covering new and previously developed software in critical and noncritical applications. IEC has a software-development document currently in development. They also have a risk analysis document (IEC 601-1-4).

Design alternatives and tradeoffs

The determination of the design and the allocation of requirements is a highly iterative process. Alternative designs are postulated that could, or are candidates to, satisfy the requirements. The determination of these designs is a fundamentally creative activity, a "cut and try" determination of what might work. The specific techniques used are numerous and call upon a broad range of skills. They include control theory, optimization, consideration of man-machine interface, use of modern control test equipment, queuing theory, communication and computer engineering, statistics, and other disciplines. These techniques are applied to factors such as performance, reliability, schedule, cost, maintainability, power consumption, weight, and life expectancy.

Some of the alternative designs will be quickly discarded, while others will require more careful analysis. The capabilities and quality of each design alternative is assessed using a set of design factors that are specific to each application and to the methods of representing the system design.

Certain design alternatives will be superior in some respects, while others will be superior in different respects. These alternatives are "traded off," one against the other, in terms of the factors that are important for the system being designed. The design ensues from a series of technology decisions, which are documented with architecture diagrams that combine aspects of data and control flow. As an iterative component of making technology decisions, the functionality expressed by the data flow and control flow diagrams from system requirements analysis is allocated to the various components of the system. Although the methods for selection of specific technology components are not a part of the methodology, the consequences of the decisions are documented in internal performance requirements and timing diagrams.

Finally, all factors are taken into account, including customer desires and political issues to establish the complete system design. The product of the system design is called an "architecture model." The architecture includes the components of the system, allocation of requirements, and topics such as maintenance, reliability, redundancy, and self-test.

Software architecture

Software architecture is the high-level part of software design, the frame that holds the more detailed parts of the design. Typically, the architecture is described in a single document called the "architecture specification." The architecture must be a prerequisite to the detailed design, because the quality of the architecture determines the conceptual integrity of the system. This, in turn, determines the ultimate quality of the system.

A system architecture first needs an overview that describes the system in broad terms. It should also contain evidence that alternatives to the final organization have been considered and the reasons why the organization used was chosen over the alternatives. The architecture should additionally contain:

- Definition of the major modules in a program. What each module does and its interface should be well defined
- Description of the major files, tables, and data structures to be used. It should describe alternatives that were considered and should justify the choices that were made
- Description of specific algorithms or reference to them
- Description of alternative algorithms that were considered and indicate the reasons why certain algorithms were chosen
- In an object-oriented system, specification of the major objects to be implemented. It should identify the responsibilities of each major object and ways in which the object interacts with other objects. It

should include descriptions of the class hierarchies, of state transitions, and of object persistence. It should also describe other objects that were considered and should give reasons for preferring the chosen organization.

- Description of a strategy for handling changes clearly. It should show that possible enhancements have been considered and that the enhancements most likely are also the easiest to implement
- Estimation of the amount of memory used for nominal and extreme cases

Software architecture contemplates two important characteristics of a computer program: The hierarchical structure of procedural components (modules); and the structure of data. Software architecture is derived through a partitioning process that relates elements of a software solution to parts of a real-world problem implicitly defined during requirements analysis. The evolution of a software structure begins with a problem definition. The solution occurs when each part of the problem is solved by one or more software elements.

Choosing a methodology

It seems there are about as many design methodologies as there are engineers to implement them. Typically, the methodology selection entails a prescription for the requirements analysis and design processes. Of the many popular methods, each has its own merit, based on the application to which the methods are applied. The tool set and methodology selection should run hand-in-hand. Tools should be procured to support established or tentative design methodology and implementation plans. In some cases, tools are purchased to support a methodology already in place. In other cases, an available tool set dictates the methodology. Ideally, the two are selected at the same time, following a thorough evaluation of need.

Selecting the right tool set and design methodology should not be based on a flashy advertisement or a suggestion from an authoritative methodology guru. It is important to understand the environment in which it will be employed and the product to which it will be applied. Among other criteria, the decision should be based on the size of the project (i.e., the number of requirements), the type of requirements (i.e., hard or soft real-time), the complexity of the end-product, the number of engineers, the experience and skill level of the engineers, the project schedules, the project budget, the reliability requirements, and the future enhancements to the product (i.e., maintenance concerns). Weight factors should be applied to the evaluation criteria. One way or another, whether the evaluation is conducted in a formal or informal way, involving one person or more, it should be done to ensure a proper fit for the organization and product.

Regardless of the approach used, the most important factor to be considered for the successful implementation of a design methodology is software-development-team buy-in. The software development team must possess the confidence that the approach is appropriate for the application and must be willing and "excited" to tackle the project. The implementation of a design methodology takes relentless discipline. Many projects have been unsuccessful as a result of lack of commitment and faith.

The two most popular formal approaches applied to the design of medical products are the Object Oriented Analysis/Design and, the more traditional, (top-down) Structured Analysis/Design. There are advantages and disadvantages to each. Either approach, if done in a disciplined and systematic manner along with the electrical system design, can provide for a safe and effective product.

Choosing a language

Programming languages are the notational mechanisms that are used to implement software products. Features available in the implementation language exert a strong influence on the architectural structure and algorithmic details of the software. Choice of language has also been shown to have an influence on programmer productivity. Industry data have shown that programmers are more productive using a familiar language than an unfamiliar one. Programmers who work with high-level languages achieve better productivity than do those who work with lower-level languages. Developers who work with interpreted languages tend to be more productive than those who work with compiled languages. In languages that are available in both interpreted and compiled forms, programs can be productively developed in the interpreted form, and then released in the better-performing, compiled form.

Modern programming languages provide a variety of features to support development and maintenance of software products. These features include:

- Strong type checking
- Separate compilation
- User-defined data types
- Data encapsulation
- Data abstraction.

The major issue in type checking is flexibility versus security. Strongly typed languages provide maximum security, while automatic type coercion provides maximum flexibility. The modern trend is to augment strong type checking with features that increase flexibility while maintaining the security of strong type checking.

Separate compilation allows retention of program modules in a library. The modules are linked into the software system, as appropriate, by the linking loader. The distinction between independent compilation and separate compilation is that type checking across compilation-unit interfaces is performed by a separate compilation facility, but not by an independent compilation facility. User-defined data types, in conjunction with strong type checking, allow the programmer to model and to segregate entities from the problem domain using a different data type for each type of problem entity. Data encapsulation defines composite data objects in terms of the operations that can be performed on them, and the details of data representation and data manipulation are suppressed by the mechanisms. Data encapsulation differs from abstract data types in that encapsulation provides only one instance of an entity.

Data abstraction provides a powerful mechanism for writing well-structured, easily modified programs. The internal details of data representation and data manipulation can be changed at will and, provided that the interfaces of the manipulation procedures remain the same, other components of the program will be unaffected by the change, except perhaps for changes in performance characteristics and capacity limits. Using a data-abstraction facility, data entities can be defined in terms of predefined types, used-defined types, and other data abstractions, thus permitting systematic development of hierarchical abstractions.

Software risk analysis

Software risk analysis techniques identify software hazards and safety-critical single- and multiple-failure sequences; determine software safety requirements; including timing requirements; and analyze and measure software for safety. While functional requirements often focus on what the system shall do, risk requirements must also include what the system shall not do, including means of eliminating and controlling system hazards and of limiting damage in case of a mishap. An important part of the risk requirements is the specification of the ways in which the software and the system can fail safely and the extent to which failure is tolerable.

Several techniques have been proposed and used in for conducting risk analysis, including:

- Software hazard analysis
- Software fault tree analysis
- Real time logic

Software hazard analysis, like hardware hazard analysis, is the process whereby hazards are identified and categorized with respect to criticality and probability. Potential hazards that must be considered include normal operating modes, maintenance modes, system failure or unusual incidents in the environment, and errors in human performance. Once hazards are identified, they are assigned a severity and probability. Severity involves a qualitative measure of the worst credible mishap that could result from the hazard. Probability refers to the frequency with which the hazard occurs. Once the probability and severity are determined, a control mode (that is, a means of reducing the probability and/or severity of the associated potential hazard) is established. Finally, a control method or methods will be selected, to achieve the associated control mode.

Real-time logic is a process whereby the system designer first specifies a model of the system in terms of events and actions. The event-action model describes the data-dependency and temporal ordering of the computational actions that must be taken in response to events in a real-time application. The model can be translated into Real Time Logic formulas. The formulas are transformed into predicates of Presburger arithmetic with uninterpreted integer functions. Decision procedures are then used to determine whether a given risk assertion is a theorem that is derivable from the system specification. If so, the system is safe with respect to the timing behavior denoted by that assertion, as long as the implementation satisfies the requirements specification. If the risk assertion is unsatisfiable with respect to the specification, then the system is inherently unsafe because successful implementation of the requirements will cause the risk assertion to be violated. Finally, if the negation of the risk assertion is satisfiable under certain conditions, then additional constraints must be imposed on the system to ensure its safety.

Software metrics

Software must be subjected to measurement in order to achieve a true indication of quality and reliability. Quality attributes must be related to specific product requirements and must be quantifiable. These aims are accomplished through the use of metrics. Software-quality metrics are defined as quantitative measures of an attribute that describes the quality of a software product or process. Using metrics for improving software quality, performance, and productivity begins with a documented software development process that will be improved incrementally. Goals are established with respect to the desired extent of quality and productivity improvements over a specified time period. These goals are derived from, and are consistent with, the strategic goals for the business enterprise.

Metrics that are useful to the specific objectives of the program, that have been derived from the program requirements, and that support the evaluation of the software consistent with the specified requirements must be selected. To develop accurate estimates, a historical baseline must be established, consisting of data collected from previous software projects. The data collected should be reasonably accurate, collected from as many projects as possible, consistent, and representative of applications that are similar to work that is to be estimated. Once the data have been collected, metric computation is possible.

Metrics that can be used to measure periodic progress in achieving the improvement goals are then defined. The metric data collected can be used as indicators of development process problem areas, and improvement actions identified. These actions can be compared and analyzed with respect to the best return on the business's investment. The measurement data provide information for investing wisely in tools for quality and productivity improvement.

A feedback mechanism must be implemented so that the metrics data can provide guidance for identifying actions to improve the software development process. Continuous improvements to the software development process result in higher-quality products and increased development team productivity. The process-improvement actions must be managed and controlled so as to achieve dynamic process improvement over time.

Requirements traceability

It is becoming increasingly apparent how important thorough requirements traceability is, during the design and development stages of a software product, especially in large projects with requirements that number in the thousands or tens of thousands. Regardless of the design and implementation methodology it is important to ensure that the design is meeting its requirements during all phases of design. To ensure that the product is designed and developed in accordance with its requirements throughout the development cycle, individual requirements should be assigned to design components. Each software requirement, as might appear in a SRS, for example, should be uniquely identifiable. Requirements that result from design decisions (i.e., implementation requirements) should be uniquely identified and tracked along with product functional requirements. This process not only ensures that all functional and safety features are built into the product as specified, but also drastically reduces the possibility of requirements "slipping through the cracks". Overlooked features can be much more expensive when they become design modifications at the tail end of development.

The RTM is generally a tabular format with requirement identifiers as rows, and design entities as column headings. Individual matrix cells are marked with file names or design-model identifiers to denote that a requirement is satisfied within a design entity. A RTM ensures completeness and consistency with the software specification, which can be accomplished by forming a table that lists the requirements from the specification versus the way each is met in each phase of the software-development process.

Software reviews

Timely and well-defined reviews are integral parts of all design processes. Each level of design should produce design-review deliverables. Software project development plans should include a list of the design phases, the expected deliverables for each phase, and a sound definition of the deliverables to be audited at each review.

Reviews of all design material have several benefits. First and foremost, authors are more compelled to elevate the quality of their work when they know that their work is being reviewed. Second, reviews often uncover design blind spots and alternative design approaches. Finally, the documentation generated by the reviews is used to acquire agency approvals for process and product.

Software reviews can take several different forms:

- Inspections, design, and code
- Code walk-throughs
- Code reading
- Dog-and-pony shows

An inspection is a specific kind of review that has been shown to be extremely effective in detecting defects, and to be relatively economical as compared to testing. Inspections differ from the usual reviews in several ways:

- Checklists focus the reviewer's attention on areas that have been problems in the past
- The emphasis is on defect detection, not correction
- Reviewers prepare for the inspection meeting beforehand and arrive with a list of the problems they have discovered
- Data are collected at each inspection and are fed into future inspections to improve them

The general experience with inspections has been that the combination of design and code inspections usually removes 60–90% of the defects in a product. Inspections identify error-prone routines early, and reports indicate that they result in 30% fewer defects per 1000 lines of code than walkthroughs do. The inspection process is systematic because of its standard checklists and standard roles. It is also self-optimizing because it uses a formal feedback loop to improve the checklists and to monitor preparation and inspection rates.

A walkthrough usually involves two or more people discussing a design or code. It might be as informal as an impromptu bull session around a whiteboard; it might be as formal as a scheduled meeting with overhead transparencies and a formal report sent to management.

Following are some of the characteristics of a walkthrough:

- The walkthrough is usually hosted and moderated by the author of the design or code under review.
- The purpose of the walkthrough is to improve the technical quality of a program, rather that to assess it.
- All participants prepare for the walkthrough by reading design or code documents and by looking for areas of concern.
- The emphasis is on error detection, not correction.
- The walkthrough concept is flexible and can be adapted to the specific needs of the organization using it.

When used intelligently, a walkthrough can produce results that are similar to those of an inspection; i.e., it typically can find between 30–70% of the errors in a program. Walkthroughs have been shown to be marginally less effective than inspections, but under some circumstances they can be preferable.

Code reading is an alternative to inspections and walkthroughs. In code reading, source code is read for errors. Readers also comment on qualitative aspects of the code, such as its design, style, readability, maintainability, and efficiency. A code reading usually involves two or more people reading code independently and then meeting with the author of the code to discuss it. To prepare for a meeting, the author hands out source listings to the code readers. Two or more people read the code independently. When the reviewers have finished reading the code, the code-reading meeting is hosted by the author of the code and focuses on problems that the reviewers have discovered. Finally, the author of the code fixes the problems that the reviewers have identified.

"Dog-and-pony shows" are reviews in which a software product is demonstrated to a customer. The purpose of these reviews is to demonstrate to the customer that the project is viable, so they are management reviews rather than technical reviews. They should not be relied upon to improve the technical quality of a program. Technical improvement comes from inspections, walkthroughs, and code reading.

Performance predictability and design simulation

The effort to predict the real-time performance of a system is a key activity of design that some software developers often overlook. During the integration phase, software designers often spend countless hours trying to fine-tune a system that has bottlenecks "designed in". Execution estimates for the system interfaces, response times for external devices, algorithm execution times, operating system context-switch time, and I/O device access times in the forefront of the design process provide essential input into software design specifications.

For single processor designs, mathematical modeling techniques such as Rate Monotonic Analysis (RMA) should be applied, to ensure that all required operations of that processing unit can be performed within the expected timeframes. System designers often fall into the trap of selecting processors before the software design has been considered, only to experience major disappointment and "finger pointing" when the product is released. It is imperative to a successful project that the processor selection comes after a processor-loading study is complete.

In a multiprocessor application, up-front system performance analysis is equally important. System anomalies can be very difficult to diagnose and resolve in multiprocessor systems with heavy interprocessor communications and functional expectations. Performance shortcomings that appear to be the fault of one processor are often the result of a landslide of smaller inadequacies from one or more of the other processors or subsystems. Person-years of integration phase defect resolution can be eliminated by front-end system design analysis and/or design simulation. Commercial tools are readily available to help perform network and multiprocessor communications analysis and execution simulation. Considering the pyramid of effort that is needed in software design, defect correction in the forefront of design yields enormous cost savings and increased reliability in the end.

Coding

For many years, the term "software development" was synonymous with coding. Today, for many software-development groups, coding is now one of the shortest phases of software development. In fact, in some cases, although very rare in the world of real-time embedded software development, coding is actually done automatically from higher-level design (mspecs) documentation by automated tools called "code generators."

With or without automatic code generators, the effectiveness of the coding stage is dependent on the quality and completeness of the design documentation generated in the immediately preceding software development phase. The coding process should be a simple transition from the module specifications, and, in particular, the pseudocode. Complete mspecs and properly developed pseudocode leave little to interpretation for the coding phase, thus reducing the chance of error.

The importance of coding style (how it looks) is not as great as the rules that facilitate comprehension of the logical flow (how it relates). In the same light, in-line code documentation (comments) should most often address "why" rather than "how" functionality is implemented. These two focuses help the code reader to understand the context in which a given segment of code is used. With precious few exceptions (e.g., high-performance-device drivers) quality source code should be recognized by its readability, and not by its raw size (i.e., the number of lines) or its ability to take advantage of processor features.

Design support tools

Software development is very labor-intensive and, is therefore, prone to human error. In recent years, commercial software-development support packages have become increasingly more powerful, less expensive, and more available to reduce the time spent doing things that computers do. Although selection of the right tools can mean up-front dedication of some of the most talented resources in a development team, it also can bring about significant long-term increase in group productivity.

Good software-development houses have taken advantage of CASE tools that reduce the time spent generating clear and thorough design documentation. The advantages of automated software-design packages are many. Formal documentation can be used as proof of product development procedure conformance for agency approvals. Clear and up-to-date design documents facilitate improved communications between engineers, thus lending to more effective and reliable designs. Standard documentation formats reduce learning curves that are associated with unique design depictions among software designers, thus leading to better and more timely design formulation. Total software life cycle costs are reduced (especially during maintenance) due to reduced ramp-up time and more efficient and reliable modifications. Finally, electronic forms of documentation can be easily backed up and stored off-site, thus eliminating a crisis in the event of an environmental disaster. In summary, the adaptation of CASE tools has an associated up-front cost that is recovered by significant improvements in software quality and development-time predictability.

Design as the basis for verification and validation

Verification is the process of ensuring that all products of a given development phase satisfy given expectations. Prior to proceeding to the next-lower level or phase of design, the product (or outputs) of the current phase should be verified against the inputs of the previous stage. A design process cannot be a "good" process without the verification process ingrained—they naturally go hand-in-hand.

Software project management plans (or software quality assurance plans (SQAP)) should specify all design reviews. Each level of design will generate documentation to be reviewed or deliverables to verify against the demands of the previous stage. For each type of review, the software-management plans should describe the purpose, materials required, scheduling rules, scope of review, attendance expectations, review responsibilities, what the minutes should look like, follow-up activities, and any other requirements that relate to company expectations.

At the code level, code reviews should ensure that the functionality implemented within a routine satisfies all expectations documented in the "mspecs." Code also should be inspected, to satisfy all coding rules.

The output of good software designs also includes implementation requirements. At a minimum, implementation requirements include the rules and expectations placed on the designers to ensure design uniformity as well as constraints, controls, and expectations placed on designs to ensure that upper-level requirements are met. General examples of implementation requirements might include rules for accessing I/O ports, timing requirements for memory accesses, semaphore arbitration, inter-task communication schemes, memory-addressing assignments, and sensor- or device-control rules. The software-verification and -validation process must address implementation requirements as well as the upper-level software requirements to ensure that the product works as it was designed to.

Verification and validation life cycle

An SVVP lays out the framework from which the verification and validation activities proceed. IEEE 1012-1986 is very detailed about development life cycle and the verification and validation products that are generated at each phase. Specifically, the standard calls out the following phases:

- Concept phase
- Requirements phase
- Design phase
- Implementation phase
- Test phase
- Installation phase
- Operation and maintenance phase

In industry, it is more typical for software verification and validation to follow a simpler model, as shown in the Figure 5.4-1.

This model is simpler in that it focuses verification and validation activities on the software that is generated,

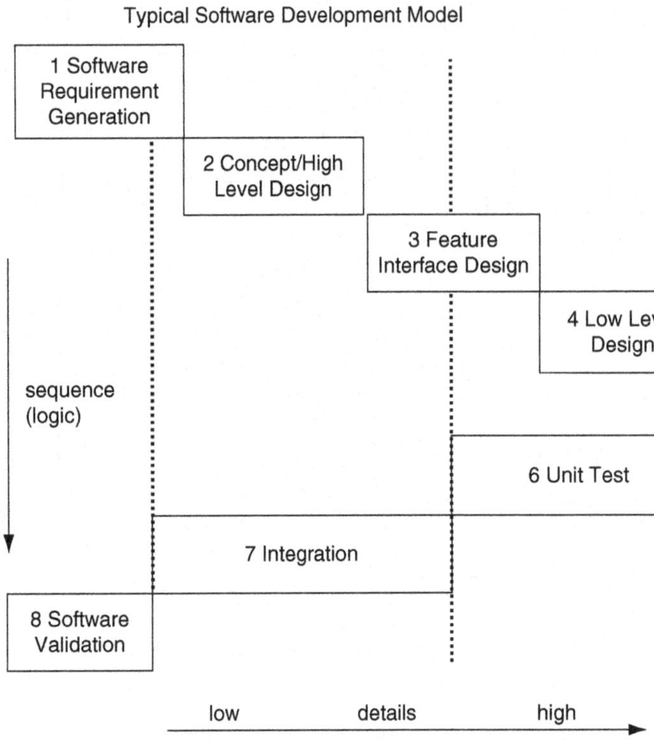

Figure 5.4-1 Typical software development model.

rather than on output of every phase of development. Another document, called the SQAP, is used to cover other verification and validation aspects of product development, such as software requirements reviews, hazard and risk analysis, and design reviews.

Using the model as shown in Figure 5.4-1, software validation testing covers validation of the software product by testing against the requirements generated in the requirements-generation phase. Software validation is analogous to black box testing. Integration testing covers verification of features defined from the requirements-generation phase, the concept/high-level design phase, and the feature-interface design phase. Integration testing tests the collection of software modules, where the software modules are joined together to provide a product feature or higher level of functionality. Finally, unit testing covers verification at the lowest level, which includes the feature interface design phase and the low-level design phase. Unit testing typically covers the verification of smaller collections of software modules.

Verification and validation overview

This overview describes many of the project management details that must be addressed in order to perform the verification and validation functions. They include the verification and validation organization; a schedule; resources; responsibilities; and any special tools, techniques, and methodologies.

The organization describes who is responsible for carrying out the various verification and validation efforts. In practice, this section is used to describe who will be responsible for maintaining the laboratory, who will manage the validation testing, who will manage verification testing, and who will be responsible for anomaly resolution.

Verification and validation scheduling is very important for planning activities in order to support the overall product schedule. Often, a higher-level schedule is entered into the SVVP, and a detailed schedule is maintained apart from the SVVP. This is generally due to different software applications for generating a schedule and for generating documentation.

The resources section describes all of the equipment, software, and hardware that are required, to perform verification and validation activities. A subsection also should be included on the validation of using any software tools, where the software tool is used as part of the verification and validation process. Per the FDA's regulations (FDA, Quality System Regulation, 21 CFR Part 820, June 1, 1997), any device that is used in a company's quality system must be validated. A test fixture or a customized software application clearly falls under this definition. Thus, it is necessary to validate the software tool.

Verification and validation life cycle activities

In the SVVP, this section typically includes the criteria, inputs/outputs, reviews, testing approach, and training for each verification and validation phase. The criteria describes the goal that each phase is defined to achieve. The inputs/outputs phase defines what things are needed as inputs into the phase and what the product of each phase are.

It is also helpful to add a subsection on the general functionality that will be tested. The IEEE standard does not call this out. However, adding this subsection is helpful to reviewers of the SVVP. It provides information on the functionality that is scheduled for testing.

Verification and validation documentation

Each verification and validation phase has its activities and documentation that is associated with it. For unit, integration, and software validation test phases, typically there are test plans, test procedures, and test results and reports.

For software validation, there usually are some additional documents. The Software validation group typically has a requirements-to-test cross-reference. The cross reference ensures that all the requirements defined in the SRS are covered in some test. It is also helpful in the cross-reference document to include a test-to-requirement cross-reference. This backward reference helps to ensure that a requirement has adequate coverage. The software validation group also generates a procedure-response document, which serves as a means to close out any issues or comments that occurred during an official validation pass. Finally, the software validation group has been responsible for writing a final report called the software verification and validation report (SVVR), which summarizes the activities from all of the previous verification and validation phases.

Verification and validation administrative procedures

This section provides additional guidance on ways that the verification and validation will be conducted. IEEE Std 1012-1986 defines this section as including anomaly reporting/resolution, task iteration policy, deviation policy, control procedures, and standards/practices/conventions.

It is also useful to include section metrics. IEEE Std 730.1-1989, IEEE Standard for Assurance Plans, includes metrics as part of a SQAP. The following subsections describe in more detail anomaly reporting, metrics collection, and reporting, and they provide some additional topics that are important to verification and validation.

Anomaly reporting and resolution

One result of verification and validation is the generation of anomalies found during testing. A process for reporting anomalies must be in place to record anomalies. Some institutions have used a lab notebook to document anomalies. More frequently, a PC-based database program is used to log and to store anomalies electronically.

In addition, a process must be in place for resolution of anomalies found against the product software. It can be a committee of people who are responsible for the product or simply the software test coordinator who is working with the primary software developer. The responsible parties determine the risk of each anomaly.

Validation metrics

There are several aspects of verification and validation in which it is helpful to generate metrics. The first and most obvious one is anomaly metrics. An example of anomaly metric tracking is shown in Figure 5.4-2. This chart shows the tracking of open severitized anomalies as a function of time.

The second aspect of test metrics is those involving test development. These metrics are used to monitor the test development process and to monitor overall test parameters for all projects. Monitoring the test development process ensures that the test development is proceeding according to schedule. An example is shown in Figure 5.4-3.

Configuration management

Just as it is important to maintain a configuration for product development code, it is equally important to practice configuration management on test protocols. Configuration of test materials includes:

- Script libraries
- Test scripts
- Manual protocol
- Test results
- Test plans and procedures
- Test system design

Protocol templates

To make the development of tests more uniform and efficient, it is advisable to have templates for the test

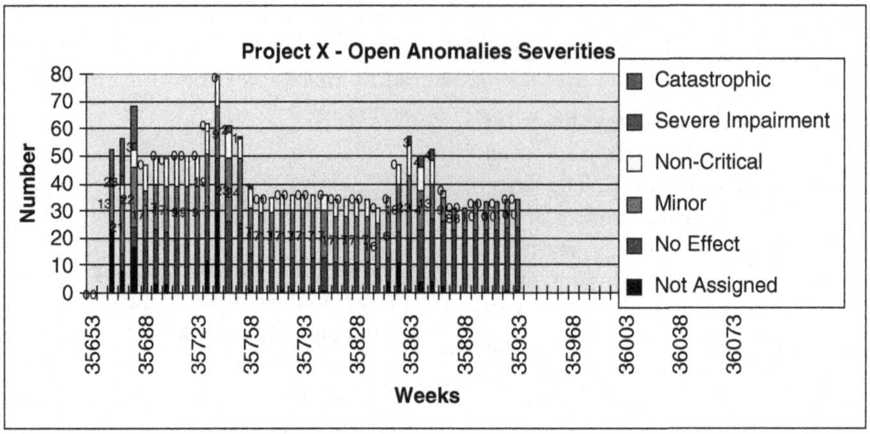

Figure 5.4-2 Metric chart showing open anomalies sorted by severity.

protocol. If test scripts are written, then a set of development guidelines is suggested.

Test execution

This section discusses approaches that are used when executing manual protocol and automated scripts.

Process improvement

No test process is perfect. There are always ways to improve the test-development process to do things more efficiently. Thus, some time should be set aside for review of the current processes that are in place. If the test team is large enough, a process team can be created to address processes and process changes in the test group. Discussions should occur among test team members to determine those processes that worked well and those that did not.

Test development

Requirements analysis and allocating requirements

Today, it is assumed that a SRS is written. The SRS describes the behavior and functionality of the software in the medical device. It should obtain its inputs from the following sources:

- Derivation of the product specification
- Control procedures from the risk and hazard analyses
- Regulatory requirements, such as safety-relevant computing

A large part of the verification and validation effort will concentrate on proving that these software requirements are fulfilled. Therefore, the requirements in the SRS

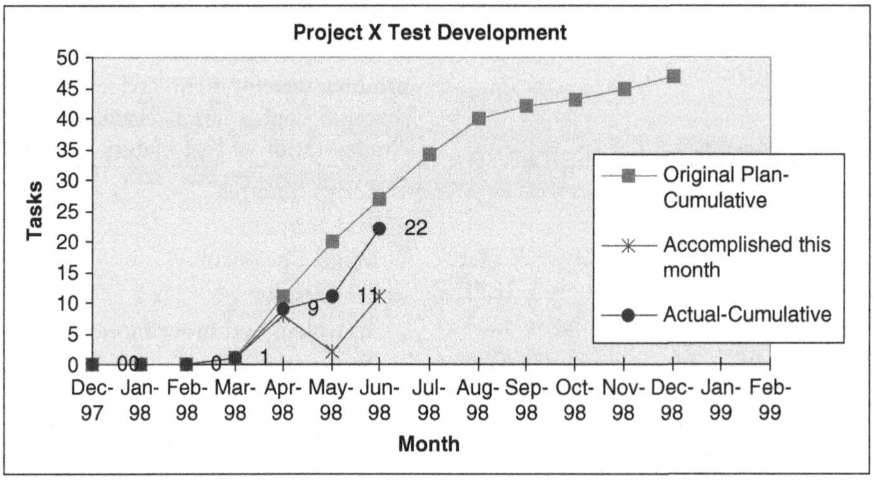

Figure 5.4-3 Monitoring the test-development process.

must be enumerated to permit proper allocation to the different test phases and the specific tests.

The initial analysis of the SRS involves determining where to test a particular requirement. Where a requirement is tested is often based on the capabilities of the test fixture used in validation. Usually, most requirements can be tested at the validation level. The balance of the requirements is then tested during integration or unit testing. In general, a requirement that has external stimulus (to the central processing unit (CPU)) and the ability to monitor externally are done at the validation level. Exceptions to this are conditions that might require very specific timing, and that timing is calculated internally to the CPU.

Requirements that are tested during integration or unit testing are typically those that cannot be tested sufficiently during validation testing. Testing a "software watchdog error" requirement is good example of a requirement tested during integration or unit testing. In this case, the software is monitoring itself, and there is no external means to cause the fault.

The second step of the requirements analysis phase is to determine the best way to group the requirements into tests. One relatively straightforward approach to doing this is to create tests based on sections in the SRS. Typically, a section in an SRS will describe a feature almost fully. By following the SRS format, the tests then are developed by feature, which is a logical approach. Of course, there will be requirements in the SRS where this approach will not apply. In such cases, further analysis must be done to assess whether additional tests should be created, or whether one of these "special" requirements can be allocated into those tests that follow the SRS.

As mentioned at the beginning of this section, control procedures from the risk and hazard analysis should be written into the SRS. If this is done, then only those risk/hazard requirements will be analyzed and allocated to tests just like all other requirements.

Storing the allocation matrix is a final aspect of requirements allocation to test. It is simple enough to keep the matrix in a spreadsheet or an electronic document. However, it can also be stored in a database. The advantage of storing it in a database is that a user can track test development and also can generate numerous reports.

Testing phases and approaches

Unit testing

Unit testing is performed on a smallest amount of code. What comprises a "unit" is often the subject of lengthy discussion and debate. No matter how a unit is defined, the intent of unit testing is to ensure that the lowest-level software modules are tested.

There are several approaches to unit testing, such as branch path analysis or module interface testing. In branch-path analysis, the developer uses a development tool to step through each path within a unit. In module-interface testing, the developer tests the unit from a "black box" perspective. Thus, the unit is only tested by varying the inputs into the software unit.

The type of unit testing that will be required will depend on many factors. In medical devices, the detailed branch path analysis testing is usually done against software units that are considered to be critical to the safety of the patient or the user. Other, lesser critical software units can be justifiably tested using the less thorough module interface testing.

Integration testing

Integration testing has several definitions. The definition covers the integration of two or more software units together. A broader definition covers the integration of physical subsystems (each with its own embedded software) to ensure that they work together.

Either type of integration testing brings separately developed entities together to ensure that they work together. The type of testing that is usually performed at these levels involves writing protocols to exercise the interface.

Validation testing

Validation testing is the process of proving that the product meets the product specification. It also can mean going the extra step and trying to ensure that the product does not do what it is not supposed to do. Thus, several approaches to testing are employed to exercise the software product. These approaches are discussed below.

Requirements-based testing

Requirements-based testing is the primary approach that is used to validate software. Essentially, the requirements from the SRS are analyzed and allocated to specific tests. Different approaches to testing, such as threshold testing or boundary testing, are used in developing these tests. The test steps are the sequential actions that must be taken to prove that the requirement has been met.

Threshold testing

Threshold testing is the process of proving that an event will occur when a specified parameter exceeds a certain value. An alarm, such as a low-battery alarm, is a good example of this on a medical device. If the device is running on a battery, and the voltage level in the battery falls below a threshold level, then an alarm is enunciated. In these types of tests, a tolerance band is placed around the threshold level. The tolerance is determined by the requirements in the SRS and also can be influenced by the accuracy of the test system. The test is then designed to vary the battery voltage to cross through the threshold and to see that the alarm is tripped within the tolerance band. Once the alarm has been tripped, the next part of the threshold test is to vary the parameter (in this case, the battery voltage) in the reverse direction to ensure that the alarm no longer enunciates.

Boundary testing

Boundary testing is the exercise of testing a parameter at its limits, and of trying to exceed those limits. A measured value of O_2 that is defined as not to exceed 110% would be a good example of this. (An O_2 reading can exceed 100% if the O_2 sensor is out of calibration or calibrated incorrectly.) The test then would be designed to prove that 110% O_2 could be displayed. Additionally, the test then would see how the device would react if the measured O_2 value were to exceed 110%.

Stress testing

Stress testing is the process of subjecting the unit to a bombardment of random inputs— often keypad presses or knob turns—to try to cause a software failure. This can be done manually or with an automated test system.

Manual stress testing and automated stress testing each have their own advantages. The advantage of manual stress testing is that the tester can test a certain aspect of the device and observe anomalous behavior in other areas of the device. The advantage of automated stress testing is that the test system can provide inputs into the device much more quickly than a human tester can.

Volume testing

Volume testing is the process of exercising the unit for an extended period of time. In our medical applications, we have found that running a volume test for a 72-hour period can uncover problems not found elsewhere.

Volume testing is almost always done with an automated test system. Volume testing does not necessarily use random fast, slam-bang key presses like stress testing does. Volume testing uses a logical approach to testing all of the paths in a software program. Once the device is in a certain state, probabilistic algorithms like Probability Density Function (PDF) are used in determining the next state.

Scenario testing

Scenario testing is the act of writing a test that emulates the actual use of the software from the user's perspective. For example, in anesthesia machines, a test could be written to cover a clinical situation. Scenario testing is helpful because it can uncover problems in the system, or it can detect flaws in the overall use of the product.

System testing

System testing occurs to validate the entire medical device, not just the software that is embedded in the device. Depending on the way a test station is instrumented for software testing, there can be significant overlap between software and system testing. For example, software embedded in a ventilator usually requires other mechanical–electrical subsystems in order to function. It is difficult to write sophisticated simulations to trick the ventilator software into thinking that all of the other subsystems of the ventilator are present. Therefore, a software test system might resemble a complete-system. It is up to the responsible parties to ensure that the degree of overlap is made known. More importantly, it is vital that everything is tested at some point in the product development.

Test execution and reporting

Test plan

For software validation, a test plan is written prior to executing a test pass. The format follows that defined in ANSI/IEEE Std 1012. The key aspects of the test plan are the description of the software changes, and the tests that will be run to prove that those software changes are implemented correctly. The test plan is written whenever a full validation or a regression test is run.

Test-configuration form

Because it is very important to be able to repeat a test, it is necessary to document the configuration of the equipment that is used to run a test against. This documentation is written on a test configuration form. The contents of these forms can vary. However, the primary fields are date, item, and item serial number, and there is a place for the test engineer to sign his name. The test configuration form becomes part of the test-documentation suite.

Executing manual protocol

Executing manual protocol for a formal test involves printing out the test procedures and going through the steps in the procedures. Because these printouts are a part of the official documentation, the test engineer must sign and date the documents. If the tester finds problems during the test, he will document the issue on the test procedures.

It is important that the handwritten notes on the paper procedures be addressed. The notes on the procedures could be problems found or procedure deviations. All of these markings must have closure. This closure is formally documented in the procedures response document. The procedures response document is covered later in this chapter.

Executing automated protocol

Typically, executing automated protocol amounts to setting up a batch job to submit a series of tests to be run. A test engineer fills out a test configuration form to have a piece of paper to start off the test. The batch job is then run overnight, and the results are analyzed the following day.

Experience has shown that a test in the batch occasionally can fail. This failure usually has been attributable either to timing issues or to the target-device feature having changed but the test not having changed. When this happens, it is usually a practical matter simply to correct the test script and to rerun it. However, in instances where correcting the test script would consume a large amount of time, it is sometimes more efficient to run that test manually to complete the test pass.

Test results

Once test results have been generated, they must be analyzed. The results from the automated tests are reviewed by running searches for keywords that indicate whether the test had problems. Additional analysis is performed by reviewing samples of the automated test results. This ensures that the tests are executed as expected. Other analysis might need to occur, such as when a large amount of data are generated during the test, after which data reduction and analysis must be performed to assess correctness.

Test results must be managed along with the other test documentation. The results from the manual test procedures must be included with the other paper documentation and stored in the formalized location. The electronic results files also must be put under configuration control. This is done with a configuration-management tool or a process of backing up the files to controlled media.

Test reports

Software validation test report

The key items in this report are a summary of what was tested, how it was tested, any problems that were encountered, and a recommendation for release.

Requirements to test cross reference

This document covers the ways in which the requirements were allocated to tests.

Procedures response document

As mentioned earlier, all handwritten marks on the manual test procedures must be reviewed and provided a closure. The results of this closure are documented in the procedures response document. Typically, each markup in a test procedure is a software problem, a procedure error, a procedure deviation, or a comment.

Further information

Bass L, Clements P, Kazman R. *Software Architecture in Practice*. Reading, MA, Addison Wesley Longman, 1998.

Bloesch A. Overview of Verification and Validation for Embedded Software in Medical Devices. *Handbook of Medical Device Design*. New York, Marcel Dekker, 2001.

Boehm BW. A Spiral Model of Software Development and Enhancement. *Computer* 5, 1988.

Boehm BW. *Software Engineering Economics*. Englewood Cliffs, NJ, Prentice Hall, 1981.

Booch G. *Object-Oriented Analysis and Design with Applications, 2nd Edition*. Redwood City, CA, Benjamin Cummings, 1994.

Booch G, Jacobson I, Rumbaugh J. *The Unified Modeling Language User Guide*. Reading, MA, Addison Wesley Longman, 1998.

Deutsch MS, Willis RR. *Software Quality Engineering—A Total Technical and Management Approach*. Englewood Cliffs, NJ, Prentice Hall, 1988.

Dyer M. The *Cleanroom Approach to Quality Software Development*. New York, Wiley, 1992.

Fairley RE. *Software Engineering Concepts*. New York, McGraw-Hill, 1985.

Food and Drug Administration, *21 CFR Part 820*. Washington, DC, U.S. Government Printing Office, 1997.

Fries RC. *Reliable Design of Medical Devices*. New York, Marcel Dekker, 1997.

Fries RC, Pienkowski P, Jorgens J. Safe, Effective, and Reliable Software Design and Development for Medical Devices. *Med Instrum* 30(2), 1996.

Hatley DJ, Pirbhai IA. *Strategies for Real-Time System Specification*. New York, Dorset House, 1987.

Humphrey WS. *Managing the Software Process*. Reading, MA, Addison-Wesley, 1989.

Institute of Electrical and Electronics Engineers. *IEEE Standard 730–IEEE Standard for Quality Assurance Plans.* New York, Institute of Electrical and Electronics Engineers, 1989.

Institute of Electrical and Electronics Engineers. *IEEE Standard 829–IEEE Recommended Practice for Software Requirements Specifications.* New York, Institute of Electrical and Electronics Engineers, 1998.

Institute of Electrical and Electronics Engineers. *IEEE Standard 830–IEEE Standard for Quality Assurance Plans.* New York, Institute of Electrical and Electronics Engineers, 1989.

Institute of Electrical and Electronics Engineers. *IEEE Standard 1012–IEEE Standard for Software Verification and Validation.* New York, Institute of Electrical and Electronics Engineers, 1986.

Institute of Electrical and Electronics Engineers. *IEEE Standard 1016–IEEE Recommended Practice for Software Design Descriptions.* New York, Institute of Electrical and Electronics Engineers, 1986.

Institute of Electrical and Electronics Engineers. *IEEE Standard 1059–IEEE Guide for Software Verification and Validation Plans.* New York, Institute of Electrical and Electronics Engineers, 1993.

Kan SH. *Metrics and Models in Software Quality Engineering*. Reading, MA, Addison-Wesley, 1995.

Leveson NG. *Safeware*. Reading, MA, Addison-Wesley, 1995.

McConnell S. *Code Complete*. Redmond, WA, Microsoft Press, 1993.

McConnell S. *Rapid Development*. Redmond, WA, Microsoft Press, 1996.

Page-Jones M. *The Practical Guide to Structured Systems Design—2nd Edition*. Englewood Cliffs, NJ, Prentice Hall, 1988.

Pressman R. *Software Engineering*. New York, McGraw-Hill, 1987.

Putnam LH, Myers W. *Measures for Excellence*. Englewood Cliffs, NJ, Prentice Hall, 1992.

Rakos JJ. *Software Project Management for Small to Medium Sized Projects*. Englewood Cliffs, NJ, Prentice Hall, 1990.

Rumbaugh J, *et al*. *Object-Oriented Modeling and Design*. Englewood Cliffs, NJ, Prentice Hall, 1991.

Rumbaugh J, Jacobson I, Booch G. *The Unified Modeling Language Reference Manual*. Reading, MA, Addison-Wesley, 1998.

Sommerville I, Sawyer P. *Requirements Engineering*. Chichester, England, Wiley, 1997.

Storey N. *Safety-Critical Computer Systems*. Harlow, England, Addison-Wesley, 1996.

Thayer RH, Dorfman M. *Software Requirements Engineering—2nd Edition*. Los Alamitos, California, IEEE Computer Society Press, 1997.

Yourdon E. *Modern Structured Analysis*. Englewood Cliffs, New Jersey, Yourden Press, 1989.

Chapter 5.5

5.5

Virtual instrumentation

Eric Rosow

Successful organizations have the ability to measure and to act on key indicators and events in real time. By leveraging the power of virtual instrumentation (VI) and open-architecture standards, multidimensional executive dashboards can empower health care organizations to make better and faster data driven decisions. This chapter will high-light ways in which user-defined virtual instruments and dashboards can connect to hospital information systems (e.g., Admission/Discharge/Transfer (ADT) systems and patient monitoring networks) and utilize statistical process control (SPC) to "visualize" information and to make timely, data driven decisions. The case studies described will illustrate enterprise-wide solutions for:

- Bed management and census control
- Operational management, data mining and business intelligence applications
- Clinical applications

Background

VI allows organizations to effectively harness the power of the PC, to access, analyze, and share information throughout the organization. With vast amounts of data available from increasingly sophisticated enterprise-level data sources, potentially useful information is often left hidden, due to a lack of useful tools. Virtual instruments can employ a wide array of technologies, such as multidimensional analyses and SPC tools, to detect patterns, trends, causalities, and discontinuities to derive knowledge and to make informed decisions.

Today's enterprises create vast amounts of raw data, and recent advances in storage technology; coupled with the desire to use these data competitively, has caused a data glut in many organizations. The health care industry, in particular, generates a tremendous amount of data. Tools such as databases and spreadsheets certainly help to manage and analyze these data, however databases, while ideal for extracting data, are generally not suited for graphing and analysis. Spreadsheets, on the other hand, are ideal for analyzing and graphing data, but this process can be cumbersome when working with multiple data files. Virtual instruments empower the user to leverage the best of both worlds by creating a suite of user-defined applications that allow the end-user to convert vast amounts of data into information that is ultimately transformed into knowledge to enable better, faster decision making.

Benefits of user-defined virtual instruments

The benefits of VI are increased performance and reduced costs. Because the user controls the technology through software, the flexibility of VI is unmatched by traditional instrumentation. The modular, hierarchical programming environment of VI is inherently reusable and reconfigurable.

In effect, VI allows the user to "morph" (i.e., replicate and/or customize) the functionality of traditional "vendor-defined" instruments (such as the oscilloscope shown in Figure 5.5-1) into virtual "user-defined" instruments on a standard computer or personal digital assistant (PDA).

VI applications have encompassed nearly every industry, including the telecommunications, automotive, semiconductor and biomedical industries. In the fields of health care and biomedical engineering, VI has empowered developers and end-users to conceive of, develop, and implement a wide variety of research-based

Biomedical Engineering Desk Reference; ISBN: 9780123746467

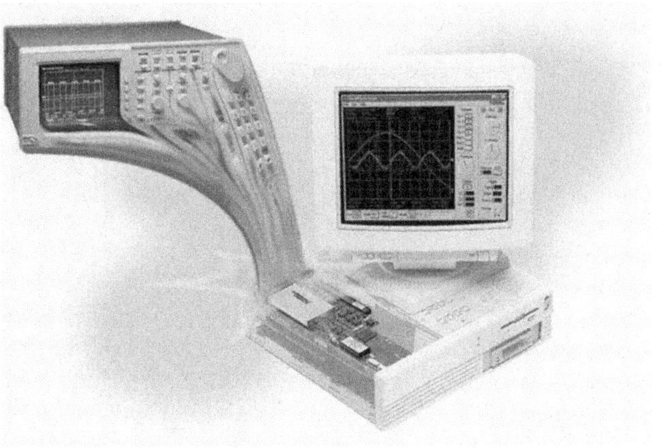

Figure 5.5-1 VI: Replicating and enhancing traditional "vendor-defined" instruments with virtual "user-defined" instruments. (Courtesy of National Instruments.)

biomedical applications and executive information tools. These applications fall into several categories, including process improvement, decision support, and clinical research. The following case studies are intended to illustrate the myriad of applications that are possible with this powerful technology.

The lab VIEW development environment

For many of its VI solutions discussed in this chapter, an application development environment (ADE) called LabVIEW has been used. LabVIEW was created by National Instruments (Austin, TX). It is an off-the-shelf graphical development environment designed specifically for developing integrated measurement and automation systems. Developers assemble user interfaces and high-level functions for data acquisition and control, signal processing and analysis, and visualization, in the same way that flowcharts are constructed. With the modularity and hierarchical structure of LabVIEW, users can quickly and easily prototype, design, deploy, and modify systems. The LabVIEW ADE is compiled for maximum execution performance, contains hundreds of advanced analysis routines, and allows developers to quickly design and build advanced applications.

This chapter presents several "real-world" virtual instrument applications and tools that have been developed to meet the specific needs of health care organizations. Particular attention is placed on the use of quality control and "performance indicators," which provide the ability to trend and forecast various metrics and improve processes. The use of SPC and modeling within virtual instruments is being demonstrated. Finally, several clinical research applications of VI are described.

Case study #1: a real-time bed-management and census-control dashboard

The problem

Most health care institutions today manage patient flow via paper, whiteboards, and phone calls. As a result, hospitals often lack precise and timely information to match bed availability with patients' clinical needs. This causes inefficient use of beds, resources, and provider time, which leads ultimately to reduced throughput; emergency department (ED) overcrowding; lost admissions; operating room (OR) delays; unhappy physicians, staff, and patients; decreased revenue; and higher expenses.

Industry trends

Concerns over patient access to hospital care are quickly becoming the largest, most urgent health care issue on the public's agenda. Many hospitals are now struggling to cope with surprising increase in the numbers of patients. Nationwide, hospitals report unprecedented inpatient and outpatient volumes, while average occupancy levels approaching 100% are not uncommon. These problems are compounded by two nationwide trends in hospitals: (1) increase in the numbers of patients; and (2) decrease in the numbers of beds.

Some hospitals are being forced to turn away ambulances and to cancel or delay elective surgeries because of a shortage of hospital beds. At other hospitals, ED patients wait hours, or even days, for rooms. Administration has no information to be able to match proactively the demand for patients with the appropriate level of nurse staffing. In addition, aging baby boomers are more likely to experience serious illness or injury, and medical advances are helping doctors to treat conditions that patients simply might have accepted in the past. There is an overall staff shortage of hospital workers, especially among nurses. With no sign of slowdown in these trends, these problems in hospital care appear set to escalate for the foreseeable future (AHA, 2001).

The solution

The Bed Management Dashboard (BMD) is a real-time process improvement and decision support product used by hospital administrators, clinicians, and managers on

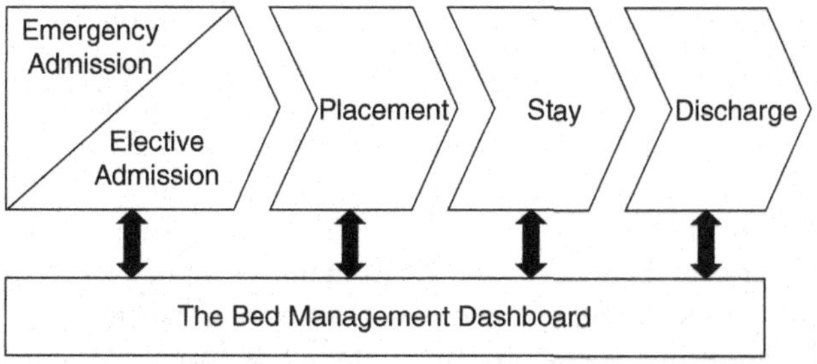

Figure 5.5-2 The BMD across the Enterprise: The BMD interacts with all departments throughout the continuum of care (e.g., admitting, ED, OR, ICU, and administration).

a constant basis. It is an enterprise-wide system that directly and indirectly interacts with all departments throughout the continuum of care (see Figure 5.5-2).

Process improvement

Process improvement is an outcome and benefit of a centralized Bed Management Process that maintains clinical input and incorporates technological support. The BMD streamlines the process of admitting, transferring, and discharging patients by:

- Optimizing patient placement processes
- Increasing staff efficiency (e.g., by reducing administrative expenses by eliminating paperwork and phone volume)
- Improving utilization of beds (e.g., by reducing inappropriate usage of inpatient beds without payer authorization, 23+ hour observation of outpatients by automatic alerts)
- Optimizing utilization of material resources (e.g., monitored beds and negative pressure rooms)
- Managing patient flow in the event of ADT system or network failure (disaster recovery)
- Enabling real-time and predictive capacity management
- Providing clinical attributes to supplement ADT data (e.g., telemetry and near nursing stations)

This results in:

- Reduced emergency room overcrowding
- Reduced OR delays
- Increased physician and staff satisfaction
- Improved patient flow
- Time savings
- Better treatment of patients and improved customer satisfaction
- Reduced diversions and lost admissions
- Increased revenues and decreased expenses
- Reduced staff

Decision support

The BMD allows administrators, clinicians, and managers to easily access, analyze, and display real-time patient and bed-availability information. It provides:

- On-demand historical, real-time, and predictive reports
- Alerts, warnings, and recommendations
- The ability to share information throughout the enterprise
- Provides point-in-time predicted bed availability
- The ability for decision-makers to move easily from "big-picture" analyses to transaction details

This results in:

- Timely data driven decisions
- Improved decisions regarding staffing levels
- Reduce reliance on third-party staffing companies to supplement in-house staff
- Improved crisis management during rapidly fluctuating census levels
- Improved ability to negotiate contracts with payers

The BMD has been running at Hartford Hospital since April 2001. It has over 900 trained users, and it can be accessed on the more than 2500 work stations throughout the organization. The system is interfaced to the Hospital's ADT Information System (Siemens/SMS), pagers, phones, and e-mail.

An air traffic control tower for beds

In many ways, the BMD is similar to an air traffic control tower. Like a real air traffic control tower, this application is real-time and mission-critical. It must handle scheduled and emergency events. The system assists with the clinical and business decision process that occur when a patient needs to be assigned to a specific bed location. Collectively, this system provides organizations with an array of enabling technologies to:

- Schedule/reserve/request patient bed assignments
- Assign and transfer patients from the ED and/or other clinical areas such as intensive care units, medical/surgical units, OR, and post-anesthesia care units
- Reduce and eliminate dependency on phone calls to communicate patient and bed requirements
- Reduce and eliminate paper processes to manage varying census levels
- Apply SPC and "Six Sigma" methodologies to manage occupancy and patient diversion

- Provide administrators, managers, and caregivers with accurate and on-demand reports and automatic alerts via pagers, e-mail, phone, and intelligent-software agents

How it works

The BMD is accessible via a web browser or via a client application. The supporting architecture of the BMD system is a standard N-tier server-based system, which, depending on the end users' needs, consists of one or more of the following: A web server, an application server, a database server, and an ADT interface server.

Key information from the hospital's ADT system is automatically fed to the BMD via a Health Level Seven (HL7) data stream. The system has been designed to accept inputs from patient monitoring and nurse-call networks. Figure 5.5-3 illustrates the flow of information from the ADT system to the user's desktop. The system typically receives up to 12,000 transaction messages per day, which are parsed into appropriate data elements by an HL7 parser and are stored on a database server.

The log-in authentication module accesses a hospital's central user log-in repository to validate the user's log-in information. This module enables single-user sign-on (SSO) capability by providing a user with the same username and password used by the other applications that authenticate to the central user log-in repository.

The integrated data-mining and report module consists of a number of standard reports that take advantage of the data warehouse nature of the BMD. These reports include, but are not limited to, a historical census report, a real-time census report, a bed manager report, a physician discharge report, and a discharge compliance report. In addition, access to the data by industry standard report writers, such as Crystal Reports and Access, give technical users the ability to create complicated or special reports as desired. For example, the BMD offers extensive ad hoc reporting capabilities ranging from length of stay (LOS) and "Care Day" metrics to asset management analyses (on beds and patient-monitor utilization), to biosurveillance reports that have been requested by state and federal organizations such as the Department of Public Health and the Centers for Disease Control and Prevention.

The utility and configuration module gives system operators the ability to administer BMD users (e.g., to add new users and to modify new or existing user security settings); to administer service, unit, rooms and beds (e.g., to add or modify clinical services, units, rooms, and beds, and their inter-relationships); to define automated alert thresholds; and to configure unit floor plan diagrams.

The embedded back-up utility module consists of a local version of the bed-management database that is

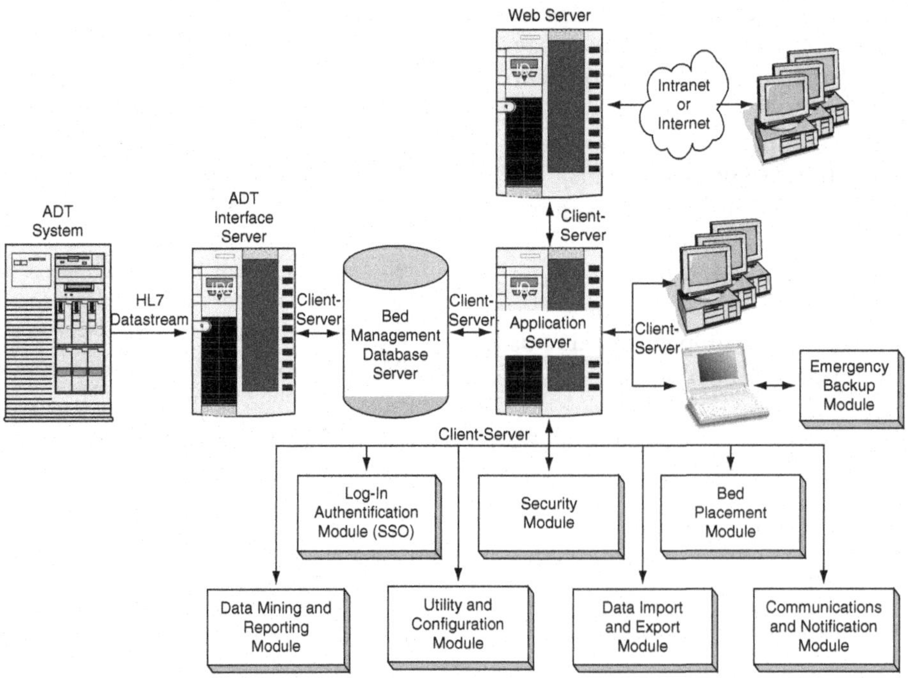

Figure 5.5-3 System diagram: The BMD consists of four major components: the ADT interface, the database server, an application server and a web server. The system has been designed to interface with any ADT system that provides an HL7 interface and utilizes TCP/IP as a communication protocol.

constantly updated. Access to this database gives users a self-contained and mobile version of the system that can be used in the event of catastrophic failure of the system hardware or network hardware or in the event of a crisis that removes the users from direct access to the hospital network.

One of the most important features of the BMD is that it reformats information from the ADT system and presents it to the clinical user in a more "user-friendly" and process-oriented manner. Dynamic and interactive graphical presentations of data are used extensively. Figure 5.5-4 illustrates the way in which all of the patients from a given admitting source, such as the ED, can be displayed and selected from a dynamically sortable "smart table."

Hospital beds are classified as having predefined "attributes," such as being "monitored" or being assigned to the "surgery" service. The needs of patients are similarly described with attributes, such as "monitor required" or "scheduled for surgery." As illustrated in Figure 5.5-5, the BMD helps to find those available beds in the hospital that meet the specific needs of a patient by guiding the clinical staff through a set of process screens that perform the match.

Decisions for patient placement can be centralized or decentralized. The dashboard allows proper communication between the appropriate parties. Status of decisions is automatically tracked, and a monitoring process can detect and notify key stakeholders of any process delays. Admitting or emergency departments can be automatically notified of decisions, if appropriate. Reporting of information is provided by online screen views of data tailored to the needs of a particular class of system user. Unit personnel can view either detailed information or summary roll-ups about their patients. Figure 5.5-6 illustrates ways in which patient information can be viewed in a dynamic and interactive floor-plan mode.

Administrators and program directors can view data over a wider scope that encompasses multiple units, services, or physicians. An example of a summary report is shown in Figure 5.5-7.

A key feature of this system is its use of "Intelligent Agents." These online agents, as shown in Figure 5.5-8, are constantly monitoring and analyzing patient and census information, and they have the ability to detect key system situations, such as high census in a unit (i.e., no available beds), excessive ED placement time for a particular patient, or delays in responses to placement requests.

The BMD allows users to run real-time queries and reports on current and future hospital census (Figure 5.5-9). These reports are stratified by inpatients, outpatients,

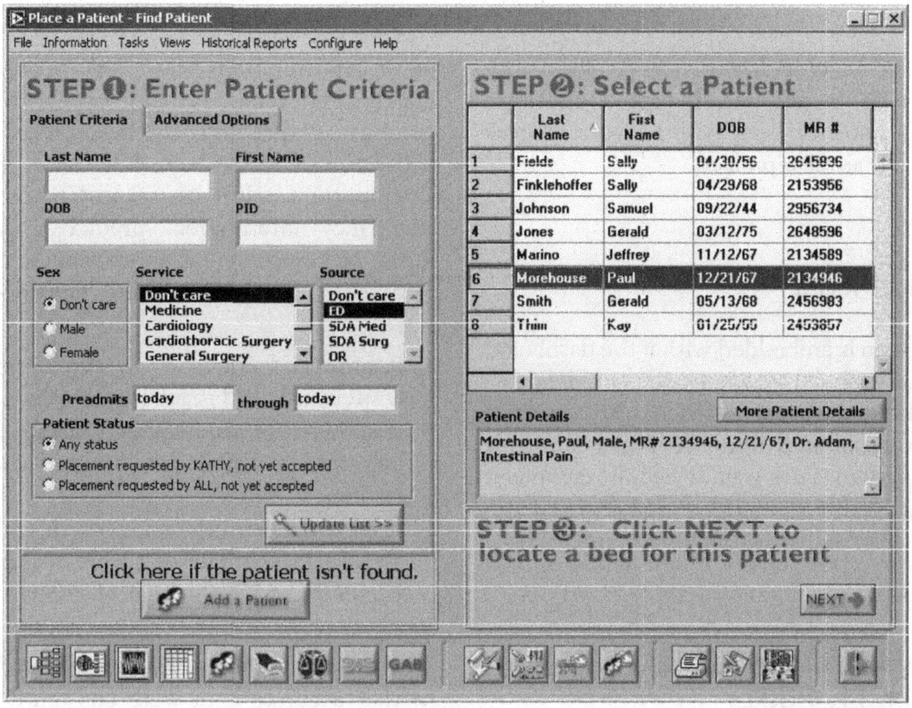

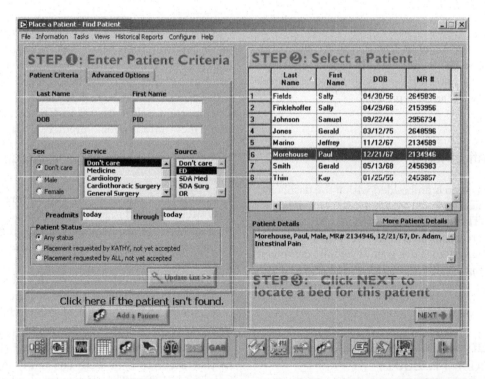

Figure 5.5-4 Find Patient: This screen is primarily used to request a bed for a patient. The user first enters various criteria to identify the patient. The system then displays all of the patients who meet the specified criteria. Finally, the user selects the patient in question and presses "NEXT" to move to another screen, where an available bed is located and requested.

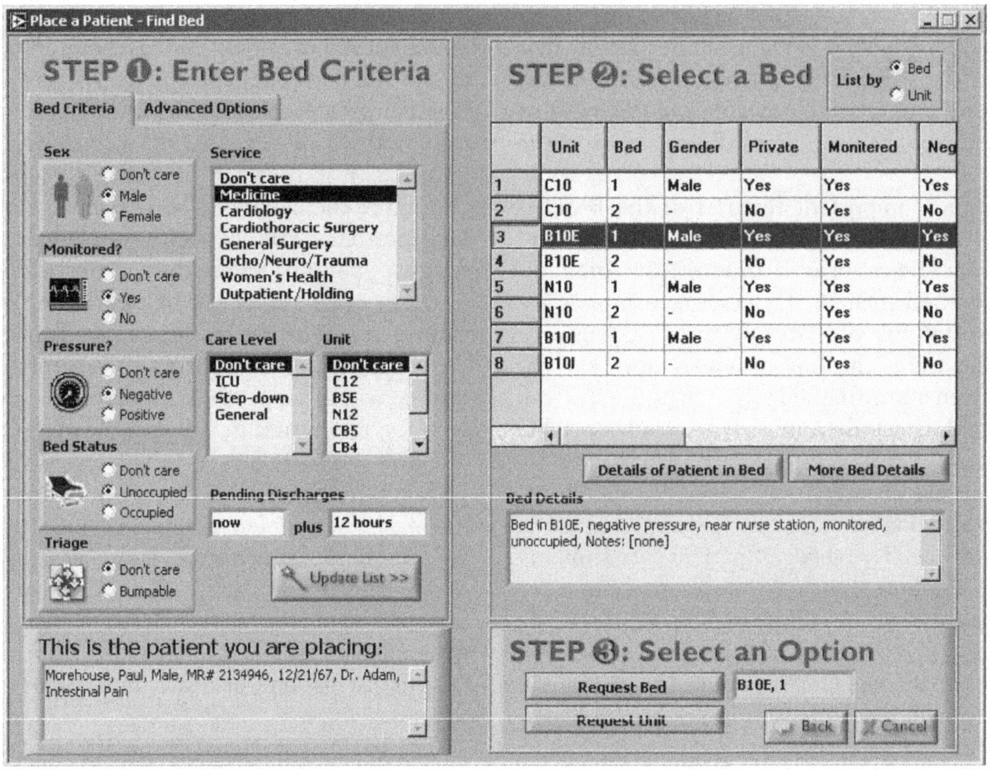

	Unit	Bed	Gender	Private	Monitored	Neg
1	C10	1	Male	Yes	Yes	Yes
2	C10	2	-	No	Yes	No
3	B10E	1	Male	Yes	Yes	Yes
4	B10E	2	-	No	Yes	No
5	N10	1	Male	Yes	Yes	Yes
6	N10	2	-	No	Yes	No
7	B10I	1	Male	Yes	Yes	Yes
8	B10I	2	-	No	Yes	No

Figure 5.5-5 Find Bed: Once a patient has been selected, this "Bed Finder" screen is used to locate a specific bed or unit for the patient. The user first enters various criteria about the type of bed that is needed (e.g., patient gender, monitor required, and negative-pressure room required). The system then displays all of the available beds that meet the specified criteria. Finally, the user selects a particular bed or unit for the previously specified patient.

acute units, non-acute units, and/or services ("collaborative management teams"). In addition, budgeted discharge, LOS, and care day statistics, along with average and peak occupancy indicators, are also profiled.

Patient confidentiality

A full-security system is embedded within the dashboard to audit user access and to assign users to definable system roles. These roles are restricted to specified processes and are prohibited from viewing or changing certain data. The system is designed to be fully compliant with the evolving Health Insurance Portability and Accountability Act (HIPAA) regulations.

The BMD at Hartford Hospital on September 11, 2001

The dashboard played a critical role at Hartford Hospital on that terrible day of September 11, 2001. When the news came in from New York (just 100 miles away) Hartford Hospital was told to expect hundreds of injured patients. At that time, the hospital had only seven open beds, two of which were ICU beds. The BMD played an important role with respect to providing real-time occupancy information, predictive capacity management, and biosurveillance reporting.

Specifically, the dashboard helped in the following ways:

- It improved efficiency by eliminating the need for whiteboards and many phone calls
- It allowed for predictive occupancy reports to help free up 150 beds in a timely and orderly fashion
- It provided an efficient way to collect and report biosurveillance data that were required by the Connecticut Department of Public Health

The key factor in this success story was that the dashboard was directly linked to the hospital census database so that accurate data were constantly available to the command center. The dashboard displays were easily, rapidly, and repeatedly revised to facilitate iterative decision making.

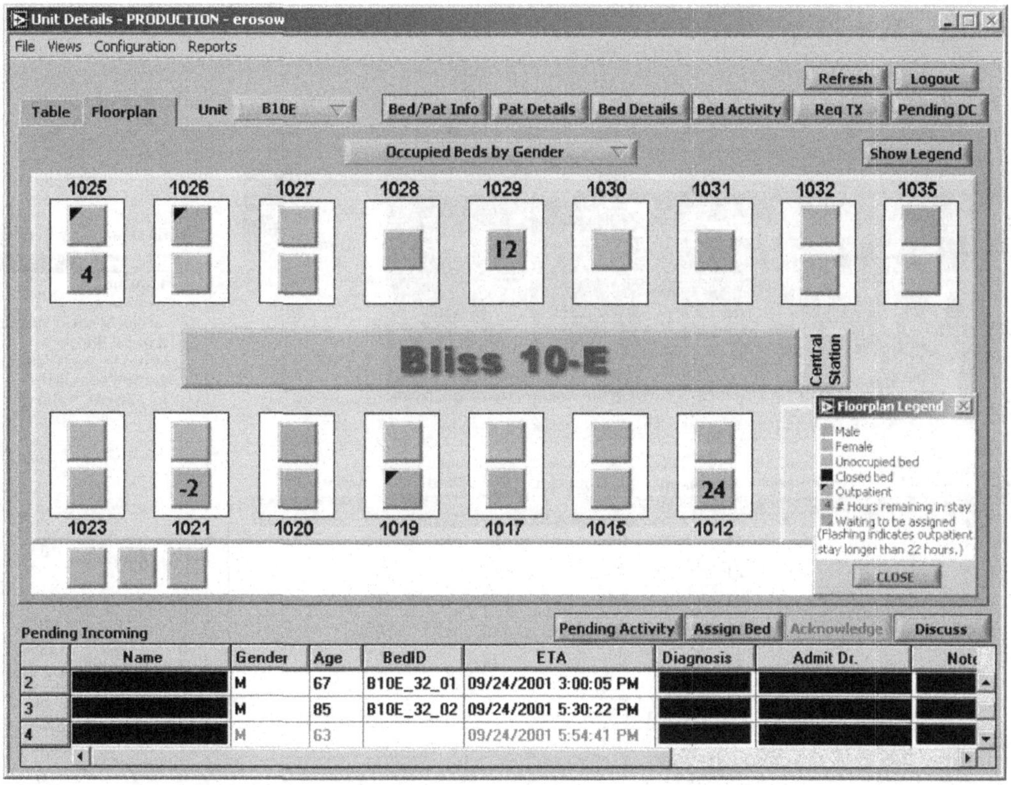

Figure 5.5-6 Unit Details (floor-plan view): This screen provides a graphical view of an intensive care unit. Each bed is presented as a square, using a simplified floor-plan view. Colors are used to indicate the selected attribute of the patient or bed. For example, the above screen indicates available beds in green, and occupied beds in red. The gray beds (which are flashing) are those with pending discharges, and the black bed is closed or inactive. Many other color-coded options are available via the pull-down selector. These include patient and bed attributes such as gender, monitored bed, negative-pressure room, and service (e.g., cardiology, surgery, and orthopedics).

Demonstrated return on investment (ROI)

Since its installation in April 2001, the system has more than paid for itself through achievement of the following indicators obtained from actual observations at Hartford Hospital.

- $200,000 annual expense avoidance by reducing the number of Bed Manager FTEs needed
- A 50% decrease in the number of phone calls per bed assignment/patient placement
- A 75–90% decrease in the overall time needed for the patient-placement process
- A 50% decrease in the number of paper forms used
- A 25% increase in throughput by alerting bed managers to discharges sooner of those that are reported to bed managers
- A 40-minute decrease in the length of the end-of-shift admission coordinator reports
- The hospital expects to save "thousands of dollars per day" by implementing the BMD outpatient alert system. (The specific dollars saved are still to be determined.)

Case study #2: PIVIT™-performance indicator virtual instrument toolkit

Most of the information management examples presented in the chapter are part of an application suite called PIVIT.™ The name is an acronym for "Performance Indicator Virtual Instrument Toolkit." PIVIT is an easy-to-use data acquisition and analysis product. PIVIT was developed specifically in response to the wide array of information and analysis needs throughout the health care setting. Figure 5.5-10 illustrates the main menu of PIVIT.

PIVIT applies virtual instrument technology to assess, analyze, and forecast clinical, operational, and financial performance indicators. Some examples include applications that profile institutional indicators (e.g., patient

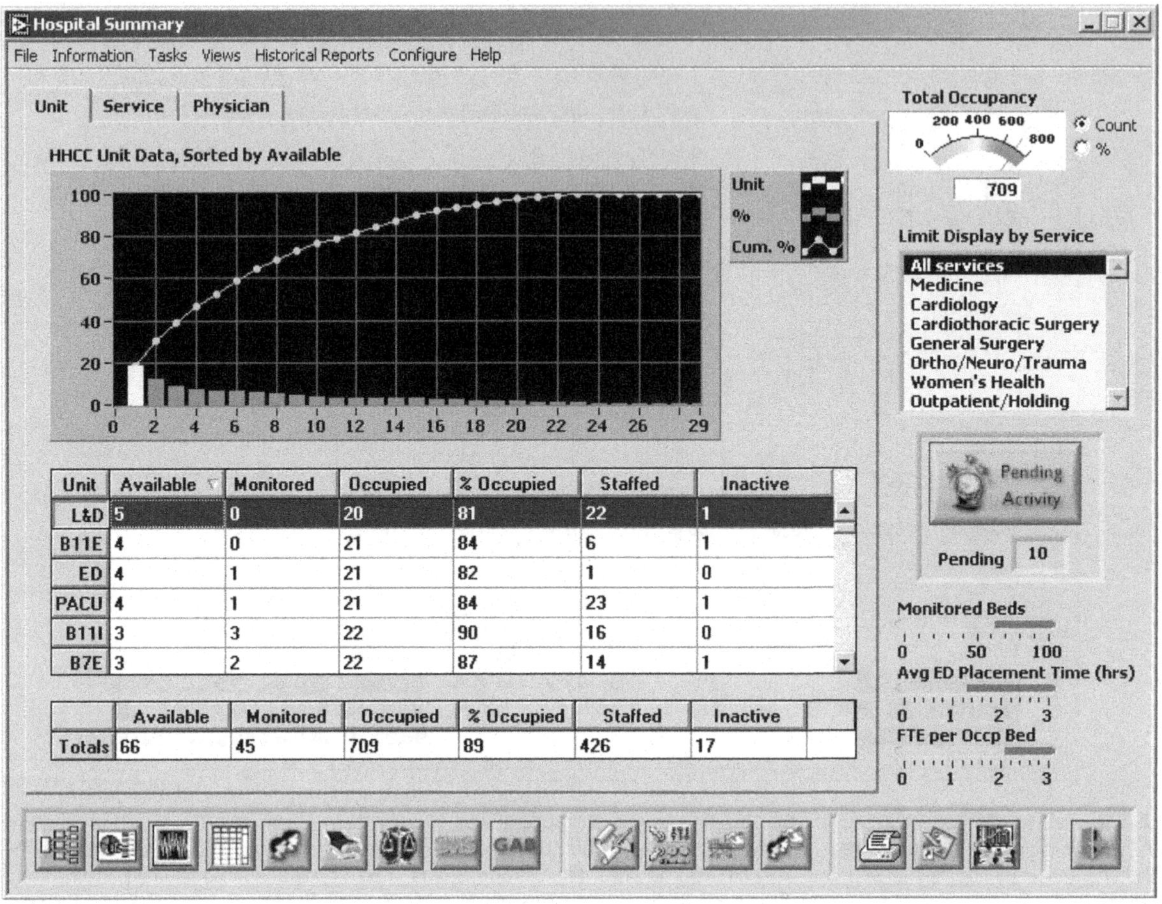

Figure 5.5-7 Hospital Summary: This screen is primarily used by administrators who need to see a global view of the hospital status. Table and Pareto charts profile the various units and the current status summaries of each unit. These patients also can be rolled up into services, or grouped by physician. In addition, patients can be aggregated in many other ways, such as time of admission, LOS, and admitting diagnosis.

days, discharges, percentage of occupancy, ALOS, revenues, and expenses), and departmental indicators (e.g., salary, non-salary, total expenses, expense per equivalent discharge, and DRGs). Other applications of PIVIT include 360-degree peer review, customer satisfaction profiling, and medical equipment risk assessment.

PIVIT can access data from multiple data sources. Virtually any parameter can be easily accessed and displayed from standard spreadsheet and database applications (e.g., Microsoft Access, Excel, Sybase, and Oracle), using Microsoft's Open Database Connectivity (ODBC) technology. Furthermore, multiple parameters can be profiled and compared in real time to any other parameter via interactive polar plots and three-dimensional displays. In addition to real-time profiling, other analyses, such as SPC, can be employed to view large data sets in a graphical format. SPC has been applied successfully for decades to help companies reduce variability in manufacturing processes. These SPC tools range from Pareto graphs to Run and Control charts. Although it will not be possible to describe all of these applications, several examples are provided below to illustrate the power of PIVIT.

Trending, relationships, and interactive alarms

Figure 5.5-11 illustrates a virtual instrument that interactively accesses institutional and department-specific indicators and profiles them for comparison. Data sets can be acquired directly from standard spreadsheet and database applications (e.g., Microsoft Access, Excel, Sybase, and Oracle). This capability has proved to be quite valuable with respect to quickly accessing and viewing large sets of data. Typically, multiple data sets contained within a spreadsheet or database had to be selected, and then a new chart of these data had to be created. Using PIVIT, the user simply selects the desired parameter from any one of the pull-down menus, and this data set is instantly graphed and compared to any other data set.

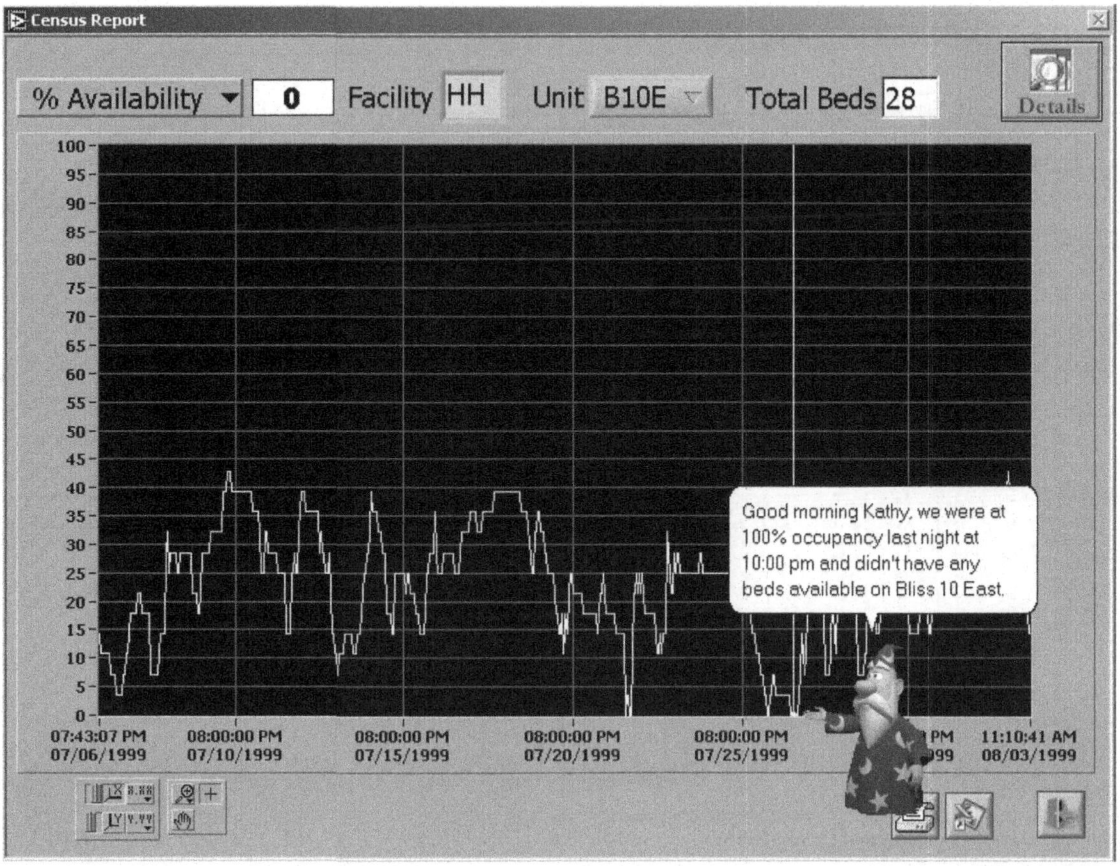

Figure 5.5-8 Intelligent Agents: The BMD can also employ online "intelligent agents" to provide assistance and to alert the user of important alarm conditions that otherwise might have gone unnoticed. Messages can be in the form of an on-screen agent (such as Merlin the Wizard) or via e-mail, pager, fax, and/or telephone.

Interactive "threshold cursors" dynamically highlight when a parameter is over and/or under a specific target. Displayed parameters can also be ratios of any measured value (e.g., "Expense per Equivalent Discharge" or "Revenue to Expense Ratio"). The indicator color will change, based on the degree to which the data value exceeds the threshold value (e.g., from green to yellow to red). If multiple thresholds are exceeded, then the entire background of the screen (normally gray) will change to red, to alert the user of an extreme condition.

Finally, multimedia has been employed by PIVIT to alert designated personnel by an audio message from the personal computer or by sending an automated message via e-mail, fax, pager, or mobile phone.

PIVIT is also able to profile historical trends and to project future values. Forecasts can be based on user-defined history (e.g., "Months for Regression"), the type of regression (linear, exponential, or polynomial), the number of days, months, or years to forecast, and whether any offset should be applied to the forecast. These features allow the user to create an unlimited number of "what-if" scenarios and to allow only the desired range of data to be applied to a forecast. In addition to the graphical display of data values, historical and projected tables are also provided. These embedded tables look and function like a standard spreadsheet.

Data modeling

Figure 5.5-12 illustrates another way in which VI can be applied to financial modeling and forecasting. This example graphically profiles the annual morbidity, mortality, and cost associated with falls within the state of Connecticut. Such an instrument has proved to be an extremely effective modeling tool due to its ability to high-light relationships and assumptions interactively, and to project the cost and/or savings of employing educational and other interventional programs.

Virtual instruments such as these are not only useful with respect to modeling and forecasting but, perhaps more importantly, they become a "knowledge base" in which interventions and the efficacy of these interventions can be statistically proved.

The example program in Figure 5.5-13 shows ways in which VI can employ standard Microsoft Windows®

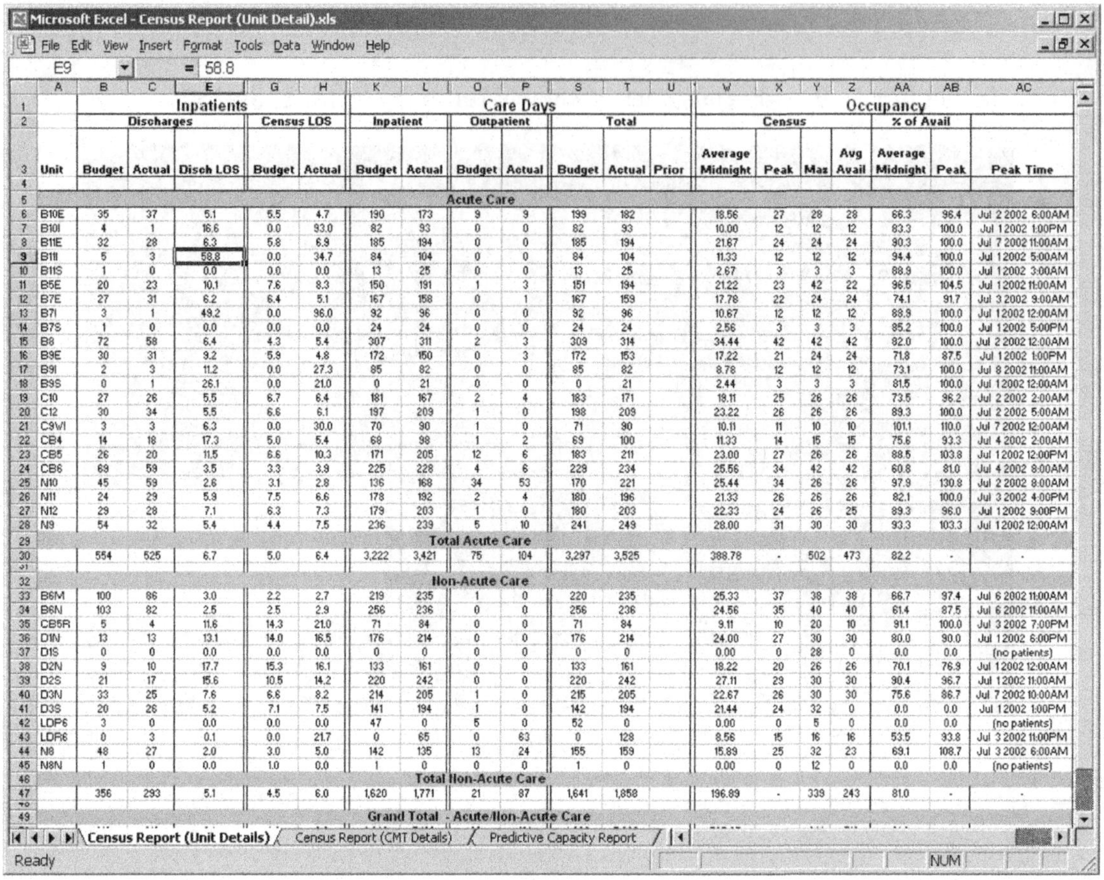

Microsoft Excel - Census Report (Unit Detail).xls

File Edit View Insert Format Tools Data Window Help

E9 = 58.8

#	Unit	Inpatients: Discharges Budget	Discharges Actual	Disch LOS	Census LOS Budget	Census LOS Actual	Care Days: Inpatient Budget	Inpatient Actual	Outpatient Budget	Outpatient Actual	Total Budget	Total Actual	Total Prior	Occupancy: Census Average Midnight	Census Peak	Census Max	Census Avg Avail	% of Avail Average Midnight	% of Avail Peak	Peak Time
5	**Acute Care**																			
6	B10E	35	37	5.1	5.5	4.7	190	173	9	9	199	182		18.56	27	28	28	66.3	96.4	Jul 2 2002 6:00AM
7	B10I	4	1	16.6	0.0	93.0	82	93	0	0	82	93		10.00	12	12	12	83.3	100.0	Jul 1 2002 1:00PM
8	B11E	32	28	6.3	5.8	6.9	185	194	0	0	185	194		21.67	24	24	24	90.3	100.0	Jul 7 2002 11:00AM
9	B11I	5	3	58.8	0.0	34.7	84	104	0	0	84	104		11.33	12	12	12	94.4	100.0	Jul 1 2002 5:00AM
10	B11S	1	0	0.0	0.0	0.0	13	25	0	0	13	25		2.67	3	3	3	88.9	100.0	Jul 1 2002 3:00AM
11	B5E	20	23	10.1	7.6	8.3	150	191	1	3	151	194		21.22	23	42	22	96.5	104.5	Jul 1 2002 11:00AM
12	B7E	27	31	6.2	6.4	5.1	167	158	0	1	167	159		17.78	22	24	24	74.1	91.7	Jul 3 2002 9:00AM
13	B7I	3	1	49.2	0.0	96.0	92	96	0	0	92	96		10.67	12	12	12	88.9	100.0	Jul 1 2002 12:00AM
14	B7S	1	0	0.0	0.0	0.0	24	24	0	0	24	24		2.56	3	3	3	85.2	100.0	Jul 1 2002 5:00PM
15	B8	72	58	6.4	4.3	5.4	307	311	2	3	309	314		34.44	42	42	42	82.0	100.0	Jul 2 2002 12:00AM
16	B9E	30	31	9.2	5.9	4.8	172	150	0	3	172	153		17.22	21	24	24	71.8	87.5	Jul 1 2002 1:00PM
17	B9I	2	3	11.2	0.0	27.3	85	82	0	0	85	82		8.78	12	12	12	73.1	100.0	Jul 8 2002 11:00AM
18	B9S	0	1	26.1	0.0	21.0	0	21	0	0	0	21		2.44	3	3	3	81.5	100.0	Jul 1 2002 12:00AM
19	C10	27	26	5.5	6.7	6.4	181	167	2	4	183	171		19.11	25	26	26	73.5	96.2	Jul 2 2002 2:00AM
20	C12	30	34	5.5	6.6	6.1	197	209	1	0	198	209		23.22	26	26	26	89.3	100.0	Jul 2 2002 5:00AM
21	C9WI	3	3	6.3	0.0	30.0	70	90	1	0	71	90		10.11	11	10	10	101.1	110.0	Jul 7 2002 12:00AM
22	CB4	14	18	17.3	5.0	5.4	68	98	1	2	69	100		11.33	14	15	15	75.6	93.3	Jul 4 2002 2:00AM
23	CB5	26	20	11.5	6.6	10.3	171	205	12	6	183	211		23.00	27	26	26	88.5	103.8	Jul 1 2002 12:00PM
24	CB6	69	59	3.5	3.3	3.9	225	228	4	6	229	234		25.56	34	42	42	60.8	81.0	Jul 4 2002 8:00AM
25	N10	45	59	2.6	3.1	2.8	136	168	34	53	170	221		25.44	34	26	26	97.9	130.8	Jul 2 2002 8:00AM
26	N11	24	29	5.9	7.5	6.6	178	192	2	4	180	196		21.33	26	26	26	82.1	100.0	Jul 3 2002 4:00PM
27	N12	29	28	7.1	6.3	7.3	179	203	1	0	180	203		22.33	24	26	25	89.3	96.0	Jul 1 2002 9:00PM
28	N9	54	32	5.4	4.4	7.5	236	239	5	10	241	249		28.00	31	30	30	93.3	103.3	Jul 1 2002 12:00AM
29	**Total Acute Care**																			
30		554	525	6.7	5.0	6.4	3,222	3,421	75	104	3,297	3,525		388.78	-	502	473	82.2	-	-
32	**Non-Acute Care**																			
33	B6M	100	86	3.0	2.2	2.7	219	235	1	0	220	235		25.33	37	38	38	66.7	97.4	Jul 6 2002 11:00AM
34	B6N	103	82	2.5	2.5	2.9	256	236	0	0	256	236		24.56	35	40	40	61.4	87.5	Jul 6 2002 11:00AM
35	CB5R	5	4	11.6	14.3	21.0	71	84	0	0	71	84		9.11	10	20	10	91.1	100.0	Jul 3 2002 7:00PM
36	D1N	13	13	13.1	14.0	16.5	176	214	0	0	176	214		24.00	27	30	30	80.0	90.0	Jul 1 2002 6:00PM
37	D1S	0	0	0.0	0.0	0.0	0	0	0	0	0	0		0.00	0	28	0	0.0	0.0	(no patients)
38	D2N	9	10	17.7	15.3	16.1	133	161	0	0	133	161		18.22	20	26	26	70.1	76.9	Jul 1 2002 12:00AM
39	D2S	21	17	15.6	10.5	14.2	220	242	0	0	220	242		27.11	29	30	30	90.4	96.7	Jul 1 2002 11:00AM
40	D3N	33	25	7.6	6.6	8.2	214	205	1	0	215	205		22.67	26	30	30	75.6	86.7	Jul 7 2002 10:00AM
41	D3S	20	26	5.2	7.1	7.5	141	194	1	0	142	194		21.44	24	32	0	0.0	0.0	Jul 1 2002 1:00PM
42	LDP6	3	0	0.0	0.0	0.0	47	0	5	0	52	0		0.00	0	5	0	0.0	0.0	(no patients)
43	LDR6	0	3	0.1	0.0	21.7	0	65	0	63	0	128		8.56	15	16	16	53.5	93.8	Jul 3 2002 11:00PM
44	N8	48	27	2.0	3.0	5.0	142	135	13	24	155	159		15.89	25	32	23	69.1	108.7	Jul 3 2002 6:00AM
45	N8N	1	0	0.0	1.0	0.0	1	0	0	0	1	0		0.00	0	12	0	0.0	0.0	(no patients)
46	**Total Non-Acute Care**																			
47		356	293	5.1	4.5	6.0	1,620	1,771	21	87	1,641	1,858		196.89	-	339	243	81.0	-	-
49	**Grand Total - Acute/Non-Acute Care**																			

Census Report (Unit Details) / Census Report (CMT Details) / Predictive Capacity Report

Ready NUM

Figure 5.5-9 Detailed census and budget report.

technology (in this case, Dynamic Data Exchange (DDE)) to transfer data to commonly used software applications such as Microsoft Access® or Microsoft Excel.® It is interesting to note that in this example, the virtual instrument can measure and graph multiple signals while sending these data to another application that could reside on the network or across the Internet.

In addition to utilizing DDE, VI can use other protocols for inter-application communication. These range from simple serial communication to TCP/IP and ActiveX.

Figure 5.5-14 illustrates the Communications Center. This application shows various methods by which the user can communicate information throughout an organization. The Communications Center can be used to simply create and print a report, or it can be used to send e-mail, faxes, or messages to a pager, or even to leave voicemail messages. This is a powerful feature in that information can be distributed easily and efficiently to individuals and groups in real time.

Additionally, Microsoft Agent technology can be used to pop up an animated help tool ("Merlin the Wizard"). Merlin then can be used to communicate a message or to indicate an alarm condition, or it can be used to help the user to solve a problem or to point out a discrepancy that otherwise might have gone unnoticed. Agents employ a "text-to-speech" algorithm to actually "speak" an analysis or alarm directly to the user or the recipient of the message. In this way, on-line help and user support can be provided in multiple languages.

In addition to real-time profiling of various parameters, more advanced analyses, such as SPC can be employed to view large data sets in a graphical format.

SPC has enormous applications throughout health care. Such perceived applications were significant drivers to embed SPC tools into PIVIT's suite of virtual instruments. For example, Figure 5.5-15 is an example of a way in which Pareto analysis can be applied to a sample trauma database of over 12,000 records. The Pareto chart could show frequency or percentage, depending on front-panel selection, and the user can select from a variety of different parameters by clicking on the "pull-down" menu. This menu can be configured to automatically display each database field directly from the database. In this example, various database fields (e.g., DRG, Principal Diagnosis, Town, and Payer), can be selected for Pareto analysis. Other tools include run charts, control charts, and process capability distributions (see, e.g. Figure 5.5-16).

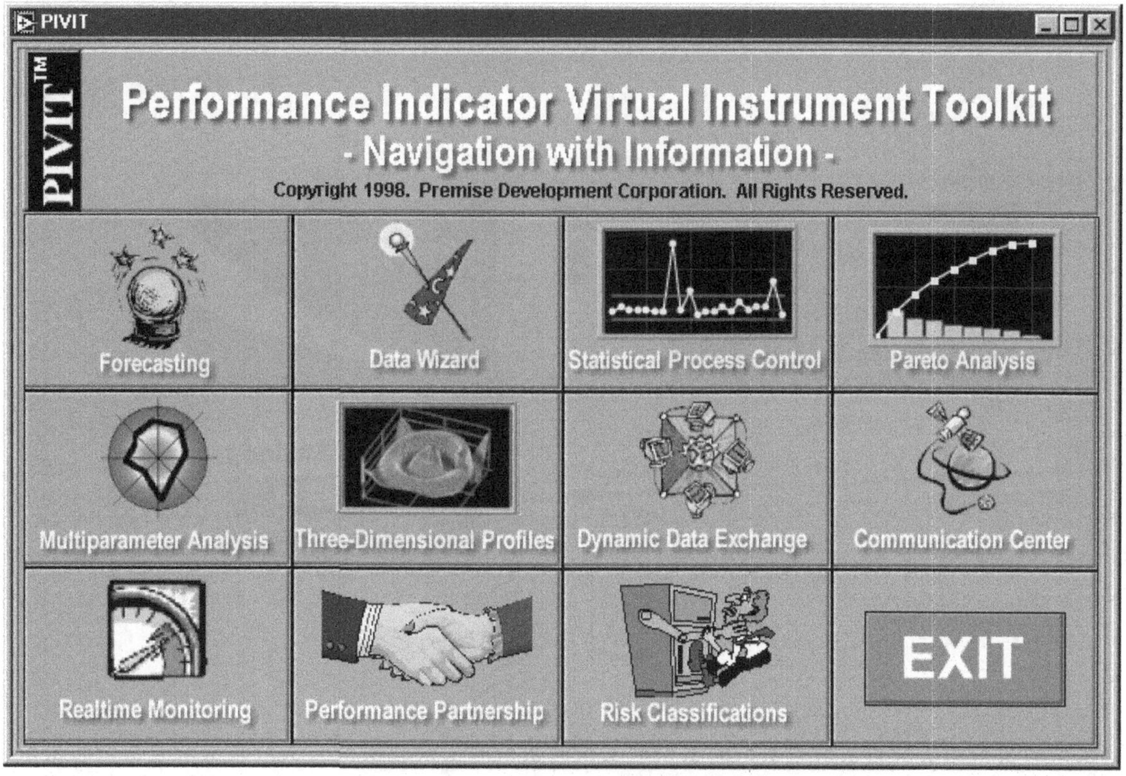

Figure 5.5-10 The Performance Indicator Virtual Instrument Toolkit (PIVIT™) main menu.

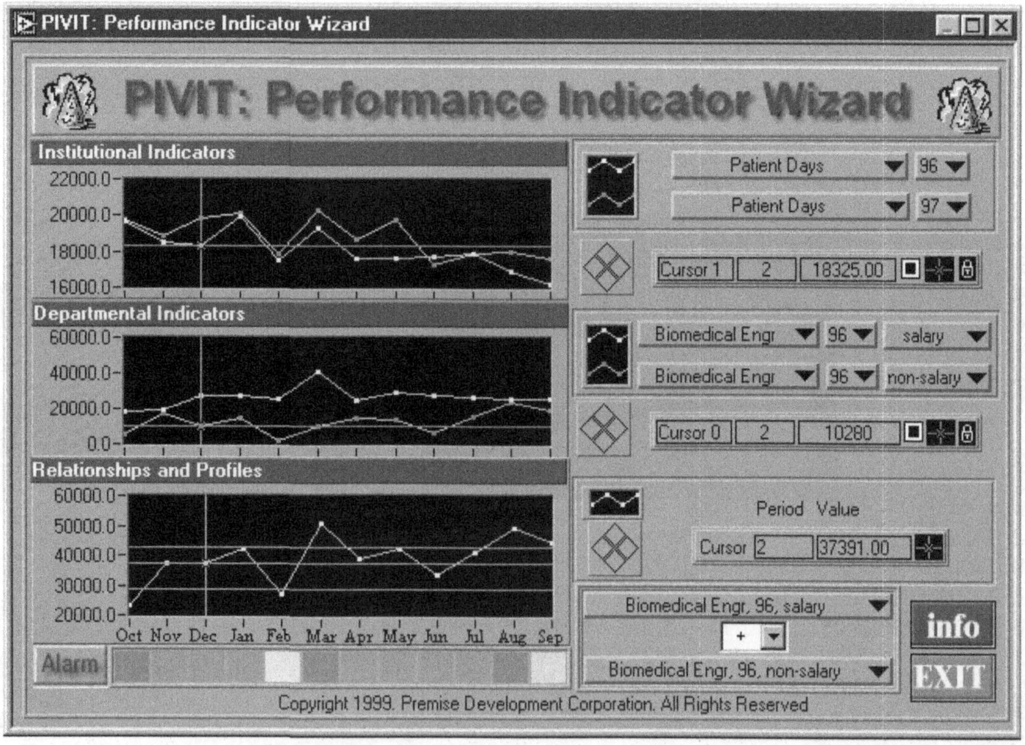

Figure 5.5-11 PIVIT–Performance Indicator Wizard Display Institutional and Departmental Indicators.

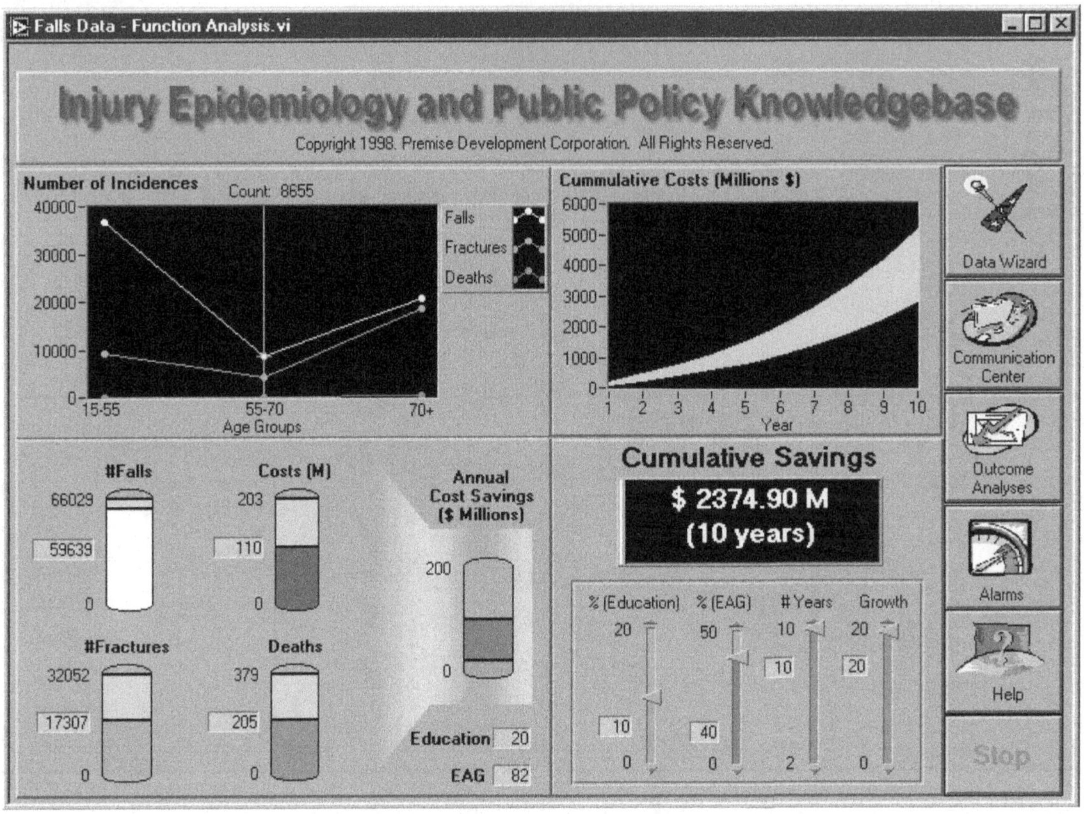

Figure 5.5-12 Injury Epidemiology and Public Policy Knowledgebase.

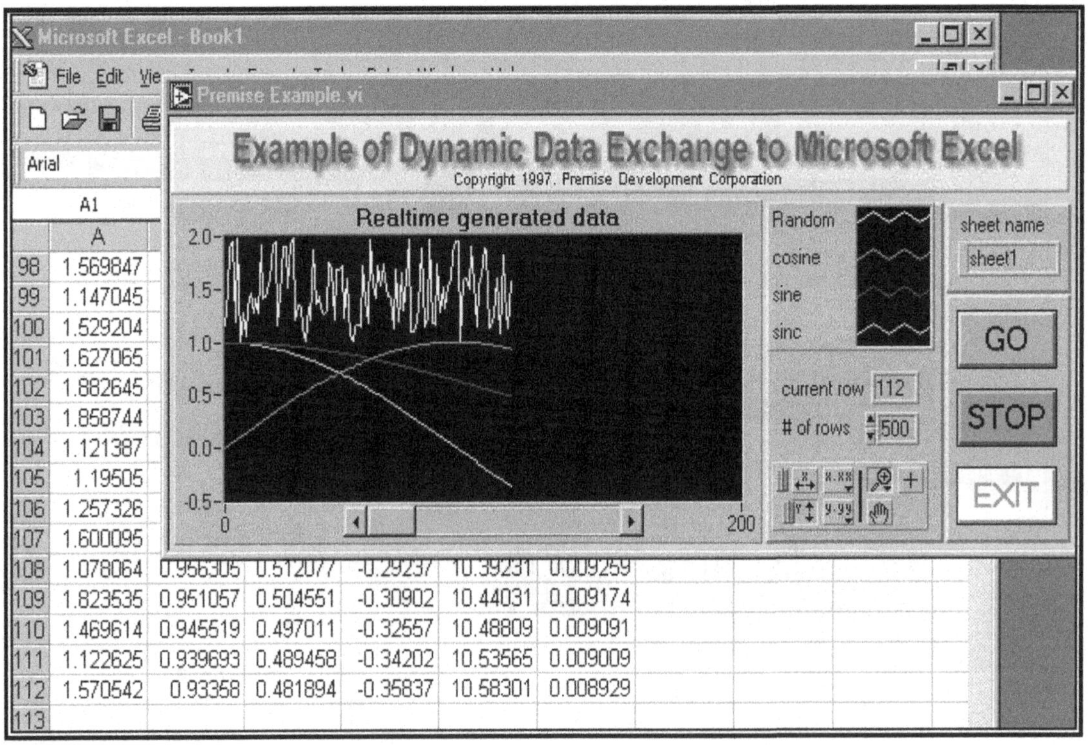

Figure 5.5-13 Example of DDE for Interapplication Communication.

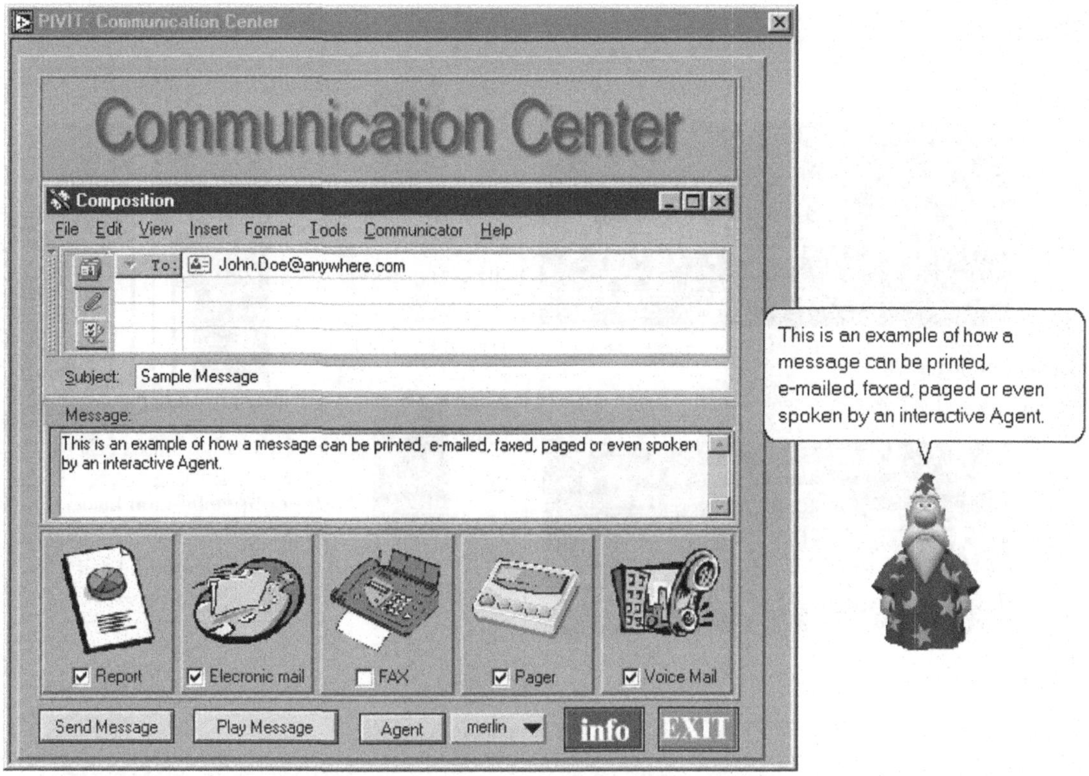

Figure 5.5-14 PIVIT's Communications Center.

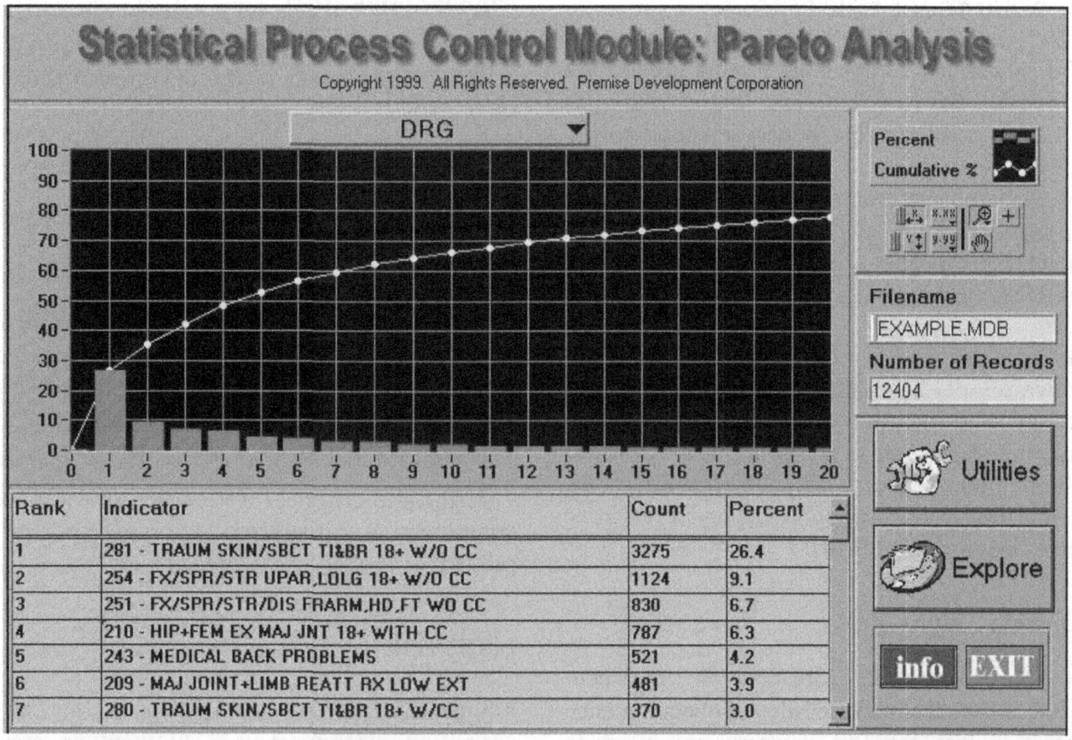

Figure 5.5-15 SPC–Pareto Ananlysis of a sample trauma registry.

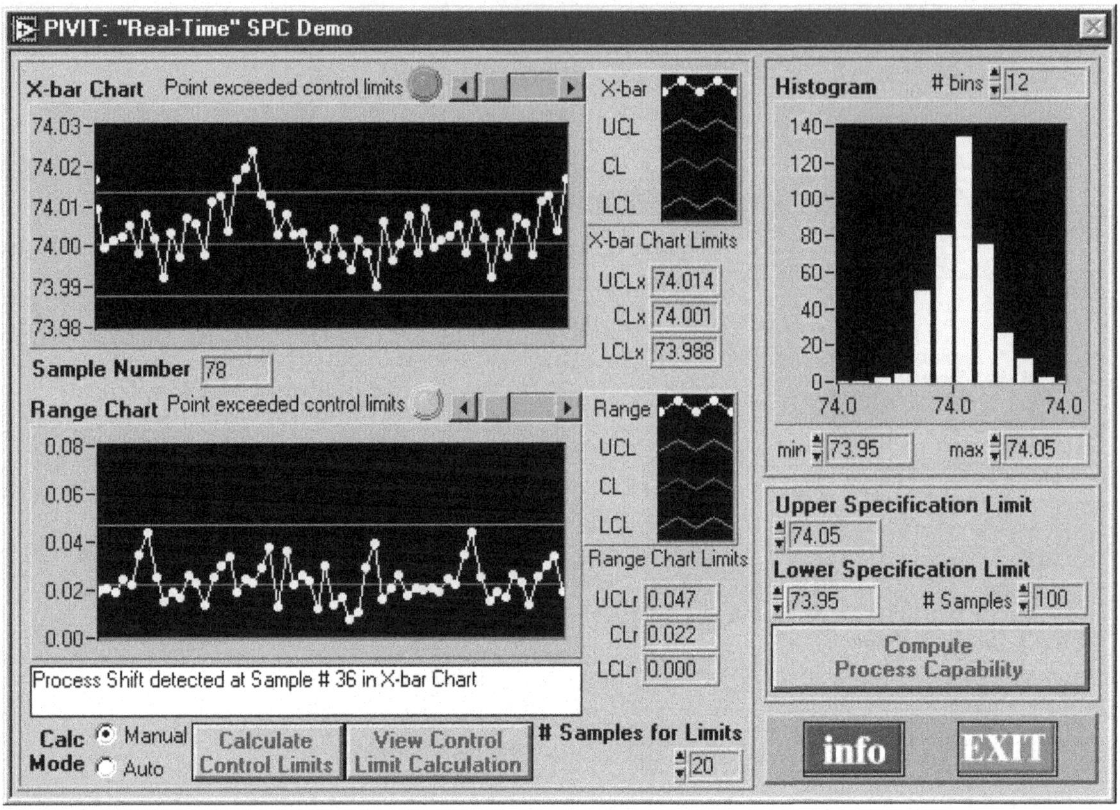

Figure 5.5-16 SPC–"Real Time" SPC Application.

Medical equipment risk criteria

Figure 5.5-17 illustrates a virtual instrument application that demonstrates the way that four "static" risk categories (and their corresponding values) are used to determine the inclusion of clinical equipment in the Medical Equipment Management Program at Hartford Hospital. Each risk category includes specific subcategories that are assigned points, which, when added together according to the formula listed below, yield a total score that will range from 4 to 25.

Considering these scores, the equipment is categorized into five priority levels (i.e., High, Medium, Low, Grey List, and Non-Inclusion into the Medical Equipment Management Program). The four static risk categories are Equipment Function (EF), Physical Risk (PR), Environmental Use Classification (EC), and Preventive Maintenance Requirements (MR).

Equipment function (EF)

Stratifies the various functional categories (i.e., therapeutic, diagnostic, analytical and miscellaneous, of equipment). The specific rankings for this category are listed in Table 5.5-1.

Physical risk (PR)

Lists the "worst-case scenario" of physical-risk potential to either the patient or the operator of the equipment (Table 5.5-2).

Environmental use classification (EC)

Lists the primary equipment area in which the equipment is used (Table 5.5-3).

Preventive maintenance requirements (MR)

Describes the level and frequency of required maintenance (Table 5.5-4). The Aggregate Static Risk Score is calculated as follows:

Aggregate Risk Score = EF + PR + EC + MR

Using the criteria's system described above, clinical equipment is categorized according to the following priority of testing and degree of risk:

High risk

Equipment that scores 18–25 points on the criteria's evaluation system. This equipment is assigned the highest risk for testing, calibration, and repair.

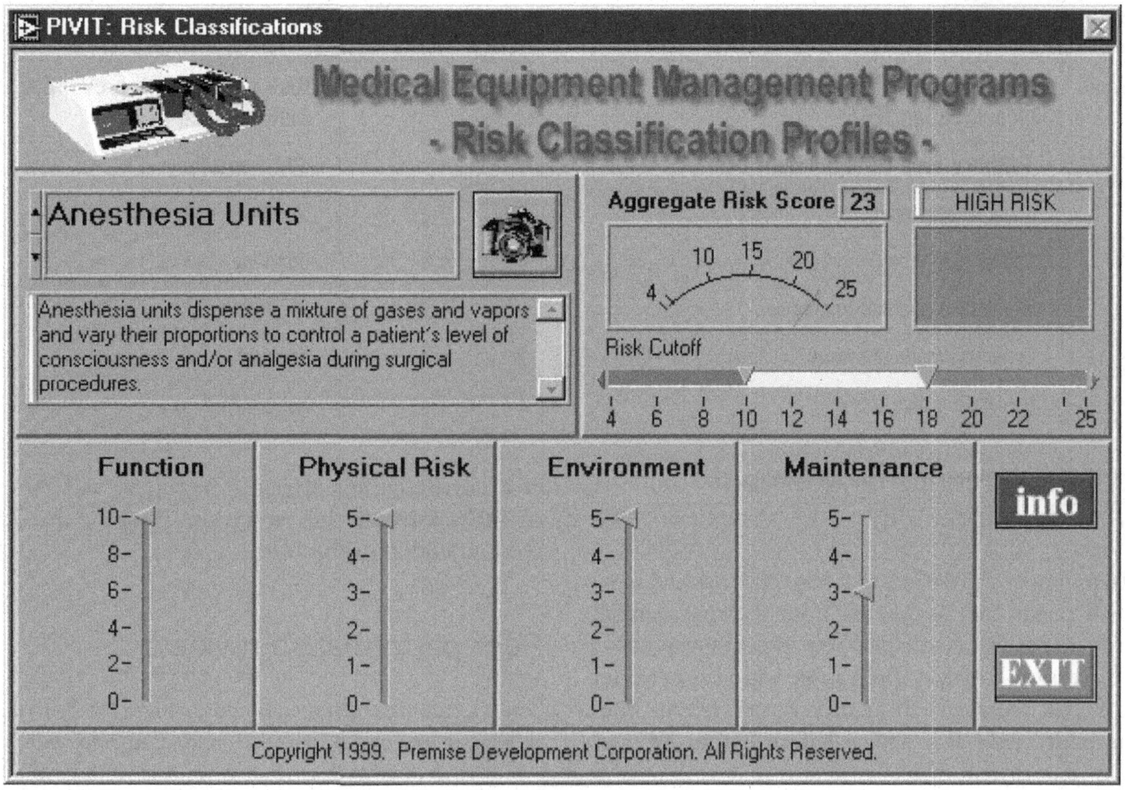

Figure 5.5-17 Medical Equipment Risk Classification Profiler.

Medium risk

Equipment that scores 15–17 points on the criteria's evaluation system.

Low risk

Equipment that scores 12–14 points on the criteria's evaluation system.

Table 5.5-1 Equipment function ranking

Point score	Risk category I: equipment function (EF) Function description
10	Therapeutic – Life Support
9	Therapeutic – Surgical or Intensive Care
8	Therapeutic – Physical Therapy or Treatment
7	Diagnostic – Surgical or Intensive Care Monitoring
6	Diagnostic – vOther physiological monitoring
5	Analytical – Laboratory analytical
4	Analytical – Laboratory Accessories
3	Analytical – Computer and related
2	Miscellaneous – Patient-related
1	Miscellaneous – Non-patient related

Hazard surveillance (gray)

Equipment that scores 6–11 points on the criteria's evaluation system are visually inspected on an annual basis during the hospital hazard-surveillance rounds.

Table 5.5-2 Physical risk ranking

Point score	Risk category II: physical risk (PR) Description of use risk
5	Potential patient death
4	Potential patient injury
3	Inappropriate therapy or mis-diagnosis
2	Equipment damage
1	No significant identified risk

Table 5.5-3 Environmental use classification ranking

Point score	Risk category IV: Environmental use classification (EC)
	Primary area of equipment use
5	Anesthetizing locations
4	Critical care areas
3	Wet locations/labs/exam areas
2	General patient care areas
1	Non-patient care areas

Table 5.5-4 Preventive maintenance ranking

Point score	Risk category III: preventive maintenance (MR)
	PM frequency
5	Monthly
4	Quarterly
3	Semi-annually
2	Annually
1	Not required

Medical equipment management program deletion

Medical equipment and devices that pose little risk and score less than 6 points can be deleted from the management program as well as from the clinical equipment inventory.

Future versions of this application will also consider "dynamic" risk factors such as user errors, mean-time-between failure (MTBF), device failure within 30 days of a preventive maintenance or repair, and the number of years beyond the American Hospital Association's recommended useful life.

Peer performance reviews

The virtual instrument shown in Figure 5.5-18 has been designed to easily acquire and compile performance

Figure 5.5-18 Performance reviews using VI.

information with respect to institution-wide competencies. It has been created to allow every member of a team or department to participate in the evaluation of a co-worker (360-degree peer review).

Upon running the application, the user is presented with a "sign-in" screen where he or she enters a username and password. The application is divided into three components. The first (top section) profiles the employee and the relevant service information. The second (middle section) indicates each competency as defined for employees, managers, and senior managers. The last (bottom section) allows the reviewer to evaluate performance by selecting one of four "radio buttons" and also to provide specific comments related to each competency. This information is then compiled (with other reviewers) as real-time feedback.

Case study #3: BioBench—a virtual instrument application for data acquisition and analysis of physiological signals

The biomedical industry relies heavily on the ability to acquire, analyze, and display large quantities of data. Whether researching disease mechanisms and treatments by monitoring and storing physiological signals, researching the effects of various drugs interactions, or teaching students in labs where they study physiological signs and symptoms, it was clear that there existed a strong demand for a flexible, easy-to-use, and cost-effective tool. In a collaborative approach, biomedical engineers, software engineers, clinicians, and researchers created a suite of virtual instruments called BioBench.™

BioBench™ (National Instruments, Austin, TX) is a new software application designed for physiological data acquisition and analysis. It was built with Lab VIEW™, the world's leading software development environment for data acquisition, analysis, and presentation.* Coupled with National Instruments data-acquisition (DAQ) boards, BioBench integrates the PC with data acquisition for the life sciences market.

Many biologists and physiologists have made major investments over time in data-acquisition hardware built before the advent of modern PCs. While these scientists cannot afford to throw out their investment in this equipment, they recognize that computers and the concept of VI yield tremendous benefits in terms of data analysis, storage, and presentation. In many cases, traditional medical instrumentation is too expensive to acquire and/or maintain. As a result, researchers and scientists are opting to create their own PC-based data-monitoring systems in the form of virtual instruments.

Other life scientists, who are just beginning to assemble laboratory equipment, face the daunting task of selecting hardware and software needed for their application. Many manufacturers in the field of life sciences focus their efforts on the acquisition of raw signals and converting them into measurable linear voltages. They do not concentrate on digitizing signals or the analysis and display of data on the PC. BioBench™ is a low-cost, turnkey package that requires no programming. BioBench is compatible with any isolation amplifier or monitoring instrument that provides an analog output signal. The user can acquire and analyze data immediately because BioBench automatically recognizes and controls the National Instruments DAQ hardware, thus minimizing configuration headaches.

Some of the advantages of PC-Based Data Monitoring include the following:

- Easy-to-use software applications
- Large memory and the PCI bus
- Powerful processing capabilities
- Simplified customization and development
- More data storage and faster data transfer
- More efficient data analysis.

Figure 5.5-19 illustrates a typical setup of a data acquisition experiment using BioBench. BioBench also features pull-down menus through which the user can configure devices. Therefore, those who have made large capital investments can easily migrate their existing equipment into the Computer Age. Integrating a combination of old and new physiological instruments from a variety of manufacturers is an important and straightforward procedure. In fact, within the clinical and research settings, it is a common requirement to be able to acquire multiple physiological signals from a variety of medical devices and instruments that do not necessarily communicate with each other. Often, this situation is compounded by the fact that end-users would like to be able to view and analyze an entire waveform and not just an average value. In order to accomplish this, the end-user must acquire multiple channels of data at a relatively high sampling rate and have the ability to manage many large data files. BioBench can collect up to 16 channels simultaneously at a sampling rate of 1000 Hz per channel. Files are stored in an efficient binary format that significantly reduces the amount of hard disk and memory requirements of the PC. During data acquisition, a number of features are available to the end-user. These features include Data Logging, Event Logging, and Alarming.

BioBench™ was developed for National Instruments (Austin, TX) by Premise Development Corporation (Hartford, CT).

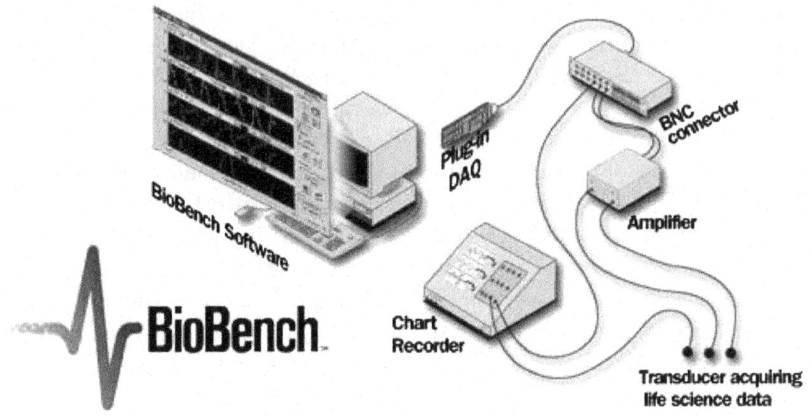

Figure 5.5-19 A typical biomedical application using BioBench.

Data logging

Logging can be enabled prior to, or during, an acquisition. The application will either prompt the user for a descriptive filename or it can be configured to automatically assign a filename for each acquisition. Turning the data-logging option on and off creates a log data-event record that can be inspected in any of the analysis views of BioBench.

Event logging

The capacity to associate and recognize user commands associated with a data file may be of significant value. BioBench has been designed to provide this capability by automatically logging user-defined events, stimulus events, and file-logging events. With user-defined events, the user can easily enter and associate date- and time-stamped notes with user actions or specific subsets of data. Stimulus events are also date- and time-stamped and provide the user information about whether a stimulus has been turned on or off. File-logging events note when data have been logged to a disk. All of these types of events are stored with the raw data when logging data to file, and they can be searched for when analyzing data.

Alarming

To alert the user about specific data values and thresholds, BioBench incorporates user-defined alarms for each signal displayed. Alarms appear on the user interface during data acquisition and notify the user if an alarm condition has occurred.

Figure 5.5-20 is an example of the Data Acquisition mode of BioBench. Once data have been acquired, BioBench can employ a wide array of easy-to-use analysis features. The user has the choice of importing recently acquired data or opening a data file that had been acquired for comparison or teaching purposes. Once a data set has been selected and opened, BioBench allows the user simply to select and highlight a region of interest and to choose the analysis options to perform a specific routine.

BioBench implements a wide array of scalar and array analyses. For example, scalar-analysis tools will determine the minimum, maximum, mean, integral, and slope of a selected data set, while the array analysis tools can employ Fast Fourier Transforms (FFTs), peak detection, histograms, and X-versus-Y plots.

The ability to compare multiple data files is important in analysis, and BioBench allows the user to open an unlimited number of data files for simultaneous comparison and analysis. All data files can be scanned using BioBench's search tools in which the user can search for particular events that are associated with areas of interest. In addition, BioBench allows the user to employ filters and transformations to their data sets, and all logged data can be easily exported to a spreadsheet or database for further analysis. Finally, any signal acquired with BioBench can be played back, thus taking lab experience into the classroom. Figure 5.5-21 illustrates the analysis features of BioBench.

Case study #4: a cardiovascular pressure–dimension analysis system

The intrinsic contractility of the heart muscle (myocardium) is the single most important determinant of prognosis in virtually all diseases affecting the heart (e.g., coronary artery disease, valvular heart disease, and cardiomyopathy). Furthermore, it is clinically important to be able to evaluate and track myocardial function in other situations, including chemotherapy, where cardiac

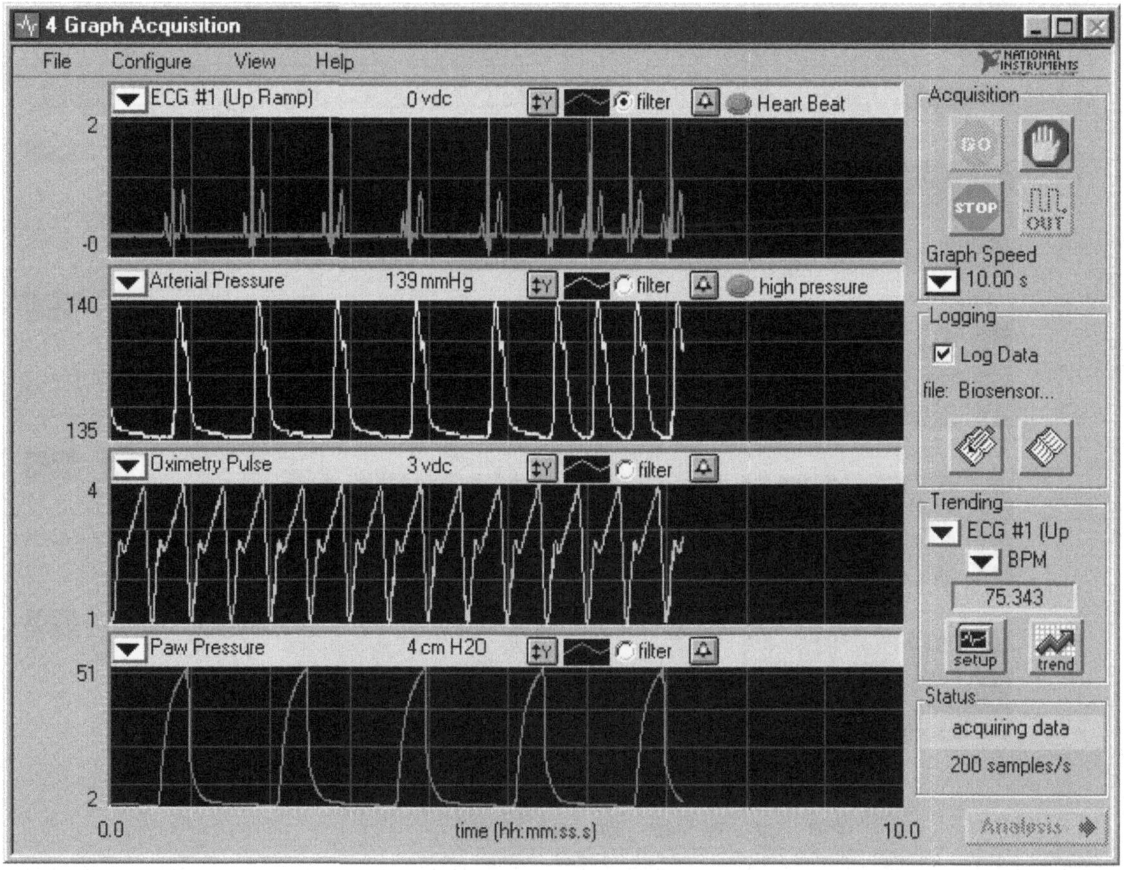

Figure 5.5-20 BioBench Acquisition mode with alarms enabled.

dysfunction could be a side effect of treatment, and liver disease, where cardiac dysfunction could complicate the disease.

The most commonly used measure of cardiac performance is the ejection fraction. Although it does provide some measure of intrinsic myocardial performance, it is also heavily influenced by other factors, such as heart rate and loading conditions (i.e., the amount of blood returning to the heart and the pressure against which the heart ejects blood).

Better indices of myocardial function based on the relationship between pressure and volume throughout the cardiac cycle (pressure–volume loops) exist. However, these methods have been limited because they require the ability to track ventricular volume continuously during rapidly changing loading conditions. While there are many techniques to measure volume under steady-state situations, or at end-diastole and end-systole (the basis of ejection fraction determinations), few have the potential to record volume during changing loading conditions.

Echocardiography can provide online images of the heart with high temporal resolution (typically 30 frames per second). Because echocardiography is radiation-free and has no identifiable toxicity, it is ideally suited to pressure–volume analyses. Until recently however, its use for this purpose has been limited by the need for manual tracing of the endocardial borders, an extremely tedious and time-consuming endeavor.

The system

Biomedical and software engineers at Premise Development Corporation (Hartford, CT), in collaboration with physicians and researchers at Hartford Hospital, have developed a sophisticated research application called the "Cardiovascular Pressure–Dimension Analysis (CPDA) System". The CPDA system acquires echocardiographic volume and area information from the acoustic quantification (AQ) port, in conjunction with ventricular pressure(s) and ECG signals to perform pressure–volume and pressure–area analyses rapidly. The development and validation of this system have led to numerous abstracts and publications at national conferences, including the American Heart Association, the American College of Cardiology, the American Society of Echocardiography, and the Association for the Advancement of Medical Instrumentation. This fully automated

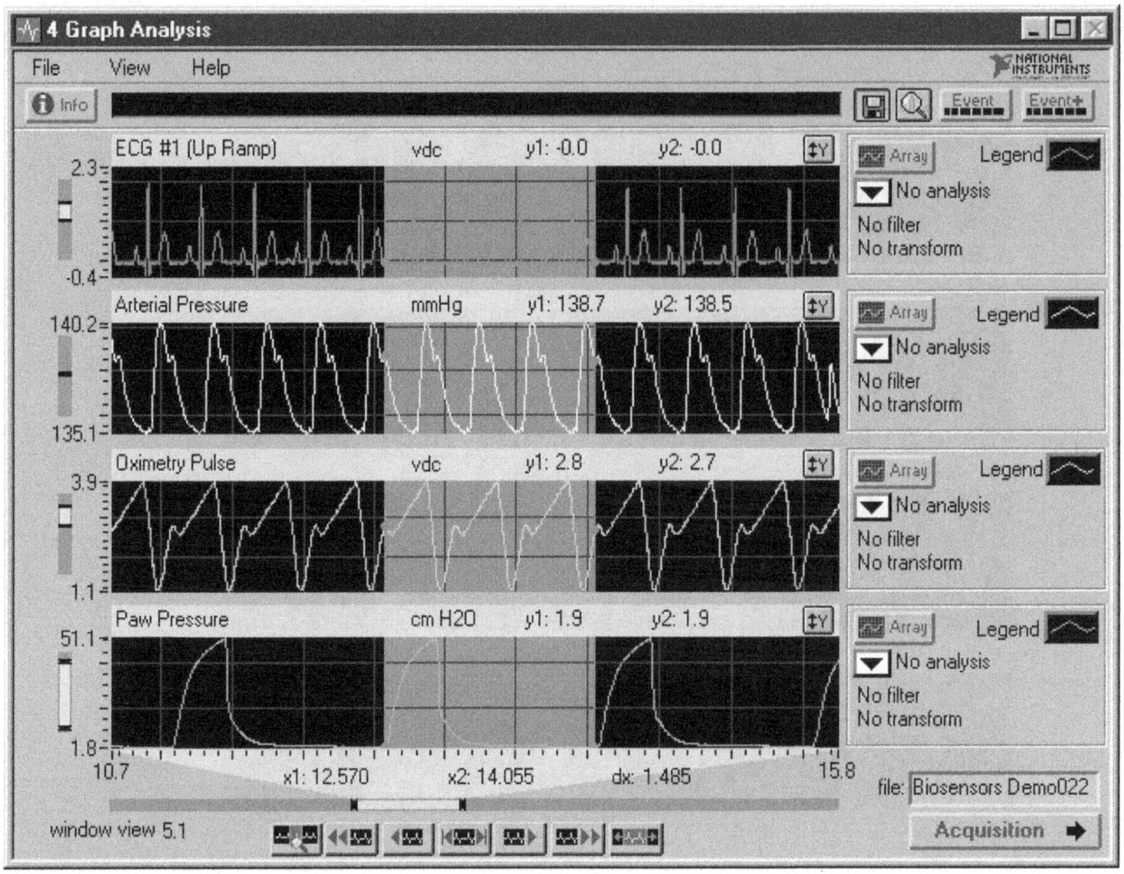

Figure 5.5-21 BioBench Analysis mode.

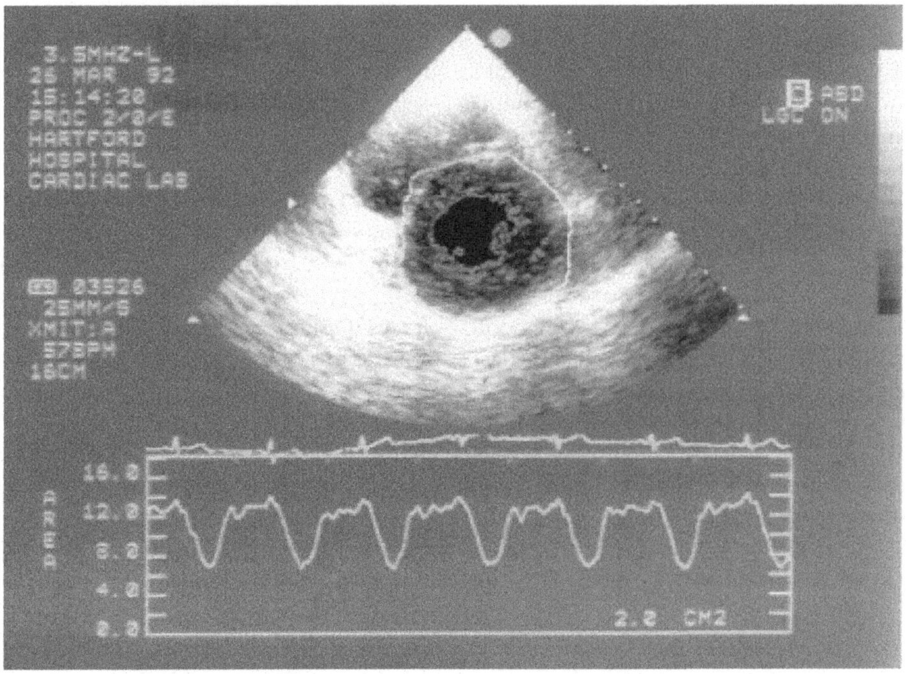

Figure 5.5-22 The Acoustic Quantification (AQ) Signal (Hewlett-Packard).

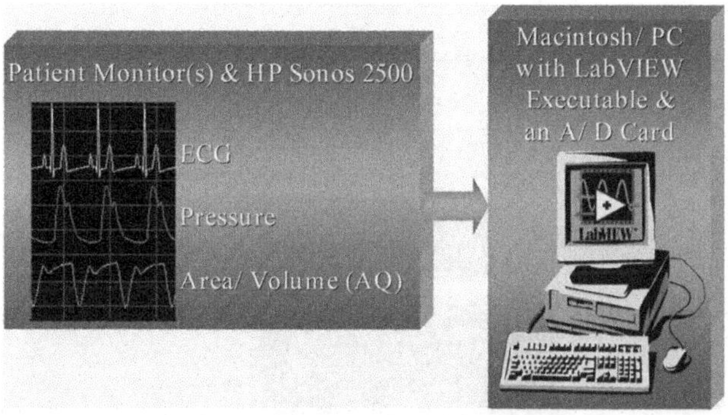

Figure 5.5-23 Schematic diagram of the CPDA system.

system allows cardiologists and researchers to perform online pressure–dimension and stroke work analyses during routine cardiac catheterizations and open-heart surgery.

The system has been designed to work with standard computer hardware. Analog signals for ECG, pressure, and area/volume (AQ) are connected to a standard BNC terminal board. Automated calibration routines ensure that each signal is properly scaled and that it allows the user to immediately collect and analyze pressure-dimension relationships.

The development of an automated, online method of tracing endocardial borders (Hewlett-Packard's AQ Technology) (Hewlett-Packard Medical Products Group, Andover, MA) has provided a method for rapid online area and volume determinations. Figure 5.5-22 illustrates this AQ signal from a Hewlett-Packard Sonos Ultrasound Machine. This signal is available as an analog voltage (−1 to +1 volts) through the Sonos Dataport option (BNC connector).

Figure 5.5-23 illustrates the measured parameters and the specific hardware used for this application. Although this application was initially developed on a Macintosh platform, the system can run on multiple platforms, including Windows 9x//NT/2000/XP. The CPDA also takes advantage of the latest hardware developments, and form-factors and can be used on either a desktop or a laptop computer.

Data acquisition and analysis

Upon launching this application, the user is presented with a dialog box that reviews the license agreement and limited warranty. Next, the main menu is displayed, allowing the user to select from one of six options as shown in Figure 5.5-24.

When conducting a test, an automated calibration sequence for the pressure and ultrasound signals (AQ) is generally performed. If the user elects not to perform a calibration, the most recent calibration values are accessed from the integrated calibration log. Pressure-calibration data are retrieved from a "lookup table" containing specific gain and offset values for a variety of different manufacturers' pressure monitors. The AQ calibration procedure involves "scaling" and "mapping" the display image signal over the −1 to +1 volt output

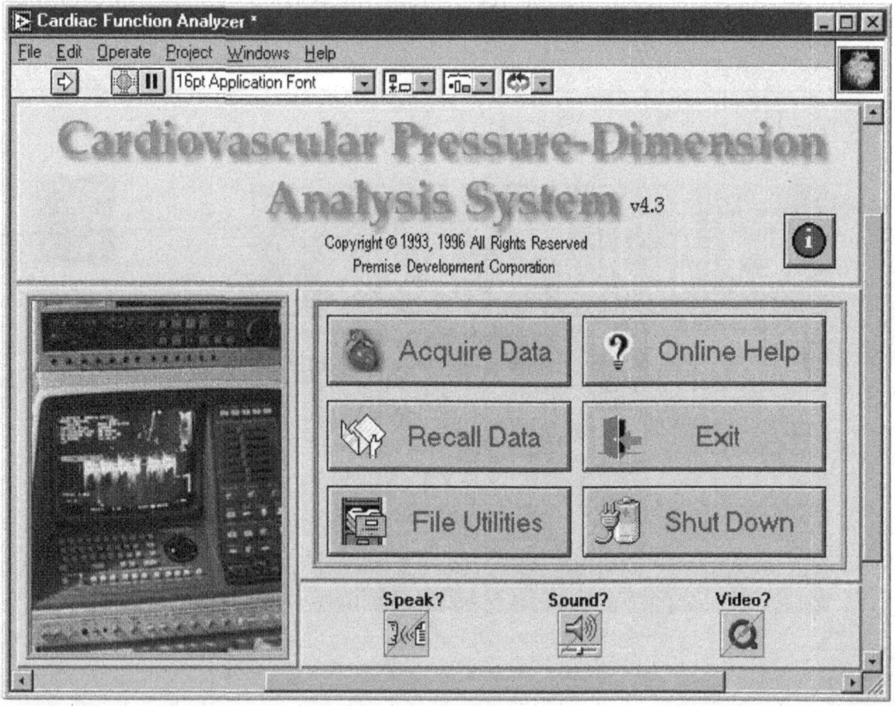

Figure 5.5-24 CPDA main menu.

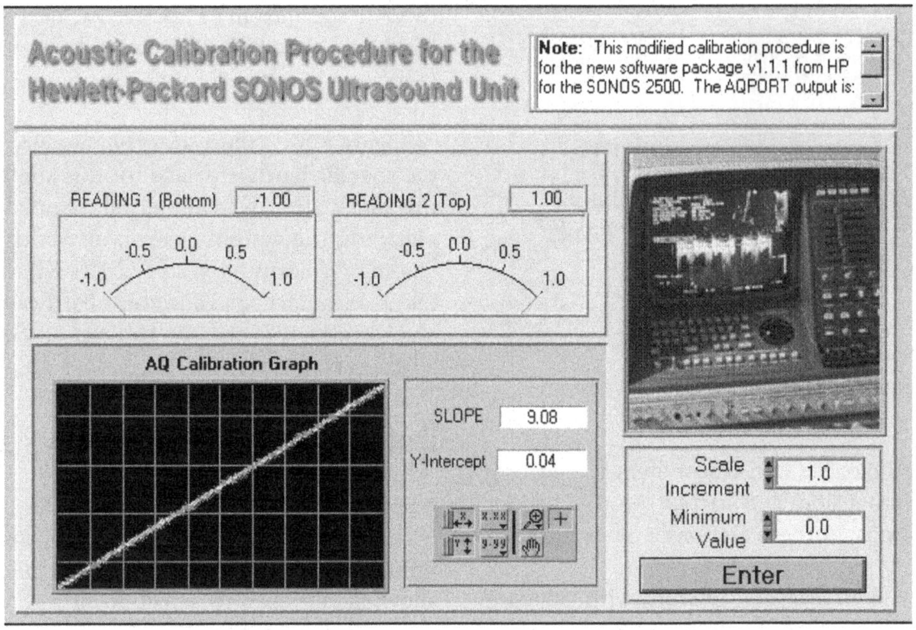

Figure 5.5-25 Hewlett-Packard's Sonos Ultrasound Machine Calibration front panel.

range. Figure 5.5-25 illustrates the front panel for the Hewlett-Packard Sonos calibration procedure. Sequential instructions in the form of a scrolling string indicator, as well as dialog boxes, are also available.

The default sampling frequency for each channel is 200 Hz. Data are typically collected for 20–60 seconds. The user is presented with a "Pre-Scan" panel to ensure that each signal is calibrated and tracking appropriately. When the user is ready to collect and store data, the "Cardiac DAQ" instrument is called (Figure 5.5-26). This sub-VI uses double-buffering to collect and display each channel for the predefined time. In order to maintain high temporal resolution, data are displayed in 10-second "sweeps." An indicator is provided to display the instantaneous and total collection time.

Once data are collected, the user is presented with the "Data Selection" sub-VI to define a particular range of data to save to a file. This option allows the user to store only the portion or subset of data that are useful among the entire collected data set (e.g., the last 25 seconds of

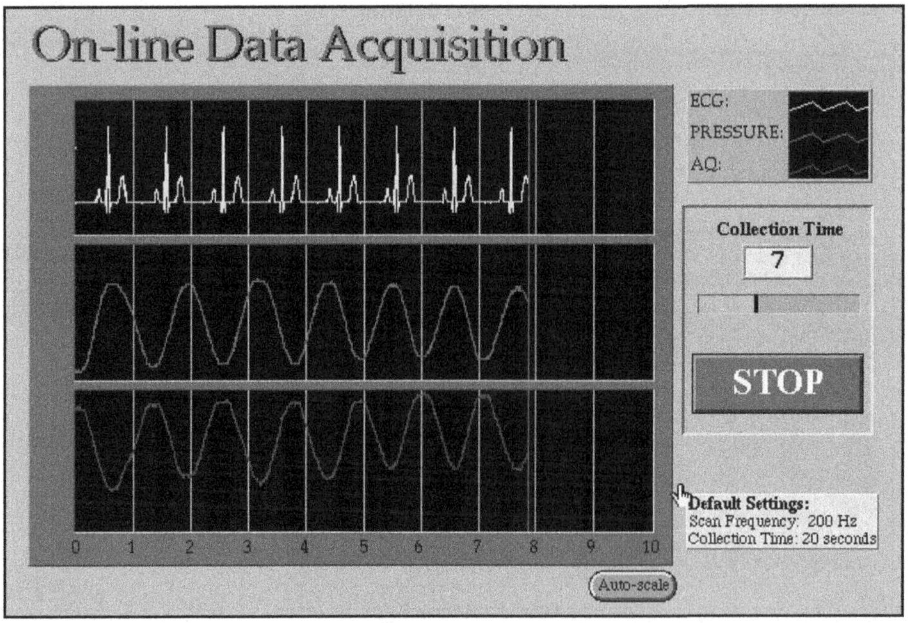

Figure 5.5-26 The Cardiac DAQ front panel.

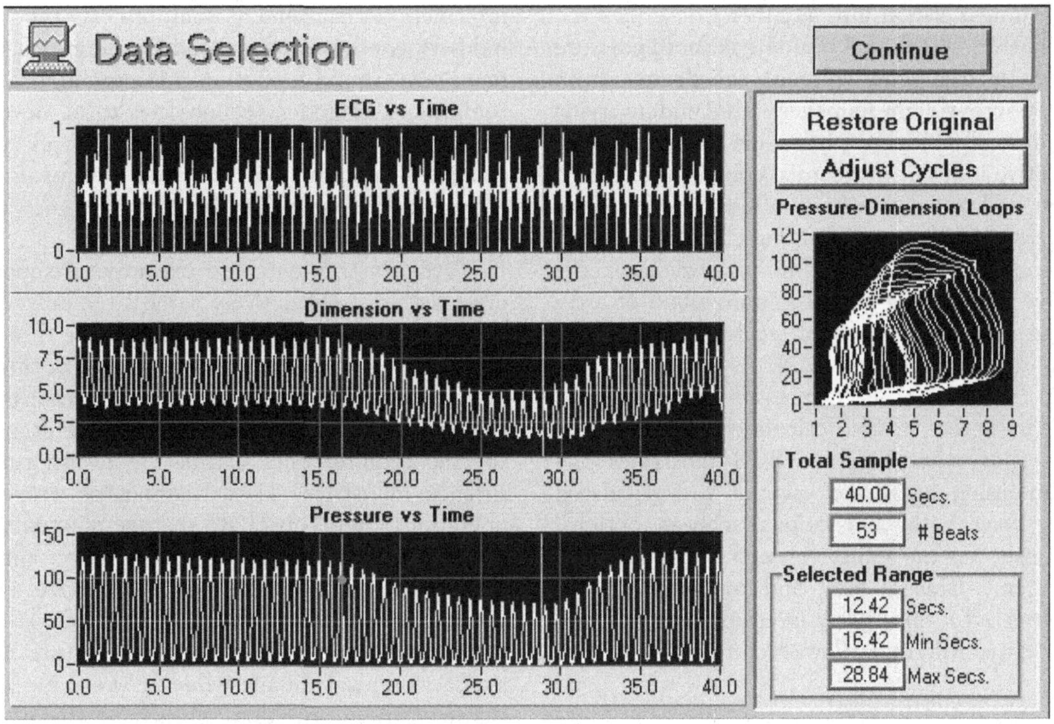

Figure 5.5-27 The Data Selection front panel.

a 60-second array). The default setting will store the entire data set. Interactive cursors are used to interactively set the initial and final indices of the data subset for analysis, as illustrated in Figure 5.5-27.

Clinical significance

Several important relationships can be derived from these signals. Specifically, a parameter called the "End-Systolic

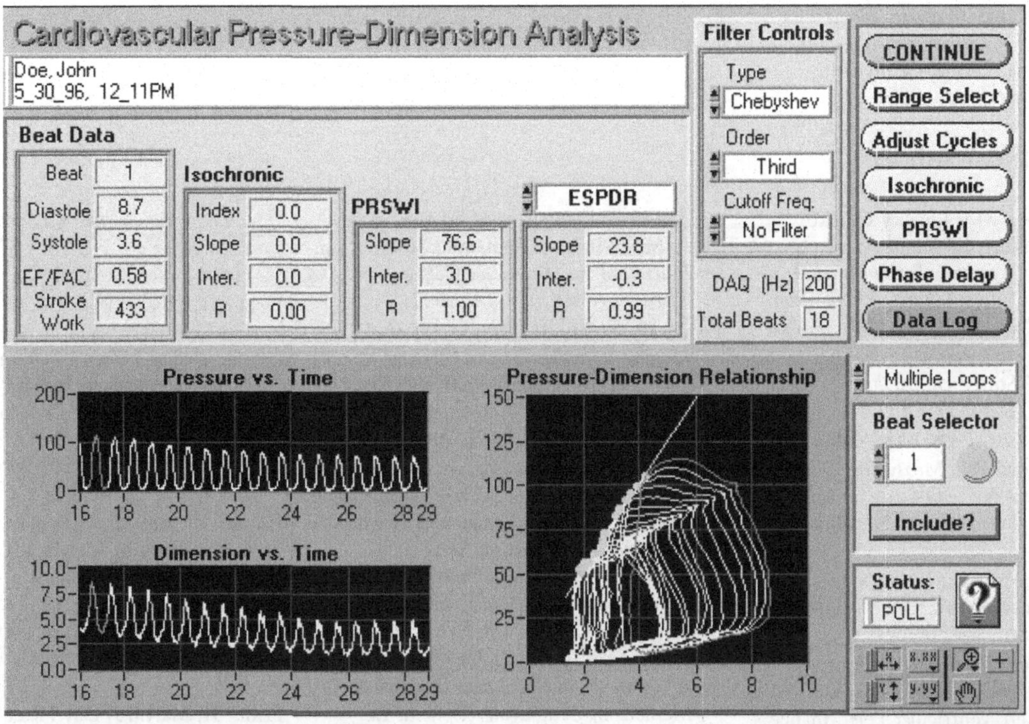

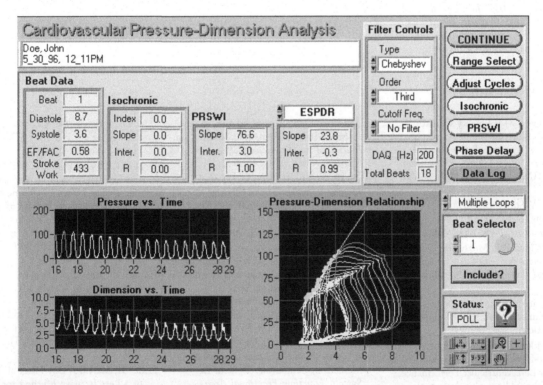

Figure 5.5-28 The Cardiac Cycle Analysis front panel.

Pressure–Volume Relationship" (ESPVR) describes the line of best fit through the peak-ratio (maximum pressure with respect to minimum volume) coordinates from a series of pressure–volume loops generated under varying loading conditions. The slope of this line has been shown to be a sensitive index of myocardial contractility that is independent of loading conditions. In addition, several other analyses, including time-varying elastance (Emax) and stroke work, are calculated. Time-varying elastance is measured by determining the maximum slope of a regression line through a series of isochronic pressure–volume coordinates. Stroke work is calculated by quantifying the area of each pressure–volume loop. Statistical parameters are also calculated and displayed for each set of data. Figure 5.5-28 illustrates the pressure–dimension loops and each of the calculated parameters along with the various analysis options. Finally, the user has the ability to export data sets into spreadsheet and database files, and to export graphs and indicators into third-party presentation software packages, such as Microsoft PowerPoint®.

Summary

VI allows the development and implementation of innovative and cost-effective biomedical applications and information-management solutions. As the health care industry continues to respond to the growing trends of managed care and capitation, it is imperative for clinically useful, cost-effective technologies to be developed and utilized. As application needs surely will continue to change, VI systems will continue to offer users flexible and powerful solutions without requiring new equipment or traditional instruments.

Virtual instruments and executive dashboards allow organizations to effectively harness the power of the PC to access, analyze, and share information throughout the enterprise. The case studies discussed in this article illustrate ways in which various institutions have conceived and developed "user-defined" solutions to meet specific requirements within the health care and insurance industries. These dashboards support general operations, help hospitals manage fluctuating patient census and bed availability, and empower clinicians and researchers with tools to acquire, analyze, and display clinical information from disparate sources. Decision-makers can easily move from big-picture analyses to transaction-level details while at the same time safely sharing this information throughout the enterprise to derive knowledge and to make timely, data driven decisions. Collectively, these integrated applications directly benefit health care providers, payers, and, most importantly, patients.

Reference

AHA. *Statistics 2001: The Clinical Advisory Board, Capacity Command Center-Best Practices for Managing a Full House.* Chicago, American Hospital Association, 2001.

Further information

American Society for Quality Control. *American National Standard: Definitions, Symbols, Formulas, and Tables for Control Charts, Publication number ANSI/ASQC A1-1987.* ANSI, 1987.

Breyfogle FW. *Statistical Methods for Testing, Development and Manufacturing.* New York, Wiley, 1982.

Carey RG, Lloyd RC. *Measuring Quality Improvement in Health care: A Guide to Statistical Process Control Applications.* 1995.

Frost, Sullivan. *Market Intelligence, File 765.* Mountain View, CA, The Dialog Corporation,

Fisher JP, Mikan JS, Rosow E, et al. Pressure–Dimension Analysis of Regional Left Ventricular Performance Using Echocardiographic Automatic Boundary Detection: Validation in an Animal Model of Inotropic Modulation. *Journal of the American College of Cardiology* 19(3):262A, 1992.

Fisher JP, McKay RG, Mikan JS, et al. *Human Left Ventricular Pressure-Area and Pressure–Volume Analysis Using Echocardiographic Automatic Boundary Detection.* Hartford, CT, American Heart Association, 1992.

Fisher JP, Mitchel JF, Rosow E, et al. *Evaluation of Left Ventricular Diastolic Pressure-Area Relations with Echocardiographic Automatic Boundary Detection.* Hartford, CT, American Heart Association, 1992.

Fisher JP, McKay RG, Mikan JS, et al. *A Comparison of Echocardiographic Methods of Evaluating Regional LV Systolic Function: Fractional Area Change Versus the End-Systolic Pressure-Area Relation.* Hartford, CT, American Heart Association, 1992.

Fisher JP, McKay RG, Rosow E, et al. On-Line Derivation of Human Left Ventricular Pressure–Volume Loops and Load-Independent Indices of Contractility Using Echocardiography with Automatic Boundary Detection: A Clinical Reality. *Circulation* 88:I–304, 1993.

Fisher JP, Chen C, Krupowies N, et al. Comparison of Mechanical and Pharmacologic Methods of Altering Loading Conditions to Determine End-Systolic Indices of Left Ventricle Function. *Circulation* 90(II):1–494, 1994.

Fisher JP, Martin J, Day FP, et al. Validation of a Less Invasive Method for

Determining Preload Recruitable Stroke Work Derived with Echocardiographic Automatic Boundary Detection. *Circulation* 92:1–278, 1995.

Fontes ML, Adam J, Rosow E, Mathew J, DeGraff AC. Non-Invasive Cardiopulmonary Function Assessment System. *J Clin Monit* 13:413, 1997.

Johnson GW. *LabVIEW Graphical Programming: Practical Applications in Instrumentation and Control, 2nd Edition.*, McGraw-Hill, 1997.

Kutzner J, Hightower L, Pruitt C. Measurement and Testing of CCD Sensors and Cameras. *SMPTE J*: 325–327, 1992.

Mathew JP, Adam J, Rosow E. Cardiovascular Pressure–Dimension Analysis System. *J Clin Monit* 13:423, 1997.

Montgomery DC. *Introduction to Statistical Quality Control, 2nd Edition.* New York, Wiley, 1992.

Rosow E, Adam J, Satlow M. *Real-time Executive Dashboards and Virtual Instrumentation: Solutions for Health care Systems*. 2002 Annual Health Information Systems Society (HIMSS) Conference and Exhibition, Georgia World Congress Center, Atlanta, GA, January 28, 2002.

Rosow E, Adam J. Virtual Instrumentation Tools for Real-Time Performance Indicators. *Biomed Instrum Technol* 34(2):99–104, 2000.

Rosow E, Olansen J. *Virtual Bioinstrumentation: Biomedical, Clinical, and Health care Applications in LabVIEW.*, Prentice-Hall, 2001.

Tufte ER. *Visual Explanations.*, Graphics Press, 1997.

Tufte ER. *Envisioning Information.*, Graphics Press, 1990.

Tufte ER. *The Visual Display of Quantitative Information.*, Graphics Press, 1983.

Walker B. *Optical Engineering Fundamentals*. New York, McGraw-Hill, 1995.

Wheeler DJ, Chambers DS. *Understanding Statistical Process Control, 2nd Edition.*, SPC Press, 1992.

Chapter 5.6

5.6

Electromagnetic interference in the hospital

W. David Paperman, Yadin David, and James Hibbetts

The density of occupancy of the electromagnetic spectrum is increasing because of the expansion of wireless services (the culprits). Despite the efforts of manufacturers to harden clinical devices to the effects of electromagnetic interference (EMI), reports of new incidents of interference to previously unaffected medical devices (the victims) appear in medical literature and anecdotally (CBS, 1994; Paperman et al, 1994). The role and the knowledge base of the clinical engineer must expand to understand and manage these ever-increasing challenges.

Electromagnetic radiation

Electromagnetic radiation occurs when an alternating current is generated. An electromagnetic field is created in the vicinity of the source. The range, or distance of the radiated electromagnetic field can be increased when coupled to a conductor. The magnetic field is further radiated by the flow of the current along this conductor. Even at a reduced amplitude, a corresponding current will be generated when another conductor is subjected to the field of the radiating conductor. The radiated field will have many characteristics. The most important of those characteristics are amplitude, periodicity, and waveform.

The amplitude of the impressed alternating current defines the energy of a radiated field. This field is generally expressed in volts per meter (V/M). Amplitude at the source point (transmitting conductor) will define, in conjunction with frequency and distance, the amplitude of the electromagnetic field induced in the secondary (receiving) conductor subject to the Inverse Square Law.

The periodicity or frequency of alternation is expressed in Hertz per second, which allows the calculation of the wavelength of the radiated electromagnetic field. Knowing the wavelength versus the physical length of both the radiator and the secondary conductor into which the radiated energy is induced provides an estimate of the potential amplitude developed by or within a device at risk. Waveforms other than linear (sinusoidal) can produce multiple and variable frequencies. The foremost example of a nonlinear waveform is the square wave. The square wave produces many multiples, or harmonics, which have a range many times that of the fundamental frequency. The majority of digital devices generate square waves.

Depending on the amplitude at the point of generation, the efficiency of the auxiliary radiator (antenna), the generated frequency, and the waveform, the potential for interference between digital devices and clinical devices can exist for many kilometers from the source of electromagnetic radiation.

EMI threats come from a multitude of sources. These are broadly divided into two categories: Devices that emit intentional electromagnetic radiation for communications and control, and devices that, as a byproduct of their operation, emit unintentional (incidental) radiation (Figure 5.6-1). Some of the major and more common sources of EMI encountered in the clinical environment follow.

Intentional radiators:

- Television broadcast stations: Analog and digital
- Commercial radio stations: Analog and digital

Biomedical Engineering Desk Reference; ISBN: 9780123746467

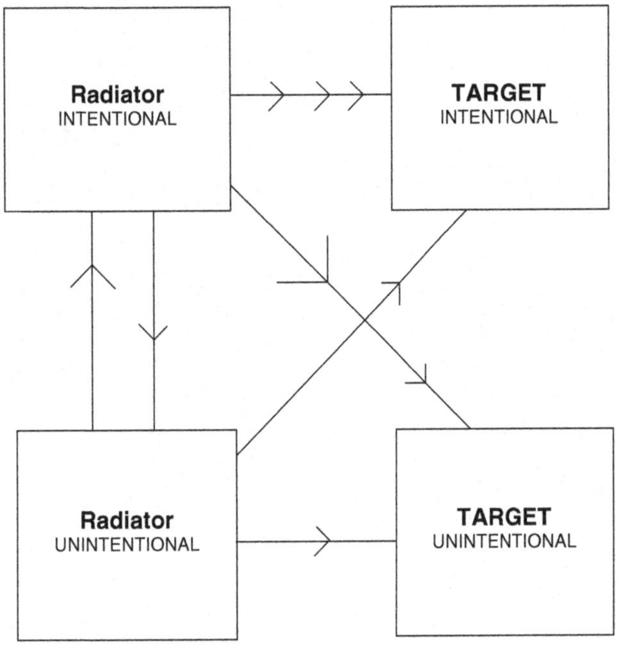

Figure 5.6-1 The complexities of EMI-radiators and targets: Culprits and victims.

- Land mobile radio: Fixed base (FB), mobile, and portable two-way radio sources (walkie talkies)*
- Paging: One-way and two-way wireless messaging service transmitters
- Cellular telephones and sites*
- Wireless personal digital assistants (PDAs)
- Wireless networking devices (an increasing threat in the 2.4 GHz industrial, scientific, and medical ISM band)
- Unlicensed and unauthorized users of two-way radio communications equipment (Pirates)*

Unintentional radiators:

- Lighting systems (especially florescent), including energy saving electronic ballasts*
- High energy control systems (HVAC controllers), especially variable-speed controllers*
- Malfunctioning electrical services*
- Universal-type electric motors*
- Pulse oximeters*
- Displays (CRT and plasma), computer, television, and instrumentation
- Wired computing networks
- "Smart" fire detection and alarm devices*
- Electrosurgical units (ESU)*
- Defibrillators

These sources of electromagnetic radiation, as seen from the previous lists, can be intentional or unintentional. The challenges and the institutional responses imposed by the presence of these EMI sources can be divided, for all intents and purposes, into four parts: Mitigation, detection, correction, and prevention. Although the mitigation and prevention phases appear to be similar, they differ in their approaches and applications.

Controlling the effect of EMI

Mitigation

In the mitigation phase, the principal concern is reducing risk to, or victimization of, clinical devices by devices emitting unintentional or intentional electromagnetic radiation. The sources of the devices may be internal or external to the institution. The devices may be intentional or unintentional radiators of electromagnetic fields, and the electromagnetic energy may be radiated or conducted. Figure 5.6-1 illustrates the electromagnetic interaction among intentional and unintentional radiators and intentional and unintentional targets.

Mitigation is accomplished in part by the careful analysis of a device, including the accessories to be used with it, and the electromagnetic environment in which it will be used. A successful approach has been implemented at Texas Children's Hospital. This approach is accomplished in two phases: Thorough analysis of the environment in which the device is to be placed, called "footprinting," and an equally thorough analysis of the clinical device itself, called "fingerprinting." The procedures of quantifying potential EMI and mitigating the effects through environment analysis are described in greater detail later in this chapter.

Another vital part of the mitigation process occurs during facility planning. Whether planning a new facility, an expansion of an existing facility, or remodeling of a facility, clear and concise communications between affected departments, architects, contractors, and the clinical engineering department are vital at all stages of the project. For example, one institution installed expensive, screened rooms at great cost. When the clinical engineer, acting on complaints of erratic EMG operation, determined that the room played a role in the culprit/victim relationship, the rooms were disassembled, also at great cost. In this case, the relocation of the affected department was not an option. The clinical engineering department had not been made aware of these rooms, or of their intended application that involved the use of known culprit devices.

* Sources (the culprits) of incidents attributed to EMI encountered at Texas Children's Hospital.

Examples of proactive mitigation include consulting services provided by the Biomedical Engineering Department and the Television Services Group during the latest expansion of Texas Children's Hospital. These services, which involved the expertise of all groups within the biomedical engineering department, included not only space design, but the design, implementation, and expansion of wireless paging and communications systems. Through the use of a distributed energy antenna systems, (i.e., leaky coax-radiax) radio frequency (RF) power levels were maintained at levels deemed safe for clinical devices while providing the required coverage area for a nurse call-specific paging system. Safe wireless telephonic communications were implemented using low-power microcellular systems. Infant abduction systems were reviewed for specifications including emission characteristics, and a system was selected based in part on its low electromagnetic radiation levels.

The use of portable radio communication devices in a large campus environment is a basic requirement for security, engineering, and guest services. As part of the mitigation process, instruction in the safe use of two-way radio equipment in the clinical environment is mandatory for all personnel who use this equipment (ECRI, 1993).

Detection

Detection is implemented upon the report of a device malfunction. A preliminary analysis of the incident may indicate that the cause of the malfunction, intermittent or permanent, may have been EMI. In the case of continuing or continuous device malfunction not otherwise attributable to defects within the device itself (e.g., no problem found [NPF] service reports), a careful investigation begins, using various test equipment, some of which may require specialized construction. This section presents a more detailed description of the basic equipment necessary to locate and identify sources of EMI and some of the basic techniques, referred to as "ghost hunting." Once the type and source of the interference is detected and analyzed, the next step is to reduce or eliminate the interference to the victim device. However, the victim device is removed in some cases, because of an intensely hostile electromagnetic environment. Replacement of the victim device with an equivalent device that may have other, less sensitive responses is an option.

The detection phase can eliminate the possibility of EMI as the culprit and cause of improper operation of a specific device. Indeed, subsequent investigation can reveal technical problems within a device that were previously attributed to EMI.

Correction

Correction of victim/culprit relationship(s) can take many forms, some practical, some impractical. Correction may be part of a process to ensure that the victim device meets all specifications that can affect its susceptibility to electromagnetic radiation. For example, hospital staff should ensure that proper case-to-case contact is made in coated conductive coatings and that the coatings have not been worn or abraded. In some instances, the victim device can and should be removed from the environment where it is at risk. If the culprit is local, removal may be one practical solution to the problem. Instances of electromagnetic radiation emanating from abandoned wiring (passive reradiation) have occurred that dictated removal of the wiring. Interference to clinical devices originating from active (nonfiberoptic) wiring, such as networking trunk conduits, mandates rerouting of the wiring. The same can apply to modern in-plant telephone systems, usually digital in nature. Experiences at Texas Children's Hospital support the work of researchers (ECRI, 1988) in finding that shielded rooms rarely correct problems when the source of electromagnetic radiation is contained within the local environment. In extreme cases, the existing shielded room must be disassembled or the victim equipment (and department), moved to other quarters within the institution.

Prevention

Sustainable EMI prevention must include institution-wide compliance with guidelines and with those policies created within an institution to limit possible sources, EMI as well as the selection and deployment of medical devices that offer more effective immunity from EMI. Overall, observance of this two-pronged policy will have the effect of reducing the risk of EMI to the proper operation of clinical devices and therefore reducing the risk to the institution. At Texas Children's Hospital, EMI prevention takes several forms (David, 1993). A proactive policies and procedures manual that, guided by the biomedical engineering department, defines allowable sources of radiation within the institution and mandates training procedures for employees of the institution that are required by job necessity to carry sources of electromagnetic radiation, i.e., intentional radiators. As an example, signage mandated by the policies and procedures manual directs that all cellular telephones be turned off. Their use within the institutional campus is not allowed. The policies and procedures manual further mandates that all employees using radio equipment, especially handheld devices, be trained in their safe use within the institution. Additionally, a stipulation that clinical devices must be EMI compatible is incorporated into the

Condition of Sale, issued and agreed to by to vendors, to define the technical conditions that must be met under the purchase contract.

Case histories

Role reversal

An unusual example occurred when the victim (i.e., a pager), normally considered a culprit, and the culprit (i.e., a physiological monitor), normally a victim, exchanged roles. Pagers had ceased receiving calls in a cardiovascular intensive care unit. Field intensity measurements, interrogation of employees, and other investigative procedures conducted by the clinical engineer yielded a scenario in which some pagers were not able to receive calls when near a well-known intensive care monitoring system.

Standards and practices relating to on-site testing for the presence of electromagnetic radiation were reviewed. A modified Open Antenna Test Site (OATS) procedure (ECRI, 1992; ECRI, 1988), was used as the measurement guideline (Bennett, 1993). As for equipment, an Empire NF-105 field intensity metering system (loaned by a staff member previously engaged in field site RF measurements), was used as the measurement device. An area characterization *(i.e., footprint)* in the area of interest was taken as well as a characterization of a representative culprit device (i.e., *fingerprint)*. The field intensity measurements of a representative unit of the monitor system showed the existence of an unintentional radiofrequency close to the operating frequency of the paging system. Once the source of the EMI was identified, the results were presented to the manufacturer. Initial responses from the manufacturer's media representatives were not encouraging. During one conversation, the manufacturer's EMI expert said that the solution was to increase the radiated power of the paging transmitters. This solution was immediately dismissed from consideration, because increasing the level of intentionally radiated electromagnetic energy to compensate for the effect of existing unintentional electromagnetic radiation would increase the potential for risk to other clinical devices. Only when the biomedical engineering department was ready to persuade the client department to cancel the purchase order for the remaining devices did the manufacturer discover a modification to reduce the severity of the interference. This modification, when installed, did not remove all of the unintentional radiation but did reduce the level of the interfering electromagnetic radiation sufficiently to relieve interference with pagers. The manufacturer essentially redistributed and dispersed the radiated energy to other points in the spectrum, which reduced the amplitude at the specific pager operating frequency.

This incident was an interesting introduction to the practical effects of device compatibility (or lack of) and EMI in the clinical environment. After a technician from the Texas Children's Hospital Biomedical Engineering Department installed the modification provided by the manufacturer, a fingerprint of the modified device was taken again.

This incident was concluded so successfully that two mutually beneficial results were obtained: The problems of EMI in the clinical environment were clearly and graphically demonstrated within the institution, and the hospital obtained funding for the purchase of modern signal analysis equipment.

An unusual source

Another "ghost hunt" occurred at one of the biomedical engineering departments' client hospitals. The wireless medical telemetry there began intermittently displaying error codes (i.e., loss of data). Staff members were carefully questioned. The times when the interference occurred were established. Further probing revealed that a recently upgraded fire alarm system was undergoing acceptance tests during the periods of interference. As this would be an unusual source of interference, cooperation was sought from plant engineering personnel and the representatives of the fire detection company. Questions posed to the fire detection engineers yielded no prior indications of interference. A spectrum analyzer and broadband antenna system were set up at the site. The spectrum analyzer showed the presence of a recurrent but not time repetitive pulse, indicating that the interference source was radiating a quasi-random pulse.

Working with the fire alarm and plant engineering personnel, circuits controlling the various alarm devices were isolated and shut down individually. When no results were obtained on the affected floor, the same shut-down procedure was implemented on the floor below. When the enunciators—the audio-visual (A/V) units that provided audible and visual warnings of a fire—were shut down one floor below the affected area, the pulse and the interference effects disappeared. The floor below the affected area was a mechanical plant floor, and so the fire alarm company had installed high-powered A/V units. These A/V units used strobe lights. On discharge, these devices emitted a fast-rising, short-duration pulse. The telemetry antennas were receiving these pulses. However, the preamplifiers in the antennas and in the receivers themselves were not saturating or being desensitized by these pulses because the pulse was of such short duration. The pulses passed right through the RF portions of the telemetry system and corrupted data bytes. Open junction boxes with excess wiring hanging in circular loops contributed to the radiation of the pulses. Properly closing the junction boxes and replacing the strobe lights with lower intensity devices within fire code guidelines resolved the situation.

Risk prevention

The following is an example of how the application of footprinting and fingerprinting may have prevented an EMI incident. The diagnostic imaging department bought new telemetry equipment. Footprinting revealed that the level of incidental emissions in that department was approximately 67 microvolts. Fingerprinting a representative telemetry transmitter showed that the transmitted energy level at the standard 1-meter distance was 12 microvolts above the background level.

The defined risk was that there was not sufficient signal-to-noise ratio to preclude intolerably long periods of loss of useable signal throughput. Testing the telemetry transmitter (fingerprinting) is done under controlled conditions at a fixed distance that can not be maintained under real world conditions. A patient wearing a telemetry transmitter cannot be expected to maintain a 1-meter distance from the receiving antennas. As the distance between a patient and the receiving antennas increases, the received signal degrades because of the various topographical conditions and the inverse square law. As the signal degrades, the level of acceptable data degrades accordingly. If the degradation is significant, there is a chance that the telemetry system will not be able to recognize the emergency if a patient is in cardiac distress.

Interference of another type

But not all ghosts are due to radiated or conducted electromagnetic fields. For example, a report was received of intermittent interference to an EMG device in the physical therapy department of the hospital. The department felt that the culprit was the MRI system located directly above the area containing the victim device. During footprinting, measurements entailing more than the normal broad spectrum procedures were indicated. In this application, the fingerprinting equipment was set up to measuring any radiated electromagnetic fields at the resonant frequency of the MRI that might leak from the shielded room environment that contains the MRI system. During a period of several hours, no leakage was detected. A broader spectrum scan showed that this department was in a remarkably quiet location with respect to RF energy.

Further interviews with the doctors and staff led the clinical engineer to perform a somewhat unorthodox series of tests. The filtration on the EMG was broadened and the leads laid out, unterminated, on the couch on which patients were placed. During this procedure the clinical engineer observed that, when pressure was applied to the couch cushion, the baseline of the EMG machine would vary synchronously. Several additional adjustments to the sensitivity (gain) of the device and the time base were made. It then appeared that any motion in the immediate vicinity of the device would cause this baseline shift. Based on the results of these tests and of an investigation that determined the material composition of the environment (vinyl cushions on the couch and highly waxed vinyl floor), the engineer concluded that the cause of the interference with the device was electrostatic, not electromagnetic. Furthermore, the intermittent appearance of the problem was attributed to the fact that hydrotherapeutic baths were located two doors away from the EMG room. Their intermittent operation would raise the humidity sufficiently to reduce the potential for the generation of intense electrostatic charges. This was a decidedly different ghost hunt.

Interference not caused by EMI

One of the recent trends related to EMI is that of a manufacturer's technical problems being attributed to EMI. Although EMI may play a significant role in the performance degradation in certain clinical devices, it is not always the cause. One example involved radiological equipment. X-ray films from one of the radiology labs displayed an artifact described by the radiologists as "chicken scratches." This artifact was present primarily during a Temporo-Mandibular Joint (TMJ) procedure. This artifact was initially attributed to either conducted or radiated EMI, specifically conducted EMI from the power lines. An outside consultant measured radiation from the power lines, but there was insufficient indication that the artifact was due to the power system. After several months, the biomedical engineering department was called in for consultation.

Initial investigation, including timing the artifacts as they appeared on the film, indicated some degree of synchronicity (timing repeatability). This implied that the interference was time-locked (synchronized) within the radiology system. Due to the characteristics of the interference, such as the timing (frequency) of the interference in relationship to the scan rate, a probe was designed, constructed, and tuned to resonate within the frequency range of the suspected interfering signal. A thorough investigation of the areas surrounding the radiology suite included probing of the three-phase electrical service panels that provided the distribution of power to the suite. No high levels of radiated electromagnetic energy were encountered within the frequency of interest. The investigation was then conducted within the radiology suite. Despite intensive probing of all devices within the suite, including video monitors, computers, and control systems, and examination of the system involved in the TMJ procedure during which these artifacts most often appeared, the source could not be attributed to EMI.

A detailed report was filed with the radiology department. The manufacturer's representative was called in. Based on the inconclusive findings, a greater in-depth analysis of the problem was conducted by the representative, the clinical engineer, and technicians from the biomedical engineering department radiology group. During the analysis, the investigators discovered a disconnected bonding strap within the camera head. Reconnecting this strap improved the performance of the system and reduced the artifact.

This incident further illustrates the benefits of cooperation to resolve interference issues, irrespective of the sources of interference. It also demonstrates the effectiveness of a proactive, competent, and supported EMI program in the clinical environment. Placing blame rarely mitigates risks.

Variability: a demonstration of the problem

Variability in the repeat cause and effect (culprit/victim) device relationship poses a problem when instituting proactive programs designed to reduce the risks attributable to EMI with the support of management.

In 1996, at the request of NHK (Japan Educational Television), the television services group of the biomedical engineering department at the Texas Children's Hospital in Houston was asked contribute a segment to an educational program about EMI issues in the hospital

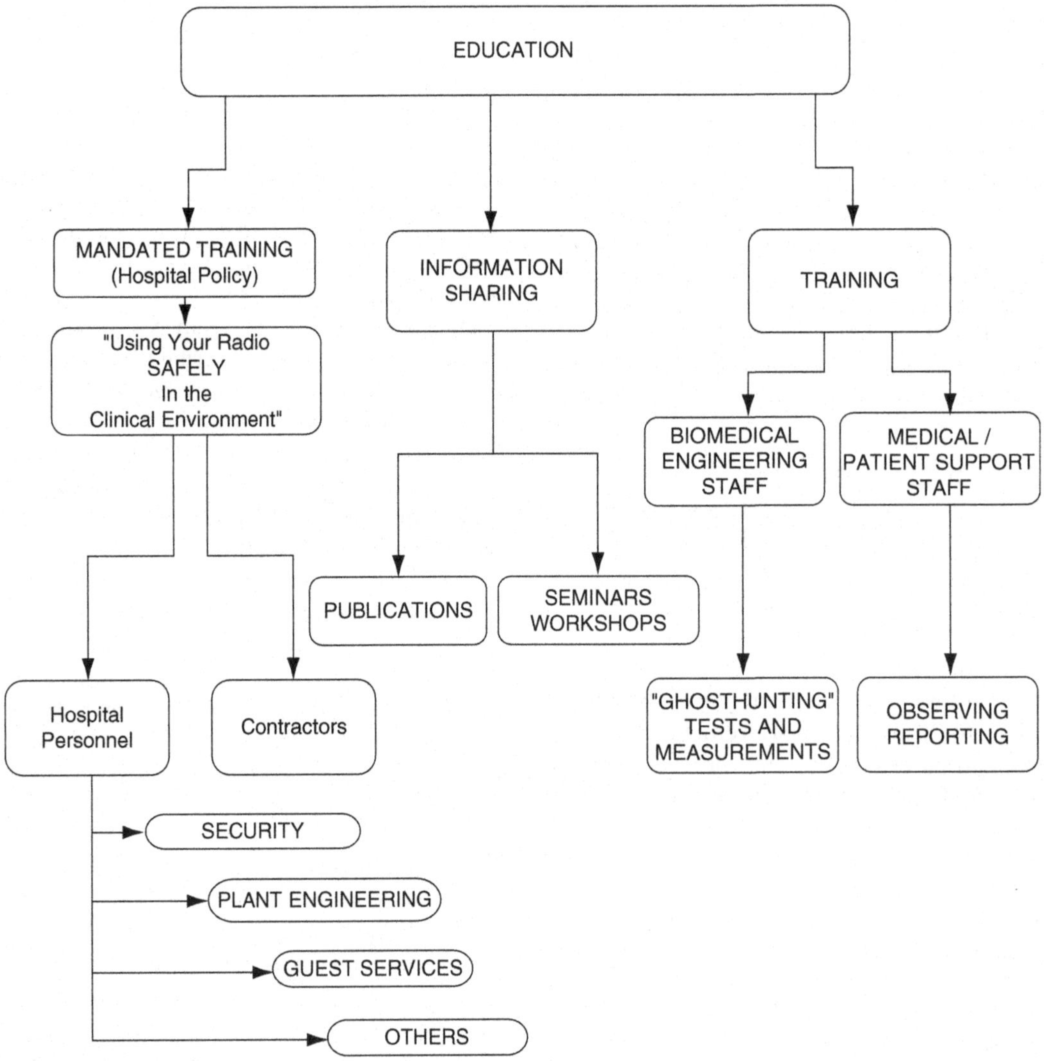

Figure 5.6-2 Flow diagram of hospital educational program in electromagnetic compatibility and interference relating to medical device technologies.

environment. The hospital administrators decided that a demonstration of EMI using actual medical equipment would meet the program's objectives of demonstrating the variability of effects of a common source radiator on identical clinical devices.

Two of the same model of hemodialysis machines from the same manufacturer were used in the demonstration. Under the supervision of the clinical engineer, the machines were prepared in such a way that accessories, calibrations, wiring positions within the devices, and location within the demonstration area were as identical as possible.

An intentional radiating source was placed in transmit mode at a distance of 1 meter from the clinical devices. The source was a walkie-talkie commonly encountered in the environment (151.625 MHz frequency band, measured power output of 4 watts at the transmitter, antenna efficiency of approximately 40%, frequency modulation at 5-KHz deviation). The results illustrated the issue of variable susceptibility. One device failed repeatedly in a noncatastrophic condition, while the other device, located nearby, was unaffected.

Demonstrations such as this, performed on demand, tend to support the many anecdotal EMI-related reports that our department receives daily from other institutions. Regrettably, many EMI incidents go unreported due to a lack of specific programs to address them.

The varied nature of EMI-related equipment malfunctions and the associated risks mandate a proactive program of EMI identification and methodology for risk reduction. An effective program relies on the cooperation of all parties potentially affected by EMI; medical staff, plant engineering, information services, biomedical engineering, and device manufacturers. Due to the highly diversified knowledge and experience required to coordinate detection and mitigation of EMI, the clinical engineering department and its personnel experienced in RF must take responsibility.

An operational protocol must be developed to address EMI issues. (See Figure 5.6-2.) This provides a defined structure to process requests for EMI investigations and a structure for processing and reporting the results of the investigations.

Programs and procedures

Testing for electromagnetic compatibility (EMC) in the clinical environment introduces a host of complex conditions not normally encountered in laboratory situations. In the clinical environment, various RF sources of EMI may be present anywhere. Isolating and analyzing the impact from the sources of interference involves a multidisciplined approach based on training in and knowledge of the following:

- Operation of medical devices and their susceptibility to EMI
- RF propagation modalities and interaction theory
- Spectrum-analysis systems and technique (preferably with signature analysis capabilities) and calibrated antennas
- Established methodology of investigating suspected EMI problems, which includes testing protocols and standards

Both standard test procedures adapted for the clinical environment and personnel trained in RF behavior increase the odds of proactively controlling EMI in the clinical environment, thus providing a safer and more effective patient care environment. The methods employed in the following procedures are variations of the OATS technique (ECRI, 1992; ECRI, 1988), a standard for open site testing (Southwick, 1992; Bennett, 1993; ANSI, 1991) and ANSI C63.4-1991 (ANSI, 1991) and ANSI C63.18-1997.

The selection of the spectrum analyzer and the options installed in it were influenced by several factors. A spectrum analyzer of the communications system test type was deemed desirable because there is no better way to characterize devices that emit RF— intentional or incidental—and the environment in which those devices operate. Broadband devices indicate relative RF activity but do not indicate the operating frequencies or modulation types. Both characteristics, independently and together, affect the susceptibility of clinical devices.

A digitally based communications analyzer was needed to archive the results of the EMI tests; both fingerprints and footprints. The flexibility of performance requirements of the device was important, and as a cost-saving benefit, Texas Children's Hospital also uses it to maintain the hospital radio communications systems.

Since no two modified OATS environments are identical, no two results obtained under the same testing parameters will be identical. Many factors affect the detailed test results, including complex absorption and reflection variables that are totally site-dependant.

Footprinting

At Texas Children's Hospital, the EMI testing program has evolved from years of experience and analysis. This program is not static. Sources of EMI and susceptibility characteristics of devices are in constant change. As new threats arise, the plan is periodically reviewed and modified to contend with them. Further modifications to the plan and procedures are based on a continuous review of wireless industry trends. They represent a proactive response to perceived future threats. At present the program consists of the procedural plan outlined above and three series of tests: (1) area characterization, footprinting (Figure 5.6-3), (2) device characterization, fingerprinting (Figure (5.6-4), and *ad hoc* susceptibility testing according to IEEE guidelines (Knudson and Bulkeley, 1994).

Footprinting an area means performing a series of spectrum/amplitude scans for electromagnetic radiation in a defined or designated area. Footprinting is an ongoing technique that defines electromagnetic radiation in a clinical facility or facilities in a multibuilding campus. Footprinting is also done upon request from a department experiencing performance degradation of clinical, diagnostic, or therapeutic devices when EMI is the suspect. The procedure involves a series of 20-MHz-wide spectrum sweeps, beginning at 2 MHz and ending at 1 GHz with antenna(s) in the horizontal plane. This procedure is repeated with the antenna(s) vertically polarized. Tunable standard antennas are adjusted for the correct resonant length for the center of each 20-MHz window.

The footprinting procedures yield two results: An overview of the radiated electromagnetic fields present at a specified location in the environment and the amplitudes and types of emission of those fields. Several incident investigations have been successful using this information. In cases where EMI has been site-originated, the source has been removed and the problem has been corrected. In some cases, the affected device required maintenance to correct the problem. An added value of footprinting is that the data obtained during the process meets the basic requirements for a site search similar to the OATS procedure for fingerprinting individual devices. The results of the footprinting scans are transferred to a storage medium and filed. They form a comparison database that is used to evaluate new devices before introducing them into a specified area.

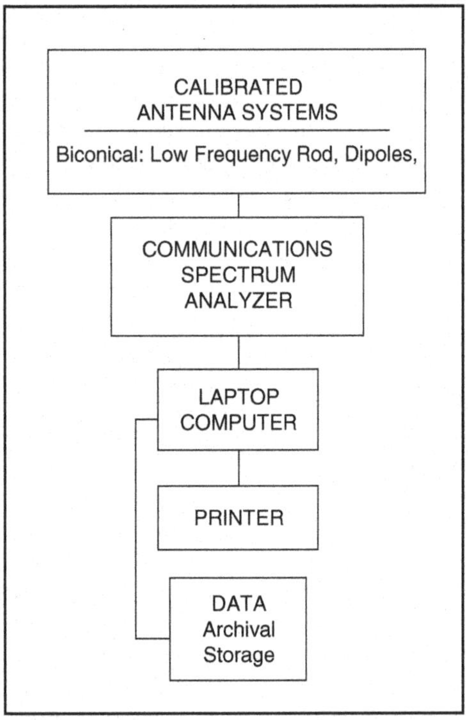

Figure 5.6-3 Typical equipment configuration for fingerprinting and footprinting.

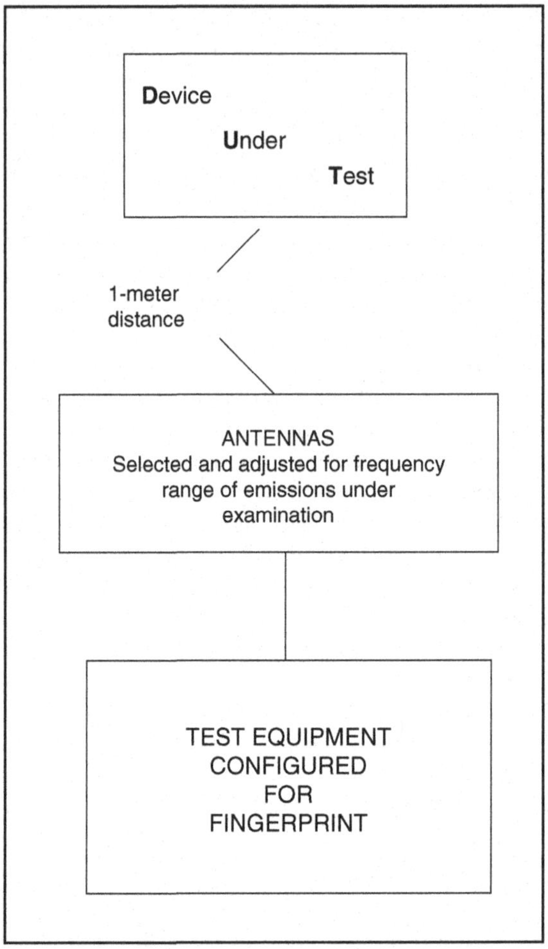

Figure 5.6-4 Fingerprinting.

Fingerprinting

The process of fingerprinting a device has several steps. Again, the spectrum analyzer and calibrated antenna system are the primary tools used to analyze the device under test (DUT) (see Figure 5.6-4). To minimize the loss of information from the DUT due to masking by other sources of electromagnetic radiation, areas within the clinical physical plant should be tested using the footprinting procedure until a relatively quiet area is found.

For the fingerprinting procedure, as in the footprinting procedures, the standards antenna should be located as far as possible from any conductive material. The standards antenna should be located at a height equal to the center of the DUT. Due to constantly changing EM fields and frequencies, they execute the footprinting procedure in the selected area immediately prior to fingerprinting the DUT. Once the data from the latest footprint has been stored, place the DUT on a nonmetallic stand located as far as possible from any conductive material. The standard antenna(s) is then mounted on a tripod and, in a horizontally polarized mode, placed 1 meter from the front of the DUT. As in the footprint procedure, a series of 20-MHz-wide spectrum sweeps are performed and the results are recorded. This procedure is repeated with the antenna(s) vertically polarized. When using tunable standard antennas, adjust them to the correct resonant length for the center of each 20-MHz window. Upon review of the data collected, any emissions attributable to the DUT should be rescanned. To increase detail in the frequency range in which the emissions were observed, the spectrum analyzer window is narrowed to a 200-KHz/division or smaller, (e.g., 5 KHz/division) sweep width.

Fingerprinting tests are conducted for two reasons. First, compliance with hospital policy on acceptance testing requires that representative samples of all devices be tested prior to entering the clinical environment for the first time. Second, both theory and experience have demonstrated the need for device testing. A device that radiates unintentionally can victimize other devices. It may also be susceptible to victimization by ingress at their egress frequencies and modulation parameters. Those devices already in the environment are fingerprinted if there is reason to believe that they have the potential to be an EMI victim or a culprit.

The procedure used to initiate and track an EMI investigation is shown in Figure 5.6-5. Members of the clinical staff are instructed to call the biomedical engineering department and television services group in any case of a device malfunction not attributable to a routine failure. In cases of a new device entering the clinical environment for the first time or a new application for an existing device, a request to test the device for compatibility is generated. This call is referred to the clinical engineer responsible for EMI investigations. When an incident might be attributable to EMI, the engineer visits the site as a part of the initial investigation. The possible victim device is viewed and tested to determine whether it might have been, or is being, affected by EMI. Interviews with the personnel responsible for the area and the operation of the device(s) must be conducted. Based on the results of this preliminary investigation, a decision is reached as to the desirability or feasibility of further investigation.

The decision to continue the investigation is based on several factors: Is there a high probability of operator error? Is this a very rare occurrence? Is this either a very old device that might be reaching the end of its reliable life cycle or a new device experiencing infant mortality. As part of the preliminary investigation process, maintenance histories of the device are reviewed. Another component involves the review of equipment added to the environment that might have increased the overall RF hostility in the area enough to cause interference. Footprinting records of the area are reviewed as are fingerprints of the victim devices(s).

The mode of device failure is an important part of the evaluation. Is the victim device alarming? Is it operating erratically? Is it changing its operational parameters either temporarily or permanently? Is it shutting down? Is there a latching change that created the alarm? The answers to these questions can all point to the criminal device or devices. If the failure mode can be duplicated and if the failure appears to have been caused by an intentional or incidental radiator, the culprit device can usually be identified. If it is within the clinical environment, it is silenced or removed. Many times, no further investigation is required. An incident report is generated and filed. A copy is provided to the department initiating the service request. A copy, if deemed necessary, is given to the appropriate reporting agency, such as the Food and Drug Administration Center for Devices and Radiological Health (FDA, CDRH).

If a complete testing procedure is required and a recent footprint of the area exists, a new set is acquired and compared to the older set. Changes in the area environment are noted and analyzed. This is also compared with any existing fingerprints of the victim device. If no fingerprints exist for a representative victim device, or when a device shows signs that its ability to resist ingress may be compromised, it will be fingerprinted. After a cursory footprint of the quiet area, to ensure that no significant changes have occurred in that environment, the victim device, if practical (size and weight can affect the test location), can be moved to this area for fingerprinting.

Generally, the fingerprinting procedure depends on the operating characteristics of the device. For example, intentional radiators are tested for emissions at their operating frequencies, modulations, and at second and third harmonic frequencies. Unintentional radiators (i.e., microprocessor controlled devices) are tested from 1/4 clock frequency to 300 MHz. There are exceptions to this guideline, such as when relatively strong emissions continue to the 300 MHz point. In these cases, readings are continued to 1 GHz. The new digital cellular telephones (GSM) are now in service. Third Generation (3G) wireless communications devices should be in service in the near future. The Biomedical Engineering Department has already received reports of interference to clinical devices by digital telephones.

As both the fingerprint and footprint tests proceed, the records presented by the spectrum analyzer are reviewed. Any RF emissions exceeding a predetermined

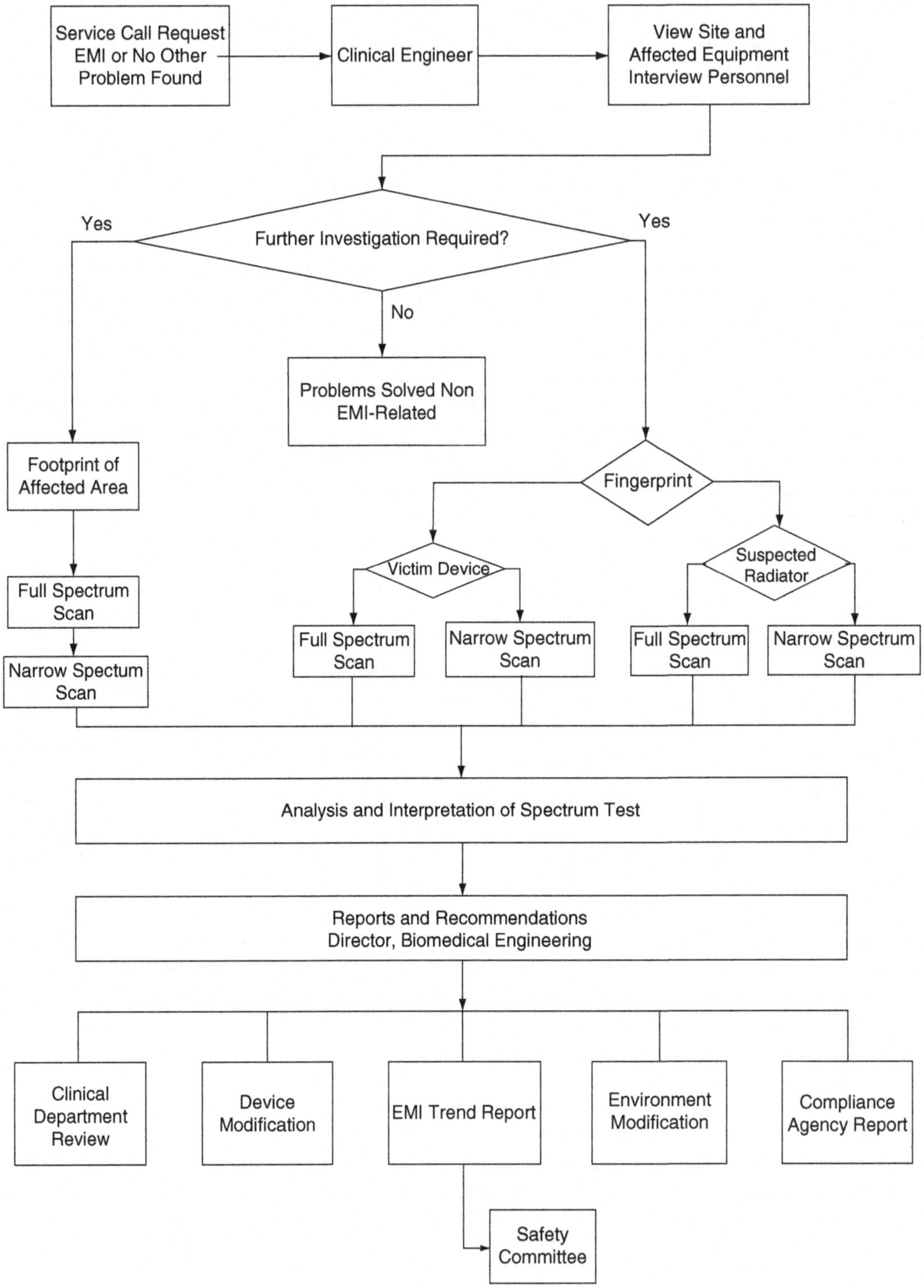

Figure 5.6-5 EMI problem resolution flowchart.

value are noted, especially frequencies not easily correlated to known sources of intentional radiation. The window of observation is then narrowed to obtain a greater resolution, or magnification, of the emission of interest. This typically identifies the type of modulation of the intentional radiator. After the series of footprinting and fingerprinting tests have been completed, both the clinical engineer and the technician performing the tests analyze the results and generate reports and recommendations. These are discussed with the Director of

Biomedical Engineering and possible resolutions of the problem are analyzed and reviewed.

There is not always an ideal solution to an interference problem. Many factors may contribute to device victimization, some of which are not easily or practically resolved. For example, if a recently added high-power paging transmitter were installed on an adjacent building, the additional energy radiated could affect devices that exhibited no previous EMI-related reactions. Historically, if the licensed transmitter is installed on a building over which the institution has no control, it will be difficult to remove. Such transmitters are licensed by the Federal Communications Commission and the medical device has no statutory protection; a one-way street.

Modifications to the environment are recommended if they are practical. These might include screening an area or relocating a device within an area. If modification of a medical device appears to be the only solution to the problem, then the manufacturer of the device is advised of the problem. The biomedical engineering department can and does advise the manufacturer. *Ad hoc* modifications of medical devices to mitigate susceptibility or egress are improper and should not be attempted. Performing such modifications would, in most cases, be a violation of FDA regulations and would increase the institution's risk for liability.

Summary

Despite gains in spectrum protection for some categories of patient monitoring devices through the development of standards, the overall electromagnetic environment is becoming increasingly hostile to the safe operation of clinical devices. The establishment and maintenance of a safe environment for the operation of clinical devices requires a multidisciplined approach. There must be a program of education involving technical, clinical, maintenance, and management personnel and staff. A clinical engineer experienced in RF and EMI, as well as in appropriate test and measurement equipment, must be available

A proactive testing and evaluation program that includes ongoing measurements of plant and incoming devices must be established. Maintained devices—especially those that have demonstrated a potential for electromagnetic waves—be spot-tested periodically using the fingerprint method.

This chapter has presented the need for a proactive EMI management program designed to limit the destructive effects of EMI on clinical devices. Some of the previously mentioned causes of EMI and methods used to test both the environment and the devices within it are based on experiences at Texas Children's Hospital.

References

ANSI. American National Standard for Methods of Measurement of Radio-Noise Emissions from Low-Voltage Electrical and Electronic Equipment in the Range of 9 KHz to 40 GHz. C63.4. New York, American National Standards Institute, 1991.

Bennett WS. Making OATS Measurements Reproducible from Site to Site. *EMC Test & Design* 4:34, 1993.

CBS. Haywire. *Eye-to-Eye with Connie Chung*. CBS, December 1, 1994.

US Government. CFR 47; Part 15.103 (c,e). Code of Federal Regulations of Telecommunications. Code of Federal Regulations, 1991.

David Y. Safety and Risk Control Issues: Biomedical Systems. In Dorf RC (ed). *The Electrical Engineering Handbook*. Boca Raton, FL, CRC Press, 1993.

ECRI. Patient-Owned Equipment. *Health Devices* 17:98, 1988.

ECRI. Ventilators, High Frequency. *Health Device Alert* p. 2, December 18, 1992.

ECRI. Guidance Article: Cellular Telephones and Radio Transmitters-Interface with Clinical Equipment. *Health Devices* 22(8,9):416, 1993.

Knudson T, Bulkeley WM. Stray Signal, Clutter on Airwaves Can Block Workings of Medical Electronics. *The Wall Street Journal*, p. 1, June 15, 1994.

Paperman WD, David Y, McKee KA. Electromagnetic Compatibility: Causes and Concerns in the Hospital Environment. *ASHE Health care Facilities Management Series*. Chicago, IL, ASHE, 1994

Southwick R. EMI Signal Measurements at Open Antenna Test Sites. *EMC Test & Design* 3:44, 1992.

Section Six

Medical imaging technology

Chapter 6.1

6.1

Fundamentals of magnetic resonance imaging

Reinaldo Perez

6.1.1 Early history of nuclear magnetic resonance

The root of nuclear magnetic resonance (NMR), or as it sometimes is called magnetic resonance imaging (MRI), goes back to World War II. In 1938 I. Rabi perfected a beam-splitting technique and successfully achieved NMR—a term that Rabi coined. The NMR experiments made use of the spin-state-dependent force that inhomogeneous magnetic fields exert on an atomic beam of silver atoms directed perpendicular to the gradient fields. For spin-$\frac{1}{2}$ nuclei, the atomic beam splits, but it reconverges when the polarity of the gradient field reverses. Rabi showed that irradiating the spins at the transition frequency, which interchanges the $m = \pm\frac{1}{2}$ states, eliminated the convergence.

The first detection of NMR in bulk matter was achieved in the mid-1940s by research groups led by Edward M. Purcell at Harvard University and Felix Bloch of Stanford University. Purcell used a resonant cavity to study the absorption of radio-frequency (RF) energy in paraffin: at resonance, the cavity output was found to be slightly reduced. By contrast, Bloch and his colleagues used what they called "nuclear induction." Bloch described the experiment as measuring an electromotive force resulting from the forced precession of the nuclear magnetization in the applied RF field.

When matter is placed in a magnetic field, the nuclear magnetic moments orient parallel to the field, leading to a paramagnetic polarization in the direction of the magnetic field—the z direction. If an oscillating magnetic field is applied in the x or y direction, the polarization vector is deflected from the z direction once the field approaches the resonance value. The resonance condition is given by

$$|\gamma|B = \omega$$

where B is the amplitude of the applied static magnetic field, ω is the nuclear precession frequency, and γ is the gyromagnetic ratio, which is a constant for a given isotope. In NMR, this rotation of the magnetic polarization vector of the nuclei in a plane perpendicular to the z axis induces an emf in a detector coil; this is the NMR signal.

Nuclear magnetic relaxation—that is, the return of the spin system to equilibrium is of great significance to imaging and was conceptualized by these early investigators. By repeatedly passing the spin system through the resonance condition (by varying the amplitude of the polarizing magnetic field) and observing the reappearance of a signal, Bloch found that for protons (that is, the hydrogen nuclei) in liquids the time constant T_1 for the return of the longitudinal magnetization was of the order of seconds. Further, he concluded from the sharpness of the resonance that the spins' phase memory time—the transverse relaxation time T_2—is of the order of hundreds of milliseconds in fluid (and, as shown later, only slightly shorter in biological tissues). It is clearly thanks to Mother Nature's good graces, or God, that NMR in human subjects is possible at all. If it took, instead of seconds, hours for the spin to repolarize, the technique would be impractical.

The commonly used detection method during the first two decades of NMR work exploited the principles of continuous wave excitation, where the field is swept while the sample is irradiated with RF energy of constant frequency. An alternative scheme, which is still in

Biomedical Engineering Desk Reference; ISBN: 9780123746467

use, consists of pulsed RF excitation followed by the detection of the resultant preprecession signal. Hence, rather than being simultaneous, in this scheme excitation and detection are performed sequentially.

A major milestone was the discovery of the chemical shift by Warren Proctor, F. C. Yu, and W. C. Dickinson. They found that in ammonium nitrate, two nitrogen-14 resonances could be observed, which they ascribe to the different chemical environments to which the nitrogen nucleus is exposed in the nitrate and ammonium ions. Similar findings were later made by others for nuclei such as fluorine, phosphorus, and hydrogen. These observations constitute the basis of modern NMR spectroscopy. A few years later, as the magnetic homogeneity that determines the frequency resolution achievable in NMR was improved further, another type of fine structure was discovered. This structure, which is due to spin–spin coupling, is fundamental to modern high-resolution spectroscopy, and together with the chemical shift provides the basic ingredients for molecular structure determination. Today, NMR is the preeminent method for determining the structures of biomolecules with molecular masses up to 100,000 Da.

The development of pulse Fourier transform by Richard Ernst and Weston Anderson was of great importance for NMR. This alternative mode of signal creation, detection, and processing led to an unprecedented enhancement in per-unit-time detection sensitivity compared with continuous wave excitation techniques. If N channels are used simultaneously in an experiment, then, provided the dominant source of noise is not the excitation, the sensitivity increases by a factor of $N^{1/2}$. Ernst and Anderson demonstrated that one can affect broadband excitation by exciting the nuclear spins with short RF pulses of a single carrier frequency.

A new breakthrough was added to NMR technology in 1973 when Paul Lauterbur at the State University of New York at Stony Brook first proposed generating spatial maps of spin distributions by what he called "NMR zeugmatography." The key to this method was the idea of superimposing magnetic field gradients onto the main magnetic field to make the resonance frequency a function of the spatial origin of the signal. In the presence of a magnetic field gradient the frequency domain signal is the equivalent of a projection of the object onto the gradient axis. By rotating the gradient in small angular increments, one obtains a series of projections from which an image can be reconstructed using back projection techniques.

In and around 1980 whole body experimental NMR scanners were in operation, and by 1981 clinicians began to explore the clinical potential of MRI. NMR has several advantages over x-ray computerized tomography (CT). First, it was noninvasive—that is, it did not require ionizing radiation or the injection of contrast material. Second, it provided intrinsic contrast far superior to that of x-ray CT. Some of this early work showed that MRI was uniquely sensitive to diseases of the white matter of the brain, such as multiple sclerosis. Further, the contrast could be controlled to a significant extent by the nature and timing relationships of the RF pulses. Third, MRI was truly multiplanar and even three dimensional (3D)—that is, it could provide images in other than the traditional transverse plane without the subject having to be repositioned. This property turned out to be of great value over x-ray CT for the study of the brain and other organs.

In 1975 Ernst introduced a new class of NMR experiments now known as two-dimensional (2D). NMR, and should be regarded as the parent of modern NMR techniques. One can understand the principle by reference to Figure 6.1-1. Suppose the spins in a point object of spin density $\rho(x, y)$ are excited by an RF pulse in the presence of a magnetic field gradient of magnitude G_y, which we call a phase-encoding gradient. These spins will resonate at a relative frequency $\omega = \gamma B(y)$, where $B(y) = G_y Y$. If the gradient is active for a period t_y, the phase at the end of the gradient period is $\phi y = \gamma G_y y t_y$. Let us then step the time t_y in equal increments, as applied by Figure 6.1-1b. We readily notice that the phase at the end of the gradient period varies cyclically with time. At time $t = t_y$, the gradient G_y is turned off and an orthogonal

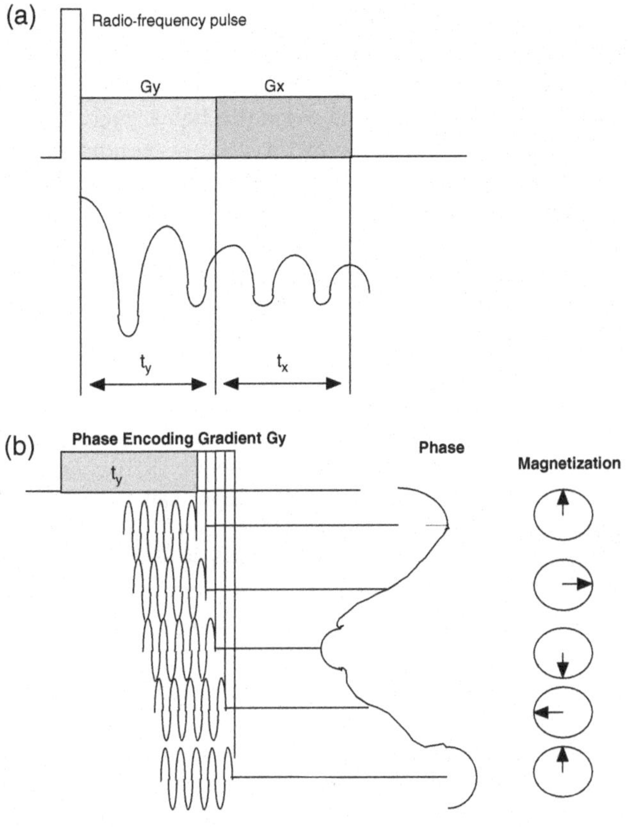

Figure 6.1-1 Fourier zeugmatography principle.

gradient G_x is applied for a duration t_x, during which the free-induction decay signal is collected. During the detection period the spins precess at a frequency $\omega x = \gamma G_x x$. One thus encodes spatial information into both the phase and the frequency of the NMR signal, whose magnitude is proportional to $\rho(x, y)$. Of course, in the case of imaging a real object, the signal has a multitude of frequency and phase components.

An obvious drawback of stepping the duration of the gradient G_y is the decrease in signal amplitude due to the irreversible decay (with relaxation time T_2) of the transverse magnetization. Because the phase shift imparted to the signal by the gradient G_y is a function of the gradient's duration as well as its amplitude, one can achieve the same effect by stepping the amplitude of the gradient while keeping its duration constant. This important modification of the technique gave rise to what is called spin-warp imaging. The term refers to the "warping" of the phase caused by the gradient. Raising the amplitude of the first gradient incrementally during each excitation and read cycle yields an $N_1 \times N_2$ array of raw data, from which one can reconstruct an image by double Fourier transformation of the signal. The resulting digital image consists of $N_1 \times N_2$ picture elements, or pixels, whose values are proportional to the amplitudes of the detected transverse magnetizations. The process of incrementally increasing a gradient in amplitude to encode spatial information into phases of processing spins is called phase encoding.

The basis of the signal and contrast in NMR is the transient nature of the signal. Spins, following excitation, return to their equilibrium state with characteristic time constants, the spin relaxation times T_1 and T_2, which for water in biological tissues are of the order of hundreds of milliseconds or even seconds. This process can be described by the phenomenological Block equations, which predict the evolution of the spin system in terms of the longitudinal and transverse components of the complex spin magnetization.

Consider a typical imaging experiment in which RF pulses are applied repeatedly at time intervals $\tau < T_1$ for the purpose of spatial encoding. Then the magnetization available for detection is attenuated by a factor $1 - e^{-\tau/T_1}$. This, in itself, would not be a sufficient mechanism for modulating the image signal were it not for the large range of relaxation times found for the water protons in mammalian tissues—from about 100 ms to several seconds.

Though the process of tissue water relaxation is not completely understood, its rate is related to the extent of binding of water to the surface of biological macromolecules. Increased binding slows molecular motion. The more closely the reorientation motion of the magnetic dipoles matches the Larmor frequency, the greater the transition probabilities between the nuclear energy levels, and thus the greater the relaxation rates. A case in point is the brain, the majority of which consists of gray and white matter adjacent to fluid-filled cavities, the ventricles. Because water is more tightly bound in white matter than in gray matter, the water molecules in white matter reorient more slowly than those in gray matter, thus more closely matching the Larmor frequency; hence T_1 wm (wm is for white matter) is less than T_1 gm (gm is for gray matter). By contrast, spinal fluid, which from the point of view of molecular motion closely parallels neat water, has much faster molecular motion, and thus T_1 sf is much greater than both T_1 wm and T_1 gm.

The equilibrium magnetization is proportional to the proton concentration in the tissues, hence the different plateau values. For short pulse-recycle times ($\tau \ll T_1$) the signal amplitudes follow the reciprocal of t_1, whereas at long recycle times ($\tau \gg T_1$) they are governed by their equilibrium magnetization—that is, they follow the proton concentrations. Contrast in MRI is therefore a continuum, and unlike in x-ray imaging, there is no universal gray scale. Further, it was recognized early on that in most diseased tissues, such as tumors, the relaxation times are prolonged. This difference provides the basis for image contrast between normal and pathological tissues.

6.1.2 General review of MRI

Protons (hydrogen nuclei) precess when placed in a magnetic field. This phenomenon is the basis for NMR. Nuclear precession occurs with a frequency directly proportional to the strength of the magnetic field, with a proportionality constant called the gyromagnetic ratio, of about 42.6 MHz/T. Typical frequencies range from 300 to 800 MHz (see Figure 6.1-2).

The precessional axis lies along the direction of the magnetic field. If an oscillating magnetic field at the precessional frequency is applied perpendicular to the static field, the protons will now precess about the axis of the oscillating field, as well as that of the static field. The condition is known as nutation. The oscillating field is generated by a tuned RF resonator, or RF coil, which usually surrounds the sample. The magnetic field of the precessing protons induces, in turn, an oscillating voltage in the RF coil, which is detected when the RF field is gated off. This voltage is then amplified and demodulated to baseband, as in a normal superheterodyne receiver, and digitized using an analog-to-digital converter (see Figure 6.1-3).

Little information of practical use would attach to the demodulated signal if all the protons exhibit identical precessional frequencies. In fact, very minor perturbations arise in the proton precessional frequency

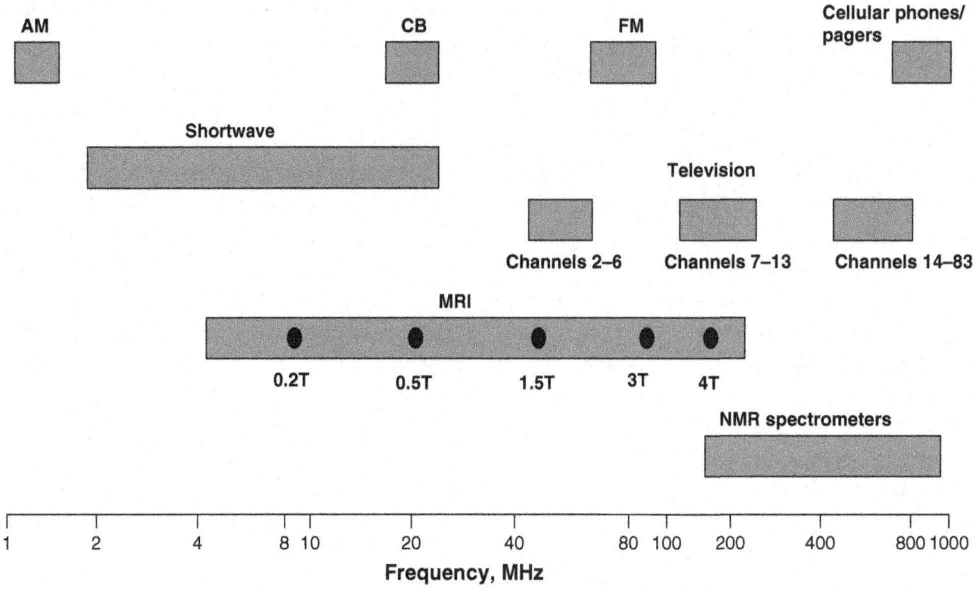

Figure 6.1-2 Magnetic resonance frequencies.

that reflect the molecular environment of the protons. In an ethanol molecule (CH_3CH_2OH), for example, the protons in the CH_3 group precess at a slightly different frequency from those in the CH_2 group or the OH group. Spectroscopy can determine structural information from the Fourier spectrum of the received signal.

To factor out the signal's dependence on the static magnetic field, NMR measurements are often given in a unitless quantity called chemical shift, which is typically measured in parts per million. It is the difference between the precession frequency of protons that are part of a particular molecular group and precession

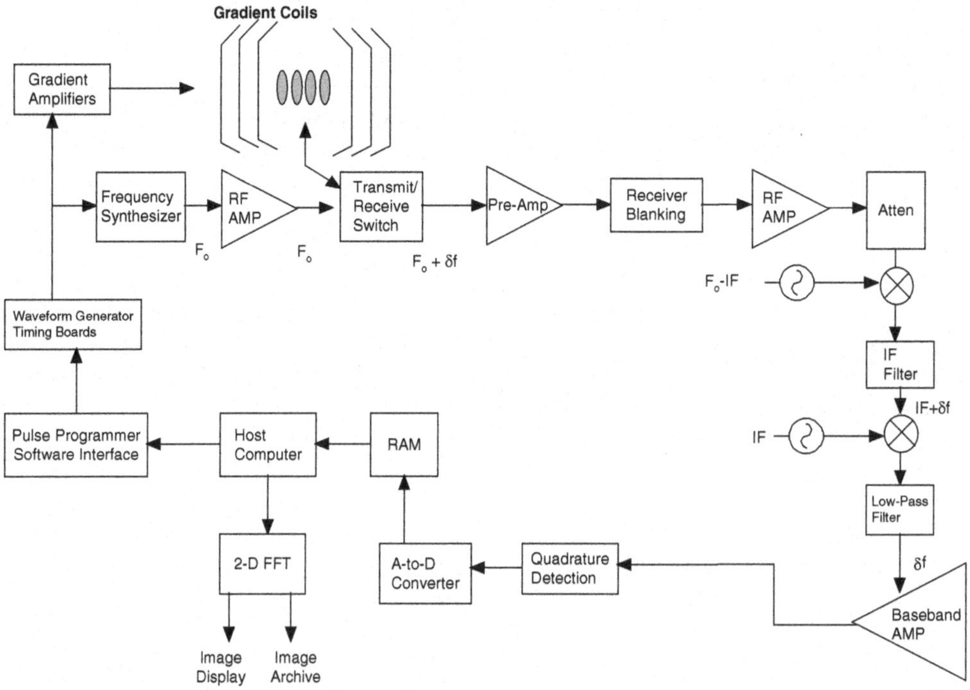

Figure 6.1-3 Block diagram of an NMR system.

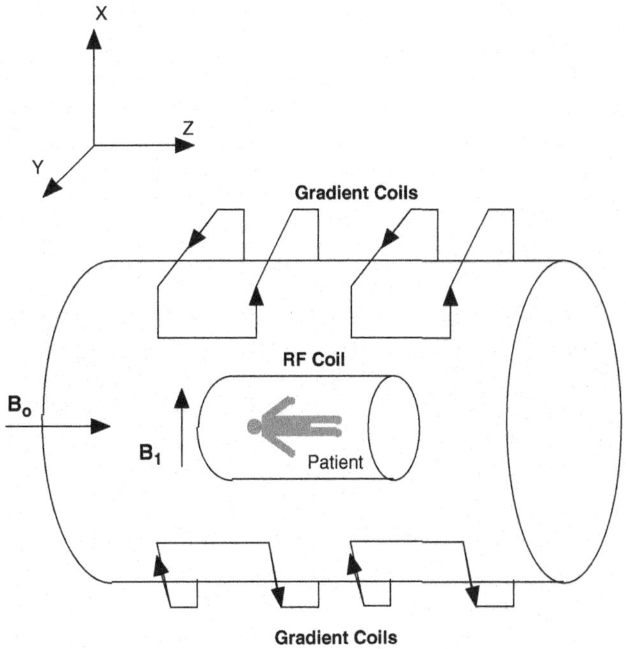

Figure 6.1-4 RF and gradient coils construction.

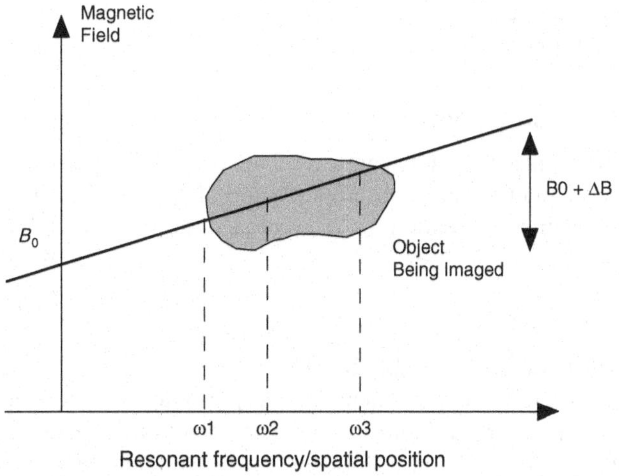

Figure 6.1-5 Spatial position caused by magnetic field.

frequency of protons in a reference compound that makes NMR possible to be measured. This application of magnetic resonance is generally referred to as high-resolution NMR spectroscopy, and is widely used in the pharmaceutical and chemical industries.

NMR imaging, better known as MRI, is one of the most important imaging technologies found in most modern hospitals. In MRI, it is the protons in the water molecules of a patient's tissue that are the source of the signal. The spatial information needed to form images from magnetic resonance is obtained by placing magnetic field gradient coils on the inside of the magnet. These coils, constructed from copper wire, create additional magnetic fields that vary in strength as a linear function of distance along the three spatial axes (Figure 6.1-4). This means that the resonant frequencies of the water protons within the patient's body are now spatially encoded.

The chief use of clinical MRI is for imaging the brain and spine. However, the recent development of rapid imaging methods has extended its application to the chest and abdomen where motion previously caused blurring of the image. The contrast in MRI images arises from differences in the number of protons in a given volume and in their relaxation times (the time taken for the magnetization of a sample to return to equilibrium after the RF pulse is turned off), which are related to the molecular environment of the protons (see Figure 6.1-5). Following excitation, each proton within the excited volume precesses at the same frequency. During detection of the echo, a gradient B_0 is applied causing a variation in the frequencies for the protons generating the echo signal. The frequency of precession ω_i for each proton depends upon its position. Frequencies measured from the echo are mapped to the corresponding position.

Brain scans are carried out to detect tumors, infarcts, aneurysms, or other pathological conditions. These are normally easy to detect since both the proton densities and relaxation times are markedly different from those of healthy tissue.

6.1.3 A more detailed overview of MRI

In clinical practice, MRI has been a great tool in the medical community. Over the years, MRI has assisted physicians in diagnosis, treatment, and presurgical treatments. The main advantage of MRI when compared with other imaging tools (e.g., x-ray computed tomography or CT scan) is that it does not require exposure of the human body to ionizing radiation; therefore it is very safe. The MRI signals are also very sensitive to different parts of tissues.

6.1.3.1 The physics of spin

Spin is a fundamental property of nature like electrical charge or mass. Spin comes in multiples of $\frac{1}{2}$ and can be + or −. Protons, electrons, and neutrons all possess spin. Individual unpaired electrons, protons, and neutrons each possesses a spin of $\frac{1}{2}$. In the deuterium atom (^2H), with one unpaired electron, one unpaired proton, and one unpaired neutron, the total electronic spin is equal to $\frac{1}{2}$ and the total nuclear spin is equal to 1. Two or more particles with spins having opposite signs can pair up to eliminate the observable manifestations of spin. An example is helium. In NMR, it is unpaired nuclear spins that are of importance.

Spin properties

When placed in a magnetic field of strength B_0, a particle with a net spin can absorb a photon of frequency υ. The frequency depends on the gyromagnetic ratio γ of the particle:

$$\upsilon = \gamma B_0 \tag{6.1.1}$$

For hydrogen, $\gamma = 42.58$ MHz/T.

Almost every element in the periodic table has an isotope with a nonzero nuclear spin. NMR can be performed only on isotopes whose natural abundance is high enough to be detected. However, some of the nuclei that are of interest in MRI are listed in Table 6.1-1.

To understand how particles with spin behave in a magnetic field, consider a proton. This proton has the property called spin. Think of the spin of this proton as a magnetic moment vector, causing the proton to behave like a tiny magnet with north and south poles. When the proton is placed in an external magnetic field, the spin vector of the particle aligns itself with the external field, just like a magnet would. There is a low-energy configuration of state where the poles are aligned N-S-N-S and a high-energy state N-N-S-S.

Transitions

A particle can undergo a transition between the energy states by the absorption of a photon. A particle in the lower energy state absorbs a photon and ends up in the upper energy state. The energy of this photon must exactly match the energy difference between the two states. The energy E of a photon is related to its frequency υ by Planck's constant ($h = 6.62 \times 10^{-34}$ J s).

$$E = h\upsilon \tag{6.1.2}$$

Table 6.1-1 Net spin of several nuclei of interest for MRI

Nuclei	Unpaired protons	Unpaired neutrons	Net spin	γ (MHz/T)
^{1}H	1	0	$\frac{1}{2}$	42.58
^{2}H	1	1	1	6.54
^{3}P	0	1	$\frac{1}{2}$	17.25
^{23}Na	2	1	$\frac{3}{2}$	11.27
^{14}N	1	1	1	3.08
^{13}C	0	1	$\frac{1}{2}$	10.71
^{19}F	0	1	$\frac{1}{2}$	40.08

In NMR and MRI, the quantity υ is called the resonance frequency and the Larmor frequency, respectively.

The energy of the two spin states can be represented by an energy level diagram. We have seen that $\upsilon = \gamma B_0$ and $E = h\upsilon$; therefore the energy of the photon needed to cause a transition between the two spin states is

$$E = h\gamma B_0 \tag{6.1.3}$$

When the energy of the photon matches the energy difference between the two spin states an absorption of energy occurs. In the NMR, the frequency of the photon is in the RF range. In NMR spectroscopy, υ is between 60 and 800 MHz for hydrogen nuclei. In clinical MRI, υ is typically between 15 and 80 MHz for hydrogen imaging.

When a group of spins is placed in a magnetic field, each spin aligns in one of the two possible orientations. At room temperature, the number of spins in a lower energy level, N^+, slightly outnumbers the number in the upper level, N^-. Boltzmann statistics tells us that

$$\frac{N^-}{N^+} = e^{-E/kT} \tag{6.1.4}$$

where E is the energy difference between the spin states; k is Boltzmann's constant, 1.3805×10^{-23} J/K; and T is the temperature in kelvin. As the temperature decreases, so does the ratio N^-/N^+. As the temperature increases, the ratio approaches one.

The signal in NMR spectroscopy results from the difference between the energy absorbed by the spins that make a transition from the lower energy state to the higher energy and the energy emitted by the spins that simultaneously make a transition from the higher energy state to the lower energy state. The signal is therefore proportional to the difference between the states. NMR is a rather sensitive spectroscopy since it is capable of detecting these very small perturbation differences. It is the resonance, or exchange of energy at a specific frequency between the spins and the spectrometer, that gives NMR its sensitivity.

It is worth noting two other factors that influence the MRI signal: the natural abundance of the isotope and biological abundance. The natural abundance of an isotope is the fraction of nuclei having a given number of protons and neutrons, or atomic weight. For example, there are three isotopes of hydrogen: ^{1}H, ^{2}H, and ^{3}H. The natural abundance of ^{1}H is 99.985%. Table 6.1-2 lists the natural abundance of some nuclei studied by MRI. The biological abundance is the fraction of one type of atom in the human body. Table 6.1-3 lists the biological abundance of some nuclei studied by MRI.

Table 6.1-2 Natural abundance of isotopes of interest to MRI

Element	Symbol	Natural abundance (%)
Hydrogen	^{1}H	99.985
Hydrogen	^{2}H	0.015
Carbon	^{13}C	1.11
Nitrogen	^{14}N	99.63
Nitrogen	^{15}N	0.37
Sodium	^{23}Na	100
Phosphorus	^{31}P	100
Potassium	^{39}K	93.1
Calcium	^{43}Ca	0.145

Spin packets

It is cumbersome to describe NMR on a microscopic scale. A microscope picture is more convenient. The first step in developing the microscopic picture is to define the spin packet. A spin packet is a group of spins experiencing the same magnetic field strength. In this example, the spins within each grid section represent a spin packet.

At any instant in time, the magnetic field due to the spins in each spin packet can be represented by a magnetization vector. The size of each vector is proportional to $(N^{+} - N^{-})$. The vector sum of the magnetization vectors from all of the spin packets is the net magnetization. To describe pulsed NMR, it is necessary to talk in terms of the net magnetization. Adapting the conventional NMR coordinate system, the external magnetic field and the net magnetization vector at equilibrium are both along the z axis.

T_1 processes

At equilibrium, the net magnetization vector lies along the direction of the applied magnetic field B_0 and is called the equilibrium magnetization M_z which equals M_0. We refer to M_z as the longitudinal magnetization. There is no transverse (M_x or M_y) magnetization here. It is possible to change the net magnetization by exposing the energy of a frequency equal to the energy difference between the spin states. If enough energy is put into the system, it is possible to saturate the spin system and make $M_z = 0$. The time constant that describes how M_z returns to its equilibrium value is called the spin lattice relaxation time (T_1). The equation governing this behavior as a function of the time t after its displacement is

$$M_z = M_0(1 - e^{-t/T_1}) \tag{6.1.5}$$

Therefore T_1 is defined as the time required to change the Z component of magnetization by a factor of e.

If the net magnetization is placed along the $-Z$ axis, it will gradually return to its equilibrium position along the $+Z$ axis at a rate governed by T_1. The equation governing this behavior as a function of the time t after its displacement is

$$M_z = M_0(1 - 2e^{-t/T_1}) \tag{6.1.6}$$

The spin-lattice relaxation time (T_1) is the time to reduce the difference between the longitudinal magnetization (M_z) and its equilibrium value by a factor of e.

If the net magnetization is placed in the XY plane it will rotate about the Z axis at a frequency equal to the frequency of the photon that would cause a transition between the two energy levels of the spin. This frequency is called the Larmor frequency.

T_2 processes

In addition to the rotation, the net magnetization starts to dephase because each of the spin packets making it up experiences a slightly different magnetic field and rotates at its own Larmor frequency. The longer the elapsed time, the greater the phase difference. Here the net magnetization vector is initially along $+Y$. For this and all dephasing examples you can think of this vector as the overlap of several thinner vectors from the individual spin packets.

The time constant that describes the return to equilibrium of the transverse magnetization M_{xy} is called the spin–spin relaxation time T_2.

$$M_{xy} = M_{xy_0} e^{-t/T_2} \tag{6.1.7}$$

T_2 is always less than or equal to T_1. The net magnetization in the XY plane goes to zero and then the longitudinal magnetization grows in until we have M_0 along the z axis.

Any transverse magnetization behaves the same way. The transverse component rotates about the direction of

Table 6.1-3 Biological elements of interest to MRI

Element	Biological abundance
Hydrogen (H)	0.63
Sodium (Na)	0.00041
Phosphorus (P)	0.0024
Carbon (C)	0.094
Oxygen (O)	0.26
Calcium (Ca)	0.0022
Nitrogen (N)	0.015

applied magnetization and dephases. Time T_1 governs the rate of recovery of the longitudinal magnetization.

In summary, the spin–spin relaxation time T_2 is the time to reduce the transverse magnetization by a factor of e. In the previous sequence, T_2 and T_1 processes are shown separately for clarity. That is, the magnetization vectors are shown filling the *XY* plane completely before growing back up along the *Z* axis. Actually, both processes occur simultaneously with the only restriction being that T_2 is less than or equal to T_1.

Two factors contribute to the decay of transverse magnetization:

1. molecular interactions (said to lead to a pure T_2 molecular effect)
2. variations in B_0 (said to lead to an inhomogeneous T_2 effect)

The combination of these two factors is what actually results in the decay of transverse magnetization. The combined time constant is called T_2^*. The relationship between the T_2 from molecular processes and that from inhomogeneities in the magnetic field is as follows.

$$\frac{1}{T_2^*} = \frac{1}{T_2} + \frac{1}{T_{2\ \text{inhomo}}} \tag{6.1.8}$$

MRI is based on the physical principle of NMR. NMR was discovered by Black and Purcell in 1946. When certain atomic nuclei are placed in a static magnetic field, it will assume either a higher energy level or a lower energy level (Figure 6.1-6). As shown in the figure, the energy between the two states is linearly dependent on the strength of the applied static field: a physical principle known as the Zeeman effect. A nucleus in a higher energy state can fall to the lower energy state. A nucleus will emit a photon when transitioning between these two energy states. The energy of the photon is equal to the energy difference between the two states. A nucleus in the lower energy state can jump to a higher energy state, absorbing a photon with energy matching the energy difference between the two states. This means that when nuclei under the effect of an applied magnetic field are irradiated by photons in the form of electromagnetic fields at a certain frequency by an RF probe, some nuclei in the lower energy states will absorb the photons and jump to a higher energy state. After the radiated energy is interrupted, the nuclei in the higher energy state will return to the lower energy state to recover to equilibrium, emitting photons on electromagnetic fields, which can be detected by an MRI probe. The frequency emitted by the electromagnetic fields is detected by the energy difference between the two states of the nuclei. The decay of the signals in time depends on the molecular environment of the nuclei. Because the energy difference between the two states of a given nucleus in an external field depends on the strength of the external field, the energy difference at any point in the body can be made different by varying the magnetic field strength from point to point. Therefore, the energy of the photons and the frequency of the electromagnetic fields absorbed and emitted by the nuclei are also different from point to point. When the signals from all nuclei are received, their frequency can be used to determine spatial information about the nuclei.

From quantum mechanics certain atomic nuclei possess what is called spin. A proton, which has a mass, can be thought of as spinning, as shown in Figure 6.1-7 with its own angular momentum. The circulating positive current creates a small magnetic field.

Neutrons can also be thought of as generating a small magnetic field since their distributed positive and negative charges are not uniformly distributed. These small magnetic fields are called magnetic moments M and are given by

$$M = \gamma J \tag{6.1.9}$$

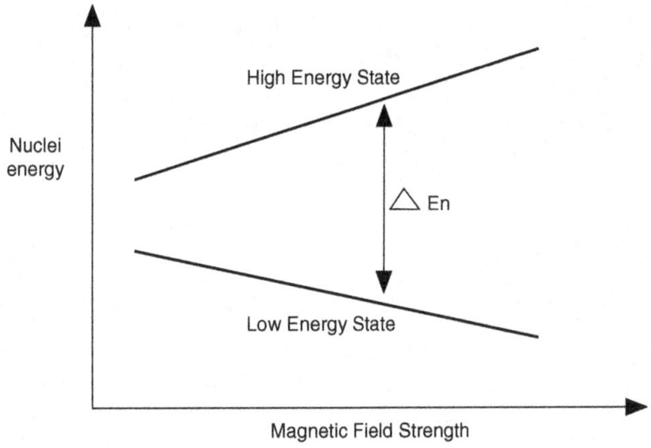

Figure 6.1-6 Energy level of nuclei under a magnetic field.

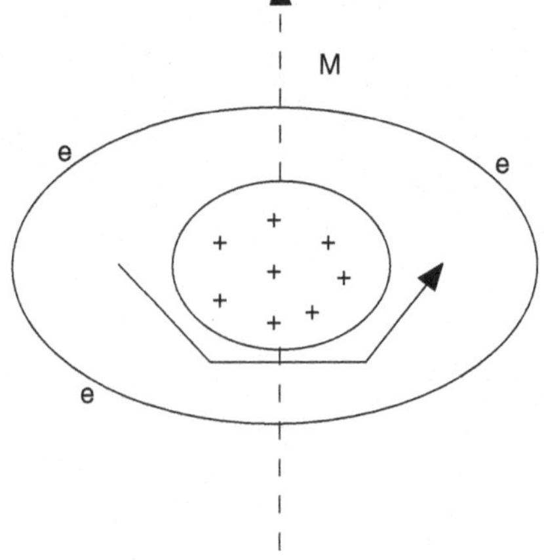

Figure 6.1-7 Magnetic field precession in magnetic fields.

where J is the angular momentum and γ is known as the geomagnetic ratio. For nuclei with an odd number of protons or an odd number of neutrons, there is never a net angular momentum equal to zero because of opposite spin states (Pauli exclusion principle that the angular momentum of each proton in a nucleus must assume opposite spin states). Many biological nuclei usually have odd numbers of protons and spins; example of such are ^1H, ^{13}C, ^{19}F, ^{23}N, and ^{31}P. Usually ^1H is the most significant nucleus in most MRI imaging because it is part of the water molecule and it has the highest NMR sensitivity with a geomagnetic ratio $\gamma/2\pi = 42.58$. Let's consider a magnetic flux B_0 in the z direction.

The proton can assume two positions: with its z component of the magnetic moment aligned with the external field or with its z component of the magnetic moment opposed to the external field. Both states are stable, but the energy associated with the latter is higher. This can be observed in Figure 6.1-8 and the angle θ_0 can be determined by the value of the magnetic moment M and its z component M_z, which are given by quantum mechanics as

$$M = \frac{\gamma h \sqrt{3}}{4\pi} \qquad M_z = \frac{\gamma h}{4\pi} \qquad (6.1.10)$$

where h is the Planck constant ($h = 6.629 \times 10^{-34}$ Js). Therefore

$$\theta_0 = \cos^{-1}\left(\frac{M_z}{M}\right) = \cos^{-1}\left(\frac{1}{\sqrt{3}}\right) \approx 54.7° \qquad (6.1.11)$$

The difference in energy between the two states

$$\Delta E = 2M_z B_0 = h\upsilon \qquad (6.1.12)$$

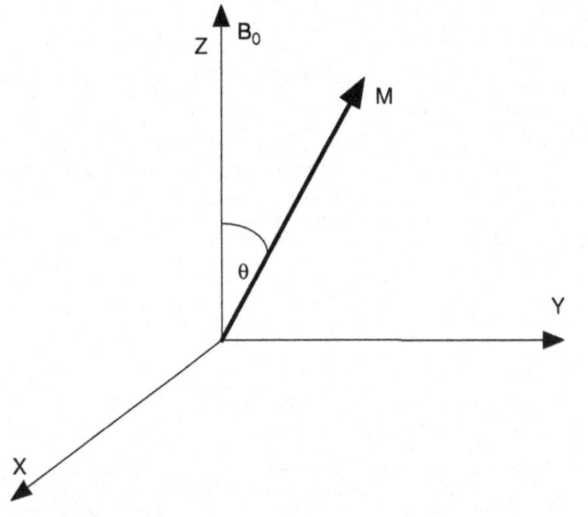

Figure 6.1-8 The magnetization vector.

Therefore the frequency υ is given by

$$\upsilon = \left[\frac{2M_z}{h}\right]B_0 \qquad (6.1.13)$$

It can be observed that the frequency is directly proportional to the magnetic field.

The angular momentum of a nucleus is linearly related to its magnetic moment

$$\frac{d\vec{M}}{dt} = \gamma\left(\vec{M} \times \vec{B_0}\right) \qquad (6.1.14)$$

Representing this equation in terms of scalar components and combining such equations we obtain

$$M(t) = a_x(M_x\cos(\gamma B_0)t + M_y\sin(\gamma B_0))$$
$$+ a_y(M_y\cos(\gamma B_0)t - M_x\sin(\gamma B_0)) + a_z M_z$$

This last expression of the magnetic moment has a frequency component given by

$$f = \frac{\upsilon B_0}{2\pi} \qquad (6.1.15)$$

which is known as the Larmor resonant frequency. It can be easily shown that the frequency of the radiation is the same as the precessional frequency of the magnetic moment.

When a magnetic field is applied to a bulk mass, each individual magnetic moment must align itself either with or against the external field. Alignment with the magnetic field involves a lower energy and is called the parallel state (α). Alignment against the magnetic field involves a higher energy and it is called the antiparallel state (β).

Boltzmann's law is given by

$$\frac{P_\alpha}{P_\beta} = \exp\left(\frac{\Delta E}{KT}\right) \qquad (6.1.16)$$

where P_α denotes the probability that a nucleus is found in the parallel state (α), P_β denotes the probability that a nucleus is found in the antiparallel state (β), K is Boltzmann's constant ($K = 1.3800 \times 10^{-23}$ J/K), T is the absolute temperature of the sample. For protons at 20°C, ΔE is on the order of 10^{-26} J and KT is on the order of 10^{-21} J. Therefore the previous equation can be expressed as

$$P_\alpha - P_\beta \approx \frac{\Delta E}{2KT} \qquad (6.1.17)$$

If we estimate, $P_\alpha \cong P_\beta \cong \frac{1}{2}$. The net magnetic moment per unit volume, known as magnetization, is given by the expression

$$M = (P_\alpha - P_\beta)nM_z\hat{a}_z \approx \frac{\Delta E}{2KT}nM_z\hat{a}_z \qquad (6.1.18)$$

where n denotes the number of protons per unit volume. Notice that as the temperature increases the magnetization is destroyed. Since ΔE is linearly dependent on B_0, the net magnetization is proportional to the magnetic field. In NMR the observation of the precession of this magnetic moment is of great importance.

The effects of RF radiation on the magnetization of the body sample in a uniformly applied B_0 field needs to be addressed. Under the influence of an RF magnetic field, directed on a given processes coordinate, the magnetization can be rotated away from its equilibrium position and the angle θ is given by

$$\theta = VB_{\mathrm{eff}}T \tag{6.1.19}$$

where B_{eff} is the value of the effective magnetic field in the rotating frame. The rotation of M will have an angular frequency

$$\omega = \gamma B_{\mathrm{eff}} \tag{6.1.20}$$

We can notice that a large gyromagnetic ratio provides a quicker perturbation of the magnetization vector. In practice it rotates M by 90° and 180°. The durations of these rotations are given by

$$T_{90} = \frac{\pi}{2\gamma B_{\mathrm{eff}}} \tag{6.1.21a}$$

$$T_{180} = \frac{M}{\gamma B_{\mathrm{eff}}} \tag{6.1.21b}$$

After the initial 90° rotation, the magnetization vector M processes in the transverse plane. The relaxation process is better explained by the Block equations described next.

$$\begin{aligned} \frac{dM_{x,y}}{dt} &= \gamma(\vec{M} \times \vec{B})_{x,y} - \frac{M_{x,y}}{T_2} \\ \frac{dM_z}{dt} &= \gamma(\vec{M} \times \vec{B})_z + \frac{M_0 - M_z}{T_1} \end{aligned} \tag{6.1.22}$$

where T_2 and T_1 are the transverse and longitudinal relaxation times, respectively, and M_0 denotes the equilibrium value of magnetization, which is assumed to lie in the z direction. The solution of Eqs. (6.1.22) is given by

$$\begin{aligned} M_x(t) &= M_0 \exp\left(-\frac{t}{T_2}\right)\cos(\gamma B_0 t) \\ M_y(t) &= -M_0 \exp\left(-\frac{t}{T_2}\right)\sin(\gamma B_0 t) \\ M_z(t) &= M_0\left[1 - \exp\left(-\frac{t}{T_1}\right)\right] \end{aligned} \tag{6.1.23}$$

The relaxation signal as it returns to equilibrium is shown in Figure 6.1-9.

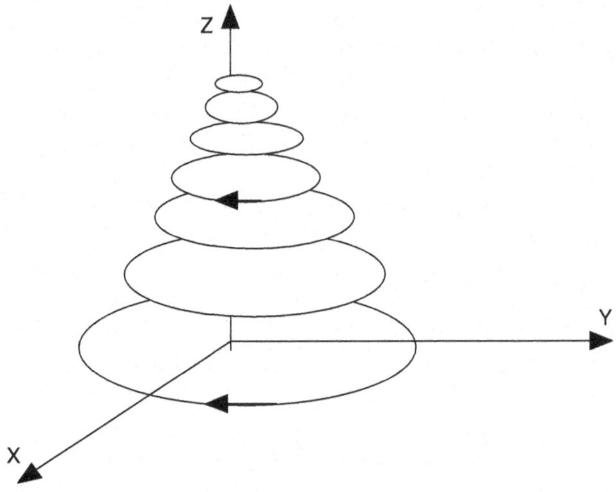

Figure 6.1-9 Relaxation signal and equilibrium.

When a magnetic field B_0 is applied in the z direction, the nuclei precess about the z axis at the Larmor frequency and with a precessional angle θ. Some of the nuclei precess around the $+z$ axis protons and others precess around the $-z$ direction and this results in a bulk magnetization, as shown in Figure 6.1-10.

Upon the application of an RF pulse phase coherence is accomplished in both the $+z$ and $-z$ precession. The RF pulse is at the resonant frequency and it stimulates the flipping between the two sections in the $+z$ and $-z$ precession. This allows energy to be imparted into the protons and nuclei migrate to the $-z$ precession. Once the population of nuclei in each $+z$ and $-z$ precession sections is the same, there will be no net axial magnetization. After a 90° pulse, as the magnetization precesses

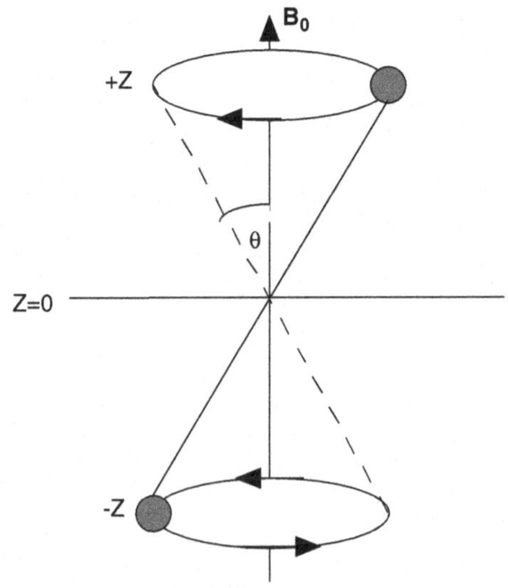

Figure 6.1-10 Bulk magnetization illustration.

in the transverse plane, two types of relaxation occur. The "longitudinal" relaxation causes the axial magnetization to be brought back to equilibrium and is given by T_1 (longitudinal relaxation time). "Transverse" relaxation causes the transverse magnetization to be reduced to zero and is given by T_2 (transverse relaxation time). In most materials $T_2 < T_1$, and T_1 is usually shorter for liquids than for solids. On the other hand, T_2 is generally greater for liquids than for solids.

6.1.4 MRI hardware design

The basic components of MRI hardware are shown in Figure 6.1-11. The magnet in MRI is responsible for generating the static magnetic field B_0. The gradient coils create a gradient which is superimposed upon the main field B_0. A strong magnetic field provides a better SNR and better resolution in both frequency spatial domains. Most clinical imaging systems have field strengths around 2 T. Some functional MRI systems have 3 T or 4 T main magnets.

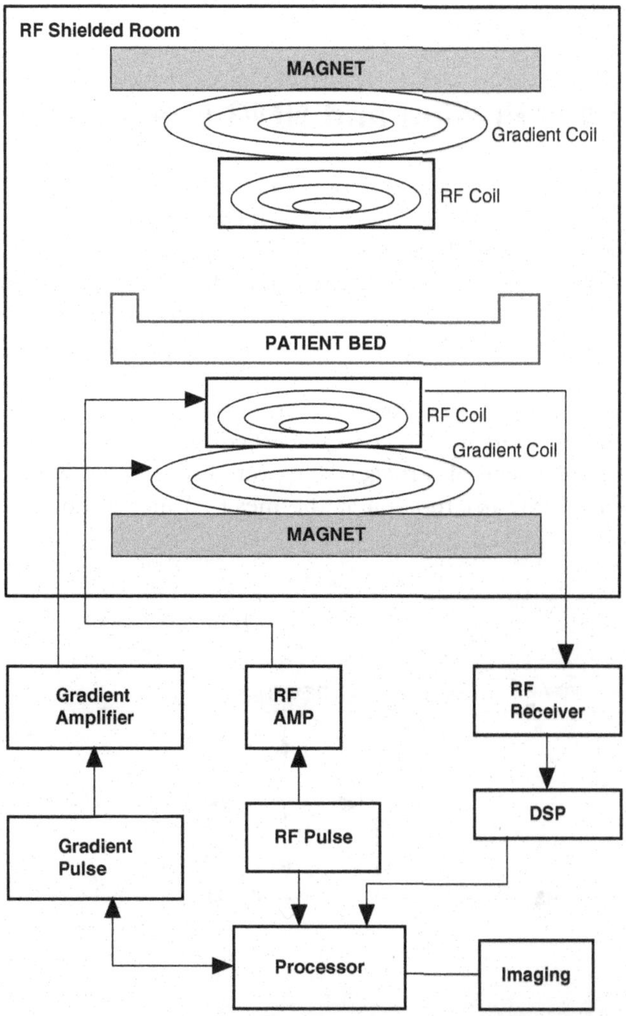

Figure 6.1-11 Hardware representation of MRI.

The primary requirements of the main magnet is that its field be uniform. In most cases, the magnetic field is not uniform and for this reason "shim" coils are frequently employed. The shim coils are a set of coils built to produce a field that is polarized in the same direction as the main field of known spatial dependence. Therefore, if the main magnet's nonuniformity is known, the shim coils can be set to carry gradients that cancel (using superposition) the inhomogeneous components of the main field.

Most magnetic fields are generated using permanent magnets. These magnets are simple and affordable. Their fringing fields are also small. Permanent magnets can also come in different sizes and shapes. In MRI applications with permanent magnets, the patient is placed between the two poles of the magnets. However, temperature drift of permanent magnets is an issue since the RF is usually not controlled by any type of feedback and would therefore not remain at the Larmor frequency if the temperature changes after calibration.

For magnetic fields greater than 0.5 T, superconducting magnets must be used. The coil windings are made from an alloy of niobium–titanium and are cooled to a temperature below 12 K using liquid helium (boiling point of 4.2 K). These fields are very strong and homogeneous.

The Larmor frequency depends on the strength of the magnetic field B_0. Strong gradients for imaging technique may require the help of gradient pulsing. The duration of these pulses must be of the order of T_2 to be effective. Because gradient coils have a natural self-inductance, they cannot be switched on or off simultaneously. Coils that have large inductances can be driven to steady state quickly at the expense of power by using other driver circuits to compensate for the power needed. Since the gradient coils can induce eddy currents in the rest of the magnet structure, to compensate, shielding must be used. Furthermore, the desired gradients, which are typically linear, must provide detailed spatial information about the sample. Gradients for most modern imaging schemes can be produced in any of the three spatial directions without physically rotating the gradient coils.

The sample under test must be irradiated with an RF field also known as the B_1 field. This is done to tip the magnetization away from the equilibrium position and generate a detectable NMR signal. The RF fields are produced by a transmitter and an RF coil. The transmitter determines pulse shape, duration, power, and timing. The RF coil is responsible for coupling the energy given by the transmitter to the nuclei. To generate an RF pulse, the transmitter uses first a frequency synthesizer to achieve the defined frequency. A waveform generator creates a user-defined pulse shape, which is then mixed with the

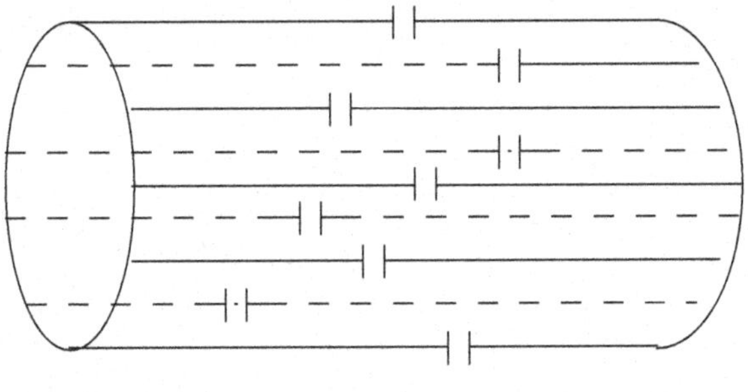

Figure 6.1-12 Lumped capacitor model for the coil in the MRI resonator.

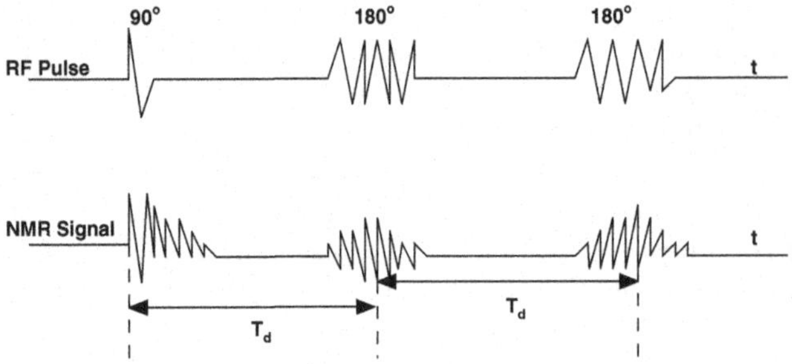

Figure 6.1-14 Spin echo sequencing.

pure tone and this creates an RF pulse. The pulse or a sequence of several pulses is repeated at a repetition rate. There are basically two types of pulses used in NMR machines: the hand pulses, which are typically broad, and rectangular pulses. The other type of pulse is known as soft pulses, which are often sinc shaped to provide frequency selectivity. In soft pulsing, it is important that the center frequency is close to the Larmor frequency.

To receive a good signal, RF coils are required to couple the energy from nuclei in precedence to a receiver. The RF coil must be responsive to frequencies within the general band of the Larmor frequency of interest. In human imaging, the wavelength is generally of the same order of magnitude as the sample. Often, a coil design known as the "birdcage" is used. This coil combines lumped capacitors with distributed inductance to form a volume resonator. If the resonance of the coil is designed well, it should increase the efficiency of the coil at the operating frequency. The lumped capacitors provide a method for the resonator to store energy, as shown in Figure 6.1-12.

The coils must generate a very uniform B_1 magnetic field. Any inhomogeneity will introduce a distortion in the images. The requirement of a uniform field places a burden on the coil design. Therefore, it is desirable to use a coil that allows quadrature excitation and detection (two RF coils are needed, one for transmitting and one for receiving). Quadrature excitation is capable of generating and receiving circularly polarized fields. To accomplish this, the transmittal power and received power must be split equally into two channels, with one of the channels having a 90° phase delay. The two channels are then fed to the two inputs of the RF coil to produce fields that are perpendicular to each other as shown in Figure 6.1-13.

6.1.5 Pulsing and NMR imaging

One "pulsing" scheme commonly used today is called "spin echo" sequences. This sequence provides a 90° pulse followed by a time delay $T_d/2$ (T_d is time delay), a 180° pulse, and another time delay, as shown in Figure 6.1-14. The term T_r denotes the length of a frame.

After the 90° pulse, the magnetization lies in the transverse plane if the pulse of 90° is applied along the x axis. It results in a magnetization aligned along the y axis and coherence has been established. The nuclei under the influence of stronger fields rotate faster and hence have some net rotation at the mean Larmor frequency.

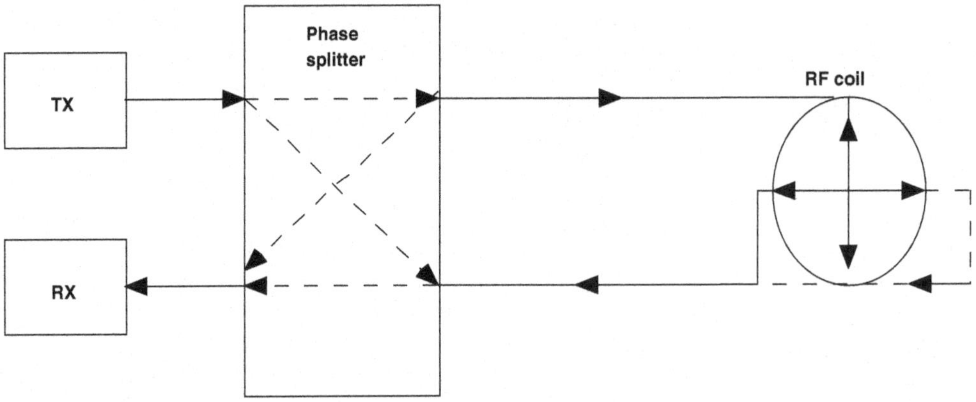

Figure 6.1-13 Pulsing RF coils with perpendicular fields.

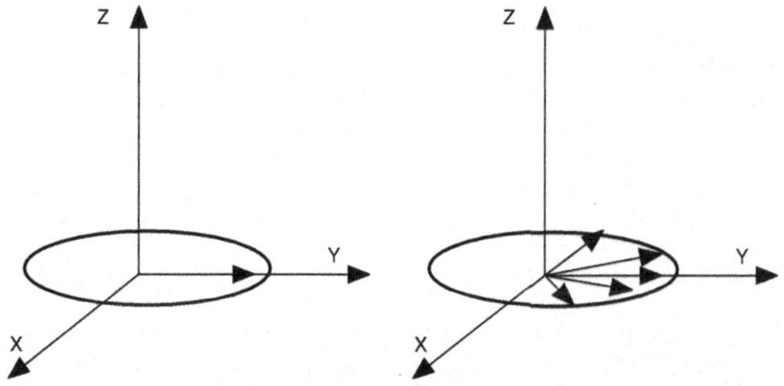

Figure 6.1-15 Nuclei under the influence of a magnetic field.

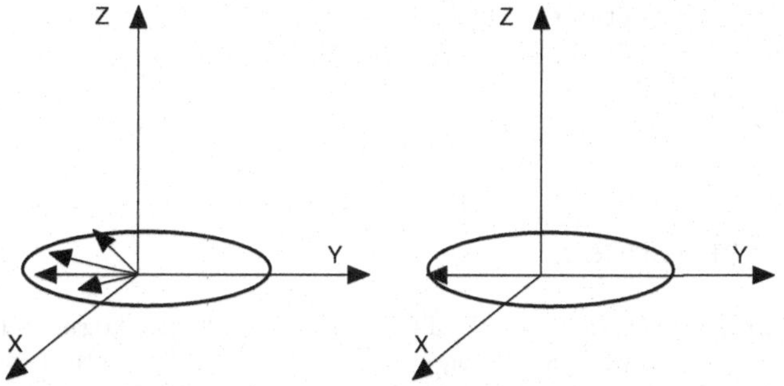

Figure 6.1-16 Nuclei under the influence of a magnetic field.

The nuclei under the influence of a weaker magnetic field will rotate slowly, one lag behind, and experience a negative sensed net rotation in the frame, as shown in Figure 6.1-15.

If we now apply a 180° pulse, also on the x axis, each individual magnetic moment will be rotated 180° about the x axis (Figure 6.1-16). Now the faster nuclei are behind the mean nuclei and the slower nuclei are ahead. This means that the faster nuclei eventually will cross the path of the slower nuclei and coherence is again established (Figure 6.1-16).

The peak value of the "echo" signal is given by

$$E_0 = (\text{free induction decay}) \exp\left(-\frac{T_d}{T_2}\right) \quad (6.1.24)$$

from which the value of T_2 can be obtained.

The inversion recovery imaging that uses the spin echo is shown in Figure 6.1-17, where G_s is the slice selection gradient, G_f is the required readout gradient, and G_ϕ is the phase-encoding gradient. The sequence starts with an RF pulse at 180° in conjunction with a slice selection gradient. The remainder of the sequence after a time T_i is the same as the spin echo imaging sequence. The spin echo sequences employ 90° and 180° RF pulses.

The inversion recovery imaging technique can be performed in conjunction with the 2D Fourier transform technique. We first define the slice to be imaged out of a 3D object by applying a steep gradient perpendicular to this slice. Next it is necessary to phase encode each of the longitudinal directions of the 2D slice. This will be accomplished by applying a gradient pulse along each of

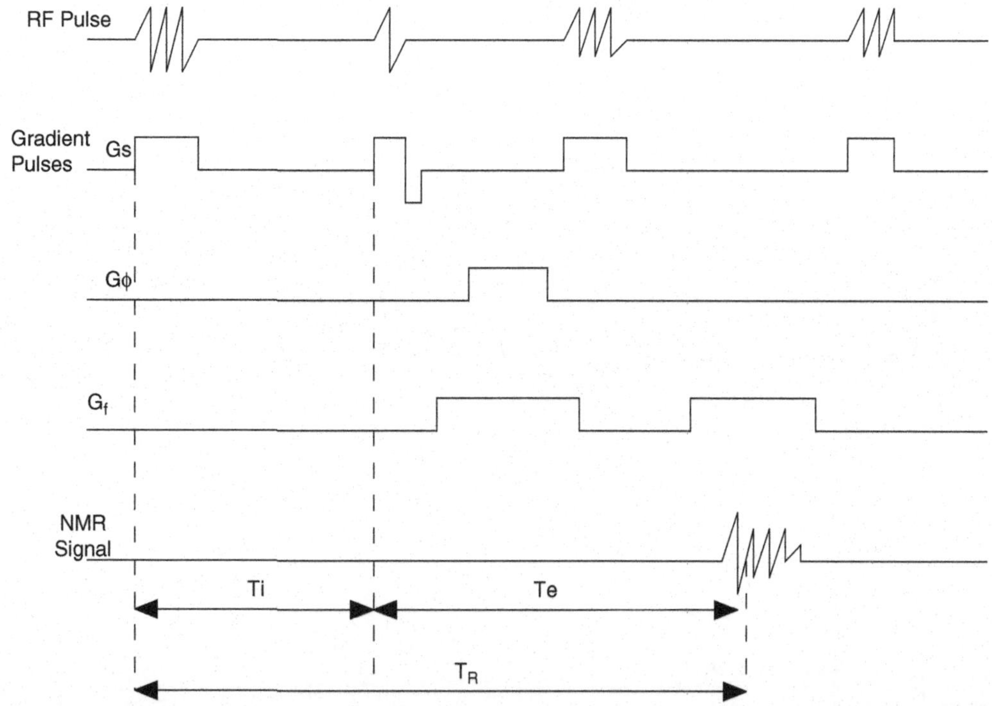

Figure 6.1-17 Recovery of imaging.

those directions. Usually, for the first acquisition, the pulse is halted once nuclei on the edges of the sample have accumulated a pulse difference of 180°. When the cycles are repeated larger phase shifts are deliberately induced in each cycle in a progressive manner. This is accomplished to increase the resolution in the longitudinal direction, which for this method depends on the number of the acquisitions.

References

Brown, Mark A., and Richard C. Semelka. 1999. "Magnetic resonance abbreviations, definitions, and description—a review." *Radiology.*

Chien, Daisy, and Robert R. Edelmen. 1991. "Ultrafast imaging using gradient echoes, magnetic resonance." *Quarterly* 7, 31–56.

De Yoe, Edgar A., Pater Bandettini, Jay Neitz, David Miller, and Paula Winans. 1994. *J. Neurosci. Meth.* 54, 171–187.

Jin, Jianming. 1999. *Electromagnetic Analysis and Design in Magnetic Resonance Imaging*, CRC.

Lauterbur, P.C. 1973. "Image formation by induced local interactions: examples employing nuclear magnetic resonances." *Nature* 242.

Mansfield, P., and P.G. Morris. 1982. *NMR Imaging in Biomedicine.* Academic Press, New York.

Slitcher, C.P. 1986. *Principles of Magnetic Resonance*, 3rd ed. Springer Verlag, New York.

Chapter 6.2

6.2

Optical sensors

Reinaldo Perez

There are basically two technologies that are widely used in optical sensors, charge-coupled devices (CCDs) and fiber optics. Both are major contributors to the development of advanced diagnostics and therapeutic techniques. In this chapter, we first address CCD and then fiber optics.

6.2.1 Charge-coupled devices

The CCD refers to a semiconductor architecture in which a charge is transferred through storage areas. The CCD architecture has three basis steps: (1) charge collection, (2) charge transfer, and (3) the conversion of charge into a measured voltage. The basic building block of the CCD is the metal semiconductor capacitor (MIS), also known as the gate. The most common MIS is the metal oxide semiconductor.

Charge generation is often considered as the initial goal of CCD. Silicon can create an electron–hole pair for each absorbed photon. The electrons and holes can be stored or transformed; charge generation occurs under an MOS capacitor. The charge created at a pixel site is proportional to the incident light. The aggregate effect of all the pixels is to produce a spatially sampled representation of the continuous picture. Pixel readout occurs by sensing the charge transfer between the capacitor at the pixel site.

When an absorbed photon creates an electron–hole pair, photodetection has occurred. The generated carriers must be stored at a site. The absorption coefficient is wavelength specific and decreases with increasing wavelength. Applying a positive voltage to the CCD causes the mobile positive holes in *p*-type silicon to migrate toward the ground electrode because like charges repel. This region, which is devoid of positive charges, is the depletion region. A photon with an energy greater than the energy gap is absorbed in the depletion region producing an electron–hole pair as shown in Figure 6.2-1.

The electron stays within the depletion region, whereas the hole moves to the ground electrode. The number of electrons collected is proportional to the applied voltage, oxide thickness, and gate electrode area.

The CCD register consists of a series of gates. The manipulation of the gate voltage in a systematic and sequential manner transfers the electrons from one gate to the next in a conveyor-belt-like fashion. For charge transfer, the depletion region must overlap. The depletion regions are gradients, and the gradients must overlap for charge transfer to occur.

Initially, a voltage is applied to gate 1 and photodetectors are collected in well 1 (Figure 6.2-2). When a voltage is applied to gate 2, electrons move to well 2 in a waterfall manner (c).

The process is rapid and then the charge quickly equilibrates in two wells (d). As the voltage is reduced on gate 1, the well potential decreases and electrons again flow in a waterfall manner into well 2 (e). The process is repeated many times until the charge is transferred through the shift register.

The CCD array is a series of column registers (Figure 6.2-3). The charge is kept within rows or columns by channel stops and the depletion regions overlap in one direction only. At the end of each column is a horizontal register of pixels. The register accumulates a line at a time, and later it transports the charge packets in a serial mode to an output amplifier. The horizontal serial register must be clocked out to be a sense node before the next line enters the serial register. Therefore, separate vertical and horizontal clocks are required for all

Biomedical Engineering Desk Reference; ISBN: 9780123746467

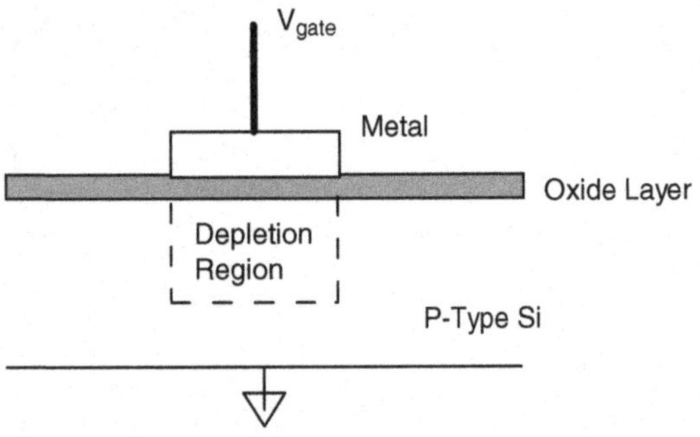

Figure 6.2-1 The depletion region in semiconductor devices.

CCD arrays. The process creates a serial data stream that represents the two-dimensional image.

Although any number of transfer gates per detector can be used, the number generally varies from two to four. With a three-phase system the charge is stored under one or two gates. This is shown in Figure 6.2-4. Only 33% of the pixel area is available for well capacity. With equal potential wells, a minimum number of three phases are required to clock out charge packets efficiently.

As the voltage is applied, the charge packet moves to that well. By sequentially varying the gate voltage, the charge moves off the horizontal shift register and onto a sense capacitor. The clock signals are identical (only one master clock is required to drive the array) for all three phases but offset in time (phase).

6.2.1.1 CCD arrays

Array architecture is dependent on the application. Full-frame arrays are used in scientific applications. Interline transfer devices are used in consumer product. In Figure 6.2-5 we can illustrate a full-frame (FFT) array. After integration the image pixels are read out line-by-line through a serial register which clocks its content onto the output sense node. All the charge must be clocked out of the serial register before the next time line can be transferred.

During the reading process, the pixels are continually illuminated, which can result in a smeared image in the direction charge flow. Data rates are limited by the amplifier bandwidth and also by the capabilities of the analog-to-digital converter. A large array can be divided into subarrays that are read out simultaneously, the effective clock rate increases by the number of subarrays. Figure 6.2-6 illustrates a large array subdivided into four subarrays. Software can reconstruct the original image, where the serial data are devoted and reformatted by a video processor.

A frame transfer image contains two almost identical arrays. One array is used for image pixel and the other one for storage. The storage cells are identical to the light-sensitive cells but are covered with a metal light shield to prevent any light exposure. Once the integration cycle is complete the charge is transferred quickly from the light-sensitive pixels to the storage cells. The transfer time is around 500 μs. The smear is limited only to the time it takes to transfer the image to the storage area.

6.2.1.2 Interline transfer

The interline transfer array consists of photodiodes separated by vertical registers that are covered by an opaque metal shield (see Fig. 6.2-7). After integration, the charge generated by the photodiodes is transferred to the vertical CCD registers very fast and the smear is then minimized. The main advantage of the interline transfer is that the transfer from the active sensors to the shield storage is quick. The shields act like a venetian blind that obscures half the information that is available in the

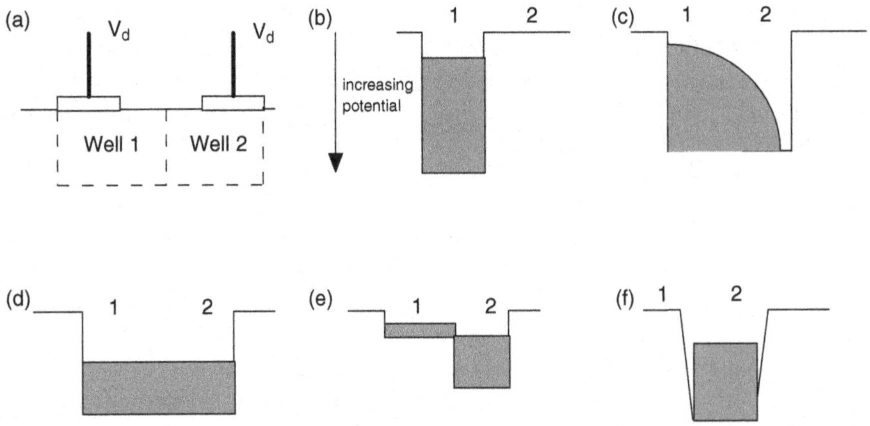

Figure 6.2-2 Movement of electrons in a potential well due to photoelectrons.

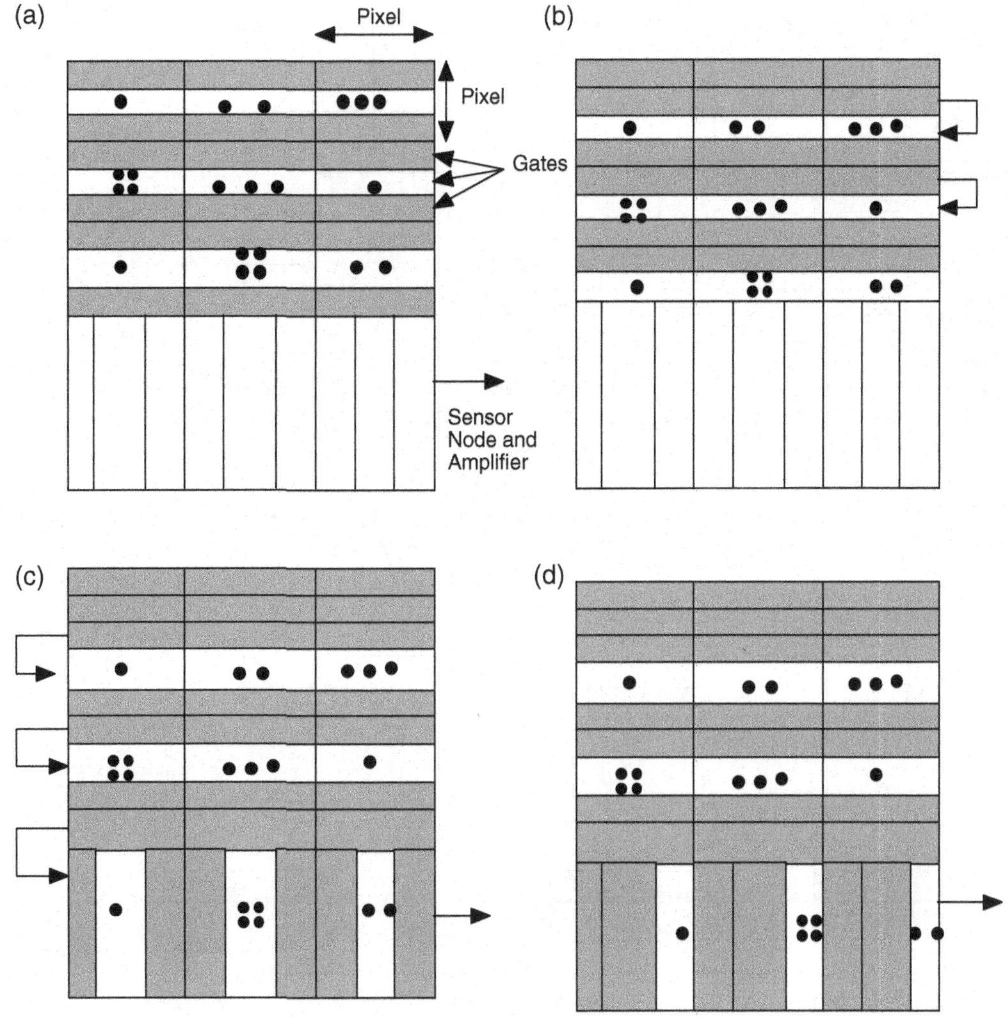

Figure 6.2-3 (a)–(d) Internal look at a CCD array and the charge transport.

scene. The area fill factor can be as low as 20%. Since the detector is only 20% of the pixel area, the output voltage is only 20% of a detector, which would fill the pixel area.

A fraction of the light can leak into vertical registers but this can be minimized by using shielded storage array. There are several types of transfer register architectures. In Figure 6.2-8 we can observe a four-phase transfer register that carries charge under two gates. The 2:1 interlace has both fields collected at the same time but are read out alternately.

6.2.2 Optical fiber

Communications using light as a signal carrier and optical fibers as the media are termed *optical fiber communications*. The applications of optical fiber communications have increased very rapidly since the first commercial installation of a fiber-optic system in 1977. Today every major telecommunications company spends millions of dollars developing and utilizing an optical fiber communications system. Fiber-optic communication systems can process information using either digital or analog modulation schemes. In the most commonly used fiber-optic communication system, voice, video, and data are converted to a coded pulse stream of light using a suitable light source. This pulse stream is carried by optical fibers to a regenerating or receiving station. At the final receiving station the light pulses are converted into the form of the original information. Optical fiber communication systems are currently used in the fields of telecommunications, data communications, and cable television. These systems are also used for many military, automotive, medical, and industrial applications.

Since the dawn of history, people have used light as a vehicle to carry information. Lanterns on ships and smoke signals or flashing mirrors on land are early examples of how light was used to communicate. It was

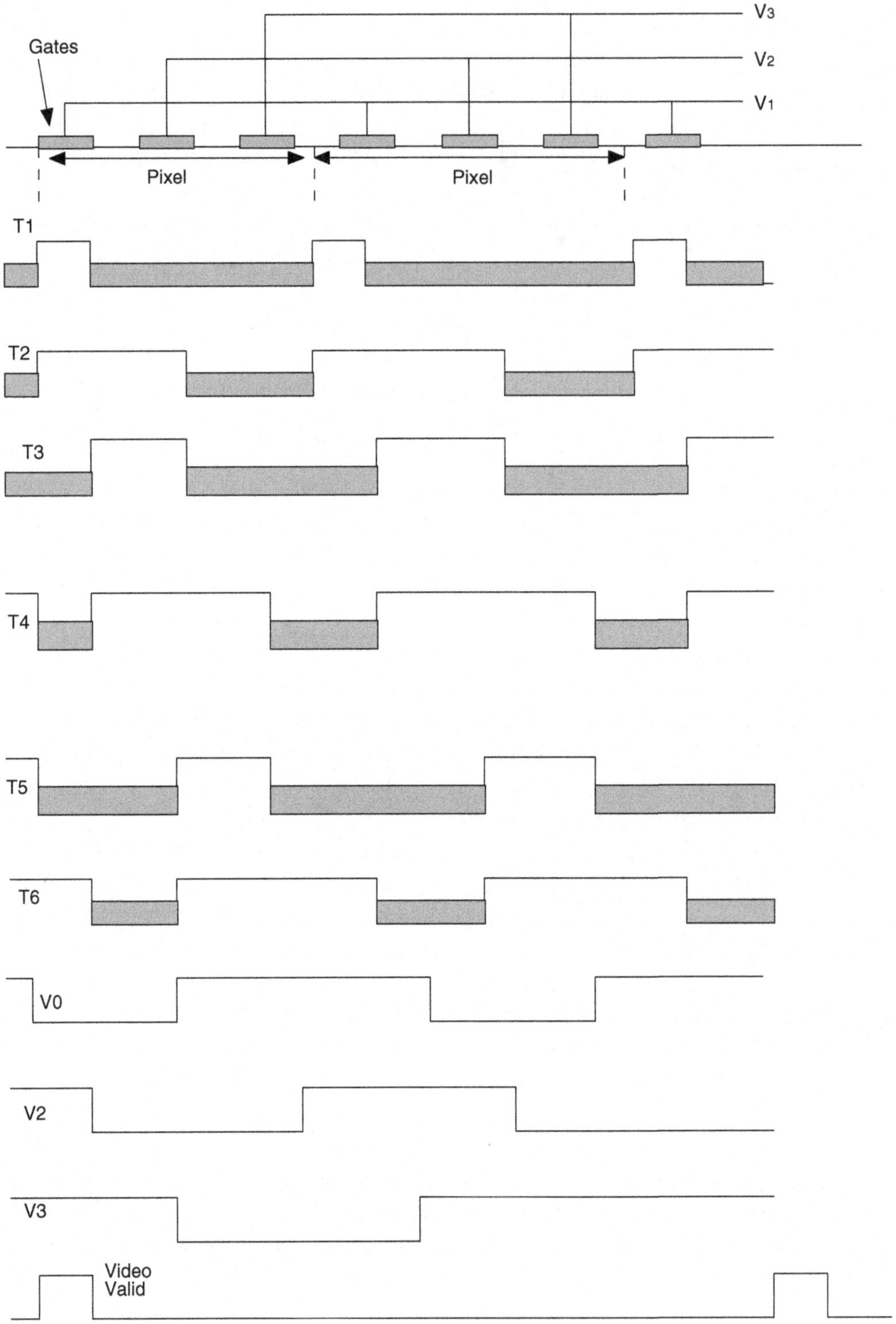

Figure 6.2-4 Charge storage in the gates of CCD.

over a hundred years ago that Alexander Graham Bell (1880) transmitted a telephone signal a distance greater than 200 m using light as a signal carrier. Bell called his invention a "photophone" and obtained a patent for it (Bruce, 1973). Bell, however, wisely gave up the photophone in favor of the electric telephone. The photophone, at the time of its invention, could not be exploited commercially because of two basic drawbacks: (1) the lack of a reliable light source and (2) the lack of dependable transmission medium.

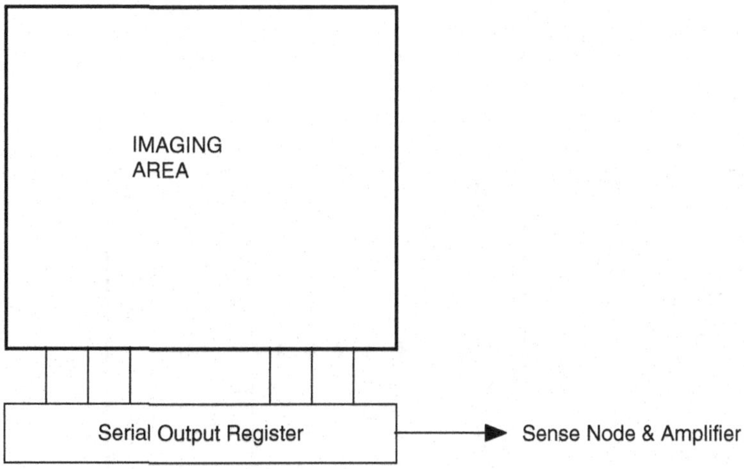

Figure 6.2-5 Interface of a CCD array and the register for data output.

Modern light-wave communication systems had their birth in the 1960s. The first demonstration of the ruby laser in 1960 (Maiman 1960) and a demonstration of laser operations in semiconductor devices in 1962 (Hall *et al.* 1962; Nathan *et al.* 1962) were early stepping stones that led to the continuous operation of room temperature, long-lifetime, semiconductor lasers that are in common use today. The laser made available a coherent optical frequency carrier of enormous communication capacity. If a communication system was built that utilized only 0.01% of the laser carrier frequency, its modulation bandwidth would be 30 GHz. In 1966 a parallel evolution of fiber technology was taking place. Although the best existing fibers at that time had attenuation greater than 1000 dB/km, researchers at Standard Telecommunication Laboratories (STC) in England (Kao and Hockham 1966) speculated that losses as low as 20 dB/km should be achievable and they further suggested that such fibers would be useful in telecommunication applications, and they were correct. In 1970 workers at Corning Glass Works (Kapron *et al.* 1970) produced the first fiber with loss under 20 dB/km. Since that time, fiber technology has advanced to the point of producing fibers with loss less than 0.25 dB/km at 1.55 μm. These fibers are approaching the Rayleigh scattering limit of the glass being used to fabricate them. The biggest advantage of an optical fiber communication system is its tremendous information-carrying capacity. There are already many systems that can carry several thousand simultaneous conversations over a pair of optical fibers that are thinner than a human hair.

6.2.2.1 Classification and features of optical fibers

Fibers that are used for optical communications are waveguides made of transparent dielectrics whose function is to guide light over long distances. An optical fiber consists of an inner cylinder of glass called the core,

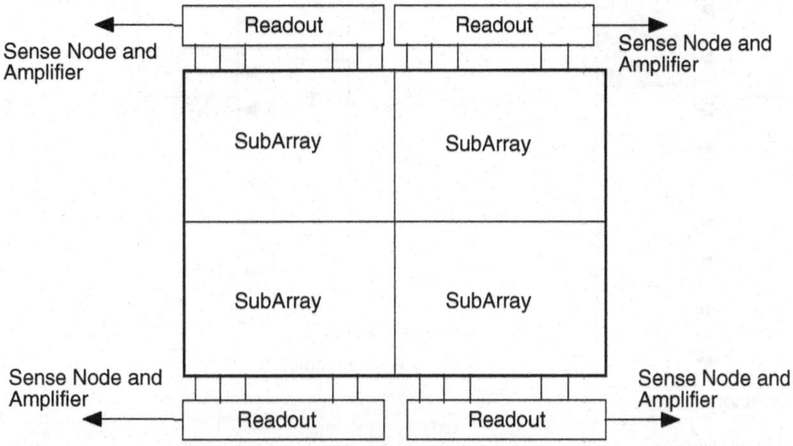

Figure 6.2-6 CCD subarrays and hardware registers for reading data out.

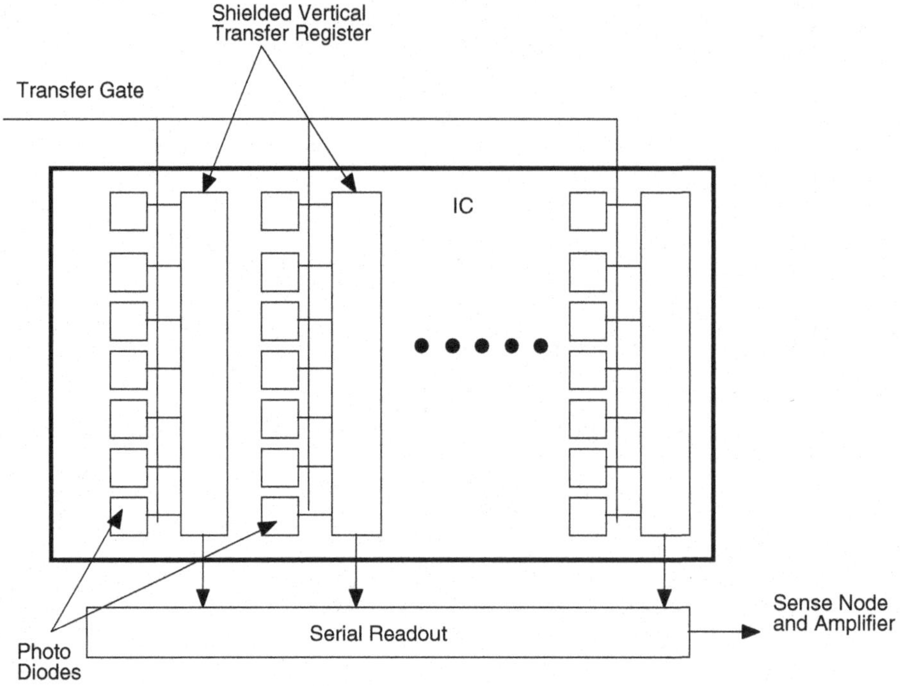

Figure 6.2-7 The interline transfer in a CCD.

surrounded by a cylindrical shell of glass of lower refractive index, called the cladding. Optical fibers may be classified in terms of the refractive index profile of the core and whether one mode (single-mode fiber) or many modes (multimode fiber) are propagating in the guide (Figure 6.2-9). If the core, which is typically made of high-silica-content glass or a multicomponent glass, has a uniform refractive index n_1, it is called a *step-index* fiber. If the core has a nonuniform refractive index that gradually decreases from the center toward the core–cladding interface, the fiber is called a *graded-index* fiber. The cladding surrounding the core has a uniform

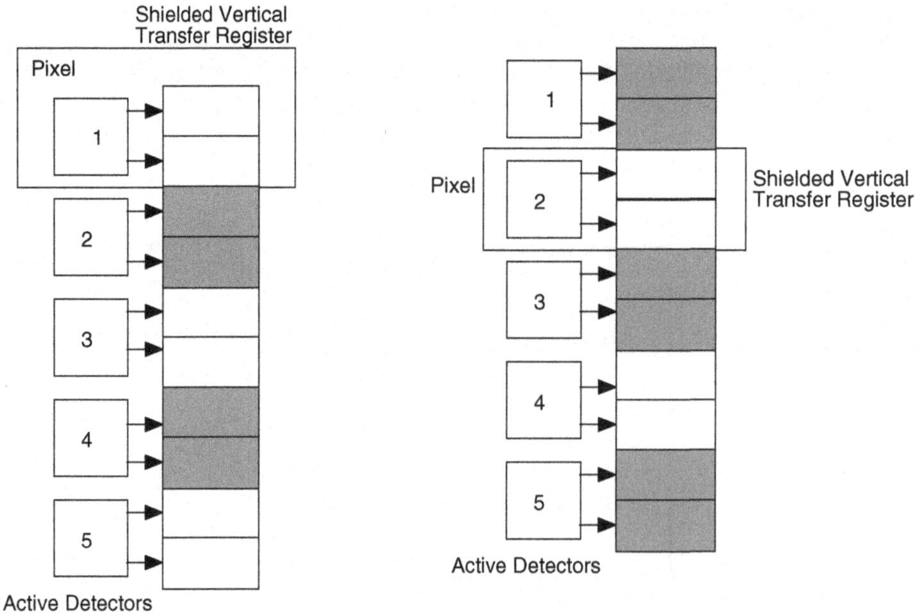

Figure 6.2-8 Detailed layout of the 2:1 interlaced array: (a) the odd field is clocked into the vertical transfer register and (b) the even field is transferred.

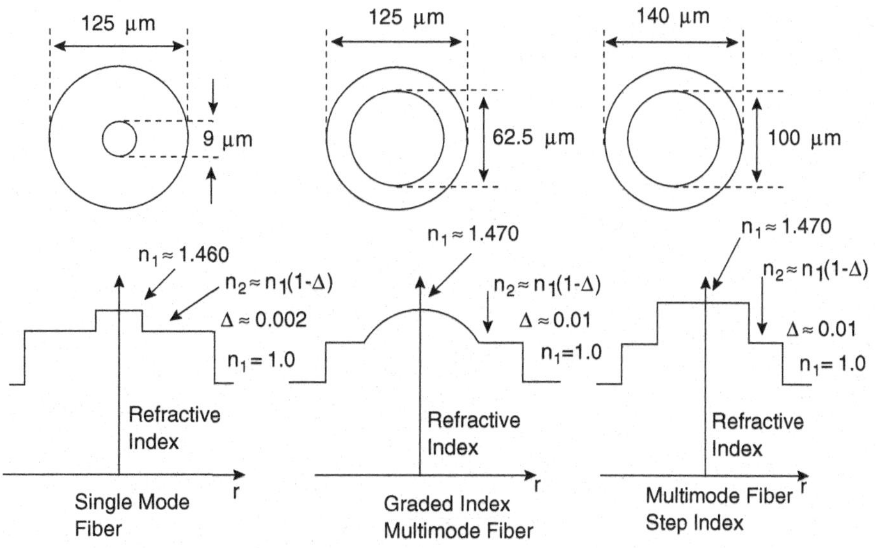

Figure 6.2-9 Dimensions and refractive indexes for commonly used optical fibers.

refractive index n_2 that is slightly lower than the refractive index of the core region. The cladding of the fiber is made of a high-silica-content glass or a multicomponent glass. Figure 6.2-9 shows the dimensions and refractive indexes of commonly used telecommunication fibers. Figure 6.2-10 enumerates some of the advantages, constraints, and applications of the different types of fibers. In general, when the transmission medium must have a very high bandwidth—for example, in an undersea or long-distance terrestrial system—a single-mode fiber is used. For intermediate system bandwidth requirements between 150 MHz-km and 2 GHz-km, such as found in local-area networks, either a single-mode or graded-index multimode fiber would be the choice. For applications such as short data links in which lower bandwidth requirements are placed on the transmission medium, either a graded-index or a step-index multimode fiber can be used.

Because of their low-loss and high-bandwidth capabilities, optical fibers have the potential for being used wherever twisted-wire pairs or coaxial cables are used as the transmission medium in a communication system. If an engineer were interested in choosing a transmission medium for a given transmission objective, he or she would tabulate the required and desired features of alternative technologies that may be available for use in the applications. With that process in mind, a summary of the attractive features and the advantages of optical fiber

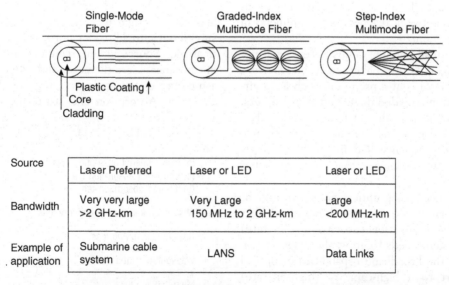

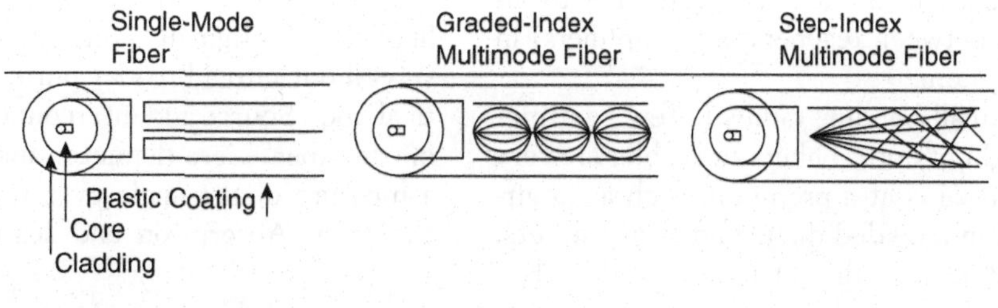

	Single-Mode Fiber	Graded-Index Multimode Fiber	Step-Index Multimode Fiber
Source	Laser Preferred	Laser or LED	Laser or LED
Bandwidth	Very very large >2 GHz-km	Very Large 150 MHz to 2 GHz-km	Large <200 MHz-km
Example of application	Submarine cable system	LANS	Data Links

Figure 6.2-10 Advantages, constraints, and applications of optical fibers.

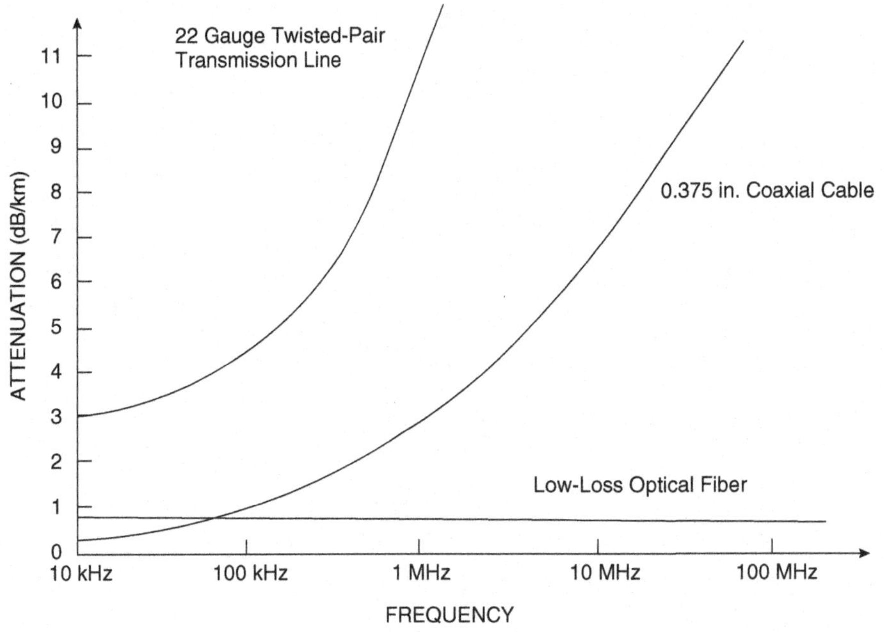

Figure 6.2-11 Signal attenuation vs frequency for different transmission media.

transmission will be given. These advantages include (1) low loss and high bandwidth; (2) small size and bending radius; (3) nonconductive, nonradiative, and noninductive; (4) light weight; and (5) providing natural growth capability.

To appreciate the low- and high-bandwidth capabilities of optical fibers, consider the curves of signal attenuation vs frequency for three different transmission media shown in Figure 6.2-11. Optical fibers have a "flat" transfer function well beyond 100 MHz. Compared with wire pairs or coaxial cables, optical fibers have far less loss for signal frequencies above a few megahertz. This is an important characteristic that strongly influences system economics, because it allows the system designer to increase the distance between regenerators (amplifiers) in a communication system.

The small size, small bending radius (a few centimeters), and light weight of optical fibers and cables are very important where space is at a premium, such as in aircraft, on ships, and in crowded ducts under city streets.

Because optical fibers are dielectric waveguides, they avoid many noise problems such as radiated interference, ground loops, and, when installed in a cable without metal, lightning-induced damage that exists in other transmission media.

Finally, the engineer using optical fibers has a great deal of flexibility. He or she can install an optical fiber cable and use it initially in a low-capacity (low-bit-rate) system. As the system needs to grow, the engineer can take advantage of the broadband capabilities of optical fibers and convert to a high-capacity (high-bit-rate) system by simply changing the terminal electronics.

The proper design and operation of an optical communication system using optical fibers as the transmission medium require a knowledge of the transmission characteristics of the optical sources, fibers, and interconnection devices (connectors, couplers, and splices) used to join lengths of fibers together. The transmission criteria that affect the choice of the fiber type used in a system are signal attenuation, information transmission capacity (bandwidth), and source coupling and interconnection efficiency. Signal attenuation is due to a number of loss mechanisms within the fiber, as shown in Table 6.2-1, and due to the losses occurring in splices and connectors. The information transmission capacity of a fiber is limited by dispersion, a phenomenon that causes light that is originally concentrated into a short pulse to spread out into a broader pulse as it travels along an optical fiber. Source and interconnection efficiency depends on the fiber's core diameter and its numerical aperture, a measure of the angle over which light is accepted in the fiber. Absorption and scattering of light traveling

Table 6.2-1 Loss mechanisms in optical fibers

1. Intrinsic material absorption loss (a) Ultraviolet absorption tail (b) Infrared absorption tail
2. Absorption loss due to impurity ions
3. Rayleigh scattering loss
4. Waveguide scattering loss
5. Microbending loss

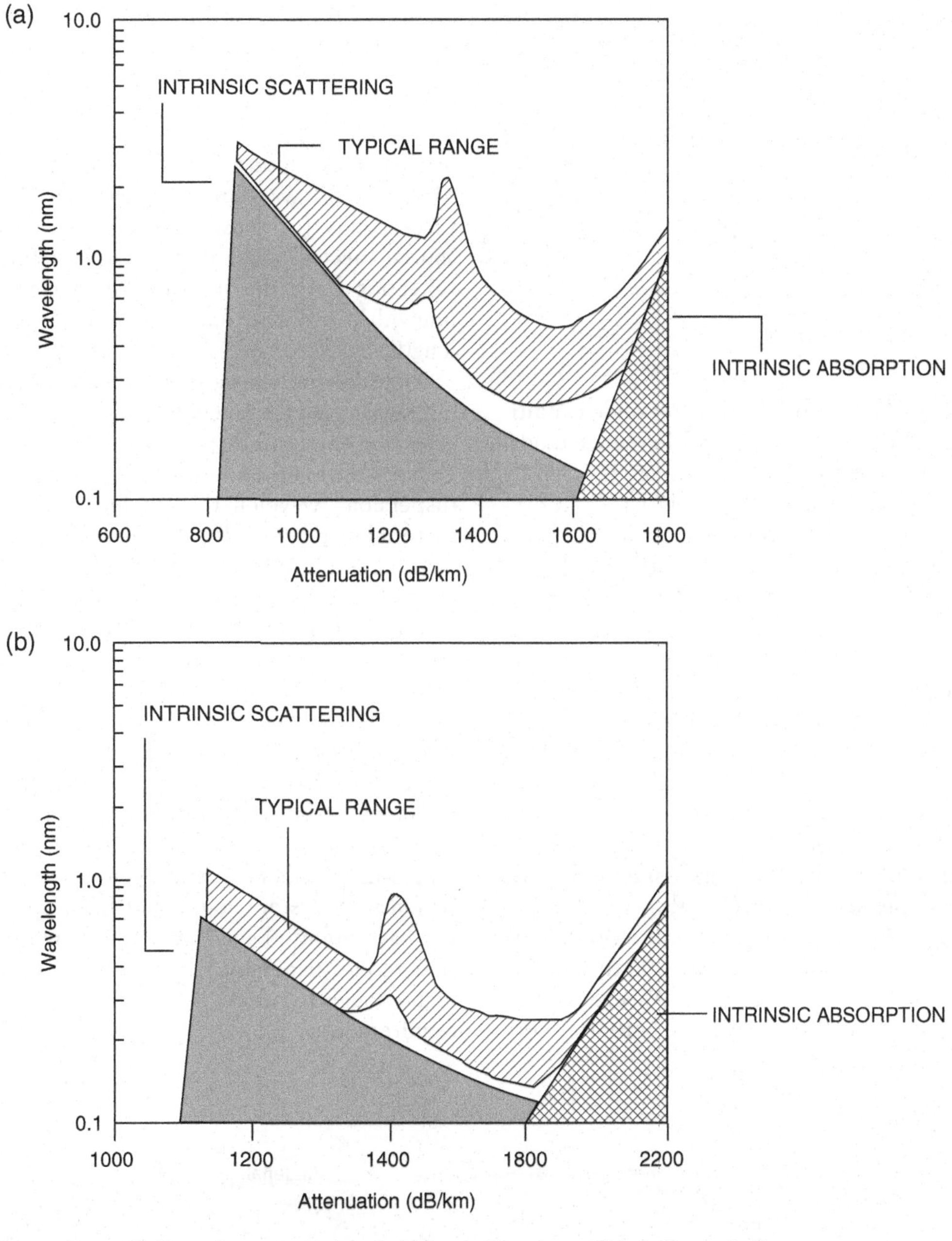

Figure 6.2-12 Dependence of attenuation on wavelength (a) for multimode and (b) single-mode fibers.

through a fiber lead to signal attenuation, the rate of which is measured in decibels per kilometer (dB/km). As can be seen in Figure 6.2-12, for both multi-mode and single-mode fibers, attenuation depends strongly on wavelength. The decrease in scattering losses with increasing wavelength is offset by an increase in material absorption such that attenuation is lowest near 1.55 μm (1550 nm).

The measured values given in Table 6.2-2 are probably close to the lower bounds for the attenuation of optical fibers. In addition to intrinisic fiber losses, extrinsic loss mechanisms, such as absorption due to impurity ions and microbending loss due to jacketing and cabling, can add loss to a fiber. The bandwidth or information-carrying capacity of a fiber is inversely related to its dispersion. The total dispersion in a fiber is a combination of three components: intermodal dispersion (modal delay distortion), material dispersion, and waveguide dispersion.

Intermodal dispersion occurs in multimode fibers because rays associated with different modes travel different effective distances through the optical fiber. This causes light in the different modes to spread out temporally as it travels along the fiber. Modal delay distortion

Table 6.2-2 Best attenuation results (dB/km) in Ge-P-SiO_2 core fibers

Wavelength (nm)	$\Delta \approx 0.2\%$ (single-mode fiber)	$\Delta \approx 1.0\%$ (graded-index multimode fiber)
850	2.1	2.20
1300	0.27	0.44
1500	0.16	0.23

can severely limit the bandwidth of a step-index multimode fiber to the order of 20 MHz/km. To reduce modal delay distortion in multimode fibers, the core is carefully doped to create a graded (approximately parabolic) refractive index profile. By carefully designing this index profile, the group velocities of the propagating modes are nearly equalized. Bandwidths of 1.0 GHz-km are readily attainable in commercially available graded-index multimode fibers. The most effective way to eliminate intermodal dispersion is to use a single-mode fiber. As only one mode propagates in a single-mode fiber, modal delay distortion between modes does not exist and very high bandwidths are possible. The bandwidth of a single-mode fiber, as mentioned previously, is limited by the combination of material and waveguide dispersion. As shown in Figure 6.2-13, both material and waveguide dispersions are dependent on wavelength.

Material dispersion is caused by the variation of the refractive index of the glass with wavelength and the spectral width of the system source. Waveguide dispersion occurs because light travels in both the core and cladding of a single-mode fiber at an effective velocity between that of the core and cladding materials. The waveguide dispersion arises because the effective velocity changes with wavelength. The amount of waveguide dispersion depends on the design of the waveguide structure as well as on the fiber material. Material and waveguide dispersions are measured in picoseconds (of pulse spreading) per nanometer (of source spectral width) per kilometer (of fiber length), reflecting both the increases in magnitude in source linewidth and the increase in dispersion with fiber length.

Material and waveguide dispersions can have different signs and effectively cancel each other's dispersive effect on the total dispersion in a single-mode fiber. In conventional germanium-doped silica fibers, the "zero-dispersion" wavelength at which the waveguide and material dispersion effects cancel each other out occurs near 1.30 μm. The zero-dispersion wavelength can be shifted to 1.55 μm, or the low-dispersion characteristics of a fiber can be broadened by modifying the refractive index profile shape of a single-mode fiber. This profile shape modification alters the waveguide dispersion characteristics of the fiber and changes the wavelength region in which waveguide and material dispersion effects cancel each other. Figure 6.2-14 illustrates the profile shapes of conventional, dispersion-shifted, and dispersion-flattened single-mode fibers. Single-mode fibers operating in their zero-dispersion region with system sources of finite spectral width do not have infinite bandwidth (and may exhibit polarization mode dispersion) but have bandwidths that are high enough to satisfy all current high-capacity system requirements.

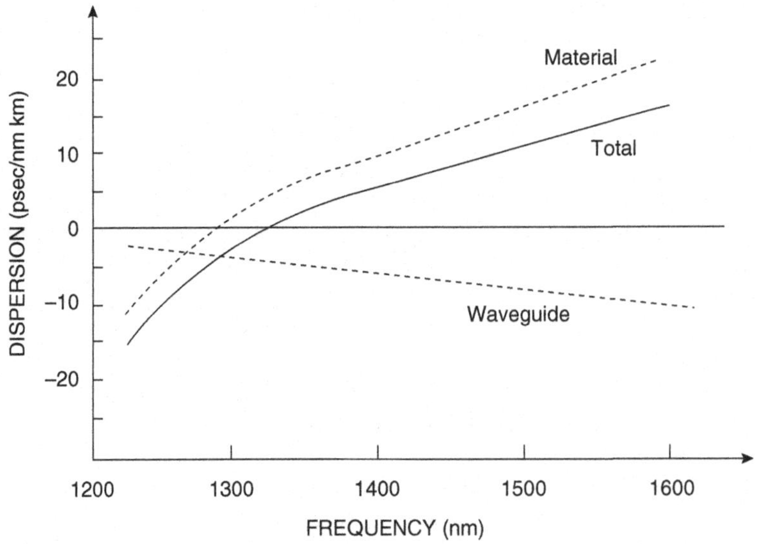

Figure 6.2-13 Dependence of waveguide dispersion on wavelength.

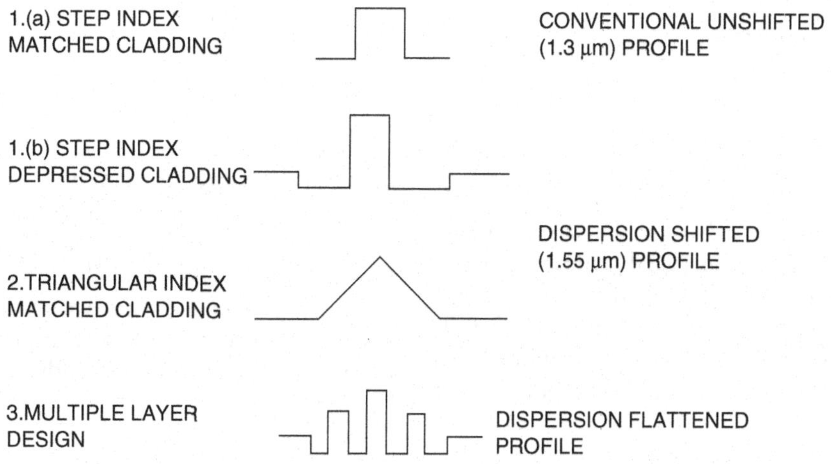

Figure 6.2-14 Profile shapes of different types of single-mode fibers.

6.2.3 Analysis of optical fibers

6.2.3.1 The step-index fiber

In this section electromagnetic field theory will be used to solve rigorously the boundary value problem of the round optical fiber with a homogeneous core (the step-index fiber) (Morse and Feshbach 1953). To simplify this analysis, we will assume that the fiber is oriented along the z axis and the radius b of the fiber cladding is large enough to ensure that the cladding field decays exponentially and approaches zero at the cladding–air interface. This allows us, as shown in Figure 6.2-15, to analyze the fiber as a tractable two-media boundary value problem. This thick-cladding assumption agrees well with the conditions that exist within a properly designed optical fiber used for communication purposes. The steps that we will follow to solve the boundary value problem of the step-index fiber are outlined in Table 6.2-3. To obtain the modes in a step-index optical fiber, one must solve the modified wave equation [Eqs. (6.2.1) and (6.2.2)] shown later for E_z and H_z in both the core and cladding regions of the fiber:

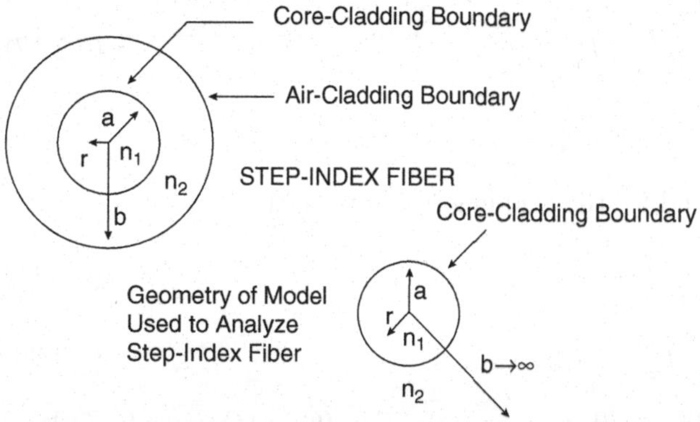

Figure 6.2-15 Physical characteristics of step-index fibers.

$$\frac{\partial^2 E_z}{\partial r^2} + \frac{1}{r}\frac{\partial E_z}{\partial r} + \frac{1}{r^2}\frac{\partial^2 E_z}{\partial \phi^2} + \kappa^2 E_z = 0 \tag{6.2.1}$$

$$\frac{\partial^2 H_z}{\partial r^2} + \frac{1}{r}\frac{\partial H_z}{\partial r} + \frac{1}{r^2}\frac{\partial^2 H_z}{\partial \phi^2} + \kappa^2 H_z = 0 \tag{6.2.2}$$

Having obtained expressions for E_z and H_z, we can directly obtain from Maxwell's equations expressions for the transverse components of the fields:

$$E_r = -\frac{j}{\kappa^2}\left(\beta\frac{\partial E_z}{\partial r} + \omega\mu\frac{1}{r}\frac{\partial H_z}{\partial \phi}\right) \tag{6.2.3}$$

$$E_\phi = -\frac{j}{\kappa^2}\left(\beta\frac{1}{r}\frac{\partial E_z}{\partial \phi} - \omega\mu\frac{\partial H_z}{\partial r}\right) \tag{6.2.4}$$

Table 6.2-3 Analysis of the step-index fiber procedures followed

1. Mathematically model the step-index fiber using the wave equation in cylindrical coordinates.
2. Use the technique of separation of variables to partition the wave equation.
3. Define the physical requirements that influence the solution of the fields in the core and cladding.
4. Select the proper functional form of the solution of the modified wave equation (Bessel's equation) in the core and cladding.
5. Apply the boundary conditions at the core–cladding interface.
6. Obtain the "characteristic" equation and its resulting modal equations.
7. Analyze the resulting modes and their cutoff conditions.

$$H_r = -\frac{j}{\kappa^2}\left(\beta\frac{\partial H_z}{\partial r} - \omega\varepsilon\frac{1}{r}\frac{\partial E_z}{\partial \phi}\right) \quad (6.2.5)$$

$$H_\phi = -\frac{j}{\kappa^2}\left(\beta\frac{1}{r}\frac{\partial H_z}{\partial \phi} + \omega\varepsilon\frac{\partial E_z}{\partial r}\right) \quad (6.2.6)$$

Because Eqs. (6.2.1) and (6.2.2) have the same mathematical form, we will solve Eq. (6.2.1), understanding that solutions obtained for it will be valid for Eq. (6.2.2). To obtain Eq. (6.2.1) we have already assumed an optical system with cylindrical symmetry. The longitudinal direction of propagation is the z axis and the dependence of the fields is of the form $e^{j(\omega t - \beta z)}$.

The technique of separation of variables will now be applied to obtain a solution of Eq. (6.2.1). We will assume that we can obtain independent solutions for E_z in ϕ and r, that is,

$$E_z(\phi, r) = A\Phi(\phi)F(r) \quad (6.2.7)$$

Since the fiber has circular symmetry, we will choose a circular function as a trial solution for $\Phi(\phi)$.

$$\Phi(\phi) = e^{j\upsilon\phi} \quad (6.2.8)$$

where υ is a positive or negative integer. Now we have

$$E_z AF(r)e^{j\upsilon\phi} \quad (6.2.9)$$

Taking the second-order derivatives of Eq. (6.2.9) with respect to r and ϕ and substituting back into Eq. (6.2.1), we obtain

$$\frac{d^2 F(r)}{dr^2} + \frac{1}{r}\frac{dF(r)}{dr} + \left(\kappa^2 - \frac{\upsilon^2}{r^2}\right)F(r) = 0 \quad (6.2.10)$$

Equation (6.2.10) is a form of Bessel's equation where κ is defined as the wave number and given by the expression $\kappa^2 = k^2 - \beta^2$, k being the complex propagation constant equal to $\omega^2\mu\varepsilon$. This well-known second-order differential equation has two independent solutions. Numerous cylinder functions satisfy Bessel's equation. Energy considerations will dictate the choice of the functions selected as solutions of Eq. (6.2.10); that is,

1. The field must be finite in the core of the fiber. Specifically the cylinder function chosen in the core of the fiber must be finite at $r = 0$.
2. The field in the cladding of the fiber must have an exponentially decaying behavior at large distances from the center of the fiber.

Because the fields must be finite at the center of the fiber core, we will choose $J_\upsilon(\kappa r)$ as the form of the solution for $r < a$. Therefore, for $r < a$,

$$E_z = AJ_\upsilon(\kappa r)e^{j\upsilon\phi} \quad (6.2.11)$$

$$H_z = BJ_\upsilon(\kappa r)e^{j\upsilon\phi} \quad (6.2.12)$$

We require that the field in the cladding of the fiber decay in the r direction and be of the form $e^{-\gamma r}$ where γ is the decay constant of the evanescent field, given by $\gamma^2 = \beta^2 - n_2^2 k_0^2$, with n_2 being the index of refraction of the cladding given by $(\varepsilon_{r2})^{1/2}$.

If we define $\kappa = j\gamma$ we can choose a modified Hankel function of the first kind to describe the decaying behavior of the field in the cladding for large r. That is, for $r > a$,

$$E_z = CH_\upsilon^{(1)}(j\gamma r)e^{j\upsilon\phi} \quad (6.2.13)$$

$$H_z = DH_\upsilon^{(1)}(j\gamma r)e^{j\upsilon\phi} \quad (6.2.14)$$

where A, B, C, and D *are* unknown constants to be determined.

To obtain the transverse fields in the core and cladding of the guide, one must use Eqs. (6.2.3) through (6.2.6). For example, to obtain E_r in Eq. (6.2.3) for both $r < a$ and $r > a$ one must differentiate the longitudinal fields (i.e., E_z) of Eqs. (6.2.11) and (6.2.13), respectively, with respect to r and ϕ and then substitute the results back into Eq. (6.2.3). After some simplification we obtain, for $r < a$,

$$E_r = -\frac{j}{\kappa^2}\left[A\beta\kappa J_\upsilon'(\kappa r)e^{j\upsilon\phi} + B(j\upsilon)(\omega\mu)\frac{1}{r}J_\upsilon(\kappa r)e^{j\upsilon\phi}\right]e^{j\upsilon\phi} \quad (6.2.15)$$

For $r > a$ we have

$$E_r = -\frac{1}{\gamma^2}\left[\beta\gamma C H_\upsilon^{(1)\prime}(j\gamma r) + \omega\mu_0 \frac{\upsilon}{r} D H_\upsilon^{(1)}(j\gamma r)\right]e^{j\upsilon\phi} \quad (6.2.16)$$

where primed terms (i.e., terms with ′) mean the first derivatives with respect to κr. Furthermore, for the core region ($r < a$)

$$\kappa^2 = k_1^2 - \beta^2 \quad (6.2.17)$$

$$k_1^2 = \omega^2\mu_0\varepsilon_1 \quad (6.2.18)$$

and for the cladding region ($r > a$)

$$\gamma^2 = \beta^2 - k_2^2 \quad (6.2.19)$$

$$k_2^2 = \omega^2\mu_0\varepsilon_2 \quad (6.2.20)$$

In a similar manner using Eqs. (6.2.4) through (6.2.6), we can obtain for the core region ($r > a$)

$$E_\phi = -\frac{j}{\kappa^2}\left[j\beta\frac{\upsilon}{r}AJ_\upsilon(\kappa r) - \kappa\omega\mu BJ'_\upsilon(\kappa r)\right]e^{j\upsilon\phi} \quad (6.2.21)$$

$$H_r = -\frac{j}{\kappa^2}\left[-j\omega\varepsilon_1\frac{\upsilon}{r}AJ_\upsilon(\kappa r) + \kappa\beta BJ'_\upsilon(\kappa r)\right]e^{j\upsilon\phi} \quad (6.2.22)$$

$$H_\phi = -\frac{j}{\kappa^2}\left[\kappa\omega\varepsilon_1 AJ'_\upsilon(\kappa r) + j\beta\frac{\upsilon}{r}BJ_\upsilon(\kappa r)\right]e^{j\upsilon\phi} \quad (6.2.23)$$

and for the cladding region ($r > a$)

$$E_\phi = -\frac{1}{\gamma^2}\left[\beta\frac{\upsilon}{r}CH_\upsilon^{(1)}(j\gamma r) - \gamma\omega\mu_0 DH_\upsilon^{(1)\prime}(j\gamma r)\right]e^{j\upsilon\phi} \quad (6.2.24)$$

$$H_r = -\frac{1}{\gamma^2}\left[-\omega\varepsilon_2\frac{\upsilon}{r}CH_\upsilon^{(1)}(j\gamma r) - \gamma\beta DH_\upsilon^{(1)\prime}(j\gamma r)\right]e^{j\upsilon\phi} \quad (6.2.25)$$

$$H_\phi = -\frac{1}{\gamma^2}\left[\gamma\omega\varepsilon_2 CH_\upsilon^{(1)\prime}(j\gamma r) + \beta\frac{\upsilon}{r}DH_\upsilon^{(1)}(j\gamma r)\right]e^{j\upsilon\phi} \quad (6.2.26)$$

The constants A, B, C, D, and β are determined by applying the boundary conditions for the two tangential components of the electric and magnetic fields at the core–cladding interface ($r = a$). The boundary conditions for the fields at the core–cladding interface can be written as

$$\begin{aligned} E_{z_1} &= E_{z_2} \\ E_{\phi_1} &= E_{\phi_2} \quad \text{for } r = a \\ H_{z_1} &= H_{z_2} \\ H_{\phi_1} &= H_{\phi_2} \end{aligned}$$

where subscripts 1 and 2 refer to the fields in the core and cladding, respectively. Applying these conditions yields four simultaneous equations for the unknown A, B, C, and D. This solution yields a determinant. The solutions for A, B, C, and D can be obtained from this determinant provided that the system determinant for the four equations vanishes. Expansion of this determinant results in what is known as the "eigenvalue" or characteristic equation of the waveguide. This equation defines the modes in the guide and yields the permissible values of β, κ, and γ associated with each mode. The resulting characteristic equation for the step-index fiber is

$$\left[\frac{\varepsilon_1}{\varepsilon_2}\frac{a\gamma^2}{\kappa}\frac{J'_\upsilon(\kappa a)}{J_\upsilon(\kappa a)} + j\gamma a\frac{H_\upsilon^{(1)\prime}(j\gamma a)}{H_\upsilon^{(1)}(j\gamma a)}\right] \times \left[\frac{a\gamma^2 J'_\upsilon(\kappa a)}{\kappa J_\upsilon(\kappa a)} + j\gamma a\frac{H_\upsilon^{(1)\prime}(j\gamma a)}{H_\upsilon^{(1)}(j\gamma a)}\right] = \left[\upsilon\left(\frac{\varepsilon_1}{\varepsilon_2} - 1\right)\frac{\beta k_2}{\kappa^2}\right]^2 \quad (6.2.27)$$

The coefficients A, B, C, and D can be written so that A is the only unknown coefficient. For example, it can be shown that

$$C = \frac{J_\upsilon(\kappa a)}{H_\upsilon^{(1)}(j\gamma a)}A \quad (6.2.28)$$

$$D = \frac{J_\upsilon(\kappa a)}{H_\upsilon^{(1)}(j\gamma a)}B \quad (6.2.29)$$

where A and B are related to each other via

$$B = j\upsilon\frac{\omega(\varepsilon_1 - \varepsilon_2)\beta J_\upsilon(\kappa a)H_\upsilon^{(1)}(j\gamma a)}{\kappa\gamma a\left[\gamma J'_\upsilon(\kappa a)H_\upsilon^{(1)}(j\gamma a) + j\kappa J_\upsilon(\kappa a)H_\upsilon^{(1)\prime}(j\gamma a)\right]}A \quad (6.2.30)$$

In general, the permissible field configurations or modes that exist in a step-index fiber have six field components. For the round fiber, hybrid modes exist as well as the transverse electric (TE) and transverse magnetic (TM) modes. The hybrid modes are denoted HE and EH modes and have both longitudinal electric and magnetic field components present. In terms of a ray analogy for the step-index fiber, the hybrid modes correspond to propagating skew rays and the TE and TM modes correspond to propagating meridional rays [$\upsilon = 0$ in Eq. (6.2.27)]. For meridional rays the right-hand side of Eq. (6.2.27) is equal to zero and one obtains two characteristic equations that define the TE and TM modes. These equations are

$$\left[\frac{a\gamma^2}{\kappa}\frac{J'_0(\kappa a)}{J_0(\kappa a)} + j\gamma a\frac{H_0^{(1)\prime}(j\gamma a)}{H_0^{(1)}(j\gamma a)}\right] = 0 \quad (6.2.31)$$

$$\left[\frac{\varepsilon_1}{\varepsilon_2}\frac{a\gamma^2}{\kappa}\frac{J'_0(\kappa a)}{(\kappa a)} + \frac{j\gamma a H_0^{(1)\prime}(j\gamma a)}{H_0^{(1)}(j\gamma a)}\right] = 0 \quad (6.2.32)$$

An important parameter for each propagating mode is its cutoff frequency. A mode is cut off when its field in the cladding ceases to be evanescent and is detached from the guide; that is, the field in the cladding does not decay. The rate of decay of the fields in the cladding is determined by the value of the constant γ. For large values of γ, the fields are tightly concentrated inside and close to the core. With decreasing values of γ, the fields reach farther out into the cladding. Finally, for $\gamma = 0$, the fields detach themselves from the guide. The frequency at which this happens is called the cutoff frequency. The mode cutoff frequency can be calculated from Eqs. (6.2.17) through (6.2.20) for $\gamma = 0$ and is given by

$$\omega_c = \frac{\kappa_c}{\sqrt{\mu_0(\varepsilon_1 - \varepsilon_2)}} \quad (6.2.33)$$

The cutoff of each mode is obtained by solving the characteristic equation for κ_c for that mode. The cutoff frequency of a mode can be zero if $\kappa_c = 0$. One and only one mode can exist in an optical fiber with $\omega_c = 0$. This mode is the hybrid HE_{11} mode and it exists for all frequencies. If the fiber is designed to operate with only the HE_{11} mode present, it will operate as a single-mode optical fiber. The single-mode fiber has a very small core diameter and small refractive index difference between its core and cladding. These parameters are chosen to ensure that all other guided modes are below their cutoff frequency. To better understand the relationship between mode cutoff and the physical parameters of a fiber, a cutoff parameter $\kappa_c a$, which is usually called the "V" number of the fiber, is defined as

$$\kappa_c a = \omega_c \sqrt{\mu_0 \varepsilon_0} \left(\sqrt{n_1^2 - n_2^2} \right) a \tag{6.2.34}$$

Noting that

$$\omega_c \sqrt{\mu_0 \varepsilon_0} = \frac{2\pi}{\lambda_0} \tag{6.2.35}$$

we finally obtain

$$V \equiv \kappa_c a = \frac{2\pi a}{\lambda_0} \sqrt{n_1^2 - n_2^2} \tag{6.2.36}$$

The number of propagating modes in a step-index fiber is proportional to the V number as shown in Figure 6.2-16. Table 6.2-4 illustrates how increasing a, n_1, n_2, or λ_0 influences the number of propagating modes in the fiber. Notice that for $V < 2.405$ it is possible to design a single-mode fiber that supports only the HE_{11} mode. Typical commercial single-mode fibers have a core diameter of approximately 8–10 μm and a refractive index difference between the core and the cladding of approximately 0.2%. These types of fibers are used for long-distance communications where their very large information-carrying capacity is needed. Single-mode fibers are used in a large majority of the fiber-optic telecommunication systems being installed today. A typical commercial step-index multimode fiber, on the other hand, will propagate hundreds of mode groups and have a relatively large core diameter and refractive index difference between the core and cladding. Step-index multimode fibers are used in short-distance data links where their lower information-carrying capacity is not an issue.

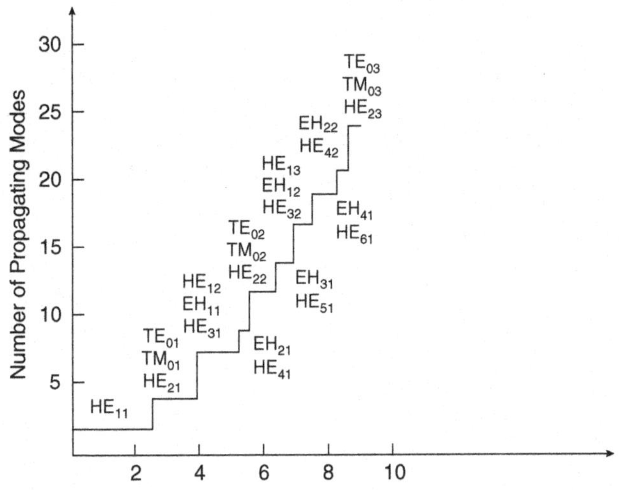

Figure 6.2-16 Types of modes in step-index fibers.

Table 6.2-4 Dependence of the number of propagating modes in a fiber on the physical properties of the fiber

Increasing physical parameters	Number of propagating modes
Core radius, a	Increases
Core refractive index, n_1	Increases
Cladding refractive index, n_2	Decreases
Source wavelength, λ_0	Decreases

6.2.4 The graded-index fiber

The graded-index fiber, because of its relatively large bandwidth and core diameter, is used in many local-area networks where moderate information-carrying capacity is needed. The multimode step-index fiber bandwidth is severely limited (less than 100 MHz-km) due to modal delay distortion. The grading of the refractive index profile of a fiber core has the effect of increasing the bandwidth of a fiber to up to 2 GHz-km by equalizing the group delays of the various propagating mode groups.

Let's consider a multimode fiber with an inhomogeneous core as shown in Figure 6.2-17. The wave equation (6.2.10) is rewritten below showing the variation of the refractive index with r:

$$\frac{d^2F(r)}{dr^2} + \frac{1}{r}\frac{dF(r)}{dr} + \left[k^2(r) - \beta^2 - \frac{v^2}{r^2} \right] F(r) = 0 \tag{6.2.37}$$

where

$$k(r) = \frac{2\pi}{\lambda_0} n(r) = k_0 n(r) \tag{6.2.38}$$

and

$$\kappa^2 = k^2(r) - \beta^2 \tag{6.2.39}$$

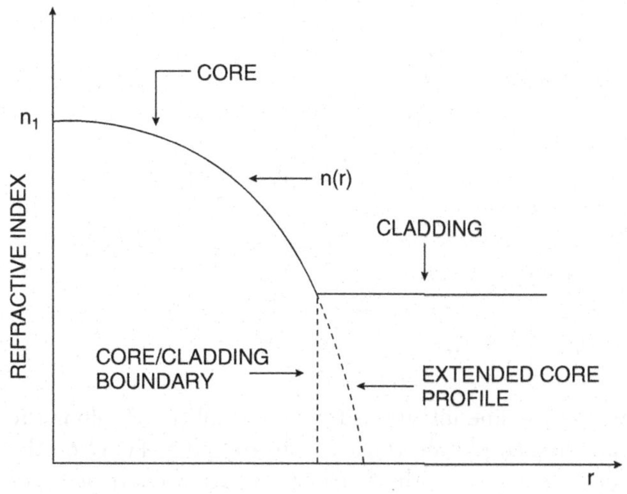

Figure 6.2-17 Physical profile of a graded-index fiber.

The solution of the wave equation leads to a "characteristic equation" for the guide which relates β to κ. To simplify the analysis we first assume that the refractive index continues to decrease in the cladding following the core profile shown by the dashed curve in Figure 6.2-17 (i.e., refractive index is circular symmetric).

Furthermore, to solve the wave equation we resort to a ray optics method of analysis based on the WKBJ method (after Wentzel, Kramers, Brillouin, and Jefferies [Morse and Feshbach 1953]). The WKJB approach is a geometric optics approximation that works whenever the refractive index of the fibers varies only slightly over distances on the order of the optical wavelength. Implementation of the WKBJ approach to a graded-index fiber yields that for a propagation mode to exist it is a necessary condition that (Cherin 1983)

$$k^2(r) - \beta^2 - \frac{\upsilon^2}{r^2} > 0 \tag{6.2.40}$$

Figure 6.2-18a illustrates $k^2\,(r)$ and υ^2/r^2 as a function of the radius r. The solid curve in Figure 6.2-18b shows $k^2(r) - \upsilon^2/k^2$ as a function of r. For a fixed value of β there exist two values of r (r_1 and r_2) such that

$$k^2(r) = \frac{\upsilon^2}{r^2} - \beta^2 = 0 \tag{6.2.41}$$

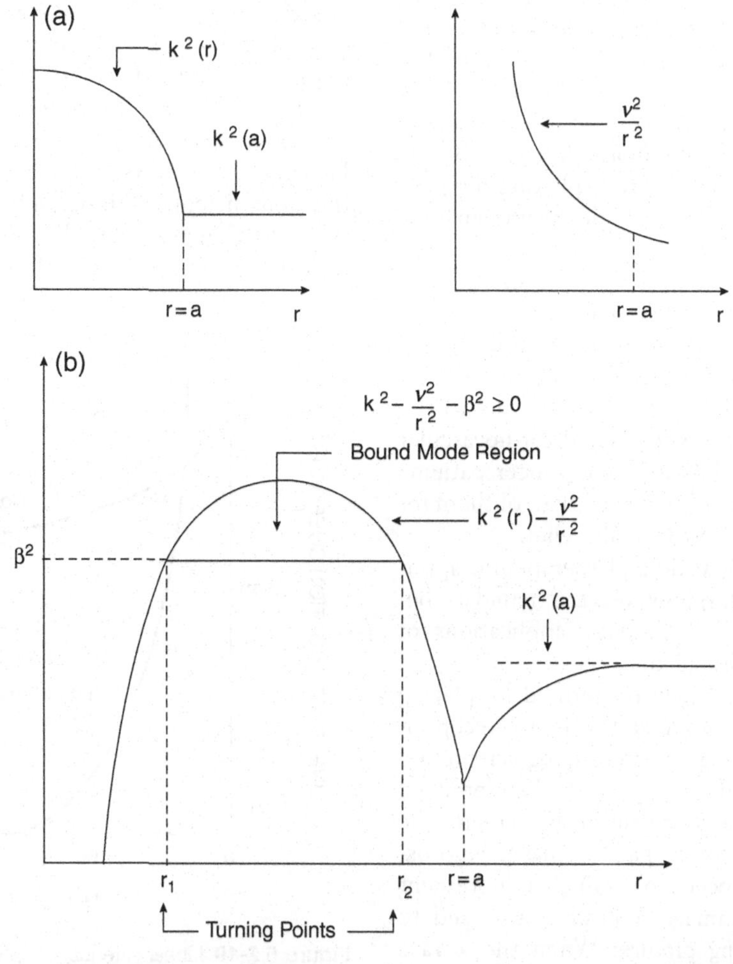

Figure 6.2-18 Wave number diagram for graded-index fiber.

It is between these two radii that the ray associated with the assumed plane-wave solution is constrained to move. Outside these two values of r, called caustics, the ray becomes imaginary, leading to decaying fields.

For a fixed value of β, as υ increases, the region between the two caustics becomes narrower. As υ is increased further a point will be reached where the caustics merge. Beyond this point the wave is no longer bound. The propagation conditions of a wave depend on the value of both β and υ. For a fixed value of υ, modes far from cutoff have large β values and correspondingly more closely spaced caustics. In general, a bound hybrid mode in a graded-index optical fiber can be represented pictorially by a skew ray spiraling down the fiber between two caustics. Both inside ($r < r_1$) and outside ($r < r_2$) the caustics, the field corresponding to the hybrid mode decays.

6.2.5 CT scanners in medicine

For almost a century, x rays have been used for medical imaging and radiation therapy. Over 100 years ago Wilhelm Röntgen, a professor of physics at the Julius Maximilian University of Wurzburg, discovered x rays while experimenting with cathode rays in a Crookes tube. Word of this discovery spread quickly, and by early 1986 the properties of x rays were under investigation in many laboratories in Europe and North America. By the turn of the century, physicians were exploiting the penetrating character of x rays to look inside the human body without cutting it open.

The usage of x rays for medical diagnosis and therapy has expanded enormously since those early years. Today, in the United States alone, over 300 million clinical x-ray examinations are performed annually for purposes ranging from static imaging of fractures and cancers to the real-time guidance of tissue biopsies and cardiovascular angioplasties. In addition, half a million cancer patients each year receive x-ray treatments, about half of them for curative purposes and the rest for pain relief.

Until recently the diagnostic and therapeutic applications were distinct. Today, however, the boundary between the diagnostic and therapeutic applications of x rays in medicine is far less distinct.

Ordinary planar x-ray images are formed by placing a patient between an x-ray tube and an image receptor, usually a cassette containing an intensifying screen and a photographic film. The film is exposed by light emitted when the transmitted x rays interact in the screen. The resulting radiograph is a static shadow image. Fluoroscopy is a variant of this procedure in which a fluorescent screen and an electronic image intensifier are used to form a continuous moving picture. When the x rays traverse the patient, they can be absorbed, scattered or transmitted undisturbed to the receptor. The scattered x rays merely interfere with the information conveyed by the shadow pattern of transmitted rays. Therefore, a mechanical grid is inserted behind the patient to prevent most of the scattered x rays from reaching the cassette. For a parallel, monoenergetic x-ray beam incident along the z axis, the distribution $N(x, y)$ of transmitted x-ray photons at the image plane is given, in the absence of scattering, by

$$N_0 A \int e^{-\mu(z)} dz$$

where the line integral is taken over all tissues along the unscattered photon trajectory to the point (x, y) on the image plane, μ is the linear attenuation coefficient for x rays of the tissue encountered at (x, y, z), and A is the x-ray energy absorption coefficient of the intensifying screen. The distribution of x rays absorbed in the screen thus forms a two-dimensional projection image of the transmission of x rays through the three-dimensional volume of tissue exposed to the x-ray beam.

The linear attenuation coefficient μ is in fact the sum of the coefficients for various types of x-ray interactions. For the range of x-ray energies employed in medical imaging, two kinds of interactions predominate: the photoelectric effect, described by the linear attenuation coefficient τ, and Compton scattering described by the linear attenuation coefficient σ. Thus $\mu = \tau + \sigma$. Figure 6.2-19 shows the x-ray energy dependencies of these coefficients in human soft tissue. The photoelectric coefficient increases with atomic number Z like Z^3, principally because x rays interact photoelectrically with

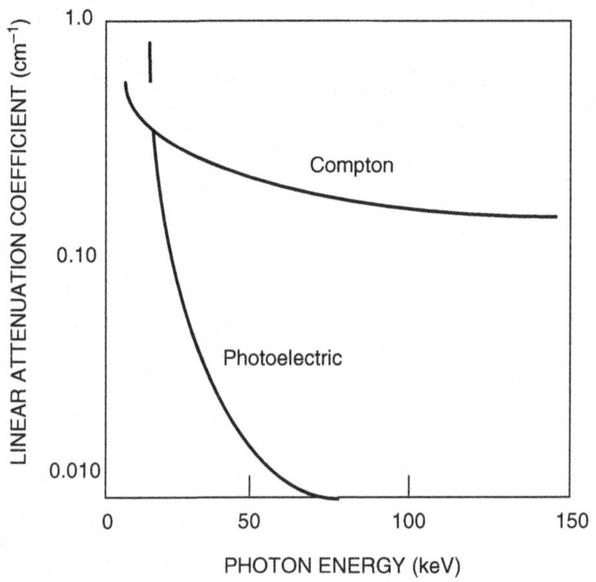

Figure 6.2-19 Linear attenuation coefficient for x rays traversing human soft tissue is the sum of two dominating contributions.

the inner, tightly bound electrons of an atom. The Compton coefficient, by contrast, is relatively independent of the atomic number of the tissue atoms, because x rays Compton-scatter almost exclusively off the outer, loosely bound atomic electrons. Both coefficients increase linearly with the tissue density. To achieve contrast between soft tissues that differ only slightly in Z, one must use low-energy x rays, because they interact predominantly by the photoelectric effect. An example is in mammography, which employs x rays in the range of 15–30 keV. For chest x-ray images, which involve tissue of greater intrinsic contrast, clinicians use x rays with energies ranging from 50 to 150 keV.

X-ray images represent a combination of four kinds of resolution: spatial, contrast, temporal, and statistical. Improving any one of these resolution factors degrades one or more of the others. The compromise among them represents a balance such that no single factor dominates the degradation of the image.

6.2.5.1 Sectional imaging

A limitation of conventional planar x-ray imaging is the projection of a three-dimensional distribution of attenuation coefficients as a shadow onto a two-dimensional detector. This type of projection discards a lot of information about tissue variation along the beam direction. For many years, techniques of analog tomography were used in attempts to overcome this limitation, but they were restricted to certain applications, and images were difficult to interpret.

An important development was achieved in 1972 with the introduction of x-ray transmission computed tomography (CAT). This technique was brought to clinical medicine through the effort of Godfrey Hounsfield and Allan Cormack, who shared the 1979 Nobel Prize in medicine.

Consider a highly collimated x-ray pencil beam in the plane of a slice of the body only a few millimeters thick. X rays transmitted all the way through the slice are measured with a collimated detector on the opposite side of the patient. The signal from the x-ray detector is converted to digital output. The tight collimation of source and detector prevents scattered radiation from degrading image contrast. The number of x-ray photons recorded by the detector at one position constitutes a single pencil-beam projection of x-ray transmission data at a specific angle through the tissue slice. This process is repeated many times at slightly different angles to create a set of multiple projections of the entire tissue slice (Figure 6.2-20).

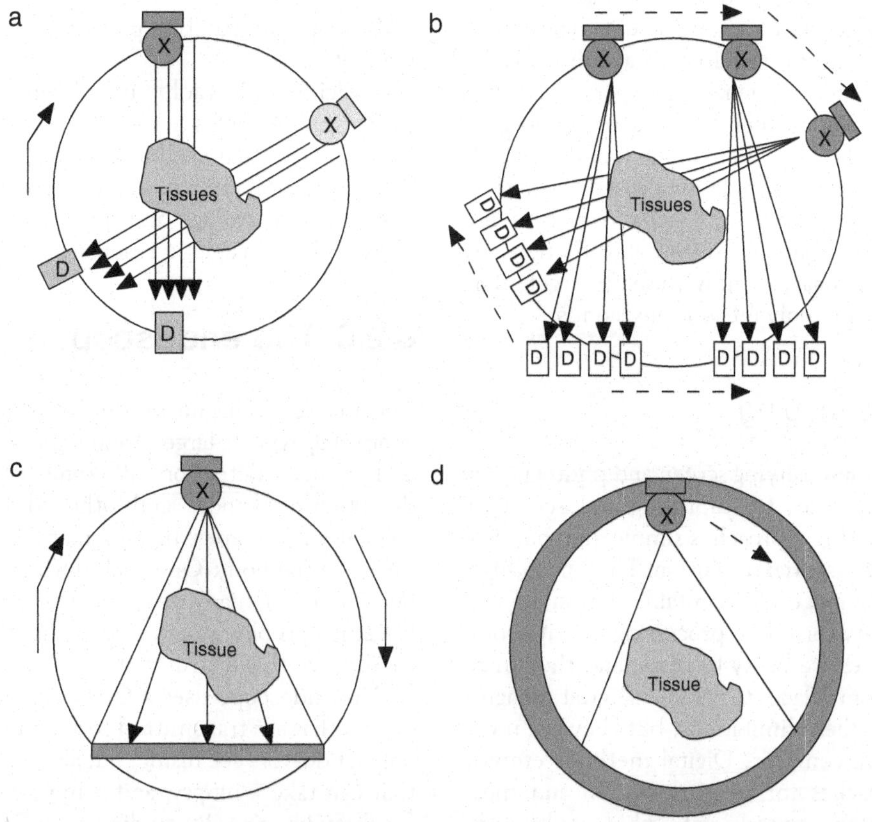

Figure 6.2-20 Evolution of geometries of x-ray CT scanners.

If the x-ray projection data are collected at a sufficient number of angles, a matrix of values of the attenuation coefficient μ for different $\delta x \delta y$ cells can be calculated by a simple back-projection technique, which yields the two-dimensional distribution $\mu(x, y)$ over the whole tissue slice. In displaying the variation of the attenuation coefficient pictorially in shades of gray, one creates an image that shows the various anatomical features of the tissue slice. In practice, the back projection is calculated by Fourier transforming the projection data into (spatial) frequency space.

The first type of CT scanners used a single collimated x-ray source and two detectors for data to be collected from two contiguous tissue slices. The source and each detector mapped out projection data in a translate-rotate geometry, one x-ray path at a time. The efforts to collect the data faster soon led to the successive generation of fan-beam scanners shown in Figures 6.2-20b, 6.2-20c, and 6.2-20d. Figure 6.2-20b shows an array of detectors that move in a translate-rotate configuration; Figure 6.2-20c shows a bank of detectors that move in a purely rotational geometry; finally Figure 6.2-20d shows a ring array of stationary detectors. Using fan-beam geometries, a whole tissue slice can be imaged in a few seconds. By combining these geometries with the patient gantry moving continuously along its long axis, one can get cross-sectional images of many slices of the patient in minimum time, a procedure called spiral scanning.

Although the evolution of x-ray scanning geometries greatly shortened the time required to acquire images, none of the three fan-beam generations permitted data acquisition in a time (less than 0.1 s) short enough to capture images of the heart and other blood-perfused organs without significant degradation caused by motion. Fifth-generation scanners employ an electron-beam gun that generates x-ray beams in different directions by scanning over a stationary concave metal target. The resulting scan times are only a few milliseconds.

6.2.5.2 Digital imaging

The combination of intensifying screen and photographic film has many advantages for capturing and recording many x-ray images. This approach is simple, inexpensive, and it yields excellent spatial resolution. This approach is limited to a narrow range of acceptable exposures and offers little flexibility for image processing or data compression. Film images are bulky to store, and they must be transported from place to place. Digital imaging methods overcome these limitations, but they are more expensive and more complex. Digital methods employ a variety of approaches for x-ray detection and measurements: fluorescent crystals with photomultipliers, semiconductor detectors, channel electron multipliers, and photostimulatable phosphors with laser-scanning readout.

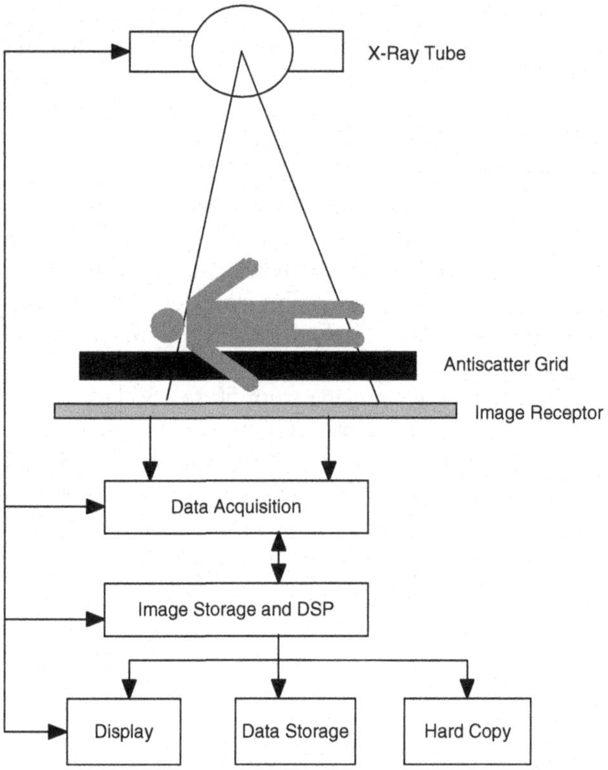

Figure 6.2-21 Digital radiography system.

A digital radiographic unit (Figure 6.2-21), the x-ray source, and receptor are computer controlled to provide digital images that can be displayed in real time on video screens. Digital images can be stored on magnetic media. Digital image storage and display are used routinely in x-ray CT and magnetic resonance imaging.

6.2.6 The endoscope

The future of diagnostic devices using photonics looks promising with a large amount of research focusing on early cancer detection. Photonic components for the detection of cancer and other diseases keep getting smaller and cheaper, making many of these instruments fit inside the body. One such device, the endoscope, has been used for many years to diagnose and treat gastrointestinal problems, and research is still being performed to improve this instrument.

The endoscope uses a CCD or optical fibers to form images that are transmitted to a monitor. Endoscopes not only let doctors see inside, but also include an instrument that can take a biopsy, and some can even dye the area for x-ray imaging. By guiding an endoscope into the gastrointestinal tract, doctors can view lesions or sources of

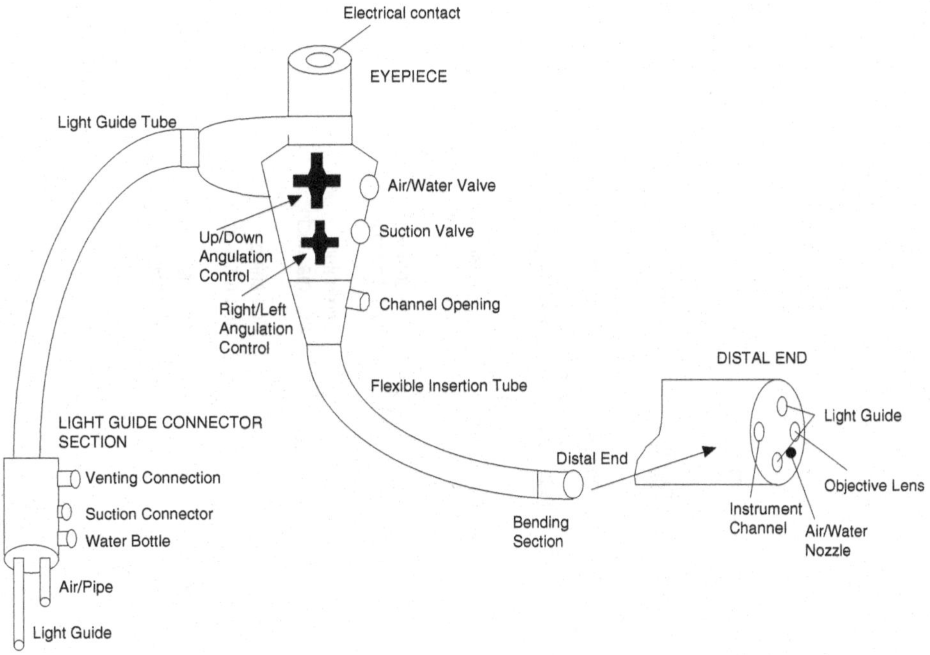

Figure 6.2-22 Flexible endoscope for looking at gastrointestinal tract without surgery.

bleeding. Endoscopes for this purpose cost around $15,000 and typically have lifetimes of about 3000 uses. The endoscope can improve its image quality and get smaller to give the patient more comfort. Furthermore, the methods of cleaning and sterilizing the reusable instruments are complicated and time consuming. Attempts to make semidisposable endoscopes have not had great success. Patients sometimes avoid the uncomfortable procedure because it requires sedation and/or local anesthetic. In addition, endoscopes cannot reach all parts of the small intestine. However, making endoscopes that can go wireless would allow doctors to take endoscopes where they could not go before and could make patients more comfortable. An endoscope is illustrated in Figure 6.2-22.

6.2.7 Digital x rays

Breast cancer remains a leading cause of cancer deaths in the world (50,000 + a year in the United States alone). Better diagnosis and treatment of breast cancer have noticeably improved the outcome of the disease, reducing death rates over the last decade by about 2% per year in the United States. A great deal of that success is due to earlier detection by the standard breast imaging technology, film-screen x-ray mammography.

The variety and sophistication of imaging technologies have increased greatly, encompassing everything from optical laser imaging to digital mammography. While these new technologies (see Table 6.2-5) have shown impressive results, a report released by the Institute of Medicine concludes that such technologies will only play a supporting role to film-screen mammography. Film mammography is the gold standard for screening for breast cancer and the technology against which all other technologies will be benchmarked.

Presently, abnormalities and lesions are discovered either by physical examination or by screening mammography, a task performed by a radiologist. Once identified, the abnormality must be diagnosed as benign or malignant by using other imaging technologies such as ultrasound or a biopsy and microscopic examination. The true tumors are biochemically characterized and categorized (staged) according to size and how much they have spread. The system is not flawless. It misses up to 20% of the tumors and many of those are found later to be benign.

Screening tools have to be highly sensitive, identifying as correctly as possible those tumors that could be malignant. Diagnostic tools must have a great specificity in order to really catch those tumors that are malignant.

Digital mammography (Figure 6.2-23) is the new technology most certain to see clinical use. With its high spatial resolution, mammography requires very small pixels and a high signal-to-noise ratio. The digital version of the technology has superior dynamic range and linearity compared to film, leading to a much better contrast resolution. It also allows the images to be manipulated and analyzed with software. This approach may lead to the discovery of more subtle features indicative of cancer

Table 6.2-5 Imaging technologies for breast cancer

Technology	Description	FDA approved	Used routinely	Used frequently	Clinical data suggest a role	Preclinical data suggest a role	Data not available
Film-screen mammography	The standard x-ray technique	Yes	Screening and diagnosis				
Digital mammography	Digital version of x-ray technique	Yes		Screening and diagnosis			
Ultrasound	Forms images by reflection of megahertz frequency sound waves	Yes	Diagnosis		Screening		
MRI	Forms images from radio emissions from nuclear spins	Yes		Diagnosis	Screening		
Scintimammography	Senses tumors from gamma ray emissions of radioactive pharmaceuticals	Yes			Diagnosis	Screening	
Thermography	Seeks tumors by infrared signature	Yes			Diagnosis	Screening	
Optical imaging	Localizes tumors by measuring scattered near infrared light				Diagnosis	Screening	
Electrical potential measurement	Identifies tumors by measuring potential at array of detectors on skin				Diagnosis	Screening	
Positron emission tomography	Forms images using emissions from annihilation of positrons from radioactive pharmaceuticals	Yes				Screening and diagnosis	
Novel ultrasound techniques	Include compound imaging, which improves resolution using sound waves from several angles; 3D and Doppler imaging					Screening and diagnosis	
Elastography	Uses ultrasound or MRI to infer the mechanical properties of tissue					Screening and diagnosis	
Magnetic resonance spectroscopy	Analyzes tissue's chemical makeup using radio emissions from nuclear spins					Screening and diagnosis	
Thermoacoustic computed tomography	Generates short sound pulses within breast using RF energy and constructs a 3D image from them						Screening and diagnosis
Microwave imaging	Views breast using scattered microwaves						Screening and diagnosis
Hall-effect imaging	Picks up sonic vibrations of charged particles exposed to a magnetic field						Screening and diagnosis
Magnetomammography	Senses magnetic contrast agents collected in tumors						Screening and diagnosis

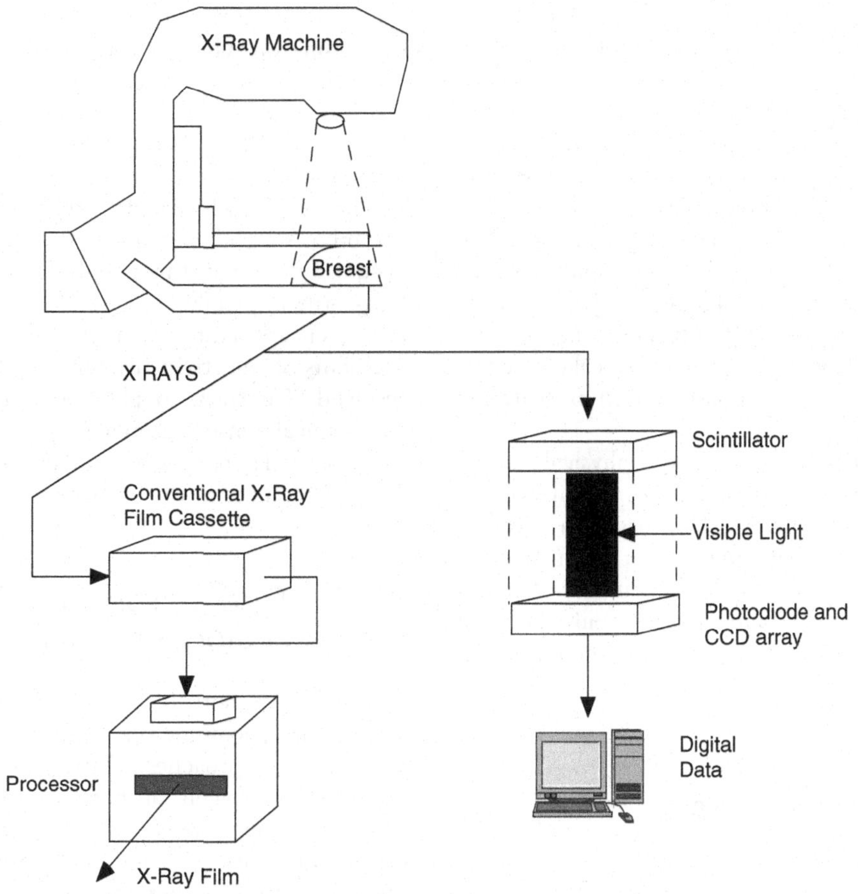

Figure 6.2-23 How digital mammography works.

and to a great ability to distinguish between potential cancers and harmless tissue abnormalities. Digital mammography may not make a difference in the numbers of cancers that are detected when compared with film mammography. However, it is also possible that digital mammography may decrease the number of follow–up tests needed for proper screening and diagnosis.

Digital mammography can employ either indirect detection, or scintillator x-ray conversion, or direct detection, or electronic x-ray conversion. In indirect detection, x rays are converted into visible light and then picked up on a solid-state detector. Direct detection, a more advanced technique aiming for higher resolution, converts x rays directly into electric signals. In the indirect detection, x rays pass through the breast and then strike a scintillator, a material that absorbs the x rays and emits visible light. The scintillator is coupled to a photodetector array or connected to tiles of CCD by tapered optical fibers. Cesium iodide can be used as a scintillator. An important design consideration in making indirect detectors is choosing the thickness of the scintillator. A thicker scintillator captures more x rays, but also leads to a loss of resolution because the photons have more chance to scatter before reaching the CCD. Direct detection omits the conversion of x rays to visible light, thus removing the loss from scattering of the light.

Clinical digital mammography provides digital files, rather than stacks of films. This allows the transfer of information more easily. Image enhancements and other software techniques will add enough functionalities for radiologists to make it worthwhile. Computer-aided detection (CAD) systems are software systems that can identify potential cancers on digital mammography images. These software systems are entering clinical practice as a way to improve radiologists' ability to detect the few cancer cases in the sea of normal-looking images they observe everyday. In a typical mammography day, a radiologist uncovers around six cancers for every 1000 images observed. Some cancers are overlooked. Therefore, allowing the easy distribution of digital images among several radiologists can significantly reduce the chances of some cancers being missed. The use of CAD software has the peculiarity that the software has to be "trained" for a particular machine to recognize tumor sizes and types.

Other well-established imaging technologies such as MRI, positron emission tomography (PET), and ultrasound can assist digital mammography in the diagnosis problem. For example, physicians frequently use ultrasound to help determine whether a lesion detected on a mammogram is a malignant mass or a harmless cyst. However, many of these techniques have real limitations. MRI specificity is highly variable, ranging from 28 to 100 percent, depending on the interpretation technique used and the patient population; PET scanners are expensive and scarce, and ultrasound has trouble seeing microcalcifications because of so-called speckle—tiny bright flecks on the image caused by scattered echoes.

The imaging principle relies on the physical characteristics of a cancer, such as its relative opacity to different wavelengths of light. This approach may need to be helped by combining the imaging approach with the usage of markers that exploit the biochemistry of cancer (e.g., compounds that will bind to tumors only) and that are picked up by imaging when such compounds react to light.

6.2.8 Medical sensors from fiber optics

Fiber optics is beginning to have great use in healthcare for intracavity imaging and safe laser delivery. It is also becoming greatly useful in monitoring physiological functions. The optical fiber technology in healthcare provides many advantages.

1. The fibers are very small and flexible and they can be inserted inside very thin catheters and hypodermic needles. These are highly noninvasive techniques useful for monitoring.
2. Fibers are nontoxic, chemically inert, and very safe for patients. They are practically immune to electromagnetic interference from other electronic sources, which is of great importance for the patients.
3. Because of interference immunity there is no crosstalk between neighboring fibers, which allows the usage of several sensors grouped together in a single catheter.

The ideal fiber-optic sensors for medical applications must have the following properties: (1) reliability, (2) automated operations, and (3) simple implantation and low-cost maintenance. The fiber-optic sensors in medical applications are based on sensing processes that are typical of other types of fiber sensors, especially in intensity and time domain modulation. Mostly, they act as point sensors, either intrinsic or extrinsic, depending on whether the modulation is produced in the fiber sensor itself, or whether there is an external transducer connected to the fiber. They can also be used as spectral sensors, where the fibers act simply as light guiders.

The design of fiber-optic probes must be such that no thrombosis or inflammatory effects of the blood vessels occurs. Of course, this is only for temporary use since long-term implantable sensors are not feasible. Proper probe encapsulation is a crucial factor, so potential problems of selectivity, hysteresis, and stability must be resolved. Contrary to physical sensors, chemical and biochemical sensors cannot usually be hermetically encapsulated. There may be problems of interference with other substances in the environments.

6.2.8.1 Fiber optics for circulatory and respiratory systems

Fiber-optic sensors in the circulatory and respiratory systems are becoming very popular in both invasive and noninvasive approaches. Optical fibers can be used to measure the oxygen saturation in the blood. This measurement is necessary to monitor the cardiovascular and cardiopulmonary systems. The simplest method of this measurement is either through reflected or absorbed light, which is collected at two different wavelengths and the oxygen saturation is calculated on the basis of isosbestic regions of Hb (hemoglobin) and OxyHb (oxygenated hemoglobin) absorption.

Optical oximeters, which calculate oxygen saturation via the light transmitted to the earlobes, toes, and fingertips, have been developed primarily for neonatal care. A difficulty in neonatal care is the importance of differentiating between the light absorption due to arterial blood and that due to all other tissues and blood in the light path, which implies the use of multiple wavelengths. This challenge can be met using an oximeter. This approach is based on the assumption that a change in the light absorbed by the tissue during systole is caused by the passage of arterial blood. Using two wavelengths, it is possible to noninvasively measure the oxygen saturation by using the pulsatile rather than the absolute level of transmitted or reflected light intensity. The detection of blood absorbance fluctuations that are synchronous with systolic heart contractions is called photoplethysmography.

Blood gases and blood pH

Monitoring of the blood pH and blood oxygen (pO_2) and carbon dioxide (pCO_2) partial pressures to determine

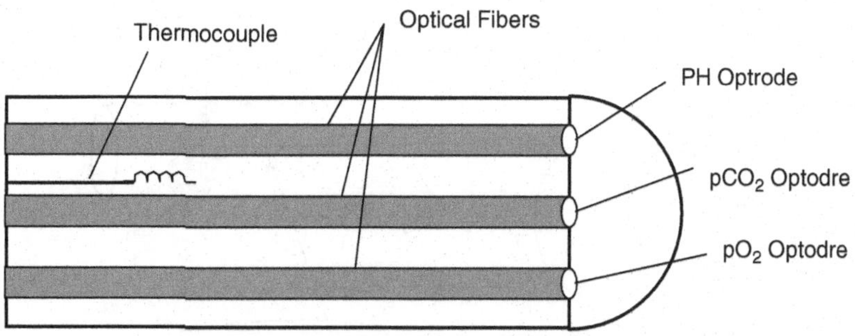

Figure 6.2-24 Fiber-optic vascular sensor for measuring blood oxygen and carbon dioxide.

the quantity of oxygen delivered to the tissues and the quality of the perfusion is of great importance. The pH is detected by a chromophore, which changes its optical spectrum as a function of the pH. The absorption-based indicators or fluorophores are usually used. The carbon dioxide is detected indirectly, since its diffusion in a carbonate solution fixed at the optical fiber tips alters the pH, which means that the CO_2 content can be determined by measuring the pH. Oxygen is measured using a separate chemical transducer; oxygen is detected via fluorescence techniques that exploit the quenching produced by oxygen on fluorophores.

The first intravascular sensor for simultaneous and continuous monitoring of the pH, pO_2, and pCO_2 is shown in Figure 6.2-24. It is composed of three optical fibers encapsulated in a polymer enclosure, including also a thermocouple used for temperature monitoring. Three fluorescent indicators are used as chromophores. The pH and pCO_2 are measured by the same fluorophore. The optoelectronics is composed of three modules, one for each sensor. A suitable filtered xenon lamp, when modulated, provides illumination for the modules. The source of light is focused through a lens onto a prism beam splitter and coupled through fiber to the sensor tip. The deflected light is collected by a reference detector for source control. The returning fluorescence is deflected by the prism beam splitter onto the signal detector.

There have been problems concerning the usage of intravascular fiberoptic sensors during clinical trials on volunteers in critical care and on surgical patients and these remain unresolved.

1. blood flow decreases due to peripheral vasoconstriction lasting for several hours after surgical operations, which can give rise to a contamination by flush solutions
2. the wall effect, which affects the oxygen count
3. the formation of a clot around the sensor tip, which can alter the value of all the analytical values.

Respiratory monitoring

In intensive care units there is a need to continuously monitor breathing condition. It is possible to accomplish this monitoring from the nurse's station in such a way that patients can be kept under observation without the need for the nurses to be physically present. An optical fiber with a moisture-sensitive cladding can be developed for that purpose. The cladding is a plastic film doped with umbelliferous dye, which is a moisture-sensitive fluorescent material when pumped with UV light. The sensitive fiber section is placed over the patient's mouth and excited with He–Cd laser on a halogen lamp. Since the water vapor content in the human exhalation exceeds that in the room, the patient's exhalation produces a fluorescent signal that is detected by an electro-optical unit at the nurse's station. This monitoring is very useful in detecting abnormal breathing in patients.

Angiology

Blood vessels that are obstructed by atherosclerotic plaques can be recanalized by means of a pulse excimer laser radiation, guided by an optical fiber (laser angioplasty). Despite its widespread uses there is the possibility of blood vessel perforation, which occurs in between 20% and 40% of patients. To minimize the risk there is the usage of laser-induced fluorescence diagnosis of the vessel wall. This is part of identifying the target under irradiation. An all-optical approach to target identification is suggested by the fact that short optical pulses, when absorbed by the tissue, generate ultrasonic thermoelastic waves. The amplitude and temporal characteristics of the acoustic signal are dependent on the target composition and can be detected by a pressure fiber-optics system. In this approach an optical fiber tipped with a Fabry–Pérot cavity is inserted in the lumbar of the artery, and the sensor is in contact with the tissue. As shown in Figure 6.2-25, two signals are guided by the optical fiber: (1) pulse light of a Nd:YAG laser used to generate the thermoelastic wave in the tissue and (2) low-power continuous wave light of a tunable laser

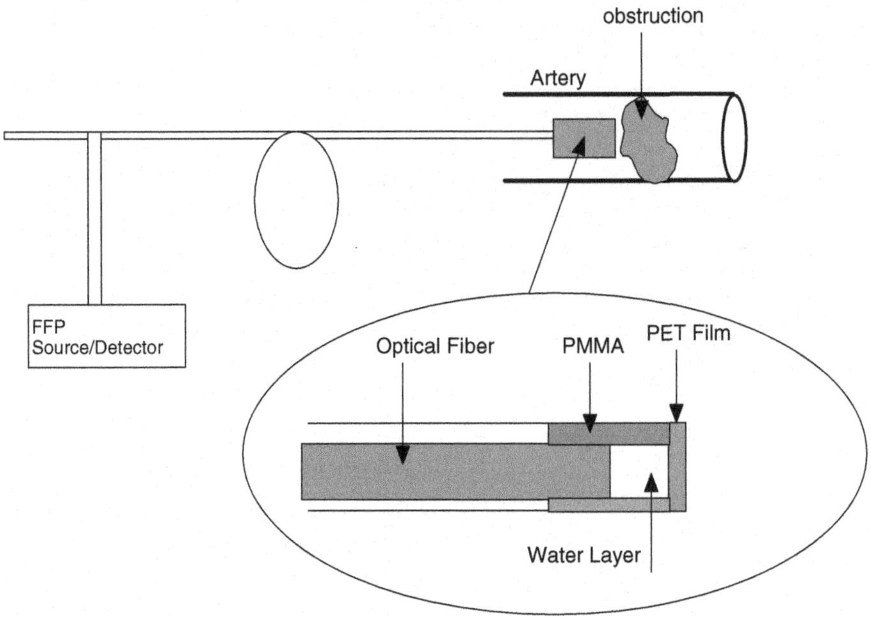

Figure 6.2-25 Angiology instrument using optical fibers.

diode used for sensor interrogation. The Fabry–Pérot cavity of the fiber tip is formed by a polyethylene-terephthalate film and the fiber end provides a fiber–polymer acoustic impedance match. As the polymer film is in contact with the tissue, the stress due to the thermoelastic wave modulates its thickness and, hence, the optical phase difference between the interfering Fresnel reflections from both sides of the film.

When the sensor head is coaxially positioned with the delivery fiber, measurements are taken from the center of the acoustic source, giving improved targeting accuracy. This kind of photoacoustic spectroscopy has been experimentally tested on postmortem human aortas.

In addition to disturbing central circulation, cardiovascular diseases may influence the peripheral circulation by affecting the microvascular perfusion in tissue. Insufficient peripheral circulation may produce chemical gastric discomfort and ulcers. The best method for assessment of microvascular perfusion is laser Doppler flow monitoring in which the use of optical fibers can improve the possibilities of both invasive and contact measurements.

The basic concept of fiber-optic laser Doppler flow is shown in Figure 6.2-26. Light from the He-Ne laser is guided by an optical fiber to the tissue or vascular network being studied. The light is diffusely scattered and partially absorbed within the illuminated volume. Light hitting a moving blood cell undergoes a small Doppler shift due to the scattering particles.

Gastroenterology

The need for fiber optic systems to monitor *in vivo* the functional aspect of the foregut is increasing. An important parameter when studying the human foregut is the gastric and esophageal pH. Monitoring gastric pH for long periods serves to analyze the physiological pattern of acidity. It provides information regarding changes in the course of the peptic ulcer and enables assessment of the effect of gastric antisecretory drugs. In the esophagus, the gastroesophageal reflux, which causes a pH decrease in the contents from 7 to 2, can determine esophagitis with possible strictures and Barrett's esophagus, which is considered a pheneoplastic lesion. In addition, in measuring the bile-containing reflux, the bile and pH should be measured simultaneously.

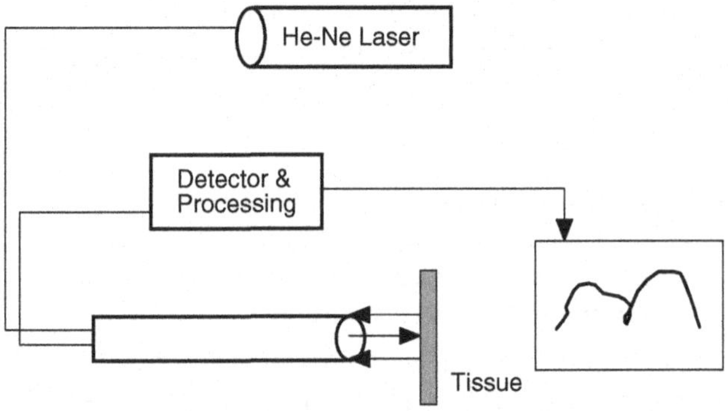

Figure 6.2-26 The fiber-optic Doppler flow meter.

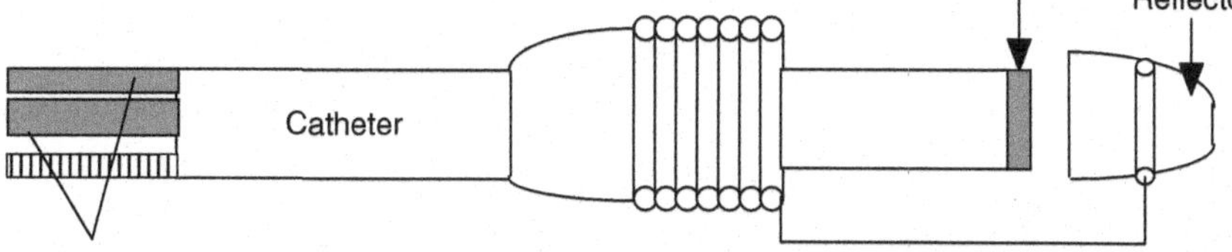

Figure 6.2-27 Gastroenterology usage of fiber optics.

The sensor shown in Figure 6.2-27 uses two dyes. Chromophores, immobilized on controlled pore glasses, are fixed at the end of plastic optical fibers. The distal end of the fibers is heated and the CPG forms a very thin pH-sensitive layer on the fiber tips. The probe has four fibers (two for each chromophore). The use of LEDs as light sources, solid-state detection, and an internal microprocessor makes this a truly portable, battery-powered sensor.

References

Bruce, R. V. 1973. *Alexander Graham Bell and the Conquest of Solitude.* Gollanez, London.

Cherin, A. H. 1983. *An Introduction to Optical Fibers.* McGraw-Hill, New York.

Hall, R. N., G. E. Fenner, J. D. Kingsley, T. J. Soltys, and R. O. Carlson. 1962. "Coherent light emissions from GaAs Junction." *Phys. Rev. Lett.* 9, 366.

Kao, K. C., and G. A. Hockham. 1966. "Dielectric fiber surface waveguides for optical waveguides." *Proc. Inst. Electr. Eng.* 133, 1151.

Kapron, F. P., D. B. Keck, and R. D. Maurer. 1970. "Radiation losses in glass optical waveguides." *Appl. Phys. Lett.* 17, 423.

Maiman, T. H. 1960. "Stimulated optical radiation in ruby." *Nature* (London) 6, 106.

Morse, P. M., and H. Feshbach. 1953. *Methods of Theoretical Physics, Part II.* McGraw-Hill, New York.

Nathan, M. I., W. P. Dumke, G. Burns, F. H. Dill, Jr., and G. Lasher. 1962. "Stimulated Emissions of Radiation from GaAs p-n Junctions." *Appl. Phys. Lett.* 1, 62.

Chapter 6.3

6.3

Fundamental enhancement techniques

Raman B. Paranjape

6.3.1 Introduction

Image enhancement techniques are used to refine a given image, so that desired image features become easier to perceive for the human visual system or more likely to be detected by automated image analysis systems [1, 13]. Image enhancement allows the observer to see details in images that may not be immediately observable in the original image. This may be the case, for example, when the dynamic range of the data and that of the display are not commensurate, when the image has a high level of noise or when contrast is insufficient [4, 5, 8, 9].

Fundamentally, image enhancement is the transformation or mapping of one image to another [10, 14]. This transformation is not necessarily one-to-one, so that two different input images may transform into the same or similar output images after enhancement. More commonly, one may want to generate multiple enhanced versions of a given image. This aspect also means that enhancement techniques may be irreversible.

Often the enhancement of certain features in images is accompanied by undesirable effects. Valuable image information may be lost or the enhanced image may be a poor representation of the original. Furthermore, enhancement algorithms cannot be expected to provide information that is not present in the original image. If the image does not contain the feature to be enhanced, noise or other unwanted image components may be inadvertently enhanced without any benefit to the user.

In this chapter we present established image enhancement algorithms commonly used for medical images. Initial concepts and definitions are presented in Section 6.3.2. Pixel-based enhancement techniques described in Section 6.3.3 are transformations applied to each pixel without utilizing specifically the information in the neighborhood of the pixel. Section 6.3.4 presents enhancement with local operators that modify the value of each pixel using the pixels in a local neighborhood. Enhancement that can be achieved with multiple images of the same scene is outlined in Section 6.3.5. Spectral domain filters that can be used for enhancement are presented in Section 6.3.6. The techniques described in this chapter are applicable to dental and medical images as illustrated in the figures.

6.3.2 Preliminaries and definitions

We define a digital image as a two-dimensional array of numbers that represents the real, continuous intensity distribution of a spatial signal. The continuous spatial signal is sampled at regular intervals and the intensity is quantized to a finite number of levels. Each element of the array is referred to as a picture element or pixel. The digital image is defined as a spatially distributed intensity signal $f(m, n)$, where f is the intensity of the pixel, and m and n define the position of the pixel along a pair of orthogonal axes usually defined as horizontal and vertical. We shall assume that the image has M rows and N columns and that the digital image has P quantized levels of intensity (gray levels) with values ranging from 0 to $P - 1$.

The histogram of an image, commonly used in image enhancement and image characterization, is defined as a vector that contains the count of the number of pixels in the image at each gray level. The histogram, $h(i)$, can be defined as

$$h(i) = \sum_{m=0}^{M-1} \sum_{n=0}^{N-1} \delta(f(m,n) - i), \quad i = 0, 1, \ldots, p - 1,$$

Biomedical Engineering Desk Reference; ISBN: 9780123746467

where

$$\delta(w) = \begin{cases} 1 & w = 0, \\ 0 & \text{otherwise.} \end{cases}$$

A useful image enhancement operation is convolution using local operators, also known as kernels. Considering a kernel $w(k, l)$ to be an array of $(2K + 1 \times 2L + 1)$ coefficients where the point $(k, l) = (0,0)$ is the center of the kernel, convolution of the image with the kernel is defined by:

$$\begin{aligned} g(m,n) &= w(k,l) * f(m,n) \\ &= \sum_{k=-K}^{K} \sum_{l=-L}^{L} w(k,l) \cdot f(m-k, n-l), \end{aligned}$$

where $g(m, n)$ is the outcome of the convolution or output image. To convolve an image with a kernel, the kernel is centered on an image pixel (m, n), the point-by-point products of the kernel coefficients and corresponding image pixels are obtained, and the subsequent summation of these products is used as the pixel value of the output image at (m, n). The complete output image $g(m, n)$ is obtained by repeating the same operation on all pixels of the original image [4, 5, 13]. A convolution kernel can be applied to an image in order to effect a specific enhancement operation or change in the image characteristics. This typically results in desirable attributes being amplified and undesirable attributes being suppressed. The specific values of the kernel coefficients depend on the different types of enhancement that may be desired.

Attention is needed at the boundaries of the image where parts of the kernel extend beyond the input image. One approach is to simply use the portion of the kernel that overlaps the input image. This approach can, however, lead to artifacts at the boundaries of the output image. In this chapter we have chosen to simply not apply the filter in parts of the input image where the kernel extends beyond the image. As a result, the output images are typically smaller than the input image by the size of the kernel.

The Fourier transform $F(u, v)$ of an image $f(m, n)$ is defined as

$$F(u,v) = \frac{1}{MN} \sum_{m=0}^{M-1} \sum_{n=0}^{N-1} f(m,n) e^{-2\pi j\left(\frac{um}{M}+\frac{vn}{N}\right)},$$
$$u = 0,1,2,\ldots,M-1, \quad v = 0,1,2,\ldots,N-1,$$

where u and v are the spatial frequency parameters. The Fourier transform provides the spectral representation of an image, which can be modified to enhance desired properties. A spatial-domain image can be obtained from a spectral-domain image with the inverse Fourier transform given by

$$f(m,n) = \sum_{u=0}^{M-1} \sum_{v=0}^{N-1} F(u,v) e^{2\pi j\left(\frac{um}{M}+\frac{vn}{N}\right)},$$
$$m = 0,1,2,\ldots,M-1, \quad n = 0,1,2,\ldots,N-1.$$

The forward or inverse Fourier transform of an $N \times N$ image, computed directly with the preceding definitions, requires a number of complex multiplications and additions proportional to N^2. By decomposing the expressions and eliminating redundancies, the fast Fourier transform (FFT) algorithm reduces the number of operations to the order of $N \log_2 N$ [5]. The computational advantage of the FFT is significant and increases with increasing N. When $N = 64$ the number of operations are reduced by an order of magnitude and when $N = 1024$, by two orders of magnitude.

6.3.3 Pixel operations

In this section we present methods of image enhancement that depend only upon the pixel gray level and do not take into account the pixel neighborhood or whole-image characteristics.

6.3.3.1 Compensation for nonlinear characteristics of display or print media

Digital images are generally displayed on cathode ray tube (CRT) type display systems or printed using some type of photographic emulsion. Most display mechanisms have nonlinear intensity characteristics that result in a nonlinear intensity profile of the image when it is observed on the display. This effect can be described succinctly by the equation

$$e(m,n) = C(f(m,n)),$$

where $f(m, n)$ is the acquired intensity image, $e(m, n)$ represents the actual intensity output by the display system, and $C()$ is a nonlinear display system operator. In order to correct for the nonlinear characteristics of the display, a transform that is the inverse of the display's nonlinearity must be applied [14, 16].

$$g(m,n) = T(e(m,n)) \cong C^{-1}(C(f(m,n)))$$
$$g(m,n) \cong f(m,n),$$

where $T()$ is a nonlinear operator which is approximately equal to $C^{-1}()$, the inverse of the display system operator, and $g(m, n)$ is the output image.

Determination of the characteristics of the nonlinearity could be difficult in practice. In general, if

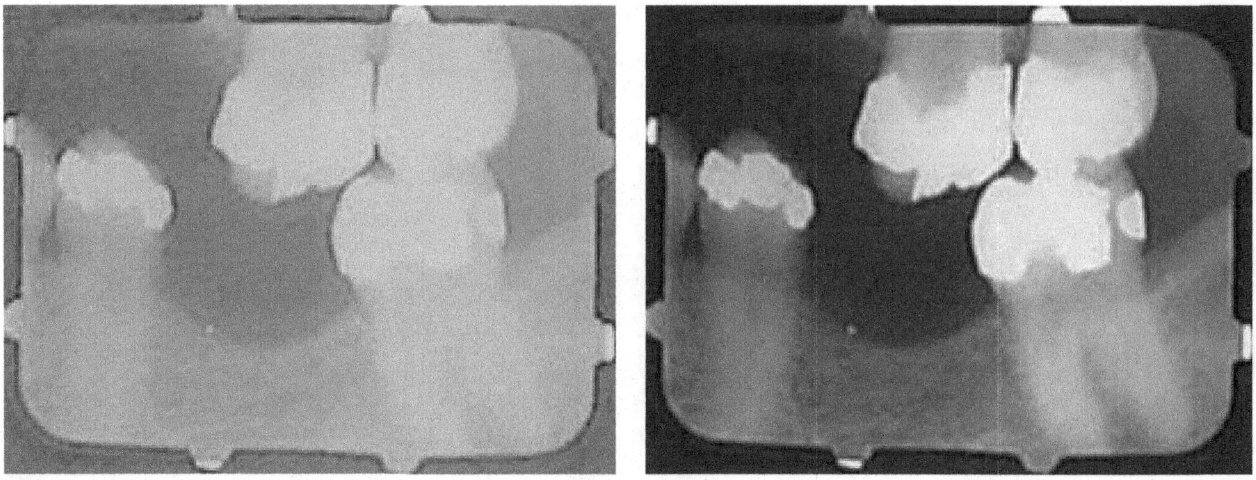

(a) (b)

Figure 6.3-1 (a) Original image as seen on a poor-quality CRT-type display. This image has poor contrast, and details are difficult to perceive—especially in the brighter parts of the image such as in areas with high tooth density or near filling material. (b) The nonlinearity of the display is reversed by the transformation, and structural details become more visible. Details within the image such as the location of amalgam, the cavity preparation liner, tooth structures, and bony structures are better visualized.

a linear intensity wedge is imaged, one can obtain a test image that captures the complete intensity scale of the image acquisition system. However, an intensity measurement device that is linear is then required to assess the output of the display system, in order to determine its actual nonlinear characteristics.

A slightly exaggerated example of this type of a transform is presented in Fig. 6.3-1. Figure 6.3-1a presents a simulated CRT display with a logarithmic characteristic. This characteristic tends to suppress the dynamic range of the image decreasing the contrast. Figure 6.3-1b presents the same image after an inverse transformation to correct for the display nonlinearity. Although these operations do in principle correct for the display, the primary mechanism for review and analysis of image information is the human visual system, which is fundamentally a nonlinear reception system and adapts locally to the intensities presented.

6.3.3.2 Intensity scaling

Intensity scaling is a method of image enhancement that can be used when the dynamic range of the acquired image data significantly exceeds the characteristics of the display system, or vice versa. It may also be the case that image information is present in specific narrow intensity bands that may be of special interest to the observer. Intensity scaling allows the observer to focus on specific intensity bands in the image by modifying the image such that the intensity band of interest spans the dynamic range of the display [14, 16]. For example, if f_1 and f_2 are known to define the intensity band of interest, a scaling transformation may be defined as

$$e = \begin{cases} f & f_1 \le f \le f_2 \\ 0 & \text{otherwise} \end{cases}$$

$$g = \left\{\frac{e - f_1}{f_2 - f_1}\right\} \cdot (f_{\max}),$$

where e is an intermediate image, g is the output image, and $f_{\max}$ is the maximum intensity of the display.

These operations may be seen through the images in Fig. 6.3-2. Figure 6.3-2a presents an image with detail in the intensity band from 90 to 170 that may be of interest to, for example a gum specialist. The image, however, is displayed such that all gray levels in the range 0 to 255 are seen. Figure 6.3-2b shows the histogram of the input image and Fig. 6.3-2c presents the same image with the 90-to-170 intensity band stretched across the output band of the display. Figure 6.3-2d shows the histogram of the output image with the intensities that were initially between 90 and 170, but are now stretched over the range 0 to 255. The detail in the narrow band is now easily perceived; however, details outside the band are completely suppressed.

6.3.3.3 Histogram equalization

Although intensity scaling can be very effective in enhancing image information present in specific intensity bands, often information is not available *a priori* to identify the useful intensity bands. In such cases, it may

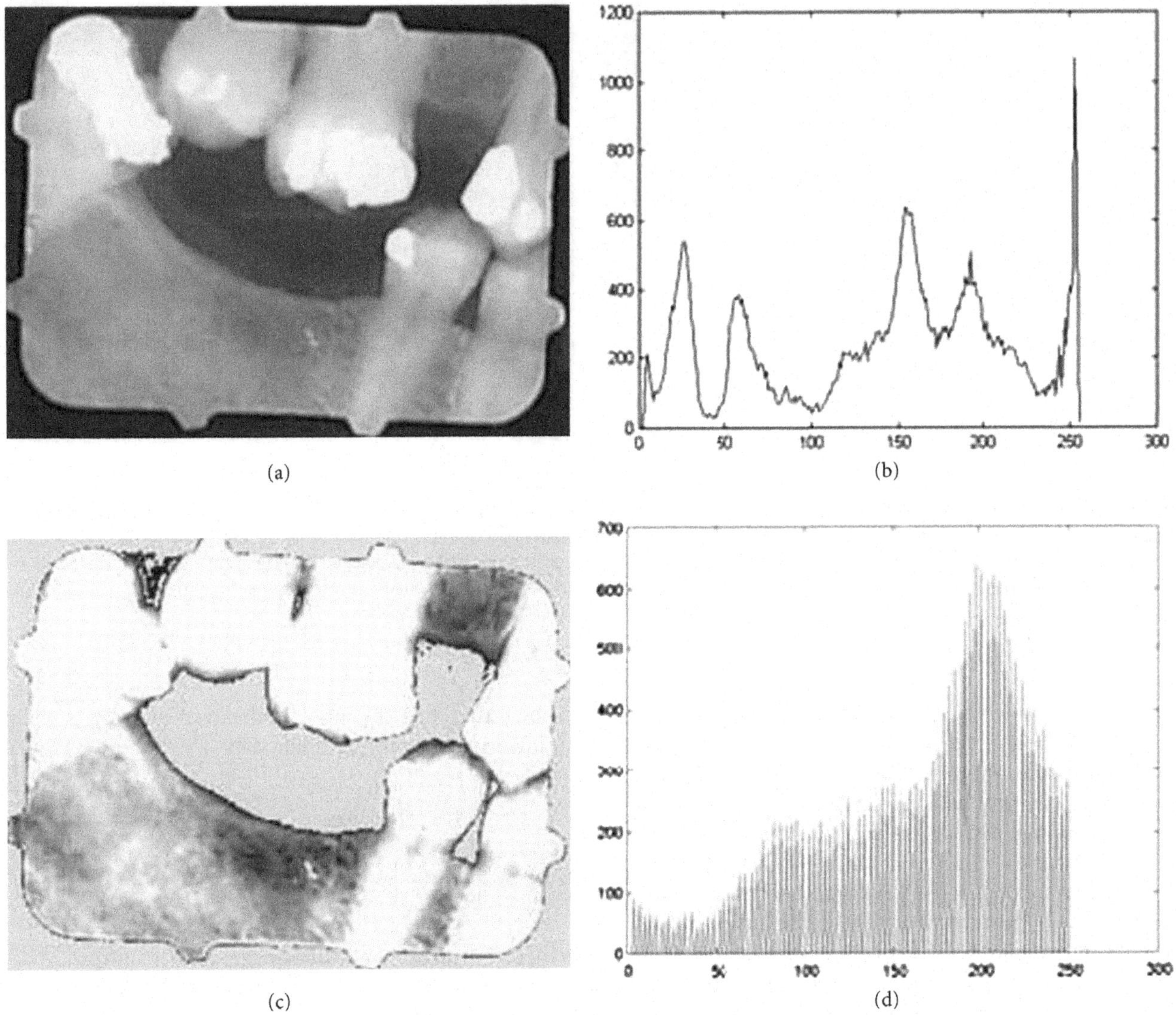

Figure 6.3-2 (a) Input image where details of interest are in the 90-to-170 gray level band. This intensity band identifies the bony structures in this image and provides an example of a feature that may be of dental interest. (b) Histogram of the input image in (a). (c) This output image selectively shows the intensity band of interest stretched over the entire dynamic range of the display. This specific enhancement may be potentially useful in highlighting features or characteristics of bony tissue in dental X-ray imagery. This technique may also be effective in focusing attention on other image features such as bony lamina dura or recurrent caries. (d) Histogram of the output image in (c). This histogram shows the gray levels in the original image in the 90-to-170 intensity band stretched over 0 to 255.

be more useful to maximize the information conveyed from the image to the user by distributing the intensity information in the image as uniformly as possible over the available intensity band [3, 6, 7]. This approach is based on an approximate realization of an information-theoretic approach in which the normalized histogram of the image is interpreted as the probability density function of the intensity of the image. In histogram equalization, the histogram of the input image is mapped to a new maximally-flat histogram.

As indicated in Section 6.3.2, the histogram is defined as $h(i)$, with 0 to $P-1$ gray levels in the image. The total number of pixels in the image, $M*N$, is also the sum of all the values in $h(i)$. Thus, in order to distribute most uniformly the intensity profile of the image, each bin of the histogram should have a pixel count of $(M*N)/P$.

It is, in general, possible to move the pixels with a given intensity to another intensity, resulting in an increase in the pixel count in the new intensity bin. On the other hand, there is no acceptable way to reduce or divide the pixel count at a specific intensity in order to reduce the pixel count to the desired $(M*N)/P$. In order to achieve approximate uniformity, the average value of the pixel count over a number of pixel values can be made close to the uniform level.

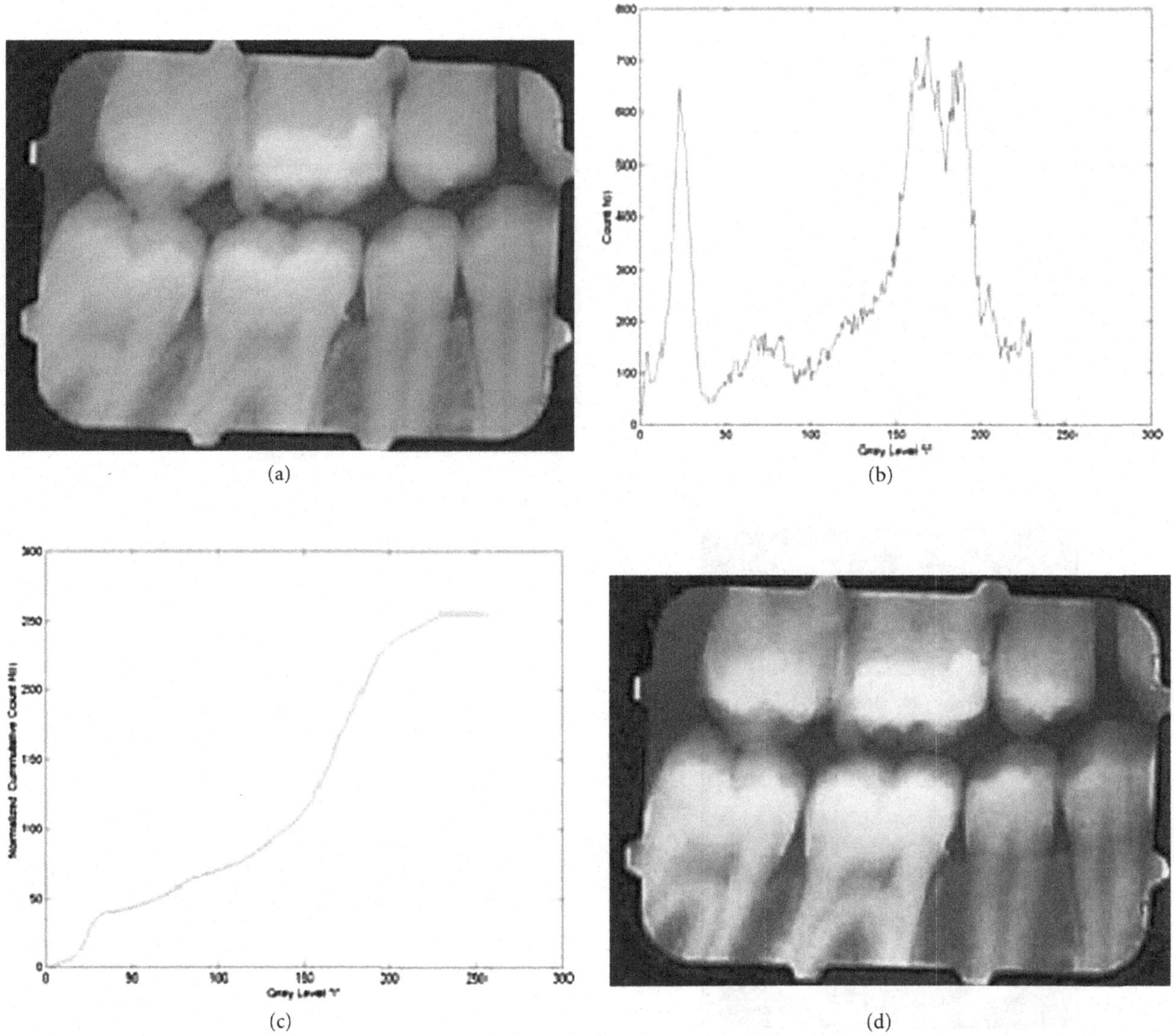

Figure 6.3-3 (a) Original image where gray levels are not uniformly distributed. Many image details are not well visualized in this image because of the low contrast. (b) Histogram of the original image in (a). Note the nonuniformity of the histogram. (c) Cumulative histogram of the original image in (a). (d) Histogram-equalized image. Contrast is enhanced so that subtle changes in intensity are more readily observable. This may allow earlier detection of pathological structures. (e) Histogram of the enhanced image in (d). Note that the distribution of intensity counts that are greater than the mean value have been distributed over a larger gray level range. (f) Cumulative histogram of the enhanced image in (d). (g) Original brain MRI image (courtesy of Dr. Christos Dzavatzikos, Johns Hopkins Radiology Department). (h) through (l) same steps as above for brain image.

A simple and readily available procedure for redistribution of the pixels in the image is based on the normalized cumulative histogram, defined as

$$H(j) = \frac{1}{M \cdot N} \sum_{i=0}^{j} h(i), \qquad j = 0, 1, \ldots, P-1.$$

The normalized cumulative histogram can be used as a mapping between the original gray levels in the image and the new gray levels required for enhancement. The enhanced image $g(m, n)$ will have a maximally uniform histogram if it is defined as

$$g(m,n) = (P-1) \cdot H(f(m,n)).$$

Figure 6.3-3a presents an original dental image where the gray levels are not uniformly distributed, while the associated histogram and cumulative histogram are shown in Figs 6.3-3b and 6.3-3c, respectively. The cumulative histogram is then used to map the gray levels of the input images to the output image shown in Fig. 6.3-3d. Figure 6.3-3e presents the histogram of Fig. 6.3-3d, and Fig. 6.3-3f shows the corresponding cumulative histogram. Figure 6.3-3f should ideally be a straight line from (0,0) to $(P-1, P-1)$, but in fact only approximates this line to the extent possible given the initial distribution of gray levels. Figure 6.3-3g

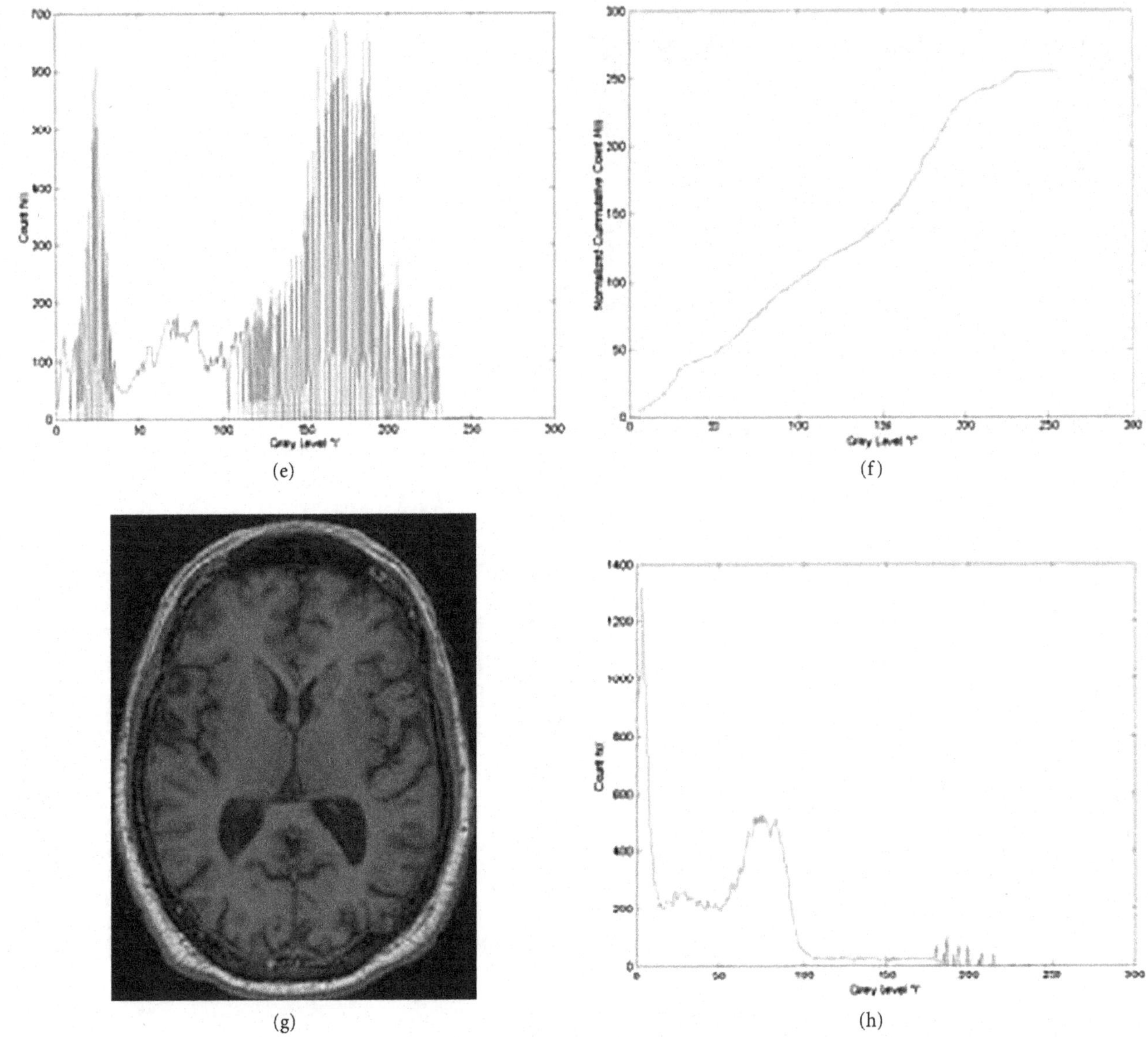

Figure 6.3-3 cont'd

through l show the enhancement of a brain MRI image with the same steps as above.

6.3.4 Local operators

Local operators enhance the image by providing a new value for each pixel in a manner that depends only on that pixel and others in a neighborhood around it. Many local operators are linear spatial filters implemented with a kernel convolution, some are nonlinear operators, and others impart histogram equalization within a neighborhood. In this section we present a set of established standard filters commonly used for enhancement. These can be easily extended to obtain slightly modified results by increasing the size of the neighborhood while maintaining the structure and function of the operator.

6.3.4.1 Noise suppression by mean filtering

Mean filtering can be achieved by convolving the image with a $(2K+1 \times 2L+1)$ kernel where each coefficient has a value equal to the reciprocal of the number of coefficients in the kernel. For example, when $L = K = 1$, we obtain

$$w(k,l) = \begin{Bmatrix} 1/9 & 1/9 & 1/9 \\ 1/9 & 1/9 & 1/9 \\ 1/9 & 1/9 & 1/9 \end{Bmatrix},$$

referred to as the 3 × 3 averaging kernel or mask. Typically, this type of smoothing reduces noise in the image,

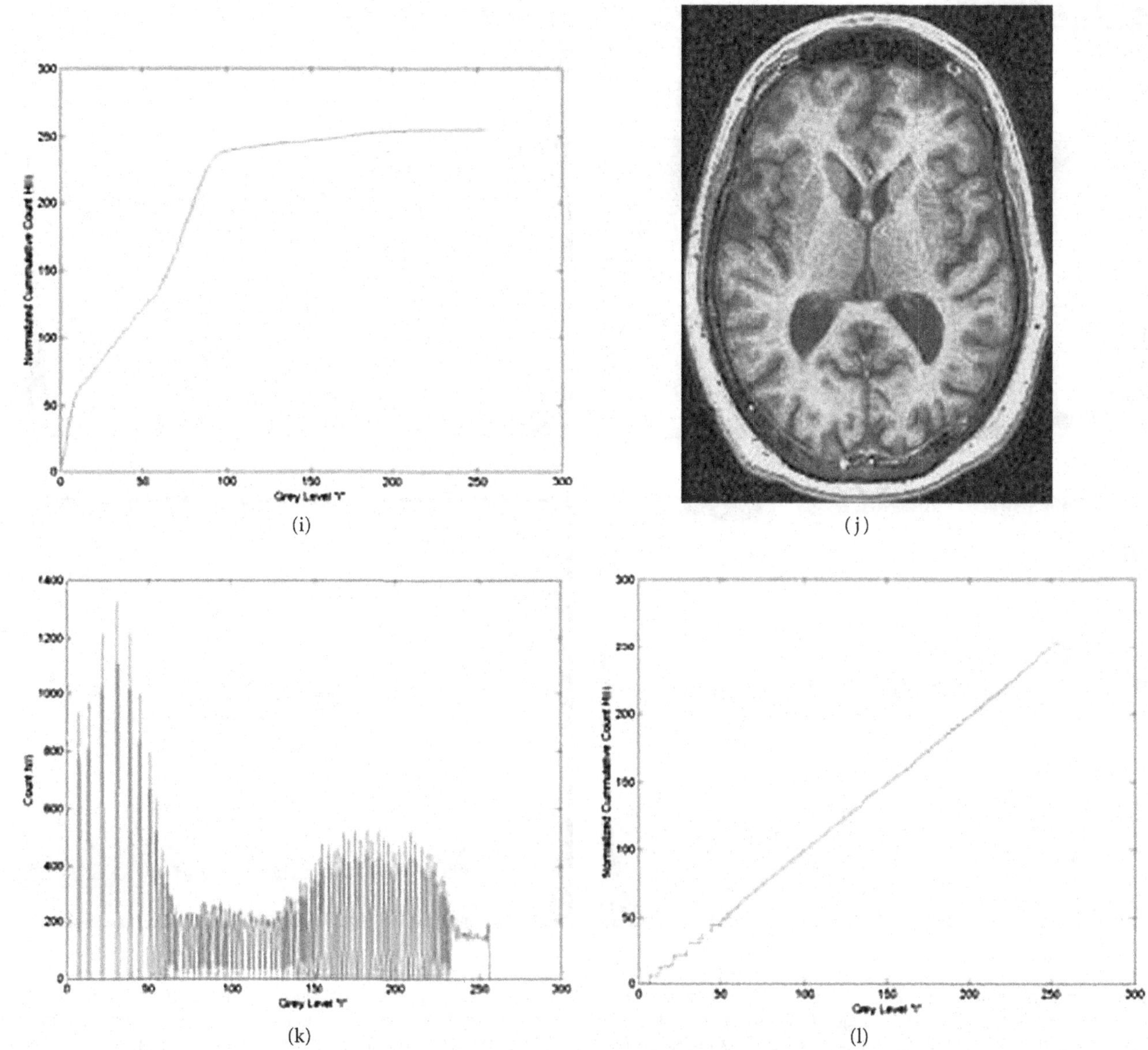

Figure 6.3-3 cont'd

but at the expense of the sharpness of edges [4, 5, 12, 13]. Examples of the application of this kernel are seen in Fig. 6.3-4a–6.3-4d. Note that the size of the kernel is a critical factor in the successful application of this type of enhancement. Image details that are small relative to the size of the kernel are significantly suppressed, while image details significantly larger than the kernel size are affected moderately. The degree of noise suppression is related to the size of the kernel, with greater suppression achieved by larger kernels.

6.3.4.2 Noise suppression by median filtering

Median filtering is a common nonlinear method for noise suppression that has unique characteristics. It does not use convolution to process the image with a kernel of coefficients. Rather, in each position of the kernel frame, a pixel of the input image contained in the frame is selected to become the output pixel located at the coordinates of the kernel center. The kernel frame is centered on each pixel (m, n) of the original image, and the median value of pixels within the kernel frame is computed. The pixel at the coordinates (m, n) of the output image is set to this median value. In general, median filters do not have the same smoothing characteristics as the mean filter [4, 5, 8, 9, 15]. Features that are smaller than half the size of the median filter kernel are completely removed by the filter. Large discontinuities such as edges and large changes in image intensity are not affected in terms of gray level intensity by the median filter, although their positions may be shifted by a few pixels. This

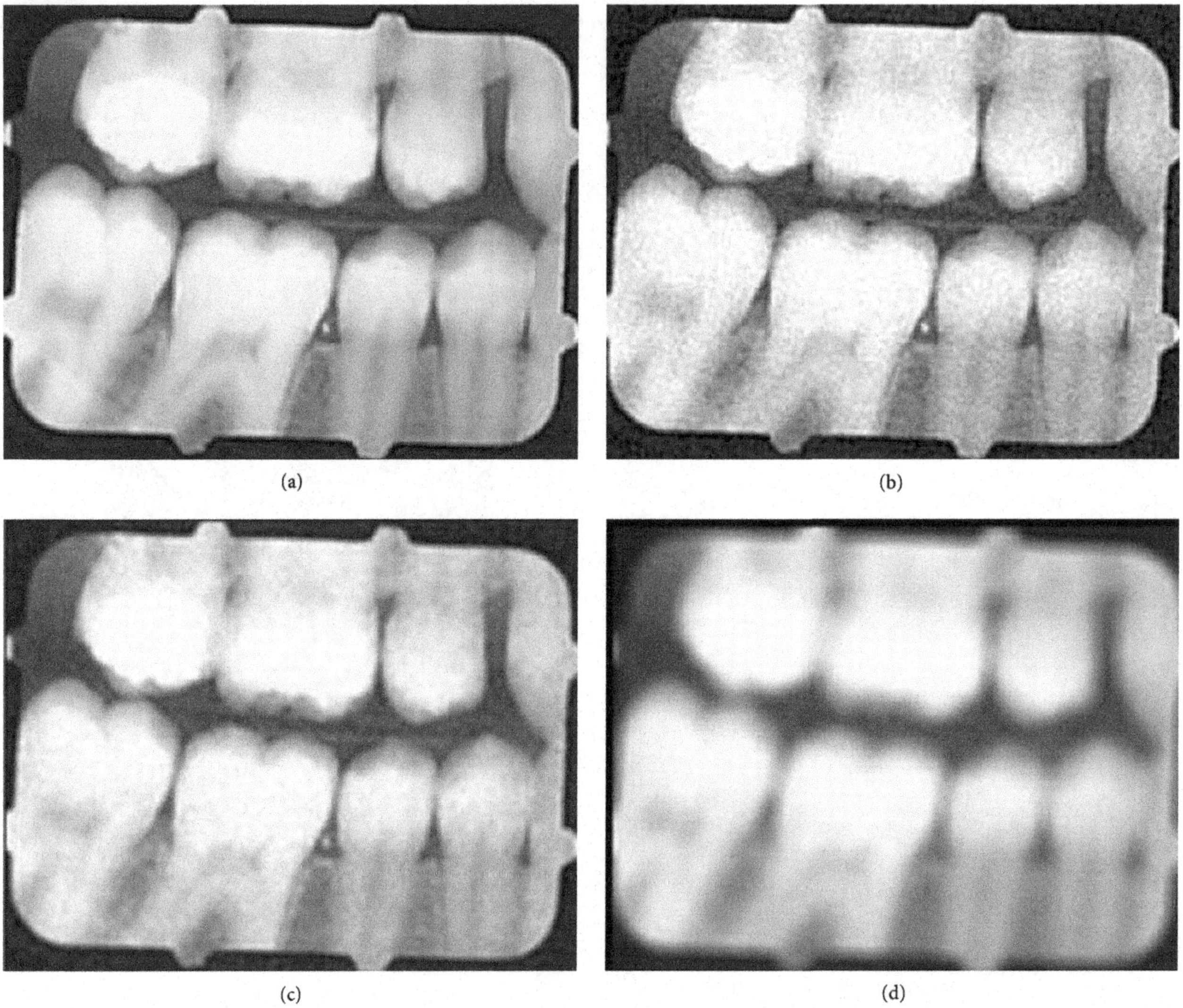

Figure 6.3-4 (a) Original bitewing X-ray image. (b) Original image in (a) corrupted by added Gaussian white noise with maximum amplitude of ±25 gray levels. (c) Image in (b) convolved with the 3 × 3 mean filter. The mean filter clearly removes some of the additive noise; however, significant blurring also occurs. This image would not have significant clinical value. (d) Image in (b) convolved with the 9 × 9 mean filter. This filter has removed almost all of the effects of the additive noise. However, the usefulness of this filter is limited because the filter size is similar to that of significant structures within the image, causing severe blurring.

nonlinear operation of the median filter allows significant reduction of specific types of noise. For example, "shot noise" may be removed completely from an image without attenuation of significant edges or image characteristics. Figure 6.3-5 presents typical results of median filtering.

6.3.4.3 Edge enhancement

Edge enhancement in images is of unique importance because the human visual system uses edges as a key factor in the comprehension of the contents of an image [2, 4, 5, 10, 13, 14]. Edges in different orientations can be selectively identified and enhanced. The edge-enhanced images may be combined with the original image in order to preserve the context.

Horizontal edges and lines are enhanced with

$$w_{H1}(k,l) = \begin{Bmatrix} 1 & 1 & 1 \\ 0 & 0 & 0 \\ -1 & -1 & -1 \end{Bmatrix} \quad \text{or}$$

$$w_{H2}(k,l) = \begin{Bmatrix} -1 & -1 & -1 \\ 0 & 0 & 0 \\ 1 & 1 & 1 \end{Bmatrix},$$

and vertical edges and lines are enhanced with

$$w_{V1}(k,l) = \begin{Bmatrix} 1 & 0 & -1 \\ 1 & 0 & -1 \\ 1 & 0 & -1 \end{Bmatrix} \quad \text{or}$$

$$w_{V2}(k,l) = \begin{Bmatrix} -1 & 0 & 1 \\ -1 & 0 & 1 \\ -1 & 0 & 1 \end{Bmatrix}.$$

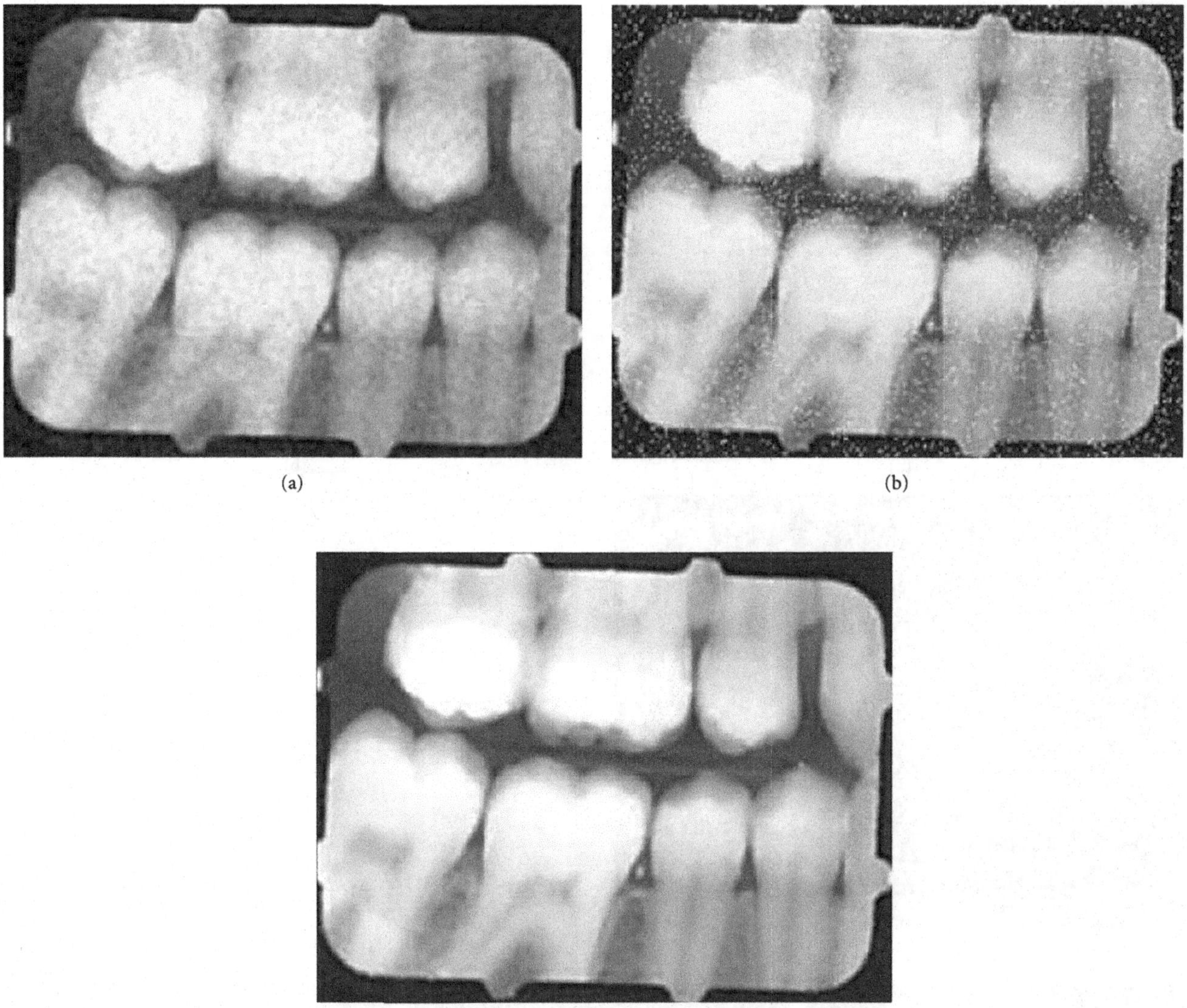

(a) (b) (c)

Figure 6.3-5 (a) Image in Fig. 6.3-4b enhanced with a 3 × 3 median filter. The median filter is not as effective in noise removal as the mean filter of the same size; however, edges are not as severely degraded by the median filter. (b) Image in Fig. 6.3-4a with added shot noise. (c) Image in Figure 6.3-5 enhanced by a 3 × 3 median filter. The median filter is able to significantly enhance this image, allowing almost all shot noise to be eliminated. This image has good diagnostic value.

The omnidirectional kernel (unsharp mask) enhances edges in all directions:

$$K_{HP}(k,l) = \begin{Bmatrix} -1/8 & -1/8 & -1/8 \\ -1/8 & 1 & -1/8 \\ -1/8 & -1/8 & -1/8 \end{Bmatrix}.$$

Note that the application of these kernels to a positive-valued image can result in an output image with both positive and negative values. An enhanced image with only positive pixels can be obtained either by adding an offset or by taking the absolute value of each pixel in the output image. If we are interested in displaying edge-only information, this may be a good approach. On the other hand, if we are interested in enhancing edges that are consistent with the kernel and suppressing those that are not, the output image may be added to the original input image. This addition will most likely result in a non-negative image.

Figure 6.3-6 illustrates enhancement after the application of the kernels w_{H1}, w_{V1}, and w_{HP} to the image in Figs. 6.3-3a and 6.3-3g. Figures 6.3-6a–6.3-6c show the absolute value of the output images obtained with w_{H1}, w_{V1}, and w_{HP}, respectively, applied to the dental image while 6.3-6d–6.3-6f show the same for the brain image. In Fig. 6.3-7, the outputs obtained with these three kernals are added to the original images of Fig. 6.3-3a and 6.3-3g. In this manner the edge

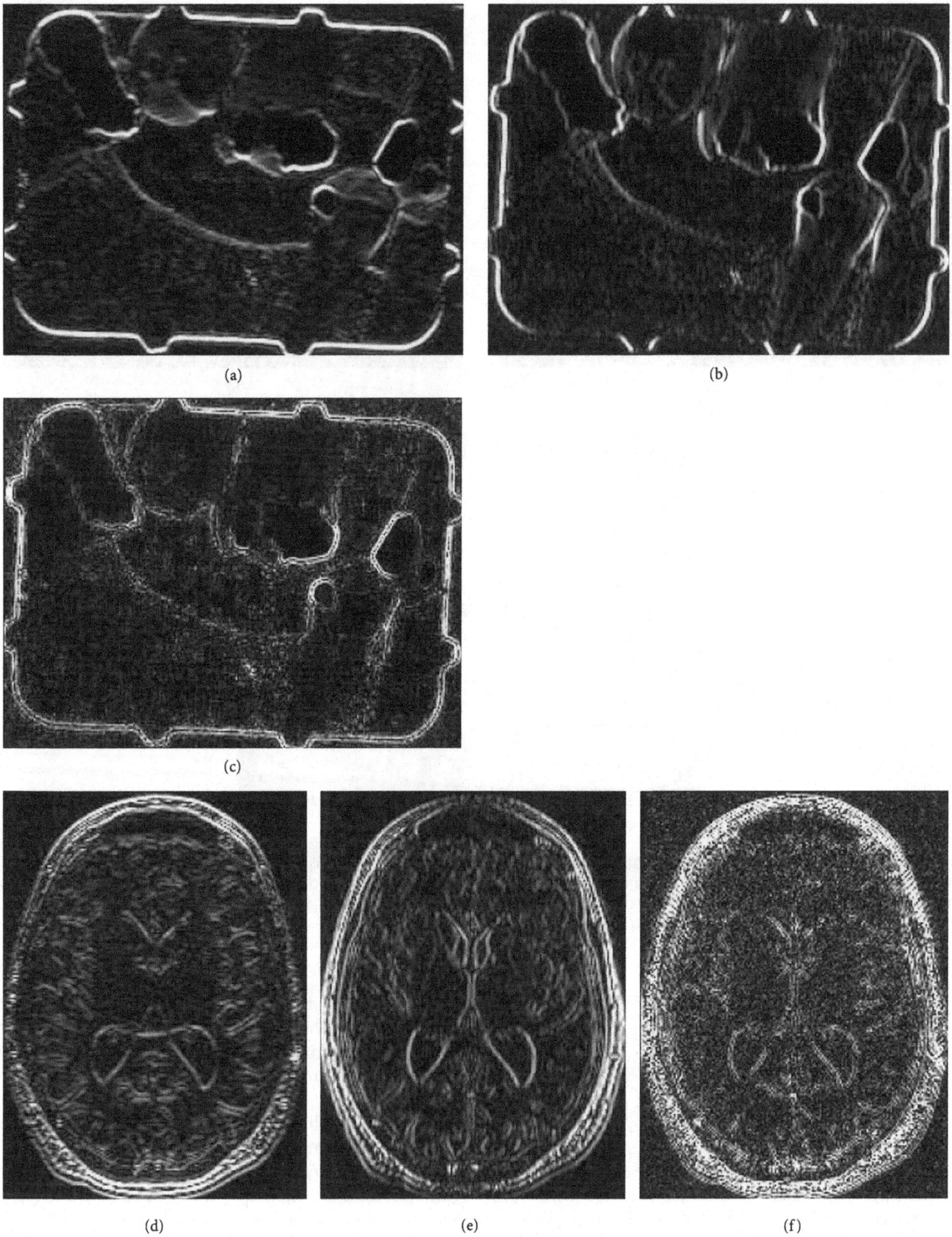

Figure 6.3-6 (a) Absolute value of output image after convolution of w_{H1} with the image in Fig. 6.3-3a. (b) Absolute value of output image after convolution of w_{V1} with the image in Fig. 6.3-3a. (c) Absolute value of output image after convolution of w_{HP}. (d through f) same as a, b, and c using image in Fig. 6.3-3g.

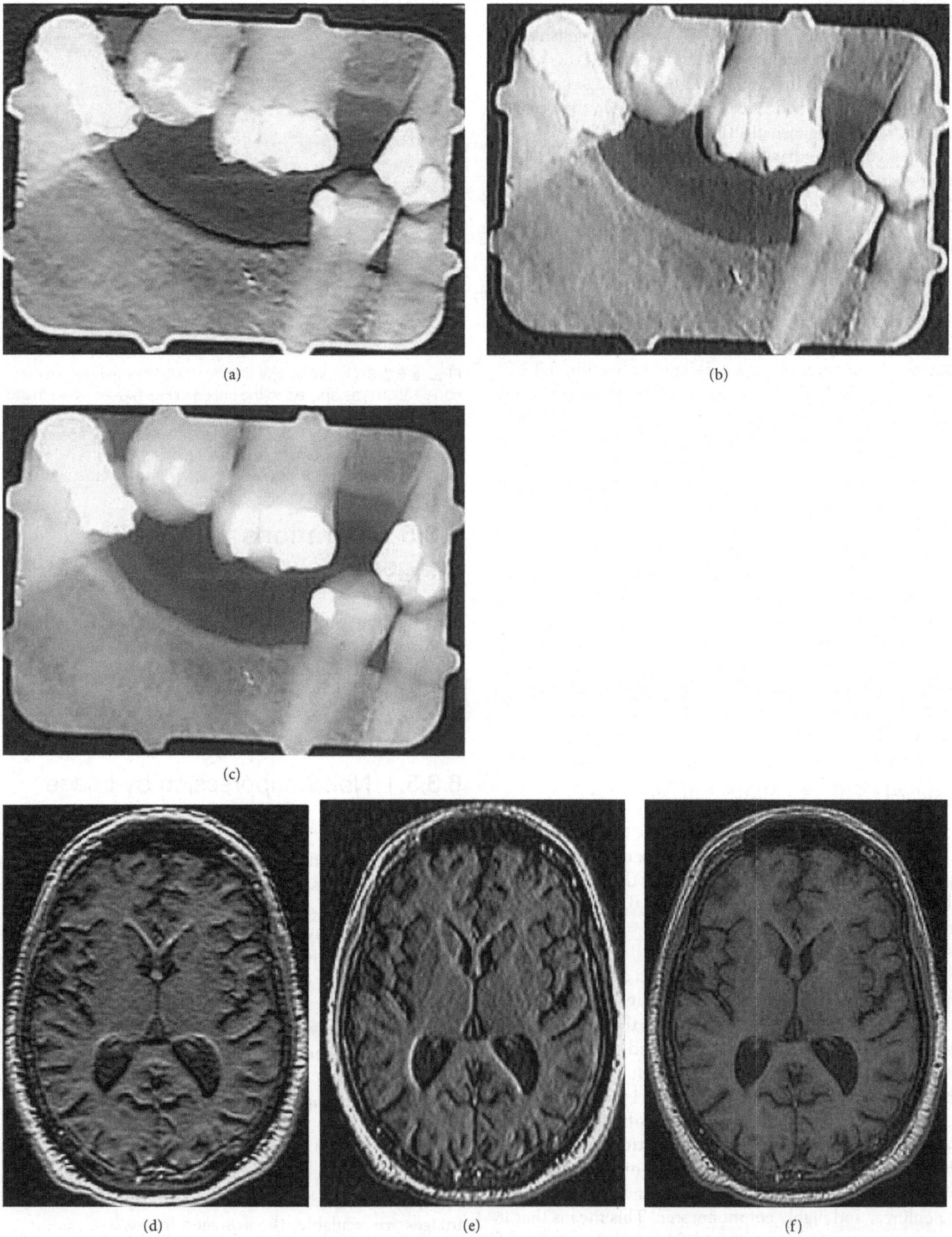

Figure 6.3-7 (a) Sum of original image in Fig. 6.3-3a and its convolution with w_{H1}, (b) with w_{V1}, and (c) with w_{HP}. (d through f) same as a, b, and c using image in Fig. 6.3-3g.

information is enhanced while retaining the context information of the original image. This is accomplished in one step by convolving the original image with the kernel after adding one to its central coefficient. Edge enhancement appears to provide greater contrast than the original imagery when diagnosing pathologies.

Edges can be enhanced with several edge operators other than those just mentioned and illustrated.

6.3.4.4 Local-area histogram equalization

A remarkably effective method of image enhancement is local-area histogram equalization, obtained with a modification of the pixel operation defined in Section 6.3.3.3. Local-area histogram equalization applies the concepts of whole-image histogram equalization to small, overlapping local areas of the image [7,11]. It is a nonlinear operation and can significantly increase the observability of subtle features in the image. The method formulated as shown next is applied at each pixel (m,n) of the input image.

$$h_{LA}(m,n)(i) = \sum_{k=-K}^{K} \sum_{l=-L}^{L} \delta(f(m+l,n+k)-i),$$
$$i = 0,1,\ldots P-1$$

$$H_{LA}(m,n)(j) = \frac{1}{(2K+1)\cdot(2L+1)}\sum_{i=0}^{j} h_{LA}(m,n)(i),$$
$$j = 0,1,\ldots P-1$$

$$g(m,n) = (P-1)\cdot H_{LA}(m,n)(f(m,n))$$

where $h_{LA}(m,n)(i)$ is the local-area histogram, $H_{LA}(m,n)(j)$ is the local-area cumulative histogram, and $g(m,n)$ is the output image. Figure 6.3-8 shows the output image obtained by enhancing the image in Fig. 6.3-2a with local-area histogram equalization using $K = L = 15$ or a 31×31 kernel size.

Local-area histogram equalization is a computationally intensive enhancement technique. The computational complexity of the algorithm goes up as the square of the size of the kernel. It should be noted that since the transformation that is applied to the image depends on the local neighborhood only, each pixel is transformed in a unique way. This results in higher visibility for hidden details spanning very few pixels in relation to the size of the full image. A significant limitation of this method is that the mapping between the input and output images is nonlinear and highly nonmonotonic. This means that it is inappropriate to make quantitative measurements of pixel intensity on the output image, as the same gray level may be transformed one way in one part of the image and a completely different way in another part.

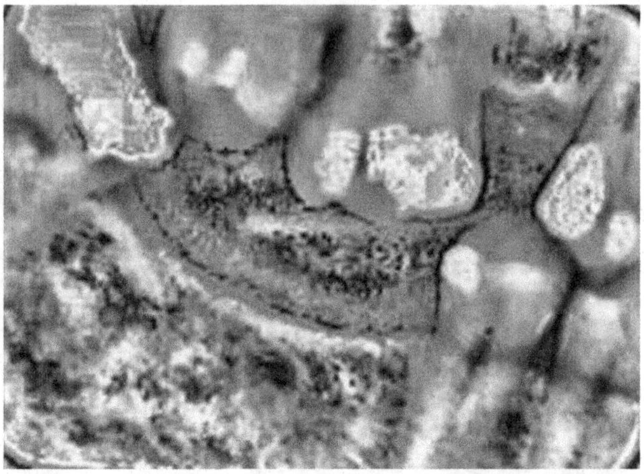

Figure 6.3-8 Output image obtained when local-area histogram equalization was applied to the image in Fig. 6.3-2a. Note that the local-area histogram equalization produces very high contrast images, emphasizing detail that may otherwise be imperceptible. This type of enhancement is computationally very intensive and it may be useful only for discovery purposes to determine if any evidence of a feature exists.

6.3.5 Operations with multiple images

This section outlines two enhancement methods that require more than one image of the same scene. In both methods, the images have to be registered and their dynamic ranges have to be comparable to provide a viable outcome.

6.3.5.1 Noise suppression by image averaging

Noise suppression using image averaging relies on three basic assumptions: (1) that a relatively large number of input images are available, (2) that each input image has been corrupted by the same type of additive noise, and (3) that the additive noise is random with zero mean value and independent of the image. When these assumptions hold, it may be advantageous to acquire multiple images with the specific purpose of using image averaging [1] since with this approach even severely corrupted images can be significantly enhanced. Each of the noisy images $a_i,(m,n)$ can be represented by

$$a_i(m,n) = f(m,n) + d_i(m,n),$$

where $f(m,n)$ is the underlying noise-free image, and $d_i(m,n)$ is the additive noise in that image. If a total of Q images are available, the averaged image is

$$g(m,n) = \frac{1}{Q}\sum_{i=1}^{Q} a_i(m,n)$$

such that

$$E\{g(m,n)\} = f(m,n)$$

and

$$\sigma_g = \frac{\sigma_d}{\sqrt{Q}},$$

where $E\{\cdot\}$ is the expected value operator, σ_g is the standard deviation of $g(m, n)$, and σ_d is that of the noise. Noise suppression is more effective for larger values of Q.

6.3.5.2 Change enhancement by image subtraction

Image subtraction is generally performed between two images that have significant similarities between them. The purpose of image subtraction is to enhance the differences between two images (1). Images that are not captured under the same or very similar conditions may need to be registered [17]. This may be the case if the images have been acquired at different times or under different settings. The output image may have a very small dynamic range and may need to be rescaled to the available display range. Given two images $f_1(m, n)$ and $f_2(m, n)$, the rescaled output image $g(m,n)$ is obtained with

$$b(m,n) = f_1(m,n) - f_2(m,n)$$

$$g(m,n) = f_{\max} \cdot \left(\frac{b(m,n) - \min\{b(m,n)\}}{\max\{b(m,n)\} - \min\{b(m,n)\}} \right)$$

where $f_{\max}$ is the maximum gray level value available, $b(m,n)$ is the unstretched difference image, and $\min\{b(m,n)\}$ and $\max\{b(m,n)\}$ are the minimal and maximal values in $b(m,n)$, respectively.

6.3.6 Frequency domain techniques

Linear filters used for enhancement can also be implemented in the frequency domain by modifying the Fourier transform of the original image and taking the inverse Fourier transform. When an image $g(m, n)$ is obtained by convolving an original image $f(m, n)$ with a kernel $w(m, n)$,

$$g(m,n) = w(m,n) * f(m,n),$$

the convolution theorem states that $G(u, v)$, the Fourier transform of $g(m, n)$ is given by

$$G(u,v) = W(u,v)F(u,v),$$

where $W(u, v)$ and $F(u, v)$ are the Fourier transforms of the kernel and the image, respectively. Therefore, enhancement can be achieved directly in the frequency domain by multiplying $F(u, v)$, pixel-by-pixel, by an appropriate $W(u, v)$ and forming the enhanced image with the inverse Fourier transform of the product. Noise suppression or image smoothing can be obtained by eliminating the high-frequency components of $F(u, v)$, while edge enhancement can be achieved by eliminating its low-frequency components. Since the spectral filtering process depends on a selection of frequency parameters as high or low, each pair (u, v) is quantified with a measure of distance from the origin of the frequency plane,

$$D(u,v) = \sqrt{u^2 + v^2},$$

which can be compared to a threshold D_T to determine if (u, v) is high or low. The simplest approach to image smoothing is the ideal low-pass filter $W_L(u, v)$, defined to be 1 when $D(u, v) \le D_T$ and 0 otherwise. Similarly, the ideal high-pass filter $W_H(u, v)$ can be defined to be 1 when $D(u, v) \ge D_T$ and 0 otherwise. However, these filters are not typically used in practice, because images that they produce generally have spurious structures that appear as intensity ripples, known as ringing [5]. The inverse Fourier transform of the rectangular window $W_L(u, v)$ or $W_H(u, v)$ has oscillations, and its convolution with the spatial-domain image produces the ringing. Because ringing is associated with the abrupt 1 to 0 discontinuity of the ideal filters, a filter that imparts a smooth transition between the desired frequencies and the attenuated ones is used to avoid ringing. The commonly used Butterworth low-pass and high-pass filters are defined respectively as

$$B_L(u,v) = \frac{1}{1 + c[D(u,v)/D_T]^{2n}}$$

and

$$B_H(u,v) = \frac{1}{1 + c[D_T/D(u,v)]^{2n}},$$

where c is a coefficient that adjusts the position of the transition and n determines its steepness. If $c = 1$, these two functions take the value 0.5 when $D(u, v) = D_T$. Another common choice for c is $\sqrt{2} - 1$, which yields 0.707 (– 3 dB) at the cutoff D_T. The most common choice of n is 1; higher values yield steeper transitions.

The threshold D_T is generally set by considering the power of the image that will be contained in the preserved frequencies. The set S of frequency parameters (u, v) that belong to the preserved region, i.e., $D(u, v) \le D_T$ for low-pass and $D(u, v) \ge D_T$ for high-pass,

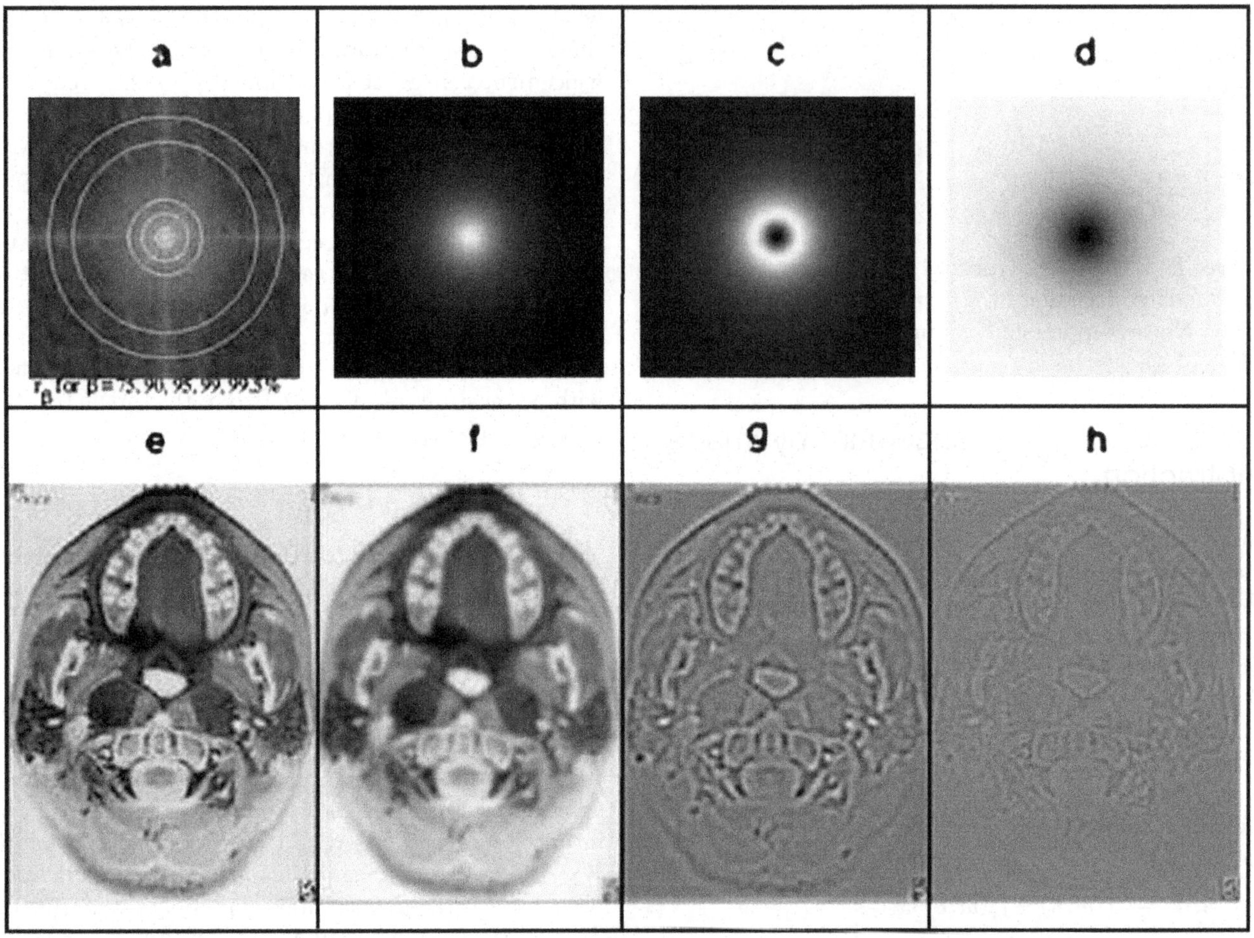

Figure 6.3-9 Filtering with the Butterworth filter. (a) Fourier transform of MRI image in (e); the five circles correspond to the β values 75, 90, 95, 99, and 99.5%. (b) Fourier transform of low-pass filter with $\beta = 90\%$ which provides the output image in (f). (c) Band-pass filter with band $\beta = 75\%$ to $\beta = 90\%$ whose output is in (g). (d) High-pass filter with $\beta = 95\%$, which yields the image in (h). (Courtesy of Dr. Patricia Murphy, Johns Hopkins University Applied Physics Laboratory.)

determines the amount of retained image power. The percentage of total power that the retained power constitutes is given by

$$\beta = \frac{\sum_{(u,v)\in S} |F(u,v)|^2}{\sum_{\forall (u,v)} |F(u,v)|^2} \times 100$$

and is used generally to guide the selection of the cutoff threshold. In Fig. 6.3-9a, circles with radii r_β that correspond to five different β values are shown on the Fourier transform of an original MRI image in Fig. 6.3-9e. The $u = v = 0$ point of the transform is in the center of the image in Fig. 6.3-9a. The Butterworth low-pass filter obtained by setting D_T equal to r_β for $\beta = 90\%$, with $c = 1$ and $n = 1$, is shown in Fig. 6.3-9b where bright points indicate high values of the function. The corresponding filtered image in Fig. 6.3-9f shows the effects of smoothing. A high-pass Butterworth filter with D_T set at the 95% level is shown in Fig. 6.3-9d, and its output in Fig. 6.3-9h highlights the highest frequency components that form 5% of the image power. Figure 6.3-9c shows a band-pass filter formed by the conjunction of a low-pass filter at 95% and a high-pass filter at 75%, while the output image of this band-pass filter is in Fig. 6.3-9g.

6.3.7 Concluding remarks

This chapter has focused on fundamental enhancement techniques used on medical and dental images. These techniques have been effective in many applications and are commonly used in practice. Typically, the techniques presented in this chapter form a first line of algorithms in attempts to enhance image information. After these algorithms have been applied and adjusted for best outcome, additional image enhancement may be required to improve image quality further. Computationally more intensive algorithms may then be considered to take advantage of context-based and object-based information in the image.

Acknowledgments

The author thanks Dr. Bill Moyer for providing access to data and for discussions. Mr. Mark Sluser and Mr. Rick Yue supported this work. NSERC and the Government of Saskatchewan, Post Secondary Education, provided financial support for this project. TR*Labs* Regina provided laboratory facilities.

References

1. de Graaf, C. N., and Viergever, M. A. (ed.), *Information Processing in Medical Imaging*. Plenum Press, New York, 1988.
2. Fong, Y. S., Pomala-Roez, C. A., and Wong, X. H.,Comparison study of non-linear filters in image processing applications. *Opti. Eng.* **28**(7), 749–760 (1989).
3. Frei, W., Image enhancement by image hyberbolization. *Comp. Graph. Image Process.* **6**, 286–294 (1977).
4. Gonzalez, R. C, and Wintz, P., *Digital Image Processing.* Addison-Wesley, Reading, MA, 1987.
5. Gonzalez, R. C, and Woods, R. E., *Digital Image Processing* Addison-Wesley, Reading, MA, 1992.
6. Hall, E. H. Almost uniform distributions from image enhancement, *IEEE Trans. Comp.* **C-23**(2), 207–208 (1974).
7. Ketchum, D. J., Real-time image enhancement techniques. *Proc. SPIE/OSA* **74**, 120–125 (1976).
8. Lewis, R., *Practical Digital Image Processing*. Ellis Horwood, West Sussex, UK, 1990.
9. Low, A., *Introduction to Computer Vision and Image Processing.* McGraw-Hill, U.K., 1991.
10. Niblack, W., *An Introduction to Digital Image Processing*. Prentice Hall, Englewood Cliffs, NJ, 1986.
11. Pizer, S. M., Amburn, P., Austin, R., Cromartie, R., Geselowitz, A., Geer, T., tar Haar Remeny, J., Zimmerman, J.B., and Zuiderveld, K., Adaptive histogram equalization and its variations. *Comp. Vision Graph. Image Process.* **39**, 355–368 (1987).
12. Restrepo, A., and Bovik, A., An adaptive trimmed mean filter for image restoration. *IEEE Trans. Acoustics, Speech, Signal Process.* **ASSP-36**(8), 8813–8818 (1988).
13. Rosenfeld, A., and Kak, A., *Digital Picture Processing.* Academic Press, New York, 1982.
14. Russ, J. *The Image Processing Handbook*, 2nd Ed. CRC Press, Boca Raton, FL 1994.
15. Sinha, P. K., and Hong, Q. H., An improved median filter. *IEE Trans. Med. Imaging* **MI-9**(3), 345–346 (1990).
16. Wahl, F., *Digital Image Signal Processing*. Artech House, Norwood, MA, 1987.
17. Watkins, C., Sadun, A., and Marenka, A., *Modern Image Processing: Warping, Morphing and Classical Techniques*. Academic Press, London, 1993.

Chapter 6.4

6.4

Fundamentals of image segmentation

Jadwiga Rogowska

6.4.1 Introduction

The principal goal of the segmentation process is to partition an image into regions (also called classes, or subsets) that are homogeneous with respect to one or more characteristics or features [11, 16, 20, 30, 36, 66, 77, 96, 107, 109]. Segmentation is an important tool in medical image processing and it has been useful in many applications. The applications include detection of the coronary border in angiograms, multiple sclerosis lesion quantification, surgery simulations, surgical planning, measuring tumor volume and its response to therapy, functional mapping, automated classification of blood cells, studying brain development, detection of micro-calcifications on mammograms, image registration, atlas-matching, heart image extraction from cardiac cineangiograms, detection of tumors, etc. [8, 14, 15, 35, 38, 41a, 61, 71, 88, 109, 115, 132].

In medical imaging, segmentation is important for feature extraction, image measurements, and image display. In some applications it may be useful to classify image pixels into anatomical regions, such as bones, muscles, and blood vessels, while in others into pathological regions, such as cancer, tissue deformities, and multiple sclerosis lesions. In some studies the goal is to divide the entire image into subregions such as the white matter, gray matter, and cerebrospinal fluid spaces of the brain [67], while in others one specific structure has to be extracted, for example breast tumors from magnetic resonance images [71].

A wide variety of segmentation techniques has been proposed (see surveys in [11, 20, 30, 41, 77, 83, 127]). However, there is no one standard segmentation technique that can produce satisfactory results for all imaging applications. The definition of the goal of segmentation varies according to the goal of the study and the type of the image data. Different assumptions about the nature of the analyzed images lead to the use of different algorithms.

Segmentation techniques can be divided into classes in many ways, depending on classification scheme:

- Manual, semiautomatic, and automatic [101].
- Pixel-based (local methods) and region-based (global methods) [4].
- Manual delineation, low-level segmentation (thresholding, region growing, etc.), and model-based segmentation (multispectral or feature map techniques, dynamic programming, contour following, etc.) [109].
- Classical (thresholding, edge-based, and region-based techniques), statistical, fuzzy, and neural network techniques [87].

The most commonly used segmentation techniques can be classified into two broad categories: (1) *region segmentation* techniques that look for the regions satisfying a given homogeneity criterion, and (2) *edge-based segmentation* techniques that look for edges between regions with different characteristics [16, 36, 77, 96, 107].

Thresholding is a common region segmentation method [25, 83, 98, 107, 127]. In this technique a threshold is selected and an image is divided into groups of pixels having values less than the threshold and groups of pixels with values greater or equal to the threshold. There are several thresholding methods: global methods based on gray-level histograms, global methods based on local properties, local threshold selection, and dynamic thresholding. *Clustering* algorithms achieve region segmentation [13, 27, 37, 54] by partitioning the image into sets or clusters of pixels that have strong similarity in the feature space. The basic operation is to examine each pixel

Biomedical Engineering Desk Reference; ISBN: 9780123746467

and assign it to the cluster that best represents the value of its characteristic vector of features of interest. *Region growing* is another class of region segmentation algorithms that assign adjacent pixels or regions to the same segment if their image values are close enough, according to some preselected criterion of closeness [77, 85].

The strategy of edge-based segmentation algorithms is to find object boundaries and segment regions enclosed by the boundaries [16, 36, 41, 72, 96]. These algorithms usually operate on edge magnitude and/or phase images produced by an edge operator suited to the expected characteristics of the image. For example, most *gradient operators* such as Prewitt, Kirsch, or Roberts operators are based on the existence of an ideal step edge. Other edge-based segmentation techniques are *graph searching* and *contour following* [6, 14, 106].

Traditionally, most image segmentation techniques use one type of images (MR, CT, PET, SPECT, ultrasound, etc.). However, the performance of these techniques can be improved by combining images from several sources (*multi-spectral* segmentation [29, 89, 117]) or integrating images over time (*dynamic* or *temporal* segmentation [71, 93, 108]).

The following sections will present some of the segmentation techniques that are commonly used in medical imaging. In Section 6.4.2 we will discuss several thresholding techniques. Section 6.4.3 will describe region growing techniques. The watershed algorithm will be reviewed in Section 6.4.4. Section 6.4.5 will present edge-based segmentation techniques. A discussion of multispectral segmentation methods will be given in Section 6.4.6.

6.4.2 Thresholding

Several thresholding techniques have been developed [16, 25, 36, 41, 51, 96–98, 107, 127]. Some of them are based on the image histogram; others are based on local properties, such as local mean value and standard deviation, or the local gradient. The most intuitive approach is global thresholding. When only one threshold is selected for the entire image, based on the image histogram, thresholding is called *global.* If the threshold depends on local properties of some image regions, for example local average gray value, thresholding is called *local.* If the local thresholds are selected independently for each pixel (or groups of pixels), thresholding is called *dynamic* or *adaptive.*

6.4.2.1 Global thresholding

Global thresholding is based on the assumption that the image has a bimodal histogram and, therefore, the object can be extracted from the background by a simple

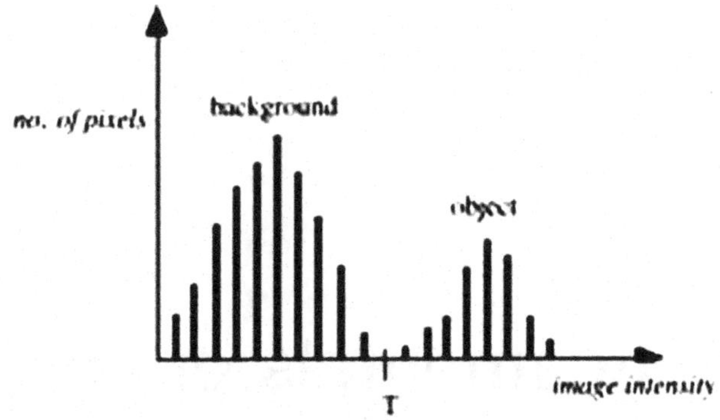

Figure 6.4-1 An example of bimodal histogram with selected threshold *T.*

operation that compares image values with a threshold value T [25, 107]. Suppose that we have an image $f(x, y)$ with the histogram shown in Fig. 6.4-1.

The object and background pixels have gray levels grouped into two dominant modes. One obvious way to extract the object from the background is to select a threshold T that separates these modes.

The thresholded image $g(x, y)$ is defined as

$$g(x,y) = \begin{cases} 1 & \text{if}(x,y) > T \\ 0 & \text{if}(x,y) \le T \end{cases} \tag{6.4.1}$$

The result of thresholding is a binary image, where pixels with intensity value of 1 correspond to objects, while pixels with value 0 correspond to the background.

Figure 6.4-2 shows the result of segmentation by thresholding. The original image (Fig. 6.4-2A) contains white cells on a black background. Pixel intensities vary between 0 and 255. The threshold $T = 127$ was selected as the minimum between two modes on a histogram (Fig. 6.4-2B), and the result of segmentation is shown in Fig. 6.4-2C, where pixels with intensity values higher than 127 are shown in white. In the last step (Fig. 6.4-2D) the edges of the cells were obtained by a 3×3 Laplacian (second-order derivative [36]; also see description in Section 6.4.5), which was applied to the thresholded image in Fig. 6.4-2C.

There are many other ways to select a global threshold. One of them is based on a classification model that minimizes the probability of error [77]. For example, if we have an image with a bimodal histogram (e.g., object and background), we can calculate the error as the total number of background pixels misclassified as object and object pixels miscalssified as background. A semi-automated version of this technique was applied by Johnson *et al.* [56] to measure ventricular volumes from 3D magnetic resonance (MR) images. In their method an operator selects two pixels—one inside an object and one in the background. By comparing the distribution of pixel

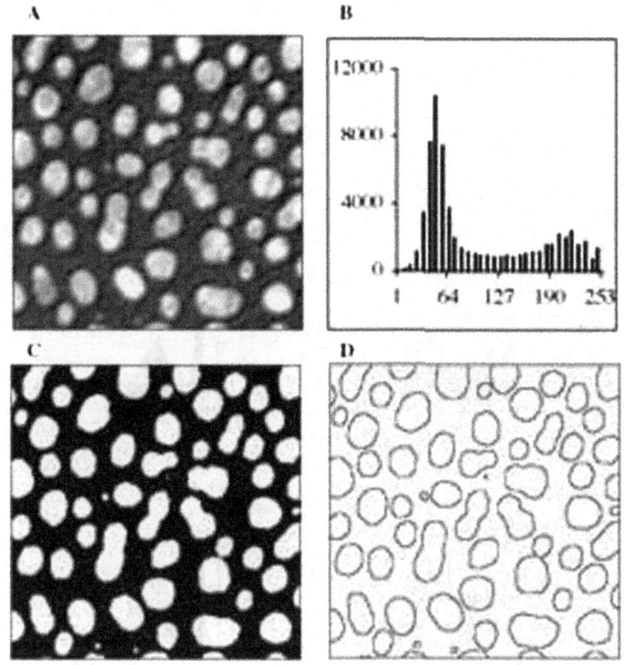

Figure 6.4-2 An example of global thresholding. (A) Original image, (B) histogram of image A, (C) result of thresholding with $T = 127$, (D) outlines of the white cells after applying a 3 × 3 Laplacian to the image shown in C.

intensities in the circular regions around selected pixels, the threshold is calculated automatically and it corresponds to the least number of misclassified pixels between two distributions. The result of the thresholding operation is displayed as a contour map and superimposed on the original image. If needed, the operator can manually modify any part of the border. The same technique was also applied to extract lymph nodes from CT images and was found to be very sensitive to user positioning of interior and exterior points [95]. Some of the threshold selection techniques are discussed in Refs. [25, 96, 127].

In many applications appropriate segmentation is obtained when the area or perimeter of the objects is minimally sensitive to small variations of the selected threshold level. Figure 6.4-3A shows the intensity profile of an object that is brighter than background, and three threshold levels for segmentation: $T1$, $T2$, and $T3$. A small variation ΔT in the lowest threshold level will cause a significant change in the area or perimeter of the segmented object. The same is true for the highest threshold level. However, a change of ΔT in the middle level will have minimal effect on the area or perimeter of the object. The object area $A(T)$ and perimeter $P(T)$ are functions of the threshold T that often exhibit the trend shown in Fig. 6.4-3B. Therefore, the threshold level that minimizes either $dA(T)/dT$ or $dP(T)/dT$ is often a good choice, especially in the absence of operator guidance and when prior information on object locations is not available.

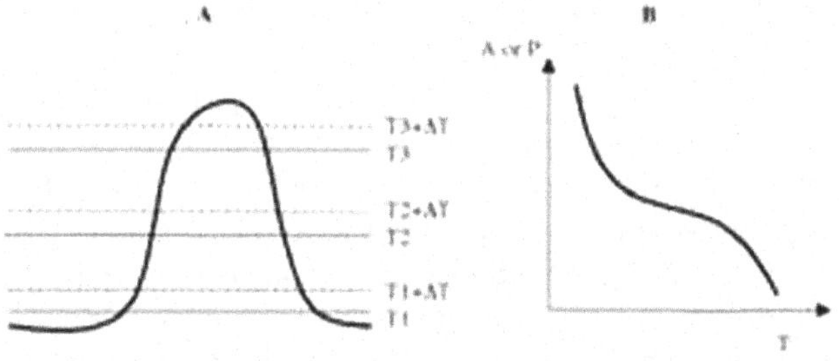

Figure 6.4-3 An example of the sensitivity of the threshold level selection. (A) Cross-sectional intensity profile of a light object on a dark background with three thresholding levels $T1$, $T2$, and $T3$, and three other levels generated by adding a small value ΔT; (B) a hypothetical plot of the area (A) or perimeter (P) versus thresholding level T.

A related technique that evaluates multiple thresholds is based on an estimate of the gradient magnitude around the segmented object edge [16]. The average gradient magnitude is given by

$$\overline{G} = \lim_{\Delta T \to 0} \frac{\Delta T \times P(T)}{\Delta A} = \frac{P(T)}{H(T)}, \qquad (6.4.2)$$

where $H(T)$ is the histogram function. The threshold that maximizes the average boundary gradient is selected.

If an image contains more than two types of regions, it may still be possible to segment it by applying several individual thresholds [96], or by using a multithresholding technique [86]. With the increasing number of regions, the histogram modes are more difficult to distinguish, and threshold selection becomes more difficult.

Global thresholding is computationally simple and fast. It works well on images that contain objects with uniform intensity values on a contrasting background. However, it fails if there is a low contrast between the object and the background, if the image is noisy, or if the background intensity varies significantly across the image.

6.4.2.2 Local (adaptive) thresholding

In many applications, a global threshold cannot be found from a histogram or a single threshold cannot give good segmentation results over an entire image. For example, when the background is not constant and the contrast of objects varies across the image, thresholding may work well in one part of the image, but may produce unsatisfactory results in other areas. If the background variations can be described by some known function of position in the image, one could attempt to correct it by using gray level correction techniques, after which a single threshold should work for the entire image.

Another solution is to apply *local (adaptive) thresholding* [6, 9, 18, 25, 41, 63, 80, 127].

Local thresholds can be determined by (1) splitting an image into subimages and calculating thresholds for each subimage, or (2) examining the image intensities in the neighborhood of each pixel. In the former method [18], an image is first divided into rectangular overlapping subimages and the histograms are calculated for each subimage. The subimages used should be large enough to include both object and background pixels. If a subimage has a bimodal histogram, then the minimum between the histogram peaks should determine a local threshold. If a histogram is unimodal, the threshold can be assigned by interpolation from the local thresholds found for nearby subimages. In the final step, a second interpolation is necessary to find the correct thresholds at each pixel.

In the latter method, a threshold can be selected using the mean value of the local intensity distribution. Sometimes other statistics can be used, such as mean plus standard deviation, mean of the maximum and minimum values [16, 25], or statistics based on local intensity gradient magnitude [25, 62].

Modifications of the above two methods can be found in Refs. [30, 41, 80, 96]. In general, local thresholding is computationally more expensive than global thresholding. It is very useful for segmenting objects from a varying background, and also for extraction of regions that are very small and sparse.

6.4.2.3 Image preprocessing and thresholding

Many medical images may contain low-contrast, fuzzy contours. The histogram modes corresponding to the different types of regions in an image may often overlap and, therefore, segmentation by thresholding becomes difficult. Image preprocessing techniques can sometimes help to improve the shape of the image histogram, for example by making it more strongly bimodal. One of the techniques is image *smoothing* by using the *mean* (*average*) or *median* filter discussed in Chapter 6.3 [53, 65, 96, 99]. The mean filter replaces the value of each pixel by the average of all pixel values in a local neighborhood (usually an N by N window, where $N = 3, 5, 7$, etc.). In the median filter, the value of each pixel is replaced by the median value calculated in a local neighborhood. Median smoothing, unlike the mean filter, does not blur the edges of regions larger than the window used while smoothing out small textural variations. Figure 6.4-4 illustrates results of preprocessing on an autoradiography image using a median filter with 7×7 and 9×9 windows. Figure 6.4-4A shows the original image and its histogram, which is unimodal and, therefore, precludes selection of an appropriate threshold. Median filtering

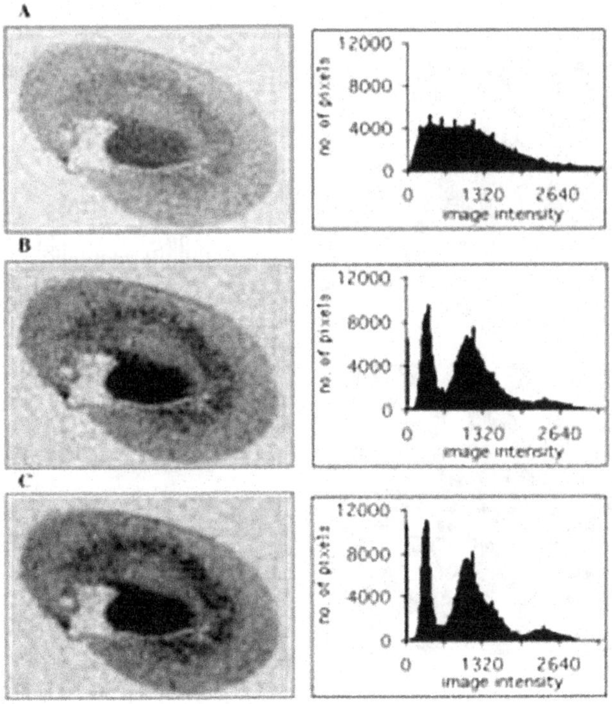

Figure 6.4-4 Median filtering as a preprocessing step for thresholding; (A) original autoradiography image, (B) result of a 7×7 median filter, (C) result of a 9×9 median filter. Corresponding image histograms are shown on the right.

sharpens the peaks on the image histogram (Figs 6.4-4B and C) and allows selection of thresholds for image segmentation.

A common smoothing filter is the *Gaussian filter*, where for each pixel $[i, j]$, the convolution mask coefficients $g[i, j]$ are based on a Gaussian function:

$$g[i,j] = \exp\left[\frac{-(i^2+j^2)}{2\sigma^2}\right], \qquad (6.4.3)$$

where σ is the spread parameter (standard deviation) that defines the degree of Gaussian smoothing: Larger σ implies a wider Gaussian filter and a greater amount of smoothing. The Gaussian filter can be approximated in digital images by an N by N convolution mask. A 7×7 Gaussian mask with $\sigma^2 = 2$ [52] is obtained with the coefficients of the following matrix:

$$\begin{array}{ccccccc}
1 & 4 & 7 & 10 & 7 & 4 & 1 \\
4 & 12 & 26 & 33 & 26 & 12 & 4 \\
7 & 26 & 55 & 71 & 55 & 26 & 7 \\
10 & 33 & 71 & 91 & 71 & 33 & 10 \\
7 & 26 & 55 & 71 & 55 & 26 & 7 \\
4 & 12 & 26 & 33 & 26 & 12 & 4 \\
1 & 4 & 7 & 10 & 7 & 4 & 1
\end{array}$$

By normalizing each coefficient with the sum of all (1115) a filter that preserves the scale of the image is obtained.

Goshtasby and Turner [38] reported that smoothing with a Gaussian filter reduced noise and helped in thresholding of the endocardial surfaces on cardiac MR images.

Preprocessing with *extremum sharpening* combined with median filtering has proven to be useful in segmenting microscopic images of blood cells [14, 65]. In this method, a minimum and maximum are calculated within an N by N window around each pixel (x, y). The value of the extremum operator is simply whichever of the two extrema is the closest to the value at pixel (x, y). When the pixel (x, y) has a value exactly midway between minimum and maximum, the operator takes the value of the pixel. The appropriate window size for the extremum sharpening has to be commensurate with the width of the image edges.

The extremum sharpening is usually followed by median filtering, which smoothes out the slightly ragged contours left by the sharpening operation. The standard procedure suggested in [65] for segmenting cells was: 9×9 median filter (noise removal), 3×3 extremum sharpening, and finally 5×5 median filter, followed by thresholding based on threshold determined from the histogram.

The median and Gaussian smoothing, as well as extremum sharpening, "improve" image histograms by producing images with strongly bimodal histograms. Additional techniques for making histogram valleys deeper are discussed in Weszka *et al.* [127].

A more elaborate approach used for specific types of images is provided by *adaptive filtering* techniques where the parameters of the algorithm are modified locally based on the pixel's neighborhood [51, 68]. If, for example, the neighborhood has relatively constant intensity, we can assume that we are within an object with constant features and we can apply an isotropic smoothing operation to this pixel to reduce the noise level. If an edge has been detected in the neighborhood, we could still apply some smoothing, but only along the edge. Adaptive filtering combines an efficient noise reduction and an ability to preserve and even enhance the edges of image structures. Westin used adaptive filtering successfully for the thresholding of bones on CT images [126].

6.4.3 Region growing

Whereas thresholding focuses on the difference of pixel intensities, the region growing method looks for groups of pixels with similar intensities. Region growing, also called region merging, starts with a pixel or a group of pixels (called seeds) that belong to the structure of interest. Seeds can be chosen by an operator, or provided by an automatic seed finding procedure. In the next step neighboring pixels are examined one at a time and added to the growing region, if they are sufficiently similar based on a uniformity test (also called a homogeneity criterion). The procedure continues until no more pixels can be added. The object is then represented by all pixels that have been accepted during the growing procedure [1, 6, 36, 77, 85, 96, 102, 104, 107, 113, 116].

One example of the uniformity test is comparing the difference between the pixel intensity value and the mean intensity value over a region. If the difference is less than a predefined value, for example, two standard deviations of the intensity across the region, the pixel is included in the region; otherwise, it is defined as an edge pixel. The results of region growing depend strongly on the selection of the homogeneity criterion. If it is not properly chosen, the regions leak out into adjoining areas or merge with regions that do not belong to the object of interest. Another problem of region growing is that different starting points may not grow into identical regions.

The advantage of region growing is that it is capable of correctly segmenting regions that have the same properties and are spatially separated. Another advantage is that it generates connected regions.

Instead of region merging, it is possible to start with some initial segmentation and subdivide the regions that do not satisfy a given uniformity test. This technique is called *splitting* [41, 96, 107]. A combination of *splitting and merging* adds together the advantages of both approaches [6, 84, 133].

Various approaches to region growing segmentation have been described by Zucker [133]. Excellent reviews of region growing techniques were done by Fu and Mui [30], Haralick and Shapiro [41], and Rosenfeld and Kak [96].

An interesting modification of region growing technique called *hill climbing* was proposed by Bankman *et al.* for detecting microcalcifications in mammograms [8]. The technique is based on the fact that in a given image $f(x, y)$, the edge of a microcalcification to be segmented is a closed contour around a known pixel (x_0, y_0), the local intensity maximum. For each pixel, a slope value $s(x, y)$ is defined as

$$s(x,y) = \frac{f(x_0,y_0) - f(x,y)}{d(x_0,y_0,x,y)}, \tag{6.4.4}$$

where $d(x_0, y_0, x, y)$ is the Euclidean distance between the local maximum pixel and pixel (x, y).

In the first step, the object's edge points are identified by radial line search emanating from the local maximum. The line search is applied in 16 equally spaced directions originating from the pixel (x_0, y_0), and for each direction,

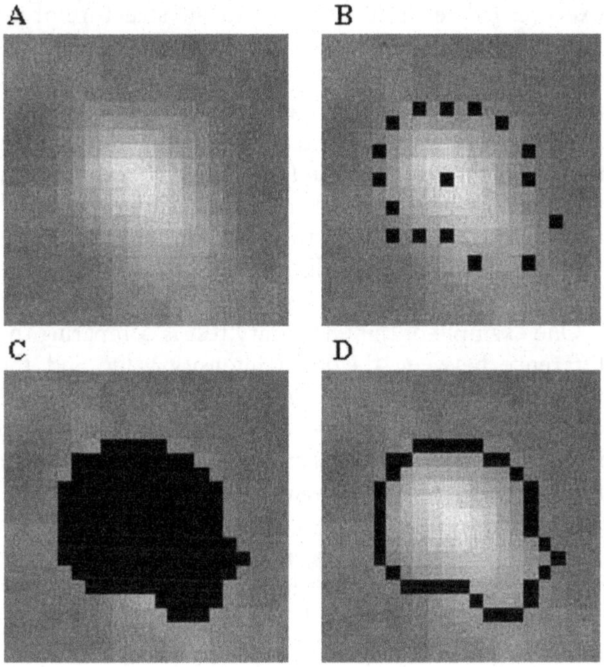

Figure 6.4-5 Steps of segmentation with the hill climbing algorithm; (A) a 0.5 × 0.5 mm image showing a subtle microcalcification, (B) 16 edge points determined by the algorithm, (C) result of region growing, (D) edges of region enclosing the segmented microcalcification. Reprinted with permission from I. N. Bankman, T. Nizialek, I. Simon, *et al.*, "Segmentation algorithms for detecting microcalcifications in mammograms", *IEEE Trans. Inform. Techn. Biomed*, vol. 1, no. *2*, pp. 141–149, 1997. ©1997 IEEE.

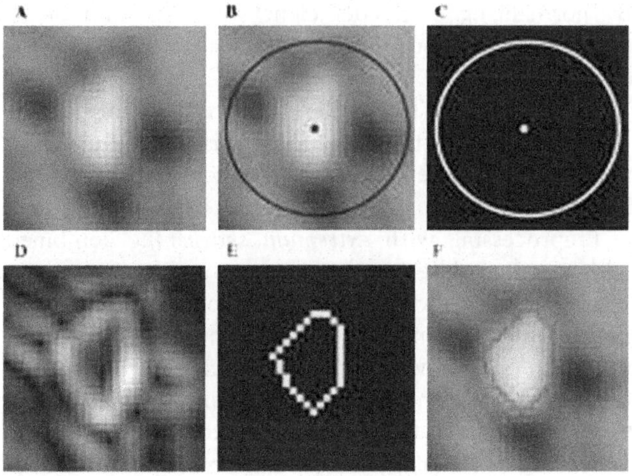

Figure 6.4-6 Image segmentation using Sobel/watershed algorithm. (A) Original image of a lymph node; (B) operator's marks: a point inside the node, and a circle enclosing the area well outside the node; (C) binary image generated from B; (D) result of a 3 × 3 Sobel edge detection operation performed on the original image A; (E) result of the watershed algorithm performed on image D using markers from image C; (F) edges of the lymph node (interior region from image E) superimposed on the original image. Reprinted with permission from J. Rogowska, K. Batchelder, G. S. Gazelle, *et al.* Quantitative CT lymphography: evaluation of selected two-dimensional techniques for computed tomography quantitation of lymph nodes. *Investigative Radiology*, vol. 31, no. 3, pp. 138–145, 1999.

a pixel is considered to be on the edge if it provides the maximal slope value. Next, the edge points are used as seeds for region growing with a spatial constraint (growing the region inward, toward local maximum) and an intensity constraint (including pixels with intensity values increasing monotonically toward the local maximum). Figure 6.4-5 shows the steps of segmentation using the hill-climbing algorithm. The technique was successfully applied to segment low-contrast microcalcification clusters on mammography images. The advantages of this algorithm are that it does not need selection of a threshold and that, because it grows the region from the edges toward the center, it circumvents excessive growth of a region.

Region growing has found many other medical applications, such as segementation of ventricles on cardiac images [104], extraction of blood vessels on angiography data [48], or extraction of brain surface [21].

6.4.4 Watershed algorithm

Watershed segmentation is a region-based technique that utilizes image morphology [16, 107]. It requires selection of at least one marker ("seed" point) interior to each object of the image, including the background as a separate object. The markers are chosen by an operator or are provided by an automatic procedure that takes into account the application-specific knowledge of the objects. Once the objects are marked, they can be grown using a morphological watershed transformation [10]. A very intuitive description of watersheds can be found in Ref. [16]. To understand the watershed, one can think of an image as a surface where the bright pixels represent mountaintops and the dark pixels valleys. The surface is punctured in some of the valleys, and then slowly submerged into a water bath. The water will pour in each puncture and start to fill the valleys. However, the water from different punctures is not allowed to mix, and therefore the dams need to be built at the points of first contact. These dams are the boundaries of the water basins, and also the boundaries of image objects.

An application of watershed segmentation to extract lymph nodes on CT images is shown in Fig. 6.4-6 [95]. In this implementation a 3 × 3 Sobel edge operator [36, 96] is used in place of the morphological gradient to extract edge strength. The original lymph node image is shown in Fig. 6.4-6A. In the first step, the operator positions a cursor inside the node (Fig. 6.4-6B). All pixels within a radius of two pixels of the mark are used as seed points for the lymph node. To mark the exterior of lymph node, the operator drags the cursor outside of the node to define a circular region, which completely encloses the node (Fig. 6.4-6C). All pixels outside this circle mark the background.

In the next step, an edge image is created using the Sobel edge operator (Fig. 6.4-6D). The edge image has high values for the pixels with strong edges. With the seed point marking the node interior, the circle marking the background (Fig. 6.4-6C), and the edge image generated by the Sobel operator (Fig. 6.4-6D), the segmentation proceeds directly with the watershed operation (Fig. 6.4-6E). The watershed operation operates on an edge image to separate the lymph node from the surrounding tissue. By using a technique called simulated immersion [119], the watershed considers whether a drop of water at each point in the edge image would flow to the interior seed point or the exterior marker. Points that drain into the interior belong to the lymph node, whereas points that drain to the exterior belong to the surrounding tissue. More formal discussions of morphological segmentation can be found in Refs. [75, 119, 120].

Watershed analysis has proven to be a powerful tool for many 2D image-segmentation applications [75]. An example of segmentation of microscopic image of human retina is included in Ref. [107]. Higgins and Ojard [43] applied a 3D extension of the watershed algorithm to cardiac volumetric images.

6.4.5 Edge-based segmentation techniques

An edge or boundary on an image is defined by the local pixel intensity gradient. A gradient is an approximation of the first-order derivative of the image function. For a given image $f(x,y)$, we can calculate the magnitude of the gradient as

$$|G| = \sqrt{\left[G_x^2 + G_y^2\right]} = \sqrt{\left[\left(\frac{\partial f}{\partial x}\right)^2 + \left(\frac{\partial f}{\partial y}\right)^2\right]} \quad (6.4.5)$$

and the direction of the gradient as

$$D = \tan^{-1}\left(\frac{G_y}{G_x}\right) \quad (6.4.6)$$

where G_x and G_y are gradients in directions x and y, respectively. Since the discrete nature of digital image does not allow the direct application of continuous differentiation, calculation of the gradient is done by differencing [36].

Both magnitude and direction of the gradient can be displayed as images. The magnitude image will have gray levels that are proportional to the magnitude of the local intensity changes, while the direction image will have gray levels representing the direction of maximum local gradient in the original image.

Most gradient operators in digital images involve calculation of convolutions, e.g., weighted summations of the pixel intensities in local neighborhoods. The weights can be listed as a numerical array in a form corresponding to the local image neighborhood (also known as a mask, window or kernel). For example, in case of a 3 × 3 Sobel edge operator, there are two 3 × 3 masks:

$$\begin{matrix} -1 & -2 & -1 \\ 0 & 0 & 0 \\ 1 & 2 & 1 \end{matrix} \qquad \begin{matrix} -1 & 0 & 1 \\ -2 & 0 & 2 \\ -1 & 0 & 1 \end{matrix}$$

The first mask is used to compute G_x while the second is used to compute G_y. The gradient magnitude image is generated by combining G_x and G_y using Eq. 6.4.5. Figure 6.4-7B shows an edge magnitude image obtained with the 3×3 Sobel operator applied to the magnetic resonance angiography (MRA) image of Fig. 6.4-7A.

The results of edge detection depend on the gradient mask. Some of the other edge operators are Roberts, Prewitt, Robinson, Kirsch, and Frei-Chen [36, 41, 53, 96, 97].

Many edge detection methods use a gradient operator, followed by a threshold operation on the gradient, in order to decide whether an edge has been found [12, 16, 25, 36, 41, 72, 96, 97, 107, 113]. As a result, the output is a binary image indicating where the edges are. Figures 6.4-7C and 6.4-7D show the results of thresholding at two different levels. Please note that the selection of the appropriate threshold is a difficult task. Edges displayed in Fig. 6.4-7C include some background pixels around

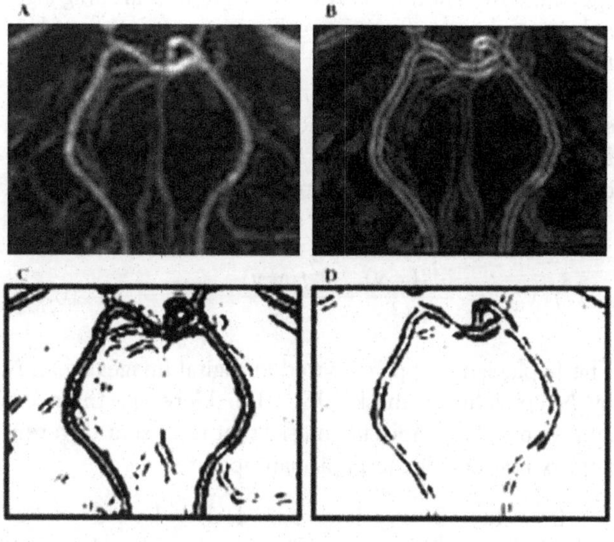

Figure 6.4-7 Edge detection using Sobel operator. (A) Original angiography image showing blood vessels, (B) edge magnitude image obtained with a 3 x 3 Sobel mask, (C) edge image thresholded with a low threshold (300), (D) edge image thresholded with a high threshold (600).

the major blood vessels, while edges in Fig. 6.4-7D do not enclose blood vessels completely.

The edge-based techniques are computationally fast and do not require a priori information about image content. The common problem of edge-based segmentation is that often the edges do not enclose the object completely. To form closed boundaries surrounding regions, a postprocessing step of linking or grouping edges that correspond to a single boundary is required. The simplest approach to edge linking involves examining pixels in a small neighborhood of the edge pixel (3×3, 5×5, etc.) and linking pixels with similar edge magnitude and/or edge direction. In general, edge linking is computationally expensive and not very reliable. One solution is to make the edge linking semiautomatic and allow a user to draw the edge when the automatic tracing becomes ambiguous. For example, Wang *et al.* developed a hybrid algorithm (for MR cardiac cineangiography) in which a human operator interacts with the edge tracing operation by using anatomic knowledge to correct errors [121]. A technique of *graph searching* for border detection has been used in many medical applications [6, 14, 64, 81, 105, 106, 112]. In this technique each image pixel corresponds to a graph node and each path in a graph corresponds to a possible edge in an image. Each node has a cost associated with it, which is usually calculated using the local edge magnitude, edge direction, and a priori knowledge about the boundary shape or location. The cost of a path through the graph is the sum of costs of all nodes that are included in the path. By finding the optimal low-cost path in the graph, the optimal border can be defined. The graph searching technique is very powerful, but it strongly depends on an application-specific cost function. A review of graph searching algorithms and cost function selection can be found in Ref. [107].

Since the peaks in the first-order derivative correspond to zeros in the second-order derivative, the *Laplacian* operator (which approximates second-order derivative) can also be used to detect edges [16, 36, 96].

The Laplace operator ∇^2 of a function $f(x, y)$ is defined as

$$\nabla^2 f(x,y) = \frac{\partial^2 f(x,y)}{\partial x^2} + \frac{\partial^2 f(x,y)}{\partial y^2}. \quad (6.4.7)$$

The Laplacian is approximated in digital images by an N by N convolution mask [96, 107]. Here are three examples of 3×3 Laplacian masks that represent different approximations of the Laplacian operator:

$$\begin{matrix} 0 & -1 & 0 \\ -1 & 4 & -1 \\ 0 & -1 & 0 \end{matrix} \qquad \begin{matrix} -1 & -1 & -1 \\ -1 & 8 & -1 \\ -1 & -1 & -1 \end{matrix} \qquad \begin{matrix} 1 & -2 & 1 \\ -2 & 4 & -2 \\ 1 & -2 & 1 \end{matrix}$$

The image edges can be found by locating pixels where the Laplacian makes a transition through zero (zero crossings). Figure 6.4-8A shows a result of a 3×3 Laplacian applied to the image in Fig. 6.4-7A. The zero crossings of the Laplacian are shown in Fig. 6.4-8C.

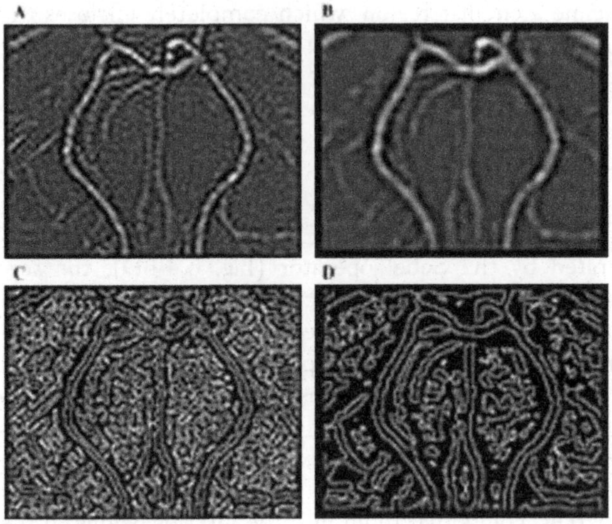

Figure 6.4-8 Results of Laplacian and Laplacian of Gaussian (LoG) applied to the original image shown in Fig. 7A. (A) 3×3 Laplacian image, (B) result of a 7×7 Gaussian smoothing followed by a 7×7 Laplacian, (C) zero-crossings of the Laplacian image A, (D) zero-crossings of the LoG image B.

All edge detection methods that are based on a gradient or Laplacian are very sensitive to noise. In some applications, noise effects can be reduced by smoothing the image before applying an edge operation. Marr and Hildreth [72] proposed smoothing the image with a Gaussian filter before application of the Laplacian (this operation is called Laplacian of Gaussian, LoG). Figure 6.4-8B shows the result of a 7×7 Gaussian followed by a 7×7 Laplacian applied to the original image in Fig. 6.4-7A. The zero crossings of the LoG operator are shown in Fig. 6.4-8D. The advantage of LoG operator compared to a Laplacian is that the edges of the blood vessels are smoother and better outlined. However, in both Figs 6.4-8C and D, the nonsignificant edges are detected in regions of almost constant gray level. To solve this problem, the information about the edges obtained using first and second derivatives can be combined [107]. This approach was used by Goshtasby and Turner [38] to extract the ventricular chambers in flow-enhanced MR cardiac images. They used a combination of zero crossings of the LoG operator and local maximum of the gradient magnitude image, followed by the curve-fitting algorithm.

The Marr–Hildreth operator was used by Bomans *et al.* [12] to segment the MR images of the head. In a study of coronary arteriograms, Sun *et al.* [110] used a directional low-pass filter to average image intensities in the direction parallel to the vessel border. Other edge-finding algorithms can be found in Refs. [24, 30, 36, 96].

6.4.6 Multispectral techniques

Most traditional segmentation techniques use images that represent only one type of data, for example MR or CT. If different images of the same object are acquired using several imaging modalities, such as CT, MR, PET, ultrasound, or collecting images over time, they can provide different features of the objects, and this spectrum of features can be used for segmentation. The segmentation techniques based on integration of information from several images are called *multispectral* or *multimodal* [20, 22, 29, 90, 103, 118].

6.4.6.1 Segmentation using multiple images acquired by different imaging techniques

In the case of a single image, pixel classification is based on a single feature (gray level), and segmentation is done in one-dimensional (single-channel) feature space. In multispectral images, each pixel is characterized by a set of features and the segmentation can be performed in multidimensional (multichannel) feature space using clustering algorithms. For example, if the MR images were collected using $T1$, $T2$, and a proton-density imaging protocol, the relative multispectral data set for each tissue class result in the formation of tissue clusters in three-dimensional feature space. The simplest approach is to construct a 3D scatter plot, where the three axes represent pixel intensities for $T1$, $T2$, and proton density images. The clusters on such a scatter plot can be analyzed and the segmentation rules for different tissues can be determined using automatic or semiautomatic methods [13, 19].

There are many segmentation techniques used in multi-modality images. Some of them are k–nearest neighbors (kNNs) [19, 55, 76], k–means [111, 118], fuzzy c–means [12, 40], artificial networks algorithms [19, 89], expectation/maximization [31, 58, 125], and adaptive template moderated spatially varying statistical classification techniques [122]. All multispectral techniques require images to be properly registered. In order to reduce noise and increase the performance of the segmentation techniques, images can be smoothed. Excellent results have been obtained with adaptive filtering [20], such as Bayesian processing, nonlinear anisotropic diffusion filtering, and filtering with wavelet transforms [32, 49, 50, 103, 124, 130].

To illustrate the advantages of using multispectral segmentation, we show in Fig. 6.4-9 the results of adaptive segmentation by Wells *et al.* [125] applied to dual-echo ($T2$-weighted and proton-density weighted) images of the brain. The adaptive segmentation technique is based on the expectation/maximization algorithm (EM) [26a] and uses knowledge of tissue properties and intensity inhomogeneities to correct and segment MR images. The technique has been very effective in segmenting brain tissue in a study including more than 1000 brain scans [125]. Figures 6.4-9A and B present the original $T2$ and proton-density images, respectively. Both images were obtained from a healthy volunteer on a 1.5-T MR scanner. Figure 6.4-9C shows a result of conventional statistical classification, using nonparametric intensity models derived from images of the same type from a different individual. The segmentation is too heavy on white matter and shows asymmetry in the gray matter thickness due to intrascan inhomogeneities. Considerable improvement is evident in Fig. 6.4-9D, which shows the result of EM segmentation after convergence at 19 iterations.

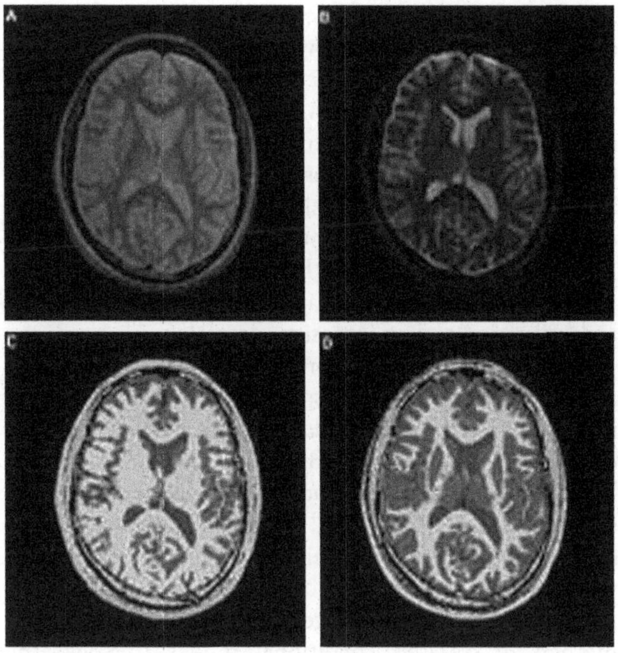

Figure 6.4-9 The results of adaptive segmentation applied to dual-echo images of the brain. (A) Original $T2$-weighted image, (B) original proton-density weighted image, (C) result of conventional statistical classification, (D) result of EM segmentation. The tissue classes are represented by colors: blue, CSF; green, white matter; gray, gray matter; pink, fat; black, background. (Courtesy of Dr. W. M. Wells III, Surgical Planning Lab, Department of Radiology, Brigham and Women's Hospital, Boston.)

Adaptive segmentation [125] is a generalization of standard intensity-based classification that, in addition to the usual tissue class conditional intensity models, incorporates models of the intra- and interscan intensity inhomogeneities that usually occur in MR images. The EM algorithm is an iterative algorithm that alternates between conventional statistical tissue classification (the "E" step) and the reestimation of a correction for the unknown intensity inhomogeneity (the "M" step).

The EM approach may be motivated by the following observations. If an improved intensity correction is available, it is a simple matter to apply it to the intensity data and obtain an improved classification. Similarly, if an improved classification is available, it can be used to derive an improved intensity correction, for example, by predicting image intensities based on tissue class, comparing the predicted intensities with the observed intensities, and smoothing. Eventually, the process converges, typically in less than 20 iterations, and yields a classification and an intensity correction.

In recent work, the algorithm has been extended in a number of directions. A spline-based modeling of the intensity artifacts associated with surface coils have been described by Gilles *et al.* [34]. The addition of an "unknown" tissue class and other refinements has been described by Guillemaud and Brady [39]. Also, Markov models of tissue homogeneity have been added to the formalism in order to reduce the thermal noise that is usually apparent in MR imagery. Held *et al.* [42] used the method of iterated conditional modes to solve the resulting combinatorial optimization problem, while Kapur *et al.* [59] used mean field methods to solve a related continuous optimization problem.

6.4.6.2 Segmentation using multiple images acquired over time

Multispectral images can also be acquired as a sequence of images, in which intensities of certain objects change with time, but the anatomical structures remain stationary. One example of such sequence is a CT image series generated after intravenous injection of a contrast medium that is carried to an organ of interest. Such an image sequence has constant morphology of the imaged structure, but regional intensity values may change from one image to the next, depending upon the local pharmacokinetics of the contrast agent.

The most popular segmentation technique that employs both intensity and temporal information contained in image sequences is the *parametric analysis technique* [44, 45, 79a, 89a]. In this technique, for each pixel or region of interest, the intensity is plotted versus time. Next, the plots are analyzed, with the assumption that the curves have similar time characteristics. Certain parameters are chosen, such as maximum or a minimum intensity, distance between maximum and minimum, or time of occurrence of maximum or minimum. The appropriate set of parameters depends on the functional characteristics of the object being studied. Then, an image is calculated for each of the chosen parameters. In such images the value of each pixel is made equal to the value of the parameter at that point. Therefore, the method is called parametric imaging. The disadvantage of the method of parametric analysis is that it assumes that all pixel intensity sequence plots have the same general pattern across the image. In fact, however, many images have pixels or regions of pixels that do not share the same characteristics in the time domain and, therefore, will have dissimilar dynamic intensity plots.

An interesting application of the parametric mapping technique to the 3D segmentation of multiple sclerosis lesions on series of MR images was proposed by Gerig *et al.* [33]. Temporal images were acquired in intervals of 1, 2, or 4 weeks during a period of 1 year. The parameters chosen for parametric maps were based on lesion characteristics, such as lesion intensity variance, time of appearance, and time of disappearance. The 3D maps displayed patterns of lesions that show similar temporal dynamics.

Another technique for temporal segmentation was introduced by Rogowska [91]. The *correlation mapping* (also called *similarity mapping*) technique identifies regions (or objects) according to their temporal similarity or dissimilarity with respect to a reference time–intensity curve obtained from a reference region of interest (ROI). Assume that we have a sequence of N spatially registered temporal images of stationary structures. The similarity map NCOR_{ij} based on normalized correlation is defined for each pixel (i, j) as

$$\mathrm{NCOR}_{ij} = \frac{\sum_{n=1}^{N}\left(A_{ij}[n]-\mu_A\right)(R[n]-\mu_R)}{\sqrt{\sum_{n=1}^{N}\left(A_{ij}[n]-\mu_A\right)^2\sum_{n=1}^{N}(R[n]-\mu_R)^2}}; \tag{6.4.8}$$

where $A_{ij}[n]$ is the time sequence of image intensity values for the consecutive N images: $A_{ij}[1], A_{ij}[2],\ldots, A_{ij}[N]$, $(i = 1,2,\ldots, I, j = 1,2,\ldots, J, n = 1,2,\ldots,N$; I is the number of image rows, J is the number of image columns), $R[n]$ is the reference sequence of mean intensity values from a selected reference ROI, μ_A is the mean value of the time sequence for pixel (i,j), and μR is the mean value of the reference sequence.

Pixels in the resulting similarity map, whose temporal sequence is similar to the reference, have high correlation values and are bright, while those with low correlation values are dark. Therefore, similarity mapping segments structures in an image sequence based on their temporal responses rather than spatial properties. In addition, similarity maps can be displayed in pseudocolor or color-coded and superimposed on one image. Figure 6.4-10 shows an application of correlation mapping technique to the temporal sequence of images acquired from a patient with a brain tumor after a bolus injection of contrast agent (Gd-DTPA) on a 1T MR scanner. The first image in a sequence of 60 MR images with the reference region of

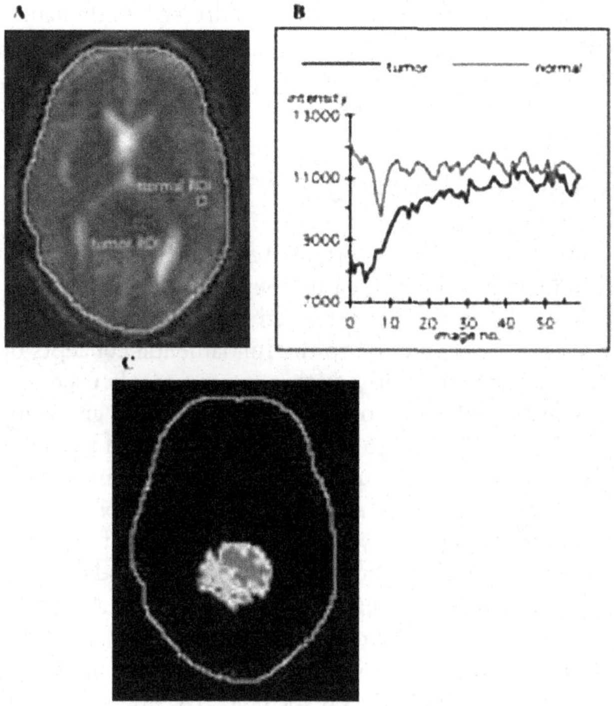

Figure 6.4-10 Image segmentation using correlation mapping. (A) First image in a sequence of 60 temporal images with 3 × 3 pixel ROIs drawn in tumor and normal area; (B) plot of the average intensity of the reference ROI (tumor) and the normal ROI for 60 images in a sequence; (C) correlation map of the tumor.

interest in the tumor area and a normal ROI is shown in Fig. 6.4-10A. Figure 6.4-10B plots the average intensities of the reference and normal ROIs. The correlation map is displayed with a pseudocolor lookup table in Fig. 6.4-10C.

The technique of correlation mapping has found numerous applications. Some of them are included in Refs [92, 93]. Other investigators have adopted this technique in brain activation studies [7], segmentation of breast tumors [71], and renal pathologies [108].

A modification of correlation mapping technique, called *delay mapping*, is also used to segment temporal sequences of images. It segments an image into regions with different time lags, which are calculated with respect to the reference [94].

Parametric maps, similarity maps, and delay maps—all are segmentation and visualization tools for temporal sequences of images. They are particularly useful for evaluation of disease processes, drug treatments, or radiotherapy results.

6.4.7 Other techniques

Combined (hybrid) strategies have also been used in many applications. Here are some examples: Kapur *et al.* [58] present a method for segmentation of brain tissue from magnetic resonance images that combines the strengths of three techniques: single-channel expectation/maximization segmentation, binary mathematical morphology, and active contours models. Masutani *et al.* [73] segment cerebral blood vessels on MRA images using a model-based region growing, controlled by morphological information of local shape.

Many segmentation techniques developed originally for two-dimensional images can be extended to three dimensions—for example, region growing, edge detection, or multispectral segmentation [12, 19, 21, 90, 125]. *3D segmentation* combined with 3D rendering allows for more comprehensive and detailed analysis of image structures than is possible in a spatially limited single-image study. A number of 3D segmentation techniques can be found in the literature, such as 3D connectivity algorithm with morphological interpolation [57], 3D matching of deformable models [70], 3D edge detection [78], coupled surfaces propagation using level set methods [131], and a hybrid algorithm based on thresholding, morphological operators, and connected component labeling [46, 100].

There has been great interest in building digital volumetric models (3D atlases) that can be used as templates, mostly for the MR segmentation of the human brain [23, 47, 61]. A *model-based segmentation* is achieved by using atlas information to guide segmentation algorithms. In the first step, a linear registration is determined for global alignment of the atlas with the image data. The linear registration establishes corresponding regions and accounts for translation, rotation and scale differences. Next, a nonlinear transform (such as elastic warping, [5]) is applied to maximize the similarity of these regions.

Warfield *et al.* [122, 123] developed a new, adaptive, template-moderated, spatially varying, statistical classification algorithm. The algorithm iterates between a classification step to identify tissues and an elastic matching step to align a template of normal anatomy with the classified tissues. Statistical classification based upon image intensities has often been used to segment major tissue types. Elastic registration can generate a segmentation by matching an anatomical atlas to a patient scan. These two segmentation approaches are often complementary. Adaptive, template moderated, spatially varying, statistical classification integrates these approaches, avoiding many of the disadvantages of each technique alone, while exploiting the combination. The algorithm was applied to several segmentation problems, such as quantification of normal anatomy (MR images of brain and knee cartilage) and pathology of various types (multiple sclerosis, brain tumors, and damaged knee cartilage). In each case, the new algorithm provided a better segmentation than statistical classification or elastic matching alone.

Figure 6.4-11 shows an example of 3D segmentation of normal and pathological brain tissues. The tumor

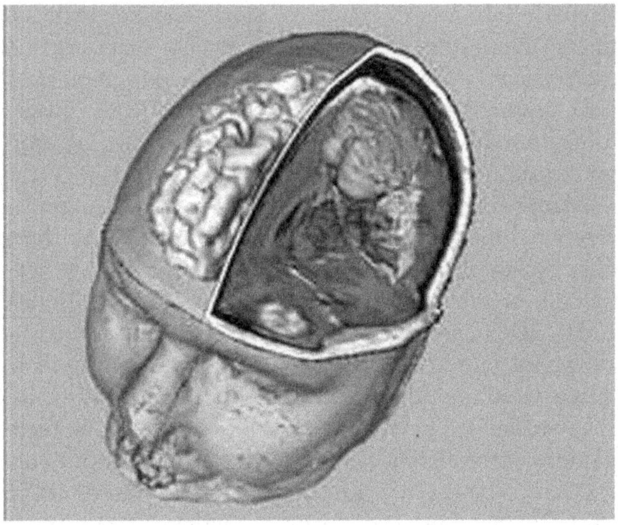

Figure 6.4-11 Rendering of 3D anatomical models and 2D MRI cross-sections of a patient with a meningioma. The models of the skin surface, the brain, and the tumor (green) are based on automatically segmented 3D MRI data. The precentral gyrus (yellow) and the corticospinal tract (blue) are based on a previously aligned digital brain atlas [61]. (Courtesy of Drs. Ron Kikinis, Michael Kaus, and Simon Warfield, Surgical Planning Lab, Department of Radiology, Brigham and Women's Hospital, Boston.)

segmentation was carried out with the algorithm of Kaus *et al.* [60]. This visualization was used to support preoperative surgical planning for tumor resection.

In some medical images, regions that have similar average intensities are visually distinguishable because they have different textures. Each pixel can be assigned a texture value and the image can be segmented using texture instead of intensity [6, 79].

Segmentation has also been addressed with neural networks in several applications [2, 28, 37, 40, 69, 82, 101, 132].

6.4.8 Concluding remarks

Segmentation is an important step in many medical applications involving measurements, 3D visualization, registration, and computer-aided diagnosis. This chapter was a brief introduction to the fundamental concepts of segmentation and methods that are commonly used.

Selection of the "correct" technique for a given application is a difficult task. Careful definition of the goals of segmentation is a must. In many cases, a combination of several techniques may be necessary to obtain the segmentation goal. Very often integration of information from many images (acquired from different modalities or over time) helps to segment structures that otherwise could not be detected on single images.

As new and more sophisticated techniques are being developed, there is a need for objective evaluation and quantitative testing procedures [17, 20, 26]. Evaluation of segmentation algorithms using standardized protocols will be useful for selection of methods for a particular clinical application.

Clinical acceptance of segmentation techniques depends also on ease of computation and limited user supervision. With the continued increases in computer power, the automated realtime segmentation of multispectral and multidimensional images will become a common tool in clinical applications.

References

1. R. Adams, L. Bischof, "Seeded region growing," *IEEE Trans. Pattern Recogn. Mach. Intell.*, Vol. 16, no. 6, pp. 641–647, 1994.
2. S. C. Amartur, D. Piraino, Y. Takefuji, "Optimization neural networks for the segmentation of Magnetic Resonance images," *IEEE Trans. Med. Imag.*, vol. 11, no. 2, pp. 215–220, 1992.
3. M. S. Atkins, B. T. Mackiewich, "Fully automatic segmentation of the brain in MRI," *IEEE Trans. Med. Imag.*, vol. 17, no. 1, pp. 98–107, 1998.
4. G. J. Awcock, R. Thomas, *Applied Image Processing.* NewYork: McGraw-Hill, Inc., 1996.
5. R. Bajcsy, S. Kovacic, "Multiresolution elastic matching," *Comp. Vision, Graphic and Image Proc.*, vol. 46, pp. 1–21, 1989.
6. D. G. Ballard, C. M. Brown, *Computer Vision*, Englewood Cliffs: Prentice Hall, 1982.
7. P. A. Bandettini, A. Jesmanowicz, E. C. Wong, J. S. Hyde, "Processing strategies for functional MRI of the human brain," *Magn. Res. Med.*, vol. 30, pp. 161–173, 1993.
8. I.N. Bankman, T. Nizialek, I. Simon, O.B. Gatewood, I.N. Weinberg, W R. Brody, "Segmentation algorithms for detecting microcalcifications in mammograms", *IEEE Trans. Inform. Techn. Biomed.*, vol. 1, no. 2, pp. 141–149.
9. J. Bernsen, "Dynamic thresholding of gray-level images,"in Proc. *8th Int. Conf. Pattern Recognition*, Paris, France, pp. 1251–55, Oct. 1986.
10. S. Beucher, "Segmentation tools in mathematical morphology," *SPIE*, vol. 1350, Image Algebra and Morphological Image Processing, pp. 70–84, 1990.
11. J.C. Bezdek, L.O. Hall, L.P. Clarke, "Review of MR image segmentation techniques using pattern recognition," *Med. Phys.*, vol. 20, no. 4, pp. 1033–1048, 1993.
12. M. Bomans, K.-H. Hohne, U. Tiede, M. Riemer, "3-D segmentation of MR images of the head for 3-D display," *IEEE Trans. Medical Imaging*, vol. 9, no. 2, pp. 177–183, 1990.
13. M.E. Bramdt, T.P. Bohan, L.A. Kramer, J.M. Fletcher, "Estimation of CSF, white and gray matter volumes in hydrocephalic children using fuzzy clustering of MR imaging,"

Computerized Med. Imag. Graphics, vol. 18, no. 1, pp. 25–34, 1994.

14. J.F. Brenner, J.M. Lester, W.D. Selles, "Scene segmentation in automated histopathology: techniques evolved from cytology automation," *Pattern Recognition*, vol. 13, pp. 65–77, 1981.
15. D. Brzakovic, X.M. Luo, P. Brzakovic, "An approach to automated detection of tumors in mammograms," *IEEE Trans. Med. Imag.*, vol. 9, no. 3, pp. 233–241, 1990.
16. K.R. Castleman, *Digital Image Processing*. Upper Saddle River: Prentice Hall, 1996.
17. V. Chalana, Y. Kim, "A methodology for evaluation of boundary detection algorithms on medical images," *IEEE Med. Imag.*, vol. 16, no. 5, pp. 634–652, 1997.
18. C.K. Chow, T. Kaneko, "Automatic boundary detection of the left ventricle from cineangiograms," *Comput. Biomed. Res.*, vol. 5, pp. 388–410, 1972.
19. L.P. Clarke, R.P. Velthuizen, S. Phuphanich, J.D.Schellenberg, J. A. Arrington, M. Silbiger, "MRI: Stability of three supervised segmentation methods," *Mag. Res. Imag.*, vol. 11, pp. 95–106, 1993.
20. L.P. Clarke, R.P. Velthuizen, M.A. Camacho, J.J. Heine, M. Vaidyanathan, L.O. Hall, R.W. Thatcher, M.L. Silbiger, "MRI segmentation: methods and applications,"*Mag. Res. Imag.*, vol. 13, no. 3, pp. 334–368, 1995.
21. H.E. Cline, C.L. Dumoulin, H.R. Hart, W.E. Lorensen, S. Ludke, "3D reconstruction of the brain from magnetic resonance images using a connectivity algorithm," *Magn. Reson. Imaging*, vol. 5, pp. 345–352, 1987.
22. H.E. Cline, W.E. Lorensen, R. Kikinis, R. Jolesz,"Three-dimensional segmentation of MR images of the head using probability and connectivity," *J. Comp. Assisted Tomography*, vol. 14, no. 6, pp. 1037–1045, 1990.
23. D. L. Collins, T.M. Peters, W. Dai, A. C. Evans, "Model based segmentation of individual brain structures from MRI data," In: R. A. Robb, Ed.: Visualization in Biomedical Computing II, *Proc. SPIE* 1808. Chapel Hill, NC, pp. 10–23, 1992.
24. L. S. Davies, "A survey of edge detection techniques," *Computer Graph. and Image Proc.*, vol.4, pp. 248–270, 1975.
25. E. R. Davies, *Machine Vision*, San Diego: Academic Press, 1997.
26. C. deGraaf, A. Koster, K. Vincken, M. Viergever, "Amethodology for the validation of image segmentation algorithms," In: *Proc. IEEE Symp. Computer-Based Medical Systems*, pp. 17–24, 1992.

26a A. P. Dempster, N. M. Laird, D. B, Rubin, Maximum likelihood from incomplete data via the EM algorithm. *J. Roy. Statist. Soc.*, vol. 39, pp. 1–38, 1977.

27. R. O. Duda, P. E. Hart, *Pattern Recognition and Scene Analysis*, New York: Wiley, 1973.
28. S. Fahlman, G. Hinton, "Connectionist architectures for artificial intelligence," *IEEE Computer*, vol. 20, no. 1, pp.100–109, 1987.
29. L. M. Fletcher, J. B. Marsotti, J. P. Hornak, "A multispectral analysis of brain tissues," *Magn. Reson. Med.*, vol. 29, pp. 623–630, 1993.
30. K. S. Fu, J. K. Mui, "A survey on image segmentation," *Pattern Recognition*, vol. 13, no. 1, pp. 3–16, 1981.
31. G. Gerig, J. Martin, R. Kikinis, O. Kuebler, M. Shenton, F. A. Jolesz, "Automating segmentation of dual-echo MR head data," In: *Information Processing in Medical Imaging* A. C. Colchester, D. J. Hawkes, Eds., Berlin: Springer-Verlag, pp. 175–185, 1991.
32. G. Gerig, O. Kubler, R. Kikinis, F. A. Jolesz, "Nonlinear anisotropic filtering of MRI data," *IEEE Trans Med. Imag.*, vol. 11, no. 2, pp. 221–232, 1992.
33. G. Gerig, D. Welti, C. Guttmann, A. Colchester, G. Szekely, "Exploring the discrimination power of the time domain for segmentation and characterization of lesions in serial MR data," *Proc. of First Intern. Conf. on Medical Image Computing and Computer-Assisted Intervention (MICCAI'98)*, pp. 469–479, in Lecture Notes in Computer Science, Eds. W.M. Wells, A. Colchester, and S. Delp, Springer Verlag, 1998.
34. S. Gilles, M. Brady, J.P. Thirion, N. Ayache, "Bias field correction and segmentation of MR images," *Proc. VBC'96, Visualization in Biomedical Computing*, in Lecture Notes in Computer Science, vol. 1131, pp. 153–158, Springer-Verlag, 1996.
35. J. E. Golston, R. H. Moss, W. V. Stoecker, "Boundary detection in skin tumor images: An overall approach and a radial search algorithm," *Pattern Recognition*, vol. 23, no. 11, pp. 1235–1247, 1990.
36. R. C. Gonzalez, R. E. Woods, *Digital Image Processing*. Reading, MA: Addison-Wesley Publishing Company, 1993.
37. E. Gose, R. Johnsonbaugh, S. Jost, *Pattern Recognition and Image Analysis*. Upper Saddle River: Prentice Hall, 1996.
38. A. Goshtasby, D. A. Turner, "Segmentation of Cardiac Cine MR Images for extraction of right and left ventricular chambers," *IEEE Trans. Med. Imag.*, vol. 14, no. 1, pp. 56–64, 1995.
39. R. Guillemaud, M. Brady, "Estimating the bias field of MR images," *IEEE Trans. Med. Imag.*, vol. 16, no. 3, pp. 238–251, 1997.
40. L. O. Hall, A. M. Bensaid, L. P. Clarke, R. P. Velthuizen, M. S. Silbiger, J. C. Bezdek, "A comparison of neural network and fuzzy clustering techniques in segmenting magnetic resonance images of the brain," *IEEE Trans. Neural Networks*, vol. 3, no. 5, pp. 672–682, 1992.
41. R. M. Haralick, L. G. Shapiro, "Survey: image segmentation techniques," *Comp. Vision Graph Image Proc.*, vol. 29, pp. 100–132, 1985.

41a. J. J. Heine, S. R. Deans, D. K. Cullers, R. Stauduhar, L. P. Clarke, "Multiresolution statistical analysis of high-resolution digital mammograms," *IEEE Trans. Med. Imag.*, vol. 16, no. 5, pp. 503–515, 1997.

42. K. Held, E. Rota Kopps, B. Krause, W. Wells, R. Kikinis, H. Muller-Gartner, "Markov random field segmentation of brain MR images," *IEEE Trans. Med. Imag.*, vol. 16, pp. 878–887, 1998.
43. W. E. Higgins, E. J. Ojard, "Interactive morphological watershed analysis for 3D medical images," *Proceedings of the Symposium on 3D Advanced Image Processing in Medicine*, Rennes, France, pp. 117–121, Nov. 2–4, 1992.
44. K. H. Höhne, M. Bohm M, G. C. Nicolae, "The processing of X-ray image sequences," In: P. Stucki, Ed., *Advances in Digital Image Processing*.

New York: Plenum Press, pp. 147–163, 1979.

45. K. H. Höhne, M. Bohm, "Processing and analysis of radiographic image sequences," In: TS Huang, Ed., *Image Sequence Processing and Dynamic Scene Analysis*. Berlin: Springer-Verlag (NATO ASI Series, Vol F2), pp. 602–623, 1983.
46. K. H. Höhne, W. A. Hanson, "Interactive 3D-segmentation of MRI and CT volumes using morphological operations," *J. Comput. Assist. Tomogr.*, vol. 16, no. 2, pp. 285–294, 1992.
47. K. H. Höhne, M. Bomans, M. Riemer, R. Schubert, U. Tiede, W. Lierse, "A 3D anatomical atlas based on a volume model," *IEEE Comput. Graphics Appl.*, vol. 12, pp. 72–78, 1992.
48. X. Hu, N. Alperin, D. N. Levin, K. K. Tan, M. Mengeot, "Visualization of MR angiography data with segmentation and volume-rendering techniques," *J. Magn. Res.Imag*, vol. 1, pp. 539–546, 1991.
49. X. Hu, V. Johnson, W. H. Wong, C. T. Chen, "Bayesean image processing in magnetic resonance imaging," *Magn. Reson. Imag.*, vol. 9, pp. 611–620, 1991.
50. H. Itagaki, "Improvements of nuclear magnetic resonance image quality using iterations of adaptive nonlinear filtering," *IEEE Trans. Med. Imag.*, vol. 12, no. 2, pp. 322–327, 1993.
51. B. Jahne, *Practical Handbook on Image Processing for Scientific Applications*, Boca Raton, FL: CRC Press, 1997.
52. R. Jain, R. Kasturi, B. G. Schunck, *Machine Vision*, NewYork: McGraw-Hill, 1995.
53. A. K. Jain, *Fundamentals of Digital Image Processing*. Englewood Cliffs, NJ: Prentice Hall, 1989.
54. A. K. Jain, P. J. Flynn, "Image segmentation using clustering," in *Advances in Image Understanding*, K. Bowyer, and N. Ahuja, Eds. Los Alamitas, CA: IEEE Computer Society Press, 1996.
55. E. F. Jackson, P. A. Narayana, J. S. Wolinsky, T. J. Doyle,"Accuracy and reproducibility in volumetric analysis of multiple sclerosis lesions," *J. Comput. Assist. Tomogr.*, vol. 17, pp. 200–205, 1993.
56. L. A. Johnson, J. D. Pearlman, C. A. Miller, T. I. Young, K. R. Thulborn, "MR quantification of cerebral ventricular volume using a semiautomated algorithm," *AJNR*, vol. 14, pp. 1313–1378, 1993.
57. M. Joliot, B. M. Mazoyer, "Three-dimensional segmentation and interpolation of magnetic resonance brain images," *IEEE Trans. Med. Imag.*, vol. 12, no. 2, pp. 269–277, 1993.
58. T. Kapur, W. E. L. Grimson, W. M. Wells, R. Kikinis, "Segmentation of brain tissue from magnetic resonance images," *Medical Image Analysis*, vol. 1, no. 2, pp. 109–127, 1996.
59. T. Kapur, W. E. L. Grimson, R. Kikinis, W. M. Wells, "Enhanced spatial priors for segmentation of magnetic resonance imagery," *Proc. of First Int.Conf. on Medical Image Computing and Computer-Assisted Intervention (MICCAI'98)*, pp. 457–468, in Lecture Notes in Computer Science, Eds. W. M. Wells, A. Colchester, and S. Delp, Springer Verlag, 1998.
60. M. R. Kaus, S. K. Warfield, A. Nabavi, E. Chatzidakis, P. M. Black, F. A. Jolesz, R. Kikinis, "Segmentation of meningiomas and low grade gliomas in MRI," *Proc. of Second Int. Conf. on Medical Image Computing and Computer-Assisted Intervention (MICCAI'99)*, pp. 1–10, in Lecture Notes in Computer Science, Eds. C. Taylor and A. Colchester, Springer Verlag, 1999.
61. R. Kikinis, M. E. Shenton, D. V. Iosifescu, R. W. McCarley, P. Saiviroonporn, H. H. Hokama, A. Robatino, D. Metcalf, C. G. Wible, C. M. Portas, R. Donnino, F. A. Jolesz, "Digital brain atlas for surgical planning, model driven segmentation and teaching,"*IEEE Trans. Visualiz. and Comp. Graph.*, vol. 2, no. 3, 1996.
62. J. Kittler, J. Illingworth, J. Foglein, "Threshold based on a simple image statistics," *Comput. Vision Graph. Image Process.*, vol. 30, pp. 125–147, 1985.
63. A. Kundu, "Local segmentation of biomedical images,"*Comput. Med. Imaging Graph.*, vol. 14, pp. 173–183, 1990.
64. L. M. Lester, H. A. Williams, B. A. Weintraub, J. F. Brenner, "Two graph searching techniques for boundary finding in white blood cell images," *Comput. Biol. Med.*,vol. 8, pp. 293–308, 1978.
65. J. M. Lester, J. F. Brenner, W. D. Selles, "Local transformsfor biomedical image analysis," *Comp. Graph. Imag. Proc.*, vol. 13, pp. 17–30, 1980.
66. Z. Liang, "Tissue classification and segmentation of MR images," *IEEE in Medicine and Biology*, vol. 12, no. 1, pp.81–85, 1993.
67. K. Lim, A. Pfefferbaum, "Segmentation of MR brain images into cerebrospinal fluid spaces, white and gray matter," *J. Comput. Assist. Tomogr.*, vol. 13, pp. 588–593, 1989.
68. J. S. Lim, *Two-Dimensional Signal and Image Processing.*Englewood Cliffs: Prentice Hall, 1990.
69. W. C. Lin, E. C. Tsao, C. T. Chen, "Constraint satisfaction neural networks for image segmentation," *Pattern Recogn.*, vol. 25, no. 7, pp. 679–693, 1992.
70. J. Lotjonen, I. E. Mangin, P-J. Reissman, J. Nenonen, T. Katila, "Segmentation of magnetic resonance images using 3D deformable models," *Proc. of First Int. Conf. on Medical Image Computing and Computer-Assisted Intervention (MICCAI'98)*, pp.1213–1221, in Lecture Notes in Computer Science, W. M. Wells, A. Colchester, and S. Delp, Eds. Springer Verlag, 1998.
71. F. A. Lucas-Quesada, U. Sinha, S. Sinha, "Segmentation strategies for breast tumors from dynamic MR images," *JMRI*, vol. 6, pp. 753–763, 1996.
72. D. Marr, E. Hildreth, "Theory of edge detection," *Proc. Roy. Soc. London*, vol. 27, pp. 187–217, 1980.
73. Y. Masutani. T. Schiemann, K-H. Hohne, "Vascular shape segmentation and structure extraction using ashape-based region-growing model," *Proc. of First Int.Conf. on Medical Image Computing and Computer-Assisted Intervention (MICCAI'98)*, pp. 1242–1249,in Lecture Notes in Computer Science, W. M. Wells, A. Colchester, and S. Delp, Eds. Springer Verlag, 1998.
74. T. Mclnerney, D. Terzopoulos, "Deformable models in medical image analysis: a survey," *Medical Image Analysis*, vol. 1, no. 2, pp. 91–108, 1996.
75. F. Meyer, S. Beucher, "Morphological segmentation,"*Journal of Visual Communication and Image Representation*, vol. 1(1), pp. 21–46, 1990.
76. J. R. Mitchell, S. J. Karlik, D. H. Lee, A. Fenster, "Computer-assisted identification and quantification of multiple sclerosis lesions in MR imaging volumes in the brain," *JMRI*, vol. 4, pp. 197–208, 1994.

77. A. Mitiche, J. K. Aggarwal, "Image segmentation by conventional and information-integrating techniques: a synopsis," *Image and Vision Computing*, vol. 3, no. 2, pp. 50–62, 1985.
78. O. Monga, N. Ayache, P. T. Sander, "From voxel to intrinsic surface features," *Image and Vision Computing*, vol. 10, no. 6, pp. 403–417, 1992.
79. R. Muzzolini, Y-H. Yang, R. Pierson, "Multiresolution texture segmentation with application to diagnostic ultrasound images," *IEEE Trans. on Med. Imag*, vol. 12, no. 1, pp. 108–123, 1993.
79a. K. Nagata, T. Asano, Functional image of dynamic computed tomography for the evaluation of cerebral hemodynamics, *Stroke*, vol. 21, pp. 882–889, 1990.
80. Y. Nakagawa, A. Rosenfeld, "Some experiments on variable thresholding," Pattern recognition, vol. 11, pp.191–204, 1979.
81. N. J. Nilsson, *Problem Solving Methods in Artificial Intelligence*. New York: McGraw-Hill, 1971.
82. M. Ozkan, B. M. Dawant, R. J. Maciunas, "Neural-network based segmentation of multi-modal medical images: A comparative and prospective study," *IEEE Trans Med. Imag.*, vol. 12, no. 3, pp. 534–544, 1993.
83. N. R. Pal, S. K. Pal, "A review on image segmentation techniques," *Pattern Recognition*, vol. 26, no. 9, pp. 1227–1249, 1993.
84. T. Pavlidis, *Structural Pattern Recognition*, Berlin:Springer-Verlag, 1977.
85. W. K. Pratt, *Digital Image Processing*. New York: JohnWiley & Sons, 1991.
86. K. Preston, M. J. Duff, *Modern Cellular Automata*. New York: Plenum Press, 1984.
87. J. C. Rajapakse, J. N. Giedd, J. L. Rapoport, "Statistical approach to segmentation of single-channel cerebral MR images," *IEEE Trans. Med. Imag.*, vol. 16, no. 2, pp. 176–186, 1997.
88. J. C. Rajapakse, J. N. Giedd, C. DeCarli, J. W. Snell, A. McLaughlin, Y. C. Vauss, A. L. Krain, S. D. Hamburger, J. L. Rapoport, "A technique for single-channel MR brain tissue segmentation: Application to a pediatric study," *Magn. Reson. Imag.*, vol. 18, no. 4, 1996.
89. W. E. Reddick, J. O. Glass, E. N. Cook, T. D. Elkin, R. J. Deaton, "Automated segmentation and classification of multispectral magnetic resonance images of brain using artificial neural networks," *IEEE Trans. Med. Imag.*, vol. 16, no. 6, pp. 911–918, 1997.
89a. G. Riediger, L. M. Gravinghoff, K. H. Hohne, E. W Keck, Digital cine angiographic evaluation of pulmonary blood flow velocity in ventricular septal defect. *Cardiovasc. Intervent. Radiol*, vol. 11, pp. 1–4, 1988.
90. R. Robb, *Three-dimensional Biomedical Imaging*. New York: VCH Publishers, Inc., 1995.
91. J. Rogowska, "Similarity methods for dynamic imageanalysis," *Proceedings of International Conf. of AMSE*, vol. 2, pp. 113–124, 1991.
92. J. Rogowska, G. L. Wolf, "Temporal correlation images derived from sequential MRI scans," *J. Comp. Assist. Tomogr.*, vol. 16, no. 5, pp. 784–788, 1992.
93. J. Rogowska, K. Preston, G. J. Hunter, L. M. Hamberg, K. K. Kwong, O. Salonen, G. L. Wolf, "Applications of similarity mapping in dynamic MRI," *IEEE Trans. on Med. Imag.*, vol. 14, no. 3, pp. 480–486, 1995.
94. J. Rogowska, B. Hoop, P. Bogorodzki, J. A. Cannillo, G. L. Wolf, "Delay time mapping of renal function with slip-ring CT," *Proc. of the RSNA Scientific Assembly and Annual Meeting*, Chicago, IL, Nov. 26–Dec. 1, 1995.
95. J. Rogowska, K. Batchelder, G. S. Gazelle, E. F. Halpern, W. Connor, G. L. Wolf, "Quantitative CT lympho-graphy: evaluation of selected two-dimensional techniques for computed tomography quantitation of lymph nodes," *Investigative Radiology*, vol. 31, no. 3, pp. 138–145, 1996.
96. A. Rosenfeld, A. C. Kak, *Digital Image Processing*. New York: Academic Press, 1982.
97. J. C. Russ, *The Image Processing Handbook*. Boca Raton: CRC Press, 1999.
98. P. K. Sahoo, S. Soltani, A. K. Wond, Y. C. Chen, "A survey of thresholding techniques," *Comput. Vision, Graph, Image Process.*, vol. 41, pp. 233–260, 1988.
99. P. Schmid, "Segmentation of digitized dermatoscopic images by two-dimensional color clustering," *IEEE Trans. on Med. Imag.*, vol. 18, no. 2, pp. 164–171, 1999.
100. T. Schiemann, J. Nuthmann, U. Tiede, and K. H. Höhne, "Segmentation of the Visible Human for high quality volume based visualization", In: Visualization in Biomedical Computing, Proc. VBC '96 (K. H. Hohne and R. Kikinis, eds.), vol. 1131 of Lecture Notes in Computer Science, pp. 13–22, Berlin: Springer-Verlag, 1996.
101. N. Shareef, D. L. Wand, R. Yagel, "Segmentation of medical images using LEGION," *IEEE Trans. Med. Imag.*,vol. 18, no. 1, pp. 74–91, 1999.
102. L. Shen, R. Rangayyan, and J. E. L. Desautels, "Detection and classification of mammographic calcifications,"*International Journal of Pattern Recognition and Artificial Intelligence*, vol. 7, pp. 1403–1416, 1993.
103. A. Simmons, S. R. Arridge, G. J. Barker, A. J. Cluckie, P. S. Tofts, "Improvements to the quality of MRI cluster analysis," *Mag. Res. Imag.*, vol. 12, no. 8, pp. 1191–1204, 1994.
104. H. R. Singleton, G. M. Pohost, "Automatic cardiac MR image segmentation using edge detection by tissue classification in pixel neighborhoods," *Mag. Res. Med.*, vol. 37, pp. 418–424, 1997.
105. M. Sonka, X. Zhang, M. Siebes, M. S. Bissing, S. C. Dejong, S. M. Collins, C. R. McKay, "Segmentation of intravascular ultrasound images: a knowledge-based approach," *IEEE Trans. Med. Imag.*, vol. 14, no. 4, pp. 719–731, 1955.
106. M. Sonka, G. K. Reddy, M. D. Winniford, S. M. Collins,"Adaptive approach to accurate analysis of small-diameter vessels in cineangiograms," *IEEE Trans Med. Imag.*, vol. 16, no. 1, pp. 87–95, 1997.
107. M. Sonka, V. Hlavac, R. Boyle, *Image Processing Analysis, and Machine Vision*. CA: PWS Publishing, Pacific Grove, 1999.
108. D. Spielman, M. Sidhu, R. Herfkens, L. Shortlife, "Correlation imaging of the kidney," *Proceedings of the International SMRM Conference*, Nice, France, p. 373, 1995.
109. P. Suetens, E. Bellon, D. Vandermeulen, M. Smet, G. Marchal, J. Nuyts, L. Mortelman, "Image segmentation: methods and applications in diagnostic radiology and nuclear medicine," *European*

Journal of Radiology, vol. 17, pp. 14–21, 1993.

110. Y. Sun, R. J. Lucariello, S. A. Chiaramida, "Directional low-pass filtering for improved accuracy and reproducibility of stenosis quantification in coronary arteriograms," *IEEE Trans. Med. Imag.*, vol. 14, no. 2, pp. 242–248, 1995.
111. T. Taxt, A. Lundervold, B. Fuglaas, H. Lien, V. Abeler,"Multispectral analysis of uterine corpus tumors in magnetic resonance imaging," *Magn. Res. Med.*, vol. 23, pp. 55–76, 1992.
112. D. R. Thedens, D. J. Skorton, S. R. Feagle, "Methods of graph searching for border detection in image sequences with applications to cardiac magnetic resonance imaging," *IEEE Trans. Med. Imag*, Vol. 14, no. 1, pp. 42–55, 1995.
113. V. Torre, T. A. Poggio, "On edge detection," *IEEE Trans PAMI*, vol. 8, pp. 147–163, 1986.
114. D. L. Toulson, J. F. Boyce, "Segmentation of MR images using neural nets," *Image Vision, Computing*, vol. 5, pp. 324-328, 1992.
115. J. K. Udupa, L. Wei, S. Samarasekera, Y. Miki, M. A. van Buchem, R. I. Grossman, "Multiple sclerosis lesion quantification using fuzzy-connectedness principles," *IEEE Trans. Med. Imag.*, vol. 16, no. 5, 1997.
116. S. E. Umbaugh, *Computer Vision and Image Processing: A Practical Approach Using CVIPtools.* Upper Saddle River, NJ: Prentice Hall, 1998.
117. M. W Vannier, R. L. Butterfield, D. Jordan, W. A. Murphy, R. G. Levitt, M. Gado, "Multispectral analysis of magnetic resonance images," *Radiology*, vol. 154, pp. 221–224, 1985.
118. M. W. Vannier, C. M. Speidel, D. L. Rickman, "Magnetic resonance imaging multispectral tissue classification," News *Physiol. Sci.*, vol. 3, pp. 148–154, 1988.
119. L. Vincen, P. Soille, "Watersheds in digital spaces: an efficient algorithm based on immersion simulations," *IEEE Trans. PAMI*, vol. 13(6), pp. 583–598, 1991.
120. L. Vincent, "Morphological grayscale reconstruction in image analysis: applications and efficient algorithms," *IEEE Trans. Image Processing*, vol. 2(2), pp. 176–201, 1993.
121. J. Z. Wang, D. A. Turner, M. D. Chutuape, "Fast,interactive algorithm for segmentation of series of relate dimages: application to volumetric analysis of MR imagesof the heart," *JMRI*, vol. 2, no. 5, pp. 575–582, 1992.
122. S. K. Warfield, M. Kaus, F. A. Jolesz, R. Kikinis, "Adaptive template moderated spatially varying statistical classification," *Proc. of First Int. Conf. on Medical Image Computing and Computer-Assisted Intervention (MICCAI'98)*, pp. 431–438, in Lecture Notes in Computer Science, Eds. W.M. Wells, A. Colchester, and S. Delp, Springer Verlag, 1998.
123. S. K. Warfield, M. K. Ferenc, A. Jolesz, R. Kikinis, "Adaptive template moderated spatially varying statistical classification," *Medical Image Analysis*, 2000 (accepted to appear).
124. J. B. Weaver, Y. Xu, D. M. Healy, L. D. Cormwell, "Filtering noise from images with wavelet transforms," *Magn. Res. Med.*, vol. 21, pp. 288–295, 1991.
125. W. M. Wells III, E. L. Grimson, R. Kikinis, F. A. Jolesz, "Adaptive segmentation of MRI data," *IEEE Trans. Med. Imag.*, vol. 15, no. 4, pp. 429–443, 1996.
126. C-F. Westin, S. Warfield, A. Bhalerao, L. Mui, J. Richolt, R. Kikinis, "Tensor controlled local structure enhancement of CT images for bone segmentation," *Proc. of First Int. Conf. on Medical Image Computing and Computer-Assisted Intervention (MICCAI'98)*, pp. 1205–1212, in Lecture Notes in Computer Science, Eds. W. M. Wells, A. Colchester, and S. Delp, Springer Verlag, 1998.
127. J. S. Weszka. "A survey of threshold selection techniques," *Computer Graphics and Image Proc.*, vol. 7, pp. 259–265, 1978.
128. A. Witkin, M. Kass, D. Terzopoulos, "Snakes: Active contour models," *International Journal of Computer Vision*, vol. 1(4), pp. 321–331, 1988.
129. A. Yezzi, S. Kichenassamy, A. Kumar, P. Olver, A. Tannenbaum, "A geometric snake model for segmentation of medical imagery," *IEEE Trans. Med. Imag.*, vol. 16, no. 2, pp. 199–209, 1997.
130. H. Yoshida, K. Doi, R. M. Nishikawa, "Automated detection of clustered microcalcifications in digital mammograms using wavelet transform techniques," *SPIE Image Processing*, vol. 2167, pp. 868–886, 1994.
131. X. Zeng, L. H. Staib, R. T. Schultz, J. S. Duncan, "Volumetric layer segmentation using coupled surfaces propagation," Proc. *IEEE Conf. on Computer Vision and Pattern Recognition*, pp. 708–715, Santa Barbara, CA, June 1998.
132. Y. Zhu, H. Yan, "Computerized tumor boundary detection using a Hopfield neural network," *IEEE Trans. Med. Imag.*, vol. 16, no. 1, pp. 55–67, 1997.
133. S. W. Zucker, "Region growing: childhood and adolescence," *Comp. Graphics Image. Proc*, vol. 5, pp. 382–399, 1976.

Chapter 6.5

6.5

Registration for image-guided surgery

Eric Grimsom and Ron Kikinis

6.5.1 Introduction

Many surgical procedures require highly precise localization, often of deeply buried structures, in order for the surgeon to extract targeted tissue with minimal damage to nearby structures. Although methods such as MRI and CT are invaluable in imaging and displaying the internal 3D structure of the body, the surgeon still faces a key problem in applying that information to the actual procedure. Since he is limited to seeing exposed surfaces within the surgical opening, he cannot easily visualize paths to targets or positions of nearby, but hidden, critical structures. In addition, the lack of visible landmarks within the surgical opening may inhibit his ability to determine his current position, and thus to navigate safe trajectories to other structures.

Because traditional clinical practice often only utilizes 2D slices of MR or CT imagery, the surgeon must mentally transform critical image information to the actual patient. Thus, there is a need for techniques to register a 3D reconstruction of internal anatomy with the surgical field. Such registered information would support image-guided surgery, by allowing the surgeon to directly visualize important structures, and plan and act accordingly. Visualization methods include "enhanced reality visualization" [14], in which rendered internal structures are overlaid on the surgeon's field-of-view, and instrument tracking, in which medical instruments acting on the patient are localized and visualized in the 3D MR or CT imagery. The benefits of image-guided surgical methods include the following:

- Accelerated migration to minimally invasive surgeries via improved hand–eye coordination and better transfer of a *priori* plans to the patient.
- Shorter procedures through increased visualization of the surgical field.
- Reduced risk of sensitive tissue damage.
- More accurate and complete tissue resection, ablation, or biopsy.

The key stages of an accurate, reliable, image-guided surgery system are as follows:

- Creating accurate, detailed, patient-specific models of relevant anatomy for the surgical procedure.
- Registering the models, and the corresponding imagery, to the patient.
- Maintaining the registration throughout the surgical procedure.
- Tracking medical instruments in the surgical field in order to visualize them in the context of the MR/CT imagery and the reconstructed models.

In this chapter, we describe the registration process used to align preoperative imagery with the actual patient position, and the process by which a surgeon visualizes and navigates through the patient using that information. We do this using an example of a neurosurgical image-guidance system, although the same issues arise in other areas as well.

6.5.2 Image-guided neurosurgery system

Neurosurgery is an ideal application for image-guided techniques, by virtue of the high precision it requires, the need to visualize nearby tissue, the need for planning of optimal trajectories to target tissue, and the need to localize visually indistinguishable, but functionally

Biomedical Engineering Desk Reference; ISBN: 9780123746467

different, tissue types. One method for aligning imagery to the patient is to use some form of extrinsic marker—a set of landmarks or other frame structures that attach to the patient prior to imaging, and that can then be used to establish a correspondence between position in the image and position on the patient. Examples include stereotactic frames, bone screws, and skin markers (e.g., [3, 11, 15, 16, 18, 19, 23–26, 28]).

In other words, by placing markers on the patient prior to imaging, and keeping them rigidly attached through the completion of surgery, one obtains a coordinate frame visible in the imagery that directly supports transference of information to the surgical field of view. Stereotactic frames, though, not only are uncomfortable for the patient, but are cumbersome for the surgeon. They are limited to guidance along fixed paths and prevent access to some parts of the head. We would like to use a frameless system both for its simplicity and generality, and for its potential for use in other parts of the body. More recently, frameless stereotaxy systems have been pursued by many groups (e.g., [1, 2, 6, 21, 29, 35]) and usually consist of two components: registration and tracking. We have added a third, initial, component to our system—reconstructed models of the patient's anatomy. The system's components are described next, with emphasis on the use of registration to align imagery with patient and surgeon's viewpoint.

The architecture of our image-guided surgery system (Fig. 6.5-1) supports frameless, nonfiducial, registration of medical imagery by matching surface data between patient and image model. The system consists of a portable cart (Fig. 6.5-2) containing a Sun UltraSPARC workstation and the hardware to drive the laser scanner and Flashpoint tracking system. On top of the cart is mounted an articulated extendable arm to which we attach a bar housing the laser scanner and Flashpoint cameras. The three linear Flashpoint cameras are inside the bar. The laser is attached to one end of the bar, and a video camera to the other. The joint between the arm and scanning bar has three degrees of freedom to allow easy placement of the bar in desired configurations.

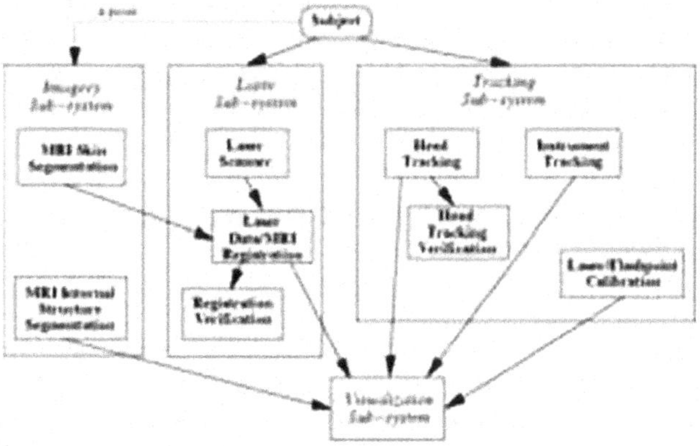

Figure 6.5-1 Image-guided surgery system architecture.

6.5.2.1 Imagery subsystem

MRI is the prime imaging modality for the neurosurgery cases we support. The images are acquired prior to surgery with no need for special landmarking strategies. To use the imagery, it is important to create detailed models of the tissues being imaged. This means that we must segment the images: identify the type of tissue associated with each voxel (or volume element) in the imagery, and then create connected geometric models of the different types of tissue. A wide range of methods (e.g., [27, 30, 33, 34, 37]) have been applied to the segmentation problem. Classes of methods include statistical classifiers (e.g., [33, 37]), which use variations in recorded tissue response to label individual elements of the medical scan, then extract surface boundaries of connected tissue regions to create structural models; deformable surface methods (e.g., [27, 30]), which directly fit boundary models to delineations between adjacent tissue types; and atlas-driven segmenters (e.g., [34]), which use generic models of standard anatomy to guide the labeling and segmentation of new scans.

Our current approach to segmentation uses an automated method to initially segment into major tissue classes while removing gain artifacts from the imagery [17,37], then uses operator-driven interactive tools to refine this segmentation. This latter step primarily relies on 3D visualization and data manipulation techniques to correct and refine the initial automated segmentation. The segmented tissue types include skin, used for registration, and internal structures such as white matter, gray matter, tumor, vessels, cerebrospinal fluid, and structures. These segmented structures are processed by the Marching Cube algorithm [20] to construct isosurfaces and to support surface rendering for visualization.

The structural models of patients constructed using such methods can be augmented with functional information. For example, functional MRI methods or transcranial magnetic stimulation methods (e.g., [9]) can be used to identify motor or sensory cortex. The key issue is then merging this data with the structural models, and to do this we use a particular type of registration method [7,8,38]. This approach uses stochastic sampling to find the registration that optimizes the mutual information between the two data sets. Optimizing mutual information makes the method insensitive to intensity differences between the two sensory modalities, and hence it can find the best alignment even if different anatomical features are highlighted in the scans.

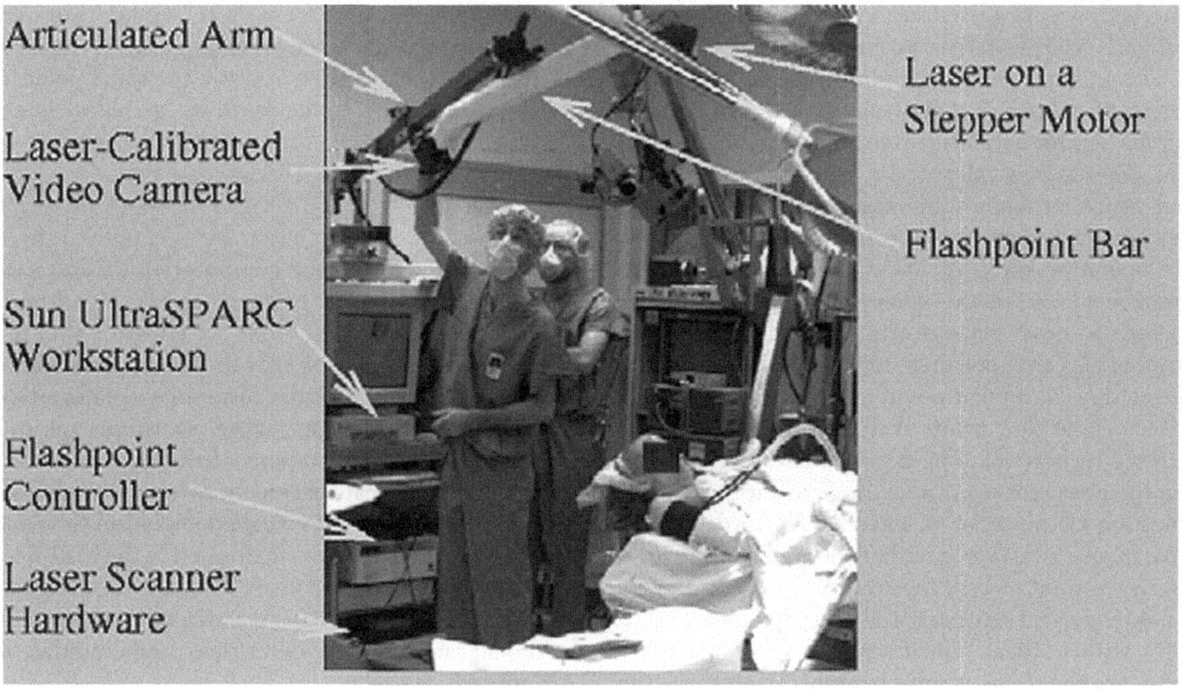

Figure 6.5-2 Physical setup for image guided surgery system.

The result is an augmented, patient-specific, geometric model of relevant structural and functional information. Examples are shown in Fig. 6.5-3.

6.5.2.2 Registration subsystem

Registration is the process by which the MRI or CT data is transformed to the coordinate frame of the patient. Excellent reviews of registration methods include [22, 24, 32].

Extrinsic forms of registration use fiducials (e.g., [1, 21, 31, 35]): either markers attached to the skin or bone prior to imaging or anatomically salient features on the head. The fiducials are manually localized in both the MR or CT imagery and on the patient, and the resulting correspondences are used to solve for the registration. Fiducial systems may not be as accurate as frame-based methods—Peters *et al.* [28] reports fiducial accuracy about an order of magnitude worse than frame-based methods, but Maciunas *et al.* [21] reports high accuracy achieved with novel implantable fiducials.

Intrinsic registration is often based on surface alignment, in which the skin surface extracted from the MRI data is aligned with the patient's scalp surface in the operating room. Ryan *et al.* [29] generates the patient's scalp surface by probing about 150 points with a trackable medical instrument. Colchester *et al.* [6] uses an active stereo system to construct the scalp surface. We also perform the registration using surface alignment [12], benefiting from its dense data representation, but use either a laser scanner to construct the patient's scalp surface or a trackable probe to obtain data points from the patient's skin surface for registration.

We have used two related methods to register the reconstructed model to the actual patient position. In the

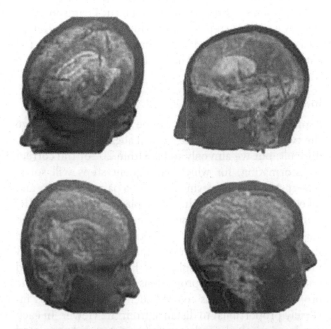

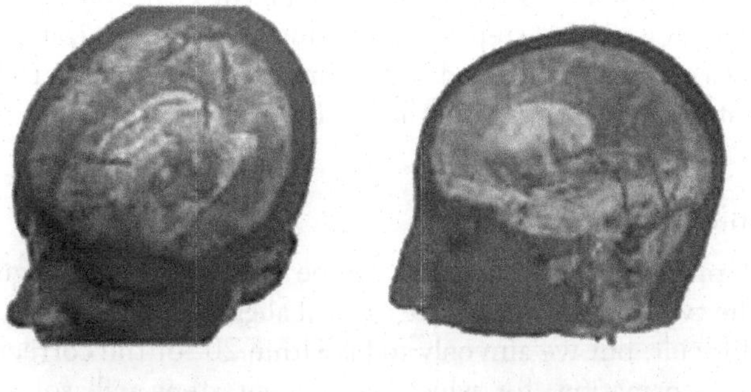

Figure 6.5-3 Examples of patient reconstructions by segmenting MRI scans into different tissue types.

first method, we use a laser scanner to collect 3D data of the patient's scalp surface as positioned on the operating table. The scanner is a laser striping triangulation system consisting of a laser unit (low-power laser source and cylindrical lens mounted on a stepper motor) and a video camera. The laser is calibrated a priori by using a calibration gauge of known dimensions to calculate the camera parameters and the sweeping angles of the laser. In the operating room the laser scanner is placed to maximize coverage of the salient bony features of the head, such as nose and eye orbits. To ensure accurate registration we can supplement the laser data with points probed with a Flashpoint pointer, similar to Ryan *et al.* [29], to include skin points that are not visible to the laser in the registration. The acquired laser data is overlaid on the laser scanner's video image of the patient for specification of the region of interest. This process uses a simple mouse interface to outline the region of the head on which we want to base the registration. This process need not be perfect—the registration is designed to deal robustly with outliers. The laser scan takes about 30 seconds once the sensor is appropriately placed above the patient.

An alternative method is to simply use a trackable probe to acquire data. In this case, we trace paths on the skin of the patient with the trackable probe, recording positional information at points along each path. These points are not landmarks, but simply replace the lines of laser data. The registration process is the same, whether matching laser data or trackable probe data to the skin surface of the MRI model.

The key to our system is the integration of a reliable and accurate data-to-MRI registration algorithm. Our registration process is described in detail in Grimson *et al.* [14]. It is a three-step process performing an optimization on a six-parameter rigid transformation, which aligns the data surface points with the MRI skin surface.

Initial alignment

A manual initial alignment can be used to roughly align the two surfaces. Accurate manual alignment can be very difficult, but we aim only to be within 20° of the correct transformation, for which subsequent steps will solve. One method for achieving this uses a pair of displays and takes about 60 seconds. In one display, the rendered MRI skin is overlaid on the laser scanner's video view of the patient, and the MRI data is rotated and translated in three dimensions to achieve a qualitatively close alignment. In the second display, the laser data is projected onto three orthogonal projections of the MRI data. The projected MRI data is colored such that intensity is inversely proportional to distance from the viewer. In each overlay view, the laser data may be rotated and translated in two dimensions to align the projections. An alternative to manual initial alignment is to record three known points using the trackable probe (e.g., tip of the nose, tip of the ear), then identify roughly the same point in the MRI model, using a mouse-driven graphical interface. This process determines a rough initial alignment of the data to the MR reconstruction and typically takes less than 5 seconds.

It is also possible to automate this process, by using search methods from the computer vision literature. In [14], we describe an efficient search algorithm that matches selected points from the patient's skin to candidate matches from the skin surface of the MRI model. By using constraints on the distance and orientation between the sets of points, these algorithms can quickly identify possible registrations of the two data sets. Applying the coordinate frame transformation defined by each match, the full set of data points from the patient's skin surface can then be transformed to the MRI frame of reference. Residual distances between the transformed data points and the MRI skin surface serve as a measure of fit and can be used to determine good candidate initial alignments.

Refined alignment

Given the initial alignment of the two data sets, we typically have registrations on the order of a few centimeters and a few tens of degrees. We need to automatically refine this alignment to a more accurate one. Ideally, we need algorithms that can converge to an optimal alignment from a large range of initial positions [12–14].

Our method iteratively refines its estimate of the transformation that aligns patient data and MRI data. Given a current estimate of the transformation, it applies that estimate to the patient data to bring it into the MRI coordinate frame. For each transformed data point, it then measures a Gaussian weighted distance between the data point and the nearest surface point in the MRI model. These Gaussian weighted distances are summed for all data points, which defines a measure of the goodness of fit of the current estimated transformation. This objective function is then optimized using a gradient descent algorithm. The role of the Gaussian weighting is to facilitate "pulling in" of one data set to the other, without needing to know the exact correspondence between data points. The process can be executed in a multiresolution manner, by first using Gaussian distributions with large spreads (to get the registration close), then reducing the spread of the distribution, and resolving in a sequence of steps.

This process runs in about 10 seconds on a Sun UltraSPARC workstation. The method basically solves for the transform that optimizes a Gaussian weighted least-squares fit of the two data sets.

Detailed alignment

Automated detailed alignment then seeks to accurately localize the best surface data to MRI transformation [12–14]. Starting from the best transformation of the previous step, the method then solves a second minimization problem. In this case it measures the least-squares fit of the two data sets under the current estimated transformation (subject to a maximum distance allowed between a transformed data point and the nearest point on the skin surface, to discount the effects of outliers in the data). This minimization can again be solved using a gradient descent algorithm.

This process runs in about 10 seconds on a Sun UltraSPARC workstation. The method basically solves a truncated least-squares fit of the two data sets, refining the transformation obtained in the previous step.

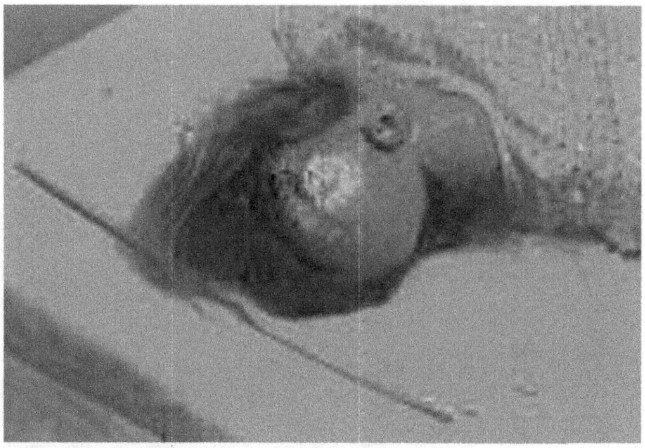

Figure 6.5-4 Example of augmented reality visualization. Tumor and ventricles have been overlaid onto live video view of patient.

Stochastic perturbation

To ensure that the solution found using this process is not a local minimum, the method arbitrarily perturbs the transformation and reruns the process. If the system converges to the same solution after several trials, the system terminates with this registration.

Camera calibration

The final stage of the process is to determine the relationship between a video camera viewing the patient, and the patient position. This can be accomplished by using a trackable probe to identify the positions of points on a calibration object in patient coordinates. By relating those coordinates to the observed positions in the video image, one can solve for the transformation relating the camera to the patient [12–14].

Augmented reality visualization

By coupling all of these transformations together, we can provide visualizations of internal structures to the surgeon. In particular, we can transform the segmented MRI model (or any portions thereof) into the coordinate frame of the patient, then render those structures through the camera transformation, to create a synthetic image of how those structures should appear in the camera. This can then be mixed with a live video view to overlay the structures onto the actual image (Fig. 6.5-4).

Verifying the registration

Three verification tools are used to inspect the registration results, as the objective functions optimized by the registration algorithm may not be sufficient to guarantee the correct solution. One verification tool overlays the MRI skin on the video image of the patient (Fig. 6.5-5), except that we animate the visualization by varying the blending of the MRI skin and video image. A second verification tool overlays the sensed data on the MRI skin by color-coding the sensed data by distance between the data points and the nearest MRI skin points. Such a residual error display identifies possible biases remaining in the registration solution. A third verification tool compares locations of landmarks. Throughout the surgery, the surgeon uses the optically tracked probe to point to distinctive anatomical structures. The offset of the probe position from the actual point in the MR volume is then observed in the display. This serves to measure residual registration errors within the surgical cavity.

6.5.2.3 Tracking subsystem

Tracking is the process by which objects are dynamically localized in the patient's coordinate system. Of particular interest to us is the tracking of medical instruments and the patient's head. The two most common methods of tracking are articulated arms and optical tracking. Articulated arms are attached to the head clamp or operating table and use encoders to accurately compute the angles of its joints and the resulting 3D position of its end point. Such devices, though, may be bulky in the operating room and, because of their mechanical nature, are not as fault tolerant as other methods. Optical trackers use multiple cameras to triangulate the 3D location of flashing LEDs that may be mounted on any object to be tracked. Such devices are generally perceived as the most accurate, efficient, and reliable localization system [2, 5]. Other methods such as acoustic or magnetic field sensing are being explored as well, but can be more sensitive to environmental effects. We use optical tracking (the Flashpoint system by IGT Inc., Boulder, CO, USA)

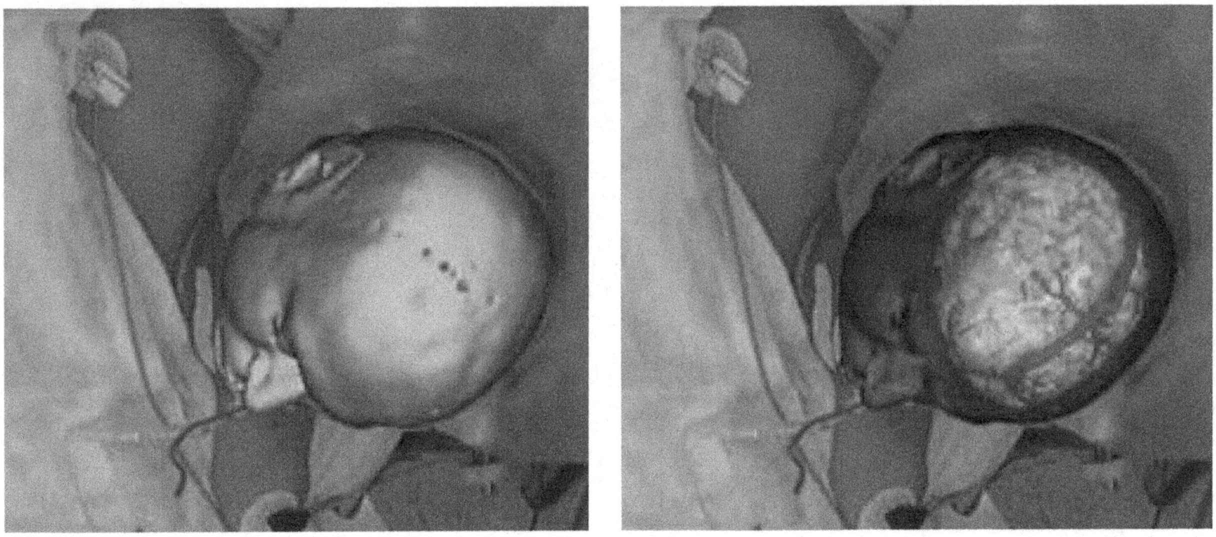

Figure 6.5-5 (*Left*) Initial registration, showing overlay of MRI skin data onto registered video image. (*Right*) Enhanced reality visualization of the patient showing hidden internal structures overlaid on the surgeon's view of the patient.

because of its accuracy and ease-of-use benefits, though magnetic tracking systems are of similar capability.

Tracking patient head motion is often necessary for a variety of reasons. The head is not always clamped to the operating table, the head may move relative to the clamp, the operating table may be moved, or the hardware performing the tracking may be moved to rearrange lights or other equipment in the operating room. Although not all image-guided surgery systems account for patient motion, [1, 2, 6, 21, 29] solve this problem by attaching trackable markers to the head or clamp. We currently utilize an optically trackable configuration of markers attached to a Mayfield clamp (Fig. 6.5-6). We have also experimented with directly attaching trackable LEDs to the skin surface of the patient. Our experience is that while in most cases this worked well, it required that the surgeon carefully plan the location of the LEDs to ensure that they did not move between initial placement and opening of the skin flap.

We require direct line-of-sight from the Flashpoint cameras to the LEDs at times when the surgeon requires image guidance. In order to maintain such line-of-sight, we can relocate the scanning bar such that it is out of the way of the surgeon but maintains visibility of the LEDs. Such dynamic reconfiguration of the scanning bar is a benefit of the head tracking process.

Instrument tracking is performed by attaching two LEDs to a sterile pointer. The two LEDs allow us to track the 3D position of the tip of the pointer as well as its orientation, up to the twist angle, which is not needed for this application. Figure 6.5-6 shows the surgeon using the trackable pointer in the opened craniotomy.

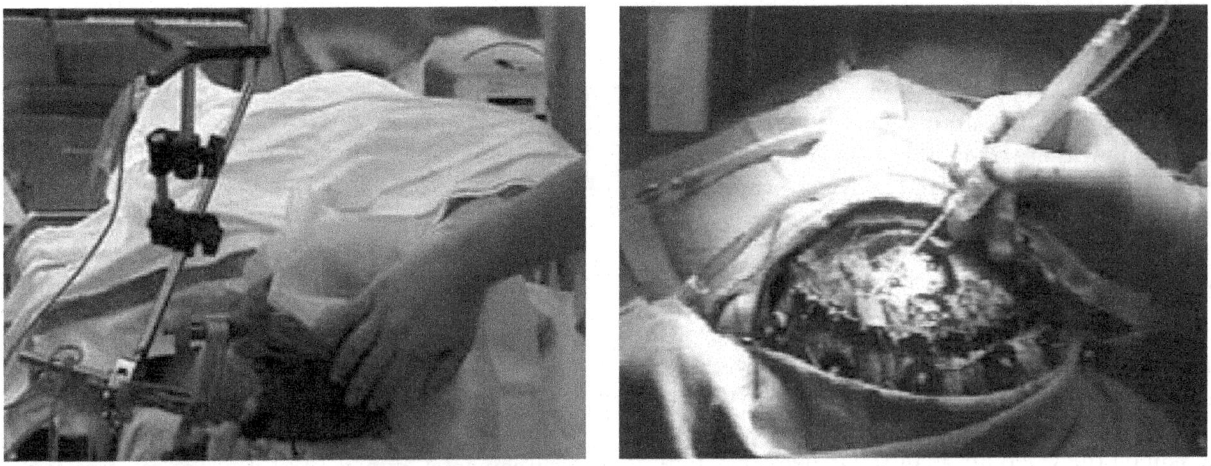

Figure 6.5-6 Trackable configuration of LEDs attached to head clamp, or to the skin flap.

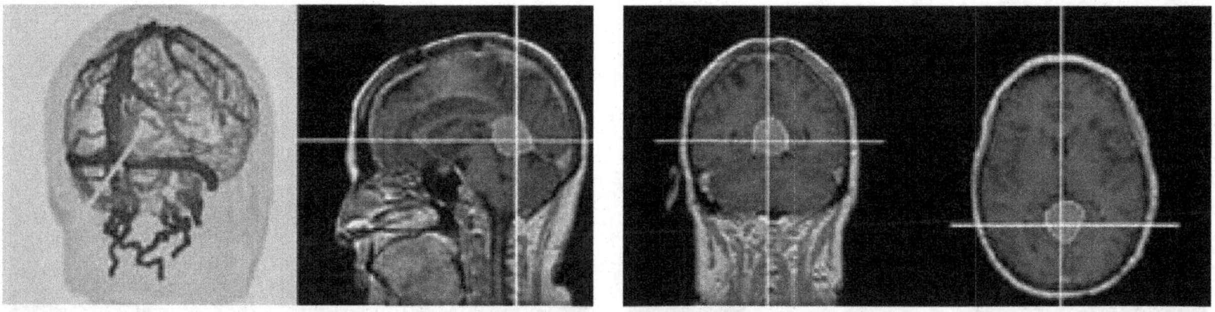

Figure 6.5-7 Pointer tracking in 3D MRI rendering and three orthogonal MRI slices.

6.5.2.4 Visualization subsystem

Two types of visualizations are provided to the surgeon on the workstation monitor. One is an enhanced reality visualization in which internal structures are overlaid on the video image of the patient. The video image is set up to duplicate the surgeon's view of the patient. Any segmented MR structures may be displayed at varying colors and opacities (see Fig. 6.5-5).

A second visualization shows the location of the pointer tip in a 3D rendering of selected MRI structures and in three orthogonal MRI slices (see Fig. 6.5-7). These visualizations are updated twice per second as the pointer is moved.

6.5.3 Operating room procedure

Using our system, as seen from the surgeon's perspective, involves the following steps:

(1) Prepare patient for surgery as per usual procedure, including clamping the head. Head is still visible.

(2) Attach a configuration of LEDs to the head clamp, and record the positions of the LEDs in the Flashpoint system.

(3) Register MRI to patient by placing our scanner bar over patient's head. The bar is generally about 1.5 m away from head. Scan patient's head by swabbing a trackable probe across the skin. Typically several swabs are used, designed to cover a wide range of positions on the patient. It is often convenient to include swabs along known paths such as across the cheeks or down the nose, as these paths will aid in inspecting the resulting registration.

(4) The Flashpoint/laser bar may be repositioned at any point to avoid interference with equipment and to maintain visibility of LEDs.

(5) Sterilize and drape patient. Any motion of the patient during this process will be recorded by movements of the LED configuration attached to the head clamp.

(6) Proceed with craniotomy and surgical procedure.

(7) At any point, use sterile Flashpoint pointer to explore structures in the MR imagery.

6.5.4 Performance analysis

To evaluate the performance of our registration and tracking subsystems, we have performed an extensive set of controlled perturbation studies [10]. In these studies, we have taken existing data sets, simulated data acquisition from the surface of the data, added noise to the simulated surface data, then perturbed the position of data and solved for the optimal registration. Since we know the starting point of the data, we can measure the accuracy with which the two data sets are reregistered.

Although extensive details of the testing are reported in Ettinger *et al.* [10], the main conclusions of the analysis are as follows:

- Accurate and stable registration is achieved for up to 45° rotational offsets of the data sets, with other perturbations.
- Accurate and stable registration is achieved for up to 75° rotational offsets of the data sets, with no other perturbations.
- Robust registration is obtained when the surface data spans at least 40% of the full range of the surface, and is generally obtained with as little as 25% coverage.
- Small numbers of outliers do not affect the registration process.

6.5.5 Operating room results

We have used the described image-guided neurosurgery system on more than 100 patients. These cases included

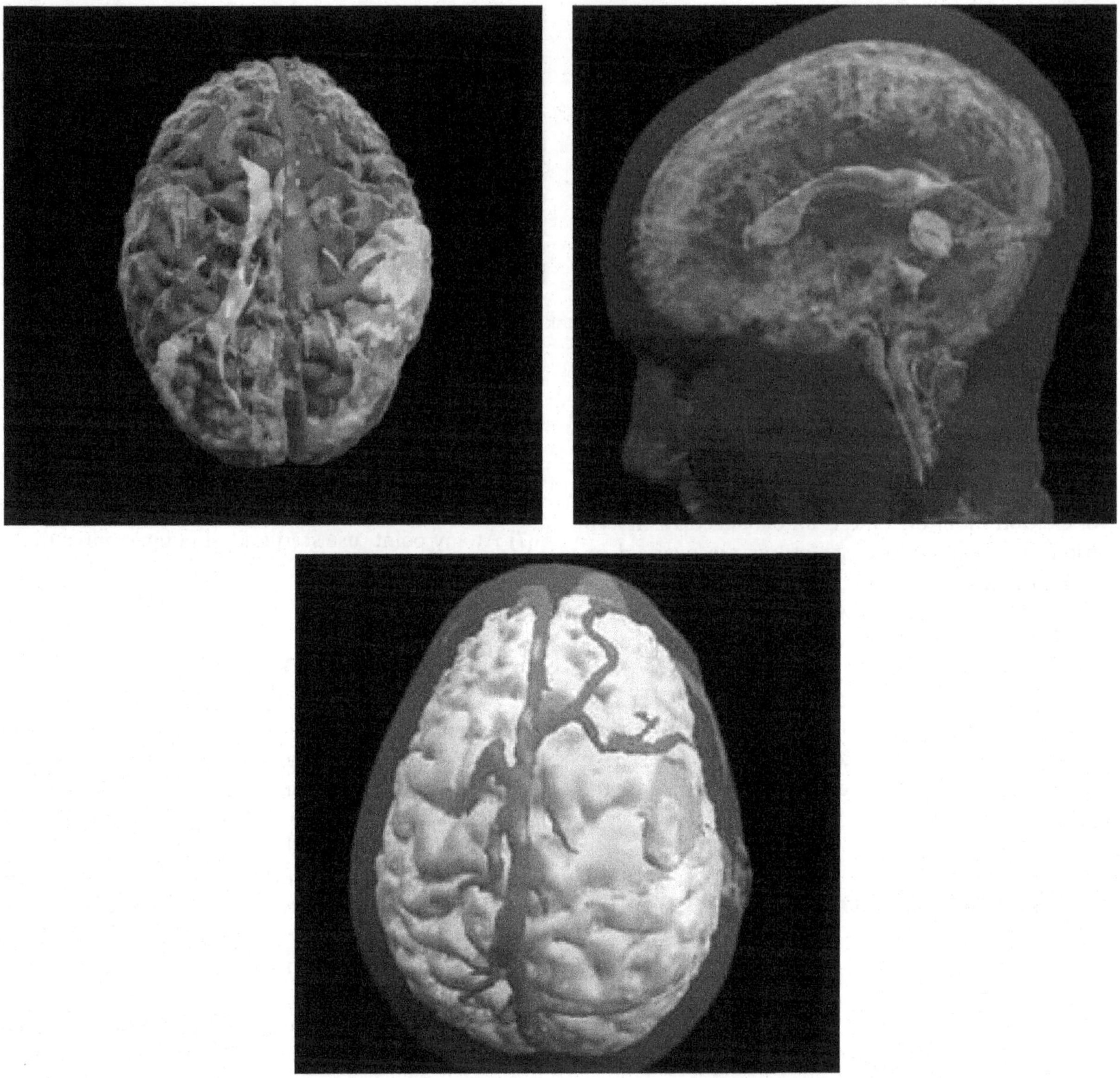

Figure 6.5-8 Examples of neurosurgical cases. The last example includes a fusion of fMRI data overlaid on top of the structure model.

high-grade and low-grade supratentorials; meningiomas; metastases; posterior fossa; meningioangiomatosis; intractable epilepsy; vascular; biopsies; and demyelinating lesion.

In all cases the system effectively supported the surgery as follows:

- By providing guidance in planning bone cap removal—this was done through the augmented reality visualization in which the surgeon could visualize paths to the critical tissue and plan an appropriate entry point.
- Identifying margins of tumor—this was done by tracing the boundaries of tissue with the trackable probe.
- Localizing key blood vessels.
- Orienting the surgeon's frame of reference.

Selected examples are shown in Fig. 6.5-8.

To qualitatively validate the system's performance, the surgeon placed the pointer on several known landmarks: skull marks from previous surgeries, ventricle tip, inner skull bones such as eye orbits, sagittal sinus, and small cysts or necrotic tissues. He then estimated their position in the MRI scan, and we compared the distance between the expected position and the system's tracked position. Typically, this error was less than two voxels (MRI resolution was 0.9375 mm by 0.9375 mm by 1.5 mm), although this does depend in some cases on the administration of drugs to control brain swelling.

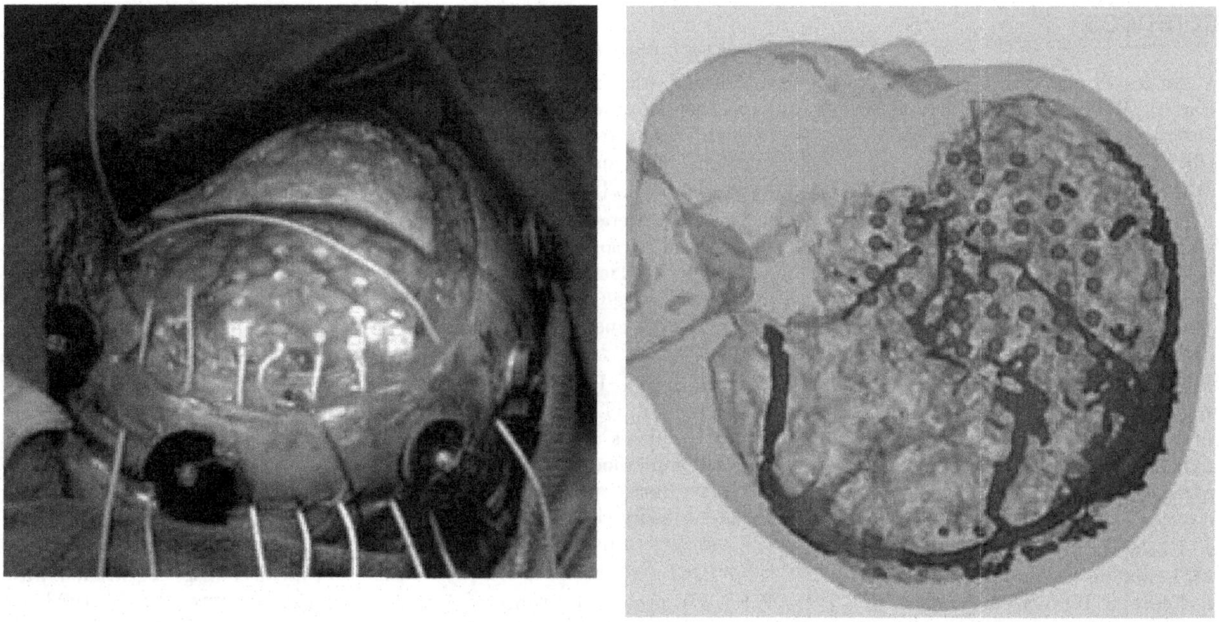

Figure 6.5-9 Grid of electrodes placed on cortical surface. Location of grid points overlaid on MR reconstruction, with focal area highlighted.

One example of the effectiveness of the system is illustrated by the following study. Twenty patients with low-grade gliomas underwent surgery with the system. The pathologies included 10 low-grade astrocytomas (grades I, II out of IV), 7 oligoastrocytomas (without anaplastic features), and 3 oligodendrogliomas. Thirteen patients underwent cortical mapping, including 7 who underwent speech and motor mapping, 2 motor alone, 1 speech alone, and 3 motor and sensory. This cortical mapping was then registered with the structural MRI model and used to provide guidance to the surgeon. In these cases, 31% had a subtotal resection; the remainder had total resection. One patient exhibited temporary left-sided weakness. Cortical mapping had represented the sensory cortex diffusely behind this patient's gross tumor. The postoperative weakness was temporary and was thought to be due to swelling. One patient showed a mild, left upper extremity proprioreceptive deficit, which was due to a vascular accident on postoperative day 1. The remaining patients were neurologically intact following the procedure.

In addition to the tumor resection cases, we have also used the system in 10 pediatric epilepsy cases [4]. In the first stage of this two-stage surgery, the patient's cortex is exposed and a grid of electrical pickups is placed on the cortical surface. A lead from each pickup is threaded out through the skin for future monitoring. In addition to registering the MRI model of the patient to his/her position, the location of each electrical contact is recorded and transformed to MRI co ordinates. The patient is then closed up and monitored for several days. During any seizure event, the activity from each cortical probe is monitored, and transformed to the MRI model. This enables the surgeon to isolate potential foci in MRI coordinates. During a second surgical procedure, the augmented MRI model is reregistered to the patient and the locations of the hypothesized foci are presented to the surgeon for navigational guidance. An example of this is shown in Fig. 6.5-9.

To see the range of cases handled by our system, we encourage readers to visit the Web site http://splweb.bwh.harvard.edu:8000/pages/comonth.html, which shows selected cases with descriptions of the use and impact of the navigation system on the case.

6.5.6 Summary

We have described an image-guided neurosurgery system, now in use in the operating room. The system achieves high positional accuracy with a simple, efficient interface that interferes little with normal operating room procedures, while supporting a wide range of cases. Qualitative assessment of the system in the operating room indicates strong potential. In addition to performing quantitative testing on the system, we are also extending its capabilities by integrating a screw-based head tracking system and improved visualization capabilities.

References

1. L. Adams, A. Knepper, D. Meyer-Ebrecht, R. Ruger, W. van der Brug. An optical navigator for brain surgery. *IEEE Computer*, 29(l):48–54, Jan. 1996.
2. D.R. Bucholz, K.R. Smith. A comparison of sonic digitizers versus light emitting diode-based localization. In R.J. Maciunas (ed.), *Interactive Image-Guided Neurosurgery*, Amer. Assoc. Neur. Surg., 1993.
3. R.D. Bucholz, K.R. Smith, J. Henderson, L. McDurmont. Intraoperative localization using a three dimensional optical digitizer. In *Medical Robotics and Computer Assisted Surgery*, pp. 283–290, 1994.
4. A. Chabrerie, F. Ozlen, S. Nakajima, M. Leventon, H. Atsumi, E. Grimson, E. Keeve, S. Helmers, J. Riviello, G. Holmes, F. Duffy, F. Jolesz, R. Kikinis, P. Black. Three- dimensional reconstruction and surgical navigation in pediatric epilepsy surgery, *Medical Image Computation and Computer Assisted Interventions*, Boston, October 1998.
5. P. Cinquin, S. Lavallee, J. Troccaz, Computer assisted medical interventions. *IEEE EMB*, 254–263, May/June 1995.
6. A.C.F. Colchester, J. Zhao, K. Holton-Tainter, C. Henri, N. Maitland, P. Roberts, C. Harris, R. Evans, Development and preliminary evaluation of VISLAN, a surgical planning and guidance system using intra-operative video imaging. *Medical Image Analysis*, l(l):73–90, March 1996.
7. A. Collignon, F. Maes, D. Delaere, D. Vandermeulen, P. Suetents, G. Marchal. Automated multimodality image registration using information theory. In Y. Bizais and C. Barillot (eds.) *Information Processing in Medical Imaging*, pp. 263–274, Kluwer Academic Publishers, 1995.
8. A. Collignon, D. Vandermeulen, P. Suetens, G. Marchal. 3D multi-modality medical image registration using feature space clustering. In Ayache, N. (ed.), *CVRMed*, Vol. 905, *Lecture Notes in Computer Science*, pp. 195–204, Springer-Verlag, 1995.
9. G. Ettinger, Hierarchical three-dimensional medical image registration, Ph.D. Thesis, MIT, 1997.
10. G. Ettinger, M. Leventon, E. Grimson, R. Kikinis, L. Gugino, W. Cote, L. Sprung, L. Aglio, M. Shenton, G. Potts, E. Alexander. Experimentation with a transcranial magnetic stimulation system for functional brain mapping. *Medical Image Analysis*, 2(2):133–142, 1998.
11. R. Galloway. Stereotactic frame systems and intraoperative localization devices. In R. Maciunas (ed.), *Interactive Image-Guided Neurosurgery*, Amer. Assoc. Neur. Surg., 1993.
12. W.E.L. Grimson, T. Lozano-Perez, W.M. Wells III, G.J. Ettinger, S.J. White, R. Kikinis. An automatic registration method for frameless stereotaxy, image guided surgery, and enhanced reality visualization. Comp. Vis. and Pattern Recognition Conference, Seattle, June 1994.
13. W.E.L. Grimson, G.J. Ettinger, S.J. White, PL. Gleason, T. Lozano-Perez, W.M. Wells III, R. Kikinis. Evaluating and validating an automated registration system for enhanced reality visualization in surgery. *First CVRMED*, Nice, France, pp. 3–12, April 1995.
14. W.E.L. Grimson, G.J. Ettinger, S.J. White, T. Lozano-Perez, W.M. Wells III, R. Kikinis. An automatic registration method for frameless stereotaxy, image guided surgery, and enhanced reality visualization. *IEEE TMI*, 15(2): 129–140, April 1996.
15. D.J. Hawkes, D.L. Hill, E.C Bracey. Multimodal data fusion to combine anatomical and physiological information in the head and heart. In J.H. Reiber and E.E. van der Wall (eds.), *Cardiovascular Nuclear Medicine and MRI*, pp. 113— 130, Kluwer Academic Publishers, 1992.
16. P.F. Hemler, P.A. van den Elsen, T.S. Sumanaweera, S. Napel, J. Drace, J.R. Adler. A quantitative comparison of residual error for three different multimodality registration techniques. In H. Bizais, C. Barillot, and R. di Paola, (eds.), *Information Processing in Medical Imaging*, pp. 389–390, Kluwer Academic Publishers, 1995.
17. T. Kapur, W.E.L. Grimson, R. Kikinis. Segmentation of brain tissue from MR images. *First CVRMED*, Nice, France, pp. 429–433, April 1995.
18. L. Lemieux, N.D. Kitchen, S.W Hughes, D.G.T. Thomas. Voxel-based localization in frame-based and frameless stereotaxy and its accuracy. *Med. Phys.* 21, 1301–1310.
19. L.D. Lunsford. *Modern Stereotactic Neurosurgery.* Martinus Nijhoff, Boston, 1988.
20. WE. Lorensen, H.E. Cline. Marching Cube: A high resolution 3-D surface construction algorithm. *Computer Graphics*, 21(3):163–169, 1987.
21. R. Maciunas, J. Fitzpatrick, R. Galloway, G. Allen. Beyond stereotaxy: Extreme levels of application accuracy are provided by implantable fiducial markers for interactive- image-guided neurosurgery. In R.J. Maciunas (ed.), *Interactive Image-Guided Neurosurgery*, Amer. Assoc. Neur. Surg., 1993.
22. J.B. Maintz, M.A. Viergever. A survey of medical image registration. *Medical Image Analysis*, 2, 1–36, 1998.
23. C.R. Maurer, J.M. Fitzpatrick. A review of medical image registration. In R.J. Maciunas (ed.), *Interactive Image- Guided Neurosurgery*, Amer. Assoc. Neur. Surg., pp. 17–44, 1993.
24. C.R. Maurer, J.J.McCrory, J.M. Fitzpatrick. Estimation of accuracy in localizing externally attached markers in multimodal volume head images. In M.H. Loew, (ed.) *Medical Imaging: Image Processing*, Vol. 1898, pp. 43–54, SPIE Press, 1993.
25. C.R. Maurer, G.B. Aboutanos, B.M. Dawant, R.A. Margolin, R.J. Maciunas, J.M Fitzpatrick. Registration of CT and MR brain images using a combination of points and surfaces. In M.H. Loew (ed.), *Medical Imaging: Image Processing, SPIE*, pp. 109–123, 1995.
26. C.R. Maurer, J.M Fitzpatrick, R.L. Galloway, M.Y. Wang, R.J. Maciunas, G.S. Allen. The accuracy of image-guided neurosurgery using implantable fiducial markers. In H. Lemke, K. Inamura, C. Jaffe, and M. Vannier (eds.), *Computer Assisted Radiology*, pp. 1197–1202, Springer- Verlag, 1995.
27. T. Mclnerney, D. Terzopoulos. Medical image segmentation using topologically adaptable snakes. *CVRMed*, 1995.
28. T. Peters, B. Davey, P. Munger, R. Comeau, A. Evans, A. Olivier. Three-dimensional multimodal image-guidance for neurosurgery. *IEEE TMI*, 15(2):121–128, April 1996.
29. M.J. Ryan, R.K. Erickson, D.N. Levin, C.A. Pelizzari, R.L. Macdonald, G.J. Dohrmann. Frameless stereotaxy with real-time tracking of patient head movement and retrospective

patient-image registration. *J. Neurosurgery,* 85:287–292, August 1996.

30. G. Szekely, A. Keleman, C. Brechbuehler, G. Gerig. Segmentation of 3D objects from MRI volume data using constrained elastic deformations of flexible Fourier surface models. *CVRMed,* 1995.
31. S. Tebo, D. Leopold, D. Long, S. Zinreich, D. Kennedy. An optical 3D digitizer for frameless stereotactic surgery. *IEEE Comp. Graph. Appl.* 16(1):55–64, Jan. 1996.
32. P.A. van den Elsen, E.J.D. Pol, M.A. Viergever. Medical image matching—a review with classification. *IEEE Eng. Med. Biol.* 12, 26–39, 1993.
33. K.L. Vincken, A.S.E. Koster, M.A. Viergever. Probabilistic hyperstack segmentation of MR brain data. *CVRMed,* 1995.
34. S. Warfield, J. Dengler, J. Zaers, C. Guttmann, W. Wells, G. Ettinger, J. Hiller, R. Kikinis. Automatic identification of grey matter structures from MRI to improve the segmentation of white matter lesions. *Journal of Image Guided Surgery,* 1(6): 326–338, June 1996.
35. E. Watanabe. The neuronavigator: A potentiometer-based localizing arm system. In R. Maciunas (ed), *Interactive Image-Guided Neurosurgery,* Amer. Assoc. Neur. Surg., 1993.
36. W.M. Wells. Adaptive segmentation of MRI data. *First CVRMED,* Nice, France, April 1995, pp. 59–69.
37. W.M. Wells, III, W.E.L. Grimson, R. Kikinis, F.A. Jolesz. Adaptive segmentation of MRI data. *IEEE Trans. Medical Imaging,* 15(4):429–442, August 1996.
38. W.M. Wells, P. Viola, H. Atsumi, S. Nakajima, R. Kikinis. Multi-modal volume registration by maximization of mutual information. *Medical Image Analysis,* 1(1):35–51, 1996.

Chapter 6.6

6.6

Visualization pathways in biomedicine

Meiyappan Solaiyappan

6.6.1 Visualization in medicine

Visualization is one of the rapidly developing areas of scientific computing. The cost of high performance computing has become increasingly more affordable in recent years. This has promoted the use of scientific visualization in many disciplines where the complex data sets are rich in quality and overwhelming in quantity. In medicine, more than cost, advancement of the physics of imaging has become a major influencing factor that has spurred the growth of image computing and visualization. The increasing cost of health care has created an awareness and demand for investigating safe and cost-effective approaches to practice diagnosis and deliver treatment. This trend has created a need for powerful ways of delivering information to physicians at every step in patient care delivery that might help the physicians to understand the problem better and faster and stay closer to the truth without leaving any room for potential instrument or human error. Traditionally, the practice of medicine requires information to be handled in a variety of ways, such as by touch, sound, appearance, and smell, in a manner comparable to a craft. The craftsmanship involved in medicine becomes more obvious when it comes to therapy or treatment procedures where the physician's hand–eye coordination is the final step that decides the outcome of what may be considered an extensive and expensive investigation. From an engineering standpoint, these aspects pose new challenges, since most of the information is qualitatively rich while quantitatively difficult to characterize. Thus, visualizing the information that describes the nature of the underlying "functional" source becomes vital in medicine. Functional visualization is a particularly effective approach because multiple functional characteristics can be mapped to different visual cues that facilitate the interpretation of multidimensional information and correlation of qualitative and quantitative information simultaneously. Some examples are three-dimensional (3D) perspective realism for representing spatial relationships effectively, animated displays for representing temporal information, and other forms of visual cues such as hues and textures for representing various quantitative or functional information. Although information visualization is one aspect of the problem, the paradigm of interaction between the physician and the information is an equally significant one for craftsmanship needs.

It is the rapid development and culmination of technology that meets such needs in graphics computing, virtual interaction, and tactile feedback systems that has created a synergy of research between physicians and engineers that has led to the development of new frontiers in visualization in medicine. The impacts these advancements are creating in biomedical visualization are many, and they can be broadly classified to different areas based on the application such as clinical diagnostic visualization, image guided therapy, procedure training simulation, pretreatment planning, and clinical hypothesis validation.

The chief purpose of this chapter is to introduce visualization concepts to the reader. Given that such concepts change in unison with technology, it is also important to present here the underlying problems, through case examples in medical visualization. This provides a better appreciation of these concepts from an application point of view and will allow the reader to follow the evolution of the concepts in the future.

Biomedical Engineering Desk Reference; ISBN: 9780123746467

6.6.1.1 The genealogy of visualization

Visualization phenotypes

Based on their visual characteristics resulting from their interaction with the user environment, visualization in medicine can be grouped into three major classes: illustrative visualization, investigative visualization, and imitative visualization (Fig. 6.6-1). The extent to which they embrace the technology of visualization can be considered as marginal, high, and maximum, respectively.

Illustrative visualization

Illustrative visualization developed over the past two decades from attempts to separate visualization into two distinctive processes: extraction of information and its presentation. Fast processing is desirable but not critical, whereas quality and accuracy are essential. The concepts of illustrative visualization form the basis of the other two classes of visualization. Illustrative systems do not rely on interactive data manipulation but present information that may have been carefully extracted from the data by other means.

The initial expectations of biomedical visualization based on this approach soon reached a plateau. As stated in a National Science Foundation (NSF) grant announcement: "In spite of enormous advances of computer hardware and data processing techniques in recent decades, the amazing and still little understood ability of human beings to 'see the big picture' and 'know where to look' when presented with the visual data is still well beyond the computer's analytic skills." Although trying to characterize humans' ability to perform such processing is a challenging scientific investigation in itself, it also became imperative in biomedical research to provide rapid visualization systems that take advantage of this human ability. This presented the motivation for investigative visualization systems.

Investigative visualization

Investigative visualization tries to focus on the explorative aspect of visualization. These techniques generally do not aim to provide a strict analytical solution; instead, they aim to provide a visual solution that the human eye might be able to interpret better. In general these approaches do not require the detailed knowledge about the data that analytical approaches require. The explorative aspect of this class of visualization has been gaining appeal because it is essential for the clinical application of new imaging methods. Some of the concepts in this class are interactive or real-time volumetric visualization, dynamic visualization, multimodality registration, functional (multidimensional) visualization, and navigational visualization. The emphasis is usually on speed, because interactive ability is essential to make these visualization tools useful in practice.

Imitative visualization

This class of visualization attempts to imitate reality through visualization. The imitation can be a visual perception as in virtual reality type systems or functional imitation as in simulation and modeling. The challenge is to provide a realistic simulation in a virtual environment. In medicine, such a challenge arises naturally in the area of pretreatment planning and training for interventional and intraoperative surgical procedures. Both the speed

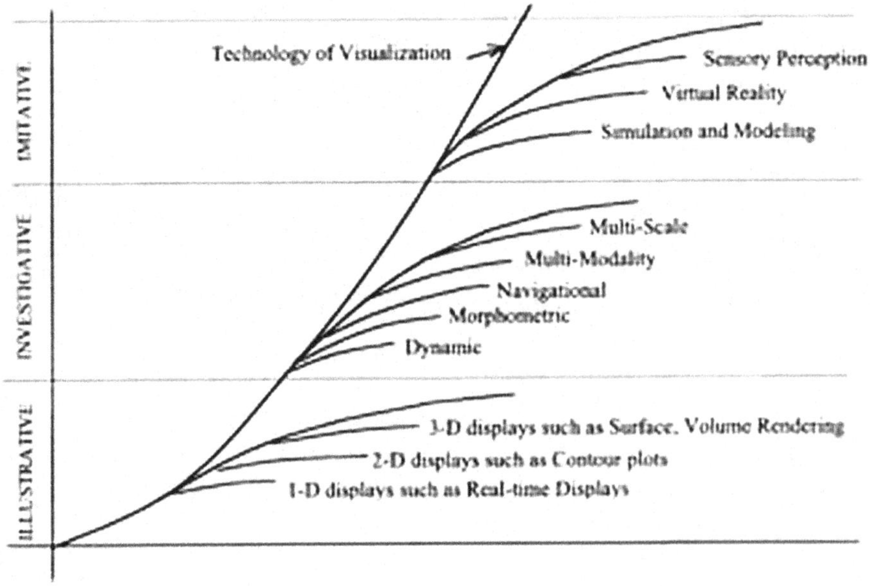

Figure 6.6-1 Visualization pathways.

and quality of visualization are critical here, and so is the paradigm of interaction. Some of the technology needed for such visualization applications may not be available today, but there is rapid progress in this area.

6.6.1.2 Visualization genotypes

In medicine, because of the inherent complexity in visualizing the information in the data, different concepts of visualization evolved as the technologies that could enable them became available. For the purpose of this discussion, the evolution of different concepts of visualization in medicine can be grouped into several generations.

First generation systems are essentially one-dimensional (ID) waveform displays such as those that appear in patient monitoring systems.

Second generation systems perform two-dimensional (2D) image processing and display. Contours and stack of contour lines that can represent the three-dimensional form of the data also were developed during this period.

Third generation systems generally involve 3D image processing and visualization. Isosurfaces, contour surfaces, shell-rendering and volume-rendering techniques were developed in this generation.

Fourth generation systems process multidimensional data such as dynamic volume data sets, sometimes called 4D data. In general the fourth dimension can be any other dimension associated with volume data.

Fifth generation systems are virtual reality type visualization systems, which combine multidimensional data with 3D (i.e., six degree of freedom) interaction.

Next generation systems represent concepts under development, such as sensory feedback techniques where the user interacting with the structures could feel the physical properties of the material and obtain valuable "visual–sensory" information in simulation type visualization systems.

The main focus of this chapter is on investigative and imitative visualization. Clinical research examples from medicine and biology are provided to illustrate the concepts of visualization and their significance.

6.6.2 Illustrative visualization

This class of visualization is the earliest one to develop and includes the first three generations, with their ID, 2D and 3D visualization displays. The concepts developed here are also applied in the other two phenotypes of visualization.

6.6.2.1 First generation systems

Real-time patient monitoring

Prior to the arrival of image-based information in radiology, the most basic forms of visualization in biomedicine were the ID waveforms that one would see on such devices as ECG monitors. Although elementary, this form of visualization is a very powerful tool for conveying the physiological state of the subject during a clinical intervention. Depending on the time scale of these waveforms, one would be able to understand both the present state of the subject and the trend condition. This information may be difficult to represent in any other form other than simple visual aid.

Subsequently, more advanced forms of these ID displays were generated to convey the combined physiological state using different signals such as ECG, blood pressure, and respiratory waveforms. These various signals were appropriately combined and presented in a form that would help clinicians rapidly observe potential problems during a clinical intervention. Thus, the "multimodality" of information helped clinicians understand the "functional" state of the physiology of the subject. These early developments already indicated the potential benefit of visualization in patient care. Anesthesiologists who operate various instruments for delivering and maintaining proper respiratory and hemodynamic state during an intraoperative procedure may need to read various displays to be careful to avoid human errors. Their task is often described to be analogous to that of a pilot inside a cockpit. As heads-up display visualization techniques helped revolutionize the organization of the cockpit, the use of high-end visualization became common for even the simplest forms of biomedical signals. Thus, the term cockpit visualization technology became popular for describing the impact of visualization in medicine.

6.6.2.2 Second generation systems

2D image processing techniques and displays formed the second generation systems. Some of the earliest 2D visualization tools were image processing techniques for enhancing image features that otherwise may have been ignored. Feature extraction techniques, expert systems, and neural networks applications were developed along with second generation visualization systems.

Interpolation

In medical images, the number of pixels might vary depending on the modality of imaging, and it is usually in the range of 128 × 128 to 512 × 512. The resolution of graphic displays is usually high, above 72 pixels/inch, making these images appear relatively small. Suitable interpolation techniques are required to enlarge these images to a proper display size. A common interpolation technique, bilinear interpolation, refers to linear interpolation along the horizontal and vertical direction. The pixel value at any display point is computed based on a weighted sum of pixel values at the corner locations of the smallest rectangular cell in the image that surrounds the point. The weighting factor is the ratio of the area the diagonally opposite corner forms with the considered display point to the area of the rectangular cell formed by the pixel image. Although bilinear interpolation may provide satisfactory results for many applications, more elaborate interpolation techniques may also be required.

2D contours and deformable models

Manipulation of the entire 2D image appeared to be a cumbersome approach when the feature of interest could be represented as contour lines delineating structures of interest. Besides, such contours could provide quantitative information such as area and perimeter, and they could also be used to build 3D models. Considering such benefits, both manual (supervised) and automatic approaches were developed. Earlier automatic contour extraction techniques suffered setbacks due to the insufficient quality of the image. Later, deformable models were developed to preserve the continuity of the contour and its topology. The user would provide an initial simple contour line that served as an initial reference. The deformable model would then shrink or expand the contour to minimize a cost function associated with its shape evaluated at each iterative step as the contour progressively approached the boundary of interest. By defining appropriate penalty values to prevent irregularities in the contour, the continuity around poor image boundaries was preserved. This approach, also called snakes or active contours, created considerable technical interest in the field.

Contour models

In early systems, contours from a stack of serial slices could be arranged to display a topographic view of the 3D form representing the boundary of structures. Later, "contour stitching" techniques were developed to sew the contours and provide a 3D surface model view of the contour stacks. Although simple in appearance, these techniques required significant attention, especially when the contour shapes changed considerably and led to aliasing problems in the rendered surface. A case of particular interest was the Y branch frequently encountered in vascular or airway trees. In these cases, one contour in a slice had to be stitched to two contours in the adjoining slice to produce the Y branch. Several approaches that could solve the problem in special cases were developed, but they did not lead to a general solution. Subsequently, isosurfaces were developed and used for this purpose, as described later. Because of the drawback of dealing with a large number of triangles to represent the branch, other approaches based on an implicit analytical representation were investigated and used in a limited number of applications. New techniques under investigation show promise, especially model-based approaches where a model representing the Y branch can be appropriately parameterized to adapt to points in the contour lines.

2D texture mapping and parametrizing images

Texture mapping is a concept introduced in computer graphics for providing high visual realism in a scene [28]. Painting the elements of a drawing with realistic effects each time they appear in the scene may be an unwarranted intensive task, especially when the purpose is to provide only a visual context, such as the background field in a game. In those cases, an image piece representing the elements, such as its photograph, could be used to create the illusion that the elements appear as if they were drawn into the scene. The simplicity and popularity of this technique enabled both graphics software developers and graphics hardware manufacturers to use it widely. The technology of texture mapping rapidly advanced and became inexpensive. In the scientific visualization field texture mapping contributes to realism and speed, but more importantly it provides a geometric representation for the image, separating the spatial information from its image-pixel fragments. This substantially simplifies the tasks involved in visualization problems. By solving the problem in its geometric formulation, which in many cases would be more straightforward, and letting the texture mapping technique take care of the pixel association with the geometry, more ingenious visualization systems could be built.

6.6.2.3 Third generation systems

With the arrival of 3D images in the biomedical field, researchers developed various methods to display volume information [17]. The effectiveness of a technique depends primarily on the source of the image. Probably the most important factor in the development of volume visualization is the fact that the data had one dimension more than the computer display. Thus, in some sense, every technique ultimately had to project the 3D information to form a 2D image, a process where information could be potentially lost or occluded. Although stereo display may have eliminated some of these problems, the fundamental problem of presenting a 3D volume of information in a form that the user can quickly interpret still remains an elusive visualization problem. In this context, it may be justified to say that the routine use of clinical visualization is waiting for a smart visualization system that can quickly determine the main interest of the user and present this information in a convenient manner. Unlike ID and 2D systems, which appear to have gained quick clinical entry, 3D systems are still used primarily in clinical research programs rather than routine clinical applications. The cost of 3D visualization systems that can be considered clinically useful is also relatively high, because of demanding volumetric computations that require very fast systems.

Surface visualization

Although presenting a 3D-volume image is a fairly complex problem, there are other ways of displaying or extracting geometrical information that have been well accepted for certain applications. The approach, which is similar to isocontour lines in topographic data, extends this concept to create a 3D surface to characterize 3D image data. The technique came to be known as isosurface extraction and was proposed by Marc Levoy and Bill Lorensen [19]. The method works very successfully for CT volume image data where the high signal-to-noise ratio allows effective classification of constituent structures.

Isosurface extraction ("marching cubes")

Volumetric images consist of a stack of 2D images and can be considered as a 3D matrix of image data points. The smallest fragment of the image is called a *voxel*, in analogy to the concept of pixel in a 2D image. The surface is extracted using a thresholding algorithm for each cube of the lattice, marching through the entire volume. In each cube, each pair of connected corners is examined for a threshold-crossover point based on linear interpolation. These points along each of the edges are linked to form the isosurface on each cube, and the process is repeated for the rest of the volume. Special interpretation is required to handle cases that correspond to multiple surfaces within the cube. The surfaces are usually represented with triangles.

The advantage of this method is its fairly detailed surface representation for objects of interest, as long as the objects are easily separable in the data. However, its computational load is high, and each time a new threshold value is selected the generation of the new surface may cause delays. The number of triangles produced by the method in a typical set of volume image data is very large, typically on the order of tens of thousands. Thus, displaying them all can be an intensive graphic task. Adaptive surface extraction techniques were later developed to address this problem using an approach that coalesces coplanar triangles to be represented by a larger polygon. This can improve performance substantially, as the number of vertices that need to be processed in the transformation pipeline is significantly reduced. Reduction can be obtained with Delaunay triangulation, also known as thin-plate techniques, where coalescing can be extended to include triangles that are approximately coplanar, within a given tolerance. This can reduce the triangle population at the expense of a negligible drop in quality.

Deformable surfaces, balloons, shrink-wrap surface

Surface extraction proves to be very effective when the signal-to-noise ratio of the data is high and structures are well segmented. However, extraction results could become unpredictable when the data are noisy or when structures cannot be segmented well, as is often the case with some MR images. Following the approach used to solve similar problems in 2D images using deformable contours [46], elastic surface approaches were proposed to solve the problem in 3D data. These surfaces are sometimes called balloons, for their expanding properties, or shrink-wrapping surfaces with elastic properties, or, in general, deformable surfaces. Like snakes, these techniques usually tend to be computationally intensive because of their iterative steps.

"Statistical" surfaces

Recent approaches that attempt to produce efficient results for noisy data are "statistical" surfaces that employ space partitioning techniques based on local statistical measures to produce a mean estimated surface within a given error deviation. This technique may not preserve the topology connectivity that deformable techniques could provide.

Wavelet surfaces

As discussed earlier, one of the problems of surface representation is the number of triangles used to represent the surface. Recently, wavelet techniques, which have inherently a multiresolution approach, have been applied to describe surfaces. One major advantage of this approach is that one can decide the desired resolution at display time; thus, during periods of interaction a low-resolution surface could be displayed, and when the interaction ceases the higher resolution display of the surface can be generated.

Volume visualization

The inherent limitation of the surface extraction method is that it represents a specific threshold value in the data and becomes restrictive or selective. Also, occasionally, false surface fragments may be produced because of the interpolation involved. Volumetric visualization methods overcome these limitations and could help visualize as much information as possible from the 3D volume image, without being restrictive. This chapter presents only a brief survey of issues in volume visualization.

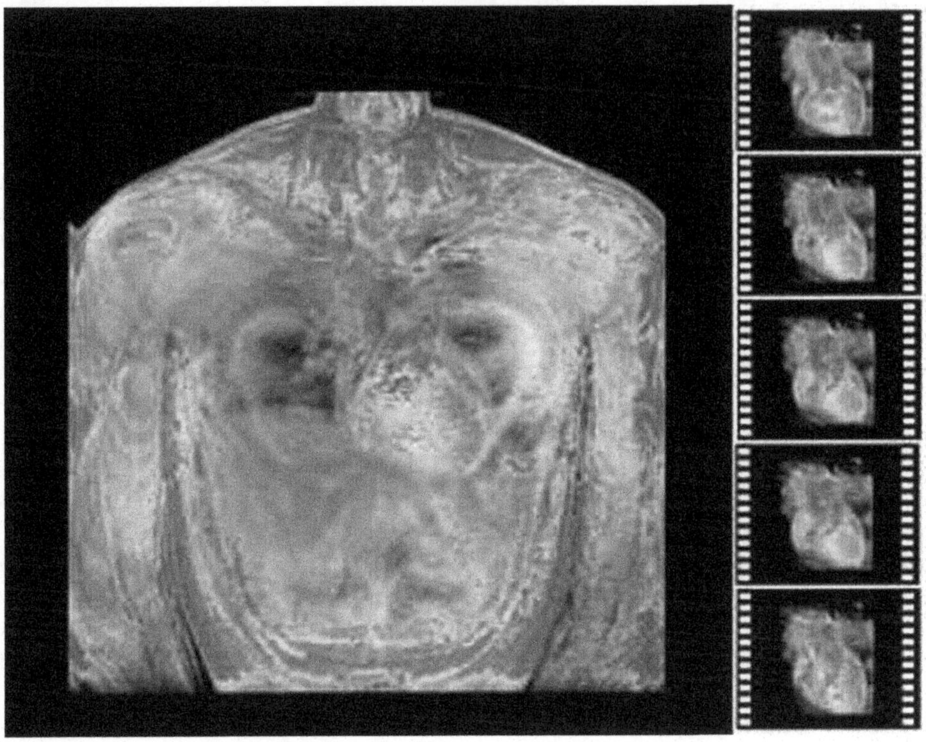

Figure 6.6-2 Visualization of beating heart: interactive dynamic volume rendering of cardiac cine-MR images. The cine frame snapshots show the contraction cycle. (Images courtesy of M. Solaiyappan, Tim Poston, Pheng Ann Heng, Elias Zerhouni, Elliot McVeigh.)

Maximum intensity projection

The earliest volume visualization method is known as maximum intensity projection (MIP). As the name suggests, in this approach, the maximum intensity value in the volume data is projected on the viewing plane along each ray of the projection. This technique is particularly useful for displaying vascular structures acquired using angiographic methods.

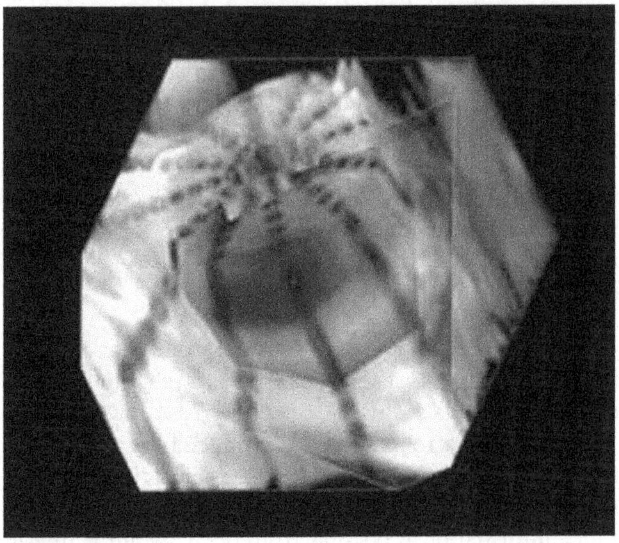

Figure 6.6-3 Visualization of left ventricle strain fields (color-coded) combined with tagged MR volume.

Volume rendering

The "ray-casting" approach was further explored to produce new visualization concepts in medicine that shared a common principle but approached the solution differently. Instead of choosing the maximum value along each ray, the general idea was to accumulate all data points along the ray with appropriate weights and to produce an aggregate value that is projected on the viewing plane. This led to the development of various volume-rendering techniques [23–27].

Various accumulation models were employed to produce different results. For instance, a simple summation of all the points along the ray can produce line-integral projections similar to X-ray images. The simple summation operation is a less intensive operation than other complex accumulation operations. This may be useful for simulating X-ray type images, but because the image will be dominated by structures that have maximum data values irrespective of their spatial distance from the viewing plane, this technique is less desirable to produce a 3D effect in the rendered image. The most popular technique turned out to be one that used a "blending" operation similar to the one used for antialiasing during scan conversion or scene composition [22]. The blending operation corresponds to a weighted mean between the data point and the background value. The weighting factor could be given a physical meaning of "opacity" of the data point, from which the transparency of the data point, i.e., 1 minus the opacity, can be computed. For generating the rendered image, each ray is traversed along the volume, from back to front with respect to the viewing plane, starting with initial background value zero. Data points are accumulated as the sum of the data value times its opacity, and the background times its transparency. The sum gives an effective blended value for the data points, which are treated as semitransparent. An interesting aspect of this technique is that hardware graphics pipelines in advanced graphics accelerated systems had blending functions that supported this operation on a per pixel basis, used for scene composition and antialiasing. The weighting factor became equivalent to the "alpha coefficient," the fourth component in the four-components-per-pixel representations (red, green, blue, alpha) of high-end graphic workstations.

Trilinear interpolation

During ray traversal through the volume, the rays need not pass through exact data points, except when the rays are parallel to the three orthogonal coordinate axes of the volume. In all other orientations the ray will pass through between data points. In this case, it becomes necessary to estimate the data point values along the ray at equal distances. One simple way would be to use the "nearest-neighbor" value, which can produce fast results. However, this gives rise to poor resolution of the feature, and the nearest neighbor can flip under even a simple rotation, causing undesirable artifacts. Thus, a smooth interpolation along the ray becomes very important, and the quality of rendering substantially improves with a good interpolation method. One popular and simple interpolation technique is the trilinear interpolation approach, similar to the well-known bilinear approach. The data value at any given location inside the image lattice is computed using the smallest cubic cell that contains the point. The interpolated value is a linear weighted sum of the contributions from all eight corners of this cell. The weighting factor for each corner is computed as the ratio of the volume of the cube whose diagonal axis is formed by the given point and the diagonally opposite corner to the volume of the cube formed by the cell.

The volume-rendering operation requires a fairly large number of data samples, to prevent the blending from being dominated by a few data values, which would produce a strong aliasing effect. When the number of sample points is low, the results may have poor quality. Also, in some applications, the number of rays needed to

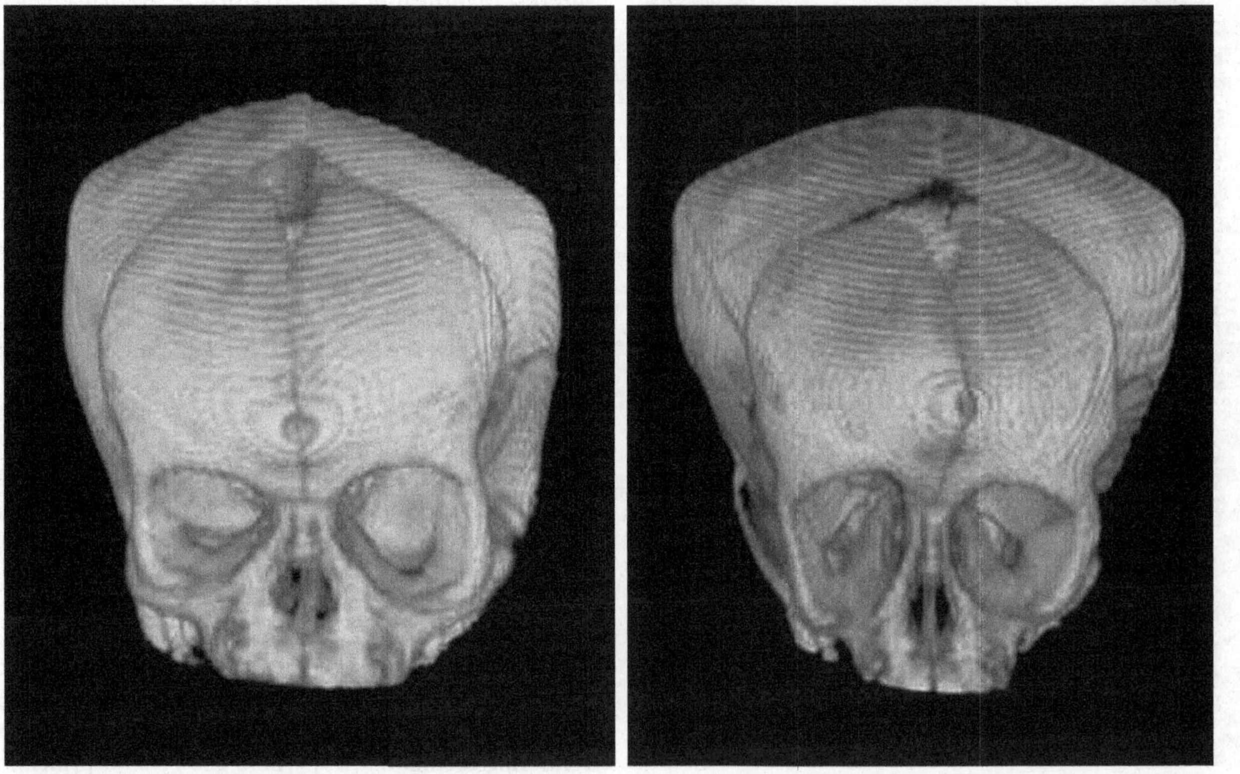

Figure 6.6-4 Visualization of volumetric morphing. Study of sagittal synostosis. Growth form derived from sagittal synostosis patients is applied on normal subject (left) to visualize the simulated synosotosis (right). (Images courtesy of Joan Richtsmeier, Shiaofen Fang, R. Srinivasan, Raghu Raghavan; M. Solaiyappan, Diana Hauser.)

produce the desired rendering may be greater than the data points can support. In such cases, interpolation becomes the single major factor that can determine the quality of the rendered image. More elaborate interpolation techniques can take into account not just eight surrounding data points but the 26 neighboring cells to compute a cubic polynomial interpolation. Such interpolation techniques give very good results, but the additional speed penalty may be quite high. When the original volume of data is fairly large (>256 × 256 × 64), trilinear interpolation may provide good results to fulfill many visualization purposes.

Lighting and shading

One of the important benefits of surface rendering compared to volume rendering is the lighting and shading effects that can improve visual cues in the 2D displayed image. This is figuratively called 2.5D image. Thus, the lack of lighting in the early volume-rendering models and in MIP represented a notable drawback. Lighting calculations were introduced in the blending equations introducing a weighting factor that represented the local gradient in the scalar 3D field data. The gradient vector was estimated using the gradient along the three principal axes of the volume or the 26 neighbors.

Lighting effects are particularly useful with CT image data where the surface features in the data are well pronounced because of the limited number of features that CT imaging resolves. However, with MR images, lighting does not always produce better results. Its contributions depend on the data contrast and noise. Noise in the data can affect gradient calculation much more strongly than the linear interpolation estimations.

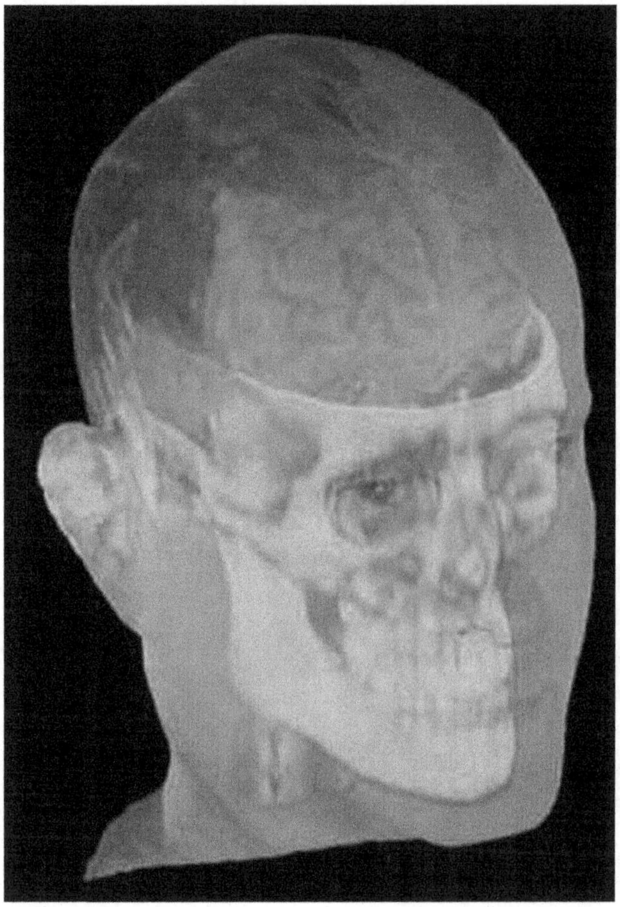

Figure 6.6-5 Multiple projection 2D photographs, back projected onto the surface obtained from CT and fused with MR of the brain. (Image courtesy of M. Solaiyappan, Nick Bryan, Pheng Ann Heng.)

Transfer functions

The hidden difficulty of volume rendering resides in the weighting factor of the accumulation equations. The weighting factor or the opacity value can be assigned for different data points in the volume to enhance or suppress their influence in the rendered image. The transfer function or the lookup table for the weighting factor serve this purpose. Since the data points are usually represented as discrete data of 8,12, or 16 bits, such table sizes are not very large. However, for floating-point representation of the data, a transfer function (a piecewise linear or a continuous polynomial representation) can be used.

Although the transfer function provides considerable flexibility for controlling transparency or opacity, it may be sometimes very difficult to enhance the feature of interest that the user wants to visualize. A small change in the transfer function may cause large differences in the final image because many interpolated samples are rendered along each ray, and this large accumulation can produce oversaturation or under-representation in the computations. Many automatic approaches to compute the transfer functions have been proposed, but none have become popular enough. Hardware-accelerated volume rendering [29–32] and dramatic improvements in the speed of computing and graphic systems led to the development of interactive approaches where the user could quickly see the result of the selected transfer function to determine to the desired one. However, a highly intelligent transfer function technique that can operate with minimal user interaction is critical for a wider clinical acceptance of volume rendering. Alternatively, more versatile rendering techniques that do not depend on sensitive input parameters may also be developed.

Shell rendering

Many techniques were investigated soon after the shortcomings of volume-rendering and surface-rendering techniques were realized. Notable among them is a shell rendering technique [20] where the general principle is to combine the strengths of surface rendering and volume rendering and suppress their weakness. Surface rendering is very selective in extracting particular

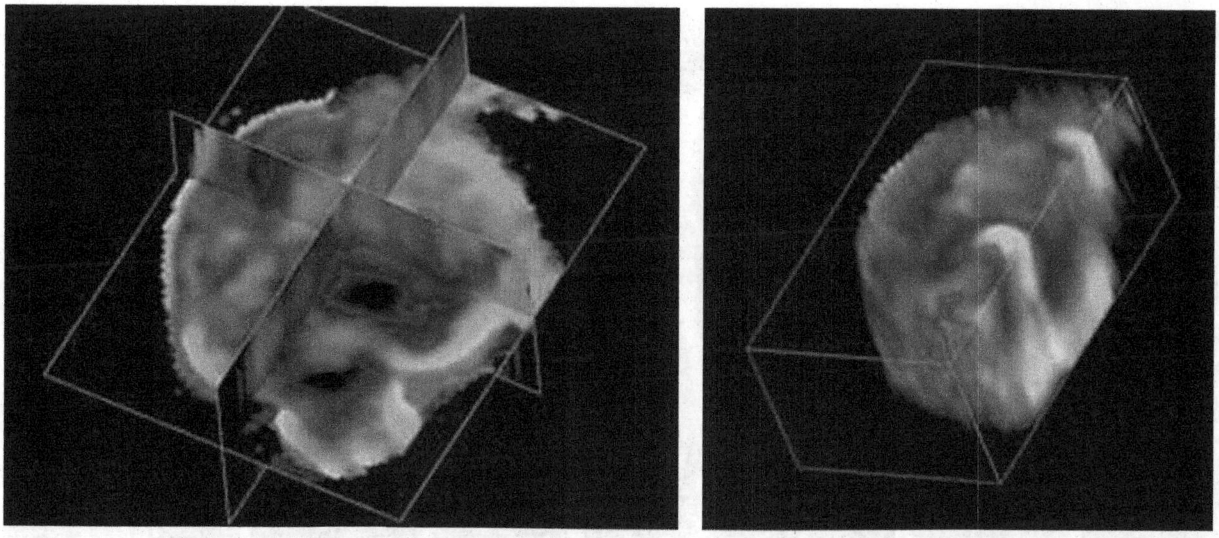

Figure 6.6-6 Visualization of the functional correlation of vascular volume and vascular permeability in tumor mass obtained *in vivo* using MR microscopy (125-micron) technique. 3D volume visualization using color space registration shows that there is little overlap between vascular volume (green) and vascular permeability (red), confirming previous observations. (Images courtesy of M. Solaiyappan, Zaver Bhujwalla, Dmitri Artemov.)

structures from the volume data. In cases where the anatomical structures of interest cannot be extracted with a unique threshold, surface rendering may be difficult to use. Volume rendering blends the data over a range suitably weighted by a transfer function. However, in its original form, it does not take into account the spatial connectivity between various structures in the data, thus making it sometimes difficult to select a particular structure of interest. Shell rendering combines spatial connectivity information addressed by surface rendering with voxel level blending of data provided by volume rendering. Thus, if surface rendering is considered as a "hard-shell" based on its specific threshold values, the shell rendering can be thought of as a "soft-shell" based on its fuzzy threshold.

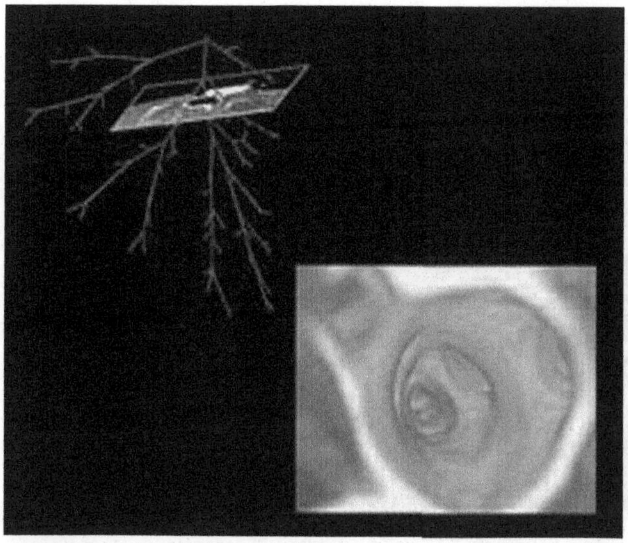

Figure 6.6-7 Visualization of breathing lung: navigational visualization (virtual bronchoscopy) using segmented bronchial tree. (Images courtesy of M. Solaiyappan, S. A. Wood, E. Zerhouni, W. Mitzner.)

Volume encoding (octrees)

During the development of various rendering techniques, encoding of volume data emerged as a research topic. When they became first available, sets of volume data were relatively large in comparison to the limited disk storage capacity of the time. Representation and processing of volume data required a great deal of memory and computing power. Octree encoding of the volume took advantage of clusters of data points that could be grouped. Appropriate grouping of these data points could result in substantial reduction in storage and yield high speed in volume rendering. A cluster of cells partitioned by octree encoding can be processed in one step instead of the several steps that may otherwise be required.

A more versatile approach would be a 3D wavelet-based approach, which could help maintain multilevel resolution of the volume. It could provide image compression and improve the speed of rendering during interaction by using low resolution.

Texture mapping for 3D images

The concept of 2D texture mapping was extended to 3D images. However, unlike 2D texture mapping, the 3D texture mapping technique does not produce a 3D rendered image directly from a texture map. It only provides a fundamental but powerful capability called multiplanar

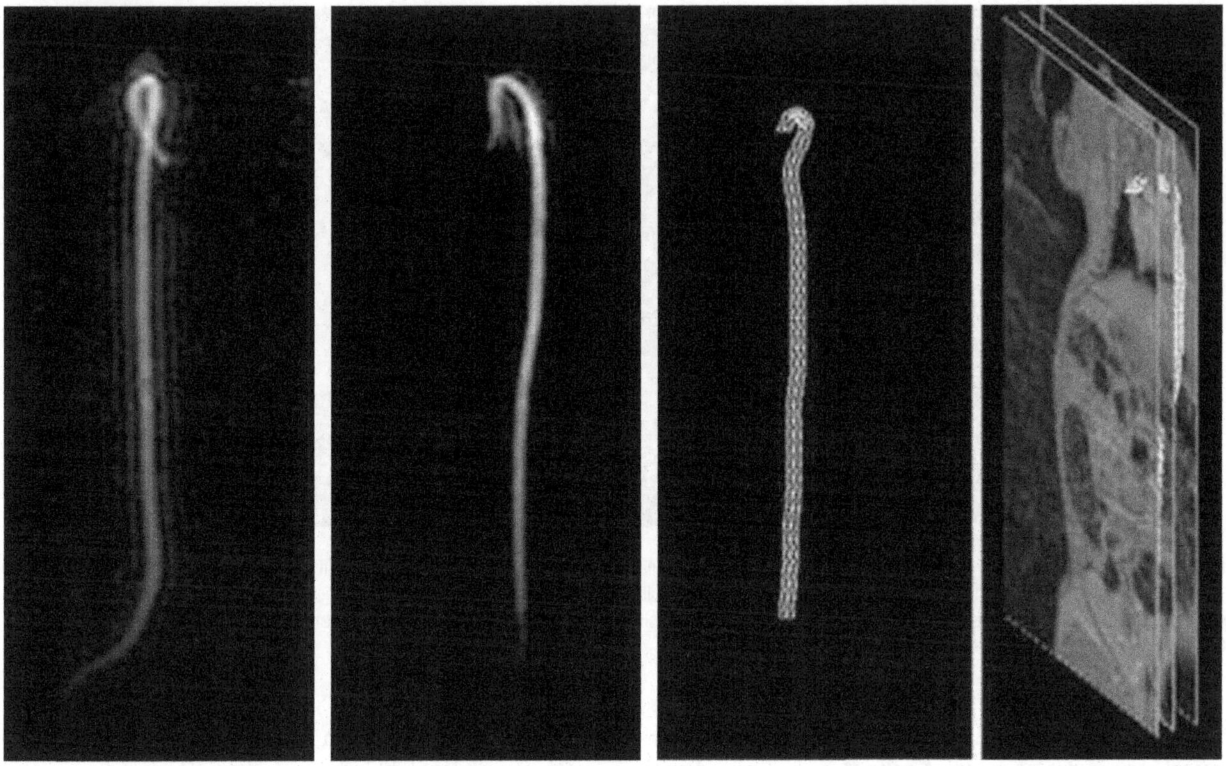

Figure 6.6-8 3D visualization of intravascular MR probes obtained using real-time MR fluoroscopy imaging. Depth projection (third image from the left) is reconstruction from coronal and sagittal projections and shown registered with MR road-map images for navigating the probe.

reformatting (MPR) that can be used for any 3D rendering application. MPR enables one to extract any arbitrary plane of image data from the 3D volume image that has been defined as a 3D texture. Applications can use this capability to produce a 3D-rendered image. For instance, if the volume is sliced into stacks of slices parallel to the viewing plane that are then blended back-to-front (with respect to the viewing plane), the volume-rendered image can be obtained. This process is also known as depth composition. The alpha component and

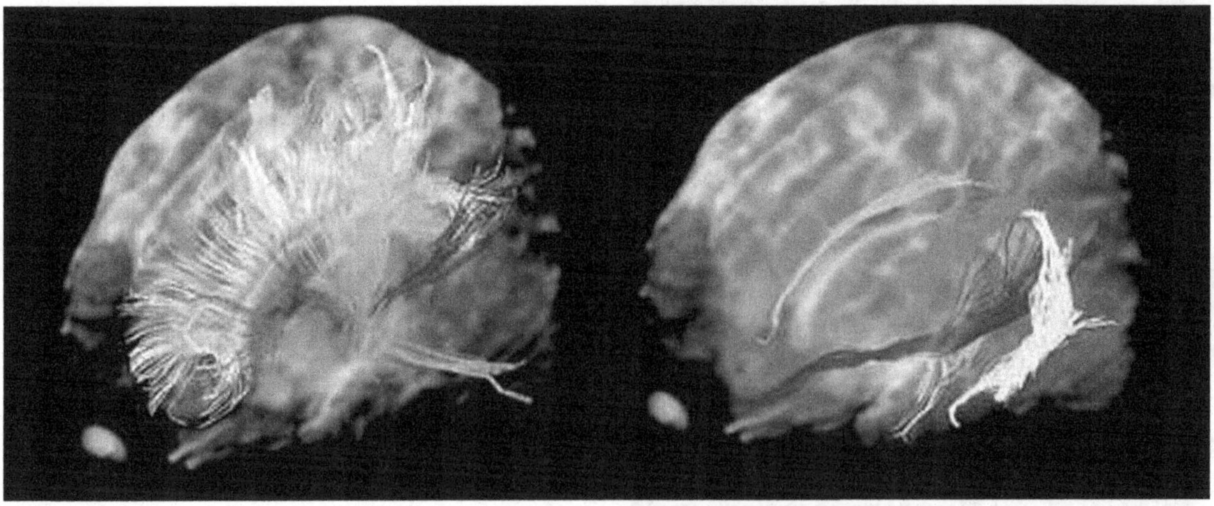

Figure 6.6-9 3D reconstruction of brain fiber from diffusion tensor imaging. (*Left*) White matter tracts that form the corona radiata are shown: corpus callosum (yellow), anterior talamic radiation (red), corticobalbar/cortiscospinal tract (green), optic radiation (blue). (*Right*) Association of fibers and tracts in the limbic system is shown: cingulum (green), fimbria (red), superior (pink) and inferior (yellow) longitudinal fasciculus, uncinate fasciculus (light blue), and inferior fronto-occipitial fasciculus (blue). (Images courtesy of Susumu Mori, R. Xue, B. Crain, M. Solaiyappan, V. P. Chacko, and P. C. M. van Zijl.)

the hardware-accelerated, per-pixel blending operation can be used for rendering.

6.6.3 Investigative visualization

As the concepts in illustrative visualization became mature enough, they became widely used in various applications. The concept of investigative visualization slowly emerged as a form of visualization for addressing specific questions that required collective use of various graphics concepts described in the earlier section on illustrative visualization. This corresponds to *fourth generation systems.*

The application of many 3D volume visualization techniques remained limited to specific purposes where the needs could not be fulfilled otherwise. The use of 3D visualization techniques was not widespread because of several potential factors: It usually required a certain amount of specialized prior knowledge about the technique, the computation time and the cost of required high-performance systems were both high. Imaging advances in medicine helped systems produce better 3D images by slicing the sampling space into sets of image planes, while the rendering techniques were tools for putting them back together to produce a 3D image that did not add any new information. Thus, volume rendering was considered to be clinically useful only in cases where the 3D morphology could provide a visualization advantage.

But soon, new forms of imaging techniques were developed that contained functional information associated with the anatomical information in the images. Such functional imaging techniques needed volume rendering to visualize the functional information that could not be visualized otherwise. Volume rendering benefited various functional visualization applications that required investigative imaging. To illustrate investigative visualization several examples are now presented. The difference between investigative visualization and illustrative visualization is not in their fundamental concepts, but in their specific applications.

Stereoscopic 3D visualization

One of the most important developments in volume visualization that demonstrated the full potential of volume rendering is stereoscopic 3D rendering. A stereo pair (left and right-eye views) of plain volume-rendered

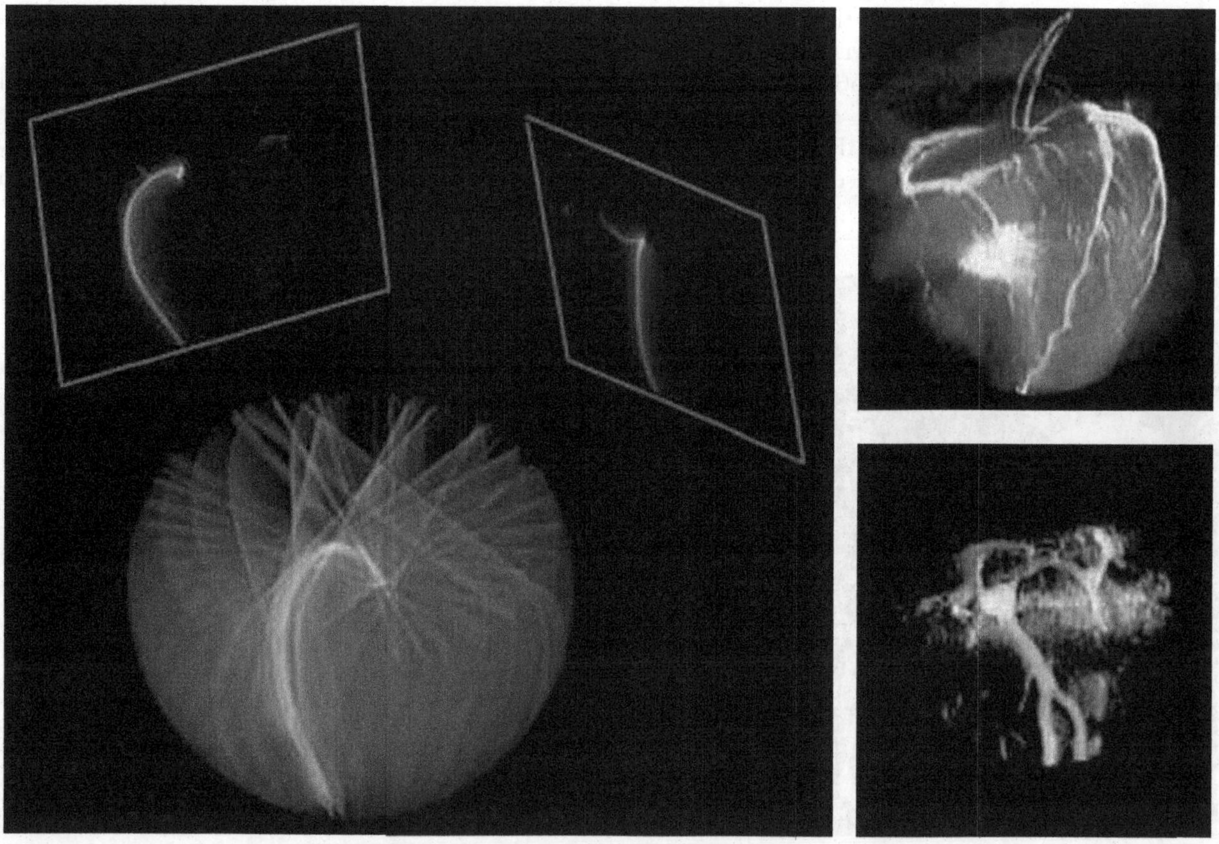

Figure 6.6-10 Visualization using back-projection reconstruction. (*Left*) Illustration of reconstruction. (*Right*) Clinical applications. (*Top*) Coronary vessel studies. (*Bottom*) Aneurysm localization studies using conventional sweep arm rotational angio systems. (Images courtesy of M. Solaiyappan, Nick Bryan, Timothy Eckel, Cliff Belden.)

images could be viewed using stereo goggles and provided the necessary left- and right-eye interleaving to produce the virtual 3D volume-rendered image. Stereoscopic visualization represents a milestone in volume visualization in terms of promoting its popularity and acceptance. The traditional display methods that present a 3D object on a 2D display do not make use of the depth perception as a powerful visual cue to convey the 3D spatial information. Another powerful advantage of stereoscopic 3D visualization is its inherent ability to let the human eye play the role of a 3D filter to suppress the effect of noise in the image. Noise in images such as MRI volume usually appears around structures of interest. When the rendered image is presented in 2D, the noise appears projected on the structures and obscures them. But when the data are presented in 3D stereo, noise is resolved by depth perception and poses little distraction when the eye focuses on the structure of interest. The computational load doubles in stereoscopic 3D rendering because it involves creating a pair of views.

Dynamic visualization

One of the simplest functional imaging examples is the cine-MR technique used to image the beating of the heart (Figs. 6.6-2 and 6.6-3). Displaying the slices as cine-frames provides the motion cues that represent the functional aspect of the beating heart. But in such a 2D display, the through-image-plane motion would be suppressed, amounting to moderation of actual motion. Thus, volume rendering as a real-time display could provide in one sequence all the necessary information that can describe the motion of the beating heart with its three-dimensional dynamic properties [35, 36].

Developmental visualization and 3D volume morphing

The study of 3D shape changes in morphology is a challenging area of investigation, but in many cases there may not be a unique way of deriving this information without making certain assumptions. Such assumptions may cause loss of generality, but in an investigative study, posing those assumptions and visualizing the corresponding results may provide a better understanding of the functional information present in the image. One such specific developmental visualization is in the field of craniofacial modeling and visualization. For example, given two different stages in the craniofacial development in children with normal and abnormal growth, it would be instructive to visualize how the normal and abnormal growth functions would exhibit themselves if they were expressed in the opposite group (Fig. 6.6-4). To describe the growth function, usually homologous landmarks that can be uniquely identified in craniofacial morphology are provided in the two stages that describe the growth. However, because there are few such homologous points (about 60), the visualization problem here would be to interpolate the growth function over the 3D space that represents the volume [48]. Using such a growth function, one can then model the morphing of the volume-rendered image to produce the developmental visualization. Thus, by applying the growth function to a different subject, one could get a qualitative

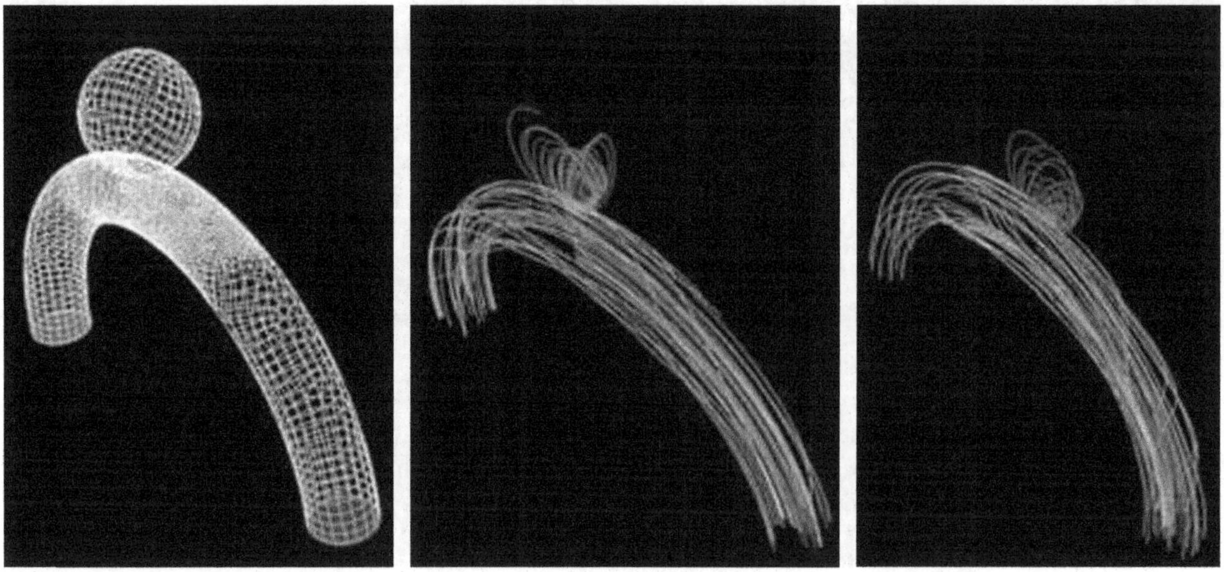

Figure 6.6-11 Modeling of the blood flow through an aneurysm and its visualization. (*Left*) Synthetic model. (*Middle*) Blood flow in the absence of coils in aneurysm. (*Right*) When there are coils to reduce the blood flow, and thereby its potential to rupture. (Images courtesy of TseMin Tsai, Jim Anderson.)

view of the effect of pseudo-craniofacial development on the subject. Thus, developmental visualization is essentially a 3D morphing technique that uses biologically appropriate control functions to describe the morphing. Also, it is important to notice that these morphing techniques attempt to morph (i.e., translate, rotate, scale) every voxel in the volume, unlike the more conventional morphing techniques used for 3D surfaces. Volumetric morphing is computationally intensive; however, it can avoid topological problems, such as self-intersection, that 3D surface morphing can easily encounter.

Multimodality visualization

3D visualization can be used for morphological correlation between different forms of volumetric data, such as those acquired using different types of imaging exams or modality, mostly on the same subject (Figs 6.6-5 and 6.6-6). Such spatial comparisons or localizations help to validate the efficacy of examinations. Often, direct correlation measures do not provide a good quantitative description of the correspondence between the data because statistical variations deteriorate the correlation measure. Multimodality volumetric visualization provides correlation information and helps the user to visualize the presence of spatial localization in the data. When the two sets of data have different spatial orientations, automated 3D registration may be difficult and interactive visualization may help explore and understand the data. The technical challenges in this class of volume rendering are not demanding, but the logic of combining two or more volumes requires attention. The data may need to be combined pixel by pixel, or pixels may need to be substituted from image to image. The

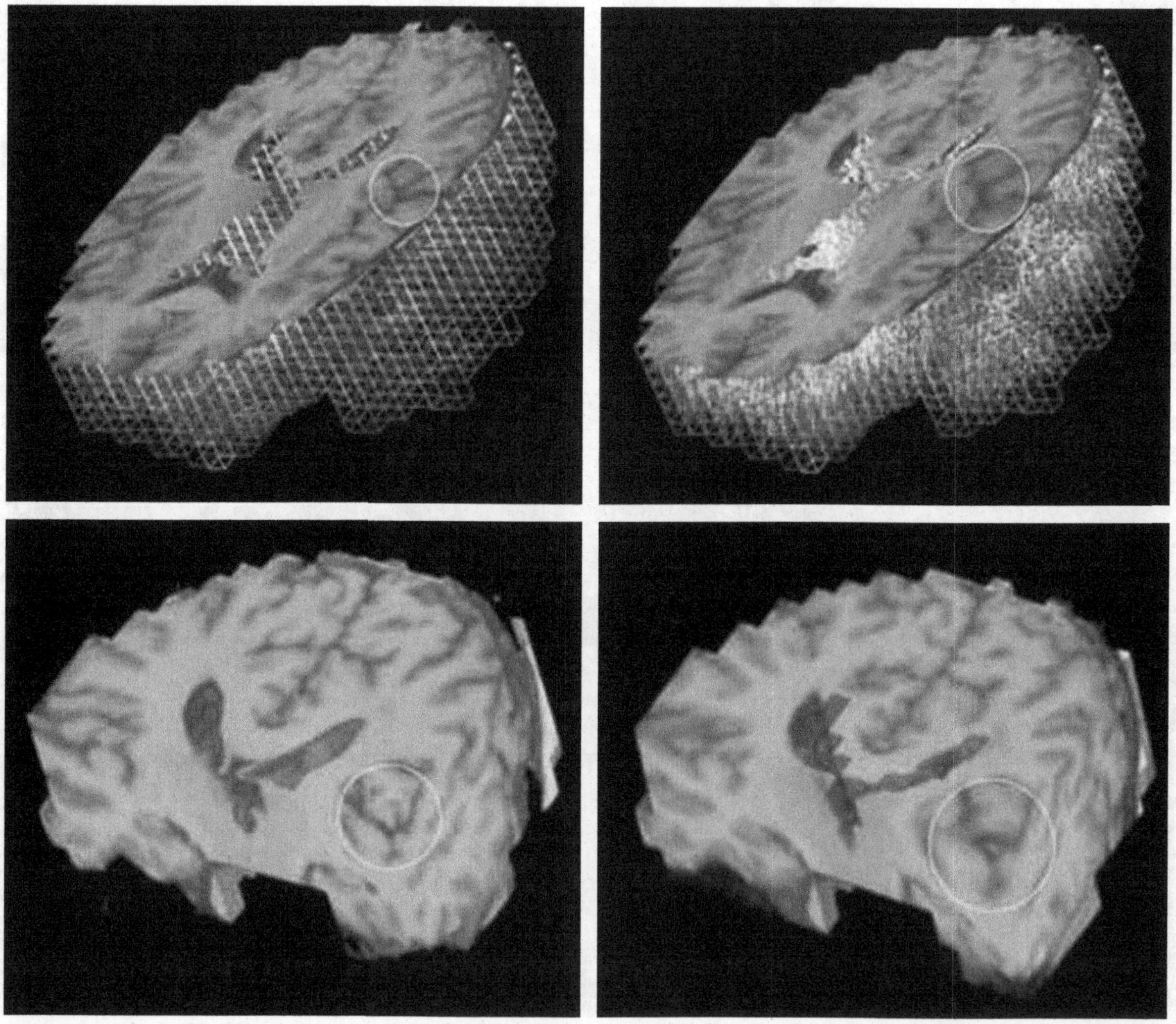

Figure 6.6-12 Finite-element modeling of tumor growth and its visualization using 3D morphing to visualize the morphological changes in the surrounding anatomy. (*Top*) FEM mesh (white: white matter, gray: gray matter, red: boundary nodes). (*Bottom*) FEM embedded 3D visualization. (*Left*) Undeformed brain. (Right) Deformed brain. (Images courtesy of Stelios Kyriacou, M. Solaiyappan, Christos Davatzikos.)

latter method is particularly useful for visualization of multimodality data that typically are complementary. A combined visualization has the potential to provide more information than the individual data sets. This occurs, for instance, when one combines a CT image, which can distinguish the skull from soft tissue, with an MRI image, which can resolve soft tissues but cannot show the skull [42, 44, 45] (Fig. 6.6-5).

Navigational visualization

Generally, visualizing structures from the outside is useful in understanding the 3D morphology. Sometimes it would be informative to look at the structures from the inside, such as in a vascular or bronchial tree or other tubular structures (Fig. 6.6-7). Such types of visualization can be called navigation visualization [55] because of their ability to provide necessary visualization for navigating through the internal structures. This class of visualization techniques became popular because of its similarity to "invitroscopic" systems such as endoscopes, bronchioscopes, or colonoscopes. In this aspect, these techniques are close to simulation-type visualization. The advantage of using navigational visualization on a 3D volume is that full spatial orientation is known at all times, unlike real endoscopes or bronchoscopes, where it

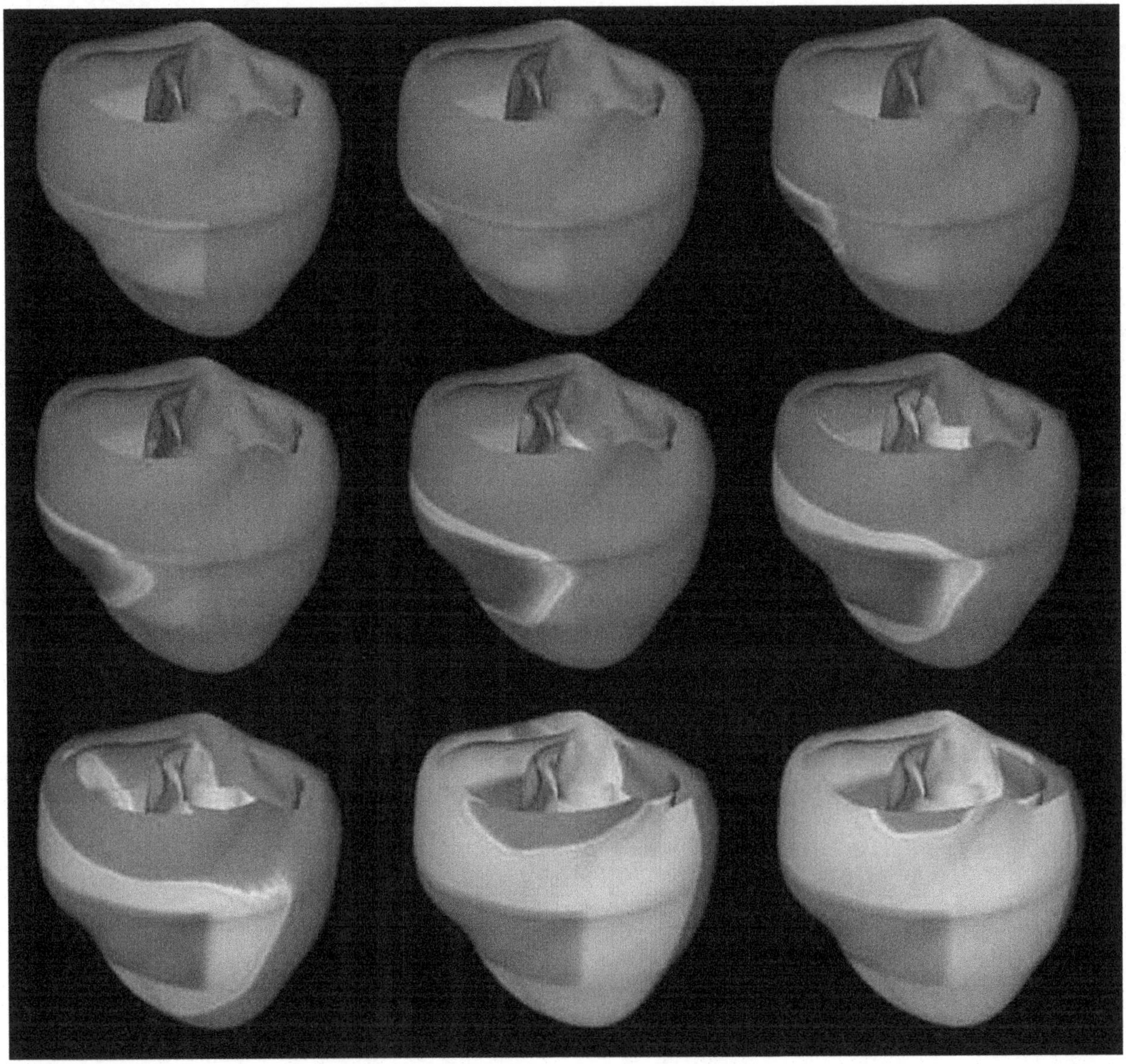

Figure 6.6-13 Finite element modeling of the electrical activity of the heart and its 3D visualization. Depolarization of the heart with both normal and abnormal cells, with the abnormal cells having ionic properties similar to those seen in patients with congestive heart failure. This simulation leads to a sustained reentrant wave of activation. The ECG of this arrhythmia is similar to those seen in CHF patients. (Images courtesy of Raimond Winslow, Dave Scollan, Prasad Gharnpure, M. Solaiyappan.)

may be very difficult to know the exact spatial location of the probe. Thus, it may be easier to get both the internal view and the surrounding spatial localization information, providing a "road-map" image [37]. The volume-rendering technique used in this type of visualization should address the strong perspective issues that are encountered.

Real-time visualization for image-guided procedures

Recent advances in imaging techniques have made it possible to acquire images in real-time during an interventional procedure (Fig. 6.6-8). During such procedures, usually, the real-time images themselves may be sufficient to provide the necessary guidance information needed for the procedure. However, there are instances when that may not be sufficient and advanced visualization becomes a powerful tool for guidance information. Information acquired in real-time and more detailed information obtained prior to the procedure are combined to provide real-time guidance. Registration is an important factor and speed is a major issue in this kind of visualization due to real-time requirements.

Flow (vector) visualization

In volume visualization, generally the data at each point in the volume lattice is a scalar quantity. Recently, new MR imaging techniques that provide diffusion images of water molecules produce information that is a vector in each data point, and new types of visualization techniques for such vector field data become necessary. For instance, one approach would try to produce continuous "flow-field-lines" through the volume that follow the vectors in the 3D space. Such visualization techniques present unique line patterns that can help to identify sources, sinks, and vortices in the volume of vector data (Fig. 6.6-9) [33, 34, 38, 39].

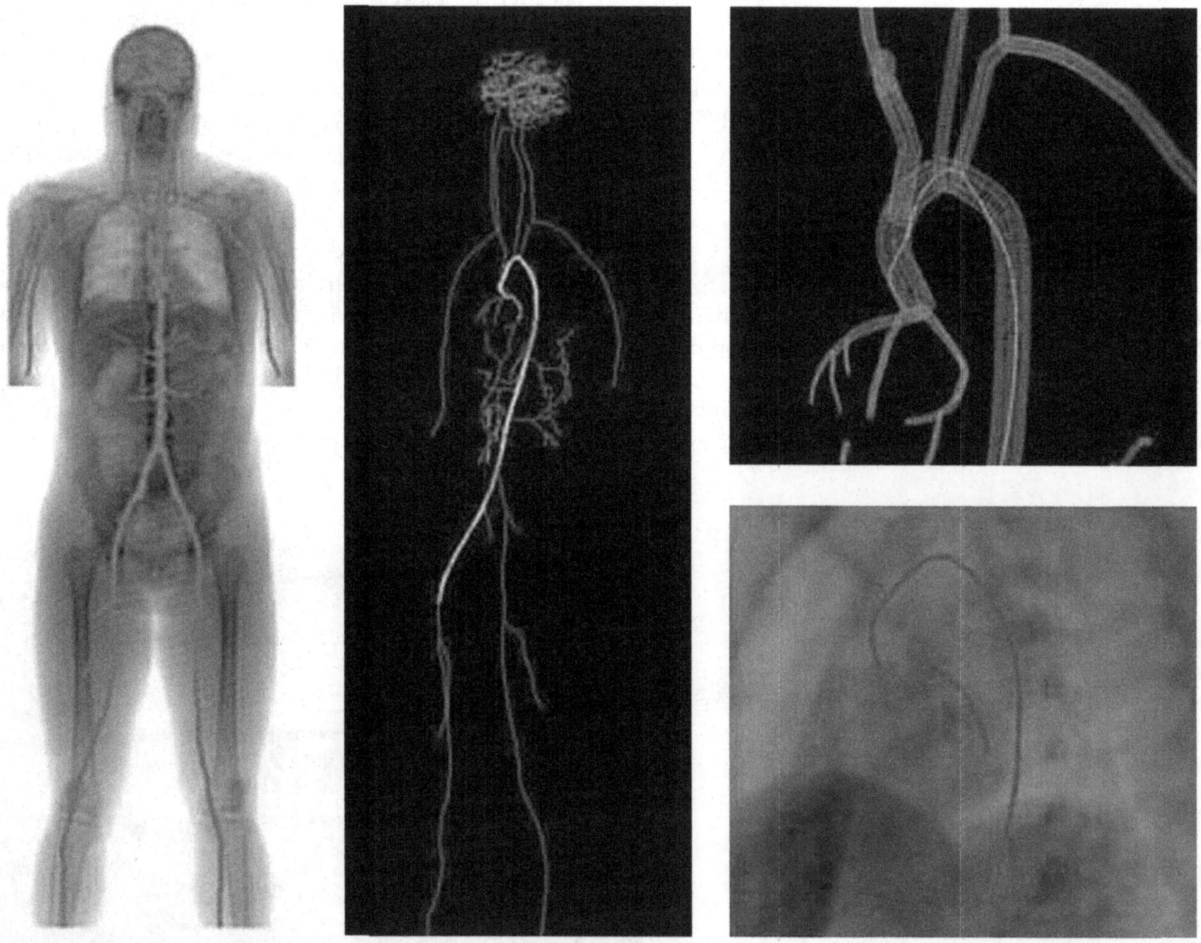

Figure 6.6-14 Real-time interactive surgical simulator. Finite-element-based modeling of the interaction of catheter and guide-wire devices with the vascular structures of the human body. Visualization provides training and planning information through the simulated fluoroscopic display. (Image courtesy of Yao Ping Wang, Chee Kong Chui, H. L. Lim, Y. Y. Cai, K. H. Mak, R. Mullick, R. Raghavan, James Anderson.)

Rapid back-projection reconstruction

Although volume visualization is an area of continuing progress, there is another need that is growing with new types of imaging techniques: visualization of a 3D structure using a set of projection images, i.e., images acquired from different orientations. Concerns of exposure to radiation and the need for high speeds make projection images a preferred mode of acquisition for some applications. Such requirements may be difficult to meet while acquiring the images in 3D mode. In this case, if the geometric nature of the object is known, a representative shape of the object can be reconstructed using image back-projection techniques. Although this principle is similar to the tomographic reconstruction used in CT, lack of sufficient accuracy and angular resolution may restrict the use of such approaches. Back-projection refers to a graphic technique in which a raster or scan line of an image is smeared, i.e., painted parallel to the direction in which the image is acquired, on an image reconstruction buffer. By back-projecting corresponding scan lines from each projection image into the reconstruction buffer, transverse-images of the desired 3D image can be obtained (Fig. 6.6-10) [29]. Recent advances in graphics hardware, such as texture mapping, enable rapid reconstruction using the back-projection technique rather than traditional generic CPU based approaches.

6.6.4 Imitative visualization

The development of illustrative and investigative visualization systems paved the way to the development of visualization systems that focus on creating realism through visualization. Such virtual realism could be exhibited in visualizing the structures or in visualizing the functions through simulation and modeling. These visualization techniques belong to two classes of fifth generation systems.

6.6.4.1 Fifth generation systems I (modeling and simulation)

Modeling represents a significant thrust in biomedical research. It attempts to build analytical formulations that describe physiological functions and involves such multidisciplinary fields as finite-element modeling and computation fluid dynamics (Figs. 6.6-11–6.6-14). Visualization needs in modeling have been minimal in the past, primarily due to the complexity of the models. The focus was on quantitative or analytical verification rather than on visual validation thought to be optional. However, over the years, the scope of models that could be built and the extent to which they can be studied became limited when all the necessary input information to define model parameters became unavailable. The new approach that was needed to address such modeling problems emerged in visualization, since it could enable researchers to qualitatively validate the model. Visualization techniques that are employed in modeling packages are more suitable for engineering applications than for biomedical applications, which deal with large volumetric images. Visualization techniques that are needed for biomedical modeling and simulation applications should be compatible with finite-element [49–51] and fluid dynamics techniques [52]. This will enable researchers to build rapid prototypes of the models for quick visual validation and determine some parameters of the model [53]. Also, this approach can enable them to build the model using actual geometry rather than the simplified representation the earlier modeling techniques preferred.

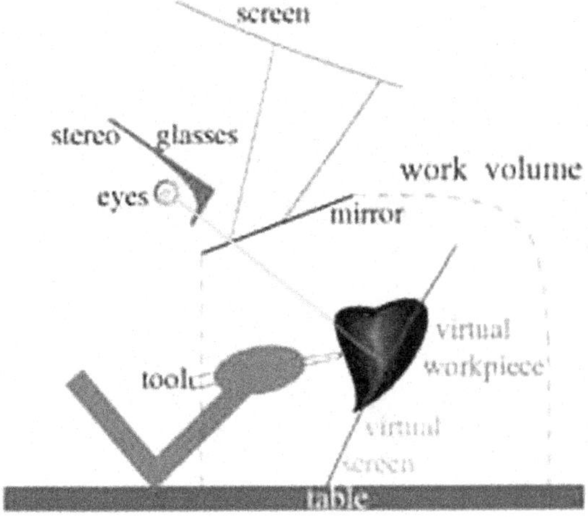

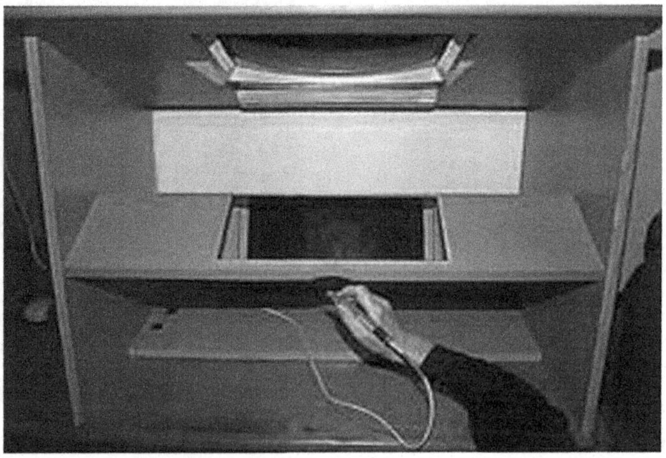

Figure 6.6-15 Virtual workbench. (*Left*) Schematic diagram illustrating the concept of reach-in work space. (*Right*) Prototype model of the workbench. (Photos courtesy of Timothy Poston, Luis Serra.)

6.6.4.2 Fifth generation systems II (virtual reality)

Virtual reality visualization represents the next milestone in biomedical visualization. While stereoscopic visualization provided the 3D perception, such a presentation also required intrinsically intuitive 3D interaction with the display. The traditional interaction paradigm using a tool or stylus such as a mouse or pointer that is away from the display lacks the intuitive control of hand–eye coordination. A more natural and intuitive interaction can be obtained by manipulating the image where it is perceived in 3D space. This can be achieved with a stylus using 3D positional sensors. However, placing such a stylus that the hand can manipulate directly on the image would obstruct the view of the image.

Some virtual systems overcome this problem by using helmet-mounted displays that present a virtual world through stereo goggles mounted to the display. This is generally called an "immersive" virtual reality paradigm. It may be suitable for representing a virtual world where the physical scale of the objects is larger than the viewers, as in architectural visualization, where they can imagine immersing themselves into such a virtual world. However, such a paradigm may not be suitable for visualizing objects of similar or smaller physical scale, as is often the case in medicine. A more suitable virtual reality paradigm proposed for such an interaction is the virtual workbench or virtual work volume concept (Figs. 6.6-15 and 6.6-16). A mirror arrangement is placed at a suitable angle to the stereo display, reflecting the 3D image to the viewer, so that the viewer, looking at the mirror through stereo goggles, perceives 3D objects behind the mirror. The virtual object that appears within the "reach-in" distance of the hand can be manipulated exactly where it appears in the virtual work volume. The 3D stylus or mouse can be replaced by a suitable visual motif functionally representing the tool in the image. Augmented reality visualization encapsulates same principles, but instead of presenting a fully synthetic virtual world, it is optically superimposed on the real world, so that the viewer perceives the virtual objects placed in a real working space. This is useful during an intraoperative procedure, combining, for instance, a 3D MRI image that can show subtle differences in the brain tissues with a real image of the brain during the surgery where the naked eye would not be able to see the differences very well. Real-time registration of the actual anatomy with

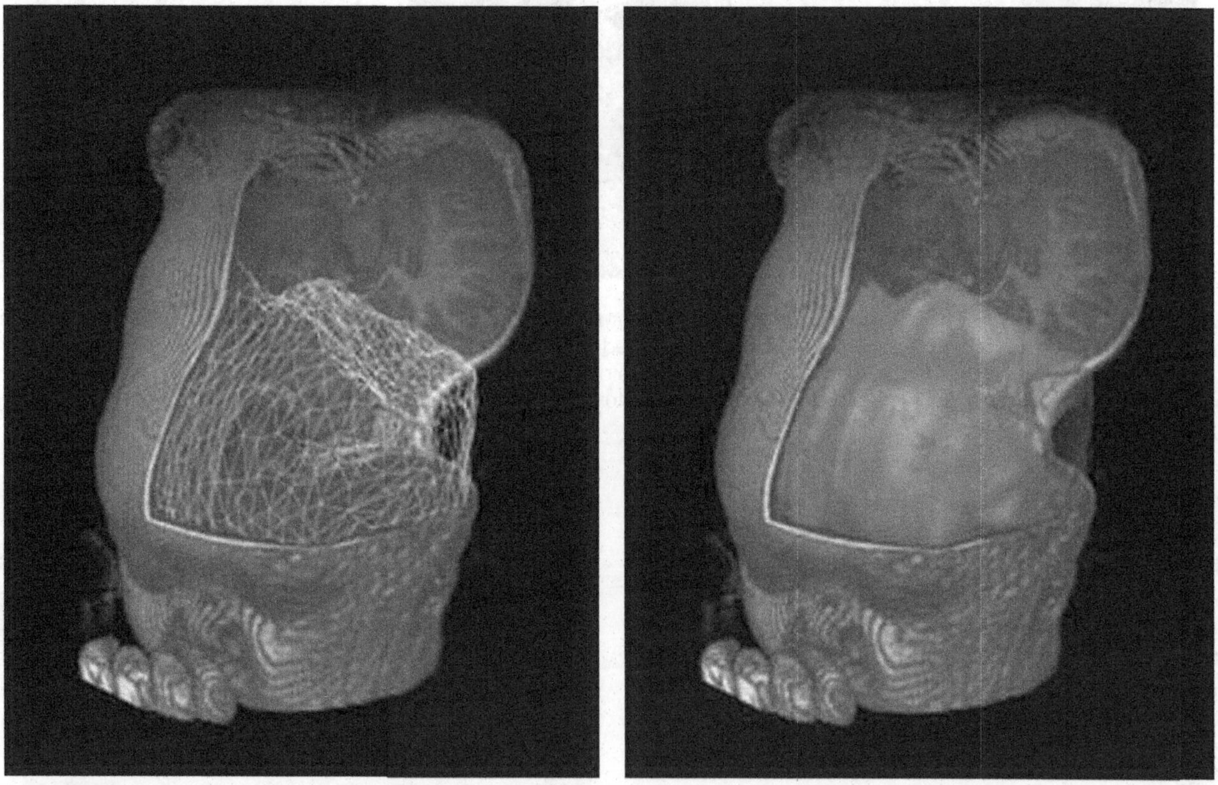

Figure 6.6-16 Presurgical planning: separation of Siamese twins. 3D visualization and 3D interactions in the virtual workspace environment using MRA and MR volumes to visualize vascular and brain morphology assisted surgeons in devising strategies in planning the extremely complex and most delicate neurosurgical procedure of separating craniopagus Siamese twins (under the notably challenging circumstance that the patient was in a distant location and could not be reached prior to surgery). Dec. 1997, Zambia. Pediatric neurosurgeon: Benjamin Carson (JHU). Software (VIVIAN): Luis Serra, Ng Hern (KRDL, Singapore). Tushar Goradia, James Anderson (JHU).

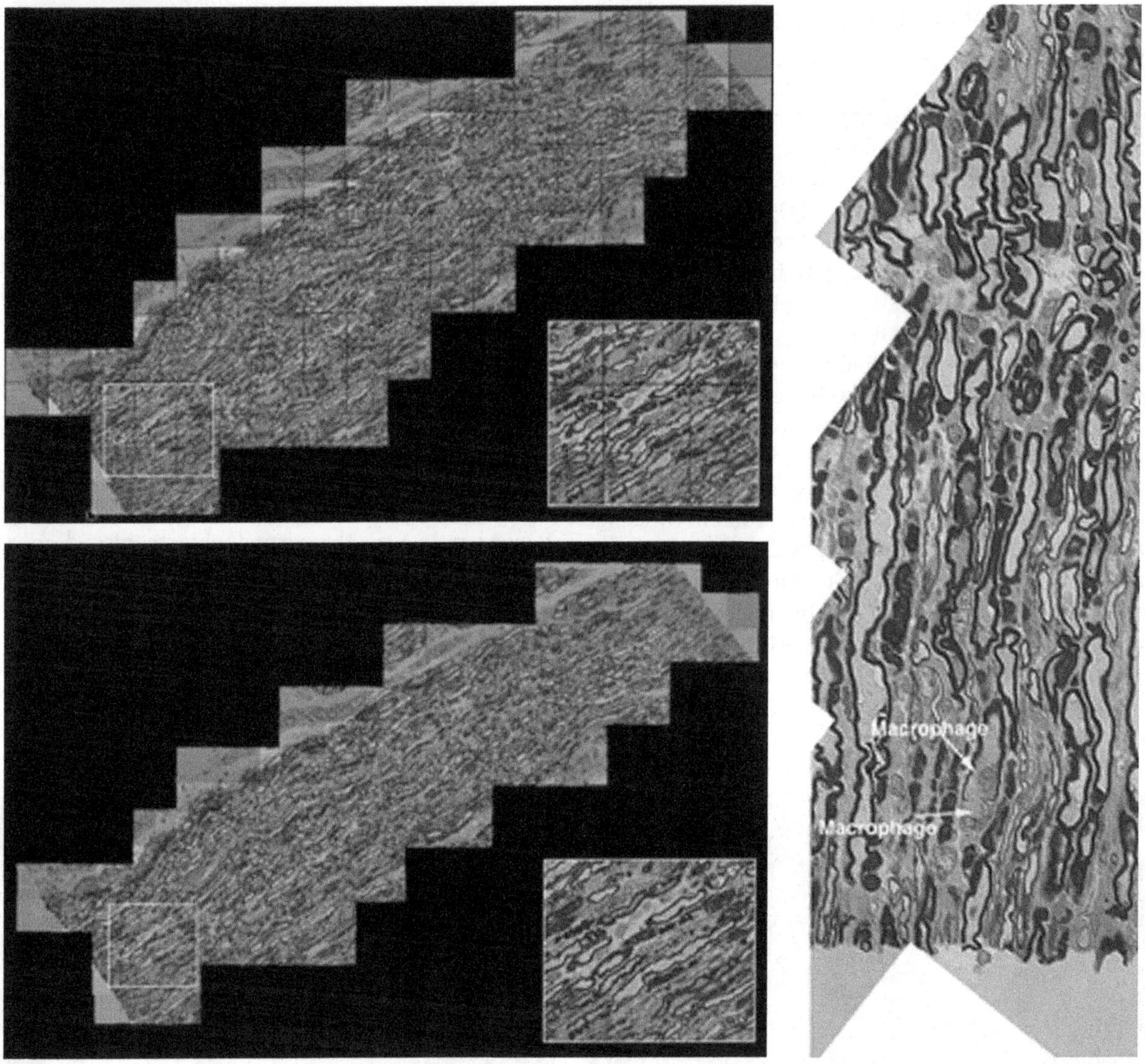

Figure 6.6-17 Virtual microscopy: visualization of nerve cells. (*Left top*) Tiles of images (~150 pcs each 640 × 480 pixels) acquired using light microscope at 1000 ×. (*Left bottom*) Seamlessly stitched virtual field of view image that can help follow single axons of the cell. (*Right*) Ventral spinal root from an individual dying of acute motor axonal neuropathy. Macrophages are seen insinuating their way into periaxonal space through the node of Ranvien. (Images courtesy of M. Solaiyappan, Tony Ho, Aline Fang-Ling Li, John Griffin, Raghu Raghavan.)

the virtual image is crucial for such integration [56–58]. For such purposes, a stereotaxic reference frame that is stationary with respect to the subject and remains physically attached to the subject during imaging and surgery is used. Another issue that needs to be addressed is the change in the shape of the brain during the craniotomy as the skull is opened for surgery.

6.6.4.3 Imitative visualization by sensory feedback

In the next generation of visualization systems, the concept of visualization will extend beyond graphical displays and include other forms of perception that convey additional information. For instance, force feedback or sensory feedback can help in understanding not only the spatial structure, but also the physical characteristics of the data, such as deformability and consistency. Multisensory visualization may increase the channels of communication for interacting with data.

6.6.5 Visualization in biology

Although visualization in biology is a less pervasive technology today than it is in medicine, there is an increasing trend toward using powerful visualization tools in biology. It is important to note that many of the

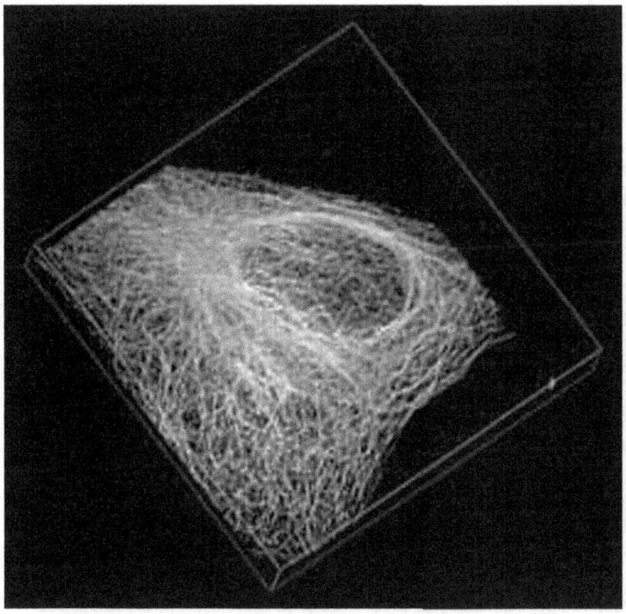

Figure 6.6-18 3D visualization of microtubules. Immunofluorescence labeling of U2OS (human osteosarcoma) cells with monoclonal antibody to tubulin and a rhodamine-labeled secondary antibody against mouse monoclonal antibody. The images were acquired using a Noran confocal microscope. (Image courtesy of Douglas Murphy, M. Solaiyappan.)

investigative research techniques in biology and medicine are pursuing similar paths. Imaging methods in medicine are increasingly becoming microscopic in scale, while imaging in biology requires addressing the functional properties of microscopic structures (Figs. 6.6-17–6.6-20). These trends may lead to the development of new types of visualization systems that try to combine the scales. Image data in medicine are functionally rich, while image datasets in biology are large because of the very high optical resolution that can be achieved. One emerging concept is virtual microscopy visualization, where tiles of microscope images acquired from different areas on a specimen are seamlessly stitched to form a single large virtual image that can be zoomed and panned as if it was acquired with a very large field of view (Fig. 6.6-17). Combining these two types of data will challenge the current visualization techniques in terms of their current limitation in handling very large data sizes.

6.6.6 Visualization in spatial biostatistics

Large collections of image databases in epidemiological studies have been growing rapidly over the years, and spatial statistical techniques are now being applied to images of large populations. Applications include relationships between such attributes as lesions and functional deficits (Fig. 6.6-21) [41]. Gene expression studies carried out with the help of imaging techniques such as *in situ* hybridization suggest future needs for correlating the information over a large number of studies in different groups. In such spatial statistical studies, each data point could be potentially represented as a multidimensional vector and presented to the viewer using visualization techniques.

6.6.7 Parametric visualization

Parametric visualization addresses an emerging trend arising from the analysis of images to characterize functional information present in the images. Traditionally, parameters that describe certain functional information are computed and stored as numeric images that can be displayed like any other image. However, it may require large amount of storage space and processing time to handle such numeric images because they are usually represented as floating-point numbers and vector quantities. To simplify the representation of such information, parametric fields are used to provide an underlying (usually smooth) model that can be evaluated at any image location using its analytical formulation [43].

One of the major advantages of a parametric field is that it can efficiently make use of graphics architectures to provide the visualization of the data, eliminating the

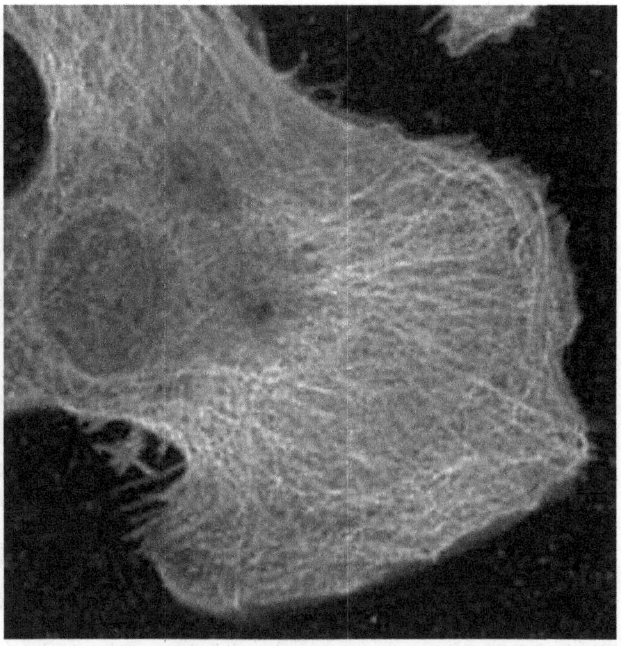

Figure 6.6-19 3D visualization of actin filaments. Reconstruction of a mouse fibroblast with antibodies microtubules and actin filaments, imaged using a Zeiss confocal microscope exciting in the FITC and TRITC fluorescence spectra.

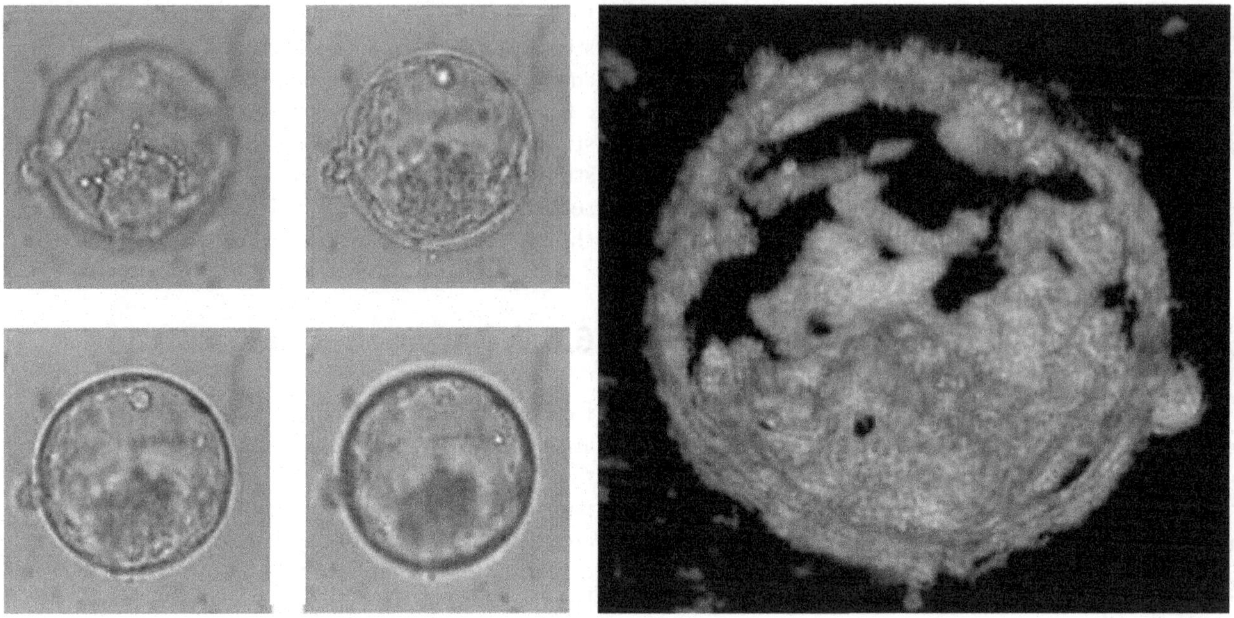

Figure 6.6-20 3D visualization of live human embryo, age 5 days (blastocyst). (*Left*) Four images from optical microscope at varying focal plane. (*Right*) 3D volume by deconvolution showing the inner cell mass. (Images courtesy of M. Solaiyappan, Fong Chui Yee, Ariffeen Bongso, Rakesh Mullick, Raghu Raghavan.)

precomputation time and the need for high-resolution storage for visualization. Furthermore, successive derivative parametric images, such as velocity or acceleration maps from the motion fields, can be displayed in real time.

The example presented here is a 4D B-spline based motion field representation of cardiac-tagged MR images (Fig. 6.6-22). A motion field with $7 \times 7 \times 7 \times 15$ control points has been shown to adequately describe the full motion of the heart during cardiac cycle. Using this field,

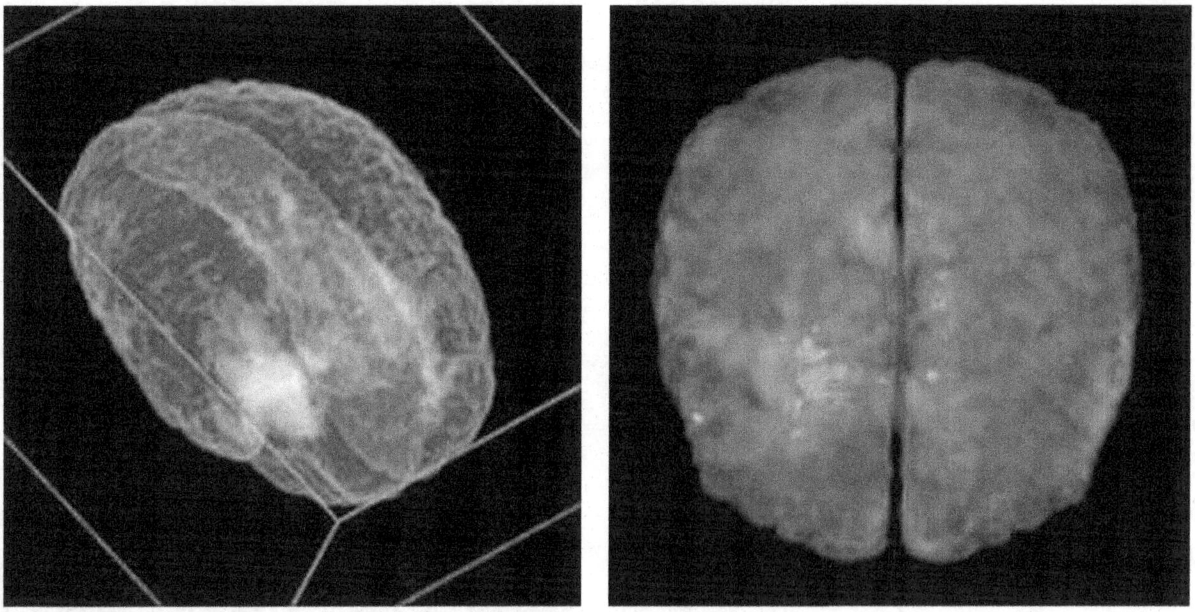

Figure 6.6-21 3D visualization of lesion–deficit associations. The development of ADHD (attention deficit hyperactivity disorder) was studied using a voxel-based approach for a spatial statistical technique (Fisher's exact test) applied to a population of children involved in frontal lobe injury. Higher intensity (left image) shows higher confidence of association in those regions. Right image shows the associated regions in a color-mapped 3D Talairach atlas registered to the volume. (Images courtesy of V. Megalooikonomou, C. Davatzikos, E. H. Herskovits, M. Solaiyappan.)

material points can be tracked over time and local mechanical properties (e.g., strain) can be computed. The visualization method presented here utilizes the similarity between the B-spline representation of the motion fields and the graphics hardware support for non-uniform rational B-spline (NURBS) display with texture mapping to achieve high-performance visualization of these parametric fields.

6.6.8 Discussion

6.6.8.1 Speed issues in visualization systems

Speed issues often become the essential factors that determine the usefulness of a visualization system. Visualization involves two types of basic graphic operations: One is related to the geometry, such as the transformation of the vertices of a polygon, and the other is associated with displaying pixels. In the graphics processing pipeline, the overhead can happen in either of these two operations and adversely affect the speed of the entire system. Thus, for instance, drawing a large number of small triangles would cause overhead in the transformation computations associated with the vertices, while the pixel filling operation for the small triangles would be fast. The opposite is true when the number of triangles is few but their size is large. Thus, an optimal balance between the two factors yields a good overall system performance. Another factor that affects speed is the geometric technique used in visualization. A simple technique may execute much faster because of its use of instruction and memory cache, whereas a more intelligent and elaborate technique that does not make effective use of instruction and memory cache will be slower. Since biomedical images are usually large, efficient use of cache capabilities of the hardware is essential.

6.6.8.2 Future issues in visualization

Many potentially important visualization concepts did not develop successfully in the past because of the lack of proper systems. These concepts had memory and processor speed requirements that were then considered unreasonable. This led to the development of logically more complex systems, which, in turn, created several other diversions. With processors and memory becoming more efficient and less expensive, and with the rapid

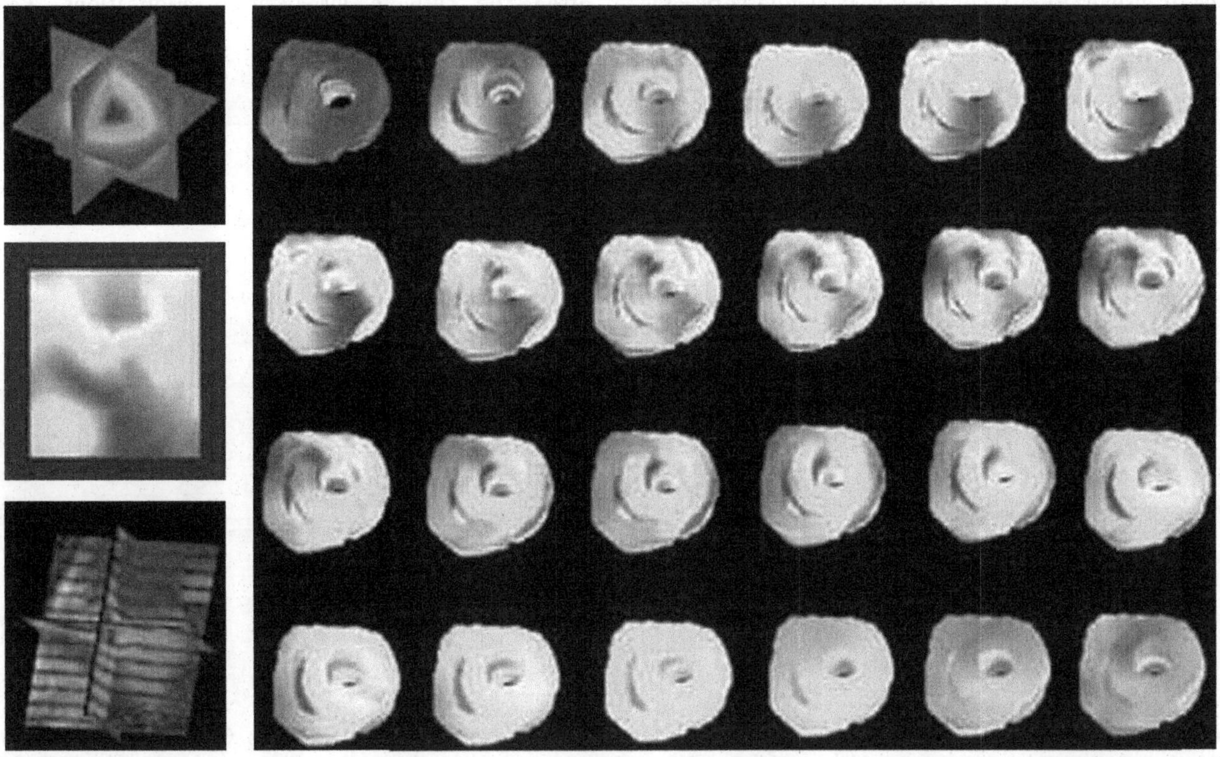

Figure 6.6-22 Visualization of the displacement field during heart motion. (*Left*) Triplanar view of a 3D parametric map (top) used for the 3D texture mapping of a NURBS representation of the displacement field (center) and modulated with image data (bottom). (*Right*) Cine-frames of volumetric rendering of the NURBS generated displacement fields during the full cardiac cycle, starting from end diastole at top left. (Frame order: left-to-right, top-to-bottom.) (Images courtesy of M. Solaiyappan, Cengizhan Ozturk, Elliot McVeigh, Albert Lardo.)

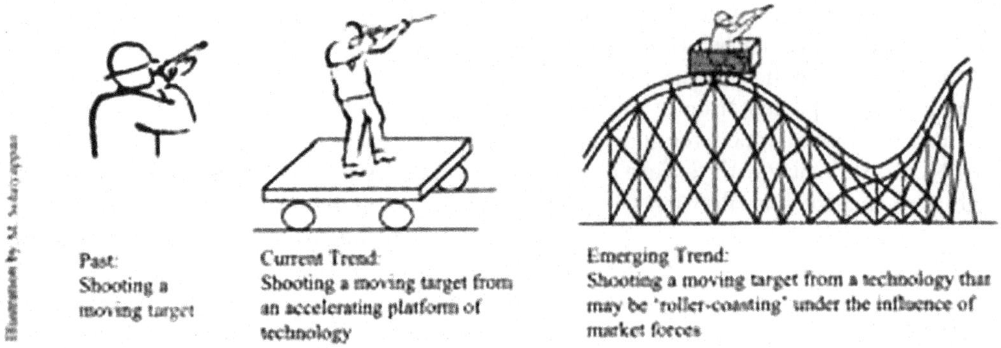

Figure 6.6-23 Targeting visualization applications.

advancement of graphics accelerators, many of those initial and simpler concepts may be revisited to provide more powerful and richer solutions. Other complex programs that required supercomputers in the past will become more affordable and useful. The synergy of these developments could produce new ideas for visualization systems. Yet, the challenges in visualization will not diminish. For instance, as the capabilities of imaging systems advance, the spatial, temporal, and channel resolutions of data will increase. A twofold increase in spatial resolution means a fourfold increase in image area and an eightfold increase in volume dimension. New types of visualization techniques that handle very large data sizes need to be developed. As the number of imaging modalities increases, system architectures for visualization may change from integrated systems to distributed visualization systems.

Targeting visualization application can be thought of as a visualization problem in itself. The rapid progress in computing and imaging technology makes the planning and coordination of new applications extremely challenging. Traditionally planning of such applications is thought of as shooting a moving target (Fig. 6.6-23). The moving target represents an application whose specification evolves during the research and development cycle, as is often the case in visualization applications. Rapid

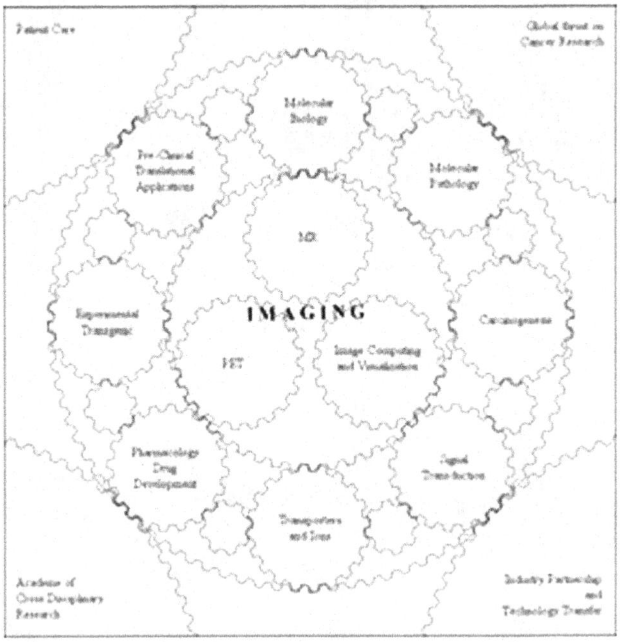

Figure 6.6-24 Synergy and synchronization form the clockwork in multicentric cross-disciplinary research. Clockwork demands both cooperative actions and precision timing of those actions. This depends on optimal coupling of the actions and inertia of each discipline involved. For instance, the illustration demonstrates the role of the *In Vivo* Cellular and Molecular Imaging Cancer Center as one such clockwork in cancer research. Imaging here represents the coupling wheel that transforms the actions from discoveries in one area to research momentum in another area with precision timing that can help generate *revolutions* in cancer research. Critical to such clockwork is also the ability to produce independent action within each unit, lack of which could mean degeneration of actions that could potentially bring the clockwork to a halt.

advances in both imaging and computing technology take the shooting game to a more challenging level, by staging it to take place on a rapidly advancing platform of technology. This emphasizes the need to foresee and understand the shrinking gap between the capability of the technology and the demands of the application, so that the development process does not undershoot or overshoot cost or performance factors when it approaches meeting the target. The newly emerging trend represents another level of complexity, where promising technology need not be the best solution if market forces are not conducive to the survival of such technology. Thus, the shooting game becomes one that takes place in a roller-coaster ride where not only is the platform of technology important, but also its stability under the influence of market forces and the strength of the supporting structure become important factors.

One of the fundamental research aims in biology and medicine is to relate structures and functions at the cell level with structures and functions at the organ or anatomical level. In this manner, diseases can be better understood and the effect of treatments on diseases can be studied more effectively. Such investigations will create needs for superscale visualization systems that allow researchers to visualize and correlate information across different scales, including molecules, cells, organs, body, and epidemiological studies in human populations.

The interaction between different disciplines presents another level of complexity in multidisciplinary research in biology and medicine. The interaction between different disciplines provides beneficial synergy. However, such synergy requires precise synchronization of information flow between different disciplines. Both the synergy and the synchronization can be facilitated by visualization systems. The clockwork arrangement shown in Fig. 6.6-24 illustrates the interaction between multidisciplinary initiatives in an imaging center for cancer research.

6.6.8.3 Software and computing systems used for figures

In-house visualization research software tools were used to generate the figures presented in this chapter. For the FEM modeling of the tumor growth and the fluid dynamics modeling of the blood flow, ABAQUS and Fluent software respectively were used. The Onyx Reality Engine system from Silicon Graphics, Inc., was used as the computing and visualization platform for the production of these images.

References

Textbooks

Scientific visualization

1. *Mathematical Visualization Algorithms, Applications and Numerics* (Hans-Christian Hege, Konrad Polthier, Eds.). Springer Publishers (1998).
2. *Data Visualization Techniques* (Chandrajit Bajaj, Ed.). John Wiley Sons, Ltd. (1999).

Medical imaging

3. *Fundamentals of Medical Imaging*. Zang Hee Cho, Joie P. Jones, Manbir Singh. John Wiley Sons (1993).
4. *Fundamentals of Digital Image Processing* A. K. Jain. Prentice-Hall (1989).
5. *Medical Imaging Systems.* A. Macovski. Prentice Hall (1983).
6. *Principles of Magnetic Resonance Imaging*. Dwight Nishimura, Dept. of Electrical Engineering, Stanford University (1996).

Space and geometry

7. *Space through the Ages. The Evolution of Geometrical Ideas from Pythagoras to Hilbert and Einstein.* Cornelius Lanczos. Academic Press (1970).
8. *Computational Geometry: An Introduction*. Franco P. Preparata, Michael Ian Shamos. Springer Verlag (1988).
9. *A Treatise on the Differential Geometry of Curves and Surfaces.* Luther Pfahler Eisenhart. Dover Publications Inc. (1909).
10. *Algorithmic Geometry.* Jean-Daniel Boissonnat, Mariette Yvinec. Cambridge University Press (1998).

Shapes and forms in biology

11. *On Growth and Form.* Sir D'Arcy Wentworth Thompson. Cambridge University Press (1961).
12. *Models for the Perception of Speech and Visual Form*: Proceedings of a Symposium, Sponsored by the Data Sciences Laboratory, Air Force Cambridge Research Laboratories, Boston, MA, Nov. 11–14, 1964 (Weiant Walthen-Dunn, Ed.). MIT Press (1967). See "A Transformation for Extracting New Descriptors of shape," Harry Blum (pp. 363–380).

Course materials/conference proceedings

13. *3D Visualization in Medicine ACM SIGGRAPH 98*. Course Organizer: Terry S. Yoo. Lecturers: Henry Fuchs, Ron Kikinis, Bill Lorensen, Andrei State, Michael Vannier.
14. *Data Visualization '99* (E. Groller, H. Loffelmann, W Ribarsky, Eds.), Proceedings of the Joint Eurographics and IEEE TCVG Symposium on Visualization, Vienna, Austria. Springer, Wien.
15. *Visualization in Scientific Computing '95*, Proceedings of the Eurographics Workshop in Chia, Italy, May 3–5, 1995 (R. Scateni, J. van Wijk, and P. Zanarini, Eds.). Springer, Wien.
16. *Advanced Graphics Programming Techniques Using OpenGL*. Organizer: Tom McReynolds. SIGGRAPH Course (1998).

Research papers

General

17. *Survey article:* Three-Dimensional Medical Imaging: Algorithms and Computer Systems. M. R. Stytz, G. Frieder, and O. Frieder. *ACM Computing Surveys* **23**(4), December (1991).
18. Medical Visualization—Why We Use CG and Does it REALLY Make

a Difference in Creating Meaningful Images? Panel Discussion, Computer Graphics Proceedings, Annual Conference Series, SIGGRAPH (1997).

Surface visualization

19. Marching Cubes: A High Resolution 3D Surface Construction Algorithm. William Lorensen and Harvey Cline. *Computer Graphics* 21:163–169 (1987).
20. Shell Rendering. Jayaram K. Udupa and Dewey Odhner. *IEEE Computer Graphics and Applications*, November, pp. 58–67 (1993).
21. Surface Reconstruction from Unstructured 3D Data. Maria-Elena Algorri and Francis Schmitt. *Computer Graphics Forum* 15: 47–60 (1996).

Scalar field, volume visualization

22. Compositing Digital Images. Thomas Porter, Tom Duff. *Computer Graphics* 18:100–106 (1984).
23. Volume Rendering. Karen A. Frenkel. *Communications of the ACM* 32(4) (1989).
24. Efficient Ray Tracing of Volume Data. Marc Levoy. Computer Graphics Proceedings, Annual Conference Series, SIGGRAPH (1993).
25. Editing Tools for 3D Medical Imaging Tools. Derek R. Ney and Elliot K. Fishmann. *IEEE Computer Graphics Applications* Nov. (1991).
26. Fast Volume Rendering Using a Shear-Warp Factorization of the Viewing Transformation. Computer Graphics Proceedings, Annual Conference, SIGGRAPH (1994).
27. Frequency Domain Volume Rendering. Takashi Totsuka, Marc Levoy. Computer Graphics Proceedings, Annual Conference Series, SIGGRAPH (1993).
28. Texture Mapping as a Fundamental Drawing Primitive. Paul Haeberli, Mark Segal. Fourth Euro-graphics Workshop on Rendering, pp. 259–266 Paris, France, June (1993).
29. Accelerated Volume Rendering and Tomographic Reconstruction Using Texture Mapping Hardware. B. Cabral, N. Cam and J. Foran. *ACM Symposium on Volume Visualization* '94, pp. 91–98 (1994).
30. Interactive Volume Rendering Using Advanced Graphics Architectures. Robert Fraser. SGI Tech Report, Nov. 1994.
31. Efficiently Using Graphics Hardware in Volume Rendering Applications. Rüdiger Westermann, Thomas Ertl. Proceedings of SIGGRAPH 98, *Computer Graphics Proceedings*, July 1998, Orlando, FL pp. 169–178.
32. The VolumePro Real-Time Ray-Casting System. Hanspeter Pfister, Jan Hardenbergh, Jim Knitte, Hugh Lauer, Larry Seiler. Computer Graphics Proceedings, Annual Conference Series, SIGGRAPH (1999).

Vector field visualization

33. Imaging Vector Fields using Line Integral Convolution. Brian Cabral, Leith (Casey) Leedom. Computer Graphics Proceedings (SIGGRAPH) Annual Conference Series (1993).
34. Conveying the 3D Shape of Smoothly Curving Transparent Surfaces via Textures. Victoria Interrante, Henry Fuchs, Stephen M. Pizer. *IEEE Transactions on Visualization and Computer Graphics*, 3(2): (1997).

Functional imaging/visualization

35. Human Heart: Tagging with MR Imaging—A Method for Noninvasive Assessment of Myocardial Motion. E. Zerhouni, D. Parish, W. Rogers, A. Yang, E. Shapiro. *Radiology* 169:59–63 (1988).
36. Interactive Visualization of Speedy Non-invasive Cardiac Assessment. Meiyappan Solaiyappan, Tim Poston, Pheng Ann Heng, Elias Zerhouni, Elliot McVeigh, Michael Guttman. *IEEE Computer, Special Issue on Surgery*, January (1995).
37. Measuring and Visualizing the Effect of Inflation and Hysteresis on Three Dimensional Lung Structure with Stereoscopic Volume Rendering. S. A. Wood, S. Meiyappan, E. Zerhouni, W. Mitzner. *American Review of Respiratory Disease* (1995).
38. Three Dimensional Tracking of Axonal projections in the Brain by Magnetic Resonance Imaging. S. Mori, B. J. Crain, V. P. Chacko, P. C. M. van Zijl *Annals of Neurology* 45:265–269 (1999).
39. In Vivo Three-Dimensional Reconstruction of Rat Brain Axonal Projections by Diffusion Tensor Imaging. R. Xue, P. C. M. van Zijl, B. J. Crain, M. Solaiyappan, and S. Mori. *Magnetic Resonance in Medicine* 42: 1123–1127 (1999).
40. Depth Reconstruction from Projection Images for 3D Visualization of Intravascular MRI Probes. Meiyappan Solaiyappan, Joanna Lee, Ergin Atalar. International Society for Magnetic Resonance in Medicine, Seventh Scientific Meeting, 22–28 May (1999).
41. Mining Lesion-Deficit Associations in a Brain Image Database. V. Megalooikonomou, C. Davatzikos, E. H. Herskovits. Proceedings of the ACM SIGKDD International Conference on Knowledge Discovery and Data Mining, San Diego, CA, Aug. 15–18 (1999).
42. Insights into Tumor Vascularization using Magnetic Resonance Imaging and Spectroscopy. Zaver M. Bhujwalla, Dmitri Artemov, Meiyappan Solaiyappan, to appear in *Experimental Oncology.*
43. Interactive Visualization of 4-D Parametric Fields. Meiyappan Solaiyappan, Cengizhan Ozturk. Medical Imaging 2000, SPIE, Feb. 2000, San Diego.

Multimodal visualization

44. Multi-Modal Volume Registration by Maximization of Mutual Information. William M. Wells, Paul Viola, Hideki Atsumi, Shin Nakajima, Ron Kikinis. *Medical Image Analysis*, Oxford University Press. 3D Visualization in Medicine course, ACM SIGGRAPH (1998).
45. Multi Modal Volume Based Tumor Neuro-surgery Planning in the Virtual Workbench. Luis Serra, Ralf A. Kockro, Chua Gim Guan, Ng Hern, Eugene C. K. Lee, Yen H. Lee, Chumpon Chan, Wieslaw L. Nowinski. First International Conference on Medical Image Computing and Computer-Assisted Intervention, MICCAI98, MIT, Cambridge MA, October, pp. 1007–1016 (1998).

Shape morphing

46. Note: On Active Contour Models and Balloons. Laurent D. Cohen. *CVGIP: Image Understanding* 5(2): 211–218 (1991).
47. Wavelet Based Volume Morphing. Taosong He, Signey Wang, Arie Kaufman. *IEEE Computer Graphics and Applications* (Reprint No. 1070-2385/94) (1994).
48. Volume Morphing and Rendering—An Integrated Approach. Shiaofen Fang, Rajagopalan Srinivasan, Raghu Raghavan, Joan T. Richtsmeier. *Computer Aided Geometric Design* 17, 59–81 (2000).

Simulation and modeling

49. Nonlinear Elastic Registration of Brain Images with Tumor Pathology Using a Biomechanical Model. S. K. Kyriacou, C. A. Davatzikos, S. J. Zinreich, R. N. Bryan. *IEEE Transactions on Medical Imaging* 18(7): 580–592 (1999).
50. A Vascular Catheterization Simulator for Training and Treatment Planning. J. Anderson, R. Raghavan. *Journal of Digital Imaging* 11:120–123 (1998).
51. Real-Time Interactive Surgical Simulator for Catheterization Procedures. Y. P. Wang, C. K. Chui, H. L. Lim, Y. Y. Cai, K. H. Mak. *Journal of Computer Aided Surgery* 3: 211–227 (1999).
52. The Hemodynamics of a Coil-Filled Saccular Aneurysm. T. M. Tsai, J. H. Anderson. Technical Report, Department of Radiology, Johns Hopkins Medical Institutions, Baltimore.
53. Electrophysiological Modeling of Cardiac Ventricular Function: From Cell to Organ. R. L. Winslow, D. F. Scollan, A. Holmes, C. K. Yung, J. Zhang, M. S. Jafri. *Ann. Rev. Biomed. Eng.*, in press.

Virtual/augmented reality systems

54. The Brain Bench: Virtual Tools for Neurosurgery. Luis Serra, Nowinski, W. L, T. Poston, Chua, B. C., Ng H., Lee, C. M., Chua, G. G. and Pillay, P. K., *Journal of Medical Image Analysis* 1(4): (1997).
55. An Interactive Visualization and Navigation Tool for Medical Volume Data. Ove Sommer, Alexander Dietz, Rudiger Westermann, Thomas Ertl. *Computers and Graphics* 23:233–244 (1999).
56. Technologies for Augmented Reality Systems: Realizing Ultrasound-Guided Needle Biopsies. Andrei State, Mark A. Livingston, William F. Garret, Gentaro Hirota, Mary C. Whitton, Etta D. Pisano, Henry Fuchs. *Computer Graphics Proceedings*, Annual Conference Series, SIGGRAPH (1996).
57. Computer Assisted Interactive Three-Dimensional Planning for Neurosurgical Procedures. Ron Kikinis, Langham, T. M. Moriarty, M. R. Moore, E. Alexander, M. Matsumae, W. E. Lorensen, P. M. Black, F. A. Jolesz. *Neurosurgery* 38(4):640–651 (1996).
58. Planning and Simulation of Neurosurgery in a Virtual Reality Environment. Ralf A. Kockro, Luis Serra, Yeo Tseng-Tsai, Chumpon Chan, Sitoh Yih-Yian, Chua Gim-Guan, Eugene Lee, Lee Yen Hoe, Ng Hern, Wieslaw L. Nowinski. *Neurosurgery* 46(1):118–137 (2000).

Section Seven

Tissue engineering

Chapter 7.1

7.1

Tissue engineering

Buddy D. Ratner, Allan Hoffman, Frederick J. Schoen, and Jack E. Lemons

7.1.1 Introduction

Frederick J. Schoen

Biomaterials investigation and development has been stimulated and informed by a logical evolution of cell and molecular biology, materials science, and engineering, and an understanding of the interactions of materials with the physiological environment. These developments have permitted the evolution of concepts of tissue–biomaterials interactions to evolve through three stages, overlapping over time, yet each with a distinctly different objective (Fig. 7.1.1-1) (Hench and Pollak, 2002). The logical and rapidly progressing state-of-the-art, called *tissue engineering*, is discussed in this chapter.

The goal of early biomaterials development and use in a wide variety of applications was to achieve a suitable combination of functional properties to adequately match those of the replaced tissue without deleterious response by the host. The "first generation" of modern biomaterials (beginning in the mid-20th century) used largely off-the-shelf, widely available, industrial materials that were not developed specifically for their intended medical use. They were selected because of a desirable combination of physical properties specific to the clinical use, and they were intended to be *bioinert* (i.e., they elicited minimal response from the host tissues). The widely used elastomeric polymer silicone rubber is prototypical. Pyrolytic carbon, originally developed in the 1960s as a coating material for nuclear fuel particles and now widely used in mechanical heart valve substitutes, exemplifies one of the first biomaterials whose formulation was studied, modified, and controlled according to engineering and biological principles specifically for medical application (Bokros, 1977).

Subsequently, technology enabled and certain applications benefited by "second-generation" biomaterials that were intended to elicit a nontrivial and controlled reaction with the tissues into which they were implanted, in order to induce a desired therapeutic advantage. In the 1980s, *bioactive* materials were in clinical use in orthopedic and dental surgery as various compositions of bioactive glasses and ceramics (Hench and Pollak, 2002), in controlled-localized drug release applications such as the Norplant hormone-loaded contraceptive formulation, and in devices such as the HeartMate® left ventricular assist device for patients with congestive heart failure, with an integrally-textured polyurethane surface that fosters a controlled thrombotic reaction to minimize the risk of thromboembolism (Rose *et al.*, 2001). Recently, drug-eluting endovascular stents have been shown to markedly limit in-stent proliferative restenosis following balloon angioplasty (Sousa *et al.*, 2003). The need for maximally effective dosing regimens, new protein-and nucleic acid-based drugs (which cannot be taken in classical pill form), and elimination of systemic toxicities have stimulated development of new implantable polymers and innovative systems for controlled drug delivery and gene therapy (LaVan *et al.*, 2002). Controlled drug delivery is now capable of providing a wide range of drugs that can be targeted (e.g., to a tumor, to a diseased blood vessel, to the pulmonary alveoli) on a one-time or sustained basis with highly regulated dosage and can regulate cell and tissue responses through delivery of growth factors and plasmid DNA containing genes that encode growth factors (Bonadio *et al.*, 1999; Richardson *et al.*, 2001).

Biomedical Engineering Desk Reference; ISBN: 9780123746467

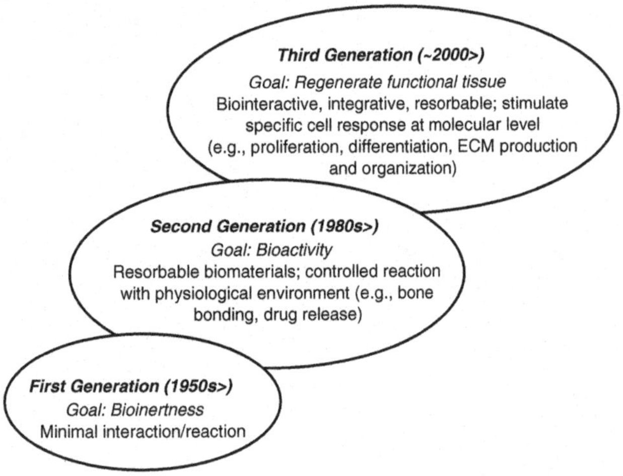

Fig. 7.1.1-1 Evolution of biomaterials science and technology. (From Rabkin, E., and Schoen, F. J. 2002. Cardiovascular tissue engineering. *Cardiovasc. Pathol.* **11:** 305.)

The second generation of biomaterials also included the development of resorbable biomaterials with variable rates of degradation matched to the requirements of a desired application. Thus, the discrete interface between the implant site and the host tissue could be eliminated in the long term, because the foreign material would ultimately be degraded by the host and replaced by tissues. A biodegradable suture composed of poly-(glycolic acid) (PGA) has been in clinical use since 1974. Many groups continue to search for biodegradable polymers with the combination of strength, flexibility, and a chemical composition conducive to tissue development (Hubbell, 1999; Griffith, 2000; Langer, 1999).

With engineered surfaces and bulk architectures tailored to specific applications, "third generation" biomaterials are intended to stimulate highly precise reactions with proteins and cells at the molecular level. Such materials provide the scientific foundation for molecular design of scaffolds that could be seeded with cells *in vitro* for subsequent implantation or specifically attract endogenous functional cells *in vivo*. A key concept is that a scaffold can contain specific chemical and structural information that controls tissue formation, in a manner analogous to cell–cell communication and patterning during embryological development. The transition from second- to third-generation biomaterials is exemplified by advances in controlled delivery of drugs or other biologically active molecules. Nanotechnology and the development of microelectromechanical systems (MEMSs) have opened new possibilities for fine control of cell behavior through manipulation of surface chemistry and the mechanical environment (Chen *et al.*, 1997; Bhatia *et al.*, 1999; Huang and Ingber, 2000; Chiu *et al.*, 2000, 2003).

Tissue engineering is a broad term describing a set of tools at the interface of the biomedical and engineering sciences that use living cells or attract endogenous cells to aid tissue formation or regeneration, and thereby produce therapeutic or diagnostic benefit. In the most frequent paradigm, cells are seeded on a scaffold composed of synthetic polymer or natural material (collagen or chemically treated tissue), a tissue is matured *in vitro*, and the construct is implanted in the appropriate anatomic location as a prosthesis (Langer and Vacanti, 1993; Fuchs *et al.*, 2001; Griffith and Naughton, 2002; Rabkin and Schoen, 2002; Vacanti and Langer, 1999). A typical scaffold is a bioresorbable polymer in a porous configuration in the desired geometry for the engineered tissue, often modified to be adhesive for cells, in some cases selective for a specific circulating cell population.

The first phase is the *in vitro* formation of a tissue *construct* by placing the chosen cells and scaffold in a metabolically and mechanically supportive environment with growth media (in a *bioreactor*), in which the cells proliferate and elaborate extracellular matrix. In the second phase, the construct is implanted in the appropriate anatomic location, where remodeling *in vivo* is intended to recapitulate the normal functional architecture of an organ or tissue. The key processes occurring during the *in vitro* and *in vivo* phases of tissue formation and maturation are (1) cell proliferation, sorting, and differentiation, (2) extracellular matrix production and organization, (3) degradation of the scaffold, and (4) remodeling and potentially growth of the tissue. The general paradigm of tissue engineering is illustrated in Fig. 7.1.1-2. Biological and engineering challenges in tissue engineering are focused on the three principal components that comprise the "cell–scaffold–bioreactor system"; control of the various parameters in device fabrication (Table 7.1.1-1) may have major impact on the ultimate result. Exciting new possibilities are opened by advances in stem cell technology (Blau *et al.*, 2001; Bianco and Robey, 2001) and the recent evidence that some multipotential cells possibly capable of tissue regeneration are released by the bone marrow and circulating systemically (Hirschi *et al.*, 2002) while others may be resident in organs such as heart and the central nervous system formally not considered capable of regeneration (Hirschi and Goddell, 2002; Grounds *et al.*, 2002; Nadal-Ginard *et al.*, 2003; Johansson, 2003; Orlic *et al.*, 2003).

Tissue-engineered configurations for skin replacement have achieved clinical use. Further examples of previous and ongoing clinical tissue engineering approaches include cartilage regeneration using autologous chondrocyte transplantation (Brittberg *et al.*, 1994) and a replacement thumb with bone composed of autologous periosteal cells and natural coral (hydroxyapatite) (Vacanti *et al.*, 2001). A key challenge in tissue engineering is to understand quantitatively how cells respond to molecular signals and integrate multiple inputs to generate a given response,

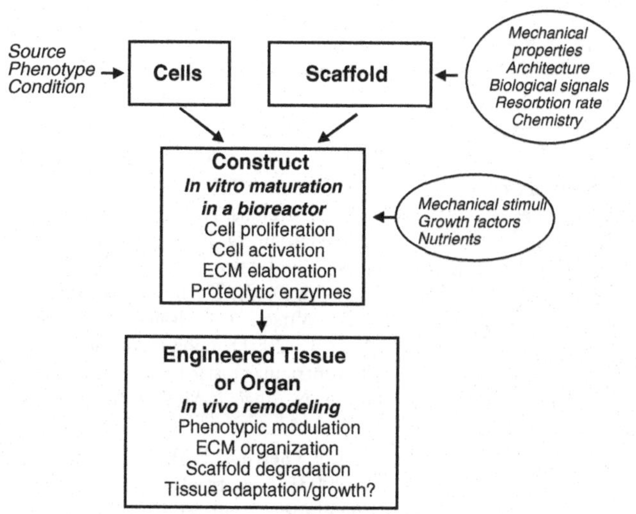

Fig. 7.1.1-2 Tissue engineering paradigm. In the first step of the typical tissue engineering approach, differentiated or undifferentiated *cells are* seeded on a bioresorbable *scaffold* and then the *construct* matured *in vitro* in a bioreactor. During maturation, the cells proliferate and elaborate extracellular matrix to form a "new" tissue. In the second step, the construct is implanted in the appropriate anatomical position, where remodeling *in vivo* is intended to recapitulate the normal tissue/organ structure and function. The key variables in the principal components—cells, scaffold, and bioreactor—are indicated. (From Rabkin, E., and Schoen, F. J., 2002, Cardiovascular tissue engineering. *Cardiovasc. Pathol.* **11:** 305.)

Table 7.1.1-1 Control of structure and function of an engineered tissue

Cells	**Biodegradable matrix/scaffold**
Source	*Architecture/porosity/chemistry*
Allogenic	Composition/charge
Xenogenic	Homogeneity/isotropy
Autologous	Stability/resorption rate
Type/phenotype	Bioactive molecules/ligands
Single versus multiple types	Soluble factors
Differentiated cells from primary or other tissue	*Mechanical properties*
	Strength
Adult bone-marrow stem cells	Compliance
Pluripotent embryonic stem cells	Ease of manufacture
Density	**Bioreactor conditions**
Viability	*Nutrients/oxygen*
Gene expression	Growth factors
Genetic manipulation	Perfusion and flow conditions
	Mechanical factors
	Pulsatile
	Hemodynamic shear stresses
	Tension/compression

(From Rabkin, E., and Schoen, F. J., 2002, Cardiovascular tissue engineering. *Cardiovasc. Pathol.* **11:** 305.)

and to control nonspecific interactions between cells and a biomaterial, so that cell responses specifically follow desired receptor–ligand interactions. Another approach uses biohybrid extracorpo-real artificial organs using functional cells that are isolated from the recipient's blood or tissues by an impermeable membrane (Colton, 1995; Humes *et al*, 1999; Strain and Neuberger, 2002). *Tissue engineering* also seeks to understand structure/function relationships in normal and pathological tissues (particularly those related to embryological development and healing) and to control cell and tissue responses to injury, physical stimuli, and biomaterials surfaces, through chemical, pharmacological, mechanical, and genetic manipulation. This is an immensely exciting field.

Bibliography

Bhatia, S. N., Balis, U. J., Yarmush, M. L., and Toner, M. (1999). Effect of cell–cell interactions in prevention of cellular phenotype: cocultivation of hepatocytes and nonparenchymal cells. *FASEB J.* **13**: 1883–1900.

Bianco, P., and Robey, P. G. (2001). Stem cells in tissue engineering. *Nature* **414**: 118–121.

Blau, H. M., Brazelton, T. R., and Weimann, J. M. (2001). The evolving concept of a stem cell: entity or function? *Cell* **105**: 829–841.

Bokros, J. C. (1977). Carbon biomedical devices. *Carbon*, **15**: 353–371.

Bonadio, J. E., Smiley, E., Patil, P., and Goldstein, S. (1999). Localized, direct plasmid gene delivery *in vivo*: prolonged therapy results in reproducible tissue regeneration. *Nat. Med.* **7**: 753–759.

Brittberg, M., Lindahl, A., Nilsson, A., Ohlsson, C., Isaksson, O., and Peterson, L. (1994). Treatment of deep cartilage defects in the knee with autologous chondrocyte transplantation. *N. Engl. J. Med.* **331**: 889–895.

Chen, C. S., Mrksich, M., Suang, S., Whitesides, G. M., and Ingber, D.E. (1997). Geometric control of cell life and death. *Science* **276**: 1425–1528.

Chiu, D. T., Jeon, N. L., Huang, S., Kane, R. S., Wargo, C. J., Choi, I. S., Ingber, D. E., and Whitesides, G. M. (2000). Patterned deposition of cells and proteins into surfaces by using three-dimensional microfluidic systems. *Proc. Natl. Acad. Sci. USA* **97**: 2408–2413.

Chiu, J-J., Chen, L-J., Lee, P-L., Lee, C-I., Lo, L-W., Usami, S., and Chien, S. (2003). Shear stress inhibits adhesion molecule expression in vascular endothelial cells induced by coculture with smooth muscle cells. *Blood* **101**: 2667–2674.

Colton, C. K. (1995). Implantable biohybrid artificial organs. *Cell Transplantation* **4**: 415–436.

Fleming, R. G., Murphy, C. J., Abrams, G. A., Goodman, S. L., and Nealey, P. F. (1999). Effects of synthetic micro- and nano-structured surfaces on cell behavior. *Biomaterials* **20**: 573–588.

Fuchs, J. R., Nasseri, B. A., and Vacanti, J. P. (2001). Tissue engineering: a 21st century solution to surgical reconstruction. *Ann. Thorac. Surg.* **72**: 577–590.

Griffith, L. G. (2000). Polymeric biomaterials. *Acta Mater.* **48**: 263–277.

Griffith, L. G., and Naughton, G. (2002). Tissue engineering—current challenges and expanding opportunities. *Science* **295**: 1009–1014.

Grounds, M. D., White, J. D., Rosenthal, N., and Bogoyevitch, M. A. (2002). The role of stem cells in skeletal and cardiac muscle repair. *J. Histochem. Cytochem.* **50**: 589–610.

Hench, L. L., and Pollak, J. M. (2002). Third-generation biomedical materials. *Science* **295**: 1014–1017.

Hirschi, K. K., and Goddell, M. A. (2002). Hematopoietic, vascular and cardiac fates of bone marrow-derived stem cells. *Gene Ther.* 9: 648–652.

Huang, S., and Ingber, D. E. (2000). Shape-dependent control of cell growth, differentiation, and apoptosis: switching between attractors in cell regulatory networks. *Exp. Cell Res.* **261**: 91–103.

Hubbell, J. A. (1999). *Bioactive biomaterials. Curr. Opin. Biotechnol.* **10**: 123–129.

Humes, H. D., Buffington, D. A., MacKay, S. M., Funk, A., and Wetzel, W. E. (1999). Replacement of renal function in uremic animals with a tissue-engineered kidney. *Nat. Biotechnol.* **17**: 451–455.

Johansson, C. B. (2003). Mechanism of stem cells in the central nervous system. *J. Cell Physiol.* **196**: 409–418.

Langer, R. (1999). Selected advances in drug delivery and tissue engineering. *J. Controlled Release* **62**: 7–11.

Langer, R., and Vacanti, J. P. (1993). Tissue engineering. *Science* **260**: 920–926.

Lauffenburger, D. A., and Griffith, L. G. (2001). Who's got pull around here? Cell organization in development and tissue engineering. *Proc. Natl. Acad. Sci. USA* **98**: 4282–4284.

La Van, D. A., Lynn, D. M., and Langer, R. (2002). Moving smaller in drug discovery and delivery. *Nat. Rev.* **1**: 77–84.

Meckstroth, K. R., and Darney, P. D. (2001). Implant contraception. *Semin. Reprod. Med.* **19**: 339–54.

Nadal-Ginard, B., Kajstura, J., Leri, A., and Anversa, P. (2003). Myocyte death, growth, and regeneration in cardiac hypertrophy and failure. *Circ. Res.* **92**: 139–150.

Orlic, D., Kajstura, J., Chimenti, S., Bodine, D. M., Leri, A., and Anversa, P. (2003). Bone marrow stem cells regenerate infracted myocardium. *Pediatr. Transplant.* 7(Suppl 3): 86–88.

Rabkin, E., and Schoen, F. J. (2002). Cardiovascular tissue engineering. *Cardiovasc. Pathol.* **11**: 305–317.

Richardson, T. P., *et al.* (2001). Polymeric delivery of proteins and plasmid DNA for tissue engineering and gene therapy. *Crit. Rev. Eukar. Gene Exp.* **11**: 47–58.

Rose, E. A., Gelijns, A. C., Muscowitz, A. J., Heitjan, D. F., Stevenson, L. W., Dembitsky, W., Long, J. W., Ascheim, D. D., Tierney, A. R., Levitan, R. G., Watson, J. T., Ronan, N. S., and Meier, P. (2001). Long-term mechanical left ventricular assist for end-stage heart failure. *N. Engl. J. Med.* **345**: 1435–1443.

Sousa, J. E., Serruys, P. W., and Costa, M. A. (2003). New frontiers in cardiology: drug-eluting stents—I and II. *Circulation* **107**: 2274–2279, 2383–2389.

Strain, A. J., and Neuberger, J. M. (2002). A bioartificial liver—state of the art. *Science* **295**: 1005–1009.

Vacanti, J. P., and Langer, R. (1999). Tissue engineering: the design and fabrication of living replacement devices for surgical reconstruction and transplantation. *Lancet* **354**: SI32–SI34.

Vacanti, C. A., Bonassar, L. J., Vacanti, M. P., and Shufflebarger, J. (2001). Replacement of an avulsed phalanx with tissue-engineered bone. *N. Engl. J. Med.* **344**: 1511–1514.

7.1.2 Overview of tissue engineering

Simon P. Hoerstrup and Joseph P. Vacanti

Introduction

The loss or failure of an organ or tissue is a frequent, devastating, and costly problem in health care, occurring in millions of patients every year. In the United States, approximately 9 million surgical procedures are performed annually to treat these disorders, and 40 to 90 million hospital days are required. The total national health-care costs for these patients exceed $500 billion per year (Langer and Vacanti, 1993,1999). Organ or tissue loss is currently treated by transplanting organs from one individual to another or performing surgical reconstruction by transferring tissue from one location in the human body to the diseased site. Furthermore, artificial devices made of plastic, metal, or fabrics are utilized. Mechanical devices such as dialysis machines or total joint replacement prostheses are used, and metabolic products of the lost tissue, such as insulin, are supplemented. Although these therapies have saved and improved millions of lives, they remain imperfect solutions.

Tissue engineering represents a new, emerging interdisciplinary field applying a set of tools at the interface of the biomedical and engineering sciences that use living cells or attract endogenous cells to aid tissue formation or regeneration (Rabkin *et al.*, 2002) to restore, maintain, or improve tissue function.

Engineered tissues using the patient's own (autologous) cells or immunologically inactive allogeneic or xenogeneic cells offer the potential to overcome the current problems of replacing lost tissue function and to provide new therapeutic options for diseases such as metabolic deficiencies.

Current therapeutic approaches for lost tissue or organ function

Transplantation

Organs or parts of organs are transplanted from a cadaveric or living-related donor into the patient suffering from lost organ function. Many innovative advances have been made in transplantation surgery during recent years and organ transplantation has been established as a curative treatment for end-stage diseases of liver, kidney, heart, lung, and pancreas (Starzl, 2001; Starzl *et al.*, 1989; Stratta *et al.*, 1994). Unfortunately, transplantation is substantially limited by a critical donor shortage. For example, fewer than 5200 liver donors are available annually for the approximately 18,500 who were on the waiting list in the year 2001 (http://www.ustransplant.org/annual.html). Besides the donor shortage, the other major problem of organ transplantation remains the necessity of lifelong immunosuppression therapy with a number of substantial and serious side effects (Keeffe, 2001).

Surgical reconstruction

Organs or tissues are moved from their original location to replace lost organ function in a different location (e.g., saphenous vein as coronary bypass graft, colon to replace esophagus or bladder, myocutaneous flaps or freegrafts for plastic surgery). Nevertheless there are a number of problems associated with this method of therapy, since the replacement tissues, consisting of a different tissue type, cannot replace all of the functions of the original tissue. Moreover, long-term complications occur, such as the development of a malignant tumor in colon tissue replacing bladder function (Kato *et al.*, 1993; Kusama *et al.*, 1989) or calcification and resulting stenosis of vascular grafts (Kurbaan *et al.*, 1998). Finally, there is also the risk of complications and surgical morbidity at the donor site.

Artificial prosthesis

The use of artificial, nonbiological materials in mechanical heart valves, blood vessels, joint replacement prostheses, eye lenses, or extracorporeal devices such as dialysis or plasmapheresis machines has improved and prolonged patients' lives dramatically. However, these methods are complicated by infection, limited durability of the material, lack of mechanism of biological repair and remodeling, chronic irritation, occlusion of vascular grafts, and the necessity of anticoagulation therapy and its side effects (Kudo *et al.*, 1999; Mow *et al.*, 1992). Regarding the pediatric patient population, not all artificial implants can provide a significant growth or remodeling potential, which often results in repeated operations associated with substantial morbidity and mortality (Mayer, 1995).

Supplementation of metabolic products of diseased tissues or organs

In the case of the loss of endocrine tissue function, hormonal products such as insulin or thyroid, adrenal, or gonadal hormones can be successfully supplemented by oral or intravenous medication. In most cases, chronic supplementation is necessary. Unfortunately, supplementation therapy cannot replace natural feedback mechanisms, frequently resulting in dysregulation of hormone levels. As a consequence, clinical conditions such as hypo- or hyperglycemic crises or the long-term complications of chronic hormonal imbalances nephropathy or microvascular disease in patients with insulin-dependent diabetes mellitus continue to occur (Orchard *et al.*, 2002, 2003).

Tissue engineering as an approach to replace lost tissue or organ function

In the most frequent paradigm of tissue engineering, isolated living cells are used to develop biological substitutes for the restoration or replacement of tissue or organ function. Generally, cells are seeded on bioabsorbable scaffolds, a tissue is matured *in vitro*, and the construct is implanted in the appropriate anatomic location as a prosthesis. Cells used in tissue engineering may come from a variety of sources including application-specific differentiated cells from the patients themselves (autologous), human donors (allogeneic) or animal sources (xenogeneic), or undifferentiated cells comprising progenitor or stem cells. The use of isolated cells or cell aggregates enables manipulation prior to implantation, e.g., transfection of genetic material or modulation of the cell surface in order to prevent immunorecognition. Three general strategies have been adopted for the creation of new tissues including cell injection, closed, or flow-through systems, and tissue engineering using biodegradable scaffolds.

Cell injection method

The cell injection method avoids the complications of surgery by allowing the replacement of only those cells that supply the needed function. Isolated, dissociated cells are injected into the bloodstream or a specific organ of the recipient. The transplanted cells will use the vascular supply and the stroma provided by the host tissue as a matrix for attachment and reorganization (Matas *et al.*, 1976). This method offers opportunities for a number of applications in replacing metabolic functions as occurs in liver disease, for example (Grossman *et al.*, 1994). However, cell mass sufficient to replace lost metabolic functions is difficult to achieve and its application for replacing functions of structural tissues such as heart valves or cartilage is rather limited. Several cell types may be used for injection, such as bone marrow cells, blood-derived progenitor cells, and muscle satellite cells (see also the later subsection on tissue engineering of muscle).

Whole bone marrow contains multipotent mesenchymal stem cells (marrow stromal cells) that are derived from somatic mesoderm and are involved in the self-maintenance and repair of various mesenchymal tissues. These cells can be induced *in vitro* and *in vivo* to differentiate into cells of mesenchymal lineage, including fat, cartilage and bone, and cardiac and skeletal muscle. The first successful allogenic bone marrow transplant in a human was carried out in 1968. More than 40,000 transplants (from bone marrow, peripheral blood, or umbilical cord blood) were carried out worldwide in 2000 (http://www.ibmtr.org/newsletter/pdf/2002Feb.pdf). The most common indications for allotransplants are acute and chronic leukemias, myelodysplasia (MDS), and nonmalignant diseases (aplastic anemia, immune deficiencies, inherited metabolic disorders). Autotransplants are generally used for non-Hodgkin's lymphoma (NHL), multiple myeloma (MM), Hodgkin's lymphoma, and solid tumors.

Experimental studies suggested that bone marrow-derived or blood-derived progenitor cells may also contribute, e.g., to the regeneration of infarcted myocardium (Orlic *et al.*, 2001a).

Closed-system method

Closed systems can be either implanted or used as extracorporeal devices. In this approach, cells are isolated from the body by a semipermeable membrane that allows diffusion of nutrients and the secreted cell products but prevents large entities such as antibodies, complement factors, or other immunocompetent cells from destroying the isolated cells. Protection is also provided to the recipient when potentially pathological (e.g., tumorigenic) cells are transplanted. Implantable systems (encapsulation systems) come in a variety of configurations, basically consisting of a matrix that cushions the cells and supports their survival and function and a surrounding porous membrane (Fig. 7.1.2-1). In vascular-type designs the transplanted secretory cells are housed in a chamber around a vascular conduit separated from the bloodstream by a semipermeable membrane. As

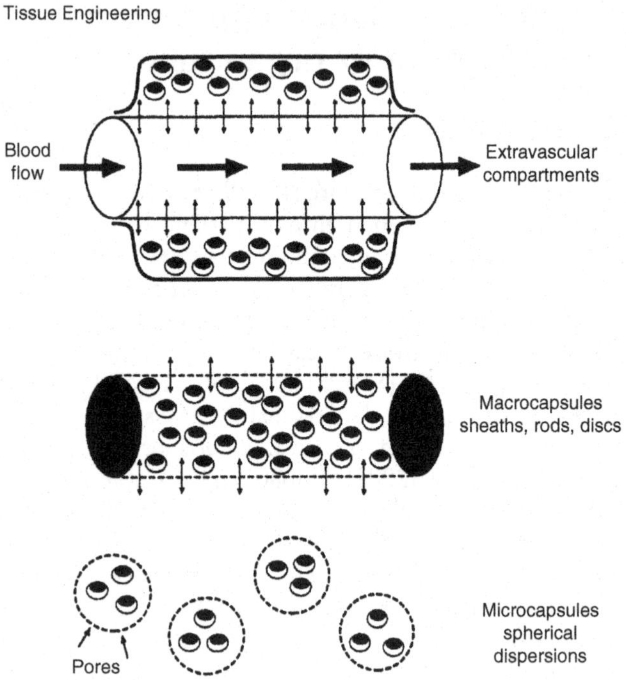

Fig. 7.1.2-1 Configurations of implantable closed-system devices for cell transplantation. (Reprinted with permission from Langer, R., and Vacanti, J. P., 1993. *Science* **260:** 920–926.)

blood flows through, it can absorb substances secreted by the therapeutic cells while the blood provides oxygen and nutrients to the cells. In macroencapsulation systems, a semipermeable membrane is used to encapsulate a relatively large (up to 50–100 million per unit) number of transplanted cells. Microcapsules are basically microdroplets of hydrogel with a diameter of less than 0.5 mm housing smaller numbers of cells. Macrocapsules are far more durable than microcapsule droplets and can be designed to be refillable in the body. Moreover, they can be retrieved, providing opportunities for more control than microcapsules. Their main limitation is the number of cells they can accommodate. In animal experiments, implantable closed-system configurations have been successfully used for the treatment of Parkinson's disease as well as diabetes mellitus (Aebischer *et al.*, 1991, 1988; Kordower *et al.*, 1995; Date *et al.*, 2000). If islets of Langerhans are used, they will match the insulin released to the concentration of glucose in the blood. This has been successfully demonstrated in small and large animals with maintenance of normoglycemia even in long-term experiments (Kin *et al.*, 2002; Lacy *et al.*, 1991; Lanza *et al.*, 1999; Sullivan *et al.*, 1991). Major drawbacks of these systems are fibrous tissue overgrowth and resultant impaired diffusion of metabolic products, nutrients, and wastes, as well as the induction of a foreign-body reaction with macrophage activation resulting in destruction of the transplanted cells within the capsule (Wiegand *et al.*, 1993).

In extracorporeal systems (vascular or flow-through designs) cells are usually separated from the bloodstream. Great progress is being made in the development of extracorporeal liver assist devices for support of patients with acute liver failure. Currently four devices that rely on allogenic or xenogenic hepatocytes cultured in hollow-fiber membrane technology are in various stages of clinical evaluation (Patzer, 2001; Rozga *et al.*, 1994).

Tissue engineering using scaffold biomaterials

Open systems of cell transplantation with cells being in direct contact to the host organism aim to provide a permanent solution to the replacement of living tissue. The rationale behind the use of open systems is based on empirical observations: dissociated cells tend to reform their original structures when given the appropriate environmental conditions in cell culture. For example, capillary endothelial cells form tubular structures and mammary epithelial cells form acini that secrete milk on the proper substrata *in vitro* (Folkman and Haudenschild, 1980). Although isolated cells have the capacity to reform their respective tissue structure, they do so only to a limited degree since they have no intrinsic tissue organization and are hindered by the lack of a template to guide restructuring. Moreover, tissue cannot be transplanted in large volumes because diffusion limitations restrict interaction with the host environment for nutrients, gas exchange, and elimination of waste products. Therefore, the implanted cells will survive poorly more than a few hundred microns from the nearest capillary or other source of nourishment (Vacanti *et al.*, 1988). With these observations in mind, an approach has been developed to regenerate tissue by attaching isolated cells to biomaterials that serve as a guiding structures for initial tissue development. Ideally, these scaffold materials are biocompatible, biodegradable into nontoxic products, and manufacturable (Rabkin *et al.*, 2002). Natural materials used in this context are usually composed of extracellular matrix components (e.g., collagen, fibrin) or complete decellularized matrices (e.g., heart valves, small intestinal submucosa). Synthetic polymer materials are advantageous in that their chemistry and material properties (biodegradation profile, microstructure) can be well controlled. The majority of scaffold-based tissue engineering concepts utilize synthetic polymers [e.g., PGA, poly(lactid acid) (PLA), or poly(hydroxy alkanoate) (PHA)]. In general, these concepts involve harvesting of the appropriate cell types and expanding them *in vitro*, followed by seeding and culturing them on the polymer matrices. The polymer scaffolds are designed to guide cell organization and growth allowing diffusion of nutrient to the transplanted cells. Ideally, the cell–polymer matrix is prevascularized or would become vascularized as the cell mass expands after implantation. Vascularization could be a natural response to the implant or be artificially induced by sustained release of angiogenic factors from the polymer scaffold (Langer and Vacanti, 1999). Since the polymer scaffold is designed to be biodegradable, concerns regarding long-term biocompatability are obviated.

Cells used in tissue engineering may come from a variety of sources including cell lines from the patients themselves (autologous), human donors (allogeneic), or animal sources (xenogeneic). However, allogeneic and xenogeneic tissue may be subjected to immunorejection. Cell-surface modulation offers a possible solution to this problem by deleting immunogenic sites and therefore preventing immunorecognition. A bank of cryopreserved cells would then be possible and genetic engineering techniques could be used to insert genes (Raper and Wilson, 1993) to replace proteins, such as the LDL receptor (Chowdhury *et al.*, 1991) or factor IX (Armentano *et al.*, 1990).

Applications of tissue engineering

Investigators have attempted to engineer virtually every mammalian tissue. In the following summary, we discuss replacement of ectodermal, endodermal, and mesodermal derived tissues.

Ectodermal derived tissue

Nervous system

Diseases of the central nervous system, such as a loss of dopamine production in Parkinsons's disease, represent an important target for tissue engineering. Transplantation of fetal dopamine-producing cells by stereotactically guided injection into the appropriate brain region has produced significant reversal of debilitating symptoms in humans (Lindvall *et al.*, 1990). Further benefit regarding survival, growth, and function has been demonstrated when implantation of dopamine-producing cells was combined with polymer-encapsulated cells continuously producing human glial cell line-derived growth factor (GDNF) (Sautter *et al.*, 1998). In the animal model PC12 cells, an immortalized cell line derived from rat pheochromocytoma, have been encapsulated in polymer membranes and implanted in the guinea pig striatum (Aebischer *et al.*, 1991) or primates (Date *et al.*, 2000; Kordower *et al.*, 1995), resulting in a dopamine release from the capsule detectable for many months. Similarly, encapsulated bovine adrenal chromafin cells have been implanted into the subarachnoid space in rats, where through their continuous production of enkephalins and catecholamines they appeared to relieve chronic intractable pain (Sagen, 1992). Finally, investigations have been undertaken to achieve brain tissue by immobilization of neuronal and glia cells in *N*-methacrylamide polymer hydrogels. These cells have shown cell viability and maintained differentiation *in vitro* (Woerly *et al.*, 1996).

Nerve regeneration is another field of current investigations. When nerve injury results in gaps that are too wide for healing, autologous nerve grafts are used as a bridge. Several laboratories have shown in animal models that artificial guiding structures composed of natural polymers (laminin, collagen, chondroitin sulfate) or synthetic polymers can enhance nerve regeneration (Valentini *et al.*, 1992). Moreover, this process can be aided by placing Schwann cells seeded in polymer membranes (Guenard *et al.*, 1992). Polymers can also be designed so that they slowly release growth factors, possibly allowing regrowth of the damaged nerve over a greater distance (Aebischer *et al.*, 1989; Haller and Saltzman, 1998). In the case of neurodegenerative diseases such as amyotrophic lateral sclerosis (ALS), progression of the motor neuron disease could be successfully delayed in the animal model by polymer encapsulation of genetically modified cells to secrete neutrotrophic factors. This suggests that encapsulated cell delivery of neutrotrophic factors may provide a general method or effective administration of therapeutic proteins for the treatment of neurodegenerative diseases (Aebischer *et al.*, 1996a, b; Tan *et al.*, 1996).

Recently, a phase I/II clinical trial has been performed in 12 ALS patients to evaluate the safety and tolerability of intrathecal implants of encapsulated genetically engineered baby hamster kidney (BHK) cells releasing human ciliary neurotrophic factor (CNTF) (Aebischer *et al.*, 1996b, Zurn *et al.*, 2000). No adverse side effects have been observed in these patients in contrast to the systemic delivery of large amount of CNTF. However, antibodies against bovine fetuin have been detected because the capsules have been kept in a medium containing fetuin before transplantation.

Micorencapsulated cells may also be used for the treatment of malignant brain tumors (Thorsen *et al.*, 2000). Genetically modified cells secrete tumor controlling/suppressing substances such as the anti-angiogenic protein endostatin (Read *et al.*, 2001).

Cornea

The cornea is a transparent window covering the front surface of the eye that protects the intraocular contents and is the main optical element that focuses light onto the retina. Worldwide, millions of people suffer from bilateral corneal blindness. Transplant donors are limited and there is a risk of infectious agent transmission. Moreover, in the case that the limbal epitethelial stem cells of the recipient are damaged (alkali burns, autoimmune conditions, or recurrent graft failures), the donor corneal epithelium desquamate and is replaced by conjuctivization and fibrovascular scarring in the repicient.

The cornea is avascular and immunologically privileged, making this tissue an excellent candidate for tissue engineering. Ideally, an artificial cornea would consist of materials that support adhesion and proliferation of corneal epithelial cells so that an intact continuous epithelial layer forms. In addition, these materials should have appropriate nutrient and fluid permeability, light transparency, low light scattering, and no toxicity. Artificial cornea has been developed that consisted of a peripheral rim of biocolonizable microporous fluorocarbon polymer (polytetrafluoroethylene, PTFE) fused to an optical core made of polydimethylsiloxane coated with polyvinylpyrrolidone (Legeais and Renard, 1998). In contrast to this "hybrid" cornea, another group used poly-(2-hydroxyethyl methacrylate) (PHEMA) for both the porous skirt (opaque sponge, 10–30 μm) and the optical core (transparent gel) (Chirila, 2001; Crawford *et al.*, 2002). PHEMA is a biomaterial with a long record of ocular tolerance in applications such as contact lenses, intraocular lenses, and intracorneal inlays. The use of this material as porous sponge allowed cellular invasion, production of collagen, and vascularization, without the formation of a foreign-body capsule. Both devices have been tested preclinically.

Furthermore, tissue-engineered implantable contact lenses could obviate the need for surgery in patients who seek convenient, reversible correction of refractive error. An onlay involves debridement of the central corneal epithelium and placement of a synthetic lenticule on the exposed stromal surface, leaving Bowman's zone intact. The anterior surface of the lenticule is then covered by the recipient eye's corneal epithelium, incorporating the lenticule to achieve the desired refractive correction by altering the curvature of the anterior corneal surface. Porous collagen-coated perfluoropolyether (PFPE) was successfully tested in cats (Evans *et al.*, 2002) and Lidifilcon A, a copolymer of methyl methacrylate and *N*-vinyl-2-pyrrolidone (Allergan Medical Optics, Irvine, CA) was implanted in monkeys (McCarey *et al.*, 1989).

The multistep procedure of corneal reconstruction has been demonstrated using corneal cells from rabbit (Zieske *et al.*, 1994) and from fetal pig (Schneider *et al.*, 1999), human cells from donor corneas (Germain *et al.*, 1999), or immortalized cell lines from the main layers of the cornea (Griffith *et al.*, 1999). In these studies collagen matrices or collagen–chondritin sulfate substrates cross-linked with glutaraldehyde have been tested. More recently, carbodiimide cross-linking and composites using urethane/urea techniques have been evaluated for biocompatibility and epithelial ingrowth (Griffith *et al.*, 2002).

Skin

Several new types of tissue transplants are being studied for treatment of burns, skin ulcers, deep wounds, and other injuries. One approach to skin grafts involves the *in vitro* culture of epidermal cells (keratinocytes). Small skin biopsies are harvested from burn patients and expanded up to 10,000-fold. This expansion is achieved, e.g., by cultivating keratinocytes on a feeder layer of irradiated NIH 3T3 fibroblasts, which, in conjunction with certain added media components, stimulates rapid cell growth. This approach allows coverage of extremely large wounds. A disadvantage is the 3- to 4-week period required for cell expansion, which may be too long for a severely burned patient. Cryopreserved allografts may help to circumvent this problem (Nave, 1992). Another promising approach uses human neonatal dermal fibroblasts grown on PGA mesh. In deep injuries involving all layers of skin the grafts are placed onto the wound bed and a skin graft is placed on top followed by vascularization of the graft. This results in formation of an organized tissue resembling dermis. Clinical trials have shown good graft acceptance without evidence of immune rejection (Hansbrough *et al.*, 1992). Fibroblasts have also been placed on hydrated collagen gel. Upon implantation, the cells migrate through the gel by enzymatic digestion of collagen, which results in reorganization of collagen fibrils (Bell *et al.*, 1979). ApliGraf, formerly known as Graftskin, is a commercially available two-layered tissue-engineered skin product composed of type I bovine collagen that contains living human dermal fibroblasts and an overlying cornified epidermal layer of living human keratinocytes. Both cell types are derived from neonatal foreskin and grow in a special mold that limits lateral contraction (Bell *et al.*, 1991a, b). ApliGraf has been investigated in a multicenter study after excisional surgery for skin cancer with good results (Eaglstein *et al.*,1999).

The artificial skin developed by Burke and Yannas (Burke *et al.*, 1981), now called Integra, consists of collagen–chondritin 6-sulfate fibers obtained from bovine hide (collagen) and shark cartilage (chondritin 6-sulfate). It has been engineered into an open matrix of uniform porosity and thickness and covered with a uniformly thick (0.1-mm) silicone sheet. This artificial skin has been studied extensively in humans (Heimbach *et al.*, 1988; Sheridan *et al.*, 1994) and was approved for use in burn patients in 1997.

Besides clinical use of artificial skin, several companies have explored the possibilities of dermal substitutes for diagnostic purposes. There is particular interest in minimizing the use of animals for topological irritation, corrosivity, and other testing (Fentem *et al.*, 2001; Portes *et al.*, 2002). Gene therapy for genodermatoses (Spirito *et al.*, 2001), junctional epidermolysis bullosa (Robbins *et al.*, 2001), and ichthyosis (Jensen *et al.*, 1993) remains a topic of great interest using either transgenic fibroblasts or keratinocytes.

Endoderm

Liver

Liver transplantation is a routine treatment for end-stage liver disease, but donor organ shortage remains a serious problem. Many patients die while waiting for a transplant and those with chronic disease often deteriorate resulting in low survival rates after transplantation. Therefore a "bridging" device that would support liver function until a donor liver became available or the patient's liver recovered is of great interest. Most liver support systems remove toxins normally metabolized by the liver through dialysis, charcoal hemoperfusion, immobilized enzymes, or exchange transfusion. However, none of these systems can offer the full functional spectrum performed by a healthy liver. Hepatocyte systems aiming at replacement of liver function by transplantation of isolated cells are being studied for both extracorporeal and implantable applications. Extracorporeal systems can be used when the patient's own organ is recovering or as a bridge to transplantation. These systems provide a good control of the medium surrounding the cell system and a minimized risk of immune rejection. Their design is primarily based on hollow-fiber, spouted-bed, or flat-bed devices (Bader *et al.*, 1995). Implantable hepatocyte systems, on the other hand, offer the possibility of permanent liver replacement (Yarmush *et al.*, 1992a). Successful hepatocyte transplantation depends on a number of critical steps. After cell harvest the hepatocytes must be cultured and expanded *in vitro* prior to transplantation. Hepatocyte morphology can be maintained by cultivating the cells on three-dimensional structures, such as sandwiching them between two hydrated collagen layers. Under these conditions the hepatocytes have been shown to secrete functional markers at physiological levels (Dunn *et al.*, 1991). Moreover the hepatocytes must be attached to the polymer substrata so that they maintain their differentiated function and viability. A sufficient mass of hepatocytes must become engrafted and remain functional to achieve metabolic replacement and vascularization, which is critical for graft survival (Yarmush *et al.*, 1992b). Finally, hepatocyte transplantation per se provides neither all cell types nor the delicate and complex structural features of the liver. Products normally excreted through bile may accumulate because of the difficulty in reconstructing the biliary tree solely from hepatocytes. However, hepatocytes placed

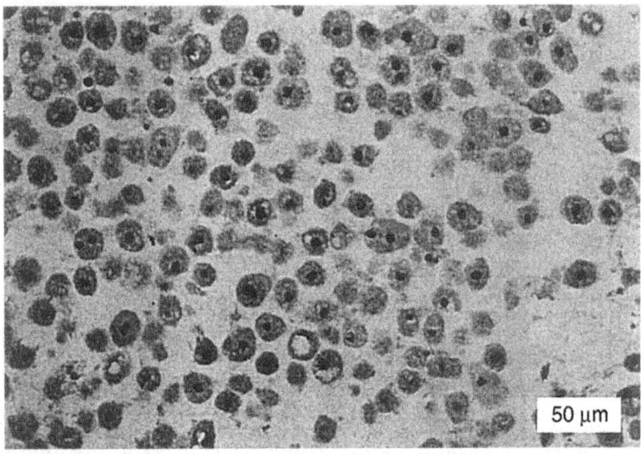

Fig. 7.1.2-2 Histologic photomicrograph demonstrating viable hepatic cells after 2 days under flow conditions (hematoxylin and eosin; original magnification ×300). (Reprinted with permission from Kim, S. S., *et al.*, 1998. *Ann. Surg.* **228:** 8–13.)

on appropriate polymers can form tissues resembling those in the natural organ and have shown evidence of bile ducts and bilirubin removal (Uyama *et al.*, 1993). More recently, model systems in which the vascular architecture is mimicked in the device have been tested using three-dimensional printing, hepatocytes, and endothelial cells (Fig. 7.1.2-2; Kim *et al.*, 1998).

Four bioartificial liver devices have entered sustained clinical trials. The device rely all on hollow-fiber membranes to isolate hepatocytes from direct contact with patient fluids. They differ in source and treatment of hepatocytes prior to patient use and in the choice of perfusate: plasma or whole blood. Three devices are perfused with the patient's plasma. The HepatAssist is filled with freshly thawed cryopreserved primary porcine hepatocytes along with collagen-coated dextran beads for cell attachment (Chen *et al.*, 1997; Rozga *et al.*, 1993; Watanabe *et al.*, 1997). The ELAD system uses a HepG2 human hepatocyte cell line that has been grown to confluence in the extracellular space (Ellis *et al.*, 1996; Sussman *et al.*, 1994). The Gerlach BELS run either with human hepatocytes (if available) or with porcine primary hepatocytes embedded in a collagen matrix in the extraluminal space (Gerlach, 1997; Gerlach *et al.*, 1997). In contrast, the bioartificial liver support system (BLSS) is perfused with whole blood. This has the advantage that a greater rate of blood concentration reduction and lower endpoint blood concentration at equivalent perfusion times is achieved compared to systems using plasma perfusion. The detoxification is performed with primary porcine hepatocytes (Mazariegos *et al.*, 2001; Patzer *et al.*, 2002, 1999).

Pancreas

Each year more than 700,000 new cases of diabetes are diagnosed in the United States and approximately 150,000 patients die from the disease and its complications. Diabetes is characterized by pancreatic islet destruction leading to more or less complete loss of glucose control. Tissue engineering approaches to treatment have focused on transplanting functional pancreatic islets, usually encapsulated to avoid immune reaction. Three general approaches have been tested in animal experiments. In the first, a tubular membrane was coiled in a housing that contained islets. The membrane was connected to a polymer graft that in turn connected the device to blood vessels. This membrane had a 50-kDa molecular mass cutoff, thereby allowing free diffusion of glucose and insulin but blocking passage of antibodies and lymphocytes. In pancreatectomized dogs treated with this device, normoglycemia was maintained for more than 150 days (Sullivan *et al.*, 1991). In a second approach, hollow fibers containing rat islets were immobilized in polysaccharide alginate. When the device was placed intraperitoneally in diabetic mice, blood glucose levels were lowered for more than 60 days and good tissue biocompatibility was observed (Lacy *et al.*, 1991). Finally, islets have been enclosed in microcapsules composed of alginate or polyacrylate. This method offers a number of distinct advantages over the use of other biohybrid devices, including greater surface-to-volume ratio and ease of implantation (simple injection) (Kin *et al.*, 2002; Lanza *et al.*, 1999, 1995). All of these transplantation strategies require a large, reliable source of donor islets. Porcine islets are used in many studies and genetically engineered cells that overproduce insulin are also under investigation (Efrat, 1999).

Tubular structures

The current concept of using tubular structures of other organs for reconstruction of bladder, ureter and urethra, trachea, esophagus, intestine, and kidney represents a major therapeutic improvement. A diseased esophagus, for example, can be treated clinically with autografts from the colon, stomach, skin, or jejunal segments. However, such procedures carry a substantial risk of graft necrosis, inadequate blood supply, infection, lack of peristaltic activity, and other complications. Copolymer tubes consisting of lactic and glycolic acid have been sutured into dogs after removal of esophageal segments, over time resulting in coverage of the polymer with connective tissue and epithelium (Grower *et al.*, 1989). Alternatively, elastin-based acellular aortic patches have been successfully used in experimental esophagus injury in the pig. While mucosal and submucosal coverage took place within 3 weeks, the majority of the elastin-based biomaterial degraded. However, the muscular layer did not regenerate (Kajitani *et al.*, 2001). In a similar approach fetal intestinal cells have been placed onto copolymer tubes and implanted in rats. Histological examination after several weeks revealed differentiated intestinal epithelium lining of the tubes and this epithelium appeared to secrete mucus (Vacanti *et al.*, 1988). Furthermore, intestinal epithelial organoid units transplanted on porous biodegradable polymer tubes have been shown to vascularize and to regenerate into complex tissue resembling small intestine (Kim *et al.*, 1999), and successful anastomosis between tissue-engineered intestine and native small bowel has been performed (Fig. 7.1.2-3; Kaihara *et al.*, 1999). Finally, Perez *et al.* demonstrated that tissue-engineered small intestine is capable of developing a mature immunocyte population and that mucosal exposure to luminal stimuli is critical to this development (Perez *et al.*, 2002). Despite these promising findings, the regeneration of the muscle layer seems to be a major problem. Autologous mesenchymal stem cells seeded onto a collagen sponge graft induced only a transient distribution of cells positive for α-smooth muscle actin (Hori *et al.*, 2002).

Tubular structures have also been used in kidney replacement. As a first step toward creating a bioartificial kidney, renal tubular cells have been grown on acrylonitrile–vinyl chloride copolymers or microporous cellulose nitrate membranes. *In vitro*, these cells transported glucose and tetraethylammonium cation in the presence of a hemofiltrate of uremic patients (Uludag *et al.*, 1990). In a further attempt to create bioartificial renal tubule, renal epithelial cells have been grown on hollow fibers and formed an intact monolayer exhibiting functional active transport capabilities (MacKay *et al.*, 1998). Finally, an extracorporeal device was developed using a standard hemofiltration cartridge containing renal tubule cells (Humes *et al.*, 1999; Nikolovski *et al.*, 1999). The pore size of the hollow fibers allows the membranes

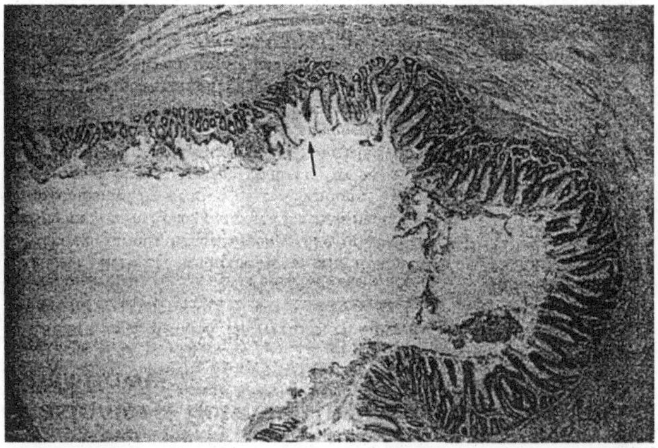

Fig. 7.1.2-3 Histology of a tissue-engineered intestine 10 weeks after implantation characterized by crypt villus structures. Arrow indicates anastomosis site; left site of the arrow is tissue-engineered intestine and right is native small bowel. (Reprinted with permission from Kaihara, S., *et al.*, 1999. *Transpl. Proc.* **31:** 661–662.)

to act as scaffolds for the cells and as an immunoprotective barrier. *In vitro* and *in vivo* studies have shown that the cells keep differentiated active transport, differentiated metabolic transport, and important endocrine processes (Humes *et al.*, 2002, 2003).

For replacement of urether, urothelial cells were seeded onto degradable PGA tubes and implanted in rats and rabbits resulting in two or three layers of urothelial cell lining (Atala *et al.*, 1992). More recently, an acellular collagen matrix from bladder submucosa seeded with cells from urethral tissue was also successfully used for tubularized replacement in the rabbit. In contrast, unseeded matrices lead to poor tissue development (de Filippo *et al.*, 2002). A neo-bladder has been created from urothelial and smooth muscle cells *in vitro* and after implantation in the animal, functional evaluation for up to 11 months has demonstrated a normal capacity to retain urine, normal elastic properties, and normal histologic architecture (Oberpenning *et al.*, 1999).

Mesoderm

Cartilage

More than 1 million surgical procedures in the United States each year involve cartilage replacement. Current therapies include cartilage transplantation and implantation of artificial polymer or metal prostheses. Unfortunately, donor tissue is limited and artificial implants may result in infection and adhesive breakdown at the host–prosthesis interface. Finally, a prosthesis cannot adapt in response to environmental stresses as does cartilage (Mow *et al.*, 1992). The need for improved treatments has motivated research aiming at creating new cartilage that is based on collagen–glycosaminglycan templates (Stone *et al.*, 1990), isolated chondrocytes (Grande *et al.*, 1989), and chondrocytes attached to natural or synthetic polymers (Cancedda *et al.*, 2003; Vacanti *et al.*, 1991; Wakitani *et al.*, 1989). It is critical that the cartilage transplant have an appropriate thickness and attachment to be mechanically functional. Chondrocytes grown in agarose gel culture have been shown to produce tissues with stiffness and compressibility comparable to those of articular cartilage (Freed *et al.*, 1993). The use of bioreactors for cultivating chondrocytes on polymer scaffolds *in vitro* enables nutrients to penetrate the center of this nonvascularized tissue, leading to relatively strong and thick (up to 0.5 cm) implants (Buschmann *et al.*, 1992). Moreover, it has been shown that the hydrodynamic conditions in tissue-culture bioreactors can modulate the composition, morphology, mechanical properties, and electromechanical function of engineered cartilage (Vunjak-Novakovic *et al.*, 1999). In other studies, chondrocytes were seeded onto PGA meshes and conditioned for several weeks on an orbital shaker. The functional cartilage was then combined with an osteoconductive support made of ceramic/collagen sponge. The composite was press-fitted in a large experimental osteochondral injury in a rabbit knee joint, where it showed good structural and functional properties (Schaefer *et al.*, 2002). With regard to the needs of reconstructive surgery, tissue-engineered autologous cartilage has been generated *in vitro* from tiny biopsies (Naumann *et al.*, 1998). Finally, some research has been undertaken to evaluate tissue engineering of cartilage even in space in order to elucidate the influence of micro/agravity on tissue formation (Freed *et al.*, 1997).

Bone

Current therapies of bone replacement include the use of autogenous or allogenic bone. Moreover, metals and ceramics are used in several forms: biotolerant (e.g., titanium), bioresorbable (e.g., tricalcium phosphate), porous (e.g., hydroxyapatite-coated metals), and bioactive (e.g., hydroxyapatite and glasses). Synthetic and natural polymers have been investigated for bone repair, but it has been difficult to create a polymer displaying optimal strength and degradation properties. Another approach involves implantation of demineralized bone powder (DBP), which is effective in stimulating bone growth. By inducing and augmenting formation of both cartilage and bone (including marrow), bone morphogenic proteins (BMPs) or growth factors such as transforming growth factor-β (TGF-β) represent other promising strategies (Toriumi *et al.*, 1991; Yasko *et al.*, 1992). Bone growth can also be induced when cells are grown on synthetic polymers and ceramics. For example, when human marrow cells are grown on porous hydroxyapatite in mice, spongious bone formation was detectable inside the pores within 8 weeks (Casabona *et al.*, 1998). Femoral shaft reconstruction has been demonstrated using bioresorbable polymer constructs seeded with osteoblasts as bridges between the bone defect (Puelacher *et al.*, 1996), and similar experiences have been reported for craniofacial applications (Breitbart *et al.*, 1998). Formation of phalanges and small joints has been demonstrated with selective placement of periosteum, chondrocytes, and tenocytes into a biodegradable synthetic polymer scaffold (Isogai *et al.*, 1999). Large bone defects in tibia of sheep were successfully reconstructed using combinations of autologous marrow stromal cells and coral (Petite *et al.*, 2000). Similar results were obtained by Kadiyala *et al.*, who have treated experimentally induced nonunion defects in adult dog femora with autologous marrow-derived cells grown on a hydroxyapatite: beta tricalcium phosphate (65: 35) scaffold (Kadiyala *et al.*, 1997). This approach was also successful in patients suffering from segmental bone defects. Abundant callus formation along the implants and good integration at the interface with the host bones was observed 2 months after surgery (Quarto *et al.*, 2001).

Muscle

The ability to generate muscle fibers has possible application regarding the treatment of muscle injury, cardiac disease, disorders involving smooth muscle of the intestine or urinary tract, and systemic muscular diseases such as Duchenne muscular dystrophy (DMD). Myoblasts from unaffected relatives have been transplanted into Duchenne patients and shown to produce dystrophin several months following the implantation. Myoblasts can migrate from one healthy muscle fiber to another (Gussoni *et al.*, 1992); thus, cell-based therapies may be useful in treating muscle atrophies. Creation of a whole hybrid muscular tissue was achieved by a sequential method of centrifugal cell packing and mechanical stress-loading resulting in tissue formation strongly resembling native muscle in terms of cell density, cell orientation, and incorporation of capillary networks. Kim and Mooney (1998) demonstrated with regard to smooth muscle cells the importance of matching both the initial mechanical properties and the degradation rate of a predefined three-dimensional scaffold to the specific tissue that is being engineered.

Loss of heart muscle tissue in the course of ischemic heart disease or cardiomyopathies is a major factor of morbidity and mortality in numerous patients. Once patients become symptomatic, their life expectancy is usually markedly shortened. This decline is mostly attributed to the inability of cardiomyocytes to regenerate after injury. Necrotic cells are replaced by fibroblasts leading to scar tissue formation and regional contractile dysfunction. In contrast, skeletal muscle has the capacity of tissue repair, presumably because of satellite cells that have regenerative capability. Satellite cells are undifferentiated skeletal myoblasts, which are located beneath the basal lamina in skeletal muscles. These cells have also been tested for myocardial repair (Chiu *et al.*, 1995; Menasche, 2003; Menasche *et al.*, 2001; Taylor *et al.*, 1998). In rats, myoblast grafts can survive for at least 1 year (Al Attar *et al.*, 2003). However, satellite cells transplanted into nonreperfused scar tissue do not transdifferentiate into cardiomyocytes but show a switch to slow-twitch fibers, which allow sustained improvement in cardiac function (Hagege *et al.*, 2003; Reinecke *et al.*, 2002).

Recent studies have suggested that bone marrow-derived or blood-derived progenitor cells contribute to the regeneration of infarcted myocardium and enhance neovascularization of ischemic myocardium (Kawamoto *et al.*, 2001; Orlic *et al.*, 2001a, b). In a pilot trial it was shown that also in patients with reperfused acute myocardial infarction, intracoronary infusion of autologous progenitor cells beneficially affected postinfarction left-ventricle remodeling processes (Assmus *et al.*, 2002).

An alternative approach to cell grafting techniques is the generation of cardiac tissue grafts *in vitro* and implanting them as spontaneously and coherently contracting tissues. As a model system, rat neonatal or embryonic chicken cardiomyocytes may be seeded on three-dimensional polymeric scaffolds (Carrier *et al.*, 1999) or collagen disks formed as a sponge (Radisic *et al.*, 2003) or by layering cell sheets three-dimensionally (Shimizu *et al.*, 2002). The latter two approaches are suitable for producing thicker cardiac tissue with more evenly distributed cells at a higher density. A principally different approach to generate engineered heart tissue was developed by Eschenhagen and colleagues. Neonatal or embryonic cardiomyocytes were mixed with freshly neutralized collagen I and cast into a cylindrical mold. After a few days the tissue patches were transferred to a stretching device, which induced hypertrophic growth and increased cell differentiation (Eschenhagen *et al.*, 2002b, 1997; Zimmermann *et al.*, 2000). Interestingly, the response to isoprenalin of stretched tissue was much more pronounced than in unstretched tissue (Eschenhagen *et al.*, 2002b; Zimmermann *et al.*, 2002).

Blood vessels

Peripheral vascular disease represents a growing health and socioeconomic burden in most developed countries (Ounpuu *et al.*, 2000). Today, artificial prostheses made of expanded polytetrafluoroethylene (ePTFE) and poly(ethylene terephthalate) (PET, Dacron) are the most widely used synthetic materials. Although successful in large diameter (>5-mm) high-flow vessels, in low-flow or smaller diameter sites they are compromised by thrombogenicity and compliance mismatch (Edelman, 1999). To circumvent these problems numerous modifications and techniques to enhance hematocompatibility and graft patency have been evaluated both *in vitro* and *in vivo*. These include chemical modifications, coatings (Gosselin *et al.*, 1996; Ye *et al.*, 2000), and surface seeding with endothelial cells (Pasic *et al.*, 1996; Zilla *et al.*, 1999). *In vitro* endothelialization of ePTFE grafts may result in patency rates comparable to state-of-the-art veinous autografts (Meinhart *et al.*, 1997). Polymer surface modifications involving protein adsorption may also be desirable. Unfortunately, materials that promote endothelial cell attachment often simultaneously promote attachment of platelets and smooth muscle cells associated with the adverse side effects of clotting and pseudointimal thickening. A possible solution has been demonstrated with polymers containing adhesion molecules (ligands) specific for endothelial cells (Hubbell *et al.*, 1991).

To overcome the limitations just mentioned, tissue engineering procedures could lead to completely biological vascular grafts. In fact, there have already been case reports regarding first human pediatric applications of tissue-engineered large-diameter vascular grafts (Naito *et al.*, 2003; Shin'oka *et al.*, 2001). As to small-caliber grafts, there are three principal approaches involving (1) synthetic biodegradable scaffolds, (2) biological scaffolds, and (3) completely autologous methods.

1. Niklason *et al.* have shown in animal models that by utilizing flow bioreactors to condition biodegradable polymers loaded with vascular cells, it is possible to generate arbitrary lengths of functional vascular grafts with significant extracellular matrix production, contractile responses to pharmacological agents, and tolerance of supraphysiologic burst pressures (Mitchell and Niklason, 2003; Niklason *et al.*, 2001, 1999). Similar *in vitro* experiments based on human vascular-derived cells seeded on PGA/PHA copolymers demonstrated the feasibility of viable, surgically implantable human small-caliber vascular grafts and the important effect of a "biomimetic" *in vitro* environment on tissue matura tion (Hoerstrup *et al.*, 2001).
2. A different approach to tissue engineering of vascular grafts comprises the use of decellularized natural matrices as initially introduced by Rosenberg *et al.* (1996). Histological examination of chemically decellularized carotid arteries revealed well-preserved structural matrix proteins. This provides an acellular scaffold that can be successfully repopulated *in vitro* prior to implantation (Teebken *et al.*, 2000). Such scaffolds have also been shown to be repopulated *in vivo* (Bader *et al.*, 2000). Recently, successful utilization of endothelial precursor cells for tissue engineering of vascular grafts based on decellularized matrices has been demonstrated (Kaushal *et al.*, 2001).
3. L'Heureux *et al.* cultured and conditioned sheets of vascular smooth muscle cells and their native extracellular matrix without any scaffold material in a flow system. Subsequently these sheets were placed around a tubular support device and after maturation the tubular support was removed and endothelial cells were seeded into the lumen. Thereby a complete scaffold-free vessel was created with a functional endothelial layer and a burst strength of more than 2000 mm Hg (L'Heureux *et al.*, 1998).

Angiogenesis (the formation of new blood vessels) is essential for growth, tissue repair, and wound healing. Therefore, many tissue-engineering concepts involve angiogenesis for the vascularization of the newly generated tissues. Unfortunately, so far advances have been compromised by the inability to vascularize thick, complex tissues, particularly those comprising large organs such as the liver, kidney, or heart. To overcome these limitations, several approaches have been investigated. Vacanti and coworkers used local delivery of basic fibroblast growth factor (bFGF) to increase angiogenesis and engraftment of hepatocytes in tissue-engineered polymer devices (Lee *et al.*, 2002). In another study sustained and localized delivery of vascular endothelial growth factor (VEGF) combined with the transplantation of human microvascular endothelial cells was used to engineer new vascular networks (Peters *et al.*, 2002). Using micromachining technologies on silicon, Kaihara *et al.* demonstrated *in vitro* generation of branched three-dimensional vascular networks formed by endothelial cells (Kaihara *et al.*, 2000).

Heart valves

For treatment of heart-valve disease, mechanical or biological valves are currently in use. The drawbacks of mechanical valves include the need for lifelong anticoagulation, the risk of thromboembolic events, prosthesis failure, and the inability of the device to grow. Biological valves (homograft, xenograft, fixed by cryopreseration of chemical treatment) have a limited durability due to their immunogenic potential and the fact that they represent nonliving tissues without regeneration capacities. All types of contemporary valve prostheses basically consist of nonliving, foreign materials, posing specific problems to pediatric applications when devices with growth potential are required for optimal treatment.

The basic concept currently used for tissue engineering of heart valve structures is to transplant autologous cells onto a biodegradable scaffold, to grow and to condition the cell-seeded scaffold device *in vitro*, and finally to implant the tissue-like construct into the donor patient.

The heart-valve scaffold may be based on either biological or synthetic materials. Donor heart valves or animal-derived valves depleted of cellular antigens can be used as a scaffold material. Removing the cellular components results in a material composed of essentially extracellular matrix proteins that can serve as an intrinsic template for cell attachment (Samouillan *et al.*, 1999). In general, nonfixed acellularized valve leaflets have shown recellularization by the host, as demonstrated in dogs (Wilson *et al.*, 1995) and sheep (Elkins *et al.*, 2001; Goldstein *et al.*, 2000). However, first clinical applications of this concept in children resulted in rapid failure of the heart valves due to severe foreign-body-type reactions associated with a 75% mortality (Simon *et al.*, 2003). In a further approach, specific biological matrix constituents can be used as scaffold material. Collagen is one of the materials that show biodegradable properties and can be used as a foam (Rothenburger *et al.*, 2002), gel or sheet (Hutmacher *et al.*, 2001), or sponge (Taylor *et al.*, 2002), and even as a fiber-based scaffold (Rothenburger *et al.*, 2001). It has the disadvantage that it is difficult to obtain from the patient. Therefore, most of the collagen scaffolds are of animal origin. Another biological material displaying good controllable biodegradable properties is fibrin. Since fibrin gels can be produced from the patient's blood to serve as autologous scaffold, no toxic degradation or inflammatory reactions are expected (Lee and Mooney, 2001).

The use of synthetic materials as scaffolds has already been broadly demonstrated for cardiovascular tissue engineering. Initial attempts to create single heart-valve leaflets were based on synthetic scaffolds, such as polyglactin, PGA, PLA, or PLGA (copolymer of PGA and PLA). To create complete trileaflet heart-valve conduits, PHA-based materials (polyhydroxyalkanoates) were used (Sodian *et al.*, 2000). These materials are thermoplastic and can therefore be easily molded into any desired three-dimensional shape. A combined polymer scaffold consisting of nonwoven PGA and P4HB (poly-4-hydroxybutyrate) has shown promising *in vivo* results (Hoerstrup *et al.*, 2000a).

In most cardiovascular tissue-engineering approaches cells are harvested from donor tissues, e.g., from peripheral arteries, and mixed vascular cell populations consisting of myofibroblasts and endothelial cells are obtained. Out of these, pure viable cell lines can be easily isolated by cell sorters (Hoerstrup *et al.*, 1998) and the subsequent seeding onto the biodegradable scaffold is undertaken in two steps. First, the myofibroblasts are seeded and grown *in vitro*. Second, the endothelial cells are seeded on top of the generated neotissue, leading to the formation of a native leaflet-analogous histological structure (Zund *et al.*, 1998).

Successful implantation of a single tissue-engineered valve leaflet has been demonstrated in the animal model (Shinoka *et al.*, 1996) and based on a novel *in vitro* conditioning protocol of the tissue-engineered valve constructs in bioreactor flow systems (pulse-duplicator) completely autologous, living heart-valves were generated (Fig. 7.1.2-4). Interestingly, these tissue-engineered valves showed good *in vivo* functionality and strongly resembled native heart valves with regard to biomechanical and morphological features (Hoerstrup *et al.*, 2000b; Rabkin *et al.*, 2002). With regard to clinical applications, several human cell sources have been investigated (Schnell *et al.*, 2001). Recently, cells derived from bone marrow or umbilical cord have been successfully utilized to generate heart valves and conduits *in vitro* (Hoerstrup *et al.*, 2002a, b). In contrast to vascular cells, these cells can be obtained without surgical interventions representing an easy-to-access cell source in a possible routine clinical scenario. Because of their good proliferation and progenitor potential, these cells are

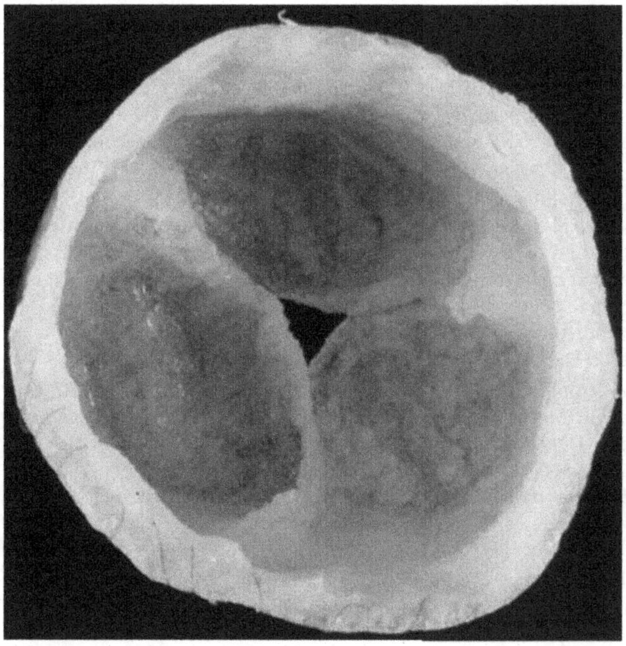

Fig. 7.1.2-4 Photograph of a living, tissue engineered heart valve after 14 days of biomimetic conditioning in a pulse–duplicator–bioreactor based on a rapidly biodegradable synthetic scaffold material. (Reprinted with permission from Hoerstrup *et al.*, 2000. *Circulation* **102:** III-44–III-49.)

expected to be an attractive alternative for cardiovascular tissue-engineering applications.

Blood

There is a critical need for blood cell substitutes since donor blood suffers from problems such as donor shortage, requirements for typing and cross-matching, limited storage time, and, even more importantly in the era of AIDS, infectious disease transmission. Oxygen-containing fluids or materials as a substitute for red blood cells offer important applications in emergency resuscitation, shock, tumor therapy, and organ preservation. Several oxygen transporters are under investigation. Hemoglobin is a primary candidate, which not only serves as the natural oxygen transporter in blood but also functions in carbon dioxide transport, as a buffer, and in regulating osmotic pressure. Early clinical trials of cell-free hemoglobin were complicated by its lack of purity, instability, high oxygen affinity, and binding nitric oxide (NO), leading to cardiovascular side effects. These problems have been subsequently addressed by various chemical modifications such as intra- and intermolecular cross-linking using diacid, glutaraldehyde, or *o*-raffinose or conjugation to dextran or polyethylene glycol. Because of the limited hemoglobin availability, genetically engineered human hemoglobin or hemoglobin from bovine sources may represent a valid alternative. Several products are now in phase II/III clinical studies (Winslow, 2000). The latest developments include nanoencapsulated genetically engineered macromolecules of poly (hemoglobin–catalase–superoxide dismutase). Biodegradable polylactides and polyglycolides are used as carriers leading to artificial red blood cells containing hemoglobin and protective enzymes (Chang, 2003). Furthermore, perfluorocarbons (PFC) may be an alternative characterized by a high gas dissolving capacity (O_2, CO_2, and others), chemical and biological inertness, and low viscosity. However, hemoglobin binds significantly more oxygen at a given partial oxygen pressure than can be dissolved in PFC. Research to create functional substitutes for platelets by encapsulating platelet proteins in lipid vesicles has also been conducted (Baldassare *et al.*, 1985). Finally, stem cells have the potential to differentiate into the various cellular elements of blood (Thomson *et al.*, 1998).

Future perspectives

Current methods of transplantation and reconstruction are among the most time-consuming and costly therapies available today. Tissue-engineering offers future promise in the treatment of loss of tissue or organ function as well as for genetic disorders with metabolic deficiencies. Besides that, tissue engineering offers the possibility of substantial future savings by providing substitutes that are less expensive than donor organs and by providing a means of intervention before patients are critically ill. Few areas of technology will require more interdisciplinary research or have the potential to affect more positively the quality and length of life. Much must be learned from cell biology, especially with regard to what controls cellular differentiation and growth and how extracellular matrix components influence cell function. Immunology and molecular genetics will contribute to the design of cells or cell transplant systems that are not rejected by the immune system. With regard to the cell source, transplanted cells may come from cell lines or primary tissue, from the patients themselves, or from other human donors, animal tissue, or fetal tissue. In choosing the cell source a balance must be found between ethical issues, safety issues, and efficacy. These considerations are particularly important when introducing new techniques in the tissue-engineering field such as the generation of histocompatible tissue by cloning (nuclear transfer) (Lanza *et al.*, 2002) or by the creation of oocytes from embryonic stem cells (Hubner *et al.*, 2003).

The materials used in tissue engineering represent a major field of research regarding, e.g., polymer processing, development of controlled-release systems, surface modifications, and mathematical models possibly predicting *in vivo* cellular events.

Bibliography

Aebischer, P., Pochon, N. A., Heyd, B., Deglon, N., Joseph, J. M., Zurn, A. D., Baetge, E. E., Hammang, J. P., Goddard, M., Lysaght, M., Kaplan, F., Kato, A. C., Schluep, M., Hirt, L., Regli, F., Porchet, F., and De Tribolet, N. (1996a). Gene therapy for amyotrophic lateral sclerosis (ALS) using a polymer encapsulated xenogenic cell line engineered to secrete hCNTF. *Hum. Gene. Ther.* 7: 851–860.

Aebischer, P., Salessiotis, A. N., and Winn, S. R. (1989). Basic fibroblast growth factor released from synthetic guidance channels facilitates peripheral nerve regeneration across long nerve gaps. *J. Neurosci. Res.* **23**: 282–289.

Aebischer, P., Schluep, M., Deglon, N., Joseph, J. M., Hirt, L., Heyd, B., Goddard, M., Hammang, J. P., Zurn, A. D., Kato, A. C., Regli, F., and Baetge, E. E. (1996b). Intrathecal delivery of CNTF using encapsulated genetically modified xenogeneic cells in amyotrophic lateral sclerosis patients. *Nat. Med.* **2**: 696–699.

Aebischer, P., Tresco, P. A., Winn, S. R., Greene, L. A., and Jaeger, C. B. (1991). Long-term cross-species brain transplantation of a polymer-encapsulated dopamine-secreting cell line. *Exp. Neurol.* **111**: 269–275.

Aebischer, P., Winn, S. R., and Galletti, P. M. (1988). Transplantation of neural tissue in polymer capsules. *Brain Res.* **448**: 364–368.

Al Attar, N., Carrion, C., Ghostine, S., Garcin, I., Vilquin, J. T., Hagege, A. A., and Menasche, P. (2003). Long-term (1 year) functional and histological results of autologous skeletal muscle cells transplantation in rat. *Cardiovasc. Res.* **58**: 142–148.

Armentano, D., Thompson, A. R., Darlington, G., and Woo, S. L. (1990). Expression of human factor IX in rabbit hepatocytes by retrovirus-mediated gene transfer: potential for gene therapy of hemophilia B. *Proc. Natl. Acad. Sci. USA* **87**: 6141–6145.

Assmus, B., Schachinger, V., Teupe, C., Britten, M., Lehmann, R., Dobert, N., Grunwald, F., Aicher, A., Urbich, C., Martin, H., Hoelzer, D., Dimmeler, S., and Zeiher, A. M. (2002). Transplantation of progenitor cells and regeneration enhancement in acute myocardial infarction (TOPCARE-AMI): *Circulation* **106**: 3009–3017.

Atala, A., Vacanti, J. P., Peters, C. A., Mandell, J., Retik, A. B., and Freeman, M. R. (1992). Formation of urothelial structures *in vivo* from dissociated cells attached to biodegradable polymer scaffolds in vitro. *J. Urol.* **148**: 658–662.

Bader, A., Knop, E., Fruhauf, N., Crome, O., Boker, K., Christians, U., Oldhafer, K., Ringe, B., Pichlmayr, R., and Sewing, K. F. (1995). Reconstruction of liver tissue *in vitro*: geometry of characteristic flat bed, hollow fiber, and spouted bed bioreactors with reference to the *in vivo* liver. *Artif. Organs* **19**: 941–950.

Bader, A., Steinhoff, G., Strobl, K., Schilling, T., Brandes, G., Mertsching, H., Tsikas, D., Froelich, J., and Haverich, A. (2000). Engineering of human vascular aortic tissue based on a xenogeneic starter matrix. *Transplantation* **70**: 7–14.

Baldassare, J. J., Kahn, R. A., Knipp, M. A., and Newman, P. J. (1985). Reconstruction of platelet proteins into phospholipid vesicles. Functional proteoliposomes. *J. Clin. Invest.* **75**: 35–39.

Bell, E., Ivarsson, B., and Merrill, C. (1979). Production of a tissue-like structure by contraction of collagen lattices by human fibroblasts of different proliferative potential in vitro. *Proc. Natl. Acad. Sci. USA* **76**: 1274–1278.

Bell, E., Parenteau, N., Gay, R., Nolte, C., Kemp, P., Bilbo, B., Ekstein, B., and Johnson, E. (1991a). The living skin equivalent: its manufacture, its organotypic properties and its responses to irritants. *Toxic. Vitro* **5**: 591–596.

Bell, E., Rosenberg, M., Kemp, P., Gay, R., Green, G. D., Muthukumaran, N., and Nolte, C. (1991b). Recipes for reconstituting skin. *J. Biomech. Eng.* **113**: 113–119.

Breitbart, A. S., Grande, D. A., Kessler, R., Ryaby, J. T., Fitzsimmons, R. J., and Grant, R. T. (1998). Tissue engineered bone repair of calvarial defects using cultured periosteal cells. *Plast. Reconstr. Surg.* **101**: 567–574; discussion 575–576.

Burke, J. F., Yannas, I. V., Quinby, W. C., Jr., Bondoc, C. C., and Jung, W. K. (1981). Successful use of a physiologically acceptable artificial skin in the treatment of extensive burn injury. *Ann. Surg.* **194**: 413–428.

Buschmann, M. D., Gluzband, Y. A., Grodzinsky, A. J., Kimura, J. H., and Hunziker, E. B. (1992). Chondrocytes in agarose culture synthesize a mechanically functional extracellular matrix. *J. Orthop. Res.* **10**: 745–758.

Cancedda, R., Dozin, B., Giannoni, P., and Quarto, R. (2003). Tissue engineering and cell therapy of cartilage and bone. *Matrix Biol.* **22**: 81–91.

Carrier, R. L., Papadaki, M., Rupnick, M., Schoen, F. J., Bursac, N., Langer, R., Freed, L. E., and Vunjak-Novakovic, G. (1999). Cardiac tissue engineering: cell seeding, cultivation parameters, and tissue construct characterization. *Biotechnol. Bioeng.* **64**: 580–589.

Casabona, F., Martin, I., Muraglia, A., Berrino, P., Santi, P., Cancedda, R., and Quarto, R. (1998). Prefabricated engineered bone flaps: an experimental model of tissue reconstruction in plastic surgery. *Plast. Reconstr. Surg.* **101**: 577–581.

Chang, T. M. (2003). Future generations of red blood cell substitutes. J. *Intern. Med.* **253**: 527–535.

Chen, S. C., Mullon, C., Kahaku, E., Watanabe, F., Hewitt, W., Eguchi, S., Middleton, Y., Arkadopoulos, N., Rozga, J., Solomon, B., and Demetriou, A. A. (1997). Treatment of severe liver failure with a bioartificial liver. *Ann. N.Y. Acad. Sci.* **831**: 350–360.

Chirila, T. V. (2001). An overview of the development of artificial corneas with porous skirts and the use of PHEMA for such an application. *Biomaterials* **22**: 3311–3317.

Chiu, R. C., Zibaitis, A., and Kao, R. L. (1995). Cellular cardiomyoplasty: myocardial regeneration with satellite cell implantation. *Ann. Thorac. Surg.* **60**: 12–18.

Chowdhury, J. R., Grossman, M., Gupta, S., Chowdhury, N. R., Baker, J. R., Jr., and Wilson, J. M. (1991). Long-term improvement of hypercholesterolemia after *ex vivo* gene therapy in LDLR-deficient rabbits. *Science* **254**: 1802–1805.

Crawford, G. J., Hicks, C. R., Lou, X., Vijayasekaran, S., Tan, D., Mulholland, B., Chirila, T. V., and Constable, I. J. (2002). The Chirila Keratoprosthesis: phase I human clinical trial. *Ophthalmology* **109**: 883–889.

Date, I., Shingo, T., Yoshida, H., Fujiwara, K., Kobayashi, K., and Ohmoto, T. (2000). Grafting of encapsulated dopamine-secreting cells in Parkinson's disease: long-term primate study. *Cell Transplant.* **9**: 705–709.

de Filippo, R. E., Yoo, J. J., and Atala, A. (2002). Urethral replacement using cell seeded tubularized collagen matrices.

J. Urol. **168**: 1789–1792; discussion 1792–1793.

Dunn, J. C., Tompkins, R. G., and Yarmush, M. L. (1991). Long-term *in vitro* function of adult hepatocytes in a collagen sandwich configuration. *Biotechnol. Prog.* **7**: 237–245.

Eaglstein, W. H., Alvarez, O. M., Auletta, M., Leffel, D., Rogers, G. S., Zitelli, J. A., Norris, J. E., Thomas, I., Irondo, M., Fewkes, J., Hardin-Young, J., Duff, R. G., and Sabolinski, M. L. (1999). Acute excisional wounds treated with a tissue-engineered skin (Apligraf). *Dermatol. Surg.* **25**: 195–201.

Edelman, E. R. (1999). Vascular tissue engineering: designer arteries. Circ. *Res.* **85**: 1115–1117.

Efrat, S. (1999). Genetically engineered pancreatic beta-cell lines for cell therapy of diabetes. *Ann. N.Y. Acad. Sci.* **875**: 286–293.

Elkins, R. C., Dawson, P. E., Goldstein, S., Walsh, S. P., and Black, K. S. (2001). Decellularized human valve allografts. *Ann. Thorac. Surg.* **71**: S428–S432.

Ellis, A. J., Hughes, R. D., Wendon, J. A., Dunne, J., Langley, P. G., Kelly, J. H., Gislason, G. T., Sussman, N. L., and Williams, R. (1996). Pilot-controlled trial of the extracorporeal liver assist device in acute liver failure. *Hepatology* **24**: 1446–1451.

Eschenhagen, T., Didie, M., Heubach, J., Ravens, U., and Zimmermann, W.H. (2002a). Cardiac tissue engineering. Transpl. *Immunol.* **9**: 315–321.

Eschenhagen, T., Didie, M., Munzel, F., Schubert, P., Schneiderbanger, K., and Zimmermann, W. H. (2002b). 3D engineered heart tissue for replacement therapy. *Basic. Res. Cardiol.* **97** (Suppl 1): I146–I152.

Eschenhagen, T., Fink, C., Remmers, U., Scholz, H., Wattchow, J., Weil, J., Zimmermann, W., Dohmen, H. H., Schafer, H., Bishopric, N., Wakatsuki, T., and Elson, E. L. (1997). Three-dimensional reconstitution of embryonic cardiomyocytes in a collagen matrix: a new heart muscle model system. *FASEB. J.* **11**: 683–694.

Evans, M. D., Xie, R. Z., Fabbri, M., Bojarski, B., Chaouk, H., Wilkie, J. S., McLean, K. M., Cheng, H. Y., Vannas, A., and Sweeney, D. F. (2002). Progress in the development of a synthetic corneal onlay. *Invest. Ophthalmol. Vis. Sci.* **43**: 3196–3201.

Fentem, J. H., Briggs, D., Chesne, C., Elliott, G. R., Harbell, J. W., Heylings, J. R., Portes, P., Roguet, R., van de Sandt, J. J., and Botham, P. A. (2001). A prevalidation study on *in vitro* tests for acute skin irritation. Results and evaluation by the Management Team. Toxicol. *Vitro* **15**: 57–93.

Folkman, J., and Haudenschild, C. (1980). Angiogenesis *in vitro*. *Nature* **288**: 551–556.

Freed, L. E., Langer, R., Martin, I., Pellis, N. R., and Vunjak-Novakovic, G. (1997). Tissue engineering of cartilage in space. *Proc. Natl. Acad. Sci. USA* **94**: 13885–13890.

Freed, L. E., Marquis, J. C., Nohria, A., Emmanual, J., Mikos, A. G., and Langer, R. (1993). Neocartilage formation *in vitro* and *in vivo* using cells cultured on synthetic biodegradable polymers. *J. Biomed. Mater. Res.* **27**: 11–23.

Gerlach, J. C. (1997). Long-term liver cell cultures in bioreactors and possible application for liver support. *Cell Biol. Toxicol.* **13**: 349–355.

Gerlach, J.C., Lemmens, P., Schon, M., Janke, J., Rossaint, R., Busse, B., Puhl, G., and Neuhaus, P. (1997). Experimental evaluation of a hybrid liver support system. *Transplant Proc.* **29**: 852.

Germain, L., Auger, F. A., Grandbois, E., Guignard, R., Giasson, M., Boisjoly, H., and Guerin, S. L. (1999). Reconstructed human cornea produced *in vitro* by tissue engineering. *Pathobiology* **67**: 140–147.

Goldstein, S., Clarke, D. R., Walsh, S. P., Black, K. S., and O'Brien, M. F. (2000). Transspecies heart valve transplant: advanced studies of a bioengineered xeno-autograft. *Ann. Thorac. Surg.* **70**: 1962–1969.

Gosselin, C., Vorp, D. A., Warty, V., Severyn, D.A., Dick, E. K., Borovetz, H. S., and Greisler, H. P. (1996). ePTFE coating with fibrin glue, FGF-1, and heparin: effect on retention of seeded endothelial cells. *J. Surg. Res.* **60**: 327–332.

Grande, D. A., Pitman, M. I., Peterson, L., Menche, D., and Klein, M. (1989). The repair of experimentally produced defects in rabbit articular cartilage by autologous chondrocyte transplantation. J. *Orthop. Res.* **7**, 208–218.

Griffith, M., Hakim, M., Shimmura, S., Watsky, M. A., Li, F., Carlsson, D., Doillon, C. J., Nakamura, M., Suuronen, E., Shi-nozaki, N., Nakata, K., and Sheardown, H. (2002). Artificial human corneas: scaffolds for transplantation and host regeneration. *Cornea* **21**: S54–S61.

Griffith, M., Osborne, R., Munger, R., Xiong, X., Doillon, C. J., Laycock, N. L., Hakim, M., Song, Y., and Watsky, M. A. (1999). Functional human corneal equivalents constructed from cell lines. *Science* **286**: 2169–2172.

Grossman, M., Raper, S. E., Kozarsky, K., Stein, E.A., Engelhardt, J. F., Muller, D., Lupien, P. J., and Wilson, J. M. (1994). Successful *ex vivo* gene therapy directed to liver in a patient with familial hypercholesterolaemia. *Nat. Genet.* **6**: 335–341.

Grower, M. F., Russell, E. A., Jr., and Cutright, D. E. (1989). Segmental neogenesis of the dog esophagus utilizing a biodegradable polymer framework. *Biomater. Artif. Cells Artif. Organs.* **17**: 291–314.

Guenard, V., Kleitman, N., Morrissey, T. K., Bunge, R. P., and Aebischer, P. (1992). Syngeneic Schwann cells derived from adult nerves seeded in semipermeable guidance channels enhance peripheral nerve regeneration. *J. Neurosci.* **12**: 3310–1320.

Gussoni, E., Pavlath, G. K., Lanctot, A. M., Sharma, K. R., Miller, R. G., Steinman, L., and Blau, H. M. (1992). Normal dystrophin transcripts detected in Duchenne muscular dystrophy patients after myoblast transplantation. *Nature* **356**: 435–438.

Hagege, A. A., Carrion, C., Menasche, P., Vilquin, J. T., Duboc, D., Marolleau, J.P., Desnos, M., and Bruneval, P. (2003). Viability and differentiation of autologous skeletal myoblast grafts in ischaemic cardiomyopathy. *Lancet* **361**: 491–492.

Haller, M. F., and Saltzman, W.M. (1998). Nerve growth factor delivery systems. *J. Controlled Release* **53**: 1–6.

Hansbrough, J. F., Cooper, M. L., Cohen, R., Spielvogel, R., Greenleaf, G., Bartel, R. L., and Naughton, G. (1992). Evaluation of a biodegradable matrix containing cultured human fibroblasts as a dermal replacement beneath meshed skin grafts on athymic mice. *Surgery* **111**: 438–446.

Heimbach, D., Luterman, A., Burke, J., Cram, A., Herndon, D., Hunt, J., Jordan, M., McManus, W., Solem, L., Warden, G., and Zanvacki, B. (1988). Artificial dermis for major burns. A multi-center randomized clinical trial. *Ann. Surg.* **208**: 313–320.

Hoerstrup, S. P., Kadner, A., Breymann, C., Maurus, C. F., Guenter, C.I., Sodian, R., Visjager, J. F., Zund, G., and Turina, M. I. (2002a). Living, autologous pulmonary artery conduits tissue engineered from human umbilical cord cells. *Ann. Thorac. Surg.* **74**: 46–52; discussion.

Hoerstrup, S. P., Kadner, A., Melnitchouk, S., Trojan, A., Eid, K., Tracy, J., Sodian, R., Visjager, J. F., Kolb, S. A., Grunenfelder, J., Zund, G., and Turina, M. I. (2002b). Tissue engineering of functional trileaflet heart valves from human marrow stromal cells. *Circulation* **106**: I143–I150.

Hoerstrup, S. P., Sodian, R., Daebritz, S., Wang, J., Bacha, E. A., Martin, D. P., Moran, A. M., Guleserian, K. J., Sperling, J. S., Kaushal, S., Vacanti, J.P., Schoen, F. J., and Mayer, J. E., Jr. (2000a). Functional living trileaflet heart valves grown *in vitro*. *Circulation* **102**: III44–III49.

Hoerstrup, S. P., Sodian, R., Sperling, J.S., Vacanti, J. P., and Mayer, J. E., Jr. (2000b). New pulsatile bioreactor for *in vitro* formation of tissue engineered heart valves. *Tissue Eng.* **6**: 75–79.

Hoerstrup, S. P., Zund, G., Schoeberlein, A., Ye, Q., Vogt, P. R., and Turina, M. I. (1998). Fluorescence activated cell sorting: a reliable method in tissue engineering of a bioprosthetic heart valve. *Ann. Thorac. Surg.* **66**: 1653–1657.

Hoerstrup, S. P., Zund, G., Sodian, R., Schnell, A. M., Grunenfelder, J., and Turina, M. I. (2001). Tissue engineering of small caliber vascular grafts. *Eur. J. Cardiothorac. Surg.* **20**: 164–169.

Hori, Y., Nakamura, T., Kimura, D., Kaino, K., Kurokawa, Y., Satomi, S., and Shimizu, Y. (2002). Experimental study on tissue engineering of the small intestine by mesenchymal stem cell seeding. *J. Surg. Res.* **102**: 156–160.

Hubbell, J. A., Massia, S. P., Desai, N. P., and Drumheller, P. D. (1991). Endothelial cell-selective materials for tissue engineering in the vascular graft via a new receptor. *Biotechnology (N.Y.)* **9**: 568–572.

Hubner, K., Fuhrmann, G., Christenson, L. K., Kehler, J., Reinbold, R., De La Fuente, R., Wood, J., Strauss, J. F., 3rd, Boiani, M., and Scholer, H. R. (2003). Derivation of oocytes from mouse embryonic stem cells. *Science* **300**: 1251–1256.

Humes, H. D., Fissell, W. H., Weitzel, W. F., Buffington, D. A., Westover, A. J., MacKay, S. M., and Gutierrez, J. M. (2002). Metabolic replacement of kidney function in uremic animals with a bioartificial kidney containing human cells. *Am. J. Kidney Dis.* 39: 1078–1087.

Humes, H. D., MacKay, S. M., Funke, A. J., and Buffington, D. A. (1999). Tissue engineering of a bioartificial renal tubule assist device: *in vitro* transport and metabolic characteristics. *Kidney Int.* **55**: 2502–2514.

Humes, H. D., Weitzel, W. F., Bartlett, R. H., Swaniker, F. C., and Paganini, E. P. (2003). Renal cell therapy is associated with dynamic and individualized responses in patients with acute renal failure. *Blood Purif.* **21**: 64–71.

Hutmacher, D. W., Goh, J. C., and Teoh, S. H. (2001). An introduction to biodegradable materials for tissue engineering applications. *Ann. Acad. Med. Singapore* **30**: 183–191.

Isogai, N., Landis, W., Kim, T. H., Gerstenfeld, L. C., Upton, J., and Vacanti, J. P. (1999). Formation of phalanges and small joints by tissue-engineering. *J. Bone. Joint. Surg. Am.* **81**: 306–316.

Jensen, T. G., Jensen, U. B., Jensen, P. K., Ibsen, H. H., Brandrup, F., Ballabio, A., and Bolund, L. (1993). Correction of steroid sulfatase deficiency by gene transfer into basal cells of tissue-cultured epidermis from patients with recessive X-linked ichthyosis. *Exp. Cell Res.* **209**: 392–397.

Kadiyala, S., Jaiswal, N., and Bruder, S. P. (1997). Culture expanded, bone marrow derived mesenchymal stem cells can regenerate a critical-sized segmental bone defect. *Tissue Eng. 3*: 173–185.

Kaihara, S., Borenstein, J., Koka, R., Lalan, S., Ochoa, E. R., Ravens, M., Pien, H., Cunningham, B., and Vacanti, J. P. (2000). Silicon micromachining to tissue engineer branched vascular channels for liver fabrication. *Tissue Eng.* **6**: 105–117.

Kaihara, S., Kim, S. S., Benvenuto, M., Choi, R., Kim, B. S., Mooney, D., Tanaka, K., and Vacanti, J.P. (1999). Anastomosis between tissue-engineered intestine and native small bowel. *Transplant Proc.* **31**: 661–662.

Kajitani, M., Wadia, Y., Hinds, M. T., Teach, J., Swartz, K. R., and Gregory, K. W. (2001). Successful repair of esophageal injury using an elastin based biomaterial patch. *ASAIO J.* **47**: 342–345.

Kato, T., Sato, K., Miyazaki, H., Sasaki, S., Matsuo, S., and Moriyama, M. (1993). The uretero-ileoceco-proctostomy (ileo-cecal rectal bladder): early experiences in 18 patients. *J. Urol.* **150**: 326–331.

Kaushal, S., Amiel, G. E., Guleserian, K. J., Shapira, O.M., Perry, T., Sutherland, F. W., Rabkin, E., Moran, A. M., Schoen, F. J., Atala, A., Soker, S., Bischoff, J., and Mayer, J. E., Jr. (2001). Functional small-diameter neovessels created using endothelial progenitor cells expanded *ex vivo*. *Nat. Med.* **7**: 1035–1040.

Kawamoto, A., Gwon, H. C., Iwaguro, H., Yamaguchi, J. I., Uchida, S., Masuda, H., Silver, M., Ma, H., Kearney, M., Isner, J. M., and Asahara, T. (2001). Therapeutic potential of *ex vivo* expanded endothelial progenitor cells for myocardial ischemia. *Circulation* **103**: 634–637.

Keeffe, E. B. (2001). Liver transplantation: current status and novel approaches to liver replacement. *Gastroenterology* **120**: 749–762.

Kim, B.-S., and Mooney, D. J. (1998). Engineering smooth muscle tissue with a predefined structure. *J. Biomed. Mater. Res.* **41**: 322–332.

Kim, S. S., Kaihara, S., Benvenuto, M. S., Choi, R. S., Kim, B. S., Mooney, D. J., Taylor, G. A., and Vacanti, J. P. (1999). Regenerative signals for intestinal epithelial organoid units transplanted on biodegradable polymer scaffolds for tissue engineering of small intestine. *Transplantation* **67**: 227–233.

Kim, S. S., Utsunomiya, H., Koski, J. A., Wu, B. M., Cima, M. J., Sohn, J., Mukai, K., Griffith, L. G., and Vacanti, J. P. (1998). Survival and function of hepatocytes on a novel three-dimensional synthetic biodegradable polymer scaffold with an intrinsic network of channels. *Ann. Surg.* **228**: 8–13.

Kin, T., Iwata, H., Aomatsu, Y., Ohyama, T., Kanehiro, H., Hisanaga, M., and Nakajima, Y. (2002). Xenotransplantation of pig islets in diabetic dogs with use of a microcapsule composed of agarose and polystyrene sulfonic acid mixed gel. *Pancreas* **25**: 94–100.

Kordower, J. H., Liu, Y. T., Winn, S., and Emerich, D. F. (1995). Encapsulated PC12 cell transplants into hemiparkinsonian monkeys: a behavioral, neuroanatomical, and neurochemical analysis. *Cell. Transplant.* **4**: 155–171.

Kudo, T., Kawase, M., Kawada, S., Kurosawa, H., Koyanagi, H., Takeuchi, Y., Hosoda, Y., and Wanibuchi, Y. (1999). Anticoag-ulation after valve replacement: a multicenter retrospective study. *Artif. Organs* **23**: 199–203.

Kurbaan, A. S., Bowker, T. J., Ilsley, C. D., and Rickards, A. F. (1998). Impact of postangioplasty restenosis on comparisons of outcome between angioplasty and bypass grafting. Coronary Angioplasty versus Bypass Revascularisation Investigation (CABRI)

Investigators. *Am. J. Cardiol.* **82**: 272–276.

Kusama, K., Donegan, W. L., and Samter, T. G. (1989). An investigation of colon cancer associated with urinary diversion. *Dis. Colon Rectum* **32**: 694–697.

Lacy, P. E., Hegre, O. D., Gerasimidi-Vazeou, A., Gentile, F. T., and Dionne, K. E. (1991). Maintenance of normoglycemia in diabetic mice by subcutaneous xenografts of encapsulated islets. *Science* **254**: 1782–1784.

Langer, R., and Vacanti, J. P. (1993). Tissue engineering. *Science* **260**: 920–926.

Langer, R. S., and Vacanti, J. P. (1999). Tissue engineering: the challenges ahead. *Sci. Am.* **280**: 86–89.

Lanza, R. P., Chung, H. Y., Yoo, J. J., Wettstein, P. J., Blackwell, C., Borson, N., Hofmeister, E., Schuch, G., Soker, S., Moraes, C. T., West, M.D., and Atala, A. (2002). Generation of histocompatible tissues using nuclear transplantation. *Nat. Biotechnol.* **20**: 689–696.

Lanza, R. P., Ecker, D. M., Kuhtreiber, W. M., Marsh, J. P., Ringeling, J., and Chick, W. L. (1999). Transplantation of islets using micro-encapsulation: studies in diabetic rodents and dogs. *J. Mol. Med.* **77**: 206–210.

Lanza, R. P., Kuhtreiber, W. M., Ecker, D., Staruk, J. E., and Chick, W. L. (1995). Xenotransplantation of porcine and bovine islets without immunosuppression using uncoated alginate microspheres. *Transplantation* **59**: 1377–1384.

Lee, H., Cusick, R.A., Browne, F., Ho Kim, T., Ma, P. X., Utsunomiya, H., Langer, R., and Vacanti, J. P. (2002). Local delivery of basic fibroblast growth factor increases both angiogenesis and engraftment of hepatocytes in tissue-engineered polymer devices. *Transplantation* **73**: 1589–1593.

Lee, K. Y., and Mooney, D. J. (2001). Hydrogels for tissue engineering. *Chem. Rev.* **101**: 1869–1879.

Legeais, J. M., and Renard, G. (1998). A second generation of artificial cornea (Biokpro II). *Biomaterials* **19**: 1517–1522.

L'Heureux, N., Paquet, S., Labbe, R., Germain, L., and Auger, F. A. (1998). A completely biological tissue-engineered human blood vessel. *FASEB. J.* **12**: 47–56.

Lindvall, O., Rehncrona, S., Brundin, P., Gustavii, B., Ast-edt, B., Widner, H., Lindholm, T., Bjorklund, A., Leenders, K. L., and Rothwell, J. C. (1990). Neural transplantation in Parkinson's disease: the Swedish experience. *Prog. Brain Res.* **82**: 729–736.

MacKay, S. M., Funke, A. J., Buffington, D. A., and Humes, H. D. (1998). Tissue engineering of a bioartificial renal tubule. *ASAIO J.* **44**: 179–183.

Matas, A. J., Sutherland, D. E., Steffes, M. W., Mauer, S. M., Sowe, A., Simmons, R. L., and Najarian, J. S. (1976). Hepatocellular transplantation for metabolic deficiencies: decrease of plasma bilirubin in Gunn rats. *Science* **192**: 892–894.

Mayer, J. E., Jr. (1995). Uses of homograft conduits for right ventricle to pulmonary artery connections in the neonatal period. *Semin. Thorac. Cardiovasc. Surg.* **7**: 130–132.

Mazariegos, G. V., Kramer, D. J., Lopez, R. C., Shakil, A. O., Rosenbloom, A. J., DeVera, M., Giraldo, M., Grogan, T. A., Zhu, Y., Fulmer, M. L., Amiot, B. P., and Patzer, J. F. (2001). Safety observations in phase I clinical evaluation of the Excorp Medical Bioartificial Liver Support System after the first four patients. *ASAIO J.* **47**: 471–475.

McCarey, B. E., McDonald, M. B., van Rij, G., Salmeron, B., Pettit, D. K., and Knight, P. M. (1989). Refractive results of hyperopic hydrogel intracorneal lenses in primate eyes. *Arch. Ophthalmol.* **107**: 724–730.

Meinhart, J., Deutsch, M., and Zilla, P. (1997). Eight years of clinical endothelial cell transplantation. Closing the gap between prosthetic grafts and vein grafts. *ASAIO J.* **43**: M515–M521.

Menasche, P. (2003). Skeletal muscle satellite cell transplantation. *Cardiovasc. Res.* **58**: 351–357.

Menasche, P., Hagege, A. A., Scorsin, M., Pouzet, B., Desnos, M., Duboc, D., Schwartz, K., Vilquin, J. T., and Marolleau, J. P. (2001). Myoblast transplantation for heart failure. *Lancet* **357**: 279–280.

Mitchell, S. L., and Niklason, L. E. (2003). Requirements for growing tissue-engineered vascular grafts. *Cardiovasc. Pathol.* **12**: 59–64.

Mow, V. C., Ratcliffe, A., and Poole, A. R. (1992). Cartilage and diarthrodial joints as paradigms for hierarchical materials and structures. *Biomaterials* **13**: 67–97.

Naito, Y., Imai, Y., Shin'oka, T., Kashiwagi, J., Aoki, M., Watanabe, M., Matsumura, G., Kosaka, Y., Konuma, T., Hibino, N., Murata, A., Miyake, T., and Kurosawa, H. (2003). Successful clinical application of tissue-engineered graft for extracardiac Fontan operation. *J. Thorac. Cardiovasc. Surg.* **125**: 419–420.

Naumann, A., Rotter, N., Bujia, J., and Aigner, J. (1998). Tissue engineering of autologous cartilage transplants for rhinology. *Am. J. Rhinol.* **12**: 59–63.

Nave, M. (1992). Wound bed preparation: approaches to replacement of dermis. *J. Burn Care Rehabil.* **13**: 147–153.

Niklason, L. E., Abbott, W., Gao, J., Klagges, B., Hirschi, K. K., Ulubayram, K., Conroy, N., Jones, R., Vasanawala, A., Sanzgiri, S., and Langer, R. (2001). Morphologic and mechanical characteristics of engineered bovine arteries. *J. Vasc. Surg.* **33**: 628–638.

Niklason, L. E., Gao, J., Abbott, W. M., Hirschi, K. K., Houser, S., Marini, R., and Langer, R. (1999). Functional arteries grown *in vitro*. *Science* **284**: 489–493.

Nikolovski, J., Gulari, E., and Humes, H.D. (1999). Design engineering of a bioartificial renal tubule cell therapy device. *Cell. Transplant.* **8**: 351–364.

Oberpenning, F., Meng, J., Yoo, J. J., and Atala, A. (1999). De novo reconstitution of a functional mammalian urinary bladder by tissue engineering. *Nat. Biotechnol.* **17**: 149–155.

Okano, T., and Matsuda, T. (1998). Muscular tissue engineering: capillary-incorporated hybrid muscular tissues *in vivo* tissue culture. *Cell Transplant.* **7**: 435–442.

Orchard, T. J., Chang, Y. F., Ferrell, R. E., Petro, N., and Ellis, D. E. (2002). Nephropathy in type 1 diabetes: a manifestation of insulin resistance and multiple genetic susceptibilities? Further evidence from the Pittsburgh Epidemiology of Diabetes Complication Study. *Kidney Int.* **62**: 963–970.

Orchard, T. J., Olson, J. C., Erbey, J. R., Williams, K., Forrest, K. Y., Smithline Kinder, L., Ellis, D., and Becker, D. J. (2003). Insulin resistance-related factors, but not glycemia, predict coronary artery disease in type 1 diabetes: 10-year follow-up data from the Pittsburgh Epidemiology of Diabetes Complications Study. *Diabetes Care* **26**: 1374–1379.

Orlic, D., Kajstura, J., Chimenti, S., Jakoniuk, I., Anderson, S. M., Li, B., Pickel, J., McKay, R., Nadal-Ginard, B., Bodine, D. M., Leri, A., and Anversa, P. (2001a). Bone marrow cells regenerate infarcted myocardium. *Nature* **410**: 701–705.

Orlic, D., Kajstura, J., Chimenti, S., Limana, F., Jakoniuk, I., Quaini, F., Nadal-Ginard, B., Bodine, D. M., Leri,

A., and Anversa, P. (2001b). Mobilized bone marrow cells repair the infarcted heart, improving function and survival. *Proc. Natl. Acad. Sci. USA* **98**: 10344–10349.

Ounpuu, S., Anand, S., and Yusuf, S. (2000). The impending global epidemic of cardiovascular diseases. *Eur. HeartJ.* **21**: 880–883.

Pasic, M., Muller-Glauser, W., von Segesser, L., Odermatt, B., Lachat, M., and Turina, M. (1996). Endothelial cell seeding improves patency of synthetic vascular grafts: manual versus automatized method. *Eur. J. Cardiothorac. Surg.* **10**: 372–379.

Patzer, J. F., 2nd (2001). Advances in bioartificial liver assist devices. *Ann. N. Y. Acad. Sci.* 944: 320–333.

Patzer, J. F., 2nd, Campbell, B., and Miller, R. (2002). Plasma versus whole blood perfusion in a bioartificial liver assist device. *ASAIO J.* **48**: 226–233.

Patzer, J. F., 2nd, Mazariegos, G. V., Lopez, R., Molmenti, E., Gerber, D., Riddervold, F., Khanna, A., Yin, W. Y., Chen, Y., Scott, V. L., Aggarwal, S., Kramer, D. J., Wagner, R. A., Zhu, Y., Fulmer, M. L., Block, G. D., and Amiot, B. P. (1999). Novel bioartificial liver support system: preclinical evaluation. *Ann. N.Y. Acad. Sci.* **875**: 340–352.

Perez, A., Grikscheit, T. C., Blumberg, R. S., Ashley, S. W., Vacanti, J. P., and Whang, E. E. (2002). Tissue-engineered small intestine: ontogeny of the immune system. *Transplantation* **74**: 619–623.

Peters, M. C., Polverini, P. J., and Mooney, D. J. (2002). Engineering vascular networks in porous polymer matrices. *J. Biomed. Mater. Res.* **60**: 668–678.

Petite, H., Viateau, V., Bensaid, W., Meunier, A., de Pollak, C., Bourguignon, M., Oudina, K., Sedel, L., and Guillemin, G. (2000). Tissue-engineered bone regeneration. *Nat. Biotechnol.* **18**: 959–963.

Portes, P., Grandidier, M. H., Cohen, C., and Roguet, R. (2002). Refinement of the Episkin protocol for the assessment of acute skin irritation of chemicals: follow-up to the ECVAM prevalidation study. *Toxicol. Vitro* **16**: 765–770.

Puelacher, W. C., Vacanti, J. P., Ferraro, N. F., Schloo, B., and Vacanti, C. A. (1996). Femoral shaft reconstruction using tissue-engineered growth of bone. *Int. J. Oral Maxillofac. Surg.* **25**: 223–228.

Quarto, R., Mastrogiacomo, M., Cancedda, R., Kutepov, S. M., Mukhachev, V., Lavroukov, A., Kon, E., and Marcacci, M. (2001). Repair of large bone defects with the use of autologous bone marrow stromal cells. *N. Engl. J. Med.* **344**: 385–386.

Rabkin, E., and Schoen, F. J. (2002). Cardiovascular tissue engineering. *Cardiovasc. Pathol.* **11**: 305–317.

Rabkin, E., Hoerstrup, S. P., Aikawa, M., Mayer, J. E., Jr., and Schoen, F. J. (2002). Evolution of cell phenotype and extracellular matrix in tissue-engineered heart valves during *in-vitro* maturation and *in-vivo* remodeling. *J. Heart Valve Dis.* **11**: 308–314; discussion 314.

Radisic, M., Euloth, M., Yang, L., Langer, R., Freed, L. E., and Vunjak-Novakovic, G. (2003). High-density seeding of myocyte cells for cardiac tissue engineering. *Biotechnol. Bioeng.* **82**: 403–414.

Raper, S. E., and Wilson, J. M. (1993). Cell transplantation in liver-directed gene therapy. *Cell. Transplant.* **2**: 381–400; discussion 407–410.

Read, T. A., Sorensen, D. R., Mahesparan, R., Enger, P.O., Timpl, R., Olsen, B. R., Hjelstuen, M. H., Haraldseth, O., and Bjerkvig, R. (2001). Local endostatin treatment of gliomas administered by microencapsulated producer cells. *Nat. Biotechnol.* **19**: 29–34.

Reinecke, H., Poppa, V., and Murry, C. E. (2002). Skeletal muscle stem cells do not transdifferentiate into cardiomyocytes after cardiac grafting. *J. Mol. Cell Cardiol.* **34**: 241–249.

Robbins, P. B., Lin, Q., Goodnough, J. B., Tian, H., Chen, X., and Khavari, P. A. (2001). *In vivo* restoration of laminin 5 beta 3 expression and function in junctional epidermolysis bullosa. *Proc. Natl. Acad. Sci. USA* **98**: 5193–5198.

Rosenberg, N., Martinez, A., Sawyer, P. N., Wesolowski, S. A., Postlethwait, R. W., and Dillon, M. L., Jr. (1996). Tanned collagen arterial prosthesis of bovine carotid origin in man. Preliminary studies of enzyme-treated heterografts. *Ann. Surg.* **164**: 247–256.

Rothenburger, M., Vischer, P., Volker, W., Glasmacher, B., Berendes, E., Scheld, H. H., and Deiwick, M. (2001). *In vitro* modelling of tissue using isolated vascular cells on a synthetic collagen matrix as a substitute for heart valves. *Thorac. Cardiovasc. Surg.* **49**: 204–209.

Rothenburger, M., Volker, W., Vischer, J. P., Berendes, E., Glasmacher, B., Scheld, H. H., and Deiwick, M. (2002). Tissue engineering of heart valves: formation of a three-dimensional tissue using porcine heart valve cells. *ASAIO J.* **48**: 586–591.

Rozga, J., Podesta, L., LePage, E., Morsiani, E., Moscioni, A.D., Hoffman, A., Sher, L., Villamil, F., Woolf, G., McGrath, M., Kong, L., Rosen, H., Lanman, T., Vierling, J., Makowka, L., and Demetriou, A.A. (1994). A bioartificial liver to treat severe acute liver failure. *Ann. Surg.* **219**: 538–544; discussion 544–546.

Rozga, J., Williams, F., Ro, M. S., Neuzil, D. F., Giorgio, T. D., Backfisch, G., Moscioni, A.D., Hakim, R., and Demetriou, A. A. (1993). Development of a bioartificial liver: properties and function of a hollow-fiber module inoculated with liver cells. *Hepatology* **17**: 258–265.

Sagen, J. (1992). Chromaffin cell transplants for alleviation of chronic pain. *ASAIO J.* **38**: 24–28.

Samouillan, V., Dandurand-Lods, J., Lamure, A., Maurel, E., Lacabanne, C., Gerosa, G., Venturini, A., Casarotto, D., Gherardini, L., and Spina, M. (1999). Thermal analysis characterization of aortic tissues for cardiac valve bioprostheses. *J. Biomed. Mater. Res.* **46**: 531–538.

Sautter, J., Tseng, J. L., Braguglia, D., Aebischer, P., Spenger, C., Seiler, R. W., Widmer, H.R., and Zurn, A. D. (1998). Implants of polymer-encapsulated genetically modified cells releasing glial cell line-derived neurotrophic factor improve survival, growth, and function of fetal dopaminergic grafts. *Exp. Neurol.* **149**: 230–236.

Schaefer, D., Martin, I., Jundt, G., Seidel, J., Heberer, M., Grodzinsky, A., Bergin, I., Vunjak-Novakovic, G., and Freed, L. E. (2002). Tissue-engineered composites for the repair of large osteochondral defects. *Arthritis. Rheum.* **46**: 2524–2534.

Schneider, A.I., Maier-Reif, K., and Graeve, T. (1999). Constructing an *in vitro* cornea from cultures of the three specific corneal cell types. *In Vitro Cell Dev. Biol. Anim.* **35**: 515–526.

Schnell, A. M., Hoerstrup, S. P., Zund, G., Kolb, S., Sodian, R., Visjager, J.F., Grunenfelder, J., Suter, A., and Turina, M. (2001). Optimal cell source for cardiovascular tissue engineering: venous vs. aortic human myofibroblasts. *Thorac. Cardiovasc. Surg.* **49**: 221–225.

Sheridan, R. L., Tompkins, R. G., and Burke, J. F. (1994). Management of burn wounds with prompt excision and immediate closure, [see comments]. *J. Intensive Care Med.* **9**: 6–17.

Shimizu, T., Yamato, M., Isoi, Y., Akutsu, T., Setomaru, T., Abe, K., Kikuchi, A., Umezu, M., and Okano, T. (2002). Fabrication of pulsatile cardiac tissue grafts using a novel 3-dimensional cell sheet manipulation technique and temperature-responsive cell culture surfaces. *Circ. Res.* 90: e40.

Shinoka, T., Imai, Y., and Ikada, Y. (2001). Transplantation of a tissue-engineered pulmonary artery. N. *Engl. J. Med.* **344**: 532–533.

Shinoka, T., Ma, P. X., Shum-Tim, D., Breuer, C. K., Cusick, R. A., Zund, G., Langer, R., Vacanti, J. P., and Mayer, J. E., Jr. (1996). Tissue-engineered heart valves. Autologous valve leaflet replacement study in a lamb model. *Circulation* **94**: II164–II168.

Simon, P., Kasimir, M. T., Seebacher, G., Weigel, G., Ullrich, R., Salzer-Muhar, U., Rieder, E., and Wolner, E. (2003). Early failure of the tissue engineered porcine heart valve SYNERGRAFT in pediatric patients. *Eur. J. Cardiothorac. Surg.* **23**: 1002–1006.

Sodian, R., Hoerstrup, S. P., Sperling, J. S., Daebritz, S., Martin, D. P., Moran, A. M., Kim, B. S., Schoen, F. J., Vacanti, J. P., and Mayer, J. E., Jr. (2000). Early *in vivo* experience with tissue-engineered trileaflet heart valves. *Circulation* **102**: III22–III29.

Spirito, F., Meneguzzi, G., Danos, O., and Mezzina, M. (2001). Cutaneous gene transfer and therapy: the present and the future. *J. Gene Med.* **3**: 21–31.

Starzl, T. E. (2001). The birth of clinical organ transplantation. *J. Am. Coll. Surg.* **192**: 431–446.

Starzl, T. E., Demetris, A. J., and Van Thiel, D. (1989). Liver transplantation (1). *N. Engl. J. Med.* **321**: 1014–1022.

Stone, K. R., Rodkey, W. G., Webber, R. J., McKinney, L., and Steadman, J. R. (1990). Future directions. Collagen-based pros-theses for meniscal regeneration. *Clin. Orthop.* 129–135.

Stratta, R. J., Taylor, R. J., Bynon, J. S., Lowell, J. A., Sindhi, R., Wahl, T. O., Knight, T. F., Weide, L. G., and Duckworth, W. C. (1994). Surgical treatment of diabetes mellitus with pancreas transplantation. *Ann. Surg.* **220**: 809–817.

Sullivan, S. J., Maki, T., Borland, K. M., Mahoney, M. D., Solomon, B. A., Muller, T. E., Monaco, A. P., and Chick, W. L. (1991). Biohybrid artificial pancreas: long-term implantation studies in diabetic, pancreatectomized dogs. *Science* **252**: 718–721.

Sussman, N. L., Gislason, G. T., and Kelly, J. H. (1994). Extra-corporeal liver support. Application to fulminant hepatic failure. *J. Clin. Gastroenterol.* **18**: 320–324.

Tan, S. A., Deglon, N., Zurn, A. D., Baetge, E. E., Bamber, B., Kato, A. C., and Aebischer, P. (1996). Rescue of motoneurons from axotomy-induced cell death by polymer encapsulated cells genetically engineered to release CNTF. *Cell Transplant.* **5**: 577–587.

Taylor, D. A., Atkins, B. Z., Hungspreugs, P., Jones, T. R., Reedy, M. C., Hutcheson, K. A., Glower, D. D., and Kraus, W. E. (1998). Regenerating functional myocardium: improved performance after skeletal myoblast transplantation. *Nat. Med.* **4**: 929–933.

Taylor, P. M., Allen, S. P., Dreger, S. A., and Yacoub, M. H. (2002). Human cardiac valve interstitial cells in collagen sponge: a biological three-dimensional matrix for tissue engineering. *J. Heart Valve Dis.* **11**: 298–306.

Teebken, O. E., Bader, A., Steinhoff, G., and Haverich, A. (2000). Tissue engineering of vascular grafts: human cell seeding of decellularised porcine matrix. *Eur. J. Vasc. Endovasc. Surg.* **19**: 381–386.

Thomson, J. A., Itskovitz-Eldor, J., Shapiro, S. S., Waknitz, M. A., Swiergiel, J. J., Marshall, V. S., and Jones, J. M. (1998). Embryonic stem cell lines derived from human blastocysts. *Science* **282**: 1145–1147.

Thorsen, F., Read, T. A., Lund-Johansen, M., Tysnes, B. B., and Bjerkvig, R. (2000). Alginate-encapsulated producer cells: a potential new approach for the treatment of malignant brain tumors. *Cell Transplant.* **9**: 773–783.

Toriumi, D. M., Kotler, H. S., Luxenberg, D. P., Holtrop, M. E., and Wang, E. A. (1991). Mandibular reconstruction with a recombinant bone-inducing factor. Functional, histologic, and biomechanical evaluation. *Arch. Otolaryngol. Head Neck Surg.* **117**: 1101–1112.

Uludag, H., Ip, T. K., and Aebischer, P. (1990). Transport functions in a bioartificial kidney under uremic conditions. *Int. J. Artif. Organs* **13**: 93–97.

Uyama, S., Kaufmann, P. M., Takeda, T., and Vacanti, J. P. (1993). Delivery of whole liver-equivalent hepatocyte mass using polymer devices and hepatotrophic stimulation. *Transplantation* **55**: 932–935.

Vacanti, C .A., Langer, R., Schloo, B., and Vacanti, J. P. (1991). Synthetic polymers seeded with chondrocytes provide a template for new cartilage formation. *Plast. Reconstr. Surg.* **88**: 753–759.

Vacanti, J. P., Morse, M. A., Saltzman, W. M., Domb, A. J., Perez-Atayde, A., and Langer, R. (1988). Selective cell transplantation using bioabsorbable artificial polymers as matrices. *J. Pediatr. Surg.* **23**: 3–9.

Valentini, R. F., Vargo, T. G., Gardella, J. A., Jr., and Aebischer, P. (1992). Electrically charged polymeric substrates enhance nerve fibre outgrowth *in vitro*. *Biomaterials* **13**: 183–190.

Vunjak-Novakovic, G., Martin, I., Obradovic, B., Treppo, S., Grodzinsky, A. J., Langer, R., and Freed, L. E. (1999). Bioreactor cultivation conditions modulate the composition and mechanical properties of tissue-engineered cartilage. *J. Orthop. Res.* **17**: 130–138.

Wakitani, S., Kimura, T., Hirooka, A., Ochi, T., Yoneda, M., Yasui, N., Owaki, H., and Ono, K. (1989). Repair of rabbit articular surfaces with allograft chondrocytes embedded in collagen gel. *J. Bone Joint Surg. Br.* **71**: 74–80.

Watanabe, F. D., Mullon, C. J., Hewitt, W. R., Arkadopoulos, N., Kahaku, E., Eguchi, S., Khalili, T., Arnaout, W., Shackleton, C. R., Rozga, J., Solomon, B., and Demetriou, A. A. (1997). Clinical experience with a bioartificial liver in the treatment of severe liver failure. A phase I clinical trial. *Ann. Surg.* **225**: 484–491; discussion 491–494.

Wiegand, F., Kroncke, K. D., and Kolb-Bachofen, V. (1993). Macrophage-generated nitric oxide as cytotoxic factor in destruction of alginate-encapsulated islets. Protection by arginine analogs and/or coencapsulated erythrocytes. *Transplantation* **56**: 1206–1212.

Wilson, G. J., Courtman, D. W., Klement, P., Lee, J. M., and Yeger, H. (1995). Acellular matrix: a biomaterials approach for coronary artery bypass and heart valve replacement. *Ann. Thorac. Surg.* **60**: S353–S358.

Winslow, R. M. (2000). Blood substitutes. Adv. Drug Deliv. *Rev.* **40**: 131–142.

Woerly, S., Plant, G. W., and Harvey, A. R. (1996). Neural tissue engineering: from polymer to biohybrid organs. *Biomaterials* **17**: 301–310.

Yarmush, M. L., Dunn, J. C., and Tompkins, R. G. (1992a). Assessment of artificial liver support technology. *Cell Transplant.* **1**: 323–341.

Yarmush, M. L., Toner, M., Dunn, J. C., Rotem, A., Hubel, A., and Tompkins, R. G. (1992b). Hepatic tissue engineering. Development of critical technologies. *Ann. N. Y. Acad. Sci.* **665**: 238–252.

Yasko, A. W., Lane, J. M., Fellinger, E. J., Rosen, V., Wozney, J. M., and Wang, E. A. (1992). The healing of segmental bone defects, induced by recombinant human bone morphogenetic protein (rhBMP-2). A radiographic, histological, and biomechanical study in rats. *J. Bone Joint Surg. Am.* **74**: 659–670.

Ye, Q., Zund, G., Jockenhoevel, S., Schoeberlein, A., Hoerstrup, S. P., Grunenfelder, J., Benedikt, P., and Turina, M. (2000). Scaffold pre-coating with human autologous extracellular matrix for improved cell attachment in cardiovascular tissue engineering. *ASAIO. J.* **46**: 730–733.

Zieske, J. D., Mason, V. S., Wasson, M. E., Meunier, S. F., Nolte, C. J., Fukai, N., Olsen, B. R., and Parenteau, N. L. (1994). Basement membrane assembly and differentiation of cultured corneal cells: importance of culture environment and endothelial cell interaction. *Exp. Cell Res.* **214**: 621–633.

Zilla, P., Deutsch, M., and Meinhart, J. (1999). Endothelial cell transplantation. *Semin. Vasc. Surg.* **12**: 52–63.

Zimmermann, W. H., Fink, C., Kralisch, D., Remmers, U., Weil, J., and Eschenhagen, T. (2000). Three-dimensional engineered heart tissue from neonatal rat cardiac myocytes. *Biotechnol. Bioeng.* **68**: 106–114.

Zimmermann, W. H., Schneiderbanger, K., Schubert, P., Didie, M., Munzel, F., Heubach, J. F., Kostin, S., Neuhuber, W. L., and Eschenhagen, T. (2002). Tissue engineering of a differentiated cardiac muscle construct. *Circ. Res.* **90**: 223–230.

Zund, G., Hoerstrup, S. P., Schoeberlein, A., Lachat, M., Uhlschmid, G., Vogt, P. R., and Turina, M. (1998). Tissue engineering: a new approach in cardiovascular surgery: seeding of human fibroblasts followed by human endothelial cells on resorbable mesh. *Eur. J. Cardiothorac. Surg.* **13**: 160–164.

Zurn, A. D., Henry, H., Schluep, M., Aubert, V., Winkel, L., Eilers, B., Bachmann, C., and Aebischer, P. (2000). Evaluation of an intrathe-cal immune response in amyotrophic lateral sclerosis patients implanted with encapsulated genetically engineered xenogeneic cells. *Cell Transplant.* **9**: 471–484.

7.1.3 Immunoisolation

Michael J. Lysaght and David Rein

Introduction

In the context of tissue engineering and cellular medicine, the terms *immunoisolation* and *encapsulation* usually refer to devices and therapies in which living cells are separated from a host by a selective membrane barrier. This barrier permits bidirectional trafficking of small molecules between host and grafted cells, and protects foreign cells from effector agents of a host's immune system. In analogy with pharmacological immunosuppression, the degree of protection afforded by immunoisolatory barriers depends upon the circumstances of application and may be total or partial, long-term or short-term. Occasional reference to the concept, which is illustrated in Fig. 7.1.3-1, can be found as early as the late 1930s and appears sporadically in the literature of the 1950s and 1960s. The approach first received serious investigational attention in the mid-1970s. Interest has expanded considerably in the past two decades. Encapsulation currently encompasses a daunting array of therapy formats, device configurations, and biomaterials.

The first modern efforts involving cell encapsulation were directed at development of a long-term implant to replace the endocrine function of a diabetic pancreas. Other investigators quickly expanded this field of study to include short-term extracorporeal replacement of the failing liver. Later applications include the use of encapsulated cells for *in situ* synthesis and local delivery of naturally occurring and recombinant cell products for the treatment of chronic pain, Parkinson's disease, macular degeneration, and similar disorders. Devices employed for encapsulated cell therapy vary in size over several orders of magnitude from small spheres, with a volume of 10^{-5} cm^3, to large extracorporeal devices with a net volume of ~ 10 cm^3. Their anticipated service life ranges from a few hours in the case of the liver to several years for other therapeutic implants. In some cases, the immunoisolatory vehicles simply serve as constitutive sources of bioactive molecules; in other cases, regulated release is required; and for still others, host detoxification is the goal. Despite such a spectrum of application parameters, devices containing immunoisolated cells share many common features and design principles: (1) Cells are rarely deployed more than 500 μm (5×10^{-2} cm) from the host; cells much farther than this critical distance either undergo necrosis or cease to synthesize and release protein. (2) Cells generally are supported on a matrix or scaffold to provide some of the functions of normal extracellular matrix and to prevent the formation of large, unvascularized cellular aggregates. (3) Separative membranes are invariably self-supporting, thus requiring a design trade-off between transport characteristics and mechanical strength. (4) Both the membrane and the matrices usually are prepared from either hydrogels or reticulated foams, themselves chosen from relatively few among the many available candidates.

In the remainder of this overview, we will describe the immunological challenge of protecting cells with barrier materials; summarize critical components of

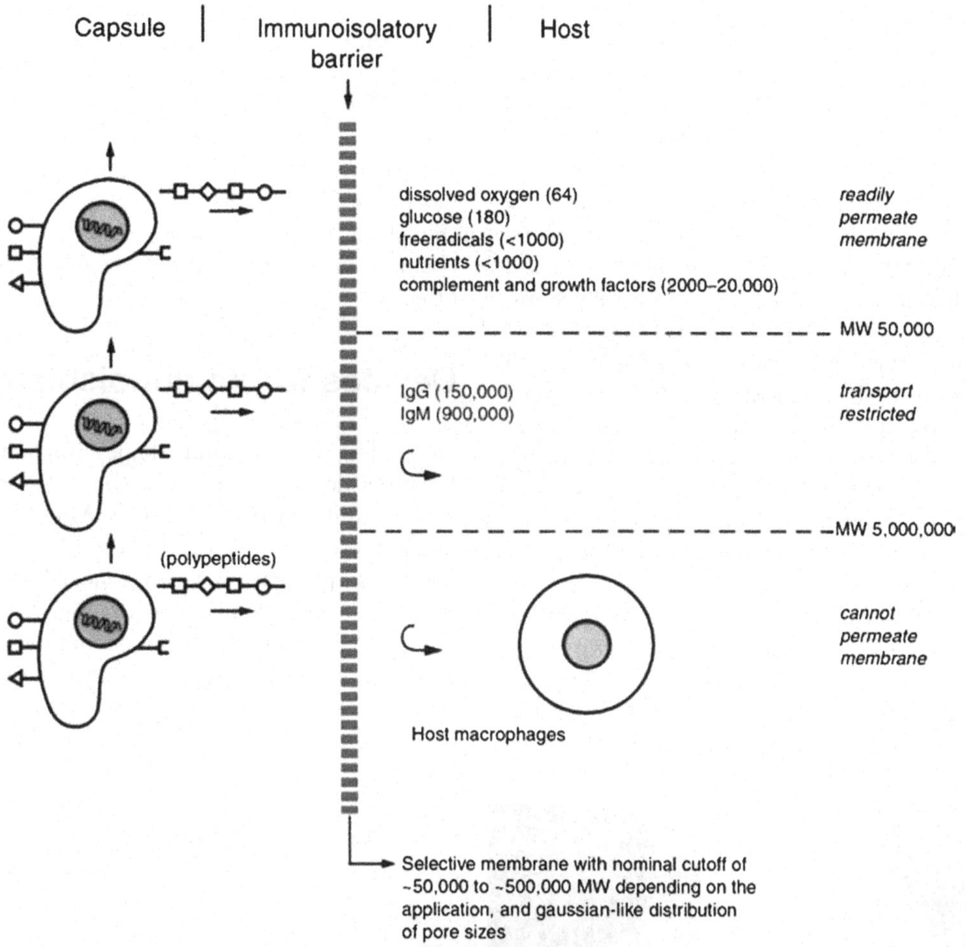

Fig. 7.1.3-1 The concept of encapsulation. A biocompatible selective membrane barrier surrounds naturally occurring or genetically modified cells. Nutrients, oxygen, and small bioactive materials freely transit the membrane but immunologically active species are too large to penetrate. Although such perfect selectivity is clearly an idealization, techniques have been successfully developed to the point of large-scale clinical evaluation.

immunoisolate devices, i.e., cells, membranes, and matrices; review the more common device configurations; and conclude with a short survey of the development status of principal applications.

The challenge of immunoisolation

The immune system is a network that deploys a complex, redundant phalanx of pathways to distinguish self from non-self and to destroy non-self. Membrane barriers have proven remarkably effective in preventing immune destruction of *allogenic* cells, i.e., cells originating from the same species as the host. Protection of allogenic cells is possible because they normally will not be subjected to immune destruction in the absence of cell–cell contact between graft and host. Even protein-permeable membranes have been found to allow long-term function and survival of allogenic cells. This happy circumstance is marred by the reality that the supply of transplantable human cells is very limited, just as is the supply of human solid organs, and thus therapies based upon transplanted human cells are not likely to have much therapeutic impact. There has been some effort to create dividing cell lines from protein secreting human cells, or to genetically engineer naturally dividing human cells (e.g., fibroblasts) to produce useful proteins. In the future, stem cells may provide unconstrained supply of tissue. In the main, however, investigators have responded to the scarcity of human cells by turning to cells of animal origins. Such so-called *xenogenic* cells are far more difficult to encapsulate and success is constrained to special cases. [Examples are devices containing xenogenic cells implanted in certain immunoprivileged sites (spinal fluid, ventricles, eyes) where the avidity of the immune response is muted or placed in contact with acellular fluid (or flowing blood) to minimize the localized inflammatory reaction.] Or, in the case of the liver, xenogenic cells are utilized for periods of time that are much shorter than that required for the development of fulminant immune responses. All such strategies have proven successful, and there is abundant evidence of survival of xenogenic cells, in these limited circumstances, for 3 to 6 months or longer. In contrast, encapsulated xenogenic cells rarely survive much beyond 14 to 28 days when implanted subcutaneously or intraperitoneally in immunocompetent hosts.

Two mechanisms are invoked to explain the inability of membranes to universally protect xenogenic implants, and the different fates of encapsulated allogeneic and xenogeneic cells. First, membranes are not ideally semipermeable and thus allow passage of small quantities of large immunomolecules, including complement and both elicited and preformed immunoglobulins. Such agents are far more active against xenogenic cells than against allogenic grafts. A second problem is that soluble antigens "leak" from cell surfaces or are released upon cell necrosis. These protein constituents are not conserved between species and are, in varying degrees, immunogenic. Their gradual release results in a localized inflammatory response in the neighborhood of the membrane, readily visualized by histology. Inflammatory cells express a number of low-molecular-weight toxins (including free radicals and cytotoxic cytokines) that pass through the membrane and attack the cells inside. Note that this mechanism requires a local nidus of inflammatory tissue and is thus unlikely to be encountered when implants are placed in cell-free fluid cavities. It is also not significant with allogenic cells whose only non-self proteins are confined to the MHC system.

Interestingly, pharmacological immunosuppression has rarely been used in conjunction with encapsulation, though on theoretical grounds the combination of mechanical and chemical agents would likely prove highly effective.

Devices for immunoisolation

Depending on size and shape, implantable immunoisolation device designs can be categorized as either microcapsules or macrocapsules. Microcapsule beads are illustrated in Fig. 7.1.3-2 (upper right), along with other materials popular for immunoisolation. A current macrocapsule design is shown in Fig. 7.1.3-3. These different designs all share the common components of a permselective membrane, an internal matrix, and the living encapsulated tissue. Macrocapsules are small (100–600 μm, 0.01–0.06 × 10^{-4} cm in diameter) spherical

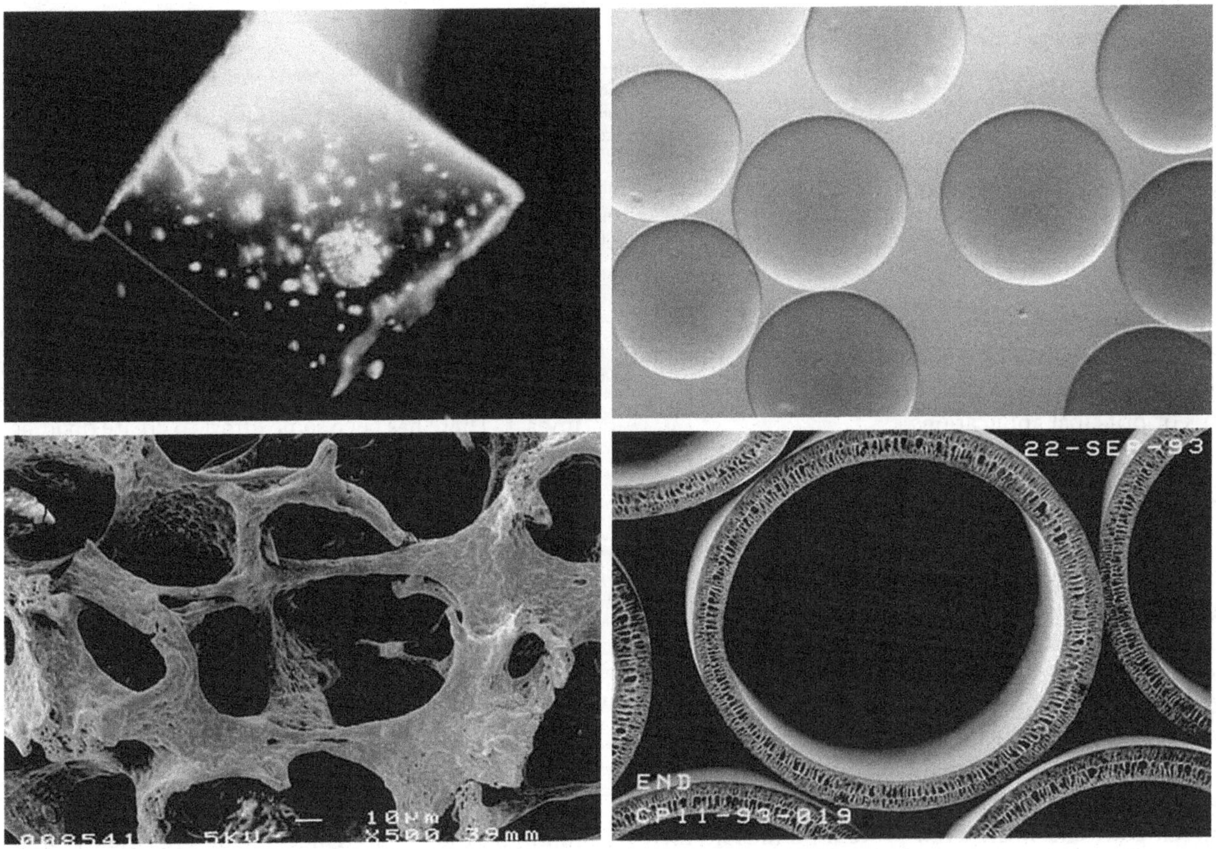

Fig. 7.1.3-2 Materials used as matrices or barrier materials. Micrographs or photomacrographs of hydrophilic materials in the form of matrices (top left) and microcapsules (top right) and of hydrophobic materials in the form of foams (bottom left) and membranes for use in macrocapsules (bottom right).

Fig. 7.1.3-3 Photograph of an implantable macrocapsule. The pencil and tweezers are included for scale.

beads containing up to several thousand cells. Typically, hundreds to thousands of microcapsules are implanted into the host to achieve a therapeutic dose. This design minimizes transport resistance, allows for easy implantation, and provides good dose control. However, microcapsules are difficult to explant and are usually quite fragile. Macrocapsules are much larger in size with the capacity to hold millions of cells, generally requiring a single device for a given therapy. These devices are implanted as tubular or flat sheet diffusion chambers with an inner diameter dimension of 0.5–2.0 mm and a length of 1–10 cm. Macrocapsules provide mechanical and chemical stability superior to those of microcapsules and are easily retrieved. A significant concern with this design is the geometric resistance to mass transport, which limits viability of encapsulated tissue. An alternative macrocapsule design involves connecting the device directly to the patient's circulatory system. The cells are contained in a chamber surrounding the macrocapsule, and the flowing blood can provide an efficient means of nutrient transport. A major challenge with this vascular design is maintaining shunt patency of the device.

Membranes

A wide variety of different materials have been used to formulate the permselective membranes for microcapsules and macrocapsules. In general, the membranes for macrocapsules have been engineered from synthetic thermoplastics, whereas those for microcapsules have been engineered using hydrogel-based materials. Table 7.1.3-1 and Fig. 7.1.3-2 illustrate the materials and appearance of hydrophilic and hydrophobic membrane materials used for immunoisolation.

The process to manufacture microcapsules typically starts with the creation of a slurry of the living cells in a dilute hydrogel solution. Next, small droplets are formed by extruding this mixture through an appropriate nozzle, followed by the cross-linking of the hydrogel to form the mechanically stable microcapsules with an immunoisolatory layer. In an alternative process, water-insoluble synthetic polymers are used in place of hydrogels to prepare the cell slurry. These microcapsules and the immunoisolatory layer are then formed upon interfacial precipitation of the polymer solution. The process to manufacture macrocapsules typically involves phase inversion of a thermoplastic polymer solution cast as a flat sheet or extruded as a hollow fiber. During phase inversion, the polymer solution is placed in controlled contact with miscible nonsolvent, resulting in the formation of the mechanically stable and immunoisolatory membrane. At a later stage, the living cells are aseptically introduced into the fiber or chamber, which is subsequently sealed.

The processes developed to manufacture macrocapsules and microcapsules are very versatile and allow for the formation of membranes with a wide variety of different transmembrane pore structures and outer surface microgeometries. Membrane selection has a strong influence on microcapsule or macrocapsule device performance and is characterized in terms of membrane chemistry, transport properties, outer surface morphology, and strength. Optimum parameters are dictated by the metabolic requirements of the encapsulated cells, the size of the therapeutic agent to be delivered, the required immunoprotection, and the desired biocompatibility. Membrane transport properties are chosen to maintain viability and functionality of the encapsulated cells and provide release of the therapeutic agent. This selection involves designing membranes that provide sufficient nutrient flux to meet the requirements of the encapsulated cells, while preventing flux of immunological species that would reject the tissue.

Biocompatibility is defined by the host reaction to the implant and has a significant impact upon device performance. Biocompatibility depends upon the nature of the encapsulated cell and both the transport properties and outer morphology of the membrane barrier.

Transport properties are routinely evaluated in combination with a physical characterization of the membrane to develop structure–property relations. Physical parameters such as inner diameter, wall thickness, wall morphology, and surface morphology can influence the transport behavior. Light micrometry is used to characterize membrane geometry, and scanning electron microscopy is used to analyze membrane morphology. The high-resolution techniques of atomic force microscopy and low-voltage scanning electron microscopy have been exploited to image the porosity and pore size of the permselective skin of ultrafiltration membranes. A wide range of membrane wall morphologies can be produced using the phase

Table 7.1.3-1 Materials commonly used in encapsulation

Hydrophilic	Hydrophobic
Membranes	
Alginate 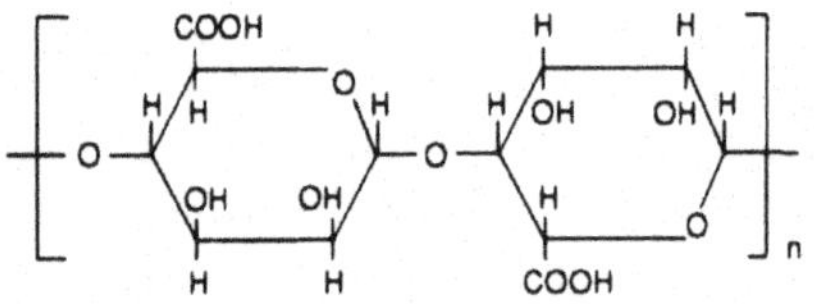	**Polysulfone** 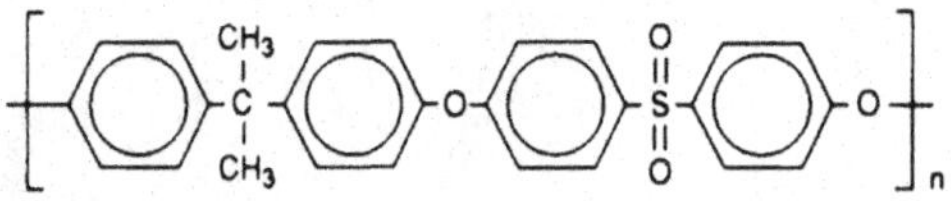
Polylysine 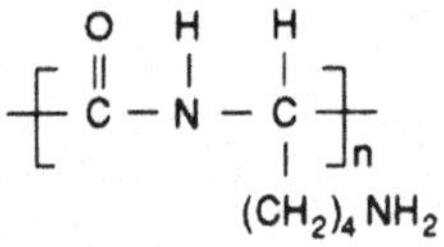	**PAN-PVC** 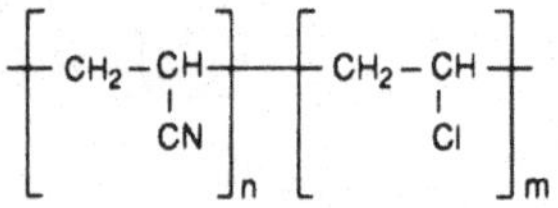
HEMA-MMA 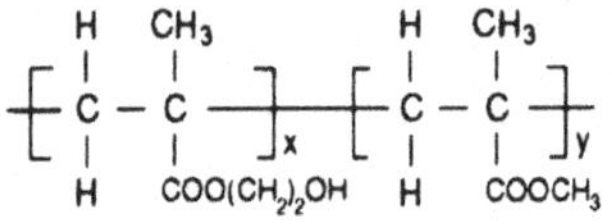	
Matrices	
Chitosan 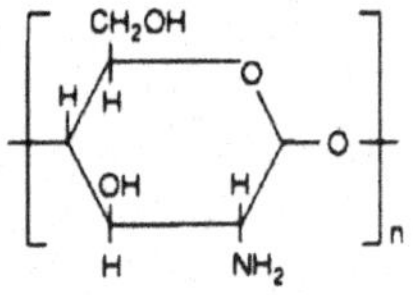	**Dacron-polyethylene terephthalate** 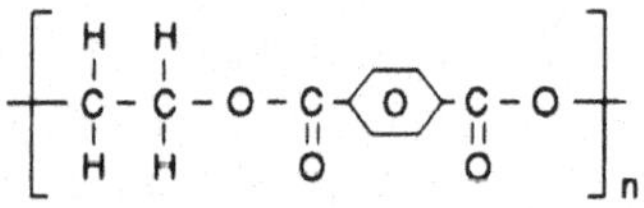
Collagen 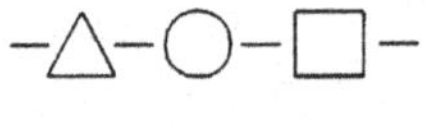 amino acid sequence	**Polyvinyl Alcohol** 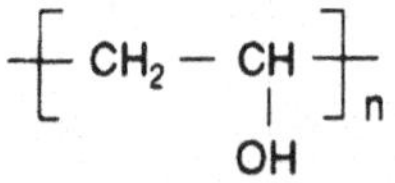
Alginate 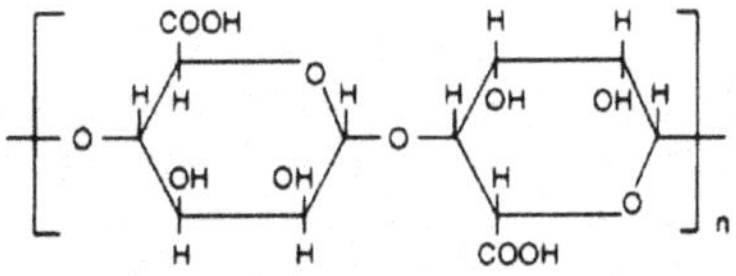	

inversion process: most common are foamlike or trabecular structures. Outer surface morphology is generally characterized as rough (microgeometries >2 μm) or smooth. Implanted into a host tissue site, rough surface will frequently evoke a significant host fibrotic reaction, whereas smooth surfaces will evoke a relatively mild reaction. In some cases, a vascularized host reaction can actually improve encapsulated device viability, by providing nutrients and oxygen to the perimembrane region.

Matrices

The second component of an immunoisolation device is the internal matrix. Hydrogels and solid scaffolds have been widely used and can be produced from synthetic or naturally derived materials (Table 7.1.3-1). Examples of hydrogel matrices include alginate, agarose, and poly(ethylene oxide), and examples of scaffold matrices include poly(ethylene terephthalate) yarn, poly(vinyl alcohol) foam, and cross-linked chitosan. This matrix serves two basic functions. The first is to provide mechanical support for the encapsulated cells in order to maintain a uniform distribution within the device. In the absence of this support, the cells often gravitate toward one region of the device and form a large necrotic cluster. The matrix also serves a biological function by stimulating the cells to secrete their own extracellular matrix, regulating cell proliferation, regulating secretory function, and maintaining the cells in a differentiated phenotype.

Selection of a matrix for a particular cell type involves several design considerations. Generally, suspension cell cultures prefer a hydrogel-based matrix, whereas anchorage-dependent cells prefer the attachment surfaces of a solid scaffold. The matrix must also be physically and chemically compatible with the permselective membrane. For example, scaffold matrices should not damage the integrity of the permselective membrane and soluble matrix components should not significantly affect the pore size. The stability of the matrix should also be considered and in general must at least match the lifetime of the device. Finally, the transport characteristics of the matrix candidates need to be considered. Certain matrices may exhibit significant resistance to the transport of small or large solutes, and thus affect overall performance.

Cells

The final component of the immunoisolation device is the encapsulated cells used to secrete the therapeutic molecules. These cells may be derived from "primary" cells, (i.e., postmitotic cells dividing very slowly if at all), continuously dividing cell lines, or genetically engineered tissue. All three cell types have been successfully encapsulated. Cell sourcing for a device begins with a definition of the desired secretory function of the implant. For example, chromaffin cells are a known source of the opoid peptide norepinephrine and have been used as a cellular delivery system to treat chronic pain. Such chromaffin cells are obtained as primary cultures from an enzymatic isolation of the bovine adrenal gland. Islets of Langerhans for the delivery of insulin to replace pancreatic function represent another widely investigated primary cell type. The PC12 rat pheochromocytoma line is an example of an immortalized cell line derived from a tumor that has been used for the delivery of L-dopa and dopamine in the treatment of Parkinson's disease. Cells engineered to secrete a variety of neurotrophic factors have been used in an encapsulated environment for the treatment of neurodegenerative diseases and include the Chinese hamster ovary (CHO) line, the Hs27 human foreskin fibroblast, and the BHK line.

Different cell types have different requirements for survival and function in a device and may result in a variety of levels of performance in any given implant site. These cell-specific considerations include metabolic requirements, proliferation rate within a device, and antigenicity and are assessed to ensure long-term device performance. For example, a highly antigenic encapsulated cell may be rapidly rejected in a nonimmunoprivileged site, such as the peritoneal cavity. This same encapsulated tissue may result in very satisfactory performance in an immunoprivileged site, such as the central nervous system. Similarly, a cell with a high nutrient requirement may provide superior performance in a nutrient-rich site, such as a subcutaneous pouch, and fail in a nutrient-poor site such as the cerebral spinal fluid.

Safety is another consideration in sourcing cells for eventual human implants. Grafts must be derived from healthy donors or from stable, contaminant-free cell lines. Before approving human clinical trials, regulatory bodies require testing for known transmittable diseases, mycoplasma, reverse transcriptase, cultivable viruses, and microbial contaminants.

Applications

At this writing (mid-1999) several applications of immunoisolated cell therapy are in clinical trials but none have reached the stage of approval by regulatory agencies and routine clinical utilization. Table 7.1.3-2 summarizes the application status of the bioartificial liver, the

Table 7.1.3-2 Application status of immunoisolation (late 1999)

Bioartificial liver	Several reports of clinical investigations in literature for bridge to transplant two successful phase I[a] trials. Two "pivotal[b] trials" underway.
Bioartificial pancreas	One case report of a single patient receiving a therapeutic dose of islets (and immunosuppression). Several reports of "survival studies" at smaller doses. Preclinical trials report outstanding success in rodents but not in dogs and nonhuman primates.
Delivery of cell and gene therapy	Pain: Successful phase I study completed; pivotal trial failed to show efficacy. ALS: Human clinical trials reported; Huntington trial is in progress. Numerous studies in primates, other large animals, and rodents.

[a] Phase I. Small trial to determine safety in ~10 patients.
[b] Pivotal. Large trial to determine efficacy. Includes control arm.

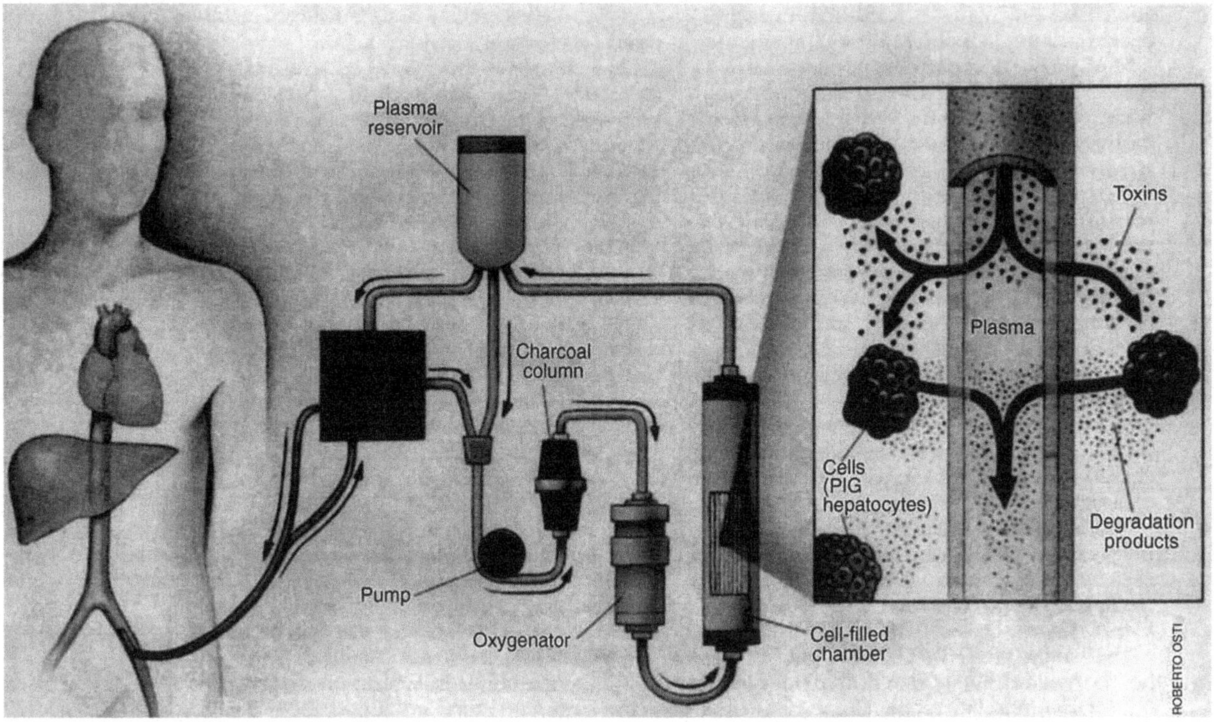

Fig. 7.1.3-4 The bioartificial liver. Several closely related versions of this intermittent extracorporeal therapy format have undergone preliminary clinical trials.

bioartificial pancreas, and the delivery of cell and gene therapy. As in all areas of tissue engineering, technology is moving rapidly and Table 7.1.3-2 should be appreciated in historical rather than current context.

The bioartificial liver currently is being evaluated as a "bridge to transplant," i.e., to extend the lifetime of patients who are medically eligible for liver transplantation until a donor organ becomes available. Several designs and treatment protocols have been proposed; one appealing format is shown in Fig. 7.1.3-4. The extracorporeal circuit is broadly similar to that used in dialysis for the treatment of kidney failure. Blood is continuously withdrawn from the patient's vasculature, at a rate of 200–300 ml/min, treated in a hollow-fiber bioartificial liver, and ultimately returned to the patient. A charcoal filter may be added to further detoxify the blood. Treatments are performed daily for 4 hours. Results in early human trials were quite encouraging and several cases of recovery without transplantation were observed.

The bioartificial pancreas has enjoyed very impressive success in rodent studies—so much so that no fewer than five reports on "proof of principle" experiments have appeared in the hallowed pages of *Science* magazine. Unfortunately, results from larger animal models and human studies have been disappointing. Investigators have not been able to reliably isolate the number of islets (500,000 to 1,000,000 or 2×10^9 cells) required for a large recipient. Moreover, species scaling of device design has proven difficult: formats that were suitable in rodents generally have been unsatisfactory in large animals. Some investigators believe that use of genetically engineered cells or their transgenic cohorts may solve the problem of cell source. Genetically engineered cells might be more productive than islets and ease some of the constraints on device design. Development of a clinically beneficial bioartificial pancreas remains an important challenge for biomedical engineering in the early 21st century.

Encapsulation is also being developed for the delivery of cell and gene therapy. Small quantities of cells producing a desired therapeutic molecule are placed inside the lumen of a sealed hollow fiber or encapsulated in microspheres. A therapeutic dose may involve a very manageable $1–10 \times 10^6$ cells (roughly two orders of magnitude fewer cells than would be required for a bioartificial pancreas). From one perspective, these devices represent a form of drug delivery providing point-source, time-constant, and site-specific delivery with the added benefit of a "regenerable" source of bioactive "drug." In another sense, when recombinant cells are involved, this approach can be considered a form of gene therapy in which the transplanted gene resides in cells housed in a capsule rather than directly in the cells of the recipient. The technical issues involved in this form of encapsulated cell therapy are largely resolved. Several successful preclinical and clinical trials have been reported.

Bibliography

Aebischer, P., and Lysaght, M. J. (1995). Immunoisolation and cellular xenotransplantation. *Xeno* **3:** 43–48.

Aebischer, P., Pochon, N.A., Heyd, B., Deglon, N., Joseph, J.M., Zurn, A.D., Baetge, E.E., Hammang, J.P., Goddard, M., Lysaght, M., Kaplan, F., Kato, A.C., Schluep, M., Hirt, L., Regli, F., Porchet, F., and DeTribolet, N. (1996a). Gene therapy for amyotrophic lateral sclerosis (ALS) using a polymer encapsulated xenogenic cell line engineered to secrete hCNTF. *Hum. Gene Ther.* **7:** 851–860.

Aebischer, P., Schluep, M., Deglon, N., Joseph, J.M., Hirt, L., Heyd, B., Goddard, M., Hammang, J.P., Zurn, A.D., Kato, A.C., Regli, F., and Baetge, E.E. (1996b). Intrathecal delivery of CNTF using encapsulated genetically modified xenogeneic cells in amyotrophic lateral sclerosis. *Nat. Med.* **2**(6): 696–699.

Avgoustiniatos, E.S., and Colton, C.K. (1997). Effect of external oxygen mass transfer on viability of immunoisolated tissue. *Ann. N.Y. Acad. Sci.* **831:** 145–167.

Brauker, J., Carr-Brendel, V., Martinson, L., Crudele, J., Johnston, W., and Johnson, R. (1995). Neovascularization of synthetic membranes directed by membrane microarchitecture. *J. Biomed. Mater. Res.* **29:** 1517–1524.

Cabasso, I. (1980). Hollow fiber membranes. in *Kirk-Othmer Encyclopedia of Chemical Technology,* Vol. 12. Wiley, New York, pp. 492–517.

Chen, S.C., Mullon, C., Kahaku, E., Watanabe, F., Hewitt, W., Eguchi, S., Middleton, Y., Arkadopoulos, N., Rozga, J., Solomon, B., and Demetriou, A.A. (1997). Treatment of severe liver failure with a bioartificial liver. *Ann. N. Y. Acad. Sci.* **831:** 350–360.

Colton, C.K. (1995). Implantable biohybrid artificial organs. *Cell Transplant.* **4**(4): 415–436.

Dionne, K.E., Cain, B.M., Li, R.H., Bell, W.J., Doherty, W.J., Rein, D.H., Lysaght, M.J., and Gentile, F.T. (1996). Transport characterization of membranes for immunoisolation. *Biomaterials* **17:** 257–266.

Emerich, D.F., Winn, S.R., Hantraye, P.M., Peschanski, M., Chen, E.Y., Chu, Y., McDermott, P., Baetge, E.E., and Kordower, J.H. (1997). Protective effect of encapsulated cells producing neurotrophic factor CNTF in a monkey model of Huntington's disease. *Nature* **386**(6623): 395–399.

Emerich, D., Lidner, M., Winn, S.R., Chen, E., Frydel, B., Koedower, J. (1996). Implants of encapsulated human CNTF producing fibroblasts prevent behavioral deficits and striatal degeneration in a rodent model of Huntington's disease. *J. Neurosci.* **16:** 5168–5181.

Inoue, K., Fujisato, T., Gu, Y.J., Burczak, K., Sumi, S., Kogire, M., Tobe, T., Uchida, K., Nakai, I., Maetani, S., and Ikada, Y. (1992). Experimental hybrid islet transplantation: application of polyvinyl alcohol membrane for entrapment of islets. *Pancreas* **7:** 562–568.

Kordower, J.H., Liu, Y., Winn, S., and Emerich, D. (1995). Encapsulated PC12 cell transplants into hemiparkinsonian monkeys: a behavioral, neuroanatomical, and neurochemical analysis. *Cell Transplant.* **4:** 155–171.

Lanza, R.P., and Chick, W.L. (1997). Transplantation of pancreatic islets. *Ann. N.Y. Acad. Sci.* **831:** 323–331.

Lanza, R.P., Cooper, D.K.C., and Chick, W.L. (1997). Xenotransplantation. *Sci. Am.* **277**(1): 54–59.

Li, R. (1998). Materials for immunoisolated cell transplantation. *Adv. Drug Dev. Rev.* **33**(1-2): 87–109.

Lysaght, M.J., and Aebischer, P.A. (1999). Encapsulated cells as therapy. *Sci. Am.* Apr. 76–83.

Roberts, T., De Boni, U., and Sefton, M.V. (1996). Dopamine secretion by PC12 cells microencapsulated in a hydroxyethyl methacrylate methyl methacrylate copolymer. *Biomaterials* **17:** 267–275.

Sagen, J., Wang, H., Tresco, P., and Aebischer, P. (1993). Transplants of immunologically isolated xenogeneic chromaffin cells provide a long-term source of pain neuroactive substances. *J. Neurosci.* **13:** 2415–2423.

Strathmann, H. (1985). Production of microporous media by phase inversion processes. In *Materials Science of Synthetic Membranes,* D.R. Lloyd (ed.). American Chemical Society, Washington, D. C.

Winn, S.R., and Tresco, P.A. (1994). Hydrogel applications for encapsulated cellular transplants. *Methods Neurosci.* **21:** 387–402.

7.1.4 Synthetic bioresorbable polymer scaffolds

Antonios G. Mikos, Lichun Lu, Johnna S. Temenoff and Joerg K. Tessmar

Scaffold design

Tissue engineering involves the development of functional substitutes to replace missing or malfunctioning human tissues and organs (Langer and Vacanti, 1993). Most strategies in tissue engineering have focused on using biomaterials as scaffolds to direct specific cell types to organize into three-dimensional (3D) structures and perform differentiated function of the targeted tissue. Synthetic bioresorbable polymers that are fully degradable into the body's natural metabolites by simple hydrolysis under physiological conditions are the most attractive scaffold materials. These scaffolds offer the possibility to create completely natural tissue or organ equivalents and thus overcome the problems such as infection and fibrous tissue formation associated with permanent implants.

These synthetic polymers must possess unique properties specific to the tissue of interest as well as satisfy some basic requirements in order to serve as an appropriate scaffold. One essential criterion is biocompatibility, i.e., the polymer scaffold should not invoke an adverse inflammatory or immune response once implanted (Babensee *et al.*, 1998). Some important factors that determine its biocompatibility, such as the chemistry, structure, and morphology, can be affected by polymer synthesis, scaffold processing, and sterilization. Toxic residual chemicals involved in these processes (e.g., monomers, stabilizers, initiators, cross-linking agents, emulsifiers, organic solvents) may be leached out from the scaffold with detrimental effects to the engineered and surrounding tissue.

The primary role of a scaffold is to provide a temporary substrate to which transplanted cells can adhere. Most organ cell types are anchorage-dependent and require the presence of a suitable substrate in order to survive and retain their ability to proliferate, migrate, and differentiate. Cell morphology correlates with cellular activities and function; strong cell adhesion and spreading often favor proliferation while a rounded cell shape is required for cell-specific function. For example, it has been demonstrated that the use of substrates with patterned surface morphologies or varied extracellular matrix (ECM) surface coatings can modulate cell shape and function (Chen *et al.*, 1998; Mooney *et al.*, 1992; Singhvi *et al.*, 1994).

For epithelial cells, cell polarity is essential for their function. Polarity refers to the distinctive arrangement, composition, and function of cell-surface and intracellular domains. This typically corresponds to a heterogeneous extracellular environment. For example, retinal pigment epithelium (RPE) cells have three major surface domains: the apical surface is covered with numerous microvilli; the lateral surface is joined with neighboring cells by junctional complexes; and the basal surface is convoluted into basal infoldings and connected to the basal lamina. The polymer scaffold for RPE transplantation should therefore provide proper surface chemistry and surface microstructure for optimal cell–substrate interaction and, along with appropriate culture conditions, be able to induce proper cell polarity (Lu *et al.*, 1998).

Besides cell morphology, the function of many organs is dependent on the 3D spatial relationship of cells with their ECM. The shape of a skeletal tissue is also critical to its function. Gene expression in cells is regulated differently by 2D versus 3D culture substrates. For instance, the differentiated phenotype of human epiphyseal chondrocytes is lost on 2D culture substrates but reexpressed when cultured in 3D agarose gels (Aulthouse *et al.*, 1989). A polymer scaffold should be easily and reproducibly processed into a desired shape that can be maintained after implantation so that it defines the ultimate shape of the regenerated tissue. A suitable scaffold should therefore act as a template to direct cell growth and ECM formation and facilitate the development of a 3D structure.

Porosity, pore size, and pore structure are important factors to be considered with respect to nutrient supply to transplanted and regenerated cells. To regenerate highly vascularized organs such as the liver, porous scaffolds with large void volume and large surface-area-to-volume ratio are desirable for maximal cell seeding, attachment, growth, ECM production, and vascularization. Small-diameter pores are preferable to yield high surface area per volume provided the pore size is greater than the diameter of a cell in suspension (typically 10 μm). However, topological constraints may require larger pores for cell growth. Previous experiments have demonstrated optimal pore sizes of 20 μm for fibroblast ingrowth, 20–125 μm for adult mammalian skin regeneration, and 200–400 μm for bone ingrowth (Boyan *et al.*, 1996; Whang *et al.*, 1995). The rate of tissue invasion into porous scaffold also depends on the pore size and polymer crystallinity (Mikos *et al.*, 1993c; Park and Cima, 1996; Wake *et al.*, 1994). Compared to isolated pore structure, an interconnected pore network enhances the diffusion rates to and from the center of the scaffold and facilitates vascularization, thus improving oxygen and nutrient supply and waste removal. The vascularization of an implant is a prerequisite for regeneration of most 3D tissues except for cartilage, which is avascular.

Mechanical properties of the polymer scaffold should be similar to the tissue or organ intended for regeneration. For load-bearing tissues such as bone, the scaffold should be strong enough to withstand physiological stresses to avoid collapse of the developing tissue. Also, transfer of load to the scaffold (stress shielding) after implantation may result in lack of sufficient mechanical stimulation to the ingrowing tissue. For the regeneration of soft tissues such as skin, the scaffolds are required to be pliable or elastic. The stiffness of the scaffold may affect the mechanical tension generated within the cell cytoskeleton, which is critical for the control of cell shape and function (Chicurel *et al.*, 1998). A more rigid surface may facilitate the assembly of stress fibers and enhance cell spreading and dividing. Scaffold compliance may also affect cell–cell contacts and aggregation (Moghe, 1996).

Understanding and controlling the degradation process of a scaffold and the effects of its degradation products on the body is crucial for long-term success of a tissue-engineered cell–polymer construct. The local drop in pH due to the release of acidic degradation products from some implants may cause tissue necrosis or inflammation. Polymer particles formed after long-term implantation of a scaffold or due to micromotion at the implantation site may elicit an inflammatory response. Microparticles of polymers have been shown to suppress initial rat-marrow stromal osteoblast proliferation in culture (Wake *et al.*, 1998). The mechanism by which the scaffold degrades should also be considered. For example the degradation products are released gradually by surface erosion, whereas during bulk degradation, the release of degradation products occurs only when the molecular weight of the polymer reaches a critical value. This late-stage burst effect may cause greater local pH drop.

The rate of scaffold degradation is tailored to allow cells to proliferate and secrete their own ECM while the polymer scaffold vanishes over a desired time period (from days to months) to leave enough space for new tissue growth. Since the mechanical strength of a scaffold usually decreases with degradation time, the degradation rate may be required to match the rate of tissue regeneration in order to maintain the structural integrity of the implant. The degradation rate of a scaffold can be affected by various factors listed in Table 7.1.4-1.

The design requirements of a tissue engineering scaffold are specific to the structure and function of the tissue to be regenerated. The polymer scaffold is typically engineered to mimic the natural ECM of the body. ECM proteins play crucial roles in the control of cell growth and function (Hay, 1993; Howe *et al.*, 1998). However, most synthetic polymer scaffolds do not possess the specific signals (ligands) that can be recognized by cell-surface receptors. It is therefore preferable that the polymer chain have chemically modifiable functional groups onto which sugars, proteins, or peptides can be attached. In addition, polymer–peptide hybrid molecules may be created or the ligand may be immobilized on the scaffold surface to generate a biomimetic microenvironment (Shakesheff *et al.*, 1998; Shin *et al.*, 2003b).

Table 7.1.4-1 Factors affecting scaffold degradation

Polymer chemistry	*Scaffold structure*
Composition	Density
Structure	Shape
Configuration	Size
Morphology	Mass
Molecular weight	Surface texture
Molecular weight distribution	Porosity
	Pore size
Chain motility	Pore structure
Molecular orientation	Wettability
Surface-to-volume ratio	Processing method and conditions
Ionic groups	Sterilization
Impurities or additives	
In vitro	*In vivo*
Degradative medium	Implantation site
pH	Access to vasculature
Ionic strength	Mechanical loading
Temperature	Tissue growth
Mechanical loading	Metabolism of degradation products
Type and density of cultured cells	Enzymes

Scaffold materials

The range of physical, chemical, mechanical, and degradative properties that may be achieved using synthetic bioresorbable polymers renders them extremely versatile as scaffold materials. Their molecular weight and chemical composition may be precisely controlled during polymer synthesis. Copolymers, polymer blends, and composites with other materials may be manufactured to give rise to properties that are advantageous over homopolymers for certain applications. Moreover, many polymers can be functionalized by converting end groups or addition of side chains with various chemical groups to obtain polymers that can be self-cross-linked or cross-linked with proteins and other bioactive molecules (Behravesh *et al.*, 1999). By choosing an appropriate processing technique, scaffolds of specific architecture and structural characteristics may be fabricated.

Not all types of currently available synthetic bioresorbable polymers can be manufactured into 3D scaffolds because of their chemical and physical properties and processability (Table 7.1.4-2). The most widely utilized scaffold materials are poly(α-hydroxy esters) such as PGA, PLA, and PLGA. They have been fabricated into thin films, fibers, porous foams, and conduits and investigated as scaffolds for regeneration of several tissues. Furthermore, the lysine groups in poly(lactic acid-*co*-lysine) provide sites for addition of cell-adhesion sequences such as RGD peptides (Barrera *et al.*, 1995; Cook *et al.*, 1997).

Poly(propylene fumarate) (PPF), an unsaturated linear polyester that can be cross-linked through its fumarate double bonds, has been investigated as a bioresorbable bone cement (Peter *et al.*, 1997a). The cross-linking, mechanical, and degradative properties of an injectable composite scaffold made of PPF and β-tricalcium phosphate have been characterized (Peter *et al.*, 1999, 1998b, 1997b). The mechanical properties of PPF scaffolds can be further modified depending on the cross-linking parameters employed. An increase in compressive modulus was observed with the use of a cross-linking agent, PPF-diacrylate, and the choice of a photo-(light-based) initiator, rather than a thermally based initiator system (Timmer *et al.*, 2003).

Poly(ethylene glycol) (PEG), although nondegradable, is often used to fabricate copolymers or polymer blends to increase the hydrophilicity, biocompatibility, and/or softness of the scaffold. Poly(propylene fumarate-*co*-ethylene glycol) [P(PF-*co*-EG)] hydrogels have been developed for cardiovascular applications (Suggs *et al.*, 1998a, 1997, 1999). When used in combination with a pore-forming agent and modified with cell-adhesion ligands, P(PF-*co*-EG) has also been used in highly porous scaffolds for bone tissue engineering (Behravesh and

Table 7.1.4-2 Scaffold materials and their applications[a]

Materials	Applications
Poly(α-hydroxy esters)	
Poly(L-lactic acid) (PLLA)	Bone, cartilage, nerve
Poly(glycolic acid) (PGA)	Cartilage, tendon, urothelium, intestine, liver, bone
Poly (D,L-lactic-*co*-glycolic acid) (PLGA)	Bone, cartilage, urothelium, nerve, RPE
PLLA-bonded PLGA fibers	Smooth muscle
PLLA coated with collagen or poly(vinyl alcohol) (PVA)	Liver
PLLA and poly(ethylene glycol) (PEG) block copolymer	Bone
PLGA and PEG blends	Soft tissue and tubular tissue
Poly(L-lactic acid-co-ε-caprolactone) (PLLACL)	Meniscal tissue, nerve
Poly(D,L-lactic acid-co-ε-caprolactone) (PDLLACL)	Vascular graft
Polyurethane/poly(L-lactic acid)	Small-caliber arteries
Poly(lysine-*co*-lactic acid)	Bone, cartilage, nerve
Poly(propylene fumarate) (PPF)	Bone
Poly(propylene fumarate-*co*-ethylene glycol) [P(PF-*co*-EG)]	Cardiovascular, bone
PPF/β-tricalcium phosphate (PPF/β-TCP)	Bone
Poly(ε-caprolactone)	Drug delivery
Polyhydroxyalkanoate (PHA)	Cardiovascular
Polydioxanone	Bone
Polyphosphates and polyphosphazenes	Skeletal tissue, nerve
Pseudo-poly(amino acids) Tyrosine-derived polyiminocarbonates Tyrosine-derived polycarbonate Tyrosine-derived polyacrylates	Bone

[a] Adapted from Babensee *et al.* (1998).

Mikos, 2003). Results indicate that this hydrogel supported the proliferation, osteogenic differentiation and matrix production from seeded bone-marrow stromal cells during 28 days of *in vitro* culture (Behravesh and Mikos, 2003).

In addition, another material including PEG, oligo [poly(ethylene glycol) fumarate] (OPF), has been developed and characterized (Jo *et al.*, 2001). Because of the chemical structure of this oligomer, it can be used to form biodegradable hydrogels with a high degree of swelling (Jo *et al.*, 2001; Temenoff *et al.*, 2003). This material may find uses in guided tissue regeneration applications because it demonstrates relatively low general cell adhesion, but at the same time, possesses an ability to be modified with peptides that could encourage adhesion of specific cell types (Shin *et al.*, 2002; Temenoff *et al.*, 2003).

Poly(ε-caprolactone) (PCL) as well as blends and copolymers containing PCL have also been studied as scaffold materials (Suggs and Mikos, 1996). Polyphosphates and polyphosphazenes have been processed into scaffolds for bone tissue engineering (Behravesh *et al.*, 1999; Renier and Kohn, 1997). Pseudo-poly(amino acids), in which amino acids are linked by both amide and nonamide bonds (such as urethane, ester, iminocarbonate, and carbonate), are amorphous and soluble in organic solvents and thus processable into scaffolds. The most studied among these polymers are tyrosine-derived polycarbonates and polyacrylates (James and Kohn, 1996). By structural modifications of the backbone and pendant chains, polymer families with systematically and gradually varied properties can be created.

Applications of scaffolds

Tissue induction

Tissue induction is the process by which ingrowth of surrounding tissue into a porous scaffold is effected (Fig. 7.1.4-1A). The scaffold provides a substrate for the migration and proliferation of the desired cell types. For example, an osteoinductive material can be used to selectively induce bone formation. This approach has been employed to regenerate several other tissues including skin and nerve.

Cell transplantation

The concept is that cells obtained from patients can be expanded in culture, seeded onto an appropriate polymer scaffold, cultured, and then transplanted (Bancroft and Mikos, 2001) (Fig. 7.1.4-1B). The time at which transplantation takes place varies with a specific application. Usually the cells are allowed to attach to the scaffold, proliferate, and differentiate before implantation. A scaffold for bone cell transplantation should be osteoconductive, meaning that it has the capacity to direct the growth of osteoblasts *in vitro* and allow the integration of the transplant with the host bone. This strategy is the most widely used in tissue engineering and has been investigated for the transplantation of many cell types including osteoblasts, chondrocytes, hepatocytes, fibroblasts, smooth muscle cells, and RPE. This method also offers the possibility that genetically modified cells could be transplanted, thereby simultaneously presenting both the cells and the bioactive factors they produce to the site of interest, with the potential of further enhancing regeneration of the injured area (Blum *et al.*, 2003). This approach may be considered a combination of the applications of cell transplantation and delivery of bioactive molecules, discussed later.

Prevascularization

The major obstacle in the development of large 3D transplants such as liver is nutrient diffusion limitation, because cells will not survive farther than a few hundred microns from the nutrient supply. Although the scaffold can be vascularized postimplantation, the rate of vascularization is usually insufficient to prevent cell death inside the scaffold. In this case, prevascularization of the scaffold may be necessary to allow the ingrowth of fibrovascular tissue or uncommitted vascular tissue such as periosteum (layer of connective tissue covering bone) before cell seeding by injection (Fig. 7.1.4-1C). The prevascularized scaffold will provide a substrate for cell attachment, growth, and function. The extent of prevascularization has to be optimized to allow sufficient nutrient diffusion as well as enough space for cell seeding and tissue growth (Mikos *et al.*, 1993c).

Some complex osseous defects created by bone tumor removal or extensive tissue damage exceed the critical size for normal healing and require a large transplant to restore function. A novel strategy is to prefabricate vascularized bone flaps by implanting a mold containing bioresorbable polymers with osteoinductive elements onto a periosteal site remote from the defect where prevascularization and ectopic bone formation can occur over a period of time as the scaffold degrades (Thomson *et al.*, 1999). The created autologous bone can then be transplanted to the defect site where vascular supply can be attached via microsurgery to existing vessels.

In situ polymerization

Injectable, *in situ* polymerizable, bioresorbable materials can be utilized to fill defects of any size and shape with minimal surgical intervention (Fig. 7.1.4-1D). For instance, PPF has been developed as an injectable bone cement that hardens within 10 to 15 minutes under physiological conditions. These materials do not require prefabrication but must meet additional requirements since polymerization or cross-linking reactions occur *in vivo*. All reagents and products must be biocompatible, and the reaction conditions such as temperature, pH, and heat release should not damage implanted cells or the surrounding tissue.

The hardened material (scaffold) must be highly porous and have interconnected pore structure in order to serve as a suitable template for guiding cell growth and differentiation. This can be achieved by combining a porogen such as sodium chloride crystals in the injectable paste that are eventually leached out, leaving a porous polymer matrix. Since the leaching step occurs *in vivo*, local high salt concentration may lead to high osmolarity and tissue damage. The amount of porogen incorporated has to be optimized to ensure biocompatibility, while enough porosity needs to be achieved to allow sufficient nutrient diffusion and vascularization.

PPF has also been developed for use in combination with cell transplantation applications through the incorporation of cells within the material during the cross-linking procedure. Because of the potentially non-cytocompatible conditions that may be present during the curing reaction, a composite material has been developed in which cells are first encapsulated in gelatin microspheres, and these are then included with the PPF during cross-linking (Payne *et al.*, 2002a, b). It has been shown that this encapsulation procedure enhances the viability, proliferation, and osteogenic differentiation of rat-marrow stromal cells (as compared to nonencapsulated cells) when cultured on cross-linking PPF *in vitro* (Payne *et al.*, 2002a, b).

Delivery of bioactive molecules

Cellular activities can be further modulated by various soluble bioactive molecules such as DNA, cytokines, growth factors, hormones, angiogenic factors, or immunosuppresant drugs (Babensee *et al.*, 2000; Holland and Mikos, 2003; Kasper and Mikos, 2003). For instance, bone morphogenetic proteins (BMPs) have been identified as a family of growth factors that regulate differentiation of bone cells (Ripamonti and Duneas, 1996). Controlled local delivery of these tissue inductive factors to transplanted and regenerated cells is often desirable. This has led to the concept of incorporation of bioactive molecules into scaffolds for implantation. These factors can be bound into polymer matrix during scaffold processing (Behravesh *et al.*, 1999; Shin *et al.*, 2003a) (Fig. 7.1.4-2A). Alternatively, bioresorbable microparticles or nanoparticles loaded with these molecules can be impregnated into the substrates (Hedberg *et al.*, 2002; Holland *et al.*, 2003; Lu *et al.*, 2000) (Fig. 7.1.4-2B). By incorporating BMPs or other osteogenic molecules into the injectable paste, PPF can also serve as a delivery vehicle for bone growth factors to induce a bone-regeneration cascade (Hedberg *et al.*, 2002). The release of bioactive molecules *in vivo* is governed by both diffusion and polymer degradation (Hedberg *et al.*, 2002; Holland *et al.*, 2003). In addition, if the molecules are

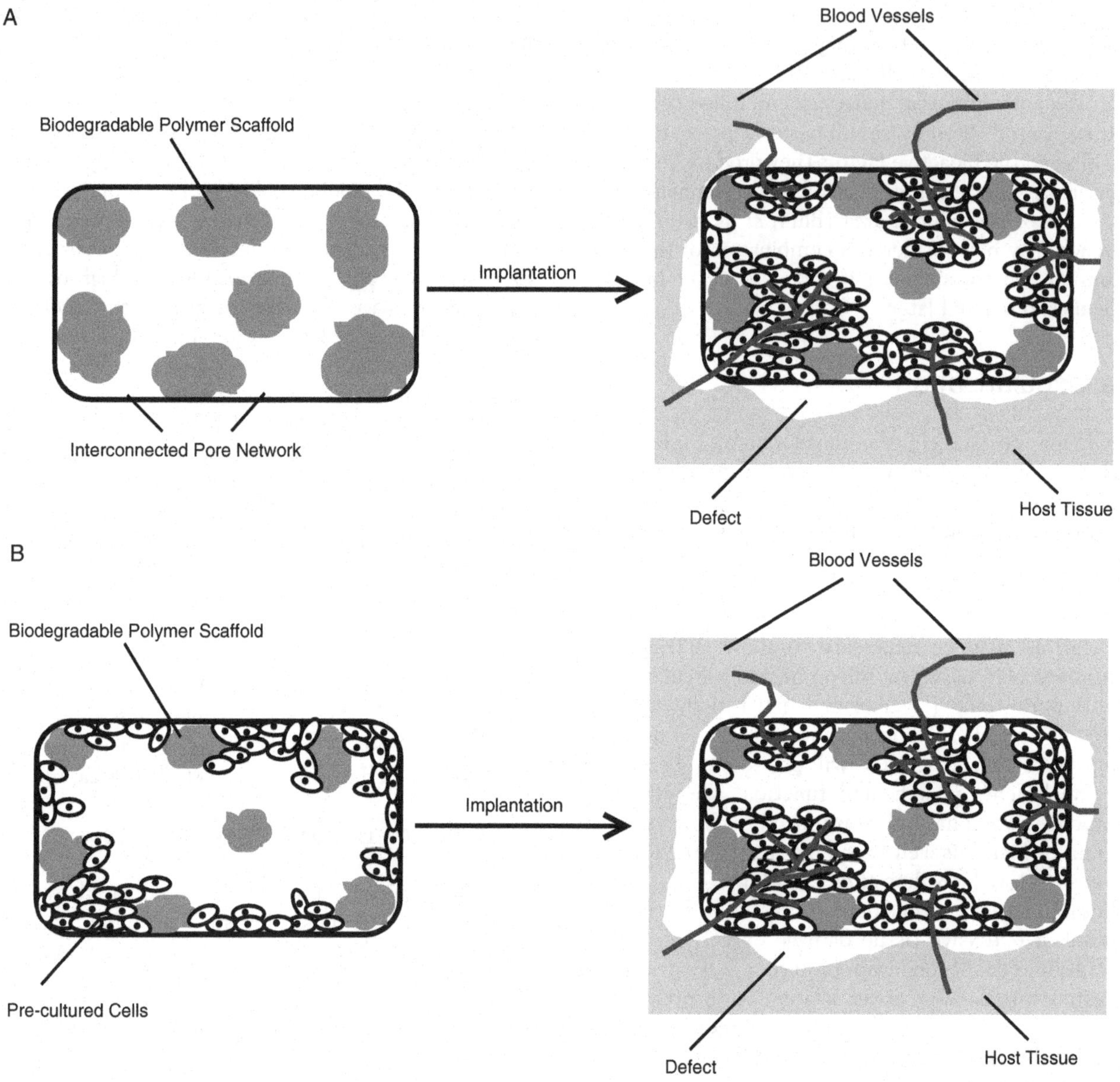

Fig. 7.1.4-1 Applications of bioresorbable polymers as porous scaffolds in tissue engineering. (A) Tissue induction. (B) Cell transplantation. (C) Prevascularization. (D) *In situ* polymerization. In all cases, the porous scaffolds allow tissue ingrowth as they degrade gradually.

encapsulated within microparticles that are degraded through enzymatic actions, such as gelatin (Holland *et al.*, 2003), the concentration and activity of these enzymes may also affect the release profile of the factors from composite scaffolds.

Scaffold processing techniques

The technique used to manufacture synthetic bioresorbable polymers into suitable scaffolds for tissue regeneration depends on the properties of the polymer and its intended application (Table 7.1.4-3). Scaffold processing usually involves (1) heating the polymers above their glass transition or melting temperatures; (2) dissolving them in organic solvents; and/or (3) incorporating and leaching of porogens (gelatin microspheres, salt crystals, etc.) in water (Temenoff and Mikos, 2000).

These processes usually result in decrease in molecular weight and have profound effects on the biocompatibility, mechanical properties, and other characteristics of the formed scaffold. Incorporation of large bioactive molecules such as proteins into the scaffolds and retention of their activity have been a major challenge.

Fiber bonding

Fibers provide a large surface-area-to-volume ratio and are therefore desirable as scaffold materials. PGA fibers in the form of tassels and felts have been utilized as

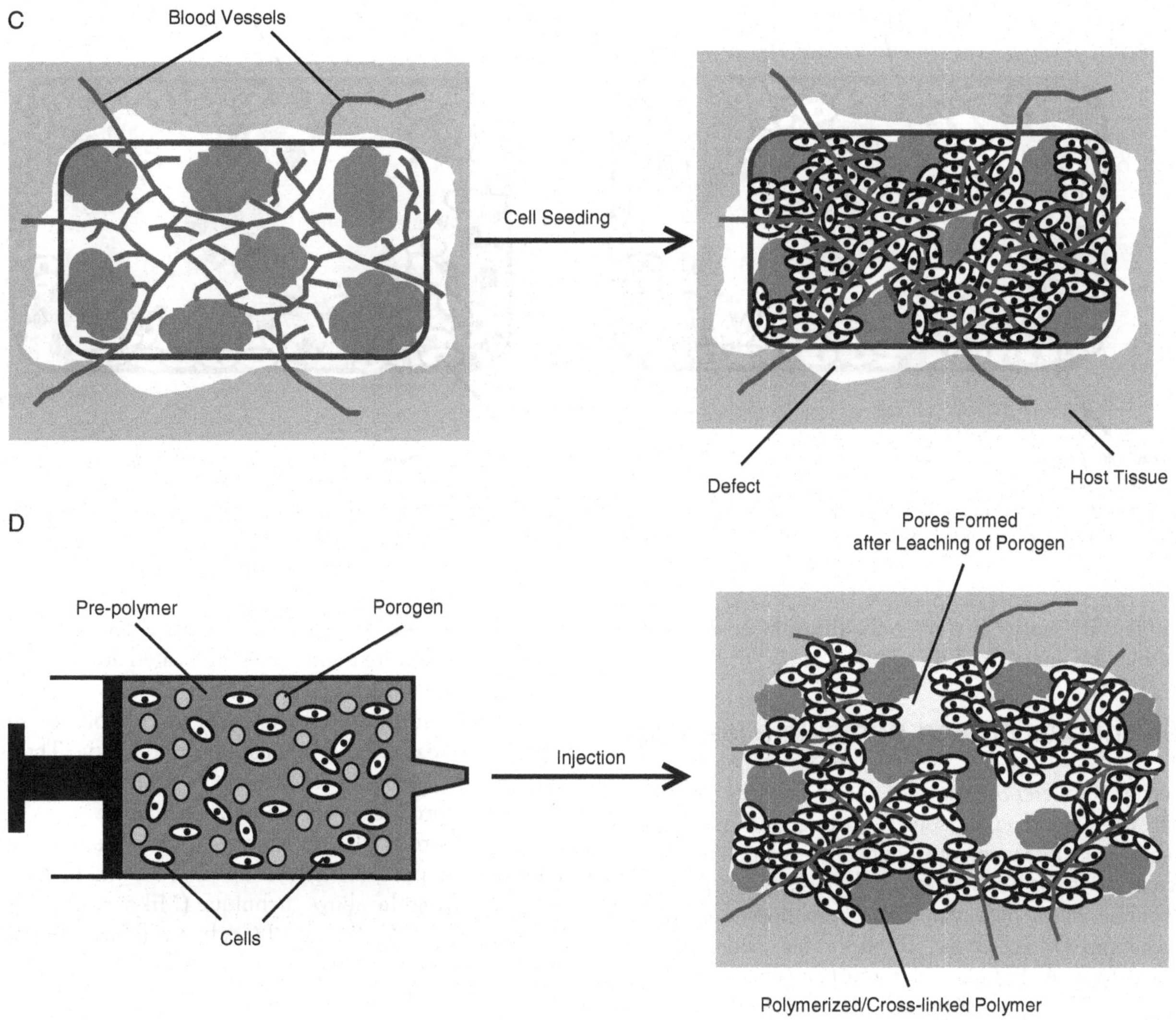

Fig. 7.1.4-1 cont'd

scaffolds to demonstrate the feasibility of organ regeneration (Cima *et al.*, 1991; Vacanti *et al.*, 1991). However, these fibers lack the structural stability necessary for *in vivo* uses, which has led to the development of a fiber bonding technique (Mikos *et al.*, 1993a). With this method, PGA fibers are aligned in the shape of the desired scaffold and then embedded in a PLA/methylene chloride solution. After evaporation of the solvent, the PLA–PGA composite is heated above the melting temperatures of both polymers. PLA is removed by selective dissolution after cooling, leaving the PGA fibers physically joined at their cross-points without any surface or bulk modifications while maintaining their initial diameter. Stipulations concerning the choice of solvent, immiscibility of the two polymers, and their relative melting temperatures restrict the general application of this technique to other polymers.

An alternative method of fiber bonding has also been developed to prepare tubular scaffolds for the regeneration of intestine, blood vessels, and ureters (Mooney *et al.*, 1996b, 1994a). In this technique, a nonwoven mesh of PGA fibers is attached to a rotating Teflon cylinder. The scaffolds are reinforced by spray casting with solutions of PLA or PLGA, which results in a thin coat that bonds the cross-points of PGA fibers. The behavior of transplanted cells is therefore determined by the PLA or PLGA coating instead of the PGA mesh. The mechanical strength of the scaffold is provided by both fibers and coating and is designed in such a way to withstand mechanical stresses or compromise degradation of PLA or PLGA. For example, PGA fiber-based matrices alone did not withstand contractile forces exerted by cultured smooth muscle cells, while scaffolds stabilized by spray-coating atomized PLA solution over the sides of the PGA matrices maintained their desired size and shape over 7 weeks in culture (Kim and Mooney, 1998). This method is very useful for fabrication of thin scaffolds; however, it does not allow the creation of

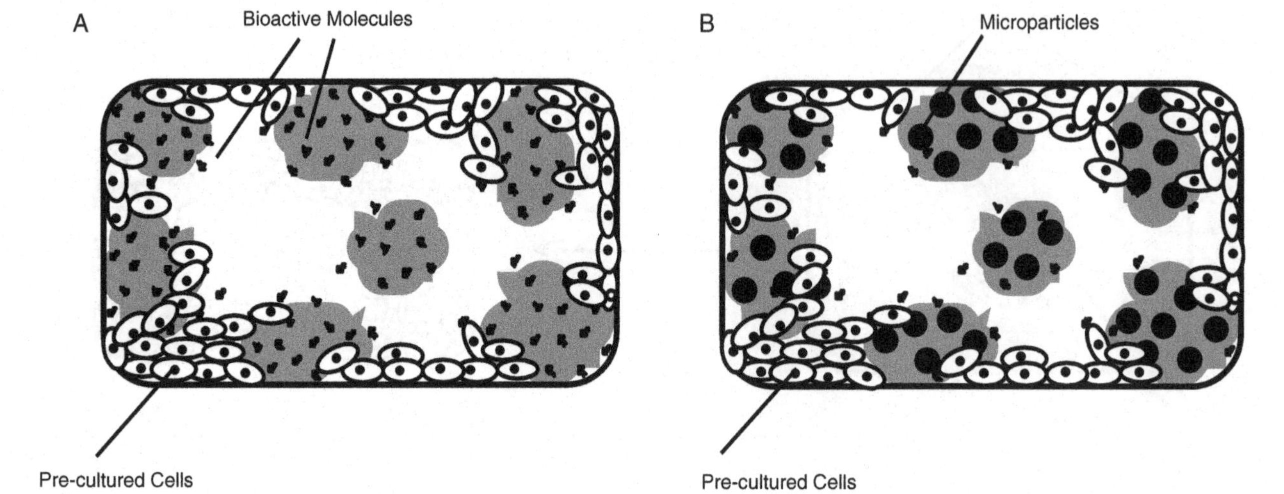

Fig. 7.1.4-2 Localized delivery of bioactive molecules from scaffolds. (A) Release directly from the supporting matrix. (B) Microparticles or nanoparticles loaded with bioactive molecules are impregnated into scaffolds and serve as delivery vehicles.

complex 3D scaffolds since only a thin layer at the surface may be engineered by coating.

Solvent casting and particulate leaching

In order to overcome some of the drawbacks associated with fiber bonding, a solvent-casting and particulate-leaching (SC/PL) technique has been developed to prepare porous scaffolds with controlled porosity, surface-area-to-volume ratio, pore size, and crystallinity for specific applications (Mikos *et al.*, 1994b). This method can be applied to PLA, PLGA, and any other polymers that are soluble in a solvent such as chloroform or methylene chloride. For example, sieved salt particles are dispersed in a PLA/chloroform solution that is used to cast a membrane onto glass petri dishes. After evaporating the solvent, the PLA/salt composite membranes are heated above the PLA melting temperature and then quenched or annealed by cooling at controlled rates to yield amorphous or semicrystalline foams with regulated crystallinity. The salt particles are eventually leached out by selective dissolution in water to produce a porous polymer matrix.

Table 7.1.4-3 Examples of scaffolds processed by various techniques

Processing technique	Examples
Fiber bonding	PGA fibers; PLA-reinforced PGA fibers
Solvent casting and particulate leaching	PLA, PLGA, PPF foams
Superstructure engineering	PLA, PLGA membranes
Compression molding	PLA, PLGA foams
Extrusion	PLA, PLGA conduits
Freeze-drying	PLGA foams
Phase separation	PLA foams
High-pressure gas foaming	PLGA, P(PF-*co*-EG) scaffolds
Solid free-form fabrication	Complex 3D PLA, PLGA structures

Highly porous PLA foams with porosities up to 93% and median pore diameters up to 500 μm have been prepared using the above technique (Mikos *et al.*, 1994b; Wake *et al.*, 1994). Porous PLGA foams fabricated by the same method have been shown to support osteoblasts growth both *in vitro* and *in vivo* (Ishaug *et al.*, 1997; Ishaug-Riley *et al.*, 1997, 1998). The porosity and pore size can be controlled independently by varying the amount and size of the salt particles, respectively. The surface-area-to-volume ratio depends on both initial salt weight fraction and particle size. In addition, the crystallinity, which affects both degradation and mechanical strength of the polymer, can be tailored to a particular application. A disadvantage of this method is that it can only be used to produce thin wafers or membranes with uniform pore morphology up to 3 mm thick (Wake *et al.*, 1996). The preparation of thicker membranes may result in the formation of a solid skin layer characteristic of asymmetric membranes. The two controlling phenomena are solvent evaporation of the surface and solvent diffusion in the bulk.

This method has been modified to fabricate tubular scaffolds (Mooney *et al.*, 1995a, 1994b). Porous PLGA membranes prepared using SC/PL are wrapped around Teflon cylinders, and the overlapping ends are fused together with chloroform. The Teflon core is then removed to leave a hollow tube. Because of the relatively brittle nature of the porous membranes used, this method is limited to tubular scaffolds with a low ratio of wall thickness to inner diameter.

To increase the pliability of the porous membranes, PEG has been blended with PLGA in the SC/PC process (Wake *et al.*, 1996). Micropores resulted from dissolution of PEG during leaching are believed to alter the structure of the pore walls and increase the pliability of the scaffold. These membranes can be rolled over into tubular scaffolds with a significantly higher ratio of wall thickness to inner diameter. The membranes fabricated from the polymer blend do not show any macroscopic damage during rolling as is observed for tubes made of PLGA alone.

Highly porous PPF scaffolds have also been formed using the SC/PL technique for both tissue induction and delivery of bioactive factors (Fisher *et al.*, 2003; Hedberg *et al.*, 2002). In this procedure, the PPF is cross-linked around the salt particles in molds of desired size. The samples are then removed from the molds and the salt is leached in water. Mechanical and degradation properties of the resulting scaffolds, with pore sizes ranging from 300 to 800 μm and porosities of 60–70%, have been characterized *in vitro* (Fisher *et al.*, 2003). These scaffolds were also found to induce a mild tissue response when implanted for up to 8 weeks either subcutaneously or in cranial defects in rabbits (Fisher *et al.*, 2002).

Superstructure engineering

Polymer scaffolds with complex 3D architecture (superstructures) can be formed by superimposing defined structural elements such as pores, fibers, or membranes in order according to stochastic, fractal, or periodic principles (Wintermantel *et al.*, 1996). This approach may provide optimal spatial organization and nutritional conditions for cells. The coherence of structural elements determines the anisotropic structural behavior of the scaffold. The major concern in engineering superstructures is the spatial organization of the elements in order to obtain desired pore sizes and interconnected pore structure.

A simple example of this technique is membrane lamination to construct foams with precise anatomical shapes (Mikos *et al.*, 1993b). A contour plot of the particular 3D shape is first prepared. Highly porous PLA or PLGA membranes with the shapes of the contour are then manufactured using SC/PL. The adjacent membranes are bonded together by coating chloroform on their contacting surfaces. The final scaffold is thus formed by laminating the constituent membranes in the proper order. It has been shown that continuous pore structures are formed with no boundary between adjacent layers. In addition, the bulk properties of the 3D scaffold are identical to those of the individual membranes.

Compression molding

Compression molding is an alternative technique of constructing 3D scaffolds. In this method, a mixture of fine PLGA powder and gelatin microspheres is loaded in a Teflon mold and then heated above the glass transition temperature of the polymer (Thomson *et al.*, 1995). The PLGA/gelatin composite is subsequently removed from the mold and gelatin microspheres are leached out. In this way, porous PLGA scaffolds with a geometry identical to the shape of the mold can be produced.

Polymer scaffolds of various shapes can be constructed by simply changing the mold geometry. This method also offers the independent control of porosity and pore size by varying the amount and size of porogen used, respectively. In addition, it is possible to incorporate bioactive molecules in either polymer or porogen for controlled delivery, because this process does not utilize organic solvents and is carried out at relatively low temperatures for amorphous PLGA scaffolds. This manufacturing technique may also be applied to PLA or PGA. However, higher temperatures are required (above the polymer melting temperatures) because these polymers are semicrystalline.

Compression molding can be combined with the SC/PL technique to form porous 3D foams. The dried PLGA/salt composites obtained by SC are broken into pieces of less than 5 mm in edge length and compression-molded into a desired 3D shape (Widmer *et al.*, 1998). The resulted composite material can then be cut into desired thickness. Subsequent leaching of the salt leaves an open-cell porous foam, with more uniform pore distribution than those obtained by SC/PL for increased thickness.

Highly porous poly(α-hydroxy ester) scaffolds, though desirable in many tissue engineering applications, may lack required mechanical strength for the replacement of load bearing tissues such as bone. Hydroxyapatite and β-tricalcium phosphate are biocompatible and osteoconductive materials and can be incorporated into these foams to improve their mechanical properties. Because the macroscopic mixing of three solid particulates (polymer powder, porogen, and ceramic) is difficult, a combined SC, compression-molding, and PL technique described earlier has been employed to fabricate an isotropic composite foam scaffold of PLGA reinforced with short hydroxyapatite fibers (15 μm in diameter and 45 μm in length) (Thomson *et al.*, 1998). Within certain range of fiber contents, these scaffolds have superior compressive strength compared to nonreinforced materials of the same porosity.

Extrusion

Various extrusion methods such as ram (piston-cylinder) extrusion, hydrostatic extrusion, or solid-state-extrusion (die drawing) have been applied to increase the orientation of polymer chains and thus produce high-strength, high-modulus materials (Ferguson *et al.*, 1996). More recently, an extrusion process has been successfully combined with the SC/PL technique to manufacture porous tubular scaffolds for guided tissue such as peripheral nerve regeneration (Widmer *et al.*, 1998). First the dry polymer/salt composite wafers obtained from SC are cut into pieces and placed in a customized piston extrusion tool (Fig. 7.1.4-3). The tool is then mounted into a hydraulic press and heated to the desired processing temperature. The temperature is allowed to equilibrate and the polymer/salt composite is then extruded by applying pressure. The extruded tubes are cut to appropriate lengths. Finally, the salt particles are leached out to yield highly porous conduits.

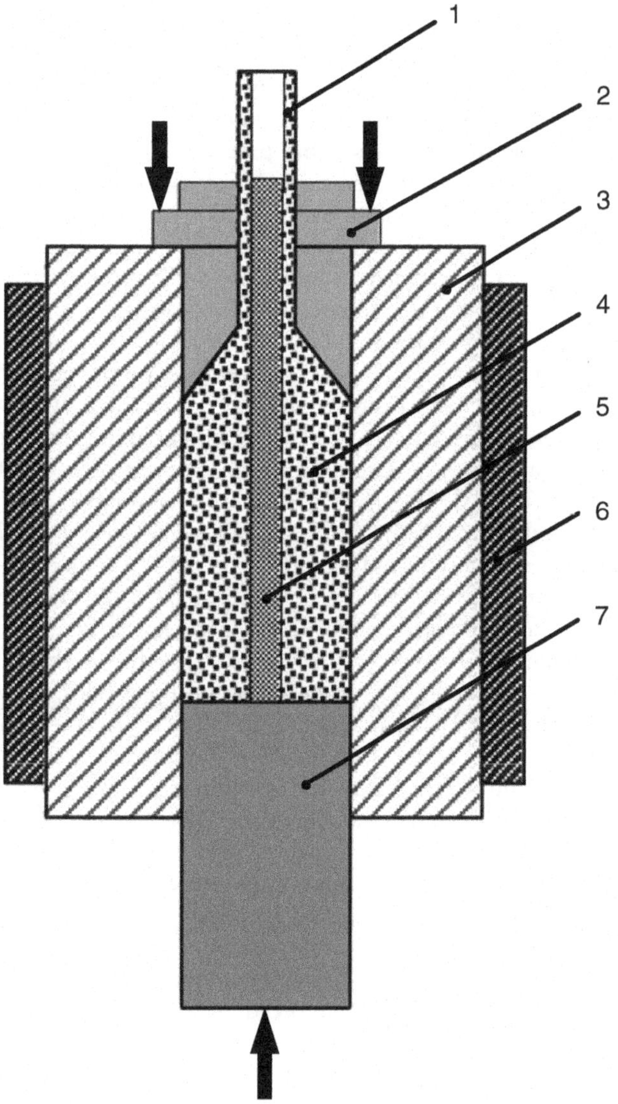

Fig. 7.1.4-3 Piston extrusion tool for the manufacture of tubular polymer/salt composite structures: 1, extruded polymer/salt construct; 2, nozzle defining the outer diameter of the tubular construct; 3, tool body; 4, melted polymer/salt mixture; 5, rod defining the inner diameter of the tubular construct; 6, heat band with temperature control; and 7, piston moving the melted polymer/salt mixture. The arrows indicate the attachment points for the forces involved in the extrusion process.

The pressure for extrusion at a constant rate is dependent on the extrusion temperature. High temperature may result in thermal degradation of the polymer. The porosity and pore size of the extruded conduits are determined by salt weight fraction, salt particle size, and processing temperatures. The fabricated conduits have an open-pore structure and are suitable for incorporation of cells or microparticles loaded with tissue inductive factors.

Freeze-drying

Low-density polymer foams have been produced using a freeze-drying technique (Hsu *et al.*, 1997). Polymer is first dissolved in a solvent such as glacial acetic acid or benzene to form a solution of desired concentration. The solution is then frozen and the solvent is removed by lyophilization under high vacuum. Several polymers including PLGA and PLGA/PPF have been prepared into porous foams with this method. The foams have either leaflet or capillary structures depending on the polymer and solvent used in fabrication. These foams are generally not suitable as scaffolds for cell transplantation. Subsequent compression of the foams by grinding and extrusion can generate matrices with varied densities. Foam density has been shown to determine the kinetics of drug release from these matrices.

An emulsion freeze-drying technique has also been developed to fabricate porous scaffolds (Whang *et al.*, 1995). In this technique, water is added to a PLGA/methylene chloride solution and the immiscible phases are homogenized. The created emulsion (water-in-oil) is then poured into a copper mold maintained in liquid nitrogen (−196°C). After quenching, the samples are freeze-dried to remove methylene chloride and water. Using this technique, PLGA foams with porosity in the rangeof 91–95% and median pore diameters of 13–35 μm with larger pores greater than 200 μm have been made by varying processing parameters such as water volume fraction, polymer weight fraction, and polymer molecular weight. Compared to the SC/PL technique, this method produces foams with smaller pore sizes but higher specific pore surface area and can produce thick (>1 cm) foams.

Phase separation

The ability to deliver bioactive molecules from a degrading polymer scaffold is desirable for tissue regeneration. However, the activity of the molecule is often dramatically decreased because of the harsh chemical or thermal environments used in some polymer processing techniques. Using a novel phase separation technique, scaffolds loaded with small hydrophilic and hydrophobic bioactive molecules have been manufactured (Lo *et al.*, 1995). The polymer is dissolved in a solvent such as molten phenol or naphthalene, followed by dispersion of the bioactive molecule in this homogeneous solution. A liquid–liquid phase separation is induced by lowering the solution temperature. The resulting bicontinuous polymer and solvent phases are then quenched to create a two-phase solid. Subsequent removal of the solidified solvent by sublimation leaves a porous polymer scaffold loaded with bioactive molecules.

The fabricated PLA foams have pore sizes up to 500 μm with relatively uniform distributions. The properties of the foams depends on the polymer type, molecular weight, concentration, and solvent used. It has been shown that proteins such as alkaline phosphatase retain as much as 75% of their activity after scaffold fabrication with the naphthalene system, but the activity is completely lost in the phenol system. Although phenol has a lower melting temperature than naphthalene, it is a more polar solvent and can interact with proteins and weaken the hydrogen bonding within the protein structure, resulting in a loss of protein activity. The phenol system may be useful for the entrapment of small drugs or short peptides instead.

Gas foaming

In one example of the gas foaming (GF) technique, solid disks of PLGA prepared by either compression molding or SC are exposed to high-pressure CO_2 (5.5 MPa, 25°C) environment to allow saturation of CO_2 in the polymer (Mooney *et al.*, 1996a). A thermodynamic instability is then created by reducing the CO_2 gas pressure to ambient level, which results in nucleation and expansion of dissolved CO_2 pores in the polymer particles. PLGA sponges with a porosity of up to 93% and a pore size of about 100 μm have been fabricated. The porosity and pore structure are dependent on the amount of CO_2 dissolved, the rate and type of gas nucleation, and the rate of gas diffusion to the pore nuclei.

The major advantage of this technique is that it involves no organic solvent or high temperature and therefore is promising for incorporating tissue induction factors in the polymer scaffolds. However, the effects of high pressure on the retention of activity of proteins still need to be assessed. In addition, this process yields mostly nonporous surfaces and a closed pore structure inside the polymer matrix, which is undesirable for cell transplantation. In an improved method, a porogen such as salt particles can be combined with the polymer to form composite disks before gas foaming (Harris *et al.*, 1998). The expansion and fusion of the polymer particles lead to the formation of a continuous matrix with entrapped salt particles, which are subsequently leached out. The GF/PL process produces porous matrices with predominately interconnected macropores (created by leaching out salt) and smaller, closed pores (created by the nucleation and growth of gas pores in the polymer particles). The fabricated matrices have a more uniform pore structure and higher mechanical strength than those obtained with SC/PL.

For injectable scaffolds, a combination of ascorbic acid, ammonium persulfate, and sodium bicarbonate has been used at atmospheric conditions to form highly porous hydrogel materials for bone tissue engineering (Behravesh *et al.*, 2002). In this case, as the hydrogel is cross-linked, carbon dioxide is produced, causing pore formation. The ratio of the three components just listed determines the final porosity (43–84%) and pore size (50–200 μm) of the scaffolds (Behravesh *et al.*, 2002). As mentioned previously, these porous [P(PF-*co*-EG)] foams supported rat-marrow stromal cell differentiation and bone matrix production during *in vitro* culture (Behravesh and Mikos, 2003).

Solid freeform fabrication

Solid freeform fabrication (SFF) refers to computer-aided design, computer-aided manufacturing (CAD/CAM) methodologies such as stereolithography, selective laser sintering (SLS), ballistic particle manufacturing, and 3D printing (3DP) for the creation of complex shapes directly from CAD models. SFF techniques, although mainly investigated for industrial applications such as rapid prototyping, offer the possibility to fabricate polymer scaffolds with well-defined architecture because local composition, macrostructure, and microstructure can be specified and controlled at high resolution in the interior of the components. These methods build complex 3D objects by material addition and fusion of cross-sectional layers (2D slices decomposed from CAD models). In addition, they allow the formation of multimaterial structures by selective deposition. Prefabricated structures can also be embedded during material buildup. By carefully controlling the processing conditions, cells, bioactive molecules, or synthetic vasculature may be included directly into layers of polymer scaffolds during fabrication.

An example of the use of stereolithography is the development of a diethyl fumarate/PPF resin as a liquid base material for a custom-designed apparatus using a computer-controlled, ultraviolet laser and suitable photoinitiator (Cooke *et al.*, 2002). In this case, the machine builds the desired structure from the bottom toward the top, with the resin allowed to wash over the sample after each layer is formed. This provides new base material to be photo-cross-linked in the desired geometry for the next layer using the computer-driven laser. The spatial resolution of such a system is 100 μm (Cooke *et al.*, 2002).

In the SLS technique, a thin layer of evenly distributed fine powder is first laid down (Bartels *et al.*, 1993; Berry *et al.*, 1997). A computer-controlled scanning laser is then used to sinter the powder within a cross-sectional layer. The energy generated by the laser heats the powder into a glassy state and individual particles fuse into a solid. Once the laser has scanned the entire cross section, another layer of powder is laid on top and the whole process is repeated.

In the 3DP process, each layer is created by adding a layer of polymer powder on top of a piston and cylinder containing a powder bed and the part being fabricated. This layer is then selectively joined where the part is to be formed by ink-jet printing of a binder material such as an organic solvent. The printed droplet has a diameter of 50–80 μm. The printhead position and speed are controlled by computer. The piston, powder bed, and part are lowered and a new layer of polymer powder is laid on top of the already processed layer and selectively joined. The layered printing process is repeated until the desired part is completed.

The local microstructure within the component can be controlled by varying the printing conditions. The resolution of features currently attainable by 3DP for degradable polyesters is about 200 μm (Griffith *et al.*, 1997). Using this technique, scaffolds with complex structures may be fabricated (Giordano *et al.*, 1996). A model drug (dye) with a concentration profile specified by a CAD model has been successfully incorporated into a scaffold during the 3DP process, demonstrating the feasibility of producing complex release regimes using a single drug-delivery device (Wu *et al.*, 1996). By mixing salt particles in the polymer powder and their subsequent leaching after 3DP process, porous PLGA scaffolds with an intrinsic network of interconnected branching channels have been fabricated for cell transplantation (Kim *et al.*, 1998). This network of channels and micropores could provide a structural template to guide cellular organization, enhance neovascularization, and increase the capacity for mass transport. Furthermore, multiple printheads containing different binder materials can be used to modify local surface chemistry and structure. Patterned PLA substrates with selective cell-adhesion domains have been fabricated by 3DP (Park *et al.*, 1998).

Characterization of processed scaffolds

Various techniques are available to characterize the fabricated polymer scaffolds (Table 7.1.4-4). The molecular weight and polydispersity index of the polymer can be measured by gel permeation chromatography (GPC). Information on chemical composition and structure can be obtained by nuclear magnetic resonance (NMR) spectroscopy, X-ray diffraction, Fourier transform infrared (FTIR), and FT-Raman (FTR) spectroscopy. The thermal properties of the polymer such as glass transition temperature (T_g), melting temperature (T_m), and crystallinity (X_c) can be determined by differential scanning calorimeter (DSC). Porosity and pore size distribution of a porous scaffold are measured by mercury intrusion porosimetry. Scanning electron microscopy (SEM) is the most common method to view the pore structure and morphology. The 3D microstructure of porous PLGA matrices has been analyzed by confocal microscopy (Tjia and Moghe, 1998). Mechanical properties of the scaffolds such as tensile strength and modulus, compression strength and modulus, compliance/hardness, flexibility, elasticity, and stress and stain at yield can be measured using mechanical testing equipment. Some tests require the processing of scaffolds into a particular shape and dimensions specified by ASTM standards.

Table 7.1.4-4 Characterization of bioresorbable polymer scaffolds

Properties	Techniques
Bulk properties	
Molecular weight	GPC
Polydispersity index	GPC
Chemical composition, structure	NMR, X-ray diffraction, FTIR, FTR
Thermal properties (T_g, T_m, X_c, etc.)	DSC
Porosity, pore size	Mercury intrusion porosimetry
Morphology	SEM, confocal microscopy
Mechanical properties	Mechanical testing
Degradative properties	*In vitro, in vivo*
Surface properties	
Surface chemistry	ESCA, SIMS
Distribution of chemistry	Imaging methods (e.g., SIMS)
Orientation of groups	Polarized IR, NEXAFS
Texture	SEM, AFM, STM
Surface energy and wettability	Contact-angle measurement

The *in vitro* degradation properties can be evaluated by placing the bioresorbable scaffolds in simulated body fluid, typically pH 7.4 phosphate-buffered saline (PBS). Changes in sample weight, molecular weight, morphology, and thermal and mechanical properties can then be measured at various time points until degradation process is completed. In addition, characterization of the chemical makeup of the degradation products may be possible through the use of GPC or high-performance liquid chromatography (HPLC) (Timmer *et al.*, 2002). However, such studies do not allow for the continuous observation of changes within the scaffolds. An *in vivo* study is often necessary to predict the degradation behavior of the scaffolds for cell transplantation (Shin *et al.*, 2003b).

Material surfaces, which are usually different from the bulk, play a crucial role in regulating cell response. Electron spectroscopy for chemical analysis (ESCA) and static secondary ion mass spectrometry (SIMS) are the most powerful tools for analyzing surface chemistry and composition. Information on the orientation of chemical groups can be obtained by polarized IR and near-edge X-ray absorption fine structure (NEXAFS). Surface morphology can be characterized by SEM, scanning probe microscopy (SPM), and atomic force microscopy (AFM). Surface wettability and energy are assessed by contact-angle measurements.

Cell seeding and culture in 3D scaffolds

The major obstacles to the *in vitro* development of 3D cell-polymer constructs for the regeneration of large organs or defects have been obtaining uniform cell seeding at high densities and maintaining nutrient transport to the cells inside the scaffolds. To achieve desired spatial and temporal distribution of cells and molecular cues affecting cellular function, cell culture conditions should provide control over hydrodynamic and biochemical factors in the cell environment.

Static culture

The conventional static cell seeding technique involves the placement of the scaffold in a cell suspension to allow the absorption of cells. However, the resulting cell distribution in the scaffold is often not uniform, with the majority of the cells attached only to the outer surfaces (Wald *et al.*, 1993). Wetting hydrophobic polymer scaffolds with ethanol and water prior to cell seeding allows for displacement of air-filled pores with water and thus facilitates penetration of cell suspension into these pores (Mikos *et al.*, 1994a). Infiltration with hydrophilic polymers or surface hydrolysis of scaffolds has also been shown to increase the cell seeding density (Gao *et al.*, 1998; Mooney *et al.*, 1995b). Seeding cells by injection or applying vacuum to ensure penetration of the cell suspension through the 3D matrix could result in uniform cell seeding initially. However, the uniformity is lost under static culture conditions because of the nutrient and oxygen diffusion limitation within the scaffold.

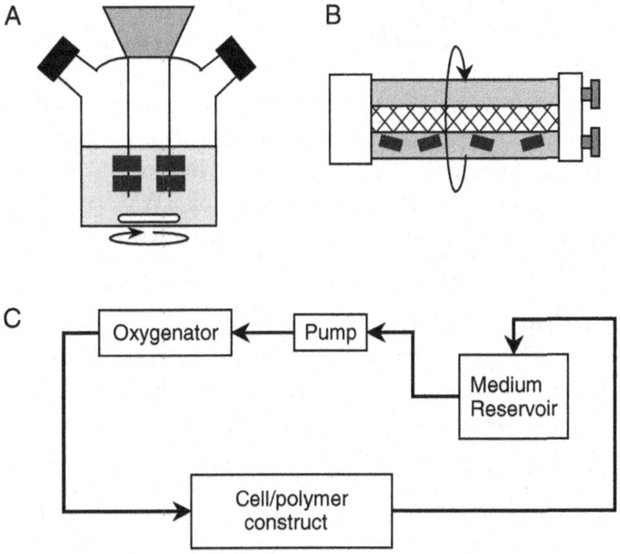

Fig. 7.1.4-4 Dynamic cell seeding and culture techniques in 3D scaffolds. (A) Spinner flask. (B) Rotary vessel. (C) Perfusion system.

Several dynamic cell seeding and culture techniques have been developed to ensure uniform cell distribution, which will lead to uniform tissue regeneration (Fig. 7.1.4-4). Compared to static culture conditions, mass transfer rates can be maintained at higher levels and cell growth is not restricted by the rate of nutrient supply under well-mixed culture conditions. These methods can be scaled up and are suitable for cell cultivation using multiple scaffolds.

Spinner flask culture

In a spinner flask, 3D polymer scaffolds are first fixed to needles attached to the cap of the flask, and then exposed to a uniform, well-mixed cell suspension (Fig. 7.1.4-4A) (Freed *et al.*, 1993). Using this method, porous PGA scaffolds have been uniformly seeded with chondrocytes at high yield and high kinetic rate (to minimize the time that cells stay in the suspension) (Vunjak-Novakovic *et al.*, 1998). Mixing has been found to promote the formation of cell aggregates with sizes of 20–32 μm. The spin rate, however, needs to be well adjusted to minimize cell damage under high shear stress. The spinner flask is also suitable for suspension culture of hepatocyte spheroids that exhibit enhanced liver function compared to monolayer culture in the long-term (Kamihira *et al.*, 1997).

Rotary vessel culture

The rotating-wall vessel (RWV) also allows enhanced mass transport and is useful for 3D cell culture

(Fig. 7.1.4-4B). The polymer scaffolds are loaded into the vessel and a uniform cell suspension is added. Vessel rotation is initiated to allow dynamic cell seeding and increased to maintain a high rate of nutrient and oxygen diffusion. Alternately, the scaffolds can be preseeded with cells under static conditions before loading (Goldstein *et al.*, 1999). Several configurations of RWV have been used in microgravity tissue engineering (Freed and Vunjak-Novakovic, 1997).

Perfusion culture

A flow perfusion culture system has been used for *in vitro* regeneration of large 3D tissues and organs (Fig.7.1.4-4C)(Bancroft *et al.*, 2002; Glowacki *et al.*, 1998; Griffith *et al.*, 1997; Kim *et al.*, 1998). The cell–polymer constructs are maintained in a continuous-flow condition. The culture medium is pumped from a reservoir through an oxygenator and the cell–polymer constructs, and recirculated back to the reservoir. The flow rate for cell survival is adjusted based on cell mass. The entire perfusion unit is maintained in normal sterile culture conditions. Compared to static culture, medium perfusion has been shown to significantly enhance cell viability and matrix production (Glowacki *et al.*, 1998). Additionally, the medium flow rate was found to influence ECM deposition and the timing of osteogenic differentiation when marrow stromal cells were cultured on three-dimensional scaffolds in a perfusion bioreactor (Bancroft *et al.*, 2002). These systems are useful for the development of complex tissue structures as well as the study of the effects of mechanical stimulation on cell viability, differentiation, and ECM production.

Other culture conditions

Ideally the culture conditions should provide all necessary signals that the cells normally experience *in vivo* for optimal tissue regeneration. For instance, mechanical stimulation plays an important role in the differentiation of mesenchymal tissues (Chiquet *et al.*, 1996; Goodman and Aspenberg, 1993). Application of well-controlled loads may stimulate bone growth into porous scaffolds. The degradation of the scaffolds can be affected by applied strain (Miller and Williams, 1984). Transwell culture systems that allow the use of different culture media for apical or basal sides are often employed to induce and maintain the polarity of epithelial cells. The growth and function of some retinal cells may be regulated by the light–dark cycle. In some cases, a gradient substrate with spatially controlled wettability or other properties may be desired (Ruardy *et al.*, 1995). Some cellular chemotactic responses may require the creation of concentration gradients of growth factors. Temporal presentation of signals is also important. For example, each phase of the differentiation of osteoblasts (proliferation, maturation of ECM, and mineralization) requires different signals (Lian and Stein, 1992; Peter *et al.*, 1998a). Coculture of several types may be preferred for *in vitro* organogenesis including angiogenesis.

Conclusions

Significant progress has been made to optimize the engineering of tissue and organ analogs. However, many challenges remain in the engineering of 3D tissues and organs for clinical use. Nevertheless, many advances have been made in synthetic polymer chemistry, scaffold processing methods, and tissue-culture techniques. These may eventually allow the generation of long-term functional complex cell–polymer constructs with precisely controlled local environment such as material microstructure, nutrient and growth factor concentration, and mechanical forces.

Bibliography

Aulthouse, A. L., Beck, M., Griffey, E., Sanford, J., Arden, K., Machado, M. A., and Horton, W. A. (1989). Expression of the human chondrocyte phenotype in vitro. *In Vitro Cell Dev. Biol.* **25**: 659–668.

Babensee, J. E., Anderson, J. M., McIntire, L. V., and Mikos, A. G. (1998). Host response to tissue engineered devices. *Adv. Drug Deliv. Rev.* **33**: 111–139.

Babensee, J. E., McIntire, L. A., and Mikos, A. G. (2000). Growth factor delivery for tissue engineering. *Pharm. Res.* **17**: 497–504.

Bancroft, G. N., and Mikos, A. G. (2001). Bone tissue engineering by cell transplantation. in *Tissue Engineering for Therapeutic Use*, Y. Ikada and N. Ohshima (eds.). Vol. 5. Elsevier Science, New York, pp. 151–163.

Bancroft, G. N., Sikavitsas, V. I., van den Dolder, J., Sheffield, T. L., Ambrose, C. G., Jansen, J. A., and Mikos, A. G. (2002). Fluid flow increases mineralized matrix deposition in 3D perfusion culture of marrow stromal osteoblasts in a dose-dependent manner. *Proc. Natl. Acad. Sci. USA* **99**: 12600–12605.

Barrera, D., Zylstra, E., Lansbury, P., and Langer, R. (1995). Copolymerization and degradation of poly(lactic-*co*-lysine). *Macromolecules* **28**: 425–432.

Bartels, K. A., Bovik, A. C., Crawford, R. C., Diller, K. R., and Aggarwal, S. J. (1993). Selective laser sintering for the creation of solid models from 3D microscopic images. *Biomed. Sci. Instrum.* **29**: 243–250.

Behravesh, E., and Mikos, A. G. (2003). Three-dimensional culture of marrow stromal osteoblasts in biomimetic poly(propylene fumarate-*co*-ethylene glycol)-based macroporous hydrogels. *J. Biomed. Mater. Res.* **66A**: 698–706.

Behravesh, E., Yasko, A. W., Engel, P. S., and Mikos, A. G. (1999). Synthetic biodegradable polymers for orthopaedic applications. *Clin. Orthop. Rel. Res.* **367**: S118–S129.

Behravesh, E., Jo, S., Zygourakis, K., and Mikos, A.G. (2002). Synthesis of in-situ cross-linkable macroporous biodegradable poly(propylene fumarate-*co*-ethylene glycol) hydrogels. *Biomacro-molecules* **3**: 374–381.

Berry, E., Brown, J. M., Connell, M., Craven, C. M., Efford, N .D., Radjenovic, A., and Smith, M. A. (1997). Preliminary experience with medical applications of rapid prototyping by selective laser sintering. *Med. Eng. Phys.* **19**: 90–96.

Blum, J. S., Barry, M. A., and Mikos, A. G. (2003). Bone regeneration through transplantation of genetically modified cells. *Clin. Plast. Surg.* **30**: 611–620.

Boyan, B. D., Hummert, T. W., Dean, D. D., and Schwartz, Z. (1996). Role of material surfaces in regulating bone and cartilage cell response. *Biomaterials* **17**: 137–146.

Chen, C. S., Mrksich, M., Huang, S., Whitesides, G. M., and Ingber, D. E. (1998). Micropatterned surfaces for control of cell shape, position, and function. *Biotechnol. Prog.* **14**: 356–363.

Chicurel, M.E., Chen, C.S., and Ingber, D.E. (1998). Cellular control lies in the balance of forces. *Curr. Opin. Cell Biol.* **10**: 232–239.

Chiquet, M., Matthisson, M., Koch, M., Tannheimer, M., and Chiquet-Ehrismann, R. (1996). Regulation of extracellular matrix synthesis by mechanical stress. *Biochem. Cell Biol.* **74**: 737–744.

Cima, L.G., Vacanti, J.P., Vacanti, C., Ingber, D., Mooney, D., and Langer, R. (1991). Tissue engineering by cell transplantation using degradable polymer substrates. *J. Biomech. Eng.* **113**: 143–151.

Cook, A.D., Hrkach, J.S., Gao, N.N., Johnson, I.M., Pajvani, U.B., Cannizzaro, S.M., and Langer, R. (1997). Characterization and development of RGD-peptide-modified poly(lactic acid-*co*-lysine) as an interactive, resorbable biomaterial. *J. Biomed. Mater. Res.* **35**: 513–523.

Cooke, M.N., Fisher, J.P., Dean, D., Rimnac, C., and Mikos, A.G. (2002). Use of stereolithography to manufacture critical-sized 3D biodegradable scaffolds for bone ingrowth. *J. Biomed. Mater. Res. Part B: Appl. Biomater.* **64B**: 65–69.

Ferguson, S., Wahl, D., and Gogolewski, S. (1996). Enhancement of the mechanical properties of polylactides by solid-state extrusion. II. Poly(L-lactide), poly(L/D-lactide), and poly(L/DL-lactide). *J. Biomed. Mater. Res.* **30**: 543–551.

Fisher, J.P., Vehof, J.W.M., Dean, D., van der Waerden, J.P.C.M., Holland, T.A., Mikos, A.G., and Jansen, J.A. (2002). Soft and hard tissue response to photocrosslinked poly(propylene fumarate) scaffolds in a rabbit model. *J. Biomed. Mater. Res.* **59**: 547–556.

Fisher, J.P., Holland, T.A., Dean, D., and Mikos, A.G. (2003). Photoinitiated cross-linking of the biodegradable polyester poly(propylene fumarate). Part II. *In vitro* degradation. *Biomacro-molecules* **4**: 1335–1342.

Freed, L.E., and Vunjak-Novakovic, G. (1997). Microgravity tissue engineering. *In Vitro Cell. Dev. Biol.-Animal* **33**: 381–385.

Freed, L.E., Vunjak-Novakovic, G., and Langer, R. (1993). Cultivation of cell-polymer cartilage implants in bioreactors. *J. Cell. Biochem.* **51**: 257–264.

Gao, J., Niklason, L., and Langer, R. (1998). Surface hydrolysis of poly(glycolic acid) meshes increases the seeding density of vascular smooth muscle cells. *J. Biomed. Mater. Res.* **42**: 417–424.

Giordano, R.A., Wu, B.M., Borland, S.W., Cima, L.G., Sachs, E.M., and Cima, M. J. (1996). Mechanical properties of dense polylactic acid structures fabricated by three dimensional printing. *J. Biomater. Sci. Polymer Ed.* **8**: 63–75.

Glowacki, J., Mizuno, S., and Greenberger, J.S. (1998). Perfusion enhances functions of bone marrow stromal cells in three-dimensional culture. *Cell Transplant.* **7**: 319–326.

Goldstein, A.S., Zhu, G., Morris, G.E., Meszlenyi, R.K., and Mikos, A.G. (1999). Effect of osteoblastic culture conditions on the structure of poly(DL-lactic-*co*-glycolic acid) foam scaffolds. *Tissue Eng.* **5**:421–433.

Goodman, S., and Aspenberg, P. (1993). Effects of mechanical stimulation on the differentiation of hard tissues. *Biomaterials* **14**: 563–569.

Griffith, L.G., Wu, B., Cima, M.J., Powers, M.J., Chaignaud, B., and Vacanti, J.P. (1997). *In vitro* organogenesis of liver tissue. *Ann. N.Y. Acad. Sci.* **831**: 382–397.

Harris, L.D., Kim, B.-S., and Mooney, D.J. (1998). Open pore biodegradable matrices formed with gas foaming. *J. Biomed. Mater. Res.* **42**: 396–402.

Hay, E.D. (1993). Extracellular matrix alters epithelial differentiation. *Curr. Opin. Cell Biol.* **5**: 1029–1035.

Hedberg, E.L., Tang, A., Crowther, R.S., Careny, D.H., and Mikos, A.G. (2002). Controlled release of an osteogenic peptide from injectable biodegradable polymeric composites. *J. Controlled Release* **84**: 137–150.

Holland, T.A., and Mikos, A.G. (2003). Review: advances in drug delivery for articular cartilage. *J. Controlled Release* **86**: 1–14.

Holland, T.A., Tabata, Y., and Mikos, A.G. (2003). *In vitro* release of transforming growth factor-β1 from gelatin microparticles encapsulated in biodegradable, injectable oligo(poly(ethylene glycol) fumarate) hydrogels. *J. Controlled Release* **91**: 299–313.

Howe, A., Aplin, A.E., Alahari, S.K., and Juliano, R.L. (1998). Integrin signaling and cell growth control. *Curr. Opin. Cell Biol.* **10**: 220–231.

Hsu, Y.Y., Gresser, J.D., Trantolo, D.J., Lyons, C.M., Gangadharam, P.R.J., and Wise, D.L. (1997). Effect of polymer foam morphology and density on kinetics of *in vitro* controlled release of isoniazid from compressed foam matrices. *J. Biomed. Mater. Res.* **35**: 107–116.

Ishaug, S.L., Crane, G.M., Miller, M.J., Yasko, A.W., Yaszemski, M.J., and Mikos, A.G. (1997). Bone formation by three-dimensional stromal osteoblast culture in biodegradable polymer scaffolds. *J. Biomed. Mater. Res.* **36**: 17–28.

Ishaug-Riley, S.L., Crane, G.M., Gurlek, A., Miller, M.J., Yasko, A.W., Yaszemski, M.J., and Mikos, A.G. (1997). Ectopic bone formation by marrow stromal osteoblast transplantation using poly(DL-lactic-*co*-glycolic acid) foams implanted into the rat mesentery. *J. Biomed. Mater. Res.* **36**: 1–8.

Ishaug-Riley, S.L., Crane-Kruger, G.M., Yaszemski, M.J., and Mikos, A.G. (1998). Three-dimensional culture of rat calvarial osteoblasts in porous biodegradable polymers. *Biomaterials* **19**: 1405–1412.

James, K., and Kohn, J. (1996). New biomaterials for tissue engineering. *Mater. Res. Soc. Bull.* **21**: 22–26.

Jo, S., Shin, H., Shung, A.K., Fisher, J.P., and Mikos, A.G. (2001). Synthesis and characterization of oligo(poly(ethylene glycol) fumarate) macromer. *Macromolecules* **34**: 2839–2844.

Kamihira, M., Yamada, K., Hamamoto, R., and Iijima, S. (1997). Spheroid formation of hepatocytes using synthetic polymer. *Ann. N.Y. Acad. Sci.* **831**: 398–407.

Kasper, F.K., and Mikos, A.G. (2003). Biomaterials and gene therapy. in *Molecular and Cellular Foundations of Biomaterials*, N. Peppas and M.V. Sefton (eds.). Academic Press, New York, pp. 131–163.

Kim, B.-S., and Mooney, D.J. (1998). Engineering smooth muscle tissue with a predefined structure. *J. Biomed. Mater. Res.* **41**: 322–332.

Kim, S.S., Utsunomiya, H., Koski, J.A., Wu, B.M., Cima, M.J., Sohn, J., Mukai, K., Griffith, L.G., and Vacanti, J.P. (1998). Survival and function of hepatocytes on a novel three-dimensional synthetic biodegradable polymer scaffold with an intrinsic network of channels. *Ann. Surg.* **228**: 8–13.

Langer, R., and Vacanti, J.P. (1993). Tissue engineering. *Science* **260**: 920–926.

Lian, J.B., and Stein, G.S. (1992). Concepts of osteoblast growth and differentiation: basis for modulation of bone cell development and tissue formation. *Crit. Rev. Oral. Biol. Med.* **3**: 269–305.

Lo, H., Ponticiello, M.S., and Leong, K.W. (1995). Fabrication of controlled release biodegradable foams by phase separation. *Tissue Eng.* **1**: 15–28.

Lu, L., Carcia, C.A., and Mikos, A.G. (1998). Retinal pigment epithelium cell culture on thin biodegradable poly(DL-lactic-*co*-glycolic acid) films. *J. Biomater. Sci. Polymer Ed.* **9**: 1187–1205.

Lu, L., Stamatas, G.N., and Mikos, A.G. (2000). Controlled release of transforming growth factor-β1 from biodegradable polymers. *J. Biomed. Mater. Res.* **50**: 440–451.

Mikos, A.G., Bao, Y., Cima, L.G., Ingber, D.E., Vacanti, J.P., and Langer, R. (1993a). Preparation of poly(glycolic acid) bonded fiber structures for cell attachment and transplantation. *J. Biomed. Mater. Res.* **27**: 183–189.

Mikos, A.G., Sarakinos, G., Leite, S. M., Vacanti, J.P., and Langer, R. (1993b). Laminated three-dimensional biodegradable foams for use in tissue engineering. *Biomaterials* **14**: 323–330.

Mikos, A.G., Sarakinos, G., Lyman, M.D., Ingber, D.E., Vacanti, J.P., and Langer, R. (1993c). Prevascularization of porous biodegradable polymers. *Biotechnol. Bioeng.* **42**: 716–723.

Mikos, A.G., Lyman, M.D., Freed, L.E., and Langer, R. (1994a). Wetting of poly(L-lactic acid) and poly(DL-lactic-*co*-glycolic acid) foams for tissue culture. *Biomaterials* **15**: 55–58.

Mikos, A.G., Thorsen, A.J., Czerwonka, L. A., Bao, Y., Langer, R., Winslow, D.N., and Vacanti, J.P. (1994b). Preparation and characterization of poly(L-lactic acid) foams. *Polymer* **35**: 1068–1077.

Miller, N.D., and Williams, D.F. (1984). The *in vivo* and *in vitro* degradation of poly(glycolic acid) suture material as a function of applied strain. *Biomaterials* **5**: 365–368.

Moghe, P.V. (1996). Soft-tissue analogue design and tissue engineering of liver. *Mater. Res. Soc. Bull.* **21**: 52–54.

Mooney, D., Hansen, L., Vacanti, J., Langer, R., Farmer, S., and Ingber, D. (1992). Switching from differentiation to growth in hepatocytes: control by extracellular matrix. *J. Cell. Physiol.* **151**: 497–505.

Mooney, D.J., Mazzoni, C.L., Organ, G.M., Puelacher, W.C., Vacanti, J.P., and Langer, R. (1994a). Stabilizing fiber-based cell delivery devices by physically bonding adjacent fibers. in *Biomaterials for Drug and Cell Delivery*, A.G. Mikos, R.M. Murphy, H. Bernstein, and N.A. Peppas (eds.). Vol. 331. Material Research Society, Pittsburgh, pp. 47–52.

Mooney, D.J., Organ, G., Vacanti, J.P., and Langer, R. (1994b). Design and fabrication of biodegradable polymer devices to engineer tubular tissues. *Cell Transplant.* **3**: 203–210.

Mooney, D.J., Breuer, C., McNamara, K., Vacanti, J.P., and Langer, R. (1995a). Fabricating tubular devices from polymers of lactic and glycolic acid for tissue engineering. *Tissue Eng.* 1:107–118.

Mooney, D.J., Park, S., Kaufmann, P.M., Sano, K., McNamara, K., Vacanti, J.P., and Langer, R. (1995b). Biodegradable sponges for hepatocyte transplantation. *J. Biomed. Mater. Res.* **29**: 959–965.

Mooney, D.J., Baldwin, D.F., Suh, N.P., Vacanti, J.P., and Langer, R. (1996a). Novel approach to fabricate porous sponges of poly(D,L-lactic-*co*-glycolic acid) without the use of organic solvents. *Biomaterials* **17**: 1417–1422.

Mooney, D.J., Mazzoni, C.L., Breuer, C., McNamara, K., Hern, D., Vacanti, J.P., and Langer, R. (1996b). Stabilized polyglycolic acid fibre-based tubes for tissue engineering. *Biomaterials* **17**: 115–124.

Park, A., and Cima, L.G. (1996). *In vitro* cell response to differences in poly-L-lactide crystallinity. *J. Biomed. Mater. Res.* **31**: 117–130.

Park, A., Wu, B., and Griffith, L.G. (1998). Integration of surface modification and 3D fabrication techniques to prepare patterned poly(L-lactide) substrates allowing regionally selective cell adhesion. *J. Biomater. Sci. Polymer Ed.* **9**: 89–110.

Payne, R.G., McGonigle, J.S., Yaszemski, M.J., Yasko, A.W., and Mikos, A.G. (2002a). Development of an injectable, *in situ* crosslinkable, degradable polymeric carrier for osteogenic cell populations. Part 2. Viability of encapsulated marrow stromal osteoblasts cultured on crosslinking poly(propylene fumarate). *Biomaterials* **23**: 4373–4380.

Payne, R.G., McGonigle, J.S., Yaszemski, M.J., Yasko, A.W., and Mikos, A.G. (2002b). Development of an injectable, *in situ* crosslinkable, degradable polymeric carrier for osteogenic cell populations. Part 3. Proliferation and differentiation of encapsulated marrow stromal osteoblasts cultured on crosslinking poly(propylene fumarate). *Biomaterials* **23**: 4381–4387.

Peter, S.J., Miller, M.J., Yaszemski, M.J., and Mikos, A.G. (1997a). Poly(propylene fumarate). in *Handbook of Biodegradable Polymers*, A.J. Domb, J. Kost and D.M. Wiseman (eds.). Harwood Academic Publishers, Amsterdam, pp. 87–97.

Peter, S.J., Nolley, J.A., Widmer, M.S., Merwin, J.E., Yaszemski, M.J., Yasko, A.W., Engel, P.S., and Mikos, A.G. (1997b). *In vitro* degradation of a poly(propylene fumarate)/β-tricalcium phosphate composite orthopaedic scaffold. *Tissue Eng.* **3**: 207–215.

Peter, S.J., Liang, C.R., Kim, D.J., Widmer, M.S., and Mikos, A.G. (1998a). Osteoblastic phenotype of rat marrow stromal cells cultured in the presence of dexamethasone, β-glycerolphosphate, and L-ascorbic acid. *J. Cell. Biochem.* **71**: 55–62.

Peter, S.J., Miller, S.T., Zhu, G., Yasko, A. W., and Mikos, A.G. (1998b). *In vivo* degradation of a poly(propylene fumarate)/ β-tricalcium phosphate injectable composite scaffold. *J. Biomed. Mater. Res.* **41**: 1–7.

Peter, S.J., Kim, P., Yasko, A.W., Yaszemski, M.J., and Mikos, A.G. (1999). Crosslinking characteristics of an injectable poly(propylene fumarate)/ β-tricalcium phosphate paste and mechanical properties of the cross-linked composite for use as a biodegradable bone cement. *J. Biomed. Mater. Res.* **44**: 314–321.

Renier, M.L., and Kohn, D.H. (1997). Development and characterization of a biodegradable polyphosphate. *J. Biomed. Mater. Res.* **34**: 95–104.

Ripamonti, U., and Duneas, N. (1996). Tissue engineering of bone by osteoinductive biomaterials. *Mater. Res. Soc. Bull.* **21**: 36–39.

Ruardy, T.G., Schakenraad, J.M., van der Mei, H.C., and Busscher, H.J. (1995). Adhesion and spreading of human skin fibroblasts on physicochemically characterized gradient surfaces. *J. Biomed. Mater. Res.* **29**: 1415–1423.

Shakesheff, K.M., Cannizzaro, S.M., and Langer, R. (1998). Creating biomimetic microenvironments with synthetic polymer-peptide hybrid molecules. *J. Biomater. Sci. Polymer Ed.* **9**: 507–518.

Shin, H., Jo, S., and Mikos, A.G. (2002). Modulation of marrow stromal osteoblast adhesion on biomimetic oligo(poly(ethylene glycol) fumarate) hydrogels modified with Arg-Gly-Asp peptides and a poly(ethylene glycol) spacer. *J. Biomed. Mater. Res.* **61**: 169–179.

Shin, H., Jo, S., and Mikos, A.G. (2003a). Review: biomimetic materials for tissue engineering. *Biomaterials* **24**: 4353–4364.

Shin, H., Ruhe, P.Q., and Mikos, A.G. (2003b). *In vivo* bone and soft tissue response to injectable, biodegradable oligo(poly(ethylene glycol) fumarate) hydrogels. *Biomaterials* **24**: 3201–3211.

Singhvi, R., Kumar, A., Lopez, G., Stephanopoulos, G.N., Wang, D.I.C., Whitesides, G.M., and Ingber, D.E. (1994). Engineering cell shape and function. *Science* **264**: 696–698.

Suggs, L.J., and Mikos, A.G. (1996). Synthetic biodegradable polymers for medical applications. in *Physical Properties of Polymers Handbook*, J.E. Mark (ed.). American Institute of Physics, Woodbury, NY, pp. 615–624.

Suggs, L.J., Payne, R.G., Yaszemski, M.J., Alemany, L.B., and Mikos, A.G. (1997). Synthesis and characterization of a block copolymer consisting of poly(propylene fumarate) and poly(ethylene glycol). *Macromolecules* **30**: 4318–4323.

Suggs, L.J., Kao, E.Y., Palombo, L.L., Krishnan, R.S., Widmer, M.S., and Mikos, A.G. (1998a). Preparation and characterization of poly(propylene fumarate-co-ethylene glycol) hydrogels. *J. Biomater. Sci. Polymer Ed.* **9**: 653–666.

Suggs, L.J., Krishnan, R.S., Garcia, C.A., Peter, S.J., Anderson, J.M., and Mikos, A.G. (1998b). *In vitro* and *in vivo* degradation of poly(propylene fumarate-*co*-ethylene glycol) hydrogels. **42**: 312–320.

Suggs, L.J., Shive, M.S., Garcia, C.A., Anderson, J.M., and Mikos, A.G. (1999). *In vitro* cytotoxicity and *in vivo* biocompatibility of poly(propylene fumarate-*co*-ethylene glycol) hydrogels. *J. Biomed. Mater. Res.* **46**: 22–32.

Temenoff, J.S., and Mikos, A.G. (2000). Formation of highly porous biodegradable scaffolds for tissue engineering. *Electr. J. Biotechnol.* **3**: http://www.ejb.org/content/vol3/issue2/full/5/index.html.

Temenoff, J.S., Steinbis, E.S., and Mikos, A.G. (2003). Effect of drying history on swelling properties and cell attachment to oligo(poly(ethylene glycol) fumarate) hydrogels for guided tissue regeneration applications. *J. Biomater. Sci. Polymer Ed.* **14**: 989–1004.

Thomson, R.C., Yaszemski, M.J., Powers, J.M., and Mikos, A.G. (1995). Fabrication of biodegradable polymer scaffolds to engineer trabecular bone. *J. Biomater. Sci. Polymer Ed.* **7**: 23–28.

Thomson, R.C., Yaszemski, M.J., Powers, J.M., and Mikos, A.G. (1998). Hydroxyapatite fiber reinforced poly(α-hydroxy ester) foams for bone regeneration. *Biomaterials* **19**: 1935–1943.

Thomson, R.C., Mikos, A.G., Beahm, E., Lemon, J.C., Satterfield, W.C., Aufdemorte, T.B., and Miller, M.J. (1999). Guided tissue fabrication from periosteum using preformed biodegradable polymer scaffolds. *Biomaterials* **20**: 2007–2018.

Timmer, M.D., Jo, S., Wang, C., Ambrose, C.G., and Mikos, A.G. (2002). Characterization of the cross-linked structure of fumarate-based degradable polymer networks. *Macromolecules* **35**: 4373–4379.

Timmer, M.D., Ambrose, C.G., and Mikos, A.G. (2003). Evaluation of thermal- and photo-crosslinked biodegradable poly(propylene fumarate)-based networks. *J. Biomed. Mater. Res.* **66A**: 811–818.

Tjia, J.S., and Moghe, P.V. (1998). Analysis of 3-D microstructure of porous poly(lactide–glycolide) matrices using confocal microscopy. *J. Biomed. Mater. Res.* **43**: 291–299.

Vacanti, C.A., Langer, R., Schloo, B., and Vacanti, J.P. (1991). Synthetic polymers seeded with chondrocytes provide a template for new cartilage formation. *Plast. Reconstr. Surg.* **88**: 753–759.

Vunjak-Novakovic, G., Obradovic, B., Martin, I., Bursac, P.M., Langer, R., and Freed, L.E. (1998). Dynamic cell seeding of polymer scaffolds for cartilage tissue engineering. *Biotechnol. Prog.* **14**: 193–202.

Wake, M.C., Patrick, C.W., Jr., and Mikos, A.G. (1994). Pore morphology effects on the fibrovascular tissue growth in porous polymer substrates. *Cell Transplant.* **3**: 339–343.

Wake, M.C., Gupta, P.K., and Mikos, A.G. (1996). Fabrication of pliable biodegradable polymer foams to engineer soft tissues. *Cell Transplant.* **5**: 465–473.

Wake, M.C., Gerecht, P.D., Lu, L., and Mikos, A.G. (1998). Effects of biodegradable polymer particles on rat marrow derived stromal osteoblasts in vitro. *Biomaterials* **19**: 1255–1268.

Wald, H.L., Sarakinos, G., Lyman, M.D., Mikos, A.G., Vacanti, J.P., and Langer, R. (1993). Cell seeding in porous transplantation devices. *Biomaterials* **14**: 270–278.

Whang, K., Thomas, C.H., Healy, K.E., and Nuber, G. (1995). A novel method to fabricate bioabsorbable scaffolds. *Polymer* **36**: 837–842.

Widmer, M.S., Gupta, P.K., Lu, L., Meszlenyi, R.K., Evans, G.R.D., Brandt, K., Savel, T., Gurlek, A., Patrick, C.W., Jr., and Mikos, A.G. (1998). Manufacture of porous biodegradable polymer conduits by an extrusion process for guided tissue regeneration. *Biomaterials* **19**: 1945–1955.

Wintermantel, E., Mayer, J., Blum, J., Eckert, K.-L., Luscher, P., and Mathey, M. (1996). Tissue engineering scaffolds using superstructures. *Biomaterials* **17**: 83–91.

Wu, B.M., Borland, S.W., Giordano, R.A., Cima, L.G., Sachs, E.M., and Cima, M. J. (1996). Solid free-from fabrication of drug delivery devices. *J. Controlled Release* **40**: 77–87.

Chapter 7.2

7.2

Scope of tissue engineering

Yoshito Ikada

In tissue engineering, a neotissue generally is regenerated from the cells seeded onto a bioabsorbable scaffold, occasionally incorporating growth factors: *cells + scaffold + growth factors → neotissue*. In theory, any tissue could be created using this basic principle of tissue engineering. However, in order to achieve successful regeneration of tissues or organs based on the tissue engineering concept, several critical elements should be deliberately considered including biomaterial scaffolds that serve as a mechanical support for cell growth, progenitor cells that can be differentiated into specific cell types, and inductive growth factors that can modulate cellular activities. The fundamentals of tissue engineering will be presented in this chapter.

7.2.1 Functions of scaffold

When a tissue is severely damaged or lost, not only large numbers of functional cells but also the matrix in tissue, generally called *extracellular matrix* (ECM), are lost. It is difficult to imagine how small-molecule drugs or even recombinant proteins would be able to restore the lost tissue and reverse the function. Because tissue represents a highly organized interplay of cells and matrices, the fabrication of replacement tissue may be facilitated by mimicking the spatial organization in tissue. To this end, we should provide an artificial or biologically derived ECM for cells to create a neotissue. Isolated cells have the capacity to form a tissue structure only to a limited degree when placed as a suspension on tissue, because they need a template that guides cell organization. In tissue engineering we designate the substitute of native ECM as "scaffold", "template", or "artificial matrix". Scaffold provides a three-dimensional (3-D) ECM analog which functions as a template required for infiltration and proliferation of cells into the targeted functional tissue or organ. If any assistance by scaffold is not required for cells, we call it "cell (or cellular) therapy" or "cell transplantation." Cell therapy avoids the complications of surgery, but allows replacement of only those cells that perform the biological functions including hormone secretion and enzyme synthesis. It would be therefore convenient to divide regenerative medicine into two subgroups, as shown in Fig. 7.2-1, depending on the scaffold requirement.

The primary function of scaffolds is to provide structure for organizing dissociated cells into appropriate tissue construction by creating an environment that enables 3-D cell growth and neotissue formation. When cells attached to a scaffold are implanted, they will be incorporated into the body. Cell attachment is the first critical element in initiating cell growth and neotissue development. Natural or synthetic biomaterials utilized for scaffold fabrication are mostly selected on the basis of their biocompatibility, bioabsorbability, and mechanical properties. Much of the secondary scaffold processing is performed to make the scaffold more porous for enhancement of cell infiltration and neotissue ingrowth. To promote cell attachment various cell adhesion molecules such as laminin (LN) have been used to coat the scaffold before cell seeding. The traditional method of seeding polymer scaffolds with cells has employed static cell culture techniques. For instance, a concentrated cell suspension is pipetted onto a collagen-coated polymer scaffold and left to incubate for variable periods of time for cells to adhere to the polymer. Dynamic cell seeding employs a method in which either the medium or the medium and scaffold are in constant motion during the incubation period.

Biomedical Engineering Desk Reference; ISBN: 9780123746467

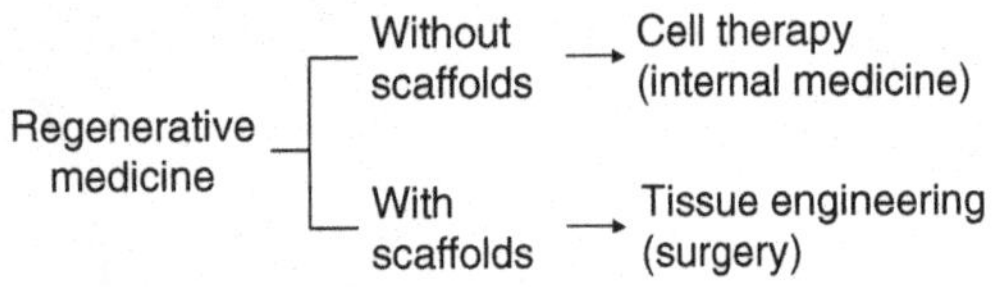

Fig. 7.2-1 Classification of regenerative medicine based on the use of scaffold.

Scaffolds should encourage the growth, migration, and organization of cells, providing support while the tissue is forming. Finally, as demonstrated in Fig. 7.2-2, the scaffolds will be replaced with host cells and a new ECM which in turn should provide functional and mechanical properties, similar to native tissue. The material and the 3-D structure of scaffolds have a significant effect on cellular activity. Depending on the tissue of interest and the specific application, the required scaffold material and its properties will be quite different. In general, a biologically active scaffold should provide the following characteristics: (1) a 3-D, well-defined porous structure to make the surface-to-volume ratio high for seeding of cells as many as possible; (2) a physicochemical structure to support cell attachment, proliferation, differentiation, and ECM production to organize cells into a 3-D architecture; (3) an interconnected, permeable pore network to promote nutrient and waste exchange; (4) a non-toxic, bioabsorbable substrate with a controllable absorption rate to match cell and tissue growth *in vitro* or *in vivo*, eventually leaving no foreign materials within the replaced tissues; (5) a biological property to facilitate vasculature network formation in the scaffold; (6) mechanical properties to support or match those of the tissue at the site of implantation and occasionally to present stimuli which direct the growth and formation of a load-bearing tissue; (7) a mechanical architecture to temporarily provide the biomechanical structural characteristics for the replacement "tissue" until the cells produce their own ECM; (8) a good carrier to act as a delivery system for bioactive agents, such as growth factors; (9) a geometry which promotes formation of the desired, anisotropic tissue structure; (10) a processable and reproducible architecture of clinically relevant size and shape; (11) sterile and stable enough for shelf life, transportation, and production; and (12) economically viable and scaleable material production, purification, and processing.

Scaffolds initially fill a space otherwise occupied by natural tissue, and then provide a framework by which a tissue will be regenerated. The architecture of 3-D scaffold can also control vascularization and tissue ingrowth *in vivo*. In this capacity, the physical and biological properties of the material are inherent in the success of the scaffold. Selection and synthesis of the appropriate scaffold material is governed by the intended scaffold application and environment in which the scaffold will be placed. For example, a scaffold designed to encapsulate cells must be capable of being gelled without damaging the cells, must allow appropriate diffusion of nutrients and metabolites to and from the encapsulated cells and surrounding tissue, and stay at the site of implantation with sufficient mechanical integrity and strength. Scaffold heterogeneity has been shown to lead to variable cell adhesion and to affect the ability of the cells to produce a uniform distribution of ECM. Tissue synthesized in a scaffold with non-uniform pore architecture may show inferior biomechanical properties compared to tissue synthesized in a scaffold with a more uniform pore structure. In scaffolds with equiaxed pores, cells aggregate into spherical structures, while in scaffolds with a more elongated pore shape, cells align with the pore axis.

Once the scaffold is produced and placed, formation of tissues with desirable properties relies on scaffold mechanical properties on both the macroscopic and the microscopic levels. Macroscopically, the scaffold must bear loads to provide stability to tissues as it forms and to fulfill its volume maintenance function. On the

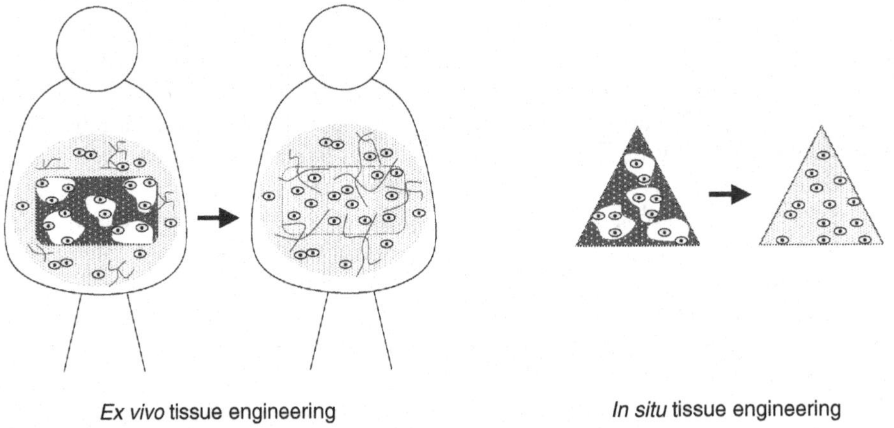

Fig. 7.2-2 Role of scaffold in tissue engineering.

microscopic level, cell growth and differentiation and ultimate tissue formation are dependent on mechanical input to the cells. As a consequence, the scaffold must be able to both withstand specific loads and transmit them in an appropriate manner to the surrounding cells and tissues. Specific mechanical properties of scaffolds include elasticity, compressibility, viscoelastic behavior, tensile strength, failure strain, and their time-dependent fatigue.

7.2.2 Absorbable biomaterials

Biomaterials are a critical enabling technology in tissue engineering, because they serve in various ways as a substrate on which cell populations can attach and migrate, as a 3-D implant with a combination of specific cell types, as a cell delivery vehicle, as a drug carrier to activate specific cellular function in the localized region, as a mechanical structure to define the shape of regenerating tissue, and as a barrier membrane to provide space for tissue regeneration along with prevention of fibroblast ingrowth into the space. In many cases a biomaterial serves a dual role as scaffold and as delivery device. Degradation and absorption of biomaterials are essential in functional tissue regeneration, unless the application is aimed at long-term encapsulation of cells to be immunologically isolated. Materials that disappear from the body after they have fulfilled their function obviate concerns about long-term biocompatibility. The by-products of degradation must be non-toxic, similar to the starting material. For a biomaterial to be accepted in the medical system, its safety and efficacy must be proven with any therapy. Ideally, the rate of scaffold degradation should mirror the rate of new tissue formation or be adequate for the controlled release of bioactive molecules. Table 7.2-1 represents naturally occurring and synthetic biomaterials that possess hydrolysable bonds in the main chain. Their current medical applications are summarized in Table 7.2-2. Clinical applications of absorbable biomaterials have a long history similar to those of non-absorbable biomaterials. Absorbable sutures like catgut and hemostatic or sealing agents like collagen, oxidized cellulose, and α-cyanoacrylate polymers have been the front runners in the medical use of absorbable biomaterials. The largest clinical application of absorbable biomaterials at present is for suturing and ligature.

The technical term "degradable" that has been used for biomaterials has several synonyms including biodegradable, absorbable, bioabsorbable, and resorbable. In general, the term "degradable" is used when the molecular weight (MW) of the polymer constructing the material does decrease over time. When such degradation takes place only in a biological environment where enzymes exist, the term "biodegradable" is preferably used because the degradation of material is a result of enzymatic, biological action. When a material disappears by being taken into another body, we call the phenomenon "absorption". In this case a decrease in MW is not a necessary condition. A good example for this is alginate which is water soluble but becomes water insoluble upon addition of divalent cations such as Ca^{2+}. This water-insoluble gel recovers to a water-soluble sol when a high concentration of monovalent, sodium ion is present. If this sol–gel transition takes place through ion exchange in the living body, one can say that the alginate–Ca^{2+} complex has been "absorbed" into the body without a decrease in MW of the starting alginate. As this example suggests, the term "absorbable" (or resorbable) seems to be more suitable than "degradable" so far as their medical use is concerned, because absorption includes both chemical degradation (either by passive hydrolytic or by enzymatic cleavage) and physicochemical absorption (through simple physical dissolution into aqueous media). Here the term "absorbable" (or "bioabsorbable") will be mostly used. The term "bioabsorbable" is used simply because the absorption proceeds in the biological environment.

Specific bulk and surface properties—including mechanical strength, absorption kinetics, wettability, and cell adhesion—are required for the biomaterials that are used for tissue engineering. Large numbers of both biologically derived and synthetic materials that meet these requirements have been extensively explored in tissue engineering. These biomaterials can be categorized according to several schemes. Here their categorization is based on their source.

7.2.2.1 Natural polymers

The origin of naturally occurring polymers is human, animals, or plants. Materials from natural sources such as collagen derived from animal tissues have been considered to be advantageous because of their inherent properties of biological recognition, including presentation of receptor-binding ligands and susceptibility to cell-triggered proteolytic degradation and remodeling. However, the biologically derived materials have concerns, especially complexities associated with purification, sustainable production, immunogenicity, and pathogen transmission. Apart from this fact, medical applications of absorbable natural polymers are limited, because their mechanical strength is not strong enough when hydrated. One exception is chitin (and chitosan) that is a crystalline polymer. Most of natural polymers are soluble in aqueous media or hydrophilic. Because water-soluble polymers are not appropriate as a 3-D scaffold, they should be converted into water-insoluble materials by physical or chemical reactions.

Table 7.2-1 Hydrolyzable bonds in bioabsorbable macromoecules and the representatives

Bond	Chemical structure	Representative polymers
Ester	$-\text{C}(=\text{O})-\text{O}-$	PGA, PLA, poly(malic acid)
Peptide	$-\text{N(H)}-\text{C}(=\text{O})-$	Collagen, fibrin, synthetic polypeptides
Glycoside	$(-\text{O},-\text{C})\text{C}-\text{O}-\text{C}(\text{C},-\text{O})$	HAc, alginate, starch, chitin, chitosan
Phosphate	$-\text{P}(=\text{O})(\text{O})-\text{O}-$	Nucleic acid
Anhydride	$-\text{C}(=\text{O})-\text{O}-\text{C}(=\text{O})-$	$+\text{C}(=\text{O})-\text{C}_6\text{H}_4-\text{O}-(\text{CH}_2)_4-\text{C}(=\text{O})-\text{O}+_n$
Carbonate	$-\text{O}-\text{C}(=\text{O})-\text{O}-$	$+\text{CH}_2-\text{CH}_2-\text{O}-\text{C}(=\text{O})-\text{O}+_n$
Orthoester	$(-\text{C},-\text{O})\text{C}(\text{O}-,\text{O}-)$	
Carbon-carbon	$-\text{CH}_2-\text{C}(\text{CN})(\text{C}(=\text{O})\text{O}-)-$	Poly(isobutyl cyabiacrylate)
Phosphazene	$-\text{P}=\text{N}-$	Polydiaminophosphazene

7.2.2.1.1 Proteins

The major source of naturally derived proteins has been bovine or porcine connective tissues from peritoneum, blood vessels, heart valves, and intestine. These connective tissues have excellent mechanical properties; reconstructed collagen sponge is incomparable to natural collagenous tissues with respect to the tensile properties. Additional drawbacks of naturally derived materials include possible risks of prion such as bovine spongiform encephalitis (BSE), immunogenicity of eventually remaining cells, their remnants, and biopolymers themselves. However, many of natural polymers can promote

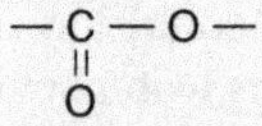

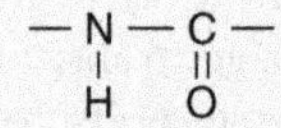

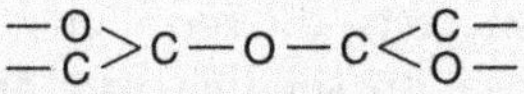

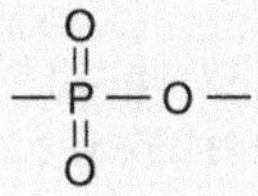

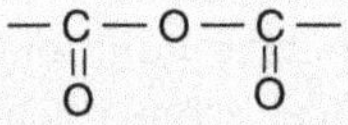

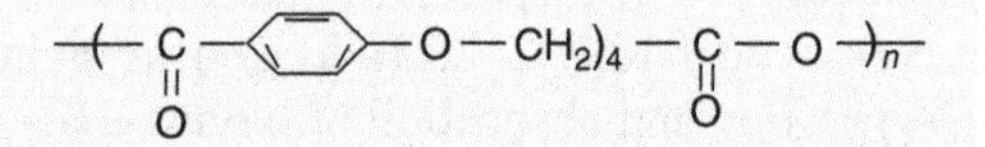

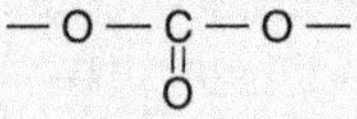

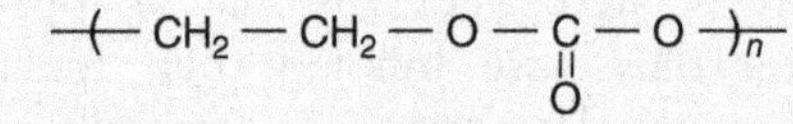

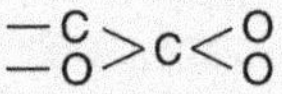

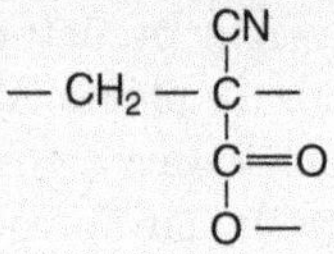

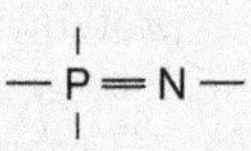

Table 7.2-2 Medical applications of bioabsorbable polymers

I. For surgical operation
1. Suturing, stapling
2. Adhesion, covering
3. Hemostasis, sealing
II. For implantation
1. Adhesion prevention
2. Bone fixation
3. Augmentation
4. Embolization, stenting
III. For drug delivery
1. Sustained release
2. Drug targeting

cell attachment owing to the presence of cell adhesion sequence.

Collagens

Collagen is the most abundant protein within the ECM of connective tissues such as skin, bone, cartilage, and tendon. At least 20 distinct types of collagen have been identified. The primary structural collagen in mammalian tissues is type I collagen (or collagen I). This protein has been well characterized and is ubiquitous across the animal and plant kingdom. Collagen contains a large number of glycine (almost 1 in 3 residues, arranged every third residue), proline and 4-hydroxyproline residues. A typical structure is -Ala-Gly-Pro-Arg-Gly-Glu-4Hyp-Gly-Pro-. Collagen is composed of triple helix of protein molecules which wrap around one another to form a three-stranded rope structure. The strands are held together by both hydrogen and covalent bonds, while collagen strands can self-aggregate to form stable fibers. Collagen is naturally degraded by metalloproteases, specifically collagenase, and serine proteases, allowing for its degradation to be locally controlled by cells present in tissues.

Allogeneic and xenogeneic, type I collagens have been long recognized as a useful scaffold source with low antigenic potential. Bovine type I collagen has perhaps been the biological scaffold most widely studied due to its abundant source and its history of successful use. Type I collagen is extracted from the bovine or porcine skin, bone, or tendon through alkaline or enzymatic procedures. Most of the telopeptide portion present at the end of collagen molecule with antigenic epitopes is removed during the extraction processes. The low mechanical stiffness and rapid biodegradation of the extracted but untreated collagen have been crucial problems that limit the use of this biomaterial as scaffold. Since crosslinking is an effective method to improve the biodegradation rate and the mechanical property of collagen, crosslinking treatments have become one of the most important issues for collagen technology. Two kinds of crosslinking methods are known for collagen: chemical and physical. The physical methods that do not introduce any potential cytotoxic chemical residues include photooxidation, UV irradiation, and dehydrothermal treatment (DHT). Chemical methods are applied generally when higher extents of crosslinking than those provided by the physical methods are needed. The reagents used in the chemical crosslinking include glutaraldehyde (GA), water-soluble carbodiimide (WSC) such as 1-ethyl-3-(3-dimethylaminopropyl)-carbodiimide (EDAC), hexamethylene diisocyanate, acyl azides, glycidyl ethers, polyepoxidic resins, and so on. A bifunctional reagent bridging amino groups between two adjacent polypeptide chains through Schiff base formation, GA, is the most predominant choice for collagen crosslinking because of its water solubility, high crosslinking efficiency, and low cost, although GA is a potentially cytotoxic aldehyde. Carbodiimides (CDIs) have been widely used for activation of carboxyl groups of natural polymers under the acidic conditions that are necessary for protonation of the CDI nitrogens, leading to nucleophilic attack of the carboxylate anion at the central carbon to form an initial O-acylisourea. The EDAC has been called a "zero-length crosslinker" because it catalyzes the intermolecular formation of peptide bonding in collagen without becoming incorporated. This crosslinking reaction results in formation of water-soluble urea as only one by-product. If both unreacted EDAC and urea are thoroughly rinsed from the material, the concern over the release of toxic residuals, commonly associated with other chemical crosslinking agents, will be reduced.

Crosslinking by UV and DHT does not introduce toxic agents into the material, but both the treatments partially give rise to denaturation of collagen. When collagen fiber scaffolds for ligament tissue engineering are crosslinked by DHT, in combination with CDI, approximately 50% of these implants rupture prior to neoligament formation, due possibly to the collagen denaturation caused by DHT. Although physical crosslinking can avoid introducing potential cytotoxic residues, most of the physical treatments do not yield high enough crosslinking to meet the demand for collagen as scaffold. Therefore, chemical treatments are applied in many cases with use of traditional GA, WSC, and other methods. Such chemical crosslinking has been shown to reduce biodegradation of collagen.

Traditional collagen crosslinking reagents may impart some degree of cytotoxicity, caused by the presence of unreacted functional groups or by the release of those groups during enzymatic degradation of the crosslinked protein. Furthermore, the chemical reaction that occurs between amine and/or carboxylic acid groups must be averted in the case of *in situ* crosslinking of cell-seeded gels. Methods have been developed that allow for collagen materials to directly crosslink without incorporation of crosslinking reagents. To date, the recognized mechanisms for strengthening collagen constructs in the presence of cells are nonenzymatic glycation and enzyme-mediated crosslinking techniques, thereby enhancing mechanical strength while remaining benign toward the cells.

A number of studies have shown that the major antigenic determinants of collagen are located within the terminal regions. In still other cases, evidence has been presented to suggest that central determinants also play a role in collagen–antibody interactions. Collagens are treated with proteolytic enzymes to remove the terminal telopeptides. However, in some cases, telopeptide remnants persisting following pepsin treatment (Fig. 7.2-3) [1] have been shown to be sufficiently large so that the antigenic activity of the pepsin-treated and native forms is almost indistinguishable. Further detailed study may be needed to characterize the human immune response to xenogeneic collagen.

Although several commercial skin products are based on bovine type I collagen which has been licensed for clinical use by the Food and Drug Administration (FDA), this would seem not to be a good long-term solution for clinical use. The possible risk of virus and prion tends to reduce the use of collagen in tissue engineering, but the use of collagen in tissue engineering likely still continues because of its excellent properties as scaffold. Development of good assay kits for virus check and reasonable consideration of risk/benefit balance are required for the safe application of collagen. A big challenge at the moment is to produce inexpensive human recombinant collagen that is completely free of any virus and prion.

Besides collagen, elastin plays a major role in determining the mechanical performance of some native tissues. Elastin fibers can extend 50–70% under physiological loads, and depending on the location of the vessel, elastin content can range from 33% to 200% that of collagen. In native tissues, elastin exists in stable fibers that resist both hydrolysis and enzymatic digestion.

Gelatin

Gelatin is obtained by denaturing collagen by heating animal tissues including bone, skin, and tendon. Alkaline pretreatment of collagen, which converts asparagine and glutamine residues to their respective acids, produces acidic gelatin with isoelectric points below 7, while extraction with diluted acid or enzymes yields basic gelatin with isoelectric points higher than 7. Gelatin is a heterogeneous mixture of single and multistranded polypeptides, each with extended left-handed proline helix conformations and containing between 300 and 4000 amino acids. The triple helix of type I collagen extracted from skin and bones is composed of two a1 (I) and one a2 (I) chains, each with MW of ~95 kDa, width of 1.4 nm, and length of 290 nm. Gelatin consists of mixtures of these strands together with their oligomers and breakdown (and other) polypeptides. Solutions undergo coil–helix transition followed by aggregation of the helices by the formation of collagen-like right-handed triple-helical

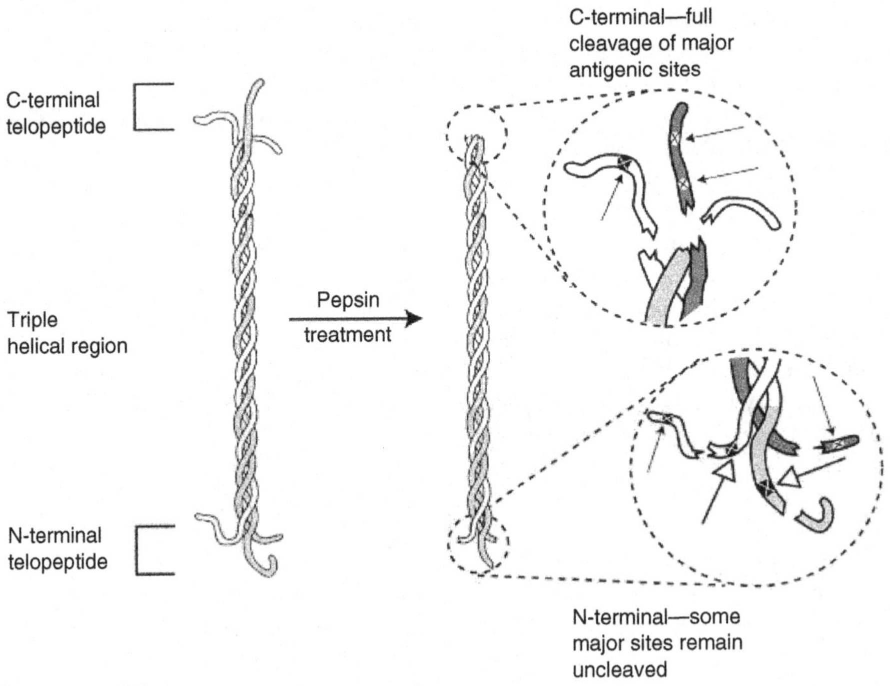

Fig. 7.2-3 Telopeptide removal from collagen via pepsin treatment.

proline/hydroxyproline(OHP)-rich junction zones. Higher levels of these pyrrolidines result in stronger gels. Each of the three strands in the triple helix requires 25 residues to complete one turn; typically there would be between one and two turns per junction zone. Gelatin films containing greater triple-helix content swell less in water and are consequently much stronger.

Gelatin has been used for a range of medical applications including adhesion prevention because of good processability, transparency, and bioabsorbability. Aqueous solution of gelatin sets to a gel through hydrogen bonding below room temperature and recovers to the sol state upon raising the temperature to destroy the hydrogen bonding. This reversible sol–gel transition facilitates the molding of gelatin into definite shapes such as block and microsphere, but chemical crosslinking is required when its dissolution in aqueous media at the body temperature should be avoided. When applied as a scaffold of cells or a carrier of growth factors, gelatin needs to be permanently crosslinked. Glutaraldehyde has most frequently been used for chemical crosslinking of gelatin to link lysine to lysine, similar to other proteins.

Fibrin

Fibrin is a product of partial hydrolysis of fibrinogen by the enzymatic action of thrombin. Upon crosslinking it converts to gel. This is called "fibrin glue" or "surgical adhesive". Human fibrin adhesives are approved and available in most major geographical regions of the world. Fibrin is applied to patients as a liquid and solidifies shortly thereafter *in situ*. Furthermore, fibrin gel can be readily infiltrated by cells, because most migrating cells locally activate the fibrinolytic cascade.

Silk fibroin

Due to their high strength, native silk proteins from silkworm have been used in the medical field as suture material for centuries. Undesirable immunological problems attributed to the sericin protein of silk limited the use of silk in the last two decades. However, purified silk fibroin, which remains after removal of sericin, exhibits low immunogenicity and retains many of the attributes of native silk fibers. This has sparked a renewed interest in the use of silk fibroin as a biomaterial. Silk fibroin has unique properties that meet many of the demands for scaffolds. Silk exhibits high strength and flexibility and permeability to water and oxygen. In addition, silk fibroin can be molded into fibers, sponges, or membranes, making silk a good substrate for biomedical applications such as implant biomaterials, cell culture scaffolds, and cell carriers.

7.2.2.1.2 Polysaccharides

In addition to proteinous materials, naturally occurring polysaccharides and their derivatives have been employed for scaffold fabrication. Among them are hyaluronic acid (HAc), chitin, chitosan, alginate, and agarose. A prominent feature common to most of polysaccharides is the lack of cell-adhesive motifs in the molecules. This makes these biopolymers suitable as biomaterials for fabrication of a scaffold whose interaction with cells should be minimized. An exception is chitosan which has basic NH_2 groups.

Hyaluronic acid

Industrially, HAc is obtained from animal tissues such as umbilical cord, cock's comb, vitreous body, and synovial fluid. Biotechnology also produces HAc on a large scale. HAc is an only non-sulfated glycosaminoglycan (GAG) that is present in all connective tissues as a major constituent of ECM and plays pivotal roles in wound healing. As shown in Fig. 7.2-4, this linear, non-adhesive polysaccharide consists of repeating disaccharide units (β-1,4-D-glucuronic acid and β-l,3-*N*-acetyl-D-glycosamine) with weight-average MWs (Mw) up to 10,000 kDa. This anionic polymer is also a major constituent of the vitreous (0.1–0.4 mg/g), synovial joint fluid (3–4 mg/ml) and hyaline cartilage, where it reaches approximately 1 mg/g wet weight. Serum HAc levels range from 10 to 100 μg/l, but are elevated during disease. Clearance of HAc from the systemic circulation results in a half-life of 2.5–5.5 min in plasma.

In solution, HAc assumes a stiffened helical configuration due to hydrogen bonding, and the ensuing coil structure traps approximately 1000-fold weight of water. The highly viscous aqueous solutions thus formed give HAc unique physicochemical and biological properties that make it possible to preserve tissue hydration, regulate tissue permeability through steric exclusion, and permit joint lubrication. In the ECM of connective tissues, HAc forms a natural scaffold for binding other large GAGs and proteoglycans (aggrecans), which are maintained through specific HAc–protein interactions. Consequently, HAc plays important roles in maintaining tissue morphologic organization, preserving extracellular space, and transporting ions, solutes, and nutrients. Along with ECM proteins, HAc binds to specific cell surface receptors such as CD44 and RHAMM. The resulting activation of intracellular signaling events leads to cartilage ECM stabilization, regulates cell adhesion and mobility, and promotes cell proliferation and differentiation. The HAc signaling takes place also during morphogenesis and embryonic development, modulation of inflammation, and in the stimulation of wound healing. In correspondence with these functions, HAc is a strong inducer of angiogenesis, although its biological activity in tissues has been shown to depend on the molecular size. High MW native-HAc (n-HAc) has been shown to inhibit angiogenesis, whereas degradation products of low MW stimulate endothelial cell proliferation and migration.

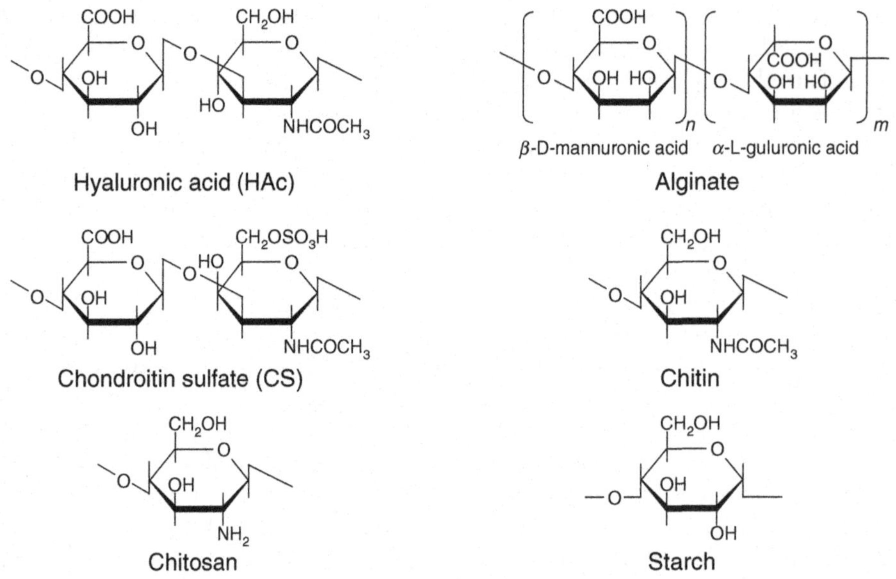

Fig. 7.2-4 Chemical structure of polysaccharides used for tissue engineering.

Oligosaccharide HAc (o-HAc) fragments have been shown to induce angiogenesis in several animal models as well as within *in vitro* collagen gels. HAc is naturally hydrolyzed by hyaluronidase, allowing cells in the body to regulate the clearance of the material in a localized manner.

Owing to its unique physicochemical properties, unmodified HAc has been widely used in the field of viscosurgery, visco-supplementation, and wound healing. However, the poor mechanical properties of this water-soluble polymer and its rapid degradation *in vivo* have precluded many clinical applications. Therefore, in an attempt to obtain materials that are more mechanically and chemically robust, a variety of covalent crosslinking via hydroxyl or carboxyl groups, esterification, and annealing strategies have been explored to produce insoluble HAc hydrogels. For example, HAc-esterified materials, collectively called "Hyaff™," are prepared by alkylation of the tetrabutylammonium salt of HAc with an alkyl or benzyl halide in dimethyl formamide solution. Crosslinked HAc has been prepared using divinyl sulfone, 1,4-butanediol diglycidyl ether, GA, WSC, and a variety of other bifunctional crosslinkers. However, the crosslinking agents are often cytotoxic small molecules, and the resulting hydrogels have to be extracted or washed extensively to remove traces of unreacted reagents and by-products.

Alginate

Alginates are linear polysaccharide derived primarily from brown seaweed and bacteria. They are block copolymers composed of regions of sequential (1-4)-linked β-D-mannuronic acid monomers (M-blocks), regions of α-L-guluronic acid (G-blocks), and regions of interspersed M and G units (Fig. 7.2-4). The length of the M- and G-blocks and sequential distribution along the polymer chain varies depending on the source of alginates. These biopolymers undergo reversible gelation in aqueous solution under mild conditions through interaction with divalent cations including Ca^{2+}, Ba^{2+}, and Sr^{2+} that can cooperatively bind between the G-blocks of adjacent alginate chains creating ionic interchain bridges. This highly cooperative binding requires more than 20 G-monomers.

Gels can also be formed by covalently crosslinking alginate with adipic hydrazide and poly(ethylene glycol) (PEG) using standard CDI chemistry. Ionically crosslinked alginate hydrogels do not specifically degrade but undergo slow uncontrolled dissolution. Mass of the alginate-Ca^{2+} is lost through ion exchange of calcium followed by dissolution of individual chains, which results in loss of mechanical stiffness over time. Alginates are easily processed into any desired shape with the use of divalent cations. One possible disadvantage of using alginates is its low and uncontrollable *in vivo* degradation rate, mainly due to the sensitivity of the gels towards calcium chelating compounds (e.g., phosphate, citrate, and lactate). Several *in vivo* studies have shown large variations in the degradation rate of calcium-crosslinked sodium alginates. Hydrolytically degradable form of alginate and an alginate derivative, polyguluronate, are oxidized alginate and poly(aldehyde guluronate), respectively.

Chondroitin sulfate

Chondroitin sulfate (CS) is composed of repeating disaccharide units of glucuronic acid and *N*-acetylgalactosamine with a sulfate group and a carboxyl group on each disaccharide (Fig. 7.2-4). It is a constituent of ECM, contributing to the functionality of the extracellular network. The cartilage ECM consists of type II collagen and proteoglycans including aggrecan, which are responsible for the tissue's compressive and tensile strength, respectively. Chondroitin sulfate forms the arms of the aggrecan molecule in cartilage.

Chitosan and chitin

Chitosan is a linear polysaccharide of (1-4)-linked D-glucosamine and *N*-acetyl-D-glucosamine residues derived from chitin, which is found in arthropod exoskeletons (Fig. 7.2-4). The degree of N-deacetylation of chitin usually varies from 50% to 90% and determines the crystallinity, which is the greatest for 0% and 100% N-deacetylation. Chitosan is soluble in dilute acids which protonate the free amino groups. Once dissolved, chitosan can be gelled by increasing the pH or extruding the solution into a non-solvent. Chitosan derivatives and blends have also been gelled via GA crosslinking, UV irradiation, and thermal variation. Chitosan is degraded by lysozyme, and the kinetics of degradation is inversely related to the degree of crystallinity. Figure 7.2-5 shows the dependence of resorption of chitin on the hydrolysis extent when partially hydrolyzed chitin (or partially acetylated chitosan) is subcutaneously implanted in rat [2]. In contrast with 100% homopolymeric chitosan, partially hydrolyzed chitin or partially acetylated chitosan and chitin are absorbable and high in the tensile strength, but it seems that clear evidence has not yet been presented regarding its safety, especially when implanted in the human body.

7.2.2.1.3 Natural composite—ECM

The native ECM provides a substrate containing adhesion proteins for cell adhesion and regulates cellular growth and function by presenting different kinds of growth factors to the cells. The ECM is a complex structural protein-based entity surrounding cells within mammalian tissues. Most normal vertebrate cells cannot survive unless they are anchored to the ECM. In tissues and organs, major ECM components are structural and functional proteins, glycoproteins, and proteoglycans arranged in a unique, specific 3-D ultrastructure, as illustrated in Fig. 7.2-6. Each tissue or organ has its own unique set and content of these biomolecules. In skin, the collagen:elastin ratio is about 9:1, whereas in an artery this ratio is 1:1 averaging all artery layers, and 1:9 when considering the lamina elastica only. In ligaments, the collagen:elastin ratio is also 1:9, and in lung about 1:1. Likewise, the amount and type of GAGs, another major ECM component, varies from matrix to matrix. For instance, in cartilage CS is the major GAG making up 20% of the dry weight. In skin, dermatan sulfate is the most abundant (about 1% of the dry weight), whereas in the vitreous body of the eye the major GAG is hyaluronate.

Natural ECMs are gels composed of various protein fibrils and fibers interwoven within a hydrated network of GAG chains. In their most elemental function, ECMs thus provide a structural scaffold that, in combination with interstitial fluid, can resist tensile (via the fibrils) and compressive (via the hydrated network) stresses. In this context it is worth mentioning just how small a proportion of solid material is needed to build mechanically quite robust structures. Structural ECM proteins include collagens—some of which are long and stiff and thus serve structural functions whereas others serve connecting and recognition functions—and elastin, which forms an extensive crosslinked network of elastic fibers and sheets. The anisotropic fibrillar architecture of natural ECMs has apparent consequences for cell behavior. Because of a tight connection between the cytoskeleton and the ECM through cell surface receptors, cells sense and respond to the mechanical properties of their environment by converting mechanical signals into chemical

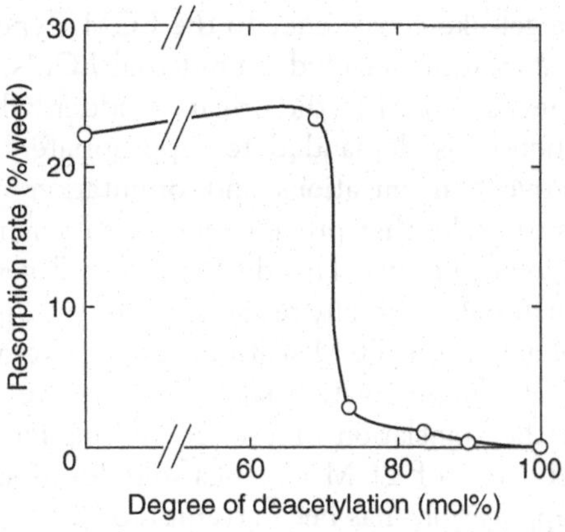

Fig. 7.2-5 Dependence of the initial resorption rate on films of chitin and its deacetylated derivatives on the degree of deacetylation.

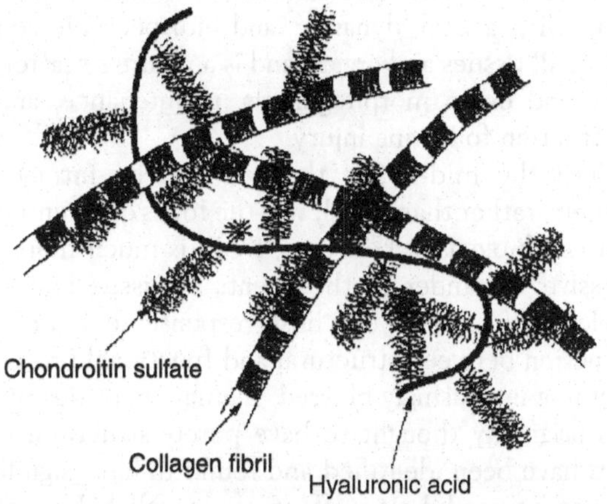

Fig. 7.2-6 Component arrangement in ECM (cartilage).

signals. Consequently, the biophysical properties of ECMs influence various cell functions, including adhesion and migration. Moreover, the fibrillar structure of matrix components brings about adhesion ligand clustering, which has been demonstrated to alter cell behavior. Structural ECM features, such as fibrils and pores, are often of a size compatible with cellular processes involved in migration, which may influence the strategy by which cells migrate through ECMs.

Natural ECMs modulate tissue dynamics through their ability to locally bind, store, and release soluble bioactive ECM effectors such as growth factors to direct them to the right place at the right time. When many growth factors bind to ECM molecules through, for example, electrostatic interactions to heparan sulfate proteoglycans, it raises their local concentration to levels appropriate for signaling, localizes their morphogenetic activity, protects them from enzymatic degradation, and in some cases may increase their biological activity by optimizing receptor–ligand interactions. Growth factors are required in only very tiny quantities to elicit a biological response.

The macromolecular components of natural ECMs are degraded by cell-secreted and cell-activated proteases, mainly by matrix metalloproteases (MMPs) and serine proteases. This creates a dynamic reciprocal response, with the ECM stimulating the cells within it and cellular proteases remodeling the ECM and releasing bioactive components from it.

With the discovery that ECM plays a role in the conversion of myoblasts to myotubes and that structural proteins such as collagen and GAGs are important in salivary gland morphogenesis it became obvious that ECM proteins serve many functions including the provision of structural support and tensile strength, attachment sites for cell surface receptors, and as a reservoir for signaling factors that modulate such diverse host processes as angiogenesis and vasculogenesis, cell migration, cell proliferation and orientation, inflammation, immune responsiveness, and wound healing. Stated differently, the ECM is a vital, dynamic, and indispensable component of all tissues and organs and is a nature's scaffold for tissue and organ morphogenesis, maintenance, and reconstruction following injury.

Until the mid-1960s the cell and its intracellular contents, rather than ECM, was the focus of attention for most cell biologists. However, ECM is much more than a passive bystander in the events of tissue and organ development and in the host response to injury. The distinction between structural and functional proteins is becoming increasingly blurred. Domain peptides of proteins originally thought to have purely structural properties have been identified and found to have significant and potent modulating effects upon cell behavior. For example, the RGD (R: arginine; G: glycine; D: aspartic acid) peptide that promotes adhesion of numerous cell types was first identified in the fibronectin (FN) molecule, a molecule originally described for its structural properties. Several other peptides have since been identified in "dual function" proteins including LN, entactin, fibrinogen, types I and VI collagen, and vitronectin. The discovery of cytokines, growth factors, and potent functional proteins that reside within the ECM characterized it as a virtual information highway between cells. The concept of "dynamic reciprocity" between the ECM and the intracellular cytoskeletal and nuclear elements has become widely accepted. The ECM is not static. The composition and the structure of the ECM are a function of location within tissues and organs, age of the host, and the physiologic requirements of the particular tissue. Organs rich in parenchymal cells, such as the kidney, have relatively little ECM. In contrast, tissues such as tendons and ligaments with primarily structural functions have large amounts of ECM relative to their cellular component. Submucosal and dermal forms of ECM reside subjacent to structures that are rich in epithelial cells (ECs) such as the mucosa of the small intestine and the epidermis of the skin. These forms of ECM tend to be well vascularized, contain primarily type I collagen and site-specific GAGs, and a wide variety of growth factors.

Collagen types other than type I exist in naturally occurring ECM, albeit in much lower quantities. These alternative collagen types each provide distinct mechanical and physical properties to the ECM and contribute to the utility of the intact ECM as a scaffold for tissue repair. Type IV collagen is present within the basement membrane of all vascular structures and is an important ligand for endothelial cells, while type VII collagen is an important component of the anchoring fibrils of keratinocytes to the underlying basement membrane of the epidermis. Type VI collagen functions as a "connector" of functional proteins and GAGs to larger structural proteins such as type I collagen, helping to provide a gel-like consistency to the ECM. Type III collagen exists within selected submucosal ECMs, such as the submucosal ECM of the urinary bladder, where less rigid structure is demanded for appropriate function. The relative concentrations and orientation of these collagens to each other provide an optimal environment for cell growth *in vivo*. This diversity of collagen within a single material is partially responsible for the distinctive biological activity of ECM scaffolds and is exemplary of the difficulty in re-creating such a composite *in vitro*, although the translation of the ECM functions to the therapeutic use of ECM as a scaffold for tissue engineering applications has been attempted.

The ECM of the basement membrane that resides immediately beneath ECs such as urothelial cells (UCs) of the urinary bladder, endothelial cells of blood vessels, and hepatocytes of the liver is comprised of distinctly different collections of proteins including LN, type IV collagen, and entactin. All ECMs share the common features of providing structural support and serving as a reservoir of growth factors and cytokines. The ECMs present these factors efficiently to resident cell surface receptors, protect the growth factors from degradation, and modulate their synthesis. In this manner, the ECM affects local concentrations and biological activity of growth factors and cytokines and makes the ECM an ideal scaffold for tissue repair and reconstruction.

The GAGs are also important components of ECM and play important roles in binding of growth factors and cytokines, water retention, and the gel properties of ECM. The heparin binding properties of numerous cell surface receptors and of many growth factors [e.g., fibroblast growth factor (FGF) family, vascular endothelial growth factor (VEGF)] make the heparin-rich GAGs extremely desirable components of scaffolds for tissue repair. The GAG components of the small intestine submucosa (SIS) consist of the naturally occurring mixture of CSs A and B, heparin, heparin sulfate, and HAc. To date it will need time-consuming labor or much money to obtain a large amount of purified components of the natural ECMs and reconstruct an ECM from the purified components.

Since ECM plays an important role in a tissue's mechanical integrity, crosslinking of ECM may be an effective means of improving the mechanical properties of tissues. The ECM crosslinking can result from the enzymatic activity of lysyl oxidase (LO), tissue transglutaminase, or nonenzymatic glycation of protein by reducing sugars. The LO is a copper-dependent amine oxidase responsible for the formation of lysine-derived crosslinks in connective tissue, particularly in collagen and elastin. Desmosine, a product of LO-mediated crosslinking of elastin, is commonly used as a biochemical marker of ECM crosslinking. The LO-catalyzed crosslinks that are present in various connective tissues within the body—including bone, cartilage, skin, and lung—are believed to be a major source of mechanical strength in tissues. Additionally, the LO-mediated enzymatic reaction renders crosslinked fibers less susceptive to proteolytic degradation.

7.2.2.2 Synthetic polymers

Before the prion shock, naturally derived materials had attracted much attention because of their natural origin which seemed to guarantee the biocompatibility. However, reports on the Creutsfeld-Jacobs disease due to the implanted sheets made from human dried dura mater diverted the focus of biomaterial scientists to non-biological materials such as synthetic polymers. Synthetic materials have long been applied for replacements of tissues and organs, fulfilling some auxiliary functions, especially improving comfort and the well-being of patients. Further possibilities exist now for synthetic materials to create tissues and organs with controlled mechanical properties and well-defined biological behavior. In the biomaterial area, there are two kinds of synthetic polymers, non-absorbable and absorbable. Non-absorbable polymers have been used as key materials for artificial organs, implants, and other medical devices. In most cases, absorbable polymers are not adequate as the major component of permanent devices, since absorption or degradation of materials in these applications has a meaning almost identical to the material deterioration which is an undesirable, negative concept for biomaterials in permanent use. Widespread clinical use of silicone, poly(ethylene terephthalate) (PET), polyethylene, polytetrafluoroethylene (PTFE), and poly(methyl methacrylate) (PMMA) as important components of artificial organs and tissues is owing to the excellent chemical stability or non-degradability in the body. If these materials undergo degradation more or less in the body, this will definitely raise a serious concern because one cannot deny that degradation by-products might evoke untoward reactions in the body. If these stable polymers exhibit deterioration over time, this might not always be due to hydrolysis but due to attack by active oxygens generated in the body as a result of inflammation.

For simplicity, bioabsorbable, synthetic polymers are here classified into three groups: poly(α-hydroxyacid)s, synthetic hydrogels, and others.

7.2.2.2.1 Poly(α-hydroxyacid)s [aliphatic α-polyesters or poly(α-hydroxyester)s]

The majority of bioabsorbable, synthetic polymers that are currently available is poly(α-hydroxyacid)s that have repeating units of-O-R-CO- (R: aliphatic) in the main chain. This is mainly because most of them have the potential to produce scaffolds with sufficient mechanical properties and some of them have been approved by the U.S. FDA for a variety of clinical applications as absorbable biomaterials with biosafe degradation by-products. By contrast, aromatic polyesters with phenyl groups in the main chain do not undergo any appreciable degradation in physiological conditions. The monomers used for synthesis of poly(α-hydroxyacid)s include glycolic acid (or glycolide), and L- and DL-lactic acid (or L- and DL-lactide) with a hydroxyl group on the α carbon. These monomers can yield not only homopolymers but also copolymers when polymerized together with other monomers such as ε-caprolactone (CL), p-dioxanone, and 1,3-trimethylene carbonate (TMC). Chemical structures of α-hydroxyacid polymers, copolymers, and their monomers are shown in Fig. 7.2-7.

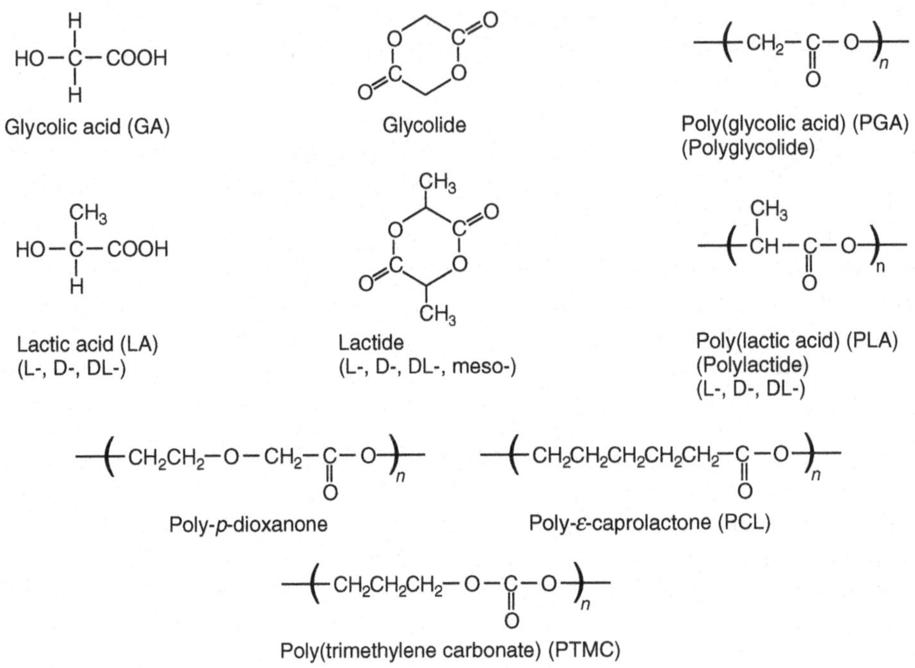

Fig. 7.2-7 Chemical structure of α-hydroxyacid polymers, copolymers, and their monomers.

Homopolymers

The most widely used absorbable sutures are made from polyglycolide (PGA) or poly(glycolide-co-lactide) (PGLA) with a glycolide (GA)/L-lactide (LLA) ratio of 90/10. This PGLA with the 90% content of GA is included in PGA here, because PGLA with a GA/LLA ratio of 90/10, which is commercially available as a multifilament suture and a Vicryl mesh (Vicryl, Ethicon, USA), exhibits properties quite similar to PGA (100% GA polymer). Poly(L-lactide) (PLLA) has been clinically used after molding into pin, screw, and mini-plate for fixation of fractured bones and maxillofacial defects of patients. Both PGA and PLLA are crystalline polymers which can provide medical devices with excellent mechanical properties, but PGA degrades mostly too quickly while PLLA degrades too slowly for use as scaffold. Nevertheless, both of them have primarily been chosen as polymers for scaffold fabrication in numerous studies worldwide.

Non-woven PGA fabrics have extensively been used as a scaffold material for cell growth in the effort to engineer many types of tissues. However, scaffolds fabricated from PGA fibers lack sufficient dimensional stability to allow molding into distinct shapes and degrade rapidly to disturb processing of this material after exposure to aqueous media. To overcome these problems, the PGA fabrics are often dipped in solution of polylactide (PLA), followed by evaporation of the solvent to deposit stiff PLA coating on the fabrics. In general, cell adhesion onto such blended materials is influenced by the polymer component existing at the outermost surface, while the degradation kinetics is the simple addition of kinetics of each component degradation.

It takes a longer time than a few years for homopolymers of LLA and ε-CL to be completely absorbed, whereas TMC homopolymer degrades too quickly in the presence of water. Homopolymers of poly(D,L-lactide) (PDLLA) are bioabsorbed at a little higher rate than PLLA, because of the absence of crystalline regions. In case there is no need to distinguish between PLLA and PDLLA, the term PLA will be used below to include PLLA and PDLLA.

Copolymers

The aliphatic copolyester that has the largest clinical application is poly(LA-*co*-GA) (PLGA) mostly with an LLA/GA ratio around 50/50. This copolymer has clinically been used as a carrier of peptide drugs for their sustained release. In many cases, copolymers are preferred for scaffold fabrication because of their more versatile, physicochemical properties. Figure 7.2-8 shows the decrease in tensile strength in phosphate buffered solution (PBS) versus the hydrolysis time for various aliphatic polyesters (Table 7.2-3) in the fiber form [3]. Copolymerization of monomers A and B offers a great potential for modifications of polymers A or B, by controlling the physical and biological properties of bioabsorbable polymers, such as degradation rate, hydrophilicity, mechanical properties, and *in vivo* shrinkage.

Assume that homopolymer A is absorbed while homopolymer B exhibits no or insignificant absorption in the body. Blending of homopolymers A and B does not

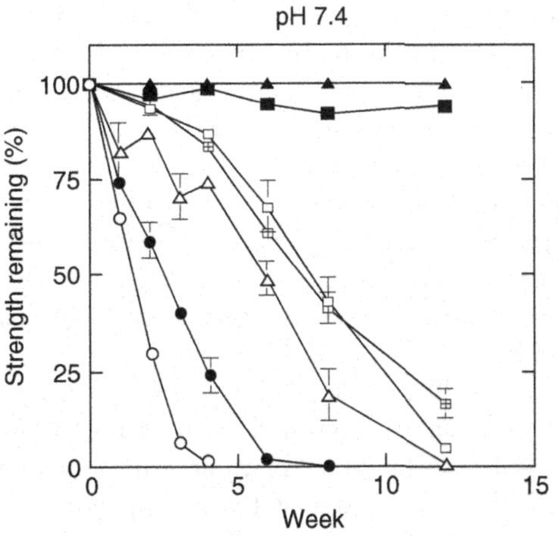

Fig. 7.2-8 Tensile strength change of monofilament sutures upon immersion in buffer solution of pH 7.4 at 37°C. Open circles, MONOCRYL; open squares, PDS II; open triangles, MAXON; solid circles, BIOSYN; crossed squares, P(LA/CL); solid squares, ETHILON; and solid triangles, PROLENE.

change the degradation kinetics of each homopolymer, but copolymerization of monomer B with monomer A converts homopolymer B to a bioabsorbable polymer component. The scheme is illustrated in Fig. 7.2-9 for the equimolar copolymerization of monomers A and B. The average continuous sequence of each monomer in the copolymer chain is governed by polymerization conditions such as initiator and temperature. If the continuous sequence of monomer B in the copolymer chain is shorter than a critical length below which oligomers B are soluble or dispersible in aqueous media, the copolymer A–B becomes bioabsorbable owing to the degradation of monomer A unit. A typical example is LA–CL copolymers that are absorbed in the body at rates higher than LA homopolymer and CL homopolymer that is virtually non-absorbable. Furthermore, this copolymerization converts the brittle LA homopolymer into much more rubber-like, tough polymer. Figure 7.2-10 shows how copolymerization of LLA with CL yields polymers with low Young's moduli [4] and high resorption rates. The DLLA copolymerization with CL will produce copolymers with properties different from those of LLA–CL copolymers, since long sequences dominating in the LLA chains will result in small crystallite formation by associating together, in marked contrast with DLLA sequences that do not have any potential to crystallize.

Copolymerization of DLLA or CL with TMC has also been attempted. The TMC homopolymer of high MW is an amorphous elastomer that shows good mechanical performance, combining high flexibility with high tensile strength, but degrades very slowly at pH 7.4 and 37°C. High MW copolymers of TMC and DLLA with 20–50 mol% of TMC are amorphous, relatively strong elastomers, can maintain mechanical properties up to 3

Table 7.2-3 Chemical structure of suture materials

Suture	Chemical composition (wt ratio)
MONOCRYL®	Glycolide:ε-caprolactone = 75:25
MAXON®	Glycolide:trimethylene carbonate = 67.5:22.5
BIOSYN®	Glycolide:dioxanone:trimethylene carbonate = 60:14:26
PDS II®	*p*-Dioxanone =100
P(LA/CL)	L-Lactide:ε-caprolactone = 80:20
ETHILON®	Caprolactam = 100
PROLENE®	Propylene =100

months at *in vitro* degradation, and are absorbed in less than a year. In contrast, copolymers of TMC and CL degrade more slowly than TMC–DLLA copolymers. The TMC–CL copolymers with high contents of CL are semicrystalline, very flexible, and tough, so that they can maintain mechanical properties for more than 1 year when incubated in buffer solution at pH 7.4 and 37°C.

In addition to A–B type copolymers, A–B–A type triblock copolymers have been actively synthesized. For instance, PEG–PLA–PEG triblock copolymers can be synthesized as follows. In the first step, diblock copolymers of methoxy(Me).PEG–PLA are prepared by ring-opening polymerization of LA in the presence of Me.PEG–OH. Then, the resultant diblock copolymer is reacted with hexamethylene diisocyanate (HMDI) at a high temperature to connect the two diblock copolymer chains. The reasoning for block copolymerization of LA with PEG is to reduce the surface hydrophobicity of PLA scaffolds or make suspension in aqueous media.

Many bioabsorbable polymers like PLA and PGA lack functional groups which facilitate further functionalization. Therefore, poly(LA-*co*-lysine) was synthesized

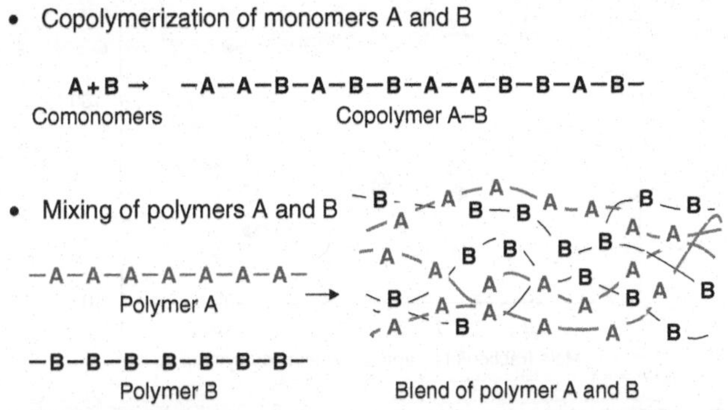

Fig. 7.2-9 Difference between copolymer A–B and blend of polymer A and polymer B.

to immobilize peptides on the pendant amino groups onto the lysine side chains, while carboxyl-functionalized PLA was prepared by copolymerizing side-chain-protected β-alkyl-α-malate and subsequent deprotection of the carboxyl groups by hydrogenation or alkaline treatment.

7.2.2.2.2 Hydrogels

Crosslinking of water-soluble polymer chains produces water-insoluble 3-D networks. The product is called "hydrogel". Several distinctive features, including tissue-like viscoelasticity, diffusive transport, and interstitial flow characteristics, make synthetic hydrogels excellent physicochemical mimetics of natural ECMs and candidates for soft tissue scaffolds. Indeed, hydrogels have a potential to efficiently encapsulate cells and high water contents to allow for nutrient and waste transport. Soft tissues of our body are also like a hydrogel because of their high water contents ranging between 70% and 90%. The structural integrity of hydrogels depends on crosslinks formed between polymer chains via physical, ionic, or covalent interactions. Synthetic hydrogels can be processed under relatively mild conditions, have structural properties similar to ECM, and can be delivered in the body in a minimally invasive manner, particularly by injection. A variety of synthetic materials including PEG, poly(vinyl alcohol) (PVA), and poly(acrylic acid) may be used to form hydrogels. These polymers are not biodegradable but water soluble unless crosslinks are introduced. Thus, once the introduced crosslinks are broken,

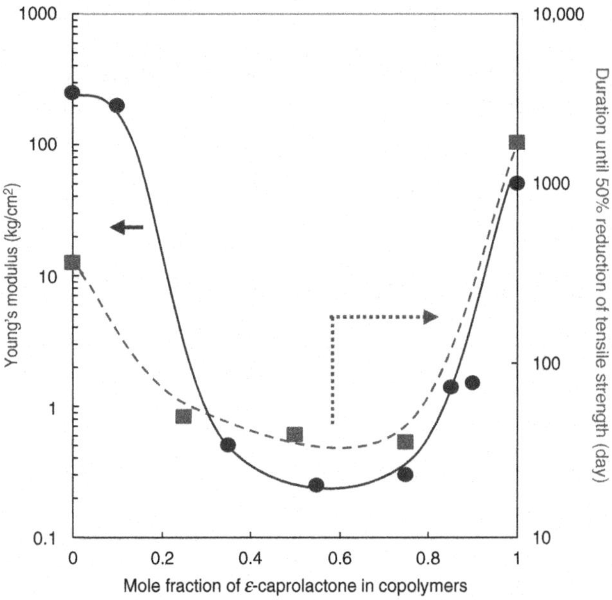

Fig. 7.2-10 Young's modulus and hydrolysis time until to 50% reduction of tensile strength for LLA-ε-CL copolymer plotted against the mole fraction of ε-CL comonomer.

the resulting polymer chains become again water soluble, being absorbed in the body, followed by excretion from kidney or bile duct, depending on the molecular size. Naturally derived polymers including collagen, gelatin, fibrinogen, albumin, polypeptides, HAc, CS, agarose, alginate, and chitosan can also provide hydrogels.

Poly(ethylene glycol) or poly(ethylene oxide) (PEO) with the same chemical structure as PEG is one of the synthetic polymers that is most commonly used for hydrogel fabrication in tissue engineering and is currently approved by FDA for several applications. Each end of PEG chains can be modified with either acrylates or methacrylates to facilitate photocrosslinking. When the modified PEG is mixed with an appropriate photoinitiator and crosslinked via UV exposure hydrogels are formed. This photo-induced method to produce hydrogels facilitates diverse, minimally invasive applications via arthroscopy/endoscopy or subcutaneous injection for tissue replacement or augmentation. Thermally reversible hydrogels have also been created from block copolymers of PEG and PLLA.

Although *in situ* forming hydrogels offer advantages as cell delivery carriers, cells do not usually adhere to highly hydrated gels because of no built-in cell-adhesive ligands and limited protein adsorption, so that the cells encapsulated inside gels are present in a "blank" environment wherein there is little to no interaction of integrins and other cell surface receptors with the gels.

7.2.2.2.3 Others

Besides aliphatic polyesters and hydrogels, a number of synthetic polymers have been used for scaffolds. The chemical structures of these polymers are shown in Fig. 7.2-11. Only polyurethanes (PUs) and polyphosphazens will be described here, because these two polymers have been studied by different research groups.

Polyurethanes

A considerable majority of tissue engineering literature has utilized a narrow array of polyester scaffolds although their mechanical properties are limited with respect to high strain, elastic capabilities. As a result, attention has been paid on developing bioabsorbable elastomers with high strain, elastic capabilities, and mechanical strength. The impact of mechanical forces on tissue development is increasingly appreciated in efforts to engineer load-bearing and mechanically responsive tissues. The polymers applicable for these tissues should address the requirement to transmit mechanical cues to the tissues over the course of tissue development, regardless of *in vitro* mechanical training or dynamic *in vivo* movement. Segmented PU elastomers potentially meet

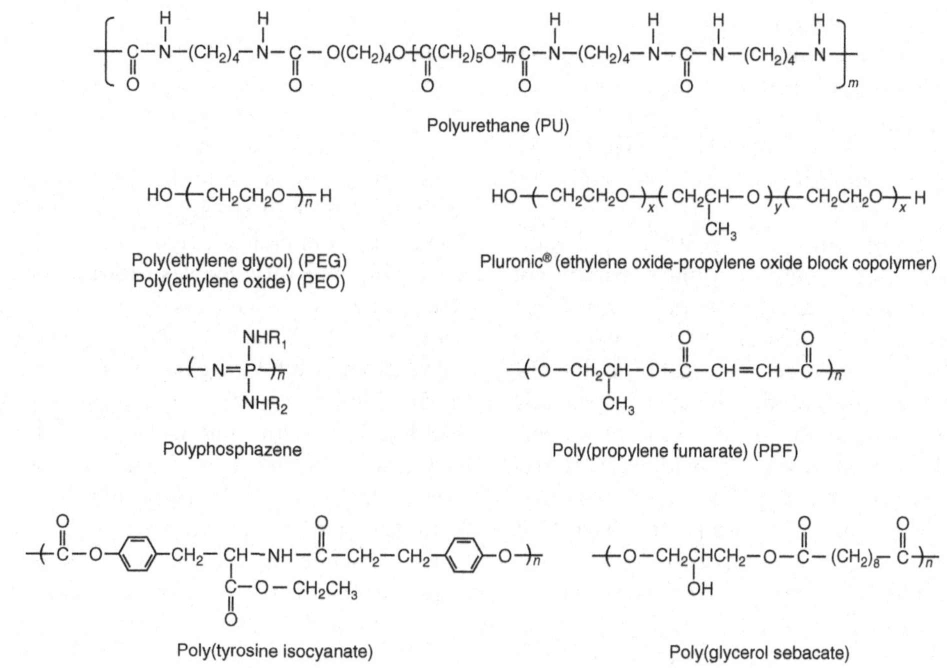

Fig. 7.2-11 Chemical structure of synthetic, absorbable polymers used for tissue engineering (except for α-hydroxyacid polymers).

these requirements, because these polymers have been used for elastic devices such as indwelling catheters, intra-aortic balloons, and left ventricular assist devices.

Segmented PUs are basically synthesized from a polymer with a terminal hydroxyl group at both chain ends (HO–P–OH) and an excess of diisocyanate (OCN–R–NCO). Upon mixing them under an excess of diisocyanate, a prepolymer having terminal isocyanate groups is produced under formation of urea bonds between –OH and –NCO. Addition of a diamine chain extender to this prepolymer increases the chain length to form segmented PU. The P portion associates to yield a soft segment while the R portion forms a hard segment. If the P portion comprises hydrolysable units, the resultant PU exhibits bioabsorption.

Polyphosphazenes

Polyphosphazenes comprise a large class of macromolecules with alternating phosphorus and nitrogen atoms in the backbone and a wide variety of organic, inorganic, or organometallic side groups. Some linear polyphosphazenes undergo rapid hydrolytic degradation, but exhibit poor mechanical properties.

7.2.2.3 Inorganic materials—calcium phosphate

Biological apatite found naturally in bone comprises a range of minerals and has osteoconductive and osteophilic properties. Inorganic minerals have been used for cell scaffolding as well. Among them are hydroxyapatite (HAp) [$Ca_{10}(PO_4)_6(OH)_2$] and β-tricalcium phosphate (β-TCP) [β-$Ca_3(PO_4)_2$]. Synthetic HAp is very slowly absorbed in the body by attack of osteoblasts, whereas β-TCP undergoes absorption probably both with and without such biological assist. A precursor of biological apatite in bone tissue is octacalcium phosphate (OCP) [$Ca_8H_2(PO_4)_6 5H_2O$]. The OCP is absorbed at a higher rate than HAp and β-TCP. A large advantage of HAp as scaffold is that this is a natural component of bones and bioactive (or bioconductive). The inherent brittleness of calcium phosphates limits their use to the exclusive graft material in non-load-bearing defects.

Coral with a chemical composition of calcium carbonate ($CaCO_3$) has an advantage of good communication between pores and is readily converted to HAp, resulting in a high affinity for cells with tissue-regenerating ability. A porous material prepared from coral skeletons is an optimal scaffold for bone regeneration, but the mechanical properties of the composite in the hydrate state are much inferior to those of the natural bone. In addition, coral plays an important role in the marine environment and has now become difficult to obtain, because its use is controlled by environmental regulations such as the Washington Treaty.

7.2.2.4 Composite materials

Inorganic scaffolds are too brittle and not amenable to trimming at operation table, when prefabricated as porous structure. This issue has led some investigators to fabrication of composites from inorganic and organic materials. A well-known example is a composite from HAp and soft collagen, which is a biomimetic from a compositional point of view. Owing to its pliability one can trim the porous composite with scissors into a desired shape.

The PLA, PGA, and some other synthetic polymers have an established safety record in humans, but are not osteoconductive. Apatite-coated polymer composites combine the osteoconductive property of bioceramics and the mechanical resilience of organic polymers. Polymeric scaffolds impregnated with apatite promote *in vitro* cellular attachment and bone nodule formation, as well as bone formation *in vivo*, providing an appropriate osteogenic environment for tissue engineering. Such bioabsorbable and bioactive composites have been developed, combining resorbable polymers with calcium phosphates, bioactive glasses, or glass-ceramics in various scaffold architectures. Besides imparting bioactivity to a polymer scaffold, the addition of bioactive phases to bioabsorbable polymer may alter polymer degradation behavior. Bioactive phases in bioabsorbable polymer allow rapid exchange of protons in water for alkali in the glass or ceramic, which may provide a pH buffering effect at the polymer surface, reducing acceleration of acidic degradation products from the polymer. A further advantage of using bioactive inorganic particles for composite preparation with absorbable polymers is increased mechanical properties owing to the filler effect.

7.2.3 Pore creation in biomaterials

The 3-D architecture of scaffolds with proper pore size, pore shape, alignment, and interconnectivity can have significant effects on regulating the tissue-specific morphogenesis of cultured cells. The rate of tissue ingrowth increases as the porosity and the pore size of the implanted devices increase. The transport of molecules through pores is a function of their size, connectivity, and tortuosity. Thus, the architecture of scaffolds will significantly affect nutrient and oxygen transport within the 3-D matrix, which may directly affect the cell motility during tissue regeneration. The porous structure of scaffold is characterized by two parameters, pore size and porosity. The average pore diameter must be large enough for cells to migrate through the pores and small enough to retain a critical total surface area for appropriate cell binding. To allow for transport of cells and metabolites the scaffold must have a high specific surface and large pore volume fraction in addition to an interconnected pore network. Scaffold pore size has been shown to influence cell adhesion, growth, and phenotype. The optimal scaffold pore size that allows maximal entry of cells as well as cell adhesion and matrix deposition varies with different cell types. Generally, the optimal pore size ranges from 100 to 500 μm and the optimal porosity is above 90%. Various bone engineering groups have noted that pore sizes of greater than 300 μm have a greater penetration of mineralized tissue in comparison with smaller pore sizes, while at pore sizes of 75 μm, hardly any mineralized tissue is found within the scaffold. High porosity enables maximal conversion of cells and tissue invasion necessary for construction, together with conduits suitable for blood vessel formation.

Scaffolds of hydrogel type such as collagen gel and fibrin glue have no geometrical pores recognizable with SEM, but nutrients and oxygen are delivered to cells via physical diffusion through the aqueous medium. The term "channel" may be more appropriate for characterizing the pore structure, because interconnected pores form a channel. However, fabrication of scaffolds with typical channel structures is extremely tedious [5].

A variety of techniques have been developed to fabricate porous scaffolds, of which some include woven and non-woven fiber-based fabrics, solvent casting, salt (or particulate) leaching, temperature-induced phase separation, gas foaming with pressurized carbon dioxide, melt molding, high-pressure processing, membrane lamination, forging, injection molding, pressing, inkjet printing, fused deposition modeling, electrospinning, rapid prototyping, among others. A simple method is to use absorbable fibers as a starting material, because fibers can yield a variety of "porous" products including woven or knitted cloth, non-woven fabric, web, mesh, felt, and fleece. Conventional spinning can produce either multifilaments or monofilaments. A representative example is products from PGA which has very few specific solvents but readily yields fibers by melt spinning. To produce aggregates of PGA fine fibers by means of electrospinning, we need expensive solvents such as trifluoropropyl alcohol for preparing the stock solution for electrospinning of PGA.

Another simple method available in small labs for porous scaffold fabrication is to use the solution of absorbable polymers. One can readily obtain a porous sheet or a 3-D block when a polymer solution is subjected to freeze drying. When a polymer solution is mixed with porogens such as NaCl particulates followed by drying and removal of the porogens from the dried material, we obtain a porous scaffold with the pore size determined by the particulate size. Supercritical liquids have also been applied to fabrication of porous materials. More sophisticated methods applied for scaffold formation are imprinting of polymers and photo-polymerization of monomers along with computer-aided machines. The 3-D printing is a novel technique and the application of this technique for tissue engineering is still limited. These modern technologies, together with computer programming, allow production of scaffolds with very regular and ultrafine structures. The control over scaffold architecture using these fabrication techniques is highly process driven and not design driven.

7.2.3.1 Phase separation (freeze drying)

A widely used method for preparation of porous scaffolds is the thermally induced phase separation, in which the solution temperature is lowered to induce phase separation of the homogeneous polymer solution. The phase separation mechanism may be liquid–liquid demixing, which generates polymer-poor and polymer-rich liquid phases. The subsequent growth and coalescence of the polymer-poor phase would develop to form pores in scaffolds. On the other hand, when the temperature is low enough to allow freeze of the solution, the phase separation mechanism would be solid–liquid demixing, which forms frozen solvent and concentrated polymer phases. By adjusting the polymer concentration, using different solvents, or varying the cooling rate, phase separation could occur via different mechanisms, resulting in the formation of scaffolds with various morphologies. "Freeze drying" is one of the most extensively used methods that produce matrices with porosity greater than 90%. The pore sizes depend on the growth rate of ice crystals during the freeze-drying process.

After removal of the liquid or frozen solvent contained in the demixed solution, the space originally occupied by the solvent would become pores in the prepared scaffolds. Obviously, in the stage of solvent removal, the porous structure contained in the solution needs to be carefully retained. Without freeze drying, a rise in temperature during the drying stage could result in remixing of the phase-separated solution or remelting of the frozen solution, leading to destruction of the porous structure. Thus, the reason of using freeze drying to remove solvent is quite obvious: to keep the temperature low enough that the polymer-rich region would not redissolve and possesses enough mechanical strength to prevent pore collapse during drying. Although freeze drying is a widely used method to prepare porous scaffolds, it is time consuming and energy consuming. For instance, it often takes 4 days to remove solvent by freeze drying. During this 4-day period, a lot of energy is consumed to keep vacuum and to maintain the low temperature needed for drying. Another problem encountered in the application of freeze drying is the occurrence of surface skin. During the freeze-drying stage, if the temperature is not controlled low enough, the polymer matrix in the demixed solution would not be rigid enough to resist the interfacial tension caused by the evaporation of solvent. Thus, the porous structure collapses and dense skin layers occur in the prepared scaffold.

In the case of hydrogels, the freeze-dry processing does not require additional chemicals, relying on the water already present in hydrogels to form ice crystals that can be sublimated from the polymer, creating a particular microarchitecture. Because the direction of growth and the size of ice crystals are a function of the temperature gradient, linear, radial, and/or random pore directions and sizes can be produced with this methodology.

For the freeze-drying process, the solvent vapor pressure at the drying temperature (usually very low) needs to be high enough to allow its removal. Dimethylcarbonate can also be used as a solvent because it exhibits high vapor pressure and melting point around 0°C which makes it suitable for sublimation. Because of a solubility parameter close to that of dioxane, this solvent might be convenient as an alternative to dioxane (potentially carcinogen) for freeze drying of poly(α-hydroxyacid)s. Dimethyl sulfoxide (DMSO) cannot be used as the solvent for PLGA for the freeze-drying process due to its low vapor pressure. This limitation of choosing solvent may be lifted when the scaffolds are prepared by a freeze-extraction method. Freeze extraction and freeze gelation are of a type of phase separation. These methods also fix the porous structure under freezing condition, but in the subsequent drying stage the freeze-drying process is not needed. The principle of the freeze-extraction method is to remove the solvent by extraction with a non-solvent. After the removal of solvent, the space originally occupied by the solvent is taken by the non-solvent and the polymer is then surrounded with the non-solvent. Under this circumstance, even at room temperature, the polymer would not dissolve. Hence, drying can be carried out at room temperature to remove the non-solvent, leaving space that becomes pores in the scaffold. For the freeze-extraction process, DMSO can be used as the solvent for preparation of PLGA scaffolds, because it can be easily extracted out by ethanol aqueous solution.

7.2.3.2 Porogen leaching

The porogen leaching method has also been widely used to prepare porous scaffolds, because of easy operation and accurately controlling pore size and porosity. The salt leaching technique consists of adding salt particulates to a polymer solution. The overall porosity and level of pore connectivity are regulated by the ratio of polymer/salt particulates and the size of the salt particulates.

7.2.3.3 Fiber bonding

Fiber mesh consists of individual fibers either woven or knitted into 3-D patterns of variable pore sizes. They can be obtained by deposition of a polymer solution over a non-woven mesh of another polymer, and the subsequent evaporation. The advantages of fiber meshes are the large surface area for cell attachments and the rapid diffusion of nutrients, but they are not suitable for the fine control of porosity.

7.2.3.4 Gas foaming

The gas-foaming technique uses high-pressure CO_2 gas processing. The porosity and pore structure depend on the amount of gas dissolved in the polymer, the rate and type of gas nucleation, and the diffusion rate of gas molecules through the polymer to the pore nuclei. However, this technique often produces a scaffold that is too compact for the cells. This may be improved by associating the gas foaming with salt leaching. Another technique derived from improvement of gas foaming consists of the substitution of high-pressure CO_2 with ammonium bicarbonate, which acts also as porogen. In this case the porosity depends only on the amount of salt particulates added, whereas the pore diameter is due to the size of the salt crystals.

7.2.3.5 Rapid prototyping

Rapid prototyping (RP) or solid free-form (SFF) fabrication refers to a group of technologies that build a physical, 3-D object in a layer-by-layer fashion. The RP is a subset of mechanical processing techniques which allow highly complex structures to be built as a series of thin two-dimensional (2-D) slices using computer-aided design (CAD) and computer-aided manufacturing (CAM) programs. These techniques essentially allow researchers to predefine properties such as porosity, interconnectivity, and pore size. The RP methodologies include stereolithography (SLA), selective laser sintering, ballistic particle manufacturing, and 3-D printing. Examples of RP applied to generate tissue engineering scaffolds include laser sintering to fabricate Nylon-6 scaffolds, fused deposition molding of poly(ε-caprolactone) (PCL) scaffolds, and SLA of PU patterns. Direct and indirect SLA have been used to make ceramic scaffolds and cancellous bone structure models. Few studies address the generation of scaffolds by RP techniques for soft tissue engineering such as 3-D hydrogel scaffolds. SLA, one of the most common types of RP, operates by selectively shining a laser beam onto a vat of liquid photopolymer. This method has been employed in numerous biomedical applications such as building models of biological structures, bone substrate scaffolds, and heart valve scaffolds. Using an RP technique to pattern not only the scaffold but also cells will accelerate and improve tissue assembly. On the basis of this idea, various RP methods have emerged, but they involve time-consuming sequential writing processes with a narrow processing window and relatively low resolution. The SFF fabrication technologies create 3-D structures in laminated fashion from numerical models. Commercially available SFF technologies, such as fused deposition modeling, SLA, and selective laser sintering have been utilized to fabricate scaffolds. Computer-aided engineering has also emerged to fabricate multifunctional scaffolds with control over scaffold composition, porosity, macroarchitecture, and mechanical properties based on optimization models. To print tissue engineering scaffolds, inkjet solvent binder is printed onto a powder bed of porogens and polymer particles. The solvent will dissolve the polymer and evaporate, and the polymer will re-precipitate to form solid structures. The final porosity is achieved after particulate leaching and solvent removal. Figure 7.2-12 shows a custom-designed fiber-deposition device [6]. Hollister reviewed how integration of computational topology design and SFF has made scaffolds with designed characteristics possible [7].

7.2.3.6 Electrospinning

Electrospinning provides a mechanism to produce nanofibrous scaffolds from synthetic and natural polymers, with high porosity, a wide distribution of pore diameter, high surface-area-to-volume ratio, and morphological similarities to natural collagen fibrils. Fiber diameters are in the range from several micrometers down to less than 100 nm. The electrospinning process is based on a fiber spinning technique driven by a high-voltage electrostatic field using a polymeric solution or liquid. Figure 7.2-13 represents an electrospinning apparatus [8]. The underlying physics of this technique relies on the application of an electrical force, especially when at the polymer droplet surface it overcomes the surface tension force, and a charged jet is ejected. As the solvent evaporates, the charge density increases on the fibers, resulting in an unstable jet, which stretches the fibers over about one million fold. The variables controlling the behavior of the electrified fluid jet during electrospinning can be divided into fluid properties and operating parameters. The relevant fluid properties are viscosity, conductivity, dielectric constant, boiling point, and surface tension. The operating parameters include flow rate, applied electric potential, and distance between the tip and the collector called "air gap."

The final product of electrospinning generally consists of randomly interconnected webs of sub-micron size fibers. Nanofibrous scaffolds formed by electro-spinning, by virtue of structural similarity to natural ECM, may represent promising structures for tissue engineering applications.

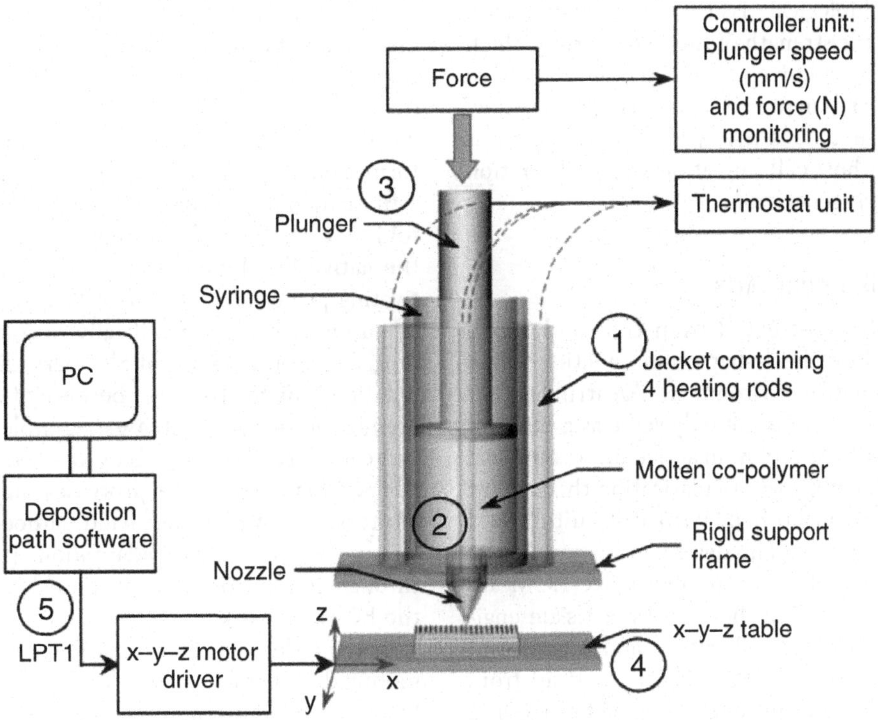

Fig. 7.2-12 The 3-D deposition device consisting of five main components: (1) a thermostatically controlled heating jacket; (2) a molton copolymer dispensing unit consisting of a syringe and nozzle; (3) a force-controlled plunger to regulate flow of molten copolymer; (4) a stepper motor driven *x–y–z* table; and (5) a positional control unit consisting of stepper-motor drivers linked to a personal computer containing software for generating fiber deposition paths.

7.2.4 Special scaffolds

7.2.4.1 Naturally derived scaffolds

Connective tissues are typically tough and pliable in contrast with synthetic, hydrophilic polymers. The high toughness and pliability of naturally derived materials in spite of their high water contents such as 70% are due largely to ordered orientation of collagen fibrils and the presence of crosslinked elastin in the connective tissue. Some researchers have attempted to fabricate scaffolds from collagen molecules for tendon and ligament tissue engineering, but the reconstructed scaffolds do not provide such good mechanical properties as natural ones because of difficulty in assembling the collagen fibers to specific orientations. When scaffolds are derived from natural tissues, they do not need the pore formation process and can provide the optimal surface for cell adhesion because of the presence of cell adhesion sites. Matrices for tissue engineering are being derived by extraction or partial purification of whole tissue, removing some components and leaving much of the 3-D matrix structure intact, likely with growth factors as well. Acellular biological tissues have therefore long been proposed to be used as scaffolds for tissue repair and tissue regeneration. The guiding principle behind such acellular grafts is that the immunogenic response associated with allografts is sufficiently reduced through removal of the cells. Once implanted, the acellular graft serves as a natural scaffold into which surrounding cells readily migrate, forming the foundation for new tissue. These biomaterials are composed of ECM proteins and polysaccharides that are conserved among different species and can serve as scaffolds for cell attachment, migration, and proliferation. In addition to the inherent

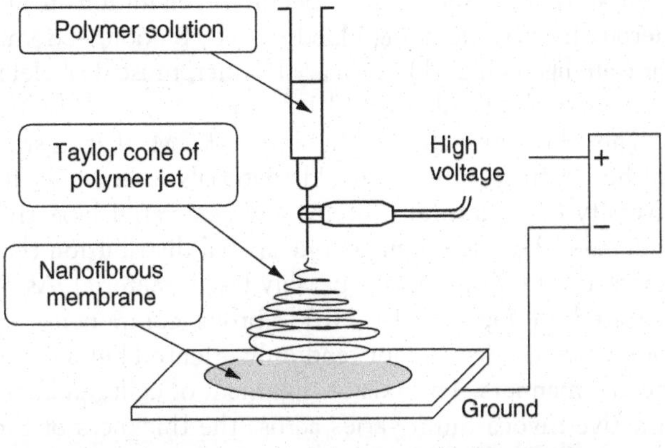

Fig. 7.2-13 Electrospinning apparatus of polymer nano-fibrous membranes.

cell compatibility, biological tissues primarily maintain desired shapes and the strength of the tissues from which the materials have been derived. This can be an advantage over synthetic materials in regard to materials processing. In addition, acellular tissues may provide a natural microenvironment for host cell migration and proliferation to accelerate tissue regeneration.

7.2.4.1.1 ECM-like scaffolds

Tissues are basically constructed from cells and ECM, and the cells will lose their shape and function when taken out from the native environment. An artificial 3-D polymeric scaffold must thus not only serve as a physical structural support, but also play an active important role in cell migration, growth, and vascularization throughout the 3-D architecture in both *in vitro* cell culture and *in vivo* tissue regeneration. The design and manufacture of 3-D polymeric scaffolds that mimic native ECM are thus thought to be crucial to the success of tissue engineering. The ECM exists in all tissues and organs, but can be harvested for use as a therapeutic scaffold from relatively few sources. Dermis of the skin, submucosa of the small intestine and urinary bladder, pericardium, basement membrane and stroma of the decellularized liver, and the decellularized Achilles tendon are all potential sources of ECM.

The host response to ECM-derived scaffolds is largely dependent upon the methods used to process the material. Chemical and non-chemical crosslinking of ECM proteins have been utilized extensively in an effort to modify the physical, mechanical, or immunogenic properties of naturally derived scaffolds. Chemical crosslinking generally involves aldehyde or CDI, and photochemical protein crosslinking has also been investigated. Although these crosslinking methods provide certain desirable physical or mechanical properties, the end result is the modification of a biologically interactive material into a relatively inert scaffold. There is abundant literature on the use of modified ECM scaffolds, especially chemically crosslinked biological scaffolds, for tissue repair and replacement. Porcine heart valves, decellularized and crosslinked human dermis (Alloderm™), and chemically crosslinked purified bovine type I collagen (Contigen™) are examples of such products currently available for use in humans. Similarly, modified ECM scaffolds have been used for the reconstitution of the cornea, skin, cartilage, bones, and nervous system.

The scaffold modifications typically result in inhibition of the scaffold degradation and cellular infiltration into the scaffold. Although there may be clinical uses for such modified biomaterials, these properties are counterintuitive to many approaches in the field of tissue engineering; especially, those approaches in which cells are seeded upon scaffolds prior to or at the time of implantation. In contrast, ECM scaffolds that remain essentially unchanged from the native ECM elicit a host response that promotes cell infiltration and rapid scaffold degradation, deposition of host-derived neomatrix, and eventually constructive tissue remodeling with minimum formation of scar tissue. The native ECM represents a fundamentally different material than the ECM scaffold that has been chemically or otherwise modified. In the native ECM, cells clear a path by secreting and activating proteases, such as MMPs, serine proteases, and hyaluronidase, that specifically degrade protein or proteoglycan components of the pericellular matrix. Degradation is highly localized because of the involvement of membrane-bound proteases, complexation of soluble proteases to cell surface receptors, and a tightly regulated balance between active proteases and their natural inhibitors. On the other hand, amoeboid migration is driven by cell shape adaptation (that is, squeezing through pre-existing matrix pores) and deformation of the ECM network.

Porcine-derived ECM scaffolds that have not been modified, except for the decellularization process and terminal sterilization, have been used for the repair of numerous body tissues. Immediately following implantation *in vivo*, there is an intense cellular infiltrate consisting of equal numbers of polymorphonuclear leukocytes and mononuclear cells (MNCs). By 3 days postimplantation, the infiltrate is almost entirely MNCs in appearance with early evidence for neovascularization. Between day 3 and 14, the number of MNCs increases, vascularization becomes intense, and there is a progressive degradation of the xenogeneic scaffold with associated deposition of host-derived neomatrix. Following day 14, the MNC infiltrate diminishes and there is the appearance of site-specific parenchymal cells that orient along lines of stress. These parenchymal cells consist of fibroblasts, smooth muscle cells, skeletal muscle cells, and™ endothelial cells (ECs) depending upon the site in which the scaffold has been placed. A devitalized collagen-based scaffold from small intestine submucosa (SIS) has been shown to induce site-specific regeneration in numerous tissues, including blood vessels, tendon, abdominal wall, ligaments, skin, urinary bladder, musculoskeletal repair, and dural substitute [9].

The characteristic of the intact ECM that distinguishes it from other scaffold materials is not only its diversity of structural proteins and associated bioactive molecules, but also their unique spatial distribution that makes the ECM an uncomparably tough material. As illustrated in Fig. 7.2-14 [10], fibrous components of connective tissues are not randomly oriented but align in specific manners. The axis of alignment of collagen fibers in native myocardium varies across the thickness of the heart wall. Such mechanical anisotropy is required also for engineered heart tissue.

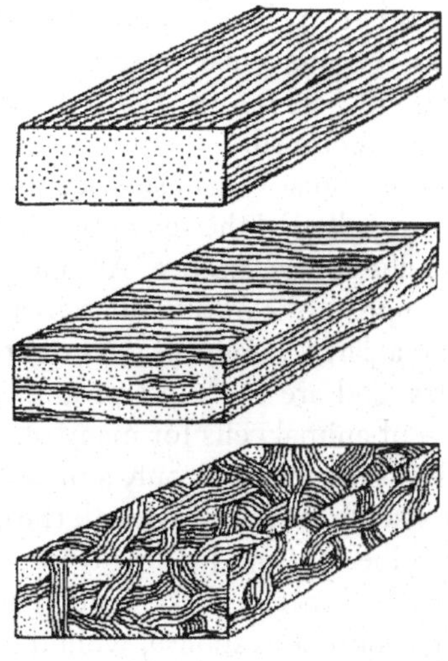

Fig. 7.2-14 Orientation of collagen fibrils in connective tissues.

7.2.4.1.2 Fibrin gel

The advantage of fibrin gels over the other gels is that it can be obtained autologous. Furthermore, cells entrapped in the fibrin gels were reported to produce more collagen and elastin than those entrapped in collagen gel. Fibrin degrades within several days by cell-associated enzymatic activities when no degradation inhibitors are used. In tissue engineering applications using cells encapsulated in fibrin gel, degradation inhibitors are often used to preserve the scaffold function of the fibrin. The effect of the inhibitor concentration on collagen formation in the gels differs among studies.

7.2.4.1.3 Matrigel™

Matrigel is commercially available from BD Biosciences (San Jose, CA, USA). This gel is a basement membrane preparation extracted from Engelbreth-Holm-Swarm mouse sarcoma and solubilized in Dulbecco's modified Eagle's medium (DMEM). This contains several components of basement membranes enriched with LN. Matrigel is liquid at 2–8°C and sets to a gel rapidly at 22–35°C.

7.2.4.1.4 Marine natural scaffold

Some natural biomaterials including coral, sea urchins, and marine sponges are much less expensive than the natural ECMs and appear to provide affordable, readily available scaffolds with a number of unique and suitable properties such as open interconnected channels.

7.2.4.2 Injectable scaffolds

Large-scale, prefabricated scaffolds require invasive surgery for implantation, are difficult to contour to the defect shape, have predetermined mechanical properties, and induce poor vascularization. Therefore, injectable biomaterials such as collagen, fibrin, gelatin, alginate, HAc, chitosan, and PEG gels emerged as candidates for scaffolds in the replacement of large-size tissue defects. Injectable hydrogel materials are advantageous because they require a minimally invasive procedure and conform to irregular shapes. However, they are mechanically weaker than soft tissues and many cells perform poorly when suspended within a bioinert hydrogel.

7.2.4.3 Soft, elastic scaffolds

For soft tissue engineering, synthetic elastomeric materials with tunable degradation properties would be preferable. For instance, for cardiovascular applications, elastomeric behavior with low modulus would allow the transmission of stresses to seeded or infiltrating cells early in the implant or culture period. Such mechanical training has been shown to be important in developing mechanically robust tissues with appropriate cellular orientation. The PGA, PLA, and their copolymers (PLGA) are relatively stiff and nonelastic and are not ideally suited for engineering of soft tissues under a mechanically demanding environment such as cardiovascular, urological, and gastrointestinal tissue, unless specifically processed. Mechanical signals are thought to be necessary for developing the cell alignment that leads to tissue structure with correct biomechanical properties and functions. Elastic scaffolds may be essential for the engineering of any tissue under conditions of cyclic mechanical strain. This is the reason for synthesis of absorbable and pliable PUs.

7.2.4.4 Inorganic scaffolds

The most abundant bioceramic HAp can be used as a scaffold for bone regeneration because HAp will be integrated in the regenerated bone so far the shape and size is acceptable, although the absorption rate is quite low unless osteoblast frequently attacks. However, HAp scaffolds whose pores are not interconnected but independent are not adequate for tissue engineering. Continuous pore interconnection is a prerequisite as scaffold. Modification of marine coral into HAp possessing porous structure suitable for tissue engineering has been applied for scaffold fabrication, but the use of coral-derived HAp should be refrained if the coral harvest violates the Washington Treaty. In contrast to HAp, β-tricalcium phosphate (β-TCP) exhibits rapid degradation and hence is a good candidate for inorganic scaffolds. To serve as the scaffold in which bone ingrowth and vascularization take place, β-TCP should be highly porous, but this inorganic material tends to produce a porous block with brittle and fragile structure. This drawback may be overcome by formation of composites from β-TCP and flexible polymers, which must be osteoconductive if portion of β-TCP is present on the surface of scaffold.

7.2.4.5 Composite scaffolds

Ceramics including dense and porous HAp, TCP ceramics, bioactive glasses, and glass-ceramics have been combined with a number of polymers including collagen, chitosan, and poly(α-hydroxyacid)s. The combination of such polymers with a bioactive component takes advantage of the osteoconducting properties (bioactivity) of HAp and bioactive glasses and of their strengthening effect on polymer matrices. The composite is expected to have superior mechanical properties than the neat (unreinforced) polymer and to improve structural integrity and flexibility over brittle glasses and ceramics for eventual load-bearing applications. Composite fabrication research has focused on developing polymer/ceramics blends, precipitating ceramic onto polymer templates, coating polymers onto ceramics, or ceramics onto polymers. Polymer processing techniques including combined solvent-casting and salt-leaching, phase separation and freeze drying, and immersion-precipitation have been used for the preparation of highly porous PLA/HAp scaffolds. *In situ* apatite formation can also be induced by a biomimetic process in which polymer foams are incubated in a simulated body fluid.

7.2.5 Surface modifications

Both bulk (e.g., strength and degradability) and surface properties of biomaterials are important in tissue engineering. Generally, bulk and surface properties are interwound. Often, materials are chosen for their favorable bulk properties, and their unfavorable biological interactions are improved by surface modifications. The success of tissue engineering depends on interactions at the cell–scaffold interface, including the cell adhesion and proliferation, expression and activity of regulatory signaling molecules, and biomechanical stimuli.

7.2.5.1 Cell interactions in natural tissues

A highly dynamic and complex array of biophysical and biochemical signals are transmitted from the outside of a cell by various cell surface receptors and integrated by intracellular signaling pathways. The signals converge to regulate gene expression and ultimately establish cell phenotype. This indicates that the ultimate decision of a cell to differentiate, proliferate, migrate, apoptose, or perform other specific functions is a coordinated response to the molecular interactions with ECM effectors. This flow of information between cells and their ECM is bidirectional.

In multicellular organisms contacts of cells with neighboring cells and the surrounding ECM are mediated by cell adhesion receptors. Among them the integrin family comprises the most numerous and versatile groups. They are a large family of heterodimeric, cell surface molecules, and are the most prominent ECM adhesion receptors of animal cells for many of the ECM adhesion molecules. They play not only a major role in linking the macromolecules of the ECM with the cell's cytoskeleton, cell–cell adhesion, and binding to proteases, but are also important in processes like embryogenesis, cell differentiation, immune response, wound healing, and hemostasis. Integrins consist of two non-covalently associated transmembrane subunits, termed α and β. To date 18α and 8β subunits are known, which form 24 different heterodimers. The combination of particular α and β subunits determines the ligand specificity of the integrin. Some integrins, however, are highly promiscuous, e.g., the αvβ3 integrin binds to vitronectin, FN, von Willebrand factor, osteopontin (OP), tenascin, bone sialoprotein, and thrombospondin. Vice versa, ECM molecules like FN are ligands for several integrins. Fibronectin is a disulfide-linked dimeric glycoprotein prominent in many ECMs and present at about 300 μg/ml in plasma. It interacts with collagen, heparin, fibrin, and cell surface receptors of the integrin family. In 1984, the tripeptide motif RGD (Fig. 7.2-15) [11] was identified by Pierschbacher and Rouslahti as a minimal essential cell adhesion peptide sequence in FN. The RGD peptides inhibit cell adhesion to FN on the one hand, and promote cell adhesion when they are immobilized on surfaces on the other hand. Since then, cell-adhesive RGD sites were identified in many other ECM proteins, including vitronectin, fibrinogen, von Willebrand factor, collagen, LN, OP, tenascin, and bone sialoprotein as well as in membrane proteins, in viral and bacterial proteins,

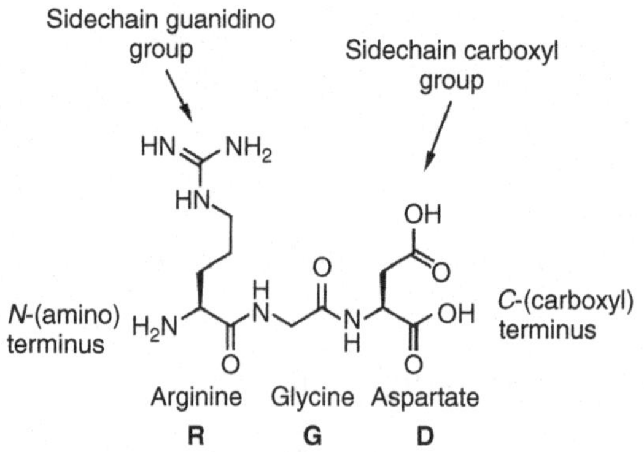

Fig. 7.2-15 The molecular formula and nomenclature of RGD.

and in snake venoms. About half of the 24 integrins have been shown to bind to ECM molecules in an RGD-dependent manner. Since other important adhesive motifs have been also identified, the RGD sequence is not the "universal cell recognition motif," but it is unique with respect to its broad distribution and usage. The conformation of the RGD-containing loop and its flanking amino acids in the respective proteins are mainly responsible for their different integrin affinity. Other factors that contribute to integrin-ligand binding affinity include the activation of integrins by divalent cations and cytoplasmatic proteins.

The process of integrin-mediated cell adhesion comprises a cascade of four different partly overlapping events: cell attachment, cell spreading, organization of actin cytoskeleton, and formation of focal adhesions. First, in the initial attachment step the cell contacts the surface and some ligand binding occurs that allows the cell to withstand gentle shear forces. Secondly, the cell body begins to flatten and its plasma membrane spreads over the substratum. Thirdly, this leads to actin organization into microfilament bundles, referred to as stress fibers. In the fourth step the formation of focal adhesions occurs, which link the ECM to molecules of the actin cytoskeleton. Focal adhesions consist of clustered integrins and more than 50 other transmembrane-associated and other cytosolic molecules. During the four steps of cell adhesion integrins are employed in physical anchoring processes as well as in signal transduction through the cell membrane.

It is well established that integrin-mediated cell spreading and focal adhesion formation trigger survival and proliferation of anchorage-dependent cells. In this context the expression of the anti-apoptotic protein Bcl-2, induced by $\alpha 5\beta 1$ integrins, and the suppression of the p53 pathway by focal adhesive kinase (FAK) are discussed. In contrast, loss of attachment causes apoptosis in many cell types, referred to as "anoikis" (a Greek word meaning homelessness). Anoikis can even be induced in the presence of immobilized ECM molecules when non-immobilized soluble ligands like RGD peptides are added, as illustrated in Fig. 7.2-16 [11]. Stable linking of RGD peptides to a surface is essential to promote strong cell adhesion, because formation of focal adhesions only occurs if the ligands withstand the cell's contractile forces. These forces are able to redistribute weakly adsorbed ligands on a surface, which leads only to weak fibrillar adhesions later on. Furthermore, cells can remove mobile integrin ligands by internalization.

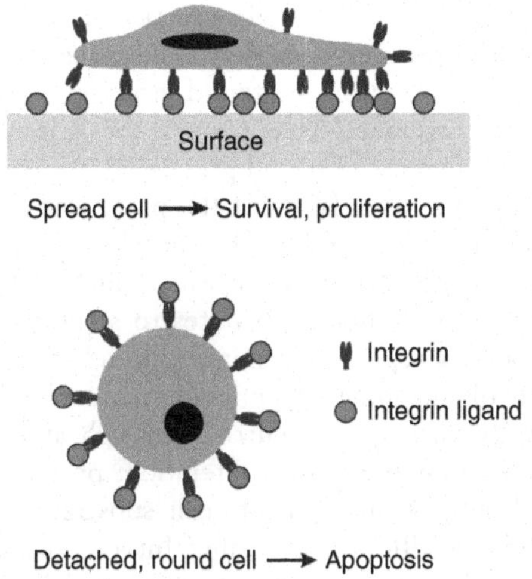

Fig. 7.2-16 Opposite effects of integrin ligands. Immobilized ligands act as agonists of the ECM, leading to cell adhesion and cell survival, while non-immobilized ligands act as antagonists, leading to cell deattachment, a round cell shape, and apoptosis.

7.2.5.2 Artificial surface in biological environment

Because common poly(α-hydroxyester)s like PLA and PGA are hydrophobic, porous scaffolds fabricated with these polymers are floating in cell culture medium. When cells in culture medium are plated on the top of the porous scaffold or injected into its interior for seeding, the majority of its pores remain empty since the scaffold does not absorb the culture medium. Because it is crucial to obtain a uniform distribution of initial seeded cells throughout the scaffold volume for the creation of a tissue with homogeneous cellularity, some approaches have been attempted to improve hydrophilicity of polymer scaffolds and thus to ensure uniform and dense cell seeding. They include treatments of the scaffolds by prewetting with ethanol, hydrolysis with NaOH, oxidation with perchloric acid solution, oxygen or ammonia plasma discharge treatment, physical or chemical coating with some hydrophilic polymers or cell-adhesive proteins, and blending with hydrophilic polymers such as Pluronics (ethylene glycol-propylene glycol block copolymers).

When foreign materials are exposed to a biological environment, ECM proteins are non-specifically adsorbed on the surface of nearly all the materials, masking its specific surface properties, and then cells indirectly interact with the material surface through the adsorbed ECM proteins [12]. A range of proteins from the culture medium come adsorbed to the substrate surface, reflecting the relative abundance of proteins in the culture medium. However, over time, weakly adsorbed proteins will leave the surface, while those bound more strongly will remain. Other protein molecules will come in and replace those that have been desorbed. The implication of this series of events is that the final nature of the attached protein layer reflects both the surface chemistry of the underlying substrate and the composition of the culture medium. This highlights the importance of optimizing the surface chemistry of the substrate to give the best performance of the cells in a particular medium. The nature of the protein layer that is adsorbed to the underlying substrate has a profound effect on the attachment and subsequent development of the cells in culture. Not only is the mass of bound protein an issue but, more importantly, the orientation of the protein with respect to the surface may be crucial. Complexity of the cell adhesion process that spans a broad range of time and length scales no doubt contributes to some of the differences among reports of cell interaction with materials. Adhesion of anchorage-dependent mammalian cells is traditionally viewed as occurring through at least four major steps that precede proliferation: protein adsorption, cell–substratum contact, cell–substratum attachment, and cell adhesion/spreading. Interactions between cells and polymers in serum are mediated by proteins that have been either preimmobilized on the material surface, absorbed from the surrounding medium, or secreted by cells during culture. Cell contact and attachment involves gravitation/sedimentation to within 50 nm or so of a surface whereupon physico-chemical forces conspire to close the cell–surface distance gap. Attached cells then spread over the surface, depending on the compatibility with the surface typically within hours.

Functional scaffolds should be capable of eliciting specific cellular responses and directing new tissue formation. This is mediated by biomolecular recognition instead of by non-specifically adsorbed proteins, similar to the native ECM. Extensive studies have been performed to render scaffolds biomimetic by surface modification, as this is the simplest way to make biomimetic scaffolds. Scaffold with an informational function, e.g., with the RGD sequence which facilitates cell attachment, must be better than non-informational synthetic polymers. These biomimetic materials potentially mimic many roles of ECM in tissues. For example, incorporation of peptide sequences renders the surface of biomaterials cell adhesive that inherently have been non-adhesive to cells. The design of biomimetic materials that are able to interact with surrounding tissues by biomolecular recognition is an attempt to make the materials such that they are capable of directing new tissue formation mediated by specific interactions, which can be manipulated by altering design parameters instead of by non-specifically adsorbed ECM proteins. For mimicking the native ECM, either long chains of ECM proteins such as FN, vitronectin, and LN or short peptide fragments composed of several amino acids along the long chain of ECM proteins can be the candidates for surface modification.

The early work on the surface modification of biomaterials with bioactive molecules has used long chains of ECM proteins for surface modification. Biomaterials can also be coated with these proteins, which usually have promoted cell adhesion and proliferation. Since the finding of the presence of signaling domains that are composed of several amino acids along the long chain of ECM proteins and primarily interact with cell membrane receptors, the short peptide fragments have been used for surface modification in numerous studies. Particularly, the use of a short peptide for surface modification is advantageous over the use of the long chain of native ECM proteins. The native ECM protein tends to be randomly folded upon adsorption to the biomaterial surface such that the receptor binding domains are not always sterically available. However, the short peptide sequences are relatively more stable during the modification process than long chain proteins such that nearly all peptides modified with spacers are available for cell binding. In addition, short peptide sequences can be massively synthesized in laboratories more economically. The biomimetic material modified with these bioactive molecules can be used as a tissue engineering scaffold that potentially serves as artificial ECM providing suitable biological cues to guide new tissue formation.

The most commonly used peptide for surface modification is RGD. Additionally, other peptide sequences such as Tyr-Ile-Gly-Ser-Arg (YIGSR), Arg-Glu-Asp-Val (REDV), and Ile-Lys-Val-Ala-Val (IKVAV) have been immobilized on various model substrates. In order to provide stable linking, RGD peptides should be covalently attached to the polymer via functional groups like hydroxyl, amino, or carboxyl groups. Simple adsorption of small RGD peptides only leads to poor cell attachment. When a polymer does not have functional groups on its surface, these have to be introduced by blending, copolymerization, or chemical or physical treatments. Different coupling techniques have been employed to ensure covalent binding of the peptides to the surface of the materials. In most cases, peptides are linked to polymers by reacting an activated surface carboxylic acid group with the nucleophilic N-terminus of the peptide, as shown in Fig. 7.2-17a [11]. First, the surface carboxyl group is converted to an active ester, that is less prone to hydrolysis, e.g., *N*-hydroxysuccinimide (NHS) ester, and, second, this is coupled to the peptide in water. Polymers that contain surface amino groups can be treated with succinic anhydride to generate surface carboxyl groups, which can be reacted with RGD peptides as described above. Amino groups can directly be converted into preactivated carboxyl groups by using an excess of bisactivated moieties like *N*,*N*′-disuccinimidyl carbonate (DSC), as shown in Fig. 7.2-17b. Surface containing hydroxyl groups can simply be preactivated with, for instance, tresyl chloride (Fig. 7.2-17c) or *p*-nitrophenyl chlorocarbonate (Fig. 7.2-17d).

A bifunctional crosslinker that has a long spacer arm can be used for the immobilization of peptides to the surface, which can enable the immobilized peptides to move flexibly in the biological environment. For polymer substrates lacking appropriate functional groups for

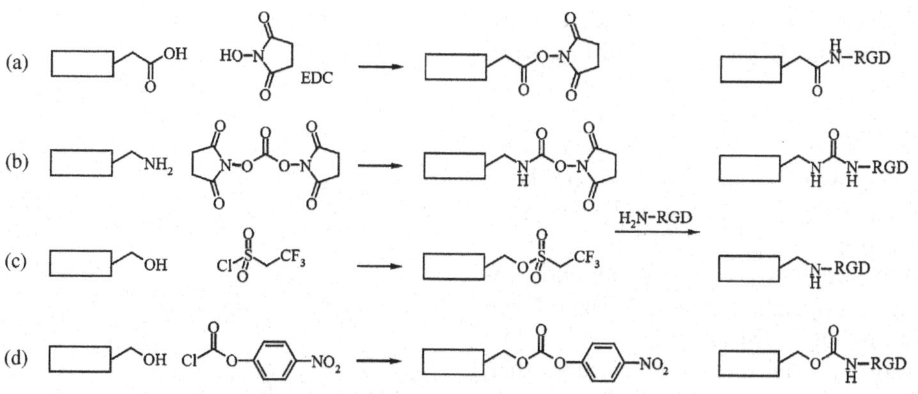

Fig. 7.2-17 RGD peptides react via the N-terminus with different groups on polymers: (a) carboxyl groups, preactivated with a carbodiimide and NHS to generate an active ester; (b) amino groups, preactivated with DSC; (c) hydroxyl groups, preactivated as tresilate; and (d) hydroxyl groups, preactivated as *p*-nitrophenyl carbonate.

a coupling reaction, a photochemical immobilization method has been utilized to graft cell-binding peptides. In order to examine that any cellular responses to the modified substrates are mediated solely by the immobilized peptides, the experiment is performed under serum-free conditions.

7.2.6 Cell expansion and differentiation

For the clinical use of tissue engineering with isolated cells, a small number of cells are initially isolated from a small biopsy from the specific body part of patients or others and then expanded in number in conventional monolayer culture before they are seeded into scaffolds. In many cases the number of harvested cells is not large enough for repairing the lost or damaged tissues, especially when used clinically in the reconstruction of large defects. Therefore, it is necessary to multiply the isolated cells for tissue engineering with a sufficient number of cells. Generally, the smaller the cell amount, the ECM formation is less. Moreover, if the density of cells seeded into scaffolds is low—in other words, the distance between the neighboring cells is long—the production of ECM such as collagen and GAG from the cells will be poor because of insufficient communication between the cells. Although the promise of tissue engineering is tremendous, it has only seldom been accomplished in humans, largely because many cells are needed to generate even small amounts of tissue. It will be often necessary to generate large amounts of tissue, starting with very few cells. To circumvent this problem, cell culture is performed to multiply cells under retention of their phenotypic characteristics either before or after seeding them into scaffold.

In cell culture the modulation of cell phenotype between the synthetic and quiescent states is important. The modulation is based on biochemical or environmental cues. A central issue in blood vessel culture is to balance the competing goals of smooth-muscle cell proliferation and ECM deposition (synthetic state or dedifferentiation), and the contractile phenotype associated with differentiation and maturation. For culture of engineered vessels *de novo*, an increased synthetic state is required, whereas at the conclusion of vessel culture, a minimally proliferative, quiescent phenotype is desired. Since cells would dedifferentiate when seeded into polymeric structures *in vitro*, in depth investigations are necessary to find out how dedifferentiation of seeded cells can be prevented.

To treat traumatic or congenital cartilage defects with tissue engineering techniques, a relatively small number of donor cells, either chondrocytes or progenitor cells, are expanded *in vitro* until sufficient cells are obtained. However, *in vitro* multiplication of chondrocytes in monolayer results in dedifferentiation of these cells. Expansion in high seeding density cultures often fails to produce sufficient chondrocytes, even after several passages. Lower seeding densities may increase cell yield, but bear the risk of decreased redifferentiation capacity. Differentiation of stem cells to target cells *in vitro* needs specific culture media.

7.2.6.1 Monolayer (2-D) and 3-D culture

Advances in cell culture techniques have culminated in the field of tissue engineering. The most common method to increase the number of cells is 2-D monolayer cell culture on a flat substrate. The 2-D culture is excellent for cell expansion but sometimes induces the loss of native functional natures of cells. In monolayer cultures, cells are forced to grow in one plane under space-limiting conditions. This results in an obviously artificial growth environment, in contrast to development *in vivo*.

Conventional monolayer cultures of chondrocytes have the disadvantage of producing matrix that differs from that produced *in vivo*, losing their typical phenotype within several days or weeks as they "dedifferentiate". Freshly isolated articular chondrocytes express cartilage-specific type II collagen and hyaline cartilage markers, aggrecan, but during prolonged culture and serial subculture these cells lose their spherical shape, begin to dedifferentiate to a fibroblast-like phenotype, and produce predominantly unspecific type I collagen. The phenotypic stability of adult human chondrocytes is lost quickly on expansion in serial monolayer cultures than that of cells of juvenile humans. This loss of phenotype in monolayer culture is reversible if chondrocytes are cultured in 3-D culture systems embedded in solid support matrices, such as collagen, agarose, or alginate gels. However, there are disadvantages to these systems, including the slow rate of proliferation and the substantial decline of matrix production in suspension cultures. Cardiomyocytes cultured in monolayers also exhibit properties different from native heart tissue, because of structural differences between 2-D and native environments, and because of the effects of cell isolation and *in vitro* cultivation.

The 3-D culture has often been stated to be preferable to 2-D culture, because most of the cells in our body are present in 3-D environments [12]. Many cellular processes including morphogenesis and organogenesis have been demonstrated to occur exclusively when cells are organized in a 3-D fashion. The fundamental process underlying most tissue engineering methodologies is 3-D culture at high cell density to enhance cell–cell interactions favorable for ECM production. The 3-D ECM culture systems have been developed to simulate natural interactions between cells and the extracellular environment. The 3-D culture maintains the cell phenotype but is poor for cell expansion. Cells in 3-D culture are surrounded with a substrate not only on one side of the cells but on many sides of the cells. Generally, the so-called "3-D scaffolds" having distinct pores of sizes much larger than the cells come in contact with cells only by their one side, as illustrated in Fig. 7.2-18a. Cells in right 3-D culture should be in contact with a substrate from multiple directions, as shown in Fig. 7.2.18b. However, the 2-D culture may change to 3-D culture once the cells begin to be surrounded by the matrix produced by the cells themselves, as shown in Fig. 7.2.18d. In culture of chondrocytes for cartilage repair the isolated cells have been expanded in monolayer culture, dedifferentiated, and then redifferentiated in a 3-D cell arrangement for new cartilage formation. A typical sign of dedifferentiation of chondrocytes is the switch from type II collagen to type I collagen synthesis. Endothelial cells grown in 2-D systems vary from 3-D model systems. A wide variety of cell types exhibit enhanced maintenance of their differentiated phenotype if cultured in 3-D systems instead of monolayers, which is attributed to associated differences in cell shape and/or increases in intercellular communication. It was shown that neonatal rat ventricular myocytes in confluent monolayers couple on average with six cells, whereas the same cells in native ventricles couple on average with nine cells [13].

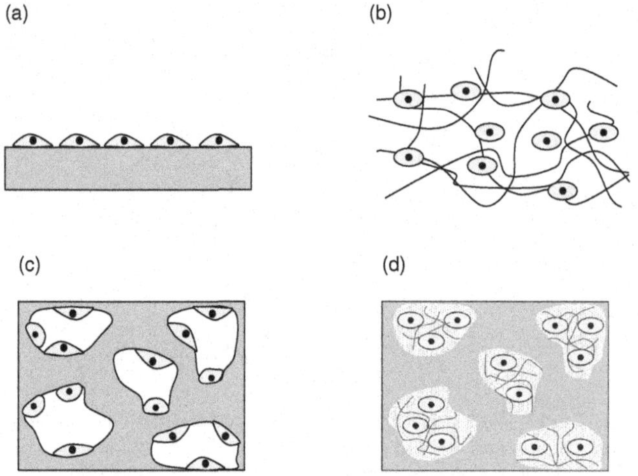

Fig. 7.2-18 Cells at 2-D and 3-D culture, and seeded on 3-D porous scaffold: (a) cells on flat dish (2-D culture); (b) cells in gel (3-D culture); (c) cells adsorbed on 3-D porous scaffold (2-D culture); and (d) cells entrapped in biosynthesized ECM.

One major constraint in the use of 3-D scaffolds has been the limitation of cell migration and tissue ingrowth within these structures. Bovine aortic endothelial cells can survive or proliferate in a 2-D model in the absence of angiogenic factors; however, they die in a 3-D collagen lattice in the absence of angiogenic factors [14]. Because cells located in the interior scaffold receive nutrients only through diffusion from the surrounding media in static culture, high cell density on the exterior of the scaffold may deplete nutrient supply before these nutrients can diffuse to the scaffold interior to support tissue growth. In addition, diffusive limitations may inhibit the efflux of cytotoxic degradation products from the scaffold and metabolic wastes produced in the scaffold interior. Some examples of possible diffusive limitations of high cell density 3-D culture are summarized in Table 7.2-4 [15]. In a series of related studies, maximum penetration depth of osseous tissues was reported to be in the range of 200–300 μm within porous 3-D PLGA scaffold after 2 months of static culture. Although several attempts have been made to alter scaffold geometry to provide adequate diffusion within 3-D constructs with some success, ingrowth limitations within 3-D scaffolds remain a pervasive problem in tissue engineering.

Some of the factors known to influence the phenotype of cells are growth factors, cell–material interactions,

Table 7.2-4 Representative examples of high-density bone cell cultivation within 3-D scaffolds

Cell type	Pore size (mm)	Pore volume (%)	Scaffold thickness (mm)	Cell density (cells/ml)	Observation
Rat marrow	300–500	90	1.9	[a]	Maximum osseous penetration of 240 ± 82 μm (day 56)
Rat calvaria	150–300	90	1.9	[a]	Maximum osseous penetration of 220 ± 40 μm (day 56)
	500–710	90	1.9	[a]	Maximum osseous penetration of 190 ± 40 μm (day 56)
SaOS-2	187	30	2.5	9.5×10^7	Preferrential cell growth on scaffold exterior (day 7)
Rat marrow	300–500[b]	78	6	5.8[b] $\times 10^{6}$[c]	Preferrential cell growth on scaffold exterior (day 7)

Notes
Refer to the original work for the references.
[a] Values of cell density within the scaffold could not be determined from the reference.
[b] An average pore diameter of 400 mm was assumed to facilitate model calculations.
[c] Reported values for cell and scaffold void volume of 3.5×10^6 and 0.60 cm^2 were used to determine cell density (cells/ml).

and mechanical stimulation, but the basic steps of cell-culturing processing, regardless of 2-D and 3-D cultures, involve sufficient nutrient transfer throughout the culture medium and sufficient removal of waste products from around the cells. Other important parameters that must be monitored are oxygen concentration and pH of the culture medium, as changing these parameters can have a positive or negative effect on the cell growth.

7.2.6.2 Cell seeding

Seeding cells into porous scaffolds using autologous or xenogeneic serum is generally the first step of tissue engineering. Achieving the optimal cell density and desired cell distribution in scaffolds is a major goal of cell seeding technologies in tissue engineering. General seeding requirements for 3-D scaffolds include (a) high yield to maximize the utilization of donor cells, (b) high kinetic rate to minimize time in suspension for anchorage-dependent cells, (c) specially uniform distribution of attached cells to provide basis for uniform tissue regeneration, (d) high initial construct cellularity to enhance the rate of tissue development, and (e) appropriate nutrient and oxygen supply to maintain cell viability during the seeding procedure. Efficiency of cell seeding is usually low and in many cases cells are found close to the surface of the scaffold, but not in the interior. In addition, time-consuming, costly, and physiologically stressful cell seeding and tissue assembly techniques would become limiting factors in the development of clinically useful tissue-engineered products.

The spatial organization of cells provides cell–cell adhesion cues that are important in directing numerous biological functions, such as embryonic tissue development, organ formation, and tissue regeneration. This spatial organization is necessary for the preservation of vital cell–cell interactions as well as cell phenotype.

7.2.6.2.1 Serum

As mentioned above, reaction of tissue-engineered constructs requires the *in vitro* propagation of human cells to increase cell numbers. The *in vitro* cell growth can be stimulated by supplementation of the growth medium with serum. So far, culture techniques have required the use of xenobiotic material—fetal calf serum (FCS). The bovine serum has been used as a rich source of mitogens for cell proliferation. However, there is a risk of viral and/or prion disease transmission when bovine serum is used in clinical cell culture. To reduce the risk of disease transmission, it is preferable to culture cells under completely defined culture conditions, without any xenobiotic or donor cell material. The ideal medium is one that contains no xenobiotic products and in which any mitogens present are recombinant.

The choice of medium for the culture of autologous cells for clinical use may be at the discretion of the consultant medical doctors in many countries. Green's medium that has been in use for some 20 years for culture of skin cells for patients with burn injuries contains xenobiotic materials. Serum for clinical use must be sourced from countries such as New Zealand or Australia, where BSE has not been found to occur. In terms of risk management, regulatory authorities would prefer that cells cultured for clinical use avoid the use of bovine and other animal-derived products, to reduce the risk of disease transmission. Thus, a "holy grail" of cell culture for clinical use is to develop an entirely defined culture system. The use of autologous human serum in tissue engineering is increasing.

Cell growth medium often contains supplements, growth factors, and serum that may have an unknown influence on cell adhesion and proliferation. For example, the inclusion of the fibroblast feeder layer for keratinocyte expansion comes from the well-documented dependence of ECs on mesenchymal cell interactions between epidermal and dermal cells via soluble factors providing important signals in regulating the re-epithelialization of wound skin. Keratinocytes regulate the expression of keratinocyte growth factor (KGF) in fibroblasts through the release of interleukin-1β (IL-1β), a proinflammatory molecule that increases fibroblast proliferation and ECM production. The effect of solid substrate on the paracrine relationship between keratinocytes and fibroblasts as modulated by KGF and IL-1β is unclear.

Although the significance of breathing ambient air (containing 21% oxygen) to our survival is obvious, physiologic "normoxia" is much lower. Therefore, the traditional paradigm of culturing cells in humidified ambient air may not be optimal for maintaining certain cell types, including stem cells. Oxygen reduction to 3–6% promotes the survival of both peripheral and central nervous system stem cells, and can influence their fate by enhancing catecholaminergic differentiation. Similarly, human hematopoietic stem cells demonstrate increased self-renewal and bone marrow repopulating capability after hypoxic treatment. Furthermore, because cartilage is a relatively avascular tissue, chondrocytes are bathed in a naturally hypoxic milieu and rely on hypoxia-induced signaling for survival. Murine marrow-derived mesenchymal stem cells (MSCs) undergo enhanced osteochondrogenic differentiation in the setting of chronic hypoxia, perhaps by returning them to a more "natural" oxygen environment. In contrast, preadipocytes do not thrive under hypoxia, and the absence of oxygen triggers a hypoxia inducible factor (HIF)-1α response that represses genes essential to adipogenic differentiation. Thus, although reduced-oxygen incubation may promote the *in vitro* expansion and/or differentiation of many cell types by mimicking *in vivo* oxygen levels, it is possible to inhibit the differentiation of others by exposing them to reduced oxygen.

7.2.6.2.2 Cell adhesion

Cell adhesion can be divided into three grades of adhesiveness: (1) *early adhesiveness*, meaning that a cell is attached but has not spread, (2) *intermediate adhesiveness* characterized by cell spreading, but which lacks stress fibers and focal adhesions, and (3) *late adhesiveness* indicating cell spreading with stress fibers and focal adhesions. Early adhesiveness is important since it is the first step of attachment to substitutes, which results in cell growth, differentiation, viability, and spreading. Figure 7.2-19 diagrams the general form and characteristics of attachment and proliferation rate assays, identifying quantitative parameters extracted by statistical fitting to data. Here, $\%I_{max}$ measures the maximum number of cells that attach to a surface from a cell suspension, expressed as percent of inoculum.

Integrins that mainly mediate the biological cell adhesion are involved in processes such as development, wound healing, tumor invasion, and inflammation. Many cells appear to be capable of attaching to implant materials through integrins, and a considerable number of proteins contain the requisite RGD and (G)RGD(S) (Gly-Arg-Asp-Ser) sequences which are recognized by integrins. Serum contains abundant proteins with unknown activities in terms of cell adhesion. A high concentration of FN in serum can serve as an immediate attachment protein and may anchor cells to implants. Cell adhesion to a substrate is dependent on the chemical properties such as material composition and wettability, and physical parameters such as porosity and roughness. With the passage of time the cell can produce the ECM, and the ECM proteins in turn will attach to the substitute surface.

Cells can also be cultured successfully on a surface coated with type I collagen, but there is a concern

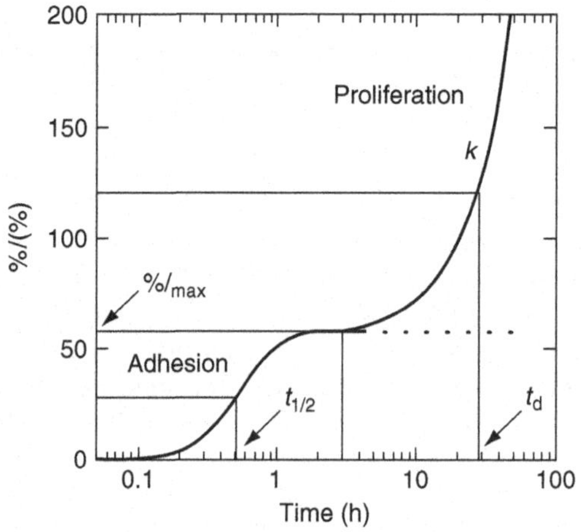

Fig. 7.2-19 Schematic illustrating cell adhesion and proliferation identifying quantitative parameters extracted from the variation of percentage of a cell inoculum (%I) with time. %I_{max} is the maximum percentage of a cell inoculum that adheres to a surface from a sessile cell suspension and $t_{1/2}$ measures half-time to %I_{max}. The proliferation rate (k) and cell-number doubling time (t_d) measure viability of attached cells.

throughout Europe that BSE cannot be detected by any *in vitro* tests and therefore it is impossible to be confident that bovine material is BSE free unless it comes from herds that have never been exposed to BSE.

7.2.6.2.3 Seeding efficiency

Although there are some techniques that successfully introduce cells into biomaterials, seeding efficiencies are not yet at optimal levels, especially at low cell concentrations. Lower seeding densities affect the amount of time and resources required to obtain scaffolds ready for implantation. The systems developed for the seeding of cells onto scaffolds include from simple techniques such as static seeding, where cells and scaffolds are brought into direct contact and allowed to sit, relatively undisturbed, with the intention of cellular attachment and migration into the scaffolds, to more elaborate techniques such as pulsatile perfusion wherein medium flows under oscillatory pressures to try to mimic the natural environments. Dynamic seeding which induces medium flow within the scaffold pores and shear stress on cells seems to be the most applicable method when relatively thin (<1 mm) scaffolds with a large 2-D surface need to be seeded.

In addition to seeding a greater number of cells into scaffolds, it is important to achieve homogeneity in cellular distribution. Achieving a confluent coating on scaffold exteriors may not be ideal because of potential problems with vascular ingrowth and nutrient/waste transport. Such an exterior cellular coating is also undesirable because of cell migration: If there is no subsequent cell migration into the scaffold interior, a bioabsorbable scaffold would ultimately collapse and the goal of quicker tissue regeneration will not be achieved.

7.2.6.2.4 Assessment of cells in scaffolds

Strategies for investigating cell growth in scaffolds include cell viability, proliferation, and metabolic active assays. In most cases, it is unknown how many cells remain viable over time in scaffolds *in vitro* and *in vivo* without assays in which the viability of the samples must be compromised to perform the specific assessment. Various assays are available for assessing cultured cell proliferation. These include (1) mitochondrial enzyme reduction of tetrazolium salts into their respective formazon by-products [MTT (3-(4,5-dimethylthiazol-2-yl)-2,5-diphenyl tetrazo-zolium bromide) and MTS]; (2) cellular redox indicators (Alamar blue); (3) ATP quantification through bioluminescence; (4) S-phase incorporation of radioactively labeled DNA precursors such as [^{3}H]thymidine and bromodeoxyuridine (BrDU); and (5) co-staining with a fluorescent DNA-specific dye for live cells (Hoechst 33258 and PicoGreen), and physical counting (hemocytometer). These assays are well established in characterizing cell proliferation in 2-D monolayer cultures of low cell densities.

The most common methods to visualize cells on biomaterials have been limited to light microscopy or use of dyes that compromises the viability of the cells. Conventional techniques, such as histomorphometry, electron microscopy, and Fourier transform infrared (FT-IR) imaging, are capable of giving information on tissue development, but they require that scaffolds be destructively sectioned. Other problems include the dissolution of polymeric scaffolds by the organic solvents used in the preparation of thin sections or the disruption of early mineral deposits by aqueous solvents. Some investigators reported on the successful application of confocal microscopy to study cells growing in the pores of scaffolds, but the depth penetration of this technique is limited and it cannot be applied to scaffolds that are optically opaque. What is required is a nondestructive technique that can provide spatially resolved, chemically specific, tissue-level information about the tissue formed within the pores of scaffolds.

DNA binding dyes coupled with confocal microscopy have been used to demonstrate the coverage of cells on materials. In a similar method, 3-D cell distribution on a material can be demonstrated with the fluorescence of enhanced green fluorescent protein (EGFP) and confocal microscopy without the need for special preparation of the cells. This approach facilitates visual assessment of cells as they exist in tissues and is not dependent on cell markers. Since EGFP is stably produced by the cells, no staining or cellular manipulation is required. Reporter genes have been utilized for a variety of applications ranging from gene expression and regulation to determination of efficiency of gene vector delivery. The technique allows the tracking of stem cells as they differentiate or become specialized. A reporter gene is inserted into a stem cell. This gene is only activated or "reports" when cells are undifferentiated and is turned off once they become specialized. Once activated, the gene directs the stem cells to produce a protein that fluoresces in a brilliant green color. Two commonly used reporter genes are EGFP and luciferase. Figure 7.2-20 illustrates the genetical modification of cell to express both EGFP and luciferase [16]. Both the fluorescent and luminescent signals from the cells follow a linear relationship as a function of cell number. These relationships provide two opportunities for quantifying cellularity from simple extracts. Both genes, when expressed in mammalian cells, will produce a molecule that can be detected by different techniques. A chromophore-containing protein, EGFP, emits fluorescent light when excited with light at a wavelength of approximately 488 nm. Qualitatively, EGFP under excitation provides an opportunity to visualize only cells expressing this protein. In contrast, the luciferase enzyme hydrolyzes its substrate, D-luciferin, in

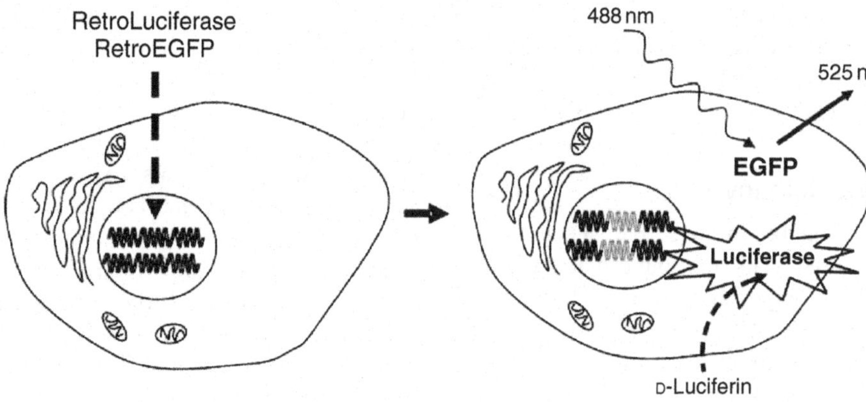

Fig. 7.2-20 Schematic depiction of cell genetically modified by retroviral vectors to express the reporter genes EGFP and luciferase. EGFP when excited with at ~488 nm will emit at ~525 nm. Luciferase in the presence of the substrate, D-luciferin, will emit light.

the presence of ATP and oxygen to produce photons of the emission spectra ranging between 400 and 620 nm. This stable retroviral transfection of cells with green fluorescent protein (GFP) offers a pathway to study tissue development, with emphasis on distinguishing between cellular components initially seeded onto a construct and those occurring as a result of cell ingrowth from surrounding tissue.

Various seeding techniques can be assessed using a DNA assay to estimate the total number of cells within a scaffold. This approach, however, yields very little information about the distribution of the cells throughout the scaffold. For destructive cellularity assays which quantify cellular DNA, the presence of ECM and/or material particulates in the sample extracts can interfere with these analyzes by making it difficult to extract the DNA. A critical issue in tissue engineering concerns whether the cellular components of tissue-engineered structures are derived from cells harvested and seeded onto an acellular scaffold or from cells originating from surrounding tissue (*e.g.*, proximal and distal anastomosis in the case of cardiovascular tissue engineering), or from circulating pluripotent stem cells. To clarify this issue, some studies have utilized fluorescent carbocyanine dyes. It was possible to identify cells labeled by this technique *in vivo* for up to 6 weeks after transplantation, but more stable long-term labeling and tracing without adverse effects, such as cell toxicity, are required. Current histomorphometric techniques evaluating rates of tissue ingrowth tend either to measure the overall tissue content in an entire sample or to depend on the user to indicate a front of tissue ingrowth. Neither method is particularly suitable for the assessment of tissue ingrowth rates, as these methods either lack the sensitivity required or are problematic when there is a tissue ingrowth gradient rather than an obvious tissue ingrowth front.

Cells interacting with scaffolds may exhibit distinct patterns of gene expression depending on the molecular nature of the surface they are contacting. Many cell types grown in 3-D culture exhibit phenotypes drastically different from those of their plate-grown counterparts. In these cases gene expression studies are beneficial.

7.2.6.2.5 Gene expression of cells

Expression of specific genes is enhanced within cell-seeded constructs or engineered tissue. For instance, in the bone tissue engineering, Runx 2, alkaline phosphatase (ALP), and osteocalcin (OCN) that are important markers of different stages of bone matrix production are expressed. Runx 2 is a transcriptional activator essential for initial osteoblast differentiation and subsequent bone formation. The ALP is expressed by pre-osteoblasts and osteoblasts before the expression of OCN. Finally, OCN is a late marker that binds HAp and is produced by osteoblasts just before and during matrix mineralization.

7.2.6.3 Bioreactors

When 3-D cellular constructs are grown in static culture, cells on the outer surface of the constructs are typically viable and proliferate readily while cells within the constructs may be less active or necrotic. In the absence of a vascular blood supply *in vitro*, nutrient delivery to cells throughout 3-D tissue engineered constructs grown in static culture must occur by diffusion. As a result, thin tissues (e.g., skin) and tissues that are naturally avascular (e.g., cartilage) have been more readily grown *in vitro* than thicker, vascular tissues such as bone. The engineering of tissues *ex vivo* in a bioreactor offers several benefits, such as better understanding of tissue development and mechanisms of disease, off-the-shelf revision of transplantable tissues, and possible scale-up for commercial production of engineered tissue. The bioreactors in tissue engineering are utilized for cell expansion on a large scale and production of 3-D tissues *in vitro*. Bioreactors offer several advantages over culture in plates and flasks. Bioreactors can be custom designed to engineer tissues with complicated 3-D geometry containing multiple cell types. More significantly, bioreactors can impart appropriate biochemical and mechanical stimuli in a controllable environment to promote cell growth, maturation, and tissue differentiation. Artificial tissue development in bioreactors generally involves less handling than static culture methods to significantly reduce contamination risk and is amenable to scale up for the generation of tissue products. Bioreactors can serve as tissue growth systems as well as packaging and shipping units that can be delivered directly to surgeons.

Bioreactors have been widely used from drug production to beer brewing to create products in an efficient manner. The basic concept of bioreactors that take advantage of microorganisms is to create an environment that is advantageous for microorganisms to the creation of a desired product, whether it be penicillin, alcohol, or ECM. Nutrient, waste, and oxygen levels must be carefully controlled to prevent the organisms from dying. There are three types for such bioreactors: batch, fed-batch, and continuous systems. In a batch system, organisms are combined with the required nutrients in a single-step process. A fed-batch process is similar to the batch reaction in that nothing is removed during the reaction. However, in contrast to the batch system, additional nutrients are added over a portion of the process to keep the reaction rate at its maximum. A continuous bioreactor adds nutrients and removes waste and products over the entire course of the reaction. Continuous reactors are the most efficient and cost-effective once the equipment is running.

The bioreactors used in engineering tissues are primarily batch processes and there is no need to run a large-scale operation. As custom design of engineered tissues is often necessary because of immunological concerns, a continuous reactor is not necessary. The bioreactors for engineering tissues, as opposed to bioprocessing, provide an environment in which cells continuously reside throughout the culturing period. Mass transfer becomes the main concern because nutrients and oxygen must be provided in sufficient amounts to tissues to grow to a usable size. The stimulus from a mechanical force can still be beneficial. Articular cartilage in the knee experiences shearing stresses during normal movement. On the surface of cartilage, a thin layer of synovial fluid provides lubrication, which reduces friction, but shearing of the tissue still occurs because of the solid-on-solid contact. Shear forces alter the phenotype of the cells seeded on a scaffold to one that is close to native cartilage. These forces can be emulated using a flowing fluid either across or through the cell-seeded scaffold. Microcarrier beads may effectively provide surface area for attachment of anchorage-dependent cells. In addition, agitation of microcarrier cultures leads to homogeneous culture conditions and improved gas–liquid oxygen transfer.

If constructs must be produced using patient-specific cells that do not produce an immune response, then continuous bioreactors are not required but large-scale bioreactors that resemble a batch process will have to be created. This makes the task much more difficult because the samples must be separated at all times and the equipment must be sterilized after each batch. A mechanical force must be applied to each scaffold in a controlled manner without any mixing of cell batches.

Tissue culture systems that provide dynamic medium flow conditions around or within tissue-engineered constructs should be designed to enhance nutrient exchange and cell growth *in vitro*. Such tissue culture systems are required as bioreactors to engineer thicker, more uniform tissues for transplantation. In addition to enhancing mass transport, bioreactor systems may be useful to deliver controlled mechanical stimuli such as flow-mediated shear stress, matrix strains, or hydrostatic pressures to tissue constructs. This general approach has been used to culture a variety of 3-D constructs including bone, cartilage, muscle, and blood vessels. Tissue culture systems that incorporate dynamic medium flow conditions for developing 3-D tissue constructs include spinner flasks, perfusion systems, and rotary cell bioreactors. In general, improved cell viability, proliferation, and ECM production have been demonstrated in dynamic systems relative to static controls. Internal fluid flow in each system is achieved in different ways. In spinner flasks, stirred culture medium is moved past the scaffolds at fixed positions within the vessel. In rotating bioreactors, the scaffolds are not fixed, but in continuous free fall within the culture medium during bioreactor rotation. In perfusion cultures, culture medium is directly perfused using a uniform fluid pressure head applied to the scaffolds and fluid. In spinner flasks, flow and mixing of culture medium is associated with turbulent shear at construct surfaces. Mass transport between the constructs and culture medium is enhanced by convection, whereas mass transport within the constructs remains controlled by diffusion. In perfused reactors, interstitial flow of culture medium enhances mass transport throughout the construct volume, and exposes all cells to laminar shear.

7.2.6.3.1 Spinner flask

One of the simplest bioreactors is the mechanically stirred flask and the most common mechanically stirred bioreactor is the spinner flask. Scaffolds seeded with cells are attached to needles hanging from the cover of the flask and suspended within a stirred suspension of cells with addition of sufficient medium to cover the scaffolds, as shown in Fig. 7.2-21 [17]. This method has produced favorable results, but potential weaknesses of the technique include the amount of time required for

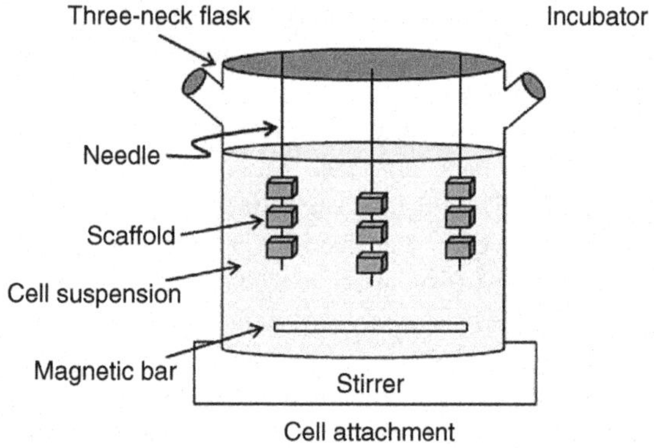

Fig. 7.2-21 Schematic view of a spinner flask designed to assist initial cell attachment. The cell-attached scaffolds are transferred onto 24-well plates for subsequent cell proliferation and differentiation assays.

seeding (typically 24 h), low efficiency at low cell concentrations, and undesirable side effects associated with mechanically stirred bioreactors of high shear rates. The degree of shear stress depends on stirring speed and the morphology of the scaffold. Cell damage has been observed at 150–300 rpm in microcarrier cultures, and although there is no apparent physical cell damage at 50 rpm, a fibrous capsule does form on the construct surface. The local shear force experienced by the cells is produced by eddies created by the turbulent flow of the impeller. The smallest turbulent eddies are on the order of several hundreds of micron with velocities of approximately 0.5 cm/sec. Cell flattening and proliferation, and formation of an outer capsule, are caused by the pressure and velocity fluctuations associated with turbulent mixing. Table 7.2-5 represents the high-shear bioreactor summary [18].

Using impellers is popular in basic cell culturing because it increases the rate of mass transfer to the cells, but impellers create many problems. Non-uniform mass transfer rates, nutrient and pH gradients, and shear gradients impart a non-uniform mechanical stimulus over the construct, resulting in the formation of tissue that is inferior to that produced by other culturing techniques. The shear force at the surface of the impeller is reported to be 10 times higher than anywhere else within the bioreactor. Because of this, cells located closer to the impeller will exhibit more of an injury response because of the high levels of shear. Cells farther away will not produce a fibrous capsule as long as the rotation rate is low enough, but they will receive fewer nutrients and experience a higher pH because of the lessened mixing.

7.2.6.3.2 Perfusion system

Mechanically stirred bioreactors are used in scaffold-seeding experiments but they might not be optimal for producing tissues such as cartilage. The most successful results in any bioreactor are based on the use of scaffolds seeded dynamically and at high densities. However, 3-D hydrogels that include cells during gel formation do not need to be seeded in this manner and cells can be distributed evenly at high densities without the use of a mechanically stirred bioreactor. For 3-D porous

Table 7.2-5 High-shear bioreactor summary

First author	Proteoglycan	Collagen	Additional notes
Gooch (construct, 2- to 4-week bovine knee, 18 $\times$ 10^6 cells/cm^3 PGA, 6 weeks)	36% decrease in GAG composition; 2- to 3-fold increase in GAG synthesis	80% increase in collagen composition	Constructs retained low levels of GAG after synthesis
Vunjak-Novakovic (construct, 2- to 3-week bovine knee, 25 $\times$ 10^6 cells/cm^3 PGA, 8 weeks)	60% increase in GAG composition	125% increase in collagen composition	Fibrous capsule formed
Smith (monolayer, adult bovine wrist, 1.5 $\times$ 10^5 cells/cm^2, 3 days)	2-fold increase in [^{35}S]-sulfate incorporation	Not evaluated	High levels of prostaglandin E_2 and metalloproteinase released
Freed (construct, 2- to 3-week bovine knee, 29 $\times$ 10^6 cells/cm^3 PGA, 6 weeks)	No significant change	50% increase in collagen composition	Primarily type I collagen
Dunkelman (construct, 4- to 8-month rabbit joints, P2, 25 $\times$ 10^6 cells/cm^3 PGA, 4 weeks)	25% dry weight GAG (15–30% in native cartilage)	15% dry weight collagen (50–73% in native cartilage)	Direct perfusion; more tissue formed on periphery than middle

Refer to the original work for the references.

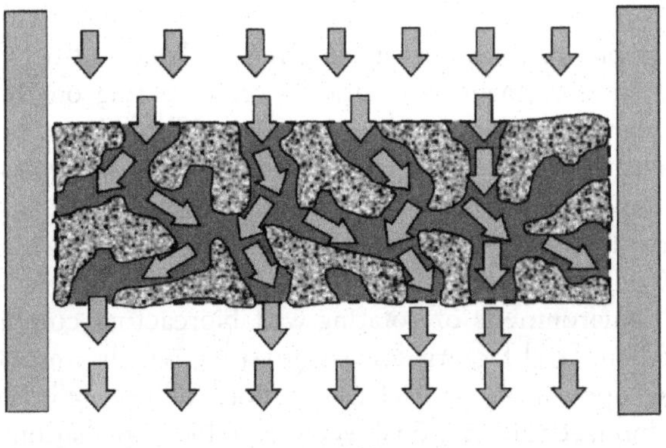

Fig. 7.2-22 Direct perfusion bioreactor.

scaffolds, a few systems that force medium through the scaffold give the most thickness-independent results. This type of system is called a "direct perfusion bioreactor," as illustrated in Fig. 7.2-22 [19]. This type of bioreactor uses a pump to perfuse medium continuously through the interconnected porous network of the scaffold, rather than around the edges. The constant availability of fresh medium, the mechanical action of shear stress, and the ability to transport nutrients through an increasingly dense mass of cells and ECM material have favored its use.

Cells within the bulk of a construct feel the shear force from the fluid flowing through the pores and may produce ECM in response to it. A modification that can be made to direct perfusion bioreactors is to mix a portion of the old medium with the fresh medium. By recycling some of the medium, beneficial proteins such as growth factors and interleukin 4 are kept in the system. Otherwise, chemical signals would be lost from the developing tissue. The tissue then develops in a medium that contains all the signal proteins that are produced, the cell-secreted collagen, and proteoglycans that are flushed out of the scaffold but are still in the medium.

One of the problems that becomes apparent when using direct perfusion bioreactors is non-uniform cellular secretions through the thickness of the construct as well as damage to some of the cells. If fluid is flowing from one side of a scaffold to the other, then the front surface will have a greater mechanical stress exerted on it because of the oncoming flow. Conversely, the back surface does not feel the force except for inside of the pores. This will cause a thicker matrix to form on the front side of the scaffold as compared with the rest of the construct. For high-shear direct perfusion devices, the fibrous response is similar to the capsule formed in spinner flasks except that it occurs only on one side of the construct. The tissue ends up having inferior mechanical characteristics compared with native tissue. The direct perfusion bioreactor must use low flow rates to fix this problem, which would nullify the effectiveness of this particular bioreactor system.

7.2.6.3.3 Rotating wall reactor

The conventional bioreactor has the disadvantage that high shear forces are generated, which damage the cells and hinder proper tissue-specific differentiation. Decreasing the stirring rate while increasing the viscosity of the culture medium might partially reduce the hydrodynamic damage, but aggregates formed under these conditions still exhibit necrotic centers. The ability of the direct perfusion bioreactors to provide a nutrient-rich environment for the cells and stimulate them simultaneously was carried over into the design of the more sophisticated rotating-wall device. The major change is mainly in the application of a mechanical force, because high or even moderate levels of them are undesirable in the formation of some tissues. A rotating fluid environment was found to be the best way to produce a low-shear bioreactor. Efforts at tissue reconstruction would benefit from a venue which promotes cell–cell association while avoiding the detrimental effects of high shear stress. Such a venue might be provided by microgravity, because it was observed many years ago that cells in suspension tended to aggregate when exposed to microgravity in space [20]. In an effort to derive the potential beneficial effects of microgravity and low fluid shear for cell culture here on the earth, scientists at the National Aeronautics and Space Agency (NASA) introduced the rotary wall vessel (RWV) bioreactor. Probably, the best overall results for cell culturing currently come from a modified version of the direct perfusion bioreactor called the "rotating-wall bioreactor" [18]. The NASA originally created it as a "microgravity" environment for cell culture, but later successfully as a low-shear, high-diffusion bioreactor for many cell types.

The horizontally rotating culture vessel was initially designed to simulate some aspects of microgravity and has proved valuable in the generation of 3-D cultures of a variety of transformed and non-transformed cells. The system allows anchorage-dependent cells to grow in 3-D and multiple layers within the framework. The NASA-developed bioreactor allows for the formation of 3-D aggregates and promotes co-spacial localization between similar and dissimilar cells under conditions of controlled access to O_2 and nutrients. As illustrated in Fig. 7.2-23, the most popular type of the bioreactor is composed of two cylinders, where microcarriers or scaffolds are placed in the annular space between the two cylinders. Gas exchange is allowed through the stationary inner cylinder while the outer cylinder is impermeable and rotates in a controlled fashion. Under carefully selected rotational rates the free falling of the scaffolds inside the bioreactor due to gravity can be balanced by the centrifugal forces due to the rotation of the outer cylinder, thus establishing microgravity-like culturing constitutions.

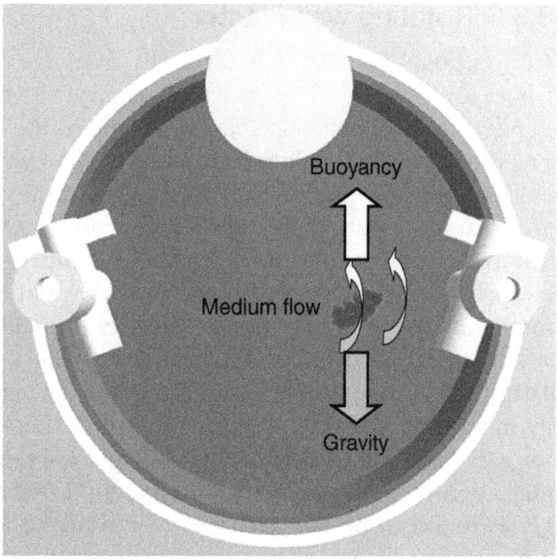

Fig. 7.2-23 Rotating-wall vessel bioreactor.

Rotating-wall bioreactors are horizontally rotated with fluid-filled culture vessels equipped with membrane diffusion gas exchange to optimize gas/oxygen supply. The initial rotational speed is adjusted so that the culture medium, the individual cells, and pre-aggregated cell constructs or tissue fragments rotate synchronously with the vessel, thus providing for an efficient low-shear mass transfer of nutrients and wastes. As the cell aggregates grow in size and exhibit increasing sedimentation velocities, the shear stress can approach up to 0.5 dyn/cm^2. These simulated microgravity conditions facilitate spatial co-location and 3-D assembly of individual cells into large aggregates. There is laminar flow of medium over the surface of constructs with fairly uniform shear stress distribution. For example, the maximum shear stress on the surface of constructs is approximately 0.3 dyn/cm^2 for a 120-ml rotating vessel containing only one construct and operated at rotational speeds of ~25 rpm. The low-shear bioreactor summary is given in Table 7.2-6 [18]. The time-averaged gravitational vector acting on these cellular assemblies is reduced to about 10^{-2} g. In a typical experiment on board a space craft or the space station cycling in near-earth orbit, the gravitational force is approximately 10^{-4}–10^{-6} g.

Cell culture conditions in the "simulated microgravity" environment of rotating-wall bioreactors combine two beneficial factors: low shear stress, which promotes close apposition (spatial co-location) of the cells; and randomized gravitational vectors. Close apposition of the cells in the absence of shear forces presumably promotes cell–cell contacts and the initiation of differentiative cellular signaling via specialized cell adhesion molecules. This process then might lead to the rapid establishment and expansion of aggregate cultures which, unlike in conventional bioreactors, are not disrupted by shear forces. In addition, the low-shear environment, in concert with randomized gravitational vectors, might restrict diffusion of mitogenic and differentiative growth factors, which are secreted by the cells. These autocrine/paracrine feedback mechanisms might further enhance the aggregation and differentiation, and contribute to the observed capability of this environment to maintain high-density cell cultures. Cultures in a 3-D matrix or scaffolds such as collagen and agarose gel largely maintain the chondrocyte phenotype, but the use of RWVs as scaffold-free bioreactors also leads to successful differentiation and the hyaline production. The most significant advantages of the RWV for a 3-D tissue culture are low shear forces, the reduced risk of cell damage, and the increased opportunity for cell–cell interaction. Thus, primarily dissociated cells can reassemble to form a 3-D tissue-like mass. The spatial orientation may be critical to the differentiation of growing aggregate cells and to the regeneration of a functional ECM.

Table 7.2-6 Low-shear bioreactor summary

First author	Proteoglycan	Collagen	Additional notes
Gooch (construct, 2- to 4-week bovine knee, 76 × 10^6 cells/cm^3 PGA, 4 weeks)	2.9-fold increase in GAG composition	1.6-fold increase in collagen composition	Supplemented with IGF-I
Martin (construct, 2- to 3-week bovine knee, 127 × 10^6 cells/cm^3 PGA, 6 weeks)	75% increase in GAG composition	39% increase in collagen composition	Equilibrium modulus and GAG composition reach native cartilage levels after 7 months *in vitro*
Freed (construct, 2- to 3-week bovine knee, 127 × 10^6 cells/cm^3 PGA, 6 weeks)	68% of native GAG levels per gram wet weight	33% of native type II collagen levels per gram wet weight	Type II collagen crosslinked in constructs

Refer to the original work for the references.

Cell aggregates generated *in vitro* (in either suspension cultures or stirred bioreactors) that exceed about 1 mm in size invariably develop necrotic cores. Culture conditions in RWV bioreactors are unique in that the fluid dynamics of the system allow for efficient mass transfer of nutrients and oxygen diffusion. In this environment, dissociated cells can assemble into macroscopic tissue aggregates several mm in size, which are largely devoid of such necrotic cores. With the possible exception of avascular tissue of low cellularity and slow metabolism, such as cartilage, all tissues, including those growing in RWV bioreactors, will eventually require internal, blood-vessel-like conduits for the delivery of oxygen and nutrients as well as for the removal of waste products. Attempts to generate 3-D constructs as replacement tissues will necessitate combining RWV technology with innovative methods for creating bioengineered blood conduits such as growing endothelial cells on the inside and outside of tubular scaffolds.

7.2.6.3.4 Kinetics

The success of scaffolds for tissue engineering is typically coupled to the appropriate transport of gases, nutrients, proteins, cells, and waste products into, out of, and/or within the scaffold. To obtain fluid flow and nutrient flux in porous, 3-D scaffolds, dynamic culture methods (Fig. 7.2-24) [15] have been employed. However, the primary mass transport property of interest, at least initially, is diffusion. In a scaffold, the rate and distance a molecule diffuses depend on both the material and the molecule characteristics and interactions. As a consequence, diffusion rates will be affected by the MW and size of the diffusion species (defined by Stokes radii) compared to the pore of scaffolds. For example, molecules such as glucose, oxygen, and vitamin B_{12}, with MWs less than 1300 Da and Stokes radii less than 1 nm, are able to freely diffuse into and from ionically cross-linked hydrogels. Gel properties such as polymer fraction, polymer size, and crosslinker concentration determine the gel nanoporous structure. However, higher MW molecules such as albumin and fibrinogen are not able to freely diffuse, and their rate of diffusion is further decreased by increases in polymer concentration, in cross-linker concentration, and/or in the extent of gelation. Ultimately, diffusion requirements and subsequent material choice depend on the scaffold application.

The mass transfer of nutrients and waste between a scaffold and the surrounding medium follows the equation below:

$$J_s = -DS\frac{dC}{dx}$$

where J_s is the diffusion of a solute, D is the diffusivity of the solute, S is the surface area normal to the direction of

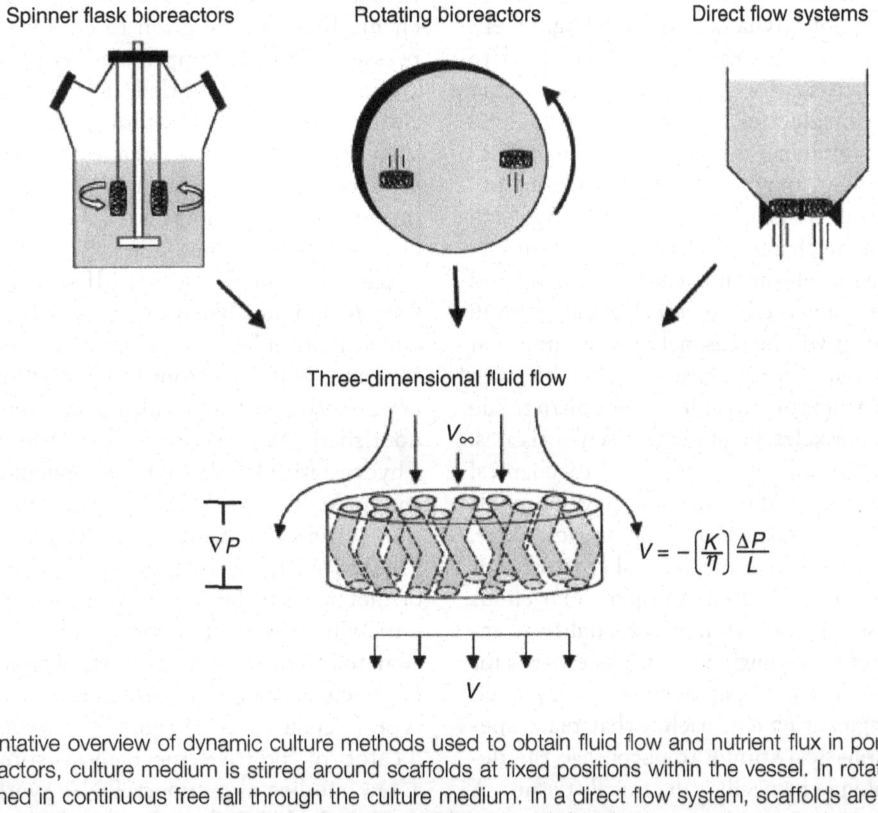

Fig. 7.2-24 Representative overview of dynamic culture methods used to obtain fluid flow and nutrient flux in porous 3-D scaffolds. In spinner flask bioreactors, culture medium is stirred around scaffolds at fixed positions within the vessel. In rotating bioreactors, scaffolds are maintained in continuous free fall through the culture medium. In a direct flow system, scaffolds are fixed in position and perfused with culture medium by a direct application of a uniform fluid pressure head.

solute diffusion, and dC/dx is the concentration gradient. However, diffusion is just one of many parameters that are important in the cell-culturing process.

To evaluate the role of nutrient diffusion and consumption within cell-seeded scaffolds in the absence of flow, Botchwey *et al.* developed a 1-D glucose diffusion model. Such a quantitative analysis will provide a basis for development of new dynamic culture methodologies to overcome the limitations of passive nutrient diffusion in 3-D cell–scaffold composite systems. However, many factors still remain to be quantified, such as the transport of oxygen, other nutrients, cell–cell signaling molecules, growth factors, metabolic wastes, and cell chemotactic factors in dynamic culture, all of which may have an effect on cell function and tissue synthesis.

7.2.6.4 Externally applied mechanical stimulation

Some tissues exist in a mechanically dynamic environment. Blood vessels are continuously exposed to mechanical forces that lead to adaptive remodeling. Although there have been many studies characterizing the responses of vascular cells to mechanical stimuli, the precise mechanical characteristics of the forces applied to cells to elicit these responses are not clear. Soft musculoskeletal tissues also adapt to immobilization and realize strengthening in response to exercise. Tendons are in a continuous state of dynamic remodeling. Cells suspended within a 3-D network of ECM respond to external changes via receptor–ligand interactions that relay signals from outside the cell to the cytoskeletal domain and thereby influence subsequent cellular function such as attachment, migration, differentiation, and apoptosis. As natural tissues, especially those which should resist against mechanical loading and pressure—such as cartilages, bones, ligaments, cardiac muscle, and blood vessels—are subjected to mechanical stimuli during development, it will be reasonable to assume that mechanical stimuli, pulsed or unpulsed, should be given to cell–scaffold constructs aimed at least at *in vitro* tissue engineering of musculoskeletal and cardiovascular tissues. Indeed, numerous studies have shown local mechanical signals to be a key factor directing the development, growth, repair, and maintenance of bone and cartilage. Since some cells can sense their mechanical environment such as endogenously generated tension, exogenous stimulus may be used as a conditioning modality to influence the efficacy of tissue engineered replacements for load-bearing tissues. Thus, researchers have employed functional tissue engineering approaches that place special focus on using physical stimuli to encourage the development of a biomechanically functional tissue. A premise of *in vitro* tissue engineering is that one may enhance the rate and quality of tissue growth by recreating *in vitro* some of the same conditions that the tissue experiences *in vivo*. Articular cartilage is amenable to such an approach, as it is a tissue defined not only by its cellular constituents but also by its dynamic physical environment. Therefore, appropriate mechanical stimulation has been applied to cells during their culture, especially in case cells in the human body live in an environment heavily influenced by mechanical forces such as in load-bearing tissues. This is the major reason why chondrocyte has been very often selected as a cell model for tissue engineering under mechanical stimulation. Another reason is that the cartilage in which chondrocytes live is an avascular tissue and receives oxygen and nutrients from the synovial fluid. Chondrocyte is known to be one of the most robust cells and to develop differently on the basis of what culturing processes are used. The presence of mechanical forces such as hydrostatic pressure or direct compression stimulates chondrocytes to secrete more ECM as compared with static culture.

There are two major methods for mechanically stimulating cells outside of their culturing environment to enhance their growth. One is the bioprocessing which uses mechanical stimuli only at intervals during the culturing period. The other is the bioreactor system which uses a constant mechanical force to stimulate the cells. The advantage of these approaches is the introduction of a mechanical stimulus and an increase in diffusion through the porous scaffold. In the bioprocess, mechanical stimuli are given to cells via either hydrostatic pressure or direct compression. When cartilage located in articular joints is loaded during walking, running, or shifting weight while standing, the force is transmitted throughout the tissue which contains water by 75–80 wt%, being absorbed primarily by the fluid. The pressure produced by the compressed fluid acts uniformly on the chondrocytes within ECM. This hydrostatic pressure ranges between 7 and 10 MPa during normal activity. One technique that uses hydrostatic pressure as a mechanical stimulus is to employ a two-step process that separates culturing from force application. The cells are kept mostly in static culture medium, where they are nourished. At prescribed times, the cells are moved to a hydrostatic chamber where a specified load is placed on them. Another technique uses a semicontinuous perfusion system that feeds the cells and applies hydrostatic pressure in the same device. In this case the cell cultures do not have to be moved as much, which reduces the possibility of contamination, and the process can be automated to run for long periods of time without any need for manual labor. The production of proteoglycans and type II collagen is currently the main experimental indicator of positive mechanical stimulation, although a few studies also report DNA synthesis or aggregate moduli. Table 7.2-7 summarizes representative results of

Table 7.2-7 Hydrostatic pressure summary

First author	Proteoglycan	Type II collagen	Additional notes
Smith (monolayer, adult bovine wrist, 10^5 cells/cm^2, 4 days)	21-fold increase in mRNA synthesis	9-fold increase in mRNA synthesis	10 MPa, 1 Hz
Carver (construct, 1-week bovine knee, 2×10^6 cells/cm^3 PGA, 5 weeks)	2-fold increase in concentration	No significant increase	3.5 MPa, 5/15 s (on/off) for 20 min every 4 h
Smith (monolayer, adult bovine wrist, 10^5 cells/cm^2, 4 h)	31% increase in mRNA synthesis	36% increase in mRNA synthesis	1% fetal bovine serum 10 MPa, 1 Hz

Refer to the original works for the references.

bioprocesses which use hydrostatic pressure as a mechanical stimulus [18]. The duration and magnitude of hydrostatic loading vary widely between studies, but the most successful studies dynamically load cartilage for longer periods of time.

In the case of engineering blood vessels, mechanical stimulation can be provided via radial distension of the silicone tube around which a tubular scaffold is sewn, as demonstrated in Fig. 7.2-25 [21]. *In vitro* fluid-induced shear stress, as occurs in agitated microcarrier cultures, can be used as a model system for mechanical stimulation of articular chondrocytes. Culture of chondrocytes on microcarriers may have beneficial effects for cartilage tissue engineering applications. Efficient expression of chondrocytes, without the loss of their ability to express the differentiated phenotype, would be highly beneficial for the engineering of cartilage for clinical purposes. In particular, such expression would exclude costly and time-consuming techniques of redifferentiation.

The thickness of constructs is significantly limited by the diffusion properties associated with porous scaffolds. Dynamic compression helps alleviate these diffusion limitations through pressure gradients within the scaffold as well as by a secondary mixing effect on the surrounding medium. In a healthy person, articular cartilage undergoes direct compression thousands of times a day without any long-term injury to the tissue. Unlike hydrostatic pressure, there is an actual solid-on-solid contact that takes place under direct compression.

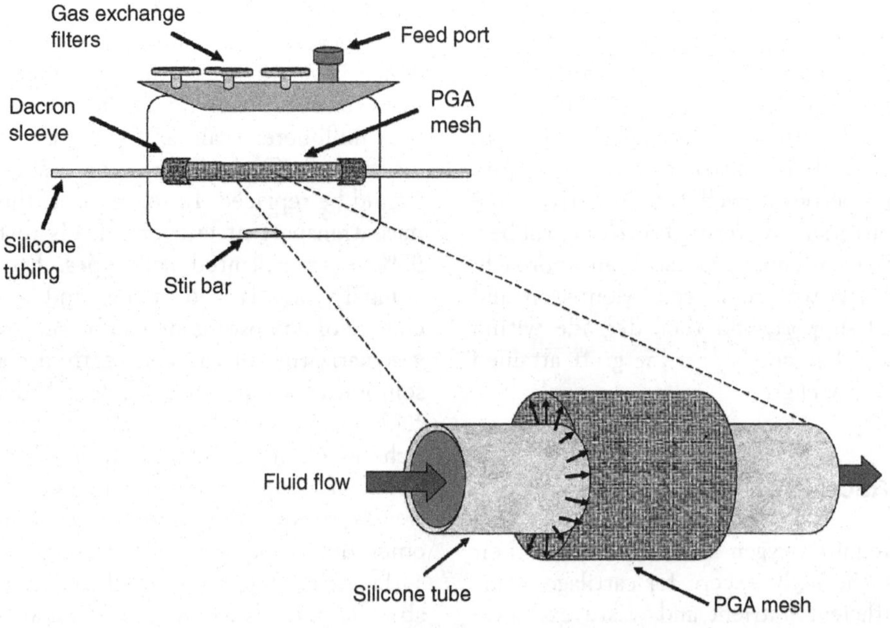

Fig. 7.2-25 Vessels made of PGA are sewn around a silicone tube and are attached to the sides of a glass bioreactor with a nondegradable Dacron sleeve. Mechanical stimulation of the construct is provided via radial distension of the silicone tube as a result of fluid flow.

Mass transfer for constructs under direct compression is expected to be better than for those cultured under hydrostatic pressure or by the static culture approach. The main parameters that must be set when using dynamic compression are the frequency of the applied load, the strain or force used, and the duration of the experiment.

New bioreactors can be developed that take advantage of several different mechanical stimuli, as well as a good culturing environment, all in one package. For example, hydrostatic pressure could be combined with a rotating bioreactor to create a stimulating environment that is self-contained. Because scaffolds are already cultured in a fluid medium, hydrostatic pressure could be applied without removing anything from the sterile environment. This reduces the chance of contamination as well as limiting the amount of time needed to transfer the scaffolds between a stimulation device and a culturing environment. If the cells are fed in sufficient amounts and are surrounded by a good environment, they will act like native cells and secrete ECM into the scaffold pores. The addition of bioactive peptides and growth factors in the scaffolds will help matrix synthesis, especially in conjunction with mechanical forces.

Most of engineered tissues are histologically similar to natural tissues, but differ greatly in mechanical strength, stiffness, and functional properties. The mechanical properties of many tissues engineered to date are inferior to those of native tissues. Skeletal muscle constructs were demonstrated to have reasonable initial histology but force generation an order of magnitude below the native tissue [22]. Generally, such poor results are observed when cells seeded onto scaffolds are cultured under static conditions. Both morphology and force generation of engineered heart tissue may be modified by cyclic loading during construct development. It is necessary to investigate the effect of mechanical conditioning of cells on cell–scaffold constructs and to quantify their mechanical properties as well as their active force capabilities. Fiber arrays made from absorbable, rubber-like materials have been chosen for such an approach. The scaffold materials will guide cell orientation and provide mechanical support, and then degrade within a certain time frame but not before the graft attained sufficient mechanical strength.

7.2.6.5 Neovascularization

Mammalian cells require oxygen and nutrients for their survival. Tissues in the body, except for cartilage, overcome issues of sufficient nutrient and waste exchange with their surroundings by containing closely spaced capillaries that provide conduits for convective transport of nutrients and waste products to and from the tissues.

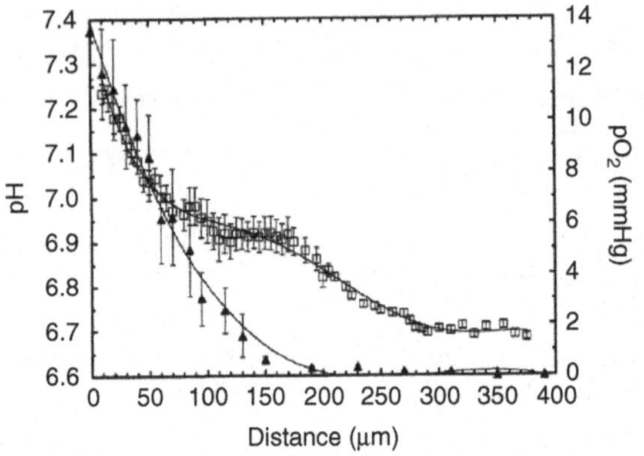

Fig. 7.2-26 Mean interstitial pH and partial pressure of oxygen (pO_2) profiles of 27-day-old tumors taken as one moves away from the nearest blood vessel. Open symbols, pH; closed symbols, pO_2.

Mammalian cells are located within 100–200 μm of blood vessels. This distance is the diffusion limit for oxygen. Figure 7.2-26 shows the relationship between the distance of tumor cells from nearby vessels and their degree of hypoxia and acidosis [23]. One of the dominant factors currently limiting clinical applications of tissue engineering is the inability of the scaffold to construct *de novo* a microvasculature. Tissue engineered constructs also require a capillary network for cell maintenance and function excluding those less than 100–200 μm thick, which may be oxygenated by diffusion. Few clinical trials in cardiac tissue engineering are due to unavailability of scaffolds with large size and excellent capability of vasculature. The lack of capillary network connected to the host tissue and the resulting poor, oxygen transport limit the thickness of generated tissue constructs to ~0.1 mm, which is generally too thin for clinical application. The mass transport issue limits the size of engineered tissues to a millimeter scale at the largest, which is clinically insufficient if a large mass of tissue or a whole organ should be replaced. In some cases, the limitation of this mass transport of nutrients leads to loss of more than 95% of transplanted cell types. Fat tissue represents a highly vascularized tissue, and a volume-persistent culture of adipose tissue can be successful only via early vascularization of the cell–scaffold construct. A simple solution to avoid considerable cell death *in vivo*, especially in the central area, is to utilize native tissues with rich vascularity. Among them is the omentum which is a fold of the peritoneum anchored to the stomach and transverse colon that drapes over the small intestine. The omentum is highly vascular, contains a relatively large surface area, and is recruited to sites of intra-abdominal abscess, perhaps to "wall off" the area and prevent diffuse peritoneal contamination. The mesentery of the small intestine that anchors the bowel to the body is also vascular and supplies blood to the small intestine.

7.2.7 Growth factors

The term "growth factors" used here include differentiation factors and angiogenic factors in addition to growth factors along with bone morphogenetic proteins. It must be recognized that the redundancy found in most biological structures is such that precise characterization of a growth factor as falling in just one category above is misleading. Many growth factors may provide a host of functions and may modulate cell attachment, cell growth (or apoptosis), cell differentiation, cell migration, neovascularization, etc., and indeed may do so differently according to the biochemical, cellular, and biomechanical context into which they are placed. Both growth factors and differentiation factors are likely essential to establishing a sufficient number and architecture of appropriately functioning cells. Specific factors favor certain cell lineage, possibly through an inhibition of certain lineage rather than promotion of a specific one.

Growth factors may be exogenously added, or the cells themselves may be induced to synthesize them in response to chemical and/or physical stresses. Although cytokines and growth factors are present within ECM in vanishingly small quantities, they act as potent modulators of cell behavior. The list of these growth factors is given in Table 7.2-8. These factors tend to exist in multiple isoforms, each with its specific biological activity. Purified forms of growth factors and biological peptides have been investigated as therapeutic means of encouraging blood vessel formation (VEGF), inhibiting blood vessel formation (angiostatin), stimulating deposition of granulation tissue (platelet-derived growth factor, PDGF), and encouraging epithelialization of wounds (KGF). However, this therapeutic approach has struggled with determination of optimal dose, sustained and localized release at the desired site, and the inability to turn the factor "on" and "off" as needed during the course of tissue repair. An advantage of utilizing the ECM in its native state as a scaffold for tissue repair is the presence of all of the attendant growth factors (and their inhibitors) in the relative amounts that exist in nature and, perhaps most importantly, in their native 3-D ultrastructure.

Growth factors must greatly contribute also to tissue engineering at various stages of cell proliferation and differentiation. Thus, numerous studies have been performed using growth factors in the field of tissue engineering, and some growth factors have produced promising results in a variety of preclinical and clinical models.

7.2.7.1 Representative growth factors

7.2.7.1.1 BMPs

The BMPs represent a family of related osteoinductive peptides akin to differentiation factors. Research involving BMPs found its beginnings more than 35 years ago, when Urist observed that demineralized bovine bone

Table 7.2-8 Growth factors used for tissue engineering

Growth factor	Abbreviation	Effects
Basic fibroblast growth factor	bFGF, FGF-2	Angiogenesis; fibroblast and osteoblast mitogen
Bone morphogenetic proteins	BMP-2 BMP-7 (OP-1)	Growth and development of some tissues; osteogenesis
Transforming growth factor-βl	TGF-β1	Proliferation and differentiation of bone forming cells; fbroblast matrix synthesis
Vascular endothelial growth factor	VEGF	Angiogenesis; proliferation and migration of endothelial cells
Platelet-derived growth factor	PDGF	Proliferation of smooth muscle cells; fibroblast mitogen and matrix synthesis
Hepatocyte growth factor	HGF	Hepatocyte mitogen; motogen and anti-apoptotic factor of cells; angiogenesis
Keratinocyte growth factor	KGF	Epithelization of wounds
Epidermal growth factor	EGF	Proliferation of epithelial; mesenchymal and fibroblast cells
Insulin-like growth factor	IGF-1	Cartilage development and homeostasis; bone formation

matrix was capable of inducing endochondral bone formation when implanted ectopically into soft tissues of experimental animals [24]. Urist and colleagues subsequently discovered that low MW glycoproteins isolated from bone were responsible for the osteogenic activity observed earlier and were capable of inducing bone formation when delivered to ectopic or orthotopic locations. Wozney *et al.* and Celeste *et al.* subsequently cloned the first bone morphogenetic proteins: BMP-2, BMP-3, and BMP-4. Ozkaynak *et al.*, using similar techniques, cloned BMP-7 and BMP-8. At least 15 BMPs have been identified, many of which can induce chondro-osteogenesis in various mammalian tissues. The BMPs are homodimeric molecules (MW: 25–30 kDa) that regulate various cellular functions such as bone induction, morphogenesis, chemotaxis, mitosis, hematopoiesis, cell survival, and apoptosis. With the exception of BMP-1, which is a protease that possesses the carboxy-terminal procollagen peptide, BMPs are part of the transforming growth factor (TGF)-β superfamily and play a major role in the growth and development of several organ systems, including the brain, eyes, heart, kidney, gonads, liver, skeleton, skeletal muscle, ligaments, tendons, and skin. Although the sequences of members of the TGF-β superfamily vary considerably, all are structurally very similar. The BMPs are further divided into subfamilies based on phylogenetic analysis and sequence similarities: BMP-2 and -4 (dpp subfamily); BMP-3; BMP-5, -6, -7, -8A, and -8B (60A subfamily). And BMP-6 is characteristically expressed in prehypertrophic chondrocytes during embryogenesis. Several investigators have reported that BMP-6 plays a role in chondrocyte differentiation both *in vivo* and *in vitro*. When involved in osteoinduction, BMPs have three main functions. First, BMPs act as a chemotactic agent, initiating the recruitment of progenitor and stem cells toward the area of bone injury. Second, BMPs function as growth factors, stimulating angiogenesis and proliferation of stem cells from surrounding mesenchymal tissues. Third, BMPs function as differentiation factors, promoting maturation of stem cells into chondrocytes, osteoblasts, and osteocytes.

7.2.7.1.2 FGFs

The function of FGFs is not restricted to cell growth. Although some of the FGFs induce fibroblast proliferation, the original FGF (FGF-2 or basic FGF) is now known also to induce proliferation of endothelial cells, chondrocytes, smooth muscle cells, melanocytes, as well as other cells. It can also promote adipocyte differentiation, induce macrophage and fibroblast IL-6 production, stimulate astrocyte migration, and prolong neural survival. The FGFs are potent modulators of cell proliferation, motility, differentiation, and survival, and play an important role in normal regeneration processes *in vivo*, such as embryonic development, angiogenesis, osteogenesis, chondrogenesis, and wound repair. The FGF superfamily consists of 23 members, all of which contain a conserved 120-amino-acid core region that contains six identical, interspersed amino acids. The superfamily members act extracellularly through four tyrosine kinase FGF receptors, with multiple specificities noted for almost all FGFs. The FGFs are considered to play substantial roles in development, angiogenesis, hematopoiesis, and tumorigenesis. Human FGF-2, otherwise known as basic FGF, HBGF-2, and EDGF, is an 18 kDa, non-glycosylated polypeptide that shows both intracellular and extracellular activity. The FGFs are stored in various sites of the body under interactions with GAGs such as heparin and heparan sulfate in the ECM. And FGF-2 binds to heparin and heparan sulfate with a high affinity. These GAGs stabilize FGF-2 by protecting from inactivation by acid and heat as well as from enzymatic degradation. Also, heparin enhances the mitogenic activity of FGF-2 and serves as a cofactor to promote binding of FGF-2 to high affinity receptors.

7.2.7.1.3 VEGF

Of many unknown angiogenic factors, VEGF is unique in that it is the only known cytokine with mitogenic effects primarily confined to endothelial cells. The VEGF is produced by a variety of normal and tumor cells. Its expression correlates with periods of capillary growth during embryologic development, wound healing, and the female reproductive cycle, as well as with tumor expansion. In addition, suppression of VEGF expression in adult mice or neutralization of VEGF receptors suppresses tumor growth, while VEGF$^{\pm}$ mice are mortal *in utero*. Consequently, VEGF is thought to be a major promoter of both physiologic and pathologic angiogenesis. Furthermore, VEGF has been shown to be an anti-apoptosis survival factor for endothelial cells even during periods of microvessel stasis. In addition, VEGF can enhance tissue secretion of a variety of pro-angiogenic proteases, including uPA, MMP-1, and MMP-2. It has also been shown that inhibition of VEGF or the VEGF-R2 receptor can suppress expression of MMP-2 and MMP-3.

The development of blood vessels includes two processes. Vasculogenesis is the embryonic formation of blood islands, the earliest vascular system, by the differentiation from mesoderm of angioblasts and hematopoietic precursor cells (HPCs). Angiogenesis is the sprouting of pre-existing vessels to form the vascular tree. In addition to a key role in embryonic development, angiogenesis is essential in such things as wound healing and tumor growth. While many growth factors exhibit angiogenic activity (FGFs, PDGF, TGF-α, HGF, and P/GF), most evidence points to a special role for VEGF. The VEGF is a dimeric glycoprotein that stimulates endothelial cells, induces angiogenesis and increases vascular permeability. There are four alternatively sliced variants of 121, 165, 189, and 206 amino acid residues. The receptors for VEGF are VEGFR-1 and VEGFR-2. Homozygous mutants of the VEGF receptors led to lack of vasculogenesis and death of mouse embryos on day 8, indicating that VEGF receptors are essential for the formation of a normal vasculature. The FGFs tend to be potent yet relatively non-specific growth factors with some angiogenic activity while the VEGF group trends to be relatively more specific to angiogenesis but with relative endothelial cell-specific, yet weaker, endothelial cell mitogenicity.

7.2.7.1.4 TGF-β1

The TGF-β1 is the superfamily of growth and differentiating factors of which BMP is a member. The MW of TGF-β1 is 25 kDa. This protein is synthesized in platelets and macrophages, as well as in some other cell types. When released by platelet degranulation or actively secreted by macrophages, TGF-β1 acts as a paracrine growth factor (i.e., growth factor secreted by one cell exerting its effect on an adjacent second cell), affecting mainly fibroblasts, marrow stem cells, and osteoblast precursors. The TGF-β1 stimulates chemotaxis and mitogenesis (increase in the cell populations of healing cells) of osteoblast precursors, promotes differentiation toward mature osteoblasts, stimulates deposition of collagen matrix, and inhibits osteoclast formation and bone resorption. Therefore, TGF-β represents a mechanism for sustaining a long-term healing and bone regeneration module.

7.2.7.1.5 PDGF

The PDGF is a glycoprotein existing mostly as a dimer of two chains of about equal size and MW (14–17 kDa). The protein that seems to be the first growth factor present in a wound is synthesized not only in platelet but also in macrophages and endothelium. It initiates connective tissue healing, including bone regeneration and repair. At the time of injury PDGF emerges from degranulating platelets and activates cell membrane receptors on target cells. The most important specific activities of PDGF include mitogenesis, angiogenesis (endothelial mitoses into functioning capillaries), and macrophage activation (debridement of the wound site and a second-phase source of growth factors for continued repair and bone formation). The PDGF is also stored in the bone matrix and released upon activation of osteoblasts, resulting in an increase of new bone formation. The PDGF is a potent stimulator of fibroblast cell migration, mitogenesis, proliferation, and matrix synthesis important in wound healing.

The four PDGF isoforms (A, B, C, and D) are characterized by a highly conserved eight-cysteine domain termed the PDGF/VEGF homology domain. The PDGF isoforms exist as disulfide-linked homodimers and heterodimers and differently bind homodimer and heterodimer combinations of two receptor tyrosine kinases, PDGF-Rα and PDGF-Rβ. The PDGF-AA is widely expressed in fibroblasts, osteoblasts, platelets, macrophages, smooth muscle cells, endothelial cells, and Langerhans cells. The PDGF-AA activity is ubiquitous, but dependent on cell expression of PDGF-Rα. The PDGF-AA plays key roles in protein synthesis, chemotaxis inhibition, embryonic neuron fiber development, and bronchial lung development. The PDGF-AB demonstrates mitogenic activity for vascular smooth muscle cells (VSMCs) as well as stimulating angiogenesis in the heart. Parallel to PDGF-AA and PDGF-BB expression, PDGF-AB is important in a wide variety of cellular processes of the immune, nervous, and cardiovascular systems. The PDGF-BB is mainly expressed in endothelial cells and participates in angiogenesis and arterialization of early organ, respiratory, and neuronal development. The PDGF-BB is also observed in platelets, neurons, macrophages, and fetal fibroblasts. The PDGF-BB, via PDGF Rβ, is involved in cellular proliferation and TGF-like activities. The PDGF-CC and PDGF–DD form a novel subgroup of the PDGF family distinguished by structural differences that include an N-terminal CUB domain. Widely expressed in multiple embryonic and adult cell, tissue, and organ types, PDGF-CC appears to be important for angiogenesis, cardiovasculature development, and tumorigenesis.

7.2.7.2 Delivery of growth factors

The use of growth factors has not always been achieved successfully *in vivo*. A major reason for this is the high diffusibility and short half-life time of growth factors *in vivo* to effectively retain their biological activities. Topical delivery of proteins remains in the site administered for a limited duration because of protein diffusion, proteolytic cleavage, and the early bioabsorption of carriers when carriers such as fibrin glue are applied. Application of growth factors in tissue engineering requires enhancement of their activities *in vivo* by means of adequate delivery systems.

The delivery methods include bolus injection; release of protein adsorbed directly on scaffold surfaces; in collagen sponge or in porous coatings; constantly delivery via osmotic pumps; and controlled release of protein trapped in an absorbable polymer. There have been reported numerous studies on the formulation of protein growth factors within absorbable polymers for use as drug delivery vehicles. Although trapping in absorbable polymers seems to yield formulations that can deliver active proteins, there is ample literature demonstrating that organic solvents can have a negative effect on protein association and function. By extension from normal

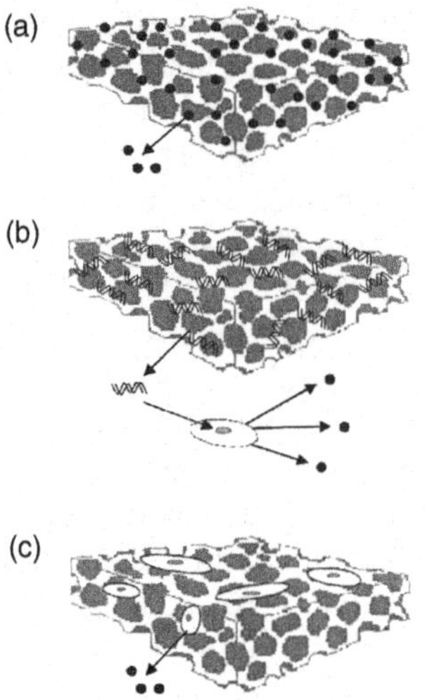

Fig. 7.2-27 Three primary polymeric growth factor delivery strategies: (a) growth factors are embedded within the polymer matrix and released; (b) genes encoding a growth factor are embedded within the polymer matrix and released, followed by cellular uptake and expression of the gene to produce growth factor; (c) growth factor is released from cells seeded on the polymer matrix that secrete the factor.

tissue formation and repair, important variables in formulations for delivery systems include the concentration, timing, and sequence in which the growth factors are introduced.

To circumvent the difficulty of sustained release, high cost, and low commercial availability of growth factors, several methods have been attempted for the sustained delivery of growth factors. As shown in Fig. 7.2-27 [25], the methods used to deliver growth factor molecules include the development of systems to deliver the protein itself, genes encoding the growth factor, or cells secreting the growth factor. Injection of recombinant plasmids and transplantation of gene-manipulated cells have been performed with an expectation that growth factors will be continuously produced from the modified cells for a certain period of time. The DNA most widely employed for such studies is those which are responsible for the biosynthesis of VEGF that induces vascularization. Plasmid vectors are relatively safe but vulnerable to nuclease attack and consequent inefficiency and expense. Viral vectors including adenoviruses, retroviruses, lentoviruses, etc., increase gene transfer efficiency but themselves have limitations including possible toxicologic and immunologic responses. Retroviruses are expressed only in proliferating cells and permanently integrate into genomic DNA.

7.2.8 Cell sources

A human body consists of approximately 60 trillion cells. Human cells differentiate from stem cells into progenitor, precursor, and mature cells, forming a tree-like

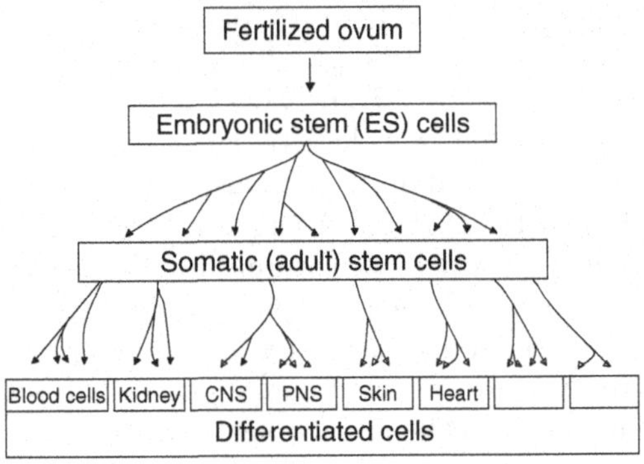

Fig. 7.2-28 From undifferentiated cells to differentiated cells.

hierarchy structure as represented in Fig. 7.2-28. The starting cell for human body development is a fertilized ovum, which then forms an embryo as a result of repeated cleavages. The modura formed after 3–4 day and the blastocyst formed after 5–7 day cleavage contains pluripotent stem cells, named "embryonic stem" (ES) cells. Circulating blood cells survive for only a short time ranging from days to months. Throughout the entire life they are replenished by hematopoietic stem cells in bone marrow which provide a continuous source of progenitors for red cells, platelets, monocytes, granulocytes, and lymphocytes. In addition to hematopoietic stem cells, adult bone marrow contains also non-hematopoietic stem cells. The stem cells for non-hematopoietic tissues are referred to either as MSCs, because of their ability to differentiate into cells that can be roughly defined as mesenchymal, or as bone marrow stromal cells (BMSCs), because they appear to arise from the complex array of supporting stromal structures found in marrow. It has often been stated that a key factor in tissue engineering is lineage-committed precursor cells, especially multilineage stem cells, but almost differentiated cells also have large applications in the current tissue engineering.

Many different types of stem cell exist, but they all are found in very small populations in the human body; in some cases 1 stem cell in 100,000 cells in circulating blood. To identify these rare types of cells found in many different cells and tissues, scientists use stem cell markers. Each cell type has a certain combination of receptors on their surface that makes them distinguishable from other kinds of cells. In many cases, a combination of multiple markers is used to identify a particular stem cell type. Table 7.2-9 lists some of the markers commonly used to identify stem cells and to characterize differentiated cell types.

7.2.8.1 Differentiated cells

Previous thinking was that many differentiated cells of the adult human have a limited capacity to divide, but

Table 7.2-9 Commonly used markers for stem cells and differentiated cell types

Tissue	Marker	Cell type	Significance
Bone marrow	$CD34^+$ $Scal^+$ Lin^- profile	Mesenchymal stem cell	Into adipocyte, osteocyte, chondrocyte, and myocyte
	CD29, CD44, CD105	Mesenchyme	A type of cell adhesion molecule
Bone	RunX		Transcriptional activatori
	ALP	Preosteoblast, osteoblast	Before the expression of OCN
	OCN	Osteoblast	Before and during matrix mineralization
Cartilage	Type II and IV collagen	Chondrocyte	Synthesized specifically by chondrocyte
	Sulfated proteoglycan	Chondrocyte	Synthesized by chondrocyte
Fat	ALBP	Adipocyte	Adipocyte lipid-binding protein
Nervous system	Nestin	Neural progenitor	A marker of neural precursors
	β-Tubulin-III	Neuron	Indicative of neuronal differentiation
Muscle	MyoD and Pax7	Myoblast, myocyte	Secondary transcriptional factors
	Myosin heavy chain	Cardiomyocyte	A component of contractile protein
	Myosin and MR4	Skeletal myocyte	Secondary transcriptional factors
Blood vessel	CD34	Endothelial progenitor	Cell surface protein
	Flkl	Endothelial progenitor, endothelial cell	Cell surface receptor protein
	VE-cadherin	Smooth muscle cell	Cell adhesion molecule

some almost differentiated cells have widely been used in clinical tissue engineering. Among them are notably fibroblasts, keratinocytes, osteoblasts, endothelial cells, chondrocytes, preadipocytes, adipocytes, and tenocytes. In the following, we will focus on chondrocytes as representative of differentiated cells.

Despite a large number of studies achieved using chondrocytes, isolation of autologous chondrocytes for human use is invasive, requiring a biopsy from a non-weight-bearing surface of a joint or a painful rib biopsy. In addition, the *ex vivo* expansion of a clinically required a number of chondrocytes from a small biopsy specimen, which may itself be diseased, is hindered by deleterious phenotypic changes in the chondrocyte. Bone marrow-derived MSCs have been reported to differentiate into multiple cell types of mesenchymal origin. Bone marrow is a reservoir of both hematopoietic and non-hematopoietic stem cells. The MSCs produce tissues such as cartilage, bone, fat, and tendon. These tissues are used daily by plastic and orthopedic surgeons for the repair and augmentation of tissue defects. For these tissues or organ repair or replacement, tissue engineers seek to manipulate cell biology and cellular environments. However, to date, few tissue-engineered systems provide an autologous, minimally invasive, and easily customizable solution for the repair or augmentation of cartilage defects using MSCs.

The activity and function of articular chondrocytes during skeletal growth differ from those found after completion of growth. Chondrogenesis begins in the central core of the developing limb end. First, cartilage is formed from undifferentiated mesenchymal cells that cluster together and synthesize cartilage collagen, proteoglycans, and noncollagenous proteins. Type II collagen forms the primary component of the cross-banded collagen fibrils. The organization of these fibrils, into a tight meshwork that extends throughout the tissue, provides the tensile stiffness and strength of articular cartilage, and contributes to the cohesiveness of the tissue by mechanically entrapping the large proteoglycans. The tissue becomes recognizable as cartilage under light microscopy, when an accumulation of matrix separates the cells and they assume spherical shape. In growing individuals, the chondrocytes produce new tissue to expand and remodel the articular surface. With skeletal maturation, the rates of cell metabolic activity, matrix synthesis, and cell division decline. After completion of skeletal growth, most chondrocytes probably never divide, but rather continue to synthesize collagens, proteoglycans, and noncollagenous proteins. This continued synthetic activity suggests that the maintenance of articular cartilage requires substantial ongoing remodeling of the macromolecular framework of the matrix, and the replacement of degraded matrix macromolecules. With aging, the capacity of the cells to synthesize some types of proteoglycans and their response to stimuli, including growth factors, decrease. These age-related changes may limit the ability of the cells to maintain the tissue, and thereby contribute to the development of degeneration of the articular cartilage.

To promote restoration of normal biomechanical function and long-term integrity of the articular cartilage, integration with surrounding cartilage and subchondral bone is important. Otherwise, the osteochondral repair construct would delaminate or subside. A lack of the integration is the reason in part why tissue engineering approaches to osteochondral defect repair have had limited success. In an attempt to enhance the integration, tissue engineering strategies have utilized heterogeneous constructs in which the upper region promotes cartilage repair while the lower region is specifically designed to encourage bone integration. The ability of chondrocyte-seeded scaffolds to promote repair of the subchondral bone has been variable, due in part to differences in the chondrocyte phenotype within tissue-engineered cartilage constructs. The cellular component of an osteochondral repair strategy may provide critical signals for enhancing bone repair that are intrinsically lacking in an acellular or devitalized implant. During growth and endochondral bone repair, chondrocytes progress to a hypertrophic phenotype and play an important role in angiogenesis, osteoblast recruitment, and mineralized matrix formation. Conversely, articular chondrocytes do not express osteoinductive factors and may inhibit bone formation by producing antiangiogenesis factors such as tissue inhibitors of metalloproteases and troponin. Therefore, chondrocytes within a tissue-engineered cartilage construct have the potential to influence adjacent bone repair response.

Chondrocytes *in vivo* respond to chemical stimuli as well. Two such biochemical regulators of matrix biosynthesis found in articular joints are TGF-β1 and IGF-I. The IGF-I enhances matrix biosynthesis and mitotic activity in chondrocytes, decreases matrix catabolism, and can enhance the tissue properties of cartilage explants in long-term culture. The TGF-β1 exerts a similar influence on matrix biosynthesis, directing production toward increased amounts of larger, more anionic proteoglycan species.

The phenotype change is well known for chondrocytes which are responsible for the hyaline cartilage consisting of type II collagen in the normal articular cartilage, but under irregular culture conditions the same cells produce the fibrous cartilage made from type I collagen which is inferior to the hyaline cartilage with respect to mechanical properties required for articular cartilage. This means that the cells lose the chondrocyte phenotype and become fibroblast-like cells.

7.2.8.2 Somatic (adult) stem cells

Mature cells, when allowed to multiply in an incubator, ultimately lose their effectiveness. Consequently, scientists are turning to other cell types. To be effective, cells must be easily procured and readily available; they must multiply well without losing their potential to generate new functional tissue; they should not be rejected by the recipient and not turn into cancer; and they must have the ability to survive in the low-oxygen environment normally associated with surgical implantation. Mature adult cells fail to meet many of these criteria. The oxygen demand of cells increases with their metabolic activity. After being expanded in the incubator for significant periods of time, they have a relatively high oxygen requirement and do not perform normally. A hepatocyte, for example, requires about 50 times more oxygen than a cell such as a chondrocyte and much attention has turned to progenitor cells and stem cells. True stem cells can turn into any type of cell, while progenitor cells are more or less committed to becoming cell types of a particular tissue or organ. Somatic adult stem cells may actually represent progenitor cells in that they may turn into all the cells of a specific tissue but not into any cell type. Somatic stem cells can be procured from the individual needing the new tissue and thus not be rejected. Since these cells are immature, they will survive a low-oxygen environment. Somatic stem cells normally reside within specific extracellular regulatory microenvironments—stem cell niches—consisting of a complex mixture of soluble and insoluble, short- and long-range ECM signals, which regulate their behavior. These multiple, local environmental cues are integrated by cells that respond by choosing self-renewal or a pathway of differentiation. Outside of their niche, adult stem cells lose their developmental potential quickly.

Somatic stem cells had been claimed to possess an unexpectedly broad differentiation potential that could be induced by exposing stem cells to the extracellular developmental signals of other lineages in mixed-cell cultures. This stem cell plasticity was thus thought to form the foundation for one of the multiple prospective uses of adult stem cells in regenerative medicine. However, experimental evidence supporting the existence of stem-cell plasticity has been refuted because stem cells have been shown to adopt the functional features of other features by means of cell-fusion-mediated acquisition of lineage-specific determinants (chromosomal DNA) rather than by signal-mediated differentiation. Data demonstrating that stem cells could fuse with and subsequently adopt the phenotypes of other cell types indicated that the very co-culture assays originally interpreted to support plasticity instead were artifacts of cell fusion.

Two types of stem cell are available: ES cells and somatic or adult autologous stem cells. The clinical use of ES cells has critical problems of cell regulation including malignant potential, allogeneic immune response, and ethical issues concerning the cell source, while the problem with autologous stem cells concerns their cell source. Bone marrow is the most suitable cell source, because it involves not only hematopoietic stem cells but also MSCs. The potential advantages of using bone marrow MSCs include low cell numbers required at the initial culture, relative simple procedure for bone marrow harvest, and the cell maintenance of high biological activity from older donors. However, the yield of MSCs obtained from aspirated bone marrow blood is too low (approximately 1 per 10^5 adherent stromal cells) to use them as clinical cell source for tissue regeneration and, therefore, an *ex vivo* cell expansion will be necessary. The frequency of long-term repopulating cells is 1 in 35,000 total epidermal cells, or in the order of 1 in 10^4 basal epidermal cells, similar to that of hematopoietic stem cells in the bone marrow [26]. Hematopoietic stem cell frequency in the bone marrow was determined to be 1 in 10,000.

Stem cells have not taken on the identity of any specific cell type and are not yet committed to any dedicated function; they can divide indefinitely and may be induced to give rise to one or more specialized cell types. It seems very likely that each tissue or organ has one somatic stem cell even in adults. A well-known somatic stem cell is the MSC that is able to differentiate into a variety of tissues including skin, cartilage, bone, muscle, and fat, as illustrated in Fig. 7.2-29 [27]. It is clearly seen how versatile this adult stem cell is for clinical application. Stem cells of embryonic as well as adult tissue origin undergo the differentiation process and eventually reach functional maturity. In the use of stem cells as part of tissue engineering, cellular behaviors including the differentiation process must be carefully monitored.

Although the potential of ES cells is enormous, the use of embryonal sources of stem cells is controversial and major ethical and political issues impede their use. Issues surrounding the rights of the unborn fetus, and subsequent government regulation and limitations on availability and applicability of embryonic tissue, have put the brakes on what appeared to be a rapidly approaching clinical reality. In contrast, recent interest has emerged in the use of bone marrow-derived stem cells for tissue engineering applications. Adult-derived precursors potentially provide ample quantities of an autologous source of regenerative tissue without these ethical and political issues. Although demonstrations of bone marrow–derived MSC plasticity have been reported and debated, widespread use of adult-derived tissue will likely require a relatively painless, convenient, and safe procurement method. Some have suggested that the skin fulfills this role, while other reports have begun to emerge suggesting that adipose tissue—which is electively aspirated in large quantities—provides a readily available autologous source. Like bone marrow, adipose is supported by a stroma whose isolation yields a significant amount of cells capable of osteogenic, adipogenic, neurogenic, myogenic, and chondrogenic differentiation.

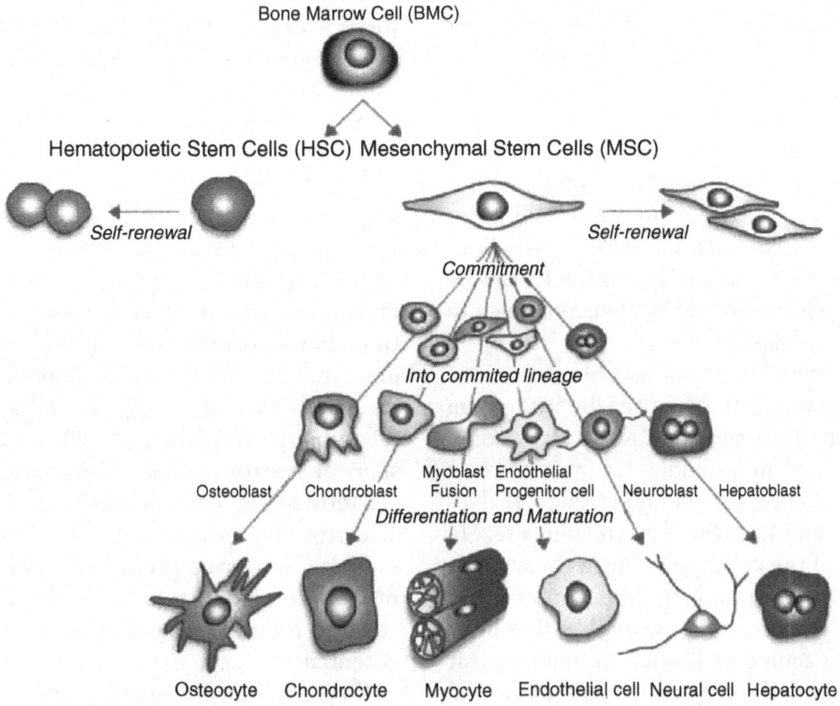

Fig. 7.2-29 MSCs differentiation cascade. There are at least two types of stem cells in bone marrow, namely hematopoietic stem cells and MSCs. One type of MSC repeats self-renewal, whereas the other type is committed to a specific cell lineage and goes through a lineage process. MSCs are reported to differentiate into a variety of cellular types, such as osteocytes, chondrocytes, myocytes, endothelial cells, neurons, hepatocytes, etc.

Given that adipose-derived mesenchymal precursors can be harvested in abundant quantities under local anesthesia with little patient discomfort, they may emerge as an important source for cell-based therapy. One can envision a scenario in which a mesenchymal cell fraction is purified from a patient's bone marrow or liposuction aspirate, is exposed to oxygen and other environmental conditions optimized for differentiation along a certain lineage, and is ultimately returned to the same patient to fill a tissue defect.

Some other tissues have been shown to be sources of somatic stem cells. Periosteum, as well as adipose tissue and peripheral blood, is a good source that removes the confounding effect of hematopoietic stem cells. Periosteum has been shown to be an effective source of cells in the repair of osteochondral defects in an animal model but again suffers from painful procurement procedures and low cell yield. On the basis of the currently available therapeutic options, the ideal reconstitutive measure would cause insignificant donor morbidity, regenerate quickly the harvested tissue, have no size limitations, be readily available, have no issues of immunogenicity, and be of low cost. Good cell volume would obviate the need for *ex vivo* expansion and easy availability would allow for expensive preliminary *in vitro* human cell testing before clinical trials.

7.2.8.2.1 MSCs

Circulating blood cells survive for only a few days or months. This means that hematopoietic stem cells in bone marrow must provide a continuous source of progenitors for blood cells. Bone marrow also contains cells that meet the criteria for stem cells of non-hematopoietic tissues. The stem-like cells for non-hematopoietic tissues are referred to as MSCs, because they can be differentiated in culture into osteoblasts, chondrocytes, adipocytes, and myoblasts. The marrow stromal cells can be isolated from other cells in marrow by their tendency to adhere to tissue culture plastic.

The presence of stem cells for non-hematopoietic cells in bone marrow was first suggested by a German pathologist Cohnheim 140 years ago [28]. He studied wound repair by injecting an insoluble dye into the veins of animals and then looking for the appearance of dye-containing cells in wounds he created at a distant site. He concluded that most of the cells appearing in the wounds came from the bloodstream and, by implication, from bone marrow. This work raised the possibility that bone marrow might be the source of fibroblasts that deposit collagen fibers as part of the normal process of wound repair.

Evidence that bone marrow contains cells that can differentiate into fibroblasts as well as other mesenchymal cells has been available since the pioneering work of Friedenstein *et al.* [29]. They placed samples of whole bone marrow in plastic culture dishes, and, after 4 h, poured off the cells that were nonadherent. In effect, they discarded most of the hematopoietic stem cells. The small number of adherent cells were heterogenous in appearance, but the most tightly adherent cells were spindle shaped and formed foci of two to four cells. The cells in the foci remained dormant for 2–4 days and then began to multiply rapidly. After passage several times in culture, the adherent cells became more uniformly spindle shaped in appearance. The cells had the ability to differentiate into colonies that resembled small deposits of bone or cartilage.

Adult or somatic MSCs, commonly referred to as BMSCs, are stem cells originated from embryonic mesoderm and are a unique class of multipotent cells that are noncommitted and remain in an undifferentiated state. When induced by the appropriate biological cues including right chemicals, hormones, and growth factors, BMSCs are capable of extensive proliferation and differentiation into several phenotypes including fibrous tissues (e.g., tendon), and hematopoiesis-supporting reticular stroma, and represent a heterogeneous cell population likely containing a range of progenitor cells. Of most importance from a tissue engineering standpoint is the fact that stem cells that have been transformed to osteoblasts are much like bone cells of a developing organism, and are capable of secreting large amounts of ECM. Moreover, the ease of isolation of MSCs makes them very attractive for tissue engineering applications; aspiration of bone marrow is only slightly more complicated than a blood donation, and MSCs can be enriched to obtain a relatively pure population of cells. The technique using bone marrow cells (BMCs) does not require cell culture with serum from other species, which might be associated with a risk of infection. In addition, as large numbers of cells can be obtained from BMCs without culture, fewer steps in the preparation of tissues would mean a lower risk of contamination and less work and time. Fewer materials for culture would mean cost benefits, and the technique can be applied even in emergency cases in a clinical setting, as cell seeding requires only a few hours. In culture of cells generated from suspensions of marrow, colonies form from a single precursor cell termed the "colony-forming-unit fibroblast" (CFU-F). The progeny of these CFU-Fs is what have been defined as BMSCs. During prolonged cultivation *ex vivo*, adult BMSCs undergo two possible interdependent procedures: replicative aging and a decline in differentiation potential.

One strategy for tissue engineering is to use these MSCs or BMSCs harvested from marrow stroma. The MSCs have attracted extensive interest due to the surprising finding that they can also commit to neural cell lineages in both animal models and cell cultures. Under certain culture conditions, MSCs were induced to assume morphology and express protein markers typical of neurons. Adult MSCs were found to migrate into the brain and develop into astrocytes. These results suggest that MSCs may be useful also in neural tissue repair and regeneration. The use of MSCs has significant advantages over other cell types. First, the use of adult MSCs means that autologous transplantation is possible, thus avoiding detrimental immune responses often caused by allogeneic transplantation. Second, ethical concerns associated with the use of ES cells or fetal tissues are eliminated. Last, unlike neural stem cells (NSCs) or other neural precursors that need to be obtained via surgery, MSCs are relatively easy not only to obtain from a small aspirate of bone marrow, but also to expand in culture under conditions in which they retain some of their potential to differentiate into multiple cell lineages. For clinical use, it might be desirable to expand and differentiate MSCs into the cell types of choice *in vitro* before transplantation, because this strategy allows reproducible and reliable generation of well-defined transplants in a precisely controlled environment. Besides making tissue-engineered constructs of a single tissue type, an important potential advantage in using MSCs is the ability to design multifunctional or composite tissue constructs, such as osteochondral grafts, from a single cell source. Interestingly, there is evidence to suggest that certain types of bone marrow-derived cells can themselves produce an angiogenic mediator VEGF. This observation raises the possibility that these mesenchymal cells could potentially regulate angiogenesis.

Several clinical trials with human MSCs, or closely related cells, have started. For most of the human trials and animal experiments, MSCs are prepared with a standard protocol in which mononucleated cells are isolated from a bone marrow aspirate with a density gradient, and then both enriched and expanded in the presence of FCS by their tight adherence to plastic tissue culture dishes. For instance, bone marrow aspirates are mixed with Hanks' balanced salt solution (HBSS) containing heparin. The same amount of high-density Ficoll solution is carefully placed beneath the HBSS mixture followed by centrifugation. The nucleated cell layer existing at the interface between the HBSS and the Ficoll is removed and placed on a dish. After removal of non-adherent cells and the subsequent continuation of cell culture, cells form colonies from which MSCs are obtained. Cultures of human MSCs, unlike murine cells, become free of hematopoietic precursors after one or two passages and can be extensively expanded before they senesce. However, when the cells are expanded under standard culture conditions, they lose their proliferative capacity and their potential to be differentiated into lineages such as adipocytes, tenocytes, and chondrocytes. Moreover, cultures of human MSCs are morphologically heterogeneous, even when cloned from single-cell-derived colonies.

Studies on human MSCs become complicated by the fact that there is no consensus as to the characteristic surface epitopes that can be used to identify the cells. A series of antibodies to surface epitopes have been employed by several investigators, but none have been into general use. Sekiya *et al.* screened over 200 antibodies but did not find any that efficiently distinguish rapidly self-renewing cells from mature MSCs [30]. It is therefore difficult to compare the results that different research groups obtained either in animal models for diseases or in clinical trials. It follows that several parameters must be considered in preparing frozen stocks of human MSCs: (1) variations in the quality and number of MSCs obtained from different bone marrow aspirates, (2) the yield of cells required as frozen stocks, (3) the quality of the cultures in terms of their content of early progenitor cells that replicate most rapidly and have the greatest potential for multilineage differentiation, and (4) the number of cell doublings the cells have undergone before they are harvested and frozen.

Autologous MSCs have advantages over ES cells that may lead to terato-carcinoma formation. However, compared with ES cells, which have an unlimited proliferative life span (period before the cells reach growth arrest in culture) and consistently high telomerase activity, MSCs have very poor replicative capacity and short proliferative longevity. Thus, an important challenge in tissue engineering is to improve the replicative capacity of MSCs, thereby to obtain a number of MSC sufficient to repair large defects. Forced expression of telomerase in MSCs markedly increased their proliferative life span and MSCs with a high telomerase activity showed osteogenic potential.

Cell expansion

The most common source of adult-derived stem cells is the bone marrow. MSC can be obtained easily and repeatedly by bone marrow aspiration, but isolation of marrow aspirates in great volume causes damage and pain, and it is difficult to isolate from the bone marrow 10^7–10^8 MSCs that are required for regeneration of large injured tissues. In addition, the heterogeneous nature of bone marrow with both hematopoietic and MSCs confounds the result of various therapies using bone marrow. The scarcity of MSC in bone marrow of older donors may impose additional requirements with respect to cell expansion and differentiation. Furthermore, a short life span of MSC and a reduction in their differentiation potential in culture have limited their clinical application. Thus, the ability to rapidly expand MSCs in culture is of obvious importance in using the cells for tissue engineering. The expansion of MSCs *in vitro* is a prerequisite for autologous cell transplantation. In other words, identification of growth factors that stimulate the proliferation of MSCs and support their multilineage differentiation potential is a critical step towards the clinical application of MSCs.

Cell differentiation

The forces driving stem cell differentiation, or maintaining stem cells in a state of suspended undifferentiation, include secreted and bound messengers or homing signals. Specific chemicals and hormones that cause the transformation of MSCs to osteoblasts, chondrocytes, and adipocytes have been elucidated. Table 7.2-10 represents the commonly used *in vitro* environments for differentiations. For instance, osteogenic differentiation occurs when MSCs are treated with dexamethasone, β-glycerophosphate, and ascorbic acid (AA), and this differentiation to osteoblasts is characterized by gene expression of osteopontin (OP) and alkaline phosphatase (ALP). Just as important are the cellular environmental sensors sensitive to oxygen, temperature, chemical gradients, mechanical forces, and others cues in the microenvironment. The notion of microenvironments affecting stem cell division and function is not new. Schofield dubbed these "niches" with respect to hematopoietic stem cells, and subsequent reports have described their presence in numerous tissues including neural, germline, skin, intestinal, and others [31].

7.2.8.2.2 Adipose-derived stem cells

A second large stromal compartment found in human subcutaneous adipose tissue has received attention because of the presence of multipotent cells named adipose-derived adult stem (ADAS) cells. Under lineage-specific biochemical and environmental conditions, ADAS cells will differentiate into osteogenic, chondrogenic, myogenic, adipogenic, and even neuronal pathways. Although it remains to be determined whether ADAS cells meet the definition of stem cells, they are multi-potential, are available in large numbers, are easily accessible, and attach and proliferate rapidly in culture, making them an attractive cell source for tissue engineering. Moreover, ADAS cells demonstrate a substantial *in vitro* bone formation capacity, equal to that of bone marrow, but are much easier to culture. It has been empirically shown that several growth factors and hormones, in combination with cellular condensation and rounded cell morphology, may promote the chondrogenic differentiation of ADAS cells.

7.2.8.2.3 Umbilical cord blood-derived cells

Human umbilical cord blood-derived cells may be alternative autologous or allogeneic cell source. Umbilical cord blood cells contain multipotent stem cells and these cells have been used to generate tissue-engineered pulmonary artery conduits in a pulsatile bioreactor [32]. Cells from umbilical cord artery, umbilical cord vein, whole umbilical cord, and saphenous vein segments were compared for their potential as cell sources for tissue-engineered vascular grafts [33]. All four cell sources generated viable myofibroblast-like cells with ECM formation including types I and III collagen and elastin. There were also $CD34^-$ umbilical cord cells which have the capacity to generate endothelial cells.

Table 7.2-10 Lineage specific differentiation induced by media supplementation

Medium	Media	Serum	Supplementation
Osteogenic	Dulbecco's minimal essential medium (DMEM)	10% Fetal calf serum (FCS)	50 μM ascorbic acid 2-phosphate, 10 mM β-glycerophosphate, 100 nM dexamethasone
Chondrogenic	High-glucose DMEM		10 ng/ml transforming growth factor (TGF)- β3,100 nM dexamethasone, 6 μg/ml insulin, 100 μM ascorbic acid 2-phosphate, 1 mM sodium pyruvate, 6 μg/ml transferrin, 0.35 mM proline, 1.25 mg/ml bovine serum albumin
Adipogenic	DMEM	10% FCS	1 μM dexamethasone, 0.2 mM indomethacin, 10 μg/ml insulin, 0.5 mM 3-isobutyl-l-methylxanthine
Myogenic	DMEM	10% FCS, 5% horse serum	10 μM 5-azacytidine, 50 μM hydrocortisone
Cardiac	Isocove's modified DMEM	20% FCS	3 μM 5-azacytidine

7.2.8.3 Cell therapy

Stem cell therapy is an emerging field, but current clinical experience is limited. One limitation of cell therapy is that donor cells often cannot functionally replace the impaired cells immediately following transplantation and this delay may impact patient survival. In the field of adult stem cell therapy, it is estimated that there are currently over 80 therapies and around 300 clinical trials underway using such cells [34]. Hematopoietic stem cell transplants are routine clinical practice and more than 300 patients with type I diabetes have now undergone transplants of islet cells using the so-called "Edmonton protocol", with a significant proportion staying off insulin injections for several years. But the field remains beset by problems of reproducibility. Often, the level of therapeutic benefit is minimal as implanted cells die because of immune attack or other problems. It is commonly not clear whether cells are expanding, fusing with recipient cells, or exerting an effect through secreted growth factors. Unraveling these problems will require progress on several fronts.

7.2.8.3.1 Angiogenesis

Adult bone marrow is a rich reservoir of tissue-specific stem and progenitor cells. Among them is a scarce population of cells known as "endothelial progenitor cells" (EPCs) that can be mobilized to the circulation and contribute to the neoangiogenic processes. Circulating endothelial progenitor cells (CEPs) have been detected in the circulation either after vascular injury or during tumor growth. The CEPs primarily originate from EPCs within the bone marrow and differ from sloughed mature, circulating endothelial cells (CECs) that randomly enter the circulation as a result of blunt vascular injury. Preclinical studies have shown that introduction of bone marrow–derived endothelial and hematopoietic progenitors can restore tissue vascularization after ischemic events in limbs, retina, and myocardium [33]. Co-recruitment of angiocompetent hematopoietic cells delivering specific angiogenic factors facilitates incorporation of EPCs into newly sprouting blood vessels. Given the morbidity associated with limb ischemia, vascular stem cell therapy provides promising adjunct therapy to current bypass surgical approaches. In preclinical studies, introduction of bone marrow–derived EPCs effectively improves collateral vessel formation, thereby minimizing limb ischemia. In patients suffering from peripheral arterial disease, placement of autologous whole bone-marrow MNCs into ischemic gastrocnemius muscle resulted in restoration of limb function. Because MNCs contain both EPCs and myeloid cells, it remains to be determined whether the improvement in these studies was due in part to the introduction of myelomonocytic cells.

Given the complexity of organ-specific microenvironmental cues essential for functional incorporation of transplanted EPCs and CEPs, numerous hurdles have to be overcome for successful stem cell therapy for tissue vascularization. As damaged tissue may lose anatomical cues for functional organ neovascularization, *in vitro* manipulation of stem cells may be essential to facilitate *in vivo* incorporation. Identification of organ-specific cytokines—including the appropriate combinations of VEGFs, FGFs, PDGFs, IGFs, angiopoietins, other as-yet-unrecognized factors and ECM components for optimal culture conditions—will provide the platform for the differentiation of stem cells, allowing delivery of a large number of angiocompetent cells.

7.2.8.3.2 Cardiac malfunction

Myocardial infarctions most commonly result from coronary occlusions, due to a thrombus overlying an atherosclerotic plaque. Because of its high metabolic rate, myocardium (cardiac muscle) begins to undergo irreversible injury within 20 minutes of ischemia, and a wavefront of cell death subsequently sweeps from the inner layers toward the outer layers of myocardium over a 3- to 6-h period. Although cardiomyocytes are the most vulnerable population, ischemia also kills vascular cells, fibroblasts, and nerves in the tissue. Myocardial necrosis elicits a vigorous inflammatory response. Hundreds of millions of marrow-derived leukocytes, initially composed of neutrophils and later of macrophages, enter the infarct. The macrophages phagocytose the necrotic cell debris and likely direct subsequent phases of wound healing. Concomitant with removal of the dead tissue, a hydrophilic provisional wound repair tissue rich in proliferating fibroblasts and endothelial cells—termed "granulation tissue"—invades the infarct zone from the surrounding tissue. Over time, granulation tissue remodels to form a densely collagenous scar tissue. In most human infarcts, this repair process requires 2 months to complete. At the organ level, myocardial infarction results in thinning of the injured wall and dilation of the ventricular cavity. These structural changes markedly increase mechanical stress on the ventricular wall and promote progressive contractile dysfunction. The extent of heart failure after a myocardial infarction is directly related to the amount of myocardium lost.

Related to myocardial repair following injury, the limited proliferative capacity of mature cardiomyocytes is the fundamental reason that numerous investigators have evaluated alternative cell sources. Cellular cardiomyoplasty to replace damaged myocardial cells has been attempted using a variety of cell types including bone-marrow precursor cells, skeletal myoblasts, satellite cells, muscle-derived stem cells, smooth muscle cells, late embryonic/fetal and neonatal cardiac cells, and ES cells. Umbilical cord blood cells have the capacity to form endothelial cells and myofibroblasts. Given this level of excitement, it is hardly surprising that the theory-to-therapy approach to the use of cells would be thrust forward into translational human studies at the earliest possible stage [34]. A prompt action came in 2001 with the publication of a paper in Nature by Orlic and colleagues, suggesting that stem cells derived from bone marrow can replace heart muscle lost as a result of heart attack, and can improve cardiac function. Injecting bone-marrow stem cells into an injured heart potentially represented a new therapy, triggering the launch of numerous clinical studies to investigate the effect of directly injecting these cells into the damaged heart muscle of patients following a heart attack. A major barrier to the long-term success of cellular cardiomyoplasty is the survival of the transplanted cells.

Recent reports suggest that hematopoietic stem cells can transdifferentiate into unexpected phenotypes such as skeletal muscle, hepatocytes, ECs, neurons, endothelial cells, and cardiomyocytes, in response to tissue injury or placement in a new environment. Although most studies suggest that transdifferentiation is extremely rare under physiological conditions, extensive regeneration of myocardial infarcts was reported after direct stem cell injection, promoting several clinical trials. Under conditions of tissue injury, myocardial replication and regeneration have been reported and a growing number of investigators have implicated adult bone marrow in this process, suggesting that marrow serves as a reservoir for cardiac precursor cells. It remains unclear which BMCs can contribute to myocardium, and whether they do so by transdifferentiation or cell fusion.

Independent studies by Murry *et al.* [35] and Balsam *et al.* [36] seriously challenged Orlic and colleagues' initial observations and the scientific underpinnings of the ongoing human studies [37]. Two strategies were used to show that bone-marrow stem cells do not take on the role of damaged heart cells. Murry *et al.* isolated and purified genetically modified bone-marrow stem cells from mice. The modification "tagged" the cells (with LacZ), enabling them to be detected in the recipient mouse heart, into which the cells were directly injected. Closer inspection of the recipient heart showed that the label could not be detected in heart muscle cells. Similar results were shown by Balsam *et al.*, although the approach was slightly different. Donor bone-marrow stem cells were transfused directly into the circulation of recipient. Again, the tag (GFP) could not be detected in heart muscle cells of the donor; indeed, the BMCs continued to differentiate into blood cells while in the heart. So, scientists are asking why there are wide discrepancies between the earlier report and the current investigations. As Murry *et al.* suggest, the differences may arise from the difficulty of tracking the *in vivo* fate of transplanted cells within an intact organ. Orlic *et al.* mainly relied on detecting unique protein constituents of bone-marrow stem cells using fluorescently tagged antibodies. Murry *et al.* and Balsam *et al.*, however, created intrinsic genetic markers that can be easily recognized without antibody staining. Owing to its high density of muscle-specific contractile proteins, intact heart muscle tends to have high inherent background fluorescence, and can also display non-specific antibody binding to the abundant muscle proteins. This makes it difficult, even for the most experienced labs using the most specialized microscopes, to track cell fate by simply using techniques that rely on fluorescent antibody staining of cardiac proteins. Various experiments using several types of stem cells support the view that transdifferentiation occurs rarely, if at all, in many organ systems, including heart muscle. Less than 2% of the transplanted or injected cells take on the *in vivo* fate of heart cells. If this is true, then the improvement in cardiac function seen by Orlic *et al.* might have arisen not because the stem cells transdifferentiated, but because new blood vessels were encouraged to grow around the injected area. Such growth of new blood vessels has been consistently found in transplantation studies of diverse cell types in the heart. Studies in large-animal models of transplanted bone-marrow-derived stem cells in the injured heart also failed to document cardiac regeneration [38]. Again, the implication is that any functional improvement seen may not be related to an increase in functioning heart muscle *per se*. A recent clinical study, in which bone-marrow stem cells were transplanted into injured hearts, was terminated because of serious cardiac side effects that threatened the blood flow to the heart. This again suggests that further experimental testing is warranted in large-animal model systems.

7.2.8.4 ES cells

The first report of a stable ES cell line derived from a human blastocyst in 1998 was by Thomson *et al.* [39]. This led to a surge of interest in ES cells as a potential cell source for tissue engineering. The ES cells are derived from the inner cell mass of the preimplantation blastocyts. They are pluripotent and can be maintained and expanded in culture in an undifferentiated state. Markers such as Oct-4, SSEA-4, and Nanog have all been used to characterize and assess the pluripotent capacity of ES lines when grown *in vitro*. However, the precise mechanisms by which the culture methods routinely employed in laboratories enable ES cells to remain pluripotent are still not fully understood and it is likely that further key genes that prevent the cell from proceeding with differentiation remain to be identified.

Human ES cell lines (hES) are pluripotent diploid cells that can proliferate in culture indefinitely and provide a unique system for studying the events in human embryonic development. The hES cells have the potential to generate all embryonic cell lineages when they undergo differentiation. Differentiation of hES can be induced in monolayer culture or by removing the cells from their feeder layer and growing them in suspension. This differentiation in suspension results in aggregation of the cells and formation of embryoid bodies (EBs), where successive differentiation steps occur.

From human ES cells we might be able to develop new transplantation therapies to replace diseased or aged cells or tissues. To this end, researchers need to develop methods with which they can derive from human ES cells their required cell types, such as cardiomyocytes or hematopoietic cells. Chemical cues provided directly by growth factors or indirectly by feeder cells can induce ES cell differentiation toward specific lineages. While ES cells show great promise for treating many diseases, such as heart disease, diabetes, and Parkinson's disease, non-matching ES cells would be rejected by patients' immune systems unless they take immunosuppressant drugs. The HLA (human histocompatibility leukocyte antigen) system has a central role in the initiation and development of immune rejection. However, hES cells and their differentiated progeny express highly polymorphic MHC (major histocompatibility complex) molecules that serve as major graft rejection antigens to the immune system of allogeneic hosts. To achieve sustained engraftment of donor cells, strategies must be developed to overcome graft rejection without broadly suppressing host immunity. One approach entails induction of donor-specific immune tolerance by establishing chimeric engraftment in hosts with hematopoietic cells derived from an existing hES cell line. To achieve best possible MHC matching we could establish large banks of HLA-defined and highly diversified hES cell lines, but this strategy might not be sufficient since minor rejection antigens are still present and difficult to define. Immunosuppressive drugs such as cyclosporine are administered to transplant recipients to prevent acute and chronic immune-mediated rejection of allogeneic bone marrow and organ transplants even with best possible MHC matching. Polymorphisms in many non-HLA histo-compatibility antigens, including highly polymorphic mitochondrial and H-Y gene products, result in rejection even in HLA-matched individuals.

7.2.8.4.1 Cell expansion and differentiation

Routine propagation of mouse ES cells in an undifferentiated state can be achieved by culture upon mitotically inactivated mouse embryonic fibroblasts (MEFs) or upon gelatin-coated dishes in the presence of the interleukin-6 family member cytokine leukemia inhibitory factor (LIF). The LIF stimulates ES cell self-renewal following binding of the LIF receptor β/gp130 heterodimer and activation of the JAK/Stat3 signaling passway. Although the mechanism by which gelatin aids in the maintenance of ES cells in an undifferentiated state is not known, it is clear that surface properties of specific substrates can exert powerful effects upon cell growth and behavior.

By controlling the culture conditions under which ES cells are allowed to differentiate, it is possible to generate cultures that are enriched for lineage-specific precursors. To this end, stem cells require an additional ability to control growth and differentiation into useful cell types. The effects of biomaterials on the behavior of stem cells have not been studied in great detail. This is due in part to the potential diversity of biomaterials and the difficulty of large-scale hES cell production.

7.2.8.4.2 Somatic cell nuclear transfer

The isolation of pluripotent hES cells and breakthroughs in somatic cell nuclear transfer (SCNT) in mammals have raised the possibility of performing human SCNT and generated potentially unlimited sources of undifferentiated cells for use in research, with potential applications in tissue repair and transplantation medicine. The SCNT concept, known as "therapeutic cloning", refers to the transfer of the nucleus of a somatic cell into an enucleated donor oocyte. In theory, the oocyte's cytoplasm would reprogram the transferred nucleus by silencing all the somatic cell genes and activating the embryonic ones. The ES cells would be isolated from the inner cell mass of the cloned preimplantation embryo. When applied in a therapeutic setting, these cells would carry the nuclear genome of the patient; therefore, after directed cell differentiation, the cells could be transplanted without immune rejection to treat degenerative disorders such as diabetes, osteoarthritis (OA), and Parkinson's disease among others.

A team led by veterinary cloning expert Woo Suk Hwang and gynecologist Shin Yong Moon of Seoul National University in South Korea showed that the cloning technique can work in humans [40]. The researchers described how they created a human ES cell line by inserting the nucleus of a human cumulus cell into a human egg from which the nucleus had been removed. (Cumulus cells surround the developing eggs in an ovary, and in mice and cattle they are particularly efficient nucleus donors for cloning.) After using chemicals to prompt the reconstructed egg to start dividing, the team allowed it to develop for a week to the blastocyst stage, when the embryo forms a hollow ball of cells. They then removed the cells that in a normal embryo are destined to become the placenta, leaving the so-called "inner-cell mass" that would develop into the fetus. When these cells are grown in culture, they can become ES cells, which reproduce indefinitely and retain the ability to form all the cell types in the body. The ES cell line the team derived seems to form bone, muscle, and immature brain cells, for example. The South Korean scientists suspect that their method of removing the egg's nucleus might have been one of the secrets of their success. Instead of sucking the nucleus out with a pipette, which in past work seemed to damage the protein machinery that controls cell division, the team nicked a small hole in the egg's membrane and gently squeezed out the genetic material. Perhaps the Korean scientists' most important advantage was the whopping 242 eggs they had to work with. The team obtained oocytes and donor cells from 16 healthy women, who underwent hormone treatment to stimulate their ovaries to overproduce maturing eggs. The women who donated specifically for the experiments were not compensated, and were informed that they would not personally benefit from the research.

In this study the adult cell (donor cell) and the egg (oocyte) came from the same person. This made difficult to prove that the embryo really was cloned. The process was also very inefficient, taking 242 eggs to create just one ES cell line. However, in 2005 the Hwang's team created 11 more ES cell lines from cloned embryos in an impressive study that answers all the criticisms of their original study. They have also greatly increased the efficiency of the process: the 11 lines came from just 185 fresh eggs donated by 18 unpaid volunteers, meaning an average of only 17 eggs was needed per ES cell line [41]. The donor "adult" cells came from patients aged 2–56, with a variety of conditions ranging from spinal injuries to an inherited immune condition. The work proves that matching ES cells can be derived via therapeutic cloning from donors of any age and sex.

References

1. A. K. Lynn, I.V. Yannas, and W. Bonfield, Antigenicity and immunogenicity of collagen, *J. Biomed. Mater. Res. B: Appl. Biomater.*, **71B**, 343 (2004).
2. K. Tomihata and Y. Ikada, *In vitro* and *in vivo* degradation of films of chitin and its deacetylated derivatives, *Biomaterials*, **18**, 567 (1997).
3. K. Tomihata, M. Suzuki, and Y. Ikada, The pH dependence of monofilament sutures on hydrolytic degradation, *J. Biomed. Mater. Res.*, **58**, 511 (2001).
4. S.-H. Hyon, W.-J. Cha, T. Nakamura *et al.*, Synthesis and properties of elastomeric lactide- caprolactone copolymers, *J. Japanese Soc. Biomater.* (in Japanese), **14**, 216 (1996).
5. V. Karageorgiou and D. Kaplan, Porosity of 3D biomaterial scaffolds and osteogenesis, *Biomaterials*, **26**, 5474 (2005).
6. T.B.F. Woodfield, J. Malda, J. de Wijn *et al.*, Design of porous scaffolds for cartilage tissue engineering using a three-dimensional fiber-deposition technique, *Biomaterials*, **25**, 4149 (2004).
7. S.J. Hollister, Porous scaffold design for tissue engineering, *Nat. Mater.*, **4**, 518 (2005).
8. K. Fujihara, M. Kotaki, and S. Ramakrishna, Guided bone regeneration membrane made of polycaprolactone/calcium carbonate composite nano-fibers, *Biomaterials*, **26**, 4139 (2005).
9. S.F. Badylak, The extracellular matrix as a scaffold for tissue reconstruction, *Semin. Cell Dev. Biol.*, **13**, 377 (2002).
10. J. Parizek, P. Mericka, J. Spacek *et al.*, Xenogeneic pericardium as a dural substitute in reconstruction of suboccipital dura mater in children, *J. Neurosurg.*, **70**, 905 (1989).
11. U. Hersel, C. Dahmen, and H. Kessler, RGD modified polymers: Biomaterials for stimulated cell adhesion and beyond, *Biomaterials*, **24**, 4385 (2003).
12. A. Abbott, Cell culture: Biology's new dimension, *Nature*, **424**, 870 (2003).
13. V. Fast and A. Kleber, Microscopic conduction in cultured strands of neonatal rat heart cells measured with voltage-sensitive dyes, *Circ. Res.*, **73**, 914 (1993).
14. M. Kuzuya, S. Satake, M.A. Ramos *et al.*, Induction of apoptotic cell death in vascular endothelial cells cultured in three-dimensional collagen lattice, *Exp. Cell Res.*, **248**, 498 (1999).
15. E.A. Botchwey, M.A. Dupree, S.R. Pollack *et al.*, Tissue engineered bone: Measurement of nutrient transport in three-dimensional matrices, *J. Biomed. Mater. Res.*, **67A**, 357 (2003).
16. J.S. Blum, J.S. Temenoff, H. Park *et al.*, Development and characterization of enhanced green fluorescent protein and luciferase expressing cell line for non-destructive evaluation of tissue engineering constructs, *Biomaterials*, **25**, 5809 (2004).
17. H.-W. Kim, H.-E. Kim, and V. Salih, Stimulation of osteoblast responses to biomimetic nanocomposites of gelatin-hydroxyapatite for tissue engineering scaffolds, *Biomaterials*, **26**, 5221 (2005).
18. E.M. Darling and K.A. Athanasiou, Articular cartilage bioreactors and bioprocesses, *Tissue Engineering*, **9**, 9 (2003).
19. G.N. Bancroft, V.I. Sikavitsas, and A.G. Mikos, Design of a flow perfusion bioreactor system for bone tissue-engineering applications, *Tissue Engineering*, **9**, 549 (2003).
20. W.C. Hymer *et al.*, Feeding frequency affects cultured rat pituitary cells in low gravity, *J. Biotechnol.*, **47**, 289 (1996).
21. A. Solan, S. Mitchell, M. Moses *et al.*, Effect of pulse rate on collagen deposition in the tissue-engineered blood vessel, *Tissue Engineering*, **9**, 579 (2003).
22. R.G. Dennis, P.E. Kosnik, M.E. Gilbert *et al.*, Excitability and contractility of skeletal muscle engineered from primary cultures and cell lines, *Am. J. Physiol. Cell Physiol.*, **280**, C288 (2000).
23. G. Helmlinger, F. Yuan, M. Dellian *et al.*, Interstitial pH and pO2 gradients in solid tumors *in vivo*: High-resolution measurements reveal a lack of correlation, *Nat. Med.*, **3**, 177 (1997).
24. M.R. Urist, Bone: Formation by autoinduction, *Science*, **150**, 893 (1965).
25. R.R. Chen and D.J. Mooney, Polymeric growth factor delivery strategies for tissue engineering, *Pharm. Res.*, **20**, 1103 (2003).

26. T.E. Schneider, C. Barland, A.M. Alex *et al.*, Measuring stem cell frequency in epidermis: A quantitative *in vivo* functional assay for long-term repopulating cells, *Proc. Nat. Acad. Sci. USA*, **100**, 11412 (2003).
27. N. Kotobuki, M. Hirose, Y. Takakura *et al.*, Cultured autologous human cells for hard tissue regeneration: Preparation and characterization of mesenchymal stem cells from bone marrow, *Artif. Organs*, **28**, 33 (2004).
28. J. Cohnheim, Ueber Entzueundung und Eiterung, *Arch. Path. Anat. Physiol. Klin. Med.*, **40**, 1 (1867).
29. A.J. Friedenstein, U. Gorskaja, and N.N. Kulagina, Fibroblast precursors in normal and irradiated mouse hematopoietic organs, *Exp. Hematol.*, **4**, 267 (1976).
30. I. Sekiya, B.L. Larson, J.R. Smith *et al.*, Expansion of human adult stem cells from bone marrow stroma: conditions that maximize the yields of early progenitors and evaluate their quality, *Stem Cells*, **20**, 530 (2002).
31. R. Schofield, The relationship between the spleen colony-forming cell and the haemopoietic stem cell, *Blood Cells*, **4**, 7 (1978).
32. S.P. Hoerstrup, A. Kadner, C. Breymann, Living, autologous pulmonary artery conduits tissue engineered from human umbilical cord cells, *Ann. Thorac. Surg.*, **74**, 46 (2002).
33. A. Kadner, G. Zund, C. Maures *et al.*, Human umbilical cord cells for cardiovascular tissue engineering: A comparative study, *Eur. J. Cardiothorac. Surg.*, **25**, 635 (2004).
34. Editorial, Focus on cell therapies. Proceed with caution, *Nat. Biotechnol.*, **23**, 763 (2005).
35. C. Murry, M.H. Soonpaa, H. Reinecke *et al.*, Haematopoietic stem cells do not transdifferentiate into cardiac myocytes in myocardial infarcts, *Nature*, **428**, 664 (2004).
36. L.B. Balsam, A.J. Wagers, J.L. Christensen *et al.*, Haematopoietic stem cells adopt mature haematopoietic fates in ischaemic myocardium, *Nature*, **428**, 668 (2004).
37. D. Orlic, J. Kajstura, S. Chimenti *et al.*, Bone marrow cells regenerate infracted myocardium, *Nature*, **410**, 701 (2001).
38. K. Wollert, G.P. Meyer, J. Lotz *et al.*, Intracoronary autologous bone-marrow cell transfer after myocardial infarction: The BOOST randomised controlled clinical trial, *Lancet*, **364**, 141 (2004).
39. J.A. Thomson, J. Itskovitz-Eldor, S.S. Shapiro *et al.*, Embryonic stem cell lines derived from human blastocysts, *Science*, **282**, 1145 (1998).
40. W.S. Hwang, Y.J. Ryu, J.H. Park *et al.*, Evidence of a pluripotent human embryonic stem cell line derived from a cloned blastocyst, *Science*, **303**, 1669 (2004).
41. W.S. Hwang, S.I. Roh, B.C. Lee *et al.*, Patient-specific embryonic stem cells derived from human SCNT blastocysts, *Science*, **308**, 1777 (2005).

Note added in proof

References 40 and 41 have been retracted as the results in them are deemed to be invalid. This retraction by Donald Kennedy was originally published in *Science Express* on 12 January 2006 and in *Science* 20 January 2006: Vol. 311, no. 5759, p. 335; DOI: 0.1126/Science. 1124926

Section Eight

Ethics in biomedical engineering

Chapter 8.1

8.1

Bioethics: a creative approach

Daniel Vallero

The Brain—is wider than the Sky—
For—put them side by side—
The one the other will contain
With ease—and You—beside—
The Brain is deeper than the sea—
For—hold them—Blue to Blue—
The one the other will absorb—
As Sponges—Buckets—do—
The Brain is just the weight of God—
For—Heft them—Pound for Pound—
And they will differ—if they do—
As Syllable from Sound—

Emily Dickinson (1830–1886)[1]

Engineers do not have to be reminded of the importance of imagination. As Dickinson poetically and eloquently elucidates, the mind is almost limitless in its ability to reason. Given enough reliable information, humans are capable of amazing feats. We can see beyond the constraints of "what is" to "what can be." Imagination is arguably the greatest asset of the engineer. Engineers design because what they are able to see does not yet exist. That is the purpose of design. Likewise, imagination is a useful device to help us to understand the nature of bioethical challenges to practicing and future engineers, as well as an important means of designing ways to avert bioethical problems before they arise. So, what is it we are up against in the decades ahead?

Here we consider bioethics questions relevant to the practice of engineering. Few, if any, of these will be answered to the complete satisfaction of every engineer. However, the very process of inquiry may help to give context and structure to some otherwise amorphous issues. So, let us begin with an essential query:

Bioethics Question: When is an action morally permissible and what kinds of behavior are morally obligatory?

Let us begin on a balmy Saturday morning in the North Carolina Piedmont, where a hundred or so new graduate students were gathered in the auditorium of Duke University's Chemistry Building. It was orientation week in late August, when Duke requires that every Ph.D.-seeking graduate student participate in a full day of ethics training.

The Responsible Conduct of Research training program developed by the Graduate School had had some inauspicious beginnings. Some years earlier, Duke, like other prominent research universities, risked losing funding from the National Institutes of Health because their research programs could not demonstrate that their researchers were receiving adequate training in ethics. So, as is often the case, Duke did not simply respond to the criticism and threat by instituting a *pro forma* program to keep the auditors happy, but decided to use this as a "teachable moment" and galvanizing event. Among the creative approaches to be adopted was to engage the students as soon as they arrive in Durham. Part of this commitment was to invite provocative speakers to evoke responses and to stir the imagination of the students. It was hoped that this first gathering would begin each student's self-directive, proactive, lifelong ethos of responsible research. On that Saturday, the students were asked to participate in two thought experiments.

In this vein, rather than to try the familiar ethical analytical approach of tediously (and often boringly) delineating, point by point, each possible bioethical aspect of engineering, we apply a tool used by philosophers to approach bioethics intuitively, using our imagination.

Biomedical Engineering Desk Reference; ISBN: 9780123746467

Thought experiments

After the students had conducted a couple of projects and were given some advice on how to avoid ethical dilemmas, Duke University philosophy professor, Alex Rosenberg,[2] presented the following thought experiments:[3]

Thought experiment 1

You are riding in a trolley car in San Francisco. You are the only passenger. Being the inquisitive type that you are, you watch the trolley driver's methods of controlling speed, changing tracks, and other aspects of handling the machinery. You ascend a hill, come to the top, and begin your descent. To your surprise and horror, during the downhill acceleration, the trolley driver jumps from the car, leaving you alone.

You move to the driver's seat and quickly find out why the driver jettisoned. The brakes have completely failed. The driver must have had a strong understanding of the laws of motion, since he jumped just before the car accelerated to the velocity at which anyone jumping from the car would suffer mortal injuries. In other words, you cannot leave the car. The horn and warning devices have also failed and do not operate. The only remaining control available to you is the lever that changes tracks.

The car continues to accelerate until you see on the track in front of you four workers (all the same size, sex, age, and demographics) who are standing on the tracks with their backs to you. There is no way that they can hear you due to the noisy equipment around them and the fact that your horn does not work. You also happen to notice that there is a switch track in front of you just before the workers that allows you to change tracks. However, a single worker (of the same size, sex, age, and demographics as the other four workers) is standing on the track in front of you to which you can switch.

What do you do?

Your ethical decision consists of only two choices. As an ethicist might ask: What is morally permissible? What is morally obligatory?

The group of students unanimously declared that switching the tracks was morally permissible since you would kill one person to save four. Most also agreed that you were morally obliged (not just permitted) to take the step.

Rosenberg then offered a second thought experiment:

Thought experiment 2

You are a respected biomedical engineering major who has gone on to become a world-renowned surgeon at St. Bob, the Scientist Hospital, in San Francisco. You have kept up your skills in engineering and blended them with your surgical expertise to design a transplant device, based on nanotechnologies, that allows you to transplant any organ from one person to another in ten minutes with absolutely no rejection. (Hey, it's a thought experiment, so we can assume such things!) The transplant device is up and running at St. Bob's.

You are also a charitable physician. In fact, the reason you chose St. Bob's is that it has two distinct wings. Wing 1 is a world-class hospital that treats the rich and famous. Wing 2 is a public clinic that allows anyone to walk in and receive treatment. You practice in both wings, including dedicating *pro bono* services in Wing 2 every Tuesday from 9:00 p.m. to 1:00 a.m. Wednesday. Many of the patients you see in Wing 2 are completely destitute and have no family or friends. They are loners. Recently, a Wing 2 patient from a local homeless shelter was brought in to see you. You find in your interview with him that he has no family, no job, and no interest in bettering himself. You gave him a checkup, including a checkup of his lungs, liver, and pancreas, and found him to be in excellent physical health.

Earlier in the evening, in Wing 1, you saw a number of patients. Four of these patients were in dire need of transplants. One needed a pancreas, one a liver, and two needed a lung. In fact, if they do not receive them by midnight, all four will die. Your transplant machine could easily save them if the organs were available from a healthy donor, such as your Wing 2 patient.

What should you do? What is morally permissible in this case?

The Ph.D. students were flummoxed. Almost all of them said that it is morally impermissible to take the organs from the one person (and in the process kill him) to save the four. But, most could not say why or put their finger on the difference between the two scenarios.

Both situations involve an interface of technology with ethics. A crude technology in the first case (the level and switch track) and an advanced technology in the second case (the nanotechnology-based transplant machine) are part of the ethical decision. In both, we can change the situation, but the bioethical question is "Should we?"

Bioethics Question: What must a professional do to be trusted?

On the face of it, these cases are identical questions about what is moral; that is, are you permitted to kill one to save four? So, what is the big difference? If we are talking about a decision concerning anything but life, the question is almost silly. It would be merely arithmetic. But, when it comes to bioethics, such questions are often complicated and debatable. But, the difference basically comes down to trust.

Teachable moment: trust

Consider the people whom you can trust. Write down at least three reasons that they can be trusted. These can be intrinsic qualities, such as, to your knowledge they have never lied to you. They may also be extrinsic constraints, such as they are duty bound by law or a professional standard (e.g., a medical doctor). Often, they are combinations of the two (e.g., a trusted family doctor).

Explain how such trust applies to the practice of engineering. Discuss the types and degrees of controls that the engineering profession has over its members (extrinsic controls), and how much the social contract with engineering depends on characteristics of the individual engineer (intrinsic controls).

The principle of double effect

It is tempting to jump in and start analyzing the student's responses for right and wrong elements. That is the typical case-based approach. That is to say, philosophers and ethicists commonly apply "casuistry," an approach to determine right and wrong decisions or behavior by analyzing cases that illustrate general ethical rules. Among the problems with this approach is that it forces us to judge. Engineers, on the other hand, are problem solvers and designers, not judges. We are more comfortable about building a solution than about sitting in judgment (see Teachable Moment: The Engineer as Agent *versus* Judge). However, to prime the pump, let us begin to think about why the students responded as they did.

Teachable moment: the engineer as agent *versus* judge

Philosopher Caroline Whitbeck is widely recognized as one of the first and strongest advocates of a design-based approach to professional ethics. She accentuated the synthetic as well as analytic elements in responses to moral problems.

In the 1980s and 1990s, she developed the parallel between ethical problems and design problems. She broke new ground in active learning methods in the teaching of engineering ethics and the responsible conduct of research, especially methods that place the learner in the position of the agent who must actually respond to the problem (rather than in the position of a judge who merely evaluates responses that have already been constructed). Among the ethical design criteria are collaboration, trust and trustworthiness, responsibility, and diligence (opposite of negligence).

Whitbeck draws four points of analogy between ethical problems and design problems:[4]

1. For interesting or substantive ethical problems, such as substantive problems of research design, there is rarely, if ever, a uniquely correct solution or response.
2. Some possible responses are unacceptable – there are wrong answers even if there is no unique right answer – and some are better than others.
3. However, solutions may have advantages of different sorts, such that where there are two candidate solutions, neither may be clearly better than the other.
4. A proposed solution must do all the following (in addition to being reasonably secure against accidents and miscarriages):
   - Achieve the desired performance or end – In the case of an ethical problem this might be to fulfill some moral responsibility, a professional responsibility, or a family responsibility.
   - Conform to given specifications or desired criteria – For an ethical problem, these specifications might include meeting the standards of care for one's profession, and not taking so much time that one fails in other particular commitments.
   - Be consistent with (usually unstated) background constraints, for example, that one not violate anyone's human rights and that one minimize the infringement of other rights.

Much like engineering design problems, ethical dilemmas are often ill-posed. Case-based approaches often erroneously assume that an engineer is confronted with a well-posed ethical problem; that is, one that is uniquely solvable (i.e., a unique solution exists) and one that is dependent upon a continuous application of data. That is to say, cases are best when a solution to the problem exists, the solution is unique, and the solution depends continuously on the data. So, if ethical cases were solved by something akin to the Laplace's equation,[5] casuistry is the way to go. By contrast, an ill-posed problem does not have a unique solution and can only be solved by discontinuous applications of data, meaning that even very small errors or perturbations can lead to large deviations in possible solutions.[6] Engineers are well aware of the complexity and challenge, as well as of the need to solve ill-posed problems. After all, most emergent

Continued

problems are ill-posed. For example, an ill-posed problem may be solved by what are known as "inverse methods," such as restricting the class of admissible solutions using *a priori* knowledge. *A priori* methods include variational regularization using a quadratic stabilizer. Usually, this requires stochastic approaches, i.e., assumptions that the processes and systems will behave in a random fashion.

By extension, small changes for good or bad can produce unexpectedly large effects in an ethical decision. A seemingly small mistake, mishap, or ethical breech can lead to some dramatic, even devastating, results. Engineers are constantly warned to pay attention to the specific details of a design. The same admonition holds for ethical issues. In fact, the famous engineer, Norman Augustine has told us that "engineers who make bad decisions often don't realize they are confronting ethical issues."[7]

The design problem model of ethical problems represents problems as characteristically possessing more than one good (i.e., wise and responsible) solution. This allows the engineer to avoid the trap of situational ethics, where perpetrators are allowed to excuse their poor ethical choices since there are no right or wrong answers. This contradicts common sense and common morality. People know that there are indeed wrong answers. We also know that in most decisions, certain answers are more ethical than others, with some clearly wrong (the worst is known as the negative paradigm). Whitbeck's approach also obviates the over-simplistic and often flawed attempt to make an ethical decision into a multiple-choice question, selecting the best among two or more choices (this is often done in professional surveys, asking that as a practicing engineer "you be the judge"). This attempt at standardizing and modularizing professional ethics flies in the face of one of engineering's most important assets, creativity. The design problem model clarifies the character of what Whitbeck calls the agent's "synthetic" or constructive task of devising and improving responses, a much-preferred approach to the "analytic" approach of the judge.

An overarching advantage of the design approach is that it places the engineer squarely within the situation. From this vantage point, we can build solutions, rather than select them. Whitbeck has argued that moral problems are frequently mischaracterized as dilemmas; the roots of this misrepresentation and its relation to the failure of applied ethics are used to illuminate how to construct responses to moral problems. She argues that in many cases, there is a tendency to examine problems retrospectively, based on the medical case method approach, as opposed to placing the engineer in his or her most comfortable position of designer. The academic approach of the philosophers seldom matches what professionals and students need.

The medical case model tends to address acts, such as whether to participate in a specific medical action (e.g., use of genetic testing, withholding treatment, performing an abortion, or removing a feeding tube). Within enumerated constraints, society has granted the physician much latitude in treating a patient. The moral decision-making process in a medical situation depends on the general circumstances that may justify or fail to corroborate such acts. In other words, the attending physician in a hospital has inherited the patient's entire history and must make specific decisions about the act of patient care at that precise moment. Conversely, engineers are more adept at looking at problems from a life cycle perspective. Indeed, like the physician, part of our fact finding should include similar instances where a design has succeeded or failed, for instance, by using event and fault tree scenarios to understand the situation. However, the engineer goes well beyond the "act" to consider how the system can be better designed not only for damage control, but to prevent similar situations in the future. Whereas an attending physician may be limited to a decision of whether to remove a feeding tube for a specific patient, the design approach of the engineer is to look at the entire situation that led to the potentially tragic consequences of this decision. What could have been done to prevent the situation from occurring, such as concrete steps toward better technologies for health care and monitoring, and even better-designed roads and vehicles so that the accident that led to the coma and brain damage could have been wholly prevented? The life cycle design perspective calls for attention to all of the functions leading to an outcome, not simply the goodness or shortcomings of products.

One of the best examples of the life cycle viewpoint has been articulated by a nonengineer. Theologian Ronald Rolheiser[8] shares a parable of a community that dutifully, carefully, and honorably pulls a continuous stream of dead bodies from the town's river. Each day the moral people of the community give a proper burial to each deceased person represented by the body, as dictated by some type of social contract. In fact, these efforts led to a well-organized system to deal with the bodies, even providing jobs sorely needed by their citizenry. However, the community never makes the effort to travel upstream to find the source of the dead bodies! The case-based approach would look at the act of each person. Many of them are behaving quite morally by providing the burials and paying their respect to the dead. The agent-based design approach, however, requires a trip upstream. What is it about our systems in contemporary society that brings us to these bioethical dilemmas? Why does the "bottom-up" approach, with each person seemingly behaving morally, not lead to an overall ethical system? Stepping back and taking the comprehensive viewpoint clearly shows an overall societal ethical transgression. So long as we have a piecemeal, myopic view, no progress can be made in eliminating the core problem.

An example of this myopia is the recent argument for stem cell research because reproductive clinics are "just going to throw the embryos away anyway, so why not get something good out of them by using them in stem cell research?" The life cycle view would require that the whole process, including the ethics of the treatment of embryos and other aspects of reproductive technologies, be part of the whole argument. In this case the question is not limited to "What is my duty?" (known as the deontological view) or "What is the best result?" (known as the consequentialist view), but is really "What is going on here?" (i.e., a rational-relationship view, which considers the entirety of the issue, including duties, consequences, and the means toward these consequences).

Amy the engineer

The design approach advantage is not limited to the big issues, but may be even more directly beneficial to the individual decisions (i.e., microethical decisions) of the practicing engineer.

As mentioned, according to Whitbeck, ethical questions are all too often incorrectly presented as "dilemmas." A dilemma is defined as a forced choice between two alternatives that are exactly and equally unfavorable. This leads to the representation of moral problems as though they were forced choices between two (or more) equally unwelcome alternatives. Again, this approach tries to make the solutions less messy by "pretending" that they are well-posed. We do this frequently in engineering, such as assuming "spherical chickens" when designing a poultry processing plant. Such a misrepresentation of moral problems as dilemmas implies that the only possible responses are the proposed courses of action (all of which are objectionable), stifling creative attempts to offer better alternatives. In a way, this is an over-prescribed design. Engineers perform best when they are not overly constrained by the client.

Whitbeck[9] shares the so-called Heinz dilemma recounted by Carol Gilligan in her book *In a Different Voice.*[10] Lawrence Kohlberg, a founder of moral development theory, posited to children a dilemma of whether a man named Heinz should steal a drug he cannot afford to save the life of his wife. Gilligan describes the performance of Jake, a child who does well by Kohlberg's criteria:

> Jake, at eleven, is clear from the outset that Heinz should steal the drug. Constructing the dilemma, as Kohlberg did, as a conflict between the values of property and life, Jake discerns the logical priority of life and uses that logic to justify his choice: For one thing, a human life is worth more than money, and if the druggist only makes $1000, he is still going to live, but if Heinz doesn't steal the drug, his wife is going to die. (Why is life worth more than money?) Because people are all different and so you couldn't get Heinz's wife again. Jake understands the game to be one of finding the covering "principles," ordering them, and cranking out a solution, and this is the abstract exercise that he has successfully performed. What is notable is that Jake has not only learned the game but also recognizes it as an abstract puzzle; he aptly describes it as "sort of like a math problem with human beings." When ethical problems are constructed as abstract math problems with human beings it is no wonder that they have nothing much to do with moral life.

Another young respondent, Amy, refused to abide by the arbitrary conditions and constraints. As a result, she did not perform well by Kohlberg's criteria because she insisted on trying to work out a better response to Heinz's problem. When Amy was asked whether Heinz should steal the drug, she proposed new alternatives:

> Well, I don't think so. I think there might be other ways besides stealing it, like if he could borrow the money or make a loan or something, but he really shouldn't steal the drug but his wife shouldn't die either.

Asked why he should not steal the drug Amy replies:

> If he stole the drug, he might save his wife then, but if he did, he might have to go to jail, and then his wife might get sicker again, and he couldn't get more of the drug, and it might not be good. So, they should really just talk it out and find some other way to make the money.

Gilligan interprets Amy's solution as "a narrative of relationships that extends over time." Like the essay question's advantage over the multiple-choice test, the narrative can express "the dynamic character of a situation with its unfolding possibilities and resolutions." This avoids the need for abstraction.

Amy's answers follow an engineering paradigm for problem solving. She looks for numerous possibilities and alternative solutions. She refuses to be pigeonholed into predetermined solutions. She is put off by the arbitrariness of the design landscape, and seeks accommodation of her client's needs. Amy deserves more credit than Gilligan offers. Her response is not simply addressing relationships, as Gilligan asserts. In fact, as Whitbeck points out, Amy did not even identify the failure of Heinz and his spouse's relationship as the result of Heinz's jail sentence. She was more concerned that Heinz's wife may need the drug again. Amy was still seeking a solution to the root problem (need for the drug) and providing a sustainable solution. As an "engineer," Amy is more concerned that the process supports acceptable outcomes on an ongoing basis than merely solving a short-term problem.

The principal lesson seems to be that cases can be arbitrary and can limit engineering (and ethics) creativity, by imposing a forced choice between two (or a few) right and wrong answers. Amy, as most good engineers do, begins with an open-ended set of possibilities and brainstorms to arrive at an acceptable means of solving the problem. Next, she looks at the feasible options and explores one or more reasonable solutions, including attempts to persuade the pharmacist to help, obtaining a loan, and other practical actions.

Continued

Amy also reminds us that engineers need to keep thinking. We must analyze each possible outcome from a number of perspectives, including those that are not so easy for technical types to grasp, such as the sociological and psychological risks and benefits. From there, the engineer can follow a critical path to an acceptable solution.

Questions

1. Consider a case of bioethical importance. The case can be a negative or a positive paradigm.
2. What are the strengths and weaknesses of this case (as per Whitbeck)?
3. How might you be more of an ethical agent than a judge in answering these questions?
4. Identity a current bioethical issue that is being approached as an abstract math problem with human beings. What are the shortcomings of this approach?

The first thought experiment introduces us to an important consideration in bioethics, the so-called principle of double effect. Think of this as an equation with two constraints. The first constraint stems from Socrates' moral challenge to "first do no harm." The second constraint was fully first articulated by the thirteenth-century Italian theologian, Saint Thomas Aquinas:

> *Good is to be done and promoted and evil is to be avoided.*[11]

In its most simple form, the doctrine says that an act that leads to negative side effects is permitted, but deliberate harm (even for good causes) is wrong. This seems similar in both scenarios, but the second thought experiment involves actively harming one person, whereas the first thought experiment involves harm as a side effect. In the trolley case, you are just redirecting the harm. However, in the second case you have to do something to the homeless man to save the four. In the first case, no person has more rights than anyone else not to be crushed, but in the second case, the homeless man has a right not to be killed, even though his organs would be put to good use.

These distinctions among professional decisions point to the fact that engineers will face risk trade-offs and double effects to some degree during their careers. For example, in designing a device, some persons (e.g., immunocompromised) may be harmed by its use. If all of a certain group are harmed at the expense of another group's benefit, this could be conceived as intentional harm. The way to mitigate this harm is to disclose fully the shortcomings of the device and to work on improvements that will decrease the likelihood of harm. The device must do inherent good (e.g., provide insulin to diabetics or deliver drugs to ill patients). In other words, the designer must not actually intend to accomplish the bad effect (harming immunocompromised people). The ill effect is simply unavoidable in the effort to do good. If another approach would avoid the bad and still accomplish the good, then such an option is the preferred and obligatory act.

Another provision of the double effect is that the good effect must be at least as directly an effect of the action as is the negative (bad) effect. In particular, the bad must not cause the good effect. Also, the benefit of the bad must not outweigh the benefit of the good. Finally, another ethically acceptable approach without the side effects must not be available.

A bioethical example of the double effect is that of the vaccine. A government and any manufacturer normally know the population risk of administering a vaccine to the public. Most are expected to benefit, with a small

Teachable moment: Who was Van Rensselaer Potter?

The word "bioethics" was coined by Van Rensselaer Potter (1911–2001), an American biochemist. He wrote two important books. The first, *Bioethics: A Bridge to the Future* (Prentice-Hall), wherein he coined the term, was written in 1970 as a call to integrate many scientific and engineering disciplines to provide for an environment that was both livable and sustainable (that term as applied to the environment did not yet exist either). Bioethics was subsequently coopted by the medical community and was redefined specifically to address the morality of medical practice (including appropriateness of treatments and technological advances). Meanwhile, Potter grew more and more respectful of the "land ethic" espoused by Aldo Leopold, and attempted to recapture his original meaning and coined another phrase, "global bioethics," in this 1988 book, *Global Bioethics: Building on the Leopold Legacy* (Michigan State University Press). For our purposes, the term also harkened a new role for biosystem engineering.

Questions

1. Consider the prefix "bio" and how it applies to both terms: bioethics and global bioethics.
2. Contrast these terms from Potter and Leopold's perspectives with those of the contemporary definition of biomedical ethics.
3. Why has the term bioethics morphed in meaning so much since Potter coined it?

number with adverse side effects. And, from this small group, a subset of vaccine recipients will die. The lives are saved as a result of the vaccine, not as a result of the deaths of those who die of side effects. The side effects do not advance any goals of the drug manufacturer. Thus, the side effects are not intended as a means to any other outcome. Finally, the proportion of lives saved is very high compared to the lives lost, satisfying the requirement that benefits outweigh the negative outcomes. It would not be permissible to produce and administer the vaccine with these side effects (deaths) if another means of preventing the disease were available.

Credat emptor

The thought experiments demonstrate, at least at some intuitive level, that trust is one of the distinguishing attributes of the professional. Engineering is no exception. Arguably, the engineer's principal client is the public. And, the clients need know little about the practice of engineering because, owing to the expertise, they have delegated the authority to the engineer. Just as society allows a patient to undergo brain surgery even if that person has no understanding of the fundamentals of brain surgery, so also does our society cede authority to engineers for design decisions. With that authority comes a commensurate amount of responsibility, and when things go wrong, culpability. The first canon of most engineering ethical codes requires that we "hold paramount" the health and welfare of the public. The public is an aggregate, not an "average." So, leaving out any segment violates this credo. Thus, even though most people favor an approach, it is still up to the engineer to ensure that any recommended approach is scientifically sound. In other words, no design, even a very popular one, is to be recommended unless it meets acceptable, scientifically sound standards. Conversely, the most scientifically sound approach may have unacceptable social side effects. The right and just decision is not a "popularity contest." Once again, the engineer is put in the position of balancing trade-offs.[12]

Bioethics Question: What is the role of values in bioethical decisions?

Recently, much debate has occurred within the legal community as to whether the use of lethal injection to execute convicted criminals is "cruel and unusual punishment." The debate includes positions across a wide spectrum, ranging from lethal injection being morally reprehensible to its being morally obligatory. The US Constitution prohibits cruel and unusual punishment, so it is not surprising that the debate reached the United States Supreme Court. In the spring of 2006, the Court refused to examine this issue, but a large controversy is brewing. A central bioethical issue important to engineering is whether the two-step system of injection may cause extreme pain if the first chemical, an anesthetic, is not properly injected. Thus, when the actual lethal dose of the second chemical is injected, the resulting pain puts the state in the position of a cruel executioner.

Perhaps another thought experiment will help us to begin to consider the debate rationally:

Thought experiment

Dante is a chemical engineer employed by InfernoChem, Inc. (ICI). Dante's expertise is in small-scale reactors. Recently, he developed a manifold system that allows two different chemicals to be injected at specific preset times. His device is particularly useful in injecting catalysts into reactors. For example, a polymerization step in one of the reactors requires that the first catalyst be injected when the reactor temperature is 50°C and the second catalyst injected 10 minutes later when the reactor temperature reaches 80°C. For the past 18 months, ICI has required that Dante continuously miniaturize the two-step injection system. In fact, the company gave Dante two monetary awards for his progress. Recently, the system was shrunk to 25 cubic centimeters of fluid volume.

Dante's perspective is that of a devout Roman Catholic. He opposes capital punishment for religious and personal reasons. He briefly mentioned these convictions to Tom, his immediate supervisor, on two occasions after executions were covered in the local news. Dante discussed this in a very matter-of-fact and dispassionate way.

Tom is aware that the reason for the short deadline to miniaturize Dante's injection system is that ICI has a contract with the State Bureau of Prisons to produce a single injection system that would ensure that prisoners receiving lethal injections get the entire dose of the anesthetic drug before the toxic dose is administered. However, based on Dante's opposition to the death penalty, Tom has chosen not to disclose the actual application of the new technology.

Was Tom morally permitted to withhold the company's plans from Dante? What is the appropriate thing for Dante to do?

One of the means of determining whether an ethical decision is sound is to see if we are being morally consistent in our values. For example, some have pointed out inconsistencies in those who support the death penalty, yet ideologically oppose abortion, embryonic stem cell research, euthanasia, and assisted suicide, as well as other acts against humans at the beginning and end of life. The counterargument is that in these other situations, innocents are being killed, while in the case of capital punishment, if due process is followed, the person being executed is being treated fairly. Again, it comes down to what is most cherished, what has the most value. If

Teachable moment: capital punishment, abortion, and the definition of human life

The American Medical Association (AMA) policy on abortion is:

> The Principles of Medical Ethics of the AMA do not prohibit a physician from performing an abortion in accordance with good medical practice and under circumstances that do not violate the law. (III, IV)[13]

And

> (1) [A]bortion is a medical procedure and should be performed only by a duly licensed physician and surgeon in conformance with standards of good medical practice and the Medical Practice Act of his state; and (2) no physician or other professional personnel shall be required to perform an act violative of good medical judgment. Neither physician, hospital, nor hospital personnel shall be required to perform any act violative of personally held moral principles. In these circumstances, good medical practice requires only that the physician or other professional withdraw from the case, so long as the withdrawal is consistent with good medical practice. (Sub. Res. 43, A-73; Reaffirmed: I-86; Reaffirmed: Sunset Report, I-96; Reaffirmed by Sub. Res. 208, I-96; Reaffirmed by BOT Rep. 26, A-97; Reaffirmed: CMS Rep. 1,I-00)[14]

Compare and contrast this policy that the AMA policy on capital punishment:

> An individual's opinion on capital punishment is the personal moral decision of the individual. A physician, as a member of a profession dedicated to preserving life when there is hope of doing so, should not be a participant in a legally authorized execution. Physician participation in execution is defined generally as actions which would fall into one or more of the following categories: (1) an action which would directly cause the death of the condemned; (2) an action which would assist, supervise, or contribute to the ability of another individual to directly cause the death of the condemned; (3) an action which could automatically cause an execution to be carried out on a condemned prisoner.

(See the rest at http://www.ama-assn.org/apps/pf_new/pf_online?f_n=browse&doc=policyfiles/-HnE/E-2.06.HTM&&s_t=&st_p=&;nth= 1 &prev_pol=policyfiles/HnE/E-1.02.HTM&nxt_ pol= policyfiles/HnE/E-2.01.HTM&.) Compare this policy to an engineering code of ethics. Identify commonalities and any inconsistencies between the two policies, especially the differences between the engineer's paramount focus on the public good and the physician's focus on the patient.

In particular, consider that for lethal injection executions, the AMA prohibits a physician aiding the execution by:

- selecting injection sites;
- starting intravenous lines as a port for a lethal injection device;
- prescribing, preparing, administering, or supervising injection drugs or their doses or types; inspecting, testing, or maintaining lethal injection devices; and
- consulting with or supervising lethal injection personnel.

Some have argued that, in an abortion, the doctor does not "kill" the unborn child intentionally, but makes the "choice" to evict the child from the womb. This position argues that the unborn child's right to life does not "entail that the child *in utero* is morally entitled to the use of the mother's body for life support."[15] Tragically, this position is tantamount to subjugating the unborn child to a "parasite."

Questions

1. Compare the technical and moral differences and similarities between the prohibitions against participating in an execution and the applications of the technologies of abortion.
2. Is the child *in utero* living? Is it human? Be specific in these definitions. If so, explain the bioethical distinction between calling abortion an act of killing and a choice to "evict" the child.

human life is sacred, as most believe (however they define it), can it ever become unsacred, as in the case of a convicted murderer?

The bottom line is drawn with values. What we perceive to be precious is protected and cherished. What we see as having less value becomes commodified. Less value translates to less protection and greater expendability.

The good engineer

> *Engineering is an active process of solving problems and building new things using computers and other technologies.*
>
> Duke University's K-PhD Program[16]

Bioethics Question: What does it mean to be a good engineer?

From the Ancient Greeks, excellence is a dichotomous phenomenon. Good was considered to be given and almost undefinable (although Socrates can be credited with starting the process of defining it). Excellence requires skill and character. Our academic and professional preparation is designed to give us the former and some of the latter. For example, the engineering curriculum continues to place high demands on the student's grasp of mathematical and scientific concepts. This has never let up. However, the curriculum of today increasingly requires an understanding of the social sciences and humanities and demands core competencies in interpersonal skills.

Engineering is active. It is a process that takes the raw materials and energy that exist at a given time and creates new things and improves things that already exist. Engineers solve problems and build. Engineers do things.[17] We are seldom satisfied merely in possessing information or mastering skills.

The real test of engineering is when we put our knowledge and aptitude into practice. I noticed this in the Duke engineering students who recently returned from Indonesia and Uganda after participating in the Engineers without Borders projects abroad. They were joyous about their opportunities to apply the theoretical information in real projects. It is this eagerness to apply what we know is a gift that has truly made the world a better place. Many of the improvements in public health, safety, and quality of life are largely attributed to engineers.

In North America and Europe, public works projects designed by engineers have given us clean water,[18] which has prevented many of the diseases responsible for the majority of premature deaths and disabilities so common 100 years ago. Pollution control equipment has allowed for cleaner air. Vehicle and transportation designs continue to improve the safety of travel. Chemical engineering advances have improved product manufacturing, leading to higher quality and safer consumer products and pharmaceuticals. And, biomedical devices and systems have lengthened and extended an improved quality of life to millions.

The engineering call will be even stronger in the future. The knowledge and creativity of the engineer will have to grow in proportion to the increased societal expectations. Kristina Johnson, Dean of Duke University's Pratt School of Engineering, characterized the engineer's future obligation to society:

> *[A]s an engineering dean, I'd argue that it is our responsibility as good citizens of the planet to solve many of our global problems, such as developing renewable energy sources, purifying water, sustaining the environment, providing low-cost health care and vaccines for infectious diseases, to mention a few. Coupled with global climate issues, transportation and urbanization, we need all the technical horsepower we can educate.*[19]

As in all aspects of engineering, however, there are challenges and obstacles. We are familiar with design challenges, such as the unique conditions of situations that make us think outside of the box. An equation or model seems to work well in most cases, but those few instances that fail can be the difference between good engineering and poor design. Or, we commonly experience the challenge of the lessons to be learned when we move from a prototype to an actual application. Slight changes in scale or complexity (i.e., effects of some unknown variable) limit the application of a design. One specific challenge, the subject of this book, is the ethical challenge, specifically the challenge of how engineering decisions and actions affect human life, for good or ill.

Feedback and enhancement of design

> *Engineering, like poetry, is an attempt to approach perfection. And engineers, like poets, are seldom completely satisfied with their creations.*
>
> Henry Petroski (1985)[20]

Engineering is not only active, but, as Petroski reminds us, it is a system filled with feedbacks. We are frequently told where we fall short, but we are sufficiently optimistic to recognize our progress. This continuous improvement was formalized by another engineer, W. Edwards Deming, who established the total quality management (TQM) process. TQM is defined as "a management approach of an organization, centered on quality, based on the participation of all its members and aiming at long-term success through customer satisfaction, and benefits to all members of the organization and to society."[21] This is a strong mandate. However, it should not imply that engineering advances are linear. In fact, growth is incremental and is often accompanied by setback, hypothetically shown in Figure 8.1-1.

Especially in biomedical and biosystematic applications, engineers push the envelope. They must. The growth in medical science is directly attributable to new technology, pharmacology, and systems. The unforeseen setbacks depicted in Figure 8.1-1 go with the territory, but the public does not take kindly to those that could have been prevented if reasonable steps had been taken. It hurts the professional and the profession because a preventable failure delays the overall advance in engineering solutions to problems, as shown in Figure 8.1-2. It is even worse if the setbacks are the results of mishaps and mistakes. And, since engineering requires intensive and extensive training and experience, most situations

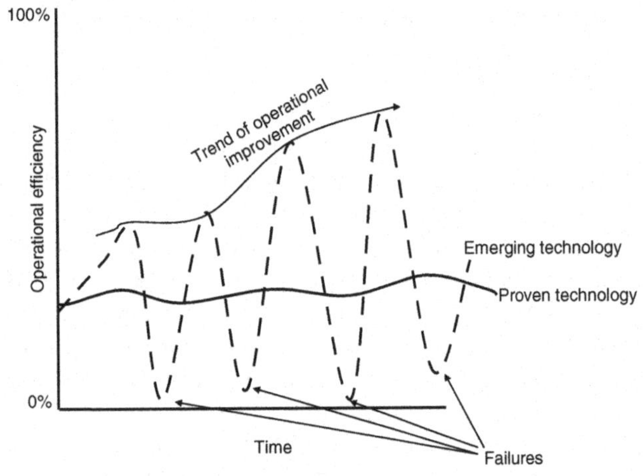

Figure 8.1-1 Hypothetical failure rate of new *versus* proven technologies.

ripe for mistakes and mishaps are only detectable to fellow engineers. This was well put by Petroski[22]:

> *Engineers... are not superhuman. They make mistakes in their assumptions, in their calculations, in their conclusions. That they make mistakes is forgivable; that they catch them is imperative. Thus it is the essence of modern engineering not only to be able to check one's own work but also to have one's work checked and to be able to check the work of others.*

Teachable moment: the good engineer

The famous physicist Freeman Dyson (born 1923) said:

> A good scientist is a person with original ideas. A good engineer is a person who makes a design that works with as few original ideas as possible. There are no prima donnas in engineering.

He also said:

> If we had a reliable way to label our toys good and bad, it would be easy to regulate technology wisely. But we can rarely see far enough ahead to know which road leads to damnation. Whoever concerns himself with big technology, either to push it forward or to stop it, is gambling in human lives.

Questions

1. Give two examples in bioengineering that demonstrate Dyson's first point and two emerging technologies in which we presently need to heed the warning of his second quote.
2. What does this tell us about risk taking, especially the optimum between necessary and irresponsible risks?
3. Give an example of when avoiding a risk leads to bigger problems than taking the risk.

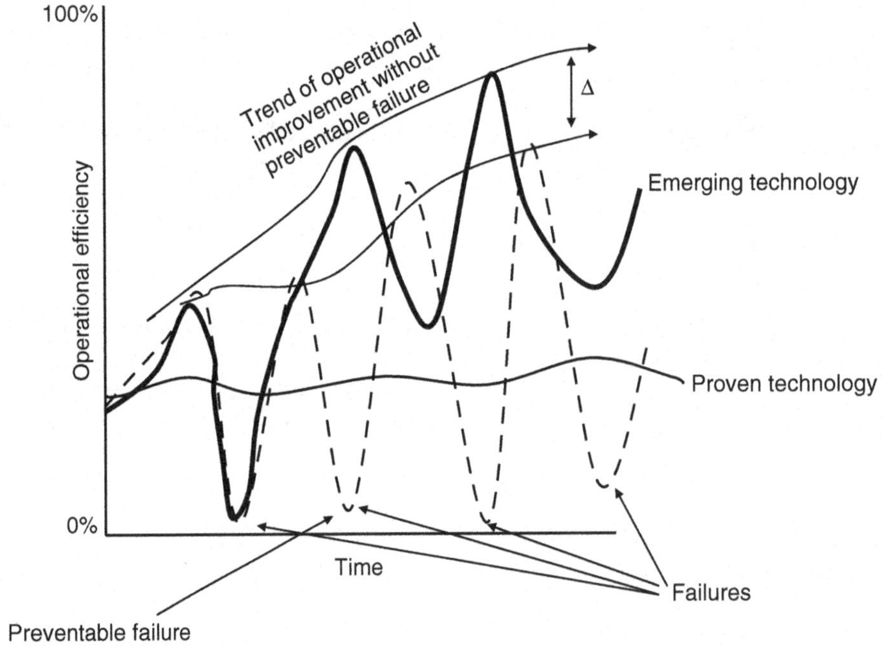

Figure 8.1-2 Hypothetical effect of preventable failure on overall advance of a new technology. Solid line is the technological advance had the preventable failure not occurred. The delta (Δ) is the difference in technology development attributed to the preventable mistake. Without the failure, the technological progress would be greater by delta.

The profession of engineering[23]

By Herbert Hoover

It is a great profession. There is the satisfaction of watching a figment of the imagination emerge through the aid of science to a plan on paper. Then it moves to realization in stone or metal or energy. Then it brings jobs and homes to men. Then it elevates the standards of living and adds to the comforts of life. That is the engineer's high privilege.

The great liability of the engineer compared to men of other professions is that his works are out in the open where all can see them. His acts, step by step, are in hard substance. He cannot bury his mistakes in the grave like the doctors. He cannot argue them into thin air or blame the judge like the lawyers. He cannot, like the architects, cover his failures with trees and vines. He cannot, like the politicians, screen his shortcomings by blaming his opponents and hope that the people will forget. The engineer simply cannot deny that he did it. If his works do not work, he is damned. That is the phantasmagoria that haunts his nights and dogs his days. He comes from the job at the end of the day resolved to calculate it again. He wakes in the night in a cold sweat and puts something on paper that looks silly in the morning. All day he shivers at the thought of the bugs which will inevitably appear to jolt his smooth consummation.

On the other hand, unlike the doctor his is not a life among the weak. Unlike the soldier, destruction is not his purpose. Unlike the lawyer, quarrels are not his daily bread. To the engineer falls the job of clothing the bare bones of science with life, comfort and hope.

No doubt as years go by people forget which engineer did it, even if they ever knew. Or some politician puts his name on it. Or they credit it to some promoter who used other people's money with which to finance it. But the engineer himself looks back at the unending stream of goodness that flows from his successes with satisfactions that few professions may know. And the verdict of his fellow professionals is all the accolade he wants.

In addition to the aesthetic aspects of design, it must be underpinned by sound science. Herbert Hoover, one of only two US presidents trained as an engineer (the other is Jimmy Carter), eloquently captured this balance in his memoirs (see The Profession of Engineering).

> *Cogito, ergo sum.* (Latin: "I think, therefore I am.")[24]
>
> René Descartes (1637)

In the broadest sense, engineering is an outgrowth of rationalism; that is, engineers apply reason to solve problems. Descartes' observation can be extended to say: "I think, therefore I design." But, on what do we base such designs, and if these are flawed, what makes an engineer choose another path? According to physicist-philosopher Thomas S. Kuhn, there are essentially two types of paradigm shifts: those that result from a discovery caused by encounters with anomaly and those that result from the invention of new theories brought about by failures of existing theories to solve problems the theory defines. In the case of a paradigm shift brought about by discovery, the first step in shifting the said paradigm is the discovery of the anomaly itself.

Engineers are constantly interpreting theory to provide realistic examples. In this regard, Louis Pasteur can be considered to be among the first "modern era engineers." He was concerned with advancing the state of the science of diseases in parallel with practicing private and public health care. Thus, engineering cannot rely on a single-minded paradigm – in this case, Pasteur's work found that the standard medical practice paradigms would fall short (e.g., in treating anthrax). He also found that a new foundational paradigm was needed (e.g., an enhancement of germ theory).

Thus, the good engineer explores the anomalies. Once the paradigm has been adjusted so that the anomalous becomes the expected, it is said that the paradigm change is complete. In the case of a paradigm shift that results from the invention of new theories caused by the failure of existing theory, the first step is the failure itself (when the system in place fails, the creation of a new system is necessary). Several things can bring about this failure: observation of discrepancies between the theory and the fact, changes in the surrounding social or cultural climate, and academic and practical criticism of the existing theory. While these problems have generally been known for a long time, Kuhn noted that on numerous occasions the scientific community has been highly resistant to change in paradigms. Kuhn also noted that in the early stages of a paradigm, it is easy to invent theoretical alternatives that can be placed on a given set of data. However, once the paradigm is better established, these theoretical alternatives are strongly resisted.[25]

Similarly, the process through which advances are made, and paradigms developed, may also be examined through the lens of the social sciences, especially economics. Although there are numerous examples of the differences in thinking between economists and engineers, this is an area of agreement. For example, economist John Maynard Keynes wanted to understand the dynamics of economics fundamentally. At the same time, he wanted to solve the problem of economic depression. Thus, Keynes was taking the engineering perspective, seeing the fusion of these goals as a joint desire to extend basic knowledge and reach applied goals.

Engineering bioethics and morality

I not only acknowledge but insist upon the fact that morality only limits the range of morally acceptable answers, it does not always provide a unique solution to a moral problem. I hold that it is very rare that any ethical theory, including mine, can resolve any controversial ethical disagreement.

Bernard Gert[26] (twentieth-century ethicist)

The more complicated the problem, the less certain and unique is the solution. Why does engineering attract the best and brightest young minds? Surely, the attraction goes beyond the mathematical and scientific accolades. It goes beyond the sorting process of high analytical and quantitative scores on college admission exams. It must have something to do with a calling. That calling is an integration of a myriad of factors and variables. This integration is an uneasy one. It is simultaneously rewarding and risky, as are all great callings. It incorporates the most basic and most complex algorithms, given that it involves science and people, respectively.

Humans are quite complex. Since engineering is a human enterprise, it should come as no surprise that applying the sciences to solve human problems is a complicated business.

Mathematics has evolved to deal with such complications. In 1902, the mathematician Jacques Hadamard defined a well-posed problem (*un problème bien pose*) as one that is uniquely solvable (*déterminé*). A year earlier he defined ill-posed problems (*questions mal posées*) as those without a unique solution; that is, such problems depend in a discontinuous way on the measurements so that tiny errors can create very large deviations in the solution. The modern rendition of this phenomenon is the "butterfly effect" (see Discussion Box: Ethics and the Butterfly Effect). Hadamard believed anything that is physically important must be well posed. We now know better that numerous engineering, medical, and physical science problems are ill-posed. It can be argued that engineering ethics cases may often be even less uniquely solvable than some of the most complicated physical and engineering problems, such as the tragic case of Jesica Santillan, where a usually well-managed variable (blood type) was mistakenly mismatched, leading to a cascade of events that ended in tragedy (rejection of the transplanted heart and subsequent death).

Discussion box: ethics and the butterfly effect

The Butterfly Effect is the name for "sensitive dependence upon initial conditions,"[27] as a postulate of chaos theory. To engineers, the effect can mean that a small change for good or bad can reap exponential rewards or costs.

Edward Lorenz, at a 1963 New York Academy of Sciences meeting, related the comments of a "meteorologist who had remarked that if the theory were correct, one flap of a seagull's wings would be enough to alter the course of the weather forever." Lorenz later revised the seagull example to that of a butterfly in his 1972 paper "Predictability: Does the Flap of a Butterfly's Wings in Brazil Set off a Tornado in Texas?" presented at a meeting of the American Association for the Advancement of Science, Washington, DC. In both instances, Lorenz argued that future outcomes are determined by seemingly small events cascading through time. Engineers and mathematicians struggle mightily to find ways to explain (and to predict) such outcomes of so-called ill-posed problems. As engineers, we generally like orderly systems so we prefer a well-posed problem; that is, one that is uniquely solvable (i.e., a unique solution exists) and one that is dependent upon a continuous application of data. By contrast, an ill-posed problem does not have a unique solution and can only be solved by discontinuous applications of data, meaning that even very small errors or perturbations can lead to large deviations in possible solutions.[28] Finding the appropriate times and places to solve ill-posed problems is a promising area of mathematical and scientific research.

By extension, small changes for good or bad can produce unexpectedly large effects. Upfront considerations of possible losses of privacy in designing information systems or possible malfunctions under certain physiological constraints when designing medical devices can prevent large problems down the road. Allowing for and dutifully considering the ideas from any member of the design team, no matter how junior, can lead to a successful study and help avoid costly mistakes. We sometimes hear after the fact how someone had noticed a disturbing trend in a laboratory study or an apparent cluster of effects in a group of early adopters, but whose questions and complaints were ignored until a larger data set eventually showed that the device or drug caused unnecessary ill effects. Many catastrophic failures, upon analysis, had at least one internal memorandum from an engineer presciently stating misgivings.

Engineers ignore this information at their peril. Ignoring small details can lead to big problems.

"Small" error and devastating outcomes

Duke University is blessed with some of the world's best physicians and medical personnel. As a research institute, it often receives some of the most challenging medical cases, as was the previously mentioned case for Jesica Santillan, a teenager in need of a heart transplant. Although the surgeon in charge had an impeccable record and the hospital is world renowned for such a surgery, something went terribly wrong. The heart that was transplanted was of a different blood type than that of the patient. The heart was rejected, and even after another heart was located and transplanted, Jesica died due to the complications brought on by the initial rejection. The logical question is how could something so vital and crucial and so easy to know as blood type be overlooked? It appears to be a systematic error. The system of checks and balances failed. And, the professional, i.e., the surgeon, is ultimately responsible and primarily accountable for this or any other failure on his watch.

What can we learn from the Santillan case? One lesson is that a system is only as good as the rigor and vigilance given to it. There is really no such thing as "auto pilot" when it comes to systems. Aristotle helps us here. He contended that the whole is greater than the sum of its parts. This is painfully true in many public health disasters. Each person or group may be doing an adequate or even superlative job, but there is no guarantee that simply adding up each of the parts will lead to success. The old adage that things "fall through the cracks" is a vivid metaphor. The first mate may be doing a great job in open waters, but may not be sufficiently trained in dire straits when the captain is away from the bridge. A first response team may be adequately trained for forest fires (where water is a very good substance for firefighting), but may not properly suited for a spill of an oxidizing agent (where applying water can make matters considerably more dangerous). Without someone with a "global view" to oversee the whole response, the perfectly adequate and even exemplary personnel may contribute to the failure.

Systems are always needed and these systems must be tested and inspected continuously. Every step in the critical path that leads to failure is important. In fact, the more seemingly "mundane" the task, the less likely people are to think a lot about it. So, these small details may be the largest areas of vulnerability. Like the butterfly effect in chaos theory, the chain of events or critical path of one's decision will ultimately determine whether it is a good one or a bad one. One must wonder how many presurgery meetings before the Santillan case had significant discussions on how to make sure that the blood type is properly labeled. One can venture that such a discussion occurs much more frequently now in pre-op meetings (as well as hospital board meetings) throughout the world. Although, it is quite likely even this focus has attenuated in the years since the tragedy. The public and our peers will judge whether we apply due diligence or whether our designs and projects are impaired by something we should have known and considered.

Technology, engineering, and economics

Although engineers are a diverse lot, most are more than a little utilitarian. They strive to provide the greatest amount of goods and services to the greatest number (however these terms are defined and constrained by the design specifications). The National Academy of Engineering recently declared that "engineers and their inventions and innovations have helped shape the changes that have made our lives more productive and fruitful."[29] But what does it mean to become more fruitful? It must be something beyond pure utilitarianism. To begin, we can consider the economic implications.

Bioethics Question: Does biosystem engineering ethics go beyond utility in defining what is "right?"

Technology is the obvious indicator of biosystem engineering. It is the "workhorse" that delivers on biosystematic designs. In economics, technology can be considered a tool of empowerment, in that it empowers producers to generate more output from given levels of the two inputs, labor and capital. In this sense, as a catalyst in chemistry is a substance that increases the rate of reaction, technology is a catalyst for production of output – as it increases the amount of output we get from given inputs.[30] Technology allows for the use of more advanced capital, which results in better and faster ways to create output. Producers are rendered more efficient, as they are able to produce more output, given the same amount of input. This results in greater profit, which results in economic growth. For example, if we consider the basic supply–demand model, we see that improvements in technology cause the supply curve to shift out to the right, meaning that producers will create more output for any given price (Figure 8.1-3). This raises equilibrium output and lowers equilibrium price. Equilibrium, in this case, is used to refer to the price level at which the aggregate supply curve (an upward sloping line that illustrates how much producers would be willing to supply at any given price) crosses the aggregate demand curve (the downward sloping line that shows how much of a good consumers would demand at any given price). This price level is called the "equilibrium" price because

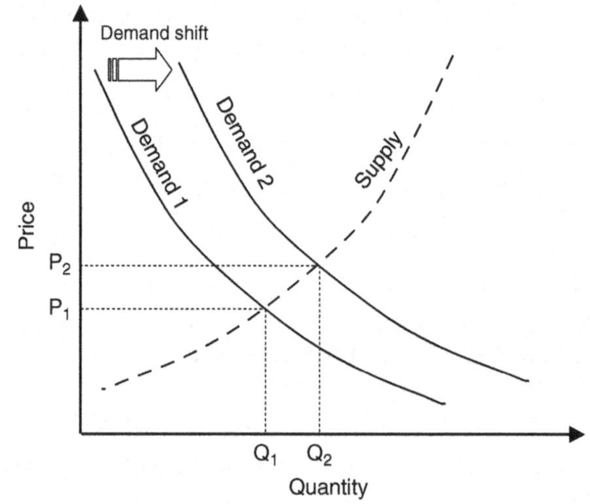

Figure 8.1-3 Supply–demand economic model. The increase in demand results in a relative increase in price and quantity needed to reach a new equilibrium point on the supply curve.

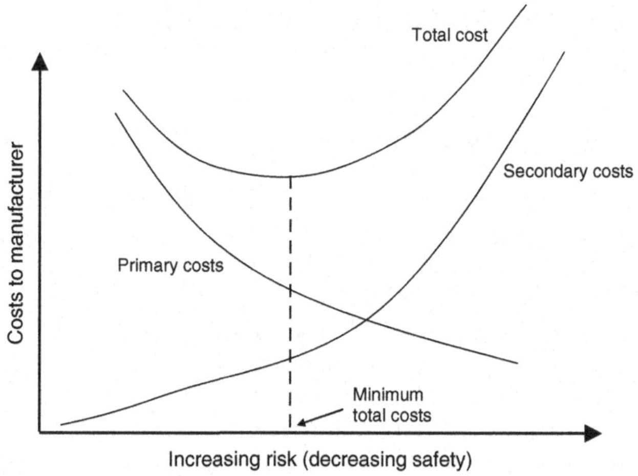

Figure 8.1-4 Safety and risks associated with primary and secondary costs. Increased safety can be gained by considering secondary costs in product and system design. Adapted from: M. Martin and R. Schinzinger's, 1996, *Ethic in Engineering*, McGraw-Hill, New York, NY.

here the amount supplied equals the amount demanded. It is a "balance," as any further increase in supply would disrupt the equilibrium. The same connotation of "balance" is seen in the usage of "equilibrium" in chemistry and thermodynamics, which is the study of work, heat, and energy on a system.

An object is said to be in thermodynamic equilibrium when it is in thermal, chemical, and mechanical equilibrium. It is observed that some properties of an object can change when the object is heated or cooled. Should two objects be brought into physical contact, an initial change in the property of both objects results. For instance, heat, a form of energy, is transferred between these two objects when they are brought into contact with one another. Eventually, this transfer of energy stops. At this point the objects are said to be in thermodynamic equilibrium.[31] Thus in both the economic and the thermodynamic definition of equilibrium, the system is said to be in balance; nothing changes unless acted upon by an exogenous force. In the case of market equilibrium, these endogenous forces act to shift the supply curve or the demand curve, subsequently changing where they cross to form the equilibrium price. Some examples of exogenous forces are increased costs for producers, changes in the regulatory environment, or changes in wealth of consumers.

This supply–demand relationship is analogous to the engineers paramount ethical demand for safety, as shown in Figure 8.1-4. Like price in the economic model, the safety of an engineered device or system translates into costs. However, this is not a linear relationship. An unsafe product is not cheap if all the factors are included in the cost calculations. Primary costs of production are only part of the equation. Premature obsolescence and failures can lead to significant costs of lawsuits, lack of public trust, and recalls and warranty costs.[32] A well-designed device may still be a failure if its useful life is too short or too uncertain.

Technological advances often improve the use of scarce resources, which are then allocated through the economic system. Therefore technology broadens the horizon through which economics operates. When firms invest, they increase capital; and increasing input means more output, i.e., more economic growth. Depreciation on capital stock yields less output. The term sustainability in economics is similar to the concept of environmental sustainability. In economics it is used to describe capital withstanding time and continuing to function to facilitate the production of output. Technology allows higher levels of sustainability for capital, including a predictable and reliable engineered system.

We must not forget that everything engineers do is to protect and to enhance life, no matter what the specific discipline. Thus, even for those of us in engineering specialties other than biomedical, we must be mindful of our work's impact on life. And, while our keen interest is usually in human life, we must be aware of how our work affects other species. In fact, some of the greatest resistance against scientific and engineering research is the result of the perceived indifference to pain and suffering in nonhuman animals. Even microbial life must be respected in our laboratories. One of the scariest scenarios is that of unchecked self-replication of unicellular species. Many engineering researchers are presently pushing the envelopes of nanotechnology, for example. This includes using microbes to manufacture complex pharmaceuticals and chemical products that either cannot be manufactured or can be manufactured much less efficiently in abiotic chemical reactors. However,

Teachable moment: the dismal scientist *versus* the technological optimist

British political economist Robert Malthus[33] (1766–1834) predicted that starvation would result as projected population growth exceeded the rate of increase in the food supply. His forecast was based on the population growth at an exponential rate whereas the food supply growth rate would be linear. These predictions greatly underestimated the role of technology in increasing the global food supply, causing many to doubt such economic models. In a later incarnation of Malthusian thinking, Paul Ehrlich's *Population Bomb* gives an exceedingly grim prognosis for the future:

> Each year food production in underdeveloped countries falls a bit further behind burgeoning population growth, and people go to bed a little hungrier. While there are temporary or local reversals of this trend, it now seems inevitable that it will continue to its logical conclusion: mass starvation.[34]

Not only does Ehrlich state that the world is headed toward calamity, he is convinced that there is nothing anyone really can do that will provide anything more than temporary abatement. Ehrlich's attitude toward technology as part of the solution to the impending problem is technological pessimism, so to speak. Ehrlich's lack of confidence in technology to deal with the problems plaguing the future is perhaps seen most explicitly in his statement: "But, you say, surely Science (with a capital "S") will find a way for us to occupy the other planets of our solar system and eventually of other stars before we get all that crowded."[35] Ehrlich was sure that "the battle to feed humanity is over." He insisted that India would be unable to provide sustenance for the 200-million-person growth in its population by 1980. He was wrong – thanks to biotechnologists like Norman Borlaug. Borlaug and his team engaged in a program that developed a special breed of dwarf wheat that was resistant to a wide spectrum of plant pests and diseases and that produced two or three times more grain mass than the traditional varieties. His team then taught local farmers in both India and Pakistan how to cultivate the new strain of wheat. This astonishing increase in the production of wheat within a few years has come to be called the green revolution, its inception credited to Borlaug.[36] Since 1968, when Ehrlich published his frightful predictions, "India's population has more than doubled, its wheat production has more than tripled, and its economy has grown nine-fold."[37] Pakistan has progressed from harvesting 3.4 million tons of wheat each year to around 18 million tons, and India has made similar impressive movement from 11 million tons to 60 million tons.[38]

Malthus' pessimistic predictive framework earned for economics the nickname of "dismal science." Conversely, engineers look upon the same problems with more of a technical optimism. Engineers "mess up" the Malthusian curve by finding ways to accomplish this (e.g., Borlaug spoiling Ehrlich's predictions). And, such positions lead to moral arguments as when Ehrlich criticizes the Pope for saying "You must strive to multiply bread so that it suffices for the tables of mankind and not, rather, favor an artificial control of birth, which would be irrational, in order to diminish the number of guests at the banquet of life."[39] Ehrlich feels that this representative bread cannot indeed be multiplied: "Can we expect great increases in food production to occur through the placing of more land under cultivation? The answer to this specific question is most definitely no."[40] The truth is, however, that Borlaug did in fact foresee ways to multiply the bread and, that engineers feel that they can continue to design technologies that help to support the growing population and the expanding needs of mankind. In *The Engineer of 2020,* the NAE describes various ways that engineers in the future will help solve the very same problems Ehrlich (and the Malthusian model in general) is concerned with. Where Ehrlich felt technology's role in solving the problem would only be seen through how "improved technology has greatly increased the potential of war as a population control device,"[41] engineers look toward technology not as a "means for self-extermination"[42] but rather as an option for supporting and improving life in the future.

Having examined Ehrlich's prognosis from 1970s and 1980s, consider now a prognosis made by engineers for a new future. According to *The Engineer of 2020,* the world's population will approach 8 billion people; much of this increase will be seen in groups that are today considered underdeveloped countries, mainly in Asia and Africa. Apparently, "by 2015, and for the first time in history, the majority of people, mostly poor will reside in urban centers, mostly in countries that lack the economic, social, and physical infrastructures to support a burgeoning population."[43] Engineers, however, see an opportunity for "the application of thoughtfully constructed solutions through the work of engineers"[44] in the challenge posed by the highly crowded and densely populated world of 2020. Likewise, engineers look upon the necessity for improved health care delivery in the world of the future with confidence. They feel that they will be able to make advanced medical technologies accessible to this ever-growing global population base. In the developed world of twenty years from now they see positive implications on human health, due to improved air quality and the control and cleanup of hazardous waste sites, and focused efforts to treat diseases like malaria and AIDS.[45]

Continued

Engineers believe that they can solve the problems posed by the future, as opposed to views like the one posed by Ehrlich who sees a future where "small pockets of *Homo sapiens* hold on for a while in the Southern Hemisphere, but slowly die out as social systems break down, radiation poisoning takes effect, climatic changes kill crops, livestock dies off, and various man-made plagues spread. The most intelligent creatures ultimately surviving this period are cockroaches."[46] Indeed, this dramatic example serves to illustrate the differences between the engineer's technical optimism and the doomsayer's technical pessimism, so to speak. Of course, this discussion has considered the extremes. Many economists are quite optimistic and many engineers are rather pessimistic. But the ethos of each discipline does travel in different directions, since the nature of engineering is to solve problems. If one has little or no hope at finding a solution, why ever bother? But what does the future call for – the proverbial idealist or the modern-day skeptic? Mankind can either resign itself to failure and deem the big problems unsolvable or it can press forward, attempting to solve the problems it faces and overcome tomorrow's challenges. While the question is subjective, it becomes clear that in order to progress, most engineers choose the latter option. Thus, it can be strongly asserted that utility is but one measure of biosystematic success albeit an important one. Reason and duty are also important, as we will investigate throughout this text.

Questions

1. Is the moniker "dismal scientist" truly applicable to the economist? Explain.
2. What are some downsides to technological optimism?
3. What is the difference between high density and crowding (or "overcrowding")?
4. Explain the bioethical aspects of population controls proposed by two recent Canadian political agendas:
   a. A private member's bill, C-407, introduced to the Canadian Parliament that proposed to allow any person, under certain conditions, to aid a person close to death or suffering from a debilitating illness to "die with dignity" if that person has expressed the free and informed consent to die.
   b. Health Canada's plans to introduce preimplantation genetic diagnosis (PGD) regulations in May 2006.

Defend your agreement or disagreement of the following statement:

Both proposals establish legally binding mandates as to when, how, and why innumerable innocent human beings shall be consigned or abandoned to death, reminiscent of the early "slippery slope" actions that led to the eugenic[47] movement embraced by many scientists in the United Kingdom, the United States, and Germany in the 1930s, which ultimately led to the heinous actions of the Nazis to produce a "master race." Thus, physicians who participate in euthanasia or in prenatal diagnosis often think in terms of "weeding out" the unfit, thus denying them their inalienable right to life. Stemming overpopulation is simply the environmental rationalization for eugenics. (Hint: Read varying views about "Social Darwinism".)

---

a number of credible researchers fear the attendant risks of such nanomachinery, including the creation of unintended toxic by-products and dangerous new strains of microbes.

The life aspects of engineering are, for the most part, the major success stories of engineering. For example, more than any other profession, engineers have prevented (in some cases, eliminated) devastating diseases with their public works projects, such as wastewater treatment, sanitary landfills, hazardous waste facilities, air quality controls, and drinking water supplies. Engineers not specifically practicing in biomedical engineering sometimes need to be reminded that their work serves life.

The "medical" part of biomedical engineering implies a strong link between the medical and the engineering professions. Like engineers, medical practitioners apply the basic sciences to achieve results. Thus, our designing of devices and structures is part of the larger health care provision. It is quite interesting to watch the growth and evolution of professions. They seem to oscillate between stages of specialization and contraction. Presently, both seem to be occurring in biomedical engineering. Medical doctors have become highly specialized, but their responsibilities have increasingly called for broader accountability. While the individual practitioner may lead one area of health care for the patient, they must build systems to ensure that all of the other specialties effectively work together to provide the best care for each patient. In this sense, medicine is part of a larger biosystem (human health).

Some of the most dramatic ethical and legal failures have resulted not from the practitioners' incompetence in their area of specialization, but in their lack of oversight and quality assurance of others who are part of the comprehensive care. Engineers are part of this system of care. In fact, some of the major advances in devices have come about through the close relationships with medical practitioners, such as the collaborations between teaching hospitals and biomedical engineering programs (like those at Duke, Johns Hopkins, and Stanford, to name a few). The engineer brings a number of assets to the team, including practicality, creativity, adaptability, and a long-term view.

The next term, ethics, has been summed up by Socrates as the way we ought to live. For engineers, this can be modified a bit. Engineering ethics is the way we ought to practice. The fundamental canons of the National Society of Professional Engineers (NSPE) code of ethics[48] captures what engineers "ought" to do. It states that engineers, in the fulfillment of their professional duties, shall:

1. Hold paramount the safety, health and welfare of the public.
2. Perform services only in areas of their competence.
3. Issue public statements only in an objective and truthful manner.
4. Act for each employer or client as faithful agents or trustees.
5. Avoid deceptive acts.
6. Conduct themselves honorably, responsibly, ethically, and lawfully so as to enhance the honor, reputation, and usefulness of the profession.

Let us consider these canons as they relate to biomedical and biosystem ethics for engineers. The canons are the professional equivalents to "morality," which refers to societal norms about acceptable (virtuous/good) and unacceptable (evil/bad) conduct. These norms are shared by members of society to provide stability as determined by consensus.[49] Philosophers consider professional codes of ethics and their respective canons to be normative ethics, which is the philosophical study of ethics concerned with classifying actions as right and wrong without bias. Normative ethics is contrasted with descriptive ethics, which is the study of what a group actually believes to be right and wrong, and how it enforces conduct. Normative ethics regards ethics as a set of norms related to actions. Descriptive ethics deals with what "is" and normative ethics addresses "what should be."

Philosopher Bernard Gert categorizes behaviors into what he calls a "common morality," which is a system that thoughtful people use implicitly to make moral judgments.[50] According to Gert, humans strive to avoid five basic harms: death; pain; disability; loss of freedom; and loss of pleasure. Arguably, the job of the engineer is to design devices, structures, and systems that mitigate against such harms in society. Similarly, Gert identifies ten moral rules of common morality:

1. Do not kill.
2. Do not cause pain.
3. Do not disable.
4. Do not deprive of freedom.
5. Do not deprive of pleasure.
6. Do not deceive.
7. Keep your promises.
8. Do not cheat.
9. Obey the law
10. Do your duty.

Most of these rules are proscriptive. Only rules 7, 9, and 10 are prescriptive, telling us what to do, rather than what not to do. The first five directly prohibit the infliction of harm on others. The next five indirectly lead to prevention of harm. Interestingly, these rules track quite closely with the tenets and canons of the engineering profession (see Table 8.1-1).

Table 8.1-1 Canons of the National Society of Professional Engineers compared to Gert's rules of morality

Engineers shall:	Most closely linked to rules of morality identified by Gert
1. Hold paramount the safety, health, and welfare of the public	• Do not kill • Do not cause pain • Do not disable • Do not deprive of pleasure • Do not deprive of freedom
2. Perform services only in areas of their competence	• Do not deceive • Keep your promises • Do not cheat • Obey the law • Do your duty
3. Issue public statements only in an objective and truthful manner	• Do not deceive
4. Act for each employer or client as faithful agents or trustees	• Do not deprive of pleasure • Keep your promises • Do not cheat • Do your duty
5. Avoid deceptive acts	• Do not deceive • Keep your promises • Do not cheat
6. Conduct themselves honorably, responsibly, ethically, and lawfully so as to enhance the honor, reputation, and usefulness of the profession	• Do your duty • Obey the law • Keep your promises

The Gert model is good news for engineering bioethics. Numerous ethical theories can form the basis for engineering ethics and moral judgment. Immanuel Kant is known for defining ethics as a sense of duty. Thomas Hobbes presented ethics within the framework of a social contract, with elements reminiscent of Gert's common morality. John Stuart Mill considered ethics with regard to the goodness of action or decision as the basis for utilitarianism. Philosophers and ethicists spend much effort and energy deciphering these and other theories as paradigms for ethical decision making. Engineers can learn much from these points of view, but in large measure, engineering ethics is an amalgam of various elements of many theories. As evidence, the American Society of Mechanical Engineers (ASME)[51] has succinctly bracketed ethical behavior into three models:

Malpractice, or Minimalist, Model – In some ways this is really not an ethical model in that the engineer is only acting in ways that are required to keep his or her license or professional membership. As such, it is more of a legalistic model. The engineer operating within this framework is concerned exclusively with adhering to standards and meeting requirements of the profession and any other applicable rules, laws, or codes. This is often a retroactive or backward-looking model, finding fault after failures, problems, or accidents happen. Any ethical breach is assigned based upon design, building, operation, or other engineering steps that have failed to meet recognized professional standards. This is a common approach in failure engineering and in ethical review board considerations. It is also the basis of numerous engineering case studies.

Reasonable Care, or Due Care, Model – This model goes a step further than the minimalist model, calling upon the engineer to take reasonable precautions and to provide care in the practice of the profession. Interestingly, every major philosophical theory of ethics includes such a provision, such as the harm principle in utilitarianism, the veil of ignorance in social contract ethics, and the categorical imperative in duty ethics. It also applies a legal mechanism, known as the reasonable person standard. Right or wrong is determined by whether the engineer's action would be seen as ethical or unethical according to a "standard of reasonableness as seen by a normal, prudent nonprofessional."[52]

Good Works Model – A truly ethical model goes beyond abiding by the law or preventing harm. An ethical engineer excels beyond the standards and codes and does the right thing to improve product safety, public health, or social welfare. An analytical tool related to this model is the net goodness model, which estimates the goodness or wrongness of an action by weighing its morality, likelihood, and importance.

This model is rooted in the moral development theories such as those expounded by Kohlberg,[53] Piaget,[54] and Rest,[55] who noted that moral action is a complex process entailing four components: moral awareness (or sensitivity), moral judgment, moral motivation, and moral character. The actor must first be aware that the situation is moral in nature; that is, at least that the actions considered would have consequences for others. Second, the actor must have the ability to judge which of the potential actions would yield the best outcome, giving consideration to those likely to be affected. Third, the actor must be motivated to prioritize moral values above other sorts of values, such as wealth or power. Fourth, the actor must have the strength of character to follow through on a decision to act morally.

Piaget, Kohlberg, and others (e.g., Duska)[56] have noted that the two most important factors in determining a person's likelihood of behaving morally are age and education; that is, of being morally aware, making moral judgments, prioritizing moral values, and following through on moral decisions. These are strong indicators of experience[57] and seem to be particularly critical regarding moral judgment: A person's ability to make moral judgments tends to increase with maturity as they pursue further education, generally reaching its final and highest stage of development in early adulthood. This theory of moral development is illustrated in Table 8.1-2.

Kohlberg insisted that these steps are progressive. He noted that in the two earliest stages of moral development, which he combined under the heading "preconventional level," a person is primarily motivated by the desire to seek pleasure and avoid pain. The "conventional level" consists of stages three and four: in stage three, the consequences that actions have for peers and their feelings about these actions; in stage four, considering how the wider community will view the

Table 8.1-2 Kohlberg's stages of moral development

Preconventional level	**1.** Punishment-obedience orientation
	2. Personal reward orientation
Conventional level	**3.** "Good boy" – "nice girl" orientation
	4. Law and order orientation
Postconventional level	**5.** Social contract orientation
	6. Universal ethical principle orientation

Source: L. Kohlberg, 1981 *The Philosophy of Moral Development (Vol. 1)*, Harper & Row, San Francisco, CA.

actions and be affected by them. Few people reach the "postconventional" stage, wherein they have an even broader perspective: Their moral decision making is guided by universal moral principles;[58] that is, by principles that reasonable people would agree should bind the actions of all people who find themselves in similar situations.

The moral need to consider the impact one's actions will have on others forms the basis for a normative model. Pursuing an activity with the goal of obeying the law has as its driving force the avoidance of punishment, and pursuing an activity with the goal of improving profitability is a goal clearly in line with stockholders' desires; presumably customers', suppliers', and employees' desires must also be met at some level. And finally, pursuing an activity with the goal of "doing the right thing," behaving in a way that is morally right and just, can be the highest level of engineering behavior. This normative model of ethical engineering can be illustrated as in Figure 8.1-5.

There is a striking similarity between Kohlberg's model of moral development (Table 8.1-2) and the model engineering professional growth. Avoiding punishment in the moral development model is similar to the need to avoid problems early in one's career. The preconventional level and early career experiences have similar driving forces.

At the second level in the moral development model is a concern with peers and community, while in the professionalism model the engineer must balance the needs of clients and fellow professionals with those of society at large. Engineering services and products must be of high quality and be profitable, but the focus is shifting away from self-centeredness and personal well-being toward external goals.

Finally at the highest level of moral development a concern with universal moral principles begins to govern actions. The driving force or motivation is trying to do the right thing on a moral (not legal or financial) basis. These behaviors set the example for the whole profession, now and in the future.

Professional growth is enhanced when engineers and technical managers base their decisions on sound business and engineering principles. Ethical content is never an afterthought, but is integrated within the business and design decision-making process: That is, the engineering exemplars recognize the broad impact their decisions may have, and they act in a way such that their action is in the best interest of not only themselves and the organization they represent, but also the broader society and even future generations of engineers.

Much of the ethics training to date has emphasized preconventional thinking; that is, adherence to codes,

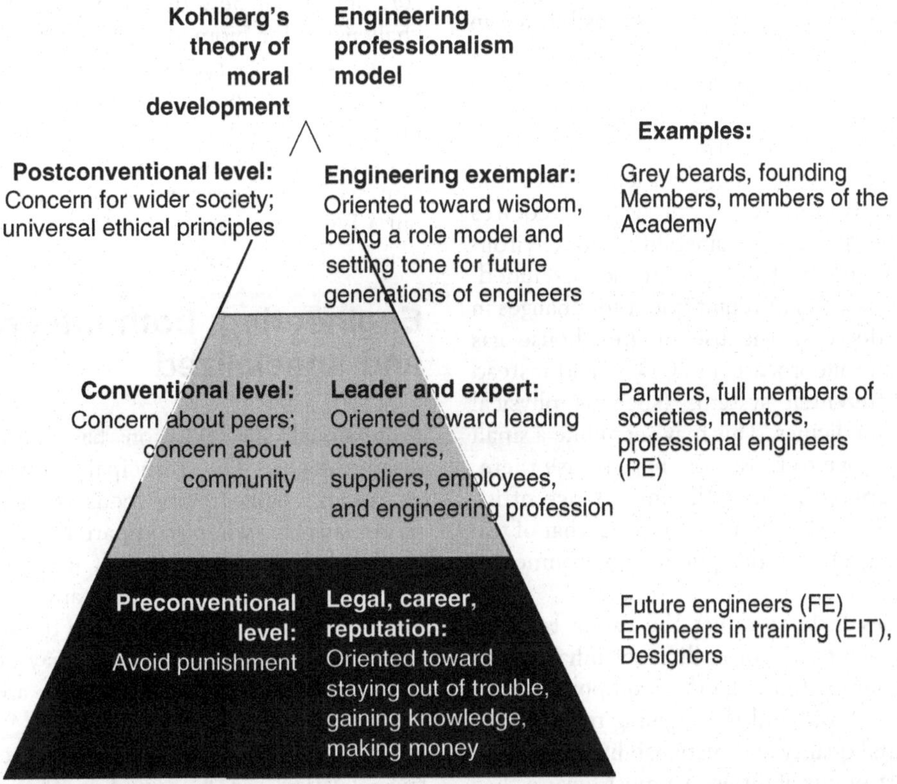

Figure 8.1-5 Comparison of Kohlberg's moral development stages to professional development in engineering.

laws, and regulations within the milieu of profitability for the organization. This benefits the engineer and the organization, but is only a step toward full professionalism, the kind needed to confront bioethical challenges. We who teach engineering ethics must stay focused on the engineer's principal client, "the public." One interpretation of the "hold paramount" provision of this ethical canon is that it has primacy over all the others. So, anything the professional engineer does cannot violate this canon. No matter how competent, objective, honest, and faithful, the engineer must not jeopardize public safety, health, or welfare. This is a challenge for such a results-oriented profession.

Bioethics Question: How can medical and engineering ethics coalesce?

In our zeal to provide the best technologies, devices, services, and plans, we cannot treat the general public as a means to such noble ends. Here is where the primary focus of the physician and that of the engineer begin to diverge. The medical practitioner's principal client is the patient, whereas the primary client of the engineer is the public. Nothing is more important to the engineer than the health and safety of the public.

The engineer, especially the biomedical engineer, must navigate both professional codes. As evidence, the Biomedical Engineering Society recently recognized this with its approval of a new code of ethics in 2004. The code recognizes that the biomedical engineer practices at the confluence of "expertise and responsibilities in engineering, science, technology, and medicine."[59] Mirroring the NSPE code, the biomedical engineering community reminds its members that "public health and welfare are paramount considerations."[60]

Public safety and health considerations affect the design process directly. Almost every design now requires at least some attention to sustainability and environmental impacts. Biomedical designs are not excluded. For example, there is a recent requirement for changes in drug delivery to decrease the use of greenhouse gas propellants like chlorofluorocarbons (CFCs) and instead using pressure differential systems (such as physical pumps) to deliver medicines. This may seem like a small thing or even a nuisance to those who have to use them, but it reflects an appreciation of the importance of incremental effects. It also combines two views, that of the patient (drug therapy) and the public (environmental quality).

One inhaler does little to affect the ozone layer or threaten the global climate, but millions of inhalers can produce enough halogenated and other compounds that the threat must be considered in designing medical devices. Environmental quality and sustainability are public virtues. To the best of our abilities, we must ensure that what we design is sustainable over its useful lifetime. This requires that the engineer think about the life cycle not only during use but when the use is complete. Such programs as "design for recycling" (DFR) and "design for disassembly" (DFD) allow the engineer to consider the consequences of various design options in space and time. They also help designers to pilot new systems and to consider scale effects when ramping up to full production of devices.

Like virtually everything else in engineering, best serving the public is a matter of optimization. The variables that we choose to give large weights will often drive the design. Treating cancer, providing devices to aid cardiovascular functioning, and delivery of efficacious drug therapies are noble and necessary ends. The engineer must continue to advance the state of the science in these high-priority areas. But, the public is a complicated entity and the human body is uniquely exquisite. Thus, any possible adverse effects must be recognized. These should be incorporated and properly weighted when we optimize benefits. We must weigh these benefits against possible hazards and societal costs.

Engineering competence

Engineering is a technical profession. It depends on scientific breakthroughs. Science and technologies are drastically and irrevocably changing. The engineer must stay abreast of new developments. This is particularly challenging for biomedical science and biosystem technology, where the scale of interest continues to decrease. It was not that long ago when organs were the most refined scale of interest, giving way to the organelles and cells. Now, the "nanoscale" is receiving the most attention, with structures and systems having design units of but a few angstroms.

Engineering: both integrated and specialized

Professional specialization has both advantages and disadvantages. The principal advantage is that the practicing engineer can focus on a specific discipline more sharply when compared to a generalist. The principal disadvantage is that integrating the different parts can be challenging. For example, in a very complex design only a few people can see the overall goals. Thus, those working in specific areas may not readily see duplication or gaps that they assume are being addressed by others.

A classic example of the shortcomings of overspecialization can be found in the video *Professional Ethics and Engineering* produced by Duke's Center for

Applied Ethics. In it, a seasoned engineer is being interviewed by a professional review panel and asked whether he knew that the foundation of a building being constructed was mismatched to the soil type. He said that he did, but it was none of his business, since it was the job of the soil engineers. The panel reminded him that people died as a result of this failure. As the Santillan case reminds us, medical scenarios can also suffer since the whole is greater than the sum of its parts. This is the essence of biosystem engineering.

The work of technical professions is both the effect and the cause of modern life. When undergoing medical treatment and procedures, people expect physicians, nurses, emergency personnel, and other health care providers to be current and capable. Society's infrastructure, buildings, roads, electronic communications, and other modern necessities and conveniences are expected to perform as designed by competent engineers and planners. But how does society ensure that these expectations are met? Much of the answer to this question is that society cedes a substantial amount of trust to a relatively small group of experts, the professionals in increasingly complex and complicated disciplines that have grown out of the technological advances that began in the middle of the twentieth century and grew exponentially in its waning decades.

Professions, including engineering, are not neatly subdivided as they once were. A visit to the hospital shows that not only do many of the physicians specialize in particular areas of medicine (e.g., neuromedicine, oncology, and geriatrics), but all of these physicians must rely on chemists, radiologists, and tomographic experts to obtain data about their patients (e.g., from serum analysis, magnetic resonance and CT scans, and sonography). In fact, many of the solutions (cures?) to health problems require an intricate cacophony among doctors, biomedical engineers, and technicians, as well as public and community health professionals, epidemiologists, and environmental engineers to prevent and control many of the diseases, making treatment unnecessary. For example, a drug delivery system requires the understanding of the biochemical needs of the patient, the fluid mechanics of the pharmacology, and the actual design of the apparatus. This is a continuum among science, engineering, and technology.

Within this highly complex, contemporary environment, practitioners must ensure that they are doing what is best for the profession and what is best for the patient and client. This best practice varies by profession and even within a single professional discipline, so the actual codified rules (codes of ethics, either explicit or implicit) must be tailored to the needs of each group. However, many of the ethical standards are quite similar for most technical professions. For example, people want to know that the professional is trustworthy. The trustworthiness is a function of how good the professional is in the chosen field and how ethical the person is in practice. Thus, the professional possesses two basic attributes: subject matter knowledge and character. Maximizing these two attributes enhances professionalism.

Who is a professional?

There is some debate about just who is a professional. I often ask the students enrolled in my Professional Ethics course to give examples of professionals. The list always includes physicians, engineers, airline pilots, and lawyers, and usually includes accountants. A few students consider clergy and military officers (not usually enlisted personnel) to be professionals. Some include businesspersons, teachers, and scientists. Only a small minority includes professional athletes, although many admit this is because the group includes the term "professional." Several other disciplines are included, but support diminishes after the first few. I approach the query quite unscientifically, simply asking their opinions, but this is interesting since I give them no criteria from which to label something as a professional; yet they are readily equipped to answer. I simply ask whom they would identify as professionals, and a list is generated.

I am often amazed by the intuitive powers of students (actually, of people in general). They usually can differentiate some very complicated subject matter (e.g., pollutant types, risk, values, and obligatory moral behavior), but they often cannot tell you why. In other words, they cannot explain their methodology, but they clearly use one. So, I delve a little more deeply by asking the students to tell me why one group is professional and another is not. They usually note readily that it is not that one is necessarily more "valuable" than the other or that ease or difficulty is a determining factor. Certain highly technical, critical, difficult, and respected "jobs," such as that of an aircraft mechanic, are not generally considered "professional."

Most of us will admit that a certain threshold of expertise is needed to ascribe the label "professional" to someone. However, as our aircraft mechanic example demonstrates, clearly expertise is a necessary but insufficient quality of professionalism. All professionals must be experts in the field, but not all experts are professionals. One distinguishing characteristic of a professional is the level of accountability and degree of responsibility. In our aircraft example, the mechanic is a highly trained expert in a particular area, but with a tightly defined span of control and realm of responsibility. However, the airline pilot is responsible for all aspects of the plane; that is, everything the mechanic does, everything the copilot does, and everything about the plane, the weather, the flying conditions, and whatever it takes to transport the plane, passenger, and cargo safely to the destination are the responsibility of the pilot. The captain is also accountable for everything that transpires on his or her "watch." Any organization, like any system, has a set of norms and mores that are distributed throughout its membership. Indeed, the mechanic who does not follow protocol will be reprimanded, usually severely, but the captain shares the blame, if for no other reason than because the "system" being led by the captain does not adequately ensure high-quality performance by the mechanic. Nor can the pilot defer and deflect blame to the airline company. No company policy or business decision should detract from the professional responsibilities of the pilot. In a word, the pilot remains responsible. The pilot is accountable for the whole flight experience.

What is technical?

With a better idea of what it means to be a professional, we can now endeavor to characterize certain professionals as "technical." The technical professional must have a mastery of technical subject matter. But, what does this mean? Is the ability to play a video game or listen to an I-Pod a technical skill? Most of us would not think so. Is the ability to run sonigraphic software and hardware a technical skill? Most would agree. However, is this ability enough to be a considered a professional? Many would say: "No, it simply means the person is a skillful technician."

Table 8.1-3 Functions that must be integrated into an engineering design

1. Baseline studies of existing conditions
2. Analyses of project alternatives
3. Feasibility studies
4. Environmental impact studies and other macro-ethical, societal considerations
5. Assistance in project planning, approval, and financing
6. Design and development of systems, processes, and products
7. Design and development of construction plans
8. Project management
9. Construction supervision and testing
10. Process design
11. Start-up operations and training
12. Assistance in operations
13. Management consulting
14. Environmental monitoring
15. Decommissioning of facilities
16. Restoration of sites for other uses
17. Resource management
18. Measuring progress for sustainable development

Source: Adapted from American Society of Mechanical Engineers, http://www.professionalpractice.asme.org/communications/sustainability/2.htm; accessed 23 May 2006.

Systematics: incorporating ethics into the design process

The key to engineering successes is ensuring that all of the right factors are considered in the design phase and that these factors are properly implemented and monitored throughout the project.

Integrated engineering approaches require that the engineer's responsibilities extend well beyond the construction, operation, and maintenance stages. Such an approach has been articulated by the ASME. One way to visualize a systematic view, such as design for the environment (DFE) recommended by the ASME is to use an integrated matrix[61] (Table 8.1-3). This allows for the engineer to see the technical and ethical considerations associated with each component of the design, as well as the relationships among these components. For example, health risks, social expectations, and environmental impacts and other societal risks and benefits associated with a device, structure, product, or activity can be visualized at various stages of the manufacturing, marketing, and application stages. This yields a number of two-dimensional matrices (Figure 8.1-6) for each relevant design component. And, each respective cell indicates both the importance of that component and the confidence (expressed as scientific certainty) that the engineer can have about the underlying information used to assess the importance (see legend in Figure 8.1-6).

The matrix approach is qualitative, but it allows comparisons of alternatives that would otherwise be incomparable, which is often the case in bioethics. To some extent, even numerical values can be assigned to each cell to compare them quantitatively, but the results are at the discretion of the analyst, who determines how different areas are weighted. The matrix approach can also focus on design for a more specific measure, such as energy efficiency or product safety, and can be extended to view corporate activities systematically.

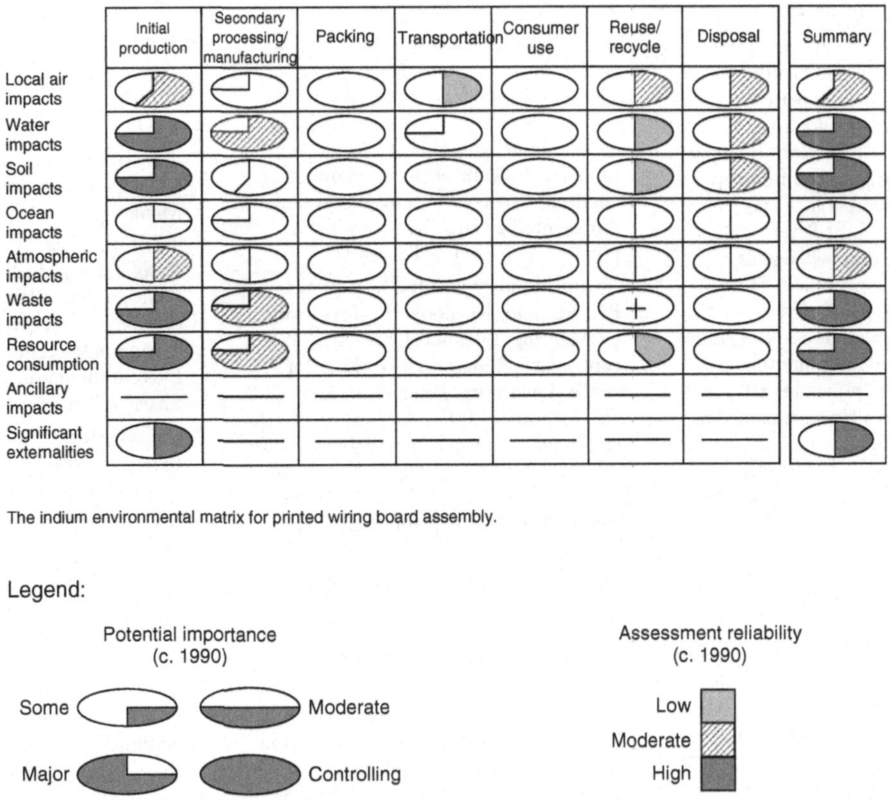

Figure 8.1-6 An example of an integrated engineering matrix; in this instance applied to sustainable designs. Adapted from: American Society of Mechanical Engineers; http://www.professionalpractice.asme.org/communications/sustainability/2.htm; accessed 25 May 2006.

Contemporary society demands many things from its professionals. For those in technical fields there are two prominent expectations: trust and competence. These expectations are built into the codes of practice and ethics of each technical discipline. They are two elements that the engineer keeps throughout his or her career.

Notes and commentary

1 E. Dickinson, 1862, *Wider than the Sky.*

2 I have modified Rosenberg's experiments a bit to make them more specifically relevant to bioethics and engineering, but the results reported are those observed in the Responsible Conduct of Research workshop, August 2004.

3 The original version of this thought experiment was conceived by philosopher Philippa Ruth Foot (born 1926). She is one of the leaders of the contemporary movement to virtue ethics. Virtue ethics centers around what a person should become; that is, the goal of ethics is "eudaimonia" (the Aristotelian concept of "success" sometimes translated as "happiness" but more correctly as "blessedness."). The twentieth-century movement is known as the "aretaic" turn, from the Greek term *arête*, meaning "excellence." This is in contrast to utilitarian and other "consequentialist" ethical models, where ethics is predominantly determined by outcome.

4 "Teaching Ethics to Scientists and Engineers: Moral Agents and Moral Problems," *Science and Engineering Ethics* **1**, no. 3 (1995): 299–308.

5 Laplace's equation on the rectangular region, $0 < x < a$, $0 < y < b$, is subject to the Dirichlet boundary conditions:

$$u(x,0) = x^2 u(x,b) = x^2 - b^2$$
$$u(0,y) = -y^2 u(a,y) = a^2 - y^2$$

The unique solution to this BVP is $u(x, y) = x^2 - y^2$, making it well posed.

6 J. Hadamard, *Lectures on the Cauchy Problem in Linear Partial Differential Equations* (New Haven, CN: Yale University Press, 1923).

7 N.R. Augustine, Ethics and the second law of the thermodynamics. *The Bridge*, 32(3) Fall, 2002.

8 R. Rolheiser, *The Holy Longing: The Search for a Christian Spirituality* (New York, NY: Double-day, 1999).

9 This case is found at http://onlineethics.org/bib/appcw-pt3.html (accessed 30 June 2006).

10 C. Gulligan, *In a Different Voice: Psychological Theory and Women's Development* (Cambridge, MA: Harvard University Press, 1993).

11 T. Aquinas, *Summa Theologica*, I-II Q94 Art 2.

12 Often, texts, manuals, and handbooks are valuable, but only when experience and good listening skills are added to the mix can wise (and ethical) decisions be made. First-century thinking linked maturity to "self-control" or "temperance" (Greek *kratos* for "strength"). St Peter, for example, considered knowledge as a prerequisite for temperance. Thus, from a professional point of view, he seemed to be arguing that one can really only understand and appropriately apply scientific theory and principles after one practices them (I realize he was talking about spirituality, but anyone who even casually studied Peter's life would see that he fully integrated the physical and the spiritual). This is actually the structure of most professions. For example, engineers who intend to practice must first submit to a rigorous curriculum (approved and accredited by the Accreditation Board for Engineering and Technology), then must sit for the Future Engineers (FE) examination. After some years in the profession (assuming tutelage by and intellectual osmosis with more seasoned professionals), the engineer has demonstrated the *kratos* (strength) to sit for the Professional Engineers (PE) exam. Only after passing the PE exam does the National Society for Professional Engineering certify that the engineer is a "professional engineer" and eligible to use the initials PE after one's name. The engineer is, supposedly, now schooled beyond textbook knowledge and knows more about why in many problems the correct answer is "It depends." In fact, the mentored engineer even has some idea of what the answer depends on (i.e., beyond "knowing that one does not know" as Socrates would say).

13 American Medical Association (AMA), 1997, Policy E-2.01 Abortion, http://www.ama-assn.org/apps/pf_new/pf_online?f_n=browse&doc=policyfiles/HnE/E-2.01.HTM&&s_t=&st_p=&nth=1&prev_pol=policyfiles/HnE/E-1.02.HTM&nxt_pol=policyfiles/HnE/E-2.01.HTM& (accessed 13 July 2006).

14 AMA, H-5.995, Abortion.

15 P. Lee and R.P. George, "The Wrong of Abortion," in *Contemporary Debates in Applied Ethics*, ed. A. Cohen and C.H. Wellman (Malden, MA: Blackwell Publishing, 2004).

16 Duke University, 2006, http://www.k-phd.duke.edu/purpose.htm (accessed 13 April 2006). According to the site, the K-PhD program "provides opportunities for children to learn to think critically and analytically while developing a passion for understanding the world and an appreciation for improving the quality of all living things." Its mission "is to increase significantly the number of children, particularly female and under-represented groups, who choose to pursue science related careers."

17 This profundity is actually a quote by P.A. Vesilind, RL Rooke Professor of Engineering, Bucknell University, at the conference, "Engineers Working for Peace, Lewisburg, Pennsylvania, 15 November 2003." Vesilind made the comment as a reminder of the practicality of the profession and the need to respect the ethos of engineers when addressing societal issues, like peace and justice.

18 The concept of "clean" is subject to debate within the engineering community. It parallels the questions about safety. When is a device or drug sufficiently "safe" to move to the production stage? Environmental engineers ask a similar question, "How clean is clean?" We wonder when we have done a sufficient job of cleaning up a spill or a hazardous waste site. It is often not possible to have nondetectable concentrations of a pollutant, especially since analytical chemistry and other scientific disciplines continue to improve. Commonly, a threshold for cancer risk to a population is one in a million excess cancers. In cleanup situations, the tolerable risk may be much higher (e.g., one in ten thousand). However, one may find that the contaminant is so difficult to remove that we almost give up on dealing with the contamination and put in measures to prevent exposures, i.e., fencing the area in and prohibiting access. This is often done as a first step in site remediation, but is unsatisfying and controversial (and usually politically and legally unacceptable). Thus, even if costs are high and technology unreliable, the engineer must find suitable and creative ways to clean up the mess and meet risk-based standards.

19 K. Johnson, "We Need to Keep Leading Students into Science, Math," Editorial in *The Durham (NC) Herald-Sun*, p. A-ll, 12 February 2006.

20 H. Petroski, *To Engineer Is Human: The Role of Failure in Successful Design* (New York, NY: St. Martin's Press, 1985).

21 T.J. Albrecht, "ISO 9002 Implementation: Lessons Learned," *Quality Digest* 14 (1994): 55–61.

22 Petroski, *To Engineer Is Human.*

23 *The Memoirs of Herbert Hoover 1874–1920: Years of Adventure*, vol. 1, Library of Congress E 802.H7 (Washington, DC: 1951), 132–3.

24 Translated from *"Je pense, donc je suis,"* in R. Descartes, 1637, *Discourse on Method.*

25 T.S. Kuhn, *The Structure of Scientific Revolutions* (Chicago, IL: University of Chicago Press, 1962). Much of the economics and engineering comparative discussion benefits from the work of Duke undergraduate student, Rayhaneh Sharif-Askary's research.

26 B. Gert, Letter to the Editor, *The Ag Bioethics Forum* (Iowa State University) 5, no. 2 (November 1993).

27 R.C. Hilborn, *Chaos and Nonlinear Dynamics* (UK: Oxford University Press, 1994).

28 Hadamard, *Lectures on the Cauchy Problem.*

29 National Academy of Engineering, *The Engineer of 2020: Visions of Engineering in the New Century* (Washington, DC: National Academy Press, 2004), 48.

30 This analogy does not hold completely to the economics of technology, since in chemistry the catalyst is a chemical substance that increases the rate of a reaction without being consumed. We know that technologies do indeed become consumed (antiquated and in need of replacement).

31 National Aeronautics and Space Administration Thermodynamic Equilibrium, 15 March 2006,http://www.grc.nasa.gov/WWW/K-12/airplane/thermoO.html.

32 M. Martin and R. Schinzinger, *Ethic in Engineering* (New York, NY: McGraw-Hill, 1996).

33 His actual given name was Thomas Robert Malthus.

34 Paul R. Ehrlich, *The Population Bomb* (New York, NY: Ballantine Books, 1968).

35 Ibid., 20.

36 Salil Singh, 17 April 2006, "Norman Borlaug: A Billion Lives Saved." A World Connected, http://www.aworldconnected.org/article.php/311.html.

37 Ibid.

38 Ibid.

39 Ehrlich, *The Population Bomb*, 95.

40 Ibid., 96.

41 Ibid., 69.

42 Ibid.

43 National Academy of Engineering, 27–8.

44 Ibid.

45 Ibid., 28–9.

46 Ehrlich, *The Population Bomb*, 78.

47 In 1883, Francis Galton, Charles Darwin's cousin, coined the term "eugenics." He reportedly objected to charity because it encouraged the poor to have more children. Such elitism is an example of social engineering run amok.

48 National Society for Professional Engineering, 2003, NSPE Code of Ethics for Engineers, http://www.nspe.org/ethics/ehl-code.asp (accessed 8 January 2006).

49 T.L. Beauchamp and J.F. Childress, "Moral Norms," in *Principles of Biomedical Ethics*, 5th ed. (New York, NY: Oxford University Press, 2001).

50 B. Gert, *Common Morality: Deciding What to Do* (New York, NY: Oxford University Press, 2004).

51 American Society of Mechanical Engineers, 2006, Professional Practice Curriculum, "Engineering Ethics," http://www.professionalpractice.asme.org/engineering/ethics/0b.htm (accessed 10 April 2006).

52 Note that this is not the "reasonable engineer standard." Thus, the reasonable person standard adds an onus to the profession, i.e., not only should an action be acceptable to one's peers in the profession, but to those outside of engineering. An action could very well be legal, and even professionally permissible, but may still fall below the ethical threshold if reasonable people consider it to be wrong.

53 L. Kohlberg, *The Philosophy of Moral Development*, vol. 1 (San Francisco, CA: Harper & Row, 1981).

54 J. Piaget, *The Moral Judgment of the Child* (New York, NY: The Free Press, 1965).

55 J.R. Rest, *Moral Development: Advances in Research and Theory* (New York, NY: Praeger, 1986); and J.D. Rest, D. Narvaez, M.J. Bebeau, and S.J. Thoma, *Postconventional Moral Thinking: A Neo- Kohlbergian Approach* (Mahwah, NJ: Lawrence Erlbaum Associates, 1999).

56 R. Duska and M. Whelan, *Moral Development: A Guide to Piaget and Kohlberg* (New York, NY: Paulist Press, 1975).

57 Hence, the engineering profession's emphasis on experience and mentorship.

58 J.A. Rawls, *A Theory of Justice* (Cambridge, MA: Harvard University Press, 1785); and I. Kant, *Foundations of the Metaphysics of Morals*, trans. L.W. Beck, 1951, (Indianapolis, IN: Bobbs-Merrill, 1959).

59 Biomedical Engineering Society, 2004, Biomedical Engineering Society, 2004, "Biomedical Engineering Society Code of Ethics," http://www.bmes.org/pdf/2004ApprovedCodeofEthicsShortForm.pdf (accessed 8 January 2006).

60 This wording is quite interesting. It omits "public safety." However, safety is added under professional obligations that biomedical engineers "use their knowledge, skills, and abilities to enhance the safety, health, and welfare of the public." The other interesting word choice is "considerations." Some of us would prefer "obligations" instead. These compromises may indicate the realities of straddling the design and medical professions. For example, there may be times when the individual patient needs supersede those of the general public and *vice versa*.

61 American Society of Mechanical Engineers, 2005, Sustainability: Engineering Tools, http://www.professionalpractice.asme.org/business_functions/suseng/1.htm (accessed 10 January 2006).

Chapter 8.2

8.2

Bioethics and the engineer

Daniel Vallero

While engineering is a rapidly evolving field that adapts to new knowledge, new technology, and the needs of society, it also draws on distinct roots that go back to the origins of civilization. Maintaining a linkage of the past with the future is fundamental to the rational and fact-based approaches that engineers use in identifying and confronting the most difficult issues.

National Academy of Engineering (2004)[1]

"Biomedical ethics" or "bioethics" has numerous definitions. Here is mine:

Bioethics is the set of moral principles and values (the ethics *part) needed to respect, to protect, and to enhance life (the* bio *part).*

Engineers, medical practitioners, and all technical professionals must be clear about meanings. A stray mark on a blueprint or a misreading of a prescription can lead to harmful outcomes, even death and destruction. So it is with each term in this book's title. Let us consider each.

Upon review, there may be a few parts of the definition that are missing, such as words like medicine, health, and biotechnologies. They are certainly embedded, but bioethics is much more. In light of the risk that my definition may appear overly simple and obvious, let us try to go back to the origins.

The term was coined by Van Rensselaer Potter II (1911–2001). Although Potter was a biochemist, he seemed to think like an engineer; that is, in a rational and fact-based manner. In fact, his original 1971 definition of bioethics was one rooted in integration and systematics. Potter considered bioethics to bridge science and the humanities to serve the best interests of human health and to protect the environment. In his own words, Potter describes this bridge:

From the outset it has been clear that bioethics must be built on an interdisciplinary or multidisciplinary base. I have proposed two major areas with interests that appear to be separate but which need each other: medical bioethics and ecological bioethics. Medical bioethics and ecological bioethics are non-overlapping in the sense that medical bioethics is chiefly concerned with short-term views: the options open to individuals and their physicians in their attempts to prolong life. . . . Ecological bioethics clearly has a long-term view that is concerned with what we must do to preserve the ecosystem in a form that is compatible with the continued existence of the human species.[2]

Biomedicine, engineering, and the development and application of emerging biotechnologies all share a common feature; they call for balance. Society demands that the state of the science be advanced as rapidly as possible and that no dangerous side effects ensue. Most engineers appreciate the value of pushing the biotechnological envelopes. Engineers are adept at optimizing among numerous variables for the best design outcomes. However, most of these emergent areas are associated with some degree of peril. A recent query of top scientists[3] addressed this very issue. Its focus was on those biotechnologies needed to help the developing countries. Thus, the study included both the societal and the technological areas of greatest potential value. Each of these international experts was asked the following questions about the specific technologies:

Biomedical Engineering Desk Reference; ISBN: 9780123746467

- *Impact.* How much difference will the technology make in improving health?
- *Appropriateness.* Will it be affordable, robust, and adjustable to health care settings in developing countries, and will it be socially, culturally, and politically acceptable?
- *Burden.* Will it address the most pressing health needs?
- *Feasibility.* Can it realistically be developed and deployed in a time frame of 5–10 years?
- *Knowledge gap.* Does the technology advance health by creating new knowledge?
- *Indirect benefits.* Does it address issues such as environmental improvement and income generation that have indirect, positive effects on health?

The top three areas require major advances in biomedical engineering (Table 8.2-1). The fourth area is within the domain of environmental and civil engineering. The fifth area is a challenge for genetic and tissue engineers. The sixth area falls within biomedical engineering research and clinical engineering. The seventh area combines the work of computer engineers and biomedical engineers, while the eighth area is a blend of agricultural and biomedical engineering with food sciences. The ninth area will require advances in biomedical, clinical, and tissue engineering, and the tenth area will call on computational pharmacological modeling (e.g., compartmental models), material sciences, biomedical engineering, and chemical engineering. This is evidence that bioethics is a growing concern for all engineering disciplines. Notably, each of these technological areas is associated with bioethical issues, but in very unique ways.

Regarding biomedical engineering, both parts of the term "biomedical" are important. Again, "bio" connotes life. The dictionary definition of this combination form denotes "life or living organisms, or systems derived from them."[4] This is an engineering-friendly definition, since it incorporates systems. In fact, the discipline of "biosystem engineering" relates to the "operation on industrial scale of biochemical processes ... and is usually now termed biochemical engineering."[5] Interestingly, this appears to be a distinction between what molecular biologists and biochemists do and what engineers do with the same information. Bioengineering is the "application of the physical sciences and engineering to the study of the functioning of the human body and to the treatment and correction of medical conditions."[6] This closely tracks with the definition of "biomedical engineering."

Thus, engineers as agents of technological progress are at a pivotal position. Technology will continue to play an increasingly important role in the future. The concomitant societal challenges require that every engineer fully understands the implications and possible drawbacks of these technological breakthroughs. Key among them will be biotechnical advances at smaller scales, well below the cell and approaching the molecular level. Technological processes at these scales require that engineers improve their grasp of the potential ethical implications. The essence of life processes is at stake.

Table 8.2-1 Ranking by global health experts of top ten biotechnologies needed to improve health in developing countries

Final ranking	Biotechnology
1	Modified molecular technologies for affordable, simple diagnosis of infectious diseases
2	Recombinant technologies to develop vaccines against infectious diseases
3	Technologies for more efficient drug and vaccine delivery systems
4	Technologies for environmental improvement (sanitation, clean water, and bioremediation)
5	Sequencing pathogen genomes to understand their biology and to identify new antimicrobials
6	Female-controlled protection against sexually transmitted diseases, both with and without contraceptive effect
7	Bioinformatics to identify drug targets and to examine pathogen–host interactions
8	Genetically modified crops with increased nutrients to counter specific deficiencies
9	Recombinant technology to make therapeutic products (e.g., insulin, interferons) more affordable
10	Combinatorial chemistry for drug discovery

Source: Data from survey conducted in: A.S. Daar, H. Thorsteinsdóttir, D.K. Martin, A.C. Smith, S. Nast, and P.A. Singer, 2002, Top Ten Biotechnologies for Improving Health in Developing Countries, *Nature Genetics*, 32, pp. 229–32.

Major bioethical areas

Engineering practice and research is deeply committed to and involved in the advancing technologies that will benefit humankind. However, this commitment and involvement calls for deliberate and serious considerations of actual and potential ethical issues. The President's Council on Bioethics[7] has summarized the dichotomy between the promise and the ethical challenges:

> *For roughly half a century, and at an ever-accelerating pace, biomedical science has been gaining wondrous new knowledge of the workings of living beings, from small to great. Increasingly, it is providing precise and*

sophisticated knowledge of the workings also of the human body and mind. Such knowledge of how things work often leads to new technological powers to control or alter these workings, powers generally sought in order to treat human disease and relieve suffering. But, once available, powers sought for one purpose are frequently usable for others. The same technological capacity to influence and control bodily processes for medical ends may lead (wittingly or unwittingly) to non-therapeutic uses, including "enhancements" of normal life processes or even alterations in "human nature." Moreover, as a result of anticipated knowledge of genetics and developmental biology, these transforming powers may soon be able to transmit such alterations to future generations.

So, let us consider some of these technological areas important to engineers and their attendant bioethical concerns.

Cloning and stem cell research

[C]loning represents a turning point in human history—the crossing of an important line separating sexual from asexual procreation and the first step toward genetic control over the next generation. It thus carries with it a number of troubling consequences for children, family, and society.

Leon R. Kass, Chair, The President's Council on Bioethics[8]

Researchers in the United States asserted in late 2001 that they were able to produce the first cloned human embryos, merely reaching a six-cell stage before the cells stopped division and died. In the meantime, a number of fertility specialists had declared a strong intent to clone human beings. In response to the technical and societal uncertainties and anxieties, the United States Congress had already begun in 1998 to consider these issues, with the House of Representative in July 2001 passing a strict ban on all human cloning, including the production of cloned human embryos. Since then, a number of cloning-related bills have been considered in the US Senate and several state legislations. Numerous nations have banned human cloning, with the United Nations considering an international convention on the issue. It suffices to say that the political and societal aspects of cloning are as challenging as the technical demands.

The biology of cloning begins with the ovum (egg). A human ovum consists of a single gamete cell, having only 23 active chromosomes. This means that the sex has not yet been determined. Once the ovum contains a complete nucleus from any species that is activated and developing – whether that has occurred by sexual fertilization or by asexual somatic cell nuclear transfer (SCNT) – an embryo of that species (*Homo sapiens*, sheep, cat, dog, etc.) is produced. Cloning is a type of reproduction wherein offspring result not from the chance of the union of ovum and sperm (sexual reproduction); rather it results from the deliberate replication of the genetic makeup of another single individual (asexual reproduction).

Human cloning is accomplished by introducing the nuclear material of a human somatic cell (donor) into an oocyte (egg) that has had its own nucleus removed or inactivated. This yields a product with a human genetic constitution virtually identical to the donor of the somatic cell. This technique is known as SCNT. Since SCNT uses human genetic material, the developing embryo is of the species *H. sapiens*.

Bioethics Question: Is research using human pluripotent stem cells ethical?

A "pluripotent" cell can be differentiated into more than one alternative type of mature cell; so it is able to produce all the cell types of the developing organism's body. Thus, a pluripotent stem cell has the same functional capacity (i.e., pluripotency) as an embryonic stem cell, though it does not necessarily share the same origin. "Stem cell research" involves isolating human embryonic stem cells from embryos at the blastocyst stage (Figure 8.2-1) or from the germinal tissue of fetuses (Figure 8.2-2). As of this writing, such harvesting kills the donor. The embryonic stem cells have been harvested from *in vitro* fertilization (IVF). Human adult stem

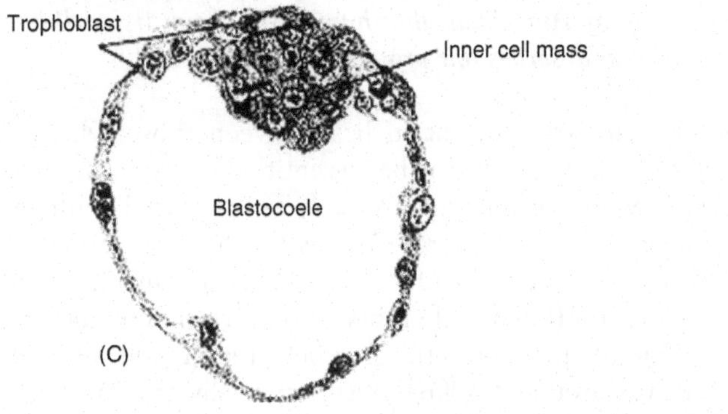

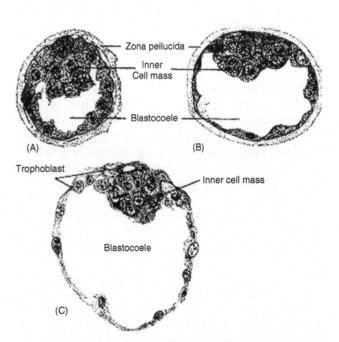

Figure 8.2-1 Three stages of blastocyst formation in the pig, drawn from sections to show the formation of the inner cell mass. Adapted from: The President's Council for Bioethics, 2004, *Monitoring Stem Cell Research*, Appendix A, Washington, DC.

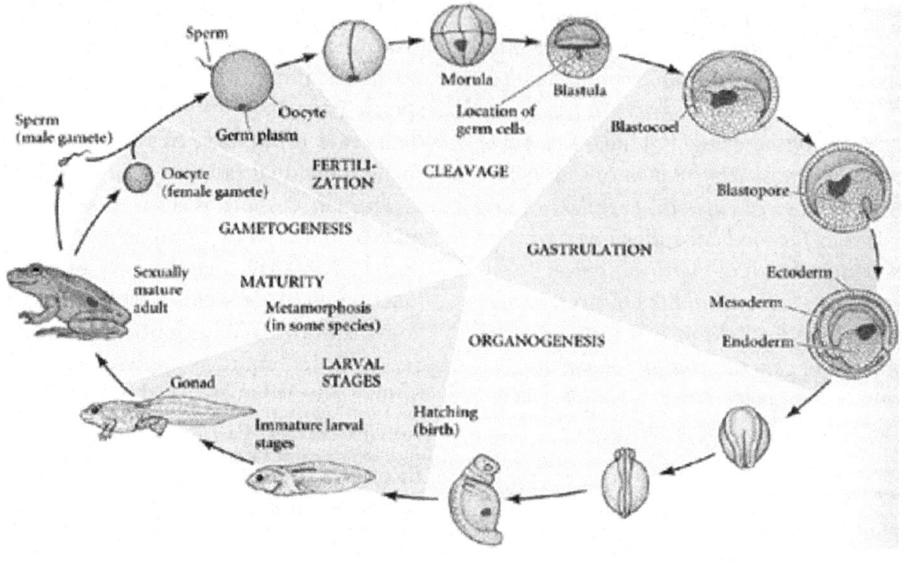

Figure 8.2-2 Stages of stem cell development in a frog. Note the continuity of germplasm.
Adapted from: The President's Council for Bioethics, 2004, *Monitoring Stem Cell Research*, Appendix A, Washington, DC.

(i.e., multipotent) cells have been isolated from a variety of tissues. These stem cell populations can be differentiated *in vitro* into various cell types, and are currently being extensively and intensively investigated for potential applications in regenerative medicine. Many scientists believe that embryonic and adult stem cells can lead to treatments for many human maladies. However, much of this is conjecture at this point.

The Chairman of the President's Council on Bioethics succinctly characterized the raging stem cell debate in a letter to the President:

> *While they may well in the future prove to be of considerable scientific and therapeutic value, new human embryonic stem cell lines cannot at present be obtained without destroying human embryos. As a consequence, the worthy goals of increasing scientific knowledge and developing therapies for grave human illnesses come into conflict with the strongly held belief of many Americans that human life, from its earliest stages, deserves our protection and respect.*[9]

The diametric opposition lies between those who see embryos as living human beings entitled to protection and those who consider them merely as "potential" humans that may be used as researchers see fit in an effort to advance the state of knowledge. This use includes the destruction (killing) of embryos to harvest stem cells.

One of the major ethical issues of stem cell research revolves around the first stage, especially the means of harvesting the embryonic stem cells by IVF. In particular, the argument contends that since fertilization has occurred, we are actually conducting research on a living human being. And, the IVF process itself is cavalier about the value of the individual, fertilized ova, destroying and discarding many in the process.

In a recent undergraduate seminar at Duke, students shared why IVF seems to "get a pass" morally, at least in comparison to the scrutiny given to stem cell and cloning research, even though all these are beginning of life (BOL) issues. One student pointed out that IVF is not new. Others noted that the technologies are more mainstream and understandable by the public. One student suggested the fact that many people know people who were conceived via IVF, so the mystique is gone. In the course of the discussion, it became obvious that most of the students do not even think about the morality of IVF. It simply exists as an alternative means of reproduction. And, the IVF industry is considered rather positively. The discarded embryos are just considered by many to be a byproduct of the process that allows people who could not ordinarily do so to have children. Thus, few seem to question the morality of IVF, so the discarding of embryos is not often seen as an immoral act of the fertility clinics. Further, those who oppose the use of embryos that would otherwise be destroyed are seen as "wasteful" and "myopic," or even as Luddites who stand in the way of progress and betterment of humankind.

As the President's Council on Bioethics puts it: "All extractions of stem cells from human embryos, cloned or not, involve the destruction of these embryos."[10] A rather common paradox is that research may either exacerbate or ameliorate ethical issues. For example, if emerging research supports the contention that embryonic stem cells are essential to provide the treatment and cures of intractable diseases, many scientists and ethicists will push the "greater good," utilitarian viewpoint, while others will see this as malicious means (sacrificing embryos and living human beings) toward the end (treatment).

Further, ethicists will be faced with the "slippery slope" dilemma. This argument sees a progression from diminished respect for embryos to a lack of respect for all unborn human beings, which leads to less respect for weaker members of the society. This last step in the progression is an argument for an incremental acceptance of eugenics, the belief that information about heredity can be used to improve the human race. The dreadful truth is that the genocide and other inhumane attempts at eugenics, such as the Nazis' attempted purging of Jewish people in the twentieth century, were clumsy and inefficient compared to emergent genetic manipulations available now and in the foreseeable future. The slippery slope advocates fear that the next Hitler will have subtle tools and a "rationale" supported by many to accomplish similar ends (i.e., eradicating selected human subpopulations).

In fact, one view extends this devaluing of certain human beings to even *after* they are born (a type of "postnatal abortion").

Peter Singer holds the title of Ira W. Decamp Chair of Bioethics at Princeton University's Center for the Study of Human Values. In his book, *Rethinking Life and Death: The Collapse of Our Traditional Ethics*, Singer asserts:

> *Human babies are not born self-aware or capable of grasping their lives over time. They are not persons. Hence their lives would seem to be no more worthy of protection than the life of a fetus.*[11]

Ironically, many pro-life advocates would agree with Singer's statement because they believe that a fetus is indeed worthy of protection as is any human person. However, they strongly oppose Singer's contention that self-awareness is an essential characteristic of personhood, so those who are not self-aware are, by Singer's definition, not persons. They may well agree with the definition for the species, but not for individuals. In other words, *H. sapiens* are indeed unique in their self-awareness, but each individual of the species varies in awareness of self and the awareness changes over the lifetime of the individual.

To Singer, those creatures at any given time that lack full self-awareness (the unborn, the newly born, those in vegetative states, the infirm, and those nearing the end of life) are merely human nonpersons. In fact, Singer, who is an emblematic advocate of animal rights, contends that unaware babies are less valuable than self-aware nonhuman animals. Singer's position points to a problem of unadulterated utilitarianism; that is, value is based entirely on the elitist's view of philosophical and scientific naturalism (materialism). In such a strident view, nonmaterial phenomena, such as a soul, are dismissed since they cannot be measured empirically (this also eliminates most human qualities, such as love, true happiness, and meaning). Such rigid scientism misrepresents the scientific data and even its own materialistic philosophy, since even the template for materialism cannot be proven by observation. That is, the premise that only that which can be measured is real is itself an unmeasurable and untestable premise.

Thus, such moral relativism is an easy and sloppy way to deal with personhood. Applying a "greatest good" defense (i.e., decisions wherein the most pleasure or desire is gained) allows for the rest of society to sacrifice certain members. Such sacrifice is easiest and scientifically erroneous if we call them "nonhuman" and it is philosophically erroneous if we call them "nonpersons." Singer and his ilk advocate both. Common sense and, hopefully, common morality would disagree. Such convenience is highly unscientific and grossly unethical. Theologian Michel Schooyans[12] has characterized this quite well:

> *Men cast doubt on the human character of certain beings whenever they sought arguments to exploit or exterminate their fellow human beings. In antiquity slaves were considered as things and barbarians as second class men. In the sixteenth century, some conquerors considered the Indians as "beasts in human appearance." The Nazis looked upon some men as "non-men", as* Unmenschen. *To these arbitrary classifications dictated by the masters corresponded real discrimination and this, in turn, "legitimized" exploitation or extermination.*

Scientific advances can wreak havoc with social values. What appears to be advanced thinking at times turns out to be retrograde attempts at dehumanization.

Advances in technology can be used to commoditize human beings. Genetic fetal testing, for example, can be used as a screen against the "unfit." Conversely, scientific advances can improve the bioethical landscape. Many of us hope that the perceived demand for embryonic stem cells will soon be obviated by scientific advances that allow the same cures and treatments using unquestionably morally acceptable tools, such as adult stem cells. Recent advances cause us to be hopefully optimistic (see Teachable Moment: Nanog).

Human enhancement

Engineering is all about enhancing our environment and ourselves. However, the tools that have emerged in recent decades are changing the nature of what we can do. Consequently, ethical issues are emerging or are becoming increasingly complicated by biomedical and biosystem breakthroughs. One area that should be discussed here, though, is "genetic engineering."

The first question that engineers must ask is whether so-called genetic engineering is really engineering. At its

Teachable Moment: Nanog

In Celtic legend, Nanog (short hand for *Tir Nan Og*) was a land of the eternally young. Fittingly, the Nanog molecule is at the threshold of what could be the unraveling of the need for embryonic stem cells to advance medical science. Scientists have argued that embryonic stem cells are unique in their ability to develop into any type of cell in the body, and further that they show great, yet still unfilled, potential to replace and to mend damaged tissue. Thus, the argument goes, embryonic stem cells are essential for such medical treatments.

Recently, researchers have become aware that the Nanog may someday allow ordinary cells to make use of that wonderful attribute.

Quite recently, working from the premise that amniotic fluid is known to contain multiple cell types derived from the developing fetus, researchers at Wake Forest University have advanced that state of science to show that these cells are able to produce an array of differentiated cells, including those of adipose, muscle, bone, and neuronal lineages.[13] The amniotic fluid-derived stem (AFS) cells are different from both adult and embryonic stem cells but share qualities of both. According to the lead researcher, Anthony Atala, AFS cells, like human embryonic stem cells, can double every one and a half days and can extensively differentiate. However, like adult cells, they do not seem to form tumors when implanted. As a scientific and potentially ethical bonus, AFS cells are readily gathered from amniotic fluid or placenta, making way for a potentially vast supply of cells available for therapeutic purposes.

Questions

1. What are some of the scientific obstacles that need to be overcome in Nanog research?
2. If the science supports this advance, what may be some of the reasons that scientists and others may still be skeptical of these findings?
3. Compare this type of skepticism to Kuhn's descriptions of the scientific community during a paradigm shift.

most basic level, genetic engineering is artificially modifying the genetic code of an organism. But what is artificial? Obviously, an active approach like inserting DNA into a cell is artificial, but what about passive techniques like choosing potential mates with certain traits as the spouse and the future parent of one's children?

Since engineering is the application of sciences to address societal needs, then genetic engineering seems to be a type of engineering. However, few engineers actually perform human genetic engineering. In fact, most of it is accomplished by biologists. And, these are mainly physicians (e.g., those in IVF clinics) and biomedical researchers (e.g., those conducting DNA work under the auspices of biochemistry). So, the lay public and clergy work in passive genetic engineering (e.g., choosing spouses) and biologists work in active genetic engineering (e.g., cellular biochemists). Where does that leave the engineering profession?

Arguably, engineers' work is affected by that of other professions. Thus, we need at least some preparation for the bioethical issues that are certain to arrive in our individual specialties.

The emergence of genetic manipulation has been rapid. Scientists made huge gains throughout the 1900s in discovering DNA and its structure. In 1977, Genetech became the first company to use recombinant DNA technology. This served as a catalyst in the industry by being the first of many opportunities and discoveries throughout the 1980s and 1990s. In 1988, the Human Genome Project started with the goal of determining the entire sequence of DNA in humans. Since then, genetic engineering has been the focus of a myriad of ethical questions and debates.

Patenting life

Bioprospecting, the search for natural substances of medicinal value, is a very divisive topic in bioethical debates. In November 1999, the US Patent and Trademark Office rescinded a patent on the plant species *Banisteriopsis caapi* held by a Californian since 1986. The plant is sacred to tribal communities living in the Amazon Basin and is the source of the hallucinogen *ayahuasca*, used in their religious rituals. In addition to being a harbinger of the complications of religious and cultural respect, it presages the looming, bitter debates about the extent to which biological materials can be "owned."

In fact, in one form or the other, humankind has been in the bioprospecting business for millennia. Like many bioethical issues, emerging technologies and research have changed the landscape (literally and figuratively) dramatically. And, powerful interests, such as pharmaceutical companies, see natural materials (including certain genes) as lucrative ventures that need to be harnessed for profit. For example, the biotechnology firm, Diversa Inc., entered into an agreement with the National Park Service to find efficacious and beneficial microbes in the geysers and springs in Yellowstone National Park. However, this met with much resistance and ultimately the agreement was suspended by a federal court ruling.

As controversial as patents on plant genetic material are, they pale in comparison to the bioethical debates surrounding that of animals. This is in part because patenting animals' genetic materials is linked to cloning. The larger bioethical issue is captured well by the Church of Scotland's Society, Religion and Technology Project:

Many people would also say that knowledge of a genetic sequence itself is part of the global commons and should be for all to benefit from. To patent parts of the human genome as such, even in the form of "copy genes," would be ethically unacceptable to many in Europe. In response it is argued that patenting is the legal assessment of patent claims, and should not be confused with ethics. But patenting is already an ethical activity, firstly in that it expresses a certain set of ethical values of our society; it is a response to a question of justice, to prevent unfair exploitation of inventions. Secondly a clause excluding inventions "contrary to public order and decency" is part of most European patent legislation – an extreme case of something like a letter bomb would be excluded as immoral. But now we have brought cancerous mice and human genetic material in the potential frame of intellectual property that ethics has moved to a much more central position, where it sits uncomfortably with the patenting profession. They do not like the role of ethical adjudicator to be thrust upon them by society.[14]

Teachable moment: patenting germplasm

Consider the following statement from Keith Douglas Warner of Environmental Studies Institute at Santa Clara University:

The privatization of germplasm formerly considered the common heritage of humankind is incompatible with notions of the common good and economic justice. The scrutiny that life industries have been receiving is well deserved, although most of this attention has been focused on the potential threats to human and ecosystem health. The economic implications of the biotechnology patent regime are less obvious because they do not impact individuals, but rather social groups. The pubic appears less interested in this dimension of the biotechnology revolution. Nevertheless, addressing this patent regime through the lens of the common good is a better strategy for critics of agricultural biotechnology, who will likely be more successful in slowing down the expansion of corporate control over germplasm by addressing economic issues.[15]

Questions

1. The biotechnical revolution has improved crop yields and has greatly increased the world's food supply in recent centuries. What have been the "human and ecosystem health" tradeoffs associated with these benefits?
2. Is it morally preferable to engage in "slowing down the expansion of corporate control over germplasm" and other genetic materials? Why or why not?
3. Compare any opportunities lost with the risks prevented if germplasm ownership by private concerns is halted.

Neuroethics

Ethicists and scientists have continuously struggled with defining just what constitutes personhood, but virtually every definition includes the human mind. So, the mind–brain connection includes elements of both ethics and neuroscience. Nanotechnology and other emergent applications of neurotechnologies affect who a person is. The medical definition of neuroethics is a bit pedestrian; that is, ethical aspects of neuromedicine. However, neuroethics is more than a subset of biomedical ethics.

Manipulations of neural tissue affect who we are, including our free will. This has intrigued ethicists for millennia. Neuroscientists Dai Rees and Steven Rose[16] consider neuroethics more broadly to include aspects of responsibility, personhood, and human agency. These issues are already upon us:

How will the rapid growth of human brain/machine interfacing – a combination of neuroscience and informatics (cyborgery) – change how we live and think? These are not esoteric or science fiction; we aren't talking about some science-fiction prospects of human cloning, but prospects and problems that will become increasingly sharply present for us and our children in the next ten to twenty years.

Every neuroethical issue is entangled with the concept of "self." Indeed, self is more than an aggregation of neurons, or even their synaptic functions. The challenge is how to characterize the mind and the self, while merging the scientific method with ethics. But this is nothing new. Rene Descartes tried but died before completing this project, followed by the rationalists (closest to Descartes' view), the empiricists (especially David Hume), until the thoroughly modern view of John Stuart Mill's utilitarianism. All of these views fall short in dealing with the self, so they also fail in providing an ethical framework for neuroethics. For example, Hume's logical positivism contended that all knowledge is based on logical inference from simple "protocol sentences" grounded in observable facts. This position is self-defeating since even this postulation cannot be observed! Perhaps, this argues that mind and brain are separate but interrelated concepts.

Thus, neuroethics must reconcile two perspectives: (1) the doctrine of psychophysical parallelism, which hold that a mental state is always accompanied by a neural activity, and (2) the concept of self, which holds that the self or person is more than the mental functions, and also includes temperament or motivation, which are affected by the physiological systems other than neural (e.g., endocrine) and, some would argue, nonphysiological factors (e.g., the soul). The first perspective is most widely held by the neuroscience community, but the second is the one that is embraced by most people, as well as many faith traditions. This difference is fodder for ethical conflict between the professional (e.g., the treating physician or hospital staff) and the patient. It may also help to explain the intense conflict in coma and other cases, such as those where a person is in a permanent vegetative state (PVS).

Organ transplantation

> *The debate over organ transplantation touches on many of the deepest issues in bioethics: the obligation of healing the sick and its limits; the blessing and the burden of medical progress; the dignity and integrity of bodily life; the dangers of turning the body, dead or alive, into just another commodity; the importance of individual consent and the limits of human autonomy; and the difficult ethical and prudential judgments required when making public policy in areas that are both morally complex and deeply important. It is no exaggeration to say that our attitudes about organ transplantation say much about the kind of society that we are, both for better and for worse.*
>
> President's Council on Bioethics, 2003[17]

Most scientists consider the human body to be like that of any other organism's body or even no different than any other system of matter. Thus, many do not see any moral relevance to organ transplants. The controversy arises when scientists and others set aside moral attitudes and strongly held convictions about the sacredness. This is in opposition to the beliefs of many people and of a number of faith traditions that see the human body as something much more than a bunch of cells. This is not much different from the psychophysical parallelism between the body and the mind, especially as it conveys a sense of self. Are we devaluing humanity if we have a free exchange of organs? The moral question asked by the President's Council on Bioethics is whether it is possible that in ignoring societal taboos have we diminished humanity, has it "lessened us, dehumanized us, and corrupted us?" This leads to the Council's more practical questions[18]:

1. What is the most ethically responsible and prudent public policy for procuring cadaver organs?
2. Should the current law be changed, modified, or preserved?

The engineer should be aware that such questions are being asked and should not assume that all, or even the majority, of people share the scientific community's perspective on organ transplantation. In fact, this may be the first time that a number of readers have heard about there even being an ethical issue associated with transplants. Suffice to say that the engineer should be aware of the diversity of opinion and of the likelihood that a number of his or her clients at a minimum are uneasy with the current system of transplants and may even be in outright opposition to the procedure.

Responsible conduct of human research

All human beings deserve respect. Those who are subjects of research must provide informed consent. Whereas philosophers for many centuries have extolled the virtues of respect, it really was not until 1979, with the publication of *the Belmont Report: Ethical Principles and Guidelines for the Protection of Human Subjects of Research,*[19] that the inviolable principles were articulated on how to treat people as research subjects. Basically this consists of three requirements: (1) respect for persons; (2) beneficence; and (3) justice.

Animal testing

Respect for animals has increasingly been integrated into the Western ethos, including rethinking the majority of the utilitarian perspectives within the medical and biological research communities. Animals have played a key role in biomedical research and technological development.

Animal research points to some of the problems of the popularly held utilitarian ethical framework, especially how the model influences compartmentalization and objectification.

Compartmentalization in thinking is not only a division of labor, but it is often a survival mechanism for many professionals engaged in having to live with an ethical decision. This was best demonstrated in my recent discussion with author and veterinarian, Richard Orzeck,[20] who candidly shared the following case:

During my second tour of duty through Cornell's pathology department, I was working late one weekday afternoon in the hospital's postmortem room performing an autopsy on a young Labrador who had died unexpectedly from no obvious cause, when one of my friends from the junior class walked into the room leading a dapple-gray, quarter horse mare. The pathology professor and I had just finished a long and detailed study of the poor dog's lungs and heart when he decided that we both needed a short break. Having nothing else to do while he was out smoking his cigarette, I went over to chitchat with my underclassman colleague.

As she stood there holding the silent and unusually well-behaved horse by its lead, we talked about all of the exciting things that vet students talk about when they're able to find the time: how our classes were going, how my rotations were going, and how demanding all of the professors were. After a couple of minutes of this small talk, I casually asked her what she was doing with the horse here in this room normally set aside for studying the dead.

Her eager and straightforward answer surprised me. She said that the animal was a donation to the college by an owner who, for whatever reason, no longer wanted it. She said that she had been working with a research professor after classes on a project focused on degenerative joint diseases in racing and performance horses, and that the animal was part of their study.

Reaching over to stroke the horse's neck, I looked into eyes of my future colleague and smiled as I congratulated her on being asked to be part of this research project. As a student research assistant, even though the job mostly involves doing all the "dirty work" such as mucking out the stalls and feeding and caring for the research subjects, you do get to interact on a higher level with the doctors and professors in charge. And it can all be pretty exciting stuff, especially if a scientific or medical breakthrough is discovered. Still curious as to why she was here in the postmortem room with this obviously healthy horse, I asked her again why they were here.

A few of seconds of awkward silence ensued as I waited for her to answer. Reaching up with her free hand to pet the horse on its muzzle, she finally told be that she was waiting for the professor to arrive. After he arrived, they would euthanize *the animal in order to harvest its healthy joint cartilage. These tissues were needed as a* positive control *in their research. And then she said no more. She seemed quite excited by the whole thing, and I remember, just briefly, being a little surprised at her enthusiasm.*

I congratulated her again, and not giving it any more thought at the time, returned to continue my autopsy of a Labrador retriever. Several minutes later I saw that the research professor had indeed arrived, and my friend and he carried out their grim task (I won't go into the details) to advance the "noble cause of medical science." My surprise at her enthusiasm was because I knew her to be—just like all my vet school colleagues—a dedicated and compassionate person who loved all animals, especially horses. She was subjecting herself to a rigorous, unrelenting, and expensive eight years of college and medical school to fulfill her life's dream of saving the lives and improving the welfare of her animal patients. But here she was, ready to end the life of a perfectly healthy horse and eager about the opportunity to do so.

Orzeck's case indicates the extent to which scientists can rationalize behaviors and decisions within a "research" context that we would not otherwise do. This is truly an example of the ends justifying the means. Orzeck continues:

I can't say for sure, but by the words that my fellow student used allowed her to justify and to absolve herself of what she was doing. "She and the researcher were going to euthanize the animal in order to harvest its healthy joint cartilage so that they could use the tissue as a positive control." Animal? Euthanize? Harvest? Positive control? To make what she was doing more bearable, she transformed this living, breathing horse into an object of exploitation through the use of these very specific words.

Animal: The creature was no longer a horse; it was now an animal object. It was easier to accomplish what she had to do to an animal or research subject than it was to a living horse.

Euthanize: Even though I and all other veterinarians (and, quite sadly nowadays, even the human medical profession) use the term all the time, the word euthanize *also is quite sneaky. The term has been bastardized and twisted and applied to many situations less ennobling. It's nicer to say "euthanize the research subject" than it is to say "kill the innocent horse."*

Harvest: The term brings to mind comforting images of strong and hardy farmers gathering up sheaves of wheat and corn, or perhaps little red-cheeked children helping their dear grandpa pick apples from well-trimmed trees in his ancient New England apple orchard. But the word also has recently been adopted by various groups of people to rationalize some despicable behavior. Credit and banking agencies "harvest" and "mine" the data that they extract and collect from their clients; various government agencies "harvest" mountains of covertly obtained information on its citizens for their so-called common good; and, most recently, scientists have sought to have the right to "harvest" stem cells from aborted babies to advance the causes of medical research. In summary, it was more pleasant to say "harvest the tissue" than it was to say, "We're exploiting this helpless horse to steal its healthy joint cartilage for our research."

Positive control: The use of positive controls is an honorable and long-used scientific technique that allows researchers to make accurate comparisons of information they are trying to obtain for whatever study they are trying to make. The best example I can think of involves human medical research and a study in which my father was asked to participate. As a frequent hospital patient suffering from heart disease, he had been asked to participate in a study of a new drug, which, if it worked, would improve the strength of his heart. On the permission sheet, he was offered free care during the length of the study, but he would not be told whether he would receive the new drug or a placebo. (A placebo is something that looks like a drug but is actually an inactive substance.) The placebo group would then be referred to as the positive controls. Their purpose was to act as a baseline, or starting point, against which any effects of the new drug could be compared.

The problems with using positive controls are many. The individuals, whether human or animal, don't receive any potential benefits of the scientific study that's being performed. In the worst case, the positive controls could be deprived of a life-giving procedure. But the biggest problem with regard to medical science and the use of positive controls is that by using the term positive control, *researchers are able to make objects out of these patients. And making an object out of living and breathing people now makes them easier to use and take advantage of, all in the name of science. It is easier to justify and to appear heroic to sacrifice a positive control than it is to put to death an innocent horse.*

I hasten to add that before we pass judgment on these otherwise compassionate and loving persons, we must be reminded that nearly all of us are guilty of objectification. For example, when clients bring their pets to my practice for euthanasia, I always ask them what they would like me to do with their remains; probably because the word remains *is easier to deal with and just sounds nicer than* dead pet's body.

Orzeck's case is yet another example of how terminology is not simply a neutral conveyance of information, but it is often steeped in ideology and perspective. Medical and engineering professionals, as Feynman reminded us, must be diligent and vigilant in using the correct terms to communicate. "Junk science" is fraught with the loose usage of language. In fact, strategically designed redefinitions and omission or selective use of actual data are common fallacies in junk science.

Is the research worth it?

Debates about animal research elucidate a number of ethical issues. First, as illustrated in the debate between Descartes and Gassendi, the difference between humans and animals is an important distinction. To most biologists, the difference is merely a continuum, as indicated by the development of the nervous system and other physiological metrics. These physiological complexities translate into sensory differences that differentiate the species' sentience (especially self-awareness), one of the variables that distinguish "humanness." In fact, sentient-centered ethics falls between "anthropocentric" (human-centered) and "biocentric" (i.e., all living creatures have inherent worth, e.g., Schweitzer's "reverence for life"[21]) ethical frameworks. That is, it calls us to appreciate the perspective of other creatures in a personal sense. With advances in the understanding of neurological processes, for example, we must assume that the more highly developed animals (and possibly even many of the less advanced) experience pain. That said, the bioethical view would cause us to want to do whatever we can to prevent or at least reduce suffering in these other species.

The ethical problem ensues when the utility of animal-derived knowledge is presented as a dichotomy. This can cause the utilitarian view to come to the fore and we are forced to choose between preventing and curing some human malady *versus* harming animals. However, this viewpoint is inherently weak, and such an argument is invalid. This illogical argument is referred to as the fallacy of *non sequitur* (Latin: "does not follow") since the outcome (cure) does not really depend on the condition (animal suffering). This particular *non sequitur* argument is known as "denying the antecedent."

1. If A then B. (Animal research leads to cures, so people benefit from animal research.)
2. Not A. (Not allowing animal research would end animal suffering.)
3. Therefore, not B. (Not allowing animal suffering would prevent cures.)

Not all animal research leads to animal suffering (at least it varies substantially). Another type of *non sequitur* is affirming the consequent:

1. If A is true, then B is true. (If animals suffer in research, cures are developed.)
2. B is stated to be true. (Animal research has resulted in cures.)
3. Therefore, A must be true. (Therefore, animals must suffer.)

In this instance, the argument derives from a false dichotomy that denies the possibility that animal research can be conducted in ways that are humane. It also denies that even if the animal research was necessary in the past (or at least beneficial to the advancement of medical knowledge), it ignores alternate approaches that can now displace such research (e.g., *in vitro* and *in silico*).

Moral concern grows with respect to what we value. For example, we are less tolerant of harm to our own pet than to a "generic" lab animal. Also, concern usually increases as an animal is considered more "human-like." This value can be seen even in people who believe that animals differ from humans both in degree (i.e., a continuum of self-awareness and cognitive processes) and in kind (i.e., inherently different, especially because many believe humans have a soul and animals do not). In both instances, people value certain animals more than others. In the first case, the species (i.e., the cat or the dog) may have more value to the pet owner than to species other than that of the pet; however, this value can be transferred to other animals. The second type of valuation increases with the complexity of the species. Greater concern may result because people ascribe human characteristics to non-human species (known as "anthropomorphism").

Primates are particularly important indicators of moral value, since they share many more "human" traits than most other animals. From a biomedical perspective, the fact that nonhuman primates have similar anatomical and physiological features makes them ripe for research. From a bioethical perspective, their similarities argue against much of the research, especially that which is invasive, painful, and unpleasant. The case of psychological researcher Edward Taub and his George Washington University student Alex Pacheco (who later became cofounder of People for the Ethical Treatment of Animals, PETA) dramatically illustrates the bioethical challenge.[22]

From 1981 to 1991, the National Institutes of Health provided $1.2 million to the Institute of Behavioral Research in Silver Spring, Maryland for Taub's attempt to regenerate severed nerves in 17 Rhesus Macaque (*Macaca mulatta*), popularly known as the Rhesus Monkey. Sensory nerves were cut in the monkeys' limbs (deafferentation), then stimuli such as electric shock, physical restraint, and food or sight deprivation were applied to compel animals to regain the use of crippled limbs. Taub's rationale for this research was to aid stroke victims and the mentally handicapped.

In 1982, Pacheco, who took a summer job at the institute, visited the laboratory and took photographs of the starved, uncared for animals. He reported the lab to the state police, who then raided the lab, removed the monkeys from the lab, and the case went to court. The monkeys were euthanized at the end of the case. Taub was convicted of six counts of animal cruelty.

The case illustrates some of the ethical aspects of animal research: definition of human life *versus* animal life, the application of precautionary principles, dignity in any research methodology, the question of whether animals have rights, and whether useful results truly justify research. On this last count, Taub's research appears to have advanced the state of neurological science. But was it worth it?

Animal research is of three types: pure research, applied research, and testing. Pure research strives to advance the understanding of biological structures, processes, and systems for the sole benefit to medical and scientific knowledge. Applied research is a research specifically done to address a biomedical need. Testing studies are special types of applied research that test the effects of a procedure, device, or drug to determine efficiencies and efficaciousness. Thus animal data provide a utilitarian purpose. Because animal researchers hold human life to be paramount, they see animals as a means to an end and view this research as justifiable based solely on possible benefits to the human race.

Viewed exclusively from this perspective, the Silver Spring monkey is a morally and ethically sound case. However, there must be other nonutilitarian aspects to the case that are immoral. Let us consider how the court may have arrived at a guilty verdict, and arguably, a pronouncement of unethical practices (Figure 8.2-3). Numerous consequences of this case are positive. Whereas the animals may have been abused and living in deplorable conditions, there were medical advances.

It appears that Pacheco made his decisions based mainly on three factors: legality (animal rights, guidelines), duty to provide health/safety of animals, and politics/public opinion. He showed less regard for the potential of these studies to aid human victims of stroke and mental retardation as Taub proposed. His position seems to have been that not only was the fruitful research not morally obligatory, but was also not morally permissible. This seems to be a more deontological view than that of Taub, who was most concerned with the benefits to science/medical research, finding cures for humans, and finance/economics of funding. In fact, many such whistle-blowing cases exhibit a duty-based view. Some are indeed utilitarian, for example, if a laboratory is doing bad science that will lead to negative consequences (e.g., dangerous device or drug), but others are in spite of good science if immoral activity is ongoing (e.g., cruelty and harassment). In the end, the courts sided with Pacheco

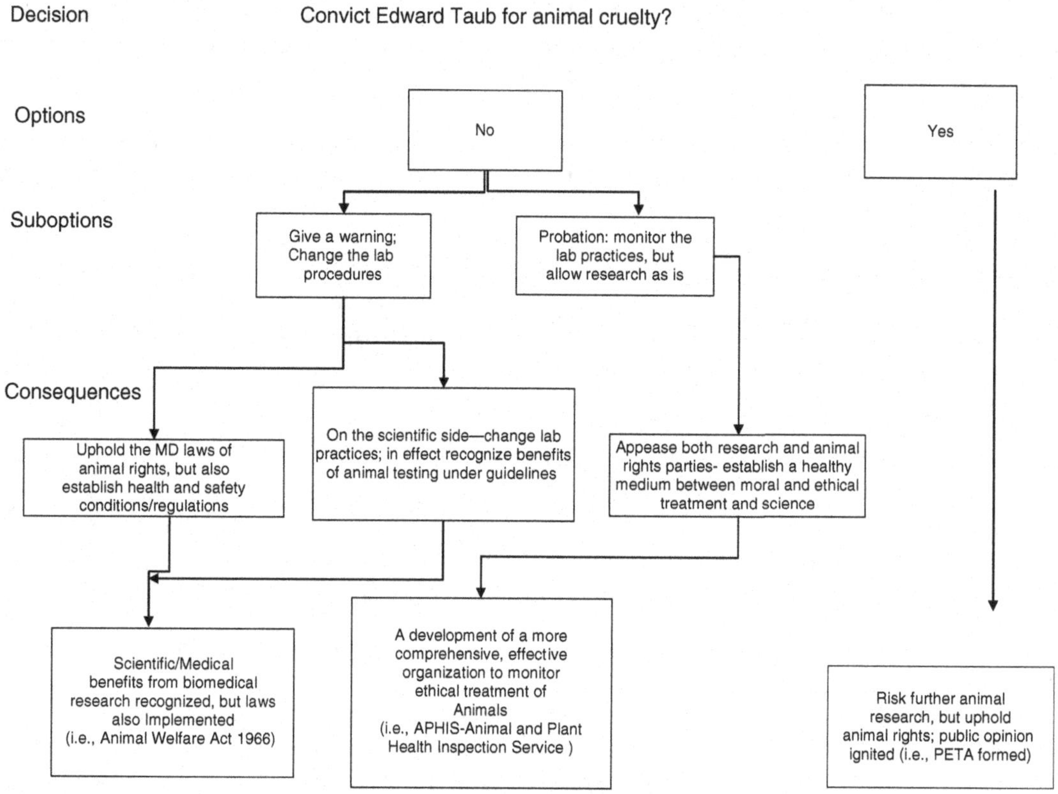

Figure 8.2-3 Event tree showing possible decision flow in Silver Spring Rhesus Monkey case.

and convicted Taub. We can show this logical progression as a syllogism:

Factual premise	Animal research is a powerful tool for scientific and medical research, and Taub's monkey-based studies were intended to study nerve regeneration and recovery of crippled limbs.
Fact-value premise	The animal subjects lived in deplorable conditions and were abused.
Evaluative premise	Even though animal research is a powerful tool for scientific and medical research, abuse of animal subjects in studies is morally wrong.
Evaluative conclusion	Therefore, Taub's studies, while intended for research, violate animals' rights and are morally wrong.

Therefore, animal research goes beyond "whether the nonhuman nature of other animals is morally relevant, and if it is relevant, what humanity and justice permit us to do with animals."[23] As we learn more about animal neurophysiology and psychology it becomes harder to argue that the values of justice and humanity are not applicable to animals, and do not merit ethical consideration. One does not have to be a hardcore member of the Animal Liberation Front to agree that some cases go beyond the pale. Few would disagree that some studies are clearly immoral, such as the studies done at the University of Pennsylvania wherein baboons were rendered brain dead from having their heads cemented to pistons that whipped the brain at accelerations up to 2,000 times the force of gravity, while researchers laughed while the primates were restrained, alert, and writhing before impact.[24] However, most animal welfare cases are not so clear.

Systematic reality check

Biomedical and biosystem engineers depend on animal studies, and they are more likely than many to support such research as a critical means toward a noble end. So, then, how can engineers be objective and realistic about matters where we are likely to have such a bias and conflicts of interests? One way is to do what engineers do best; use the systematic approach.[25] Start by gathering the pertinent facts and identifying the uncertainties and data gaps. Ethical problem solving is fraught with ambiguities and assumptions, so it is crucial that decisions are based on the best available factual information and that the decision maker be honest about what is not known.

Next, use the gathered information to identify realistic options and alternate solutions. Conduct a sensitivity analysis, i.e., compare how each alternative performs in terms of how sensitive it is to slight changes in scenarios needed to reach the decision. For example, one alternative may produce excellent outcomes (high on the utility scale), but in the process it leads to injustices or disproportional harm (low on the fairness scale). The alternative of withholding scientific findings that may promote bioterrorism (i.e., "dual use"), for instance, could be seen as one where one utility is optimized (i.e., security), but at the expense of another value (i.e., scientific freedom). Based on these analyses, a number of plans for addressing the problem can be considered.

All the reasonable plans can be compared and assessed against moral metrics, such as those embodied in the moral theories (e.g., utilitarianism, Rawlsism, deontology, and rational models). Key to these assessments is characterizing all the potentially affected parties and the stake each has in the decision. What are the risks and the benefits to each party from each option?

Finally, make a well-informed decision. However, even after the decision is made, keep seeking feedback and revisiting the options and alternatives to ensure that the decision continues to be the right one. Ethical decision making can be messy and chaotic. Unanimous decisions are the exception. Even consensus can be difficult. Sometimes, the right decision is made in the face of a resistant majority. So, the process is never static and the actions may need to be adapted as new information becomes available. For example, the constraint may disappear. The choice of keeping some lead (Pb) in gasoline even though it was known to have neurotoxic effects eventually went away in the 1980s when alternative fuel additives became available and when engines were redesigned. Or, when a sufficient repository of data becomes available, the need for even important animal testing can be replaced by computational methods (i.e., *in silico* studies and informatics). Thus, we need to continue to look for alternatives to animal research.

Animal welfare is sometimes categorized as the "3 Rs:"

1. *Reduction.* Methods that result in the use of fewer animals to obtain scientifically valid information.
2. *Refinement.* Methods to reduce stress or discomfort to the animals involved and to improve animals' overall well-being and environment.
3. *Replacement.* Methods other than animal studies that can provide robust biomedical information and modeling.

The bioengineering researcher must constantly rethink the research to adhere to these three methods.

Genetically modified organisms

Genetic modification of organisms is a very old endeavor. Humans have changed the characteristics of numerous plants and animals through selective breeding techniques, beginning with attempts to encourage offspring from organisms with favorable traits, such as size, color, texture, and taste. However, in recent decades the process has become supercharged with the onset of direct genetic manipulations of DNA and RNA. Like a number of bioethical issues, genetic manipulation is associated with the slippery slope fear.[26] The slippery slope occurs when allowing an act makes it "impossible to hold the line and prevent extension to a less justifiable situation."[27] The goal of genetic modification is to delete specific phenotypic characteristics from or introduce new characteristics to an organism's progeny. Prior to the late twentieth century, this was done externally, but now it is increasingly accomplished internally within the cell's genetic material. The resulting progeny is known as a "transgenic organism." Transgenesis occurs when DNA from another organism is introduced using artificial gene transfer techniques.

Such techniques allow researcher to understand the interactions of certain genes more completely. The major ethical issues involved in genetically modified organisms (GMO) often center on animal welfare, and risks to human health and environment. For example, what if a new creature is so different in kind that it has such a competitive advantage (i.e., no effective predators) or an ability to self-replicate that it would pose risks to public health and welfare, in violation of the engineer's first ethical canon. And, how many animals' lives are worth an important discovery? Are we decreasing the genetic diversity of our wildlife or destroying the habitats of other animals?

Such concerns come from within and outside of the scientific community and have at least a basis in utilitarianism (i.e., disagreement about the utility *versus* risks). Others oppose the research based on religious and moral concerns, arguing that the researchers and the biotechnological companies are immorally attempting to "play God" by creating entirely new beings and unnaturally altering the genetic makeup of progeny.

Furthermore, GMOs are generally supported due to the dual effects principal. First, it is believed that such research could lead to more profitable or productive animals. Second, scientists hope that this experimentation could aid humans by developing treatments to deadly diseases or methods to assist in the creation of tissues and organs.

Transgenic species

An organism that has been genetically modified (GM) is known as a "transgenic species." There are at least two ethical considerations of transgenic species. First, what

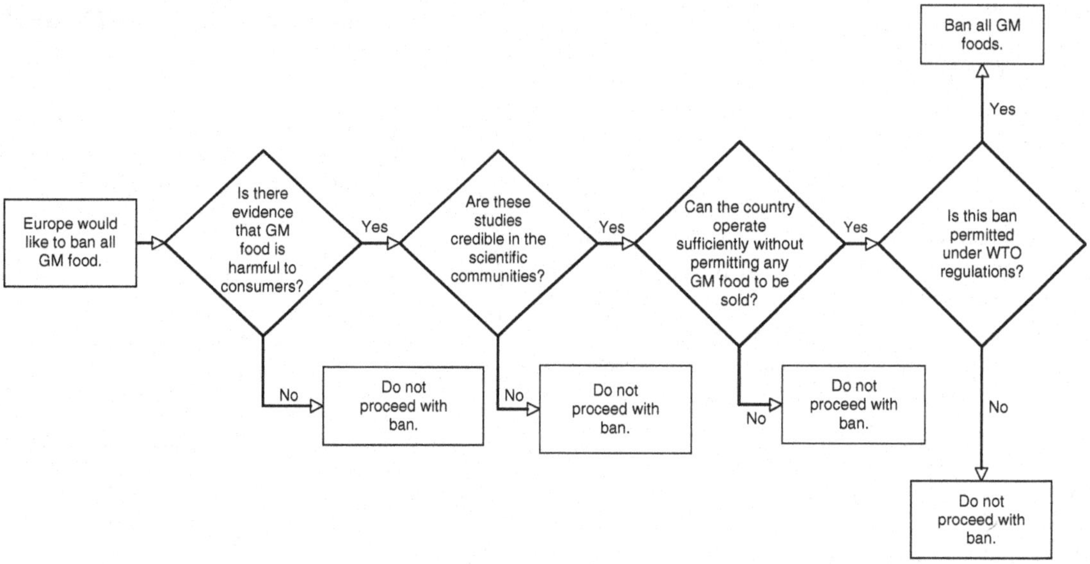

Figure 8.2-4 Decision flow chart for GM food in Europe.

impact might research and marketing on these species have on society, such as threats to health? Second, what are the ethical considerations needed on behalf of the creatures themselves? We address the first issue next in discussing GM food. The second is predominantly about animals. For example, is it ethical to modify a monkey so that its fur glows in the dark?

One issue with transgenic species is about who benefits. For example, Harvard University modified the genetic structure of a mouse, known as the "oncomouse," to produce a human cancer in the animal. From an anthropocentric viewpoint, this can be justified so long as it helps to advance cancer research, but it does not stand against a biocentric view or even a deontological view if one considers the protection of lower sentient species obligatory. It is also an example of bioprospecting, as Harvard has sought protection of their ownership of the oncomouse genetic information.

Food

The European Union has taken a number of recent actions, many precautionary, to address GMOs. These actions can be treated as a bioethical case. In 1999, a four-year ban on all GM crops was the result of several studies indicating that GM food could be harmful to humans. One study of the effects of GM potatoes on the digestive tract of rats found that there were large differences between the intestines of rats fed GM potatoes and those fed unmodified potatoes.[28] Other studies have indicated that introducing a new gene to a plant could create new allergens or cause reactions in susceptible people (e.g., a gene from peanuts transferred to soybeans could cause those allergic to peanuts to have reactions when they eat soybeans).

Based on these studies, the European Union announced a ban on all GM crops. At the end of 2002, the ban was lifted, but producers were forced to label all GM goods with a special DNA sequence that identified the origin of the crops. This made it easier for regulators to spot contaminated crops and feed. Later, in June 2003, the European Parliament agreed upon a UN biosafety protocol regulating the international trade of GM food. This protocol allows countries to ban the importation of GM organisms. However, there must be credible evidence about the dangers of the product. This protocol, therefore, states that goods developed from new technologies should be regulated based on the precautionary principle.

A number of factors surround GMOs. The safety and welfare of consumers appear to be the biggest concern. The European nations feared that their inhabitants would experience sicknesses similar to the mad cow epidemic (bovine spongiform encephalopathy). However, GM crops could lead to future discoveries in other fields. It is possible that certain technologies could be first tested on crops before animals, or that methods for altering plant genetics could be transferable to humans and/or animals. Opportunity costs, such as precautions slowing possible improvements in health, need to be considered. Enhanced nutritional value, disease prevention, and other health-related positive outcomes should be considered. Another possible negative outcome that should be investigated is the effects of the crops on the environment. Will new plant species interrupt predator–prey balances or be opportunistic, nuisance species?

A flow chart (Figure 8.2-4) shows the decision-making process in this case. At the beginning, Europe would like to ban the GM food. So, they approach their first question: Is there evidence to support such an action? The countries then must determine if there is sufficient evidence to support such an action. If they do have the evidence, Europe must next ensure that the studies meet all scientific requirements (i.e., they are replicable, etc.). If the studies are not sufficient, the ban cannot be established until further research has been conducted.

If all the scientific evidence is credible, then the countries encounter the feasibility issues. Can the countries be sustained without GM foods? The majority of the food produced has some form of genetic alteration, so such a ban would be dramatic.

This question also entails a political aspect. Moreover, with such a ban, it is undeniable that many of Europe's trading partners will be infuriated. So, will all the nations be able to handle the political backlash of such a regulation?

By answering yes to the above questions, Europe is able to move onto the final question – would this regulation be permitted under World Trade Organization guidelines? If the answer is yes, then they ban the food. If the answer is no, then they have no choice but to allow the goods to be sold in their nations.

The initial decision posed was whether Europe should ban GM foods. Europe had two options, to ban or not to ban the food. If they chose to ban, they had two suboptions. First, they could attempt to get studies and research to support the decision. Second, they could avoid the research step and proceed to ban the GM food. This demonstrates that the GMOs have macroethical and geopolitical ramifications, so no matter how comfortable the GMO researcher may feel, there will be opposition to the research.

Environmental health: the ethics of scale and the scale of ethics

The ultimate measure of a man is not where he stands in moments of comfort and convenience, but where he stands at times of challenge and controversy.

Martin Luther King, Jr. (1963)[29]

Environmental problems are often characterized by scale, a concept familiar to most engineers. In fact, we often describe phenomena by their dimensions and by when they occur; that is, by their respective spatial and temporal scales. Engineers are comfortable with dimensional analysis. But, can we "measure" ethics in a similar way? Of course, physical analogies do not completely hold for metaphysical phenomena, but they can be instructive. King's advice is that we can measure ethics, especially in our behavior during worst cases. How well can we stick to our principles and duties when things get tough? Philosophers and teachers of ethical philosophy at the university level frequently subscribe to one classical theory or another for the most part, but most concede the value of other models. However, they all agree that ethics is a rational and reflective process of deciding how we ought to treat each other.

Temporal aspects of bioethical decisions: environmental case studies

An important consideration in making bioethical decisions is the amount and type of effects that will result from an action. This is the beginning of rational ethics, that is, forming a factual premise. A particularly difficult aspect of a decision or an activity is predicting the cascade of events and their future impacts.

From a teleological perspective, an event can represent a means or an end. Indeed, an end can actually be a means toward another end. So let us consider a few bioethical problems of various scales. Manufacturing and commercial decisions about material use, such as metallic pigments in paint, lead-based fuel additives, and industrial processes that generate carcinogenic byproducts, can have lasting effects for generations. Such decisions are quite complex. A case in point is the comparison of short- and long-term effects of using coal *versus* nuclear fission to generate electricity. Combusting coal releases particle matter and damaging compounds like sulfur dioxide and toxic substances like mercury. Nuclear power presents a short-term concern about potential accidental releases of radioactive materials and long-lived radioactive wastes (sometimes with half-lives of hundreds of thousands of years). Nuclear events have also been extremely influential in our perception of pollution and threats to public health. Most notably, the cases of Three Mile Island, Dauphin County, Pennsylvania in 1979 and the Chernobyl nuclear power plant disaster in the Ukraine in 1986 have had an unquestionable impact on not only nuclear power, but on other aspects of environmental policy, such as community "right to know" and the importance of risk assessment, management, and communication.

Similarly, decisions regarding armed conflict must consider not only the tactical warfare, but the geopolitical and public health changes wrought by the conflict. Furthermore, the psychological and medical effects on combatants and noncombatants must be taken into account. Prominent cases of these effects include the use of the defoliant Agent Orange in Vietnam and decisions to prescribe drugs and to use chemicals in the Persian Gulf War in the 1990s. The World War II atomic bombings on the Japanese cities of Hiroshima and Nagasaki in August 1945 not only served the purpose of accelerating the end of the war in the Pacific arena, but also ushered in the continuing threat of nuclear war. In addition, the bombings were the world's first entrees to the linkage of chronic illness and mortality (e.g., leukemia and radiation disease) that could be directly linked to radiation exposure. Following are a few cases where the effects did not become apparent until after a protracted lag period.

Agent orange

The use of Agent Orange during the Vietnam War (used between 1961 and 1970) dramatically demonstrates the concept of "latency period," where possible effects may not be manifested until years or decades after exposure. Agent Orange is a defoliant and weed-killing chemical that was used by the US military during the Vietnam War. It was sprayed to remove leaves from the trees behind which the enemy troops would hide. Agent Orange was dispersed by airplanes, helicopters, trucks, and backpack sprayers. In the 1970s, years after the tours of duty in Vietnam, some veterans became concerned that exposure to Agent Orange might be the cause of delayed health effects. One of the chemicals in Agent Orange contained small amounts of the highly toxic compound 2,3,7,8-tetrachlorodibenzo-*para*-dioxin (TCDD).

The US Department of Veteran Affairs (VA) has listed a number of diseases, which could have resulted from exposure to herbicides like Agent Orange. The law requires that some of these diseases be at least 10% disabling under VA's rating regulations within a deadline that began to run the day a person left Vietnam. If there is a deadline, it is listed in parentheses after the name of the disease as follows:

- Chloracne or other acneform disease consistent with chloracne (must occur within one year of exposure to Agent Orange)
- Chronic lymphocytic leukemia
- Diabetes mellitus, type II
- Hodgkin's disease
- Multiple myeloma
- Non-Hodgkin's lymphoma
- Acute and subacute peripheral neuropathy (for the purpose of this section, the term acute and subacute peripheral neuropathy means temporary peripheral neuropathy that appears within weeks or months of exposure to an herbicide agent and resolves within two years of the date of onset)
- Porphyria cutanea tarda (must occur within one year of exposure to Agent Orange)
- Prostate cancer
- Respiratory cancers (cancer of the lung, bronchus, larynx, or trachea)
- Soft-tissue sarcoma (other than osteosarcoma, chondrosarcoma, Kaposi's sarcoma, or mesothelioma)

The issue is international. After all, if it is true that US soldiers exposed to Agent Orange show symptoms often associated with dioxin exposure, then there is also likely to be residual dioxin contamination in the treated areas of Vietnam. Dioxin is highly persistent.

Scientists from Vietnam, the United States, and 11 other countries have discussed the state of the science of research in the health effects of dioxin. The Vietnamese and US have agreed to a plan that addresses the need for direct research on human health outcomes from exposure to dioxin and research on the environmental and ecological effects of dioxin and Agent Orange.

Chlorinated dioxins have 75 different forms and there are 135 different chlorinated furans, depending on the number and arrangement of chlorine atoms in the molecules. The compounds can be separated into groups that have the same number of chlorine atoms attached to the furan or to the dioxin ring. Each form varies in its chemical, physical, and toxicological characteristics (Figure 8.2-5) The primary concerns in Vietnam from prolonged exposure to dioxin are reproductive and developmental disorders that may be occurring in the general population.

Dioxins are created only unintentionally during chemical reactions, especially combustion processes; that is, they have never been synthesized for any other reason than for scientific investigation, for example, to make analytical standards for testing. The most toxic form is the TCDD isomer, which is a byproduct when certain

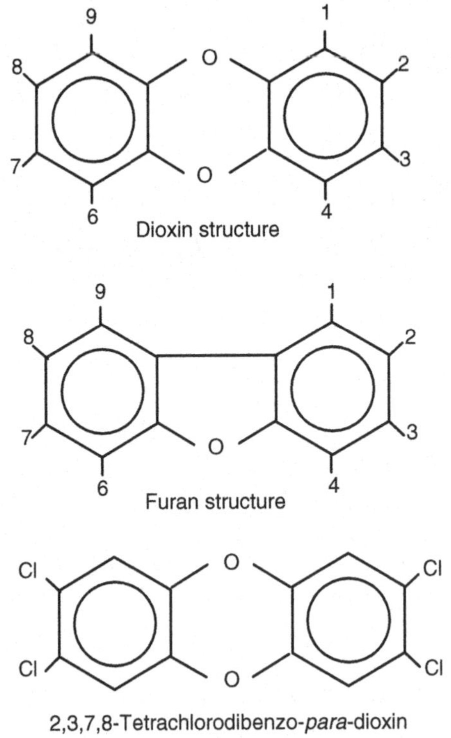

Figure 8.2-5 Molecular structures of dioxins and furans. Bottom structure is of the most toxic dioxin congener, tetrachlorodibenzo-*para*-dioxin (TCDD), formed by the substitution of chlorine for hydrogen atoms at positions 2, 3, 7, and 8 in the molecule.

pesticides, such as those used in the Vietnam defoliants, are produced. Other isomers, which may have been present in the formulations along with the 2,3,7,8 configurations, are also considered to have higher toxicity than the dioxins and furans with different chlorine atom arrangements.

Dioxin contaminants of Agent Orange have persisted in the environment in Vietnam for over thirty years. In addition to a better understanding of the outcomes of exposure, an improved understanding of residue levels and rates of migration of dioxin and other chemicals into the environment is needed. "Hot spots" containing high levels of dioxin in soil have been identified and others are presumed to exist but have yet to be located.

Dioxin has migrated through soil and has been transported through natural processes such as windblown dust and erosion into the aquatic environment. Contamination of soil and sediments provides a reservoir source of dioxin for direct and indirect exposure pathways for humans and wildlife. Movement of dioxin through the food web results in bioconcentration and biomagnification with potential ecological impacts and continuing human exposure. Research is needed to develop approaches for more rapid and less expensive screening of dioxin residue levels in soil, sediment, and biological samples, which can be applied in Vietnam.

Actually, a number of defoliating agents were used in Vietnam, including those listed in Table 8.2-2. Most of the formulations included the two herbicides 2,4-D and 2,4,5-T. The combined product was mixed with kerosene or diesel fuel and dispersed by aircraft, vehicle, and hand spraying. An estimated 80 million liters of the formulation was applied in South Vietnam during the war.[30]

Table 8.2-2 Formulations of defoliants used in the Vietnam war

Agent	Formulation
Purple	2,4-D and 2,4,5-T used between 1962 and 1964
Green	Contained 2,4,5-T and was used during 1962–1964
Pink	Contained 2,4,5-T and was used during 1962–1964
Orange	A formulation of 2,4-D and 2,4,5-T used between 1965 and 1970
White	A formulation of picloram and 2,4-D
Blue	Contained cacodylic acid
Orange II	2,4-D and 2,4,5-T used in 1968 and 1969 (also sometimes referred to as "Super Orange")
Dinoxol	2,4-D and 2,4,5-T. Small quantities were tested in Vietnam between 1962 and 1964
Trinoxol	2,4,5-T. Small quantities tested in Vietnam during 1962–1964

Source: Agent Orange website: http://www.lewispublishing.com/orange.htm; accessed on 22 April 2005.

The Agent Orange problem illustrates the problem of uncertainty in characterizing and enumerating effects. There is little consensus on whether the symptoms and disorders suggested to be linked to Agent Orange are sufficiently strong and well documented; that is, provide weight of evidence, to support cause and effect. This complicates bioethical decision making since the factual premises are fraught with uncertainties.

Japanese metal industries

As in military decisions, industrial decisions also have significant long-term health considerations.

Minamata mercury case

One of the most telling cases of improper bioethical decision making was that of Minamata, a small factory town on Japan's Shiranui Sea. Minamata means "nitrogen," emblematic of the town's production of commercial fertilizer by the Chisso Corporation for decades, beginning in 1907.[31] Beginning in 1932, the company produced pharmaceutical products, perfumes, and plastics and processed petrochemicals. Chisso became highly profitable, notably because it became the only Japanese source of a high-demand primary chemical, diotyl phthalate (DOP), a plasticizing agent. These processes needed the reactive organic compound, acetaldehyde, which is produced using mercury. The residents of Minamata payed a huge price for this industrial heritage. Records indicate that from 1932 to 1968, the company released approximately 27 tons of mercury compounds into the adjacent Minamata Bay. This directly affected the dietary intake of toxic mercury by fisherman, farmers, and their families in Kumamoto, a small village about 900 kilometers from Tokyo. The consumed fish contained extremely elevated concentrations of a number of mercury compounds, including the highly toxic methylated forms (i.e., monomethyl mercury and dimethyl mercury), leading to classic symptoms of methyl mercury poisoning. In fact, the symptoms were so pronounced that the syndrome of these effects came to be known as the "Minamata disease."

In the middle of the 1950s, residents began to report what they called the "strange disease," including the classic form of mercury toxicity; that is, disorders of the central nervous system (CNS) and peripheral nervous systems (PNS). Diagnoses included numbness in lips and limbs, slurred speech, and constricted vision. A number of people engaged in uncontrollable shouting. Pets and domestic animals also demonstrated mercury toxicity, including cat suicides and birds dying in flight. These events were met with panic by the townspeople.

Physician, Hajime Hosokawa from the Chisso Corporation Hospital, reported in 1956 that "an unclarified disease of the central nervous system has broken out." Hosokawa correctly associated the fish dietary exposure with the health effects. Soon after this initial public health declaration, government investigators linked the dietary exposures to the bay water. Chisso denied the linkages and continued the chemical production, but within two years, they moved their chemical releases upstream from Minamata Bay to the Minamata River, with the intent of reducing the public outcry. The mercury pollution became more widespread. For example, towns along the Minamata River were also contaminated. Hachimon residents also showed symptoms of the "strange disease" within a few months. This led to a partial ban by the Kumamoto Prefecture government, which responded by allowing fishermen to catch, but not to sell, fish from Minamata Bay. The ban did not reduce the local people's primary exposure, since they depended on the bay's fish for sustenance. However, the ban did acquit the government from further liability.

Some three years after the initial public health declaration, in 1959, Kumamoto University researchers determined that the organic forms of mercury were the cause of the "Minamata disease." A number of panels and committees, which included Chisso Corporation membership, studied the problem. They rejected the scientific findings and any direct linkages between the symptoms and the mercury-tainted water. After Hosokawa performed cat experiments that dramatically demonstrated the effects of mercury poisoning, Chisso managers no longer allowed him to conduct such research and his findings were concealed from the public.[32] Realizing that the links were true, the Chisso Corporation began to settle with the victims. The desperate and relatively illiterate residents signed agreements with the company for payment, which released the company from any responsibility. The agreement included the exclusion: "... if Chisso Corporation were later proven guilty, the company would not be liable for further compensation." Notwithstanding these setbacks, Minamata also represents one of the first cases of environmental activism. Residents began protests in 1959, demanding monetary compensation. However, these protests led to threats and intimidation by Chisso; so the victims settled for fear of losing even the limited compensation.

Chisso installed a mercury removal device at the outfall, known as a "cyclator," but it omitted a key production phase so the removal was not effective. Finally, in 1968, the company stopped releasing mercury compounds into the Minamata River and Bay. Ironically, the decision was neither an environmental one, nor an engineering solution. The decision was made because the old mercury production method had become antiquated. Subsequently, the courts found that the Chisso Corporation repeatedly and persistently contaminated Minamata Bay from 1932 to 1968.

Victim compensation has been slow. About 4000 people have either been officially recognized as having "Minamata Disease" or are in the queue for verification from the board of physicians in Kumamoto Prefecture. Fish consumption from the bay has never stopped, but mercury levels appear to have dropped, since cases of severe poisoning are no longer reported.

Cadmium and *Itai Itai* disease

The mines of central Japan located near the Toyama Prefecture have been removing metals from the surrounding mountains since as early as 710 A.D. Gold was the first metal to be mined from the area, followed by silver in 1589, and shortly thereafter lead, copper, and zinc. At the start of the twentieth century, the Mitsui Mining and Smelting Co. Ltd controlled the production of these mines. As a result of the Russo-Japanese War, World War I, and World War II, a surge in the demand for metals in the use of weapons and military equipment caused massive increases in the mines' production that was aided with the advent of new European technologies in mining.

Along with the huge increase in mining production came a significant increase in pollution produced from the mines. Liquid and solid wastes were dumped into the surrounding waters, including the Jinzu River that flows into the Sea of Japan, and the five major tributaries that flow into the Jinzu River. The Jinzu River water system supplies water to the surrounding city of Toyama, 30 kilometers downstream from the main mining operations.[33] This water was primarily used by the surrounding areas for irrigation of the rice paddies. In addition, water provided a source for drinking, washing, and fishing.

Large amounts of cadmium were released into the Jinzu River Basin from 1910 to 1945. Cadmium was extracted from the earth's crust during the production of other metals like zinc, lead, and copper that were being mined near the Toyama area. Cadmium is a naturally occurring element that does not corrode easily, enters the air during mining, can travel long distances, and then falls into the ground or water only to be taken up by fish, plants, animals, or humans from the environment.[34] The cadmium from the mines deposited in the river and land of the Jinzu River Basin was absorbed by the surrounding plants and animals causing fish to die and the rice to grow poorly. Furthermore, humans living in that area consumed poisoned water and rice.

As a result of the ingestion of cadmium, a previously undiagnosed disease specific to the Toyama Prefecture appeared in 1912. Initially, the symptoms were not well understood and suspected to stem from either a regional or a bacterial disease or the result of lead poisoning. However, in 1955, cadmium was linked to the disease which came to be known as *itai itai*. In 1961, the

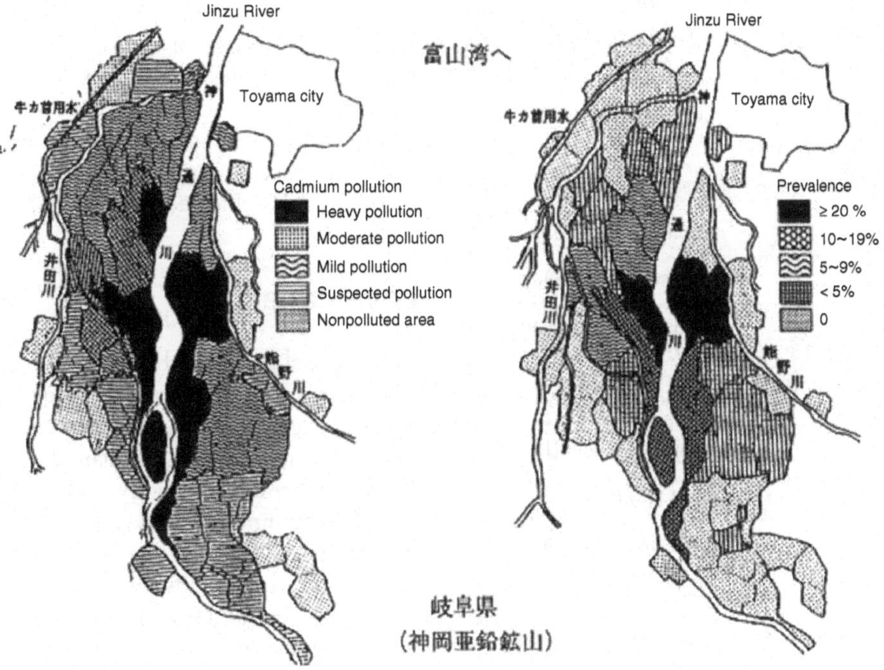

Figure 8.2-6 The co-occurrence of cadmium contamination with the prevalence of itai itai disease in woman over fifty years of age in Toyama Prefecture (c. 1961). Adapted from: Kanazawa Medicine, 1998.

Kamioka Mining Station of Mitsui Mining and Smelting Co. Ltd was linked as the direct source of cadmium poisoning, and the Toyama Prefecture, 30 kilometers downstream, was designated the worst cadmium-contaminated area (Figure 8.2-6). The concentrations of cadmium in this prefecture were orders of magnitude higher than that found in background levels and were well above other industrialized locations (Table 8.2-3).

Cadmium exposure results in major health-related problems of *itai itai* disease, irreversible kidney damage, and bone disease (*itai itai* is Japanese for "Ouch Ouch"). After exposure to high levels of cadmium, the kidneys have decreased ability to remove acids from the blood due to proximal tubular dysfunction resulting in hypophosphatemia (low-phosphate blood levels), gout (arthritic disease), hyperuricemia (elevated uric acid levels in the blood), hyperchloremia (elevated chloride blood levels), and kidney atrophy (as much as 30%). Following kidney dysfunction, victims of the disease develop osteomalacia (soft bones), loss of bone mass, and osteoporosis leading to severe joint and back pains and increased risk of fractures ATSDR (1999).

Scale is more than size

As the previous cases indicate, the timing of exposure, not just the amount of time, is of the essence. Corporations and governments have changed the health baseline of the world. The background and baseline of contaminants in the human body has increased dramatically in a relatively short time (see Table 8.2-3). Theo Colburn, who is known for her publications on the growth and effects of environmental endocrine disrupting compounds, sums up the problem of synthetic chemicals and the moral responsibility of corporations that gained much attention in the 1960s and 1970s:

> *Every one of you sitting here today is carrying at least 500 measurable chemicals in your body that were never in anybody's body before the 1920s.... We have dusted the globe with man-made chemicals that can undermine the development of the brain and behavior, and the endocrine, immune and reproductive systems, vital systems that assure perpetuity.... Everyone is exposed. You are not exposed to one chemical at a time, but a complex mixture of chemicals that changes day by day, hour by hour, depending on where you are and the environment you are in.... In the United States alone it is estimated that over 72,000 different chemicals are used regularly. Two thousand five hundred new chemicals are introduced annually—and of these, only 15 are partially tested for their safety. Not one of the chemicals in use today has been adequately tested for these intergenerational effects that are initiated in the womb.*[35]

Table 8.2-3 Estimates of average daily dietary intake of cadmium based on food analysis in various countries

Country	Estimates (μg Cd per day)	Reference
Areas of normal exposure		
Belgium	15	Buchet et al. (1983)
Finland	13	Koivistoinen (1980)
Japan	31	Yamagata & Iwashima (1975)
Japan	48	Suzuki & Lu (1976)
Japan	49	Ushio & Doguchi (1977)
Japan	35	Iwao (1977)
Japan	49	Ohmomo & Sumiya (1981)
Japan	59	Iwao et al. (1981a)
Japan	43.9 (males), 37.0 (females)	Watanabe et al. (1985)
New Zealand	21	Guthrie & Robinson (1977)
Sweden	10	Wester (1974)
Sweden	17	Kjellström (1977)
United Kingdom	10–20	Walters & Sherlock (1981)
USA	41	Mahaffey et al. (1975)
Areas of elevated exposure		
Japan	211–245	Japan Public Health Association (1970)
Japan	180–391	
Japan	136	Iwao et al. (1981a)
United Kingdom	36	Sherlock et al. (1983)
United Kingdom	29	Sherlock et al. (1983)
USA	33	Spencer et al. (1979)

Source: International Programme on Chemical Safety, 1992, Environmental Health Criteria: 134 (Cadmium), World Health Organization, Geneva, Switzerland.

The corporate track record in the twentieth century was not good. Toxic substances sprang up at Love Canal in New York, Times Beach in Missouri, and the Valley of the Drum in Kentucky. Soon numerous other sites throughout the United States were found to be contaminated, leading to a progression of environmental laws, especially the Comprehensive Environmental Response, Compensation and Liability Act, better known as "Superfund," beginning in 1980. Much of the previous legal precedence for environmental jurisprudence had been more on the order of nuisance laws. With the greater recognition of public health risks associated with toxic substances like those found in these hazardous waste cases, the public and the environmental professionals called for a more aggressive and scientifically based approach. This changed the bioethical framework for engineering.

Love canal

The seminal and arguably the most infamous case is the contamination in and around Love Canal, New York. The beneficent beginnings of the case belie its infamy. In the nineteenth century, William T. Love had an opportunity to generate electricity from Niagara Falls and the potential for industrial development. To achieve this, Love planned to build a canal that would also allow ships to pass around the Niagara falls and travel between the two Great Lakes, Erie and Ontario. The project started in the 1890s, but soon floundered due to inadequate financing and also due to the development of alternating electrical current, which made it unnecessary for industries to locate near a source of power production. Hooker Chemical Company purchased the land adjacent to the canal in the early 1990s and constructed a production facility. In 1942, Hooker Chemical began disposing its industrial waste in the canal. This was wartime in the United States, and there was little concern for possible environmental consequences. Hooker Chemical (which later became Occidental Chemical Corporation) disposed of over 21 000 tons of chemical wastes including halogenated pesticides, chlorobenzenes, and other hazardous materials into the old Love Canal. The disposal continued until 1952 at which time the company covered the site with soil and deeded it to the City of Niagara Falls, which wanted to use it for a public park. In the transfer of the deed, Hooker specifically stated that the site was used for the burial of hazardous materials, and the company warned the city that this fact should govern future decisions on the use of the land. Everything Hooker Chemical did during those years was seemingly legal.

About this time, the Niagara Falls Board of Education was looking for a place to construct a new elementary school, and the old Love Canal seemed like a perfect spot. This area was a growing suburb, with densely packed single-family residences on streets paralleling the old canal. A school on this site seemed like a perfect solution and so it was built.

In the 1960s, the first complaints began, and intensified during the early 1970s. The groundwater table rose during those years and brought some of the buried chemicals to the surface. Children in the school playground were seen playing with strange 55-gallon drums that popped out of the ground. The contaminated liquids started to ooze into the basements of the nearby residents, causing odor and complaints of various health problems. More importantly, perhaps, the contaminated liquid was found to have entered the storm sewers and was being discharged upstream of the water intake for the Niagara Falls water treatment plant.

The situation reached a crisis point and President Jimmy Carter declared an environmental emergency in 1978, resulting in the evacuation of 950 families in an area of 10 square blocks around the canal. But the solution presented a difficult engineering problem. Excavating the waste would have been a dangerous undertaking, and would probably have caused the death of some of the workers. Digging up the waste would also have exposed it to the atmosphere resulting in uncontrolled toxic air emissions. Finally, there was the question as to what would be done with the waste. Since it was mixed, no single solution such as incineration would have been appropriate. The US Environmental Protection Agency (EPA) finally decided that the only thing to do with this dump was to isolate it and continue to monitor and to treat the groundwater. The contaminated soil on the school site was excavated, detoxified, and stabilized and the building itself was razed. All the sewers were cleaned, removing 62 000 tons of sediment that had to be treated and moved to a remote site. At the present time, the groundwater is still being pumped and treated, thus preventing further contamination.

The cost is staggering, and a final accounting is still not available. Occidental Chemical paid $129 million and continues to pay for oversight and monitoring. The rest of the funds are from the Federal Emergency Management Agency and from the US Army, which was found to have contributed waste to the canal.

The Love Canal story had the effect of galvanizing the American public into understanding the problems of hazardous waste, and was the impetus for the passage of several significant pieces of legislation such as the Resource Conservation and Recovery Act, the Comprehensive Environmental Response, Compensation, and Liability Act, and the Toxic Substances Control Act. It also ushered in a new bioethical perspective: "toxics."

Times beach

The Time's Beach story is an example of a confluence of events that can lead to difficult bioethical decisions. It had inauspicious beginnings. Times Beach was a popular resort community along the Meramec River, about 17 miles west of St Louis. With few resources, the roads in the town were not paved and dust on the roads was controlled by spraying oil. For two years, 1972 and 1973, the contract for road spraying went to a waste oil hauler, Russell Bliss. The roads were paved in 1973 and the spraying ceased.

Bliss obtained his waste oil from the Northeastern Pharmaceutical and Chemical Company in Verona, Missouri, which manufactured hexachlorophene, a bactericidal chemical. In the production of hexachlorophene, considerable quantities of dioxin-laden waste had to be removed and disposed. A significant amount of the dioxin was contained in the "still bottoms" of chemical reactors. Incineration of the wastes was very costly. The company was taken over by Syntex Agribusiness in 1972, and the new company decided to contract with Bliss to haul away the still bottom waste without telling Bliss what was in the oily substance. Bliss mixed it with other waste oils and he used it to oil the roads in Times Beach, unaware that the oil contained high concentrations of dioxin (greater than 2000 ppm), including the most toxic congener, TCDD.

Bliss oiled roads and sprayed oil to control dust, especially in horse arenas. He used the dioxin-laden oil to spray the roads and horse runs in nearby farms. In fact, it was the death of horses in these farms that first alerted the Center for Disease Control to sample the soil at the farms. They found dioxin, but did not make the connection with Bliss. Finally in 1979, the US EPA became aware of the problem when a former employee of the company told them about the sloppy practices in handling the dioxin-laden waste. The EPA converged on Times Beach in "moon suits" and panic set in among the populace. The situation was not helped by the message from the EPA to the residence of the town. "If you are in town it is advisable for you to leave and if you are out of town do not go back." In February 1983, on the basis of an advisory from the Centers for Disease Control, the EPA permanently relocated all the residents and businesses at a cost of $33 million. Times Beach was by no means the only problem stemming from the contaminated waste oil. Twenty-seven other sites in Missouri were also contaminated with dioxins. Most of the dioxin contamination has since been cleaned up.

The contamination of Times Beach, Missouri, while affecting much of the town, was not the key reason for the national attention. The event occurred shortly after the Love Canal hazardous waste problem was identified and people were wondering just how extensively dioxin and other persistent organic compounds were going to be found in the environment. Times Beach also occurred at the time when scientists and engineers were beginning to get a handle on how to measure and even how to treat (i.e., by incineration) contaminated soil and water.

Other events, such as the worries about DDT and its effect on eagles and other wildlife, cryptosporidium outbreaks, and Legionnaire's Disease, also seem to have received greater attention due to their timing. This illustrate the importance of timing in bioethical decision making. One would be hard-pressed to identify a single event that caused the public concern about the metal lead. In fact, numerous incremental steps brought the world to appreciate lead toxicity and risk. For example, studies following lead reductions in gasoline and paint showed marked improvements in lead levels in blood in many children. Meanwhile, scientific and medical research was linking lead to numerous neurotoxic effects in PNS and CNS, especially of children. Similarly, stepwise progressions of the knowledge of environmental risk occurred for polychlorinated biphenyls (PCBs), numerous organochlorine, organophosphate, and other pesticides, depletion of the stratospheric ozone layer by halogenated (especially chlorinated) compounds, and even the effect of releases of carbon dioxide, methane, and other "greenhouse gases" on global warming (though more properly called global climate change).

Teachable moment: the whole is greater than the sum of its parts

Aristotle (384 BC–322 BC) is generally credited with the famous philosophical principle, "the whole is greater than the sum of its parts." To engineers and scientists, this is synergy. We also know that in some instances the whole is less than the sum of its parts, which is antagonism. Such principles also hold for ethics. These cases have demonstrated that the combination of a few unethical decisions can lead to dramatic consequences.

The engineering profession is one of creative problems solving and optimization. Decisions lead to events that lead to consequences, which in turn, lead to other decisions, events, and consequences. At each decision point, the engineer is presented options. The design is successful when it meets the needs of the client, and it provides for the optimal outcome in terms of the public's health, safety, and welfare. Thus, every design decision is to some degree an ethical decision. And, every engineering project is an ethics-laden project. The design must lie between timidity and recklessness.

> *The most exciting phrase to hear in science, the one that heralds new discoveries is not "Eureka," (I found it!) but "That's funny... ."*
>
> Isaac Asimov (1920–1992)

One hard and fast rule is that engineers are purveyors of science; so we do not have the prerogative of messing with the facts and scientific laws. We must respect factual information, yet we must be careful not to label something prematurely as "correct," especially in light of possible contradicting information. Just because we are comfortable with the *status quo* is not a sufficient reason to hold to wrong and invalid arguments. In fact, such stubbornness is quite unscientific. Ironically, scientists are considered to be searchers of truth no matter where the journey takes us, yet like most people we have a comfort zone. Such xenophobia translates into scientific and ethical myopia. We do not like change and resist it reflexively, particularly if it means undoing some of our cherished tenets.

Surprise, even when unwelcome, as Asimov seemed to say, is a necessary part of the discovery process in science and in ethics. Women and men of science, both researchers and practitioners, must simultaneously keep an open mind and an eye toward better ways of doing things and must maintain scientific rigor. Thus, engineering is an intellectually and morally active pursuit.

Questions

1. What do the Agent Orange, Japanese metal, Love Canal and Times Beach cases have in common?
2. How has ethical responsibility changed since the early 1900s?

Active engineering

An important tenet of ethics and engineering communications is that we be clear in what we say and mean. The first step in bioethical analysis is to reach an understanding about the facts of the matter. So, then, what is meant when we say that engineering is an "active" process? The first definition in the dictionary[36] is "characterized by action, rather than by contemplation or speculation." This connotation is interesting and valuable from a number of perspectives. Obviously, the noun "action" drives the adjective "active." But among the definitions of the noun that best fit the adjective's definition, one seems to stand out; that is, "the bringing about of an alteration by force or natural agency." Alteration means that something has changed. Engineers hope and expect the change to be better than what existed before. However, if this is not the case, it is a type of failure. Failure in itself is not unethical. Only some failures are rooted in ethical breaches.

Another interesting aspect of the first dictionary definition of "active" is the contrast between an action and contemplation or speculation. Contemplation should precede any action. Prudence dictates that "you should look before you leap." And, the proper sequence of ethical or any decision making is "ready, aim, shoot," although many of us frequently "ready, shoot, aim." In other words, action-oriented people can have a natural proclivity to act, even before much or any thinking. This can lead to addressing symptoms of problems, but not solving the problems themselves. The definition is not a value judgment, but simply recognizes that the two steps, contemplation and action, are unique and sequential.

The other contrast between action and speculation, however, is value-laden. There are times, deservedly or not, when decision makers appear to suffer from "analysis paralysis." Decisions are always made with some degree of uncertainty. The key dilemma for the designer is to know when a sufficient amount of information about the risks and the benefits of a design has been ascertained. The sufficiency is a function of the severity of the outcome (costs of being wrong) and the loss of a benefit (opportunity risk). In some cases, a "50/50" flip of the coin decision-making approach is sufficient, such as whether a bridge between Chapel Hill and Durham should be painted Carolina or Duke blue (although such decisions carry great local import in the North Carolina Piedmont). Conversely, a decision as to whether to set a drinking health standard at 10 parts per million (ppm) or 15 ppm for a pollutant can account for a margin of waterborne diseases and even allow for increased mortality if the higher concentration is applied. Speculation in the former case may lead to some unhappy fans, but speculation in the latter case can translate into increased morbidity and mortality.

Another way to think about activity is to consider what one means by the opposite of active. At least two very different antonyms must be considered. If something or someone is "inactive" they are idle. Another antonym of active is passive. An active solution is one that requires added energy, whereas a passive solution is one that occurs as a matter of course. Sometimes, the passive solution is preferable, as when a noninvasive, homeopathic treatment is used instead of an invasive, pharmaceutical approach to treat a similar malady. And, at the other time, professional judgment dictates the need to "do something."

Yes, engineers do things, and what we do should give us great pride. Engineers are, by nature, a thoughtful and outwardly directed lot. We are sympathetic to the needs of others, beginning with our clients. In fact, designers, in general, are optimistic and future-oriented. We see things long before they take on physical reality. Thus, engineers are highly suited to take a long view of things. Some faith traditions may characterize us as having much "faith." Our code calls us to be "faithful agents." While this is a statement about the trust that our clients place in us, it is also a statement about our confidence and faith in ourselves as professionals. This is the engineering perspective. We "expect" the medical device, or building, or water supply to become real, to become "substance."[37] Even those engineers who are called in to fix things, such as biomedical engineers finding ways to ameliorate human suffering, environmental engineers who remediate hazardous waste sites or failure analysts who look for ways to prevent future disasters, always expect things to improve with action.

Engineers are willing to stand behind our work (e.g., we build in feedback in the operation and maintenance (O&M) process, we check progress continuously on biomedical devices, install monitoring wells to ensure waste cleanup is going as planned, and we work closely with inspectors to incorporate lessons learned from failure analyses). Hence, in addition to being faithful, we adapt. In fact, adapting is a big part of design. The dynamism of biomedicine has been a fertile ground for engineering for many centuries. A classic example was the 1592 visit of Galileo to one of the first medical institutions in Padua, Italy.[38] Galileo, acting as an engineer, took the opportunity to lecture the future medical practitioners and researchers on mathematical principles including his own theories, as well as their applications (notably the pendulum, thermoscope, and telescope). Such applications led to an enhanced understanding of physiology, such as studies by Galileo's student, Sanctorius, of human body temperature and pulse rates. One student at Padua during Galileo's visit was William Harvey, renowned for characterizing the circulatory system, which he based on the mechanical and motion laws expounded by Galileo. Such interplay between medicine and engineering, while often subtle and indirect, has evolved into the more formal collaborations we see today. Duke, Johns Hopkins, and other world class biomedical engineering programs are co-located and intellectually intertwined with leading medical schools. But, the dynamic design processes of the contemporary engineers make for complicated ethical challenges and none more so than those related to biomedicine.

The stakes are very high in biomedicine. It is truly and literally a matter of life and death. Not advancing the state of biomedical science is simply not an ethical option. Consequently, the typical precautions that may hold for other areas of engineering need special considerations when addressing biomedical challenges. For example, the "no action" alternative is seldom satisfying to biomedical practitioners. Not looking for a better device or system that can improve the quality of a patient's life is simply not acceptable. Risk assessors refer to this as "opportunity risk." That is, if we simply seek the refuge of no added risk, we may lose an opportunity to really improve things. For example, nanotechnologies (using systems that are only a few hundred molecules in size) are fraught with risks, such as the chance that working at this level may cause changes to self-replication and other cellular signals, which could cause unknown damage. However, if we do not seek the uses of nanotechnologies, such as the application of highly efficient, biologically inspired processes for drug delivery, we may lose the opportunity to make the drugs more efficacious. If we can use biologically inspired nanomanufacturing to synthesize and deliver tumor-reducing drugs, it may be possible to treat cancers that have heretofore been untreatable. So, we must be bold in applying nascent sciences and simultaneously take great care to ensure that we are not opening some awful "Pandora's box."[39] Finding this balance of acceptable and reasonable risk goes beyond science, engineering, and technology and enters the realms of ethics.

In a way, the advancement of science and its applications must follow a type of ethical index. Indices have weighted values for each of the important variables. Some more sophisticated indices have operators that can shut the index down, such as a water quality index that automatically goes to zero if certain levels of dissolved oxygen are not available (even if all the other values are fine). Thus, engineers need some type of working index for biomedical ethics. They must be able to tell when something is going awry. This is a tall order, since most of us are well trained in the physical sciences and mathematics, but few in the intricacies of biology associated with health care. The exception, of course, is the biomedical engineer, but even their perspective of biomedicine varies from that of others in the health care profession.

Choosing the right "ethics index" requires some knowledge of the common ethical models used to evaluate engineering decisions and actions.

Ethical theories: a primer

Engineers are more familiar with the works of Newton, Boyle, Einstein, and Bohr than they are with those of Hammarabi, Socrates, Hippocrates, Aristotle, Hobbes, Mill, Kant and Rawls. However, it was only four centuries ago that philosophy was central to all studies. In fact, physical science was considered to be within the realm of "natural philosophy." The Ancient and Renaissance scientists did not distinguish ethics and other aspects of philosophy from physics and other underpinning sciences of engineering. Later Descartes, Humes, and Mill tried to turn things around by attempting to place ethics under the scientific method. Although we cannot return to such thinking completely, it may be useful to consider our careers as engineers a bit more comprehensively as a starting point for understanding bioethics.

The comprehensive view allows us to consider some topics not often covered in engineering texts; concepts like good *versus* evil, moral *versus* immoral acts, and obligations *versus* prohibitions are understood by most professionals at some intuitive level, but unlike our colleagues in the humanities and social sciences, we are more likely to avoid considering them theoretically. However, reading the classical works of Aristotle, Aquinas, Kant, *et al.* makes the case for life being a mix of virtues and vices available to humans. In fact, our previous discussion about active engineering sets the stage for ethical engineering. Ethicists sometimes refer to goodness to be the result of "right action." Virtue can be defined as the power to do good or a habit of doing good. In fact, one of Aristotle's most memorable lines is "Excellence is habit." So, if we do good, we are more likely, according to Aristotle, to keep doing good. Conversely, vice is the power and the habit of doing evil.

Truth

The subjectivity or relational nature of good and evil, however, leads to some discomfort in scientific and engineering circles, where meanings of certainty and consistency of definition are crucial to problem solving. Actually, this is consistent with the perspective of philosophers, especially ethicists. Scientific facts are a crucial part of any bioethical case analysis, and so are other factors. To wit, philosophers tell us that we can determine whether a moral argument is valid (not necessarily correct) by parsing the argument into a "syllogism," which consists of four parts:

1. The factual premise
2. The connecting premise (i.e., factual to evaluative)
3. The evaluative premise
4. The moral conclusion

For example, the facts may show that exposing people to a chemical at a certain dosage (e.g., one part per million or ppm) leads to a specific form of cancer in one in every ten thousand people. We also know that, from a public health perspective, allowing people to contract cancer as a result of some human activity is morally wrong. Thus, the syllogism would be the following:

1. Factual premise: exposure to chemical X at 1 ppm leads to cancer
2. Connecting premise: A company is releasing 10 kg of chemical X per day which leads to 1 ppm exposure to people living near an industrial plant
3. Evaluative premise: decisions that allow industrial releases that lead to cancer are morally wrong
4. Moral conclusion: therefore, corporate executives who decide to release 10 or more kilograms of chemical X from their plants are morally wrong.

Science provides us with the factual premise and part of the connecting premise, but social mores and norms provide us with the evaluative premise and drive the moral conclusion. However, if we are uncertain about the facts, the validity of the argument is disrupted. Scientific uncertainties are brought about both by variability and error.[40] Variability is ever present in space and time. Every case has a unique set of factors, dependent variables, situations, and scenarios, so that what occurred will never be completely repeated again. Every cubic centimeter of soil is different from every other cubic centimeter. The same goes for a sample of water, sediment, air, and organic tissue. And, these all change with time. Taking a sample in the winter is different from that in the summer. Conditions in 1977 are different in so many ways from conditions in 2007. And, of course, there are errors. Some are random in that the conditions that led to the cases in this book are partially explained by chance and things that are neither predictable nor correctable, although we can explain (or at least try to explain) them statistically, (e.g., with normal distributions).

Other error is systematic, such as those of my own bias. I see things through a prism different from anyone else's. This prism, like yours, is the result of my own experience and expertise. This prism is my perception of what is real and what is important. My bias is heavily weighted in sound science, or at least what I believe to be sound science (as opposed to "junk science").[41] Sound science requires sufficient precision and accuracy in presenting the facts. "Precision" describes how refined and repeatable an operation can be performed, such as the exactness in the instruments and the methods used to obtain a result. It is an indication of the uniformity or reproducibility of a result. This can be likened to shooting arrows,[42] with each arrow representing a data point. Targets A and B in Figure 8.2-7 show equal precision. Assuming that the center of the target; that is, the bull's-eye, is the "true value," data set B is more accurate than A. If we are consistently missing the bull's-eye in the same direction at the same distance, this is an example of bias or systematic error. The good news is that if we are aware that we are missing the bull's-eye (e.g., by comparing our results with those of known standards when using our analytical equipment), we can calibrate and adjust the equipment. To stay with our archery analogy, the archer would move her sight up and to the right.

Thus, "accuracy" is an expression of how well a study conforms to some defined standard (the true value). So,

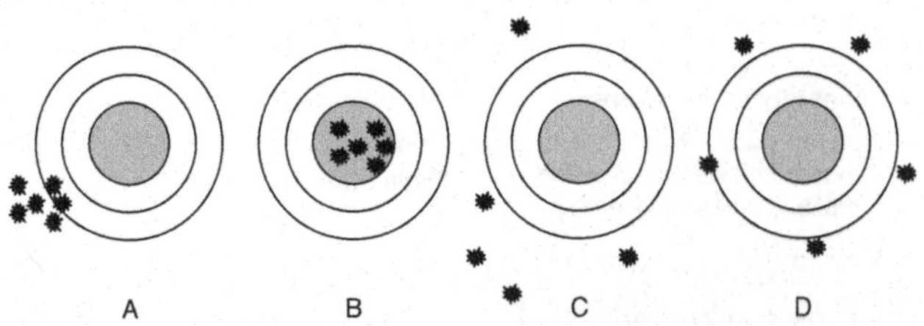

Figure 8.2-7 Precision and accuracy. The bull's-eye represents the true value. Targets A and B demonstrate data sets that are precise, targets B and D demonstrate data sets that are accurate, and targets C and D demonstrate data sets that are imprecise. Target B is the ideal data set, which is precise and accurate.

accuracy expresses the quality of what we find, while precision expresses the quality of the operation by which we obtained our finding. So, the other two scenarios of data quality are shown in targets C and D. Thus, the four possibilities are that our data is precise but inaccurate (target A), precise and accurate (target B), imprecise and inaccurate (target C), and imprecise and accurate (target D).

At first blush, target D may seem unlikely, but it is really not all that uncommon. The difference between targets B and D is simply that D has more "spread" in the data. For example, the variance and standard deviation of D is much larger than that of B. However, their measures of central tendency, i.e., the means, are nearly the same. So, both data sets are giving us the right answer, but almost all of the data points in B are near the true value. None of the data points in D is near the true value, but the mean (average location) is near the center of the bull's-eye; so, it has the same accuracy as target B, but with much less precision. The key is that precision and accuracy of the facts surrounding a case must be known.

Even if we agree on the facts, we all will not agree on which of the virtues and vices are best or even whether something is a virtue or a vice (e.g., loyalty). But one concept does seem to come to the fore in most major religions and moral philosophies: empathy. Putting oneself in another's situation is a good metric for virtuous acts. The "Golden Rule" is at the heart of Immanuel Kant's "categorical imperative." With apologies to Kant, here is a simplified way to describe the categorical imperative: When deciding whether to act in a certain way, ask if your action (or inaction) will make for a better world if all others in your situation acted in the same way. An individual action's virtue or vice is seen in a comprehensive manner. It is not whether one should cut corners and use deficient materials when designing a device, it is whether everyone in the same situation should do likewise. A corollary to this concept is what Elizabeth Kiss, formerly of Duke's Kenan Center for Ethics and now president of Agnes Scott College calls the "Six O'clock News" imperative. That is, when deciding whether an action is ethical or not, consider how my friends and family would feel if they heard about all of its details on tonight's TV news. That may cause me to consider more fully the possible externalities and consequences of my decision (and maybe even tempt me to "overdesign").

Psychological aspects of ethics

Engineers are familiar with conditions and constraints. For example, we know from chaos theory the importance of initial conditions on outcomes, but also know these can be changed by boundary conditions and constraints imposed by the engineer (i.e., interventions). These concepts are analogous to moral development. At the most basic level, a person's path toward moral decision making is a continuum. Since much of an engineer's ethics is influenced by his or her own "conscience," a basic understanding of the theory of moral development is in order.[43]

"Conscience" is something that develops rather rapidly in human development, but is honed constantly throughout life. As a child learns to communicate, especially with the first caregivers, the conscience is controlled and informed by external sources (i.e., we learn to do what we are told by an authority figure, or we learn by watching and imitating others). However, as we mature, the conscience to a greater extent is controlled internally. A mature conscience is the result of internalization. In other words, we start to depend completely on outside sources, grow to become independent, but eventually behave as interdependent members of a community.

> *Another doctrine repugnant to Civil Society is, that whatsoever a man does against his Conscience is Sin; For a man's Conscience, and his Judgment is the same thing; and as the Judgment, so also the Conscience may be erroneous; once a man lives in a commonwealth, the Law is the public Conscience, by which he hath already undertaken to be guided. Otherwise in such diversity, as there is of private Consciences, which are but private opinions, the Commonwealth must needs be distracted, and no man dare to obey the Sovereign Power, farther than it shall seem good in his own eyes.*
>
> Thomas Hobbes (1660)[44]

Individual, personal conscience is the building block for what the seventeenth century natural philosopher Thomas Hobbes considered to be needed in a "social contract." Hobbes argued that humans are egoistic and self-serving by nature, so an efficient society first instructs its individual members to behave in ways that support the community. Hobbes thought that society had to overcome the "state of nature" where selfish, brutish individual desires would lead to anarchy. This argument sees the role of conscience as twofold: to benefit the individual member (private conscience) and to benefit the society as a whole (the commonwealth).

Character. As mentioned in Chapter 8.1, Jean Piaget, Lawrence Kohlberg, and other educational psychologists have argued that moral development takes a predictable and stepwise progression. The development is the result of social interactions over time. For example, Kohlberg[45] identified six stages in three levels (see Table 8.1-2.)

Kohlberg insisted that these steps are progressive. Every person must pass through the preceding step before advancing to the next. Thus, a person first behaves according to authority (stages 1 and 2), then according to approval (stages 3 and 4), before finally maturing to the point where they are genuinely interested in the welfare of others. My experience has been gratifying in that most of my colleagues and the majority of engineering students in my courses have indicated moral development well within the postconventional level.

We can apply the Kohlberg model directly to the engineering profession (Figure 8.2-8). The most basic (bottom tier) actions are preconditional; that is, engineering decisions are made solely to stay out of trouble. While proscriptions against unethical behavior at this level are effective, the training, mentorship, and other

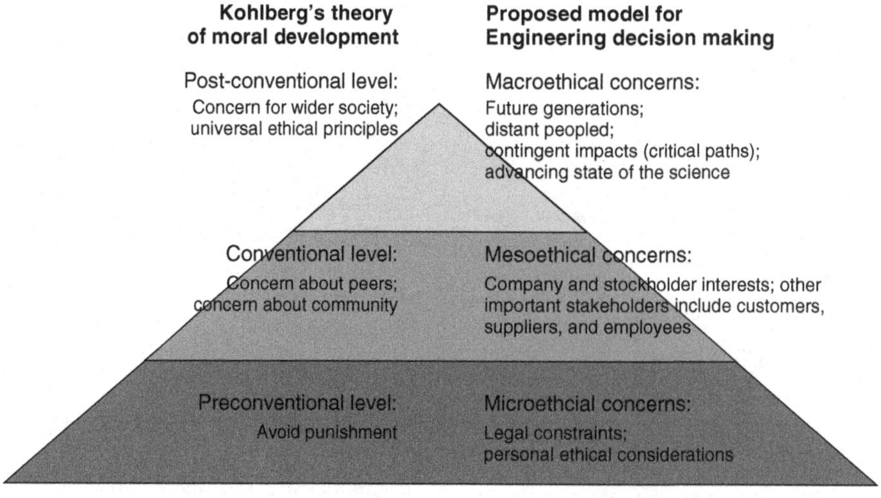

Figure 8.2-8 Adaptation of Kohlberg's stages of moral development to the ethical expectations and growth in the engineering profession.

opportunities for professional growth push the engineer to higher ethical expectations. This is the normative aspect of professionalism. In other words, with experience as guided by observing and emulating ethical role models, the engineer moves to conventional stages. The engineering practice is the convention, as articulated in our codes of ethics.

Above the conventional stages, the truly ethical engineer makes decisions based on the greater good of the society, even at personal costs. In fact, the "payoff "for the engineer in these cases is usually for people he or she will never meet and may occur in future he or she will not share personally. The payoff does provide benefits to the profession as a whole, notably that we as a profession can be trusted. This top-down benefit has incremental value for every engineer. Two common sayings come to mind about top-down benefits. Financial analysts often say about the effect of a growing economy on individual companies: "A rising tide lifts all ships." Likewise, environmentalists ask us: "To think globally, but to act locally." In a like manner, the individual engineer is an emissary of the whole profession.

Bioethical considerations introduce a number of challenges that must be approached at all three ethical levels. At the most basic, microethical level, laws, rules, regulations, and policies dictate certain behaviors. For example, cloning and blastocyst research, especially that which receives federal funding, is controlled by rules overseen by federal and state agencies. Such rules are often proscriptive; that is, they tell what "not to do," but are less clear on what actually "to do."

At the next level, beyond legal considerations, the engineer is charged with being a loyal and faithful agent to the clients. Researchers are beholden to their respective universities and institutions. Engineers working in companies and agencies are required to follow mandates to employees (although never in conflict with their obligations to the engineering profession). Thus, engineers must stay within budget, use appropriate materials, and follow best practices as they concern their respective designs. For example, if an engineer is engaged in work that would benefit from a collaboration with another company working with similar genetic material, the engineer must take precautionary steps to avoid breaches in confidentiality, such as trade secrets and intellectual property.

The highest level, the macroethical perspective, has a number of bioethical aspects. Many of the research and development projects are in areas that could greatly benefit society but may lead to unforeseen costs. The engineer is called to consider possible contingencies. For example, if an engineer is designing "nanomachinery" at the subcellular level, is there a possibility that self-replication mechanisms in the cell could be modified to lead to potential adverse effects, such as generating mutant pathological cells, toxic byproducts, or changes in genetic structure not previously expected? Thus, this highest level of professional development is often where "risk tradeoffs" must be considered. In the case of our example, the risk of adverse genetic outcomes must be weighed against the loss of advancing the state of medical science (e.g., finding nanomachines that manufacture and deliver tumor-destroying drugs efficiently).

Ongoing cutting-edge research (such as the efficient manufacturing of chemicals at the cellular scale, the development of cybernetic storage, and data transfer systems using biological or biologically inspired processes, etc.) will create new solutions to perennial human problems by designing more effective devices and improving computational methodologies. Nonetheless, in our zeal to push the envelopes of science, we must not ignore some of the larger, societal repercussions of our research; that is, we must employ new paradigms of "macroethics."

William A. Wulf, President of the National Academy of Engineering, introduced the term macroethics, defining it as a societal behavior that increases the intellectual pressure "to do the right thing" for the long-term improvement of society. Balancing the potential benefits of the advances in nanotechnology to society while also avoiding negative societal consequences is a type of macroethical dilemma.[46] Macroethics asks us to consider the broad societal impact of science in shaping research agendas and priorities. At the same time, "microethics" is needed to ensure that researchers and practitioners act in accordance with scientific and professional norms, as dictated by standards of practice, community standards of excellence, and codes of ethics.[47] The engineering profession and engineering education standards require attention to both the macro- and microdimensions of ethics. Criterion 3, "Program Outcomes and Assessment," of the Accreditation Board for Engineering and Technology (ABET), Inc. includes a basic microethical requirement for engineering education programs, identified as "(f) an understanding of professional and ethical responsibility," along with the macroethical requirements that graduates of these programs should have "(h) the broad education necessary to understand the impact of engineering solutions in a global and societal context" and "(j) a knowledge of contemporary issues."[48]

Personal and organizational ethics and morality are affected by psychology. Attitudes are complex mental processes that influence how a person processes information and that motivate behavior.[49] They have been explored in depth in the psychological literature, where "attitude" has been defined as:

Teachable moment: the physiome project: the macroethics of engineering toward health

In *The Bridge,* the Journal of the National Academy of Engineering, James B. Bassingthwaighte Vol. 32, No. 3 – (Fall 2002), identified and described some of the most important *engineering tools and techniques that can be used to advance health care.*

According to Bassingthwaighte, the new tools fall into four categories:

1. **Informatics and information flow**
The problem in medicine and biology is that much relevant information is either irretrievable or undiscovered. Even a complete human genome cannot define human function. In fact, it is only a guide to the possible ingredients. The genetically derived aspects of the genome (i.e., proteins) are much more numerous than the genes. To get an idea of the magnitude of the problem, consider that yeast has about three proteins per gene, and humans have about 10 proteins per gene. Pretranslational selection from different parts of the DNA sequence, the posttranslational slicing out of parts of the protein, the splicing of two or more proteins together, and the combining of groups of proteins into functional, assembly-line-like complexes, all contribute to the variety of the products of gene expression. Even a completely identified proteome, which is still beyond the scientific horizon, will be like a list of the types of parts of a jumbo jet with no indication of how many should be used or where they go. The concentration of proteins, the balance between synthesis and decay in each cell type, is governed by environment, by behavior, and by the dynamic relationships among proteins, substrates, ionic composition, energy balance, and so on and, thus, cannot be predicted on the basis of the genome.
2. **The combinatorial dilemma**
Sorting out the genome will leave us with a huge number of proteins to think about. The estimates of the number of genes have come down by about half from earlier estimates of 60 000 to 100 000; because new ones are also being found, 50 000 is a reasonable estimate. The level of complexity in mammalian protein expression far exceeds that of *C. elegans,* which has 19 536 genes and 952 cells. Humans might only have two or three times as many genes, but probably have a much higher ratio of proteins per gene. Assuming 10 proteins per gene, we have on the order of a half million proteins in widely varied abundance, and each protein has several possible states. If a protein in a given state interacts with only five other proteins (e.g., exchanging substrates with neighbors in a pathway or modifying the kinetics of others in a signaling sequence), then it may "connect" to any other protein through only a few links, a kind of "six degrees of separation" from any other protein. Moreover, cells contain not just proteins, but also substrates and metabolites, and they are influenced by their environments. Given the possible permutations and combinations of linkages and the many multiples further in the dynamics of their interactions, the combinatorial explosion would appear to preclude predictions.
3. **Managing complexity**
The complexity ... briefly described provides a basis for functionality that cannot be predicted from knowledge about each of the components. "Emergent" behavior is the result of interactions among proteins, subcellular systems, and aggregates of cells, tissues, organs, and systems within an organism. Physiological systems are highly nonlinear, higher order systems, and dynamics are often chaotic. Chaotic systems are only predictable over the short term; but they have a limited operating range. Even when Bernard (1927) defined the stability of the "milieu interieure," he meant a mildly fluctuating state rather than a stagnant "homeostasis." Biological systems are "homeodynamic"; they fluctuate, but under control, and they are neither "static" nor "randomly varying."
4. **Bioinformatics**
This vast array of information must be linked into a consistent whole. The databases must be well curated and easily accessible, and they must provide a substrate for behaviorally realistic models of physiological systems. The arguments for building large databases to capture biological data are fairly new. Federal funds support genomic and proteomic databases, but not databases of higher level physiological information. Organ and systems data acquired over the past century have not been collected in databases and are poorly indexed in the print literature. Providing searchable texts of articles online will help but will not be a substitute for organized databases. The Visible Human Project, the National Library of Medicine's effort to preserve anatomic information is a part of the morphome (which we define as providing anatomic and morphometric information), analogous to the genome and the proteome.

Bassingthwaighte continues:

All of this information must then be captured in a comprehensive, consistent, conceptual framework; that is, a model of the system that conveys understanding, and for this we will need to use engineering approaches. Understanding complicated systems, modeling them, and learning the tricks for reducing their complexity to attain computability, are in the engineering domain, and bioengineering-trained investigators will be the integrators of the future.

Of course, all models are incomplete. They come in a variety of forms, such as sketches of concepts, diagrams of relationships, schemas of interactions, mathematical models defined by sets of equations, and computational models (from analytical mathematical solutions or from numerical solutions to differential or algebraic equations). The behavior of a well developed, well documented computer model can give us some insight into the behavior of the real system.

We must do our utmost to predict well, not just the direct results of a proposed intervention, but also the secondary and long-term effects. Thus, databasing, the development, archiving, and dissemination of simple and complex systems models, and the evaluation (and rejection or improvement) of data and of models – are all part of the moral imperative. They are the tools necessary to thinking in depth about the problems that accompany, or are created by, interventions in human systems or ecosystems.

As a step toward providing these tools, Bassingthwaighte has initiated the Physiome and the Physiome Project:

A physiome can be defined as the quantitative description of the functional state of an organism. A quantitative model is a way of removing contradictions among observations and concepts and creating a consistent, reproducible representation of a system. Like the genome, the physiome can be defined for each species and for each individual within the species. The composite and integrated system behavior of the living organism is described quantitatively in hierarchical sets of mathematical models defining the behavior of the system. The models will be linked to databases of information from a multitude of studies. Without data, there is nothing to model; and without models, there is no source of deep predictive understanding.

The Physiome Project provides one response to the macroethical imperative to minimize risk while advancing medical science and therapy. The project is an effort to define the physiome, through databasing and modeling, of individual species, from bacteria to man. The project began with collaborations among groups of scientists in a few fields and is developing spontaneously as a multinational collaborative effort. Investigators first defined goals and then proceeded to put pieces together into impressive edifices. Via iteration with new experimentation, models can remove contradictions and demonstrate emergent properties. These models are part of the tool kit for the "reverse engineering" of biology. The scale of the models, like the scale of models for weather prediction, presents computational grand challenges.

The Physiome Project is not likely to result in a virtual human being as a single computational entity. Instead, small models linked together will form large integrative systems for analyzing data. There is a growing appreciation of the importance, indeed the necessity, of modeling for analysis and for prediction in biological systems as much as in physical and chemical systems.

The hierarchical nature of biological systems is being used as a guide to the development of hierarchies of models. Models at the molecular level can be based on biophysics, chemistry, energetics, and molecular dynamics, but it is obviously not practical to use molecular dynamics in describing the fluxes through sets of biochemical pathways, just as it is not practical to use the full set of biochemical reactions when describing force–velocity relationships in muscle, or to use the details of myofilament crossbridge reactions when describing limb movement and athletic performance. One cannot build a truck out of quarks.

Biological models can be defined at many hierarchical levels from gene to protein to cell to organ to intact organism. Practical models comprised of sets of linked component models, each somewhat simplified, represent one level of the hierarchy. The strategy is to avoid computing the details of underlying events and to capture, at the higher level, the essence of their dynamic behavior. But monohierarchical models are not necessarily built to adapt to changes in conditions. Handling transients is like using adjustable time steps in systems of stiff equations, but more complicated; the lower level model must be used to correct the higher level representation. Once we have very good models that extend from gene regulation to the functions of the organism, they can be used to predict the short-term and long-term efficacy and side effects of various therapies.

Questions:

1. Like genomic information, should biological information and models be put in the public domain?
2. What types of risks and benefits can be expected from this project?
3. Is this a model for medical and engineering collaboration? Why or why not?

[A] psychological tendency that is expressed by evaluating a particular entity {the object of the attitude} with some degree of favour or disfavour.[50]

Attitudes are inferred by observing an individual's response to a situation (a stimulus); but they cannot be measured directly (Figure 8.2-1)[51]. For example, a doctor holding a pro-life attitude, when confronted by a patient requesting a termination of pregnancy, might respond by refusing to act as the patient wishes, or by explaining his or her beliefs to the patient, or both. But if, in spite of the doctor's beliefs, he or she agrees to arrange the termination of pregnancy, then the patient might not be able to infer that the doctor has a pro-life attitude, and a knowledgeable observer might conclude that the doctor's pro-life attitude is not strongly held compared with competing pressures to act in a counter-attitude manner.

Attitude affects empathy, which is central to justice. Justice is the virtue that enables us to give others what is due to them as our fellow human beings. This means we must not only avoid hurting others by our actions, but we ought to safeguard the rights of others in what we do and what we leave undone.

The categorical imperative is emblematic of empathy. Kant uses this maxim to underpin duty ethics (so-called deontology) with empathetic scrutiny. However, empathy is not the exclusive domain of duty ethics. In consequentialism, also known as teleological ethics, empathy is one of the palliative approaches to deal with the problem of "ends justifying the means." Other philosophers also incorporated the empathic viewpoint into their frameworks. In fact, Mill's utilitarianism axiom of "greatest good for the greatest number of people" is moderated by his "harm principle," which, at its heart, is empathetic. That is, even though an act can be good for the majority, it may still be unethical if it causes undue harm to even one person. Empathy also comes into play in contractarianism, as articulated by Hobbes' social contract theory. For example, John Rawls has moderated the social contract with the "veil of ignorance" as a way to consider the perspective of the weakest, one might say "most disenfranchised," members of society. Finally, the rationalist frameworks incorporate empathy into all ethical decisions when they ask the guiding question of "what is going on here?" In other words, what benefit or harm, based on reason, can I expect from actions brought about by the decision I am about to make? One calculus of this harm or benefit is to be empathetic to all others, particularly the weakest members of society, those with little or no "voice."

The word "empathy" has an interesting beginning. It originally comes from the German word *einfühlung*, which means the ability to project oneself into a work of art, like a painting. Psychologists at the beginning of the 1900s searched for a word that meant the projection of oneself into another person, and chose the German word, translated into English as "empathy." The concept itself was known, such as the Native Americans' admonition to walk in another's moccasins, but it needed a construction. The earlier meaning of empathy was thus the ability to project oneself into another person, to imitate the emotions of that person by physical actions. For example, watching someone prick a finger would result in a visible winching on the part of the observer because the observer would know how it feels.

From that notion of empathy, it was natural to move to more cognitive role taking, imagining the other person's thoughts and motives. From here, empathy began to be thought of as the response that a person has for another's situation. Psychologists and educators, especially Jean Piaget,[52] began to believe that empathy develops throughout childhood, beginning with the child's first notion of others who might be suffering personal stress. The child's growing cognitive sense eventually allows him or her to experience the stress in others. Because people are social animals, this understanding of the stress in others, according to the psychologists, eventually leads to true compassion for others.

A problem with this notion of empathy development is that some experiments have shown that the state of mind of a person is very important in that person's ability to empathize. Apparently, small gifts or compliments significantly increase the likelihood that a person will show empathy toward third parties. A person in a good mood tends to be more understanding of others. If this is true, then empathy is (at least partly) independent of the object of the empathy, and empathy becomes a characteristic of the individual.[53]

Charles Morris defines empathy as:[54]

> *The arousal of an emotion in an observer that is a vicarious response to the other person's situation ... Empathy depends not only on one's ability to identify someone else's emotions but also on one's capacity to put oneself in the other person' place and to experience an appropriate emotional response. Just as sensitivity to non-verbal cues increases with age, so does empathy: The cognitive and perceptual abilities required for empathy develop on as a child matures.*

Such definitions of empathy seem to be widely accepted in the moral psychology field. But there are some serious problems with such a definition.

First, we have no way of knowing if the emotion triggered in the observer is an accurate representation of the stress in the subject. We presume that a pin prick would be felt in a similar way because we have had this done to us and we know what it feels like. But what about the stress caused by a broken promise? How can an observer know that he or she is on the same wavelength as the subject when the stress is emotional?[55]

If a subject says that she is sad, the observer would know what it is like to be sad, and would share in the sadness; that is, the observer would empathize with the subject's sadness and be able to tell the subject what is being felt. But is the observer really feeling what the subject is feeling? There is no way to define or measure "sadness," and thus there is no way to prove that the observer is actually feeling the same sadness that the subject is feeling.[56] An existentialist and empiricist might say that this is true for everything, even physical realities, but that is beyond the scope of this discussion.

The second problem relates to nonhuman animals. Psychologists have studied empathy exclusively as a human–human interaction, and yet many nonhuman animals can exhibit empathy. Witness the actions of a dog when its master is sick. You can read the caring and sympathy and hopefulness in the dog's eyes.[57] Since sentience and pain management are important to biomedical engineering, these uncertainties are no small matter.

Humans also have strong emotional feelings toward nonhuman animals. The easiest to understand in these terms is the empathy we feel when animals are in pain. We do not know for sure that they are in pain of course, since they cannot tell us, but they act in ways similar to the way humans behave when they are in pain and there is every reason to believe that they feel pain in the same way. Anatomical studies on animals confirm that many of their nervous systems do not differ substantially from those of humans, indicating that they feel pain. Indirect measures, such as tomography, also support the contention that animals feel pain in ways similar to humans.

More problematic are the lower animals and plants. There is some evidence that trees respond physiologically when they are damaged, but this is far from certain. The response may not be pain at all, but some other sensation (if we can even suggest that trees have sensations). And yet many of us are loathe to cut down a tree, believing that the tree ought to be respected for what it is, a center of life. This idea was best articulated by Albert Schweitzer in his discussions on the "reverence for life," or the idea that all life is sacred.

Empathy toward the non-human world cannot be based solely on sentience. Something else is going on. When a person does not want to cut down a tree because of caring for the tree, this is certainly some form of empathy, but it does not come close to the definitions used by the psychologists.

The third problem with this definition of empathy is that there is a huge disconnect between empathy and sympathy. If an observer watches a subject getting a finger pricked, the observer may know exactly what it feels like, having had a similar experience in the past. So there is great empathy. But there might be little sympathy for the subject. The observer might actually be glad that the subject is being hurt, or it might be funny to the observer to watch the subject suffer.

Years ago on the popular television show "Saturday Night Live," there was an occasional bit where a clay figure, Mr Bill, suffered all manner of horrible disasters and ended up being cut, mangled, crumbled, and squashed. Watching this may have elicited some empathy on the part of the observers, but certainly there was no sympathy for the destruction of the little clay man. His destruction was meant to be funny.

We could argue that a lack of sympathy might indicate that there must be a lack of empathy also. How is it possible for someone to empathize with another person getting a finger pricked, but think it to be humorous? Perhaps there has been no empathy at all. Or perhaps we have conditioned ourselves to laugh at others when they get hurt as a defense mechanism (e.g., "whistling in the dark") to somehow separate the violence from our own experience. Or we have learned from and have become desensitized by mass media and video games to destroy others without regret.

Fairness

Empathy is not a moral value in the same way that loyalty, truthfulness, and honesty are moral values. Each of us can choose to tell the truth or to lie in any particular circumstance, and a moral person will tell the truth (unless there is an overwhelming reason not to, such as to save a life). But it is not possible to choose to have or not to have empathy. One either has empathy or one does not. One either cares for those in need or one does not.

Because we believe that empathy is worthwhile, and respect and admire people who have empathy, we tend to assign moral worth to this characteristic, and we believe that people with empathy are virtuous. On the other hand, we do not condemn those who do not have empathy. For example, people who contribute to various relief organizations such as CARE and Catholic Charities do so because they have empathy for those in need, but many people choose not to contribute. They lack some measure of empathy for others in need, but this does not make them bad people. They simply choose not to contribute.

Can engineers not have empathy and still do good engineering? That is, is empathy necessary for good engineering? Certainly on a personal level, engineers read the same newspapers and hear the same TV news as everyone else, and thus their lack of empathy ought not to be any more or less criticized than the lack of empathy by anyone else. But the truth is that responsibility of professional engineers is supererogatory to everyday ethics. Engineering ethics is a different layer on top of everyday common morality, and engineers share many responsibilities not required of nonengineers. By virtue of their training and skills engineers serve others and have certain responsibilities that relate to their place in society. The oft-quoted first canon in many codes of engineering ethics

> *The engineer shall hold paramount the health, safety, and welfare of the public*

is very clear. It states that the engineer has responsibility to the "public," not to a segment of the public that fits the design paradigm, or that segment that employs the engineer, or that segment that has power and money. The engineer is responsible to the public. Full stop. And in so doing, the engineer must help that segment of the public least able to look out for themselves. There is a *noblesse oblige* in engineering, the responsibility of the "nobles" to care for the less fortunate.

Thus, to be an effective and "good" engineer requires that we be able to put ourselves in the place of those who have given us their trust. The implications for justice are that it is much easier to export "canned" answers and solutions to problems from our vested viewpoints. This view must span time and space. What will the product performance look like in ten years if the project is implemented? What happens if some of the optimistic assumptions are realized? The users will be left with the consequences. It is much better, but much more difficult, to see the problem from the perspective of those with the least power to change things. We are empowered as professionals to be agents of change. So, as agents of change and justice, engineers must strive to hold paramount the health, safety, and welfare of all the public, we must be competent, and we must be fair.

Value as a bioethical and engineering concept

In engineering, one might consider "value" through the idea of value engineering. This concept was created at General Electric Co. during World War II. As the war caused shortages of labor and materials, the company was forced to look for more accessible substitutes. Through this process, they saw that the substitutes often reduced costs or improved a product.[58] Consequently, they turned the process into a systematic procedure called "value analysis."

Value engineering consists of assessing the value of goods in terms of "function." Value, as a ratio of function to cost, can be improved in various ways. Oftentimes, value engineering is done systematically through the four basic steps:[59] information gathering, alternative generation, evaluation, and presentation. In the information gathering step, engineers consider what the requirements for the object are. Part of this step includes function analysis, which attempts to determine what functions or performance characteristics are important. In the next step, i.e., alternative generation, value engineers consider the possible alternative ways of meeting the requirements. Next, in evaluation, the engineers assess the alternatives in terms of functionality and cost-effectiveness. Finally, in the presentation stage, the best alternative is chosen and presented to the client for the final decision.[60]

In the realm of economics, value is considered to be the worth of one commodity in terms of other commodities (or currency). There are three main value theories in economics. The first, an intrinsic theory of value, holds that the value of an object, good, or service is contained in the item itself. These theories tend to consider the costs associated with the process of producing an item when assigning the item value. For example, the labor theory of value, a model developed by David Ricardo, holds that the value of a good is derived from the effort of its production, reduced to the two inputs in the production frontier, labor, and capital.[61] In this model, if a lamp is produced in five hours by 3 people, then the lamp is worth $3 \times 5 = 15$ man-hours. On the other hand, the subjective theory of value holds that goods have no "intrinsic" value, outside the desire of individuals to have the items. Here, value becomes a function of how much an individual is willing to give up in order to have that item.[62]

Similarly, the marginal theory of value accounts for both the scarcity and the desirability of a good, holding that the utility rendered by the last unit consumed determines the total value of a good. The main difference between the labor theory of value (and the concept of intrinsic value) and marginal theory of value is that the former accomodates a form of value derived from utility – from satisfying human desire.[63]

Furthermore, the common understanding of the term "value" generally means that the item is of importance for one reason or the other. At the same time, to say something is "valuable" would mean that it costs a lot of money, meaning that a high demand for the item in society has driven the prices up. Therefore, person A might value her beaded necklace because she derives pleasure from it, but her diamond necklace would be considered valuable, because of the large number of other members of the society that also feel they would derive pleasure from it (which causes a high demand that increases the price).

The so-called diamond – water paradox, is a noteworthy example of the role of scarcity in economic value theory. Diamonds, which have relatively little use, i.e., only aesthetic value, have an extremely high price when compared to water, which is essential to life itself. The example illustrates the importance of scarcity in the economic value. Here, as diamonds are far scarcer than is water, they have the higher price. However, this is situational. In the middle of the desert, where water is extremely scarce, someone would almost certainly be willing to pay more money for water than for diamonds. Thus, that person would value water more highly.

Since utility is a common metric for value in biomedicine, the differing definitions should give us pause. Often, the design's value as perceived by decision makers is what drives ethics. However, we may be defining value in substantially different ways.

Technical optimism versus dismal science

The National Academy of Engineering has declared:

> *Engineers and their inventions and innovations have helped shape the changes that have made our lives more productive and fruitful.*[64]

But what does it mean to become more fruitful? One way to determine such success is to consider the economic implications.

In economics, technology can be considered a tool of "empowerment," in that it empowers producers to generate more output from given levels of the two inputs, labor, and capital. A catalyst in chemistry is a substance that increases the rate of reaction; in this context, one might say that technology is a "catalyst" for the production of output – as it increases the amount of output we get from given inputs. Technology allows for the use of more advanced capital, which results in better and faster ways to create output. Producers are rendered more efficient, as they are able to produce more output, given the same amount of input. This results in greater profit, which results in economic growth.

Technology improves utility as it allows for better use of scarce resources, which are then allocated through the economic system, facilitating the achievement of Pareto optimality. That is, given a decision of whether to take an action, the selected alternative must improve the lot of at least one member of the affected group and cannot worsen the plight of any other member (known as a Pareto improvement). Thus, economics would drive the decision toward improved utility, at least for that one person.

Therefore, technology broadens the horizon through which economics operates. When firms invest, they increase capital; increasing our input means more output and more economic growth. And when capital is depreciating, it is less productive, yielding less output. Technology also allows higher levels of sustainability for capital.

As discussed in Chapter 8.1 the economist, Malthus did not realize that technology could increase food supply. And his modern day disciple Paul Ehrlich, gives an exceedingly grim prognosis for the future: "Each year food production in underdeveloped countries falls a bit further behind burgeoning population growth, and people go to bed a little hungrier. While there are temporary or local reversals of this trend, it now seems inevitable that it will continue to its logical conclusion: mass starvation."[65] Not only does Ehrlich state that the world is headed toward calamity, he is convinced that there is nothing anyone can really do that will provide anything more than temporary abatement. To focus on Ehrlich's attitude towards technology as part of the solution to the impending problem is to see Ehrlich's "technological pessimism," so to speak. Ehrlich's lack of confidence in technology to deal with the problems plaguing the future is seen in his statement: "But, you say, surely Science (with a capital "S") will find a way for us to occupy the other planets of our solar system and eventually of other stars before we get all that crowded."[66] Ehrlich was sure that "the battle to feed humanity is over." He insisted that India would be unable to provide sustenance for the 200 million person influx in its population by 1980. He was wrong – Ehrlich did not count on the "green revolution."

As this predictive framework earned the title of "dismal science" for economics, engineers look upon the same problems with more of a technical optimism. Engineers bump up the Malthusian curve by finding ways to improve conditions. In *The Engineer of 2020*, one finds descriptions of the various ways engineers in the future will help to solve the very same problems about which Ehrlich (and the Malthusian model in general) is concerned. Where Ehrlich considered technology's role in solving the problem would only be seen through how "improved technology has greatly increased the potential of war as a population control device,"[67] engineers look towards technology not as Ehrlich's "means for self-extermination"[68] but rather they opt to use it to support and improve life in the future.

According to National Academy of Engineering, the world's population will approach 8 billion people; much of this increase will be seen in groups that are today considered underdeveloped countries, mainly in Asia and Africa. Apparently, "by 2015, and for the first time in history, the majority of people, mostly poor will reside in urban centers, mostly in countries that lack the economic, social, and physical infrastructures to support a burgeoning population."[69] However, engineers see in the challenge posed by the highly crowded and densely populated world of 2020 an opportunity for "the application of thoughtfully constructed solutions through the work of engineers."[70] Likewise, engineers look upon the necessity for improved health care delivery in the world of future with confidence. The key word is confidence, and not "arrogance." Engineers must make advanced technologies accessible to this evergrowing global population base. In the next twenty years, positive implications on human health, will be possible due to improved air quality and the control and clean up of hazardous waste sites, and focused efforts to treat diseases like malaria and AIDS.[71]

Engineers believe that they can solve the problems posed by the future, as opposed to views like the one posed by Ehrlich who sees a future where "small pockets of *Homo sapiens* hold on for a while in the Southern Hemisphere, but slowly die out as social systems break down, radiation poisoning takes effect, climatic changes kill crops, livestock dies off, and various man-made plagues spread. The most intelligent creatures ultimately surviving this period are cockroaches."[72] Indeed, this dramatic example serves to illustrate the differences between engineer's technical optimism and doomsayer's technical pessimism, so to speak. But what does the future call for – the proverbial idealist or the modern-day skeptic? Human kind can either resign itself to failure and deem the problems it will come to face unsolvable, or it can press forward, attempting to solve the problems it faces and overcome tomorrow's challenges. While the question is subjective, it becomes clear that in order to have progress, most engineers choose creativity and action.

Notes and commentary

1 National Academy of Engineering, *The Engineer of 2020: Visions of Engineering in the New Century* (Washington, DC: National Academy Press, 2004), 49.

2 V.R. Potter II,"What Does Bioethics Mean?"*The Ag Bioethics Forum* 8, no.1 (1996): 2 – 3.

3 A.S. Daar, H. Thorsteinsdóttir, D.K. Martin, A.C. Smith, S. Nast, and P.A. Singer, "Top Ten Biotechnologies for Improving Health in Developing Countries," *Nature Genetics* 32(2002): 229 – 32.

4 *Oxford Dictionary of Biochemistry and Molecular Biology* (New York, NY: Oxford University Press, 1997).

5 Ibid.

6 Ibid.

7 The President's Council on Bioethics, "Working Paper 1, Session 4: Human Cloning 1: Human Procreation and Biotechnology," (17 January 2002).

8 L.R. Kass, The President's Council on Bioethics, Transmittal Memo to "Human Cloning and Human Dignity: An Ethical Inquiry," (Washington, DC: 2002).

9 L.R. Kass, 2005, Letter of Transmittal to President George W. Bush, *Alternative Sources of Human Pluripotent Stem Cells*, a White Paper of the President's Council on Bioethics (Washington, DC, 10 May 2005).

10 The President's Council on Bioethics, Executive Summary, "Human Cloning and Human Dignity: An Ethical Inquiry," (Washington, DC: 2002), xxvii.

11 P. Singer, *Rethinking Life and Death: The Collapse of Our Traditional Ethics* (New York, NY: St. Martin's Griffin, 1994).

12 M. Schooyans, *Bioethics and Population: The Choice of Life* (St Louis, MO: Central Bureau, Community Center for Vital Aging, 1996).

13 P. De Coppi, G. Bartsch Jr., M.M. Siddiqui, T. Xu, C.C. Santos, L. Perin, G. Mostoslavsky, A.C. Serre, E.Y. Synder, J.J. Yoo, M.E. Furth, S. Soker, and A. Atala "Isolation of Amniotic Stem Cell Lines with Potential for Therapy" *Nature Biotechnology* 25(1) (2007): 100–6.

14 Church of Scotland, 2006, Society, Religion and Technology Project, Patenting Life: An Introduction to the Issues, http://www.srtp.org.uk/scsunpat.shtml (accessed 17 September 2006).

15 K.D. Warner, "Are Life Patents Ethical? Conflict between Catholic Social Teaching and Agricultural Biotechnology's Patent Regime," *Journal of Agricultural and Environmental Ethics* 14, no. 3 (2002): 301–19.

16 D. Rees and S. Rose, *The New Brain Sciences: Perils and Prospects* (Cambridge, UK: Cambridge University Press, 2004).

17 President's Council on Bioethics, Staff Background Paper: Organ Transplantation: Ethical Dilemmas and Policy Choices (Washington, DC, 2003).

18 Ibid.

19 US Department of Health, Education and Welfare, National Commission for the Protection of Human Subjects of Biomedical and Behavioral Research, *The Belmont Report: Ethical Principles and Guidelines for the Protection of Human Subjects of Research*, 18 April 1979.

20 R.V. Orzeck shared this case that will appear in his book, *So Now You'll Know*, to be published in 2007.

21 One of the common themes of this book, along with a systematic approach and the need for professional trust, is that of empathy. Schweitzer's reverence for life has been characterized as a "bioempathetic" viewpoint. See: A. Sweitzer, *Out of My Life and Thought* (translated by A.B. Lemke) (Henry Holt & Co., New York, NY, 1990), 157.

22 This discussion draws upon the ideas of Diana Chang, who conducted undergraduate research in an independent study course that I facilitated at Duke.

23 F.B. Orlans, T.L. Beauchamp, R. Dresser, D.B. Morton, and J.P. Gluck, *The Human Use of Animals: Case Studies in Ethical Choice* (New York: Oxford University Press, 1998).

24 Ibid.

25 T.F. Budinger and M.D. Budinger refer to this approach as the "four As:" (1) acquire facts; (2) alternatives; (3) assessment; and (4) action (T.F. Budinger and M.D. Budinger, *Ethics of Emerging Technologies: Scientific Facts and Moral Challenges)* (Hoboken, NJ: John Wiley & Sons, 2006).

26 This discussion draws upon the ideas of Zach Abrams, who conducted undergraduate research on GMOs in my Ethis in professions course of Duke.

27 G. Tulloch, *Euthanasia: Choice and Death* (Edinburgh, UK: Edinburgh University Press, 2005).

28 S. Ewen and A. Pusztai, "Effect of Diets Containing Genetically Modified Potatoes Expressing *Galanthus nivalis* Lectin on Rat Small Intestine," *The Lancet*, 354 (1999), 9187.

29 M.L. King Jr., *Strength to Love*, Fortress Edition (May 1981) (Minneapolis, MN: Augsburg Fortress Publishers, 1963).

30 Agent Orange website: http://www.lewispublishing.com/orange.htm (accessed 22 April 2005).

31 A principal source for the Minamata case is the Trade & Environment Database, developed by James R. Lee, American University, The School of International Service, http://www.american.edu/TED/(accessed 19 April 2005).

32 This is an all too common professional ethics problem, i.e., lack of full disclosure. It is often, in retrospect, a very costly decision to withhold information about a product, even if the consequences of releasing the information would adversely affect the "bottom line." Ultimately, as has been seen in numerous ethical case studies, the costs of not disclosing are severe, such as bankruptcy and massive class action lawsuits, let alone the fact that a company's decision may have led to the death and disease of the very people they claim to be serving, their customers and workers!

33 International Programme on Chemical Safety, United Nations Environmental Programme, "Cadmium." Environmental Health Criteria (EHC134), Geneva, Switzerland, 1992.

34 Agency for Toxic Substances and Disease Registry, US Department of Health and Human Services "Toxicological Profile for Cadmium," Washington, DC, 1999.

35 T. Colburn, Speech at *the State of the World Forum* (San Francisco, CA: 1996).

36 *Webster's Ninth New Collegiate Dictionary* (Springfield, MA: Merriam-Webster, 1990).

37 Christian Scripture has defined faith as "the substance of things hoped for, the evidence of things not seen" (Hebrews 11:1). This is an extension of the Judaic outlook for the "promised land" (a metaphor to the engineer's expectation to see the designs reach the build phase, to the medical researcher's search for cures of obdurate maladies, or to the city planner's long-range plan, envisioning green spaces and copious public amenities). This optimistic view lends a temporal dimension to faithfulness. For example, the concept of sustainability in terms of public health and environmental quality is a requirement to take (or to avoid) actions based on their impact on future generations. This means that engineers must be faithful agents to distant and future people. An action that is most expedient for the present may not be the best if it has severe effects in the long run.

38 This account can be found in J. Enderle, S. Blanchard, and J. Bronzino, ed., *Introduction to Biomedical Engineering* (Burlington, MA: Elsevier Academic Press, 2005).

39 The Pandora's box is an interesting and useful ethical metaphor that conveys the potential of unintended, negative outcomes from a seemingly innocuous or beneficial act. The term comes from the Greek myth about a box left by Mercury with Epimetheus and Pandora for safekeeping. Epimetheus warned Pandora not to open the box, but eventually upon hearing voices asking to be freed, Pandora's curiosity got the best of her. She opened the lid and out came diseases, vices, and other ills to humanity. Hence, the myth has been used as a warning not to unadvisedly or prematurely rush into the unknown (e.g., viral research, neurotechnologies, nanotechnologies, and new drug therapies). However, an often forgotten part of the story is that Pandora opened the box a second time and "hope" was released. Biomedicine and engineering are modern manifestations of this hope.

40 Another way to look at uncertainty is that it is a function of variability and ignorance. This has been well articulated by L. Ginzburg in his review of ecological case studies in US Environmental Protection Agency, 1994, Peer Review Workshop Report on Ecological Risk Assessment Issue Papers, Report Number EPA/630/R-94/008. According to Ginzburg, "variability includes stochasticity arising from temporal and spatial heterogeneity in environmental factors and among exposed individuals. Ignorance includes measurement error, indecision about the form of the mathematical model, or appropriate level of abstraction." Thus, variability can be lessened by increased attention, e.g., empirical evidence, and "translated into risk (i.e., probability) by the application of a probabilistic model," but ignorance cannot. Ignorance simply translates into confidence intervals, or "error bounds" on any statement of risk.

41 See, for example, Physical Principles of Unworkable Devices, http://www.lhup.edu/~dsimanek/museum/physgal.htm. Donald E. Simanek's humorous but informative site on why perpetual motion machines cannot work, i.e., their inventors assumed erroneous "principles." This site is instructive to biomedical decision makers to beware of "junk science." Sometimes a good way to learn why something works the way it does is to consider all the reasons that it fails to work.

42 My apologies to the originator of this analogy, who deserves much credit for this teaching device. The target is a widely used way to describe precision and accuracy.

43 It is probably safe to say that most engineers lack a substantial amount of formal training in psychology and the behavioral sciences, so no previous background in these areas is needed this text. However, I have observed many engineers who are gifted in "people skills." Also, most have gained knowledge and have read extensively in these areas after their baccalaureate education. An interesting and valuable change in engineering education has been greater emphasis on the student's grasp of the social sciences and the humanities.

My own educational interests were a bit ahead of this trend (by three decades), not due to prescience or an unusual sense of what I would need to prepare for the challenges of a very rewarding career. No, my second major in psychology was because I met a beautiful young woman in a sophomore course. She happened to be a psychology major, so I found myself taking an inordinate amount of social science courses, which eventually translated into a major. Oh, by the way, my insistence paid off. I married her and we recently celebrated our thirtieth anniversary. I have never drawn a critical path or fault tree for contingent probabilities of this outcome, and the benefits derived (at least by me) but perhaps, I shall. Analyze that!

44 T. Hobbes, 1660, *Leviathan*.

45 Lawrence Kohlberg, *Child Psychology and Childhood Education: A Cognitive-Developmental View* (New York, NY: Longman Press, 1987).

46 J.B. Bassingthwaighte, "The Physiome Project: The Macroethics of Engineering toward Health," *The Bridge* 32, no. 3 (2002): 24–9.

47 J.R. Herkert, "Microethics, Macroethics, and Professional Engineering Societies," in *Emerging Technologies and Ethical Issues in Engineering* (Washington, DC: The National Academies Press, 2004), 107–14.

48 Accreditation Board for Engineering and Technology (ABET), Inc., *Criteria for Accrediting Engineering Programs: Effective for Evaluations during the 2004-2005 Accreditation Cycle* (Baltimore, MD: ABET, 2003).

49 I. Ajzen, *Attitudes, Personality, and Behaviour* (Milton Keynes: Open University Press, 1988).

50 A.H. Eagly and S. Chaiken, *The Psychology of Attitudes* (Orlando, FL: Harcourt Brace & Company, 1993).

51 Ibid.

52 J. Piaget, *The Moral Judgment of the Child* (New York, NY: The Free Press, 1965).

53 S. Vaknin, *Malignant Self Love: Narcissism Revisited* (Macedonia: Lidija Rangelovska Narcissus Publications, 2005).

54 C.G. Morris, *Psychology – An Introduction*, 9th ed. (Englewood Cliffs, NJ: Prentice-Hall, 1996).

55 This is one of the problems with B.F. Skinner's brand of behaviorism, as articulated in *Beyond Freedom and Dignity* (Hackett Publishing, 1971). Certainly, persons act out on what they have learned and that learning is an aggregate of their responses to stimuli. However, human emotions and empathy are much more than this. Empathy is a very high form of social and personal development. So, although one might be able to "train" an ant or a bee to respond to light stimuli, or a pigeon to "play ping-pong" (as Skinner did), even these lower animals have overriding social complexities. At the heart of humanity are freedom and dignity, in spite of what some behaviorists tell us.

56 Vaknin, *Malignant Self Love*.

57 The concept may be innate and extended to other animals, such as elephants sensing "awe" for their ancestral graveyards.

58 "Value Engineering," 24 February 2006. DOD Value Engineering Program, http://ve.ida.org/ve/ve.html (accessed 16 March 2006).

59 "What Is the Value Method?", 5 May 2006. Systemic Analytic Methods and Innovations, http://www.value-engineering.com (accessed 5 May 2006).

60 "The Value Engineering (VE) Process," 11 March 2005, US Department of Transportation Federal Highway Administration, http://www.fhwa.dot.gov/ve/veproc.htm (accessed 16 March 2006).

61 "Value in Economics" The Columbia Encyclopedia, 6th ed. 2001–2005, http://www.bartleby.com/65/va/value2.html (accessed 2 May 2006).

62 David Ricardo, *On the Principles of Political Economy and Taxation* (John Murray, London: 1821), http://www.econlib.org/library/Ricardo/ricP.html (accessed 15 March 2006)

63 Value in Economics.

64 National Academy of Engineering, *The Engineer of 2020: Visions of Engineering in the New Century* (Washington, DC: National Academy Press, 2004), 48.

65 P. Ehrlich, *The Population Bomb* (New York, NY: Ballantine Books, 1968).

66 Ibid., 20.

67 Ibid., 69.

68 Ibid.

69 National Academy of Engineering, 27–8.

70 Ibid.

71 Ibid., 28–9.

72 Ehrlich, The Population Bomb, 78.

Index

Information on source books

S. Grimnes and O.G. Martinsen
Bioimpedance and Bioelectricity Basics
9780123032607
2000
£84.00/$136.00
The complete text for aerospace engineering and maintenance students and professionals

- One complete source and reference guide to a complex and disparate field
- For graduate-level students, researchers and practitioners in physics, biophysics, instrumentation, and biomedical engineering
- Highly illustrated, with an in-depth explanation of all mathematics

Academic Press

Edited by Buddy D. Ratner, Allan S. Hoffman, Frederick J. Schoen, Jack E. Lemons
Biomaterials Science
9780125824637
2004
£54.99/$108.00
Completely revised and expanded update of the best-selling classic text/reference which defined an entire subject field

- Provides comprehensive coverage of principles and applications of all classes of biomaterials
- Integrates concepts of biomaterials science and biological interactions with clinical science and societal issues including law, regulation, and ethics
- Discusses successes and failures of biomaterials applications in clinical medicine and the future directions of the field

Academic Press

Daniel A. Vallero
Biomedical Ethics for Engineers
9780750682275
03/2007
£29.99/$49.95
Explore important biomedical ethical issues with the only book available to discuss them!

- Working tool for biomedical engineers in the new age of technology
- Numerous case studies to illustrate the direct application of ethical techniques and standards
- Ancillary materials available online for easy integration into any academic program

Academic Press

John Semmlow
Circuits, Signals and Systems for Bioengineers
9780120884933
2005
£54.99/$108.00
The only book that relates important electrical engineering concepts to biomedical engineering and biological studies

- Translates important electrical engineering tools such as Fourier Transform, Laplace Transform, analog modeling, systems modeling, and other linear systems analysis techniques for bioengineering students
- Includes MATLAB examples and problems
- Includes CD-Rom with PowerPoint presentations, extra examples, figures, and support routines

Academic Press

Joseph F. Dyro
Clinical Engineering Handbook
9780122265709
2004
£82.00/$142.00
The world's first comprehensive reference for technology, best practice and future prospects in clinical engineering with a fully international perspective

- For graduate-level students, researchers and practitioners in physics, biophysics, instrumentation, and biomedical engineering
- Highly illustrated, with an in-depth explanation of all mathematics
- Helps Clinical Engineers to be safety and quality facilitators in medical facilities

Academic Press

Richard R. Kyle Jr. and W. Bosseau Murray
Clinical Simulation
9780123725318
12/2007
£59.95/$99.95
The ultimate guide to prevention, containment, treatment, and procedure in a risk-free healthcare setting

- A step-by-step manual to developing successful simulation programs
- Learn how to design, construct, outfit and run simulation facilities for clinical education and research
- Benefit from a wealth of hard-won experiences and lessons learned from leading international practitioners

Academic Press

Reinaldo J. Perez
Design of Medical Electronic Devices
9780125507110
2002
£68.99/$122.00
Bridges the gap between the medical professional and the engineer in designing medical devices

- An essential book for graduate students as well as professionals involved in the design of medical equipment
- Covers every stage of the process, from design to manufacturing to implementation
- Topics covered include analogue/digital conversions, data acquisition, signal processing, optics, and reliability and failure

Academic Press

Editor-in-Chief Isaac Bankman
Handbook of Medical Imaging
9780120777907
2002
£100.00/$185.00
The first comprehensive compilation of the concepts and techniques used to analyze and manipulate medical images after they have been generated or digitized

- Containing work from Internationally renowned authors, from Johns Hopkins, Harvard, UCLA, Yale, Columbia and UCSF
- Includes imaging and visualization
- Contains over 60 pages of stunning, four-color images

Academic Press

Stanley M. Dunn, Alkis Constantinides and Prabhas V. Moghe
Numerical Methods in Biomedical Engineering
9780121860318
2005
£54.99/$102.00
Provides bioengineers with vital tools to be able to read, understand and implement numerical modeling

- Supported by Whitaker Foundation Teaching Materials Program; ABET-oriented pedagogical layout
- MATLAB problem sets and examples available electronically; UNIX, Windows, Mac OS compatible
- Extensive hands-on homework exercises

Academic Press

Yoshito Ikada
Tissue Engineering: Fundamentals and Applications
9780123705822
2006
£105.00/$175.00
A base for directing future research of tissue engineering toward revolutionising healthcare

- Discover the key issues required in promotion of clinical trials in tissue engineering
- Focus on the fundamentals (biomaterial, scaffold, cell culture, bioreactor, animal model etc.), recent animal and human trials, and the future prospect regarding the tissue engineering
- Gain up-to-date knowledge of the most important technological developments in the field

Academic Press